ATLAS OF CLINICAL FUNGI

2nd edition

G.S. de Hoog, J. Guarro, J. Gené & M.J. Figueras

Centraalbureau voor Schimmelcultures
Utrecht, The Netherlands

Universitat Rovira i Virgili
Reus, Spain

This atlas is dedicated to
André Botter, who kept the torch
of Dutch medical mycology
burning over many years

ISBN 90-70351-43-9.

Ordering information: Centraalbureau voor Schimmelcultures, Padualaan 8, Utrecht, NL-3584 CT The Netherlands, and Institut d'Estudis Avancats, Facultad de Medicina i Ciencies de la Salut, Universitat Rovira i Virgili, Carrer St. Llorenç, 21, Reus, E-43201 Spain.

Cover illustration: SEM of ascomal hairs of *Chaetomium funicola*.

CONTENTS

INTRODUCTION

Fungi are gaining importance with the increased incidence of chronic, often fatal mycoses in immunocompromised patients. This Atlas means to stimulate interest in the identification of pathogenic and opportunistic agents. Increasing precision of identification can augment adequate and timely therapy. We hope the user will publish cases of new mycoses and deposit voucher specimens in recognized culture collections, so that a more balanced picture of the diversity of clinical fungi can be built up.

Those particularly at risk are patients with leukemia, solid tumours, AIDS or ketoacidotic diabetes. Hospitalised patients who have undergone major surgery or suffer from circulation failure or extended burn wounds and those subjected to prolonged radiotherapy, corticosteroids, cytostatics or antibiotics are also susceptible to mycotic infection. The number of uncommon infections in bone marrow transplant (BMT) patients are increasing rapidly. Mycoses decrease the chance of recovery of these patients and can seriously hamper the improvement of their quality of life.

The taxonomic diversity of etiologic agents seems to be increasing. In addition to long established mycoses caused by dermatophytes, systemic dimorphic fungi, *Candida, Cryptococcus* and *Aspergillus*, moulds and yeasts are encountered which were formerly only known to be saprobic. A surprisingly large number of these species are now seen to survive in human tissue. However, whether or not new fungi are emerging as clinical agents is difficult to verify with certainty. Comparison with older literature is often impossible, because many fungi have been described under names which have not been clarified. In the period between 1920 and 1940 a wide variety of purported pathogens were described whose identity is still uncertain. Therefore the same fungi may later have been described again under a different name. An example is *Pseudallescheria boydii*. This species is now emerging as an agent of systemic mycoses. However, the same fungus, under the anamorph name *Scedosporium apiospermum*, was already an established agent of mycetoma in the 1920s. Even earlier, before 1900, the name *Glenospora graphii* had been used for this fungus, then known as an agent of otitis externa. It is likely that the etiologic agents of many of these infections are frequently misdiagnosed as aspergillosis.

The vast number of reports of unverified names actually indicates that fungi were clinically relevant long before the emergence of the 'modern' hospital mycoses. Comparison of historic and recent fungal overviews will be possible when obsolete names have been clarified. This Atlas contains a full nomenclature of each species, and is supplemented by a checklist of every fungal name which has appeared in medical literature.

Proper recognition of rare agents of disease necessitates the availability of a broad taxonomic skill and experience in clinical laboratories, because the majority of fungal species are still identified by morphological criteria. Incorrect identification has a negative impact on the growth of knowledge and understanding of these mycoses. Uncommon agents are often identified by comparing the fungus at hand to the limited species lists published of major handbooks in medical mycology. But many of these lists are incomplete. When cases are 'recognized' on the basis of such insufficient data, the supposed restricted etiology of the respective mycosis becomes self-fulfilling. Erroneous identifications tend to persist in the literature. In a number of opportunistic clinical settings, such as in mycetoma, the spectrum of agents is probably much larger than generally supposed.

Organization of the Atlas

The main aim of the Atlas is to illustrate the diversity of fungal agents. For detailed clinical data we refer to specialised handbooks, of which Badillet *et al.* (1986), Rippon (1988) and Kwon-Chung & Bennett (1992) are particularly recommended. Excellent overviews of histopathology of major mycoses are those of Chandler *et al.* (1980) and Salfelder (1990); of the latter a new edition will be published shortly. The rare opportunists, with which most clinicians are unfamiliar, are dealt with in detail in this Atlas.

Each fungus is described with **colony characteristics** on one of the recommended identification media after 14 days incubation at room temperature (unless stated otherwise). Colony diameters may vary but fall roughly into three categories: rapidly expanding (over 4 cm), moderate (2-4 cm) or slow, restricted (less than 2 cm). **Microscopy** of anamorph and teleomorph is occasionally described on separate pages. If so, cross-references are given. **Physiological** profiles are provided for all yeasts, yeast-like fungi and dermatophytes.

Molecular diagnostics essentially comprises restriction maps of parts of the ribosomal operon, digested with fixed panel enzymes. In a number of groups phylogenetic trees are provided to give insight into relationships with other fungi. Selected sequences are available on Internet. The persistent uniform resource locator (PURL) is http://purl.oclc.org/net/ridom that is currently associated with the following URL http://www.ridom.hygiene.uni-wuerzburg.de.

In addition, literature data are provided. Occasionally a brief **differential diagnosis** is given to segregate the species from all its neighbouring species including saprobic taxa. This has systematically been done with the yeasts, using physiological characteristics; data for this comparison have been taken from Barnett *et al.* (1990). **Pathogenicity** of established species is briefly described in the chapter on clinical pathology; essential references and exceptional cases are listed under each species. With each species a BSL attribution is given according to the most widely accepted classification (de Hoog, 1997). The normal ecological niche of opportunistic fungi is supposed to be in environments outside the human or animal body. Data on **antimycotic susceptibility** are partly taken from the literature. Finally, with each fungus some selected **references** for taxonomic data and further reading are given, plus full **nomenclature** of both ana- and teleomorph.

Several fundamental problems were encountered during compilation. Reported new fungal agents should match the first description of that species. Unfortunately the state of taxonomy in many groups is insufficient to be certain of such matching. Species occurring in nature may be very similar but not identical to their pathogenic counterparts. In the case of rare mycoses we have used type or representative strains of that species. Often the strain from the case report is no longer available, thus it sometimes remains uncertain whether the actual opportunist being described.

Acknowledgements

All those who have helped us in collecting the data used in the Atlas are gratefully acknowledged: M.J. Figge, A.H.G. Gerrits van den Ende, K. Luijsterburg, G. Poot, H.J. Roeijmans, F. Snippe-Claus, M. Vermaas, R. Verwoerd-Kuyt, D. Zimmerman and all staff members who provided us with media, equipment and facilities. A. Botter, E.S. Hoekstra, P. Kager, A.F.A. Kuijpers, J.G.F.M. Meis, R. Rüchel, R.A. Samson, M.Th. Smith, C.S. Tan, K. Tintelnot, C. de Vroey and R.G.F. Wintermans presented specialized lectures at CBS courses, and their information is partly used in the Atlas. Molecular data were kindly provided by D. Attili, Y. Gräser, G. Haase, R. Horré, N. Poonwan and K. Sterflinger; numerous sequences were taken from the public domain. Antifungal susceptibility data were provided by G. Quindós, B. Fernández-Torres, M. Ortoneda and I. Pujol and some figures were kindly provided by L. Zaror, R. Negroni and C. da Silva Lacaz. Data on yeasts were donated by E. Guého, M.-L. Kerkmann, K.J. Kwon-Chung, M.Th. Smith and D. Yarrow; E.G. Simmons commented on Dematiaceae. We thank H.A. van der Aa, A. Aptroot, E. Dei-Cas, E. Göttlich, E. Guého, R.A. Humber, C.A.N. Jacobs-van Oorschot, M.R. McGinnis, F.C. Odds, E. Presterl, M.A.A. Schipper, D.J. Sullivan and R.C. Summerbell for their comments on parts of the manuscript. D. Harmsen is involved in the Internet version of the Atlas. We thank T. van den Berg-Visser, M. de la Croix, M. Pizarroso and E. Aramburu for excellent secretarial help and P.J. Hoody for correcting the English text.

We are much indebted to the Curators of the following culture collections for supplying cultures: AMMRL, ATCC, CDC, FMC, IFM, IFO, IMI, IP, NCPF, NHL, RV, TRTC, UAMH, UTHSC and WCLR. Most of the scanning micrographs were made using the facilities of the Centres of Electron Microscopy of the University of Barcelona and of the University Rovira i Virgili (M. Moncusí). Part of the work was supported by the Institut d'Estudis Avançats of the University Rovira i Virgili.

Utrecht / Reus, 1 September 2000   The authors

CLASSIFICATION

Fungi are treated as a separate kingdom of living organisms. They are **heterotrophic**, i.e., lacking chlorophyll, and are dependent on organic carbon compounds for their nutrition. They feed by the excretion of enzymes into the substrate and by subsequent take-up of digested compounds through the cell wall. The major cell wall constituents are mostly glucan and chitin, exceptionally other macromolecules such as cellulose or chitosan.

Fungi are **eukaryotes** because the genome is organized in a nucleus surrounded by a membrane. This membrane is continuous with the endoplasmic reticulum. Cellular division is accompanied by meiosis or mitosis. Cell organelles such as mitochondria, ribosomes, vacuoles, lipid bodies and other storage inclusions are present.

Of the over 100,000 described species of fungi, about 100 are known to be regularly involved in human and animal mycoses. A few hundred species occur as opportunists. These clinically significant fungi are found at several locations in the fungal kingdom. Pathogenicity is a complicated process, and fungi have acquired capacities to live in association with or to penetrate warm-blooded animals more than once in the course of evolution. But fungi that are truly pathogenic, i.e., dependent on living hosts, are only found in a few taxonomic groups. The system of classification presented below is incomplete in that it treats the main taxa only inasmuch as they have members that are of medical relevance.

Nomenclature

The kingdom is subdivided in a hierarchical manner, each rank being named with, and recognizable by, a particular ending:

Division	:-mycota
Class	:-mycetes
Order	:-ales
Family	:-aceae

A family is composed of genera, and these contain species. A species name is a **binomial**, consisting of the generic name with a specific epitheton. In scientific publications also the name of the author who introduced that name is added, e.g., *Histoplasma capsulatum* Darling. If this name is introduced in a publication of another author, this author may also be mentioned, e.g., *Trichophyton ochropyrraceum* Muijs, *in* Papegaaij. All Latin names of biological groups are written in *Italics*, e.g., *Ascomycota*. If a single member is indicated, an Anglicised noun is used, e.g., an ascomycete. Artificial groups (see below) are not taken to have official Latin names and thus are written in Roman, e.g., Coelomycetes.

Each species is standardized by its first description. A dried sample is maintained in an official herbarium, for later comparative study: the **type specimen**. If it appears that the same species was described twice under two different names, the older name stands and the newer name becomes a synonym.

Many fungi pass through a characteristic life cycle, and thus can be found as separate, independently sporulating organisms of quite different morphology (**pleomorphism**). The biochemical make-up and the physiological properties of these separate organisms remain largely identical. An important type of pleomorphism is the change of sexual and asexual generations. Each of these generations may propagate independently, forming sexual and asexual thalli, respectively. The sexually propagating thallus is known as the **teleomorph**, and the asexual thallus is the **anamorph**. The total organism is referred to as the **holomorph**. Several fungi display various types of independently propagating anamorphs, which are then called **synanamorphs**. Note that teleo- and (syn)anamorphs may all have their own names. The species *Pseudallescheria boydii*, for example, is also known under its synanamorph names *Graphium eumorphum* and *Scedosporium apiospermum*; these three names all refer to one and the same organism.

The fungal system

A group frequently treated in handbooks on medical mycology is the Actinomycetes. Similar to the fungi and bacteria, they are heterotrophic organisms. Actinomycetes are bacteria, because they are prokaryotes: the DNA is

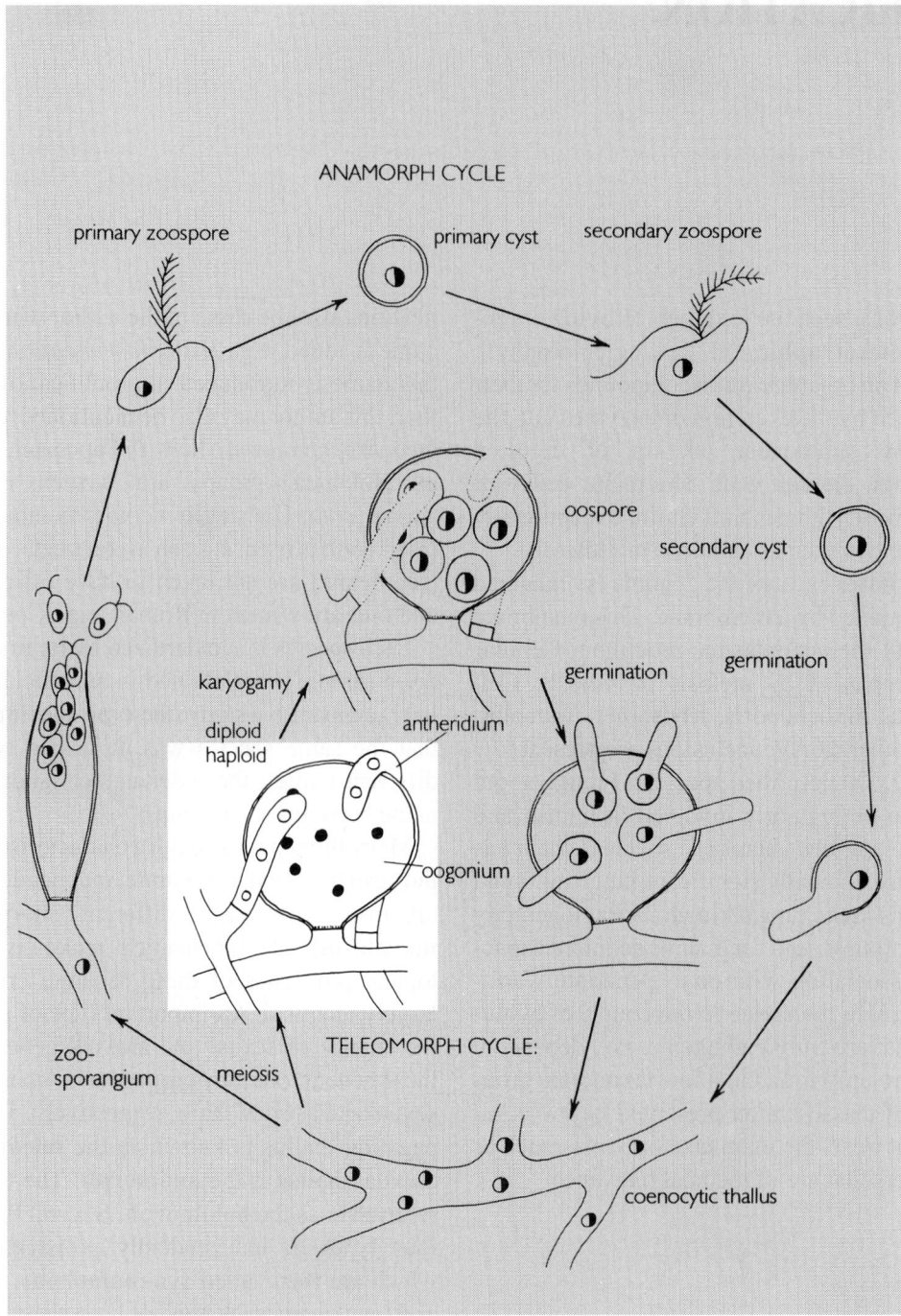

Fig. 1. Diagram of the life cycle of *Oomycota*. The inner ring displays the sexual series of events, while in the outer ring the asexual route is depicted. The diploid part of the life cycle is shaded in grey.

distributed freely in the cell rather than being organized in a nucleus, as in the fungi.

Until recently, six fungal divisions were recognized. On the basis of molecular phylogenetic data, some groups are no longer accepted as fungi (Guarro *et al.*, 1999c), while, in contrast, other organisms appear to have unexpected relationships to the fungi. The *Mycetozoa* and the *Mesomycetozoa* are purported members of the kingdom *Protista;* the *Oomycota* belong to the kingdom *Chromista*. Because of their fungus-like

appearance the are indicated as 'Pseudofungi'. For the phylogenetic position of these organisms in the general tree of life is referred to the following website: http://www.evolutionsbiologie.uni-konstanz.de/peer-lab/crown_500.html. Recently also the protozoan genus *Microsporidium* was supposed to have close affinity to the fungi (Keeling *et al.*, 2000). Members of the kingdom fungi (*Eumycota*) are currently restricted to four divisions: the *Chytridiomycota, Zygomycota,*

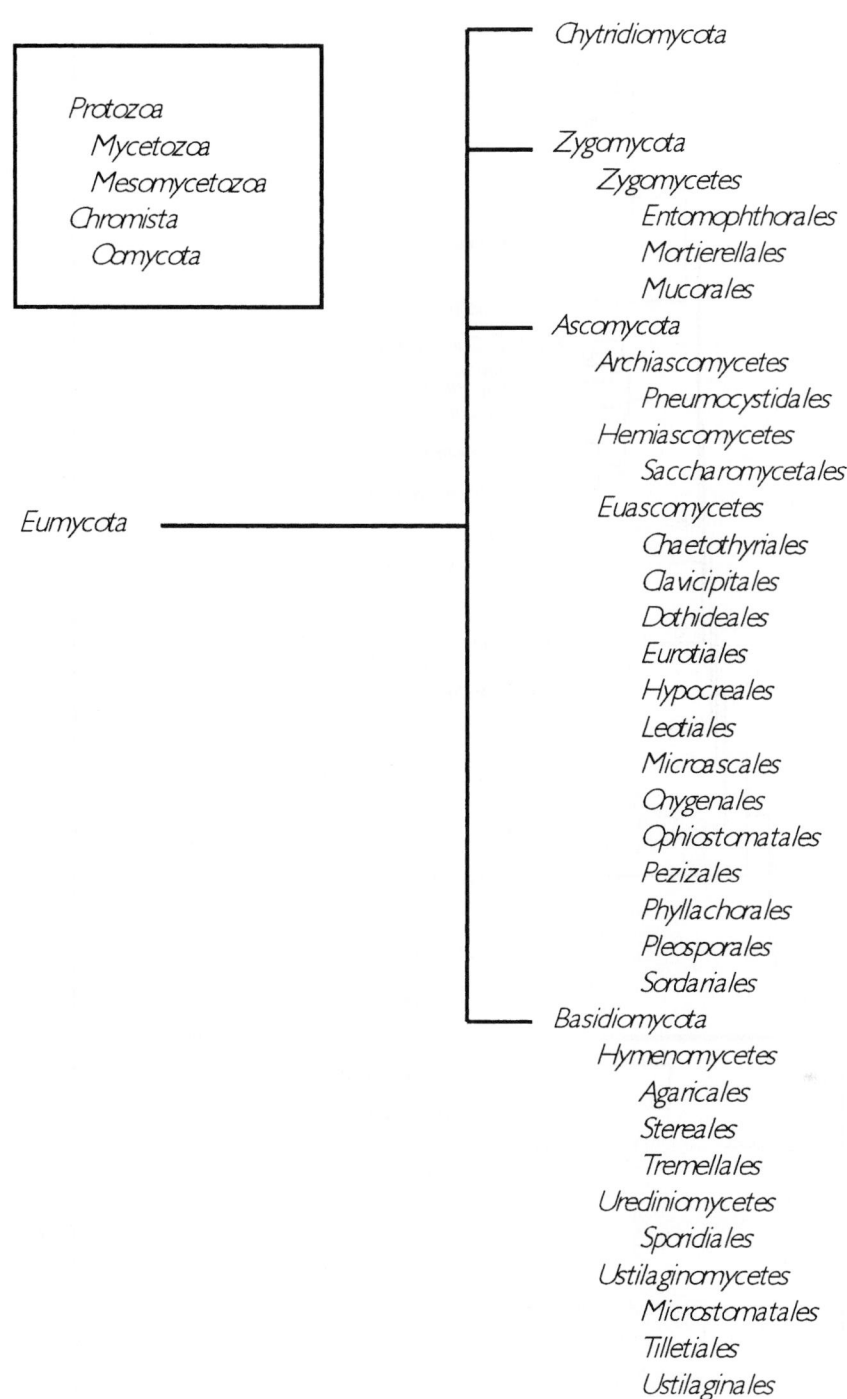

```
                                          ┌──── Chytridiomycota
                                          │
                                          │
                                          ├──── Zygomycota
                                          │         Zygomycetes
                                          │            Entomophthorales
┌─────────────────────────┐              │            Mortierellales
│   Protozoa              │              │            Mucorales
│     Mycetozoa           │              ├──── Ascomycota
│       Mesomycetozoa     │              │         Archiascomycetes
│     Chromista           │              │            Pneumocystidales
│       Oomycota          │              │         Hemiascomycetes
└─────────────────────────┘              │            Saccharomycetales
                                          │         Euascomycetes
                                          │            Chaetothyriales
Eumycota ──────────────────────────┤            Clavicipitales
                                          │            Dothideales
                                          │            Eurotiales
                                          │            Hypocreales
                                          │            Leotiales
                                          │            Microascales
                                          │            Onygenales
                                          │            Ophiostomatales
                                          │            Pezizales
                                          │            Phyllachorales
                                          │            Pleosporales
                                          │            Sordariales
                                          └──── Basidiomycota
                                                    Hymenomycetes
                                                       Agaricales
                                                       Stereales
                                                       Tremellales
                                                    Urediniomycetes
                                                       Sporidiales
                                                    Ustilaginomycetes
                                                       Microstomatales
                                                       Tilletiales
                                                       Ustilaginales
```

Fig. 2. Summary of main groups of fungi and pseudofungi, with accent on groups with medical significance.

Ascomycota and *Basidiomycota*. The *Zygomycota* are regarded as lower organisms in an evolutionary sense. They generally have non-septate mycelia, whereas the higher fungy: have regularly septate thalli.

The spectrum of clinically relevant fungi is expanding steadily, particularly in hospitalised patients with severe immunodeficiency. It is sometimes supposed that practically any fungus is a potential opportunist. However, the 397 species described in this Atlas comprise only a small fraction of the about 100,000 described species of *Eumycota*. The treated BSL-2 (BioSafety Level) species are classified in 18 orders (Table 1), while currently about 104 orders are known, some containing thousands of species which never have been observed as etiologic agents of disease in humans or animals. The taxonomic overview of this Atlas thus displays only a part of the *Eumycota,* namely those groups that contain species which have been implicated in infections of warm-blooded animals. Ascomycetes classified in BSL-3 (see p. 39) are

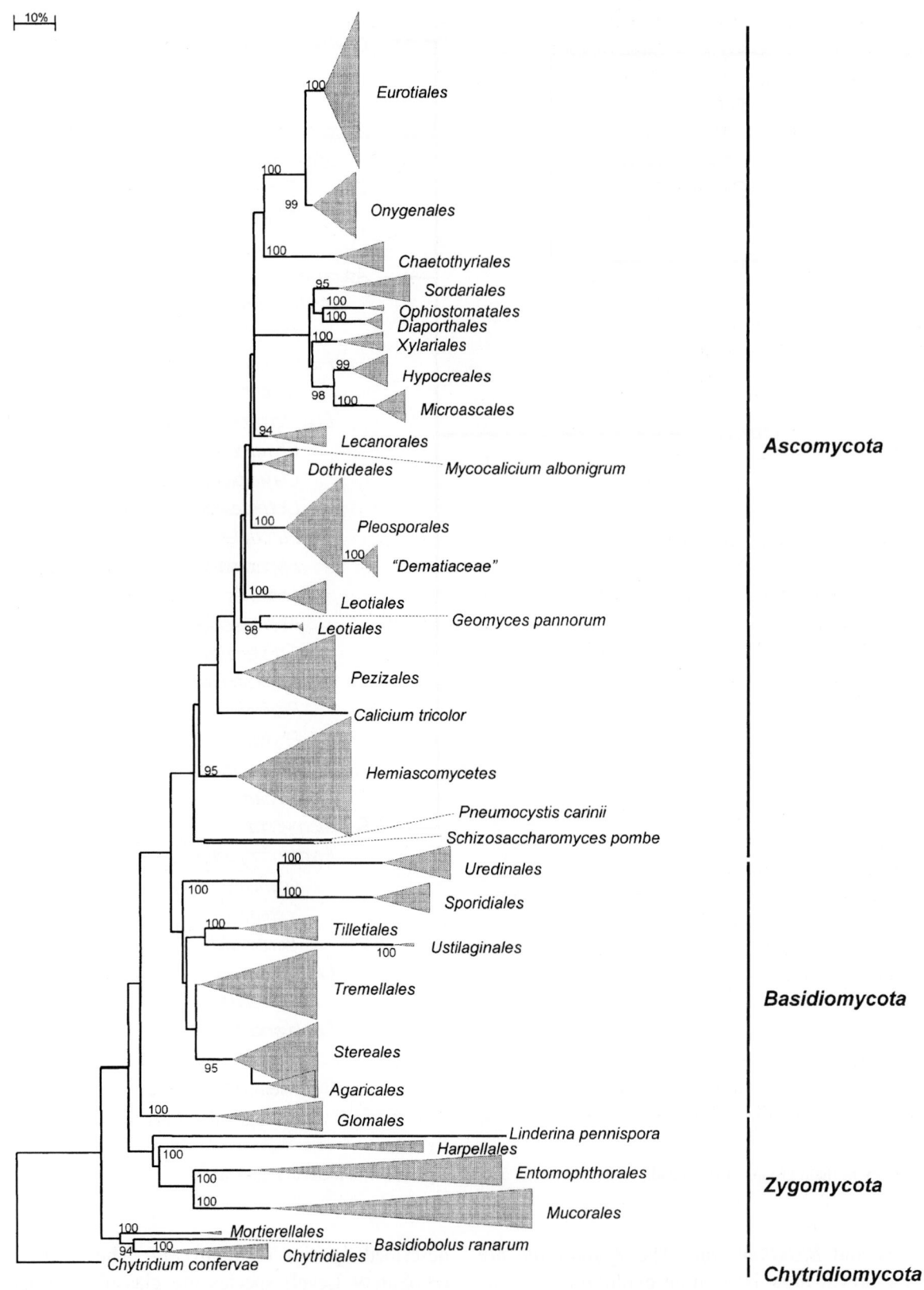

Fig. 3. Phylogenetic tree of the *Eumycota*, mainly showing the members which have been implicated in diseases of human and animals. The tree is based on 414 near-complete SSU rDNA sequencesaligned by Y. van de Peer, using Neighbor joining algorithm with Kimura correction. Bootstrap values >90 from 100 resampled datasets are shown. The tree is highly unbalanced, since of large groups with high biodiversity, e.g., *Dothideales* and *Pezizales*, only relatively few members have been sequenced; well-known order such as the *Eurotiales* are over-represented. Some orders, like the *Pezizales*, are heterogeneous so that they appear several times in the tree; others could not be resolved, e.g., the *Agaricales* being paraphyletic to the *Stereales*. A few individual taxa are attributed with difficulty to any order, such as *Geomyces pannorum*, *Pneumocystis carinii* and *Basidiobolus ranarum*. Note that in the *Zygomycota* the number of species is low but the mutual distances are enormous (compare also Fig. 13 on p. 61).

limited to only three orders: the *Onygenales*, the *Eurotiales* and the *Chaetothyriales*. It is remarkable, that these orders are phylogenetically closely interrelated (Fig. 3), which suggests that pathogenicity is an evolutionary very rare phenomenon.

1. Kingdom *Protozoa*, division *Mycetozoa*

The thallus consists of a multinuclear plasmodium without a cell wall. It creeps and feeds with pseudopodia. Later it forms large, characteristically shaped sporangia in which haploid resting spores are produced after meiosis. The sporangium contains numerous capillitium threads which aid in the dispersal of the spores. The spores germinate with myxamoebae (zoospores) which bear two, unequal flagellae. After copulation of two myxamoebae, a diploid plasmodium develops. There are no members of *Mycetozoa* with clinical significance.

2. Kingdom *Protozoa*, division *Mesomycetozoa*

This is a small group of pathogens associated with aquatic habitats, phylogenetically located between animals and fungi. They have endospore-like parasitic stages and cannot be grown on artificial media. A species with clinical significance is *Rhinosporidium seeberi*.

3. Kingdom *Chromista*, division *Oomycota*

The life cycle of a prototypical oomycete is depicted in Fig. 1. The thallus is coenocytic and the cell walls contain cellulose and glucan. Asexual propagation is through motile zoospores, having two types of flagellae: one tinsel and the other whiplash. Sexual reproduction occurs between antheridia and oogonia, which are morphologically different (**anisogamy**); these sex organs develop after meiosis. Nuclei of the antheridium are brought into the oogonium; one or a few nuclei of opposite sex fuse, while the remaining nuclei deteriorate. Non-motile **oospores** develop in the oogonium which becomes characteristically thick-walled. The oospores germinate without meiosis. Thus in the *Oomycota*, in contrast to other groups of fungi, only the gametes (i.e., the nuclei of antheridium and oogonium) are haploid, while the remaining thallus is diploid. Two orders are distinguished. The order *Saprolegniales* comprises aquatic species, some of which are pathogenic to fish. Representatives of the order *Peronosporales* mostly occur in soil or on plants; zoosporangia may be dispersed by wind. The latter order includes one species of clinical significance, *Pythium insidiosum*.

4. Kingdom *Eumycota*

I. Division *Chytridiomycota*

The thalli are small, often unicellular. Most species live in a watery environment, and are probably closely related to the Algae. Gametes and asexual spores are motile, having a single flagella at their ends. After copulation, a new thallus is formed which produces either zoospores or motile gametes. There are no chytridiomycetes with clinical significance. Occurrence in cold-blooded animals is known through Berger *et al.* (1998) and Longcore *et al.* (1999) who described zoonoses in amphibians by *Batrachochytrium dendrobatidis*.

II. Division *Zygomycota*

The life cycle of a prototypical zygomycete is depicted in Fig. 4. The thallus is basically coenocytic; the cell walls are composed of chitosan and chitin. Asexual reproduction is by non-motile spores. Sporangia contain numerous nuclei, which develop into sporangiospores after cleavage of the cytoplasm. Sexual reproduction is **isogamic**, i.e., with gametangial cells which are indistinguishable from each other. Cellular fusion leads to a **zygospore** with a thickened, characteristically ornamented wall. The following orders are treated:

* Order *Mucorales*
Asexual spores are produced endogenously in stalked sporangia. Sporangiospores are passively distributed by air. Sexual zygospores are formed after mating of suitable partners. Interfertility of heterothallic strains is an important criterion for identification.
* Order *Mortierellales*
Rapid, scant growth; columellae are absent from sporangia.
* Order *Entomophthorales*
Species are difficult to grow in culture. They produce exogenous spores on spore-bearing cells; spores are forcibly discharged by a turgor mechanism. Several species are insect pathogens. Hyphae are sometimes septate.

III. Division *Ascomycota*

The life cycle of a prototypical ascomycete is depicted in Fig. 5. The thallus is septate; septa mostly have simple pores with Woronin bodies, or they have micropores. Asexual propagules (conidia) are of widely divergent morphology. Most species reproduce with an asexual thallus, which in culture grows as yeast, as a coelomycete or as a hyphomycete (see p. 7). The teleomorph cycle is as follows. Haploid gametangia are mostly formed in a fruit body. The antheridium and ascogonium contain multiple nuclei. Soon the two cells fuse, the ascogonium

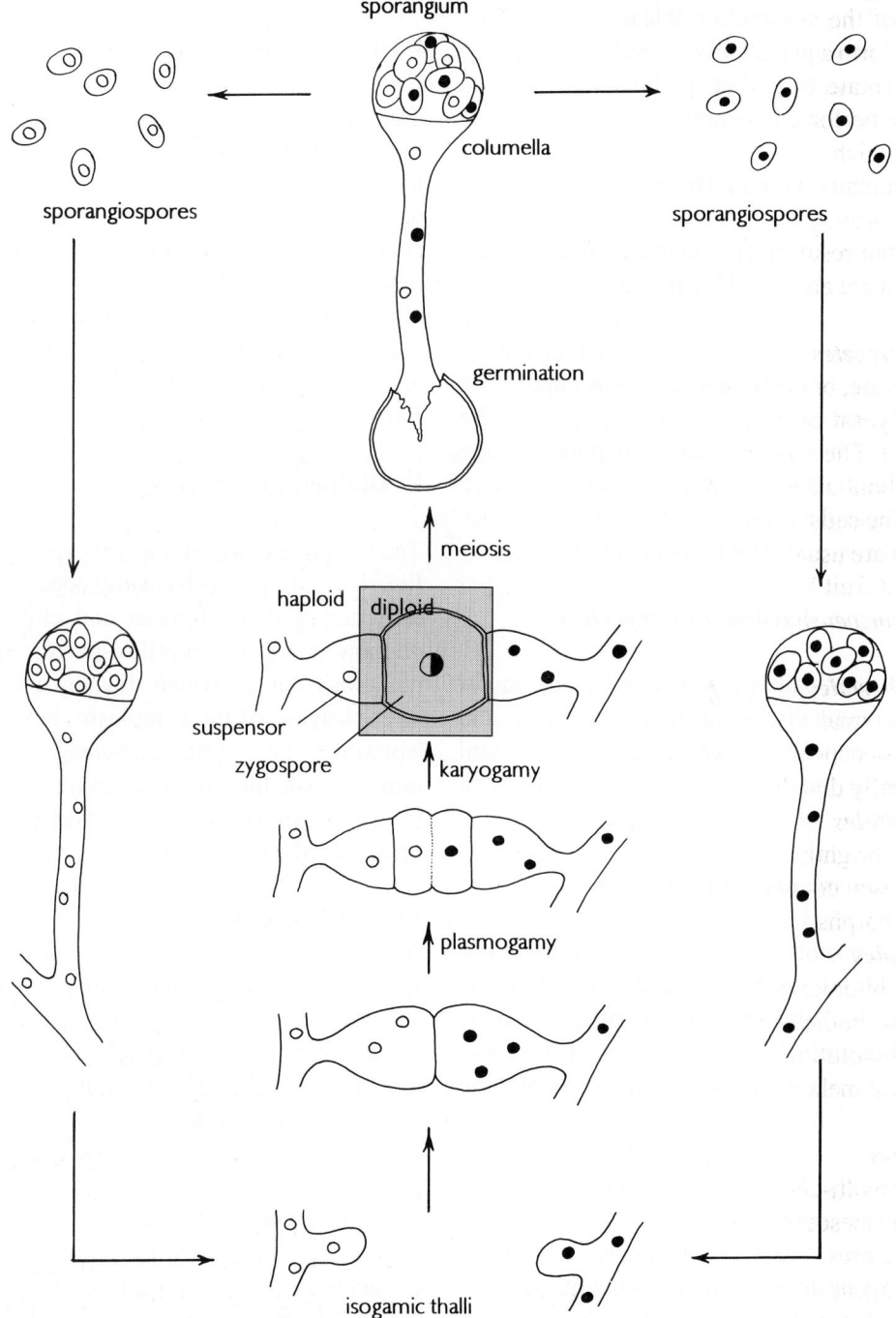

Fig. 4. Diagram of life cycle of *Zygomycota*. Central portion displays the sexual series of events, while on the outside the asexual route is depicted. The diploid part of the life cycle is within the grey box.

then containing numerous nuclei of opposite sex. From this cell short, dikaryotic hyphae arise, the two nuclei in each cell being derived from one of the parents. The hyphal tip curves to form a crozier, in which mitosis takes place. Subsequently one of the pairs of nuclei fuses, the other deteriorates. After karyogamy and meiosis, haploid spores are produced endogenously in an **ascus**. The resulting ascospores (mostly 8 or a multiple of 4) germinate forming a haploid **mycelium** (= thallus consisting of filaments) or **yeast** (= discrete budding

cells). Budding cells and hyphal thallus may again reproduce independently, giving rise to an asexual life cycle or anamorph. Three classes of *Ascomycota* will be treated:

a. Class ***Archiascomycetes***
This contains a single species, *Pneumocystis carinii*, formerly treated as a member of *Protozoa*.

b. Class ***Hemiascomycetes (Endomycetes)***

Most of these fungi are unicellular, and they comprise the main group of the yeasts. Individual budding cells often cohere and form a pseudomycelium. Some species, however, form septate hyphae. In most cases the septa then appear to be perforated by micropores of less than 10 nm width, which are concentrically arranged or dispersed; occasionally only one, central micropore is present. Sexual propagation is through isogamy, the gametangia and the resulting ascus often being discrete cells. Fruit bodies are absent.

c. Class *Euascomycetes*

The thallus is septate, or unicellular during some parts of the life cycle. If yeast cells are preponderant, these are usually melanised. The septa are perforated by a simple, central pore of about 50 nm width, often occluded by a Woronin body. The cells mostly contain a single, haploid nucleus. The asci are usually produced in or on a more or less differentiated fruit body. Representatives of twelve orders of *Euascomycetes* will be treated:

* Order *Chaetothyriales*
Characterized by small ascomata, bitunicate asci and dark, muriform ascospores. Anamorphs are melanized, and morphologically diverse.

* Order *Clavicipitales*
Characterized by brightly coloured, stromatic ascomata and cylindrical asci containing multi-septate, filiform ascospores. Anamorphs show phialidic conidiogenesis.

* Order *Dothideales*
Characterized by bitunicate asci often containing septate ascospores. Fruit bodies are thick, closed, the asci emerging after dissolution of the central portion. Anamorphs are mostly melanised, and are morphologically diverse.

* Order *Eurotiales*
Characterized by multi-celled, closed fruit bodies containing spherical, evanescent asci with often ornamented, unicellular ascospores. Anamorphs show phialidic or annellidic conidiogenesis, mostly with dry conidia.

* Order *Hypocreales*
Characterized by fleshy, brightly coloured perithecia with cylindrical asci containing hyaline ascospores of variable shape. Anamorphs show phialidic conidiogenesis.

* Order *Leotiales*
Characterized by small, often brightly coloured apothecia surrounded by seta-like hairs; asci small, containing hyaline ascospores. Anamorphs absent or poorly differentiated.

* Order *Microascales*
Characterized by membranaceous fruit bodies containing thin-walled, ellipsoidal or spherical asci. Ascospores are one-celled, mostly brown. Anamorphs generally show annellidic conidiogenesis.

* Order *Onygenales*
Characterized by fruit bodies mostly composed of loosely interwoven hyphae and by small, spherical, evanescent asci with unicellular, hyaline ascospores. Many species are keratinophilic. Anamorphs show thallic conidiogenesis.

* Order *Ophiostomatales*
Characterized by perithecia with long necks, and small, evanescent, spherical asci with hyaline, unicellular ascospores possessing thin, slimy sheaths. Anamorphs are often sympodial or yeast-like.

* Order *Pezizales*
Characterized by apothecial fruit bodies with slender asci and large, often ornamented, unicellular ascospores. Anamorphs are often with large, synchronously produced conidia.

* Order *Phyllachorales*
Characterised by black stromata mostly immersed in plant tissue, perithecia containing cylindrical asci with non-septate ascospores. Anamorphs mostly coelomycetous.

* Order *Sordariales*
Characterised by cylindrical, unitunicate asci with mostly 8 ascospores. The asci are in parallel arrangement in a fruit body with a preformed opening. Anamorphs are often hyaline, phialidic, with slimy conidia.

IV. Division *Basidiomycota*

The life cycle of a prototypical basidiomycete is depicted in Fig. 6. The thallus consists of a septate, dikaryotic mycelium, often provided with **clamp connections**; these are bridges between adjacent cells to provide each of them with a nucleus derived from one of the parent strains. The septa are perforated by a single, central pore. The wall of the pore canal is often characteristically swollen; such a structure is called a **dolipore**. Spores each produce a short-lived, haploid mycelium. Cells of suitable mating type show plasmogamy, but karyogamy is postponed. Consequently a heterokaryon with clamp connections is formed. This condition is maintained during the major part of the life cycle, including fruit body production. Karyogamy, immediately followed by meiosis, takes place in the **basidium**, which produces meiospores (basidiospores) exogenously, the spores often being forcibly discharged. Three classes are distinguished within the *Basidiomycota*:

a. Class *Urediniomycetes*

Rust fungi, having a complicated life cycle as plant pathogens. A few members grow as red yeasts in culture; these are classified in the order *Sporidiales*.

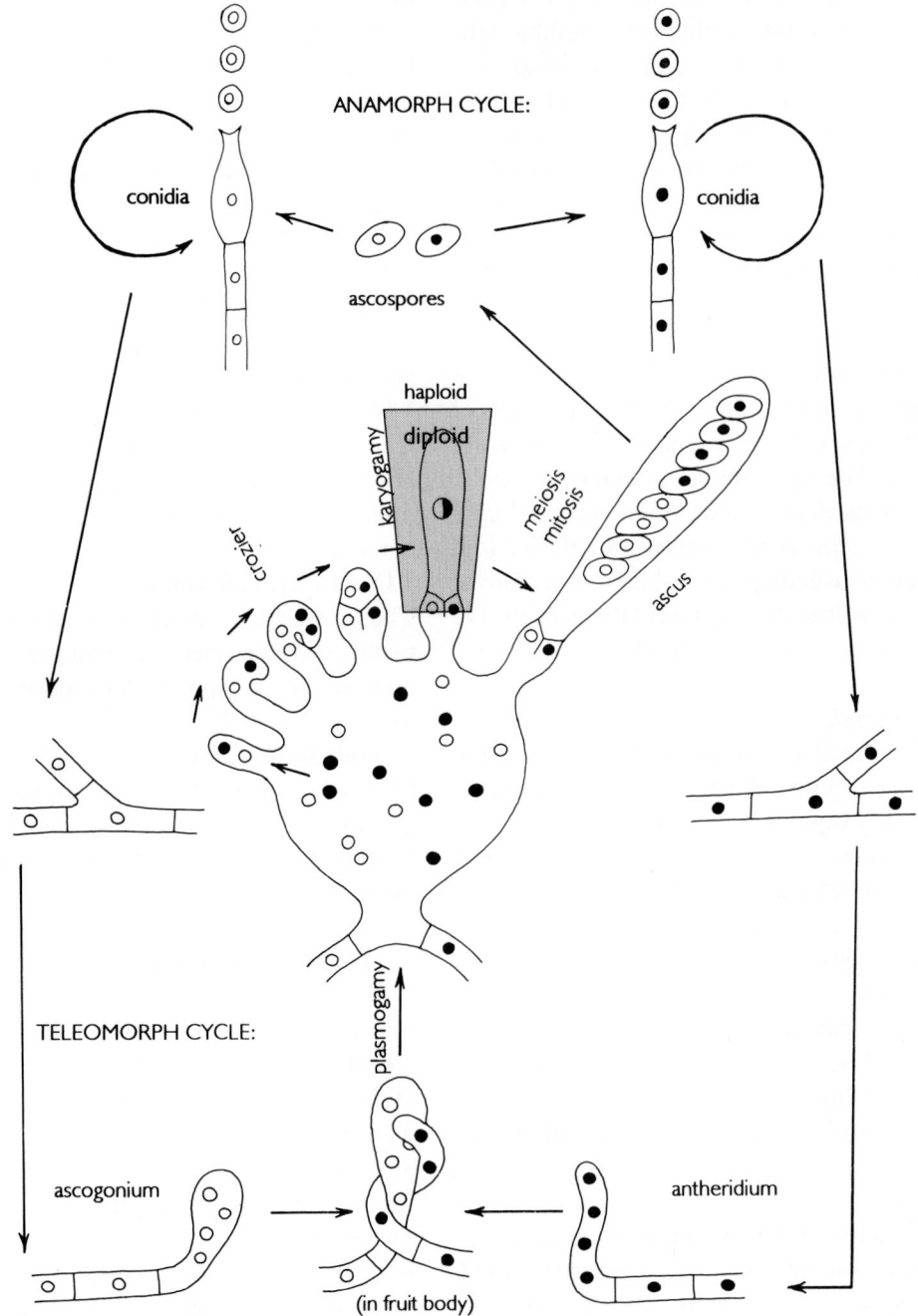

Fig. 5. Diagram of life cycle of *Ascomycota*. Lower half displays the sexual series of events (teleomorph cycle); in the upper half the asexual route (anamorph cycle) is depicted. The diploid part of the life cycle is within the grey box.

b. Class *Ustilaginomycetes*
Smut fungi, occurring as highly specialized plant pathogens. The group is phylogenetically highly diverse, containing a large number of orders. In this Atlas some yeast-like members of *Microstomatales, Tilletiales* and *Ustilaginales* are treated.

c. Class *Hymenomycetes*
Most species are characterized by a unicellular basidium, often with forcible spore discharge. Basidia are arranged in a dense palisade, which is borne on a large, macroscopically visible fruit body. The class compromises most mushrooms and toadstools. In culture, undifferentiated anamorphs with arthroconidia may be produced; these are listed under filamentous basidiomycetes. Representa-

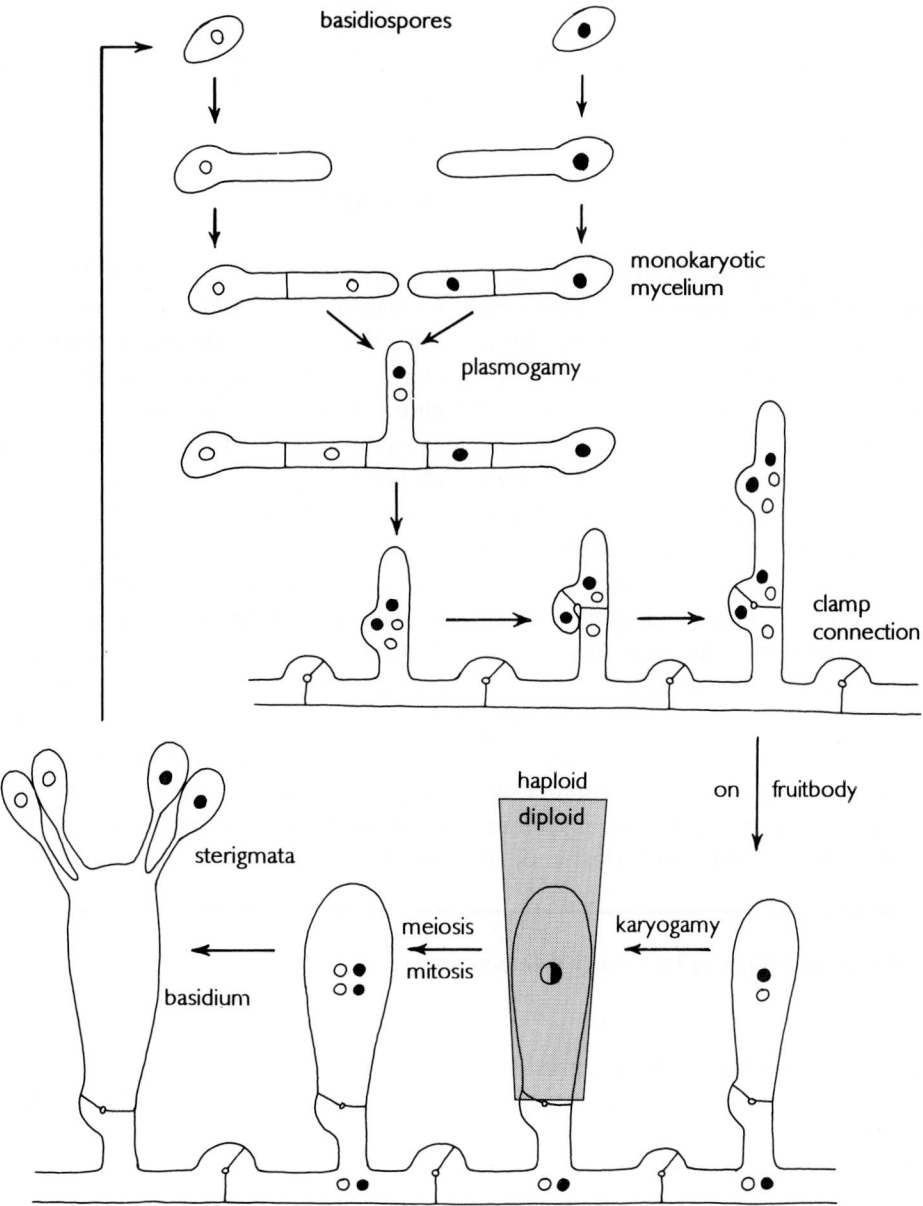

Fig. 6. Diagram of life cycle of *Basidiomycota*. The diploid part of the life cycle is within the grey box. An asexual cycle is generally absent, or can be yeast-like.

tives of two orders have very rarely been implied in medical mycology, viz. the *Agaricales* (toadstools) and *Stereales* (shelf fungi). In addition, a group with gelatinous, cerebriform fruit bodies, the order *Tremellales*, is classified under the *Hymenomycetes*. In clinical species, basidiospores are not forcibly discharged. In culture they are white yeasts (*Cryptococcus* and relatives).

5. Anamorphic fungi

Of many asexually reproducing thalli, no teleomorph is known. Most of them are supposed to have a teleomorph in the *Ascomycota;* a few belong to the *Basidiomycota*. In the absence of a teleomorph, this taxonomic affinity can

be established by ultrastructural characters: ascomycetes have 2-layered cell walls, while walls of basidiomycetes are multilamellar. The anamorphic fungi are artificially classified according to their form of growth and the production of asexual fruit bodies, as follows:

* Yeasts

When fungi have no hyphae but just consist of loose budding cells only, they are treated under the yeasts (for reviews, see Kurtzman & Fell, 1998 and Barnett *et al.*, 2000). The term 'yeasts' is a descriptive term and stands for any fungus which reproduces by budding. Yeasts are either white or red. Criteria for identification of yeasts are fundamentally different from those of hyphomycetes, viz. largely physiological and largely morphological, respectively, regardless of their biological relationship. Yeast-like fungi with melanised cell walls are so-called **black yeasts**. These are anamorphs of several orders of *Euascomycetes*. They are nearly always able to produce an additional true mycelium and are therefore treated under the Hyphomycetes.

* Hyphomycetes

The mycelium consists of septate hyphae; variously shaped conidia are produced on more or less differentiated branches. Conidial fruit bodies are absent (for a review, see Carmichael *et al.*, 1980). Criteria for identification of hyphomycetes are details in the formation of conidia (conidiogenesis). Types of conidiogenesis broadly correspond with the orders of *Euascomycetes* listed on p. 7. Sometimes, however, several types of asexual propagation can be found next to each other in the same strain. In nature, the various types of conidiation have different ecological roles in dispersing the fungus through different micro-habitats. The great majority of Hyphomycetes are of ascomycetous affinity.

* Coelomycetes

The mycelium consists of septate hyphae; conidia are produced in fruit bodies, while the rest of the mycelium remains sterile. Fruit bodies are either spherical with an apical opening (pycnidia), or flat, cup-shaped (acervuli). Except for morphological characters of fruit bodies and conidia, the process of conidiogenesis is also of significance for the taxonomy of Coelomycetes. Nearly all members of Coelomycetes are of ascomycetous affinity.

Identification

Criteria for the recognition of taxa are mostly different from the fundamental characters for biological classification; these are the **primary characters** treated in this chapter. Instead, species are identified with **secondary characters**, or **key features** (Fig. 7). This discrepancy is caused by the fact that stages in pleomorphic life cycles of fungi propagate as independent organisms. Key features may be artificial, such as colony diameter, while taxonomic criteria are mostly biologically meaningful, such as sexuality, and indicate the position of a species in the fungal kingdom. In the laboratory, most fungi exhibit their asexual state only; observation of the sexual process usually requires the application of special methods.

The main groups of fungi are roughly distinguished using the simplified key below. A general treatment of fungi sporulating in pure culture is that of von Arx (1981a). The taxonomic position of all fungi described in the Atlas, as far known, is given in Table 1.

Simplified key to the main groups of medically relevant fungi:

1a. Fungus not culturable on routine media; present with cyst-like cells in tissue . . . .
                         ***Mesomycetozoa* (56), *Archiascomycetes* (130)**
1b. Fungus culturable ➡ 2
2a. Loose budding cells abundant ➡ 3
2b. Thallus entirely consisting of filaments ➡ 4
3a. Colonies white or pink . . . . **Yeasts (125)**
3b. Colonies black . . . . **Black yeasts and relatives (374)**
4a. Mycelium regularly septate ➡ 5
4b. Mycelium nearly aseptate ➡ 8
5a. Clamp connections present, at least on some hyphae . . . . **Filamentous basidiomycetes (238)**
5b. Clamp connections absent ➡ 6
6a. Fruit bodies absent . . . . **Hyphomycetes (361)**
6b. Fruit bodies present ➡ 7
7a. Fruit bodies containing spores in asci . . . . ***Euascomycetes* (244)**
7b. Fruit bodies containing loose conidia . . . . **Coelomycetes (312)**
8a. Sporulation abundant . . . . ***Zygomycota* (58)**
8b. Sporulation absent; zoospores formed in water cultures . . . . ***Oomycota* (55)**

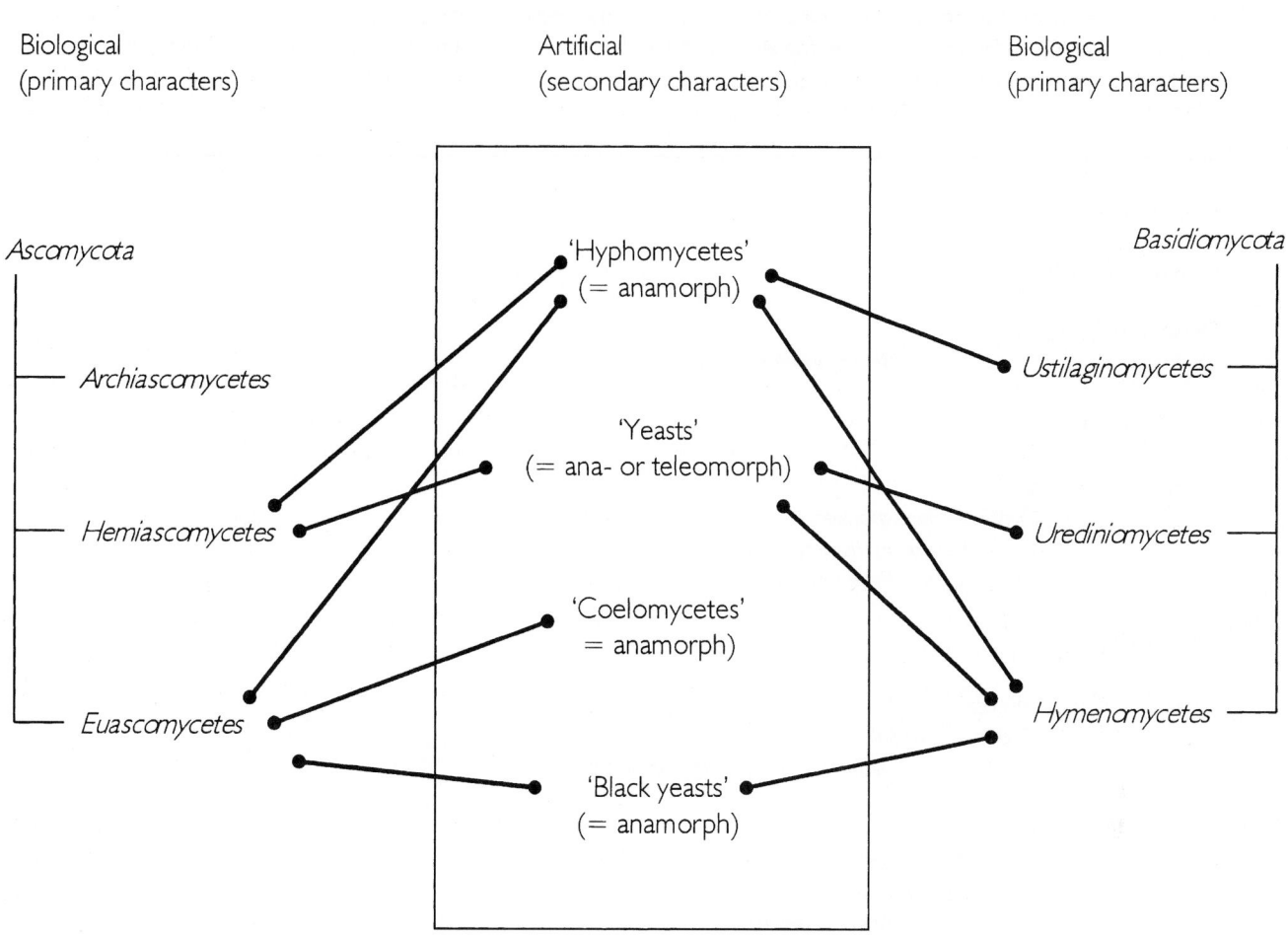

Fig. 7. Generally applied main groups of *Ascomycota* and *Basidiomycota*, being based on primary and secondary characters (biological or artificial).

Table 1. Proven or supposed phylogenetic classification of the clinically relevant fungi, with their recommended BioSafety Level (BSL) classification. Under the *Asco-* and *Basidiomycota,* current teleomorph names are listed at the left hand side; (syn)anamorph names are listed at the right. Note that some positions are highly speculative, as neither teleomorph nor phylogenetic data are currently available.

	BSL
PROTOZOA:	
Division: *Mycetozoa*	
Division: *Mesomycetozoa*	
Rhinosporidium seeberi	2
CHROMISTA:	
Division: *Oomycota*	
Order: *Peronosporales*	
Family: *Pythiaceae*	
Pythium insidiosum	2
EUMYCOTA:	
Division: *Chytridiomycota*	
Order: *Chytridiales*	
Batrachochytrium dendrobatidis	-
Division: *Zygomycota*	
Order: *Mucorales*	
Family: *Mucoraceae*	2
Absidia coerulea	2
Absidia corymbifera	2
Apophysomyces elegans	2
Chlamydoabsidia padenii	1
Mucor amphibiorum	2
Mucor circinelloides	1
Mucor hiemalis	1
Mucor indicus	1
Mucor racemosus	1
Mucor ramosissimus	1
Rhizomucor miehei	1
Rhizomucor pusillus	2
Rhizomucor variabilis	2
Rhizopus azygosporus	2
Rhizopus microsporus	2
Rhizopus oryzae	1
Rhizopus schipperae	2
Rhizopus stolonifer	1
Family: *Thamnidiaceae*	
Cokeromyces recurvatus	1
Family: *Cunninghamellaceae*	
Cunninghamella bertholletiae	2
Family: *Syncephalastraceae*	
Syncephalastrum racemosum	2
Family: *Saksenaeaceae*	
Saksenaea vasiformis	2
Order: *Mortierellales*	
Family: *Mortierellaceae*	
Mortierella polycephala	1
Mortierella wolfii	2

Table 1 (Contd.)   BSL

Order: *Entomophthorales*
  Family: *Basidiobolaceae*
    Basidiobolus ranarum   2
  Family: *Ancylistaceae*
    Conidiobolus coronatus   2
    Conidiobolus incongruus   2
    Conidiobolus lamprauges   I

Division: *Ascomycota*
  Class: *Archiascomycetes*
    Order: *Pneumocystidales*
      Family: *Pneumocystidaceae*
        Pneumocystis carinii   - -   -
  Class: *Hemiascomycetes*
    Order: *Saccharomycetales*
      Family: *Saccharomycetaceae*
        Issatchenkia orientalis   - *Candida krusei*   2
        Arxiozyma telluris   - *Candida pintolopesii*   I
        Saccharomyces cerevisiae   - unnamed   GRAS
      Family: *Endomycetaceae*
        Stephanoascus ciferrii   - *Candida ciferrii*   2
        Debaryomyces hansenii   - *Candida famata*   I
        Pichia guilliermondii   - *Candida guilliermondii*   I
        Hansenula anomala   - *Candida pelliculosa*   I
        Pichia norvegensis   - *Candida norvegensis*   I
        Pichia jadinii   - *Candida utilis*   I
      Family: *Metschnikowiaceae*
        Clavispora lusitaniae   - *Candida lusitaniae*   2
        Metschnikowia pulcherrima   - *Candida pulcherrima*   I
      Family: *Dipodascaceae*
        Galactomyces geotrichum   - *Geotrichum candidum*   I
        Dipodascus capitatus   - *Geotrichum capitatum*   2
        -                              - *Geotrichum clavatum*   2
      Family: *Lipomycetaceae*
        Kluyveromyces marxianus   - *Candida kefyr*   I
      Family: *Ascoideaceae*
        Yarrowia lipolytica   - *Candida lipolytica*   I

  Class: *Euascomycetes*
    Order: *Onygenales*
      Family: *Arthrodermataceae*
        -                              - *Epidermophyton floccosum*   2
        -                              - *Keratinomyces ceretanicus*   I
        Arthroderma borellii   - *Microsporum amazonicum*   2
        -                              - *Microsporum audouinii*   2
        Arthroderma otae   - *Microsporum canis*   2
        Arthroderma cajetani   - *Microsporum cookei*   I
        -                              - *Microsporum ferrugineum*   2
        Arthroderma fulvum   - *Microsporum fulvum*   I
        Arthroderma grubyi   - *Microsporum gallinae*   2
        Arthroderma gypseum   - *Microsporum gypseum*   I
        Arthroderma incurvatum   - *Microsporum gypseum*   I
        Arthroderma obtusum   - *Microsporum nanum*   2
        Arthroderma persicolor   - *Microsporum persicolor*   2
        -                              - *Microsporum praecox*   2
        Arthroderma racemosum   - *Microsporum racemosum*   I
        Arthroderma corniculatum   - *Microsporum* sp.   I

Table 1 (Contd.)   BSL

	Arthroderma uncinatum	- *Trichophyton ajelloi*	1
	-	- *Trichophyton concentricum*	2
	-	- *Trichophyton duboisii*	2
	Arthroderma benhamiae	- *Trichophyton erinacei*	2
	Arthroderma flavescens	- *Trichophyton flavescens*	1
	Arthroderma gloriae	- *Trichophyton gloriae*	1
	-	- *Trichophyton mentagrophytes*	2
	Arthroderma vanbreuseghemii	- *Trichophyton interdigitale*	2
	-	- *Trichophyton phaseoliforme*	1
	-	- *Trichophyton rubrum*	2
	-	- *Trichophyton schoenleinii*	2
	Arthroderma simii	- *Trichophyton simii*	2
	Arthroderma insingulare	- *Trichophyton terrestre*	1
	Arthroderma lenticulare	- *Trichophyton terrestre*	1
	Arthroderma quadrifidum	- *Trichophyton terrestre*	1
	-	- *Trichophyton thuringiense*	1
	-	- *Trichophyton tonsurans*	2
	Arthroderma gertleri	- *Trichophyton vanbreuseghemii*	1
	-	- *Trichophyton verrucosum*	2
	-	- *Trichophyton violaceum*	2
Family: *Onygenaceae*			
	Ajellomyces dermatitidis	- *Blastomyces dermatitidis*	3
	-	- *Chrysosporium inops*	1
	Aphanoascus keratinophilus	- *Chrysosporium keratinophilum*	1
	-	- *Chrysosporium pannicola*	1
	Aphanoascus fulvescens	- *Chrysosporium* sp.	2
	Nannizziopsis vriesii	- *Chrysosporium* sp.	1
	Uncinocarpus queenslandicus	- *Chrysosporium queenslandicum*	1
	-	- *Chrysosporium tropicum*	2
	Uncinocarpus orissi	- *Chrysosporium zonatum*	1
	-	- *Coccidioides immitis*	3
	Ajellomyces crescens	- *Emmonsia crescens*	2
	-	- *Emmonsia parva*	2
	-	- *Emmonsia pasteuriana*	2
	Ajellomyces capsulatus	- *Histoplasma capsulatum*	3
	Neoarachnotheca keratinophila	- *Myriodontium keratinophilum*	1
	-	- *Paracoccidioides brasiliensis*	3
Family: *Gymnoascaceae*			
	Narasimhella hyalinospora	- -	1
	Gymnoascus dankaliensis	- -	2
	-	- *Malbranchea pulchella*	1
	-	- *Ovadendron sulphureo-ochraceum*	1
	Arachnomyces nodosetosus	- *Onychocola canadensis*	2
Order: *Eurotiales*			
Family: *Eremomycetaceae*			
	Eremomyces langeronii	- *Arthrographis kalrae*	2
Family: *Monascaceae*			
	Monascus ruber	- *Basipetospora rubra*	1
Family: *Pseudeurotiaceae*			
	Pseudeurotium ovale	- *Sporothrix* sp.	1
Family: *Thermoascaceae*			
	Thermoascus crustaceus	- *Paecilomyces crustaceus*	1
Family: *Trichocomaceae*			
	Petromyces alliaceus	- *Aspergillus alliaceus*	1
	-	- *Aspergillus avenaceus*	1
	-	- *Aspergillus caesiellus*	1
	-	- *Aspergillus candidus*	1

Table 1 (Contd.) BSL

		BSL
-	- *Aspergillus carneus*	I
Eurotium chevalieri	- *Aspergillus chevalieri*	I
-	- *Aspergillus clavato-nanicus*	I
-	- *Aspergillus clavatus*	I
-	- *Aspergillus conicus*	I
-	- *Aspergillus deflectus*	I
Neosartorya fischeri	- *Aspergillus fischerianus*	I
Fennellia flavipes	- *Aspergillus flavipes*	I
-	- *Aspergillus flavus*	2
-	- *Aspergillus fumigatus*	2
-	- *Aspergillus granulosus*	I
Eurotium herbariorum	- *Aspergillus glaucus*	I
Eurotium amstelodami	- *Aspergillus hollandicus*	I
-	- *Aspergillus janus*	I
-	- *Aspergillus japonicus*	I
Emericella nidulans	- *Aspergillus nidulans*	I
-	- *Aspergillus niger*	I
Fennellia nivea	- *Aspergillus niveus*	I
-	- *Aspergillus ochraceus*	I
-	- *Aspergillus oryzae*	I
Eurotium repens	- *Aspergillus reptans*	I
-	- *Aspergillus restrictus*	I
-	- *Aspergillus sclerotiorum*	I
-	- *Aspergillus sydowii*	I
-	- *Aspergillus tamarii*	I
-	- *Aspergillus terreus*	2
Emericella quadrilineata	- *Aspergillus tetrazonus*	I
Neosartorya pseudofischeri	- *Aspergillus thermomutatus*	2
Neosartorya spinosa	- *Aspergillus spinosus*	I
Emericella unguis	- *Aspergillus unguis*	I
-	- *Aspergillus ustus*	I
-	- *Aspergillus versicolor*	I
-	- *Paecilomyces lilacinus*	I
-	- *Paecilomyces marquandii*	I
-	- *Paecilomyces puntonii*	I
-	- *Paecilomyces variotii*	2
-	- *Paecilomyces viridis*	I
-	- *Penicillium aurantiogriseum*	I
-	- *Penicillium brevicompactum*	I
-	- *Penicillium chrysogenum*	I
-	- *Penicillium citrinum*	I
-	- *Penicillium commune*	I
-	- *Penicillium decumbens*	I
-	- *Penicillium expansum*	I
-	- *Penicillium griseofulvum*	I
-	- *Penicillium marneffei*	3
-	- *Penicillium piceum*	I
-	- *Penicillium purpurogenum*	I
-	- *Penicillium rugulosum*	I
-	- *Penicillium spinulosum*	I
-	- *Penicillium verruculosum*	I
-	- *Polypaecilum insolutum*	I
Order: *Chaetothyriales*		
Family: *Herpotrichiellaceae*		
-	- *Anthopsis deltoidea*	I
-	- *Cladophialophora arxii*	2
-	- *Cladophialophora bantiana*	3

15

Table 1 (Contd.)

			BSL
	-	- *Cladophialophora boppii*	2
	-	- *Cladophialophora carrionii*	2
	-	- *Cladophialophora devriesii*	2
	-	- *Cladophialophora emmonsii*	2
	-	- *Cladophialophora modesta*	2
	-	- *Exophiala bergeri*	2
	-	- *Exophiala castellanii*	2
	-	- *Exophiala dermatitidis*	2
	-	- *Exophiala jeanselmei*	2
	-	- *Exophiala lecanii-corni*	2
	-	- *Exophiala moniliae*	2
	-	- *Exophiala pisciphila*	I
	-	- *Exophiala salmonis*	I
	-	- *Exophiala spinifera*	2
	-	- *Fonsecaea compacta*	2
	-	- *Fonsecaea pedrosoi*	2
	Capronia semiimmersa	- *Phialophora americana*	2
	-	- *Phialophora bubakii*	I
	-	- *Phialophora europaea*	2
	-	- *Phialophora reptans*	I
	-	- *Phialophora repens*	I
	-	- *Phialophora richardsiae*	2
	-	- *Phialophora verrucosa*	2
	-	- *Ramichloridium mackenziei*	3
	-	- *Ramichloridium schulzeri*	I
	-	- *Rhinocladiella aquaspersa*	2
	-	- *Rhinocladiella atrovirens*	I
	-	- *Sarcinomyces phaeomuriformis*	2
Order: *Clavicipitales*			
	-	- *Beauveria bassiana*	I
	-	- *Engyodontium album*	I
	-	- *Paecilomyces fumosoroseus*	I
	-	- *Paecilomyces javanicus*	I
Order: *Microascales*			
Family: *Microascaceae*			
	Microascus cinereus	- *Scopulariopsis cinereus*	I
	Microascus cirrosus	- *Scopulariopsis paisii*	I
	Microascus manginii	- *Scopulariopsis candida*	I
	-	- *Scopulariopsis acremonium*	I
	-	- *Scopulariopsis asperula*	I
	Microascus brevicaulis	- *Scopulariopsis brevicaulis*	2
	-	- *Scopulariopsis brumptii*	2
	-	- *Scopulariopsis flava*	I
	-	- *Scopulariopsis fusca*	I
	-	- *Scopulariopsis koningii*	I
	Petriella setifera	- *Scedosporium* sp.	I
	Petriella setifera	- *Graphium* sp.	I
	-	- *Scedosporium prolificans*	2
	-	- *Graphium putredinis*	I
	Pseudallescheria boydii	- *Scedosporium apiospermum*	2
	Pseudallescheria boydii	- *Graphium eumorphum*	2
Order: *Ophiostomatales*			
Family: *Ophiostomataceae*			
	Ophiostoma stenoceras	- *Sporothrix* sp.	I
	-	- *Sporothrix schenckii*	2
Order: *Dothideales*			
Family: *Didymosphaeriaceae*			
	Neotestudina rosatii	- -	2

16

Table 1 (Contd.) BSL

			BSL
Family: *Dothioraceae*			
Discosphaerina fulvida	-	*Aureobasidium pullulans*	1
-	-	*Cyphellophora laciniata*	1
-	-	*Cyphellophora pluriseptata*	1
Sydowia polyspora	-	*Hormonema dematioides*	1
-	-	*Hortaea werneckii*	1
-	-	*Nattrassia mangiferae*	2
-	-	*Pseudomicrodochium suttonii*	2
Famiily: *Botryosphaeriaceae*			
Botryosphaeria rhodina	-	*Lasiodiplodia theobromae*	1
Botryosphaeria subglobosa	-	*Sphaeropsis subglobosa*	1
Family: *Mycosphaerellaceae*			
-	-	*Cladosporium cladosporioides*	1
-	-	*Cladosporium elatum*	1
Mycosphaerella tassiana	-	*Cladosporium herbarum*	1
-	-	*Cladosporium oxysporum*	1
-	-	*Cladosporium sphaerospermum*	1
Family: *Piedraiaceae*			
Piedraia hortai	-	-	1
Family: *Lophiostomataceae*			
-	-	*Madurella grisea*	2
-	-	*Madurella mycetomatis*	2
-	-	*Pseudochaetosphaeronema larense*	2
-	-	*Pyrenochaeta mackinnonii*	2
-	-	*Pyrenochaeta romeroi*	2
-	-	*Pyrenochaeta unguis-hominis*	2
Order: *Pleosporales*			
Family: *Leptosphaeriaceae*			
Leptosphaeria coniothyrium	-	*Coniothyrium fuckelii*	1
Leptosphaeria senegalensis	-	-	1
Leptosphaeria thompkinsii	-	-	1
-	-	*Microsphaeropsis olivacea*	1
Family: *Pleosporaceae*			
Cochliobolus australiensis	-	*Bipolaris australiensis*	1
Cochliobolus hawaiiensis	-	*Bipolaris hawaiiensis*	1
-	-	*Bipolaris papendorfii*	1
Cochliobolus spiciferus	-	*Bipolaris spicifera*	1
-	-	*Alternaria alternata*	1
-	-	*Alternaria chlamydospora*	1
-	-	*Alternaria dianthicola*	1
Lewia infectoria	-	*Alternaria infectoria*	1
-	-	*Alternaria longipes*	1
-	-	*Alternaria tenuissima*	1
-	-	*Botryomyces caespitosus*	2
-	-	*Corynespora cassiicola*	1
Cochliobolus geniculatus	-	*Curvularia geniculata*	1
-	-	*Curvularia brachyspora*	1
-	-	*Curvularia clavata*	1
Cochliobolus lunatus	-	*Curvularia lunata*	1
Cochliobolus pallescens	-	*Curvularia pallescens*	1
-	-	*Curvularia senegalensis*	1
Cochliobolus verruculosus	-	*Curvularia verruculosa*	1
-	-	*Dichotomophthora portulacae*	1
-	-	*Dichotomophthoropsis nymphaearum*	1
-	-	*Dissitimurus exedrus*	2
-	-	*Drechslera biseptata*	1

Table 1 (Contd.)   BSL

	-	- *Exserohilum longirostratum*	1
		- *Exserohilum mcginnisii*	1
	Setosphaeria rostrata	- *Exerohilum rostratum*	1
	-	- *Mycocentrospora acerina*	1
	-	- *Papulaspora equi*	1
	-	- *Phaeosclera dematioides*	1
	-	- *Phaeotrichoconis crotalariae*	1
	-	- *Phoma cruris-hominis*	2
	-	- *Phoma dennissii* v. *oculo-hominis*	2
	-	- *Phoma eupyrena*	1
	-	- *Phoma glomerata*	1
	-	- *Phoma herbarum*	1
	-	- *Phoma minutella*	1
	-	- *Phoma minutispora*	1
	-	- *Phoma sorghina*	1
	-	- *Polycytella hominis*	2
	-	- *Ulocladium botrytis*	1
	-	- *Ulocladium chartarum*	1

Order: *Polystigmatales*
    Family: *Polystigmataceae*

	-	- *Colletotrichum coccodes*	1
	-	- *Colletotrichum dematium*	1
	Glomerella cingulata	- *Colletotrichum gloeosporioides*	2
	Glomerella tucumanensis	- *Colletotrichum graminicola*	1

    Family: *Phyllachoraceae*
       Plectosphaerella cucumerina - *Plectosporium tabacinum*   1

Order: *Sordariales*
    Family: *Lasiosphaeriaceae*
       Arnium leporinum

	Arnium leporinum	- -	1
	-	- *Arthrinium phaeospermum*	1
		- *Nigrospora sphaerica*	1

    Family: *Chaetomiaceae*

	Thielavia terrestris	- *Acremonium alabamense*	1
	Achaetomium strumarium	- -	1
	Ascotricha chartarum	- *Dicyma ampullifera*	1
	Chaetomium atrobrunneum	- -	1
	Chaetomium funicola	- -	1
	Chaetomium globosum	- -	1
	Chaetomium murorum	- -	1
	Corynascus heterothallicus	- *Myceliophthora thermophila*	1
	-	- *Phaeoisaria clematidis*	1
	-	- *Staphylotrichum coccosporum*	1

    Family: *Coniochaetaceae*

	-	- *Acrophialophora fusispora*	1
	Coniochaeta ligniaria	- *Lecythophora hoffmannii*	1
	-	- *Lecythophora mutabilis*	1
	-	- *Mycoleptodiscus indicus*	1
	-	- *Phaeoacremonium inflatipes*	1
	-	- *Phaeoacremonium parasiticum*	2
	-	- *Phaeoacremonium rubrigenum*	2
	-	- *Phialemonium curvatum*	2
	-	- *Phialemonium obovatum*	2

    Family: *Myxotrichaceae*

	Myxotrichum deflexum	- -	1
	-	- *Oidiodendron cerealis*	1
	-	- *Geomyces pannorum*	1

    Family: *Sordariaceae*
       Neurospora sitophila   - *Chrysonilia sitophila*   1

Table 1 (Contd.)   BSL

Order: *Leotiales*
  Family: *Leotiaceae*
                    -                          - *Pleurophoma cava*   1
                    -                          - *Pleurophoma pleurospora*   1
                    -                          - *Pleurophomopsis lignicola*   1
                    -                          - *Ochroconis constricta*   1
                    -                          - *Ochroconis gallopava*   2
                    -                          - *Ochroconis humicola*   2
                    -                          - *Ochroconis tshawytschae*   2
Order: Pezizales
  Family: *Ascodesmiaceae*
                    -                          - *Cephaliophora irregularis*   1
Order: *Hypocreales*
  Family: *Hypocreaceae*
    Neocosmospora vasinfecta   - *Acremonium* sp.   1
                    -                          - *Acremonium atrogriseum*   1
                    -                          - *Acremonium blochii*   1
                    -                          - *Acremonium curvulum*   1
                    -                          - *Acremonium falciforme*   2
                    -                          - *Acremonium hyalinulum*   1
                    -                          - *Acremonium kiliense*   2
                    -                          - *Acremonium potronii*   1
                    -                          - *Acremonium recifei*   2
                    -                          - *Acremonium roseogriseum*   1
                    -                          - *Acremonium strictum*   1
                    -                          - *Acremonium spinosum*   1
                    -                          - *Cylindrocarpon cyanescens*   2
    Nectria radicicola   - *Cylindrocarpon destructans*   1
                    -                          - *Cylindrocarpon lichenicola*   1
                    -                          - *Fusarium antophilum*   1
    Cosmospora episphaerica   - *Fusarium aquaeductuum*   1
                    -                          - *Fusarium chlamydosporum*   1
                    -                          - *Fusarium dimerum*   1
                    -                          - *Fusarium incarnatum*   1
                    -                          - *Fusarium napiforme*   1
    Gibberella nygamai   - *Fusarium nygamai*   1
                    -                          - *Fusarium oxysporum*   2
    Gibberella fujikuroi   - *Fusarium proliferatum*   1
    Gibberella fujikuroi v. *subglutinans*   - *Fusarium subglutinans*   1
    Nectria haematococca v. *breviconia*   - *Fusarium solani*   2
    Gibberella moniliformis   - *Fusarium verticillioides*   2
                    -                          - *Metarhizium anisopliae*   1
                    -                          - *Trichoderma harzianum*   1
    Hypocrea koningii   - *Trichoderma koningii*   1
                    -                          - *Trichoderma longibrachiatum*   1
    Hypocrea pseudokoningii   - *Trichoderma pseudokoningii*   1
    Hypocrea rufa   - *Trichoderma viride*   1

Division: *Basidiomycota*
  Class: *Ustilaginomycetes*
    Order: *Microstromatales*
                    -                          - *Cerinosterus cyanescens*   2
    Order: *Tilletiales*
                    -                          - *Tilletiopsis minor*   1

  Class: *Urediniomycetes*
    Order: *Sporidiales*

Family: *Sporidiobolaceae*

		BSL
Rhodosporidium diobovatum	- *Rhodotorula glutinis*	1
Rhodosporidium sphaerocarpum	- *Rhodotorula glutinis*	1
Rhodosporidium toruloides	- *Rhodotorula glutinis*	1
-	- *Rhodotorula minuta*	1
-	- *Rhodotorula mucilaginosa*	1
Sporidiobolus johnstonii	- *Sporobolomyces salmonicolor*	1

Class: *Hymenomycetes*
     Order: *Tremellales*
         Family: *Filobasidiaceae*

		BSL
-	- *Cryptococcus albidus*	1
-	- *Cryptococcus ater*	1
-	- *Cryptococcus curvatus*	2
-	- *Cryptococcus humicola*	1
-	- *Cryptococcus laurentii*	1
-	- *Cryptococcus macerans*	1
Filobasidiella neoformans	- *Cryptococus neoformans* v. *grubii*	2/3 *
Filobasidiella neoformans	- *Cryptococcus neoformans* v. *neoformans*	2/3 *
Filobasidiella bacillispora	- *Cryptococcus neoformans* v. *gattii*	2/3*
Filobasidium uniguttulatum	- *Cryptococcus uniguttulatus*	1
-	- *Malassezia furfur*	2
-	- *Malassezia globosa*	2
-	- *Malassezia obtusa*	2
-	- *Malassezia pachydermatis*	2
-	- *Malassezia restricta*	2
-	- *Malassezia slooffiae*	2
-	- *Malassezia sympodialis*	2
-	- *Moniliella suaveolens*	1
-	- *Trichosporon asahii*	2
-	- *Trichosporon asteroides*	2
-	- *Trichosporon cutaneum*	2
-	- *Trichosporon inkin*	2
-	- *Trichosporon mucoides*	2
-	- *Trichosporon ovoides*	2

     Order: *Stereales*
         Family: *Corticiaceae*

		BSL
Phanerochaete chrysosporium	- *Sporotrichum pruinosum*	1

         Family: *Bjerkanderaceae*

		BSL
Bjerkandera adusta	- -	1

         Family: *Schizophyllaceae*

		BSL
Schizophyllum commune	- -	1

     Order: *Agaricales*
         Family: *Coprinaceae*

		BSL
Coprinus cinereus	- *Hormographiella aspergillata*	1
-	- *Hormographiella verticillata*	1

*Clinical strains are yeast-like and are classified as BSL-2. For mating experiments, where basidiospores are generated, safety conditions at BSL-3 level are recommended.

CLINICAL PATHOLOGY

Traditionally four main categories of fungal infections, based on localization, are recognized, viz. superficial colonization (without tissue response), (muco)cutaneous infection (skin, eyes, sinuses, oropharynx, external ears, vagina), subcutaneous infection and deep infection, the latter being either localized or disseminated. Clinical pictures are treated in most general handbooks on medical mycology, of which Rippon (1988), Kwon-Chung & Bennett (1992), and the condensed overviews Richardson & Warnock (1997) and Richardson & Johnson (2000) are particularly recommended. Nomenclature of mycotic diseases is used largely according to Odds *et al.* (1992). In recent times we have witnessed an increase of opportunistic, systemic mycoses in immunocompromised patients. The patients may have been impaired as a result of the development of neoplasia, leukemia, metabolic diseases (diabetes mellitus), drug therapy (cytostatics, antibiotics, corticosteroids), intravenous drug abuse or infections (AIDS). Histopathologic responses to fungal invasion can differ considerably, depending on the species, and on the site and duration of the infection (McKenzie, 1988; Table 2). For reviews of histopathology see Chandler *et al.* (1980) and Salfelder (1990).

SUPERFICIAL MYCOSES

This category concerns fungi which living on compounds available on the human or animal body, skin surface, such as hair or lipids, without provoking any immune response.

1. **Piedra.** This is a colonization of the hair shaft, characterized by the presence of superficial nodules, leaving the hairs largely undamaged. The fungal elements are cemented together and firmly attached to the hair. Hard, black nodules are formed in the case of **black piedra**, caused by *Piedraia hortae*, or soft, white nodules in

Table 2. Principal histopathological responses to subcutaneous and systemic mycotic invasion.

	purulence	granuloma	fibrosis	necrosis	other
phaeohyphomycosis	–	+	–	–,w	
mycetoma	+	+	+	+	sinus tracts, grains
chromoblastomycosis	–	+	+	–	hyperkeratosis, acanthosis
sporotrichosis	+	+	+	+	asteroid bodies (rare)
aspergillosis	+	+	–	+	
blastomycosis	+	+	+	–	
paracoccidioidomycosis	+	+	+	+	
coccidioidomycosis	+	+	+	–	
histoplasmosis	–	+	+	–	calcification, coin lesions
zygomycosis	+	–	–	+	oedema
entomophthoromycosis	–	+	+	–	oedema
candidiasis	+	+	–	+	
cryptococcosis	–	+	–	–	

white piedra, caused by *Trichosporon* species. In white piedra, several causative species have a characteristic localisation, i.e., on capital, axillary or crural hairs.

2. **Tinea nigra.** Brown or black, superficial skin lesions, mostly on the palm of the hand, sometimes on the sole. Pigmentation is most intense near the border. The fungus has a predilection for affecting females, persons under 20 years of age, and hyperhydrotic individuals. Probably hydrophobic adhesion of an osmophilic fungus is concerned. Causative agent: *Hortaea werneckii.*

3. **Pityriasis versicolor.** Yeasts of the genus *Malassezia* are common saprobes on the skin in lipid excretions. Overabundant growth of the fungus may lead to seborrhoeic dermatitis, characterized by redness of the skin, itching and desquamation, caused in part by a hypersensitivity-reaction to the yeast. Colonization of the stratum corneum can also lead to local, confluent lesions, mostly on the chest and back. Young lesions are hypopigmented, older ones become brown and show more intense scaling. In more severe forms, e.g., in immunocompromised patients, folliculitis may be noted.

4. **Otitis externa.** This is the result of overabundant colonization of the external ear by *Aspergillus* (mostly *A. niger* or *A. fumigatus*), *Malassezia* spp. or *Pseudallescheria boydii*, generally promoted by an earlier, bacterial disorder or by local use of broad spectrum antibiotics. A coexisting hypersensitivity reaction is suspected.

5. **Superficial dermatophytoses of hair.** Various types of hair, nail and adjacent skin infection are known to be caused by keratinophilic dermatophytes. A distinction is generally made between colonization of dead parts of skin and of hairs on the one hand, and growth in cutaneous skin layers on the other.

a. **Ectothrix.** Arthroconidia are formed on the outside of hair shafts. The cuticle of the hair is damaged.

b. **Endothrix.** Arthroconidia (5-8 μm diam) are formed inside the hair. The cuticle remains intact; the hair may burst, break, crumble and curl. This leads to black dots. Main causative agents: *Trichophyton tonsurans, T. violaceum.*

CUTANEOUS MYCOSES

This category contains those disorders in which living parts of outermost skin layers (epidermis, stratum spinosum, stratum corneum), mucocutaneous membranes, genitalia or external ears are affected by fungi. Also adjacent dead skin or hairs may be involved.

Dermatophytoses of skin

1. **Ringworm.** Skin lesions are characterized by circular lesions with slightly raised, red margins, containing numerous scales and surrounded by reddish, itching skin. The fungus remains restricted to the stratum corneum. As a result of keratinolytic activity, metabolites are produced, which provoke inflammation. The horny layer is deteriorated and becomes scaly. The affected areas expand gradually and tissue may heal at the centre. Classification of ringworm lesions is according to their localisation on the human body:

a. **Tinea pedis.** Dermatophyte infection of the feet and the interdigital spaces of the toes. The macerated, moist skin between toes is infected, the horny layer is damaged, and painful fissures may develop. Main causative agents: *Trichophyton rubrum, Epidermophyton floccosum.*

b. **Tinea manuum.** Small lesions between the fingers later may affect the whole hand. The fungus affects the epidermis and causes erythema, which is particularly noted in the wrinkles of the palm. The skin becomes scaly. Main causative agent: *Trichophyton rubrum.*

c. **Tinea corporis.** Lesions on exposed, hairless skin, e.g., on the face, arms or trunk, occurring particularly in children. Usually such infections have been transferred from animals to humans. Main causative agents: *Microsporum canis, Trichophyton interdigitale, T. verrucosum, T. tonsurans, T. violaceum.*

d. **Tinea imbricata.** This is a special form of tinea corporis. It is common in Polynesia, and is characterized by regularly concentric rings on the skin. Causative agent: *Trichophyton concentricum.*

e. **Tinea cruris.** Pruritic lesions on hairy skin around the genitalia. Secondary bacterial involvement is not uncommon. Causative agents: *Epidermophyton floccosum, Trichophyton interdigitale, T. rubrum.*

f. **Tinea barbae.** Occurs on hairy parts of the face, often provoking a granulocytous reaction. Causative agents: *Trichophyton rubrum, T. verrucosum.*

g. **Tinea capitis.** Occurs on the scalp and eyebrows. Hair invasion is of the endothrix, ectothrix or favus (with crusty lesions) type, as described above. Metabolite production during hair infection causes more severe, inflammatory reactions than on hairless skin. Suppurating reactions (folliculitis) are common. This is seen with *Trichophyton verrucosum* and *Microsporum canis* in particular. Epilation may be necessary. Causative agents: *Trichophyton verrucosum* (from cattle), *T. mentagrophytes* (from horses or dogs), *T. tonsurans, T. schoenleinii, T. violaceum, Microsporum audouinii, M. canis* (from cats / dogs), *M. ferrugineum.*

2. **Favus.** Hyphae create air spaces in the hair during decomposition. The hair does not break, but is lost

completely and permanently due to degeneration of the follicle. Subsequent loss of hair causes a sharply delimited bald spot on the haired skin. The adjacent skin is invaded, crusty and scarred. Causative agent: *Trichophyton schoenleinii*.

3. **Onychomycosis.** This is a chronic infection of the nail bed and nail, most commonly toenails, which expands from the periphery towards the centre and through which the nail may eventually be released. Lateral hyperkeratosis may lift the nail distally. The distal part gradually breaks and crumbles. Adjacent nails usually remain unaffected. Main causative agents: *Trichophyton rubrum*, *T. interdigitale*.

Non-dermatophyte cutaneous infections

1. **Onychomycosis.** Described above, may also be caused by non-dermatophytes such as *Scopulariopsis brevicaulis*. *Candida* species may occur on finger nails, with inflammation of the surrounding skin (paronychium).

2. **Hyperkeratosis.** Extended scaling areas on hands and feet may be caused by the non-dermatophyte fungus *Nattrassia mangiferae*.

3. **Intertrigo.** A fungus or yeast infection of folds of the skin, strongly associated with humid maceration. Often a bacterial secondary infection is noted with Gram-positive microorganisms. The skin shows a vaguely delimited, moist, reddish, thickened area with necrotic skin. Caused by *Candida* species, mostly *C. albicans*. A special form is **perleche**, an infection of the corner of the mouth, mostly in elderly people.

4. **Mucocutaneous candidiasis.** The yeast *Candida albicans* is an extremely common colonizer of mucous membranes. Invasion is promoted by slight or severe decrease of immunocompetence. The following main clinical pictures can be distinguished:
a. **Thrush.** This is an abundant growth of *Candida* species, mostly *C. albicans*, in the oral cavity. A whitish, removable layer covers the reddish, eroded, easily bleeding mucosa. The disorder may extend into the oesophagus and is strongly connected with prolonged use of broad spectrum antibiotics and impaired T-cell immunity. It is one of the first indications of a development of AIDS.
b. **Vulvo-vaginitis.** Caused by *Candida* species, mostly *C. albicans*. A whitish, dry and crumbling fluor is produced with a sweetish smell. The yeast is easily observed in KOH-preparations. Often erythema or irritation occurs, mostly noted on the skin of the vulva, and occasionally with dysuria caused by urethritis. Invasion is promoted by temporal changes in hormone balance, e.g.,

during pregnancy, periods and with use of birth control pills. The comparable infection on the glans of the penis is known as **balanitis**.

5. **Keratitis.** Colonization or infiltration of corneal epithelium after trauma, surgery, use of corticosteroids, or careless application of contact lenses. In addition, many eye drops contain antimicrobial preservatives, and thus may stimulate fungal growth. Affected eyes usually show ulceration and scarring. Diagnosis of fungal involvement is often hampered by the difficulty of taking adequate clinical samples. Mostly *Aspergillus* and *Fusarium* species are concerned, but many otherwise saprobic fungi have been reported as causative agents.

6. **Chromoblastomycosis.** The disease occurs mostly on the extremities and is characterized by localized, slowly expanding lesions. Superficial, warty to cauliflower-like tumours and deformations develop, due to hyperkeratosis and acanthosis. The lesions are greyish, crusted and dry in appearance. Lymphogenous metastasis may occur. The fungus occurs as dark, muriform cells in the tissue. Causative agents: *Fonsecaea compacta*, *F. pedrosoi*, *Cladophialophora carrionii*, *Phialophora verrucosa*, *Rhinocladiella aquaspersa*.

SUBCUTANEOUS MYCOSES

Local, traumatic, primary infection of the subcutis, provoking a leukocytic or eosinophilic response and leading to cysts or granuloma. Most of these are chronic, sometimes disfiguring infections which may last for decades without being life-threatening. They should not be confused with secondary skin infections caused by systemic fungi treated below.

Non-ulcerative infections by diverse fungi

These concern local, non-ulcerative infections caused by a variety of fungi. The fungus is present in the form of septate hyphae or hyphal elements. In the case of **hyphomycotic cysts** the affected area is surrounded by a fibrous, collagenous secretion of the host, and hence no inflammation occurs. According to the appearance of the causative agent as seen in native preparations, the following subdivision is made:
a. **Hyalohyphomycosis.** The fungal elements are colourless.
b. **Phaeohyphomycosis.** The fungal elements are melanised (special staining often necessary).

Eumycetoma

Necrotic, suppurative lesions are formed with cavities in tissue, draining through multiple fistulae to the surface of the skin. The exudate contains compact grains, which are hyphal microcolonies. The morphology of the grains is more or less characteristic for the species. The affected area becomes swollen by oedema and granulation. In final stages the bones are severely affected. The infection is mostly found in one of the lower extremities. The disease is common in tropical regions. According to the appearance of the grains as seen in native preparations, the following subdivision is made:
a. **White-grain mycetoma.** The grains are colourless. Main causative agents: *Acremonium* spp., *Neotestudina rosatii, Pseudallescheria boydii.*
b. **Black-grain mycetoma.** The grains are black. Main causative agents: *Exophiala jeanselmei* and species of *Madurella, Leptosphaeria* and *Pyrenochaeta.*

Sporotrichosis

An infection caused by *Sporothrix schenckii* after traumatic implantation into the skin. In the classical form, a small, asymptomatic swelling develops into an ulcer with purulent excretion. The margin of the ulcer is red and raised. Subsequently, characteristic series of lesions involving the lymph nodes develop through lymphogenous spread of fungal cells. Rarely primary pulmoneous sporotrichosis is observed. Four types of sporotrichosis are distinguished (Table 3).

Primary subcutaneous blastomycosis

A rare chronic, granulomatous inflammation of the skin with ulceration, caused by *Blastomyces dermatitidis.* The disorder may disseminate; see under systemic mycoses.

Lobomycosis

This is a chronic, elastic granuloma, sometimes with ulceration. The disorder occurs in the northern parts of South America. The causative agent, *Lacazia loboi,* may be a fungus, but has thus far not been cultured.

Subcutaneous conidiobolomycosis

A disease characterized by a slow, progressive infection of the subcutis, with granulated, swollen, woody hard, lens-shaped tissue which is fixed to the muscles and not to skin and bones. The cellular reaction mainly concerns eosinophilic leukocytes. Mostly legs, arms and buttocks are involved. Later the thorax, abdomen, neck and face can be infected. The portal of entry is traumatic inoculation. It is caused by *Conidiobolus* and *Basidiobolus* species.

Table 3. Clinical types of sporotrichosis.

1. primary cutaneous after trauma

        - immunocompetent: cutaneo-lymphatic
        - immunocompromised: chronic progressive

2. primary pulmonary after inhalation

        - immunocompetent: asymptomatic, hypersensitivity
        - immunocompromised: pulmonary dissemination

3. respiratory reinfection in hypersensitized persons: minor, transient lesions

4. secondary sporotrichosis: haematogenous dissemination

        - immunocompromised patients and chronic alcoholics

Paranasal conidiobolomycosis

This is an infection of the submucosa of the paranasal sinuses caused by *Conidiobolus coronatus.* Progression is slow but severe, with formation of granulated tissue with particularly eosinophilic leukocytes and fibrosis. Blood-vessels and bone are not invaded but growth is *per continuitatum.*

Rhinosporidiosis

This is an infection of primarily the nasal mucosa caused by *Rhinosporidium seeberi*. The organism has thus far not been cultured. A granulomatous inflammation is seen with the formation of polyps.

Zygomycotic rhinitis

The mucosa gains a greyish-black appearance, resembling clotted blood. As the fungus invades through the arteries it causes thrombosis. Cerebral involvement leads to infarction of the brain and meningitis. Characteristically unilateral blindness through loss of function of 5th and 7th cranial nerves is noted. The most common agents are *Rhizopus* species.

Other types of mycotic sinusitis

Acute infection of the nasal mucosa with allergic reactions to local colonization. It is caused by normally non-pathogenic fungi, mostly *Mucorales* or *Aspergillus* species, are found primarily in debilitated hosts. Sinusitis can also be caused by plant pathogenic *Curvularia, Bipolaris* and *Exserohilum* species and is then particularly found in cattle. Humans infected by the same fungi present with recurrent nasal polyps and often develop chronic, allergic rhinitis. In immunocompromised patients this may lead to invasion and fatal cerebral involvement.

DEEP MYCOSES

Deep mycoses in immunosuppressed patients are acquired either by inhalation, entering internal organs via the pulmonary route, through paranasal sinuses, by enteric invasion, through colonization and mucosal invasion. They can also emerge as a complication of traumatic inoculation, e.g., major surgery, catheter-related infection or after use of contaminated inoculation needles. The mycosis may remain localized in deep tissue and organs (**systemic mycosis**) or may spread (**disseminated mycosis**) through the blood vessels (**sepsis**) or the lymphatic system. In the latter case, frequently dissemination to the skin takes place (**secondary cutaneous mycosis**). Such overall fungal dissemination often indicates that the effectiveness of the immune system has decreased dramatically.

Systemic mycoses

a. *Onygenales*

The following mycoses are caused by a group of closely related fungi, all belonging to the order *Onygenales*. Most of them are restricted to particular endemic areas. Patients who have been in contact with the organism mostly do not develop clinical symptoms, but develop a positive skin reaction within 14 d after exposure. Initiation of infection is through inhalation of fungal propagules. Acute reactions are either asymptomatic or symptomatic pneumonia, which may heal or develop into chronic disease. Endogenous reactivation may be observed many years after exposure. Occasionally chronic disease may lead to extrapulmonary dissemination. Events are summarized in Fig. 8.

1. **Blastomycosis.** This is a granulomatous disease, caused by *Blastomyces dermatitidis*. It mostly originates as a pulmonary infection and disseminates via the blood, initially leading to osseous and cutaneous involvement. Later the nervous system and visceral organs are affected. The gastro-intestinal tract is not involved. If untreated, the disease is usually fatal.

2. **Paracoccidioidomycosis.** Caused by *Paracoccidioides brasiliensis*. This chronic disease starts in the lungs with the fungus disseminating to the mucosal areas of the mouth and later often causing painful, erosive stomatitis with loss of teeth. The fungus has a predilection for mucosa. Oronasal lesions mostly soon develop and massive enlargement of the lymph nodes in the neck is an early sign of the presence of the disease. Later, spleen, intestines and liver are involved. Ulcerative skin lesions may finally develop, particularly in the face. When untreated, the disease is nearly always fatal.

3. **Coccidioidomycosis.** Primary infection is through inhalation of airborne propagules of *Coccidioides immitis*, leading to non-specific fever with broncho-pneumonia. Further, erythema and conjunctivitis may occur; mostly the disease resolves spontaneously. Dissemination to the skin, bones, joints, subcutaneous tissue and visceral organs may occur through haematogenous spread of endoconidia. In that case the disease is often fatal.

4. **Adiaspiromycosis.** Adiaspores are liberated conidia of *Emmonsia* species, which after inhalation enlarge in the alveoli of the host, thus hindering normal pulmonary functioning. The adiaspore remains at its implantation site and apparently does not reproduce, eventually becoming calcified. Only a slight host reaction is noted. It is common in rodents but extremely rare in humans.

5. **Histoplasmosis.** The species *Histoplasma capsulatum* causes an intracellular mycosis of the monocyte-macrophage system. Characteristically, monocytes or

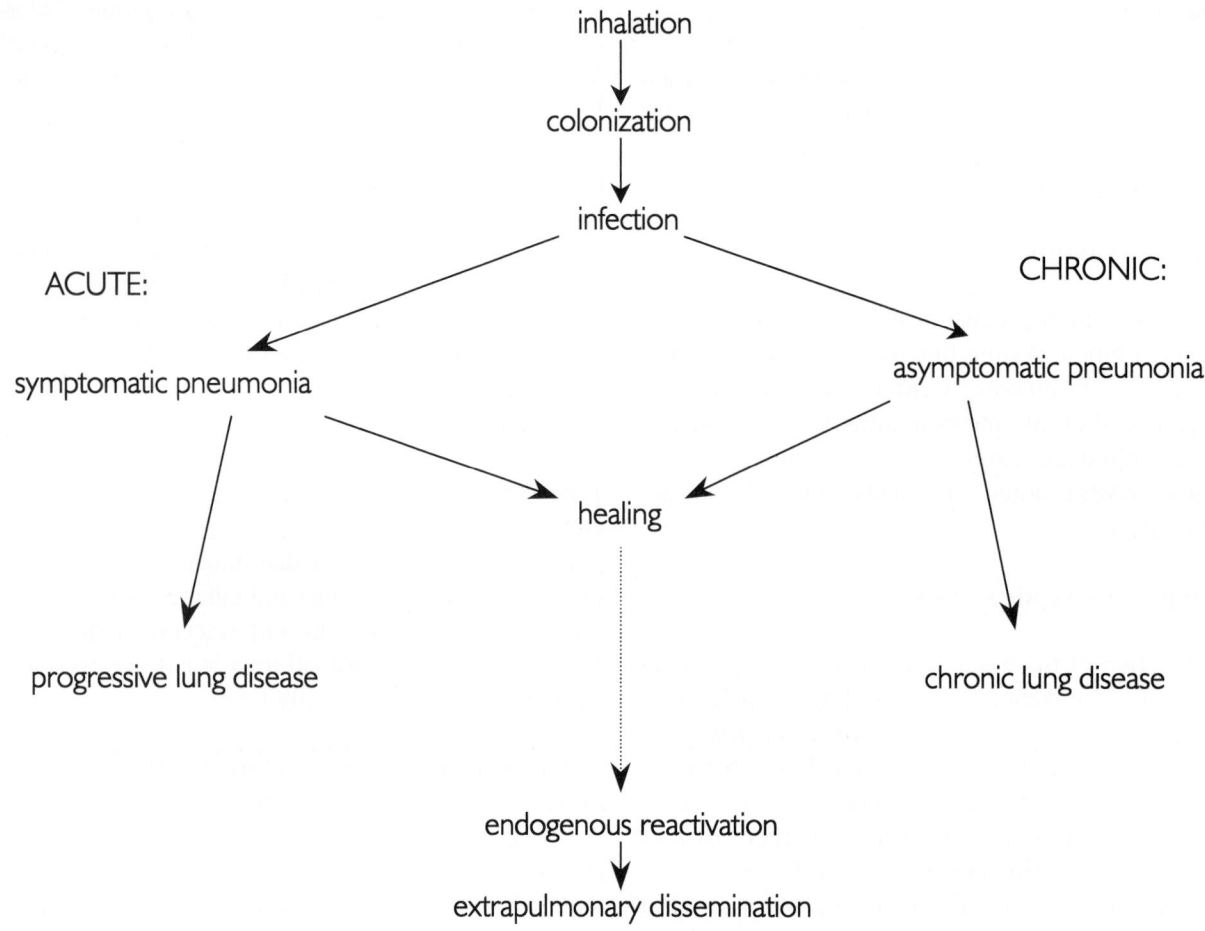

inhalation

colonization

infection

ACUTE:

CHRONIC:

symptomatic pneumonia

asymptomatic pneumonia

healing

progressive lung disease

chronic lung disease

endogenous reactivation

extrapulmonary dissemination

Fig. 8. Diagram of possible clinical courses in agents of sytemic mycoses caused by dimorphic *Onygenales*.

macrophages containing *Histoplasma* cells multiply in great numbers and give rise to lymphogenous or haematogenous metastases. Reaction is through granulomatous inflammation which later necrotizes and calcifies. Little neutrophilic and eosinophilic reaction is noted. The disseminated form is frequently fatal. The following types can be distinguished:

a. **Asymptomatic infection.** This is the case in over 99% of the exposed patients. Only a positive skin reaction and seroconversion are noted.

b. **Acute pneumonia.** This may be epidemic or the result of endogenous reactivation. Acute epidemic pneumonia is particularly found with visitors to natural caves. Within 12-14 days fever is noted with cough and pain in the chest. After healing, calcifications are seen. Acute reactivation is noted in persons who have been exposed earlier. Within 5-10 days bilateral infiltrates are formed in the lungs. Regional lymph nodes are often enlarged and infected by the organism. This is frequently seen in AIDS patients.

c. **Chronic pneumonia.** This is particularly found in the USA in males who smoke. Most have underlying lung disorders. Symptoms initially comprise cough and fever,

later accompanied by loss of weight and general fatigue. Unilateral, later bilateral, triangular lesions are produced with fibrosis and cavity formation. Later, characteristic single or multiple coin lesions are formed. Such lesions resemble carcinoma but show calcification.

d. **Disseminated histoplasmosis.** In immunocompromised persons, elderly as well as children, acute lymphatic dissemination may occur from initial foci in the lungs towards bone marrow, liver and spleen. More rarely, gastrointestinal involvement is noted, leading to encapsulated nodules. This results in fever, loss of weight and anaemia. In case of slow progression, oral ulcers may develop. In AIDS patients, involvement of the CNS is prevalent.

e. **Primary cutaneous histoplasmosis.** These are mostly laboratory-acquired infections. Local, self-limiting ulcers develop with involvement of one or more adjacent lymph nodes.

b. Other fungi

Penicilliosis

26

An infection caused by *Penicillium marneffei*, a species endemic in Southeast Asia. Fungal propagules are inhaled and may remain dormant over several decades. With impairment of the cellular immune system, particularly with AIDS, the fungus multiplies in the monocyte-macrophage system. This intracellular mycosis, similar to histoplasmosis, develops rapidly and may be fatal if untreated.

Aspergillosis

A conglomerate of clinical pictures mainly in immunocompromised patients (Table 4), caused by allergic reaction to, or colonization or infiltration by various *Aspergillus* species (Table 5).

a. **Aspergilloma.** This is a saprobic colonization of a preexisting cavity by an *Aspergillus* species. It may involve one of the sinuses or a secondary cavity in the lungs caused by a previous disease such as tuberculosis, bronchiectasis or bullae. The fungus forms a compact mycelial ball and does not invade adjacent tissues. The inflammation may expand towards the pleura and lead to formation of bronchopleural fistulae. Cultures from sputum are positive only when the cavity is connected to the bronchial system.

b. **Invasive aspergillosis.** Mostly due to *Aspergillus flavus* or *A. fumigatus*. Mycelium grows between epithelial cells and disseminates through blood. Primary localization is mostly pulmonary. Generally acute pneumonia is observed with high fever, dyspnoea and minor infiltrates on X-ray; it is usually fatal within 1-3 weeks. An essential precondition is a general or local, severe decrease in immunocompetence. Fungal hyphae in tissue can be recognized by dichotomous branching at angles of 45° (other hyphomycetes may be similar). If untreated, the mortality rate is high.

c. **Toxic aspergillosis.** An acute dyspnoea is observed after alveolar reaction towards massive inhalation of *Aspergillus* conidia. The reaction is short in duration and abates without residual symptoms.

Table 4. Predisposing factors for the development of invasive aspergillosis.

neutropenia (less than 0.5×10^9 cells/mL)
acute leukemia
immunosuppression
organ transplant
antibiotics
alcohol abuse, impairment of liver functions
chronic debilitating disease

Table 5. Diagnostic review of clinical symptoms of various forms of aspergillosis.

	dyspnoea	asthma	haemorrhages	IgG	IgE	eosinophilic reaction	X-ray	culture
aspergilloma:	−	−	(+)	+++	+	−	++	(+)
invasive aspergillosis:	+	−	−	(+)	−	−	+	(+)
toxic aspergillosis:	+	−	−	−	−	−	−	−
asthmatic aspergillosis:	−	+	−	−	+	−	−	−
alveolar infiltrative aspergillosis:	+	−	−	++	−	+	+	+
allergic bronchopulmonary aspergillosis:	+	−	−	++	+	−	+	+

d. **Asthmatic aspergillosis.** This is an acute, allergic bronchial constriction as a result of repeated exposure to *Aspergillus* conidia and leading to formation of specific IgE and eosinophilia. The reaction is short in duration but recurrent on repeated exposure.

e. **Alveolar infiltrative aspergillosis.** In this case dyspnoea is caused by (multi)local outgrowth of inhaled conidia, without tissue infiltration but with local inflammatory reactions which be visible on X-ray. Cultures from sputum are positive and specific IgG can be demonstrated serologically.

f. **Allergic bronchopulmonary aspergillosis.** This is a combination of d and e with asthmatic dyspnoea, specific IgG and IgE, eosinophilia, positive cultures and infiltrates on X-ray.

g. **Cerebral aspergilloma.** Rarely neurological disorders are noted, usually as a complication of sinusitis, with the fungus entering the brain via arteries. Usually the lesion develops extracerebrally.

Cryptococcosis

Invasion of tissues by *Cryptococcus neoformans* follows decreased T-lymphocyte immunity, often leading to subclinical pneumonia. About 5-10% of the AIDS patients in Western countries and 30% in African countries develop cryptococcosis. The fungus may rapidly penetrate the blood vessels and disseminate haematogenously. Its preference for the nervous system frequently leads to meningitis. The following clinical types can be distinguished:

a. **Pulmonary cryptococcosis.** The fungus is inhaled into the lung, but rarely provokes primary pulmonary symptoms. In the case of an underlying pulmonary disease, benign lesions may develop. Lesions in the lungs are whitish, gelatinous, without necrosis or cavitation.

b. **Disseminated cryptococcosis.** Dissemination may occur in immunocompromised hosts, usually leading to chronic meningitis. Multiple, slimy foci with a soap-bubble appearance are noted mainly on the meninges at the base of the brain. Cryptococcomas are sharply delimited, but not encapsulated. The disease may also involve other organs. Nodular, eventually ulcerating skin lesions may develop.

c. **Cerebral cryptococcosis.** Infection in the brain parenchyma, frequently in otherwise healthy patients, is mostly due to *C. neoformans* var. *gattii*.

d. **Primary cutaneous cryptococcosis.** Rarely granulomatous lesions have been reported after inoculating trauma.

Systemic candidiasis

Candidiasis is mainly caused by *Candida albicans*, although other species, such as *C. tropicalis* and *C. glabrata*, are emerging. The species is commonly present as a saprophyte in human mucosa. It may become invasive as a result of local or systemic decrease of defence mechanisms of the host. Deficiency in T-cell number or function, as occurs, e.g., in AIDS, leads to (muco)cutaneous candidiasis, whereas neutropenia is an important predisposing factor to systemic candidiasis. In neonates natural resistance is low, and hence systemic candidiasis can develop within a few days. Candidiasis may be acute or chronic. Acute candidiasis, e.g., in neonates, is the development of superficial, pseudo-membraneous growth in the oral cavity. After antibacterial therapy erythematous atrophy is seen on the tongue. A review of predisposing factors for candidiasis is given in Table 6. The following types can be distinguished:

a. **Fungaemia.** Either associated with indwelling catheters or with systemic infection in immunocompromised hosts.

b. **Systemic infection.** Involving deep tissues or organs, promoted by local or systemic impairment of innate immunity.

c. **Disseminated infection.** This is observed in hosts with impaired innate immunity, particularly neutropenia and leukemia, and develops when the patient is in an advanced state of deterioration. The fungus may invade several tissues by haematogenous proliferation and causes a general fatigue with fever. The disease is usually fatal when untreated.

Pulmonary colonization by yeast-like fungi

A whitish, floccose or smooth growth is formed on tracheal or broncheal surfaces. It is frequently associated with asthma or other infectious pulmonary diseases. Main causative agents: *Candida albicans, Geotrichum candidum, G. capitatum, G. clavatum*. In cystic fibrosis, *Exophiala dermatitidis* can be found in addition to *C. albicans*.

Disseminated infection by yeast-like fungi

Haematogenous dissemination of *Trichosporon, Candida* or *Geotrichum* species is frequently observed in patients with impaired innate immunity, particularly neutropenia or leukemia. Disseminated *Malassezia* infections are known in neonates fed with lipid-rich diets through a catheter.

Table 6. Predisposing factors for the development of candidiasis.

1. underlying conditions:	malignancy
	congenital immunodeficiencies
	viral infections
	diabetes mellitus
	endocrinopathy
	immune deficiency (e.g., in neonates, AIDS)
	neutropenia
	hormonal changes (pregnancy, oral contraceptives)
2. physiological factors:	unbalanced, carbohydrate-rich diets
	hypovitaminosis
3. mechanical trauma:	skin damage (wounds, burns, maceration)
	deep trauma, surgery
	indwelling catheters / prostheses
4. chemical causes:	heroin addiction
	iatrogenic factors (therapy with corticosteroids, antibiotics, cytostatics, irradiation)

Zygomycosis

An opportunistic infection caused by various species of *Mucorales*, particularly *Rhizopus* species. Acute infection is characterized by inflammation. Mostly little cellular reaction is noted, but oedema and necrosis soon develop. The fungi show invasive growth, preferentially in the walls and lumina of blood vessels, causing vascular thrombosis and infarction. At the same time, haemorrhages occur due to destruction of blood vessel walls. In animals growth in the placenta frequently leads to abortion. In humans, ketoacidotic patients are at risk due to the presence of free iron in their blood. The following types of zygomycosis can be distinguished:

a. **Rhinocerebral zygomycosis.** Fulminating lesions of orbital and nasal tissues and mucosa are noted, often with cerebral involvement. The disease is particularly found in patients on diabetes mellitus or on desferroxamine therapy. It mostly takes a rapidly fatal course.

b. **Pulmonary zygomycosis.** Pulmonary lesions are seen particularly in patients with impaired innate immunity or on corticosteroid therapy, or it may develop after aspiration.

c. **Gastrointestinal zygomycosis.** Extended peritonitis occasionally occurs in patients with poor general health due to malnutrition or immunosuppression.

d. **Primary cutaneous zygomycosis.** Local oedematous (sub)cutaneous gangrene occurs frequently in patients with extended skin burns or trauma.

Cerebral phaeohyphomycosis

Deep mycoses caused by melanised fungi often show involvement of the central nervous system. *Cladophialophora bantiana* and *Ramichloridium mackenziei* in particular characteristically produce lesions in the brain, in which melanized hyphal elements can be seen. The disorder is usually fatal. Also, several *Exophiala* species are neurotropic, either in humans or in fish, as is *Ochroconis gallopava*, which is primarily found in birds. Portal of entry is mostly the lungs. Brain tissue invasion is rarely encountered as a complication of sinusitis caused by *Bipolaris, Curvularia* or *Exserohilum* species.

Systemic *Pseudallescheria* infection

Pseudallescheria boydii occasionally colonizes pulmonary mucosa, particularly in patients with cystic fibrosis. With impairment of innate immunity, as may accompany circulation failure, coma or advanced leukemia, progression can result in chronic abscesses in which septate hyphae can be seen in the CSF. If untreated, the mycosis is mostly fatal.

Systemic hyalohyphomycosis

Infections are deep, caused by various, non-melanized fungi. They are localized and are mostly acquired after major surgery. Liver and spleen are frequently involved. Endocarditis may result from growth on mitral valves. *Fusarium* species are relatively common.

NATURAL ECOLOGY

Introduction

Natural ecology of a fungus explains the behaviour of the organism by focussing on its **ecological niche**, that is the environment where it produces its assimilative thallus and where it generates its progeny. Ecology therefore focusses on the understanding of the causative agent of disease, the fungus; this is in contrast to the medical approach, outlined in the previous chapter, which is primarily interested in the protection of its victim, the human host. These two approaches cause a significantly different usage of terminology, particularly concerning the notion 'pathogenicity'. Reduced virulence in the course of evolution of human-associated infectious agents has been reported, which leads to the paradoxical situation that the 'pathogens' with the highest degree of adaptation hardly cause disease. This chapter tries to circumvent confusion by using medical terms in their classical sense and adding ecological terminology to indicate the different viewpoints.

In medical microbiology, the clinician is used to treat the etiological agent of a particular disease as the pathogen, and hence to speak of strains rather than of species. This approach is indeed meaningful in bacteriology, where strains may differ considerably from each other in virulence and antibiotic resistance, due to, e.g., the occurrence of pathogenicity islands. However, fungal strains belonging to a single species or other taxonomic entity mostly are closely similar to each other in their genetic make-up. The behaviour of the *etiologic agent*, the strain, can be coincidental when the remaining members of that species live outside the human body. Species are virtual entities; only species can be pathogens. They have a potential behaviour, like pathogenicity, which needs not to be displayed under all circumstances. In contrast, the strains are concrete and have an actual behaviour.

It is likely that some of the pathogens have another natural warm-blooded host apart from humans. Therefore in the text below a distinction has been made between 'humans and animals' and 'mammals', the latter term being meant to include humans.

Do fungi have species-specific behaviour?

In practice the location of the ecological niche of a fungus can be distinguished from its non-optimal environment by the fact that a species is significantly more often isolated from its natural niche than from other sources. However, many fungi are able to endure non-optimal growth conditions outside their natural niche (Pitt & Hocking, 1997). Such species are *euryoecous*, having a wide *ecological amplitude*, i.e., being able to inhabit wide variety of environments. The presence of mechanisms that allow the conversion into a metabolically inactive state, often with a distinctive morphology such as thick-walled chlamydospores, isodiametrically enlarging thalli, or endoconidia, is a widely distributed feature in the fungal kingdom. This is illustrated by the fact that extremophily is polyphyletic, such fungi being encountered in very different orders (Sterflinger *et al.*, 1999).

There is a widely held belief that fungi all behave rather unspecifically and would lack ecological preferences. This misconception may be rooted in the fact that most fungi we know are the easily culturable, euryoecous species. Many fungi are *stenoecous*, i.e., highly specialized, and require special methods for isolation. Part of the clinically significant taxa belong to this category. Of the 73,000 currently described and accepted species of fungi and the possibly over one million existing taxa (Hawksworth, 1991), only 396 have been implicated in human mycoses as listed in this Atlas. In medical mycology many authors deny preferred human-association in claiming that immunocompromised patients may be invaded by practically any fungus (Khardori, 1989). Some of them (e.g., Ajello, 1980; Perfect, 1996) even regard the established pathogenic fungi as having their natural niche in the environment and provoking mycoses only when they are coincidentally implanted into a human host. The rationale for this view lies in the observation that, in contrast to pathogenic bacteria, the agents of deep mycoses are not transmitted directly from host to host (Steele, 1991). Thus, indeed very few of them are obligatory pathogens. Morphologically similar fungi may differ considerably in in their optimal environment, as can be illustrated by the example of *Cladosporium cladosporioides* and *Cladophialophora bantiana*. Both are melanized hyphomycetes and produce hydrophobic conidia in chains.

Table 7. Possible ecology of pathogenic fungi.

Name:	Saprobic niche:	Animal reservoir:
Blastomyces dermatitidis	wood ?	dog
Coccidioides immitis	desert soil	rodent
Cryptococcus neoformans var. *neoformans*	wood	pigeon
Cryptococcus neoformans var. *gattii*	wood	koala
Emmonsia parva	desert soil	rodent
Histoplasma capsulatum	bat dung, guano	?
Paracoccidioides brasiliensis	wood ?	armadillo
Penicillium marneffei	soil	bamboo rat
Sporothrix schenckii	wood	?

Phylogenetically they are widely separated. *Cladosporium cladosporioides* is an extremely common air-borne saprobe, of which only very few clinical cases have been described (Annessi *et al.*, 1992). In contrast, *Cladophialophora bantiana* is extremely rare, nearly all strains known to date originating from cerebral infections in humans (Horré & de Hoog, 1999). Thus, the two species must have different ecological preferences, resulting in these divergent isolation statistics.

As will be explained below, some fungi have the ability or even a preference to make use of the mammalian body during a part of their life cycle, but nevertheless they probably all are able to grow and reproduce in the environment. Thus they have two ecological niches, as their life cycle is subdivided into two radically different parts. Fungi therefore at most are **facultative pathogens**. Even species that do not grow in culture, such as *Pneumocystis carinii*, can be found outside the human body, perhaps in a dormant state. The essential observation is that they are nearly always able to grow and reproduce in the host. Very few species are exclusively found on mammals; these are **obligatory pathogens**. Completion of the entire life cycle on the human host is found by some non-invasive, commensal species such as *Malassezia* and *Piedraia*.

Some basic requirements for adaptation to the animal host

Among the factors contributing to the development of mycoses, two types should be distinguished which are explained below: vitality factors and virulence factors.

Properties which enable a fungus to endure hostile environmental conditions like heat or drought are **vitality factors** (Morschhäuser *et al.*, 1996). They are **unspecific**, i.e., not specially designed for growth in animal tissue. Among the vitality factors are the presence of melanin (Dixon *et al.*, 1991b), carotenes (Geis & Szaniszlo, 1984), thick cell walls (Emmons & Jellison, 1960) and meristematic growth (Staley *et al.*, 1982). However, it is

remarkable that some of the extremely resistant, apparently ubiquitous fungi growing on sun-exposed stone (Sterflinger *et al.*, 1998), for which we would expect frequent human infection through traumatic inoculation, are not known as etiological agents of mycoses. Apparently vitality factors alone are insufficient to cause a mycosis and do not contribute towards adaptation to mammal hosts. Resistant saprobes, when inoculated into living animal tissue mostly produce hyphal clumps or sterile elements. These are merely highly reduced forms which have lost the properties that they need to overcome the conditions of their normal environment. Therefore the fungus does not have effective means to return to its natural habitat after the host has died; it is likely that it dies along with the host. Host passage would then be a poor vehicle for dispersal for the fungus and does not seem to be advantageous in an evolutionary sense. Evolutionary adaptation to prevailing host conditions via natural selection is not possible, because the thallus is vegetative.

To allow adaptation to enhanced maintenance in living tissue, generations of fungal cells must be subjected to evolutionary pressure. Thalli present within host tissue or body fluids and consisting of independently propagating generations of fungal cells, are termed **zoodemes**. Zoodemes can be regarded as a prerequisite for the evolutionary development of pathogenicity. They are produced in yeasts which multiply by budding cells (Swerdloff *et al.*, 1993) or arthroconidia (Schiemann *et al.*, 1998). Some hyphomycetes, like *Acremonium, Fusarium* and *Paecilomyces*, are able to develop their reproductive structures in tissue. The ascomycete *Pseudallescheria boydii* may sporulate in blood with its anamorph (Wilson *et al.*, 1990). This **adventitious sporulation** may enhance invasive growth, e.g., vascular penetration, and explains the high rate of positive blood cultures of such species (Schell, 1995; Liv *et al.*, 1998)

Pathogenic fungi: evolutionary advantage of animal-association

31

In the case of mammal-associated fungi there is a balance between fungal vitality and host response. The fungus aims to maximize its reproduction and dispersal so that it has optimal evolutionary *fitness*, and the host tries to minimize the damage of the infection. Initial confrontation usually is a clash: vigorous fungal growth leading to severe inflammation. It is of interest to both parties to minimize the amount of mutual damage; they then have a *shared agenda* (Dawkins, 1990). A higher degree of adaptation to host-association thus may lead to a lower degree of virulence, or to the development of a benign condition. Because of this paradox, such low-virulence organisms are indicated here as *adapted infectious agents*, while the term 'pathogen' is restricted to those organisms causing visible damage to the host.

Many of the systemic, mammal-associated fungi are *dimorphic*, i.e., they develop two morphologically distinct vegetative forms: an invasive zoodeme (tissue form) and an additional, hyphal form in the environment. Among the systemic *Onygenales*, and in *Penicillium marneffei*, the requirements of host *vs.* environment are extremely divergent. The tissue and environmental forms are correspondingly different from each other. The tissue form is not indispensible, since the fungi can grow and propagate in the environment. The ability to exhibit such a unique structural and physiologic tissue form must have arisen through a prolonged succession of evolutionary adaptations. Therefore it is unlikely that this would be just coincidental. Rather, some kind of evolutionary advantage of animal association seems probable. This implies that the survival of the species is enhanced when an appropriate host is utilized transiently. Such optimization of propagation and survival can be considered as an *ecological strategy* of the fungus. Only properties enhancing penetration and survival in host animals within the scope of an ecological strategy are *virulence factors*. The term '*pathogenic*' is here applied in an evolutionary sense, i.e., restricted to fungi which have an ecological strategy based on utilization of an animal host and which therefore exhibit virulence factors. Since fungi have a high rate of ecological specialization, their clinical pictures and localization are rather invariant: the ecological behaviour is characteristic for each species.

A proposed model dimorphic life cycle of pathogenic *Onygenales* can be illustrated by using *Coccidioides immitis* and *Histoplasma capsulatum*. The saprobic stages found in the environmental niche of both species are filamentous. They are found in endemic areas with rather particular or extreme (micro)climatic conditions, such as desert-like soil (*Coccidioides;* Maddy, 1957) or sheltered soil enriched with guano (*Histoplasma;* DiSalvo *et al.*, 1970). The fungi possibly show low competitive ability with other saprobes in adjacent niches. Using an extreme environment as a refugium has been termed an *S-strategy*, in which S stands for stress-toler-

ance (Dix & Webster, 1995). More specifically, the afore mentioned dimorphic fungi produce zoodemes and hence have a *P-strategy*, in which P stands for pathogenicity.

For *Coccidioides*, the endemic area coincides with the range of a corresponding host animal (Emmons, 1942). At maturity, *Coccidioides* passively releases arthroconidia from the vegetative mycelium at the site of growth. The conidia are persistent and may be inhaled by an appropriate animal and then develop inside the host as spherules containing endospores. In *Histoplasma*, inhalation leads to phagocytosis without killing in macrophages, triggering the propagation of the intracellular yeast phase. The tissue forms of both fungi are equipped with factors that enable the cells to modulate phagolysosome action. The hosts often remain asymptomatic after the initial exposure, or may develop mild, transient symptoms. The fungus is able to survive, probably in a dormant stage, over extended periods (Livas *et al.*, 1995). The host animal serves as a *reservoir* of the fungus, that is, a location where it hides without causing disease and from which it may colonize a new part of its environmental niche. In the case of reservoir relations there is infection but no disease. *Histoplasma* hides in the monocyte-macrophage system; for *Coccidioides* the site of dormancy is not known. When phagocytosis and killing becomes less efficient due to impairment of Th1-cell functions, the fungi disseminate. This process is known as *endogenous reactivation*. Subsequently the host may die, and, in the case of a rodent in the *Coccidioides* endemic area, it may be supposed that the fungus develops a saprobic phase by contaminating a new environmental site from the dead body.

The two systemic members of the *Onygenales* briefly described above exhibit different types of host relations. In *Coccidioides*, as well as in the related species *Emmonsia parva*, the animal host resides in the endemic area where the frequency of infected animals may be high (Emmons & Ashburn, 1942; Hubálek *et al.*, 1995). The fungus is partly harboured and sheltered by the animal host. Apparently the rodents infected by *Coccidioides* and *Emmonsia* contribute to maintenance and dispersal of the fungus within its endemic area, where the fungus has a *continuous distribution*, i.e., where the sites of occurrence are potentially interconnected.

Histoplasma capsulatum does not infect the birds or bats that dwell in the niche of its saprobic phase. These animals produce the dung on which the saprobic phase of the species survives (Gugnani *et al.*, 1994). In contrast to *Coccidioides*, *Histoplasma* has a *disjunct distribution* since it occurs on sheltered places like grottos and indoor roosts within an endemic area. Infection rates in animals entering endemic foci are considerably higher. This is also seen in humans entering a contaminated site (Suzaki *et al.*, 1995). If non-resident animals function as a vehicle for dispersal of the fungus from one focus to the next, they can be regarded as ***vectors***.

The fungus' ecological strategy requires that the host animal functioning as a reservoir is not immediately sickened or killed. Hence the fungus, after having entered the host nearly asymptomatically, remains dormant over an extended period, during which the animal carries the fungus to another site. This tends to favour a gentle rate of virulence (Ewald, 1996; Lederberg, 1999). The trigger for endogenous reactivation is impairment of acquired cellular immunity, probably near the end of the life of the host animal when its immune system deteriorates. In humans we witness reactivation of such endogenously carried fungi, inhaled earlier while travelling through endemic areas, at the onset of development of AIDS (Livas *et al.*, 1995). This may help to explain why fungi such as the systemic members of the *Onygenales* and *Penicillium marneffei* are among the principal pathogens of AIDS patients (Eidbo *et al.*, 1993; Heath *et al.*, 1994; Chinn *et al.*, 1995; Wheat, 1995; Ampel, 1999).

In several of the examples described above we note that humans are unlikely to be the primary host of the fungus. Rather, the fungi may have balanced strategies with known or unknown animal species. Since the fungus is insufficiently adapted to the human host, its virulence in humans may be higher than that in the natural host. From an evolutionary point of view, human infections are useful only when the fungus is transmitted from the dead body into its environmental niche, such as a grotto. In our modern society this is rarely the case, and thus most human infections can be regarded as evolutionary dead ends.

Opportunistic fungi: coincidental agents of human disease

Fungi that are inoculated traumatically generally have their natural niche in the environment. In tissue of their otherwise healthy hosts they are immediately confronted with the innate cellular immune system and are phagocytized or inactivated by polymorphonuclear neutrophils (PMNs) and macrophages. The same holds true for fungi that are inhaled and do generally not expand beyond the organ where they were deposited, such as the sinus or the lung. Several hundreds of species appear able to persist under the conditions of living animal tissue, although at a low level of vitality. A strong unspecific immune reaction is required for their control, which may lead to considerable inflammation. The persistence of the fungus despite appropriate immune response is responsible for the chronic nature of the mycosis. In contrast to the systemic *Onygenales*, the fungus is probably unable to re-colonize the environment from the body of the host, be it living or dead. This can be illustrated by

the fact that such fungi are often difficult to grow in primary culture. The infection thus does not contribute to the evolutionary fitness of the species and must be regarded as a ***spill-over*** infection (Bourhy *et al.*, 1999).

Fungal vitality and human defence are more or less in balance, but when innate cellular immunity becomes impaired, for example due to myeloid leukemia or the use of corticoids, the balance shifts towards fungal advantage and sepsis may take place. The agents are then referred to as ***opportunists***. It should be stressed that the opportunistic fungi of today are not really new to mycology, but most of them are known for over 100 years from traumatic mycoses.

The classical example of a traumatic mycosis by a saprobe in an otherwise healthy host is mycetoma, a mycosis characterized by tumefaction and draining sinuses. The innate immune response causes extended tissue necrosis. The fungus produces hyphae which quickly organize into aggregates that are very similar to the ones formed *in vitro* by the same fungus in submerged shake culture. As these aggregates mature and become associated with host tissue they evolve into sclerotia.

The highly resistant agents of mycetoma-like infections are able to cause disorder in human beings which are otherwise in apparent good health. First reports possibly date back for thousands of years (McGinnis, 1996), long before hospital populations of severely immunocompromis ed patients existed. It is logical that fungi resistant enough to produce chronic mycoses in otherwise healthy persons will be even more successful when the host's immunity or other defence mechanisms are impaired.

A patient without any cellular immunity can be viewed as a virgin substratum. Fungi that rapidly colonize new substrates, subsequently reproduce at high speed and abundance, and disappear before other organisms come in, have an ***R-strategy***, in which R stands for ruderal (Dix & Webster, 1995). Among these are members of *Mucorales*, such as *Absidia* and *Rhizomucor*. Their mycoses often are devastating. Other fungi in nature are able to colonize substrates which are already occupied by other microorganisms, achieving a high level of competitive ability by the production of extracellular antibiotics and toxic metabolites. They have a ***C-strategy***, in which C stands for combative (Dix & Webster, 1995). Among these are members of the genera *Aspergillus* and *Fusarium*, which in compromised humans mostly cause more chronic mycoses than *Mucorales*.

Commensals and superficial pathogens: single-factor association and mild pathogenicity

A ***contaminant*** on the skin is a fungus which has simply become associated with the animal by landing on its body. It does not grow, nor it uses the host or host products for nutrient. Fungi may utilize specific compounds produced by mammals, such as excretions of the skin or keratinous materials, without needing or damaging the living animalitself. Such fungi are referred to as ***commensals.*** Invasion of adjacent living tissue may occur, but is coincidental.

Piedraia hortae is a unique commensal because it completes its total life cycle, including propagation with meiospores, on the hair of vertebrate hosts. It forms ascomata with asci and ascospores on the outside of the hair shaft (Figueras *et al.*, 1997) and occurs quite frequently in tropical regions on humans (Coimbra & Santos, 1989) as well as on the pelts of apes (Kaplan, 1959). The fungus has never been isolated from the environment. In this context ***colonization*** of a host is defined as growth on animal tissue without invasion of living tissue and without triggering a host response. There is no infection and no disease.

Despite its dependence on primates, *P. hortae* is strictly non-pathogenic. The fungus can be viewed as an ***autochtonous commensal***, i.e., a commensal for which the vertebrate host is its natural ecological niche (Seebacher & Blaschke-Hellmessen, 1990). Similarly, *Malassezia* species are restricted to the vertebrate host (Guillot *et al.*, 1994). They are colonizers of human and animal skin, living on compounds associated with the dead cells of the stratum corneum. But *Malassezia* species, when introduced into the circulatory system by a catheter containing lipid-rich fluid, are able to cause transitory fungemia (Shparago, 1995). Their potential ability to produce zoodemes within the host makes them more prone to pathogenic adaptation than *Piedraia*.

Hortaea werneckii can be regarded as an ***allochtonous commensal*** (Seebacher & Blaschke-Hellmessen, 1990). It is found colonizing stratum corneum of feet and hands, which results in brownish patches owing to constitutive dihydroxynaphthalene melanin. The fungus is non-invasive and no host reaction is noted (Rippon, 1988). It is believed to adhere to human epithelial cells by an hydrophobic interaction where it maintains itself by the assimilation of lipidic excretions (Göttlich *et al.,* 1995). Its natural niche, however, is in salt-saturated pools and tidal waters (Zalar *et al.*, 1999). A high salt content is supposed to be the common factor between salterns and sweaty, sun-dried human hands. This strongly suggests that commensalism is indeed determined by a single or very few characteristics.

In this sense, dermatophytes may primarily be regarded as commensals, because they are dependent on the single nutrient keratin as their main source of carbon and nitrogen. For geo- and zoophilic *Microsporum* species no living host tissue is needed, but presence of keratinous remains of hairs, feathers or nails is sufficient

(Chabasse & Contet-Audonneau, 1994). They are frequently found in the immediate vicinity of birds and mammals, e.g., in nests, burrows and adjacent soil. Fungal propagules have cell wall projections that allow them to remain between fur and feathers, which contributes to their efficient dispersal. Macroconidia typically have rough walls, while teleomorphic fruit bodies are often covered with appendages in the form of stiff coils, combs or hooks (Currah, 1985). Anthropophilic *Trichophyton* species such as *T. rubrum* colonize hairless skin and thus do not require such ornamentations (Rippon, 1985). Some pathogenic species are highly reduced, being nearly unable to produce macro- or microconidia. Transmission is then through skin flakes containing infectious arthroconidia.

The genus *Trichophyton* thus seems to be phylogenetically recent, as is also exemplified by the coevolution of some dermatophyte species with human demographic movements (Rippon, 1985). This is underlined by the fact that some anthropophilic taxa seem to consist of a single clone (Gräser *et al.*, 1999c).

Because keratin occurs as an integral interface next to living tissue, the anthropophilic dermatophytes are more likely to become confronted with the cellular immune system. An evolutionary tendency towards development of immunomodulating properties by the fungus might then be expected, i.e., development in a direction leading towards more adaptation to the host. Indeed the dermatophytes have developed some ability to surpass the innate immune reaction, the infection then being held in check by acquired cellular immunity (Smith & Griffin, 1995). The anthropophilic dermatophytes are unique in producing antigens that trigger both Th1 and Th2 cell responses (Slunt *et al.*, 1996). Since Th2 cells produce cytokines which stimulate eosinophil proliferation, leading to inflammation of the airways, this reaction must be regarded as inappropriate. Th1 and Th2 cells are in balance through opposite inhibition by interferon-γ and interleukin-4, respectively. A mannan is produced by the fungus that interferes with phagocytosis (Blake *et al.*, 1991) and thus hampers Th1 proliferation, thus giving room for the ineffective Th2 response. Through this mechanism the fungus is insufficiently cleared and a chronic, recurrent infection is provoked. Zoophilic dermatophytes on humans generally give rise to strong inflammation and subsequent clearance. *Cryptococcus neoformans* possesses a mechanism with different means but with the same outcome. After interaction with capsulated yeast cells, IL-10 production is induced and thus the appropriate Th1 response is suppressed (Vecchiarelli *et al.*, 1996). In addition, selectins are stimulated to be shed from leukocytes which leads to decreased adhesion to endothelial cells (Murphy, 1999).

The occurrence of an acquired immune response suggests that there is a relatively long-standing relationship between human defence mechanisms and the fungus, which may have led to a fungal ability to evade the innate immune system without being killed. This is a prime character of a pathogenic species or an adapted infectious agent. For such fungi the acquired immunity has become particularly significant. In the dermatophytes, the difference between commensal and pathogenic fungi is gradational. Species that normally occur asymptomatically on dead keratinous material, e.g., *Microsporum canis* in the fur of animals, tend to provoke a vigorous immune response when implanted into the hairless human skin. Vigorous inflammation and disease might thus indicate a low level of adaptation, whereas reduced virulence suggests a high degree of adaptation.

Endosaprobes: delicate balance with the immune system

A comparable situation is found in fungi inhabiting the digestive tract, primarily living on half-way digested products that pass by, rather than utilizing host tissue.

The fungal flora on faeces, dung and guano is quite diverse (Seifert *et al.*, 1983). Several genera of *Ascomycota*, such as *Sporormia, Sordaria* and *Podospora* are common on mammalian dung, and hence some of them are found in the close vicinity to animals. It might be supposed that such fungi have a greater opportunity to come into contact with the particular animal whose dung is colonised. However, mycoses caused by typical dung-fungi such as *Arnium leporinum* are extremely rare (Restrepo *et al.*, 1984). Spores of several asco- and zygomycetes must be stimulated by passage through the animal digestive tract before they are able to germinate on the faeces (Webster, 1970). The ascospores are dormant during this passage through the host. This is in contrast to *Candida albicans*, which lives and propagates within the digestive tract. Its ecological niche inside the animal makes it an **endosaprobe** (Vanbreuseghem *et al.*, 1978). Even though *Candida* can be isolated from faeces, its cells have a short survival capacity in this environment, mainly due to their low competitive ability (Bernhardt *et al.*, 1995). Because the fungus dwells inside the animal throughout its life cycle, it has developed abilities to defend itself in this unique environment against the host as well as against bacterial hostilities. For example, some Gram negative bacteria are antagonistic towards *C. albicans* (Kennedy & Volz, 1985). With respect to host response, *C. albicans* is the only fungus which is able to take advantage, depending on prevailing conditions, of a deficiency of innate as well as acquired immunity. Human immunodeficiency virus (HIV) infection predisposes the AIDS patient to mucocutaneous candidiasis, whereas neutropenia, e.g., in leukemic patient populations, leads to disseminated candidiasis

(Wheat, 1995). This suggests that in healthy hosts the fungus lives in a delicate balance with the human immune system and other host factors.

Pseudallescheria boydii may have similar behaviour in water-associated animals. It occurs in environments rich in nutrients and with reduced oxygen tension, e.g., agricultural soils (Bell, 1976) or heavily polluted water (Cooke & Kahler, 1955). In human tissue it has a remarkable survival capacity (Bell, 1978). This may be explained by its microaerophily (de Hoog *et al.*, 1994a): *in vitro* in submersion it shows normal growth and conidiation. In humans, the fungus behaves as an opportunist. The clinical spectrum of infections is highly diverse (Warnock & Johnson, 1991). The main mycoses caused by *P. boydii* are mycetoma following traumatic inoculation into subcutaneous tissue (Ajello, 1952) and pulmonary infections which may evolve into neurotropic mycoses in the CSF (Berenguer *et al.*, 1989). Several patients have acquired the fungus through aspiration of polluted water (Fisher *et al.*, 1982). The diversity in the clinical spectrum suggests that this species is not adapted as a pathogen, but rather, through capacities needed for its successful survival in its natural ecological niche, it is able to survive in human tissue.

Host responses: recognition of pathogens and saprobes

Schaffner (1989) reproduced the host reaction to fungal invasion in laboratory animals. He described two fundamentally different types of kinetics of fungal elimination after intravenous inoculation. Strictly **non-pathogenic fungi** are eradicated immediately by the animal's unspecific cellular immunity. The decrease in numbers of fungal cells is unaltered when mice are athymic, demonstrating that the host response to such fungi is independent from acquired cellular immunity. Exponential increase of cell numbers is seen in immunosuppressed mice only. In contrast, cell numbers of **pathogenic fungi** initially increase exponentially when inoculated into normal mice, and with some delay growth of the fungus is controlled by acquired cellular immunity (Beaman *et al.*, 1977). Immunosuppression does not have any effect on this growth curve. In these fungi there is no exponential increase in athymic mice. The rare special case of the low-virulence adapted infectious agents has not been investigated from this viewpoint.

Schaffner (1989) indeed touched some basic issues in medical mycology. Although the bipartition is sometimes ambiguous, for the most part pathogens and opportunists can be distinguished by their interaction with the host's cellular immune system, i.e., acquired *vs.* unspecific, respectively. From an evolutionary perspective a pathogen can thus be viewed as an organism which has developed the capacity to evade the host's first, unspecific line of defence. Various kinds of association with mammals have been developed independently in the course of evolution, as pathogens are found in phylogenetically very different groups, e.g., in *Histoplasma capsulatum, Candida albicans* and *Penicillium marneffei*. It can be supposed that species which dwell in the vicinity of animals are statistically more likely to be confronted with their immune system, and thus may have developed ways to escape natural immunity. This leads to the conclusion that those species that are primarily controlled by acquired immunity must be the ones that have reached a higher level of pathogenic adaptation. Th1-cell controlled fungi are members of the *Onygenales, P. marneffei, C. albicans, Cryptococcus neoformans, Pneumocystis carinii*, and to a less extend *Sporothrix schenckii,* and perhaps some members of *Chaetothyriales*. Their abilities specifically developed to evade natural immunity, which make them pathogens, should thus be regarded as virulence factors. Pathogenicity is polyphyletic, virulence factors are unlikely to be homologous, but nevertheless the concentration of pathogens in the closely related orders *Eurotiales, Onygenales* and *Chaetothyriales* is remarkable.

In phylogenetic trees, systemic pathogens are sometimes found separated from each other by less pathogenic species, or even by species with quite different ecology. For example, Bowman *et al.* (1995) found the geophilic species *Uncinocarpus reessii* among the systemic *Onygenales*, and Masclaux *et al.* (1995) found neurotropic species in clearly distinct branches within the *Herpotrichiellaceae*. The specialized pathogens are close together, but not members of a single clade. The most likely evolutionary model is the assumption of separate specialization of each of these species, on the basis of more general factors which just provide the opportunity for such further specialization via separate lines. This is indeed what we observe both in the *Onygenales* and the *Chaetothyriales:* the systemic *Onygenales* each have a remarkably different pathogenic phase, whereas the neurotropic *Herpotrichiellaceae* differ significantly from each other by a deviating predilection for patient groups (Horré & de Hoog, 1999). The more general factors probably are the undirected vitality factors, from which an additional virulence factor may or may not emerge. Thus, primary pathogens are expected to be phylogenetically located in groups that have acquired a number of suitable vitality factors. This implies the occurrence of 'hot spots' in the fungal Kingdom, where silent opportunism has been developed and true pathogenicity can be expected. In the phylogenetic tree the vitality factors are present as *plesiomorphs* (original characters), whereas specific virulence factors develop as *apomorphs* (derived characters).

The apomorphs enabling further specialization are likely to be very different between unrelated groups of fungi, and the groups may have achieved quite different levels of pathogenic specialisation. It seems that the *Onygenales*, containing several families of animal-associated fungi, represent the highest level of pathogenic adaptation known in the fungal kingdom. A tendency to take advantage of defects of acquired immunity is prevalent in the entire group. However, the *Herpotrichiellaceae* seem to have taken some first steps towards true pathogenicity, while members of the *Microascaceae*, showing spontaneous healing after restoration of unspecific immunity (Bouza *et al.*, 1996), can be regarded as having coincidental vitality factors enabling opportunism, but do not show true pathogenicity. The plesiomorphic vitality factors, as the driving forces of pathogenic evolution and development of apomorphic virulence factors, have been supposed to include keratin assimilation in *Onygenales* (Currah, 1985), hydrophobicity allowing adhesion to nervous tissue in *Herpotrichiellaceae* (de Hoog, 1997), extracellular proteinases in *Candida* (Monod *et al.*, 1994), capsule and melanin in *Cryptococcus* (Buchanan & Murphy, 1998) and anaerobic growth in *Microascaceae* (de Hoog *et al.*, 1994a).

Fungi supposed to be systemic human pathogens are subjected to considerable ecological pressure during their residence inside the host. Subsequent evolutionary adaptation probably would lead to an even more successful pathogen, where **adaptation** is defined as the acquisition of an ability to grow and reproduce under a new set of conditions. With a high degree of pressure, survival rates of random mutants probably are low. This is reflected in the taxonomy of systemic *Onygenales*, each genus containing only a very few species. Such a group of clearly separate species, forming a **discrete taxonomic structure**, has a shared phylogeny with nearly monotypic branches (**clades**), which are clearly different from each other and have led to large morphological and behavioural gaps between the entities. In the systemic *Onygenales*, each taxon seems to have found a different way to modulate innate immunity. When evolutionary pressure is lower, due to the fact that the environment is more permissive with respect to random mutations, it may be expected that survival rates of mutants are higher. This seems the case with anthropophilic dermatophytes (Gräser *et al.*, 1999a). A clade which contains a large number of species with minute differences is termed a **radiate taxonomic structure.**

Non-adapted agents of mycosis:

Agents of mycosis with various degrees of adaptation:

Saprobe:	Commensal:	Endosaprobe:	Facultative pathogen:	Obligatory pathogen:	Human-adapted infectious agent:
Niche outside human	Niche on human	Niche inside human	Strategy: transmission mammal / environment	Strategy: transmission mammal / mammal	Strategy: transmission human / human
Healthy host: strong non-specific defence; compromised host: opportunism	No confrontation with immune system	Healthy host: little confrontation with immune system; compromised host: opportunism	Healthy human: moderate virulence	Healthy human: high virulence	Healthy human: low virulence
Example: *Aspergillus fumigatus*	Example: *Malassezia furfur*	Example: *Candida albicans*	Example: *Histoplasma capsulatum*	Example: *Trichophyton verrucosum*	Example: *Trichophyton rubrum*

Mycosis does not contribute to fitness;
Vitality factors

Fitness is increased by use of mammal vector;
Virulence factors

Table 8. Diagrammatic representation of host relationships in fungi.

Conclusions

The different relations between fungus and mammalian host are summarized in Table 8.

Despite the fact that nearly all fungi do occur in the environment, there are reasons to believe that a small number of them have become adapted to life inside the mammalian body and thus are facultative pathogens. In current bacteriological literature, a pathogen is defined as a microorganism that is capable of causing disease in the immunocompetent host; an opportunist is unable to enter the immunocompetent host, causing disease only after an impairment of immunity. The situation in the fungi is, however, somewhat less clear-cut. Fungi tend to have more ability to survive in the immunocompetent host, particularly after traumatic inoculation. Practically all fungi, including baker's yeast, seem able to do so (de Hoog, 1997), albeit extremely rarely for some species.

Species adapted to life in association with mammals have the advantage of being carried around by their hosts over extended periods. Consequently a low degree of virulence would be predicted. Some such species are able to enter the immunocompetent host through routes other than trauma. In an ecological sense only the fungi that have increased fitness by infecting a mammalian host can be defined as pathogens. Most dimorphic pathogens – i.e., those with a specialized tissue phase – live in endemic areas, due to conditions set by the requirements of their saprobic phase. It is of little use to the fungus to be distributed outside such areas. The fungus' advantage is being distributed by an animal precisely within the area where the saprobic phase can maintain itself. It is logically deducible that only advantageous ecological and evolutionary behaviour can constitute an ecological strategy. The natural host of the pathogenic phase of the fungus is thus most likely to be the animal species that has the same geographical limitation as the fungus, not a human being.

Being carried around by the host during the entire life cycle of the fungus can be viewed as the highest degree of pathogenic adaptation. Infections are then transmitted from host to host. Such fungi are adapted infectious agents to particular species of mammals or to humans. They cause highly inflammatory infections when they are transmitted to other host species; such strains fit the definition of an obligatory pathogen. However, as it is in the interest of both host and fungus, mechanisms have been developed to minimize the damage caused by the immune response. Therefore obligatory pathogenicity is evolutionary an unsuccessful approach, and will quickly lead to benign infection. In fungi this situation may have been reached by some anthropophilic dermatophytes such as as *Trichophyton rubrum*.

For all other clinically relevant fungi, in which the occurrence in the living host is an accidental encounter rather than an ecologically strategic move, the term opportunist should be applied.

The above paragraphs can be summarized in Table 9, displaying a number of possible characteristics of pathogens and adapted infectious agents *vs.* opportunists. Given the fact that diverse evolutionary lines may lead to refinement of pathogenicity in fungi, not all 'pathogens' as defined above will display all characters listed in the table.

Table 9. Summary of potential differences between pathogens and opportunists.

Pathogen/adapted infectious agent:	Opportunist:
1. Tissue phase different from saprobic growth.	1. Tissue phase identical to saprobic growth, or sterile.
2. Zoodemes formed.	2. Zoodemes absent.
3. Clinical spectrum limited and more or less characteristic for the species.	3. Clinical spectrum variable with infection site and route.
4. Ecological strategy using an animal as vector or reservoir.	4. Inoculation coincidental, without ecological strategy.
5. Environmental re-colonization from the host body; infection increases evolutionary fitness of the species.	5. Environmental re-colonization from the host body unlikely; spill-over infection.
6. Reduced virulence due to modulated immunity.	6. Highly inflammatory due to absence of immune modulation.
7. Endogenous reactivation with impaired acquired cellular immune system.	7. Opportunism with impaired innate cellular immune system.
8. Gauss-like kinetics of proliferation after experimental infection.	8. First order kinetics of proliferation after experimental infection.
9. Deep or disseminated mycosis in otherwise healthy patient.	9. Local mycosis in immunocompetent patient, deep or disseminated mycosis only in immunocompromised patient.
10. Discrete taxonomic structure.	10. Radiate taxonomic structure.
11. Virulence factors present.	11. Vitality factors present.

GENERAL TECHNIQUES

Laboratory conditions

Health risks. – All fungi that have been mentioned in the medical literature as potential etiologic agents of disease and hence are listed in this Atlas are attributed to one of four categories according to their occupational health risks (for fungi the highest category is not applied). These are GRAS (Generally Regarded As Safe) and BioSafety Levels 1-3. Reasons for attribution were explained by de Hoog (1996). The categories are defined as follows:

GRAS: Harmless organisms widely applied in food production.

BSL-1: Saprobes or plant pathogens occupying non-vertebrate ecological niches, or commensals. Infections are coincidental, superficial, and non-invasive or mild.

BSL-2: Species principally occupying non-vertebrate ecological niches, but with a relatively pronounced ability to survive in vertebrate tissue. In severely immunocompromis ed patients they may cause deep, opportunistic mycoses. Also pathogens causing superficial infections are in this category.

BSL-3: Pathogens potentially able to cause severe, deep mycoses in otherwise healthy patients.

The handling of fungi listed in each category is prescribed as follows:

GRAS and *BSL-1:* no special precautions needed.

BSL-2: small quantities can be carefully manipulated on normal benches; for larger quantities and for fungi which are easily dispersed by aerosols the use of a safety cabinet is highly recommended. Petri-dishes should be sealed with parafilm.

BSL-3: all manipulations must be carried out in a safety cabinet, air being filtered through HEPA filter; no recirculation is allowed. The cabinet must have its own equipment and must be desinfected after use. No organisms other than BSL-3 class organisms are allowed. It is recommended to leave the air flow always on at half strength. The room should be decontaminated after use. It must be lockable for complete decontamination in case of an accident. Access to the lab is restricted to qualified personnel. It is recommended to work with tubes rather than Petri-dishes.

Discarding slides and specimens. – Slides and specimens can be 'killed off' by autoclaving (30 min at 121°C, 1 Atm) or by dipping into desinfectans, such as chlorate, ethanol 70%, 1% Halacid or glutaraldehyde-containing compounds.

Desinfection. – Cleansing of surfaces is done with quad's, e.g., Tego 51/15 DL (Goldschmidt) or ethanol 70%. Chronically infested incubators may eventually be decontaminated by intensive cleansing with 1% Halacid, 20% benzylbenzoate or by formalin when allowed. Hands are washed with ethanol- or propanol-containing desinfecting agents (Amphisept 80, Goldschmidt).

Mite infestation. – Cultures maintained at room temperature for more than two weeks are frequently infested by mites. Petri-dishes with slowly growing cultures should therefore be sealed with parafilm. To avoid mite infestation in tubes, 1% mercury chloride should be applied on the cotton plug. Infected cultures may be decontaminated from mites by transferring them onto Lindane agar, and infested incubators by applying 20% benzylbenzoate in water.

Collection and pre-treatment of specimens

Appropriate collection of specimens is essential to any mycological study. The quantity should be sufficient for direct microscopic examination as well as for isolation. Transport should preferably take less than 4 hours, or if longer, the specimens should be cooled to prevent overgrowth.

A Wood's lamp may be used to localize mycotic infections, such as favus or ectothrix in cats and dogs. Affected areas show a light yellowish-green fluorescence.

Superficial locations

Skin. – Skin lesions should be superficially decontaminated with a 70% alcohol solution. Small scales are scraped off from the margin of the lesion using a rounded scalpel and are collected on a sterilized object glass. Note that the causative agent in the central portion of

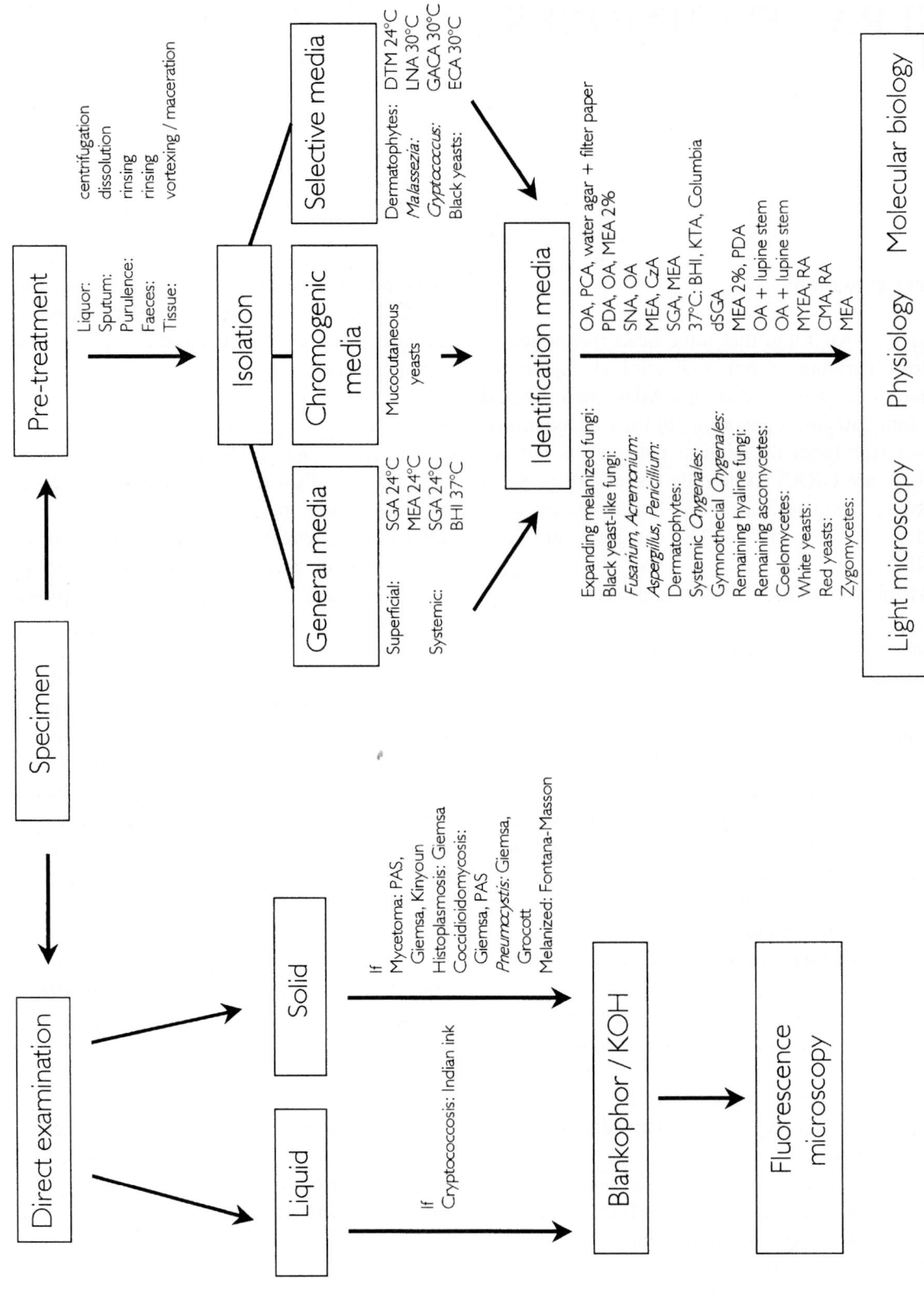

Fig. 9. Overview of procedures for identification of clinical specimens.

the lesion has mostly died off. Strongly macerated skin between the toes can be removed with a forceps. Cotton swabs can be used only on moist skin areas, mucous membranes and external ears. For the latter, also ear washings can be used.

Hair. – Affected or fluorescent hairs are collected by removing them completely using depilation forceps. Fungi are particularly grown from the hair bases. Alternatively, a (massage) brush is rubbed vigorously on the scalp and pressed into a Petri-dish with agar medium. Adhesive tape can also be used to collect the fungi which are transferred directly onto growth media.

Nails. – Superficial layers are polished, affected areas are then filed off and small pieces from the basis of the infection are collected.

Deep locations

Cerebro-spinal fluid and aspirates. – CSF and aspirates can be centrifuged to concentrate fungal particles. Larger liquid samples may be sieved with the aid of a membrane filter, pore size 0.2 μm. For isolation, the filter is placed on an agar medium.

Sputum and bronchial lavage. – Flakes from sputum are rinsed thrice with 0.9% NaCl and cultivated. Bronchial lavages are mostly processed without pretreatment. Mucus of patients with cystic fibrosis may be dissolved first by sputolysin. Centrifugation can be used for concentration of fungal particles.

Purulence. – Purulence can be rinsed and centrifuged to concentrate grains. Grains are collected, rinsed thrice with 0.9% NaCl and cultivated; see also under *tissue*.

Faeces. – Faeces are suspended in 0.9% NaCl and subsequently cultivated.

Blood. – Blood cultures are made in commercially available lysis-centrifugation systems.

Urine. – Cultures are done quantitatively with untreated urine. CFU's over $10^3 \cdot$/mL^{-1} fungal cells are indicative of an infection.

Tissue. – Tissue is rinsed with 0.9% NaCl, cut into small pieces which are placed in a drop of sterile water and cultivated on an agar medium. Alternatively, specimens are gently vortexed with some glass beads with or without trypsine in 0.9% NaCl. Sterile specimens from deep loca-

tions may be cultured on agar or in liquid, shaken growth media at 35°C.

Direct examination

Direct examination of all clinical specimens is recommended by addition (1:1, v/v) of a solution of Calcofluor white (0.1%, w/v) or Blankophor in 20% KOH and observed under fluorescence at 420 nm. Fungal staining is also possible with Grocott (Gomori's methenamine-silver). It is also recommended to use Fontana Masson's staining to distinguish melanized pathogens from, e.g., *Aspergillus*. Solid specimens are digested with 20% KOH and incubated for some minutes at 56°C. For general treatment and staining of tissues refer to Isenberg (1992), and for interpretation of histopathology to Chandler *et al.* (1980), Salfelder (1990), Powers (1998) and others.

Skin, nails and tissue. – Small pieces are incubated in 20% KOH. Optionally, the specimen is gently heated over a small Bunsen flame or incubated for 2 hours (skin) or overnight (nails) in order to soften and clear the material. Staining is usually not necessary. Eventually Parker 51 (Blue-Black Ink permanent / 60% KOH, 1:1) may be used, staining fungal elements deep blue. Methylene Blue can be used for the examination of thin scales from cases of pityriasis and erythrasma. Care should be taken to distinguish fungal elements from the artefact 'mosaic-fungus'. Fungi grow as septate hyphae or as chains of cells, straight through the host tissue regardless of cellular outlines, while 'mosaic-fungus' is unseptate and follows the host cell structure.

Hairs. – Hairs are examined in Amann's Chloral-lactophenol after gentle heating. Action of this fixative is less strong than that of KOH, and artefacts such as 'mosaic fungi' do not occur.

Respiratory specimens, exudates and body fluids. – Calcofluor white stained, wet body fluids may be examined microscopically after centrifugation. If cryptococcosis is suspected, Indian Ink (1:1) preparations should be made to demonstrate capsules around yeasts cells present in spinal fluids. Exudates should be examined macroscopically for granules; when present, these should be stained with Gram and PAS. Kinyoun acid-fast stain is used to recognize actinomycotic granules. Giemsa stain is useful when histoplasmosis or *Pneumocystis carinii* is suspected.

Isolation

Isolation is done on general-purpose, unspecific media when no particular causative agent is suspected. To avoid bacterial contamination, the following antibiotics are added:

* chloramphenicol 40 µg·mL^{-1} (can be autoclaved), or
* penicillin 20 units·mL^{-1} + streptomycin 40 units·mL^{-1} (can not be autoclaved).

In case of mixed fungal populations, cycloheximide (0.05%) can be used to suppress contaminants. However, some pathogenic fungi are sensitive to this compound.

In general, primary cultures of all fungi other than yeasts should be incubated for at least 10 days. For example, dermatophytes are maintained for at least 3 weeks prior to identification.

For a number of fungi, selective media are necessary for a reliable isolation. These are listed in Fig. 9; recipes are given in Table 14.

- Dermatophyte Test Medium (DTM) is a selective medium for the isolation of dermatophytes. It excludes most bacteria and many contaminating fungi. The phenol-red indicator changes from yellow to red as a result of the ammonia production by dermatophytes.

- Systemic, dimorphic *Onygenales* are grown additionally on BHI at 37°C, which reveals their pathogenic stage.

- *Cryptococcus* colonies can be recognized by their tyrosinase activity expressed on GACA medium, staining the colony brown.

- Black yeasts can be isolated by erythritol-containing ECA medium, this carbon source not being utilized by the majority of competing fungi.

- *Malassezia* species are lipophilic or lipid dependent and are thus isolated on media with olive oil or Tweens.

- Yeasts are isolated on commercially available chromogenic agars, enabling the separation of mixed yeast populations. Similar media enable rapid identification of *Candida albicans* using the primary culture; see this chapter under Identification.

Identification

Cultivation and physiology

Cultivation. – The use of Petri-dishes is recommended, except for specimens suspected of BSL-3 agents (*Cladophialophora bantiana, Histoplasma, Coccidioides, Paracoccidioides, Blastomyces, Penicillium marneffei* and *Ramichloridium mackenziei*), which, for reasons of safety, are grown in tubes. Tubes plugged with cotton-wool are preferred over screw-capped bottles.

In general, sporulation of Hyphomycetes is enhanced on poor media such as PCA or SNA. Sterile mycelial growth is reduced when the culture is incubated under near-UV light. Optimal media for the various fungal groups are shown in the diagram of Fig. 9 and their recipes in Table 14; for additional media refer to Atlas & Parks (1993).

Hair perforation. – Some dermatophytes are able to perforate hairs *in vitro*. The fungus is inoculated into 25 mL H$_2$O containing 2-3 drops 10% yeast extract and sterilized human hair, preferably blond children's hair. Hairs are examined microscopically after 3 weeks incubation.

Amino-acid requirements. – Various dermatophytes are heterotrophic for amino acids. Key features are the growth with inositol, thiamine, nicotinic acid and histidine. These are tested against media consisting of casein agar and ammonium nitrate agar (see Table 14: '*Trichophyton* agars' T1-T7 and p. 392).

Nitrite assimilation. – This test (NaNO$_2$, 0.26‰) is particularly suitable for recognizing the most pathogenic black yeast, *Exophiala dermatitidis*, from less pathogenic species of the genus. It is performed in 5 mL liquid medium in test tubes with positive (ammoniumsulphate) and negative controls; background growth may be significant.

Fermentation of carbon compounds. – This is detected by examining a Durham insert tube in a test tube with liquid medium for the presence of gas over a period of 3 weeks. The small inverted tube is placed in 4.5 mL of a 2% solution of the test sugar. After autoclaving, the insert should have completely filled with liquid. The organism is added as 0.5 mL suspension in 7 mL yeast-nitrogen base. D-Glucosamine can be used as source of carbon (C) as well as nitrogen (N). The scoring system used in this Atlas is as follows:

+	positive	+,w	positive or weak
−	negative	−,w	negative or weak
w	weak	s	slow or delayed
v	positive or negative		

Growth with carbon or nitrogen compounds. – The ability to grow on various substances as the sole source of carbon or nitrogen is tested in either liquid or on solid media, or with commercially available strips. The liquid medium test is the most reliable but rather slow. The ID 32C strip method gives good results for yeast diagnostics within 2 days; a variety of miniaturized methods is commercially available. These methods can also be applied to filamentous fungi or black yeasts, but due to a higher degree of background growth and frequent delayed assimilation, the results should be interpreted with care.

For the auxanographic method, tubes with nitrogen base

(for carbon growth tests) and carbon base (for nitrogen growth tests) media are melted and cooled to 45°C in a water bath. A few mL of a suspension of the test organism in water are poured into a sterile Petri-dish, a tube of the appropriate medium is added, and the whole is gently swirled round to mix thoroughly. When the agar has set, a small amount of the test substance is deposited at 4 to 6 sites (depending on the size of the dishes used) around the perimeter of each dish. In most cases powdered or crystaline chemical is used and transferred by the tip of a small spathula. In the case of ethylamine and nitrite the tip of an inoculating needle is dipped into a saturated solution of the chemical and then touched to the surface of the agar; this avoids over-dosing to toxic levels with these substances. Another way is to use small discs soaked with the substance to be tested and transferred to dishes. The Petri-dishes are inspected daily for 4 days for an opaque area of growth around the respective carbon and nitrogen compounds.

Chromogenic methods for rapid identification of Candida albicans. – Chromogenic agars such as ID Albicans® or CandiSelect® enable direct and specific identification of *Candida albicans* in primary cultures.

Cycloheximide tolerance. – This is tested at concentrations of 0.01% and 0.1% in liquid nitrogen base medium supplemented with 0.5% glucose. The tubes are inspected for growth for up to 7 days before being recorded as negative. The test is useful for a number of basidiomycetous yeasts. The less precise testing with commercial Mycosel agar (0.04%) can be used to distinguish, e.g., *Scedosporium* species.

Benomyl tolerance. – Strains are inoculated near the edge of a Petri-dish with modified Leonian's benomyl agar (LBA) and colony diameters are compared to those of control cultures on medium without benomyl. This is useful for the recognition of basidiomycetes.

Urease activity. – Many fungi can hydrolyze urea to ammonia, but specific differences are found when high urea concentrations are used. Christensen's urea medium is commercially available. A loopful of cells from a 3-day-old culture is suspended in urea broth and incubated at 37°C (even strains that do not grow at this temperature). For yeasts the test is performed in liquid medium and inspected at intervals of approximately 30 min for the change of colour (note that after 4 hours all organisms may be positive). Otherwise, the test is performed at 24°C and incubated for up to 7 days (note that after 10 days all organisms may be positive).

Extracellular DNAse. – The production of extracellular DNAse can be tested on a commercially available medium (Difco). The test organisms are inoculated onto Petri-dishes containing the test medium. After incubation and growth the Petri-dishes are flooded with ethanol 96%. After 1 hours a positive result can be observed as a halo around the colony, while the rest of the medium turns opaque.

Enzymes. – The APIZYM system is a semiquantitative micromethod consisting of 20 microcupules, 19 of which contain dehydrated chromogenic substrates for detecting 19 preformed-enzyme activities. The system includes peptidases and glycosidases plus other biochemical reactions. The test strips are inoculated and incubated aerobically at 37°C for 4 hours, and then two reagents are added to develop the chromogenic substrates. The resultant colorimetric reactions are indicative of the degree of enzyme activity and are graded on a scale of 0 to 5 in comparison with the control well and a colour chart (Head & Ratnam, 1988). These test strips are used to distinguish between *Conidiobolus* species.

Temperature tolerance. – Strains are precultured on slants and subsequently transferred to Petri-dishes and incubated at 25, 35, 37, 40 and 42°C. Diameters of colonies are compared.

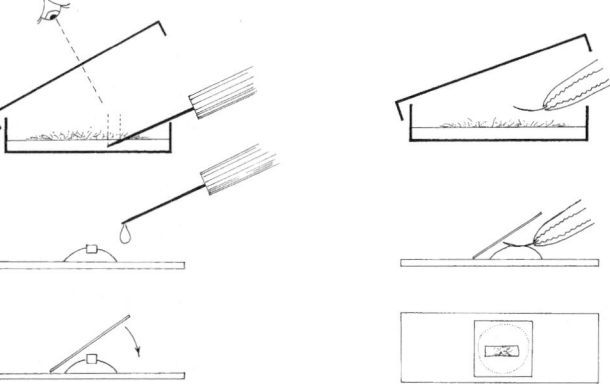

Fig. 10. Diagram of slide preparation. a-c. Sectioning method; d-f. adhesive tape method.

Microscopy

Direct observation. – Colonies of fungi producing large structures (fruit bodies, synnemata, conidial chains) can be observed under a stereo microscope prior to light microscopy. When grown on transparent media, they can be examined through the agar in inverted Petri-dishes under the light microscope at low magnification.

For light microscopic identification, most observations should be done using 100× oil immersion objective and eyepiece micrometer. Slides are made by opening the disk slightly, looking through the lid, in several ways (Fig. 10).

- A small portion of the colony with a straight needle (Fig. 10a-c) . Slides can be made in 0.9% NaCl or in specific dyes (Table 10). A coverslip is provided gently from one side to prevent air bubbles. In case of heavily sporulating cultures it is recommended to make preparations near the colony margin and then, if necessary, move towards the centre.

- Alternatively, a small piece of adhesive tape is pressed onto the colony and subsequently placed over a drop of dye, and a cover slip is placed on top of this (Fig. 10d-f). The latter technique is especially recommended with fungi producing delicate structures or chains of conidia.

Fungi with a strongly hygrophobic conidial system can be best examined in lactic acid/cottonblue after gentle heating. Alternatively, a drop of 96% ethanol can be put upon the specimen prior to covering with the coverslip. Note, however, that delicate structures may shrink, and thus all measurements should be performed in water.

Table 10. Dyes used for microscopic examination.

Shear's fluid:	K-acetate	3 g
	glycerol	60 mL
	methanol	90 mL
	dH_2O	50 mL
Lactic acid:	DL-lactic acid	85%
Lactic acid-cotton blue:	DL-lactic acid 85% aniline blue	100 mL
		0.1 g
Lactophenol:	phenol	20 g
	lactic acid	16 mL
	glycerol	31 mL
	dH_2O	20 mL
Lactophenol-cottonblue:	lactophenol	100 mL
	aniline blue	0.05 g

Undisturbed slide cultures of filamentous fungi are made by bringing a small piece of a colony onto a sterilized slide, and placing a sterilized cover slip on top of it. The slide is incubated in a moist chamber (Fig. 11). For examination, the cover slip is lifted aseptically and lowered onto a slide with a drop of mounting medium. The agar block is removed, and the slide is mounted similarly.

Yeast morphology is observed on rice agar (RA; Table 14). Up to six strains can be streaked onto a single plate; a sterilized cover slip is placed on top of the cells directly after inoculation. Plates are incubated in inverted position at 22°C for 48 hours and observed directly under the microscope. Micromorphology is optimally displayed near the margin of the slide.

Molecular detection

Although widely applied, DNA-based methods are still in a developmental phase, due to insufficent availability of reference data. Moreover, the dignostic applicability is often unclear because the degree of variability of most species for the particular gene analyzed is unknown. The data that are currently collected are used for several purposes, which are partly fundamental and partly practical (Table 15). First, the organisms have to be classified *(taxonomy)* in a evolutionary representation of the fungal kingdom *(phylogeny)*. Then, the units of evolution, the species, are recognized using simple markers (*diagnostics*). Thirdly the spread of the infection can be monitored (*epidemiology*). Each of these areas has its own set of optimal methods (Fig. 15).

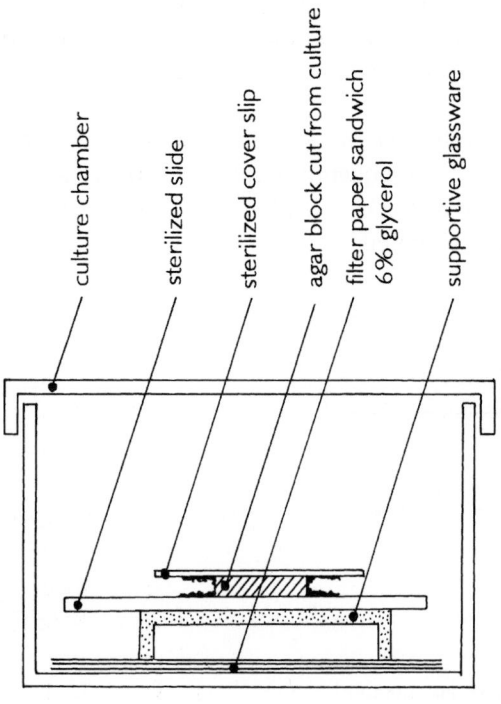

culture chamber
sterilized slide
sterilized cover slip
agar block cut from culture
filter paper sandwich
6% glycerol
supportive glassware

Fig. 11. Diagram of simplified culture chamber for observation of conidiogenesis in undisturbed cultures.

Commercial probes. – DNA-probes for the recognition of the main agents of systemic mycosis, viz. *Blastomyces dermatitidis, Coccidioides immitis, Cryptococcus neoformans* and *Histoplasma capsulatum* are commercially available. The test is based on the ability to release RNA from the fungus and subsegment reassociation with the target strain, thus forming stable hybrids with complementary strands of DNA of the probe (Accuprobe: Gen-Probe, San Diego). The probe is labeled with chemiluminescent acridinium ester molecules which are detected by their emission of light. Portions of 1-2 mm^2 colony are required. These are transferred into lysing tubes with 200 μL of hybridization buffer. Tubes are vortexed and sonicated for 15 min to break the cells and release RNA. Temperature is then raised to 95°C to inactivate the fungi. 100 μL solution is transferred into the probe-tube with labeled DNA. Capped tubes are incubated for 15 min at 60°C to allow hybridization. Non-bound nucleic acids are washed away with selection reagent. The emitted light is recorded with a luminometer (Gen-Probe). Reactions are listed as positive when over 50,000 relative light units (RLU) are recorded. For further methodology is referred to Maresca & Kobayashi (1994).

Fluorescent in situ *hybridization* (*FISH*). – The technique of whole cell FISH is regularly used in bacterial classification. In mycology first attempts have been made. Lischewski *et al.* (1996) identified *Candida* species with specific 18S rRNA-targeted oligonucleotide probes, Sandhu *et al.* (1995) experimented with filamentous fungi and Sterflinger & Hain (1999) established a specific probe for *Coniosporium*. The largest problem in mycological FISH is the rigid cell wall. To preserve cell morphology, fungal material should be fixed in (para)formaldehyde or methanol:acetic acid (3:1, v/v), but fixation decreases cell wall permeability and increases the formation of protein-DNA complexes. Treatment with 0.1N HCl or ß-glucanase (Sterflinger & Hain, 1999) improves permeability, where proteinase K or pepsin treatment destroys the protein-DNA complexes, thus improving penetration and annealing of the probe. Since ITS is spliced out during rRNA maturation, ITS-targeted probes cannot be used. A third problem, especially with black yeasts, is the intensity of the fluorescent signal. Probes bearing several fluorescent labels per molecule are recommended. False positive results occur by autofluorescence of the cell wall and/or organelles. Hybridization is therefore performed with negative controls. In short, the procedure is as follows: young cells are fixed in methanol:acetic acid (3:1, v/v) overnight at 4°C. Following fixation, the cells are washed twice in 70% ethanol and stored at -20°C in 70% ethanol until hybridization. If fixed in freshly prepared 4% paraformaldehyde in phosphate-buffered saline, the cells should be washed

twice for 10-20 min intervals in diethyl pyrocarbonate (0.1%)-treated, autoclaved distilled water (Li *et al.*, 1997) and stored as described above. Pretreated cells are transferred to gelatine or poly-L-lysin coated slides and air-dried. Cell material is dehydrated in three consecutive steps of 70, 85 and 96% ethanol 3-5 min each and air-dried. Hybridization buffer (HB; Table 13) is added and a pre-hybridization is carried out during 1 hour at 46°C in a moist chamber saturated with hybridization buffer. Following pre-hybridization, probes (final amount: 5 ng·10 μL^{-1} HB) are added to the samples and hybridized during 2 hour at 46°C in a moist chamber using HB. Next to specific probes, a universal fungal probe should be used as a positive control and a non-fungal probe to check for unspecific binding. Slides are washed with post-hybridization washbuffer, 20 min at 48°C. After a short rinse with distilled water, the slides are quickly air-dried. The samples are mounted in Vectashield or Citifluor AF1 to prevent the fluorescent signal from bleaching. Results should be observed with a epifluorescence microscope containing the appropriate filtercombinations for the fluorescent dye used. The hybridization temperature varies according to the G+C% of the probe-DNA hybrid. The stringency of the hybridization can be controlled by varying formamide concentration in the hybridization buffer. To optimize hybridization and signal intensity a series of formamide concentrations ranging from 5-25% with 5% intervals should be tested.

rDNA-based identification. – Molecular diagnostics is largely based on comparison of more or less conserved target stretches of ribosomal genes. These can be sequenced, or detected by blotting or by PCR-mediated amplification and subsequent digestion with restriction enzymes. Hopfer *et al.* (1993) and Makimura *et al.* (1994) used a highly conserved 310 bp fragment of SSU DNA; Haynes *et al.* (1995) applied LSU universal primers and subsequent nested PCR. Sandhu *et al.* (1995) listed a large number of LSU target sequences for species identification. For primer and target sequences, see Table 12. Below the methods applied in this Atlas are described.

DNA extraction. – 300 μL of CTAB buffer is added to a 1 mL Eppendorf tube with 40-100 mg Kieselgel-2. Fungal fragments (spathula-tip volume) are added and cells are disrupted mechanically with a tight-fit pestle for 1 min. Add 200 μL CTAB buffer, vortex, and incubate at 65°C for 10 min. Add 500 μL chloroform and mix. Centrifuge at 14,000 rpm for 5 min. Transfer upper layer to a new Eppendorf tube. Add 2 vols (about 800 μL) cold absolute ethanol, mix gently and precipitate DNA at −20°C for 30 min or longer (overnight). Centrifuge at 14,000 rpm for 5-10 min. Decant and wash pellet with cold ethanol 70% and dry for 10 min *in vacuo*. Resuspend in 97.5 µL TE buffer with 2.5 µL RNAse.

Amplification. – 10-100 ng nuclear DNA is added to 50 µL PCR buffer. The buffer should contain Mg^{2+} because most DNA polymerases are Mg-dependant. An improvement of the PCR can be achieved by adding Tween 20, Triton X-100 and/or gelatine. Add 50 pmol of each primer and 0.5-2 U Taq DNA polymerase, and 2 drops of mineral oil. Mineral oil can be omitted when a PCR machine with heated lid is used. PCR amplification is performed as follows: denaturing at 94°C 2-5 min, followed by 30 cycles: 94°C 0.5 min, 45-60°C 1 min annealing, 72°C 1-2 min elongation, with a final extension of 2-5 min at 72°C. The annealing temperature depends on the primer combination used. The melting temperature (T_m) of the primer should be around 50-80°C, and the difference between T_m of the primer combination should not exceed 5-10°C. The advised annealing temperature is 5-10°C below the lowest T_m of any of the primers, but it is recommended to determine the actual annealing temperature empirically. The elongation time depends on the expected length of the amplicon obtained. Normally elongation takes 1 min per Kb at 72°C.

Restriction analysis. – Add 2 U of each restriction enzyme to 8 µL amplicon solution and incubate for at least 2 hours. Enzymes used are *Hin*fI, *Hae*III, *Rsa*I, *Nde*II, *Hha*I, *Dde*I, *Taq*I and *Hpa*II (Table 12). Digests are electrophoresed on 1.2% (SSU and LSU) or 1.5% (ITS) agarose gels during 2-3 hours at 4-6 V·cm⁻¹. For ITS-RFLP an ITS1-ITS4 amplicon is used and a NS1-NS24 or NS1-NS8 amplicon for SSU-RFLP. Each digestion will give a specific banding pattern. The lengths of the expected bands are given with each species in the form of a restriction map, which start and end at both primer annealing sites, thus reflecting actual band lengths. Note that bands <100 bp generally cannot be seen on agarose gels. In general, separate maps are given for SSU, ITS and LSU rDNAs. If maps are combined, essential primer locations are given by an arrow.

Sequencing. – For diagnostic purpose, parts of the ribosomal RNA are sequenced, particularly LSU and ITS. Although there are several sequencing strategies, BigDye terminator cycle sequencing (Perkin Elmer) is convenient and results in a high performance with little background.

Table 11. Universal fungal primes (see also Gargas & DePriest, 1996; Sterflinger *et al.*, 1997).

primer sequence (5'→3')*		position corresponding to *S. cerevisiae*		
Oli4:	CTggTTgATYCTgCCAgT→	18S:	4-22	(Hendriks *et al.*, 1989)
NS1:	gTAgTCATATgCTTgTCT→	18S:	20-38	(White *et al.*, 1990)
Oli7:	AgggYTCgAYYCCggAgA→	18S:	369-387	(Hendriks *et al.*, 1989)
Oli8:	←TCTCCggRRTCgARCCCT	18S:	387-369	(Hendriks *et al.*, 1989)
NS2:	←ggCTgCTggCACCAgACTTgC	18S:	573-553	(White *et al.*, 1990)
NS3:	gCAAgTCTggTgCCAgCAgCC→	18S:	553-573	(White *et al.*, 1990)
Oli10:	←TggYRAATgCTTTCgC	18S:	951-935	(Hendriks *et al.*, 1989)
Oli11:	TTRATCAAgAACgAAAgT→	18S:	962-980	(Hendriks *et al.*, 1989)
NS4:	←CTTCCgTCAATTCCTTTAAg	18S:	1150-1131	(White *et al.*, 1990)
NS5:	AATCTTAAAggAATTgACggAAg→	18S:	1128-1150	(Hopfer *et al.*, 1993)
NS6:	←gCATCACAgACCTgTTATTgCCTC	18S:	1436-1412	(Hopfer *et al.*, 1993)
NS7:	gAggCAATAACAggTCTgTgATgC→	18S:	1413-1436	(White *et al.*, 1990)
V9D:	TTAAgTCCCTgCCCTTTgTA→	18S:	1609-1627	(de Hoog & Gerrits van den Ende,1998)
NS8:	←TCCgCAggTTCACCTACggA	18S:	1788-1769	(White *et al.*, 1990)
NS24:	←AAACCTTgTTACgACTTTTA	18S:	1769-1750	(Gargas & Taylor, 1992)
ITS1:	TCCgTAggTgAACCTgCgg→	18S:	1769-1787	(White *et al.*, 1990)
ITS2:	←gCTgCgTTCTTCATCgATgC	5.8S:	50-31	(White *et al.*, 1990)
ITS3:	gCATCgATgAAgAACgCAgC→	5.8S	31-50	(White *et al.*, 1990)
5.8S-R:	TCgATgAAgAACgCAgC→	5.8S:	34-51	(Vilgalys & Hester, 1990)
ITS4:	←TCCTCCgCTTATTgATATgC	26S:	41-60	(White *et al.*, 1990)
ITS5:	ggAAgTAAAAgTCgTAACAAgg→	18S:	1745-1767	(White *et al.*, 1990)
P1Gu:	←gCTgCATTCCCAAACAACTCgACTC	26S:	283-266	(Guadet *et al.*, 1995)
LS266:	←gCATTCCCAAACAACTCgACTC	26S:	287-266	(Masclaux *et al.*, 1995)
U1:	gTgAAATTgTTgAAAgggAA→	26S:	403-422	(Sandhu *et al.*, 1995)
U2:	←gACTCCTTggTCCgTgTT	26S:	645-662	(Sandhu *et al.*, 1995)
LR7:	←TACTACCACCACCAAgATCT	26S:	1448-1422	(Vilgalys & Hester, 1990)

*R = purines A or G; Y = pyrimidines C or T.

The ITS ribosomal DNA is amplified with primers V9D and LS266 (annealing temperature 50-55°C), using methods as detailed by Gerrits van den Ende & de Hoog (1999). Amplicons are cleaned using Microspin S-300 HR colums (Pharmacia) and sequenced with ITS5 or ITS1 and ITS4 according to the manufacturer's protocol. The material is collected by precipitation and run on an ABI (PE Biosystems) automatic sequencer.

Table 12. Restriction enzymes used.

Enzyme:	Recognition site:	Enzyme:	Recognition site:
*Dde*I	C / TNAG	*Msp*I	C / CGG
*Hae*III	GG / CC	*Nde*II (= *Mbo*I = Sau3AI)	/ GATC
*Hha*I (= *Cfo*I)	GCG / C	*Rsa*I	GT / AC
*Hpa*II	C / CGG	*Taq*I	T / CGA
*Hinf*I	G / ANTC		

Table 13. Recipes of solutions for molecular detection of fungi.

RNAse
pancreatic RNAse (20 U·mg^{-1})	10 mg
Na-acetate 0.01 M (pH 5.2)	1 mL
boil 15 min	
Tris.HCl (1 M, pH 7.4)	

Pepsin
pepsin	100 mg
HCl 1N	1 mL
dH$_2$O	9 mL

Cetyltrimethylammonium bromide(CTAB) buffer
Tris	2.42 g
NaCl	8.2 g
Na-EDTA	0.74 g
CTAB	2.0 g
dH$_2$O	100 mL
pH 7.5 with 1 N HCl	
autoclave 15 min 121°C	
store at room temperature	

Phosphate-buffered saline (PBS 3×)
NaCl	11.4 g
Na$_2$HPO$_4$	1.1 g
NaH$_2$PO$_4$	1.2 g
dH$_2$O 500 mL	
pH 7.2 with 1 N HCl	
autoclave 15 min 121°C	
store at room temperature	

Proteinase K
proteinase K	10 U·mL^{-1}
Tris (pH 7.4)	0.24 g
CaCl$_2$·2H$_2$O	3.0 mg
dH$_2$O	100 mL
filter sterilize	
store in aliquots at −20°C	

Paraformaldehyde fixative
demiwater 65°C	65 mL
paraformaldehyde	4 g
NaOH 2M	1 drop
stir until solution is clarified	
PBS 3×	33 mL
filter sterilize 0.2 μm	
quick chill to 4°C	
store at 4°C	
(Note: use paraformaldehyde solution within 1 week)	

Hybridization buffer (HB)
NaCl 5M	360 μL
Tris 1M (pH 7.2)	40 μL
dH$_2$O	1600-X μL
SDS 10%	4 μL
formamide	X μL

(X depends on the final concentration of formamide, e.g., 200 μl formamide gives a 10% final concentration)

Tris-EDTA buffer (TE)
Tris	0.12 g
Na-EDTA	0.04 g
dH$_2$O	100 mL
pH 8.0 with 1 N HCl	
autoclave 15 min at 121°C	
store at room temperature	

Post-hybridization wash buffer
Tris 1M (pH 7.2)	1 mL
NaCl 5M	4.5 mL
dH$_2$O to a total volume of	50 mL
SDS 10%	100 μL

(Note: the amount of NaCl depends on the formamide concentration used during hybridization)

PCR buffer
Tris (pH 8.3)	10 mM
KCl	50 mM
MgCl$_2$	1.5 mM
gelatin	0.01%
each deoxynucleotide triphosphate	200 μm

Antifungal susceptibility testing

In recent years the National Committee for Clinical Laboratory Standards (NCCLS) of the USA has published two reference methods for antifungal susceptibility, one for yeasts (NCCLS, 1997) and one for filamentous fungi (NCCLS, 1998). Although only few species and not all the available antifungal drugs are included in these documents, these techniques are highly reproducible and provide reliable results.

Yeasts - macrodilution method

Medium. – RPMI 1640 with L-glutamine and without sodium bicarbonate, buffered at pH 7.0 with 0.165 M morpholinepropanesulfonic acid (MOPS).

Procedure. Antifungal stock solutions are to be prepared at concentrations of at least 1280 mg·mL^{-1}, or ten times the highest concentration to be tested . For antifungal agents that cannot be prepared as stock solutions in water, such as ketoconazole, amphotericin B, itraconazole or fluconazole, a dilution series of the agents should be prepared first at 100× final strength in dimethylsulfoxide (DMSO). Each of these solutions should be diluted tenfold in RPMI 1640 broth. Small volumes of the sterile stock solutions are dispensed into sterile polypropylene or polyethylene vials, carefully sealed, and stored at –60°C or –20°C. For preparing diluted antifungal agents, sterile, 12 × 75 mm plastic tubes are used. As there will be a 1:10 dilution of the drugs when combined with the inoculum, the working antifungal solutions concentrations are tenfold those of the final concentration. For each experiment, strains are subcultured onto Sabouraud's dextrose agar or peptone dextrose agar at 35°C for 48 hour for *Candida* species and 72 hours for *Cryptococcus neoformans*. The inoculum suspension is prepared by picking five colonies of at least 1 mm diam and suspending the material in 5 mL of sterile 0.85% NaCl. The turbidity of the cell suspension is adjusted by spectrophotometer at 530 nm to obtain the turbidity of a 0.5 McFarland standard. This procedure will yield a yeast stock suspension of 1-5 × 10^6 cells·mL^{-1}. A working suspension is made by a 1:100 dilution followed by a 1:20 dilution of the stock suspension with RPMI medium, which results in 0.5-2.5 × 10^3 cell·mL^{-1}. Before adjusting the inoculum, 0.1 mL of the various antifungal concentrations are placed in the tubes to bring the drug dilutions to the final test drug concentrations (0.00313 to 16 mg·mL^{-1} for amphotericin B, ketoconazole and itraconazole and 0.125 to 64 mg·mL^{-1} for flucytosine and fluconazole). Growth control tubes (0.9 mL of inoculum suspension and 0.1 mL of drug-free medium) are included for each isolate tested. In addition, 1 mL of uninoculated, drug-free medium is included as a sterility control. The quality control organisms [*Candida*

parapsilosis CBS 604 (= ATCC 22019), *C. krusei* CBS 578 (= ATCC 6258), *C. albicans* ATCC 90028 and 24433, *C. parapsilosis* ATCC 90018 and *C. tropicalis* CBS 94 (= ATCC 750)] are tested in the same manner and are included each time a test is performed. All tubes are incubated without agitation at 35°C for 46 to 50 hours for *Candida* species and for 70 to 74 hours for *Cryptococcus neoformans*, and readings are taken when the turbidity of the growth control tube is evident. For flucytosine, ketoconazole, itraconazole and fluconazole MICs are defined as the lowest drug concentration which shows a visual turbidity less than or equal to 80% inhibition compared with that produced by the growth control tube (1:5 dilution of corresponding growth control). Amphotericin B MIC is defined as the lowest drug concentration at which there is an absence of growth.

Yeasts - microdilution method

The NCCLS microdilution test is essentially identical to the macrodilution reference method, with the exception of the volume of the dilution used and the fact that the test is performed using sterile, disposable, multiwell microdilution plates (96 U-shaped wells). The 10-fold drug dilutions described for the broth macrodilution procedure should be diluted 1:5 with RPMI. The stock inoculum suspensions are prepared and adjusted as described for the macrodilution test. The stock yeast suspension is diluted 1:50, and further diluted 1:20 with medium to obtain the two times test inoculum (1 × 10^3 to 5 × 10^3 CFU·mL^{-1}). The inoculum is diluted 1:1 when the wells are inoculated and the desired final inoculum size is achieved (0.5 × 10^3 to 2.5 × 10^3 CFU·mL^{-1}). The microdilution plates are incubated at 35°C and read at 48 hours (72 hours for most *C. neoformans*). The microdilution wells are scored with the aid of a reading mirror; the growth in each well is compared with that of the growth control (drug-free) well. A numerical score, which ranges from 0 to 4, is given to each well using the following scale: 0, optically clear; 1, slightly hazy; 2, prominent decrease in turbidity; 3, slight reduction in turbidity; and 4, no reduction of turbidity. The MIC for amphotericin B is defined as the lowest concentration at which a score of 0 is observed and, for flucytocine and the azoles, as the concentration at which a score of 2 is observed.

Filamentous fungi - microdilution method

Medium. – RPMI 1640 with L-glutamine and without sodium bicarbonate, and buffered at pH 7.0 with 0.165 M MOPS.

Procedure. – The solutions in DMSO will be diluted 1:50 in test medium. Small volumes of the sterile stock solutions are dispensed into sterile polypropylene or polyethylene vials, carefully sealed, and stored at –60ºC or –20ºC. For preparing diluted antifungal agent use sterile, 12 × 75 mm plastic tubes. As there will be a 1:2 dilution of the drugs when combined with the inoculum, the working antifungal solutions are two times concentrated than the final concentrations. For each experiment, strains are subcultured onto potato dextrose agar slants at 35ºC for 48 hours to 7 days depending of the species. Colonies will be covered with approximately 1 mL of sterile 0.85% saline, and the suspensions are made by gently probing the colonies with the tip of a Pasteur pipette. The resulting mixture of conidia or sporangiospores and hyphal fragments is withdrawn and transferred to a sterile tube. After heavy particles are allowed to settle for 3 to 5 min, the upper homogeneous suspension is collect and mixed. The densities of the suspensions are read and adjusted by spectrophotometer at 530 nm at different transmittances depending of the species. The 1:50 inoculum dilutions will correspond to $2\times$ the density needed of approximately 0.4×10^4 to 5×10^4 CFU·mL^{-1}. The test is performed by using sterile, disposible, multiwell microdilution plates (96 U-shaped wells). Each well will be inoculated with 0.1 mL of the $2\times$ inoculum suspension. This step will dilute the drug concentrations, and inoculum densities, to the final desired test concentrations. The growth control wells will contain 0.1 mL of the corresponding diluted inoculum suspension and 0.1 mL of the drug diluent without antifungal agent. The quality control organisms used in the NCCLS document are *Candida parapsilosis* ATCC

22019, *C. krusei* ATCC 6258, and one strain of *Aspergillus flavus* and one of *A. fumigatus* not numered; al-though other authors use also the strain *Paecilomyces variotii* ATCC 22319. These strains are tested in the same way and are included each time a test is performed. All microdilution trays are incubated at 35ºC without agitation.

Interpretation of results

The predictive value of the MICs obtained using these techniques is unknown because studies of correlation *in vitro - in vivo* are scarce and have only involved a few species and a few antifungals drugs so concise breakpoints are not avaible. However, the following tentative breakpoints (mg·mL^{-1}) have been recently published and although they have to be taken with cautionthey can provide guidelines for antifungal therapy (Sutton *et al.*,1998, 1999a):

Amphotericin B (AMB):	< 1 susceptible	> 2 resistant
Clotrimazole (CTZ):	not determined	
5-Fluorocytosine (5FC):	< 16 susceptible	> 32 resistant
Griseofulvine (GRF):	not determined	
Ketoconazole (KTZ):	< 8 susceptible	> 16 resistant
Fluconazole (FLZ):	< 32 susceptible	> 64 resistant
Miconazole (MCZ):	< 8 susceptible	> 16 resistant
Itraconazole (ITZ):	< 0.5 susceptible	> 1 resistant
Terbinafine (TBF):	not determined	
Voriconazole (VCZ):	not determined	

Table 14. Recipes of recommended media (all recipes based on 1 litre medium).

BCPCG: Bromocresol purple casein glucose agar

skimmed milk	80	g
bromocresol purple 1% in ethanol	2	mL
glucose	40	g
agar	30	g
pH 6.8		

BHI: Brain-heart infusion agar

commercial Bacto BHI agar	52	g
final pH 7.4		

CEA: Casamino acids erythritol agar

casamino acids (Bacto)	3	g
$MgSO_4$	0.1	g
KH_2PO_4	1.8	g
meso-erythritol	10	g
albumin	10	mL
agar	15	g
final pH 6.8		

CDBT: Creatine dextrose bromothymol blue thymine agar

Solution A:

creatinine	1	g
glucose	0.5	g
KH_2PO_4	1	g
$MgSO_4 \cdot 7H_2O$	0.5	g
thymine	0.1	g
agar	20	g
dH_2O	980	mL
pH 5.7		
autoclave 15 lbs		

Solution B:

bromothymol blue	0.4	g
0.01 N NaOH	64	mL
dH_2O	36	mL
store at room temperature and use within 6 weeks		

Medium:

solution A	980	mL
solution B	20	mL
autoclave 15 lbs		

CGB: Canavanine glycine bromothymol blue medium

Solution A:

glycine	10	g
KH_2PO_4	1	g
$MgSO_4$	1	g
L-canavanine sulphate	30	mg
thiamine-HCl	1	mg
dH_2O	100	mL
final pH 6; filter-sterilize		

Solution B:

bromothymol blue	0.4	g
0.01 N NaOH	64	mL
dH_2O	36	mL

Medium:

dH_2O	880	mL
solution B	20	mL

agar	15	g
autoclave 15 lbs		
add solution A	100	mL

Christensen's: Christensen's urea agar

peptone	1	g
glucose	1	g
NaCl	5	g
KH_2PO_4	2	g
phenolred	12	mg
agar	20	g
final pH 6.8		
autoclave 15 lbs		
add 100 mL urea solution (20% in water; filter-sterilize) to 900 mL medium (50°C)		

CMA: Cornmeal agar

ground corn	60	g
glucose	10	g
agar	15	g

CMTA: Cornmeal Tween agar

ground corn	60	g
glucose	10	g
Tween 80	10	mL
agar	15	g

Columbia: Columbia CNA agar

commercial BBL Columbia agar	42.5	g
defibrinated sheep blood	50	g

CzA: Czapek agar

saccharose	30	g
$NaNO_3$	3	g
K_2HPO_4	1	g
KCl	0.5	g
$MgSO4 \cdot 7H_2O$	0.5	g
$FeSO_4 \cdot 7H_2O$	0.01	g
agar	15	g
final pH 6.5		

DRYES: Dichloran rose bengal yeast extract sucrose agar

yeast extract	20	g
sucrose	150	g
dichloran (0.2% in ethanol)	1.0	mL
rose bengal (5%, w/v)	0.5	mL
agar	20	g
final pH 5.6		

dSGA: Diluted Sabouraud's glucose agar

	1:6		1:8	
peptone	1.67	g	1.25	g
glucose	3.33	g	2.5	g
$MgSO_4 \cdot 7H_2O$	1	g	1	g
KH_2PO_4	1	g	1	g
agar	20	g	20	g

*Phenolred solution: 0.5 g in 100 mL 0.06% NaOH.

Table 14 (Contd.)

DTM: Dermatophyte test medium

phytone	10	g
glucose	10	g
phenolred* solution	40	mL
0.8 N HCl	6	mL
cycloheximide	0.5	g
gentamycin sulphate	0.1	g
chlortetracycline·HCl	0.1	g
agar	20	g
final pH 5.5		

ECA: Erythritol chloramphenicol agar

yeast nitrogen base	6.7	g
meso-erythritol	10	g
chloramphenicol	0.05	g
agar	25	g

GACA: Guizotia abyssinica-creatinine agar

G. abyssinica seed	50	g
glucose	1	g
KH$_2$PO$_4$	1	g
creatinine	1	g
agar	15	g

GCP: glycine cycloheximide phenolred agar

cycloheximide	1.6	g
yeast nitrogen base (Difco)	6	g
glycine	10	g
filter-sterilize		
phenolred* solution	30	mL
agar (final pH 5.6)	20	g

Gorodkowa agar

glucose	1	g
peptone	10	g
NaCl	5	g
agar	20	g

KTA: Kane's Tween agar

Tween 80	0.2	mL
K$_2$SO$_4$	0.5	g
Mg$_3$C$_6$H$_5$O$_7$.14H$_2$O	1.5	g
asparagine	5	g
albumin	5	g
glucose	7	g
NaCl	0.85	g
glycerol	20	mL
casamino acids	3	mL
agar	20	g

LNA: Leeming & Notman agar

peptone	10	g
glucose	5	g
yeast extract	0.1	g
ox bile, desiccated	8	g
glycerol	1	mL
glycerol monostearate	0.5	g
Tween 60		0.5mL

whole fat cow's milk	10	mL
agar	12	g

LBA: Modified Leonian's / benomyl agar

malt extract	6.25	g
maltose	6.25	g
KH$_2$PO$_4$	1.25	g
yeast extract	1	g
MgSO$_4$·7H$_2$O	0.625	g
Bacto peptone	0.625	g
agar	20	g
benomyl (Benlate)	10	g

Littman: Littman's oxgall agar

peptone	10	g
glucose	10	g
ox bile, desiccated	10	g
crystal violet	0.01	g
agar	20	g

LIN: Lindane agar
1 mL acetone containing 75 mg lindane dissolved in 1 litre MEA

LA: Lactrimel agar

skimmed milk	200	mL
wheat flour	20	g
honey	10	g
chloramphenicol	0.05	g
agar	15	g

Löwenstein

KH$_2$PO$_4$	2.4	g
MgSO$_4$·7H$_2$O	0.24	g
magnesium citrate	0.6	g
asparagine	3.6	g
potato flour	30	g
glycerol	2	g
malachite green 2%	20	mL
homogenized whole eggs	1000	mL

mDixon: Modified Dixon's agar

malt extract	36	g
peptone 0.6%	6	g
ox bile, desiccated	20	g
Tween 40	10	mL
glycerol	2	mL
oleic acid	2	mL
agar	12	g
final pH 6.0		

McClary's: McClary's acetate agar

potassium acetate	9.8	g
D-glucose	1	g
NaCl	1.2	g
MgSO$_4$·7H$_2$O	0.7	g
yeast extract	2.5	g
agar	15	g

Table 14 (Contd.)

MEA: Malt extract agar	4%	2%
malt extract (10% sugar)	400 mL	200 mL
agar	15 g	15 g
final pH 5.5		

MEYA: Malt extract / yeast-extract agar

malt extract	10	g
yeast extract	4	g
glucose	4	g
agar	15	g
final pH 7.3		

MM: Milk medium

Tween 80	10	mL
agar	15	g
autoclave, cool to 50°C		
fresh milk	10	mL

OA: Oatmeal agar

filtered oat flakes	30	g
agar	15	g

PCA: Potato carrot agar

carrots (boiled & filtered)	20	g
potatoes	20	g
agar	15	g

PDA: Potato dextrose agar

potatoes (boiled & filtered)	75	g
glucose (dextrose)	20	g
agar	15	g

RA: Rice agar

rice extract (ground rice, filtered, autoclaved 60 min, 121° C)	65	g
agar	15	g

SEA: Soil extract agar

filtered garden soil	10	g
agar	15	g

SGA: Sabouraud's glucose agar

peptone	10	g
glucose	10	g
agar	15	g
(or 65 g commercial SGA medium; pH 5.6)		

SNA: Synthetic nutrient agar

KH_2PO_4	1	g
KNO_3	1	g
$MgSO_4 \cdot 7H_2O$	0.5	g
KCl	0.5	g
glucose	0.2	g

saccharose	0.2	g
agar	20	g

Trichophyton-agar 1 (casein agar; basal medium of *Trichophyton* agars T2-T5)

casein (10%, vitamin free)	25	mL
glucose	40	g
$MgSO_4 \cdot 7H_2O$	0.1	g
KH_2PO_4	1.8	g
agar	20	g
final pH 6.8		

Trichophyton agar 2 (inositol-casein agar)

stock inositol[2] solution	2	mL
casein agar	100	mL

Trichophyton agar 3 (thiamine-inositol-casein agar)

stock inositol[2] solution	2	mL
stock thiamine[3] solution	2	mL
casein agar	100	mL

Trichophyton agar 4 (thiamine-casein agar)

stock thiamine[3] solution	2	mL
casein agar	100	mL

Trichophyton agar 5 (nicotinic acid-casein agar)

stock nicotinic acid[4] solution	2	mL
casein agar	100	mL

Trichophyton agar 6 (ammonium nitrate agar, basal medium for *Trichophyton agar* T7)

glucose	40	g
NH_4NO_3	1.5	g
$MgSO_4 \cdot 7H_2O$	0.1	g
KH_2PO_4	1.8	g
agar	20	g

Trichophyton agar 7 (histidine-ammonium nitrate agar)

stock histidine[5] solution	2	mL
ammonium nitrate agar	100	mlL

V8: V8-agar

V8-juice	200	mL
$CaCO_3$	3	g
agar	15	g

YPGA: Yeast peptone glucose agar

glucose	20	g
mycological peptone (Oxoid)	5	g
yeast extract (Oxoid)	10	g
agar	15	g

[2]Stock inositol solution: 2.5 g·L[-1]. [3]Stock thiamine-hydrochloride solution: 10 mg·L[-1].
[4]Stock nicotinic acid solution: 0.1 g·L[-1].
[5]Stock histidine solution: 1.5 g·L[-1]. *Phenolred solution: 0.5 g in 100 mL 0.06% NaOH.

Table 15. Overview of approximate effective application of comparative techniques in mycology.

	Use	Strain	Variety	Species	Genus	Family	Order	Class	Keyref
Ultrastructure:									
Cell wall	Tax							■	Kreger & Veenhuis, 1971
Pore	Tax							■	Moore, 1987
Karyology	Tax						■		Takeo & de Hoog, 1991
Co-Q	Tax			■	■				Yamada et al., 1987
Carbohydrate pattern	Tax				■	■			Weijman & Golubev, 1987
Physiology:									
Classical	Tax			■	■				Yarrow, 1998
API 32C	Diag			■					Guého et al., 1994b
API-Zym	Diag			■					Fromentin et al., 1981
mole% G+C	Tax			■	■				Guého et al., 1992b
rDNA:									
SSU sequencing / RFLP	Tax				■	■	■	■	Sterflinger et al., 1999
LSU sequencing	Diag			■	■	■			Kurtzman & Robnett, 1998
5.8S sequencing	Tax				■	■			de Hoog et al., 1999c
ITS sequencing / RFLP	Diag			■					Lieckfeldt & Seifert, 2000
IGS sequencing	Epid	■	■						Diaz & Fell, 2000
Sequencing coding genes:									
Tubulin	Tax			■	■	■	■		Keeling et al., 2000
Actin	Tax			■	■	■	■		Donnelly et al., 1999
Chitin synthase	Tax				■	■	■	■	Karuppayil et al., 1996
NASBA	Diag			■	■				Compton, 1991
nDNA Homology	Tax	■	■						Smith et al., 2000
RAPD / UP-PCR	Epid	■	■						Voigt et al., 1995
Isoenzymes (MLEE)	Epid	■	■						Barr et al., 1997
G+C Melting curve	Tax								Guého et al., 1997
Classical /									
PCR-Fingerprinting:									
M-13	Tax	■	■	■					Weising et al., 1995
T3B	Epid/Tax	■	■						Gräser et al., 2000a
Microsatellite	Epid	■	■						Weising et al., 1995
mtDNA RFLP	Epid/Tax	■	■	■					de Cock 1994
AFLP	Epid	■							Savelkoul et al., 1999
Karyotyping	Epid	■							Franzot et al., 1998

AFLP = Amplified Fragment Length Polymorphism; Co-Q = Coenzyme Q; IGS = InterGeneric Spacer; ITS = Internal Transcribed Spacer; LSU = Large SubUnit; MLEE = MultiLocus Enzyme Electrophoresis; NASBA = Nucleic Acid Sequence-Based Amplification; RAPD = Random Amplified Polymorphic DNA; RFLP = Restriction Fragment Length Polymorphism; SSU = Small SubUnit rDNA; UP-PCR = Universally Primed PCR.
Diag = Diagnostics; Epid = Epidemiology; Tax = Taxonomy.

PSEUDOFUNGI: *OOMYCOTA*

General remarks. These groups of lower, fungus-like organisms comprise very few clinically relevant species. Both groups contain preponderantly plant- and animal-inhabiting species, with environmental phases in aquatic environments. On the basis of SSU rDNA sequence data the groups are only distantly related to the true fungi (Bowman *et al.*, 1992).

The *Oomycota* are filamentous, coenocytic. There are two main orders. Representatives of the order *Saprolegniales* occur in water and are fish pathogens. Members of the *Peronosporales* cause rots in watery plant substrates; numerous species are serious plant pathogens. On the basis of SSU rDNA sequence data, Bowman *et al.* (1992) found the *Oomycota* close to the algae rather than to the fungi. Therefore they are classified in a separate Kingdom, the *Chromista*. Kwon-Chung (1994) stressed that despite this phylogenetic position, the clinical picture closely resembles that of fungi. Only a single, opportunistic pathogen in warm-blooded animals has been described, *Pythium insidiosum*.

Chromista, Oomycota, Peronosporales, Pythiaceae. Genus: *PYTHIUM*

Generic description. Colonies spreading; mycelium coenocytic. Zoosporangia filamentous or obovoidal, liberating biflagellate zoospores through discharge tubes. Oogonia of variable shape; antheridia present or absent. Oospores usually single.
Reference. Van der Plaats-Niterink (1981).

Pythium insidiosum de Cock *et al.*

Colony characteristics. Cultures (CMA) expanding, white, flat, submerged.
Microscopy. Hyphae 4-6 µm wide, irregularly branched (branches 2.5-4.0 µm diam), sparsely septate in wider hyphae, locally disarticulating. Club-shaped appressoria present. Zoosporangia (water cultures) undifferentiated, filamentous. Zoospores with two lateral flagella.
Sexual organs. Oogonia intercalary, subspherical, 23-30 µm wide. Antheridia produced from adjacent hyphae, clavate, terminally up to 10 µm wide.
Physiology. Optimal development at 35°C, maximum growth temperature 45°C.
Differential diagnosis. *P. insidiosum* can be recognized by local, double septa formed in hyphae on various agar media, and by thick-walled fertilization tubes (de Cock *et al.*, 1987).
General remarks. An immunodiffusion test was developed by Prachharktam *et al.* (1991). Antigenic relationships were further studied by Mendoza & Marin (1989). A method to induce zoospore formation was introduced by Chaiprasert *et al.* (1990); the life cycle was described by Mendoza *et al.* (1993).
Pathogenicity. BSL-2. This is the causative agent of 'swamp cancer' in horses (Purcell *et al.*, 1994), cattle (Santurio *et al.*, 1998) and dogs (Dykstra *et al.*, 1999). The disorder is characterized by tumoural masses which develop mostly in the extremities and are ulcerative, with numerous sinus tracts; bones may be affected. It may also occur in cats (Bissonnette *et al.*, 1991), rarely in humans (Sathapayavongs *et al.*, 1989; Virgile *et al.*, 1993; Wanachiwanawim *et al.*, 1993). Shenep *et al.* (1998) reported a facial infection in a child.
Distribution. The species is restricted to tropical regions (Mendoza *et al.*, 1988; Meireles *et al.*, 1993).

References. De Cock *et al.* (1987), Mendoza *et al.* (1993, 1996), Imwidthaya (1994), Chaffin *et al.* (1995).

Nomenclature. *Hyphomyces destruens* Bridges & Emmons - J. Am. Vet. Med. Assoc. 138: 588, 1961 (invalid) ≡ *Pythium destruens* Shipton - J. Med. Vet. Mycol. 25: 150, 1987.

Pythium insidiosum de Cock, Mendoza, Padhye, Ajello & Kaufman - J. Clin. Microbiol. 25: 345, 1987.

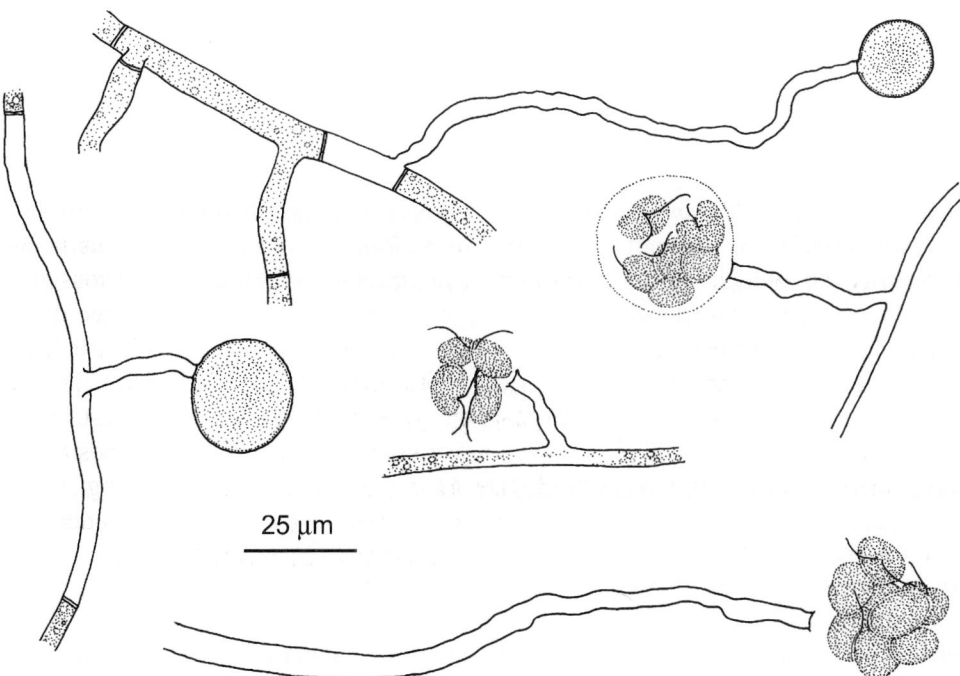

Pythium insidiosum, CBS 673.85. Filamentous zoosporangia releasing zoospores. Redrawn, courtesy A.W.A.M. de Cock.

PSEUDOFUNGI: *MESOMYCETOZOA*

General remarks. The division *Mesomycetozoa* was created by Herr *et al.* (1999) for a small group of protozoa-like organisms, of which *Rhinosporidium* *seeberi* was classified until recently in the fungi. However, *Rhinosporidium* has also been suggested to be a cyanobacterium (Ahluwalia, 1999).

Protista, Mesomycetozoa. **Genus**: *RHINOSPORIDIUM*

Rhinosporidium seeberi (Wernicke) Seeber

Description *in vivo*. The fungus penetrates the mucosal epithelium. Spherical, 6-10 µm wide cysts are formed in subepithelial tissue, prevalently in the nasal mucosa. The sporangia grow out to about 100 µm diam; the up to 5 µm thick wall has a chitinous outer layer and a cellulosic inner layer. After repeated nuclear division, cleavage of the cytoplasm occurs and the spherule swells to 250 µm diam. Several thousands of spherical endospores 7-9 µm diam, each containing multiple, indistinct, spherical bodies, are liberated through a rupture in the wall at a pre-formed thinner wall region. Liberated spores are encapsulated in mucoid, granular material and remain in the epithelium. They are surrounded by tissue, and the cycle is repeated. Massive growth of the fungus leads to the formation of large, friable, polyp-like structures. The polyps may become polymorphous and pedunculated, and develop from pink to finally deep red. At close examination the spherules can be seen macroscopically as whitish spots.
Differential diagnosis. The sporangia stain with mucicarmine, unlike the spherules of *Coccidioides immitis*. For histopathological diagnosis, see Kamal *et al.* (1995).
Molecular diagnostics. SSU restriction map based on NCBI AF118851:

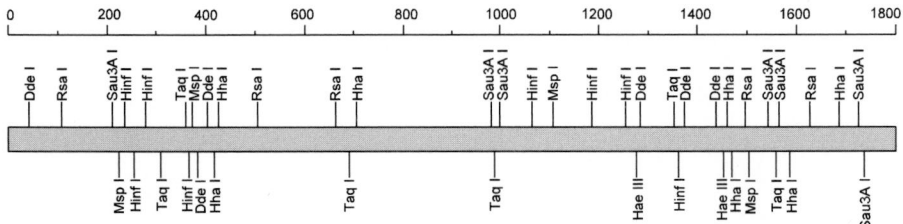

Pathogenicity. BSL-2. The disorder is referred to as rhinosporidiosis (p. 25). There is an association of the disease with patients working in stagnant fresh water. Probably some initial trauma of the mucosa is a predisposing factor. Large polyps may obstruct respiration. They may be recurrent after surgical removal. In dry, dusty areas, most cases are ocular rather than nasal (Pe'er *et al.*, 1996); conjunctival cases were reported by Mukhtar (1999). Swollen, freely movable growth is formed in the lacrymoidal duct (Moses *et al.*, 1990). Cutaneous cases were reported by Ramanan & Ghorpade (1996) and Thappa *et al.* (1998). The disease is chronic, and usually painless. The infrequent cases of infection of trachea and bronchus may be fatal due to obstruction of air passage. Dissemination is extremely rare. The disease is also frequently found in horses and cattle (Niño & Freire, 1964; Pal, 1995a). An epidemic in captive swans on a lake in Florida was reported by Kennedy *et al.* (1995).
Distribution. The species occurs mainly on the Indian subcontinent, in Nepal, South East Asia, East Africa, and tropical South America. Vukovic *et al.* (1995) reported on an outbreak in the Balkan. The species may be underdiagnosed (Snidvongs *et al.*, 1998).

General remarks. The fungus cannot be grown on routine laboratory media. *In vitro* maintenance concerns culture on epithelial cell lines. Ahluwalia (1999) suggested that a *Microcystis* species (*Cyanobacteria*) living in pond water is concerned.

References. Rippon (1988), Thianprasit & Thagerngpol (1989), Kwon-Chung & Bennett (1992), Gigase & Kastelyn (1993), Gori & Scasso (1994), Azadeh *et al.* (1994), Kennedy *et al.* (1995), Fadl *et al.* (1995), Date *et al.* (1995), Jacyk *et al.* (1997).

Nomenclature. *Coccidium seeberi* Wernicke, *in* Belou - Trat. Parasitol. p. 62, 1903 ≡ *Rhinosporidium seeberi* (Wernicke) Seeber - B. Aires, 1912.

Rhinosporidium kinealyi Minchin & Famtham - Quart. J. Microsc. Sci. 49: 521, 1905.

Rhinosporidium equi Zschokke - Schweiz. Arch. Tierheilk. 55: 641, 1913.

Rhinosporidium ayyari Allen & Dave - Ind. Med. Gaz. 71: 376-395, 1936.

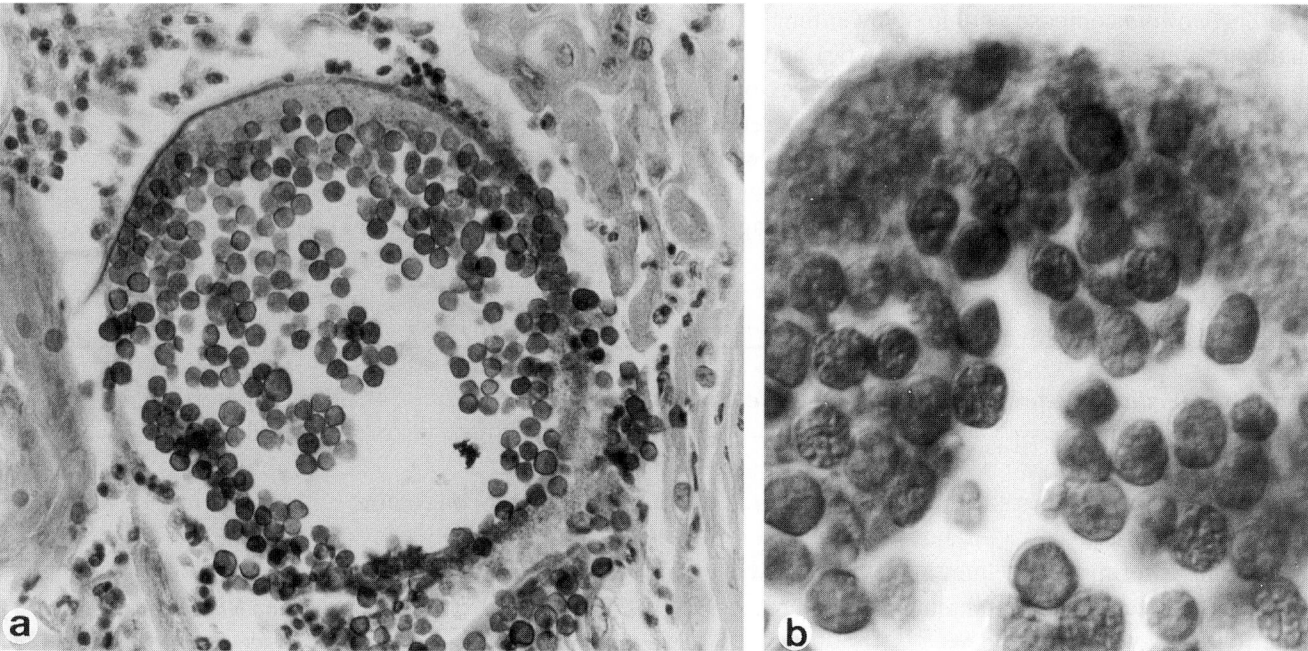

Rhinosporidium seeberi in human tissue. Mature sporangium and endospores. a. ×320; b.×1600.

ZYGOMYCOTA: EXPLANATORY CHAPTER, KEY TO THE GENERA, AND DESCRIPTIONS

Criteria for distinction of *Zygomycota*

The *Zygomycota* compose a group of lower fungi whose thalli are generally unseptate (coenocytic) and which produce zygospores after fusion of isogamic sex organs (gametangia). Three orders contain clinically relevant speices, viz. the small group of *Entomophthorales*, which have forcibly discharged spores, and the *Mucorales* and *Mortierellales* in which the spores arise by cleavage of the sporangial plasma and are passively liberated.

In tissue, members of *Zygomycota* are recognized by 10-20 μm wide, hyaline, haphazardly branched hyphal elements without septa. The fungi often stain with difficulty in PAS, H&E or GMS. Cultures are expanding and woolly (*Mucorales*), expanding and thin floccose (*Mortierellales*) or smooth and grooved (*Entomoph-thorales*), grey or white.

A 26S rDNA phylogeny of clinically relevant species was provided by Voigt *et al.* (1999).
Reference. O'Donnell (1979).

Key to the medically relevant genera of Zygomycota:

1a. Fungus forming cysts with endosporulation in the host; not growing on routine media → **2**
1b. Fungus forming mycelium in the host; growth on most media rapid and abundant → **3**
2a. Cysts prevailing in nasal mucosa . . . . see ***Rhinosporidium* (57)**
2b. Cysts exclusively in lungs . . . . see ***Pneumocystis* (131)**
3a. Spores single, borne on tubular sporophores, forcibly discharged, each with large, inflated scar at the base → (***Entomophthorales***) **4**
3b. Spores arising in sporangia which are mostly multi-spored, passively discharged, with insignificant scars → (***Mucorales, Mortierellales***) **5**
4a. Thallus hyphal, often becoming septate; sporophores elongate; spores with rounded tips ***Conidiobolus* (118)**
4b. Thallus often yeast-like; sporophores short; spores often conical . . . . ***Basidiobolus* (116)**
5a. Sporangia 1-spored or few-spored on small denticles on swollen cells → **6**
5b. Sporangia multi-spored → **8**
6a. Sporangia cylindrical, with spores in rows (merosporangia) . . . . ***Syncephalastrum* (113)**
6b. Sporangia with 1 or few spores (sporangiola) → **7**
7a. Sporangiola on recurved stalks . . . . ***Cokeromyces* (72)**
7b. Sporangiola borne on the surface of large swellings . . . . ***Cunninghamella* (74)**
8a. Sporangia flask-shaped . . . . ***Saksenaea* (112)**
8b. Sporangia spherical or pyriform → **9**
9a. Sporangia usually pyriform due to large apophyses of at least one fourth of the sporangium → **10**
9b. Sporangia spherical; apophyses, when present, less than one fifth of the sporangium → **12**
10a. Dark, septate chlamydospores formed in aerial mycelium . . . . ***Chlamydoabsidia* (70)**
10b. Dark, septate chlamydospores absent → **11**
11a. Apophyses funnel-shaped . . . . ***Absidia* (62)**
11b. Apophyses vase-shaped . . . . ***Apophysomyces* (68)**
12a. Sporangia without columellae . . . . ***Mortierella* (76)**
12b. Sporangia with columellae → **13**

13a. Sporangia with apophyses; sporangiophores in groups, unbranched .... ***Rhizopus* (101)**
13b. Above characteristics not combined ➞ **14**
14a. Rhizoids present .... ***Rhizomucor* (94)**
14b. Rhizoids absent .... ***Mucor* (81)**

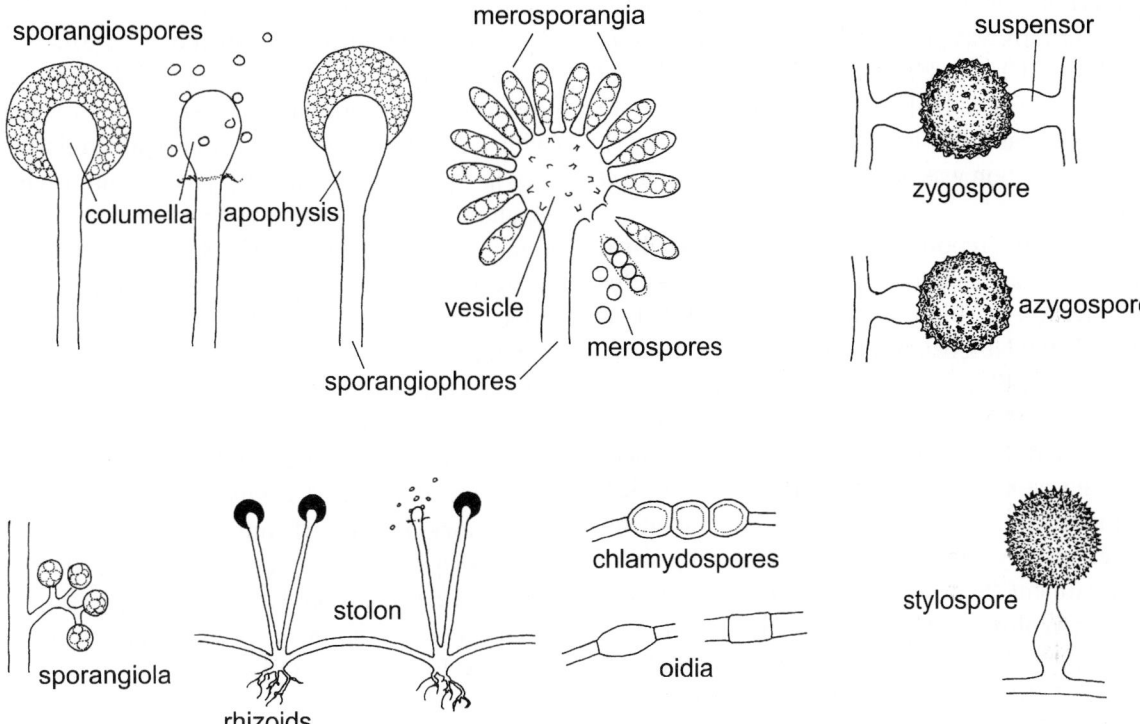

Fig. 12. Diagram of used terminology for the identification of *Mucorales* and *Mortierellales*.

Zygomycota: *Mucorales* and *Mortierellales*

Criteria for distinction of *Mucorales* and *Mortierellales*

Members of *Mucorales* and *Mortierellales* are widely distributed saprophytes in food, soil and air. Several species are thermophilic. Serodiagnosis was elaborated by Kaufman *et al*. (1990). A review of the medical significance of the groups was presented by Espinel-Ingroff *et al*. (1987), Kwon-Chung & Bennett (1992) and Ribes *et al*. (2000); an extensive bibliography of older literature on zygomycoses was published by Ader & Dodd (1979). Experimental infection was performed by van Cutsem *et al*. (1988).

Most commonly reported clinical **rhinocerebral mycoses** pictures are in patients with diabetes mellitus (e.g., Ferry & Abedi, 1983; Galetta *et al*., 1990; Lellouche *et al*., 1993; Nussbaum & Hall, 1994; Yohai *et al*., 1994; Lee *et al*., 1998), with transplants (Nenoff *et al*., 1998; Oo *et al*., 1998) or with HIV infection (van den Saffele & Boelaert, 1996). Diabetic keto-acidosis reduces the Fe-binding capacity of serum (Artis *et al*., 1982); freely available iron stimulates the growth of *Mucorales* in blood vessels. Zygomycoses are promoted by deferoxamine treatment (Seeverens *et al*., 1992; Boelaert *et al*. 1993). **Gastrointestinal** infections also occur relatively frequently (Mooney & Wanger, 1993; Reimund & Ramos, 1994). Patients are generally severely weakened; recently such infections have been noted particularly after organ transplant (Singh *et al*., 1995). Gastric disorders are frequent in insufficiently fed cattle and wild ruminants (Naglič *et al.,* 1986). **Pulmonary** cases were reviewed by Tedder *et al*. (1994). Further, **cutaneous** and **systemic** cases are known, nearly always in patients with severely impaired innate immunity.

Asexual reproduction of members of *Mucorales* and *Mortierellales* takes place by means of sporangiospores produced in **sporangia** (multi-spored), in **sporangiola** (with 1 or few spores) or in **merosporangia** (spores in rows). Sporangia generally have central **columella**, which may extend and be visible below the sporangium as an inflation, the **apophysis**. Several species creep over the agar surface with **stolons**; the substrate can be penetrated by means of **rhizoids**. Hyphae are generally non-septate, or some scattered septa are formed. Thick-walled **chlamydospores** or thin-walled **oidia** may be produced. Sexual reproduction occurs by fusion of two morphologically identical gametangia and results in dark, thick-walled, often ornamented **zygospores** which are borne by **suspensors** (Fig. 12). If similar structures are borne on a single suspensor, they are called **azygospores**. Some *Mortierella* species produce stalked **stylospores**.

References. Zycha *et al*. (1969), Hanlin (1973), Hesseltine & Ellis (1973), O'Donnell (1979).

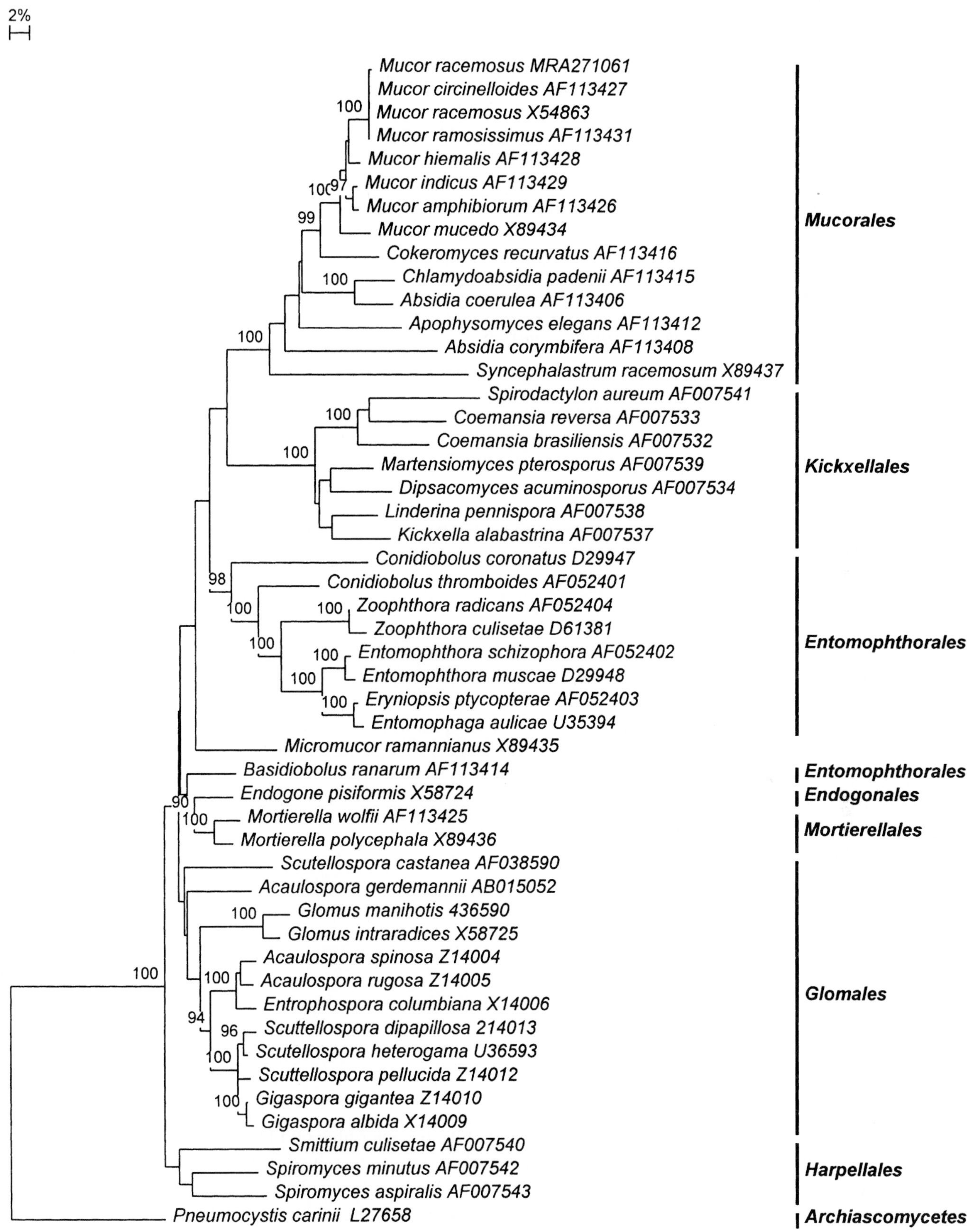

2%

Fig. 13. Phylogeny of *Zygomycota* based on SSU rDNA sequences. Large sections could not be aligned with confidence and have been omitted from the comparison; hence the actual distances between main groups are much larger. The tree was made using the Neighbor joining algorithm with Kimura correction. Bootstrap values >90 from 100 resampled datasets are shown. The orders *Kickxellales, Endogonales, Glomales* and *Harpellales*, which do not contain any clinically significant members, are included to show that the *Entomophthorales* are polyphyletic. *Pneumocystis carinii,* which in the past has been suggested to be a zygomycete, is used as outgroup.

Zygomycota, Mucorales, Mucoraceae. Genus: *ABSIDIA*

Generic description. Colonies growing rapidly, lanose. Stolons and rhizoids present. Sporangiophores arising more or less verticillately from stolons or aerial mycelium. Sporangia terminal, multi-spored, spherical to pyriform; columella with a conspicuous conical apophysis and with one or more projections at the tip. Sporangiospores spherical to ovoidal, smooth-walled or rarely echinulate. Zygospores usually surrounded by finger-shaped appendages arising from one, or from both suspensors.

References: Ellis & Hesseltine (1965), Hesseltine & Ellis (1966), Zycha *et al.* (1969), Scholer *et al.* (1983), Schipper (1990).

Key to the treated species of Absidia and Apophysomyces:

1a.	Sporangiophores greyish-brown; apophyses vase-shaped	*Apophysomyces elegans* (68)
1b.	Sporangiophores hyaline or pale violet-blue; apophyses funnel-shaped → 2	
2a.	No growth at 45°C; colonies blue to violet-blue on PDA	*Absidia coerulea* (62)
2b.	Growth at 45°C; colonies white to grey on PDA	*Absidia corymbifera* (65)

Absidia coerulea Bain.

Colony characteristics. Colonies (MEA) expanding, grey; blue to violet-blue on PDA.

Microscopy. Sporangiophores arising from stolons in whorls of two to five, hyaline to pale violet-blue. Sporangia hyaline, pyriform, 18-65 µm diam; columella 12-32 µm diam, hyaline, hemispherical above the apophysis, with a single apical projection; apophysis conical, separated from the sporangiophore by a septum. Sporangiospores hyaline, smooth-walled, spherical, 3-5 µm diam.

Zygospores. Zygospores spherical, 83-156 (-180) µm diam, coarsely roughened to verrucose, brown to black. Suspensors smooth-walled, typically unequal with finger-like appendages. Heterothallic.

Physiology. No growth at 37°C.

Molecular diagnostics. SSU restriction map based on NCBI AF113406:

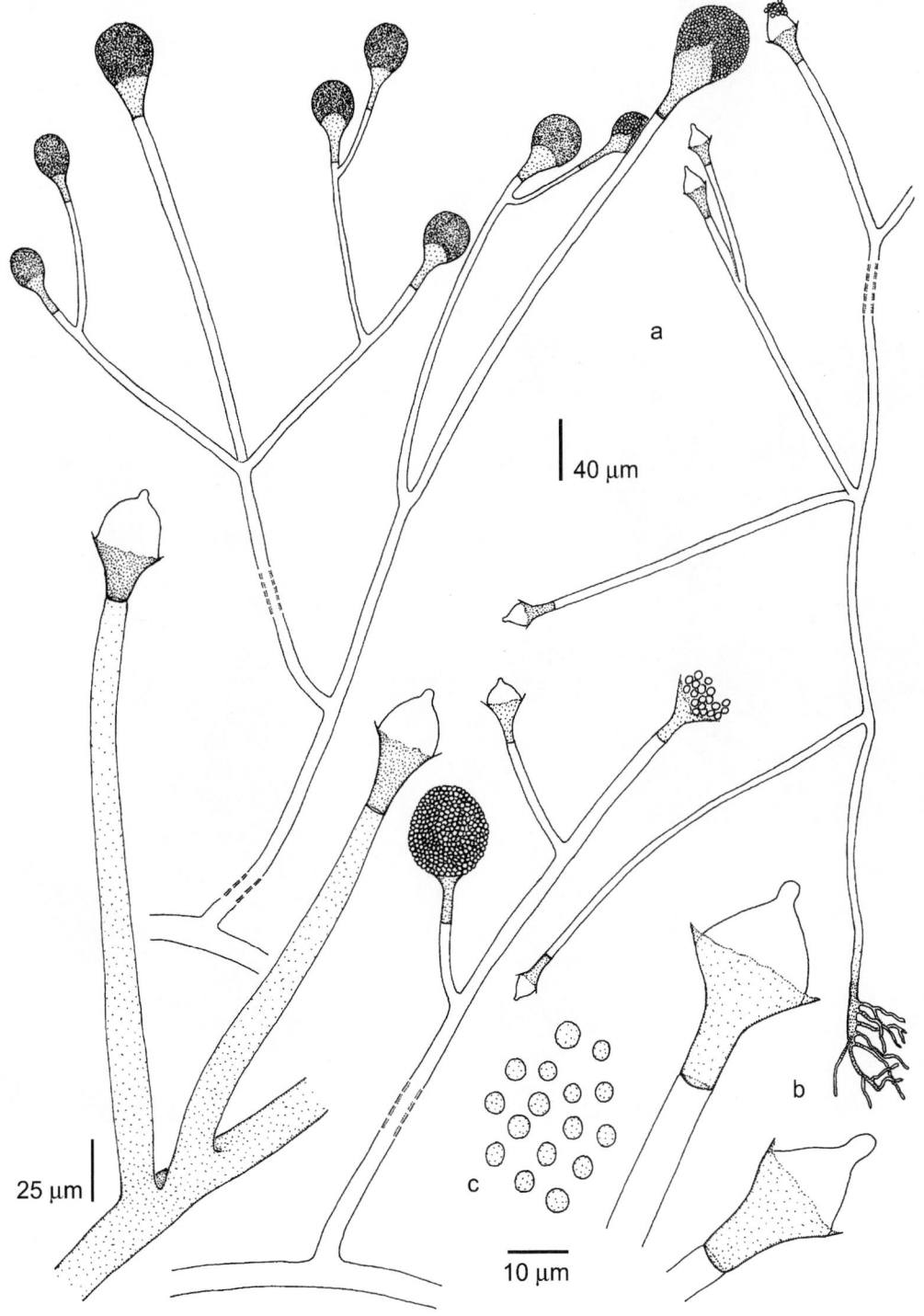

Pathogenicity. BSL-2. Pulmonary infections (Podsiadio & Halweg, 1988).
References. Ellis & Hesseltine (1965).
Nomenclature. *Absidia coerulea* Bainier - Bull. Soc. Bot. Fr. 36: 184, 1889.

Absidia coerulea, CBS 104.08. a. Sporangiophores; b. columellae; c. sporangiospores.

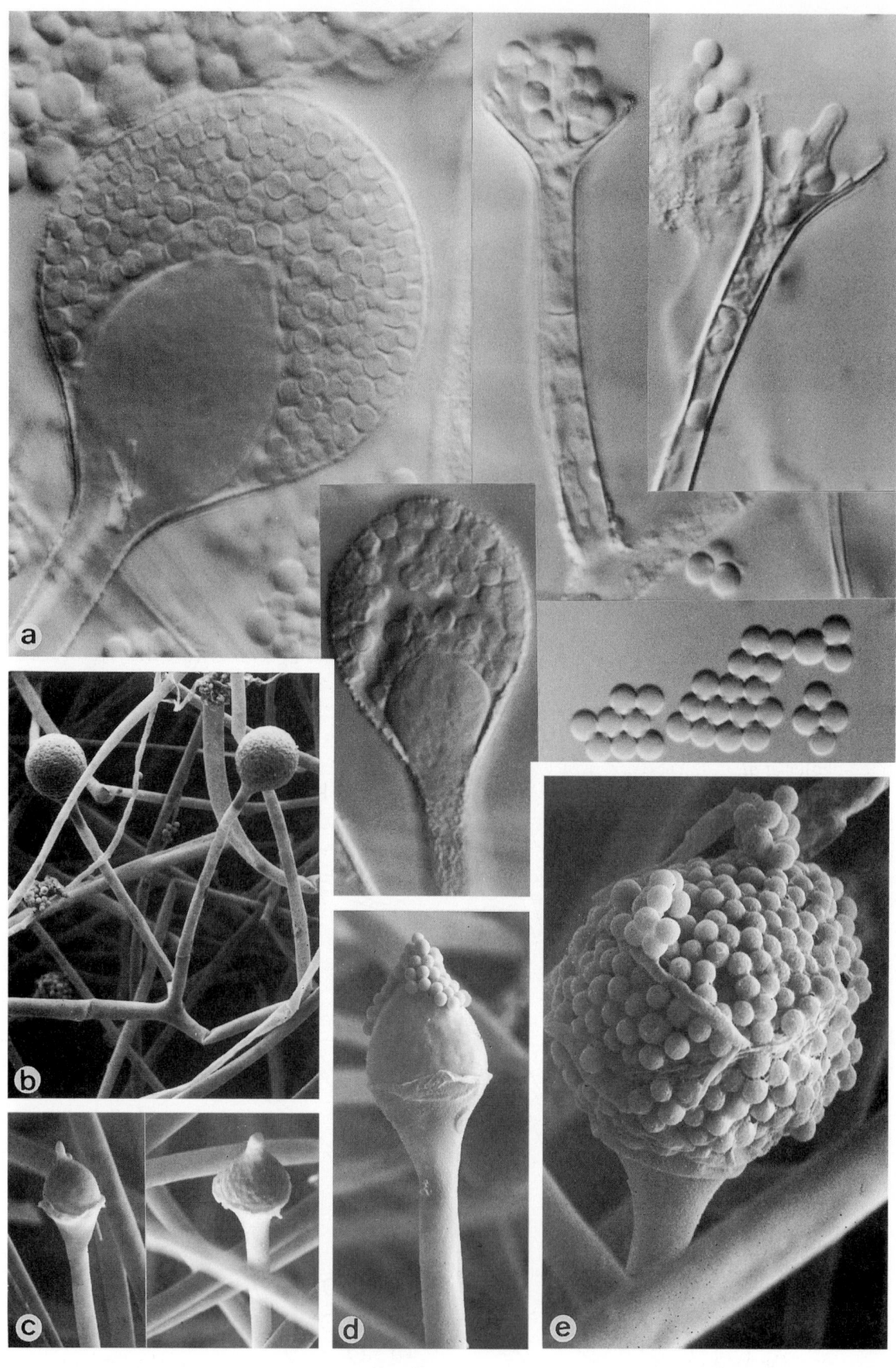

Absidia coerulea, CBS 104.08. Sporangiophores, columellae and sporangiospores. a. ×1600, b. ×435; c. ×740; d. ×980; e. ×1850.

Absidia corymbifera (Cohn) Sacc. & Trott.

Colony characteristics. Colonies (MEA, 30°C) expanding, white to greyish-brown; mycelium profusely branched, with stolons and rhizoids.

Microscopy. Sporangiophores solitary or in groups, arising from aerial hyphae, (sub)hyaline. Sporangia spherical to pyriform, 100-120 (-150) µm; columella comprising 40-60% of the sporangium, with conspicuous conical apophysis, hemispherical, with one or more apical projections. Spores smooth-walled, spherical, 3-4 µm diam, or long-ellipsoidal, 4-5 × 2.5 µm.

Zygospores. Zygospores short-ellipsoidal between suspensors, 60-100 × 45-80 µm, with 1-3 equatorial ridges, thick-walled with flat projections, reddish-brown. Suspensors rough-walled, slightly unequal or equal. Heterothallic.

Physiology. No fermentation; thiamin dependent; max. growth temperature 48-52°C.

Molecular diagnostics. SSU restriction map based on NCBI AF113408:

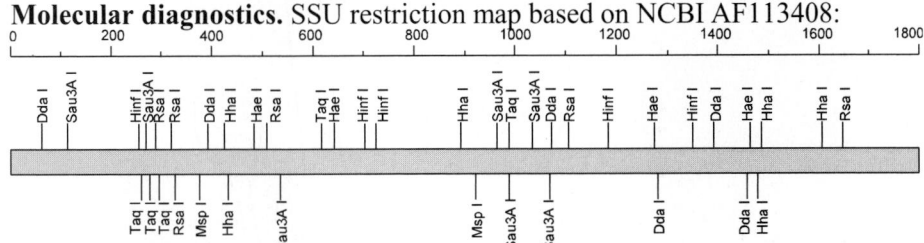

Pathogenicity. BSL-2. Uncommon agent of human (sub)cutaneous mycosis (Hopwood *et al.*, 1992; Jensen, 1992; Lopes *et al.*, 1995c; Kawasaki *et al.*, 1997; Piens *et al.*, 1999). Several cases of invasive infections in AIDS patients have been published (Smith *et al.*, 1989; Chavanet *et al.*, 1990; Hopwood *et al.*, 1992), but also in neutropenic (Manso *et al.*, 1994) or transplant (Jantunen *et al.*, 1996; Leong *et al.*, 1997) patients. Facial zygomycosis in a leukemic patient was reported by Gebhard *et al.* (1995) and a keratitis by Marshall *et al.* (1997). It is the most frequent agent of bovine mycotic abortion. Experimental inoculation was conducted by Jensen *et al.* (1995).

References. Scholer *et al.* (1983), Samson *et al.* (1996), Schipper (1990).

Antifungal susceptibility.

Antifungal	MICs range	MIC 90	Strains	Reference
AMB	0.39-0.78	0.78	3	Wildfeuer *et al.* (1998)
AMB	0.06-0.25	0.5	10	Johnson *et al.* (1999)
ITZ	0.05-0.39	0.39	3	Wildfeuer *et al.* (1998)
ITZ	0.25-0.5	0.25	10	Johnson *et al.* (1999)
KTZ	0.39-0.78	0.78	3	Wildfeuer *et al.* (1998)
VCZ	3.13	3.13	3	Wildfeuer *et al.* (1998)

Nomenclature. *Mucor corymbifera* Cohn, *in* Lichtheim - Z. Klin. Med. 7: 147, 1884 ≡ *Lichtheimia corymbifera* (Cohn) Vuillemin - C.R. Acad. Sci., Paris 136: 516, 1903 ≡ *Absidia corymbifera* (Cohn) Saccardo & Trotter - Syll. Fung. 21: 825, 1912.

Mucor ramosus Lindt - Arch. Exp. Path. Pharmacol. 21: 269, 1886 ≡ *Lichtheimia ramosa* (Lindt) Vuillemin - Arch. Parasit. 8: 570, 1904 ≡ *Absidia ramosa* (Lindt) Lendner - Mat. Fl. Crypt. Suisse 3: 144, 1908.

Mucor corymbifera Cohn var. *lichtheimi* Lucet & Costantin - Arch. Parasit. 4: 380, 1901 ≡ *Absidia lichtheimi* (Lucet & Costantin) Lendner - Mat. Fl. Crypt. Suisse 3: 143, 1908.

Mucor truchisi Lucet & Costantin - Arch. Parasit. 4: 362, 1901 ≡ *Absidia truchisi* (Lucet & Costantin) Lendner - Mat. Fl. Crypt. Suisse 3: 146, 1908 ≡ *Mucor corymbifera* Cohn var. *truchisi* (Lucet & Costantin) Sartory - Champ. Paras. Homme Anim. 3: 91, 1921.

Mucor regnieri Lucet & Costantin - Arch. Parasit. 4: 362, 1901 ≡ *Lichtheimia regnieri* (Lucet & Costantin) Vuillemin - C.R. Acad. Sci., Paris 136: 516, 1903 ≡ *Absidia regnieri* (Lucet & Costantin) Lendner - Mat. Fl. Crypt. Suisse 3: 146, 1908 ≡ *Mucor corymbifera* Cohn var. *regnieri* (Lucet & Costantin) Sartory - Champ. Paras. Homme Anim. 3: 92, 1921.

Mucor cornealis Cavara & Saccardo, *in* Cavara - Ann. Ottalmol. 42: 668, 1913 ≡ *Absidia cornealis* (Cavara & Saccardo) C.W. Dodge - Med. Mycol. p. 114, 1935 ≡ *Lichtheimia cornealis* (Cavara & Saccardo) Naumov - Clés Mucorin. p. 77, 1939.

Lichtheimia italiana Costantin & Perin - Bull. Soc. Med. Chir. Pavia 35: 5, 1922 ≡ *Lichtheimia italica* Pollacci & Nannizzi - Miceti Pat. Uomo Anim. 3: 26, 1924 (name change) ≡ *Absidia italiana* (Costantin & Perin) C.W. Dodge - Med. Mycol. p. 112, 1935.

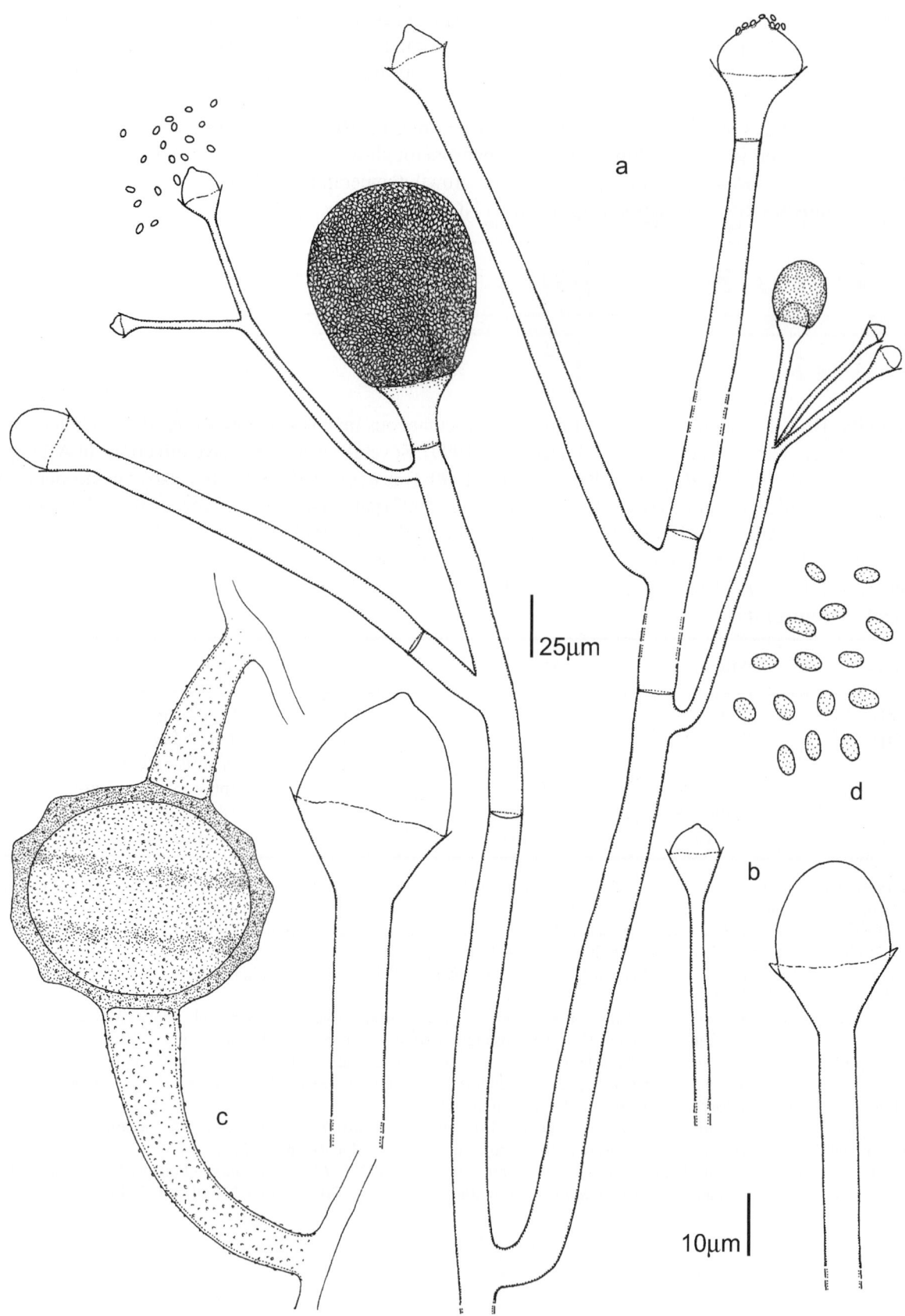

Absidia corymbifera, a, b. CBS 100.31; c, d. CBS 271.65 × 270.65. a. Sporangiophores; b. columellae; c. zygospore, d. sporangiospores.

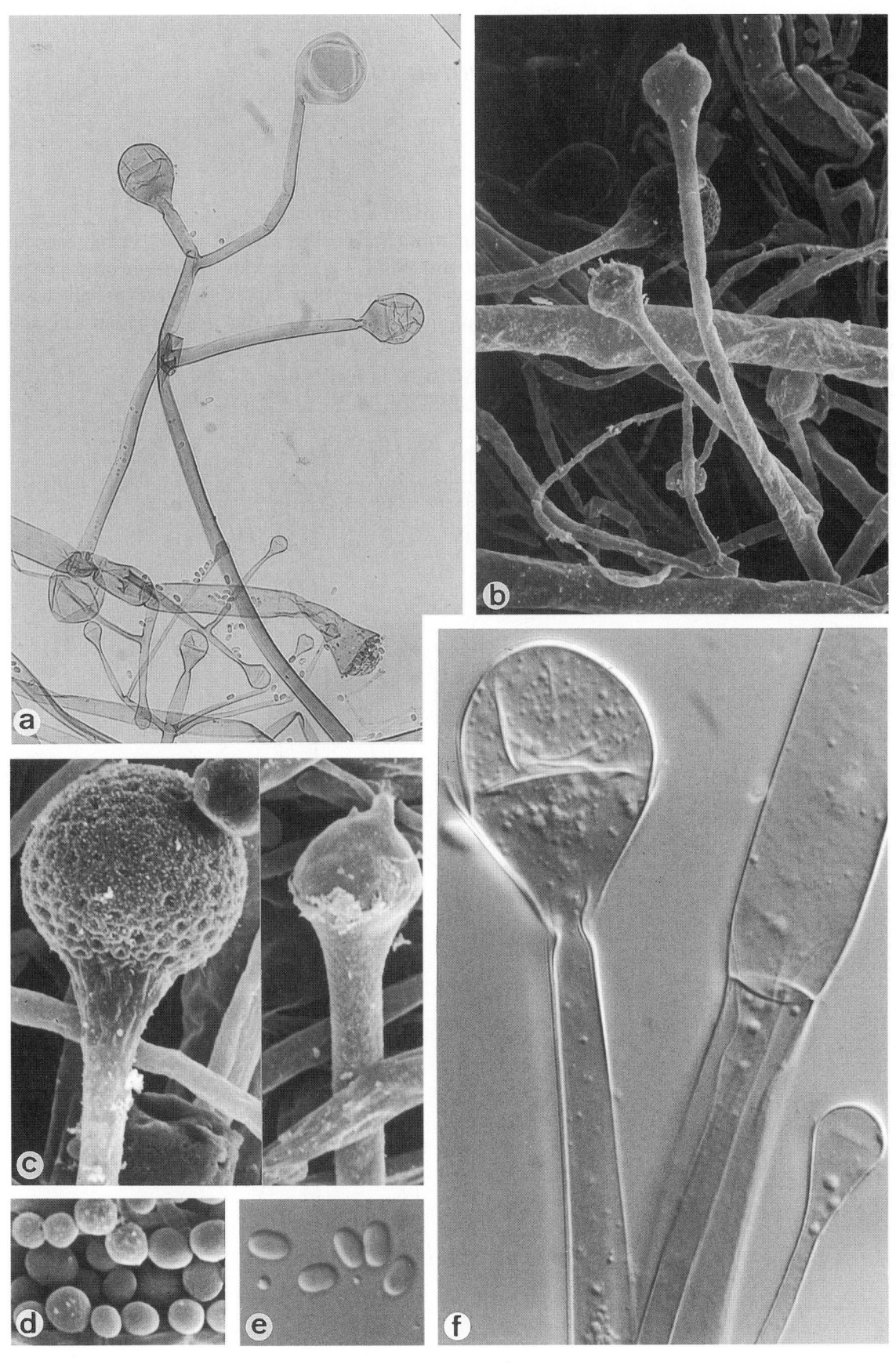

Absidia corymbifera, CBS 100.31. Sporangiophores, columellae and sporangiospores. a. ×512; b. ×1050; c. ×2375; d. ×3000; e,f. ×1600.

Zygomycota, Mucorales, Mucoraceae. Genus: *APOPHYSOMYCES*

Apophysomyces elegans Misra *et al.*

Colony characteristics. Colonies (MEA, 30°C) growing rapidly, brownish-grey.

Microscopy. Sporangiophores generally singly, emerging from aerial hyphae, straight or curved, unbranched, slightly tapering towards the apex, light greyish-brown, up to 540 μm long, 3.4-5.7 μm wide. Sporangia produced terminally and singly, pyriform, with distinct apophyses and columellae, 20-58 μm diam; apophyses vase- or bell-shaped, 10-46 × 11-40 μm; columellae hemispherical, 18-28 μm diam. Sporangiospores subspherical to cylindrical, subhyaline, smooth-walled, 5.4-8.0 × 4.0-5.7 μm.

Molecular diagnostics. SSU restriction map based on NCBI AF113412:

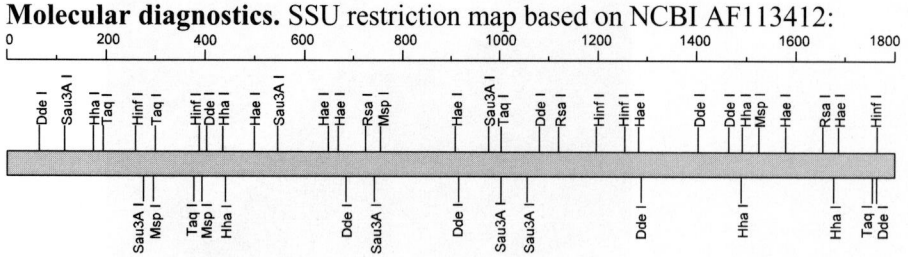

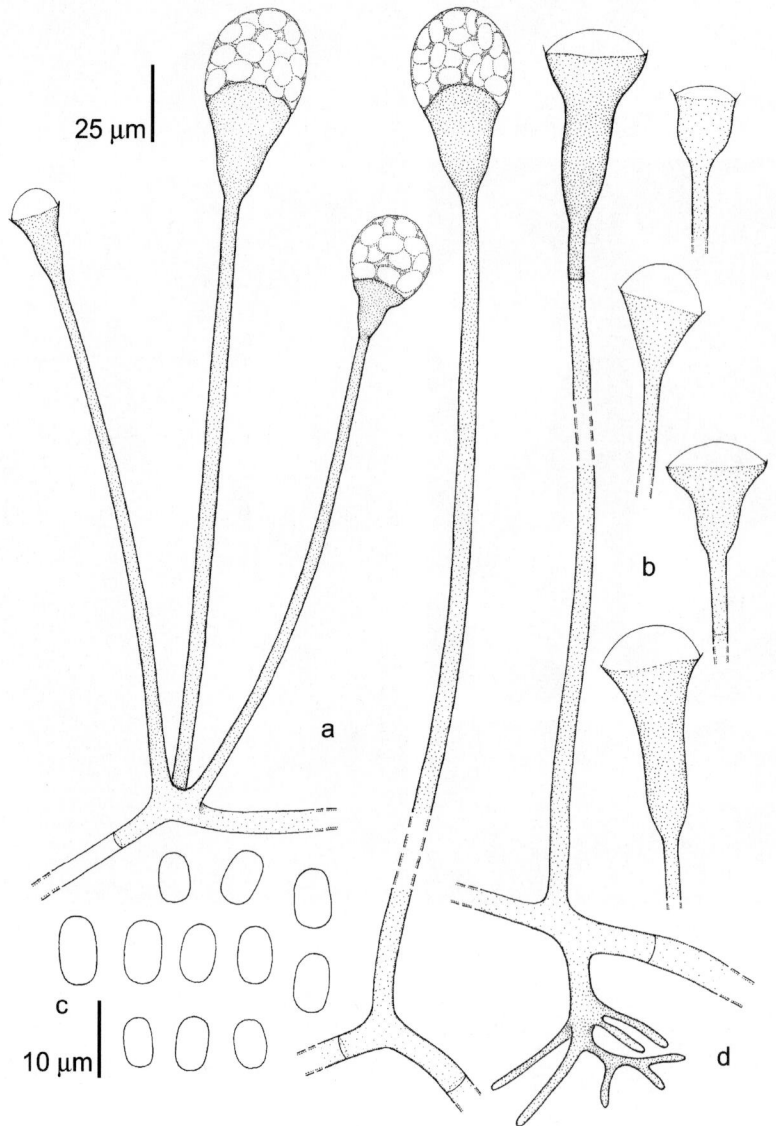

Apophysomyces elegans, CBS 476.78. a. Sporangiophores; b. columellae; c. sporangiospores; d. rhizoids.

68

Pathogenicity. BSL-2. Huffnagle *et al.* (1992) described four cases. Two of these were in patients with diabetes and the other two developed after traumatic inoculation. The case reported by Burrell *et al.* (1998) developed after a cactus spine injury. Lakshmi *et al.* (1993) and McGinnis *et al.* (1993) reported fatal cases of necrotizing fasciitis. Cooter *et al.* (1990) reported a burn wound infection, Mathews *et al.* (1997) an infection after surgery, Naguib *et al.* (1995) a traumatic infection in a patient under immunosuppressive therapy and Chakrabarti *et al.* (1997a) and Cáceres *et al.* (1997) cutaneous infections. In general, cases emerge after severe, extended injury to the cutaneous barrier (Kimura *et al.*, 1999). Systemic cases were presented by Lawrence *et al.* (1986) and Wieden *et al.* (1985). A case of osteomyelitis was presented by Eaton *et al.* (1994). Rhinocerebral mycoses were reported by Radner *et al.* (1995) and Chakrabarti *et al.* (1997b) and a renal infection by Chugh *et al.* (1996). Some of these were in otherwise healthy patients.

References. Misra *et al.* (1979), McGinnis *et al.* (1993), Holland (1997).

Nomenclature. *Apophysomyces elegans* Misra, Srivastava & Lata - Mycotaxon 8: 378, 1979.

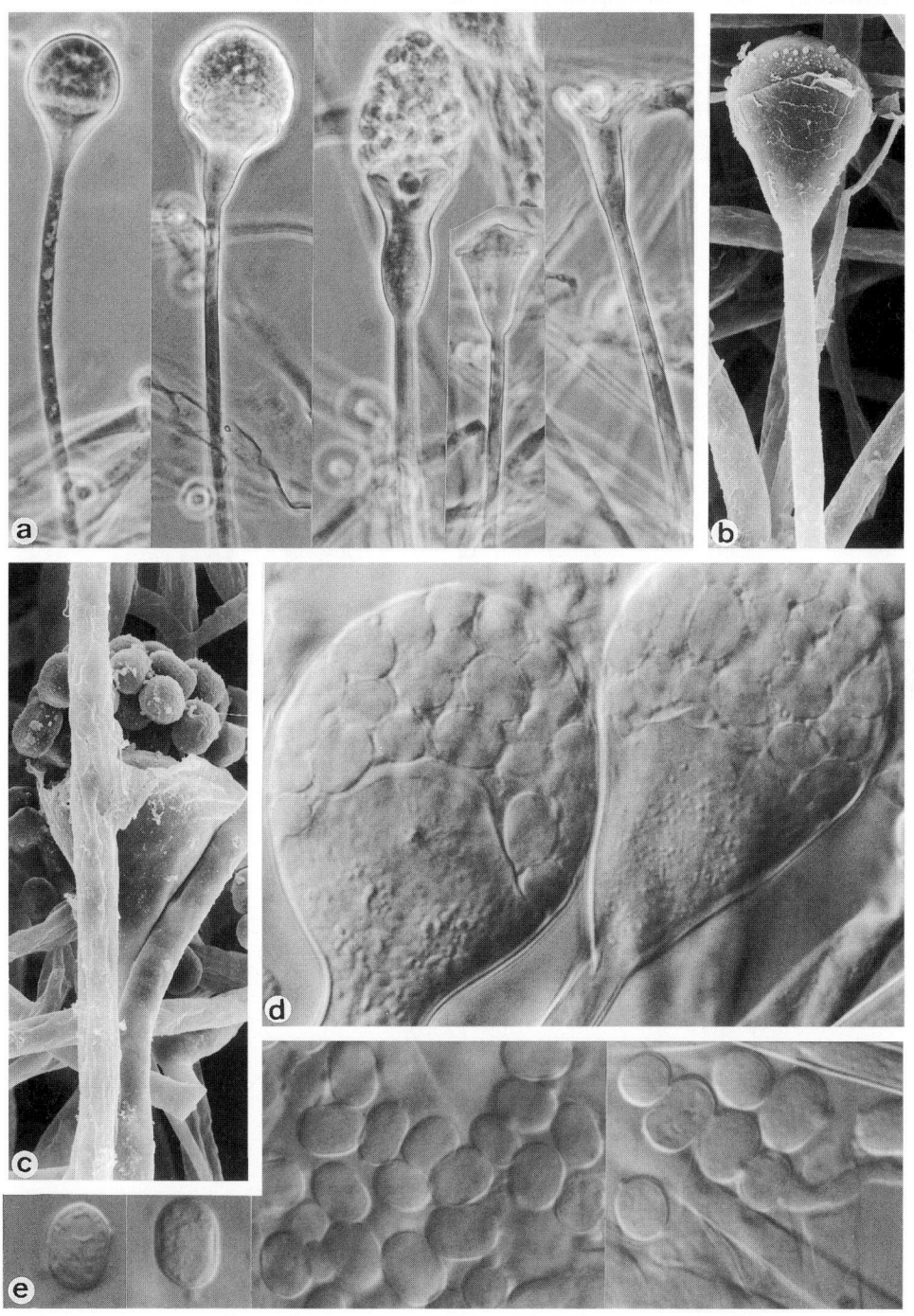

Apophysomyces elegans, CBS 476.78. a. Sporangiophores; b-d. Young and mature sporangia; e. sporangiospores. a. ×510; b. ×1270; c. ×1350; d, e. ×1600.

Zygomycota, Mucorales, Mucoraceae. Genus: *CHLAMYDOABSIDIA*

Chlamydoabsidia padenii Hesseltine & Ellis

Colony characteristics. Colonies (MEA) expanding, greyish-brown.

Microscopy. Sporangiophores erect, arising singly from stolons, alternating with rhizoids and also directly from creeping mycelium, hyaline, minutely striate, usually simple but sometimes sympodially branched two or three times, each successive sporangium borne at a higher level. Sporangia 12-40 μm wide, pyriform, pale yellow, multi-spored; columella 6.5-25.0 μm diam, ovoidal, often with a collar, with conspicuous apophysis. Sporangiospores spherical, thin-walled, 3-5-6.6 μm diam, hyaline. Chlamydospores solitary, terminal, clavate, septate, 26-27 μm wide, thick-walled, brown to black.

Molecular diagnostics. SSU restriction map based on NCBI AF113415:

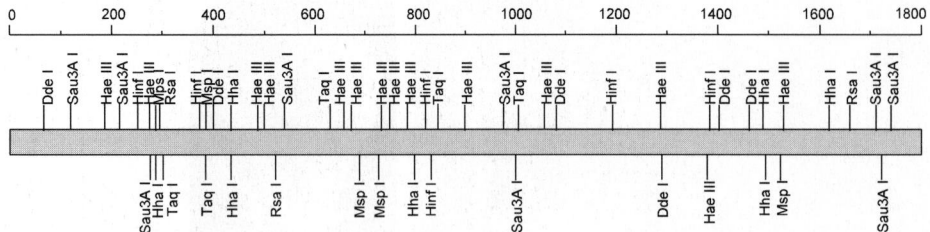

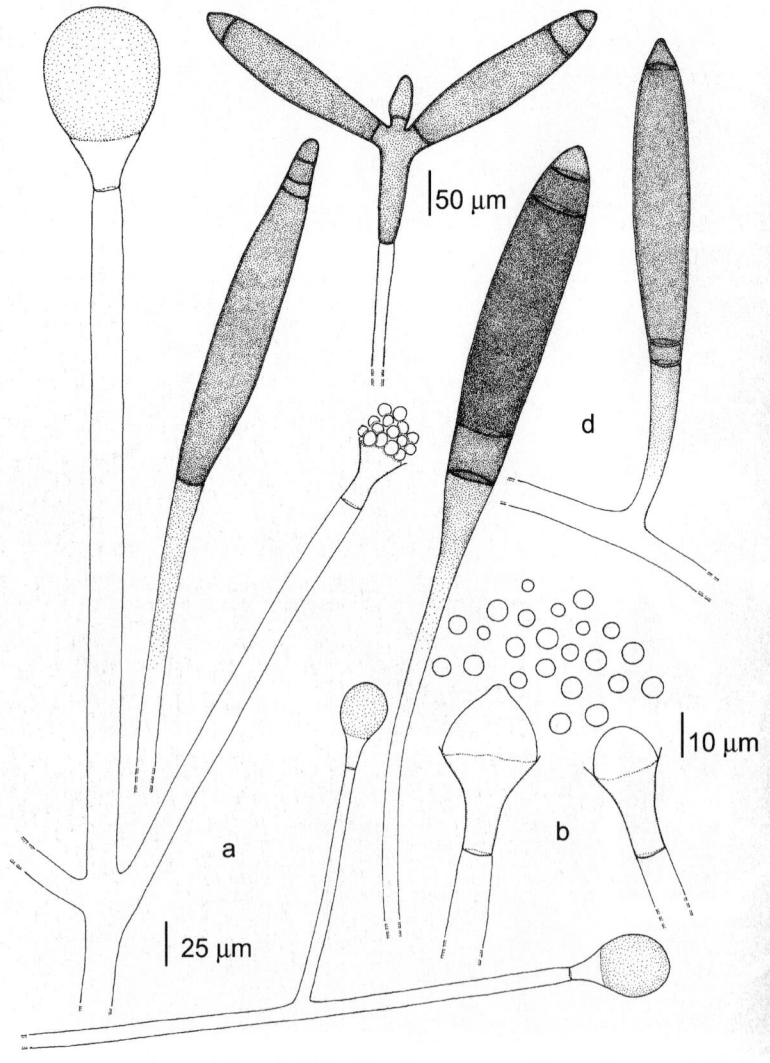

Chlamydoabsidia padenii, CBS 172.67. a. Sporangiophores; b. columellae; c. sporangiospores; d. chlamydospores.

Pathogenicity. BSL-1. A keratitis has been reported by Cambon *et al.* (1992), but this strain, CBS 101040, was reidentified as *Absidia corymbifera* (M.A.A. Schipper, pers. comm.). The clinical relevance of this species thus is doubtful.

Reference. Hesseltine & Ellis (1966).

Nomenclature. *Chlamydoabsidia padenii* Hesseltine & Ellis - Mycologia 58: 763. 1963.

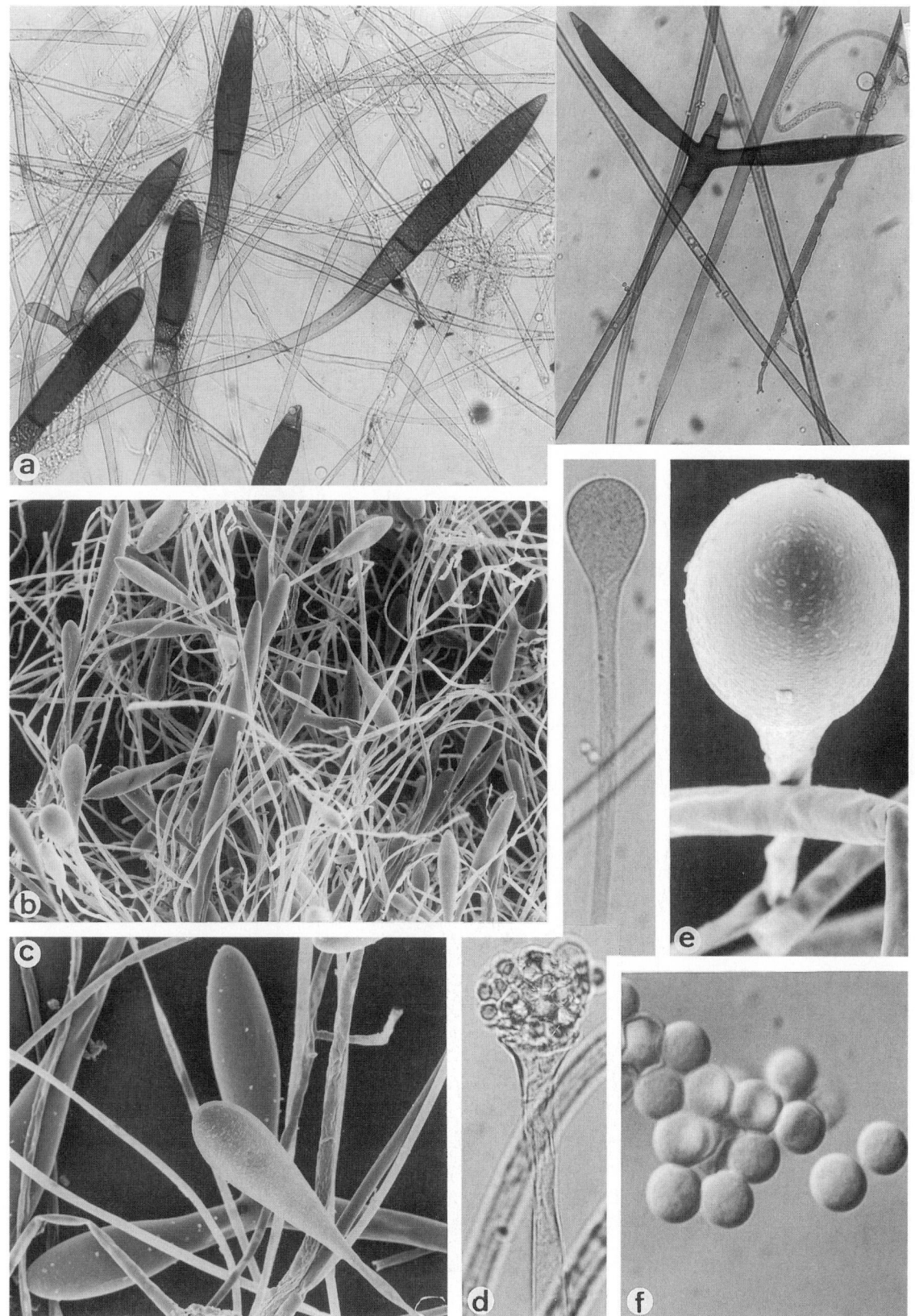

Chlamydoabsidia padenii, CBS 172.67. a-c. Chlamydospores; d, e. sporangia; f. sporangiospores. a. ×128; b. ×80; c. ×230; d. ×512; e. ×950; f. ×1600.

Zygomycota, Mucorales, Thamnidiaceae. Genus: *COKEROMYCES*

Cokeromyces recurvatus Poitras

Colony characteristics. Colonies (MEA) with slow to moderate growth, thin, greyish to brown.

Microscopy. Sporangiophores arising from vegetative hyphae, without rhizoids, mostly unbranched, 100-500 × 9 μm. Sporangiola with few spores, spherical to obovoidal, with columella, without apophysis, 8.4-12.6 μm, borne terminally on long, recurved and contorted stalks that arise from fertile vesicles formed at the apical end of sporangiophores. Sporangiospores smooth-walled, ovoidal to ellipsoidal, 2.5 × 4.5 μm.

Zygospores. Zygospores between opposite suspensors, spherical, brown, with pointed projections, 33-55 μm diam. Homothallic.

Molecular diagnostics. SSU restriction map based on NCBI AF113416:

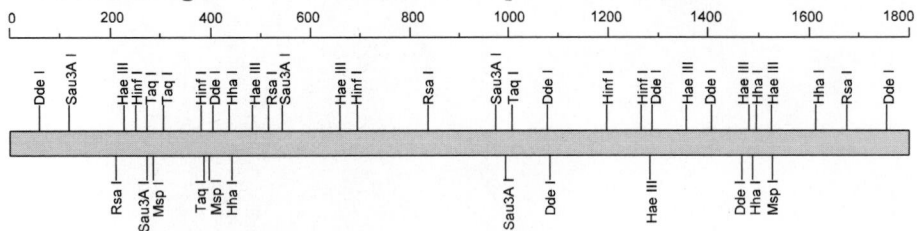

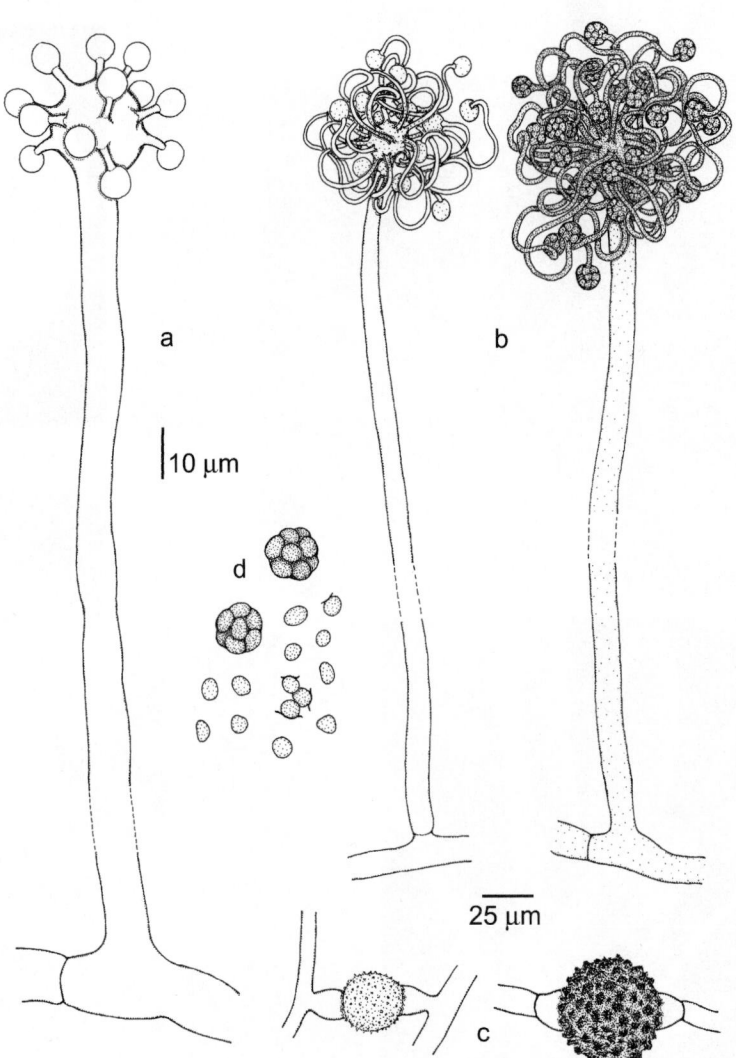

Cokeromyces recurvatus, CBS 158.50. a. Young sporangiophore with sporangiolum initiation; b. sporangiophores; c. zygospores; d. sporangiola and sporangiospores.

Pathogenicity. BSL-1. Two cases of genitary tract infection probably caused by this fungus were reported by McGough *et al.* (1990). It was isolated from brain tissue by Kemna *et al.* (1993). Alvarez *et al.* (1995) and Tsai *et al.* (1997) reported cases in transplant patients. A fatal peritonitis in an alcoholic was reported by Munipalli *et al.* (1996).
References. Zycha *et al.* (1969), Kemna *et al.* (1994).
Nomenclature. *Cokeromyces recurvatus* Poitras - Mycologia 42: 272, 1950.

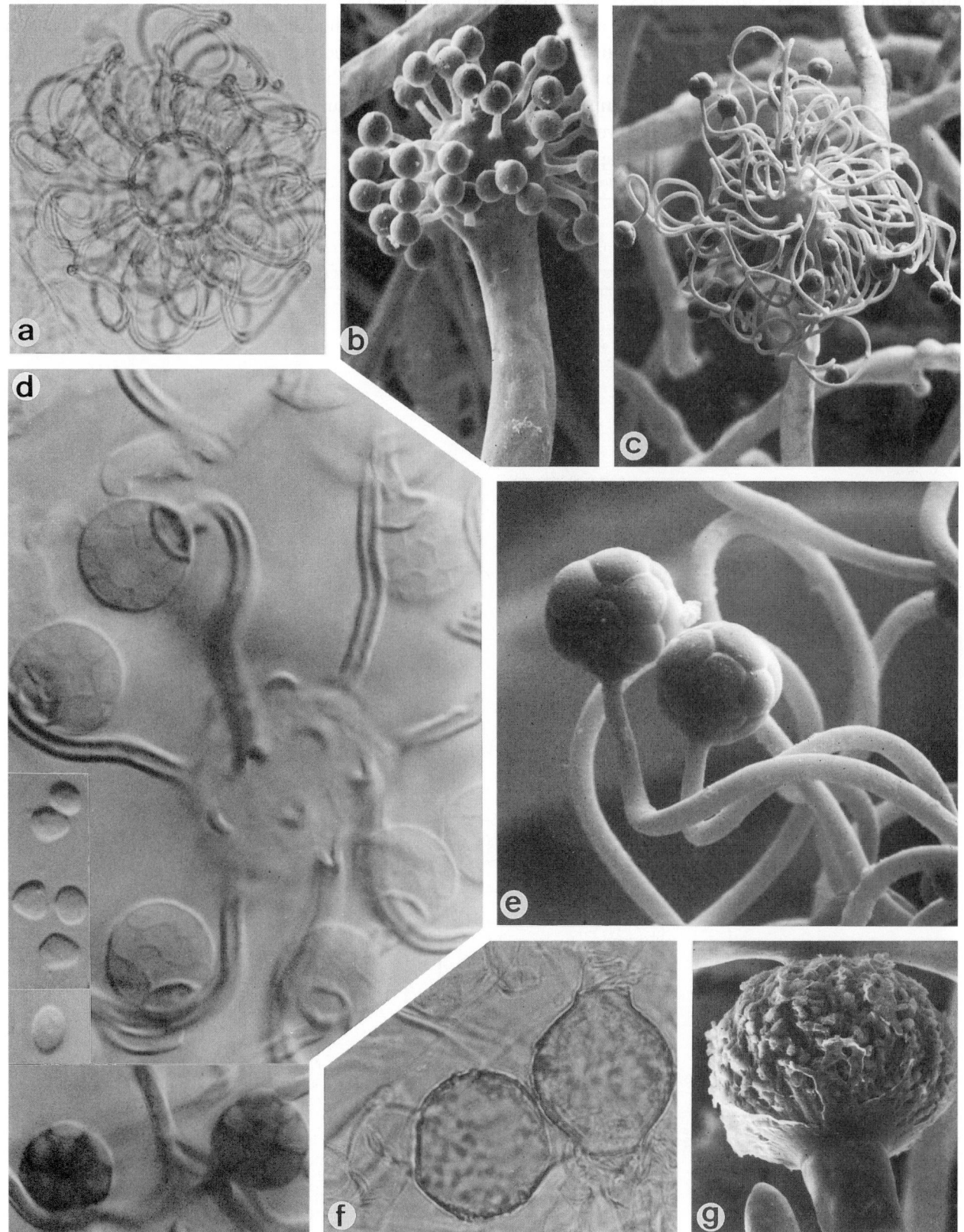

Cokeromyces recurvatus, CBS 158.50. a, c-e. Sporangiola borne on long and recurved stalks; b. sporangiophore with sporangiolum initiation; f,g. zygospores. a. ×512; b. ×880; c. ×490; d. ×1600; e. ×2700; f. ×512; g. ×1250.

Zygomycota, Mucorales, Cunninghamellaceae. **Genus:** *CUNNINGHAMELLA*

Cunninghamella bertholletiae Stadel

Colony characteristics. Colonies (MEA) expanding, with abundant floccose mycelium, tannish-grey.
Microscopy. Sporangiophores erect, in the apical region with a whorl of short lateral branches, each branch ending in a swollen vesicle up to 40 µm diam, with 1-spored sporangiola (each becoming a sporangiospore) covering the entire surface. Sporangiospores spherical to ovoidal, 7-11 µm diam, smooth-walled, sometimes finely echinulate.
Zygospores. Zygospores spherical, 25-55 µm diam, brownish, with tuberculate projections. Heterothallic.
Physiology. The species is thermophilic, growing at 45°C.
Molecular diagnostics. SSU restriction map based on NCBI AF113421:

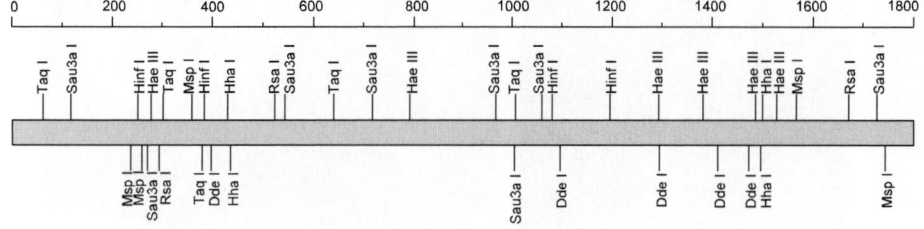

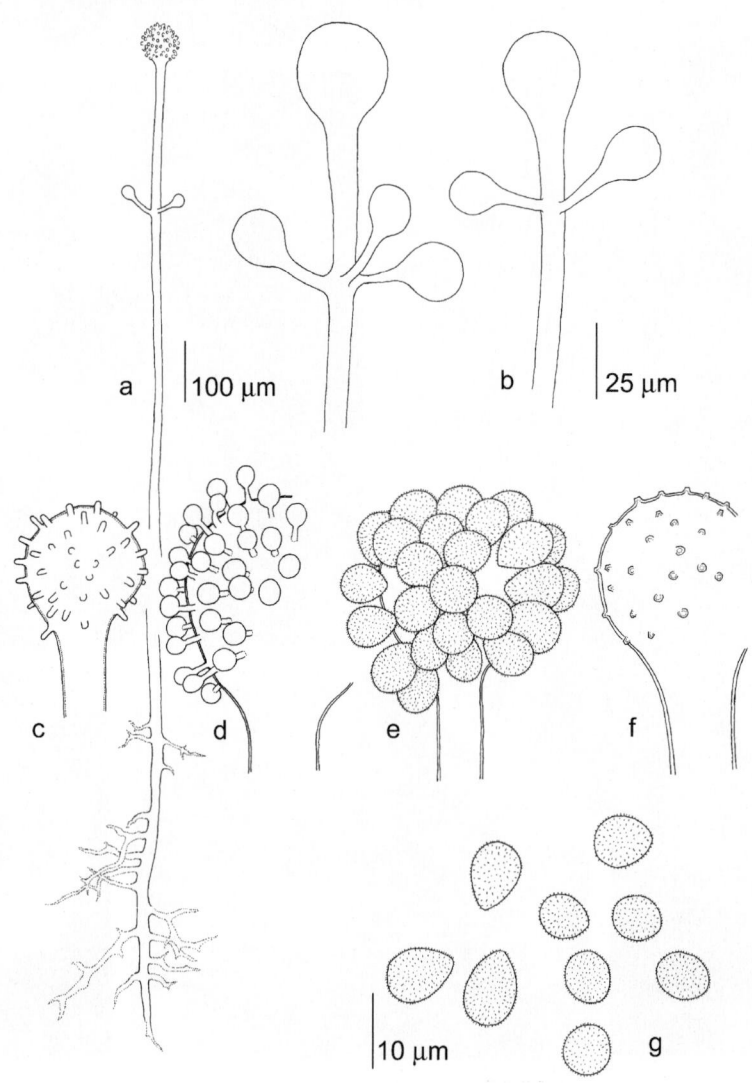

Cunninghamella bertholletiae, CBS 187.84. a, b. Sporangiophores; c-f. development of 1-spored sporangiola; g. sporangiospores.

Differential diagnosis. The non-pathogenic species *Cunninghamella elegans* differs by purely grey colonies and absence of growth at 45°C.

Pathogenicity. BSL-2. The species is known as an occasional opportunist (Weitzman & Christ, 1979; Maloisel *et al.*, 1991), mostly after traumatic inoculation (Boyce *et al.*, 1981). Rex *et al.* (1988) and Maloisel *et al.* (1991) reported on patients under deferoxamine therapy and Brennan *et al.* (1983) on a diabetic patient. Infection in an AIDS patient was reported by Mostaza *et al.* (1989), but severe infections are mostly seen in patients with leukemia (Kiehn *et al.*, 1979; McGinnis *et al.*, 1982b; Lopes *et al.*, 1996c). Disseminated cases in renal (Kolbeck *et al.*, 1985) and liver (Nimmo *et al.*, 1988) transplant recipients were described. Pulmonary cases are rare (Lopes *et al.*, 1996; Mazade *et al.*, 1998) and have poor prognosis (Cohen-Abbo *et al.*, 1993; Dermoumi, 1993; Kontoyianis *et al.*, 1994).

References. Samson (1969), Weitzman & Christ (1979, 1980), Sands *et al.* (1985), Iwatsu *et al.* (1990), Su *et al.* (1999).

Nomenclature. *Cunninghamella bertholletiae* Stadel - Űber Neuen Pilz C. bertholl., Thesis, Kiel, p. 1, 1911.

Cunninghamella bertholletiae, CBS 187.84. Sporangiophores with single-spored sporangiola. a. ×490; b. ×2140; c. ×1225.

Zygomycota, Mortierellales, Mortierellaceae. Genus: *MORTIERELLA*

Generic description. Colonies restricted or spreading in overlapping 'waves' or lobes, often with garlic-like odour. Sporangia spherical, multi-spored, without columella, developing at the tips of sporangiophores. Mycelium mostly dichotomously branched. Sporangiospores produced in variable numbers, colourless, brownish or pink. Chlamydospores, supported by short aerial stalks and with ornamented walls (stylospores), may be present.
References. Zycha *et al.* (1969), Domsch *et al.* (1980).

Key to the treated species of Mortierella:

1a. Stylospores present; sporangiospores ovoidal to irregular, up to 11 µm wide . . . . . . **M. polycephala (76)**
1b. Stylospores absent; sporangiospores ellipsoidal to reniform, 2.5 × 5.5 µm wide . . . . **M. wolfii (79)**

Mortierella polycephala Coemans

Colony characteristics. Colonies (PDA) with sparse growth.
Microscopy. Sporangiophores single or in tufts, 300-500 µm long, 12.5-20.0 µm wide at the base and narrowed to 3.5-5.0 µm at the tip, with branches mostly in whorls in the upper part. Sporangia spherical, 37-75 µm diam, without columella, leaving a collar at the base after dehiscence. Sporangiospores smooth-walled, ovoidal to irregular in shape, 5.5-13.2 × 5.5-11.0 µm. Stylospores spherical, 15-25 µm diam, on long stalks, occasionally with a swollen apex, verrucose to echinulate.
Zygospores. Zygospores between slightly unequal suspensors, embedded in a dense hyphal network. Homo- or heterothallic.
Molecular diagnostics. SSU restriction map based on NCBI X89436:

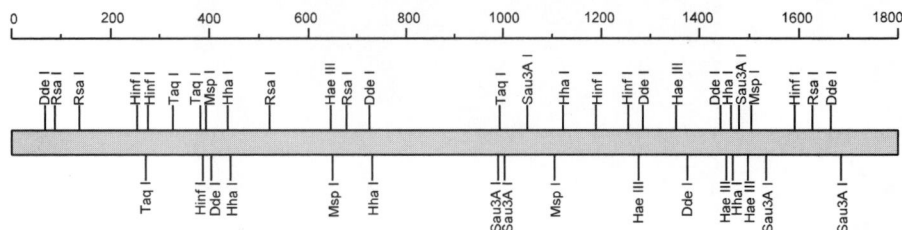

Pathogenicity. BSL-1. Pulmonary infections in cattle (Scholz & Meyer, 1965).
References. Zycha *et al.* (1969), Domsch *et al.* (1980).
Nomenclature. *Mortierella polycephala* Coemans - Bull. Acad. Belg., Sér. 2 , 15: 536, 1908.

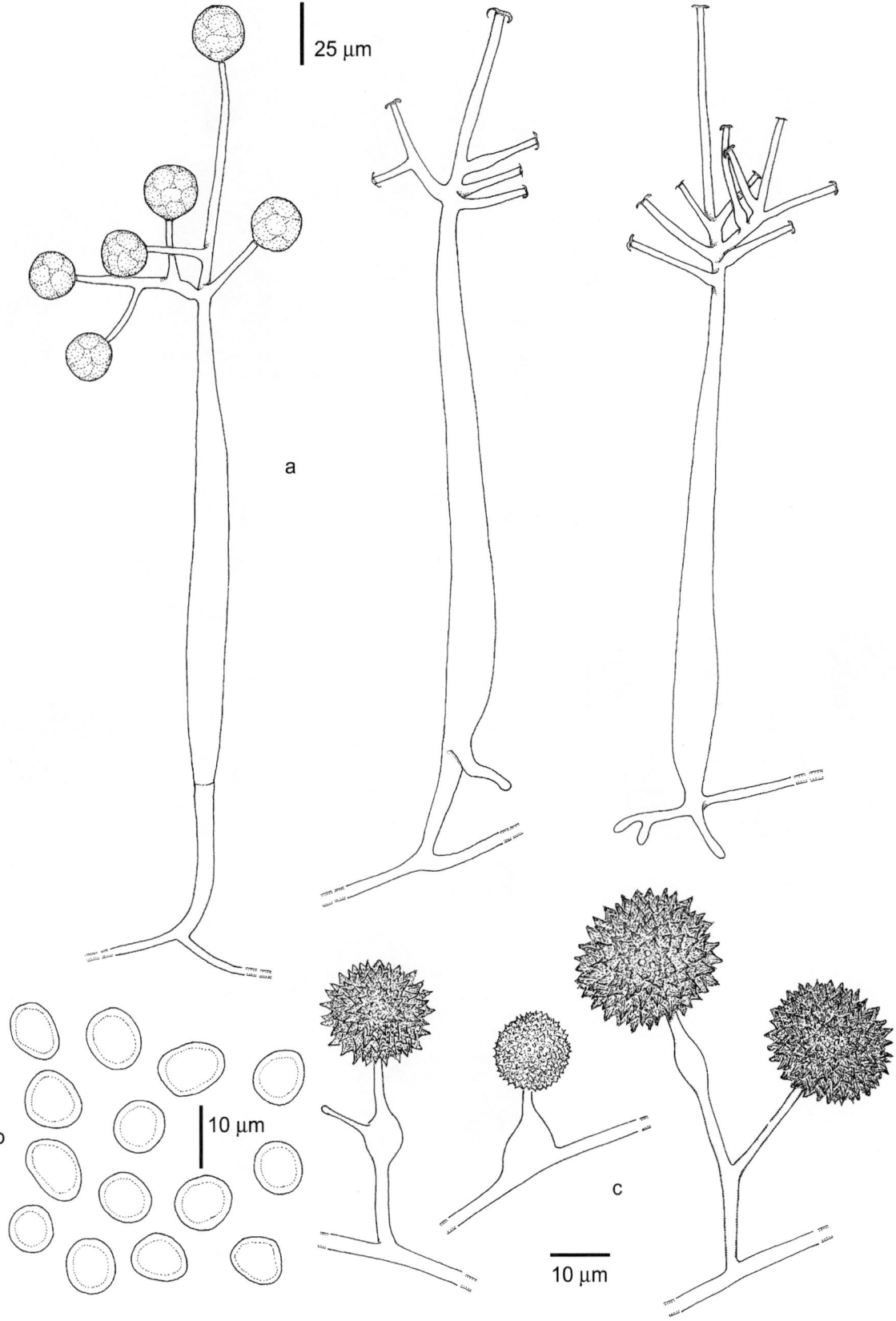

Mortierella polycephala, CBS 327.72. a. Sporangiophores; b. sporangiospores; c. stylospores.

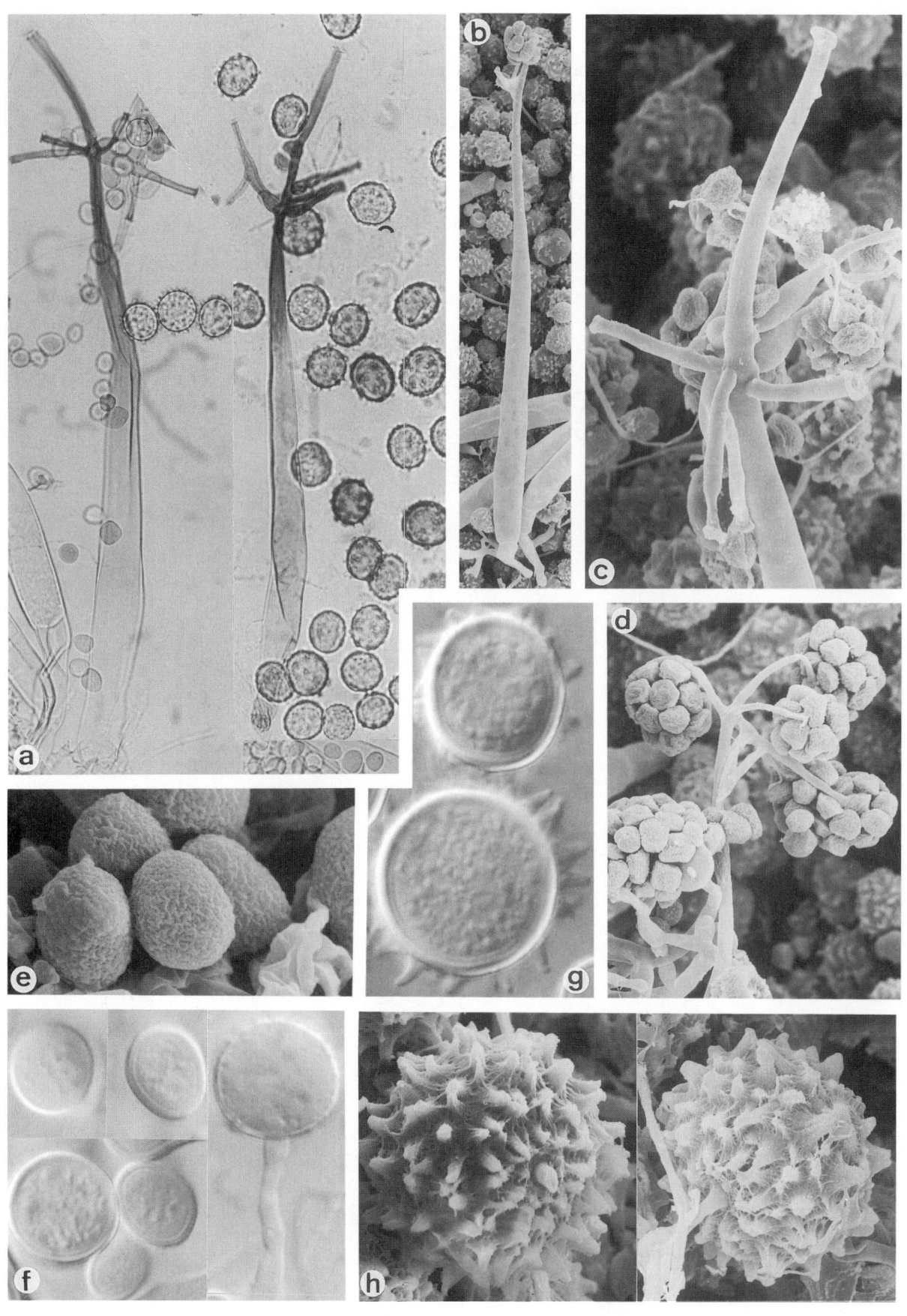

Mortierella polycephala, CBS 327.72. a-c. Sporangiophores; d. sporangia; e-h. sporangiospores and stylospores. a. ×280; b. ×350; c. ×1100; d. ×800; e. ×3700; f, g. ×2000; h. ×2300.

Mortierella wolfii Mehrotra & Baijal

Colony characteristics. Colonies (SEA, 30°C) with sparse growth, flat, whitish.

Microscopy. Sporangiophores often arising from rhizoids, 90-350 µm long, 11-20 µm wide at the base and tapering towards the tip, with a subterminal whorl of sporangium-bearing branches. Sporangia spherical, 20-50 µm diam, leaving a collarette-like frill at the base after dehiscence. Sporangiospores smooth-walled, ellipsoidal to reniform, 4-12 × 2.5-5.5 µm. Chlamydospores with amoeboid appendages often present.

Molecular diagnostics. SSU restriction map based on NCBI AF113425:

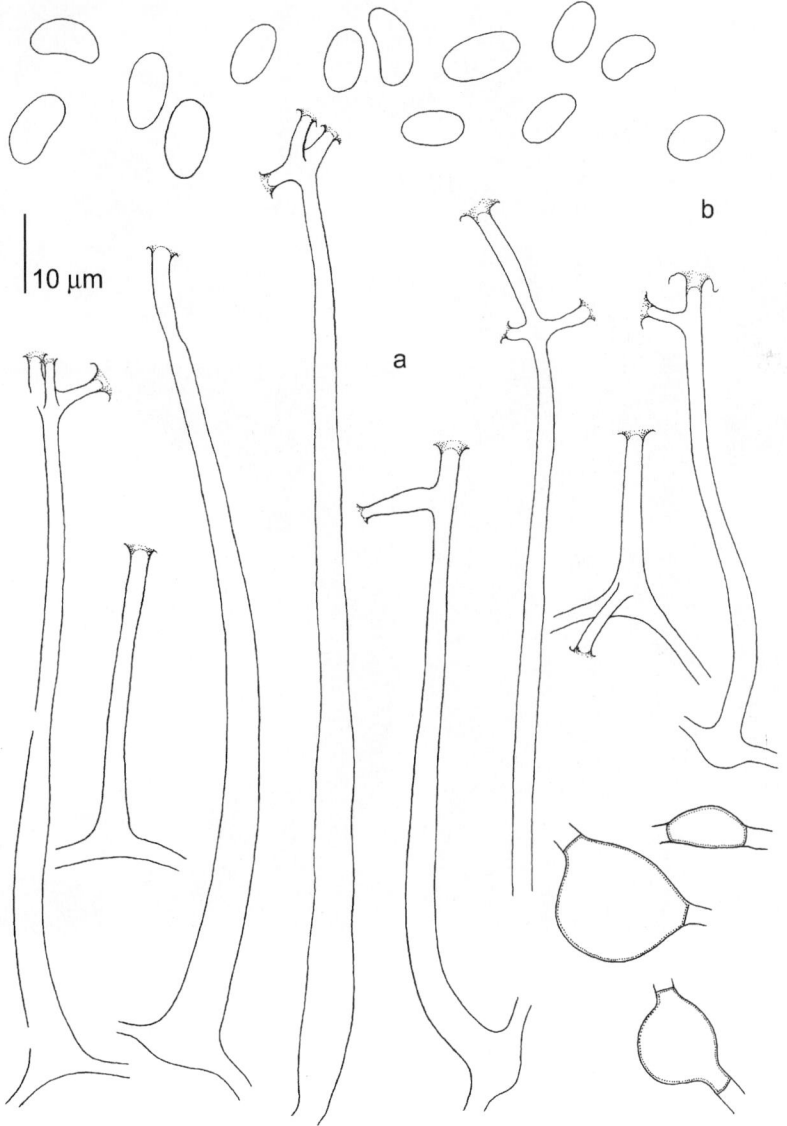

Pathogenicity. BSL-2. Common agent of bovine mycotic abortion in Australia and New Zealand, sometimes occurring elsewhere (Ellis, 1994).

References. Mehrotra & Baijal (1963), Komoda *et al.* (1988).

Nomenclature. *Mortierella wolfii* Mehrotra & Baijal - Mycopath. Mycol. Appl. 20: 51, 1963.

Mortierella wolfii, CBS 611.70. a. Sporangiophores with several collars after liberating spores; b. liberated sporangiospores; c. chlamydospores.

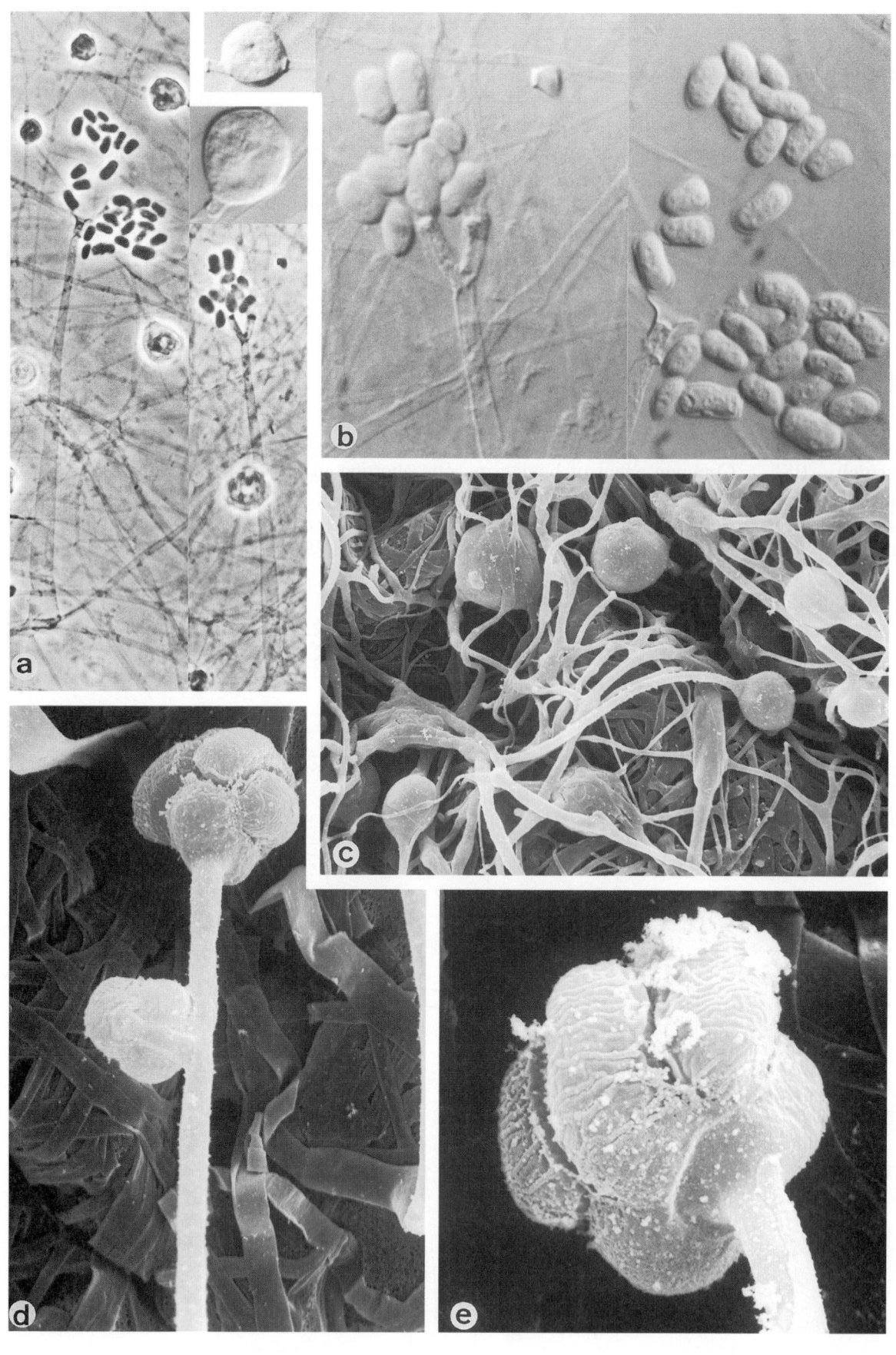

Mortierella wolfii, CBS 611.70. a. Sporangiophores and liberated sporangiospores; b-e. chlamydospores, detail of sporangia and liberated sporangiospores. a.×640; b. ×1600; c. ×2100; d. ×2850; e. ×8500.

80

Zygomycota, Mucorales, Mucoraceae. Genus: *MUCOR*

Generic description. Colonies fast-growing, whitish to greyish, often several cm high, usually forming a thick mat due to abundant production of erect sporangiophores. Sporangiophores without basal rhizoids, not originating from stolons, unbranched or irregularly branched, terminally bearing multi-spored sporangia. Sporangia spherical, without apophyses, with large columellae; wall deliquescent or persistent and rupturing at maturity, often with adhering calcium oxalate crystals. Zygospores without appendages on suspensors.

General remarks. *Mucor* species are rather frequently isolated as culture contaminants. Only few, thermolerant species may have clinical significance.

References. Zycha *et al.* (1969), Schipper (1978a), Domsch *et al.* (1980).

Key to the treated species of Mucor:

1a. Sporangiophores unbranched or weakly sympodially branched → **2**
1b. Sporangiophores repeatedly branched → **3**
2a. Sporangiospores spherical, 3.5-5.5 µm diam .... *M. amphibiorum* **(82)**
2b. Sporangiospores ellipsoidal, 5.7-8.7 × 2.7-5.4 µm .... *M. hiemalis* **(86)**
3a. Growth at 40°C .... *M. indicus* **(88)**
3b. Poor or no growth at 37°C → **4**
4a. Sporangiospores mostly about 3.5-6 µm diam; chlamydospores in hyphae rare → **5**
4b. Sporangiospores up to 8-10 µm diam; chlamydospores in hyphae abundant, occasionally
     in the columella .... *M. racemosus* **(90)**
5a. Colonies with restricted growth; columellae applanate; no assimilation of ethanol *M. ramosissimus* **(92)**
5b. Colonies with expanding growth; columellae spherical to ellipsoidal; assimilation
     of ethanol .... *M. circinelloides* **(84)**

Mucor amphibiorum Schipper

Colony characteristics. Colonies (MEA) expanding, greyish-brown, slightly aromatic.

Microscopy. Sporangiophores hyaline, up to 25 mm high, up to 20 μm wide, erect and unbranched, rarely sympodial-ly branched. Sporangia up to 75 (-100) μm diam, slightly flattened, dark brown, with diffluent membranes; columellae cylindrical-ellipsoidal, up to 60 × 50 μm, with small collars. Sporangiospores smooth-walled, spherical, 3.5-5.5 μm diam.

Zygospores. Zygospores spherical to slightly compressed, up to 70 × 60 μm, with stellate projections. Suspensors unequal. Heterothallic.

Physiology. Maximum growth temperature 36°C. No assimilation of ethanol and nitrate.

Pathogenicity. BSL-2. This species has been implicated in a rare mycosis in various amphibia (Berger *et al.*, 1997). Main characteristic was the production of spherules in several internal organs (Frank *et al.*, 1974; Schipper, 1978a). It also caused ulcerative skin lesions in a free-living platypus (Obendorf *et al.*, 1993).

Reference. Schipper (1978a).

Nomenclature. *Mucor amphibiorum* Schipper - Stud. Mycol. 17: 14, 1978.

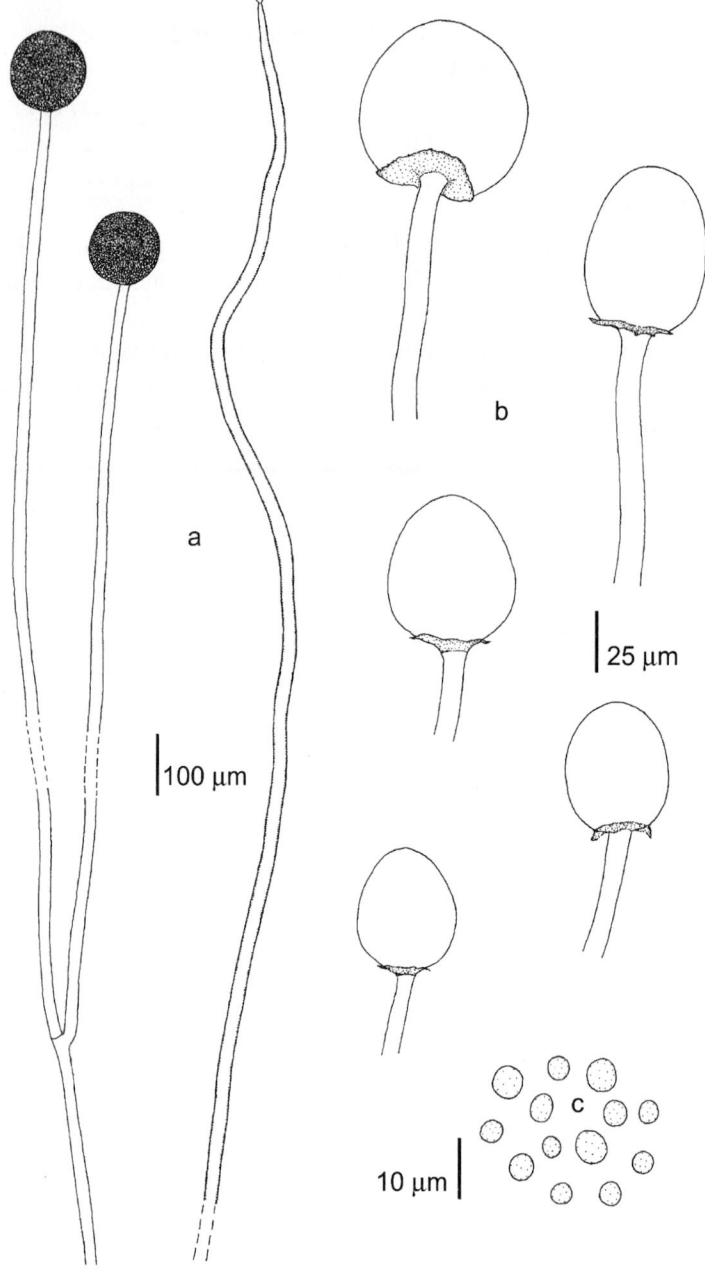

Mucor amphibiorum, CBS 763.74. a. Sporangiophores; b. columellae; c. sporangiospores.

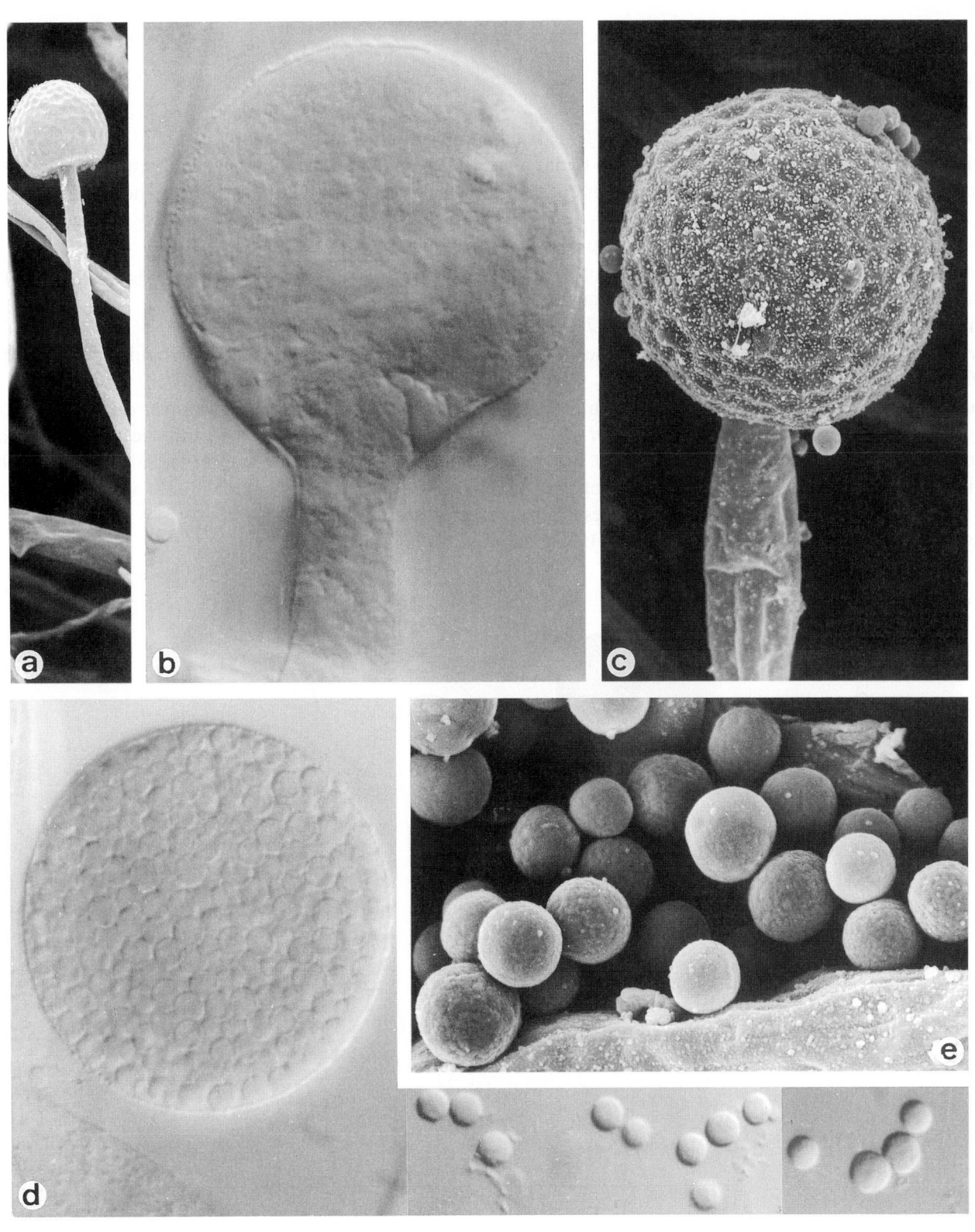

Mucor amphibiorum, CBS 763.74. Sporangiophores, columellae and sporangiospores. a. ×700; b. ×1600; c. ×1700; d. ×1600; e. ×4900.

Mucor circinelloides v. Tiegh.

Colony characteristics. Colonies (MEA, 24°C) expanding, floccose, light greyish-brown.

Microscopy. Sporangiophores hyaline, up to 6 mm high, 17 μm wide, repeatedly branched, forming two layers of different height: longer branches erect, shorter ones often recurved. Sporangia 20-80 μm diam, membrane diffluent in larger but persistent in smaller sporangia; columellae spherical to ellipsoidal, about 50 μm wide. Sporangiospores smooth-walled, ellipsoidal, 4.5-7.0 × 3.5-5.0 μm. Chlamydospores absent or scant.

Zygospores. Zygospores spherical to slightly compressed, up to 100 μm wide, with stellate spines, reddish-brown to dark brown. Suspensors equal to slightly unequal. Heterothallic.

Physiology. Maximum growth temperature 37°C. Assimilates ethanol and nitrate.

Remark. Four varieties have been distinguished on the basis of shapes of columellae and sporangiospores (Schipper, 1976).

Pathogenicity. BSL-1. Frequently reported infecting animals such as cattle and swine (Morquer *et al.*, 1965), fowl (Porges *et al.*, 1935), ganders (Marjanková *et al.*, 1978), platypus (Stewart *et al.*, 1999) and occasionally humans (Fetchick *et al.*, 1986; Berenguer *et al.*, 1988). Particularly ketoacidotic patients are at risk (del Palacio *et al.*, 1999). A human cutaneous case was reported by Wang *et al.* (1990) as *Mucor lusitanicus*, which is now supposed to belong to the same species as it occasionally produces zygospores in crosses (Schipper, 1976).

References. Schipper (1976), Scholer *et al.* (1983), Samson *et al.* (1996).

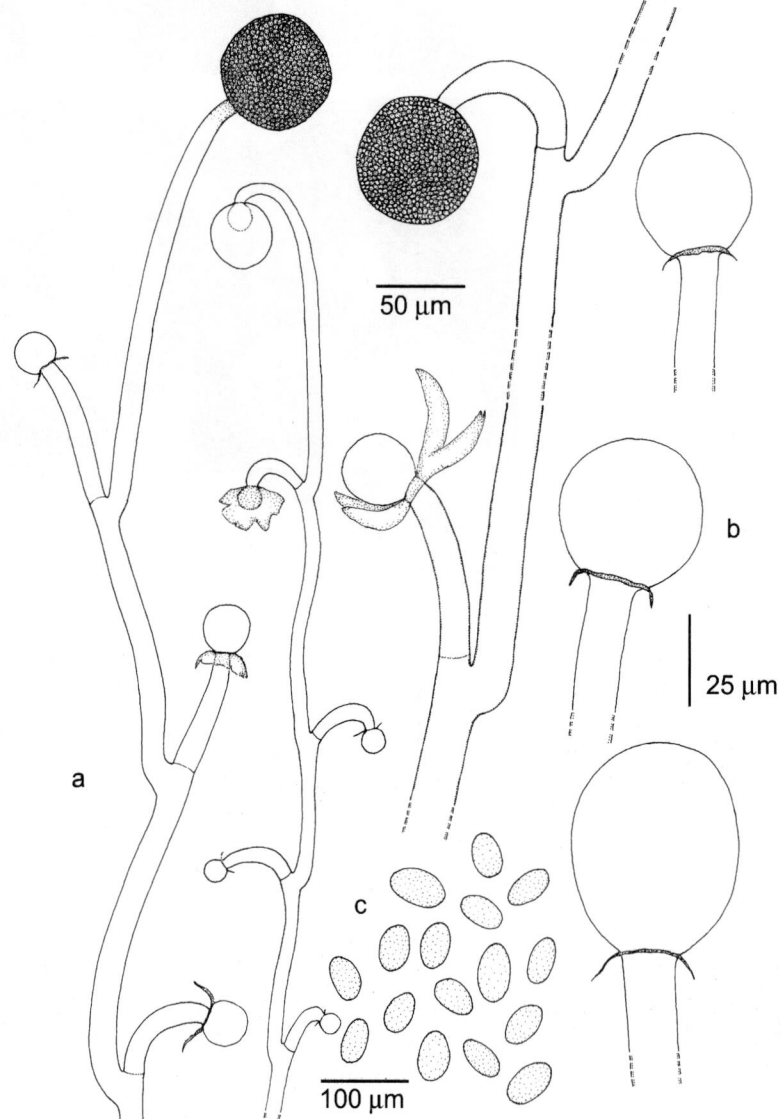

Mucor circinelloides, IMI 200.943. a. Short, recurved sporangiophores; b. columellae; c. sporangiospores.

Antifungal susceptibility.

Antifungal	Mean MICs	MIC 90	Strains	Reference
AMB	1.11	1.56	2	Wildfeuer *et al.* (1998)
ITZ	8.84	12.5	2	Wildfeuer *et al.* (1998)
KTZ	6.25	6.25	2	Wildfeuer *et al.* (1998)
VCZ	35.4	50	2	Wildfeuer *et al.* (1998)

Nomenclature. *Mucor circinelloides* van Tieghem - Annls Sci. Nat. 1: 94, 1875 ≡ *Calyptromyces circinelloides* (van Tieghem) Sumstine - Mycologia 2: 148, 1910.

Mucor lusitanicus Bruderlein - Bull. Soc. Bot. Genève 8: 276, 1916 ≡ *Mucor circinelloides* van Tieghem f. *lusitanicus* (Bruderlein) Schipper - Stud. Mycol. 12: 9, 1976.

Mucor paronychius Sutherland-Campbell & Plunkett - Arch. Derm. 30: 656, 1934.

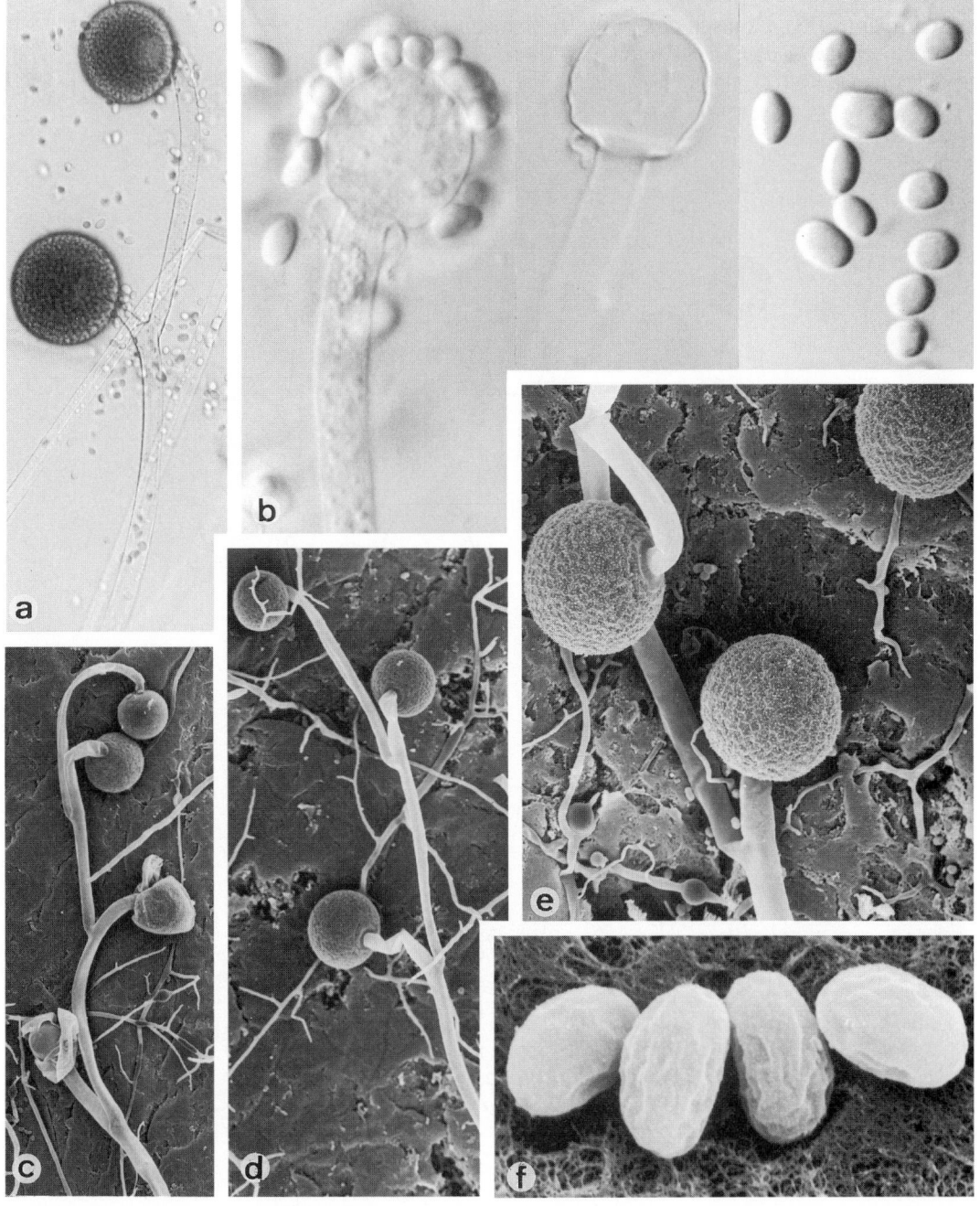

Mucor circinelloides, CBS 192.68. a, c, d. Sporangiophores; b. columellae and sporangiospores; e. sporangia; f. sporangiospores. a. ×250; b. ×1600; c. ×205; d. ×255; e. ×575; f. ×6750.

Mucor hiemalis Wehmer

Colony characteristics. Colonies (MEA) expanding, greyish-ochraceous.

Microscopy. Sporangiophores hyaline, up to 15 mm high, up to 14 µm wide, erect, unbranched at first, later sparingly branched. Sporangia yellowish at first, becoming dark brown, up to 80 µm diam, with diffluent membranes; columellae ellipsoidal, spherical when young, up to 38 × 30 µm. Sporangiospores ellipsoidal, sometimes flattened at one side, 5.7-8.7 × 2.7-5.4 µm, smooth-walled. Oidia present in substrate hyphae.

Zygospores. Zygospores up to 100 µm diam, roughened with long spines, blackish-brown. Suspensors equal or nearly equal. Heterothallic.

Physiology. Maximum growth temperature 30°C. Nitrate negative. Mostly no thiamin required.

Pathogenicity. BSL-1. This species was isolated in a case of zygomycosis (Neame & Rayner, 1960). Recently it was also reported as the causative agent of a primary cutaneous mycosis in an otherwise healthy girl (Prevoo *et al.*, 1991). A subcutaneous case was published by Costa *et al.* (1990) as *M. hiemalis* f. *luteus*, a slightly different taxon which, however, sometimes produces zygospores in mating (Schipper, 1973).

References. Schipper (1973), Domsch *et al.* (1980), Samson *et al.* (1996).

Nomenclature. *Mucor hiemalis* Wehmer - Annls Mycol. 1: 37, 1902.

Mucor luteus Linnemann - Beitr. Fl. Mucor. Marb., 1936 ≡ *Mucor hiemalis* Wehmer f. *luteus* (Linnemann) Schipper - Stud. Mycol. 4: 33, 1973.

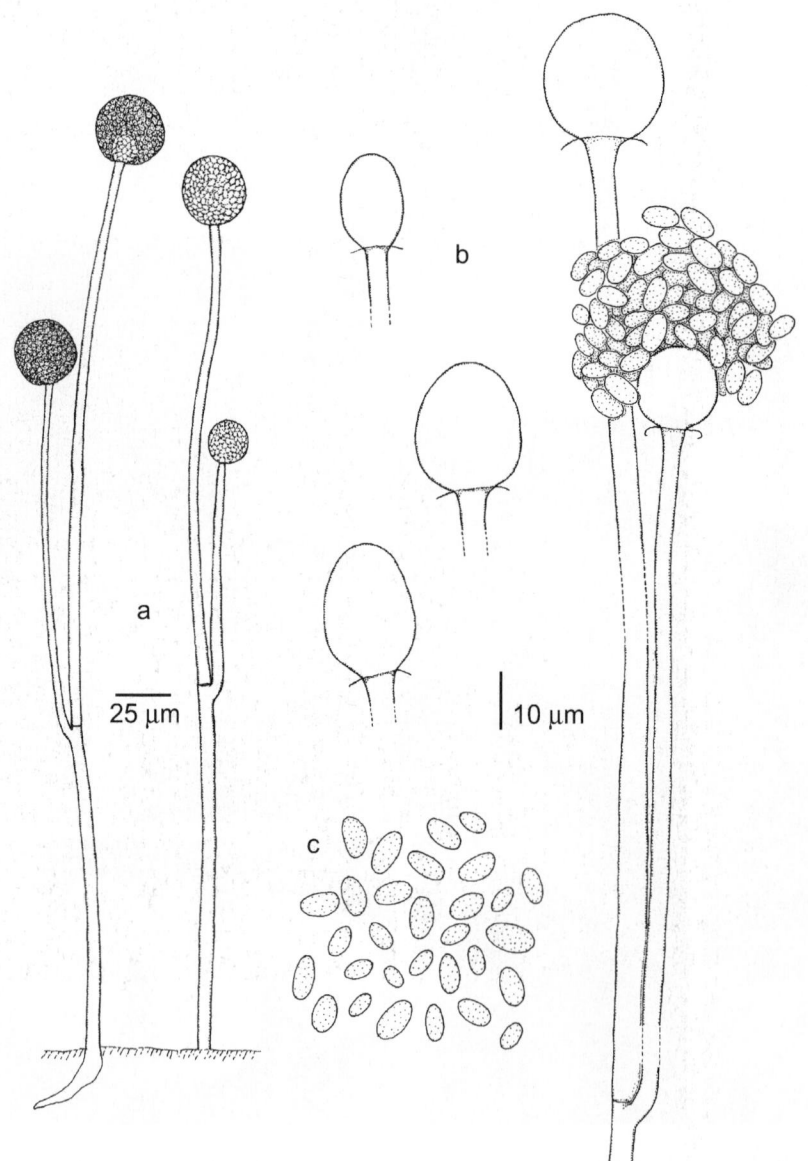

Mucor hiemalis, IMI 21217. a. Sporangiophores; b. mature sporangia and columellae; c. sporangiospores.

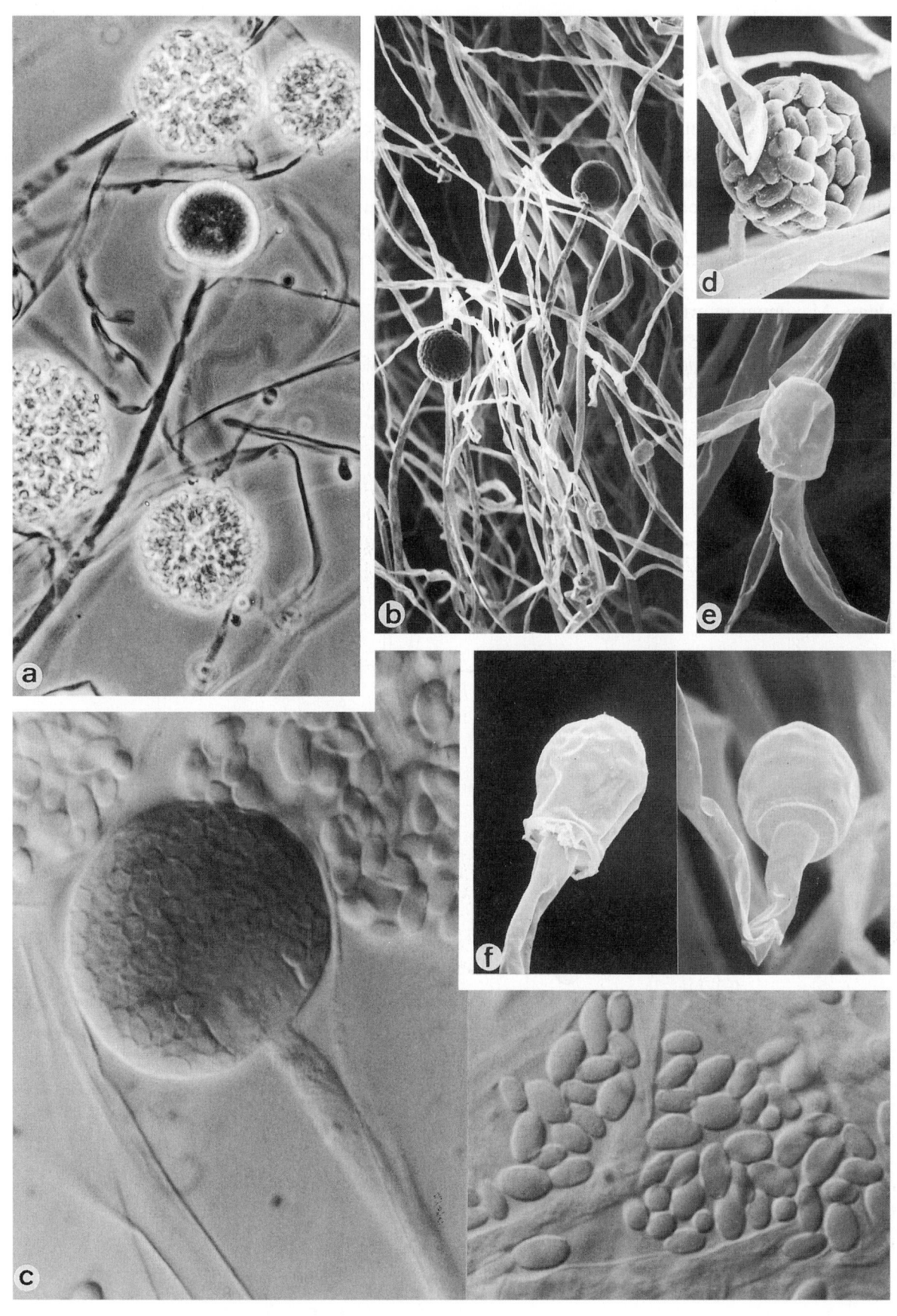

Mucor hiemalis, IMI 21217. a, b. Sporangiophores; c, d. sporangia and sporangiospores; e, f. columellae. a. ×640; b. ×460; c. ×1600; d. ×1900; e. ×1600; f. ×2500.

Mucor indicus Lendner

Colony characteristics. Colonies (MEA, 30°C) expanding, deep yellow, aromatic.

Microscopy. Sporangiophores hyaline to yellowish, up to 10 mm high, 14 μm wide, repeatedly sympodially branched, with long branches. Sporangia yellow to brown, up to about 75 × 40-50 μm, with diffluent, transparent walls; columellae subspherical to applanate, rarely elongated, up to 56 × 53 μm. Sporangiospores smooth-walled, subspherical to ellipsoidal, 5.4-5.7 × 4.4 μm. Chlamydospores abundant in light.

Zygospores. Zygospores spherical to slightly compressed, up to 100 μm diam, with stellate spines, black. Suspensors unequal. Heterothallic.

Physiology. Maximum growth temperature 42°C. Growth with ethanol; nitrate negative; thiamin dependent. Sporulation strongly influenced by light.

Pathogenicity. BSL-1. *M. indicus* was found in human cases of gastric (Douvin *et al.*, 1975) and pulmonary (Krasinski *et al.*, 1985) infections.

Reference. Schipper (1978a).

Nomenclature. *Mucor indicus* Lendner - Bull. Soc. Bot. Genève 21: 258, 1930.

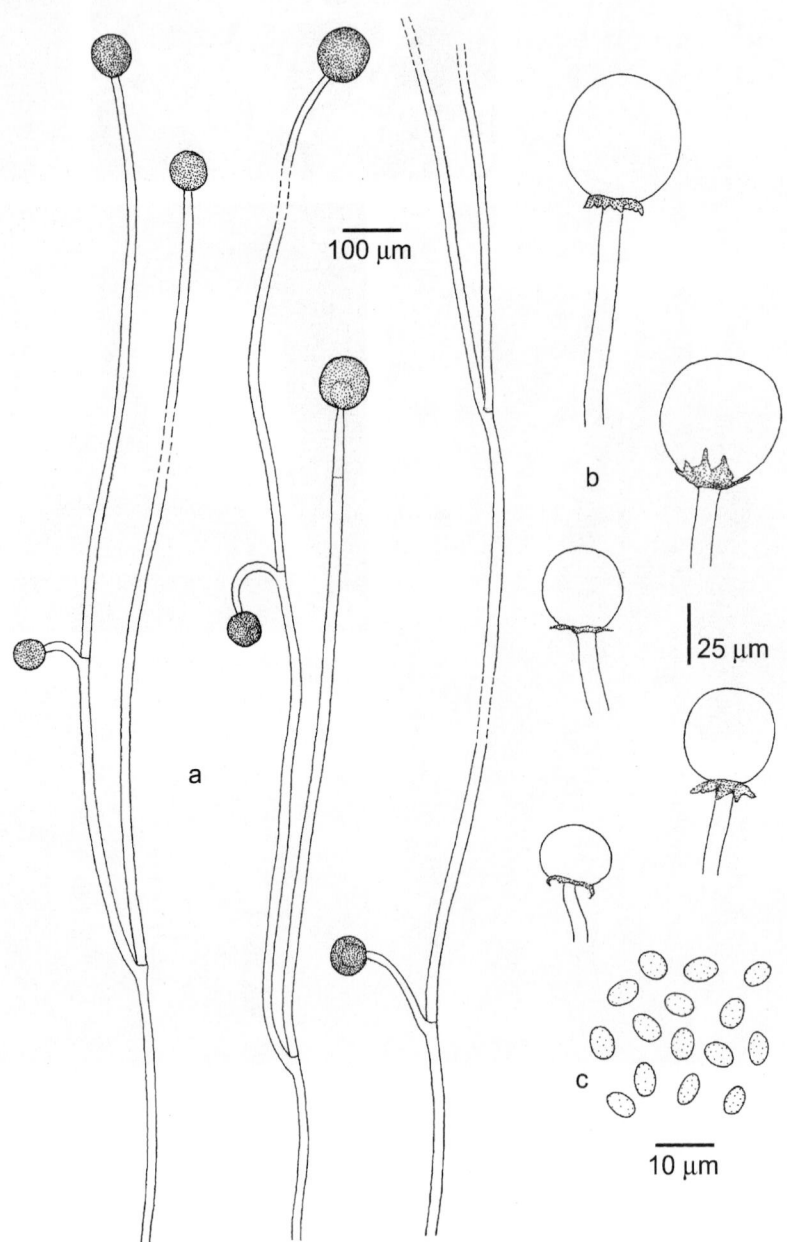

Mucor indicus, CBS 226.29. a. Sporangiophores; b. columellae; c. sporangiospores.

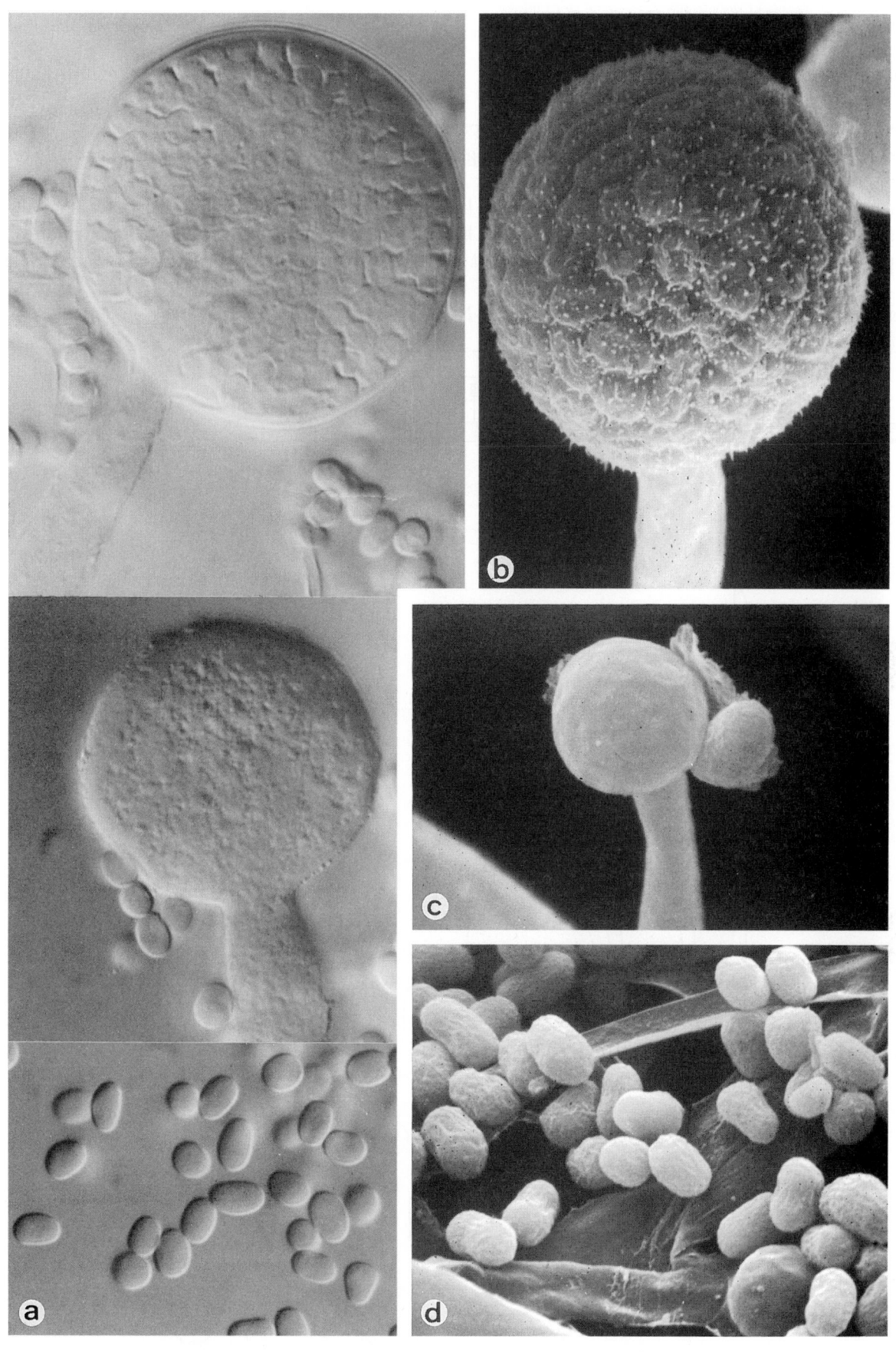

Mucor indicus, CBS 226.29. Sporangiophores, columellae and sporangiospores. a. ×1600; b. ×4300; c. ×5400; d. ×3700.

Mucor racemosus Fres.

Colony characteristics. Colonies (MEA) expanding, pale greyish-brown.

Microscopy. Sporangiophores hyaline, up to 20 mm high, 14-17 μm wide, sympodially and monopodially branched, the short monopodial branches often being recurved. Sporangia brownish, up to 80 (-90) μm diam; columellae subspherical to pyriform, often with truncate bases, light brown, with collars. Sporangiospores smooth-walled, spherical to broadly ellipsoidal, up to 8-10 μm diam. Chlamydospores mostly occurring in sporangiophores.

Zygospores. Zygospores up to 110 μm diam, with short spines, brown. Suspensors equal or nearly equal. Heterothallic.

Molecular diagnostics. SSU restriction map based on NCBI X54863:

Mucor racemosus, IMI 35565. a. Sporangiophores; b. columellae; c. part of a sporangium; d. sporangiospores; e. chlamydospores.

Physiology. Maximum growth temperature 32°C. Assimilation of sucrose.

Pathogenicity. BSL-1. Some reports on this species being involved in animal and human mycoses exist, but its pathogenicity is questionable (Scholer *et al.*, 1983).

References. Schipper (1976), Samson *et al.* (1996).

Antifungal susceptibility.

Antifungal	Mean MICs	MIC 90	Strains	Reference
AMB	0.66	0.78	4	Wildfeuer *et al.* (1998)
ITZ	3.13	3.13	4	Wildfeuer *et al.* (1998)
KTZ	3.72	6.25	4	Wildfeuer *et al.* (1998)
VCZ	10.5	12.5	4	Wildfeuer *et al.* (1998)

Nomenclature. *Mucor racemosus* Fresenius - Beitr. Mykol. 1: 12, 1850.

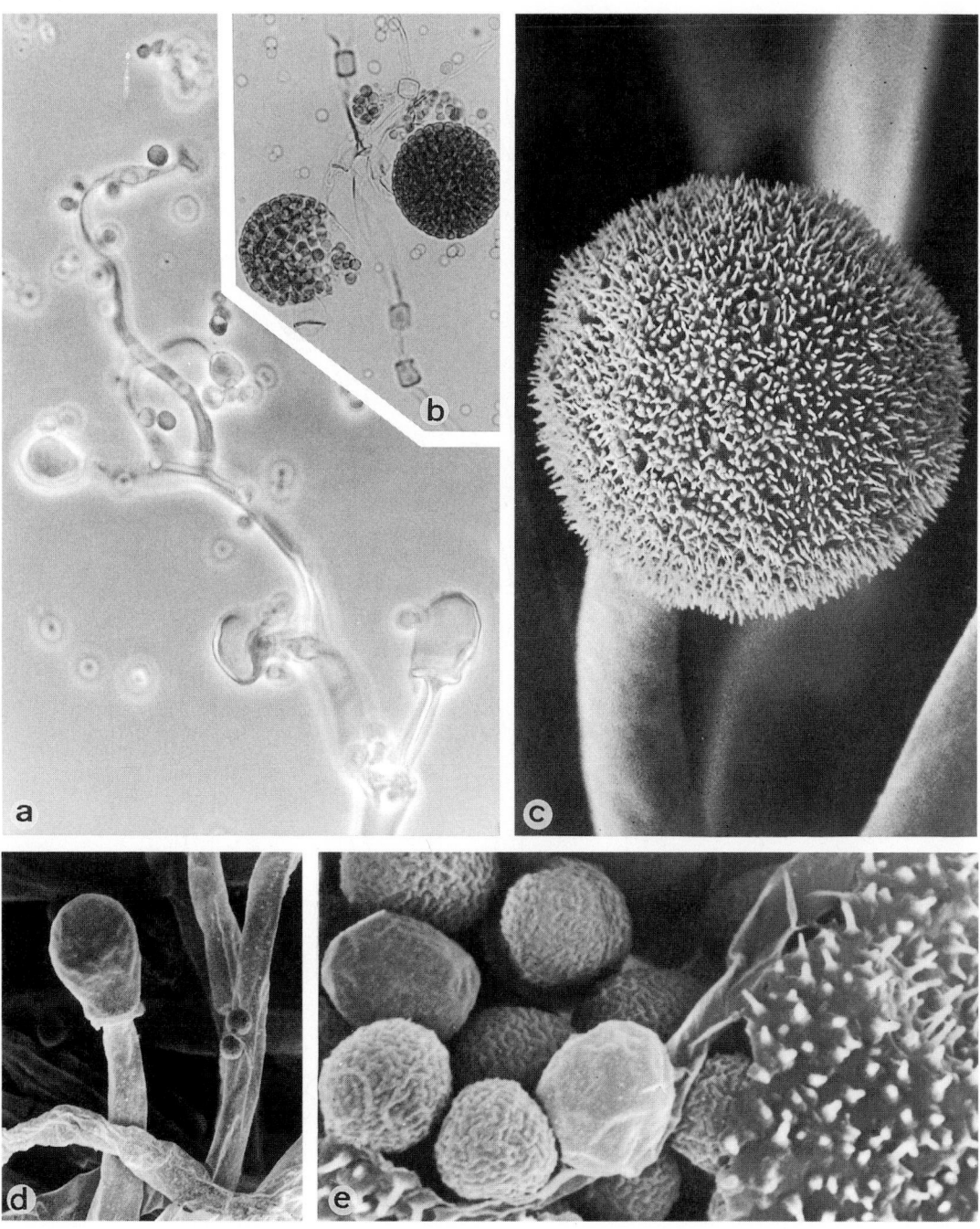

Mucor racemosus, IMI 35565. a. Sporangiophores; b, c. sporangia and chlamydospores; d. columella; e. sporangiospores. a. ×512; b. ×256; c. ×1750; d. ×1425; e. ×6000.

Mucor ramosissimus Samutsevich

Colony characteristics. Colonies (MEA) restricted, greyish.

Microscopy. Sporangiophores hyaline, up to 2 mm high, 18 µm wide, slightly roughened, tapering towards the apex, repeatedly sympodially branched. Sporangia blackish, spherical to dorsiventrally flattened, up to 80 µm diam, with persistent walls; columellae applanate, up to 40-50 µm, absent in small sporangia. Sporangiospores smooth-walled, spherical to broadly ellipsoidal, 5-8 × 4.5-6.0 (-7) µm. Oidia present in substrate hyphae. Chlamydospores absent.

Physiology. Maximum growth temperature 36°C. Assimilation of ethanol negative and of nitrate positive.

Pathogenicity. BSL-1. The species has been implicated in chronic mucocutaneous infection with destruction of the soft tissues of the face (Vignale *et al.*, 1964), a cutaneous lesion (Weitzman *et al.*, 1993), rhinocerebral mycosis (Bullock *et al.*, 1974), and septic arthritis in a neonate (Sharma *et al.*, 1994, 1995).

Reference. Schipper (1976).

Nomenclature. *Mucor ramosissimus* Samutsevich - Mat. Mikol. Fitopatol. 6: 210, 1927.

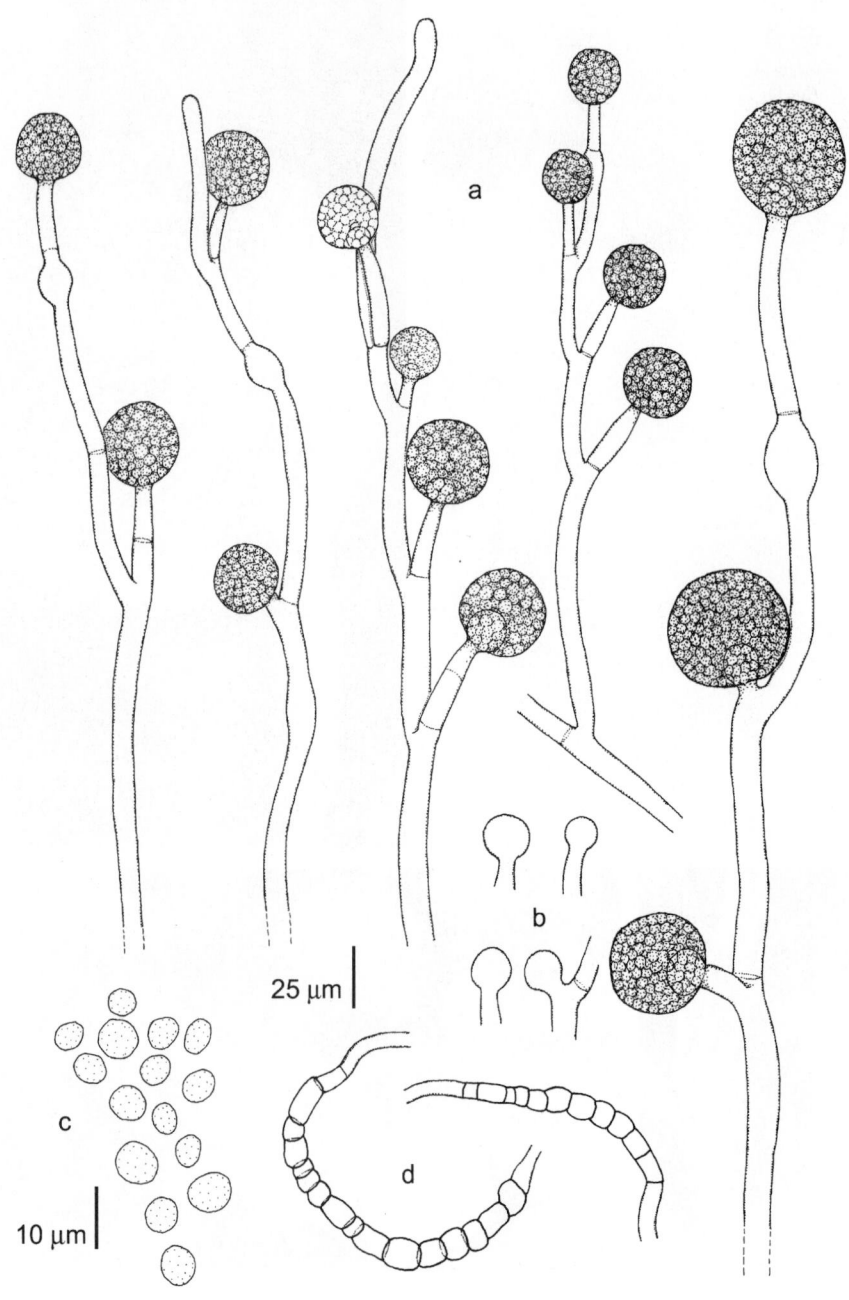

Mucor ramosissimus, AAMRL 40.1. a. Sporangiophores; b. columellae; c. sporangiospores; d. oidia.

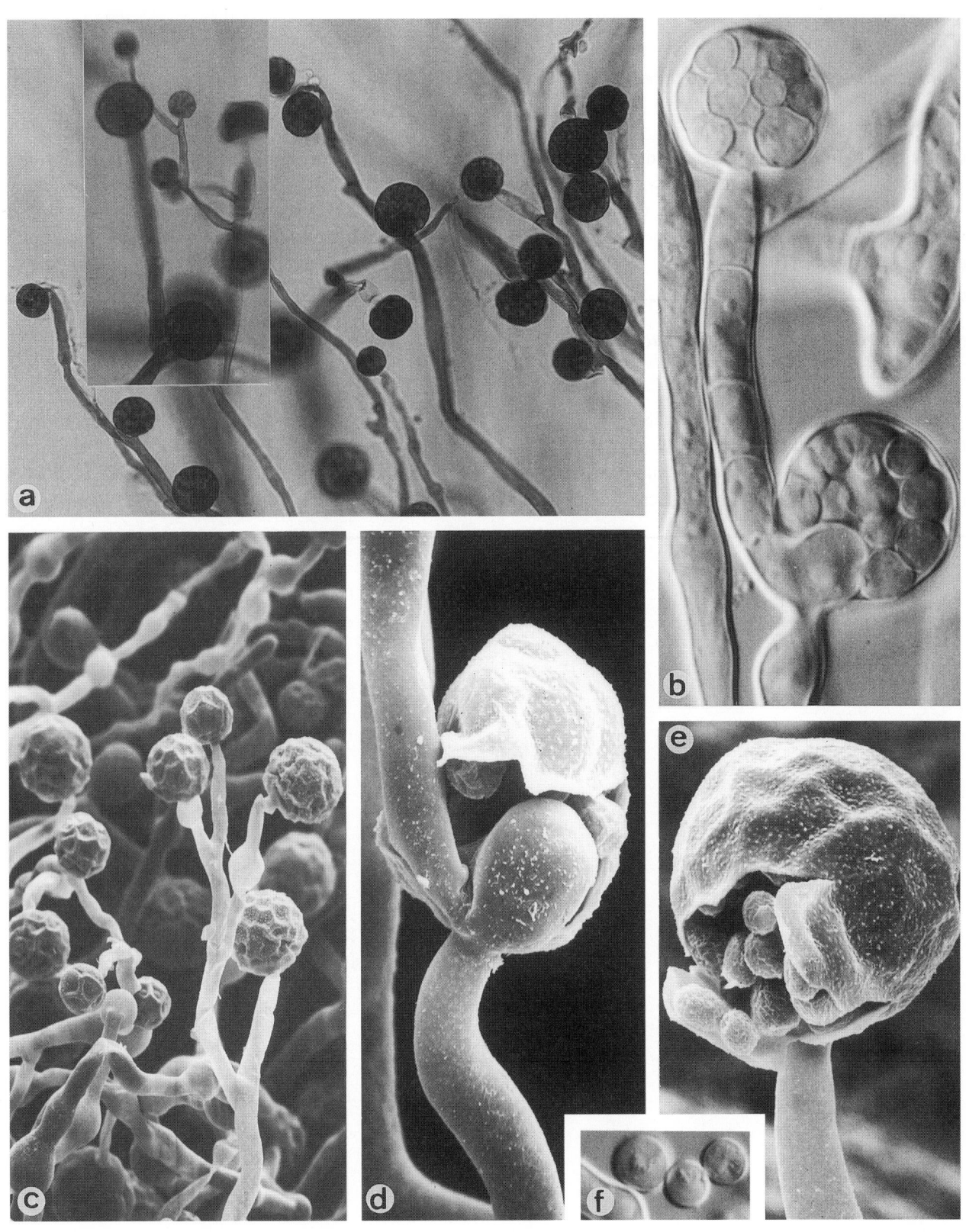

Mucor ramosissimus, AAMRL 40.1. Sporangiophores and sporangiospores. a. ×320; b. ×1600; c. ×700; d. ×3400; e. ×2400; f. ×1600.

Zygomycota, Mucorales, Mucoraceae. Genus: *RHIZOMUCOR*

Generic description. Colonies expanding, grey to deep olive. Sporangiophores originating either from short aerial hyphae or from distinct stolons, both with simple or weakly branched rhizoids. Sporangia spherical to subspherical, brown, multi-spored, with columellae, without apophyses, borne terminally on simple or branched sporangiophores. Sporangiospores subspherical. Zygospores spherical, covered with blunt outgrowths. Most species are thermophilic. **Reference.** Schipper (1978b).

Key to the treated species of Rhizomucor:

1a. Maximum growth temperature 38°C .... *R. variabilis* (99)
1b. Good growth at 45°C ➝ **2**
2a. Sporangia up to 100 µm diam; sucrose assimilated; not thiamin dependent .... *R. pusillus* (97)
2b. Sporangia up to 60 µm diam; sucrose not assimilated; thiamin dependent .... *R. miehei* (95)

Rhizomucor miehei (Cooney & Emerson) Schipper

Colony characteristics. Colonies (MEA, 30°C) expanding, olive-grey.

Microscopy. Sporangiophores borne on stolons with weakly developed rhizoids, up to 1 mm high and 7 μm diam, sympodially branched, brownish. Sporangia spherical to subspherical, up to 60 μm diam; membranes diffluent; columellae obovoidal to slightly pyriform, up to 28 × 25 μm, with small protrusions. Sporangiospores hyaline, smooth-walled; subspherical to ellipsoidal, 3-4 μm diam.

Zygospores. Zygospores spherical to slightly compressed, up to 50 μm, reddish- to blackish-brown, with equal suspensors. Homothallic.

Physiology. No assimilation of sucrose, thiamin dependent. Maximum growth temperature 57°C.

Pathogenicity. BSL-1. This species probably caused two cases of bovine mastitis (Scholer *et al.*, 1983). Compare also *Rhizomucor variabilis* (p. 99), which has been suggested to be a synonym (M.A.A. Schipper, pers. comm.).

Reference. Schipper (1978b).

Nomenclature. *Mucor miehei* Cooney & Emerson - Thermoph. Fungi p. 26, 1964 ≡ *Rhizomucor miehei* (Cooney & Emerson) Schipper - Stud. Mycol. 17: 58, 1978.

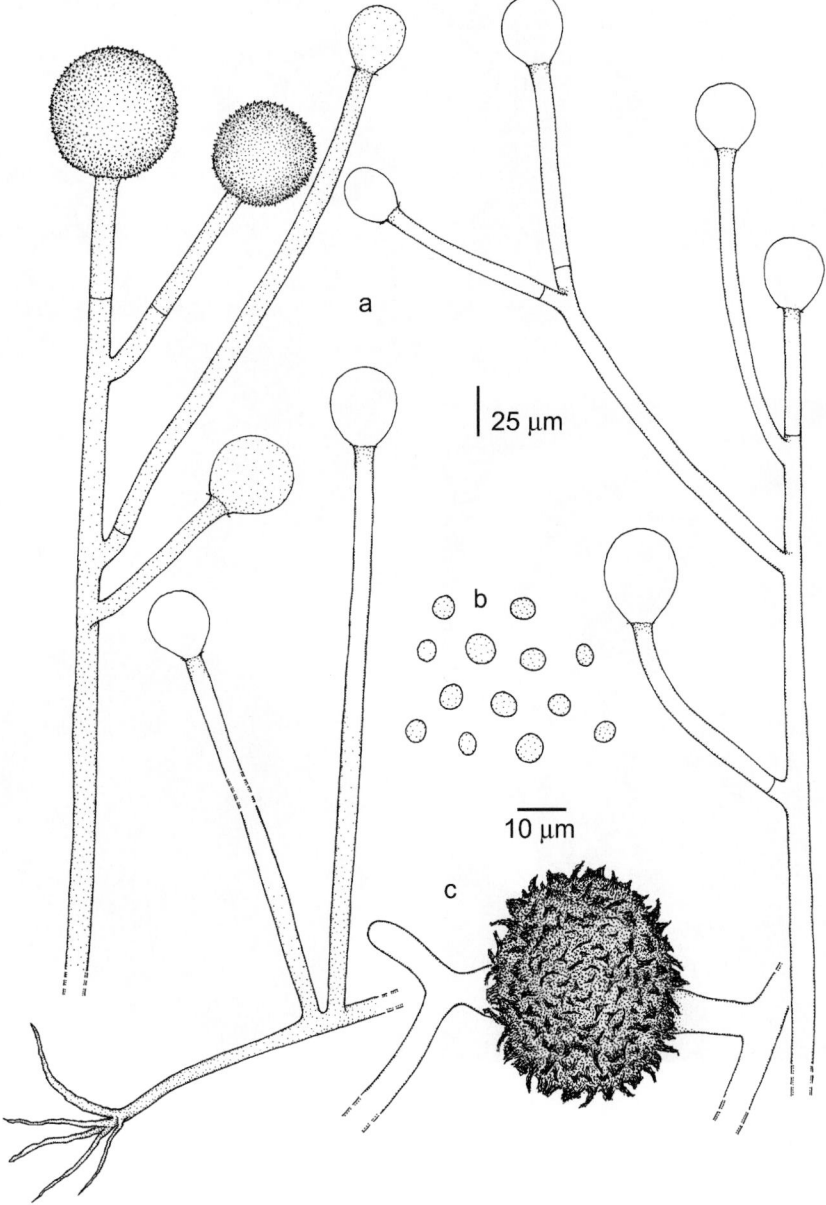

Rhizomucor miehei, IMI 240410. a. Sporangiophores; b. sporangiospores; c. zygospore.

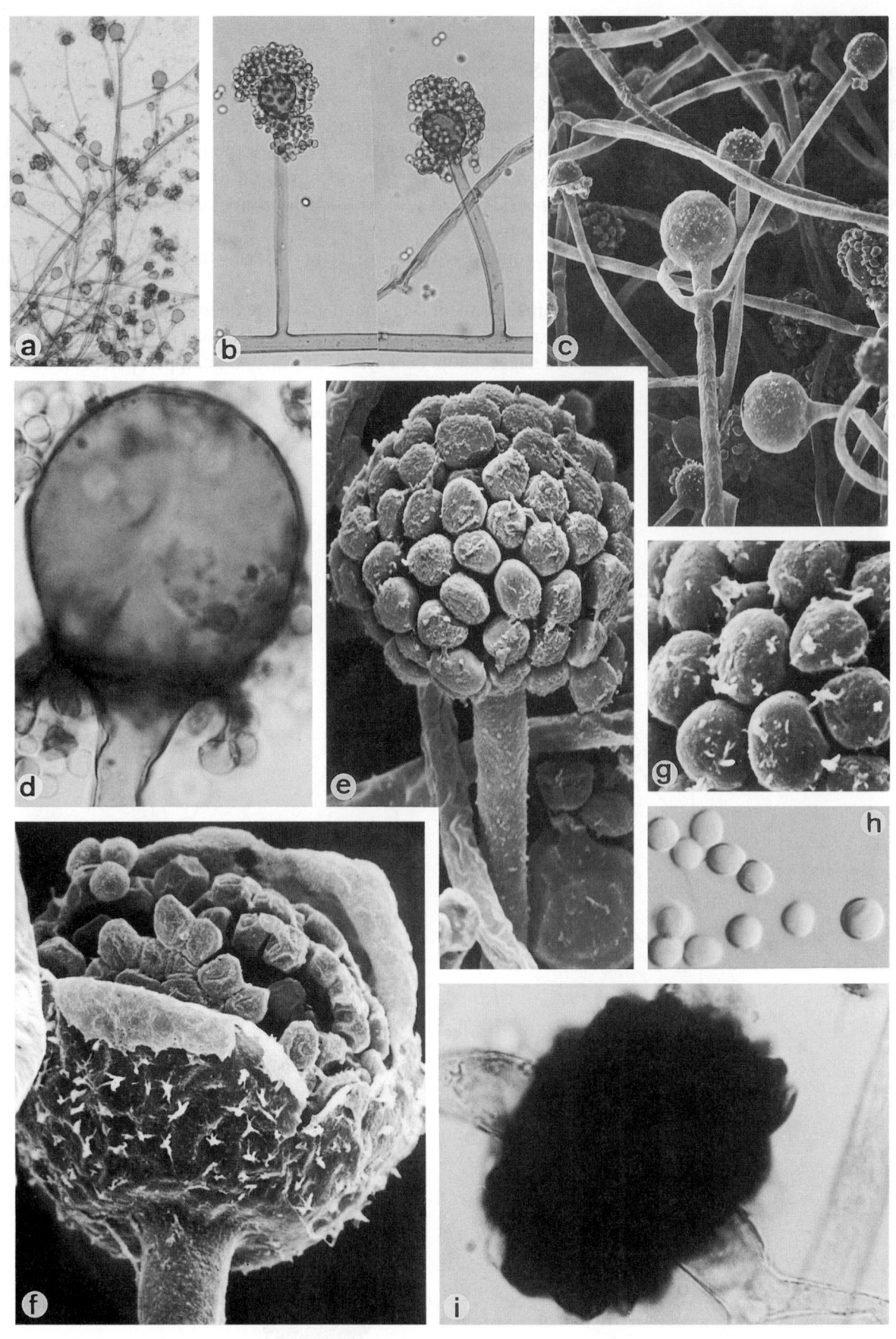

Rhizomucor miehei, IMI 240410. a-c. Sporangiophores; d. columella; e, f. mature sporangia; g, h. sporangiospores; i. zygospore. a. ×76; b. ×256; c. ×520; d. ×1600; e. ×3050; f. ×2450; g. ×5000; h, i. ×1600.

96

Rhizomucor pusillus (Lindt) Schipper

Colony characteristics. Colonies (MEA, 30°C) expanding, brownish.

Microscopy. Sporangiophores originating from aerial hyphae or from stolons, with branched, thin-walled rhizoids. Sporangiophores brownish, 11-15 µm wide, apical branches each terminating with a sporangium. Sporangia spherical, up to 100 µm diam; membrane diffluent; columella (sub)spherical to pyriform, about 40 µm wide, without apophysis. Sporangiospores hyaline, smooth-walled, (sub)spherical, 3-4 µm diam.

Zygospores. Zygospores spherical to slightly compressed, up to 70 µm wide, dark brown to blackish-brown, with stellate warts. Suspensors equal. Homo- or heterothallic.

Physiology. Growth with sucrose; not thiamin dependent; thermophilic, maximum growth temperature 55°C.

Pathogenicity. BSL-2. Rare cause of human zygomycosis, particularly in leukemic patients (St-Germain *et al.*, 1993). Cases of systemic (Erdos *et al.*, 1972) and disseminated (Kramer *et al.*, 1977) infections have been reported. A localized infection was seen in a diabetic patient (Wickline *et al.*, 1989). It is a frequent agent of bovine mycotic abortion.

References. Schipper (1978b), Vastag *et al.* (1998).

Nomenclature. *Mucor pusillus* Lindt - Arch. Exp. Path. Pharmakol. 21: 272, 1886 ≡ *Rhizomucor pusillus* (Lindt) Schipper - Stud. Mycol. 17: 54, 1978.

Mucor parasiticus Lucet & Costantin - C.R. Hebd. Acad. Sci., Paris 129: 1033, 1899 (≡ *Rhizomucor parasiticum* auctt.) ≡ *Rhizopus parasiticus* (Lucet & Costantin) Lendner - Matér. Fl. Crypt. Suisse 3: 115, 1908.

Mucor septatus Bezold, *in* Siebenmann - Schimmelmyc. Menschl. Ohres p. 97, 1889 ≡ *Rhizomucor septatus* (Bezold) Lucet & Costantin - Arch. Parasit. 4: 362, 1901.

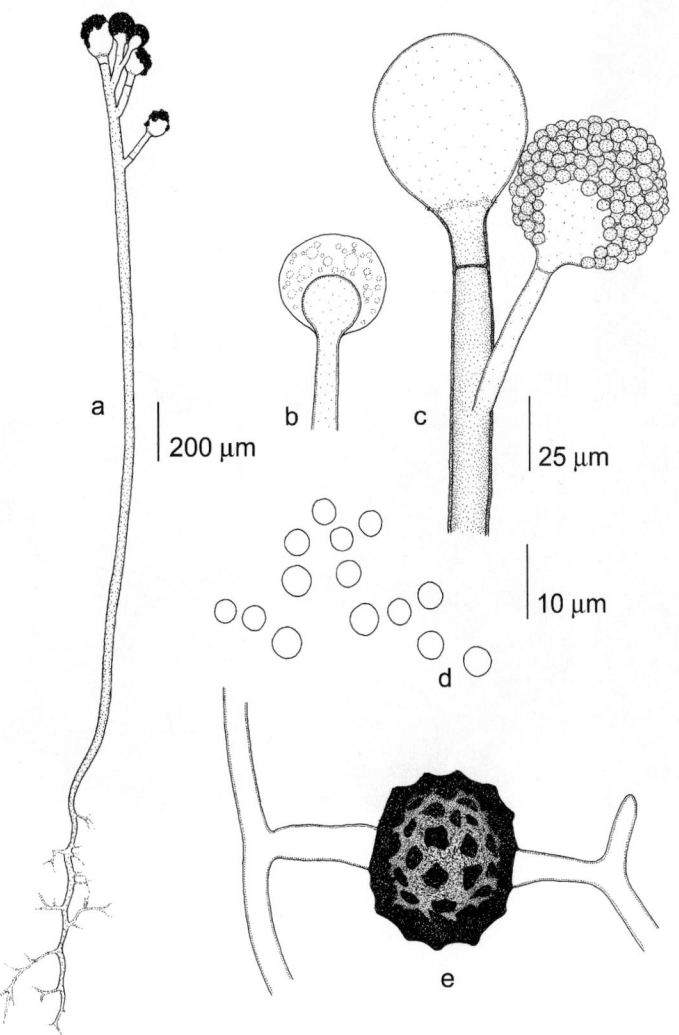

Rhizomucor pusillus, CBS 245.58. a. Sporangiophore with columellae and clusters of sporangiospores; b. section of young sporangium; c. columellae; d. sporangiospores; e. zygospore.

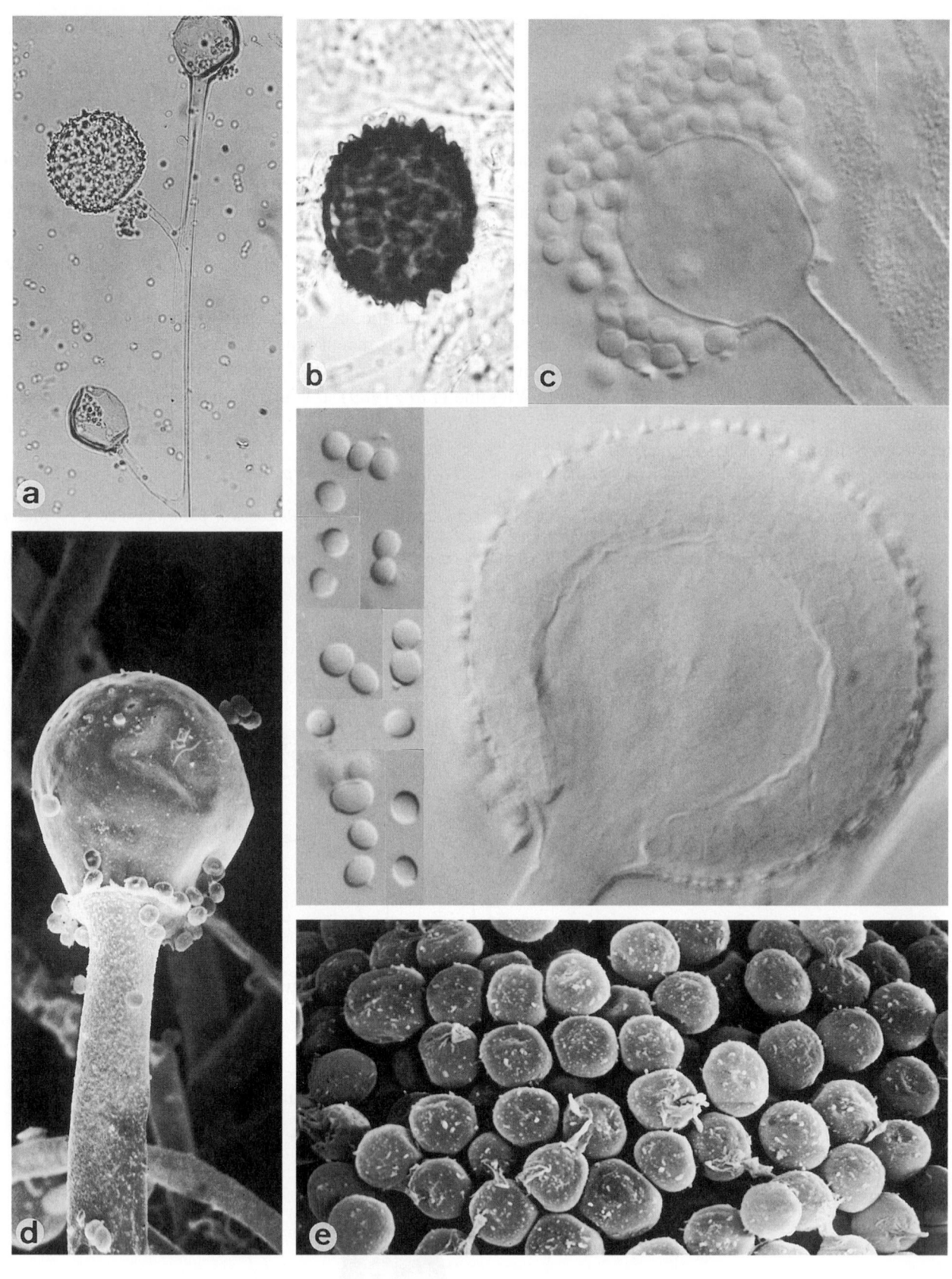

Rhizomucor pusillus, IMI 226172. a. Sporangiophore; b. zygospore; c. young and mature sporangia, and sporangiospores; d. columella with liberated sporangiospores; e. sporangiospores. a. ×320; b. ×640; c. ×1600; d. ×1450; e. ×4700.

Rhizomucor variabilis Zheng & Chen

Colony characteristics. Colonies (MEA, 30°C) expanding, lanose to hairy, whitish to ochraceous, with buff-ochre reverse.

Microscopy. Sporangiophores arising from hyphae or from stolons; rhizoids abundant. Sporangiophores hyaline, up to 2 mm long, 9-23 μm wide, simple or once branched, with branches terminating at a higher level than the main stems; branches all ending in a sporangium. Sporangia (sub)spherical, up to 100 μm diam; membrane diffluent; columella spherical, ellipsoidal to cylindrical, about 40 μm wide, sometimes lobed, with or without apophysis. Sporangiospores hyaline, smooth-walled, very variable, mostly subspherical to ellipsoidal, 3-11 × 2-7 μm. Chlamydospores abundant.

Zygospores. Zygospores unknown.

Physiology. Maximum growth temperature 38°C.

Differential diagnosis. The species may concern degenerate cultures of *Mucor hiemalis* (p. 86; M.A.A. Schipper, pers. comm.). Phylogenetically it proved to be very close to that species (Voigt *et al.*, 1999).

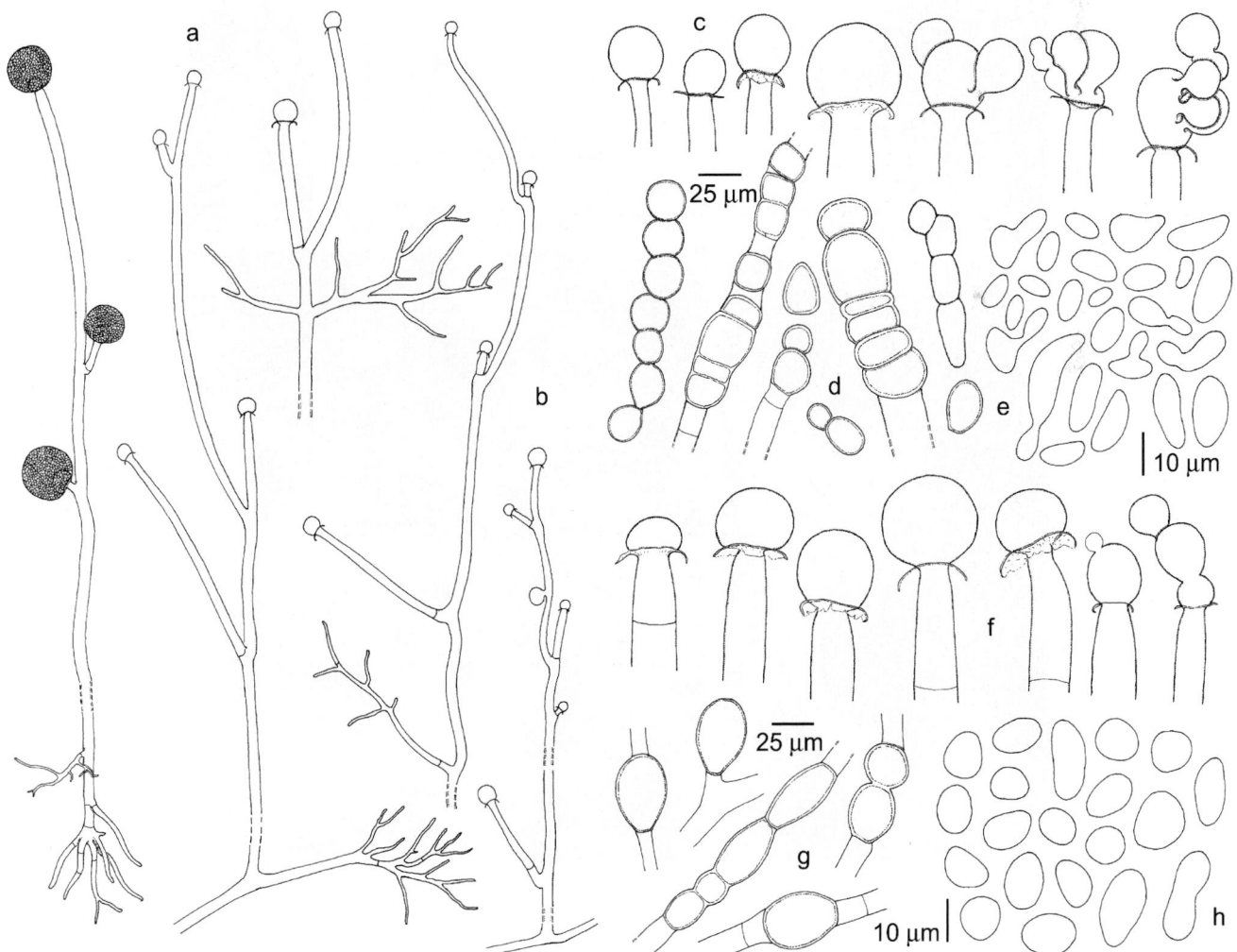

Rhizomucor variabilis var. *variabilis*, a, c, d, e, CBS 103.93; *Rhizomucor variabilis* var. *regularior*, b, f, g, h, CBS 384.95. a, b. Sporangiophores; c, f. columellae of various shapes; d, g. chlamydospores; e, h. sporangiospores.

Varieties. The var. *regularior* Zheng & Chen was distinguished on the basis of more profusely branched sporophores which mostly had a septum at the point of branching, and of some additional, gradational morphological differences (Zheng & Chen, 1993).

Pathogenicity. BSL-2. Known from a chronic cutaneous lesion in a human patient (Zheng & Chen, 1991). The var. *regularior* was reported from a disfiguring facial mycosis (Zheng & Chen, 1993). Both patients were without underlying disease.

References. Zheng & Chen (1991, 1993).

Nomenclature var. *regularior*. *Rhizomucor variabilis* Zheng & Chen var. *regularior* Zheng & Chen - Mycosystema 6: 2, 1993.
Nomenclature var. *variabilis*. *Rhizomucor variabilis* Zheng & Chen - Mycosystema 4: 47, 1991.

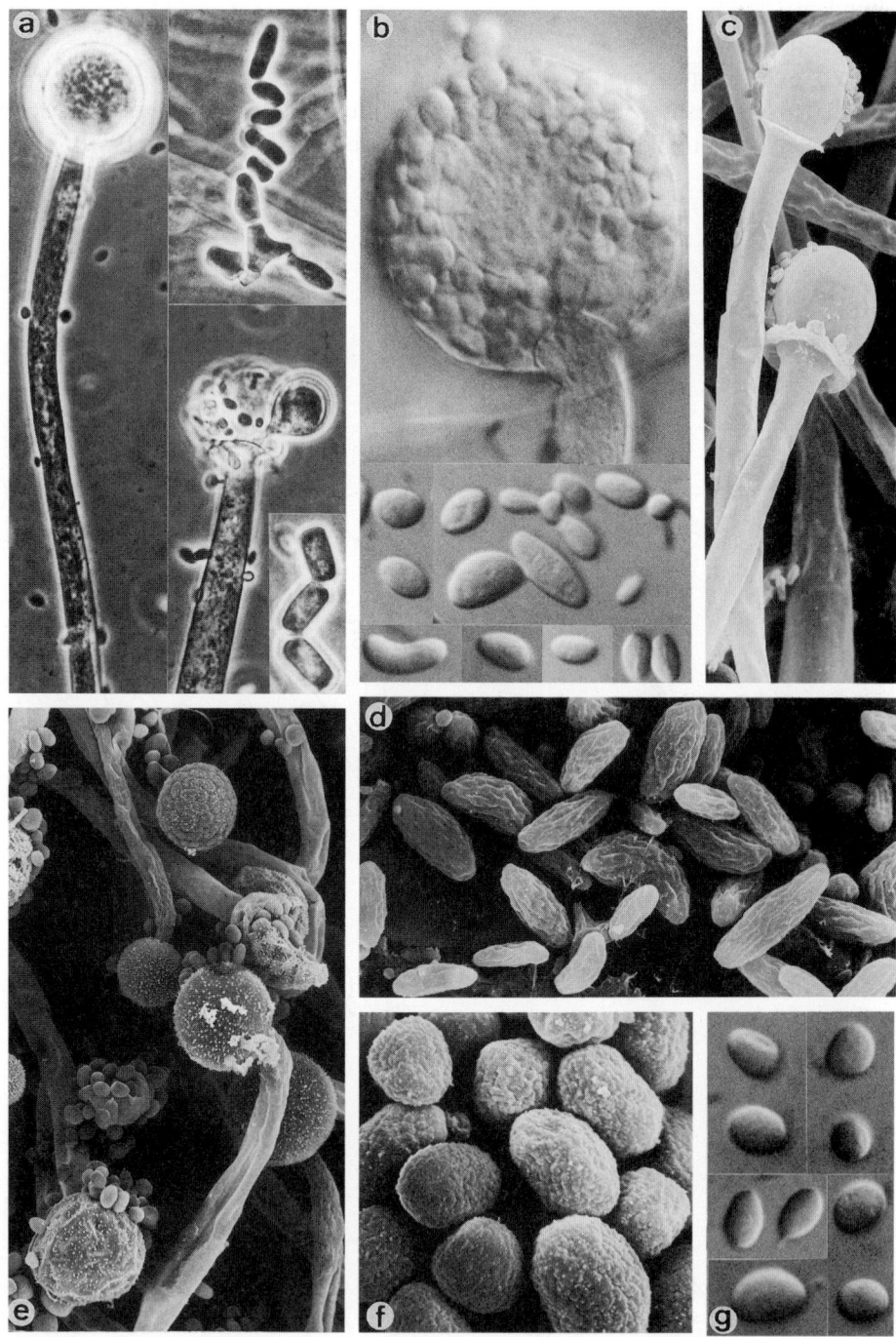

Rhizomucor variabilis var. *variabilis*, a-d. CBS 103.93; *Rhizomucor variabilis* var. *regularior*, e-g. CBS 384.95. a. Sporangiophore, chlamydospores and an irregular columella; b, e. sporangia and sporangiospores; c. columellae; d-g. sporangiospores. a. ×485; b. ×1215; c. ×645; d. ×2735; e. ×570; f. ×3500; g. ×1215.

Zygomycota, Mucorales, Mucoraceae. Genus: *RHIZOPUS*

Generic description. Colonies expanding, hairy. Stolons and rhizoids present. Sporangiophores arising from swellings just above rhizoids. Sporangia terminal, multi-spored, with apophysis and columella; columellae subspherical to slightly ellipsoidal. Sporangiospores angular or (sub)spherical, ornamented. Zygospores reddish, between unequal, spherical suspensors.

General remarks. Experimental infection was performed by van Cutsem *et al.* (1988). The GC% of DNA of medical species were listed by Frye & Reinhardt (1993).

References. Domsch *et al.* (1980), Schipper (1984), Schipper & Stalpers (1984).

Key to the treated species of Rhizopus:

1a.	Azygospores abundant	***R. azygosporus* (102)**
1b.	Azygospores infrequent **→ 2**	
2a.	Sporangiophores mostly not exceeding 0.8 mm in height; sporangia up to 100 µm diam; growth at 45°C, though sometimes poor **→3**	
2b.	Sporangiophores often higher than 1 mm; sporangia mostly over 100 µm diam; no growth at 45°C **→ 7**	
3a.	Sporangiospores angular	***R. microsporus* var. *chinensis* (105)**
3b.	Sporangiospores broadly ellipsoidal to subspherical **→ 4**	
4a.	Sporangiospores rarely exceeding 6 µm in diam and 6.5 µm in length **→ 5**	
4b.	Sporangiospores often exceeding 6 µm in diam and 9 µm in length	***R. microsporus* var. *oligosporus* (105)**
5a.	Sporangiospores distinctly striate **→ 6**	
5b.	Sporangiospores minutely spinulose	***R. microsporus* var. *rhizopodiformis* (105)**
6a.	Colonies hairy, dark greyish-brown; good growth at 45°C	***R. microsporus* var. *microsporus* (104)**
6b.	Colonies thinly floccose, greyish-white; restricted growth at 45°C	***R. schipperae* (108)**
7a.	Sporangia up to 275 µm diam; no growth at 37°C	***R. stolonifer* (110)**
7b.	Sporangia not exceeding 250 µm diam; growth at 37°C	***R. oryzae* (106)**

Rhizopus azygosporus Yuan & Jong

Colony characteristics. Colonies (MEA, 30°C) expanding; hairy, greyish-ochraceous, becoming dark grey with age, up to 10 mm high. Colony reverse yellowish-brown.

Microscopy. Sporangiophores single or in small groups, brown, up to 500 µm high, 6-14 µm wide. Sporangia spherical, up to 100 µm diam, greyish-black; columella spherical to conical, comprising 80% of the sporangium. Sporangiospores hyaline, spherical to ovoidal, 4-7 µm in length with faint striations. Chlamydospores present.

Zygospores. Zygospores absent. Azygospores abundant, terminal or lateral, spherical, golden to dark brown, 15-70 µm in diam, with stellate projections.

Physiology. Maximum growth temperature 45-50°C.

Pathogenicity. BSL-2. The fungus was found as the agent of three fatal cases of gastrointestinal infections in premature infants in Australia (Schipper *et al.*, 1996).

Nomenclature. *Rhizopus azygosporus* Yuan & Jong - Mycotaxon 20: 398, 1984.

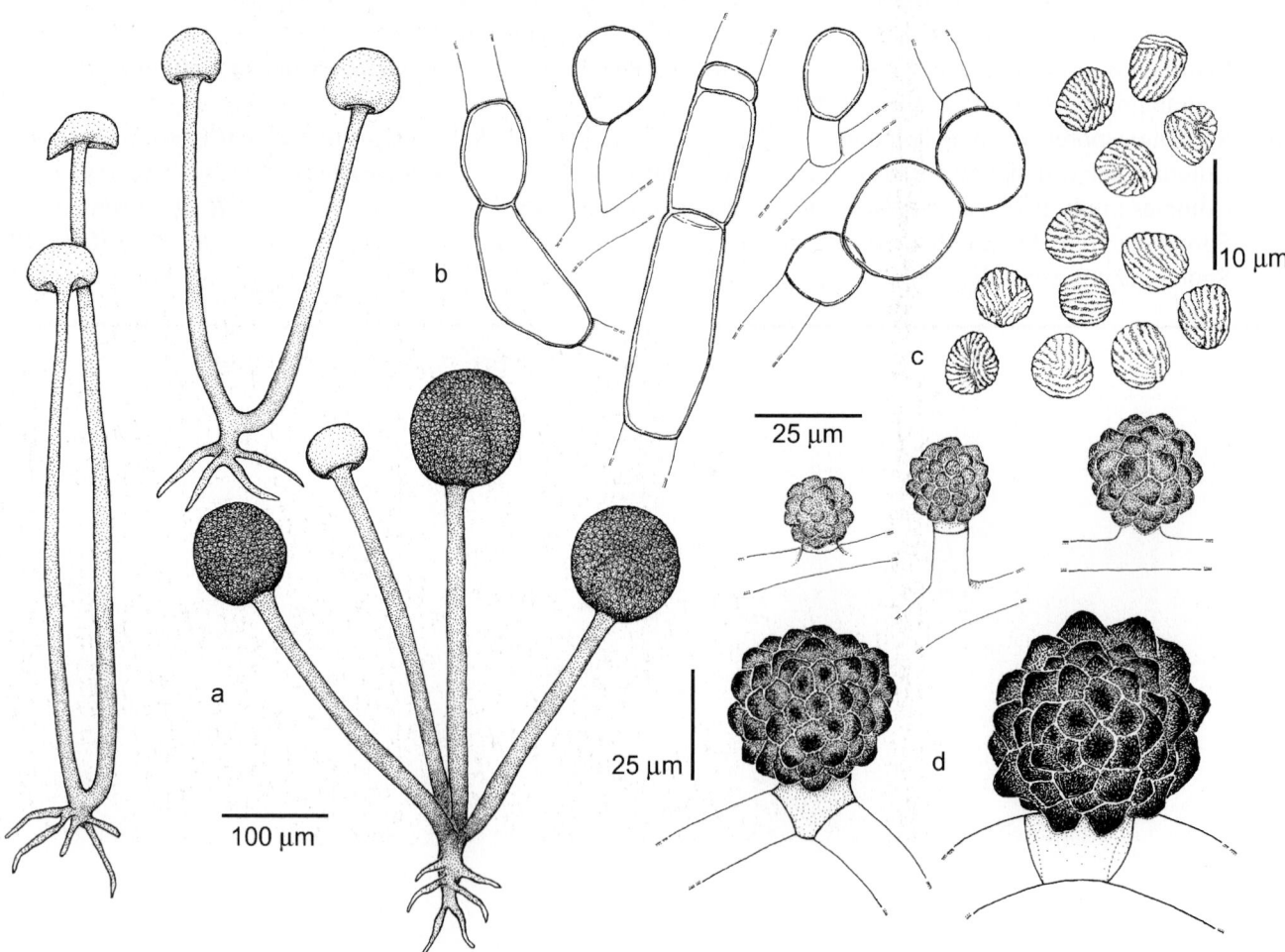

Rhizopus azygosporus, CBS 357.93. a. Sporangiophores; b. chlamydospores; c. sporangiospores; d. azygospores.

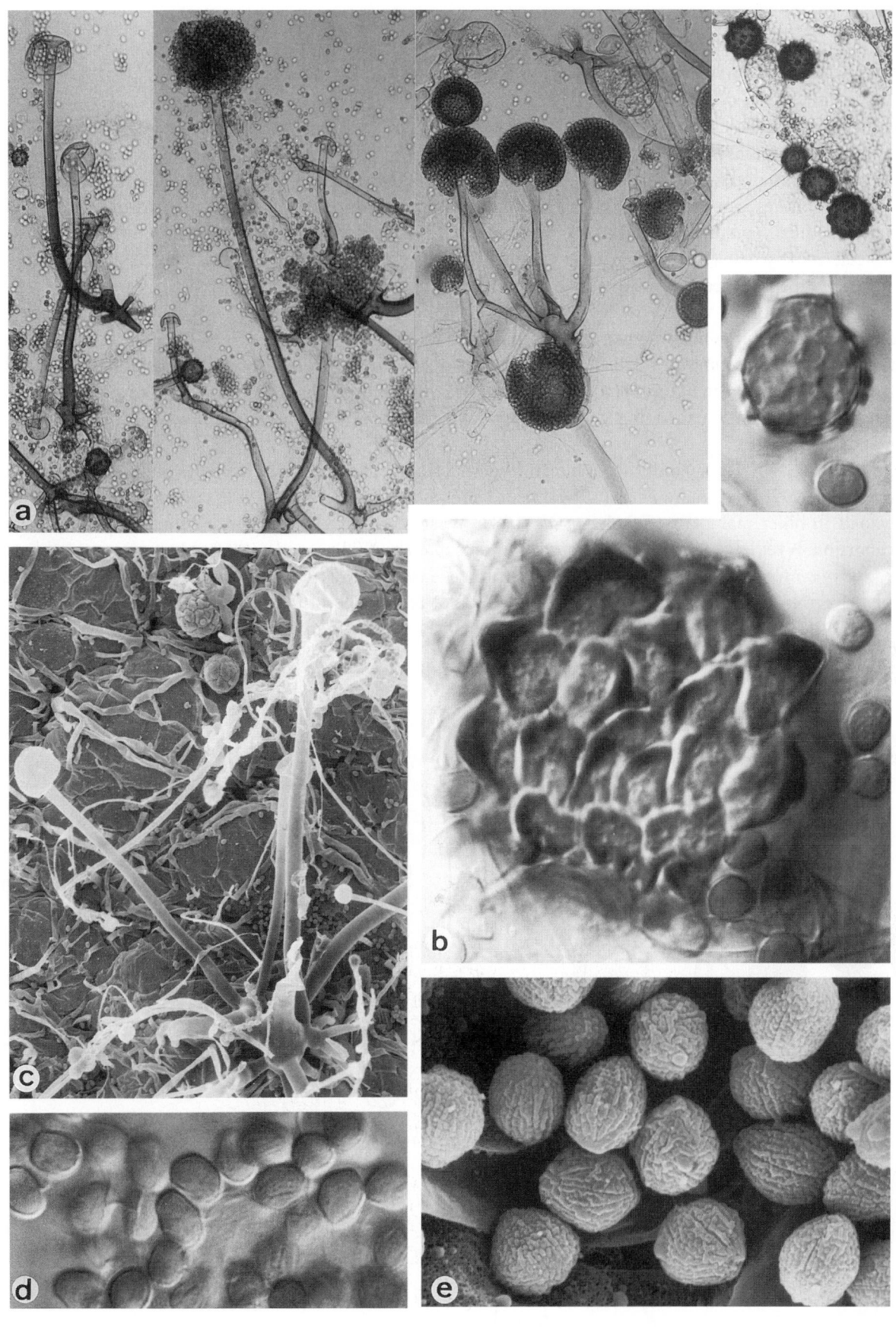

Rhizopus azygosporus, CBS 357.93. a, c. Sporangiophores, chlamydospores and azygospores; b, d, e. azygospores and sporangiospores. a. ×128; b, d. ×1600; c. ×240; e. ×3800.

103

Rhizopus microsporus v. Tiegh.

Colony characteristics. Colonies (MEA, 30°C) expanding, hairy, dark greyish-brown, up to 10 mm high. Colony reverse yellowish-brown.

Microscopy. Sporangiophores often in pairs, brownish, up to 400 μm high, 8-10 μm wide; usually mats of macro- and microsporangiophores can be distinghuished. Sporangia spherical, up to 100 μm diam, greyish-black; columella spherical to conical, comprising 80% of the sporangium. Sporangiospores hyaline, angular, broadly ellipsoidal or subspherical, mostly between 6 and 9 μm in length, distinctly striate. Chlamydospores may be present.

Zygospores. Zygospores reddish-brown, up to 100 μm diam, with stellate projections, usually borne between unequal suspensors. Heterothallic.

Physiology. Thermophilic: maximum growth temperature 50-52°C.

Varieties. Var. *rhizopodiformis* (Cohn) Schipper & Stalpers differs from the typical variety by (sub)spherical sporangiospores which are minutely spinulose, var. *oligosporus* (Saito) Schipper & Stalpers by irregularly ornamented sporangiospores frequently exceeding 9 μm in length and var. *chinensis* (Saito) Schipper & Stalpers by somewhat angular sporangiospores.

Pathogenicity. BSL-2. The species causes human zygomycosis, particularly the cutaneous and gastrointestinal forms (Kimura *et al.,* 1995). Nakamura *et al.* (1989) described a fatal infection in a CAPD patient. It is the main agent of zygomycosis in other mammals (Scholer *et al.,* 1983; West *et al.,* 1983), especially being involved in swine and bovine abortion. A review of the recent literature was presented by Waller *et al.* (1993). Most cases probably concern the var. *rhizopodiformis* (West *et al.,* 1995). Confirmed cases by var. *microsporus* were necrotic subcutaneous infections (Kerr *et al.,* 1988; West *et al.,* 1995). Tintelnot & Nitsche (1989) reported a wound infection by var. *oligosporus.* Van Beyma's (1931) strain from cow foetus is now known to be var. *chinensis.*

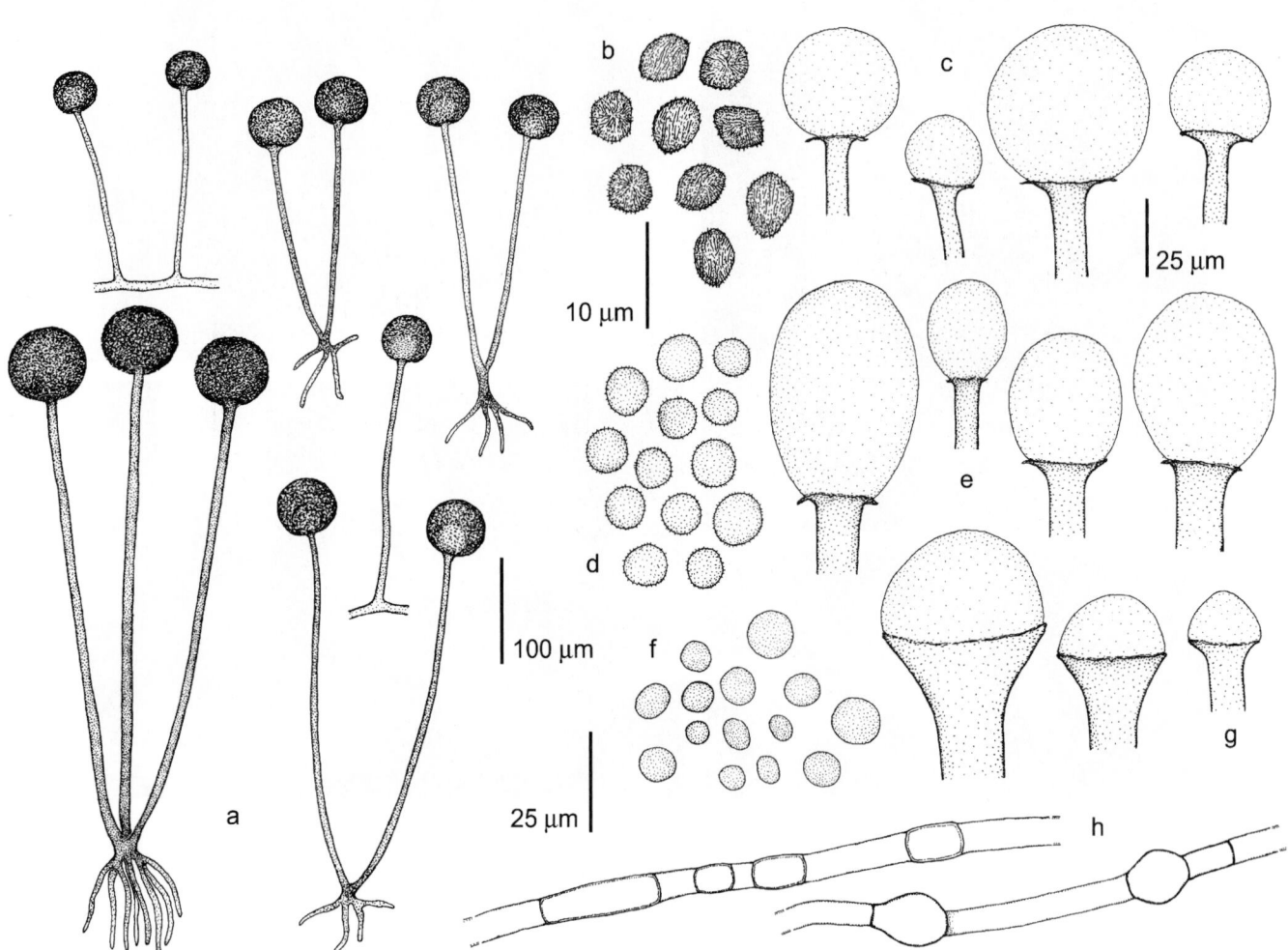

Rhizopus microsporus var. *microsporus*, a-c. CBS 308.87; *Rhizopus microsporus* var. *rhizopodiformis*, d, e. CBS 102277; *Rhizopus microsporus* var. *oligosporus*, f-h. CBS 338.62. a. Sporangiophore; b, d, f. sporangiospores; c, e, g. columellae; h. chlamydospores.

References. Bottone *et al.* (1979), Schipper & Stalpers (1984), Ellis (1986), Polonelli *et al.* (1988), Waller *et al.* (1993).

Nomenclature var. *chinensis*. *Rhizopus chinensis* Saito - Zentbl. Bakt. Parasitkde, Abt. 2, 13: 156, 1904 ≡ *Rhizopus microsporus* van Tieghem var. *chinensis* (Saito) Schipper & Stalpers - Stud. Mycol. 25: 31, 1984.

   Rhizopus bovinus van Beyma - Verh. K. Ned. Akad. Wet., Ser. C, 29: 38, 1931.

Nomenclature var. *microsporus*. *Rhizopus microsporus* van Tieghem - Ann. Sci. Nat., Sér. 6, 1: 83, 1875.

Nomenclature var. *oligosporus*. *Rhizopus oligosporus* Saito - Zentbl. Bakt. Parasitkde, Abt. 2, 14: 626, 1905 ≡ *Rhizopus microsporus* van Tieghem var. *oligosporus* (Saito) Schipper & Stalpers - Stud. Mycol. 25: 31, 1984.

Nomenclature var. *rhizopodiformis*. *Mucor rhizopodiformis* Cohn, *in* Lichtheim - Z. Klin. Med. 7: 140, 1884 ≡ *Rhizopus cohnii* Berlese & de Toni, *in* Saccardo - Syll. Fung. 7: 213, 1888 (name change) ≡ *Rhizopus rhizopodiformis* (Cohn) Zopf, *in* Schenk - Handb. Bot. 4: 587, 1890 ≡ *Rhizopus microsporus* van Tieghem var. *rhizopodiformis* (Cohn) Schipper & Stalpers - Stud. Mycol. 25: 30, 1984.

   Rhizopus equinus Costantin & Lucet - Bull. Soc. Mycol. Fr. 19: 200, 1903.

   Rhizopus pusillus Naumov - Opred. Mucor. p. 74, 1935.

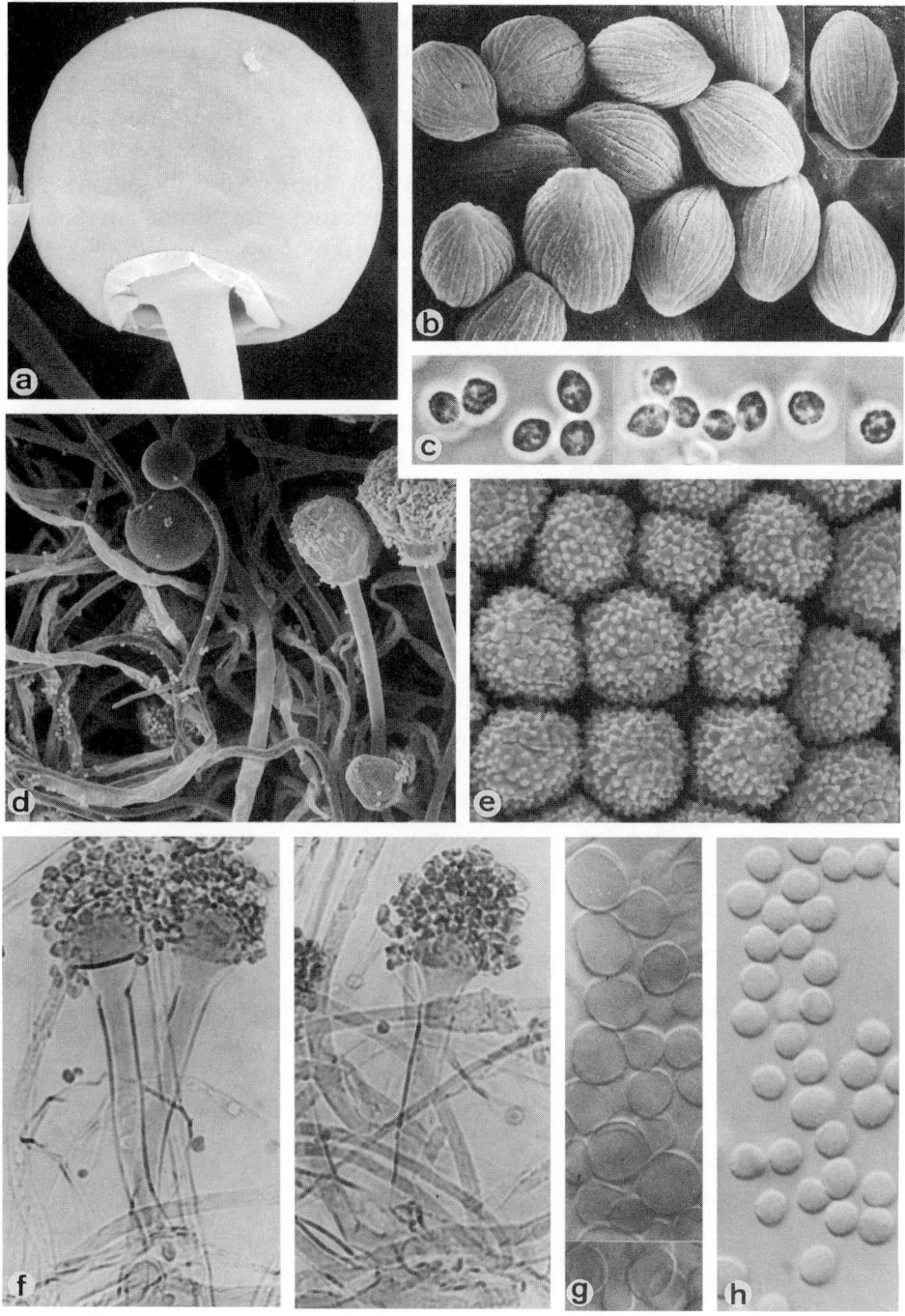

Rhizopus microsporus var. *microsporus*, a-c. CBS 308.87; *Rhizopus microsporus* var. *rhizopodiformis*, d, e, h. CBS 102277; *Rhizopus microsporus* var. *oligosporus*, f, g. CBS 338.62. a. Columella; b, c, e, g, h. sporangiospores; d, f. sporangiophores. a. ×1015; b. ×3740; c. ×1090; d. ×160; e. ×4590; f. ×350; g, h. ×1090.

Rhizopus oryzae Went & Prinsen Geerligs

Colony characteristics. Colonies (MEA, 30°C) expanding, up to 1 cm high, whitish to greyish-brown.

Microscopy. Sporangiophores singly or in tufts, brown, 1-2 mm high, 18 μm wide, mostly unbranched, sometimes with brownish swellings up to 50 μm diam. Rhizoids sparingly branched, up to 250 μm long, brownish. Sporangia spherical, 50-250 μm diam, brownish-grey to black; columella comprising 50-70% of sporangium, spherical; apophysis short, 3-12 μm high. Sporangiospores greyish-green, angular, subspherical to ellipsoidal, longitudinally striate, 6-8 × 4.5-5.0 μm. Chlamydospores single or in chains, spherical to ovoidal, 10-35 μm diam, hyaline, smooth-walled.

Zygospores. Zygospores red to brown, spherical or laterally flattened, 60-140 μm, with flat projections. Suspensors unequal, spherical and conical. Heterothallic.

Physiology. Thermotolerant: growth at 40°C, no growth at 45°C.

Pathogenicity. BSL-1. Important agent of human mucormycosis. The great majority of these are rhinocerebral infections (Scholer *et al.*, 1983; Siddiqi & Freedman, 1994; Attapattu, 1995; Belhadj *et al.*, 1997; Adler *et al.*, 1998). (Sub)cutaneous infections cause severe local necrosis (Liao *et al.*, 1995; Toro *et al.*, 1998). A gastric ulceration in a renal transplant patient was reported by Winkler *et al.* (1966), a fatal pulmonary infection in a heart transplant recipient by Muhm *et al.* (1996), a genitourinary infection in a diabetic patient by Williams *et al.* (1995) and cutaneous infections by Hurle *et al.* (1996) and Linder *et al.* (1998). The case fatality rate in such infections is high. Most cases are associated with diabetes mellitus; AIDS-related infections are uncommon (Sanchez *et al.*, 1994). A pulmonary mycosis in an otherwise healthy patient was described by Yokoi *et al.* (1999). The species is rarely reported from animals (Perelman & Kuttin, 1992).

References. Schipper (1984), Ellis (1985), Oliveri *et al.* (1988).

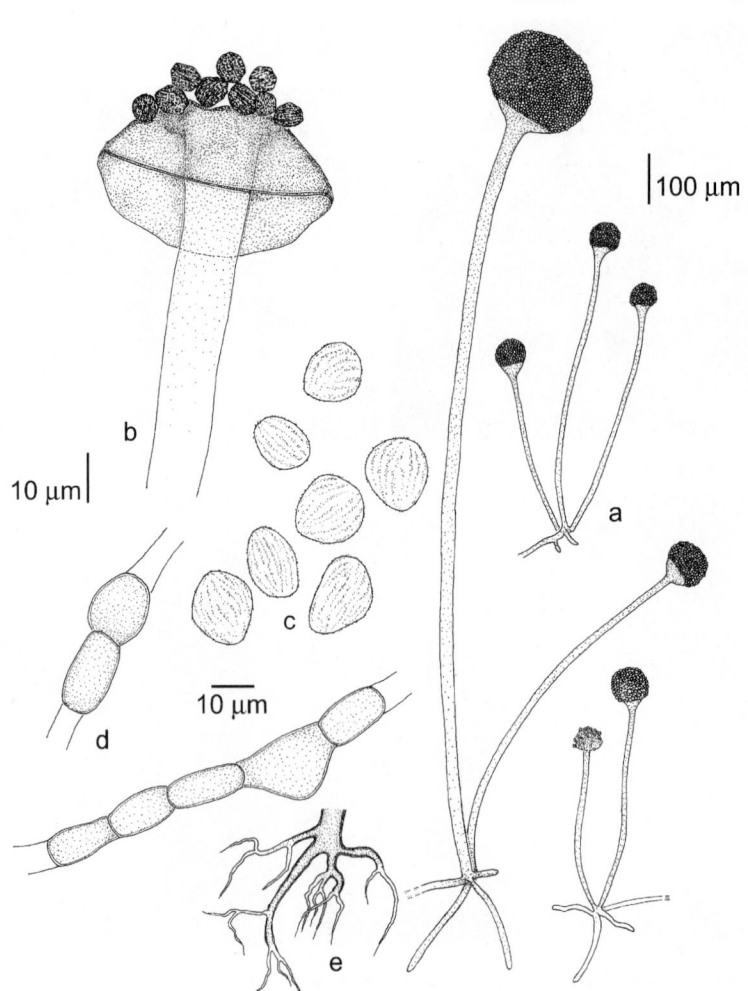

Rhizopus oryzae. a-d. IMI 215407; e. CBS 110.17. a . Habit sketch; b. sporangiophore; c. sporangiospores; d. chlamydospores; e. rhizoids..

Antifungal susceptibility.

Antifungal	mean MICs	MIC 90	Strains	Reference
AMB	1.24	1.56	3	Wildfeuer *et al.* (1998)
AMB	0.5	nd	5	Espinel-Ingroff *et al.* (1995)
FLZ	64	nd	5	Espinel-Ingroff *et al.* (1995)
ITZ	3.13	6.25	3	Wildfeuer *et al.* (1998)
KTZ	1.56	1.56	3	Wildfeuer *et al.* (1998)
KTZ	8	nd	5	Espinel-Ingroff *et al.* (1995)
VCZ	2.48	12.5	3	Wildfeuer *et al.* (1998)

Nomenclature. (?) *Rhizopus arrhizus* Fischer, *in* Rabenhorst - Krypt. Fl. 1: 233, 1892.
   Rhizopus oryzae Went & Prinsen Geerligs - Verk. K. Akad. Wet., Sect. 2, 4: 16, 1895.
   Rhizopus nodosus Namyslowski - Bull. Acad. Sci. Cracovie 1906: 682, 1906.
   Rhizopus norvegicus Hagem - Unters. Norw. Mucorin. p. 39, 1907/08.
   Rhizopus suinus Nielsen - Virchow's Arch. Path. Anat. 273: 859, 1929.

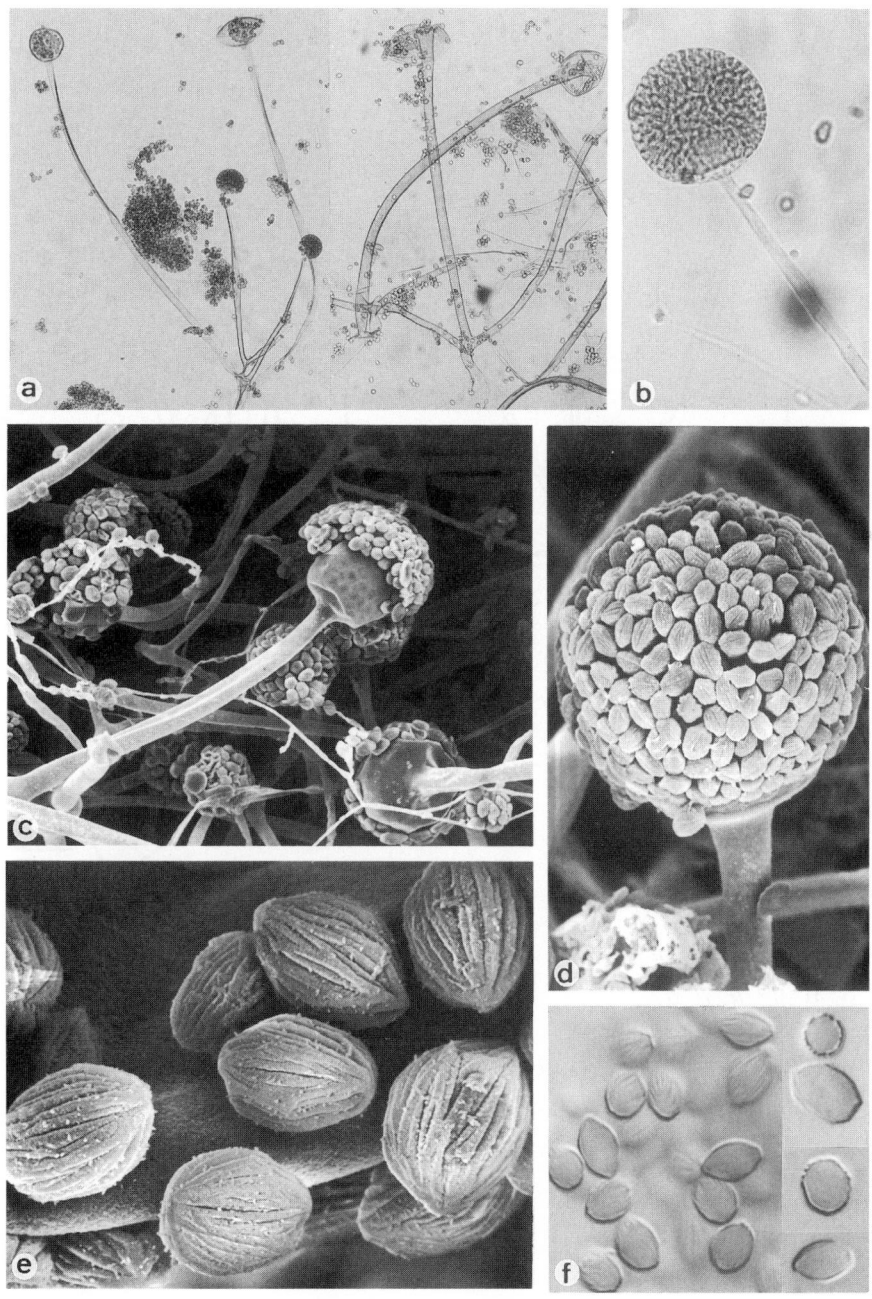

Rhizopus oryzae, IMI 215407. a-d. Sporangiophores and sporangia; e, f. sporangiospores. a. ×160; b. ×415; c. ×320; d. ×860; e. ×3350; f. ×1040.

Rhizopus schipperae Weitzman *et al.*

Colony characteristics. (MEA 30°C) expanding, thinly floccose, greyish-white.
Microscopy. Sporangiophores arising with groups from rhizoids, brown, 100-460 μm high, 5-15 μm wide. Sporangia spherical, up to 80 μm diam, greyish black; columellae subspherical to somewhat conical. Sporangiospores hyaline, subspherical to ovoidal, 5.5-7.0 × 4.8-5.8 μm, distinctly striate. Chlamydospores abundant, terminally or intercalary, spherical up to 20 μm diam.
Zygospores. Zygospores unknown.
Physiology. Thermotolerant: optimal growth at 30°C, very restricted growth at 45°C.
Differential diagnosis. Characteristically, optimal sporulation is obtained on CzA, with production of dense clusters of sporangiophores on rhizoids.
Pathogenicity. BSL-2. The species is known from a case of pulmonary infection in a patient with AML (Weitzman *et al.*, 1996) and from a disseminated infection (Amstead *et al.*, 1999).
Reference. Weitzman *et al.* (1996).
Nomenclature. *Rhizopus schipperae* Weitzman, McGough, Rinaldi & Dalla-Latta - Mycotaxon 59: 220, 1996.

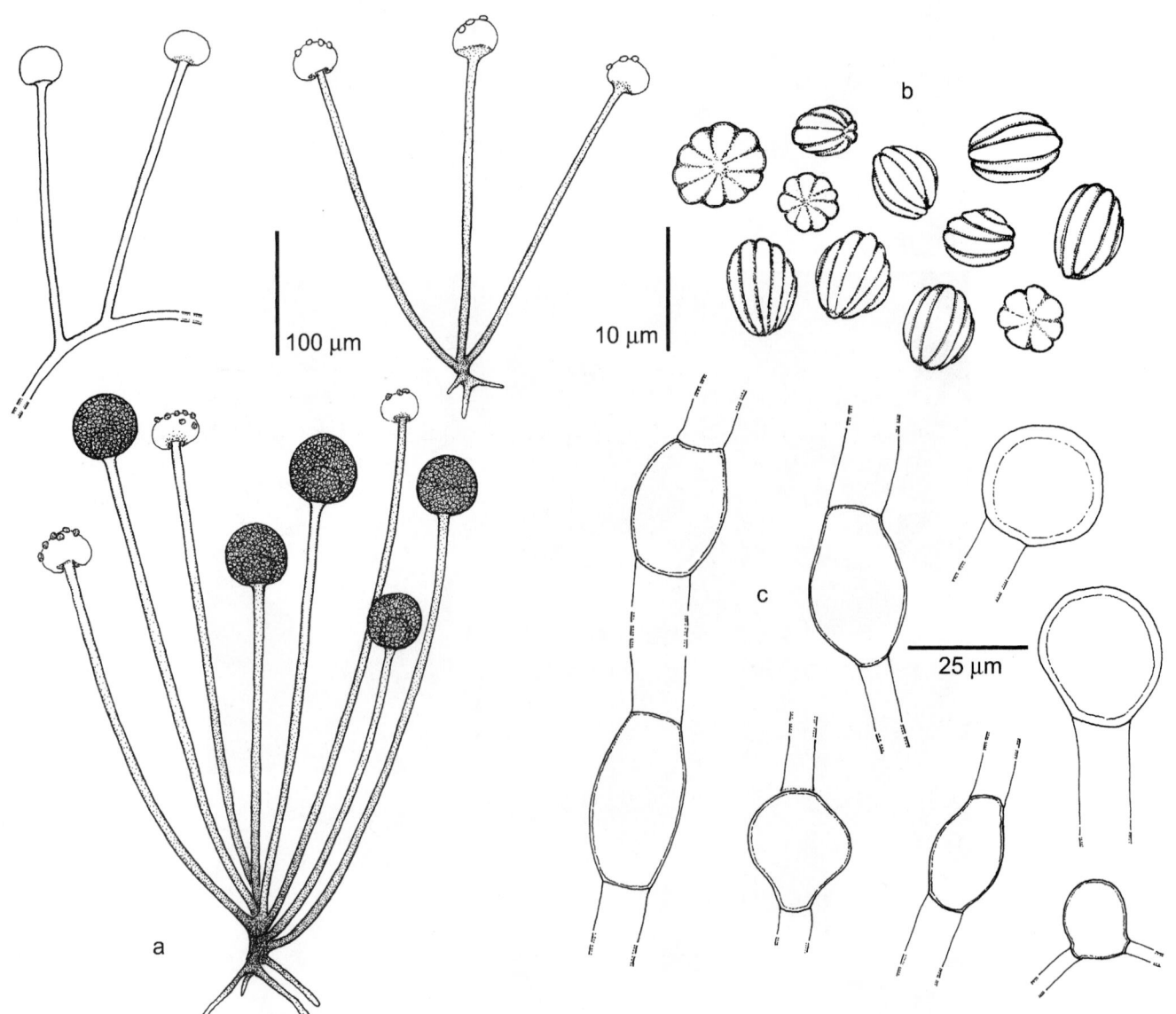

Rhizopus schipperae, CBS. 138.95. a. Sporangiophores; b. sporangiospores; c. chlamydospores.

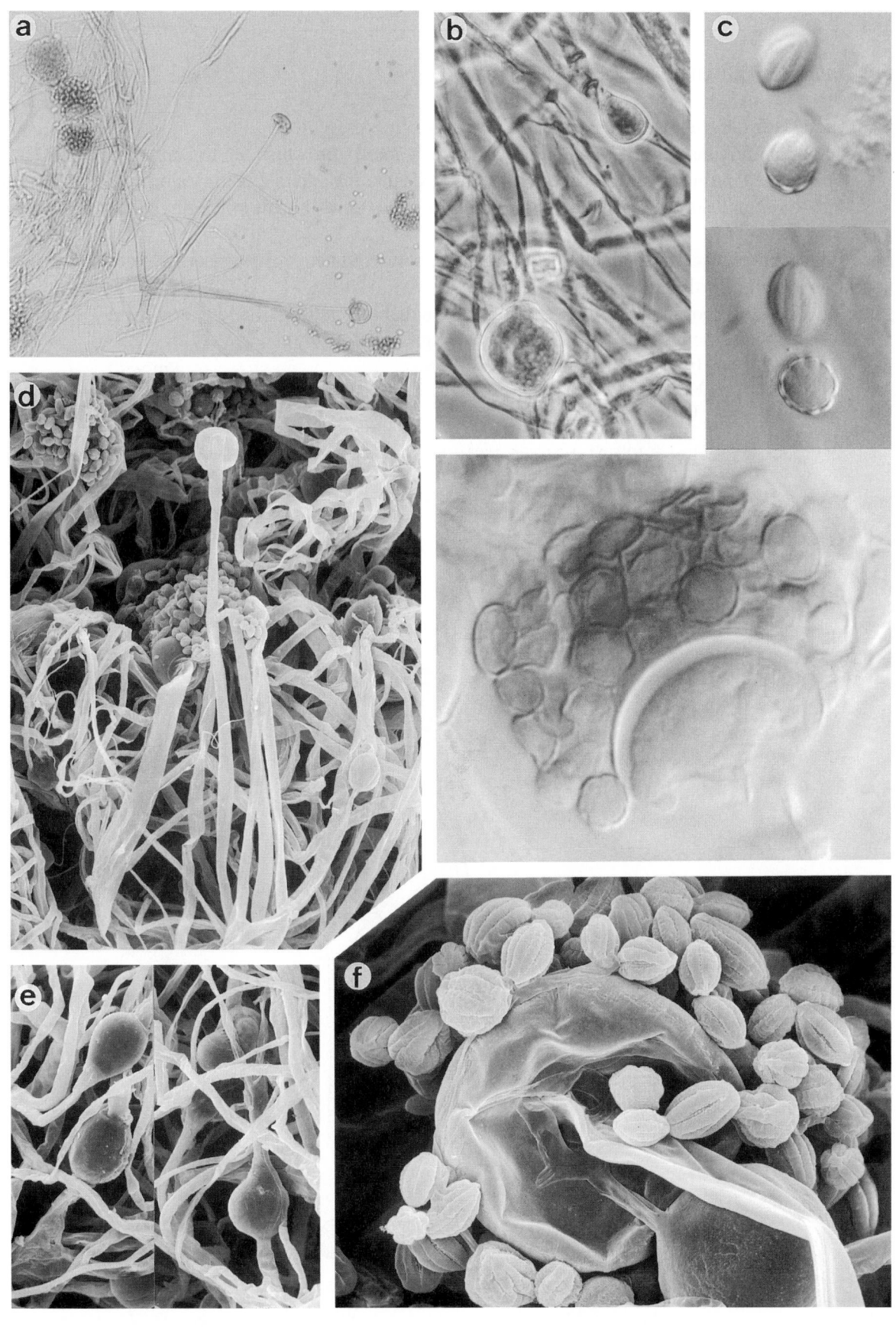

Rhizopus schipperae, CBS. 138.95. a, d. Sporangiophores; b, e. chlamydospores; c, f. columellae and sporangiospores. a. ×250; b. ×512; c. ×1600; d. ×425; e. ×575; f. ×2100.

Rhizopus stolonifer (Ehrenb.:Fr.) Vuill.

Colony characteristics. Colonies (MEA) expanding, whitish.

Microscopy. Sporangiophores with 1-3 together (occasionally more), brownish, up to 2 mm high and 20 µm wide; rhizoids well developed, profusely branched. Sporangia black, spherical, up to 275 µm diam; columellae conical, up to 140 µm high. Sporangiospores angular-spherical to ellipsoidal, up to 13 µm in length, striate. Chlamydospores absent.

Zygospores. Zygospores black, warted, up to 200 µm diam, between unequal suspensors. Heterothallic.

Physiology. Maximum growth temperature 30-32°C.

Pathogenicity. BSL-1. Rhinocerebral mycoses particulary in diabetic patients (Sandler *et al.*, 1997; Ferry & Abedi, 1983). A gastrointestinal case in cattle was described by Naglie *et al.* (1986).

References. Domsch *et al.* (1980), Schipper & Stalpers (1984).

Nomenclature. *Mucor stolonifer* Ehrenberg - Silvae Mycol. Berol. p. 25, 1818 ≡ *Rhizopus nigricans* Ehrenberg - Mycetogenesi p. 198, 1820 (name change) ≡ *Mucor stolonifer* Ehrenberg:Fries - Syst. Mycol. 3: 321, 1829 ≡ *Rhizopus stolonifer* (Ehrenberg:Fries) Vuillemin - Revue Mycol. 24: 54, 1992.

Mucor niger Ciaglinski & Hewelke - Z. Klin. Med. 22: 626, 1893 ≡ *Rhizopus niger* (Ciaglinski & Hewelke) Barthelat - Arch. Parasit. 7: 46, 1903.

Rhizopus artocarpi Raciborski - Paras. Algen Pilze Java's 1: 11, 1900.

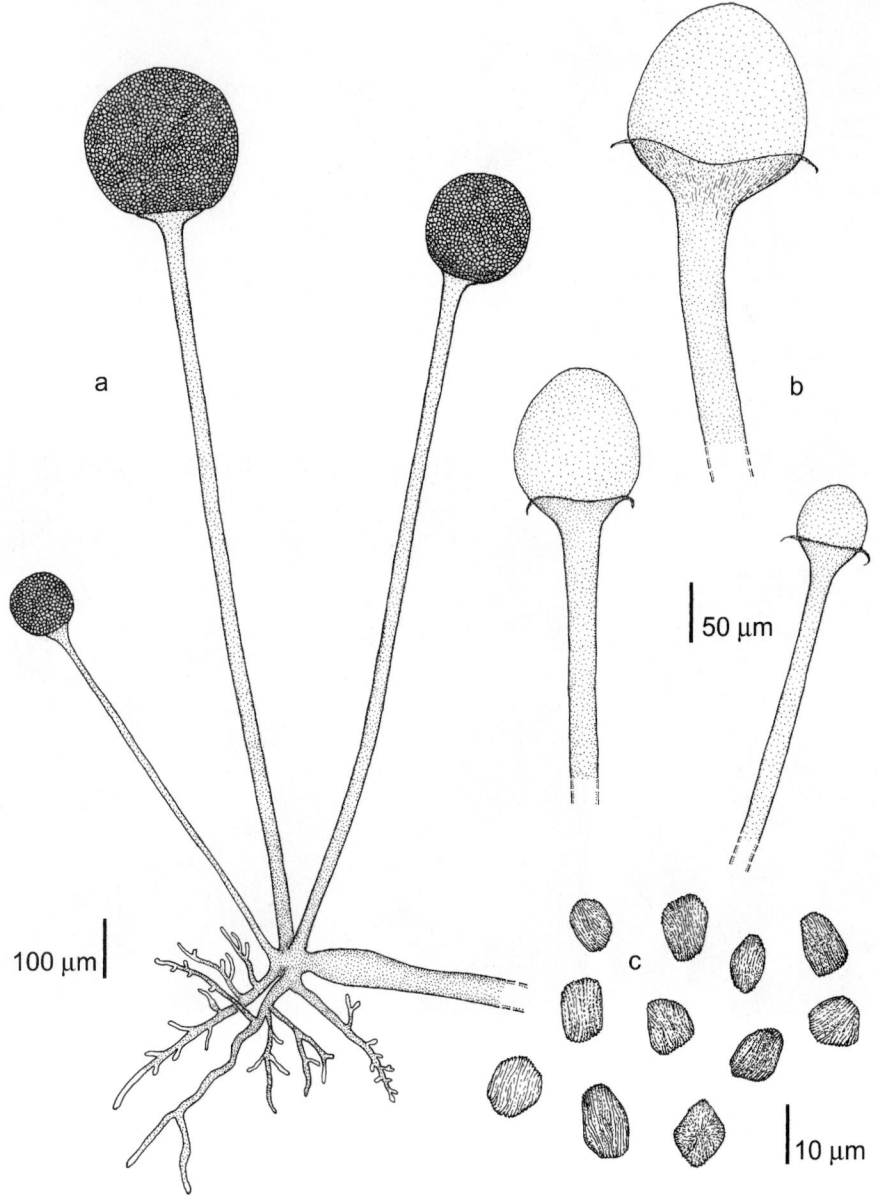

Rhizopus stolonifer, IMI 57762. a. Sporangiophores with rhizoids; b. columellae; c. sporangiospores.

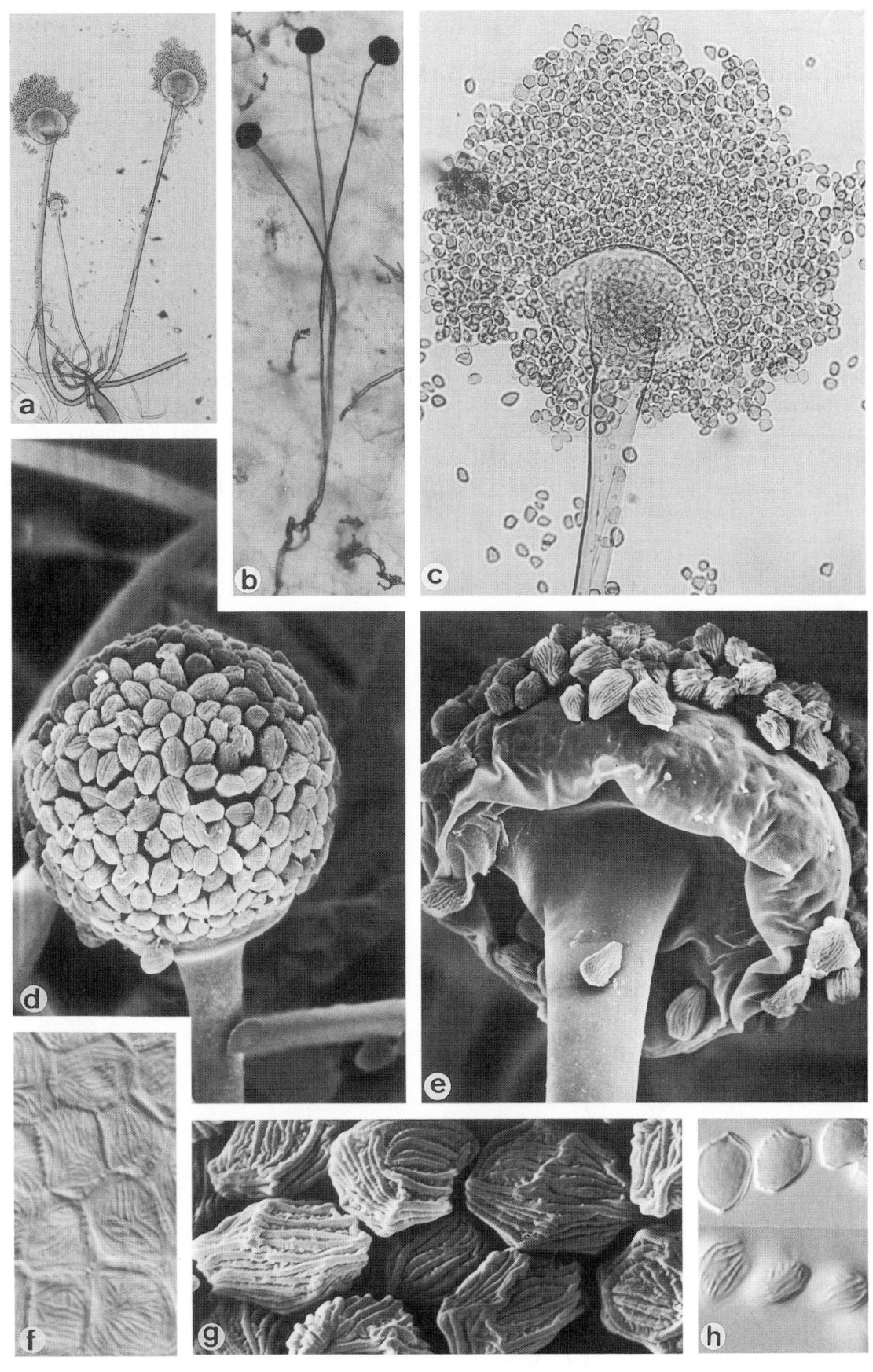

Rhizopus stolonifer, IMI 57762. a,b. Sporangiophores; c-e. mature sporangia; f-h. sporangiospores. a. ×50; b. ×256; c. ×1280; d. ×1325; e. ×1450; f. ×1600; g. ×6000; h. ×1600.

Zygomycota, Mucorales, Saksenaeaceae. Genus: *SAKSENAEA*

Saksenaea vasiformis Saksena

Colony characteristics. Colonies (OA, 30°C) expanding, grey.

Microscopy. Sporangiophores single, unbranched, 25-60 × 6-9 μm, with dichotomously branched, darkly pigmented rhizoids. Sporangia single, terminal, flask-shaped, up to 50 μm long, with basis up to 20 μm wide, multi-spored; columella hemispherical, 11-15 μm diam. Sporangiospores smooth-walled, ellipsoidal to cylindrical, 3-4 × 1.5-2.0 μm.

Physiology. Maximum growth temperature 44°C.

Serology. Exoantigen tests using immunodiffusion were described by Lombardi *et al.* (1989).

Pathogenicity. BSL-2. About a dozen cutaneous cases (Al-Hedaithy, 1998) have been described, whereby the portal of entry was traumatized skin (Chakrabarti *et al.*, 1997a; Bearer *et al.*, 1994; Wilson *et al.*, 1998). The fungus is able to attack immunocompetent hosts and the infection is often fatal (Padhye *et al.*, 1988b). Kaufman *et al.* (1988) and Gonis & Starr (1997) described fatal rhinocerebral infections.

References. Tauphaichitr *et al.* (1990), Chien *et al.* (1992), Mathews *et al.* (1993), Hye *et al.* (1996), Holland (1997).

Nomenclature. *Saksenaea vasiformis* Saksena - Mycologia 45: 434, 1953.

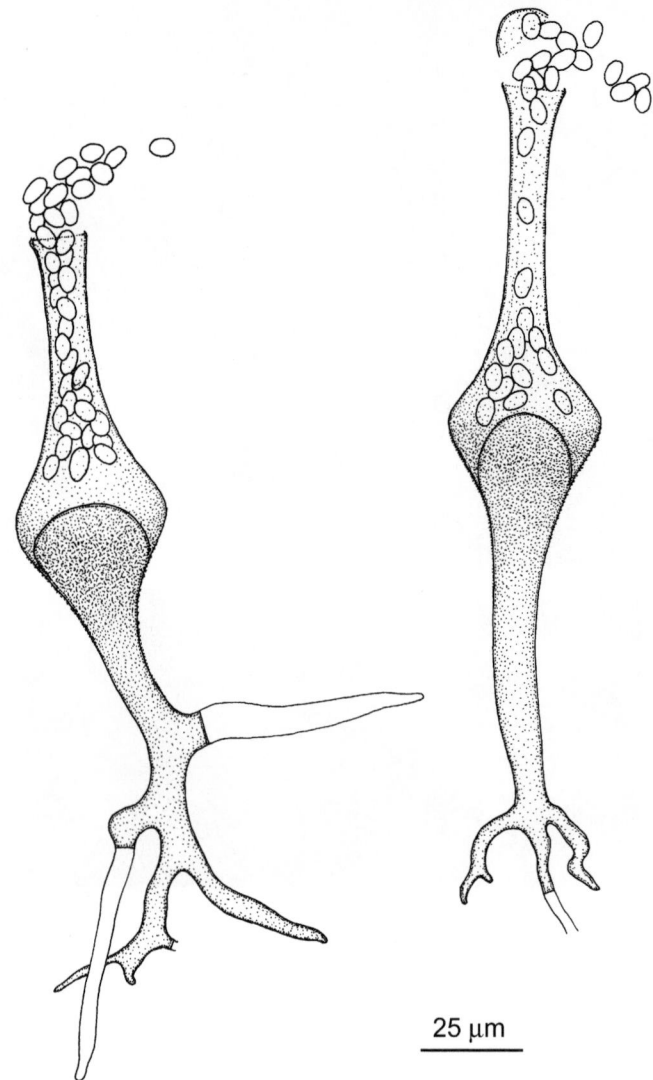

25 μm

Saksenaea vasiformis, CBS 133.90. Sporangia liberating sporangiospores.

Zygomycota, Mucorales, Syncephalastraceae. Genus: *SYNCEPHALASTRUM*

Syncephalastrum racemosum Cohn

Colony characteristics. Colonies (MEA) expanding, with abundant aerial mycelium, greyish.

Microscopy. Sporangiophores arising from rhizoids, irregularly branched, 10-25 µm wide, each branch bearing a terminal vesicle up to 80 µm diam, which produces merosporangia over the entire surface. Merosporangia grey, cylindrical, up to 33 × 4 µm, containing 3-18 spores in a single row. Merospores smooth-walled, pale brown, spherical to ovoidal, 3-7 µm diam.

Zygospores. Zygospores between equal suspensors, black, spherical, 50-90 µm diam, with conical projections. Heterothallic.

Molecular diagnostics. SSU restriction map based on NCBI X89437:

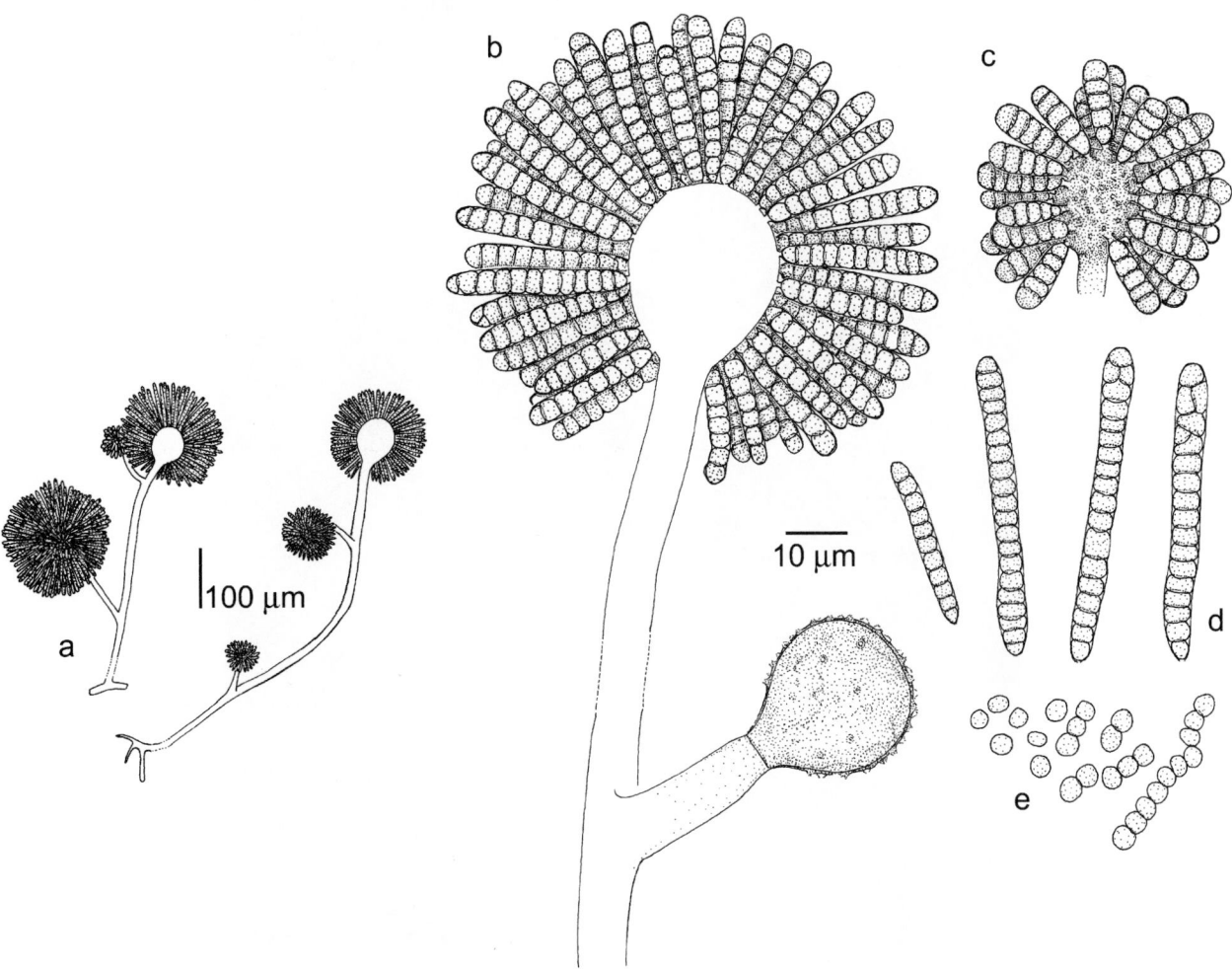

Syncephalastrum racemosum, IMI 7760. a. Habit sketch; b. sporangiophore; c. fertile vesicle bearing cylindric, mature merosporangia; d. merosporangia with uniseriate merospores; e. merospores.

113

Physiology. Maximum growth temperature 40°C.

Pathogenicity. BSL-2. The species is a saprobe, rarely occurring on humans. A case of cutaneous infection in a human (Kalamam & Thambiah, 1980) and a bovine mycotic abortion have been reported (Turner, 1964).

References. Zycha *et al.* (1969), Samson et al. (1996).

Antifungal susceptibility.

Antifungal	Mean MICs	MIC 90	Strains	Reference
AMB	0.52	0.78	10	Wildfeuer *et al.* (1998)
ITZ	0.84	1.56	10	Wildfeuer *et al.* (1998)
KTZ	0.36	0.78	10	Wildfeuer *et al.* (1998)
VCZ	3.13	6.25	10	Wildfeuer *et al.* (1998)

Nomenclature. *Syncephalastrum racemosum* Cohn - Kryptfl. Schles. 3-1: 217, 1886.

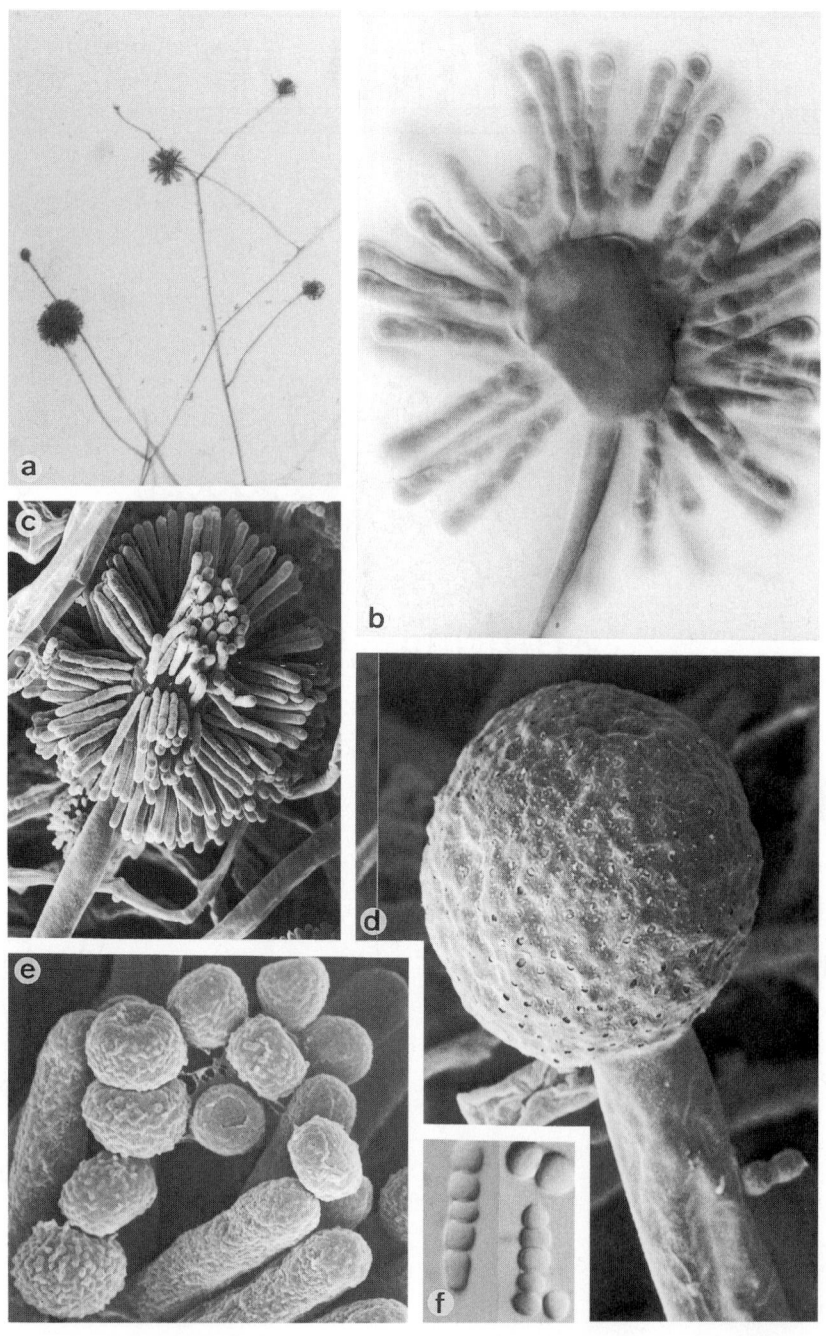

Syncephalastrum racemosum, IMI 7760. a. Sporangiophores; b, c. fertile vesicle bearing cylindrical, immature merosporangia; d. fertile vesicle without merosporangia; e. merospores; f. mature merosporangia. a. ×140; b. ×1120; c. ×490; d. ×1400; e. ×3920; f. ×1120.

Zygomycota: Entomophthorales

Criteria for distinction of *Entomophthorales*

Most members of this order are pathogens of insects and other invertebrates, but species can also be found in soil and on dung. For an overview of the group the reader is referred to Ben-Ze'ev & Kenneth (1982). The order is characterized by the presence of forcibly discharged propagules. **Primary conidia** are produced directly from the thallus; they are actively shot off by a turgor mechanism. They germinate with a germ tube, or, alternatively, produce **secondary conidia** which are morphologically identical to the primary conidium, or slightly smaller. This cycle may be repeated several times. Older conidia may develop hair-like appendages and become **villose conidia**. **Microconidia** are passively discharged. In some species, **capilliconidia** are present, which are elongate cells often with a terminal adhesive knob, borne by a thin supporting hypha. **Zygospores** are usually intercalary in creeping hyphae; they may have lateral protuberances of gametangial remains (**beaks**).
References. Keller (1987), Humber (1989), Nagahama *et al.* (1998), Ribes *et al.* (2000).

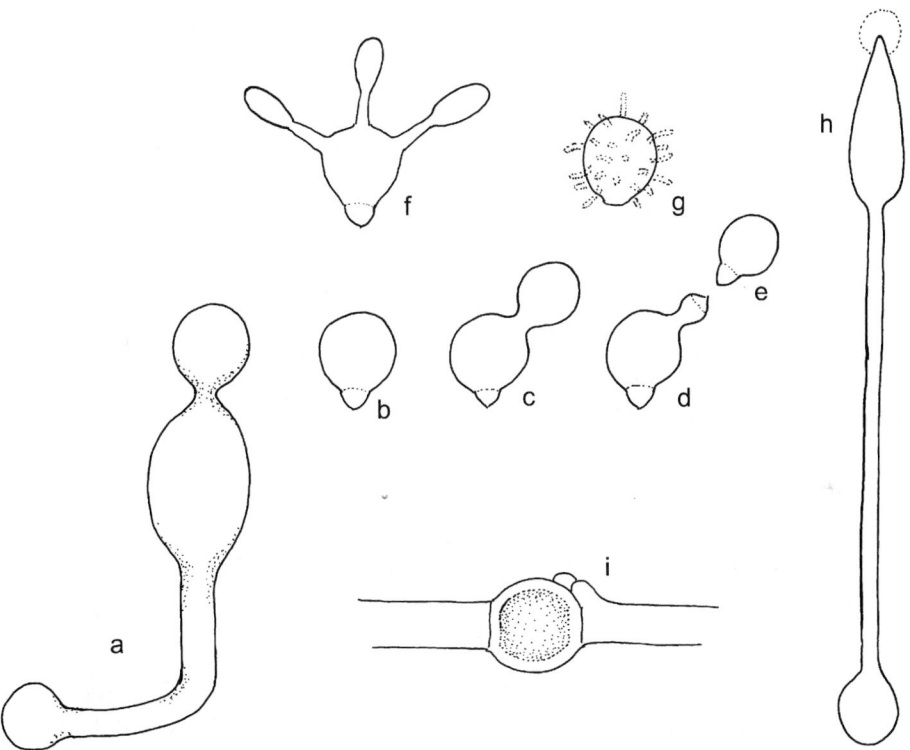

Fig. 14. Diagram of used terminology for the identification of *Entomophthorales*. a-e. Production of forcibly discharged conidia. a. Primary conidiophore; b-d. primary conidia; e. secondary conidium. f. Passively discharged microconidia. g. Villose conidium. h. Capilliconidium with adhesive tip. i. Zygospore with beaks.

Key to the genera:

1a.   Thallus in mature colonies hyphal, locally septate; cells multinucleate (but nuclei
       cryptic) . . . . *Conidiobolus* **(118)**
1b.   Thallus in mature colonies consisting of short elements and / or budding cells;
       cells prominently uninucleate . . . . *Basidiobolus* **(116)**

Zygomycota, Entomophthorales, Basidiobolaceae. Genus: *BASIDIOBOLUS*

Basidiobolus ranarum Eidam

Colony characteristics. Colonies (PDA) expanding, subhyaline, waxy, without aerial mycelium, with cerebriform centre and deep radial fissures in the outer zone.

Microscopy. Primary conidiophores short, continuous, with swollen ends. Primary conidia spherical, forcibly discharged and with a large papilla after dehiscence. Secondary conidia pyriform. Cells unambiguously uninucleate. Capillispores with adhesive knobs may be present. Zygospores intercalary in hyphae, thick-walled, subhyaline, with lateral protuberances of gametangial remains (beaks).

Physiology.

Alkaline phosphatase	+	β-Glucuronidase	–	Elastase	+
Esterase	+	α-Glucosidase	+	Keratinase	–
Esterase lipase	+	β-Glucosidase	–	Amylase	–
Lipase	–	α-Mannosidase	v	DNAse	v
Leucine arylamidase	+	α-Fucosidase	–	Urease	+
Valine arylamidase	v	Chitinase	–,w	Phosphatase	v
Cystine arylamidase	–	Hyaluronidase	v	*N*-Acetyl-D-glucosaminidase	+
Acid phosphatase	+	Lipase	+	β-Glucosidase	–
Phosphoamidase	+	Lecithinase	+	Trypsin	+
α-Galactosidase	–	Gelatinase	+	Chymotrypsin	v
β-Galactosidase	–	Collagenase	+	Growth at 37°C	+

Differential diagnosis. Species distinction in *Basidiobolus* has insufficiently been resolved; described cases in the literature were often identified without culturing. All three species reported from humans were supposed to be identical with rDNA RFLP (Kwon-Chung & Bennett, 1992), but *B. meristosporus* proved to be antigenically different (Yangco *et al.*, 1986).

Molecular diagnostics. SSU restriction map based on NCBI D29946:

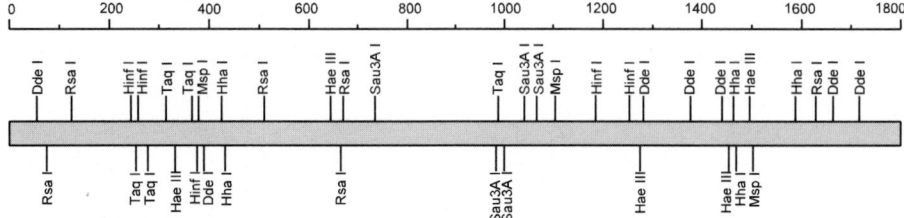

Pathogenicity. BSL-2. The species is found in decaying vegetation and on dung of reptiles and amphibians and is supposed to be a normal component of the intestinal flora of these animals (Okafor *et al.*, 1984). Rarely infection occurs in these animals (Groff *et al.*, 1991). Also in humans gastrointestinal infection is known (Kahn *et al.*, 1998), but mostly the fungus occurs as an agent of subcutaneous mycosis (conidiobolomycosis, p. 24) and is particularly found in boys below the age of ten (Dasgupta *et al.*, 1976; Bittencourt *et al.*, 1979; Vismer *et al.*, 1980; Ravisse, 1987; Scholtens & Harrison, 1994; Davis *et al.*, 1994). Limbs, buttocks and perineum are frequently involved (Burkitt *et al.*, 1964; Nazir *et al.*, 1997). Gastrointestinal infections are very rare in humans (Pasha *et al.*, 1997). An infection in a horse was reported by Owens *et al.* (1984).

Distribution. Moist, tropical regions, mainly in Africa and Asia.

References. Greer & Friedman (1966), Fromentin & Ravisse (1977), Krishnan *et al.* (1998), Feio *et al.* (1999).

Antifungal susceptibility.

Antifungal	MICs range	MIC 90	Strains	Reference
AMB	0.5-16	4	9	Guarro *et al.* (1999a)
5FC	64-256	256	9	Guarro *et al.* (1999a)
FLZ	2-128	128	9	Guarro *et al.* (1999a)
ITZ	0.25-32	32	9	Guarro *et al.* (1999a)
KTZ	0.25-32	4	9	Guarro *et al.* (1999a)
MCZ	0.5-8	8	9	Guarro *et al.* (1999a)

Nomenclature. *Basidiobolus ranarum* Eidam - Beitr. Biol. Pfl. 4: 194, 1886.
  Basidiobolus haptosporus Drechsler - Bull. Torrey Bot. Club 74: 411, 1947.
  Basidiobolus meristosporus Drechsler - J. Wash. Acad. Sci. 45: 50, 1955.

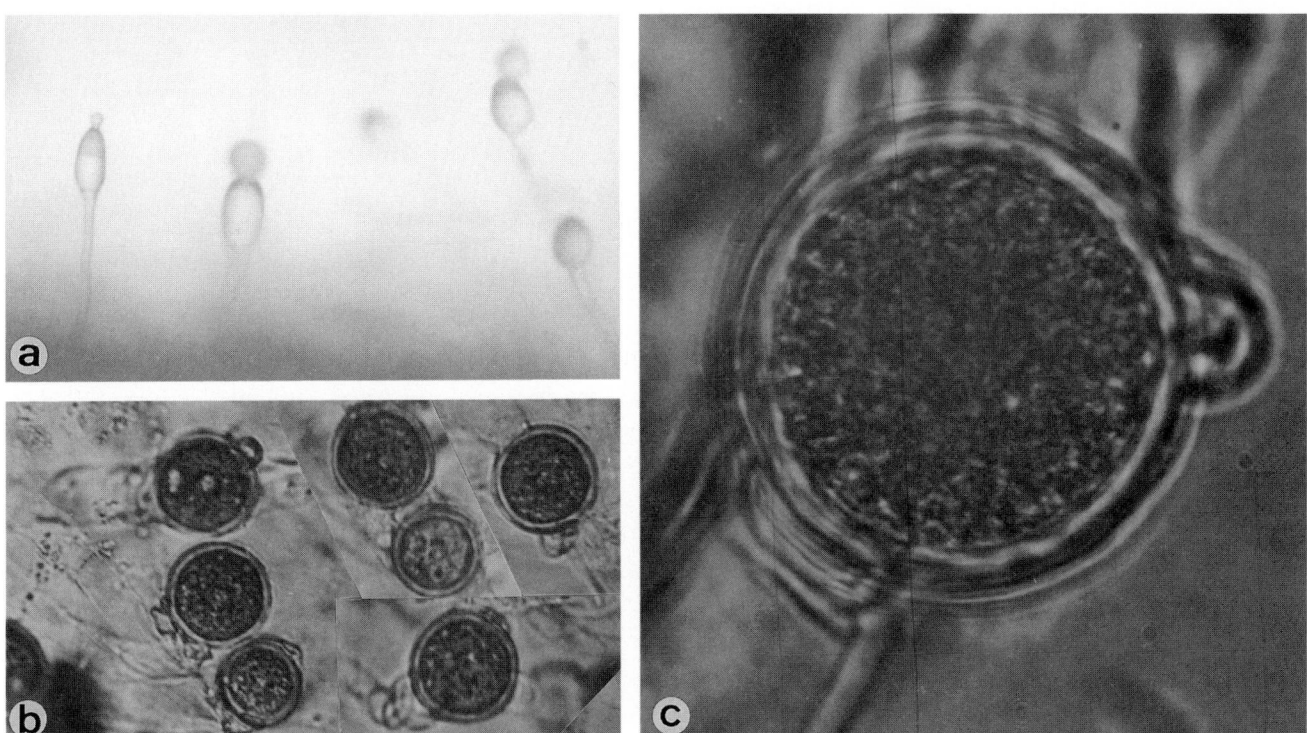

Basidiobolus ranarum., strain R. Nelson #85. a. Capilliconidia on the lid of the Petri-dish; b, c. zygospores. a. ×150; b. ×400; c. ×1800. Courtesy R. Nelson.

Zygomycota, Entomophthorales, Anylistaceae. Genus: *CONIDIOBOLUS*

Generic description. Colonies waxy, mycelium wide, coenocytic or irregularly septate. Conidiophores simple, bearing a single terminal conidium which is forcibly discharged; a papillar scar remains. Conidia may reproduce repetitively. Cells consistently multinucleate, nuclei small and cryptic. Zygospores intercalary, thick-walled, (sub)hyaline, formed after conjugation of adjacent cells, without beaks.
References. Fromentin (1982), Humber (1989).

Key to the treated species of Conidiobolus:

1a. Colonies (PDA) attaining more less than 30 mm in 3 days; forcibly discharged conidia with or without secondary conidia; villose conidia remaining absent **→ 2**

1b. Colonies (PDA) attaining more than 40 mm in 3 days; forcibly discharged conidia producing multiple secondary conidia; some conidia in older cultures become villose .... *C. coronatus* (**119**)

2a. Conidia 10-20 μm diam, never producing forcibly discharged secondary conidia ..... *C. lamprauges* (**123**)

2b. Conidia usually over 20 μm diam, producing multiple forcibly discharged secondary conidia .... *C. incongruus* (**121**)

Conidiobolus coronatus (Cost.) Batko

Colony characteristics. Colonies (PDA) expanding, hyaline, soon with irregularly radial blooms.
Microscopy. Hyphae 6-15 μm diam. Conidiophores 60-90 μm high, slightly tapering towards the tip. Primary conidia, about 40 μm diam, with prominent, papillate base, in older cultures forming hair-like appendages (villose conidia); conidia forcibly discharged, replicative microconidia often produced, passively discharged.
Physiology.

Alkaline phosphatase	+	Cystine arylamidase	v	α-Glucosidase	v
Esterase	v	Acid phosphatase	+	β-Glucosidase	+
Esterase lipase	v	Phosphoamidase	+	α-Mannosidase	+
Lipase	–	α-Galactosidase	–	Trypsin	–
Leucine arylamidase	+	β-Galactosidase	–	Chymotrypsin	+
Valine arylamidase	+	β-Glucuronidase	–	*N*-Acetyl-D-glucosaminidase	+

Molecular diagnostics. SSU restriction map based on NCBI D29947:

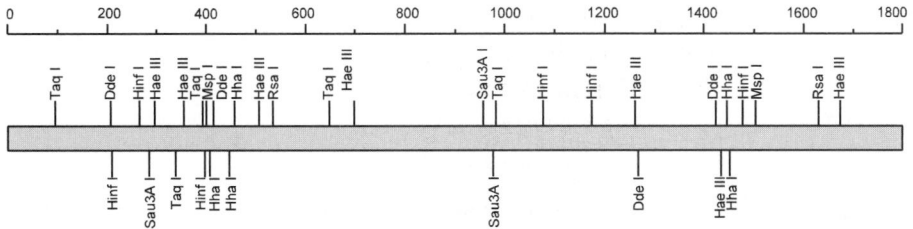

Pathogenicity. BSL-2. Agent of nasal granuloma in man (Ng *et al.*, 1991; Fournier *et al.*, 1995; Mukhopadhyay *et al.*, 1995; Ochoa *et al.*, 1997) and other higher mammals (Moll *et al.*, 1992). Facial subcutaneous tissues and paranasal sinuses are involved, leading to nasal obstruction. Finally firm, subcutaneous nodules are formed. The disorder is found particularly in otherwise healthy outdoor workers in tropical rainforests of West Africa, where the fungus is normally found in soil and rotten plant material, and also in patients with underlying disease such as Burkitt lymphoma (Pecarrere *et al.*, 1994). A nasal infection in a dog was reported by Lemairé *et al.* (1994) and in a horse by Zamos *et al.* (1996).
Distribution. Central America, equatorial Africa, India.
References. Emmons & Bridges (1961), Fromentin & Ravisse (1977), Testa *et al.* (1987), Costa *et al.* (1991), Gugnani (1992).
Antifungal susceptibility.

Antifungal	MICs range	MIC90	Strains	Reference
AMB	0.5-4	4	6	Guarro *et al.* (1999a)
5FC	256	256	6	Guarro *et al.* (1999a)
FLZ	128	128	6	Guarro *et al.* (1999a)
ITZ	0.25-32	32	6	Guarro *et al.* (1999a)
KTZ	4-32	32	6	Guarro *et al.* (1999a)
MCZ	4-16	16	6	Guarro *et al.* (1999a)

Nomenclature. *Boudierella coronata* Costantin - Bull. Soc. Mycol. Fr. 13: 40, 1897 ≡ *Delacroixia coronata* (Costantin) Saccardo & P. Sydow - Syll. Fung. 14: 457, 1899 ≡ *Entomophthora coronata* (Costantin) Kevorkian - J. Agric. Univ. P. Rico 21: 191, 1937 ≡ *Conidiobolus coronatus* (Costantin) Batko - Entomophaga, Mém. Hors. 2: 129, 1964.
   Conidiobolus villosus Martin - Bot. Gaz. 80: 311, 1925.

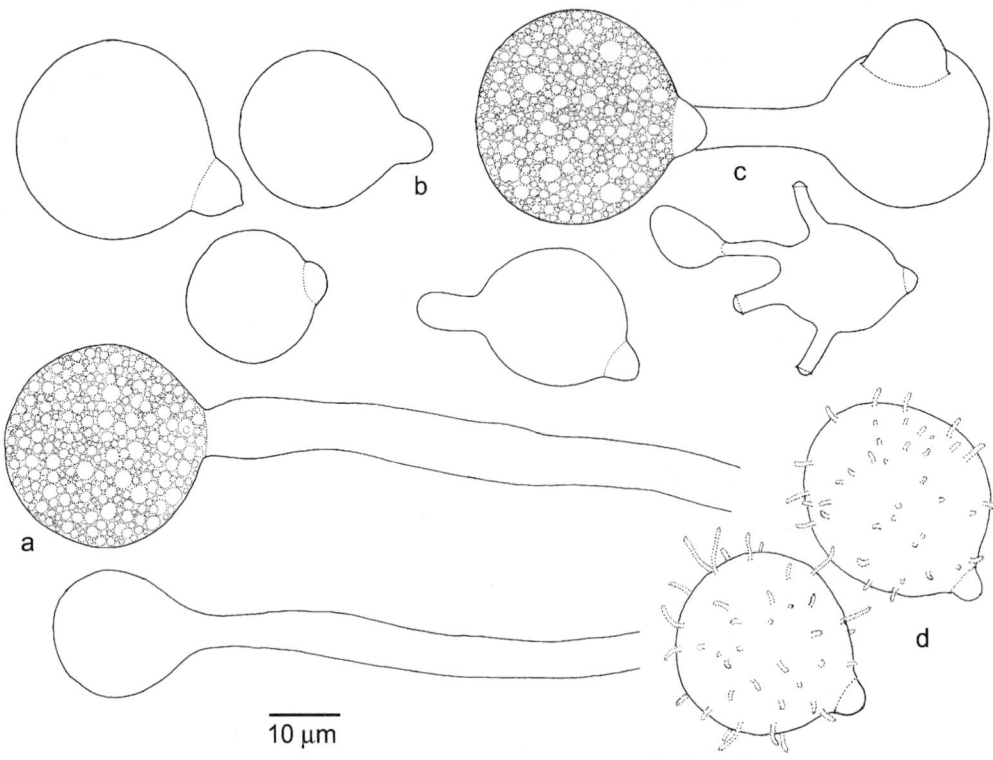

Conidiobolus coronatus, CBS 176.55. a. Conidiophores; b. primary conidia; c. microconidia; d. villose conidia.

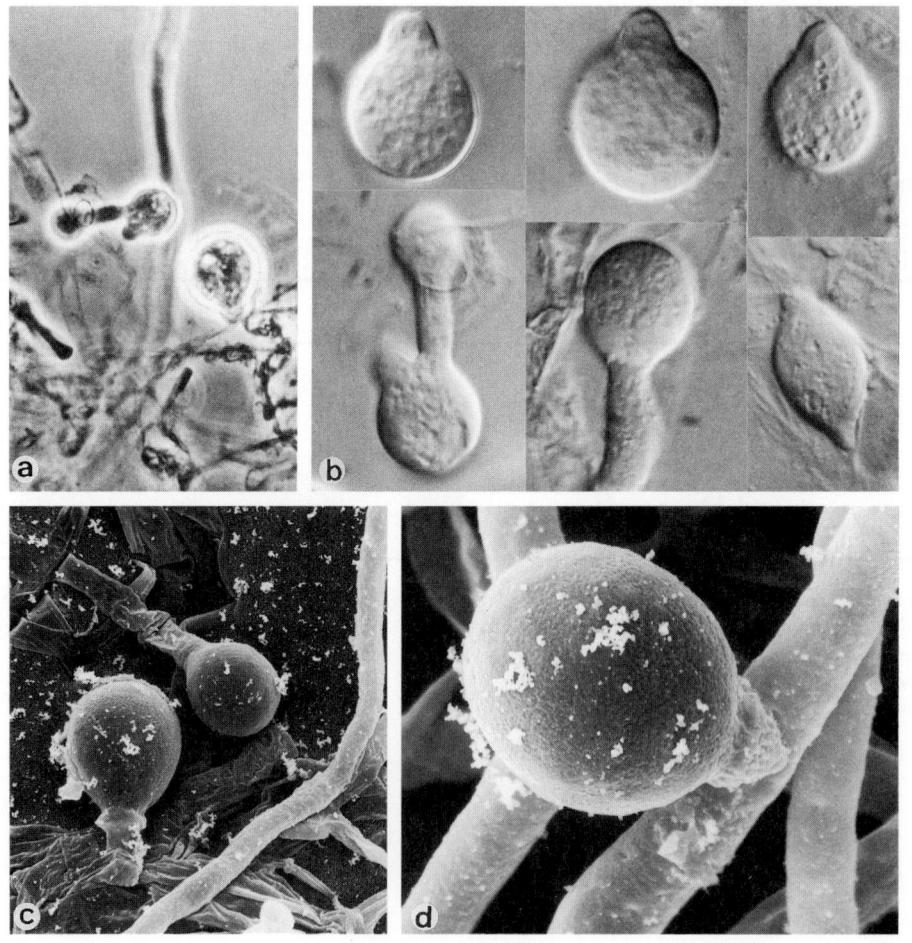

Conidiobolus coronatus, CBS 140.26. Conidiophores and primary conidia. a. ×430; b. ×1070; c. ×1405; d. ×3850.

Conidiobolus incongruus Drechsler

Colony characteristics. Colonies (PDA) expanding, initially flat with sparse aerial mycelium, becoming folded irregularly or radially, white, with age becoming tan to brown.

Microscopy. Hyphae 5-9 µm wide. Conidiophores hypha-like, not swollen at their apices. Primary conidia forcibly discharged, spherical to pyriform, 11-37 × 12-42 µm, with a single, 2 µm wide, tapering basal papilla about 10 µm in length. Microconidia passive, spherical, 7-15 µm diam, formed simultaneously on sterigmata on primary conidia.

Zygospores. Zygospores usually spherical to elongate, 15-25 µm diam, thick-walled, formed by gametangia. Homothallic.

Physiology.

Alkaline phosphatase	+	Acid phosphatase	+	α-Mannosidase	+
Esterase	v	Phosphoamidase	+	α-Fucosidase	–
Esterase lipase	v	α-Galactosidase	–	Trypsin	–
Lipase	–	β-Galactosidase	–	Chymotrypsin	+
Leucine arylamidase	+	β-Glucuronidase	–	N-Acetyl-D-glucosaminidase	+
Valine arylamidase	+	α-Glucosidase	+		
Cystine arylamidase	+	β-Glucosidase	+		

Pathogenicity. BSL-2. The species has been described as an extremely rare agent of systemic mycosis with inhalative portal of entry. An orbitofacial case was reported by Al-Hajjar *et al.* (1996), and fatal disseminated cases in a patient without underlying disease was described by Busapakum *et al.* (1983), in a drug abuser by Jaffay *et al.* (1990) and in a renal transplant patient byWalker *et al.* (1992). A pulmonary and pericardial mycosis in a granulocytopenic patient was reported by Walsh *et al.* (1994). Koshi *et al.* (1972) described three subcutaneous cases. A disseminated infection in a deer was reported by Stephans & Gibson (1997).

References. King & Jong (1976), King (1977), Fromentin *et al.* (1981).

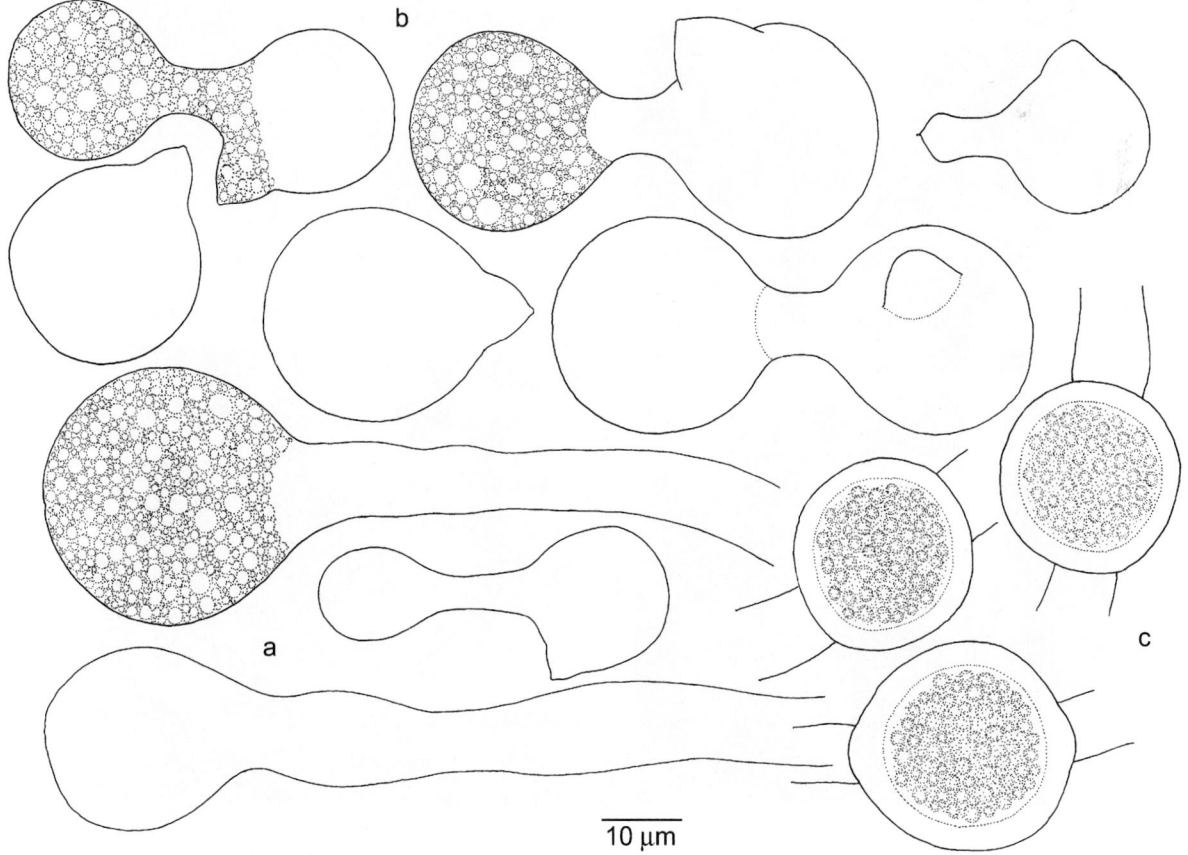

10 µm

Conidiobolus incongruus, CBS 108.84. a. Conidiophores with primary conidia; b. replicative conidia; c. zygospores.

121

Antifungal susceptibility.

Antifungal	MICs	Strains	Reference
AMB	4	1	Guarro *et al.* (1999a)
5FC	256	1	Guarro *et al.* (1999a)
FLZ	128	1	Guarro *et al.* (1999a)
ITZ	32	1	Guarro *et al.* (1999a)
KTZ	32	1	Guarro *et al.* (1999a)
MCZ	16	1	Guarro *et al.* (1999a)

Nomenclature. *Conidiobolus incongruus* Drechsler - Am. J. Bot. 47: 370, 1960.

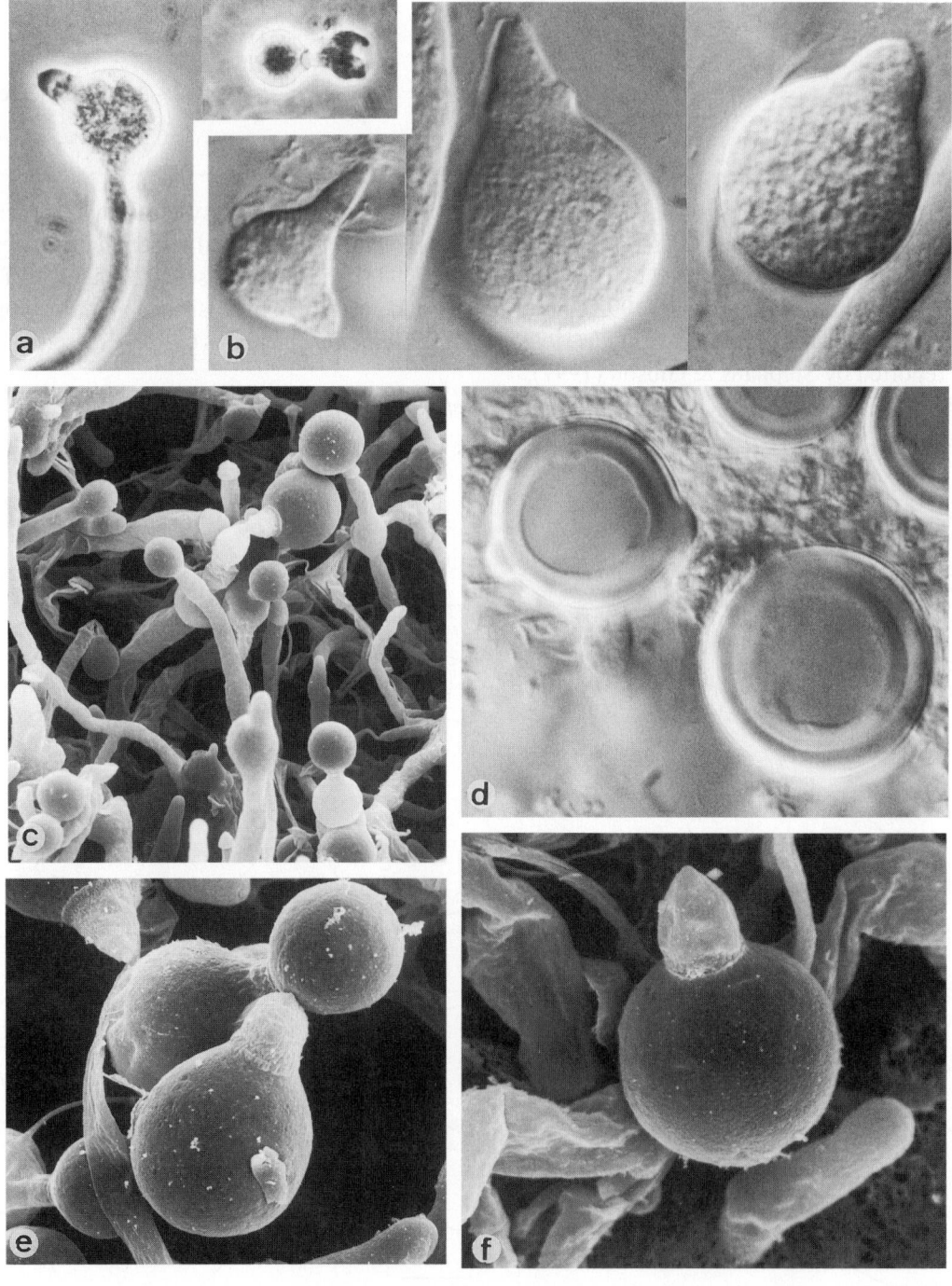

Conidiobolus incongruus, CBS 108.84. a-c, e. Conidiophores, primary and replicative conidia; d. zygospores. a. ×545; b. ×1360; c. ×830; d. ×1360; e. ×1785; f. ×3825.

Conidiobolus lamprauges Drechsler

Colony characteristics. Colonies (SGA) expanding, flat, somewhat farinose near the centre, with shallow radial fissures, whitish.

Microscopy. Hyphae 3-8 μm wide, irregular. Conidiophores hypha-like, irregular, not swollen at their apices. Primary conidia forcibly discharged, densely granular when immature, spherical, thin-walled, with one to several flat papilla after liberation, 13-22 μm diam.

Zygospores. Zygospores spherical, 12-18 μm in diam, thick-walled, with a large oil globule near the centre, formed intercalarily on undifferentiated hyphae, easily liberated, with two protuding lateral appendages (beaks).

Pathogenicity. BSL-1. Humber *et al.* (1989) described a large, fibrosing, granulomatous mass in a horse.

References. Drechsler (1953), King & Jong (1976).

Antifungal susceptibility.

Antifungal	MICs	Strains	Reference
AMB	8	1	Guarro *et al.* (1999a)
5FC	256	1	Guarro *et al.* (1999a)
FLZ	32	1	Guarro *et al.* (1999a)
ITZ	32	1	Guarro *et al.* (1999a)
KTZ	8	1	Guarro *et al.* (1999a)
MCZ	8	1	Guarro *et al.* (1999a)

Nomenclature. *Conidiobolus lamprauges* Drechsler - J. Wash. Acad. Sci. 43: 35, 1953.

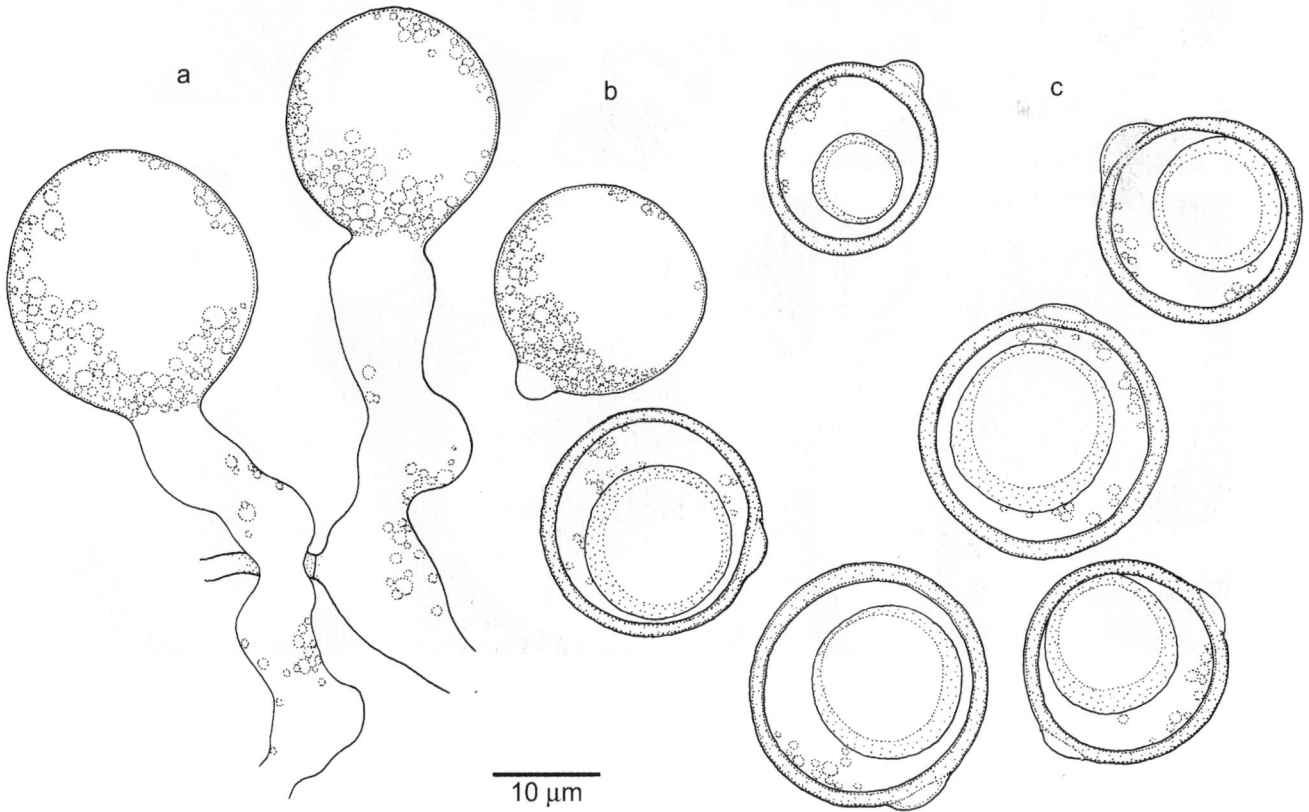

Conidiobolus lamprauges, CBS 728.97. a. Young conidiophores; b. primary conidium; c. zygospores.

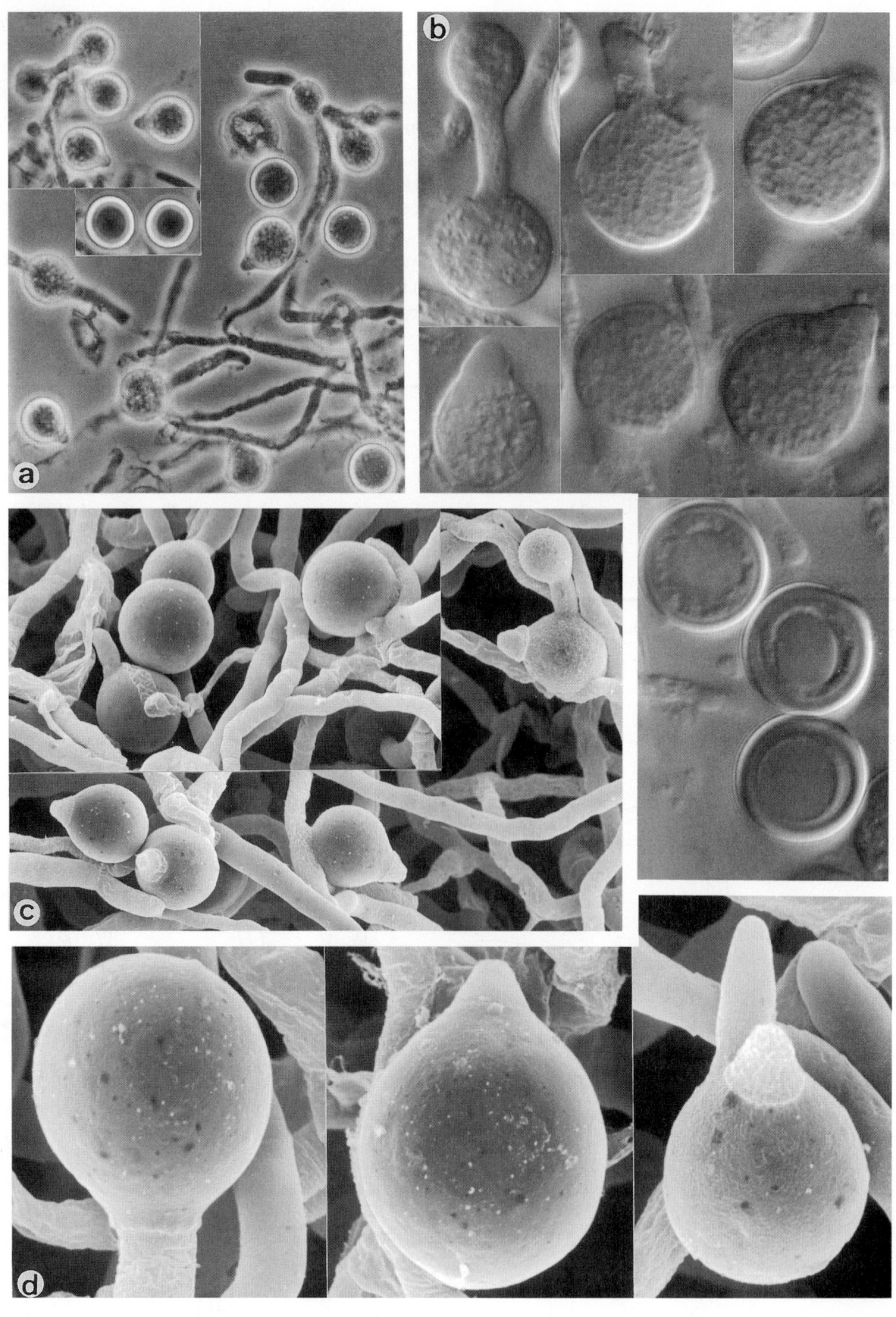

Conidiobolus lamprauges, CBS 153.56. a-d. Primary, replicative conidia and zygospores; a. ×640; b. ×1600; c. ×1450; d. ×4000.

YEASTS AND YEAST-LIKE FUNGI: EXPLANATORY CHAPTER, KEY TO THE GENERA, AND DESCRIPTIONS

General remarks. Yeasts are defined artificially as unicellular, non-melanized fungi. More or less coherent chains of cells (pseudomycelium) are common. Some groups even have strictly hyphal thalli, without budding cells. Instead, they produce arthroconidia. They are treated here under the yeasts because they are phylogenetically similar to either one of the main groups of the yeasts.

Prevalence of unicellular growth is known in the *Basidiomycota* as well as in the *Ascomycota*. Hence the 'yeasts' comprise two totally unrelated groups (Fig. 15). A subdivision of the *Basidiomycota* into three classes is adopted here (see p. 9). Most basidiomycetes are mushroom-like in nature, filamentous in culture and are classified in the subclass *Hymenomycetes* (p. 10). Two basidiomycetous groups in culture reproduce prevalently by budding cells (rarely arthroconidia): the *Urediniomycetes* and the *Ustilaginomycetes*. The few members of *Hymenomycetes* and *Urediniomycetes* that grow in culture with budding cells are treated in this Atlas under the heading **basidiomycetous yeasts** (p. 130).

The **ascomycetous yeasts** (class *Hemiascomycetes*; p. 178) are one of the main groups of the *Ascomycota*, thus at the same taxonomic level as the *Archiascomycetes* (with *Pneumocystis* as one of its members; p. 176) and the *Euascomycetes*, comprising the filamentous ascomycetes (p. 244). The *Hemiascomycetes* characteristically lack fruit bodies, the asci being produced on hyphae or they arise after conversion or conjunction of budding cells.

The above main groups of yeasts can easily be separated with the urease tests, basidiomycetous yeasts being positive and ascomycetous yeasts negative. Despite the taxonomic heterogeneity of the yeasts, the diagnostic methodology of these organisms is similar, and therefore the biologically unrelated groups are treated together in the text below, as is done in all handbooks on fungal taxonomy. Yeasts are routinely identified by nutritional physiology. Molecular recognition for many clinical species includes 26S rDNA sequencing and PCR-RFLP. Primers for a number of species have been selected (Sandhu *et al.*, 1995).

References. Two extensive monographs of the yeasts are available, *viz.* Kurtzman & Fell (1997, 1998) with emphasis on taxonomy and Barnett *et al.* (2000) with emphasis on diagnostics. For extended LSU sequence data and phylogeny is referred to Kurtzman & Robnett (1998). See also de Hoog *et al.* (1987) for related organisms. The genus *Prototheca*, which grows yeast-like in culture, has been proven to concern chorophyll-less mutants of the alga *Chlorella* (Huss & Sogin, 1990). In the unicellular, melanized ascomycetes (black yeasts) hyphae are mostly prevalent. They are anamorphs of *Chaetothyriales* (*Euascomycetes*) and are therefore treated under the hyphomycetes (p. 361).

Morphological criteria for distinction of yeasts

Forms of thalli

Multicellular thalli composed of filaments are formed by a number of species; this property is often strain-dependent. The filaments are of two types: (1) true septate hyphae and (2) pseudohyphae. **True hyphae** grow by elongation of the tip followed by the formation of a septum. The formation of septa lags slightly behind the growth so that the apical cell is mostly longer than the ones lower down. True hyphae are not constricted at the site of the septum. **Pseudohyphae** are filaments formed by elongated budding cells which do not separate. The apical cell is shorter than the ones lower down, and there are constrictions at the site where the cells are connected. A composite network of pseudohyphae is a **pseudomycelium**.

Vegetative reproduction

Budding cells. – Budding occurs when part of the cell wall inflates to give rise to a daughter cell. The bud may detach from the mother cell while it is still quite small, or it may not detach until the two cells are about the same size. Sometimes several buds or generations of buds may remain attached to one another forming a cluster of cells. Buds may be produced anywhere on the cell (**multilateral budding**), at both poles of the cell (**bipolar budding**), or at a single site on the cell (**monopolar budding**). Bud scars may produce only one bud (**single**) or they may be able to produce buds repetitively (**percurrent**).

Fission cells. – A cell is divided by one or more cross-walls, each segment elongating longitudinally to become a separate cell.

Arthroconidia. – Hyphae, either specialized or non-specialized, may become closely septate and disarticulate into separate, mostly rectangular cells.

Sterigmatoconidia. – Daughter cells are formed on short stalks. The cells are detached passively when the stalk breaks either in the middle or close to the bud. The stalks are more or less cylindrical.

Ballistoconidia. – These conidia are also formed on stalks, but they are forcibly discharged by a droplet mechanism. The stalks are narrower towards the apex. Descriptive terminology is illustrated below (Fig. 16).

Generative reproduction

Basidiomycetous yeasts are heterothallic, but mate with difficulty. A clamped mycelium is then produced. Thick-walled teliospores may be formed, which may germinate with fragile basidia producing basidiospores, or the basidia arise directly on the mycelium. The basidium can have cross walls (*phragmobasidium*) or is unseptate *(holobasidium)* illustrated below (Fig. 16). Note that all basidiospores are sessile, not forcibly discharged. **Ascomycetous yeasts** are homo- or heterothallic. They form naked asci in which the ascospores are borne. Three different types of fertilization are known: the ascus can be formed (1) by direct transformation of a vegetative cell (unconjugated ascus), (2) by 'mother-bud' conjugation, or (3) by conjugation between independent single cells. Ascospores can be smooth- or rough-walled; they are variously shaped, e.g., spherical, hat-shaped or reniform. Note that most clinically relevant yeasts reproduce vegetatively.

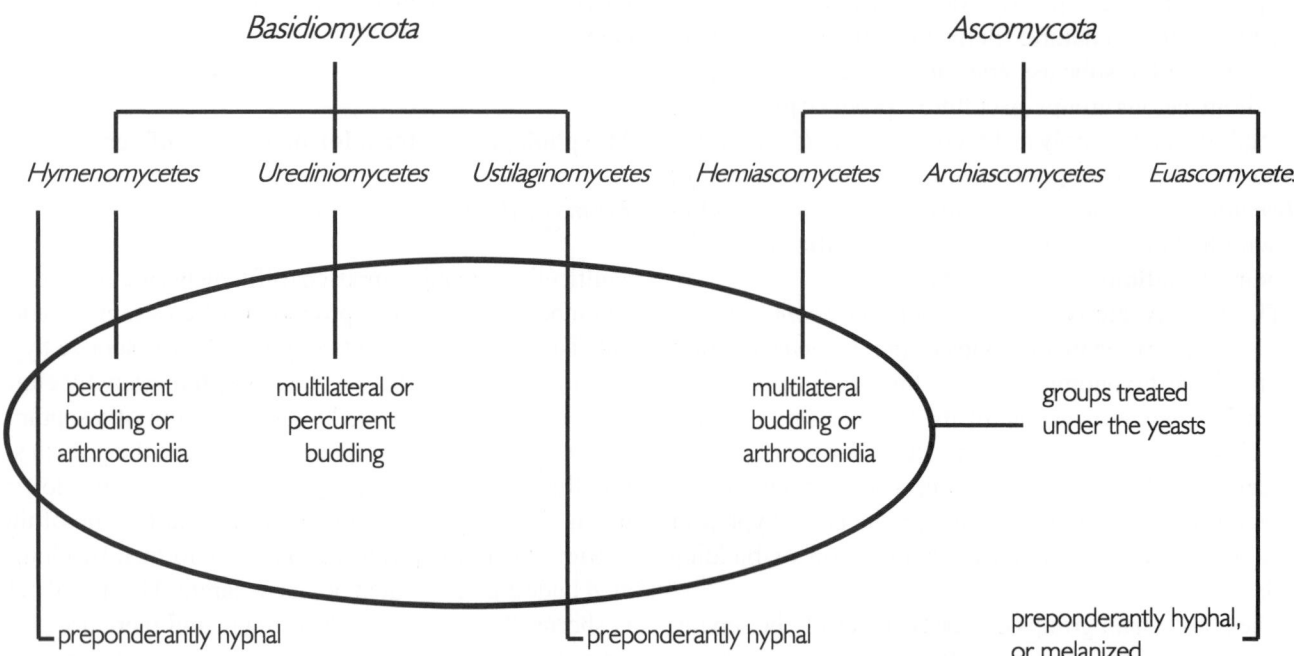

Fig. 15. Diagram of phylogenetic coherence of groups treated under the yeasts. Note that basidiomycete classes mentioned under the yeasts contain many other, non-yeast species.

126

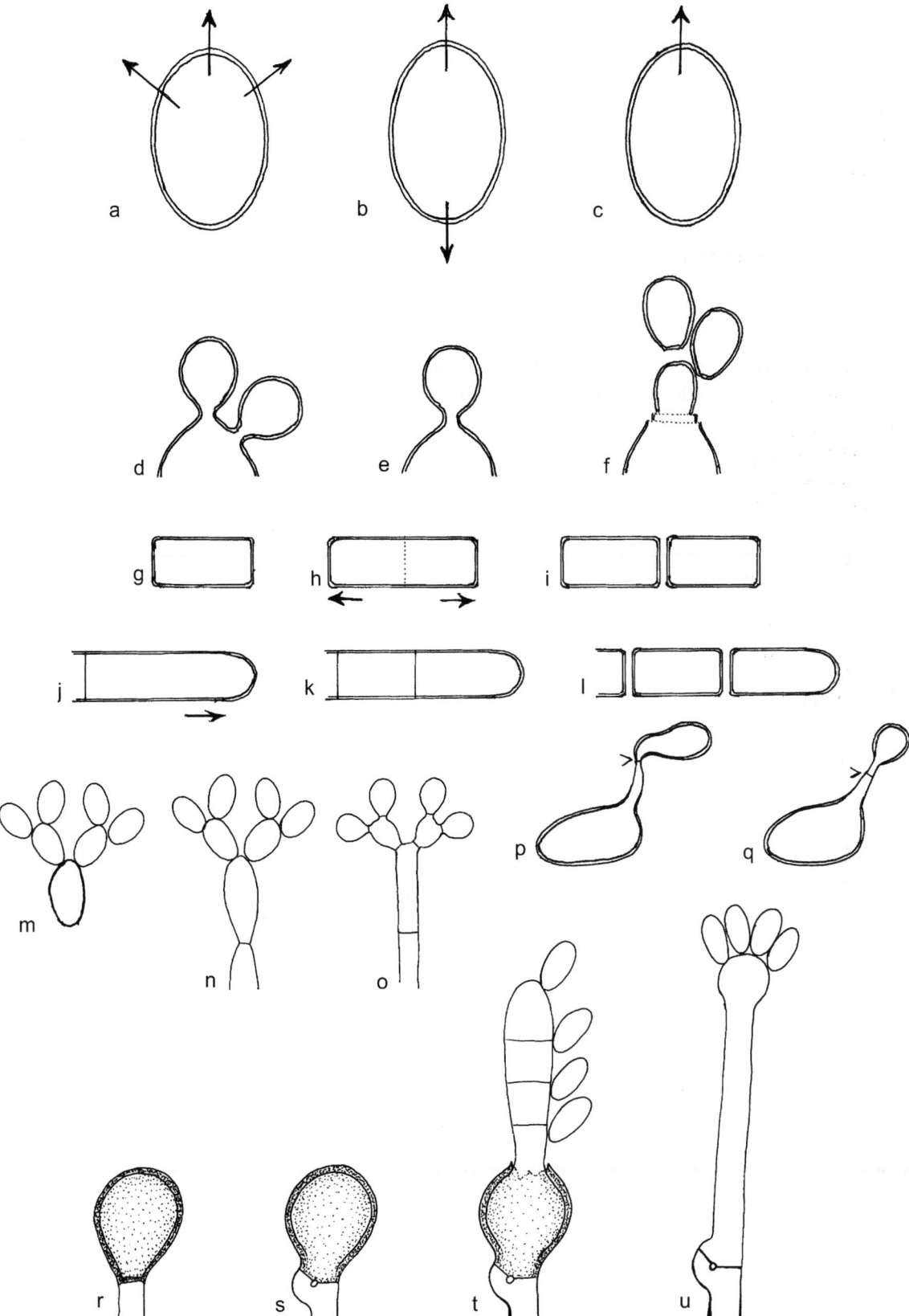

Fig. 16. Diagram of terminology of yeast morphology. a-c. Localization of buds. a. Multilateral; b. bipolar; c. monopolar; d-f. Types of budding. d. Single (multilateral); e. solitary; f. percurrent. g-i. Development of fission cells. g. Initial cell; h. bilateral elongation; i. fission. j-l. Development of arthroconidia. j. Initial hypha; k. septum formation; l. arthric disarticulation. m-o. Types of coherence of cells. m. Coherent budding cells; n. pseudomycelium; o. true mycelium. p, q. Types of stalked conidia. p. Ballistoconidium; q. sterigmatoconidium. r-u. Types of survival and meiotic structures. r. Chlamydospore; s. teliospore; t. phragmobasidium; u. holobasidium.

Simplified key to the clinically relevant genera of yeasts:

1a. Meristematic cell clumps preponderant .... *Trichosporon* **(164)**
1b. Budding cells, (pseudo)hyphae or arthroconidia preponderant ➔ **2**
2a. Colonies with pink colours ➔ **3**
2b. Colonies white, cream-coloured or ochraceous ➔ **4**
3a. Ballistoconidia present .... *Sporobolomyces* **(162)**
3b. Ballistoconidia absent .... *Rhodotorula* **(156)**
4a. Regular, true hyphae present, often preponderant ➔ **5**
4b. Budding cells preponderant; pseudomycelium may be present ➔ **7**
5a. Arthroconidia present ➔ **6**
5b. Arthroconidia absent .... *Yarrowia* **(236)**
6a. Urease present .... *Trichosporon* **(164)**
6b. Urease absent .... *Geotrichum* **(227)**
7a. Colonies with slow growth, lipophilic or lipid-dependent .... *Malassezia* **(145)**
7b. Colonies with rapid growth, not stimulated by lipids ➔ **8**
8a. Urease present .... *Cryptococcus* **(131)**
8b. Urease absent ➔ **9**
9a. Fermentation strong .... *Saccharomyces* **(234)**
9b. Fermentation absent or moderate .... *Candida* **(180)**

Table 16. Anamorph - teleomorph connections of yeasts.

Basidiomycetous yeasts		Ascomycetous yeasts	
anamorph:	- teleomorph:	anamorph:	- teleomorph:
Hymenomycetes: Tremellales		*Hemiascomycetes: Saccharomycetales*	
Cryptococcus	- *Filobasidiella*	*Candida*	- *Arxiozyma*
Malassezia	- unknown	*Candida*	- *Clavispora*
Trichosporon	- unknown	*Candida*	- *Debaryomyces*
		Candida	- *Issatchenkia*
Urediniomycetes: Sporidiales		*Candida*	- *Kluyveromyces*
		Candida	- *Metschnikowia*
Rhodotorula	- *Rhodosporidium*	*Candida*	- *Pichia*
Sporobolomyces	- *Sporidiobolus*	*Candida*	- *Stephanoascus*
		Geotrichum	- *Dipodascus*
		Geotrichum	- *Galactomyces*
		unnamed	- *Saccharomyces*
		Candida	- *Yarrowia*

2%

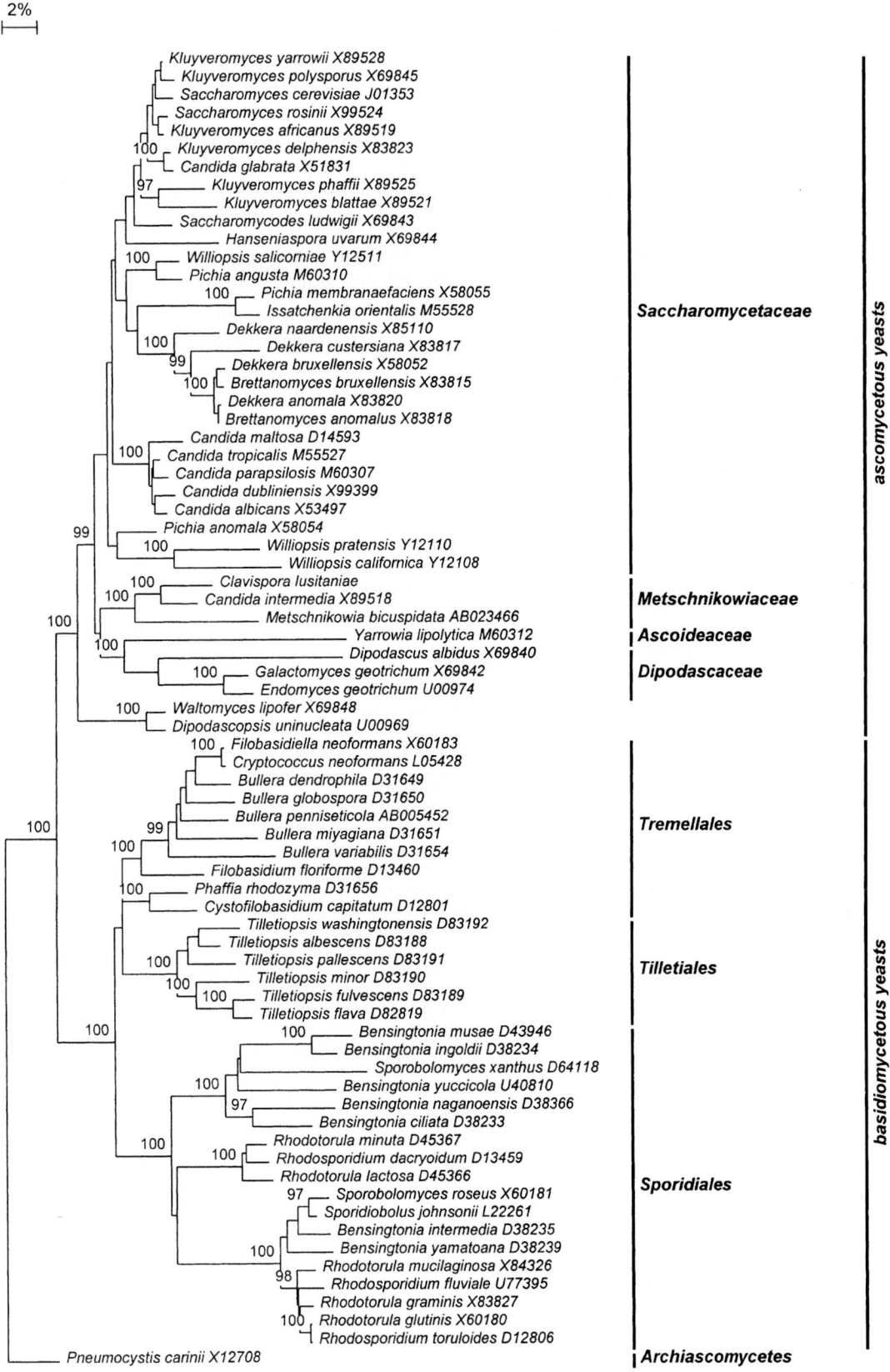

Fig. 17. Phylogenetic tree of the *Archiascomycetes* and the asco- and basidiomycetous yeasts, based on confidently aligned, near-complete SSU rDNA sequences using Neighbor joining algorithm with Kimura correction. Bootstrap values >90 from 100 resampled datasets are shown. *Pneumocystis carinii* was chosen as outgroup. All ascomycetous yeasts (*Hemiascomycetes*) are classified in a single order, the *Saccharomycetales*. In the basidiomycetous yeasts the diversity is larger, which has led to the distinction of several classes (see Table 1 on p. 19). All basidiomycetous orders contain numerous hyphal representatives.

Basidiomycetous yeasts

General remarks. Basidiomycetous yeasts are anamorphs of members of jelly fungi (*Hymenomycetes; Tremellales*) or of smuts (*Ustilaginomycetes*, *Ustilaginales*). They are recognized by presence of urease and extracellular DNAse, and by the less widely used Diazonium Blue B (DBB) staining reaction, which is also positive. In addition, mostly extracellular starch-like compounds are produced, inositol is mostly assimilated, and sugars are not fermented or only in amounts that are not detected by standard methods. Bud formation mostly percurrent. Generative reproduction is mostly produced after mating of suitable partners. A clamped mycelium with thick-walled, brown teliospores is formed, which eventually germinate with a non-septate basidium (holobasidium) or a septate basidium (phragmobasidium), bearing sessile basidiospores. Ultrastructure: cell walls are multilamellar; septa have dolipores or simple pores.

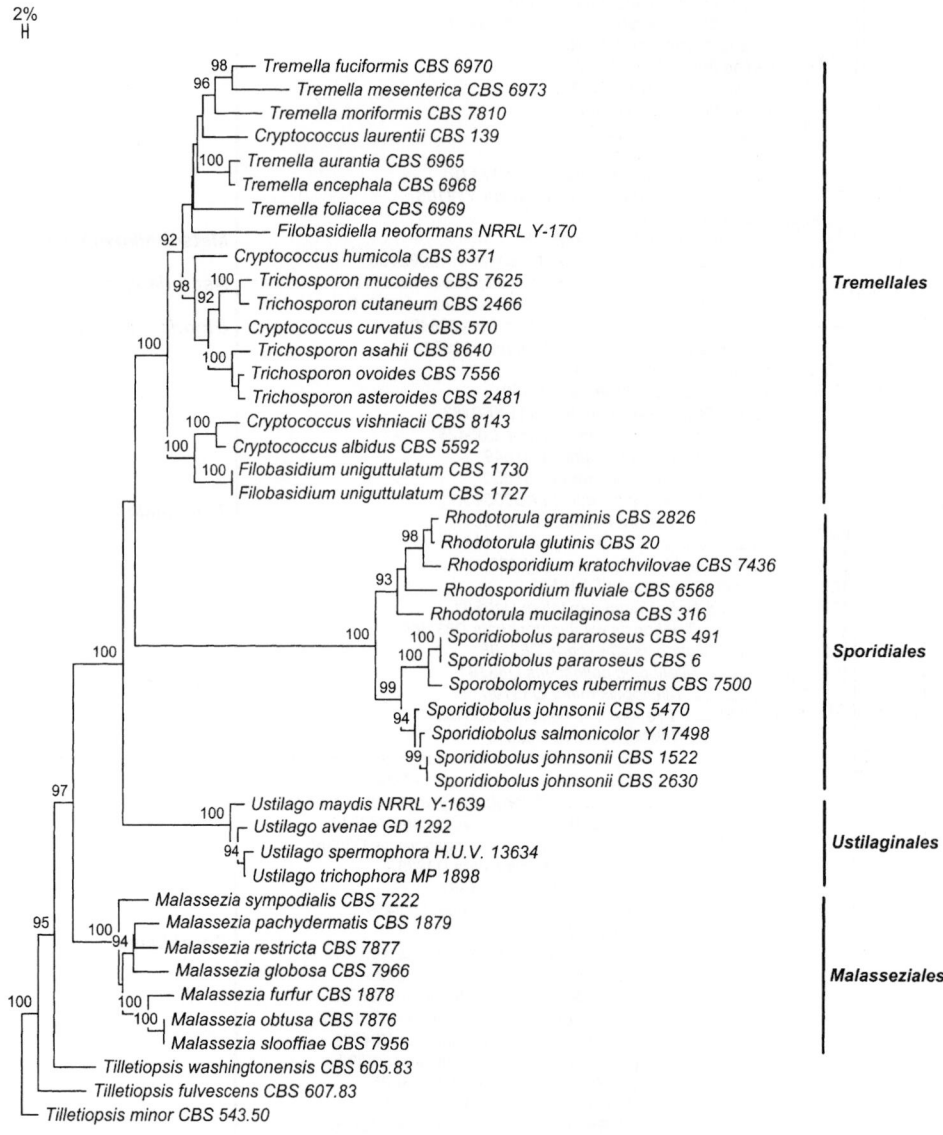

Fig. 18. Phylogenetic tree of basidiomycetous yeasts based on confidently aligned, D1/D2 domains of LSU rDNA, using Neighbor joining algorithm with Kimura correction. Bootstrap values >90 from 100 resampled datasets are shown. *Tilletiopsis minor* was chosen as outgroup. Note that numerous orders of basidiomycetous yeasts are known, which do not contain any clinically significant representatives. Medical fungi are mainly found in two orders, the *Tremellales* and the *Sporidiales*, which are clearly apart from each other. *Malassezia* takes an isolated position and therefore Begerow *et al.* (2000) maintained the order *Malasseziales* for this group.

Basidiomycetous yeasts. Genus: *CRYPTOCOCCUS*

Generic description. Pseudomycelium mostly absent; percurrent budding from spherical yeast cells; capsule may be present; no fermentation; cell walls containing xylose; DBB +, inositol +, urease +, extracellular starch produced; coenzyme Q-10.

Teleomorphs. *Filobasidium* Olive, *Filobasidiella* Kwon-Chung (*Basidiomycota, Hymenomycetes, Tremellales: Filobasidiaceae*).

General remarks. *Cryptococcus* is a large genus with species of diverse relationships. The pathogenic species *C. neoformans* is phylogenetically rather well delimited from the remaining species (Fell *et al.*, 1992; Guého *et al.*, 1993; Kwon-Chung & Chang, 1994). It is probably the only agent of cryptococcosis; the pathogenic role of non-*neoformans* strains was questioned by Krajden *et al.* (1991).

References. Barnett *et al.* (2000), Guého *et al.* (1993).

Key to the treated (sub)species of Cryptococcus:

1a.	Colonies beige to brownish → **2**	
1b.	Colonies red	*C. macerans* (**138**)
2a.	Brown pigment on GACA present → **3**	
2b.	Brown pigment on GACA absent → **5**	
3a.	Blue pigment on CGB absent; no assimilation of D-proline → **4**	
3b.	Blue pigment on CGB present; assimilation of D-proline	*C. neoformans* var. *gattii* (**141**)
4a.	No growth on CDBT agar	*C. neoformans* var. *grubii* (**141**)
4b.	Growth on CDBT agar	*C. neoformans* var. *neoformans* (**139**)
5a.	Nitrate assimilated	*C. albidus* (**132**)
5b.	Nitrate not assimilated → **6**	
6a.	Melibiose assimilated → **7**	
6b.	Melibiose not assimilated → **8**	
7a.	*N*-Acetyl-D-glucosamine assimilated	*C. humicola* (**136**)
7b.	*N*-Acetyl-D-glucosamine not assimilated	*C. laurentii* (**137**)
8a.	Lactose assimilated → **9**	
8b.	Lactose not assimilated	*C. uniguttulatus* (**143**)
9a.	Glycerol assimilated	*C. curvatus* (**135**)
9b.	Glycerol not assimilated	*C. ater* (**134**)

Table 17. Variable physiological characters among treated *Cryptococcus* species.

	albidus	ater	curvatus	humicola	laurentii	macerans	neoformans	uniguttulatus
L-Sorbose	v	−	−,s	+	v	v	−	−
Cellobiose	+	+	+	+	+	+	+,w	−
Lactose	v	+	+	+	+	v	−	−
Melibiose	v	−	−	+	+	−	−	−
Inulin	−	−	−	w,−	−	−	v	+
Soluble starch	v	−	+	v	v	v	+	v
D-Arabinose	v	v	−	+	+	+	+	w
D-Ribose	v	v	+	+	+	v	v	−
D-Glucosamine	−	+	s	+	−	−	v	−
N-Acetyl-D-glucosamine	−	+	+	+	−	−	v	+
Ethanol	+	−	+	+	v	v	w	−
Glycerol	v	−	+	+	v	v	−	w
meso-Erythritol	v	−	+	+	v	+	−	−
Ribitol	v	−	+	+	+	−	v	v
Galactitol	v	−	−	+	+	−	+	−
α-Methyl-D-glucoside	v	+	v	+	+	−	+	s
DL-Lactate	v	−	v	+	v	v	−	v
Nitrate	+	−	−	−	−	+	−	−
Vitamin-free	−	−	w	−	v	−	+	−
D-Glucarate	+	+	−	s	+	−	−	+
10% NaCl/5% glucose	−	−	−	−	−	−	+	−
Gelatin liquefaction	−	−	−	−	−	+	−	+

Cryptococcus albidus (Saito) Skinner

Colony characteristics. Colonies (MEA) glossy, slimy, smooth, cream-coloured. No brown pigment on GACA.
Microscopy. Budding (RA) percurrent. Yeast cells rather thick-walled, spherical or ovoidal, 5-10 × 3.5-8.0 μm, with thin capsule visible in Indian Ink. Hyphae or pseudohyphae absent.
Physiology.

Fermentation:	−	L-Arabinose	+	DL-Lactate	v
		D-Arabinose	v	Succinate	+
Growth:		D-Ribose	v	Citrate	+
Glucose	+	L-Rhamnose	v	*myo*-Inositol	+
Galactose	v	D-Glucosamine (C)	−	Hexadecane	−
L-Sorbose	v	N-Acetyl-D-glucosamine	−	Nitrate	+
Sucrose	+	Methanol	−	Vitamin-free	−
Maltose	+	Ethanol	+,w	2-Keto-D-gluconate	+
Cellobiose	+	Glycerol	v	5-Keto-D-gluconate	v
α,α-Trehalose	+,w	*meso*-Erythritol	v	D-Glucarate	+
Lactose	v	Ribitol	v	D-Glucuronate	+
Melibiose	v	Galactitol	v	10% NaCl/5% glucose	−
Raffinose	+	D-Mannitol	+	Urease	+
Melezitose	+	D-Glucitol	+	Starch formation	+,w
Inulin	−	α-Methyl-D-glucoside	v	Gelatin liquefaction	−
Soluble starch	v	Salicin	+	Growth at 30°C	+
D-Xylose	+	D-Gluconate	+	Growth at 37°C	v

Differential diagnosis. Species signature: sucrose +; maltose +; cellobiose +; 5-keto-D-gluconate –; butane-2,3-diol –; nitrite +; creatinine –; glucosamine (N) –; 37°C –; 0.1% cycloheximide –; 16% NaCl –; urease +;. Physiologically indistinguishable from a large number of basidiomycetous yeasts.

Molecular diagnostics. SSU restriction map based on NCBI D31655:

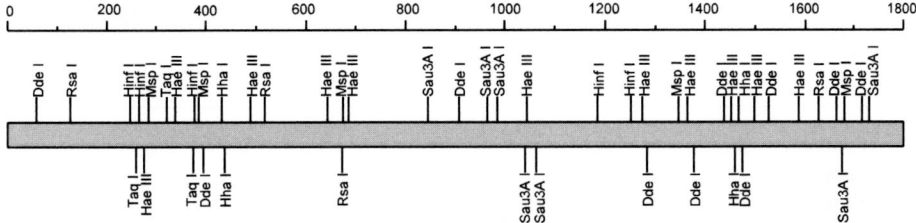

Pathogenicity. BSL-1. Cases of meningitis have been reported (Cunha & Lusins, 1973; Wieser, 1974; Melo *et al.*, 1980; Yasin *et al.*, 1988; Kordossis *et al.*, 1998). Horowitz & Blumberg (1993) reported an infection in a dialysis patient, Chabasse *et al.* (1995) sepsis in AIDS, and Wells *et al.* (1998) a mixed pulmonary infection. The identity of a strain causing a pulmonary infection (Krumholz, 1972) was questioned by Gordon & Weitzman (1972). In older literature the species was repeatedly reported from onychomycoses and pulmonary disorders; cases of which the identity of the etiologic agent was proven are Wolfram & Zach (1934, onychomycosis) and Castellani (1963, balanitis).

References. Olive (1968), Barnett *et al.* (2000), Fonseca *et al.* (2000)

Nomenclature. *Torula albida* Saito - Jpn. J. Bot. 1: 1-54, 1922 ≡ *Torulopsis albida* (Saito) Lodder - Anaskosp. Hefen 1: 163, 1934 ≡ *Torulopsis neoformans* (Sanfelder) Redaelli var. *albida* (Saito) W. Kaufman - Zbl. Bakt. Parasitkde, Abt. 2, 106: 442, 1944 ≡ *Cryptococcus albidus* (Saito) Skinner - Am. Midl. Nat. 43: 249, 1950.

Torulopsis diffluens Zach, *in* Wolfram & Zach - Arch. Derm. Syph. 170: 690, 1934 ≡ *Cryptococcus diffluens* (Zach) Lodder & Kreger-van Rij - The Yeasts p. 391, 1952 ≡ *Cryptococcus albidus* (Saito) Skinner var. *diffluens* (Zach) Phaff & Fell, *in* Lodder - The Yeasts, ed. 2, p. 1099, 1970.

Hansenula amylofaciens Dietrichson - Annls Parasit. Hum. Comp. 29: 472, 1954.

Cryptococcus neoformans (Sanfelice) Vuillemin var. *innocuous* Benham - Trans. N.Y. Acad. Sci., Ser. 2, 17: 418-429, 1955.

Cryptococcus genitalis Castellani - Derm. Trop. 2: 140, 1963.

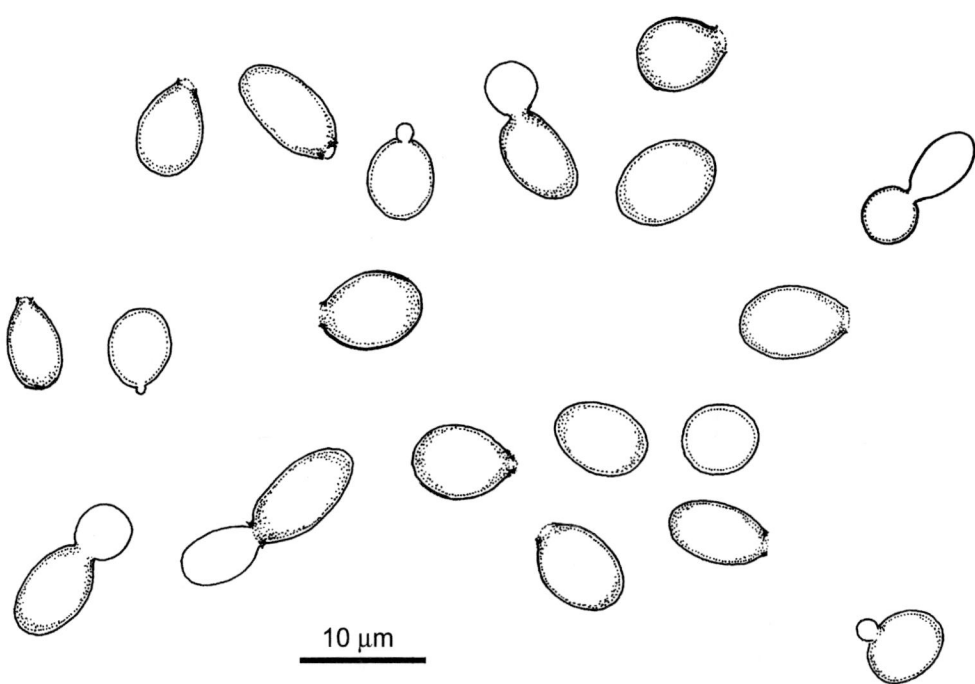

10 µm

Cryptococcus albidus, Strain M.L. Kerk.mann. Budding cells.

133

Cryptococcus ater (Castellani ex W.B. Cooke) Rodrigues de Miranda

Colony characteristics. Colonies (MEA) mucoid, cream-coloured, tan or dark.
Microscopy. Budding (RA) percurrent or multilateral. Yeast cells rather thick-walled, (sub)spherical to broadly ellipsoidal, 4-7 µm long, up to 7 µm diam, often cohering in short strings, finally with broad, flaring conidial scars. Capsule absent. Hyphae or pseudohyphae absent.
Physiology.

Fermentation:	–	L-Arabinose	+	DL-Lactate	–
		D-Arabinose	v	Succinate	+
Growth:		D-Ribose	v	Citrate	+,w
Glucose	+	L-Rhamnose	+	*myo*-Inositol	+
Galactose	+,w	D-Glucosamine (C)	+	Hexadecane	–
L-Sorbose	–,s	*N*-Acetyl-D-glucosamine	+	Nitrate	–
Sucrose	+	Methanol	–	Vitamin-free	–
Maltose	+	Ethanol	–	2-Keto-D-gluconate	+
Cellobiose	+	Glycerol	–	5-Keto-D-gluconate	+
α,α-Trehalose	+	*meso*-Erythritol	–	D-Glucarate	+
Lactose	+	Ribitol	–	D-Glucuronate	+
Melibiose	–	Galactitol	–	10% NaС1/5% glucose	–
Raffinose	s	D-Mannitol	+	Urease	+
Melezitose	v	D-Glucitol	+	Starch formation	+
Inulin	–	α-Methyl-D-glucoside	+	Gelatin liquefaction	–
Soluble starch	–	Salicin	+,w	Growth at 30°C	+
D-Xylose	+	D-Gluconate	+	Growth at 37°C	–

Differential diagnosis. Species signature: L-rhamnose +, melibiose –, lactose +, D-glucono-δ-lactone –, propane-1,2-diol –, quinic acid +, nitrite –, ethylamine +, starch formation +. Differs from *C. humicola* by absence of filaments.
Pathogenicity. BSL-1. The species was described by Castellani (1960) from an ulcer on the leg of a human patient.
Reference. Barnett *et al.* (2000).

Nomenclature. *Cryptococcus ater* Castellani - J. Trop. Med. Hyg. 63: 27, 1960 (invalid) ≡ *Melanocryptococcus ater* (Castellani) Della Torre & Ciferri - Atti Ist. Bot. Univ. Lab. Crittogam. Pavia, Ser. 5, 21: 11, 1964 ≡ *Cryptococcus laurentii* (Kufferath) Skinner var. *magnus* Lodder & Kreger-van Rij f. *ater* Castellani ex W.B. Cooke - Mycopath. Mycol. Appl. 30: 349, 1966 ≡ *Cryptococcus ater* (W.B. Cooke) Rodrigues de Miranda, *in* Kreger-van Rij - The Yeasts, ed. 3, p. 852, 1984.

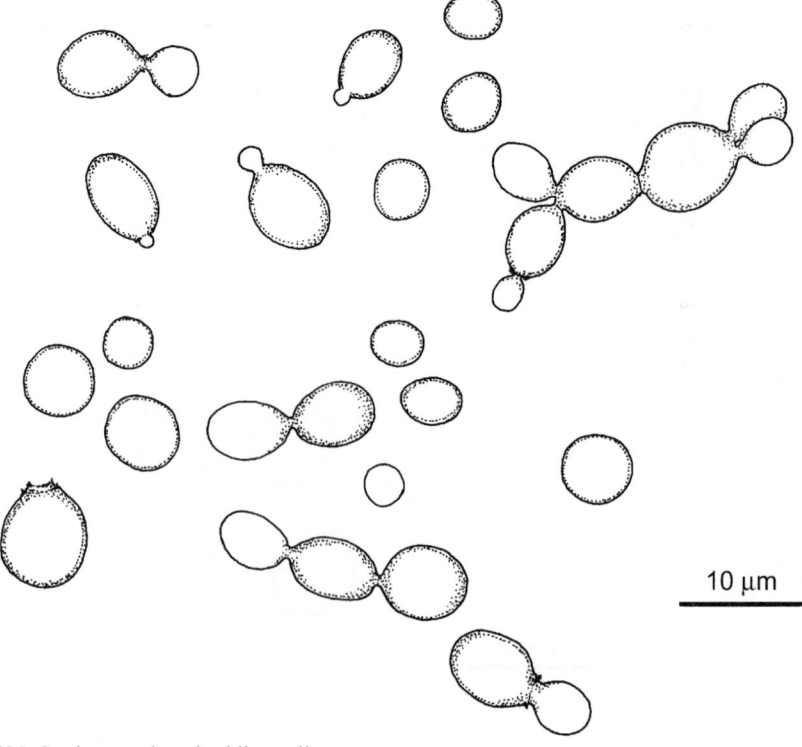

10 µm

Cryptococcus ater, CBS 4685. Stationary phase budding cells.

Cryptococcus curvatus (Diddens & Lodder) Golubev

Colony characteristics. Colonies (MEA) mucoid, wrinkled, yellowish to brownish.
Microscopy. Budding (RA) percurrent from a rather broad base. Yeast cells ovoidal to reniform, 2.5-4.5 × 3.5-9.5 μm.
Physiology.

Fermentation:	–	L-Arabinose	v	DL-Lactate	v
		D-Arabinose	–	Succinate	v
Growth:		D-Ribose	+	Citrate	v
Glucose	+	L-Rhamnose	+,w	*myo*-Inositol	+,w
Galactose	+	D-Glucosamine (C)	s	Hexadecane	–
L-Sorbose	–,s	N-Acetyl-D-glucosamine	+	Nitrate	–
Sucrose	+	Methanol	–	Vitamin-free	–,w
Maltose	+,w	Ethanol	+,w	D-Glucarate	–
Cellobiose	+	Glycerol	+	D-Glucuronate	+
α,α-Trehalose	+	*meso*-Erythritol	+,w	10% NaC1/5% glucose	–
Lactose	+	Ribitol	+,w	Urease	+
Melibiose	–	Galactitol	–	Starch formation	+
Raffinose	+	D-Mannitol	v	Gelatin liquefaction	–
Melezitose	v	D-Glucitol	v	Growth at 30°C	+
Inulin	–	α-Methyl-D-glucoside	v	Growth at 37°C	–
Soluble starch	+	Salicin	+		
D-Xylose	+	D-Gluconate	+		

Differential diagnosis. Species signature: raffinose +, D-glucuronate +, 5-keto-D-gluconate +, cadaverine +, D-glucosamine (N) –, w/o biotin –, growth at 30°C +, urease +. Physiologically indistinguishable from *Trichosporon asahii, T. cutaneum, T. ovoides* and *Rhodotorula minuta*.

Pathogenicity. BSL-2. The species is an uncommon agent of neurological disorders in man (Dromer *et al.*, 1993) or animals (Herceq *et al.*, 1977) and can be isolated from urine and sputum.

Nomenclature. *Candida heveanensis* (Groenewege) Diddens & Lodder var. *curvata* Diddens & Lodder - Aanaskosp. Hefen p. 310, 1942 ≡ *Candida curvata* (Diddens & Lodder) Lodder & Kreger-van Rij - The Yeasts p. 576, 1952 ≡ *Cryptococcus curvatus* (Diddens & Lodder) Golubev - Mikol. Fitopat. 15: 467, 1981.

Cryptococcus humicola (Daszewska) Golubev

Colony characteristics. Colonies (MEA) mucoid or dry, then finely cerebriform, whitish to cream-coloured.
Microscopy. Budding cells (RA) irregular in shape and size, broadly ellipsoidal to fusiform, often intermingled met pseudomycelial elements, then often with some elevated scars alongside the cells. Hyphae and pseudohyphae may be present.
Physiology.

Fermentation:	–	L-Arabinose	+	DL-Lactate	+
		D-Arabinose	+	Succinate	+
Growth:		D-Ribose	+	Citrate	+
Glucose	+	L-Rhamnose	+	*myo*-Inositol	+
Galactose	+	D-Glucosamine (C)	+	Hexadecane	–
L-Sorbose	+	*N*-Acetyl-D-glucosamine	+	Nitrate	–
Sucrose	+	Methanol	–	Vitamin-free	–
Maltose	+	Ethanol	+	D-Glucarate	s
Cellobiose	+	Glycerol	+	D-Glucuronate	+
α,α-Trehalose	+	*meso*-Erythritol	+	10% NaC1/5% glucose	–
Lactose	+	Ribitol	+	Urease	+
Melibiose	+	Galactitol	+	Starch formation	+
Raffinose	v	D-Mannitol	+	Gelatin liquefaction	–
Melezitose	+	D-Glucitol	+	Growth at 30°C	+
Inulin	w,–	α-Methyl-D-glucoside	+	Growth at 37°C	–
Soluble starch	v	Salicin	+		
D-Xylose	+	D-Gluconate	+		

Differential diagnosis. Species signature: sucrose +, cellobiose +, inulin –, *myo*-inositol +, nitrate –, 5-keto-D-gluconate +, ethylamine +, w/o biotin +, growth at 25°C but not at 37°C, 50% glucose –, urease +.
Pathogenicity. BSL-1. A case of ophthalmitis after traumatic inoculation was reported by Nitzulescu & Niculescu (1975) and melanonychia of toenails by Velez *et al*. (1996).
Nomenclature. *Torula humicola* Daszewska - Bull. Soc. Bot. Genève, Sér. 2, 4: 255-316, 1912 ≡ *Candida humicola* (Daszweska) Diddens & Lodder - Anaskospor. Hefen 2: 268, 1942 ≡ *Cryptococcus humicola* (Daszewska) Golubev - Mikol. Fitopat. 15: 467, 1981.

Cryptococcus humicola, CBS 571. Pseudomycelium and budding cells.

Cryptococcus laurentii (Kufferath) Skinner

Colony characteristics. Colonies (MEA) moist, slimy, cream-coloured, becoming pale ochraceous with age. No brown pigment on GACA.

Microscopy. Budding (RA) percurrent. Yeast cells spherical or ellipsoidal, 3-7 × 2.0-5.5 µm, with thin capsule visible in Indian Ink. Cells finally inflating and developing irregular protrusions. Hyphae or pseudohyphae absent.

Physiology.

Fermentation:	–	L-Arabinose	+	DL-Lactate	v
		D-Arabinose	+	Succinate	+
Growth:		D-Ribose	+	Citrate	+,w
Glucose	+	L-Rhamnose	+	*myo*-Inositol	+
Galactose	+	D-Glucosamine (C)	–	Hexadecane	–
L-Sorbose	v	*N*-Acetyl-D-glucosamine	–	Nitrate	–
Sucrose	+	Methanol	–	Vitamin-free	v
Maltose	+	Ethanol	v	2-Keto-D-gluconate	+
Cellobiose	+	Glycerol	v	5-Keto-D-gluconate	+
α,α-Trehalose	+	*meso*-Erythritol	v	D-Glucarate	+
Lactose	+	Ribitol	+	D-Glucuronate	+
Melibiose	+	Galactitol	+	10% NaCl/5% glucose	–
Raffinose	+	D-Mannitol	+	Urease	+
Melezitose	+	D-Glucitol	+	Starch formation	+
Inulin	–	α-Methyl-D-glucoside	+	Gelatin liquefaction	–
Soluble starch	v	Salicin	+,w	Growth at 30°C	+
D-Xylose	+	D-Gluconate	+	Growth at 37°C	–

Differential diagnosis. Species signature: lactose +, melibiose +, nitrate –, D- gluconate +, 5-keto-D-gluconate +, D-glucuronate +, butane-1,2-diol –, w/o biotin +, growth at 25°C but not at 37°C, urease +.

Molecular diagnostics. LSU restriction map based on NCBI AF075469:

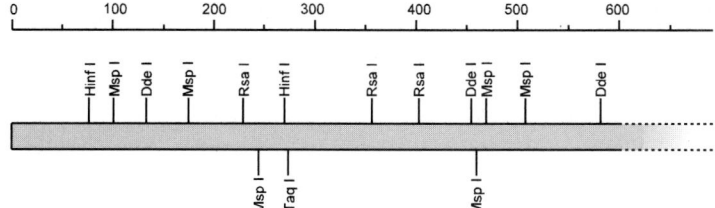

Pathogenicity. BSL-1. Species occurring on phyllosphere. A pulmonary abscess was reported by Lynch *et al*. (1981), but the identity of the agent was questioned by Krajden *et al*. (1991). A CAPD-related peritonitis was reported by Sinnott *et al*. (1989), subcutaneous nodules in a drug abuser by Johnson *et al*. (1998) and a meningitis in an AIDS patient by Kordossis *et al*. (1998). Buracco & Gallo (1990) reported a bone infection mixed with *Leishmania* in a dog.

References. Kurtzman (1973), Barnett *et al*. (2000).

Nomenclature. *Torula laurentii* Kufferath - Annls Bull. Séanc. Soc. R. Sci. Nat. Brûx. 74: 16-46, 1920 ≡ *Cryptococcus laurentii* (Kufferath) Skinner - Am. Midl. Nat. 43: 249, 1950.

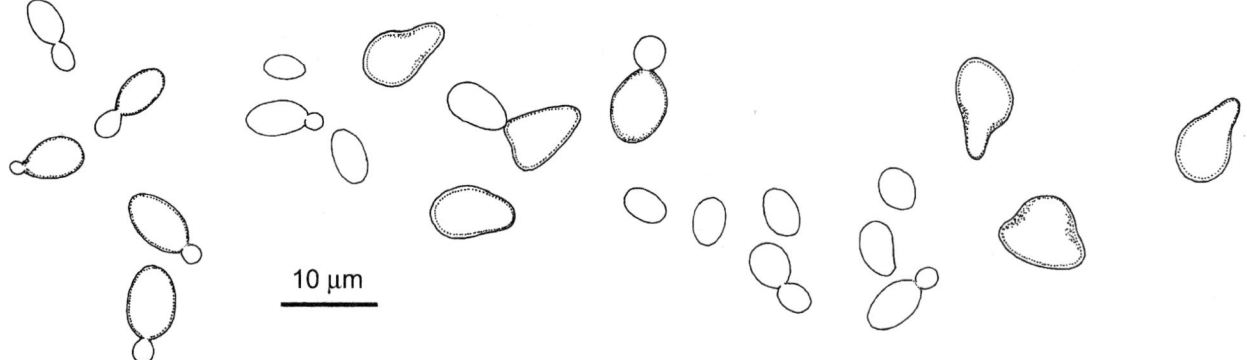

Cryptococcus laurentii, CBS 139. Budding cells intermingled with irregular, inflated cells.

Cryptococcus macerans (Frederiksen) Phaff & Fell

Colony characteristics. Colonies (MEA) pasty, pink to bright red, smooth with some rugose areas.
Microscopy. Budding (RA) percurrent. Yeast cells ellipsoidal, 5-15 × 3.0-4.5 μm.
Physiology.

Fermentation:	–	L-Arabinose	+	DL-Lactate	v
		D-Arabinose	+	Succinate	+
Growth:		D-Ribose	v	Citrate	+
Glucose	+	L-Rhamnose	v	*myo*-Inositol	+
Galactose	+,w	D-Glucosamine (C)	–	Hexadecane	–
L-Sorbose	v	*N*-Acetyl-D-glucosamine	–	Nitrate	+
Sucrose	+	Methanol	–	Vitamin-free	–
Maltose	+	Ethanol	v	2-Keto-D-gluconate	+
Cellobiose	+	Glycerol	v	5-Keto-D-gluconate	+
α,α-Trehalose	+	*meso*-Erythritol	+	D-Glucarate	–
Lactose	v	Ribitol	–	D-Glucuronate	+
Melibiose	–	Galactitol	–	10% NaC1/5% glucose	–
Raffinose	+	D-Mannitol	+	Urease	+
Melezitose	+,w	D-Glucitol	+	Starch formation	+
Inulin	–	α-Methyl-D-glucoside	–	Gelatin liquefaction	+
Soluble starch	v	Salicin	+,w	Growth at 25°C	+
D-Xylose	+	D-Gluconate	s	Growth at 30°C	–

Differential diagnosis. Species signature: D-arabinose +, *meso*-erythritol +, α-methyl-D-glucoside –, *myo*-inositol +, D-glucarate –, galactonate –, nitrate +, 2-keto-D-gluconate +, urease +, growth at 30°C –.
Pathogenicity. BSL-1. Lindsberg *et al.* (1997) reported a case of human meningoencephalitis. It should be noted, however, that the species is unable to grow at elevated temperatures.
Reference. Fell & Statzell (1998).
Nomenclature. *Rhodotorula macerans* Frederiksen - Friesia 5: 237, 1956 ≡ *Cryptococcus macerans* (Frederiksen) Phaff & Fell, *in* Lodder - The Yeasts, ed. 2, p. 1127, 1970.

Cryptococcus neoformans (Sanfelice) Vuill.

Colony characteristics. Colonies (MEA) glossy, slimy, cream-coloured to yellowish-brown, with brown pigment on GACA. Margin entire.

Microscopy. Budding (RA) mulilateral. Budding cells rather firm-walled, spherical or ellipsoidal, 3.5-7.5 × 3-7 μm, with thick capsule visible in Indian Ink. Hyphae and pseudohyphae absent.

Histopathology. Staining of tissue with H & E reveals infected areas as hyaline foci due to mucoid capsular material. In the direct preparations stained with Indian Ink cells are visible as hyaline vesicles; with GMS the cells are blackish. With mucicarmine stain, stellate deposits are visible on the cell surface, the capsule remaining partly unstained.

Teleomorphs. *Filobasidiella neoformans* Kwon-Chung, *Filobasidiella bacillispora* Kwon-Chung (*Basidiomycota, Hymenomycetes, Tremellales: Filobasidiaceae*).

At mating, dikaryotic hyphae with clamp connections are formed, bearing slender, aseptate basidia with terminal swelling. *F. neoformans* has spherical basidiospores; in *F. bacillispora* the basidiospores are reniform and smooth.

Physiology.

Fermentation:	–	L-Arabinose	+,w	DL-Lactate	–
		D-Arabinose	+	Succinate	v
Growth:		D-Ribose	v	Citrate	v
Glucose	+	L-Rhamnose	+	*myo*-Inositol	+
Galactose	+	D-Glucosamine (C)	v	Hexadecane	–
L-Sorbose	–	*N*-Acetyl-D-glucosamine	v	Nitrate	–
Sucrose	+	Methanol	–	Creatinine	+
Maltose	+	Ethanol	w	D-Proline	v
Cellobiose	+,w	Glycerol	–	D-Tryptophan	v
α,α-Trehalose	+	*meso*-Erythritol	–	Vitamin-free	–
Lactose	–	Ribitol	v	Splitting of arbutin	+
Melibiose	–	Galactitol	+	D-Glucarate	v
Raffinose	+,w	D-Mannitol	+	Urease	+
Melezitose	+	D-Glucitol	+	Starch formation	+
Inulin	v	α-Methyl-D-glucoside	+	0.1% Cycloheximide	–
Soluble starch	+	Salicin	+	Growth at 37°C	+
D-Xylose	+	D-Gluconate	+	Growth at 40°C	v

Serology. Serotyping using a capsular polysaccharide specific monoclonal antibody was performed by Dromer *et al.* (1993). A dot enzyme assay was described by Belay *et al.* (1996). Antigen detection in cerebrospinal fluid (Frank *et al.*, 1993) and serum (Hamilton & Goodley, 1993) has been described.

Differential diagnosis. Species signature: lactose –, melezitose +, *myo*-inositol +, nitrite –, w/o biotin +, growth at 37°C +, 0.1% cycloheximide –, 16% NaCl –.

Molecular diagnostics. Species-specific DNA-probes and primers based on ribosomal ITS sequences (Mitchell *et al.*,

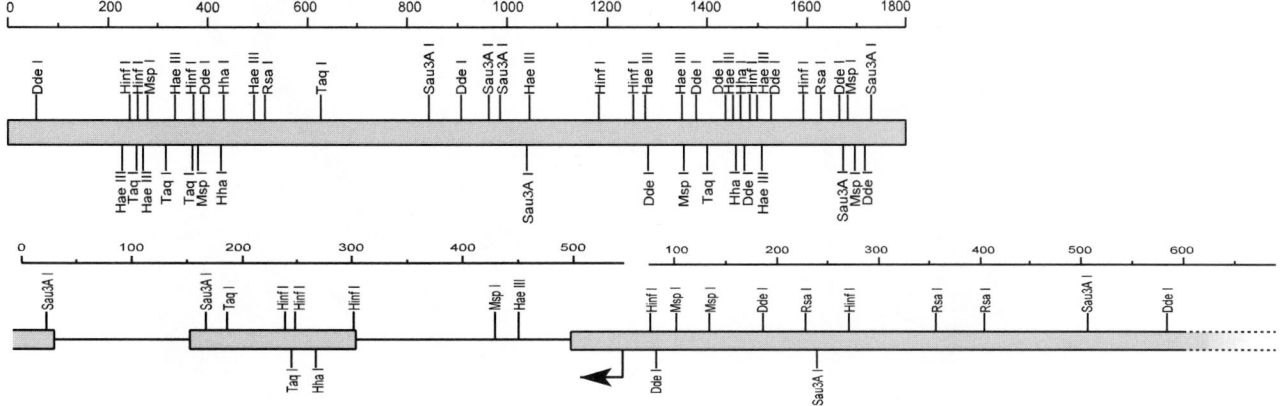

1994) and 18S sequences (Prariyachatigul *et al.*, 1996) have been developed. The species can also be characterized by PCR-amplification of rDNA fragments and subsequent digestion with restriction enzymes (Vilgalys & Hester, 1990).

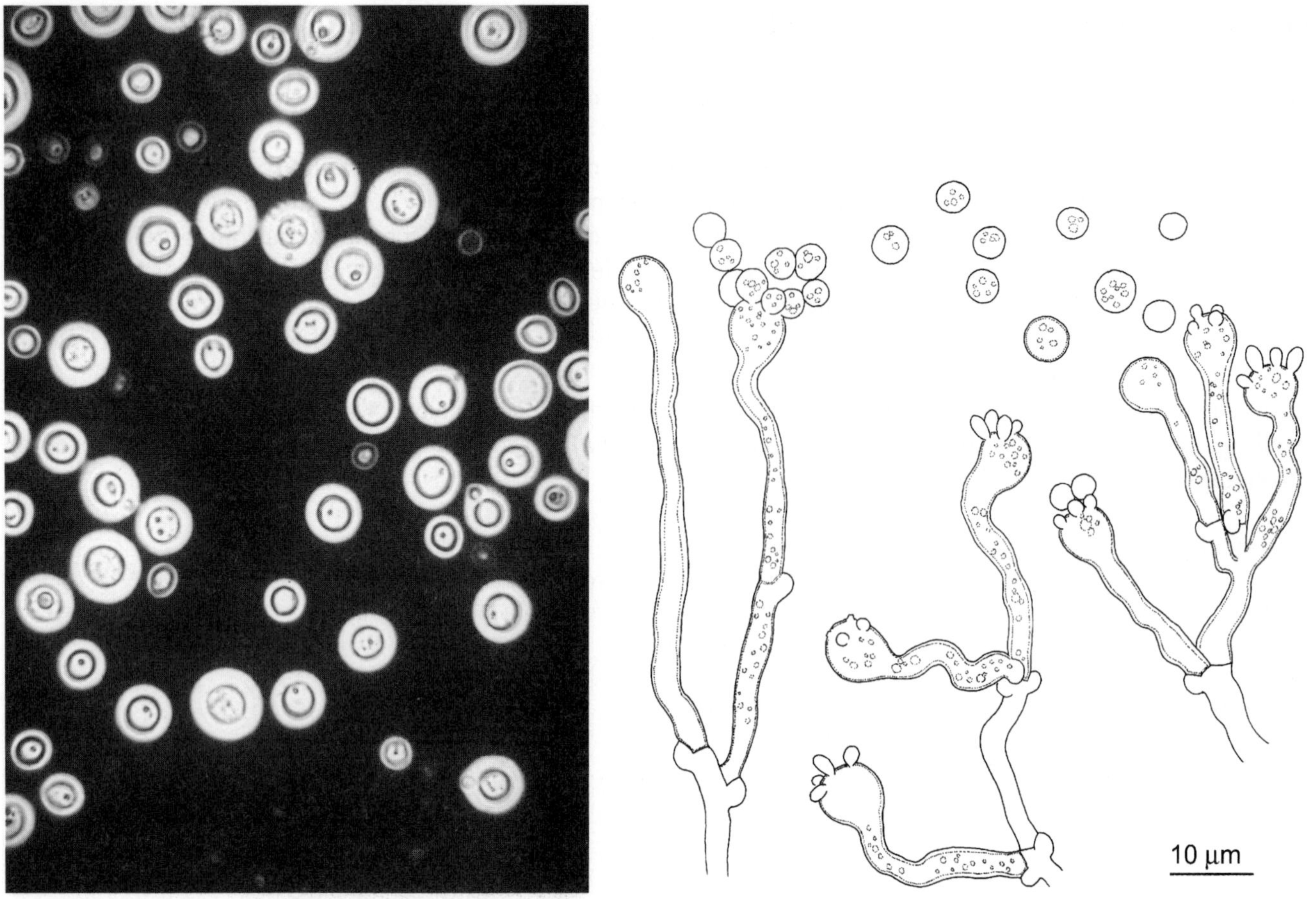

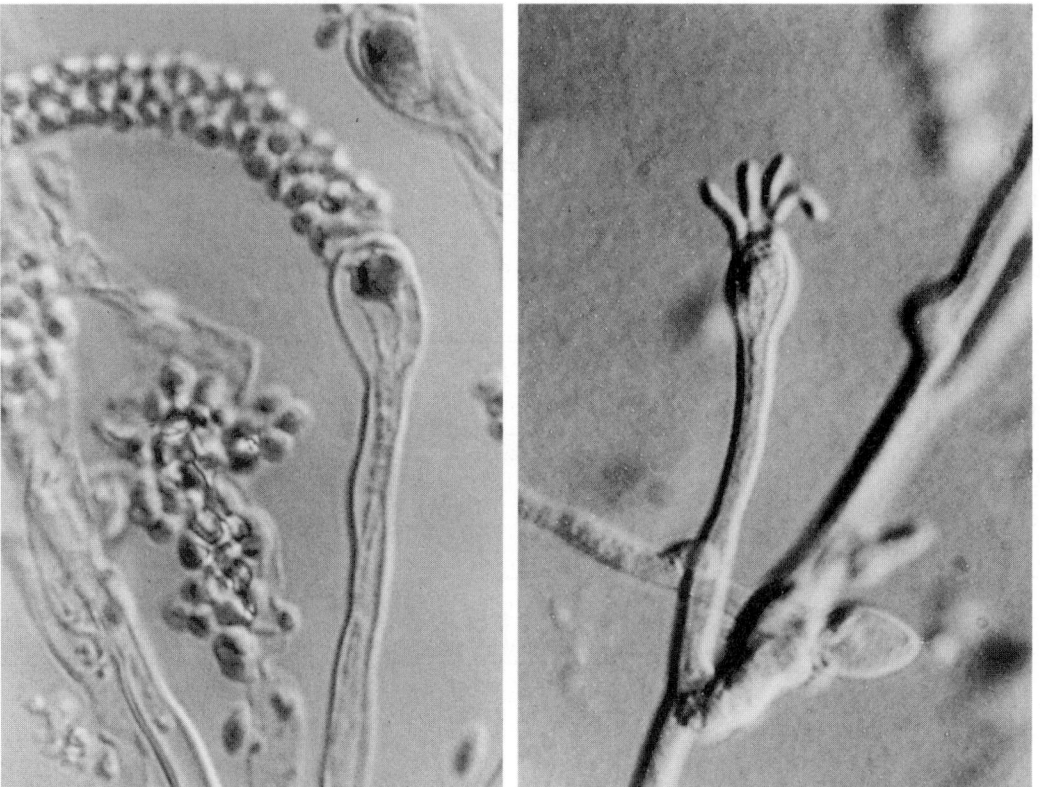

Cryptococcus neoformans, (left) capsules in Indian Ink-stained CSF; (right) CBS 132 × 6886. Clamped hyphae with terminal spore heads. Liberated basidiospores are converted into yeast cells.

Filobasidiella neoformans (*Cryptococcus neoformans*), teleomorphs with basidia producing basidiospores. a. *F. neoformans* var. *neoformans,* B-3501 × B-3502; b. *F. neoformans* var. *bacillispora,* NIH 444 × NIH 191. Courtesy K.J. Kwon-Chung. a, b. ×2400.

With these molecular methods the two varieties of *C. neoformans* were found to be identical; this was confirmed on the basis of ribosomal ITS sequences (Mitchell *et al.*, 1992). They also proved to have a high rate of DNA-DNA homology (Aulakh *et al.*, 1981). The phylogenetic position of the species was established with ribosomal genes by Guého *et al.* (1993), Fan *et al.* (1994), Yamada *et al.* (1990) and Kwon-Chung & Chang (1994) and with actin genes by Cox *et al.* (1995). 26S-based probes, covering both varieties, were developed by Sandhu *et al.* (1995). Nested PCR detection was developed by Tanaka *et al.* (1996). The SSU, ITS and LSU restriction maps based on NCBI L05428, L14068 and U94941 are given below.

Pathogenicity. BSL-2/3 (greater safety precautions are recommended when working with mated strains). Sukroong-reung *et al.* (1998) suggested that basidiospores are the most infectious propagules. The varieties of *C. neoformans* are markedly different in their clinical behaviour (Mitchell *et al.*, 1995). Cryptococcosis due to the varieties *neoformans* and *grubii* is an inhalation-mycosis, occurring almost exclusively in patients with impaired T-cell function (Sugar, 1991); association with Hodgkin's disease was reviewed by Korfel *et al.* (1998). The variety *gattii* occurs in non-AIDS patients. Systemic cases after immunosuppression are very rare (Schepelmann *et al.*, 1993; Tumietto *et al.*, 1995). Pleural effusion leading to pulmonary infiltrates (Chechani & Kamholz, 1990; Meyohas *et al.*, 1995) is a first indicator of AIDS (de Lalla *et al.*, 1993). Dissemination leads to chronic meningitis (Powderly, 1993; Yu, 1996) and later meningoencephalitis (Lee *et al.*, 1996), patients sometimes becoming disoriented (Sa'adah *et al.*, 1995). Cryptococcosis in AIDS patients probably results from endogenous reactivation of earlier infections (Spitzer & Spitzer, 1992). Secondary cutaneous manifestations are frequently observed (Murakawa *et al.*, 1996), but rarely in otherwise healthy patients (Mirza, 1996). The infection is usually fatal when untreated. Infections in other parts of the body are rare (Kerma & Graham, 1995). Primary cutaneous infections are mostly caused by serotype D strains (Naka *et al.*, 1995; Vogelaers *et al.*, 1997). Animal infections include those in shrews (Tell *et al.*, 1997) and in a dog (Kerwin *et al.*, 1998).

Table 18. Specific and varietal characters of *C. neoformans*.

	var. *neoformans*	var. *grubii*	var. *gattii*
Sucrose	+	+	+
Lactose	–	–	–
myo-Inositol	+	+	+
Nitrate	–	–	+
Creatinine	+	+	+
D-Proline	–	–	+
D-Tryptophan	–	–	+
37°C	+	+	w
GACA	+	+	+
CGB	–	–	+
GCP	–	–	+
CDBT[1]	–	orange	blue-green
Killer sensitivity[2]	–	–	+
Budding cell shape	spherical	spherical	spherical, ellipsoidal
Bud formation	multilateral	multilateral	multilateral
Teleomorph	var. *neoformans*	unknown	var. *bacillispora*
Basidiospore shape	(sub)spherical	(sub)spherical	reniform
Serotype	D	A	B, C
Distribution	Northern Europe	world-wide	Central Africa, Australia, California, Central America
Prevalent clinical	AIDS	cutaneous	no underlying disease

[1]Test results may be ambiguous.
[2]Sensitive to *Cryptococcus laurentii*, CBS 139 (Boekhout & Scorzetti, 1997).

Varieties. Three varieties of *C. neoformans* are known in part to be connected with different teleomorphs; see Table 18. The variety *gattii* also occurs in otherwise healthy patients or patients having non-AIDS-related diseases (López-Martínez *et al.*, 1996), while hosts of the var. *neoformans* nearly always have a compromised specific immunity (Speed & Dunt, 1995), although exceptions are known (Wendisch *et al.*, 1996). These varieties are probably sufficiently separate to warrant their recognition at the species level (Boekhout *et al.*, 1997). The var. *grubii* has been introduced for serotype A strains.

Distribution. *Cryptococcus neoformans* occurs subclinically in the nasal cavity of animals, particularly koalas (Connolly *et al.*, 1999). The var. *neoformans* has a world-wide distribution in dry excrements in bird shelters, e.g. pigeon roosts (Pal, 1997) or zoo animals (Irokanulo *et al.*, 1997). However, the serotypes are unevenly distributed: serotype A is common in Europe but almost absent from North America (Dromer *et al.*, 1996). A natural habitat in hollow trees has been supposed (Lazéra *et al.*, 1996). The var. *gattii* is restricted to Central Africa, Australasia, California, (sub)tropical America (López-Martínez *et al.*, 1996) and Southern Europe. This variety was also linked to mammals which carry the fungus asymptomatically, such as koalas (Connolly *et al.*, 1999) and bats (Lazéra *et al.*, 1998). It has a natural association with *Eucalyptus* (Pfeiffer & Ellis, 1993; Montagna *et al.*, 1997a, b), and rarely with other trees (Lazéra *et al.*, 1998) such as almond (Callejas *et al.*, 1998).

References. Casadevall & Perfect (1998). Var. *neoformans:* Staib (1987), Kwon-Chung (1987a), Swinne & Kayembe (1987), Kwon-Chung *et al.* (1990), Sato *et al.* (1990), Ellis & Pfeiffer (1992), Müller (1994). Var. *gattii:* Kwon-Chung *et al.* (1982), Dufait *et al.* (1987), Kwon-Chung (1987a), Ellis & Pfeiffer (1990), Mitchell & Perfect (1995). Var. *grubii:* Franzot *et al.* (1999).

Antifungal susceptibility.

Antifungal	MICs range	MIC 90	Strains	Reference
AMB	1-2	1	12	Espinel-Ingroff (1998)
AMB	0.12-0.5	0.25	20	Davey *et al.* (1998)
AMB	≤0.03-0.25	0.25	15	Karlowsky *et al.* (1997)
5FC	1-16	8	20	Davey *et al.* (1998)
5FC	0.02-16	8	15	Karlowsky *et al.* (1997)
FLZ	2-16	1	12	Espinel-Ingroff (1998)
FLZ	2-8	8	20	Davey *et al.* (1998)
FLZ	1-4	4	15	Karlowsky *et al.* (1997)
ITZ	0.06-1	1	12	Espinel-Ingroff (1998)
ITZ	0.06-0.5	0.5	20	Davey *et al.* (1998)
KTZ	≤0.12-1	0.5	20	Davey *et al.* (1998)
KTZ	0.004-0.5	0.25	15	Karlowsky *et al.* (1997)
MCZ	≤0.12-1	0.5	20	Davey *et al.* (1998)
VCZ	≤0.03-0.25	0.12	12	Espinel-Ingroff (1998)

Nomenclature var. *neoformans*. *Saccharomyces neoformans* Sanfelice - Ann. Igiene Sperim. 5: 239-262, 1895 ≡ *Cryptococcus neoformans* (Sanfelice) Vuillemin - Revue Gén. Sci. Pures Appl. 12: 747, 1901 ≡ *Torula neoformans* (Sanfelice) Redaelli & Ciferri - Riv. Biol. 13: 171-235, 1931 ≡ *Torulopsis neoformans* (Sanfelice) de Almeida - Anais Fac. Med. Univ. S. Paulo 9: 76, 1933 ≡ *Debaryomyces neoformans* (Sanfelice) Redaelli, Ciferri & Giordano - Boll. Sez. Ital. Soc. Int. Microbiol. 9: 24, 1937 ≡ *Lipomyces neoformans* (Sanfelice) Ciferri - Man. Micol. Med. 1960.

Saccharomyces lithogenes Sanfelice - Centbl. Bakt. Parasitkde, Abt. 1, 18: 521, 1895 ≡ *Blastomyces lithogenes* (Sanfelice) Sakawa - Zentbl. Bakt. Parasitkde, Abt. 1, 88: 274, 1922 ≡ *Torulopsis lithogenes* (Sanfelice) de Almeida - Anais Fac. Med. Univ. S. Paulo 9: 76, 1933.

Saccharomyces hominis Busse - Virchow's Arch. 140: 23-46, 1895 ≡ *Cryptococcus hominis* (Busse) Vuillemin - Revue Gén. Sci. Pures Appl. 12: 735, 1901 ≡ *Atelosaccharomyces busse-buschki* de Beurmann & Gougerot - Nouv. Mycoses p. 29, 1909 (name change) ≡ *Atelosaccharomyces hominis* (Busse) Verdun - Précis Parasitol., 1912 ≡ *Torulopsis hominis* (Busse) Castellani & Jacono - J. Trop. Med. Hyg. 36: 297, 1933 ≡ *Debaryomyces hominis* (Busse) Todd & Hermann - J. Bact. 32: 36,1936.

Saccharomyces hominis Costantin - Bull. Soc. Mycol. Fr. 17: 145, 1901 ≡ *Cryptococcus costantinii* Froilano de Mello & Fernandes - Arq. Hig. Pat. Exot. 6: 294, 1918 (name change) ≡ *Torulopsis costantinii* (Froilano de Mello & Fernandes) de Almeida - Anais Fac. Med. Univ. S. Paulo 9: 76, 1933 ≡ *Cryptococcus hominis* (Busse) Vuillemin var. *costantinii* (Froilano de Mello & Fernandes) C.W. Dodge - Med. Mycol. p. 338, 1935.

Saccharomyces plimmeri Costantin - Bull. Soc. Mycol. Fr. 17: 147, 1901 ≡ *Torula plimmeri* (Costantin) Weis - J. Med. Res. 7: 280-311, 1902 ≡ *Cryptococcus plimmeri* (Costantin) Neveu-Lemaire - Parasitol. Anim. Domest. p. 60, 1912 ≡ *Torulopsis plimmeri* (Costantin) de Almeida - Anais Fac. Med. Univ. S. Paulo 9: 76, 1933.

Atelosaccharomyces breweri Verdun - Précis Parasitol. 1912 ≡ *Cryptococcus breweri* (Verdun) Castellani & Chalmers - Man. Trop. Med., ed. 2, p. 771, 1913 ≡ *Saccharomyces breweri* (Verdun) Neveu-Lemaire - Précis Parasitol. Hum., ed. 5, p. 97, 1921 ≡ *Torulopsis breweri* (Verdun) de Almeida - Anais Fac. Med. Univ. S. Paulo 9: 76, 1933.

Torula histolytica Stoddard & Cutler - Monr. Rockefeller Inst. Med. Res. 6: 1, 1916 ≡ *Cryptococcus histolyticus* (Stoddard & Cutler) Castellani - Am. J. Trop. Med. 8: 393, 1928 ≡ *Torulopsis histolytica* (Stoddard & Cutler) Castellani - J. Trop. Med. Hyg. 33: 31, 1933 ≡ *Rhodotorula histolytica* (Stoddard & Cutler) Krassilnikov, *in* Kursanova - Opred. Nizsh. Rast. 3: 133, 1954.

Blastomyces neoformans Arzt - Arch. Derm. Syph. 145: 311, 1924.

Torula nasalis Harrison - Trans. R. Soc. Can., Biol. Sci., 22: 207, 1928 ≡ *Cryptococcus nasalis* (Harrison) C.W. Dodge - Med. Mycol. p 332, 1935 ≡ *Torulopsis neoformans* (Sanfelice) de Almeida race *nasalis* (Harrison) Lodder - Anaskosp. Hefen 1: 176, 1934.

Cryptococcus psicrofilus Niño, *in* Speroni, Llambias, Parodi & Niño - Bol. Inst. Clín. Quirúrg. Univ. B. Aires 5: 94, 1930.

Torulopsis hominis (Busse) Castellani & Jacono var. *honduriana* Castellani - Med. Press Circ. 136: 438, 1933 ≡ *Cryptococcus hondurianus* (Castellani) C.W. Dodge - Med. Mycol. p. 334, 1935.

Cryptococcus meningitidis C.W. Dodge - Med. Mycol. p. 333, 1935.

Filobasidiella neoformans Kwon-Chung - Mycologia 67: 1199, 1975.

Nomenclature var. *gattii*. *Cryptococcus neoformans* (Sanfelice) Vuillemin var. *gattii* Vanbreuseghem & Takashio - Annls Soc. Belge Méd. Trop. 50: 701, 1970.

Filobasidiella bacillispora Kwon-Chung - Mycologia 68: 945, 1976 ≡ *Filobasidiella neoformans* Kwon-Chung var. *bacillispora* (Kwon-Chung) Kwon-Chung, Bennett & Rhodes - Antonie van Leeuwenhoek 48: 35, 1982.

Cryptococcus bacillisporus Kwon-Chung & Bennett, *in* Kwon-Chung, Bennett & Theodore - Int. J. Syst. Bact. 28: 618, 1978.

Cryptococcus neoformans (Sanfelice) Vuillemin var. *shanghaiensis* Liao, Shao, Wu, Zhang & Li - Chin. Med. J. 96: 287, 1983.

Nomenclature var. *grubii*. *Cryptococcus neoformans* (Sanfelice) Vuillemin var. *grubii* Franzot, Salkin & Casadevall - J. Clin. Microbiol. 37: 839, 1999.

Cryptococcus uniguttulatus (Zach) Phaff & Fell

Colony characteristics. Colonies (MEA) glossy, slimy, cream-coloured; margin entire.

Microscopy. Budding (RA) percurrent. Budding cells spherical to ellipsoidal, 3.5-7.0 × 3.0-5.5 μm, frequently cohering in clusters.

Teleomorph. *Filobasidium uniguttulatum* Kwon-Chung (*Basidiomycota, Hymenomycetes, Tremellales: Filobasidiaceae*).

Physiology.

Fermentation:	–	L-Arabinose	+,w	DL-Lactate	v
		D-Arabinose	w,–	Succinate	v
Growth:		D-Ribose	–	Citrate	v
Glucose	+	L-Rhamnose	v	*myo*-Inositol	+
Galactose	v	D-Glucosamine (C)	–	Hexadecane	w
L-Sorbose	–	*N*-Acetyl-D-glucosamine	+	Nitrate	–
Sucrose	+	Methanol	–	Vitamin-free	–
Maltose	+	Ethanol	–	2-Keto-D-gluconate	+
Cellobiose	–	Glycerol	w	D-Glucarate	–
α,α-Trehalose	+,w	*meso*-Erythritol	–	D-Glucuronate	+
Lactose	–	Ribitol	v	L-Lysine	+,w
Melibiose	–	Galactitol	–	Ethylamine	–
Raffinose	s	D-Mannitol	s	Urease	+
Melezitose	+	D-Glucitol	+,w	Starch formation	+
Inulin	+	α-Methyl-D-glucoside	s	Growth at 30°C	+
Soluble starch	v	Salicin	+,w	Growth at 35°C	–
D-Xylose	+	D-Gluconate	s		

Differential diagnosis. Species signature: sucrose +, lactose –, melibiose –, *myo*-inositol +, D-glucarate –, D-glucuronate +, ethylamine –, growth at 35°C –, L-arabinitol –, D-glucono-δ-lactone –, nitrite –, w/o biotin +.

Molecular diagnostics. LSU restriction map based on NCBI AF075468:

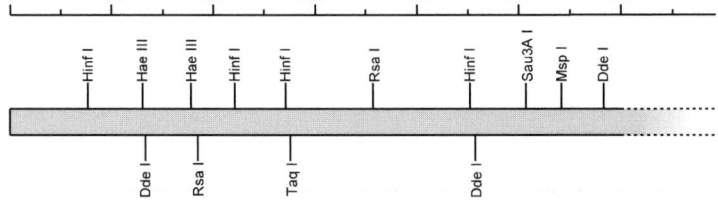

Pathogenicity. BSL-1. The species was originally described from onychomycosis (Wolfram & Zach, 1934). Additional strains in the CBS collection mostly originate from humans. It should be noted that the species is unable to grow at 35°C.

Reference. Kwon-Chung (1977).

Nomenclature. *Eutorulopsis uniguttulata* Zach, *in* Wolfram & Zach - Arch. Derm. Syph. 170: 688, 1934 ≡ *Cryptococcus neoformans* (Sanfelice) Vuillemin var. *uniguttulatus* (Zach) Lodder & Kreger-van Rij - The Yeasts p. 378, 1952 ≡ *Cryptococcus uniguttulatus* (Zach) Phaff & Fell, *in* Lodder - The Yeasts, ed. 2, p. 1140, 1970.

Filobasidium uniguttulatum Kwon-Chung - Int. J. Syst. Bact. 27: 293, 1977.

Basidiomycetous yeasts. Genus: *MALASSEZIA*

Generic description. Lipophilic. Resistant to benomyl. Budding monopolar from a broad base, percurrent. Cell wall ultrastructure multilamellar with spiral plasma membrane invaginations (Kreger-van Rij & Veenhuis; 1970; Takeo & Nakai, 1986).

General remarks. The species are common saprobes in human and animal excretions on the body surface. It has been suggested that several species can be present on the same host (Raabe *et al.*, 1998).

References. Rippon (1988), Guého & Meyer (1989), Midgley (1993), Guillot & Guého (1995), Guého *et al.* (1996), Mayser *et al.* (1997), Boekhout *et al.* (1998), Makimura *et al.* (2000), Gupta *et al.* (2000).

Key to the treated species of Malassezia:

| 1a. | Growth on SGA | .... | *M. pachydermatis* (150) |

1a. Growth on SGA .... ***M. pachydermatis* (150)**
1b. No growth on SGA ➝ **2**
2a. Catalase reaction positive ➝ **3**
2b. No catalase reaction .... ***M. restricta* (152)**
3a. Growth on SGA with Tween 40 ➝ **4**
3b. No growth SGA with Tween 40 ➝ **5**
4a. Growth on SGA with Tween 80 ➝ **6**
4b. No growth on the above medium .... ***M. slooffiae* (153)**
5a. Cells long, cylindrical .... ***M. obtusa* (149)**
5b. Cells spherical .... ***M. globosa* (147)**
6a. Growth on SGA with cremophor EL .... ***M. furfur* (145)**
6b. No growth on the above medium .... ***M. sympodialis* (155)**

Table 19. Summary of diagnostics of *Malassezia* species:

	Buds	SGA	40°C	Cremophor EL	Tween 80	Tween 40	Tween 20	Esculine	Catalase
M. furfur	wide	–	+	+	+	+	+	w	+
M. globosa	narrow	–	–	–	–	–	–	–	+
M. pachydermatis	wide	+	+	–,w	+	+	–,w	v	v
M. obtusa	wide	–	–	–	–	–	–	+	+
M. restricta	narrow	–	–	–	–	–	–	–	–
M. slooffiae	wide	–	+	–	+,w	+	+	–	+
M. sympodialis	narrow	–	+	–,w	+	+	+	+	+

Malassezia furfur (Robin) Baillon

Colony characteristics. Lipid dependent. Growth (LNA, 30°C) cream-coloured to yellowish, convex or slightly wrinkled, glistening or dull; margin entire or lobed.

Microscopy. Budding percurrent; buds nearly as wide as the mother cell. After 3 days at 37°C two forms can be recognized. The oval form is most abundant with ovoidal, ellipsoidal or cylindrical cells, 2.0-6.5 × 1.5-4.5 µm. The sparse orbicular form consists predominantly of spherical cells measuring 2.5-4.5 µm diam. Transition from one form to the other may occur.

Physiology. It is the only *Malassezia* species that assimilates cremophor EL (Mayser *et al.*, 1997).

Catalase	+	Tween 80	+	Esculine	w
SGA	–	Tween 40	+		
37°C	+	Tween 20	+		
40°C	+	Cremophor EL	+		

Molecular diagnostics. LSU restriction map based on NCBI AF063214:

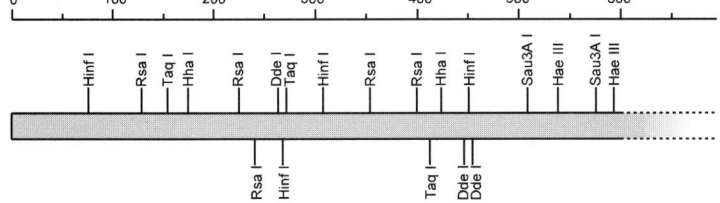

Pathogenicity. BSL-2. Reports of particular diseases in the literature often cannot be linked to the new classification of Guého and coworkers; the references below should therefore be interpreted with care. The fungus is dimorphic on skin, producing true hyphae in addition to yeast cells. It is present on skin and in ears of healthy individuals, and is acquired at very early age (Leeming *et al.*, 1995). Hyperhydrotic individuals develop an asymptomatic disorder known as pityriasis versicolor (p. 22), dark spots on the skin probably being caused by the conversion of L-tryptophan to brown indole derivatives (Mayser *et al.*, 1998). Excessive fungal growth causes seborrheic hyperkeratosis. Neonatal pustulosis may be due to this fungus (Rapelanoro *et al.*, 1996). The disease is somewhat more common in adolescent, white males, usually disappearing in their early twenties. The species occasionally causes white piedra (Lopes *et al.*, 1994a). Blepharitis after keratoplasty was reported by Toth *et al.* (1996). Severe cases involving pruritic folliculitis are particularly seen in immunocompromised patients (Helm & Lookingbill, 1993; Sandin *et al.*, 1993a). Seborrhoeic dermatitis occurs significantly more often in AIDS patients (Schechtman *et al.*, 1995). Disseminated infections are seen in neutropenic children (Schoepfer *et al.*, 1995) or neonates (Shekh *et al.*, 1989), occasionally in adults (Barber *et al.*, 1993; Shparago *et al.*, 1995) on parenteral nutrition with lipids. Clinical manifestation mostly concerns pulmonary infiltrates with yeast cells growing in arteries.

References. Ingham & Cunningham (1993), Hernandez-Molina (1993), Guého *et al.* (1996), Mayser & Pape (1998).

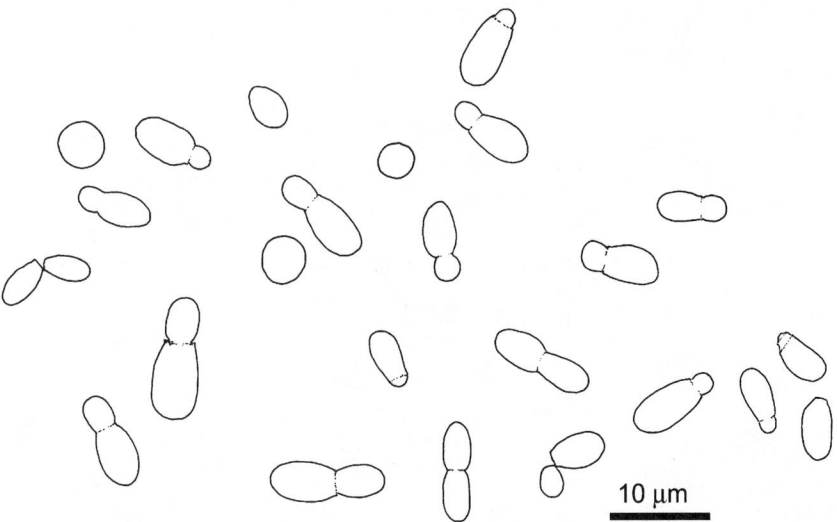

Malassezia furfur, CBS 7019. Budding cells, oval and orbicular forms.

Nomenclature. *Microsporum furfur* Robin - Hist. Nat. Vég. Paras. Homme Anim. p. 436, 1853 = *Sporotrichum furfur* (Robin) Saccardo - Syll. Fung. 4: 100, 1886 = *Malassezia furfur* (Robin) Baillon - Traité Bot. Méd. Crypt. p. 243, 1889 = *Oidium furfur* (Robin) Zopf - Die Pilze p. 257, 1890 = *Monilia furfur* (Robin) Vuillemin - Champ. Paras. Myc. Homme p. 89, 1931 = *Pityrosporum furfur* (Robin) Emmons, Binford & Utz - Med. Mycol., ed. 2, p. 159, 1970.

Cryptococcus psoriasis Rivolta - Paras. Veget. p. 469, 1883 = *Torulopsis psoriasis* (Rivolta) de Almeida - Anais Fac. Med. Univ. S. Paulo 9: 76, 1933.

Saccharomyces ovalis Bizzozero - Virchow's Arch. Path. Anat. 98: 441, 1884 = *Pitryosporum ovale* (Bizzozero) Castellani & Chalmers - Man. Trop. Med., ed. 2, p. 836, 1913 = *Malassezia ovalis* (Bizzozero) Acton & Panja - Ind. Med. Gaz. 62: 603, 1927 = *Torulopsis ovalis* (Bizzozero) de Almeida - Anais Fac. Med. Univ. S. Paulo 9: 76, 1933.

Saccharomyces capillitii Oudemans & Pekelharing - Ned. Tijdschr. Geneesk. 21: 997-1005, 1885.

Pityrosporum malassezii Sabouraud - Malad. Cuir Chev. 640, 1904 = *Dermatophyton malassezii* (Sabouraud) Don - Parasitology 3: 279, 1910 = *Cryptococcus malassezii* (Sabouraud) Benedek - Zentbl. Bakt. Parasitkde, Abt. 1, 116: 317-332, 1930.

Microsporum tropicum Castellani - Br. Med. J. 2: 1271, 1905 = *Malassezia tropica* (Castellani) Schmitter - J. Trop. Med. Hyg. 26: 194, 923.

Pityrosporum cantliei Castellani - J. Cutan. Dis. 26: 393-399, 1908.

Malassezia macfadyeni Castellani - J. Cutan. Dis. 26: 393-399, 1908.

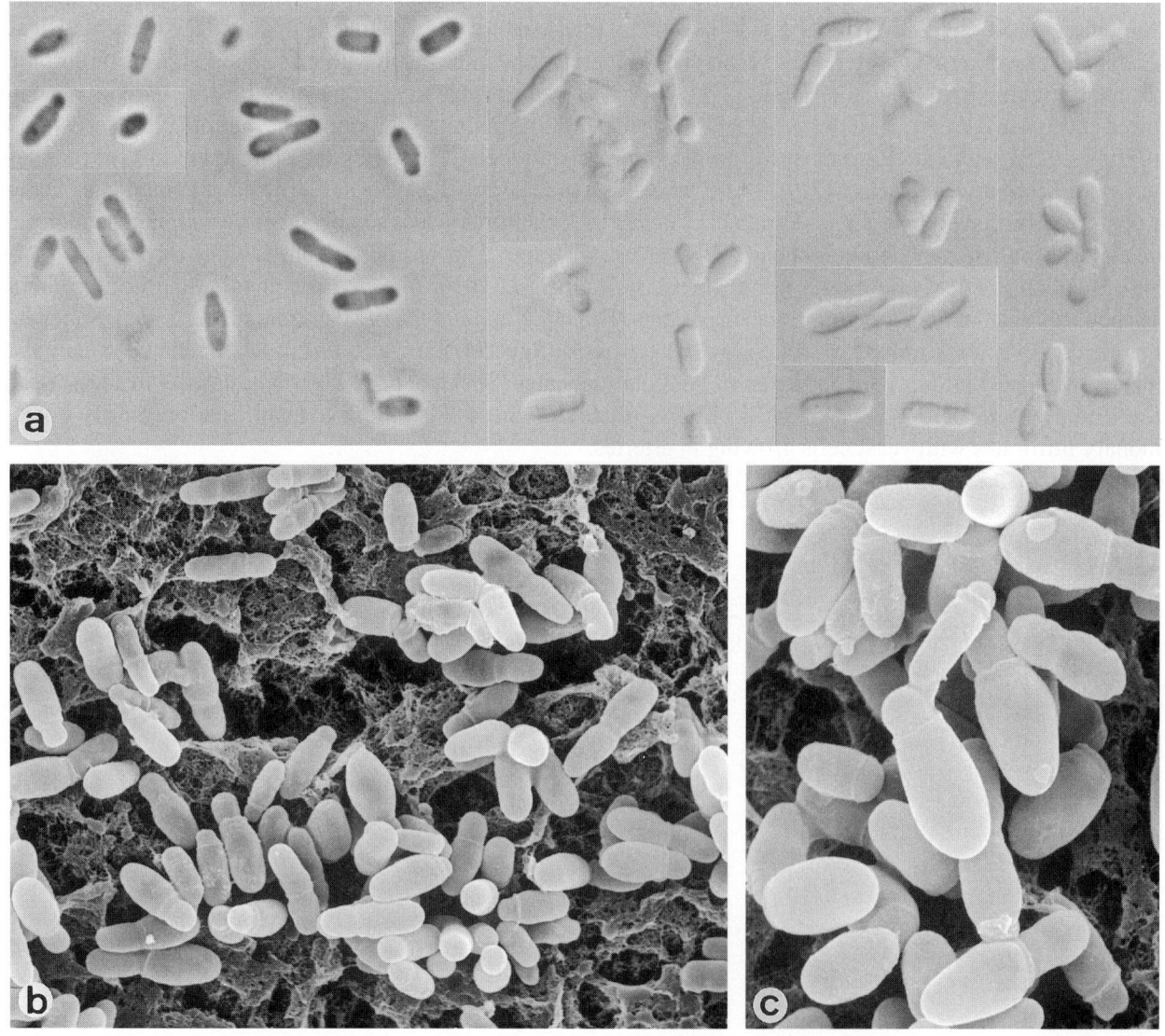

Malassezia furfur, CBS 1878. Budding cells. a. ×1600; b. ×3050; c. ×5250.

146

Malassezia globosa Midgley *et al.*

Colony characteristics. Lipid dependent. Colonies (LNA, 30°C) cream-coloured, raised, folded and wrinkled, coarse, with finely-lobed margin.
Microscopy. Budding percurrent; buds considerably narrower than the mother cell. Yeast cells spherical, 2.5-8.0 μm diam; short filaments may be present.
Physiology.

Catalase	+	Tween 80	–	Esculine	–
SGA	–	Tween 40	–	Serovar	B
37°C	w,–	Tween 20	–		
40°C	–	Cremophor EL	–		

Molecular diagnostics. LSU restriction map based on NCBI AF064025:

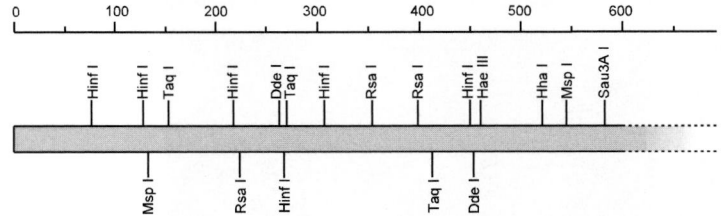

Pathogenicity. BSL-2. The species is found on healthy human skin and may cause pityriasis versicolor (Crespo Erchiga *et al.*, 1999).
Reference. Cunningham *et al.* (1990), Guého *et al.* (1996).
Nomenclature. *Malassezia globosa* Midgley, Guého & Guillot, *in* Guého, Midgley & Guillot - Antonie van Leeuwenhoek 69: 347, 1996.

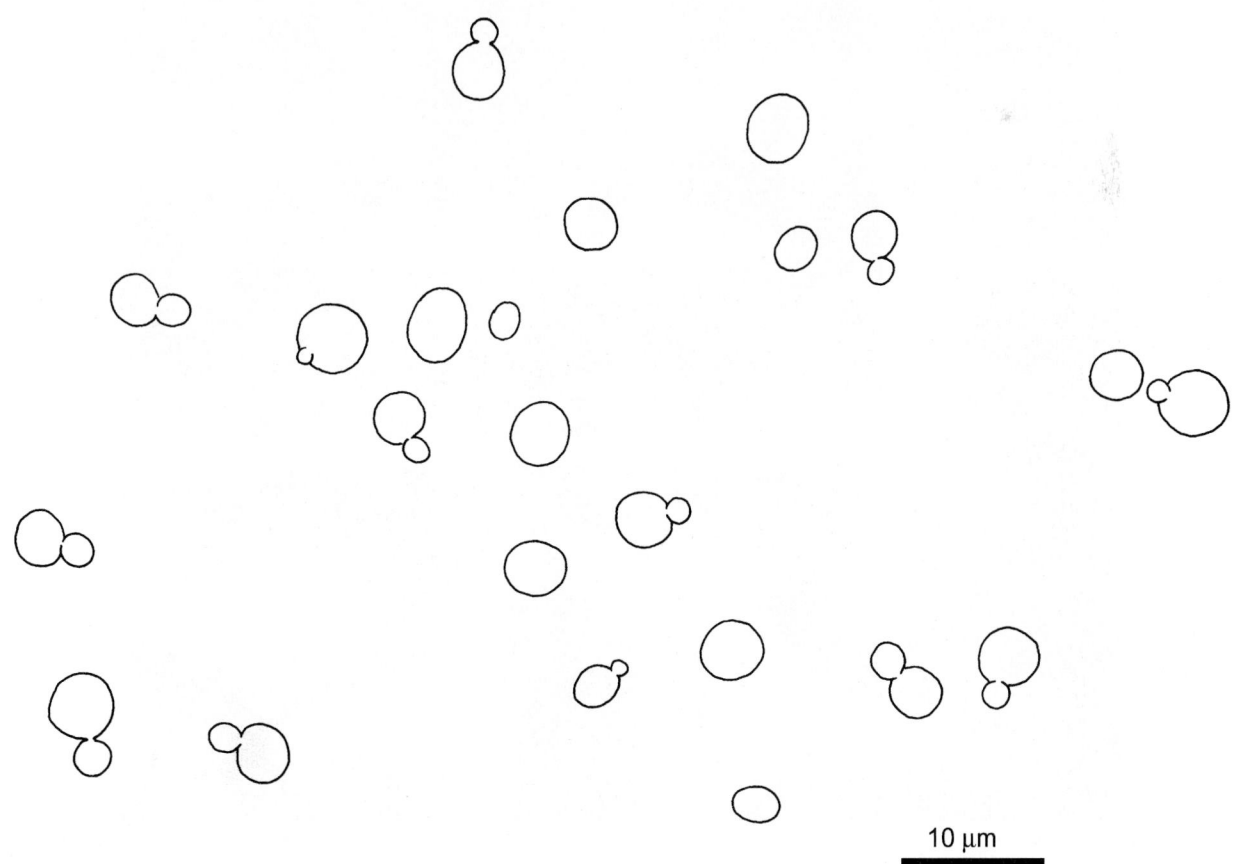

10 μm

Malassezia globosa, CBS 7966. Budding cells.

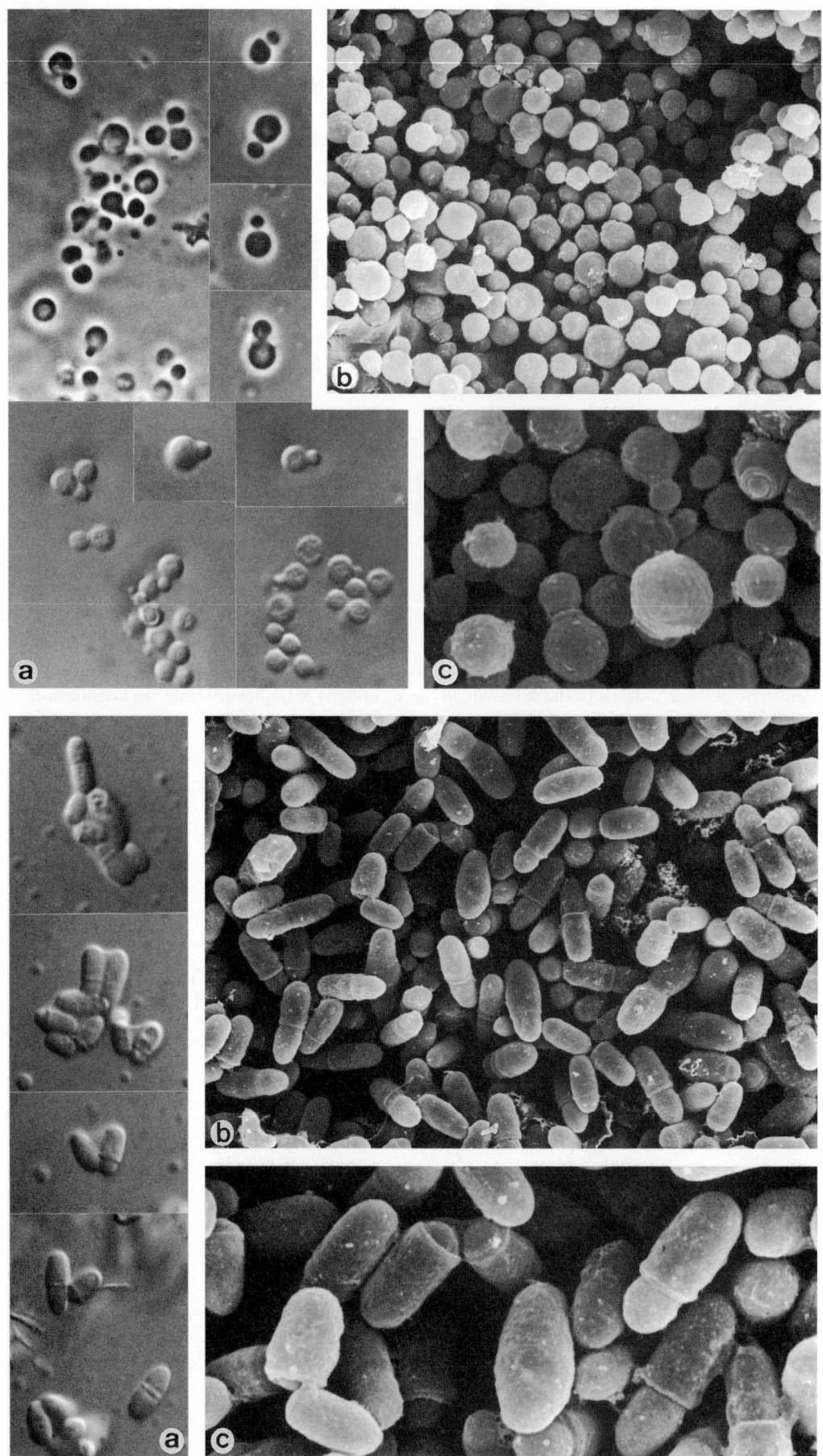

Upper plate: *Malassezia globosa,* CBS 7966. Percurrent budding cells. a. ×1600; b. ×2600; c. ×5500.
Lower plate: *Malassezia obtusa,* CBS 7876. Percurrent budding cells. a. ×1600; b. ×2600; c. ×6000.

Malassezia obtusa Midgley *et al.*

Colony characteristics. Lipid dependent. Colonies (LNA, 30°C) cream-coloured, flat, smooth, glistening, sticky.
Microscopy. Budding percurrent; buds as wide as the mother cell. Yeast cells cylindrical, 4-6 × 1.5-2.0 μm. Filaments may be present.
Physiology.

Catalase	+	Tween 80	–	Esculine	+
SGA	–	Tween 40	–		
37°C	w,–	Tween 20	–		
40°C	–	Cremophor EL	–		

Molecular diagnostics. LSU restriction map based on NCBI AF064027:

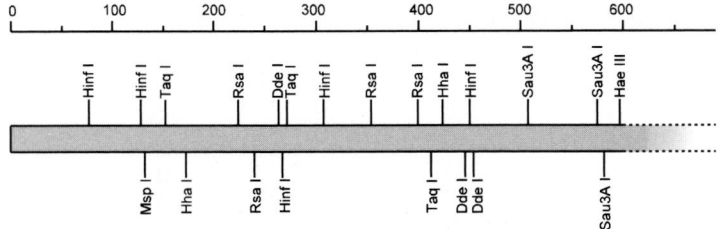

Pathogenicity. BSL-2. Rare species occurring on human skin.
Reference. Guého *et al.* (1996).
Nomenclature. *Malassezia obtusa* Midgley, Guillot & Guého, *in* Guého, Midgley & Guillot - Antonie van Leeuwenhoek 69: 349, 1996.

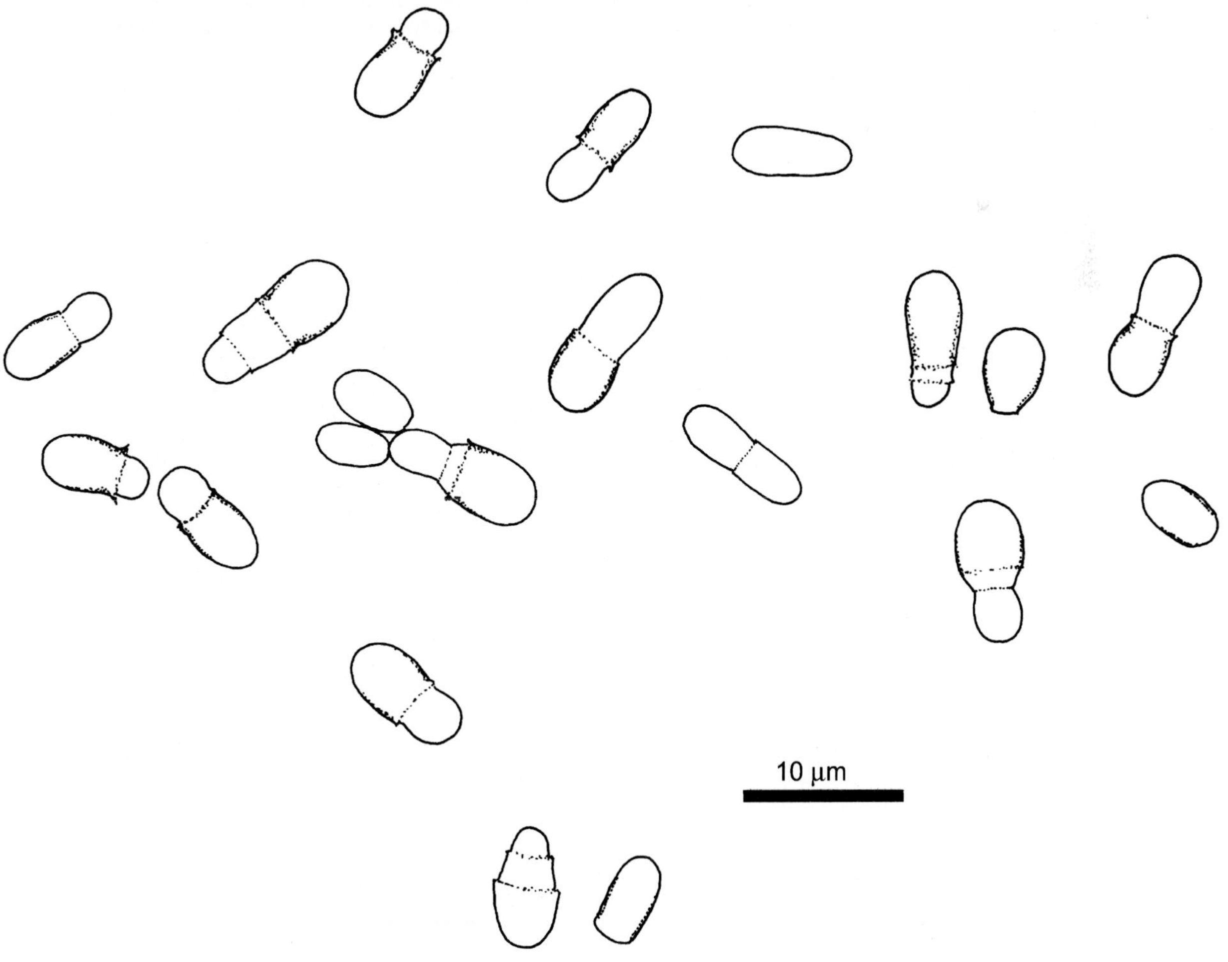

Malassezia obtusa, CBS 7876. Percurrent budding cells.

Malassezia pachydermatis (Weidman) C.W. Dodge

Colony characteristics. Not lipid dependent. Colonies (SGA, 30°C) cream-coloured, convex, soft; margin entire or slightly lobed.

Microscopy. Budding percurrent; buds as wide as the mother cells. Yeast cells subspherical to ovoidal, 3.0-6.5 × 2.5 μm.

Physiology.

Catalase	v	Tween 80	+,w	Esculine	v
SGA	+	Tween 40	+		
37°C	+	Tween 20	v		
40°C	+	Cremophor EL	–,w		

Molecular diagnostics. LSU restriction map based on NCBI AF063215:

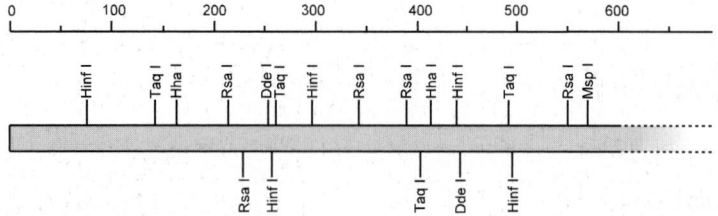

Pathogenicity. BSL-2. Zoophilic. It is the most common species in canine otitis externa (Huang *et al.*, 1993) and seems to be prevalent on carnivores (Guillot *et al.*, 1994). Disseminated infections in humans are known in IV-fed neonates (Guého *et al.*, 1987b; Mickelsen *et al.*, 1988; Welbel *et al.*, 1994).

References. Guého & Meyer (1989), Guého & Guillot (1999), Guillot & Bond (1999).

Nomenclature. *Pityrosporum pachydermatis* Weidman, *in* Fox - Rept Lab. Mus. Comp. Path. Zool. Soc. Philad. 36, 1925 ≡ *Cryptococcus pachydermatis* (Weidman) Nannizzi - Tratt. Micopat. Um. 4: 345, 1934 ≡ *Malassezia pachydermatis* (Weidman) C.W. Dodge - Med. Mycol. p. 340, 1935.

    Pityrosporum canis Gustafson - Nord. Vet. Med. 6: 434-442, 1954.

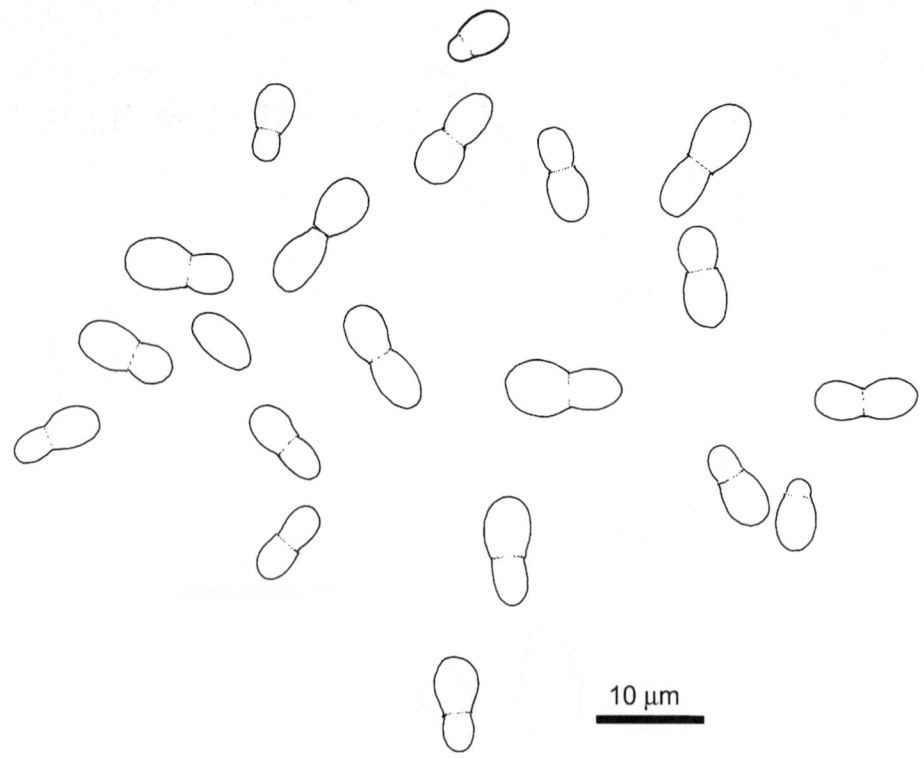

Malassezia pachydermatis, CBS 1879. Budding cells.

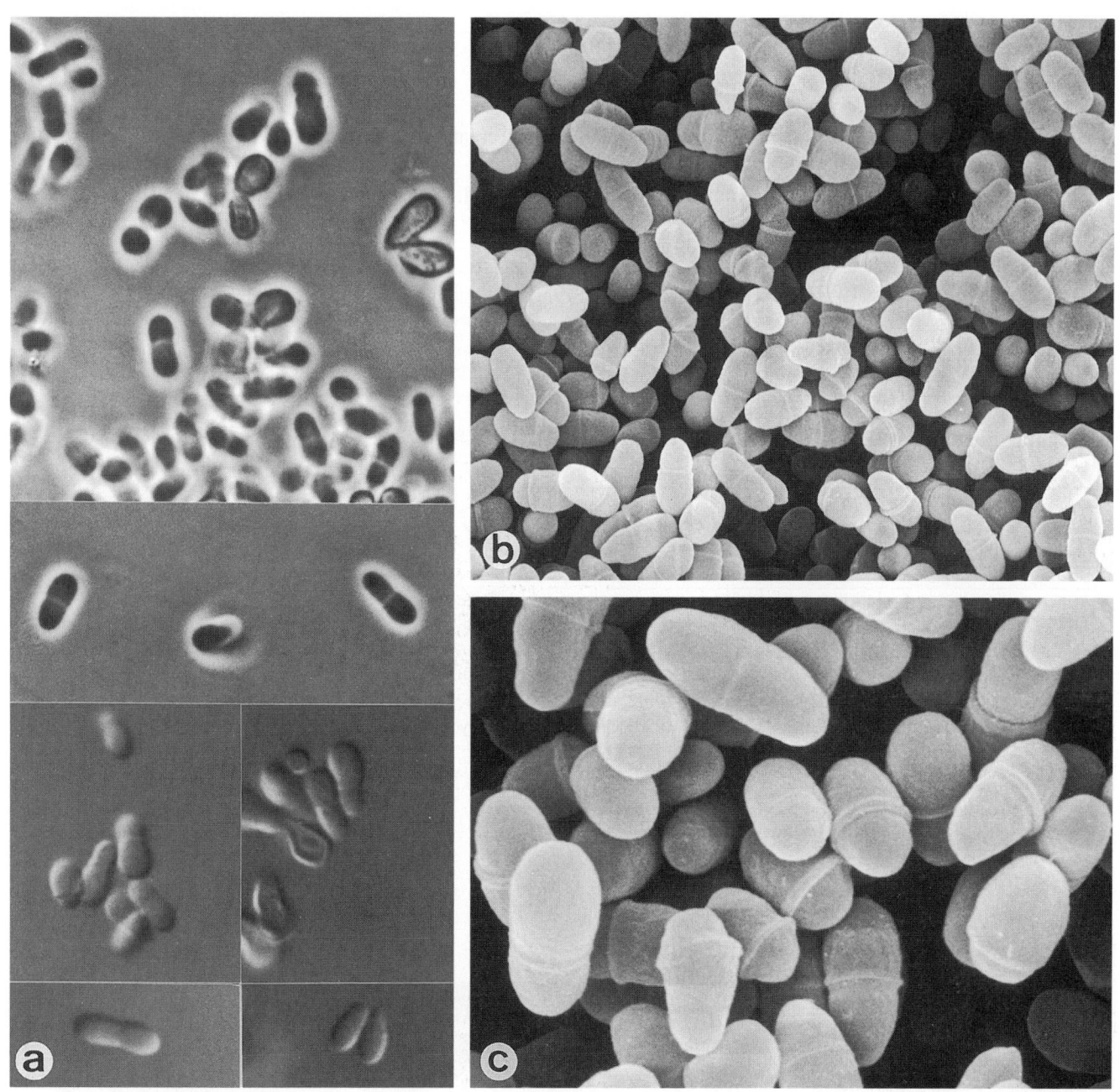

Malassezia pachydermatis, CBS 1879. Budding cells. a. ×1600; b. ×2600; c. ×6000.

Malassezia restricta Guého *et al.*

Colony characteristics. Lipid dependent. Colonies (LNA, 30°C) cream-coloured, smooth or rough, firm, with regular or finely-lobed margin.

Microscopy. Budding percurrent; buds narrower than the mother cell. Yeast cells spherical to ovoidal, 2.5-4.0 × 1.5-2.0 μm.

Physiology.

Catalase	–	Tween 80	–	Esculine	–
SGA	–	Tween 40	–	Serovar	C
37°C	–,w	Tween 20	–		
40°C	–	Cremophor EL	–		

Molecular diagnostics. LSU restriction map based on NCBI AF064026:

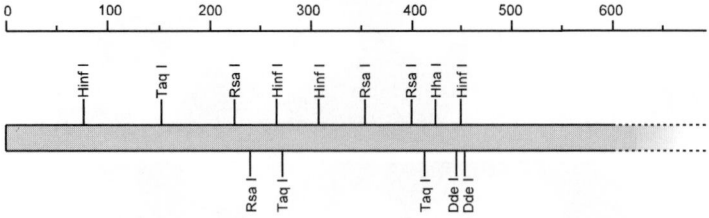

Pathogenicity. BSL-2. Species mostly found on normal human skin.

Reference. Cunningham *et al.* (1990), Guého *et al.* (1996).

Nomenclature. *Malassezia restricta* Guého, Guillot & Midgley, *in* Guého, Midgley & Guillot - Antonie van Leeuwenhoek 69: 349, 1996.

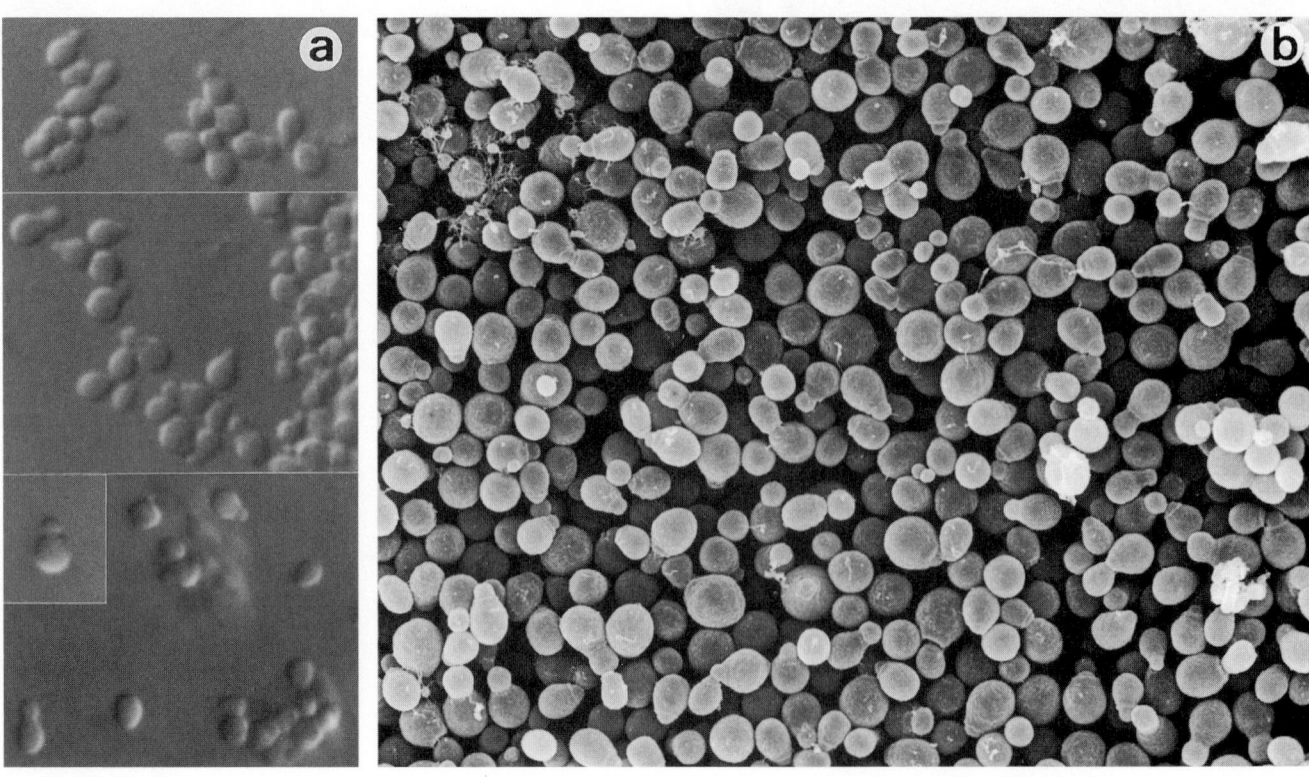

Malassezia restricta, CBS 7966. Percurrent budding cells. a. 1600; b. ×2600.

Malassezia slooffiae Guillot *et al.*

Colony characteristics. Lipid dependent. Colonies (LNA, 30° C) cream-coloured, rough with fine radial grooves.
Microscopy. Budding percurrent; buds as wide as the mother cell. Yeast cells short cylindrical, 1.5-4.0 × 1.0-2.0 μm.
Physiology.

Catalase	+	Tween 80	–	Esculine	–	
SGA	–	Tween 40	+			
37°C	+	Tween 20	+,w			
40°C	+	Cremophor EL	–			

Molecular diagnostics. LSU restriction map based on NCBI AF064028:

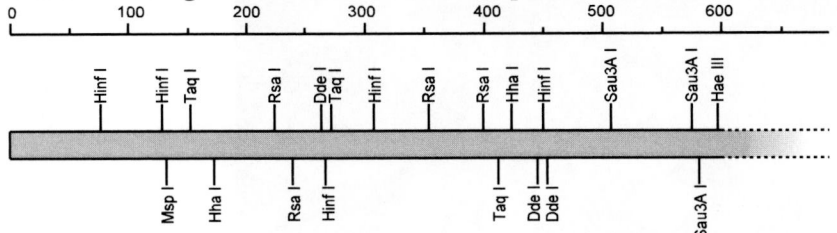

Pathogenicity. BSL-2. Anthropo- and zoophilic species.
Reference. Guého *et al.* (1996).
Nomenclature. *Malassezia slooffiae* Guillot, Midgley & Guého, *in* Guého, Midgley & Guillot - Antonie van Leeuwenhoek 69: 351, 1996.

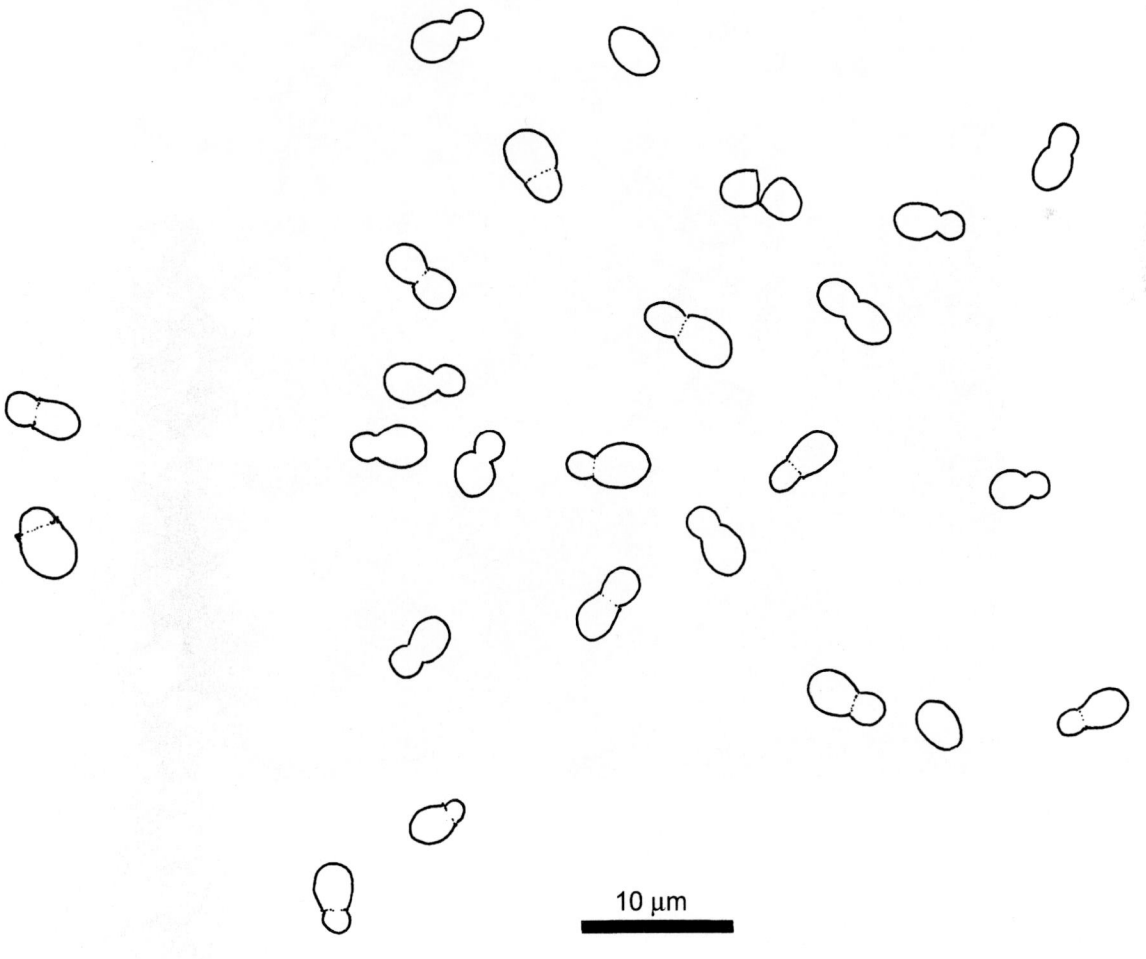

10 μm

Malassezia slooffiae, CBS 7956. Budding cells.

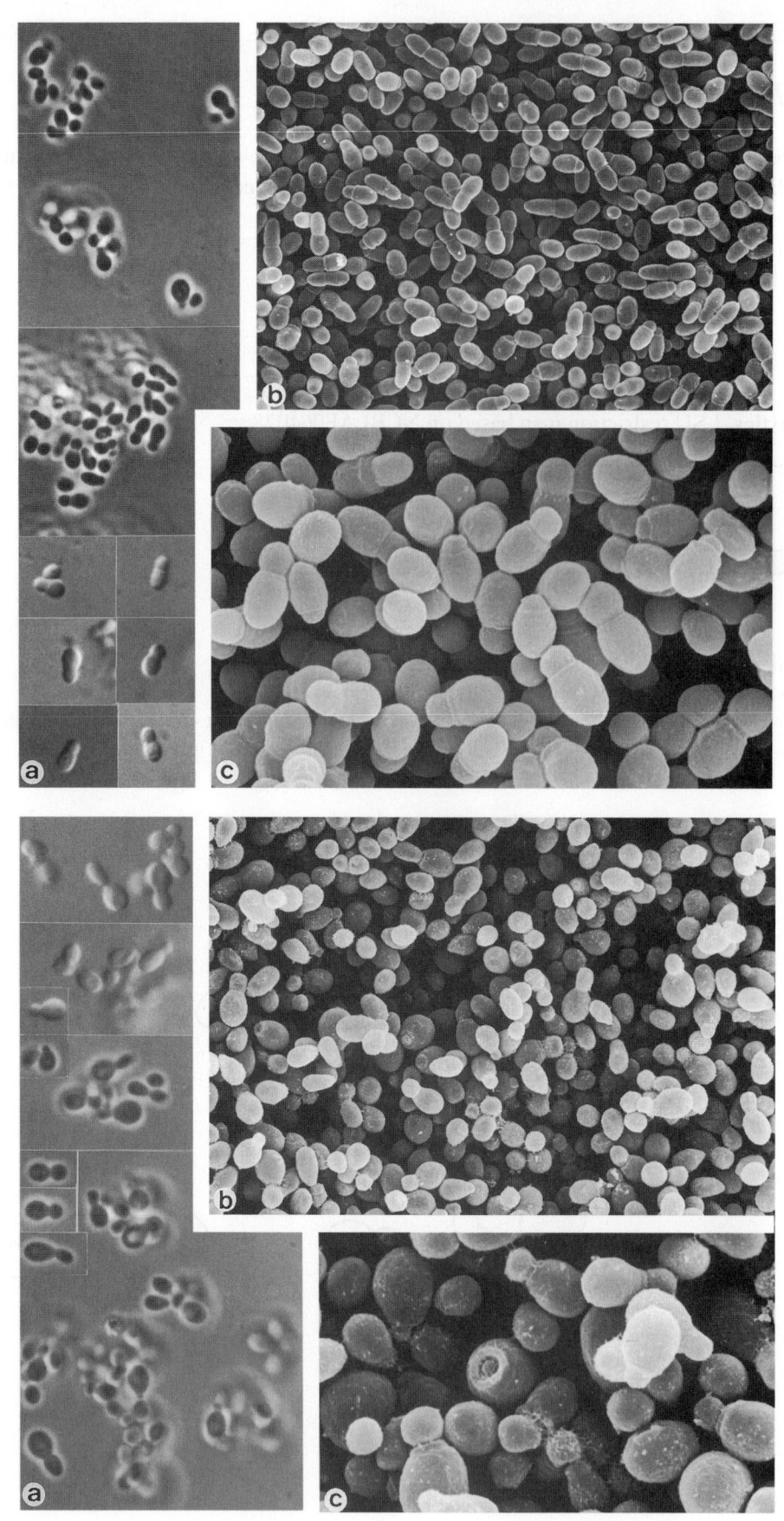

Upper plate: *Malassezia slooffiae,* CBS 7956. Percurrent budding cells. a. ×1600; b. ×2600; c. ×5500.
Lower plate: *Malassezia sympodialis,* CBS 7222. Percurrent budding cells often cohering next to each other. a. ×1600; b. ×2600; c. ×6000.

154

Malassezia sympodialis Simmons & Guého

Colony characteristics. Lipid dependent. Colonies (LNA, 30°C) cream-coloured, flat with a slight central elevation, glistening, smooth.
Microscopy. Budding percurrent, often cohering next to each other; buds narrower than the mother cell. Yeast cells spherical to ovoidal, 2.5-6.0 × 1.5-2.5 µm.
Physiology.

Catalase	+	Tween 80	+	Esculine	+
SGA	–	Tween 40	+	Serovar	A
37°C	+	Tween 20	–		
40°C	+	Cremophor EL	–		

Molecular diagnostics. LSU restriction map based on NCBI AF064024:

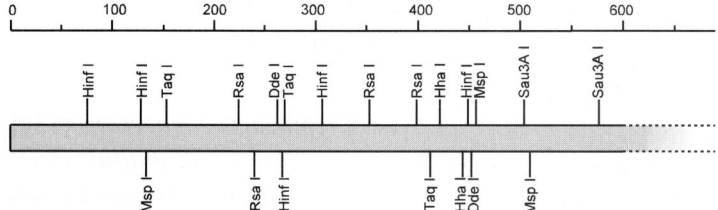

Pathogenicity. BSL-2. The species is found on healthy human skin and may cause pityriasis versicolor. In neonates cephalic pustulosis is noted (Niamba *et al.*, 1998). Chai *et al.* (2000) reported otitis externa.
Reference. Cunningham *et al.* (1990), Guého *et al.* (1996).
Nomenclature. *Malassezia sympodialis* Simmons & Guého - Mycol. Res. 94: 1147, 1990.

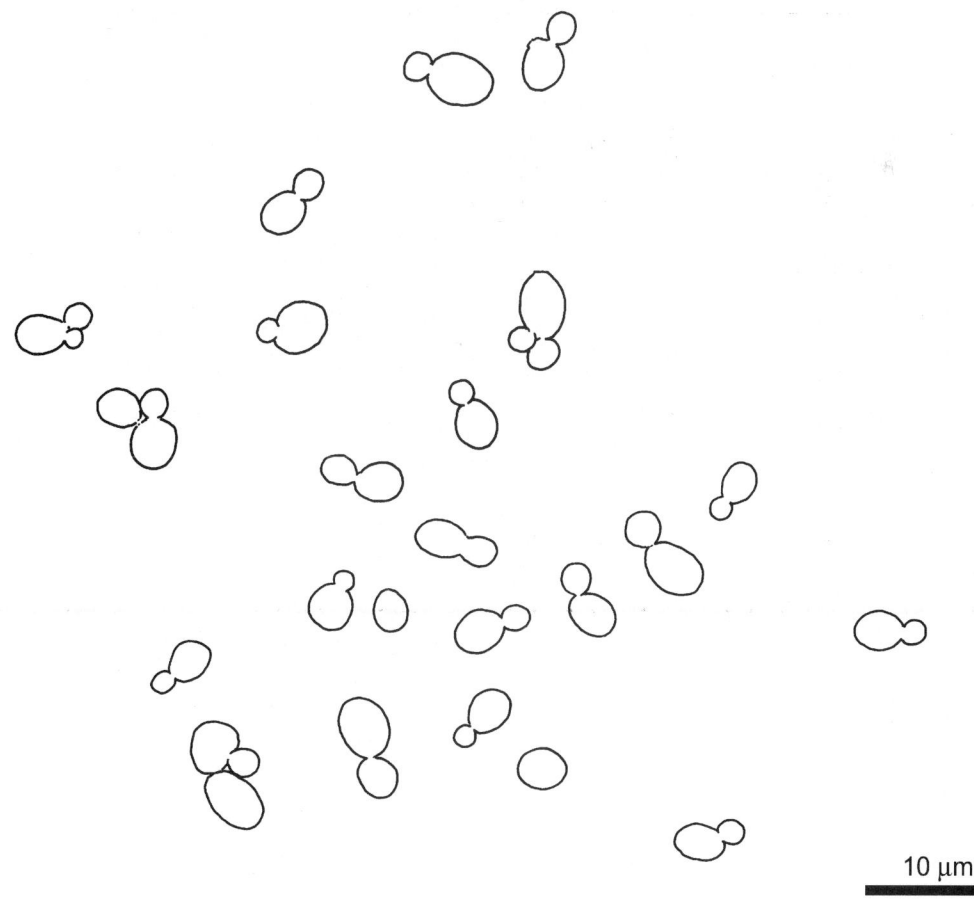

Malassezia sympodialis, CBS 7222. Budding cells.

Basidiomycetous yeasts. Genus: *RHODOTORULA*

Generic description. Colonies pink, reddish or yellowish, slimy or dry. Budding cells spherical to ellipsoidal; bud scars narrow. Ballistoconidia absent. Hyphal elements may be present. *myo*-Inositol not assimilated, fermentation absent, DBB +, urease +, extracellular starch not produced.
Teleomorph. *Rhodosporidium* Banno (*Basidiomycota, Urediniomycetes, Sporidiales: Sporidiobolaceae*).
General remarks. The genus *Rhodotorula* contains numerous saprobes, many of which are psychrophiles in lakes and ocean water or are phyllosphere inhabitants.
Reference. Fell & Statzell-Tallman (1998).

Key to the treated species of Rhodotorula and other yeasts with red colonies:

1a. Nitrate assimilated **→ 2**
1b. Nitrate not assimilated **→ 3**
2a. *meso*-Erythritol assimilated .... ***Cryptococcus macerans* (138)**
2b. *meso*-Erythritol not assimilated .... ***R. glutinis* (157)**
3a. Raffinose assimilated .... ***R. mucilaginosa* (160)**
3b. Raffinose not assimilated **→ 4**
4a. Urease present .... ***R. minuta* (159)**
4b. Urease absent .... ***Candida albicans* (184)**

Table 20. Variable physiological characters among treated *Rhodotorula* species, compared with a red *Cryptococcus* species, *C. macerans* (p. 138), and red strains of *Candida albicans* (p. 184).

	glutinis	*minuta*	*mucilaginosa*	*macerans*	*C. albicans*
Maltose	+	−	v	+	+
Raffinose	v	−	+	+	−
N-Acetyl-D-glucosamine	−	+	−	−	+
meso-Erythritol	−	−	−	+	−
Citrate	v	−	v	+	+
myo-Inositol	−	−	−	+	−
Hexadecane	+	−	−	−	s
Nitrate	+	−	−	+	−
D-Glucuronate	−	+	v	+	−
Gelatin liquefaction	−	−	−	+	nd
Urease	+	+	+	+	−

Rhodotorula glutinis (Fres.) Harrison

Colony characteristics. Colonies (MEA) coral red to salmon or slightly orange; surface smooth to wrinkled, often with fine transverse striations, glossy, later dull. Texture slimy to pasty or slightly tough.

Microscopy. Budding cells ellipsoidal, often with rudimentary pseudomycelium. Budding percurrent on rich media, multilateral with increasing coherence on poor media. Budding cells ellipsoidal.

Teleomorphs. *Rhodosporium diobovatum* Newell & Hunter, *R. sphaerocarpum* Newell & Fell, *R. toruloides* Banno (*Basidiomycota, Urediniomycetes, Sporidiales: Sporidiobolaceae*).

Heterothallic. Mycelium with clamps formed after mating, bearing dark brown, thick-walled, variously shaped teliospores.

Physiology.

Fermentation:	–	L-Arabinose	v	DL-Lactate	v
		D-Arabinose	v	Succinate	+
Growth:		D-Ribose	v	Citrate	v
Glucose	+	L-Rhamnose	v	*myo*-Inositol	–
Galactose	v	D-Glucosamine (C)	–	Hexadecane	+
L-Sorbose	v	*N*-Acetyl-D-glucosamine	–	Nitrate	+
Sucrose	+	Methanol	–	Vitamin-free	v
Maltose	+	Ethanol	v	2-Keto-D-gluconate	v
Cellobiose	v	Glycerol	v	5-Keto-D-gluconate	–
α,α-Trehalose	+	*meso*-Erythritol	–	D-Glucarate	–
Lactose	–	Ribitol	v	D-Glucuronate	–
Melibiose	–	Galactitol	v	10% NaCl/5% glucose	w
Raffinose	v	D-Mannitol	v	Urease	+
Melezitose	+	D-Glucitol	v	Starch formation	–
Inulin	–	α-Methyl-D-glucoside	v	Gelatin liquefaction	–
Soluble starch	–	Salicin	+,w	Growth at 30°C	+
D-Xylose	v	D-Gluconate	+	Growth at 37°C	v

Differential diagnosis. Species signature: lactose –, melibiose –, D-glucosamine (C) –, *meso*-erythritol –, *myo*-inositol –, starch formation –, nitrite +, w/o biotin +.

Molecular diagnostics. SSU and LSU restriction maps based on NCBI X69853 and AF070430:

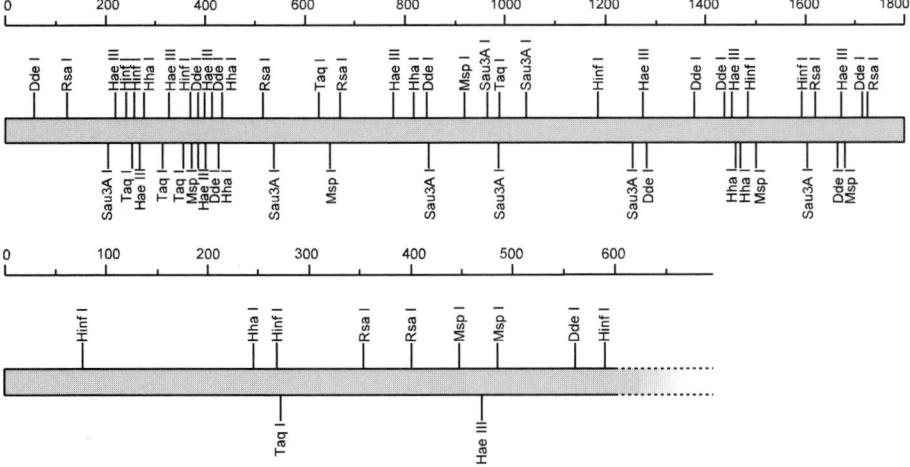

Pathogenicity. BSL-1. The species is a common saprophyte on various substrates, and its pathogenicity is generally estimated as low. Fungemia in patients with compromised innate immunity does occur, although clinical symptoms may remain absent (Fanci *et al.*, 1997). Sepsis in relation to the use of indwelling catheters has repeatedly been reported (Kiehn *et al.*, 1992; Braun & Kaufmann, 1992). Bertoli *et al.* (1992), Guerra *et al.* (1992) and Casolari *et al.* (1992) described cases of keratitis and Block & McCormick (1994) a case of dacryoadenitis.

Reference. Banno (1967).

Nomenclature. *Cryptococcus glutinus* Fresenius - Beitr. Mykol. 2: 77, 1852 ≡ *Rhodotorula glutinis* (Fresenius) Harrison - Trans. R. Soc. Can., Sect. 5, 22: 187, 1928 ≡ *Torulopsis glutinis* (Fresenius) C.W. Dodge - Med. Mycol. p. 351, 1935.

Torulopsis bronchialis Ciferri & Redaelli - Atti Ist. Bot. R. Crittog. Lab. Univ. Pavia, Ser. 3, 2: 245, 1925 ≡ *Cryptococcus bronchialis* (Ciferri & Redaelli) Nannizzi - Tratt. Micopat. Um. 4: 304, 1934 ≡ *Rhodotorula bronchialis* (Ciferri & Redaelli) Lodder - Anaskosp. Hefen 1: 91, 1934.
    Rhodosporidium toruloides Banno - J. Gen. Appl. Microbiol. 13: 193, 1967.
    Rhodosporidium diobovatum Newell & Hunter - J. Bact. 104: 504, 1970.
    Rhodosporidium sphaerocarpum Newell & Fell - Mycologia 62: 276, 1970.

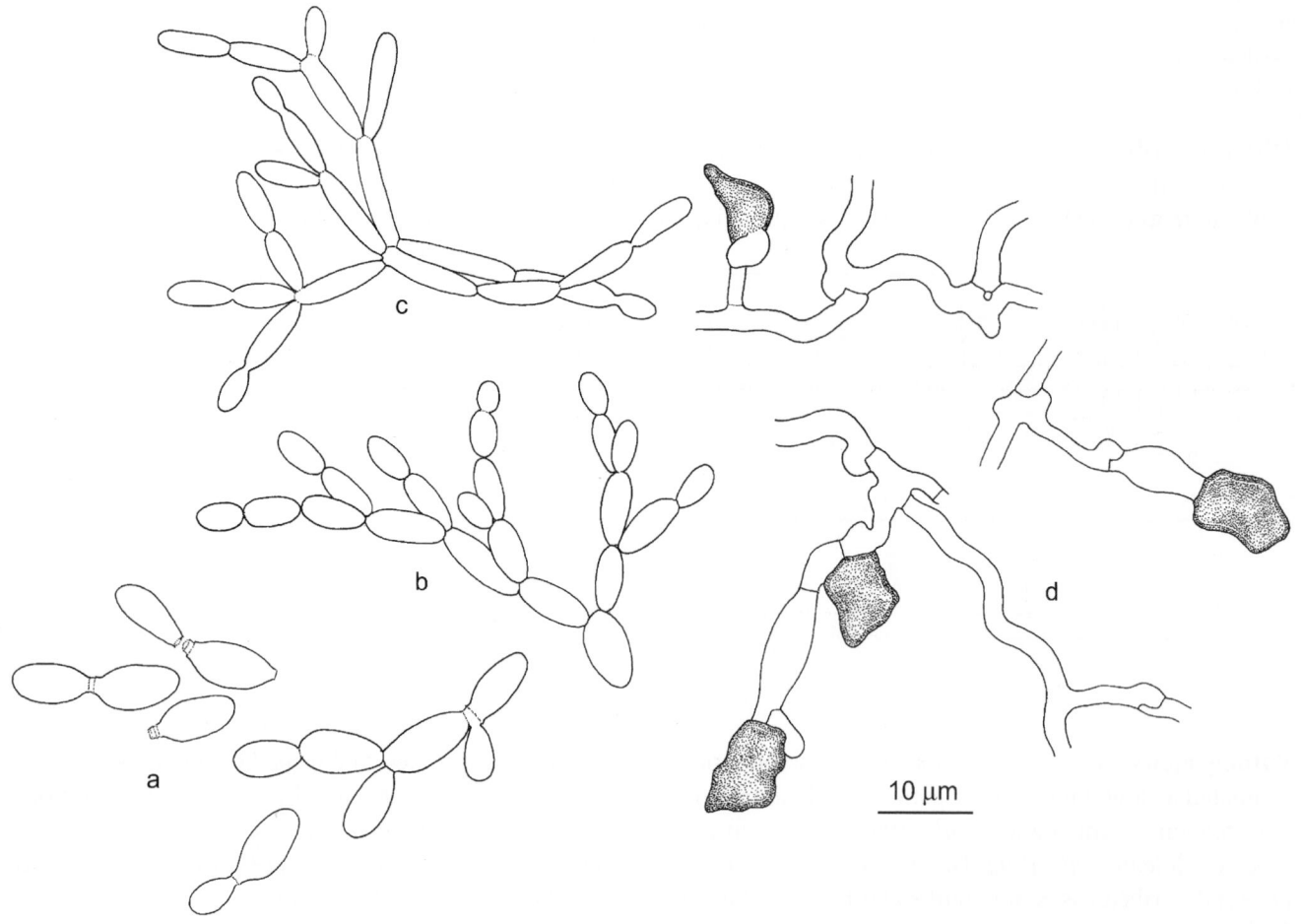

Rhodotorula glutinis (*Rhodosporium toruloides*), CBS 14 ×349. a. Percurrent budding with broadly ellipsoidal cells on YPGA; b. multilateral budding; c. pseudomycelium formation on PCA; d. mycelium with clamp connections and teliospores.

Rhodotorula minuta (Saito) Harrison

Colony characteristics. Colonies (MEA) pink, smooth, glistening, soft; margin straight.
Microscopy. Budding cells (RA) spherical to broadly ellipsoidal, 3.5-6.5 × 2.5-4.5 µm. Pseudomycelium absent.
Physiology.

Fermentation:	–	L-Arabinose	+	DL-Lactate	v
		D-Arabinose	s	Succinate	+
Growth:		D-Ribose	v	Citrate	–
Glucose	+	L-Rhamnose	–	*myo*-Inositol	–
Galactose	v	D-Glucosamine (C)	–	Hexadecane	–
L-Sorbose	v	*N*-Acetyl-D-glucosamine	+	Nitrate	–
Sucrose	+	Methanol	–	Vitamin-free	–
Maltose	–	Ethanol	v	2-Keto-D-gluconate	+
Cellobiose	v	Glycerol	+	5-Keto-D-gluconate	+
α,α-Trehalose	+	*meso*-Erythritol	–	D-Glucarate	–
Lactose	v	Ribitol	v	D-Glucuronate	+
Melibiose	–	Galactitol	–	10% NaC1/5% glucose	–
Raffinose	–	D-Mannitol	v	Urease	+
Melezitose	+	D-Glucitol	v	Starch formation	–
Inulin	–	α-Methyl-D-glucoside	–	Gelatin liquefaction	–
Soluble starch	–	Salicin	v	Growth at 30°C	v
D-Xylose	+	D-Gluconate	+	Growth at 37°C	v

Differential diagnosis. Species signature: maltose –, melibiose –, soluble starch –, L-rhamnose –, glycerol +, *meso*-erythritol –, galactitol –, *myo*-inositol –, nitrate –, starch formation –, nitrite –, w/o pyridoxine +.
Molecular diagnostics. SSU restriction map based on NCBI D45367:

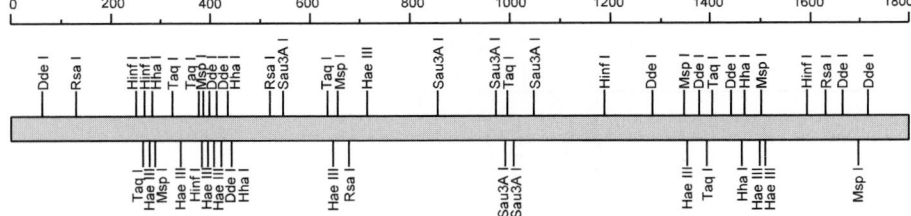

Pathogenicity. BSL-1. A case of postoperative endophthalmitis was reported by Gregory & Haller (1992) and isolation from bronchoscopy specimens by Hagan *et al.* (1994). Goldani *et al.* (1995) described a fungemia after central venous catheter infection in an AIDS patient. Ciferri & Redaelli (1925) reported the species from a systemic mycosis in a white rat. Rusthoven *et al.* (1984) described a systemic infection in a leukemic patient.

Nomenclature. *Torula minuta* Saito - Jpn. J. Bot. 1: 48, 1922 ≡ *Torulopsis minuta* (Saito) Ciferri & Redaelli - Atti Ist. Bot. Lab. Crittogam. Univ. Pavia, Ser. 1, 2: 261, 1925 ≡ *Rhodotorula minuta* (Saito) Harrison - Trans. R. Soc. Can., Sect. 5, 22: 196, 1928.

Mycotorula muris Ciferri & Redaelli - Atti Ist. Bot. Lab. Crittogam. Univ. Pavia, Ser. 3, 2 : 245, 1925 ≡ *Proteomyces muris* (Ciferri & Redaelli) C.W. Dodge - Med. Mycol. p. 208, 1935.

Rhodotorula pallida Lodder - Anaskosp. Hefen 1: 97, 1934.

Rhodotorula marina Phaff, Mrak & Williams - Mycologia 44: 436, 1952.

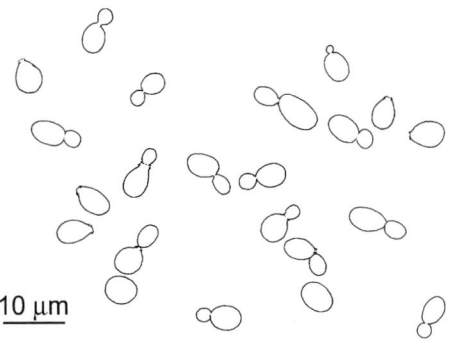

Rhodotorula minuta, CBS 319. Budding cells.

Rhodotorula mucilaginosa (Jörgensen) Harrison

Colony characteristics. Colonies (MEA) coral red to pink, glistening, smooth or rough, soft, mucous.
Microscopy. Budding cells (RA) ellipsoidal, 2.5-6.5 × 2.0-5.5 µm, unipolar; rudimentary pseudomycelium may be present.
Physiology.

Fermentation:	–	L-Arabinose	v	DL-Lactate	v
		D-Arabinose	v	Succinate	+
Growth:		D-Ribose	v	Citrate	v
Glucose	+	L-Rhamnose	v	*myo*-Inositol	–
Galactose	v	D-Glucosamine (C)	v	Hexadecane	–
L-Sorbose	v	*N*-Acetyl-D-glucosamine	–	Nitrate	–
Sucrose	+	Methanol	–	Vitamin-free	v
Maltose	v	Ethanol	v	2-Keto-D-gluconate	–
Cellobiose	v	Glycerol	v	5-Keto-D-gluconate	–
α,α-Trehalose	+	*meso*-Erythritol	–	D-Glucarate	–
Lactose	–	Ribitol	v	D-Glucuronate	v
Melibiose	–	Galactitol	v	10% NaCl/5% glucose	v
Raffinose	+	D-Mannitol	v	Urease	+
Melezitose	v	D-Glucitol	v	Starch formation	–
Inulin	–	α-Methyl-D-glucoside	v	Gelatin liquefaction	–
Soluble starch	–	Salicin	v	Growth at 37°C	+
D-Xylose	+	D-Gluconate	+		

Molecular diagnostics. SSU and LSU restriction maps based on NCBI X84326 and AF070432:

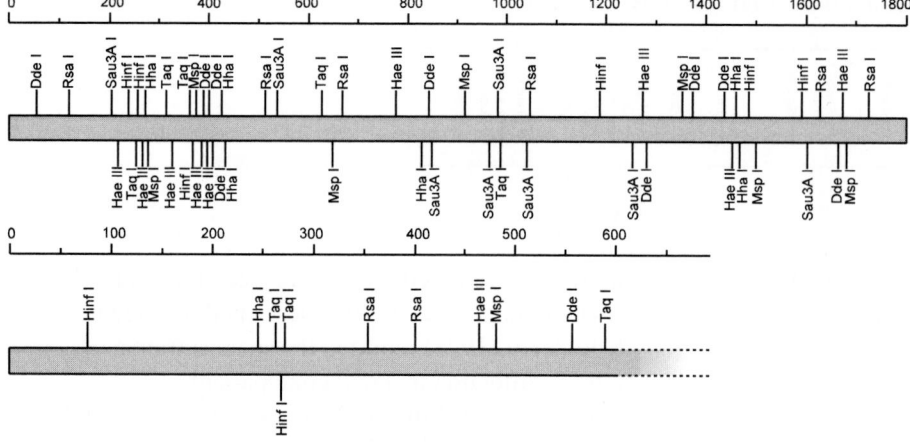

Pathogenicity. BSL-1. In older literature the species was repeatedly encountered in clinical specimens (see nomenclature below), but their etiologic role is uncertain. Recent reports are those of Benevent *et al.* (1985) from a CAPD peritonitis, Muralidhar & Sulthana (1995) from a chronic dacryocystitis and Lui *et al.* (1998) and Kiraz *et al.* (2000) from fungemia in AIDS resp. Hodgkin's patients. An infection in a neutropenic patient was listed by Nucci *et al.* (1995). Gyaurgieva *et al.* (1996) reported a meningitis in a HIV-positive patient.
Reference. Fell & Statzell-Tallman (1998).
Antifungal susceptibility.

Antifungal	MICs range	MIC 90	Strains	Reference
AMB	0.5-1	0.5	5	Espinel-Ingroff (1998)
FLZ	0.5->64	>64	5	Espinel-Ingroff (1998)
ITZ	0.25-4	2	5	Espinel-Ingroff (1998)
VCZ	0.25-4	4	5	Espinel-Ingroff (1998)

Nomenclature. *Saccharomyces ruber* Demme - Ann. Microgr. 2: 555, 1889 ≡ *Torulopsis rubra* (Demme) de Almeida - Anais Fac. Med. Univ. S. Paulo 9: 76, 1933 ≡ *Rhodotorula rubra* (Demme) Lodder - Anaskosp. Hefen 1: 69, 1934.

Torula mucilaginosa Jörgensen - Mikroorg. Gärungsind., ed. 5, p. 402, 1909 ≡ *Rhodotorula mucilaginosa* (Jörgensen) Harrison - Trans. R. Soc. Can., Sect. 5, 22: 191, 1928.

Cryptococcus mena Fontoynont & Boucher - Annls Derm. Syph., Sér. 6, 4: 213, 1923 ≡ *Torulopsis mena* (Fontoynont & Boucher) C.W. Dodge - Med. Mycol. p. 349, 1935.

Blastodendron carbonei Ciferri & Redaelli - Atti Ist. Bot. Univ. Lab. Crittogam. Pavia, Ser. 3, 2: 147, 1925 ≡ *Torulopsis mucilaginosa* (Jörgensen) Ciferri & Redaelli var. *carbonei* (Ciferri & Redaelli) C.W. Dodge - Med. Mycol. p. 354, 1935.

Mycotorula pulmonalis Ciferri & Redaelli - Atti Ist. Bot. Univ. Lab. Crittogam Pavia, Ser. 3,2: 205, 1925 ≡ *Geotrichum pulmonale* (Ciferri & Redaelli) C.W. Dodge - Med. Mycol. p. 217, 1935.

Torulopsis sanniei Ciferri & Redaelli, *in* Redaelli - Micet. Assoc. Microb. Tuberc. Polmon. Cavit. p. 61, 1925.

Cryptococcus pararoseus Castellani - Arch. Derm. Syph. 16: 402, 1927 ≡ *Rhodotorula mucilaginosa* (Jörgensen) Harrison var. *pararosea* (Castellani) Lodder - Anaskosp. Hefen 1: 102, 1934 ≡ *Torulopsis macilaginosa* (Jörgensen) Ciferri & Redaelli var. *pararosea* (Castellani) C.W. Dodge - Med. Mycol. p. 353, 1935.

Cryptococcus rubrorugosus Castellani - Arch. Derm. Syph. 16: 403, 1927 ≡ *Rhodotorula mucilaginosa* (Jörgensen) Harrison var. *rubrorugosa* (Castellani) Lodder - Anaskosp. Hefen 1: 105, 1934 ≡ *Rhodotorula rubrorugosa* (Castellani) Krasil'nikov - Opred. Niszh. Rast. 3: 129-136, 1954.

Cryptococcus corallinus Sartory, R. Sartory, Hufschmitt & Meyer - C.R. Soc. Biol. 82: 1316, 1930.

Cryptococcus radiatus Sartory, R. Sartory & Meyer - C.R. Soc. Biol. 106: 597, 1931.

Rhodotorula mucilaginosa (Jörgensen) Harrison var. *plicata* Lodder - Anaskosp. Hefen 1: 109, 1934.

Torulopsis aurantia Zach, *in* Wolfram & Zach - Arch. Derm. Syph. 170: 689, 1934.

Rhodotorula pilimanae Hedrick & Burke - Mycopath. Mycol. Appl. 6: 94, 1951.

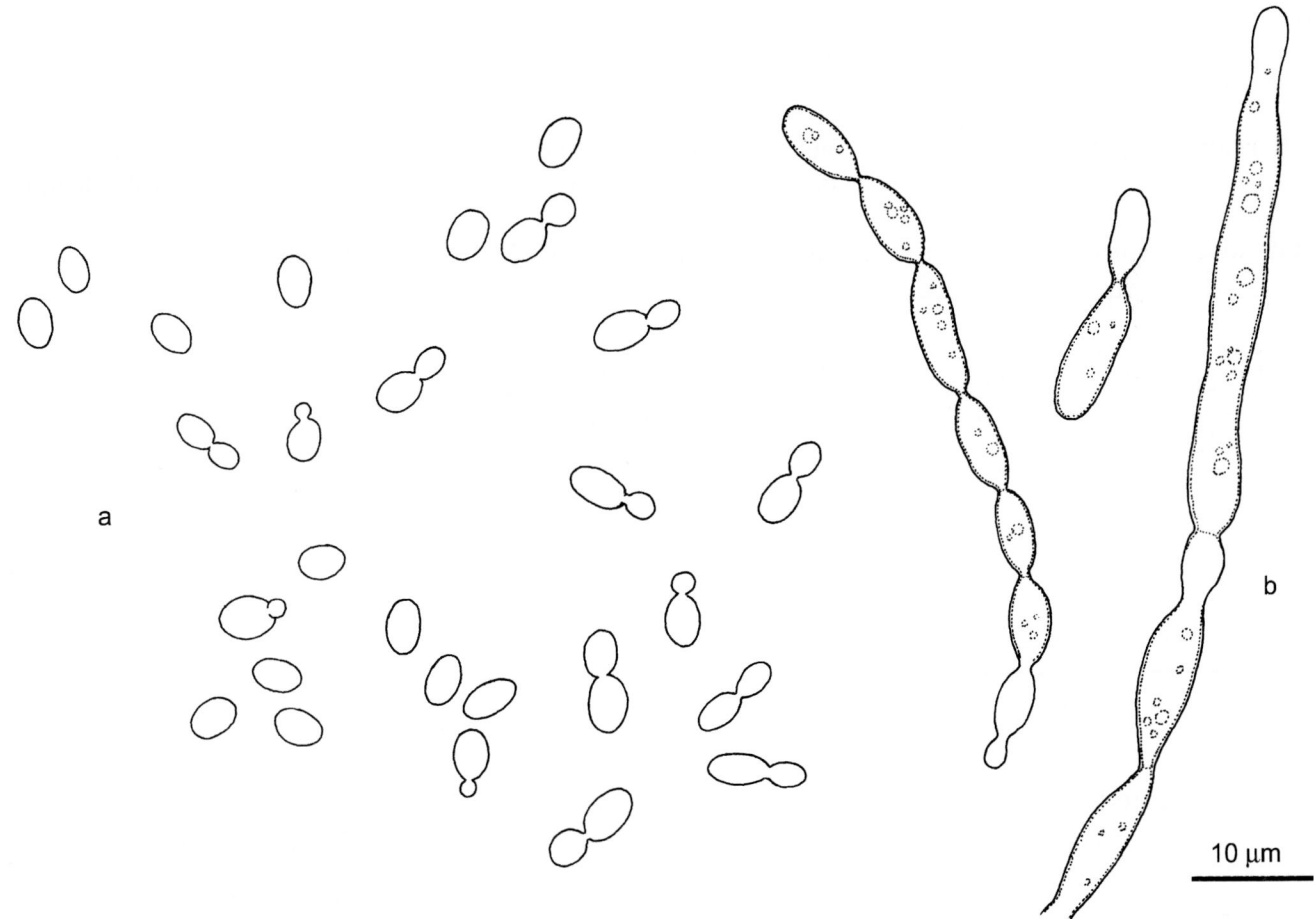

Rhodotorula mucilaginosa, Kerkmann. a. Budding cells; b. pseudomycelium.

Basidiomycetous yeasts. Genus: *SPOROBOLOMYCES*

Sporobolomyces salmonicolor (Fischer & Brebeck) Kluyver & van Niel

Colony characteristics. Colonies (MEA) butyrous, soft, salmon pink.
Microscopy. Budding cells variable, ellipsoidal to subcylindrical, 8-25 × 2-5.5 μm; budding percurrent. Pseudohyphae present. Ballistoconidia on large sterigmata, ellipsoidal to allantoid, 6-18 × 2.5-7.0 μm, unilaterally flattened.
Teleomorph. *Sporidiobolus johnsonii* Nyland (*Basidiomycota, Urediniomycetes, Sporidiales: Sporidiobolaceae*). Dikaryotic hyphae formed after mating, with clamp connections. Teliospores (sub)spherical, 9-15 μm diam, thick-walled, brown. Homo- or heterothallic.
Physiology.

Fermentation:	–	D-Xylose	s	Salicin	+,w
		L-Arabinose	v	D-Gluconate	+
Growth:		D-Arabinose	+	DL-Lactate	v
Glucose	+	D-Ribose	s	Succinate	+
Galactose	v	L-Rhamnose	–	Citrate	v
L-Sorbose	s	D-Glucosamine (C)	–	*myo*-Inositol	–
Sucrose	+	N-Acetyl-D-glucosamine	–	Hexadecane	s
Maltose	+	Methanol	–	Nitrate	+
Cellobiose	+	Ethanol	+	Vitamin-free	+
α,α-Trehalose	+	Glycerol	+	Urease	+
Lactose	–	*meso*-Erythritol	–	Starch formation	–,w
Melibiose	–	Ribitol	s	Gelatin liquefaction	–
Raffinose	–	Galactitol	–	Growth at 30°C	+
Melezitose	+	D-Mannitol	+	Growth at 37°C	v
Inulin	–	D-Glucitol	+		
Soluble starch	v	α-Methyl-D-glucoside	v		

Differential diagnosis. Species signature: inulin –, galactitol –, D-glucitol +, *myo*-inositol –, urease +, growth at 30°C +, D-glucono-δ-lactone +, 2-keto-D-gluconate –, D-glucarate –, cadaverine –. Physiologically indistinguishable from *S. roseus*.

Molecular diagnostics. SSU and LSU restriction maps based on NCBI L22261 and AF070435:

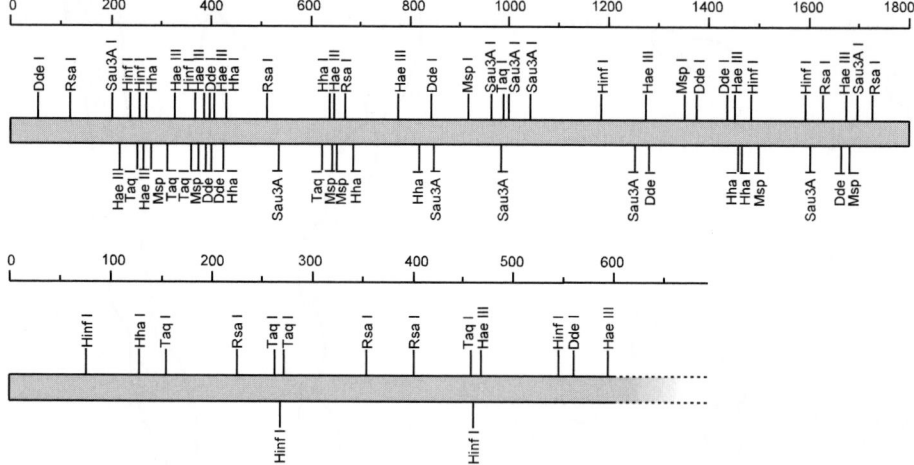

Pathogenicity. BSL-1. The species is a phyllosphere fungus. Morris *et al.* (1991) and Plzas *et al.* (1994) reported AIDS-related infections, Bergman & Kauffman (1984) a dermatitis and Misra & Randhawa (1976) a cerebral infection.
Reference. Boekhout (1991).

Antifungal susceptibility.

Antifungal	MICs range	Strains	Reference
AMB	0.5-1	3	Espinel-Ingroff (1998)
FLZ	8->64	3	Espinel-Ingroff (1998)
ITZ	1-2	3	Espinel-Ingroff (1998)
VCZ	0.25-4	3	Espinel-Ingroff (1998)

Nomenclature. *Blastoderma salmonicolor* Fischer & Brebeck - Morph. Biol. Syst. Kahm-Pilze p. 47, 1894 ≡ *Sporobolomyces salmonicolor* (Fischer & Brebeck) Kluyver & van Niel - Centbl. Bakt. ParasitKde, Abt. 2, 63: 19, 1924.

Sporobolomyces holsaticus Windisch - Arch. Mikrobiol. 14: 289, 1948 (invalid) ≡ *Sporobolomyces holsaticus* Windisch ex Yarrow & Fell - Mycotaxon 12: 253, 1980.

Sporidiobolus johnsonii Nyland - Mycologia 41: 687, 1949.

Sporobolomyces salmonicolor (Fischer & Brebeck) var. *fischeri* V.C. Misra & Randhawa - Arch. Mikrobiol. 108: 141, 1976.

Sporidiobolus salmonicolor Fell & Tallman - Curr. Microbiol. 5: 80, 1981.

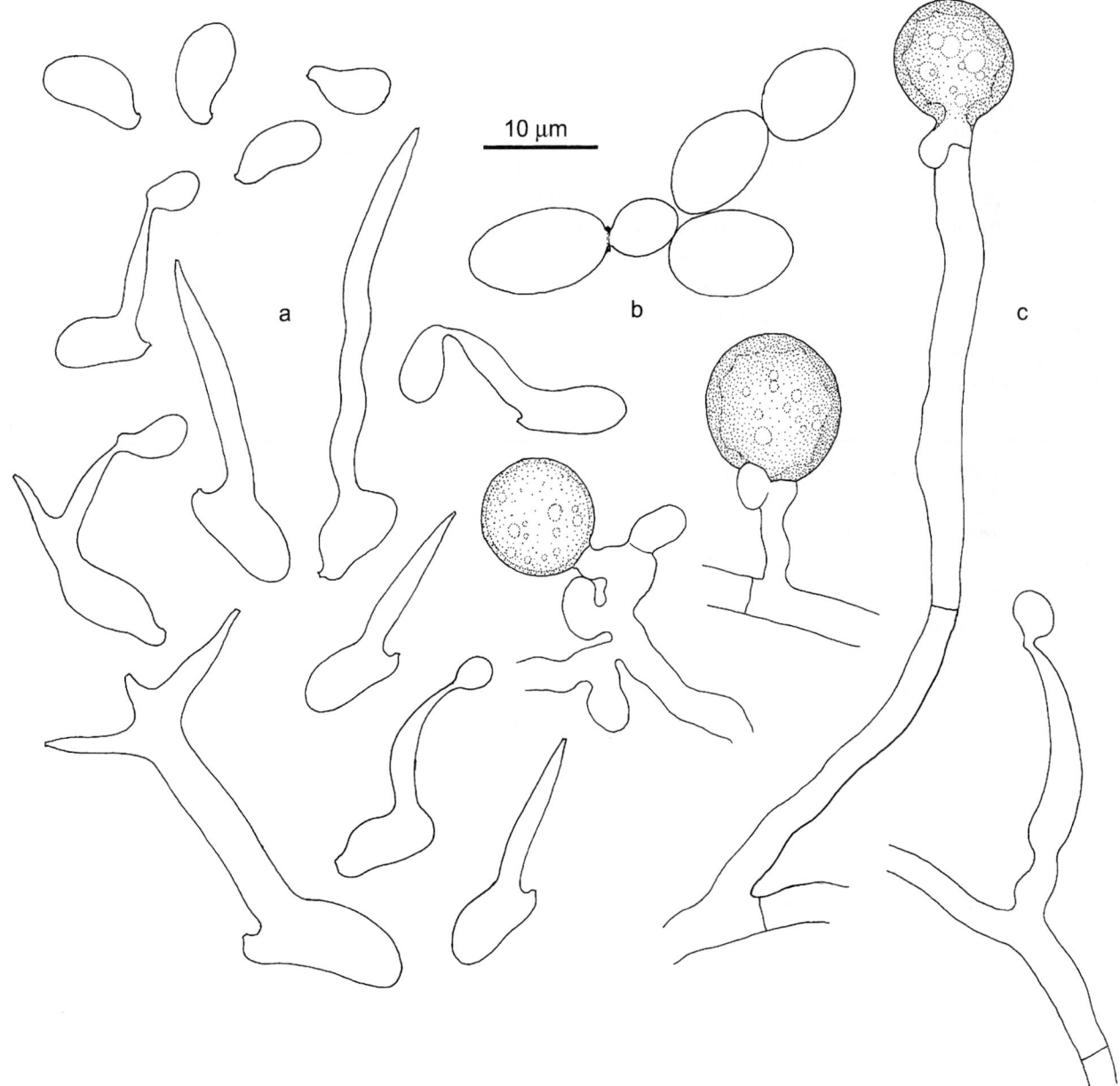

Sporobolomyces salmonicolor (*Sporidiobolus johnsonii*), CBS 5470. a. Ballistoconidia, producing secondary sterigmata; b. budding cells; c. teliospores.

Basidiomycetous yeasts. Genus: *TRICHOSPORON*

Generic description. Colonies initially yeast-like, later becoming dry. Budding absent or present. Arthroconidia abundant; lateral conidia may be present. Appressoria or meristematic cells often formed. Fermentation absent; many carbon sources assimilated. Nitrate –, urease +, DBB +, cell walls multilamellar; septa with rudimentary dolipores.
Differential diagnosis. *Cryptococcus* (p. 132) differs by presence of capsules and absence of arthroconidia. *Geotrichum* species (p. 227) are urease negative and assimilate only a limited number of carbon sources.
Molecular diagnostics. A generic primer system based on 18S rDNA genes was developed by Sugita *et al.* (1998b). Nested PCR based on 26S rDNA sequences for the diagnosis of some species was developed by Nagai *et al.* (1999). Molecular phylogeny showed the genus to be related to *Cryptococcus* (Guého *et al.*, 1992b, 1993; Sugita & Nakase, 1998).
General remarks. The genus is treated in most older handbooks with only a limited number of variable species. However, molecular studies (Guého *et al.*, 1992b) have shown that many microspecies can be distinguished, each occupying narrowly circumscribed ecological niches. Some species are somewhat psychrophilic living in soil, or are found in association with animals. Five species are of clinical significance. Localized systemic as well as disseminated infections are known in patients with acute leukemia; agents are mostly *T. asahii* (p. 168) or *T. mucoides* (p. 175). Other species are particularly involved in superficial mycoses. For a review of the clinical spectrum of *Trichosporon*, see Anaissie *et al.* (1989b), Herbrecht *et al.* (1993) and Vartivarian *et al.* (1993). The modern taxonomy of the genus has rendered a vast amount of earlier literature almost unusable, because it is unclear which species were dealt with. Among recent reports under the name '*Trichosporon beigelii*' or '*T. cutaneum*', there are several cases of septicaemia (Grauer *et al.*, 1994; Miro *et al.*, 1994; Still *et al.*, 1994; Vicek *et al.*, 1995; Yoss *et al.*, 1997; Hsu *et al.*, 1998), chronic meningitis (Mathews & Prabhakar, 1995), disseminated infection in AIDS (Lascaux *et al.*, 1998) and a bloodstream infection in burn patients (Hajjeh & Blumberg, 1995). Systemic cases in leukemic patients were presented by Hung *et al.* (1995) and Kataoka-Nishimura *et al.* (1998). Catheter-related infections in surgical patients were reported by Woodhouse *et al.* (1995) and Wang & Lin (1999). A review of infections in patients with hematological malignancies was presented by Tashiro *et al.* (1995). An infection through an implanted medical device in a captive monkey was reported by Dannemiller *et al.* (1995).
References. Guého *et al.* (1992ab, 1994, 1998), Douchet *et al.* (1994), Sugita *et al.* (1994, 1998).

Key to the treated species of Trichosporon:

1a.	Growth with melibiose ➞ **2**
1b.	No growth with melibiose ➞ **3**
2a.	Tolerant to cycloheximide . . . . *T. mucoides* (173)
2b.	Not tolerant to cycloheximide . . . . *T. cutaneum* (170)
3a.	Growth with *myo*-inositol, no growth with L-arabinose . . . . *T. inkin* (172)
3b.	No growth with *myo*-inositol, growth with L-arabinose ➞ **4**
4a.	Colony with very slow growth; thallus consisting of clumps of meristematic cells . . . . *Fissuricella filamenta,* see *T. asteroides* (168)
4b.	Colonies and microscopy otherwise ➞ **5**
5a.	Appressoria present in slide cultures . . . . *T. ovoides* (175)
5b.	Appressoria absent in slide cultures ➞ **6**
6a.	Arthroconidia barrel-shaped; thallus not meristematic . . . . *T. asahii* (166)
6b.	Arthroconidia elongate, or thallus meristematic . . . . *T. asteroides* (168)

Table 21. Summary of key characters of treated species of Trichosporon. Physiological parameters based on testing with ID 32C.

	asahii	asteroides	cutaneum	inkin	mucoides	ovoides
L-Arabinose	+	+	+	−	+	v
Sorbitol	−	−	+	−	+	−
Melibiose	−	−	+	−	+	−
myo-Inositol	−	−	+	+	+	−
37°C	+	v	−	+	+	v
0.1% Cycloheximide	+	v	−	v	+	+
Appressoria	−	−	−	+	−	+

Trichosporon asahii Akagi ex Sugita *et al.*

Colony characteristics. Colonies (SGA) moderately expanding, dry, pustular with white, farinose covering and broad, deeply fissured marginal zone.
Microscopy. Budding cells and lateral conidia absent. Arthroconidia barrel-shaped. Appressoria absent.
Physiology.

Fermentation:	−	L-Arabinose	v	DL-Lactate	v
		D-Arabinose	+	Succinate	+
Growth:		D-Ribose	+	Citrate	+
Glucose	+	L-Rhamnose	+	*myo*-Inositol	v
Galactose	+	D-Glucosamine (C)	+	Nitrate	−
L-Sorbose	v	N-Acetyl-D-glucosamine	+	Vitamin-free	−
Sucrose	v	Methanol	−	2-Keto-D-gluconate	+
Maltose	+	Ethanol	+	5-Keto-D-gluconate	+
Cellobiose	+	Glycerol	v	D-Glucuronate	+
α,α-Trehalose	v	*meso*-Erythritol	+	10% NaC1/5% glucose	+
Lactose	+	Ribitol	v	0.01% Cycloheximide	+
Melibiose	−	Galactitol	−	0.1% Cycloheximide	v
Raffinose	−	D-Mannitol	v	Urease	+
Melezitose	v	D-Glucitol	v	Starch formation	+
Inulin	−	α-Methyl-D-glucoside	+	Growth at 37°C	+
Soluble starch	v	Salicin	+	Growth at 42°C	−
D-Xylose	v	D-Gluconate	+		

Differential diagnosis. The species assimilates L-arabinose but not melibiose; growth at 37°C.
Pathogenicity. BSL-2. Most confirmed cases showed hematogenous dissemination in patients with impaired innate immunity (Guého, 1994a; Grauer *et al.*, 1994; Sugita *et al.*, 1995; Itoh *et al.*, 1996). The species has also been reported from skin lesions. It may cause white piedra on furry mammals.
References. Herbrecht *et al.* (1993), Guého *et al.* (1994b).

Antifungal susceptibility.

Antifungal	Mean MICs	Strains	Reference
AMB	0.06	6	Guého *et al.* (1994)
5FC	22.2	6	Guého *et al.* (1994)
FLZ	3	6	Guého *et al.* (1994)
ITZ	0.09	6	Guého *et al.* (1994)
KTZ	0.28	6	Guého *et al.* (1994)

Nomenclature. *Trichosporon asahii* Akagi - Jpn. J. Derm. Urol. 29: 733, 1929 (invalid) ≡ *Trichosporon asahii* Akagi ex Sugita, Nishikawa & Shinoda - J. Gen. Appl. Microbiol. 40: 405, 1994.

Proteomyces infestans Moses & Vianna - Mem. Inst. Oswaldo Cruz 5: 192, 1913 ≡ *Sporotrichum infestans* (Moses & Vianna) Sartory - Champ. Parasit. Homme Anim. p. 655, 1922 ≡ *Mycoderma infestans* (Moses & Vianna) da Fonseca & de Area Leão - Brasil Med. 43: 667, 1929 ≡ *Trichosporon infestans* (Moses & Vianna) Ciferri & Redaelli - Arch. Mikrobiol. 6: 58, 1935 ≡ *Geotrichum infestans* (Moses & Vianna) Brumpt - Précis Parasit., ed. 5, p. 1738, 1936.

Trichosporon figueriae Batista & Silveira - Mycopathologia 12: 196, 1960.

Trichosporon loboi Batista, Campos & Oliveira - Publçoes Inst. Micol. Recife 207: 5, 1963.

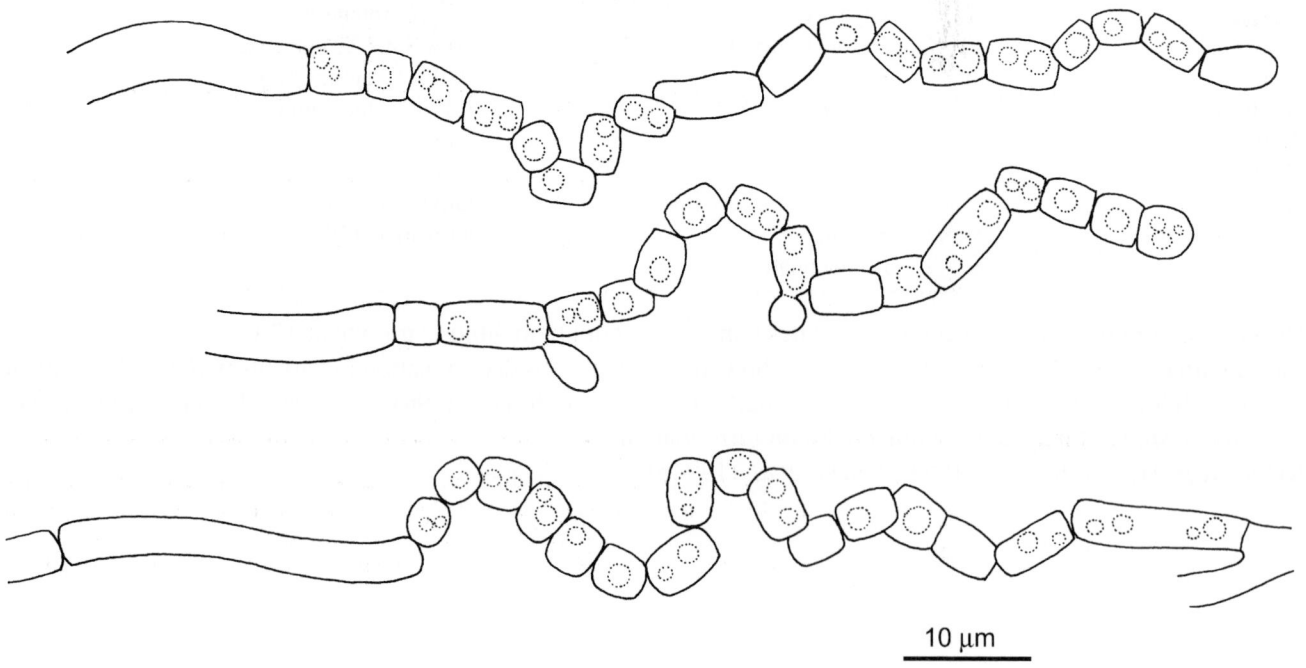

10 µm

Trichosporon asahii, CBS 2479. Disarticulating hyphae with liberated, cubic arthroconidia.

Trichosporon asteroides (Rischin) Ota

Colony characteristics. Colonies (SGA) restricted, dry, cream-coloured, cerebriform, with a radially furrowed outer zone. The meristematic form is punctiform and brownish.
Microscopy. Budding cells and lateral conidia absent. Arthroconidia elongate; hyphae often present. Appressoria absent. The meristematic form consists of hyphae which swell, become multiseptate and fall apart into smaller cell packets.
Physiology.

Fermentation:	–	L-Arabinose	+	DL-Lactate	+
		D-Arabinose	+	Succinate	+
Growth:		D-Ribose	+	Citrate	+
Glucose	+	L-Rhamnose	+	*myo*-Inositol	–
Galactose	+	D-Glucosamine (C)	v	Nitrate	–
L-Sorbose	v	*N*-Acetyl-D-glucosamine	+	Vitamin-free	–
Sucrose	+	Methanol	–	2-Keto-D-gluconate	+
Maltose	+	Ethanol	+	5-Keto-D-gluconate	+
Cellobiose	+	Glycerol	+	D-Glucuronate	+
α,α-Trehalose	+	*meso*-Erythritol	+	10% NaC1/5% glucose	+
Lactose	+	Ribitol	v	0.01% Cycloheximide	v
Melibiose	–	Galactitol	–	0.1% Cycloheximide	v
Raffinose	–	D-Mannitol	v	Urease	+
Melezitose	+	D-Glucitol	v	Starch formation	w
Inulin	–	α-Methyl-D-glucoside	+	Growth at 37°C	v
Soluble starch	+	Salicin	v	Growth at 42°C	–
D-Xylose	+	D-Gluconate	+		

Differential diagnosis. The species assimilates L-arabinose but not *myo*-inositol. It cannot reliably be distinguished physiologically from *T. asahii*.
Molecular diagnostics. LSU restriction map based on NCBI AF075513:

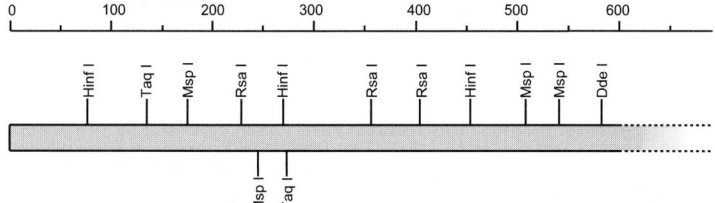

Pathogenicity. BSL-2. Very rare agent of superficial infections (Rischin, 1921).
References. Guého *et al.* (1992b, 1994).
Antifungal susceptibility.

Antifungal	Mean MICs	Strains	Reference
AMB	0.02	3	Guého *et al.* (1994)
5FC	24.95	3	Guého *et al.* (1994)
FLZ	0.45	3	Guého *et al.* (1994)
ITZ	0.02	3	Guého *et al.* (1994)
KTZ	0.03	3	Guého *et al.* (1994)

Nomenclature. *Parendomyces asteroides* Rischin - Arch. Derm. Syph. 134: 242, 1921 ≡ *Trichosporon asteroides* (Rischin) Ota - Annls Parasit. Hum. Comp. 4: 12, 1926 ≡ *Geotrichoides asteroides* (Rischin) Langeron & Talice - Annls Parasit. Hum. Comp. 10: 68, 1932 ≡ *Proteomyces asteroides* (Rischin) C.W. Dodge - Med. Mycol. p. 210, 1935.

Prototheca filamenta Arnold & Ahearn - Mycologia 64: 270, 1972 ≡ *Fissuricella filamenta* (Arnold & Ahearn) Pore, d'Amato & Ajello - Sabouraudia 15: 71, 1977.

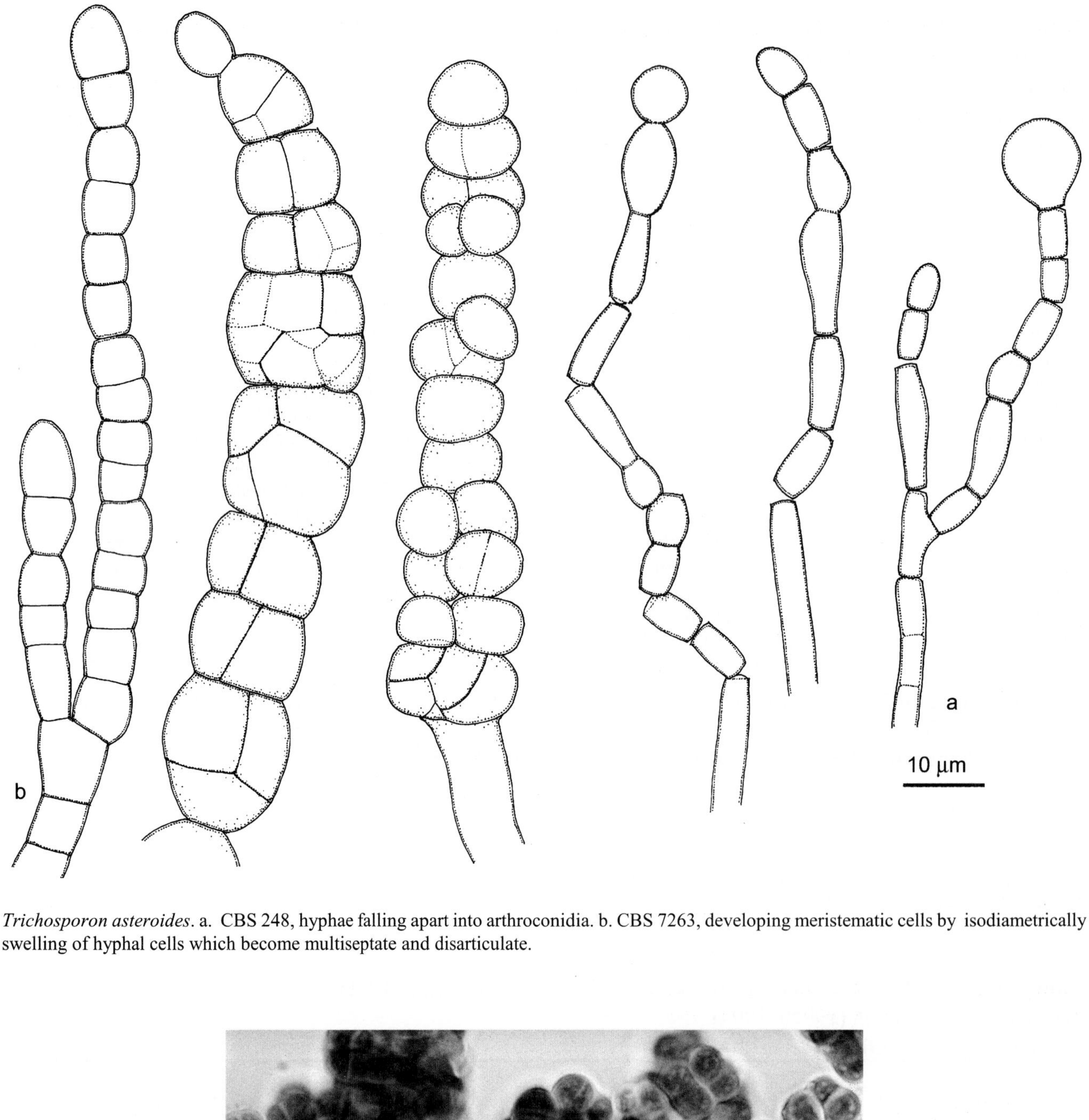

Trichosporon asteroides. a. CBS 248, hyphae falling apart into arthroconidia. b. CBS 7263, developing meristematic cells by isodiametrically swelling of hyphal cells which become multiseptate and disarticulate.

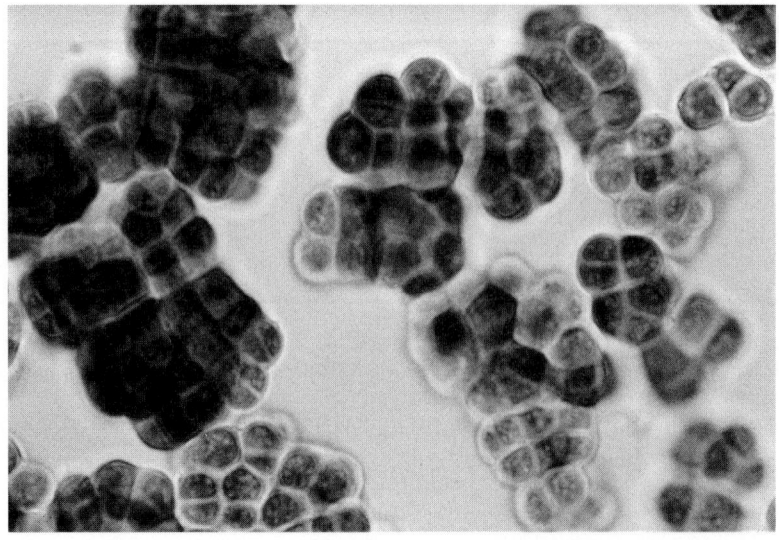

Trichosporon asteroides, CBS 7623, mature meristematic growth form. ×3040. Courtesy E. Guého.

168

Trichosporon cutaneum (de Beurmann *et al.*) Ota

Colony characteristics. Colonies (SGA) moderately expanding, cerebriform, shiny, not becoming farinose with age, with broad, fissured marginal zone.

Microscopy. Budding cells abundant in primary cultures; hyphae developing after repeated transfer. Arthroconidia cylindrical to ellipsoidal.

Physiology.

Fermentation:	–	L-Arabinose	+	DL-Lactate	+
		D-Arabinose	v	Succinate	+
Growth:		D-Ribose	+	Citrate	+
Glucose	+	L-Rhamnose	+	*myo*-Inositol	+
Galactose	+	D-Glucosamine (C)	v	Nitrate	–
L-Sorbose	v	*N*-Acetyl-D-glucosamine	+	Vitamin-free	–
Sucrose	+	Methanol	–	2-Keto-D-gluconate	+
Maltose	+	Ethanol	+	5-Keto-D-gluconate	+
Cellobiose	+	Glycerol	+	D-Glucuronate	+
α,α-Trehalose	+	*meso*-Erythritol	+	10% NaC1/5% glucose	+
Lactose	+	Ribitol	+	0.01% Cycloheximide	v
Melibiose	+	Galactitol	–	0.1% Cycloheximide	–
Raffinose	+	D-Mannitol	+	Urease	+
Melezitose	+	D-Glucitol	+	Starch formation	+
Inulin	–	α-Methyl-D-glucoside	+	Growth at 30°C	+
Soluble starch	+	Salicin	+	Growth at 35°C	–
D-Xylose	+	D-Gluconate	+		

Molecular diagnostics. SSU and LSU restriction maps based on NCBI X60182 and AF075483:

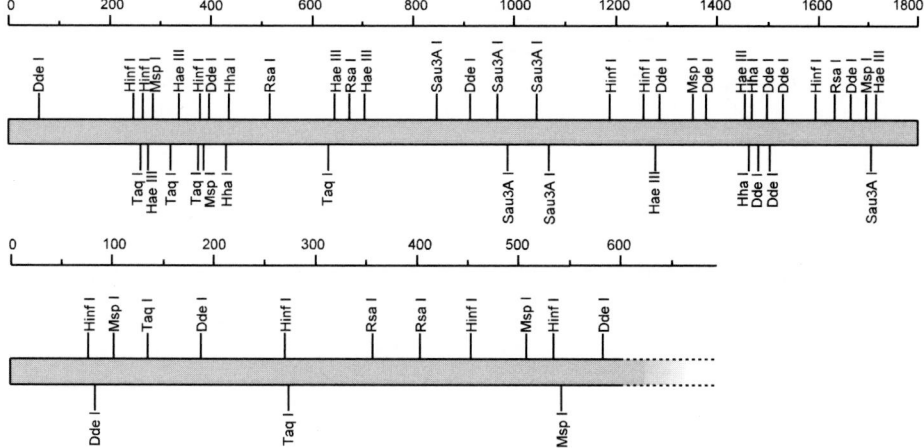

Differential diagnosis. Melibiose assimilated; no growth at 35°C; 0.1% cycloheximide not tolerated.

Pathogenicity. BSL-2. The species is rare; most reports in the literature under this name probably concerned other taxa. It causes skin lesions or white piedra.

References. Herbrecht *et al.* (1993), Guého *et al.* (1994).

Antifungal susceptibility.

Antifungal	Mean MICs	Strains	Reference
AMB	0.012	3	Guého *et al.* (1994)
5FC	6.25	3	Guého *et al.* (1994)
FLZ	2.2	3	Guého *et al.* (1994)
ITZ	0.01	3	Guého *et al.* (1994)
KTZ	0.01	3	Guého *et al.* (1994)

Nomenclature. *Oidium cutaneum* de Beurmann, Gougerot & Vaucher - Bull. Mém. Soc. Hôp., Paris 3: 256, 1909 ≡ *Mycoderma cutanea* (de Beurmann, Gougerot & Vaucher) Brumpt - Précis Parasitol., ed. 2, p. 932, 1913 ≡ *Monilia cutanea* (de Beurmann, Gougerot & Vaucher)

Castellani & Chalmers - Man. Trop. Med., ed. 2, p. 830, 1913 ≡ *Oospora cutanea* (de Beurmann, Gougerot & Vaucher) Saccardo - Syll. Fung. 22: 1243, 1913 ≡ *Trichosporon cutaneum* (de Beurmann, Gougerot & Vaucher) Ota - Annls Parasit. Hum. Comp. 4: 12, 1926 ≡ *Geotrichoides cutaneus* (de Beurmann, Gougerot & Vaucher) Langeron & Talice - Annls Parasit. Hum. Comp. 10: 67, 1932 ≡ *Geotrichum cutaneum* (de Beurmann, Gougerot & Vaucher) de Almeida - Anais Fac. Med. Univ. S. Paulo 9: 78, 1933 ≡ *Proteomyces cutaneus* (de Beurmann, Gougerot & Vaucher) C.W. Dodge - Med. Mycol. p. 211, 1935.

Trichosporon minus de Arêa Leâo - Mem. Inst. Oswaldo Cruz 35: 742, 1940.

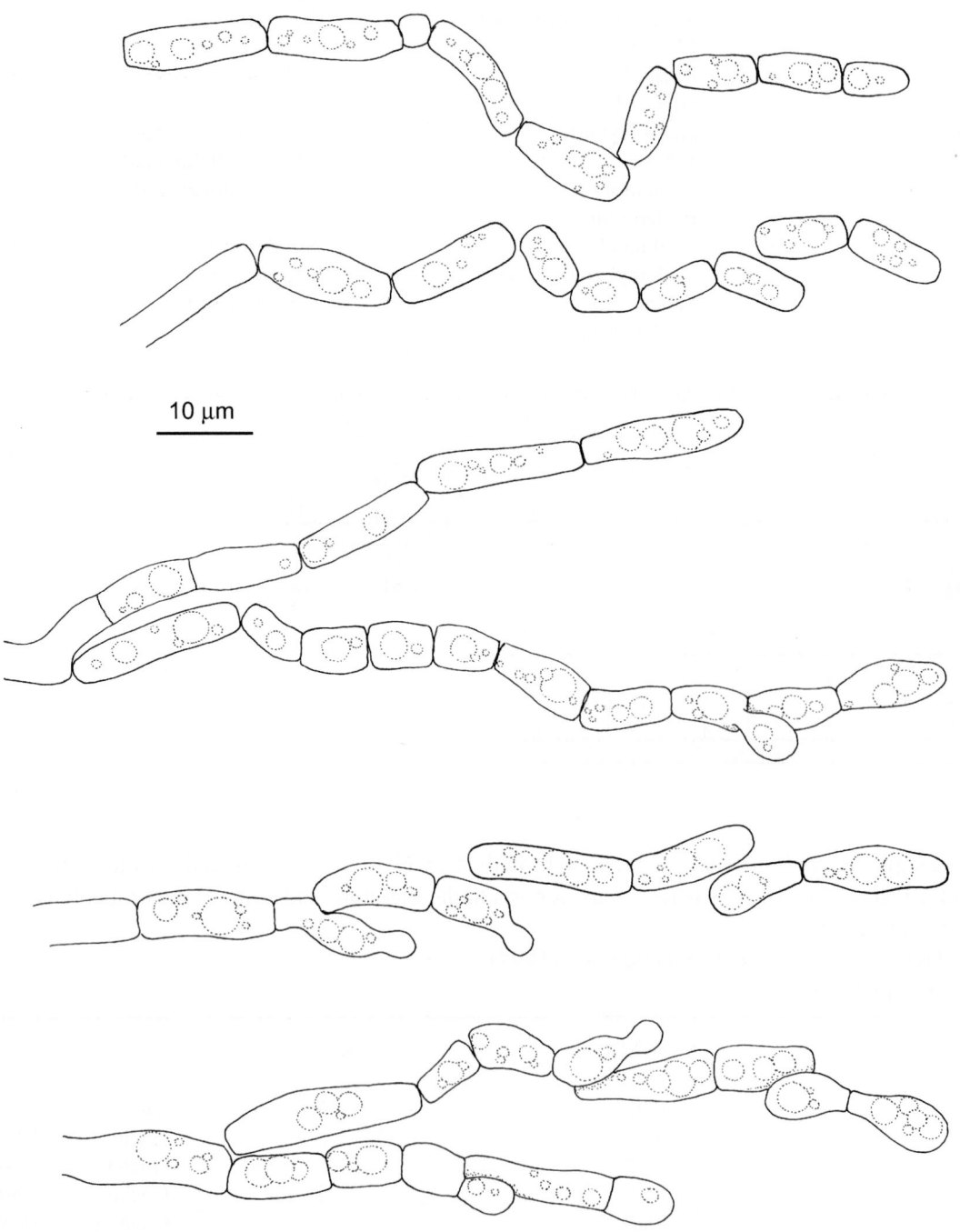

10 μm

Trichosporon cutaneum, CBS 2466. Disarticulating hyphae in strain after repeated transfer. In the lower series of arthroconidia ome poorly differentiated budding cells are produced.

Trichosporon inkin (Oho ex Ota) do Carmo-Sousa & van Uden

Colony characteristics. Colonies (SGA) restricted, finely cerebriform, white, farinose, without marginal zone, often cracking the agar medium.

Microscopy. Budding cells and lateral conidia absent. Arthroconidia long cylindrical. Appressoria present in slide cultures. Sarcinae present on media with high sugar-content.

Physiology.

Fermentation:	–	L-Arabinose	v	DL-Lactate	+
		D-Arabinose	v	Succinate	+
Growth:		D-Ribose	+	Citrate	+
Glucose	+	L-Rhamnose	–	*myo*-Inositol	+
Galactose	v	D-Glucosamine (C)	v	Nitrate	–
L-Sorbose	v	*N*-Acetyl-D-glucosamine	+	Vitamin-free	–
Sucrose	+	Methanol	–	2-Keto-D-gluconate	+
Maltose	+	Ethanol	+	5-Keto-D-gluconate	+
Cellobiose	+	Glycerol	v	D-Glucuronate	+
α,α-Trehalose	+	*meso*-Erythritol	+	10% NaC1/5% glucose	+
Lactose	+	Ribitol	–	0.01% Cycloheximide	+
Melibiose	–	Galactitol	–	0.1% Cycloheximide	v
Raffinose	–	D-Mannitol	v	Urease	+
Melezitose	+	D-Glucitol	–	Starch formation	+
Inulin	–	α-Methyl-D-glucoside	+	Growth at 37°C	+
Soluble starch	+	Salicin	v	Growth at 40°C	v
D-Xylose	+	D-Gluconate	+		

Differential diagnosis. The species grows with *myo*-inositol but not with melibiose. Sarcinae are also known in *Prototheca filamenta* Arnold & Ahearn (Pore *et al.*, 1977), which was found to be a *Trichosporon* species, probably *T. asteroides*, by Guého *et al.* (1992). The latter species is different from *T. inkin* by assimilation of inositol and D-mannitol.

Pathogenicity. BSL-2. The species is almost exclusively isolated from the human crural area, where it causes white piedra on pubic hairs (Thérizol-Ferly *et al.* 1994). Occasional disseminated cases have been described (Mussa *et al.*, 1998); Chaumentin *et al.* (1996) reported an endocarditis and Lopes *et al.* (1997) a CAPD-related peritonitis.

References. King & Jong (1975), Herbrecht *et al.* (1993), Guého *et al.* (1994).

Antifungal susceptibility.

Antifungal	Mean MICs	Strains	Reference
AMB	0.04	6	Guého *et al.* (1994)
5FC	31.4	6	Guého *et al.* (1994)
FLZ	3.9	6	Guého *et al.* (1994)
ITZ	0.03	6	Guého *et al.* (1994)
KTZ	0.02	6	Guého *et al.* (1994)

Nomenclature. *Sarcinomyces inkin* Oho - Kyoto Ikagu Zasshi 16: 15, 1919 (invalid) ≡ *Sarcinomyces inkin* Oho ex Ota - Jpn. J. Derm. Urol. 26: 137, 1926 ≡ *Trichosporon inkin* (Oho ex Ota) do Carmo Sousa & van Uden - Mycologia 59: 653, 1967 ≡ *Sarcinosporon inkin* (Oho ex Ota) King & Jong - Mycotaxon 3: 93, 1975.

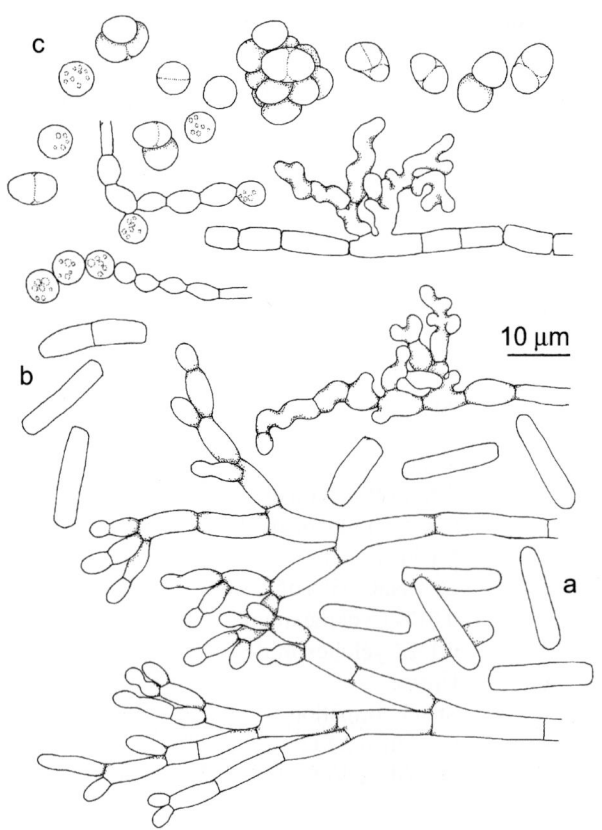

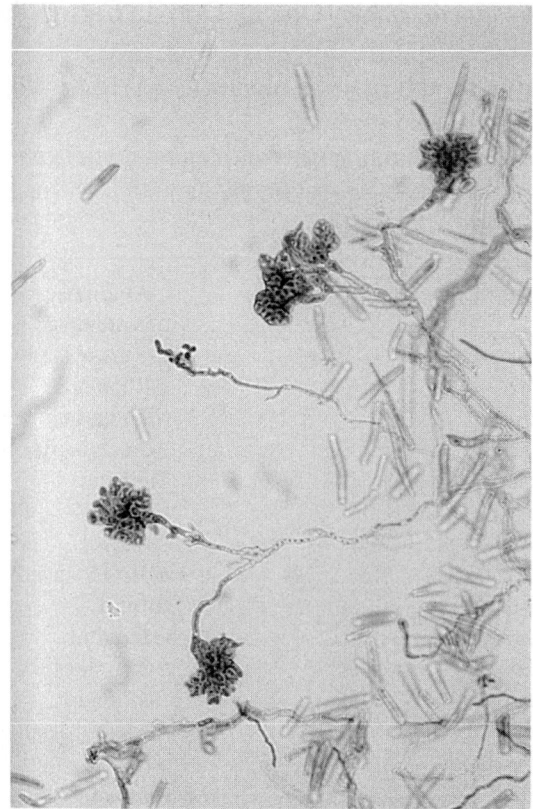

Trichosporon inkin, CBS 5585. a. Disarticulating hyphae with appressoria; b. cylindrical arthroconidia; c. sarcinae on media rich in sugar.

Trichosporon inkin, CBS 5585. Appressoria. ×2645. Courtesy E. Guého.

Trichosporon mucoides Guého & M.Th. Smith

Colony characteristics. Colonies (SGA) moderately expanding, moist and shiny, elevated, with deep, narrow radial fissures.

Microscopy. Budding cells present in primary cultures. Broadly clavate, terminal or lateral blastoconidia often present, at maturity developing a thick cell wall. Arthroconidia barrel-shaped.

Physiology.

Fermentation:	–	L-Arabinose	+	DL-Lactate	+
		D-Arabinose	+	Succinate	+
Growth:		D-Ribose	+	Citrate	+
Glucose	+	L-Rhamnose	+	*myo*-Inositol	+
Galactose	+	D-Glucosamine (C)	+	Nitrate	–
L-Sorbose	+	*N*-Acetyl-D-glucosamine	+	Vitamin-free	–
Sucrose	+	Methanol	–	2-Keto-D-gluconate	+
Maltose	+	Ethanol	+	5-Keto-D-gluconate	+
Cellobiose	+	Glycerol	+	D-Glucuronate	+
α,α-Trehalose	+	*meso*-Erythritol	+	10% NaC1/5% glucose	+
Lactose	+	Ribitol	+	0.01% Cycloheximide	+
Melibiose	+	Galactitol	+	0.1% Cycloheximide	v
Raffinose	+	D-Mannitol	+	Urease	+
Melezitose	+	D-Glucitol	+	Growth at 37°C	+
Inulin	–	α-Methyl-D-glucoside	+	Growth at 42°C	–
Soluble starch	+	Salicin	+		
D-Xylose	+	D-Gluconate	+		

Differential diagnosis. Growth with melibiose and at 37°C; tolerant to 0.1 % cycloheximide.

Molecular diagnostics. LSU restriction map based on NCBI AF075515:

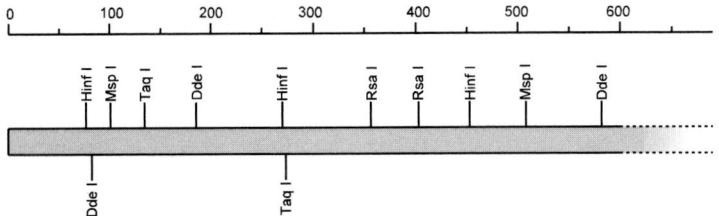

Pathogenicity. BSL-2. The species is fairly common on superficial locations, where it causes pubic white piedra (Thérizol-Ferly *et al.*, 1994) or is involved in onychomycosis. It may cause disseminated infections in patients with impaired innate immunity (Herbrecht *et al.*, 1993).

Reference. Herbrecht *et al.* (1993), Guého *et al.* (1994).

Antifungal susceptibility.

Antifungal	Mean MICs	Strains	Reference
AMB	0.02	4	Guého *et al.* (1994)
5FC	50	4	Guého *et al.* (1994)
FLZ	10.5	4	Guého *et al.* (1994)
ITZ	0.05	4	Guého *et al.* (1994)
KTZ	0.05	4	Guého *et al.* (1994)

Nomenclature. *Trichosporon mucoides* Guého & M.Th. Smith, *in* Guého, Smith, de Hoog, Billon-Grand, Christen & Batenburg-van der Vegte - Antonie van Leeuwenhoek 61: 312, 1992.

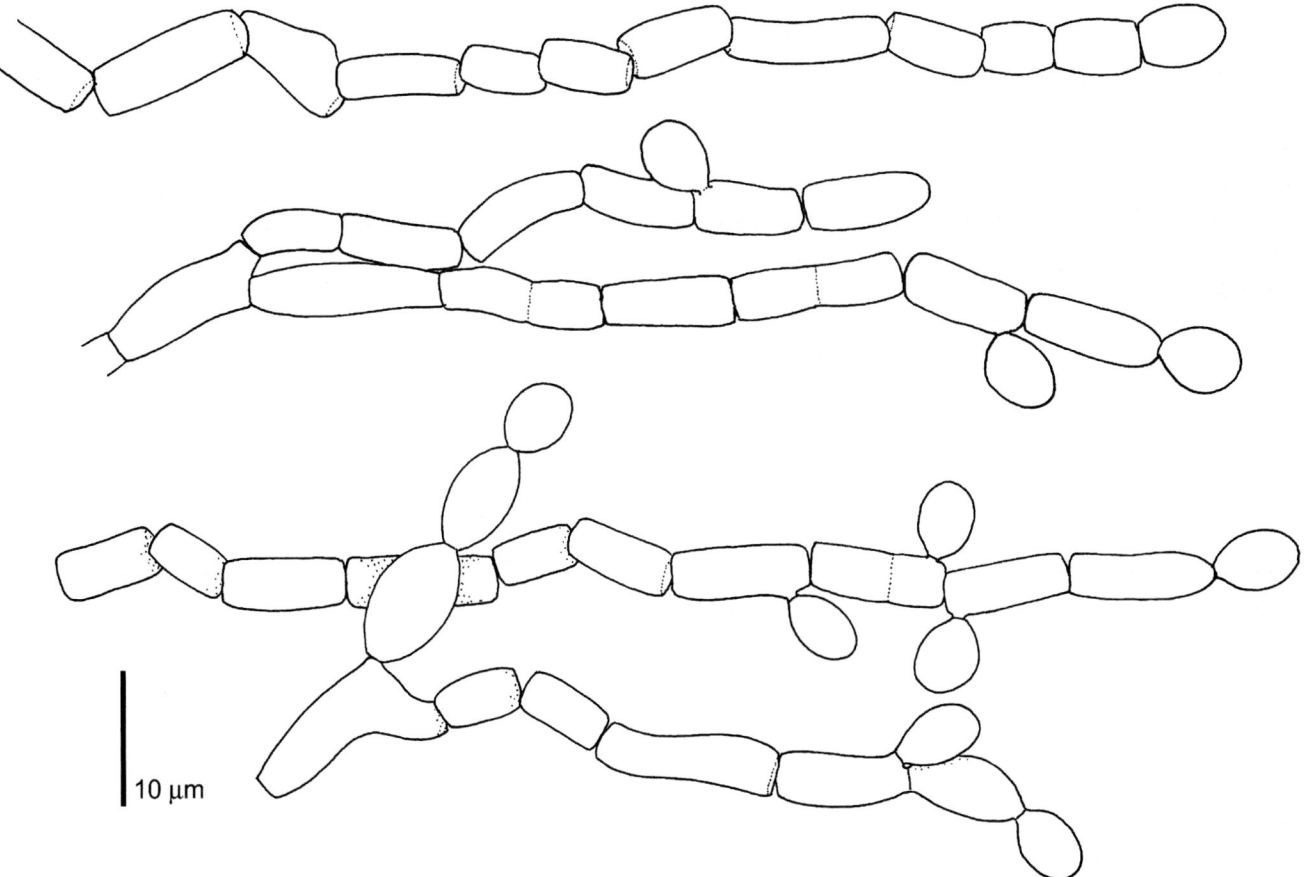

Trichosporon mucoides, CBS 7625. Hyphae with formation of arthroconidia, and lateral and terminal blastoconidia.

Trichosporon ovoides Behrend

Colony characteristics. Colonies (SGA) restricted, white, farinose, irregularly folded at the centre, with a flat marginal zone.

Microscopy. Budding cells and lateral conidia absent. Arthroconidia cylindrical. Appressoria present in slide cultures.

Physiology.

Fermentation:	–	L-Arabinose	v	DL-Lactate	+
		D-Arabinose	v	Succinate	+
Growth:		D-Ribose	+	Citrate	v
Glucose	+	L-Rhamnose	+	*myo*-Inositol	+
Galactose	+	D-Glucosamine (C)	v	Nitrate	–
L-Sorbose	v	*N*-Acetyl-D-glucosamine	+	Vitamin-free	–
Sucrose	+	Methanol	–	2-Keto-D-gluconate	+
Maltose	+	Ethanol	+	5-Keto-D-gluconate	+
Cellobiose	+	Glycerol	v	D-Glucuronate	+
α,α-Trehalose	v	*meso*-Erythritol	+	10% NaC1/5% glucose	+
Lactose	+	Ribitol	–	0.01% Cycloheximide	+
Melibiose	–	Galactitol	–	0.1% Cycloheximide	–
Raffinose	v	D-Mannitol	+	Urease	+
Melezitose	v	D-Glucitol	v	Growth at 37°C	v
Inulin	–	α-Methyl-D-glucoside	+	Growth at 42°C	–
Soluble starch	+	Salicin	+		
D-Xylose	+	D-Gluconate	+		

Differential diagnosis. The species does not grow with melibiose but tolerates 0.1% cycloheximide.

Molecular diagnostics. LSU restriction map based on NCBI AF075523:

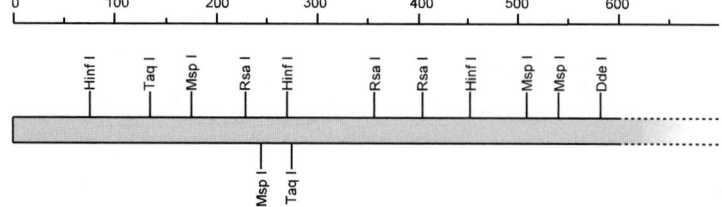

Pathogenicity. BSL-2. Infrequently involved in superficial mycoses, particularly capital white piedra. The species may play a role in hypersensitivity pneumonitis (Sugita *et al.*, 1998a).

References. Herbrecht *et al.* (1993), Guého *et al.* (1994).

Antifungal susceptibility.

Antifungal	Mean MICs	Strains	Reference
AMB	0.03	2	Guého *et al.* (1994)
5FC	35.3	2	Guého *et al.* (1994)
FLZ	4.4	2	Guého *et al.* (1994)
ITZ	0.02	2	Guého *et al.* (1994)
KTZ	0.03	2	Guého *et al.* (1994)

Nomenclature. *Trichosporon ovoides* Behrend - Berl. Klin. Wschr. 27: 467, 1890.

Geotrichum amycelicum Redaelli & Ciferri - Arch. Mikrobiol. 6: 60, 1935 ≡ *Oospora amycelica* (Redaelli & Ciferri) Delitsch - Syst. Schimmelp. p. 52, 1943.

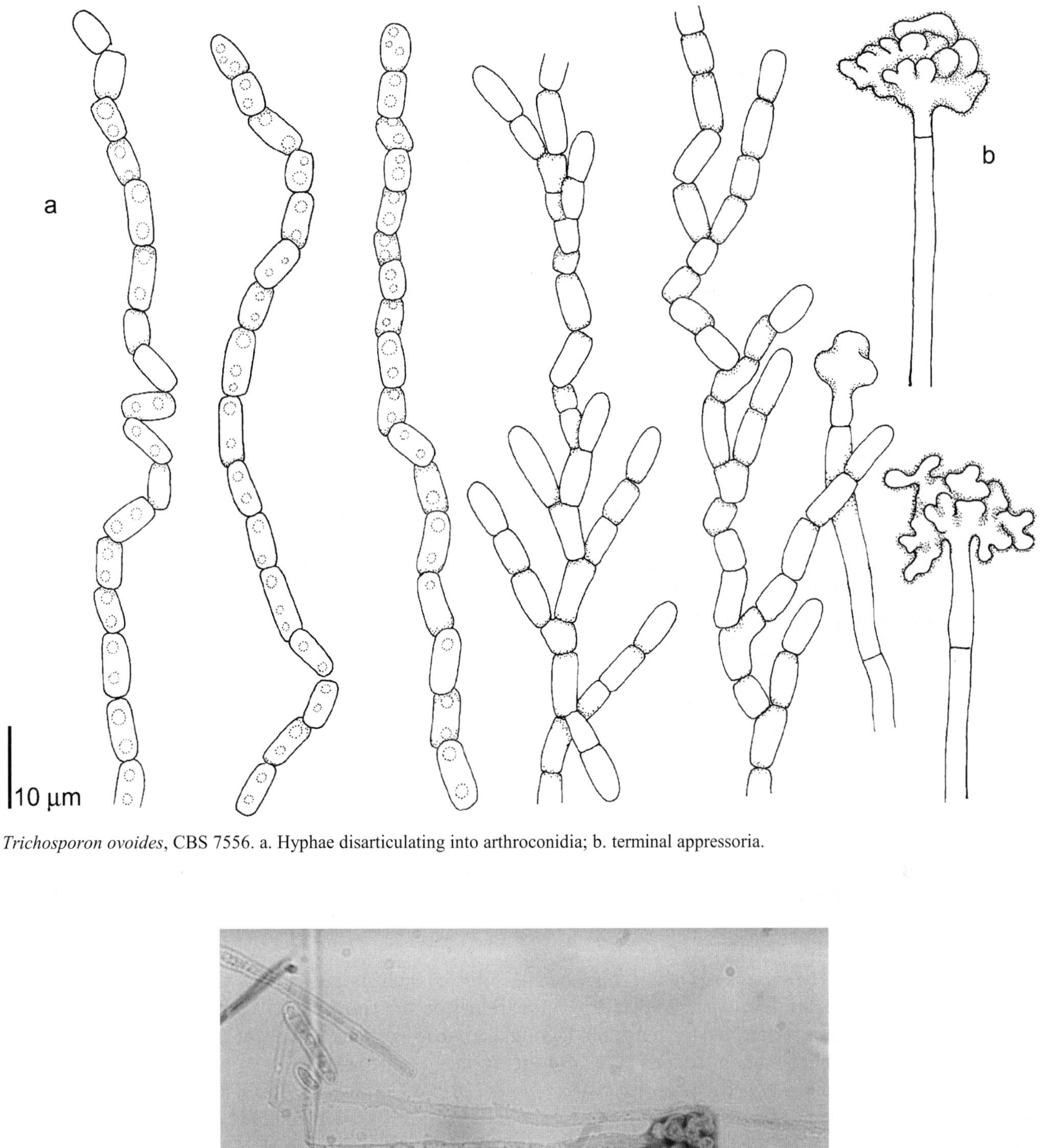

10 µm

Trichosporon ovoides, CBS 7556. a. Hyphae disarticulating into arthroconidia; b. terminal appressoria.

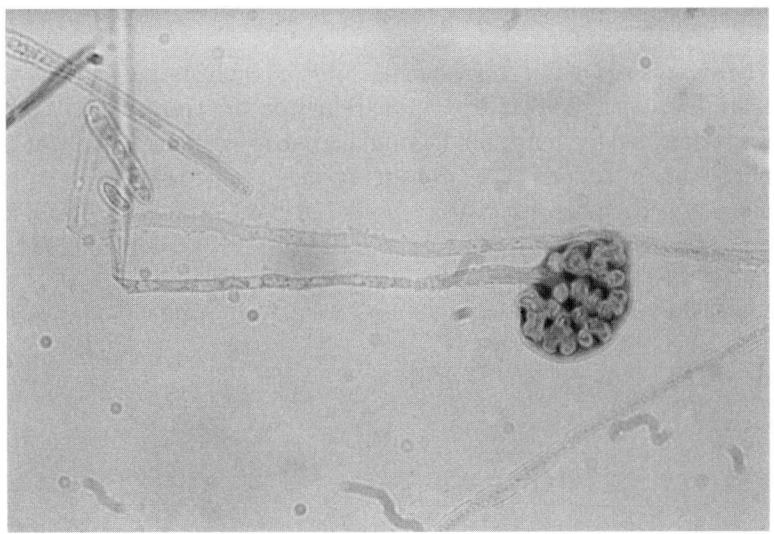

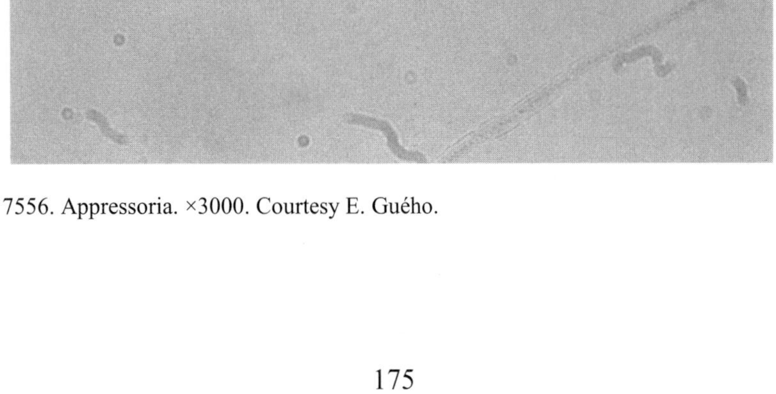

Trichosporon ovoides, CBS 7556. Appressoria. ×3000. Courtesy E. Guého.

Archiascomycetes

Ascomycota, Archiascomycetes, Pneumocystidales. Genus: *PNEUMOCYSTIS*

Pneumocystis carinii Delanoe & M. Delanoe

Description *in vivo*. Thick-walled, polymorphous cysts, 5-7 μm diam and stainable by GMS, develop in alveolar tissue. With Giemsa stain, up to 8 internal bodies can be seen. Crescent-shaped cysts are supposed to have liberated these endospores. In alveolar lumen a foamy exudate develops.

General remark. *Pneumocystis* cannot be cultivated on routine laboratory media (Cushion & Ebbets, 1990). It was longtime considered to be a *Protozoa*, but on the basis of similarities in 5S and 18S ribonucleic acids it is now supposed to be a fungus (Watanabe *et al.*, 1989). The species seems intermediate between *Basidiomycota* (Wakefield *et al.*, 1992) and *Ascomycota* (Taylor & Bowman, 1993) and takes a rather isolated position (Fan *et al.*, 1994). It was recently attributed to the *Archiascomycetes* (Nishida *et al.*, 1995), together with *Schizosaccharomyces;* compare also the phylogenetic tree (Fig. 3) of the fungal kingdom on p. 4.

Molecular diagnostics. Detection is enhanced by the use of calcofluor staining (Baselski *et al.*, 1990), or immunofluorescence staining using monoclonal antibodies (Ng *et al.*, 1990). PCR-Detection using 5S rDNA was introduced by Kitada *et al.* (1991) and Ribes *et al.* (1997) and with thymidylate synthase genes by Olsson *et al.* (1993). A comparison of ribosomal ITS-based amplicons with other molecular detection methods was performed by Lu *et al.* (1995). Chouaid *et al.* (1995) presented an evaluation of PCR diagnosis in sputum of AIDS patients. Ribes *et al.* (1997) reported PCR detection in BAL, while Lautenschlager *et al.* (1996) applied immunofluorescence. SSU and ITS restriction maps based on L27658:

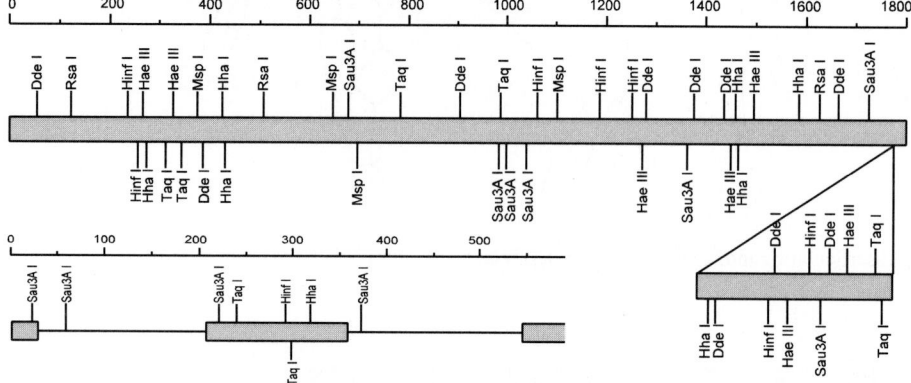

Pathogenicity. BSL-2. The organism is rather common in debilitated humans and animals as a mild infection against which residual antibodies are developed. Epidemic pneumonia may occur in institutional housing, such as orphanages, under conditions of overcrowding and malnutrition. During the last 15 years the organism has emerged as one of the most common cause of pulmonary infections in AIDS patients; the presence of *Pneumocystis* has become one of the first indications for the disease. A cutaneous case was reported by Sandler *et al.* (1996).

References. Young (1984), McKenzie (1990), Dei-Cas *et al.* (1992, 1998), Blumenfeld *et al.* (1992), Cailliez *et al.* (1996), Mazars & Dei-Cas (1998), Durand-Joly (2000).

Nomenclature. *Pneumocystis carinii* Delanoë & M. Delanoë - C.R. Hebd. Séanc. Acad. Sci., Paris 155: 658-660, 1912.
Pneumocystis jiroveci Frenkel - Nat. Cancer Inst. Monogr. 43: 13-30, 1976.

Varieties. The following *formae speciales* are currently recognized (E. Dei-Cas, pers. comm.), which were confirmed by phylogeny based on superoxide dismutase genes (Denis *et al.*, 2000):

Name	Host species	Authority
P. carinii f. sp. *carinii*	*Rattus norvegicus*	Anon. (1994)
P. carinii f. sp. *rattus*	*Rattus norvegicus*	Anon. (1994)
P. carinii f. sp. *hominis*	*Homo sapiens*	Anon. (1994)
P. carinii f. sp. *muris*	*Mus musculi*	Anon. (1994)
P. carinii f. sp. *mustelae*	*Mustela furo*	Anon. (1994)
P. carinii f. sp. *oryctolagi*	*Oyctolagus cuniculi*	Anon. (1994)
P. carinii f. sp. *equi*	*Equus caballi*	Anon. (1994)
P. carinii f. sp. *suis*	*Sus scrofa*	Anon. (1994)
P. carinii f. sp. *mustelae* A	*Mustela furo*	Banerji *et al.* (1995)
P. carinii f. sp. *mustelae* B	*Mustela furo*	Banerji *et al.* (1995)
P. carinii f. sp. *sorex*	*Sorex araneus*	Laakkonen & Sukura (1997)
P. carinii f. sp. *macacae*	*Macaca mulatta*	Durand-Joly *et al.* (2000)

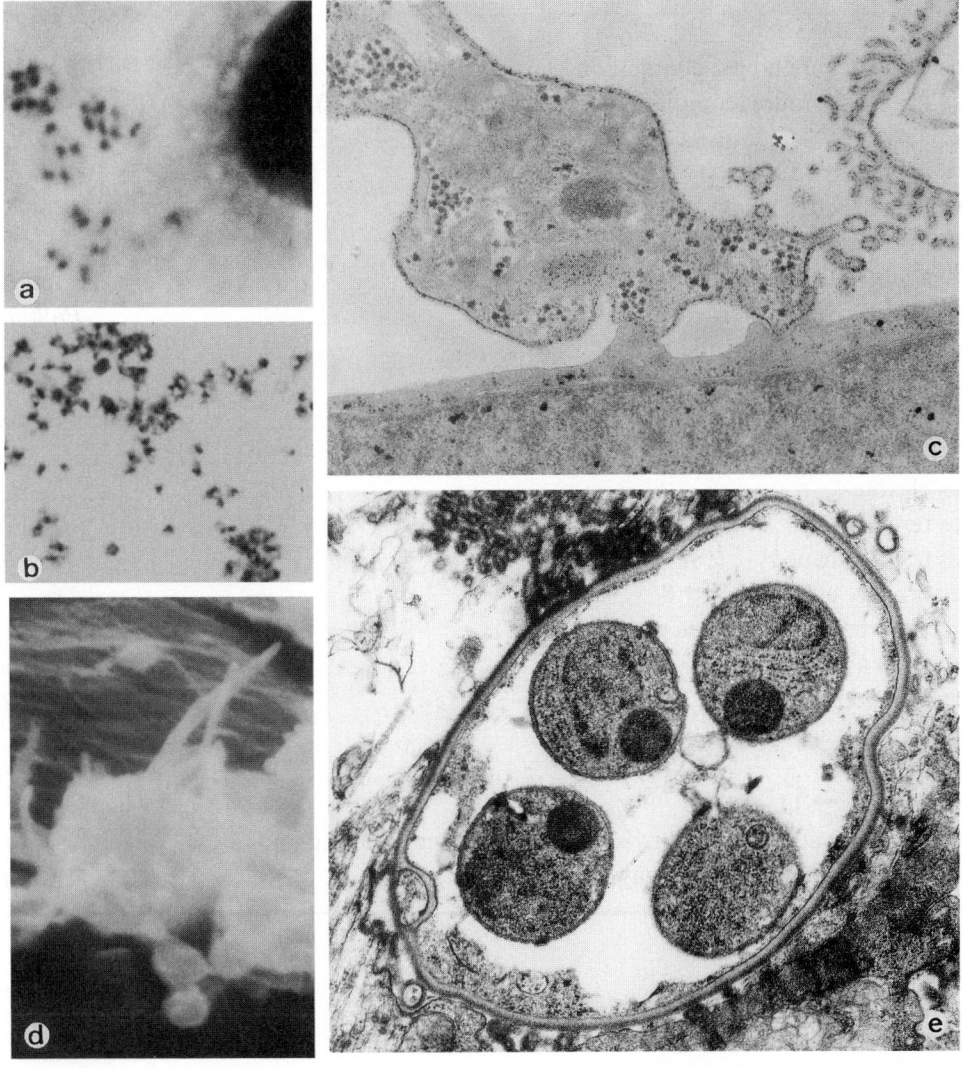

Pneumocystis carinii, a. ATCC CCL-81; b. ATCC CCL-149; c-e. ATCC CCL-185. a. Rat-derived, developing *in vitro* on Vero cell monolayers. Cystic form containing eight mononuclear intracystic bodies is observed at the top. At the right, the nucleus of a Vero cell is seen. Methanol Giemsa staining. b. Rat-derived trophozoites *in vitro* on L2 rat lung epithelial-like host cell monolayers. Methanol Giemsa staining,. c. Rat-derived trophozoite (T) attached *in vitro* to an A549 human lung epithelial cell (EC). The arrow indicates an attachment zone. Transmission electron microscopy. d. Rat-derived trophozoides developing *in vitro* on L2 cell monolayers. Arrow: filopodia. Scanning electron microscopy. e. Mature cyst containing intracystic bodies. Mi: mitochondrion; N: nucleus. Transmission electron microccopy. a. ×1250; b. ×1000; c. ×7500; d. ×7000; e. ×11,700. a. Reproduced with permission from Cailliez *et al.* - Med. Hypoth. 43: 167-171, 1995; b, c, d. courtesy E.M. Aliouat and E. Dei-Cas; e. reproduced with persmission from Dei-Cas & Cailliez - FEMS Immunol. Med. Microbiol. 22: 1-4, 1998.

Ascomycetous yeasts

General remarks. Members of the order *Saccharomycetales,* belonging to the ascomycete class *Hemiascomycetes* (formerly *Endomycetes*) are ascus-forming fungi without fruit bodies, close relatives of baker's yeast (*Saccharomyces*). Six families of ascomycetous yeasts are treated, most of which have anamorphs with budding cells (*Candida*), while the *Dipodascaceae* mostly have anamorphs with arthroconidia (*Geotrichum*). Phylogeny based on partial 26S rDNA sequences was presented by Kurtzman & Robnett (1997, 1998).

Most *Saccharomycetales* occur in liquids rich in saccharides, such as fruit juices or polluted water. Species are homo- or heterothallic. They are usually fermentative; a limited number of carbohydrates are assimilated. Budding multilateral; urease –; extracellular DNAse –; DBB –; cell wall 3-layered, without xylose or fucose; septa with micropores.

Simplified key to the treated genera of ascomycetous yeasts and their teleomorphs:

1a. Budding cells mostly abundant; conidiation blastic ➞ **2**
1b. Budding cells absent or rare; conidiation with arthroconidia ➞ **10**
2a. Ascospores absent . . . . **Candida (180)**
2b. Ascospores present ➞ **3**
3a. Ascospores hat-shaped, with unilateral brim ➞ **4**
3b. Ascospores of other shape, without brim ➞ **5**
4a. Asci firm-walled, with sterile cell on top . . . . **Stephanoascus (192)**
4b. Asci deliquescent, without sterile cell . . . . **Pichia (211, 215, 223)**
5a. Ascospores spherical, rough-walled ➞ **6**
5b. Ascospores of other shape ➞ **7**
6a. Growth with galactose, sucrose and maltose . . . . **Debaryomyces (196)**
6b. No growth with galactose, sucrose and maltose . . . . **Arxiozyma (179)**
7a. Ascospores clavate . . . . **Clavispora (209)**
7b. Ascospores reniform . . . . **Kluyveromyces (204)**
7c. Ascospores needle-shaped . . . . **Metschnikowia (217)**
7d. Ascospores broadly ellipsoidal ➞ **8**
8a. Pseudomycelium or true mycelium present
8b. Pseudomycelium absent . . . . **Saccharomyces (234)**
9a. Asci formed in budding cells . . . . **Issatchenkia (206)**
9b. Asci formed on hyphae . . . . **Yarrowia (236)**
10a. Ascospores absent . . . . **Geotrichum (227)**
10b. Ascospores present ➞ **11**
11a. Ascospores smooth-walled, with slimy sheath of even width throughout . . . . **Dipodascus (231)**
11b. Ascospores rough-walled, lacking slimy sheath . . . . **Galactomyces (228)**

Ascomycota, Hemiascomycetes, Saccaromycetales, Saccharomycetaceae. Genus: *ARXIOZYMA*

Arxiozyma telluris (van der Walt) van der Walt & Yarrow

Cultural characteristics. Colonies (YPGA) buttery, cream coloured.
Microscopy. Budding cells (RA) ellipsoidal, initially up to about 3 μm long, then swelling to subspherical, 6 × 5 μm and containing a single, spherical, rough-walled ascospore 2.5-3.0 μm diam.
Physiology.

Fermentation:		Melibiose	–	D-Mannitol	–
Glucose	+	Raffinose	–	D-Glucitol	–
Galactose	–	Melezitose	–	α-Methyl-D-glucoside	–
Sucrose	–	Inulin	–	Salicin	–
Maltose	–	Soluble starch	–	D-Gluconate	–
Lactose	–	D-Xylose	–	DL-Lactate	v
Raffinose	–	L-Arabinose	–	Succinate	v
α,α-Trehalose	–	D-Arabinose	–	Citrate	–
		D-Ribose	–	*myo*-Inositol	–
Growth:		L-Rhamnose	–	Hexadecane	–
Glucose	+	D-Glucosamine (C)	–	Nitrate	–
Galactose	–	*N*-Acetyl-D-glucosamine	–	Vitamin-free	–
L-Sorbose	–	Methanol	–	2-Keto-D-gluconate	–
Sucrose	–	Ethanol	w,–	5-Keto-D-gluconate	–
Maltose	–	Glycerol	–	D-Glucarate	–
Cellobiose	–	*meso*-Erythritol	–	10% NaC1/5% glucose	–
α,α-Trehalose	–	Ribitol	–	Gelatin liquefaction	w,–
Lactose	–	Galactitol	–	Growth at 40°C	+

Differential diagnosis. Species signature: fermentation glucose + but sucrose, trehalose and cellobiose –; assimilation: D-mannitol –, ethylamine –, lysine –, growth at 40°C +, 0.01% cycloheximide –. Physiologically indistinguishable from *Saccharomyces cerevisiae* (p. 234).
Molecular diagnostics. SSU and LSU restriction maps based on NCBI Y15809 and U72158, resp.:

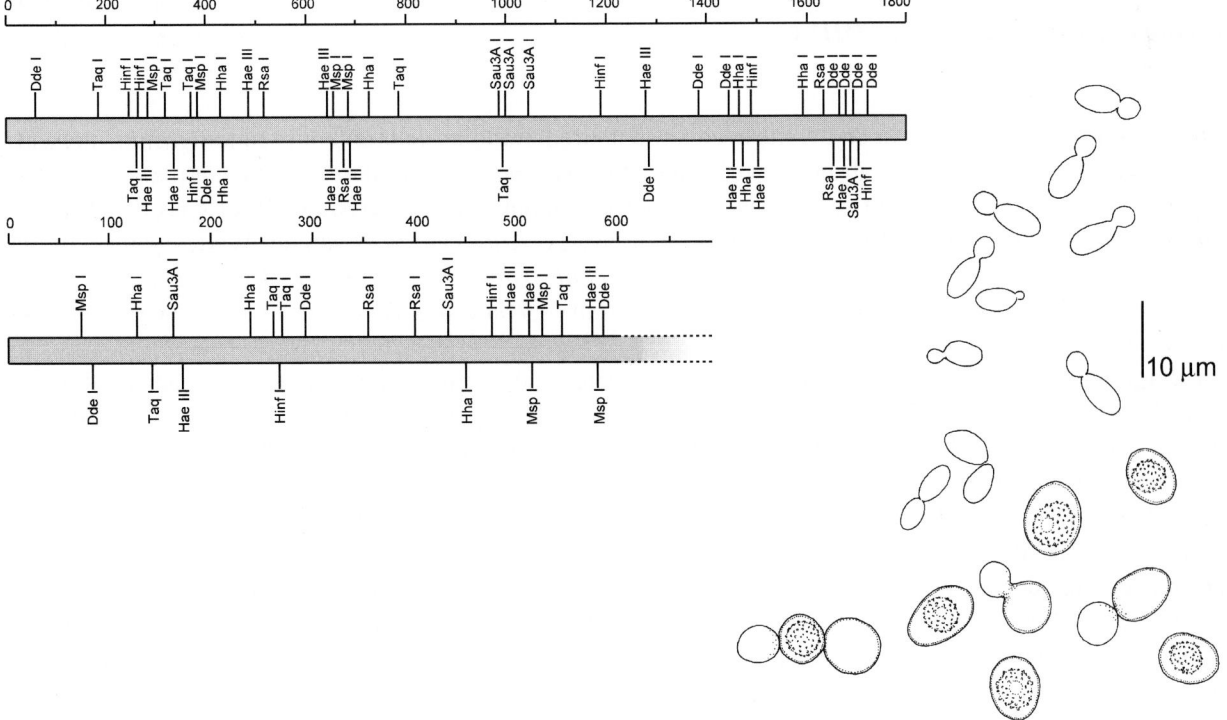

*Arxiozyma telluris,*CBS 6265. Young, elongate budding cells and mature, subspherical asci with ascospores.

Pathogenicity. BSL-1. The species is a rare endobiont in the intestinal tract of cattle (van Uden, 1952). It was listed among potential causes of gastric bezoar after surgery (Perttala *et al.*, 1975). Kunstýř *et al.* (1980) reported fatal enteritis in guinea pigs.

Reference. Kurtzman (1998)

Nomenclature. *Torulopsis pintolopesii* van Uden - Arch. Mikrobiol. 17: 207, 1952 = *Candida pintolopesii* (van Uden) S.A. Meyer & Yarrow, *in* Yarrow & Meyer - Int. J. Syst. Bact. 28: 613, 1978.

   Saccharomyces telluris van der Walt - Antonie van Leeuwenhoek 23: 27, 1957 = *Arxiozyma telluris* (van de Walt) van der Walt & Yarrow - S. Afr. J. Bot. 3: 340-342, 1984.

   Candida bovina van Uden & do Carmo Sousa - J. Gen. Microbiol. 16: 390, 1957.

Ascomycetous yeasts. Genus: *CANDIDA*

Generic description. Colonies slimy or dry, white to cream-coloured. Budding cells and/or pseudomycelium present. Budding multilateral, holoblastic. Asci absent. *myo*-Inositol and creatine not assimilated. Fermentative.

Teleomorphs. *Arxiozyma* van der Walt & Yarrow, *Citeromyces* Santa Maria, *Clavispora* Rodrigues de Miranda, *Debaryomyces* Lodder & Kreger-van Rij, *Issatchenkia* Kudryavtsev, *Kluyveromyces* van der Walt, *Pichia* Hansen, *Saccharomyces* Meyen ex Reess, *Stephanoascus* M.Th. Smith *et al.*, *Torulaspora* Lindner, *Wickerhamiella* van der Walt, *Yarrowia* van der Walt & von Arx, *Zygoascus* M.Th. Smith (*Ascomycota, Hemiascomycetes, Saccharomycetales: Saccharomycetaceae*).

Molecular diagnostics. Non-culture diagnosis of *Candida* species in disseminated infections was made by Reiss & Morrison (1993). PCR-Ribotyping for species recognition was described by Niesters *et al.* (1993) and Maiwald *et al.* (1994). Probing on the basis of V3 LSU rDNA was perfomed by Haynes & Westerneng (1996), of chitin synthase genes by Jordan (1994), on ITS sequences by Fujita *et al.* (1995), Williams *et al.* (1995) and Chen *et al.* (2000) and on lanosterol-demethylase genes by Morace *et al.* (1997). Thanos *et al.* (1996) introduced species recognition using non-specific primers (RAPD).

General remarks. Members of the genus, particularly *C. albicans*, are among the major agents of mycoses in mucous membranes and urogenital systems, and are frequently observed in deep infections in patients with impairment of either natural or acquired immunity. The widescale use of antimycotics is probably responsible for a shift towards the occurrence of non-*albicans* strains, which is noted in recent years (Odds, 1996; Nguyen *et al.*, 1996). For sequence data and phylogenetic relationships of *Candida* species, see Kurtzman & Robnett (1998).

References. Barnett *et al.* (1990), Meyer *et al.* (1998).

Key to the treated species of Candida:

1a. Germtube test positive → **2**
1b. Germtube test negative → **3**
2a. Growth at 45°C .... *C. albicans* **(184)**
2b. No growth at 45⁰C .... *C. dubliniensis* **(194)**
3a. Sucrose assimilated → **10**
3b. Sucrose not assimilated → **4**
4a. Galactose assimilated → **5**
4b. Galactose not or slowly assimilated → **7**
5a. 0.01% Cycloheximide tolerated → **6**
5b. 0.01% Cycloheximide not tolerated .... *C. rugosa* **(218)**
6a. Nitrate assimilated .... *C. catenulata* **(189)**
6b. Nitrate not assimilated .... *C. tropicalis* **(220)**
7a. Trehalose assimilated .... *C. zeylanoides* **(225)**
7b. Trehalose not assimilated → **8**
8a. D-Glucosamine assimilated .... *C. krusei* **(206)**
8b. D-Glucosamine not assimilated → **9**
9a. Cellobiose assimilated .... *C. norvegensis* **(211)**
9b. Cellobiose not assimilated .... *C. glabrata* **(198)**
10a. *myo*-Inositol assimilated → **11**
10b. *myo*-Inositol not assimilated → **12**
11a. D-Glucosamine assimilated .... *C. chiropterorum* **(190)**
11b. D-Glucosamine not assimilated .... *C. ciferrii* **(192)**
12a. Nitrate assimilated → **13**
12b. Nitrate not assimilated → **14**
13a. *meso*-Erythritol assimilated .... *C. pelliculosa* **(215)**
13b. *meso*-Erythritol not assimilated .... *C. utilis* **(223)**
14a. Maltose assimilated → **15**
14b. Maltose not assimilated .... *C. kefyr* **(204)**
15a. Cellobiose not assimilated → **16**
15b. Cellobiose assimilated or slowly assimilated → **18**
16a. D-Mannitol assimilated → **17**
16b. D-Mannitol not assimilated .... *Saccharomyces cerevisiae* **(234)**
17a. α-Methyl-D-glucoside assimilated .... *C. parapsilosis* **(212)**
17b. α-Methyl-D-glucoside not assimilated .... *C. haemulonii* **(202)**
18a. Raffinose assimilated → **19**
18b. Raffinose not assimilated → **21**
19a. Glucose fermented → **20**
19b. Glucose not or weakly fermented .... *C. famata* **(196)**
20a. Lactose assimilated .... *C. intermedia* **(203)**
20b. Lactose not assimilated .... *C. guilliermondii* **(200)**
21a. 0.1% Cycloheximide tolerated .... *C. tropicalis* **(220)**, *C. viswanathii* **(224)**
21b. 0.1% Cycloheximide not tolerated → **22**
22a. Growth at 40°C .... *C. lusitaniae* **(209)**
22b. No growth at 40°C .... *C. pulcherrima* **(217)**

Table 22. Variable physiological characters among treated *Candida* species, compared with *Saccharomyces cerevisiae* (p. 234).

	cerevisiae	zeylanoides	viswanathii	utilis	tropicalis	rugosa	pulcherrima	pelliculosa	parapsilosis	norvegensis	lusitaniae	krusei	kefyr	intermedia	haemulonii	guilliermondii	glabrata	famata	dubliniensis	ciferrii	chiropterorum	catenulata	albicans
Fermentation:																							
Glucose	+	s,–	+	+	+	–	+	+	+	s	+	+	+	+	+	+	+	–,w	+	–	–	>	+
Galactose	v	–	s	–	+	–	w,–	>	>	–	>	–	s	s,+	–	>	–	–,w	>	–	–	s,–	>
Sucrose	+	–	>	+	>	–	s,–	+	s,–	–	>	–	+	+	+	+	–	–,w	–	–	–	–	s,–
Maltose	v	–	+	–	+	–	s,–	>	s,–	–	>	–	–	>	+	–	–	–	+	–	–	>	+
Raffinose	+	–	–	w	–	–	–	w,–	–	–	–	–	+	>	–	+	–	–,w	–	–	–	–	–
Trehalose	–	s,–	s,–	–	s,+	–	–	–	s,–	–	>	–	–	s,–	s,+	+	>	–,w	>	–	–	–	>
Growth:																							
Galactose	v	s,–	+	–	+	+	+	+	+	–	+	–	s	+	s	+	–	+	+	+	+	+	+
L-Sorbose	–	s,+	>	–	>	>	+	–	s,+	–	+	–	>	+	–	>	–	>	–	+	+	–	>
Sucrose	+	–	+	+	>	–	+	+	+	–	+	–	+	+	+	+	–	+	+	+	+	–	>
Maltose	+	+	+	+	+	–	+	+	+	–	+	–	–	+	+	+	–	+	+	+	+	>	+
Cellobiose	–	s,–	+	+	s,+	–	+	+	–	–	+	–	>	+	–	+	–	+	–	>	+	–	s,+
Trehalose	+	+	+	+	+	–	+	+	+	–	+	–	w,–	+	+	+	>	+	+	+	+	s,+	+
Lactose	–	–	–	–	–	–	–	–	–	–	–	–	>	+	–	–	–	>	–	–	–	–	–
Melibiose	v	–	–	–	–	–	–	–	–	–	–	–	–	–	–	+	–	>	>	>	+	–	–
Raffinose	+	–	–	+	–	–	–	+	–	–	–	–	+	+	s,+	+	–	+	–	+	s	–	–
Melezitose	v	–	+	+	>	–	+	+	+	–	+	–	–	+	s,+	+	–	>	+	–	–	–	>
Inulin	–	–	–	w	–	–	–	–	–	–	–	–	+	s,–	–	–	–	>	–	–	+	–	–
Soluble Starch	–	–	+	–	+	–	+	+	+	–	+	–	–	>	>	+	–	>	+	>	+	>	+
D-Xylose	–	–	+	+	+	>	–	>	+	–	+	–	s	+	s,–	+	–	+	>	+	+	>	+
L-Arabinose	–	–	s,–	–	–	–	+	>	–	–	>	–	>	s,–	s,–	+	–	w,+	–	+	s	–	>
D-Arabinose	–	–	–	–	s,–	–	–	+	–	–	–	–	–	s,–	s,–	+	–	>	–	+	s	–	>
D-Ribose	–	–	s,–	–	–	s,–	–	>	>	–	>	–	>	s,–	s,–	>	–	>	–	>	+	s,–	s,–
L-Rhamnose	–	–	–	–	–	–	w,–	–	–	–	–	–	–	>	s,+	+	–	>	–	+	+	–	–
D-Glucosamine	–	s,–	s,–	–	>	–	–	–	>	–	+	–	+	>	s,+	+	>	s,+	s,–	–	–	s,–	>
Ethanol	+	v	+	+	+	+	+	+	+	–	+	+	+	+	s	+	+	+	+	>	+	+	+
Glycerol	–	+	+	+	>	+	+	+	+	–	+	+	s	s,–	s,+	+	s,+	+	s,+	+	+	+	>

Table 22. (Contd.)

Species	meso-Erythritol	Ribitol	Galactitol	D-Mannitol	D-Glucitol	α-Methyl-D-glucoside	Salicin	D-Gluconate	DL-Lactate	Succinate	Citrate	myo-Inositol	Vitamin-free	2-Keto-D-gluconate	Nitrate	0.01% Cycloheximide	0.1% Cycloheximide	Growth at 37°C
cerevisiae	–	–	–	–	–	v	v	v	v	v	–	–	–	nd	–	–	–	v
zeylanoides	–	v	–	+	+	–	v	s,+	–	+	+	–	v	+	–	+	+	w,–
viswanathii	+	+	–	+	+	+	s,	s,	–	+	+	–	–	+	–	+	+	+
utilis	–	–	–	+	+	+	+	+	+	–	+	–	+	w,–	+	–	–	+
tropicalis	–	s,+	–	+	+	>	>	>	>	+	s,+	–	>	+	–	+	+	+
rugosa	–	s,–	–	s,+	s,+	–	–	>	s,+	>	>	–	–	–	–	–	–	+
pulcherrima	–	>	–	+	+	+	+	w,–	+	>	–	–	+	–	–	–	–	>
pelliculosa	+	>	–	+	+	+	+	>	+	+	+	–	+	–	+	–	–	>
parapsilosis	–	s,+	–	+	+	+	–	s,+	–	+	+	–	+	–	–	>	>	+
norvegensis	+	+	–	–	–	–	+	–	w	+	w	–	–	–	–	–	–	+
lussitaniae	–	s	–	+	+	>	+	s	w,+	+	>	–	–	+	–	s	–	+
krusei	–	–	–	–	–	–	–	–	+	+	w,+	–	+	+	–	–	–	+
kefyr	–	s	–	>	>	–	>	–	+	+	>	–	–	–	–	+	+	+
intermedia	–	s,+	>	+	+	s,+	+	s,–	–	+	+	–	s,–	+	–	>	>	–
haemulonii	–	s	s,–	+	+	–	–	+	–	+	s,+	–	–	+	–	+	+	+
guilliermondii	–	+	>	+	+	+	+	>	>	+	>	–	–	+	–	+	+	+
glabrata	–	–	–	–	–	–	–	+	>	–	–	–	–	>	–	–	–	+
famata	–	>	+	>	+	w,+	+	w,+	w,+	>	+	>	–	+	–	>	>	+
dubliniensis	–	s,+	–	+	+	s,+	–	s,–	+	+	+	–	s,–	+	–	+	+	+
ciferrii	+	+	+	+	+	>	>	+	+	>	>	+	–	nd	+	+	+	+
chiropterorum	+	+	+	+	+	+	+	s	–	+	+	–	+	–	+	+	+	+
catenulata	–	s,+	–	+	s,+	–	–	s,–	+	+	+	–	–	>	+	+	+	>
albicans	–	>	–	+	s,+	>	–	s,–	+	+	+	–	s,–	+	–	+	+	+

Candida albicans (Robin) Berkhout

Colony characteristics. Colonies (YPGA) cream-coloured, glistening or somewhat waxy, soft and usually smooth. Some strains may become wrinkled and have a mycelial border.
Microscopy. Budding cells (RA) (sub)spherical, 3-8 × 2-7 μm. Pseudomycelium present, with blastoconidia in dense grape-like arrangment. Dark, spherical chlamydospores mostly terminal, often on a slightly swollen subtending cell. Germ tubes formed with serum.
Physiology.

Fermentation:		Soluble starch	+	*myo*-Inositol	–
Glucose	+	D-Xylose	+	Hexadecane	s
Galactose	v	L-Arabinose	v	Vitamin-free	–,s
Sucrose	–,s	D-Arabinose	v	2-Keto-D-gluconate	+
Maltose	+	D-Ribose	–,s	D-Glucuronate	–
Lactose	–	L-Rhamnose	–	Xylitol	+
Raffinose	–	D-Glucosamine (C)	v	L-Arabinitol	–,s
Trehalose	v	*N*-Acetyl-D-glucosamine	+	Arbutin	–
		Methanol	–	Nitrate	–
Growth:		Ethanol	+	Nitrite	–
Glucose	+	Glycerol	v	Cadaverine	+
Galactose	+	*meso*-Erythritol	–	Creatinine	–
L-Sorbose	v	Ribitol	v	L-Lysine	+
Sucrose	v	Galactitol	–	Ethylamine	+
Maltose	+	D-Mannitol	+	D-Glucosamine (N)	v
Cellobiose	–	D-Glucitol	+,s	10% NaC1/5% glucose	v
Trehalose	+,s	α-Methyl-D-glucoside	v	Starch formation	–
Lactose	–	Salicin	–	Urease	–
Melibiose	–	D-Gluconate	–,s	0.1% Cycloheximide	+
Raffinose	–	DL-Lactate	+	Growth at 42°C	+
Melezitose	v	Succinate	+	Growth at 45°C	+
Inulin	–	Citrate	+		

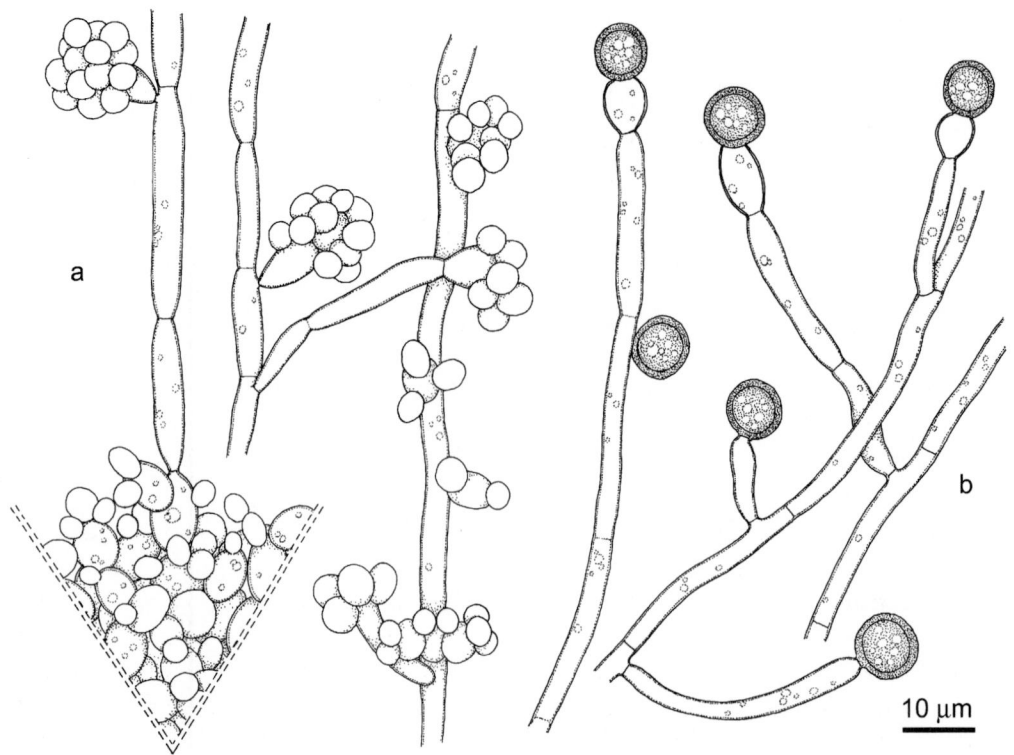

Candida albicans, strain M.-L. Kerkmann. a. Pseudomycelium emerging from cellular clumps; b. chlamydospores.

184

Differential diagnosis. Diagnostic is the production of chlamydospores on rice/Tween-80 agar and germ tubes with human serum or C/T agar (Berardinelli & Opheim, 1985), fermentation of glucose, and the following physiological characters: D-xylose +, arbutin –, soluble starch + and growth at 42°C and 45°C. *Candida dubliniensis* (Sullivan *et al.*, 1995; p. 194) differs by more abundant chlamydospore production, absence of growth at 45°C, and a dark-green rather than bluish green colour on CHROMagar (Odds & Bernaerts, 1994). A hexosamine medium for rapid identification of *C. albicans* was evaluated by Lipperheide *et al.* (1993). Variants with negative germ tube test, known as *C. claussenii*, have proven to be synonymous with *C. albicans*. Sucrose-negative variants, described as *C. stellatoidea*, also proved identical to *C. albicans* (Mahrous *et al.*, 1992). Very rarely red pigmented strains are encountered (Kerkmann *et al.*, 1999b), which might easily be mistaken for *Rhodotorula* species (p. 156); these are urea positive.

Molecular diagnostics. Molecular probes have been developed by Wickes *et al.* (1992), Holmes *et al.* (1992), del Castillo *et al.* (1993), Miyakawa *et al.* (1993) and Miyakawa & Mabuchi (1994). Niesters *et al.* (1993) used ribotyping of V4 SSU RNA for recognition of pathogenic *Candida* species and PCR-based detection in blood (van Deventer *et al.*, 1995), and Lischewski *et al.* (1996) fluorescent *in situ* hybridization in tissue. Recognition with nested-PCR of aspartate protease was performed by Reichard *et al.* (1993), the same gene on which specific probes were also developed (Mahrous *et al.*, 1992; Kanaizuka *et al.*, 1992; Sugita *et al.*, 1993). Williams *et al.* (1995) applied ribosomal ITS RFLP. Buchman *et al.* (1990), Burgener-Kairuz *et al.* (1994) and Wildfeuer *et al.* (1996) used segments of cytochrome encoding genes for detection of *C. albicans* and similar tests were developed on the basis of ribosomal genes (Holmes *et al.*, 1994), heat shock proteins (Camprin & Matthews, 1993), actin genes (Kan, 1993), aspartic protease genes (Flahaut *et al.*, 1997) and benomyl resistance genes (Weissman *et al.*, 1995). Lott *et al.* (1993) used ribosomal ITS2 sequences. A 26S-based specific probe was developed by Sandhu *et al.* (1995). Sullivan *et al.* (1995) described the closely related species *C. dubliniensis* (p. 194) on the basis of 18S V3 and RAPD differences. SSU, ITS and LSU restriction maps based on NCBI M60302, X71088 and U45776, resp., are given below. Note that the depicted 26S map is identical for the two species.

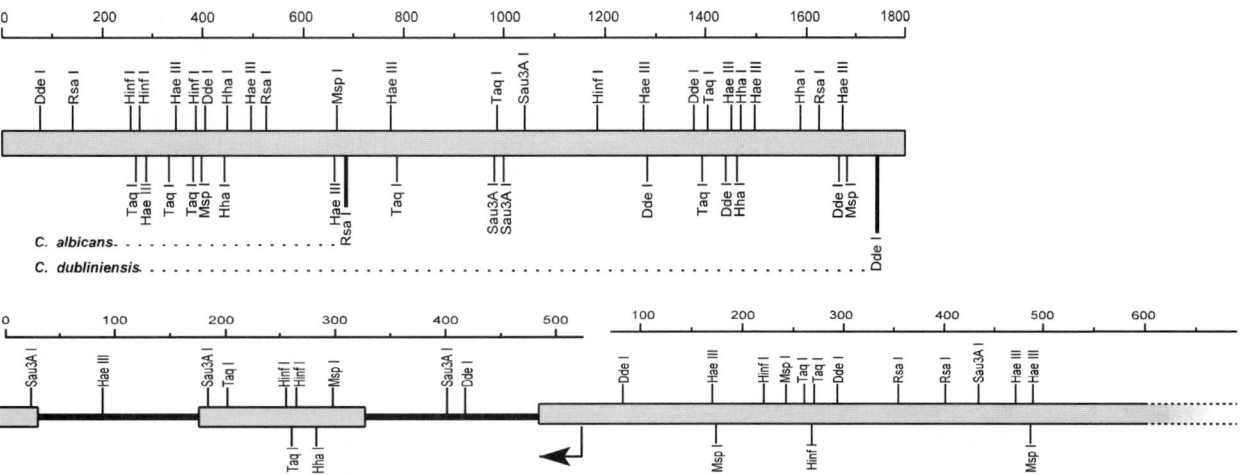

Serology. Quindos *et al.* (1997) evaluated a rapid, highly specific latex agglutination test for *C. albicans*. Discrimination between invasive candidiasis and colonization proved possible by using enolase antibodies (van Deventer *et al.*, 1994). Serology of systemic candidiasis was investigated by Navarro *et al.* (1993). Disseminated infections can be monitored by enzymatic measurement of D-arabinitol in human serum (Switchenko *et al.*, 1994; Walsh *et al.*, 1994). Larsson *et al.* (1994) diagnosed disseminated candidiasis by GC-determined D-arabinitol/L-arabinitol ratios in urine. An immunoassay for cytoplasmic *Candida* antigen diagnosis was developed by Morhart *et al.* (1994). Obayashi *et al.* (1995) introduced the possibility of diagnosis via ß-1,3-glucan limulus tests, which can potentially be used also for disseminated mycoses by other ascomycetous fungi. A review of antigen detection methods was presented by Matthews (1996).

Pathogenicity. BSL-2. The species commonly occurs in the digestive tract; candidiasis is one of the most important mycoses (see p. 23, 28). Vaginal candidiasis is extremely frequent. Mucocutaneous candidiasis occasionally leads to osteomyelitis (Mateev *et al.*, 1993). In debilitated patients, the species may become invasive through catheters (Hantschke, 1989) or lesions or perforations in mucous membranes, leading to peritonitis (Kujath *et al.*, 1990). Similarly, shunt infections are reported occasionally (Sánchez-Portocarrero *et al.*, 1994). Systemic candidiasis is frequent in neonates with low birth weight (Mcdonnell & Isaacs, 1995; Huang *et al.*, 1998). Thoracic surgery may lead to pericarditis (Schrank & Dooley, 1995). Gilbert *et al.* (1996) and Kuehnert *et al.* (1998) reported prosthetic valve endocarditis. Deep infections may originate in empyemas (Duffner *et al.*, 1997). Disseminated infections occur in cases of neutropenia (Swerdloff *et al.*, 1993) or haematologic malignancies, in patients receiving antimicrobial therapy or chemotherapy, in those who underwent abdominal surgery and in patients with severe burns (Still *et al.*, 1995). Mortality in such cases is high. Jonnalagadda *et al.* (1996) reported an osteomyelitis in a transplant patient. *Candida* pneumonia has been reviewed by Haron *et al.* (1993). Other clinical pictures, such as endophthalmitis (Towler *et al.*, 1995) or otitis (McDonald & Saulsbury, 1997) are rare. In immunocompromised patients, serotype B is twice as common as A, while in other patients both serotypes are equally distributed (Brawner *et al.*, 1992; Ollert *et al.*, 1995). Oral candidiasis is often a first sign of a HIV infection (Coleman *et al.*, 1993; Nielsen *et al.*, 1994), serotype B becoming more frequent (Torssander *et al.*, 1996).

References. Odds (1979, 1987), Bodey & Fainstein (1985), Saltarelli (1989), Blechschmidt & Meinhof (1989), Samaranayake & MacFarlane (1990), Tümbay *et al.* (1991), Cole *et al.* (1993), Thomas (1993), Bodey (1993), Scully *et al.* (1994).

Antifungal susceptibility.

Antifungal	MICs range	MIC 90	Strains	Reference
AMB	0.25-1	1	186/178	Pfaller *et al.* (1997/1998b)
AMB	0.5-2	1	206	Marco *et al.* (1998)
AMB	0.06-0.5	0.5	413	Zhanei *et al.* (1998)
AMB	≤0.03-1	1	97	Karlowsky *et al.* (1997)
AMB	0.5-4	2	32	Espinel-Ingroff (1998)
AMB	0.12-0.5	0.5	50	Davey *et al.* (1998)
5FC	0.06-128	4/2	186/178	Pfaller *et al.* (1997/1998b)
5FC	≤0.06->128	4	206	Marco *et al.* (1998)
5FC	≤0.03->64	0.5	413	Zhanei *et al.* (1998)
5FC	≤0.02-4	0.5	97	Karlowsky *et al.* (1997)
FLZ	0.12->128	2/16	186/178	Pfaller *et al.* (1997, 1998b)
FLZ	≤0.125->128	2	206	Marco *et al.* (1998)
FLZ	0.06->64	8	181	Kauffman & Zarins (1998)
FLZ	≤0.03-8	0.25	413	Zhanei *et al.* (1998)
FLZ	0.06-2	1	97	Karlowsky *et al.* (1997)
ITZ	0.015->8	0.25/1	186/178	Pfaller *et al.* (1997, 1998b)
ITZ	0.01->4	0.12	181	Kauffman & Zarins (1998)
ITZ	0.03-8	0.25	206	Marco *et al.* (1998)
ITZ	0.03-1	0.25	413	Zhanei *et al.* (1998)
ITZ	≤0.03-1	0.5	32	Espinel-Ingroff (1998)
KTZ	≤0.004-2	0.06	97	Karlowsky *et al.* (1997)
KTZ	≤0.03-1	0.13	413	Zhanei *et al.* (1998)
KTZ	≤0.12–8	1	50	Davey *et al.* (1998)
MCZ	≤0.12-2	1	50	Davey *et al.* (1998)
VCZ	0.007-4	0.12	181	Kauffman & Zarins (1998)
VCZ	≤0.0.15->16	0.06	206	Marco *et al.* (1998)
VCZ	≤0.03-4	0.5	32	Espinel-Ingroff (1998)

Nomenclature. *Oidium albicans* Robin - Hist. Nat. Vég. Paras. p. 488, 1853 ≡ *Saccharomyces albicans* (Robin) Reess - Sitzber. Phys. Med. Soc. Erlang. 8: 195, 1877 ≡ *Dematium albicans* (Robin) Laurent - Bull. Séanc. Soc. Belge Microsc. 16: 14, 1889 ≡ *Monilia albicans* (Robin) Zopf - Die Pilze p. 478, 1890 ≡ *Endomyces albicans* (Robin) Vuillemin - C.R. Hebd. Séanc. Acad. Sci., Paris 127: 630, 1898 ≡ *Endomyces vuilleminii* Landrieu - Myc. Ocul. p. 106, 1912 (name change) ≡ *Parasaccharomyces albicans* (Robin) Froilano de Mello, *in* Froilano de Mello & Fernandes - Arq. Hig. Pat. Exot. 6: 271, 1918 ≡ *Candida albicans* (Robin) Berkhout - Schimmelges Monilia, Oidium, Oospora, Torula p. 44, 1923 ≡ *Endomycopsis albicans* (Robin) Stelling-Dekker - Die Sporog. Hefen p. 265, 1931 ≡ *Mycotorula albicans* (Robin) Langeron & Talice - Annls Parasit. Hum. Comp. 18: 44, 1932 ≡ *Guilliermondella vuilleminii* (Landrieu) C.W. Dodge - Med. Mycol. p. 136, 1935 ≡ *Syringospora albicans* (Robin) C.W. Dodge - Med. Mycol. p. 274, 1935 ≡ *Procandida albicans* (Robin) Novák & Zsolt - Acta Bot. Acad. Sci. Hung. 7: 133, 1961.

Syringospora robinii Quinquad - Arch. Physiol. Norm. Path. 1: 290-305, 1868.

Saccharomyces tumefaciens albus Foulerton - J. Path. Bact. 6: 37-63, 1900 ≡ *Monilia tumefaciens albus* (Foulerton) Ota, *in* Motta - Atti R. Accad. Fisiocrit. Siena, Ser. 9, 17: 639-654, 1926 ≡ *Myceloblastanon tumefaciens album* (Foulerton) Ota - Jpn. J. Derm. Urol. 28: 4, 1928.

Parendomyces albus Querat & Laroche - Bull. Méd. Soc. Méd. Hôp. Paris, Sér. 3, 28: 111-136, 1909 ≡ *Cryptococcus albus* (Querat & Laroche) Castellani & Chalmers - Man. Trop. Med., ed. 3, p. 1080, 1919 ≡ *Monilia alba* (Querat & Laroche) Sartory - Champ. Paras. Homme Anim. p. 707, 1923 ≡ *Candida alba* (Querat & Laroche) de Almeida - Anais Fac. Med. Univ. S. Paulo 9: 11, 1933.

Endomyces faecalis Castellani - Br. Med. J. 2: 1209, 1912 ≡ *Monilia faecalis* (Castellani) Castellani - J. Trop. Med. Hyg. 17: 307, 1914 ≡ *Myceloblastanon faecale* (Castellani) Ota - Jpn. J. Derm. Urol. 27: 178, 1927 ≡ *Castellania faecalis* (Castellani) C.W. Dodge - Med. Mycol. p. 254, 1935.

Endomyces pinoyi Castellani - Br. Med. J. 2: 1210, 1912 ≡ *Monilia pinoyi* (Castellani) Castellani & Chalmers - Man. Trop. Med., ed. 2, p. 826, 1913 ≡ *Myceloblastanon pinoyi* (Castellani) Ota - Jpn. J. Derm. Urol. 27: 178, 1927 ≡ *Candida pinoyi* (Castellani) Basgal - Contrib. Estud. Blastom. Pulmon p. 50, 1931 ≡ *Blastodendrion pinoyi* (Castellani) Langeron & Talice - Annls Parasit. Hum. Comp. 10: 62, 1932 ≡ *Mycotorula pinoyi* (Castellani) Saggese - Riv. Clin. Pediatr. 32: 941, 1934.

Endomyces pulmonalis Castellani - Arch. Parasit. 16: 187, 1913 ≡ *Monilia pulmonalis* (Castellani) Castellani & Chalmers - Man. Trop. Med, ed. 2, p. 826, 1913 ≡ *Myceloblastanon pulmonale* (Castellani) Ota - Jpn. J. Derm. Urol. 27: 178, 1927 ≡ *Candida pulmonalis* (Castellani) Basgal - Contrib. Estud. Blastom. Pulmon. p. 49, 1931 ≡ *Castellania pulmonalis* (Castellani) C.W. Dodge - Med. Mycol. p. 255, 1935.

Monilia decolorans Castellani & Low - J. Trop. Med. Hyg. 16: 33, 1913 ≡ *Myceloblastanon decolorans* (Castellani & Low) Ota - Jpn. J. Derm. Urol. 27: 178, 1927 ≡ *Castellania decolorans* (Castellani & Low) C.W. Dodge - Med. Mycol. p. 253, 1935.

Monilia metchnikoffii Castellani - Annls Inst. Pasteur 30: 144, 1916 ≡ *Castellania metchnikoffii* (Castellani) C.W. Dodge - Med. Mycol. p. 250, 1935.

Monilia psilosis Ashford - Am. J. Med. Sci. 154: 157, 1917 ≡ *Myceloblastanon psilose* (Ashford) Ota - Jpn. J. Derm. Urol. 27: 176, 1927 ≡ *Mycotorula psilose* (Ashford) Langeron & Talice - Annls Parasit. Hum. Comp. 10: 47, 1932 ≡ *Candida psilosis* (Ashford) de Almeida - Anais Fac. Med. Univ. S. Paulo 9: 77, 1933 ≡ *Syringospora psilosis* (Ashford) C.W. Dodge - Med. Mycol. p. 279, 1935.

Parasaccharomyces ashfordii Anderson - J. Infect. Dis. 21: 341-387, 1917 ≡ *Myceloblastanon ashfordii* (Anderson) Ota - Jpn. J. Derm. Urol. 27: 170, 1927.

Monilia bethaliensis Pijper, *in* Castellani & Chalmers - Man. Trop. Med., ed. 3, p. 1091, 1919 ≡ *Myceloblastanon bethaliense* (Pijper) Ota - Jpn. J. Derm. Urol. 27: 171, 1927 ≡ *Candida bethaliensis* (Pijper) C.W. Dodge - Med. Mycol. p. 231, 1935.

Monilia metalondinensis Castellani & Chalmers - Man. Trop. Med., ed. 3, p. 1082, 1919 ≡ *Myceloblastanon metalondinensis* (Castellani & Chalmers) Ota - Jpn. J. Derm. Urol. 27: 178, 1927 ≡ *Candida metalondinensis* (Castellani & Chalmers) Basgal - Contrib. Estud. Blastom. Pulmon. p. 49, 1931 ≡ *Castellania metalondinensis* (Castellani & Chalmers) C.W. Dodge - Med. Mycol. p. 256, 1935 ≡ *Candida albicans* (Robin) Berkhout var. *metalondinensis* (Castellani & Chalmers) Ciferri - Man. Micol. Med., ed. 2, p. 252, 1960.

Monilia nabarroi Castellani & Chalmers - Man. Trop. Med., ed. 3, p. 1090, 1919 ≡ *Myceloblastanon nabarroi* (Castellani & Chalmers) Ota - Jpn. J. Derm. Urol. 27: 178, 1927 ≡ *Castellania nabarroi* (Castellani & Chalmers) C.W. Dodge - Med. Mycol. p. 253, 1935.

Monilia pseudolondinoides Castellani & Chalmers - Man. Trop. Med., ed. 3, p. 1082, 1919 ≡ *Castellania pseudolondinoides* (Castellani & Chalmers) C.W. Dodge - Med. Mycol. p. 255, 1935.

Endomyces actoni Vuillemin - C.R. Hebd. Séanc. Acad. Sci., Paris 170: 788-790, 1920 ≡ *Monilia actoni* (Vuillemin) Vuillemin - Champ. Paras. Myc. Homme p. 84, 1931.

Monilia pseudoalbicans Neveu-Lemaire - Précis Parasitol. Hum., ed. 5, p. 77, 1921 ≡ *Myceloblastanon pseudoalbicans* (Neveu-Lemaire) Ota - Jpn. J. Derm. Urol. 27: 173, 1927 ≡ *Mycoderma pseudoalbicans* (Neveu-Lemaire) C.W. Dodge - Med. Mycol. p. 229, 1935 ≡ *Geotrichum pseudoalbicans* (Neveu-Lemaire) de Almeida & Lacaz - Folia Clín. Biol., S. Paulo 12: 46, 1940.

Cryptococcus copellii Neveu-Lemaire - Précis Parasitol. Hum. p. 79, 1921 ≡ *Myceloblastanon copellii* (Neveu-Lemaire) Ota - Jpn. J. Derm. Urol. 27: 171, 1927 ≡ *Torulopsis copellii* (Neveu-Lemaire) de Almeida - Anais Fac. Med. Univ. S. Paulo 9: 76, 1933 ≡ *Castellania copellii* (Neveu-Lemaire) C.W. Dodge - Med. Mycol. p. 252, 1935.

Endomyces molardii Salvat & Fontoynont - Bull. Soc. Path. Exot. 15: 320, 1922 ≡ *Zymonema molardii* (Salvat & Fontoynont) C.W. Dodge - Med. Mycol. p. 173, 1935.

Cryptococcus laryngitidis Sartory, Petges & Claqué - C.R. Soc. Biol. 84: 179, 1923 ≡ *Monilia laryngitidis* (Sartory, Petges & Claqué) Vuillemin - Champ. Paras. Myc. Homme p. 84, 1931 ≡ *Torulopsis laryngitidis* (Sartory, Petges & Claqué) de Almeida - Anais Fac. Med. Univ. S. Paulo 9: 76, 1933 ≡ *Atelosaccharomyces laryngitidis* (Sartory, Petges & Claqué) C.W. Dodge - Med. Mycol. p. 344, 1935.

Monilia butantanensis Gomes de Cisneros - Ann. Paulist. Med. Cirurg. 15: 246, 1924 ≡ *Candida butantanensis* (Gomes de Cisneros) Langeron & Talice - Annls Parasit. Hum. Comp. 10: 54, 1932 ≡ *Parendomyces butantanensis* (Gomes de Cisneros) C.W. Dodge - Med. Mycol. p. 244, 1935.

Myceloblastanon cutaneum Ota - Derm. Wschr. 78: 232, 1924 ≡ *Monilia cutanea* (Ota) Nannizzi - Tratt. Micopat. Um. 4: 356, 1934 ≡ *Blastodendrion cutaneum* (Ota) C.W. Dodge - Med. Mycol. p. 287, 1935.

Myceloblastanon favrei Ota - Annls Parasit. Hum. Comp. 3: 181, 1925 ≡ *Monilia favrei* (Ota) Brumpt ≡ Précis Parasitol., ed. 4, p. 1241, 1927 ≡ *Cryptococcus favrei* (Ota) Pollacci & Nannizzi - Miceti Patog. Uomo Anim. 9: 83, 1929 ≡ *Blastodendrion favrei* (Ota) Langeron & Talice - Annls Parasit. Hum. Comp. 10: 62, 1932 ≡ *Candida favrei* (Ota) de Almeida - Anais Fac. Med. Univ. S. Paulo 9: 77, 1933.

Myceloblastanon gifuense Taniguchi - Jpn. J. Med. Sci., Ser. 13, Dermatol. Urol. 1: 74, 1926 ≡ *Monilia gifuense* (Taniguchi) Vuillemin - Champ. Paras. Myc. Homme p. 274, 1931 ≡ *Blastodendrion gifuense* (Taniguchi) C.W. Dodge - Med. Mycol. p. 288, 1935.

Monilia richmondii Shaw - Science 64: 300, 1926 ≡ *Castellania richmondii* (Shaw) C.W. Dodge - Med. Mycol. p. 256, 1935.

Monilia bucalis Niño & Puglisi - Semana Méd., B. Aires 34: 229, 1927 ≡ *Zymonema bucale* (Niño & Puglisi) C.W. Dodge - Med. Mycol. p. 176, 1935.

Monilia aldoi Pereira Filho - J. Trop. Med. Hyg. 30: 10, 1927 ≡ *Mycotoruloides aldoi* (Pereira Filho) Langeron & Talice - Annls Parasit. Hum. Comp. 10: 48, 1932 ≡ *Candida aldoi* (Pereira Filho) Castellani & Jacono - J. Trop. Med. Hyg. 36: 317, 1933.

Monilia fioccoi Pollacci & Nannizzi - Micet. Patog. Uomo. Anim., Fasc. 8, No. 76, 1928.

Endomyces albicans Okabe - Centbl. Bakt. Parasitkde, Abt. 1, 3: 186, 1929 ≡ *Zynonema albicans* (Okabe) C.W. Dodge - Med. Mycol. p. 174, 1935.

Monilia mannitofermentans Castellani - Proc. Soc. Exp. Biol. Med. 26: 544, 1929 ≡ *Monosporium tulanense* (Castellani) Agostini var. *mannitofermentans* (Castellani) Castellani - J. Trop. Med. Hyg. 36: 306, 1933 ≡ *Castellania mannitofermentans* (Castellani) C.W. Dodge - Med. Mycol. p. 255, 1935.

Monilia periunguealis Niño - Boln Inst. Clín. Quirúrg. Univ. B. Aires 5: 270-283, 1930 ≡ *Parendomyces periunguealis* (Niño) C.W. Dodge - Med. Mycol. p. 239, 1935.

Monilia alvarezsotoi Mazza & Niño, *in* Mazza, Niño, Quintana & Bernasconi - Boln Inst. Clín. Quirúrg., B. Aires 6: 180, 1931 ≡ *Zymonema alvarezsotoi* (Mazza & Niño) C.W. Dodge - Med. Mycol. p. 177, 1935 ≡ *Mycotorula alvarezsotoi* (Mazza & Niño) Niño - Boln Inst. Clín. Quirúrg, B. Aires 14: 591-1042, 1938.

Monilia vaginalis Mazza & Los Ríos - Boln. Inst. Clín. Quirúrg., B. Aires 6: 216-225, 1931 ≡ *Parendomyces vaginalis* (Mazza & Los Ríos) C.W. Dodge - Med. Mycol. p. 243, 1935.

Monilia inexorabilis Mazza & Palamedi - Reun. Soc. Argent. Patol. Reg. Norte Tucumán 7: 424-467, 1932 ≡ *Syringospora inexorabilis* (Mazza & Palamedi) C.W. Dodge - Med. Mycol. p. 242, 1935.

Mycotoruloides triadis Langeron & Talice - Annls Parasit. Hum. Comp. 10: 48, 1932 ≡ *Candida triadis* (Langeron & Talice) Langeron & Guerra - Annls Parasit. Hum. Comp. 16: 452, 1938.

Mycotoruloides ovalis Langeron & Talice - Annls Parasit. Hum. Comp. 10: 49, 1932.

Blastodendrion oosporioides Zach, *in* Wolfram & Zach - Arch. Derm. Syph. 169: 102, 1933 ≡ *Parasaccharomyces oosporioides* (Zach) C.W. Dodge - Med. Mycol. p. 266, 1935.

Cryptococcus pinoysimilis Castellani - Med. Press Circ. 136: 441, 1933 ≡ *Candida pinoysimilis* (Castellani) Castellani - J. Trop. Med. Hyg. 36: 312, 1933.

Candida mycotoruloidea Ciferri & Redaelli - Arch. Mikrobiol. 6: 50, 1935.

Candida desidiosa Ciferri & Redaelli - Arch. Mikrobiol. 6: 65, 1935.

Zynonema album C.W. Dodge - Med. Mycol. p. 175, 1935.

Parasacharomyces colardi C.W. Dodge - Med. Mycol. p. 267, 1935.

Syringospora cutanea C.W. Dodge - Med. Mycol. p. 278, 1935.

Syringospora negronii C.W. Dodge - Med. Mycol. p. 280, 1935.

Syringospora hasegawae C.W. Dodge - Med. Mycol. p. 282, 1935.

Monilia stellatoidea Jones, Martin & Durham - Am. J. Obstet. Gynecol. 35: 105, 1938 ≡ *Candida stellatoidea* (Jones, Martin & Durham) Langeron & Guerra - Annls Parasit. Hum. Comp. 17: 257, 1939 ≡ *Procandida stellatoidea* (Jones, Martin & Durham) Novák & Zsolt - Acta Bot. Acad. Sci. Hung. 7: 93-145, 1961.

Candida claussenii Lodder & Kreger-van Rij - The Yeasts p. 578, 1952.

Candida langeroni Dietrichson - Annls Parasit. Hum. Comp. 29: 479, 1954 (invalid) ≡ *Candida langeroni* Dietrichson ex van Uden & Buckley, *in* Lodder - The Yeasts, ed. 2, p. 989, 1970

Candida intestinalis Batista & J. Silveira - Hospital, Rio de Janeiro 56: 293-299, 1959.

Candida genitalis Batista & J. Silveira - Publçoes Inst. Micol. Recife 170: 11, 1962.

Candida nouvelii Saëz - Bull. Soc. Mycol. Fr. 89: 82, 1973.

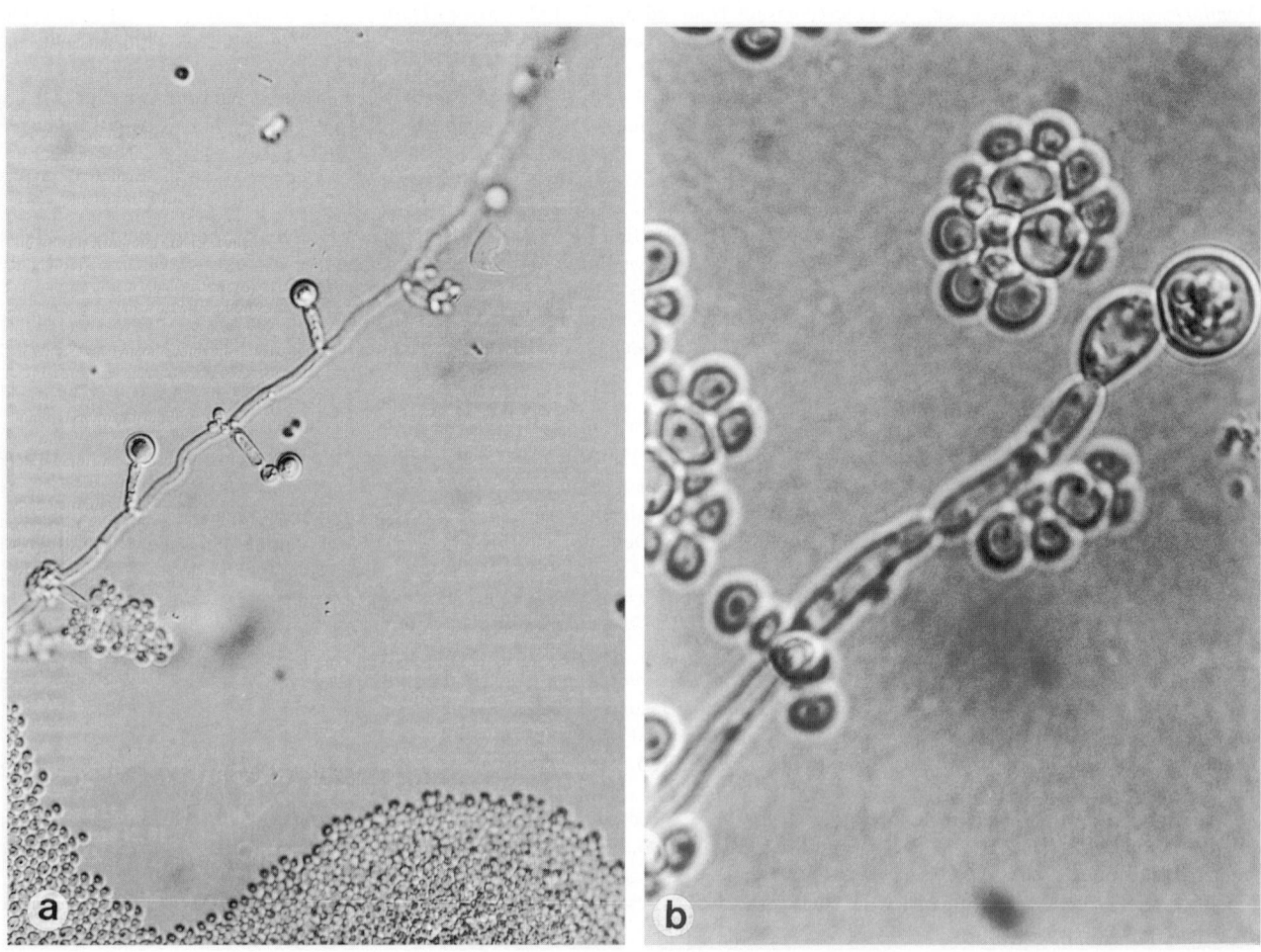

Candida albicans, strain M.-L. Kerkmann. Budding cells and chlamydospores. a. ×600; b. ×1500.

Candida catenulata Diddens & Lodder

Colony characteristics. Colonies (YPGA) whitish to cream-coloured, buttery, finely cerebriform.
Microscopy. Budding cells ellipsoidal, 3.0-7.0 × 3.0-3.5 µm. Pseudomycelium may be present.
Physiology.

Fermentation:		Soluble starch	v	*myo*-Inositol	–
Glucose	v	D-Xylose	v	Hexadecane	–
Galactose	–,s	L-Arabinose	–	Vitamin-free	–
Sucrose	–	D-Arabinose	–	2-Keto-D-gluconate	v
Maltose	–,s	D-Ribose	–,s	D-Glucuronate	–
Lactose	–	L-Rhamnose	–	Xylitol	v
Raffinose	–	D-Glucosamine (C)	–,s	L-Arabinitol	–
α,α-Trehalose	–	*N*-Acetyl-D-glucosamine	–	Arbutin	–
		Methanol	–	Nitrate	+
Growth:		Ethanol	+	Nitrite	–
Glucose	+	Glycerol	+	Cadaverine	+
Galactose	+	*meso*-Erythritol	–	Creatinine	–
L-Sorbose	–	Ribitol	+,s	L-Lysine	+
Sucrose	–	Galactitol	–	Ethylamine	v
Maltose	v	D-Mannitol	+	D-Glucosamine (N)	–
Cellobiose	–	D-Glucitol	+,s	10% NaC1/5% glucose	–
α,α-Trehalose	+,s	α-Methyl-D-glucoside	–	Starch formation	–
Lactose	–	Salicin	–	Urease	–
Melibiose	–	D-Gluconate	–,s	0.01% Cycloheximide	+
Raffinose	–	DL-Lactate	+	0.1% Cycloheximide	+,s
Melezitose	–	Succinate	+	Growth at 37°C	v
Inulin	–	Citrate	+	Growth at 40°C	–

Differential diagnosis. Species signature: D-galactose +, L-sorbose –, sucrose –, cellobiose –, *meso*-erythritol –, 5-keto-D-gluconate –, DL-lactate –, citrate +, ethanol +, nitrate –, cadaverine +, w/o vitamins –, 10% NaCl +, urease –. Physiologically indistinguishable from *C. rugosa* (p. 218), *C. tropicalis* (p. 220) and *C. zeylanoides* (p. 225).
Molecular diagnostics. SSU and LSU restriction map based on NCBI AB013539 and U45714:

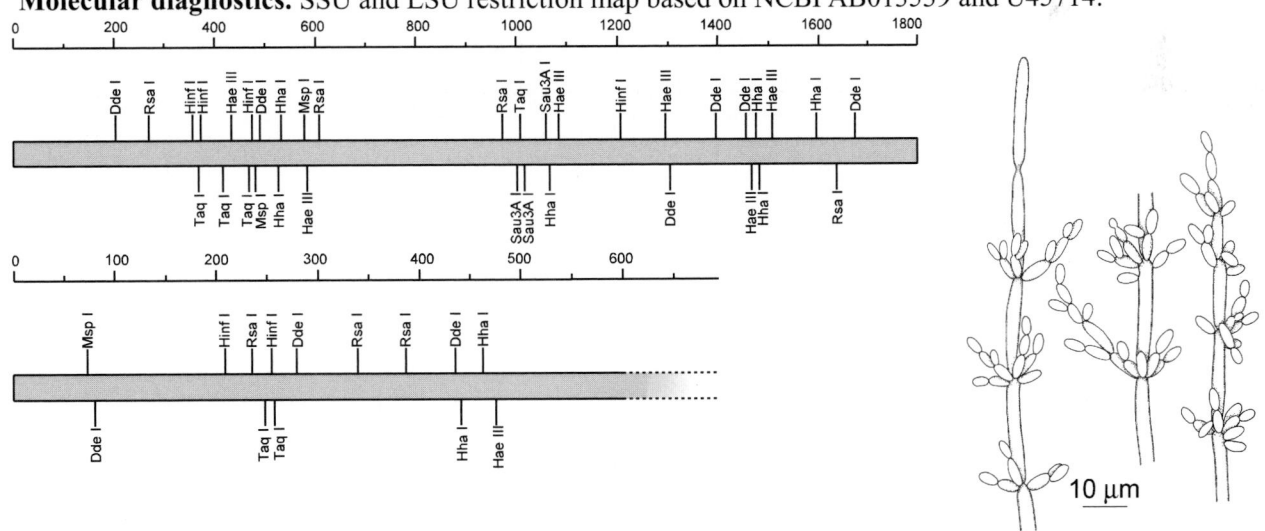

Candida catenulata, CBS 565. Pseudomycelium.

Pathogenicity. BSL-1. Mentioned as one of the fungi occurring in cancer patients (Smolyanskaya *et al*., 1996; Radosavljevic *et al*., 1999). Crozier & Coast (1977) reported an onychomycosis.
Nomenclature. *Blastodendrion brumpti* Guerra - Rôle Levures Dermatol. p. 44, 1935 (invalid) ≡ *Candida brumpti* (Guerra) Langeron & Guerra - Annls Parasit. Hum. Comp. 16: 496, 1938 ≡ *Mycotorula brumpti* (Guerra) Krassilnikov - Opred. Nizsh. Rast. 3: 145, 1954.
Candida ravauti Guerra - Rôle Levures Dermatol. p. 45, 1935 (invalid).
Candida catenulata Diddens & Lodder - Anaskosp. Hefen p. 486, 1942.

Candida chiropterorum Grose & Marinkelle

Colony characteristics. Colonies (YPGA) whitish to cream-coloured, buttery, later with small marginal patches of filaments.

Microscopy. Budding cells abundant, strongly cohering in poorly branched chains of up to 20 cells. Hyphae septate, branched, producing small groups of ellipsoidal to spherical, catenulate conidia on small, grouped denticles or on large, solitary pegs, just below hyphal septa.

Physiology

Fermentation:	–	L-Rhamnose	+	2-Keto-D-gluconate	+
		D-Glucosamine (C)	+	D-Glucarate	–
Growth:		N-Acetyl-D-glucosamine	+	D-Glucuronate	+
Glucose	+	Methanol	–	Xylitol	–
Galactose	+	Ethanol	–	L-Arabinitol	–
L-Sorbose	+	Glycerol	+	Arbutin	+
Sucrose	+	*meso*-Erythritol	+	Nitrate	–
Maltose	+	Ribitol	+	Nitrite	–
Cellobiose	+	Galactitol	+	Cadaverine	+
α,α-Trehalose	+	D-Mannitol	+	Creatinine	–
Lactose	–	D-Glucitol	+	L-Lysine	+
Melibiose	–	α-Methyl-D-glucoside	+	Ethylamine	+
Raffinose	+	Salicin	+	10% NaCl/5% glucose	+
Melezitose	s	D-Gluconate	+	Starch formation	–
Inulin	–	DL-Lactate	s	Urease	–
Soluble starch	+	Succinate	–	0.01% Cycloheximide	+
D-Xylose	+	Citrate	+	0.1% Cycloheximide	s
L-Arabinose	+	*myo*-Inositol	+	Growth at 40°C	+
D-Arabinose	s	Hexadecane	–		
D-Ribose	s	Vitamin-free	–		

Differential diagnosis. Species signature: ethanol –, *myo*-inositol +, nitrate –, 10% NaCl +.

Molecular diagnostics. SSU and LSU restriction maps based on NCBI AB013591 and U45822:

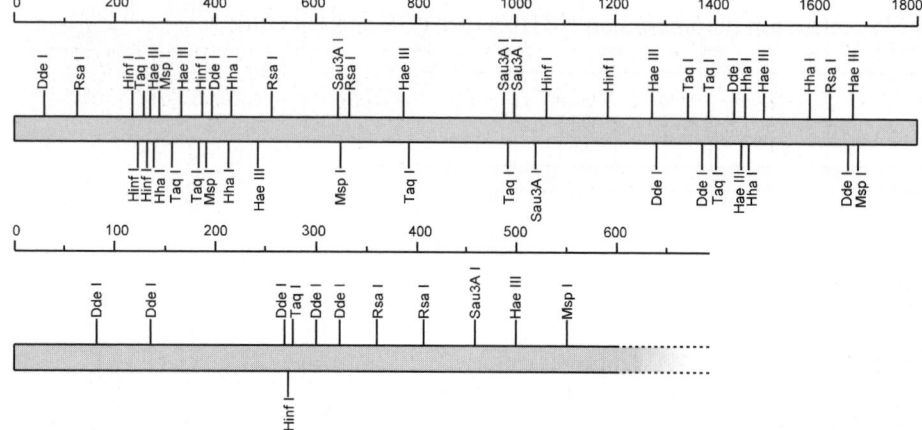

Pathogenicity. BSL-2. The species was originally described from bat liver (Grose & Marinkelle, 1968). Furman & Ahearn (1983) found it in human dialysis fluid.

Reference. Grose & Marinkelle (1968).

Nomenclature. *Candida chiropterorum* Grose & Marinkelle - Mycopath. Mycol. Appl. 36: 225, 1968.

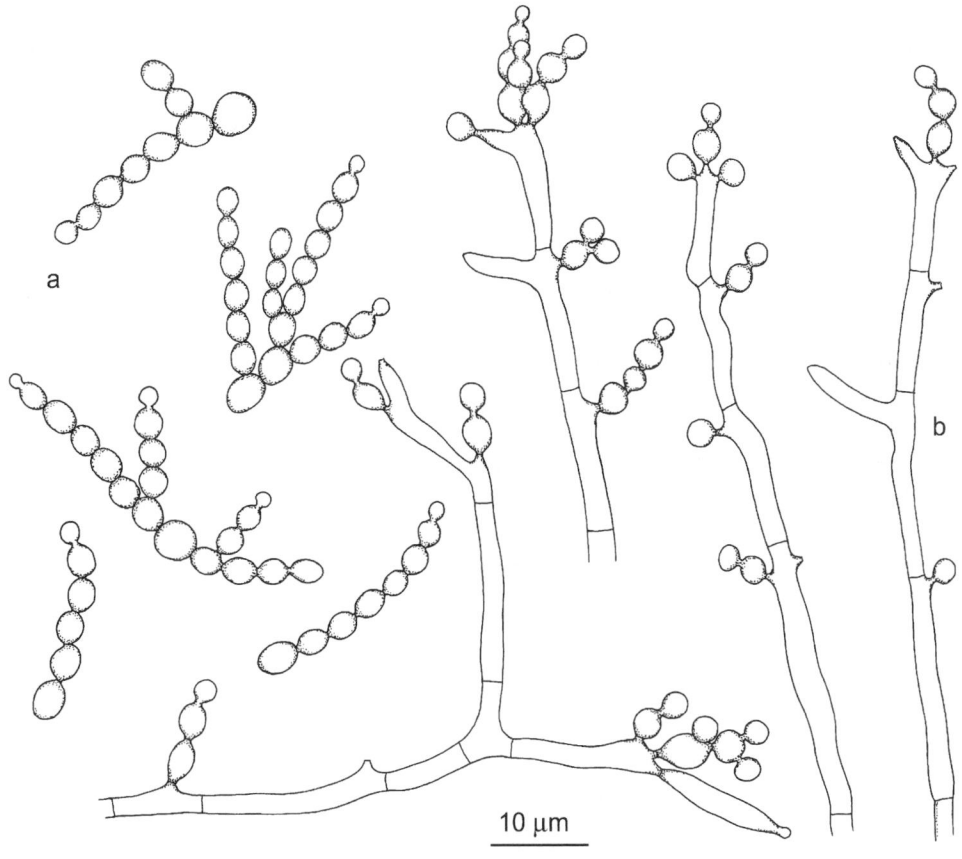

Candida chiropterorum, CBS 6064. a. Catenate budding cells; b. true mycelium with conidia on denticles.

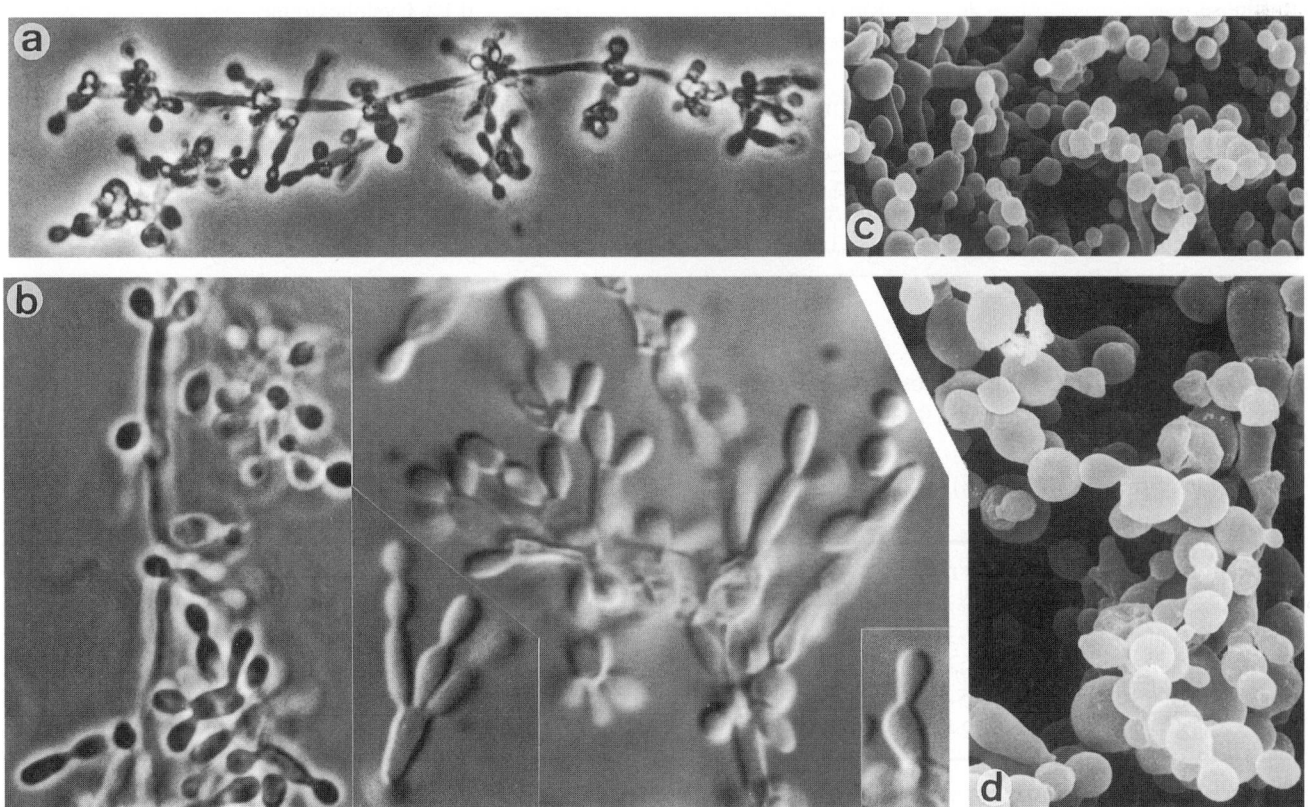

Candida chiropterorum, CBS 6064. Catenate budding cells and true mycelium with conidia on denticles. a. ×640; b, c. ×1600; d. ×2600.

191

Candida ciferrii Kreger-van Rij

Colony characteristics. Colonies (YPGA) whitish to cream-coloured; hairy projections may be present giving the colony a rough, feathery appearance; colonies convex with fringed margin.

Microscopy. Budding cells abundant, often catenate. On CMA septate hyphae are formed on which cylindrical conidiogenous cells are borne apically with a swollen head with distinct denticles bearing clusters of conidia. Conidia tear-shaped. Note that some strains are yeast-like, whereas others may be entirely hyphal.

Teleomorph. *Stephanoascus ciferrii* M.Th. Smith *et al.* (*Ascomycota, Hemiascomycetes, Saccharomycetales: Dipodascaceae*).

Asci formed after mating, spherical with firm walls and sterile top-cell, containing 1-4 hat-shaped ascospores. Heterothallic.

Differential diagnosis. Species signature: L-rhamnose +, ethanol +, D-glucuronate +, w/o thiamine –, D-glucosamine (N) –, 0.1% cycloheximide +, 10% NaCl +, urease –. Some strains are strictly yeast-like, whereas in others hyphae are preponderant. Hyphal colonies can be mistaken for *Sporothrix schenckii* (p. 925), but the colonies of *C. ciferrii* are soft and buttery, and the heads with conidium-bearing denticles are distinctly swollen. *Cerinosterus cyanescens* (p. 541) has unswollen conidial heads, produces a red/blue pigment and is urease positive.

Physiology.

Fermentation:	–	Inulin	–	D-Mannitol	+
		Soluble starch	v	D-Glucitol	+
Growth:		D-Xylose	+	α-Methyl-D-glucoside	v
Glucose	+	L-Arabinose	+	Salicin	v
Galactose	+	D-Arabinose	+	D-Gluconate	+
L-Sorbose	+	D-Ribose	v	DL-Lactate	+
Sucrose	+	L-Rhamnose	+	Succinate	v
Maltose	+	D-Glucosamine (C)	–	Citrate	v
Cellobiose	v	Methanol	–	*myo*-Inositol	+
α,α-Trehalose	+	Ethanol	+	Nitrate	–
Lactose	–	Glycerol	+	Vitamin-free	–
Melibiose	v	*meso*-Erythritol	+	Arbutin	v
Raffinose	+	Ribitol	+	0.1% Cycloheximide	+
Melezitose	–	Galactitol	+	Growth at 37°C	+

Molecular diagnostics. LSU restriction map based on NCBI U40138:

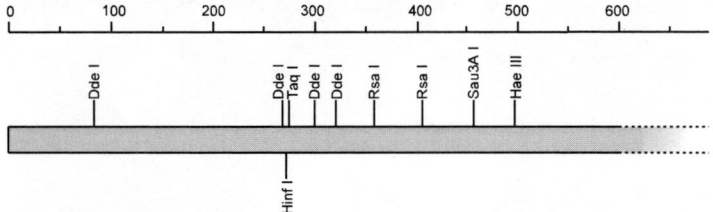

Pathogenicity. BSL-2. The species is occasionally found in association with cattle. It is rarely isolated from clinical specimens (Furman & Ahearn, 1983) and has been reported as an agent of human onychomycosis (de Gentile *et al.*, 1991).

Reference. Smith *et al.* (1976).

Antifungal susceptibility.

Antifungal	MICs range	MIC 50	Strains	Reference
AMB	1-2	1	6	Espinel-Ingroff (1998)
FLZ	32->64	64	6	Espinel-Ingroff (1998)
ITZ	0.5-2	1	6	Espinel-Ingroff (1998)
VCZ	0.12-0.5	0.25	6	Espinel-Ingroff (1998)

Nomenclature. *Candida ciferrii* Kreger-van Rij - Mycopath. Mycol. Appl. 26: 50, 1965.
  Stephanoascus ciferrii M.Th. Smith, van der Walt & Johannsen - Antonie van Leeuwenhoek 42: 125, 1976.
  Sporothrix catenata de Hoog & Constantinescu - Antonie van Leeuwenhoek 47: 367, 1983.

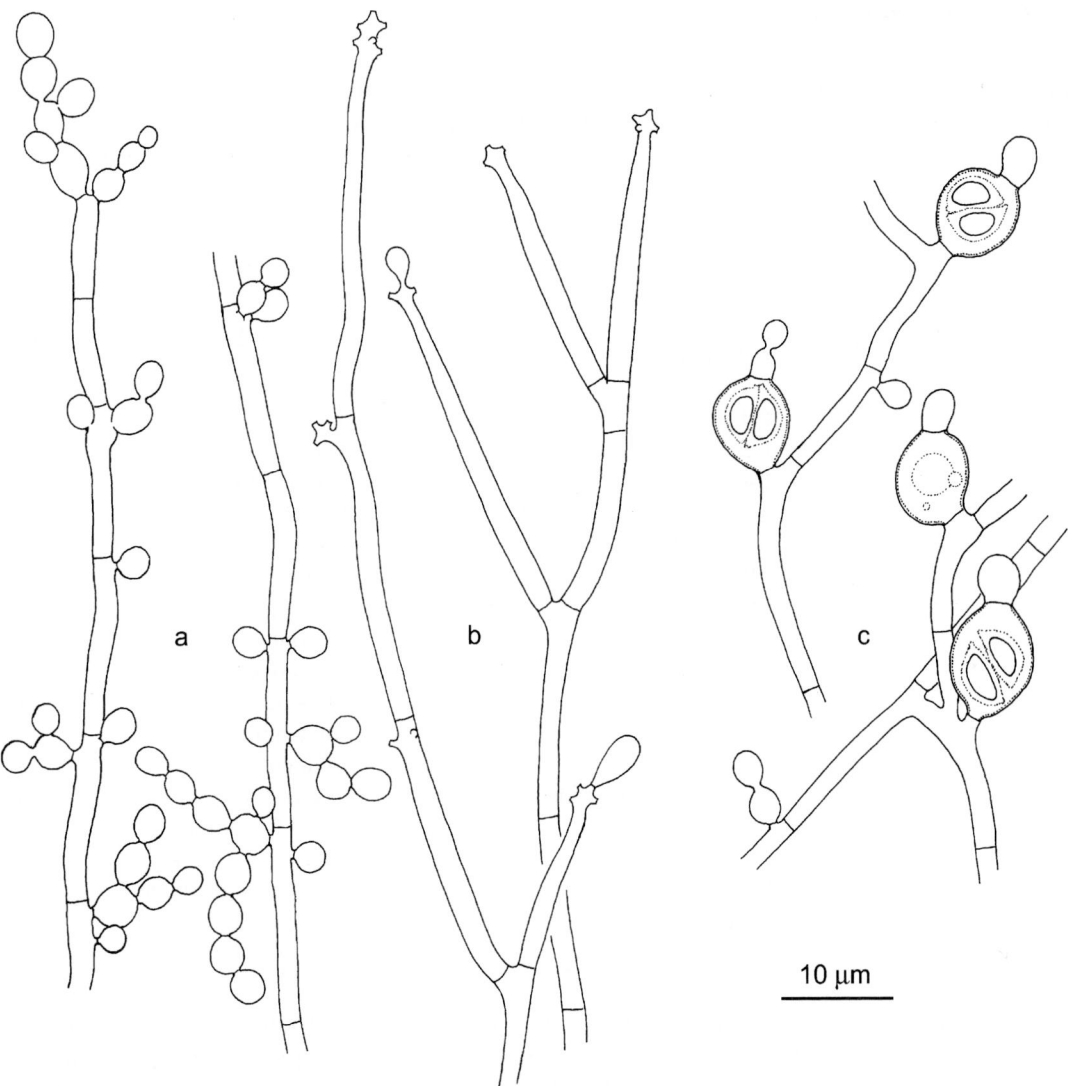

Candida ciferrii (*Stephanoascus ciferrii*), CBS 5295 × 7409. a. Pseudomycelium; b. *Sporothrix*-like conidiogenous cells; c. asci with ascospores.

Candida dubliniensis Sullivan *et al.*

Colony characteristics. Colonies (YPGA) cream-coloured, glistening or somewhat waxy, soft and usually smooth. **Microscopy**. Budding cells (RA) (sub)spherical, 3-8 × 2-7 μm. Pseudomycelium present, with blastoconidia in dense grape-like arrangement. Dark, spherical chlamydospores mostly terminally in chains of 1-3 on a slightly swollen subtending cell. Germ tubes formed with serum.
Physiology.

Fermentation:		Soluble starch	+	*myo*-Inositol	–
Glucose	+	D-Xylose	v	Vitamin-free	–,s
Galactose	v	L-Arabinose	–	2-Keto-D-gluconate	+
Sucrose	–	D-Arabinose	–	D-Glucarate	–
Maltose	+	D-Ribose	–	D-Glucuronate	–
Lactose	–	L-Rhamnose	–	Xylitol	+
Raffinose	–	D-Glucosamine (C)	–,s	L-Arabinitol	–
α,α-Trehalose	v	*N*-Acetyl-D-glucosamine	+	Arbutin	–
		Methanol	–	Nitrate	–
Growth:		Ethanol	+	Nitrite	–
Glucose	+	Glycerol	+,s	Cadaverine	+
Galactose	+	*meso*-Erythritol	–	Creatine	–
L-Sorbose	–	Ribitol	+,s	Creatinine	–
Sucrose	+	Galactitol	–	L-Lysine	+,w
Maltose	+	D-Mannitol	+	Ethylamine	+
Cellobiose	–	D-Glucitol	+	D-Glucosamine (N)	–
α,α-Trehalose	+	α-Methyl-D-glucoside	+,s	10% NaC1/5% glucose	v
Lactose	–	Salicin	–	Starch formation	–
Melibiose	v	D-Gluconate	–,s	Urease	–
Raffinose	–	DL-Lactate	+	0.1% Cycloheximide	+
Melezitose	+	Succinate	+	Growth at 40°C	+
Inulin	–	Citrate	+	Growth at 45°C	–

Differential diagnosis. The characteristic chlamydospores are particularly produced on GACA, on which medium they are absent in *C. albicans* (Staib & Morschhäuser, 1999). The physiological pattern is similar to that of *C. albicans* (p. 184), except for absence of growth at 45°C (Pinjon *et al.*, 1998). The species can further be distinguished by initial colonies being dark green on CHROMagar (Kilpatrick *et al.*, 1998), by absence of fluorescence of colonies on methyl-blue Sabouraud's agar under Wood's light (Schoofs *et al.*, 1997), the reduction of tetrazolium salt (Velagraki & Logotheti, 1998), and by absence of ß-glucosidase activity (Boerlin *et al.*, 1995). Using assimilation velocities within 48h in miniaturized test systems, Pincus *et al.* (1999) noted that glycerol (mostly +), methyl-α-D-glucoside (–), trehalose (–) and D-xylose (–) were suitable. In combination with absence of growth at 45°C this provides reliable identification (Gales *et al.*, 1999). Note that the species is physiologically indistinguishable from *C. guilliermondii* (p. 200) and. *C. tropicalis* (p. 220).
Serology. Bikandi *et al.* (1998) developed immunofluorescence for distinction from *C. albicans*. A monoclonal antibody specific for *C. albicans* germ tubes is available (Marot-Leblond *et al.*, 2000).
Molecular diagnostics. The 18S rDNA V3 variable region was considered to diverge sufficiently to warrant the separation of this species from *Candida albicans* (Sullivan *et al.*, 1995; Morschhäuser *et al.*, 1999). This was confirmed by ACT1 sequencing data (Donnelly *et al.*, 1999). The species also shows different fingerprint (Diaz Guerra *et al.*, 1999) and karyotype patterns (Sullivan *et al.,* 1993). Joly *et al.* (1999) and Tamura *et al.* (2000) selected random diagnostic sequences and Kurzai *et al.* (1999) introduced a PCR test on the basis of PHR-1 sequences for *C. albicans* which does not detect *C. dubliniensis*. With the panel of restriction enzymes used in this Atlas, the difference is a *Rsa*I site at position 684 which is absent in *C. dubliniensis* and a *Dde*I site at position 1741, which is absent from *C. albicans* (p. 184). SSU and LSU restriction maps based on NCBI X99399 and U57685, provide diagnostic differences with *C. albicans* only (p. 185).
Pathogenicity. BSL-2. The species occurs in the oral cavity of mostly HIV-positive individuals, particularly those suffering from recurrent episodes of infection (Coleman *et al.*, 1997; Meiller *et al.*, 1999), and is rarely isolated from other parts of the body. In HIV it may be underdiagnosed (Jabra-Rizk *et al.*, 1999, 2000). It also emerges with bloodstream infections in transplant patients with neutropenia (Meis *et al.*, 1999), patients undergoing broad-spectrum

antibiotic therapy (Polachek *et al.*, 2000), suffering from gastric carcinoma (Kamei *et al.*, 2000) or having other constitutional disorders, among them HIV infection (Brandt *et al.*, 2000).

References. Coleman *et al.* (1997), Schoofs *et al.* (1997), Sullivan & Coleman (1998), Pinjon *et al.* (1998), Gifillan *et al.* (1998), Odds *et al.* (1998b).

Antifungal susceptibility.

Antifungal	MICs range	MIC 90	Strains	Reference
AMB	0.05-0.38	0.2	71	Pfaller *et al.* (1999)
AMB	≤0.03-0.25	nd	22	Kilpatrick *et al.* (1998)
5FC	≤0.12	≤0.12	71	Pfaller *et al.* (1999)
FLZ	0.12-64	8.0	71	Pfaller *et al.* (1999)
FLZ	≤0.25-64	nd	22	Kilpatrick *et al.* (1998)
ITZ	0.015-0.5	0.25	71	Pfaller *et al.* (1999)
ITZ	0.03-0.25	nd	22	Kilpatrick *et al.* (1998)
VCZ	≤0.008-0.5	0.3	71	Pfaller *et al.* (1999)
VCZ	≤0.125-2	nd	22	Kilpatrick *et al.* (1998)

Nomenclature. *Candida dubliniensis* Sullivan, Westerneng, Haynes, Bennett & Coleman - Microbiology, Reading 141: 1519, 1995.

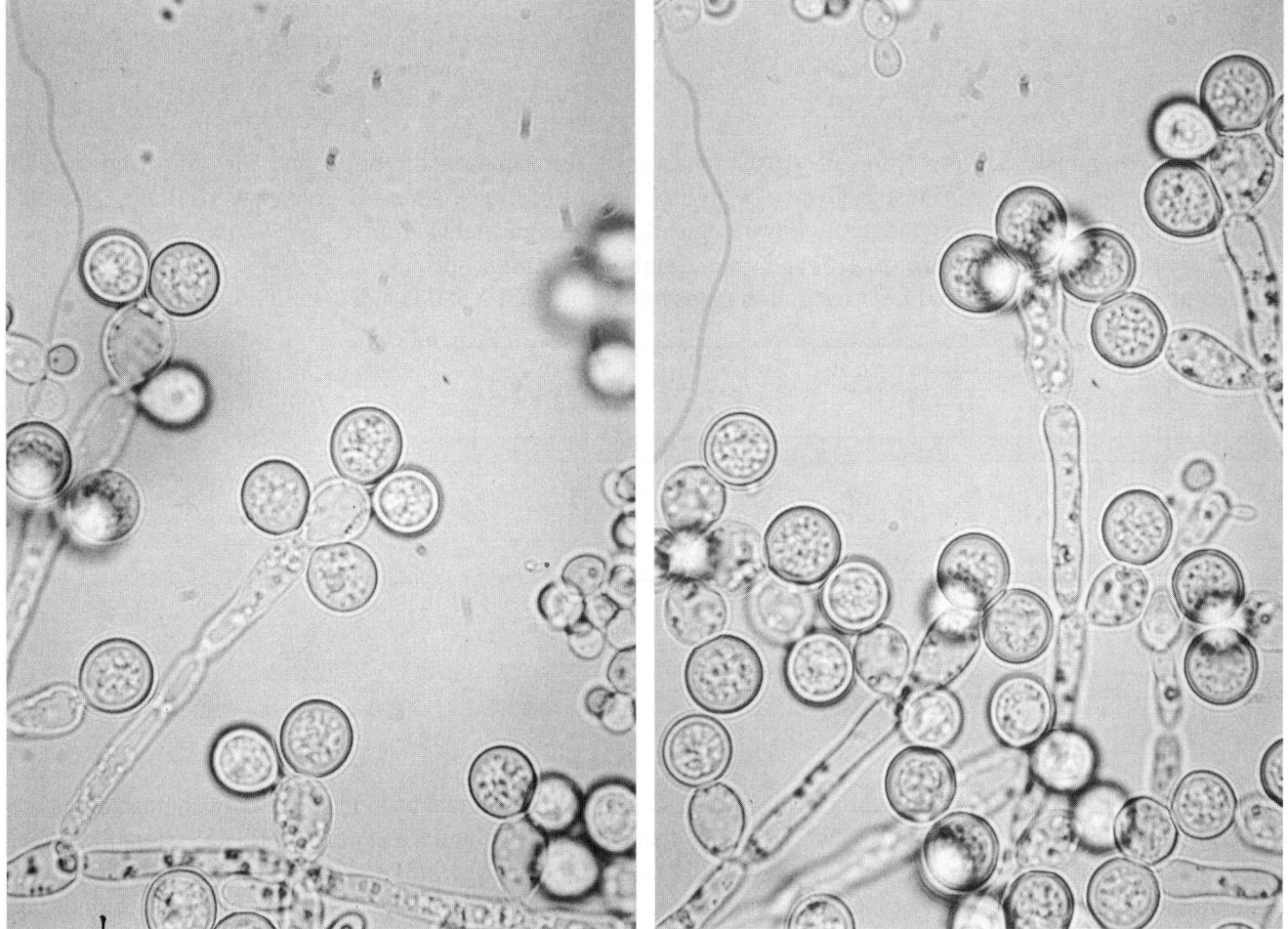

Candida dubliniensis, CD33. Chlamydospores and budding cells. Courtesy D.C. Coleman and D.J. Sullivan.

Candida famata (Harrison) S.A. Meyer & Yarrow

Colony characteristics. Colonies (YPGA) white to cream-coloured, butyrous.
Microscopy. Budding cells broadly ellipsoidal, 3.5-5.0 × 2-3.5 μm. Pseudomycelium absent.
Teleomorph. *Debaryomyces hansenii* (Zopf) Lodder & Kreger-van Rij (*Ascomycota, Hemiascomycetes, Saccharomycetales: Endomycetaceae*).
Asci spherical, persistent, containing 1-2 spherical ascospores with rough walls. Homothallic.
Physiology.

Fermentation:		Soluble starch	v	*myo*-Inositol	–
Glucose	–,w	D-Xylose	+	Hexadecane	v
Galactose	–,w	L-Arabinose	+,w	Vitamin-free	–
Sucrose	–,w	D-Arabinose	v	2-Keto-D-gluconate	+
Maltose	–	D-Ribose	v	5-Keto-D-gluconate	v
Lactose	–	L-Rhamnose	v	D-Glucarate	–
Raffinose	–,w	D-Glucosamine (C)	v	D-Glucuronate	v
α,α-Trehalose	–,w	*N*-Acetyl-D-glucosamine	v	Xylitol	+
		Methanol	–	L-Arabinitol	+
Growth:		Ethanol	+,w	Arbutin	v
Glucose	+	Glycerol	+	Nitrate	–
Galactose	+	*meso*-Erythritol	v	Nitrite	v
L-Sorbose	v	Ribitol	+	Cadaverine	v
Sucrose	+	Galactitol	v	Creatine	v
Maltose	+	D-Mannitol	+	Creatinine	v
Cellobiose	+	D-Glucitol	+,w	L-Lysine	+
α,α-Trehalose	+	α-Methyl-D-glucoside	+	Ethylamine	+
Lactose	v	Salicin	+,w	Urease	–
Melibiose	v	D-Gluconate	+,w	0.01% Cycloheximide	v
Raffinose	+	DL-Lactate	v	0.1% Cycloheximide	v
Melezitose	v	Succinate	+	Growth at 37°C	+
Inulin	v	Citrate	v	Growth at 40°C	+,w

Differential diagnosis. Differs from all yeasts by lack of fermentation of inulin and the following growth characteristics: galactose +, raffinose +, glycerol +, xylitol +, *myo*-inositol –, 2-keto-D-gluconate +, citrate +, butane-2,3-diol –, 50% glucose + and urease –. Physiologically indistinguishable from *C. guilliermondii* (p. 200)*, C. intermedia* (p. 203), *C. pelliculosa* (p. 215) and several saprobic *Candida* species.
Molecular diagnostics. SSU and LSU restriction maps based on NCBI X58053 and U45808:

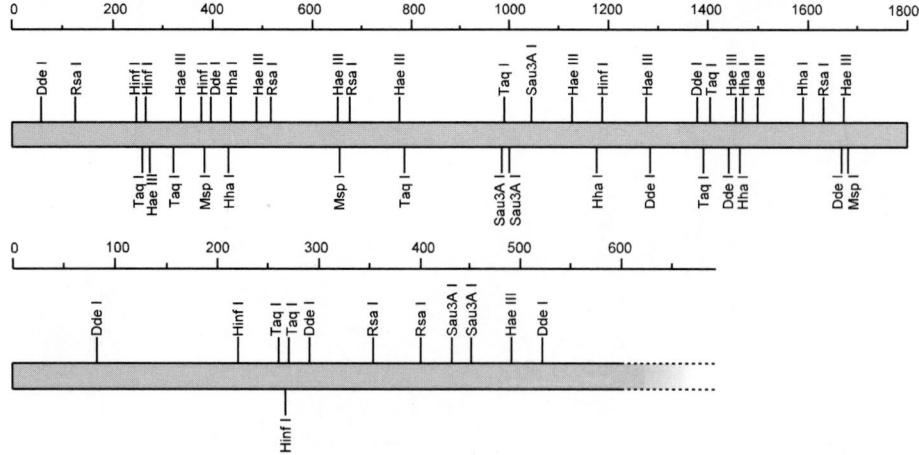

Pathogenicity. BSL-1. *Candida famata* var. *famata* is a common contaminant of food stuffs. Most human cases are caused by the var. *flareri*. Wong *et al.* (1982) described an osteomyelitis and Rao *et al.* (1991) an endophthalmitis. Disseminated cases were reported by Vazquez *et al.* (1993), Nicand *et al.* (1993) and Quindos *et al.* (1994), usually in debilitated patients (Carrega *et al.*, 1997). A cutaneous infection in a reindeer was described by Rehbinder & Mattson (1994).

Varieties. The var. *flareri* was described with ability to grow at 37°C, whereas the maximum growth temperature of the var. *famata* is 35°C (Nakase & Suzuki, 1985).

References. Nakase & Suzuki (1985), Ellis (1994), Nishikawa *et al.* (1996).

Nomenclature var. *famata.* *Saccharomyces hansenii* Zopf - Ber. Dt. Bot. Ges. 7: 94, 1889 ≡ *Debaryomyces hansenii* (Zopf) Lodder & Kreger-van Rij - The Yeasts p. 280, 1952.

Debaryomyces matruchotii Grigorakis & Péju - C.R. Soc. Biol. 35: 459, 1921.

Atelosaccharomyces hudeli de Beurmann & Gougerot - Les Nouv. Myc., 1918 ≡ *Debaryomyces hudeli* (de Beurmann & Gougerot) da Fonseca - Brasil Med. Bull. Inst. Pasteur 21: 651, 1923 ≡ *Debaryomyces kloeckeri* Guilliermond & Péju var. *hudeli* (de Beurmann & Gougerot) Stelling-Dekker - Sporog. Hefen p. 479, 1931.

Blastomyces hildegaardii Sasakawa - Centbl. Bakt. Parasitkde, Abt. 1, 88: 273, 1922 ≡ *Debaryomyces hildegaardii* (Sasakawa) Ota - Annls Parasit. Hum. Comp. 1: 124, 1923 ≡ *Debaryomyces matruchotii* Grigorakis & Péju race *hildegaardii* (Sasakawa) Stelling-Dekker - Sporog. Hefen p. 471, 1931.

Debaryomyces leopoldii Ota - Annls Parasit. Hum. Comp. 1: 128, 1923.

Debaryomyces lundsgaardii Ota - Annls Parasit. Hum. Comp. 1: 130, 1923.

Debaryomyces laedegaardii Ota - Annls Parasit. Hum. Comp. 1: 132, 1923.

Debaryomyces gruetzii Ota - Derm. Wschr. 78: 284, 1924.

Debaryomyces emphysematosus Ota - Derm. Wschr. 78: 289, 1924.

Mycotorula famata Harrison - Trans. R. Soc. Can., Biol. Sci., 22: 216, 1928 ≡ *Torulopsis famata* (Harrison) Lodder & Kreger-van Rij - The Yeasts p. 417, 1952 ≡ *Candida famata* (Harrison) S.A. Meyer & Yarrow - Int. J. Syst. Bact. 28: 612, 1978.

Debaryomyces mucosus Sartory, R. Sartory, Hufschmitt & J. Meyer - C.R. Hebd. Séanc. Acad. Sci., Paris 191: 281, 1930.

Saccharomyces sternoni Sartory, R. Sartory, Sternon & J. Meyer - Bull. Acad. Méd. Paris, Sér. 3, 107: 120, 1932.

Cryptococcus minor Pollacci & Nannizzi in Nannizzi - Tratt. Micopat. Um. 4: 317, 1934 ≡ *Torulopsis minor* (Pollacci & Nannizzi) Lodder - Anaskosp. Hefen 1: 178, 1934 ≡ *Parendomyces minor* (Pollacci & Nannizzi) C.W. Dodge - Med. Mycol. p. 244, 1935.

Nomenclature var. *flareri.* *Blastodendrion flareri* Ciferri & Redaelli - Arch. Mikrobiol. 6: 53, 1935 ≡ *Parendomyces flareri* (Redaelli & Ciferri) C.W. Dodge & M. Moore - Ann. Mo. Bot. Gdn 23: 143, 1936 ≡ *Candida flareri* (Ciferri & Redaelli) Langeron & Guerra - Annls Parasit. Hum. Comp. 16: 499, 1938 ≡ *Candida famata* (Harrison) S.A. Meyer & Yarrow var. *flareri* (Ciferri & Redaelli) Nakase & Suzuki - J. Gen. Appl. Microbiol., Tokyo 31: 83, 1985.

Debaryomyces fabryi Ota - Derm. Wschr. 78: 287, 1924 ≡ *Debaryomyces hansenii* (Zopf) Lodder & Kreger-van Rij var. *fabryi* (Ota) Nakase & Suzuki - J. Gen. Appl. Microbiol., Tokyo 31: 83, 1985.

Eutorulopsis subglobosa Zach, *in* Wolfram & Zach - Arch. Derm. Syph. 170: 688, 1934.

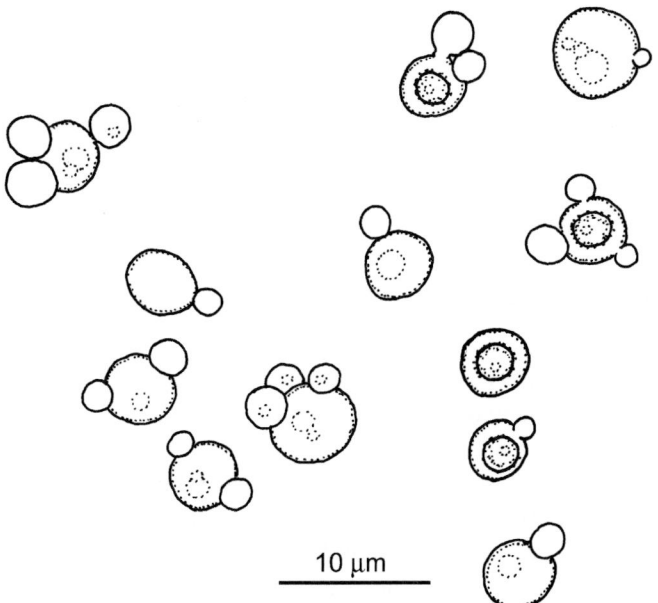

Candida famata var. *flareri* (*Debaryomyces hansenii*), CBS 1962. Budding cells, some containing ascospores.

Candida glabrata (Anderson) S.A. Meyer & Yarrow

Colony characteristics. Colonies (YPGA) cream-coloured, soft, glossy and smooth.
Microscopy. Pseudomycelium (RA) absent. Some strains may form a few branched chains of ovoidal cells. Budding unipolar; cells regularly ellipsoidal, about 3.4 × 2.0 μm, typically arranged in dense groups. Chlamydospores absent.
Physiology.

Fermentation:		Soluble starch	–	*myo*-Inositol	–
Glucose	+	D-Xylose	–	Hexadecane	–
Galactose	–	L-Arabinose	–	Vitamin-free	–
Sucrose	–	D-Arabinose	–	2-Keto-D-gluconate	v
Maltose	–	D-Ribose	–	5-Keto-D-gluconate	–
Lactose	–	L-Rhamnose	–	D-Glucuronate	–
Raffinose	–	D-Glucosamine (C)	–	Xylitol	–
α,α-Trehalose	v	*N*-Acetyl-D-glucosamine	–	L-Arabinitol	–
		Methanol	–	Arbutin	–
Growth:		Ethanol	v	Nitrate	–
Glucose	+	Glycerol	+,s	Nitrite	–
Galactose	–	*meso*-Erythritol	–	Cadaverine	+
L-Sorbose	–	Ribitol	–	Creatine	–
Sucrose	–	Galactitol	–	Creatinine	–
Maltose	–	D-Mannitol	–	L-Lysine	w,–
Cellobiose	–	D-Glucitol	–	Ethylamine	–
α,α-Trehalose	–	α-Methyl-D-glucoside	–	D-Glucosamine (N)	–
Lactose	–	Salicin	–	10% NaCl/5% glucose	+
Melibiose	–	D-Gluconate	+	Starch formation	–
Raffinose	–	DL-Lactate	v	Urease	–
Melezitose	–	Succinate	–	0.01% Cycloheximide	–
Inulin	–	Citrate	–	Growth at 42°C	+

Differential diagnosis. Species signature: sucrose –, ethylamine –, w/o niacin –, growth at 40°C +, 10% NaCl +. A colorimetric dipstick test based on assimilation of trehalose was introduced by Peltroche-Llascahuanga *et al.* (1999); rapid diagnostics using trehalose fermentation was elaborated by Fenn *et al.* (1999).
Molecular diagnostics. Species-specific nested PCR using cytochrome gene fragments was reported by Burgener-Kairuz *et al.* (1994). A 26S-based specific probe was developed by Sandhu *et al.* (1995). SSU, ITS and LSU restriction maps based on NCBI X51831 and NCBI U44808:

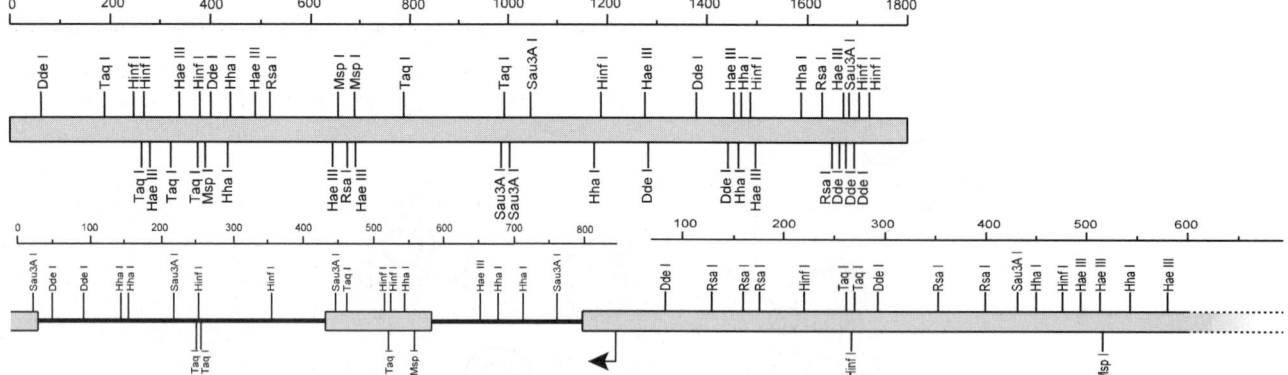

Pathogenicity. BSL-2. The species occurs as a saprophyte in the human body (Vennewald *et al.*, 1998). It is often involved in urogenital infections (Kauffman & Tan, 1984; Redondo-Lopez *et al.*, 1990; Spinillo *et al.*, 1995; Toy *et al.*, 1995). A spinal epidural abscess in an elderly patient was reported by Bonomo *et al.* (1996) and a spinal osteomyelitis by Curran & Lenke (1996). It may also be involved in deep infections (Morris & McAllister, 1993), e.g. in heart (Nishida *et al.*, 1994), pancreas (Escalante-Glorsky *et al.*, 1995), prostheses (Nayeri *et al.*, 1997), occasionally with sepsis (Hickley *et al.*, 1983) and osteomyelitis (Rubin & Sanfilippo, 1990; Owen *et al.*, 1992). A fatal pneumonia was reported by Srivastava *et al.* (1996). It is emerging as an opportunist in neonates (Glick *et al.*, 1993; Reich *et al.*, 1997). Localized infections have also been reported (Reed *et al.*, 1993; Pujol *et al.*, 1995). A catheter-related sepsis was reported by Spapen *et al.* (1995). The species may emerge after anti-*C. albicans* therapy (Millon *et al.*, 1994;

Hoppe *et al.*, 1994); cases of azole-resistant strains in transplant patients were reported by Fortun *et al.* (1997). Hemorrhagic necrosis in the forestomach of a calf due to *C. glabrata* colonization was reported by Wada *et al.* (1994).
References. Berkowitz *et al.* (1979), Lindahl & Limbird (1987), Komshian *et al.* (1989), Fidel *et al.* (1999).
Antifungal susceptibility.

Antifungal	MICs range	MIC 90	Strains	Reference
AMB	0.5-1	1	67	Pfaller *et al.* (1997)
AMB	0.06-2	2	17	Karlowsky *et al.* (1997)
AMB	1-2	2	77	Marco *et al.* (1998)
AMB	0.06-1	1	275	Zhanei *et al.* (1998)
AMB	0.25-1	0.5	50	Davey *et al.* (1998)
AMB	1-2	2	14	Espinel-Ingroff (1998)
5FC	0.06-1	0.12	67	Pfaller *et al.* (1997)
5FC	≤0.02-0.5	0.13	17	Karlowsky *et al.* (1997)
5FC	≤0.06-16	0.25	77	Marco *et al.* (1998)
5FC	≤0.03-2	0.06	275	Zhanei *et al.* 1998)
5FC	≤0.12->64	≤0.12	50	Davey *et al.* (1998)
FLZ	0.25->128	128	67	Pfaller *et al.* (1997)
FLZ	2-32	32	17	Karlowsky *et al.* (1997)
FLZ	0.25->128	64	77	Marco *et al.* (1998)
FLZ	0.5-64	32	275	Zhanei *et al.* (1998)
FLZ	1->64	32	124	Kauffman & Zarins (1998)
FLZ	4->64	64	50	Davey *et al.* (1998)
FLZ	4-64	64	14	Espinel-Ingroff (1998)
ITZ	0.02->8	2	67	Pfaller *et al.* (1997)
ITZ	0.06->8	4	77	Marco *et al.* (1998)
ITZ	≤0.03-2	1	275	Zhanei *et al.* (1998)
ITZ	0.06->4	4	124	Kauffman & Zarins (1998)
ITZ	0.12->16	8	50	Davey *et al.* (1998)
ITZ	0.5->16	2	14	Espinel-Ingroff (1998)
KTZ	≤0.004-2	1	17	Karlowsky *et al.* (1997)
KTZ	≤0.03-1	0.5	275	Zhanei *et al.* (1998)
KTZ	≤0.12-8	4	50	Davey *et al.* (1998)
MCZ	≤0.12-2	1	50	Davey *et al.* (1998)
VCZ	0.03-8	1	77	Marco *et al.* (1998)
VCZ	0.06->4	2	124	Kauffman & Zarins (1998)
VCZ	0.06-4	4	14	Espinel-Ingroff (1998)

Nomenclature. *Cryptococcus glabratus* Anderson - J. Infect. Dis. 21: 379, 1917 ≡ *Torulopsis glabrata* (Anderson) Lodder & de Vries - Mycopathologia 1: 102, 1938 ≡ *Candida glabrata* (Anderson) S.A. Meyer & Yarrow - J. Int. Syst. Bact. 28: 612, 1978.

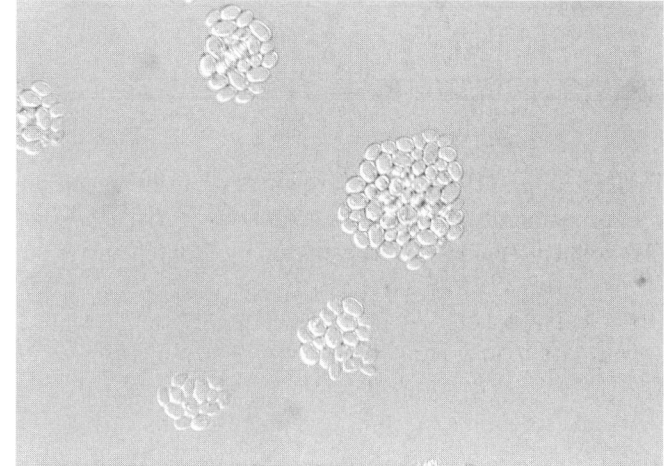

Candida glabrata, strain M.-L. Kerkmann. Budding cells. ×370.

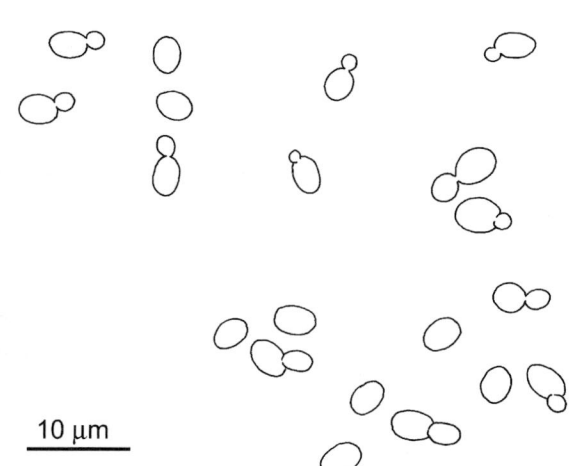

10 µm

Candida glabrata, strain M.-L. Kerkmann. Budding cells.

Candida guilliermondii (Castell.) Langer. & Guerra

Colony characteristics. Colonies (YPGA) white to cream-coloured, butyrous.
Microscopy. Budding cells (RA) (sub)spherical to broadly ellipsoidal, 3-6 × 2-4 µm. Pseudomycelium may be present, radiating from the centre of masses of budding cells; hyphae not produced.
Teleomorph. *Pichia guilliermondii* Wickerham (*Ascomycota, Hemiascomycetes, Saccharomycetales: Saccharomycetaceae*).
Asci evanescent, containing 1-4 hat-shaped ascospores, produced in low numbers. Heterothallic.
Physiology.

Fermentation:		Raffinose	+	α-Methyl-D-glucoside	v
Glucose	+	Melezitose	**v**	Salicin	v
Galactose	v	Inulin	v	D-Gluconate	v
Sucrose	+	Soluble starch	−	DL-Lactate	v
Maltose	−	D-Xylose	+	Succinate	+
Lactose	−	L-Arabinose	**v**	Citrate	+
Raffinose	v	D-Arabinose	v	*myo*-Inositol	−
α,α-Trehalose	+	D-Ribose	+	Hexadecane	+
		L-Rhamnose	v	Vitamin-free	v
Growth:		D-Glucosamine (C)	+	2-Keto-D-gluconate	+
Glucose	+	*N*-Acetyl-D-glucosamine	+	5-Keto-D-gluconate	−
Galactose	+	Methanol	−	D-Glucarate	−
L-Sorbose	v	Ethanol	+	Nitrate	−
Sucrose	+	Glycerol	+	10% NaC1/5% glucose	+
Maltose	+	*meso*-Erythritol	−	0.1% Cycloheximide	v
Cellobiose	v	Ribitol	+	Starch formation	−
α,α-Trehalose	+	Galactitol	v	Gelatin liquefaction	+
Lactose	−	D-Mannitol	v	Growth at 37°C	v
Melibiose	**v**	D-Glucitol	v		

Differential diagnosis. The species is taxonomically insufficiently defined. Its physiological pattern is variable and overlaps with numerous other species; see Barnett *et al.* (2000).
Molecular diagnostics. A 26S-based specific probe was developed by Sandhu *et al.* (1995). SSU, ITS and LSU restriction maps based on NCBI AB013587, L47110 and U45709:

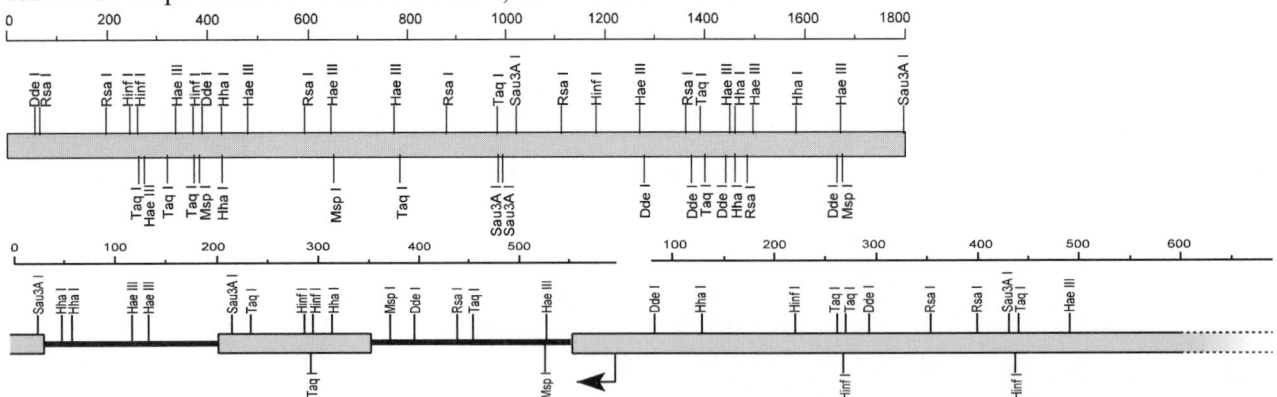

Pathogenicity. BSL-1. Disseminated cases were reported by Dick *et al.* (1985) and Vazquez *et al.* (1995), and an osteomyelitis by Tietz *et al.* (1999). Otherwise the species is occasionally involved in cutaneous (Ellis, 1994) or subcutaneous (Graham & Frost, 1973) infections. Mild catheter-related infections were reported by Montenegro *et al.* (1995).
Reference. Barnett *et al.* (1990).

Antifungal susceptibility.

Antifungal	MICs range	MIC 50	Strains	Reference
AMB	0.5-1	1	8	Espinel-Ingroff (1998)
AMB	0.25-1	0.5	9	Pfaller *et al.* (1997)
5FC	0.06-0.25	0.12	9	Pfaller *et al.* (1997)
FLZ	0.5-8	2	8	Espinel-Ingroff (1998)
FLZ	2-64	4	9	Pfaller *et al.* (1997)
ITZ	0.12-0.5	0.25	8	Espinel-Ingroff (1998)
ITZ	0.12-1	0.5	9	Pfaller *et al.* (1997)
VCZ	<0.03-0.06	0.06	8	Espinel-Ingroff (1998)

Nomenclature. *Endomyces guilliermondii* Castellani - Br. Med. J. 2: 1210, 1912 ≡ *Monilia guilliermondii* (Castellani) Castellani & Chalmers - Man. Trop. Med., ed. 2, p. 826, 1913 ≡ *Myceloblastanon guilliermondii* (Castellani) Ota - Jpn. J. Derm. Urol. 27: 178, 1927 ≡ *Blastodendrion guilliermondii* (Castellani) Guerra - Rôle Levures Dermatol. p. 25, 1935 ≡ *Mycotorula guilliermondii* (Castellani) Langeron & Guerra - Proc. 6th. Int. Congr., Amsterdam 2: 165, 1935 ≡ *Castellania guilliermondii* (Castellani) C.W. Dodge - Med. Mycol. p. 257, 1935 ≡ *Candida guilliermondii* (Castellani) Langeron & Guerra - Annls Parasit. Hum. Comp. 16: 468, 1938.

Endomyces negrii Castellani - Br. Med. J. 2: 1210, 1912 ≡ *Monilia negrii* (Castellani) Castellani & Chalmers - Man. Trop. Med., ed. 2, p. 822, 1913 ≡ *Castellania negrii* (Castellani) C.W. Dodge - Med. Mycol. p. 253, 1935.

Monilia lustigi Castellani & Chalmers - Man. Trop. Med., ed. 2, p. 828, 1913 ≡ *Castellania lustigi* (Castellani & Chalmers) C.W. Dodge - Med. Mycol. p. 258, 1935.

Monilia pseudoguilliermondii Castellani & Chalmers - Man. Trop. Med., ed. 3, p. 1088, 1919 ≡ *Myceloblastanon pseudoguilliermondii* (Castellani & Chalmers) Ota - Jpn. J. Derm. Urol. 27: 178, 1927 ≡ *Castellania pseudoguilliermondii* (Castellani & Chalmers) C.W. Dodge - Med. Mycol. p. 256, 1935.

Torula fermentati Saito - Jpn. J. Bot. 1: 1-54, 1922.

Cryptococcus krausi Ota - Derm. Wschr. 78: 229, 1924 ≡ *Blastodendrion krausi* (Ota) Redaelli & Ciferri - Annls Mycol. 27: 269, 1929 ≡ *Monilia krausi* (Ota) Nannizzi - Tratt. Micopat. Um. 4: 363, 1934 ≡ *Parendomyces krausi* (Ota) C.W. Dodge - Med. Mycol. p. 244, 1935 ≡ *Mycotorula krausi* (Ota) Ciferri & Redaelli - Atti Ist. Bot. Univ. Lab. Crittogam. Pavia, Ser. 5, 3: 36, 1943.

Blastodendrion arzti Ota - Derm. Wschr. 78: 232, 1924 ≡ *Monilia arzti* (Ota) Nannizzi - Tratt. Micopat. Um. 4: 350, 1934 ≡ *Myceloblastanon arzti* (Ota) Rippon - Med. Mycol. p. 570, 1988.

Monilia muhira Mattlet - Ann. Soc. Belge Méd. Trop. 6: 23, 1926 ≡ *Castellania muhira* (Mattlet) C.W. Dodge - Med. Mycol. p. 257, 1935 ≡ *Mycotorula guilliermondii* (Castellani) Langeron & Guerra var. *muhira* (Mattlet) Redaelli & Ciferri - Atti Ist. Bot. Univ. Lab. Crittogam. Pavia, Ser. 5, 3: 31, 1943.

Trichosporon appendiculare Batista, J. Silveira & G. Silveira - Rev. Assoc. Méd. Brasil 5: 351, 1959.

Pichia guilliermondii Wickerham - J. Bact. 92: 1269, 1966.

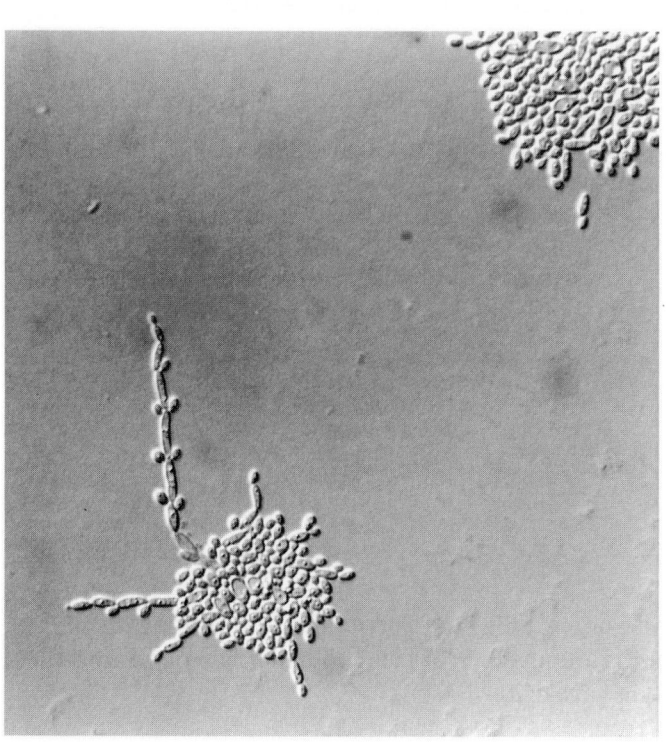 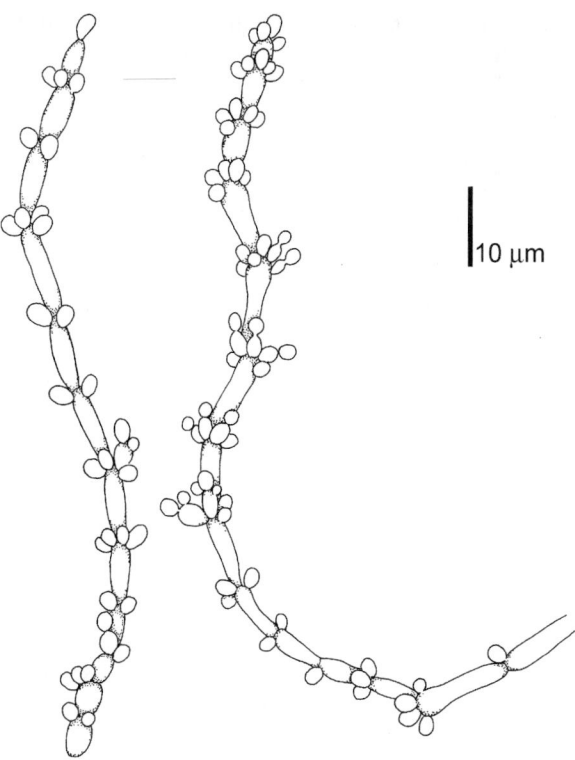

10 μm

Candida guilliermondii, CBS 2030. Pseudomycelium.

Candida haemulonii (van Uden & Kolipinski) S.A. Meyer & Yarrow

Colony characteristics. Colonies (MEYA) white to cream-coloured, butyrous.
Microscopy. Budding cells (RA) (sub)spherical, 3.0-6.5 × 3-5 μm. Pseudomycelium absent, or fragmentary pseudomycelium composed of clavate cells present.
Physiology.

Fermentation:		Soluble starch	v	*myo*-Inositol	–
Glucose	+	D-Xylose	–,s	Hexadecane	+,s
Galactose	–	L-Arabinose	–,s	Vitamin-free	–
Sucrose	+	D-Arabinose	–,s	2-Keto-D-gluconate	+
Maltose	–	D-Ribose	–,s	5-Keto-D-gluconate	–
Lactose	–	L-Rhamnose	+,s	D-Glucuronate	–
Raffinose	–	D-Glucosamine (C)	+,s	Xylitol	v
α,α-Trehalose	+,s	*N*-Acetyl-D-glucosamine	+	L-Arabinitol	–
		Methanol	–	Arbutin	v
Growth:		Ethanol	s	Nitrate	–
Glucose	+	Glycerol	+,s	Nitrite	–
Galactose	s	*meso*-Erythritol	–	Cadaverine	+
L-Sorbose	–	Ribitol	s	Creatinine	–
Sucrose	+	Galactitol	–,s	L-Lysine	+
Maltose	+	D-Mannitol	+	Ethylamine	–
Cellobiose	–	D-Glucitol	+	D-Glucosamine (N)	–
α,α-Trehalose	+	α-Methyl-D-glucoside	–	10% NaCl/5% glucose	+
Lactose	–	Salicin	–	Starch formation	–
Melibiose	–	D-Gluconate	+	Urease	–
Raffinose	+,s	DL-Lactate	–	0.1% Cycloheximide	+
Melezitose	+,s	Succinate	+	Growth at 37°C	+
Inulin	–	Citrate	+,s	Growth at 40°C	–

Differential diagnosis. Species signature: cellobiose –, *meso*-erythritol –, α-methyl-D-glucoside –, 2-keto-D-gluconate +, 0.1% cycloheximide +, w/o biotin –, 16% NaCl +. Physiologically not reliably distinguishable from *Candida famata* (p. 196), *C. guilliermondii* (p. 200) and *C. parapsilosis* (p. 212).
Molecular diagnostics. SSU and LSU restriction maps based on NCBI AB013572 and U44812:

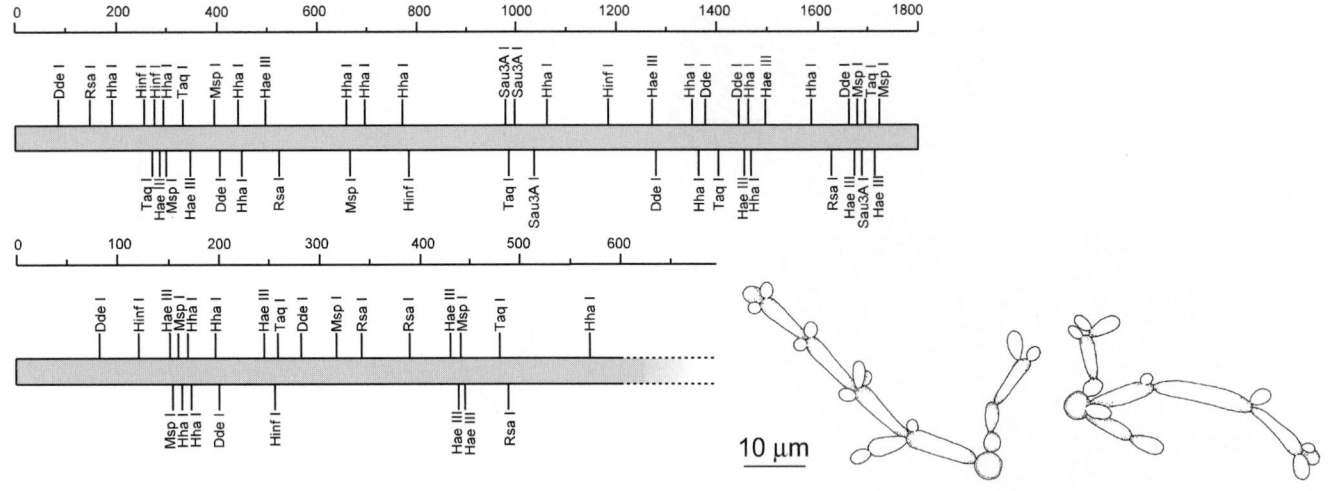

Candida haemulonii, CBS 5149. Fragmentary pseudomycelium.

Pathogenicity. BSL-1. A case of fungemia was reported by Majoret (1986) and a peritonitis by Lavarde *et al*. (1984). The species was originally described from fish (van Uden & Kolipinski, 1962) and dolphin (de Vries & Laarman, 1973). It is occasionally isolated from clinical specimens (Gargeya *et al*., 1991).
Reference. Van Uden & Kolipinski (1962).
Nomenclature. *Torulopsis haemulonii* van Uden & Kolipinski - Antonie van Leeuwenhoek 28: 78, 1962 ≡ *Candida haemulonii* (van Uden & Kolipinski) S.A. Meyer & Yarrow, *in* Yarrow & Meyer - Int. J. Syst. Bact. 28: 612, 1978.

Candida intermedia (Cif. & Ashford) Langer. & Guerra

Colony characteristics. Colonies (MEYA) cream-coloured, butyrous.
Microscopy. Budding cells (RA) ellipsoidal, 4-6 × 2-3 µm, coherent, with slender, short-celled pseudomycelium.
Physiology.

Fermentation:		Inulin	–,s	Citrate	+
Glucose	+	Soluble starch	v	*myo*-Inositol	–
Galactose	+,s	D-Xylose	+	Vitamin-free	–,s
Sucrose	+	L-Arabinose	–,s	2-Keto-D-gluconate	+
Maltose	v	D-Arabinose	–,s	D-Glucuronate	–
Lactose	–	D-Ribose	–,s	Xylitol	v
Raffinose	v	L-Rhamnose	v	L-Arabinitol	–
α,α-Trehalose	s,–	D-Glucosamine (C)	v	Arbutin	+
		Methanol	–	Nitrate	–
Growth:		Ethanol	+	Nitrite	–
Glucose	+	Glycerol	–,s	Cadaverine	+
Galactose	+	*meso*-Erythritol	–	Creatinine	–
L-Sorbose	+	Ribitol	+,s	L-Lysine	+
Sucrose	+	Galactitol	v	Ethylamine	+
Maltose	+	D-Mannitol	+	D-Glucosamine	–
Cellobiose	+	D-Glucitol	+	Starch formation	–
α,α-Trehalose	+	α-Methyl-D-glucoside	+,s	Urease	–
Lactose	+	Salicin	+	0.01% Cycloheximide	v
Melibiose	–	D-Gluconate	–,s	0.1% Cycloheximide	–
Raffinose	+	DL-Lactate	–	Growth at 35°C	v
Melezitose	+	Succinate	+	Growth at 37°C	–

Differential diagnosis. Species signature: fermentation of sucrose + and assimilation: L-sorbose +, lactose +, melibiose –, *meso*-erythritol –, 2-keto-D-gluconate +, 5-keto-D-gluconate –, nitrate –. Physiologically indistinguishable from *C. famata* (p. 196).
Molecular diagnostics. SSU and LSU restriction maps based on NCBI X89518 and U44809:

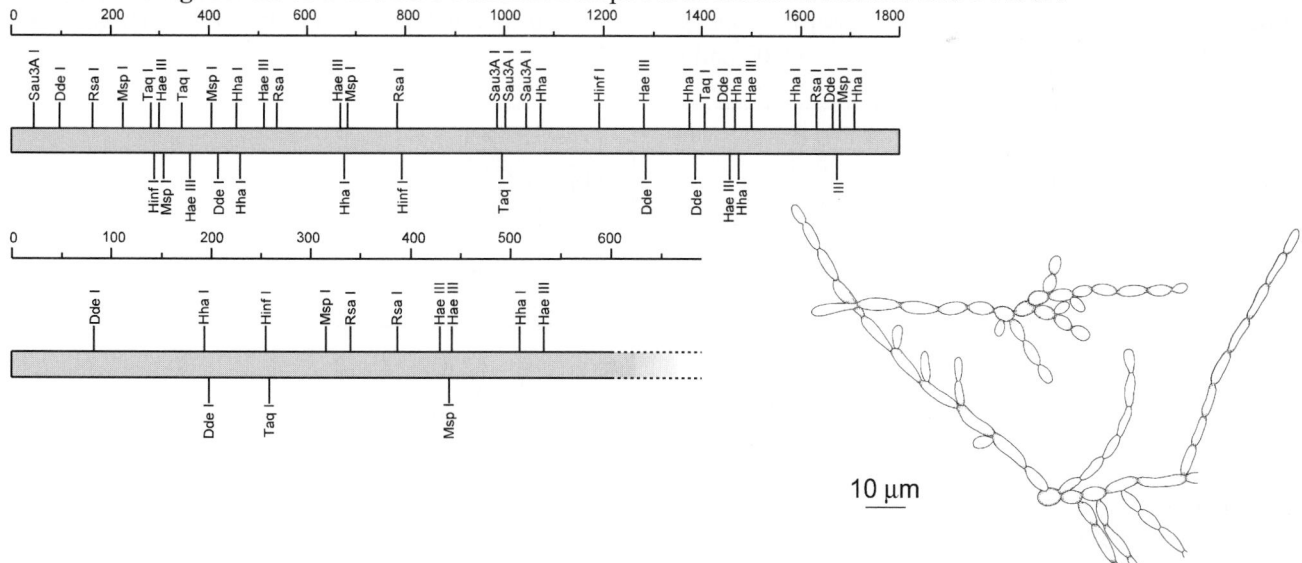

Candida intermedia, CBS 572. Pseudomycelium.

Pathogenicity. BSL-1. In older literature this species has repeatedly been reported from humans (e.g. Blank, 1951; Raab, 1964) but no recent, confirmed reports are available.
Reference. Meyer *et al.* (1998).
Nomenclature. *Blastodendrion intermedius* Ciferri & Ashford - Puerto Rico J. Publ. Health Trop. Med. 5: 103, 1929 ≡ *Cryptococcus intermedius* (Ciferri & Ashford) Nannizzi - Tratt. Micopat. Um. 4: 326, 1934 ≡ *Candida intermedia* (Ciferri & Ashford) Langeron & Guerra - Annls Parasit. Hum. Comp. 16: 461, 1938.

Candida kefyr (Beijerinck) van Uden & Buckley

Colony characteristics. Colonies (YPGA) white, cream-coloured or pink, butyrous.

Microscopy. Budding cells (RA) ellipsoidal to cylindrical, 6-10 × 3-6 µm. Hyphae absent; pseudomycelium mostly present, fragile, emerging in a volcano-shaped arrangement with parallel cells.

Teleomorph. *Kluyveromyces marxianus* (Hansen) van der Walt (*Ascomycota, Hemiascomycetes, Saccharomycetales: Saccharomycetaceae*).

Asci evanescent, containing 1-4 smooth-walled, ellipsoidal to reniform ascospores. Homothallic.

Physiology.

Fermentation:		Melibiose	–	D-Glucitol	v
Glucose	+	Raffinose	+	α-Methyl-D-glucoside	–
Galactose	s	Melezitose	–	Salicin	v
Sucrose	+	Inulin	–	D-Gluconate	–
Maltose	–	Soluble starch	–	DL-Lactate	+
Lactose	v	D-Xylose	s	Succinate	+
Raffinose	+	L-Arabinose	v	Citrate	v
α,α-Trehalose	–	D-Arabinose	–	*myo*-Inositol	–
Inulin	s	D-Ribose	v	Hexadecane	–
		L-Rhamnose	–	Vitamin-free	–
Growth:		D-Glucosamine (C)	–	2-Keto-D-gluconate	–
Glucose	+	*N*-Acetyl-D-glucosamine	–	Cadaverine	+
Galactose	s	Methanol	–	L-Lysine	+
L-Sorbose	v	Ethanol	+	Nitrate	–
Sucrose	+	Glycerol	s	Ethylamine	+
Maltose	–	*meso*-Erythritol	–	Gelatin liquefaction	–
Cellobiose	v	Ribitol	s	0.1% Cycloheximide	+
α,α-Trehalose	w,–	Galactitol	–	Growth at 40°C	+
Lactose	v	D-Mannitol	v		

Differential diagnosis. Species signature: raffinose +, 2-keto-D-gluconate –, ethylamine +, 0.1% cycloheximide +, growth at 37°C +, w/o niacin +. Physiologically indistinguishable from *Kluyveromyces lactis*.

Molecular diagnostics. A 26S-based specific probe was developed by Sandhu *et al.* (1995). SSU, ITS and LSU restriction maps based on NCBI M60303, L47107 and U94924:

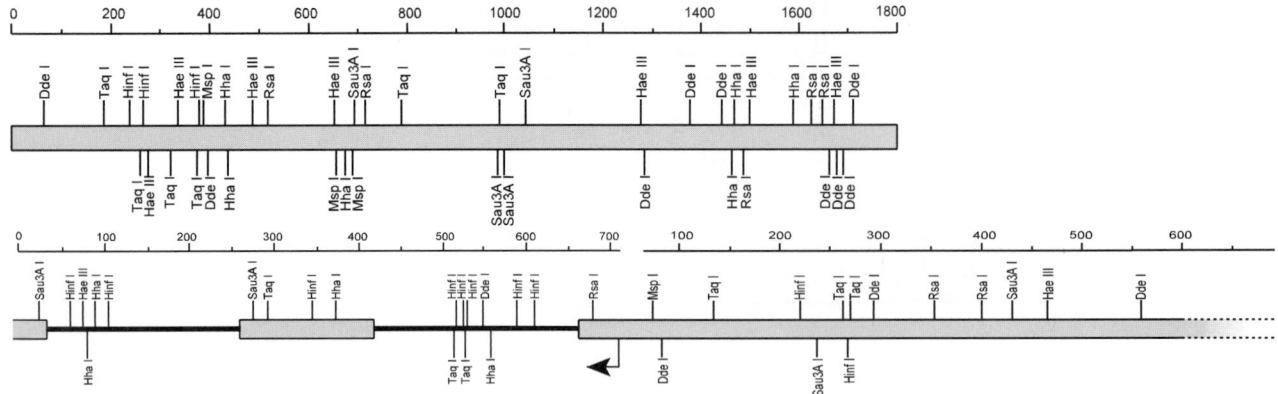

Pathogenicity. BSL-1. The species is occasionally involved in superficial candidiasis (Hernandez-Molina *et al.*, 1994). A cardiac transplant patient with pulmonary infection was described by Lutwick *et al.* (1980), an oesophagitis in a patient with carcinoma of the oropharynx by Listemann *et al.* (1998), and a mixed pulmonary infection with *Aspergillus flavus* in a leukemic patient by Lopes *et al.* (1996a). Frequent reports from older literature.

Reference. Wickes *et al.* (1992).

Antifungal susceptibility.

Antifungal	MICs range	MIC 50	MIC 90	Strains	Reference
AMB	1	1	1	10	Davey *et al.* (1998)
AMB	0.5-2	1	nd	5	Espinel-Ingroff (1998)
FLZ	0.25-1	0.25	0.25	10	Davey *et al.* (1998)
FLZ	0.5-4	2	nd	5	Espinel-Ingroff (1998)
ITZ	≤0.03-0.25	0.06	0.12	10	Davey *et al.* (1998)
ITZ	0.06-0.25	0.25	nd	5	Espinel-Ingroff (1998)
KTZ	≤0.12-1	≤0.12	≤0.12	10	Davey *et al.* (1998)
MCZ	≤0.12-1	≤0.12	≤0.12	10	Davey *et al.* (1998)
VCZ	<0.03-0.06	<0.03	nd	5	Espinel-Ingroff (1998)

Nomenclature. *Saccharomyces marxianus* Hansen - C.R. Trav. Lab. Carlsberg 2: 143-167, 1888 ≡ *Kluyveromyces marxians* (Hansen) van de Walt - Bothalia 10: 417, 1971.

Saccharomyces kefyr Beijerinck - Arch. Neerl. Sci. Exact. Nat. 23: 212, 1889 ≡ *Geotrichoides kefyr* (Beijerinck) Langeron & Talice - Annls Parasit. Hum. Comp. 10: 68, 1932 ≡ *Candida kefyr* (Beijerinck) van Uden & Buckley, *in* S.A. Meyer & Ahearn - Mycotaxon 17: 297, 1983.

Endomyces pseudotropicalis Castellani - Centbl. Bakt. Parasitkde, Abt. 1, 58: 237, 1911 ≡ *Monilia pseudotropicalis* (Castellani) Castellani & Chalmers - Man. Trop. Med., ed. 2, p. 825, 1913 ≡ *Atelosaccharomyces pseudotropicalis* (Castellani) Froilano de Mello, *in* Froilano de Mello & Fernandes - Arq. Hig. Patol. Exot. 6: 264, 1918 ≡ *Myceloblastanon pseudotropicalis* (Castellani) Ota - Jpn. J. Derm. Urol. 28: 178, 1928 ≡ *Candida pseudotropicalis* (Castellani) Basgal - Contrib. Estud. Blastom. Pulmon. p. 49, 1931 ≡ *Castellania pseudotropicalis* (Castellani) C.W. Dodge - Med. Mycol. p. 259, 1935 ≡ *Mycocandida pseudotropicalis* (Castellani) Ciferri & Redaelli - Arch. Mikrobiol. 6: 16, 1935 ≡ *Mycotorula pseudotropicalis* (Castellani) Redaelli & Ciferri - Mycopath. Mycol. Appl. 4: 37, 1947.

Cryptococcus sulfureus Beauverie & Lesieur - J. Physiol. Pathol. Gén. 14: 996, 1912 ≡ *Monilia sulfurea* (Beauverie & Lesieur) Vuillemin - Champ. Paras. Myc. Homme p. 84, 1931 ≡ *Mycoderma sulfurea* (Beauverie & Lesieur) C.W. Dodge - Med. Mycol. p. 227, 1935.

Monilia macedoniensis Castellani, *in* Castellani & Chalmers - Man. Trop. Med. p. 1087, 1919 ≡ *Candida macedoniensis* (Castellani) Berkhout - Schimmelges. Monilia, Oidium, Oospora Torula p. 44, 1923 ≡ *Myceloblastanon macedoniense* (Castellani) Ota - Jpn. J. Derm. Urol. 27: 178, 1927 ≡ *Castellania macedoniensis* (Castellani) C.W. Dodge - Med. Mycol. p. 259, 1935 ≡ *Mycotoruloides macedoniensis* (Castellani) C.W. Dodge & M. Moore - Annls Mo. Bot. Gdn 23: 141, 1936.

Saccharomyces cavernicula Redaelli - Micet. Assoc. Microb. Tuberc. Polmon. Cavit. p. 54, 1925.

Monilia macedonensoides Castellani & Taylor - J. Trop. Med. Hyg. 28: 244, 1925 ≡ *Candida macedoniensis* (Castellani) Berkhout var. *macedonensoides* (Castellani & Taylor) Westerdijk - CBS List Cult. p. 14, 1932 ≡ *Castellania macedonensoides* (Castellani) C.W. Dodge - Med. Mycol. p. 253, 1935.

Candida mortifera Redaelli - Micet. Assoc. Microb. Tuberc. Pulmon. Cavit. p. 21, 1925 ≡ *Mycocandida mortifera* (Redaelli) Langeron & Talice - Annls Parasit. Hum. Comp. 10: 58, 1932 ≡ *Monilia mortifera* (Redaelli) Nannizzi - Tratt. Micopat. Um. 4: 367, 1934.

Cryptococcus kartulisii Castellani - Amer. J. Trop. Med. 8: 413, 1928 ≡ *Myceloblastanon kartulisii* (Castellani) Ota - Jpn. J. Derm. Urol. 28: 4, 1928 ≡ *Castellania kartulisii* (Castellani) C.W. Dodge - Med. Mycol. p. 260, 1935.

Blastodendrion procerum Zach, *in* Wolfram & Zach - Arch. Derm. Syph. 170: 685, 1934.

Pseudomycoderma mazzae C.W. Dodge - Med. Mycol. p. 237, 1935.

Saccharomyces chevalieri Guilliermond var. *atypica* Dietrichson - Annls Parasit. Hum. Comp. 29: 460, 1954.

Kluyveromyces cicerosporus van der Walt, Nel & van Kerken - Antonie van Leeuwenhoek 32: 395, 1966.

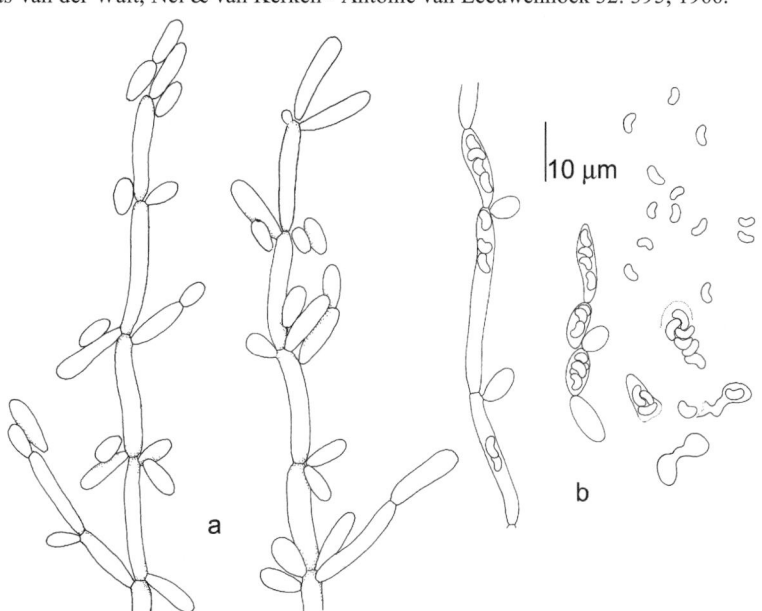

Candida kefyr (*Kluyveromyces marxianus*), CBS 6556. a. Pseudomycelium; b. asci and ascospores.

Candida krusei (Castellani) Berkhout

Colony characteristics. Colonies (YPGA) cream-coloured to tannish-white, butyrous, dry.

Microscopy. Budding cells (RA) ellipsoidal to cylindrical, 4-5 × 2-5 μm, with flat, well-circumscribed scars. Pseudomycelium often present, robust; cells liberated and arranged parallel to the main axis.

Teleomorph. *Issatchenkia orientalis* Kudryavtsev (*Ascomycota, Hemiascomycetes, Saccharomycetales: Saccharomycetaceae*).

Asci persistent, containing 1-2 rough- or smooth-walled, spherical ascospores. Homothallic.

Physiology.

Fermentation:		Raffinose	–	α-Methyl-D-glucoside	–
Glucose	+	Melezitose	–	Salicin	–
Galactose	–	Inulin	–	D-Gluconate	–
Sucrose	–	Soluble starch	–	DL-Lactate	+
Maltose	–	D-Xylose	–	Succinate	+
Lactose	–	L-Arabinose	–	Citrate	+,w
Cellobiose	–	D-Arabinose	–	*myo*-Inositol	–
α,α-Trehalose	–	D-Ribose	–	Nitrate	–
		L-Rhamnose	–	Hexadecane	–
Growth:		D-Glucosamine (C)	+	Vitamin-free	+
Glucose	+	*N*-Acetyl-D-glucosamine	+	2-Keto-D-gluconate	–
Galactose	–	Methanol	–	5-Keto-D-gluconate	–
L-Sorbose	–	Ethanol	+	D-Glucarate	–
Sucrose	–	Glycerol	+	D-Glucuronate	–
Maltose	–	*meso*-Erythritol	–	Nitrate	–
Cellobiose	–	Ribitol	–	10% NaC1/5% glucose	+
α,α-Trehalose	–	Galactitol	–	0.01% Cycloheximide	–
Lactose	–	D-Mannitol	–	Starch formation	–
Melibiose	–	D-Glucitol	–	Growth at 40°C	+

Differential diagnosis. Species signature: fermentation of glucose +, cellobiose –, and assimilation: galactose –, D-xylose –, D-mannitol –, DL-lactate +, citrate +(w), nitrate –, D-glucosamine –, 0.01% cycloheximide –, lysine +. Physiologically indistinguishable from *Issatchenkia occidentalis*.

Molecular diagnostics. Specific probes were developed by Manavathu *et al.* (1996) based on a mitochondrial proteinase, by Sandhu *et al.* (1995) based on 26S rDNA and by Carlotti *et al.* (1996) based on *Eco*RI digested nDNA. Confident distiction from *C. inconspicua* is possible with RAPD (Baily *et al.,* 1997). Nho *et al.* (1997) described specific RFLP's of rDNA amplicons. SSU, ITS and LSU restriction maps based on NCBI M55528, L47113 and U76347:

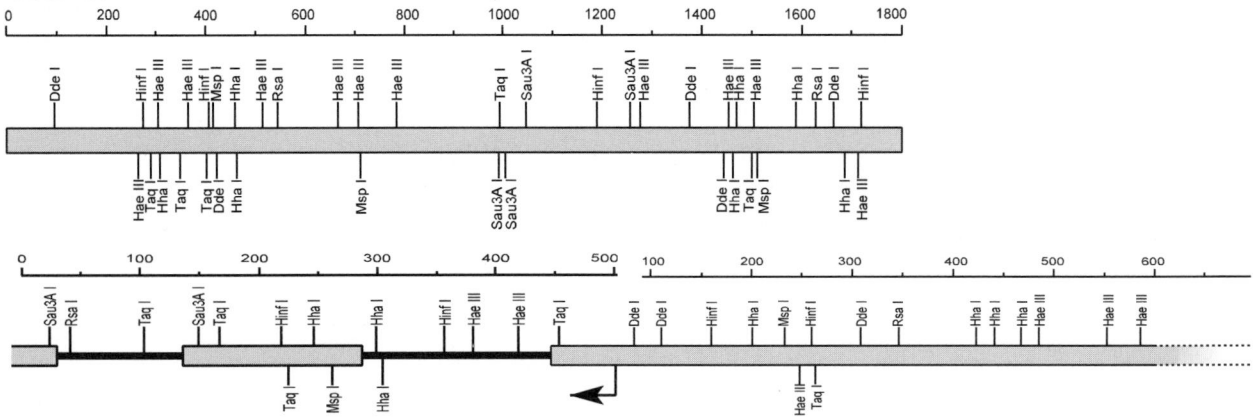

Pathogenicity. BSL-2. The species is occasionally involved in fatal systemic candidiasis, usually in patients with impaired innate immunity (Gordon *et al.*, 1980; Merz *et al.*, 1986; Wingard *et al.*, 1991; Goldman *et al.*, 1993; Iwen *et al.*, 1993, 1995; Sandin *et al.*, 1993b), e.g., transplant patients (Krueger *et al.*, 1996) or leukemia (Gregory *et al.* 1999), rarely with AIDS (Genet *et al.,* 1995). A rectal infection was reported by Cuozzo *et al.* (1995) and a catheter-associated infection by Verduyn Lunel *et al.* (1996). It may emerge after the use of antimycotics (Akova *et al.*, 1991; Millon *et al.*, 1994).

References. Goldman *et al.* (1993), Samaranayake & Samaranayake (1994).
Antifungal susceptibility.

Antifungal	MICs range	MIC 90	Strains	Reference
AMB	0.12-1	0.5	20	Davey *et al.* (1998)
AMB	0.5-1/0.5-2	1/2	36/50	Pfaller *et al* (1997/1998)
AMB	0.25-1	1	12	Espinel-Ingroff (1998)
AMB	1-2	2	17	Marco *et al.* (1998)
5FC	2-16	16	20	Davey *et al.* (1998)
5FC	1-32/16-32	16/32	36/50	Pfaller *et al.* (1997/1998)
5FC	16-64	64	17	Marco *et al.* (1998)
FLZ	16->64	>64	20	Davey *et al.* (1998)
FLZ	0.25-128/16-64	64/64	36/50	Pfaller *et al.* (1997/1998)
FLZ	16-32	16	12	Espinel-Ingroff (1998)
FLZ	16->64	>64	20	Davey *et al.* (1998)
FLZ	32-128	128	17	Marco *et al.* (1998)
ITZ	0.12-1	0.5	20	Davey *et al.* (1998)
ITZ	0.03-1/0.25-1	0.5/2	36/50	Pfaller *et al.* (1997/1998)
ITZ	0.12-1	1	12	Espinel-Ingroff (1998)
ITZ	0.12->4	4	20	Davey *et al.* (1998)
ITZ	0.5-2	2	17	Marco *at al.* (1998)
KTZ	≤0.12-2	2	20	Davey *et al.* (1998)
MCZ	1-4	4	20	Davey *et al.* (1998)
VCZ	0.12-0.5	0.5	12	Espinel-Ingroff (1998)
VCZ	0.25-2	2	20	Kauffman & Zarins (1998)
VCZ	0.25-1	1	17	Marco *et al.* (1998)

Nomenclature. *Saccharomyces krusei* Castellani - Philip. J. Sci., Ser. B, Med. Sci. 5: 202, 1910 ≡ *Endomyces krusei* (Castellani) Castellani - Br. Med. J. 2: 1210, 1912 ≡ *Monilia krusei* (Castellani) Castellani & Chalmers - Man. Trop. Med., ed. 2, p. 826, 1913 ≡ *Candida krusei* (Castellani) Berkhout - Schimmelges. Monilia, Oidium, Oospora Torula p. 44, 1923 ≡ *Myceloblastanon krusei* (Castellani) Ota - Jpn. J. Derm. Urol. 28: 178, 1928 ≡ *Geotrichoides krusei* (Castellani) Langeron & Talice - Annls Parasit. Hum. Comp. 10: 67, 1932 ≡ *Trichosporon krusei* (Castellani) Ciferri & Redaelli - Arch. Mikrobiol. 6: 19, 1935 ≡ *Mycotoruloides krusei* (Castellani) Langeron & Guerra - Proc. 6th Inst. Bot. Congr. 2: 167, 1935.

Euantiothamnus braulti Pinoy, *in* Brault & Masselot - Annls Derm. Syph., Sér. 5, 2: 592-602, 1911 ≡ *Monilia braultii* (Pinoy) Vuillemin - Champ. Paras. Myc. Homme p. 87, 1931 ≡ *Blastodendrion braultii* (Pinoy) Langeron & Talice - Annls Parasit. Hum. Comp. 10: 62, 1932 ≡ *Syringospora braultii* (Pinoy) C.W. Dodge - Med. Mycol. p. 277, 1935.

Endomyces intestinalis Castellani - Br. Med. J. 2: 1209, 1912 ≡ *Monilia intestinalis* (Castellani) Brumpt - Précis Parasitol., ed. 2, p. 925, 1913 ≡ *Myceloblastanon intestinale* (Castellani) Ota - Jpn. J. Derm. Urol. 27: 178, 1927 ≡ *Castellania intestinalis* (Castellani) C.W. Dodge - Med. Mycol. p. 253, 1935.

Endomyces nitidus Castellani - Br. Med. J. 2: 1210, 1912 ≡ *Monilia nitida* (Castellani) Castellani & Chalmers - Man. Trop. Med., ed. 2, p. 826, 1913 ≡ *Myceloblastanon nitidus* (Castellani) Ota - Jpn. J. Derm. Urol. 27: 178, 1927 ≡ *Candida nitida* (Castellani) Basgal - Contrib. Estud. Blast. Pulmon. p. 50, 1931 ≡ *Castellania nitida* (Castellani) C.W. Dodge - Med. Mycol. p. 261, 1935.

Cryptococcus lesieuri Beauverie, *in* Beauverie & Lesieur - J. Physiol. Path. Gén. 14: 994, 1912 ≡ *Castellania lesieuri* (Beauverie) C.W. Dodge - Med. Mycol. p. 249, 1935.

Endomyces blanchardii Castellani - Arch. Parasit. 16: 187, 1913 ≡ *Monilia blanchardii* (Castellani) Castellani & Chalmers - Man. Trop. Med., ed. 2. 829, 1913 ≡ *Parendomyces blanchardii* (Castellani) C.W. Dodge - Med. Mycol. p. 243, 1935.

Monilia balcanica Castellani & Chalmers - Man. Trop. Med., ed. 3, p. 1090, 1919 ≡ *Myceloblastanon balcanicum* (Castellani & Chalmers) Ota - Jpn. J. Derm. Urol. 27: 178, 1927 ≡ *Castellania balcanica* (Castellani & Chalmers) C.W. Dodge - Med. Mycol. p. 249, 1935.

Monilia londinensis Castellani & Chalmers - Man. Trop. Med., ed. 3, p 1084, 1919 ≡ *Myceloblastanon londinense* (Castellani & Chalmers) Ota - Jpn. J. Derm. Urol. 27: 178, 1927 ≡ *Castellania londinensis* (Castellani & Chalmers) C.W. Dodge - Med. Mycol. p. 252, 1935 ≡ *Candida albicans* (Robin) Berkhout var. *londinensis* (Castellani & Chalmers) Ciferri - Man. Micol. Med., ed. 2, p. 248, 1960.

Monilia parabalcanica Castellani & Chalmers - Man. Trop. Med., ed. 3, p. 1090, 1919 ≡ *Myceloblastanon parabalcanicum* (Castellani & Chalmers) Ota - Jpn. J. Derm. Urol. 27: 178, 1927 ≡ *Castellania parabalcanica* (Castellani & Chalmers) C.W. Dodge - Med. Mycol. p. 249, 1935.

Monilia parakrusei Castellani & Chalmers - Man. Trop. Med., ed. 3, p. 1092, 1919 ≡ *Myceloblastanon parakrusei* (Castallani & Chalmers) Ota - Jpn. J. Derm. Urol. 27: 178, 1927 ≡ *Castellania parakrusei* (Castellani & Chalmers) C.W. Dodge - Med. Mycol. p. 261, 1935 ≡ *Candida parakrusei* (Castellani & Chalmers) Langeron & Guerra - Annls Parasit. Hum. Comp. 16: 36-179, 1938.

Monilia africana Macfie - Annls Trop. Med. Parasitol. 15: 277, 1921 ≡ *Castellania africana* (Macfie) C.W. Dodge - Med. Mycol. p. 262, 1935.

Mycoderma bordetii Kufferath - Annls Soc. Méd. Nat. Brux. 74: 38, 1920.

Monilia enterocola Macfie - Annls Trop. Med. Parasit. 15: 278, 1921 ≡ *Myceloblastanon enterocola* (Macfie) Ota - Jpn. J. Derm. Urol. 27: 172, 1927 ≡ *Parendomyces enterocola* (Macfie) C.W. Dodge - Med. Mycol. p. 245, 1935.

Monilia tonge Mattlet - Annls Soc. Belge Méd. Trop. 6: 22, 1926 ≡ *Syringospora tonge* (Mattlet) C.W. Dodge - Med. Mycol. p. 278, 1935

Monilia dissocians Mattlet - Annls Soc. Belge Méd. Trop. 6: 24, 1926 ≡ *Castellania dissocians* (Mattlet) C.W. Dodge - Med. Mycol. p. 250, 1935.

Monilia inexpectata Mazza, Niño & Egües - Reunión Soc. Argent. Patol. Reg. Norte 5: 284, 1929 ≡ *Mycocandida inexpectata* (Mazza, Niño & Egües) Talice & Mackinnon - Reunión Soc. Argent. Patol. Reg. Norte 8: 164, 1934 ≡ *Pseudomonilia inexpectata* (Mazza, Niño & Egües) C.W. Dodge - Med. Mycol. p. 296, 1935.

Myceloblastanon tokioense Fujii - Jpn. J. Derm. Urol. 31: 970, 1931 ≡ *Schizoblastosporion tokioense* (Fujii) C.W. Dodge - Med. Mycol. p. 235, 1935.

Trichosporon dendriticum Ciferri & Redaelli - Arch. Mikrobiol. 6: 58, 1935 ≡ *Candida dendritica* (Ciferri & Redaelli) C.W. Dodge & M. Moore - Annls Mo. Bot. Gdn 23: 145, 1936.

Castellania orticonii C.W. Dodge - Med. Mycol. p. 261, 1935.

Monilia krusoides Castellani - J. Trop Med. Hyg. 40: 293 - 307, 1937.

Issatchenkia orientalis Kudryavtsev - Syst. Hefen p. 162, 1963.

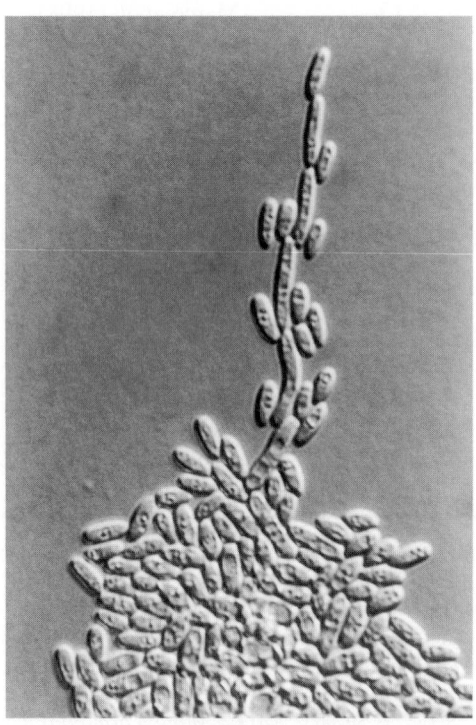
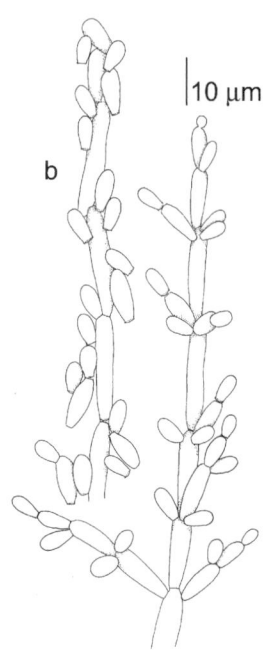
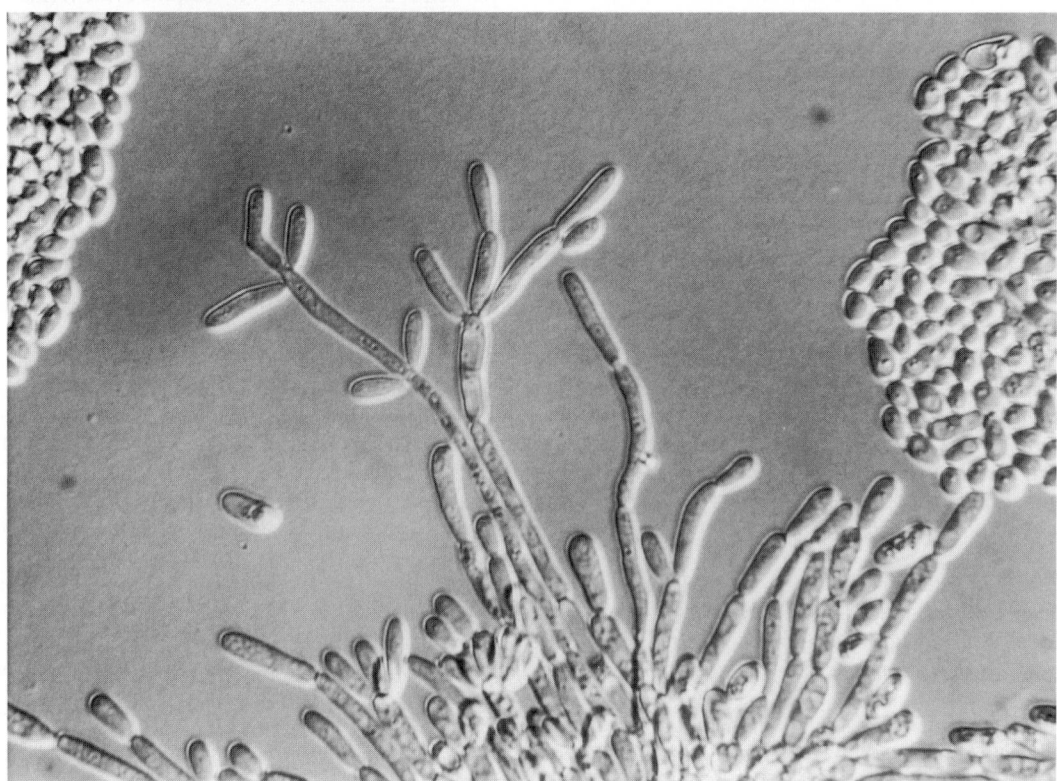

Candida krusei, CBS 573. Pseudomycelium. a, c. Courtesy M.-L. Kerkmann.

Candida lusitaniae van Uden & do Carmo-Sousa

Colony characteristics. Colonies (YPGA) white to cream-coloured, glistening, soft and smooth.
Microscopy. Budding cells (RA) ellipsoidal. Hyphae and germ tubes absent; well-developed pseudomycelium often present, strongly flexuose, growing down into the medium.
Teleomorph. *Clavispora lusitaniae* Rodrigues de Miranda (*Ascomycota, Hemiascomycetes, Saccharomycetales: Metschnikowiaceae*).
Asci ellipsoidal, evanescent, containing 4 smooth-walled, clavate ascospores. Heterothallic.
Physiology.

Fermentation:		Raffinose	–	α-Methyl-D-glucoside	v
Glucose	+	Melezitose	+	Salicin	+
Galactose	v	Inulin	–	D-Gluconate	s
Sucrose	v	Soluble starch	–	DL-Lactate	+,w
Maltose	v	D-Xylose	+	Succinate	+
Lactose	–	L-Arabinose	v	Citrate	v
Raffinose	–	D-Arabinose	–	*myo*-Inositol	–
α,α-Trehalose	v	D-Ribose	–	Hexadecane	w,–
		L-Rhamnose	v	Vitamin-free	–
Growth:		D-Glucosamine (C)	–	2-Keto-D-gluconate	+
Glucose	+	*N*-Acetyl-D-glucosamine	+	Nitrate	–
Galactose	+	Methanol	–	Ethylamine	+
L-Sorbose	+	Ethanol	+	L-Lysine	+
Sucrose	+	Glycerol	+	0.01% Cycloheximide	–,w
Maltose	+	*meso*-Erythritol	–	0.1% Cycloheximide	–
Cellobiose	+	Ribitol	s	Gelatin liquefaction	–
α,α-Trehalose	+	Galactitol	–	Growth at 42°C	+
Lactose	–	D-Mannitol	+		
Melibiose	–	D-Glucitol	+		

Differential diagnosis. Species signature: sucrose +, soluble starch –, *meso*-erythritol –, D-mannitol +, 2-keto-D-gluconate +, nitrate –, w/o pyridoxine +, growth at 42°C + .
Molecular diagnostics. A 26S-based specific probe was developed by Sandhu *et al*. (1995). SSU, ITS and LSU restriction maps based on NCBI M55526, AF009215 and U44817:

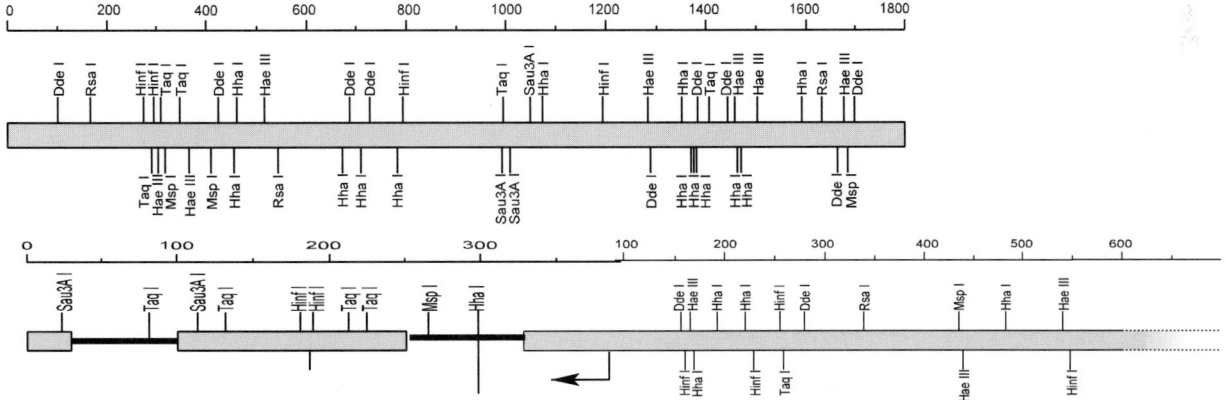

Pathogenicity. BSL-2. The species is emerging as an opportunist (Baker *et al*., 1984; Libertin *et al*., 1985; Christenson *et al*., 1987; Michel-Nguyen *et al*., 1996), particularly in humans with leukemia (Holschu *et al*., 1979), in patients undergoing cytotoxic chemotherapy (Blinkhorn *et al*., 1989), prednisone (Behar & Chertow, 1998) or receiving broad-spectrum antibiotics, in major surgery and transplant patients (Alcomar *et al*., 1989; Dilorenzo *et al.,* 1997) and in neonates (Sanchez & Cooper, 1987; Yinnon *et al*., 1992). Malignancy is frequently the underlying disease. Systemic infections are often fatal (Sarma *et al*., 1993). Fatal peritonitis after perforating appendicitis was reported by Guinet *et al*. (1983) and an endocarditis by Wendt *et al*. (1998). The fungus normally occurs as a saprophyte in the gastrointestinal tracts of warm-blooded animals. A mastitis in a cow was reported by Mós *et al*. (1978).
References. Holschu *et al*. (1979), Merz (1984), Christenson *et al*. (1987), Hadfield *et al*. (1987), Blinkhorn *et al*. (1989), Gargeya *et al*. (1990), Roder *et al*. (1991), Sanchez *et al*. (1992).

Antifungal susceptibility.

Antifungal	MICs range	MIC 90	Strains	Reference
AMB	0.5-2	1	10	Davey *et al.* (1998)
AMB	0.5-2	1	12	Pfaller *et al.* (1997)
AMB	0.06-0.5	0.5	35	Favel *et al.* (1997)
AMB	0.25-2	2	17	Espinel-Ingroff (1998)
5FC	≤0.12->64	>64	10	Davey *et al.* (1998)
5FC	0.06-128	128	12	Pfaller *et al.* (1997)
5FC	≤0.03-≥64	≥64	35	Pavel *et al.* (1997)
FLZ	≤0.12-16	1	10	Davey *et al.* (1998)
FLZ	0.12-4	4	12	Pfaller *et al.* (1997)
FLZ	0.12-4	2	35	Favel *et al.* (1997)
FLZ	0.12-4	2	17	Espinel-Ingroff (1998)
ITZ	≤0.03-0.25	0.12	10	Davey *et al.* (1998)
ITZ	0.007-0.25	0.25	12	Pfaller *et al.* (1997)
ITZ	0.06-0.5	0.5	35	Favel *et al.* (1997)
ITZ	<0.03-0.5	0.25	17	Espinel-Ingroff (1998)
KTZ	≤0.12-0.25	≤0.12	10	Davey *et al.* (1998)
KTZ	0.03-0.12	0.06	35	Favel *et al.* (1997)
MCZ	≤0.12-1	0.25	10	Davey *et al.* (1998)
VCZ	<0.03-0.06	0.06	17	Espinel-Ingroff (1998)

Nomenclature. *Candida parapsilosis* (Ashford) Langeron & Talice var. *obtusa* Dietrichson - Annls Parasit. Hum. Comp. 29: 483, 1954 (invalid) ≡ *Candida obtusa* (Dietrichson) van Uden & do Carmo Sousa - Port. Acta Biol., Ser. B, 6: 249, 1959 (invalid) ≡ *Candia obtusa* (Dietrichson) van Uden & do Carmo Sousa ex van Uden & Buckley, *in* Lodder - The Yeasts, ed. 2, p. 1018, 1970.
 Candida lusitaniae van Uden & do Carmo Sousa - Port. Acta. Biol., Ser. B, 6: 251, 1959.
 Clavispora lusitaniae Rodrigues de Miranda - Antonie van Leeuwenhoek 45: 480, 1979.

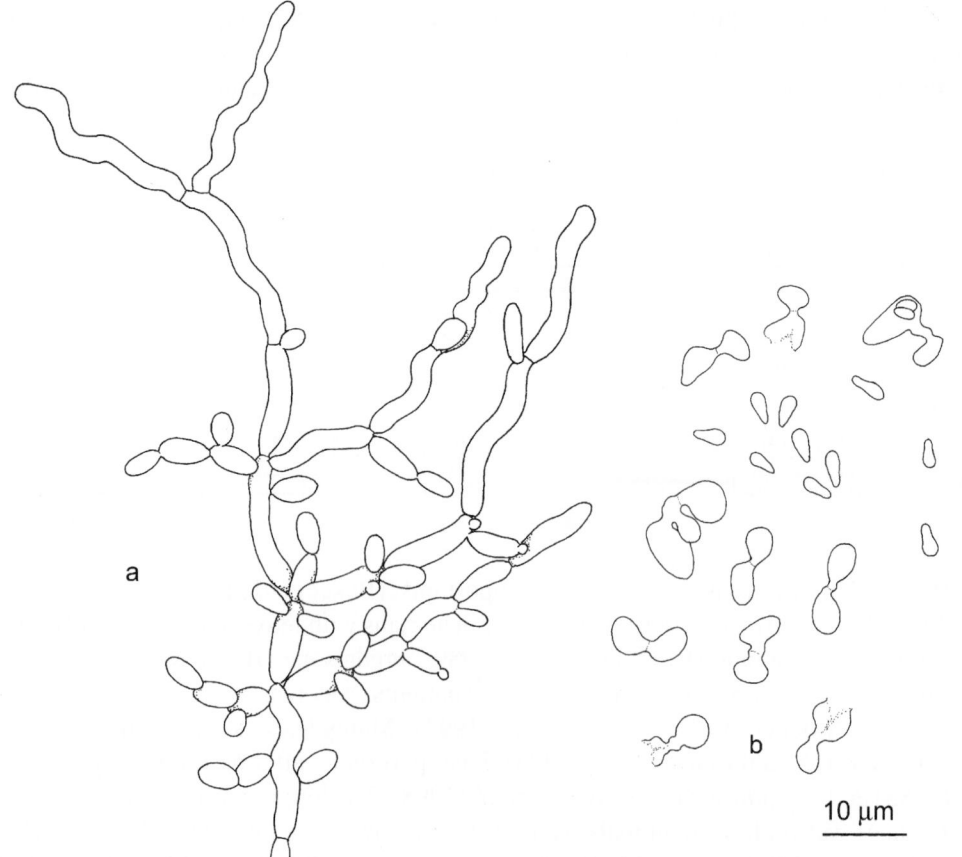

Candida lusitaniae (*Clavispora lusitaniae*), CBS 1944 × 5901. a. Pseudomycelium; b. asci and ascospores.

Candida norvegensis Dietrichson ex van Uden & Buckley

Colony characteristics. Colonies (YPGA) white to cream-coloured, butyrous.
Microscopy. Budding cells broadly ellipsoidal, 3.5-5.0 × 2.0-3.5 µm. Some pseudomycelium rarely present.
Teleomorph. *Pichia norvegensis* Leask & Yarrow (*Ascomycota, Hemiascomycetes, Saccharomycetales: Saccharomycetaceae*).
Asci evanescent, containing 1-4 hat-shaped ascospores. Homothallic.
Physiology.

Fermentation:		Raffinose	–	α-Methyl-D-glucoside	–
Glucose	s	Melezitose	–	Salicin	+
Galactose	–	Inulin	–	D-Gluconate	–
Sucrose	–	Soluble starch	–	DL-Lactate	w
Maltose	–	D-Xylose	–	Succinate	+
Lactose	–	L-Arabinose	–	Citrate	w
Raffinose	–	D-Arabinose	–	*myo*-Inositol	–
α,α-Trehalose	–	D-Ribose	–	Hexadecane	–
		L-Rhamnose	–	Vitamin-free	–
Growth:		D-Glucosamine (C)	+	2-Keto-D-gluconate	–
Glucose	+	*N*-Acetyl-D-glucosamine	–	5-Keto-D-gluconate	–
Galactose	–	Methanol	–	D-Glucarate	–
L-Sorbose	–	Ethanol	+	Nitrate	–
Sucrose	–	Glycerol	+	D-Glucosamine (N)	+
Maltose	–	*meso*-Erythritol	–	10% NaCl/5% glucose	+
Cellobiose	+	Ribitol	–	0.1% Cycloheximide	–
α,α-Trehalose	–	Galactitol	–	Gelatin liquefaction	–
Lactose	–	D-Mannitol	–	Starch formation	–
Melibiose	–	D-Glucitol	–	Growth at 37°C	+

Differential diagnosis. Species signature: D-glucosamine (C, N) +, D-mannitol –, D-tryptophan +, growth at 40°C +.
Molecular diagnostics. Nho *et al.* (1997) introduced specific *Hha*I digestion of rDNA amplicons. LSU restriction map based on NCBI U75730:

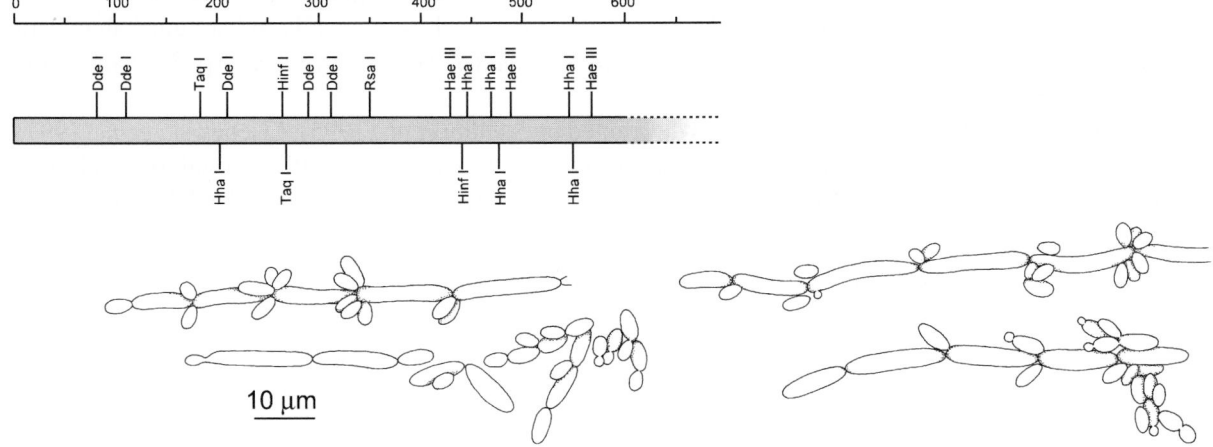

Candida norvegensis, CBS 6564. Pseudomycelium.

Pathogenicity. BSL-1. The species was described by Dietrichson (1954) from human sputum and by Leask & Yarrow (1976) from the vagina of a pregnant woman. Further isolations from clinical samples were listed by Hood *et al.* (1997). A systemic infection in a CAPD patient was reported by Nielsen *et al.* (1990). Nielsen & Stenderup (1996) described further disseminated cases in severely immunocompromised patients.
References. Leask & Yarrow (1976), Sandven *et al.* (1997).
Nomenclature. *Candida zeylanoides* (Castellani) Langeron & Guerra var. *norvegensis* Dietrichson - Annls Parasit. Hum. Comp. 29: 490, 1954 (invalid) ≡ *Candida norvegensis* Dietrichson ex van Uden & Buckley, *in* Lodder - The Yeasts, ed. 2, p. 1016, 1970.
Candida mycoderma (Reess) Lodder & Kreger-van Rij var. *annulata* Dietrichson - Annls Parasit. Hum. Comp. 29: 481, 1954 (invalid).
Candida trigonopsoides Dietrichson - Annls Parasit. Hum. Comp. 29: 488, 1954.
Pichia norvegensis Leask & Yarrow - Sabouraudia 14: 61, 1996.

Candida parapsilosis (Ashford) Langeron & Talice

Colony characteristics. Colonies (YPGA) cream-coloured to yellowish, glistening and soft, mostly smooth or partly or entirely wrinkled.

Microscopy. Pseudomycelium (RA) present, mostly abundant, consisting of branched chains of elongate cells in more or less christmastree-like arrangement, lateral branches gradually becoming shorter towards the hyphal apex.

Physiology.

Fermentation:		Soluble starch	–	*myo*-Inositol	–
Glucose	+	D-Xylose	+	Hexadecane	+,s
Galactose	v	L-Arabinose	+	Vitamin-free	–
Sucrose	–,s	D-Arabinose	–	2-Keto-D-gluconate	+
Maltose	–,s	D-Ribose	v	5-Keto-D-gluconate	–,s
Lactose	–	L-Rhamnose	–	D-Glucuronate	–
Raffinose	–	D-Glucosamine (C)	v	Xylitol	+,s
α,α-Trehalose	–,s	*N*-Acetyl-D-glucosamine	+	L-Arabinitol	–,s
		Methanol	–	Arbutin	–
Growth:		Ethanol	+	Nitrate	–
Glucose	+	Glycerol	+	Nitrite	–
Galactose	+	*meso*-Erythritol	–	Cadaverine	+
L-Sorbose	+,s	Ribitol	+,s	Creatinine	–
Sucrose	+	Galactitol	–	L-Lysine	+
Maltose	+	D-Mannitol	+	Ethylamine	+
Cellobiose	–	D-Glucitol	+	D-Glucosamine (N)	–
α,α-Trehalose	+	α-Methyl-D-glucoside	+	10% NaCl/5% glucose	+
Lactose	–	Salicin	–	Starch formation	–
Melibiose	–	D-Gluconate	+,s	Urease	–
Raffinose	–	DL-Lactate	–	0.01% Cycloheximide	v
Melezitose	+	Succinate	+	0.1% Cycloheximide	–
Inulin	–	Citrate	+	Growth at 37°C	+

Differential diagnosis. Species signature: fermentation of glucose +, and assimilation: cellobiose –, raffinose –, melebiose –, melezitose +, soluble starch –, D-xylose +, salicin –, arbutin –, 5-keto-D-gluconate – (but may be slowly positive), nitrate –, growth at 37°C +, D-tryptophan (N) –, w/o thiamine +. Physiologically indistinguishable from *Candida guilliermondii* (p. 200), *C. haemulonii* (p. 202), *C. pulcherrima* (p. 217), *C. tropicalis* (p. 220) and some saprobic species.

Molecular diagnostics. A 26S-based specific probe was developed by Sandhu *et al.* (1995). Lischewski *et al.* (1997) applied fluorescent *in situ* hybridization in tissue on the basis of an 18S rDNA probe. RAPD recognition in clinical samples was developed by Gautret *et al.* (1998). SSU, ITS and LSU restriction maps based on NCBI M60307 and L47109, resp.:

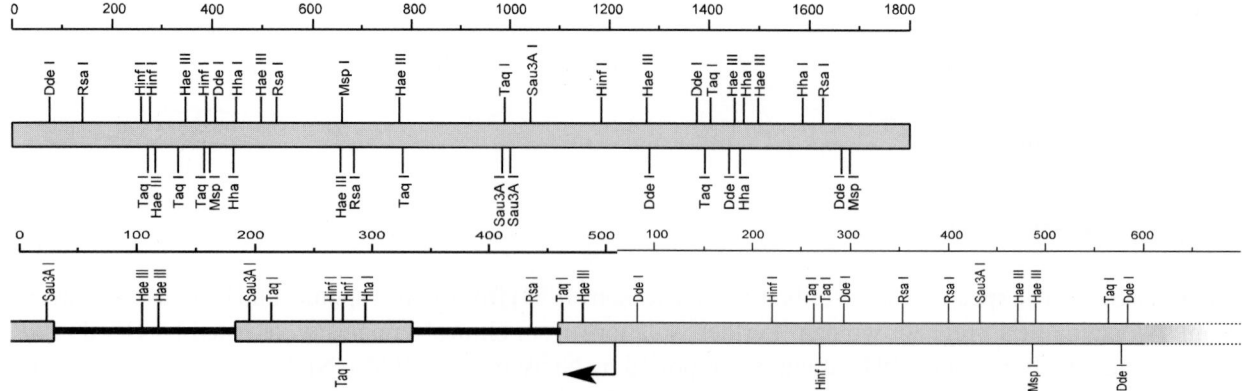

Pathogenicity. BSL-1. The species is occasionally involved as an opportunist in systemic mycoses (Amon *et al.*, 1990), particularly in patients with impaired natural immunity due to leukemia (d'Antonio *et al.*, 1992; Martino *et al.*, 1993; Girmenia *et al.*, 1996), in cancer patients (Krcmery *et al.*, 1998), after major surgery (Lopez-Jimenez *et al.*, 1994) and in children (Levy *et al.*, 1998). Nosocomial bloodstream infections are relatively frequent were reported by Solomon *et al.* (1984) and Welbel *et al.* (1996), and a pericarditis by McNamee *et al.* (1998). Plouffe *et al.* (1977)

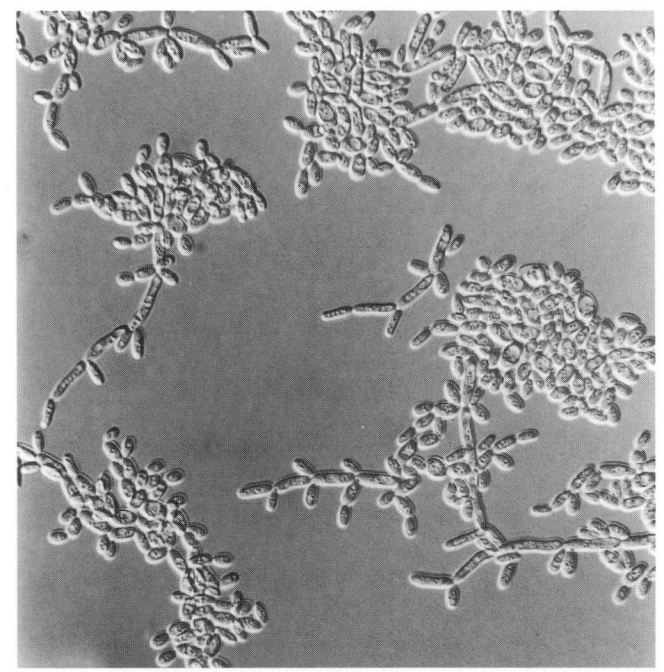

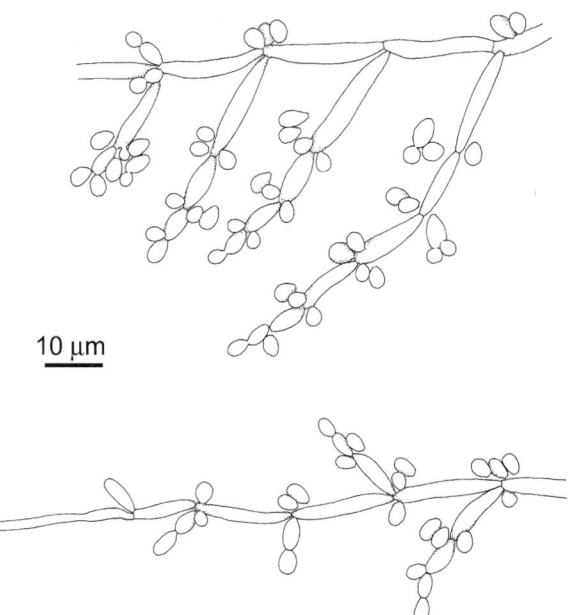

10 µm

Candida parapsilosis, strain M.-L. Kerkmann. Pseudomycelium. Courtesy M.-L. Kerkmann.

Candida parapsilosis, strain M.-L. Kerkmann. Pseudomycelium.

and Still *et al.* (1995) described infections in postoperative and burn patients. The species forms slimy biofilms on tubing systems and prosthetic materials (Branchini *et al.*, 1994; Montenegro *et al.*, 1995; Levin *et al.*, 1998); a prosthetic joint infection was described by White & Goetz (1995). Outbreaks among low birth weight neonates were reported by Saxen *et al.* (1993), Mcdonnell & Isaacs (1995) and Huang *et al.* (1999). Endocarditis was reported by Czwerwiec *et al.* (1993), Johnston *et al.* (1994), Kontou-Kastellanou *et al.* (1990) and Zahid *et al.* (1994); Diekema *et al.* (1997) traced cases of prosthetic valve endocarditis back to infection during surgery. Endophthalmitis was described by McCray *et al.* (1986), Borne *et al.* (1993) and Wong *et al.* (1997), fatal pancreatitis by Ibanez & Serrano Heranz (1999), and arthritis by MacGregor *et al.* (1979). Cases of sepsis in patients with burns were reviewed by Green *et al.* (1994). An AIDS-related case was reported by Tumbarello *et al.* (1996).

Reference. Barnett *et al.* (1990).

Nomenclature. *Monilia onychophila* Pollacci & Nannizzi, *in* Marengo - Archiv Biol. 3: 25-36, 1926 ≡ *Mycocandida onychophila* (Pollacci & Nannizzi) Langeron & Talice - Annls Parasit. Hum. Comp. 10: 58, 1932 ≡ *Mycotorula onychophila* (Pollacci & Nannizzi) Ciarrochi - Giorn. Ital. Derm. 74: 415-429, 1933.

Monilia parapsilosis Ashford - Am. J. Trop. Med. 8: 518, 1928 ≡ *Candida parapsilosis* (Ashford) Langeron & Talice - Annls Parasit. Hum. Comp. 10: 54, 1932 ≡ *Mycocandida parapsilosis* (Ashford) C.W. Dodge - Med. Mycol. p. 294, 193 ≡ *Mycotorula parapsilosis* (Ashford) Ciferri & Redaelli - Atti Inst. Bot. Univ. Lab. Crittogam. Pavia 5: 47, 1943.

Mycotorula vesica Harrison - Trans. R. Soc. Can., Sect. 5, 22: 219, 1928 ≡ *Pseudomycoderma vesica* (Harrison) C.W. Dodge - Med. Mycol. p. 236, 1935.

Blastodendrion intestinale Mattlet var. *epidermicum* Ciferri & Alfonseca - Centbl. Bakt. Parastikde, Abt. 2, 83: 274, 1931 ≡ *Castellania epidermica* (Ciferri & Alfonseca) C.W. Dodge - Med. Mycol. p. 251, 1935.

Blastodendrion globosum Zach, *in* Wolfram & Zach - Arch. Derm. Syph. 169: 99, 1933 ≡ *Schizoblastosporion globosum* (Zach) C.W. Dodge - Med. Mycol. p. 234, 1935.

Blastodendrion gracile Zach, *in* Wolfram & Zach - Arch. Derm. Syph. 169: 103, 1933 ≡ *Schizoblastosporion globosum* (Zach) C.W. Dodge - Med. Mycol. p. 234, 1935.

Saccharomyces vossii Dietrichson - Annls Parasit. Hum. Comp. 29: 467, 1954.

Saccharomyces verticillatus Dietrichson - Annls Parasit. Hum. Comp. 29: 463, 1954.

Antifungal susceptibility.

Antifungal	MICs range	MIC 90	Strains	Reference
AMB	1-2	1	18	Espinel-Ingroff (1998)
AMB	0.5-1	1	28	Pfaller *et al.* (1997)
AMB	0.25-1	1	18	Zhanei *et al.* (1998)
AMB	0.5-2	2	40	Marco *et al.* (1998)
AMB	0.25-1	1	11	Karlowsky *et al.* (1998)
AMB	0.25-0.5	0.25	20	Davey *et al.* (1998)
5FC	≤0.03-0.5	0.13	18	Zhanei *et al.* (1998)
5FC	≤0.06->128	1	40	Marco *et al.* (1998)
5FC	0.03-0.13	0.13	11	Karlowsky *et al.* (1998)
5FC	≤0.12-0.5	0.25	20	Davey *et al.* (1998)
FLZ	0.25-8	2	18	Espinel-Ingroff (1998)
FLZ	0.25-12	1	28	Pfaller *et al.* (1997)
FLZ	0.06-1	0.5	18	Zhanei *et al.* (1998)
FLZ	0.25-16	8	40	Marco *et al.* (1998)
FLZ	0.25-4	2	11	Karlowsky *et al.* (1998)
FLZ	0.25-1	1	20	Davey *et al.* (1998)
ITZ	0.03-0.5	0.25	18	Espinel-Ingroff (1998)
ITZ	0.02-0.25	0.12	28	Pfaller *et al.* (1997)
ITZ	≤0.03-0.5	0.25	18	Zhanei *et al.* (1998)
ITZ	0.13-2	0.5	40	Marco *et al.* (1998)
ITZ	≤0.03-0.12	0.12	20	Davey *et al.* (1998)
KTZ	≤0.004-0.008	≤0.004	11	Karlowsky *et al.* (1998)
KTZ	≤0.12-0.25	≤0.12	20	Davey *et al.* (1998)
KTZ	≤0.03-0.25	0.25	18	Zhanei *et al.* (1998)
MCZ	≤0.12-1	0.5	20	Davey *et al* (1998)
VCZ	<0.03-0.5	0.25	18	Espinel-Ingroff (1998)
VCZ	≤0.02-1	0.25	40	Marco *et al.* (1998)

Candida pelliculosa Redaelli

Colony characteristics. Colonies (YPGA) cream-coloured, glistening, moist, butyrous.
Microscopy. Budding cells (RA) ellipsoidal, 2-6 × 2-4 μm. Pseudomycelium may be present.
Teleomorph. *Pichia anomala* (Hansen) Kurtzman (*Ascomycota, Hemiascomycetes, Saccharomycetales: Saccharomycetaceae*).
Diploid budding cells contain 1-4 hat-shaped ascospores; ascus wall evanescent. Heterothallic.
Physiology.

Fermentation:		Raffinose	+	α-Methyl-D-glucoside	+
Glucose	+	Melezitose	+	Salicin	+
Galactose	v	Inulin	–	D-Gluconate	v
Sucrose	+	Soluble starch	+	DL-Lactate	+
Maltose	v	D-Xylose	v	Succinate	+
Lactose	–	L-Arabinose	v	Citrate	+
Raffinose	w,–	D-Arabinose	–	*myo*-Inositol	–
α,α-Trehalose	–	D-Ribose	v	Hexadecane	–
		L-Rhamnose	–	Vitamin-free	+
Growth:		D-Glucosamine (C)	–	2-Keto-D-gluconate	–
Glucose	+	*N*-Acetyl-D-glucosamine	–	5-Keto-D-gluconate	–
Galactose	v	Methanol	–	D-Glucarate	–
L-Sorbose	–	Ethanol	+	Nitrate	+
Sucrose	+	Glycerol	+	10% NaC1/5% glucose	+
Maltose	+	*meso*-Erythritol	+	0.01% Cycloheximide	–
Cellobiose	+	Ribitol	v	Starch formation	–
α,α-Trehalose	+	Galactitol	–	Gelatin liquefaction	–
Lactose	–	D-Mannitol	+	Growth at 37°C	v
Melibiose	–	D-Glucitol	+		

Differential diagnosis. Species signature: glucose fermentation +, and assimilation: melibiose –, inulin –, galactitol –, salicin +, DL-lactate +, w/o vitamins +, 10% NaCl +, D-tryptophan (N) –, 0.1% cycloheximide –. Physiologically indistinguishable from *C. guilliermondii* and similar saprobic yeasts.
Molecular diagnostics. SSU and LSU restriction maps based on NCBI X58054 and U74592:

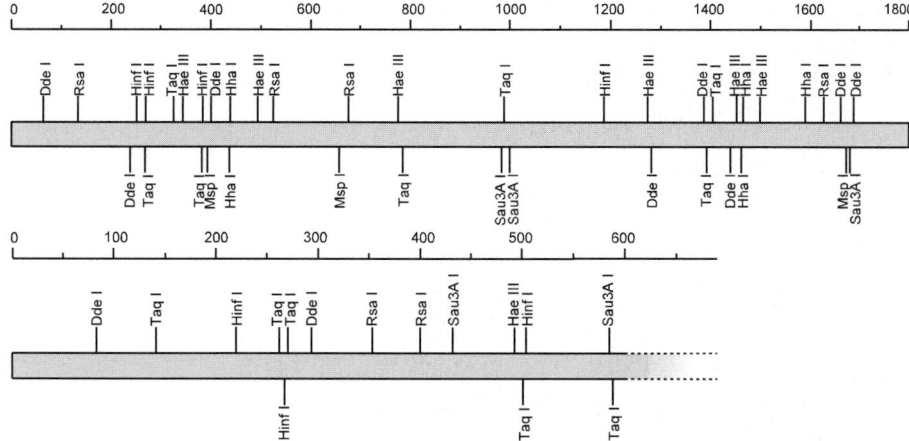

Pathogenicity. BSL-1. Neumeister *et al.* (1992) described a case of fatal fungemia in the pancreas of an immunocompetent patient. Endocarditis was reported by Nohinek *et al.* (1987). García-Martos (1994) described two cases of sepsis in debilitated patients and Kunová *et al.* (1996) a fungemia in a leukemic patient. Catheter-related infections are relatively frequent (Klein *et al.*, 1988; Haron *et al.*, 1988; Muñoz *et al.*, 1989; Dickensheets, 1989; Makimura *et al.*, 1993; Goss *et al.*, 1994; Yamada *et al.*, 1995). Thuler *et al.* (1997) described an epidemy of 24 catheter-related cases in a cancer hospital. The species was also diagnosed in AIDS patients (Salesa *et al.*, 1991) and neonates (Murphy *et al.*, 1986). The species may emerge under therapy of antimycotics to which it is less sensitive (Alter & Farley, 1994).
References. Kurtzman (1984, 1998), Barnett *et al.* (1990).

Antifungal susceptibility.

Antifungal	MICs range	MIC 50	Strains	Reference
AMB	0.25-1	1	5	Espinel-Ingroff (1998)
FLZ	2-4	2	5	Espinel-Ingroff (1998)
ITZ	0.25-1	0.5	5	Espinel-Ingroff (1998)
VCZ	0.12-0.25	0.25	5	Espinel-Ingroff (1998)

Nomenclature. *Saccharomyces anomalus* Hansen - Annls Microgr. 3: 467, 1891 ≡ *Willia anomala* (Hansen) Hansen - Centbl. Bakt. Parasitkde, Abt. 2, 12: 529, 1904 ≡ *Hansenula anomala* (Hansen) H. & P. Sydow - Annls Mycol. 17: 44, 1919 ≡ *Pichia anomala* (Hansen) Kurtzman - Antonie van Leeuwenhoek 50: 212, 1984.

Candida pelliculosa Redaelli - Micet. Assoc. Microbiol. Tuberc. Pulm. Cavit. p. 69, 1925 ≡ *Monilia pelliculosa* (Redaelli) Nannizzi - Tratt. Micopat. Um. 4: 373, 1934 ≡ *Mycocandida pelliculosa* (Redaelli) Guerra - Rôle Levures Dermatol. p. 25, 1935.

Candida beverwijkiae Novák & Vitéz - Zentbl. Bakt. Parasitkde, Abt. 1, 193: 129, 1964.

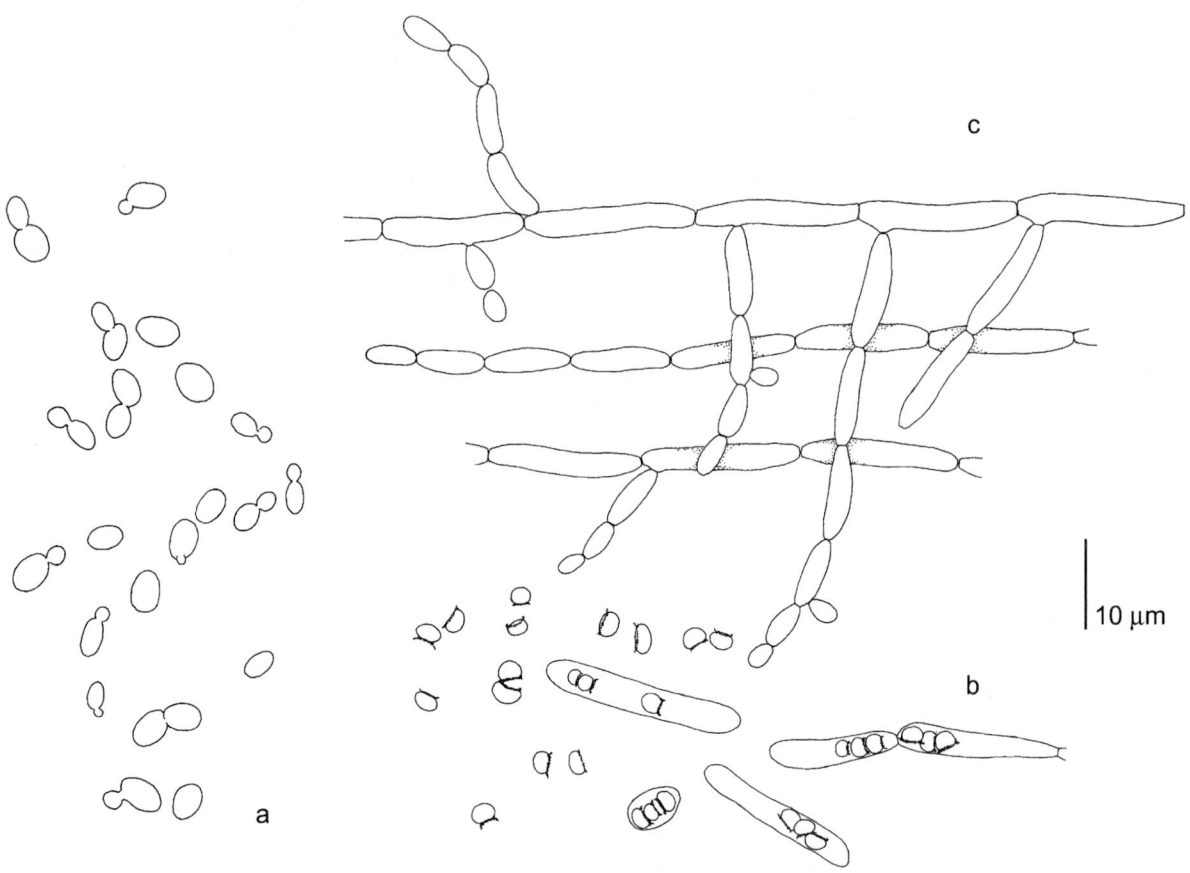

Candida pelliculosa (*Pichia anomala*), CBS 110. a. Budding cells; b. asci and ascospores; c. pseudomycelium.

Candida pulcherrima (Lindner) Windisch

Colony characteristics. Colonies (YPGA) whitish to cream-coloured, later with faint reddish-brown tinges, smooth, waxy.

Microscopy. Budding cells (RA) subspherical to broadly ellipsoidal, up to about 6 × 4 µm, containing some small oil droplets. Chlamydospores (sub)spherical, thick-walled, about 6 µm diam, containing a single, large oil drop.

Teleomorph. *Metschnikowia pulcherrima* Pitt & Miller (*Ascomycota, Hemiascomycetes, Saccharomycetales: Metschnikowiaceae*).

Physiology.

Fermentation:		Inulin	–	Succinate	+
Glucose	+	Soluble starch	–	Citrate	v
Galactose	w,–	D-Xylose	+	*myo*-Inositol	–
Sucrose	–	L-Arabinose	–	Hexadecane	+
Maltose	–	D-Arabinose	–	Vitamin-free	–
Lactose	–	D-Ribose	w,–	2-Keto-D-gluconate	+
Raffinose	–	L-Rhamnose	–	5-Keto-D-gluconate	–
α,α-Trehalose	–	D-Glucosamine (C)	–	Nitrate	–
		N-Acetyl-D-glucosamine	+	Cadaverine	+
Growth:		Methanol	–	L-Lysine	+
Glucose	+	Ethanol	+	Ethylamine	+
Galactose	+	Glycerol	+	D-Glucosamine (N)	–
L-Sorbose	+	*meso*-Erythritol	–	10% NaC1/5% glucose	+
Sucrose	+	Ribitol	v	Starch formation	–
Maltose	+	Galactitol	–	Urease	–
Cellobiose	+	D-Mannitol	+	Gelatin liquefaction	–
α,α-Trehalose	+	D-Glucitol	+	0.01% Cycloheximide	–
Lactose	–	α-Methyl-D-glucoside	+	Growth at 30°C	+
Melibiose	–	Salicin	+	Growth at 37°C	v
Raffinose	–	D-Gluconate	+		
Melezitose	+	DL-Lactate	w,–		

Differential diagnosis. Species signature: L-rhamnose –, melibiose –, lactose –, raffinose –, melezitose +, *meso*-erythritol –, xylitol +, L-arabinitol –, D-mannitol +, 5-keto-D-gluconate –, DL-lactate –, nitrate –, urease –, w/o thiamine +, 0.01% cycloheximide –, 50% glucose +. Physiologically not reliably distinguishable from *Candida guilliermondii* (p. 200), *C. parapsilosis* (p. 212) and many saprobic yeasts.

Molecular diagnostics. LSU restriction map based on NCBI U45736:

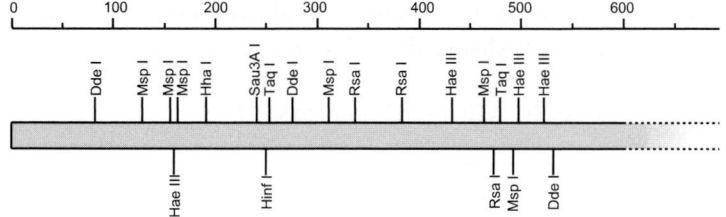

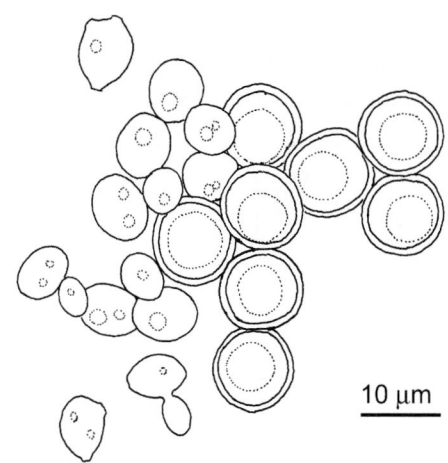

Candida pulcherrima (*Metschnikowia pulcherrima*), strain Kerkmann. Budding cells and chlamydospores.

Pathogenicity. BSL-1. Occasionally reported from cases of onychomycosis (Pospísil, 1989) and from blood cultures after infection of catheters (Weber & Kolb, 1986) or implantates (Mohl *et al.*, 1998).

Reference. Meyer *et al.* (1978).

Nomenclature. *Torula pulcherrima* Lindner - Mikrosk. Betriebskontr., ed. 3, 1901 ≡ *Candida pulcherrima* (Lindner) Windisch - Arch. Mikrobiol. 11: 368-390, 1940.

Monilia castellanii Re - J. Trop. Med. Hyg. 28: 217, 1925 ≡ *Cryptococcus castellanii* (Re) Castellani - Arch. Derm. Syph. 16: 401, 1927 ≡ *Torulopsis castellanii* (Re) Castellani & Jacano - J. Trop. Med. Hyg. 36: 312, 1933 ≡ *Castellania castellanii* (Re) C.W. Dodge - Med. Mycol. p. 250, 1935.

Cryptococcus interdigitalis Pollacci & Nannizzi - Micet. Patog. Uomo Anim., Fasc. 5, No. 44, 1926 ≡ *Mycotorula interdigitalis* (Pollacci & Nannizzi) Redaelli, *in* Flarer - Arch. Ital. Derm. Sif. Venereol. 7: 438, 1931 ≡ *Syringospora interdigitalis* (Pollacci & Nannizzi) C.W. Dodge - Med. Mycol. p. 277, 1935.

Candida rugosa (Anderson) Diddens & Lodder

Colony characteristics. Colonies (YPGA) cream-coloured, butyrous.
Microscopy. Budding cells (RA) slender, ellipsoidal to subcylindrical, 6-12 × 2.0-3.5 μm, sometimes coherent, pseudomycelium-like short hyphae are present.
Physiology.

Fermentation:	–	L-Rhamnose	–	2-Keto-D-gluconate	–
		D-Glucosamine (C)	–	5-Keto-D-gluconate	–
Growth:		*N*-Acetyl-D-glucosamine	s	D-Glucuronate	–
Glucose	+	Methanol	–	D-Glucarate	–
Galactose	+	Ethanol	+	Xylitol	–,s
L-Sorbose	v	Glycerol	+	L-Arabinitol	–
Sucrose	–	*meso*-Erythritol	–	Arbutin	–
Maltose	–	Ribitol	–,s	Nitrate	–
Cellobiose	–	Galactitol	–	Nitrite	–
α,α-Trehalose	–	D-Mannitol	+,s	Cadaverine	+
Lactose	–	D-Glucitol	+,s	Creatinine	–
Melibiose	–	α-Methyl-D-glucoside	–	L-Lysine	+
Raffinose	–	Salicin	–	Ethylamine	+
Melezitose	–	D-Gluconate	v	D-Glucosamine (N)	–
Inulin	–	DL-Lactate	+,s	Starch formation	–
Soluble starch	–	Succinate	v	Urease	–
D-Xylose	v	Citrate	v	0.01% Cycloheximide	–
L-Arabinose	–	*myo*-Inositol	–	Growth at 37°C	+
D-Arabinose	–	Hexadecane	s	Growth at 40°C	v
D-Ribose	–	Vitamin-free	–		

Differential diagnosis. Species signature: fermentation –, and assimilation: sucrose –, maltose –, cellobiose –, trehalose –, ethanol +, *meso*-erythritol –, 2-keto-D-gluconate –, nitrate –, L-lysine +, D-glucosamine (N) –, w/o thiamine –, 60% glucose –, 10% NaCl +. Physiologically indistinguishable from *Candida catenulata* (p. 189), *Geotrichum capitatum* (p. 231) and several saprobic yeast species.
Molecular diagnostics. SSU and LSU restriction maps based on NCBI AB013502 and U45727:

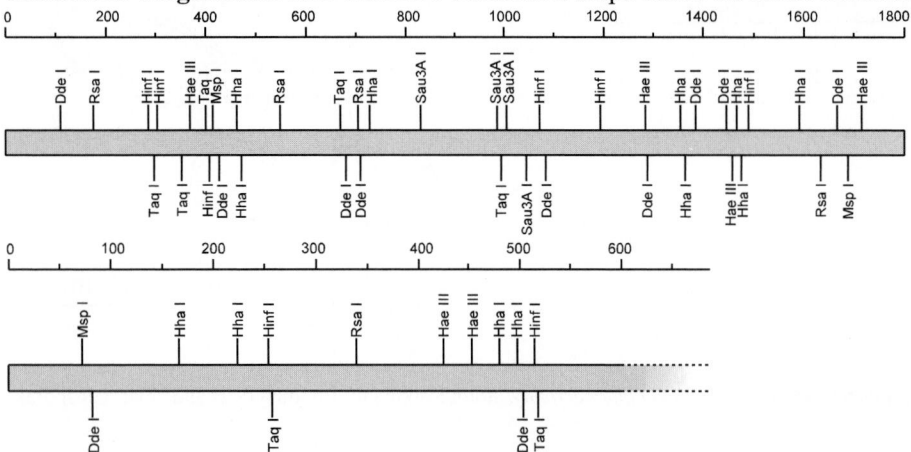

Pathogenicity. BSL-1. Catheter-associated fungemia was reported by Reinhardt *et al.* (1985), Frappaz *et al.* (1990) and Shenoy *et al.* (1996) and colonization in burn patients by Dube *et al.* (1994). An uterus infection in a horse was described by Giorgi *et al.* (1986), and a mastitis in a cow by Dixon & Dukes (1982). Its significance as an opportunist was reviewed by Sugar & Stevens (1985).
Reference. Redkar *et al.* (1996).

Antifungal susceptibility.

Antifungal	MICs range	MIC 50	Strains	Reference
AMB	0.5-1	1	7	Pfaller *et al.* (1997)
AMB	1-2	1	5	Espinel-Ingroff (1998)
AMB	0.58-1.16	nd	10	Dubé *et al.* (1994)
5FC	0.12-1	0.5	7	Pfaller *et al.* (1997)
5FC	0->323	nd	10	Dubé *et al.* (1994)
FLZ	1-8	1	7	Pfaller *et al.* (1997)
FLZ	4-16	8	5	Espinel-Ingroff (1998)
FLZ	2.5-20	nd	10	Dubé *et al.* (1994)
ITZ	0.03-0.12	0.03	7	Pfaller *et al.* (1997)
ITZ	0.06-0.5	0.25	5	Espinel-Ingroff (1998)
KTZ	0.1-0.8	nd	10	Dubé *et al.* (1994)
VCZ	<0.03-0.06	0.06	5	Espinel-Ingroff (1998)

Nomenclature. *Mycoderma rugosa* Anderson - J. Infect. Dis. 21: 341, 1917 ≡ *Candida rugosa* (Anderson) Diddens & Lodder - Anaskospor. Hefen p. 280, 1942 ≡ *Azymocandida rugosa* (Anderson) Novák & Zsolt - Acta Bot. Acad. Sci. Hung. 7: 134, 1961.

Candida rugosa (Anderson) Diddens & Lodder var. *elegans* Dietrichson - Annls Parasit. Hum. Comp. 29: 485, 1954.

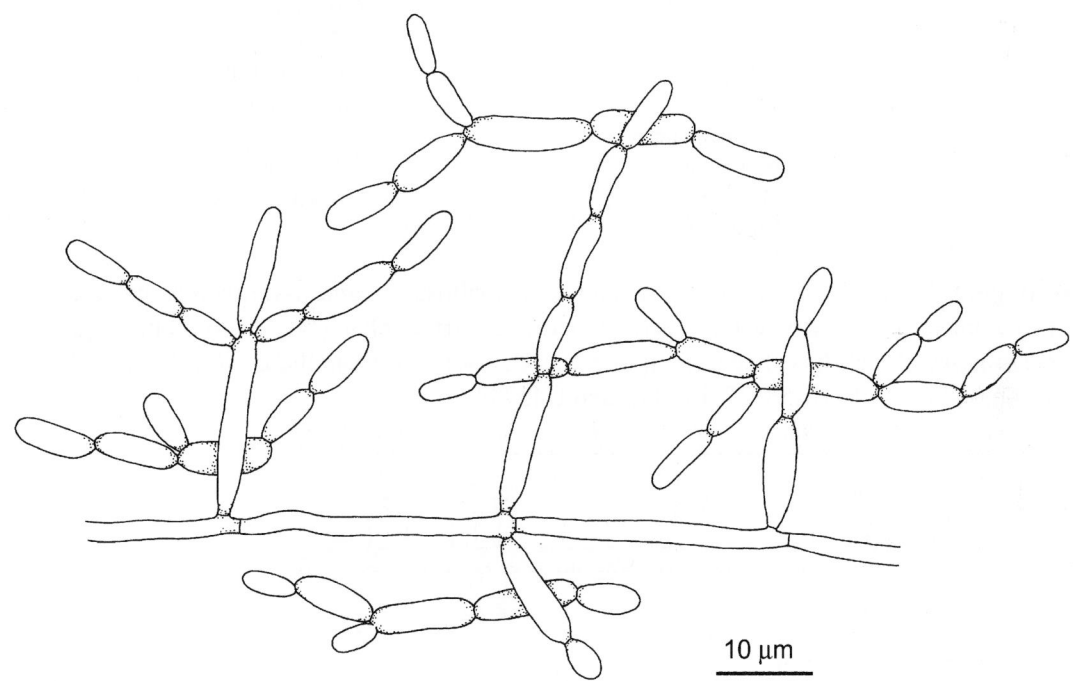

10 µm

Candida rugosa, CBS 613. Hyphal fragment and pseudomycelial chains of budding cells.

Candida tropicalis (Castellani) Berkhout

Colony characteristics. Colonies (YPGA) cream-coloured, off-white, soft, smooth and creamy or wrinkled near the margin.

Microscopy. Budding cells (RA) ellipsoidal. Pseudomycelium abundant, consisting of long, poorly branched elements, often narrowed towards a sterile apex; conidia arranged in small groups around the middle of each cellular element.

Physiology.

Fermentation:		Soluble starch	+	*myo*-Inositol	–
Glucose	+	D-Xylose	+	Hexadecane	+
Galactose	+	L-Arabinose	–	Vitamin-free	v
Sucrose	v	D-Arabinose	–	2-Keto-D-gluconate	+
Maltose	+	D-Ribose	–, s	5-Keto-D-gluconate	+
Lactose	–	L-Rhamnose	–	D-Glucuronate	–
Raffinose	–	D-Glucosamine (C)	v	Xylitol	+,s
α,α-Trehalose	+,s	*N*-Acetyl-D-glucosamine	+	L-Arabinitol	–
		Methanol	–	Arbutin	+
Growth:		Ethanol	+	Nitrate	–
Glucose	+	Glycerol	v	Nitrite	–
Galactose	+	*meso*-Erythritol	–	Cadaverine	+
L-Sorbose	v	Ribitol	+,s	L-Lysine	+
Sucrose	v	Galactitol	–	Ethylamine	+
Maltose	+	D-Mannitol	+	Creatinine	–
Cellobiose	+,s	D-Glucitol	+	D-Glucosamine (N)	–
α,α-Trehalose	+	α-Methyl-D-glucoside	v	10% NaCl/5% glucose	s
Lactose	–	Salicin	v	Starch formation	–
Melibiose	–	D-Gluconate	v	Urease	–
Raffinose	–	DL-Lactate	v	0.1% Cycloheximide	+
Melezitose	v	Succinate	+	Growth at 40°C	+
Inulin	–	Citrate	+,s		

Differential diagnosis. Species signature: fermentation of maltose, +, and assimilation: galactose +, lactose –, raffinose –, L-rhamnose –, *meso*-erythritol –, *myo*-inositol –, D-tryptophan (N) –, w/o biotin –, growth at 40°C +.

Molecular diagnostics. A 26S-based specific probe was developed by Sandhu *et al.* (1995). SSU, ITS and LSU restriction maps based on NCBI M55527, L47112 and U45749:

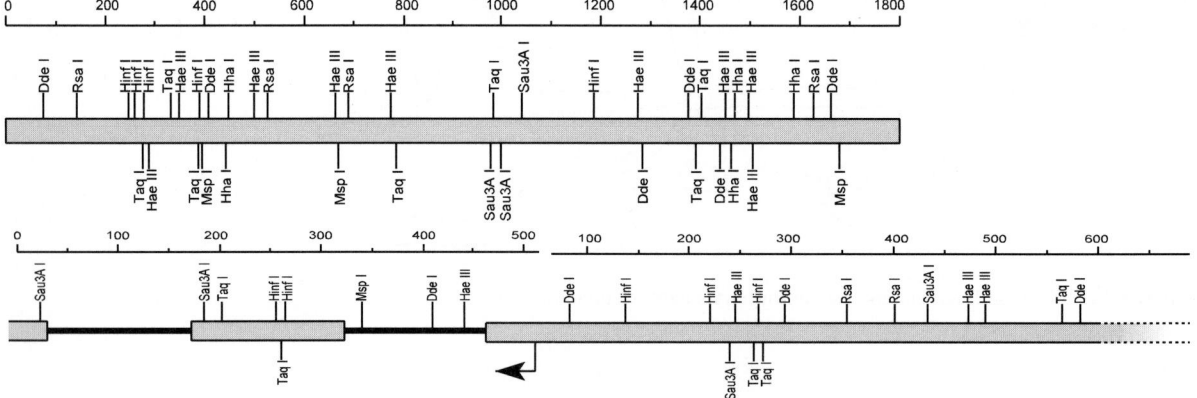

Pathogenicity. BSL-2. The species is emerging as an opportunist in patients with impaired innate immunity (Wingard *et al.*, 1979; Sandin *et al.*, 1993b; Huang *et al.*, 1993; Sakata *et al.*, 1995; Speers *et al.*, 1995; Weers-Pothof *et al.*, 1997) or neoplastic disease (Prats *et al.*, 1995). Comiter *et al.* (1996) reported a fungal obstruction and subsequent bladder rupture, and Faure-Fontenla *et al.* (1997) a gastric perforation in an otherwise healthy, 3-years old girl. Disseminated infections in neonates were reported by Finkelstein *et al.* (1993) and major surgery infections by Isenberg *et al.* (1989). Ferra *et al.* (1994) reported two cases of osteomyelitis in bone marrow transplant patients and a pulmonary infection by Prats *et al.* (1995). Endocarditis was reported by Mansur *et al.* (1996), Shmuely *et al.* (1997) and Gerritsen *et al.* (1998). Multidrug-resistance is known (Jandourek *et al.*, 1999).

References. Wingard *et al.* (1979), Schlitzer & Ahearn (1982), Cohen & Montgomerie (1993).

Antifungal susceptibility.

Antifungal	MICs range	MIC 90	Strains	Reference
AMB	0.5	0.5	20	Davey *et al.* (1998)
AMB	0.5-8	1	16	Espinel-Ingroff (1998)
AMB	0.25-1/0.5-1	1/1	58/38	Pfaller *et al.* (1997/1998)
AMB	0.5-1	1	15	Zhanei *et al.* (1998)
AMB	0.5-2	2	54	Marco *et al.* (1998)
AMB	0.25-2	1	11	Karlowsky *et al.* (1997)
5FC	≤0.12->64	0.5	20	Davey *et al.* (1998)
5FC	0.06-1/0.06->128	1/128	58/38	Pfaller *et al.* (1997/1998)
5FC	0.06-64	0.25	15	Zhanei *et al.* (1998)
5FC	≤0.06->128	1	54	Marco *et al.* (1998)
5FC	0.13-8	0.5	11	Karlowsky *et al.* (1997)
FLZ	0.25-64	8	20	Davey *et al.* (1998)
FLZ	0.25->64	32	16	Espinel-Ingroff (1998)
FLZ	0.12->128/0.12->128	1/2	58/38	Pfaller *et al.* (1997/1998)
FLZ	0.06->64	4	15	Zhanei *et al.* (1998)
FLZ	0.25->128	2	54	Marco *et al.* (1998)
FLZ	0.13->128	128	11	Karlowsky *et al.* (1997)
ITZ	≤0.03->16	0.5	20	Davey *et al.* (1998)
ITZ	0.06->16	0.5	16	Espinel-Ingroff (1998)
ITZ	0.02->8/0.02->8	0.12/0.25	58/38	Pfaller *et al.* (1997/1998)
ITZ	0.06-0.25	0.25	15	Zhanei *et al.* (1998)
ITZ	0.03->8	0.5	54	Marco *et al.* (1998)
KTZ	≤0.12-8	4	20	Davey *et al.* (1998)
KTZ	≤0.03-2	0.12	15	Zhanei *et al.* (1998)
KTZ	≤0.004-4	4	11	Karlowsky *et al.* (1997)
MCZ	≤0.12-4	2	20	Davey *et al.* (1998)
VCZ	<0.03->16	0.25	16	Espinel-Ingroff (1998)
VCZ	≤0.015->16	0.125	54	Marco *et al.* (1998)

Nomenclature. *Saccharomyces linguae-pilosae* Lucet - Arch. Parasit. 4: 262, 1901 ≡ *Cryptococcus linguae-pilosae* (Lucet) Castellani & Chalmers - Man. Trop. Med., ed. 2, p. 770, 1913 ≡ *Myceloblastanon linguae-pilosae* (Lucet) Ota - Jpn. J. Derm. Urol. 27: 175, 1927 ≡ *Torulopsis linguae-pilosae* (Lucet) de Almeida - Anais Fac. Med. Univ. S. Paulo 9: 76, 1933 ≡ *Castellania linguae-pilosae* (Lucet) C.W. Dodge - Med. Mycol. p. 260, 1935.

Oidium tropicalis Castellani - Br. Med. J. 2: 868, 1910 ≡ *Endomyces tropicalis* (Castellani) Castellani - Centbl. Bakt. Parasitkde, Abt. 1, 58: 236, 1911 ≡ *Monilia tropicalis* (Castellani) Castellani & Chalmers - Man. Trop. Med., ed. 3, p. 1086, 1918 ≡ *Atelosaccharomyces tropicalis* (Castellani) Froilano de Mello, *in* Froilano de Mello & Fernandes - Arq. Hig. Patol. Exot. 6: 263, 1918 ≡ *Candida tropicalis* (Castellani) Berkhout - Schimmelges. Monilia, Oidium, Oospora Torula p. 43, 1923 ≡ *Myceloblastanon tropicale* (Castellani) Ota - Jpn. J. Derm. Urol. 27: 178, 1927 ≡ *Castellania tropicalis* (Castellani) C.W. Dodge - Med. Mycol. p. 258, 1935 ≡ *Mycotorula tropicalis* (Castellani) Ciferri & Redaelli - Atti Ist. Bot. Univ. Pavia 5: 48, 1943 ≡ *Candida albicans* (Robin) Berkhout var. *tropicalis* (Castellani) Ciferri - Man. Micol. Med., ed. 2, p. 252, 1960 ≡ *Procandida tropicalis* (Castellani) Novák & Zsolt - Acta Bot. Acad. Sci. Hung. 7: 133, 1961.

Endomyces paratropicalis Castellani - Centbl. Bakt. Parasitkde, Abt. 1, 58: 237, 1911 ≡ *Monilia paratropicalis* (Castellani) Castellani & Chalmers - Man. Trop. Med., ed. 2, p. 826, 1913 ≡ *Atelosaccharomyces paratropicalis* (Castellani) Froilano de Mello, *in* Froilano de Mello & Fernandes - Arq. Hyg. Pat. Exot. 6: 264, 1918 ≡ *Myceloblastanon paratropicalis* (Castellani) Ota - Jpn. J. Derm. Urol. 27: 178, 1927 ≡ *Candida paratropicalis* (Castellani) Basgal - Contrib. Estud. Blastom. Pulmon. p. 49, 1931 ≡ *Castellania paratropicalis* (Castellani) C.W. Dodge - Med. Mycol. p. 264, 1935.

Endomyces bronchialis Castellani - Br. Med. J. 2: 1210, 1912 ≡ *Monilia bronchialis* (Castellani) Castellani & Chalmers - Man. Trop. Med., ed. 2, p. 826, 1913 ≡ *Myceloblastanon brochiale* (Castellani) Ota - Jpn. J. Derm. Urol. 27: 178, 1927 ≡ *Castellania bronchialis* (Castellani) C.W. Dodge - Med. Mycol. p. 254, 1935.

Endomyces burgesii Castellani - Arch. Parasit. 16: 187, 1912 ≡ *Monilia burgesii* (Castellani) Castellani & Chalmers - Man. Trop. Med., ed. 2, p. 828, 1913 ≡ *Castellania burgesii* C.W. Dodge - Med. Mycol. p. 255, 1935.

Endomyces cruzi Froilano de Mello & Paes - Arq. Hig. Pat. Exot. 6: 51 1918 ≡ *Zynonema cruzi* (Froilano de Mello & Paes) C.W. Dodge - Med. Mycol. p. 176, 1935.

Endomyces entericus Castellani - Br. Med. J. 2: 1209, 1912 ≡ *Monilia enterica* (Castellani) Castellani & Chalmers - Man. Trop. Med., ed. 2, p. 827, 1913 ≡ *Myceloblastanon entericum* (Castellani) Ota - Jpn. J. Derm. Urol. 27: 178, 1927 ≡ *Candida enterica* (Castellani) de Almeida - Anais Fac. Med. Univ. S. Paulo 9: 77, 1933 ≡ *Castellania enterica* (Castellani) C.W. Dodge - Med. Mycol. p. 264, 1935.

Endomyces insolitus Castellani - Br. Med. J. 2: 1210, 1912 ≡ *Monilia insolita* (Castellani) Castellani & Chalmers - Man. Trop. Med., ed. 2, p. 827, 1913 ≡ *Myceloblastanon insolitum* (Castellani) Ota - Jpn. J. Derm. Urol. 27: 178, 1927 ≡ *Candida insolita* (Castellani) Basgal - Contrib. Estud. Blast. Pulmon. p. 49, 1931 = *Castellania insolita* (Castellani) C.W. Dodge - Med. Mycol. p. 258, 1935.

Endomyces niveus Castellani - Br. Med. J. 2: 1210, 1912 ≡ *Monilia nivea* (Castellani) Castellani & Chalmers - Man. Trop. Med., ed. 2, p. 826, 1913 ≡ *Myceloblastanon niveus* (Castellani) Ota - Jpn. J. Derm. Urol. 27: 178, 1927 ≡ *Candida nivea* (Castellani) Basgal - Contrib. Estud. Blastom. Pulmon. p. 49, 1931 ≡ *Castellania nivea* (Castellani) C.W. Dodge - Med. Mycol. p. 235, 1935.

Endomyces perryi Castellani - Arch. Parasit. 16: 187, 1913 ≡ *Monilia perryi* (Castellani) Castellani & Chalmers - Man. Trop. Med., ed. 2., p. 829, 1913 ≡ *Parendomyces perryi* (Castellani) C.W. Dodge - Med. Mycol. p. 243, 1935.

Endomyces tropicalis Acton - Ind. J. Med. Res. 6: 598, 1919 ≡ *Actonia tropicalis* (Acton) C.W. Dodge - Med. Mycol. p. 146, 1935.

Monilia metatropicalis Castellani & Chalmers - Man. Trop. Med., ed. 3, p. 1087, 1919 ≡ *Castellania metatropicalis* (Castellani & Chalmers) C.W. Dodge - Med. Mycol. p. 258, 1935.

Monilia accraensis Macfie & Ingram - Annls Trop. Med. Parasit. 15: 274, 1921 ≡ *Candida accraensis* (Macfie & Ingram) Basgal - Contrib. Estud. Blastom. Pulmon. p. 49, 1931 ≡ *Castellania accraensis* (Macfie & Ingram) C.W. Dodge - Med. Mycol. p. 264, 1935.

Monilia issavi Mattlet - Annls Soc. Belge Méd. Trop. 6: 21, 1926 ≡ *Syringospora issavi* (Mattlet) C.W. Dodge - Med. Mycol. p. 280, 1935.

Candida vulgaris Berkhout - Schimmelges. Monilia, Oidium, Oospora Torula p. 42, 1923 ≡ *Geotrichum vulgaris* (Berkhout) Langeron & Talice - Annls Parasit. Hum. Comp. 10: 68, 1932.

Blastodendrion irritans Mattlet - Annls Soc. Belge Med. Trop. 6: 16, 1926 ≡ *Parasaccharomyces irritans* (Mattlet) C.W. Dodge - Med. Mycol. p. 268, 1935.

Blastodendrion kayongosi Mattlet - Annls Soc. Belge Med. Trop. 6: 16, 1926 ≡ *Monilia kyongosi* (Mattlet) Brumpt - Précis Parasitol., ed. 4, p. 1241, 1927 ≡ *Cryptococcus kayongosi* (Mattlet) Nannizzi - Tratt. Micopat. Um. 4: 327, 1934.

Monilia argentina Vivoli, Avellaneda & de Bardessi - Reunión Soc. Argent. Patol. Reg. Norte Tucumán 7: 239, 1932 ≡ *Mycotoruloides argentina* (Vivoli, Avellaneda & de Bardessi) C.W. Dodge - Med. Mycol. p. 291, 1935.

Monilia aegyptiaca Khouri - Bull. Soc. Pathol. Exot. 26: 9, 1933 ≡ *Castellania aegyptiaca* (Khouri) C.W. Dodge - Med. Mycol. p. 254, 1935.

Mycotorula dimorpha Redaelli & Ciferri in Ciferri & Redaelli - Arch. Mikrobiol. 6: 46, 1935 ≡ *Syringospora dimorpha* (Redaelli & Ciferri) C.W. Dodge & M. Moore - Ann. Mo. Bot. Gdn 23: 140, 1936.

Mycotorula trimorpha Redaelli & Ciferri, *in* Ciferri & Redaelli - Arch. Mikrobiol. 6: 9-72, 1935 ≡ *Mycotoruloides trimorpha* (Redaelli & Ciferri) C.W. Dodge & M. Moore - Ann. Mo. Bot. Gdn 23: 142, 1936.

Parasaccharomyces talicei C.W. Dodge - Med. Mycol. p. 269, 1935.

Candida benhamiae Novák & Vitéz - Zentbl. Bakt. Parasitkde, Abt. 1, 193: 127, 1964.

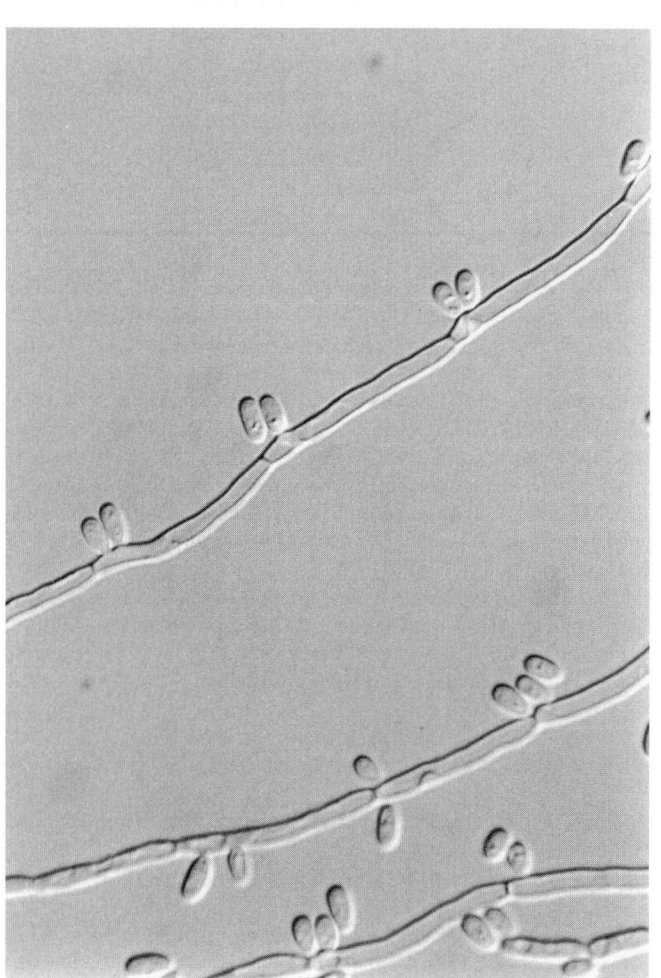
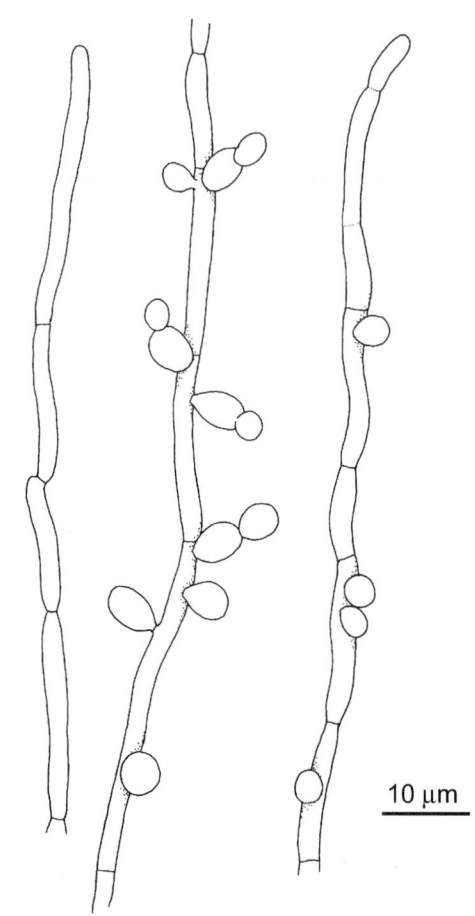

10 µm

Candida tropicalis, strain M.-L. Kerkmann. Poorly branched pseudomyclium with lateral conidia.

Candida utilis (Henneberg) Lodder & Kreger-van Rij

Colony characteristics. Colonies (YPGA) cream coloured, smooth, moist.
Microscopy. Budding cells (RA) slender, ellipsoidal to cylindrical, 7-10 × 2-3 μm, later swelling to broadly ellipsoidal. Pseudomycelium absent; budding cells may initially cohere.
Teleomorph. *Pichia jadinii* (Sartory *et al.*) Kurtzman (*Ascomycota, Hemiascomycetes, Saccharomycetales: Saccharomycetaceae*).
Physiology.

Fermentation:		Raffinose	+	α-Methyl-D-glucoside	+
Glucose	+	Melezitose	+	Salicin	+
Galactose	–	Inulin	+,w	D-Gluconate	+
Sucrose	+	Soluble starch	–	DL-Lactate	+
Maltose	–	D-Xylose	+	Succinate	+
Lactose	–	L-Arabinose	–	Citrate	+
Raffinose	w	D-Arabinose	–	*myo*-Inositol	–
α,α-Trehalose	–	D-Ribose	–	Hexadecane	–
		L-Rhamnose	–	Vitamin-free	+
Growth:		D-Glucosamine (C)	–	2-Keto-D-gluconate	w,–
Glucose	+	N-Acetyl-D-glucosamine	–	5-Keto-D-gluconate	–
Galactose	–	Methanol	–	D-Glucarate	–
L-Sorbose	–	Ethanol	+	Nitrate	+
Sucrose	+	Glycerol	+	10% NaCl/5% glucose	–
Maltose	+	*meso*-Erythritol	–	0.01% Cycloheximide	–
Cellobiose	+	Ribitol	–	Starch formation	–
α,α-Trehalose	+	Galactitol	–	Gelatin liquefaction	–
Lactose	–	D-Mannitol	+	Growth at 37°C	+
Melibiose	–	D-Glucitol	+		

Differential diagnosis. Species signature: fermentation of glucose +, and assimilation: galactose –, raffinose +, citrate +, w/o vitamins +, 10% NaCl –, growth at 37°C +. Physiologically indistinguishable from *Williopsis saturnus*.
Molecular diagnostics. LSU restriction map based on NCBI U73570:

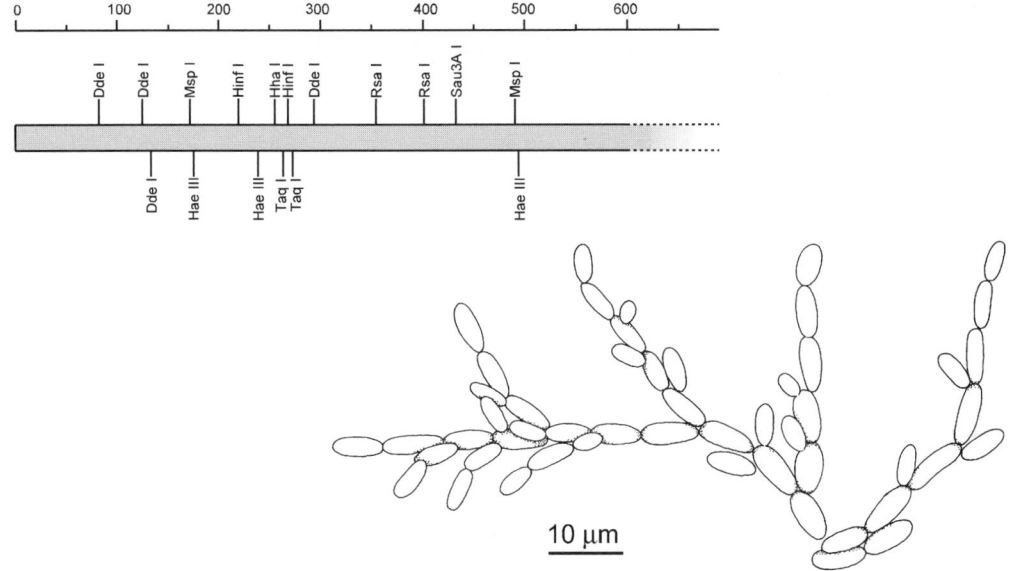

Candida utilis, CBS 621. Cohering budding cells.

Pathogenicity. BSL-1. Cases of transient hematogenous infection were presented by Alsina *et al.* (1988) and Bougnoux *et al.* (1993). Hazen *et al.* (1999) reported a chronic urinary tract infection and Shih *et al.* (1999) a keratitis.
Reference. Meyer *et al.* (1978).

Nomenclature. *Torula utilis* Henneberg - Handb. Gärungsbakt., ed. 2, 1926 ≡ *Candida utilis* (Henneberg) Lodder & Kreger-van Rij - The Yeasts p. 546, 1952.
Saccharomyces jadinii Sartory, R. Sartory, Weill & J. Meyer - C.R. Hebd. Séanc. Acad. Sci., Paris 194: 1688, 1932 ≡ *Pichia jadinii* (Sartory, R. Sartory, Weill & J. Meyer) Kurtzman - Antonie van Leeuwenhoek 50: 213, 1984.

Candida viswanathii Viswanathan & H.S. Randhawa ex R.S. Sandhu & H.S. Randhawa

Colony characteristics. Colonies (YPGA) white to cream-coloured, smooth, moist.

Microscopy. Budding cells (RA) obovoidal to ellipsoidal, 4-12 × 2.5-7.0 μm; pseudomycelium present, poorly branched, bearing few, obovoidal conidia with truncate scars.

Physiology.

Fermentation:		Soluble starch	+	*myo*-Inositol	–
Glucose	+	D-Xylose	+	Hexadecane	+
Galactose	s	L-Arabinose	–,s	Vitamin-free	–
Sucrose	v	D-Arabinose	–	2-Keto-D-gluconate	+
Maltose	+	D-Ribose	–, s	5-Keto-D-gluconate	+
Lactose	–	L-Rhamnose	–	D-Glucuronate	–
Raffinose	–	D-Glucosamine (C)	–,s	Xylitol	+,s
α,α-Trehalose	+,s	*N*-Acetyl-D-glucosamine	+	L-Arabinitol	–,s
		Methanol	–	Arbutin	+
Growth:		Ethanol	+	Nitrate	–
Glucose	+	Glycerol	+	Nitrite	–
Galactose	+	*meso*-Erythritol	–	Cadaverine	+
L-Sorbose	v	Ribitol	+	Creatinine	–
Sucrose	+	Galactitol	–	L-Lysine	+
Maltose	+	D-Mannitol	+	Ethylamine	+
Cellobiose	+	D-Glucitol	+	D-Glucosamine (N)	–
α,α-Trehalose	+	α-Methyl-D-glucoside	+	10% NaC1/5% glucose	+
Lactose	–	Salicin	+,s	Starch formation	–
Melibiose	–	D-Gluconate	+,s	Urease	–
Raffinose	–	DL-Lactate	–	0.1% Cycloheximide	+
Melezitose	+	Succinate	+	Growth at 37°C	+
Inulin	–	Citrate	+	Growth at 40°C	v

Differential diagnosis. Species signature: fermentation of maltose +, and assimilation: galactose + , cellobiose +, raffinose –, soluble starch +, L-rhamnose –, 5-keto-D-gluconate +, 0.1% cycloheximide +, growth at 37°C +. Physiologically indistinguishable from *Candida albicans* (p. 184), *C. tropicalis* (p. 220) and some saprobic yeasts.

Molecular diagnostics. SSU and LSU restriction maps based on NCBI M60309 and U45752:

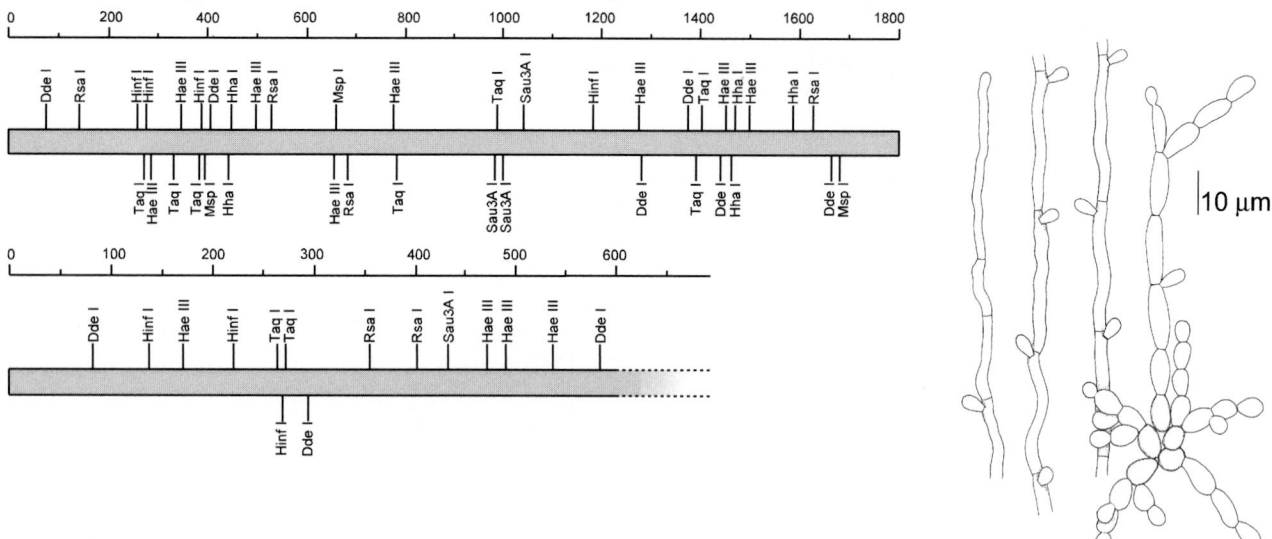

Candida viswanathii, CBS 4024. Pseudomycelium.

Pathogenicity. BSL-1. The species is a rare agent of meningitis (Viswanathan & Randhawa, 1959; Sandhu *et al.*, 1976).

Reference. Sandhu & Randhawa (1962).

Nomenclature. *Candida viswanathii* Viswanathan & H.S. Randhawa - Sci. Cult. 25: 86, 1962 (invalid) ≡ *Candida viswanathii* Viswanathan & H.S. Randhawa ex R.S. Sandhu & H.S. Randhawa - Mycopath. Mycol. Appl. 18: 179, 1962.

Candida zeylanoides (Castell. *et al.*) Langer. & Guerra

Colony characteristics. Colonies (YPGA) cream coloured, moist, glassy, slightly cerebriform.
Microscopy. Budding cells (RA) fusiform, often slightly narrowed to the end with lateral cells inserted slightly away from the apex; cells measuring about 8-12 × 3 μm.
Physiology.

Fermentation:		Soluble starch	–	*myo*-Inositol	–
Glucose	–,s	D-Xylose	–	Hexadecane	–
Galactose	–	L-Arabinose	–	Vitamin-free	v
Sucrose	–	D-Arabinose	–	2-Keto-D-gluconate	+
Maltose	–	D-Ribose	–	5-Keto-D-gluconate	+
Lactose	–	L-Rhamnose	–	D-Glucuronate	–
Cellobiose	–	D-Glucosamine (C)	–,s	Xylitol	–
α,α-Trehalose	–,s	*N*-Acetyl-D-glucosamine	+,s	L-Arabinitol	–
		Methanol	–	Arbutin	v
Growth:		Ethanol	v	Nitrate	–
Glucose	+	Glycerol	+	Nitrite	v
Galactose	–,s	*meso*-Erythritol	–	Cadaverine	v
L-Sorbose	+,s	Ribitol	v	Creatinine	–
Sucrose	–	Galactitol	–	L-Lysine	v
Maltose	–	D-Mannitol	+	Ethylamine	v
Cellobiose	–,s	D-Glucitol	+	D-Glucosamine (N)	–
α,α-Trehalose	+	α-Methyl-D-glucoside	–	10% NaC1/5% glucose	v
Lactose	–	Salicin	v	Starch formation	–
Melibiose	–	D-Gluconate	+,s	Urease	–
Raffinose	–	DL-Lactate	–	0.1% Cycloheximide	+
Melezitose	–	Succinate	+	Growth at 35°C	v
Inulin	–	Citrate	+	Growth at 37°C	w,–

Differential diagnosis. Species signature: fermentation of cellobiose –, and assimilation: sucrose –, maltose –, melibiose –, *meso*-erythrtitol –, D-mannitol +, 2-keto-D-gluconate +, xylitol –, nitrate –, 0.1% cycloheximide +, propane-1,2-diol –, w/o thiamine +. Physiologically indistinguishable from *Candida albicans* (p. 184), *C. catenulata* (p. 189) and some saprobic yeast species.
Molecular diagnostics. SSU and LSU restriction maps based on NCBI AB013509 and U45832:

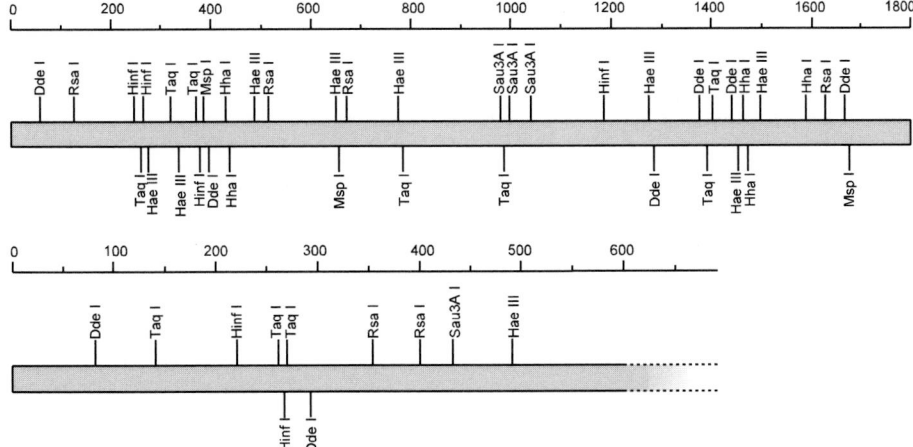

Pathogenicity. BSL-1. An uncommon commensal. Whitby *et al.* (1996) reported an endocarditis in a HIV positive patient, and Liao *et al.* (1993) a crural infection.
Reference. Meyer *et al.* (1998).

Nomenclature. *Monilia macroglossiae* Castellani - J. Trop. Med. Hyg. 28: 219, 1925 ≡ *Cryptococcus macroglossiae* (Castellani) Castellani - J. Trop. Med. Hyg. 28: 220, 1925 ≡ *Torulopsis macroglossiae* (Castellani) Castellani & Jacono - J. Trop. Med. Hyg. 36: 314, 1933 ≡ *Parendomyces macroglossiae* (Castellani) C.W. Dodge - Med. Mycol. p. 241, 1935.

Monilia zeylanoides Castellani, Douglas & Thompson - J. Trop. Med. Hyg. 28: 258, 1925 ≡ *Parendomyces zeylanoides* (Castellani, Douglas & Thompson) C.W. Dodge - Med. Mycol. p. 240, 1935 ≡ *Pseudomonilia zeylanoides* (Castellani, Douglas & Thompson) C.W. Dodge & M. Moore - Annls Mo. Bot. Gdn 23: 144, 1936 ≡ *Candida zeylanoides* (Castellani, Douglas & Thompson) Langeron & Guerra - Annls Parasit. Hum. Comp. 16: 501, 1938.

Cryptococcus uvae Pollacci & Nannizzi, *in* Motta - Atti R. Accad. Fisiocrit. Siena, Ser. 9, 17: 636, 1925 ≡ *Monilia uvae* (Pollacci & Nannizzi)
Vuillemin - Champ. Paras. Myc. Homme p. 84, 1931 ≡ *Syringospora uvae* (Pollacci & Nannizzi) C.W. Dodge - Med. Mycol. p. 28, 1935.
   Blastodendrion canis von Szilvinyi - Zentbl. Bakt. Parasitkde, Abt. 2, 89: 284, 1933.
   Pichia dubia Dietrichson - Annls Parasit. Hum. Comp. 29: 469, 1954.

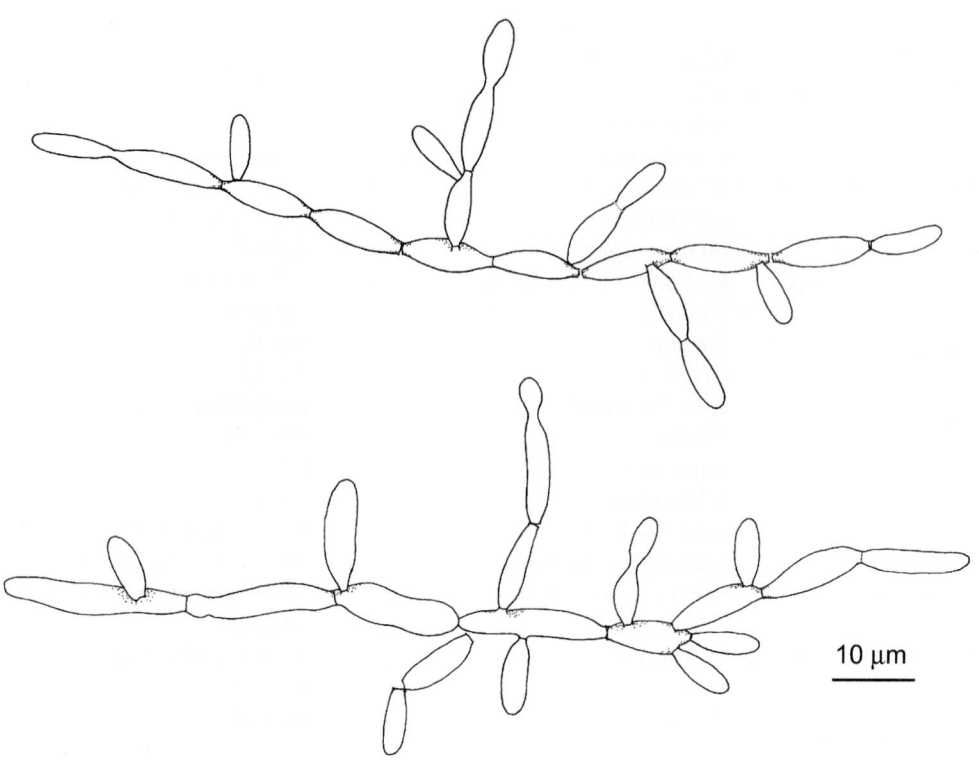

10 μm

Candida zeylanoides, CBS 619. Pseudomycelium.

Ascomycetous yeasts. Genus: *GEOTRICHUM*

Generic description. Colonies white, farinose, hairy or butyrous, usually dry. Hyphae hyaline, falling apart into rectangular arthroconidia. Cells without capsules or extracellular starch. Conidia occasionally sympodial or percurrent; chlamydospores and endoconidia often present. Cell walls 3-layered, without xylose or fucose; septa with micropores; urease –, nitrate –, DBB –.

Teleomorphs. *Dipodascus, Galactomyces* (*Ascomycota, Hemiascomycetes, Saccharomycetales: Dipodascaceae*).
References. De Hoog *et al.* (1986), Smith *et al.* (2000). Phylogenetic relations within *Geotrichum* and its teleomorphs were presented by Kurtzman & Robnett (1998).

Morphological key to the treated species of Geotrichum:

1a. Colonies rapidly expanding; marginal hyphae up to 12 µm wide, often with dichotomous branching .... **G. candidum (228)**
1b. Colonies with moderate growth; marginal hyphae up to 4 µm wide, without dichotomous branching **➜ 2**
2a. Sympodial conidia abundantly produced from cicatrized rachids .... **G. capitatum (231)**
2b. Sympodial conidia absent .... **G. clavatum (233)**

Physiological key to the treated species of Geotrichum:

1a. D-Xylose assimilated .... **G. candidum (231)**
1b. D-Xylose not assimilated **➜ 2**
2a. Cellobiose, salicin and arbutin assimilated .... **G. clavatum (234)**
2b. Cellobiose, salicin and arbutin not assimilated .... **G. capitatum (236)**

Table 23. Variable physiological characters among treated *Geotrichum* species.

	candidum	capitatum	clavatum
Cellobiose	–	–	+
D-Xylose	+	–	–
Salicin	–	–	+
Vitamin-free	+	–	–
Arbutin	–	–	+
Growth at 37°C	v	+	+

Geotrichum candidum Link:Fr.

Colony characteristics. Colonies (MEYA) rapidly expanding, white, flat or with cottony aerial mycelium and somewhat fimbriate margin. Odour insignificant or fruity.

Microscopy. Expanding hyphae often with di- or trichotomous branching; main branches 7-12 µm wide, walls firm, bearing much narrower (2.5-4.0 µm) lateral branches at acute or nearly right angles. Lateral branches soon disarticulating into short-cylindrical cells which slightly inflate, finally measuring about 5-17 × 4-6 µm.

Teleomorph. *Galactomyces geotrichum* (Butler & Petersen) Redhead & Malloch (*Ascomycota, Hemiascomycetes, Saccharomycetales: Dipodascaceae*).

Asci subhyaline, (sub)spherical, 6-9 × 7-10 µm, containing one, rarely two ascospores. Ascospores broadly ellipsoidal, 6-9 × 7-10 µm, pale yellowish brown, with an echinate inner wall (spines up to 1 µm long)and an irregular outer wall, often with a hyaline equatorial furrow. Heterothallic.

Physiology.

Fermentation:		Lactose	–	Galactitol	–
Glucose	v	Melibiose	–	D-Mannitol	v
Galactose	v	Raffinose	–	D-Glucitol	+
Sucrose	–	Melezitose	–	α-Methyl-D-glucoside	–
Maltose	–	Inulin	–	Salicin	–
Lactose	–	Soluble starch	–	D-Gluconate	–
Raffinose	–	D-Xylose	+	DL-Lactate	v
		L-Arabinose	–	Succinate	v
Growth:		D-Arabinose	–	Citrate	v
Glucose	+	D-Ribose	v	*myo*-Inositol	–
Galactose	+	L-Rhamnose	–	Vitamin-free	+
L-Sorbose	+	D-Glucosamine (C)	–	2-Keto-D-gluconate	–
Sucrose	–	Ethanol	+	Arbutin	–
Maltose	–	Glycerol	+	Nitrate	–
Cellobiose	–	*meso*-Erythritol	–	Growth at 37°C	v
α,α-Trehalose	–	Ribitol	v	Growth at 40°C	–

Differential diagnosis. The species is characterized by up to 12 µm wide hyphae with dichotomous branching at the colony margin, and the following growth characteristics: cellobiose –, xylose +, and no growth at 40°C. The closely related species *Galactomyces citri-aurantii* Butler differs by absence of growth without vitamins.

Molecular diagnostics. SSU and LSU restriction maps based on NCBI AB000652 and U40118:

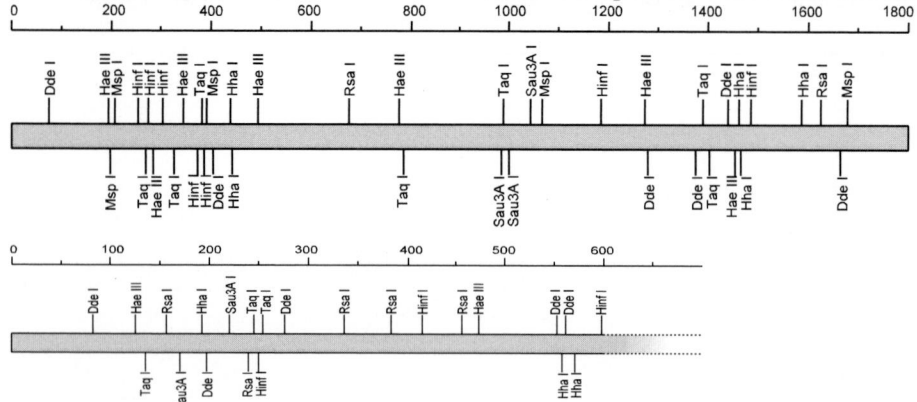

Pathogenicity. BSL-1. The main human disorders caused by *G. candidum* are colonization of the intestinal tract (Vasei & Imanieh, 1999) and bronchial or pulmonary infections, in humans as well as in other mammals (Rhyan *et al.*, 1990). The fungus causes disseminated, irregular infiltrates; the sputum is slimy and contains large quantities of arthroconidia. Blood-stream infection with sepsis is occasionally noted (Kassamili *et al.*, 1987). Hardy *et al.* (1995) and Choi *et al.* (1999) reported traumatic infections.

References. Morenz (1963, 1970), de Hoog *et al.* (1986), Smith *et al.* (1995).

Antifungal susceptibility.

Antifungal	MICs range	MIC 90	Strains	Reference
AMB	0.2-1.6	1.56	22	Wildfeuer *et al.* (1998)
ITZ	0.01-6.3	3.13	22	Wildfeuer *et al.* (1998)
KTZ	0.2-3.1	1.56	22	Wildfeuer *et al.* (1998)
VCZ	0.1-0.8	0.39	22	Wildfeuer *et al.* (1998)

Nomenclature. *Geotrichum candidum* Link - Mag. Naturf. Freunde 9: 17, 1809 ≡ *Geotrichum candidum* Link:Fries - Syst. Mycol. 3: 420, 1832.

Monilia asteroides Castellani - J. Trop. Med. Hyg. 17: 307, 1914 ≡ *Oidium asteroides* (Castellani) Castellani & Chalmers - Man. Trop. Med., ed. 3, p. 1095, 1919 ≡ *Mycoderma asteroides* (Castellani) Brumpt - Précis Parasitol., ed. 3, p. 1076, 1922 ≡ *Geotrichum asteroides* (Castellani) Basgal - Contrib. Estud. Blastom. Pulmon. p. 48, 1931.

Oidium matalense Castellani - Lect. Higher Fungi Rel. Hum. Path., Lond. 1915 ≡ *Oospora matalensis* (Castellani) Berkhout - Schimmelges. Monilia, Oidium, Oospora Torula p. 46, 1923 ≡ *Mycoderma matalensis* (Castellani) Brumpt - Précis Parasitol., ed. 3, p. 1084, 1922 ≡ *Pseudomycoderma matalensis* (Castellani) Ciferri - Arch. Protistenkde 71: 436, 1930 ≡ *Geotrichum matalense* (Castellani) Castellani - J. Trop. Med. Hyg. 35: 278, 1932 ≡ *Pseudomonilia matalensis* (Castellani) C.W. Dodge - Med. Mycol. p. 295, 1935 ≡ *Endomyces lactis* Windisch var. *matalensis* (Castellani) Windisch - Beitr. Biol. Pfl. 28: 123, 1951 ≡ *Trichosporon matalense* (Castellani) Ciferri - Anais Soc. Biol. Pernambuco 30: 140, 1955.

Geotrichum matalense (Castellani) Castellani var. *chapmanii* Castellani - J. Trop. Med. Hyg. 35: 279, 1932.

Sporotrichum cutaneum Schabinski - Grundriss Med. Mykol. p. 82, 1960.

Geotrichum pseudocandidum Saéz - Mycopath. Mycol. Appl. 34: 363, 1968.

Endomyces geotrichum Butler & Petersen - Mycologia 64: 367, 1972 ≡ *Galactomyces geotrichum* (Butler & Petersen) Redhead & Malloch - Can. J. Bot. 55: 1708, 1977.

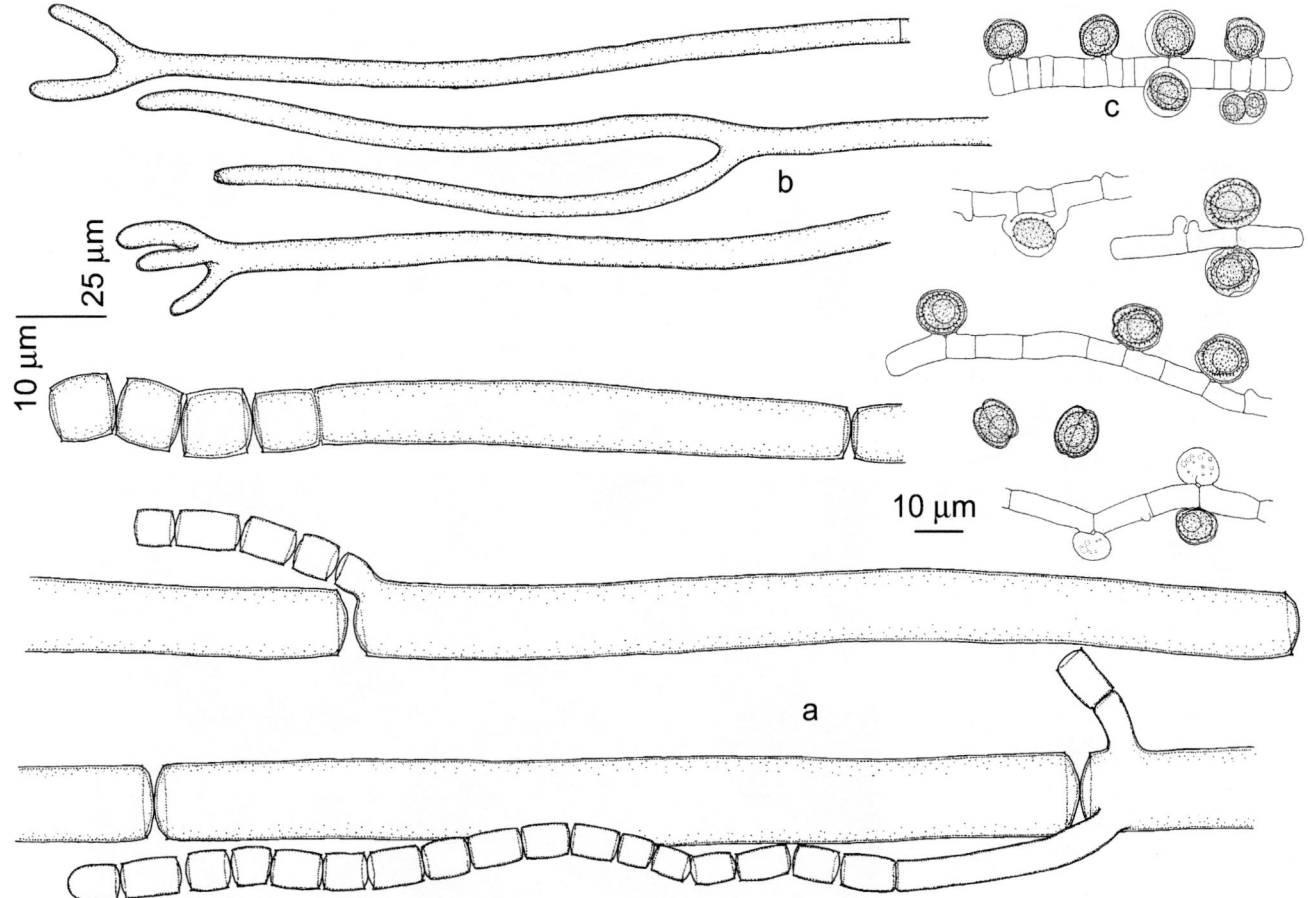

Geotrichum candidum, CBS 772.71. a. Main and lateral branches with arthroconidia; b. di- and trichotomously branched expanding hyphae; c. bipodal asci with single, ornamented ascospore (c, reproduced with permission from Stud. Mycol. 29: 81, 1986).

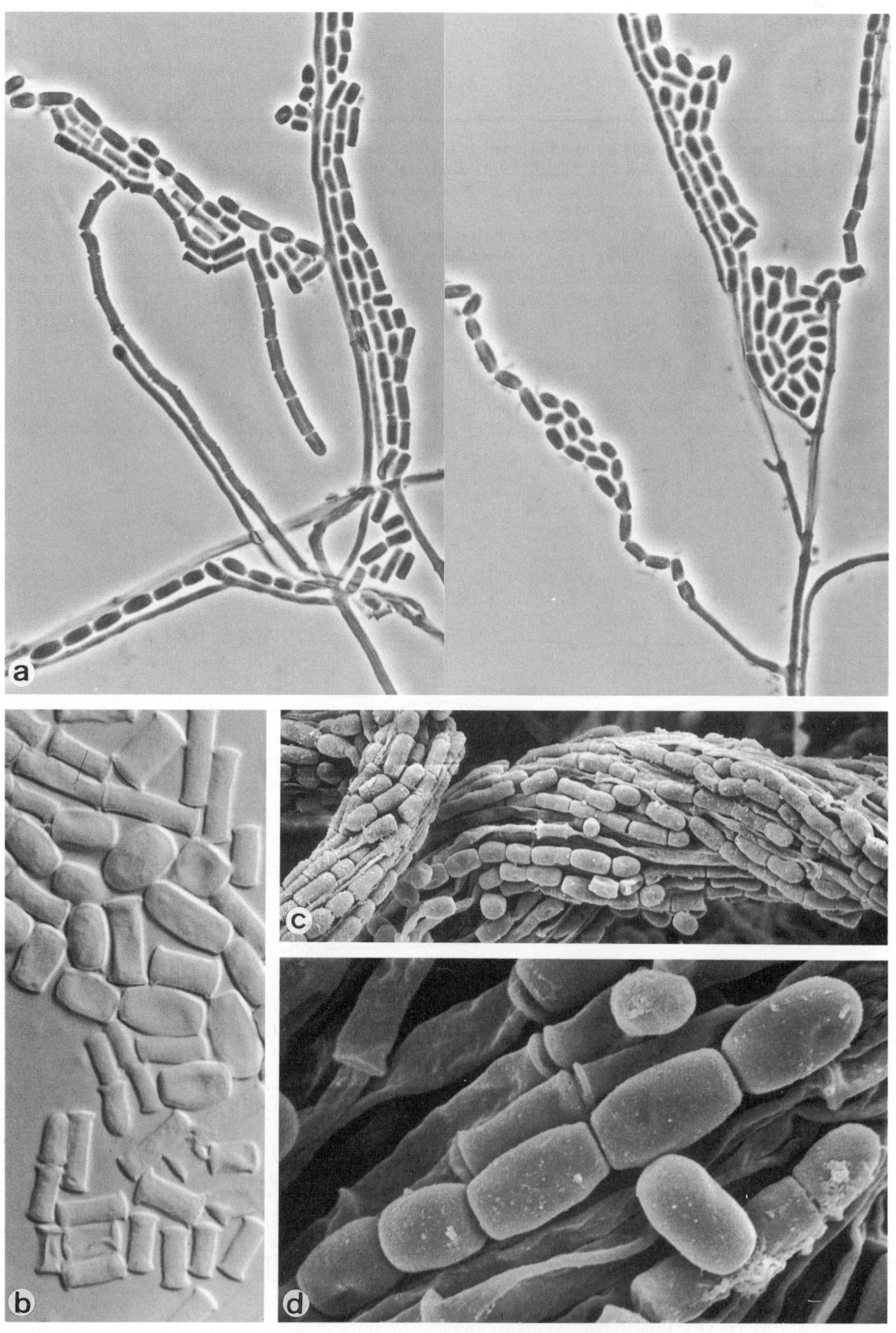

Geotrichum candidum, UAMH 77. Hyphae and arthroconidia. a. ×512; b. ×1600; c. ×1040; d. ×4300.

Geotrichum capitatum (Diddens & Lodder) v. Arx

Colony characteristics. Colonies (MEYA) with moderate growth, whitish, butyrous.
Microscopy. Conidiophores creeping or ascending, profusely branched at acute angles, 180-500 μm long, with terminal and intercalary conidiogenous cells which form long, cicatrized rachids on which the conidia are borne. Conidia cylindrical-clavate, with rounded apex and flat base, hyaline, 1-celled, 7-10 × 2.5-3.5 μm. Rectangular arthroconidia often also present. Endoconidia occasionally present.
Teleomorph. *Dipodascus capitatus* de Hoog *et al.* (*Ascomycota, Hemiascomycetes, Saccharomycetales: Dipodascaceae*).
Asci infrequently formed after mating, bipodal, broadly ellipsoidal, 7.5-11.0 × 7-9 μm, rupturing at the apex and liberating 4 ascospores which are hyaline, ellipsoidal, 5.5-6.5 × 2.5-3.2 μm, with persistent slimy sheath. Heterothallic.
Physiology.

Fermentation:	–	Inulin	–	D-Glucitol	–
		Soluble starch	–	α-Methyl-D-glucoside	–
Growth:		D-Xylose	–	Salicin	–
Glucose	+	L-Arabinose	–	D-Gluconate	–
Galactose	+	D-Arabinose	–	DL-Lactate	+
L-Sorbose	v	D-Ribose	v	Succinate	+
Sucrose	–	L-Rhamnose	–	Citrate	v
Maltose	–	D-Glucosamine (C)	–	*myo*-Inositol	–
Cellobiose	–	Ethanol	+	Vitamin-free	–
α,α-Trehalose	–	Glycerol	+	2-Keto-D-gluconate	–
Lactose	–	*meso*-Erythritol	–	Arbutin	–
Melibiose	–	Ribitol	–	Nitrate	–
Raffinose	–	Galactitol	–	0.1% Cycloheximide	+
Melezitose	–	D-Mannitol	–	Growth at 40°C	+

Differential diagnosis. The species can be recognized by its arthroconidia and by absence of growth with D-xylose, sucrose, cellobiose, but growth at 40°C and with 0.1% cycloheximide. The closely related species *Geotrichum clavatum* differs by being cellobiose +, arbutin + and salicin +.
Molecular diagnostics. SSU and LSU restriction maps based on NCBI AB000650 and NCBI U40084:

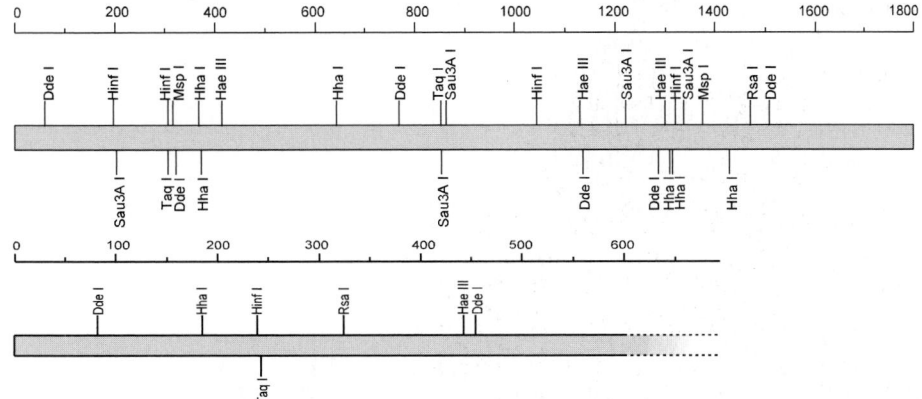

Pathogenicity. BSL-2. *Geotrichum capitatum* occurs quite frequently in human sputum (Gemeinhardt, 1965). In nearly all cases it was a secondary invader. In immunosuppressed, debilitated, neutropenic (Plum *et al.*, 1996) or particularly leukemic hosts (Oelz *et al.*, 1983; Baird *et al.*, 1985; Barbor *et al.*, 1995; d'Antonio *et al.*, 1996; Listemann *et al.*, 1996; Pagano *et al.*, 1996b) dissemination may occur (Winston *et al.*, 1977). Positive blood cultures are obtained in later stages of disease (Schiemann *et al.*, 1998). Amft *et al.* (1996) reported a disseminated case in a patient with non-Hodgkin's lymphoma and Ortiz *et al.* (1998) spondylodiscitis in a BMT patient. Also endocarditis (Polachek *et al.*, 1992), encephalitis (Deicke & Gemeinhardt, 1980), osteomyelitis (d'Antonio *et al.*, 1994) and onychomycosis (d'Antonio *et al.*, 1999) have been noted. A case of an abortion in a cow was reported by Hellman & Raethel (1964).
References. De Hoog *et al.* (1986), Guého *et al.* (1987a), Smith & Poot (1998).

231

Antifungal susceptibility.

Antifungal	MICs Range	MIC50	Strains	Reference
AMB	0.15-0.62	nd	15	Martino *et al.* (1990)
AMB	1-2	1	5	Espinel-Ingroff (1998)
5FC	0.04->100	nd	15	Martino *et al.* (1990)
FLZ	16-64	5	5	Espinel-Ingroff (1998)
ITZ	0.25-0.5	5	5	Espinel-Ingroff (1998)
KTZ	0.04-50	15	15	Martino *et al.* (1998)
VCZ	0.06-0.25	5	5	Espinel-Ingroff (1998)

Nomenclature. *Trichosporon capitatum* Diddens & Lodder - Anaskosp. Hefen 2: 488, 1942 ≡ *Geotrichum capitatum* Diddens & Lodder) von Arx, *in* von Arx, Rodrigues de Miranda, M.Th. Smith & Yarrow - Stud. Mycol. 14: 32, 1977 ≡ *Ascotrichosporon capitatum* (Diddens & Lodder) Kocková-Kratochvílová, Slaviková, Zemek & Kuniak - Proc. 5th Int. Spec. Symp. Yeasts, Bratislava p. 9, 1977 ≡ *Blastoschizomyces capitatus* (Diddens & Lodder) Salkin, Gordon, Samsonoff & Rieder - Mycotaxon 22: 378, 1985.

Sporotrichum spicatum Delitsch - Syst. Schimmelp. p. 106, 1943.

Geotrichum linkii Vöros-Felkai - Acta Microbiol. Hung. 8: 95, 1961.

Blastoschizomyces pseudotrichosporon Salkin, Gordon, Samsonoff & Rieder - Mycotaxon 14: 503, 1982.

Dipodascus capitatus de Hoog, M.Th. Smith & Guého - Stud. Mycol. 29: 51, 1986.

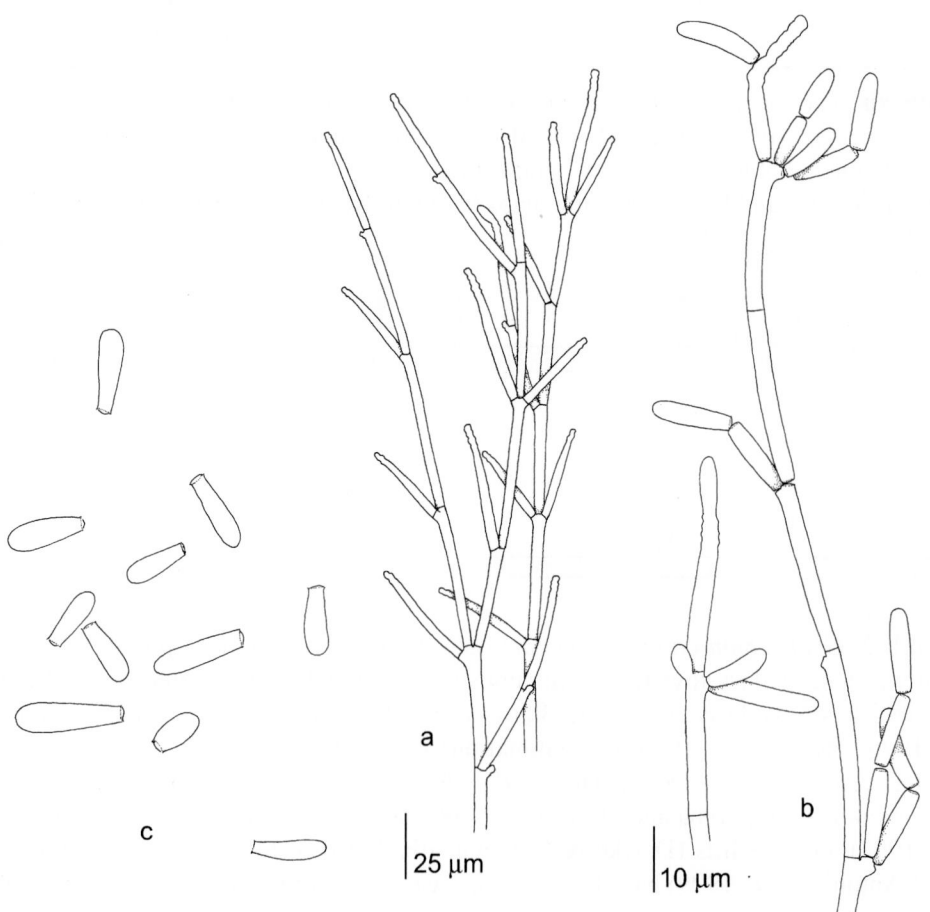

Geotrichum capitatum, CBS 580.82. a. Branching pattern; b. sympodial and arthric conidiogenesis; c. sympodial conidia.

Geotrichum clavatum de Hoog *et al.*

Colony characteristics. Colonies (MEYA) with moderate growth, whitish, butyrous.
Microscopy. True hyphae abundant, soon falling apart into rectangular arthroconidia. Sympodial conidiogenesis occasionally present. Terminal parts of hyphae may swell and become thick-walled.
Physiology.

Fermentation:	–	Inulin	–	D-Glucitol	–
		Soluble starch	–	α-Methyl-D-glucoside	–
Growth:		D-Xylose	–	Salicin	+
Glucose	+	L-Arabinose	–	D-Gluconate	–
Galactose	+	D-Arabinose	–	DL-Lactate	v
L-Sorbose	+	L-Rhamnose	–	Succinate	+
Sucrose	–	D-Glucosamine (C)	–	Citrate	+
Maltose	–	*N*-Acetyl-D-glucosamine	–	*myo*-Inositol	–
Cellobiose	+	Ethanol	+	Vitamin-free	–
α,α-Trehalose	–	Glycerol	+	2-Keto-D-gluconate	–
Lactose	–	*meso*-Erythritol	–	Arbutin	+
Melibiose	–	Ribitol	–	Nitrate	–
Raffinose	–	Galactitol	–	Growth at 37°C	+
Melezitose	–	D-Mannitol	–		

Differential diagnosis. The species lacks sympodial conidiogenesis but is D-xylose –, cellobiose +, salicin + and arbutin +. *Dipodascus spicifera* de Hoog *et al.*, having the same physiological pattern but showing abundant sympodial conidiogenesis, was once reported from a disseminated infection (Mahul *et al.*, 1989).
Molecular diagnostics. LSU restriction map based on NCBI U40112:

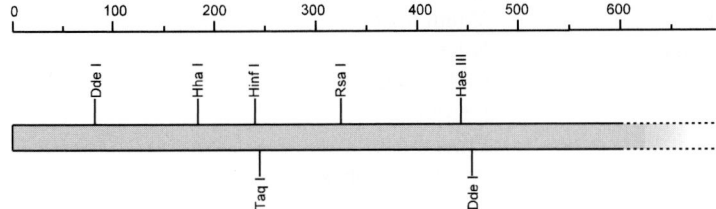

Pathogenicity. BSL-2. The species is occasionally involved in respiratory disorders (de Hoog *et al.*, 1986).
References. De Hoog *et al.* (1986), Smith & Poot (1998).
Nomenclature. *Geotrichum clavatum* de Hoog, M.Th. Smith & Guého - Stud. Mycol. 29: 57, 1986.

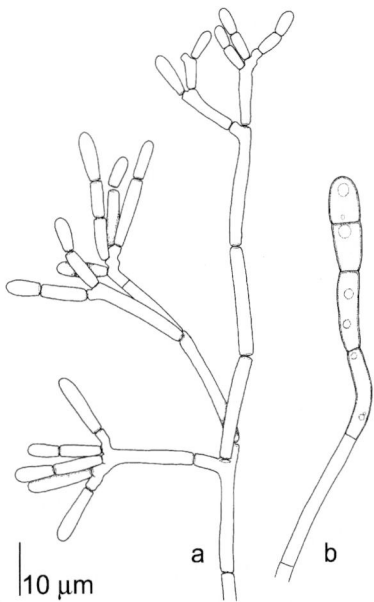

Geotrichum clavatum, CBS 576.82. a. Hyphal system and arthroconidia; b. chlamydospore-like cells.

Ascomycota, Hemiascomycetes, Saccharomycetales, Saccharomycetaceae.

Genus: *SACCHAROMYCES*

Saccharomyces cerevisiae Meyen ex Hansen

Colony characteristics. Colonies (MEYA) restricted, cream-coloured, moist.

Microscopy. Budding cells broadly ellipsoidal, with multilateral bud formation only. Some pseudomycelium may be present. Ascospores formed with 1-4 inside budding cells, spherical to broadly ellipsoidal, smooth-walled.

General remark. Some closely related taxa are distinguished within *Saccharomyces*, which cannot be recognized with confidence using classical data. ITS-RFLP was applied by McCullough *et al.* (1998a), and a PCR for RPL2 domains by Ryu *et al.* (1998).

Physiology.

Fermentation:						
Glucose	+	Lactose	–	Galactitol	–
Galactose	v	Melibiose	v	D-Mannitol	–
Sucrose	+	Raffinose	+	D-Glucitol	–
Maltose	v	Melezitose	v	α-Methyl-D-glucoside	v
Lactose	–	Inulin	–	Salicin	–
Raffinose	+	Soluble starch	–	D-Gluconate	–
α,α-Trehalose	–	D-Xylose	–	DL-Lactate	v
Melibiose	v	L-Arabinose	–	Succinate	v
		D-Arabinose	–	Citrate	–
		D-Ribose	–	*myo*-Inositol	–
Growth:		L-Rhamnose	–	Hexadecane	–
Glucose	+	D-Glucosamine (C)	–	Vitamin-free	–
Galactose	v	*N*-Acetyl-D-glucosamine	–	Nitrate	–
L-Sorbose	–	Methanol	–	Cadaverine	–
Sucrose	+	Ethanol	+	L-Lysine	–
Maltose	+	Glycerol	–	Ethylamine	–
Cellobiose	–	*meso*-Erythritol	–	0.01% Cycloheximide	–
α,α-Trehalose	+	Ribitol	–	Growth at 37°C	v

Differential diagnosis. Species signature: vigorous fermentation of glucose, and assimilation: D-xylose -, salicin -, D-gluconate -, L-lysine -, ethylamine -, 0.01% cycloheximide -, xylitol -, 2-keto-D-gluconate -, D-tryptophane -, w/o niacin +, growth at 30°C +, 60% glucose -, urease -. Physiologically indistinguishable from *Arxiozyma telluris* (p. 179) and many saprobic yeasts.

Molecular diagnostics. SSU, ITS and LSU restriction maps based on NCBI Z75578, D89886 and U44806:

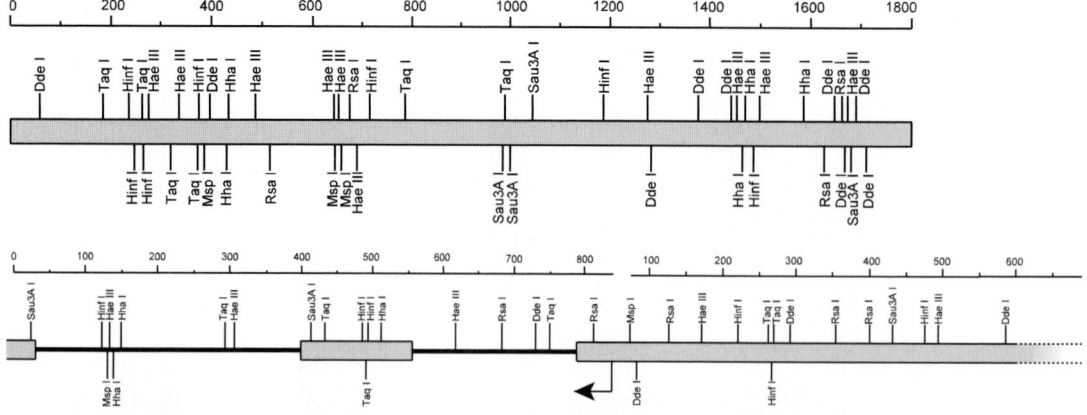

Pathogenicity. GRAS. However, this status was questioned by Murphy & Kavanagh (1999). Baker's yeast is widely applied in the production of wines, beers and other food stuffs. It can be transmitted from food materials and then be involved in superficial mycosis (Nyirjesy *et al.,* 1995). The species is commonly found in stool samples of BMT-patients (Zerva *et al.,* 1998). Infections in AIDS patients have rarely been reported (Sethi & Mandell, 1988; Tawfik *et al.*, 1989). Deep fungemia in a transplant patient was described by Cairoli *et al.* (1995) and sepsis in a leukemic patient by Oriol *et al*. (1993). Pletincx *et al*. (1995) reported a fungemia in a one-year-old child and Senneville *et al*. (1996) an uretral obstruction. *Saccharomyces boullardii*, applied for treatment of intestinal disorders (Lewis & Freedman, 1998), is a synonym of *S. cerevisiae* (McCullough *et al*., 1998b). Overload may lead to fungemia (Bassetti *et al*., 1998; Fredenucci *et al*., 1998).

References. Barnett *et al*. (1990), Vaughan-Martini & Martini (1998).

Antifungal susceptibility.

Antifungal	MICs Range	MIC 50	Mic 90	Strains	Reference
AMB	0.12-2	1	1	74	Zerva *et al*. (1998)
AMB	0.5-2	1	nd	5	Espinel-Ingroff (1998)
AMB	0.5-1	1	1	22	Pfaller *et al*. (1997)
AMB	0.13-1	1	1	24	Zhanei *et al*. (1998)
5FC	0.25-32	0.25	0.25	74	Zerva *et al*. (1998)
5FC	0.06-0.12	0.06	0.12	22	Pfaller *et al*. (1997)
5FC	≤0.03-0.5	0.06	0.25	24	Zhanei *et al*. (1998)
FLZ	0.12-16	2	8	74	Zerva *et al*. (1998)
FLZ	1-4	2	nd	5	Espinel-Ingroff (1998)
FLZ	0.5-16	2	16	22	Pfaller *et al*. (1997)
FLZ	≤0.03-4	0.5	2	24	Zhanei *et al*. (1998)
ITZ	0.5-1	0.5	nd	5	Espinel-Ingroff (1998)
ITZ	0.03-0.5	0.5	0.5	22	Pfaller *et al*. (1997)
ITZ	≤0.03-4	0.25	2	24	Zhanei *et al*. (1998)
KTZ	0.02-1	0.5	1	74	Zerva *et al*. (1998)
KTZ	≤0.03-2	0.25	1	24	Zhanei *et al*. (1998)
VCZ	0.06-0.25	0.12	nd	5	Espinel-Ingroff (1998)

Nomenclature. *Saccharomyces cerevisiae* Meyen - Weigman's Arch. Naturges. 4: 100, 1838 ≡ *Saccharomyces cerevisiae* Meyen ex Hansen - Medd. Carlsberg Lab. 2: 67, 1883.

Saccharomyces pulmonalis Redaelli - Micet. Assoc. Microb. Tuberc. Polmon. Cavit. p. 41, 1925.

Saccharomyces annulatus Negroni - Annls Parasit. Hum. Comp. 7: 303, 1929.

Saccharomyces boulardii Seguela, Bastide & Massot - 6th Int. Symp. Yeasts, Montpellier p. XI-II-P, 1984.

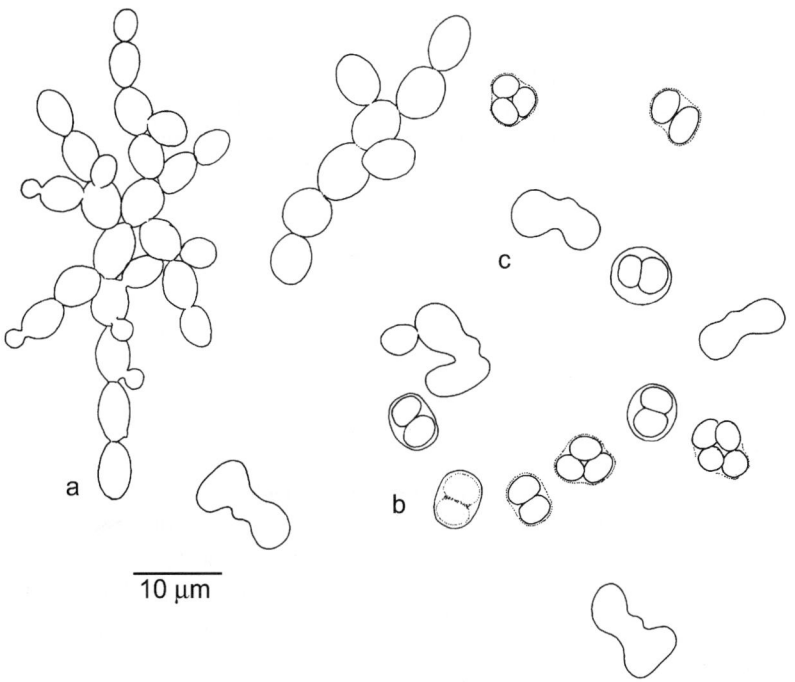

Saccharomyces cerevisiae, CBS 1171. a. Diploid vegetative multilateral budding; b. asci containing haploid ascospores; c. fusion of ascospores and subsequent budding of fusion product.

Ascomycota, Hemiascomycetes, Saccharomycetales, Ascoideaceae. Genus: *YARROWIA*

Yarrowia lipolytica (Wickerham *et al.*) van der Walt & v. Arx

Colony characteristics. Colonies (YPGA) butyrous, tannish-white.

Microscopy. Hyphae (RA) straight and stiff, 2-4 µm wide. Asci sessile with 1-3 on hyphae near the septa, ellipsoidal, 6-15 × 5-8 µm, containing 2(-4) ascospores. Ascospores subspherical to ellipsoidal, with irregular ornamentation, 3.5-6 × 3-4 µm. Homothallic.

Anamorph. *Candida lipolytica* (Harrison) Diddens & Lodder.

Ellipsoidal to cylindrical conidia produced on hyphae, 4-18 × 3-6 µm. Budding cells present or absent.

Physiology.

Fermentation:	–	D-Xylose	–	Salicin	w,–
		L-Arabinose	–	D-Gluconate	v
Growth:		D-Arabinose	–	DL-Lactate	+
Glucose	+	D-Ribose	v	Succinate	+
Galactose	v	L-Rhamnose	–	Citrate	+
L-Sorbose	v	D-Glucosamine (C)	–	*myo*-Inositol	–
Sucrose	–	*N*-Acetyl-D-glucosamine	+	Hexadecane	+
Maltose	–	Methanol	–	Vitamin-free	–
Cellobiose	w,–	Ethanol	+	2-Keto-D-gluconate	–
α,α-Trehalose	–	Glycerol	+	5-Keto-D-gluconate	–
Lactose	–	*meso*-Erythritol	+	D-Glucarate	–
Melibiose	–	Ribitol	v	Nitrate	–
Raffinose	–	Galactitol	–	10% NaC1/5% glucose	–
Melezitose	–	D-Mannitol	+	Starch formation	–
Inulin	–	D-Glucitol	+	Gelatin liquefaction	+
Soluble starch	–	α-Methyl-D-glucoside	–	Growth at 37°C	v

Differential diagnosis. Species signature: no fermentation, and assimilation: sucrose –, trehalose –, L-rhamnose –, methanol –, glycerol +, *meso*-erythritol +, α-methyl-D-glucoside –, growth at 30°C +, 0.1% cycloheximide +, 60% glucose –.

Molecular diagnostics. SSU and LSU restriction map based on NCBI M60312 and U40080:

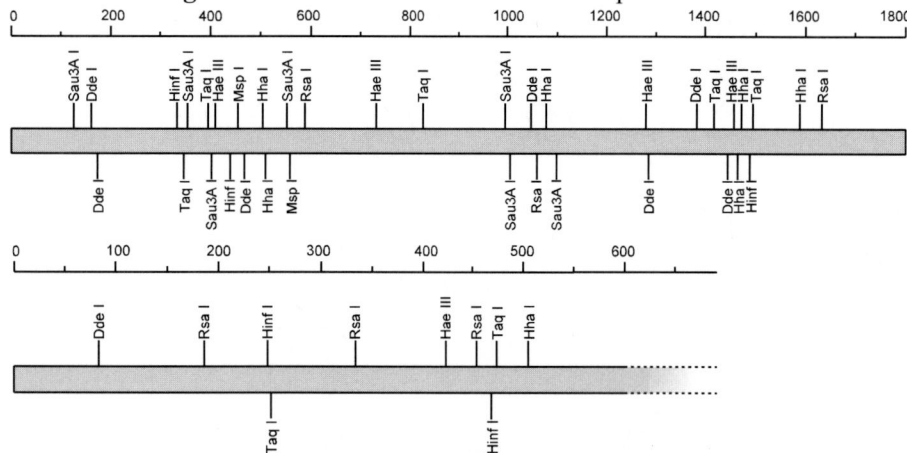

Pathogenicity. BSL-1. This is an industrial fungus producing lipases. A catheter-related infection of a patient on parenteral nutrition was reported by Ninin *et al.* (1997). Glowacka *et al.* (2000) mentioned frequent occurrence of alimentary tract disorder by this species. Disseminated infections were reported by Wehrspann & Füllbrandt (1984) and García-Martos *et al.* (1993). On the basis of animal studies, Walsh *et al.* (1989) concluded that it is only a low virulence pathogen.

Reference. Barth & Gaillardin (1997).

Antifungal susceptibility.

Antifungal	MICs range	MIC 50	Strains	Reference
AMB	1	1	5	Espinel-Ingroff (1998)
FLZ	4-32	8	5	Espinel-Ingroff (1998)
KTZ	0.12-1	0.25	5	Espinel-Ingroff (1998)
VCZ	<0.03-0.25	0.06	5	Espinel-Ingroff (1998)

Nomenclature. *Mycotorula lipolytica* Harrison - Trans. R. Soc. Can., Ser. 3, Sect. 5, 22: 187, 1928 ≡ *Candida lipolytica* (Harrison) Diddens & Lodder - Anaskosp. Hefen 2: 324, 1942.

Monilia cornealis Nannizzi, *in* Bencini & Federici - Atti R. Accad. Fisiocrit. Siena, Ser. 10, 3: 748-766, 1929 ≡ *Proteomyces cornealis* (Nannizzi) C.W. Dodge - Med. Mycol. p. 215, 1935.

Pseudomonilia deformans Zach, *in* Wolfram & Zach - Arch. Derm. Syph. 170: 686, 1934 ≡ *Candida deformans* (Zach) Langeron & Guerra - Annls Parasit. Hum. Comp. 16: 503, 1938.

Endomycopsis lipolytica Wickerham, Kurtzman & Herman - Spectr. Monogr. Ser. Arts Sci., Georgia State Univ. 1: 90, 1970 ≡ *Yarrowia lipolytica* (Wickerham, Kurtzman & Herman) van der Walt & von Arx - Antonie van Leeuwenhoek 46: 519, 1980.

10 μm

Yarrowia lipolytica, CBS 6124. Hyphae with asci and conidia.

FILAMENTOUS BASIDIOMYCETES: EXPLANATORY CHAPTER, AND KEY TO THE GENERA

Criteria for distinction of filamentous basidiomycetes

Only a few members of the large division *Basidiomycota* have clinical significance. The most important taxa produce budding cells and are treated under the yeasts (p. 125). Strictly filamentous basidiomycetes mostly are wood-rotters or occur in wood, having mushroom-like fruit bodies or being obligate plant pathogens.

In culture basidiomycetes generally produce rapidly growing, white colonies with mono- or dikaryotic hyphae. Clamp connections are very rare. Mating is often necessary to obtain fruit bodies. They are generally urease +, mostly tolerant to cycloheximide, tolerant to benomyl, and show a positive Diazonium Blue B reaction. With ultrastructure, the cell walls are multilamellar and the compartments are connected by variously-shaped dolipores. Molecular diagnostics of the most common clinically significant filamentous basidiomycetes was pre-

sented by de Hoog & Gerrits van den Ende (1998a).

Cultures may produce conidia, and some of them have therefore been attributed hyphomycete names. Among these is *Cerinosterus cyanescens* (p. 541; teleomorph unknown, member of order *Microstomatales*), *Tilletiopsis minor* (p. 940, teleomorph unknown, member of order *Tilletiales*), *Hormographiella aspergillata* (p. 712; teleomorph *Coprinus,* order *Agaricales*) and *Sporotrichum pruinosum* (p. 928; teleomorph *Phanerochaete*, order *Stereales*). Of the non-sporulating taxa, *Rhizoctonia* was reported to be involved in a keratitis (Srivastava *et al.*, 1977). Clinical strains have also been identified as *Bjerkandera* sp. (p. 240; order *Stereales*), but as yet no convincing case report has been published. The most commonly reported filamentous basidiomycetes is *Schizophyllum commune* (p. 242; order *Stereales*).

References. Lacaz *et al.* (1996), Sigler & Abbott (1997), de Hoog & Gerrits van den Ende (1998a).

Simplified key to the treated genera of filamentous basidiomycetes:

1a.	Cultures rapidly expanding, white or cream coloured, dry **→ 2**
1b.	Not combining above characteristics **→ 5**
2a.	Cultures farinose; conidia terminal with broad bases on densely branched systems . . . ***Sporotrichum* (928)**
2b.	Cultures cottony or hairy; conidia arthric or absent **→ 3**
3a.	Arthroconidia absent; hyphal pegs frequently present; no growth on Mycosel agar . . . . ***Schizophyllum* (242)**
3b.	Arthroconidia present; hyphal pegs absent; growth on Mycosel agar **→ 4**
4a.	Arthroconidia produced from lateral cells which are often conidiophore-like . . . . . ***Hormographiella* (712)**
4b.	Arthroconidia produced from creeping, undifferentiated hyphae . . . . ***Bjerkandera* (240)**
5a.	Cultures often with olivaceous tinges, producing a pleasant, fruit-like odour . . . . ***Moniliella* (767)**
5b.	Cultures otherwise **→ 6**
6a.	Cultures pinkish to brown; eventually a dull brown pigment is exuded into the agar . . . . ***Tilletiopsis* (940)**
6b.	Cultures whitish or cream coloured, eventually exuding a blue or purple pigment into the agar **→ 7**
7a.	Cultures thin, soft; hyphae narrow, easily fragmenting into separate cells; conidia arthric or sympodial . . . . ***Cerinosterus* (541)**
7b.	Not combining above characteristics **→ 8**
8a.	Cultures with slow growth, hard or slimy, often cracking the agar, cerebriform or folded; conidia predominantly arthric . . . . ***Trichosporon* (164)**
8b.	Cultures otherwise; budding cells preponderant . . . . **Basidiomycetous yeasts (130)**

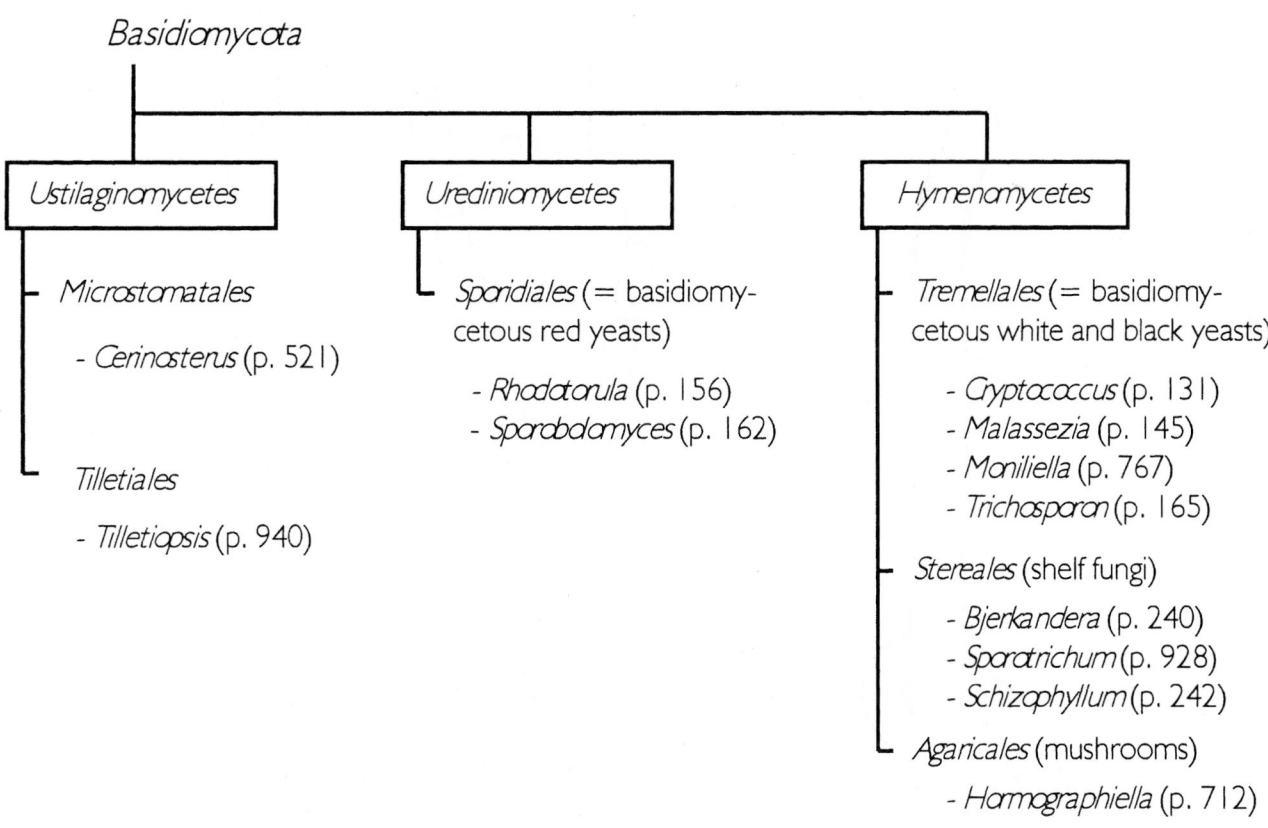

Fig. 19. Diagram of main relationships of treated members of *Basidiomycota*.

Basidiomycota, Hymenomycetes, Stereales, Coriolaceae. Genus: *BJERKANDERA*

Bjerkandera adusta (Willd.: Fr.) Karst.

Colony characteristics. Colonies (MEA 2%) spreading, cottony to woolly, white, finally cream-coloured. Reverse bleached yellow. Producing a strong smell.

Microscopy. Hyphae hyaline, 2-6 μm wide, poorly branched with narrower hyphae. Clamp connections present or absent. Arthroconidia rectangular, up to 12 μm long. Chlamydospores ellipsoidal, up to 10 μm in length. Fruit bodies absent.

Pathogenicity. BSL-1. Shelf fungus occurring on deciduous wood. Sigler & Abbott (1997) mentioned its common occurrence in clinical samples, but, as yet, no case report has been published.

References. Stalpers (1978), Phillips (1981), Breitenbach & Kränzlin (1986), Ryvarden & Gilbertson (1993).

Nomenclature. *Boletus adustus* Willdenow - Fl. Berl. Prod. p. 392, 1787 ≡ *Polyporus adustus* Willdenow:Fries - Syst. Mycol. 1: 363, 1821 ≡ *Bjerkandera adusta* (Willdenow:Fries) Karsten - Medd. Soc. Fauna Fl. Fenn. 5: 38, 1879.

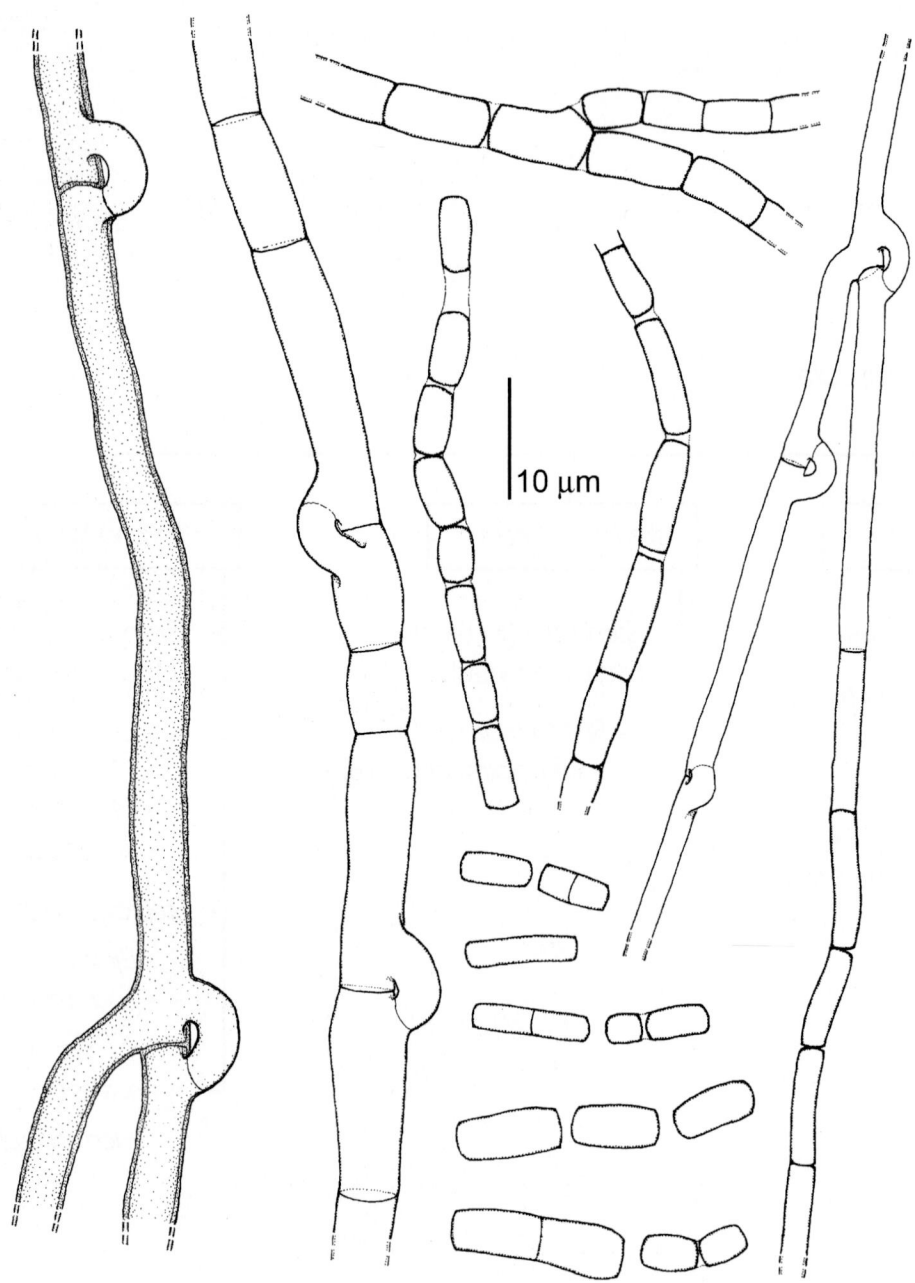

10 μm

Bjerkandera adusta, CBS 230.93. Arthroconidia and hyphae with clamp connections.

241

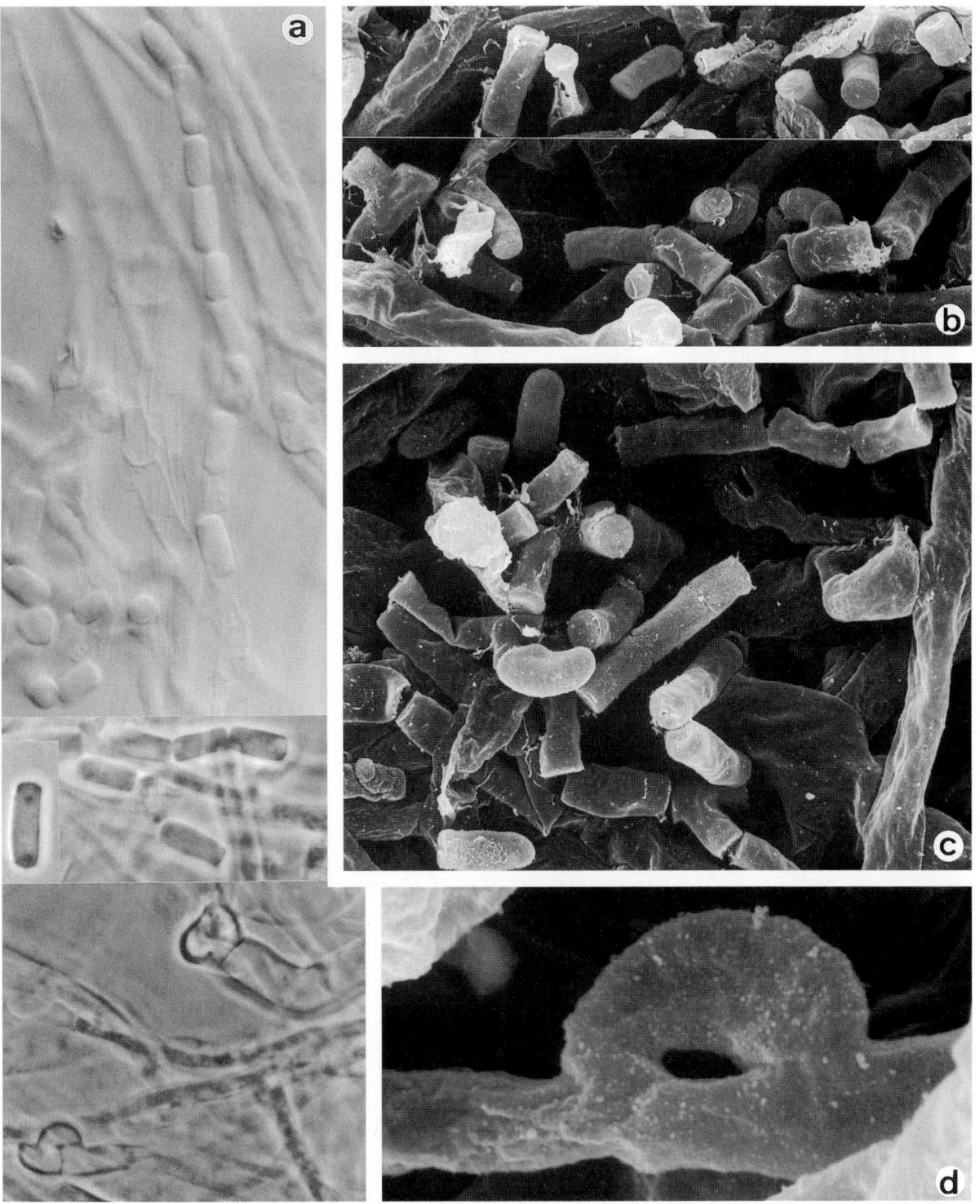

Bjerkandera adusta, CBS 230.93. Hyphae with arthroconidium formation and clamp connections. a. ×1600; b. ×2900; c. ×3500; d. ×16000.

Basidiomycota, Hymenomycetes, Stereales, Schizophyllaceae. Genus: *SCHIZOPHYLLUM*

Schizophyllum commune Fr.

Colony characteristics. Colonies (MEA 2%) spreading, woolly, pale greyish-brown, soon forming macroscopically visible fruit bodies in concentric zones. Fruit bodies sessile, kidney-shaped or medusoid, lobed with split gills on the lower side. Fungus produces an unpleasant odour.

Microscopy. Hyphae hyaline, wide, of rather diverse structure, often with clamp connections or curved lateral pegs (spicules). Conidia absent. Basidia arranged in dense palissade on the lower sides of fruit bodies, each bearing 4 basidiospores on erect sterigmata. Basidiospores hyaline, smooth- and thin-walled, elongate with lateral scar at the lower end, 6-7 × 2-3 μm.

Physiology. Tolerant to benomyl; no growth on Mycosel agar.

Differential diagnosis. Normal, dikaryotic isolates can be recognized by clamp connections, spicules and the formation of fruit bodies. Monokaryotic cultures may lack these features. Such strains show rapid, fluffy growth, produce an unpleasant smell, and the hyphae are without sporulation.

Molecular diagnostics. SSU and ITS restriction maps based on NCBI X54865 and UAMH 7695 + NCBI AF062633:

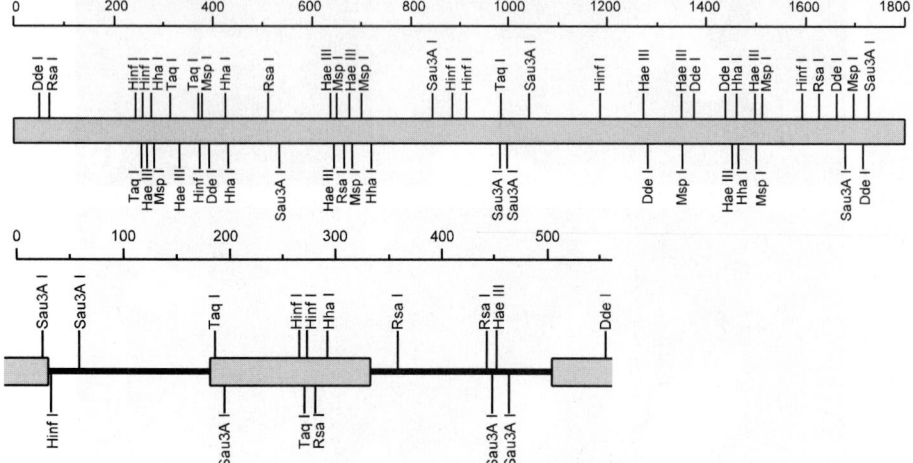

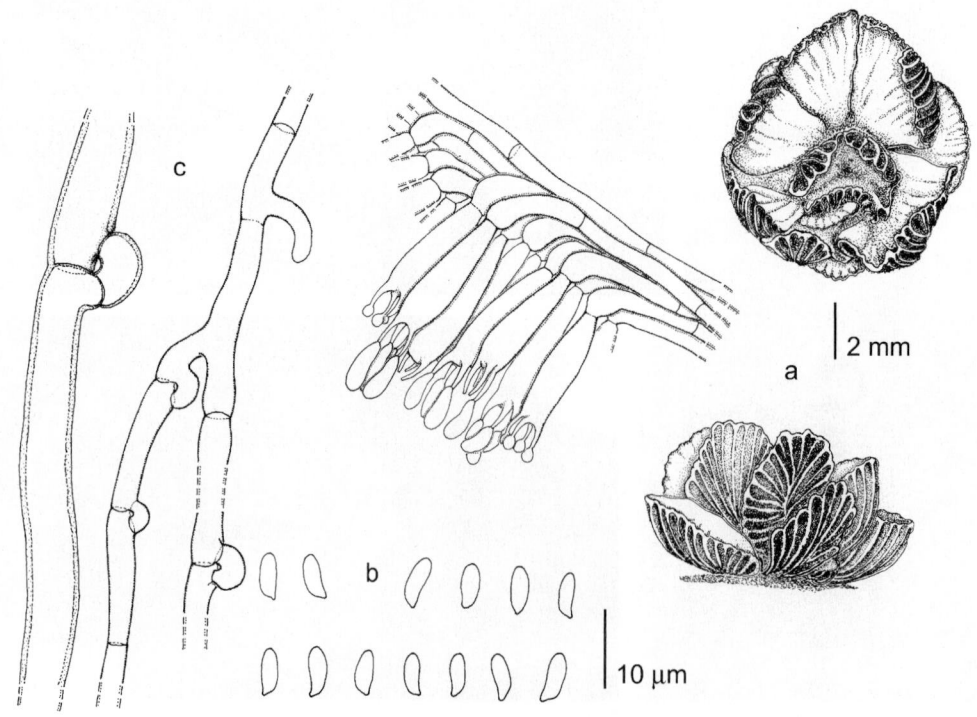

Schizophyllum commune, CBS 103.20. a. Basidiocarp; b. basidia and basidiospores; c. hyphae with clamp connections.

Pathogenicity. BSL-1. This shelf fungus is a common invader of rotten wood. Cases of sinusitis have been described (Ciferri *et al.*, 1956; Kern & Uecker, 1986; Catalano *et al.*, 1990; Rosenthal *et al.*, 1992; Sigler *et al.*, 1997a, 1999a). It was also reported to be involved in cases of allergic bronchopulmonary mycosis (Kamei *et al.*, 1994; Amitani *et al.*, 1996), pulmonary fungus ball (Sigler *et al.*, 1995), meningitis (Batista *et al.*, 1955), brain abscess (Rihs *et al.*, 1996) and onychomycosis (Kligman, 1950). Ulceration of the palate was reported by Restrepo *et al.* (1973).

References. Raper & Krongelb (1958), Watling & Sweeney (1971), Sigler *et al.* (1995).

Nomenclature. *Schizophyllum commune* Fries - Syst. Mycol. 1: 330, 1821.

Schizophyllum commune, CBS 103.20. a-c. Basidiocarps; d. cross section through hymenium; e, f. basidia and basidiospores; g. hyphae with clamp connections. a. ×64; b. ×43; c. ×30; d. ×1500; e. ×4300; f, g. ×1600.

ASCOMYCETES PRODUCING FRUIT BODIES IN CULTURE: EXPLANATORY CHAPTER, KEY TO GENERA, AND DESCRIPTIONS

Criteria for distinction of ascomycetes

Most clinically relevant fungi belong to the phylum *Ascomycota*. However, fruit bodies (**ascomata**) with asci and ascospores are rarely produced in primary cultures. Usually a conidial anamorph is prevalent, which is treated in this Atlas under the hyphomycetes (p. 361) or the yeasts (p. 125). Below, a small number of ascomycetes are treated which either lack conidia, or produce ascomata in relative abundance under routine laboratory conditions. In addition, some teleomorph genera of clinically relevant *Onygenales* are treated, i.e., of dermatophytes and dimorphic systemic pathogens. These *Onygenales* are mostly heterothallic, teleomorphs being produced only after mating of suitable partners.

An ascomycete is readily recognizable when it possesses fruit bodies containing asci with ascospores. Fruit-bodies are of four different types (Fig. 20):
a. composed of a network of hyphae (peridial hyphae) which centrally contain asci (**gymnothecium**);

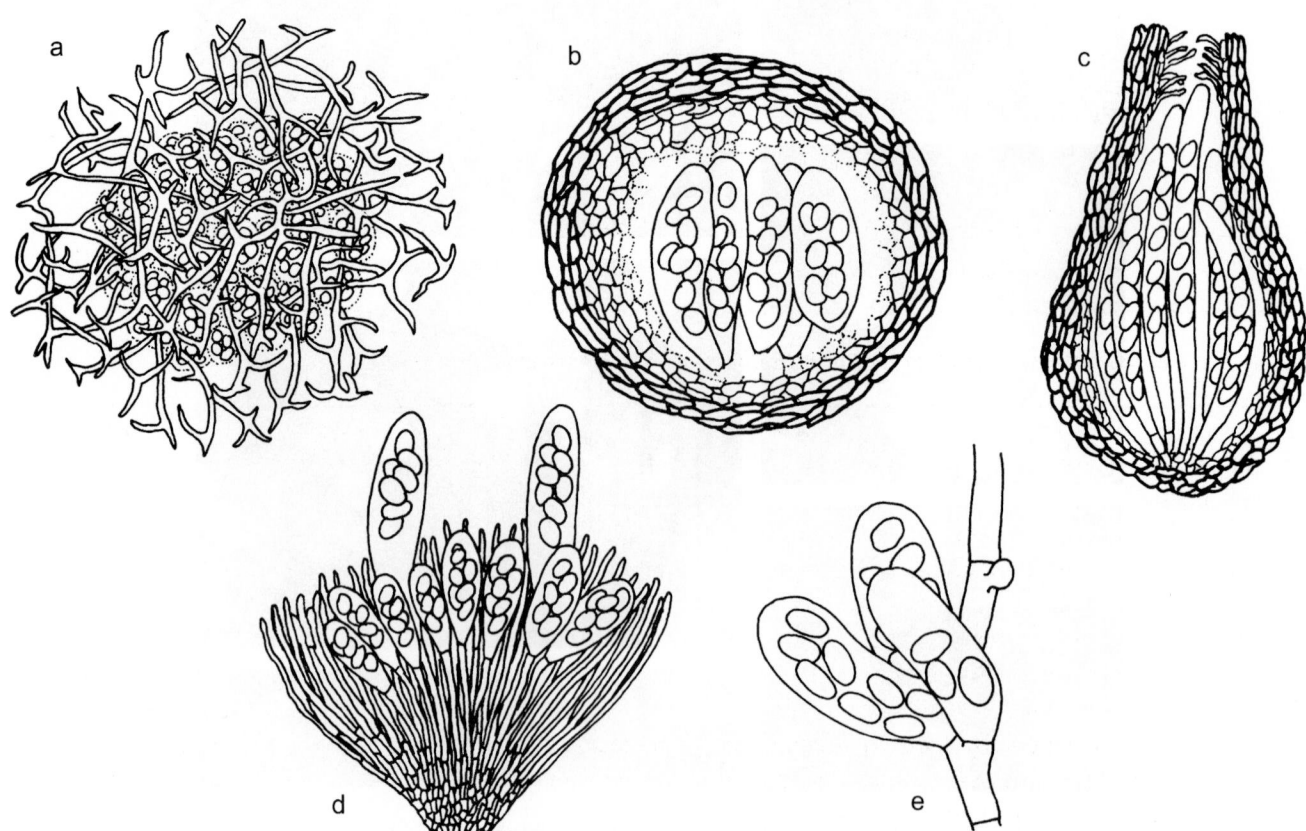

Fig. 20. Diagram of fruit body types in *Ascomycota*. a. Gymnothecium; b. cleistothecium; c. perithecium; d. apothecium; e. naked asci.

b. spherical, closed, at maturity the centrum being dissolved and developing asci which are liberated by rupture of the fruit body (**cleistothecium**);

c. spherical to pyriform, with an apical, preformed opening (ostiolum) and a preformed cavity in which the asci arise in parallel arrangement (**perithecium**);

d. flat, consisting of a saucer-like palissade of cells, some of which develop into asci (**apothecium**);

e. fruit bodies may be absent, the asci developing alongside undifferentiated hyphae (**naked asci**).

The fruit body wall (**peridium**; Fig. 21) can be composed of densely interwoven hyphae (***textura intricata***), of rounded ellipsoidal cells (***textura globulosa***), of angular cells (***textura angularis***) or of irregular, jig-saw-shaped cells (***textura epidermoidea***).

Asci can have a single wall (**unitunicate**) or a double wall (**bitunicate**). The latter is microscopically visible as a thick wall, especially in the apical part of the ascus. Ascospores are liberated by an expanding inner wall which bursts through the terminal end of the outer wall. Unitunicate asci are cylindrical or spherical and can have thin, evanescent walls, i.e., they disappear at maturation of the ascospores. Consequently mature spores are free in the fruit body cavity. Other unitunicate ascomycetes have persistent asci with more or less developed apical structures to enhance spore liberation. In order to prevent confusion with Coelomycetes, which produce conidia instead of ascospores, young fruit bodies should be squashed to see the development of asci.

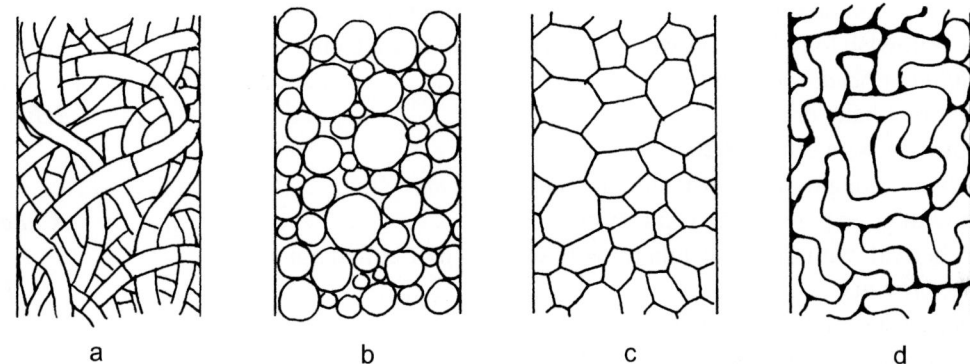

Fig. 21. Diagram of types of fruit body wall structures in *Ascomycota*. a. *Textura intricata;* b. *textura globulosa;* c. *textura angularis;* d. *textura epidermoidea.*

Key to the treated genera of Ascomycota on the basis of their fruit bodies:

1a. Ascomata composed of loosely interwoven hyphae, often bearing coils or spines ➜ **2**
1b. Ascomata with distinct, cellular walls, which may be stromatic ➜ **9**
2a. Ascomata without spines or coiled appendages ➜ **3**
2b. Ascomata with spines or coiled appendages ➜ **6**
3a. Ascospores discoid, lenticular or bivalvate, smooth-walled ➜ **4**
3b. Ascospores (sub)spherical, puntate-reticulate ➜ **5**
4a. Ascospores hyaline, bivalvate . . . . *Narasimhella* (**287**)
4b. Ascospores pigmented, discoid or lenticula . . . . *Gymnoascus* (**271**)
5a. Ascospores surrounded by a sheath; anamorph *Myriodontium* . . . . *Neorachnotheca*[1]
5b. Above characteristics not combined; anamorph *Chrysosporium* . . . . *Nannizziopsis*[2]
6a. Ascomata bearing branched or comb-shaped appendages . . . . *Myxotrichum* (**285**)
6b. Ascomata bearing coiled appendages ➜ **7**
7a. Ascospores yellow-brown to reddish-brown . . . . *Uncinocarpus*[3]

7b. Ascospores hyaline ➡ **8**

8a. Peridial hyphae composed of swollen cells; ascospores (sub)spherical . . . . *Ajellomyces* **(250)**

8b. Peridial hyphae composed of ossiform cells; ascospores ellipsoidal or lenticular . . . . *Arthroderma* **(259)**

9a. Fungus forming stromata around human hairs . . . . *Piedraia* **(303)**

9b. Fungus not stromatic ➡ **10**

10a. Ascospores brown, rhomboidal, two-celled . . . . *Neotestudina* **(297)**

10b. Ascospores otherwise ➡ **11**

11a. Ascospores with at least 3 transversal septa and with hyaline slime sheaths . . . . *Leptosphaeria* **(273)**

11b. Ascospores otherwise ➡ **12**

12a. Ascospores with slimy appendages at both ends . . . . *Arnium* **(257)**

12b. Ascospores without appendages ➡ **13**

13a. Ascospores with longitudinal striations . . . . *Neurospora*[4]

13b. Ascospores without longitudinal striations ➡ **14**

14a. Ascomata with preformed opening at the top ➡ **15**

14b. Ascomata without preformed opening ➡ **21**

15a. Ascospores hyaline, yellowish, pale brown or straw-coloured, never dark brown ➡ **16**

15b. Ascospores dark brown ➡ **19**

16a. Ascomata with a very long neck and with a ring of hyphae around ostiole;
ascospores hyaline; anamorph *Sporothrix*-like . . . . *Ophiostoma* **(299)**

16b. Above characteristics not combined ➡ **17**

17a. Ascospores usually asymmetrial, reniform, heart-shaped or triangular,
smooth-walled and with germ pores ➡ **18**

17b. Ascospores usually symmetrical, with thick and ornamented walls . . . . *Neocosmospora* **(289)**

18a. Ascospores straw coloured with a single germ pore; anamorph *Scopulariopsis* . . . . *Microascus* **(276)**

18b. Ascospores yellowish to reddish-brown, with a germ pore at each end;
anamorph *Graphium* or *Scedosporium* . . . . *Petriella* **(301)**

19a. Ascomata without hairs or setae . . . . *Achaetomium* **(248)**

19b. Ascomata hairy or with setae ➡ **20**

20a. Ascospores with a longitudinal germ slit; anamorph *Dicyma* . . . . *Ascotricha* **(260)**

20b. Ascospores with germ pores; anamorph absent . . . . *Chaetomium* **(262)**

21a. Ascomata spherical, with long appendages with contorted or coiled apices;
ascospores oblate, brownish . . . . *Arachnomyces*[5]

21b. Above characteristics not combined ➡ **22**

22a. Ascospores with germ pores ➡ **23**

22b. Ascospores without germ pores ➡ **25**

23a. Ascospores with an apical germ pore . . . . *Thielavia*[6]

23b. Ascospores with a germ pore at each end ➡ **24**

24a. Ascospores dark-brown; anamorph *Myceliophthora* . . . . *Corynascus*[7]

24b. Ascospores pale yellow to golden brown anamorph *Scedosporium* and
synnematous *(Graphium)* . . . . *Pseudallescheria* **(305)**

25a. Ascospores smooth, without equatorial furrows ➡ **26**

25b. Ascospores ornamented, or if smooth with equatorial furrows ➡ **27**

26a. Ascospores brown; anamorph *Sporothrix*-like . . . . *Pseudeurotium* **(310)**

26b. Ascospores hyaline; ascomata borne on a short erect hyphal stalk; anamorph *Basipetospora*   *Monascus* **(283)**

27a. Anamorph *Paecilomyces* or *Chrysosporium* ➡ **28**

27b. Anamorph *Aspergillus* ➡ **29**

28a. Ascospores finely verrucose; anamorph *Paecilomyces* . . . . *Thermoascus*[8]

28b. Ascospores with reticulate ornamentation; anamorph *Chrysosporium* . . . . *Aphanoascus* **(254)**

29a. Anamorph *Aspergillus* with biseriate condiogenous cells; ascomata surrounded by dense
layers of Hülle cells ➡ **30**

29b. Anamorph *Aspergillus* with uniseriate conidiogenous cells; ascomata
not surrounded by Hülle cells ➡ **31**

30a. Ascomata greenish or yellow; ascospores pale . . . . *Fennellia*[9]

30b. Ascomata purple; ascospores red or blue-violet . . . . *Emericella*[10]

31a. Colonies spreading; conidial heads strongly columnar; ascospores with well
    developed equatorial crests . . . . **Neosartorya (291)**
31b. Colonies restricted, conidial heads radiate to loosely columnar; ascospores
    usually with a pronounced furrow and/or inconspicuous crests . . . . **Eurotium**[11]

[1-11]Treated under the respective anamorphs: [1]*Myriodontium* (775), [2]*Chrysosporium* (556), [3]*Chrysosporium* (552), [4]*Chrysonilia* (543), [5]*Onychocola* (790), [6]*Acremonium* (396), [7]*Myceliophthora* (769), [8]*Paecilomyces* (795), [9]*Aspergillus* (468, 492), [10]*Aspergillus* (486, 512, 514), [11]*Aspergillus* (458, 476, 480, 499).

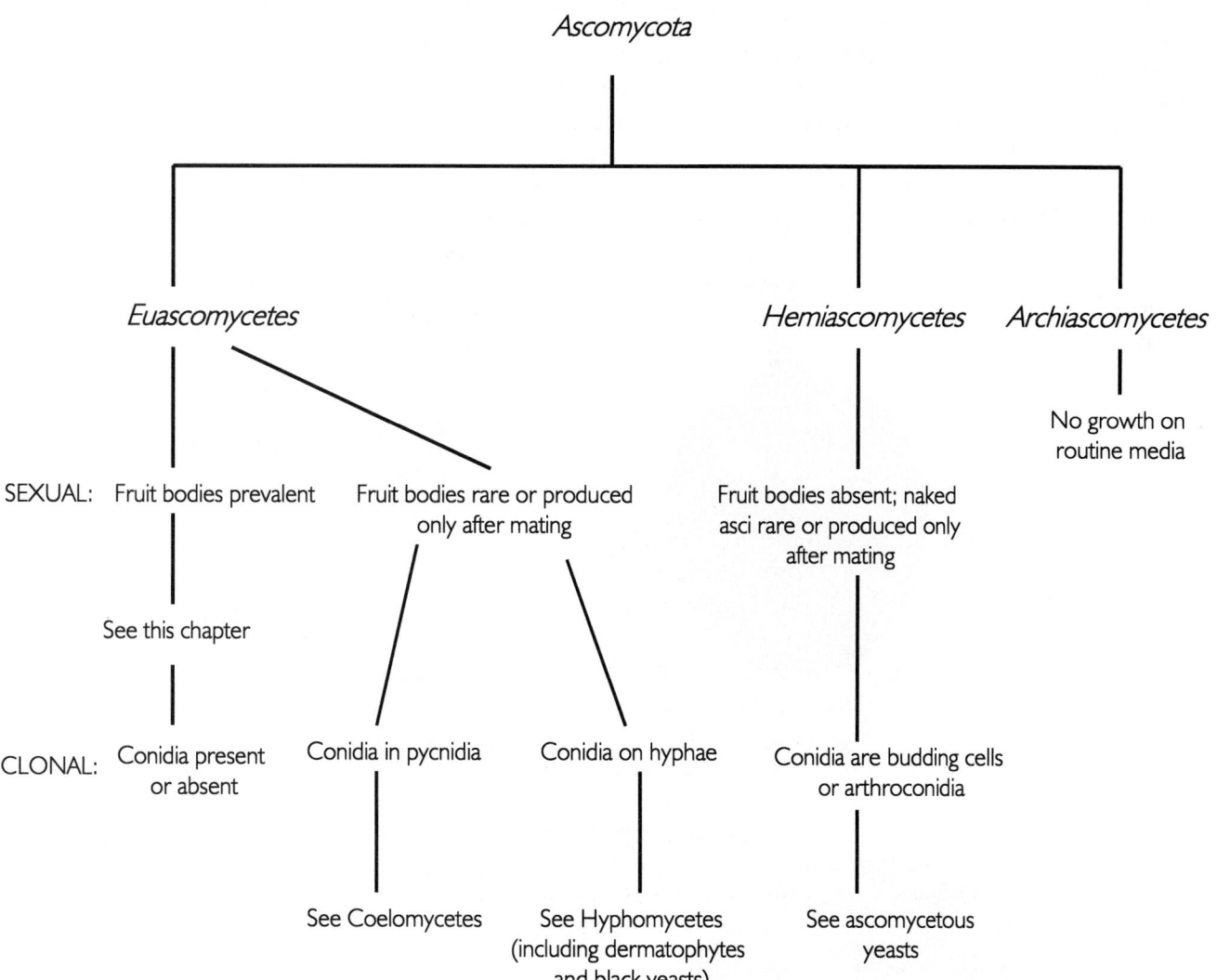

Fig. 22. Diagram of *Ascomycota* as maintained in this Atlas, in relation to anamorphic groups.

Ascomycota, Euascomycetes, Sordariales, Chaetomiaceae. Genus: *ACHAETOMIUM*

Achaetomium strumarium Rai *et al.*

Colony characteristics. Colonies (OA) growing rapidly, pale, white or yellow, later with a red exudate; reverse dark reddish-brown.

Microscopy. Ascomata formed in groups, ampulliform to ovoidal, ostiolate, usually with a conical beak, 150-270 μm diam, yellow green, tomentose. Peridium pseudoparenchymatous, composed of small, often indistinct hyphal cells (*textura intricata*). Asci narrow cylindrical, 8-spored, 60-90 × 8-11 μm. Ascospores ellipsoidal or obovoidal, attenuated and rounded at the ends, dark brown, 10-13 × 6-8 μm, with an apical germ pore.

Physiology. Optimal growth at 35-42°C; intolerant to 0.4% cycloheximide and benomyl.

Pathogenicity. BSL-1. Soil fungus. The species was reported as causative agent of brain lesions in a human (McAleer, 1988; Abbott *et al.*, 1995).

References. Von Arx *et al.* (1988), Abbott *et al.* (1995)

Nomenclature. *Achaetomium strumarium* Rai, Tewari & Mukerji - Can. J. Bot. 42: 684, 1964 ≡ *Chaetomium strumarium* (Rai, Tewari & Mukerji) Cannon - Trans. Br. Mycol. Soc. 87: 64, 1986.

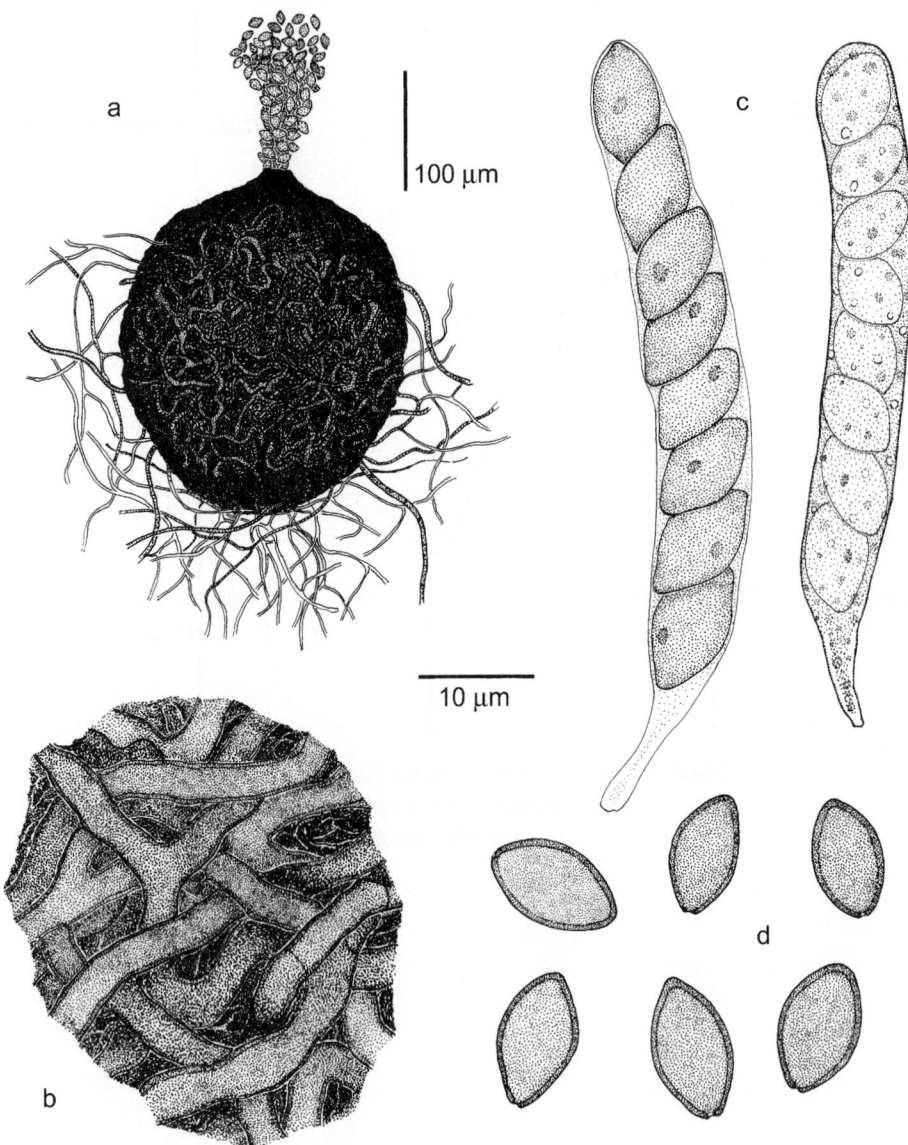

Achaetomium strumarium, CBS 758.83. a. Ascoma; b. part of peridium; c. asci; d. ascospores.

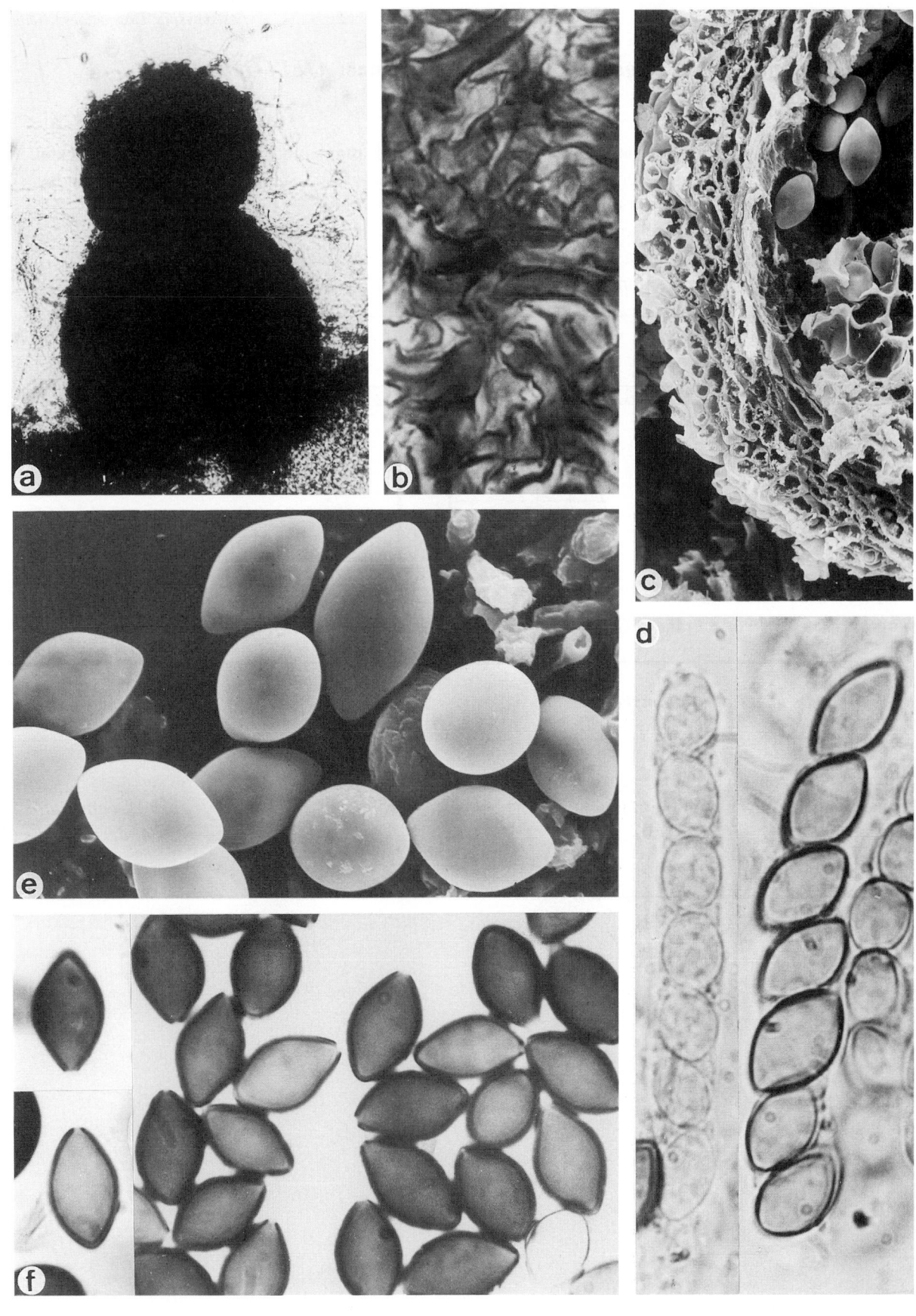

Achaetomium strumarium, CBS 758.83. a. Ascoma; b. peridium in face view; c. peridium in cross section; d-f. asci and ascospores. a. ×200; b. ×1280; c. ×1250; d-f. ×1280. Reproduced with permission from Beih. Nova Hedwigia 94: 63, 1988.

Ascomycota, Euascomycetes, Onygenales, Onygenaceae. Genus: *AJELLOMYCES*

Generic description. Colonies butyrous or cottony. Gymnothecia pale pigmented, initially spherical, later stellate due to continuing growth of appendages. Peridial hyphae smooth- and thin-walled, with spirally twisted appendages. Asci subspherical, evanescent. Ascospores (sub)spherical, hyaline. Heterothallic.

General remarks. This is the main teleomorph genus of dimorphic systemic fungi; anamorph genera are *Blastomyces, Emmonsia* and *Histoplasma*. Ascomata are obtained after mating of suitable partners on diluted SGA.

Molecular diagnostics. SSU restriction map based on NCBI X58572 and M63096:

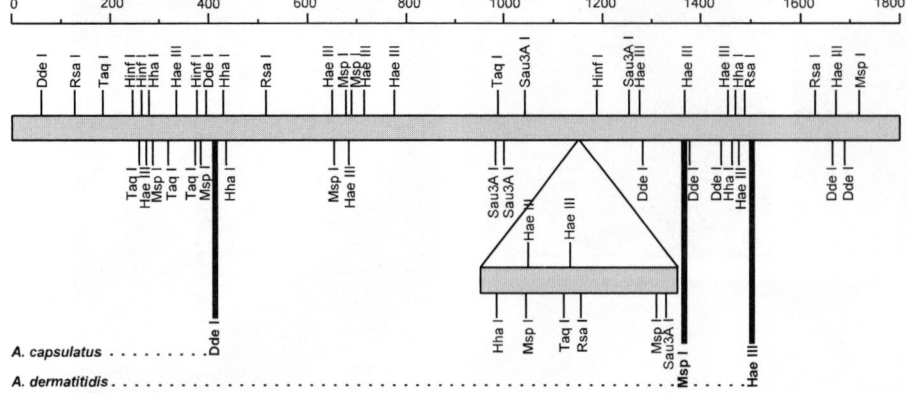

References. Glick & Kwon-Chung (1973), Kwon-Chung (1973), Currah (1985), Sigler (1996).

Key to the treated species of Ajellomyces:

1a. Peridial hyphae constricted at the septa; anamorph other than *Histoplasma* ➞ **2**
1b. Peridial hyphae of even width throughout; anamorph *Histoplasma* . . . . *A. capsulatus* (251)
2a. Anamorph *Blastomyces* . . . . *A. dermatitidis* (252)
2b. Anamorph *Emmonsia* . . . . *A. crescens* (635)

Ajellomyces capsulatus (Kwon-Chung) McGinnis & Katz

Colony characteristics. Colonies (1:10 SGA, 30°C) whitish, finally becoming brownish, cottony; reverse initially cream-coloured, becoming brownish with age.

Microscopy. Gymnothecia yellowish to tan, initially spherical, later stellate due to continuing growth of spirally twisted appendages. Peridial hyphae smooth- and thin-walled, branched, composed of cells with parallel walls, with spirally twisted appendages originating from the centre of the ascoma, having firm walls. Asci subspherical, evanescent. Ascospores subspherical, 1.2-1.5 μm diam. Heterothallic.

Anamorph. *Histoplasma capsulatum* Darling (p. 708).

Molecular diagnostics. ITS restriction map based on NCBI AF038354:

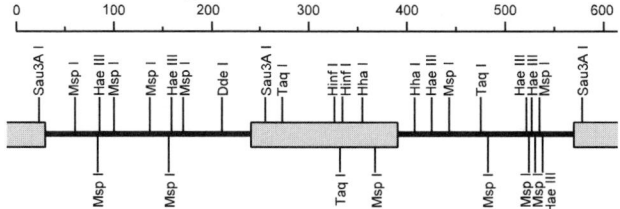

Pathogenicity. BSL-3. Agent of histoplasmosis (p. 25).

Reference. Currah (1985).

Nomenclature. *Emmonsiella capsulata* Kwon-Chung - Science 177: 368, 1972 ≡ *Ajellomyces capsulatus* (Kwon-Chung) McGinnis & Katz - Mycotaxon 8: 158, 1979.

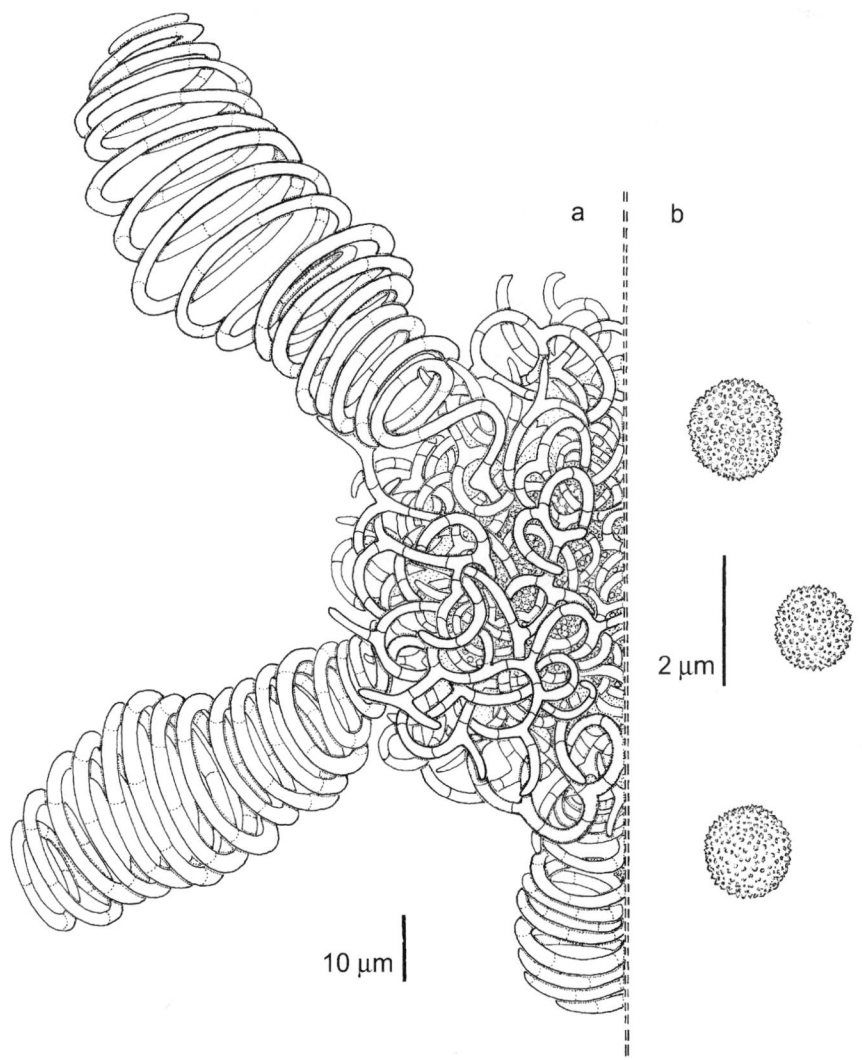

Ajellomyces capsulatus. a. Part of gymnothecium with appendages; b. ascospores. Redrawn.

Ajellomyces dermatitidis McDonough & Lewis

Colony characteristics. Colonies (1:10 SGA, 30°C) whitish, butyrous to felty; reverse cream-coloured.

Microscopy. Gymnothecia white to buff, initially spherical, later stellate due to continuing growth of spirally twisted appendages. Peridial hyphae hyaline, smooth- and thin-walled, branched, composed of swollen cells and strongly constricted at the septa, with spirally twisted appendages originating from the centre of the ascoma, having thickened inner walls. Asci subspherical, evanescent. Ascospores hyaline, spherical, 1.3-1.6 μm diam. Heterothallic.

Anamorph. *Blastomyces dermatitidis* Gilchrist & Stokes (p. 535).

Molecular diagnostics. ITS restriction map based on NCBI U18364:

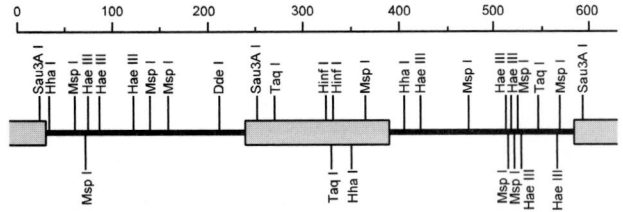

Pathogenicity. BSL-2. Agent of blastomycosis (p. 25).

References. McDonough & Lewis (1968), Kwon-Chung (1973), Currah (1985).

Nomenclature. *Ajellomyces dermatitidis* McDonough & Lewis - Mycologia 60: 77, 1968.

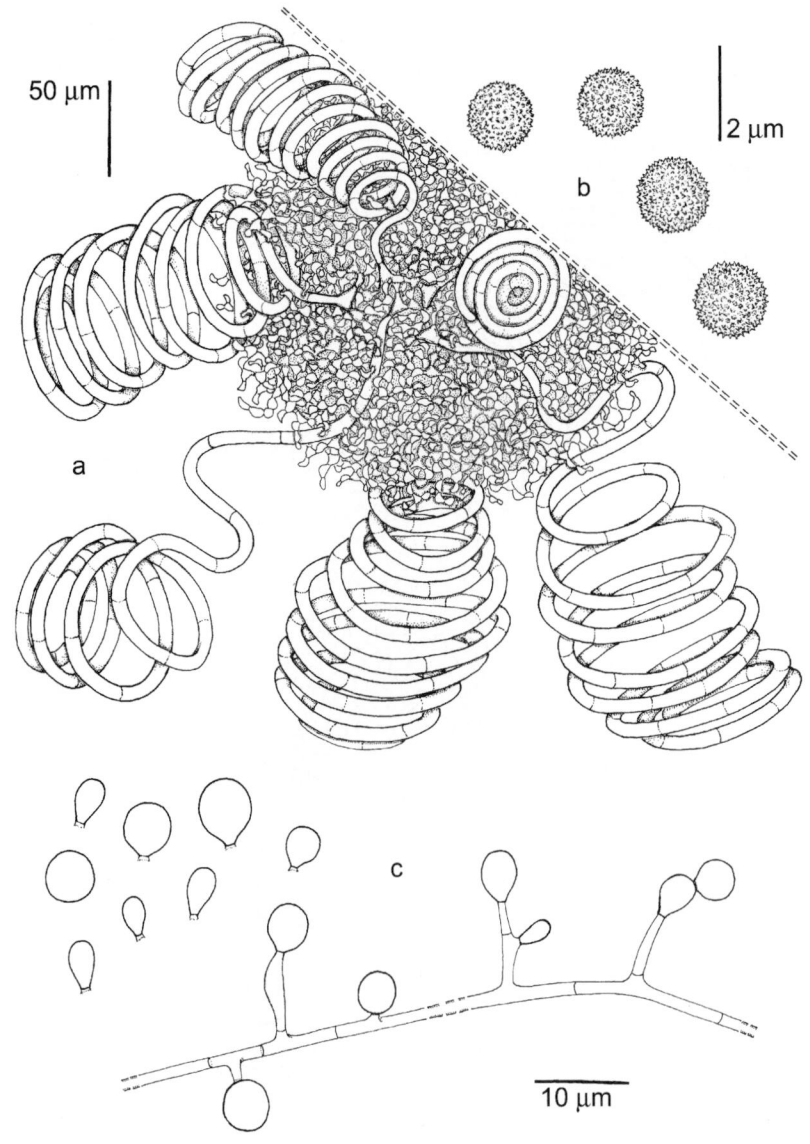

Ajellomyces dermatitidis (*Blastomyces dermatitidis*), CBS 673.68 × 674.68. a. Part of gymnothecium; b. ascospores; c. conidia.

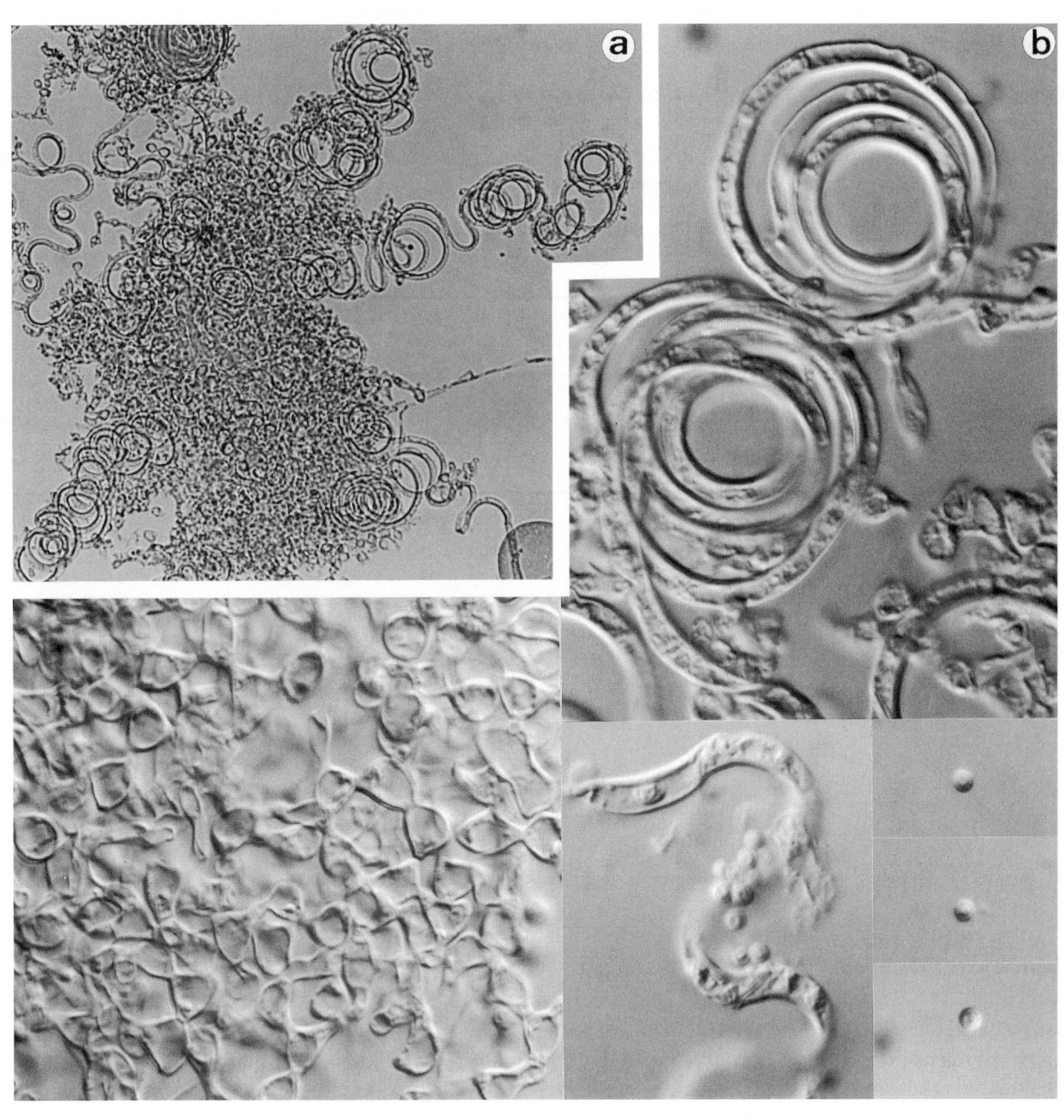

Ajellomyces dermatitidis, FMR 6636. a. Gymnothecium. b. Spirally twisted appendages and peridial hypha of the gymnothecium, and ascospores. a. ×320; b. ×1600.

Ascomycota, Euascomycetes, Onygenales, Onygenaceae. Genus: *APHANOASCUS*

Generic description. Colonies cottony, dry, white to pale pigmented. Cleistothecia pseudoparenchymatous, with subspherical asci; ascospores 1-celled, ornamented.
Anamorph. *Chrysosporium.*
Reference. Cano & Guarro (1990).

Key to the treated species of Aphanoascus:

1a. Ascospores 6.5-8.5 × 4.5-6 µm; conidia 8.5-13 × 5.5-9 µm . . . . *A. keratinophilus* (549)
1b. Ascospores 3.5-4.7 × 2.5-3.5 µm; conidia 11-15 × 4-5.5 µm . . . . *A. fulvescens* (254)

Aphanoascus fulvescens (Cooke) Apinis

Colony characteristics. Colonies (OA) growing moderately rapidly, white to tan, flat.
Microscopy. Ascomata spherical, non-ostiolate, buff to light brown, 290-500 µm diam. Peridium pseudoparenchymatous. Asci subspherical to ellipsoidal, 8-spored, 9.5-11.0 × 7-9 µm. Ascospores light brown, yellowish to pale brown in mass, irregularly reticulate, lens-shaped, 3.5-4.7 × 2.5-3.5 µm.
Physiology. Intolerant to benomyl.
Anamorph. *Chrysosporium* sp.
Fertile hyphae hyaline, not differentiated. Terminal and lateral conidia sessile or on short, unswollen protusions, solitary, hyaline, smooth- and moderately thick-walled, clavate, 15.0-17.5 × 3.7-6.0 µm, with conspicuous basal scars. Intercalary conidia frequent, cylindrical, 11-15 × 4.0-5.5 µm.
Molecular diagnostics. ITS restriction map based on NCBI AF038357:

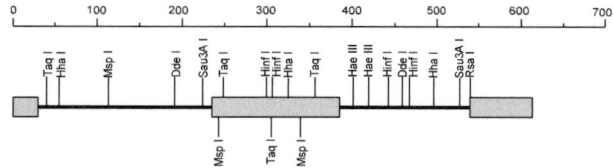

Pathogenicity. BSL-2. Keratinolytic species. Several cases of skin infection in man (Guého *et al.*, 1985; Marín & Campos, 1984) and in animals (Vanbreuseghem & de Vroey, 1979; Pal, 1995b) have been reported.
References. Guého *et al.* (1985), Cano & Guarro (1990).
Nomenclature. *Badhamia fulvescens* Cooke - Grevillea 4: 69, 1875 ≡ *Aphanoascus fulvescens* (Cooke) Apinis - Mycopath. Mycol. Appl. 35: 99, 1968 ≡ *Anixiopsis fulvescens* (Cooke) de Vries - Mykosen 12: 120, 1969.

   Eurotium stercorarium Hansen - Videnskab. Medd. Naturh. For. Kjöbenhavn 1876: 310, 1876 ≡ *Anixiopsis stercoraria* (Hansen) Hansen - Bot. Ztg. 55: 127, 1897 ≡ *Anixiopsis fulvescens* (Cooke) de Vries var. *stercoraria* (Hansen) de Vries - Mykosen 12: 121, 1969.

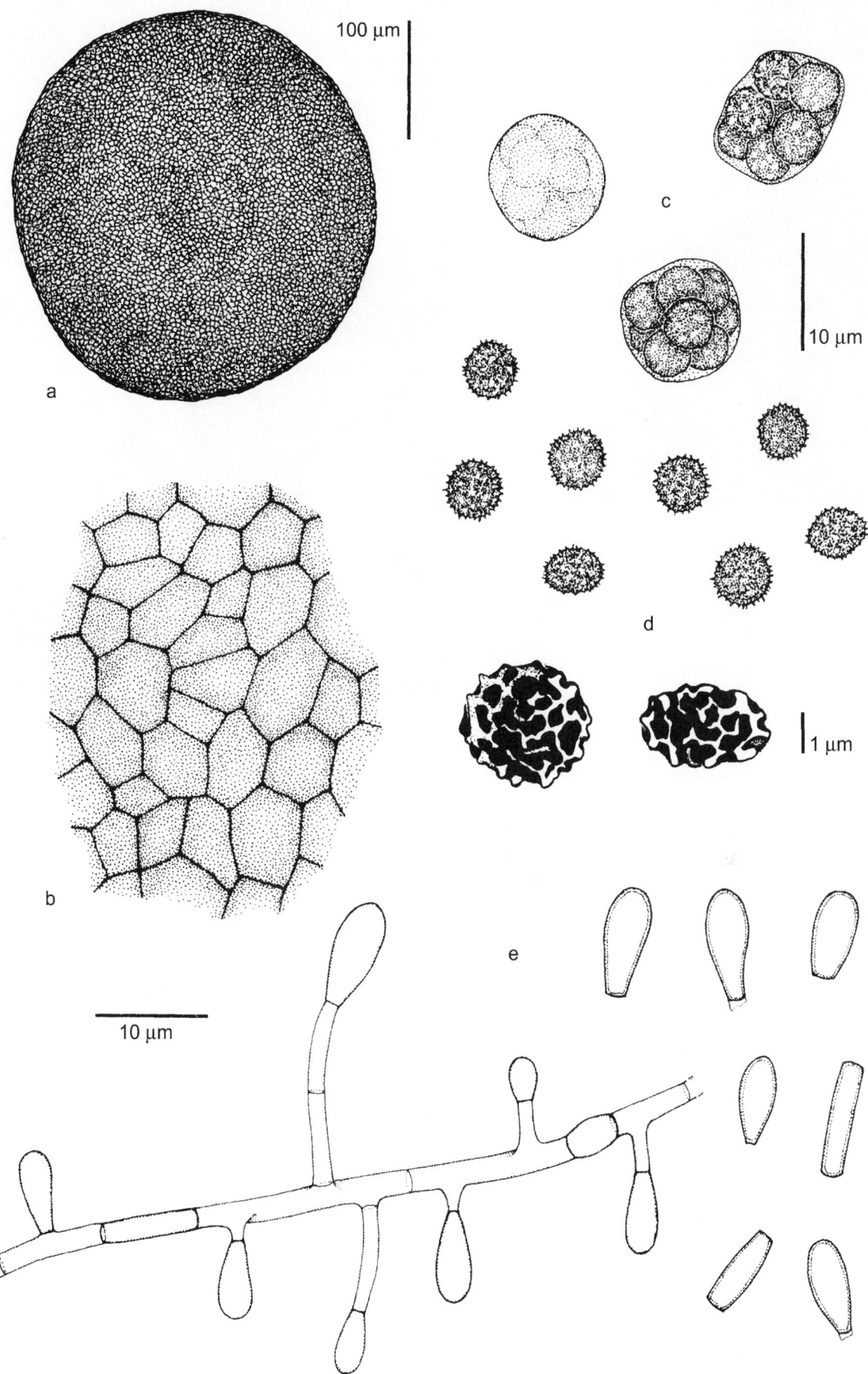

Aphanoascus fulvescens, FMR 3946. a. Ascoma; b. part of peridium; c. asci; d. ascospores; e. fertile hyphae and conidia.

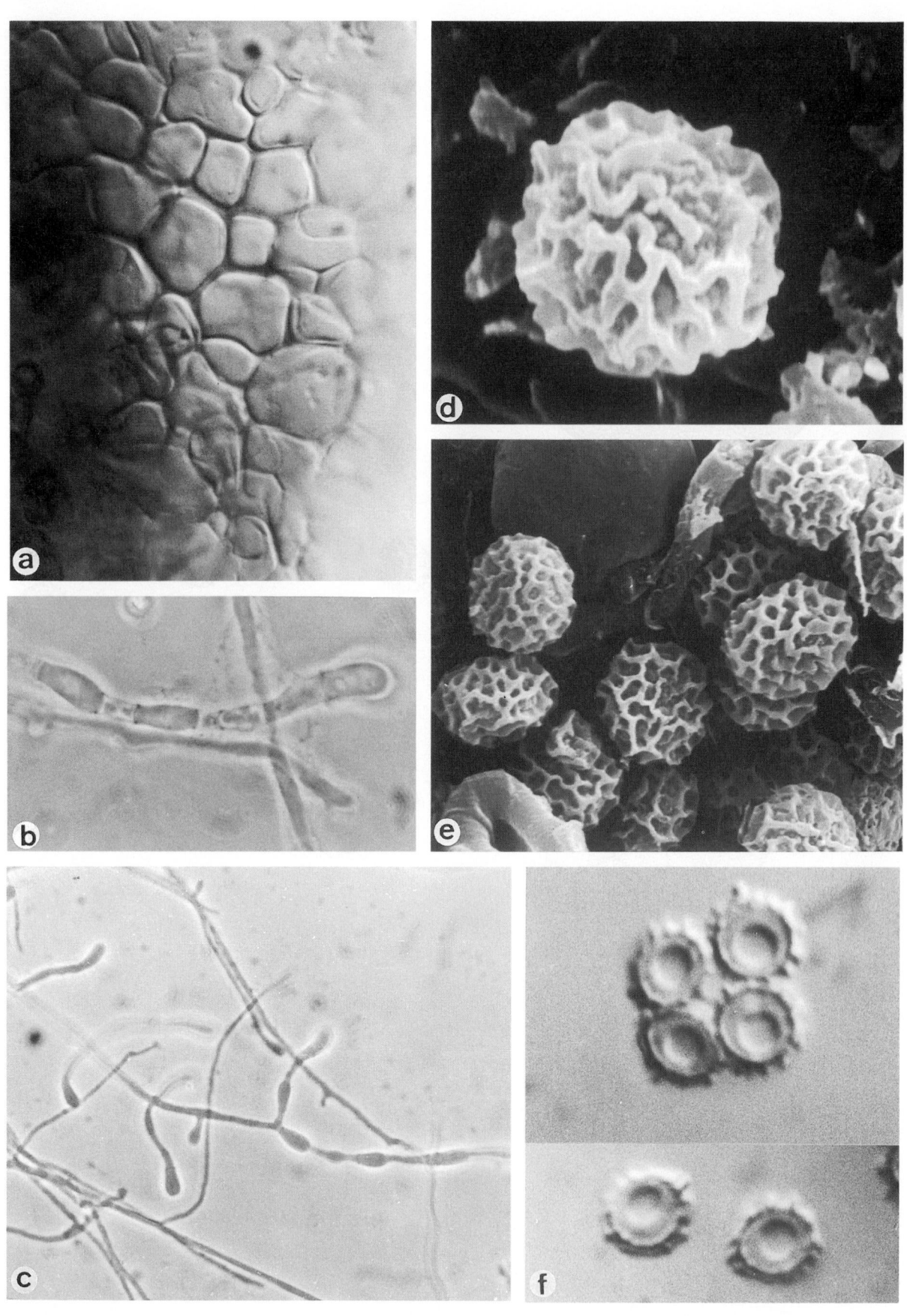

Aphanoascus fulvescens, FMR 3946. a. Part of peridium; b, c. anamorph; d-f, ascospores. a, b. ×1600; c. ×640; d. ×13500; e. ×6000; f. ×2000.

Ascomycota, Euascomycetes, Sordariales, Lasiosphaeriaceae. Genus: *ARNIUM*

Arnium leporinum (Cain) Lundqvist & Krug

Colony characteristics. Colonies (OA) spreading moderately, brownish-grey, plane, arachnoid.
Microscopy. Ascomata blackish, pyriform, 625-1000 × 300-625 μm, covered with flexuose hairs; neck cylindrical to conical. Peridium membranaceous, semi-transparent. Asci clavate, 275-350 × 46-75 μm, with a long stipe and a broad rounded apex, containing 128 spores. Ascospores ellipsoidal to broadly fusiform, 1-celled, 18-22 × 9.5-12 μm, at first hyaline, becoming olivaceous to dark brown, smooth-walled, with slimy appendages at each end, with an apical germ pore.
Pathogenicity. BSL-1. Fungus occurring on dung. A case of endocarditis has been reported (Restrepo *et al.*, 1984).
Reference. Lundqvist (1972).
Nomenclature. *Sordaria leporina* Cain - Univ. Toronto Stud., Biol. Ser., 38: 30, 1934 ≡ *Podospora leporina* (Cain) Cain - Can. J. Bot. 40: 460, 1962 ≡ *Arnium leporinum* (Cain) Lundqvist & Krug - Symb. Bot. Upsal. 20: 234, 1972.

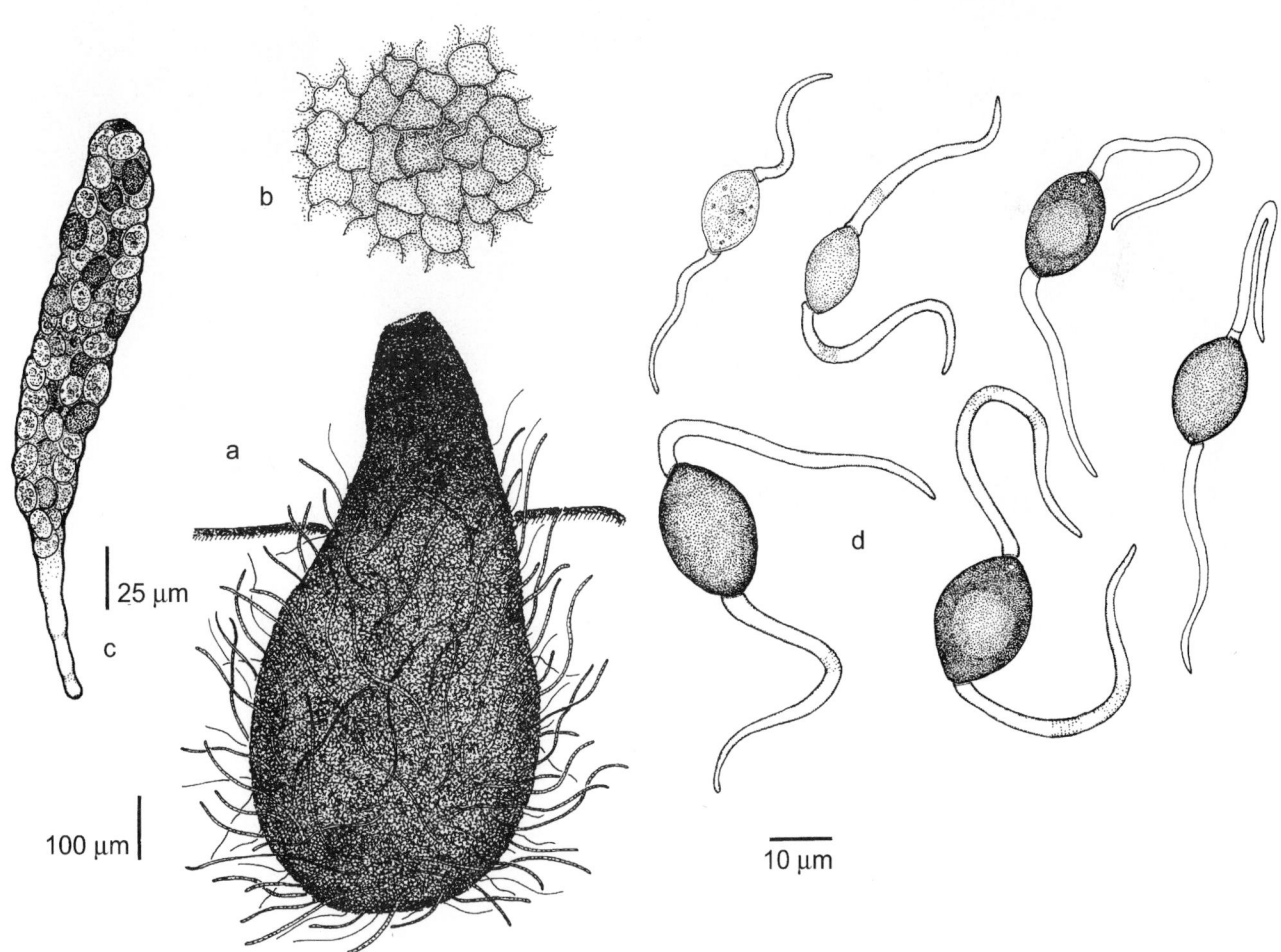

Arnium leporinum, TRTC 5197. a. Ascoma; b. part of peridium; c. ascus; d. ascospores.

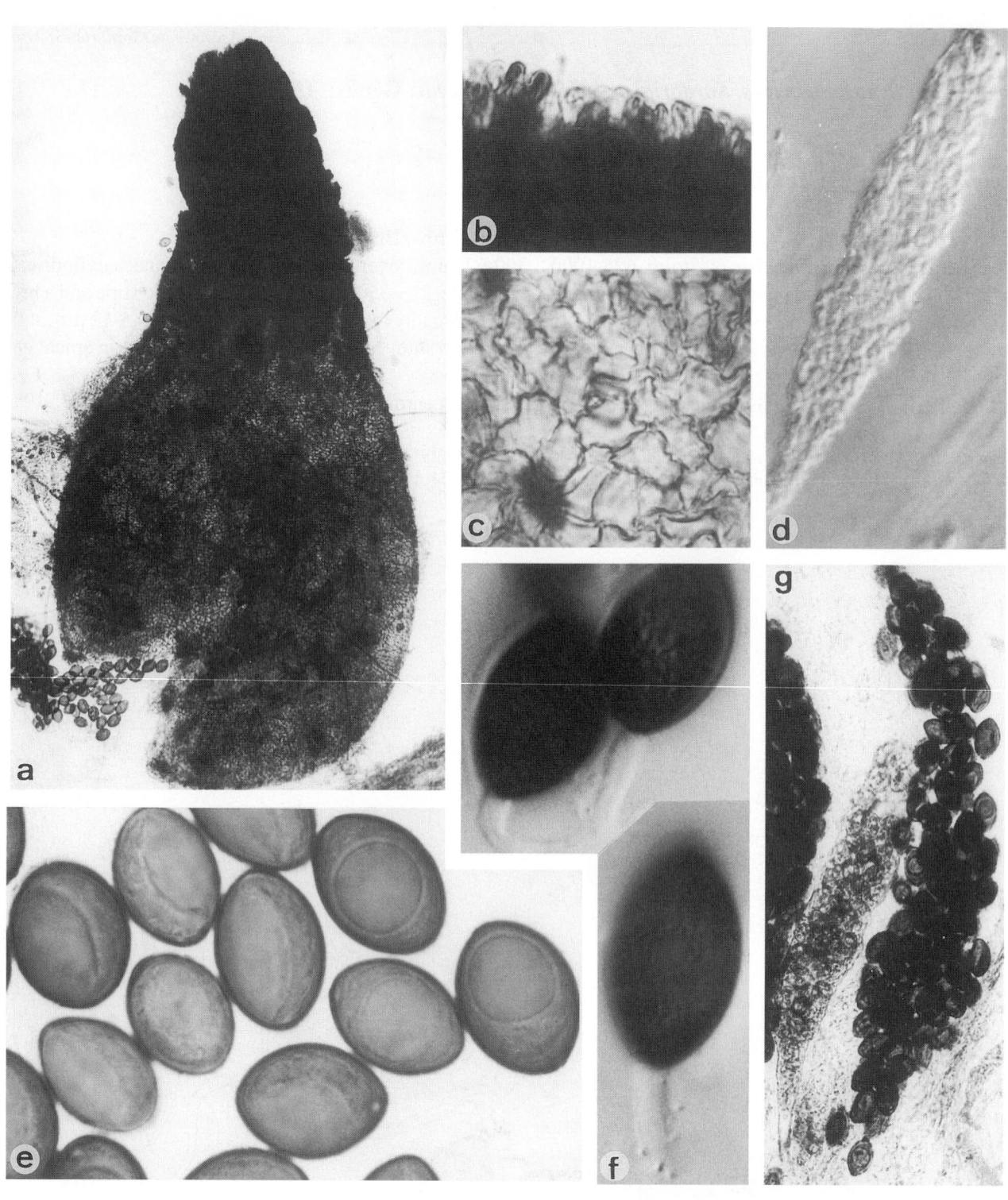

Arnium leporinum, TRTC 5197. a. Ascoma; b, c. detail of the upper part of the ascoma neck and of peridium; d, g. asci; e, f. ascospores. a. ×128; b, c. ×1280; d. ×1600; e. ×1280; f. ×1600; g. ×1280.

Ascomycota, Euascomycetes, Onygenales, Arthrodermataceae. Genus: *ARTHRODERMA*

Arthroderma simii Stockdale *et al.*

Colony characteristics. Colonies (1:10 SGA) white, cottony, with white to buff gymnothecia.

Microscopy. Gymnothecia composed of a network of unilaterally branched hyphae (peridial hyphae), consisting of ossiform, thick- and rough-walled cells terminating with a hyaline, thin-walled, spirally twisted appendage. Asci subspherical to ellipsoidal, thin-walled, evanescent, 8-spored, up to 6 μm wide. Ascospores subhyaline, lenticular, about 3 μm wide.

Anamorph. *Trichophyton simii* (Pinoy) Stockdale *et al.* (p. 979; see there for pathogenicity).

Remarks. The presently described species is an example of the large diversity of teleomorphs known in dermatophytes (genera *Microsporum* and *Trichophyton*). Gymnothecia are produced after mating of suitable partners on 1:10 diluted SGA. The genus *Nannizzia* has been reduced to synonymy with *Arthroderma* (Weitzman *et al.*, 1986).

Molecular diagnostics. ITS restriction map based on NCBI Z98017, see under anamorph (p. 979).

References. Padhye & Carmichael (1971), Kwon-Chung (1972), Currah (1985), Kawasaki *et al.* (1990).

Nomenclature. *Arthroderma simii* Stockdale, McKenzie & Austwick - Sabouraudia 4: 113, 1965.

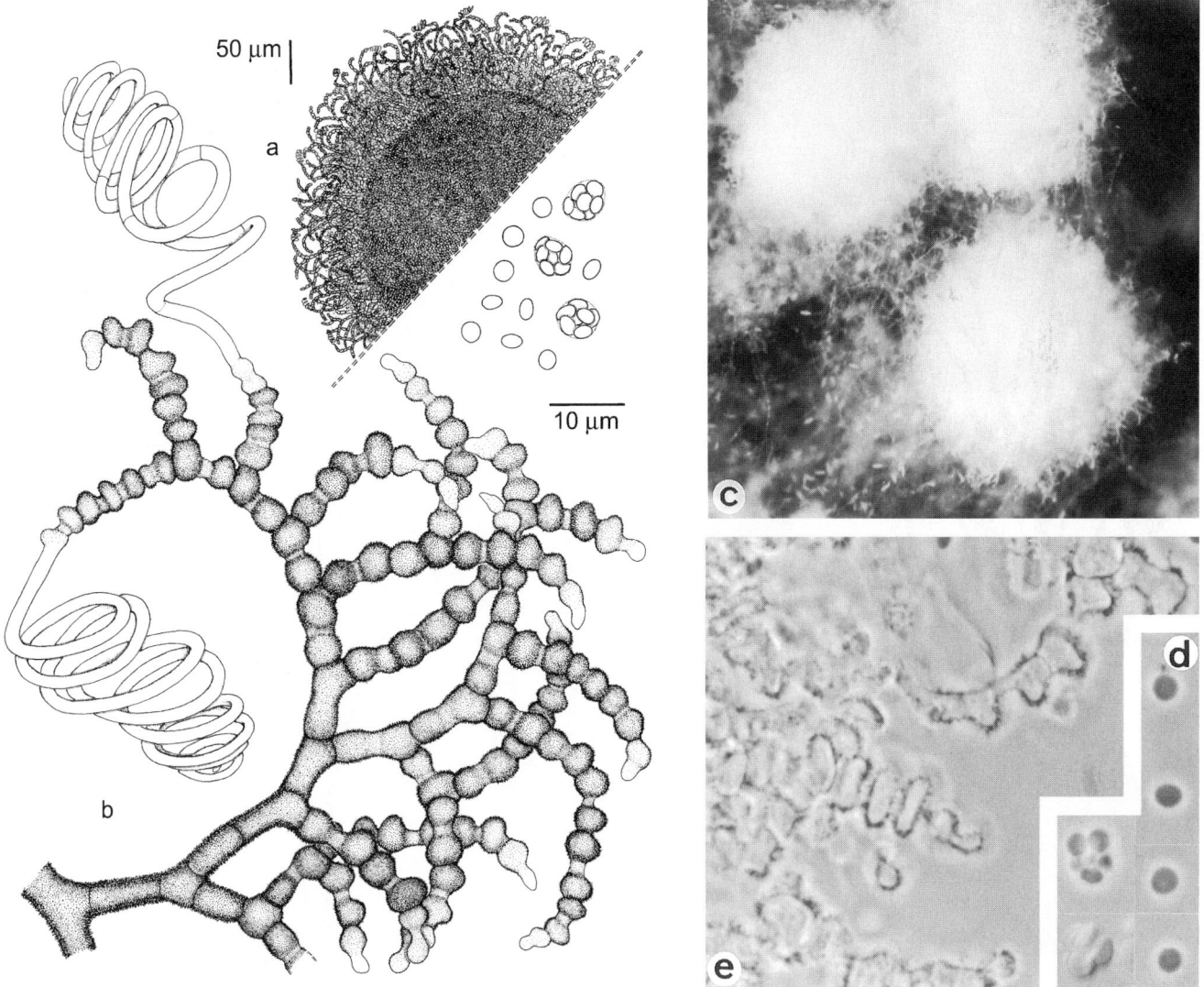

Arthroderma simii, CBS 448.65 × 417.65. a, c, d. Gymnothecia, asci and ascospores; b, e. peridial hyphae. c. ×70; d, e. ×1600.

Ascomycota, Euascomycetes, Sordariales, Chaetomiaceae. Genus: *ASCOTRICHA*

Ascotricha chartarum Berk.

Colony characteristics. Colonies (OA) growing moderately rapidly, dull blackish-brown, with dark grey patches of conidiation.

Microscopy. Perithecia black, pear-shaped with apical opening, 130-230 µm wide, discharching a black spore cirrhus, bearing olivaceous-brown, erect, stiff setae, geniculate with pale, thin-walled vesicles at the bends. Asci cylindrical, 8-spored, 65 × 10 µm. Ascospores one-celled, olivaceous-brown to black, smooth-walled, lenticular, 8 µm diam in face view, 5 µm wide in lateral view, with a distinct equatorial germ slit.

Anamorph. *Dicyma ampullifera* Boul.

Conidiophores straight, stiff, profusely branched, up to 1 mm tall, 3.5-5.5 µm wide, with pale, thin-walled vesicles at the bends. Conidiogenous cells terminal and lateral, cylindrical, bearing clusters of conidia on denticles. Conidia verrucose, yellowish-olivaceous, (sub)spherical, 5-7 × 3-6 µm.

Pathogenicity. BSL-1. Cellulolytic saprophyte. A maxillary sinusitis was reported by Singh *et al.* (1996).

References. Ames (1963), Hawksworth (1971).

Nomenclature. *Ascotricha chartarum* Berkeley - Ann. Nat. Hist. 1: 257, 1838.

  Dicyma ampullifera Boulanger - Revue Gén. Bot. 9: 25, 1897.

  Ascotrichum chartarum Berkeley var. *orientalis* Castellani & Jacono - J. Trop. Med. Hyg. 37: 362, 1934.

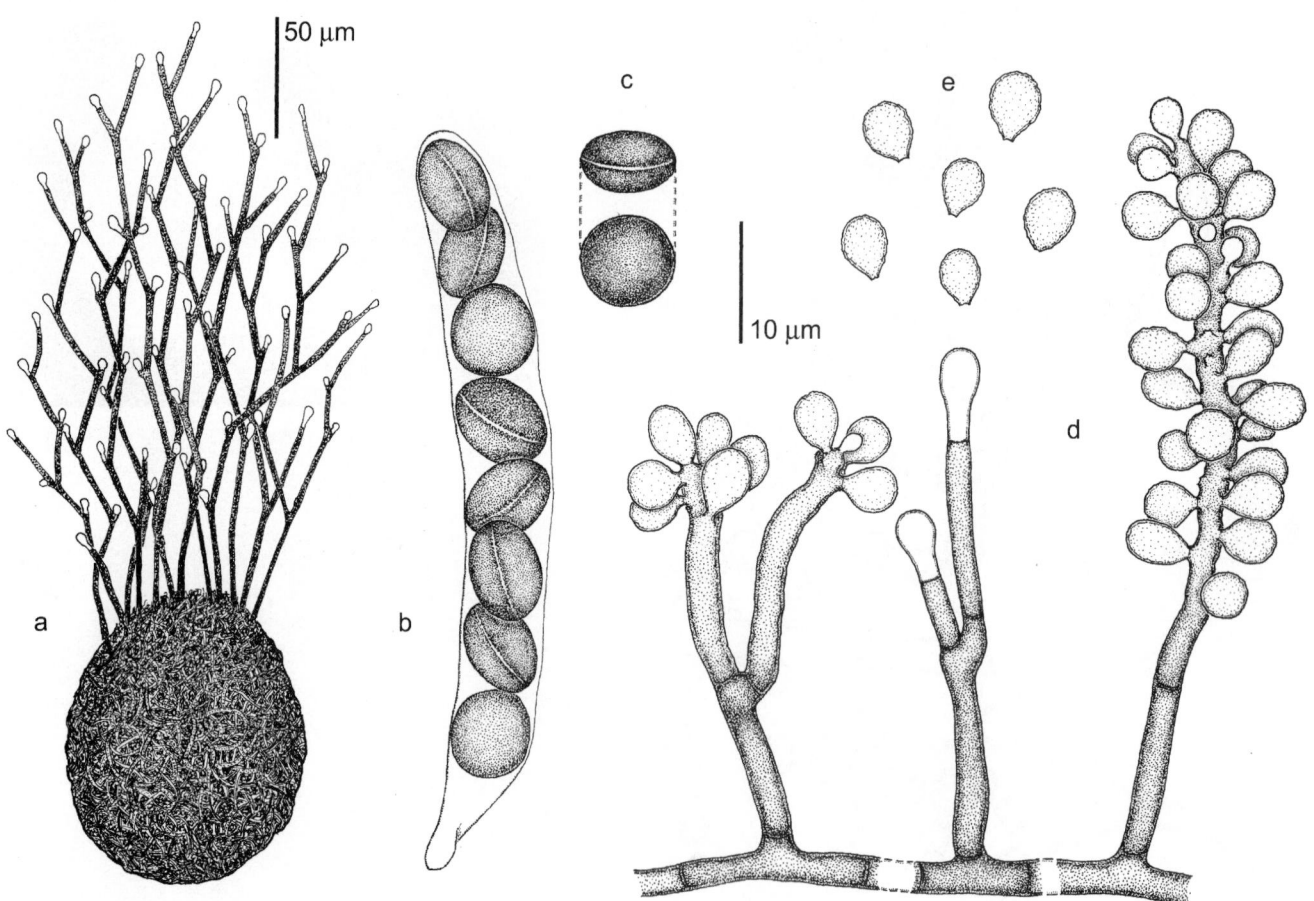

Ascotricha chartarum, CBS 657.95. a. Ascoma; b. ascus; c. oblate ascospore; d, e. anamorph with conidia.

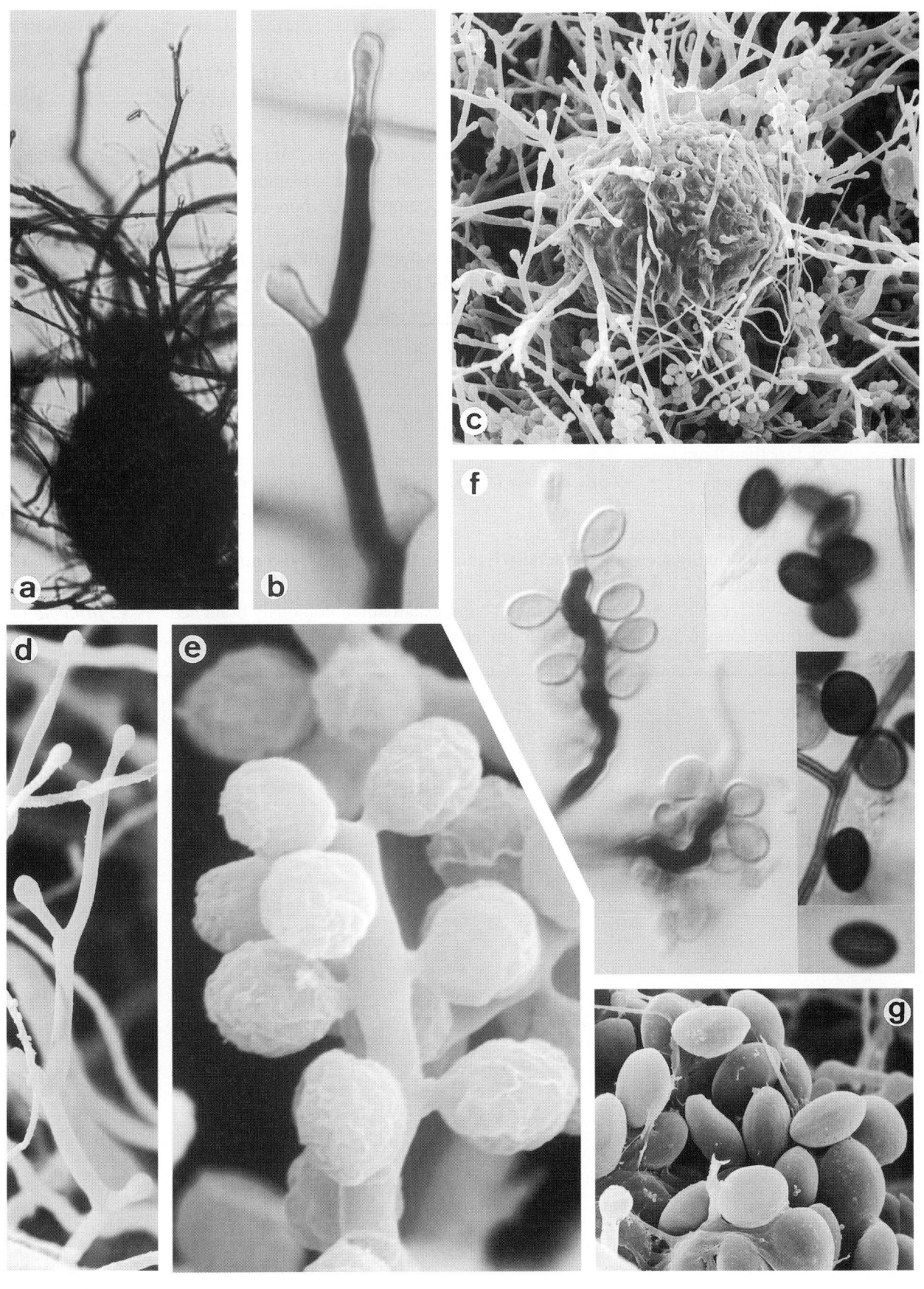

Ascotricha chartarum, CBS 657.95. a, c. Ascomata; b, d. peridial hairs; e. conidia attached to sympodial conidiogenous cells; f. conidiophores and ascospores; g. ascospores. a. ×250; b. ×1600; c. ×650; d. ×2000; e. ×5750; f. ×1600; g. ×2000.

Ascomycota, Euascomycetes, Sordariales, Chaetomiaceae. Genus: *CHAETOMIUM*

Generic description. Colonies expanding. Ascomata spherical to pyriform, attached to the substrate by rhizoidal hyphae, ostiolate, usually covered with characteristic hairs or setae, which are branched or unbranched and are often undulate or spirally coiled. Peridium pseudoparenchymatous, composed of more or less flattened hyphal cells. Asci mostly 8-spored, cylindrical, obovoidal or clavate, stalked, with evanescent walls. Ascospores aseptate, without appendages, smooth-walled, pigmented, with one or two germ pores, becoming liberated in a dark cirrhus or mass. **References.** Von Arx *et al.* (1986). Phylogeny was studied by Lee & Hanlin (1999).

Key to the treated species of Chaetomium:

1a. Ascospores mostly longer than 9 μm **→ 2**
1b. Ascospores less than 9 μm long **→ 4**
2a. Ascospores measuring 13-17 × 7-9 μm, ellipsoidal; ascomatal hairs undulate or wavy, often circinate at the apex . . . . **C. murorum (269)**
2b. Ascospores less than 13 μm long **→ 3**
3a. Ascospores limoniform in face view, bilaterally flattened, 9-12 × 8-10 × 6-8 μm . . . . **C. globosum (267)**
3b. Ascospores fusiform, 9-11 × 4.5-6 μm . . . . **C. atrobrunneum (263)**
4a. Ascospores ovoidal, 6-7.5 × 4-5.5 μm; ascomatal hairs usually dichotomously branched . . **C. funicola (265)**
4b. Ascospores limoniform, 6.5-8.5 × 6-7.5 × 5-6 μm; chlamydospores usually present . . **C. homopilatum (1024)**

Chaetomium atrobrunneum Ames

Colony characteristics. Colonies (OA) growing rapidly, dark brown.

Microscopy. Ascomata subspherical, olivaceous or pale green, 70-150 µm diam. Peridium with *textura angularis*. Ascomatal setae arising around the ostiolar pore and lower down on fruit body, long, tapering, mostly unbranched, 3-4 µm broad at the base, smooth-walled. Asci clavate, 8-spored. Ascospores fusiform or elongate pyriform, greyish, 9-11 × 4.5-6.0 µm, with a distinct, often slightly subapical germ pore.

Pathogenicity. BSL-1. A case of fatal systemic mycosis in a patient with leukemia (Rinaldi *et al.*, 1991) and a cerebral infection in a BMT patient (Guppy *et al.*, 1998; Thomas *et al.*, 1999) have been reported.

Antifungal susceptibility.

Antifungal	MICs range	Strains	Reference
AMB	0.58-1.16	4	Guarro *et al.* (1995)
5FC	>645.5	4	Guarro *et al.* (1995)
FLZ	40-80	4	Guarro *et al.* (1995)
ITZ	0.04-0.07	4	Guarro *et al.* (1995)
KTZ	0.2-0.8	4	Guarro *et al.* (1995)
MCZ	0.04-0.07	4	Guarro *et al.* (1995)

Reference. Von Arx *et al.* (1986).

Nomenclature. *Chaetomium atrobrunneum* Ames - Mycologia 41: 641, 1949.

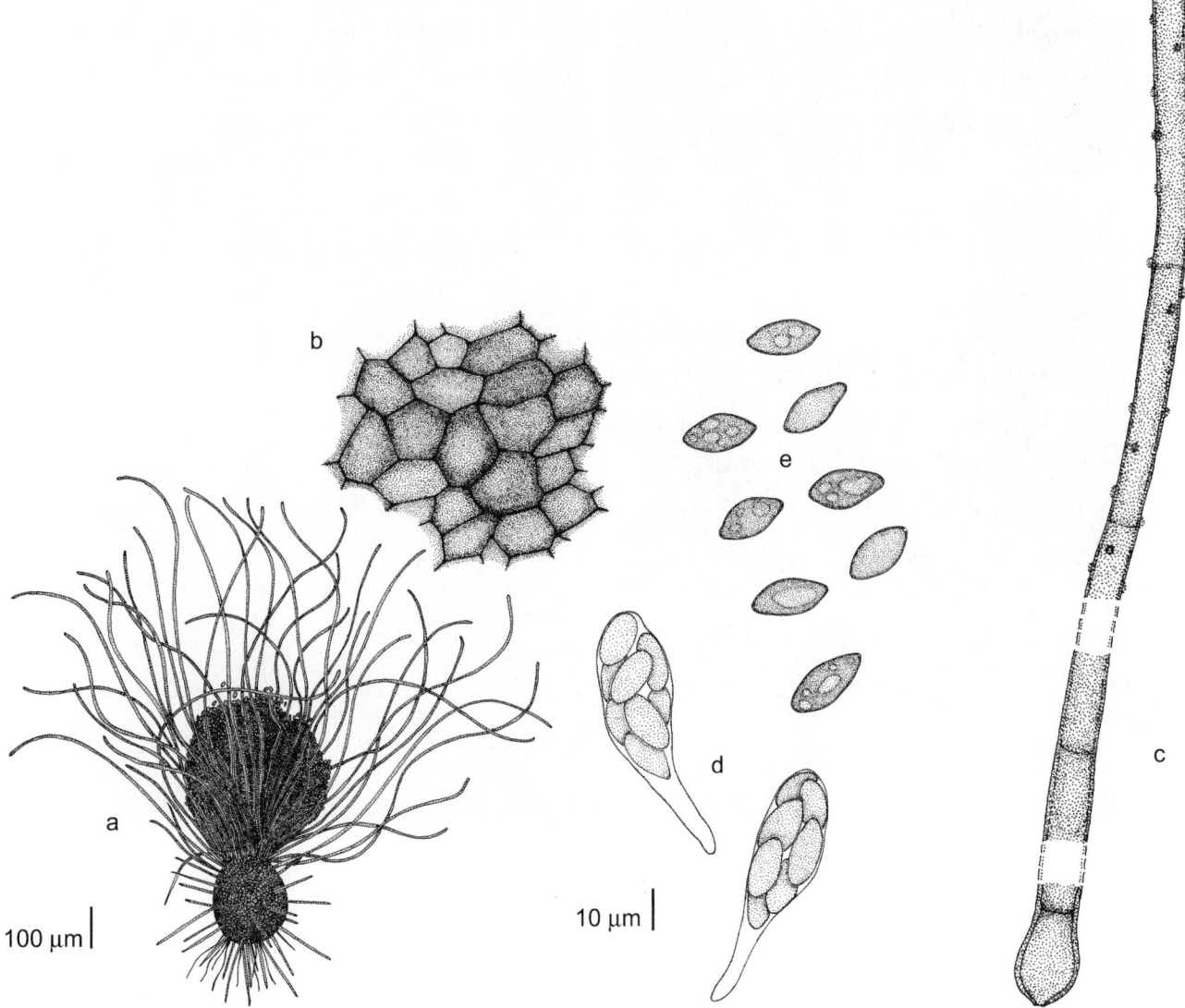

Chaetomium atrobrunneum, CBS 110.63. a. Ascoma; b. part of peridium; c. ascomal seta; d. asci; e. ascospores.

Chaetomium atrobrunneum, CBS 110.63. a-c. Ascomata; d. peridial hairs; e, f. ascospores. a. ×130; b. ×330; c. ×360; d. ×1650; e. ×1600; f. ×2400. Reproduced with permission from Beih. Nova Hedwigia 84: 75, 1986.

Chaetomium funicola Cooke

Colony characteristics. Colonies (OA) growing rapidly, with a white or pale aerial mycelium and a yellow reverse.
Microscopy. Ascomata spherical or ovoidal, dark olivaceous or grey, 120-200 μm. Peridium with *textura angularis*. Ascomal hairs straight and stiff; in part repeatedly dichotomously branched, pale or brown, verrucose. Asci 8-spored, clavate, 22-32 × 8-12 μm. Ascospores brown, ovoidal, 6.0-7.5 × 4.0-5.5 μm, attenuated at one or both ends, often slightly apiculate, at the most attenuated end with a distinct germ pore, the opposite end often with a paler spot suggesting a second germ pore.
Pathogenicity. BSL-1. This species was reported as probably responsible of subcutaneous lesions (Koch & Haneke, 1965).
Antifungal susceptibility.

Antifungal	MICs	Strains	Reference
AMB	1.6	1	Guarro *et al.* (1995)
5FC	>645.5	1	Guarro *et al.* (1995)
FLZ	80	1	Guarro *et al.* (1995)
ITZ	0.15	1	Guarro *et al.* (1995)
KTZ	1.6	1	Guarro *et al.* (1995)
MCZ	0.15	1	Guarro *et al.* (1995)

Reference. Von Arx *et al.* (1986).
Nomenclature. *Chaetomium funicola* Cooke - Grevillea 1: 176, 1873.

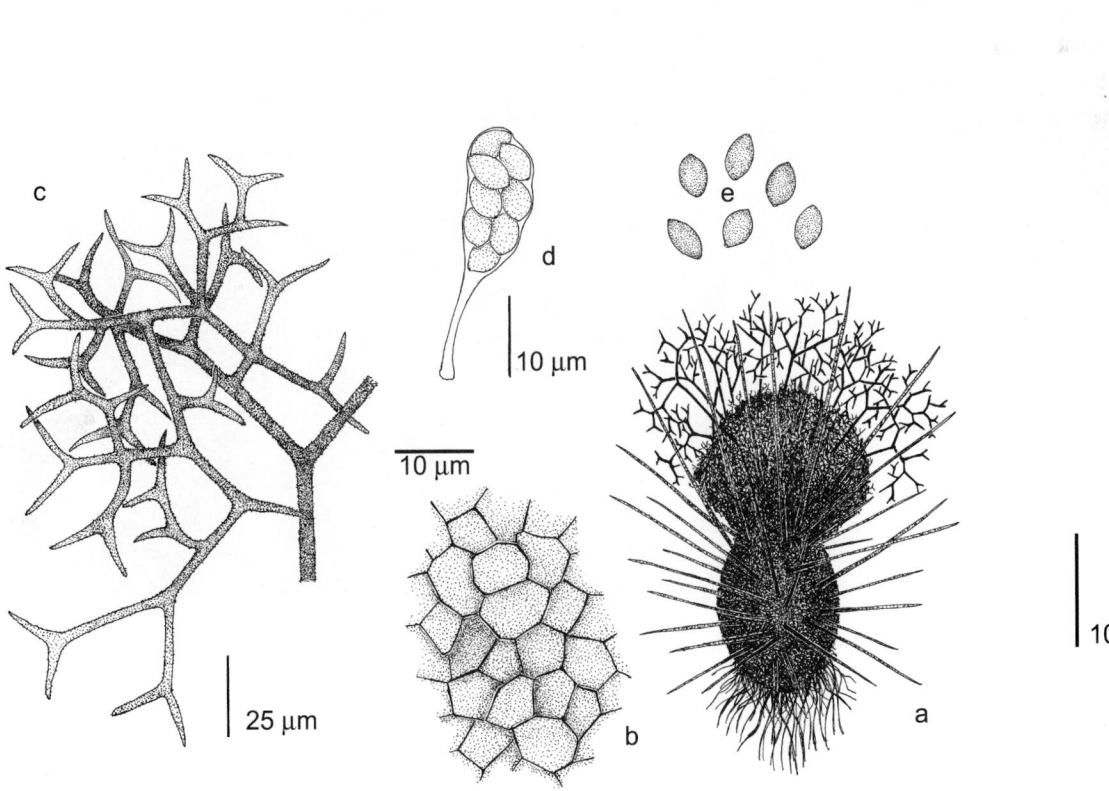

Chaetomium funicola, CBS 158.52. a. Ascoma; b. part of peridium; c. ascomal hairs; d. ascus; e. ascospores.

265

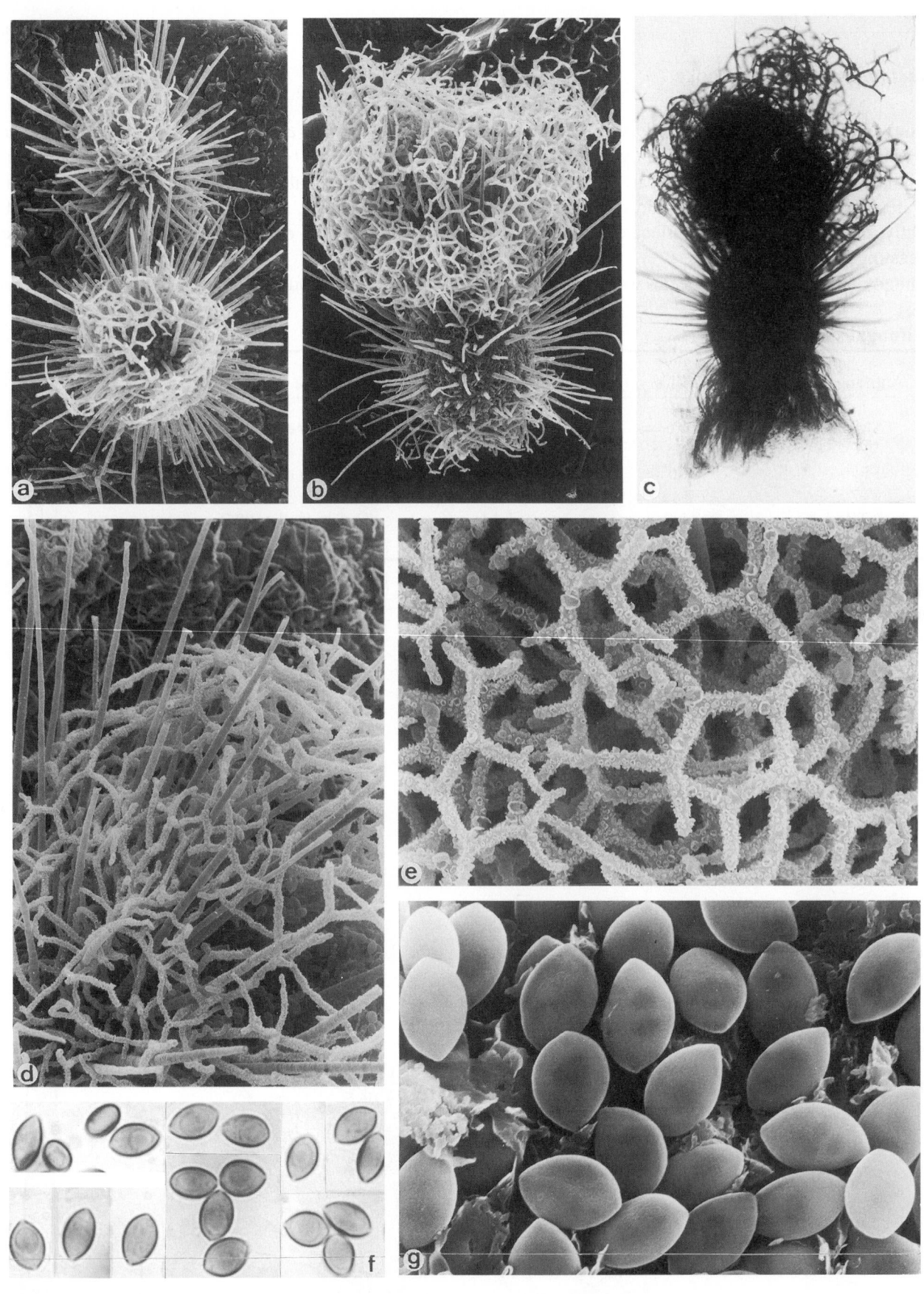

Chaetomium funicola, CBS 158.52. a-c. Ascomata; d,e. ascomal hairs; f,g. ascospores. a. ×168; b. ×147; c. ×210; d. ×310; e. ×750; f. ×2440; g. ×916. Reproduced with permission from Beih. Nova Hedwigia 84: 96, 1986.

266

Chaetomium globosum Kunze

Colony characteristics. Colonies (OA) growing rapidly, with a pale or olivaceous aerial mycelium and often with yellow, greyish-green, green or red exudates.

Microscopy. Ascomata spherical, ovoidal or obovoidal, 175-280 µm diam. Peridium brown, composed of *textura intricata*. Ascomal hairs numerous, usually unbranched, flexuose, undulate or coiled, septate, brownish, up to 500 µm long. Asci clavate, 30-40 × 11-16 µm, 8-spored. Ascospores limoniform in face view, bilaterally flattened, usually brownish, 9-12 × 8-10 × 6-8 µm, with an apical germ pore.

Physiology. No growth at 42°C.

Pathogenicity. BSL-1. Reported cases comprise onychomycosis (Rippon, 1988; Naidu *et al.*, 1991; Stiller *et al.*, 1992; Hattori *et al.*, 2000), cutaneous lesions (Costa *et al.*, 1988b; Wang *et al.*, 1998). Contamination of dialysis fluid was reported by Febré *et al.* (1999) and peritonitis by Barthez *et al.* (1984). A case of pleural effusion after autologous BMT was described by Lesire *et al.* (1999). The species was also repeatedly isolated from the pleural fluid of a patient with leukemia (Hoppin *et al.*, 1983). The identity of a cerebral case reported by Anandi *et al.* (1989) was questioned by Abbott *et al.* (1995). A pneumonia in a leukemic patient, reported by Yeghen *et al.* (1996), was probably caused by another fungus (Guarro, 1998), as was a case of sinusitis reported by Aru *et al.* (1997) (J. Guarro, unpublished data).

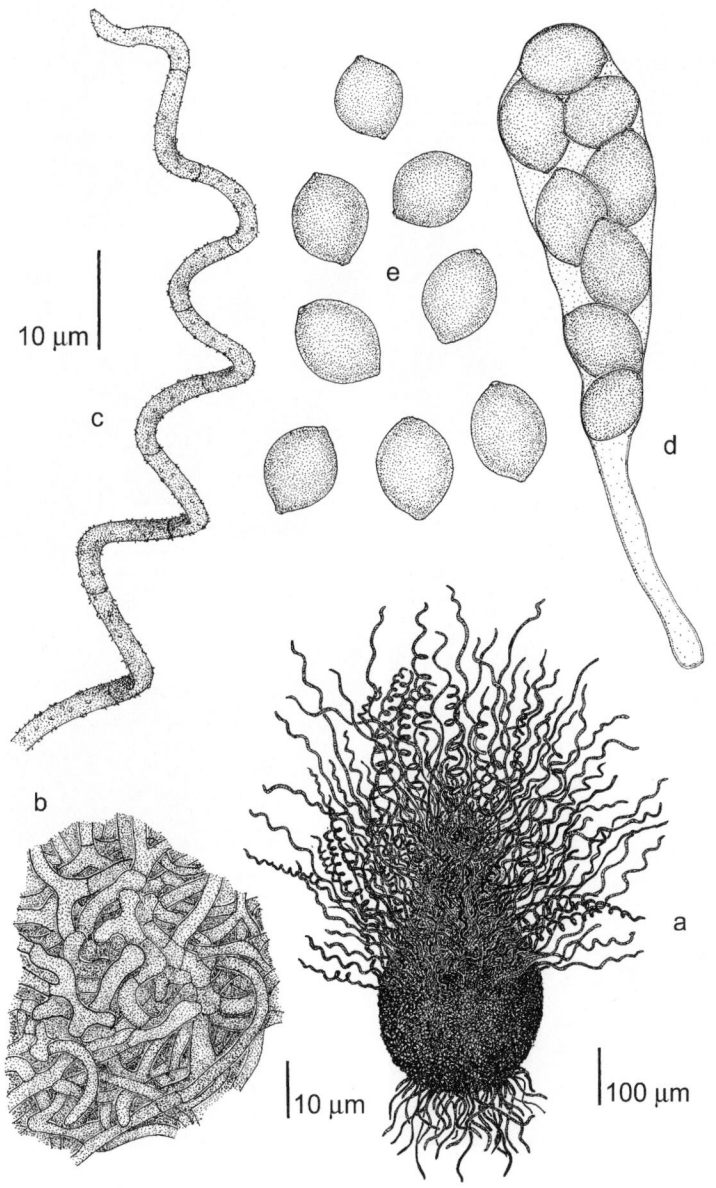

Chaetomium globosum, CBS 545.83. a. Ascoma; b. part of peridium; c. ascomal hair; d. ascus; e. ascospores.

267

Antifungal susceptibility.

Antifungal	MICs range	Strains	Reference
AMB	0.58-2.31	16	Guarro *et al.* (1995)
5FC	>645.5	16	Guarro *et al.* (1995)
FLZ	40-160	16	Guarro *et al.* (1995)
ITZ	0.035-0.3	16	Guarro *et al.* (1995)
KTZ	0.8-1.6	16	Guarro *et al.* (1995)
MCZ	0.075-0.15	16	Guarro *et al.* (1995)

Reference. Von Arx *et al.* (1986).

Nomenclature. *Chaetomium globosum* Kunze - Mykol. Hefte 1: 16, 1817.
Chaetomium cochliodes Palliser - North Amer. Fl. 3: 61, 1910.

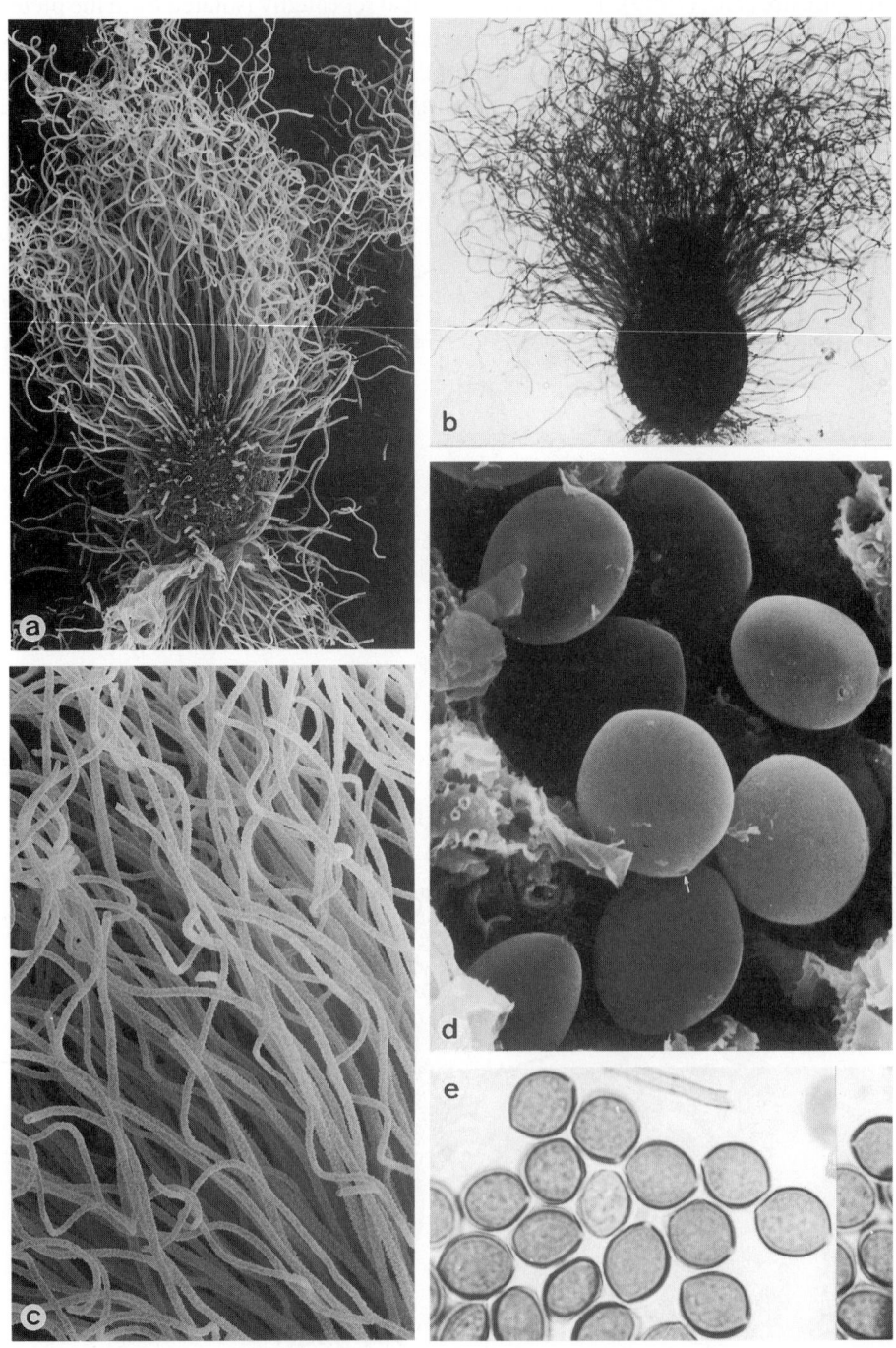

Chaetomium globosum. a, c, d. CBS 161.52; b, e. CBS 147.51 a, b. Ascoma; c. ascomal hairs; d, e. ascospores. a. ×175; b. ×128; c. ×600; d. ×3275; e. ×1280. Reproduced with permission from Beih. Nova Hedwigia 84: 103, 1986.

Chaetomium murorum Corda

Colony characteristics. Colonies (OA) growing rapidly, grey to olivaceous brown.
Microscopy. Ascomata spherical or obovoidal, reddish-brown or violet-blue. Peridium with *textura epidermoidea* or *intricata*. Ascomatal setae long, flexuous or undulate, often recurved or circinate at the apex, smooth, verrucose or spinulose. Asci clavate, 8-spored. Ascospores ellipsoidal to broadly fusiform, often attenuated at the ends, brown, 13-17 × 7-9 μm, with a distinct apical germ pore and usually with darker, longitudinal bands.
Pathogenicity. BSL-1. An infection of the dermis and subcutaneous tissue of the chest and abdominal region of a patient was reported from China (Lin & Li, 1995).
Reference. Von Arx *et al.* (1986).
Nomenclature. *Chaetomium murorum* Corda - Icon. Fung. 1: 24, 1837.

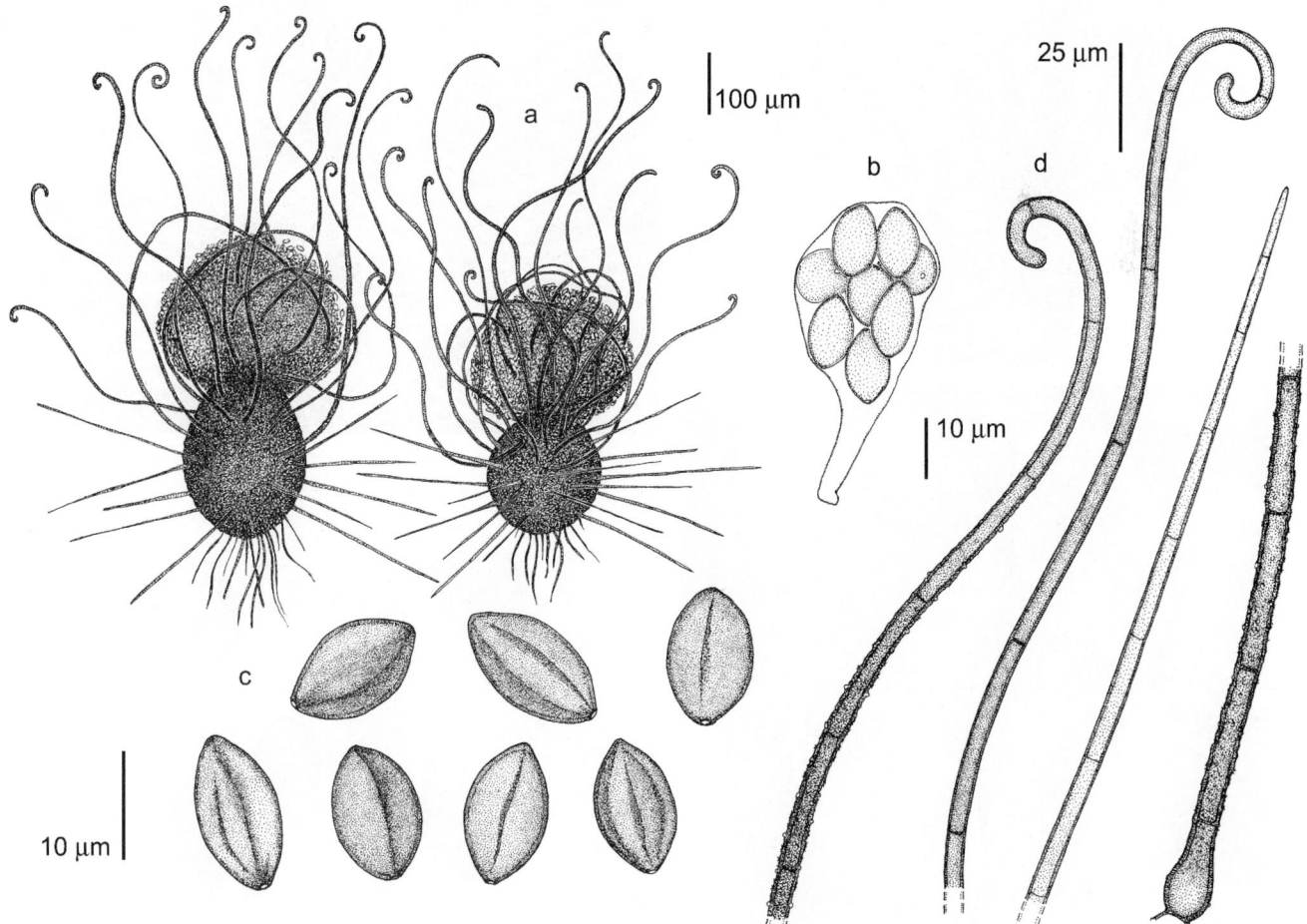

Chaetomium murorum, CBS 566.85. a. Ascomata; b. ascus; c. ascospores; d. ascomatal setae.

Chaetomium murorum, CBS 566.85. a, b. Ascomata; c, d. ascospores. a. ×150; b. ×210; c. ×2400; d. ×1450.

Ascomycota, Euascomycetes, Onygenales, Gymnoascaceae. Genus: *GYMNOASCUS*

Gymnoascus dankaliensis (Castell.) v. Arx

Colony characteristics. Colonies (OA) growing moderately rapidly, reddish-yellow or orange.

Microscopy. Ascomata are clumps of ascospores surrounded by undifferentiated peridial hyphae, 80-600 µm diam, orange-yellow to brownish-orange. Asci 8-spored, spherical to subspherial, 8-12 µm diam. Ascospores lenticular with equatorial and polar thickenings, (3.0-) 4.3-5.3 × 6-8 µm, yellow to orange or red-brown.

Pathogenicity. BSL-2. This species was implicated in cases of otomycosis (Batista *et al.*, 1960b), skin lesions (Vanbreuseghem & de Vroey, 1966; Sanyal *et al.*, 1971) and in onychomycosis (Summerbell *et al.*, 1989a).

Reference. Currah (1985).

Nomenclature. *Trichophyton dankaliense* Castellani - J. Trop. Med. Hyg. 40: 313, 1937 ≡ *Arachniotus dankaliense* (Castellani) van Beyma - Antonie van Leeuwenhoek 8: 105, 1942 ≡ *Gymnascella dankaliense* (Castellani) Currah - Mycotaxon 24: 77, 1985 ≡ *Gymnoascus dankaliensis* (Castellani) von Arx - Persoonia 13: 177, 1986.

    Pseudoarachniotus roseus Kuehn - Mycologia 49: 695, 1957.

    Arachniotus flavoluteus Kuehn & Orr - Mycologia 51: 864, 1959.

    Waldemaria pernambucensis Batista, Maia & Cavalcanti - Atas Inst. Micol., Recife 1: 6, 1960.

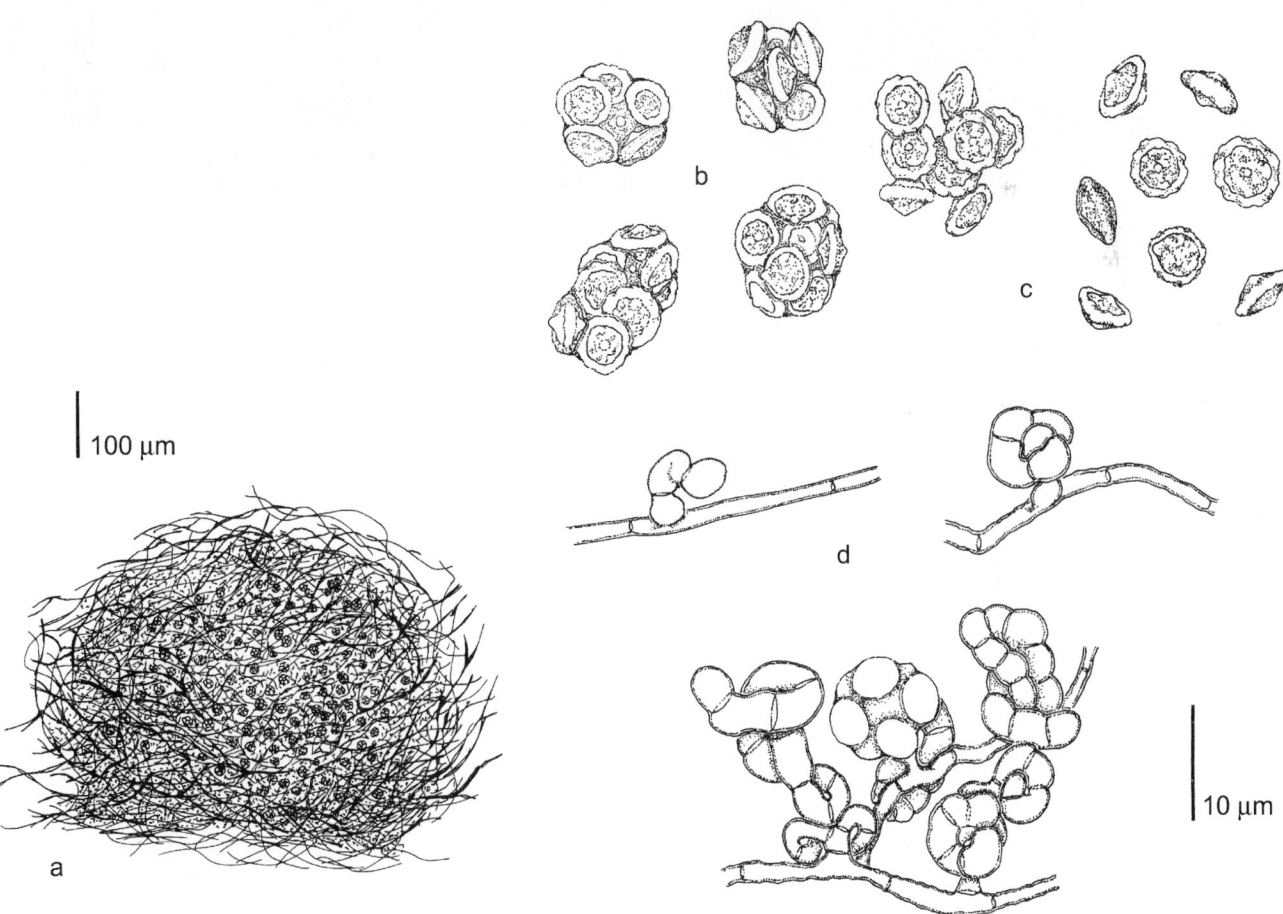

Gymnoascus dankaliensis, UAMH 3552. a. Ascoma; b. asci; c. ascospores; d. ascoma initials.

271

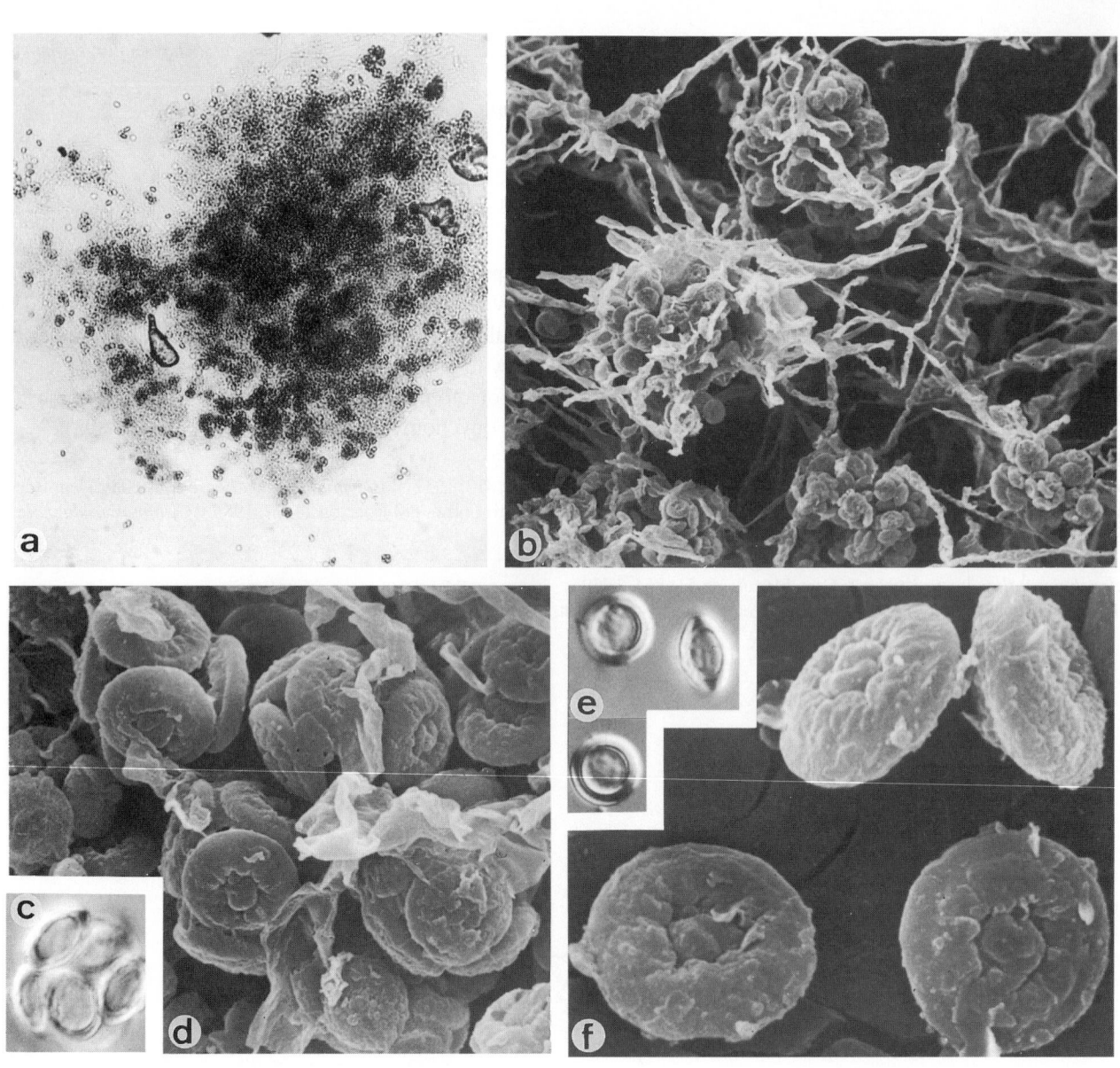

Gymnoascus dankaliensis, UAMH 3552. Ascomata, asci and ascospores. a. ×128; b. ×630; c. ×1600; d. ×2750; e. ×1600; f. ×5300.

Ascomycota, Euascomycetes, Pleosporales, Leptosphaeriaceae. Genus: *LEPTOSPHAERIA*

Generic description. Colonies growing rather slowly. Ascomata sperical, without ostiole, dark brown to black; peridium pseudoparenchymatous, composed of several layers of thick-walled cells. Paraphyses present. Asci elongate, bitunicate. Ascospores (sub)hyaline, with several transverse septa, with a slime sheath at liberation.
References. Shoemaker (1984), Khashnobish & Shearer (1996).

Key to the treated species of Leptosphaeria:

1a. Ascospores mostly 4-septate, with rounded ends . . . . **L. senegalensis (275)**
1b. Ascospores mostly 6-septate, with pointed ends . . . . **L. thompkinsii (273)**

Leptosphaeria thompkinsii El-Ani

Colony characteristics. Colonies (OA) growing slowly, woolly, dark olivaceous-grey; reverse blackish.
Microscopy. Ascomata developing after 1-2 months incubation, solitary, black, (sub)spherical, 220-530 μm diam. Peridium composed of brown, thick-walled, isodiametric cells, covered with hyphae in the upper part. Asci between paraphyses, clavate to cylindrical, 80-115 μm, bitunicate, 8-spored. Ascospores hyaline, fusiform, 32-45 × 9-11 μm, usually 6-7 septate, slightly constricted at septa, with thick, slimy sheath.
Pathogenicity. BSL-2. Described as a rare agent of human mycetoma (El-Ani, 1966).
Reference. El-Ani (1966).
Nomenclature. *Leptosphaeria thompkinsii* El-Ani - Mycologia 58: 409, 1966.

Leptosphaeria senegalensis Segretain *et al.*

Colony characteristics. Colonies (OA) growing slowly, woolly, dark olive with grey margin; reverse nearly black.
Microscopy. Ascomata developing after 1-2 months of incubation, solitary, black, (sub)spherical, 100-300 µm diam. Peridium composed of brown, thick-walled, isodiametric cells, covered with hyphae in the upper part. Asci between paraphyses, clavate, 80-160 × 17-28 µm, bitunicate, 8-spored. Ascospores subhyaline, obovoidal to ellipsoidal, 23-36 × 8.0-13.5 µm, usually 4-septate, constricted at septa, with thick, slimy sheath.
Pathogenicity. BSL-2. The species is one of the agents of human mycetoma in Central Africa (Segretain *et al.*, 1959). Grains are black with pale centre, soft, irregular in shape, about 1 mm diam.
Reference. El-Ani & Gordon (1965).
Nomenclature. *Leptosphaeria senegalensis* Segretain, Baylet, Darasse & Camain - C.R. Hebd. Séanc. Acad. Sci., Paris 248: 3732, 1959.

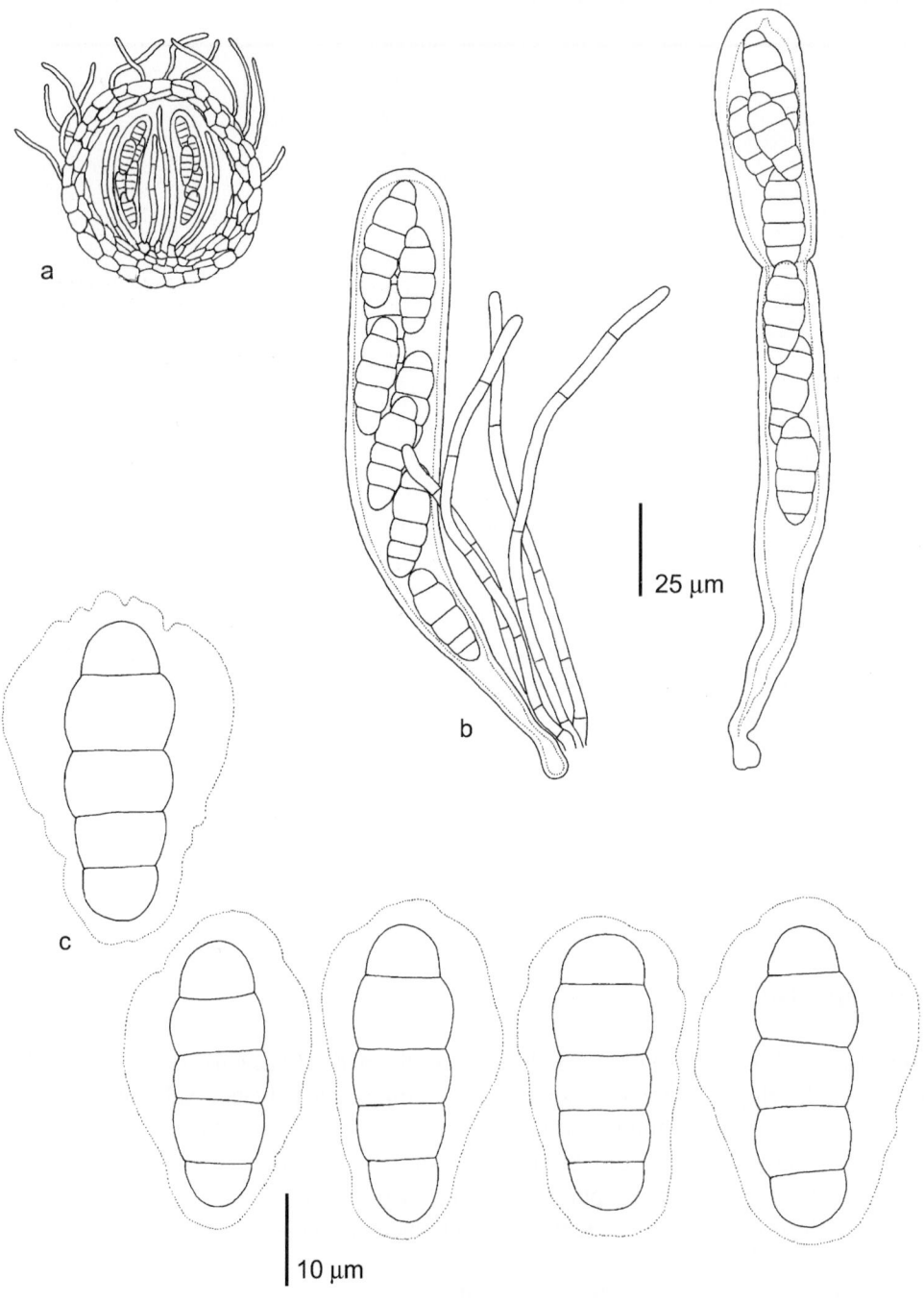

Leptosphaeria senegalensis, CBS 196.79. a. Diagram of cleistothecium; b. asci with paraphyses, right ascus liberating ascospores; c. ascospores.

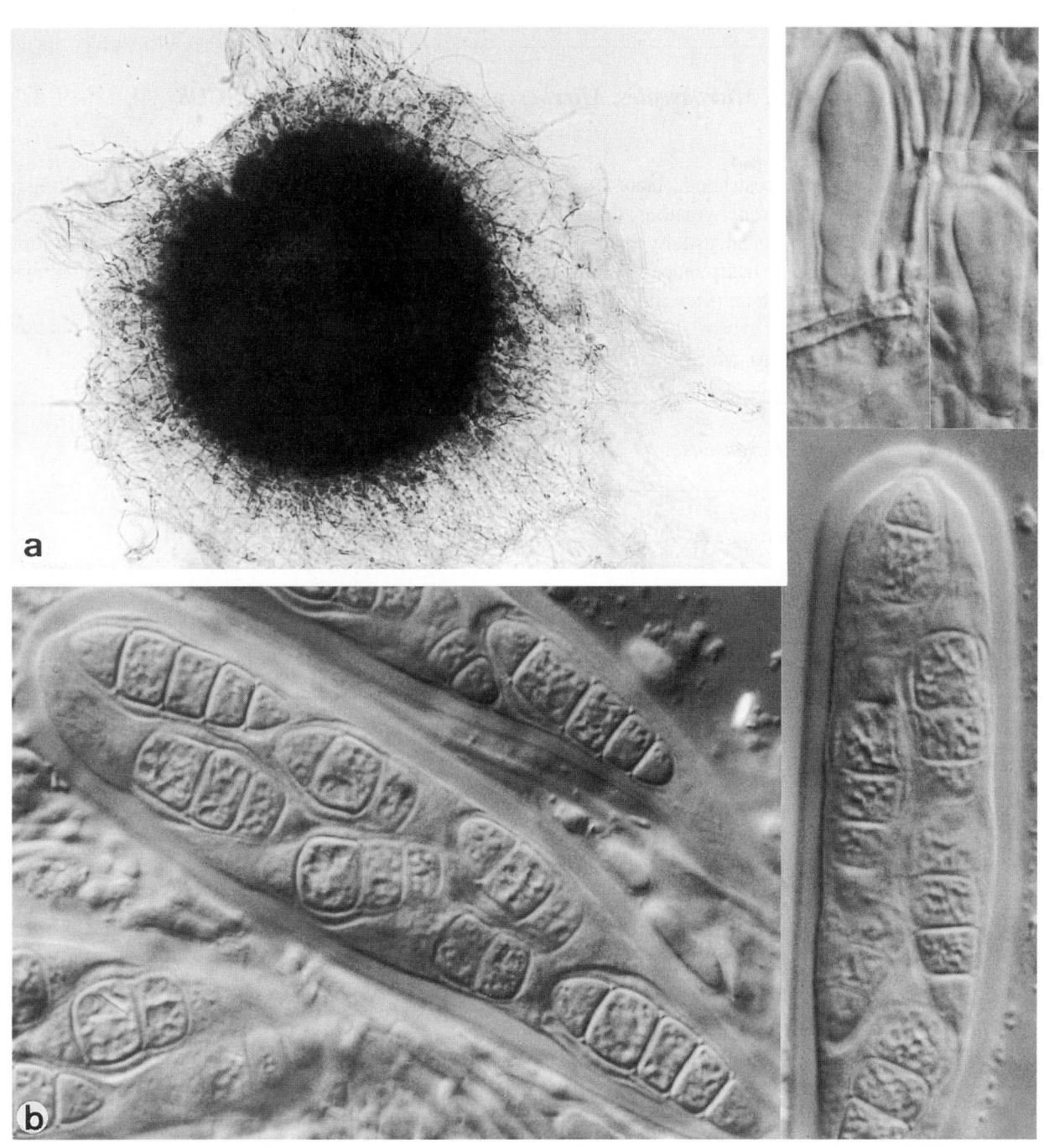

Leptosphaeria senegalensis, CBS 196.79. Cleistothecium and asci in different states of development. a. ×128; b. ×1600.

Ascomycota, Euascomycetes, Microascales, Microascaceae. Genus: *MICROASCUS*

Generic description. Colonies restricted. Ascomata spherical or ampulliform, with a cylindrical, short or long neck, dark brown; peridium pseudoparenchymatous, dark brown, composed of thick-walled, slightly flattened cells. Asci obovoidal, barrel-shaped or spherical, usually formed in basipetal rows, 8-spored, evanescent. Ascospores asymmetrical, short, navicular, reniform or heart-shaped in lateral view, dextrinoid when young, straw-coloured when mature, non-septate, smooth-walled, with an often inconspicuous germ pore.
Anamorphs. Genera *Scopulariopsis* (p. 902), *Wardomyces*, *Wardomycopsis*.
References. Barron *et al.* (1961a), Morton & Smith (1963), von Arx *et al.* (1988).

Key to the treated species of Microascus:

1a. Ascospores about twice as long as wide (5-7 × 2.5-4.0 µm), usually navicular
in lateral view . . . . *M. cinereus* (277)
1b. Ascospores less than twice as long as wide, usually heart-shaped or broadly reniform → 2
2a. Conidia 3-5 µm diam . . . . *M. cirrosus* (279)
2b. Conidia 5-6 µm diam . . . . *M. manginii* (281)

Microascus cinereus (Emile-Weil & Gaudin) Curzi

Colony characteristics. Colonies (OA) restricted, greyish, brownish or olivaceous.

Microscopy. Ascomata spherical or ampulliform, black, 170-340 µm diam, with a short-cylindrical ostiolar beak. Asci obovoidal, barrel-shaped or nearly spherical, 8-spored, 11-15 × 7-11 µm. Ascospores navicular in lateral view, pale reddish-brown when mature, 5-7 × 2.5-4.0 µm, with an apical germ pore.

Anamorph. *Scopulariopsis cinereus* Emile-Weil & Gaudin.

Conidiogenous cells cylindrical or slightly tapering, annellidic, 6-13 × 2.0-3.5 µm. Conidia pale yellowish or brownish, catenate, obovoidal to clavate, 3.5-5.0 × 3-4 µm, with truncate base.

Pathogenicity. BSL-1. Although this species has been isolated from clinical sources such as cutaneous lesions and cases of onychomycosis, chromoblastomycosis and mycetoma, its pathogenicity is questionable (Lacey, 1986). A confirmed case of suppurative cutaneous granuloma was published by Marques *et al.* (1995), and a secondary endocarditis by Celard *et al.* (1999), and a brain abscess in a BMT recipient by Baddley *et al.* (2000). .

Reference. Von Arx *et al.* (1988).

Nomenclature. *Scopulariopsis cinereus* Emile-Weil & Gaudin - Archs Méd. Exp. Anat. Path., Paris 28: 452, 1919 ≡ *Microascus cinereus* (Emile-Weil & Gaudin) Curzi - Boll. Staz. Pathol. Veg. Roma 11: 60, 1931.

    Microascus lunasporus Jones - Mycologia 28: 503, 1934.
    Scopulariopsis lunasporus Jones - Mycologia 28: 504, 1934.
    Microascus pedrosoi Fuentes & Wolf - Mycologia 48: 63, 1956.

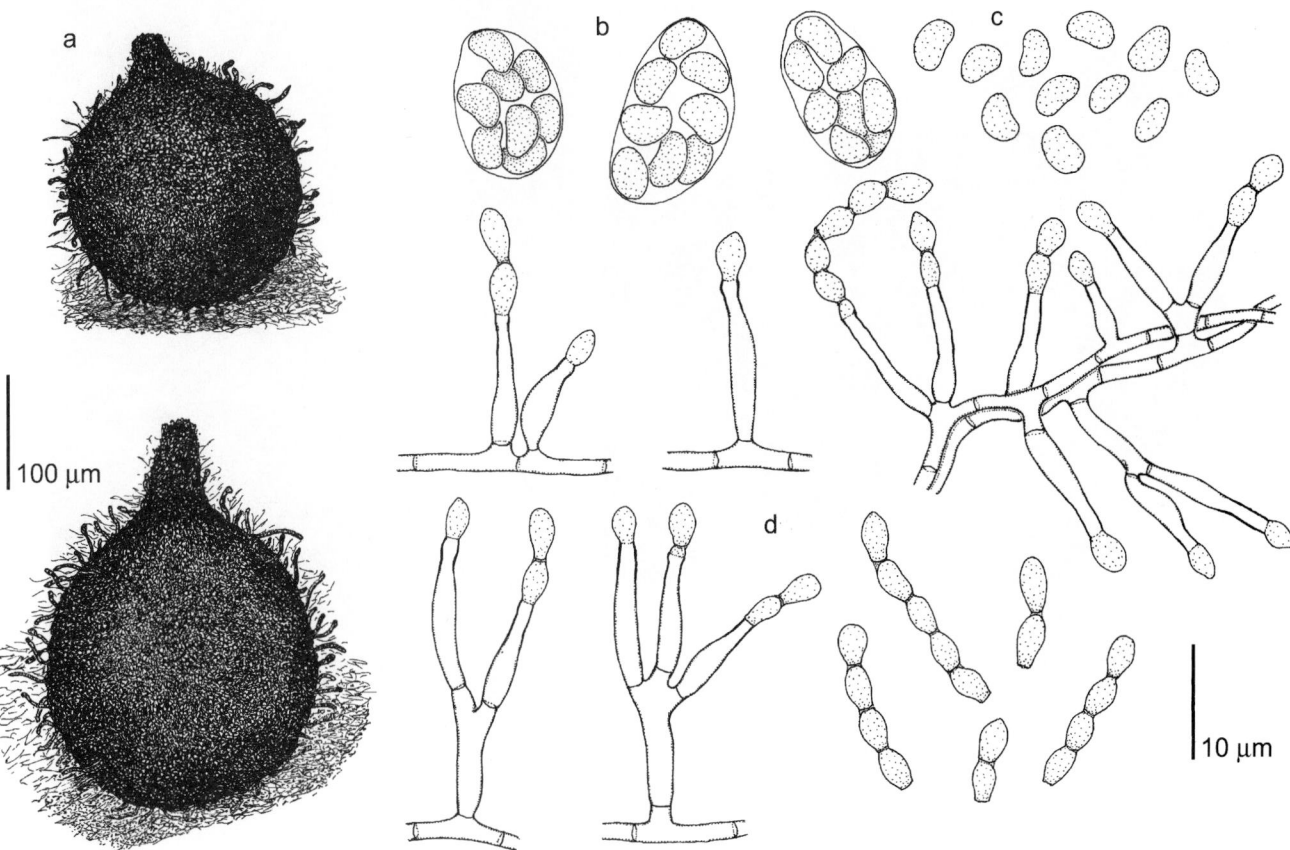

Microascus cinereus, CBS 664.71. a. Ascomata; b. asci; c. ascospores; d. *Scopulariopsis* anamorph: annellated conidiogenous cells, some on conidiophores, and conidia.

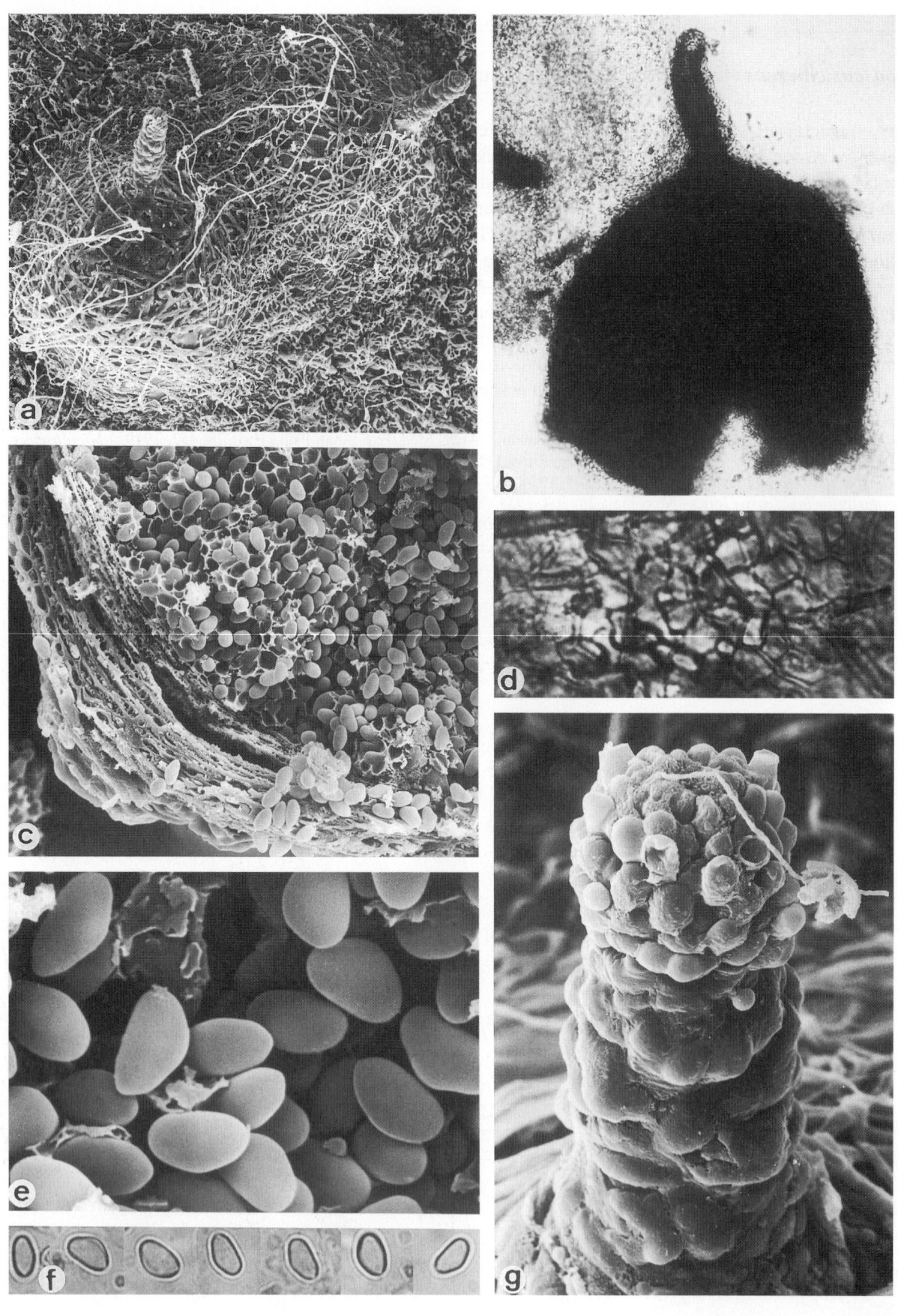

Microascus cinereus, CBS 807.73. a, b. Ascomata; c. cross-section of peridium; d. part of peridium; e, f. ascospores; g. detail of ascomal neck. a. ×245; b. ×200; c. ×1050; d. ×1280; e. ×4700; f. ×1280; g. ×1950. Reproduced with permission from Beih. Nova Hedwigia 94: 94, 1988.

278

Microascus cirrosus Curzi

Colony characteristics. Colonies (OA) restricted, brownish-grey.
Microscopy. Ascomata subspherical with a relatively long, cylindrical or tapering beak, bearing a whorl of apical hairs, 140-230 µm diam. Asci obovoidal or nearly spherical, 8-spored, 9-12 × 8-11 µm. Ascospores broadly reniform, straw-coloured when mature, 5-6 × 3-4 µm, with an inconspicuous germ pore.
Anamorph. *Scopulariopsis paisii* (Poll.) Nann.
Conidiogenous cells cylindrical, 10-20 × 2.3-5 µm, annellidic. Conidia in long, dry chains, broadly clavate, 4-6 × 3.5-5 µm, with a truncate base, pale yellow when mature.
Molecular diagnostics. SSU restriction map based on NCBI M89994:

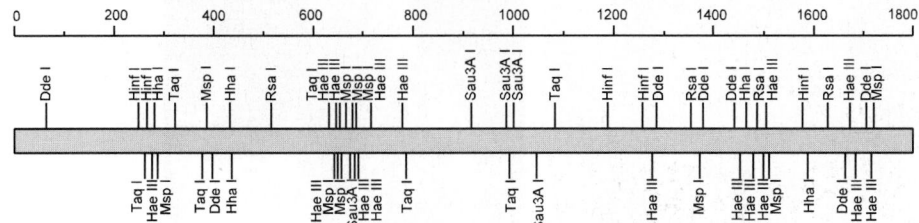

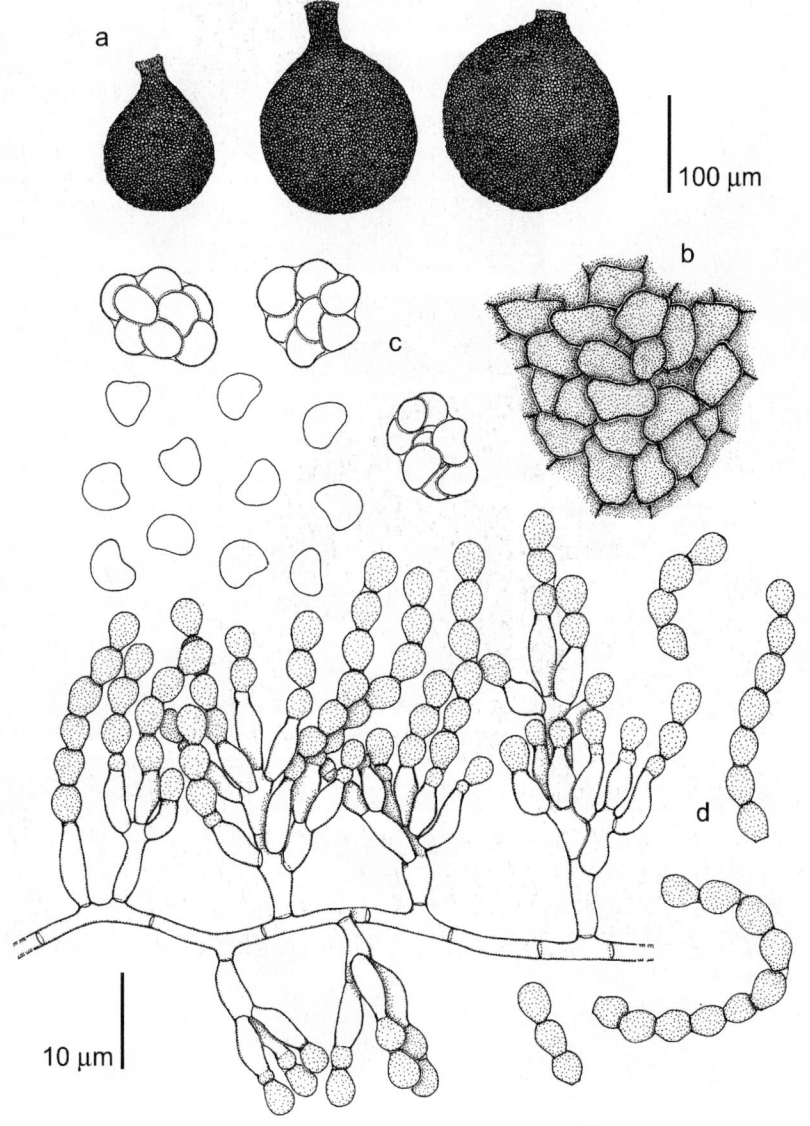

Microascus cirrosus, CBS 217.31. a. Ascomata; b. part of peridium; c. asci and ascospores; d. *Scopulariopsis* anamorph: conidiophores with annellated conidiogenous cells and conidia.

279

Pathogenicity. BSL-1. This species has been reported as the etiologic agent in cases of onychomycoses (Schönborn & Schmoranzer, 1970; de Vroey *et al.*, 1992) and also in connection with cattle abortion (Austwick & Venn, 1962). Krisher *et al.* (1995) and Toubas *et al.* (2000) reported systemic infections in immunocompromised patients

Reference. Von Arx *et al.* (1988).

Nomenclature. *Microascus cirrosus* Curzi - Boll. Staz. Pat. Veg., Roma 10: 302, 1910.

Peristomium desmosporum Lechmère - Bull. Trimest. Soc. Mycol. Fr. 39: 307, 1913 ≡ *Microascus desmosporus* (Lechmère) Curzi - Boll. Staz. Pat. Veg., Roma 11: 60, 1931.

Torula paisii Pollacci - Atti Ist. Bot. Univ. Lab. Crittogam. Pavia, Ser. 2, 18: 130, 1921 ≡ *Phaeoscopulariopsis paisii* (Pollacci) Ota - Jpn. J. Derm. Urol. 28: 405, 1928 ≡ *Scopulariopsis paisii* (Pollacci) Nannizzi - Tratt. Micopat. Um. 4: 259, 1934.

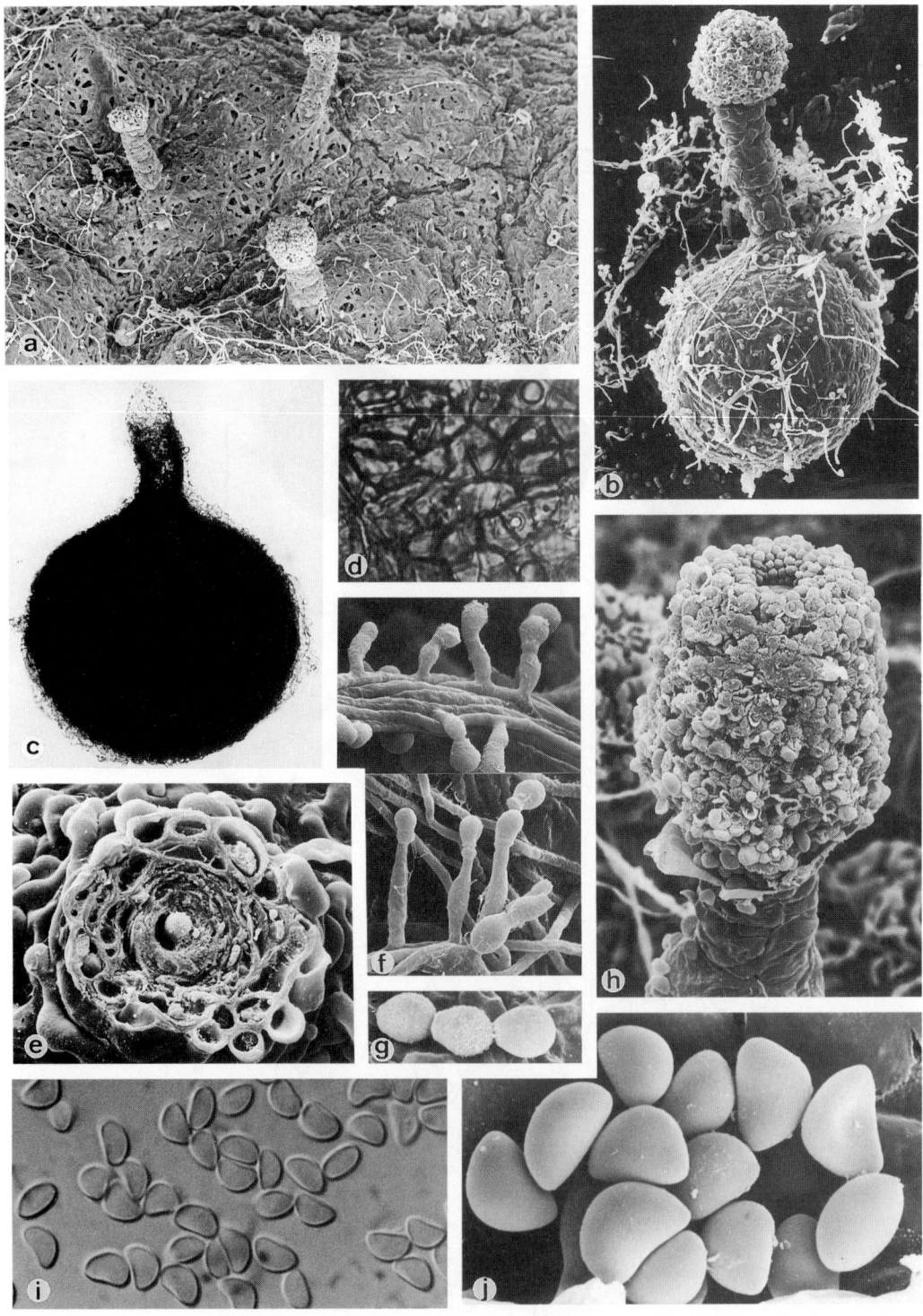

Microascus cirrosus, CBS 217.31. a. Erumpent beaks of ascomata; b, c. ascomata; d. peridium in surface view; e, h. apical parts of ostiolar beaks; f, g. annellated conidiogenous cells and conidia of the *Scopulariopsis* anamorph; i, j. ascospores. a. ×80; b. ×165; c. ×155; d. ×650; e. ×1225; f. ×1610; g. ×1695; h. ×500; i. ×820; j. ×1635. Reproduced with permission from Beih. Nova Hedwigia 94: 95, 1988.

Microascus manginii (Loubière) Curzi

Colony characteristics. Colonies (OA) restricted, pale.
Microscopy. Ascomata spherical, black, with pore in the apical papilla (not beaked), smooth-walled, 100-190 μm diam. Asci obovoidal, broadly clavate or nearly spherical, 8-spored, 11-16 × 8-13 μm. Ascospores heart-shaped in lateral view, pale straw-coloured when mature, 5-6 × 4.5-5.0 μm, with an inconspicuous germ pore.
Anamorph. *Scopulariopsis candida* (Guéguen) Vuill.
Conidiogenous cells borne on aerial hyphae or on short conidiophores, cylindrical, later with indistinct annellations, 11-22 × 2-4 μm. Conidia in dry chains, obovoidal, with a truncate base, smooth-walled, hyaline, 6-8 × 5-6 μm.
Pathogenicity. BSL-1. Isolated in several cases of onychomycoses (Frágner, 1966; Krempl-Lamprecht, 1970) and in a case of disseminated infection in a leukemic patient who had undergone bone narrow transplantation (Neglia *et al.*, 1987, identified as *S. candida* by Anaissie *et al.*, 1989).
Reference. Von Arx *et al.* (1988).
Antifungal susceptibility.

Antifungal	Mean MICs	Strains	Reference
AMB	2	1	Aguilar *et al.* (1998b)
5FC	256	1	Aguilar *et al.* (1998b)
FLZ	128	1	Aguilar *et al.* (1998b)
ITZ	32	1	Aguilar *et al.* (1998b)
KTZ	2	1	Aguilar *et al.* (1998b)
MCZ	4	1	Aguilar *et al.* (1998b)

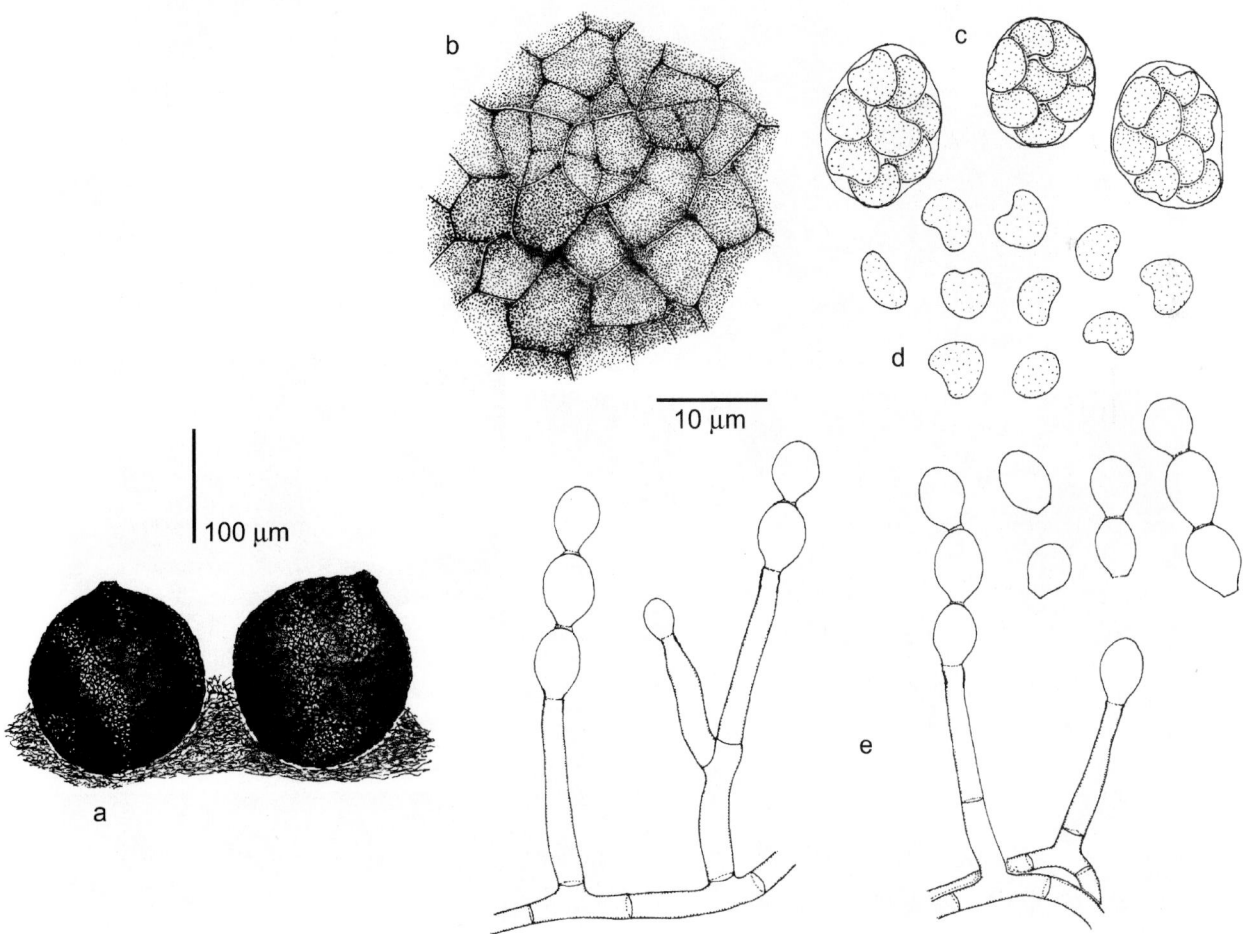

Microascus manginii (*Scopulariopsis candida*), CBS 667.71. a. Ascomata; b. part of peridium; c. asci; d. ascospores; e. anamorph: annellated conidiogenous cells and conidia.

281

Nomenclature. *Monilia candida* Guéguen - Bull. Soc. Mycol. Fr. 15: 271, 1899 ≡ *Scopulariopsis candida* (Guéguen) Vuillemin - Bull. Soc. Mycol. Fr. 27: 143, 1911.

Nephrospora manginii Loubière - C.R. Hebd. Séanc. Acad. Sci., Paris 177: 211, 1923 ≡ *Microascus manginii* (Loubière) Curzi - Bull. Staz. Pat. Veg., Roma 11: 60, 1931.

Chrysosporium keratinophilum (Frey) Carmichael var. *denticola* C. Moreau - Mycopath. Mycol. Appl. 37: 37, 1969 ≡ *Basipetospora denticola* (C. Moreau) C. Moreau - Bull. Soc. Mycol. Fr. 87: 43, 1971.

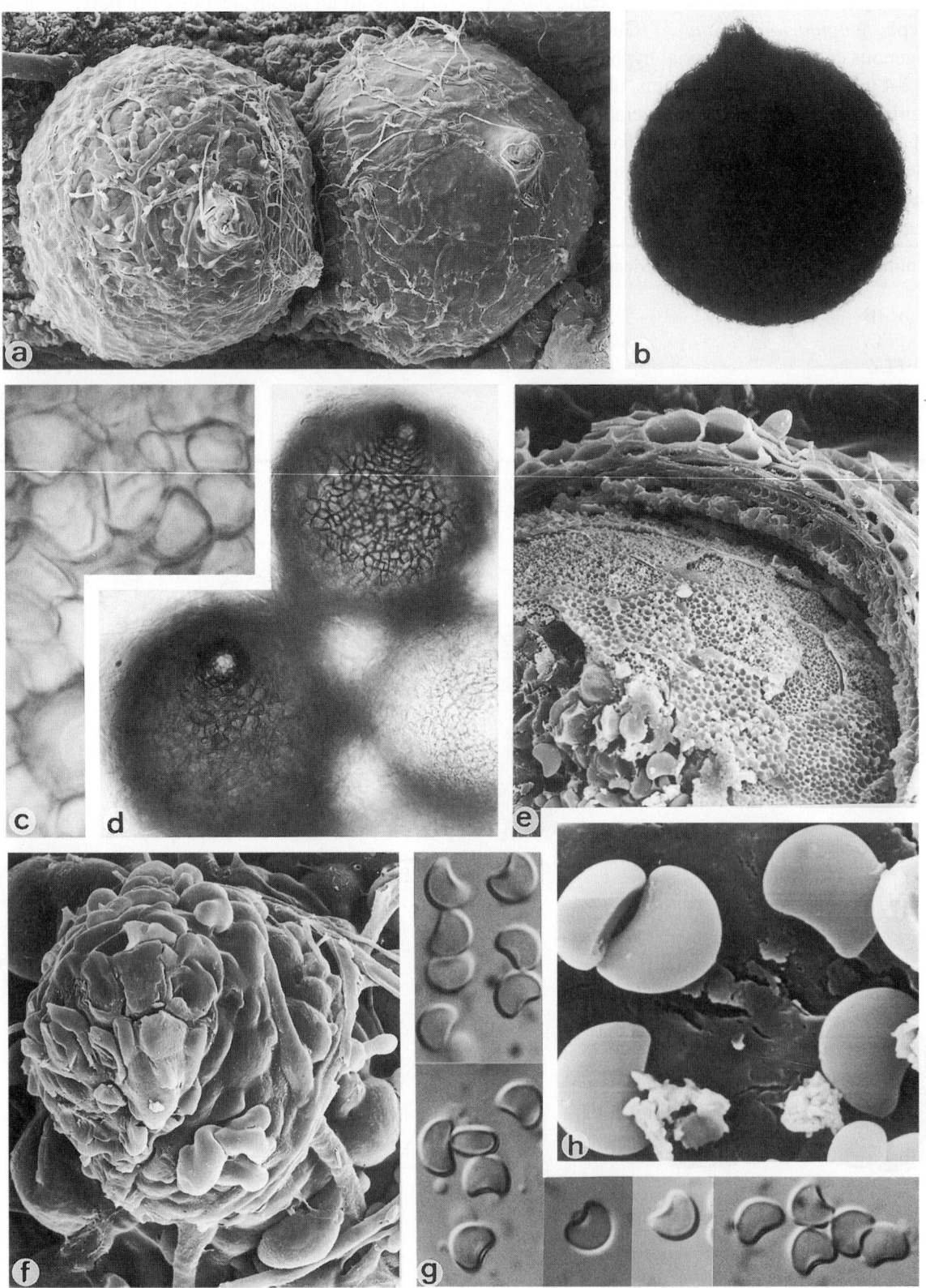

Microascus manginii, CBS 816.73. a, b, d. Ascomata; c. peridium in face view; e. peridium in cross section; f. ascomal papilla; g, h. ascospores. a. ×350; b. ×320; c. ×1600; d. ×320; e. ×1400; f. ×1500; g. ×1600; h. ×5000. Reproduced with permission from Beih. Nova Hedwigia 94: 96, 1988.

Ascomycota, Euascomycetes, Eurotiales, Monascaceae. Genus: *MONASCUS*

Monascus ruber v. Tiegh.

Colony characteristics. Colonies (MEA) expanding, flat, brownish.

Microscopy. Cleisthothecia borne terminally on an up to 150 µm long stalk, spherical, 30-60 µm diam, subhyaline to reddish-brown, with a wall composed of interwoven, flattened hyphal elements (*textura intricata*). Asci (sub)spherical, 7.5-10.0 µm in diam, evanescent. Ascospores yellowish, smooth-walled, hyaline, ovoidal to ellipsoidal, 5-6 × 4.0-4.5 µm.

Anamorph. *Basipetospora rubra* Cole & Kendrick.

Conidia thin-walled, hyaline, borne basipetally in chains, obpyriform with truncate base, 6-8 × 5-6 µm, finally inflating and becoming thick-walled. Additional arthroconidia frequently present.

Physiology. Growth at 37°C.

Pathogenicity. BSL-1. Species mostly occurring in food-stuffs and in soil. Sigler *et al.* (1999b) reported a kidney infection after surgery.

References. Cole & Kendrick (1968), Hawksworth & Pitt (1983).

Nomenclature. *Monascus ruber* van Tieghem - Bull. Soc. Bot. Fr. 31: 227, 1884.

   Basipetospora rubra Cole & Kendrick - Can. J. Bot. 46: 991, 1968.

   Coccidioides rosea Batista, Maia, Shome & Oliveira - Revta Fac. Med. Univ. Ceará 3: 71, 1963.

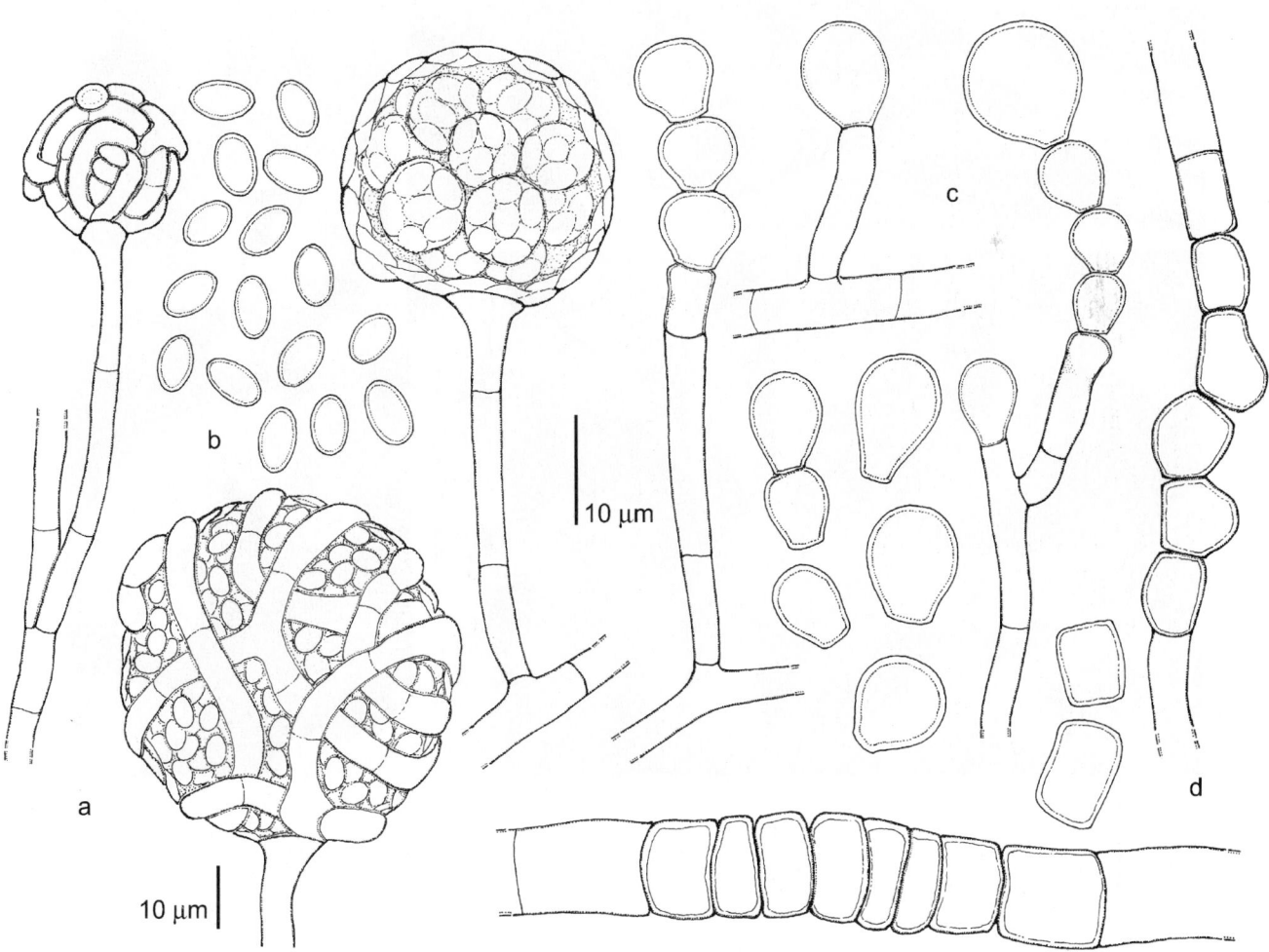

Monascus ruber, CBS 135.60. a. Pedunculate ascomata; b. ascospores; c. blastic conidia; d. arthric conidia.

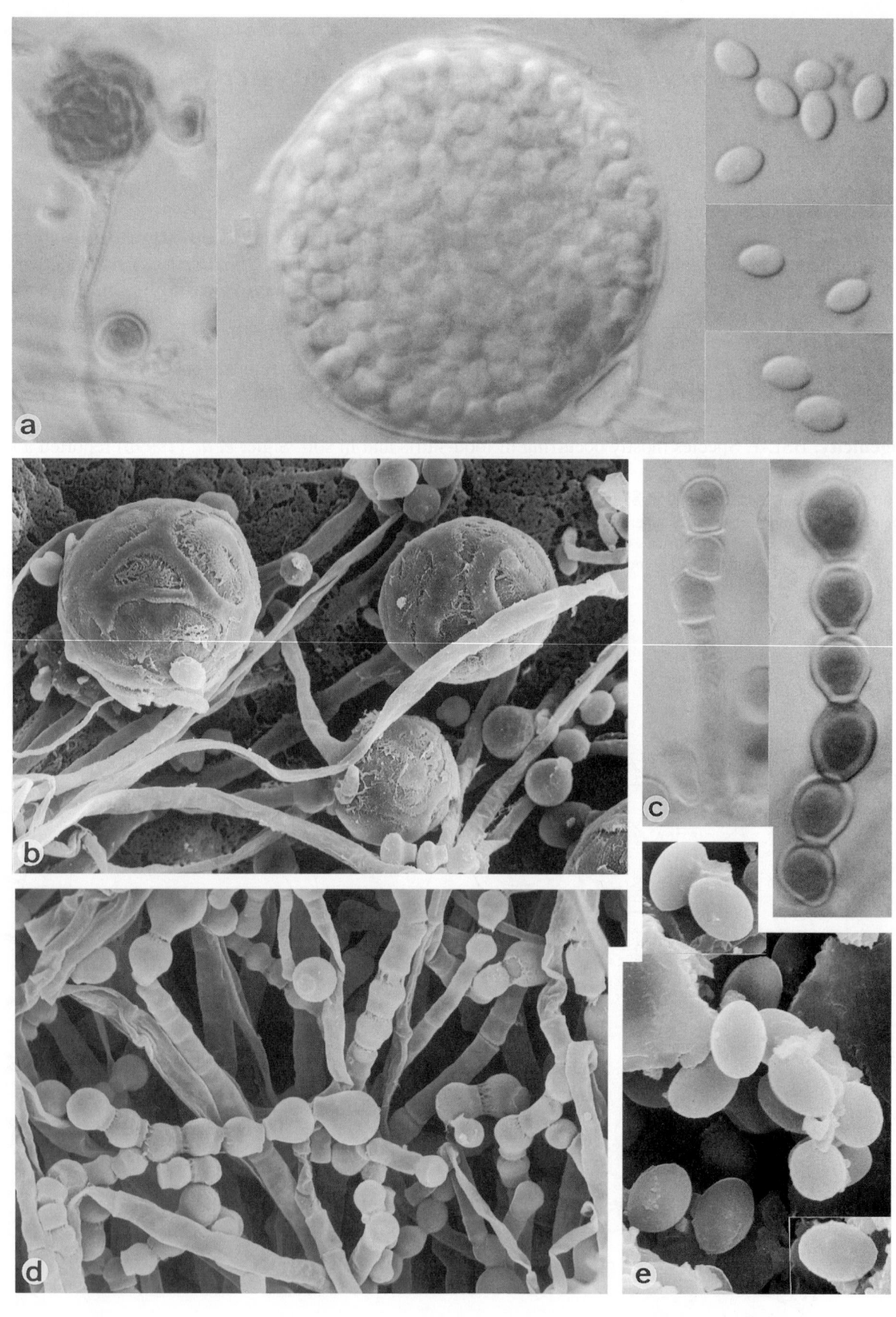

Monascus ruber, CBS 135.60. a, b. Ascomata and ascospores; c, d. blastic conidiogenous cells and conidia; e. ascospores. a. ×1600; b. ×1250; c. ×1600; d. ×1450; e. ×3400.

Ascomycota, Euascomycetes, Sordariales, Myxotrichaceae. Genus: *MYXOTRICHUM*

Myxotrichum deflexum Berk.

Colony characteristics. Colonies (OA) growing rather slowly, floccose, greenish; reverse light purple.

Microscopy. Gymnothecia dark brown to black, spherical, 200-450 μm including appendages. Peridial hyphae dark brown, branched, septate, thick- and smooth-walled. Appendages dark brown, with stiff, monopodial branching, up to 200 μm long, with lateral branches; apices acute and hyaline. Asci hyaline, 8-spored, clavate, 14-22 × 7-8 μm. Ascospores hyaline to pale yellow, ovoidal to ellipsoidal, longitudinally striate, 3.6-5.5 × 2.4-2.9 μm.

Physiology. Intolerant to benomyl.

Molecular diagnostics. Sugiyama *et al.* (1999), using rDNA phylogeny, proved that the genus is unrelated to the *Onygenales* where it was maintained earlier, but its precise position is as yet unclear. ITS restriction map based on NCBI AF062814:

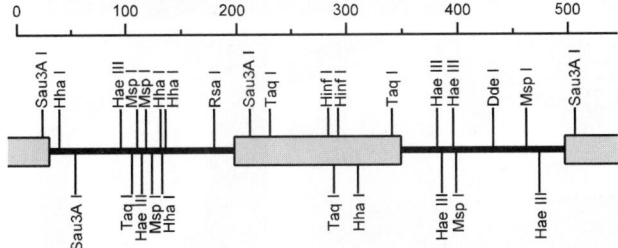

Pathogenicity. BSL-1. This species was reported as the probable cause of a case of onychomycosis (de Vroey, 1976). It is keratinophilic and occurs in soil.

References. Orr *et al.* (1963), Currah (1985).

Nomenclature. *Myxotrichum deflexum* Berkeley - Ann. Nat. Hist. 1: 260, 1838.

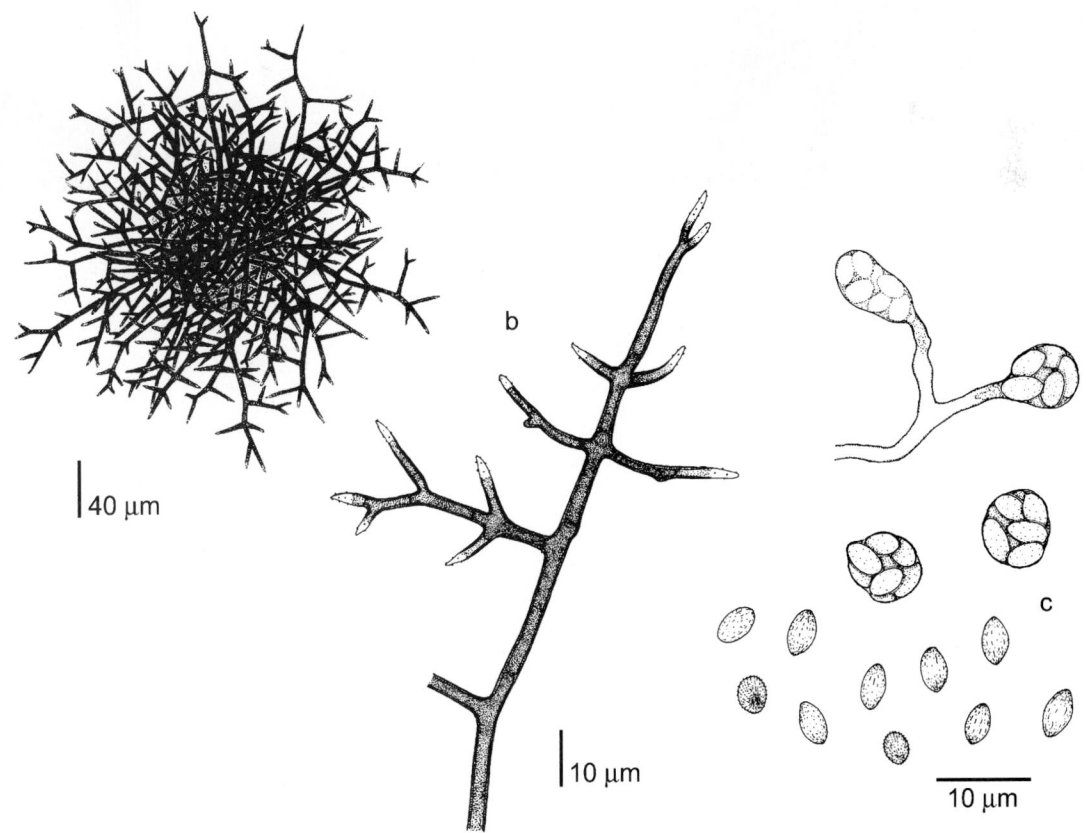

Myxotrichum deflexum, ATCC 15686. a. Ascoma; b. ascomatal appendage; c. asci and ascospores.

285

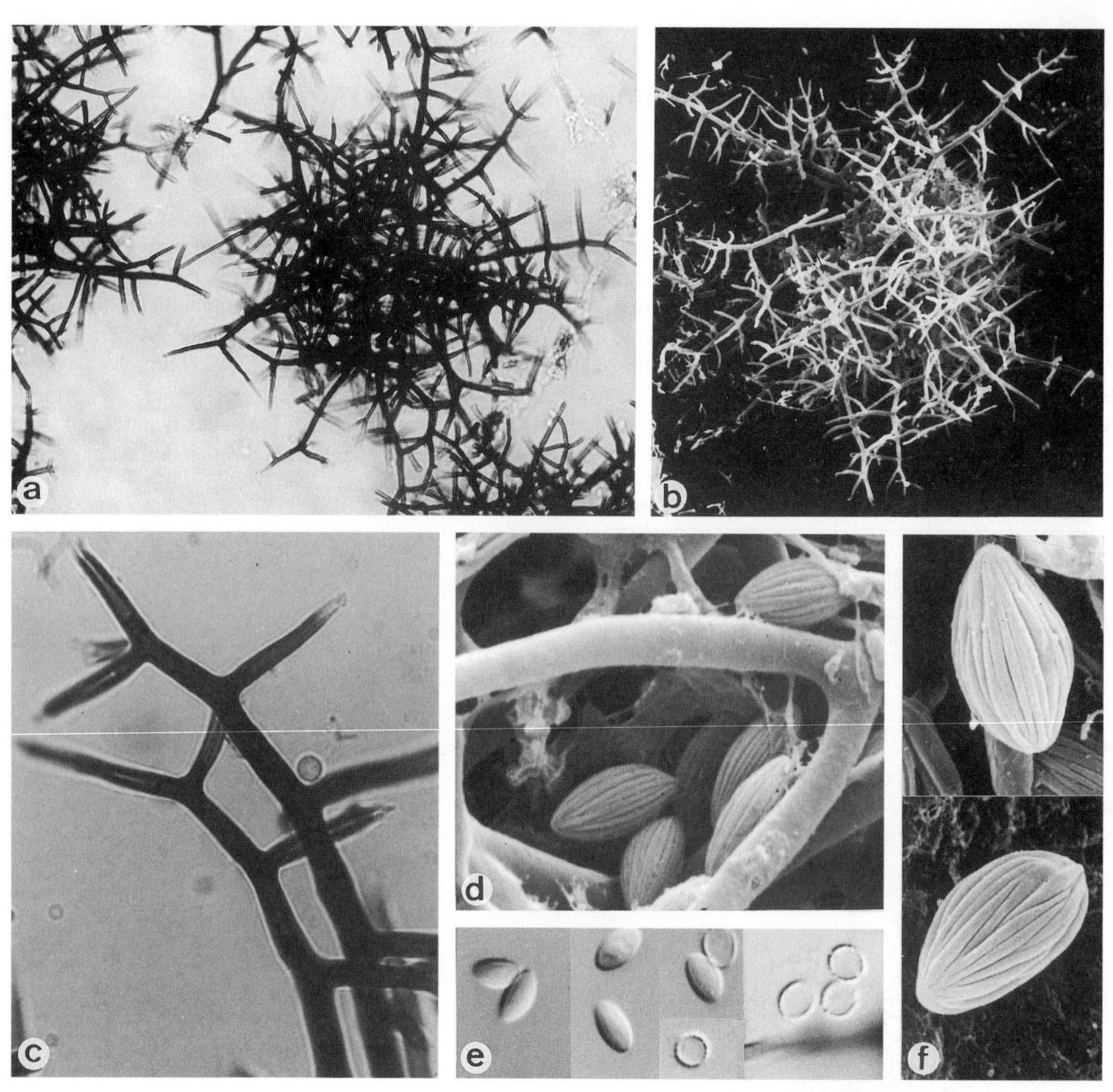

Myxotrichum deflexum, ATCC 15686. a, b. Ascomata; c. peridial appendages; d-f. asci and ascospores. a. ×256; b. ×340; c. ×1600; d. ×5600; e. ×1600; f. ×6000.

Ascomycota, Euascomycetes, Onygenales, Gymnoascaceae. Genus: *NARASIMHELLA*

Narasimhella hyalinospora (Kuehn *et al.*) v. Arx

Colony characteristics. Colonies (OA) expanding, yellowish to pale green, cottony, often with hyphal tufts.

Microscopy. Ascomata absent; spherical asci aggregated in dense groups which are covered by undifferentiated hyphae. Ascospores hyaline to pale yellow, ovoidal to oblate, with slight ornamentation, 2.5-3.0 × 2.2-2.5 µm. Anamorph absent.

Pathogenicity. BSL-1. Species mainly occurring on reptile and rodent dung. A pulmonary infection in a patient with AML was reported by Iwen *et al.* (1999a).

References. Kuehn *et al.* (1961), Currah (1985).

Nomenclature. *Pseudoarachniotus hyalinosporus* Kuehn, Orr & Ghosh - Mycopath. Mycol. Appl. 14: 217, 1961 ≡ *Narasimhella hyalinospora* (Kuehn, Orr & Ghosh) von Arx - Persoonia 6: 374, 1971 ≡ *Gymnascella hyalinospora* (Kuehn, Orr & Ghosh) Currah - Mycotaxon 24: 84, 1985.

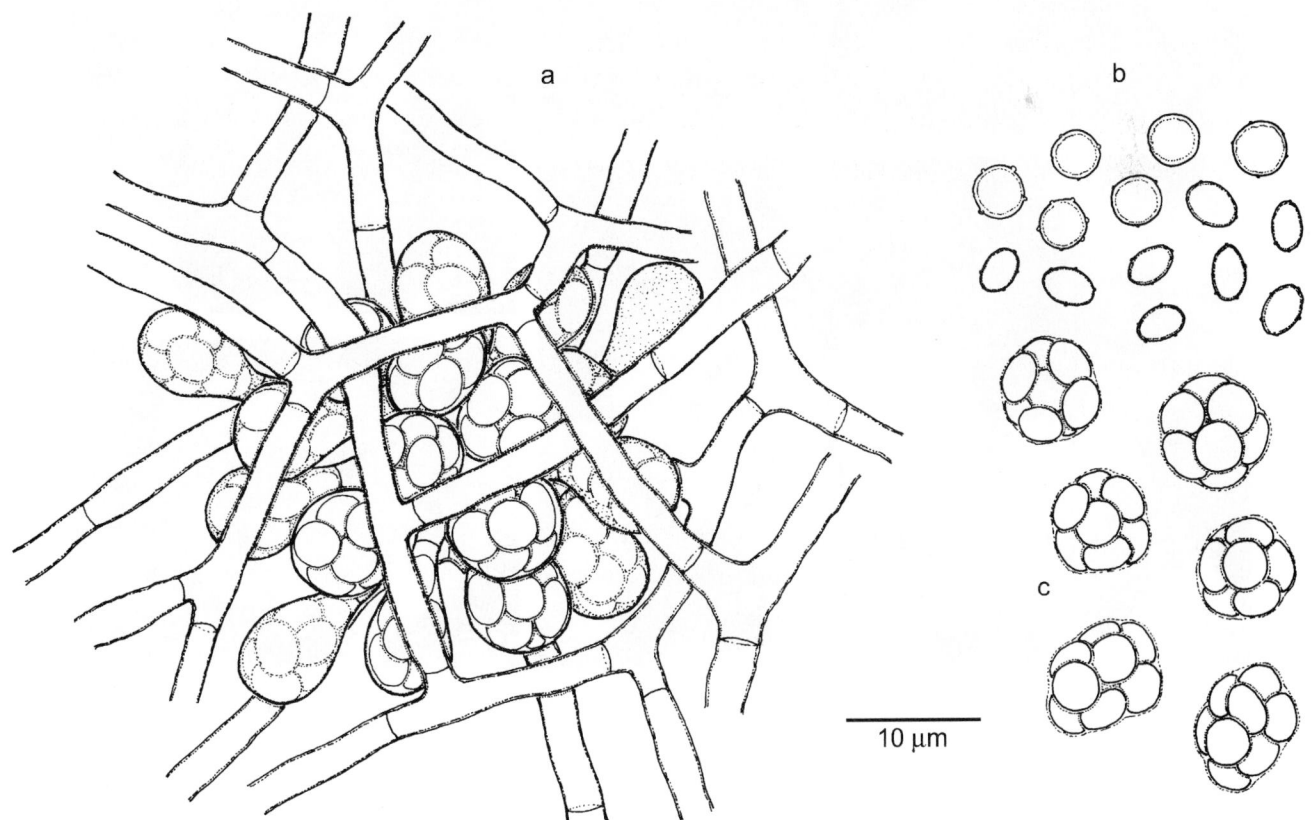

Narasimhella hyalinospora, FMR 6612. a. Part of an ascoma composed of peridial hyphae and young asci; b. ascospores; c. asci.

Narasimhella hyalinospora, FMR 6612. a Tufts of ascomata; b, d-g. asci and ascospores; c. ascomata. a. ×96; b. ×1600; c. ×575; d. ×2250; e. ×5000; f. ×5750; g. ×6000.

Ascomycota, Hypocreales, Hypocreaceae. Genus: *NEOCOSMOSPORA*

Neocosmospora vasinfecta E.F. Smith

Cultural characteristics. Colonies (SNA) expanding, flat and thin, transparent, punctate by the production of ascomata.

Microscopy. Ascomata (sub)spherical, orange-brown to red, with apical pore, smooth-walled, 280-480 µm in diam. Ascoma wall composed of several layers of *textura angularis*. Periphysoids evanescent. Asci cylindrical, 8-spored, 80-100 × 11-15 µm diam. Ascospores 1-celled, uniseriate, yellowish-brown, spherical to ellipsoidal, 10.0-15.5 × 7.5-12.0 µm, with thick walls and roughened without germ pores.

Anamorph. *Acremonium* sp.

Conidiogenous cells cylindrical, 30-100 µm long, up to 2 µm diam, arising from undifferentiated hyphae. Conidia hyaline, elongate to cylindrical, 5-13 × 2.0-3.5 µm, aggregating in slimy heads. Chlamydospores present.

Molecular diagnostics. ITS restriction map based on NCBI L36627:

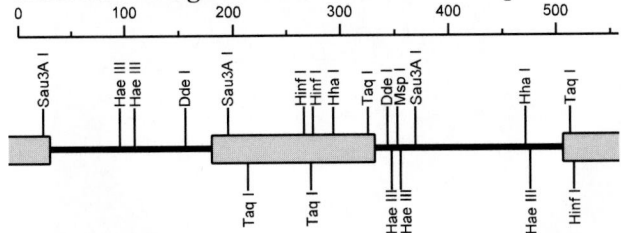

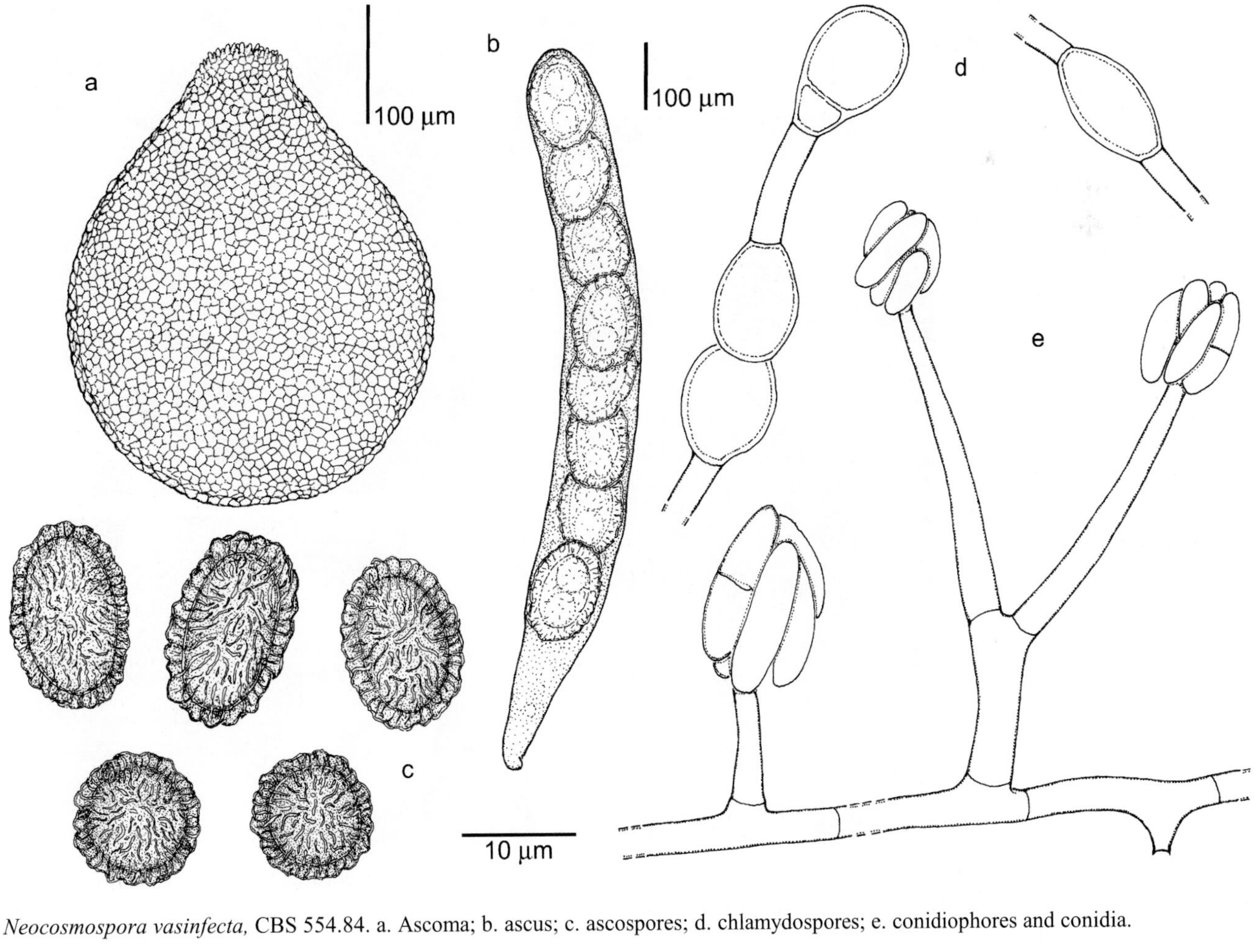

Neocosmospora vasinfecta, CBS 554.84. a. Ascoma; b. ascus; c. ascospores; d. chlamydospores; e. conidiophores and conidia.

Pathogenicity. BSL-1. This is a common soil-borne fungus in (sub)tropical climates. A systemic infection in a renal transplant recipient was reported by Chandenier *et al.* (1993) and Ben Hamida et al. (1993) and an osteoarthritis of traumatic origin by Kac *et al.* (1999).

References. Cannon & Hawksworth (1984), Udagawa *et al.* (1989).

Nomenclature. *Neocosmospora vasinfecta* E.F. Smith - Bull. U.S. Dep. Agric. 17: 45, 1889.

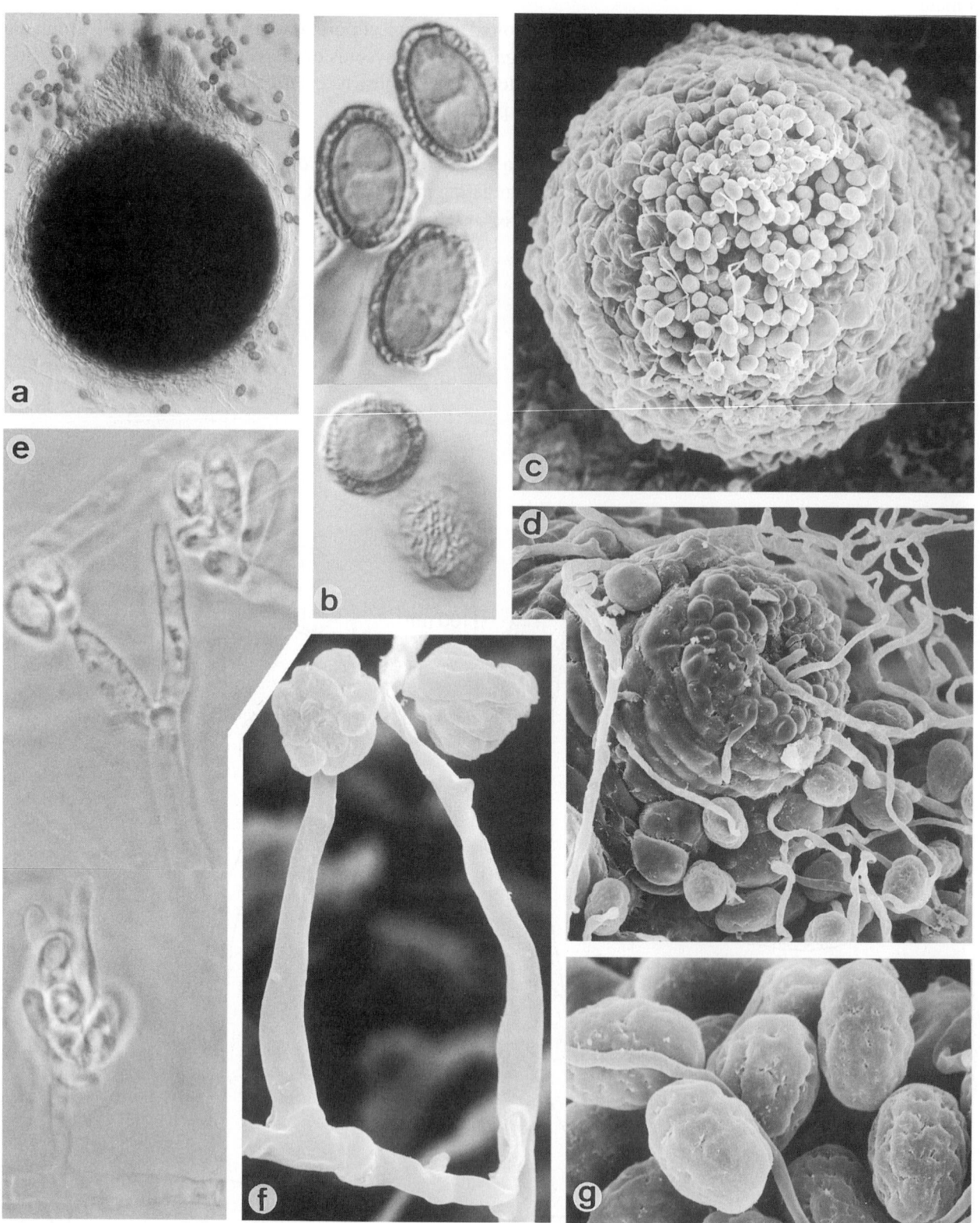

Neocosmospora vasinfecta, CBS 554.94. a-c, g. Ascomata and ascospores; d. detail of an ostiole; e, f. conidiophores and conidia. a.× 400; b. ×1600; c. ×330; d. ×900; e. ×1600; f. ×2400; g. ×2200.

Ascomycota, Euascomycetes, Eurotiales, Trichocomaceae. Genus: *NEOSARTORYA*

Generic description. Colonies expanding. Ascomata spherical, non-ostiolate, white, rarely orange, usually surrounded by a loose convering of hyaline to pale yellowish-brown hyphae. Peridium pseudoparenchymatous, composed of several layers of thin-walled, flattened cells. Asci 8-spored, spherical, irregularly disposed inside the ascoma, thin-walled, evanescent. Ascospores aseptate, without appendages, oblate or lenticular, hyaline to pale yellowish-brown, thick-walled, smooth-walled or verruculose to tuberculate, with two to four equatorial crests. Thermophilic.

Anamorph. *Aspergillus.*

Remark. The genus is phylogenetically close to *Aspergillus fumigatus* and its relatives (Girardin *et al.*, 1995; Varga *et al.*, 2000).

References. Raper & Fennell (1965), Kozakiewicz (1989).

Key to the treated species of Neosartorya:

1a.	Hemispheres of the ascospores reticulate or irregularly ribbed	*N. fischeri* **(292)**
1b.	Hemispheres of the ascospores spinulose, verruculose or tuberculate ➙ **2**	
2a.	Hemispheres of the ascospores strongly and irregularly ribbed	*N. pseudofischeri* **(294)**
2b.	Hemispheres of the ascospores spinulose	*N. spinosa* **(296)**

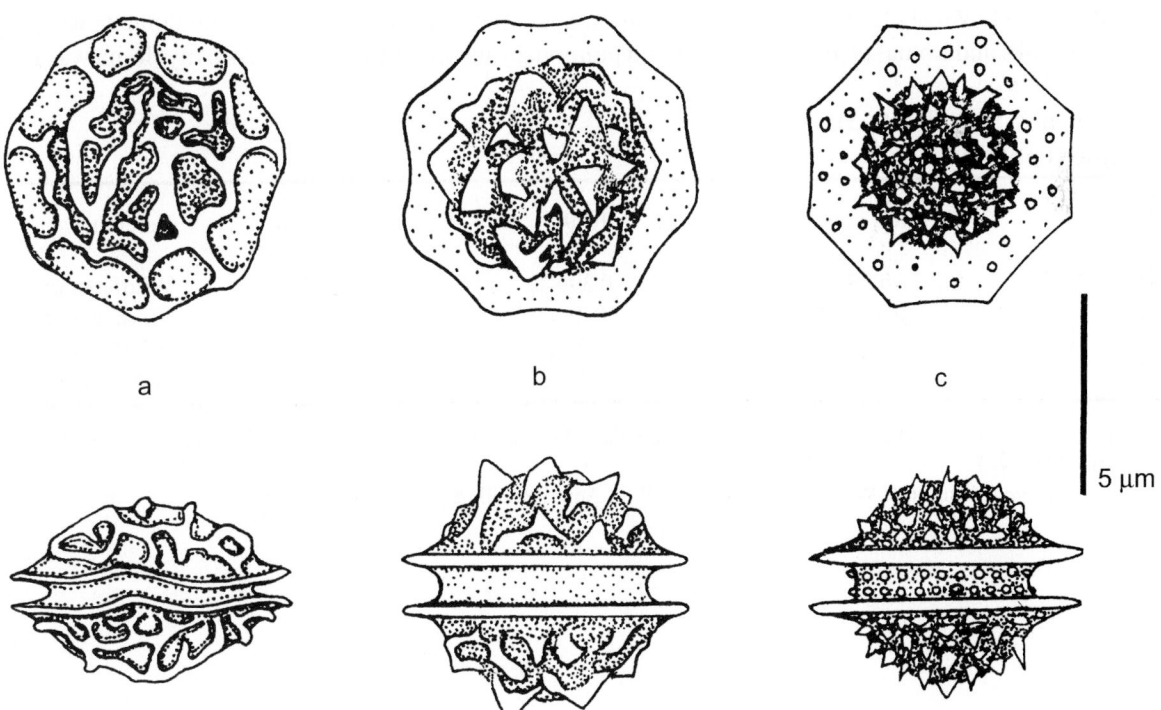

a   b   c

5 µm

Fig. 23. Diagram of ascospore shapes and ornamentations of *Neosartorya* species, based on scanning electron microscope data. a. *N. fischeri;* b. *N. pseudofischeri;* c. *N. spinosa.*

291

Neosartorya fischeri (Wehmer) Malloch & Cain

Colony characteristics. Colonies (OA) growing rapidly, farinose to felty, whitish, with greyish-blue shades, somewhat pinkish around ascomata.; reverse off-white to pinkish.

Microscopy. Ascomata borne singly or in small clusters within a loose hyphal envelope, spherical, pale pink, up to 400 μm diam; peridium pseudoparenchymatous, thin. Asci spherical to subspherical, 8-spored, 10-12 × 8-10 μm. Ascospores subspherical, with two prominent, flexuose, equatorial crests, about 5 μm diam (excluding the crests), hyaline, hemispheres bearing anastomosing ridges (reticulate).

Anamorph. *Aspergillus fischerianus* Samson & W. Gams.

Conidiophores smooth-walled, hyaline to pale green, 300-500 μm. Conidial heads columnar, terminally somewhat radiate, pale blue-grey, uniseriate. Vesicles flask-shaped, up to 18 μm diam. Phialides hyaline to pale greyish-green. Conidia subspherical, verruculose, 2-3 × 2.0-2.5 μm, pale greyish-green.

Physiology. Good growth at 37°C.

Molecular diagnostics. SSU and ITS restriction maps based on NCBI U21299 and U18355, resp.:

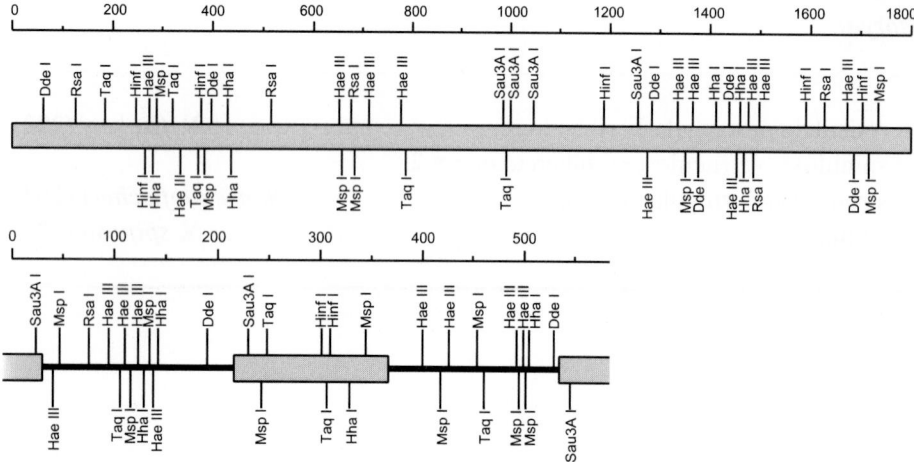

Pathogenicity. BSL-1. Mostly found in canned foodstuffs. Systemic infections in transplant recipients were reported by Gori *et al.* (1998b) and Lonial *et al.* (1997). A mixed pulmonary infection in a patient with myeloma was described by Chim *et al.* (1998).

Reference. Raper & Fennell (1965).

Antifungal susceptibility.

Antifungal	MICs range	MIC 90	Strains	Reference
AMB	0.39	0.39	2	Wildfeuer *et al.* (1998)
ITZ	0.1-0.2	0.2	2	Wildfeuer *et al.* (1998)
KTZ	0.1-0.2	0.2	2	Wildfeuer *et al.* (1998)
VCZ	0.1	0.1	2	Wildfeuer *et al.* (1998)

Nomenclature. *Aspergillus fischeri* Wehmer - Centbl. Bakt. Parasitkde, Abt. 2, 18: 390, 1907 ≡ *Neosartorya fischeri* (Wehmer) Malloch & Cain - Can. J. Bot. 50: 2621, 1973.

  Aspergillus fischerianus Samson & W. Gams, *in* Samson & Pitt - Adv. Penicill. Asp. Syst. p. 39, 1985.

Neosartorya pseudofischeri S.W. Peterson

Colony characteristics. Colonies (OA) growing rapidly, farinose to felty, cream coloured ; reverse pale ochraceous.
Microscopy. Ascomata borne singly or in small clusters within a loose hyphal envelope, spherical, pale pink, 150-300 µm diam; peridium pseudoparenchymatous, thin. Asci spherical to subspherical, 8-spored, 10-12 × 8-10 µm. Ascospores subspherical, with two prominent, flexuose, equatorial crests, about 5 µm diam (excluding the crests), hyaline; hemispheres irregularly ribbed, bearing triangular projections or elongate ridges.
Anamorph. *Aspergillus thermomutatus* (Paden) S.W. Peterson.
Conidiophores smooth-walled, hyaline to pale yellowish, 200-300 µm. Conidial heads initially radiate, later somewhat columnar, pale olivaceous grey, uniseriate. Vesicles subspherical, 10-17 µm wide. Phialides hyaline to pale greyish-green. Conidia subspherical, smooth-walled, about 4 × 3 µm, pale greyish-green.

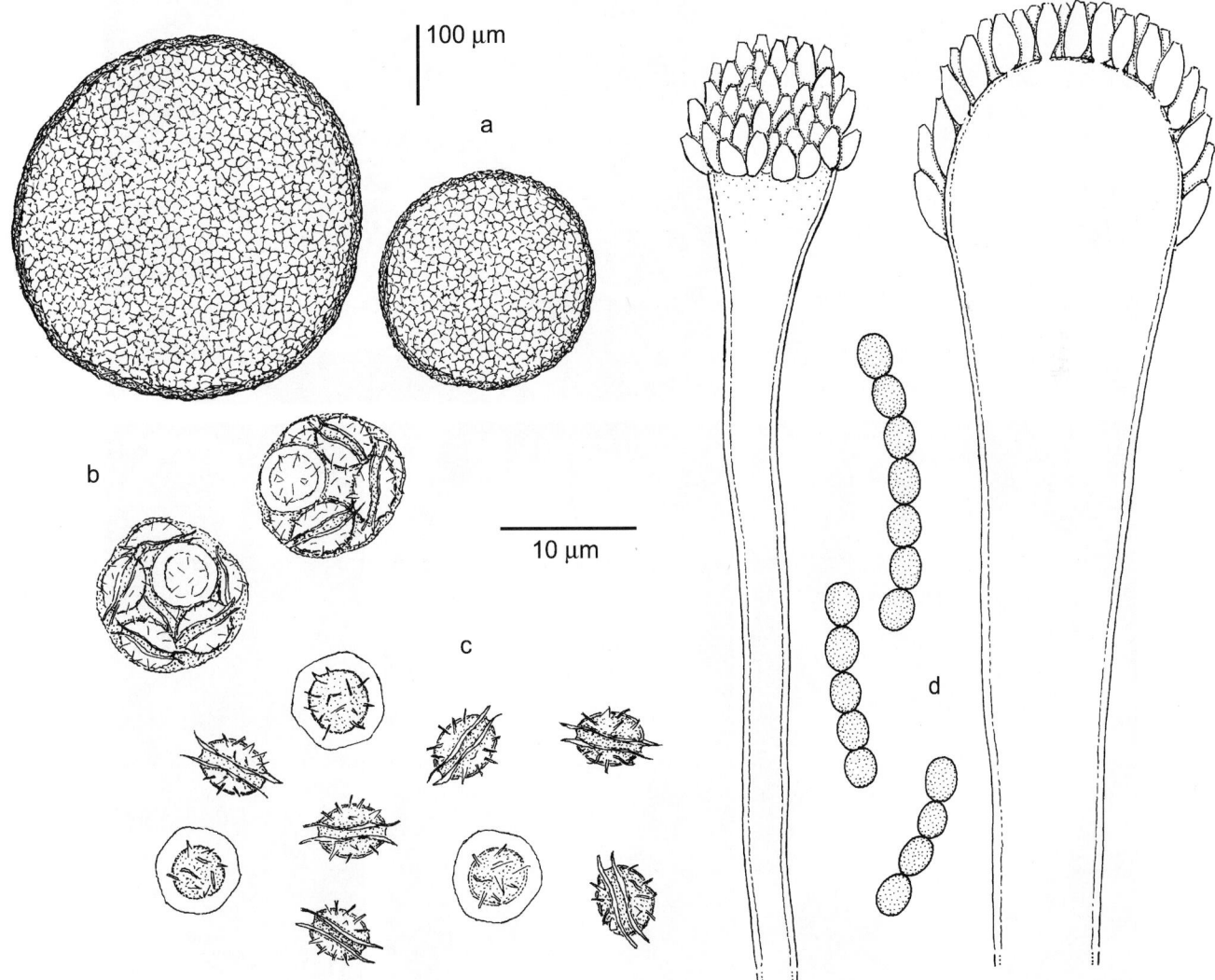

Neosartorya pseudofischeri (Aspergillus thermomutatus), CBS 207.92. a. Ascomata; b. asci; c. ascospores; d. conidiophores and conidia.

Pathogenicity. BSL-2. The species has been reported from osteomyelitis (Padhye *et al.,* 1994b), endocarditis (Summerbell *et al.,* 1992) and keratitis (Coriglione *et al.,* 1990).

Reference. Peterson (1992).

Nomenclature. *Neosartorya pseudofischeri* S.W. Peterson - Mycol. Res. 96: 549, 1992.

    Aspergillus fischeri Wehmer var. *thermomutatus* Paden - Mycopath. Mycol. Appl. 36: 161, 1968 ≡ *Aspergillus thermomutatus* (Paden) S.W. Peterson - Mycol. Res. 96: 549, 1992.

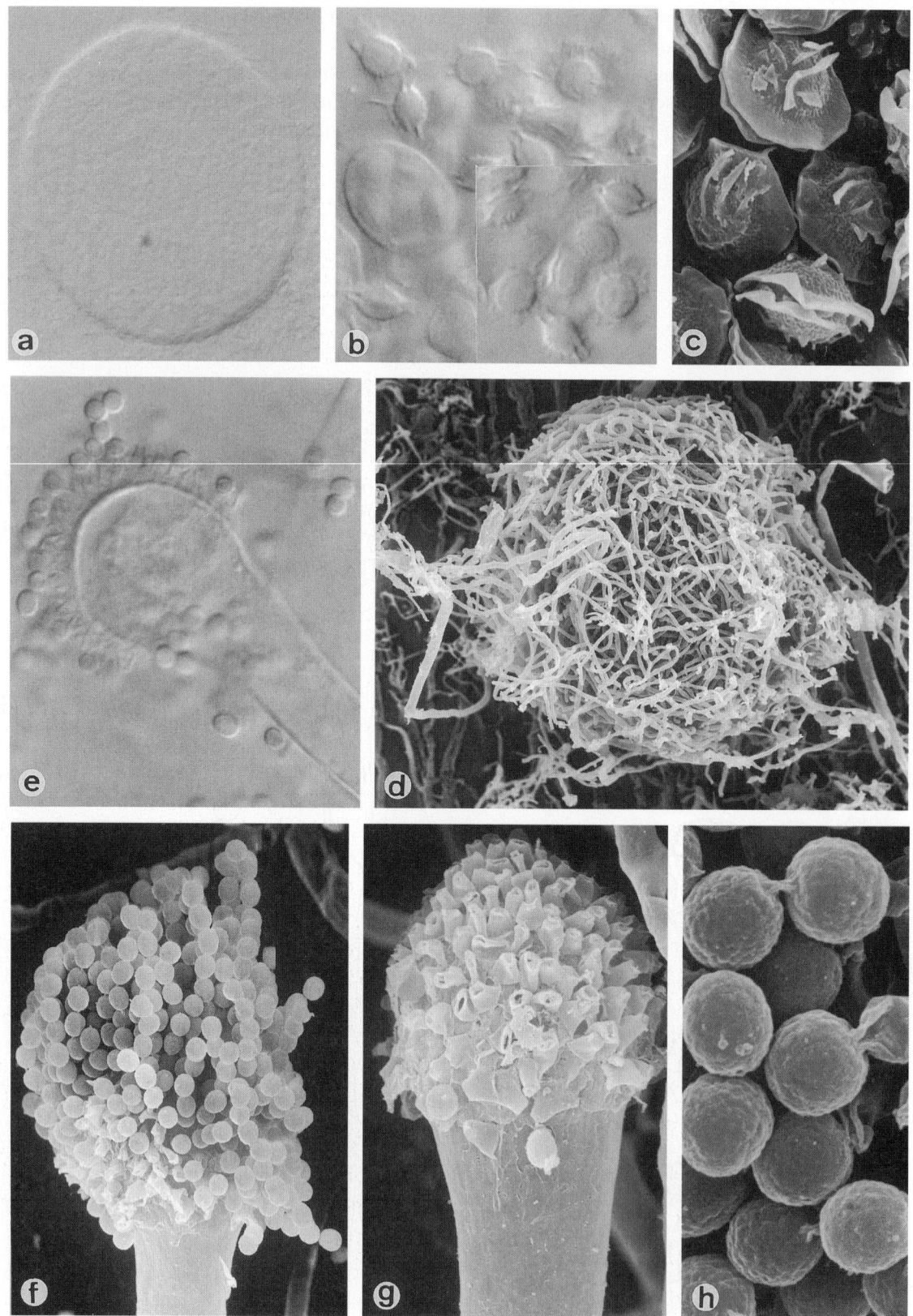

Neosartorya pseudofischeri (Aspergillus thermomutatus), CBS 207.92. a, d. Ascomata; b, c. asci and ascospores; e-h. conidiophores and conidia. a. ×128; b. ×1600; c. ×4650; d. ×345; e. ×1600; f. ×2100; g. ×3100; h. ×9500.

Neosartorya spinosa (Raper & Fennell) Kozakiewicz

Colony characteristics. Colonies (OA) growing rapidly, floccose, whitish, with pale blue-green shades.
Microscopy. Ascomata borne singly or in small clusters within a loose hyphal envelope, spherical, pale yellow, 200-300 µm; peridium pseudoparenchymatous, thin. Asci spherical to subspherical, 8-spored, 10-12 × 8-10 µm. Ascospores spherical to subspherical, with two prominent, flexuose, equatorial crests, 4-5 µm diam (excluding the crests), hyaline, hemispheres bearing large spines (spinulose).
Anamorph. *Aspergillus spinosus* Kozakiewicz.
Conidiophores smooth-walled, hyaline to pale green, 300-500 µm. Conidial heads columnar, pale blue-grey, uniseriate. Vesicles flask-shaped, up to 18 µm diam. Phialides hyaline to pale greyish-green. Conidia spherical to subspherical, smooth-walled, 2.0-4.5 × 2.4-3.5 µm, pale greyish-green.
Physiology. Good growth at 37°C.

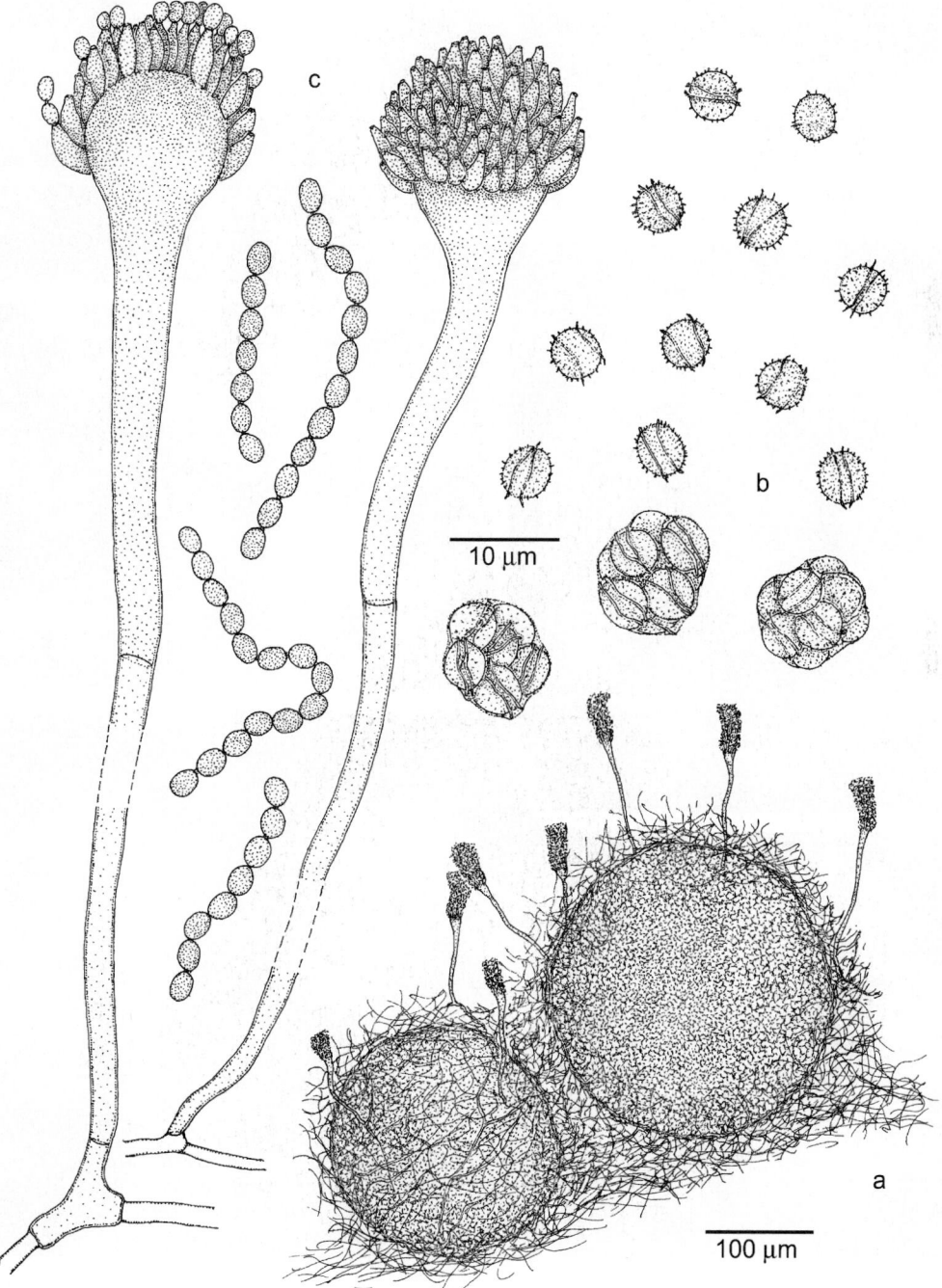

Neosartorya spinosa (*Aspergillus spinosus*), FMR 3722. a. Ascomata; b. asci and ascospores; c. anamorph: conidiophores and conidia.

Pathogenicity. BSL-1. Cases of pulmonary infection (Gerber *et al.*, 1973) has been reported.

Reference. Raper & Fennell (1965), Kozakiewicz (1989), Samson *et al.* (1990).

Nomenclature. *Aspergillus fischeri* Wehmer var. *spinosus* Raper & Fennell - Gen. Asperg. p. 256, 1965 ≡ *Neosartorya fischeri* (Wehmer) Malloch & Cain var. *spinosa* (Raper & Fennell) Malloch & Cain - Can. J. Bot. 50: 2621, 1965 ≡ *Neosartorya spinosa* (Raper & Fennell) Kozakiewicz - Mycol. Pap. 161: 58, 1989.

Aspergillus spinosus Kozakiewicz - Mycol. Pap. 161: 58, 1989.

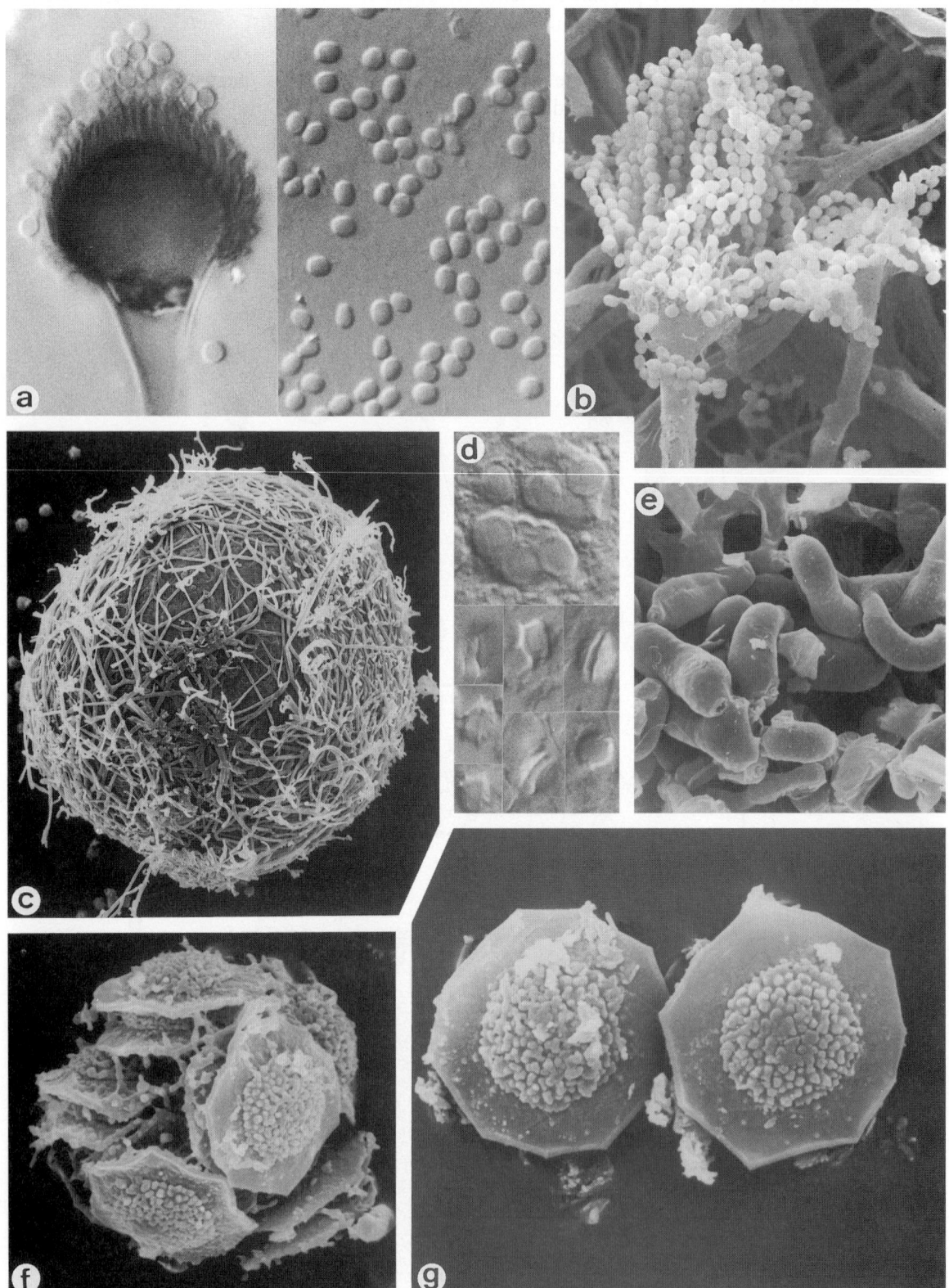

Neosartorya spinosa (*Aspergillus spinosus*), CBS 448.75. a, b. Conidiophores and conidia; c. ascoma; d, f, g. asci and ascospores; e. peridial cells. a. ×1600; b. ×1250; c. ×350; d. ×1600; e. ×2000; f. ×5000; g. ×6500.

Ascomycota, Dothideales, Didymosphaeriaceae. Genus: *NEOTESTUDINA*

Neotestudina rosatii Segretain & Destombes

Colony characteristics. Colonies (OA) restricted, dark brown to nearly black, with radial grooves; reverse dark brown.

Microscopy. Cleistothecia black, spherical, 180-900 μm. Peridium pseudoparenchymatous, divided into plates which are composed of radially arranged cells. Asci bitunicate, spherical to broadly ellipsoidal, 15-20 × 12-15 μm, evanescent. Ascospores, 2-celled, brown, smooth- and thick-walled, rhomboidal, 9.0-12.5 × 4.5-8.0 μm, with a germ pore at each end.

Anamorph. None.

Physiology. Maximum growth temperature 40°C. Cellulolytic.

Differential diagnosis. The genus *Ulospora* D. Hawksw. *et al.* is similar but has ascospores with longitudinal furrows.

Molecular diagnostics. ITS restriction map based on CBS 217.75:

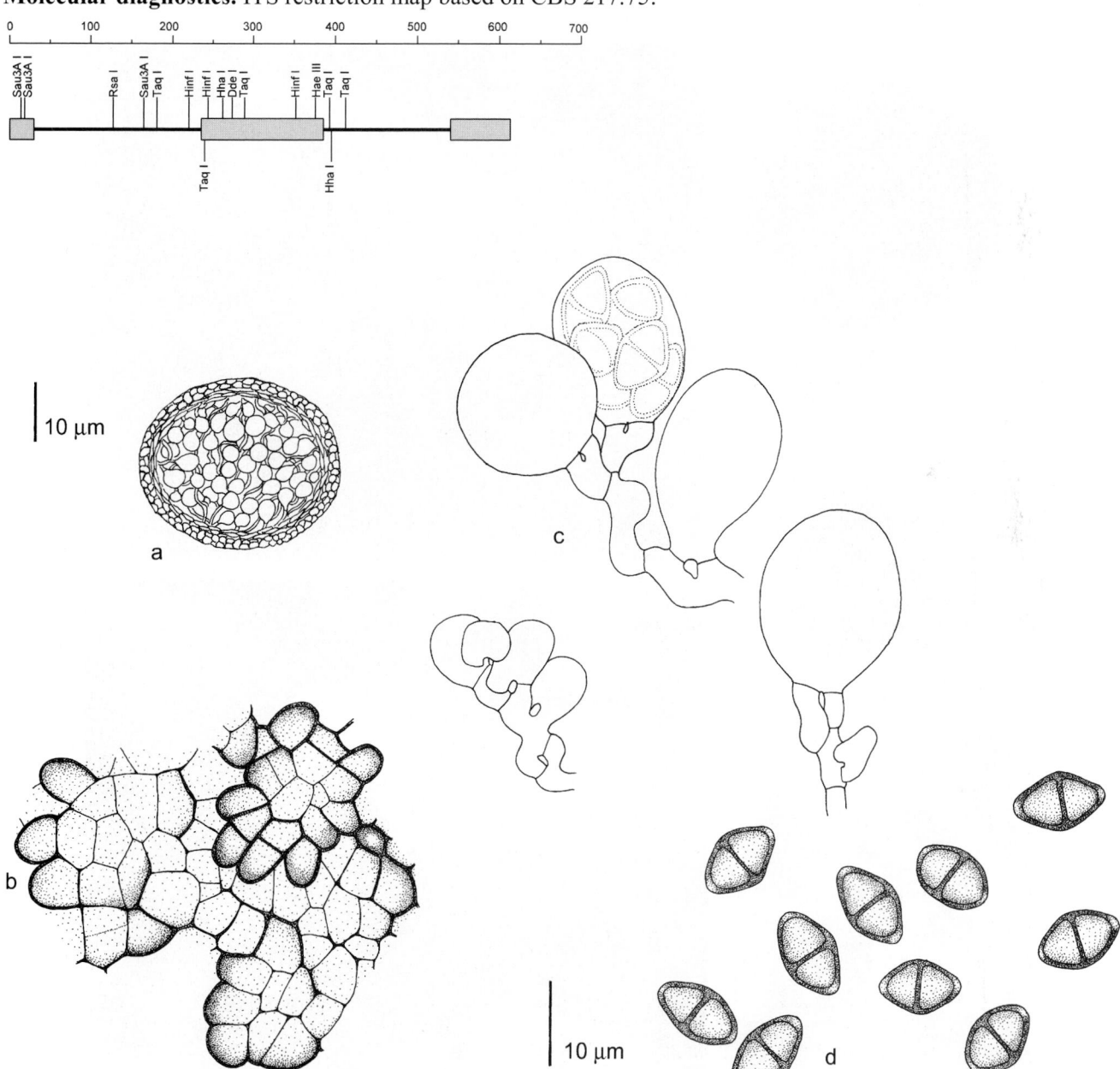

Neotestudina rosatii, CBS 331.78. a. Section through cleistothecium; b. detail of cleistothecium wall; c. young asci; d. mature ascospores.

Pathogenicity. BSL-2. The species causes mycetoma in humans (Segretain & Destombes, 1961). Grains are whitish but eventually may become dark; they are soft, irregular in shape, often with angular edges, and measure 0.5-1.0 mm in diam. The species occurs in soil in tropical countries.

References. Aue *et al.* (1969), Hawksworth & Booth (1974), Hawksworth (1979a), Sivanesan (1991).

Nomenclature. *Neotestudina rosatii* Segretain & Destombes - C.R. Hebd. Séanc. Acad. Sci., Paris 253: 2579, 1961 ≡ *Zopfia rosatii* (Segretain & Destombes) D. Hawksworth & Booth - Mycol. Pap. 135: 27, 1974.

   Pseudophaeotrichum sudanense Aue, E. Müller & Stoll - Nova Hedwigia 17: 84, 1969.

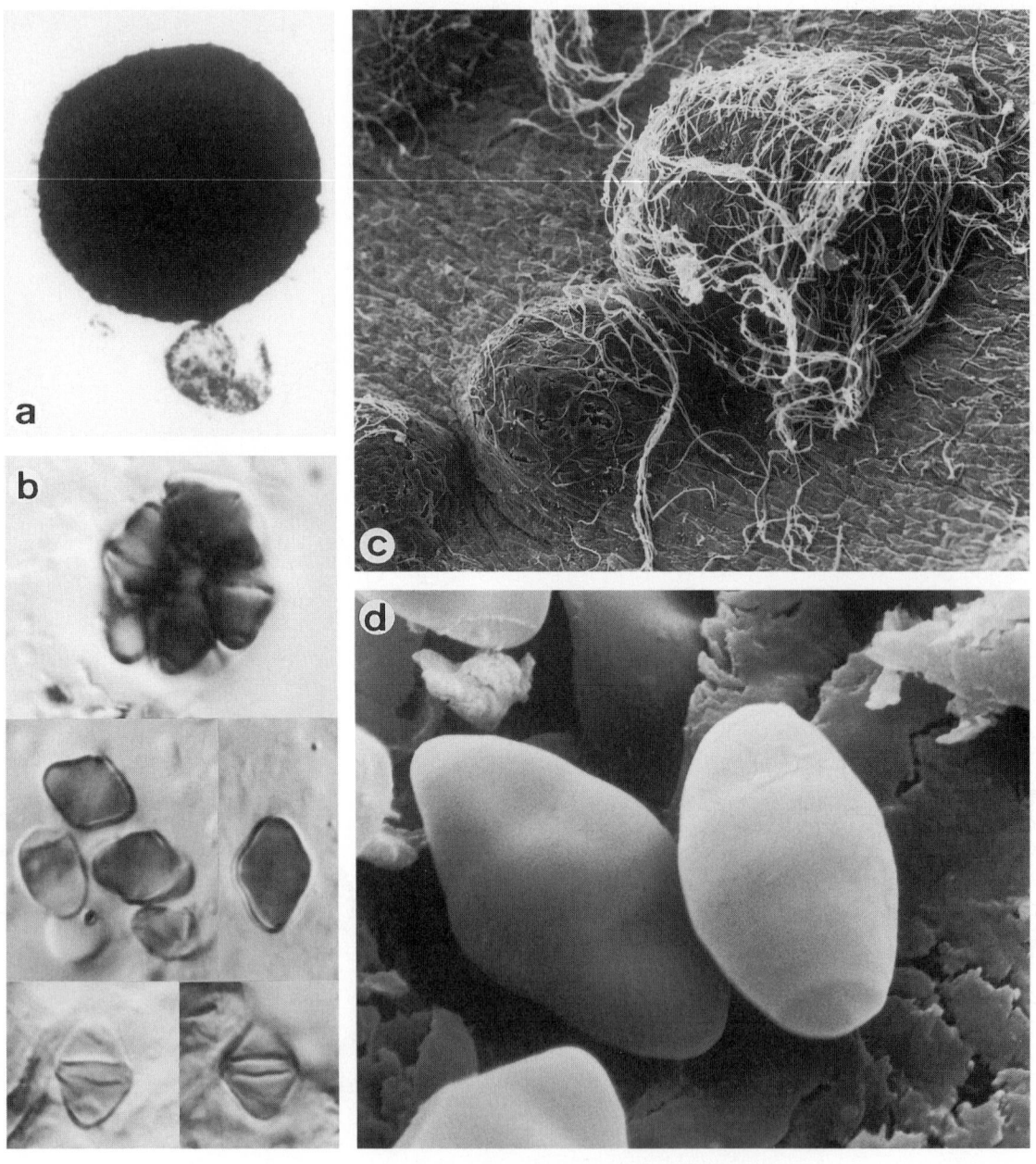

Neotestudina rosatii, a, b. IMI 210762ii; c, d. CBS 105.75. a, c. Ascomata; b, d. ascus and ascospores. a. ×128; b. ×1600; c. ×140; d. ×6500.

Ascomycota, Euascomycetes, Ophiostomatales, Ophiostomataceae. Genus: *OPHIOSTOMA*

Ophiostoma stenoceras (Robak) Melin & Nannf.

Colony characteristics. Colonies (OA) spreading, white, dry, floccose; reverse yellowish or brownish.

Microscopy. Hyphae hyaline. Perithecia scattered, black, composed of a spherical basal part, 80-180 μm diam, and a long (up to 2 mm), filiform neck which bears a whorl of pale brown setae, 18-35 (-55) μm long, at the ostiolum. Asci spherical, evanescent. Ascospores hyaline, bean-shaped, (2.0-) 2.5-4.5 × 1.0-1.5 μm, extruded in a slimy cirrhus.

Anamorph. *Sporothrix* sp.

Hyphae producing groups of navicular conidia terminally or laterally in dense clusters on acute denticles.

Differential diagnosis. The anamorph is similar to *Sporothrix schenckii* (p. 925). That species can, however, be distinguished by flat, restricted colonies which often turn brown, and by assimilation of starch. Molecular separation of the two taxa was proven by Suzuki *et al.* (1988) using mtDNA RFLP.

Physiology. Tolerant to cycloheximide; no assimilation of starch; growth at 37°C but no yeast conversion on BHI. Intolerant to benomyl.

Molecular diagnostics. SSU restriction map based on EMBL M85054:

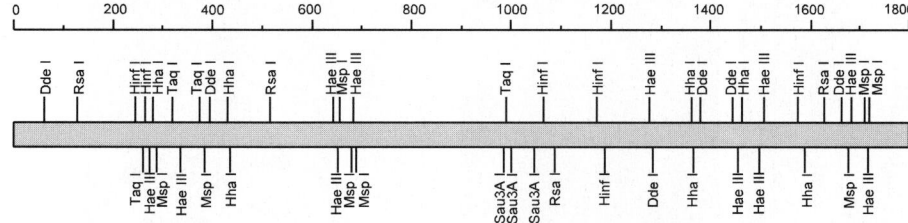

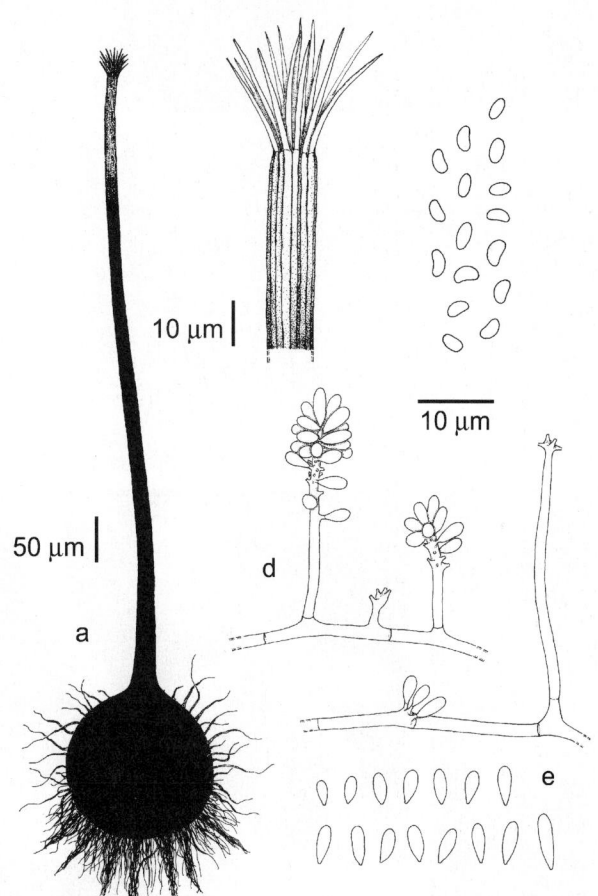

Ophiostoma stenoceras, CBS 360.71. a. Ascoma; b. end of perithecial neck with ostiolar hyphae; c. ascospores; d. conidiogenous cells of *Sporothrix* anamorph; e. conidia.

Pathogenicity. BSL-1. Fungus associated with bark insects. The cutaneous lesion of the scalp reported by Mariat (1971) probably concerned a mixed culture. Summerbell *et al.* (1993a) mentioned some strains from onychomycoses.
References. Upadhyay (1981), Summerbell *et al.* (1993).

Nomenclature. *Ceratostomella stenoceras* Robak - Nyt Mag. Naturvid. 71: 207, 1932 ≡ *Ophiostoma stenoceras* (Robak) Melin & Nannfeldt - Svenska Skogsvför. Tidskr. 32: 408, 1932.

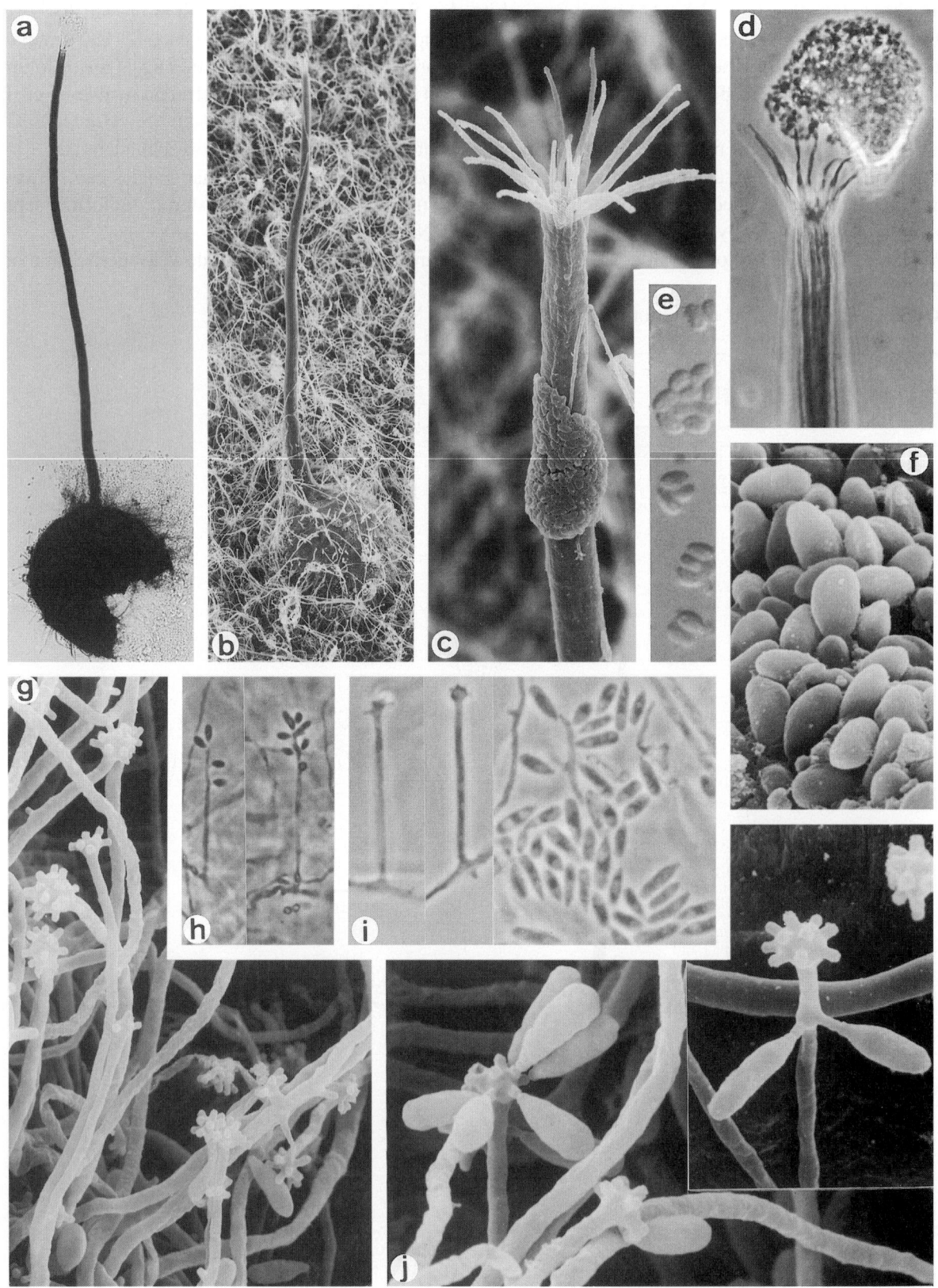

Ophiostoma stenoceras, CBS 360.71. a, b. Ascomata; c, d. end of perithecial neck with ostiolar hyphae; e, f. ascospores; g-j. conidiogenous cells of *Sporothrix* anamorph. a. ×128; b. ×200; c. ×1100; d. ×640; e. ×1600; f. ×6000; g. ×3000; h. ×512; i. ×1600; j. ×5500.

Ascomycota, Euascomycetes, Microascales, Microascaceae. Genus: *PETRIELLA*

Petriella setifera (Schmidt) Curzi

Colony characteristics. Colonies (OA) funiculose, dirty white with dark grey spots, soon becoming granular due to perithecium production.

Microscopy. Perithecia immersed, membranaceous, spherical, pale to dark brown, with a central ostiole in the short apical neck, covered by stiff hairs, smooth-walled, 75-125 µm diam. Peridium composed of angular cells. Asci ovoidal to broadly clarate, 8-spored, 21-25 × 12-15 µm, evanescent. Ascospores 1-celled, broadly fusiform, slightly flattened at one side, reddish-brown, 8-10.5 × 5-6 µm, with two inconspicuous germ pores.

Anamorphs. Conidiogenous cells arising from undifferentiated hyphae (*Scedosporium* sp.), cylindrical, producing slimy heads of 1-celled, smooth-walled, hyaline, obovoidal conidia, 7-14 × 3-4 µm, which often have a somewhat protruding basal scar. A *Graphium*-like synanamorph is occasionally present.

Molecular diagnostics. For a phylogenetic overview, see under *Pseudallescheria boydii* (p. 305). SSU and ITS restriction maps based on NCBI U43908 and AF043596, resp.:

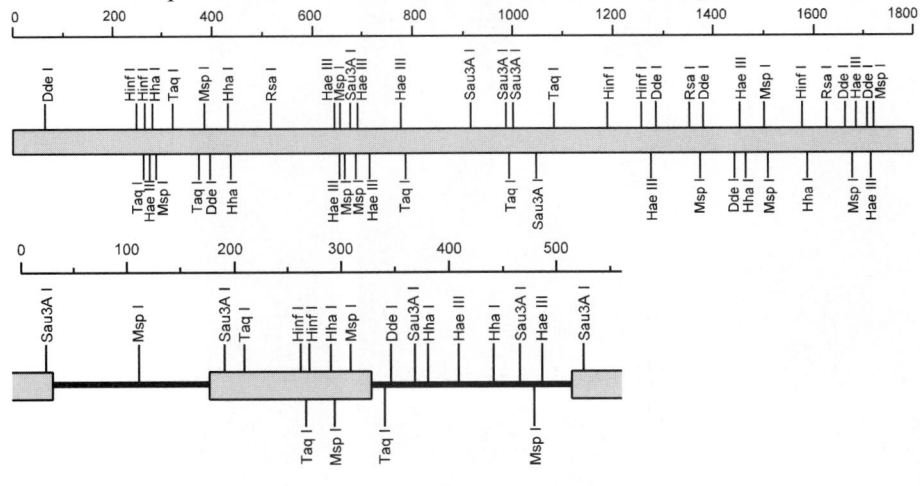

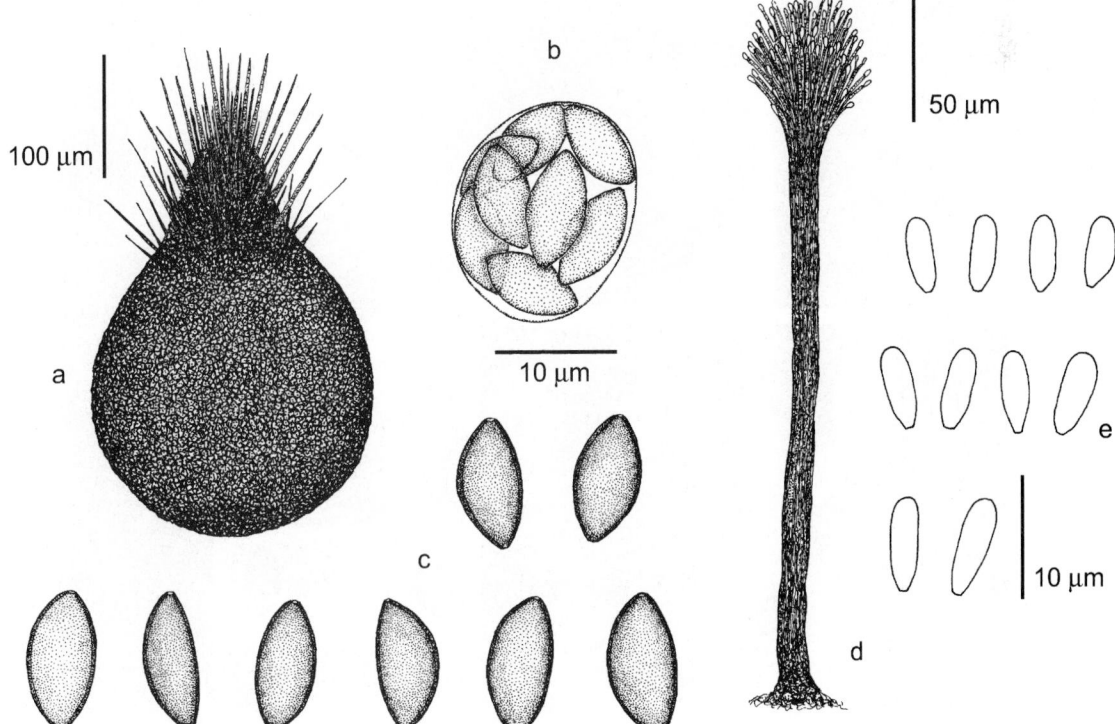

Petriella setifera, CBS 559.80. a. Ascoma; b. ascus; c. ascospores; d. synnema; e. conidia.

301

Pathogenicity. BSL-1. A mixed subcutaneous infection in a captive dolphin was reported by Poelma *et al.* (1974).
Reference. Barron *et al.* (1961).
Antifungal susceptibility.

Antifungal	Mean MICs	Strains	Reference
AMB	5.64	2	Guarro *et al.* (1995)
5FC	256	2	Guarro *et al.* (1995)
FLZ	128	2	Guarro *et al.* (1995)
ITZ	3.99	2	Guarro *et al.* (1995)
KTZ	3.99	2	Guarro *et al.* (1995)
MCZ	7.99	2	Guarro *et al.* (1995)

Nomenclature. *Microascus setifer* Schmidt - Thesis, Breslau, 1912 ≡ *Petriella setifera* (Schmidt) Curzi - Boll. Staz. Pal. Veg., Roma 10: 380-423, 1930.

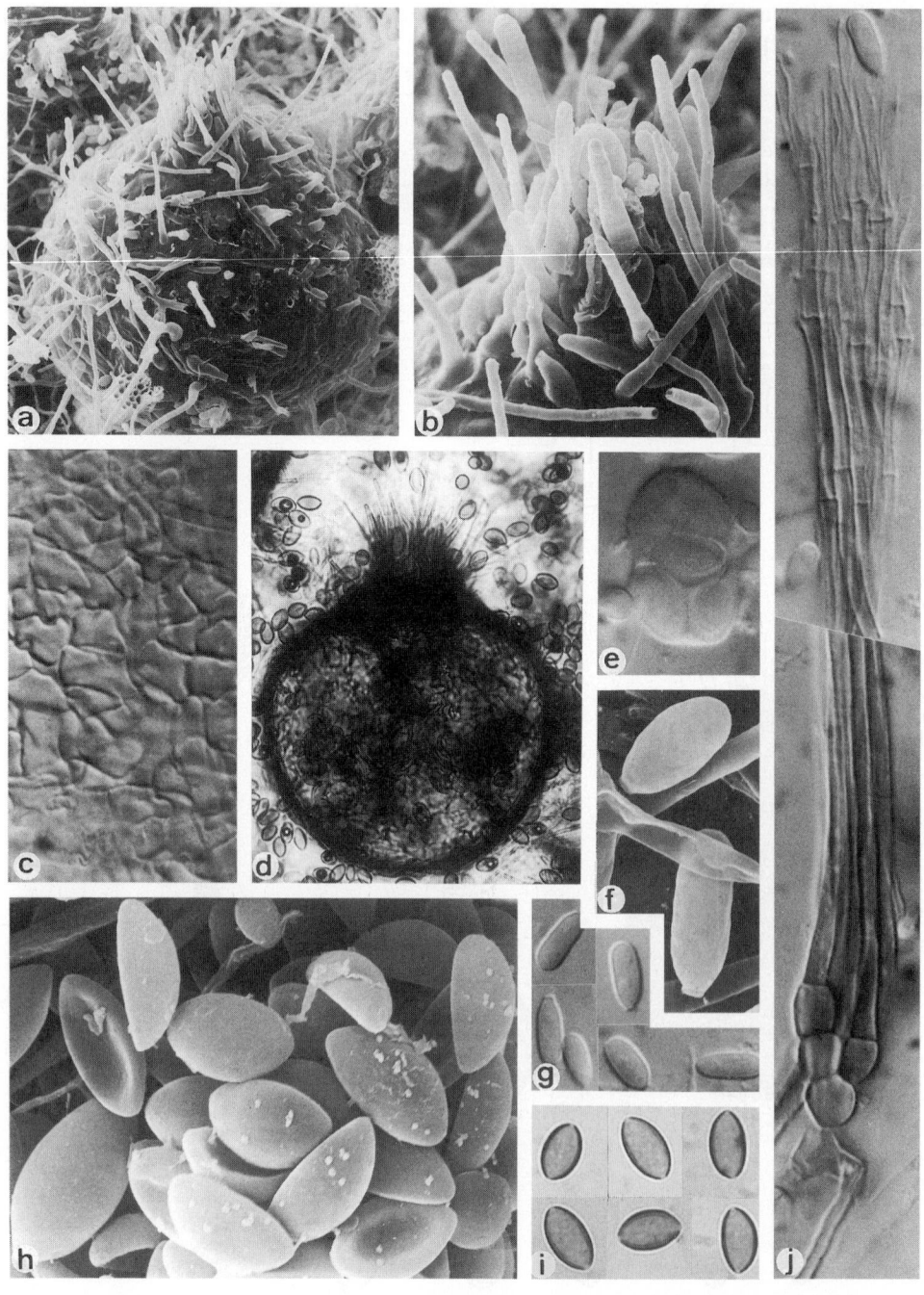

Petriella setifera, CBS 559.80. a, d. Ascomata; b. ascoma neck; c. peridium; e. ascus; f, g. conidia; h, i. ascospores; j. synnema. a. ×720; b. ×1800; c. ×1600; d. ×512; e. ×1600; f. ×4000; g. ×1600; h. ×2900; i. 1×280; j. ×1600.

Ascomycota, Euascomycetes, Dothideales, Piedraiaceae. Genus: *PIEDRAIA*

Piedraia hortae (Brumpt) da Fonseca & de Arêa Leão

Colony characteristics. Colonies (SGA + glycerine) with slow growth, restricted, folded, velvety, dark brown to black, mostly remaining sterile.

Microscopy. Ascostromata on hair compact, pseudoparenchymatous, subspherical to elongate, up to 1 mm long and 0.3 mm wide, black, with irregularly distributed loculi each mostly containing a single ascus; cavities without a proper wall, opening to the surface by an inconspicuous pore. Asci bitunicate, ellipsoidal, with 2-8 ascospores and readily dissolving walls. Ascospores (sub)hyaline, 1-celled, spindle-shaped, curved, 30-45 × 5.5-10 µm, tapering towards both ends to form whip-like extensions.

Physiology. Keratinolytic.

Differential diagnosis. A species lacking ascospore appendages was described as *P. quintanilhae* van Uden *et al.* It occurs on chimpanzees in Central Africa (Takashio & de Vroey, 1975). Agents of white piedra (p. 21) are basidiomycetous yeasts (*Trichosporon;* p. 165).

Pathogenicity. BSL-1. The fungus causes black piedra (p. 21) on hairs of the scalp. Firmly attached, black, hard nodules composed of cemented fungal cells are formed around the hair shaft (Figueras & Guarro, 1997). The hairs are weakened by fungal growth and may break easily. The disorder is found on primates including man in humid, tropical regions, and is particularly seen in South America in those populations where hair care is by treatment with oily substances (Adam *et al.*, 1977; Coimbra & Santos, 1989). When untreated the disorder may remain for years. Mixed infections with *Trichosporon* species may occur. The ultrastructure of the fungus was studied by Almeida *et al.* (1991) and the course of hair invasion by Figueras *et al.* (1996, 1997).

References. Takashio & Vanbreuseghem (1971), Takashio (1973), Figueras *et al.* (1996, 1997), Figueras & Guarro (2000).

Nomenclature. *Trichosporon hortae* Brumpt - Précis Parasitol. p. 951, 1913 ≡ *Piedraia hortae* (Brumpt) da Fonseca & de Arêa Leão - Mem. Inst. Oswaldo Cruz, Suppl. 4: 124, 1928.

    Trichosporon guayo Delamare & Gatti - C.R. Soc. Biol. 99: 1425, 1928.

    Piedraia sarmentoi Pereira - Rev. Med. Cirurg. Brasil 38: 49, 1929 ≡ *Trichosporon sarmentoi* (Pereira) Nannizzi - Tratt. Micopat. Um. 4: 294, 1934.

    Piedraia venezuelensis Brumpt & Langeron - Annls Parasit. Hum. Comp. 12: 155, 1934.

    Piedraia javanica Boedijn & Verbunt - Mycopathologia 1: 196, 1938.

    Piedraia surinamensis C.W. Dodge - Med. Mycol. p. 133, 1935.

    Piedraia malayi Green & Mankikar - Trans. R. Soc. Med. 43: 525, 1950.

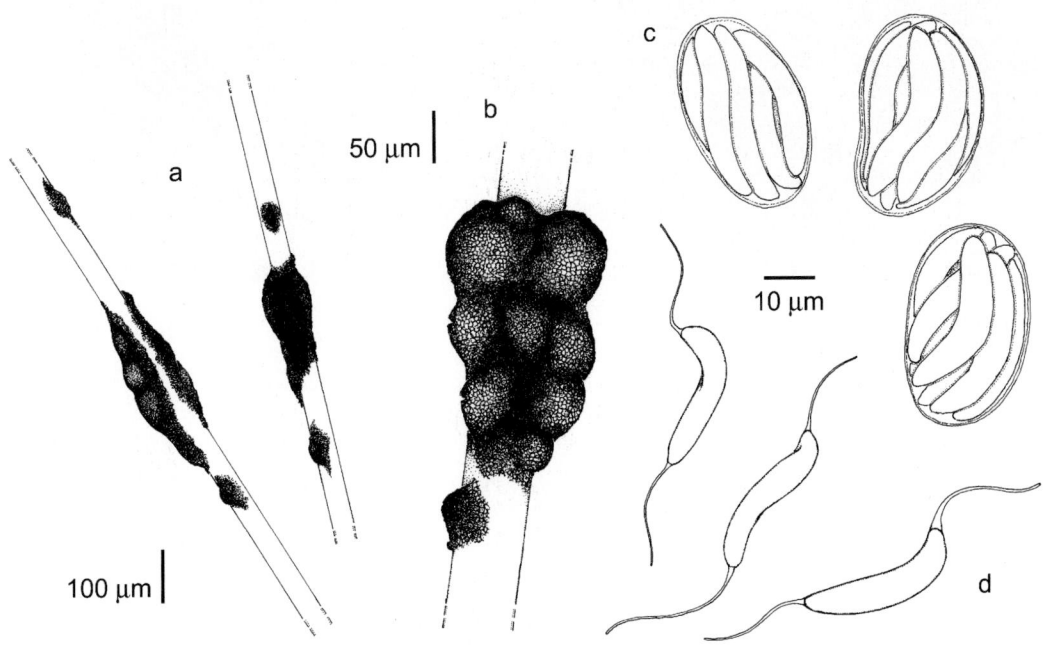

Piedraia hortae, FMR 5023. a, b. Ascostromata on hair; c. asci; d. ascospores.

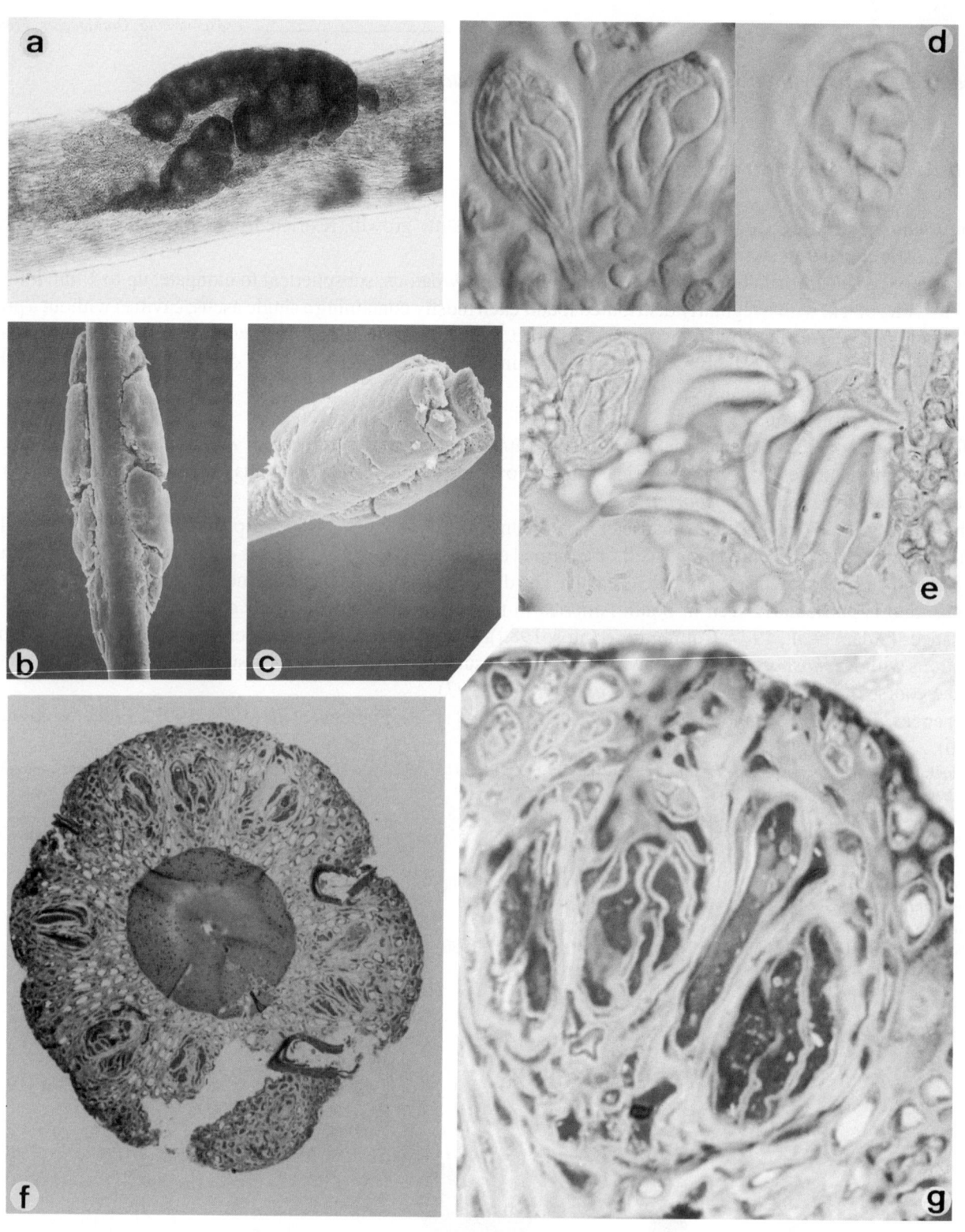

Piedraia hortae, FMR 5023. a-c. Nodules of black piedra developed on the surface of hair; d, e. asci and ascospores; f. section of a nodule showing the pseudoparenchymatous organzation of the ascomata, with well-developed asci; g. detail of asci. a. ×120; b. ×90; c. ×120; d. ×1600; e. ×1174; f. ×260; g. ×1280.

Ascomycota, Euascomycetes, Microascales, Microascaceae. Genus: *PSEUDALLESCHERIA*

Pseudallescheria boydii (Shear) McGinnis *et al.*

Colony characteristics. Colonies (OA, 30°C) growing rapidly, cottony to lanose, initially dirty white, becoming pale smoky brown.

Microscopy. Cleistothecia immersed, membranaceous, spherical, light brown to black, 140-200 μm diam. Peridium rather thin, composed of jig-saw-shaped cells. Asci (sub)spherical, 8-spored, evanescent. Ascospores 1-celled, lemon-shaped, 6-7 × 4.0-4.5 μm, smooth-walled, pale yellow to golden brown, with two terminal germ pores.

Anamorphs. Erect synnemata [synanam. *Graphium eumorphum* Sacc.] may be present, producing broadly clavate, subhyaline or pale brown conidia, 6-12 × 3.5-4.0 μm, from percurrently elongating cells. Conidiogenous cells also arising from undifferentiated hyphae [synanam. *Scedosporium apiospermum* (Sacc.) Sacc.], cylindrical, producing slimy heads of 1-celled, smooth-walled, subhyaline to brown, subspherical to elongate conidia, 6-12 × 3.5-6.0 μm, which swell and become brown and thick-walled after liberation; conidia sometimes sessile.

Physiology.

Fermentation:		Salicin	+	DL-Lactate	–
D-Glucose	–	Melibiose	v	Succinate	–
		Lactose	v	Citrate	–
Growth:		Raffinose	–	Methanol	–
D-Glucose	+	Melezitose	–	Ethanol	+
D-Galactose	v	Inulin	–	Nitrate	+
L-Sorbose	–	Starch	+	Nitrite	+
D-Glucosamine (C)	v	Glycerol	w	Ethylamine	+
Ribose	v	*meso*-Erythritol	–	L-Lysine	+
D-Xylose	+	Ribitol	+	Cadaverine	+
L-Arabinose	+	Xylitol	+	Creatine	–
D-Arabinose	w,–	L-Arabinitol	+	Creatinine	–
L-Rhamnose	v	D-Glucitol	v	10% NaCl	–
Sucrose	+	D-Mannitol	v	0.1% Cycloheximide	v
Maltose	+	Inositol	–	Mycosel	+,w
α,α-Trehalose	+	Glucono-δ-lactone	–	Urease	+
α-Methyl-D-glucoside	–	D-Gluconate	–	Growth at 37°C	+
Cellobiose	+	D-Galacturonate	–		

Differential diagnosis. Clinical strains rarely produce ascomata; they then are recognized by cylindrical conidiogenous cells with olivaceous, thick-walled conidia. Conidia are produced in slimy heads, in contrast to the dry conidia of *Scopulariopsis*. A related species is *Scedosporium prolificans* (p. 899); it has darker cultures, inflated conidiogenous cells, slightly wider conidia and does not assimilate ribitol, xylitol and L-arabinitol.

Molecular diagnostics. SSU sequencing was performed by Issakainen *et al.* (1997, 1999). A phylogenetic tree showing related taxa is given on p. 308. A 26S-based probe was developed by Sandhu *et al.* (1995) and ITS-based primers by Wedde *et al.* (1998). The species shows considerable infraspecific variability (Rainer *et al.*, 2000). SSU and ITS restriction maps based on NCBI U43913, U43914 and CBS 101.22, resp.:

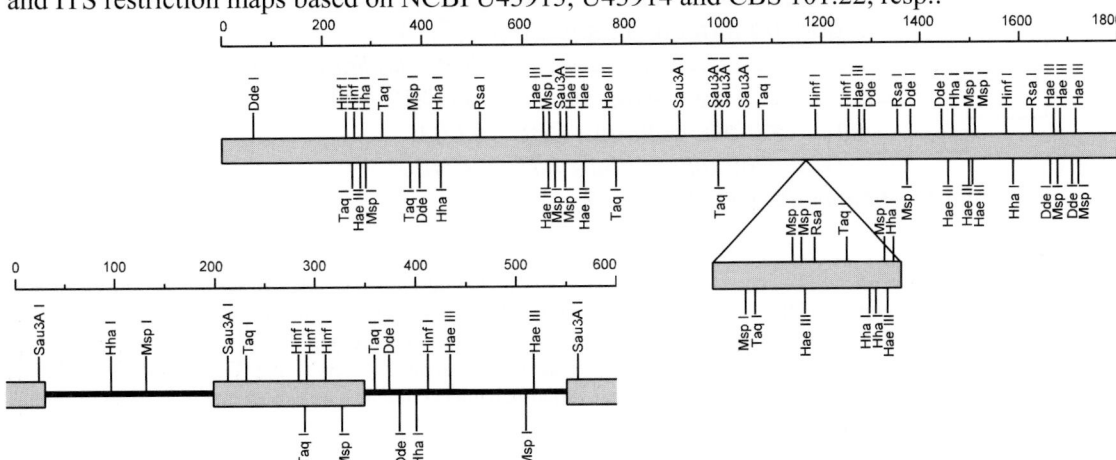

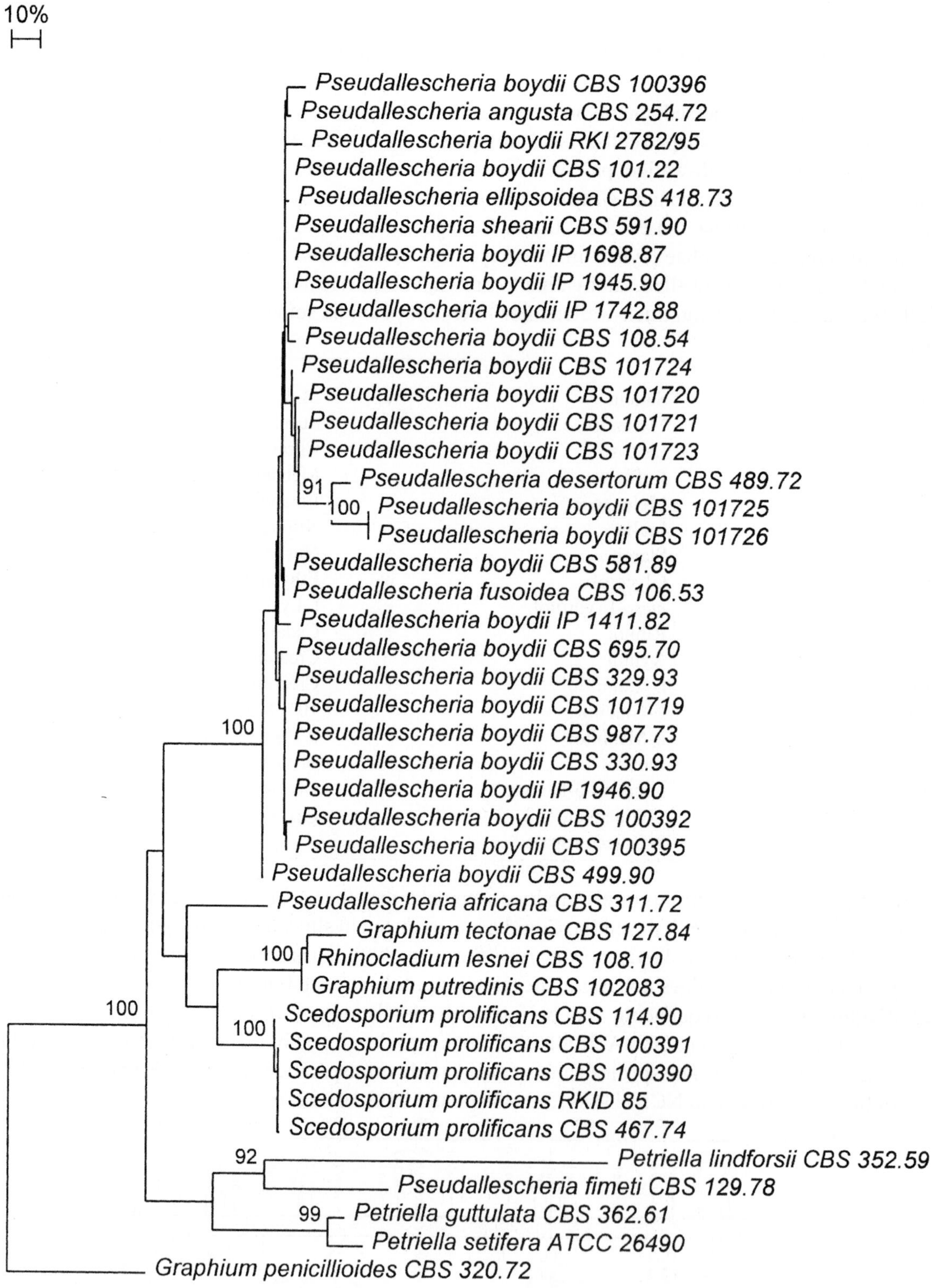

10%

Fig. 23. Phylogenetic tree of of *Pseudallescheria boydii* and related *Microascaceae* including clinically relevant *Graphium, Petriella* and *Scedosporium* species, based on confidently aligned ITS1-2 rDNA sequences, using Neighbor joining algorithm with Kimura correction. Bootstrap values >90 from 100 resampled datasets are shown. *Graphium penicillioides* is selected as outgroup. *Pseudallescheria boydii* shows a relatively wide molecular diversity; conversely, some morphologically deviating taxa are found have closely similar ITS sequences. Although the specific borderline of the species is statistically well-supported, the internal variation hampers diagnostics by sequencing or fingerprinting. For further data, see Rainer *et al.* (2000).

Pathogenicity. BSL-2. The species is frequently encountered as a saprobe in agricultural soil, manure and polluted water. In older literature it has frequently been reported as an agent of white-grain mycetoma, occurring also in temperate climates (Ajello, 1952). The species is frequently involved in arthritis (Ginter *et al.*, 1995; Willemsen *et al.*, 1997) and otitis (Rippon & Carmichael, 1976; del Palacio *et al.*, 1999a). Also cutaneous (Lemerle *et al.*, 1998) and ophthalmic cases (Orr *et al.*, 1993) have been reported. The species is an emerging opportunist in immunocompromised hosts, e.g., in leukemic, transplant (Garcia-Arata *et al.*, 1996; Ginter *et al.*, 1999) or CGD patients (Jabado *et al.*, 1998), producing rather variable clinical pictures (Tadros *et al.*, 1998). Cases are known after severe trauma due to traffic accidents (Horré *et al.*, 2000), or after aspiration of polluted water due to near-drowning (Fisher *et al.*, 1982; Dworzack *et al.*, 1989). In the disseminated cases the fungus shows a marked predilection for the central nervous system (Lopez *et al.*, 1998; Montero *et al.*, 1998). Also endocarditis may occur (Sobottka *et al.*, 1999). Without treatment, systemic mycosis mostly leads to death (Berenguer *et al.*, 1989). Further the species occurs in the respiratory tract where it may cause allergic reactions, sinusitis or pneumonia and it is frequently encountered subclinically in lungs of CF patients. An animal case was described by Käufer & Weber (1977). Cutaneous infections have been reviewed by Miyamoto *et al.* (1998).

Antifungal susceptibility.

Antifungal	MICs range	MIC 90	Strains	Reference
AMB	0.25-2	2	21	Walsh *et al.* (1995)
AMB	0.5-16	nd	5	Espinel-Ingroff *et al.* (1995)
AMB	1-8	8	13	Hennequin *et al.* (1997)
AMB	2->16	nd	11	Guarro *et al.* (2000)
FLZ	2-32	nd	5	Espinel-Ingroff *et al.* (1995)
FLZ	2-32	32	21	Walsh *et al.* (1995)
FLZ	32->64	>64	11	Guarro *et al.* (2000)
ITZ	≤0.03-4	0.25	21	Walsh *et al.* (1995)
ITZ	0.03-1	nd	5	Espinel-Ingroff *et al.* (1995)
ITZ	0.125-8	0.5	13	Hennequin *et al.* (1997)
ITZ	0.03->16	>16	11	Guarro *et al.* (2000)
KTZ	0.03-2	nd	5	Espinel-Ingroff *et al.* (1995)
KTZ	4-16	16	11	Guarro *et al.* (2000)
MCZ	0.125-0.5	0.5	21	Walsh *et al.* (1995)
MCZ	0.06-1	nd	5	Espinel-Ingroff *et al.* (1995)
MCZ	0.06-1	0.12	13	Hennequin *et al.* (1997)
MCZ	0.03-0.06	nd	11	Guarro *et al.* (2000)

References. Hironaga & Watanabe (1980), McGinnis *et al.* (1982a), Campbell & Smith (1982), Travis *et al.* (1985), Dykstra *et al.* (1989), de Hoog *et al.* (1994a), Rainer *et al.* (2000).

Nomenclature. *Verticillium graphii* Harz & Bezold, *in* Siebenmann - Schimmelmyk. Menschl. Ohres p. 95, 1889 ≡ *Glenospora graphii* (Harz & Bezold) Vuillemin - C.R. Acad. Sci., Paris 154: 141, 1912 ≡ *Trichosporium graphii* (Harz & Bezold) C.W. Dodge - Med. Mycol. p. 793, 1935 ≡ *Glenosporopsis graphii* (Harz & Bezold) da Fonseca - Parasitol. Méd. 1: 727, 1943.

Actinomyces albus Tarozzi - Archivio Sci. Med. 33: 535-632, 1909 (nom. inval.) ≡ *Streptothrix tarozzii* Miescher - Arch. Derm. Syph. 124: 297-442, 1917 ≡ *Actinomyces tarozzii* (Miescher) C.W. Dodge - Med. Mycol. p. 735, 1935.

Monosporium apiospermum Saccardo - Annls Mycol. 9: 254, 1911 ≡ *Scedosporium apiospermum* (Saccardo) Castellani & Chalmers - Man. Trop. Med., ed. 3, p. 1122, 1919 ≡ *Aleurisma apiospermum* (Saccardo) Maire, *in* Montpellier & Gouillon - Bull. Soc. Path. Exot. 14: 290, 1921.

Monosporium sclerotiale Pepere - Soc. Cult. Sci. Med. Nat. Cagl. 68: 543, 1914 ≡ *Scedosporium sclerotiale* (Pepere) Neveu-Lemaire - Précis Parasitol. Hum., éd. 5, p. 86, 1921 ≡ *Monosporium apiospermum* Saccardo var. *peperei* Sartory - Champ. Paras. Homme Anim. p. 681, 1922 (name change).

Cephalosporium boydii Shear - Mycologia 14: 242, 1922 ≡ *Glenospora boydii* (Shear) Pollacci & Nannizzi - Tratt. Micopat. Um. 4: 246, 1934.

Dendrostilbella boydii Shear - Mycologia 14: 242, 1922.

Allescheria boydii Shear - Mycologia 14: 242, 1922 ≡ *Petriellidium boydii* (Shear) Malloch - Mycologia 62: 738, 1970 ≡ *Pseudallescheria boydii* (Shear) McGinnis, Padhye & Ajello - Mycotaxon 14: 97, 1982.

Glenospora clapieri Catanei, *in* Montpellier, Catanei & Clapier - Bull. Path. Exot. 20: 502, 1927 ≡ *Trichosporium clapieri* (Catanei) C.W. Dodge - Med. Mycol. p. 794, 1935 ≡ *Glenosporopsis clapieri* (Catanei) da Fonseca - Parasit. Med. 1: 724, 1943 ≡ *Madurella clapieri* (Catanei) Redaelli & Ciferri - Mycopath. Mycol. Appl. 3: 196, 1943.

Indiella americana Delamare & Gatti - C.R. Hebd. Séanc. Acad. Sci., Paris 188: 1264, 1929 ≡ *Madurella americana* (Delamare & Gatti) Vuillemin - Champ. Paras. Myc. Homme p. 155, 1931.

Acremoniella lutzi Arêa Leao & Lobo - Acta Med., Rio de Janeiro 4: 219, 1939.

Pseudallescheria sheari Negroni & I. Fischer - Revta Inst. Bact., B. Aires 12: 201, 1944.

Acremonium suis Bakai, *in* Gladenko - Bolezni Svinei, Kiev p. 198, 1967.

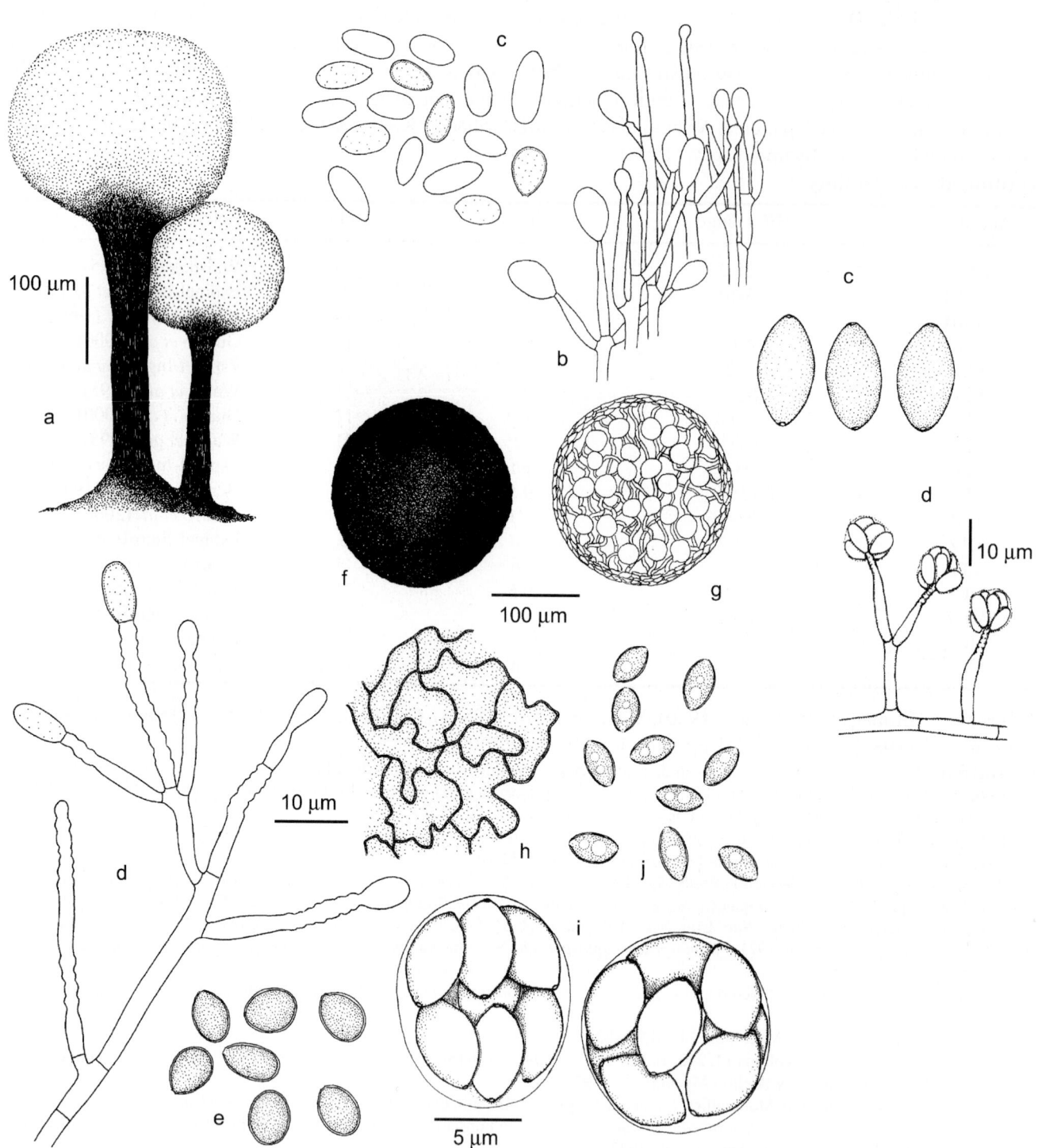

Pseudallescheria boydii, a, f, j. FMR 6697; b-e, g, h, j. CBS CBS 254.66. a-c. Synnematous (*Graphium*) synanamorph. a. Synnema; b. detail of conidiogenous cells; c. conidia. d, e. *Scedosporium* synanamorph. d. Conidiogenous cells; e. conidia. f-j. Teleomorph. f. Cleistothecium; g. diagram of sectioned cleistothecium; h. detail of cleistothecium wall; i. asci at very large magnification; j. liberated ascospores at lower magnification.

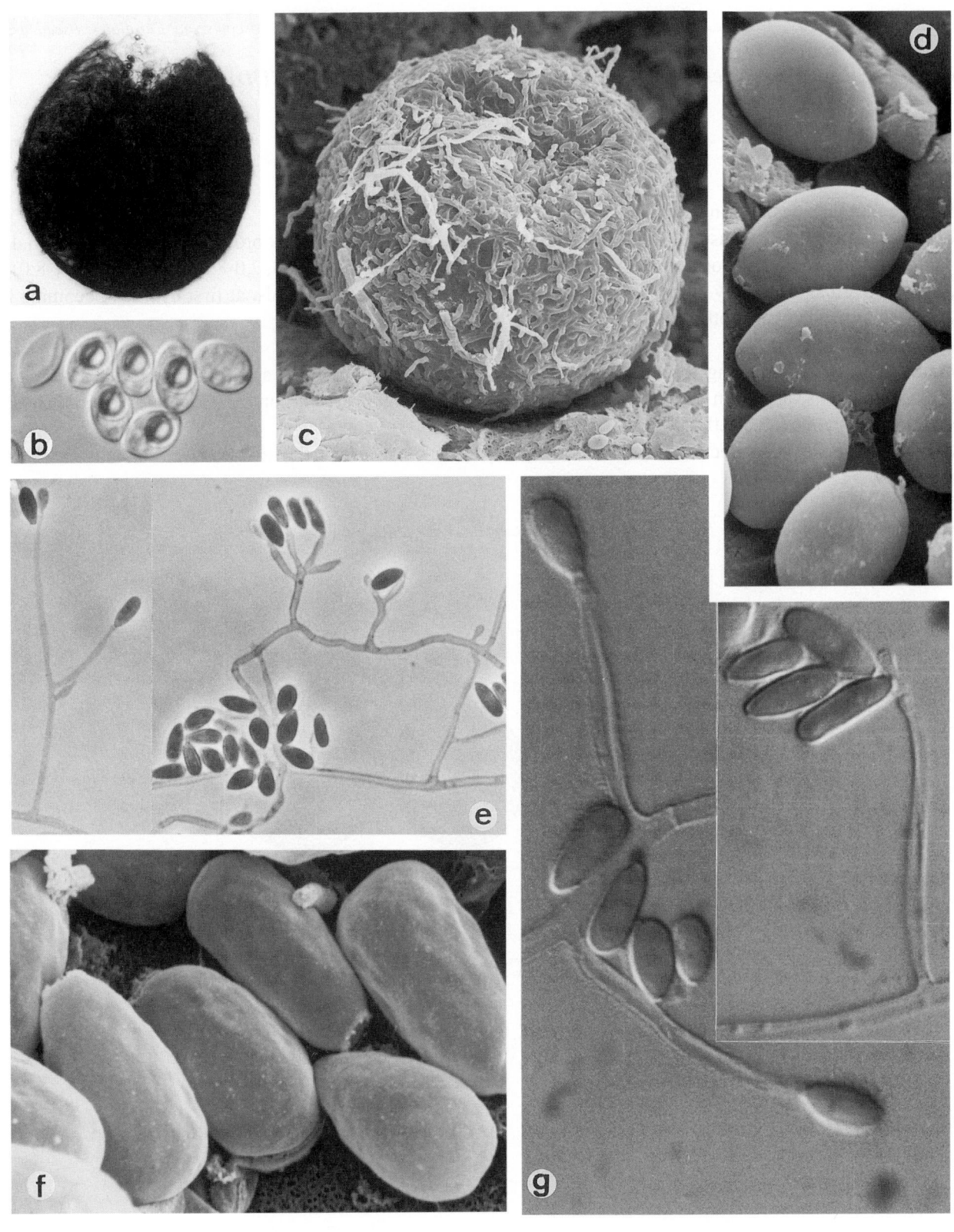

Pseudallescheria boydii, a-d, f. FMR 6425; e, g. AMMRL 28.14. a, c. Ascomata; b, d. ascospores; e-g. *Scedosporium apiospermum* anamorph, conidiogenous cells and conidia. a. ×250; b. ×1600; c. ×600; d. ×5000; f. ×7000; g. ×1600..

309

Ascomycota, Euascomycetes, Eurotiales, Pseudeurotiaceae. Genus: *PSEUDEUROTIUM*

Pseudeurotium ovale ok

Colony characteristics. Colonies (OA) spreading broadly, light brownish-grey.

Microscopy. Ascomata in clusters, spherical, smooth-walled, non-ostiolate, dark brown, 100-250 (-300) μm diam. Peridium membranaceous to coriaceous. Asci 8-spored, subspherical to ellipsoidal, 7.0-8.0 (-9.5) × 6.5-7.0 (-8.0) μm. Ascospores ovoidal to broadly ellipsoidal, 4-5 × 2.5-3.0 μm, rounded at both ends, at first hyaline, becoming light olive brown at maturity, smooth-walled, without germ pores.

Anamorph. *Sporothrix* sp.

Conidiophores short, emerging from aerial hyphae, 15-60 × 1-2 μm, smooth-walled, hyaline, with tapering tips. Conidiogenous cells variable in size, cylindrical, apically with a cluster of sessile conidia. Conidia sympodial, hyaline, subspherical to ovoidal, 3.5-8.0 × 2-5 μm, smooth-walled.

Pathogenicity. BSL-1. Isolated in two cases of onychomycoses, but its role in the infection was not clarified (English *et al.*, 1967).

Reference. Stolk (1955).

Nomenclature. *Pseudeurotium ovale* Stolk - Antonie van Leeuwenhoek 21: 65, 1922.

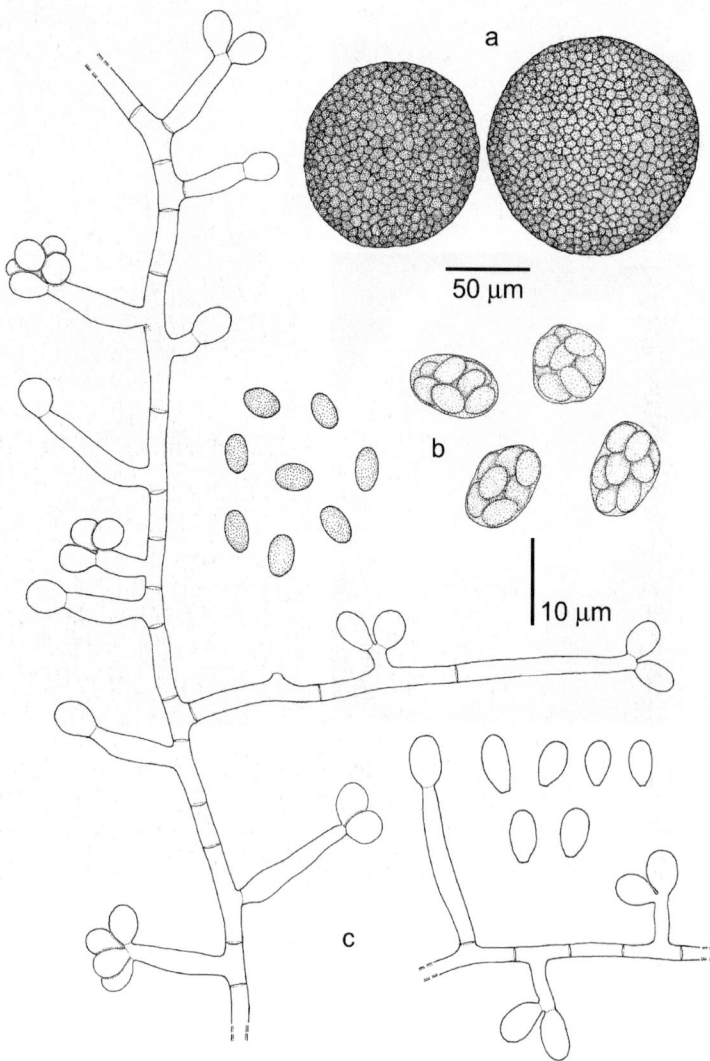

Pseudeurotium ovale, CBS 454.62. a. Ascomata; b. asci and ascospores; c. *Sporothrix*-like anamorph: conidiogenous cells and conidia.

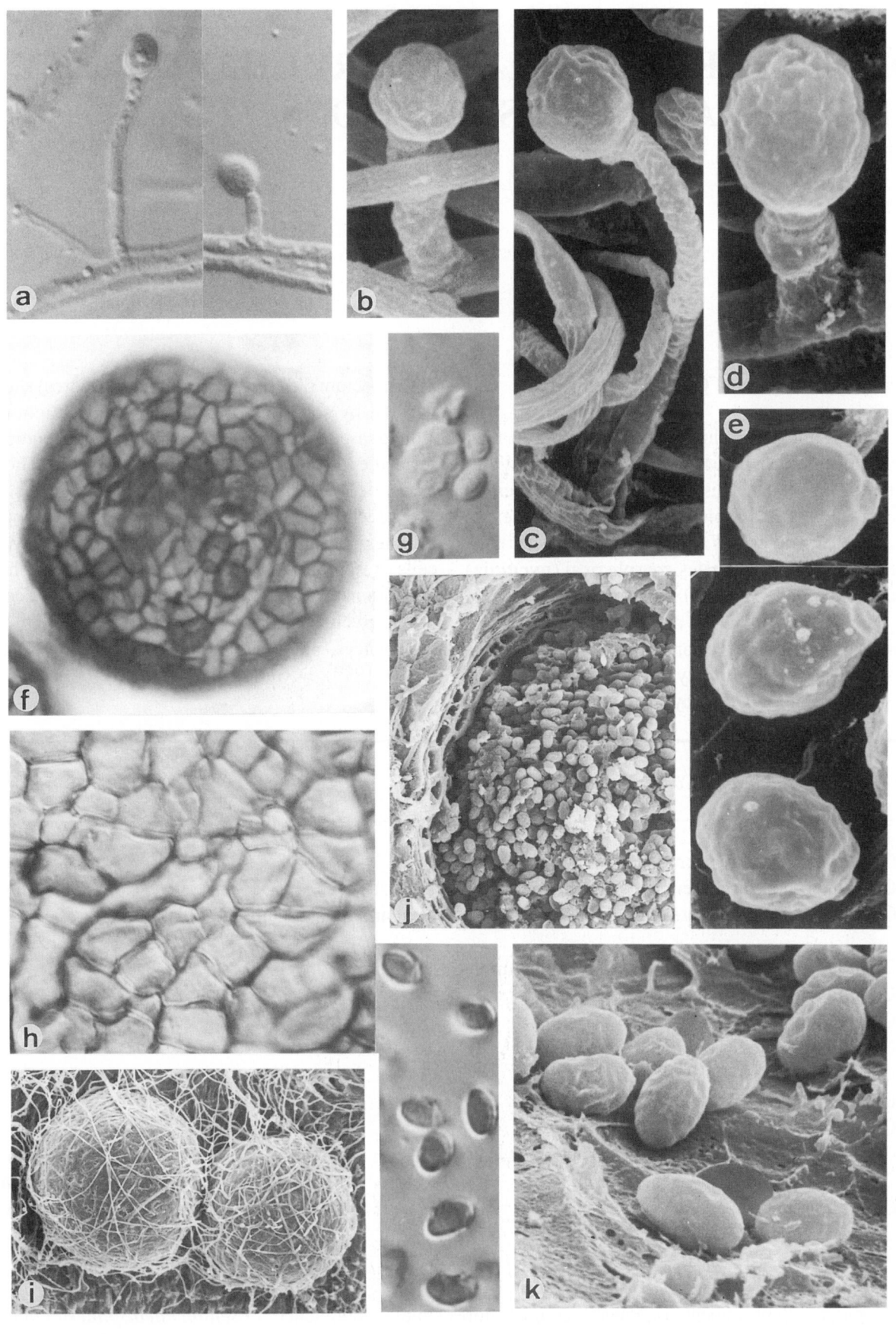

Pseudeurotium ovale, CBS 454.62. a-e. Conidiogenous cells and conidia; f, i. ascomata; g. ascus; h. part of peridium and ascospores; j. cross section of ascoma; k. ascospores. a. ×1600; b. ×8800; c. ×7000; d. ×1100; e. ×8500; f. ×512; g. ×1600; h. ×1600; i. ×280; j. ×720; k. ×4400.

311

COELOMYCETES: EXPLANATORY CHAPTER, KEY TO THE GENERA, AND DESCRIPTIONS

Criteria for distinction of Coelomycetes

Coelomycetes are fungi which produce conidia in fruit bodies (**conidiomata**). In such fruit bodies the propagules are mitotic, non-sexual. This is in contrast to the meiotic spores of Ascomycetes, which have arisen after a sexual process; sexual fruit bodies are referred to as ascomata.

Fruit bodies of Coelomycetes are spherical (**pycnidia**) with an apical opening (**ostiole**), with conidiogenous cells lining the inner cavity wall, or are open, cup-shaped (**acervuli**), in which case the conidiogenous cells compose a palisade on the fruit body surface. The difference between a pycnidium (Coelomycetes: with conidia) and

a perithecium (Ascomycetes: with ascospores) should be verified by squashing a fruit body under a cover slip. Young ascospores are inside asci, while conidia are formed exogenously on conidiogenous cells. The difference between an acervulus of Coelomycetes and a sporodochium of hyphomycetes lies in the conidiogenous cells being inserted on ellipsoidal, pseudoparenchymatous cells, resp. being part of a bundle of erect hyphae. Conidiomata may occur singly, or are united in a compact tissue (**pycnostroma**).

References. Punithalingam (1969, 1979), B. Sutton (1980, 1999), D. Sutton (1999).

Key to the treated genera of Coelomycetes:

1a. Conidia produced in pycnidia or pycnostromata → **2**
1b. Conidia produced on acervuli .... *Colletotrichum* (314)
2a. Conidia (1-)2 (-5)-septate, with darker central cell; mycelium, breaking up into arthroconidia ....
.... *Nattrassia* (329)
2b. Conidia 0-1-septate; mycelium not breaking up into arthroconidia → **3**
3a. Conidia with longitudinal striations .... *Lasiodiplodia* (325)
3b. Conidia without striations → **4**
4a. Conidia mostly over 4 μm wide → **5**
4b. Conidia mostly less than 4 μm wide → **8**
5a. Conidia hyaline → **6**
5b. Conidia brown → **7**
6a. Conidia with gelatinous appendages at the ends .... *Phyllosticta*[1]
6b. Conidia without appendages .... *Dothiorella*[2]
7a. Conidia 9-12 × 6-9 μm .... *Sphaeropsis* (359)
7b. Conidia 5.5-6.5 × 3.2-3.4 μm .... *Microsphaeropsis* (327)
8a. Pycnidia with stiff apical setae → **9**
8b. Pycnidia without setae → **10**
9a. Pycnidia with prominent, long necks; conidiophores simple, unbranched *Pseudochaetosphaeronema* (353)
9b. Pycnidia with short necks; conidiophores usually septate and branched *Pyrenochaeta* (355)
10a. Conidia (sub)hyaline → **11**
10b. Conidia pigmented .... *Coniothyrium* (323)
11a. Conidia of two types: fusiform and filiform .... *Phomopsis*[3]
11b. Conidia all of the same type → **12**
12a. Conidiogenous cells discrete → **13**

312

12b. Conidiogenous cells integrated in septate conidiophores ⟶ **14**

13a. Pycnidia superficial . . . . ***Chaetophoma***[4]

13b. Pycnidia at least partly immersed . . . . ***Phoma* (332)**

14a. Conidiophores filiform, septate and branched; conidiogenous cells all integrated . . . . . ***Pleurophoma* (349)**

14b. Conidiophores comparatively short; discrete conidiogenous cells also present . . . . . ***Pleurophomopsis* (351)**

[1]An unidentified *Phyllostictina* species was reported by Gip & Paldrok (1967) from onychomycosis. Note that *Phyllostictina* is a synonym of *Phyllosticta* (Van der Aa, 1973).

[2]Strains from human nails were listed by Punithalingam (1979), but no proven cases have been published. Note that *Dothiorella* is a synonym of *Fusicoccum* (Sutton, 1980).

[3]An unidentified *Phomopsis* species was reported by Sutton *et al.* (1999b) from osteomyelitis.

[4]*Chaetophoma dermo-unguis* Singh & Barde was reported from a nail infection (Singh & Barde, 1986). However, the type material was too poor to establish its identity.

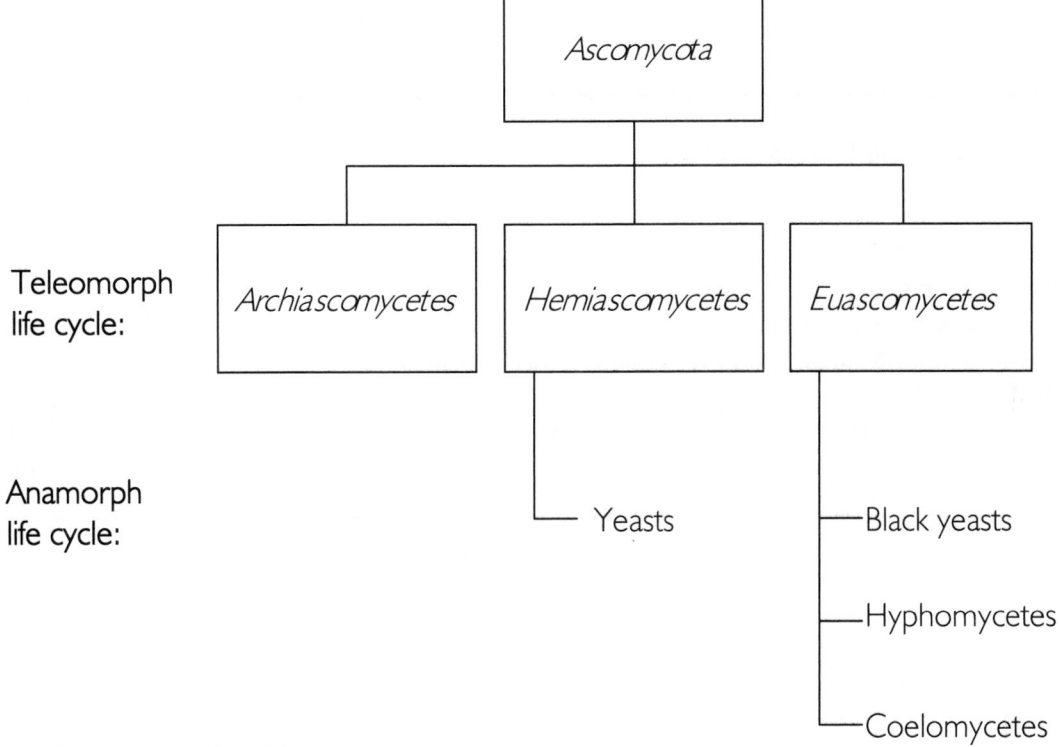

Fig. 25. Diagram of taxonomic position of the Coelomycetes.

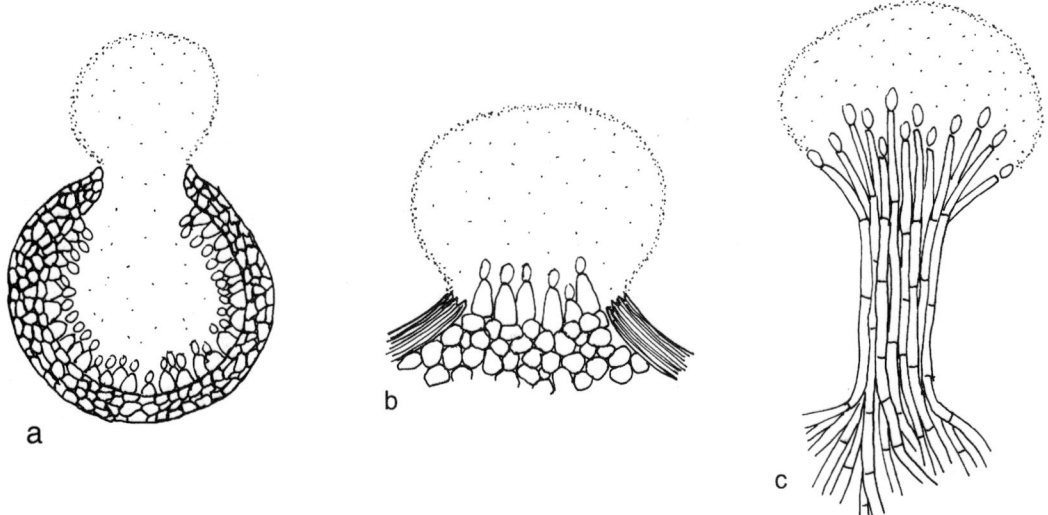

Fig. 26. Diagram of asexual fruit bodies of Coelomycetes. a. Pycnidium; b. acervulus, often breaking through the outer layer of host plant tissue; c. sporodochium of hyphomycetes.

Coelomycetes. Genus: *COLLETOTRICHUM*

Generic description. Colonies effuse, grey to dark brown or olive-green, sometimes with pinkish shades. Conidiomata stromatic, acervular, with a wall of isodiametric or elongate cells, usually intermingled with or surrounded by stiff, pigmented setae; phialides in dense layers, cylindrical or tapering; conidia formed in basipetal order, falcate, fusiform, cylindrical or ellipsoidal, non-septate, hyaline, forming pigmented appressoria when germinating. Sclerotia sometimes present in culture, dark brown to black, occasionally setose, often confluent.

Teleomorph. *Glomerella* (*Ascomycota, Euascomycetes, Polystigmatales: Polystigmataceae*).

General remark. *Colletotrichum* species are plant pathogens, only exceptionally being involved in human mycoses (Midha *et al.*, 1996).

References. Von Arx (1957, 1970), Sutton (1980), Bailey & Jeger (1992), Johnston & Jones (1997), Buddie *et al.* (1999).

Key to the treated species of Colletotrichum:

1a.	Conidia straight → **2**	
1b.	Conidia falcate → **3**	
2a.	Colonies dark; sclerotia present	*C. coccodes* **(315)**
2b.	Colonies with pinkish patches and usually dark brown to vinaceous reverse; sclerotia mostly absent	*C. gloeosporioides* **(319)**
3a.	Conidia 2.0-3.5 µm wide; appressoria more or less clavate	*C. dematium* **(317)**
3b.	Conidia 3.5-5.0 µm wide; appressoria irregular	*C. graminicola* **(312)**

Colletotrichum coccodes (Wallr.) S. Hughes

Colony characteristics. Colonies (OA) dark brown, consisting of numerous black sclerotia, occasionally with sparse, white aerial mycelium; conidial masses honey-coloured; reverse dark brown.

Microscopy. Sclerotia abundant, setose, spherical, often confluent. Conidia straight, fusiform, attenuated at the ends, 16-22 × 3-4 μm. Appressoria common, clavate, brown, 11.0-16.5 × 6.0-9.5 μm.

Molecular diagnostics. ITS restriction map based on NCBI Z32929, Z32930 and X73813:

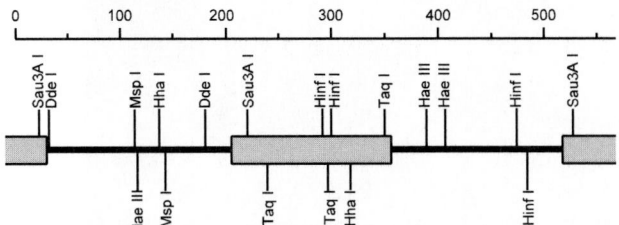

Pathogenicity. BSL-1. Keratitis (Liesegang & Forster, 1980).

References. Mordue (1967), Sutton (1980).

Antifungal susceptibility.

Antifungal	Mean MICs	MICs range	Strains	Reference
AMB	0.13	<0.03->16	5	Guarro *et al.* (1998)
5FC	256	>128	5	Guarro *et al.* (1998)
FLZ	24.2	8->64	5	Guarro *et al.* (1998)
ITZ	0.65	0.06->16	5	Guarro *et al.* (1998)
KTZ	1.52	0.125-16	5	Guarro *et al.* (1998)
MCZ	1.52	0.5->16	5	Guarro *et al.* (1998)

Nomenclature. *Chaetomium coccodes* Wallroth - Fl. Crypt. Germ. 2: 265, 1833 ≡ *Colletotrichum coccodes* (Wallroth) S. Hughes - Can. J. Bot. 36: 754, 1958.

Vermicularia atramentarium Berkeley & Broome - An. Mag. Nat. Hist. 5: 978, 1850 ≡ *Colletotrichum atramentarium* (Berkeley & Broome) Taubenheim - Mem. N.Y. Bot. Gdn. 6: 549, 1916.

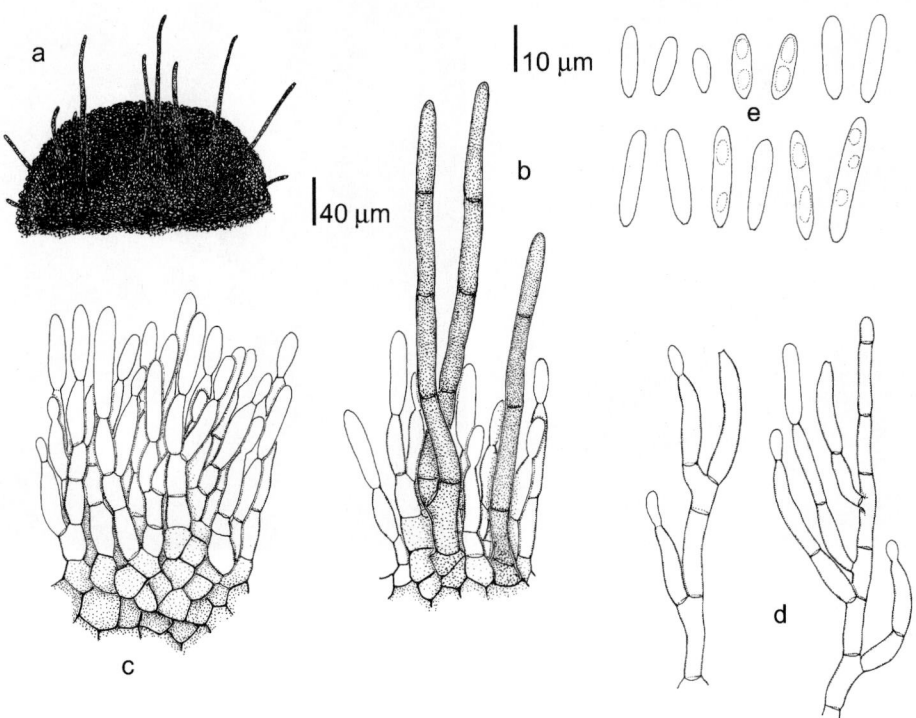

Colletotrichum coccodes, IMI 136601. a. Part of a setose sclerotium; b, c. section of a conidioma with conidiophores and setae; d. conidiophores; e. conidia.

315

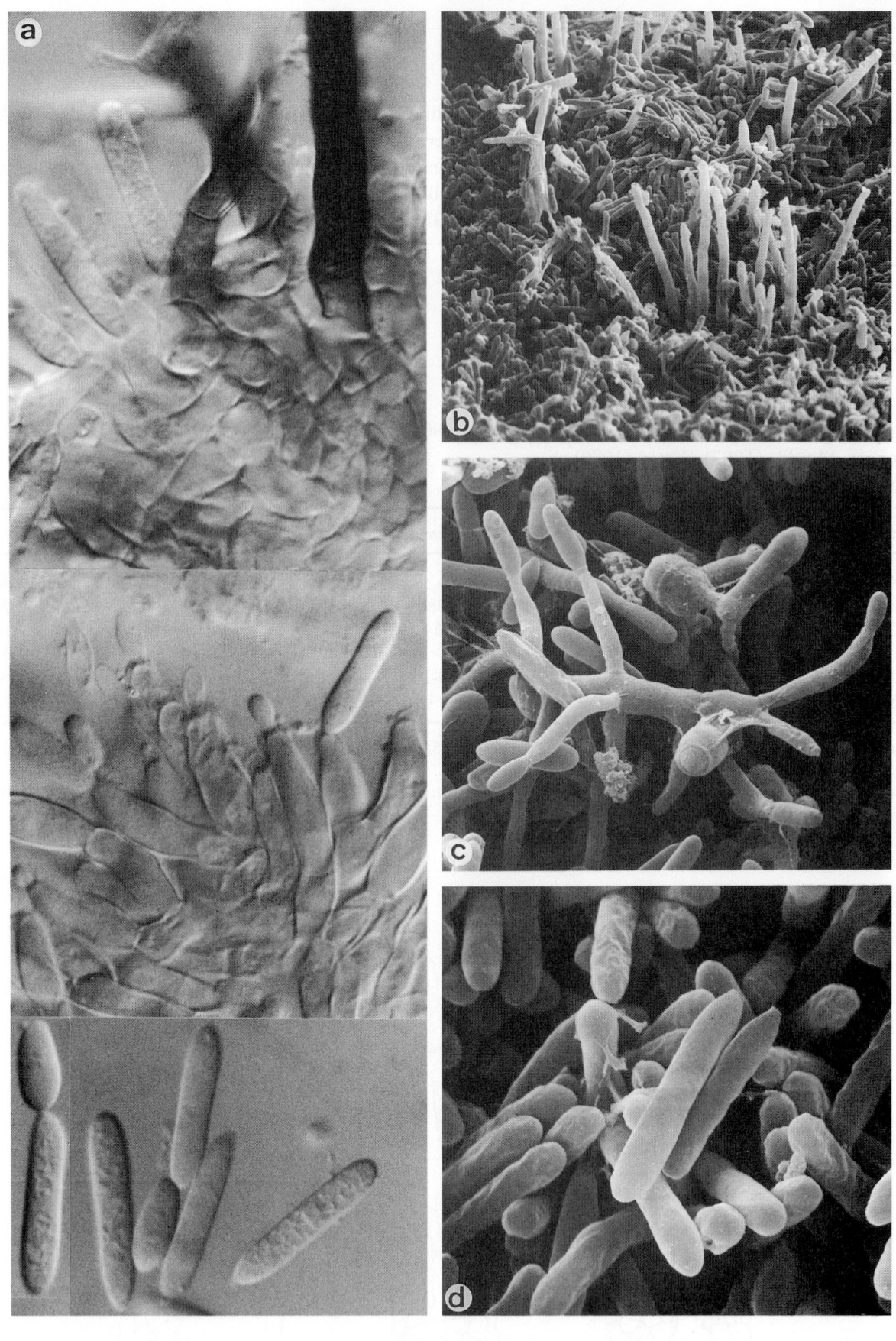

Colletotrichum coccodes, IMI 136601. a, c, d. Conidiophores and conidia; b. part of a setose conidioma. a. ×1600; b. ×450; c. ×5800; d. ×6000.

Colletotrichum dematium (Pers.: Fr.) Grove

Colony characteristics. Colonies (OA) mouse grey, with woolly aerial mycelium, becoming felt-like; reverse dark brown.

Microscopy. Sclerotia and setae abundant. Conidia falcate, with acute apex, 20-30 × 2.0-3.5 μm. Appressoria abundant, brown, clavate to circular, 8.0-11.5 × 6.5-8.0 μm.

Molecular diagnostics. ITS restriction map:

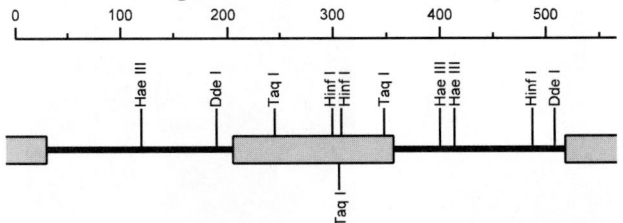

Pathogenicity. BSL-1. Keratitis (Liao *et al.*, 1983).
References. Sutton (1980), Bailey & Jeger (1992).
Antifungal susceptibility.

Antifungal	Mean MICs	MICs range	Strains	Reference
AMB	0.25	0.06-1	4	Guarro *et al.* (1998)
5FC	128	16->128	4	Guarro *et al.* (1998)
FLZ	64	8->64	4	Guarro *et al.* (1998)
ITZ	9.51	0.25->16	4	Guarro *et al.* (1998)
KTZ	4.76	0.25-16	4	Guarro *et al.* (1998)
MCZ	4.0	2-8	4	Guarro *et al.* (1998)

Nomenclature. *Sphaeria dematium* Persoon - Syn. Meth. Fung. p. 88, 1801 ≡ *Vermicularia dematia* Persoon:Fries - Summa Veg. Scand. p. 420, 1849 ≡ *Colletotrichum dematium* (Persoon:Fries) Grove - J. Bot., Lond. 56: 341, 1918.

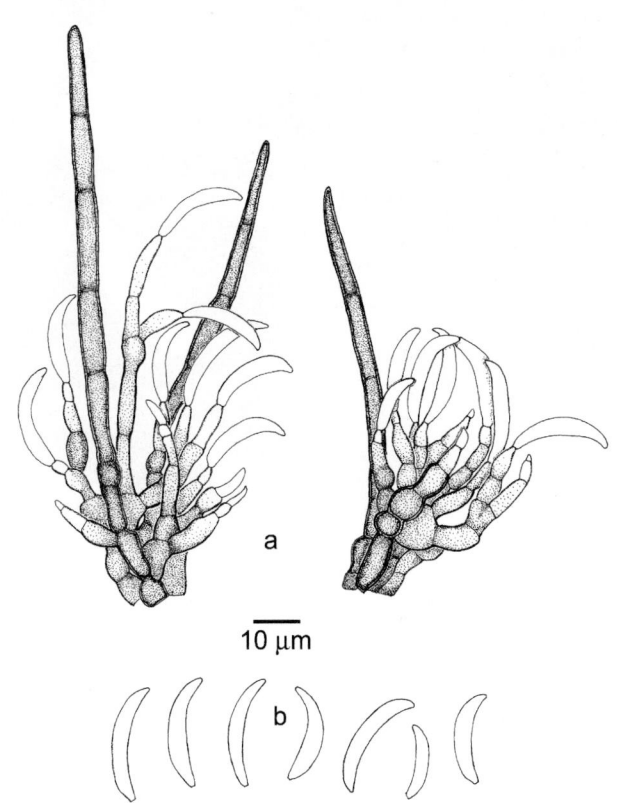

Colletotrichum dematium, IFO 6704. a. Section of conidioma with conidiophores, conidia and setae; b. conidia.

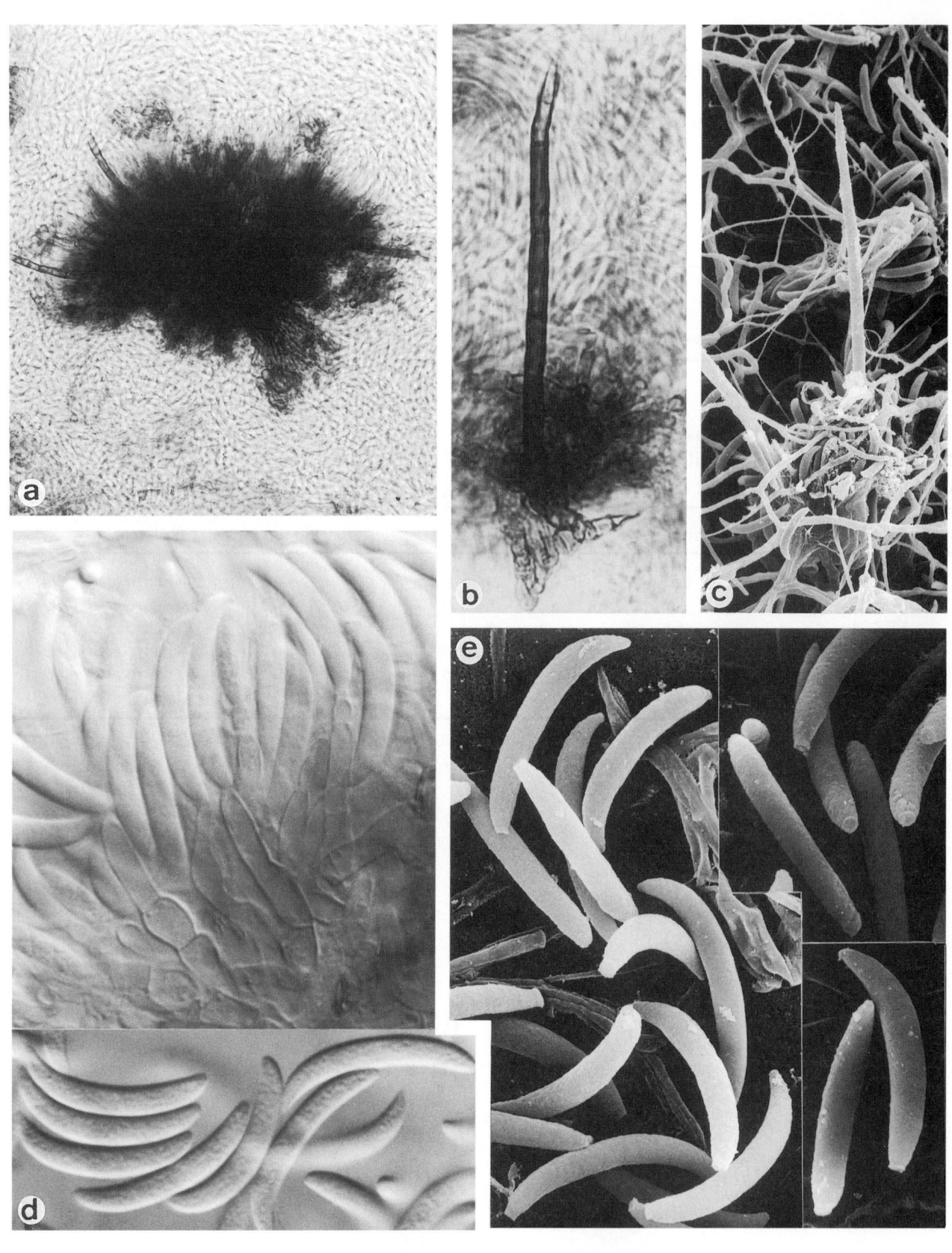

Colletotrichum dematium, IFO 6704. a-c. Conidioma with conidiophores, conidia and setae; d,e. conidiophores and conidia. a. ×320; b. ×512; c. ×800; d. ×1600; e. ×2850.

Colletotrichum gloeosporioides (Penz.) Sacc.

Colony characteristics. Colonies (OA) extremely variable, effuse, grey to brown, with pinkish patches; reverse dark brown with vinaceous stains.

Microscopy. Stromata generally absent in culture. Conidia borne on elongated phialides in acervular conidiomata, or, in early stages of development, on solitary fertile hyphae and appressoria. Conidia straight, cylindrical, obtuse at the apex, 9-24 × 3.0-4.5 µm. Appressoria 6-20 × 4-12 µm, clavate or irregular.

Teleomorph. *Glomerella cingulata* (Stoneman) Spaulding & v. Schrenk (*Ascomycota, Euascomycetes, Polystigmatales: Polystigmataceae*).

Molecular diagnostics. ITS restriction map based on NCBI Z32949, Z32948 and X73811:

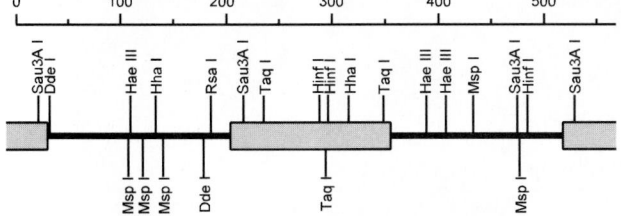

Pathogenicity. BSL-2. Keratitis (Shukla *et al.*, 1983; Matsuzaki *et al.*, 1988), subcutaneous infection (Guarro *et al.*, 1998). Another isolate that recently was reported to cause a subcutaneous infection in a renal transplant recipient (Castro *et al.*, 2000) was reidentified as *Colletotrichum crassipes* (Speg.) v. Arx (J. Guarro, unpublished data).

References. Mordue (1971), Sutton (1980).

Antifungal susceptibility.

Antifungal	Mean MICs	MICs range	Strains	Reference
AMB	0.08	0.03-0.25	7	Guarro *et al.* (1998)
5FC	116	16->128	7	Guarro *et al.* (1998)
FLZ	35.3	4->64	7	Guarro *et al.* (1998)
ITZ	0.6	<0.03->16	7	Guarro *et al.* (1998)
KTZ	0.74	0.125-4	7	Guarro *et al.* (1998)
MCZ	2.0	0.25-4	7	Guarro *et al.* (1998)

Nomenclature. *Vermicularia gloeosporioides* Penzig - Michelia 2: 450, 1880 ≡ *Colletotrichum gloeosporioides* (Penzig) Saccardo - Fung. Agrum. 2: 6, 1882.

Gloeosporium cingulatum Stoneman - Bot. Gaz. 26: 69, 1898 ≡ *Glomerella cingulata* (Stoneman) Spaulding & von Schrenk - Science, Ser. 2, 17: 751, 1903.

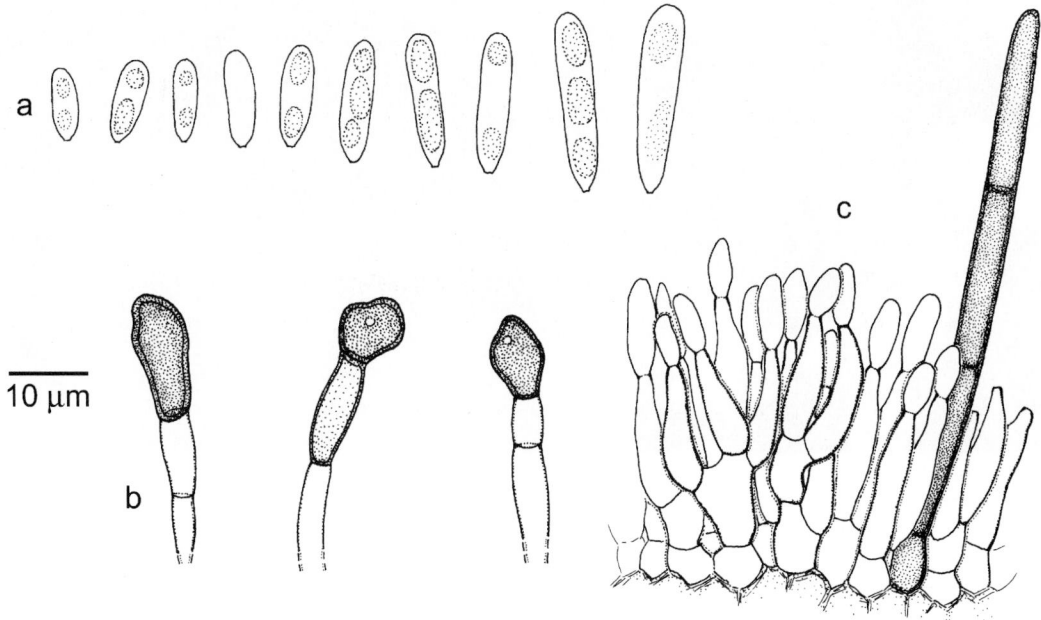

Colletotrichum gloeosporioides, CBS 465.83. a. Conidia; b. appressoria; c. Part of a conidioma.

319

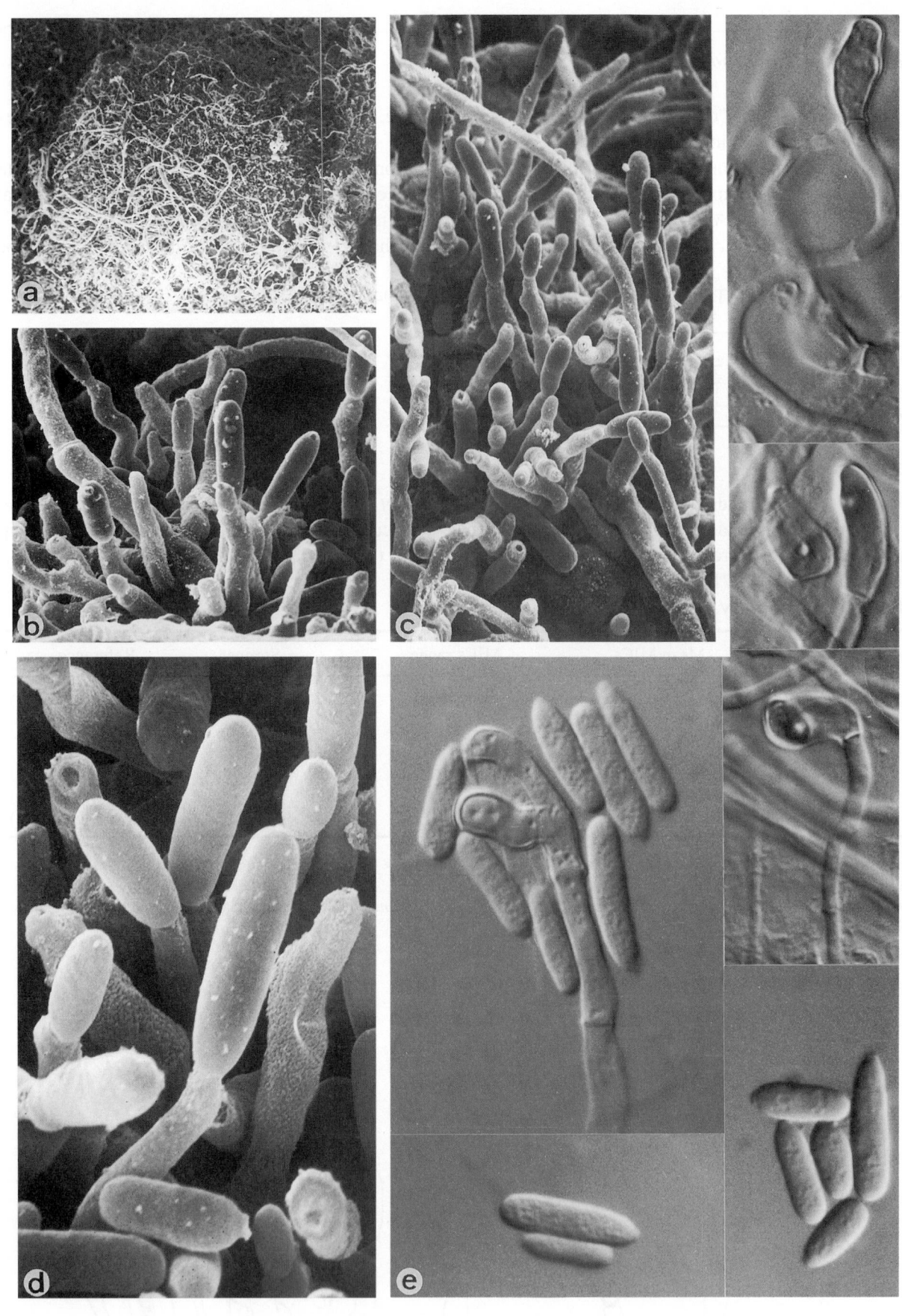

Colletotrichum gloeosporioides, CBS 465.83. a. Conidioma; b-d. conidiophores; e. appressoria and conidia. a. ×95; b. ×2000; c. ×1800; d. ×5000; e. ×1600.

320

Colletotrichum graminicola (Ces.) Wilson

Colony characteristics. Colonies (OA) effuse, grey to brown, with salmon patches of conidial slime; reverse vinaceous to purple.

Microscopy. Stromata formed from dendroid, brown hyphae. Setae abundant. Conidia borne on elongated phialides in acervular conidiomata. Conidia fusiform to falcate, 23-29 × 3.5-5.0 µm. Appressoria 17-20 × 12-14 µm, irregular.

Teleomorph. *Glomerella tucumanensis* (Speg.) v. Arx & E. Müller (*Ascomycota, Euascomycetes, Polystigmatales: Polystigmataceae*).

Molecular diagnostics. ITS restriction map based on NCBI AF059676:

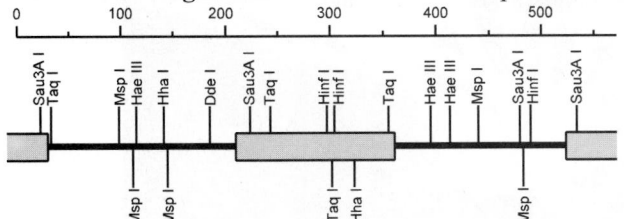

Pathogenicity. BSL-1. Plant pathogen on mais. A keratitis was reported by Ritterband *et al.* (1997).

References. Von Arx & Müller (1954), Sutton (1980).

Nomenclature. *Dicladium graminicola* Cesati - Flora 35: 398, 1852 ≡ *Colletotrichum graminicola* (Cesati) Wilson - Phytopathology 4: 110, 1914.

Psysalospora tucumanensis Spegazzini - Revta Agr. Vet., La Plata p. 228, 1896 ≡ *Glomerella tucumanensis* (Spegazzini) von Arx & E. Müller - Beitr. KryptFl. Schweiz 11-1: 195, 1954.

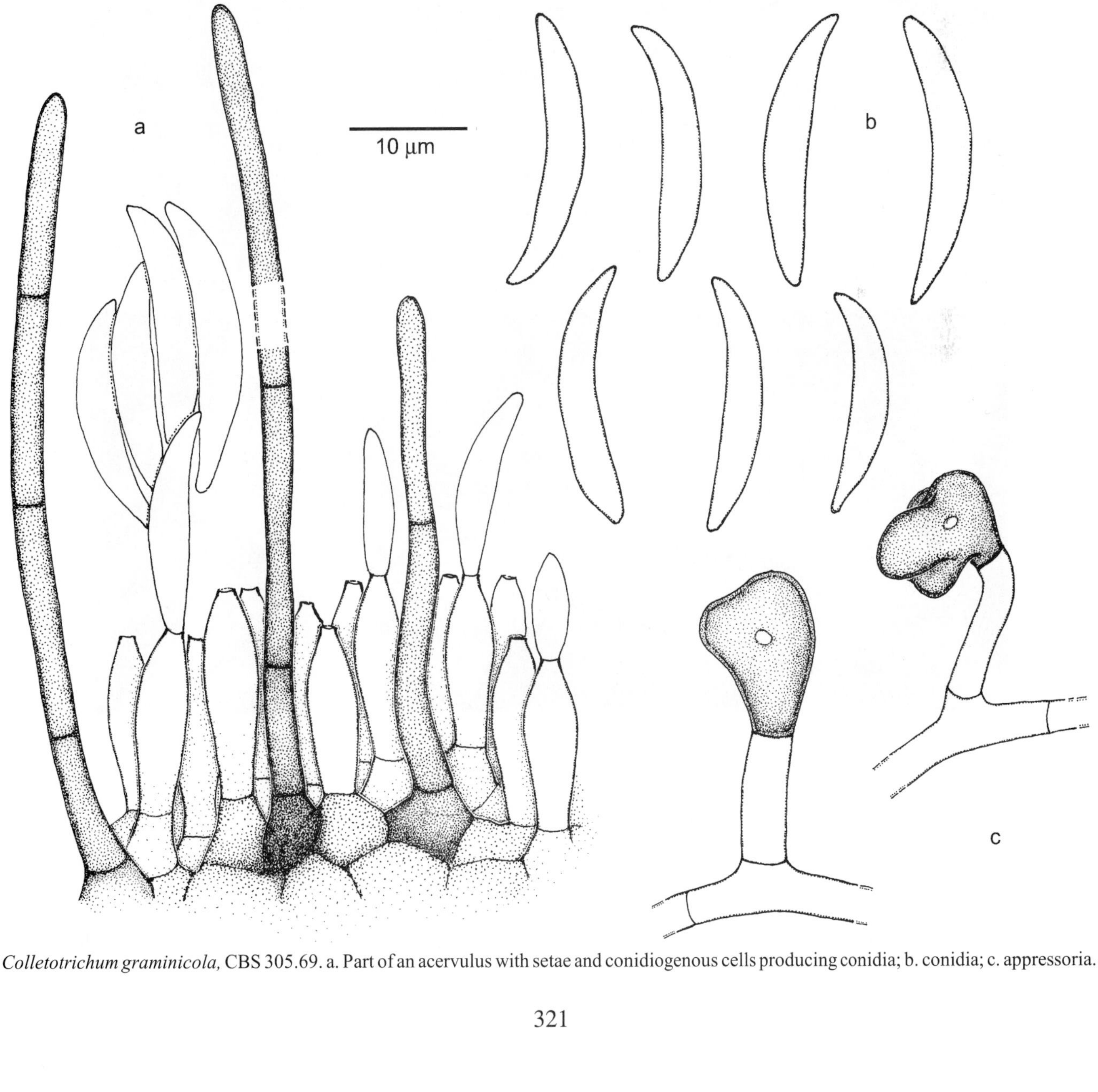

Colletotrichum graminicola, CBS 305.69. a. Part of an acervulus with setae and conidiogenous cells producing conidia; b. conidia; c. appressoria.

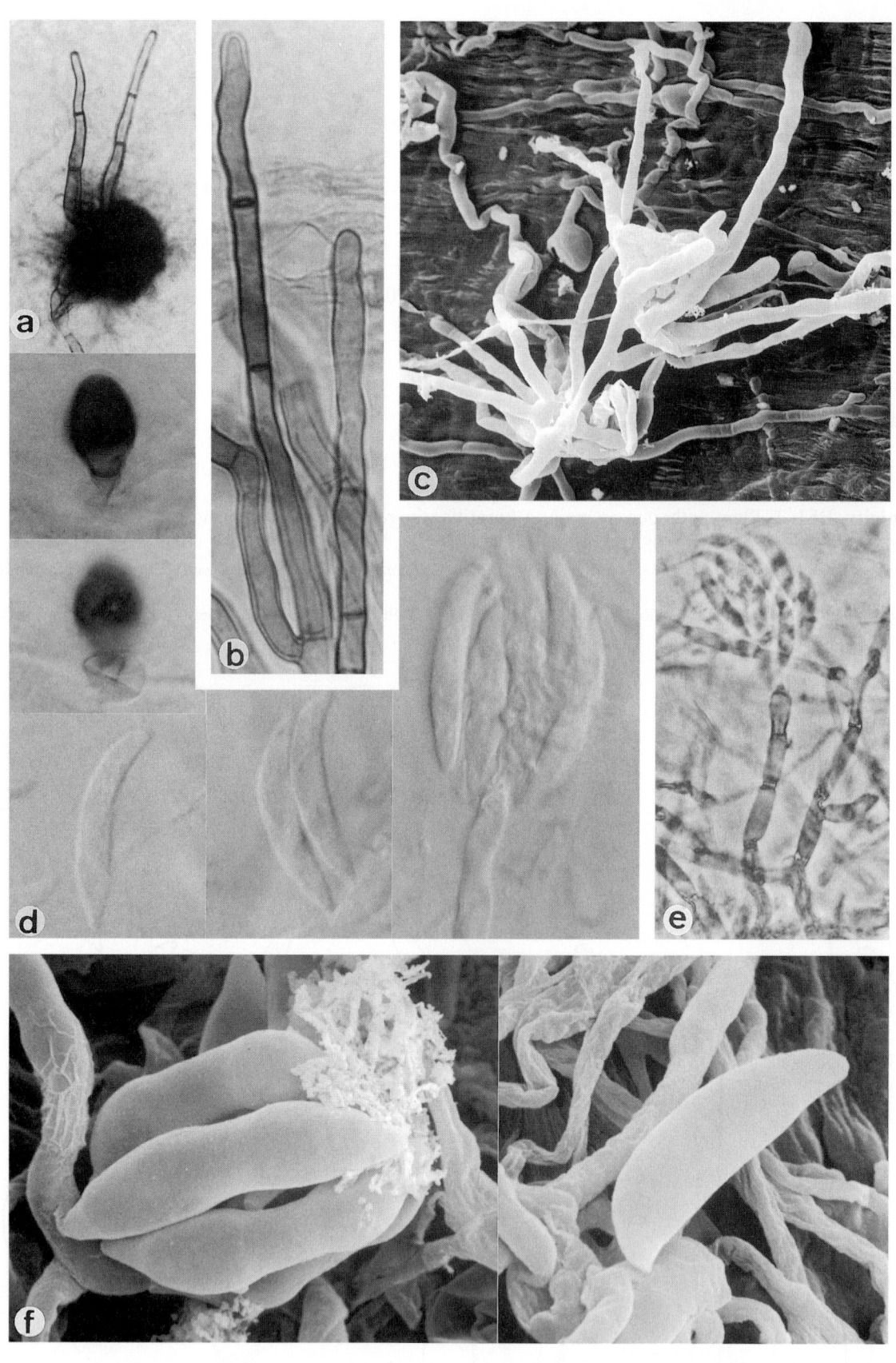

Colletotrichum graminicola, CBS 305.69. a-c. Setae; d. appressoria, conidia, and a conidiogenous cell producing conidia; e. conidiophore; f. conidia. a. ×640; b. ×1280; c. ×900; d. ×1600; e. ×640; f. ×4250.

Coelomycetes. Genus: *CONIOTHYRIUM*

Coniothyrium fuckelii Sacc.

Colony characteristics. Colonies (OA) moderately fast-growing, tan; reverse dark brown.

Microscopy. Pycnidia brown, solitary, occasionally aggregated, subspherical, 180-300 µm diam. Pycnidial wall pseudoparenchymatous of *textura angularis*. Conidiogenous cells hyaline, ampulliform, phialidic, lining the internal wall of the pycnidia. Paraphyses abundant, hyaline, cylindrical. Conidia pale brown, shortly cylindrical with rounded ends, unicellular, smooth-walled, 2.5-4.0 × 1.5-2.0 µm.

Teleomorph. *Leptosphaeria coniothyrium* (Fuckel) Sacc. (*Ascomycota, Euascomycetes, Pleosporales: Leptosphaeriaceae*).

Remark. The genus *Coniothyrium* needs revision, the species are as yet poorly circumscribed. Pycnidial species with 1-celled conidia, including *C. fuckelii*, should be reclassified in *Microsphaeropsis* (H.A. van der Aa, pers. comm.).

Pathogenicity. BSL-1. The species is a plant-pathogen, particularly found on *Rosaceae*. A case of liver infection in a human patient with acute myelogenous leukemia has been reported (Kiehn *et al.*, 1987). A second human case was reported by Schell & Perfect (1993), and further cases were announced by Paterson *et al.* (1997). In a tortoise, a superficial infection was observed (Austwick, 1983).

Antifungal susceptibility.

Antifungal	Mean MICs	Strains	Reference
AMB	0.14	5	McGinnis & Pasarell (1998b)
ITZ	0.38	5	McGinnis & Pasarell (1998b)
VCZ	0.22	5	McGinnis & Pasarell (1998b)

References. Wollenweber & Hochapfel (1937), Müller (1950).
Nomenclature. *Sphaeria coniothyrium* Fuckel - Symb. Mycol. p. 115, 1869 ≡ *Leptosphaeria coniothyrium* (Fuckel) Saccardo - Nuovo Giorn. Bot. Ital. 7: 317, 1875 ≡ *Coniothyrium fuckelii* Saccardo - Nuovo Giorn. Bot. Ital. 7: 317, 1875 (name change).

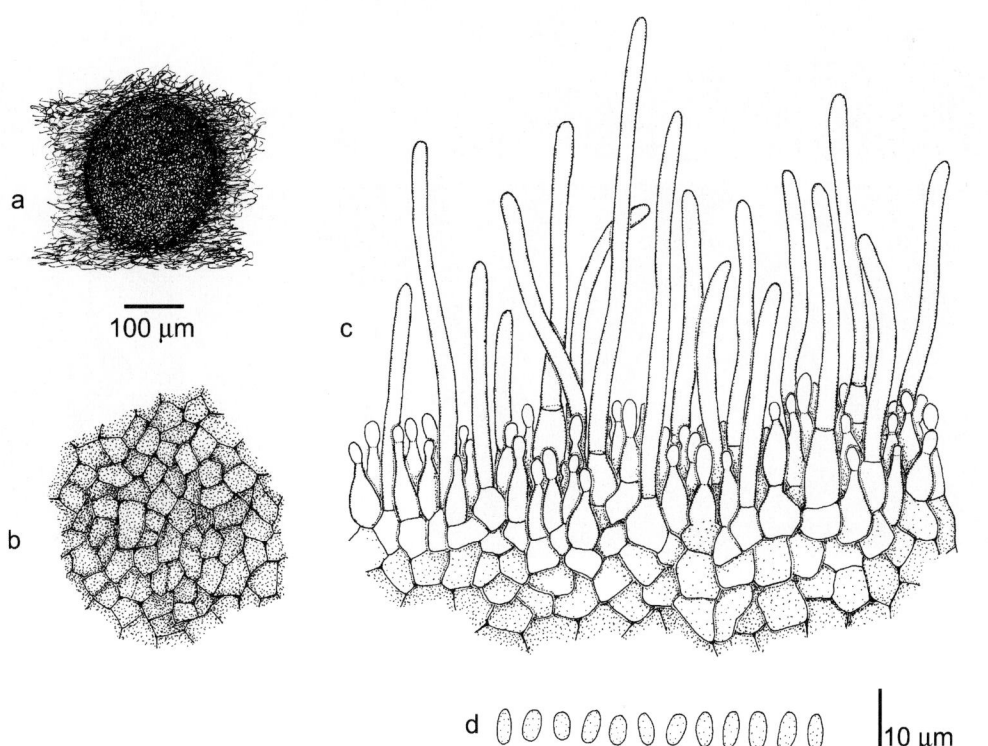

Coniothyrium fuckelii, RV 42525. a. Pycnidium; b. part of pycnidium wall; c. section of a pycnidium with conidiogenous cells and paraphyses lining the inner wall of the pycnidium; d. conidia.

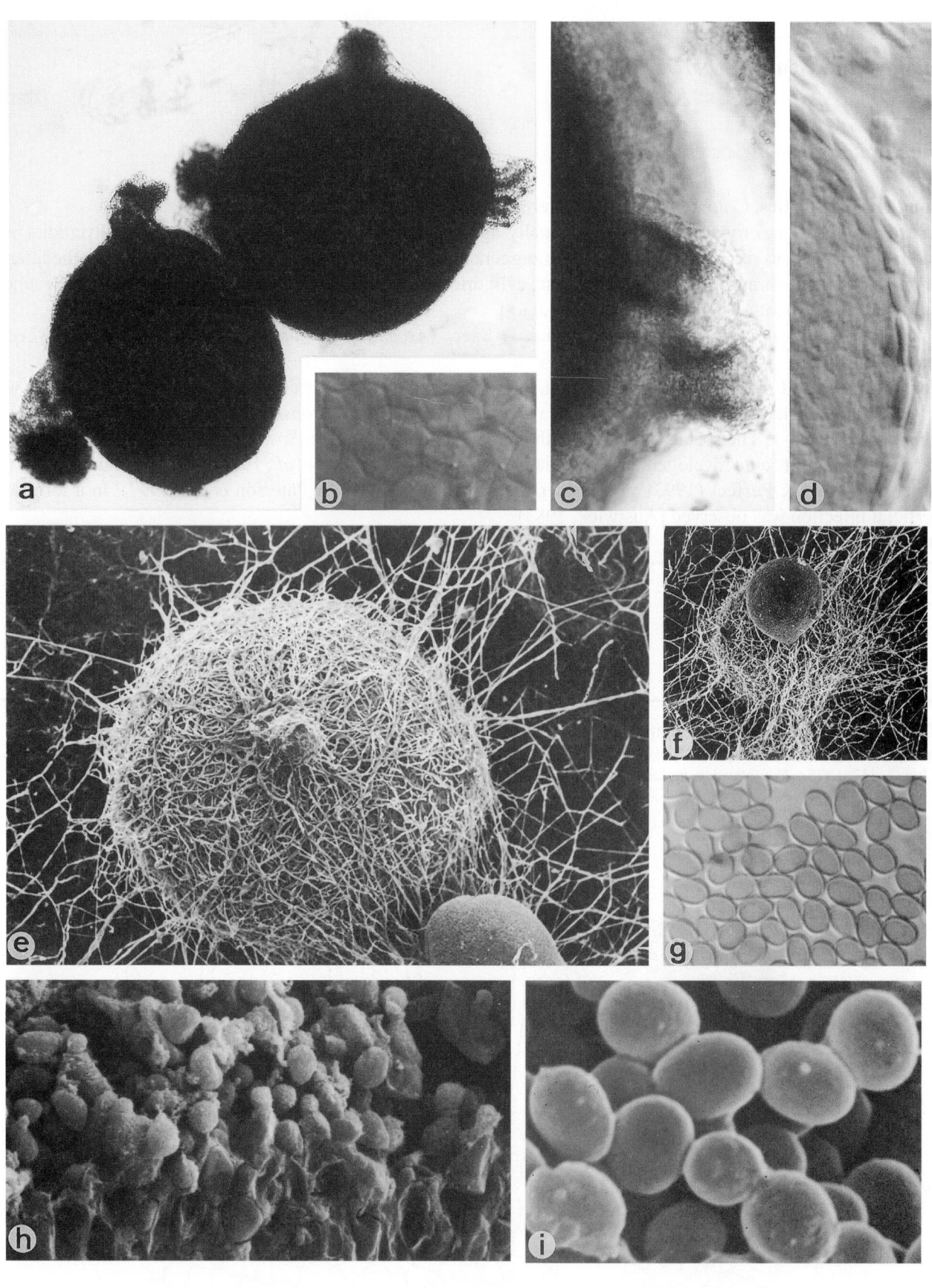

Coniothyrium fuckelii, RV 42525. a. Pycnidia; b. face view of pycnidium wall; c. detail of ostiole; d. pycnidium wall in side view; e, f. pycnidia with conidial masses; g, i. conidia; h. conidiogenous cells and conidia. a. ×140; b. ×1600; c. ×1280; d. ×1600; e. ×190; f. ×57; g. ×1600; h. ×2675; i. ×5200.

Coelomycetes. Genus: *LASIODIPLODIA*

Lasiodiplodia theobromae (Pat.) Griffon & Maublanc

Colony characteristics. Colonies (OA) grey to brown or black, with abundant aerial mycelium; reverse black.

Microscopy. Pycnostromata frequently hairy, up to 5 mm diam; occasionally solitary pycnidia are formed. Coniogenous cells hyaline, simple, cylindrical to somewhat obpyriform, holoblastic, 5-15 × 3 μm. Conidia maturing slowly, initially hyaline, later becoming 1-septate, dark brown, thick-walled, ellipsoidal, truncate at the base, 20-30 × 10-15 μm, with longitudinal striations. Paraphyses cylindrical, hyaline, septate.

Teleomorph. *Botryosphaeria rhodina* (Berk. & Curt.) v. Arx (*Ascomycota, Euascomycetes, Dothideales: Botryosphaeriaceae*).

Molecular diagnostics. SSU and ITS restriction maps based on NCBI U42476 and AF027760, resp.:

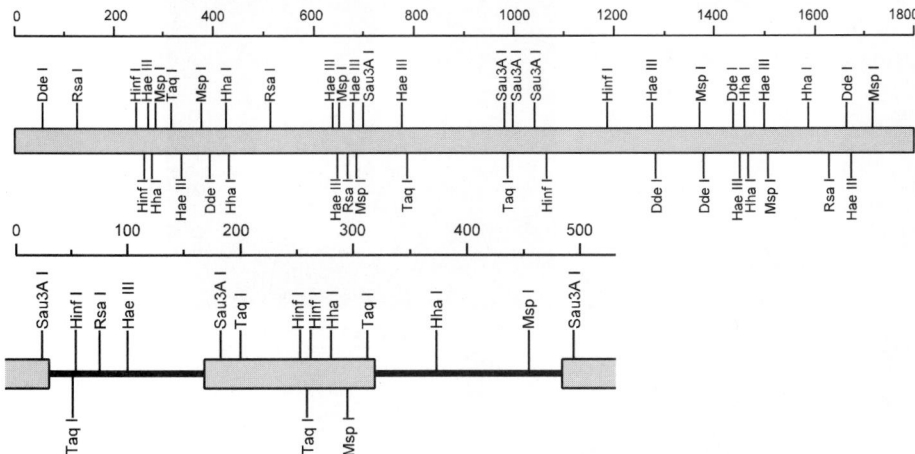

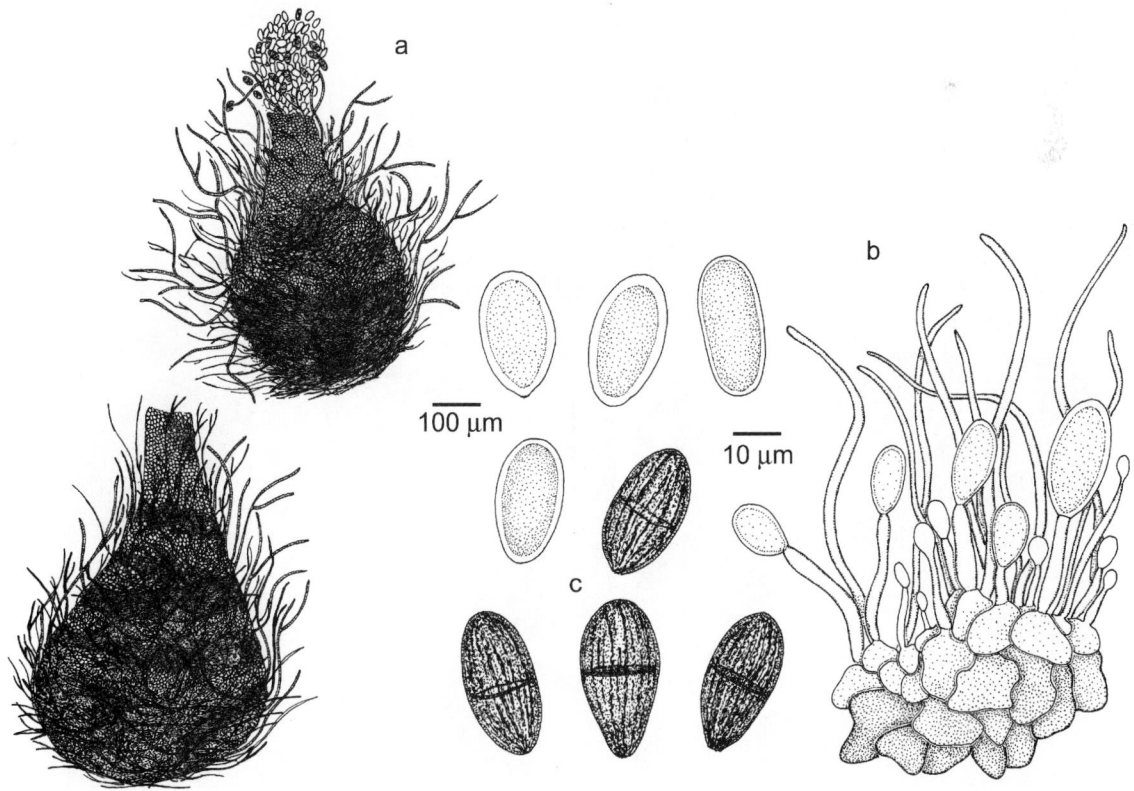

Lasiodiplodia theobromae, IFO 30028. a. Pycnidia; b. section of a pycnidium with conidiogenous cells and paraphyses lining the inner wall of the pycnidium; c. young and mature conidia.

Pathology. BSL-1. Frequently reported as agent of keratitis (Valenton *et al.*, 1975; Rebell & Forster, 1976; Slomovic *et al.*, 1985; Thomas *et al.*, 1991; Pasarell *et al.*, 1994; Borelli, 1995) and endophthalmitis (Borderie *et al., 1977)*. Also known from onychomycosis (Restrepo *et al.*, 1976; Borelli, 1995) and phaeohyphomycosis (Summerbell *et al.*, 1993b; Maslen *et al.*, 1996).

Reference. Punithalingam (1979).

Nomenclature. *Botryodiplodia theobromae* Patouillard, *in* Patouillard & Lagerheim - Bull. Trimest. Soc. Mycol. Fr. 8. 136, 1892 ≡ *Lasiodiplodia theobromae* (Patouillard) Griffon & Maublanc - Bull. Trimest. Soc. Mycol. Fr. 25: 57, 1909.

Physalospora rhodina Berkely & Curtis - Grevillea 17: 92, 1889 ≡ *Botryosphaeria rhodina* (Berkeley & Curtis) von Arx - Gen. Fungi Sporul. Pure Cult. p. 143, 1970.

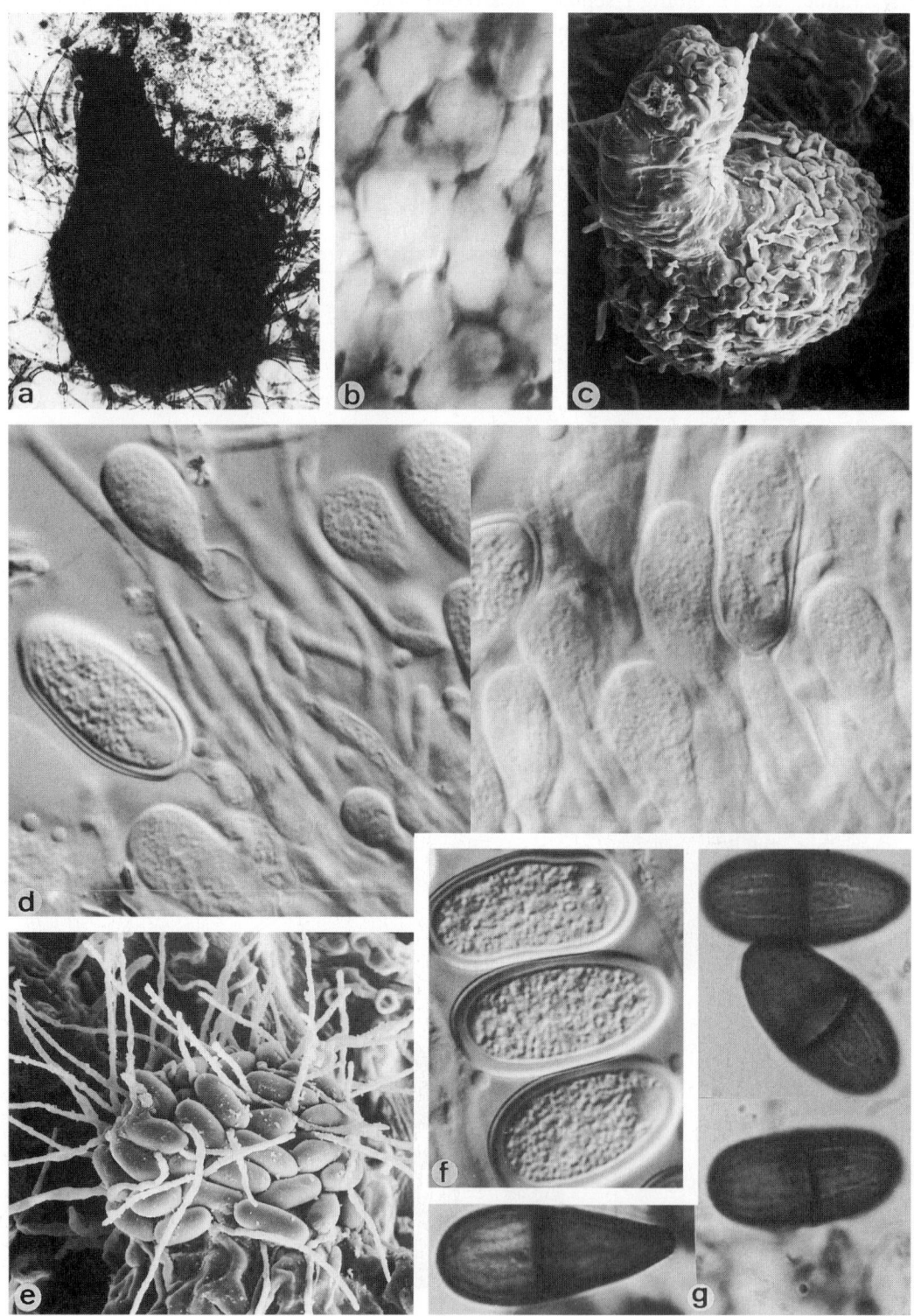

Lasiodiplodia theobromae, IFO 30028. a, c. Pycnidia; b. part of pycnidium wall; d. conidiogenous cells; e. conidial masses; f, g. conidia. a. ×128; b. ×1600; c. ×410; d. ×1600; e. ×620; f. ×1600; g. ×1280.

Coelomycetes. Genus: *MICROSPHAEROPSIS*

Microsphaeropsis olivacea Bon.

Colony characteristics. Colonies (PDA) expanding, velvety, white at the centre, becoming brown due to production of pycnidia, with concentric rings; reverse whitish.

Microscopy. Pycnidia (sub)spherical, brown, 200-240 × 160-200 µm; wall composed of *textura angularis*. Conidiogenous cells phialidic. Conidia 1-celled, pale brown, thick- and smooth-walled, ellipsoidal to cylindrical, 5.5-6.5 × 3.2-3.4 µm.

Pathogenicity. BSL-1. Plant pathogenic species. A skin infection in an otherwise healthy human patient was reported by Guarro *et al.* (1999e).

Reference. Guarro *et al.* (1999e).

Nomenclature. *Coniothyrium olivaceum* Bonorden, *in* Fuckel - Jahrb. Nassau. Ver. Naturk. 23-24: 377, 1870 ≡ *Microsphaeropsis olivacea* (Bonorden) Höhnel - Hedwigia 59: 267, 1917.

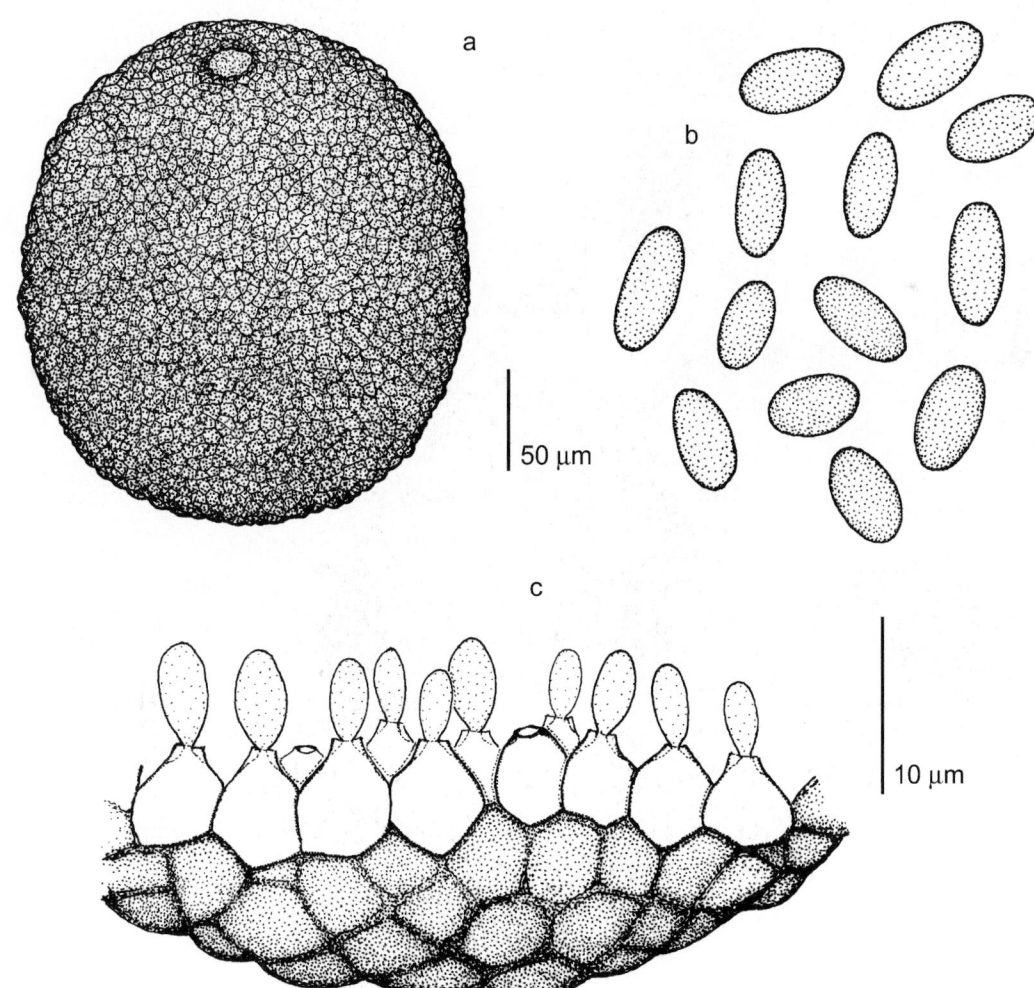

Microsphaeropsis olivacea, IMI 378167. a. Pycnidium; b. conidia; c. section of pycnidium with conidia produced from conidiogenous cells lining the inner cavity wall.

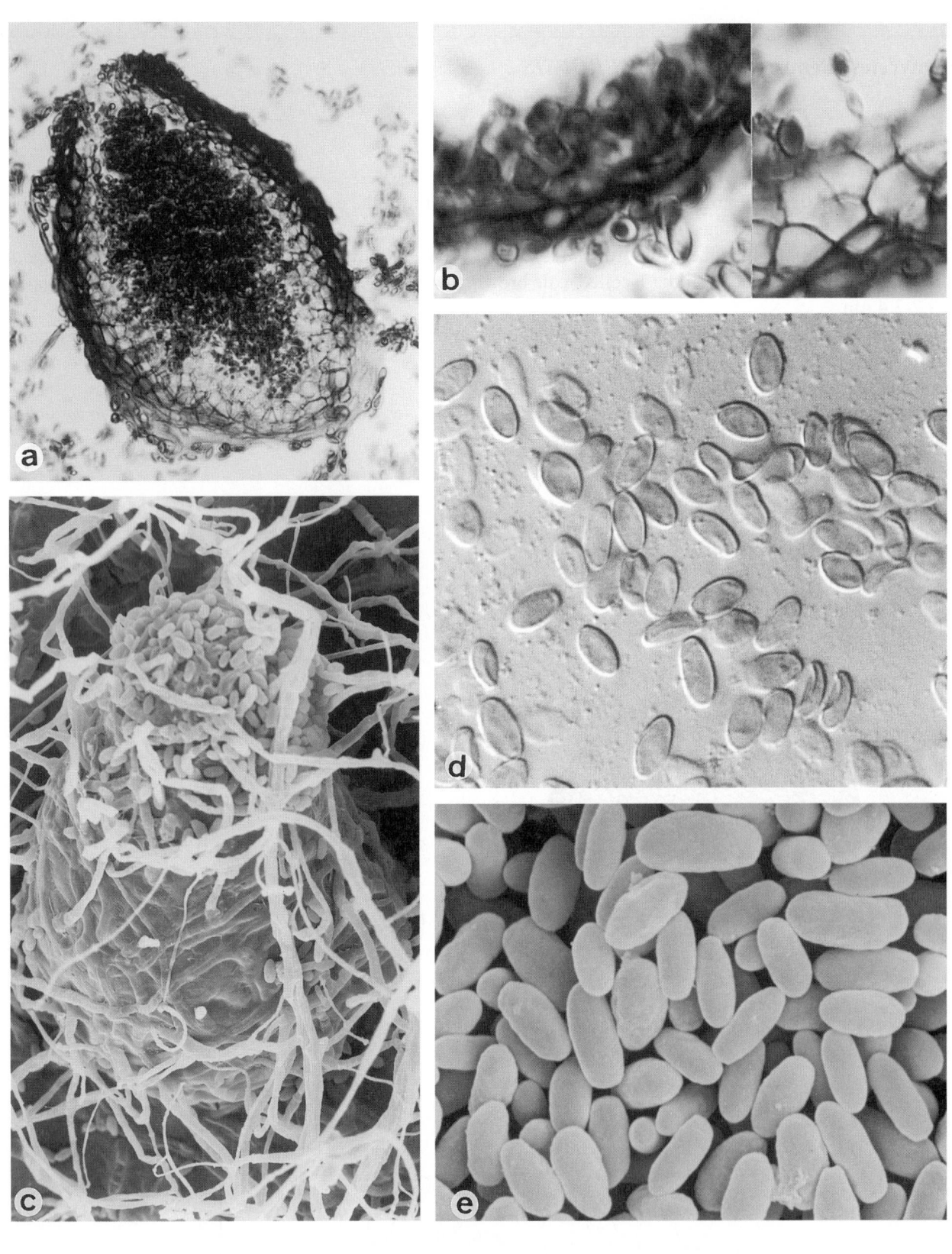

Microsphaeropsis olivacea, IMI 378167. a. Section of pycnidium; b. detail of conidiogenous cells and conidia; c. pycnidia; d, e. conidia. a. ×512; b. ×1400; c. ×1760; d. ×770; e. ×3500.

Coelomycetes. Genus: *NATTRASSIA*

Nattrassia mangiferae (H. Syd. & Syd.) Sutton & Dyko

Colony characteristics. Colonies (PDA) effuse, hairy, dark grey to blackish-brown, or white to greyish, then with a cream-coloured to deep ochraceous-yellow colony reverse.

Microscopy. Chains of arthroconidia develop from undifferentiated, broad (4-7 µm), brown hyphae, often with some slimy exudate, or from narrower branches. Arthroconidia (synanam. *Scytalidium dimidiatum*) continuous or septate, smooth- and often thick-walled, initially cylindrical, soon becoming ellipsoidal, 3.5-5.0 × 6.5-12.0 µm. Pycnidia (synanam. *Nattrassia mangiferae*) occasionally formed in old cultures, black, spherical, 110-300 µm diam. Pycnidia uni- or multilocular, then confluent. Conidiogenous cells densely packed, hyaline, somewhat tapering, 8-10 × 4 µm, holoblastic. Conidia smooth- and thin-walled, 1-celled and hyaline when young, (1-) 2 (-5)-septate when older, then central cell dark brown, 12-20 × 4-8 µm. Colourless mutants (*Scytalidium hyalinum*) often occur.

Physiology. Ribose, melibiose, lactose and *myo*-inositol assimilated; creatine and creatinine not assimilated; urease absent (Oyeka & Gugnani, 1989, 1991).

Molecular diagnostics. ITS restriction map:

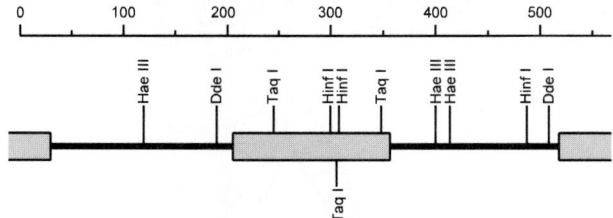

Differential diagnosis. The hyaline mutant resembles cultural states of mushrooms (*Basidiomyccota: Agaricales*; see p. 238), but these are urease positive. For further distinction of arthroconidial fungi, see under generic discussion of *Scytalidium* (p. 917).

Pathogenicity. BSL-2. The species, until recently generally known under the pycnidial synanamorph name as *Hendersonula toruloidea*, occurs on a wide range of tropical fruit trees, sometimes causing branch rots. It is frequently found causing hyperkeratotic skin and nail infections in people resident or formerly resident in the tropics (Hay & Moore, 1984; Moore, 1988). In tropical regions the fungus is very common as an infection of the sole in those patients who regularly go bare-footed (Kombila *et al.*, 1990). Also cases of onychomycosis are common (Elewski, 1996). Occasionally subcutaneous infections occur (Moore, 1992), all with traumatic portal of entry in patients with underlying disease (Sigler *et al.*, 1997b); endophthalmitis was reported by Al-Rajhi *et al.* (1993). Disseminated cases are extremely rare in patients with impaired natural immunity (Benne *et al.*, 1993).

References. Campbell (1974), Sigler & Carmichael (1976), Moore (1986, 1988, 1992), Badillet *et al.* (1987), Frankel & Rippon (1989), Gugnani & Oyeka (1989), Sutton & Dyko (1989), Oyeka & Gugnani (1991), Elewski & Greer (1991), Roeijmans *et al.* (1997).

Antifungal susceptibility.

Antifungal	Mean MICs	MICs range	MIC90	Strains	Reference
AMB	0.14	0.06-1	1	13	Guarro *et al.* (2000b)
AMB	0.5	nd	nd	10	McGinnis & Pasarell (1998b)
5FC	13.5	4-32	32	9	Unpublished data
FLZ	14.6	0.5-64	32	9	Guarro *et al.* (2000b)
ITZ	0.65	nd	nd	10	McGinnis & Pasarell (1998b)
KTZ	1.3	≤0.03-16	8	9	Guarro *et al.* (2000b)
MCZ	0.91	≤0.03-8	4	13	Unpublished data
VCZ	0.1	nd	nd	10	McGinnis & Pasarell (1998b)
VCZ	2.9	≤0.03->16	>16	13	Guarro *et al.* (2000b)

Nomenclature. *Torula dimidiata* Penzig - Michelia 2: 466, 1882 ≡ *Scytalidium dimidiatum* (Penzig) Sutton & Dyko - Mycol. Res. 93: 484, 1989.

Dothiorella mangiferae H. Sydow & Sydow, *in* H. Sydow, Sydow & Butler - Annls Mycol. 14: 1921, 1916 ≡ *Nattrassia mangiferae* (H. Sydow & Sydow) Sutton & Dyko - Mycol. Res. 93: 484, 1989.

Hendersonula toruloidea Nattrass - Trans Br. Mycol. Soc. 18: 197, 1933.
Scytalidium hyalinum Campbell & Mulder - Sabouraudia 15: 163, 1977.

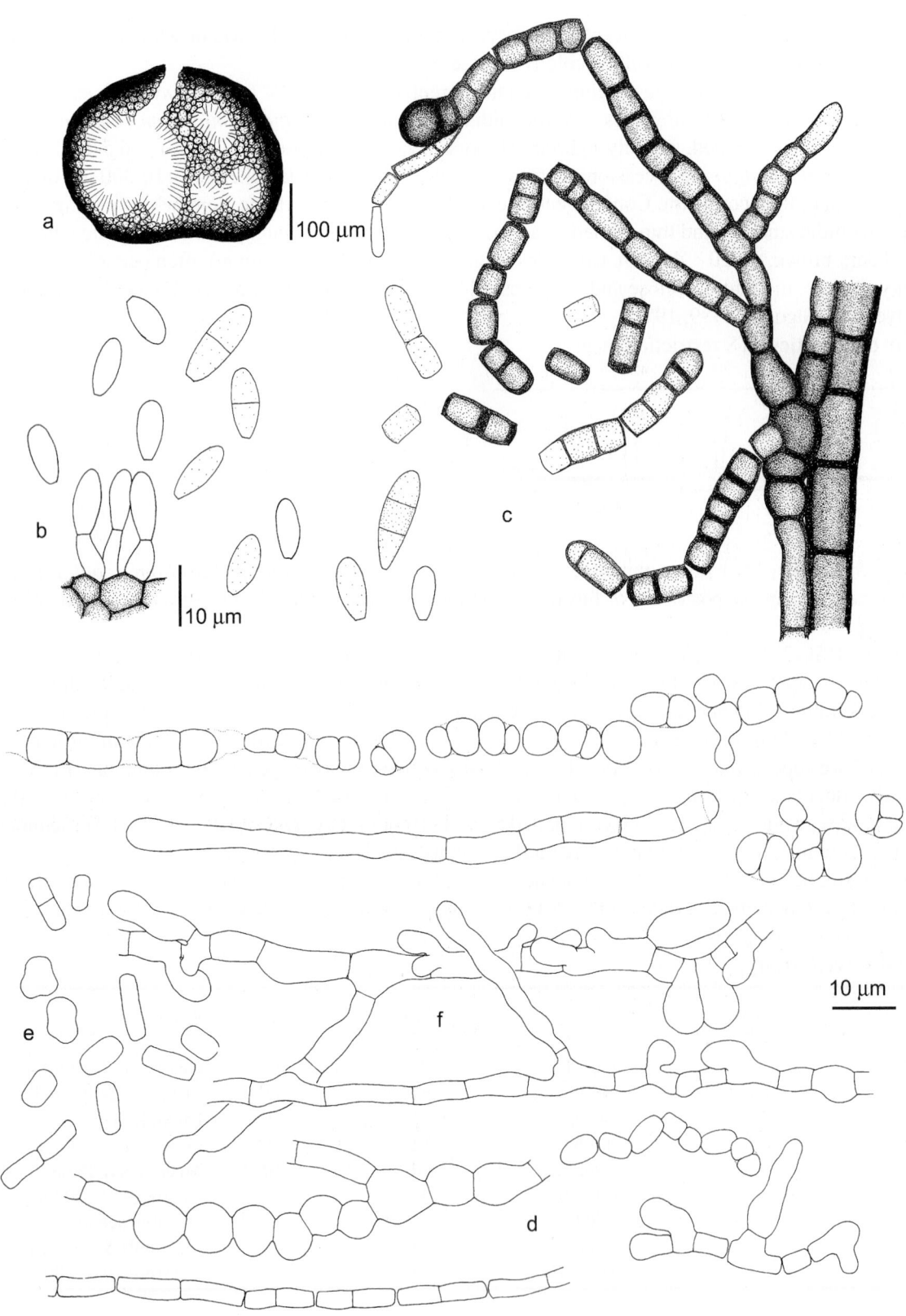

Nattrassia mangiferae, CBS 637.89. a, b. *Nattrassia mangiferae* synanamorph. a. Sketch of pycnidium; b. Conidiogenous cells and conidia. c. *Scytalidium dimidiatum* synanamorph, wild type., hyphae with arthroconidia. d-g. Hyaline mutant ('*Scytalidium hyalinum*'). d, e. CBS 145.78. d. Disarticulating hyphae; e. arthroconidia. f, g.. CBS 619.84. f. Hyphae with swollen cells; g. subdividing arthroconidia.

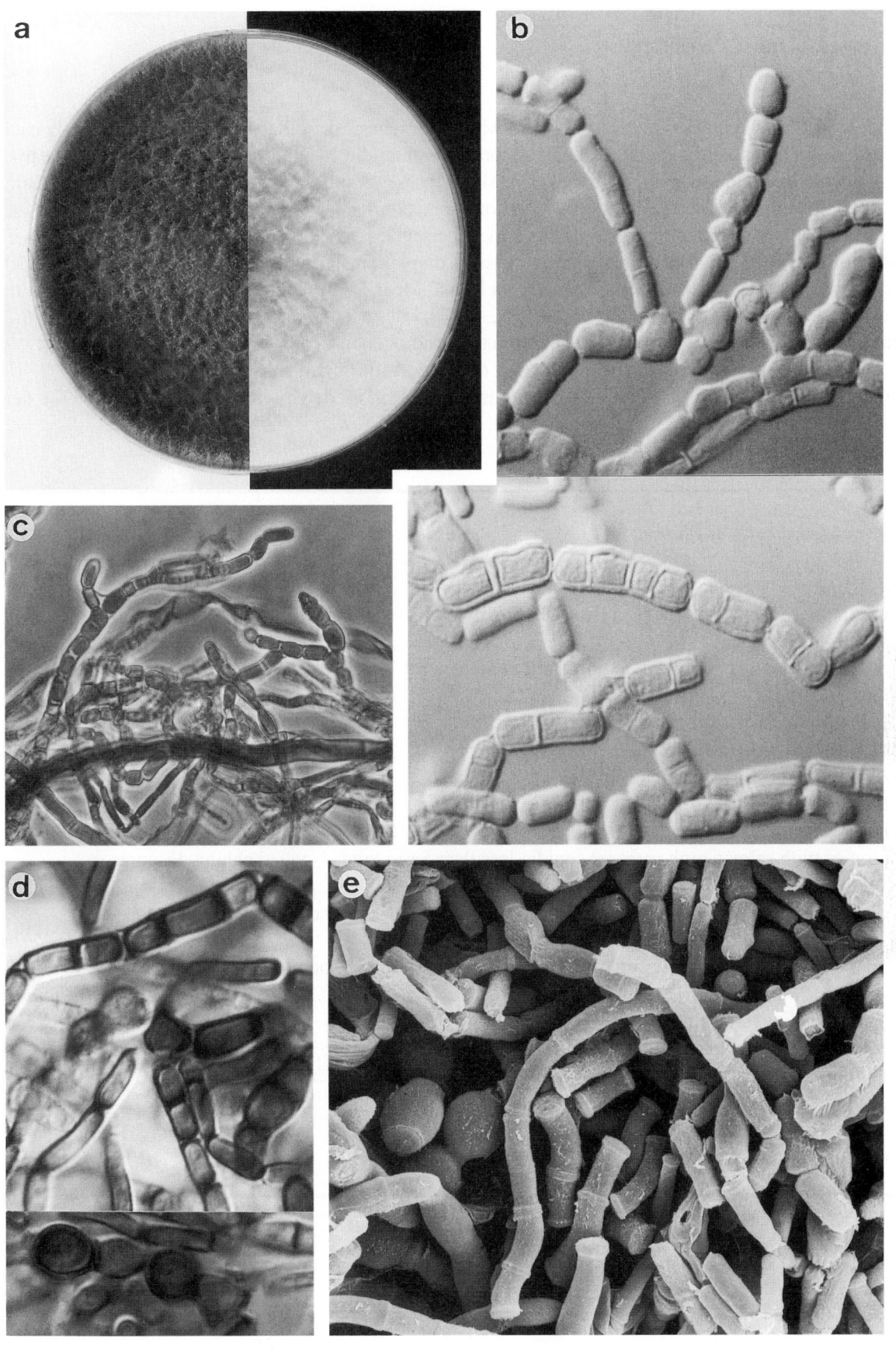

Nattrassia mangiferae, a, c-e. CBS 466.81, wild type (*Scytalidium dimidiatum* synanamorph); a, b. CBS 637.89, hyaline mutant (*Scytalidium hyalinum*). a. Black and white colony types; b-e. hyphae with arthroconidia and chlamydospores. b. ×1600; c. ×640; d. ×1600; e. ×2300.

Coelomycetes. Genus: *PHOMA*

Generic description. Colonies with dark pycnidia which are spherical, solitary or aggregated, usually each with a single ostiole, sometimes multi-ostiolate. Conidiogenous cells phialidic, hyaline, lining the inner wall of the pycnidium, producing abundant conidia which are extruded in slimy masses. Conidia hyaline to pale coloured, small, mostly unicellular, ellipsoidal, cylindrical, fusiform, pyriform or spherical. Chlamydospores with muriform septation, resembling the conidia of *Alternaria*, are produced in some species.

General remark. *Phoma* species are ubiquitous saprobes on plant material. Proven human cases are rare. Often the etiologic agent is not identified down to the species level). Most of these cases are (sub)cutaneous and traumatic (Gordon *et al.*, 1975; Young *et al.*, 1973; Arrese *et al.*, 1997), but systemic infections do occur (Morris *et al.*, 1995b).

Differential diagnosis. *Phoma* species have discrete conidiogenous cells lining the pycnidial cavity, whereas conidiogenous cells of *Pleurophoma* species are part of multicellular filaments.

References. Dorenbosch (1970), Sutton (1980), Domsch *et al.* (1980), de Gruyter & Noordeloos (1992), de Gruyter *et al.* (1993), Boerema (1993), Boerema *et al.* (1996).

Key to the treated species of Phoma:

1a.	Conidia at least partly borne laterally and apically on filiform, septate conidiophores	*Pleurophoma cava* (349)
1b.	Conidia borne on ampulliform to doliiform phialides → 2	
2a.	Chlamydospores present → 3	
2b.	Chlamydospores absent → 6	
3a.	Colonies dull red; conidia 2.0-4.5 × 1.5-1.9 µm	*P. minutispora* (345)
3b.	Not combining above characters → 4	
4a.	Chlamydospores strictly 1-celled	*P. eupyrena* (337)
4b.	Chlamydospores frequently multi-celled → 5	
5a.	Colonies with salmon pink patches; conidia less than 5 µm long; 1-celled chlamydospores present	*P. sorghina* (347)
5b.	Colonies without pinkish tinges; conidia over 5 µm long; 1-celled chlamydospores absent	*P. glomerata* (339)
6a.	Conidia hyaline to pale straw yellow, with variable morphology, occasionally 1-septate	*P. dennissii* var. *oculo-hominis* (335)
6b.	Conidia hyaline, uniform in morphology → 7	
7a.	Colonies with red pigmentation; conidia ellipsoidal, straight, over 1 µm wide → 8	
7b.	Colonies lacking red pigmentation; conidia allantoid and slightly curved, less than 1 µm wide	*P. minutella* (343)
8a.	Conidia 4-5.5 × 1.5-2.0 µm	*P. herbarum* (341)
8b.	Conidia 2-4 × 1.2-2.0 µm	*P. cruris-hominis* (333)

Phoma cruris-hominis Punithalingam

Colony characteristics. Colonies (OA) growing slowly, pale greenish-grey, becoming reddish, fluffy.

Microscopy. Pycnidia dark brick, spherical to pyriform, up to 180 µm diam. Conidia unicellular, hyaline, obovoidal, 2-4 × 1.2-2.0 µm, guttulate.

Pathogenicity. BSL-2. This species was isolated from scrapings and pus from a human skin lesion (Punithalingam, 1979). A subcutaneous case was reported by Carrière *et al.* (1997).

Reference. Punithalingam (1979).

Nomenclature. *Phoma cruris-hominis* Punithalingam - Nova Hedwigia 31: 135, 1979.

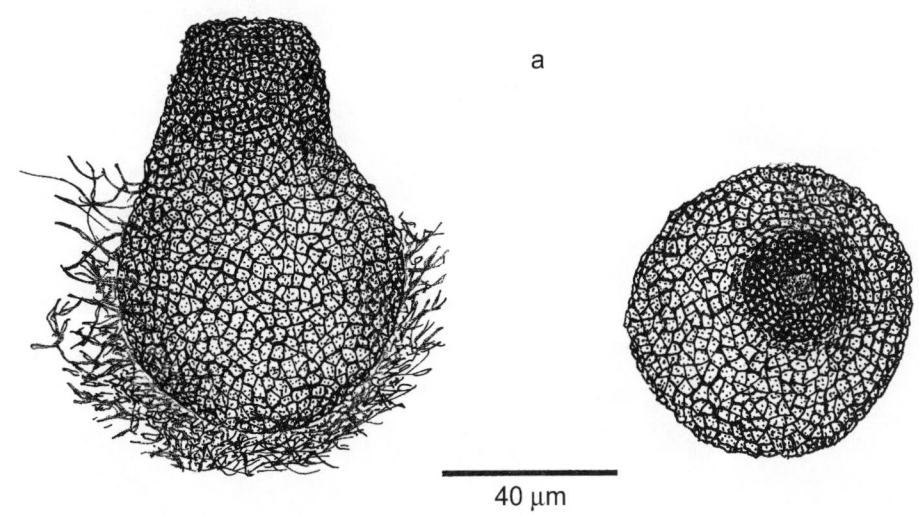

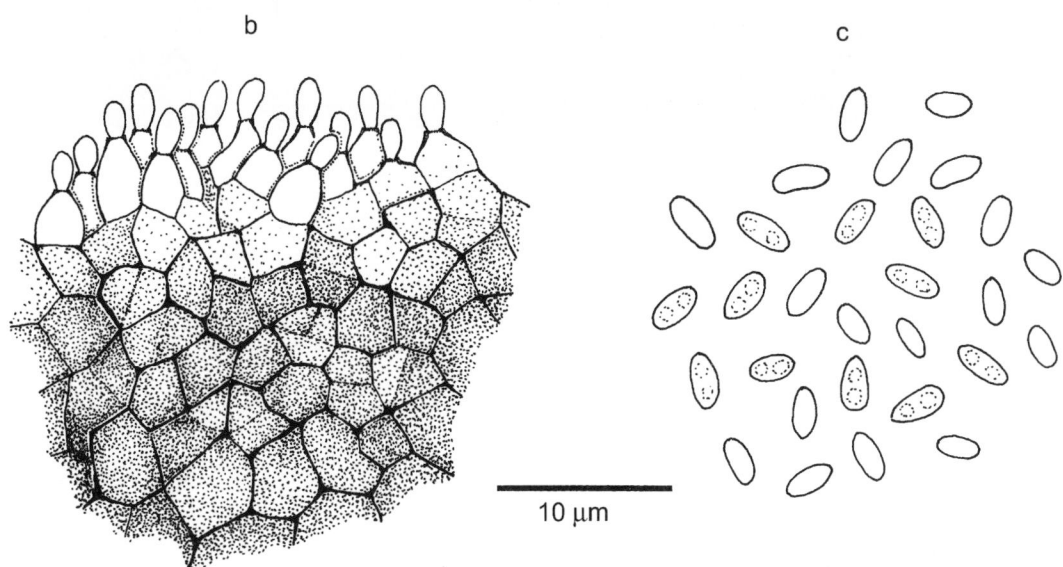

Phoma dennisii var. *cruris-hominis*, IMI 213845. a. Pycnidia; b. section of a pycnidium with conidia produced from conidiogenous cells lining the inner wall; c. conidia.

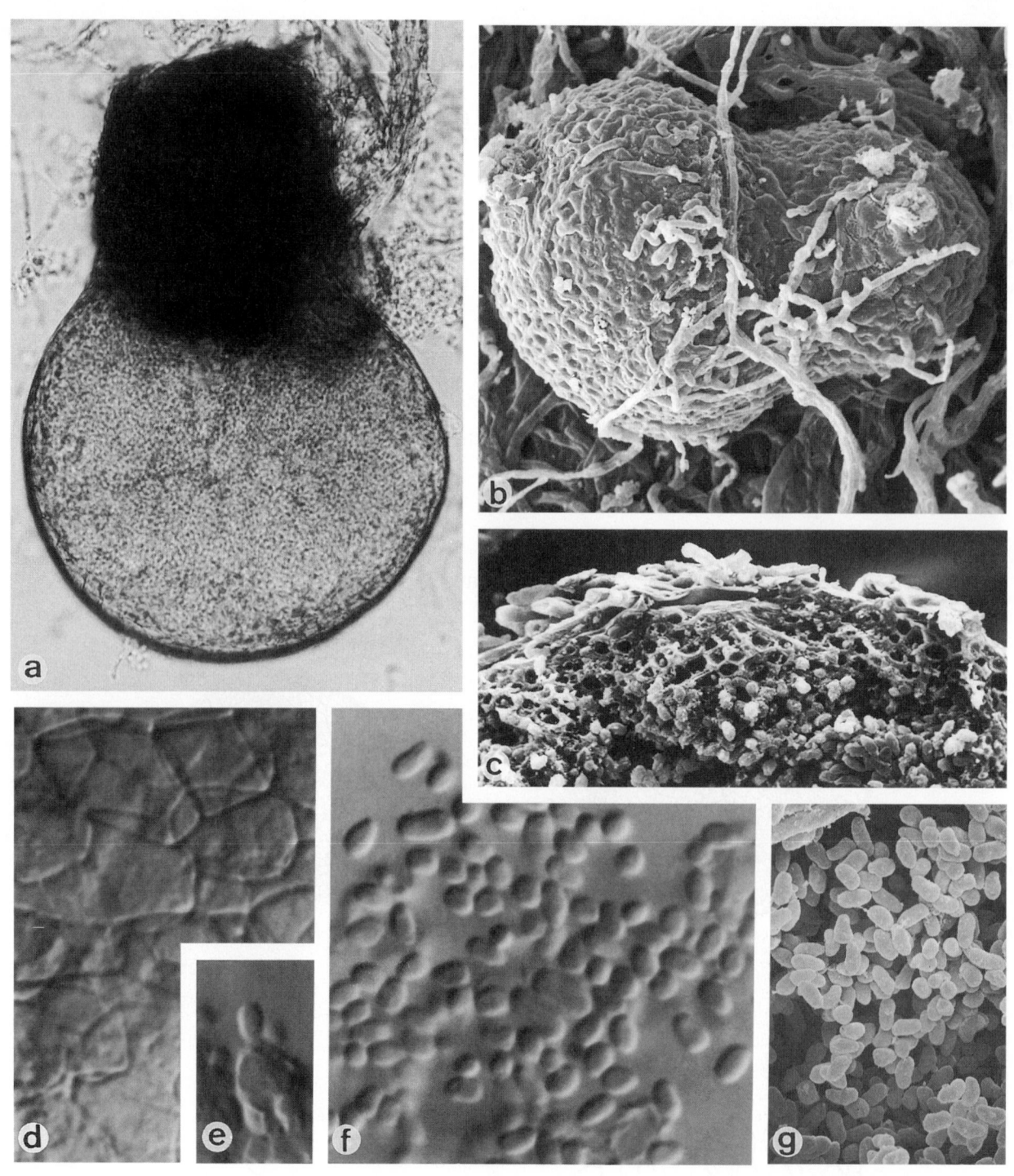

Phoma cruris-hominis, IMI 213845. a,b. Pycnidia; c. section of pycnidium wall; d. face view of part of pycnidium wall; e-g. conidiogenous cells and conidia. a. ×512; b. ×580; c. ×1350; d-f. ×1600; g. ×2100.

Phoma dennissii Boerema var. *oculo-hominis* (Punithalingam) Boerema *et al.*

Colony characteristics. Colonies (OA) growing slowly, brown to black, cottony; reverse brown.

Microscopy. Pycnidia brown-vinaceous to fuscous-black, subspherical to obpyriform, 120-200 μm diam. Conidia hyaline, straw-yellow or brown, unicellular, occasionally with a septum, short cylindrical, sometimes slightly curved, guttulate. Unicellular conidia 3-7 × 1-2 μm; septate conidia 9-16 × 3.0-4.5 μm.

Pathogenicity. BSL-2. This species was isolated from ulcerated human cornea (Punithalingam, 1979).

References. Punithalingam (1979), Boerema *et al.* (1997).

Nomenclature. *Phoma oculo-hominis* Punithalingam - Trans Br. Mycol. Soc. 67: 142, 1976 ≡ *Phoma dennisii* Boerema var. *oculo-hominis* (Punithalingam) Boerema, de Gruyter & Noordeloos - Persoonia 16: 351, 1997.

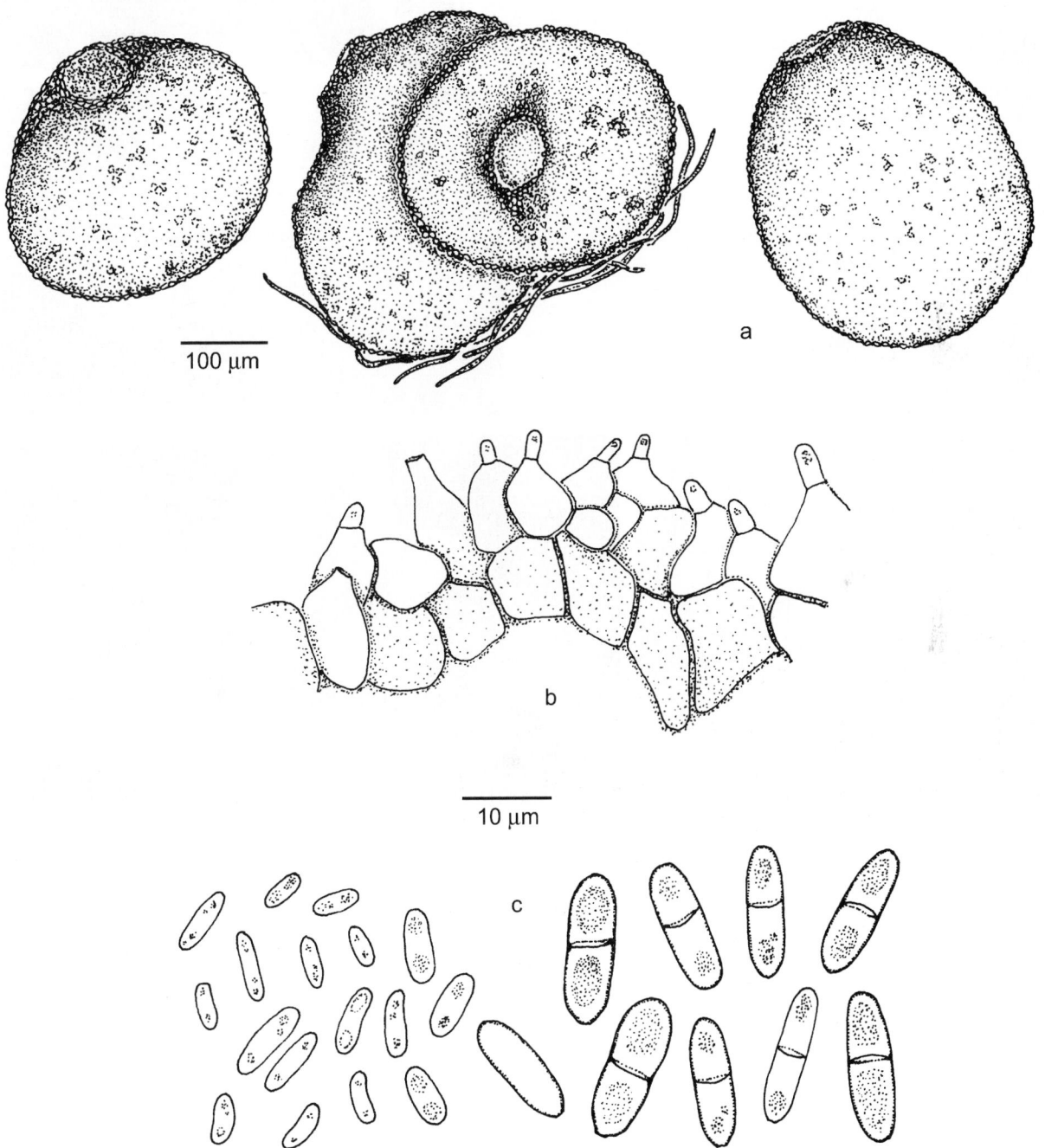

100 μm

10 μm

a

b

c

Phoma oculo-hominis, IMI 193307. a. Pycnidia; b. section of a pycnidium with conidia produced from conidiogenous cells lining the inner wall; c. conidia.

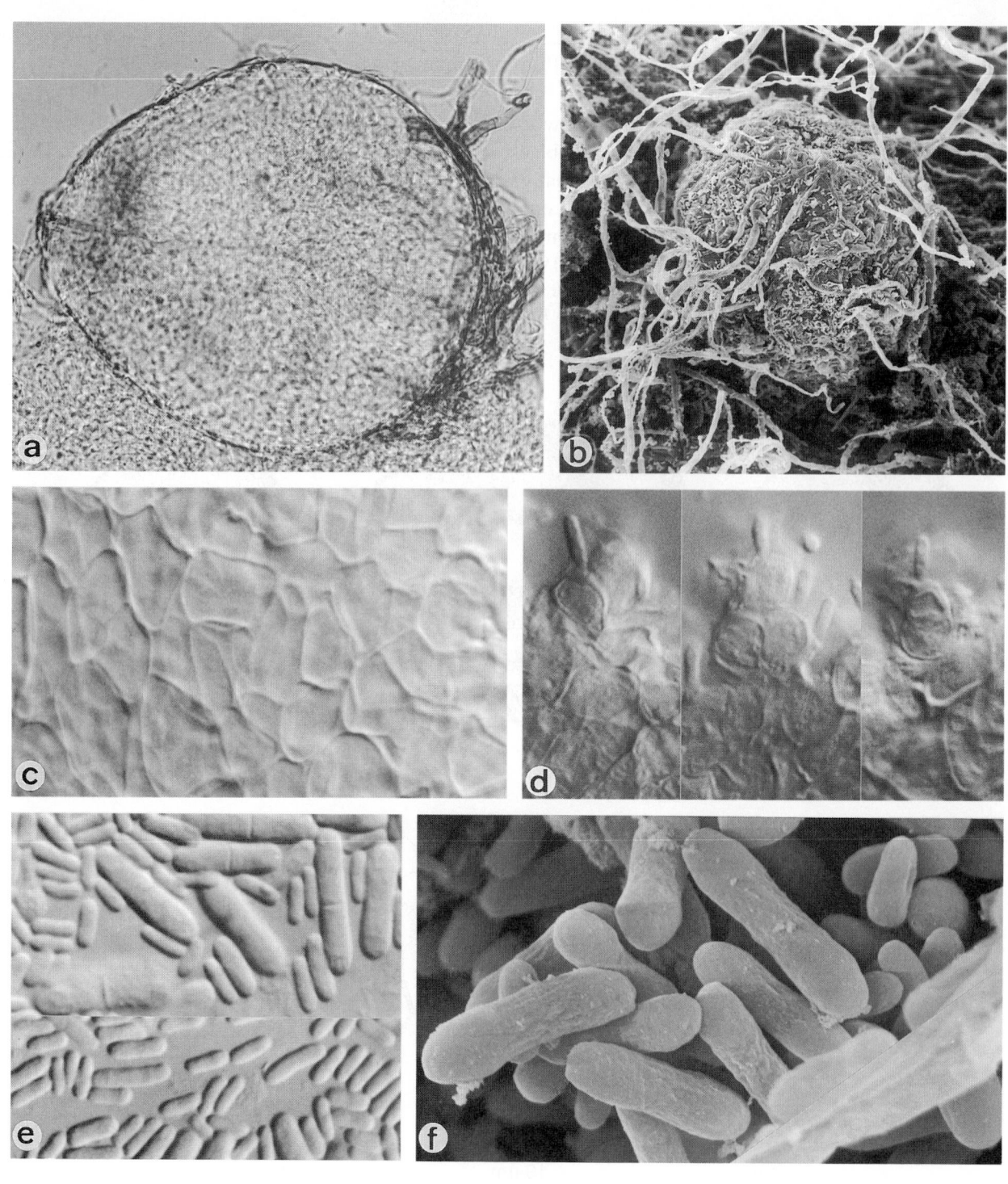

Phoma dennissii var. *oculo-hominis*, IMI 193307. a,b. Pycnidia; c. part of pycnidium wall; d-f. conidiogenous cells and conidia. a. ×256; b. ×250; c. ×1600; d. ×1600; e. ×1600; f. ×5000.

Phoma eupyrena Sacc.

Colony characteristics. Colonies (OA) with restricted growth, dark brown, covered with a dense, dark green mycelial felt; reverse dark brown.

Microscopy. Pycnidia spherical to subspherical, occasionally with a short neck, dark brown, deeply pigmented around the ostiole, up to 400 μm diam. Conidia 3.5-6.0 × 1.5-3.0 μm, cylindrical or ellipsoidal, straight or slightly curved, often biguttulate, hyaline. Small, ellipsoidal, unicellular, chlamydospores abundantly present, pale brown, smooth-walled, terminal or intercalary, often catenate.

Pathogenicity. BSL-1. The species was isolated from a human skin lesion (Bakerspigel *et al.*, 1981).

Reference. Sutton (1980).

Nomenclature. *Phoma eupyrena* Saccardo - Michelia 1: 525, 1879.

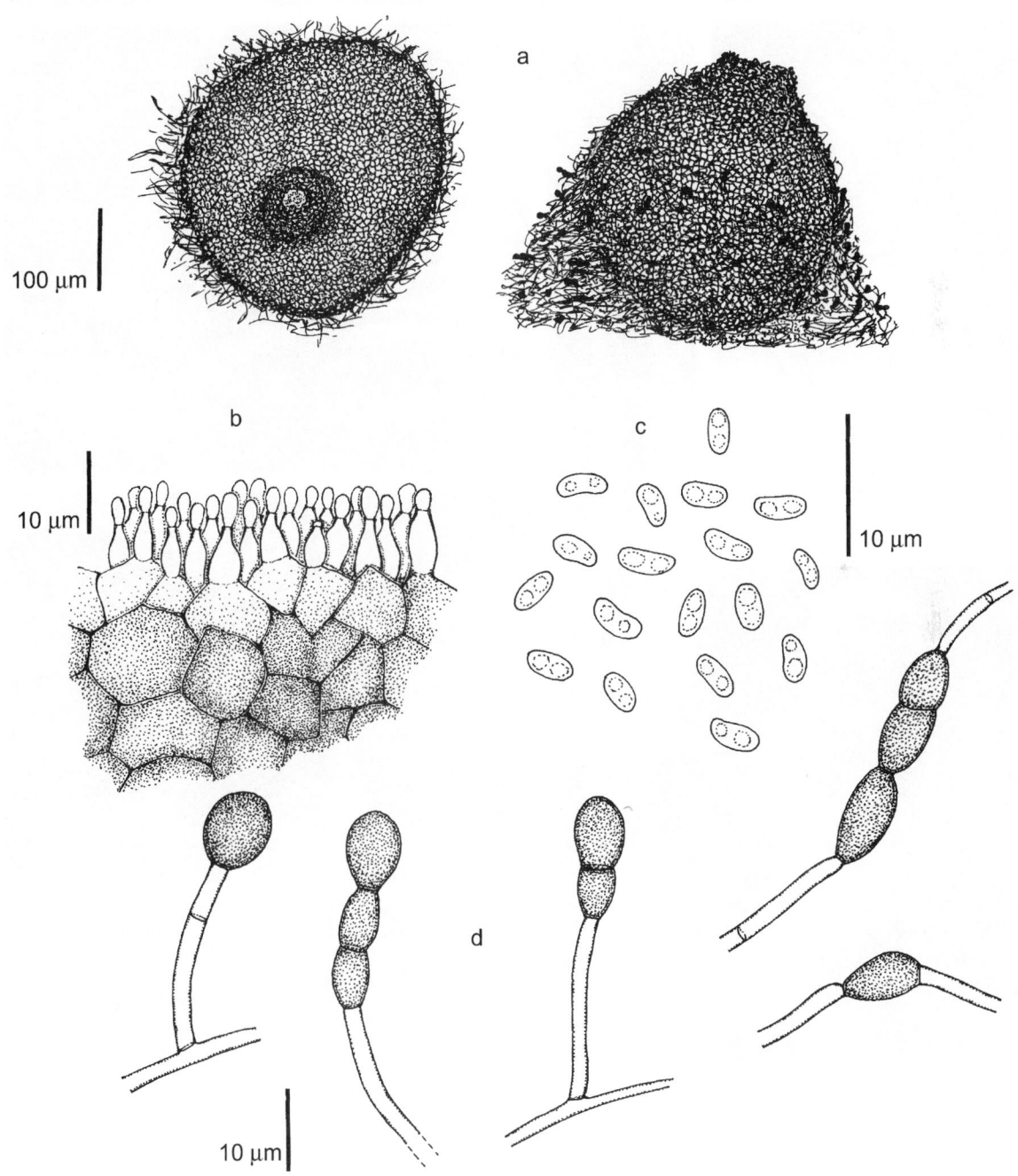

Phoma eupyrena, CBS 832.89. a. Pycnidia; b. section of a pycnidium with conidia produced from conidiogenous cells lining the inner wall; c. conidia; d. chlamydospores.

337

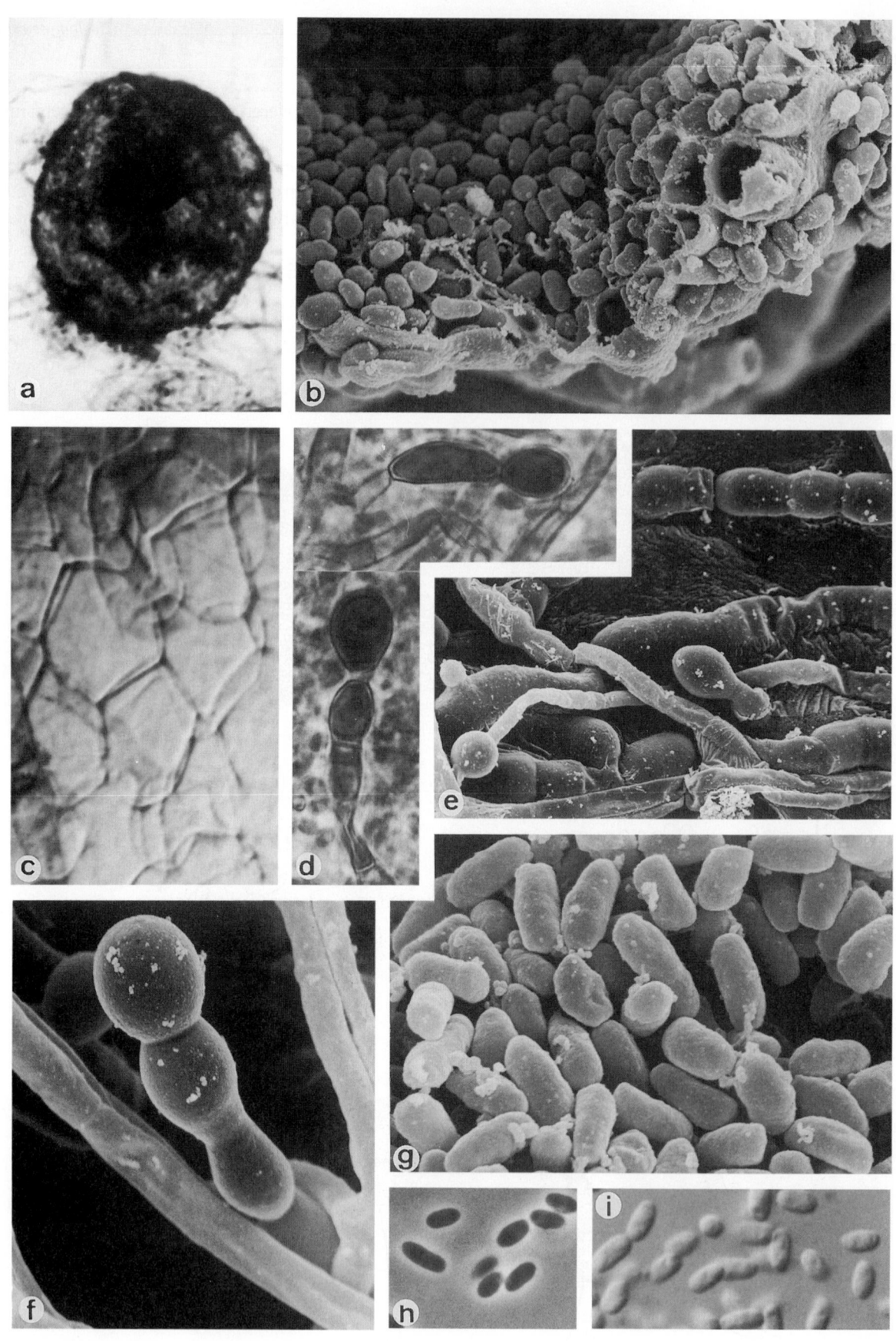

Phoma eupyrena, CBS 832.89. a. Pycnidium; b. cross section of pycnidium wall; c. part of pycnidium wall; d-f. chlamydospores; g-i. conidia. a. ×1280; b. ×2700; c. ×1600; d. ×1280; e. ×1600; f. ×3000; g. ×6000; h, i. ×1600.

Phoma glomerata (Corda) Wollenw. & Hochapfel

Colony characteristics. Colonies (OA) growing rapidly, grey, green or brown, with sparse aerial mycelium; reverse brown to black.

Microscopy. Pycnidia spherical to subspherical, ostiolate, frequently with a neck, light-coloured to black, deeply pigmented around the ostiole, up to 400 μm diam. Conidia 1-celled, occasionally 2-celled, mostly ovoidal to ellipsoidal, hyaline to pale brown, 5-10 × 2.5-3.0 μm, usually biguttulate. Chlamydospores in branched or unbranched chains, brown, with longitudinal, oblique and transverse septa (muriform).

Molecular diagnostics. ITS restriction map based on NCBI AF126819:

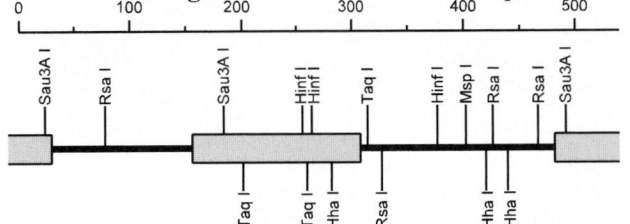

Pathogenicity. BSL-1. *P. glomerata* has been implicated with several mycotic diseases in humans (Punithalingam, 1979) and possibly also in animals (Dawson & Lepper, 1970).

References. Boerema *et al.* (1965), Punithalingam (1979), Boerema (1993).

Nomenclature. *Coniothyrium glomeratum* Corda - Ic. Fung. 4: 39, 1840 ≡ *Phoma glomerata* (Corda) Wollenweber & Hochapfel - Z. Parasitkde 8: 592, 1936 ≡ *Peyronellaea glomerata* (Corda) Goidánich - Rc. Accad. Lincei 1: 455, 1946.

Phoma hominis Agostini & Tredici, *in* Pollacci - Atti Ist. Bot. Univ. Lab. Crittogam. Pavia, Ser. 4, 6: 154, 1935 ≡ *Alternaria hominis* (Agostini & Tredici) Agostini - Atti Ist. Bot. Univ. Lab. Crittogam. Pavia, Ser. 4, 9: 187, 1937 ≡ *Peyronellaea hominis* Goidánich - Rc. Accad. Lincei 1: 455, 1946.

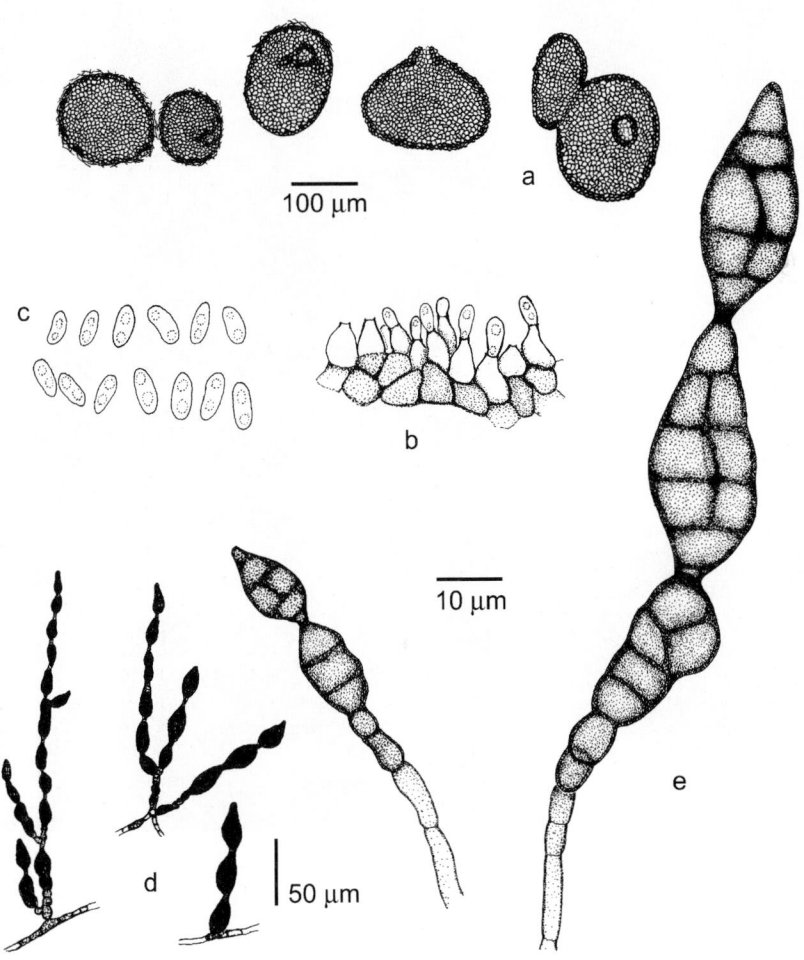

Phoma glomerata, IMI 294572. a. Pycnidia; b. section of a pycnidium with conidia produced from conidiogenous cells lining the inner wall; c. conidia; d. habit sketch of chlamydospores; e. chlamydospores.

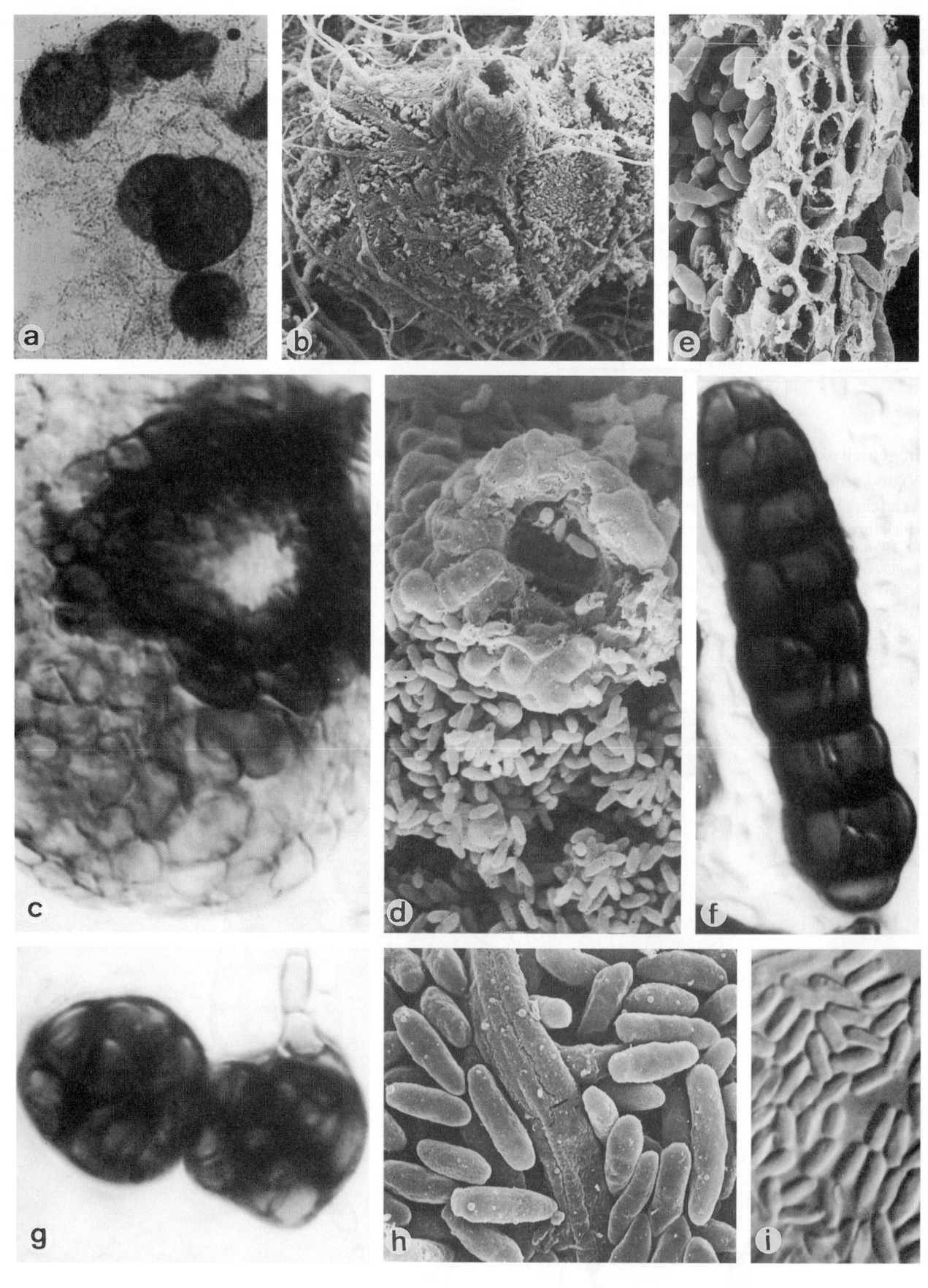

Phoma glomerata, IMI 294572. a, b. Pycnidia; c, d. details of the ostioles; e. cross section of pycnidium wall; f, g. chlamydospores; h, i. conidia. a. ×128; b. ×410; c. ×1600; d. ×1525; e. ×2300; f, g. ×1600; h. ×4800; i. ×1600.

340

Phoma herbarum Westend.

Colony characteristics. Colonies (OA) relatively slow-growing, usually a reddish pigment being exuded into the agar, with sparse grey-green aerial mycelium, turning purplish-blue instantaneously after application of a drop of 1N NaOH.
Microscopy. Pycnidia spherical, 100-200 μm diam, with distinct, rounded ostioles. Conidia hyaline, in mass hyaline to pinkish, oblong to cylindrical, unicellular, straight, 4-5 × 1.5-2.0 μm.
Pathogenicity. BSL-1. This species was implicated in human skin lesions in Canada (Bakerspigel, 1970).
Reference. Sutton (1980).
Nomenclature. *Phoma herbarum* Westendorp - Bull. Acad. Belg., Cl. Sci., 19: 118, 1852.
   Phoma hibernica Grimes, O'Connor & Cummins - Trans Br. Mycol. Soc. 17: 100, 1932.

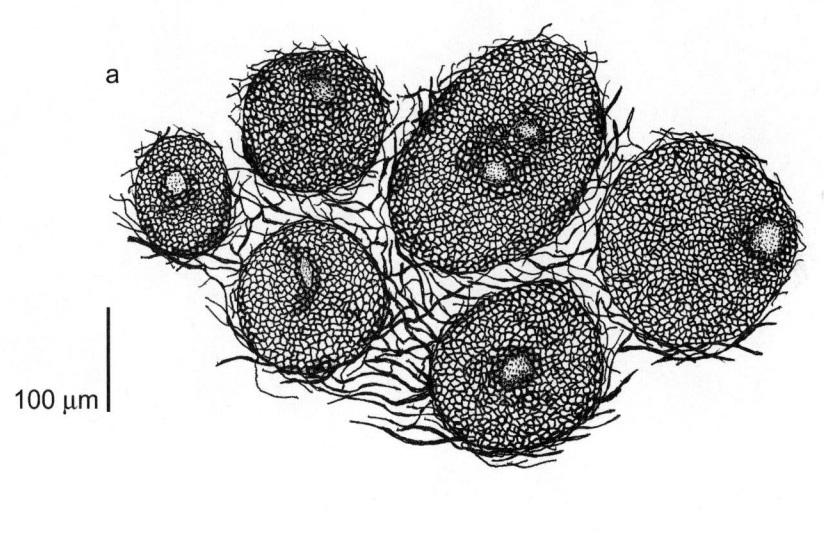

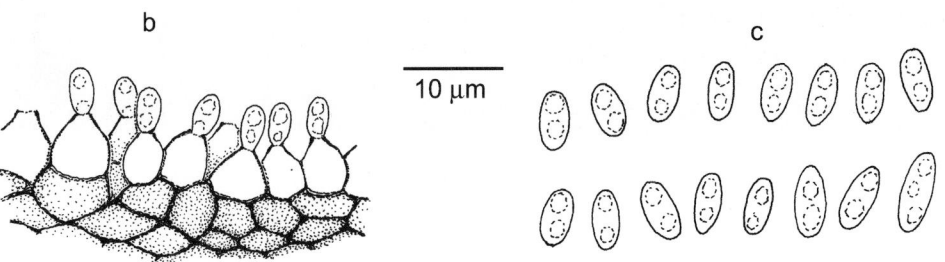

Phoma herbarum, CBS 615.75. a. Pycnidia; b. section of a pycnidium with conidia produced from conidiogenous cells lining the inner wall; c. conidia.

341

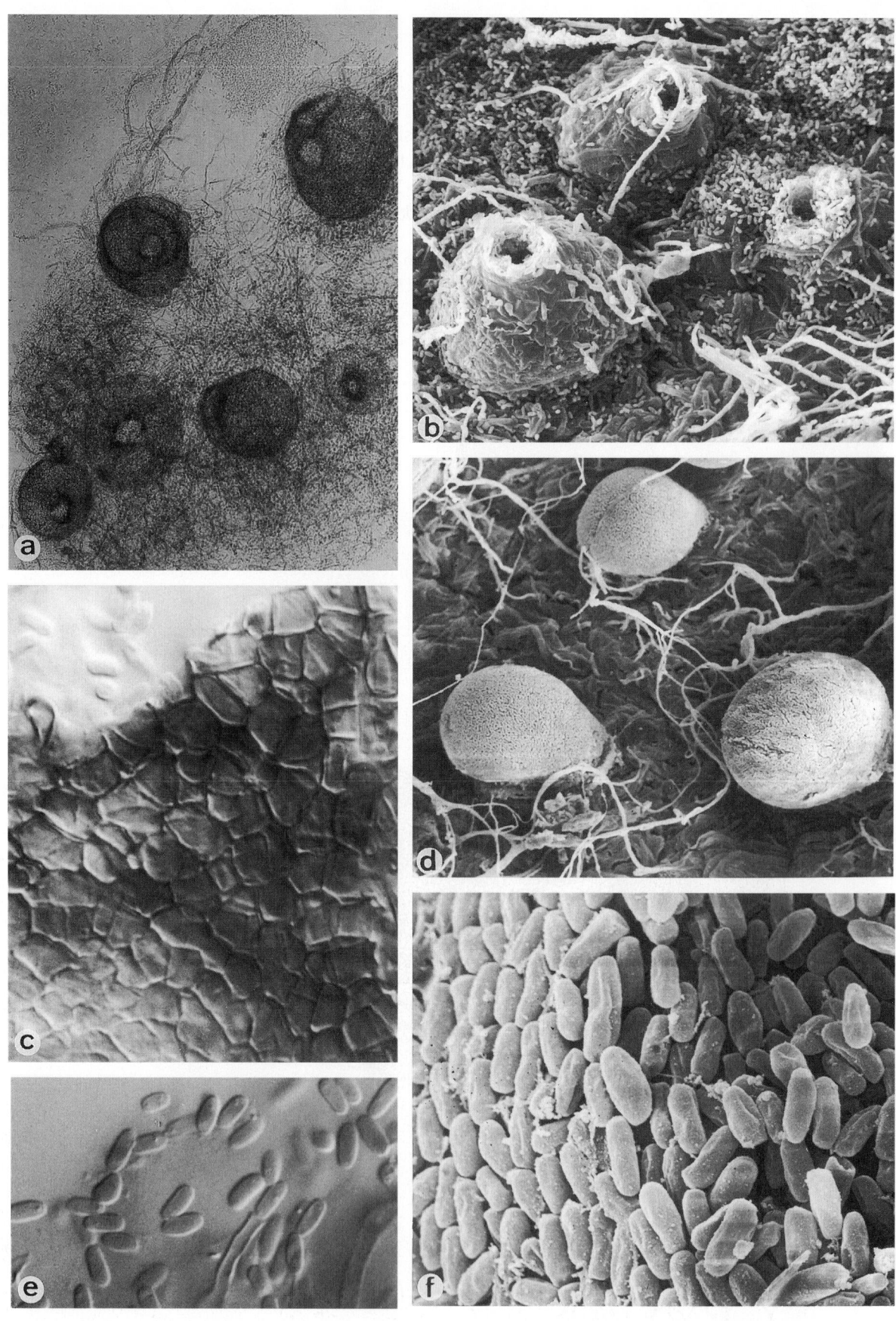

Phoma herbarum, CBS 615.75. a, b. Pycnidia; c. part of pycnidium wall; d. conidial masses; e, f. conidia. a. ×320; b. ×410; c. ×1600; d. ×250;
e. ×1600; f. ×3400.

Phoma minutella Sacc. & Penz.

Colony characteristics. Colonies (OA) growing slowly, olive-grey, with dense aerial mycelium; reverse greyish-brown.

Microscopy. Pycnidia spherical to pyriform, partly immersed. Conidia hyaline, allantoid, 2-3 × 0.5-1.0 μm.

Pathogenicity. BSL-1. The species was implicated in a subcutaneous inflammatory process on the human foot (Baker *et al.*, 1987).

Reference. Baker *et al.* (1987).

Nomenclature. *Phoma minutella* Saccardo & Penzig - Michelia 2: 618, 1882.

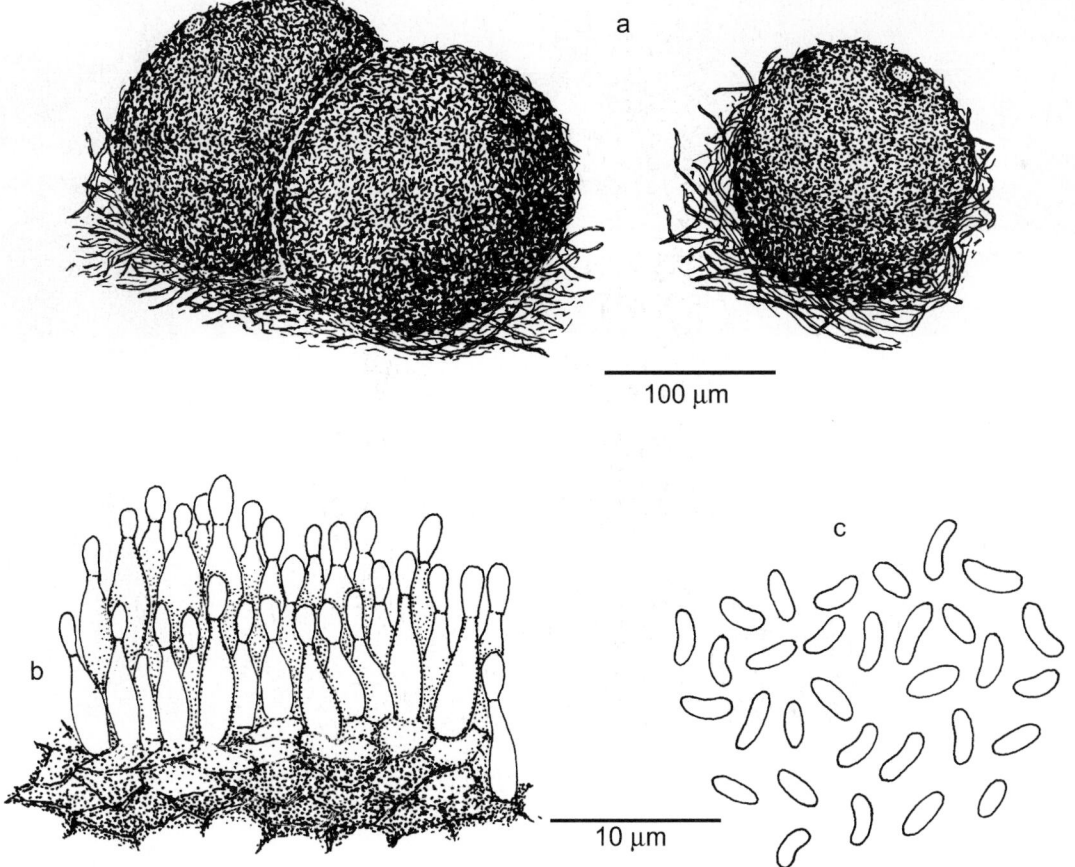

Phoma minutella, WCLR 63-86. a. Pycnidia; b. section of a pycnidium with conidia produced from conidiogenous cells lining the inner wall; c. conidia.

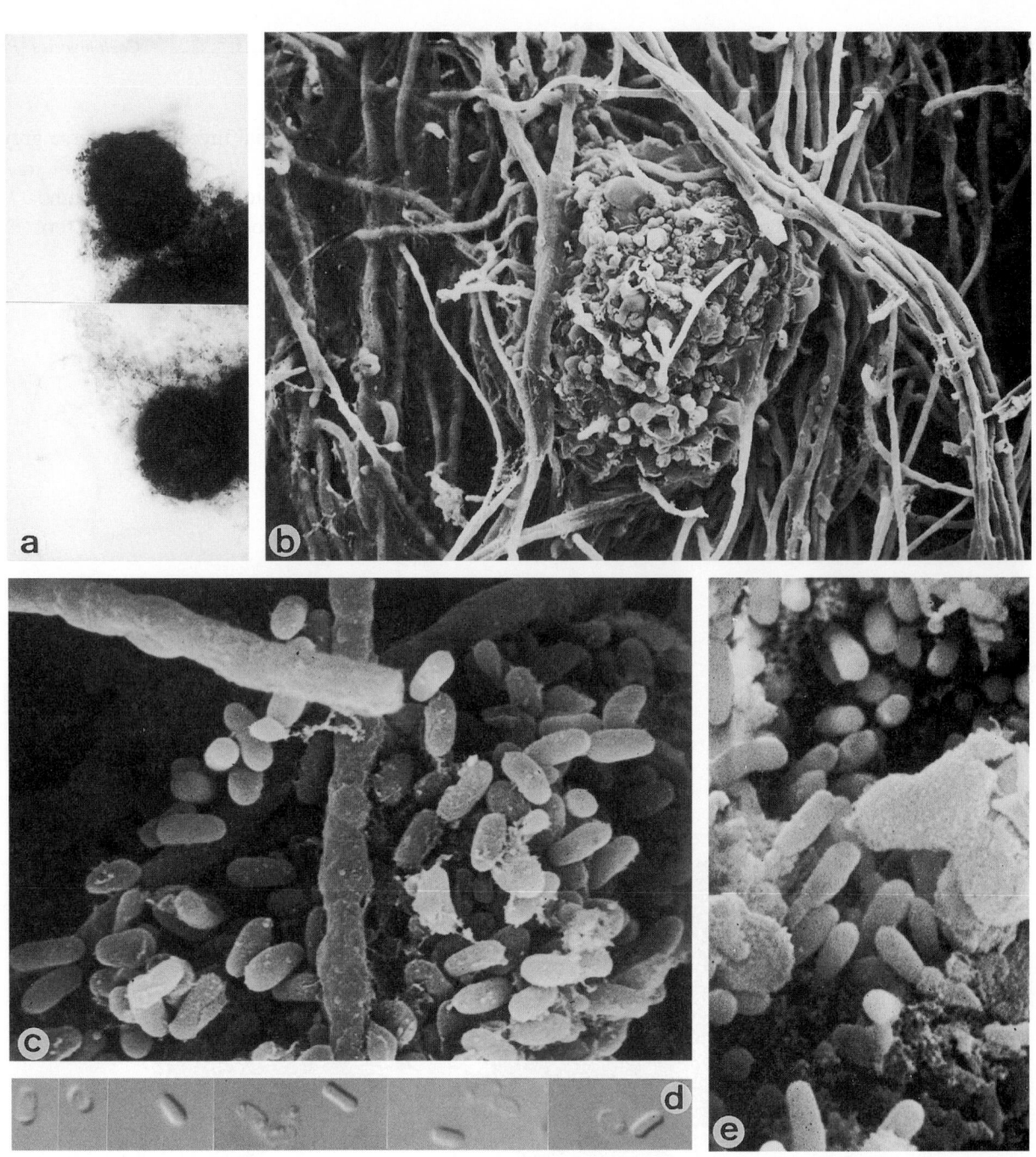

Phoma minutella, WCLR 63-86. a, b. Pycnidia; c-e. conidiogenous cells and conidia; a. ×128; b. ×680; c. ×4500; d. ×1600; e. ×5800.

Phoma minutispora P.N. Mathur

Colony characteristics. Colonies (OA) growing moderately rapidly, dull red, with little aerial mycelium.

Microscopy. Pycnidia spherical to subspherical, yellow, 60-175 μm diam. Conidia hyaline, unicellular, ovoidal, occasionally ellipsoidal, 2.0-4.5 × 1.5-1.9 μm. Chlamydospores usually present, 6-15 μm diam, thick-walled, usually finely warted, mostly formed terminally on hyphae.

Pathogenicity. BSL-1. This species was repeatedly isolated from skin lesions of two patients in India and an experimental infection in animals was positive (Shukla *et al.*, 1984b).

Reference. Mathur (1967), de Gruyter & Noordeloos (1992).

Nomenclature. *Phoma oryzae* Cooke & Massee - Grevillea 16: 15, 1887 ≡ *Phoma minutispora* P.N. Mathur, *in* de Gruyter & Noordeloos - Persoonia 15: 75, 1992 (name change).

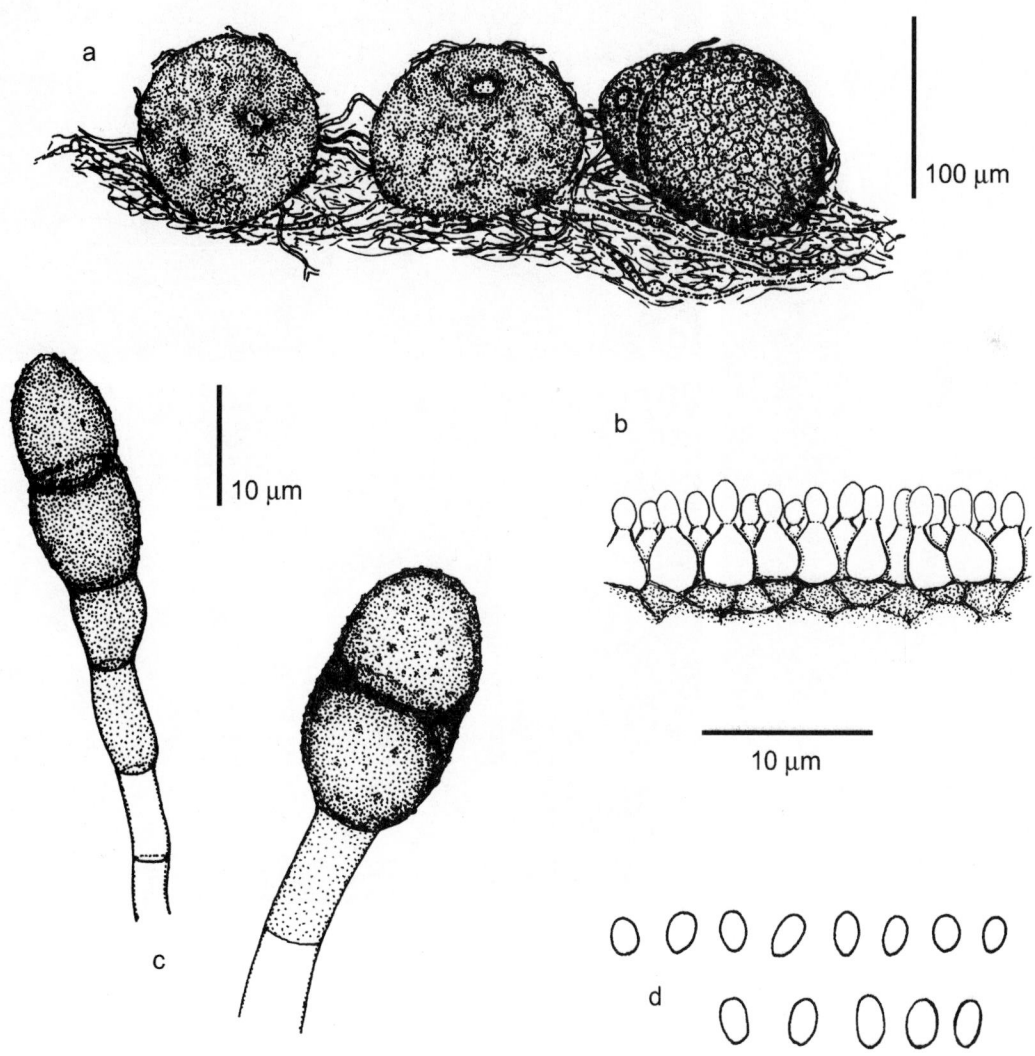

Phoma minutispora, CBS 711.76. a. Pycnidia; b. conidiogenous cells producing conidia; c. chlamydospores; d. liberated conidia.

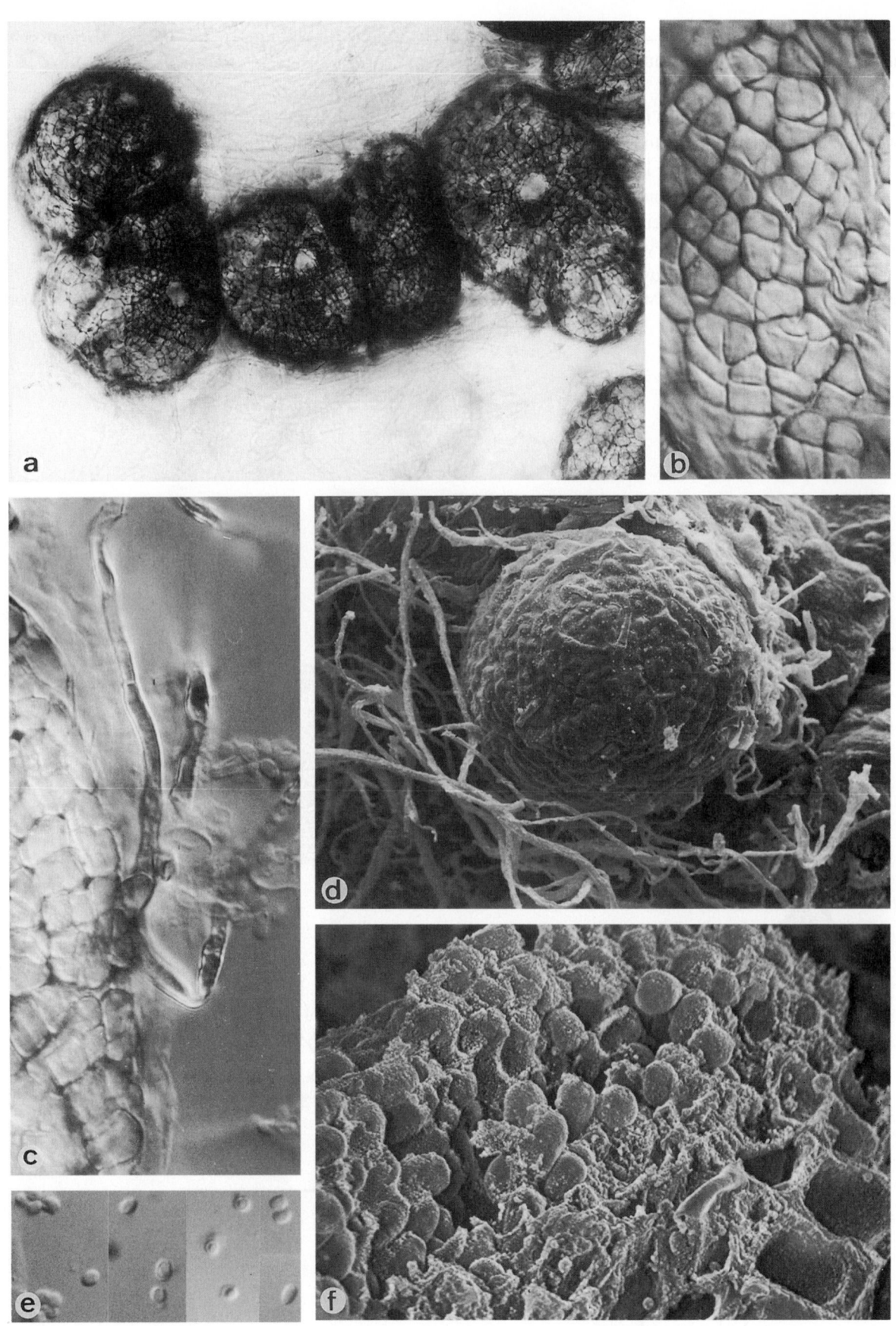

Phoma minutispora, CBS 711.76. a, d. Pycnidia; b, c. pycnidium in surface view; e, f. conidia. a. ×512; b, c. ×1600; d. ×800; e. ×1600; f. ×7400.

Phoma sorghina (Sacc.) Boerema *et al.*

Colony characteristics. Colonies (OA, 18°C) growing rapidly, with a fluffy to dense aerial mycelium, grey-green with salmon-pink patches.

Microscopy. Pycnidia dark brown, ampulliform, occasionally with a pronounced neck, 80-200 × 75-200 µm. Conidia ellipsoidal, 4-5 × 2.0-2.5 µm, without oil droplets, hyaline to pale brown. Chlamydospores 1-celled or with muriform septation, brown, 15-50 × 7-25 µm.

Pathogenicity. BSL-1. The species is a common plant pathogen. It was repeatedly isolated from two patients with skin lesions; experimental inoculation of animals proved successful (Rai, 1989a). *Chaetophoma dermo-unguis* S.M. Singh & Barde, described from onychomycosis, probably is a synonym (H.A. van der Aa, pers. comm.).

Reference. Sutton (1980), Boerema (1993).

Nomenclature. *Phyllosticta sorghina* Saccardo - Michelia 1: 140, 1878 ≡ *Phoma sorghina* (Saccardo) Boerema, Dorenbosch & van Kesteren - Persoonia 7: 139, 1973.

Chaetophoma dermo-unguis S.M. Singh & Barde - Mykosen 29: 276, 1986.

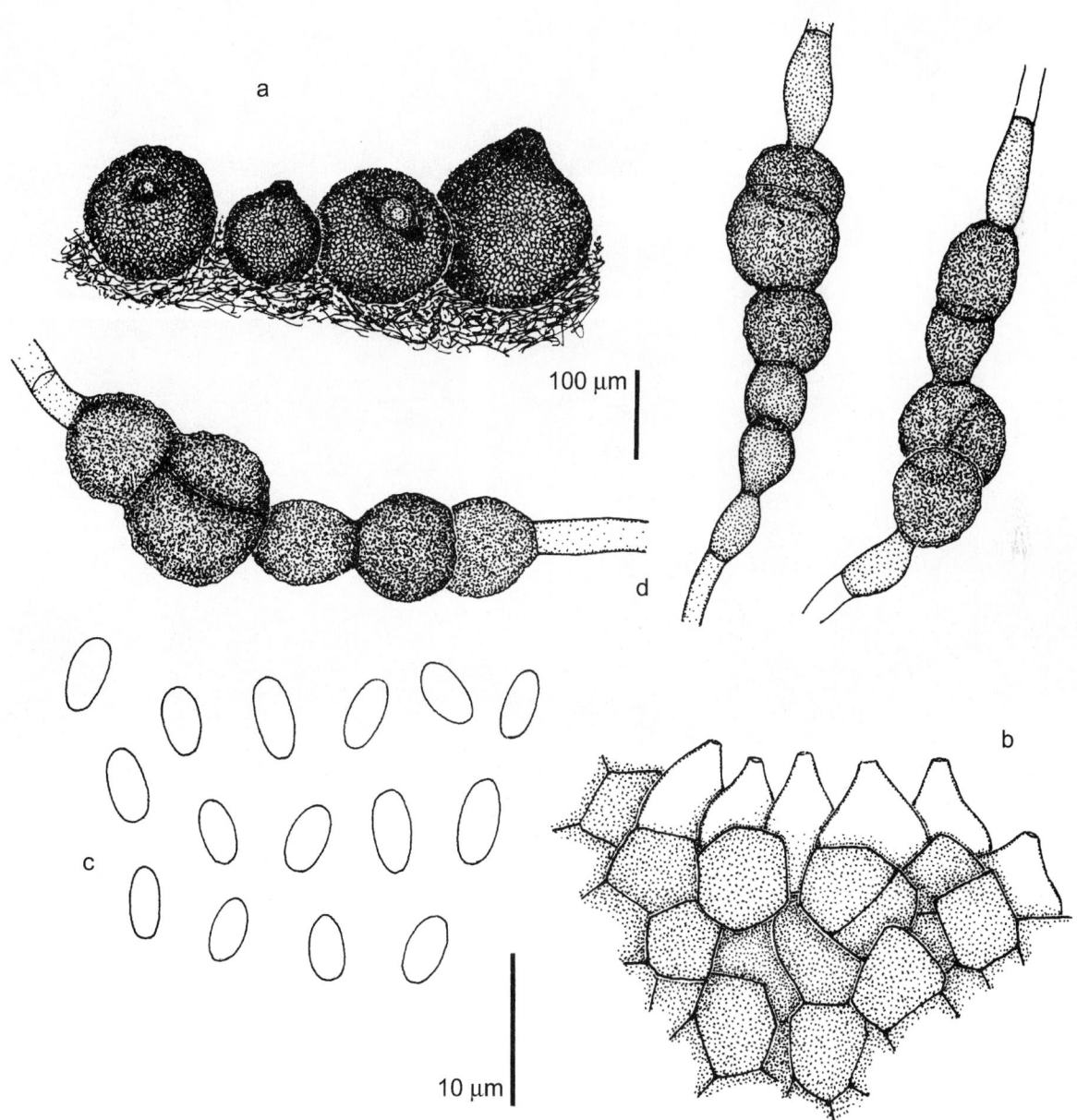

Phoma sorghina, CBS 181.80. a. Pycnidia, b. section of a pycnidium with conidia produced from conidiogenous cells lining the inner wall; c. conidia; d. chlamydospores.

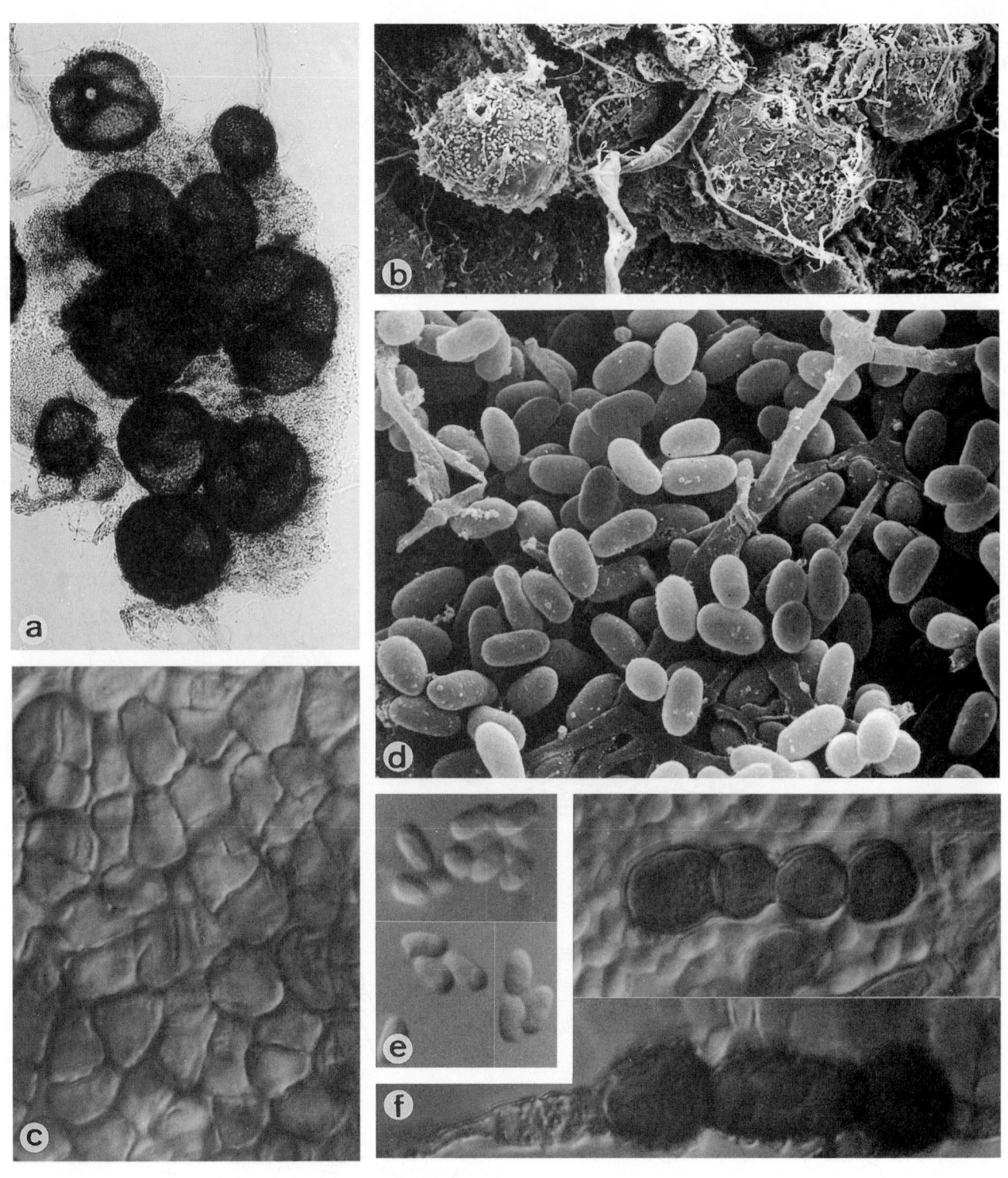

Phoma sorghina, CBS 181.80. a, b. Pycnidia; c. part of pycnidium wall; d, e. conidia; f. chlamydospores. a. ×128; b. ×235; c. ×1600; d. ×3300; e, f. ×1600.

Coelomycetes. Genus: *PLEUROPHOMA*

Generic description. Colonies with dark pycnidia which are spherical, usually each with an ostiole. Conidiogenous cells phialidic, hyaline, integrated in filiform, septate conidiophores, producing abundant conidia which are extruded in slimy masses. Conidia hyaline, small, unicellular, ellipsoidal.
Reference. Sutton (1980).

Key to the treated species of Pleurophoma:

1a.	Conidia cylindrical, often slightly curved	***P. cava* (349)**
1b.	Conidia narrow ellipsoidal, straight	***P. pleurospora* (1030)**

Pleurophoma cava (Schulz.) Boerema *et al.*

Colony characteristics. Colonies (OA) growing moderately rapidly, olivaceous grey, with sparse aerial mycelium; reverse brownish.
Microscopy. Pycnidia spherical to subspherical, brown, 190-250 μm diam; ostioles lined by compacted periphyses; cells of the pycnidium wall separated by dark, encrusted material. Conidiogenous cells integrated in filiform, septate conidiophores. Conidia 3-4 × 1-2 μm, cylindrical, straight or slightly curved, occasionally guttulate.
Pathogenicity. BSL-1. Dermatitis in a deer (Gordon *et al.*, 1975). A subcutaneous infection in an immunosuppressed patient was reported by Záitz *et al.* (1997).
Reference. Boerema & Dorenbosch (1973), Sutton (1980).
Nomenclature. *Phoma cava* Schulzer - Verh. Zool.-Bot. Ges. Wien 21: 1248, 1871 ≡ *Pleurophoma cava* (Schulzer) Boerema, Loerakker & Hamers - Persoonia 16: 172, 1996.

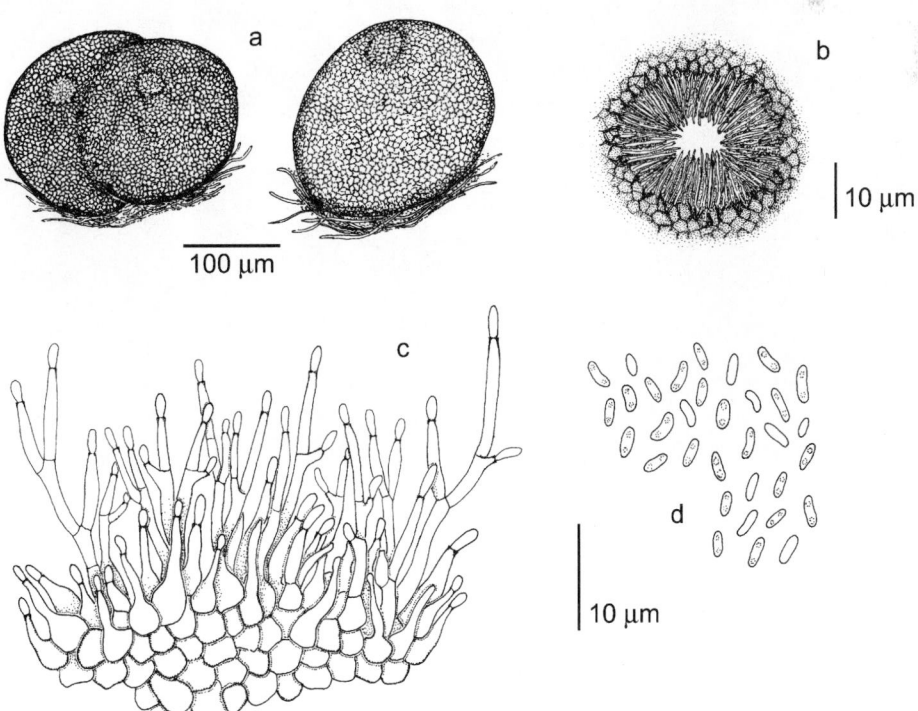

Pleurophoma cava, IMI 163515. a. Pycnidia; b. section of a pycnidium with conidiophores lining the inner wall; c. conidia.

Pleurophoma cava, IMI 163515. a. Pycnidia; b. part of pycnidium wall; c, d, f-h. conidiogenous cells and conidia; e. cross section of a pycnidium with conidiophores lining the inner wall. a. ×128; b. ×1600; c. ×1600; d. ×2050; e. ×4800; f. ×4600; g. ×8900; h. ×10750.

Coelomycetes. Genus: *PLEUROPHOMOPSIS*

Pleurophomopsis lignicola Petrak

Colony characteristics. Colonies (OA) expanding, velvety, dark greyish-brown.

Microscopy. Pycnidia dark brown, subspherical to pear-shaped, 50-300 µm wide, with a thick wall consisting of *textura angularis*, bearing ostiolum more or less pronounced, up to 150 µm in length. Conidiogenous cells arranged in a dense palissade, discrete or integrated in short conidiophores, acicular, conidia being produced from terminal phialide openings. Conidia ellipsoidal, hyaline, 2-3 × 0.5-1.0 µm.

Pathogenicity. BSL-1. A subcutaneous infection was reported by Chabasse *et al*. (1995a), a case of maxillary sinusitis by Padhye *et al*. (1997) and a systemic infection in a transplant patient by Farina *et al*. (1997).

Reference. Petrak (1924).

Antifungal susceptibility.

Antifungal	MICs	Strains	Reference
AMB	0.5	1	Unpublished data
5FC	0.5	1	Unpublished data
FLZ	32	1	Unpublished data
ITZ	0.06	1	Unpublished data
KTZ	0.06	1	Unpublished data
MCZ	0.06	1	Unpublished data

Nomenclature. *Pleurophomopsis lignicola* Petrak - Annls Mycol. 22: 165, 1924.

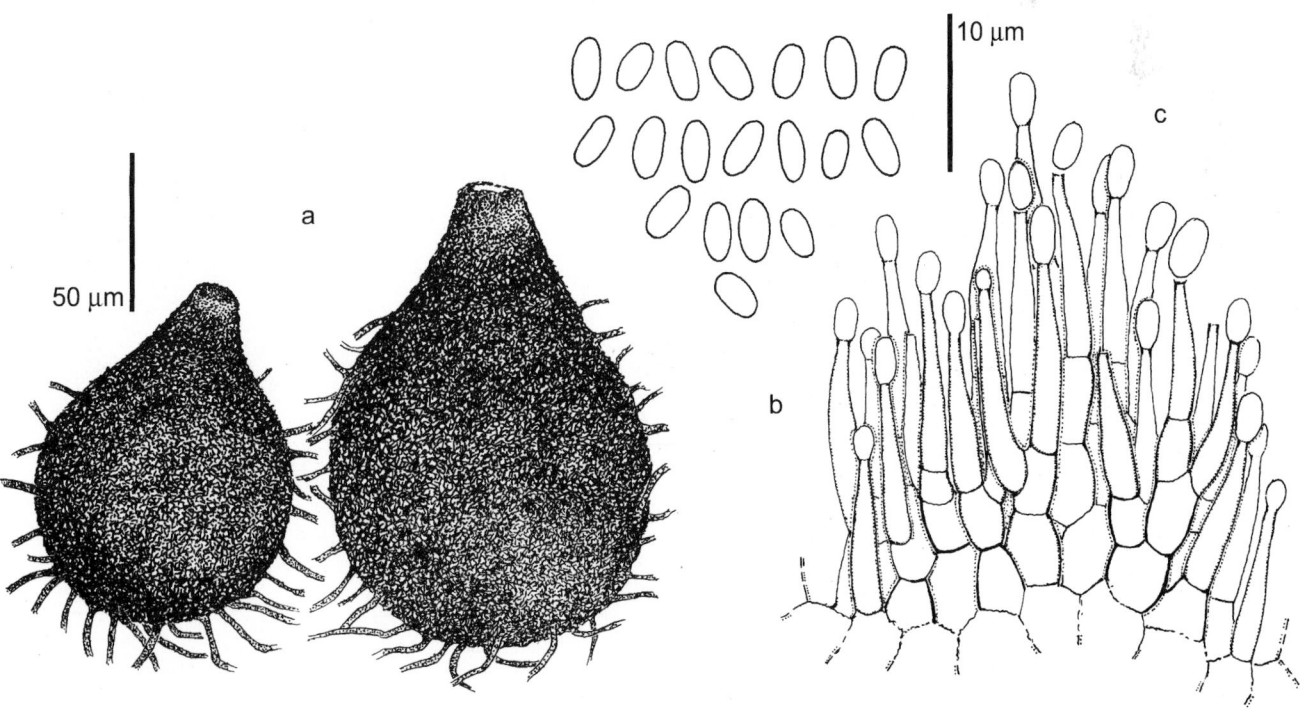

Pleurophomopsis lignicola, CBS 727.97. a. Pycnidia; b. conidia; c. section of a pycnidium with conidia produced from conidiogenous cells lining the inner cavity wall.

351

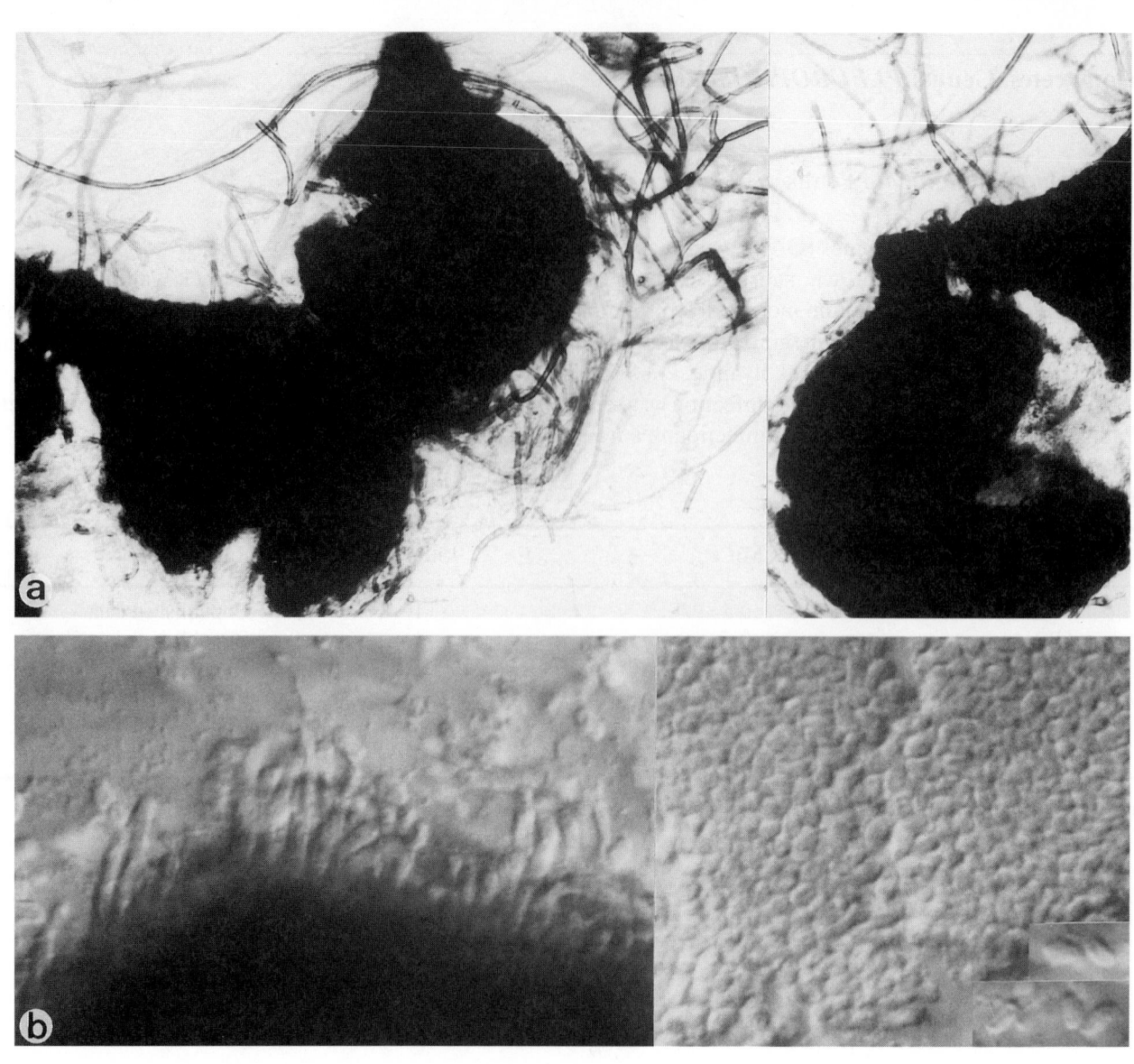

Pleurophomopsis lignicola, CBS 727.97. a. Pycnidia; b. section of a pycnidium with conidia produced from conidiogenous cells lining the inner cavity wall, and liberated conidia. a. ×250; b. ×1600.

Coelomycetes. Genus: *PSEUDOCHAETOSPHAERONEMA*

Pseudochaetosphaeronema larense (Borelli & Zamora) Punithalingam

Colony characteristics. Colonies (OA) growing slowly, olivaceous-black; reverse greyish.

Microscopy. Pycnidia sparse, solitary, occasionally in chains of two (one located at the apex of another), blackish, obpyriform, 300-500 × 150-300 µm. Pycnidial wall pseudoparenchymatous. Conidiogenous cells hyaline, ampulliform to cylindrical, phialidic, lining the internal wall of the pycnidium. Conidia hyaline, short-cylindrical, with rounded ends, 2-3 × 1.0-1.8 µm, with one or two guttules.

Pathogenicity. BSL-2. The fungus was isolated from a mycetoma of the foot of an agricultural worker (Borelli *et al.*, 1976).

Reference. Punithalingam (1979).

Nomenclature. *Chaetosphaeronema larense* Borelli & Zamora - Bull. Mens. Soc. Venez. Derm. 6: 17, 1973 ≡ *Pseudochaetosphaeronema larense* (Borelli & Zamora) Punithalingam - Nova Hedwigia 31: 127, 1979.

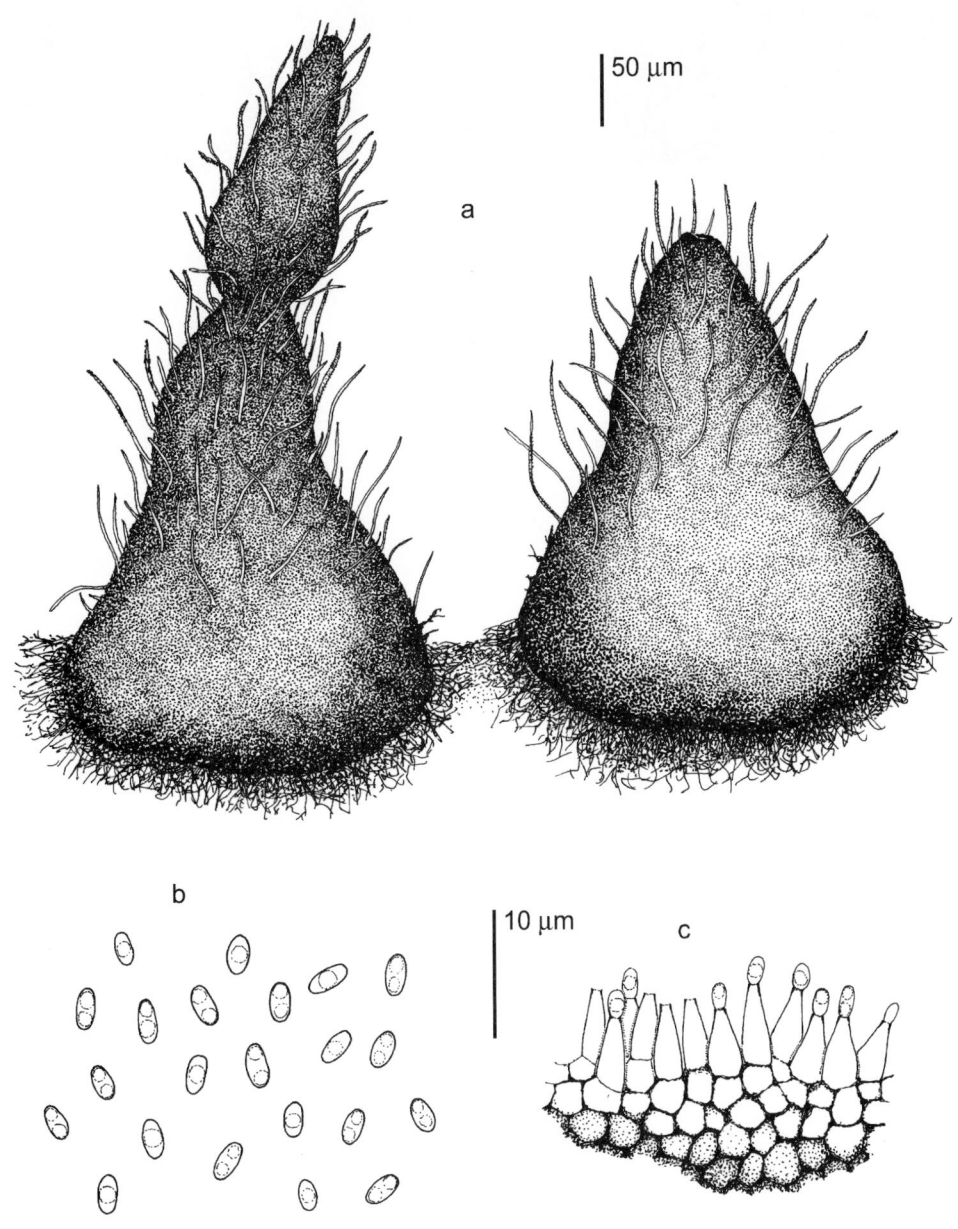

Pseudochaetosphaeronema larense, CBS 640.73. a. Pycnidia; b. conidia; c. conidiogenous cells lining the inner wall of the pycnidium.

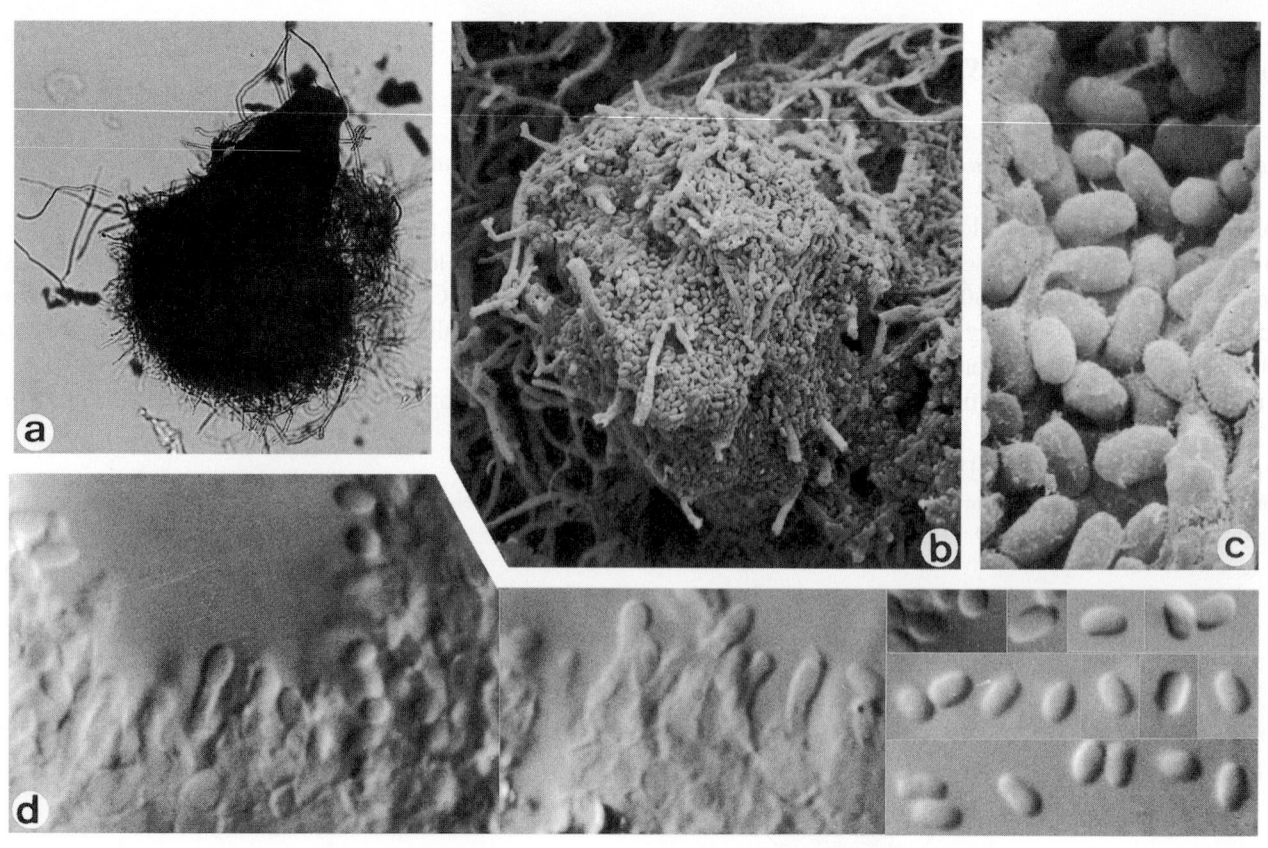

Pseudochaetosphaeronema larense, CBS 640.73. a. Pycnidium; b. detail of ostiolum; c. conidia; d. conidiogenous cells and conidia. a. ×128; b. ×630; c. ×4500; d. ×1600.

Coelomycetes. Genus: *PYRENOCHAETA*

Generic description. Colonies restricted, velvety, dark olive grey; reverse olivaceous.

Microscopy. Pycnidia solitary, rarely aggregated, spherical to subspherical, ostiolate, setose, usually brown. Setae abundant around the ostiole, dark brown, thick- and smooth-walled, septate, usually tapering towards the tip. Conidiophores ampulliform or filiform, hyaline, septate, branched at the base. Conidiogenous cells phialidic, hyaline, smooth-walled. Conidia unicellular, hyaline, smooth-walled, cylindrical to ellipsoidal.

Reference. Schneider (1979), Sutton (1980).

Key to the treated species of Pyrenochaeta:

1a. Colonies raised or wrinkled . . . . *P. mackinnonii* (355)
1b. Colonies flat, evenly velvety or floccose → **2**
2a. Colonies vinaceous-brown . . . . *P. unguis-hominis* (357)
2b. Colonies greyish-sepia to fuscous-black . . . . *P. romeroi* (355)

Pyrenochaeta romeroi Borelli

Colony characteristics. Colonies (OA) restricted, flat, evenly velvety or floccose, greyish-sepia to fuscous-black; reverse olivaceous-black.

Microscopy. Pycnidia submerged, ostiolate, black, spherical to pyriform, 85-100 µm diam, with thick walls, often covered with stiff, dark hyphae. Conidia produced from ampulliform phialides lining the innermost pycnidial wall, oozing out in slimy drops, hyaline, ellipsoidal to bacilliform, 2.5-4.0 × 1.3-1.5 µm.

Differential diagnosis. Colonies often lack pycnidia and then consist of sterile mycelium similar to *Madurella grisea* (p. 730), which is connected to the same clinical features. Another *Pyrenochaeta* species, *P. mackinnonii* Borelli, described from cases of mycetoma (Borelli, 1976; Serrano *et al.,* 1998), has smaller conidia and more restricted colonies.

Pathogenicity. BSL-2. The species causes mycetoma in humans. Grains are soft, irregular in shape, black with subhyaline centre, about 1 mm diam. The species normally occurs as a saprobe on plant debris and in soil in tropical countries. *P. romeroi* should be classified in *Phoma* rather than in *Pyrenochaeta* (Schneider, 1979); it may be identical to *Phoma leveillei* Boerema & Bollen.

References. Young *et al.* (1973), Punithalingam (1979).

Nomenclature. *Pyrenochaeta romeroi* Borelli - Derm. Venez. 1: 325-327, 1959.

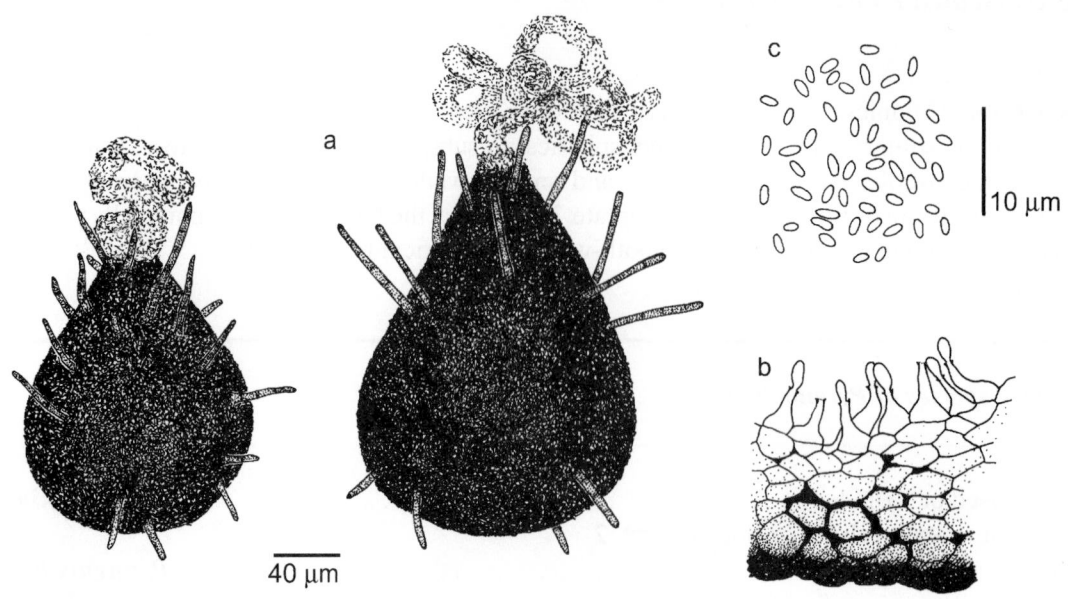

Pyrenochaeta romeroi. a, c. RV 45935; b. CBS 252.60. a. Pycnidia; b. conidiogenous cells lining the inner cavity wall; c. conidia.

Pyrenochaeta romeroi, CBS 252.60. a. Pycnidia; b, c. conidia. a. ×320; b. ×1600; c. ×10,000.

Pyrenochaeta unguis-hominis Punithalingam & English

Colony characteristics. Colonies (OA) floccose, brown-vinaceous to fawn, with abundant aerial mycelium.
Microscopy. Pycnidia solitary or aggregated, brown to black, subspherical, up to 500µm diam, ostiolate, setose; setae sparse, up to 45 µm long, 4 µm wide, obtuse at the apices. Conidiophores 15-20 × 2 µm, with 2-3 septa, conidia being produced from lateral and terminal phialide openings. Conidia unicellular, hyaline, usually short cylindrical, sometimes slightly curved, 2.0-3.5 × 1.0-1.5 µm, guttulate.
Pathogenicity. BSL-2. Repeatedly isolated from human nails, but its etiologic role in onychomycosis is questionable (Punithalingam, 1979).
Antifungal susceptibility.

Antifungal	Mean MICs	Strains	Reference
AMB	1.41	2	Unpublished data
5FC	4.0	2	Unpublished data
FLZ	128	2	Unpublished data
ITZ	32	2	Unpublished data
KTZ	5.65	2	Unpublished data
MCZ	5.65	2	Unpublished data

Reference. Punithalingam (1979), Schneider (1979).
Nomenclature. *Pyrenochaeta unguis-hominis* Punithalingam & English - Trans. Br. Mycol. Soc. 64: 539, 1975.

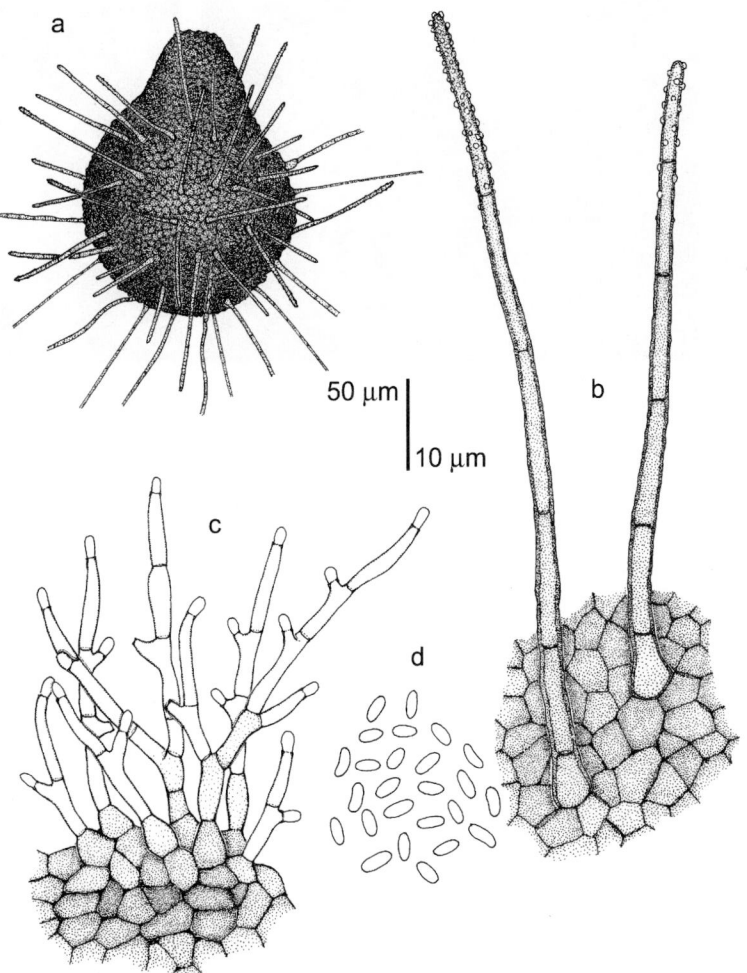

Pyrenochaeta unguis-hominis, CBS 378.92. a. Pycnidium; b. part of pycnidium wall; c. section of a pycnidium with conidia produced from conidiophores lining the inner wall of the pycnidium; d. conidia.

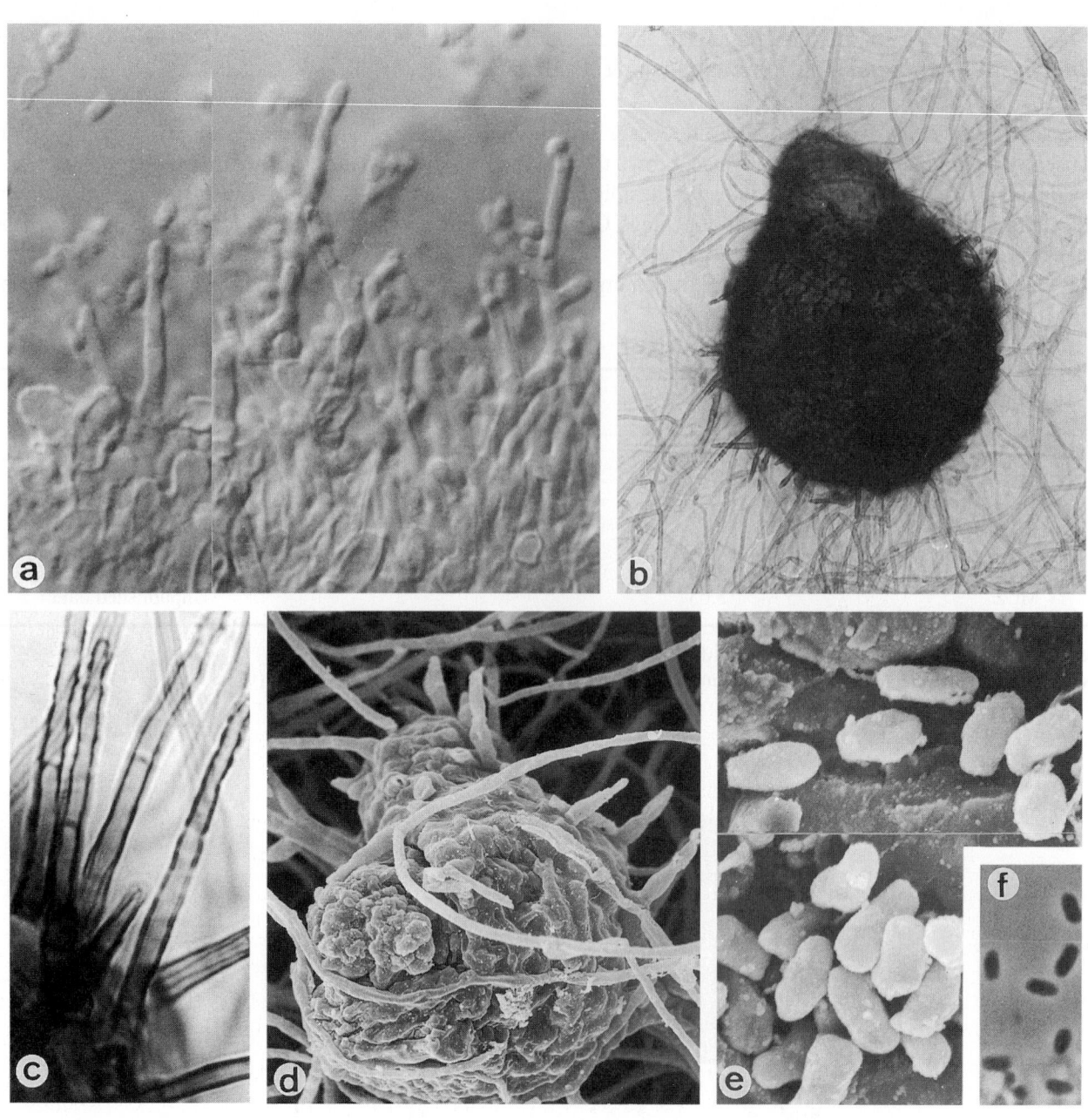

Pyrenochaeta unguis-hominis, CBS 378.92. a. Conidiophores; b, d. pycnidia; c. part of pycnidium wall; e, f. conidia. a. ×1600; b. ×256; c. ×1280; d. ×950; e. ×6900; f. ×1280.

Coelomycetes. Genus: *SPHAEROPSIS*

Sphaeropsis subglobosa Cooke

Colony characteristics. Colonies (OA) growing rapidly, floccose, dark grey to black.

Microscopy. Conidiomata stromatic, dark brown to black, solitary or aggregated, subspherical to obpyriform, 300-600 × 320-550 μm; wall composed of several layers of laterally compressed, brown cells. Conidiogenous cells holoblastic, hyaline, cylindrical with swollen base, arising from the innermost layer of cells lining the central cavity. Conidia unicellular, obovoidal, subspherical or spherical, with truncate base, thick-walled, brown, sometimes with a lighter pigmented longitudinal band, 9-12 × 6-9 μm.

Teleomorph. *Botryosphaeria subglobosa* (C. Booth) v. Arx & E. Müller (*Ascomycota, Euascomycetes, Dothideales: Botryosphaeriaceae*).

Pathogenicity. BSL-1. Keratitis (Kirkness *et al.*, 1991).

Reference. Punithalingam (1969).

Nomenclature. *Sphaeropsis subglobosa* Cooke - Grevillea 7: 95, 1879.

    Neodeightonia subglobosa C. Booth, *in* Punithalingam - Mycol. Pap. 119: 19, 1969 ≡ *Botryosphaeria subglobosa* (C. Booth) von Arx & E. Müller - Stud. Mycol. 9: 15, 1975.

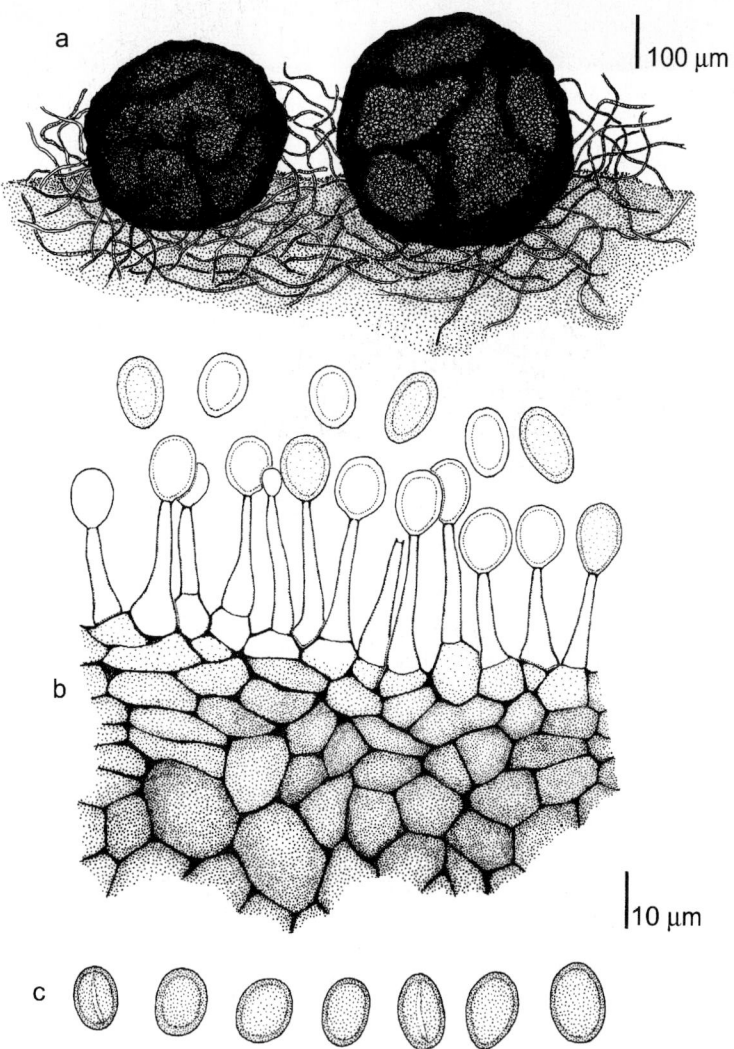

Sphaeropsis subglobosa, IMI 287616. a. Pycnidia; b. section of a pycnidium with conidia produced from conidiogenous cells lining the inner wall of the pycnidium; c. conidia.

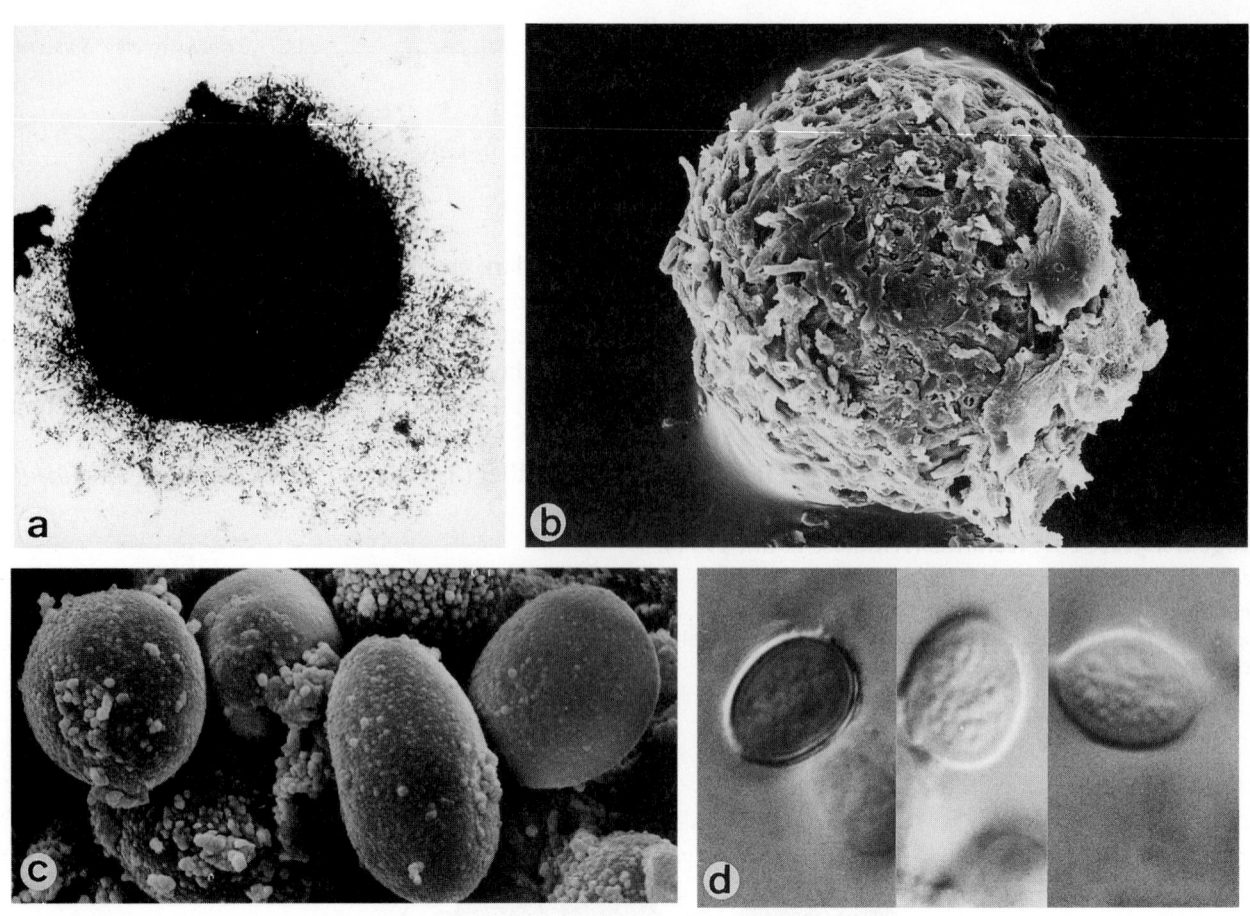

Sphaeropsis subglobosa, IMI 287616. a, b. Pycnidia; c, d. conidia. a. ×256; b. ×370; c. ×3550; d. ×1600.

HYPHOMYCETES: EXPLANATORY CHAPTERS, AND KEYS TO THE GENERA

'Hyphomycetes' form the major part of the moulds encountered in the laboratory. They are composed of regularly septate hyphae, and produce asexual propagules directly on the hyphae, without fruit bodies. Most are cultural states of Ascomycetes, occasionally of Basidiomycetes, showing mitotic propagation only. Asco- and basidiomycetes are classified on the basis of the sexual part of their life cycle (see chapter 1), and hence mitotic fungi (anamorphs) are difficult to attribute to known teleomorph species. Of many, the sexual part of the life cycle is very rare, or unknown, or it may even be non-existent. The consistently asexually reproducing fungi have been given separate anamorph names and are treated artificially under the 'hyphomycetes'. In others the complete life cycle is known and the teleomorph may even be prevalent. Such species are better treated under their teleomorph, despite the presence of conidia. This is the case with the hemiascomycetous genus *Geotrichum*, with Ascomycetes producing fruit bodies in culture, and with several mushrooms (*Basidiomycota*). Below (Fig. 27) an alternative representation of Fig. 7 on p. 11 is given, showing the biological and artificial systems as they are maintained in this Atlas, demonstrating the position of the 'hyphomycetes' in the *Eumycota*. **References.** Carmichael *et al.* (1980), von Arx (1981).

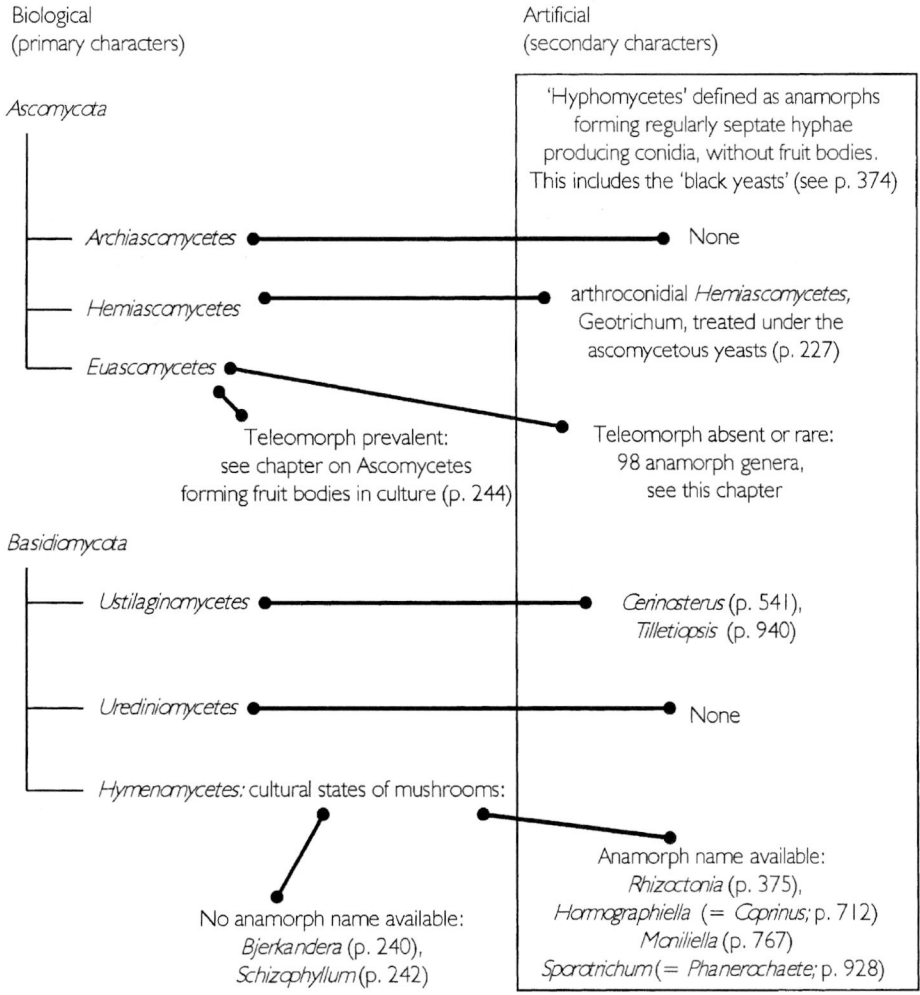

Fig. 27. Diagram of relationships between *Ascomycota, Basidiomycota,* and 'hyphomycetes'.

361

HYPHOMYCETES: EXPLANATORY CHAPTERS

1. Conidiogenesis

Criteria for distinction of hyphomycetes using production of conidia

For routine identification their biological (sexual) relationship usually do not need to be determined, but, instead, accent is laid upon the mode of conidium formation. The criteria of **conidiogenesis** (Fig. 28) relevant to the identification of medical fungi are sum-marized below; for further reading on the subject is referred to Cole & Samson (1979).

Cells which produce conidia are **conidiogenous cells.** Often a differentiated structure is present which bears one or more conidiogenous cells; this structure is the **conidiophore.**

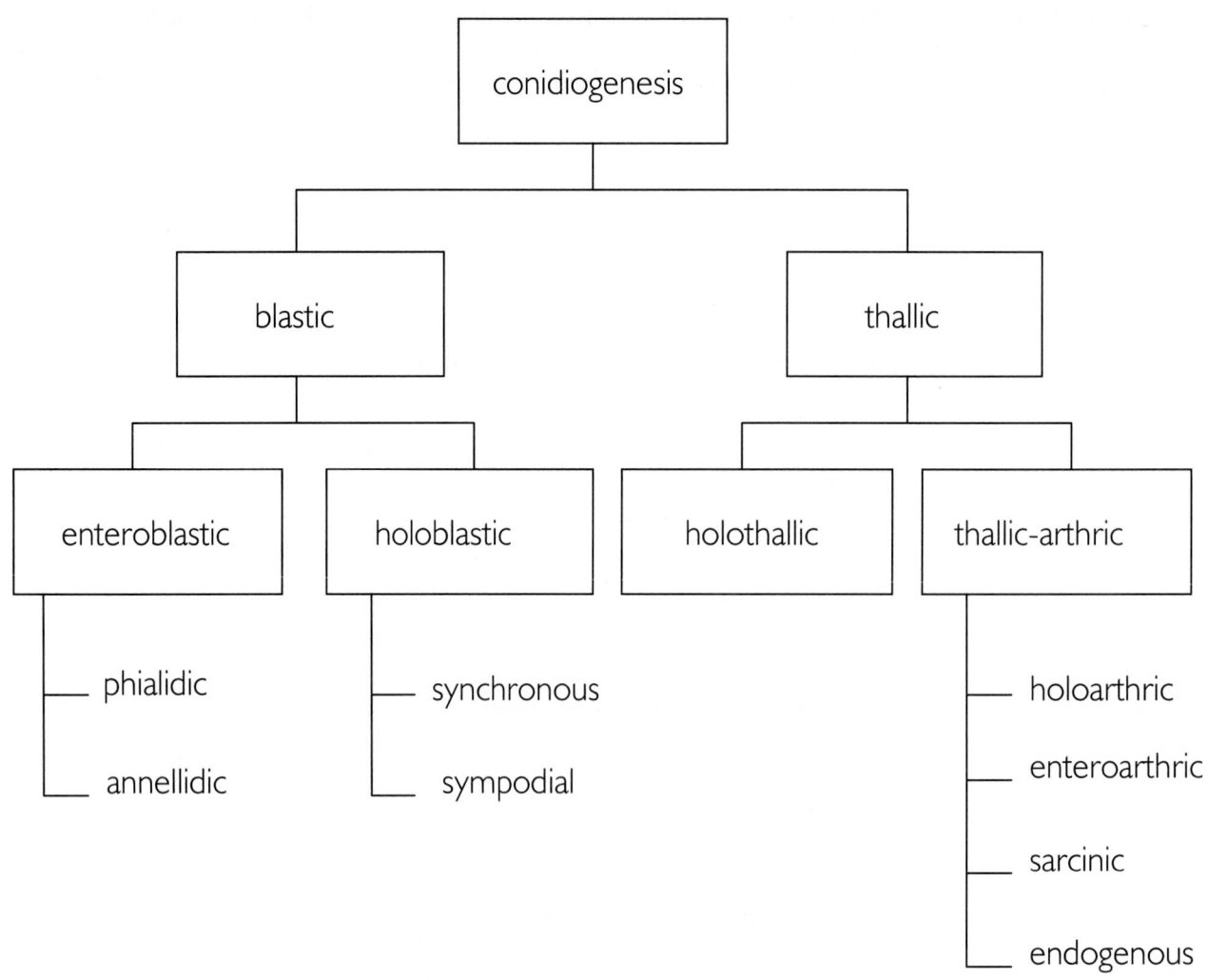

Fig. 28. Diagram of main types of conidiogenesis in clinically relevant hyphomycetes.

362

Two basic modes of conidiogenesis are distinguished, viz. blastic and thallic (Fig. 29). In blastic conidiogenesis there is a small spot on the conidiogenous cell from which the conidia are produced. In thallic conidiogenesis the entire conidiogenous cell is converted into one or more conidia.

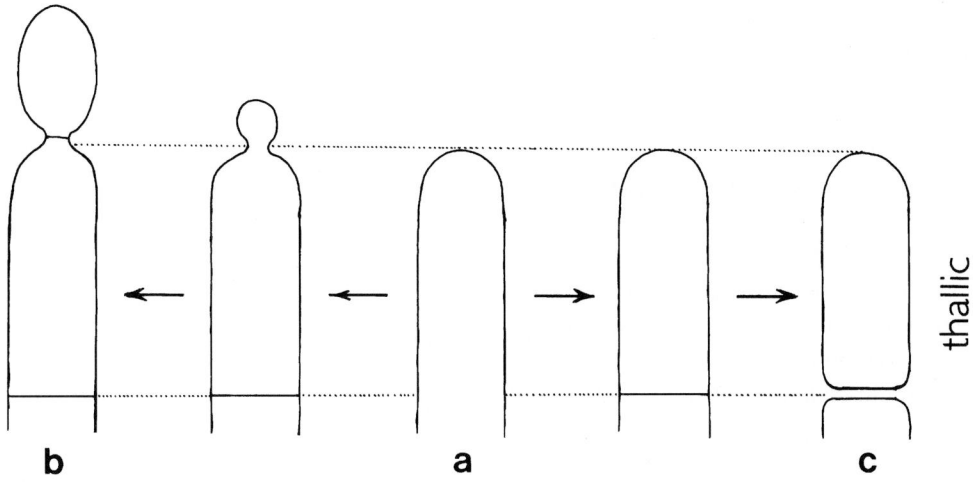

Fig. 29. Two basic possible modes of development of a hyphal element (a): blastic (b) or thallic (c).

Blastic conidiogenesis

The newly formed conidium differentiates by blowing out from a weakened spot on a cell. The conidiogenous cell has taken a particular shape and size, on which the conidium initial is clearly recognizable as an extension. If the conidiogenous cell is loose and not part of a mycelial system, as is the case in yeasts, this process is referred to as **budding.**

There are two basic modes of blastic conidium formation (Fig. 30). The blown-out end can be a local, balloon-like inflation of the cell wall **(holoblastic)**. In holoblastic development all wall layers are involved in blowing out. Alternatively, the wall is rigid, breaks open with a scar and the conidium initial, surrounded by a new wall, is forced through the opening **(enteroblastic)**. Note that these concepts are largely based on light and scanning electron microscopy, and are not always concordant with transmission electron microscopic data.

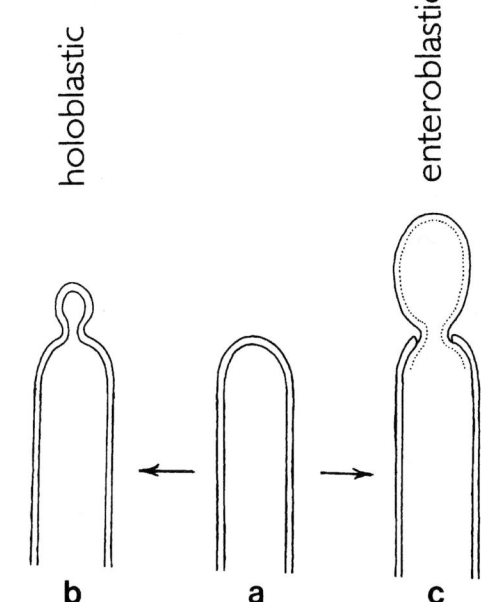

Fig. 30. Two basic possible modes of blastic development starting from (a) the young, undetermined conidiogenous cell: either (b) local balloon-like inflation of the entire cell wall (= holoblastic), or (c) through a hole in a rigid wall (= enteroblastic).

Holoblastic. – Holoblastic conidiogenous cells are often undifferentiated, or are part of a specialized conidiophore. Thus formed conidia may again form conidia in the same way, and then a chain of conidia is formed, which elongates at its tip (**acropetal**; Fig. 33a). Two main modes of holoblastic conidiogenesis are relevant:

1. *Synchronous*. – All conidia from a single conidiogenous cell blow-out at the same time (Fig. 31). A typical example of this kind of conidiogenesis is found in the genus *Cephaliophora* (p. 539).

2. *Sympodial*. – A single conidium is formed, and the conidiogenous cell grows out laterally to form a new one (Fig. 32). This may be repeated many times, so that a long outgrowth is formed, which after conidial secession bears denticles or flat scars. This is termed a **rachis**, which may be part of a single cell, or becomes septate and integrated in a multi-celled conidiophore. Typical examples of sympodial development are found in *Tritirachium* (single-celled rachis; p. 995) and *Bipolaris* (multi-celled rachis; p. 526).

A special form of sympodial development is found in fungi that have darkened scars. The conidia seem to be forced through a hole in a rigid wall and have therefore been called 'poroconidia'. However, these have proven to be holoblastic, the 'pores' as seen by light microscopy mainly being caused by differences in pigmentation.

In synchronous as well as in sympodial conidiogenesis, each conidium may produce a new conidium at its tip, by the same kind of holoblastic conidiogenesis. Subsequent conidia repeat this processand an acropetal chain of conidia is formed (Fig. 33a).

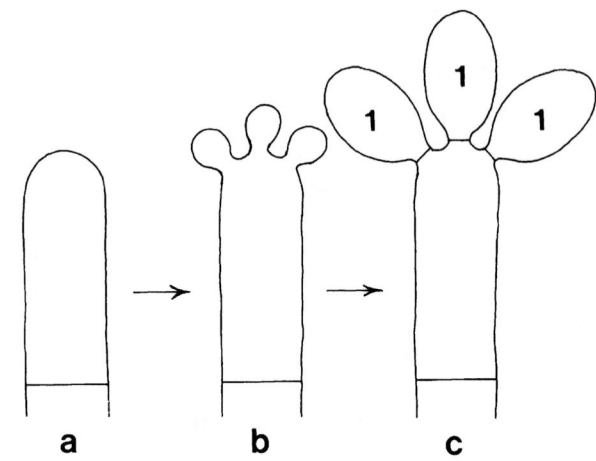

Fig. 31. Diagram of synchronous conidiogenesis. All conidium initials are formed at the same time.

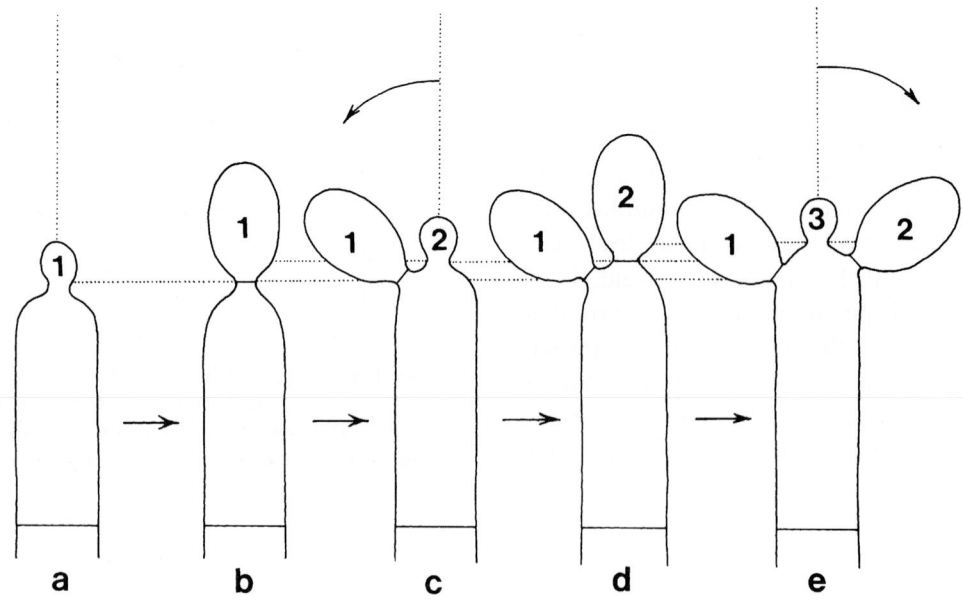

Fig. 32. Diagram of sympodial conidiogenesis. Conidium initials are formed in succession; each subsequent conidium pushes the previous one aside.

364

Enteroblastic. – Enteroblastic conidiogenous cells are mostly differentiated, often flask-shaped, clearly distinghuishable from the rest of the mycelium. Usually many conidia are produced after each other in **basipetal** succession (Fig. 33b, c), i.e., the lower conidium being the youngest. There are two main modes of enteroblastic conidiogenesis (Fig. 28):

1. *Phialidic.* – A first blown-out cell breaks open at its tip and remains as a **collarette**, from the inside of which conidia are produced successively (Fig. 34). The collarette may be wide and cup-shaped, or is hardly visible as a tiny frill at the apex of the conidiogenous cell which is termed a **phialide.**

The conidia may cohere in slimy balls close to the phialide tip, or form fragile, non-coherent chains (Fig. 34e). The arrangement of conidia is seen when a Petri-dish with a poor, transparent culture medium is examined upside down under the light microscope. Each conidium initial forms a new wall which protrudes through the scar of the previous conidium.

A typical example of phialidic development with **non-catenate** (Fig. 33c) slimy conidia is found in *Phialophora* (p. 864).

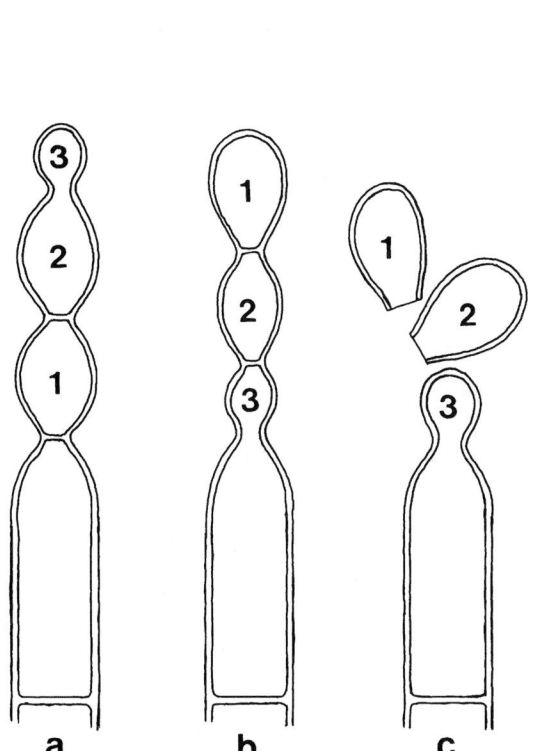

Fig. 33. Diagram of repetitive formation of conidia from a single conidiogenous cell. a. catenate, acropetal; b. basipetal, catenate; c. basipetal, non-catenate.

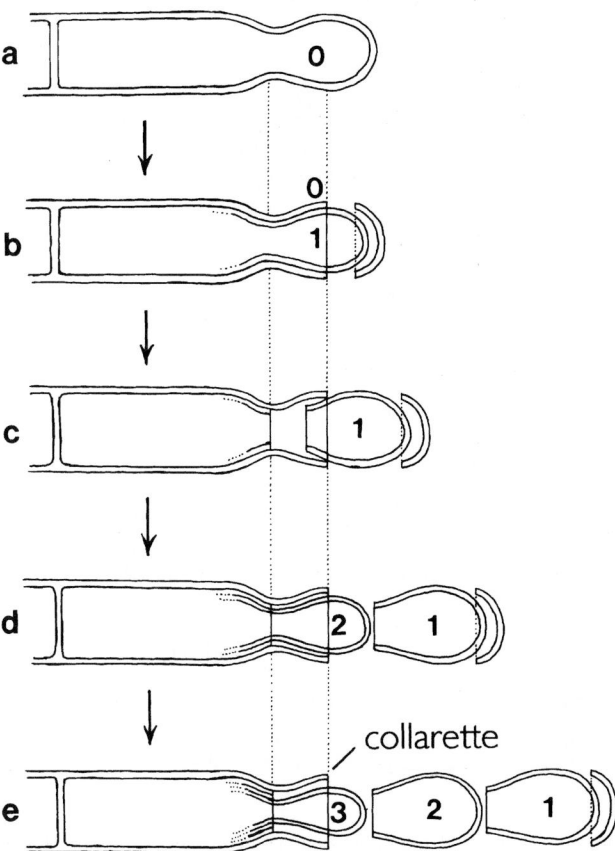

Fig. 34. Diagram of phialidic mode of conidium development with slimy, catenate or mostly non-catenate conidia. The conidium initial of the phialide is converted into a collarette. Each conidium initial forms a new wall.

In many other phialidic fungi, chains are formed by a single element which is pushed out of the phialide opening and differentiates into conidia. Such conidia are catenate and coherent (Fig. 35); chains are usually dry and hygrophobic. The chains elongate perpetually, with the youngest conidium being at the base of the chain (**basipetal**; Fig. 33b).

A typical example of phialidic development with dry, catenate conidia is found in *Aspergillus* (p. 442).

2. *Annellidic.* – The first blown-out cell is converted into a conidium, and subsequent conidia are formed by blowing out through the scar of the previous one (Fig. 36). Often each later-formed scar is located at a slightly higher level, in which case an **annellated zone** is formed. A typical example of annellidic development is found in *Scopulariopsis* (p. 902). As in phialides, a distinction may be made between slimy conidia aggregating in clusters (non-catenate) and dry conidia arising in chains (catenate). Both types of conidium production are basipetal (Fig. 33b, c).

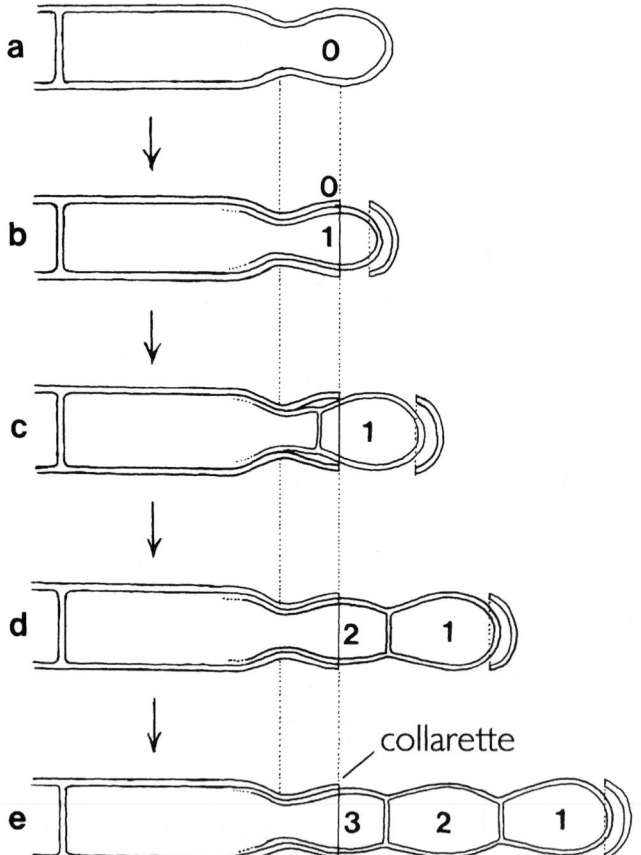 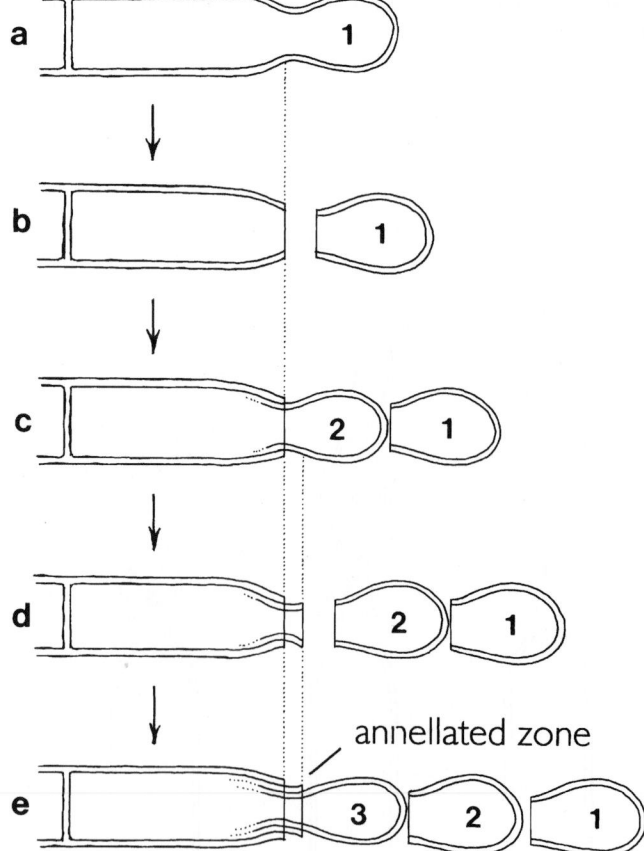

Fig. 35. Diagram of phialidic mode of conidium development with dry, catenate conidia. The conidium initial of the phialide is converted into a collarette. A cell is pushed out of the collarette and maturates to become a series of conidia.

Fig. 36. Diagram of annellidic mode of conidium development. The conidial scars are at slightly higher levels, so that an annellated zone is formed.

Thallic conidiogenesis

The newly formed conidium differentiates by conversion of a pre-existing element. A terminal or intercalary hyphal cell has become shut off by septa, and differentiates into a conidium. Two basic types of thallic development can be distinguished (Fig. 37):

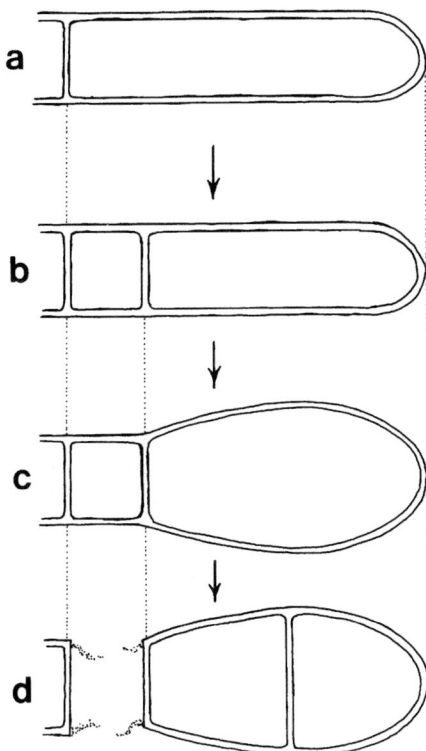

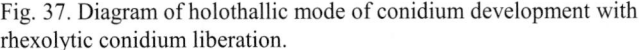

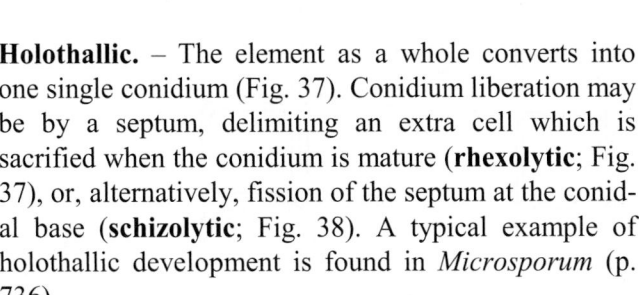

Fig. 37. Diagram of holothallic mode of conidium development with rhexolytic conidium liberation.

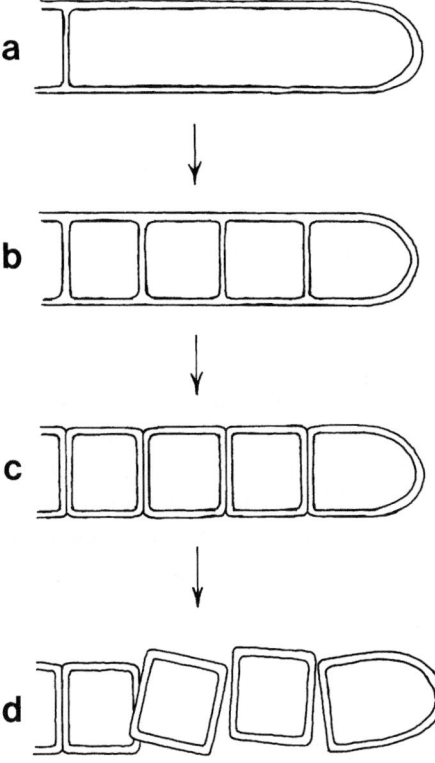

Fig. 38. Diagram of holoarthric mode of conidium development with schizolytic conidium liberation.

Holothallic. – The element as a whole converts into one single conidium (Fig. 37). Conidium liberation may be by a septum, delimiting an extra cell which is sacrificed when the conidium is mature (**rhexolytic**; Fig. 37), or, alternatively, fission of the septum at the conidal base (**schizolytic**; Fig. 38). A typical example of holothallic development is found in *Microsporum* (p. 736).

A rather general, unspecialized type of holothallic development is found in survival structures, i.e. cells that become thick-walled, mostly melanized, often encapsulated; such cells are generally referred to as **chlamydospores**. They are usually present in low abundance, in addition to conidia which serve dispersal.

Thallic-arthric. – The element is converted into a series of conidia, which are liberated at maturation, the ripened structure falling apart into single cells. Four types of thallic-arthric conidiogenesis can be distinguished:

1. *Holoarthric.* – The pre-existing hyphal element simply falls apart into a series of conidia (Fig. 38). These are liberated by splitting of the septum between cells (**schizolytic**; Fig. 38). The order of septum formation is at random; mostly the hyphal elements are perpetually subdivided into smaller units until the final conidial length is attained. A typical example of holoarthric development is found in *Geotrichum* (p. 227).

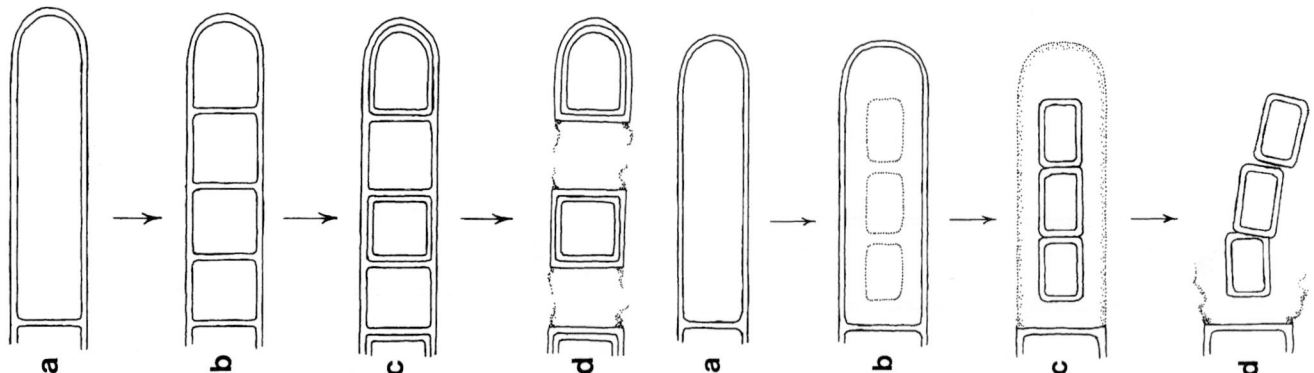

Fig. 39. Diagram of enteroarthric mode of conidium development with rhexolytic conidium liberation.

Fig. 40. Diagram of endogenous mode of conidium development.

2. *Enteroarthric.* – The pre-existing hyphal element develops two kinds of alternating cells: one kind acquires thick cell walls and becomes conidia, while the other kind dies and is sacrified to liberate the conidia (Fig. 39). This type of conidial liberation by dissolution of intermediate cells is known as **rhexolytic** cell separation. A typical example of enteroarthric conidiogenesis is found in *Coccidioides* (p. 593).

3. *Endogenous.* – The hyphal element forms internal wall layers which surround the nuclei and each become an individual conidium. The original cell wall later breaks open and the internal conidia are liberated (Fig. 40). Typical endoconidia are found in old cultures of *Aureobasidium* (p. 520).

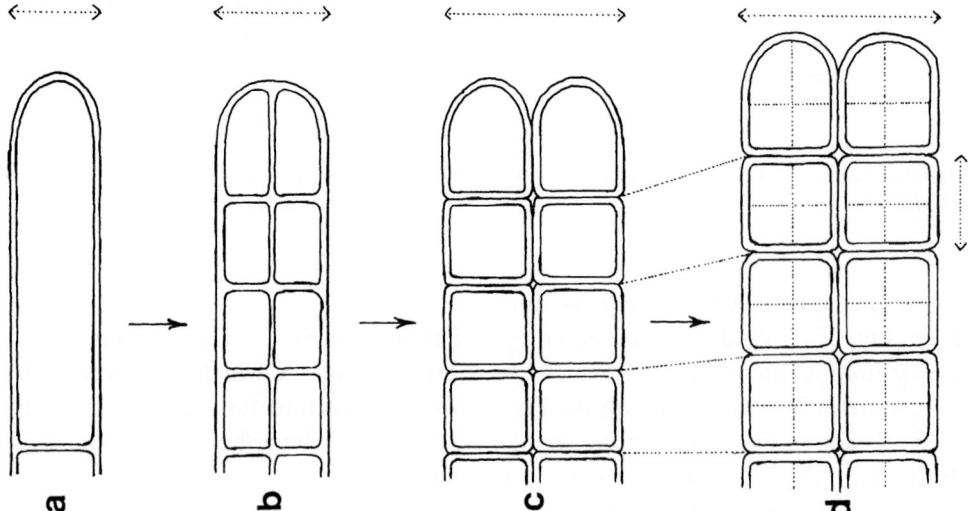

Fig. 41. Diagram of sarcinic mode of conidium development. Dotted arrows indicate expansion of cellular size.

4. *Sarcinic.* – The hyphal element gradually increases in width and becomes transversely as well as longitudinally septate (Fig. 41). Such an isodiametrically growing cell packet is a **meristem**, and subsequent conidial liberation is termed meristematic. Each cell is converted into a conidium, which later swells and becomes cruciately septate again, after which the same series of events is repeated. If the cell walls are in all directions, the conidiogenesis is called sacrcinic. A typical example of sarcinic development is found in *Botryomyces* (p. 537).

Key to the medically relevant genera of Hyphomycetes:

1a. Conidia absent → **2**
1b. Conidia present → **9**
2a. Thallus consisting of meristematically enlarging clumps of cells; mycelium sparse or absent → **3**
2b. Thallus filamentous → **5**
3a. Cell clumps in chains . . . . *Phaeosclera* **(855)**
3b. Cell clumps in dense clusters → **4**
4a. Meristematic clumps of cells initially hyaline, obliquely septate . . . . *Botryomyces* **(537)**
4b. Meristematic clumps of cells brown, cruciately septate . . . . *Sarcinomyces* **(897)**
5a. Sclerotia or multicellular bodies present → **6**
5b. Sclerotia or multicellular bodies absent → **7**
6a. Sclerotia homogeneous, usually with a dark outer wall . . . . *Rhizoctonia*[1]
6b. Sclerotia composed of some large central and many, smaller surrounding cells . . . . *Papulaspora* **(810)**
7a. Colonies black . . . . *Madurella* **(730)**
7b. Colonies of other colour → **8**
8a. Synchronous budding at 37°C . . . . *Paracoccidioides* **(812)**
8b. No budding at 37°C . . . . *Microsporum* **(736)**, *Trichophyton* **(954)**
9a. Conidia lens-shaped, dark to blackish-brown → **10**
9b. Conidia otherwise → **11**
10a. Conidia densely aggregated on specialized hyphae . . . . *Arthrinium* **(438)**
10b. Conidia singly on ampulliform cells . . . . *Nigrospora* **(777)**
11a. Thallus largely consisting of meristematically enlarging clumps of cells; mycelium and conidia sparse or absent . . . . *Sarcinomyces* **(897)**
11b. Thallus filamentous → **12**
12a. Ballistoconidia present . . . . *Tilletiopsis* **(940)**
12b. Ballistoconidia absent → **13**
13a. Conidia with appendages at one or at both ends → **14**
13b. Conidia without appendages → **15**
14a. Conidia hyaline, one-celled, with filiform appendages at both ends . . . . *Mycoleptodiscus* **(773)**
14b. Conidia brown, septate, with several apical spines . . . . *Tetraploa* **(936)**
15a. Arthroconidia abundant → **16**
15b. Arthroconidia absent or rarely present in low abundance → **29**
16a. Blastoconidia present in addition to arthroconidia; colonies soft and buttery, with olivaceous tinges . . . . *Moniliella* **(767)**
16b. Not combining above characteristics → **17**
17a. Colonies white to cream-coloured → **20**
17b. Colonies brown to olivaceous → **18**
18a. Colonies pinhead-shaped, reddish-brown; conidia short cylindrical, falling apart into cubic or spherical cells . . . . *Wallemia* **(1006)**
18b. Above characters not combined → **19**
19a. Erect conidiophores present . . . . *Oidiodendron* **(788)**
19b. Conidia arising from undifferentiated hyphae . . . . *Scytalidium* **(917)**
20a. Colonies cerebriform, restricted, with radial fissures, moist or tough; arthroconidial chains easily disarticulating, arthroconidia soon swelling to broadly ellipsoidal . . . . **see** *Trichosporon* **(164)**
20b. Colonies and conidia otherwise → **21**
21a. Colonies soft, somewhat moist, buttery . . . . **see** *Geotrichum* **(227)**
21b. Colonies dry, cottony or hairy → **22**
22a. Spherical blastoconidia present; colonies dry, eventually becoming canary yellow; conidia often formed in tufts . . . . *Arthrographis* **(440)**
22b. Not combining above characteristics → **23**
23b. Stiff, straight or curved, differentiated hyphae present; colonies usually with bright colours . . . . *Malbranchea* **(732)**

369

23b. Not combining above characteristics ➡ **24**

24a. Conidia formed on more or less differentiated, erect conidiophores, in rather dense heads . . . . *Hormographiella* (712)

24b. Conidia formed on undifferentiated, creeping hyphae ➡ **25**

25a. Colonies expanding , with strong smell; conidia produced by disarticulation of creeping hyphae . . . . see *Bjerkandera* (240)

25b. Not combining above characteristics ➡ **26**

26b. Conidia schizolytic, without separating cells; conidia arising in regular, strongly coherent chains . . . . *Onychocola* (790)

26b. Conidia rhexolytic, separated by alternating sterile cells; conidial chains otherwise ➡27

27a. Erect conidiophores present; conidia barrel-shaped to pyriform . . . . *Geomyces* (706)

27a. Not combining above characteristics ➡ **28**

28b. Conidia barrel-shaped, wider than the supporting hypha . . . . *Ovadendron* (792)

28a. Conidia rectangular, of the same width as the supporting hypha . . . . *Coccidioides* (593)

29a. Conidia arising in basipetal succession (phialidic or annellidic) ➡ **30**

29b. Conidia not borne in basipetal succession, single on each scar or in acropetal chains (sympodial, synchronous or thallic) ➡ **56**

30a. Conidiophores absent or poorly differentiated ➡ **31**

30b. Conidiophores present and well differentiated ➡ **42**

31a. Hyphae wide, hyphal cells often cubical; conidia produced in clumps from inconspicuous scars . . . . *Hormonema* (717)

31b. Not combining above characteristics ➡32

32a. Fertile hyphae densely branched; phialides intercalary; with abundant sessile collarettes; sclerotia frequently present . . . . *Cladorrhinum* (580)

32b. Not combining above characteristics ➡ **33**

33a. Colonies purely grey, brown or olivaceous black ➡ **34**

33b. Colonies white or with yellowish or pinkish tinges ➡ **39**

34a. Conidiogenous cells phialidic ➡ **35**

34b. Conidiogenous cells annellidic ➡ **38**

35a. Conidia septate . . . . *Pseudomicrodochium* (886), *Cyphellophora* (620)

35b. Conidia one-celled ➡ **36**

36a. Conidia spherical to triangular . . . . *Anthopsis* (436)

36b. Conidia of other shape ➡ **37**

37a. Phialides with flaring collarettes; conidia hyaline or pigmented, usually not allantoid . . *Phialophora* (864)

37b. Phialides with narrow collarettes; conidia hyaline, usually allantoid . . . . *Phaeoacremonium* (846)

37c. Phialides with inconspicuous collarettes; conidia pigmented . . . . *Acremonium* (394)

38a. Annellated zones conspicuous, often over 1 μm wide (if conidia dry and hydrophobic, compare *Scopulariopsis*) . . . . *Hortaea* (720)

38b. Annellated zones inconspicuous, maximally 1 μm wide . . . . *Exophiala* (645)

39a. Collarettes conspicuous, cylindrical or funnel-shaped ➡ **40**

39b. Collarettes inconspicuous, seeming a single scar at the apex of a tapering cell ➡ **41**

40a. Phialides are intercalary cells; collarettes sessile . . . . *Lecythophora* (725)

40b. Phialides differentiated, with collarettes at their apex . . . . *Phialophora* (864)

41a. Phialides with basal septum, long, slender, tapering towards the apex . . . . *Plectosporium* (880), *Acremonium* (394)

41b. Phialides intercalary or lateral, then without basal septum . . . . *Phialemonium* (859)

42a. Conidia in chains ➡ **43**

42b. Conidia not in chains; when in chains, then more than one type of conidia present ➡ **47**

43a. Conidiophores with apical swelling . . . . *Aspergillus* (442)

43b. Conidiophores without apical swelling ➡ **44**

44a. Conidiogenous cells annellidic; conidia dry, with truncate base . . . . *Scopulariopsis* (902)

44b. Not combining above characteristics ➡ **45**

45a. Phialides in dense, broom-like, penicillate arrangement . . . . *Penicillium* (814)

45b. Phialides in loose, penicillate or verticillate arrangement ➡ **46**

46a. Conidia ornamented . . . . **Acrophialophora (420)**

46b. Conidia smooth-walled . . . . **Paecilomyces (794)**

47a. Sporodochia present ➡ **48**

47b. Sporodochia absent ➡ **50**

48a. Sporodochia convex, without differentiated marginal hyphae . . . . **Fusarium (681), Volutella (1004)**

48b. Sporodochia cupulate, or surrounded by differentiated marginal hyphae ➡ **49**

49a. Sporodochia dark green, compact, cupulate, surrounded by white marginal hyphae;
hyaline setae sometimes present . . . . **Myrothecium**[2]

49b. Not combining above characteristics . . . . **Gliocladium**[3]

50a. Conidia purely green ➡ **51**

50b. Conidia of other colour ➡ **52**

51a. Conidia in chains, forming compact, dry columns . . . . **Metarrhizium (734)**

51b. Conidia not in chains, forming slimy heads . . . . **Trichoderma (943)**

52a. Conidia brownish, with rather firm walls, formed in slimy heads . . . . **Scedosporium (899)**

52b. Not combining above characteristics ➡ **53**

53a. Conidia one-celled ➡ **54**

53b. Conidia septate, falcate or cylindrical, additional microconidia may be present ➡ **55**

54a. Conidia spherical, formed on dichotomously branched conidiogenous cells . . . . **Polypaecilum (884)**

54b. Conidia usually allantoid, not formed on dichotomously branched conidiogenous cells . . . .
. . . . **Phaeoacremonium (846)**

55a. Conidia asymmetrical, usually lunate or falcate . . . . **Plectosporium (880), Fusarium (681)**

55b. Conidia symmetrical, cylindrical, with rounded ends . . . . **Cylindrocarpon (613)**

56a. Conidia arising sympodially or synchronously; if single, in acropetal chains ➡ **57**

56b. Conidia solitary on a conidiogenous cell, not in chains ➡ **98**

57a. Colonies restricted, white, at 37°C forming blastoconidia . . . . **Blastomyces (535)**

57b. Colonies otherwise ➡ **58**

58a. Conidia arising synchronously ➡ **59**

58b. Conidia arising sympodially or in acropetal chains ➡ **63**

59a. Conidia borne directly on hyphae ➡ **60**

59b. Conidia borne on swollen cells ➡ **61**

60a. Colonies mucoid, becoming dark; conidia sessile, showing budding after liberation . **Aureobasidium (520)**

60b. Colonies dry, white or cream; conidia formed on prominent denticles, without budding **Myriodontium (775)**

61a. Conidia septate . . . . **Cephaliophora (539)**

61b. Conidia one-celled ➡ **62**

62a. Conidia usually more than 5 μm in length; colonies spreading . . . . **Myceliophthora (769)**

62b. Conidia less than 5 μm in length; colonies restricted at 37°C forming large spherical
cells with thick walls . . . . **Emmonsia (633)**

63a. Conidia in mature colonies often in chains of 2 or more ➡ **64**

63b. Conidia never in chains ➡ **76**

64a. Hyphae and conidiophores hyaline; conidia swollen, with connectives . . . . **Chrysonilia (543)**

64b. Not combining above characteristics ➡ **65**

65a. Conidiogenous cells with an apical cluster of denticles ➡**66**

65b. Conidiogenous cells otherwise ➡**71**

66a. Thallus and conidia melanized . . . . **Fonsecaea (676)**

66b. Thallus hyaline, conidia hyaline or coloured ➡ **67**

67a. Colonies tough, dirty white, becoming dark . . . . **Sporothrix (924)**

67b. Colonies otherwise ➡ **68**

68a. Colonies soft, buttery, white to cream-coloured . . . . **see Candida (180)**

68b. Colonies farinose or cottony ➡ **69**

69a. Budding cells present; filaments easily falling apart . . . . **Cerinosterus (541)**

69b. Budding cells absent; hyphae not falling apart ➡ **70**

70a. Conidia borne on minute denticles and with acute base . . . . **anamorph of Ophiostoma (299)**

70b. Conidia sessile, with truncate base . . . . **Sporotrichum (928)**

71a. Conidia with transverse and longitudinal septa . . . . **Alternaria (422)**

71b. Conidia one-celled or with transverse septa only ➡ **72**

72a. Conidia generally septate ➡ **73**

72b. Conidia generally one-celled ➡ **75**

73a. Conidia partly distoseptate, smooth-walled or verruculose . . . . *Dissitimurus* (629)

73b. Conidia with true septa (euseptate), distinctly verrucose ➡ **74**

74a. Conidiogenous cells with a single conidiogenous locus . . . . *Taeniolella* (934)

74b. Conidiogenous cells with several conidiogenous loci . . . . *Stenella* (932)

75a. Conidia with dark brown scars . . . . *Cladosporium* (582)

75b. Conidial scars not conspicuously darkened . . . . *Cladophialophora* (560)

76a. Conidia one-celled ➡ **77**

76b. Conidia septate ➡ **83**

77a. Synnemata present ➡ **78**

77b. Synnemata absent ➡ **79**

78a. Conidiogenous cells annellidic; conidia aggregated in slimy masses . . . .
. . . . *Graphium* anamorph of *Petriella* (301) or *Pseudallescheria* (305)

78b. Conidiogenous cells sympodial, with denticles; conidia dry . . . . *Phaeoisaria* (853)

79a. Colonies white or yellowish ➡ **80**

79b. Colonies ochraceous, rosa or olivaceous ➡ **81**

80a. Conidiogenous cells in whorls along ascendent hyphae, with acicular basal part . . . . . *Engyodontium* (640)

80b. Conidiogenous cells in dense clusters, with spherical or ellipsoidal basal part . . . . *Beauveria* (523)

81a. Colonies ochraceous or rosa . . . . *Tritirachium* (995)

81b. Colonies olivaceous brown or olivaceous black ➡ **82**

82a. Conidiophores erect, unbranched, arising at right angles from creeping hyphae . . . . *Ramichloridium*[4] (888)

82b. Conidiophores otherwise, arising at acute angles from ascending hyphae . . . . *Rhinocladiella* (893)

83a. Conidiophores undifferentiated; conidia cylindrical, septate, hyaline except the upper cells which
is pale brown . . . . *Polycytella* (882)

83b. Not combining above characteristics ➡ **84**

84a. Conidiophores long, erect, with flat scars; conidiogenous cells sympodial; conidia
usually 2-celled . . . . *Veronaea* (1002)

84b. Not combining above characteristics ➡ **85**

85a. Conidiogenous cells with conical denticles . . . . *Ochroconis* (779)

85b. Conidiogenous cells without denticles ➡ **86**

86a. Conidia with a long and narrow, hyaline beak ➡ **87**

86b. Conidia without beak ➡ **88**

87a. Conidia with a lateral branch near the base . . . . *Mycocentrospora* (771)

87b. Conidia without a lateral branch at the base . . . . *Phaeotrichoconis* (857)

88a. Conidiogenous cells often elongating through the terminal scar ➡ **89**

88b. Conidiogenous cells otherwise ➡ **90**

89a. Conidia with transverse and longitudinal septa . . . . *Stemphylium*[5]

89b. Conidia with transverse septa only . . . . *Corynespora* (596)

90a. Conidiogenous cells lobate ➡ **91**

90b. Conidiogenous cells not lobate ➡ **92**

91a. Conidia coiled . . . . *Dichotomophthoropsis* (627)

91b. Conidia cylindrical or ellipsoidal . . . . *Dichotomophthora* (625)

92a. Conidiogenous cells straight, with small, porous, hyaline scars; conidia
obclavate . . . .*Helminthosporium*[6]

92b. Conidiogenous cells sympodial, geniculate ➡ **93**

93a. Conidia curved, euseptate with transverse septa only . . . . *Curvularia* (598)

93b. Conidia straight, or if curved distoseptate ➡ **94**

94a. Conidia with prominent scar at the base . . . . *Exserohilum* (669)

94b. Conidia without prominent basal scar ➡ **95**

95a. Conidia distoseptate, with transverse septa only ➡ **96**

95b. Conidia euseptate, with transverse and often also with longitudinal septa ➡ **97**

96a. Conidia straight, germinating from all cells . . . . *Drechslera* (631)

96b. Conidia straight or curved, germinating from polar cells only . . . . *Bipolaris* (526)

97a. Young conidia rounded at the base; mature conidia obclavate and rostrate . . . . *Alternaria* (422)

97b. Young conidia attenuated at the base; mature conidia broadly ellipsoidal . . . . *Ulocladium* (997)

98a. Conidiophores well differentiated, long, apically branched; conidia spherical, solitary, thick-walled . . . . *Staphylotrichum* (930)

98b. Not combining above characteristics ➝ **99**

99a. Conidia olivaceous, blackish-brown or black ➝ **100**

99b. Conidia hyaline to pale brown ➝ **104**

100a. Conidia septate ➝ **101**

100b. Conidia one-celled ➝ **102**

101a. Conidia two-celled, constricted at the septum, with verrucose walls *Ochroconis* (779) *Trichocladium* (941)

101b. Conidia muriform . . . . *Monodictys*[7]

102a. Conidia black, smooth-walled . . . . *Nigrospora* (777)

102b. Conidia brown, distinctly ornamented ➝ **103**

103a. Conidia coarsely verrucose . . . . *Thermomyces* (938)

103b. Conidia (macroconidia or chlamydospores) tuberculate . . . . *Histoplasma* (708)

104a. Conidia long, septate; additional one-celled microconidia often present (**dermatophytes**) ➝ **107**

104b. Conidia all short, one-celled ➝ **105**

105a. Conidia with narrow base, slimy . . . . *Scedosporium* (899)

105b. Conidia with truncate base, dry ➝ **106**

106a. Erect, profusely branched conidial apparatus present . . . . *Sporotrichum* (928)

106b. Conidia produced on undifferentiated hyphae . . . . *Chrysosporium* (545), *Trichophyton* (954)

107a. Macroconidia rough-walled . . . . *Microsporum* (736)

107b. Macroconidia smooth-walled ➝ **108**

108a. Macroconidia thick-walled . . . . *Keratinomyces* (723)

108b. Macroconidia thin-walled ➝ **109**

109a. Microconidia absent ➝ **110**

109b. Microconidia present, often predominant . . . . *Trichophyton* (954)

110a. Colonies with yellowish or brownish tinge; conidia clavate . . . . *Epidermophyton*[8] (642)

110b. Colonies white; conidia fusiform . . . . *Polycytella* (882)

[1-3]Case reports in lit. mentioned without identified species: [1]*Rhizoctonia* sp.: keratitis (Srivastava *et al.*, 1977), [2]*Myrothecium* sp.: keratitis (Liesegang & Forster, 1980), [3]*Gliocladium* sp.: keratitis (Kwon-Chung & Bennett, 1992).

[4]If conidia rhexolytic, with frilled, truncate bases, see *Nodulisporium* and *Geniculosporium*. An unnamed *Nodulisporium* sp. was reported from a sinusitis (Cox *et al.*, 1994). The *Geniculosporium* sp. Reported by Suzuki *et al.* (1998) from a human phaeohyphomycosis has been reidentified as *Ramichloridium* (G.S. de Hoog, unpublished data).

[5]*Stemphylium macrosporoideum*: mixed infection in antromycosis (Bassiouny *et al.*, 1982).

[6]*Helminthosporium*: the numerous published cases of keratitis and sinusitis probably concerned *Bipolaris* species (see p. 526).

[7]*Monodictys indica*: skin lesion (Singh & Barde, 1985), but the single strain of this species has not been preserved.

[8]Occasionally *Trichophyton* species lack microconidia; try also the key on p. 954.

HYPHOMYCETES, EXPLANATORY CHAPTERS

2. Black yeasts and their relatives

Black yeasts are defined as asexual fungi potentially able to produce melanized budding cells (a yeast phase) in any stage of their life cycle. This condition occurs in phylogenetically highly diverse fungi (Fig. 42), namely in some basidiomycetes (1) and in members of the ascomycete orders *Chaetothyriales* (2) and *Dothideales* (3). Often also the strictly filamentous relatives of pathogenic black yeasts are comprised under this group; then the indication 'black yeast-like fungi' is applied. Occasionally black yeast-like fungi reproduce by isodiametric enlargement and subsequent meriste-matic development, e.g., in *Sarcinomyces* (p. 897). Melanized asexual fungi more in general are frequently referred to as '*Dematiaceae*' ('Schwärzepilze'; p. 379). This artificial term also covers the large-conidial ana-morphs of *Pleosporales*, such as *Alternaria* (p. 422) and *Curvularia* (p. 598). The black meristematic genus *Botryomyces* (p. 537) was proven to be a member of the *Pleosporales* and thus is unrelated to the black yeasts.

Groups containing melanized budding cells are the following:
1. The basidiomycetous black yeasts are classified in the genus *Moniliella* (p. 767); the precise phylogenetic position of this group is, as yet, unknown. The species can be recognized by their ability to ferment glucose and by the frequent occurrence of rectangular arthroco-nidia. Most of these fungi are of industrial significance and are rarely seen in clinical practice.

2. In the ascomycete order *Chaetothyriales*, mainly comprising the family *Herpotrichiellaceae*, the genus *Exophiala* is one of the most commonly encountered anamorphs. The order contains numerous human patho-gens, with a wide spectrum of clinical pictures. The order is related to the *Eurotiales* and *Onygenales* rather than to the dothidealean black yeasts (Fig. 3, p. 4). The ascomycete order *Dothideales* (anamorph genus *Aureo-basidium* and its relatives) mainly comprises plant-associated fungi, which are only exceptionally involved in human disease.

In the two ascomycete orders the anamorphs are morphologically diverse, both with a similar range of diversity. Given the difference in clinical significance of members of the two orders, it is crucial to attribute each strain to the right order by identifying it with the right genus. The potential pathology largely determines whether the strain under consideration is the possible etiologic agent, or whether it should be discarded as a probable contaminant. For example, the genus *Clado-phialophora (Chaetothyriales)* contains highly virulent agents of systemic disease, whereas its morphologically similar counterpart, *Cladosporium (Dothideales)*, contains ubiquitous saprophytes on plants. The *Chaeto-thyriales* are still relatively unexplored; many new, clinically relevant taxa are still to be described.
Reference. De Hoog (1999).

Key to the genera of black yeasts and their relatives:

1a. Conidiogenesis predominantly annellidic ➞ **2**
1b. Conidiogenesis otherwise ➞ **3**
2a. Annellated zones about 1-2 µm wide, often nearly as wide as the supporting cell .... ***Hortaea* (720)**
2b. Annellated zones maximally 1 µm wide, significantly narrower than half the width of the supporting cell .... ***Exophiala* (645)**
3a. Nearly exclusively budding cells and / or meristematic cells present ➞ **4**
3b. Hyphae present in relative abundance ➞ **9**
4a. Isodiametric, septate cell clumps present ➞ **5**
4b. Budding cells remaining non-septate ➞ **7**
5a. Cell clumps falling apart into smaller packets ➞ **6**
5b. Cell clumps not disintegrating .... ***Phaeosclera* (855)**

6a. Cell clumps initially hyaline; nitrate assimilated .... *Botryomyces* **(537)**

6b. Cell clumps olivaceous black from the beginning; nitrate not assimilated .... *Sarcinomyces* **(897)**

7a. Budding strictly multilateral .... *Moniliella* **(767)**

7b. Budding frequently annellidic ➡ **8**

8a. Tolerant to 10% NaCl .... *Aureobasidium* **(520)**

8b. Intolerant to 10% NaCl .... *Exophiala* **(645)**

9a. Colonies initially pink; hyphae of irregular width, up to 6 μm in diam ➡ **10**

9b. Colonies grey to olivaceous black; hyphae of regular width, up to 2 μm in diam ➡ **12**

10a. Budding cells present; conidiophores absent ➡ **11**

10b. Budding cells absent; erect, brown conidiophores present .... *Ramichloridium* **(888)**

11a. Synchronous conidia (>2 per cell) present in young cultures; methyl-α-D-glucoside assimilated .... *Aureobasidium* **(520)**

11b. Maximally 2 synchronous conidia per cell present; methyl-α-D-glucoside not assimilated .... *Hormonema* **(717)**

12a. Conidiogenesis predominantly phialidic ➡ **13**

12b. Conidiogenesis otherwise; phialides may be present in low abundance ➡ **14**

13a. Conidia exclusively produced on stiff, spine-like, brown phialides .... see *Phaeoacremonium* **(846)**

13b. Phialides otherwise, or conidia produced on several types of conidiogenous cells ...... *Phialophora* **(864)**

14a. Conidia produced sympodially, each denticle bearing a single conidium ➡ **15**

14b. At least part of the conidia produced in short (2-3 cells) or long chains ➡ **16**

15a. Conidiophores dark brown, erect, unbranched, arising at right angles from creeping hyphae .... *Ramichloridium* **(888)**

15b. Conidiophores brown, arising at acute angles in a branched system .... *Rhinocladiella* **(893)**

16a. Conidial scars blackish-brown, darker than the conidial body; conidia brown .... *Cladosporium* **(582)**

16b. Conidial scars not conspicuously darkened; conidia subhyaline to pale brown ➡ **17**

17a. Conidia in chains of up to 4 .... *Fonsecaea* **(676)**

17b. Conidial chains significantly longer .... *Cladophialophora* **(560)**

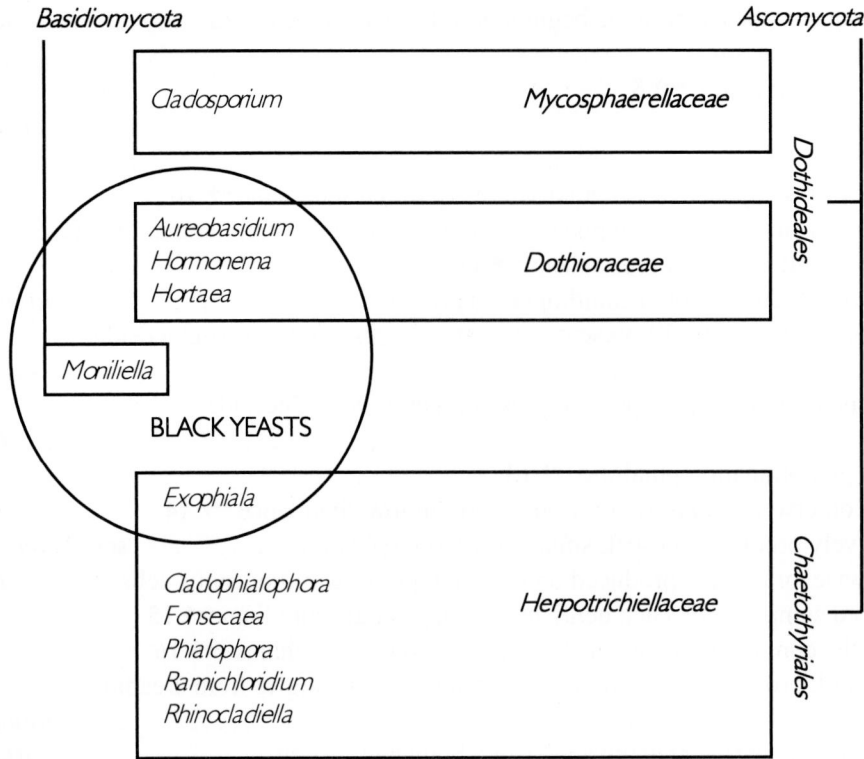

Fig. 42. Diagram of relationships of fungi with melanized budding cells (black yeasts).

Table 24. Taxonomic overview of black yeasts and their relatives, with the prevalent forms of growth.

Basidiomycota

        Moniliella   - yeast (filamentous when degenerate)

Ascomycota

  Chaetothyriales

    Herpotrichiellaceae

Cladophialophora	- filamentous
Exophiala	- filamentous with yeast
Fonsecaea	- filamentous
Phialophora	- filamentous
Ramichloridium	- filamentous with or without yeast
Rhinocladiella	- filamentous with yeast
Sarcinomyces	- meristematic

  Dothideales

    Dothioraceae

Aureobasidium	- filamentous with yeast
Hormonema	- filamentous with yeast
Hortaea	- yeast with or without filaments

    Mycosphaerellaceae

Cladosporium	- filamentous

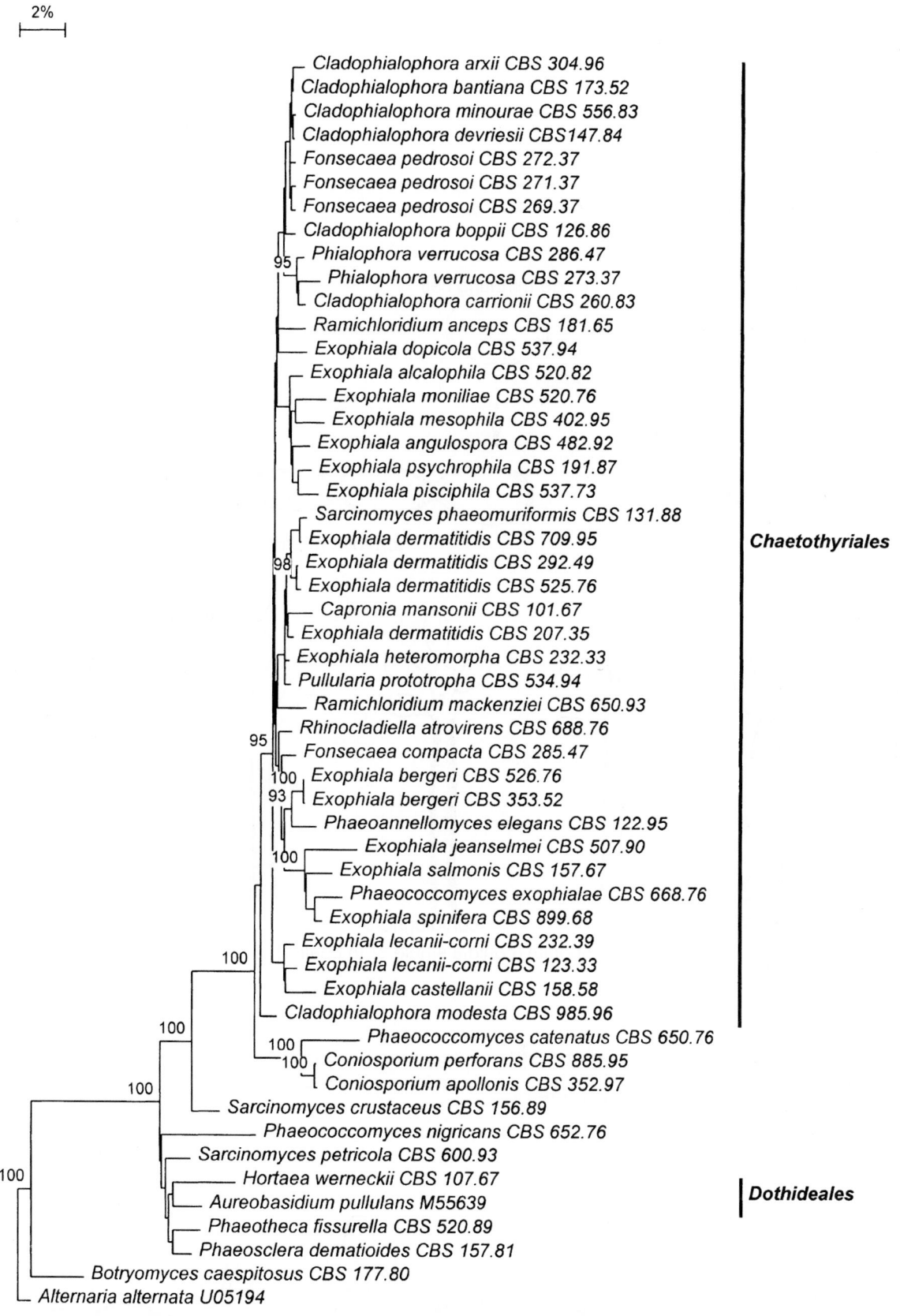

Fig. 43. Phylogenetic tree of black yeasts based on confidently aligned, near-complete SSU rDNA sequences using Neighbor joining algorithm with Kimura correction. Bootstrap values >90 from 100 resampled datasets are shown. The tree shows the two main orders of ascomycetous black yeasts: *Dothideales* and *Chaetothyriales*. The *Dothideales* is an extremely diverse group, containing thousands of species of saprobic and plant pathogenic fungi (not shown). The *Chaetothyriales* contains numerous clinically relevant species. Note that within the *Chaetothyriales* hardly any subdivision can be made. The anamorph genera maintained in the Atlas (Table 24) are based on morphology, but do not have any evolutionary significance. The yeast-like genus *Exophiala* is found intermingled with representatives of the filamentous genera *Cladophialophora, Fonsecaea, Phialophora, Ramichloridium* and *Rhinocladiella*. *Capronia* is a teleomorph genus.

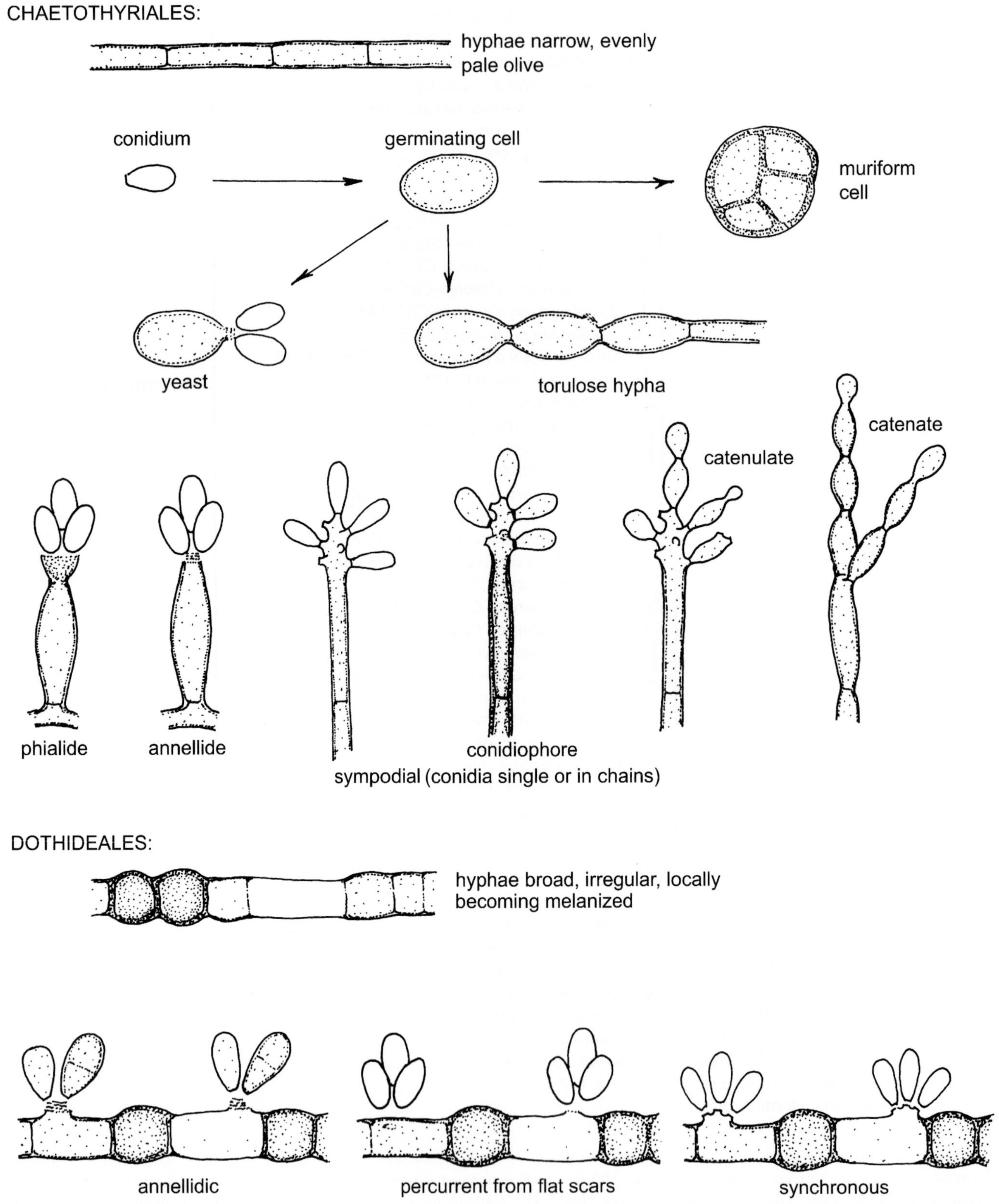

CHAETOTHYRIALES:

hyphae narrow, evenly pale olive

conidium → germinating cell → muriform cell

yeast

torulose hypha

catenulate

catenate

phialide   annellide

conidiophore

sympodial (conidia single or in chains)

DOTHIDEALES:

hyphae broad, irregular, locally becoming melanized

annellidic

percurrent from flat scars

synchronous

Fig. 44. Diagram of descriptive terminology in the two most significant orders black yeasts and their relatives, with synoptic distinction of main groups.

378

Table 24. Overview of prevalent ecology of black yeasts and their relatives, with recommended BSL classification. In brackets are exceptional clinical cases. Note that a relatively large number of members of *Chaetothyriales* in BSL 2 or 3.

Moniliella suaveolens	- industrial (cutaneous)	1
Cladophialophora arxii	- systemic	3
Cladophialophora bantiana	- primary cerebritis	3
Cladophialophora boppii	- cutaneous	2
Cladophialophora carrionii	- chromoblastomycosis	2
Cladophialophora devriesii	- disseminated	3
Cladophialophora emmonsii	- traumatic phaeohyphomycosis	2
Cladophialophora modesta	- secondary cerebritis	2
Exophiala bergeri	- cutaneous	2
Exophiala castellanii	- cutaneous	2
Exophiala dermatitidis	- pulmonary, neurotropic	2
Exophiala jeanselmei	- mycetoma	2
Exophiala lecanii-corni	- cutaneous	2
Exophiala moniliae	- saprophyte (phaeohyphomycosis)	1
Exophiala pisciphila	- opportunistic in fish	1
Exophiala salmonis	- neurotropic in fish	1
Exophiala spinifera	- cutaneous, disseminated	2
Fonsecaea compacta	- chromoblastomycosis	2
Fonsecaea pedrosoi	- chromoblastomycosis	2
Phialophora americana	- saprophyte on wood (traumatic)	2
Phialophora bubakii	- saprophyte on wood (traumatic)	1
Phialophora europaea	- cutaneous, onychomycosis	2
Phialophora repens	- saprophyte on wood (traumatic)	1
Phialophora reptans	- cutaneous	2
Phialophora richardsiae	- subcutaneous cyst	2
Phialophora verrucosa	- saprophyte on wood (traumatic)	2
Ramichloridium mackenziei	- primary cerebritis	3
Ramichloridium schulzeri	- saprophyte on wood (opportunistic)	1
Rhinocladiella aquaspersa	- chromoblastomycosis?	1
Rhinocladiella atrovirens	- saprophyte on wood (opportunistic)	1
Sarcinomyces phaeomuriformis	- cutaneous	2
Aureobasidium pullulans	- phyllosphere (opportunistic)	1
Hormonema dematioides	- plant opportunist (opportunistic)	1
Hortaea werneckii	- halophilic (tinea nigra)	1
Cladosporium spp.	- saprophytes on plant material (cutaneous)	all 1

HYPHOMYCETES, EXPLANATORY CHAPTERS

3. Dematiaceae - anamorphic Pleosporales

Criteria for distinction of dematiaceae

The artificial term 'dematiaceae' generally indicates large-conidial, melanized fungi. Teleomorphs, where known, are in *Cochliobolus, Lewia, Pyrenophora* or *Setospheria,* all members of a single family, the *Pleosporaceae* in the order *Pleosporales.* They all have hairy, black, rapidly expanding colonies, which show optimal conidium production on media poor in nutrients, such as PCA. Conidia are then mostly visible with the aid of a stereo microscope.

Conidiophores arise vertically from creeping hyphae and are clearly differentiated, dark brown, with rather thick walls, and are usually multiseptate. The apical portion develops conidia in sympodial order, and then becomes septate, so that the conidiogenous cells become **integrated** in the flexuose conidiophore (Fig. 45a). The conidial scars are dark brown to black. The conidia consist of several compartments showing different types of septation. True septa are found in those conidia where outer wall and septum are continuous; such conidia are **euseptate**. False septa are observed in those conidia where only the inner wall layers are involved in septation and outer wall forms a sac-like structure around the individual cells; such conidia are **distoseptate**, also termed 'pseudoseptate'. The latter type of septa can easily be distinguished from true septa by crushing the conidia under the cover slip in such a way that a hole in the outer wall liberates or dislocates the inner cells. Alternatively, with fixative mounts, such as lactic acid, the cytoplasma of distoseptate conidia contracts and becomes angular (Fig. 45e). Many species have with narrowed apex (**beaks**). If a new conidium is produced on top of this beak, it functions as a secondary conidiophore (**false beak**).

The kind of septation is used as a key feature for generic distinction. A further criterion is the type of germination, observed under low magnification by streaking a conidial suspension on water agar.

Germination may take place from all conidial cells, or is confined to the polar cells (Fig. 45h). Finally, the shape (straight *vs.* curved) and location of septa (transverse alone *vs.* transverse and longitudinal) are main criteria for distinction of anamorph genera. Conidia may be single or in chains, which can also be observed under the stereo microscope. Strains sometimes tend to degenerate, demonstrated by the appearance of narrow conidia which become structurally similar to the conidiophores on which they are borne. Possibly *Dissitimurus* is a degenerate counterpart of a *Bipolaris* species.

Ecologically two approximate main groups are distinguished. Most *Bipolaris, Curvularia, Drechslera* and *Exserohilum* species are pathogens on a restricted range of grass species. They are commonly found in air. Due to their large size, they remain in the sinus after inhalation and then a chronic, allergic sinusitis may be provoked. In contrast, *Alternaria* and *Ulocladium* contain a group of ubiquitous saprobes on rotten plant material and in soil. Clinically they are more often encountered as traumatic mycoses. It should be noted that cutaneous mycoses by such fungi are characterized by brownish hyphal elements in tissue, whereas subcutaneous mycoses often consist of large, hyaline, yeast-like cells. The two ecological groups with teleomorph coherence is displayed in the distance tree on p. 382.

Some extremotolerant saprobic taxa, particularly in *Alternaria*, have a strong tendency to produce meristematic clumps of cells. Some fungi exclusively producing this morphology, e.g., *Botryomyces* and *Phaeosclera*, are probably meristematic counterparts of *Alternaria*-like taxa. The possibility exists that occasional reports of mixed infections by these species actually concern conidial and meristematic synanamorphs of a single fungus.

References. Ellis (1971, 1976), Sivanesan (1987), Berbee *et al.* (1999).

Key to the treated genera of dematiaceae:

1a. Conidiophores absent, propagation being confined to clumps of meristematic cells → **2**
1b. Conidiophores present; occasionally additional clumps of meristematic cells are present → **3**
2a. Thallus exclusively consisting of meristematically enlarging cell clumps . . . . *Botryomyces* **(537)**
2b. Multi-celled clumps borne on undifferentiated hyphae . . . . *Phaeosclera* **(855)**
3a. Conidia euseptate → **4**
3b. Conidia preponderantly distoseptate → **6**
4a. Conidia curved, non-catenate, with transverse septa only . . . . *Curvularia* **(598)**
4b. Conidia straight, often catenate, mostly with additional longitudinal septa → **5**
5a. Conidia with no or very short apical beak, blackish-brown . . . . *Ulocladium* **(997)**
5b. Conidia with clearly distinguishable beak, medium brown to golden brown . . . . *Alternaria* **(422)**
6a. Conidia with marked hilum at the base . . . . *Exserohilum* **(669)**
6b. Conidial base without protruding hilum → **7**
7a. Conidia catenate . . . . *Dissitimurus* **(629)**
7b. Conidia non-catenate → **8**
8a. Conidia straight, germinating from all cells . . . . *Drechslera* **(631)**
8b. Conidia curved, germinating from polar cells only . . . . *Bipolaris* **(526)**

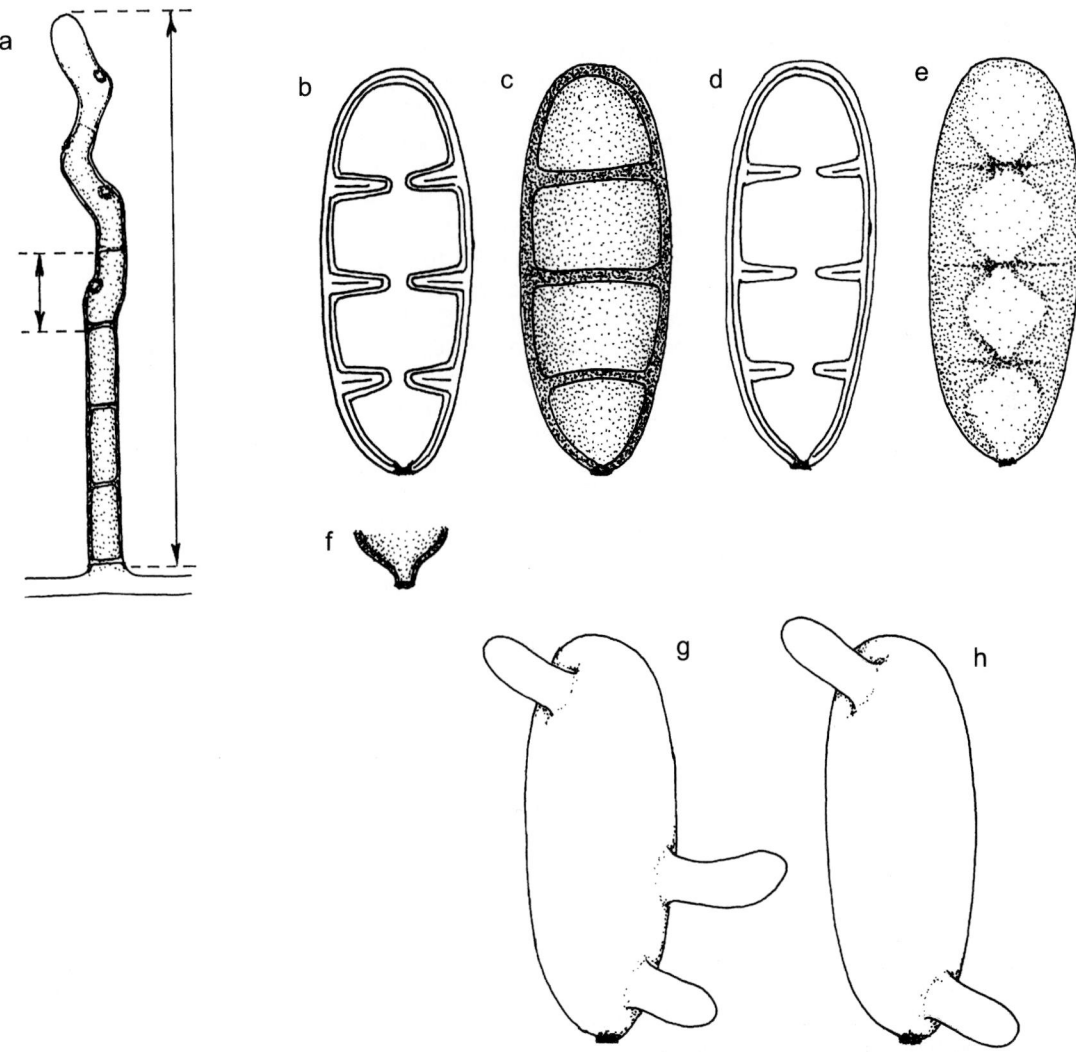

Fig. 45. Criteria for distinction of dematiaceae. a. Terminology of conidiophores, right arrow indicating the entire conidiophore, left arrow indicating the integrated conidiogenous cell. b-e: Conidial septation. b, d. Sectioned; c, e. light microsopic appearance. f. Hilum. g, h. Types of germination. g. Random; h. polar.

381

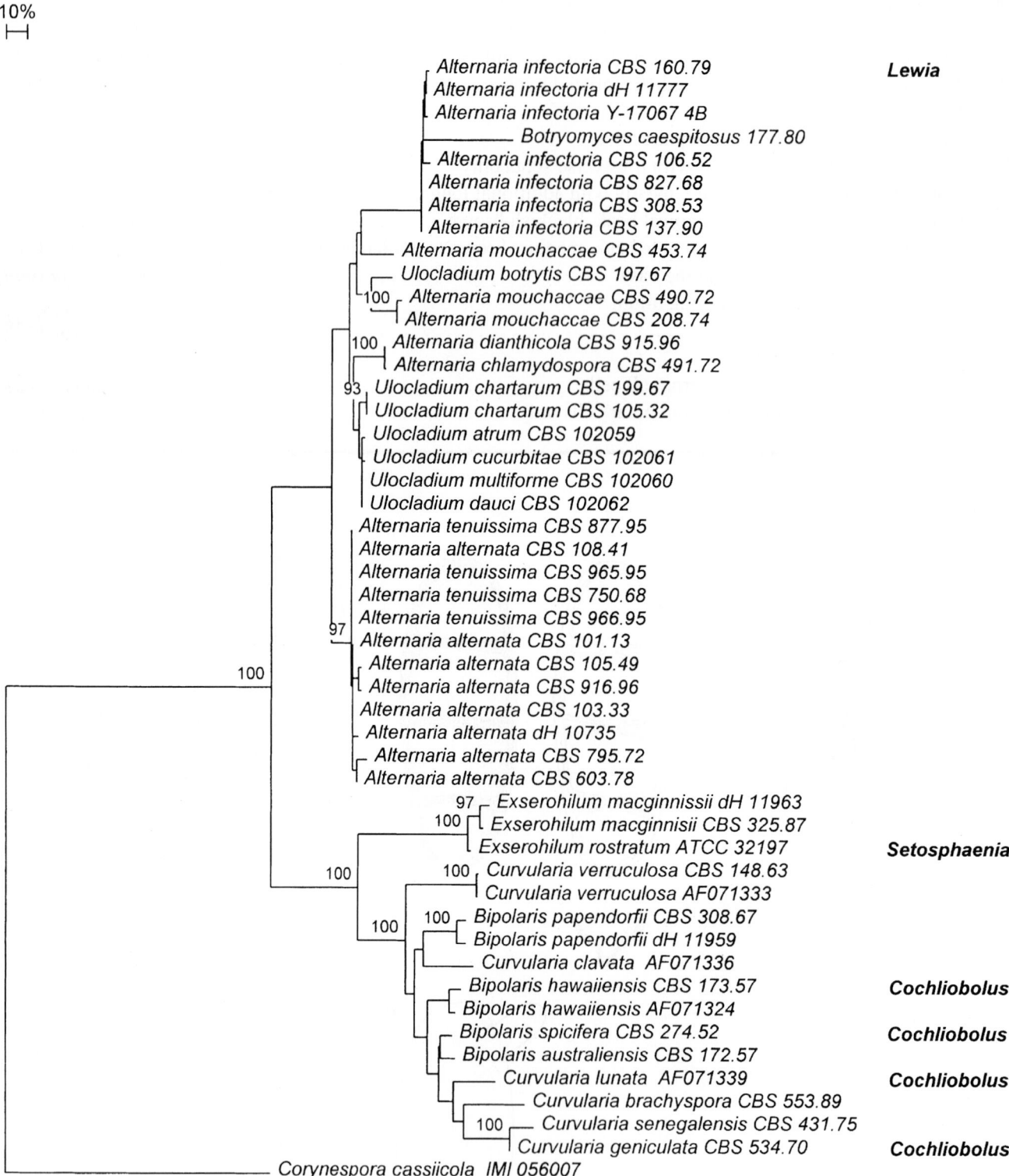

10%

Lewia

Alternaria infectoria CBS 160.79
Alternaria infectoria dH 11777
Alternaria infectoria Y-17067 4B
Botryomyces caespitosus 177.80
Alternaria infectoria CBS 106.52
Alternaria infectoria CBS 827.68
Alternaria infectoria CBS 308.53
Alternaria infectoria CBS 137.90
Alternaria mouchaccae CBS 453.74
Ulocladium botrytis CBS 197.67
100 — Alternaria mouchaccae CBS 490.72
Alternaria mouchaccae CBS 208.74
100 Alternaria dianthicola CBS 915.96
Alternaria chlamydospora CBS 491.72
93 Ulocladium chartarum CBS 199.67
Ulocladium chartarum CBS 105.32
Ulocladium atrum CBS 102059
Ulocladium cucurbitae CBS 102061
Ulocladium multiforme CBS 102060
Ulocladium dauci CBS 102062
Alternaria tenuissima CBS 877.95
Alternaria alternata CBS 108.41
Alternaria tenuissima CBS 965.95
Alternaria tenuissima CBS 750.68
Alternaria tenuissima CBS 966.95
97 Alternaria alternata CBS 101.13
Alternaria alternata CBS 105.49
Alternaria alternata CBS 916.96
Alternaria alternata CBS 103.33
Alternaria alternata dH 10735
Alternaria alternata CBS 795.72
Alternaria alternata CBS 603.78

100

97 — Exserohilum macginnissii dH 11963
100 Exserohilum macginnisii CBS 325.87
Exserohilum rostratum ATCC 32197   **Setosphaenia**
100 Curvularia verruculosa CBS 148.63
100 Curvularia verruculosa AF071333
100 Bipolaris papendorfii CBS 308.67
100 Bipolaris papendorfii dH 11959
Curvularia clavata AF071336
Bipolaris hawaiiensis CBS 173.57   **Cochliobolus**
Bipolaris hawaiiensis AF071324
Bipolaris spicifera CBS 274.52   **Cochliobolus**
Bipolaris australiensis CBS 172.57
Curvularia lunata AF071339   **Cochliobolus**
Curvularia brachyspora CBS 553.89
100 Curvularia senegalensis CBS 431.75
Curvularia geniculata CBS 534.70   **Cochliobolus**
Corynespora cassiicola IMI 056007

Fig. 46. Distance tree of dematiaceae based on confidently aligned ITS rDNA sequences, with the exception of *Corynespora cassiicola* which was taken as outgroup. The tree was made using the Neighbor joining algorithm with Kimura correction. Bootstrap values >90 from 100 resampled datasets are shown. Teleomorph genera are listed where such connections are known. Two main groups can be recognized, viz. the *Alternaria / Ulocladium* complex, where hardly any teleomorphs are known, and *Curvularia* with its relatives. The latter group comprises a large number of mainly plant-pathogenic species which are not shown; this explains why the branches in the tree are relatively wide apart. The *Alternaria / Ulocladium* complex in the tree mainly displays the saprobic species of the two genera (*Alternaria* additionally contains numrous plant pathogens). The resolution within the complex is poor; morphological species like *Alternaria alternata* and *A. tenuissima* cannot be distinguished from each other. *Botryomyces caespitosus* seems to be a meristematic segregant of *Alternaria infectoria*.

382

HYPHOMYCETES, EXPLANATORY CHAPTERS

4. Dermatophytes, dimorphic Onygenales and their relatives

One of the main, classically known groups of clinical significance is the ascomycete order *Onygenales*. It is somewhat separate from the majority of *Euascomycetes* and phylogenetically found adjacent to the orders *Eurotiales* and *Chaetothyriales* (compare Fig. 3 on p. 4). These orders also contain numerous potential pathogens, such as members of the genus *Aspergillus*, and *Exophiala* (black yeasts) and its relatives, respectively. The order *Onygenales* in main traits is characterized by gymnothecia containing spherical, evanescent asci, and by the presence of thallic conidia. A taxonomic overview of the order has been published by Currah (1985).

Members of three families of *Onygenales* are treated, although recent molecular studies do not support this classification (Vidal *et al.*, 2000; Guarro *et al.*, 2000b). A fourth family, the *Myxotrichaceae,* was recently proven to be unrelated (Sugiyama *et al.*, 1999).
1. One of the families, the **Gymnoascaceae**, is clinically nearly insignificant. Members are keratinophilic soil fungi which only exceptionally invade living mammals. Two genera listed in the Atlas, *Gymnoascus* (p. 271) and *Narasimhella* (p. 287) are ascigerous fungi lacking anamorphs.
2. A main group of keratinophilic *Onygenales* is the family **Arthrodermataceae**. This group contains the dermatophytes (p. 386) and some genera of soil fungi. Dermatophytes are those *Arthrodermataceae* that are potentially found as agents of human or animal disorders. They usually have two types of conidia (synanamorphs): one-celled **microconidia** as well as multi-celled **macroconidia.** Clinical strains may be highly reduced in morphology, showing no sporulation at all. For the dermatophytes producing conidia, the genera *Epidermophyton, Keratinomyces, Microsporum* and *Trichophyton* are available. Modern molecular revisions of the *Arthrodermataceae* have been presented by Gräser and co-workers (Gräser *et al.*, 1998, 1999, 2000). The microconidial synanamorph of dermatophytes morphologically fits the anamorph genus *Chrysosporium* (p. 546), but if the fungus can be readily identified as an invader of a human or an animal, this name is generally not applied to dermatophytes. *Chrysosporium* is restricted to geophilic fungi having one-celled, thallic microconidia only. Note that some members of the *Onygenaceae* also have *Chrysosporium*-like anamorphs. Teleomorphs of dermatophytes have spherical, evanescent asci containing 8 ascospores; the ascoma wall is often a loose network of hyphae with complicated branching and ornamentation. Ascomata are generally produced on special media after mating.
3. The third family treated is the **Onygenaceae**. This group contains the 'classical' agents of systemic mycoses (see p. 25). Three further genera treated, *Neoarachnotheca, Aphanoascus* and *Arachnomyces*, are listed under their anamorphs: *Myriodontium* (p. 775), *Onychocola* (p. 790), and *Chrysosporium* (p. 546), respectively. Also *Nannizziopsis* has a *Chrysosporium* anamorph (p. 557). These fungi are phylogenetically related to the dermatophytes; the family affinity has been confirmed by partial sequencing of the ribosomal gene (Bowman & Taylor, 1993; Leclerc *et al.*, 1994; Pan *et al.*, 1994).

All systemic *Onygenales* are supposed to have a natural association with animals, although in some cases this vector is as yet unknown. They have an environmental form, which is filamentous and infectious, and a specialized tissue form, which is yeast-like, consists of adiaspores or spherules with endoconidia. In culture at room temperature they grow as hyphomycetes, but they are **dimorphic**; the tissue form can be reproduced *in vitro* by incubation at 37°C. In healthy animals the infection is mostly subclinical. In healthy humans the infection is mostly transitory, but can be endogenously reactivated at impairment of acquired cellular immunity (see p. 37). The genera of systemic, dimorphic *Ony-genales* are morphologically widely apart from each other, both in environmental and in tissue forms. Each genus contains only a small number of species, which, within the same genus, all have comparable virulence.

References. Currah *et al.* (1985), Gräser *et al.* (1998).

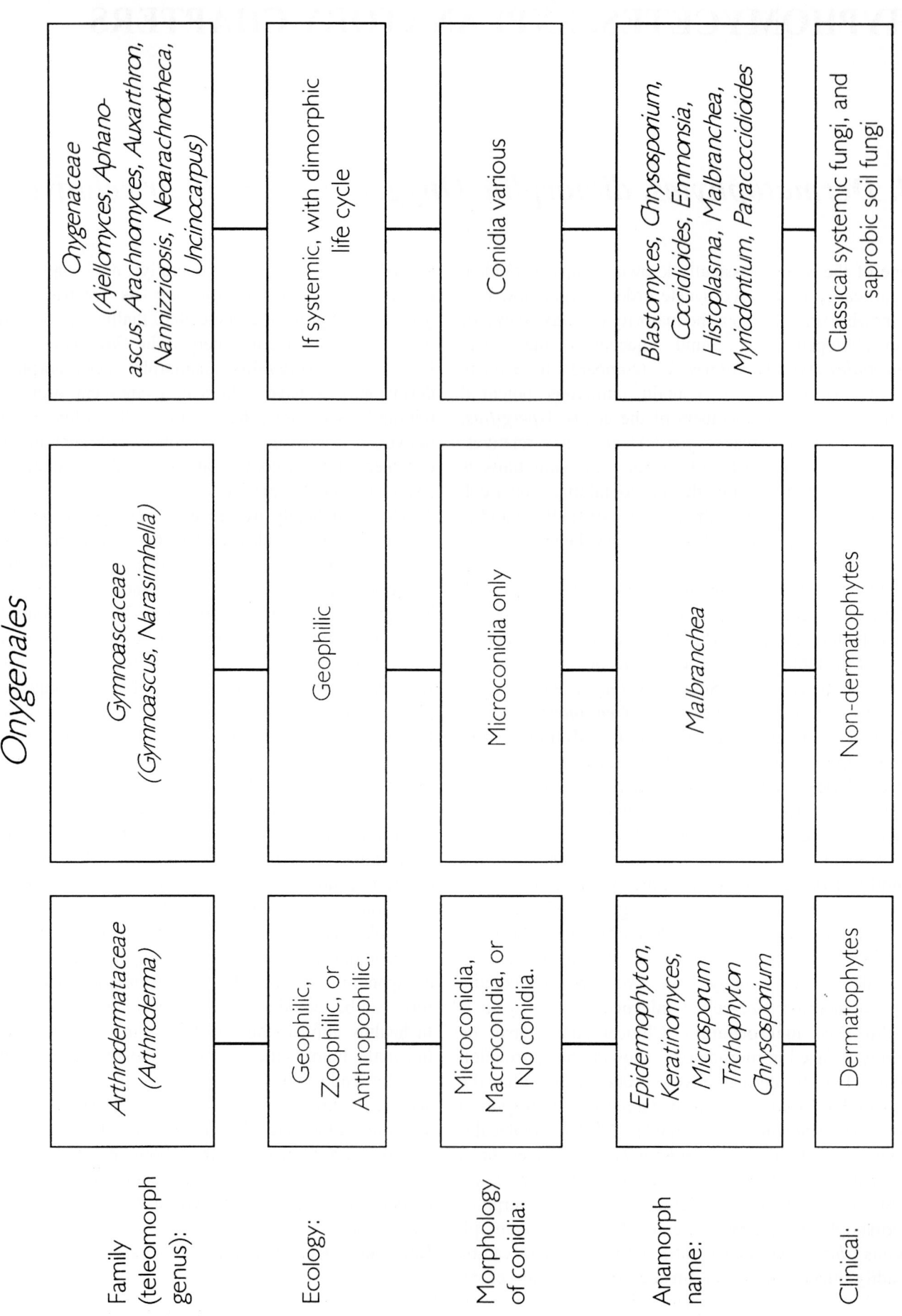

Fig. 47. Diagramatic summary of taxonomic relationships, morphology and ecology of *Onygenales*.

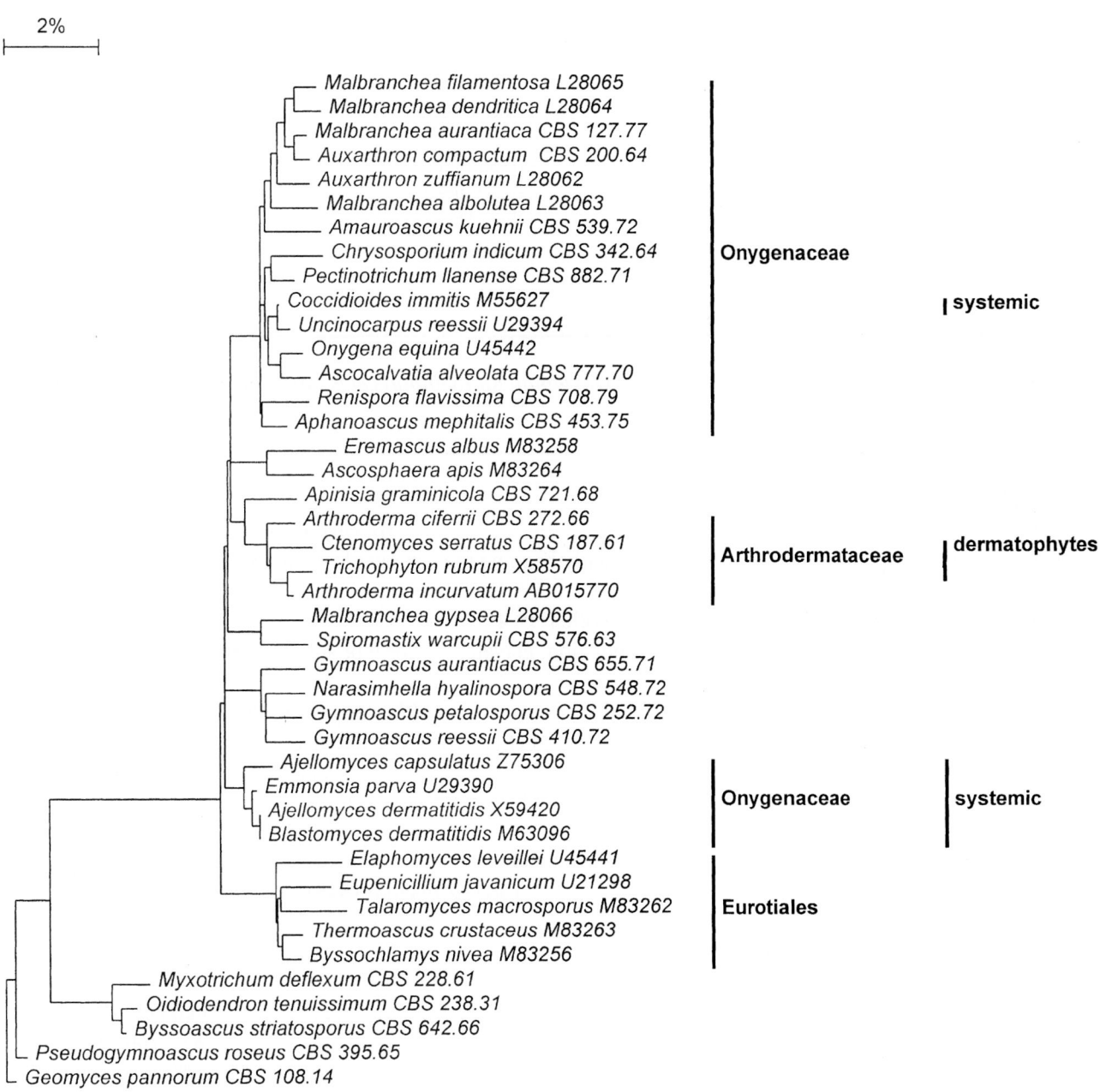

2%

Fig. 48. Phylogenetic tree of *Onygenales* based on confidently aligned, near-complete SSU rDNA sequences using Neighbor joining algorithm with Kimura correction. Bootstrap values >90 from 100 resampled datasets are shown. *Geomyces pannorum* is selected as outgroup; the order *Eurotiales* is clearly separate. Also *Myxotrichum*, until recently regarded as a member of the *Onygenales*, is found at a large distance. The two families with main clinical significance are indicated, *viz.* the *Onygenaceae*, containing agents of systemic mycosis, and the *Arthrodermataceae*, containing the dermatophytes. However, the known families within the *Onygenales* are not supported: supposed members of the *Onygenaceae* appear in two different groups. The systemic *Onygenales* do not compose a single cluster, since *Coccidioides* is found at a considerable distance from *Histoplasma* and its allies.

Table 25. Summary of generic names in *Onygenales*.

Family:	Anamorph:	Teleomorph:
Arthrodermataceae:	*Epidermophyton*	-
	Microsporum	*Arthroderma*
	Trichophyton	*Arthroderma*
	Keratinomyces	-
Gymnoascaceae:	*Malbranchea*	-
	-	*Gymnoascus*
	-	*Narasimhella*
Onygenaceae:	*Blastomyces*	-
	Chrysosporium	*Aphanoascus*
	Chrysosporium	*Uncinocarpus*
	Coccidioides	-
	Emmonsia	*Ajellomyces*
	Histoplasma	*Ajellomyces*
	Malbranchea	*Auxarthron*
	Myriodontium	*Neoarachnotheca*
	Onychocola	*Arachnomyces*
	Paracoccidioides	–

Morphological key to clinically relevant genera of Onygenalean hyphomycetes based on cultures grown at room temperature:

1a. Conidia present ➡ **2**
1b. Conidia remaining absent . . . . **Microsporum (736)**, **Trichophyton (954)**
2a. Conidia arthric, in coherent chains which gradually become wider towards the apex . . . . **Onychocola (790)**
2b. Conidia otherwise ➡ **3**
3a. Conidia all intercalary, cubic, rectangular or swollen ➡ **4**
3b. Conidia predominantly in lateral or terminal position, clavate, fusiform or spherical ➡ **5**
4a. Conidia rectangular; colonies expanding . . . . **Coccidioides (593)**
4b. Conidia irregularly swollen; colonies restricted . . . . **Blastomyces (535)**, **Paracoccidioides (812)**
5a. Large (>15 µm diam), brown, thick-walled, one-celled chlamydospore-like macroconidia
present in addition to small conidia . . . . **Histoplasma (708)**
5b. One-celled chlamydospore-like conidia, if present, not brown ➡ **6**
6a. Conidia all one-celled ➡ **7**
6b. Additional septate conidia present (dermatophytes) ➡ **8**
7a. Conidia clavate, with truncate base . . . . **Chrysosporium (545)**, see also dermatophytes ➡ **8**
7b. Conida spherical, with narrow base . . . . **Emmonsia (633)**
8a. Septate macroconidia present only ➡ **9**
8b. Micro- as well as macroconidia present ➡ **10**
9a. Macroconidia thin-walled . . . . **Epidermophyton (642)**
9b. Macroconidia thick-walled . . . . **Keratinomyces (723)**
10a. Macroconidia rough-walled . . . . **Microsporum (736)**
10b. Macroconidia smooth-walled . . . . **Trichophyton (954)**

Dermatophytes (superficial *Onygenales*)

Criteria for the distinction of dermatophytes

The species produce a smell reminiscent of ammonia due to decomposition of keratin. Species living on dead keratinous material are **geophilic**. Many species occur on animals and are principally **zoophilic** (Table 27), but some of them are easily transmitted to man. Others are prevalently **anthropophilic**, i.e., inhabiting living humans; several of these have a restricted geographical distribution (Rippon, 1985; Table 29). A gradual evolutionary adaptation to life on the human host is supposed (Chabasse & Contet-Audonneau, 1994).

Dermatophyte-specific rDNA fragments were amplified with primers TR1 and TR2 (Bock *et al.*, 1994). SSU phylogeny was presented by Leclerc *et al.* (1994) and Harmsen *et al.* (1995), and ITS 1+2 sequencing by Gräser *et al.* (1998, 1999a). The species form a closely interrelated evolutionary entity, as is demonstrated, e.g., by a narrow range of % G+C of DNA (Davison *et al.*, 1980), a frequent occurrence of high DNA homology levels (Davison *et al.*, 1980; Davison & McKenzie, 1984) and similarities in mtDNA RFLP patterns (Kawasaki *et al.*, 1995). Using a combination of ITS-sequencing, PCR fingerprinting, AFLP and classical data, the main traits of taxonomy of zoo- and anthropophilic species has now been resolved (Gräser *et al.*, 1999a-c, 2000a, b).

The site of infection as well as the type of hair invasion may be characteristic for the species (Table 26). *In vivo*, spores are produced outside the hair-shafts (**ectothrix**) or inside the hair (**endothrix**). Infections are nearly always superficial or cutaneous, only rarely deeper tissues being involved. Extended lesions are particularly seen in patients with T-cell immunodeficiency (Lowinger-Seoane *et al.*, 1992; Porro *et al.*, 1997), occasionally after immuno-depression (Grau Salvat *et al.*, 1998). Immune reactions were reviewed by Smith & Griffin (1995); dermatophytes are serologically similar (Pier *et al.*, 1995). Serology was discussed by Kaufman & López (1980).

A number of physiological tests are in use in species identification, viz. growth on Christensen's urea agar, *in vitro* hair perforation (see also Padhye *et al.*, 1980), growth on polished rice, alkaline production on bromo-cresol purple medium (Summerbell *et al.*, 1988b), sorbitol assimilation (Rezusta *et al.*, 1991) and growth on agars lacking essential vitamins or aminoacids (Table 32). For extended description of methods, see Kane *et al.* (1997).

References. Badillet (1982), Meinhof (1990), Weitzman & Summerbell (1995), Weitzman & Padhye (1996), Kane *et al.* (1997), Gräser *et al.* (1998, 1999a).

Key to the genera of dermatophytes:

1a.	Macroconidia[1,2] rough-walled	*Microsporum* (**736**)
1b.	Macroconidia smooth-walled ➔ **2**	
2a.	Macroconidia thick-walled	*Keratinomyces* (**723**)
2b.	Macroconidia thin-walled ➔ **3**	
3a.	Microconidia absent, macroconidia present	*Epidermophyton* (**642**)
3b.	Microconidia and macroconidia mostly present, though sporulation sometimes sparse .	*Trichophyton* (**954**)

---

[1] If abundant microconidia only, compare *Chrysosporium* (p. 545).
[2] Cultures may be degenerate, nearly lacking conidia; then try both *Microsporum* and *Trichophyton*.

Table 26: Forms of parasitism on human hair (after Rebell & Taplin, 1970).

1. Ectothrix: cells outside hair shaft.

    a. Small, spherical cells (2-3 μm) in masses. Hair usually brightly fluorescent.
>M. audouinii
>M. canis
>M. ferrugineum

    b. Large, spherical cells, 5-8 μm diam in sparse chains inside and outside hair. Hair not fluorescent.
>M. gallinae
>M. gypseum/M. fulvum
>M. nanum
>M. vanbreuseghemii

    c. Large, spherical cells in chains. Hair not fluorescent.
>Megaspore type (5-10 μm):
>>T. verrucosum (Fig. 49)
>Microides type (3-5 μm):
>>T. interdigitale
>Intermediate type:
>>T. rubrum

2. Endothrix: cells (5-8 μm) inside hair shaft. Hairs break off short and curl.
>T. tonsurans
>T. violaceum (Fig. 50)

3. Favus: hyphae and air spaces but usually no spores in hair. Hairs remain long; dull fluorescence.
>T. schoenleinii

4. No parasitism of hair.
>T. concentricum
>M. persicolor
>E. floccosum

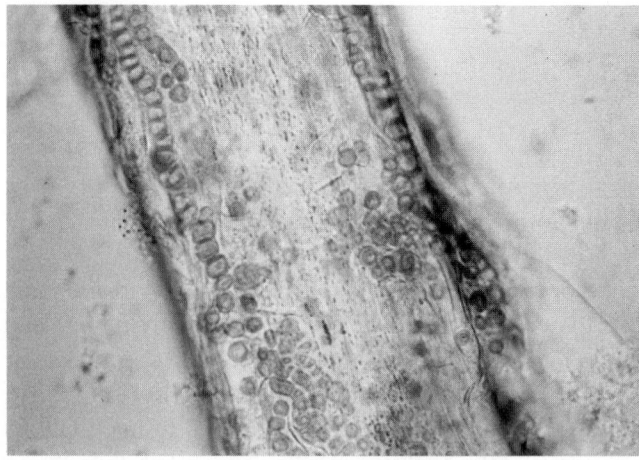

Fig. 49. Ectothrix by *Trichophyton verrucosum*.

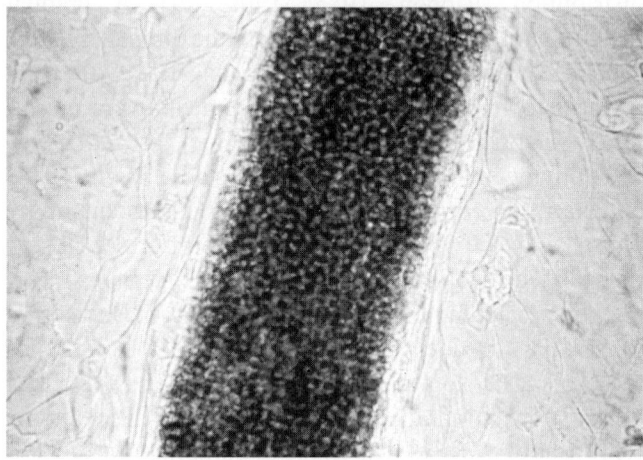

Fig. 50. Endothrix by *Trichophyton violaceum*.

Table 27. Zoophilic dermatophytes with their prevalent hosts.

M. amazonicum	rat
M. canis:	cat, dog
M. gallinae:	fowl
M. nanum:	pig
M. persicolor:	vole
M. praecox:	horse
T. erinacei:	hedgehog
T. mentagrophytes:	rodent, camel
T. simii:	monkey, fowl
T. tonsurans:	horse
T. verrucosum:	cattle

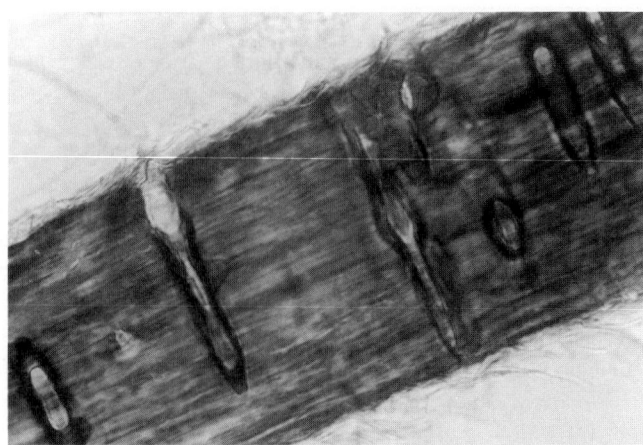

Fig. 51. *In vitro* hair perforation test: positive result.

Table 28. Dermatophytes listed in Atlas of Clinical Fungi 1ˢᵗ and 2ⁿᵈ edition.

1ˢᵗ edition	2ⁿᵈ edition
Epidermophyton floccosum	id.
Keratinomyces ajelloi	*Trichophyton ajelloi*
Keratinomyces ceretanicus	id.
Microsporum amazonicum	id.
Microsporum audouinii	id.
Microsporum boullardii	*Microsporum fulvum*
Microsporum canis	id.
Microsporum cookei	id.
Microsporum distortum	*Microsporum canis*
Microsporum equinum	*Microsporum canis*
-	*Microsporum duboisii*
Microsporum ferrugineum	id.
Microsporum fulvum	id.
Microsporum gallinae	id.
Microsporum gypseum	id.
Microsporum nanum	id.
Microsporum persicolor	id.
Microsporum praecox	id.
Microsporum racemosum	id.
Trichophyton concentricum	id.
Trichophyton equinum	*Trichophyton tonsurans*
Trichophyton equinum var. *autotrophicum*	*Trichophyton tonsurans*
Trichophyton fischeri	*Trichophyton rubrum*
Trichophyton flavescens	id.
Trichophyton gloriae	id.
Trichophyton gourvilii	*Trichophyton violaceum*
Trichophyton kanei	*Trichophyton rubrum*
Trichophyton krajdenii	*Trichophyton interdigitale*
Trichophyton longifusum	*Microsporum fulvum*
Trichophyton megninii	*Trichophyton rubrum*
Trichophyton mentagrophytes var. *erinacei*	*Trichophyton erinacei*
Trichophyton mentagrophytes var. *interdigitale*	*Trichophyton interdigitale*
Trichophyton mentagrophytes var. *mentagrophytes*	*Trichophyton mentagrophytes*
Trichophyton mentagrophytes var. *nodulare*	*Trichophyton interdigitale*
Trichophyton mentagrophytes var. *quickeanum*	*Trichophyton mentagrophytes*
Trichophyton phaseoliforme	id.
Trichophyton raubitschekii	*Trichophyton rubrum*
Trichophyton rubrum	id.
Trichophyton schoenleinii	id.
Trichophyton simii	id.
Trichophyton soudanense	*Trichophyton violaceum*
Trichophyton terrestre	id.
-	*Trichophyton thuringiense*
Trichophyton tonsurans	id.
Trichophyton vanbreuseghemii	id.
Trichophyton verrucosum	id.
Trichophyton violaceum	id.
Trichophyton yaoundei	*Trichophyton violaceum*

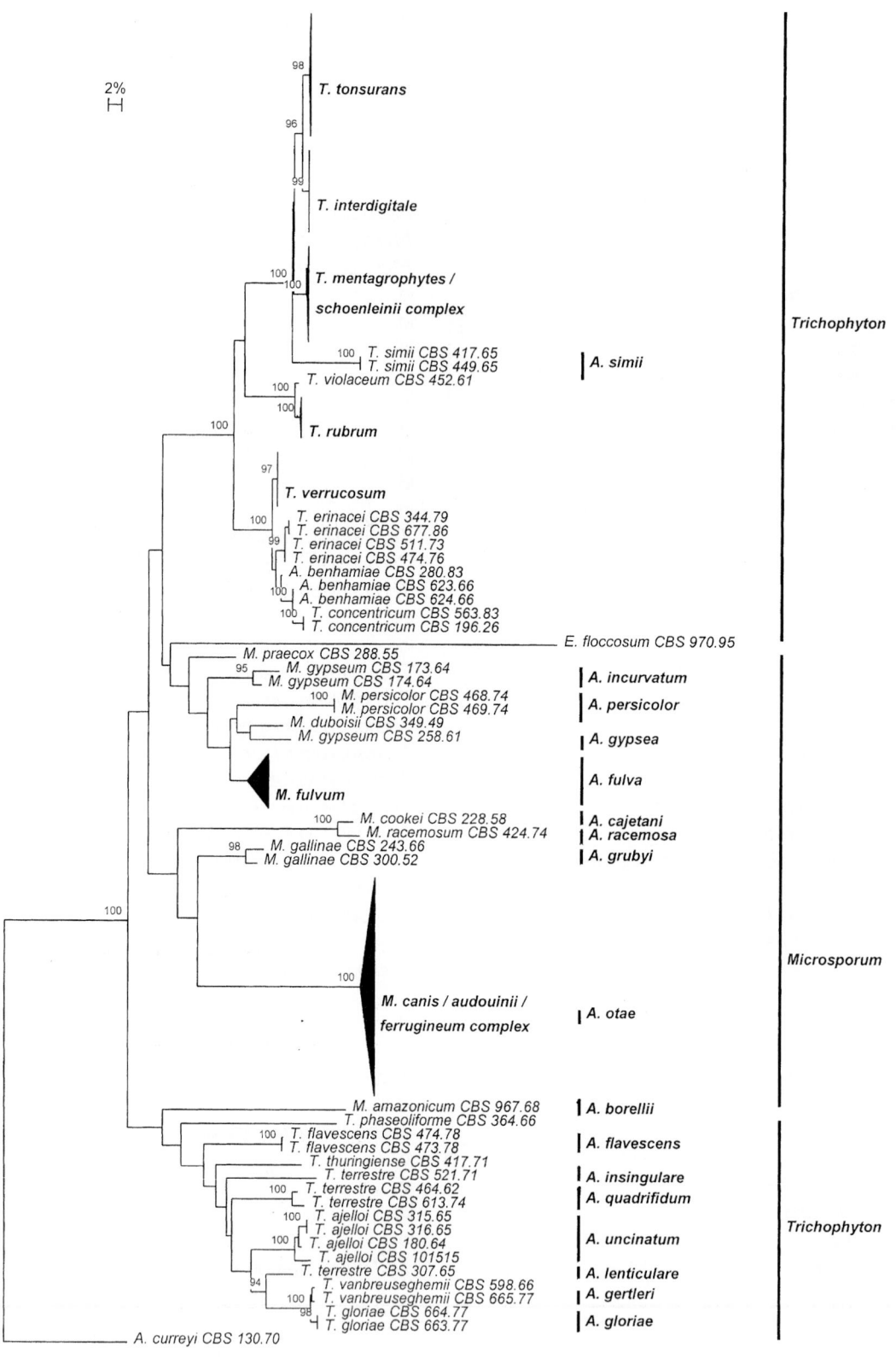

Fig. 52. Phylogenetic tree of the *Arthrodermataceae* based on confidently aligned, partial ITS rDNA sequences using Neighbor joining algorithm with Kimura correction. Bootstrap values >90 from 100 resampled datasets are shown; data were provided by Y. Gräser. *Arthroderma curreyi* is selected as outgroup; *Epidermophyton floccosum* was aligned with difficulty. The classical separation of the two major anamorph genera, *Microsporum* and *Trichophyton*, is not supported. For further details, see under *Microsporum* (p. 736) and *Trichophyton* (p. 954).
A = Arthroderma, M = Microsporum, T = Trichophyton.

Table 29. Main occurrence of dermatophytes with restricted distribution (partly after Rippon, 1985).

M. amazonicum:	Brazil, Africa
M. ferrugineum:	Asia, Eastern Europe
M. persicolor:	Europe
T. concentricum:	Pacific
T. flavescens:	Australia
T. gloriae:	North America
T. schoenleinii:	Eurasia, North Africa
T. simii:	India
T. tonsurans:	Europe, North America, Pacific
T. violaceum:	North and Equatorial Africa, Eastern Europe

Table 30. Anamorph - teleomorph connections in dermatophytes, with their prevalent pathology (recommended BSL-level).

M. amazonicum	- *A. borellii*	- zoophilic	2
M. canis	- *A. otae*	- zoophilic	2
M. cookei	- *A. cajetani*	- geophilic	1
M. fulvum	- *A. fulvum*	- geophilic	1
M. gallinae	- *A. grubyi*	- zoophilic	2
M. gypseum	- *A. gypseum*	- geophilic	1
M. gypseum	- *A. incurvatum*	- geophilic	1
M. nanum	- *A. obtusum*	- zoophilic	2
M. persicolor	- *A. persicolor*	- zoophilic	2
M. racemosum	- *A. racemosum*	- geophilic	1
T. ajelloi	- *A. uncinatum*	- geophilic	1
T. erinacei	- *A. benhamiae*	- zoophilic	2
T. flavescens	- *A. flavescens*	- geophilic	1
T. gloriae	- *A. gloriae*	- geophilic	1
T. interdigitale	- *A. vanbreuseghemii*	- anthropophilic	2
T. simii	- *A. simii*	- zoophilic	2
T. terrestre	- *A. insingulare*	- geophilic	1
T. terrestre	- *A. quadrifidum*	- geophilic	1
T. terrestre	- *A. lenticulare*	- geophilic	1
T. vanbreuseghemii	- *A. gertleri*	- geophilic	1

Table 31. Predominantly clonal species of dermatophytes, with their prevalent pathology (recommended BSL-level).

E. floccosum	- anthropophilic	2
K. ceretanicus	- geophilic	1
M. audouinii	- anthropophilic	2
M. duboisii	- anthropophilic	2
M. ferrugineum	- anthropophilic	2
M. praecox	- zoophilic	2
T. concentricum	- anthropophilic	2
T. mentagrophytes	- zoophilic	2
T. phaseoliforme	- geophilic	1
T. rubrum	- anthropophilic	2
T. schoenleinii	- anthropophilic	2
T. thuringiense	- geophilic	2
T. tonsurans	- zoophilic / anthropophilic	2
T. violaceum	- anthropophilic	2

Table 32. Physiological tests for Epidermophyton, Microsporum and Trichophyton.

	BCPCG	Urease	Hair	1	2	3	4	5	6	7
E. floccosum	+	w	−	+	+	+	+	+	w	w
M. amazonicum	nd	+	−	+	+	+	+	+	+	+
M. audouinii	+	v	−	+	+	+	+	+	+	+
M. canis	−	w,+	+	++	++	++	++	++	++	++
M. cookei	−	+	+	+	+	+	+	+	+	+
M. duboisii	nd	s	+	++	++	++	++	++	++	++
M. ferrugineum	−	−	−	+	+	+	++	+	+	+
M. fulvum	−	+	+	+	+	+	+	+	+	+
M. gallinae	nd	w	+	+	++	++	++	++	++	++
M. gypseum	−	+	+	+	+	+	+	+	+	+
M. nanum	nd	+	+	+	+	+	+	++	+	+
M. persicolor	−	+	+	+	+	+	+	+	+	+
M. praecox	−	+	−	+	+	+	+	+	+	+
M. racemosum	−	+	+	+	+	+	+	+	+	+
T. ajelloi	nd	+	+	+	+	+	+	+	+	+
T. concentricum	nd	−	−	−	+	++	+	+	−	w
T. erinacei	nd	+,w	+,−	++	++	++	++	++	++	++
T. flavescens	nd	w	+	++	++	++	++	++	++	++
T. interdigitale	+	+,−	+	++	++	++	++	++	++	++
T. mentagrophytes	nd	+	+,−	++	++	++	++	++	++	++
T. phaseoliforme	nd	+	+	+	+	+	+	+	+	+
T. rubrum	−	v	−	++	++	++	++	++	+,−	++
T. schoenleinii	+	v	−	+	+	+	++	+	+	+
T. simii	+	+,w	+	++	++	++	++	++	++	++
T. terrestre	+	+	+	+	+	+	+	+	+	+
T. thuringiense	nd	+	−	+	+	+	+	+	+	+
T. tonsurans	v	−,+	−	+,−	+,−	+,−	++,−	+	+,−	w,−
T. vanbreuseghemii	nd	+	+	++	++	++	++	++	++	++
T. verrucosum	+	−	−	−	w	++	+	−	−	−
T. violaceum	+	−,+	−	+,−	+	+	++	+	+	+

BCPCG = bromocresol purple casein glucose agar; T1 = basal medium for Trichophyton agars T2-T5; T2 = T1 + inositol; T3 = T1 + inositol + thiamine; T4 = T1 + thiamine; T5 = T1 + nicotinic acid; T6 = basal medium for Trichophyton agar T7; T7 = T6 + histidine.
v = variable; w = weak; nd = not determined.

Dimorphic fungi (systemic *Onygenales*)

Criteria for the distinction of dimorphic *Onygenales*

Five clearly different genera can be distinguished, which each comprise only a very few species and all are pathogenic. In their ecology and pathogenicity, a basic pattern is shared by all taxa. They have a natural association with warm-blooded animals. Humans are infected by inhalation of dry propagules or by trauma. In healthy persons, symptoms are mild and mostly heal spontaneously. The infectious agent remains dormant inside the human body, and can endogenously be reactivated, particularly when T-cell immunity of the host is impaired. The fungi are emerging as opportunistic systemic infections in AIDS patients (VandenBossche *et al.*, 1990; Wheat, 1995). Animal studies have shown that protection can be transmitted by transfer of T-cells (Longley & Cozad, 1979).

For identification, organisms are cultivated on SGA (and optionally on MEA) at 24°C to obtain the saprophytic synanamorph in optimally sporulating condition. Cultivation on BHI at 37° and 40°C is also necessary for a reliable identification, using characters of the yeast phase. Strains that show poor conversion may be grown on KTA or Columbia agar.

Immunohistologic diagnosis was reviewed by Kaufman (1992). Serological and DNA probes for recognition of the main agents of systemic mycosis, viz. *Blastomyces dermatitidis*, *Coccidioides immitis* and *Histoplasma capsulatum*, are commercially available.

All species should be carefully handled in a safety cabinet.

Reference. Currah (1985), VandenBossche *et al.* (1990).

Key to the systemic Onygenales grown at 24 and 37°C:

1a. Regular, short cylindrical arthroconidia produced at 24°C; no yeast phase at 37°C . . . . . . ***Coccidioides* (593)**
1b. Sporulation at 24°C otherwise; if arthroconidium-like, a yeast phase is produced at 37 °C ➡ **2**
2a. At 24°C brown macroconidia with blunt ornamentation present; at 37°C yeast cells small, with few buds on narrow bases . . . . ***Histoplasma* (708)**
2b. At 24°C ornamented macroconidia absent; yeast cells at 37°C absent or otherwise ➡ **3**
3a. At 24°C regular, spherical conidia produced on ampulliform swellings . . . . ***Emmonsia* (633)**
3b. At 24°C conidia absent or ellipsoidal, not on ampulliform swellings ➡ **4**
4a. At 24°C conidia, when present, ellipsoidal, stalked; at 37°C broad-based yeast cells produced . . . . ***Blastomyces* (535)**
4b. At 24°C conidia absent; at 37°C yeast cells with numerous synchronous, narrow-based buds produced . . . . ***Paracoccidioides* (812)**

393

HYPHOMYCETES: DESCRIPTIONS

Hyphomycetes. Genus: *ACREMONIUM*

Generic description. Colonies moderately fast growing, flat, occasionally raised at the centre, velvety to membranaceous and glabrous, whitish, yellowish or pinkish. Hyphae often forming dense fascicles. Phialides hyaline, erect, mostly arising singly from creeping hyphae, acicular, gradually tapering towards the apex, with usually inconspicuous collarettes. Conidia normally 1-celled, hyaline, subhyaline, or rarely pigmented, usually smooth- and thin-walled, spherical to cylindrical, accumulating in slimy heads at the phialide tips, occasionally formed in dry chains. Chlamydospores may be present.
Teleomorphs. *Emericellopsis* (*Ascomycota, Euascomycetes, Eurotiales: Trichocomaceae)*, *Nectria* (*Ascomycota, Euascomycetes, Hypocreales: Hypocreaceae)*, *Thielavia* (*Ascomycota, Euascomycetes, Sordariales: Chaetomiaceae*).
Differential diagnosis. *Acremonium* species are recognized by the occurrence, though sometimes in low abundance, of acicular phialides which have a septum at the base. Species which have short, tapering phialides, mostly lacking a basal septum are classified in *Phialemonium* (p. 859). Species in which conidial heads are formed directly on creeping hyphae may be classified in *Lecythophora* (with sessile phialidic collarettes; p. 738) or in *Hormonema* (without collarettes; p. 717).
General remarks. Teleomorphs are known in a few species only; they mostly belong to the ascomycete orders *Hypocreales* and *Sordariales*. This matches with molecular phylogeny on the basis of partial 18S rDNA sequences (Glenn *et al.*, 1996). The polyphyletic origin of the genus is shown below (Fig. 53). Many *Acremonium* species have been described, most of them occurring as saprobes on rotten plant material and as mycoparasites.
Pathogenicity. Clinical cases mostly concern traumatic inoculations. Etiologic agents have generally been attributed to the few species which are listed in current medical handbooks as agents of mycetoma. However, the spectrum of species involved in mycetoma might be considerably wider (de Hoog *et al.*, 1993). Two reviews of *Acremonium* infections in humans have been published (Fincher *et al.*, 1991; Guarro *et al.*, 1997a).
Reference. Gams (1971).

Key to the treated species of Acremonium:

1a. Conidia with truncate base, straight, broadly rounded at the tip, borne in slimy heads; thermotolerant to thermophilic species . . . . ***A. alabamense* (396)**
1b. Conidia otherwise; species if thermophilic with curved conidia → **2**
2a. Conidia becoming greyish → **3**
2b. Conidia remaining hyaline → **4**
3a. Conidia obovoidal, initially in long chains . . . . ***A. atrogriseum* (398)**
3b. Conidia tear-shaped, always in slimy heads . . . . ***A. roseogriseum* (414)**
4a. Conidia curved, reniform to falcate → **5**
4b. Conidia straight, fusiform, cylindrical, (ob)ovoidal or (sub) spherical → **7**
5a. Conidia 1-3-celled, usually falcate, over 6 μm long; thermotolerant . . . . ***A. falciforme* (404)**
5b. Conidia otherwise; not thermotolerant → **6**
6a. Conidiophores branched; phialides with slightly thickened apex . . . . ***A. recifei* (412)**
6b. Conidiophores simple; phialides without thickened apex . . . . ***A. curvulum* (402)**
7a. Conidia arising in chains → **8**
7b. Conidia grouped in mucilaginous heads → **9**
8a. Conidia subspherical to ovoidal, less than 4 μm long . . . . ***A. blochii* (400)**
8b. Conidia fusiform with rounded apex, over 4 μm long . . . . ***A. hyalinulum* (406)**
9a. Conidia mostly cylindrical → **10**
9b. Conidia otherwise → **11**
10a. Chlamydospores present in 3-wk-old colonies . . . . ***A. kiliense* (408)**
10b. Chlamydospores absent from 3-wk-old colonies . . . . ***A. strictum* (418)**
11a. Conidia obovoidal, smooth-walled; colonies pink . . . . ***A. potronii* (410)**
11b. Conidia (sub)spherical, verruculose; colonies whitish to pale ochre . . . . ***A. spinosum* (416)**

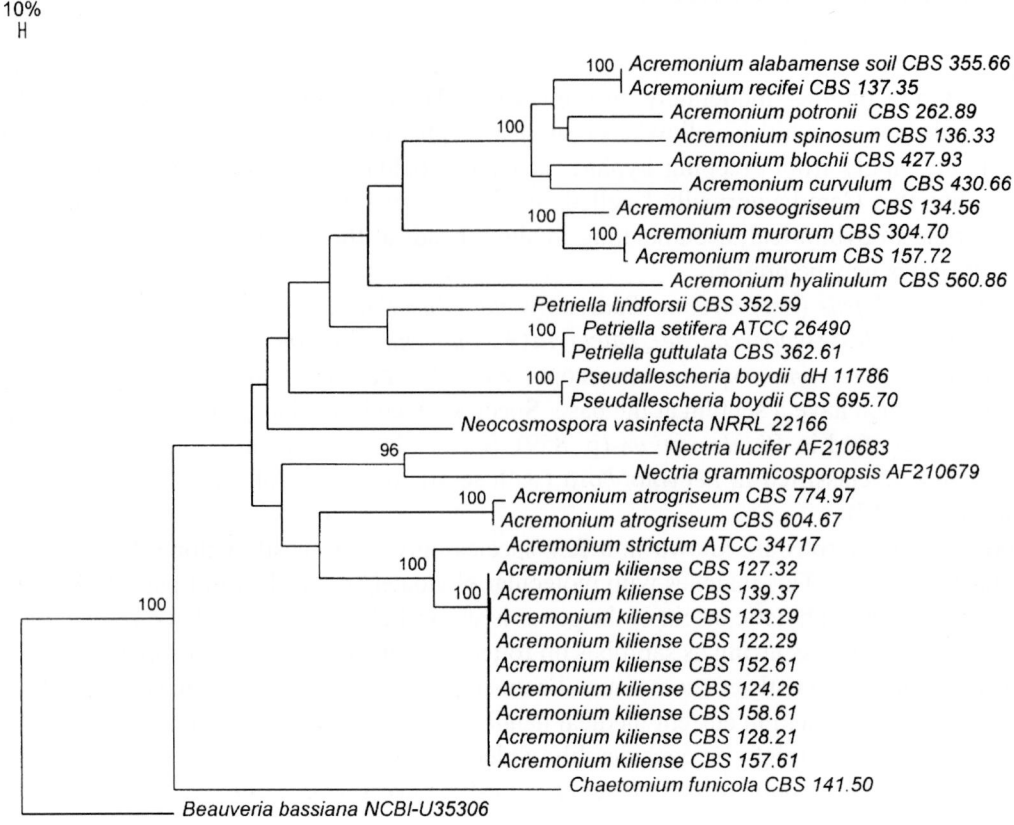

Fig. 53. Distance tree of *Acremonium* based on ITS rDNA sequences using Neighbor joining alalgorithm with Kimura correction. Bootstrap values >90 from 100 resampled datasets are shown. The mutual distances between main groups (*A. kiliense* group / *A. alabamense* group) very large, so that the majority of positions of ITS 1 and 2 could not be used in the comparison; hence the tree is mainly generated by the 5.8S rDNA gene. The actual distances are much larger. The tree shows that *Acremonium*-like morphology is found in very different groups of the fungal kingdom. *Neocosmospora vasinfecta* (p. 289) has an *Acremonium* anamorph. It belongs, with *Nectria,* to the order *Hypocreales. Acremonium alabamense* is the anamorph of *Thielavia terrestris,* which is a supposed member of the family *Chaetomiaceae,* order *Sordariales.* However, its purported relative *Chaetomium funicola* is found at a large distance. The *Petriella* and *Pseudallescheria* species included belong to the order *Microascales. Beauveria bassiana* (*Clavicitipitales*) is used as outgroup. The 5.8S gene correctly indicates that the two main groups of *Acremonium* are highly divergent, but cannot be used for phylogeny.

Acremonium alabamense Morgan-Jones

Colony characteristics. Colonies (MEA 2%, 30°C) growing rapidly, white at first, becoming pale ochraceous, floccose to powdery.

Microscopy. Conidiophores simple, cylindrical, lateral. Phialides subulate, hyaline, smooth-walled, with a distinct, thickened collarette at the apex, 8-75 × 1.0-1.5 µm. Conidia in slimy heads, obovoidal, clavate to pyriform, truncate at the base, hyaline, smooth-walled, 3-6 × 2-3 µm.

Physiology. The species is thermophilic, tolerating up to 45°C.

Teleomorph. *Thielavia terrestris* (Apinis) Malloch & Cain (*Ascomycota, Euascomycetes, Sordariales: Chaetomiaceae*).

Ascomata spherical, 70-120 µm diam, non-ostiolate. Peridium pale brown, *textura epidermoidea*, thin. Asci obovoidal or broadly clavate, 8-spored, evanescent, 26-40 × 15-18 µm. Ascospores ellipsoidal to broadly fusiform, brown, 11-16 × 609 µm, with a distinct apical germ pore.

Molecular diagnostics. ITS restriction map based on CBS 355.66:

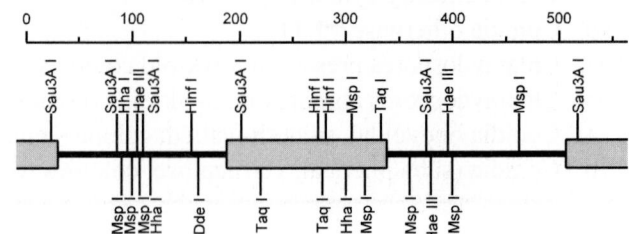

Pathogenicity. BSL-1. Brain infection in a drug abuser (Welti *et al.*, 1984).

References. Morgan-Jones (1974), von Arx (1975).

Nomenclature. *Allescheria terrestris* Apinis - Nova Hedwigia 5: 68, 1963 ≡ *Thielavia terrestris* (Apinis) Malloch & Cain - Can. J. Bot. 50: 66, 1972.

    Acremonium alabamense Morgan-Jones - Can. J. Bot. 52: 429, 1974.

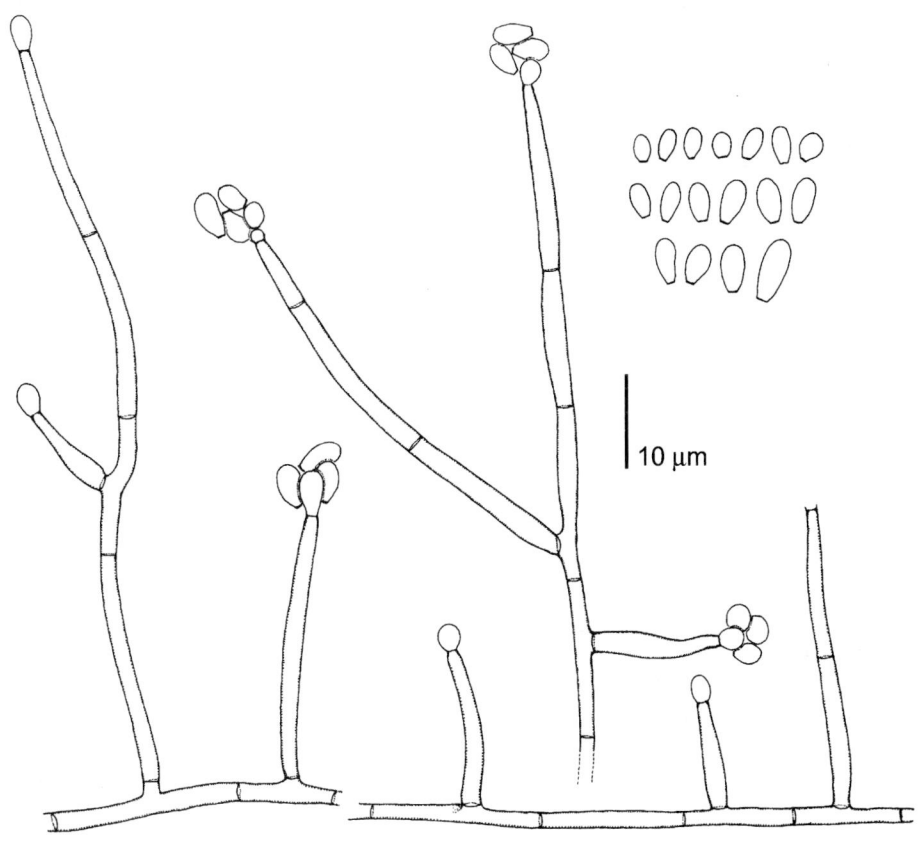

Acremonium alabamense, CBS 351.90. Phialides and conidia.

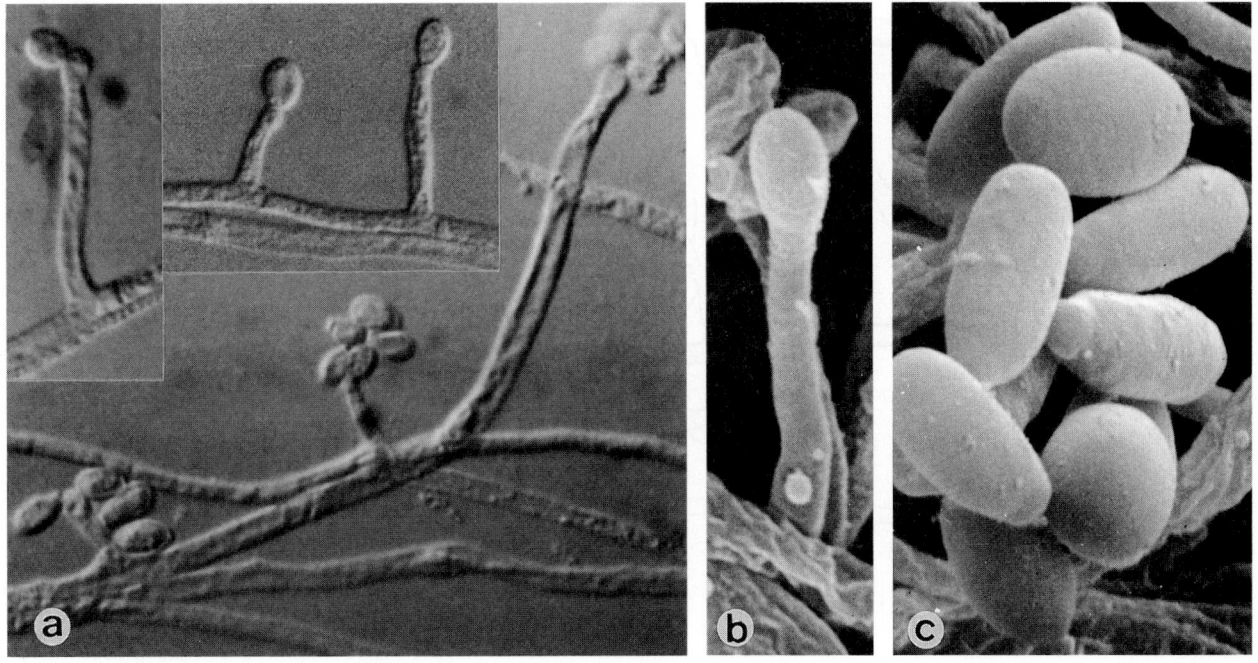

Acremonium alabamense, CBS 351.90. Phialides and conidia. a.×1600; b. ×8200; c. ×6700.

Acremonium atrogriseum (Panasenko) W. Gams

Colony characteristics. Colonies (MEA 2%) growing moderately rapidly, flat, soft, velvety, dark olivaceous-grey.
Microscopy. Phialides arising at right angles, singly or in dense groups, from brown, rather thick-walled, creeping hyphae, often in small whorls on top of a hyaline lateral supporting cell. Phialides with inflated basal part and long tapering neck, ending with a barely visible collarette, 1.0-1.2 µm in width, smooth-walled, greyish-brown, obovoidal; basal septa are frequently absent. Conidia initially in long chains, later in clumps or parallel packets, subhyaline, ellipsoidal, 3.5-4.8 × 1.8-2.1 µm.
Molecular diagnostics. ITS restriction map based on CBS 604.67:

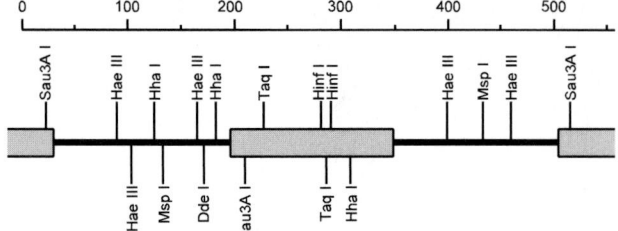

Differential diagnosis. *Acremonium rosegriseum* (p. 414) has tear-shaped conidia. The species is easily distinguished by *Dde*I×ITS. *Phialophora repens* (p. 872) has larger phialides with more gradually tapering necks, and produces reniform conidia.
Pathogenicity. BSL-1. CBS 774.97 was the agent of a human systemic infection (A. Haas, pers. comm.). CBS 306.85 was repeatedly isolated from the lower lung of a human patient (A.A. Padhye, pers. comm.).
Reference. Gams (1971).

Nomenclature. *Phaeoscopulariopsis atrogisea* Panasenko - Mycologia 56: 60, 1964 ≡ *Acremonium atrogriseum* (Panasenko) W. Gams - Cephalosp. Schimmelp. p. 89, 1971.

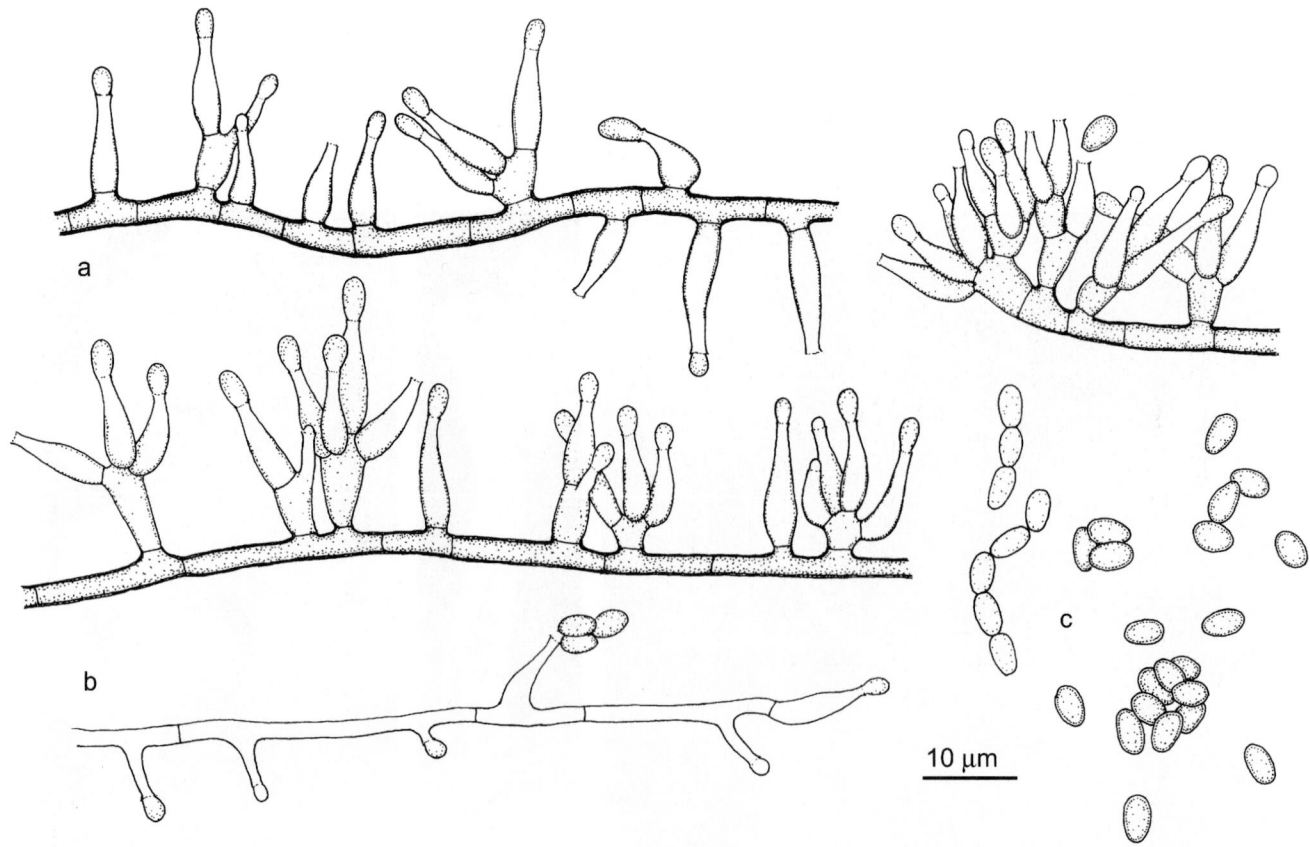

Acremonium atrogriseum, CBS 774.97, a. Hyphae with phialides; b. thin-walled submerged hyphae; c. catenate and non-catenate conidia.

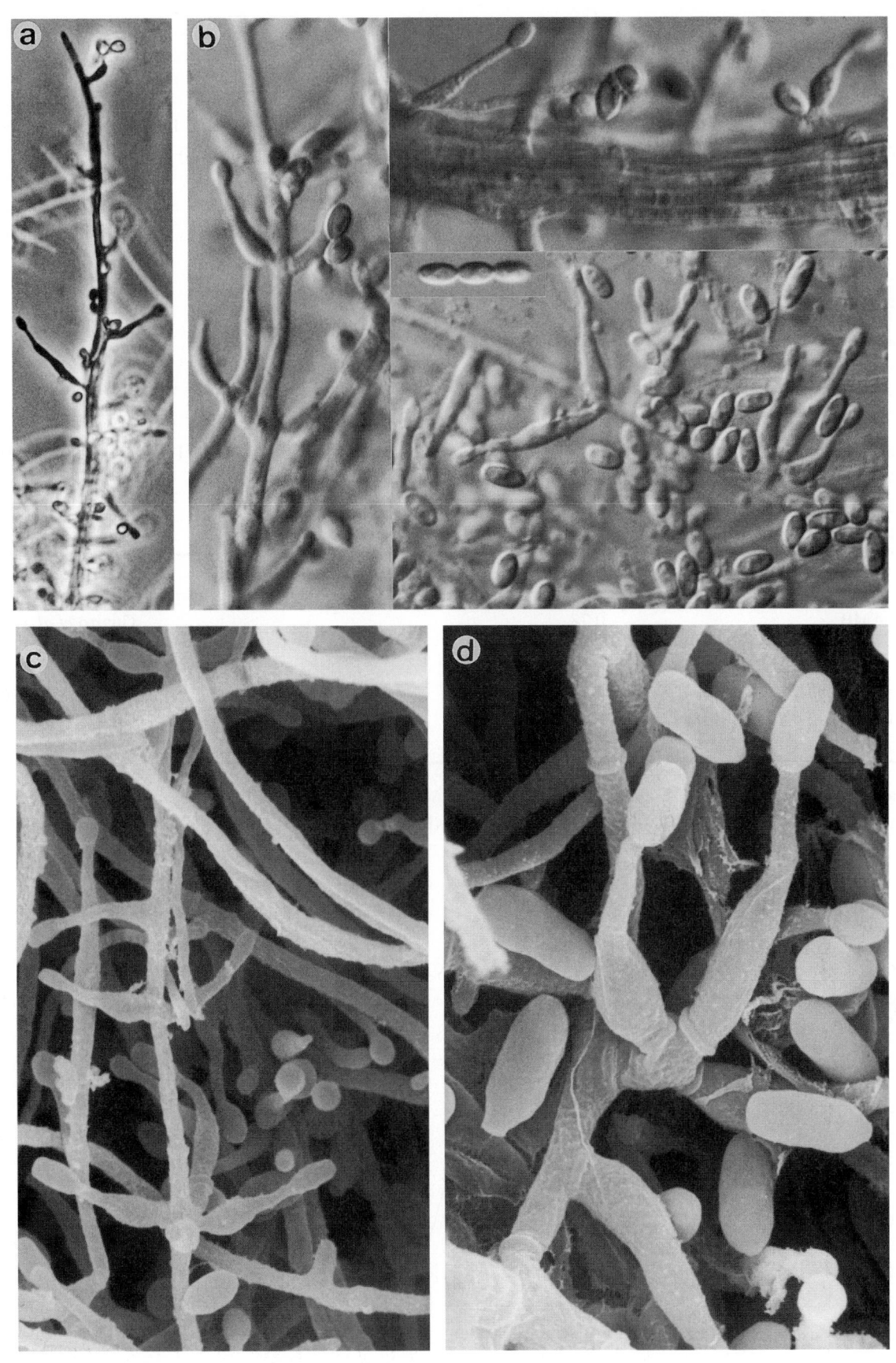

Acremonium atrogriseum, CBS 774.94. Hyphae with phialides, and catenate and non-catenate conidia. a. ×640; b. ×1600; c. ×3500; d. ×6750.

Acremonium blochii (Matruchot) W. Gams

Colony characteristics. Colonies (MEA 2%) growing slowly, white to pinkish, moist, flat, felty.
Microscopy. Phialides arising from undifferentiated hyphae, subulate, hyaline, smooth-walled, 8-25 × 1.0-1.5 μm. Conidia in chains, subspherical to ovoidal, hyaline, smooth-walled, 3.0-3.5 × 2-2.5 μm.
Molecular diagnostics. ITS restriction map based on CBS 427.93:

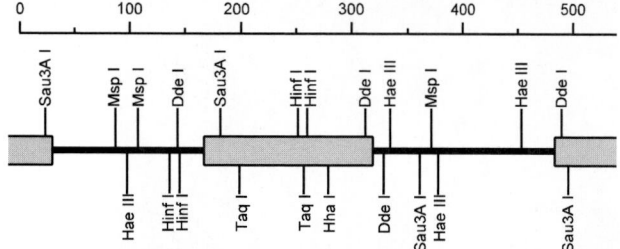

Pathogenicity. BSL-1. Cutaneous lesions in buffaloes (Chatterjee *et al.*, 1987). Gams (1971) described strains from human skin and referred to older medical literature on this species.
Reference. Gams (1971).
Antifungal susceptibility.

Antifungal	Mean MICs	Strains	Reference
AMB	0.78	3	Guarro *et al.* (1997a)
5FC	323	3	Guarro *et al.* (1997a)
FLZ	160	3	Guarro *et al.* (1997a)
ITZ	4.93	3	Guarro *et al.* (1997a)
KTZ	8.06	3	Guarro *et al.* (1997a)
MCZ	31.75	3	Guarro *et al.* (1997a)

Nomenclature. *Mastigocladium blochii* Matruchot - C.R. Hebd. Séanc. Acad. Sci., Paris 152: 325, 1911 ≡ *Scopulariopsis blochii* (Matruchot) Vuillemin - Bull. Soc. Mycol. Fr. 27: 148, 1911 ≡ *Acremonium blochii* (Matruchot) W. Gams - Cephalosp. Schimmelp. p. 78, 1971.

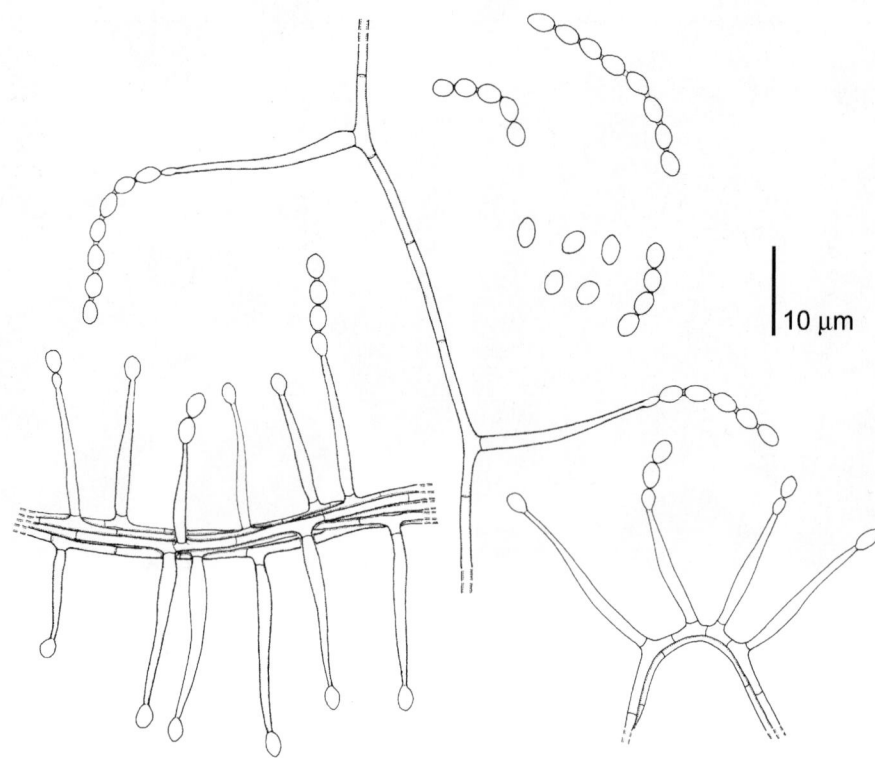

Acremonium blochii, CBS 993.69. Phialides and conidia.

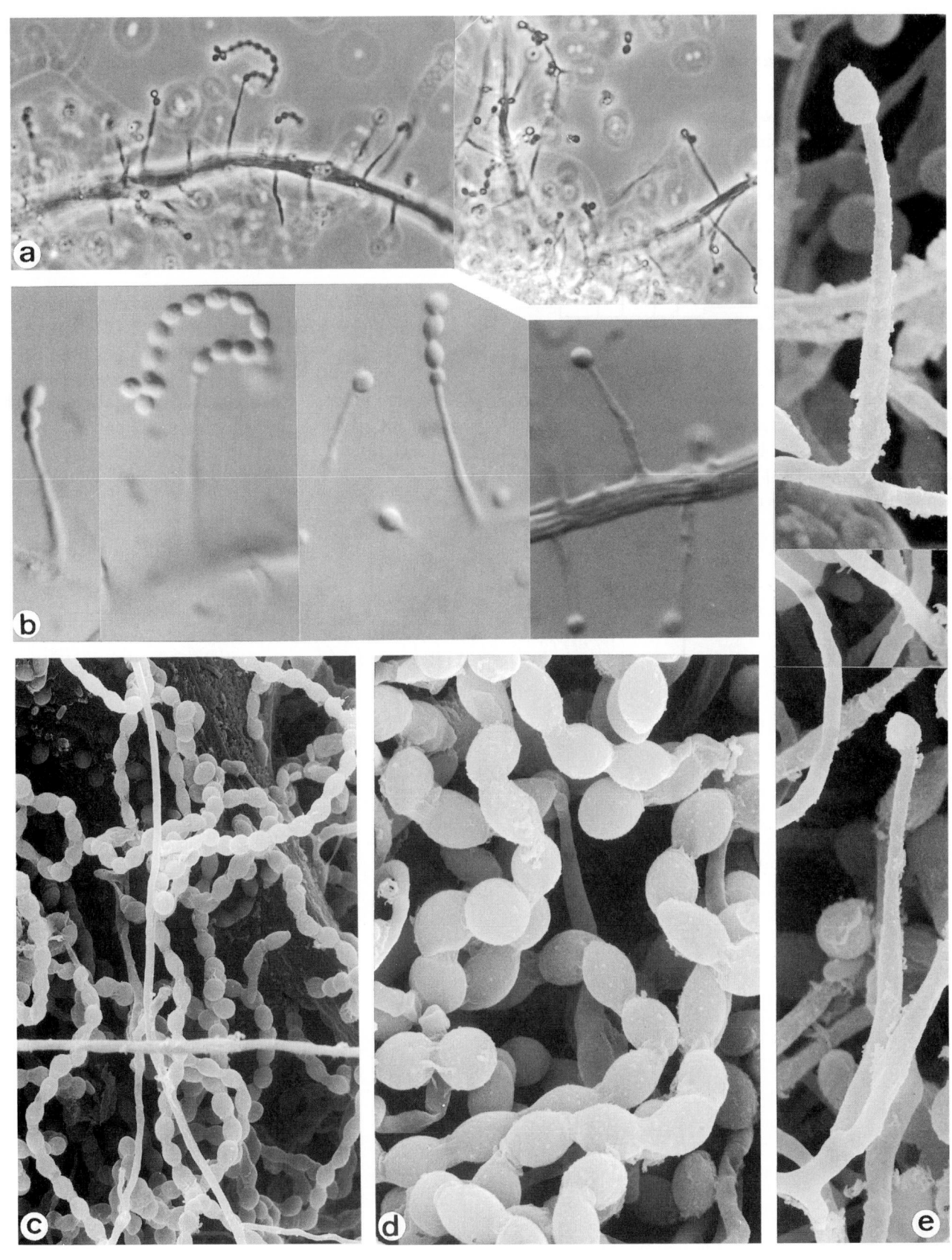

Acremonium blochii, CBS 993.69. Phialides and conidia. a. ×512; b. ×1600; c. ×1900; d,e. ×5500.

Acremonium curvulum W. Gams

Colony characteristics. Colonies (MEA 2%) growing moderately rapidly, yellow-orange, downy; reverse pale brown.
Microscopy. Conidiophores usually simple. Phialides slender, erect, arising from substrate mycelium or from aerial hyphae, 25-60 × 1.7-3.0 μm. Conidia falciform, 1-celled, sometimes 2-celled, hyaline, 4.0-6.7 × 2.0-2.5 μm.
Molecular diagnostics. ITS restriction map based on CBS 430.66:

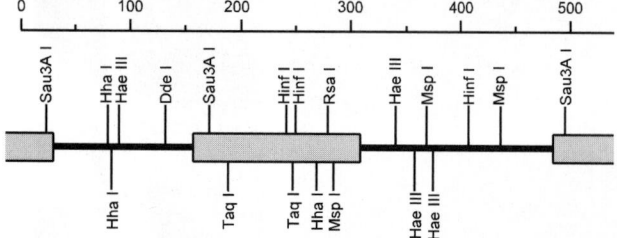

Differential diagnosis. *Acremonium recifei* (p. 412) has a somewhat more profusely branched conidial system. The two species further differ by *Dde*I and *Taq*I restriction sites in the 5.8S rDNA gene.
Pathogenicity. BSL-1. Endophthalmitis (Pflugfelder *et al.*, 1988).
Reference. Gams (1971).
Nomenclature. *Acremonium curvulum* W. Gams - Cephalosp. Schimmelp. p. 57, 1971.

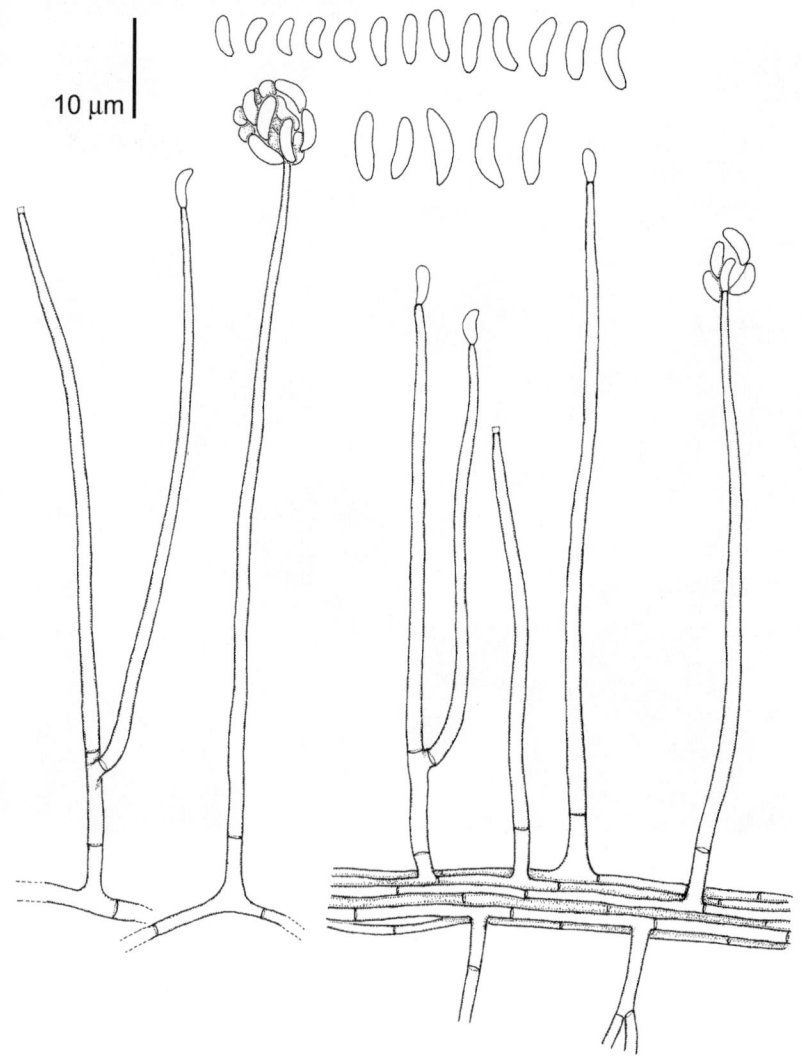

10 μm

Acremonium curvulum, CBS 430.66. Conidiophores, phialides and conidia.

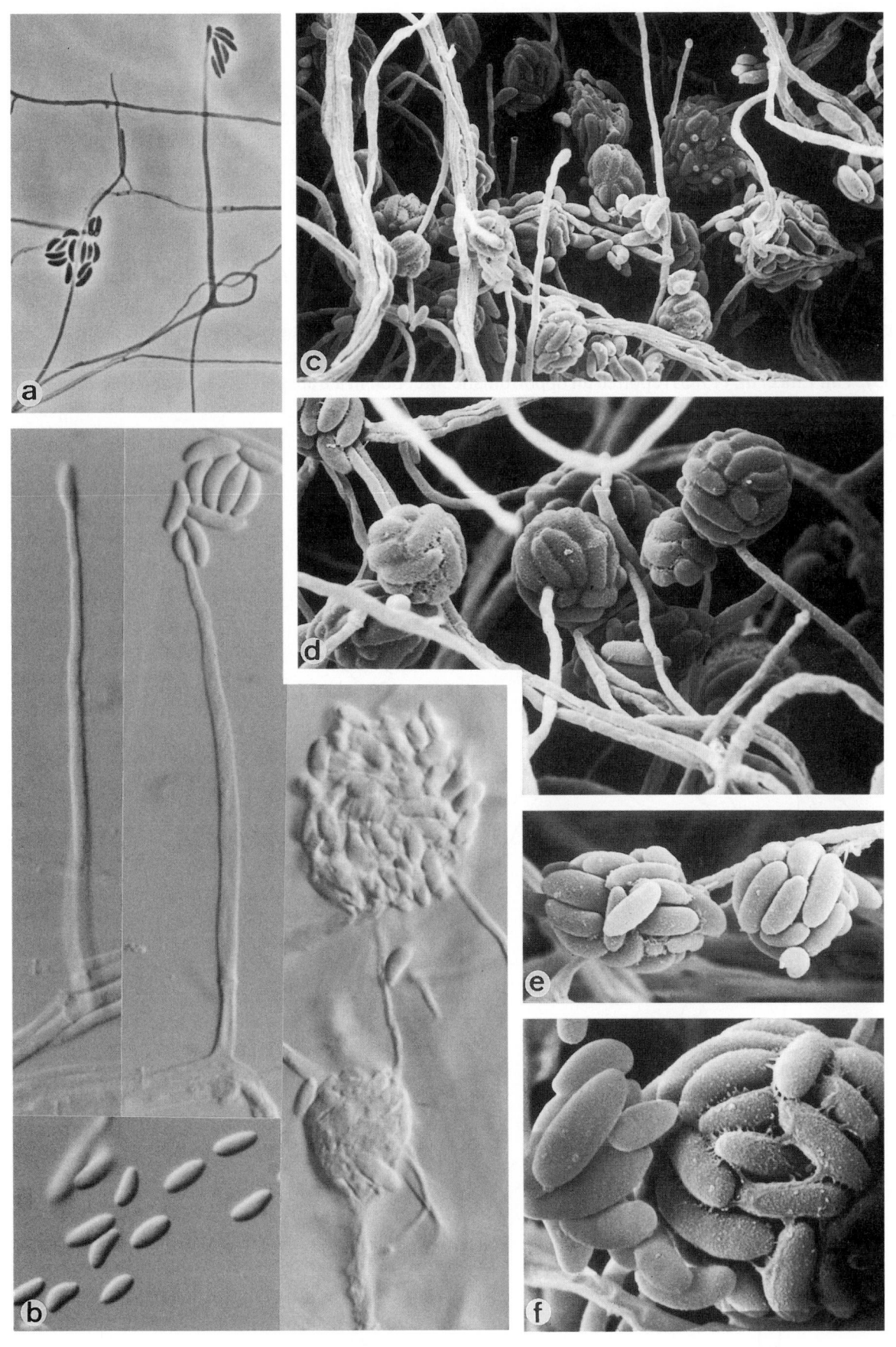

Acremonium curvulum, CBS 430.66. Conidiophores, phialides and conidia. a. ×512; b. ×1600; c. ×1650; d. ×3300; e. ×3800; f. ×6200.

Acremonium falciforme (Carrión) W. Gams

Colony characteristics. Colonies (MEA 2%) restricted, smooth or somewhat felty, brown, poorly sporulating. Colony reverse on SGA violet.

Microscopy. Conidiophores long, undifferentiated. Phialides cylindrical, often intercalary. Conidia 1-3-celled, cylindrical to falcate, reniform, with flat, hyaline basal scars, 7-10 × 2.8-3.2 µm. Pale brown chlamydospores often present.

Physiology. The species is thermotolerant.

Differential diagnosis. There is some resemblance to poorly differentiated *Fusarium* strains, but its slower growth is a distinguishing feature of this species.

Pathogenicity. BSL-2. It has been described as a causative agent of white-grain mycetoma of the extremities (Carrión 1939, 1951; MacKinnon, 1951; Avram, 1964; Halde *et al.*, 1976; McCormack *et al.*, 1987; Záitz *et al.*, 1988; Lee *et al.*, 1995). Van Etta *et al.* (1983) reported an opportunistic infection in a transplant patient, and Cameron *et al.* (1996) and Noble *et al.* (1997) cases of endophthalmitis.

References. Avram (1967), Gams (1971), Williams (1987).

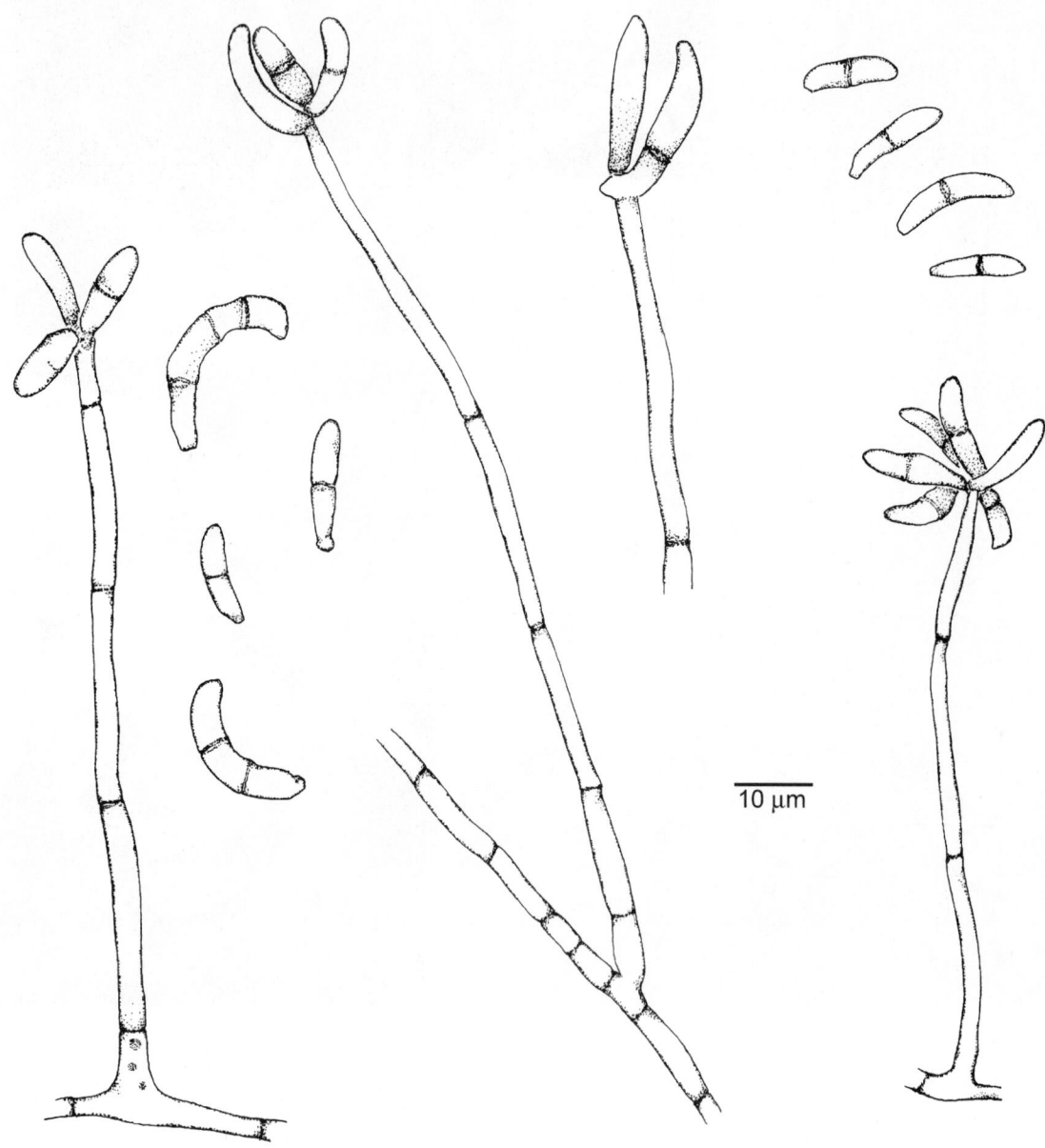

Acremonium falciforme, IMI 97567. Phialides and conidia.

Antifungal susceptibility.

Antifungal	MIC	Strains	Reference
AMB	1	1	Unpublished data
5FC	256	1	Unpublished data
FLZ	128	1	Unpublished data
ITZ	32	1	Unpublished data
KTZ	32	1	Unpublished data
MCZ	16	1	Unpublished data

Nomenclature. *Cephalosporium falciforme* Carrión - Mycologia 43: 523, 1951 ≡ *Acremonium falciforme* (Carrión) W. Gams - Cephalosp. Schimmelp. p. 139, 1971.

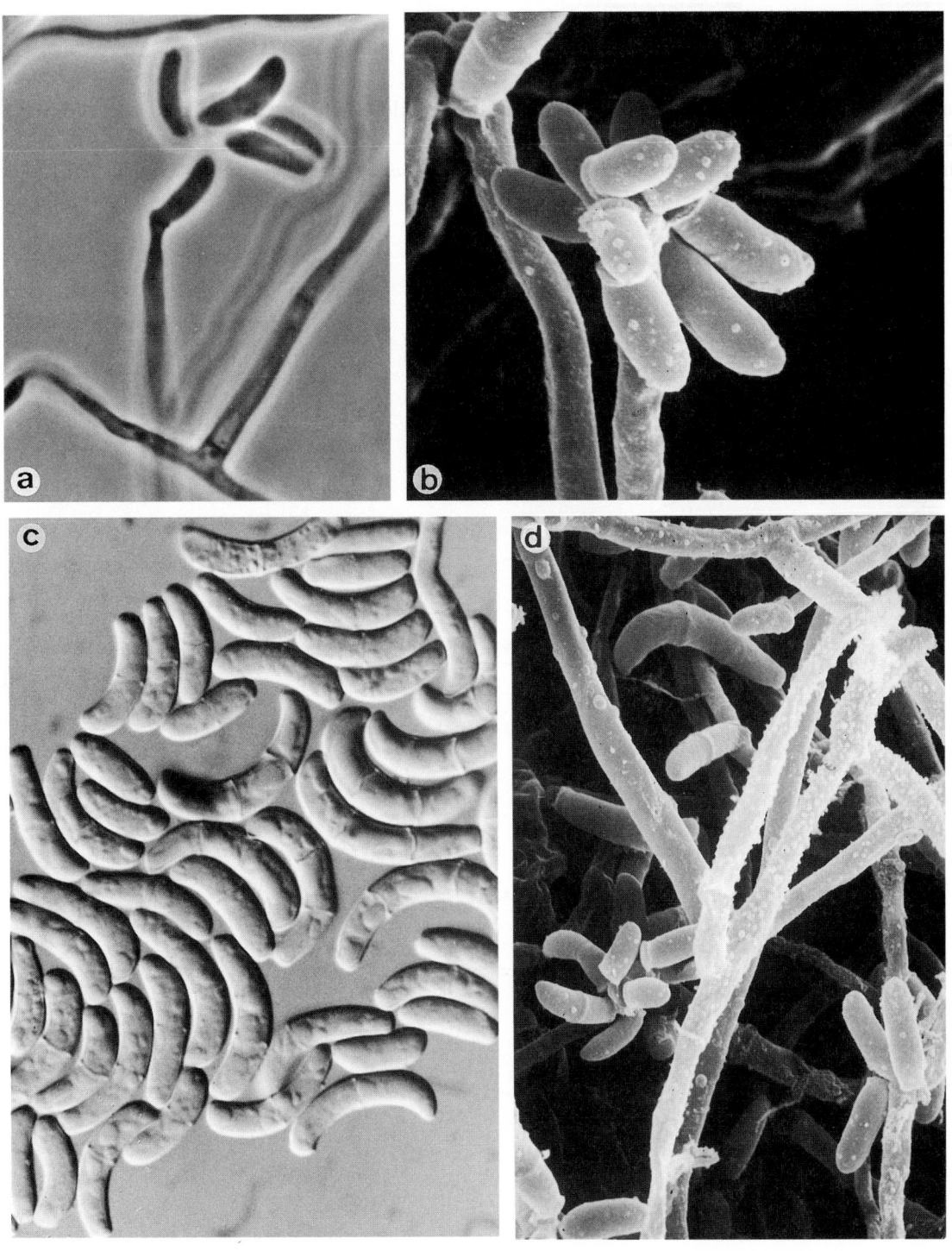

Acremonium falciforme, IMI 97567. Phialides and conidia. a. ×1600; b. ×4500; c. ×1600; d. ×2200.

Acremonium hyalinulum (Sacc.) W. Gams

Colony characteristics. Colonies (MEA 2%) growing rapidly, white to ochraceous, powdery to slightly floccose; reverse ochraceous to greyish-brown.

Microscopy. Conidiophores simple or repeatedly branched. Phialides in whorls on very short side branches, developing short, apical, polyphialidic branches with age, 15-40 × 1.5-3.0 μm; tips with localized wall thickenings. Conidia in chains, spindle-shaped with rounded upper ends, hyaline, relatively thick-walled, 4.2-6.9 × 1.6-2.5 μm.

Molecular diagnostics. ITS restriction map based on CBS 560.86:

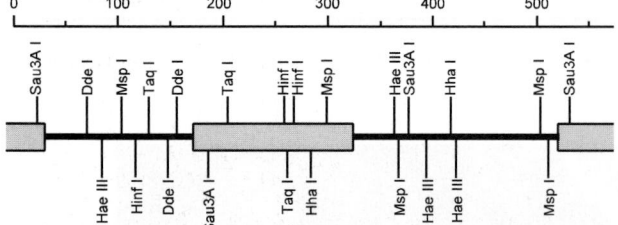

Pathogenicity. BSL-1. A case of a mycetoma-like infection in a dog (Hay *et al.*, 1978).

Reference. Gams (1971).

Nomenclature. *Torula hyalinula* Saccardo - Michelia 1: 265, 1878 ≡ *Acremonium hyalinulum* (Saccardo) W. Gams - Cephalosp. Schimmelp. p. 104, 1971.

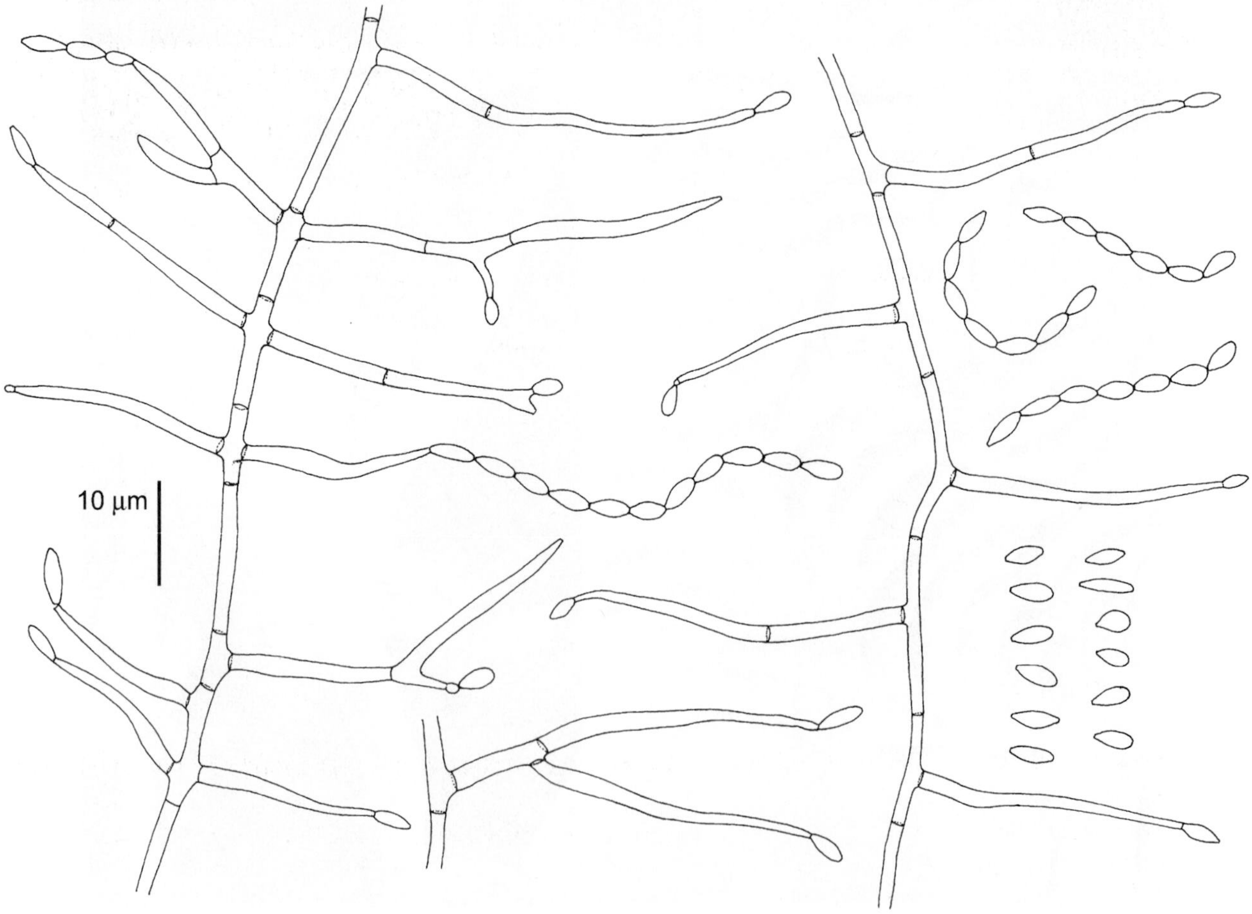

10 μm

Acremonium hyalinulum, FMR 3962. Conidiophores, some with sympodially proliferating phialides, and conidia.

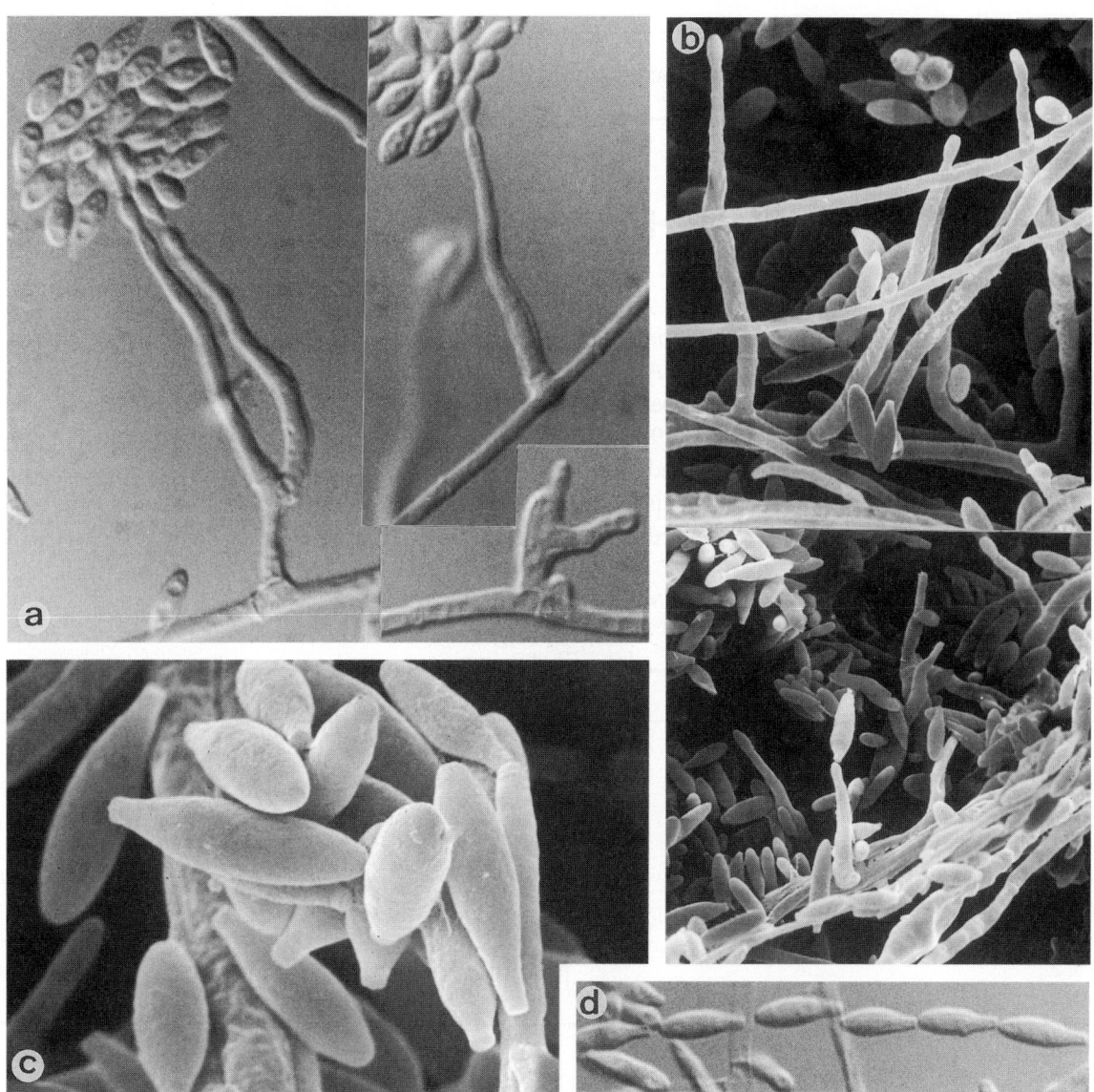

Acremonium hyalinulum, FMR 3962. Conidiophores and conidia. a. ×1600; b. ×1762; c. ×6600; d. ×1600.

Acremonium kiliense Grütz

Colony characteristics. Colonies (MEA 2%) restricted, smooth, greyish-orange or ochraceous, with thin hyphal fascicles; reverse brown on SGA.

Microscopy. Hyphae narrow, fragile. Phialides mostly on undifferentiated hyphae, thin-walled, cylindrical and slightly tapering, 25-45 μm long when having a basal septum, or intercalary, then producing a lateral phialidic outgrowth of variable length. Conidia ellipsoidal to short-cylindrical, 3-6 × 1.5 μm, hyaline, accumulating in slimy heads. Unicellular chlamydospores present, terminal or intercalary.

Molecular diagnostics. ITS restriction map based on CBS 122.29:

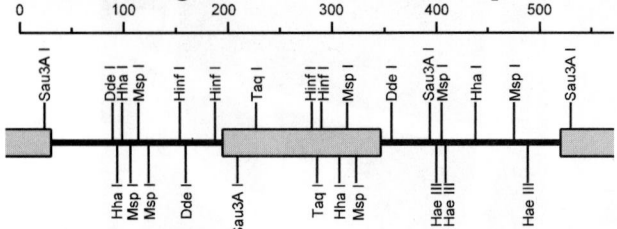

Differential diagnosis. The species is rather variable; main characteristics are the brown colony reverse on SGA and the presence of chlamydospores on OA and MEA.

Pathogenicity. BSL-2. This common saprobe has rather frequently been described as causing ulcerative, nodulose hyalohyphomycosis in humans (Lacroix *et al.*, 1988). Previously it was particularly known as an agent of mycetoma (Lacaz *et al.*, 1979). Lopes *et al.* (1995b) reported a kerion-like lesion on the scalp. Also cases of keratitis are known (Lund *et al.*, 1993); Fridkin *et al.* (1996) and Weissbold *et al.* (1996) reported cases of endophthalmitis after cataract extraction. Lacaz *et al.* (1981) reported endocarditis by this species, and Lopes *et al.* (1995a) two cases of CAPD-associated peritonitis.

References. Arievich *et al.* (1966), Gams (1971), Simon *et al.* (1991).

Antifungal susceptibility.

Antifungal	Mean MICs	MIC90	Strains	Reference
AMB	1.33	nd	5	Guarro *et al.* (1997a)
AMB	1.34	6.25	9	Wildfeuer *et al.* (1998)
5FC	645	nd	5	Guarro *et al.* (1997a)
FLZ	106	nd	5	Guarro *et al.* (1997a)
ITZ	0.91	25	9	Wildfeuer *et al.* (1998)
KTZ	4.85	nd	5	Guarro *et al.* (1997a)
KTZ	1.56	3.13	9	Wildfeuer *et al.* (1998)
MCZ	22.8	nd	5	Guarro *et al.* (1997a)
VCZ	0.39	0.78	9	Wildfeuer *et al.* (1998)

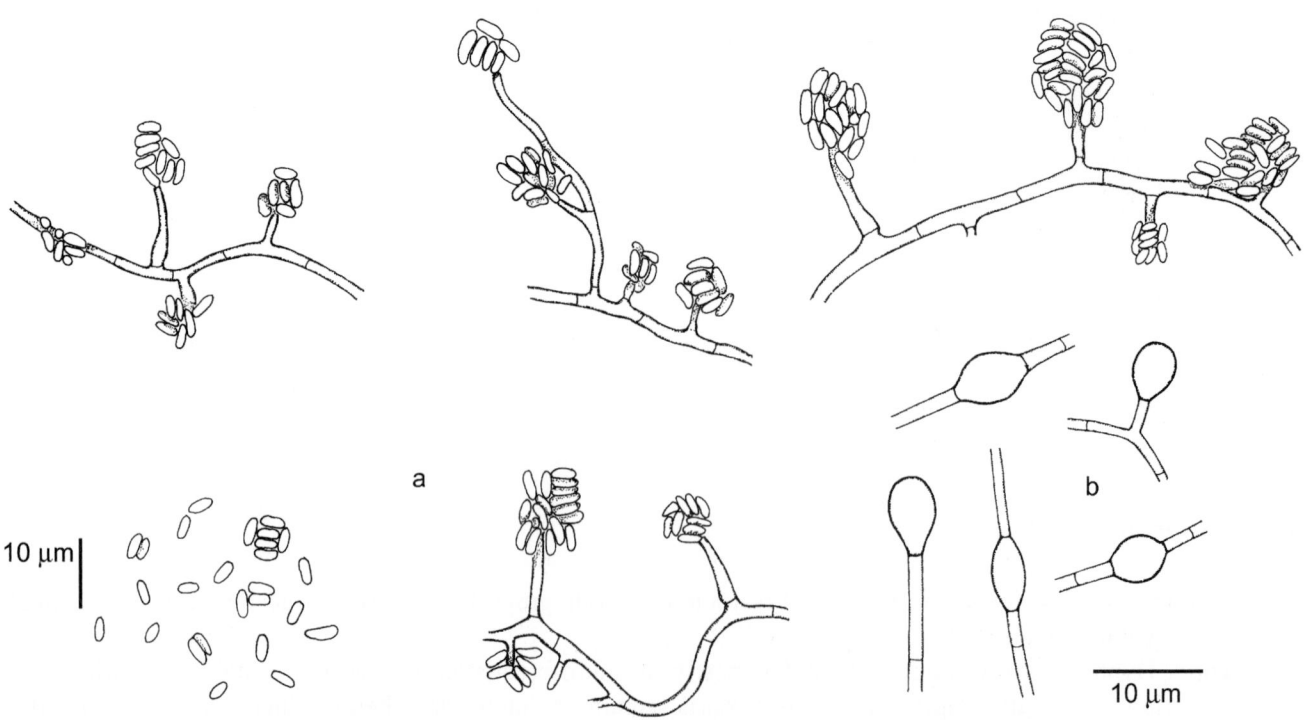

Acremonium kiliense, CBS 953.96. a. Phialides with slimy heads of conidia, b. chlamydospores.

408

Nomenclature. *Acremonium kiliense* Grütz - Derm. Wschr. 80: 774, 1925 = *Cephalosporium kiliense* (Grütz) Hartmann - Derm. Wschr. 82: 569, 1926 = *Cephalosporium asteroides griseum grützii* Benedek - Arch. Derm. Syph. 154: 166, 1928 (name change).

   Cephalosporium niveolanosum Benedek - Arch. Derm. Syph. 154: 166, 1928 = *Hyalopus niveolanosus* (Benedek) Barbosa - Subs. Estud. Hyalopus, Recife p. 46, 1941.

   Acremonium keio Nakamura & Takatsuki - Jpn. J. Derm. 32: 1100-1108, 1932.

   Cephalosporium pseudofermentum Ciferri - Arch. Protistenkde 78: 227, 1932 = *Hyalopus pseudofermentum* (Ciferri) Barbosa - Subs. Estud. Hyalopus, Recife p. 17, 1941.

   Cephalosporium stuehmeri Schmidt & van Beyma - Zentbl. Bakt. Parasitkde, Abt. 1, 130: 102, 1933 = *Hyalopus stuehmeri* (Schmidt & van Beyma) Barbosa - Subs. Estud. Hyalopus, Recife p. 43, 1941.

   Cryptomyces pleomorpha Gruner - Can. Med. Ass. J. 32: 15, 1935.

   Cephalosporium madurae Padhye, Sukapure & Thirumalachar - Mycopath. Mycol. Appl. 16: 318, 1962.

   Cephalosporium infestans Gaind & Thirumalachar - Sabouraudia 1: 230, 1962.

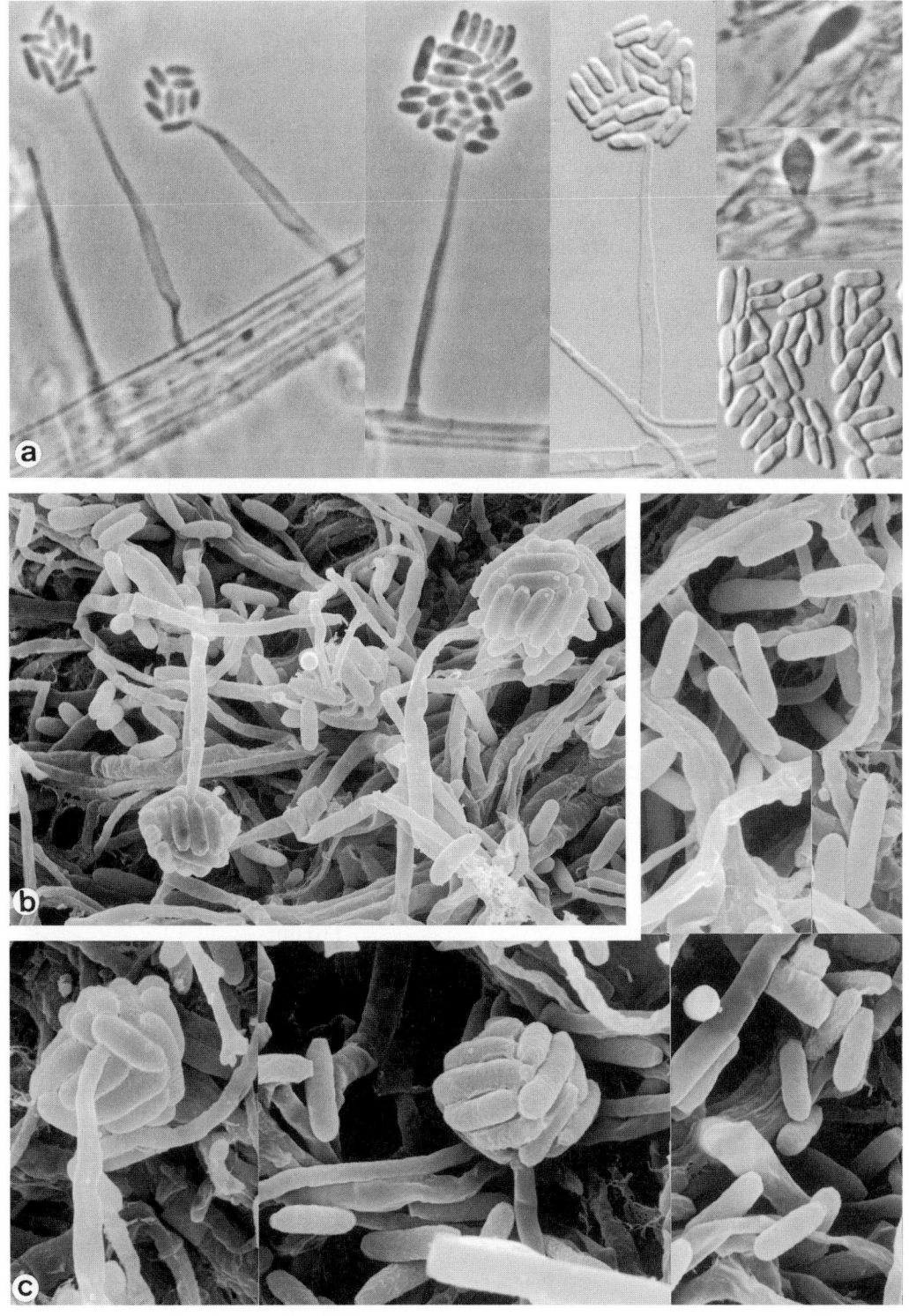

Acremonium kiliense, CBS 158.61. Phialides, conidia and chlamydospores. a. ×1265; b. ×2765; c. ×4740.

Acremonium potronii Vuill.

Colony characteristics. Colonies (MEA 2%) restricted, pink, slimy, becoming powdery to granulose; reverse colourless.

Microscopy. Phialides simple, erect, arising from creeping hyphae. Conidia aggregating in slimy heads, obovoidal or tear-shaped, smooth-walled, 2.1-4.0 × 1.3-2.5 μm.

Molecular diagnostics. ITS restriction map based on CBS 262.89:

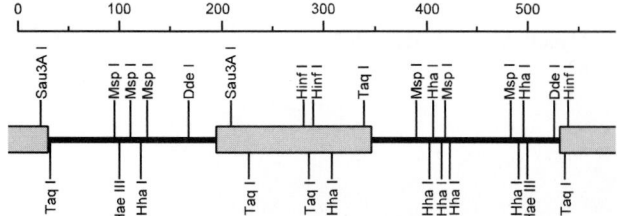

Pathogenicity. BSL-1. Human cases of keratitis (Forster & Rebell, 1975; Forster *et al.*, 1975a) and mycetoma (Lacaz & Netto, 1954) are known. In addition, an infection in the oral cavity of a cat has been reported (van den Arker, 1952). The original publication (Potron & Noisette, 1911) concerned a widely extended cutaneous case. Gams (1971) listed several strains from human skin and nails.

Reference. Gams (1971).

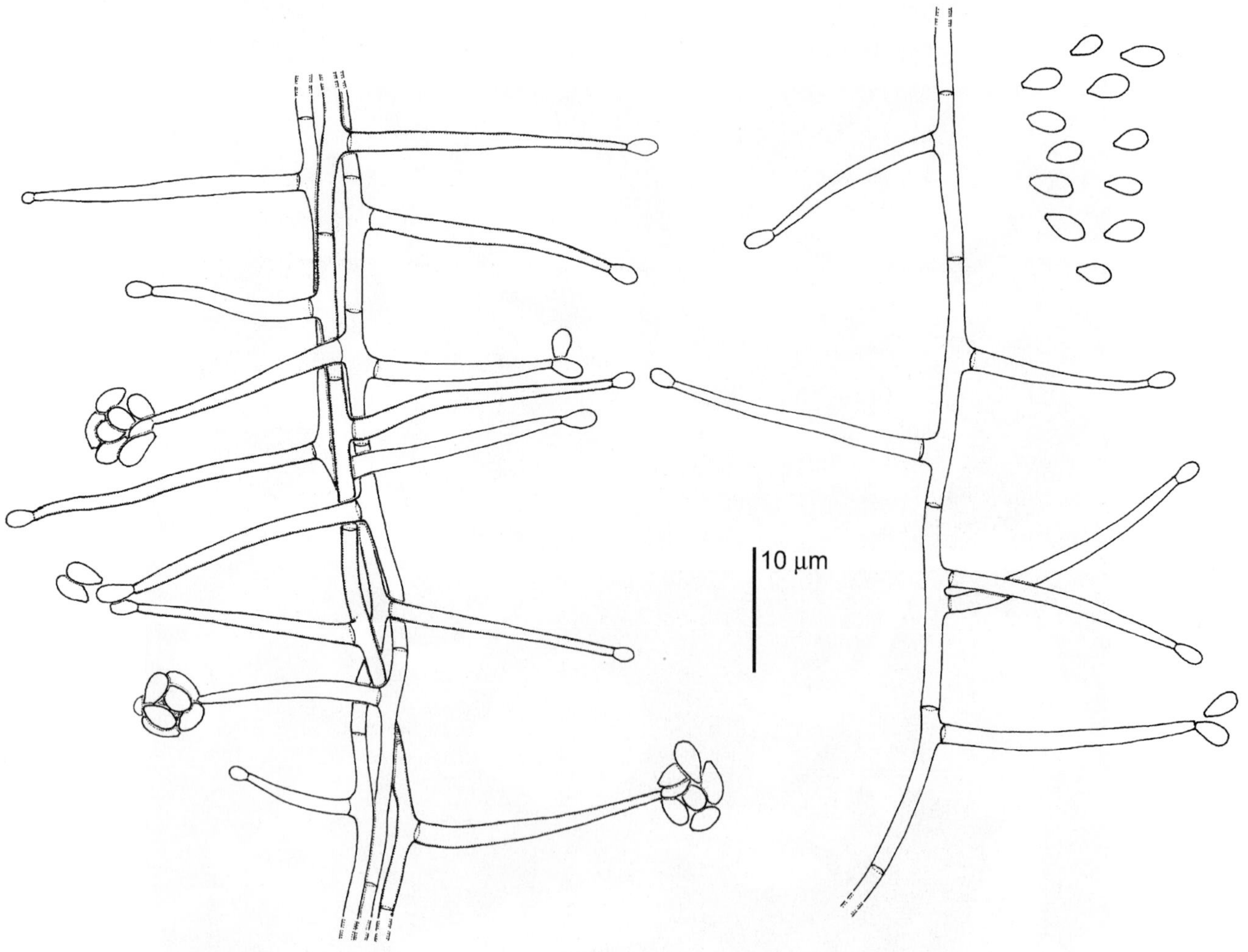

Acremonium potronii, CBS 262.89. Phialides and conidia.

410

Antifungal susceptibility.

Antifungal	Mean MICs	Strains	Reference
AMB	1.46	3	Guarro *et al.* (1997a)
5FC	645	3	Guarro *et al.* (1997a)
FLZ	160	3	Guarro *et al.* (1997a)
ITZ	3.15	3	Guarro *et al.* (1997a)
KTZ	8.06	3	Guarro *et al.* (1997a)
MCZ	6.3	3	Guarro *et al.* (1997a)

Nomenclature. *Acremonium potronii* Vuillemin - Bull. Soc. Sci. Nancy 1910: 19, 1910.

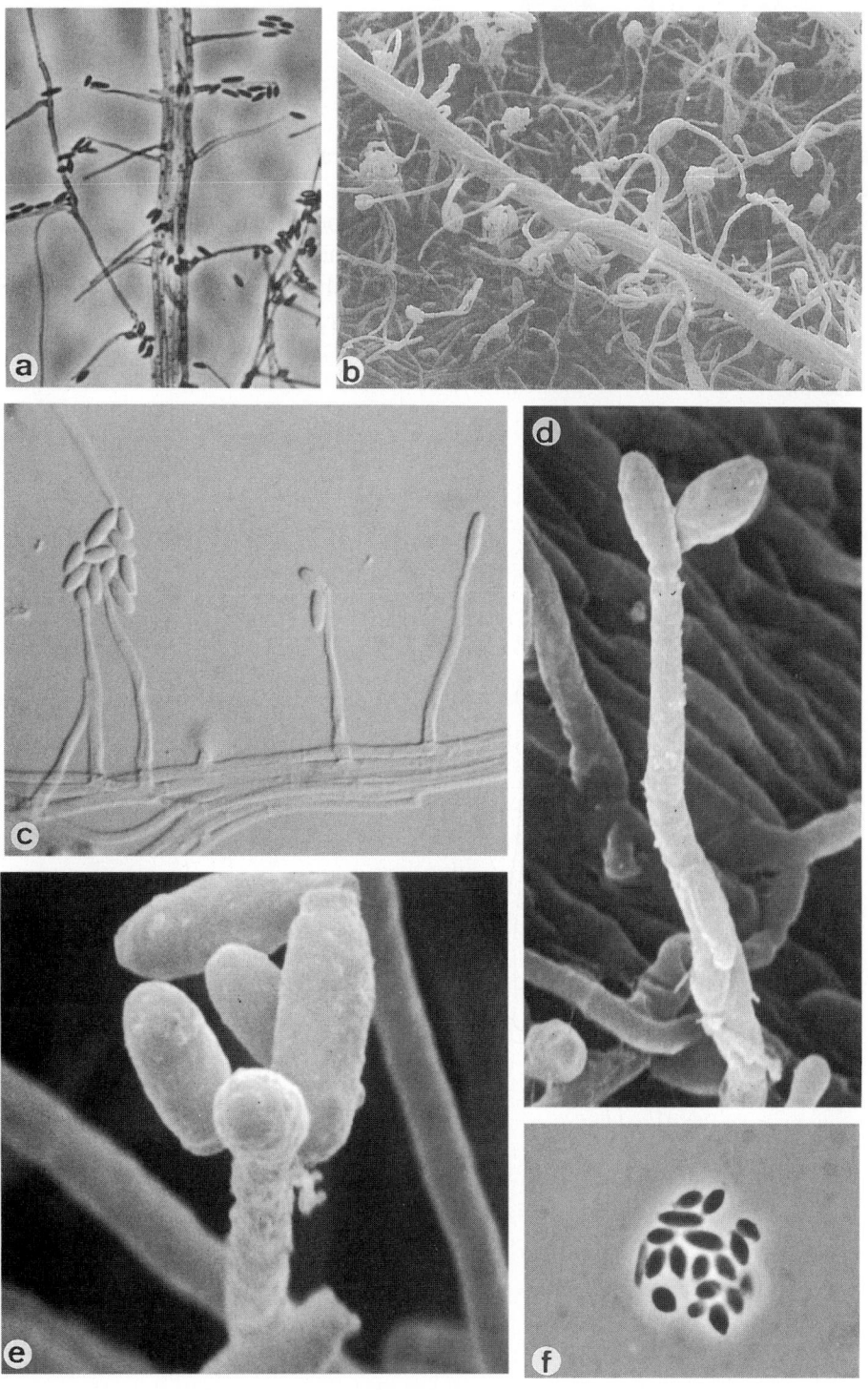

Acremonium potronii, CBS 262.89. a-c. Hyphal strands with phialides; d-f. phialides and conidia. a. ×640; b. ×1400; c. ×1600; d. ×6500; e. ×10900; f. ×1280.

Acremonium recifei (Arêa Leão & Lobo) W. Gams

Colony characteristics. Colonies (MEA 2%) restricted, smooth or felty, greenish-yellow or pale pink; reverse ochraceous-brown.

Microscopy. Conidiophores erect, short, branched in the lower part. Phialides narrow-acicular, 15-55 µm long, often in small groups on short, rather firm lateral branches, with slightly thickened apex. Conidia reniform, 4-6 × 1.3-2.0 µm. In older cultures rather thin-walled chlamydospores are present.

Molecular diagnostics. ITS restriction map based on CBS 137.35:

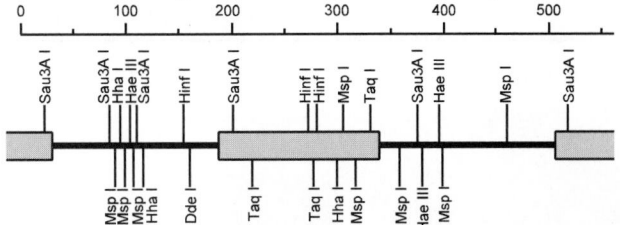

Differential diagnosis. *Acremonium curvulum* (p. 402) has a less branched conidial system. The two species further differ by *Dde* I and *Taq*I restriction sites in the 5.8S rDNA gene.

Pathogenicity. BSL-2. The species is listed in most handbooks (Rippon, 1988; Kwon-Chung & Bennett, 1992) as one of the major agents of white-grain mycetoma in humans, but actually very few confirmed cases other than the authentic description by Arêa Leâo & Lobo (1934) have been published. Cases from India were reported by Koshi *et al.* (1979). Záitz *et al.* (1995) reported a hyalohyphomycosis by this species. A catheter-related fungemia in a BMT patient was reported by Moulias *et al.* (1998).

Reference. Gams (1971).

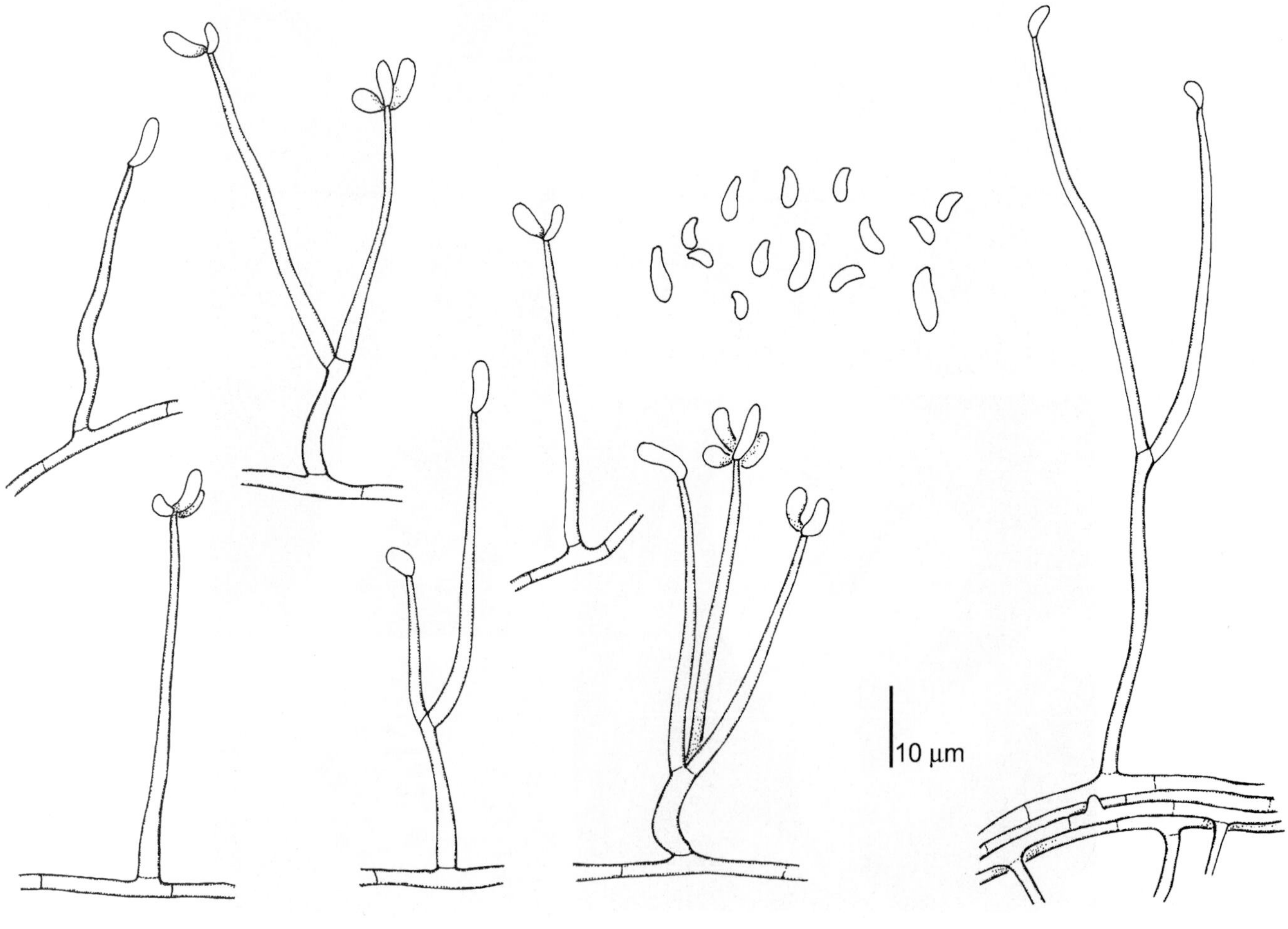

Acremonium recifei, IMI 225009. Phialides and conidia.

412

Antifungal susceptibility.

Antifungal	Mean MICs	Strains	Reference
AMB	0.58	3	Guarro *et al.* (1997a)
5FC	323	3	Guarro *et al.* (1997a)
FLZ	160	3	Guarro *et al.* (1997a)
ITZ	20	3	Guarro *et al.* (1997a)
KTZ	8.06	3	Guarro *et al.* (1997a)
MCZ	6.3	3	Guarro *et al.* (1997a)

Nomenclature. *Cephalosporium recifei* Arêa Leão & Lobo - C.R. Soc. Biol. R. Janeiro 117: 205, 1934 ≡ *Hyalopus recifei* (Arêa Leão & Lobo) Barboso - Subs. Estud. Hyalopus, Recife p. 39, 1941 ≡ *Acremonium recifei* (Arêa Leão & Lobo) W. Gams - Cephal. Schimmelp. p. 133, 1971.

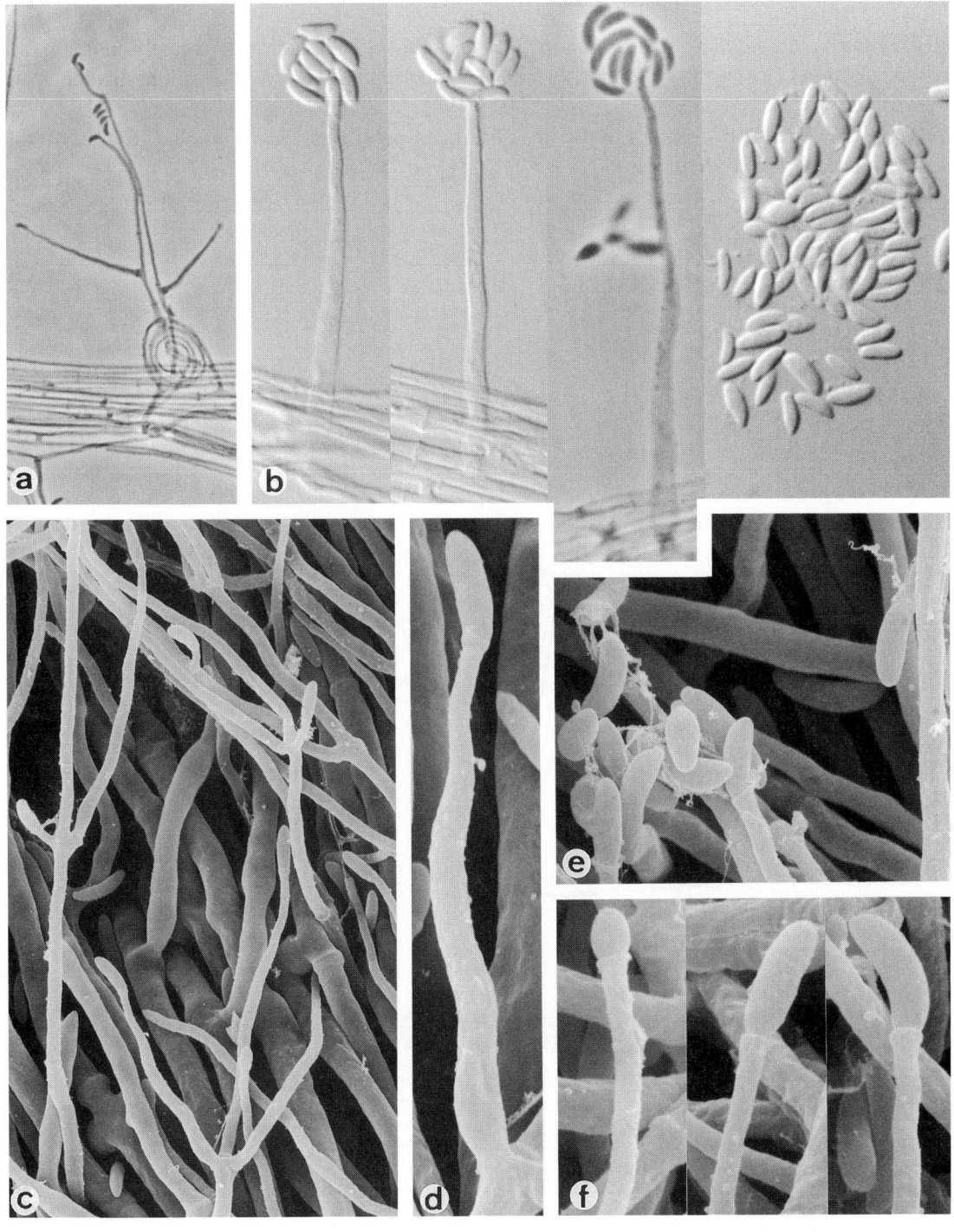

Acremonium recifei, CBS 485.77. Phialides and conidia. a. ×640; b. ×1600; c. ×1900; d. ×5500; e. ×4500; f. ×6000.

Acremonium roseogriseum (S.B. Saksena) W. Gams

Colony characteristics. Colonies (MEA 2%) growing moderately rapidly, reddish-brown, becoming olivaceous-black in areas of sporulation, often with hyphae aggregated in rope-like strands.

Microscopy. Phialides subulate, erect, arising from strands of aerial hyphae, 25-35 × 2.0-2.5 µm. Conidia in slimy heads, mostly tear-shaped, 4.9-6.5 × 2.6-4.1 µm, greyish-black, smooth-walled or finely warted.

Molecular diagnostics. ITS restriction map based on CBS 134.56:

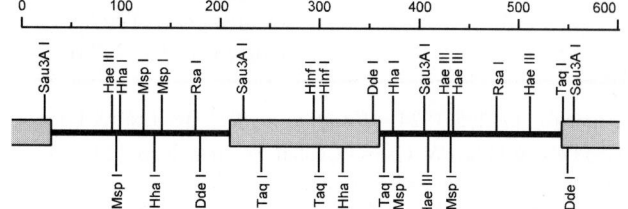

Differential diagnosis. *Acremonium atrogriseum* (p. 398) has ellipsoidal conidia and is easily distinguished by *Dde*I × ITS. *A. roseogriseum* is close to the common species *A. murorum* (Corda) W. Gams, but differs by tear-shaped rather than broadly ellipsoidal conidia.

Pathogenicity. BSL-1. A case of arthritis in the knee has been published (Ward *et al.*, 1961), but it is uncertain whether this strain was correctly identified.

Reference. Gams (1971).

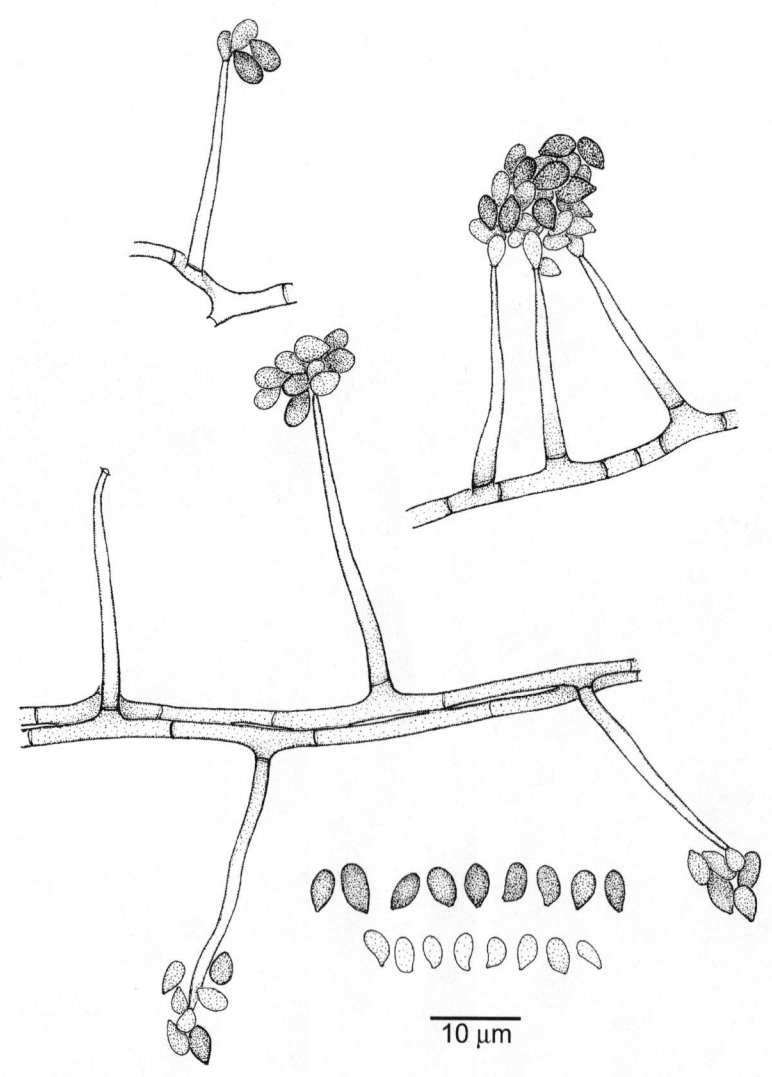

Acremonium roseogriseum, CBS 134.56. Phialides and conidia.

Antifungal susceptibility.

Antifungal	MICs	Strains	Reference
AMB	4.62	1	Guarro *et al.* (1997a)
5FC	>323	1	Guarro *et al.* (1997a)
FLZ	>80	1	Guarro *et al.* (1997a)
ITZ	2.5	1	Guarro *et al.* (1997a)
KTZ	>12.8	1	Guarro *et al.* (1997a)
MCZ	5	1	Guarro *et al.* (1997a)

Nomenclature. *Cephalosporium roseogriseum* S.B. Saksena - Mycologia 47: 895, 1955 ≡ *Acremonium roseogriseum* (S.B. Saksena) W. Gams - Cephalosp. Schimmelp. p. 87, 1976.

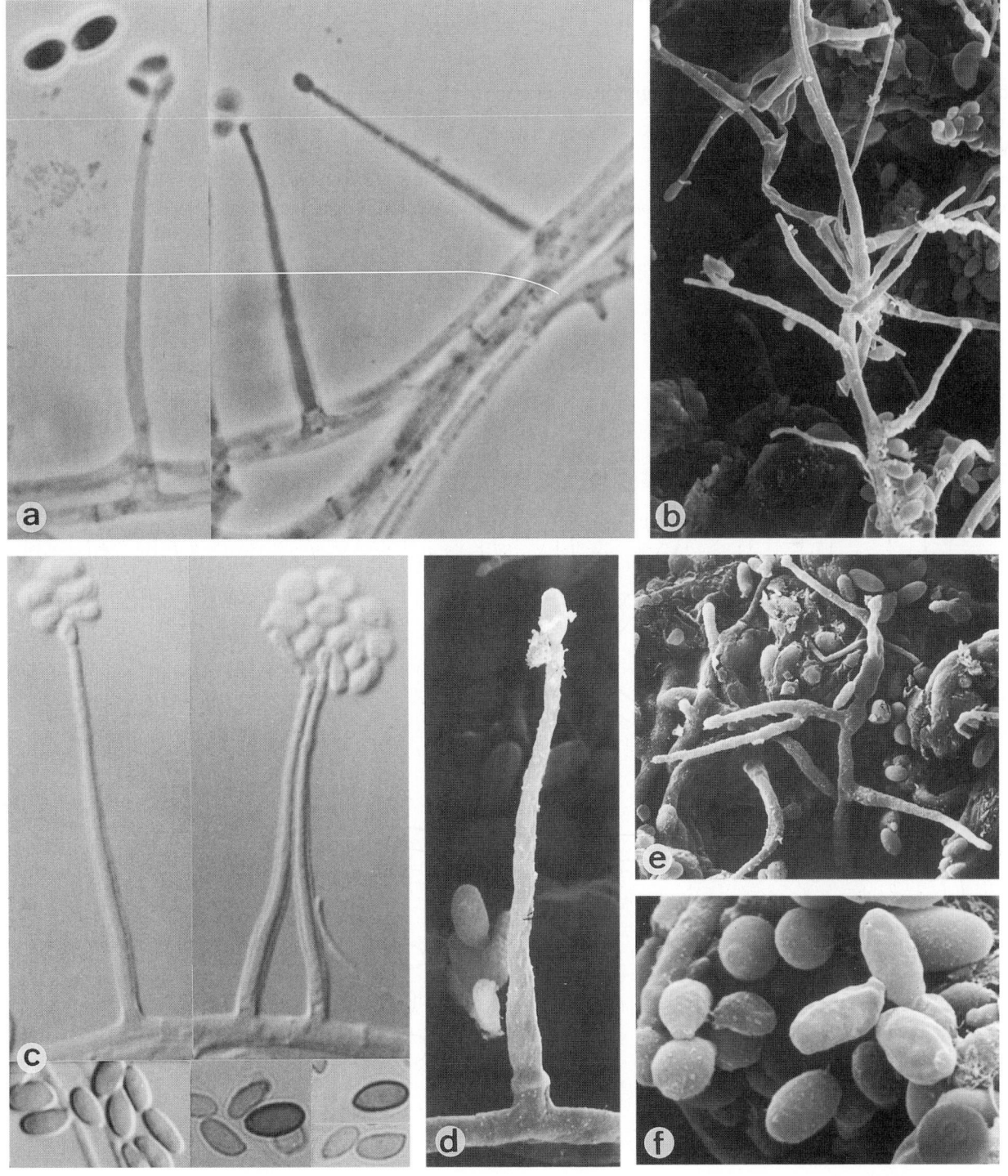

Acremonium roseogriseum, CBS 626.85. Phialides and conidia. a. ×1280; b. ×1300; c. ×1600; d. ×3400; e. ×1500; f. ×5400.

Acremonium spinosum (Negroni) W. Gams

Colony characteristics. Colonies (MEA 2%) restricted, smooth or felty, whitish to pale ochre; reverse uncoloured or pale ochraceous-brown.

Microscopy. Mycelium very tough. Sporulation poor. Phialides arising from creeping hyphae. narrow-acicular, 8-30 μm long. Conidia produced in mucilaginous heads, rarely also in short chains, hyaline, subspherical, 2.5-3.5 × 2.5 μm, verruculose. Chlamydospores absent.

Molecular diagnostics: ITS restricton map based on CBS 136.33:

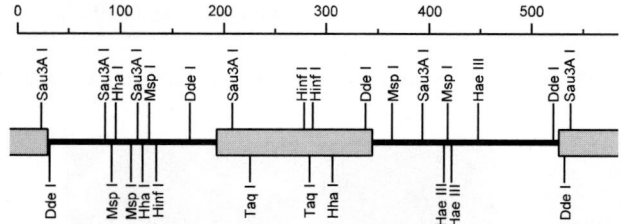

Pathogenicity. BSL-1. The species was originally described from human onychomycosis (Negroni, 1933), but has not been reported since.

Reference. Gams (1971).

Nomenclature. *Cephalosporium spinosum* - Negroni - C.R. Séanc. Soc. Biol., B. Aires 113: 480, 1933 ≡ *Hyalopus spinosus* (Negroni) Barbosa - Sub. Estud. Gen. Hyalopus, Recife. 1941 ≡ *Acremonium spinosum* (Negroni) W. Gams - Cephalosp. Schimmelp. p. 78, 1971.

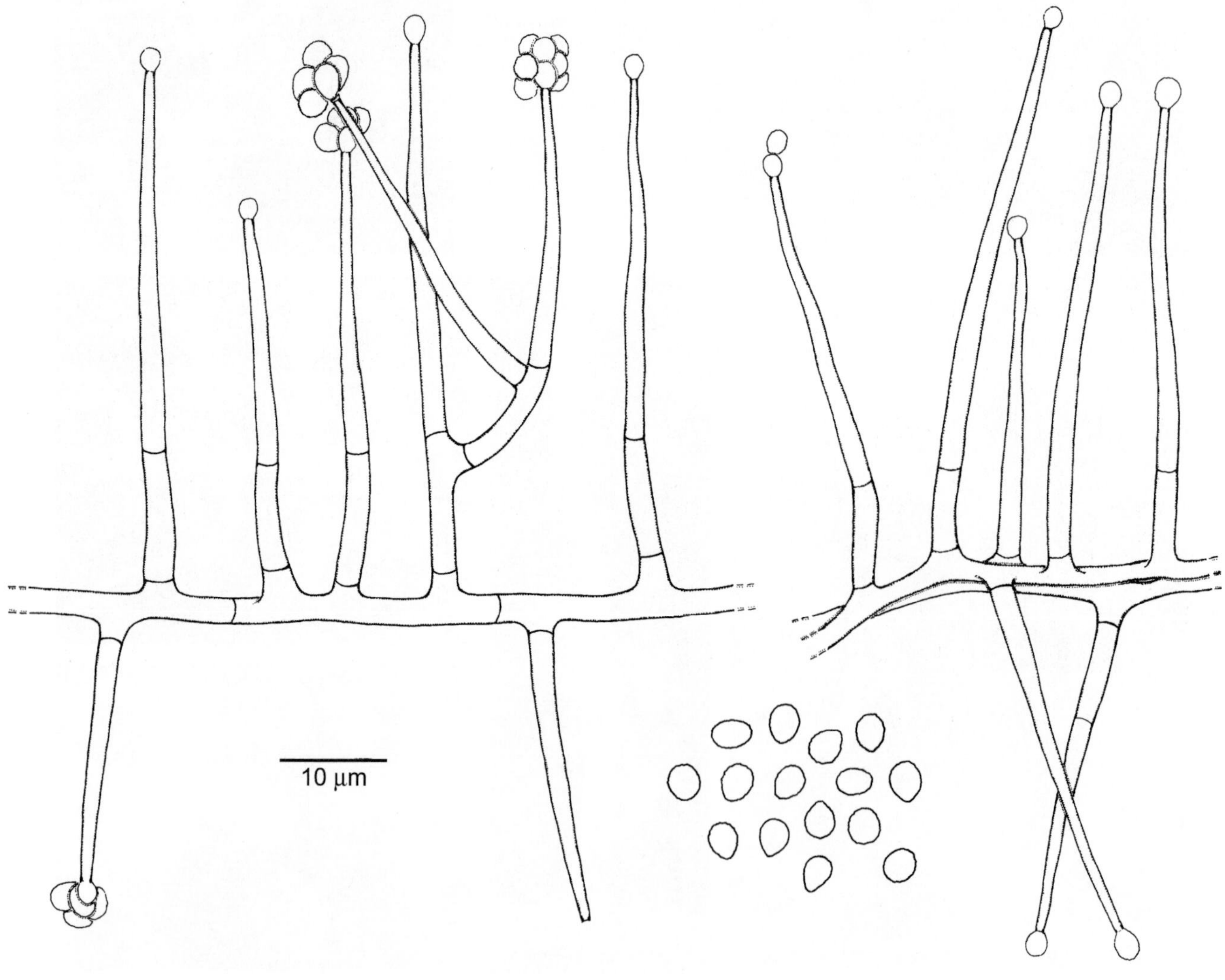

Acremonium spinosum, CBS 136.33. Conidiogenous cells and conidia.

416

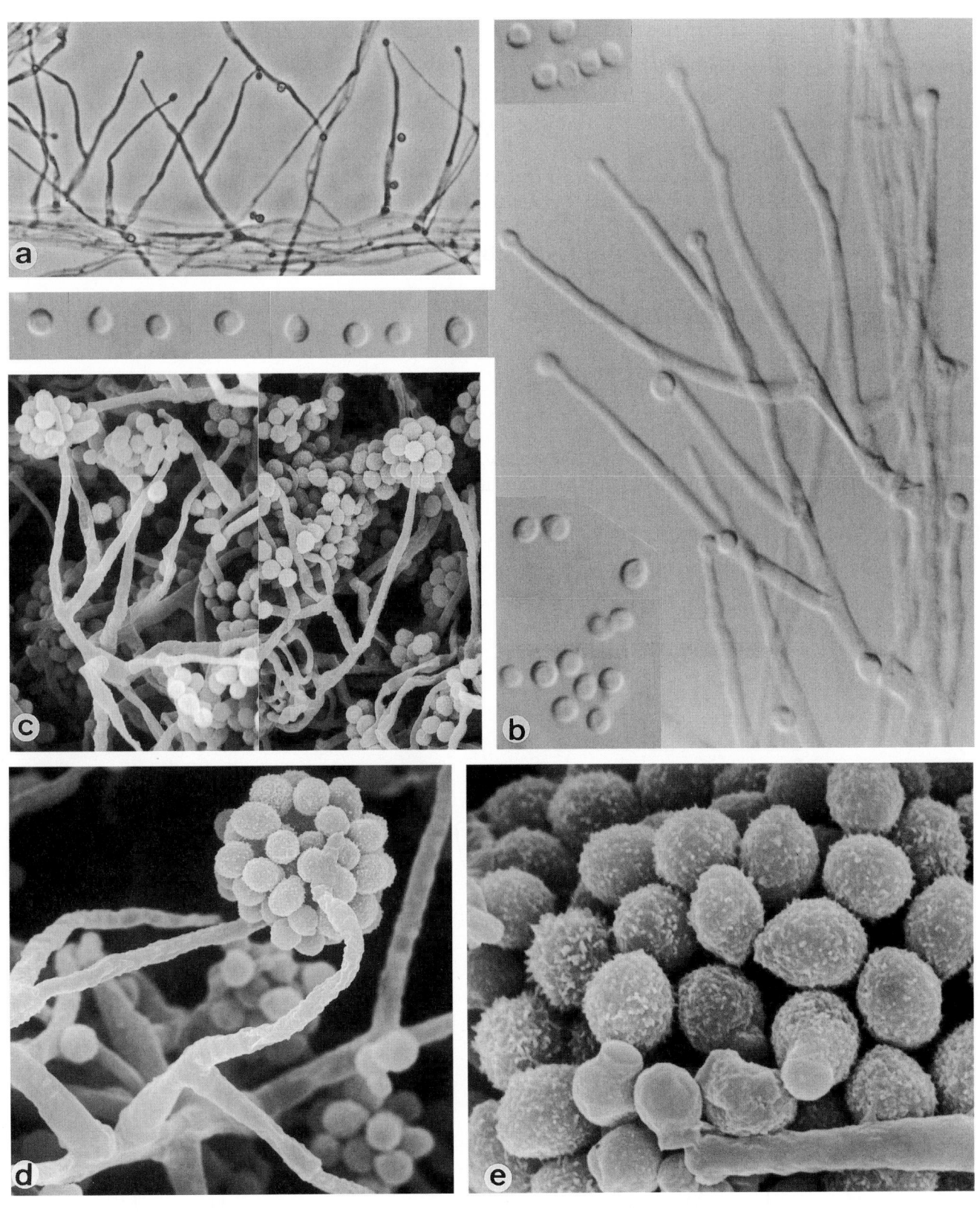

Acremonium spinosum, CBS 136.33. Phialides and conidia. a. ×640; b. ×1600; c. ×1500; d. ×3000; e. ×750.

Acremonium strictum W. Gams

Colony characteristics. Colonies (MEA 2%) growing rapidly, moist to slimy, pink or orange; reverse remaining colourless or turning pink to orange.

Microscopy. Conidiophores simple, occasionally branched. Phialides slender, arising from submerged or slightly fasciculate aerial hyphae, 20-65 × 1.4-2.5 µm. Submerged sporulation frequent from reduced phialides. Conidia grouped in slimy heads, cylindrical or ellipsoidal, 3.3-5.5 × 0.9-1.8 µm, hyaline.

Molecular diagnostics. ITS restriction map based on NCBI U57671:

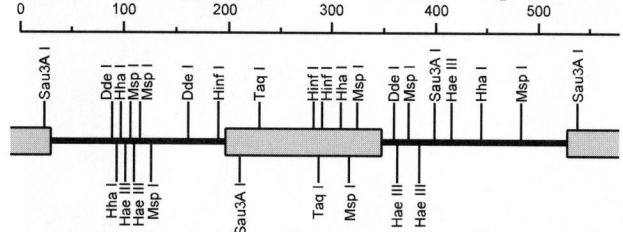

Pathogenicity. BSL-1. Invasive infections in neutropenic patients (Morin *et al.*, 1988; Schell & Perfect, 1996); pulmonary infection in a patient with chronic granulomatous disease (Boltansky *et al.*, 1984). A CAPD-related peritonitis was reported by Koc *et al.* (1998).

References. Gams (1971), Domsch *et al.* (1980), Samson *et al.* (1996).

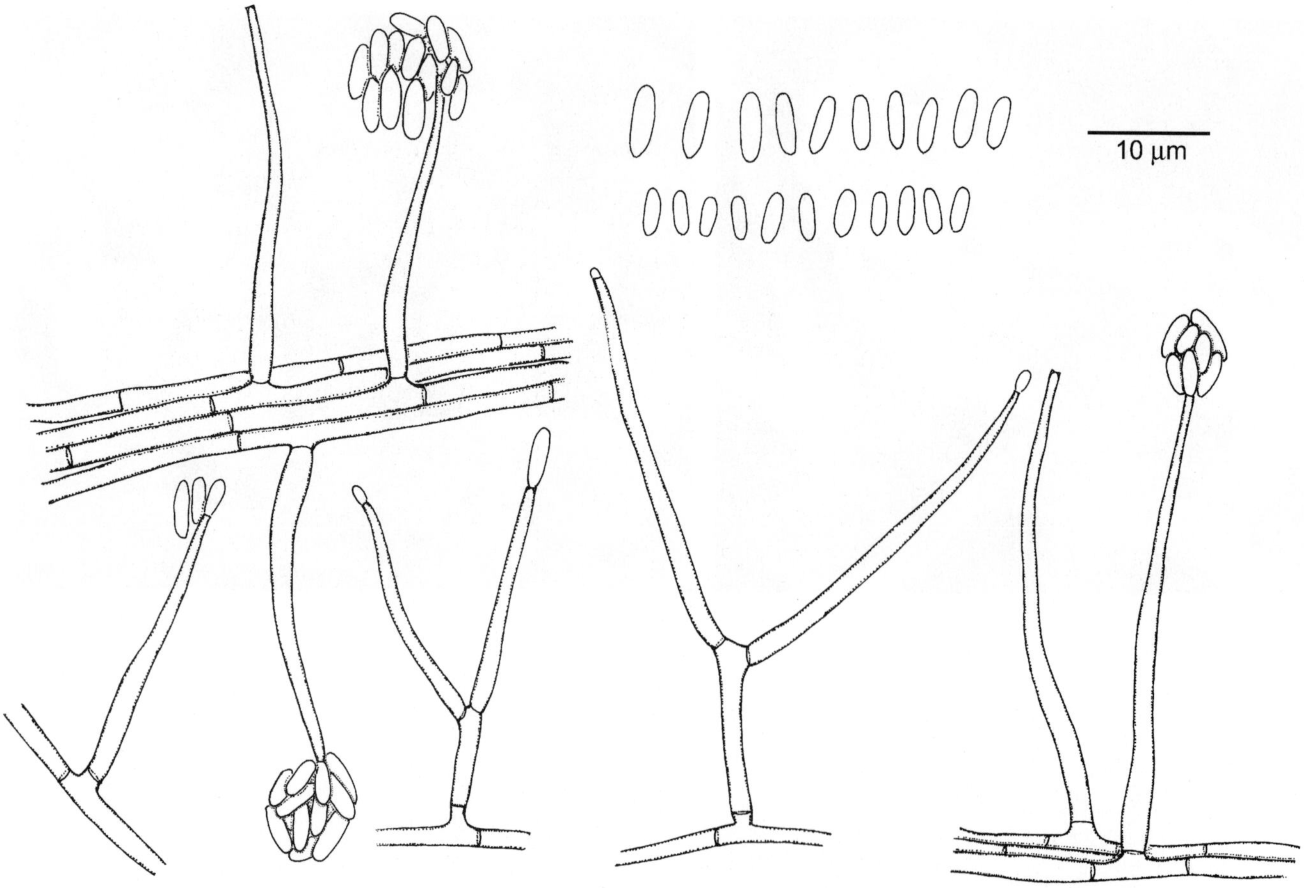

10 µm

Acremonium strictum, CBS 654.96. Conidiophores, phialides and conidia.

418

Antifungal susceptibility.

Antifungal	MICs	Strains	Reference
AMB	1.16	1	Guarro *et al.* (1997a)
5FC	>323	1	Guarro *et al.* (1997a)
FLZ	>80.0	1	Guarro *et al.* (1997a)
ITZ	>10.0	1	Guarro *et al.* (1997a)
KTZ	>12.8	1	Guarro *et al.* (1997a)
MCZ	>40.0	1	Guarro *et al.* (1997a)

Nomenclature. *Cephalosporium acremonium* Corda *ss. auctt.* ≡ *Acremonium strictum* W. Gams - Cephalosp. Schimmelp. 42, 1971.

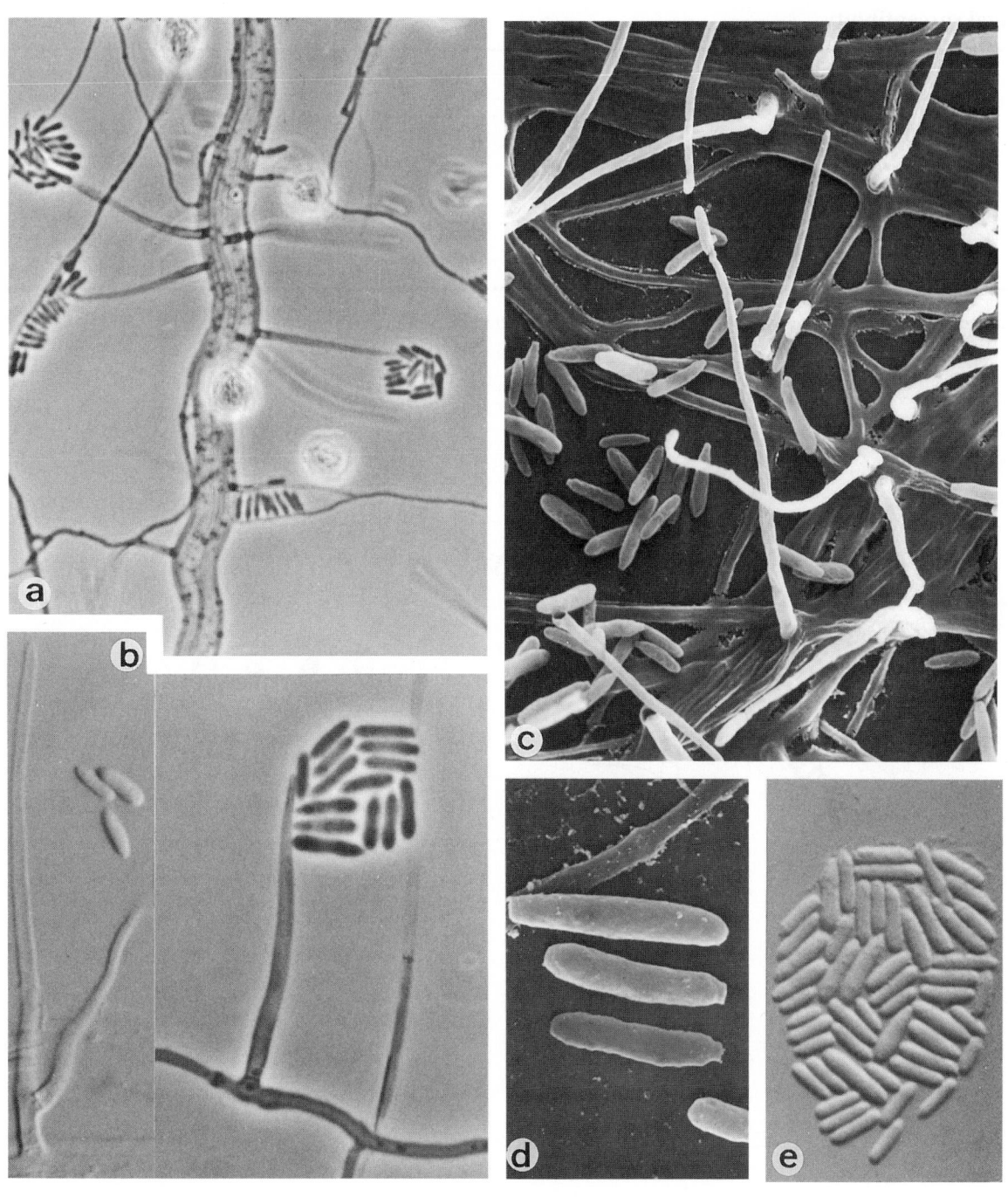

Acremonium strictum, CBS 654.96. Conidiophores and conidia. a. ×640; b. ×1600; c. ×2100; d. ×5500; e. ×1600.

Hyphomycetes. Genus: *ACROPHIALOPHORA*

Acrophialophora fusispora (S.B. Saksena) Samson

Colony characteristics. Colonies (MEA 2%) growing rapidly, greyish-brown; reverse almost black.
Microscopy. Hyphae pale brown, 1.5-3.5 μm wide. Conidiophores arising singly, terminally and laterally on the hyphae, erect, straight or slightly flexuose, tapering towards the apex, pale brown, rough-walled, up to 1.5 μm long, 2-5 μm wide, with whorls of phialides in the upper part. Phialides flask-shaped with a long and narrow neck, hyaline, smooth-walled or echinulate, 9-15 × 3.0-4.5 μm in the broadest part. Conidia arising in long chains, limoniform, one-celled, pale brown, 6-10 × 3.5-5.0 μm, finely echinulate with distinct spiral bands. Thermotolerant.
Pathogenicity. BSL-1. Keratitis (Shukla *et al.*, 1983); pulmonary infection (Sutton *et al.*, 1997), pulmonary colonization in a child with CF (González-Escalada *et al.*, 2000); disseminated infection (J. Guarro, unpublished data).
References. Samson & Mahmood (1970), Ellis (1971), Kirk (1991a).
Antifungal susceptibility.

Antifungal	MICs	Strains	Reference
AMB	2	1	Unpublished data
5FC	256	1	Unpublished data
FLZ	16	1	Unpublished data
ITZ	0.25	1	Unpublished data
KTZ	1	1	Unpublished data
MCZ	1	1	Unpublished data

Nomenclature. *Acrophialophora fusispora* (S.B. Saksena) Samson - Acta Bot. Neerl. 19: 805, 1970.

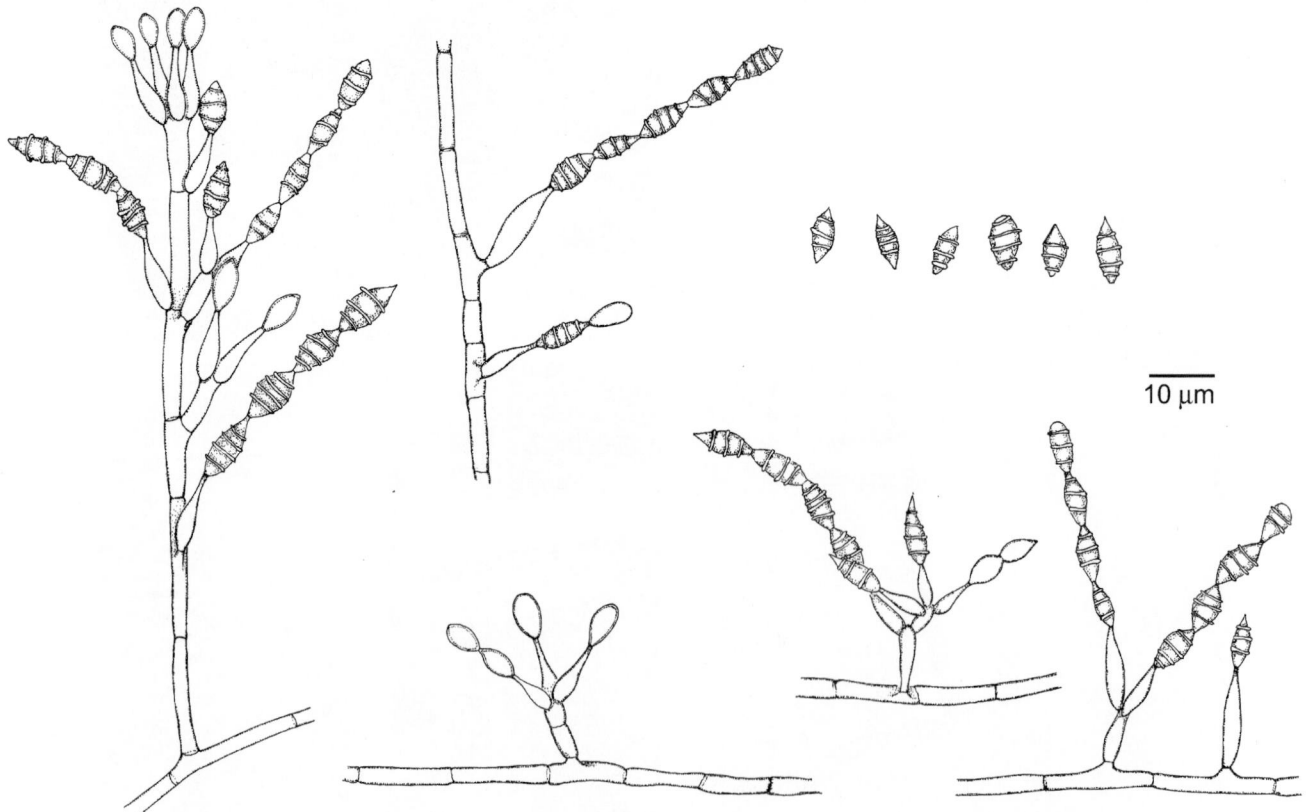

Acrophialophora fusispora, CBS 380.55. Conidiophore, phialides and conidia.

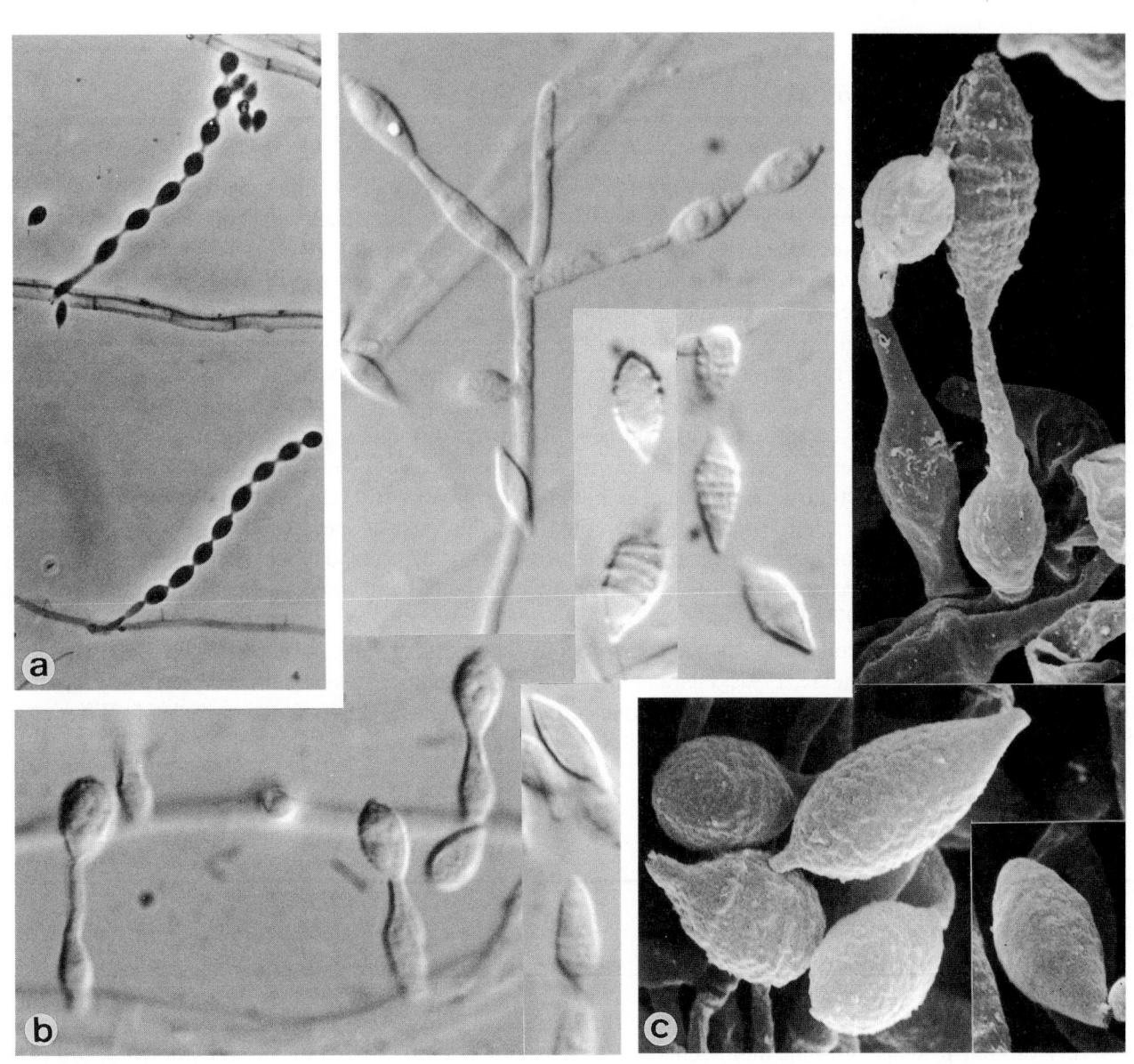

Acrophialophora fusispora, CBS 380.55. Conidiophores, phialides and conidia. a. ×512; b. ×1600; c. ×5900.

Hyphomycetes. Genus: *ALTERNARIA*

Generic description. Colonies expanding, grey to olivaceous, powdery or woolly. Conidiophores erect, brown, multi-celled, producing conidia in sympodial order; conidial scars flat, brown. Conidia brown, smooth-walled or verruculo-se, with rounded base and beaked apex, with muriform septation, catenate or single.

Teleomorphs. *Pleospora, Lewia* (*Ascomycota, Euascomycetes, Pleosporales: Pleosporaceae*).

Differential diagnosis. The genus is different from *Ulocladium* (p. 997) by its conidia having rounded bases and having more or less elongated tips. For a discussion and phylogenetic overview of related genera, see p. 380. Arrangement of conidia, either in chains or singly, can best be observed on poor media in Petri-dishes under a stereo microscope.

General remarks. The genus comprises a large number of mostly saprobic or plant-pathogenic species. Strains encountered in clinical laboratories are in the saprobic group and nearly always have conidia in chains. The majority of clinical cases are skin infections after trauma. The etiologic agent is mostly referred to as *A. alternata*, but it is likely that mostly other, closely related species are concerned, particularly those species forming chlamydospore-like structures. In such species, conversion to meristematic growth may be a response to stress. Compare also the meristematic fungus *Botryomyces caespitosus* (p. 537), which on the basis of molecular data is closely related to *Alternaria* (de Hoog *et al.*, 1997b; Fig. 42 on p. 382).

Pathogenicity. Cases mainly concern (sub)cutaneous phaeohyphomycoses (Male & Pehamberger, 1985; Viviani *et al.*, 1986; Badillet, 1991; de Bièvre, 1991; Vieira *et al.*, 1998) or onychomycoses (Wadhwani & Srivastava, 1985). *Alternaria* species are emerging as pathogens in immunocompromised hosts (Levy-Klotz *et al.*, 1985; Wiest *et al.*, 1987; Morrison & Weisdorf, 1993). *In vitro* susceptibility tests have been performed by Pujol *et al.* (2000).

References. Joly (1964), Ellis (1971, 1976).

Key to the treated species of Alternaria:

1a.	Conidia at first obclavate or obpyriform, later swelling and becoming irregular and very variable in shape, forming multi-celled chlamydosporic bodies . . . . *A. chlamydospora* (426)
1b.	Conidia slender, not as above, without chlamydosporic bodies ➞ 2
2a.	Conidia usually with transverse septation only . . . . *A. dianthicola* (428)
2b.	Conidia with transverse and longitudinal and/or oblique septa ➞ 3
3a.	Conidia often with long, multi-septate secondary conidiophores; conidial bodies of variable size, often becoming nearly tubular; sporulation poor . . . . *A. infectoria* (430)
3b.	Conidia mostly without secondary conidiophores; conidial bodies more regular in width ➞ 4
4a.	Conidia in chains of 10 or more; conidia medium to dark brown ➞ 5
4b.	Conidia usually in chains of 3-5, tapering gradually into a beak of half the conidial length; conidia pale golden brown . . . . *A. tenuissima* (434)
5a.	Conidia medium brown, mostly finely punctate, with short beak . . . . *A. alternata* (423)
5b.	Conidia dark brown, mostly smooth-walled, often with longer beak . . . . *A. longipes* (432)

Alternaria alternata (Fr.) Keissl.

Colony characteristics. Colonies (PCA) expanding, grey to olivaceous, powdery or felty.

Microscopy. Conidiophores mostly unbranched, with one or few conidial scars, up to 50 μm long, 3-6 μm wide. Conidia obclavate to ellipsoidal, with a short, cylindrical beak, 23-56 × 8-17 μm, medium brown, rugulose with muriform septation, with a single scar at the tip, arising in mostly unbranched chains of ten or more.

Differential diagnosis. *Alternaria infectoria* (p. 430) has darker and shorter conidia which arise in repeatedly branched chains, conidia having long secondary conidiophores which soon become tubular at degeneration; cultures of *A. infectoria* tend to become sterile and are white on DRYES. Possibly most clinical cases concern *A. infectoria* rather than *A. alternata*. *Ulocladium chartarum* (p. 1000) has spindle-shaped rather than obclavate conidia.

Physiology. Tolerant to benomyl. Colonies pigmented on DRYES agar.

Molecular diagnostics. ITS restriction map based on NCBI U05195:

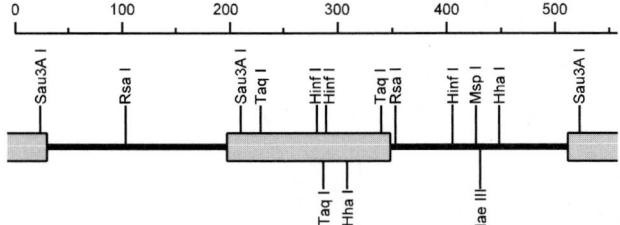

Pathogenicity. BSL-1. The species is a saprobe on dead plant material, but may cause skin lesions after trauma (Chabasse *et al.*, 1982; Verret *et al.*, 1982; Ackland *et al.*, 1998), often in immunocompromised patients (del Palacio *et al.*, 1996). Endophthalmitis following eye surgery was reported by Rummelt *et al.* (1991), and an onychomycosis by Pritchard & Muir (1987). Infections may heal spontaneously when the underlying disease is cured (Chung *et al.*, 1999). In cutaneous infections, brown hyphal elements can be observed in tissue, but subcutaneous lesions mostly show hyaline, ellipsoidal fungal elements. Rarely cases of systemic infection have been reported (Ohashi, 1960), including cases of invasive mycoses in AIDS patients (Levy-Klotz *et al.*, 1985; Wiest *et al.*, 1987).

References. Ellis (1971), Male & Pehamberger (1985), Simmons (1990, 1995a, b), Badillet (1991).

Antifungal susceptibility.

Antifungal	Mean MICs	Strains	Reference
AMB	0.33	8	Wildfeuer *et al.* (1998)
5FC	256	4	Unpublished data
FLZ	64	4	Unpublished data
ITZ	0.18	8	Wildfeuer *et al.* (1998)
KTZ	0.43	8	Wildfeuer *et al.* (1998)
MCZ	3.36	4	Unpublished data
VCZ	0.63	3	McGinnis & Pasarell (1998b)

Nomenclature. *Alternaria tenuis* C.G. Nees - Syst. Pilze Schwämme p. 72, 1817.

Torula alternata Fries - Syst. Mycol. 3: 500, 1832 ≡ *Alternaria alternata* (Fries) Keissler - Beih. Bot. Zentbl. 29: 434, 1912.
Alternaria geophila Daszewska - Bull. Soc. Bot. Genève, Ser. 2, 4: 294, 1912.
Alternaria stemphylioides Bliss - Mycologia 36: 538, 1944.

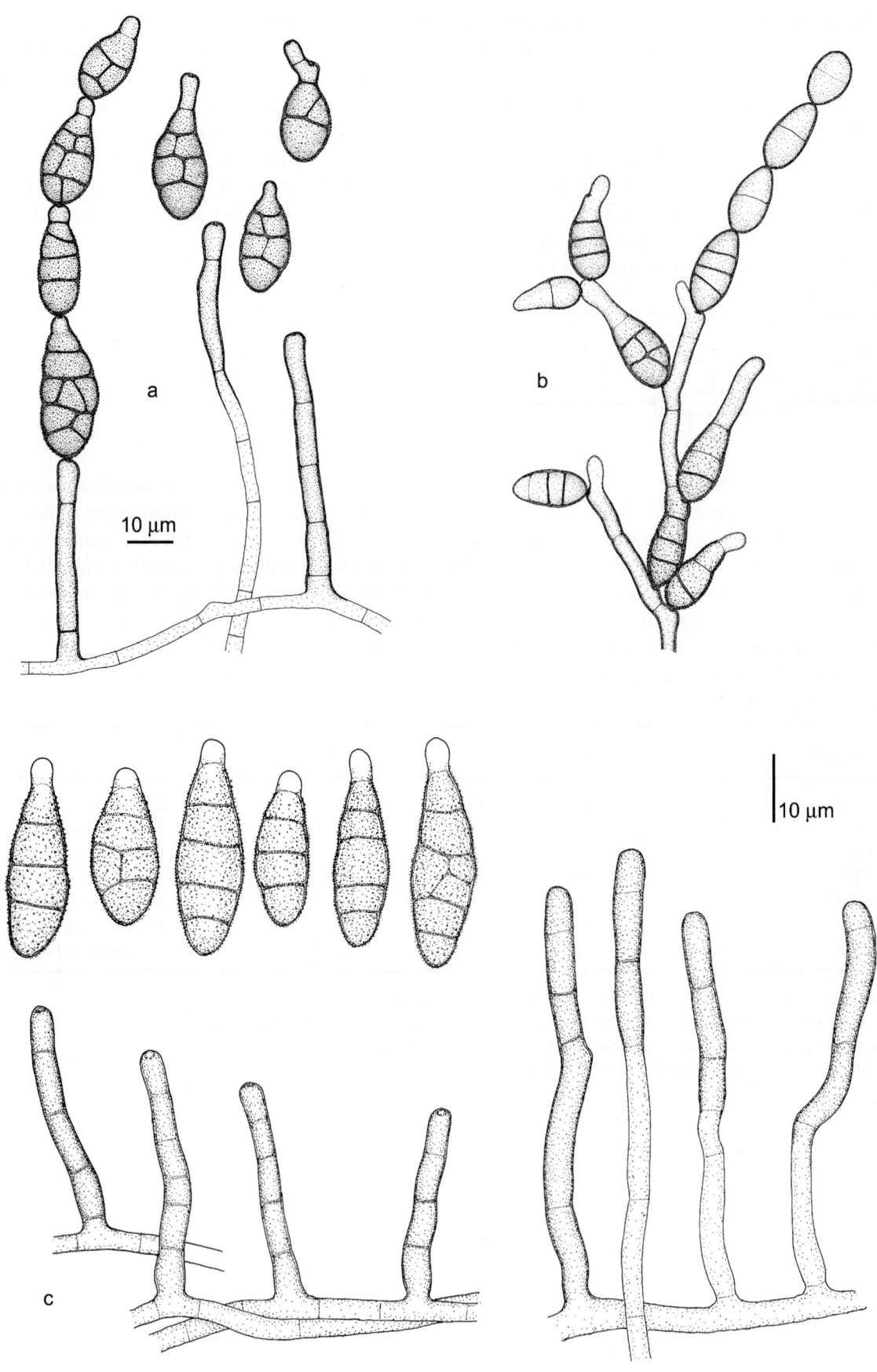

Alternaria alternata, CBS 603.78. a. Conidiophores, part of a conidial chain, and liberated conidia; b. degenerate subculture on DRYES agar; c. mature conidiophores and liberated conidia.

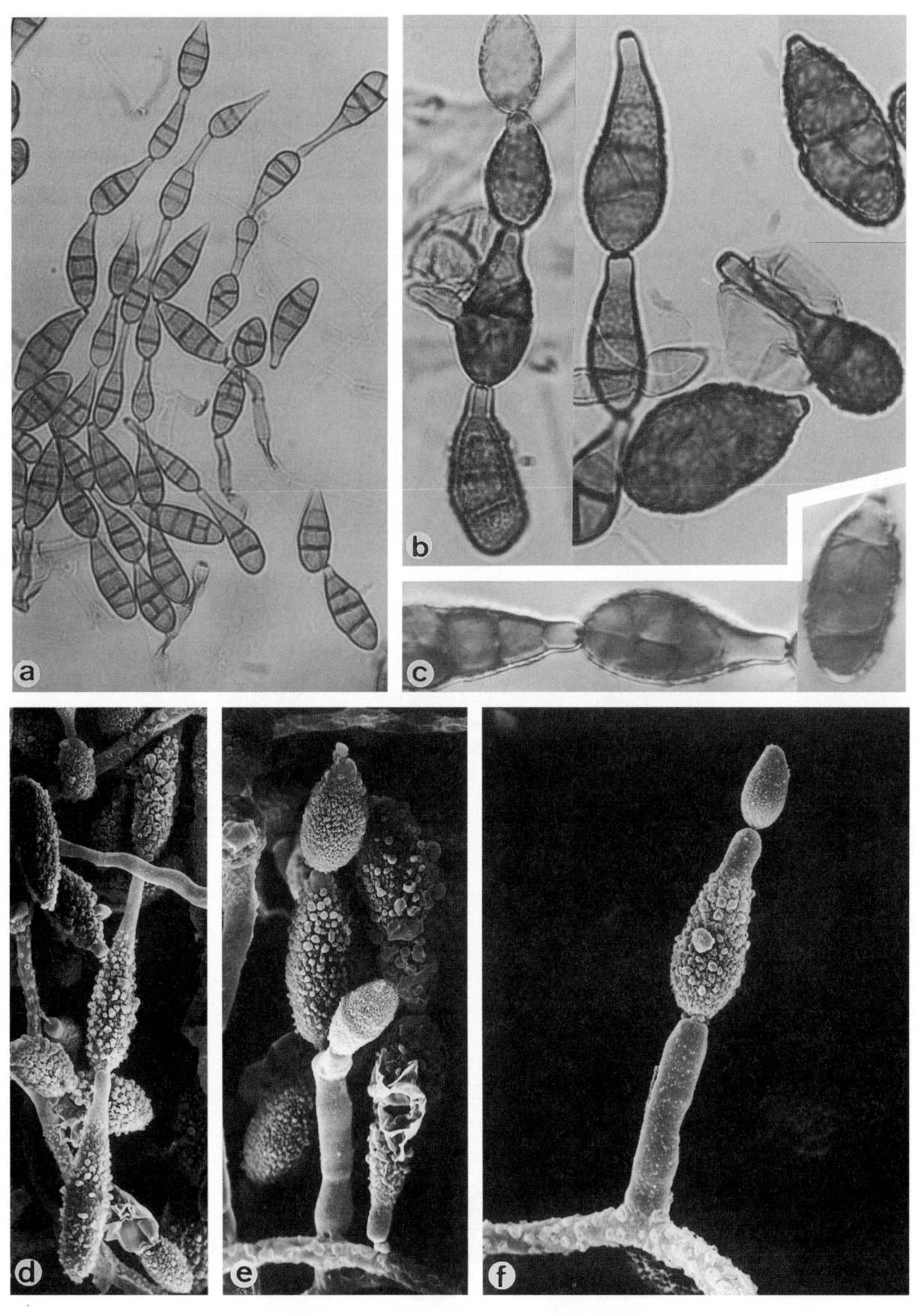

Alternaria alternata, CBS 603.78. a-d. Conidial chains; e, f. conidiophores bearing conidia. a. ×512; b. ×280; c. ×1600; d. ×1100; e. ×1900; f. ×2450.

425

Alternaria chlamydospora Mouchacca

Colony characteristics. Colonies (PCA) growing rapidly, olivaceous-black, fluffy; reverse fuscous-black.

Microscopy. Conidiophores up to 150 µm long, 3-6 µm wide, pale brown. Conidia initially obclavate or obpyriform, later swelling and becoming very irregular in shape, multi-celled, often in chains, golden-brown, smooth-walled to slightly verrucose, 26-70 × 8-48 µm, with a short, pale beak 2-6 µm wide. Similar conidia, produced directly on hyphae, are chlamydospore-like.

Molecular diagnostics. ITS restriction map based on CBS 491.72:

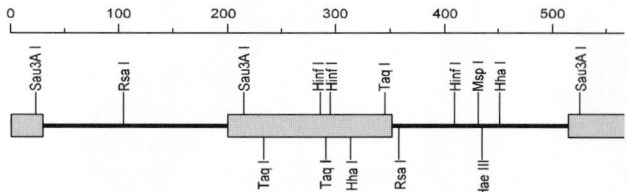

Pathogenicity. BSL-1. The species is primarily found in salty soil (Mouchacca, 1973). Onychomycoses and cutaneous infections (Singh *et al.*, 1990a; Chartois-Léauté *et al.*, 1995; Bartolome *et al.*, 1999) have been reported. It should be mentioned that the strain published by Singh *et al.* (1990a) under this name has an ITS sequence identical to *A. alternata*.

References. Mouchacca (1973), Ellis (1976), Simmons (1981).

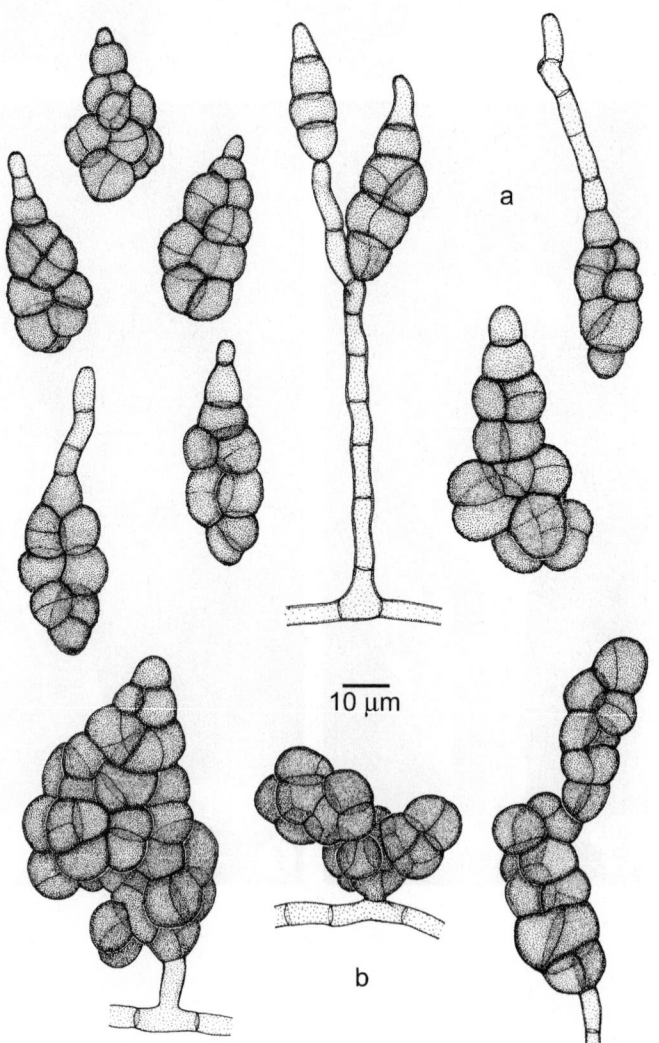

Alternaria chlamydospora, IMI 326463. a. Conidiophore and conidia; b. chlamydospores.

426

Antifungal susceptibility.

Antifungal	Mean MICs	Strains	Reference
AMB	2.8	2	Unpublished data
5FC	256	2	Unpublished data
FLZ	22.6	2	Unpublished data
ITZ	0.35	2	Unpublished data
KTZ	2	2	Unpublished data
MCZ	5.7	2	Unpublished data

Nomenclature. *Alternaria chlamydospora* Mouchacca - Mycopath. Mycol. Appl. 50: 217, 1973.

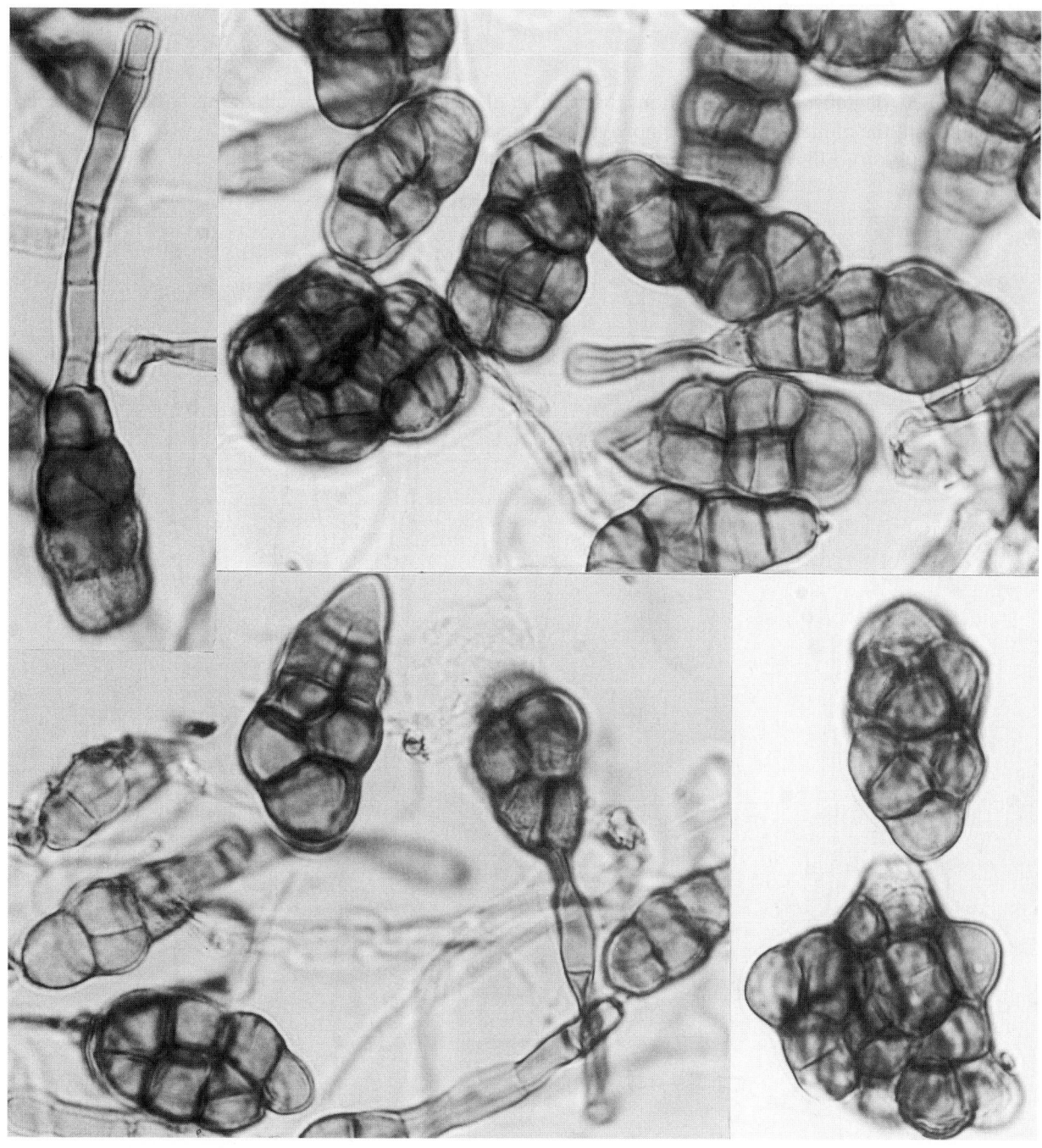

Alternaria chlamydospora, IMI 326463. Chlamydospores and conidia. ×1280.

Alternaria dianthicola Neergaard

Colony characteristics. Colonies (PCA) growing rapidly, dark brown with white to pale grey patches, usually without conidial production.

Microscopy. Conidiophores arising singly or in groups, erect or ascending, usually simple, occasionally branched, straight or flexuose, pale olivaceous brown, up to 150 μm long, 4-6 μm wide. Conidia usually in chains of 4-5, straight or curved, obclavate or almost cylindrical, rostrate, pale olivaceous-brown, smooth-walled, with up to 14 transverse septa, occasionally with few longitudinal or oblique septa, constricted at the septa, 58-120 × 9-14 μm, with a pronounced beak which is sometimes inflated at the tip.

Molecular diagnostics. With ITS RFLP the species is indistinguishable from *A. alternata* (p. 423). ITS restriction map based on CBS 915.96:

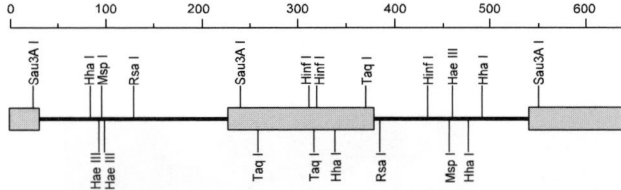

Pathogenicity. BSL-1. Subcutaneous infections (Mitchell *et al.*, 1983). The type strain of this species was described as a plant pathogen; the clinical strains were not available for verification.

References. Joly (1964), Ellis (1971), David (1991).

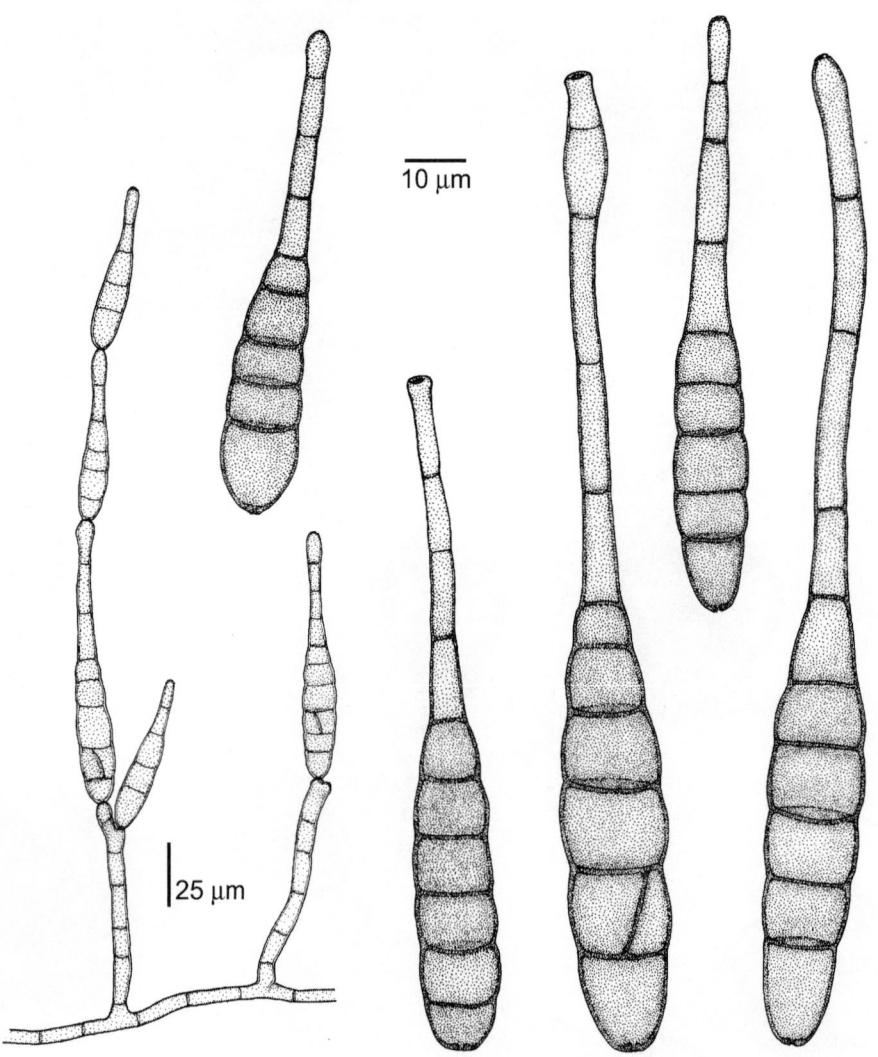

Alternaria dianthicola, CBS 112.38. Conidiophores and conidia.

Antifungal susceptibility.

Antifungal	Mean MICs	Strains	Reference
AMB	0.5	2	Unpublished data
5FC	256	2	Unpublished data
FLZ	64	2	Unpublished data
ITZ	0.5	2	Unpublished data
KTZ	1.4	2	Unpublished data
MCZ	1	2	Unpublished data

Nomenclature. *Alternaria dianthicola* Neergaard - Danish Spec. Alt. Stemph. p. 190, 1945.

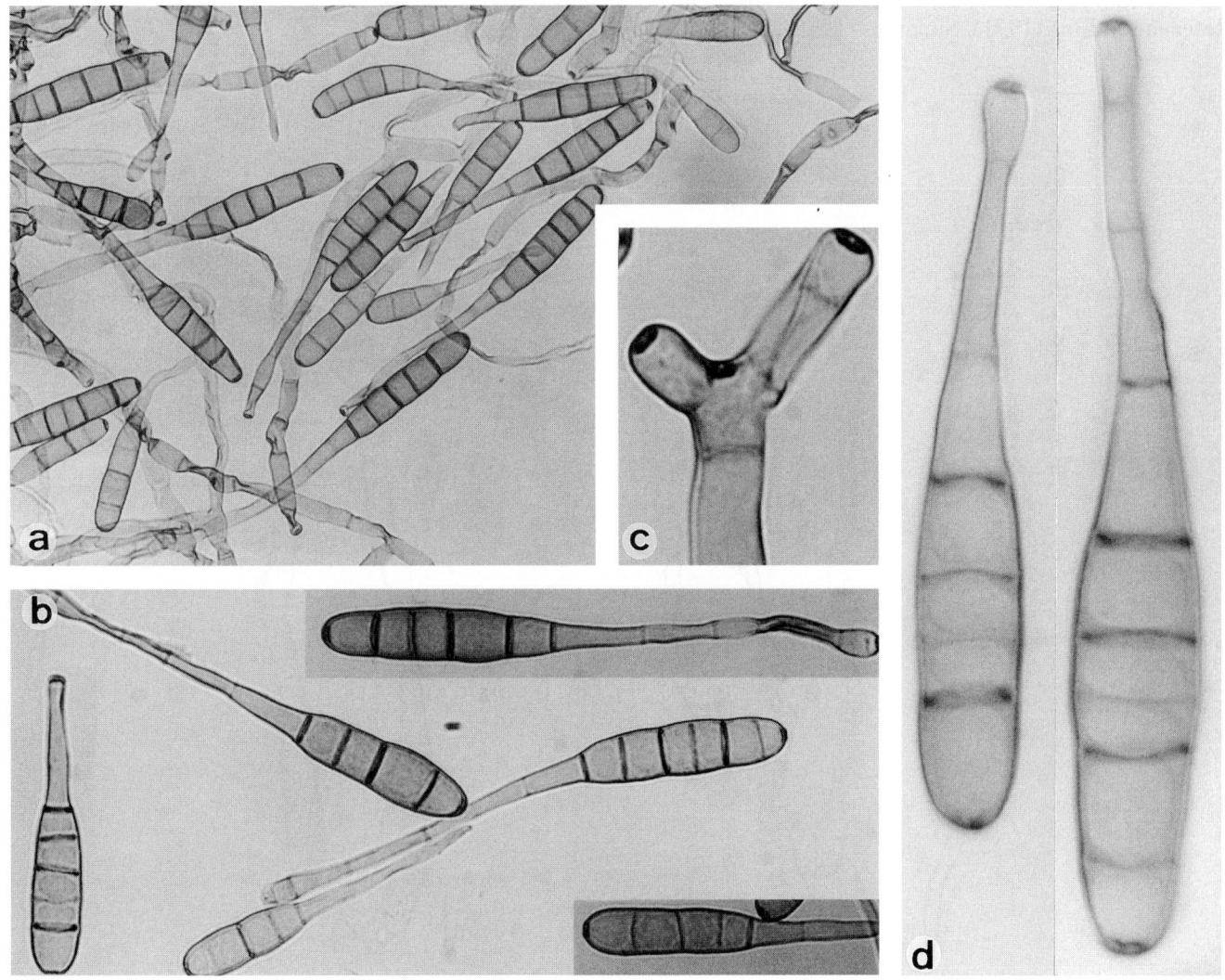

Alternaria dianthicola, CBS 112.38. Conidia and detail of the conidiophores. a. ×320; b. ×512; c, d. ×1600.

Alternaria infectoria Simmons

Colony characteristics. Colonies (PCA) expanding, grey or to olivaceous, powdery or felty, often with poor sporulation which is enhanced on water agar.

Microscopy. Conidiophores mostly unbranched, flexuose, elongating considerably, with conidial scars at irregular distances at the bends, up to 80 μm long, 3-6 μm wide, dark brown at maturity. Conidia arising in strongly branched chains, ovoidal, often with very short apical beak which extends into a secondary conidiophore, 12-30 × 7-9 μm exclusive of the beak, dark brown, smooth-walled, finally becoming rugulose, with transverse and sometimes oblique septation.

Teleomorph. *Lewia infectoria* (Fuckel) M.E. Barr & Simmons (*Ascomycota, Euascomycetes, Pleosporales: Pleosporaceae*).

Differential diagnosis. Distinct from *A. alternata* (p. 423) by poorly sporulating cultures which are white on DRYES agar and by frequent occurrence of nearly tubular conidia.

Molecular diagnostics. ITS restriction map based on NCBI Y17067:

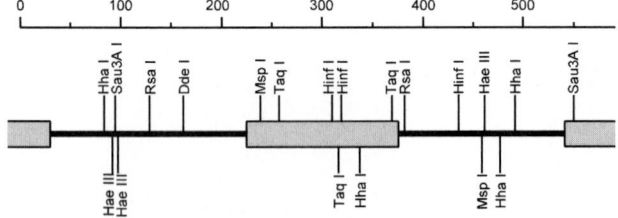

Pathogenicity. BSL-1. The species is a common saprobe on plant material. A case of phaeohyphomycosis in a cat was reported by Roosje *et al.* (1993). With ITS sequencing, most clinical strains cluster in *A. infectoria* (Fig. 42 on p. 381).

References. Ellis (1971), Simmons (1995a, b), Anderson & Thrane (1996).

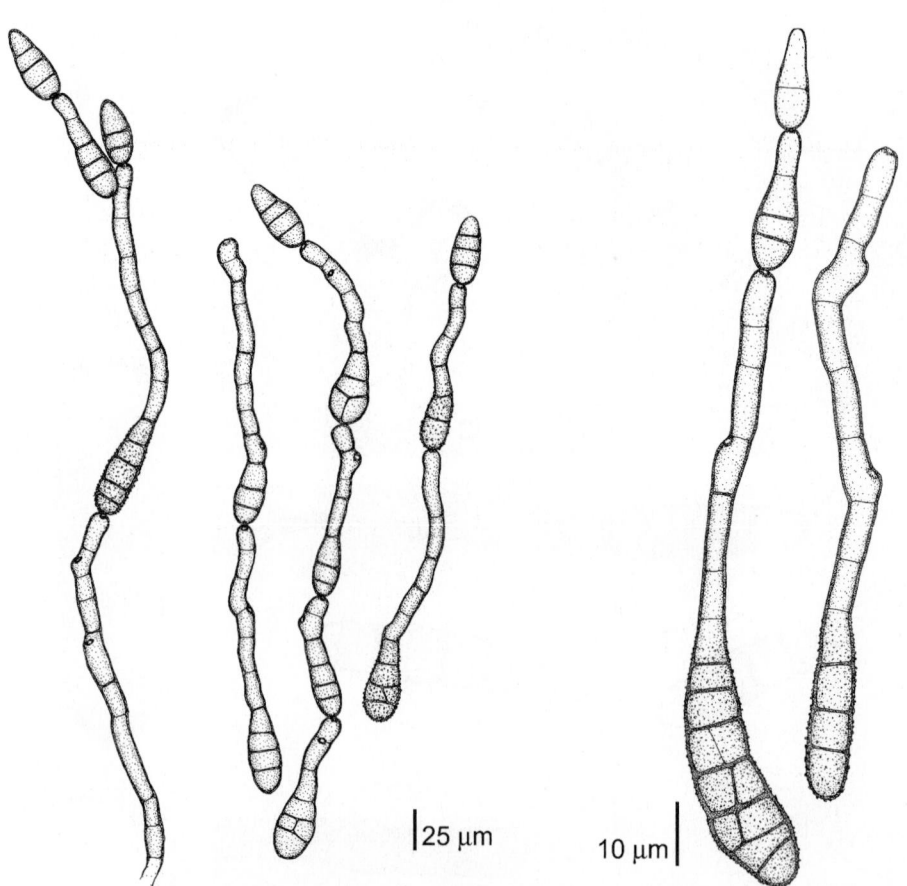

Alternaria infectoria, CBS 210.86. Conidiophore and conidial chains. a. Secondary conidiophores ('false beaks'); b. conidia.

Antifungal susceptibility.

Antifungal	Mean MICs	Strains	Reference
AMB	1.4	2	Unpublished data
5FC	256	2	Unpublished data
FLZ	32	2	Unpublished data
ITZ	2.8	2	Unpublished data
KTZ	1.4	2	Unpublished data
MCZ	4	2	Unpublished data

Nomenclature. *Alternaria infectoria* Simmons - Mycotaxon 25: 298, 1986.

Pleospora infectoria Fuckel - Jahrb. Nassau. Ver. Naturk. 23-24: 132, 1870 ≡ *Lewia infectoria* (Fuckel) M.E. Barr & Simmons, *in* Simmons - Mycotaxon 25: 296, 1986.

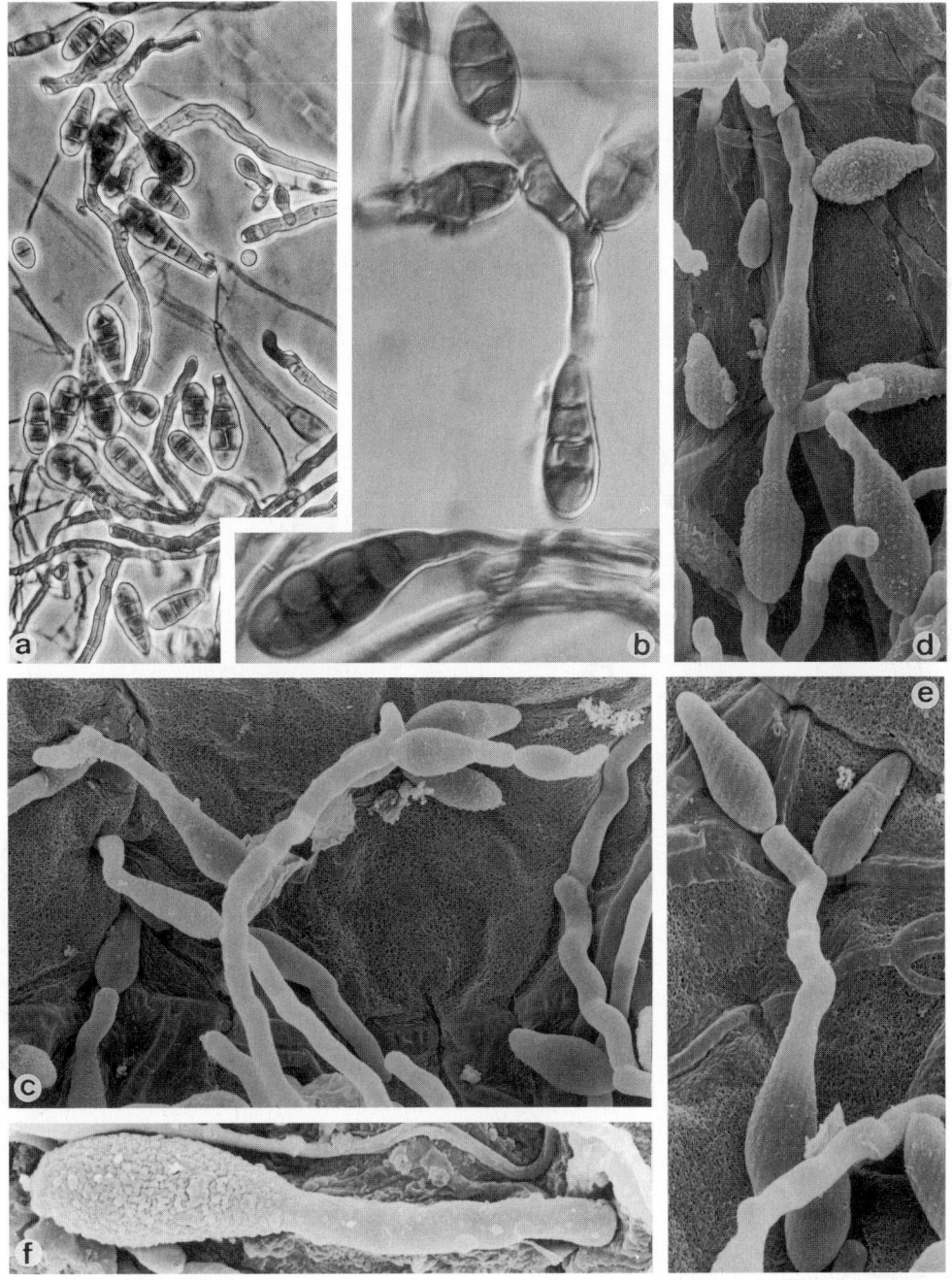

Alternaria infectoria, CBS 210.86. Conidiophores, conidial chains and liberated conidia. a. ×510; b. ×1280; c. ×1200; d. ×1360; e. ×1760; f. ×2400.

Alternaria longipes (Ellis & Everh.) Mason

Colony characteristics. Colonies (PCA) expanding, dark grey to olivaceous black, powdery.

Microscopy. Conidiophores mostly unbranched, with one or few conidial scars, up to 80 μm long, 3-5 μm wide. Conidia obclavate, mostly with a beak up to one third of the length of the conidial body, 35-110 × 11-21 μm, dark brown, smooth-walled or finely warted, with muriform septation, with a single scar at the tip, arising in mostly unbranched chains of ten or more.

Differential diagnosis. The species has the same ITS profile as *A. alternata*, but differs by darker conidia with longer beaks; conidia are mostly smooth-walled.

Pathogenicity. BSL-1. A cutaneous mycosis was reported by Gené *et al.* (1995).

Reference. Simmons (1981).

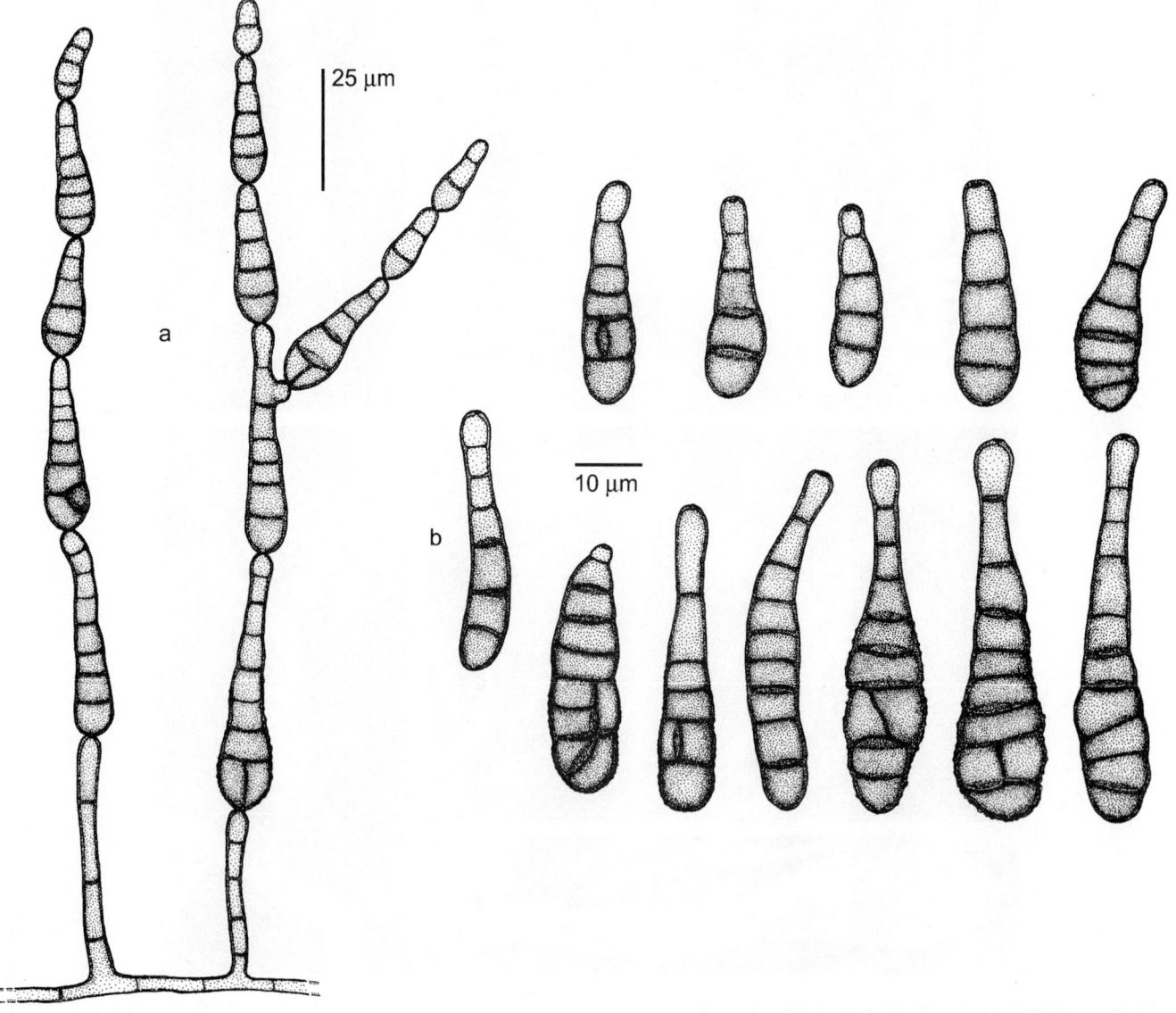

Alternaria longipes, CBS 655.96. a. Conidiophores bearing conidial chains; b. conidia.

Antifungal susceptibility.

Antifungal	Mean MICs	Strains	Reference
AMB	1	5	Unpublished data
5FC	256	5	Unpublished data
FLZ	42.5	5	Unpublished data
ITZ	0.35	5	Unpublished data
KTZ	2	5	Unpublished data
MCZ	3.48	5	Unpublished data

Nomenclature. *Macrosporium longipes* Ellis & Everhart - J. Mycol. 7: 134, 1892 ≡ *Alternaria longipes* (Ellis & Everhart) Mason - Mycol. Pap. 2: 19, 1928.

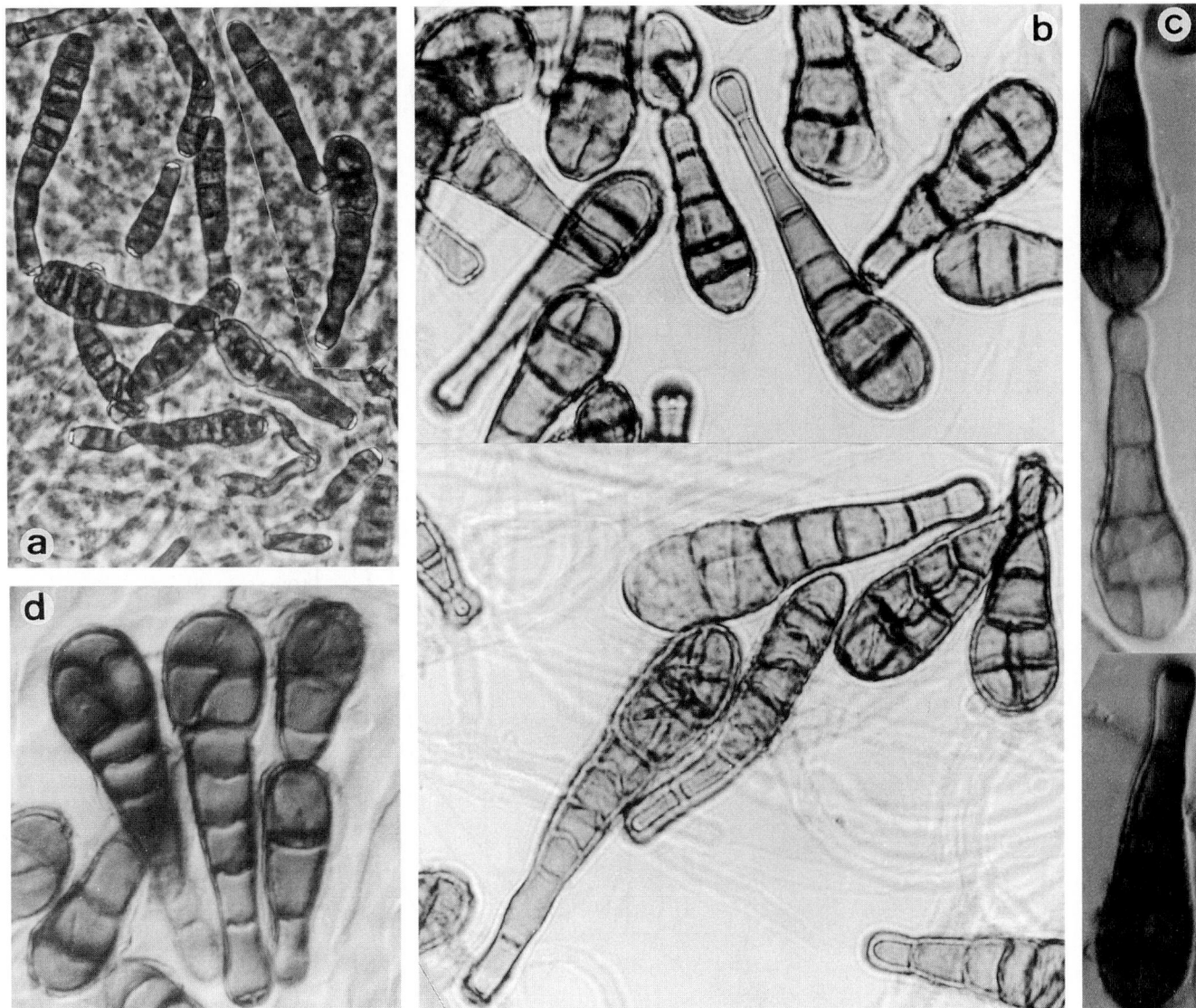

Alternaria longipes, CBS 655.96. Conidia. a. ×640; b. ×1280; c, d. ×1600.

Alternaria tenuissima (Kunze:Pers.) Wiltsh.

Colony characteristics. Colonies (PCA) growing rapidly, velvety.

Microscopy. Conidiophores solitary or in groups, simple or branched, straight or flexuose, pale golden brown, smooth-walled, up to 115 μm long, 4-6 μm wide. Conidia arising in unbranched chains of 3-5, straight or curved, obclavate or tapering gradually into a beak which is up to half the length of the conidium, golden brown, smooth-walled or slightly verruculose, generally with 4-7 transverse and one or two longitudinal septa, 22-95 × 8-19 μm.

Differential diagnosis. It has more slender conidia than *Alternaria alternata*. It cannot be distinguished using ITS sequencing data.

Molecular diagnostics. For ITS restriction map, see under *A. alternata* (p. 423).

Pathogenicity. BSL-1. It is a cosmopolitan species, occurring on a wide range of plants as a secondary invader. Cutaneous infections were reported by Camenen *et al.* (1988), Badillet (1991), Panagiotidou *et al.* (1991), Castanet *et al.* (1995) and Romano *et al.* (1996). Lesions are mostly papulonodular, and the organism often occurs with large ellipsoidal cells in tissue (Viviani *et al.*, 1986). It is generally found in patients with underlying disease (Contet-Audonneau *et al.*, 1991a; Drouhet *et al.*, 1991b).

References. Ellis (1971), Domsch *et al.* (1980), Simmons (1995).

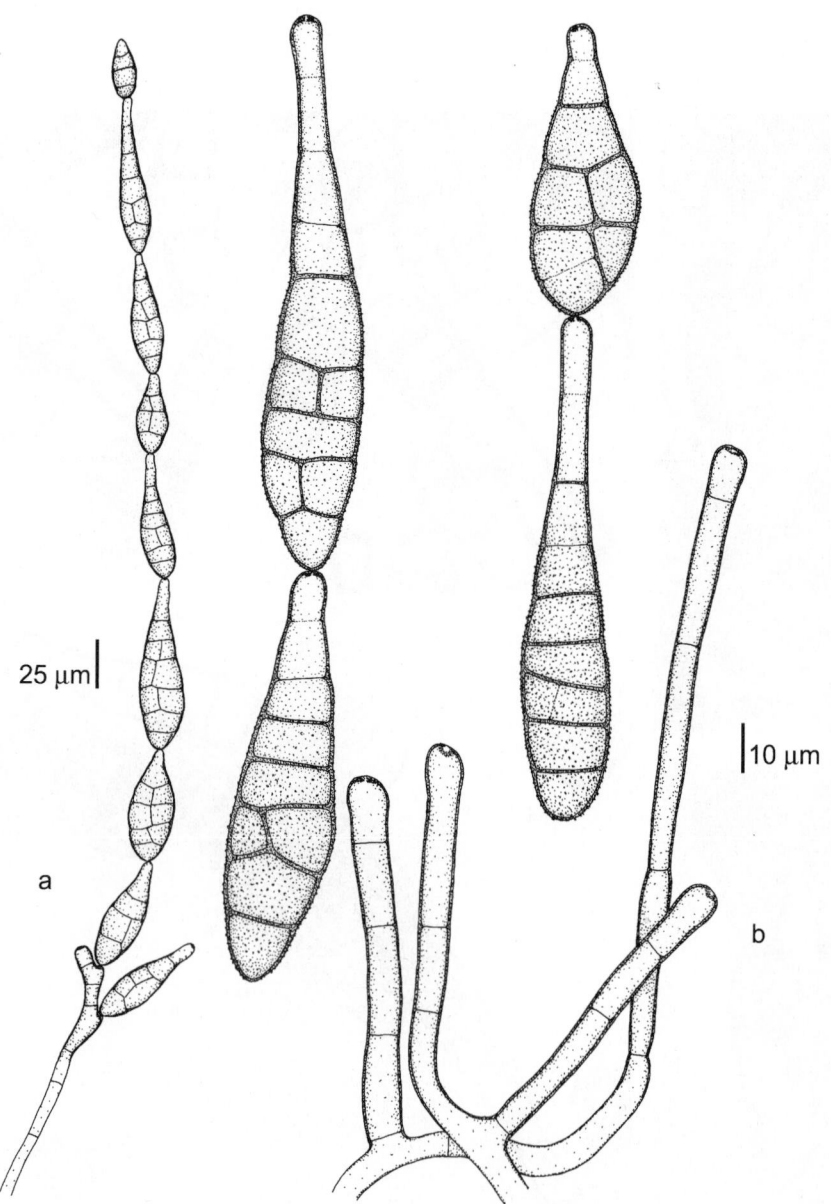

Alternaria tenuissima, CBS 878.95. a. Chain of conidia at low magnification; b. conidiophores and conidia.

Antifungal susceptibility.

Antifungal	Mean MICs	Strains	Reference
AMB	1.54	8	Unpublished data
5FC	256	8	Unpublished data
FLZ	58.6	8	Unpublished data
ITZ	0.7	8	Unpublished data
KTZ	2.38	8	Unpublished data
MCZ	4	8	Unpublished data

Nomenclature. *Helminthosporium tenuissimum* Kunze, *in* C.G. Nees & T.F.L. Nees - Nova Acta Acad. Caesar. Leop. Carol. 9: 242, 1818 ≡ *Helminthosporium tenuissimum* Kunze:Persoon - Mycol. Eur. 1: 18, 1822 ≡ *Alternaria tenuissima* (Kunze:Persoon) Wiltshire - Trans. Br. Mycol. Soc. 18: 157, 1933.

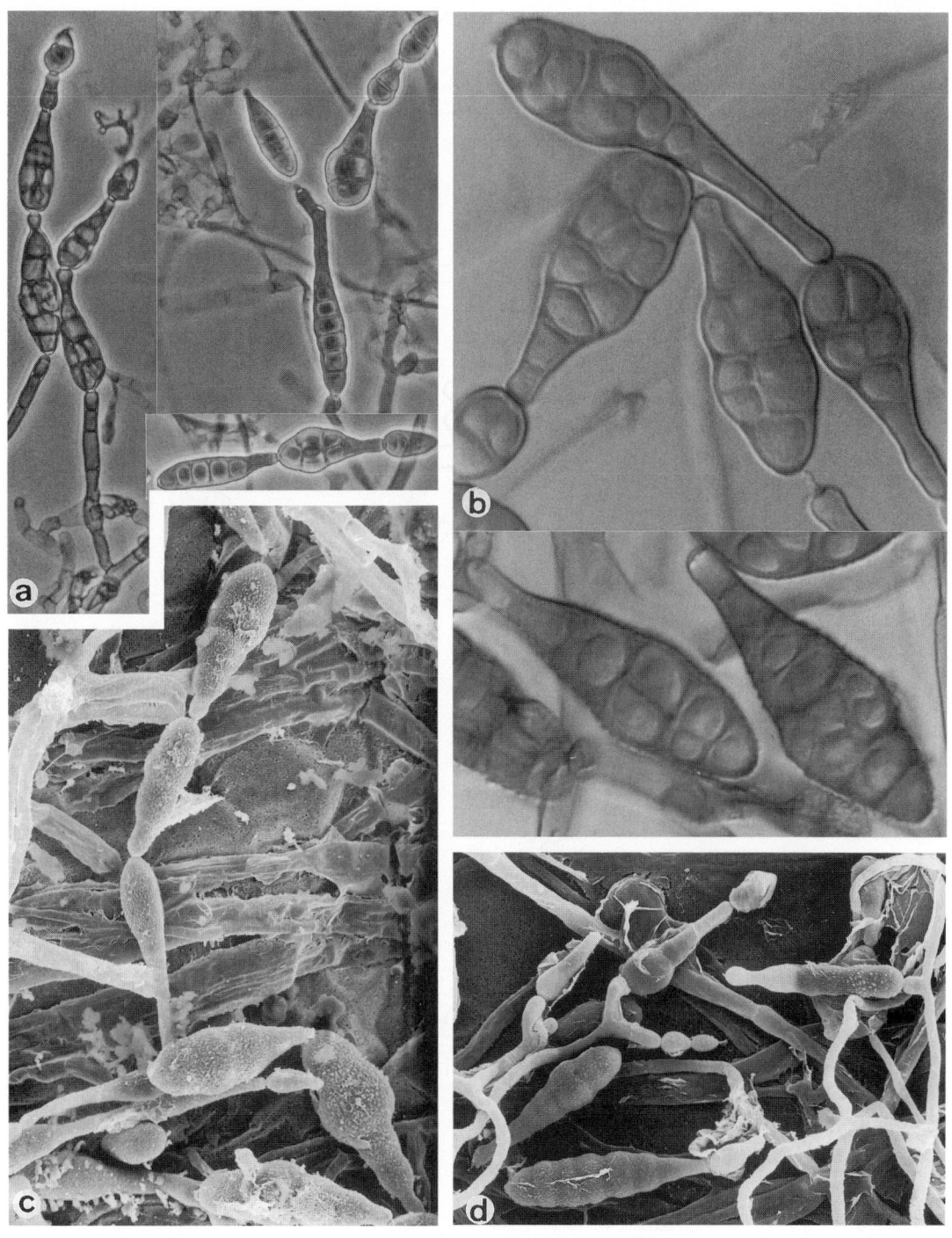

Alternaria tenuissima, CBS 878.95. Conidiophores and conidia. a. ×640; b. ×1600; c. ×800; d. ×1250; e. ×950; f. ×870.

Hyphomycetes. Genus: *ANTHOPSIS*

Anthopsis deltoidea Filipello Marchisio *et al.*

Colony characteristics. Colonies (PDA) growing slowly, velvety to lanose, olivaceous-grey to mouse grey; reverse black.

Microscopy. Phialides ovoidal, ellipsoidal, subspherical or ampulliform, 5-8 × 2-3 μm, forming compact lateral clusters on undifferentiated hyphae; generally the distinctive collarette is located at the base of the phialide. Conidia triangular, smooth-walled, 2.0-3.5 μm long, usually adhering in dense masses.

Pathogenicity. BSL-1. Olecranon bursitis (Kwon-Chung & Droller, 1984), subcutaneous infection (Badillet *et al.*, 1986).

References. Filipello Marchisio *et al.* (1976), Kwon-Chung & Droller (1984).

Nomenclature. *Anthopsis deltoidea* Filipello Marchisio, Fontana & Luppi Mosca - Can. J. Bot. 55: 117, 1976.

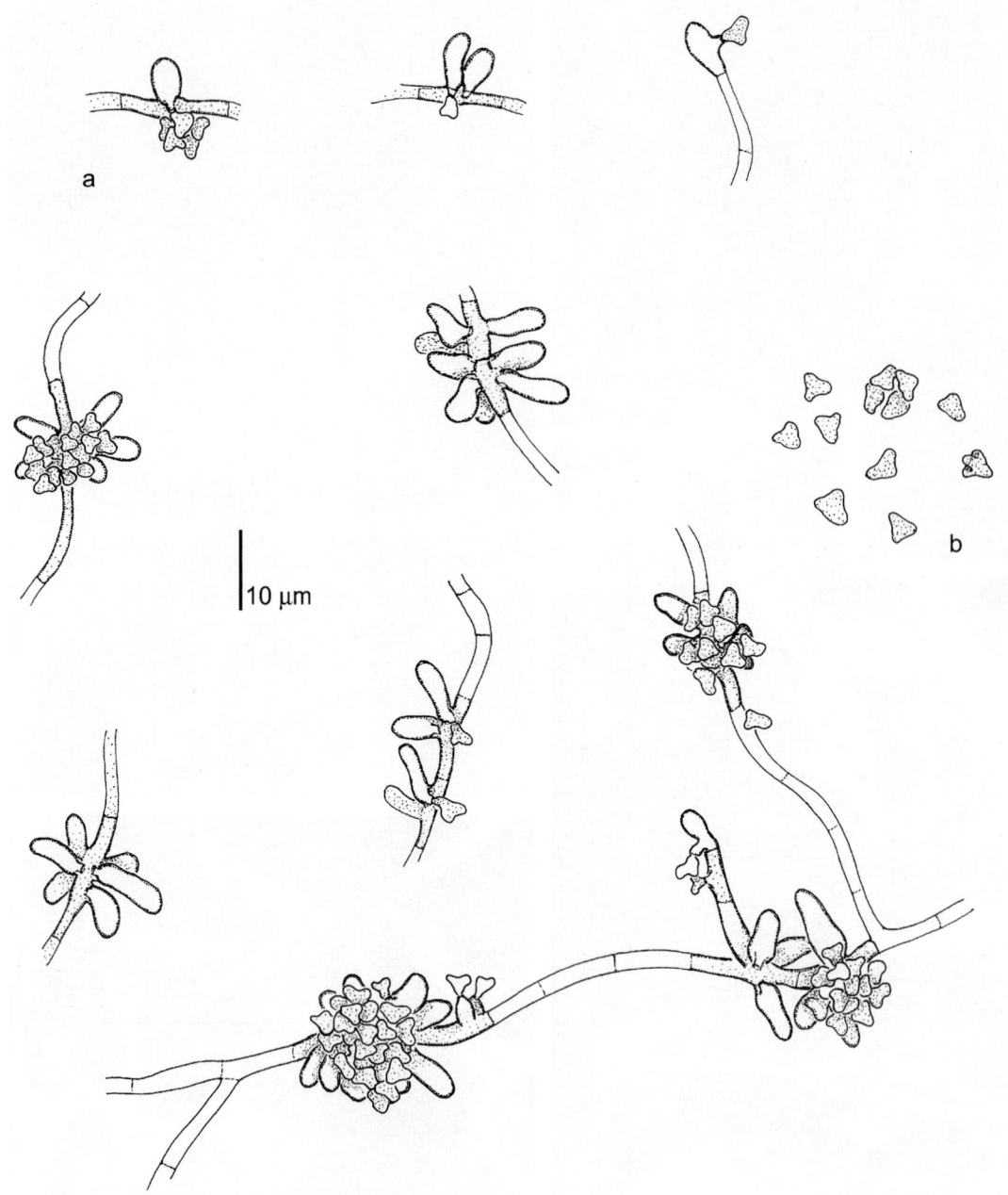

Anthopsis deltoidea, CBS 263.77. a. Conidiogenous cells; b. conidia.

436

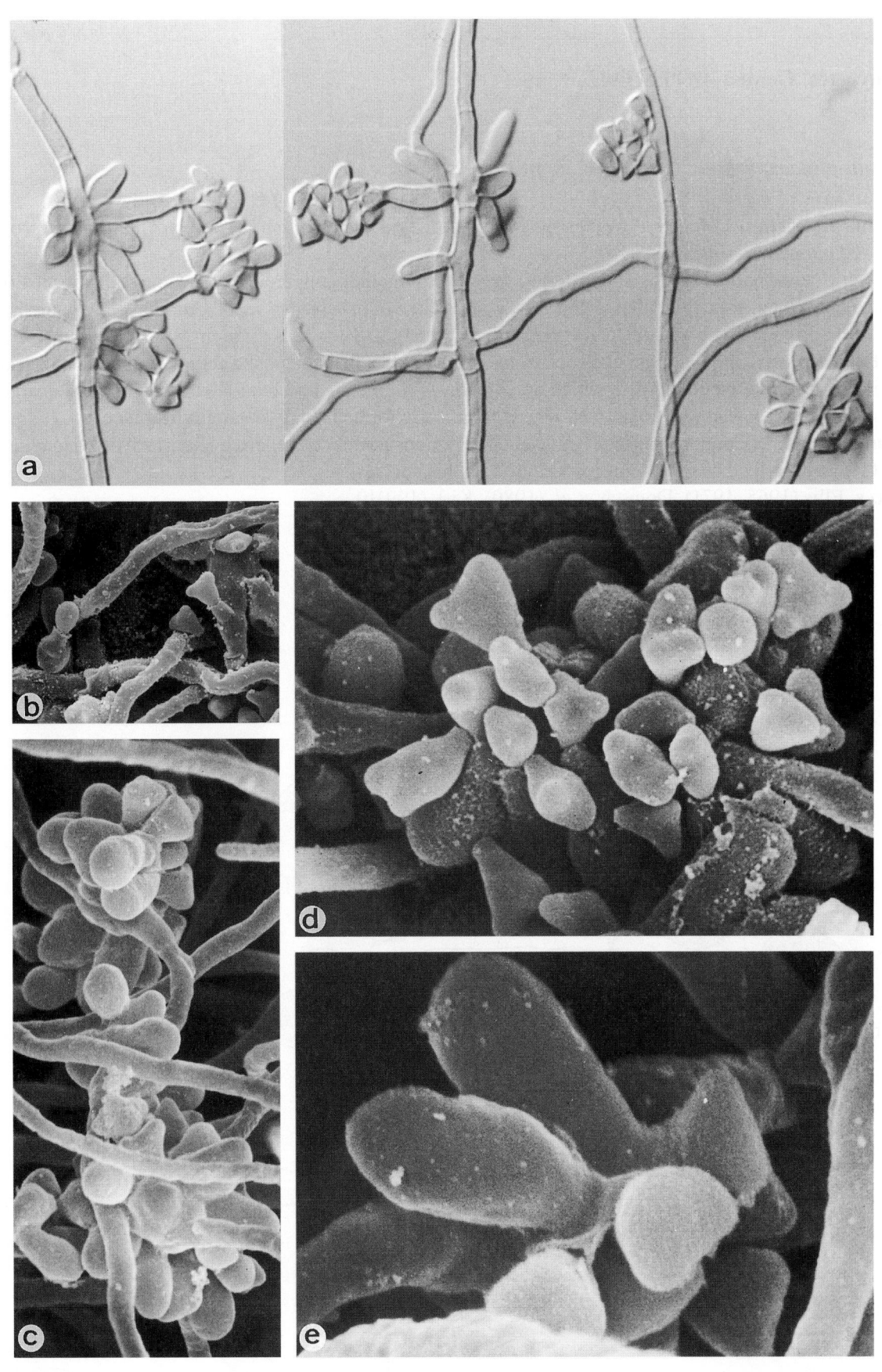

Anthopsis deltoidea, CBS 263.77. Conidiogenous cells and conidia. a. ×1600; b. ×2200; c. ×4350; d. ×7200; e. ×14500.

Hyphomycetes. Genus: *ARTHRINIUM*

Arthrinium phaeospermum (Corda) M.B. Ellis

Colony characteristics. Colonies (OA) fast-growing, with floccose, whitish aerial mycelium, often producing a diffusible red pigment.

Microscopy. Conidiophores arising singly from a lageniform mother cell, erect or ascending, flexuose, 1.0-1.5 μm wide, up to 65 μm long, smooth-walled, colourless, with hyaline or pale brown septa. Conidia lenticular, dark brown, 8-12 × 5-7 μm, provided with an equatorial germ slit.

Differential diagnosis. Sporulation mostly occurs massively, so that the diagnostic conidiophores are not easily seen. The hyaline, thread-like conidiophores which arise from swollen cells and bear lateral conidia, are characteristic. Cultures may resemble *Nigrospora* (p. 777), where conidia are formed singly on ampulliform cells.

Pathogenicity. BSL-1. Common saprobe. Cutaneous infections have been reported (Rai, 1989b; Zhao *et al.*, 1990). Agostini (1932) described an onychomycosis caused by this species.

References. Ellis (1965, 1971), Domsch *et al.* (1980), Kirk (1991b).

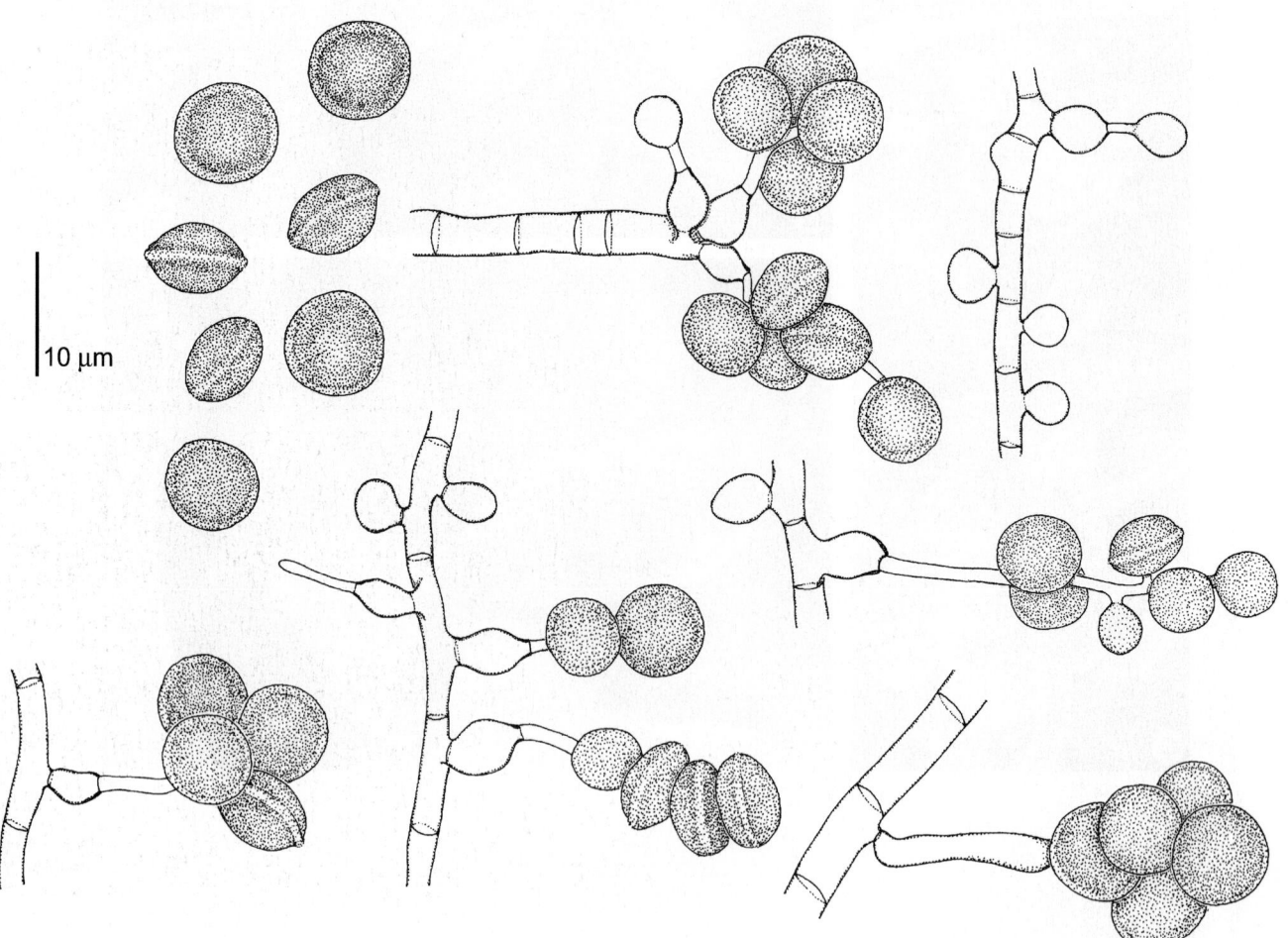

10 μm

Arthrinium phaeospermum, IMI 253199. Conidiophores and conidia.

Antifungal susceptibility.

Antifungal	MICs	Strains	Reference
AMB	0.13	1	Unpublished data
5FC	256	1	Unpublished data
FLZ	128	1	Unpublished data
ITZ	0.5	1	Unpublished data
KTZ	2	1	Unpublished data
MCZ	2	1	Unpublished data

Nomenclature. *Gymnosporium phaeospermum* Corda - Ic. Fung. 1: 1, 1837 ≡ *Arthrinium phaeospermum* (Corda) M.B. Ellis - Mycol. Pap. 103: 8, 1965.
    Coniosporium onychophilum Agostini - Atti Ist. Bot. Lab. Crittog. Univ. Paria, Ser. 4a, 3: 35, 1932.
    Arthrinium phaeospermum (Corda) M.B. Ellis var. *indicum* K.R. Khan & Sullia - Acta Bot. Ind. 8: 103, 1980.

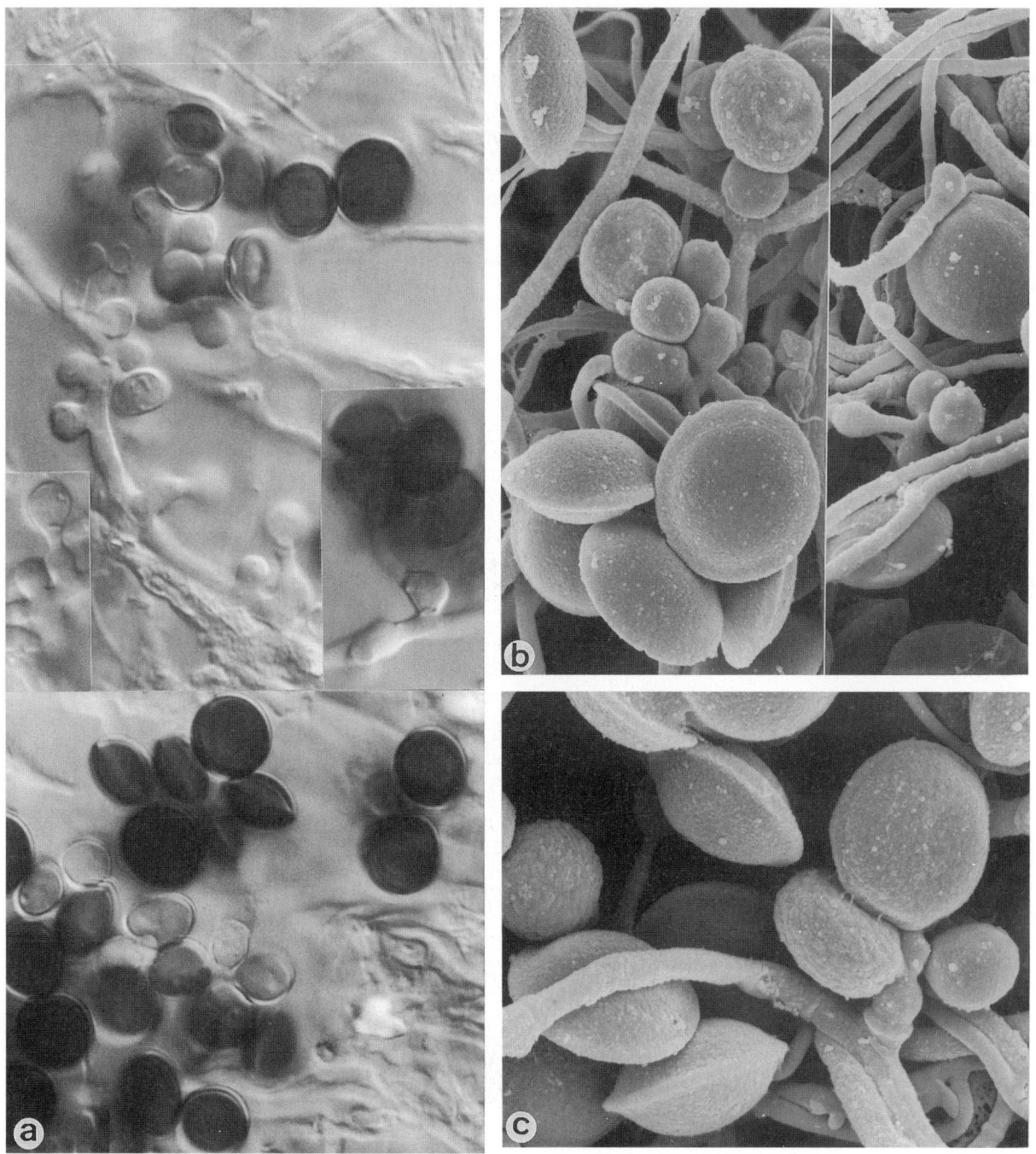

Arthrinium phaeospermum, IMI 253199. Conidiophores and conidia. a. ×1600; b. ×2600; c. ×3500.

Hyphomycetes. Genus: *ARTHROGRAPHIS*

Arthrographis kalrae (Tewari & Macpherson) Sigler & Carmichael

Colony characteristics. Colonies (SGA) spreading, cream-coloured to tan.

Microscopy. Conidiophores (sub)hyaline, narrow, branched, often in bundles, occasionally forming whitish, 0.5 cm large, linear synnemata. Arthroconidia 1-celled, hyaline, smooth-walled, cylindrical, 3-7 × 2.0-3.5 μm, dry. In addition, hyaline, spherical, 3.5-5.5 μm diam, sessile blastoconidia are formed on undifferentiated hyphae.

Teleomorph. *Eremomyces langeronii* (von Arx) Malloch & Sigler (*Ascomycota, Euascomycetes, Eurotiales: Eremomycetaceae*).

Cleistothecia spherical, 75-160 μm diam, with hyphal outer wall layer and pseudoparenchymatous inner wall layer, initially yellowish, finally black. Asci formed in dense rows, ellipsoidal, thin-walled, rather persistent, 8-13 × 4-7 μm. Ascospores ellipsoidal-fusiform, unilaterally flattened, golden brown, 3.5-5.0 × 2.0-2.5 μm.

Remark. Molecular data do not support the connection *E. langeronii* / *A. kalrae;* two different species may be concerned (Gené *et al.*, 1996b).

Physiology. Keratinolytic; *in vitro* hair perforation positive; growth at 37°C. Tolerant to cycloheximide, intolerant to benomyl.

Pathogenicity. BSL-2. Cochet (1939) reported the fungus from a case of onychomycosis, but Tewari & Macpherson (1968) observed neurotropism in artificial inoculation. A keratitis was reported by Perlman & Binns (1997) and a mycetoma by Degrave *et al.* (1997). Two infections in immunocompromised patients were presented by Delage *et al.* (1998).

References. Tewari & Macpherson (1971), Sigler & Carmichael (1976, 1983), Malloch & Sigler (1988), Gené *et al.* (1996b).

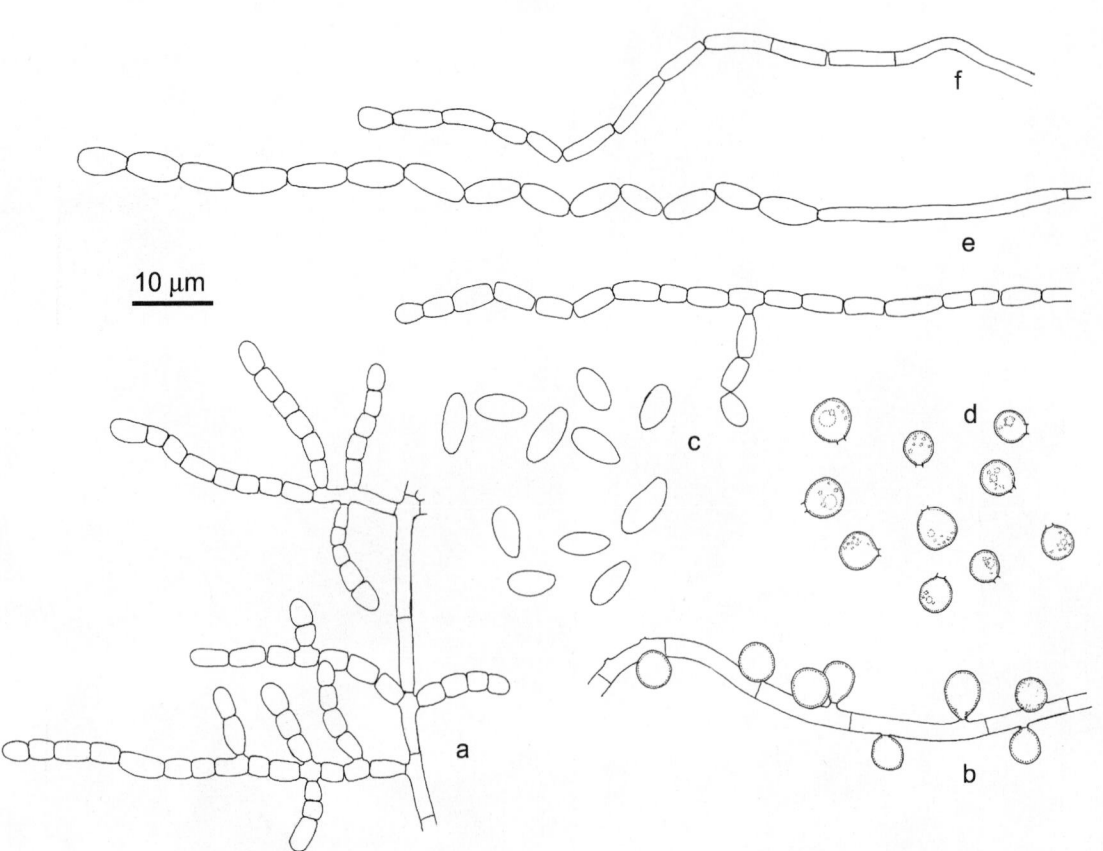

Arthrographis kalrae, a-d. CBS 693.77; e. CBS 203.78; f. CBS 112.39. Fresh culture. a. branching pattern; b. lateral sessile conidia; c. mature arthroconidia; d. liberated lateral conidia. e, f. cultures after repeated transfer.

Nomenclature. *Arthrographis langeronii* Cochet - Annls Parasit. Hum. Comp. 17: 97, 1939.

Oidiodendron kalrae Tewari & Macpherson - Mycologia 63: 603, 1971 ≡ *Arthrographis kalrae* (Tewari & Macpherson) Sigler & Carmichael - Mycotaxon 4: 360, 1976.

Pithoascus langeronii von Arx - Persoonia 10: 24, 1978 ≡ *Pithoascina langeronii* (von Arx) Valmaseda, Martínez & Barrasa - Can. J. Bot. 65: 1805, 1987 ≡ *Eremomyces langeronii* (von Arx) Malloch & Sigler - Can. J. Bot. 66: 1931, 1988.

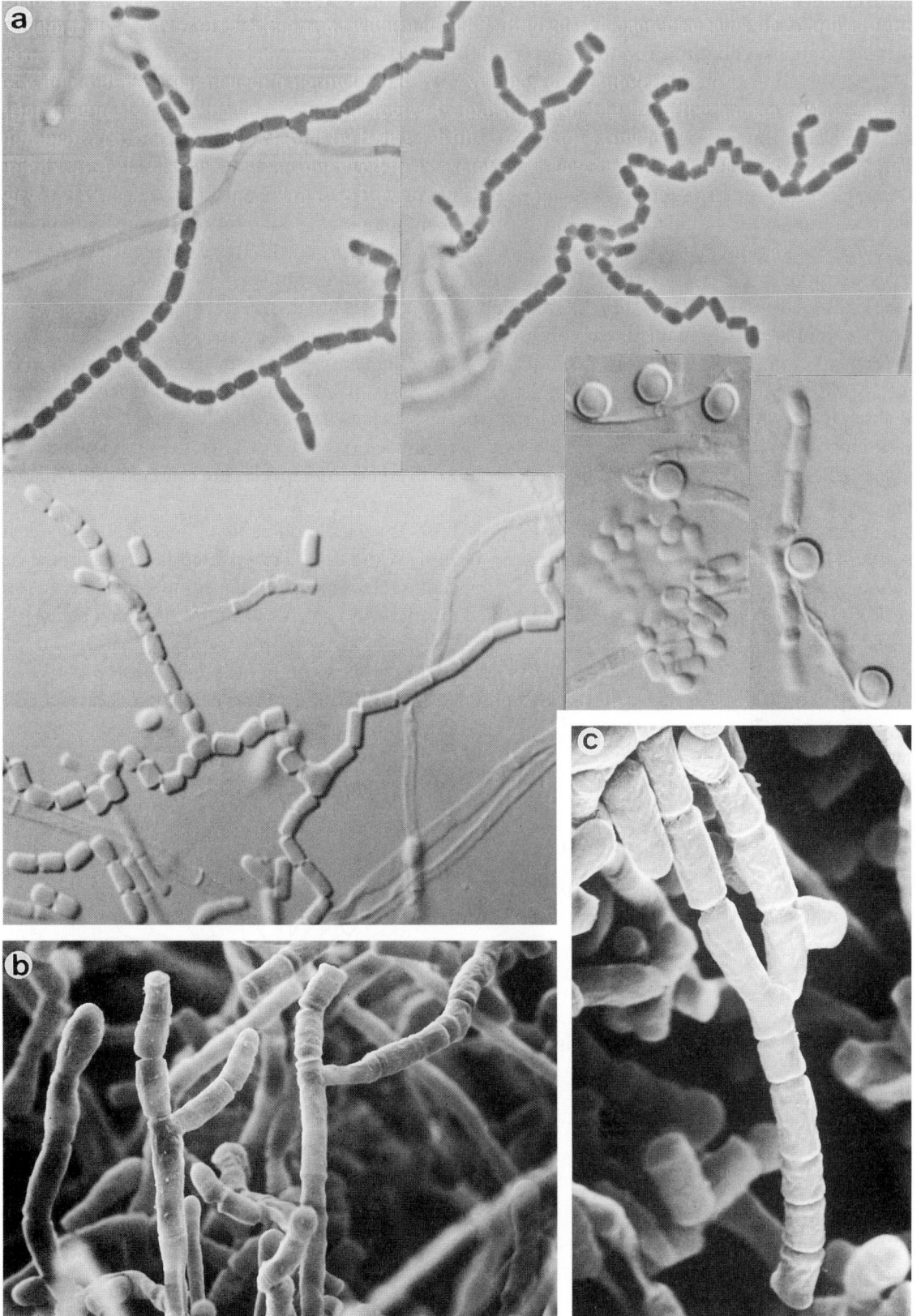

Arthrographis kalrae, UAMH 4472. Fertile hyphae forming arthroconidia, and lateral, sessile blastoconidia. a. ×600; b. ×3000; c. ×5100.

Hyphomycetes. Genus: *ASPERGILLUS*

Generic description. Colonies usually growing rapidly, powdery, white, green, yellowish, brown or black. Conidiophores erect, unbranched, with swollen apical vesicles and without septum at the base; the part of the supporting hypha which is continuous with the stipe is known as the **foot cell**. Phialides borne directly on the vesicle **(uniseriate)** or on subtending metulae **(biseriate)**; phialide openings without collarettes. Conidia produced in dry chains, forming columns **(columnar)** or diverging **(radiate)**, from flask-shaped phialides, 1-celled, smooth-walled or ornamented, (sub)hyaline or pigmented. Cells with extremely thick walls **(Hülle cells)** or sclerotia may be present. Intolerant to benomyl.

Teleomorphs. *Chaetosartorya, Dichlaena, Eurotium, Emericella, Fennellia, Hemicarpenteles, Neosartorya, Petromyces, Sclerocleista, Warcupiella* (*Ascomycota, Euascomycetes, Eurotiales: Trichocomaceae*).

General remarks. This is a large genus of common contaminants, containing about 175 species. Quite a diversity of species has been implicated in human mycoses (Pitt, 1994), although these fungi normally inhabit other ecological niches. A few species, particularly *A. flavus* (p. 470) and *A. fumigatus* (p. 473) are frequent causative agents of pulmonary aspergillosis, either allergic, or invasive or a combination of the two; for a description of the clinical patterns, see p. 27. Also cases of sinusitis are frequent. Many species are thermotolerant and their one-celled, dry, hydrophobic conidia are easily inhaled. Cerebral infections are often caused by species other than *A. flavus* or *A. fumigatus* (Pagano *et al.*, 1996); it should be noted that other filamentous fungi may also be concerned which are histologically indistinguishable. Culturing of the etiologic agent is mandatory. Below the *Aspergillus* species with published case reports are treated. Some osmotolerant species, such as *A. niger* (p. 489) and *A. terreus* (p. 509), are relatively commonly isolated as secondary invaders in cases of bacterial otitis externa; their pathogenic role has not clearly been established. Terminology used for identification of *Aspergillus* species is explained in the diagram below (Fig. 54).

Molecular diagnostics. Recognition of species using V7 to V9 variable domains of 18S rRNA was designed by Melchers *et al.* (1994) and mtDNA by Bretagne *et al.* (1995). Wang *et al.* (1998) used the cytochrome-b gene for differentiation of some clinically relevant species. Phylogenetic relationships of the entire genus were presented by Peterson (2000). Brenier-Pinchart *et al.* (1999) reviewed recent achievements of molecular diagnostics of invasive aspergillosis.

References. Raper & Fennell (1965), Samson (1979), VandenBossche *et al.* (1988), Samson *et al.* (1996), Samson & Pitt (2000), Seifert (2000). An *Aspergillus* web site is available on Internet: www.aspergillus.man.ac.uk.

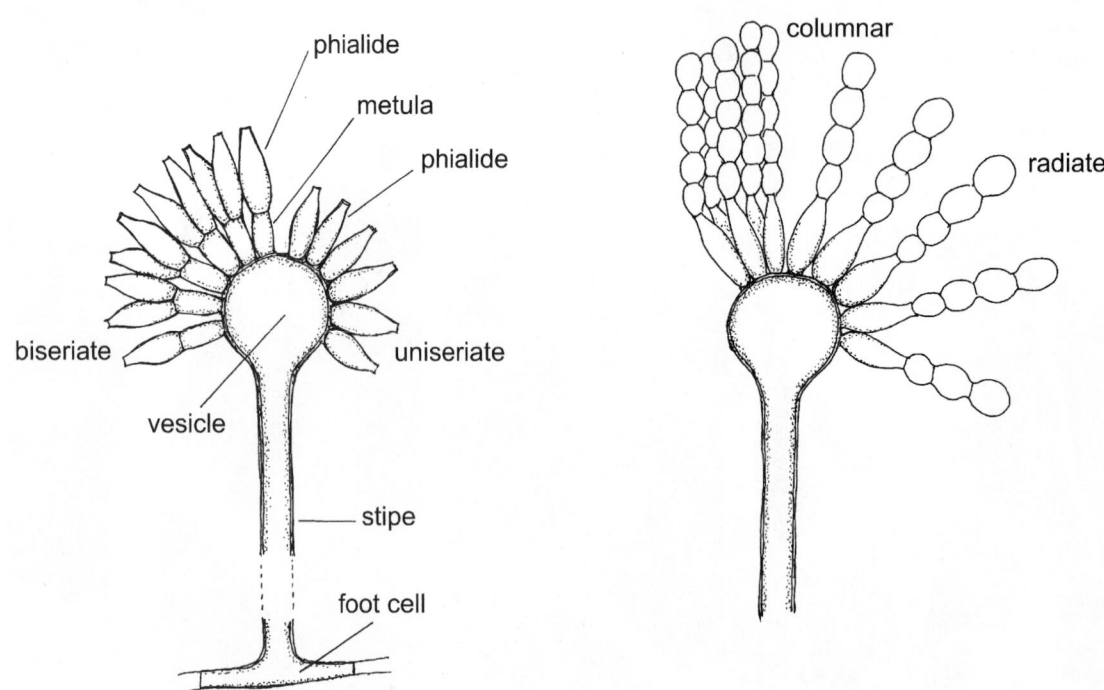

Fig. 54. Diagram of morphological terminology in *Aspergillus*.

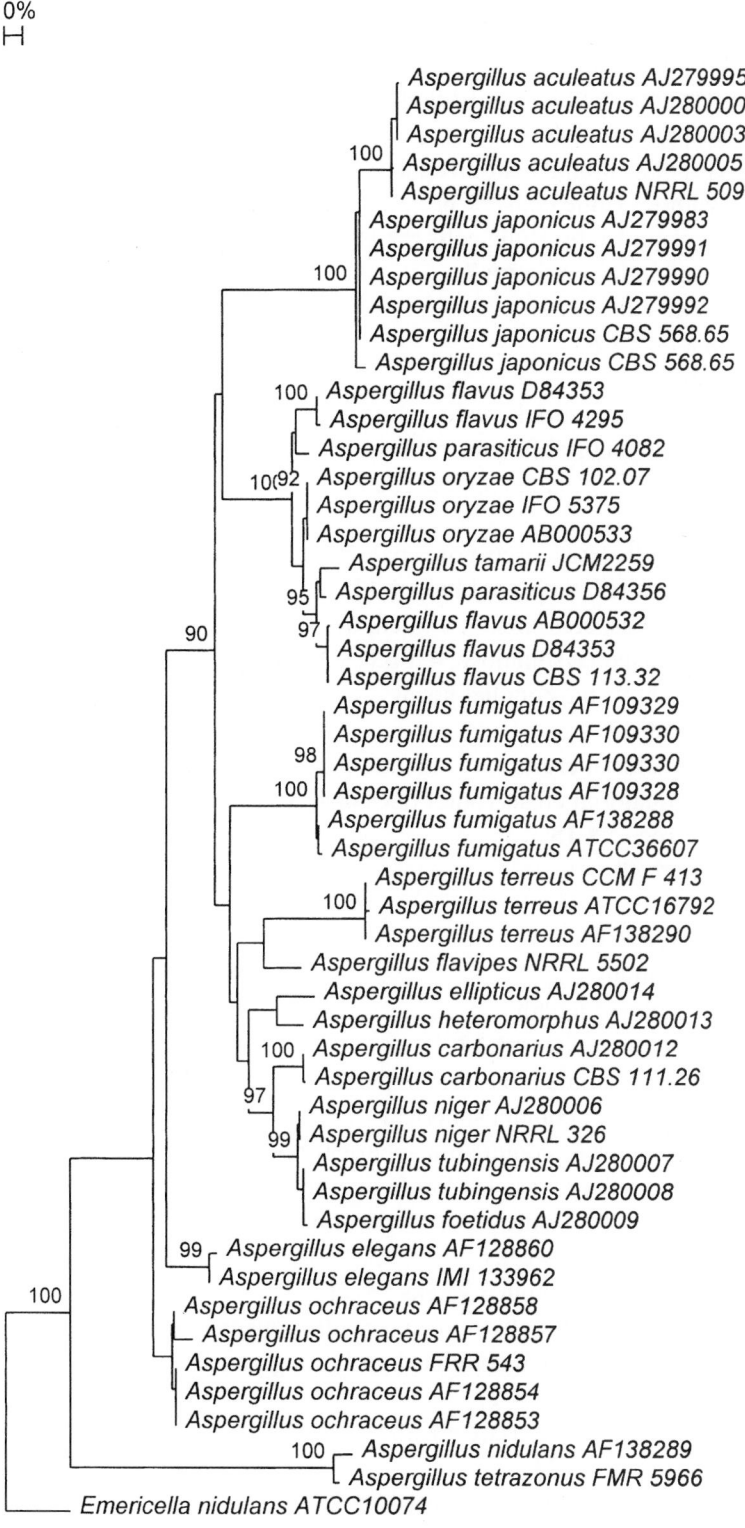

Fig. 55. Phylogenetic tree of *Aspergillus.*species based on confidently aligned ITS rDNA sequences using Neighbor joining algorithm with Kimura correction. Bootstrap values >90 from 100 resampled datasets are shown. Many species are clearly delimited from each other, which makes ITS rDNA a suitable target for species recognition.

Key to the treated species of Aspergillus:

1a. Ascomata present in culture ➞ **2**
1b. Ascomata lacking in culture ➞ **14**
2a. Colonies spreading; conidial heads strongly columnar; conidiogenous cells uniseriate; ascomata not surrounded by Hülle cells; ascospores with well developed equatorial crests ➞ **3**
2b. Above characteristics not combined ➞ **5**
3a. Convex surface of ascospores reticulate . . . . *Neosartorya fischeri* (292)
3b. Convex surface of ascospores spinulose or irregularly ribbed ➞ **4**
4a. Convex surface of ascospores spinulose . . . . *Neosartorya spinosa* (295)
4b. Convex surface of ascospores strongly and irregularly ribbed . . . . *Neosartorya pseudofischeri* (293)
5a. Wall of the ascomata without Hülle cells ➞ **6**
5b. Wall of the ascomata surrounded by Hülle cells ➞ **10**
6a. Ascospores less than 6 µm in long axis ➞ **7**
6b. Ascospores 6-7 µm in long axis . . . . *A. glaucus* (476)
7a. Ascospores up to 5 µm long ➞ **8**
7b. Ascospores 4.5-6 µm long . . . . *A. hollandicus* (480)
8a. Ascospores with pronounced equatorial crest . . . . *A. chevalieri* (458)
8b. Ascospores with a distinct equatorial furrow ➞ **9**
9a. Colonies red or brick-red . . . . *A. rubrobrunneus*[1]
9b. Colonies dull-green . . . . *A. reptans* (499)
10a. Ascomata yellow or green, clustered; ascospores pale ➞ **11**
10b. Ascomata purple; ascospores red, violet or blue ➞ **12**
11a. Conidiophores brownish; ascospores smooth-walled, 6-8 × 4.5-6.0 µm . . . . *A. flavipes* (468)
11b. Conidiophores hyaline; ascospores spinulose, 4.5-6.0 × 3.2-4.8 µm . . . . *A. niveus* (492)
12a. Spine-like hyphae arising above the conidial heads . . . . *A. unguis* (514)
12b. Spine-like hyphae absent ➞ **13**
13a. Ascospores with 2 thin equatorial crests . . . . *A. nidulans* (486)
13b. Ascospores with 4 short equatorial crests (outer ones observed with difficulty) . . . . *A. tetrazonus* (512)
14a. Colonies with shades of green ➞ **15**
14b. Colonies of other colour ➞ **30**
15a. Vesicles clavate ➞ **16**
15b. Vesicles mostly not clavate ➞ **17**
16a. Conidiophores 2-4 mm in length . . . . *A. clavatus* (462)
16b. Conidiophores less than 1 mm in length . . . . *A. clavato-nanicus* (460)
17a. Conidiogenous cells strictly uniseriate ➞ **18**
17b. Conidiogenous cells strictly biseriate or at least in part biseriate ➞ **22**
18a. Conidial heads densely columnar; conidia spherical . . . . *A. fumigatus* (473)
18b. Conidial heads radiate or columnar; conidia not spherical ➞ **19**
19a. Conidial heads radiate when young; vesicles fertile over upper half to two-thirds . . . . *A. penicilloides*[2]
19b. Conidial heads columnar; vesicles fertile on distal part only ➞ **20**
20a. Colonies attaining 4-5 cm diam in 3 weeks . . . . *A. caesiellus* (452)
20b. Colonies attaining less than 4 cm diam in 3 weeks ➞ **21**
21a. Conidial heads long and twisted, dark olive-green; conidia conspicuously roughened . . . . *A. restrictus* (464)
21b. Conidial heads shorter, pale grey-green; conidia slightly roughened . . . . *A. conicus* (466)
22a. Conidiogenous cells strictly biseriate ➞ **23**
22b. Conidiogenous cells biseriate and uniseriate ➞ **28**
23a. Conidiophore stipes dull-brown ➞ **24**
23b. Conidiophore stipes not or only slightly pigmented ➞ **25**
24a. Vesicles forming right angles to the conidiophore stipes . . . . *A. deflectus* (466)
24b. Without the above characteristic; rough-walled, encrusted spine-like hyphae present . . . . . . *A. unguis* (514)
25a. Colonies evenly yellow-green or blue-green ➞ **26**
25b. Colonies green and white . . . . *A. janus* (482)

444

26a. Colonies orange-yellow to yellow-green ➡ **27**
26b. Colonies blue-green with reddish-brown shades . . . . *A. sydowii* **(505)**
27a. Conidia 2.0-3.5 µm in diam; conidiophores hyaline or yellowish . . . . *A. versicolor* **(518)**
27b. Conidia 3.5-5.5 µm in diam; conidiophores tan to light brown . . . . *A. granulosus* **(478)**
28a. Colonies with dull brown shades with age . . . . *A. oryzae* **(496)**
28b. Colonies yellow-green or yellowish-olive ➡ **29**
29a. Conidia echinulate . . . . *A. flavus* **(470)**
29b. Conidia smooth-walled . . . . *A. avenaceus* **(450)**
30a. Conidiogenous cells strictly uniseriate; conidial heads radiate, black ➡ **31**
30b. Conidiogenous cells strictly biseriate or biseriate and uniseriate ➡ **32**
31a. Conidia (sub)spherical; vesicles 15-45 µm (commonly 20-35 µm) diam . . . . *A. japonicus* **(484)**
31b. Conidia subspherical to ellipsoidal; vesicles 35-100 µm (commonly 60-80 µm) diam . . *A. aculeatus* **(446)**
32a. Conidiogenous cells strictly biseriate ➡ **33**
32b. Conidiogenous cells biseriate and uniseriate ➡ **38**
33a. Colonies black . . . . *A. niger* **(489)**
33b. Colonies not black ➡ **34**
34a. Conidia conspicuously roughened . . . . *A. ustus* **(516)**
34b. Conidia smooth-walled to finely roughened ➡ **35**
35a. Conidial heads densely columnar . . . . *A. terreus* **(509)**
35b. Conidial heads radiate to loosely columnar ➡ **36**
36a. Colonies pale tan to pink; conidiophore stipes smooth-walled . . . . *A. carneus* **(456)**
36b. Colonies yellow to ochraceus; conidiophore stipes roughened or echinulate ➡ **37**
37a. Sclerotia pink to vinaceous-purple . . . . *A. ochraceus* **(494)**
37b. Sclerotia white to cream-coloured . . . . *A. sclerotiorum* **(503)**
38a. Colonies white . . . . *A. candidus* **(454)**
38b. Colonies not white ➡ **39**
39a. Colonies pale buff to ginger-brown; conidia smooth-walled . . . . *A. alliaceus* **(448)**
39b. Colonies dark brown; conidia echinulate . . . . *A. tamarii* **(507)**

[1]This species (as *Aspergillus sejunctus*) was implicated in human corneal ulcers (Shukla *et al.*, 1983).
[2]See under *A. restrictus* (p. 501).

Aspergillus aculeatus Iizuka

Colony characteristics. Colonies (CzA) growing rapidly, felty, purplish-black.

Microscopy. Conidial heads spherical to radiate, splitting into well-defined, divergent columns, black. Conidiophore stipes smooth-walled, hyaline or slightly pigmented at the apex, up to 4.5 mm high. Vesicles brown, subspherical, (35-)60-80 (-100) μm diam. Conidiogenous cells uniseriate. Conidia hyaline to brown, conspicuously echinulate, subspherical to ellipsoidal, 4-5 × 3.5-4.5 μm.

Molecular diagnostics. ITS restriction map based on NCBI U65309:

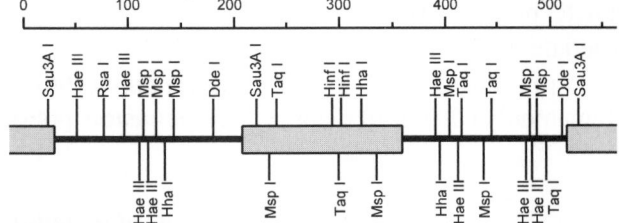

Differential diagnosis. The species is very close to *A. japonicus* (p. 484) but differentiated by larger vesicles and somewhat more elongate conidia. Its separation as an independent taxon was confirmed by Pařenicová *et al.* (2000) using molecular and metabolite data.

Pathogenicity. BSL-1. The species was isolated from scrapings of the tongue of a patient with respiratory illness (Williams *et al.*, 1984).

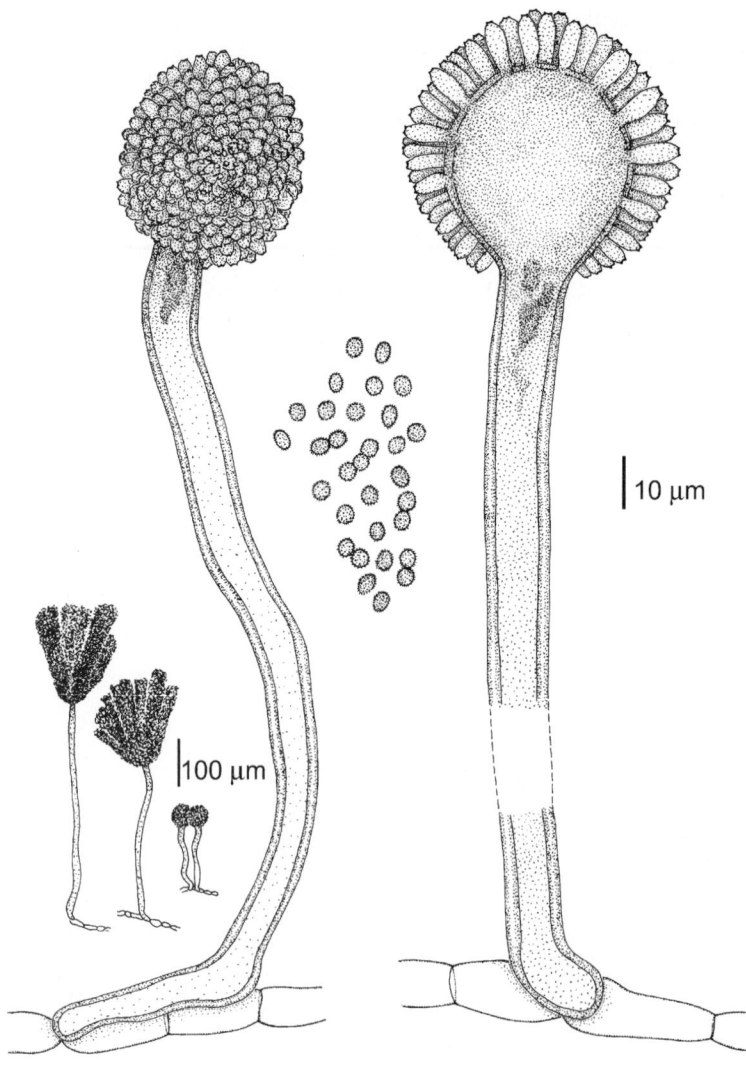

Aspergillus aculeatus, CBS 287.89. Conidiophores and conidia.

Antifungal susceptibility.

Antifungal	MICs	Strains	Reference
AMB	0.03	1	Unpublished data
5FC	4	1	Unpublished data
FLZ	4	1	Unpublished data
ITZ	0.25	1	Unpublished data
KTZ	0.5	1	Unpublished data
MCZ	2	1	Unpublished data

References. Al-Musallam (1980), Pařenicová *et al.* (2000).

Nomenclature. *Aspergillus aculeatus* Iizuka - J. Agric. Chem. Soc. Jpn. 27: 807, 1953 ≡ *Aspergillus japonicus* Saito var. *aculeatus* (Iizaki) Al-Musallam - Rev. Black Asp. Species p. 30, 1980.

Aspergillus aculeatus, CBS 287.89. Conidiophores and conidia. a. ×385; b. ×345; c. ×1010; d, e. ×960; f. ×2925; g. ×1200; h. ×1650.

Aspergillus alliaceus Thom & Church

Colony characteristics. Colonies (CzA) pale buff to ginger-brown, with grey to black sclerotial bodies.

Microscopy. Conidial heads radiate or columnar. Conidiophore stipes smooth-walled, up to 1.2 mm long, hyaline. Vesicles spherical, occasionally somewhat elongate. Conidiogenous cells biseriate on large vesicles, but on small vesicles often uniseriate. Metulae or phialides covering at least the upper half of the vesicle. Conidia ovoidal to subspherical, smooth-walled, yellow, 2.5-4.0 × 2.0-3.5 µm. Sclerotia, when present, black with white tips, ovoidal to ellipsoidal, 500-700 µm wide.

Teleomorph. *Petromyces alliaceus* Malloch & Cain (*Ascomycota, Euascomycetes, Eurotiales: Trichocomaceae*).

Pathogenicity. BSL-1. This species was the etiologic agent in a chronic otitis externa after surgery (Koenig *et al.*, 1985).

References. Raper & Fennell (1965).

Nomenclature. *Aspergillus alliaceus* Thom & Church - The Aspergilli p. 163, 1926.
  Petromyces alliaceus Malloch & Cain - Can. J. Bot. 50: 2623, 1972.

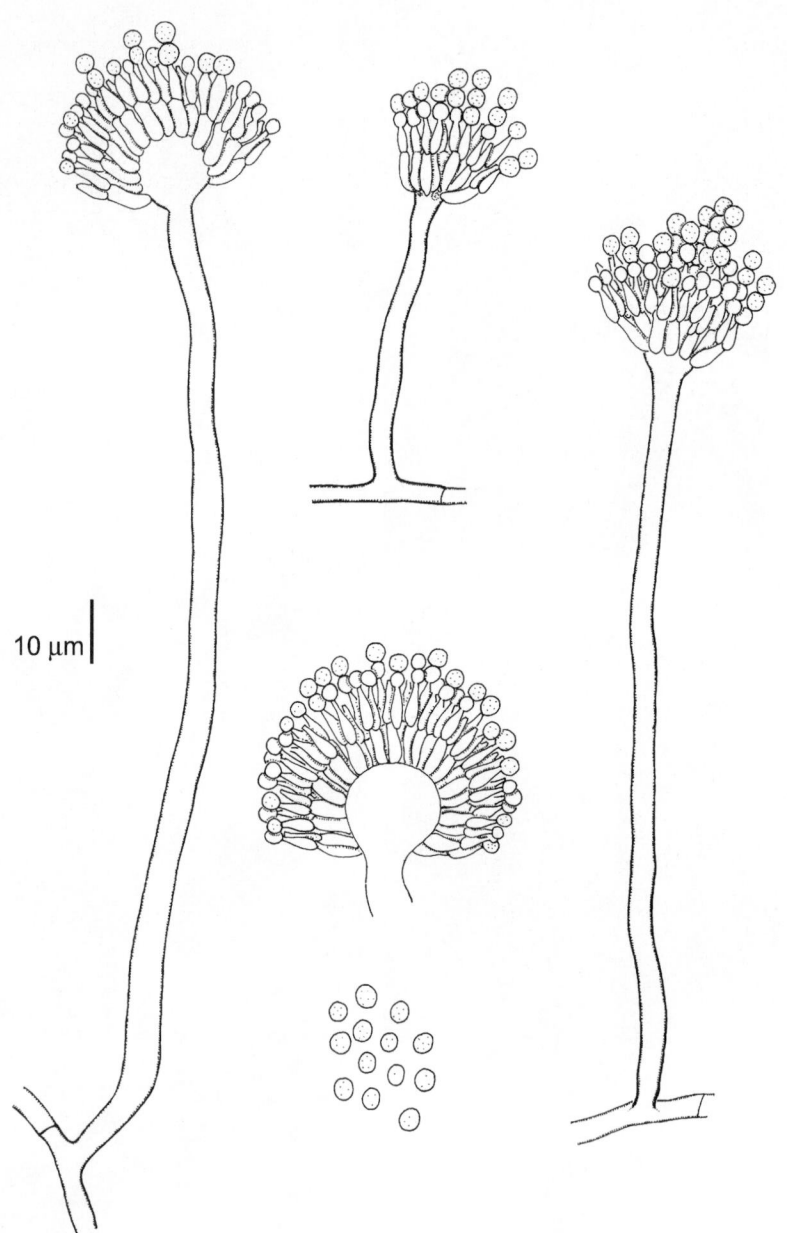

10 µm

Aspergillus alliaceus, IMI 87209. Conidiophores and conidia.

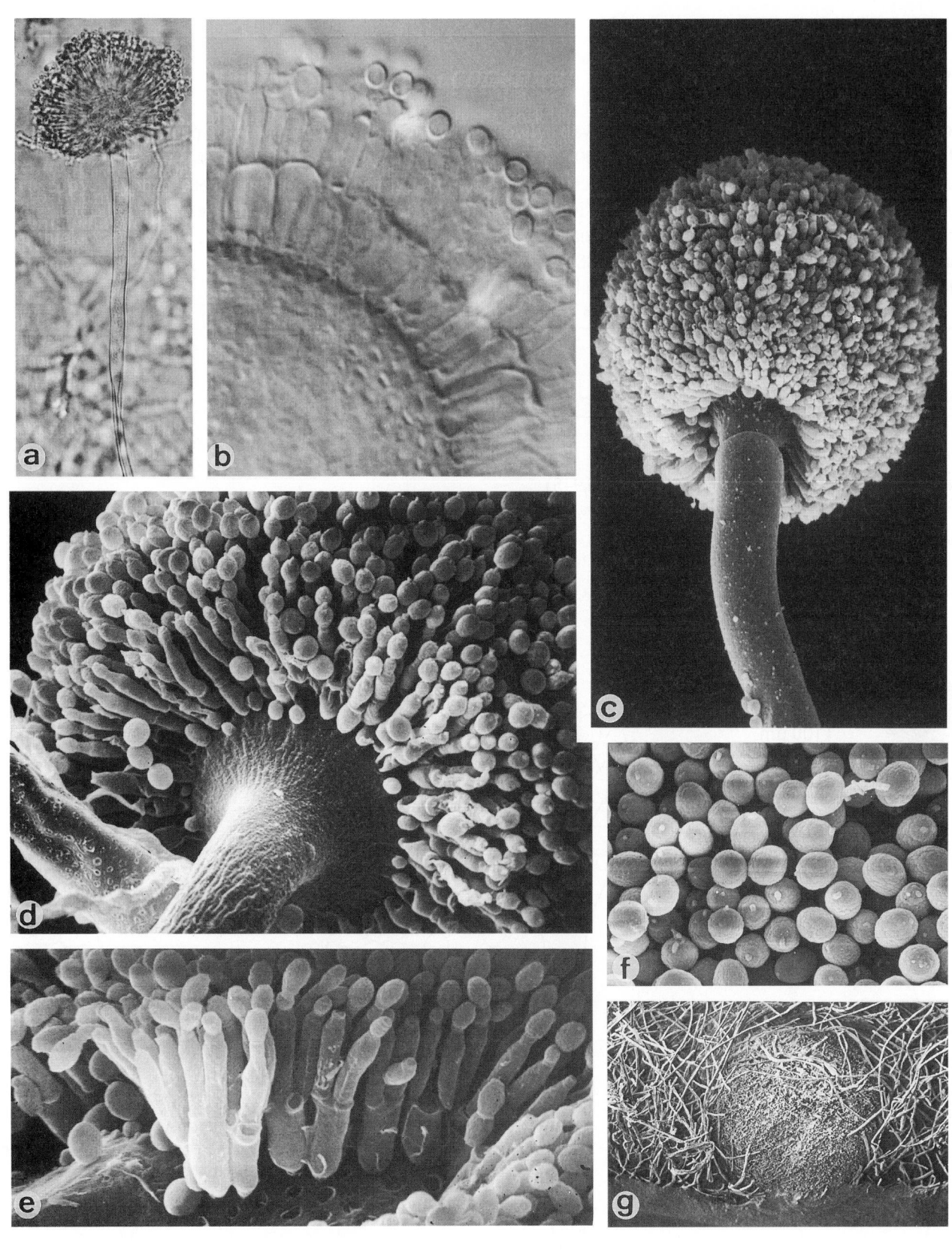

Aspergillus alliaceus, IMI 87209. a-f. Conidiophores and conidia; g. sclerotia. a. ×512; b. ×1600; c. ×1030; d. ×2050; e. ×3100; f. ×3300. g. ×55.

449

Aspergillus avenaceus G. Smith

Colony characteristics. Colonies (CzA) olivaceous yellow to olive, zonate; reverse with pinkish tinge.

Microscopy. Conidial heads radiate, up to 600 μm diam, becoming columnar. Conidiophore stipes smooth-walled to verrululose, up to 5 mm long, hyaline. Vesicles spherical or slightly flattened. Conidiogenous cells may be uniseriate but are mostly biseriate. Metulae up to 50 μm long, covering the entire surface of the vesicle. Conidia smooth-walled, yellow, ellipsoidal, 4-6 × 3.2-4.0 μm. Sclerotia greyish-brown to black, of irregular shape, up to 2-3 μm long.

Pathogenicity. BSL-1. Washburn *et al.* (1988) presented a case of sinusitis.

Molecular diagnostics. SSU restriction map based on NCBI AB008395:

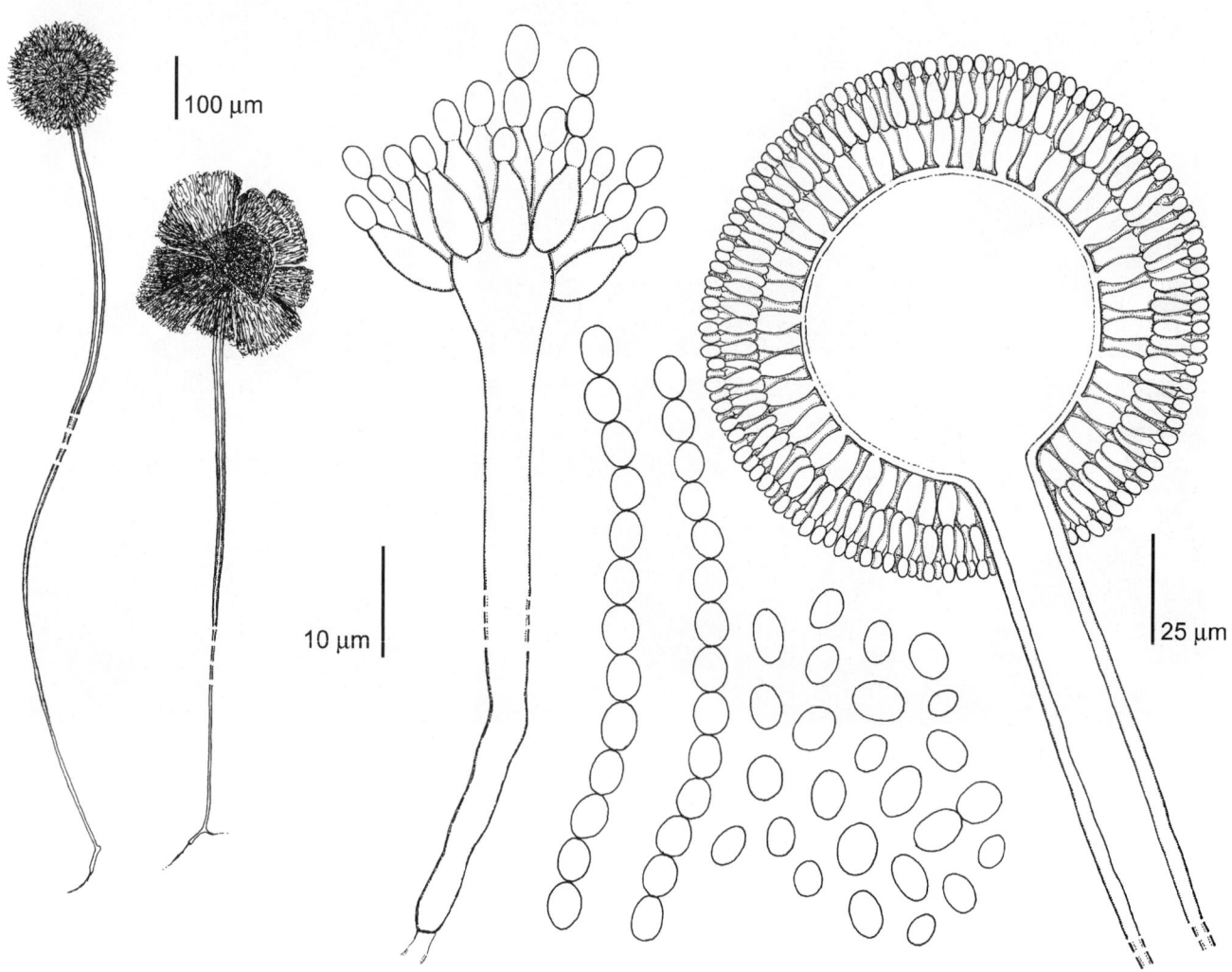

References. Smith (1943), Thom & Raper (1945).

Nomenclature. *Aspergillus avenaceus* G. Smith - Trans. Br. Mycol. Soc. 26: 24, 1943.

Aspergillus avenaceus, CBS 109.46. a. Habit sketch; b. conidiophore with uniseriate conidial head; c. conidia; d. conidiophore with biseriate conidial head.

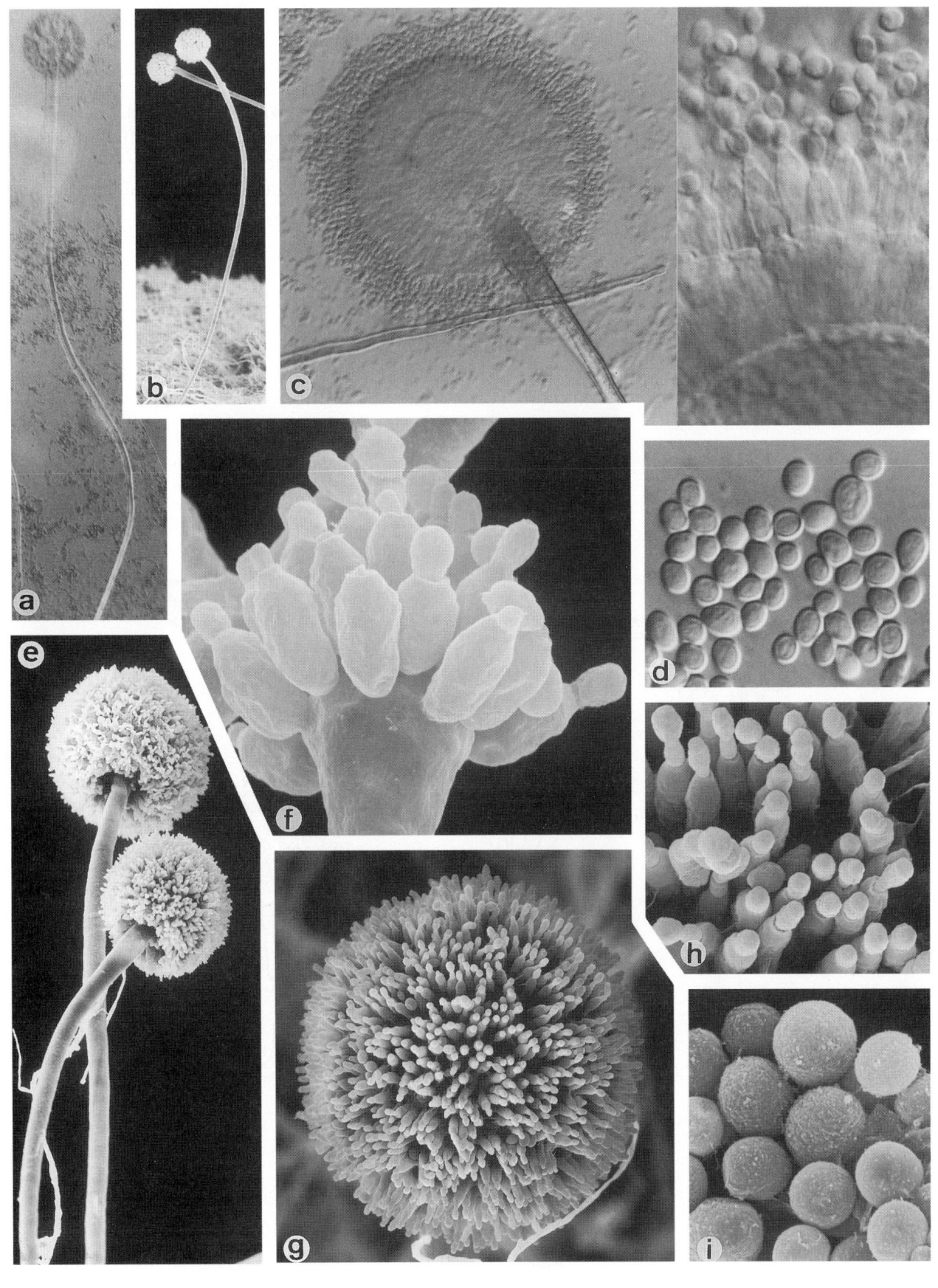

Aspergillus avenaceus, CBS 109.46. Conidiophores and conidia. a. ×128; b. ×80; c. 640; d. ×1600; e. ×310; f. ×4500; g. ×800; h. ×1867; i. ×4000.

Aspergillus caesiellus Saito

Colony characteristics. Colonies (CzA) growing rather slowly, white, mottled with green patches.

Microscopy. Conidial heads strongly columnar, columns often twisted. Conidiophore stipes 50-100 µm in length, smooth-walled, hyaline to slightly greenish. Vesicles small, drumstick-shaped, fertile on distal part only, 5-12 µm diam. Conidiogenous cells uniseriate. Conidia coarsely roughened, at first hyaline and cylindrical, later becoming ellipsoidal to pyriform, flattened at both ends, 4.5-5.5 × 3.0-3.5 µm, forming very long chains.

Pathogenicity. BSL-1. Pulmonary aspergilloma (Otcenásek *et al.*, 1976).

References. Raper & Fennell (1965), Kozakiewicz (1989).

Nomenclature. *Aspergillus caesiellus* Saito - J. Fac. Sci. Coll. Imp. Univ. Tokyo 18: 49, 1904.

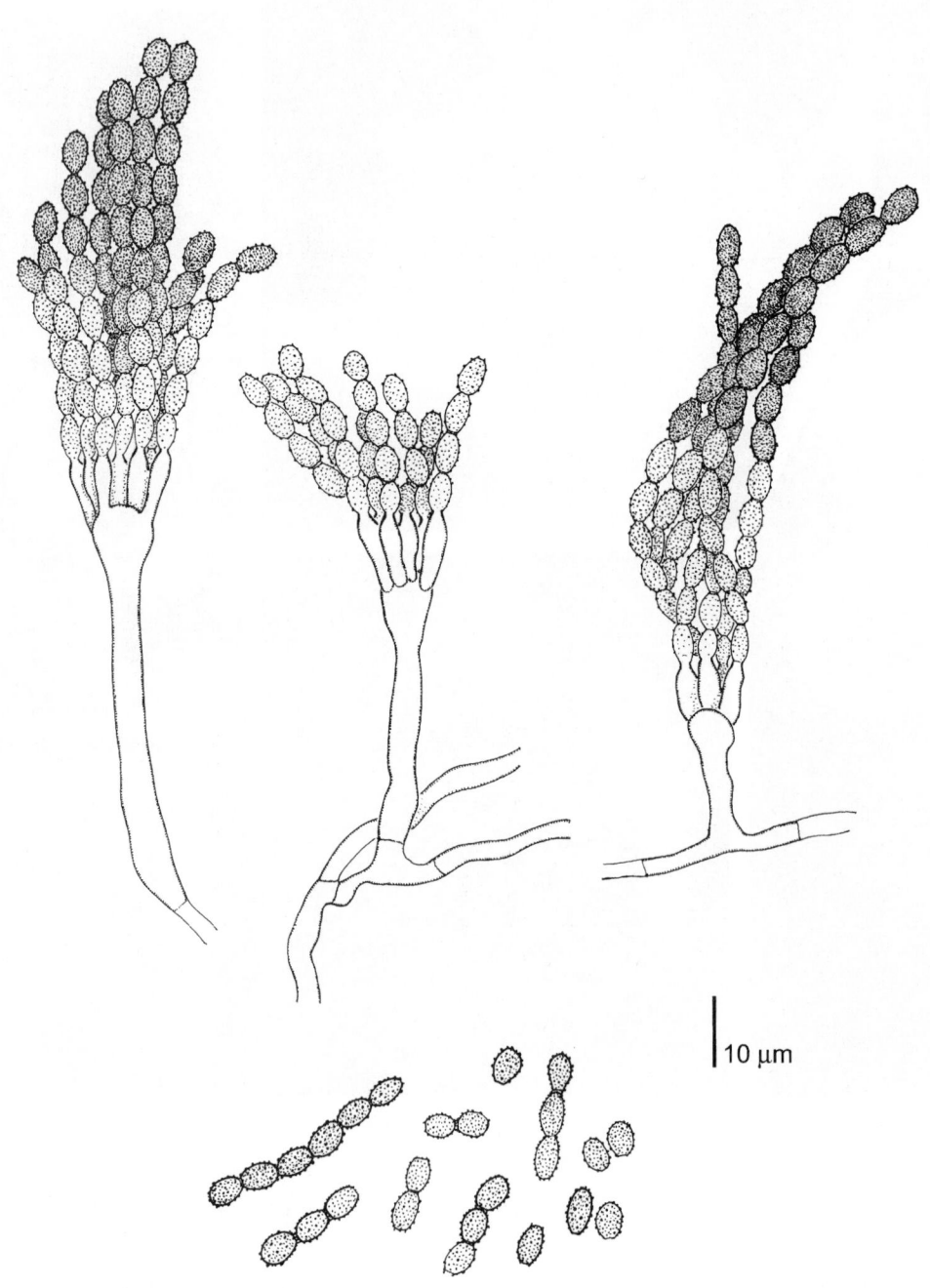

Aspergillus caesiellus, IMI 254821. Conidiophores and conidia.

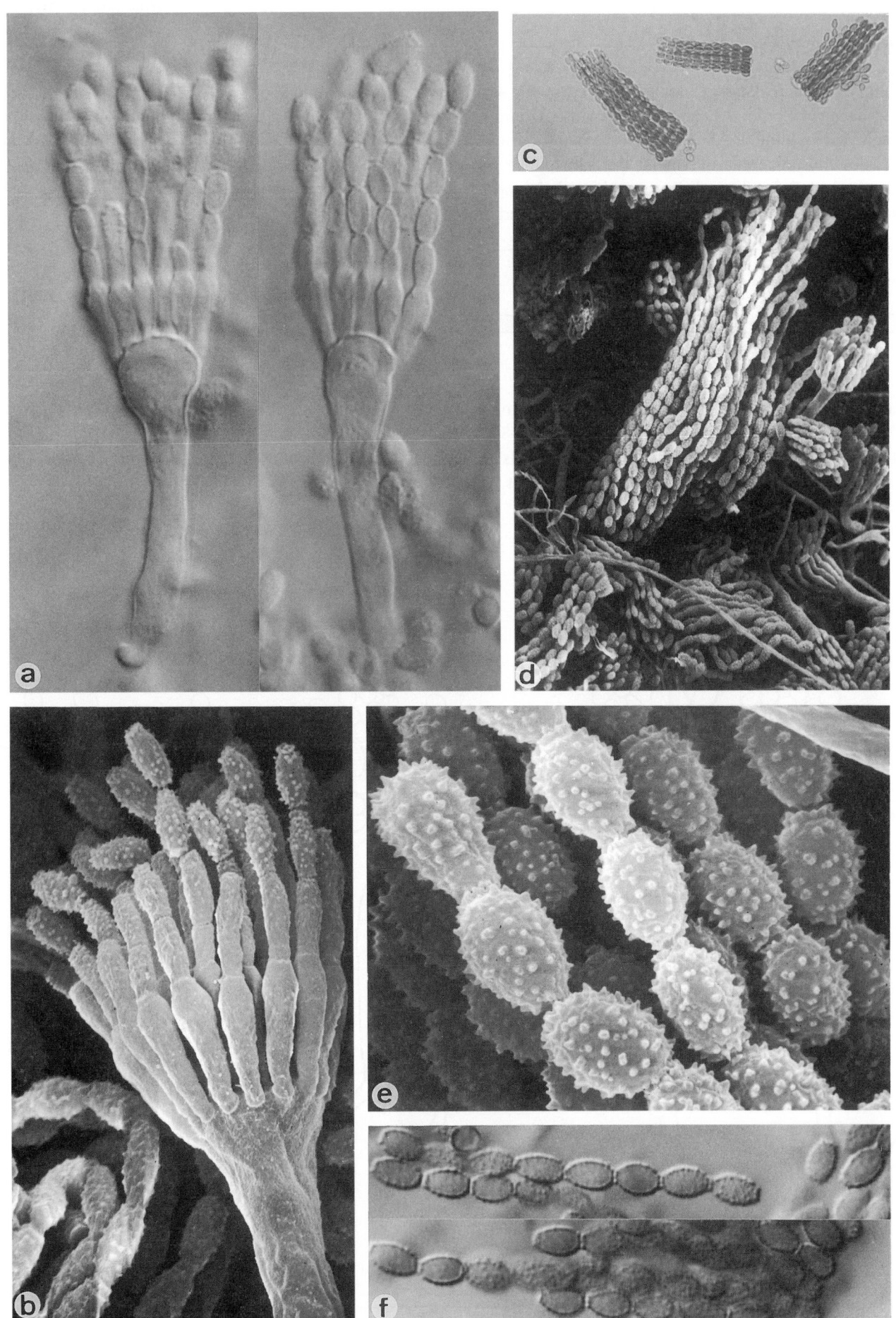

Aspergillus caesiellus, IMI 254821. Conidiophores and conidia. a. ×1600; b. ×3300; c. ×256; d. ×660; e. ×6400; f. ×1280.

453

Aspergillus candidus Link

Colony characteristics. Colonies (CzA) growing slowly, white to pale yellow.

Microscopy. Conidial heads radiate, white. Conidiophore stipes 200-500 µm in length, smooth-walled to finely roughened, hyaline. Vesicles spherical to subspherical, 10-40 µm diam. Conidiogenous cells biseriate in larger heads, occasionally uniseriate in smaller heads. Metulae covering the entire surface of the vesicle. Conidia spherical to subspherical, 2.5-4.0 µm diam, smooth-walled, hyaline. Sclerotia when present, reddish-purple to black.

Molecular diagnostics. SSU restriction map based on NCBI AB008396:

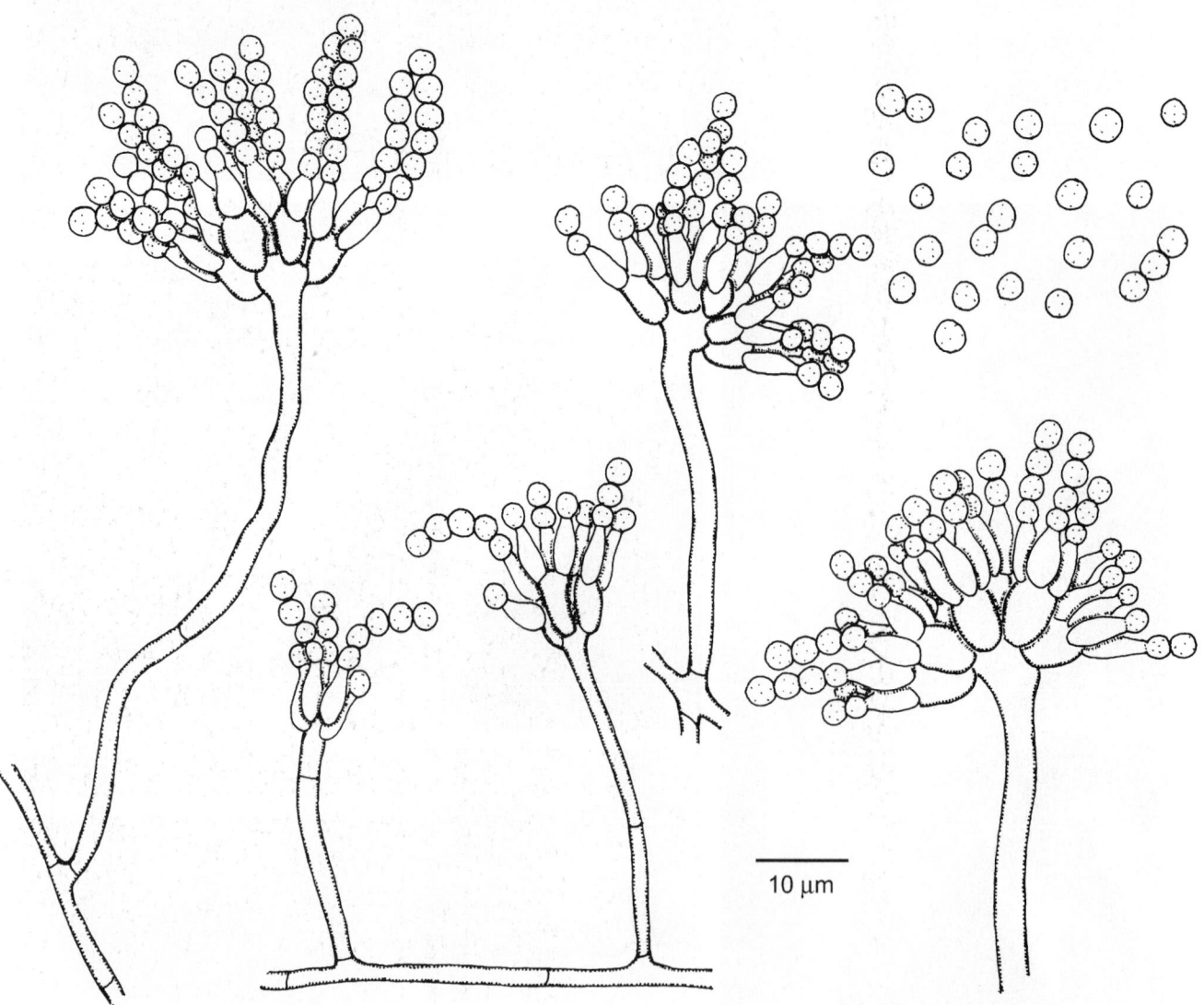

Pathogenicity. BSL-1. This species was involved in a wide range of human infections: invasive aspergillosis (Rippon, 1988), otomycosis (Yasin *et al.*, 1978; Falser, 1983), onychomycosis (Schönborn & Schmoranzer, 1970; Zaror & Moreno, 1980). It has also been isolated from birds, either healthy or with lesions (Saëz, 1970; Sharma *et al.*, 1971).

References. Raper & Fennell (1965), Klich & Pitt (1988), Kozakiewicz (1989).

Aspergillus candidus, IFO 8816. Conidiophores and conidia.

Antifungal susceptibility.

Antifungal	MICs range	MIC 90	Strains	Reference
AMB	0.39	0.39	3	Wildfeuer *et al.* (1998)
ITZ	0.03-0.2	0.2	3	Wildfeuer *et al.* (1998)
KTZ	0.2	0.2	3	Wildfeuer *et al.* (1998)
VCZ	0.05-0.1	0.1	3	Wildfeuer *et al.* (1998)

Nomenclature. *Aspergillus candidus* Link - Mag. Ges. Naturf. Freunde 3: 16, 1809.

Aspergillus candidus, IFO 8816. Conidiophores and conidia. a. ×1600; b. ×3200; c. ×6600.

Aspergillus carneus (v. Tiegh.) Blochwitz

Colony characteristics. Colonies (CzA) spreading, pale tan to pink.
Microscopy. Conidial heads radiate to loosely columnar, 150-200 × 25-35 µm. Conidiophore stipes variable in length, up to 1 mm, smooth-walled, hyaline to very pale brown. Vesicles hemispherical, 5.5-10.0 µm diam. Conidiogenous cells biseriate. Metulae covering the upper third to half of the vesicle. Conidia 2.5-3.5 µm diam, spherical, smooth-walled.
Pathogenicity. BSL-1. This species was involved in a pulmonary infection in man (Morquer & Enjalbert, 1957) and in a disseminated aspergillosis in a dog (Charles, 1989). Its experimental pathogenicity in mice was proven (Pore & Larsh, 1968).
References. Raper & Fennell (1965), Klich & Pitt (1988).
Nomenclature. *Sterigmatocystis carnea* van Tieghem - Bull. Soc. Bot. Fr. 24: 183, 1877 ≡ *Aspergillus carneus* (van Tieghem) Blochwitz - Annls Mycol. 31: 81, 1933.

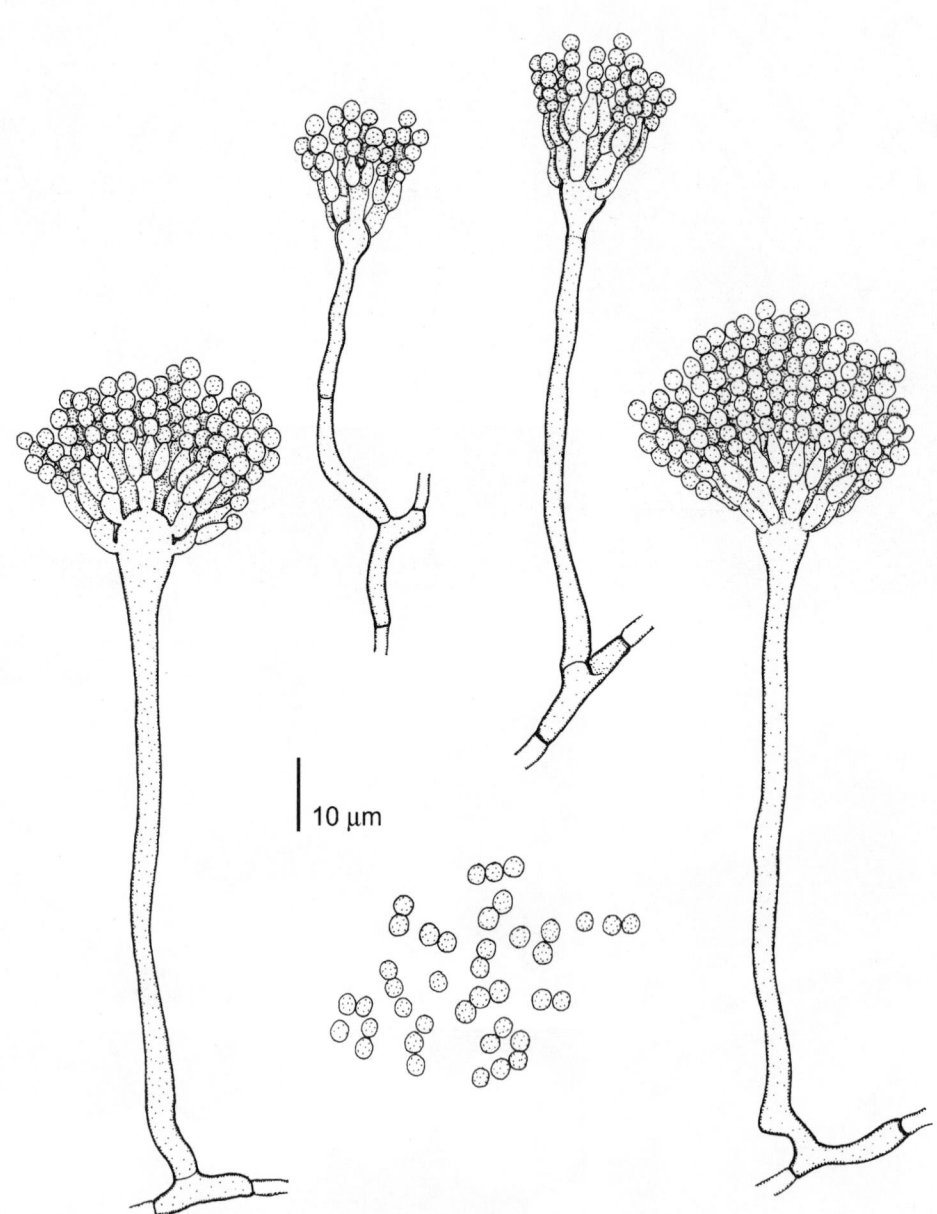

Aspergillus carneus, IFO 30898. Conidiophores and conidia.

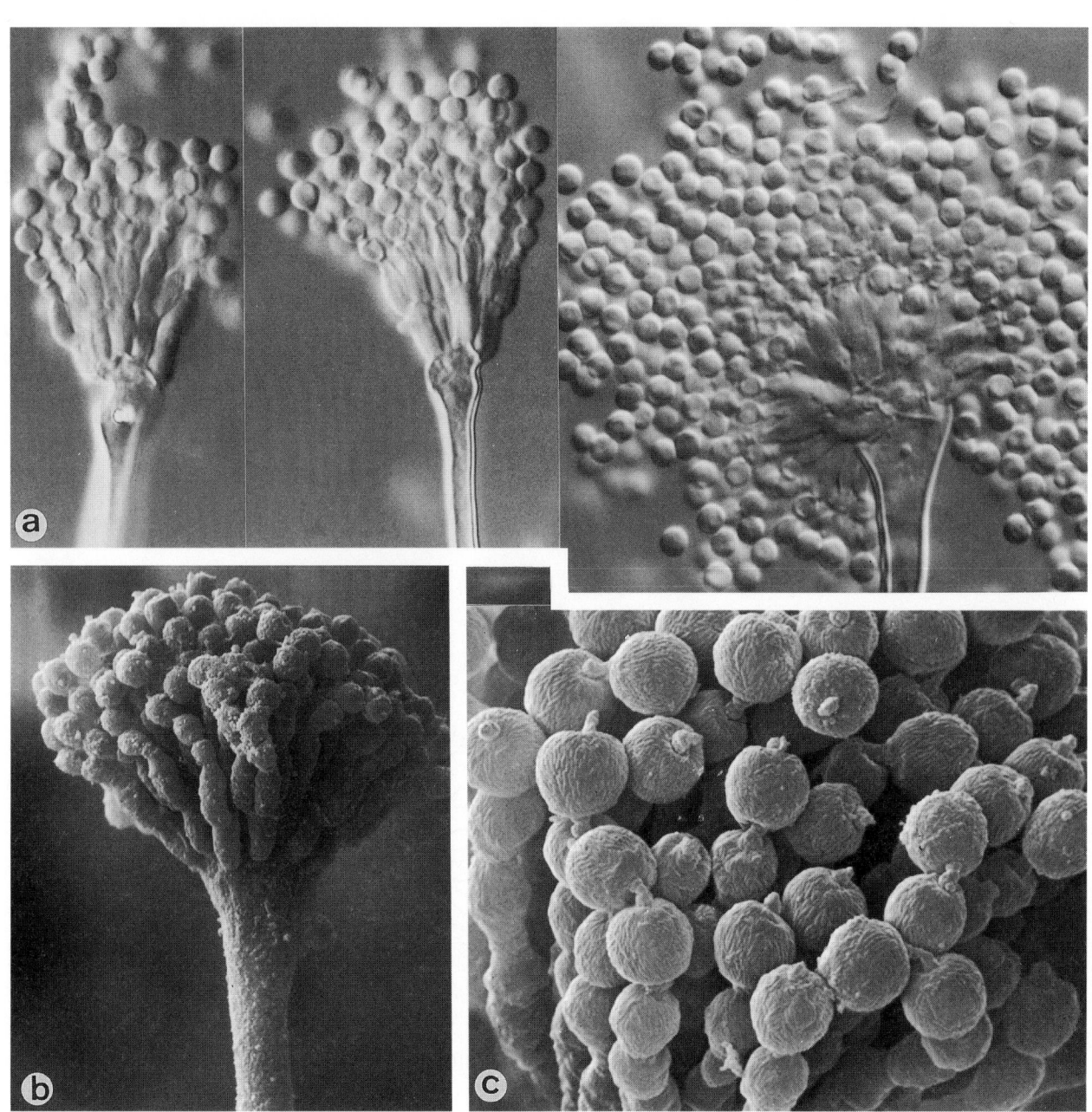

Aspergillus carneus, IFO 30898. Conidiophores and conidia. a. ×1600; b. ×2350; c. ×6000.

Aspergillus chevalieri Mangin

Colony characteristics. Colonies (CzA) with restricted growth, flat, bluish-grey.

Microscopy. Conidial heads radiate, grey-green, 125-175 μm diam. Conidiophores 700-850 μm in length, hyaline. Vesicles spherical, 25-35 μm diam. Conidiogenous cells uniseriate, 5-7 × 3.0-3.5 μm. Conidia ovoidal to ellipsoidal with flattened ends, spinulose, 4.5-5.5 μm in length, hyaline.

Teleomorph. *Eurotium chevalieri* Mangin (*Ascomycota, Euascomycetes, Eurotiales: Trichocomaceae*). Ascomata spherical to subspherical, non-ostiolate, 100-140 μm diam, yellow to orange, with pseudoparenchymatous wall composed of flattened cells, surrounded by a dense network of orange-red hyphae. Asci 8-spored, 10-13 μm diam. Ascospores lenticular, 4.6-5.0 × 3.4-3.8 μm, smooth-walled or very faintly roughened and with prominent equatorial crests, hyaline to subhyaline.

Pathogenicity. BSL-1. Three cases of primary cutaneous aspergillosis were reported by Naidu & Singh (1994). In cases of otitis media no pathogenicity could be proven (Wadhwani & Srivastava, 1984).

Reference. Blaser (1975).

Nomenclature. *Eurotium chevalieri* Mangin - Annls Sci. Nat., Bot., Sér. 9, 10: 361, 1909.
Aspergillus chevalieri Mangin - Annls Sci. Nat., Bot., Sér. 9, 10: 362, 1909.

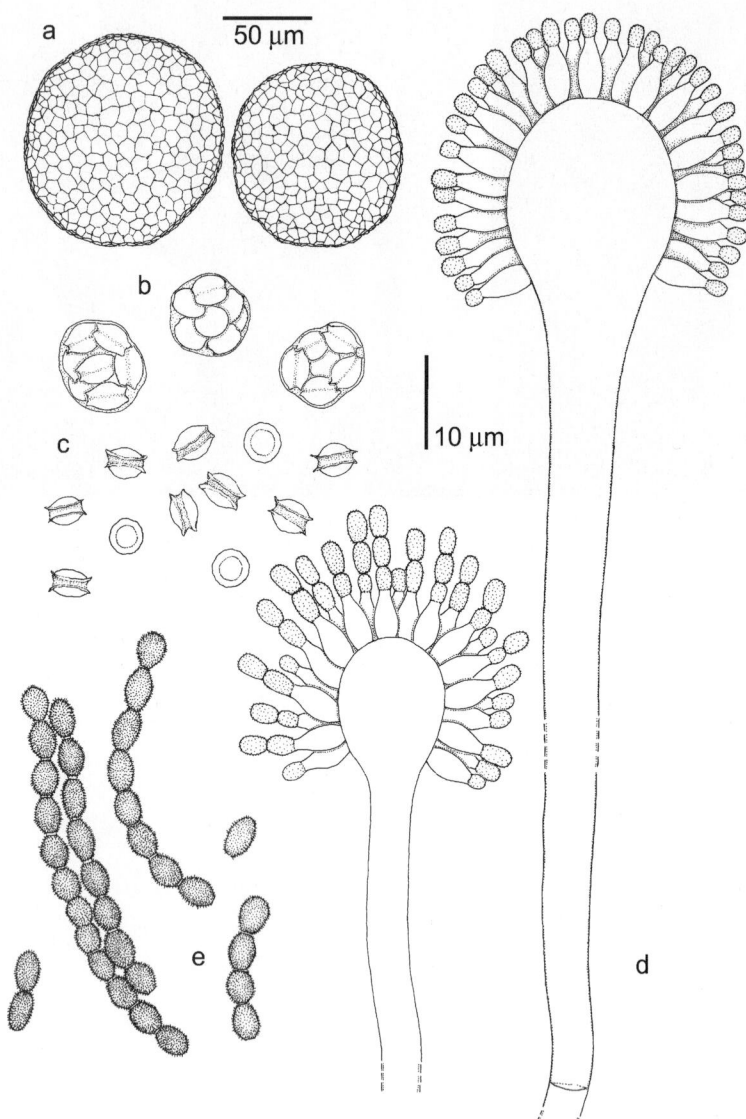

Aspergillus chevalieri (Eurotium chevalieri), CBS 522.65. a. Ascomata; b. asci; c. ascospores; d. conidiophores; e. conidia.

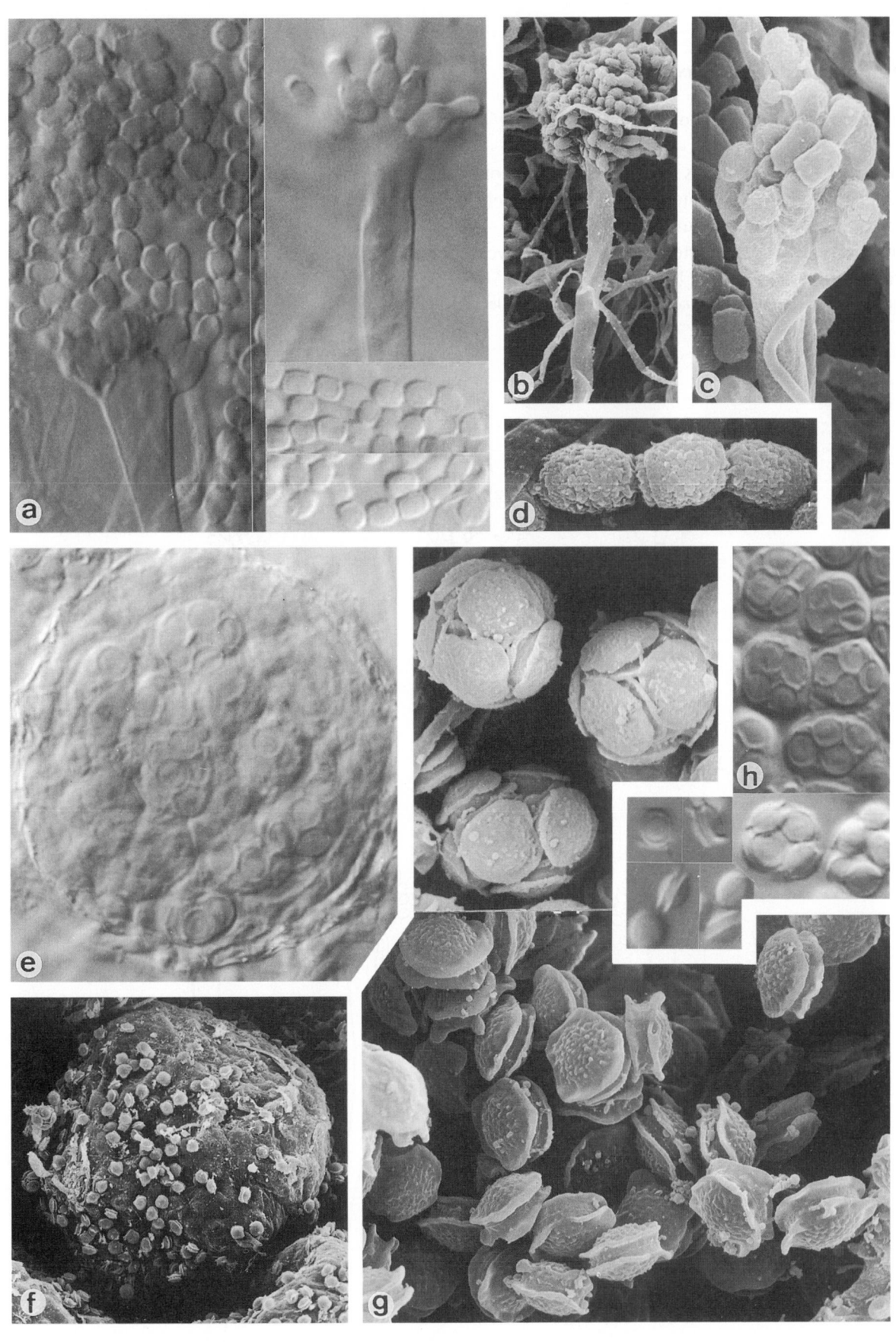

Aspergillus chevalieri (Eurotium chevalieri), CBS 522.65. a-d. Conidiophores and conidia; e-h. ascomata, asci and ascospores. a. ×1600; b. ×1100; c. ×3600; d. ×6500; e. ×1600; f. ×600; g. ×3700; h. ×1600.

459

Aspergillus clavato-nanicus Batista *et al.*

Colony characteristics. Colonies (CzA) growing rapidly, greyish blue-green, floccose; conidial heads evenly distributed.

Microscopy. Conidial heads radiate. Conidiophores 22-125 × 5-22 μm. Vesicles clavate, 5-22 μm wide. Conidiogenous cells uniseriate, covering the vesicle entirely. Conidia smooth-walled, yellowish, ellipsoidal or cylindrical, 3.8-5.0 × 2.8-3.5 μm.

Pathogenicity. BSL-1. Reported from a case of onychomycosis in Brazil (Batista *et al.*, 1955).

Reference. Raper & Fennell (1965).

Nomenclature. *Aspergillus clavato-nanicus* Batista, Maia & Alecrim - Ann. Fac. Med. Univ. Recife 15: 197, 1955.

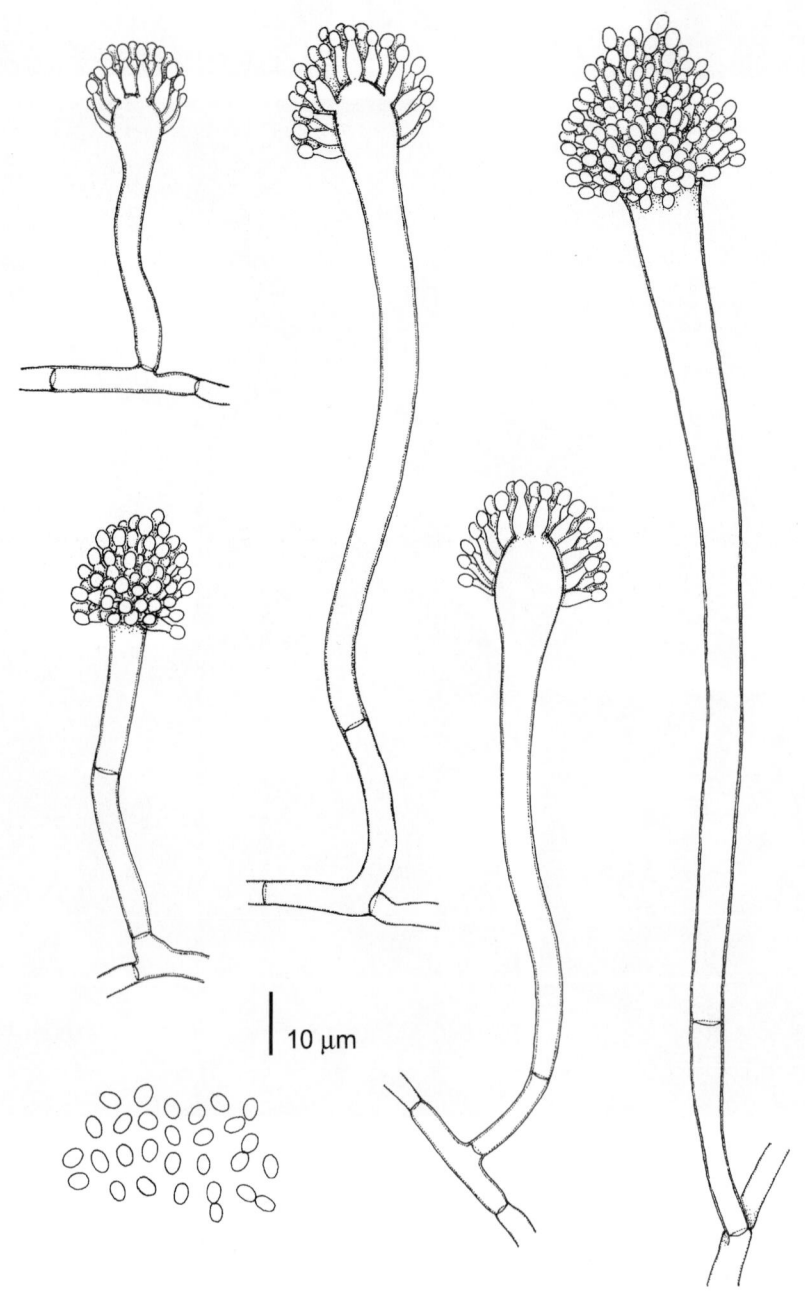

Aspergillus clavato-nanicus, CBS 474.65. Conidiophores and conidia.

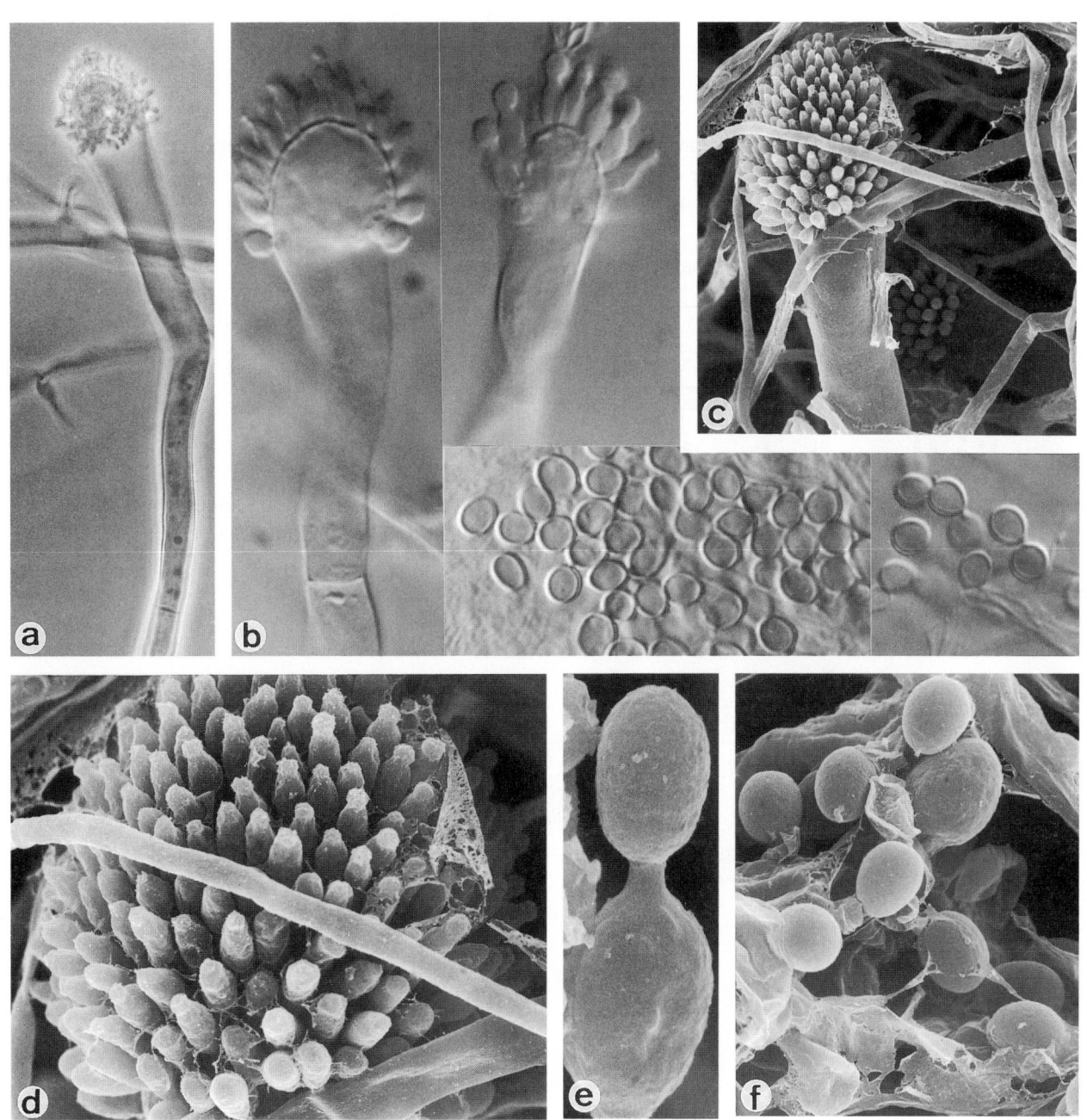

Aspergillus clavato-nanicus, CBS 475.65. Conidiophores and conidia. a. ×512; b. ×1600; c. ×1450; d. ×3700; e. ×10000; f. ×4200.

Aspergillus clavatus Desm.

Colony characteristics. Colonies (CzA) growing rapidly, bluish-green, consisting of a dense felt of conidiophores.
Microscopy. Conidial heads radiate, later splitting into several columns. Conidiophores 2-4 mm in length; stipes smooth-walled, hyaline. Vesicles clavate, 40-60 µm diam. Conidiogenous cells uniseriate. Conidia smooth-walled, pale green, ellipsoidal, 7-8 × 2-3 µm.
Molecular diagnostics. SSU restriction map based on NCBI AB008398:

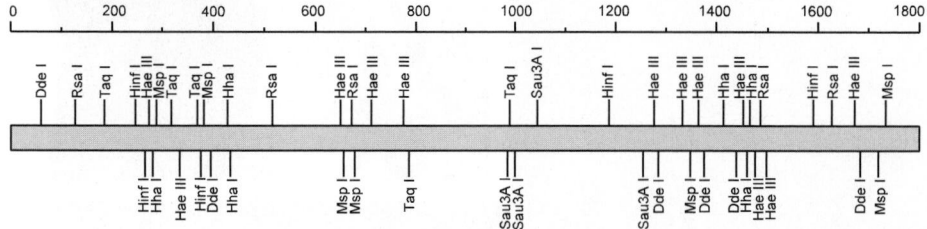

Pathogenicity. BSL-1. Agent of allergic aspergillosis (Rippon 1988). It has been implicated in various pulmonary infections (Van der Werff, 1951). In addition, a case of otomycosis was reported (Yasin *et al.*, 1978). A neurotoxicosis in sheep was reported by Shlosberg *et al.* (1991).
Reference. Raper & Fennell (1965).
Nomenclature. *Aspergillus clavtus* Desmazières - Ann. Sci. Nat., Sér. Bot., 2: 71, 1834.

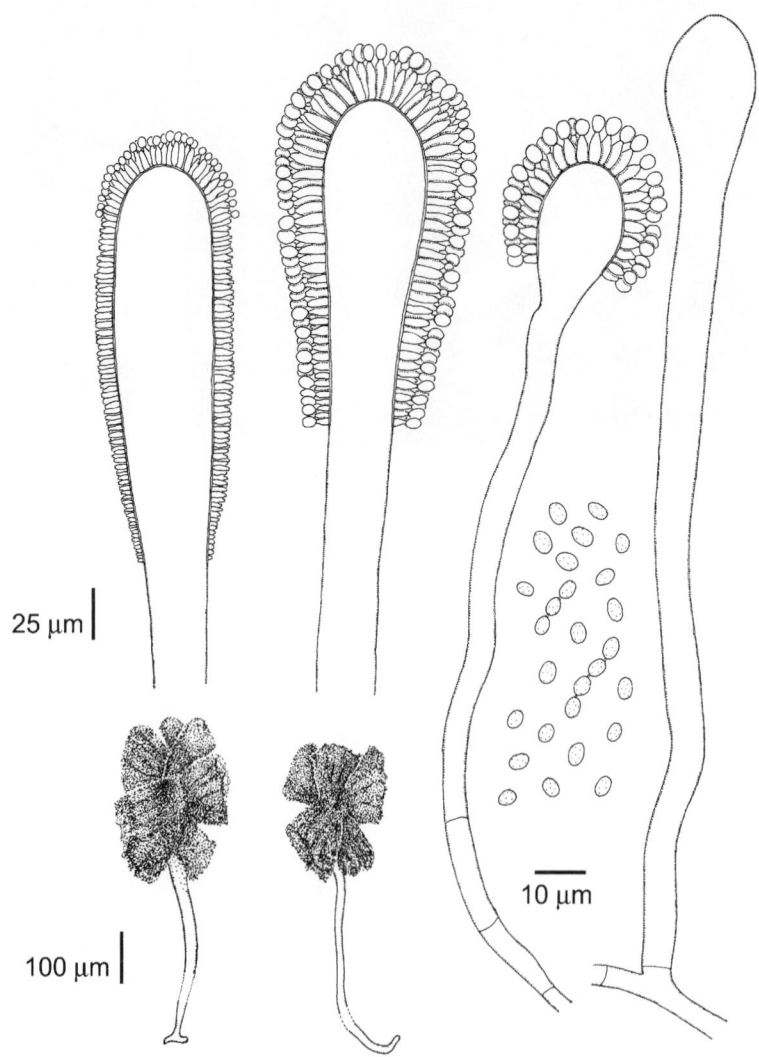

Aspergillus clavatus, IFO 31830. Conidiophores and conidia.

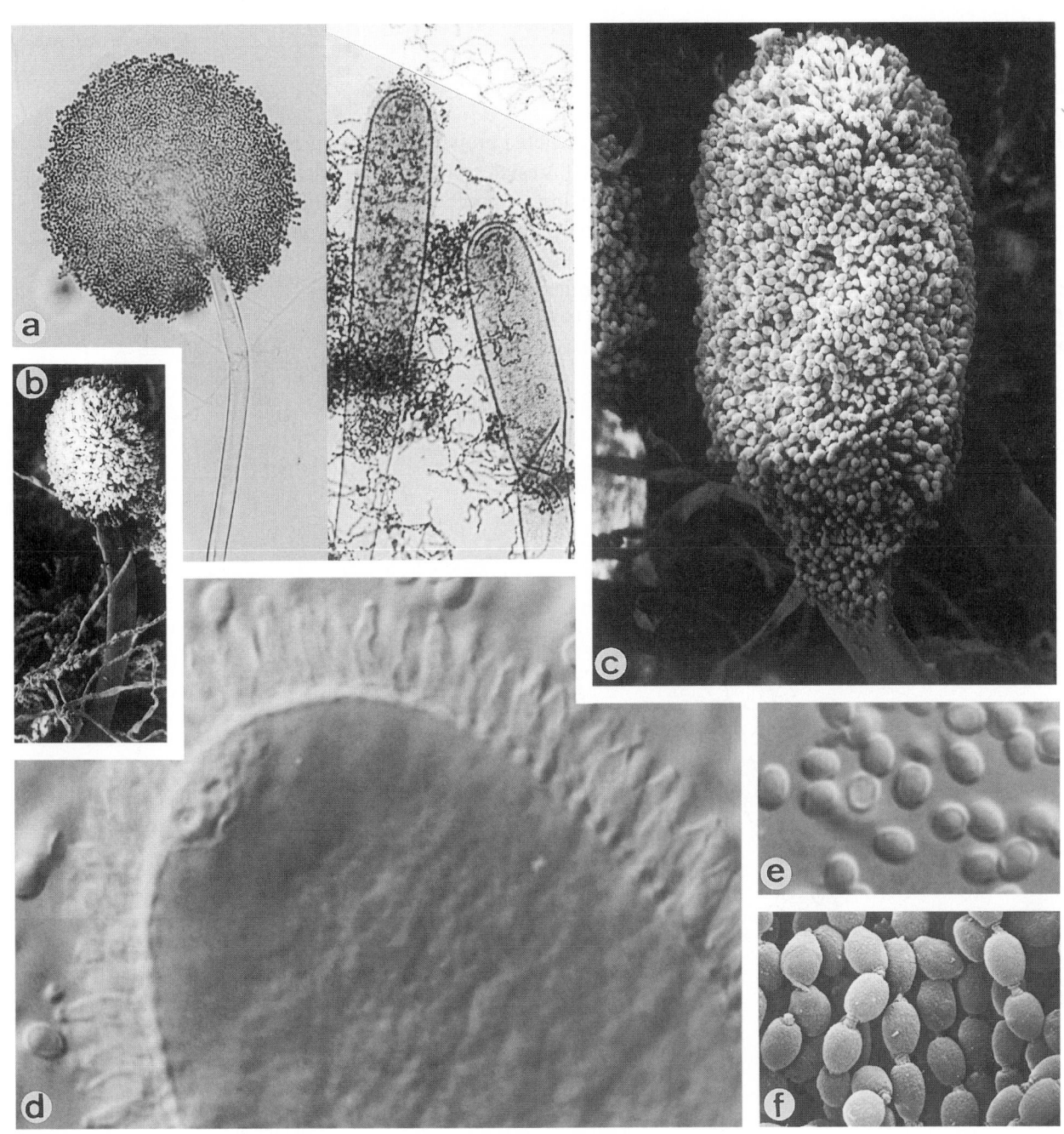

Aspergillus clavatus, IFO 31830. Conidiophores and conidia. a. ×128; b. ×300; c. ×640; d, e. ×1600; f. ×2840.

Aspergillus conicus Blochwitz

Colony characteristics. Colonies (CzA) with very restricted growth, dull grey-green.

Microscopy. Conidial heads columnar, pale grey-green. Vesicles small, subspherical, 3.0-7.5 µm diam. Conidiophore stipes arising from submerged hyphae, up to 50-75 µm long, smooth-walled but slightly roughened below the vesicle. Conidiogenous cells uniseriate, borne on the uppermost part of the vesicle only. Conidia slightly roughened, aculeate, 3.5-4.0 µm diam, olive-green in mass.

Pathogenicity. BSL-1. Reported in a case of postsurgical endophthalmitis (Jones, 1977).

References. Raper & Fennell (1965), Kozakiewicz (1989).

Nomenclature. *Aspergillus conicus* Blochwitz, *in* Dale - Annls Mycol. 12: 38, 1914.

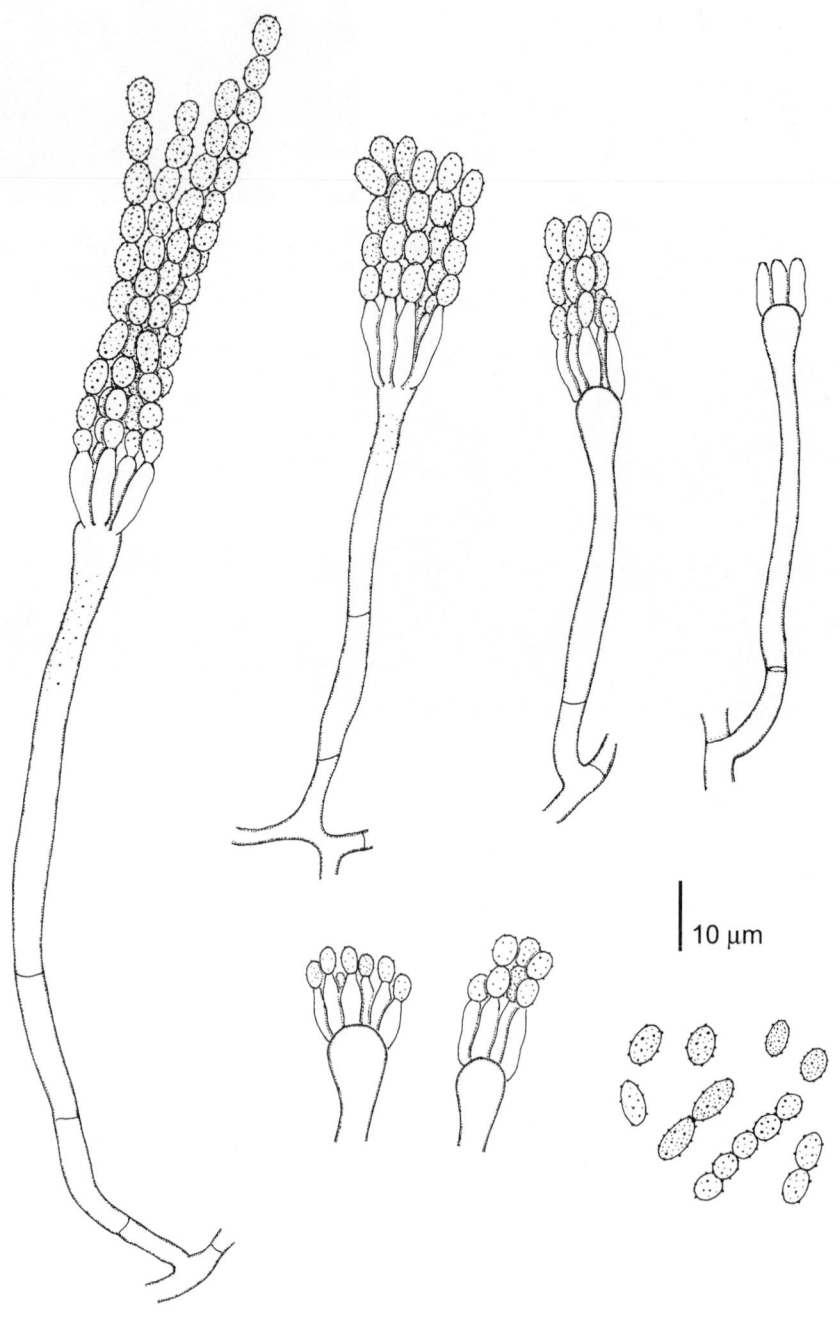

10 µm

Aspergillus conicus, IFO 4046. Conidiophores and conidia.

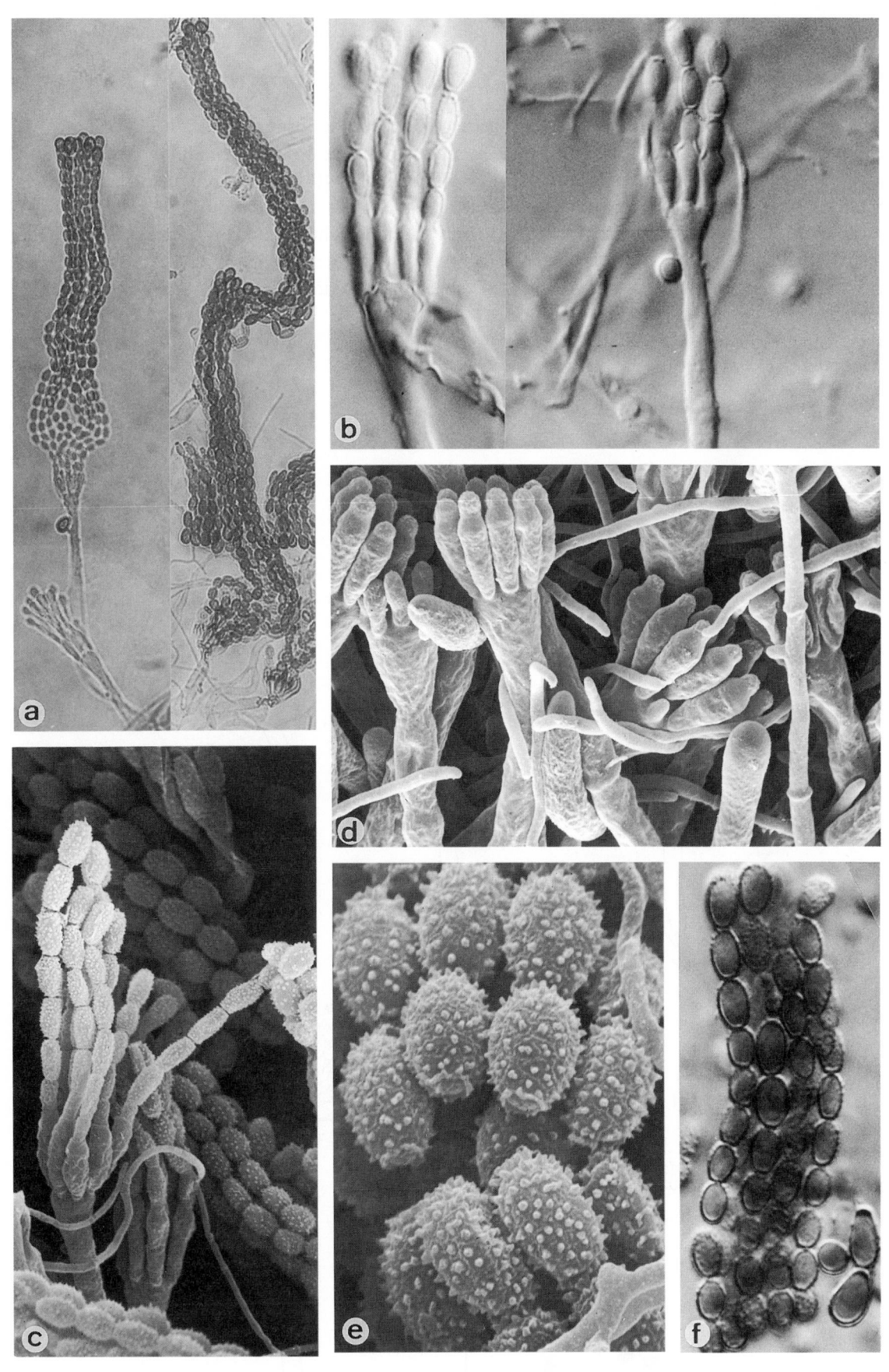

Aspergillus conicus, IFO 4046. Conidiophores and conidia. a. ×512; b. ×1600; c. ×1850; d. ×2550; e. ×6400; f. ×1600.

Aspergillus deflectus Fennell & Raper

Colony characteristics. Colonies (CzA) with very restricted growth, mouse-grey.
Microscopy. Conidial heads broadly columnar, 25-30 µm diam, borne on short conidiophores from aerial hyphae. Conidiophore stipes apically bent downward, smooth-walled, usually 40-50 µm, rarely up to 125 µm long, reddish-brown. Vesicles spherical to flask-shaped, at right angles to the main stipe. Conidiogenous cells 5.5-6.5 µm diam, biseriate. Metulae on the uppermost part of the vesicle only. Conidia spherical to subspherical, with lobate-reticulate walls, 3.0-3.5 µm diam.
Pathogenicity. BSL-1. Several cases of disseminated aspergillosis in dogs were reported (Jang *et al.*, 1986; Kahler *et al.*, 1990).
References. Raper & Fennell (1965), Kozakiewicz (1989).
Nomenclature. *Aspergillus deflectus* Fennell & Raper - Mycologia 47: 82, 1955.

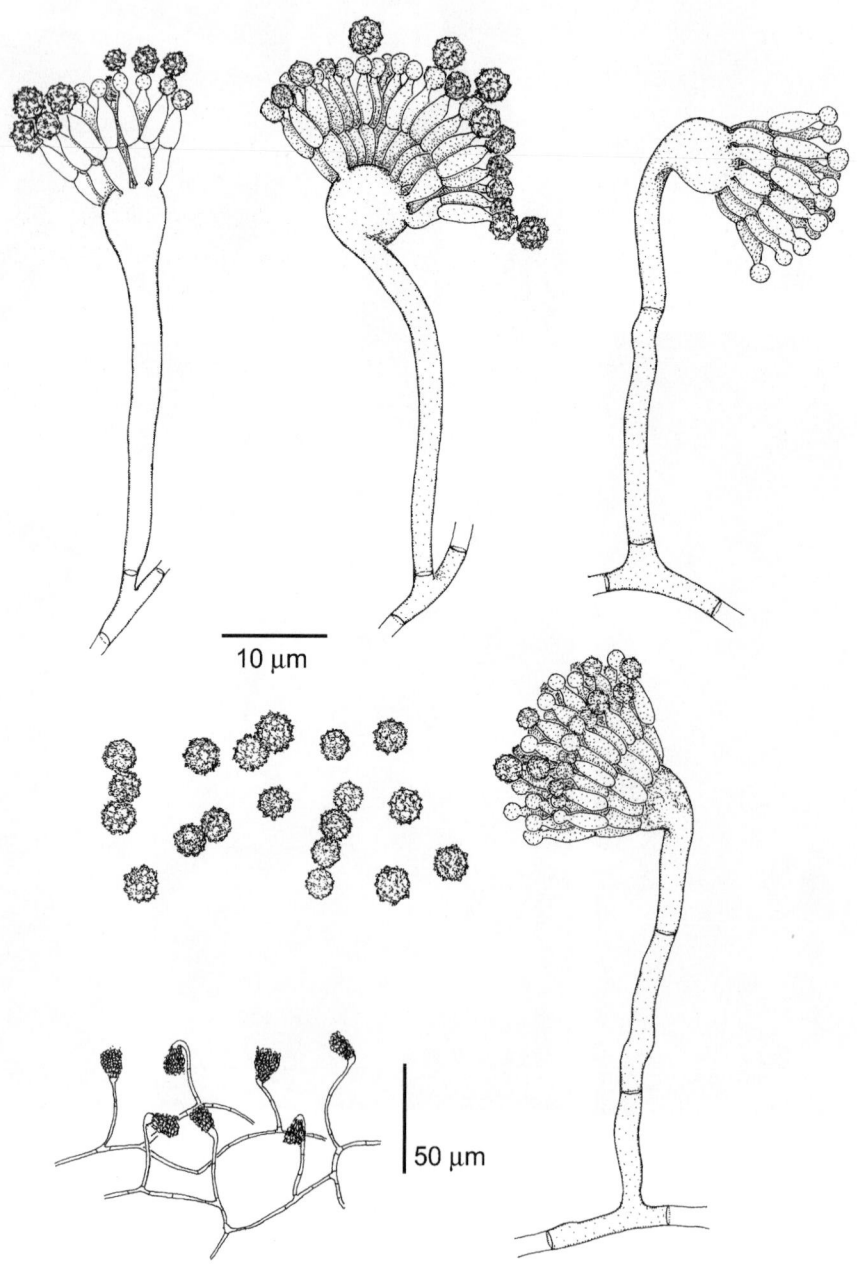

10 µm

50 µm

Aspergillus deflectus, CBS 109.55. a. Habit sketch; b, c. conidiophores and conidia.

466

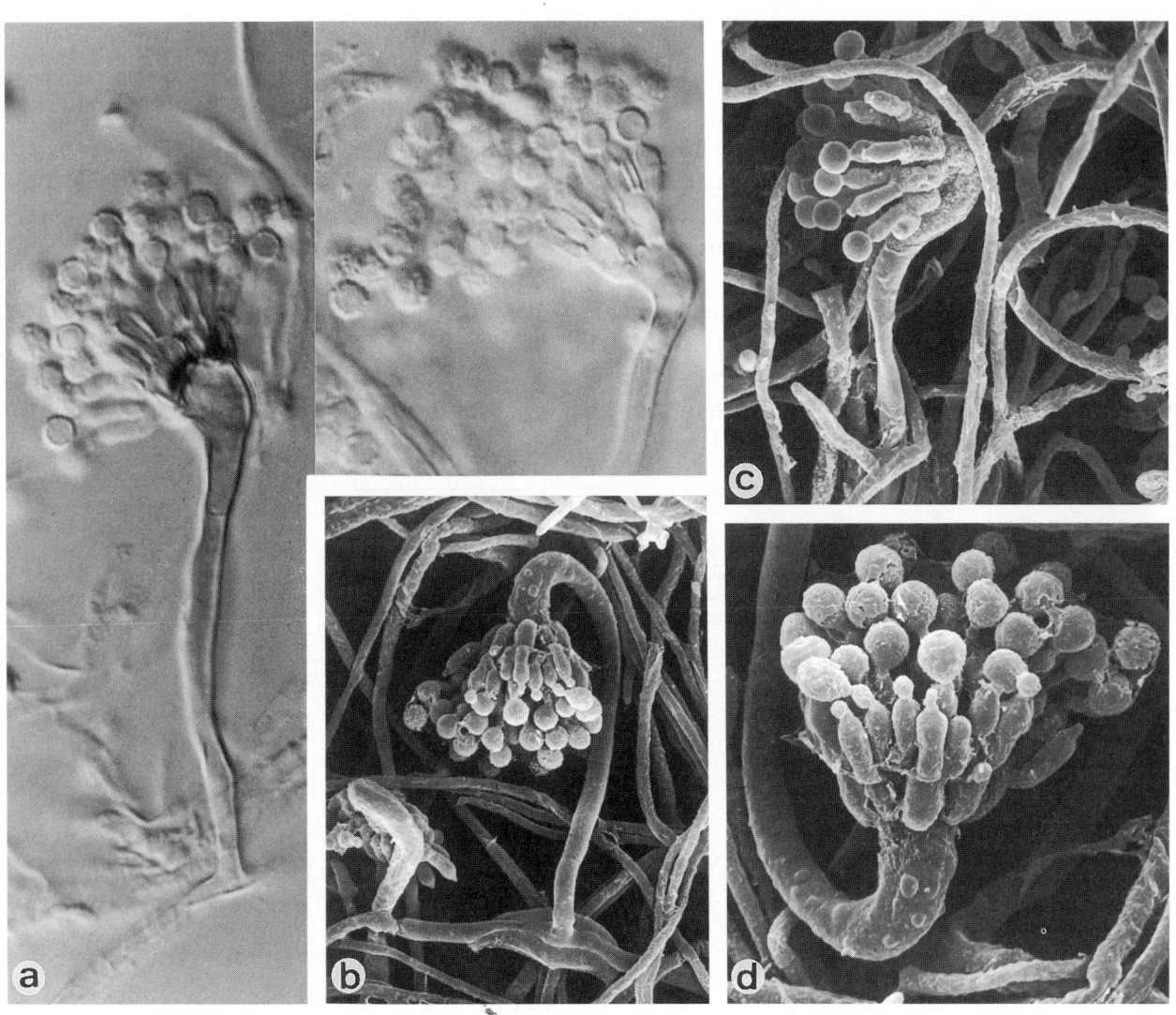

Aspergillus deflectus, CBS 109.55. Conidiophores and conidia. a, b. ×1600; c. ×2200; d. ×3300.

Aspergillus flavipes (Bain. & Sart.) Thom & Church

Colony characteristics. Colonies (CzA) growing rather slowly, yellowish, buff or brownish.

Microscopy. Conidial heads sparse, loosely columnar to radiate, white to pale buff. Conidiophore stipes smooth-walled, 2.4-3.2 μm wide, yellow to light brown. Vesicles subspherical to vertically elongate. Conidiogenous cells biseriate. Metulae covering at least the upper third of the vesicle. Conidia spherical to subspherical, smooth-walled, 2-3 μm diam, colourless.

Teleomorph. *Fennellia flavipes* Wiley & Simmons (*Ascomycota, Euascomycetes, Eurotiales: Trichocomaceae*). Ascomata surrounded by masses of Hülle cells, 600-800 μm diam, yellowish. Peridium composed of several layers of thin-walled, hypha-like cells. Asci subspherical to pyriform, 12-16 × 10-14 μm. Ascospores hyaline to pale yellow, smooth-walled, spherical to broadly ovoidal, 6-8 × 4.5-6.0 μm.

Molecular diagnostics. SSU restriction map based on NCBI AB008400:

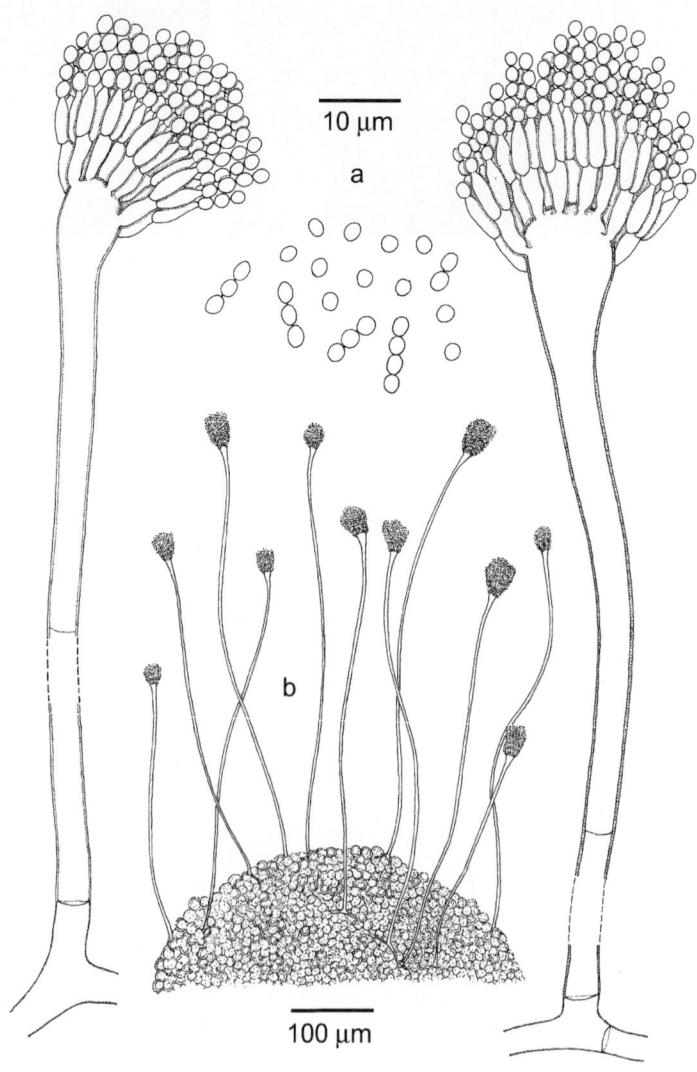

Aspergillus flavipes (Fennellia flavipes), CBS 444.75. a. Conidiophores and conidia; b. part of ascoma wall with Hülle cells and conidiophores.

Pathogenicity. BSL-1. This species was implicated in a case of cutaneous aspergillosis in an immunocompromised child (Barson & Ruymann, 1986) and in a case of osteomyelitis (Tack *et al.*, 1982).

References. Raper & Fennell (1965), Wiley & Simmons (1973), Klich & Pitt (1988), Yaguchi *et al.* (1994).

Nomenclature. *Sterigmatocystis flavipes* Bainier & Sartory - Bull. Trim. Soc. Mycol. Fr. 27: 90, 1911 ≡ *Aspergillus flavipes* (Bainier & Sartory) Thom & Church - The Aspergilli p. 155, 1926.

Fennellia flavipes Wiley & Simmons - Mycologia 65: 937, 1973.

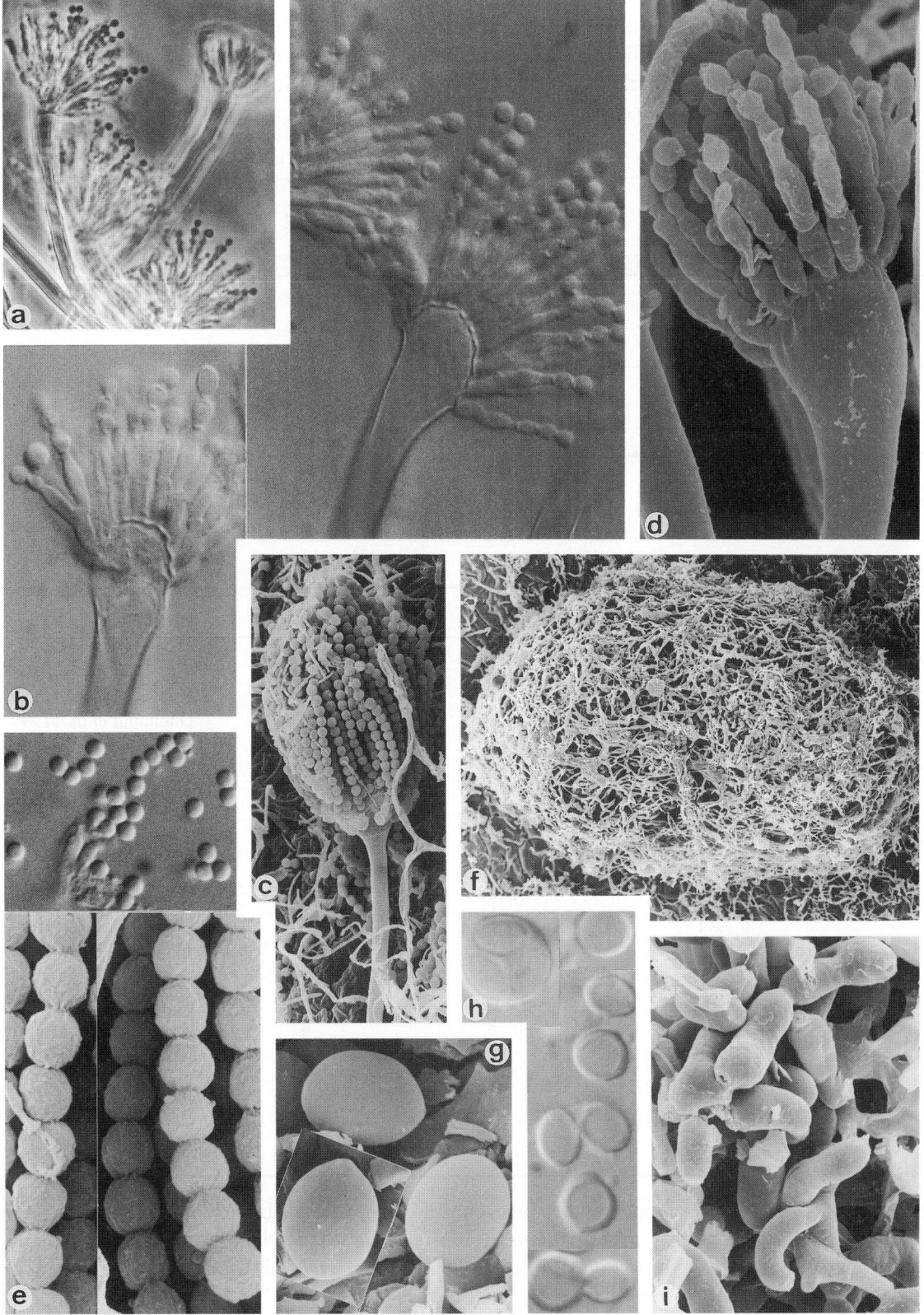

Aspergillus flavipes (*Fennellia flavipes*), CBS 444.75. a-d. Conidiophores; e. conidia; f. ascoma; g, h. asci and ascospores; i. hypha-like cells of the peridium. a. ×415; b. ×1295; c. ×690; d. ×3000; e. ×4050; f. ×155; g. ×3240; h. ×1295; i. ×1620.

Aspergillus flavus Link:Fr.

Colony characteristics. Colonies (CzA) yellowish-green, consisting of a dense felt of conidiophores.

Microscopy. Conidial heads radiate, conidiogenous cells uni- and biseriate. Conidiophore stipes rough-walled, hyaline. Vesicles spherical, 25-45 µm diam. Conidia echinulate, (sub)spherical, 3.5 µm diam. Sclerotia may be present.

Differential diagnosis. The species is easily distinguished from *A. fumigatus* (p. 473) by the characters mentioned in the key, and by mature vesicles bearing phialides over their entire surface.

Molecular diagnostics. PCR-amplification of partial alkaline protease genes and subsequent Southern blotting was developed by Tang *et al.* (1993). A 26S rDNA-based specific probe was developed by Sandhu *et al.* (1995). Verweij *et al.* (1995) compared complete 18S sequences.SSU and ITS restriction maps based on NCBI D63696 and NCBI AB008414:

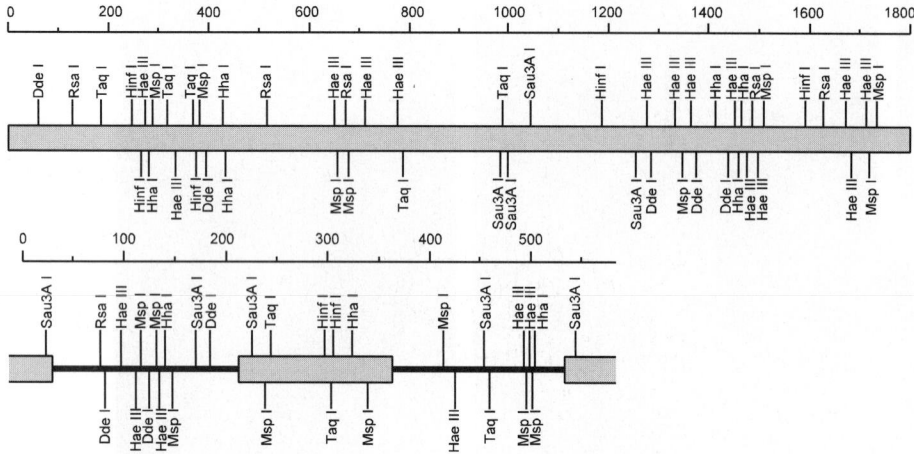

Antifungal susceptibility.

Antifungal	MICs range	MIC 90	Strains	Reference
AMB	0.78-1.56	1.56	4	Wildfeuer *et al.* (1998)
AMB	0.25-1	0.5	10	Hennequin *et al.* (1997)
AMB	0.5-2	2	22	Dannaoui *et al.* (1999)
AMB	0.5-1	1	10	Johnson *et al.* (1998)
AMB	0.5-1	1	9	Clancy & Nguyen (1998)
AMB	0.5-1	1	13	Tawara *et al.* (2000)
AMB	0.5-32	nd	10	Moore *et al.* (2000)
FLZ	8->64	>64	13	Tawara *et al.* (2000)
ITZ	0.39-0.78	0.78	4	Wildfeuer *et al.* (1998)
ITZ	0.06-0.13	0.12	10	Hennequin *et al.* (1997)
ITZ	0.12-1	0.5	22	Dannaoui *et al.* (1999)
ITZ	0.12-0.25	0.25	10	Johnson *et al.* (1998)
ITZ	0.125-0.5	0.5	13	Tawara *et al.* (2000)
ITZ	0.25->16	nd	10	Moore *et al.* (2000)
KTZ	0.78	0.78	4	Wildfeuer *et al.* (1998)
TBF	0.03	nd	12	Jessup *et al.* (2000)
TBF	0.025-0.5	0.4	29	Ryder *et al.* (1999)
VCZ	0.39-0.78	0.78	4	Wildfeuer *et al.* (1998)
VCZ	0.5-1	1	10	Johnson *et al.* (1998)
VCZ	0.5-2	2	9	Clancy & Nguyen (1998)

Pathogenicity. BSL-2. The species is one of the main agents of human allergic bronchial aspergillosis (p. 27) and of pulmonary infections in immunocompromised patients (Robinson *et al.*, 1995), where its pathology may be aggravated by the production of mycotoxins (Mori *et al.*, 1998). It also occurs in external ears and may be involved in otitis (Jesenska *et al.*, 1992), occasionally with severe complications (Harley *et al.*, 1995). Single-organ systemic infections occur in leukemic patients (Shitara *et al.*, 1993), e.g., in the kidney (Khan *et al.*, 1995). An endocarditis after application of broad-spectrum antibiotics was reported by Khan *et al.* (1997). The species is also one of the common agents of mycotic sinusitis (Drakos *et al.*, 1993; El-Shoura, 1993; Chang *et al.*, 1996). Cutaneous aspergillosis is rare (Harmon *et al.*, 1993); traumatic cases are mostly associated with underlying diseases such as diabetes (Witzig *et al.*, 1996). Also cases of aspergillosis in other vertebrates have been reported (Barton *et al.*, 1992).

Variety. The var. *columnaris* Raper & Fennell, with consistently columnar conidia on MEA, was reported from sinusitis by Mahgoub (1971).

References. Kurtzman *et al.* (1986), VandenBossche *et al.* (1988), Kozakiewicz (1995), Samson *et al.* (1996).

Nomenclature var. *columnaris*. *Aspergillus flavus* Link:Fries var. *columnaris* Raper & Fennell - Gen. Aspergillus p. 366, 1965.

Nomenclature var. *flavus*. *Aspergillus flavus* Link - Mag. Ges. Naturf. Freunde 3: 16, 1809 ≡ *Aspergillus flavus* Link:Fries - Syst. Mycol. 3: 386, 1832.

Sterigmatocystis lutea van Tieghem - Bull. Soc. Bot. Fr. 24: 103, 1877 ≡ *Aspergillus luteus* (van Tieghem) C.W. Dodge - Med. Mycol. p. 625, 1925.

Aspergillus flavus Link:Fr. var. *proliferans* Anguli, Rajam, Thirumalachar, Rangiah & Ramam - Ind. J. Microbiol. 5: 94, 1965.

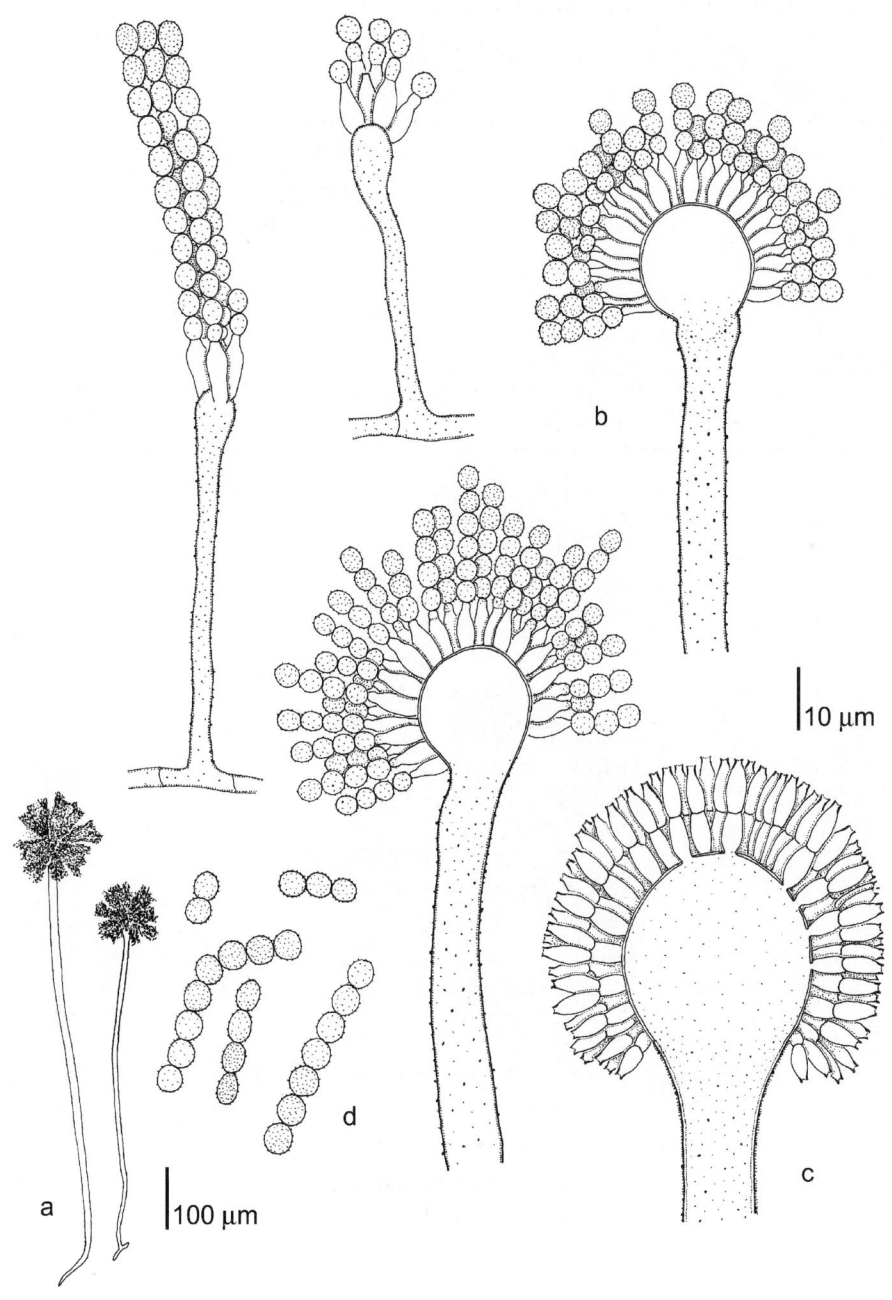

Aspergillus flavus, IFO 30107. a. Conidiophores, habit sketch; b. uniseriate conidiophores; c. biseriate conidiophore; d. conidia.

471

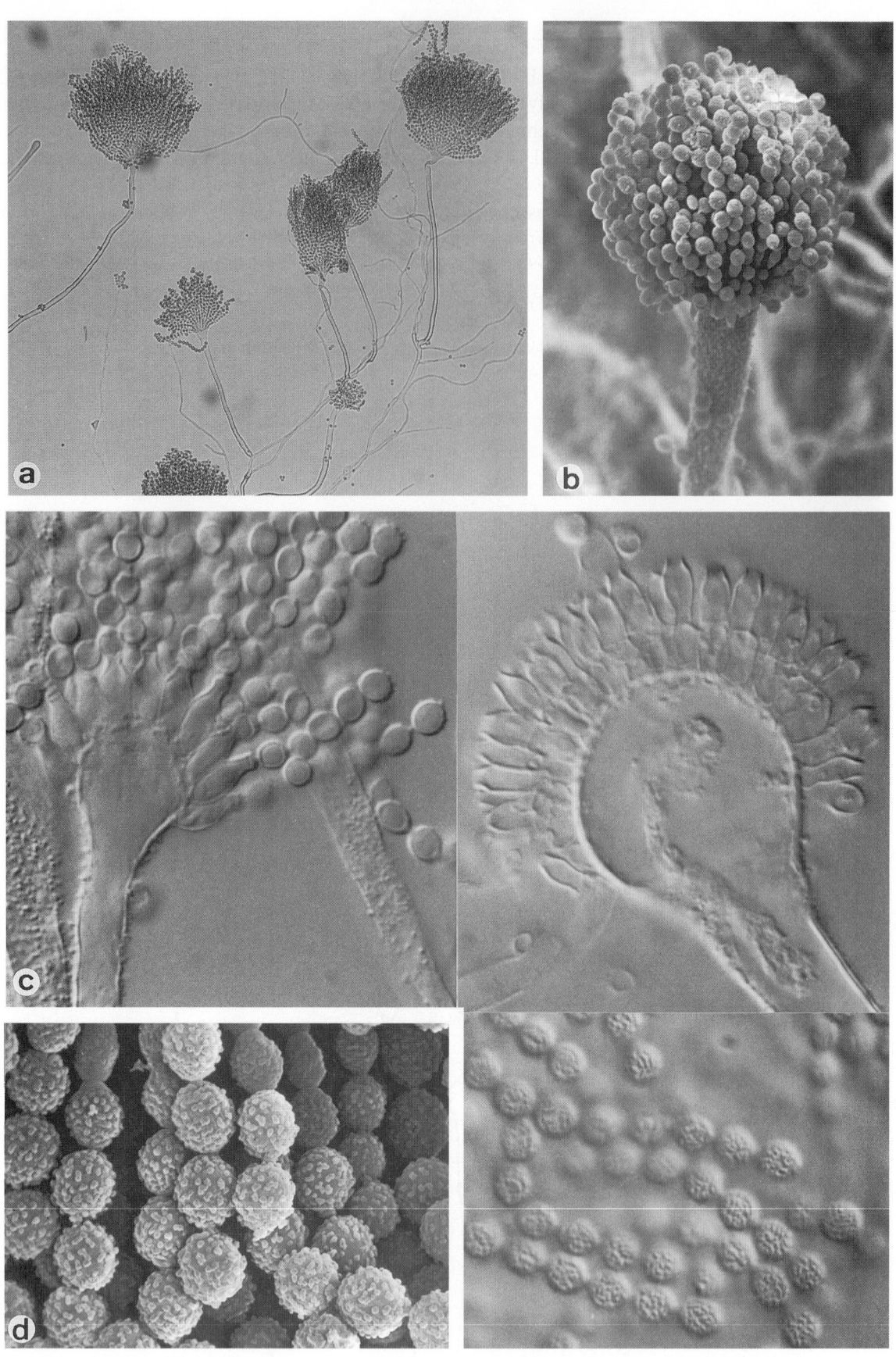

Aspergillus flavus, IFO 30107. Uniseriate and biseriate conidiophores, and conidia. a. ×130; b. ×940; c. ×1600; d. ×3600.

Aspergillus fumigatus Fres.

Colony characteristics. Colonies (CzA) dark blue-green, consisting of a dense felt of conidiophores, intermingled with aerial hyphae.

Microscopy. Conidial heads columnar; conidiogenous cells uniseriate. Conidiophore stipes smooth-walled, often green in the upper part. Vesicles subclavate, 20-30 μm wide. Conidia verrucose, (sub)spherical, 2.5-3.0 μm diam.

Differential diagnosis. The species is easily distinghuished from *A. flavus* (p. 470) by the characters mentioned in the key, and by mature vesicles bearing phialides which are upward directed. Young conidial heads are radiate.

Molecular diagnostics. PCR-based molecular detection using a fragment of 26S rDNA was described by Spreadbury *et al.* (1993) and using interrepeat PCR by van Belkum *et al.* (1993). Nested PCR of V7 to V9 variable regions of 18S rDNA was developed for detection in serum by Yamakami *et al.* (1996). Relationships to other species were revealed by partial β-tubulin and hydrophobin sequences (Geiser *et al.*, 1998). The species is close to the anamorph of *Neosartorya fischeri* (p. 292). Intermediate DNA/DNA homology values are found (Peterson, 1992), as well as similar mycotoxin profiles (Samson *et al.*, 1990) and *Eco*RI-digested DNA fingerprints (Girardin *et al.*, 1995). A 26S rDNA-based specific probe was developed by Sandhu *et al.* (1995). PCR-amplified fragments of alkaline protease genes and subsequent Southern blotting were applied by Tang *et al.* (1993) to differentiate the species from *A. flavus*. SSU and ITS restriction maps based on NCBI M60300 and AF078890:

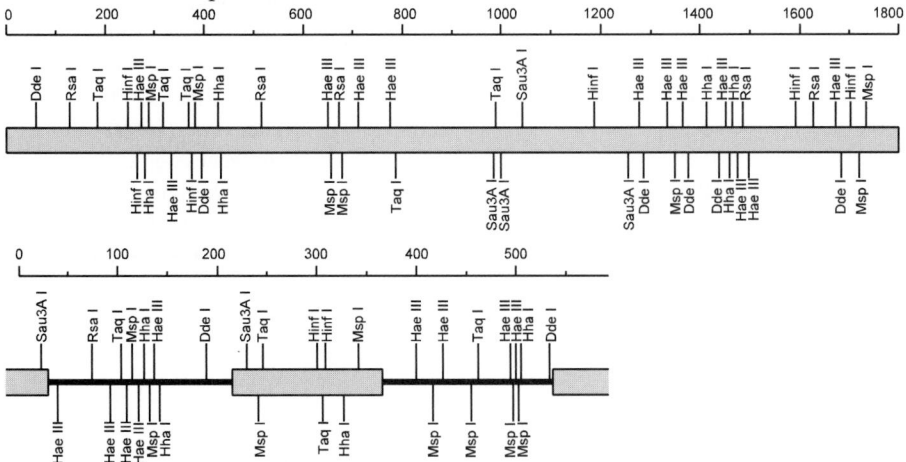

Pathogenicity. BSL-2. It is the main agent of aspergillosis in patients with impaired natural immunity (p. 27; Johnson, 1987; Bodey & Vartivarian, 1989; Vogeser *et al.*, 1997). The species causes a typical inhalation mycosis, whereby colonization and invasion are generally accompanied by allergic reactions. Particularly neutropenic patients (Moreau *et al.*, 1993), transplant patients (Waller *et al.*, 1991a; Egan *et al.*, 1996; Taillandier *et al.*, 1997), patients under intensive care (le Conte *et al.*, 1995) and patients with chronic granulomatous disease (Mouy *et al.*, 1995) are at risk, but systemic cases are also reported in patients without apparent underlying disease (Karim *et al.*, 1997). Pulmonary aspergillosis occurs among patients with <50 CD4 cells·μL^{-1} (Mylonakis *et al.*, 1998), but those with CD4 <100 cells·μL^{-1} are at risk (Addrizzo-Harris *et al.*, 1997). AIDS-associated cases are uncommon (Mylonakis *et al.*, 1996; Kuemmerk & Wedler, 1998) and are mostly secondary, due to steroid administration and neutropenia (Miller *et al.*, 1994; Tumbarello *et al.*, 1994), but may be underdiagnosed (Manfredi *et al.*, 1998). Disseminated infections in patients without apparent immune disorder are extremely rare (Orem *et al.*, 1998). Traumatic mycoses are rare, and are restricted to immunodeficient patients (Barson & Ruyman, 1986; Perzigian & Faix, 1993; Bretagne *et al.*, 1997), or wound-contamination is concerned (Bryce *et al.*, 1996). Non-pulmonary cavities can also be colonized (Matsuo *et al.*, 1995). The species occurs commonly in external ears and is one of the prevalent agents of fungal sinusitis (Rowe-Jones, 1993; Min *et al.*, 1996; Klossek *et al.*, 1996; Vennewald *et al.*, 1999). Sinusitis may be allergic, lead to fungus balls (Eloy *et al.*, 1996; Ferreiro *et al.*, 1997) or become invasive (Rieske *et al.*, 1998). In the latter case, complication may occur which may be cranial and life-threatening (Naim-ur-Rahman *et al.*, 1996). Invasion of adjacent tissues and dissemination is generally linked to innate immunosuppression (Katz *et al.*, 1993), e.g., in transplant patients (Grigg & Clouston, 1995) or after near-drowning (ter Maaten *et al.*, 1995). Cases of cerebral aspergillosis have been described (Kim *et al.*, 1993; Kerkmann *et al.*, 1994; Iemmolo *et al.*, 1998). Catheter-related cases are uncommon (Girmenia *et al.*, 1995; Berner *et al.*, 1996). Animal cases were described by Moore *et al.* (1993) and Pastor *et al.* (1993). The species natural occurrence is in rotten plant material at higher temperatures; it is abundant in the air during biological waste treatment (Göttlich, 1996). Nosocomial infections in immunocompromised patients may be due to hygiene problems (Anderson *et al.*, 1996) or construction works (Opal *et al.*, 1986; Dewhurst *et al.*, 1990).

Variety. The var. *ellipticus* was described with conidia 3.5-4.5 × 2.2-2.8 µm (Raper & Fennell, 1965; Latgé, 1999).
References. Kozakiewicz (1995), Brakhage *et al.* (1995), Samson *et al.* (1996), Latgé *et al.* (2000).
Antifungal susceptibility.

Antifungal	MICs range	MIC 90	Strains	Reference
AMB	0.39-1.56	1.56	25	Wildfeuer *et al.* (1998)
AMB	0.125-1	0.5	73	Hennequin *et al.* (1997)
AMB	0.25-1	1	20	Clancy & Nguyen (1998)
AMB	0.125-1	1	29	Tawara *et al.* (2000)
AMB	0.12-2	1	156	Dannaoui *et al.* (1999)
AMB	0.13-0.5	nd	50	Moore *et al.* (2000)
AMB	0.25-2	1	130	Verweij *et al.* (1998)
FLZ	8->64	>64	29	Tawara *et al.* (2000)
ITZ	0.5	1.56	25	Wildfeuer *et al.* (1998)
ITZ	0.03-0.12	0.06	73	Hennequin *et al.* (1997)
ITZ	0.016-1	0.5	29	Tawara *et al.* (2000)
ITZ	0.12-16	2	156	Dannaoui *et al.* (1999)
ITZ	0.13->16	nd	50	Moore *et al.* (2000)
ITZ	0.06-1	0.25	130	Verweij *et al.* (1998)
KTZ	0.39-3.13	3.13	25	Wildfeuer *et al.* (1998)
TBF	0.03-1	nd	14	Jessup *et al.* (2000)
TBF	0.05-4	1.6	82	Ryder *et al.* (1999)
VCZ	0.10-0.78	0.39	25	Wildfeuer *et al.* (1998)
VCZ	0.125-2	0.5	20	Clancy & Nguyen (1998)
VCZ	0.25-4	0.5	130	Verweij *et al.* (1998)

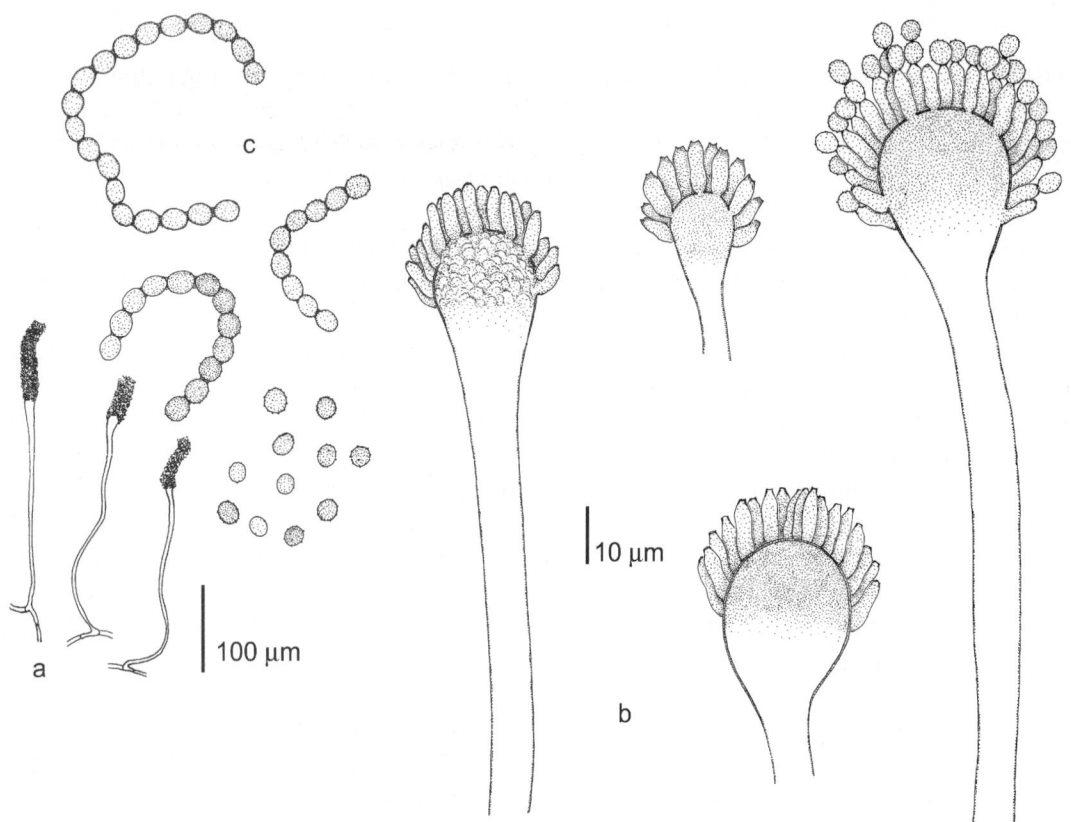

Aspergillus fumigatus, CBS 542.75. a. Conidiophores habit sketch; b. uniseriate conidiophores; c. conidia.

Nomenclature var. *ellipticus.* *Aspergillus fumigatus* Fresenius var. *ellipticus* Raper & Fennell - Gen. Aspergillus p. 246, 1965.

Nomenclature var. *fumigatus.* *Aspergillus fumigatus* Fresenius - Beitr. Mykol. p. 81, 1850.

  Aspergillus bronchialis Blumentritt - Ber. Deutsch. Bot. Ges. 19: 442, 1901.

  Aspergillus fumigatus Fresenius var. *minimus* Sartory - Bull. Acad. Méd. Paris, Sér. 3, 82: 304, 1919.

  Aspergillus septatus A. Sartory & R. Sartory - C.R. Hebd. Séanc. Acad. Sci., Paris 216: 428, 1943.

  Aspergillus phialoseptus Kwon-Chung - Mycologia 67: 771, 1975.

  Aspergillus fumigatus Fresenius var. *fulvi-ruber* Rai, Tewari & Agarwal - Beih. Nova Hedw. 47: 492, 1974 ≡ *Aspergillus arvii* Aho, Horie, Nishimura & Miyaji - Mycoses 37: 389, 1994 (name change).

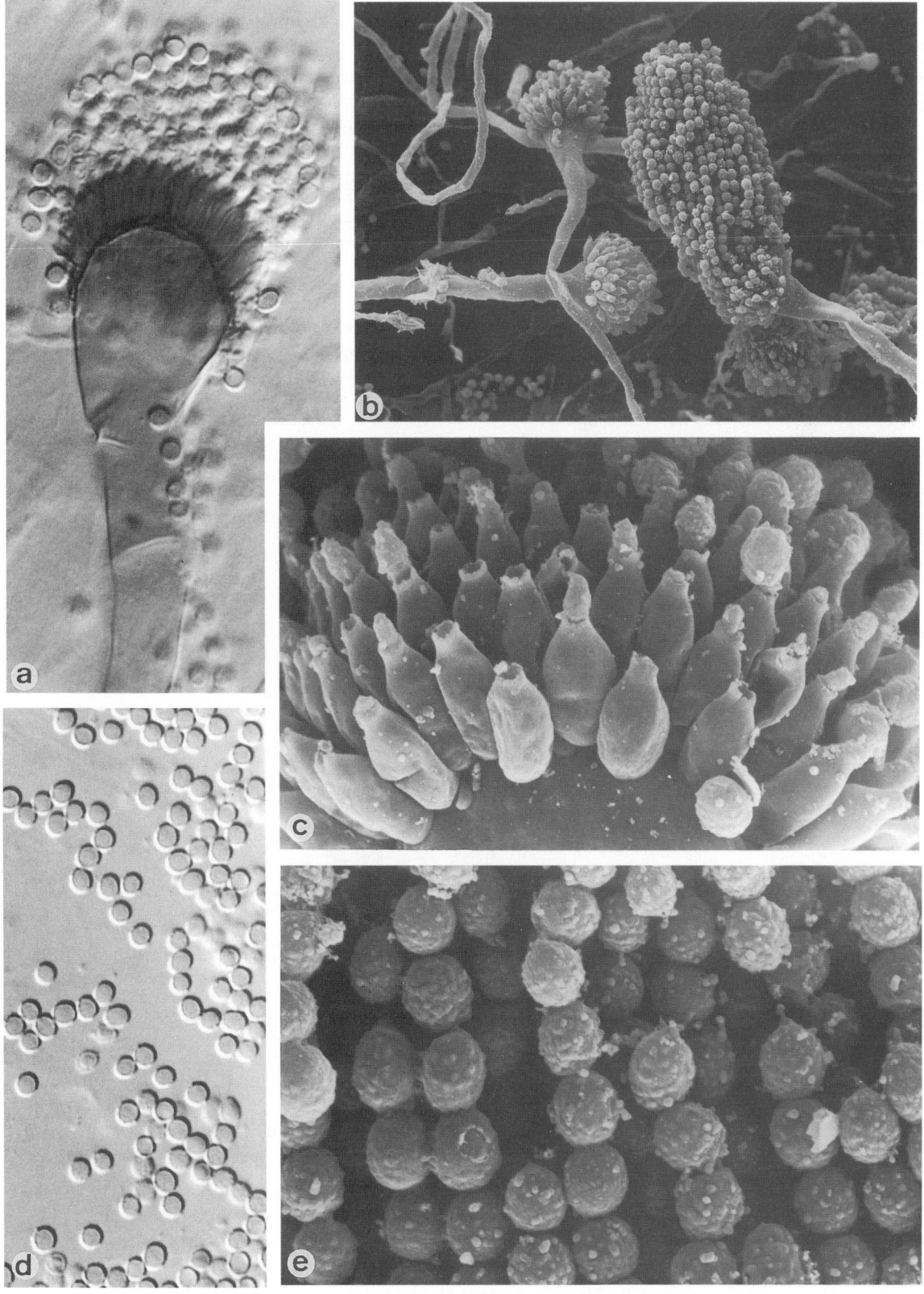

Aspergillus fumigatus, CBS 542.75. Uniseriate conidiophores and conidia. a. ×1600; b. ×800; c. ×5000; d. ×1600; e. ×6000.

Aspergillus glaucus Link

Colony characteristics. Colonies (CzA) spreading broadly, flat, dull green to grey-green.

Microscopy. Conidial heads sparse, radiate, 500-1000 μm diam, pale blue-green. Vesicles spherical. Conidiophores 700-800 × 2-3 μm, hyaline, smooth-walled. Conidiogenous cells uniseriate, 5.0-7.5 × 3-4 μm. Conidia ovoidal or aculeate, echinulate, hyaline, 4.5-7.5 μm diam.

Teleomorph. *Eurotium herbariorum* (Wiggers) Link (*Ascomycota, Euascomycetes, Eurotiales: Trichocomaceae*). Ascomata covered with red hyphae, yellow, spherical to subspherical, 75-125 μm diam. Asci 8-spored, 10-12 μm diam. Ascospores smooth-walled or occasionally slightly roughened, with a pronounced equatorial furrow, hyaline to subhyaline, lenticular, 6-7 × 3.2-5.1 μm.

Molecular diagnostics. A 26S rDNA-based specific probe was developed by Sandhu *et al.* (1995). SSU restriction map based on NCBI AB008402:

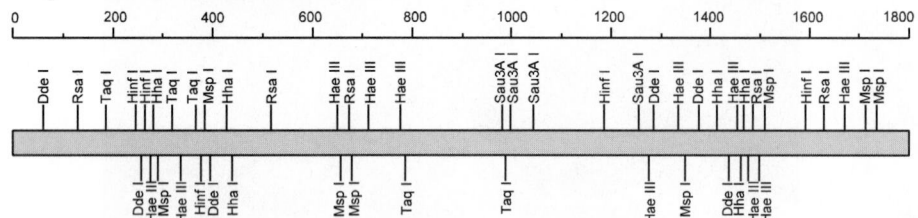

Pathogenicity. BSL-1. The species has been reported from cases of onychomycosis (Bereston & Waring, 1946). More recently, otitis (Bambule *et al.*, 1982) and orofacial infections (Dreizen *et al.*, 1985) have been reported, as well as cerebral, cardiovascular and visceral infections (Rippon, 1988).

Reference. Blaser (1975).

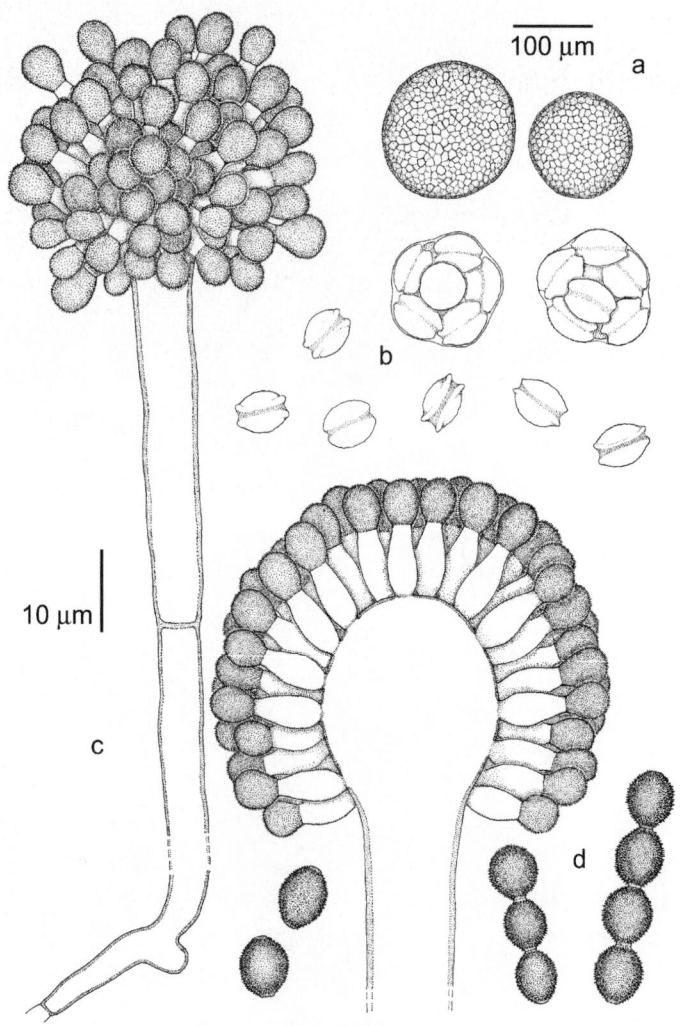

Aspergillus glaucus (*Eurotium herbariorum*), CBS 758.74. a. Ascomata; b. asci and ascospores; c. conidiophores; d. conidia.

Antifungal susceptibility.

Antifungal	MICs range	MIC 90	Strains	Reference
AMB	0.39-1.56	1.56	8	Wildfeuer *et al.* (1998)
ITZ	0.39	0.39	8	Wildfeuer *et al.* (1998)
KTZ	0.39-1.56	1.56	8	Wildfeuer *et al.* (1998)
VCZ	0.2-0.78	0.78	8	Wildfeuer *et al.* (1998)

Nomenclature. *Mucor herbariorum* Wiggers - Prim. Fl. Holst. p. 111, 1780 ≡ *Eurotium herbariorum* (Wiggers) Link - Mag. Ges. Naturf. Freunde 3: 31, 1809.

Aspergillus glaucus Link - Mag. Ges. Naturf. Freunde 3: 16, 1809.

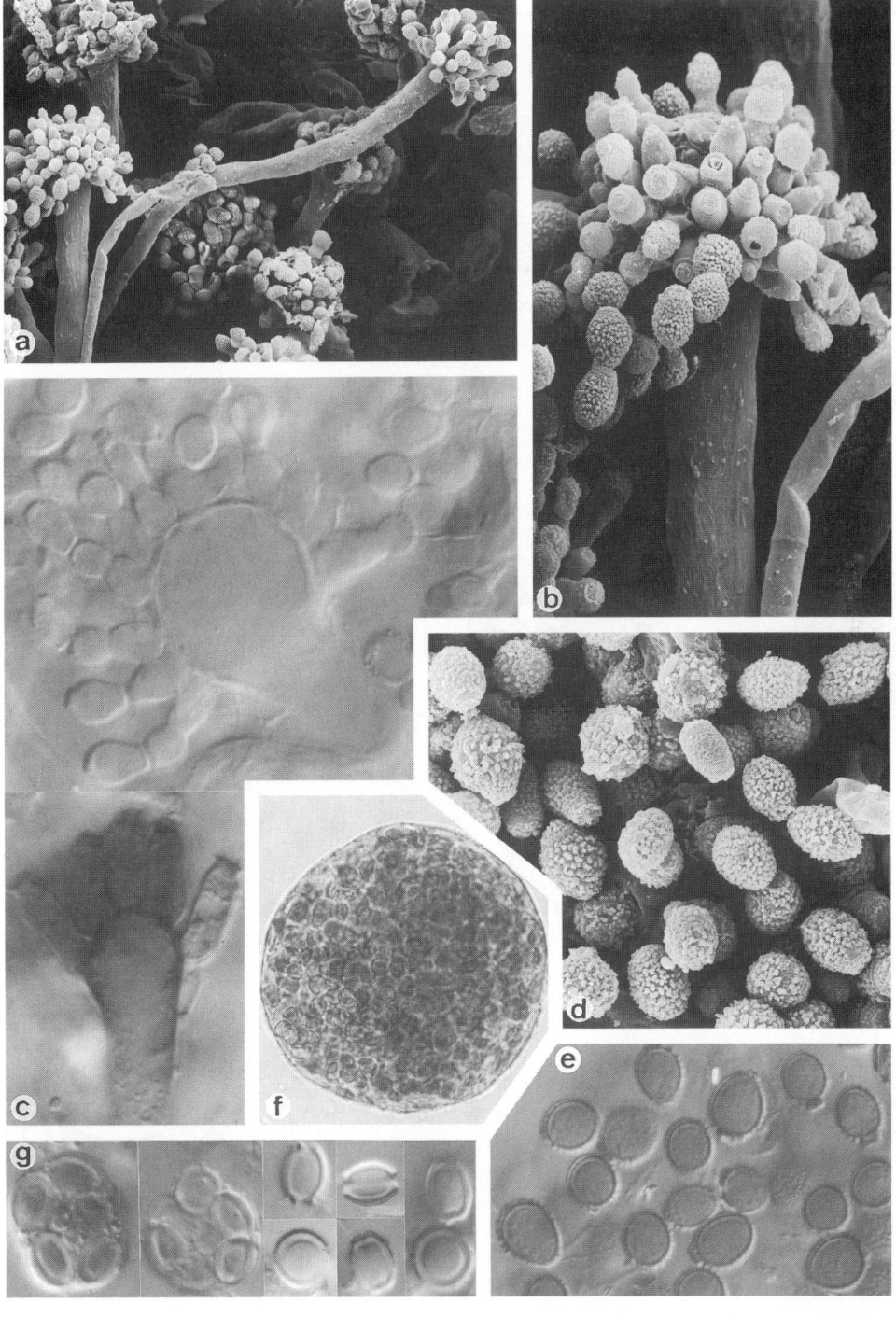

Aspergillus glaucus (Eurotium herbariorum), CBS 758.74. a-e. Conidiophores and conidia; f, g. ascomata, asci and ascospores. a. ×1405; b. ×1610; c. ×1265; d. ×4740; e. ×1265; f. ×405; g. ×1265.

Aspergillus granulosus Raper & Thom

Colony characteristics. Colonies (CzA) growing rapidly, irregularly furrowed, farinose, olivaceous-buff to purplish-brown.

Microscopy. Conidial heads more or less radiate, pale green. Conidiophores smooth-walled, up to 300 μm long; stipes tan to light brown. Vesicles ovoidal, 12-18 μm wide and 15-25 μm long. Conidiogenous cells biseriate, 3.5-5.0 μm, covering the entire surface of the vesicle. Conidia verruculose, 3.5-5.5 μm diam. Hülle cells abundant, spherical to ellipsoidal, 12-30 × 4-5 μm.

Pathogenicity. BSL-1. Soil-borne species. A disseminated infection in a heart tranplant patient was reported by Fakih *et al*. (1995).

Reference. Raper & Fennell (1965).

Antifungal susceptibility.

Antifungal	MICs	Strains	Reference
AMB	4	1	Unpublished data
5FC	256	1	Unpublished data
FLZ	128	1	Unpublished data
ITZ	1	1	Unpublished data
KTZ	4	1	Unpublished data
MCZ	4	1	Unpublished data

Nomenclature. *Aspergillus granulosus* Raper & Thom - Mycologia 36: 565, 1944.

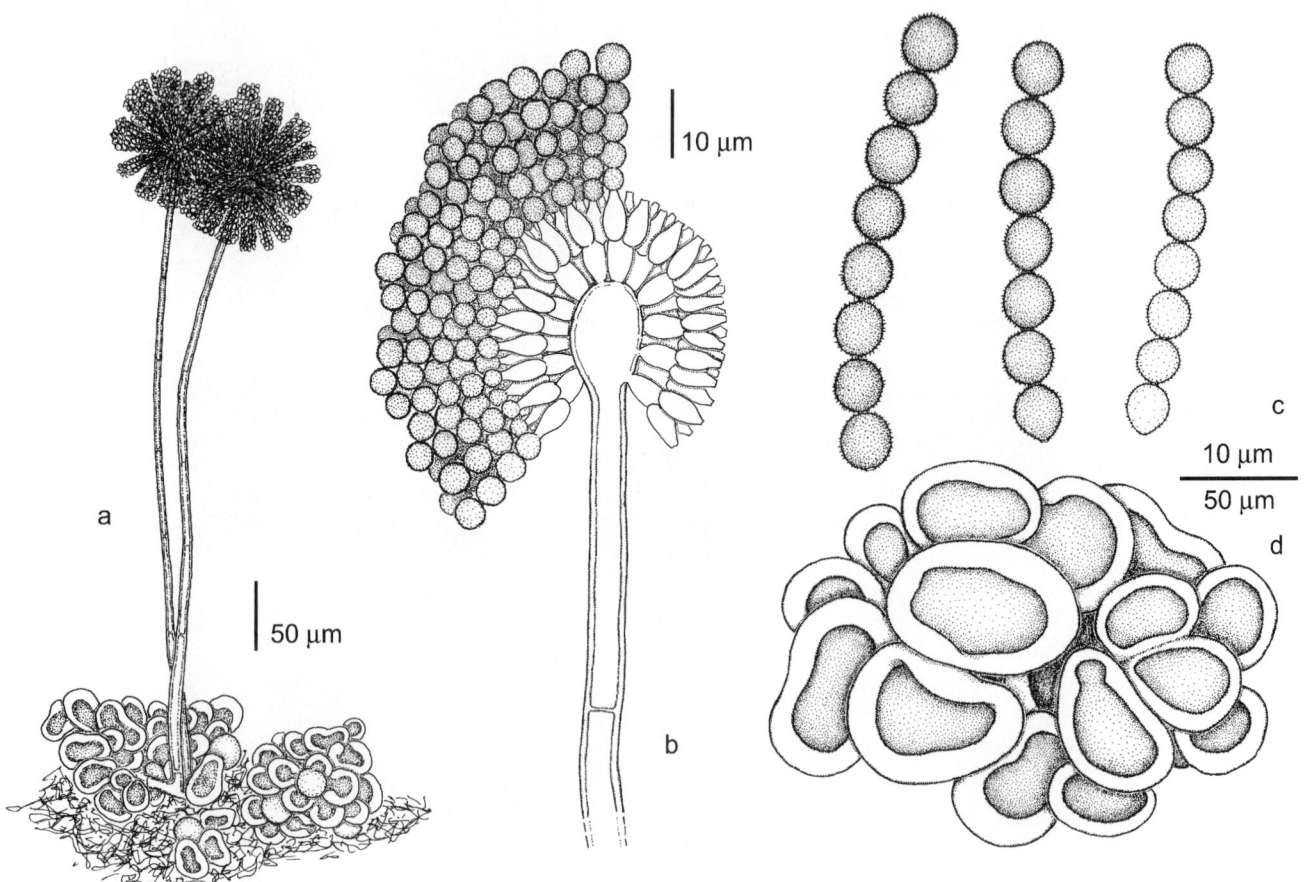

Aspergillus granulosus, CBS 119.58. a. Habit sketch; b. conidial head; c. conidia; d. cluster of Hülle cells.

Aspergillus granulosus, CBS 119.58. a, b, e, f. Conidiophores and conidia; c. cluster of Hülle cells; d. conidiophores and a cluster of Hülle cells. a. ×128; b. ×1600; c. ×640; d. ×500; e. ×2000; f. ×5500.

479

Aspergillus hollandicus Samson & W. Gams

Colony characteristics. Colonies (CzA) restricted, growing rapidly on media with 20% additional sucrose, yellow to dull yellow-grey.

Microscopy. Conidial heads radiate to loosely columnar, olive-green. Conidiophores smooth-walled, 275-350 μm long. Vesicles spherical, 18-25 μm diam. Conidiogenous cells uniseriate, 5-8 × 3.4-5.0 μm, covering at least the upper two-third of the vesicle. Conidia finely roughened to densely spinulose, spherical to subspherical, 3-5 μm diam.

Teleomorph. *Eurotium amstelodami* Mangin (*Ascomycota, Euascomycetes, Eurotiales: Trichocomaceae*).
Ascomata clustered, forming a dense layer, spherical, bright yellow, 120-160 μm diam. Asci 8-spored, spherical to subspherical, 10-12 μm diam. Ascospores pale, rough-walled, with a V-shaped equatorial furrow, lenticular, 4.5-6.0 × 3.5-4.0 μm.

Pathogenicity. BSL-1. The species has been isolated from a wide range of human mycoses: otitis (Wadhwani & Srivastava, 1984), mycetoma (Fonseca, 1930; Lacaz & Netto, 1954), dermatomycosis (Janke, 1954), cerebral abscess (David *et al.*, 1951), onychomycosis (Grigoriu & Grigoriu, 1975), keratitis (Shukla *et al.*, 1985) and pulmonary infections (Young *et al.*, 1972).

References. Raper & Fennell (1965), Blaser (1975).

Nomenclature. *Eurotium amstelodami* Mangin - Annls Sci. Nat., Bot., Sér. 9, 10: 360, 1909 ≡ *Aspergillus amstelodami* (Mangin) Thom & Church - The Aspergilli p. 113, 1926.

Aspergillus vitis Novobranova - Nov. Sist. Nizš. Rast. 9: 175, 1975. On the basis of a neotype, this would be the valid name for the anamorph (Kozakiewicz, 1989).

Aspergillus hollandicus Samson & W. Gams, *in* Samson - Adv. Pen. Asp. Syst. p. 33, 1985.

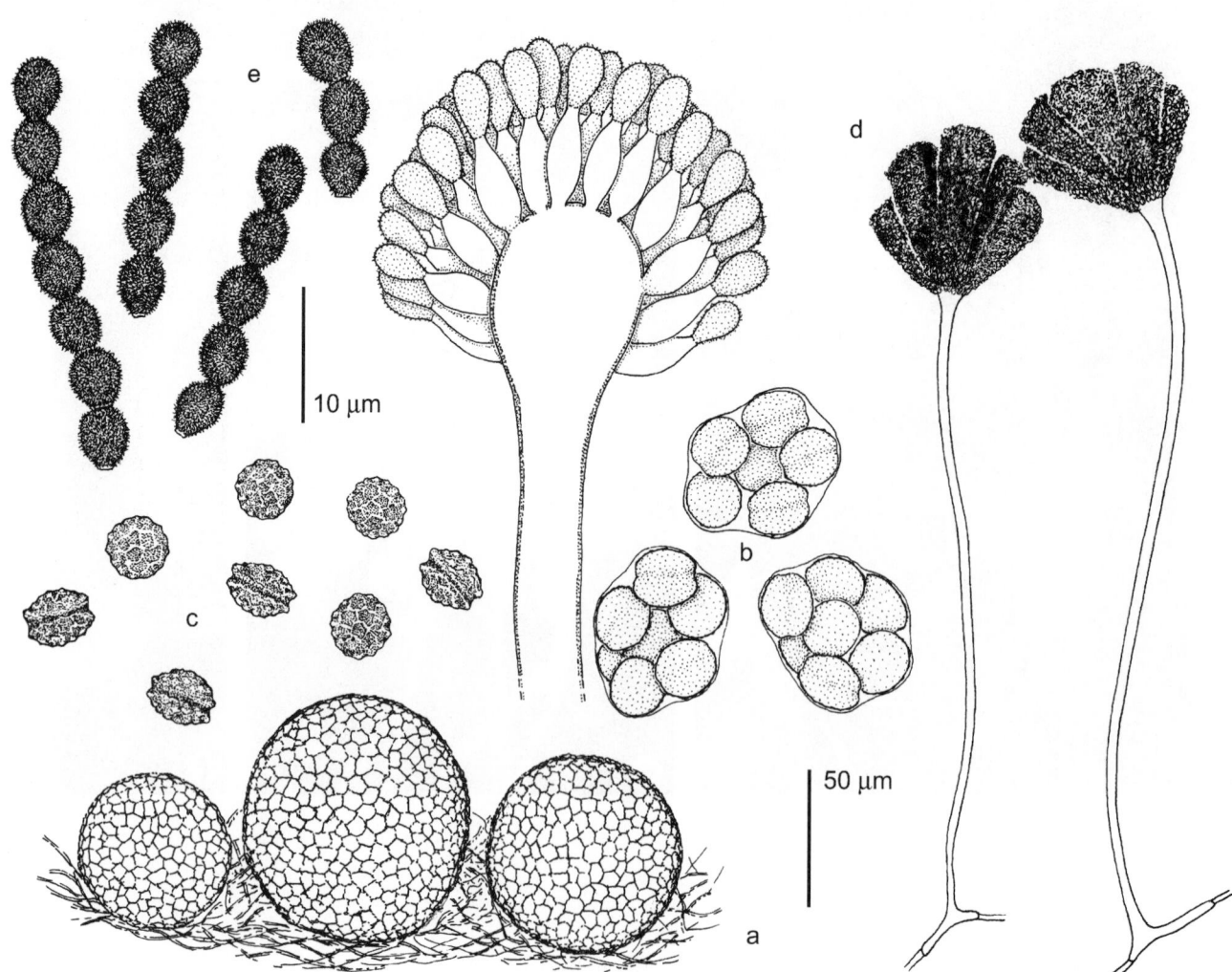

Aspergillus hollandicus (Eurotium amstelodami), CBS 518.65. a. Ascomata; b. asci; c. ascospores; d. conidiophores; e. conidia.

Aspergillus hollandicus (Eurotium amstelodami), CBS 518.65. a-d. Conidiophores and conidia; e, f. ascomata; g. part of ascoma wall; h, i. asci and ascospores. a. ×1600; b. ×890. c. ×3500; d. ×6500; e. ×490; f. ×512; g. ×1600; h. ×1600; i. ×6500.

Aspergillus janus Raper & Thom

Colony characteristics. Colonies (CzA) spreading irregularly, pale yellow-buff with dark green and white zones.
Microscopy. Conidial heads of two types: green and columnar, and white and radiate. (1) Conidiophores with green heads, smooth-walled, 300-400 µm in length; stipes hyaline to slightly yellow; vesicles ovoidal; conidiogenous cells biseriate with metulae all over. Conidia spherical, dark green in mass, spinulose, 2.5-3.5 µm diam. (2) Conidiophores with white heads, smooth-walled, 2.0-2.5 mm in length, stipes hyaline to slightly yellow; vesicles clavate, entire surface covered by metulae; conidiogenous cells biseriate. Conidia hyaline, minutely verrucose (smooth-walled with light microscopy), (sub)spherical, 2.0-2.5 µm diam.
Pathogenicity. BSL-1. Keratitis (Neuhann, 1976).
References. Raper & Fennell (1965), Kozakiewicz (1989).
Nomenclature. *Aspergillus janus* Raper & Thom - Mycologia 36: 556, 1944.

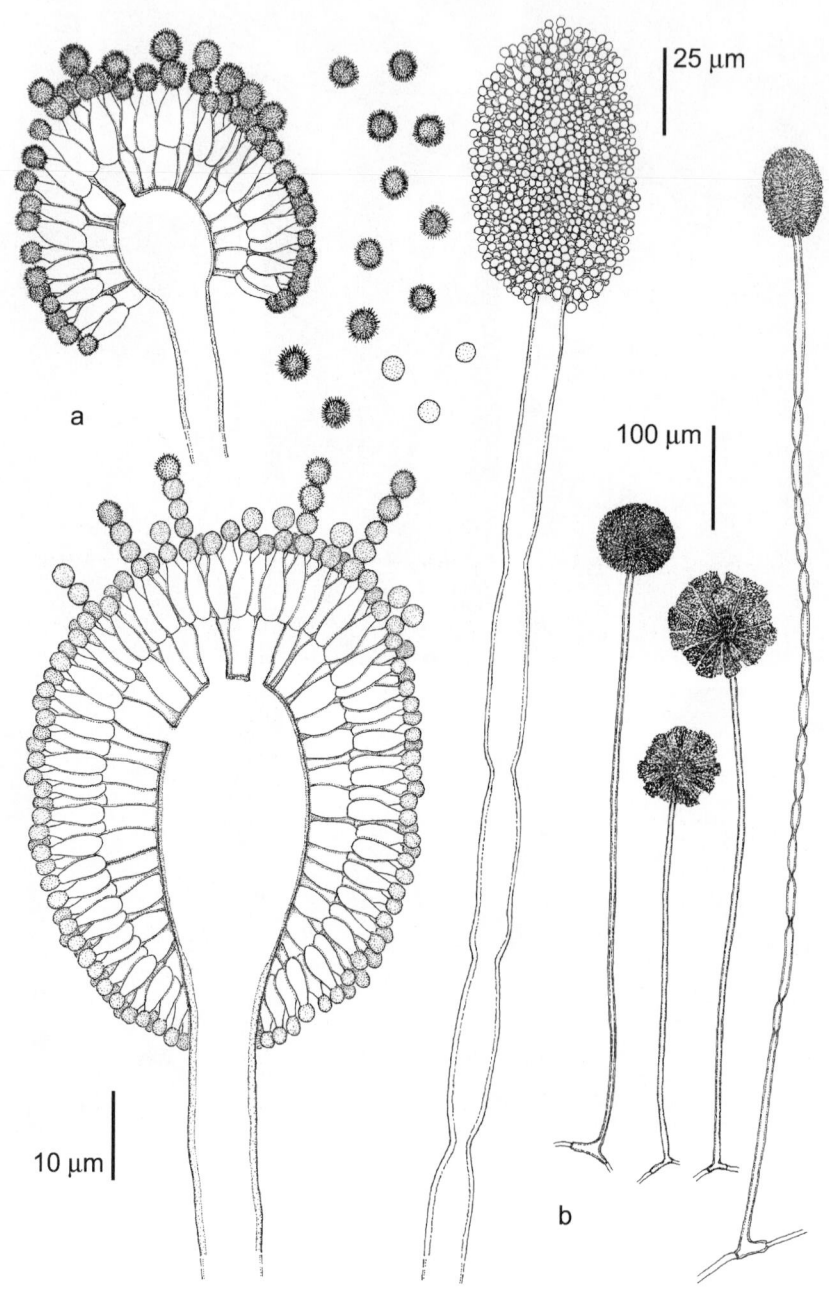

Aspergillus janus, CBS 118.45. a. Conidial heads and conidia; b. conidiophores.

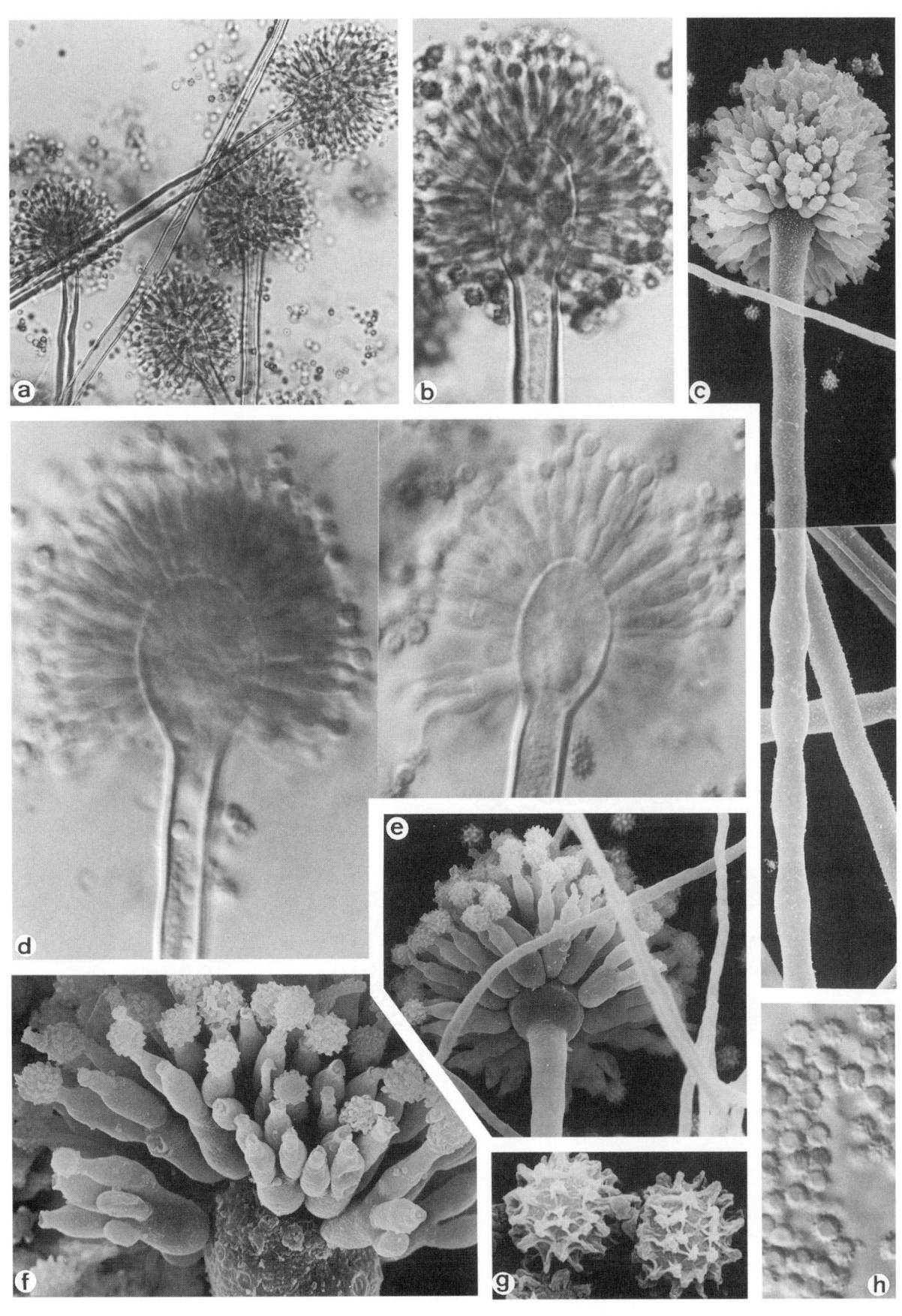

Aspergillus janus, CBS 118.45. Conidiophores and conidia. a. ×270; b. ×430; c. ×1130; d. ×1340; e. ×1675; f. ×3100; g. ×6280; h. ×1340.

Aspergillus japonicus Saito

Colony characteristics. Colonies (CzA) purplish-black.

Microscopy. Conidial heads spherical to radiate, splitting into well-defined, divergent columns, black. Conidiophore stipes smooth-walled, hyaline or slightly pigmented at the apex. Vesicles brown, subspherical, 15-45 μm (commonly 20-35 μm) diam. Conidiogenous cells uniseriate. Conidia hyaline to brown, conspicuously echinulate, subspherical to ellipsoidal, 4.5-6.0 × 4-5 μm. Sclerotia produced in some strains.

Molecular diagnostics. ITS restriction map based on NCBI U65308:

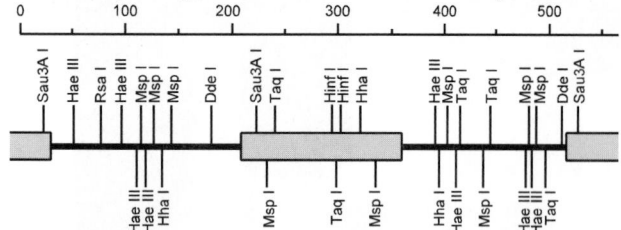

Pathogenicity. BSL-1. This fungus was isolated from cases of otitis (Wadhwani & Srivastava, 1985).

References. Raper & Fennell (1965), Al-Musallam (1980).

Nomenclature. *Aspergillus japonicus* Saito - Bot. Mag., Tokyo 20: 61, 1906.
  Aspergillus japonicus Saito var. *capillatus* Nakazawa, Takeda & Suematsu - J. Agric. Chem. Soc. Jap. 8: 12, 1932.

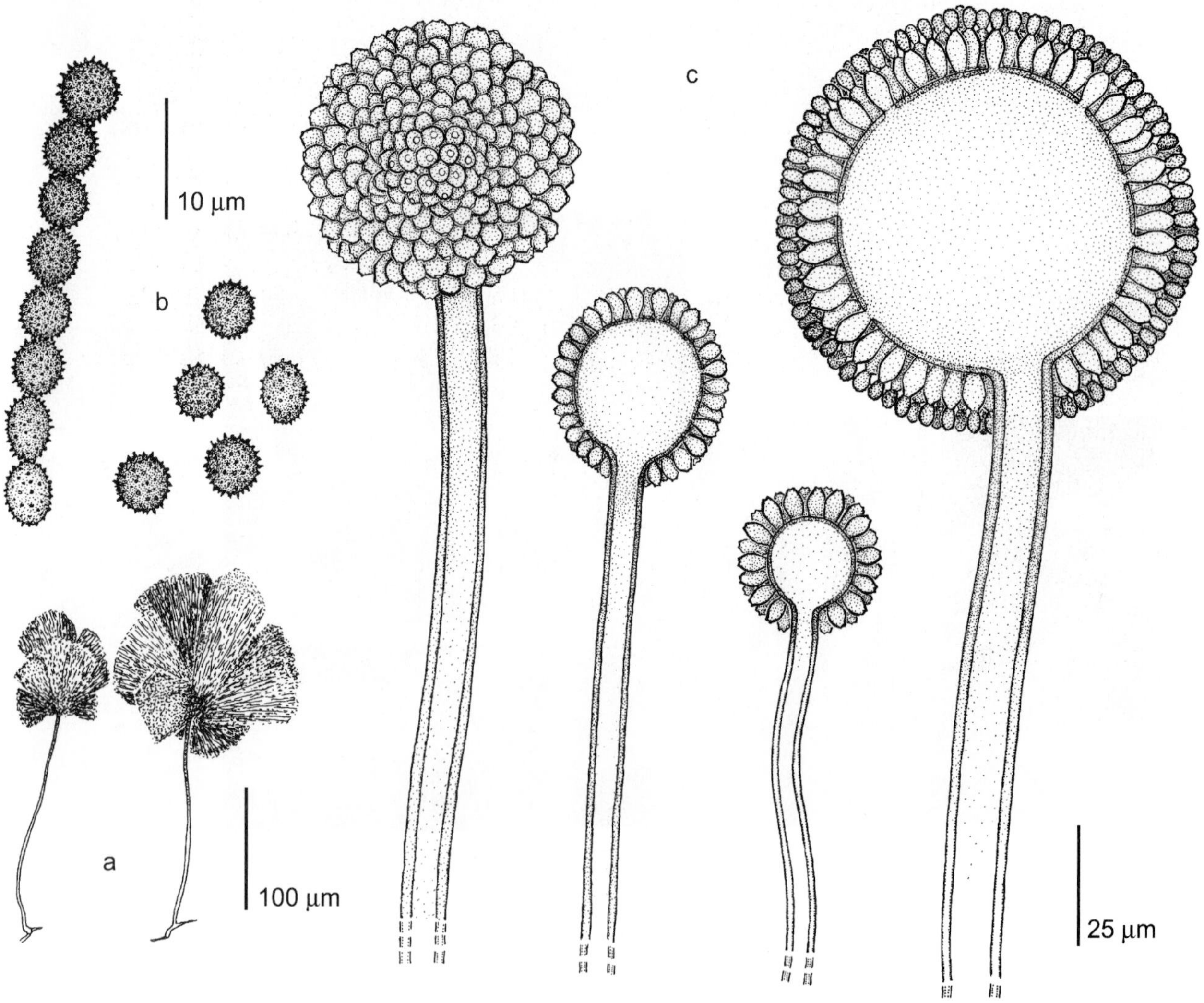

Aspergillus japonicus, CBS 114.51. a. Habit sketch; b. conidia; c. conidiophores with uniseriate conidial heads.

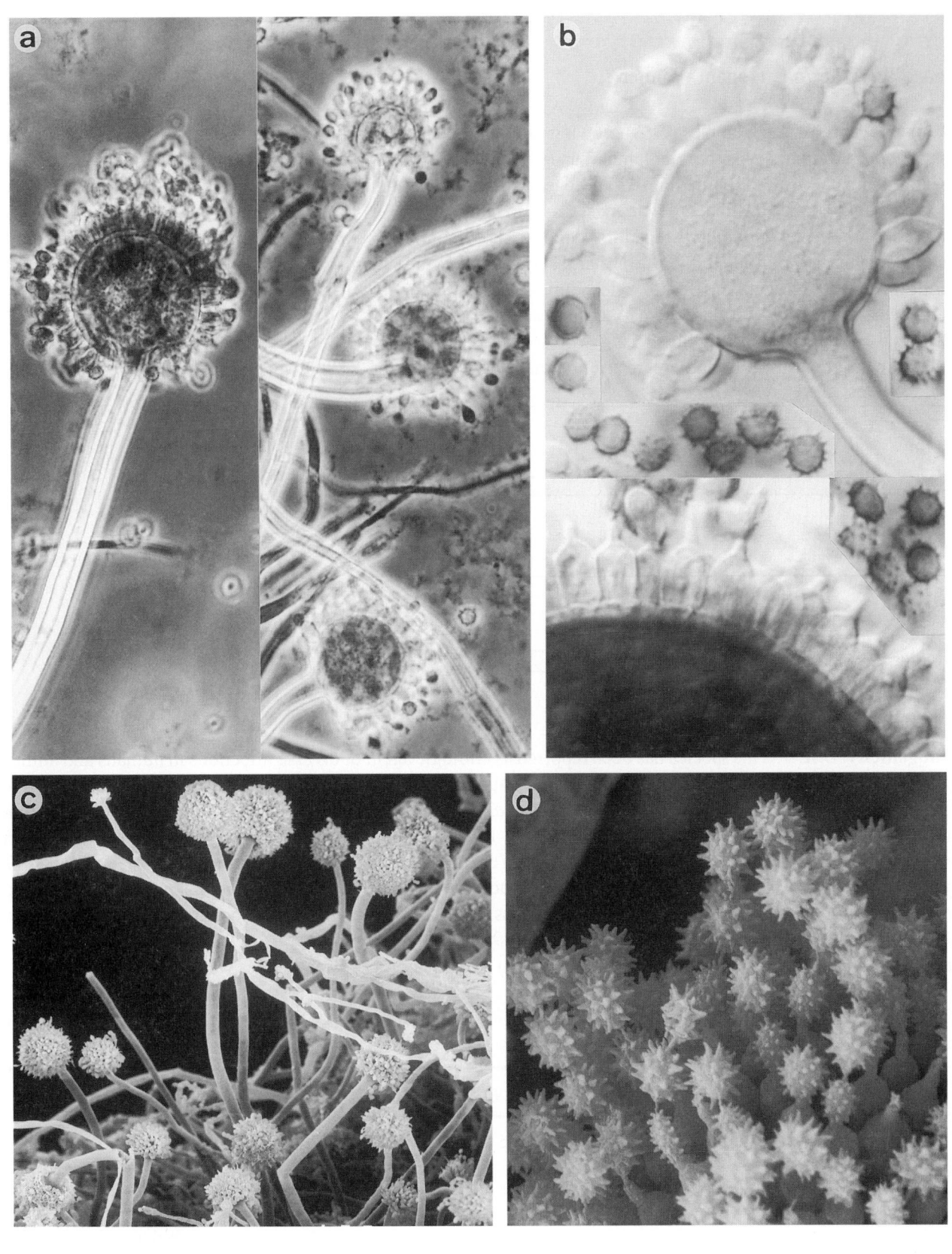

Aspergillus japonicus, CBS 114.51. a, c. Conidiophores; b, d. details of conidial heads and conidia. a. ×640; b. ×1600; c. ×255; d. ×3100.

Aspergillus nidulans (Eidam) Winter

Colony characteristics. Colonies (CzA) growing rapidly, green, cream-buff or honey-yellow; reverse dark purplish.
Microscopy. Conidial heads short, columnar, up to 80 µm long. Conidiophore stipes brownish, 60-130 × 2.5-3.0 µm. Vesicles hemispherical, 8-10 µm diam. Conidiogenous cells biseriate, 5-9 × 2-3 µm. Metulae 5-6 × 2.3 µm. Conidia spherical, rugulose, subhyaline, green in mass, 3-4 µm diam.
Teleomorph. *Emericella nidulans* (Eidam) Winter (*Ascomycota, Euascomycetes, Eurotiales: Trichocomaceae*). Ascomata spherical, purple, 100-200 µm diam, surrounded by a yellowish to cinnamon layer of scattered hyphae bearing a dense aggregation of pale yellow, thick-walled, spherical to subspherical Hülle cells. Asci 8-spored, spherical to subspherical, 7-12 µm diam. Ascospores purple-red, smooth-walled, with two equatorial crests, lenticular, 3.8-4.5 × 3.5-4.0 µm (excluding the crests).
Molecular diagnostics. SSU and ITS restriction maps based on NCBI X78539 and AF078899, resp.:

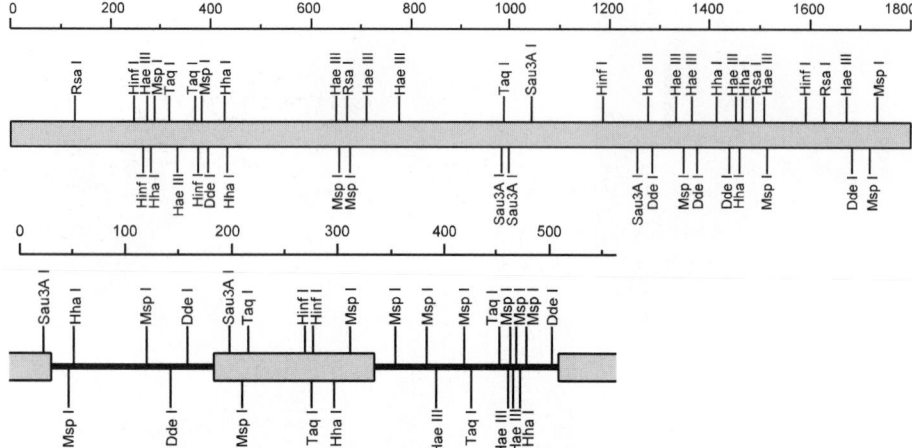

Pathogenicity. BSL-1. This species has been reported as etiologic agent of diverse infections in humans and animals, either alone or in association with other opportunistic fungi, among them pulmonary infections (Segretain & Vien, 1957; Shao *et al.*, 1983; Mizuki *et al.*, 1994; Neyens *et al.*, 1990), sinusitis (Doby & Kombila-Favry, 1978; Mitchell *et al.*, 1987), endophthalmitis (Biasoli & de Bracalenti, 1986), osteomyelitis (Redmond *et al.*, 1965) and superficial (Beverton & Waring, 1946) as well as disseminated (Welsh & Buchness, 1955) infections. Van 't Hek *et al.* (1998) reported a disseminated case of a CGD patient and Lucas *et al.* (1999) a catheter-related skin infection. A mixed cerebral and pulmonary infection was reported by Morris *et al.* (1995). Animal cases concern ducklings (Deka & Rao, 1988) and horses (Weiler *et al.*, 1991; Guillot *et al.*, 1997).
Variety. *A. nidulans* var. *dentatus* with ascospores with two equatorial rows of spines was described from a nail infection (Sandhu & Sandhu, 1963).
References. Verweij *et al.* (1995), Raper & Fennell (1965), Kozakiewicz (1995).
Antifungal susceptibility.

Antifungal	MICs range	MIC 90	Strains	Reference
AMB	0.78-6.2	6.25	7	Wildfeuer *et al.* (1998)
AMB	0.5-2	2	17	Dannaoui *et al.* (1999)
ITZ	0.2-0.78	0.78	7	Wildfeuer *et al.* (1998)
ITZ	0.12->16	4	17	Dannaoui *et al.* (1999)
KTZ	0.03-3.13	3.13	7	Wildfeuer *et al.* (1998)
VCZ	0.05-0.78	0.78	7	Wildfeuer *et al.* (1998)

Nomenclature var. *nidulans*. *Sterigmatocystis nidulans* Eidam, *in* Cohn - Beitr. Biol. Pfl. 3: 392, 1883 ≡ *Aspergillus nidulans* (Eidam) Winter, *in* Rabenhorst - Krypt Fl., 1-2: 62, 1883 ≡ *Diplostephanus nidulans* (Eidam) Neveu-Lemaire - Précis Parasitol. Hum., éd. 5, p. 101, 1921 ≡ *Emericella nidulans* (Eidam) Vuillemin - C.R. Hebd. Séanc. Acad. Sci., Paris 184: 137, 1927.

   Sterigmatocystis nidulans Eidam var. *nicollei* Pinoy - Arch. Parasit. 10: 437, 1906 ≡ *Aspergillus nidulans* (Eidam) Winter var. *nicollei* (Pinoy) Nicolle & Pinoy - C.R. Hebd. Séanc. Acad. Sci., Paris 144: 396, 1907.

   Aspergillus nidulans (Eidam) Winter var. *cesarii* Pinoy - Bull. Soc. Path. Exot. 8: 11, 1915.

   Aspergillus nidulellus W. Gams & Samson, *in* Samson & Pitt - Adv. Pen. Asp. Syst. p. 44, 1985.

Nomenclature var. *dentatus*. *Aspergillus nidulans* (Eidam) Winter var. *dentatus* D.K. Sandhu & R.S. Sandhu - Mycologia 55: 297, 1963 ≡ *Emericella nidulans* (Eidam) Winter var. *dentata* (D.K. Sandhu & R.S. Sandhu) Subramanian - Curr. Sci. 41: 758, 1972.

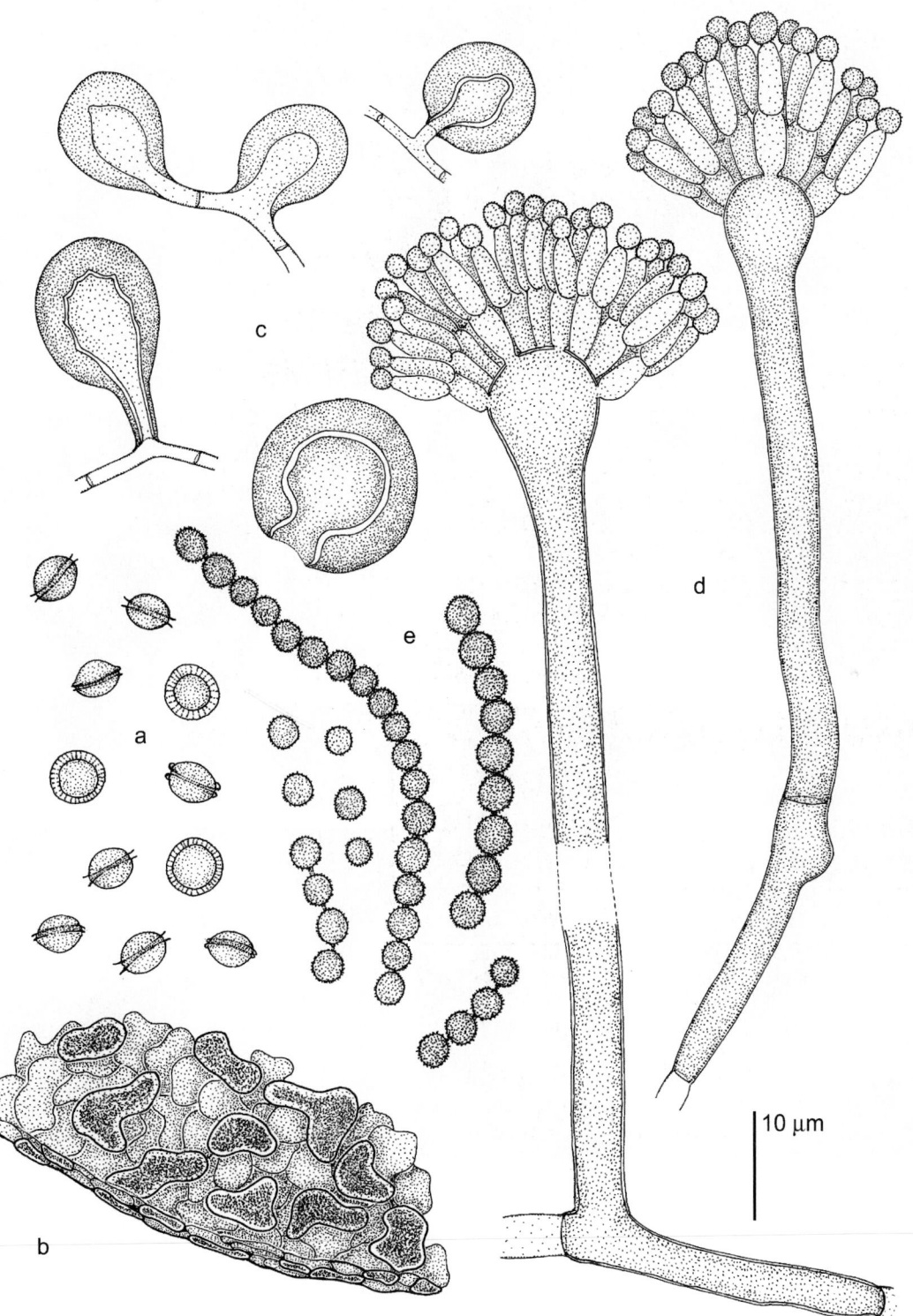

10 μm

Aspergillus nidulans (Emericella nidulans), CBS 100.20. a. Ascospores; b. part of ascoma wall; c. Hülle cells; d. conidiophores; e. conidia.

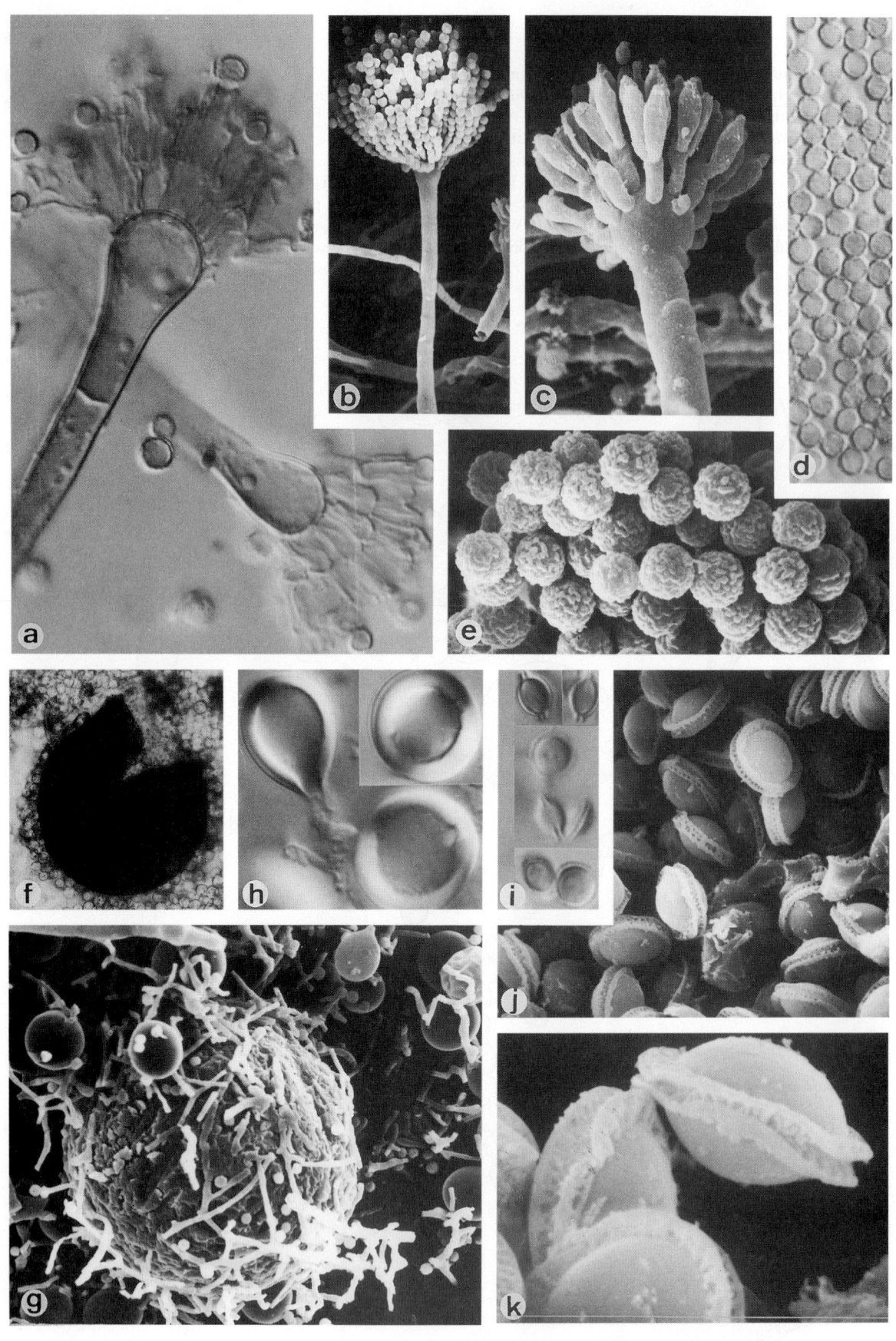

Aspergillus nidulans (Emericella nidulans), CBS 100.20. a-e. Conidiophores and conidia; f-h. ascomata and Hülle cells; i-k. ascospores. a. ×1600; b. ×780; c. ×2150; d. ×1600; e. ×4200; f. ×128; g. ×860; h. ×1600; i. ×1600; j. ×3400; k. ×9600.

Aspergillus niger v. Tiegh.

Colony characteristics. Colonies (CzA) black, consisting of a dense felt of conidiophores.

Microscopy. Conidial heads radiate. Conidiophore stipes smooth-walled, hyaline or pigmented. Vesicles subspherical, 50-100 μm diam. Conidiogenous cells biseriate. Metulae twice as long as the phialides. Conidia brown, ornamented with warts and ridges, subspherical, 3.5-5.0 μm diam.

Molecular diagnostics. 18S rDNA sequence data were published by Verweij *et al.* (1995). A 26S rDNA-based specific probe was developed by Sandhu *et al.* (1995). SSU and ITS restriction maps based on NCBI X78538 and U65306, resp.:

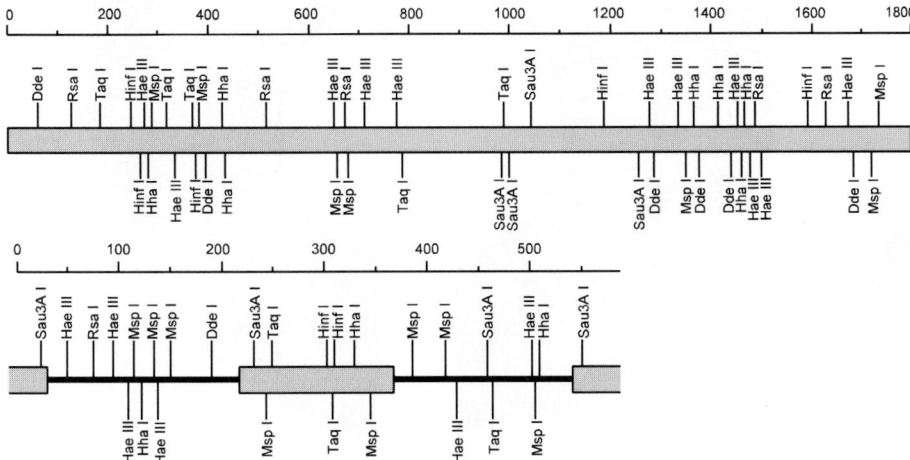

Pathogenicity. BSL-1. This species is frequently isolated subclinically (Beer & Taine, 1990) or clinically (Landry & Parkins, 1993) from human external ears; Bayó *et al.* (1994) regard the species as the prime etiologic agent of otomycosis. The species occasionally causes disseminated aspergillosis after major surgery (Yeldandi *et al.*, 1995). A cutaneous infection in a bone narrow transplant recipient was reported by Johnson *et al.* (1993) and an endophthalmitis by Krzystolik *et al.* (1997). A pulmonary aspergilloma was described by Korzeniowska-Kosela *et al.* (1990); pulmonary cases in diabetic patients have poor prognosis (Severo *et al.*, 1997). Peritonitis was reported by Bibashi *et al.* (1993) and endocarditis by Vivas (1998). Cases in immunocompromised patients are clearly associated with environmental hygiene (Loudon *et al.*, 1996). Onychomycoses were reported by Tosti & Piraccini (1998).

Variety. The variety *A. niger* var. *awamori* (Nakazawa) Al-Musallam, with predominantly olivaceous conidial heads and conidia with smaller warts, has been reported as responsible for a case of subcutaneous infection (Paldrok, 1965).

Reference. Raper & Fennell (1965).

Antifungal susceptibility.

Antifungal	MICs range	MIC 90	Strains	Reference
AMB	0.39-1.56	1.56	15	Wildfeuer *et al.* (1998)
AMB	0.5	0.5	15	Tawara *et al.* (2000)
AMB	0.12-0.5	0.5	15	Dannaoui *et al.* (1999)
AMB	0.13-0.5	nd	10	Moore *et al.* (2000)
FLZ	64->64	>64	15	Tawara *et al.* (2000)
ITZ	0.2-1.56	1.56	15	Wildfeuer *et al.* (1998)
ITZ	0.5-1	1	15	Tawara *et al.* (2000)
ITZ	0.12-1	1	15	Dannaoui *et al.* (1999)
ITZ	0.5-4	nd	10	Moore *et al.* (2000)
KTZ	0.39-3.13	1.56	15	Wildfeuer *et al.* (1998)
TBF	0.025-2.5	1.2	36	Ryder *et al.* (1999)
VCZ	0.1-0.39	0.39	15	Wildfeuer *et al.* (1998)

Nomenclature var. *awamori*. *Aspergillus awamori* Nakazawa - Rept Inst. Gov. Res. Formosa 1, 1907 ≡ *Aspergillus niger* van Tieghem var. *awamori* (Nakazawa) Al-Musallam - Rev. Black Asp. Spec. p. 60, 1980.

Nomenclature var. *niger*. *Aspergillus niger* van Tieghem - Annls Sci. Nat., Bot., Sér. 5, 8: 240, 1867 ≡ *Sterigmatocystis nigra* (van Tieghem) van Tieghem - Bull. Soc. Bot. Fr. 24: 102, 1877.

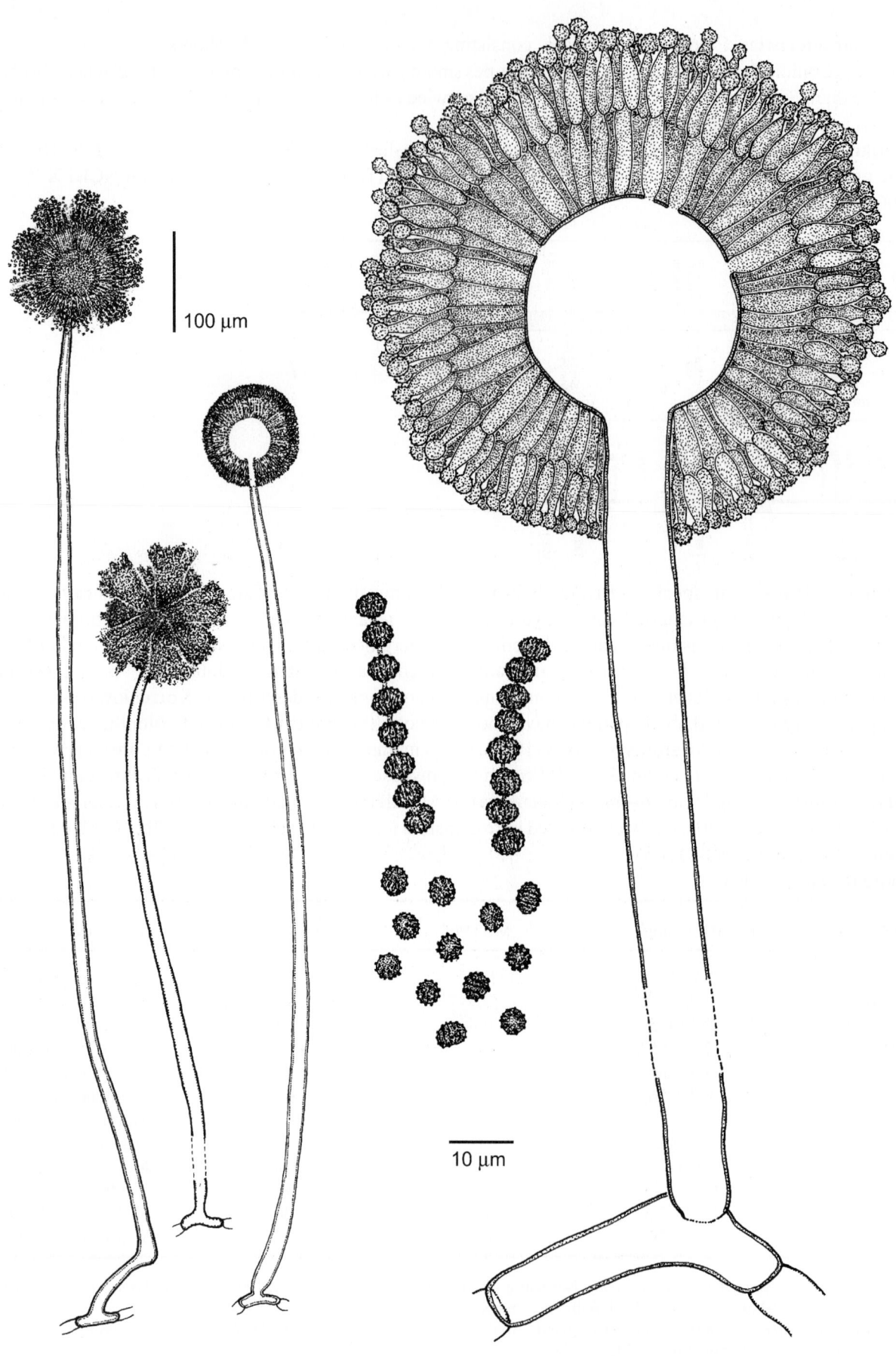

100 μm

10 μm

Aspergillus niger, FMR 2249. Conidiophores and conidia.

Aspergillus niger, FMR 2249. Conidiophores and conidia. a. ×320; b. ×242; c. ×1280; d. ×5500.

Aspergillus niveus Blochwitz

Colony characteristics. Colonies (CzA) growing rather slowly, white.

Microscopy. Conidial heads small, radiate to loosely columnar, white, becoming dull ivory with age. Vesicles Conidiophore stipes smooth-walled, hyaline, up to 1000 μm long. hemispherical, 8-15 μm diam. Conidiogenous cells biseriate. Metulae covering the upper one- to two-thirds of the vesicle. Conidia spherical, hyaline, 2.0-2.5 μm diam, smooth-walled.

Teleomorph. *Fennellia nivea* (Wiley & Simmons) Samson (*Ascomycota, Euascomycetes, Eurotiales: Trichocomaceae*).

Ascomata yellowish, spherical, 80-130 μm diam, surrounded by masses of Hülle cells. Asci 8-spored, subspherical, 9.6-11.2 μm. Ascospores spinulose, hyaline to yellowish, lenticular, 4.5-6.0 × 3.2-4.8 μm, with inconspicuous longitudinal grooves and two very low longitudinal crests.

Pathogenicity. BSL-1. Cited as responsible of otitis media (Wadhwani & Srivastava, 1984).

References. Raper & Fennell (1965), Klich & Pitt (1988).

Nomenclature. *Aspergillus niveus* Blockwitz - Annls Mycol. 27: 205, 1929.
    Aspergillus niveus Blochwitz var. *bifidus* Maia & Alecrim - Anais Fac. Med. Univ. Recife 15: 189, 1955.
    Emericella nivea Wiley & Simmons - Mycologia 65: 934, 1973 ≡ *Fennellia nivea* (Wiley & Simmons) Samson - Stud. Mycol. 18: 5, 1979.

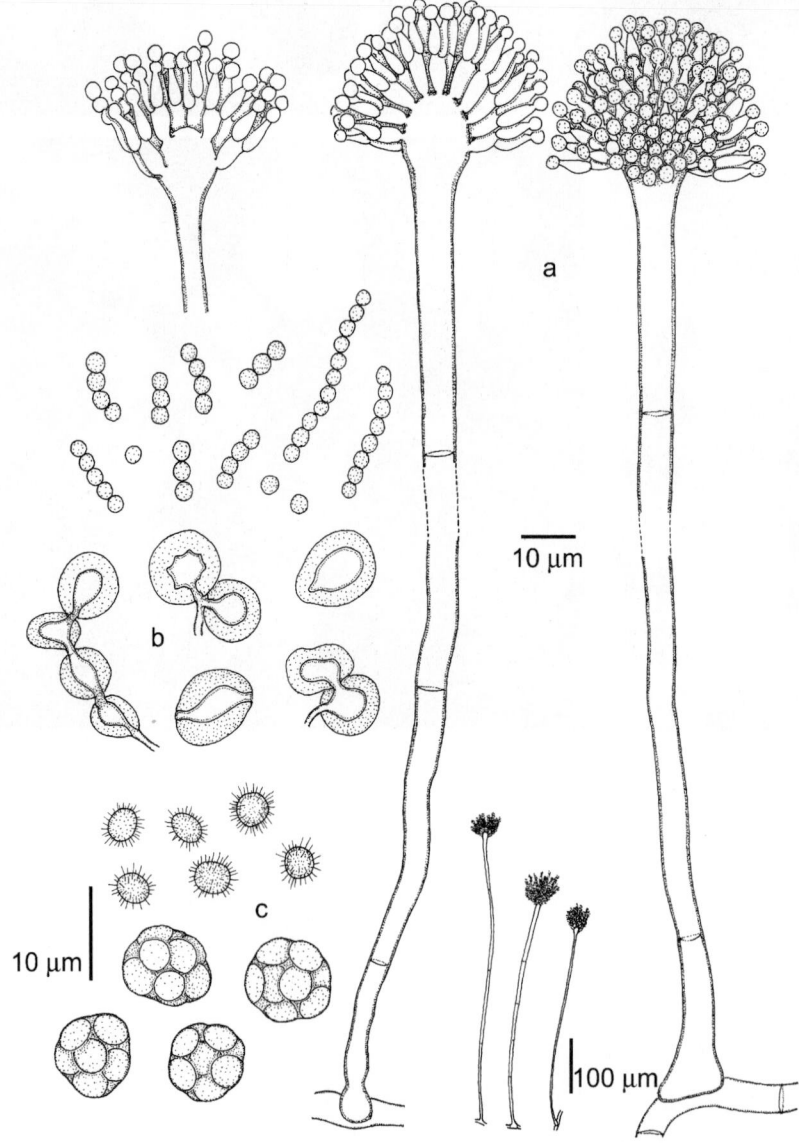

Aspergillus niveus (Fennellia nivea), CBS 262.73. a. Conidiophores and conidia; b. Hülle cells; c. asci and ascospores.

492

Aspergillus niveus (Fennellia nivea), CBS 262.73. a-d. Conidiophores and conidia; e, f. Hülle cells; g. ascomata and Hülle cells; h, i. ascoma wall, asci and ascospores. a. ×256; b. ×1600; c. ×390; d. ×2700; e. ×340; f. ×1600; g. ×256; h. ×1600; i. ×6000.

Aspergillus ochraceus Wilhelm

Colony characteristics. Colonies (CzA) with restricted growth, yellow-orange, ochraceous or buff.

Microscopy. Conidial heads radiate, splitting into several columns with age. Conidiophore stipes brownish, commonly 1.0-1.5 mm in length, with roughened walls. Vesicles spherical, thin-walled, hyaline, 35-50 µm diam. Conidiogenous cells biseriate. Metulae covering the entire vesicle. Conidia spherical to subspherical, 2.5-3.5 µm diam, smooth-walled to finely roughened. Sclerotia pink to vinaceous-purple, irregular in shape, up to 1 mm diam.

Molecular diagnostics. SSU restriction map based on NCBI AB008405:

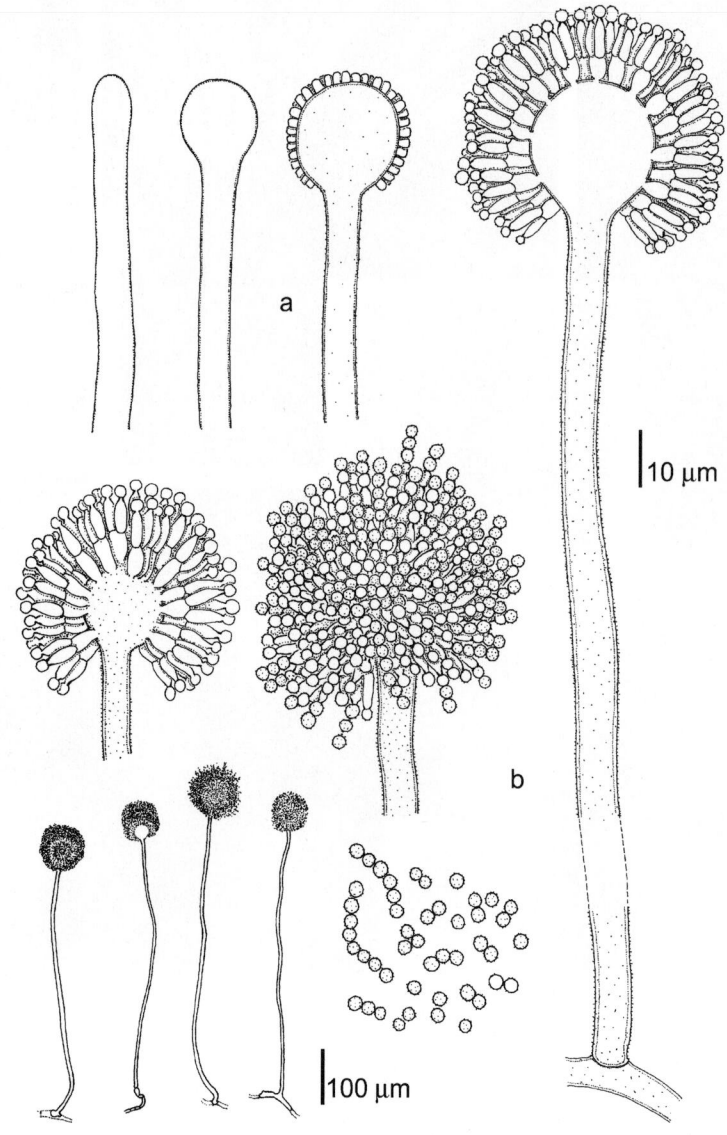

Pathogenicity. BSL-1. This species was implicated in a case of antromycosis in humans (Bassiouny *et al.*, 1982) and in a case of mycotic placentitis in a cow (Muñoz *et al.*, 1989).

References. Raper & Fennell (1965), Klich & Pitt (1988).

Aspergillus ochraceus, CBS 123.55. a. Habit sketch; b. young conidiophores; c. mature conidiophores and conidia.

Antifungal susceptibility.

Antifungal	MICs range	MIC 90	Strains	Reference
AMB	1.56-3.13	3.13	4	Wildfeuer *et al.* (1998)
ITZ	0.39-0.78	0.78	4	Wildfeuer *et al.* (1998)
KTZ	0.39-1.56	1.56	4	Wildfeuer *et al.* (1998)
VCZ	0.1-0.04	0.39	4	Wildfeuer *et al.* (1998)

Nomenclature. *Aspergillus ochraceus* Wilhelm - Rabenh. Fl. Europ. No. 2361, 1877.
  Aspergillus alutaceus Berkeley & Curtis - Grevillea 3: 108. 1875.
  Aspergillus ochraceo-petaliformis Batista & Maia - Anais Soc. Biol. Pernambuco 15: 213, 1957.

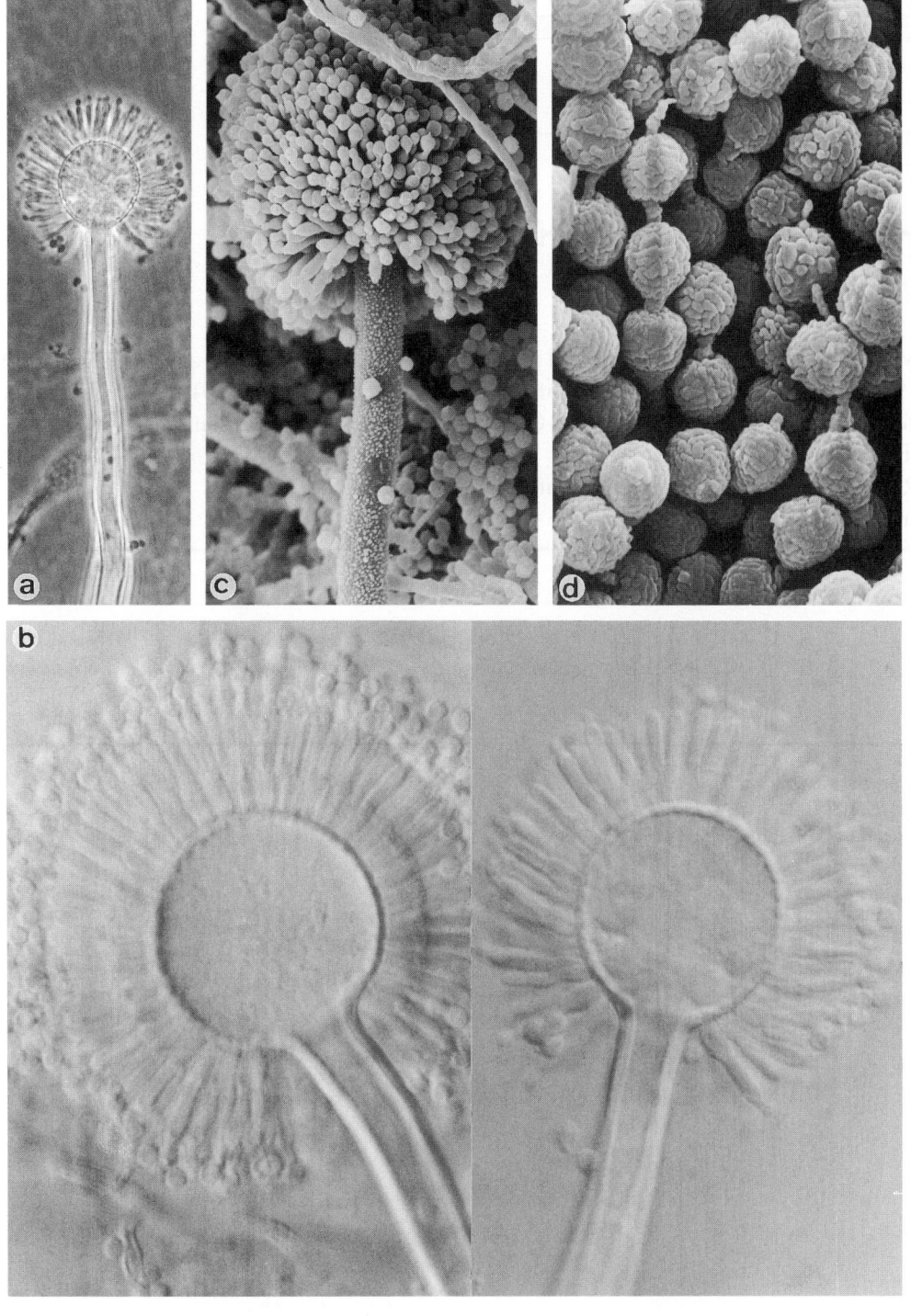

Aspergillus ochraceus, CBS 123.55. Conidiophores and conidia. a. ×390; b. ×1215; c. ×1170; d. ×6000.

Aspergillus oryzae (Ahlburg) Cohn

Colony characteristics. Colonies (CzA) growing rapidly, pale greenish-yellow, olive-yellow or with different shades of green, typically with dull brown shades with age.

Microscopy. Conidial heads radiate to loosely columnar, 150-300 μm diam. Conidiophore stipes hyaline, up to 4-5 mm in length. Vesicles subspherical, up to 75 μm diam. Conidiogenous cells uniseriate and biseriate. Metulae or phialides covering the entire surface or the upper three-fourths of the vesicle. Conidia (sub)spherical to ovoidal, 4.5-8.0 (-10.0) × 4.5-7.0 μm, smooth-walled to roughened, greenish to brownish.

Molecular diagnostics. SSU and ITS restriction maps based on NCBI D63698 and NCBI AB008417, resp.:

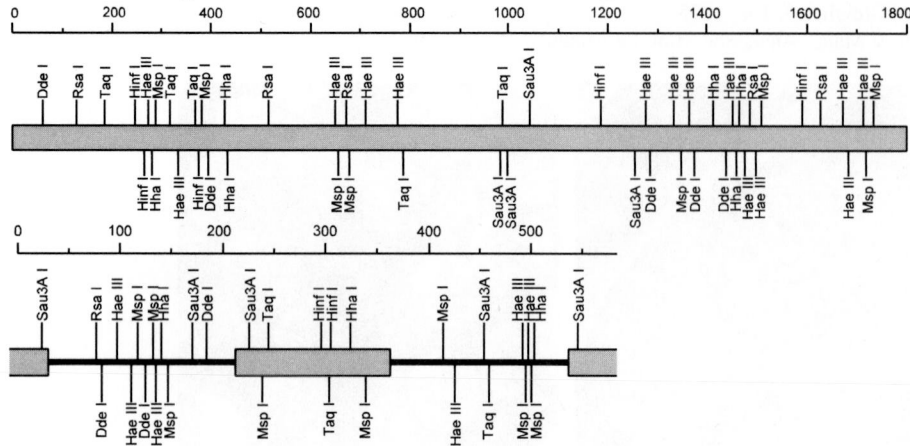

Pathogenicity. BSL-1. This species was reported as responsible for invasion of paranasal sinuses (Green *et al.*, 1969; Byard *et al.*, 1986) and was implicated in cases of meningitis (Gordon *et al.*, 1976), cerebritis (Ziskind *et al.*, 1958), pulmonary infections (Retamal *et al.*, 1984; Wan-Qing, 1988) and scleritis (Stenson *et al.*, 1982). Ota (1923) reported onychomycosis.

References. Raper & Fennell (1965), Klich & Pitt (1988).

Antifungal susceptibility.

Antifungal	MICs range	MIC 90	Strains	Reference
AMB	0.78-1.56	1.56	5	Wildfeuer *et al.* (1998)
ITZ	0.1-0.39	0.39	5	Wildfeuer *et al.* (1998)
KTZ	0.1-0.78	0.78	5	Wildfeuer *et al.* (1998)
VCZ	0.1-0.2	0.20	5	Wildfeuer *et al.* (1998)

Nomenclature. *Eurotium oryzae* Ahlburg, *in* Korschelt - Dingl. Polytech. J. 230: 330, 1878 ≡ *Aspergillus oryzae* (Ahlburg) Cohn - Jahresb. Schles. Ges. Vaterl. Kult. 61: 226, 1884.

   Aspergillus jeanselmei Ota - Annls Parasit. Hum. Comp. 1: 146, 1923 ≡ *Sterigmatocystis jeanselmei* (Ota) Nannizzi - Tratt. Micopat. Um. 4: 229, 1934.

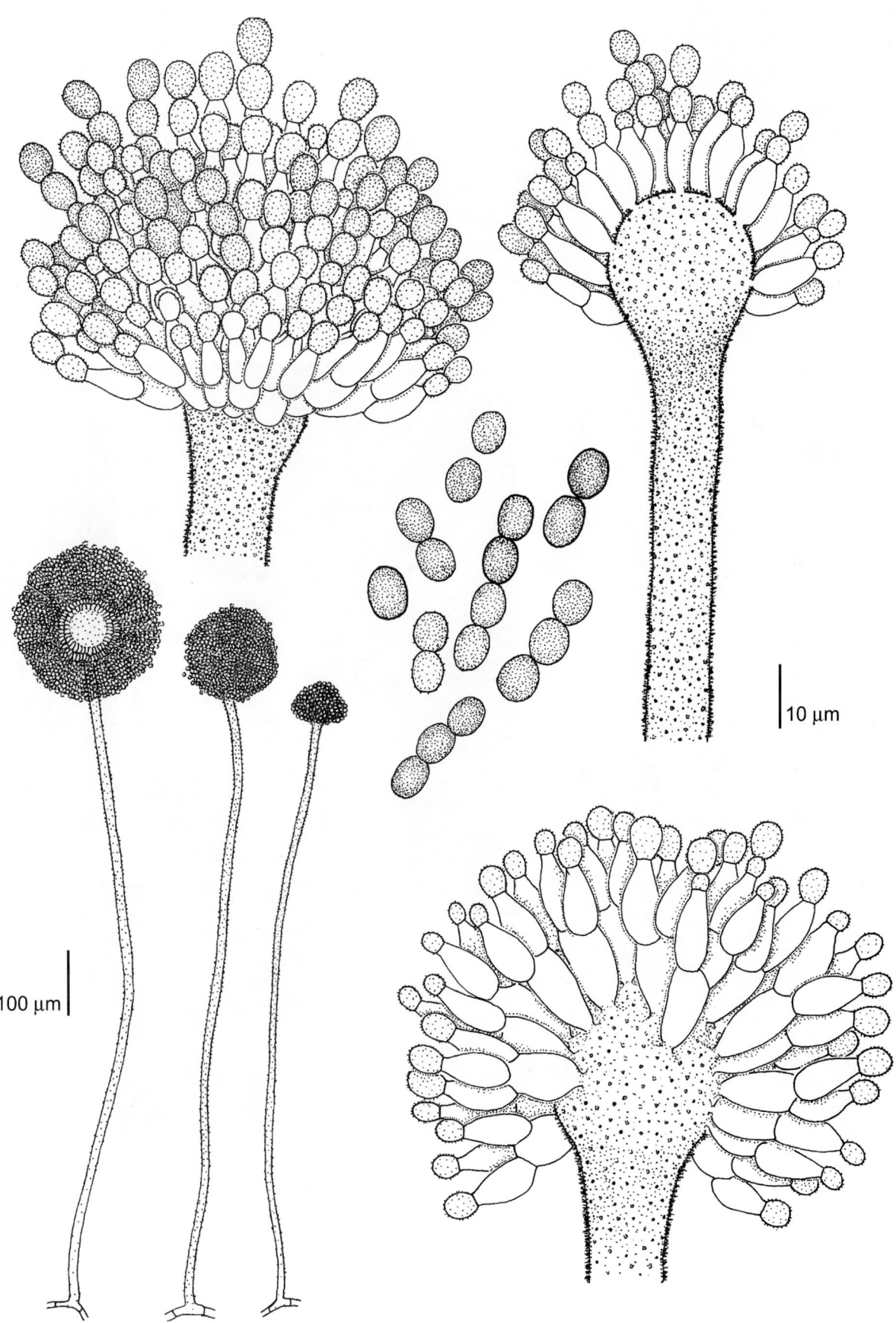

Aspergillus oryzae, CBS 819.72. Conidiophores and conidia.

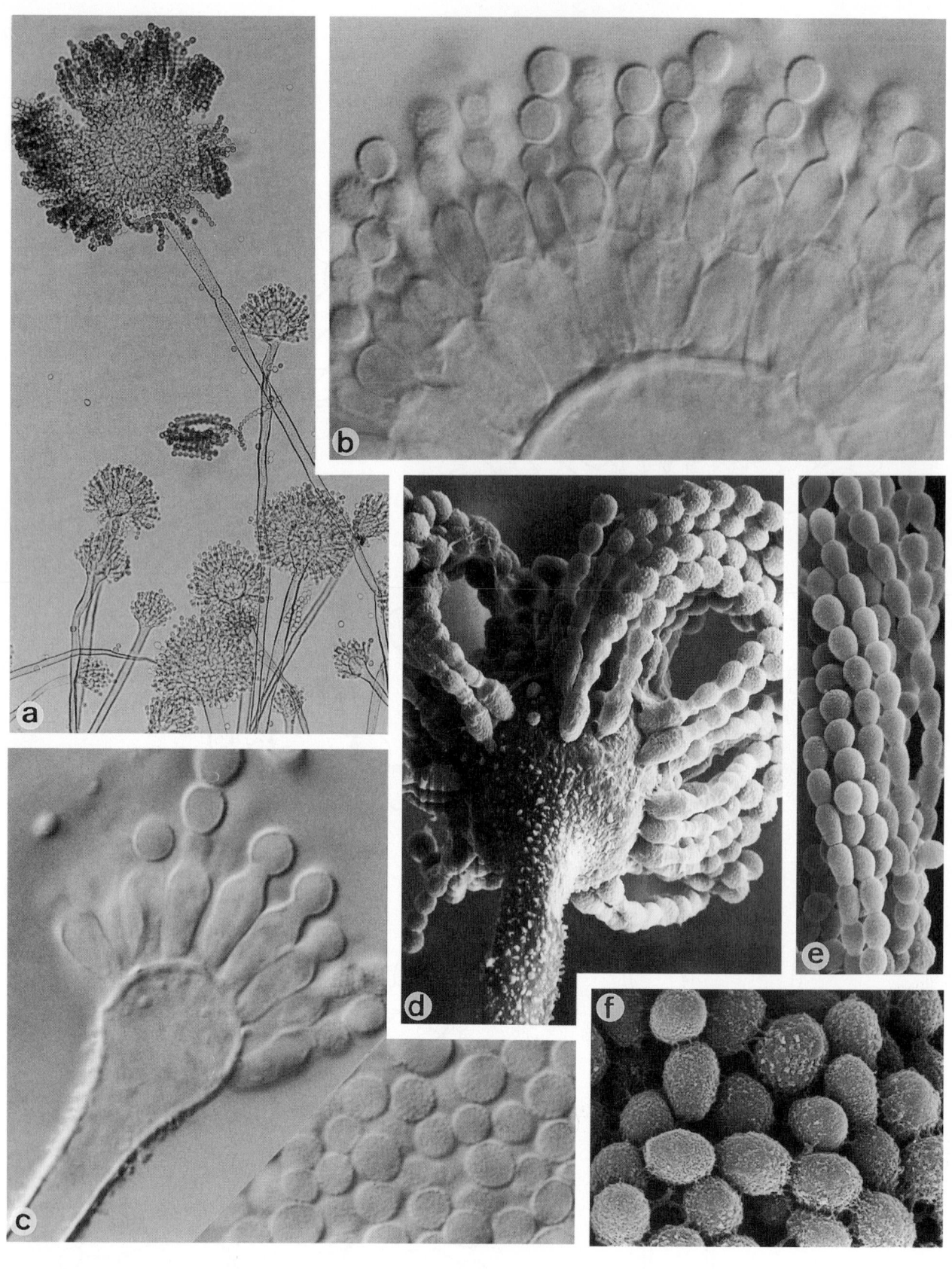

Aspergillus oryzae, CBS 819.72. Conidiophores and conidia. a. ×160; b, c. ×1600; d. ×1350; e. ×1450; f. ×2900.

Aspergillus reptans Samson & W. Gams

Colony characteristics. Colonies (CzA) with restricted growth, dull-green.

Microscopy. Conidial heads radiate to loosely columnar. Conidiophore stipes smooth-walled, hyaline, 500-1000 μm in length. Vesicles hemispherical. Conidiogenous cells uniseriate. Conidia spherical to subspherical or ovoidal, spinulose, 5.0-6.5 (-8.0) μm long.

Teleomorph. *Eurotium repens* de Bary (*Ascomycota, Euascomycetes, Eurotiales: Trichocomaceae*).

Ascomata abundant, spherical to subspherical, non-ostiolate, 75-100 μm diam, yellow, borne in loose networks of yellow to orange-red hyphae. Asci 8-spored, 10-12 μm diam. Ascospores pale, lenticular, 4.8-5.0 × 3.8-4.4 μm, smooth-walled, with a sistinct equatorial furrow.

Pathogenicity. BSL-1. This species was associated, together with *Microascus cinereus* (p. 277), in a case of sinusitis maxillaris, but its pathogenicity was not proven (Aznar *et al.*, 1989).

References. Raper & Fennell (1965), Kozakiewicz (1989, 1995).

Nomenclature. *Aspergillus glaucus* Link var. *repens* Corda - Ic. Fung. 5: 53, 1839 ≡ *Aspergillus reptans* Samson & W. Gams, *in* Samson & Pitt - Adv. Pen. Asp. Syst. p. 48, 1985 (name change).

　Eurotium repens de Bary - Abh. Senckenberg. Naturf. Ges. 7: 53, 1839 ≡ *Aspergillus repens* (de Bary) Fischer, *in* Engler & Prantl - Natürl. Pflfam. 1: 302, 1897.

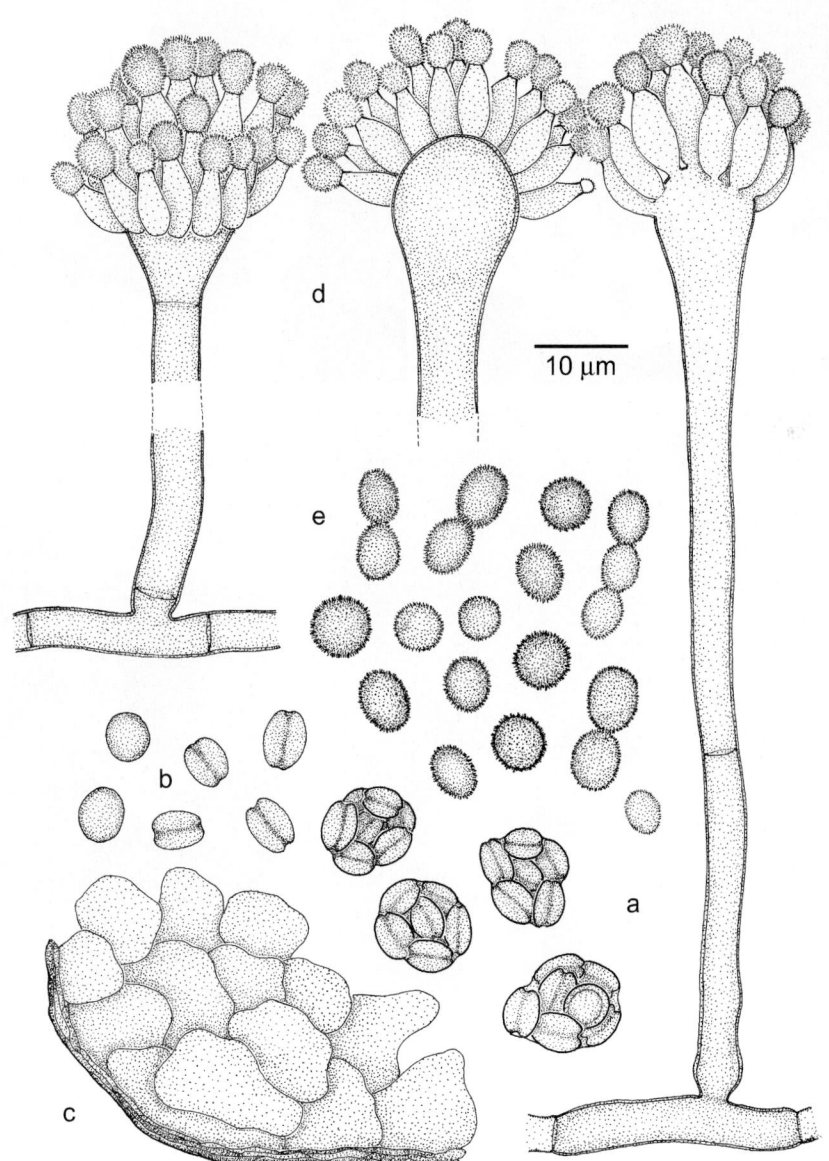

Aspergillus reptans (Eurotium repens), CBS 531.65. a. Asci; b. ascospores; c. part of ascoma wall; d. conidiophores; e. conidia.

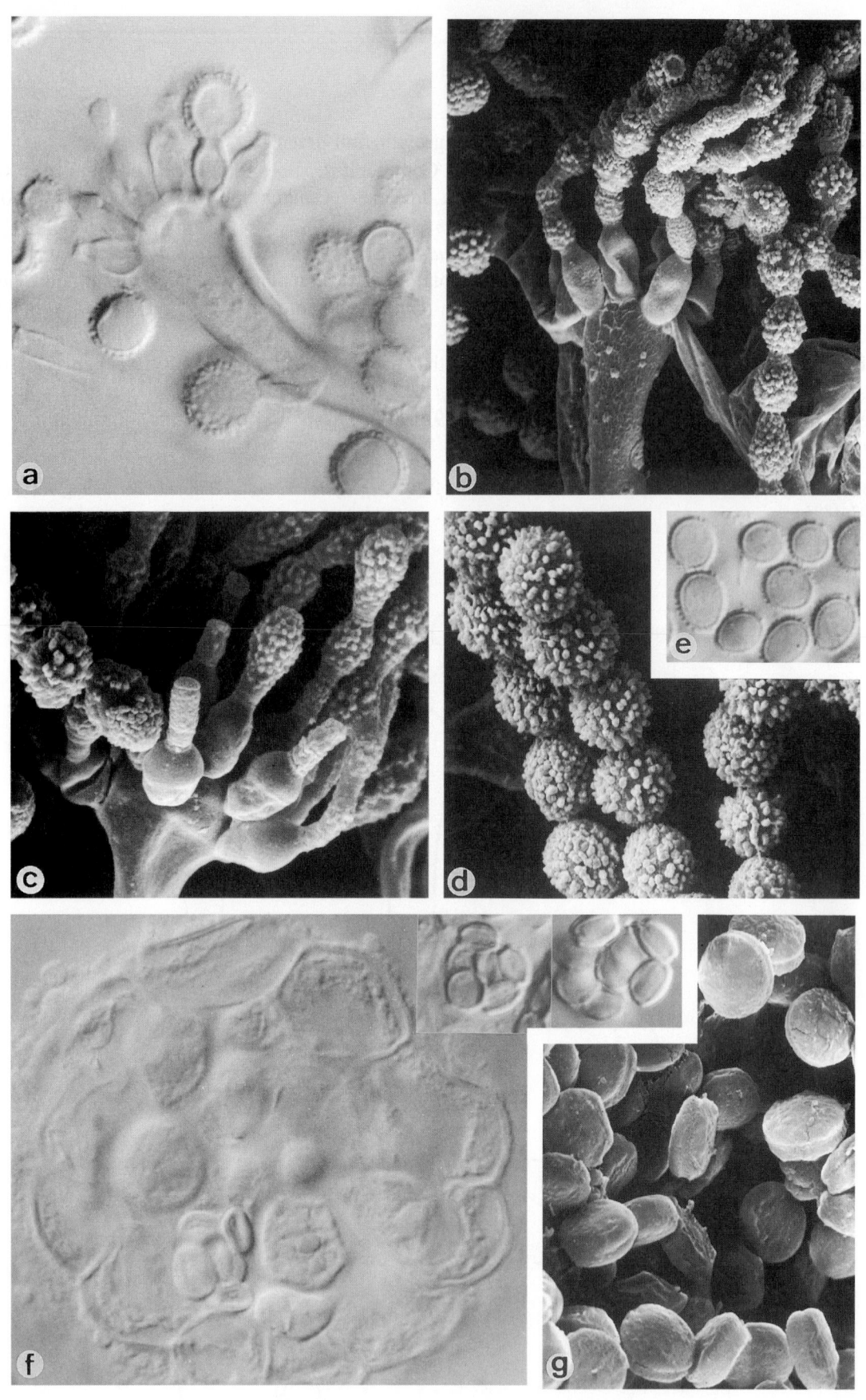

Aspergillus reptans (Eurotium repens), CBS 531.65. a-e. Conidiophores and conidia; f, g. ascoma, asci and ascospores. a. ×1600; b. ×1800; c. ×2700; d. ×3150; e, f. ×1600; g. ×2300.

Aspergillus restrictus G. Smith

Colony characteristics. Colonies (CzA) with very restricted growth, coloured in different shades of green.
Microscopy. Conidial heads columnar, often twisted, dark olive-green. Conidiophore stipes hyaline, smooth-walled or finely roughened. Vesicles flask-shaped to hemispherical, 6-12 µm diam. Conidiogenous cells uniseriate, covering the upper third of the vesicle. Conidia roughened, dark green in mass, almost cylindrical when young, ellipsoidal or pyriform at maturity, 4-7 (-10) × 3-4 (-6) µm.
Molecular diagnostics. SSU restriction map based on NCBI AB008407:

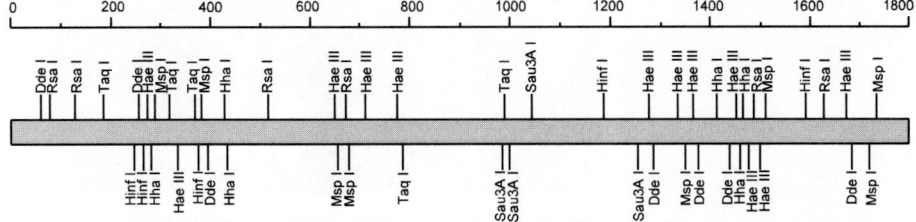

Pathogenicity. BSL-1. This species was reported as responsible for endocarditis after heart valve replacement (Mencl *et al.*, 1985). It is also known from onychomycosis (Schönborn & Schmoranzer, 1970) and from aspergilloma (Estrader *et al.*, 1972). A very closely related species, *A. penicilloides* Speg., was probably implicated in a case of generalized pulmonary aspergillosis (Maršálek *et al.*, 1960).
References. Raper & Fennell (1965), Klich & Pitt (1988).
Nomenclature. *Aspergillus restictus* G. Smith - J. Textile Inst. 22: 115, 1931.

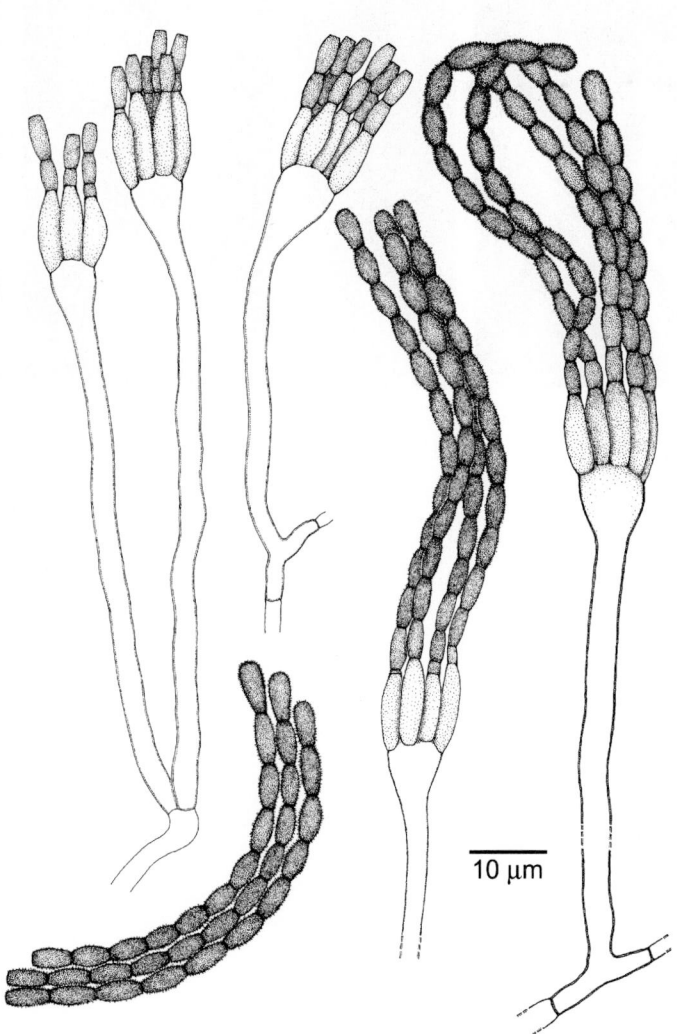

Aspergillus restrictus, CBS 542.65. Conidiophores and conidia.

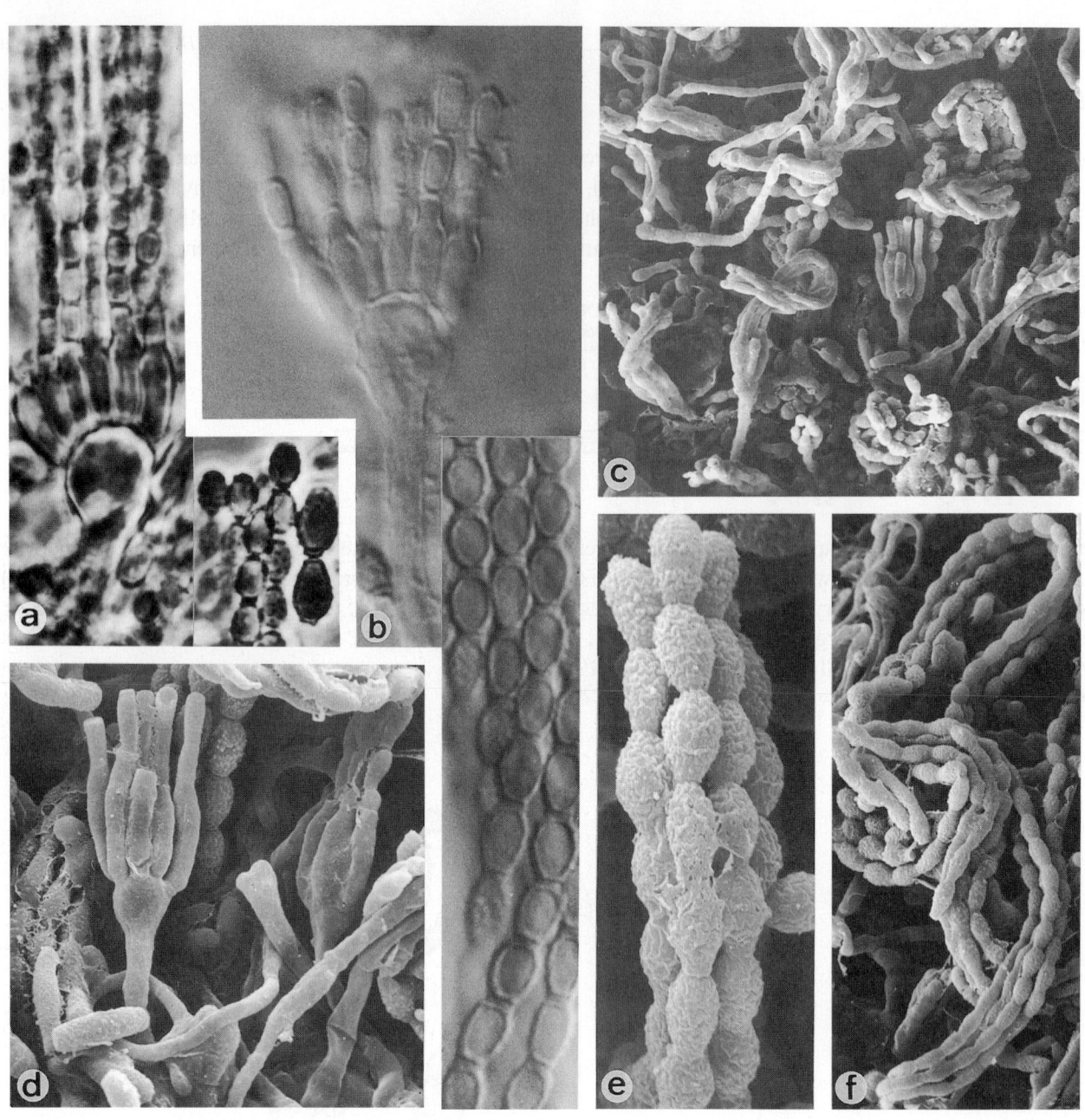

Aspergillus restrictus, CBS 542.65. Conidiophores and conidia. a. ×1280; b. ×1600; c. ×900; d. ×1950; e. ×3650; f. ×1500.

Aspergillus sclerotiorum Huber

Colony characteristics. Colonies (CzA) growing rather slowly, pale yellow.

Microscopy. Conidial heads radiate, occasionally splitting into two or more compact, divergent columns. Conidiophore stipes up to 1.2 mm long, light yellow, with thick and echinulate walls. Vesicles spherical, 17-35 μm diam. Conidiogenous cells biseriate. Metulae covering the entire vesicle surface. Conidia smooth-walled to finely roughened, spherical, 2.0-3.5 μm diam. Sclerotia spherical to subspherical, 1.0-1.5 mm diam, white to cream-coloured.

Pathogenicity. BSL-1. The species was isolated in a case of onychomycosis (Feuilhade de Chauvin & de Bièvre, 1985).

Referenes. Raper & Fennell (1965), Klich & Pitt (1988).

Nomenclature. *Aspergillus sclerotiorum* Huber - Phytopathology 23: 306, 1933.

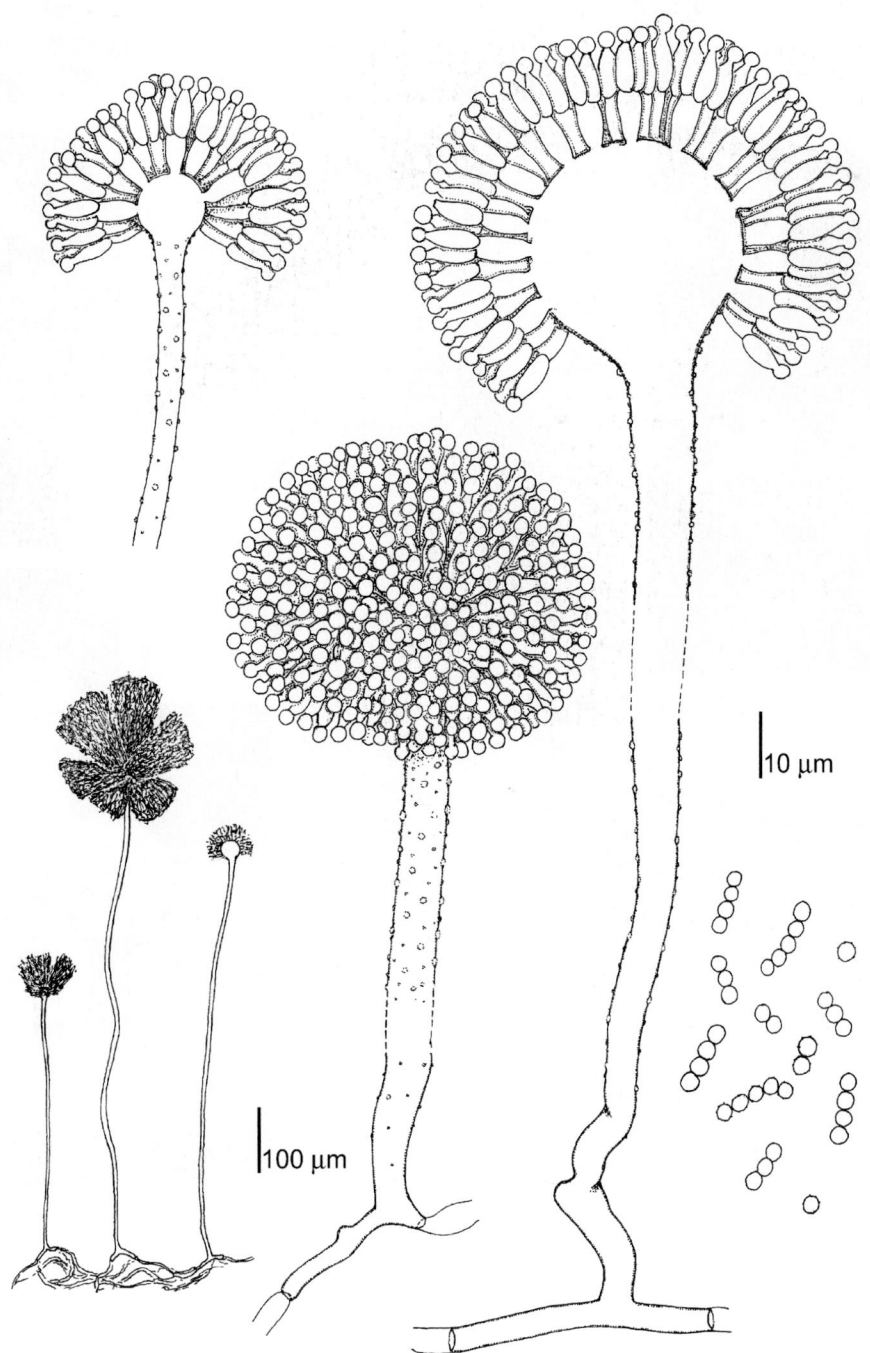

Aspergillus sclerotiorum, CBS 628.67. Conidiophores and conidia.

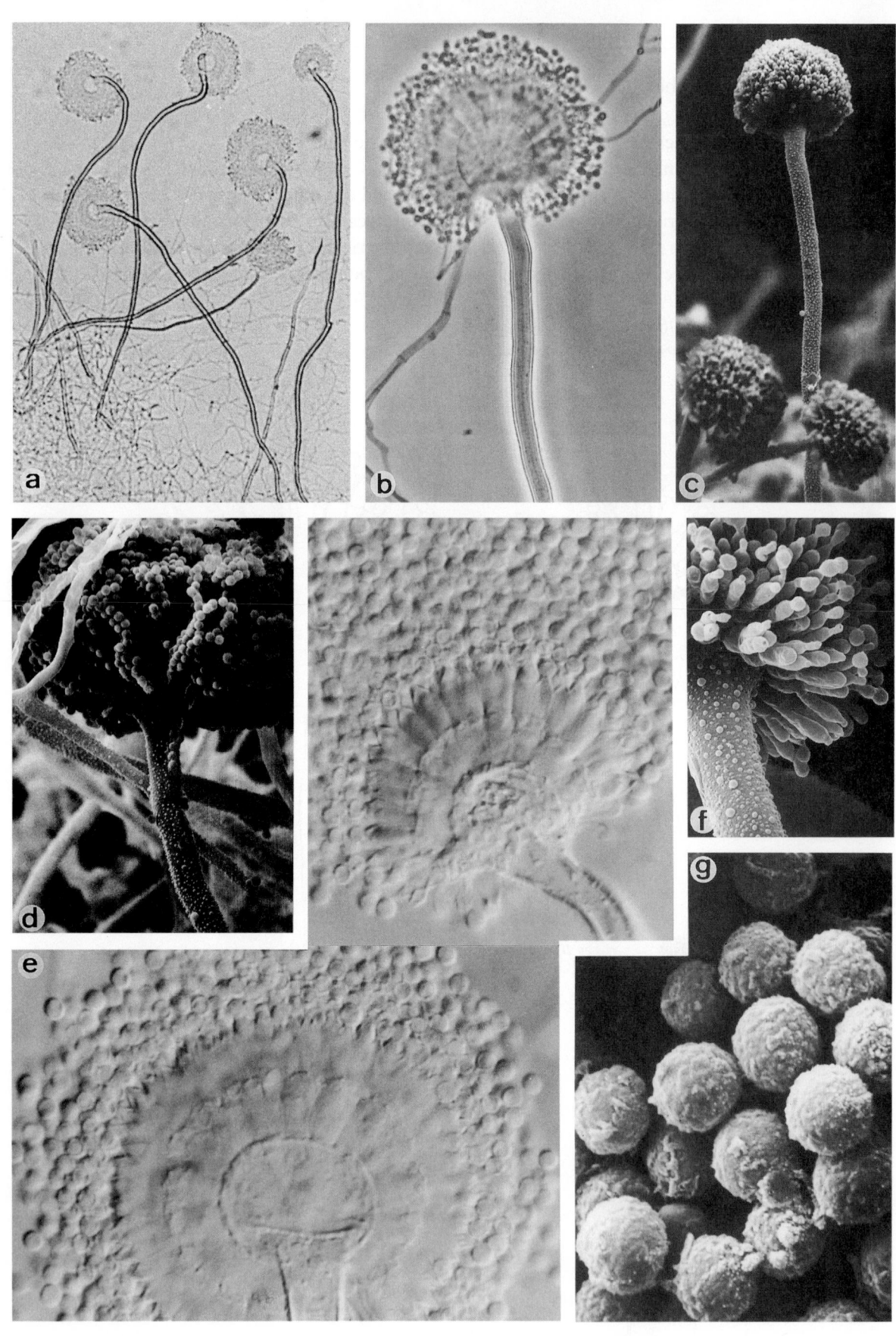

Aspergillus sclerotiorum, CBS 628.67. Conidiophores and conidia. a. ×102; b. ×512; c. ×560; d. ×840; e. ×1600; f. ×1150; g. ×8900.

Aspergillus sydowii (Bain. & Sart.) Thom & Church

Colony characteristics. Colonies (CzA) spreading, blue-green, with straw-coloured to reddish-brown shades, often with abundant exudate; reverse usually reddish.

Microscopy. Conidial heads radiate. Conidiophore stipes up to 500 μm in length, hyaline, smooth-walled. Vesicles spherical to subspherical, fertile over almost the entire surface. Conidiogenous cells biseriate. Conidia echinulate, green in mass, spherical to subspherical, 2.5-4.0 μm diam. Spherical Hülle cells may be present.

Pathogenicity. BSL-1. This fungus is a frequent agent of invasive aspergillosis (Rippon, 1988), onychomycosis (Negroni, 1943; Walshe & English, 1966; Schönborn & Schmoranzer, 1970; Achten *et al.*, 1979; Wadhwani & Srivastava, 1985) and keratomycosis (Shukla *et al.*, 1985). Also isolated from BAL (Magalhaes *et al.*, 1996)

References. Raper & Fennell (1965), Klich & Pitt (1988), Kozakiewicz (1989).

Nomenclature. *Sterigmatocystis sydowii* Bainier & Sartory - Annls Mycol. 11: 25, 1913 ≡ *Aspergillus sydowii* (Bainier & Sartory) Thom & Church - The Aspergilli p. 147, 1926.

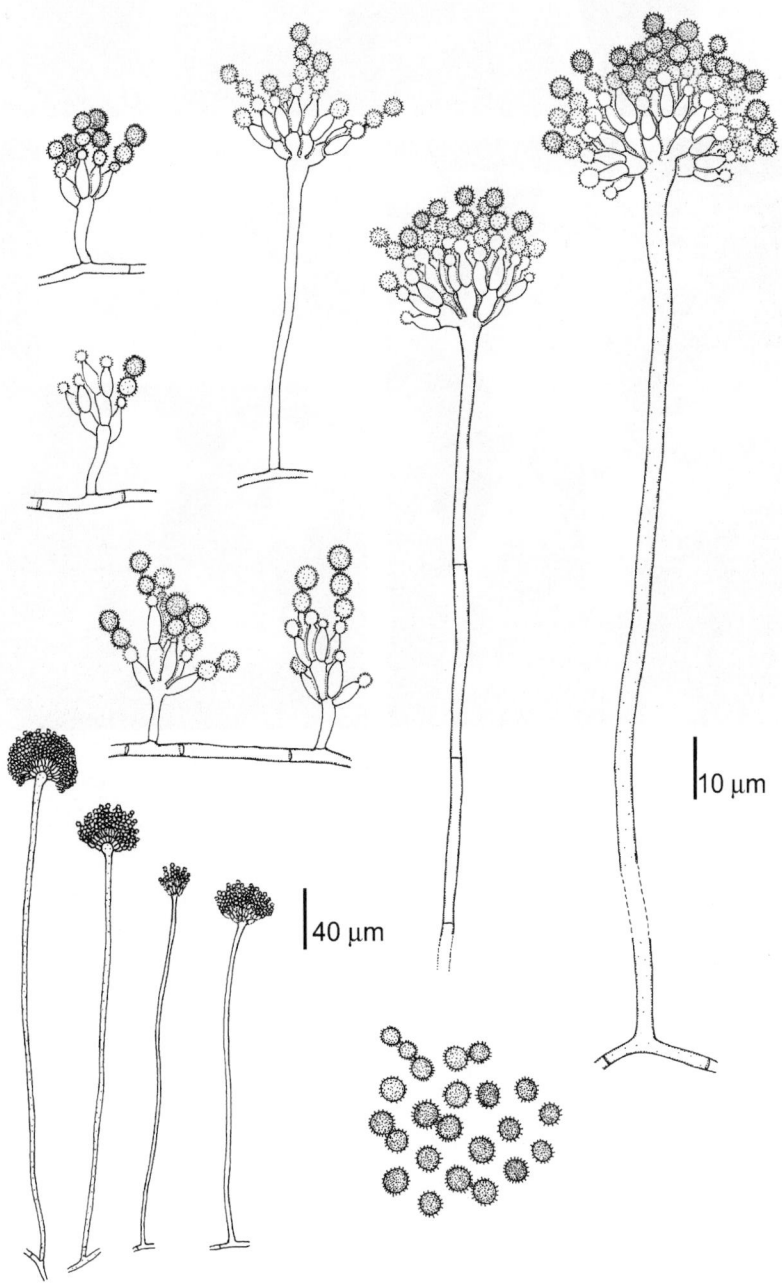

Aspergillus sydowii, CBS 129.55. Conidiophores and conidia.

Aspergillus sydowii, CBS 129.55. Conidiophores and conidia. a. ×512; b. ×1600; c. ×1750; d. ×3100; e. ×2800; f. ×1600; g. ×5700.

Aspergillus tamarii Kita

Colony characteristics. Colonies (CzA) growing rapidly, dark brown.

Microscopy. Conidial heads compact and spherical or loosely radiate, 500-600 µm diam. Conidiophore stipes usually 1-2 mm in length, hyaline, usually roughened. Vesicles spherical, 10-50 µm diam. Conidiogenous cells uniseriate and biseriate. Metulae or phialides covering the entire surface of the vesicle. Conidia echinulate to tuberculate, subspherical, 5-8 µm diam.

Molecular diagnostics. SSU and ITS restriction maps based on NCBI D63701 and AB008420:

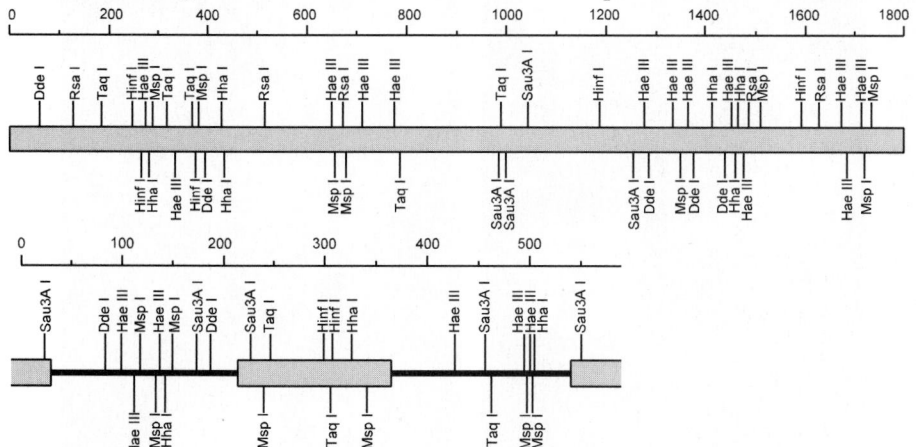

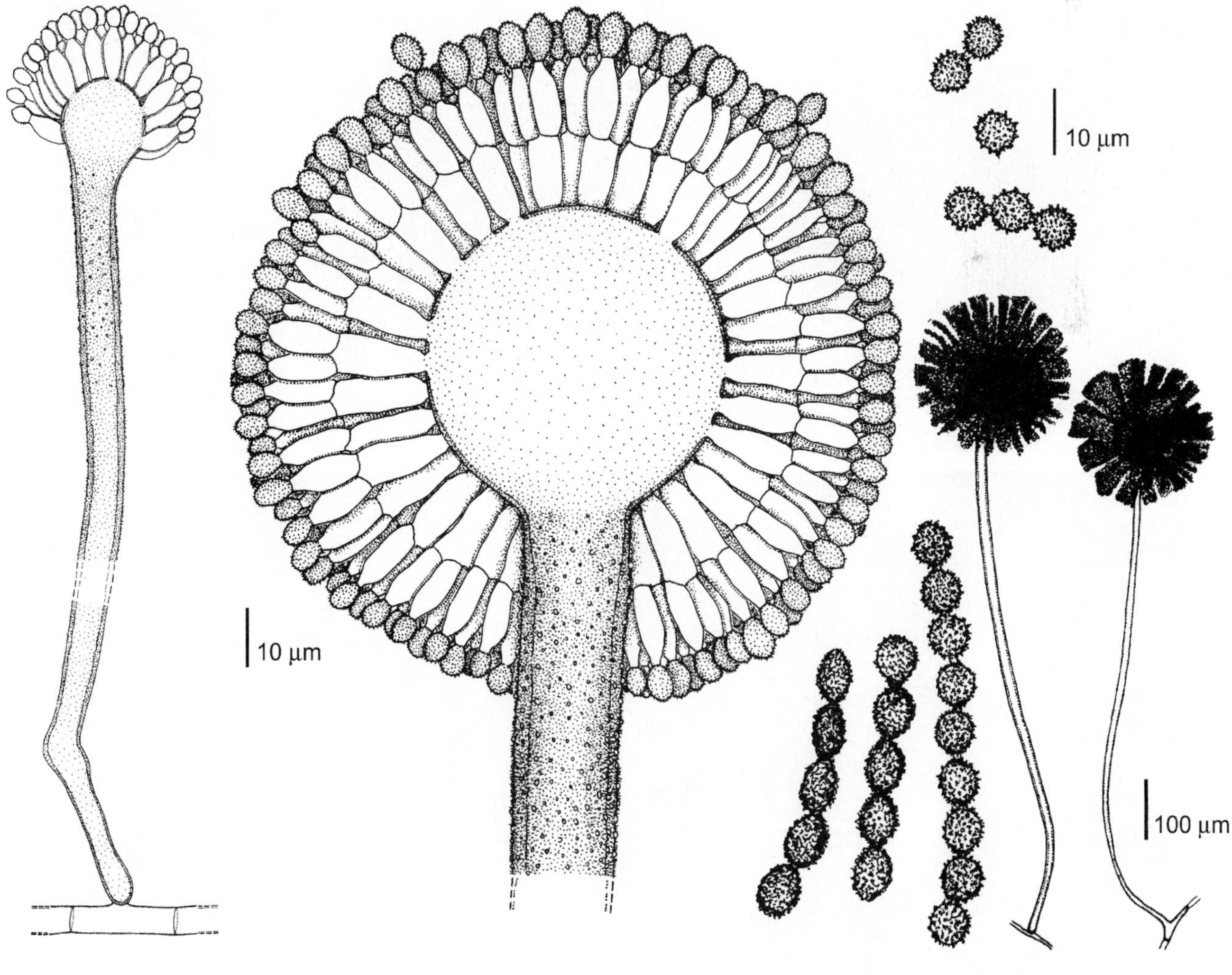

Aspergillus tamarii, CBS 104.13. Conidiophores and conidia.

Pathogenicity. BSL-1. The species was implicated in a case of eyelid infection (Degos *et al.*, 1970).

References. Raper & Fennell (1965), Klich & Pitt (1988), Kozakiewicz (1989).

Nomenclature. *Aspergillus tamarii* Kita - Centbl. Bakt. Parsitkde, Abt. 2, 37: 433, 1913.

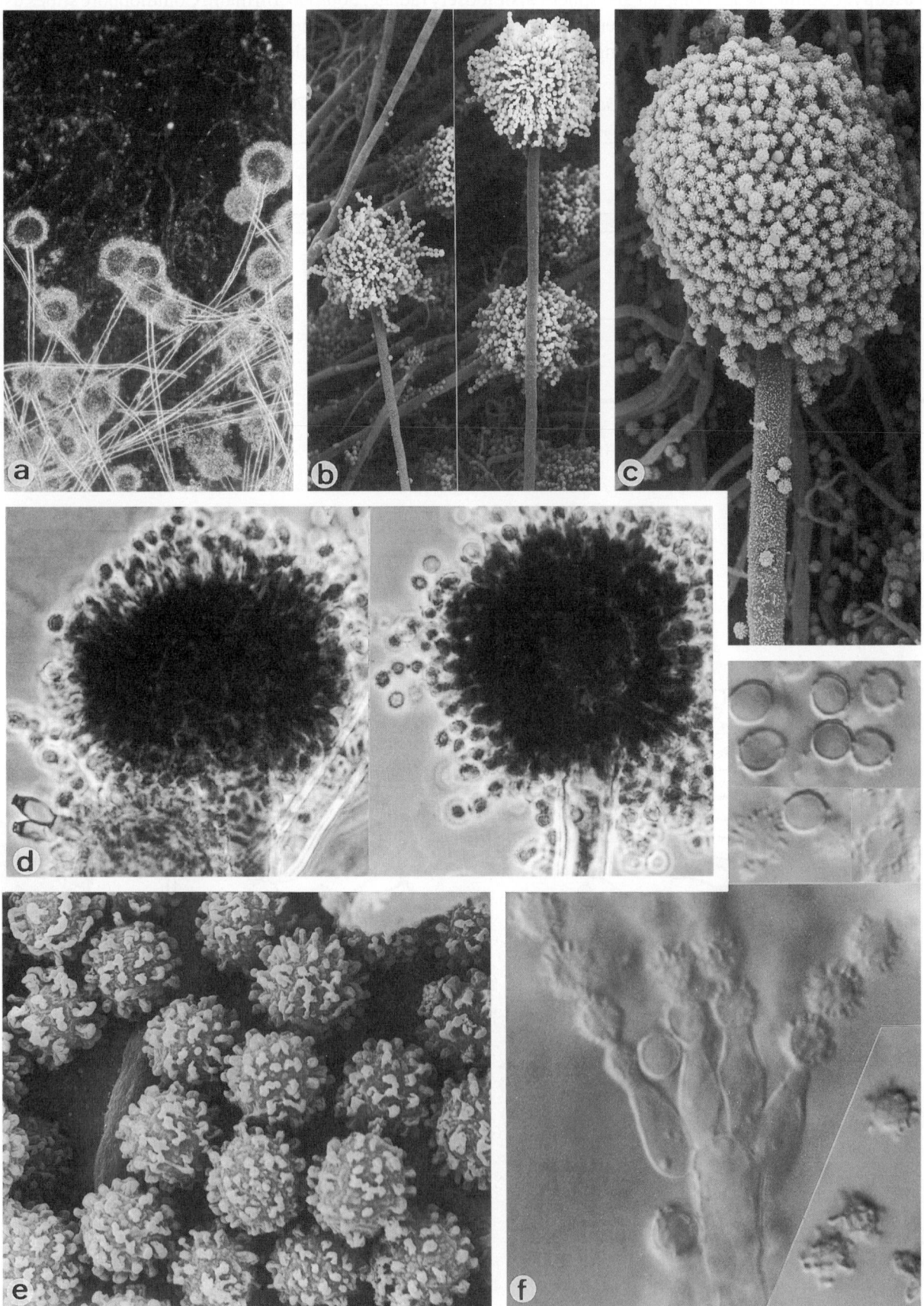

Aspergillus tamarii, CBS 104.13. Conidiophores and conidia. a. ×64; b. ×200; c. ×550; d. ×512; e. ×4900; f. ×1600.

Aspergillus terreus Thom

Colony characteristics. Colonies (CzA) yellowish-brown to cinnamon-brown, consisting of a dense felt of conidiophores.

Microscopy. Conidial heads densely columnar. Conidiophore stipes smooth-walled, hyaline. Vesicles subspherical, 10-20 μm diam. Conidiogenous cells biseriate. Metulae as long as the phialides. Conidia smooth-walled, striate with SEM, spherical to broadly ellipsoidal, 1.5-2.5 μm, hyaline.

Molecular diagnostics. 18S rDNA sequences were published by Verweij *et al.* (1995). A 26S rDNA-based specific probe was developed by Sandhu *et al.* (1995). SSU and ITS restriction maps based on NCBI AB008409 and NCBI AJ001368, resp.:

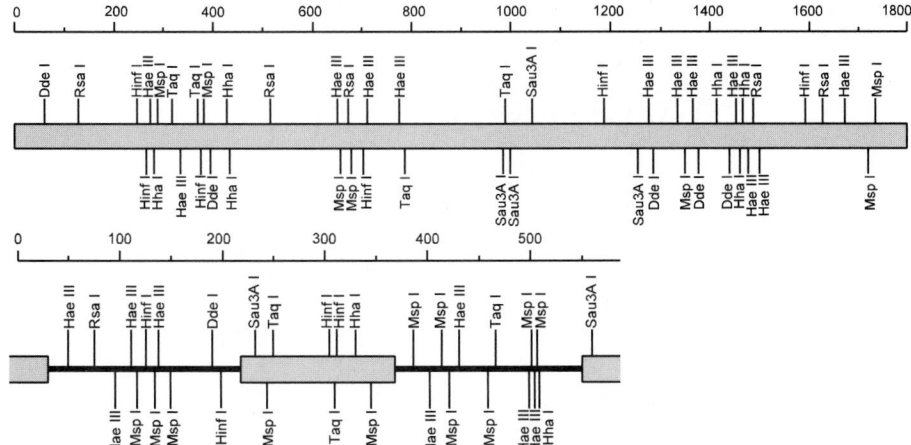

Pathogenicity. BSL-2. The species causes allergic or invasive bronchopulmonary aspergillosis (Moore *et al.*, 1988). About 10% of cases of aspergillosis may be ascribed to this species (Iwen *et al.*, 1998). Nosocomial infections have been reported (Flynn *et al.*, 1993). It is regularly implicated in a wide variety of infections in humans, such as cutaneous, ophthalmic (Das *et al.*, 1993), pulmonary and disseminated infections (Waller *et al.*, 1991b). Keratitis was reported by Singh *et al.* (1990b), arthritis by Steinfeld *et al.* (1997), spondylodiscitis by Penn *et al.* (1998) and suppurative otitis by Tiwari *et al.* (1995). It has also been reported from other vertebrates, such as pigeons (Pal, 1992; Tritz & Woods, 1993) and dogs (Kelly *et al.*, 1995; Berry & Leisewitz, 1996).

References. Raper & Fennell (1965), Klich & Pitt (1988), Samson *et al.* (1996), Kozakiewicz (1989).

Antifungal susceptibility.

Antifungal	MICs range	MIC90	Strains	Reference
AMB	3.37	nd	101	Sutton *et al.* (1999a)
AMB	1.5	nd	1	Wildfeuer *et al.* (1998)
AMB	0.063-0.5	0.5	7	Tawara *et al.* (2000)
AMB	0.25-2	2	20	Dannaoui *et al.* (1999)
AMB	2-8	nd	10	Moore *et al.* (2000a)
FLZ	4->64	>64	7	Tawara *et al.* (2000)
ITZ	0.03-0.13	0.13	7	Tawara *et al.* (2000)
ITZ	0.25-1	1	20	Dannaoui *et al.* (1999)
ITZ	0.25-0.5	nd	10	Moore *et al.* (2000)
KTZ	0.78	nd	1	Wildfeuer *et al.* (1998)
VCZ	0.22	nd	51	Sutton *et al.* (1999)
VCZ	0.78	nd	1	Wilfeuer *et al.* (1998)

Nomenclature. *Aspergillus terreus* Thom, *in* Thom & Church - Am. J. Bot. 5: 85, 1918.

Sterigmatomyces hortai Langeron - Bull. Soc. Pathol. Exot. 15: 383, 1922 ≡ *Aspergillus hortai* (Langeron) C.W. Dodge - Med. Mycol. p. 628, 1935.

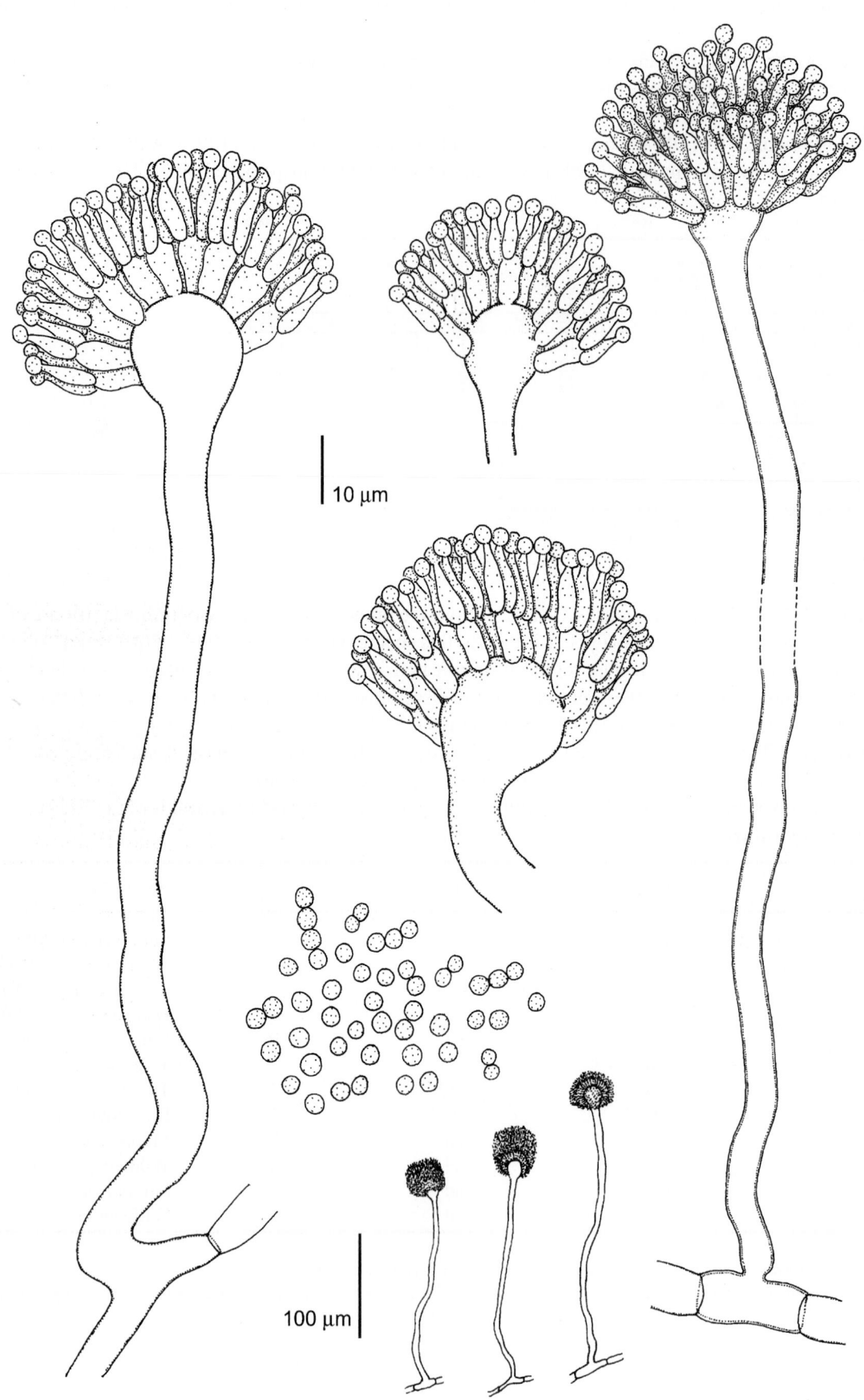

Aspergillus terreus, FMR 2207. Conidiophores and conidia.

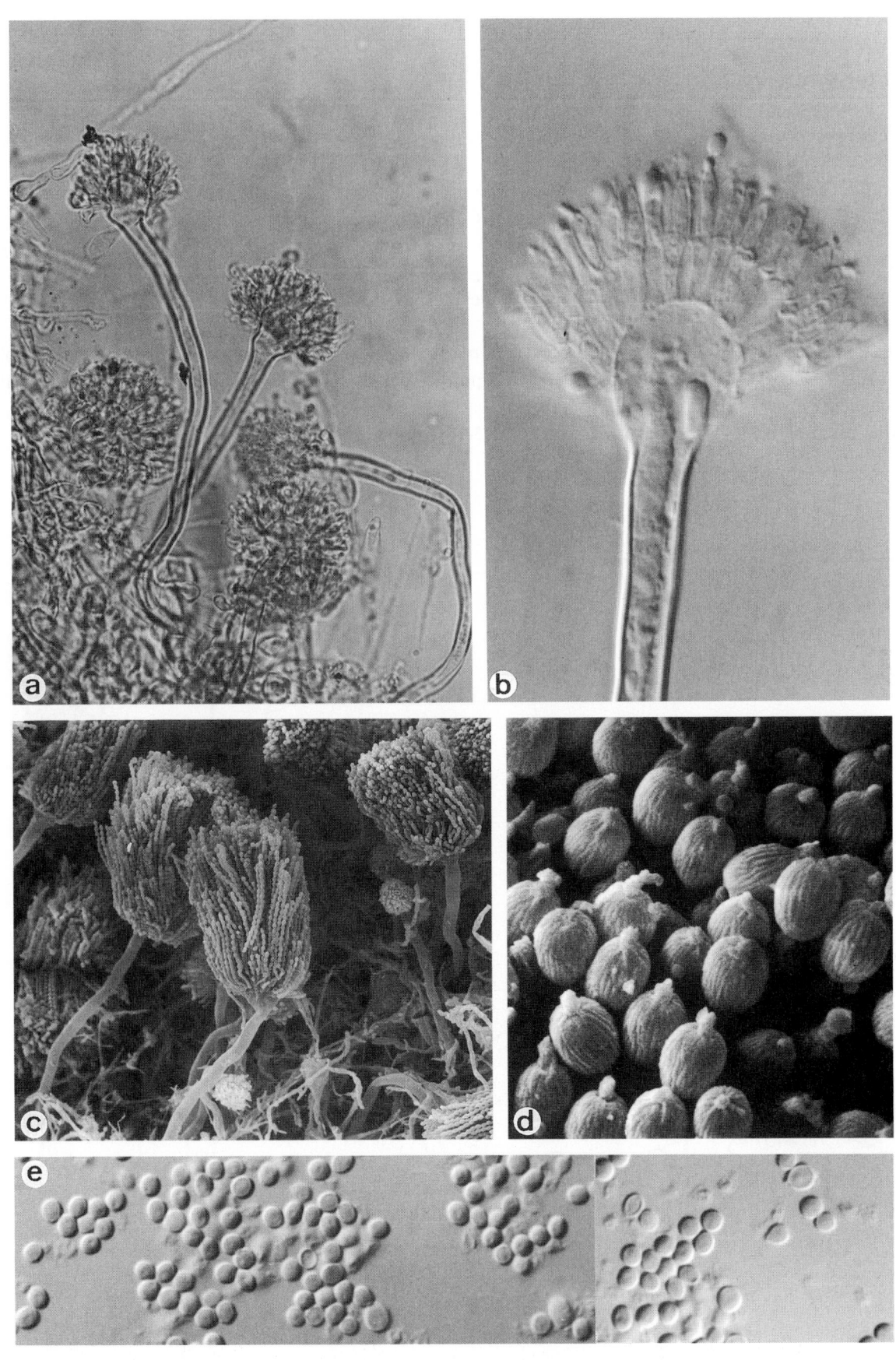

Aspergillus terreus, FMR 2207. Conidiophores and conidia. a. ×512; b. ×1600; c. ×440; d. ×8750; e. ×1600.

Aspergillus tetrazonus Samson & W. Gams

Colony characteristics. Colonies (CzA) spreading, grey with purplish shades at the centre and dull-green towards the margin.

Microscopy. Conidial heads short columnar, green. Conidiophore stipes 25-200 × 4-6 µm, brown, smooth-walled. Vesicles flask-shaped to hemispherical, 8-15 µm diam. Conidiogenous cells biseriate. Metulae 4-7 × 3-4 µm, covering only the upper half of the vesicle. Phialides 5-8 × 2.0-3.5 µm. Conidia spherical, 2.5-3.5 µm diam, finely roughened, green in mass.

Teleomorph. *Emericella quadrilineata* (Thom & Raper) C.R. Benjamin (*Ascomycota, Euascomycetes, Eurotiales: Trichocomaceae*).

Ascomata surrounded by Hülle cells, light brownish to purple, spherical, 125-150 µm diam. Asci 8-spored, spherical, 10-12 µm diam. Ascospores lenticular, smooth-walled, purple-red, 4.5-6.0 × 3.5-4.0 µm, with 4 short equatorial crests, two of which are seen with difficulty using light microscopy.

Molecular diagnostics. ITS restriction map based on NCBI AJ000934:

Aspergillus tetrazonus (Emericella quadrilineata), CBS 426.77. a. Ascospores; b. part of ascoma wall; c. Hülle cells; d. conidiophores; e. conidia.

Pathogenicity. BSL-1. The species has been isolated from sheep with mycotic dermatitis (Singh & Singh, 1970) and was the causative agent of a human case of fungal sinusitis (Polacheck *et al.*, 1992).

References. Raper & Fennell (1965), Klich & Pitt (1988), Stchigel *et al.* (1999).

Nomenclature. *Aspergillus quadrilineatus* Thom & Raper - Mycologia 31: 660, 1939 ≡ *Emericella quadrilineatus* (Thom & Raper) C.R. Benjamin - Mycologia 47: 680, 1955.

Aspergillus tetrazonus Samson & W. Gams, *in* Samson & Pitt - Adv. Pen. Asp. Syst. p. 48, 1985.

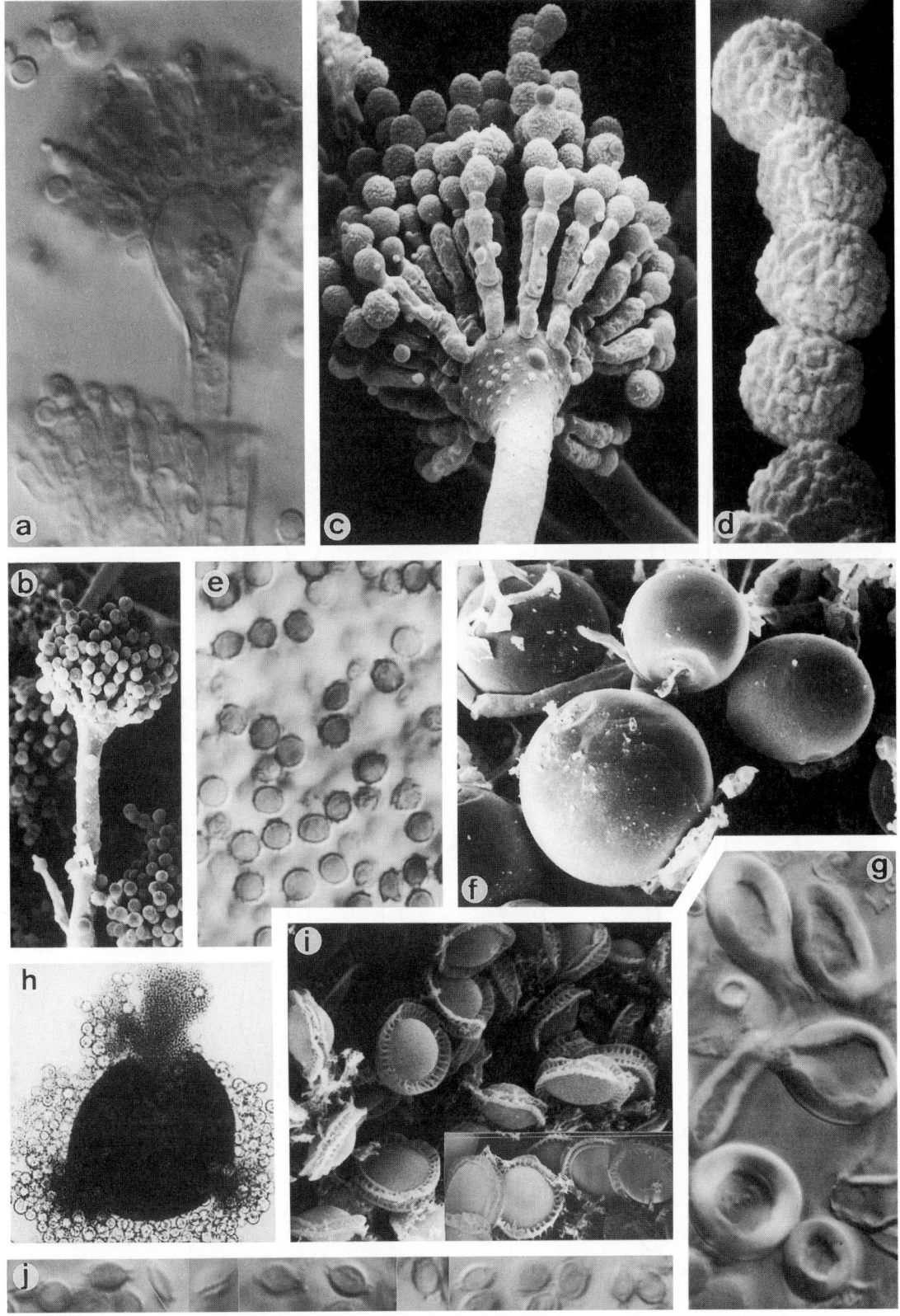

Aspergillus tetrazonus (Emericella quadrilineata), CBS 426.77. a-e. Conidiophores and conidia; f, g. Hülle cells; h-j. ascoma and ascospores. a. ×1600; b. ×1040; c. ×2700; d. ×9800; e. ×1280; f. ×2475; g. ×1600; h. ×128; i. ×4300; j. ×1600.

Aspergillus unguis (Emile-Weil & Gaudin) C.W. Dodge

Colony characteristics. Colonies (CzA) restricted, yellowish-green to dark green.

Microscopy. Conidial heads radiate to loosely columnar depending on the medium used. Conidiophore stipes smooth-walled, dull-brown. Thick- and rough-walled, seta-like hyphae arising from foot cells usually present. Vesicles spathulate, 9-12 μm diam. Conidiogenous cells biseriate. Metulae covering the upper part of the vesicle. Conidia smooth-walled or somewhat rough-walled, dull-green, spherical, 2.5-3.5 μm diam.

Teleomorph. *Emericella unguis* Malloch & Cain (*Ascomycota, Euascomycetes, Eurotiales: Trichocomaceae*). Ascomata purple, spherical to subspherical, 200-250 μm diam, surrounded by a layer of spherical Hülle cells. Asci 8-spored, subspherical, 9.5-10.5 μm diam. Ascospores purplish-red, with two equatorial crests and convex walls, smooth-walled, lenticular, 4.5-5.5 × 3.2-2.5 μm.

Pathogenicity. BSL-1. The species is known from cases of human onychomycosis (Schönborn & Schmoranzer, 1970; Grigoriu & Grigoriu, 1975).

References. Raper & Fennell (1965).

Nomenclature. *Sterigmatocystis unguis* Emile-Weil & Gaudin - Arch. Méd. Exp. Anat. Path. 28: 463, 1919 ≡ *Aspergillus unguis* (Emile-Weil & Gaudin) C.W. Dodge - Med. Mycol. p. 637, 1935.
    Emericella unguis Malloch & Cain - Can. J. Bot. 50: 62, 1972.

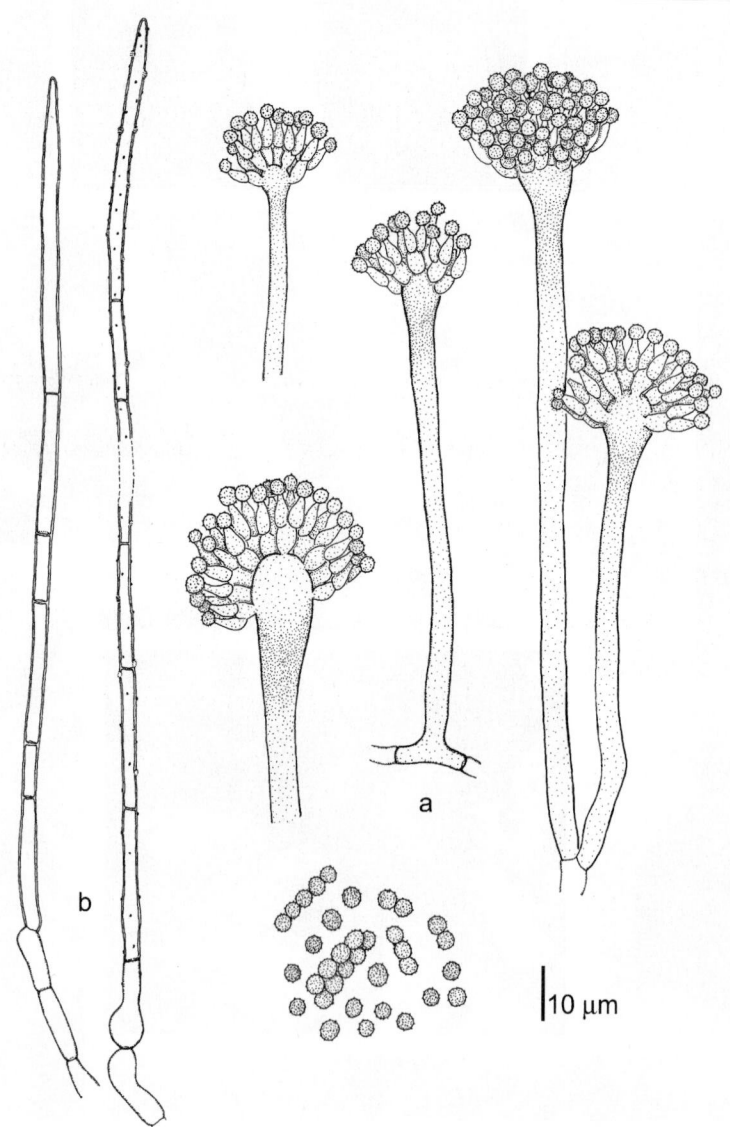

Aspergillus unguis (Emericella unguis), IMI 60313. a. Conidiophores and conidia; b. sterile seta-like hyphae.

Aspergillus unguis (Emericella unguis), IMI 60313. a, c-f. Conidiophores and conidia; b. sterile seta-like hyphae and conidiophores. a. ×320; b. ×512; c. ×1600; d. ×3600; e. ×750; f. ×3000.

Aspergillus ustus (Bain.) Thom & Church

Colony characteristics. Colonies (CzA) drab olive to dull brown, sometimes with dark purple exudate.

Microscopy. Conidial heads radiate to loosely columnar with age, commonly splitting into more or less well-defined columns. Conidiophore stipes smooth-walled, coloured in brown shades. Vesicles hemispherical to subspherical, 7-15 µm diam. Conidiogenous cells biseriate. Metulae covering upper half to upper third quarter of the vesicle. Conidia with very rough walls, spherical, 3.0-4.5 µm diam, dark yellow-brown. Irregular to elongate Hülle cells sometimes present.

Molecular diagnostics. SSU restriction map based on NCBI AB008410:

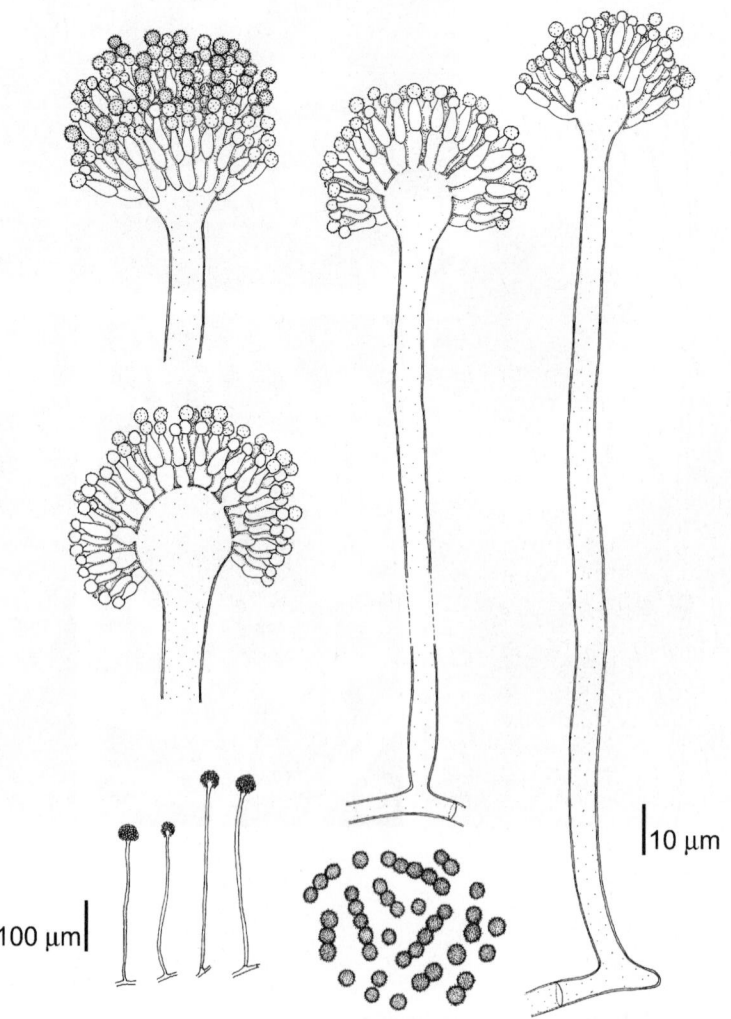

Pathogenicity. BSL-1. This species causes disseminated (Weiss & Thiemke, 1983; Iwen *et al.*, 1998) or pulmonary infections (Verweij *et al.*, 1999) in BMT patients. It was also isolated from cases of otitis media (Wadhwani & Srivastava, 1984), infected skin burns (Sandner & Schönborn, 1973) and cutaneous infections (Stiller *et al.*, 1994; Ricci *et al.*, 1998).

Reference. Raper & Fennell (1965).

Aspergillus ustus, CBS 261.67. Conidiophores and conidia.

Antifungal susceptibility.

Antifungal	Mean MICs	MIC 90	Strains	Reference
AMB	2	2	11	Verweij *et al.* (1999)
ITZ	2.24	4	11	Verweij *et al.* (1999)
VCZ	3.55	8	11	Verweij *et al.* (1999)
TBF	0.37	0.5	11	Verweij *et al.* (1999)
TBF	0.06-0.5	0.5	14	Ryder *et al.* (1999)

Nomenclature. *Sterigmatocystis usta* Bainier - Bull. Soc. Bot. Fr. 28: 78, 1881 ≡ *Aspergillus ustus* (Bainier) Thom & Church - The Aspergilli p. 152, 1926.

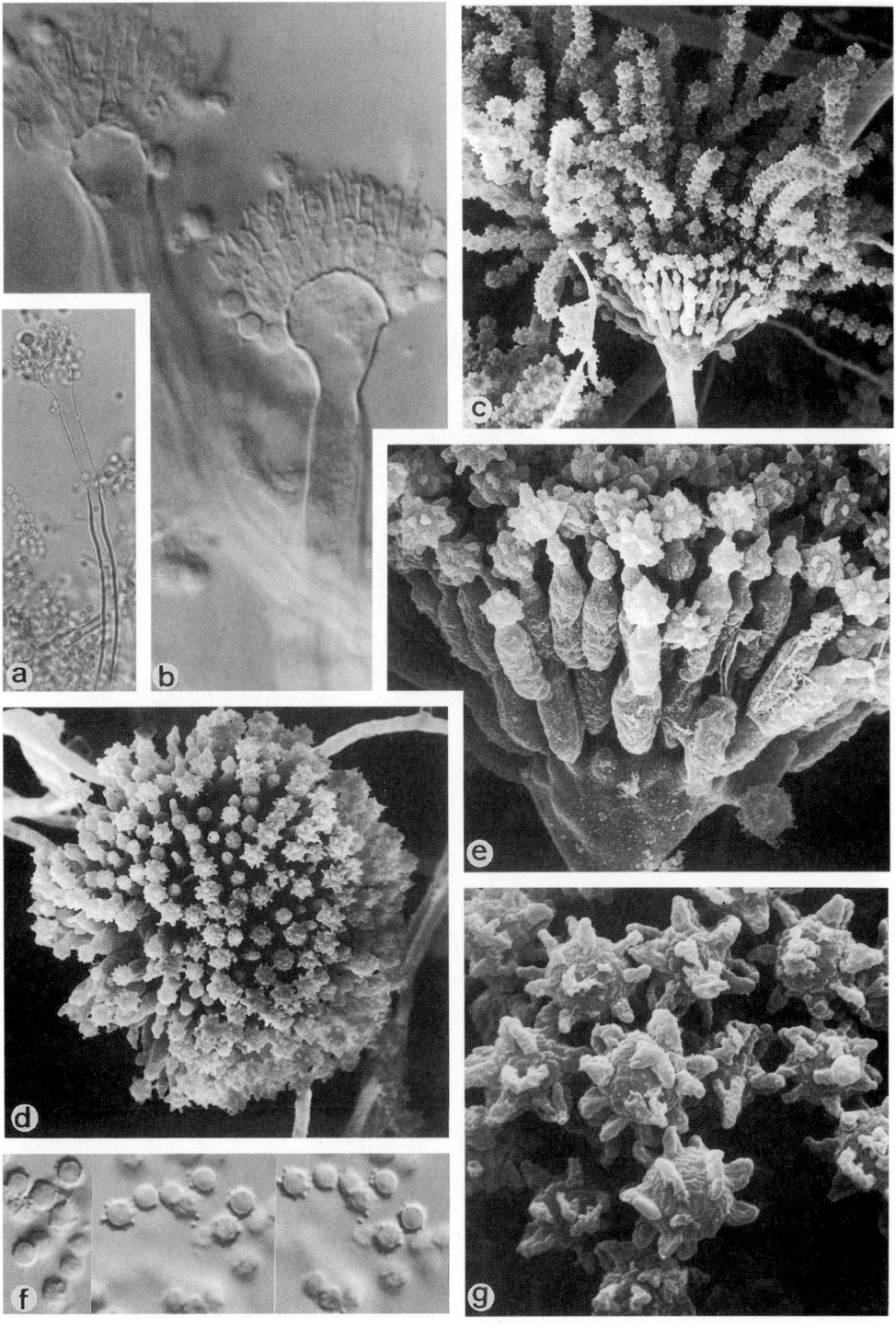

Aspergillus ustus, CBS 261.67. Conidiophores and conidia. a. ×320; b. ×1600; c. ×1100; d. ×2050; e. ×3900; f. ×1600; g. ×7800.

517

Aspergillus versicolor (Vuill.) Tiraboschi

Colony characteristics. Colonies (CzA) yellow, orange-yellow to yellow-green or pale pink, often poorly sporulating.
Microscopy. Conidial heads radiate. Conidiophore stipes smooth-walled, hyaline. Vesicles subspherical to ellipsoidal, 12-16 μm diam. Conidiogenous cells biseriate, with phialides longer than or as long as the metulae. Conidia echinulate, spherical, 2-3 μm diam. Hülle cells may be present in some isolates.
Molecular diagnostics. SSU restriction map based on NCBI AB008411:

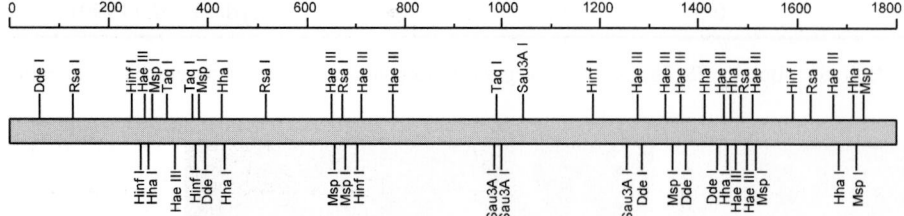

Pathogenicity. BSL-1. Cited as the causative agent of different human mycoses (Rippon, 1988). A recent case is that of Liu *et al.* (1995), who reported an osteomyelitis.
Reference. Raper & Fennell (1965).

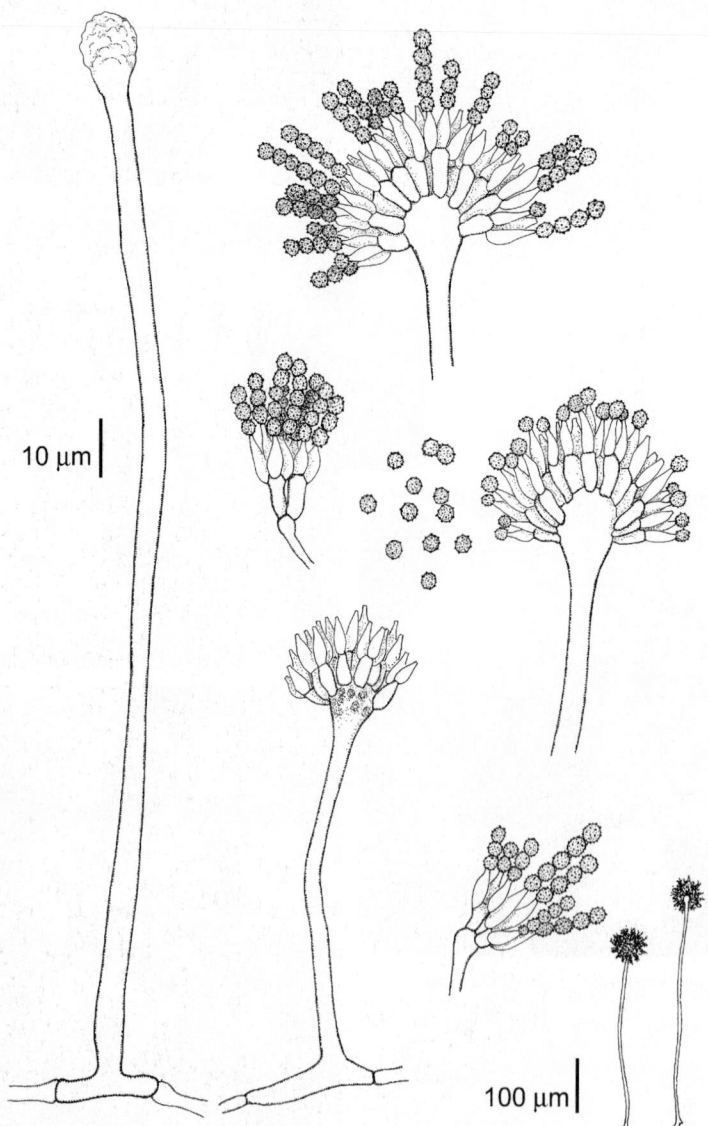

Aspergillus versicolor, FMR 2227. Conidiophores and conidia.

Antifungal susceptibility.

Antifungal	Mean MICs	Strains	MIC90	Reference
AMB	1.56	1	nd	Wildfeuer *et al.* (1998)
AMB	21.1	12	>64	Torrez-Rodríguez *et al.* (1998)
FLZ	89.1	12	>64	Torrez-Rodríguez *et al.* (1998)
GRF	118.8	12	>64	Torrez-Rodríguez *et al.* (1998)
ITZ	1.6	12	4	Torrez-Rodríguez *et al.* (1998)
ITZ	0.20	1	nd	Wildfeuer *et al.* (1998)
KTZ	2.5	12	4	Torrez-Rodríguez *et al.* (1998)
KTZ	0.39	1	nd	Wildfeuer *et al.* (1998)
TBF	0.2	2.5	1	Torrez-Rodríguez *et al.* (1998)
VCZ	0.78	1	nd	Wildfeuer *et al.* (1998)

Nomenclature. *Sterigmatocystis versicolor* Vuillemin, *in* Mirsky - Erreur Dét. Asp. Paras. Homme p. 15, 1903 ≡ *Aspergillus versicolor* (Vuillemin) Tiraboschi - Annls Bot., Roma 7: 9, 1908.

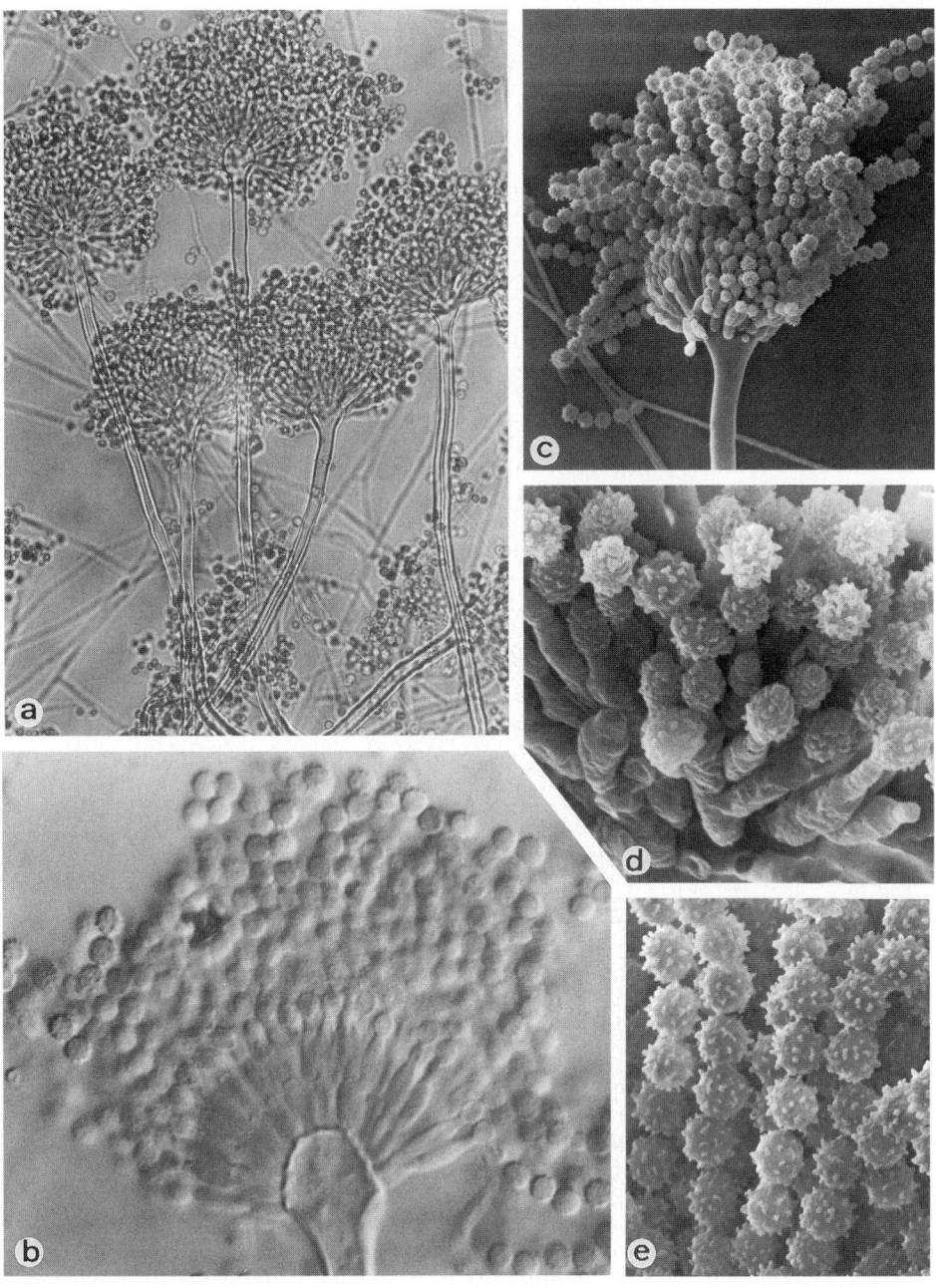

Aspergillus versicolor, FMR 2227. Conidiophores and conidia. a. ×345; b. ×1070; c. ×735; d. ×3150; e. ×2110.

Hyphomycetes, black yeasts and relatives. Genus: *AUREOBASIDIUM*

Aureobasidium pullulans (de Bary) Arn.

Colony characteristics. Colonies (MEA) growing rapidly, appearing smooth, soon covered with a slimy exudate, cream-coloured or pink, later mostly becoming brown or black.

Microscopy. Vegetative hyphae 3-12 µm wide, hyaline, locally converting into blackish-brown, thick-walled chlamydospores; expanding hyphae with irregularly dichotomous branching. Conidiogenous cells undifferentiated, mostly intercalary in hyaline hyphae. Conidia produced synchronously in dense groups from small denticles, later formed percurrently and adhering in slimy heads. Conidia hyaline, ellipsoidal, very variable in shape and size, (7.5-) 9.0-11.0 (-16.0) × (3.5-) 4.0-5.5 (-7.0) µm, 1-celled, often with an indistinct hilum. Budding frequently observed. Endoconidia often present in intercalary cells.

Teleomorph. *Discosphaerina fulvida* (F.R. Sanderson) Sivanesan (*Ascomycota, Euascomycetes, Dothideales: Dothioraceae*).

Physiology.

Growth:		Lactose	+	DL-Lactate	w
D-Glucose	+	Raffinose	+	Succinate	+,w
D-Galactose	+	Melezitose	+	Citrate	+,w
L-Sorbose	v	Inulin	w	Methanol	–
D-Glucosamine	w	Soluble starch	+	Ethanol	+
D-Ribose	+	Glycerol	+	Nitrate	+
D-Xylose	+	*meso*-Erythritol	+	Nitrite	+
L-Arabinose	+	Ribitol	+,w	Ethylamine	+
D-Arabinose	w	Xylitol	+	L-Lysine	+
L-Rhamnose	+	L-Arabinitol	+	Cadaverine	+
Sucrose	+	D-Glucitol	+	Creatine	–
Maltose	+	D-Mannitol	+	Creatinine	–
α,α-Trehalose	+	Galactitol	v	10% NaCl	+
methyl-α-D-Glucoside	+	*myo*-Inositol	+	0.1% Cycloheximide	–
Cellobiose	+	Glucono-δ-lactone	+,w	Mycosel	–
Salicin	+	D-Gluconate	+	Urease	+
Arbutin	+	D-Glucuronate	+	Benomyl	–
Melibiose	+	D-Galacturonate	+	Growth at 37°C	v

Molecular diagnostics. RFLP for distinction of *A. pullulans* from related species was developed by Yurlova *et al.* (1996) and ITS sequencing by Yurlova *et al.* (1999). Li *et al.* (1996) developed a species-specific probe based on SSU rDNA sequences. For a phylogenetic comparison with other black yeasts, see p. 374. SSU and ITS restriction maps based on NCBI M55639 and AF013229, resp.:

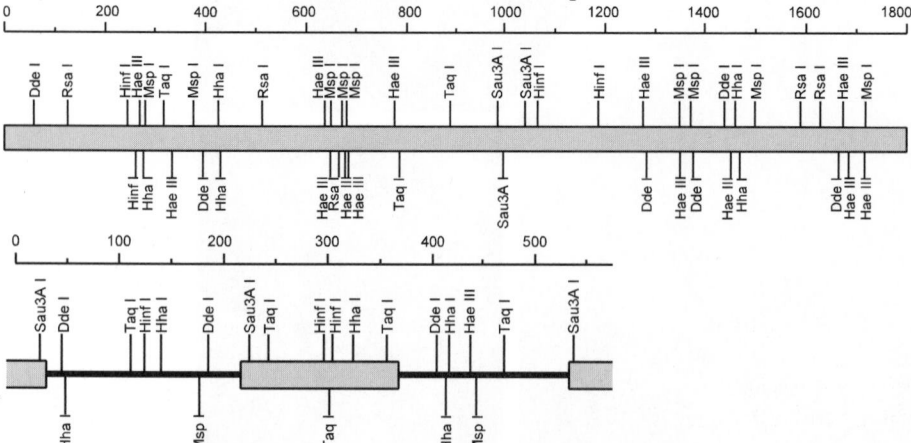

Pathogenicity. BSL-1. The fungus occurs very commonly in somewhat osmotic environments, such as in food and on plant leaves. It is a frequent colonizer of damp stone and glass and commonly found in the clinical lab as a contaminant. Infections are mostly by traumatic inoculation (Koppang *et al.,* 1991; Romand *et al.,* 1995; Hirsch *et al.*, 1996). The following human cases have been reported: keratitis (Ashikaga, 1920; Jones & Christensen, 1974), pulmonary infections (Akagi *et al.*, 1958; Tan *et al.*, 1997), systemic infections (Salkin *et al.*, 1986; Kaczmarski *et al.*, 1986; Girardi *et al.*, 1993), cutaneous infection (Vermeil *et al.*, 1971), peritonitis (Pritchard & Muir, 1987), CAPD peritonitis (Clark *et al.*, 1995; Caporale *et al.*, 1996) and invasive mycosis in an AIDS patient (Yarrish *et al.*, 1991). Salkin *et al.* (1976) described a cutaneous infection in a wild porcupine.

References. Hermanides-Nijhoff (1977), Domsch *et al.* (1980), de Hoog & Yurlova (1994), Yurlova *et al.* (1999).

Antifungal susceptibility.

Antifungal	Mean MICs	Strains	Reference
AMB	0.38	5	McGinnis & Pasarell (1998b)
AMB	0.25	2	Tintelnot *et al.* (1999)
AMB	0.25	3	Wildfeuer *et al.* (1998)
5FC	8	3	Wildfeuer *et al.* (1998)
ITZ	0.05	5	McGinnis & Pasarell (1998b)
ITZ	0.25	3	Wildfeuer *et al.* (1998)
KTZ	0.25	2	Tintelnot *et al.* (1999)
VCZ	0.14	5	McGinnis & Pasarell (1998b)
VCZ	0.12	3	Wildfeuer *et al.* (1998)

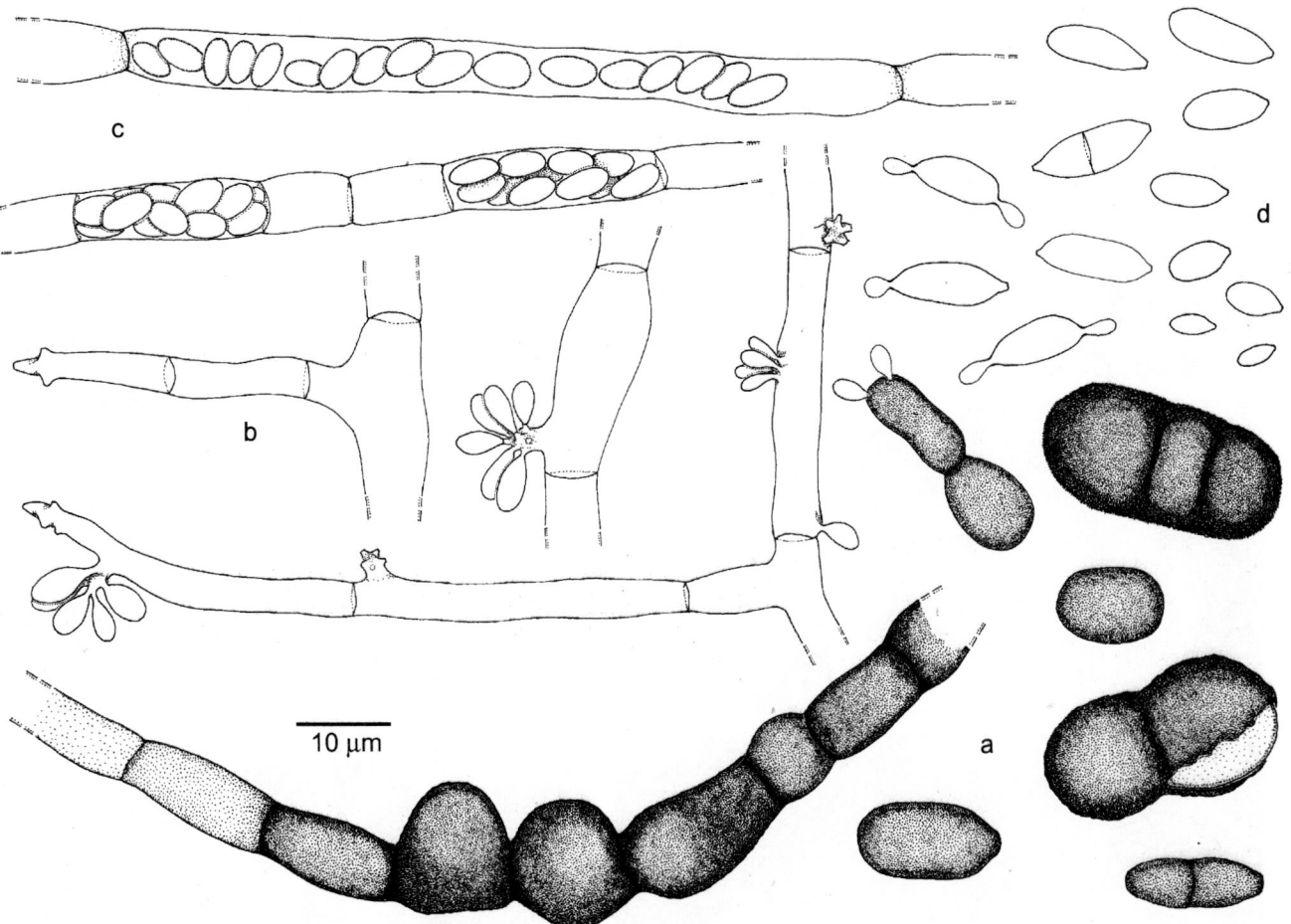

Aureobasidium pullulans, CBS 584.75. a. Intercalary and discrete chlamydospores; b. hyphae with synchronous conidia; c. endoconidia; d. conidia and budding cells.

Nomenclature. *Dematium pullulans* de Bary - Morphol. Physiol. Pilze p.182, 1886 ≡ *Aureobasidium pullulans* (de Bary) Arnaud - Ann. Ec. Agric. Montpellier, N. Sér. 16: 39, 1918.

Guignardia fulvida F.R. Sanderson - N. Zl. J. Agric. Res. 8: 139, 1965 ≡ *Discosphaerina fulvida* (F.R. Sanderson) Sivanesan - Bitunicat. Ascom. p. 148, 1984.

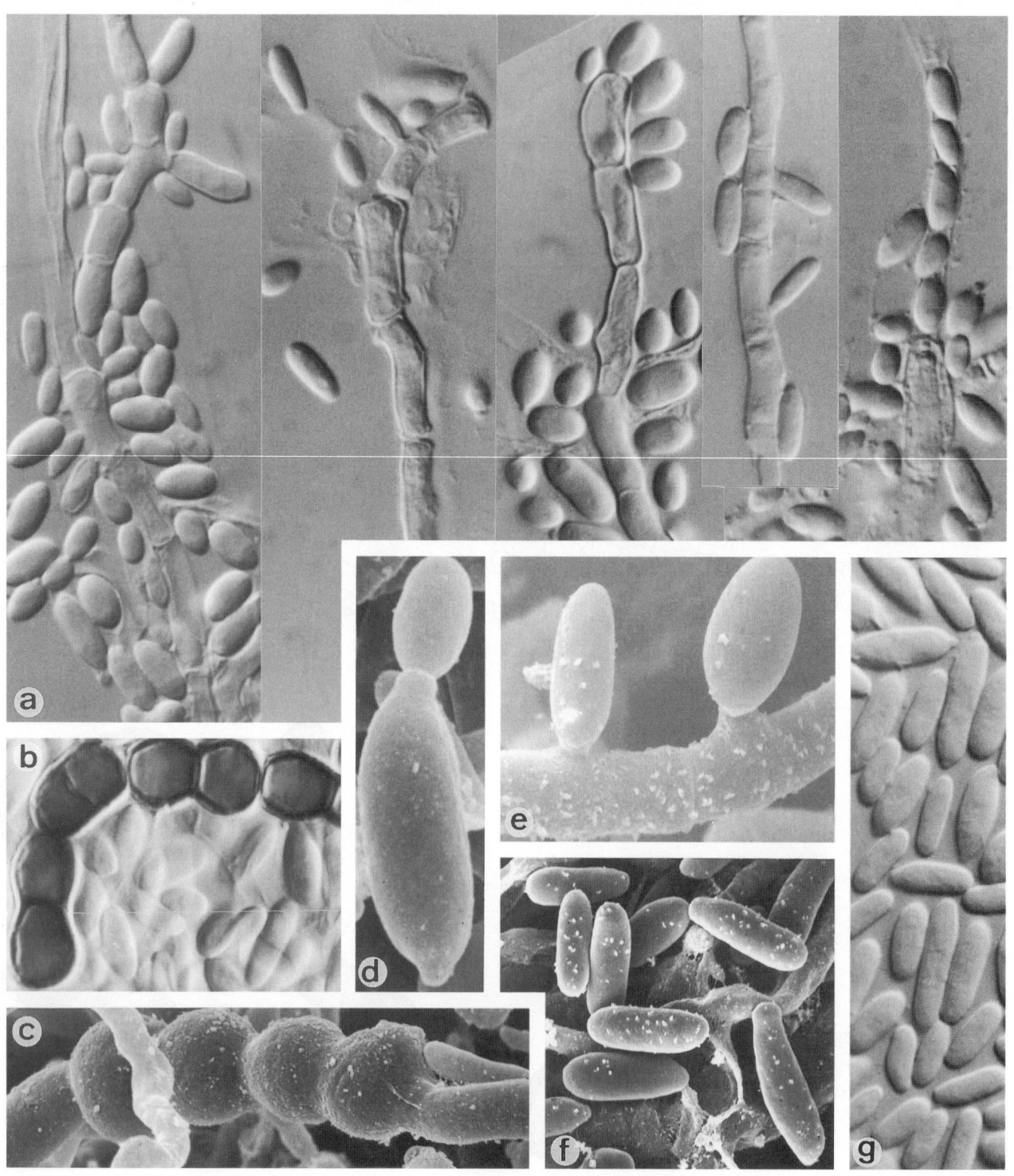

Aureobasidium pullulans, CBS 626.85. a, d-g. Conidia; b, c. dark hyphae with chlamydospores. a, b. ×1600; c. ×3500; d. ×7800; e. ×6200; f, g. ×2300.

Hyphomycetes. Genus: *BEAUVERIA*

Beauveria bassiana (Bals.) Vuill.

Colony characteristics. Colonies (OA) growing moderately rapidly, lanose, floccose, velvety to powdery, sometimes funiculose, at first white, often becoming pale yellow later.

Microscopy. Conidiogenous cells in fresh cultures aggregated in dense clusters alongside wide hyphae or in small synnemata, swollen, subspherical or ampulliform, 3-6 × 2.5-3.5 µm, with a peg-like, narrow rachis which often elongates in zig-zag manner, denticulate, up to 25 µm long and 1 µm wide. Conidia single, spherical or subspherical, hyaline, smooth-walled, 2-3 µm diam.

Serology. Exoantigen testing for species recognition was developed by Sekhon *et al.* (1997).

Physiology. Intolerant to benomyl. Physiological diagnostics within the genus was developed by Todorova *et al.* (1998).

Growth:		Arbutin	+	D-Fructose	+
D-Glucose	+	Melibiose	–	D-Mannose	+
Galactose	+	Lactose	+	D-Glucitol	+
L-Sorbose	–	D-Raffinose	+	Sorbitol	+
Ribose	+	Melezitose	w	*N*-Acetyl-D-glucosamine	+
D-Xylose	+	Inulin	–	Amygdalin	+
L-Xylose	–	Soluble starch	+	Aesculin	+
L-Arabinose	–	Glycerol	+	Glycogen	+
D-Arabinose	–	*meso*-Erythritol	+	β-Gentibiose	+
Rhamnose	–	Xylitol	–	D-Turanose	+
Sucrose	+	D-Mannitol	+	D-Lyxose	–
Maltose	+	*myo*-Inositol	+	D-Tagatose	–
α,α-Trehalose	+	Gluconate	–	D-Fucose	–
α-methyl-D-Mannoside	–	Adonitol	+	L-Fucose	–
α-methyl-D-Glucoside	–	2-Keto-D-gluconate	–	D-Arabitol	+
Cellobiose	+	5-Keto-D-gluconate	–	L-Arabitol	w
Salicin	+	β-Methyl-D-xyloside	–		

Molecular diagnostics. Specific primers were proposed by Hegedus & Khachatourians (1996). ITS restriction map based on CBS 337.72:

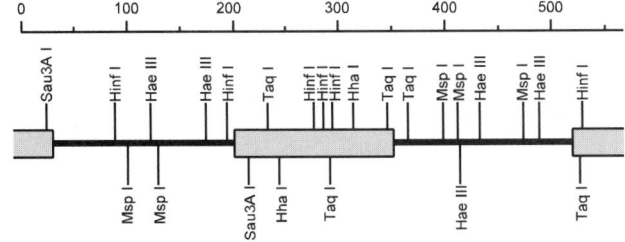

Pathogenicity. BSL-1. The fungus is a virulent insect pathogen. Infections in humans are very rare. Cases of keratitis were described by Sachs *et al.* (1985) and Ishibashi *et al.* (1987). A pulmonary infection presented by Freour *et al.* (1966) probably concerned an *Acrodontium* species. A pulmonary infection was also reported in a captive American alligator (Fromtling *et al.*, 1979) and in a tortoise (González-Cabo *et al.*, 1995).

References. De Hoog (1972), Domsch *et al.* (1980).

Nomenclature. *Botrytis bassiana* Balsamo - Linnaea 10: 611, 1835 ≡ *Beauveria bassiana* (Balsamo) Vuillemin - Bull. Soc. Bot. Fr. 59: 40, 1912.

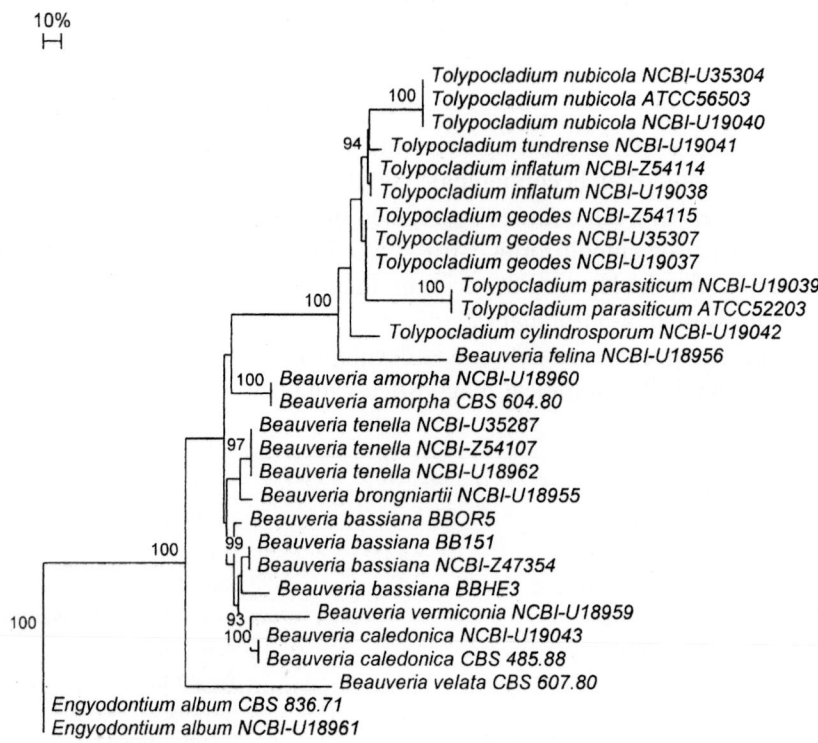

10%

Tolypocladium nubicola NCBI-U35304
100 *Tolypocladium nubicola ATCC56503*
Tolypocladium nubicola NCBI-U19040
94 *Tolypocladium tundrense NCBI-U19041*
Tolypocladium inflatum NCBI-Z54114
Tolypocladium inflatum NCBI-U19038
Tolypocladium geodes NCBI-Z54115
Tolypocladium geodes NCBI-U35307
Tolypocladium geodes NCBI-U19037
100 *Tolypocladium parasiticum NCBI-U19039*
100 *Tolypocladium parasiticum ATCC52203*
Tolypocladium cylindrosporum NCBI-U19042
Beauveria felina NCBI-U18956
100 *Beauveria amorpha NCBI-U18960*
Beauveria amorpha CBS 604.80
Beauveria tenella NCBI-U35287
97 *Beauveria tenella NCBI-Z54107*
Beauveria tenella NCBI-U18962
Beauveria brongniartii NCBI-U18955
Beauveria bassiana BBOR5
100 99 *Beauveria bassiana BB151*
Beauveria bassiana NCBI-Z47354
Beauveria bassiana BBHE3
100 93 *Beauveria vermiconia NCBI-U18959*
100 *Beauveria caledonica NCBI-U19043*
Beauveria caledonica CBS 485.88
Beauveria velata CBS 607.80
Engyodontium album CBS 836.71
Engyodontium album NCBI-U18961

Fig. 56. Phylogenetic tree of *Beauveria bassiana* and its relatives based on confidently aligned ITS rDNA sequences using Neighbor joining algorithm with Kimura correction. Bootstrap values >90 from 100 resampled datasets are shown. *Engyodontium album* (p. 640) is selected as outgroup. The phialidic genus *Tolypocladium* is closely related.

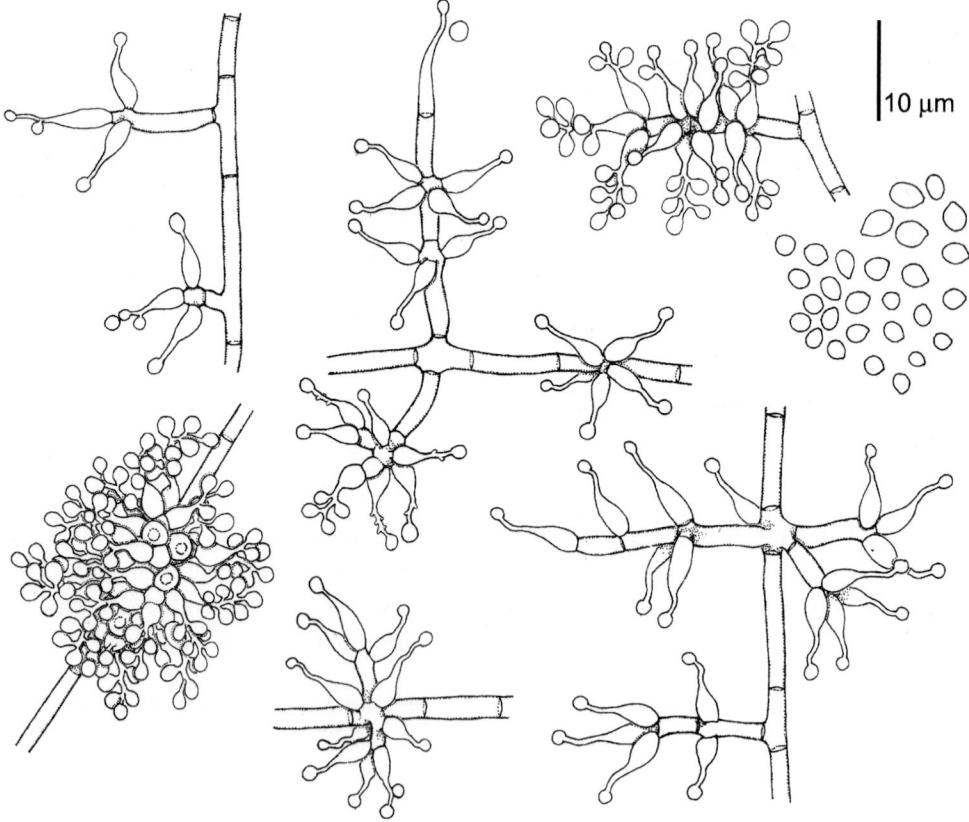

10 μm

Beauveria bassiana, CBS 337.32. Conidiogenous cells and conidia.

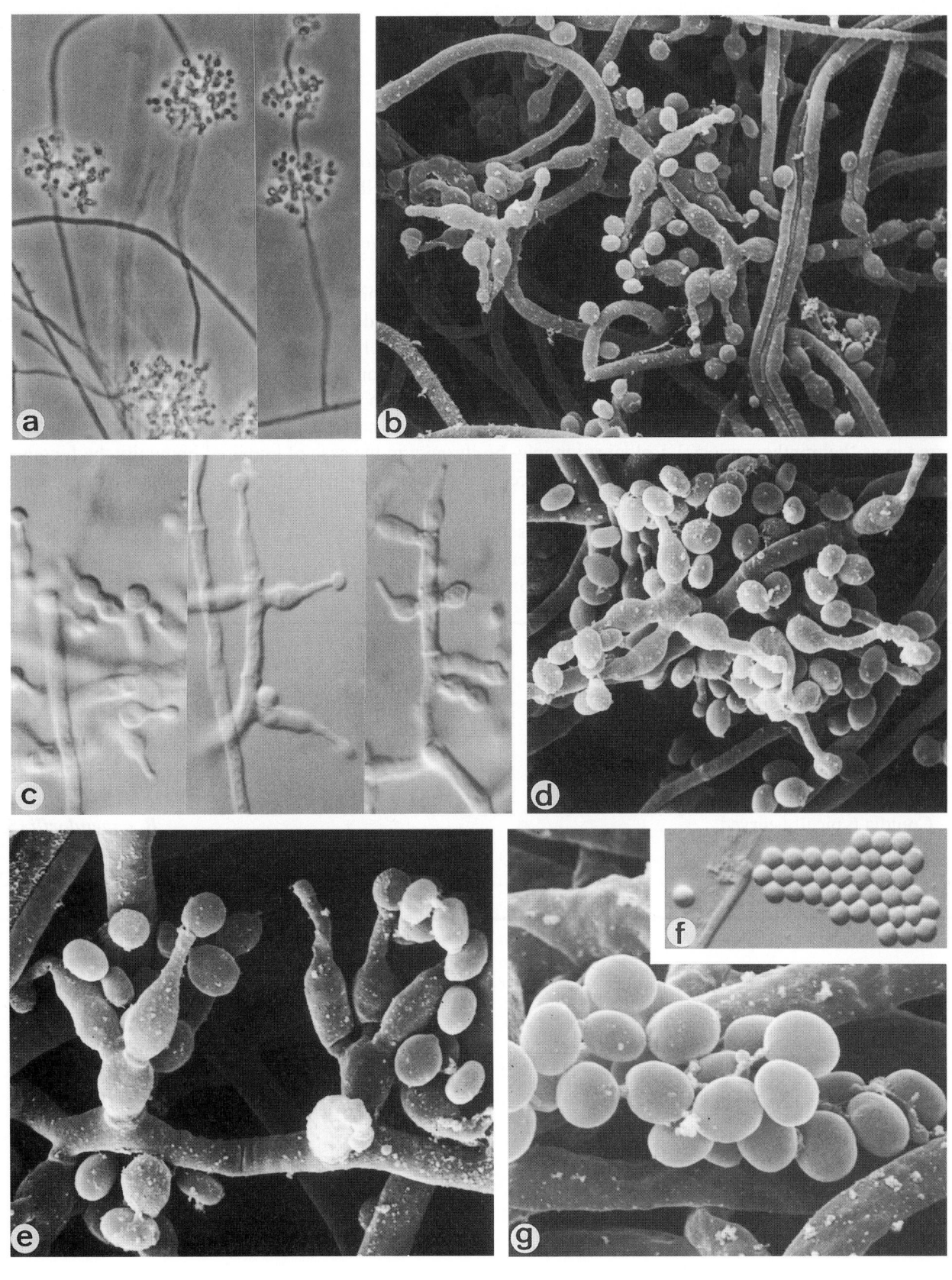

Beauveria bassiana, CBS 337.32. Conidiogenous cells and conidia. a. ×640; b. ×2850; c. ×1600; d. ×3850; e. ×6000; f. ×8600; g. ×1600.

Hyphomycetes, Dematiaceae. Genus: *BIPOLARIS*

Generic description. Colonies black, hairy, expanding. Conidiophores brown, erect, multi-cellular, producing conidia in sympodial order; conidial scars dark brown, flat. Conidia ellipsoidal, straight or curved, compartimented by distosepta. Tolerant to benomyl.

Teleomorph. *Cochliobolus* (*Ascomycota, Euascomycetes, Pleosporales: Pleosporaceae*).

Differential diagnosis. *Dissitimurus* has reduced, catenate conidia. *Curvularia* (p. 598) differs by having truly septate conidia. *Bipolaris* species were treated in the first edition of this Atlas under *Drechslera*, but species of this genus (p. 631) have straight conidia. *Exserohilum* (p. 669) is differentiated by protuberant conidial hila. Serodiagnosis of *Bipolaris* and similar fungi was developed by Pasarell *et al.* (1990).

General remarks. Nearly all species of this large genus are pathogenic to grasses. Some are common saprobes on dead plant material and in soil. These are occasionally found in humans, mainly causing allergic sinusitis (Washburn *et al.*, 1988). This disorder is more commonly found in cattle, and in older literature has mostly been referred to as 'helminthosporiosis'. Colonization may be chronic; in the case of impaired immunity the fungus may become invasive and bone erosion may take place, the fungus eventually forming lesions in the brain (Killingsworth & Wetmore, 1990).

Molecular diagnostics. Shimizu *et al.* (1998) presented a phylogeny of the genus based on a reductase gene involved in malanin biosynthesis. An ITS tree is given in Fig. 42 (p. 382).

References. Ellis (1971, 1976), Alcorn (1983), McGinnis *et al.* (1986b), Sivanesan (1987).

Key to the treated species of Bipolaris:

1a. Conidia mostly 3-septate ➞ **2**
1b. Conidia usually 3-7-septate . . . . **B. hawaiiensis (529)**
2a. Conidia straight ➞ **3**
2b. Conidia typically curved . . . . **B. papendorfii (531)**
3a. Conidia 6-11 µm wide . . . . **B. australiensis (527)**
3b. Conidia 9-14 µm wide . . . . **B. spicifera (533)**

Bipolaris australiensis (M.B. Ellis) Tsuda & Ueyama

Colony characteristics. Colonies (PCA) spreading, grey to blackish-brown, velvety.

Microscopy. Conidiophores brown, solitary, flexuose or geniculate, smooth-walled, up to 150 µm long, mostly 3-7 µm wide. Conidiogenous nodes verruculose. Conidia straight, rounded at the ends, pale brown to mid reddish-brown, mostly 3-, rarely 4-5-distoseptate, ellipsoidal or oblong, 14-40 × 6-11 µm, smooth-walled to finely roughened.

Teleomorph. *Cochliobolus australiensis* (Tsuda & Ueyama) Alcorn (*Ascomycota, Euascomycetes, Pleosporales: Pleosporaceae*).

Molecular diagnostics. ITS restriction map based on CBS 172.57:

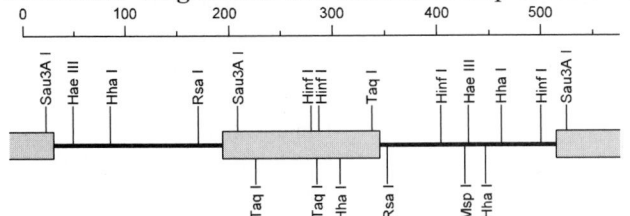

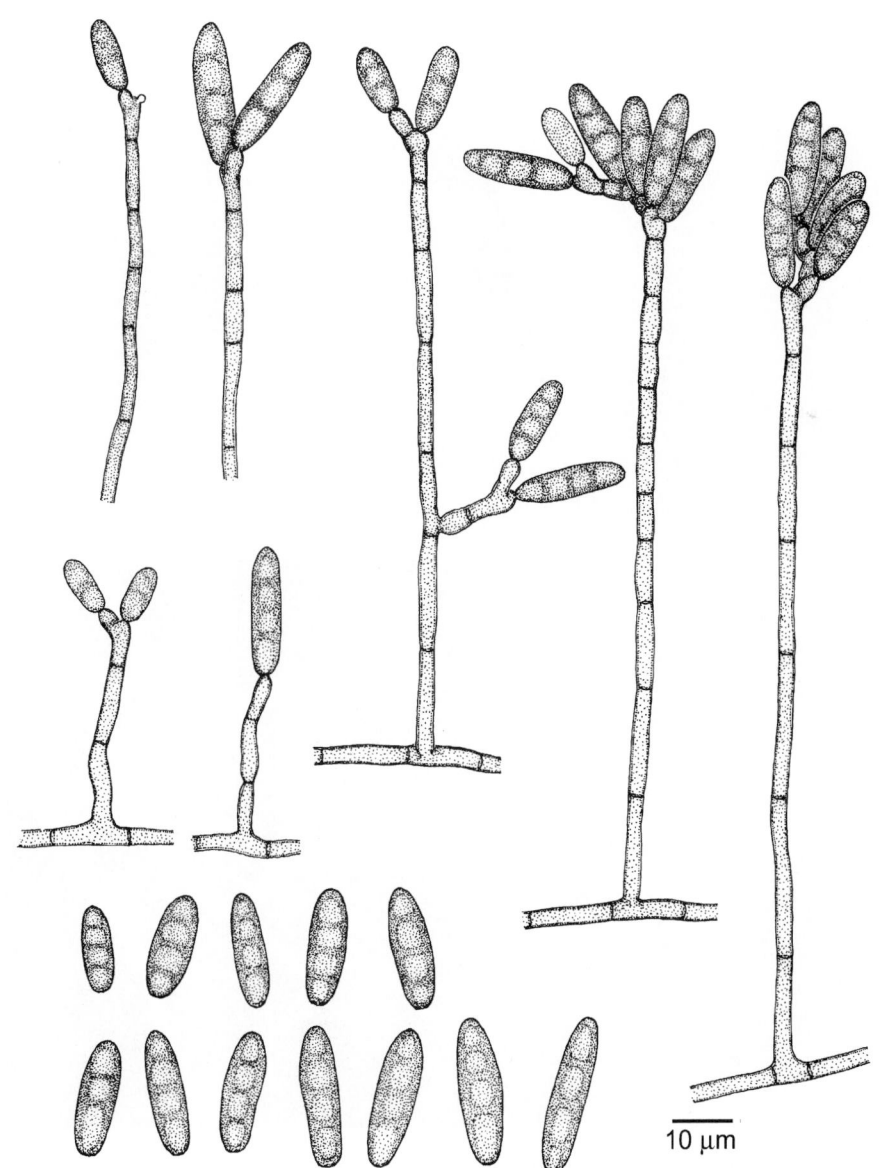

Bipolaris australiensis, CBS 172.57. Conidiophores and conidia.

Pathogenicity. BSL-1. A common saprobe. Human cases: peritonitis (Mousdale *et al.*, 1981), subcutaneous (McGinnis *et al.*, 1986b) infections, sinusitis (Harpster *et al.*, 1985), fungus ball in paranasal sinus (Pritchard & Muir, 1987). A disseminated infection in an immunocompetent patient was described by Flanagan & Bryceson (1997).

References. Ellis (1971), McGinnis *et al.* (1986b), Sivanesan (1987).

Antifungal susceptibility.

Antifungal	Mean MICs	Strains	Reference
AMB	0.25	17	McGinnis & Pasarell (1998b)
5FC	128	1	Unpublished data
FLZ	8	1	Unpublished data
ITZ	0.07	17	McGinnis & Pasarell (1998b)
KTZ	0.5	1	Unpublished data
MCZ	0.03	1	Unpublished data
TBF	0.12	3	McGinnis & Pasarell (1998a)
VCZ	0.25	17	McGinnis & Pasarell (1998b)

Nomenclature. *Helminthosporium australiense* Bugnicourt - Revue Gén. Bot. 62: 242, 1955 ≡ *Drechslera australiensis* (Bugnicourt) Subramanian & Jain - Curr. Sci. 35: 354, 1966 ≡ *Drechslera australiensis* (Bugnicourt) Subramanian & Jain ex M.B. Ellis - Demat. Hyphom. p. 413, 1971 ≡ *Bipolaris australiensis* (M.B. Ellis) Tsuda & Ueyama - Mycologia 73: 90, 1981.

Pseudocochliobolus australiensis Tsuda & Ueyama - Mycologia 73: 93, 1981 ≡ *Cochliobolus australiensis* (Tsuda & Ueyama) Alcorn - Mycotaxon 16: 373, 1983.

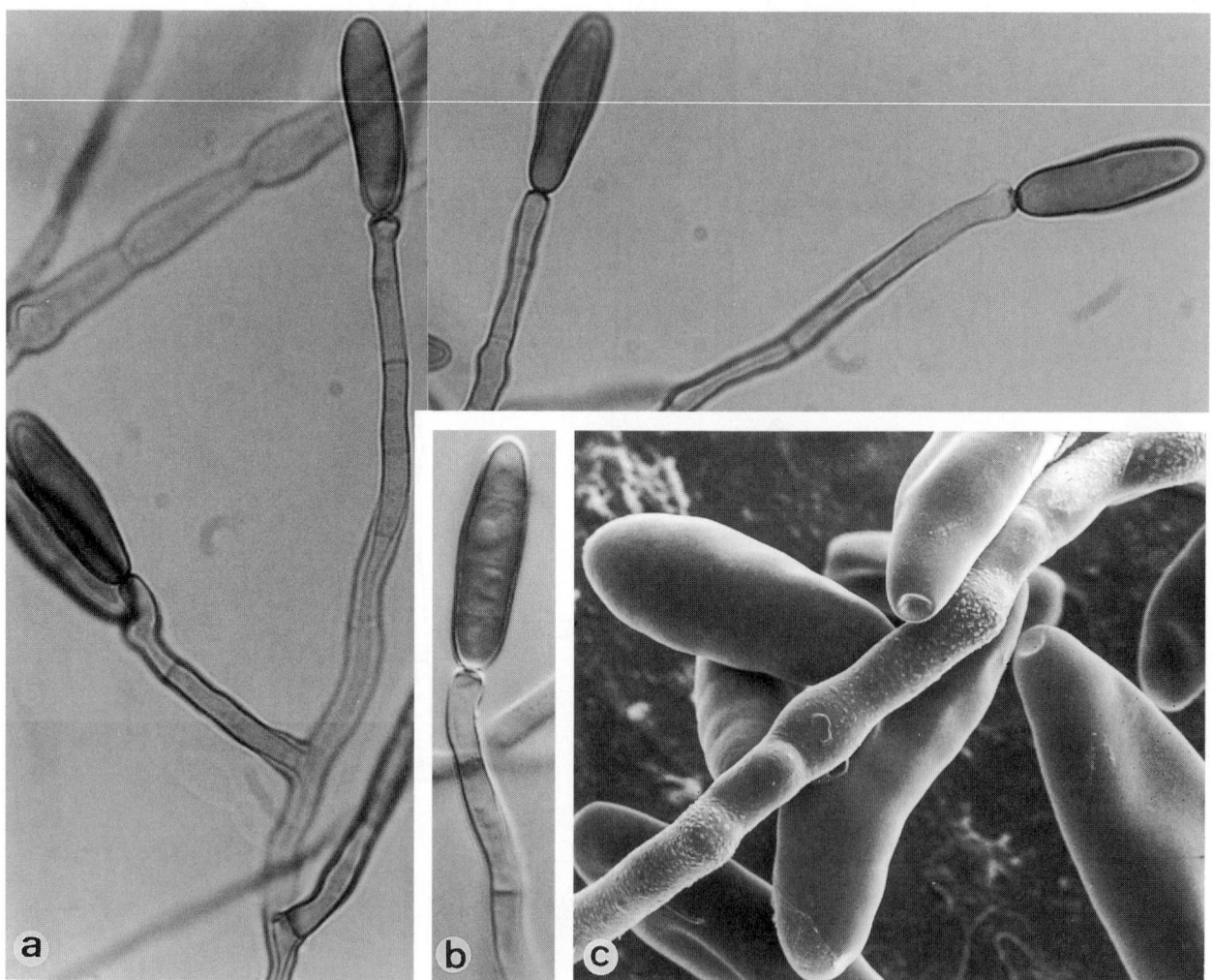

*Bipolaris australiensis,*CBS 172.57. Conidiophores and conidia. a. ×1280; b. ×1600; c. ×3600.

Bipolaris hawaiiensis (M.B. Ellis) Uchida & Aragaki

Colony characteristics. Colonies (PCA) spreading, powdery to hairy, black.

Microscopy. Conidiophores erect, unbranched, septate, apically flexuose with flat conidial scars on the edges, up to 80 μm long. Conidia smooth- and rather thick-walled, brown, with (3-) 5 (-7) distosepta, cylindrical to cigar-shaped, 18-35 × 6-9 μm.

Teleomorph. *Cochliobolus hawaiiensis* Alcorn (*Ascomycota, Euascomycetes, Pleosporales: Pleosporaceae*).

Molecular diagnosics. ITS restriction map based on CBS 173.57:

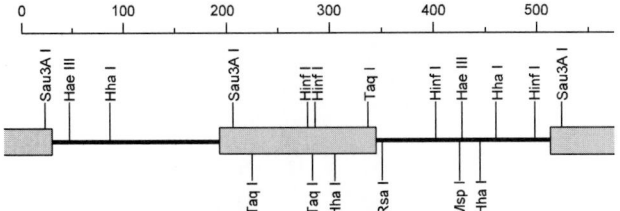

Pathogenicity. BSL-1. The species is a common saprophyte on plant material. Among the human cases, sinusitis is prevalent (Young *et al.*, 1978; Koshi *et al.*, 1987; Fryen *et al.*, 1999). Also cases of pulmonary (Koenig *et al.*, 1984) and cerebral (Morton *et al.*, 1986; Ruben *et al.*, 1987) mycosis have been reported. A fatal case of meningoencephalitis was described by Fuste *et al.* (1973). Keratitis has occasionally been reported (Krachmer *et al.*, 1978; Anandi *et al.*, 1988), as well as endophthalmitis (Pavan & Margo, 1993). A catheter-related infection was reported by Gadallah *et al.* (1995).

References. Ellis (1971), de Hoog (1983), Sivanesan (1987).

Antifungal susceptibility.

Antifungal	Mean MICs	Strains	Reference
AMB	0.25	17	McGinnis & Pasarell (1998b)
ITZ	0.07	17	McGinnis & Pasarell (1998a)
TBF	0.14	17	McGinnis & Pasarell (1998a)
VCZ	0.15	17	McGinnis & Pasarell (1998b)

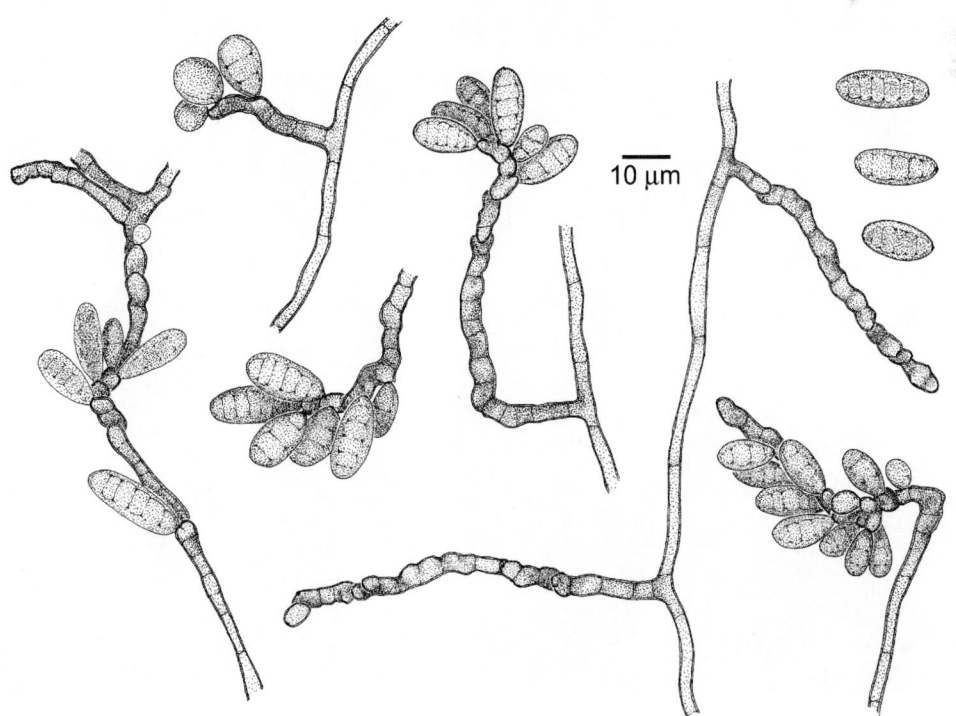

Bipolaris hawaiiensis, WCLR M. 384.86. Conidiophores and conidia.

529

Nomenclature. *Helminthosporium hawaiiense* Bugnicourt - Revue Gén. Bot. 62: 238, 1955 ≡ *Drechslera hawaiiensis* (Bugnicourt) Subramanian & Jain - Curr. Sci. 35: 354, 1966 ≡ *Drechslera hawaiiensis* (Bugnicourt) Subramanian & Jain ex M.B. Ellis - Demat. Hyphom. p. 415, 1971 ≡ *Bipolaris hawaiiensis* (M.B. Ellis) Uchida & Aragaki - Phytopathology 69: 1115, 1979.
　　Cochliobolus hawaiiensis Alcorn - Trans. Br. Mycol. Soc. 70: 64, 1978.

Bipolaris hawaiiensis, WCLR 384.86. Conidiophores and conidia. a. ×512; b. ×1280; c. ×850; d. ×1900; e. ×2800.

Bipolaris papendorfii (van der Aa) Alcorn

Colony characteristics. Colonies (PCA) effuse, velvety, grey to black.

Microscopy. Conidiophores solitary, straight or flexuose, geniculate, pale to medium brown, up to 200 μm long, 4-9 μm wide, smooth-walled, septate, with prominent, dark scars. Conidiogenous nodes verruculose, distinct, swollen, up to 5 μm wide. Conidia curved, kidney-shaped or obpyriform, usually broadest at the second cell from the base, medium brown to dark olivaceous brown, paler at the ends, smooth-walled, 3-distoseptate, 30-40 × 17-30 μm.

Molecular diagnostics. ITS restriction map based on CBS 308.67:

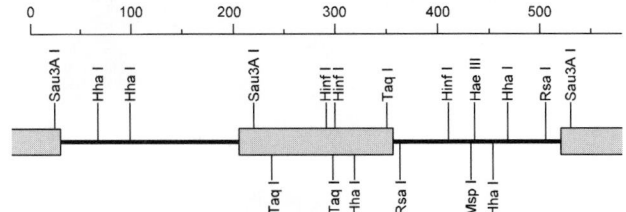

Pathogenicity. BSL-1. Reported as a possible causal agent in a case of keratitis (Pritchard & Muir, 1987).

References. Ellis (1971), Sivanesan (1987).

Nomenclature. *Curvularia papendorfii* van der Aa - Persoonia 5: 45, 1967 ≡ *Drechslera papendorfii* (van der Aa) M.B. Ellis - Demat. Hyphom. p. 413, 1971 ≡ *Bipolaris papendorfii* (van der Aa) Alcorn - Mycotaxon 17: 68, 1983.

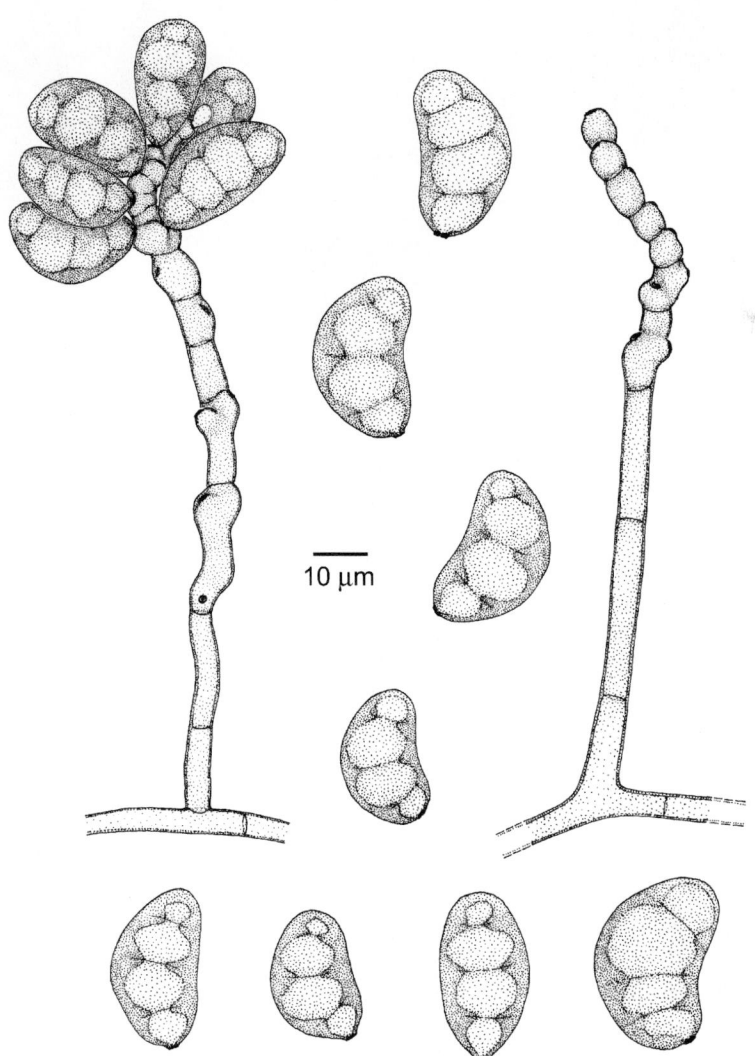

Bipolaris papendorfii, IMI 337278. Conidiophores and conidia.

531

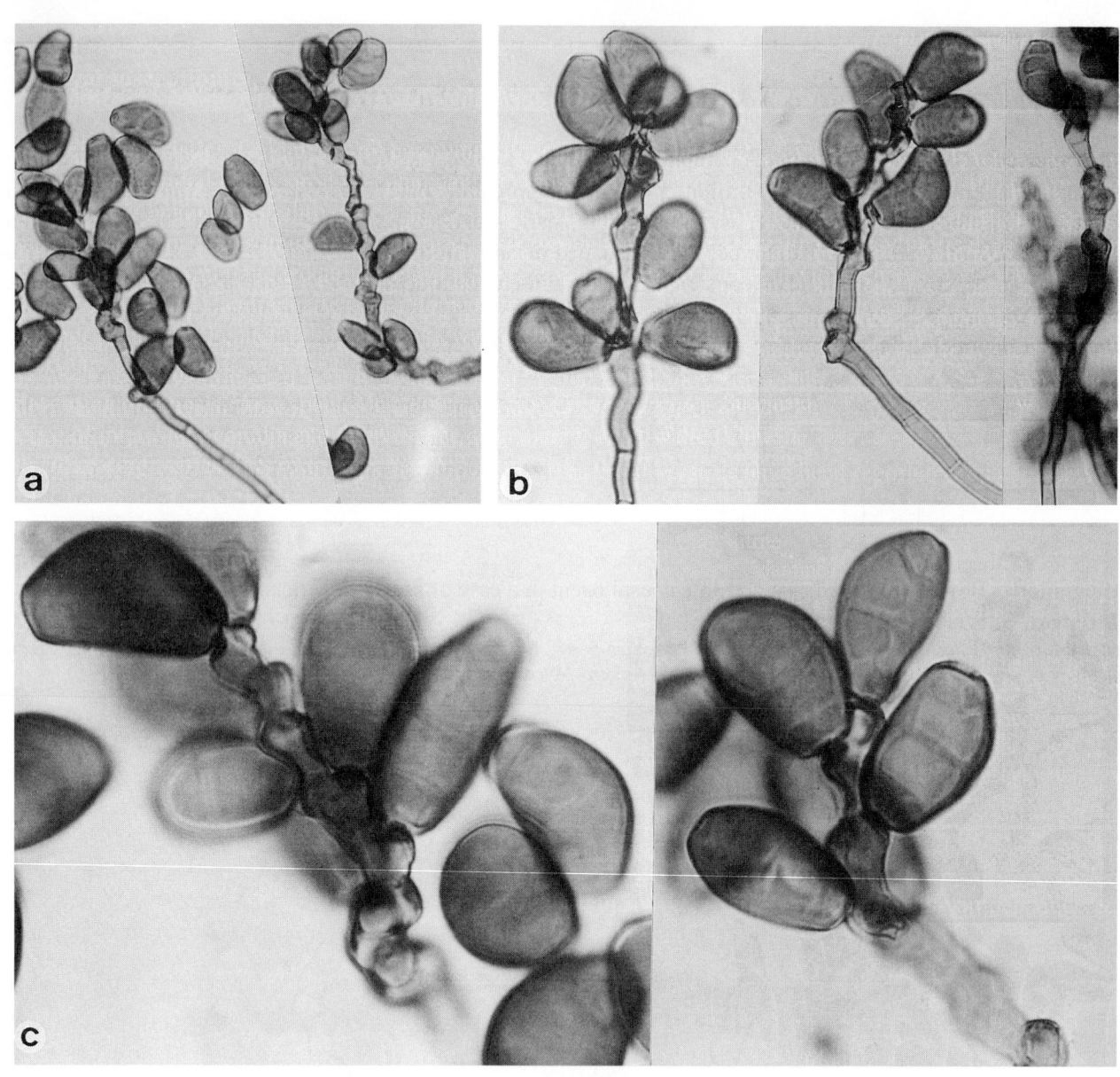

Bipolaris papendorfii, IMI 337278. Conidiophores and conidia. a. ×400; b. ×720; c. ×1280.

Bipolaris spicifera (Bain.) Subram.

Colony characteristics. Colonies (PCA) expanding, appearing glassy with sooty powder of conidia, or hairy if sporulation is poor.

Microscopy. Conidiophores erect, unbranched, septate, up to 250 μm long and 4-8 μm wide, regularly zig-zagged in the apical part, with flat, dark brown scars on the edges. Conidia brown, cylindrical with rounded ends, medium brown except for narrow subhyaline spots at the extremites, 20-40 × 9-14 μm, with 3 distosepta.

Teleomorph. *Cochliobolus spiciferus* Nelson (*Ascomycota, Euascomycetes, Pleosporales: Pleosporaceae*).

Molecular diagnostics. ITS restriction map based on CBS 274.52:

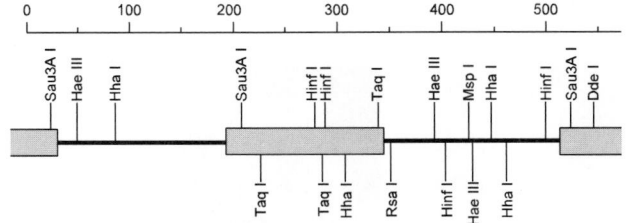

Pathogenicity. BSL-1. The species is a common saprophyte on plant material and in soil. It is frequently encountered as an agent of human or animal sinusitis (Sobol *et al.*, 1984; Drouhet *et al.*, 1985; Rolston *et al.*, 1985; Adam *et al.*, 1986; McGinnis *et al.*, 1992; Schubert & Goetz, 1998), eventually with cerebral involvement (Young *et al.*, 1978). Biggs *et al.* (1986) described a fatal cerebral case after head injury. Also cases of keratitis (Zapater *et al.*, 1975; Forster *et al.*, 1975b) and cutaneous phaeohyphomycosis (Estes *et al.*, 1977) have been reported. Karim *et al.* (1993) described a pulmonary case and Pauzner *et al.* (1997) an endocarditis. Animal cases involved horses (Schauffer, 1972; Kaplan *et al.*, 1975), cats (Müller *et al.*, 1975), cows (Patton, 1977) and dogs (Hall, 1965; Kwochka *et al.*, 1984).

References. Ellis (1971), de Hoog (1983), Douer *et al.* (1987), Sivanesan (1987).

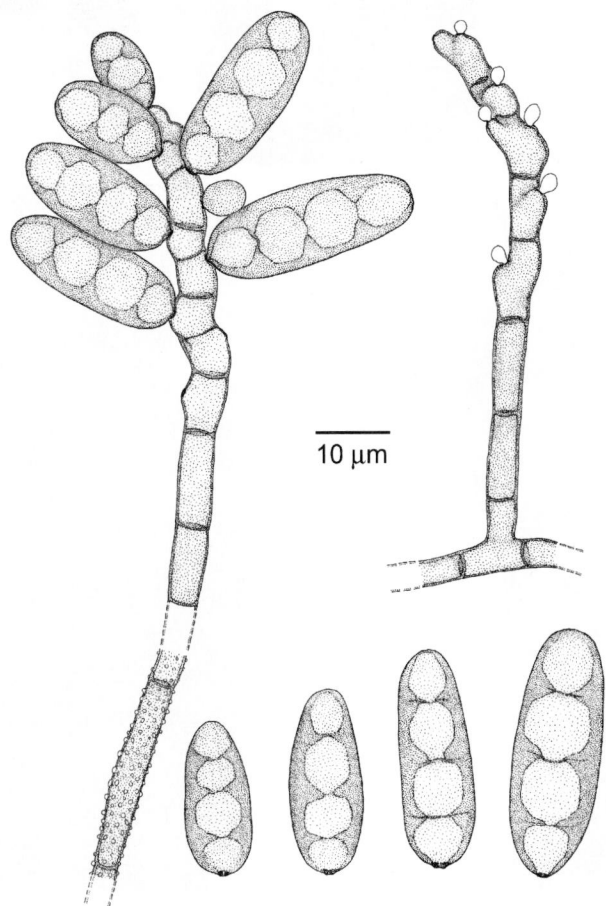

Bipolaris spicifera, CBS 586.80. Conidiophores and conidia.

Antifungal susceptibility.

Antifungal	Mean MICs	Strains	Reference
AMB	0.25	24	McGinnis & Pasarell (1998b)
ITZ	0.09	22	McGinnis & Pasarell (1998a)
TBF	0.20	22	McGinnis & Pasarell (1998a)
VCZ	0.29	24	McGinnis & Pasarell (1998b)

Nomenclature. *Brachycladium spiciferum* Bainier - Bull. Trim. Soc. Mycol. Fr. 24: 81, 1908 ≡ *Drechslera spicifera* (Bainier) von Arx - Gen. Fungi. Sporul. Pure Cult. p. 222, 1970 ≡ *Bipolaris spicifera* (Bainier) Subramanian - Hyphomycetes p. 756, 1971.
    Cochliobolus spiciferus Nelson - Mycologia 56: 198, 1964.

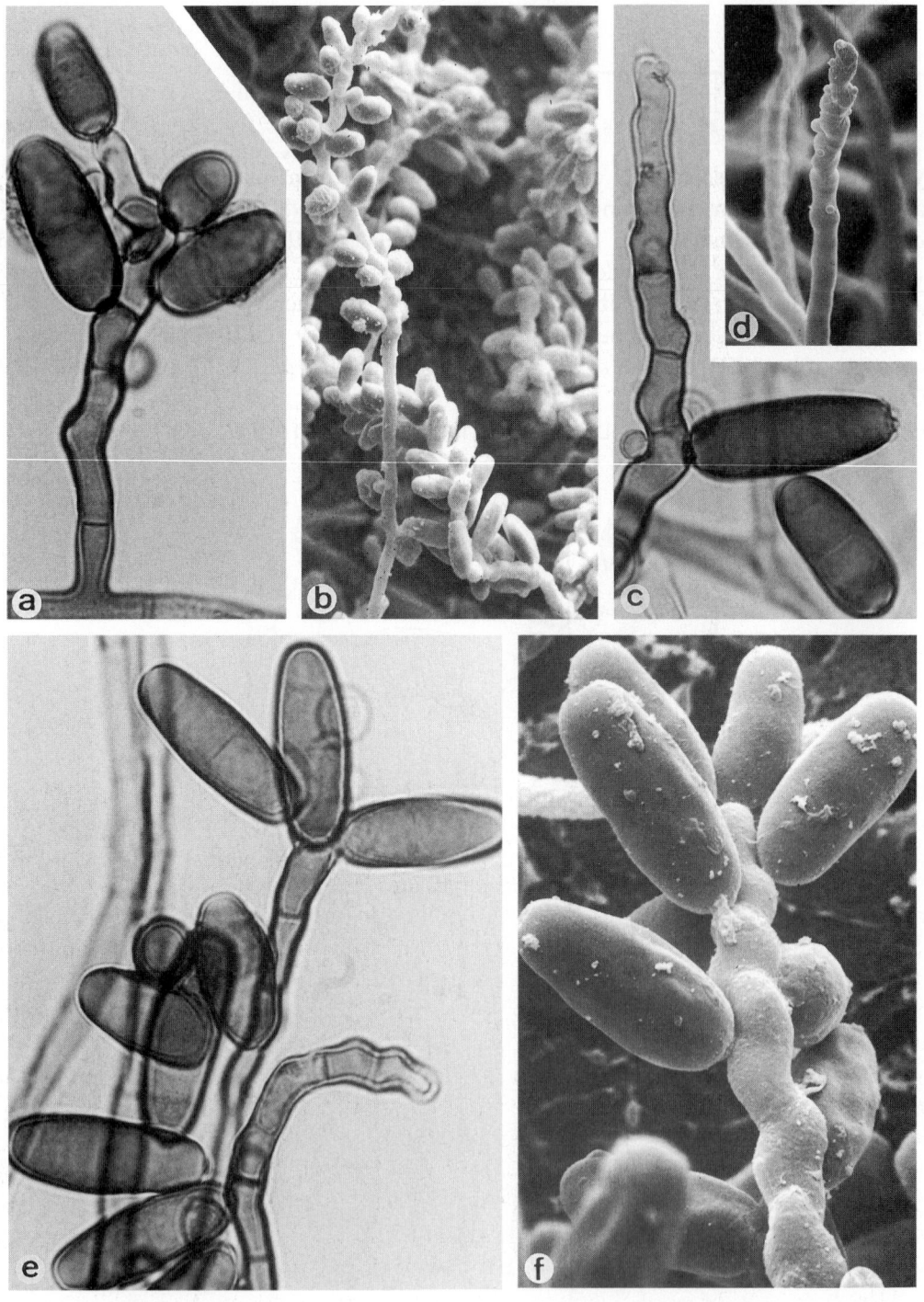

Bipolaris spicifera, CBS 586.80. Conidiophores and conidia. a. ×1280; b. ×660; c. ×1280; d. ×1100; e. ×1280; f. ×2850.

Hyphomycetes, dimorphic *Onygenales.* Genus: *BLASTOMYCES*

Blastomyces dermatitidis Gilchrist & Stokes

Colony characteristics. Colonies (SGA, 24°C) restricted, whitish, felty; reverse cream-coloured to brownish; colonies on KTA at 37°C butyrous.

Microscopy. Conidia at 24°C sessile or on short, swollen stalks, occasionally 1-2 conidia borne on ampulliform swellings, hyaline, smooth-walled, rarely echinulate, thin-walled, (sub)spherical, 1-celled, 2-7 × 2.0-4.5 μm. At 37°C on BHI yeast cells are produced, which are hyaline, smooth- and thick-walled, irregular to (sub)spherical, up to 15 μm diam, daughter bud being connected to parent cell by a broad base up to 4-5 μm.

Teleomorph. *Ajellomyces dermatitidis* McDonough & Lewis (*Ascomycota, Euascomycetes, Onygenales*: *Onygenaceae;* p. 252).

Physiology. Intolerant to benomyl.

Differential diagnosis. *In vivo*: the pathogenic form of growth consists of multinuclear, thick-walled budding cells, 8-15 μm diam. The buds are attached with a broad base, which is a distinguishing feature from *Histoplasma capsulatum* var. *duboisii* (p. 709): a variety which has uninuclear, narrow-based buds. *In vitro*: stalked conidia at 24°C and broad-based yeast cells at 37°C.

Molecular diagnostics. A DNA probe for recognition of the species is commercially available (Scalarone *et al.*, 1992; Stockman *et al.*, 1993; Padhye *et al.* 1994d). Fraser *et al.* (1991) used an *Eco*RI digested ribosomal DNA repeat as a probe in an epidemiological study. A 26S rDNA-based specific probe was developed by Sandhu *et al.* (1995). Serodiagnostic kits are commercially available. SSU and ITS restriction maps based on ATCC 26199:

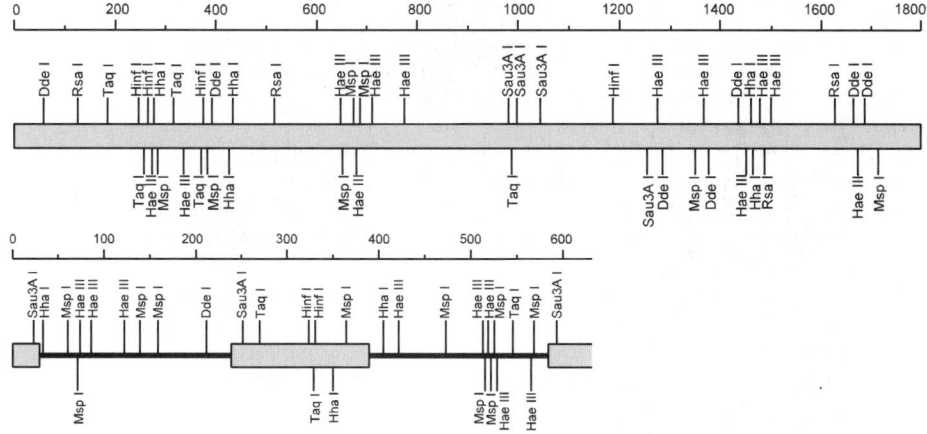

Pathogenicity. BSL-3. Agent of chronic, granulomatous blastomycosis (p. 25) of the skin, mostly orginating as a pulmonary infection (Failla *et al.*, 1995; Kuzo & Goodman, 1996) or possibly also from trauma (Yen *et al.*, 1994), rarely as a primary infection in other organs (Montes *et al.*, 1998). With hematogenous dissemination, bones and skin become affected (Gottlieb *et al.*, 1995); vertebral cases were described by Saccente *et al.* (1998). In a final stage the nervous system and visceral organs become involved (Guccion *et al.*, 1996). A cerebral case in an otherwise healthy patient was reported by Wylen & Nanda (1999). In contrast to *H. capsulatum*, the gastro-intestinal tract is not affected. A genital case was described by Mouzin & Beilke (1996). The disseminated form is frequently fatal. Outside endemic areas the species is emerging as an agent of disseminated mycoses in patients with impaired T-cell function, e.g., in AIDS (Pappas *et al.*, 1992; Tan *et al.*, 1993) and Hodgkin's disease (Winquist *et al.*, 1993) occasionally in transplant recipients (Serody *et al.*, 1993; Winkler *et al.*, 1996), CF patients (Gershan *et al.*, 1994) and tuberculosis patients (Li *et al.*, 1998). In addition to human cases, canine epizootics are known (Sarosi *et al.*, 1979; Baumgardner *et al.*, 1995). Marcelin-Little *et al.* (1996) and Gortel *et al.* (1999) reported systemic infections in dogs. The fungus probably has a reservoir in dogs (Garma-Avina, 1995; Bloom *et al.*, 1996; Arceneaux *et al.*, 1998), possibly also in bats (Randhawa *et al.*, 1985). Its environmental occurrence is mainly in abandoned, formerly man- or animal-used shelters (Baumgardner & Paretsky, 1999).

Distribution. Endemic areas are mainly in the Eastern part of the United States, South America and in Africa. American and African isolates differ in yeast-conversion rates (Lombardi *et al.*, 1988) and may represent different taxa (Guého *et al.*, 1997). Autochtonous cases outside endemic regions are extremely rare (Chodorowska & Lecewicz-Torún, 1996).

References. McDonough & Lewis (1968), MacKinnon (1970), Kwon-Chung (1973), van Oorschot (1980), Carman *et al.* (1989), Al-Doory & DiSalvo (1992), Winer-Muram & Rubin (1992), Frean *et al.* (1993), Meyer *et al.* (1993), Mercantini *et al.* (1995).

Antifungal susceptibility.

Antifungal	Mean MICs	MIC90	Strains	Reference
AMB	≤0.03-0.5	0.25	29	Karlowsky *et al.* (1997)
5FC	>64	>64	100	Li *et al.* (2000)
FLZ	12	nd	18	Chapman *et al.* (1998)
FLZ	2-32	16	29	Karlowsky *et al.* (1997)
ITZ	≤0.02	nd	18	Chapman *et al.* (1998)
ITZ	≤0.03->16	0.13	100	Li *et al.* (2000)
KTZ	0.25	nd	18	Chapman *et al.* (1998)
KTZ	≤0.003-<0.25	0.13	29	Karlowsky *et al.* (1997)
VCZ	≤0.03-16	0.25	100	Li *et al.* (2000)

Nomenclature. *Blastomyces dermatitidis* Gilchrist & Stokes - J. Exp. Med. 3: 76, 1898 ≡ *Oidium dermatitidis* (Gilchrist & Stokes) Ricketts - J. Med. Res. 6: 373, 1901 ≡ *Cryptococcus gilchristi* Vuillemin, *in* Guéguen - Champ. Paras. p. 108, 1904 (name change) ≡ *Zymonema gilchristi* (Vuillemin) de Beurmann & Gougerot - Trib. Méd., Paris 42: 503, 1909 ≡ *Mycoderma gilchristi* (Vuillemin) Jannin - Les Mycodermas p. 71, 1913 ≡ *Cryptococcus dermatitidis* (Gilchrist & Stokes) Castellani & Chalmers - Man. Trop. Med., ed. 2, p. 769, 1913 ≡ *Mycoderma dermatitidis* (Gilchrist & Stokes) Brumpt - Précis Parasitol. p. 934, 1913 ≡ *Blastomycoides dermatitidis* (Gilchrist & Stokes) Castellani - Am. J. Trop. Med. 8: 383, 1928 ≡ *Geotrichum dermatitidis* (Gilchrist & Stokes) Basgal - Contrib. Estudo Blastom. Pulmon. p. 156, 1931 ≡ *Torulopsis dermatitidis* (Gilchrist & Stokes) de Almeida - Anais Fac. Med., S. Paulo 9: 76, 1933 ≡ *Endomyces dermatitidis* (Gilchrist & Stokes) Moore - Annls Mo. Bot. Gdn 20: 106, 1933 ≡ *Gilchristia dermatitidis* (Gilchrist & Stokes) Redaelli & Ciferri - J. Trop. Med. Hyg. 37: 281, 1934 ≡ *Zymonema dermatitidis* (Gilchrist & Stokes) C.W. Dodge - Med. Mycol. p. 168, 1935 ≡ *Chrysosporium dermatitidis* (Gilchrist & Stokes) Carmichael - Can. J. Bot. 40: 1154, 1962.

Glenospora gammelii Pollacci & Nannizzi, *in* Blankenhorn & Gammel - J. Clin. Invest. 4: 483, 1927 ≡ *Acladium gammelii* (Pollacci & Nannizzi) Ota - Jpn. J. Derm. Urol. 28: 413, 1928 ≡ *Trichosporium gammelii* (Pollacci & Nannizzi) C.W. Dodge - Med. Mycol. p. 792, 1935 ≡ *Glenosporopsis gammelii* (Pollacci & Nannizzi) da Fonseca - Parasitol. Méd. 1: 725, 1943.

Scopulariopsis americana Ota - Jpn. J. Derm. Urol. 28: 400, 1928.

Blastomycoides tulanensis Castellani - Proc. R. Soc. Med. 21: 451, 1928 ≡ *Monosporium tulanense* (Castellani) Agostini - J. Trop. Med. Hyg. 35: 268, 1932 ≡ *Aleurisma tulanense* (Castellani) Ota & Kawatsure - Arch. Derm. Syph. 169: 156, 1933.

Endomyces capsulatus C.W. Dodge & Ayers, *in* Rewbridge, Dodge & Ayers - Am. J. Path. 5: 356, 1929 ≡ *Monilia capsulata* (C.W. Dodge & Ayers) Vuillemin - Champ. Paras. Myc. Homme p. 283, 1934 ≡ *Zymonema capsulatum* (C.W. Dodge & Ayers) C.W. Dodge - Med. Mycol. p. 167, 1935.

Glenospora brevis Castellani - J. Trop. Med. Hyg. 36: 311, 1933 ≡ *Aleurisma brevis* (Castellani) C.W. Dodge - Med. Mycol. p. 790, 1935 ≡ *Glenosporopsis brevis* (Castellani) da Fonseca - Parasitol. Méd. 1: 724, 1943.

Ajellomyces dermatitidis McDonough & Lewis - Mycologia 60: 77, 1968.

Endomyces capsulatus C.W. Dodge & Ayers var. *isabellinus* Moore, *in* MacBride & Thompson - Arch. Derm. Syph., Chicago 27: 61, 1933.

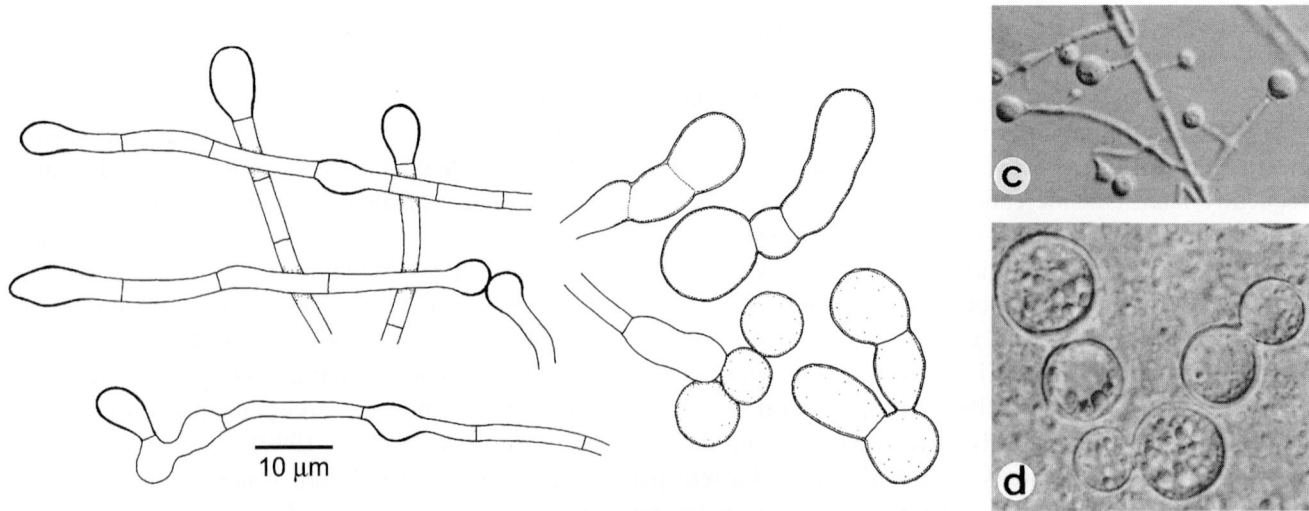

Blastomyces dermatitidis, CBS 673.68. a, c. Young conidiophores and conidia at 24°C; b. yeast-like cells at 37°C; d. spherical cells in pus smear. c. ×750; d. ×950. c, d. Reproduced with permission from Kwon-Chung & Bennett (1992).

Hyphomycetes. Genus: *BOTRYOMYCES*

Botryomyces caespitosus de Hoog & Rubio

Colony characteristics. Colonies (MEA) restricted, heaped, cauliflower-like, initially pale brown, becoming blackish-brown with age.

Microscopy. Hyphae and budding cells absent. Thallus entirely composed of clumps of irregularly septate, thick-walled cells which are subhyaline, become dark brown with age and disarticulate into smaller cell packets. Blastic conidia occasionally present.

Physiology.

Growth:						
D-Glucose	+	Melibiose	+	D-Glucuronate	–
D-Galactose	+	Lactose	+	D-Galacturonate	–
L-Sorbose	+	Raffinose	+	DL-Lactate	–
D-Glucosamine	–	Melezitose	+	Succinate	+
D-Ribose	+	Inulin	+	Citrate	+
D-Xylose	+	Soluble starch	w	Methanol	–
L-Arabinose	+	Glycerol	+	Ethanol	+
D-Arabinose	+	*meso*-Erythritol	+	Nitrate	+
L-Rhamnose	+	Ribitol	+	Nitrite	+
Sucrose	+	Xylitol	+	Ethylamine	+
Maltose	+	L-Arabinitol	+	L-Lysine	+
α,α-Trehalose	+	D-Glucitol	+	Cadaverine	+
α-methyl-D-glucoside	+	D-Mannitol	+	Creatine	+
Cellobiose	+	Galactitol	+	Creatinine	w
Salicin	+	*myo*-Inositol	+	10% NaCl	–
Arbutin	+	Glucono-δ-lactone	–	Urease	+
		D-Gluconate	–	Growth at 37°C	w

Differential diagnosis. The species differs from *Sarcinomyces phaeomuriformis* (p. 897) by young colonies being pink. Note that a meristematic synanamorph of an *Alternaria* species is probably concerned (de Hoog *et al.*,1997b; compare Fig. 42 on p. 381).

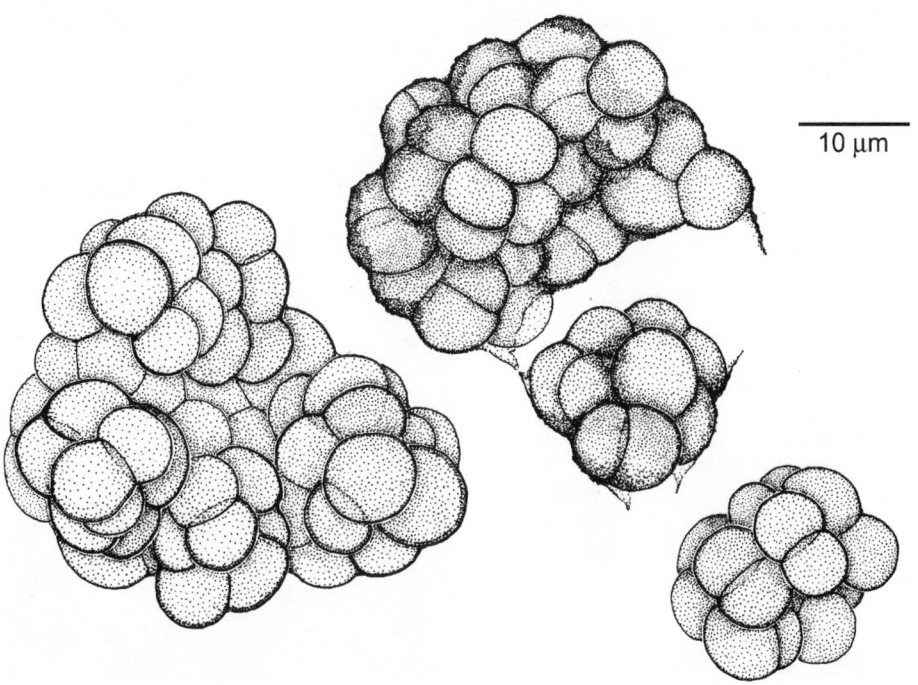

10 μm

Botryomyces caespitosus, CBS 177.80. Meristematic clumps of cells.

537

Molecular diagnostics. SSU and ITS restriction maps based on NCBI Y18695 and CBS 177.80, resp.:

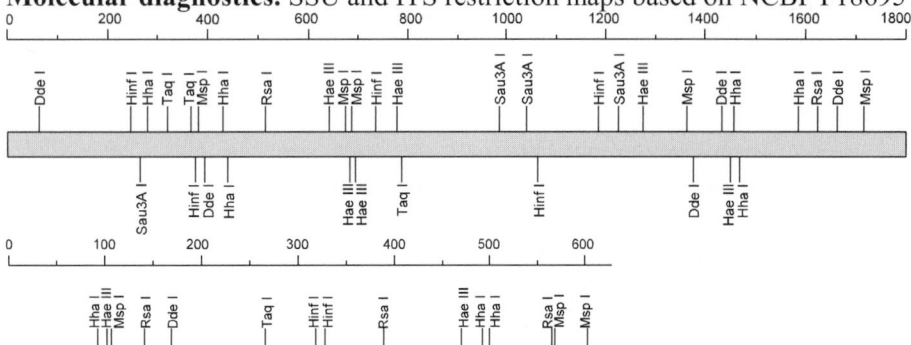

Pathogenicity. BSL-2. A chromoblastomycosis-like infection was reported by Benoldi *et al.* (1991).

Reference. De Hoog & Rubio (1982).

Antifungal susceptibility.

Antifungal	Mean MICs	Strains	Reference
AMB	0.25	1	McGinnis & Pasarell (1998b)
ITZ	0.25	1	McGinnis & Pasarell (1998b)
VCZ	2	1	McGinnis & Pasarell (1998b)

Nomenclature. *Botryomyces caespitosus* de Hoog & Rubio - Sabouraudia 20: 19, 1982.

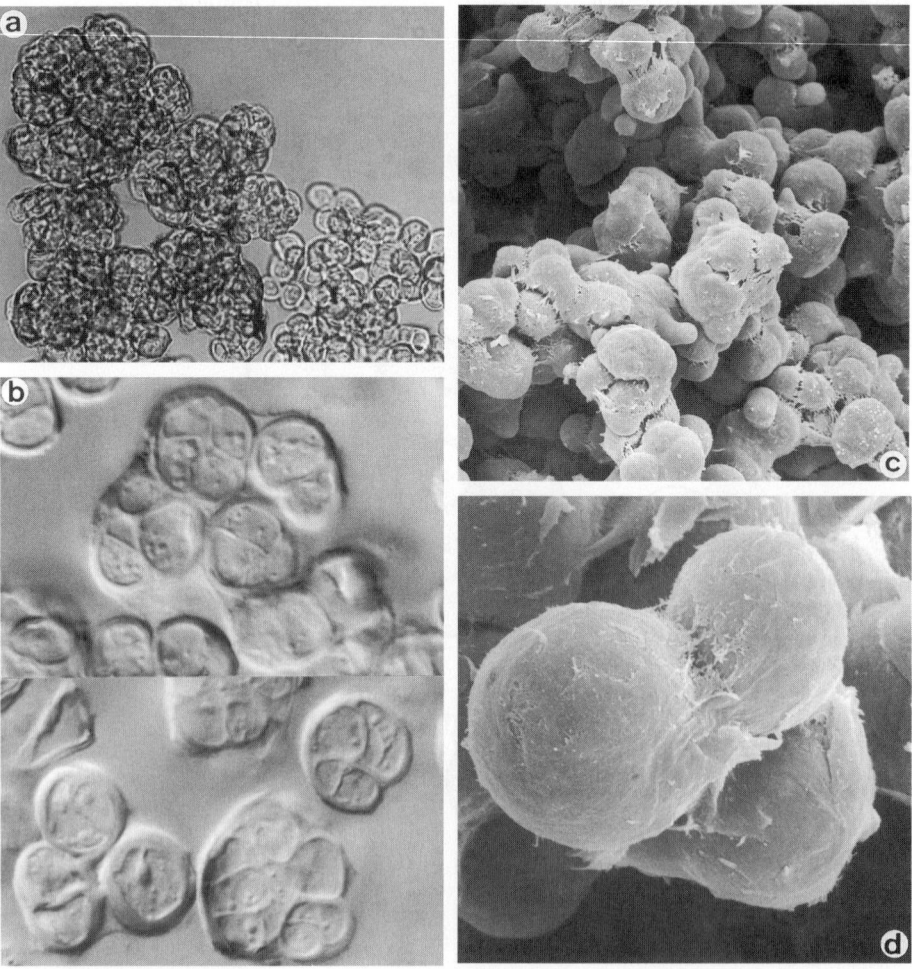

Botryomyces caespitosus, CBS 177.80. Meristematic cell clumps. a. ×325; b. ×1010; c. ×505; d. ×2520.

Hyphomycetes. Genus: *CEPHALIOPHORA*

Cephaliophora irregularis Thaxter

Colony characteristics. Colonies (PDA) growing rapidly, floccose, white to cream-coloured.

Microscopy. Hyphae hyaline, 2.5-7.0 μm diam. Conidiophores straight or flexuose, pale brown, up to 110 μm long. Conidia borne in heads on the inflated tips of conidiophores, arising synchronously next to each other, pyriform or obtriangular, smooth-walled, pedicellate, pale brown, 1-2-septate, 21-36 × 12-25 μm.

Pathogenicity. BSL-1. A case of keratitis from India (Thomas & Kuriakose, 1990, as *Arthrobotrys oligospora*). A second case, also from India, was reported by Mathews & Kuriakose (1995). Guarro & Stchigel (1999) reported a *Cephaliophora* species from monkeys originally attributed to *Trichophyton simii*.

References. Ellis (1971), Guarro *et al.* (1991c), Tanabe *et al.* (1999).

Nomenclature. *Cephaliophora irregularis* Thaxter - Bot. Gaz. 35: 158, 1903.

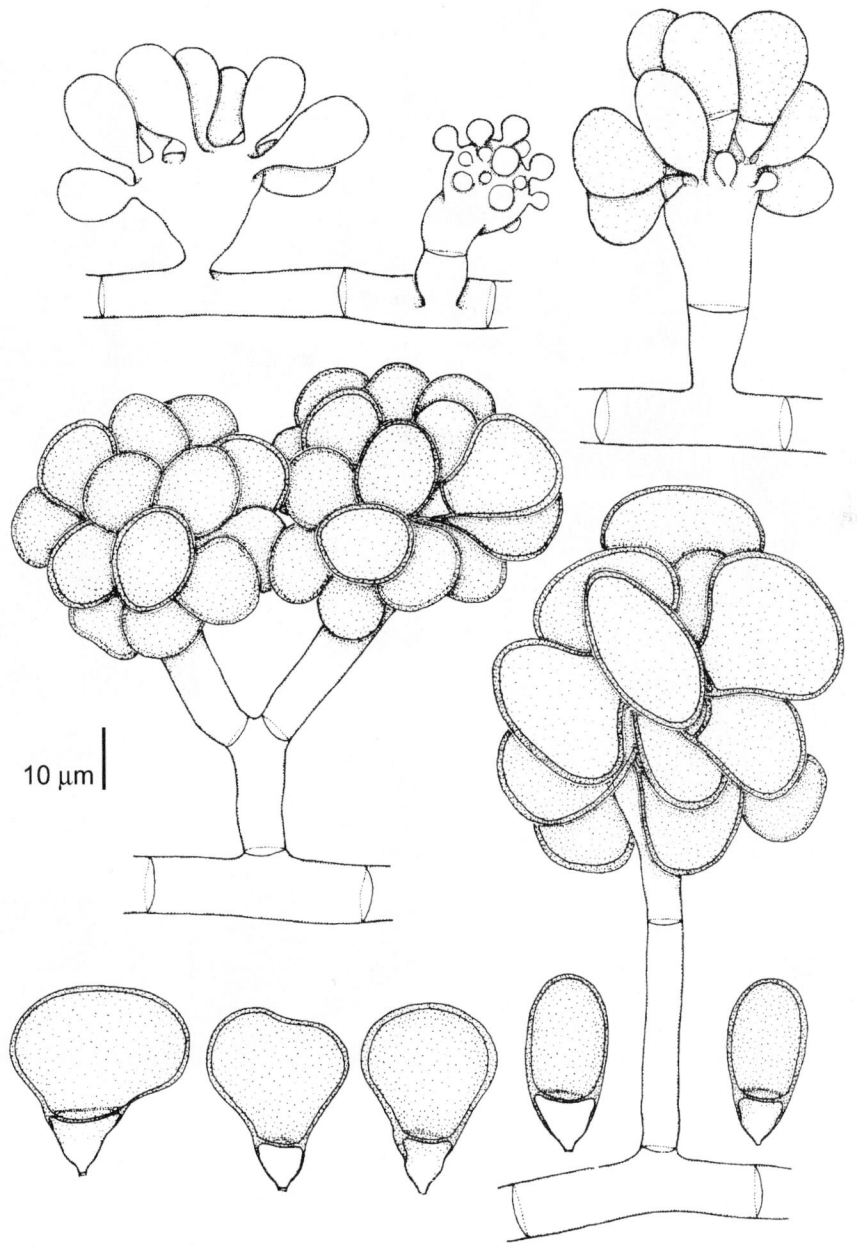

Cephaliophora irregularis, FMR 3784. Conidiophores and conidia.

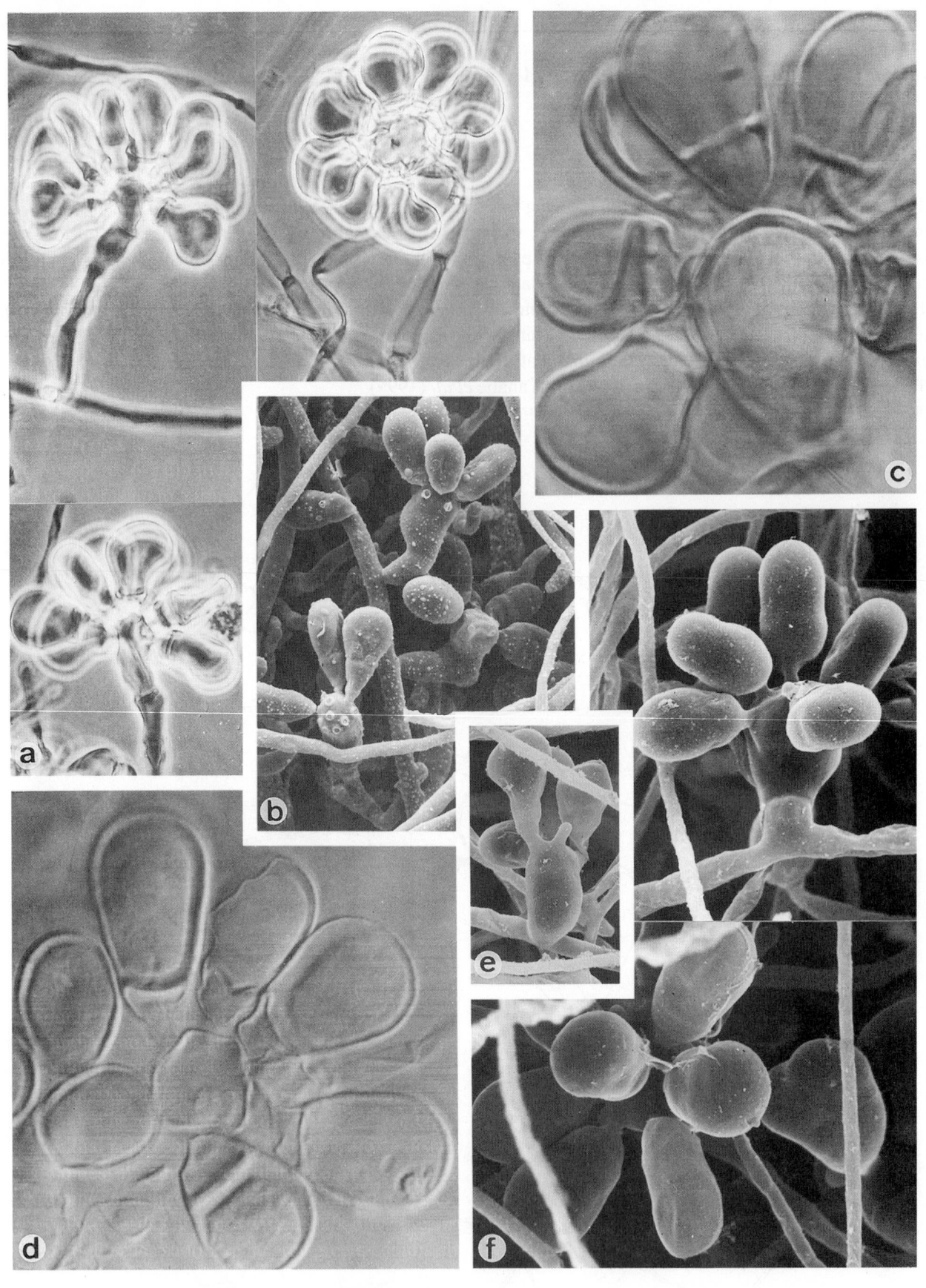

Cephaliophora irregularis, FMR 3784. Conidiophores and conidia. a. ×512; b. ×680; c, d. ×1600; e. ×1100; f. ×3250.

Hyphomycetes, filamentous basidiomycetes. Genus: *CERINOSTERUS*

Cerinosterus cyanescens (de Hoog & de Vries) R.T. Moore

Colony characteristics. Colonies (OA) restricted, farinose or velvety, often compact and somewhat cerebriform, snow-white, later often exuding a pH-dependent, deep blue/violet pigment into the agar.

Microscopy. Conidiogenous cells undifferentiated, cylindrical, of variable size (1.5-3.0 μm wide), apically with a cluster of small denticles, the cluster often repeatedly proliferating and forming similar clusters. Conidia hyaline, smooth-walled or finely verrucose, obovoidal, 3-4 μm long, somewhat larger (3.5-6.5 μm long) when bearing secondary conidia.

Physiology.

Fermentation:	–	Maltose	+	D-Mannitol	+
		α,α-Trehalose	+	Succinate	+,–
Growth:		α-methyl-D-Glucoside	+	Citrate	+,–
D-Glucose	+	Cellobiose	–	Nitrate	+
D-Galactose	–	Salicin	–	Creatine	–
L-Sorbose	–	Melibiose	–,+	Creatinine	–
D-Ribose	–	Lactose	+	Cycloheximide	–
D-Xylose	–,+	Raffinose	+	Benomyl	+
L-Arabinose	+	Melezitose	+	Urease	+
D-Arabinose	–	Soluble starch	–	Growth at 37°C	+
L-Rhamnose	–	Glycerol	+	Growth at 40°C	+
Sucrose	+	*meso*-Erythritol	+		

Differential diagnosis. The species is *Sporothrix*-like, but conidial scars are very small and cultures are thin, fragile. In fresh cultures the diffusable pigment is characteristic; *Sporothrix schenckii* (p. 925) forms tough colonies which finally become blackish-brown. *Candida ciferrii* (p. 192) is urease negative. *Cerinosterus cyanescens* is of basidiomycetous affinity (Smith & Batenburg-van der Vegte, 1985).

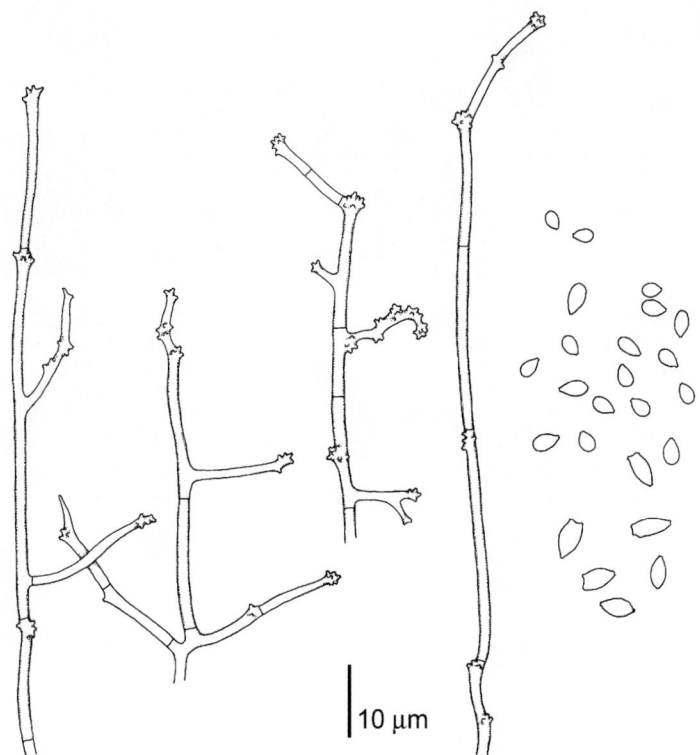

Cerinosterus cyanescens, CBS 549.93. Conidiogenous cells and conidia.

541

Pathogenicity. BSL-2. The species has occasionally been isolated from human skin and blood and was involved in nosocomial infections in patients with pneumonia (Jackson *et al.*, 1990). A pulmonary case in a transplant patient was reported by Tambini *et al.* (1996) and a fungemia in a neonate by Schmidt *et al.* (2000). Experimental inoculation showed low virulence (Sigler *et al.*, 1990).

References. De Hoog & de Vries (1973), Sigler *et al.* (1990), Middelhoven *et al.* (2000).

Nomenclature. *Sporothrix cyanescens* de Hoog & de Vries - Antonie van Leeuwenhoek 39: 515, 1973 ≡ *Cerinosterus cyanescens* (de Hoog & de Vries) R.T. Moore - Stud. Mycol. 30: 216, 1987.

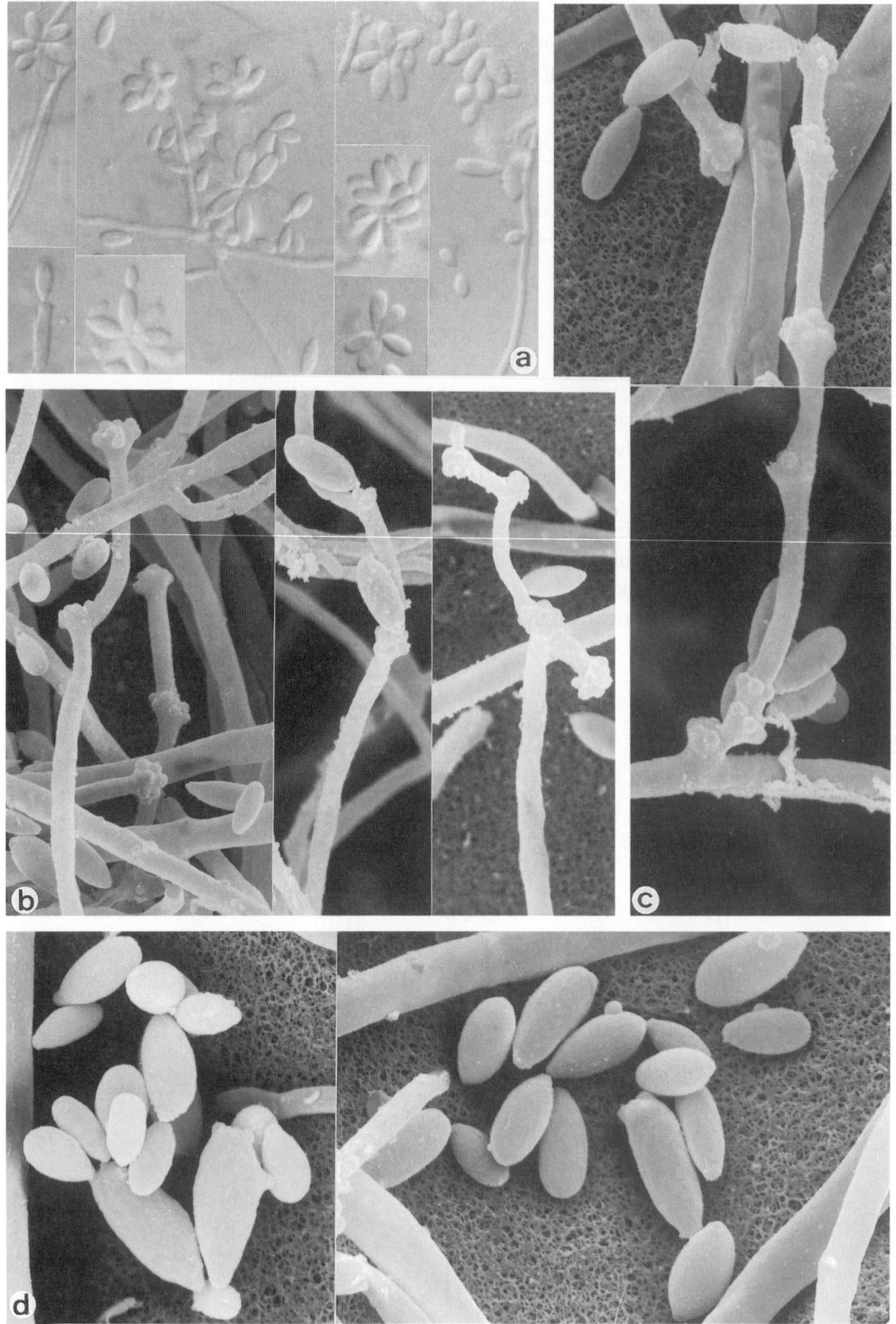

Cerinosterus cyanescens, CBS 549.93. Conidiogenous cells and conidia. a. ×1600; b. ×4500; c, d. ×6500.

Hyphomycetes. Genus: *CHRYSONILIA*

Chrysonilia sitophila (Mont.) v. Arx

Colony characteristics. Colonies (OA) very fast growing, locally forming loose, floccose masses having a pale pink or salmon pink colour.

Microscopy. Conidiogenous hyphae more or less ascending, smooth-walled, septate, with lateral branches which form chains of conidia. Conidia are arthroconidia connected by disjunctors, 1-celled, usually adhering in irregular masses, smooth-walled, ovoidal to ellipsoidal, 10-15 × 5-10 μm, hyaline, pinkish to orange in mass.

Teleomorph. *Neurospora sitophila* Shear & Dodge (*Ascomycota, Euascomyctes, Sordariales: Sordariaceae*).

Pathogenicity. BSL-1. This is the 'bread mould' which is frequently found on foodstuffs. A human case was an endophthalmitis following cataract extraction (Theodore *et al.*, 1961). Radix *et al.* (1996) reported a peritonitis in a dialysis patient.

References. Von Arx (1981b), Samson *et al.* (1996).

Nomenclature. *Penicillium sitophilum* Montagne - Annls Sci. Nat., Bot., Sér. 2, 20: 377, 1843 ≡ *Monilia sitophila* (Montagne) Saccardo - Michelia 2: 359, 1881 ≡ *Chrysonilia sitophila* (Montagne) von Arx - Sydowia 34: 16, 1981.

   Neurospora sitophila Shear & Dodge - J. Agric. Res. 34: 2016, 1927.

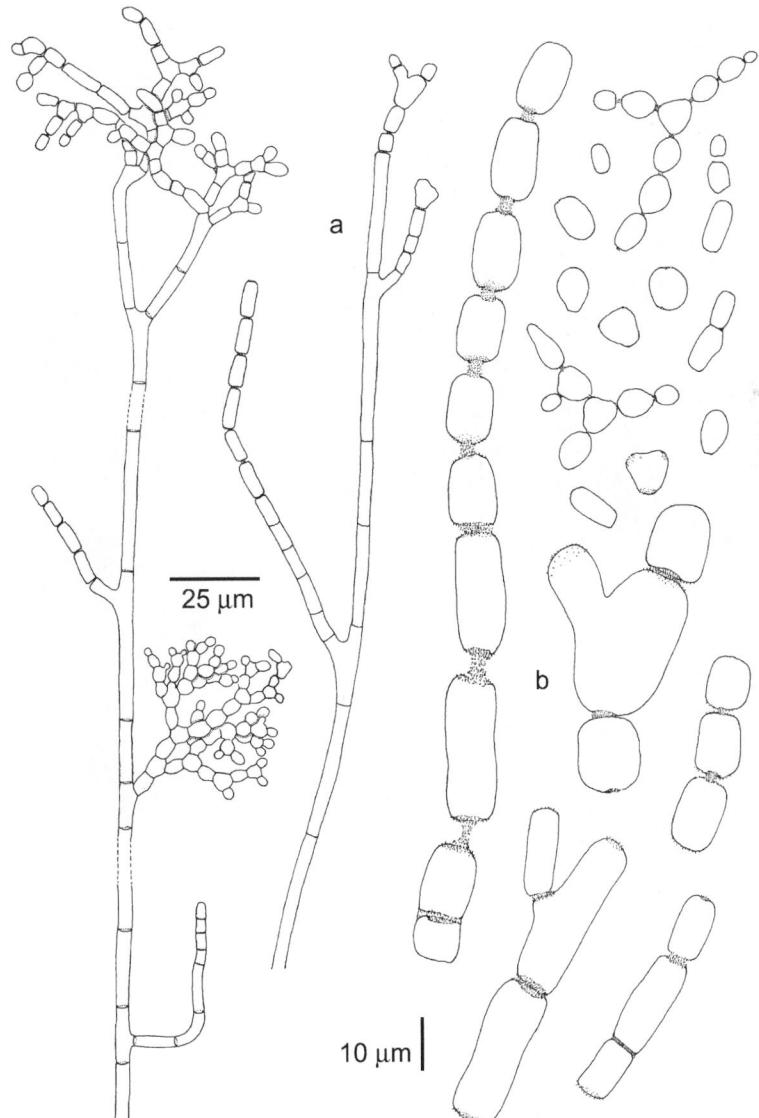

Chrysonilia sitophila, RV 66948. a. Conidial apparatus; b. arthroconidia with disjunctors between the cells.

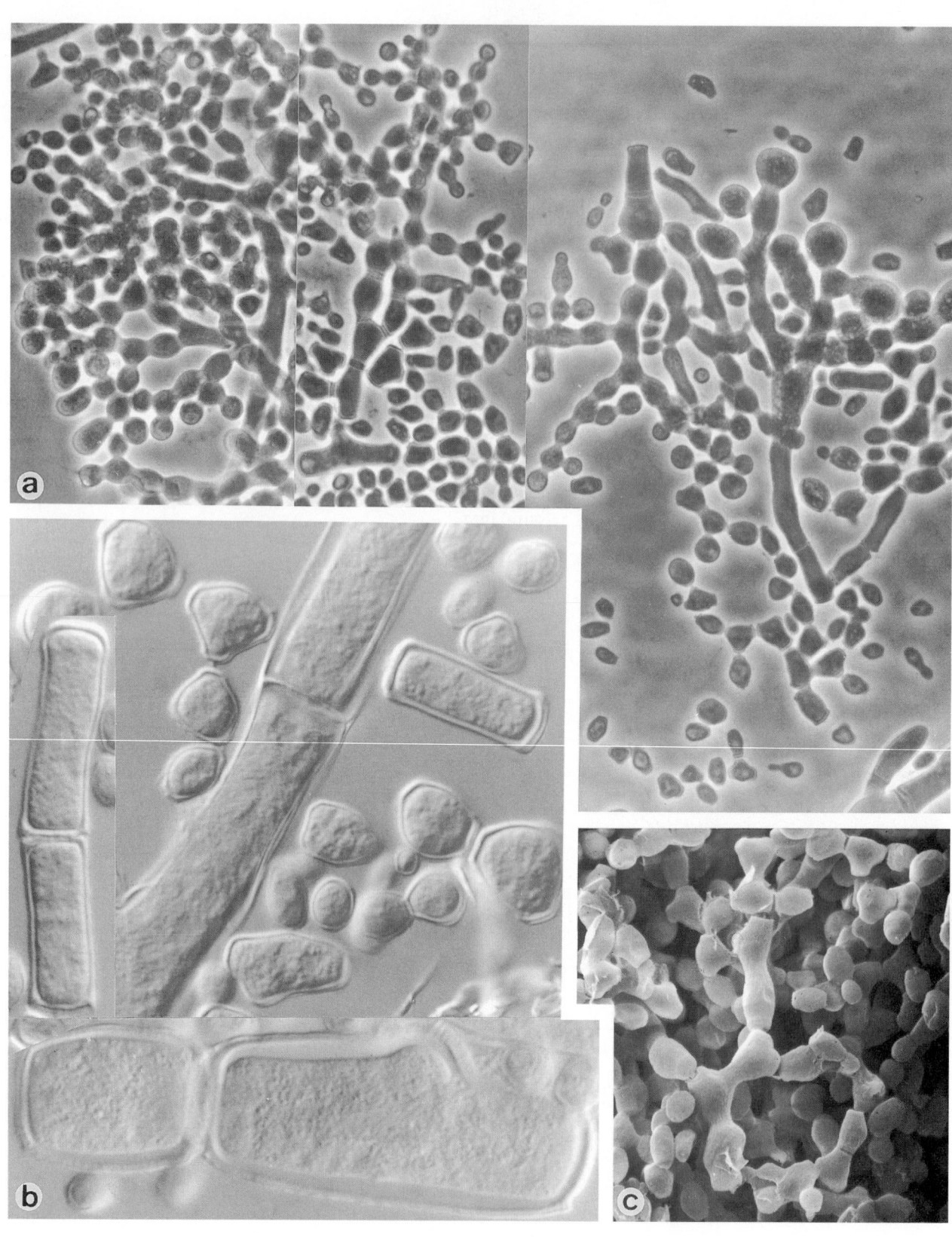

Chrysonilia sitophila, RV 66948. Conidiophores and arthroconidia. a. ×1280, b. ×1600, c. ×1345.

Hyphomycetes. Genus: *CHRYSOSPORIUM*

Generic description. Colonies usually spreading, whitish, or pale pigmented. Hyphae mostly hyaline, smooth-walled, branched. Fertile hyphae with terminal and lateral conidia. Conidia sessile or on short protrusions or side branches, subhyaline or pale yellow, thin- or thick-walled, subspherical, clavate, pyriform or obovoidal, usually one-celled, occasionally two-celled. Intercalary conidia sometimes present. Chlamydospores occasionally present.

Teleomorphs. *Amaurascopsis, Amauroascus, Aphanoascus, Arthroderma, Ctenomyces, Neogymnomyces, Pectinotrichum, Renispora, Uncinocarpus* (*Ascomycota, Euascomycetes, Onygenales: Onygenaceae*), *Nannizziopsis* (*Ascomycota, Euascomycetes, Onygenales: Gymnoascaceae*), *Bettsia* (*Ascomycota, Euascomycetes, Ascosphaerales: Ascosphaeraceae*).

General remarks. Most *Chrysosporium* species are keratinophilic fungi, living on remains of hairs and feathers in soil. *Microsporum* (p. 736) and *Trichophyton* (p. 954) species have *Chrysosporium*-like microconidia, but they are phylogenetically different and mostly can be distinguished by having prevalently lateral rather than intercalary microconidia. The polyphyletic origin of *Chrysosporium* was demonstrated by Vidal *et al.* (2000) on the basis of ITS sequences. Eight species occasionally occurring on humans are treated in this Atlas.

References. Carmichael (1962), van Oorschot (1980).

Key to the treated species of Chrysosporium:

1a.	Colonies butyrous or tough; sporulation poor, often irregularly swollen hyphal elements present	*C. inops* (546)
1b.	Colonies dry, powdery or cottony; swollen hyphal elements absent ➝ 2	
2a.	Conidia large, mostly over 8 μm long ➝ 3	
2b.	Conidia smaller ➝ 4	
3a.	Conidia thick-walled; intercalary conidia very rare	*C. keratinophilum* (548)
3b.	Conidia thin-walled; intercalary conidia abundant	**C. anam. of** *Aphanoascus fulvescens* (254)
4a.	Conidia echinulate to verrucose ➝ 5	
4b.	Conidia smooth-walled ➝ 6	
5a.	Colonies white; conidia borne on swollen stalks; no growth at 37°C	*C. pannicola* (550)
5b.	Colonies darkening after 2 wk, finally brownish-orange; conidia borne on non-swollen stalks; growth at 37°C	*C. zonatum* (558)
6a.	Conidia 2-3 μm wide	**C. anam. of** *Nannizziopsis vriesii* (556)
6b.	Conidia over 3 μm wide ➝ 7	
7a.	Intercalary absent or very rare	*C. tropicum* (554)
7b.	Intercalary conidia frequent	*C. queenslandicum* (552)

Chrysosporium inops Carmichael

Colony characteristics. Colonies (SGA) restricted, butyrous or tough, heaped, smooth, cream-coloured, with irregular margin.

Microscopy. Hyphae hyaline, irregular in width, often with swollen cells up to 10 μm in diam, with poor sporulation. Conidia sessile alongside undifferentiated hyphae, subhyaline, smooth- and rather thick-walled, 1-celled, obovoidal to ellipsoidal, 6.5-12.0 × 5-9 μm, with truncate base, scar being up to 2 μm wide.

Pathogenicity. BSL-2. The species was previously known from two strains from human skin (Agostini, 1930; Carmichael, 1962), but is now regarded as a xerophilic species occurring in foodstuffs (Kinderlerer, 1995).

Reference. Van Oorschot (1980).

Nomenclature. *Glenosporella dermatitidis* Agostini - Atti Ist. Bot. Univ. Lab. Crittog. Pavia 4: 98, 1930 ≡ *Aleurisma dermatitidis* (Agostini) C.W. Dodge - Med. Mycol. p. 789, 1935 ≡ *Chrysosporium inops* Carmichael - Can. J. Bot. 40: 1156, 1962 (name change).

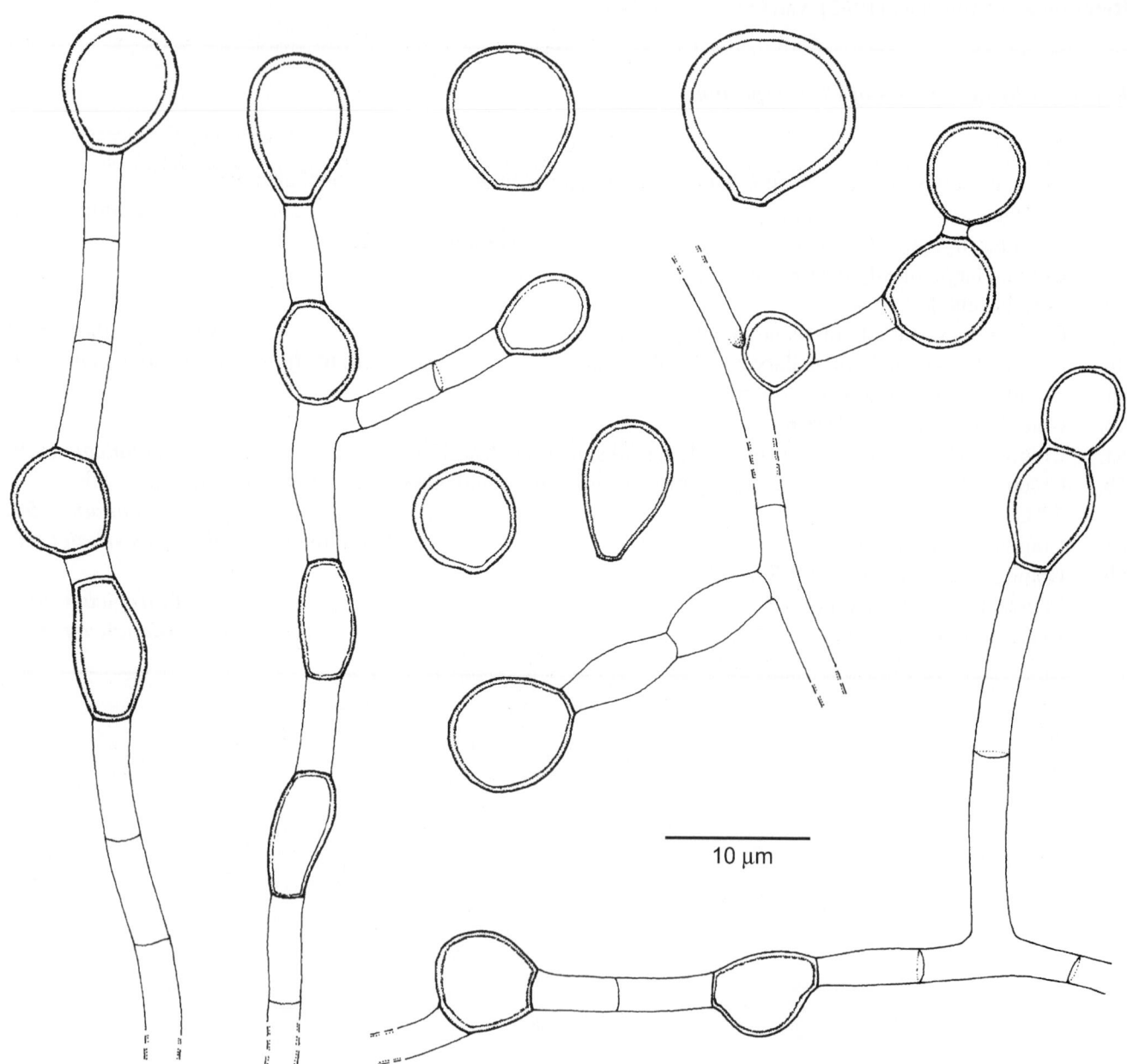

10 μm

Chrysosporium inops, CBS 132.31. Fertile hyphae and conidia.

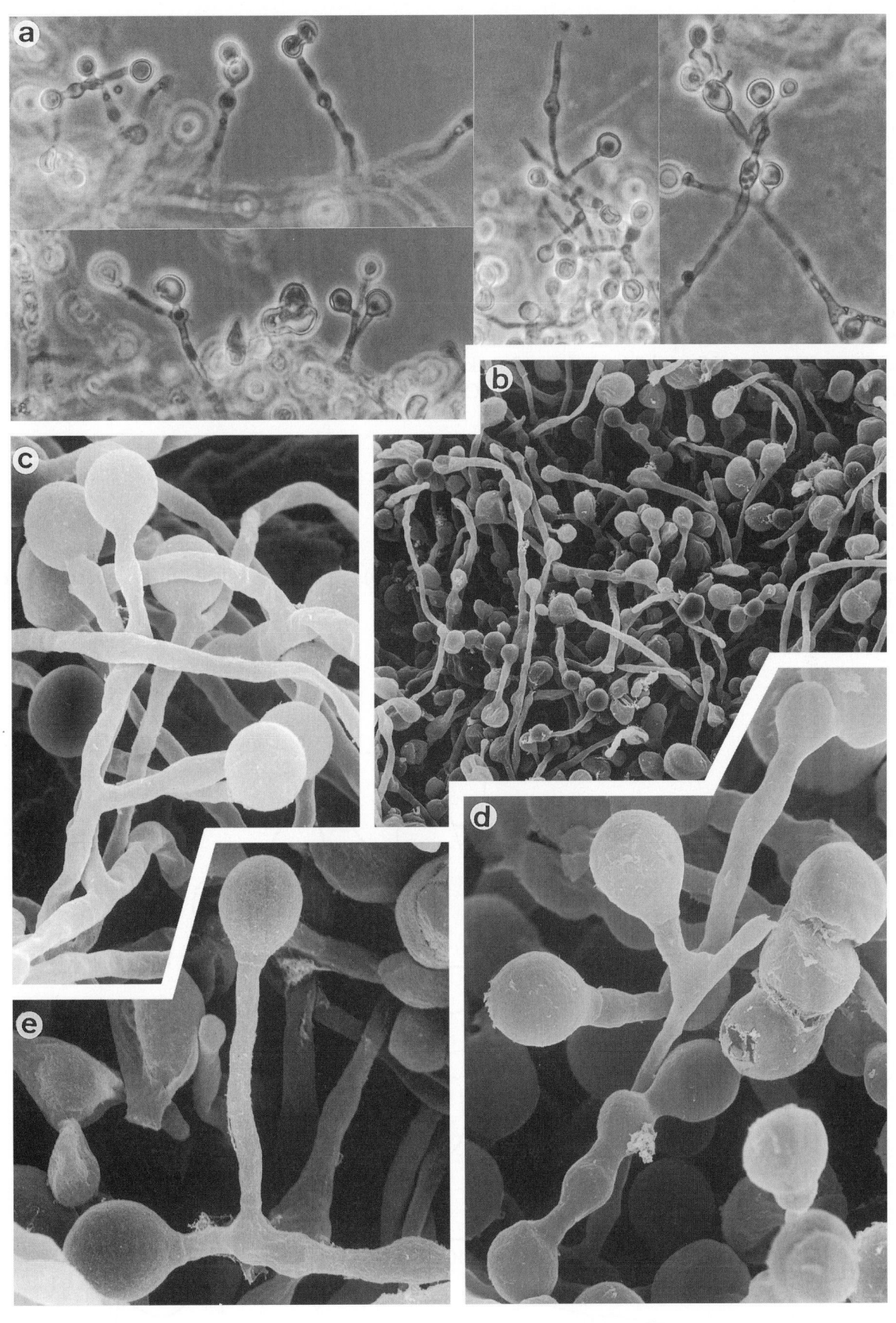

Chrysosporium inops, CBS 132.31. Fertile hyphae and conidia. a. ×640; b. ×700; c. ×2800; d. ×3100; e. ×3500.

Chrysosporium keratinophilum (Frey) Carmichael

Colony characteristics. Colonies (PYE) growing moderately rapidly, cream-coloured to yellowish, dense and powdery at the centre.

Microscopy. Fertile hyphae hyaline, not differentiated. Terminal and lateral conidia sessile or on short, unswollen protrusions, subhyaline, smooth-walled or echinulate, thick-walled, obovoidal to clavate, 8.5-13.0 × 5.5-9.0 μm, with conspicuous basal scars. Intercalary conidia uncommon.

Teleomorph. *Aphanoascus keratinophilus* Punsola & Cano (*Ascomycota, Euascomycetes, Onygenales: Onygenaceae*).

Physiology. Keratinolytic.

Differential diagnosis. The species differs from the *Chrysosporium* anamorph of *Aphanoascus fulvescens* by absence of intercalary conidia and the presence of markedly thick-walled conidia.

Molecular diagnostics. ITS restriction map based on CBS 104.62:

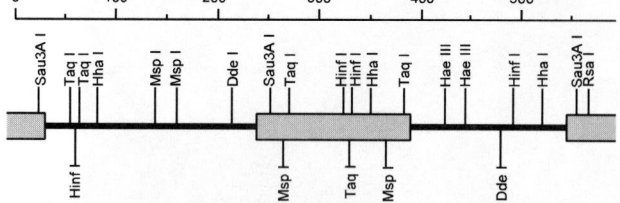

Pathogenicity. BSL-1. Geophilic species (Cano & Guarro, 1990), repeatedly isolated from onychomycoses and superficial infections (Reboux *et al.*, 1995).

Reference. Cano & Guarro (1990).

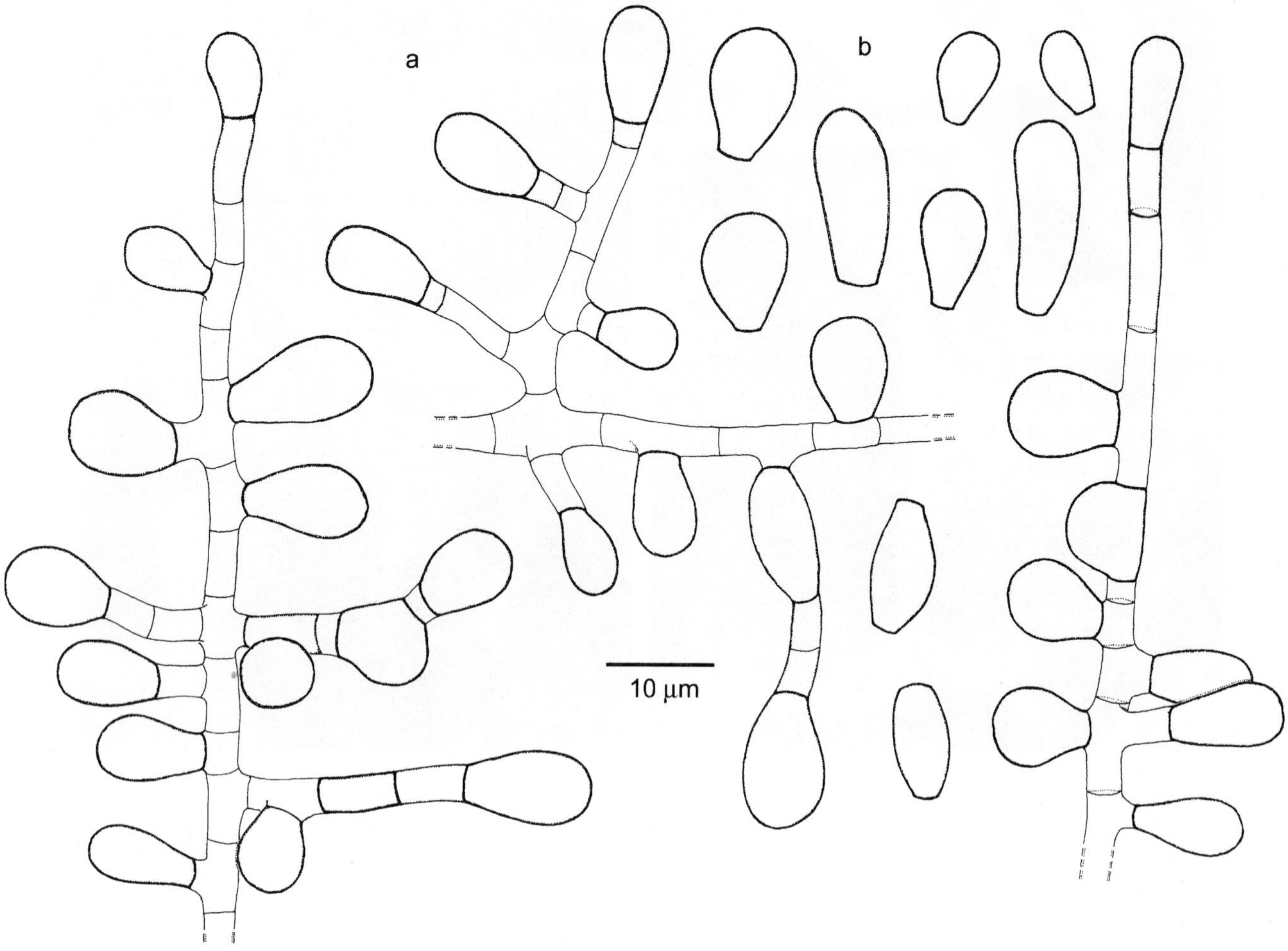

Chrysosporium keratinophilum, CBS 305.67. a. Fertile hyphae bearing conidia; b. liberated conidia.

Antifungal susceptibility.

Antifungal	Mean MICs	MIC 90	Strains	Reference
AMB	0.34	2.34	10	Wildfeuer *et al.* (1998)
AMB	≤0.08	nd	3	Guarro *et al.* (2000d)
5FC	>128	nd	3	Guarro *et al.* (2000d)
FLZ	6.3	nd	3	Guarro *et al.* (2000d)
ITZ	0.12	0.39	10	Wildfeuer *et al.* (1998)
KTZ	0.13	0.39	10	Wildfeuer *et al.* (1998)
MCZ	1.25	nd	3	Guarro *et al.* (2000d)
VCZ	0.08	0.29	10	Wildfeuer *et al.* (1998)

Nomenclature. *Aleurisma keratinophilum* Fry - Mycologia 51: 641, 1961 ≡ *Chrysosporium keratinophilum* (Frey) Carmichael - Can. J. Bot. 40: 1197, 1962.

Aphanoascus keratinophilus Punsola & Cano, *in* Cano & Guarro - Mycol. Res. 94: 358, 1990.

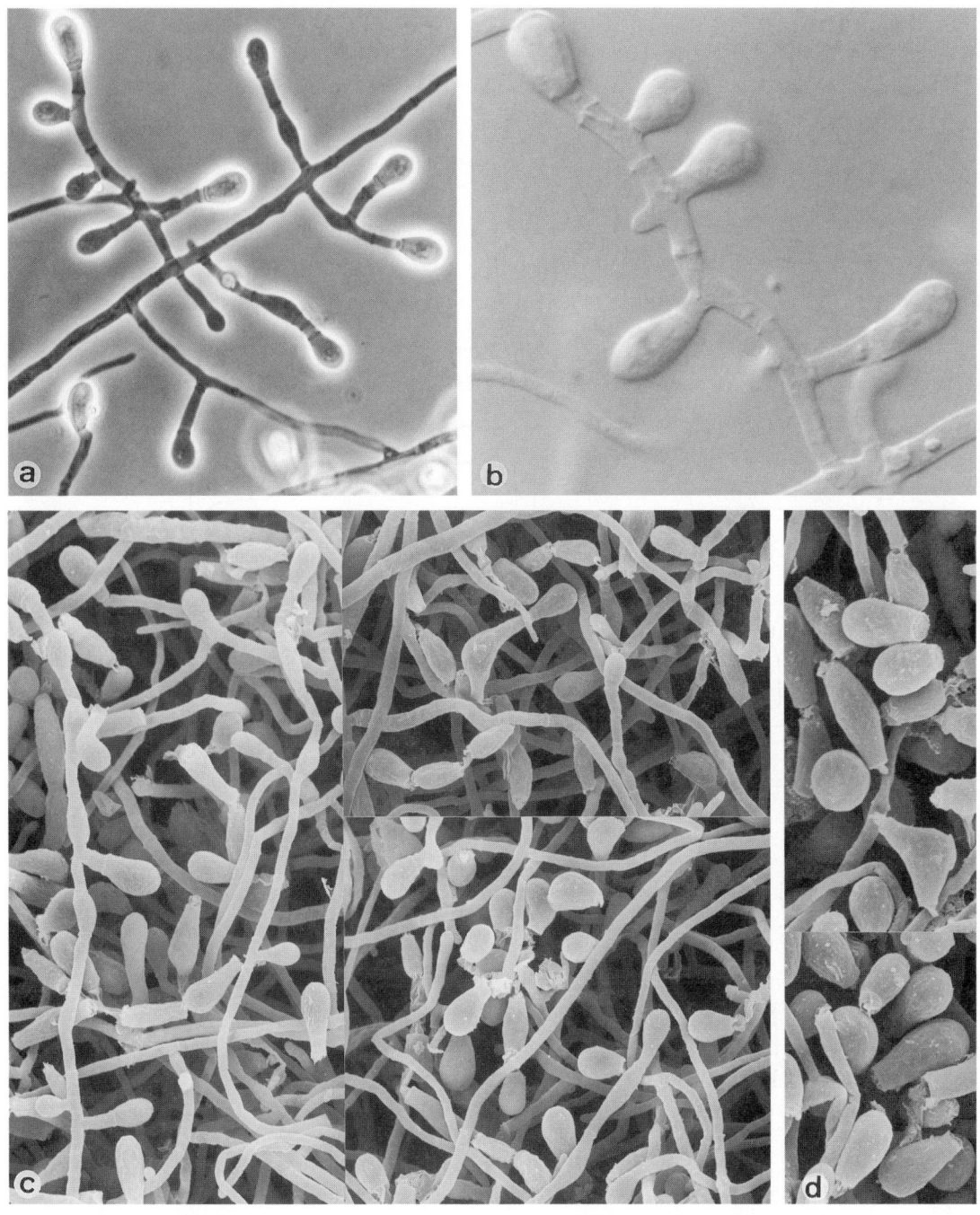

Chrysosporium keratinophilum, CBS 305.67. Fertile hyphae and conidia. a. ×640; b. ×1600; c. ×1350; d. ×2050.

Chrysosporium pannicola (Corda) v. Oorschot & Stalpers

Colony characteristics. Colonies (PYE) growing moderately rapidly, white, dry, powdery to cottony, dense and fluffly at the centre.

Microscopy. Fertile hyphae hyaline, not differentiated, repeatedly branched. Terminal and lateral conidia sessile or on short and swollen protrusions, subhyaline, echinulate, thick-walled, obovoidal to clavate, 6-11 × 3.5-4.5 µm, with conspicuous basal scars. Intercalary conidia also present.

Pathogenicity. BSL-1. Human cases concerned superficial skin infections (Ghosh *et al.*, 1976). Also skin infections in dogs are known (Hajsig *et al.*, 1974). The species is keratinophilic.

References. Van Oorschot (1980), Kane *et al.* (1997). These authors claim that *C. pannicola* and *C. evolceanei* are different species.

Nomenclature. *Capillaria pannicola* Corda - Ic. Fung. 1: 10, 1837 ≡ *Chrysoporium pannicola* (Corda) van Oorschot & Stalpers - Stud. Mycol. 20: 43, 1980.

    Trichophyton evolceanui Randhawa & Sandhu - Mycopath. Mycol. Appl. 20: 225, 1963 ≡ *Chrysosporium evolceanui* (Randhawa & Sandhu) Garg - Sabouraudia 4: 262, 1966.

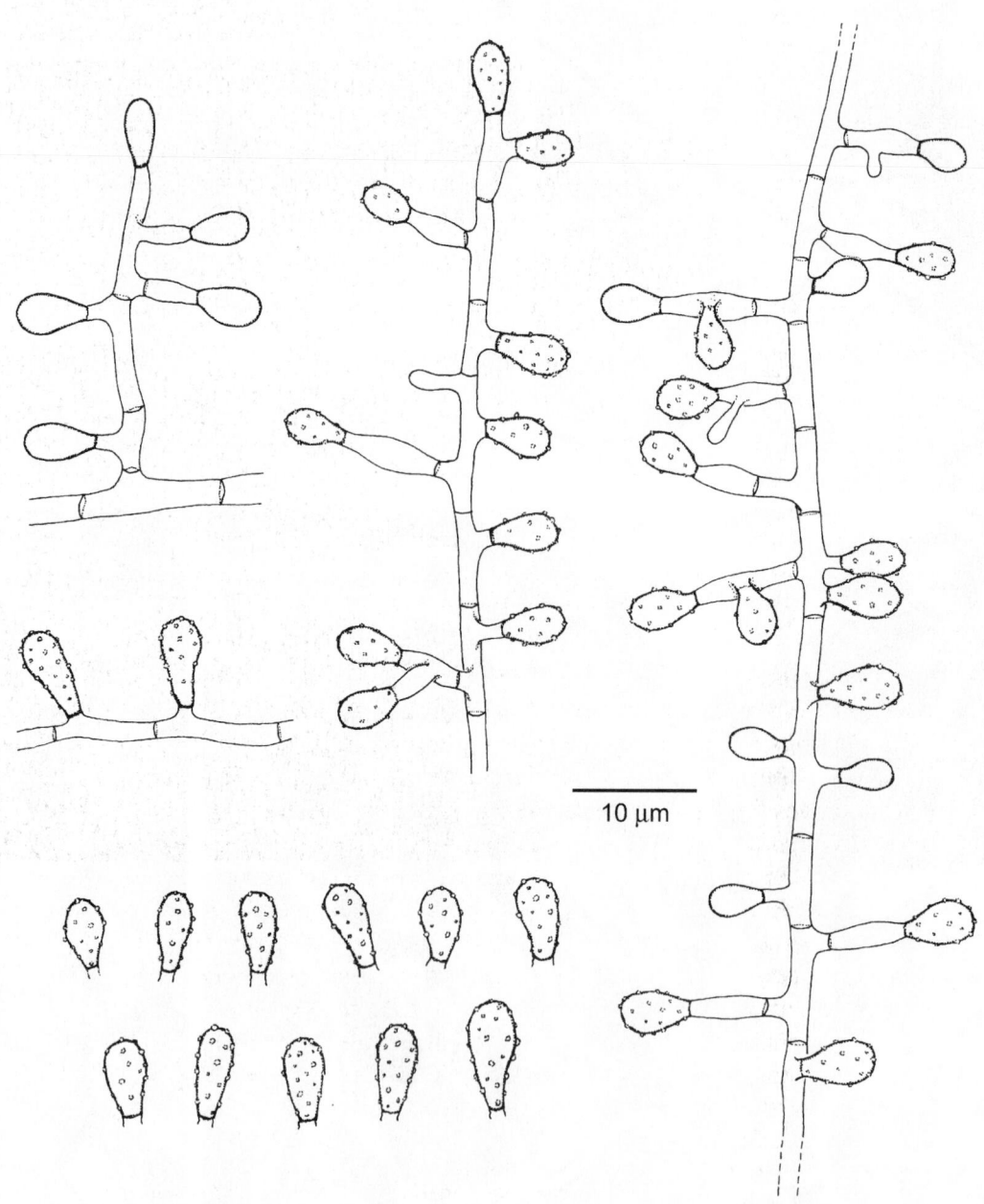

Chrysosporium pannicola, RV 12857. Fertile hyphae and conidia.

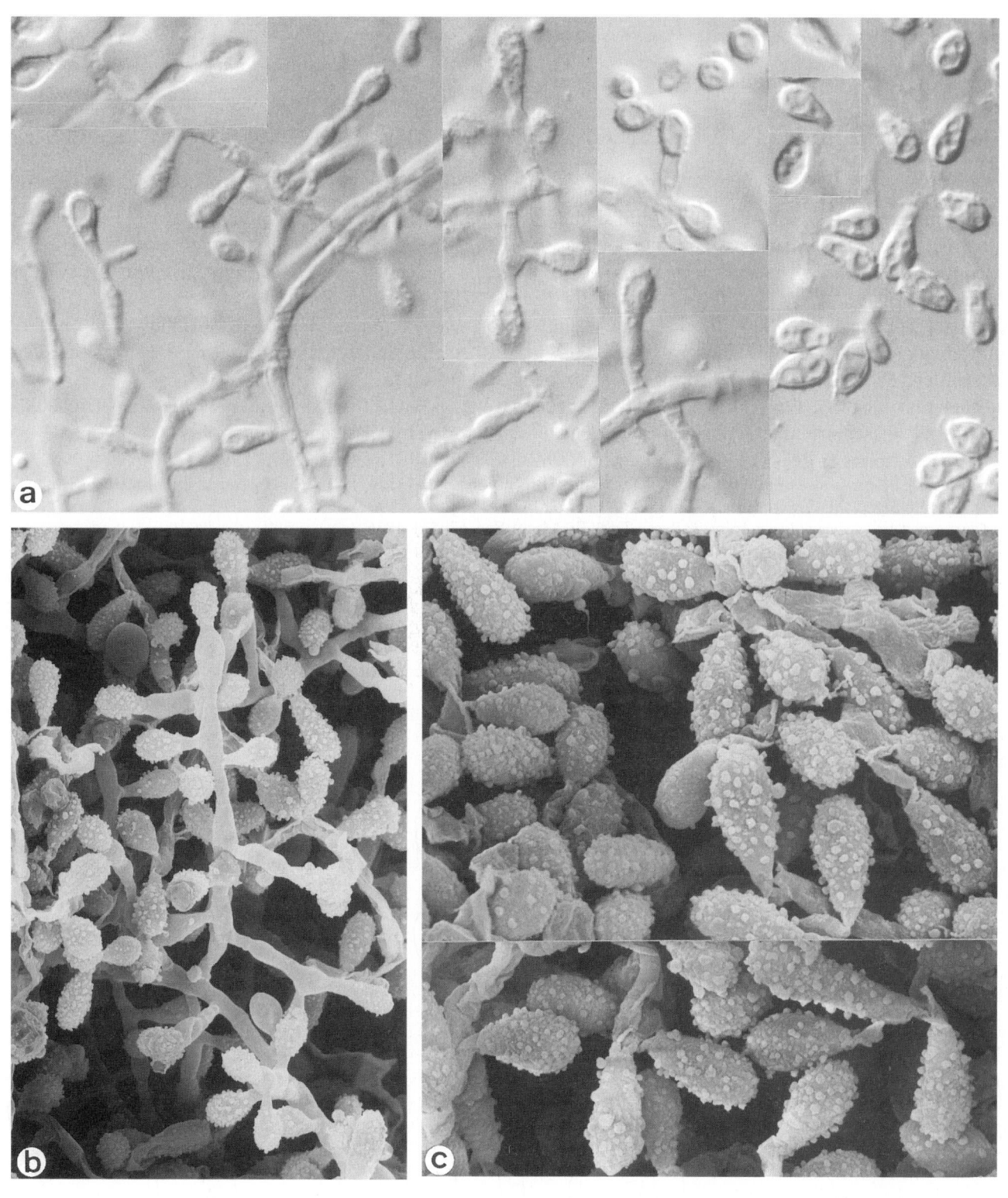

Chrysosporium pannicola, RV 12857. Fertile hyphae and conidia. a. ×1600; b. ×2000; c. ×3700.

Chrysosporium queenslandicum Apinis & Rees

Colony characteristics. Colonies (PYE) growing moderately rapidly, white, flat, dry, cottony; reverse cream-coloured to pinkish.

Microscopy. Fertile hyphae hyaline, not differentiated. Terminal and lateral conidia sessile or on short, unswollen protrusions, subhyaline, smooth-walled, with slightly thickened walls, obovoidal to broadly clavate, 3-6 × 3-5 µm, with broad basal scars. Intercalary conidia common.

Teleomorph. *Uncinocarpus queenslandicus* (Apinis & Rees) Sigler (*Ascomycota, Euascomycetes, Onygenales: Onygenaceae*).

Ascomata spherical, 350-800 µm in diam, yellowish-brown. Peridium composed of interwoven, hyaline, undifferentiated hyphae with loosely coiled appendages. Asci ovoidal to subspherical, 8-14 µm in diam. Ascospores yellowish- to reddish-brown, broadly ovoidal, 6-7 × 4-6 µm.

Pathogenicity. BSL-1. Geophilic, keratinolytic species. It was isolated from skin and nail infections (Reboux *et al.*, 1995). A disseminated infection in a captive snake was reported by Vissiennon *et al.* (1999).

References. Apinis & Rees (1976), van Oorschot (1980), Currah (1985), Sigler *et al.* (1998).

Nomenclature. *Apinisia queenslandica* Apinis & Rees - Trans. Br. Mycol. Soc. 67: 524, 1976 ≡ *Uncinocarpus queenslandicus* (Apinis & Rees) Sigler, *in* Sigler, Flis & Carmichael - Can. J. Bot. 76: 1632, 1998.

    Chrysosporium queenslandicum Apinis & Rees - Trans Br. Mycol. Soc. 67: 524, 1976.

    Brunneospora reticulata Guarro & Punsola, *in* Guarro, Punsola & Figueras - Persoonia 13: 387, 1987.

    Orromyces spiralis B. Sur & G.R. Ghosh, *in* Ghosh & Sur - Kavaka 12: 63, 1985.

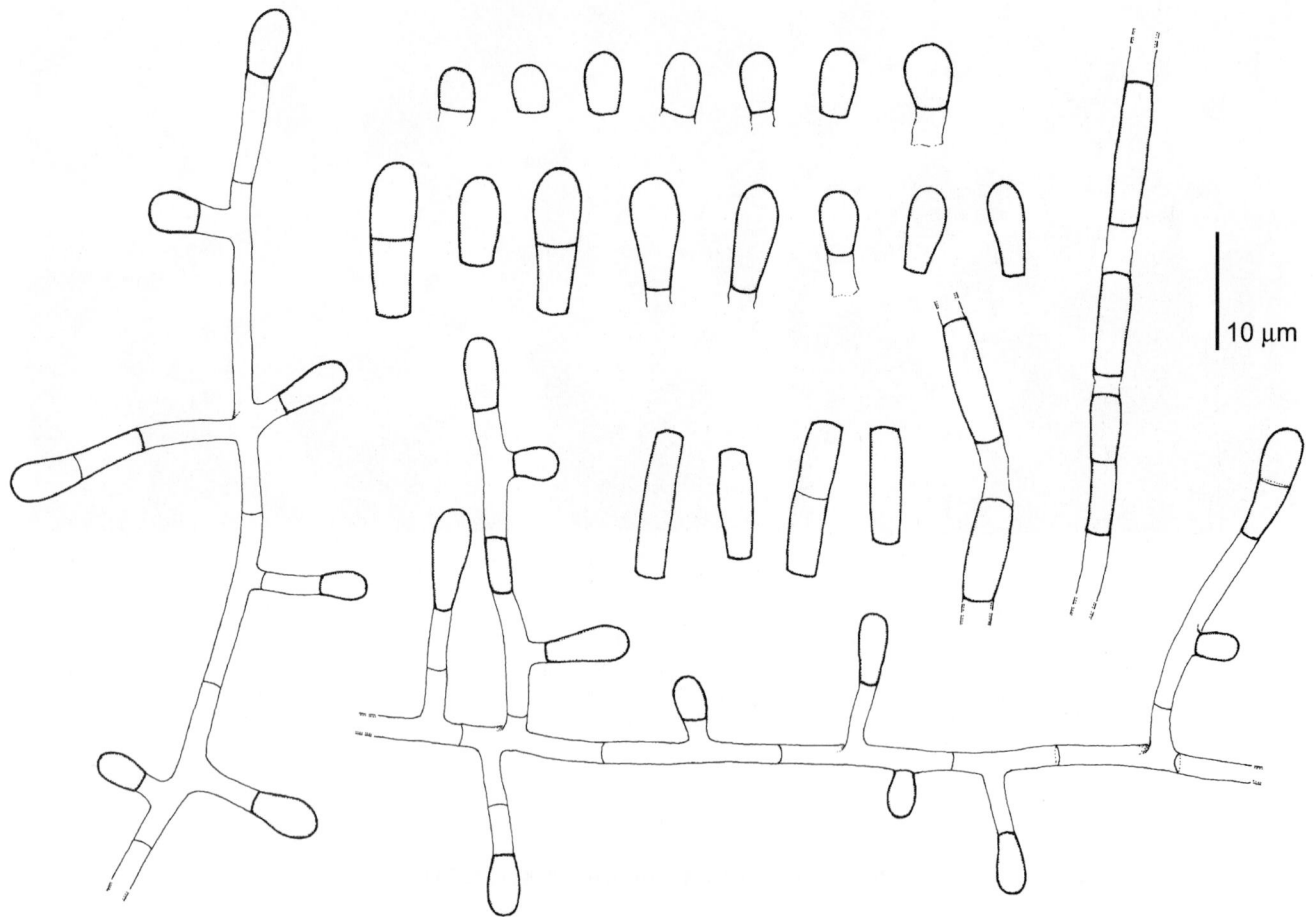

Chrysosporium queenslandicum, CBS 280.77. Fertile hyphae and conidia.

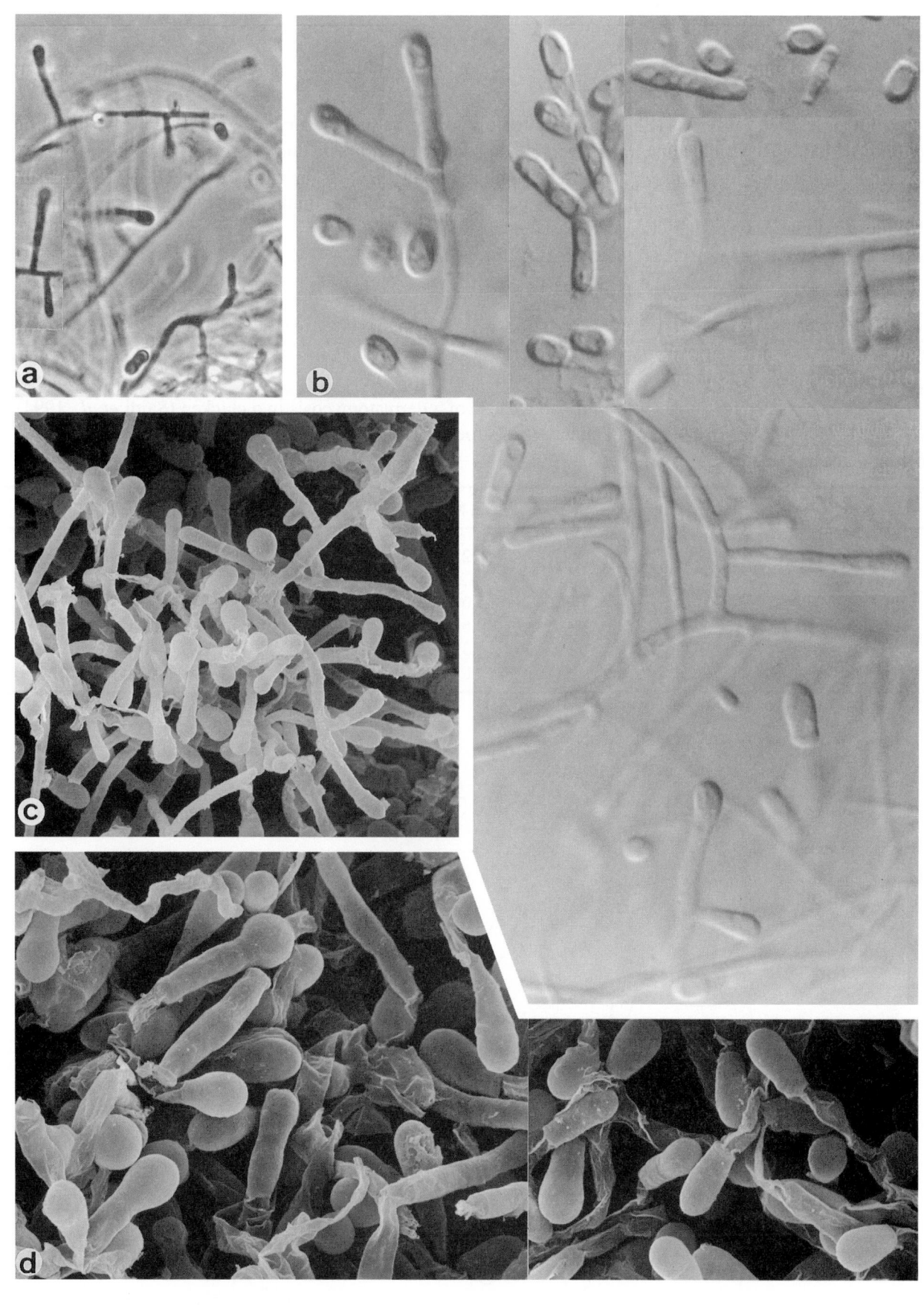

Chrysosporium queenslandicum, CBS 280.77. Fertile hyphae and conidia. a. ×640; b. ×1600; c. ×1800; d. ×2700.

Chrysosporium tropicum Carmichael

Colony characteristics. Colonies (PYE) growing rapidly, white, felty, powdery.
Microscopy. Fertile hyphae hyaline, undifferentiated. Terminal and lateral conidia sessile or on short protrusions or side branches, solitary, subhyaline, smooth- and rather thick-walled, obovoidal to clavate, 3.5-7.5 (-10.0) × 3.0-4.5 µm, 1-celled, occasionally 2-celled, with wide basal scars. Intercalary conidia occasionally present.
Molecular diagnostics. ITS restriction map based on UAMH 691:

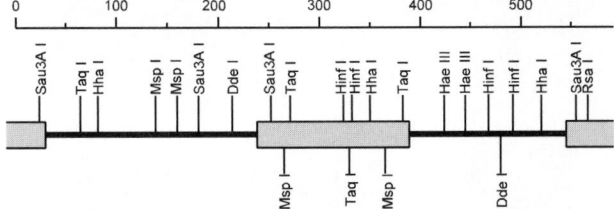

Pathogenicity. BSL-2. The fungus was a probable cause of a dermatomycosis in a chicken (Saidi *et al.*, 1994) and was isolated from skin and nail lesions in humans (de Vroey, 1976; Reboux *et al.*, 1995).
Antifungal susceptibility.

Antifungal	MICs range	Mean MICs	Strains	Reference
AMB	0.12-0.5	0.25	3	Guarro *et al.* (2000d)
5FC	>128	>128	3	Guarro *et al.* (2000d)
FLZ	2-8	5.01	3	Guarro *et al.* (2000d)
ITZ	0.5-1	0.63	3	Guarro *et al.* (2000d)
KTZ	0.25-0.5	0.39	3	Guarro *et al.* (2000d)
MCZ	1	1	3	Guarro *et al.* (2000d)

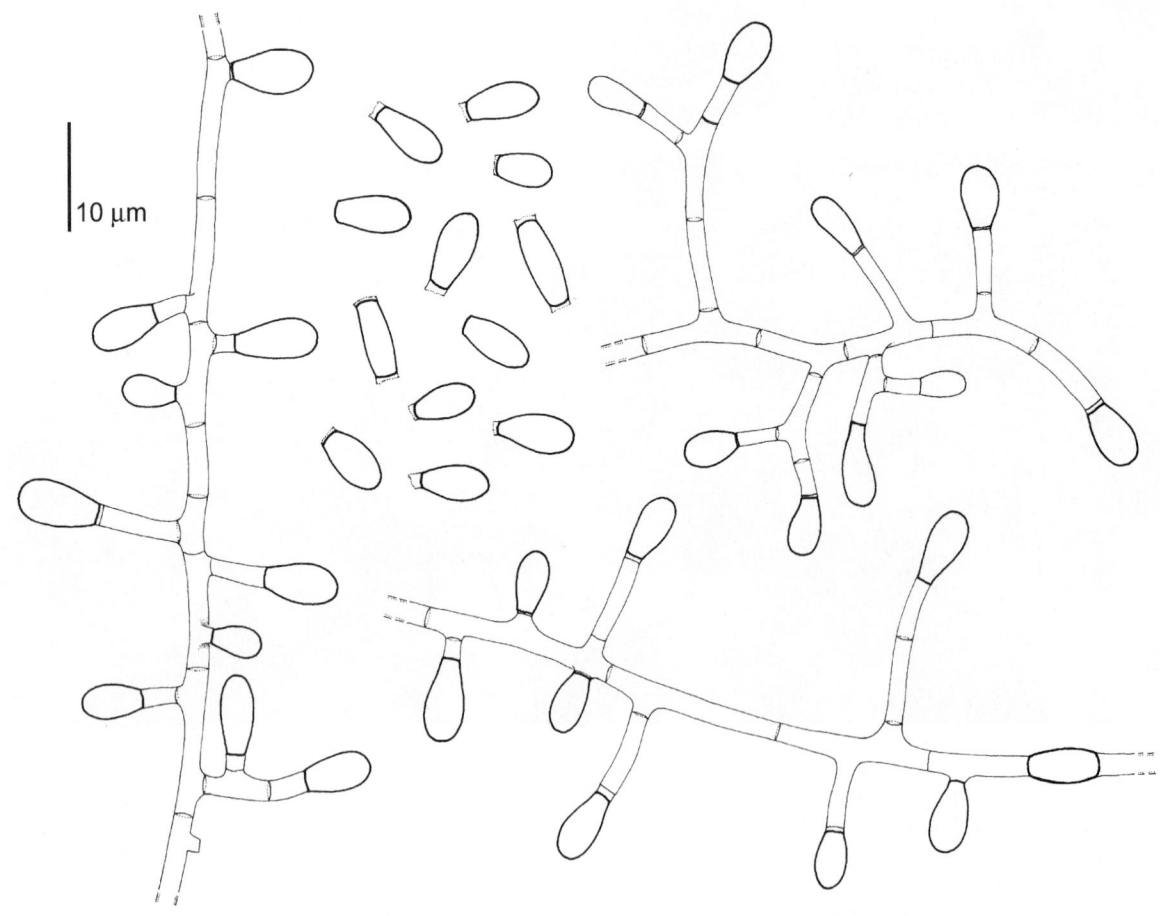

Chrysosporium tropicum, FMR 3567. Fertile hyphae and conidia.

Reference. Van Oorschot (1980).

Nomenclature. *Chrysosporium tropicum* Carmichael - Can. J. Bot. 40: 1170, 1962.

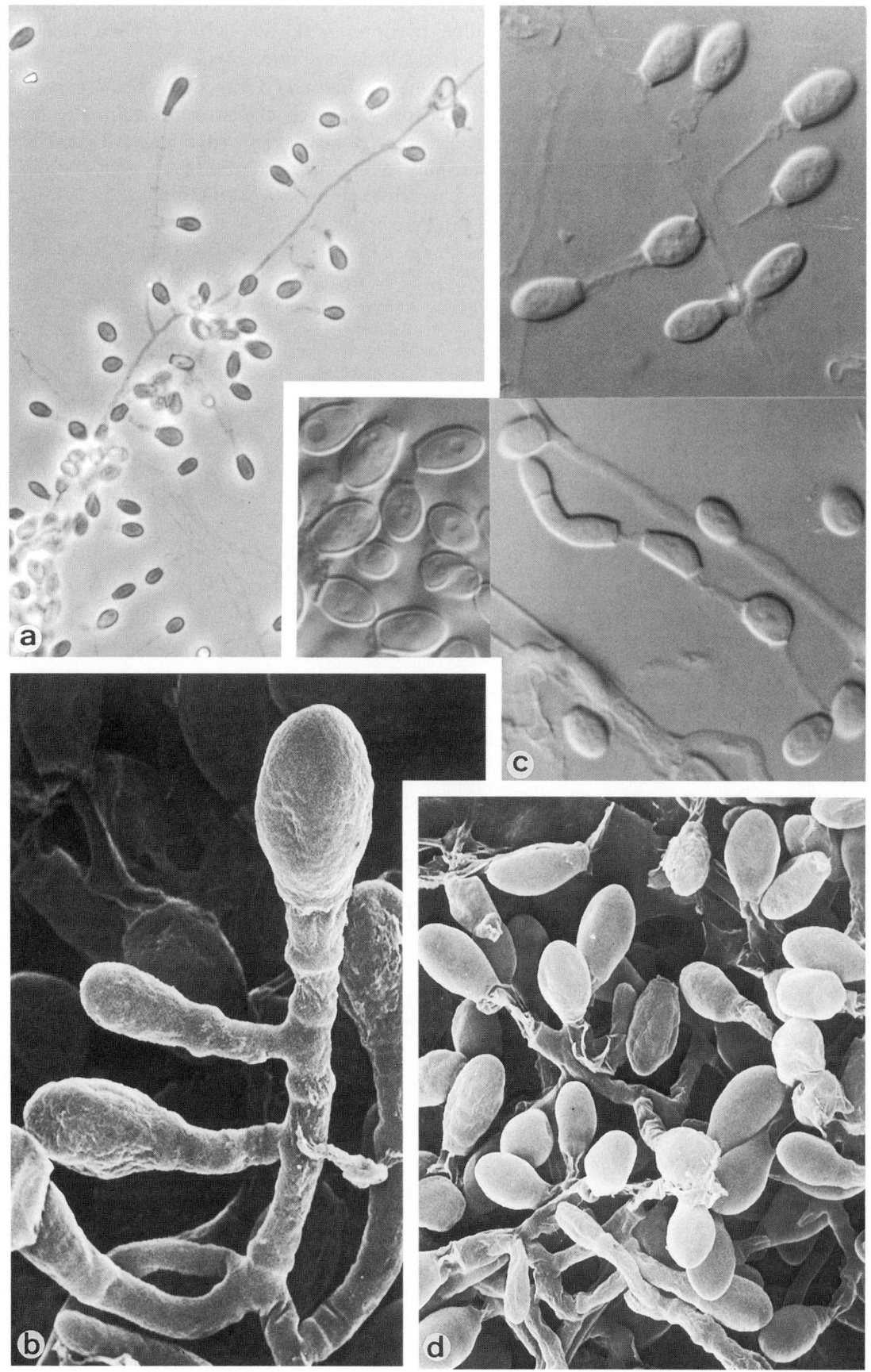

Chrysosporium tropicum, FMR 3567. Fertile hyphae and conidia. a. ×512; b. ×6400; c. ×1600; d. ×2750.

Chrysosporium anamorph of *Nannizziopsis vriesii* (Apinis) Currah

Colony characteristics. Colonies (PYE) growing moderately rapidly, felty or powdery, yellowish-white.

Microscopy. Fertile hyphae hyaline, undifferentiated. Terminal and lateral conidia sessile, usually on side branches, solitary, subhyaline, smooth- and thin-walled, pyriform or clavate, 3-10 × 2-3 µm, 1-celled, rarely 2-celled, with broad basal scars. Intercalary conidia very rare. Chlamydospores absent.

Teleomorph. *Nannizziopsis vriesii* (Apinis) Currah (*Ascomycota, Euascomycetes, Onygenales: Onygenaceae*).
Ascomata spherical, white; peridium composed of a network of loosely interwoven, hyaline hyphae which are verrucose and constricted at the septa. Asci spherical, 8-spored. Ascospores spherical, 2-3 µm diam,. hyaline, thick-walled, smooth-walled with light microscopy but spiny to reticulate under SEM.

Pathogenicity. BSL-1. Known from skin infections in chameleons (Paré *et al.*, 1997).

References. Van Oorschot (1980), Guarro *et al.* (1991a), Paré *et al.* (1997).

Nomenclature. *Rollandina vriesii* Apinis - Trans. Br. Mycol. Soc. 24: 164, 1985 ≡ *Arachnotheca vriesii* (Apinis) Samson, *in* von Arx - Gen. Fungi Sporul. Pure Cult. p. 130, 1981 ≡ *Nannizziopsis vriesii* (Apinis) Currah - Mycotaxon 24: 164, 1985.

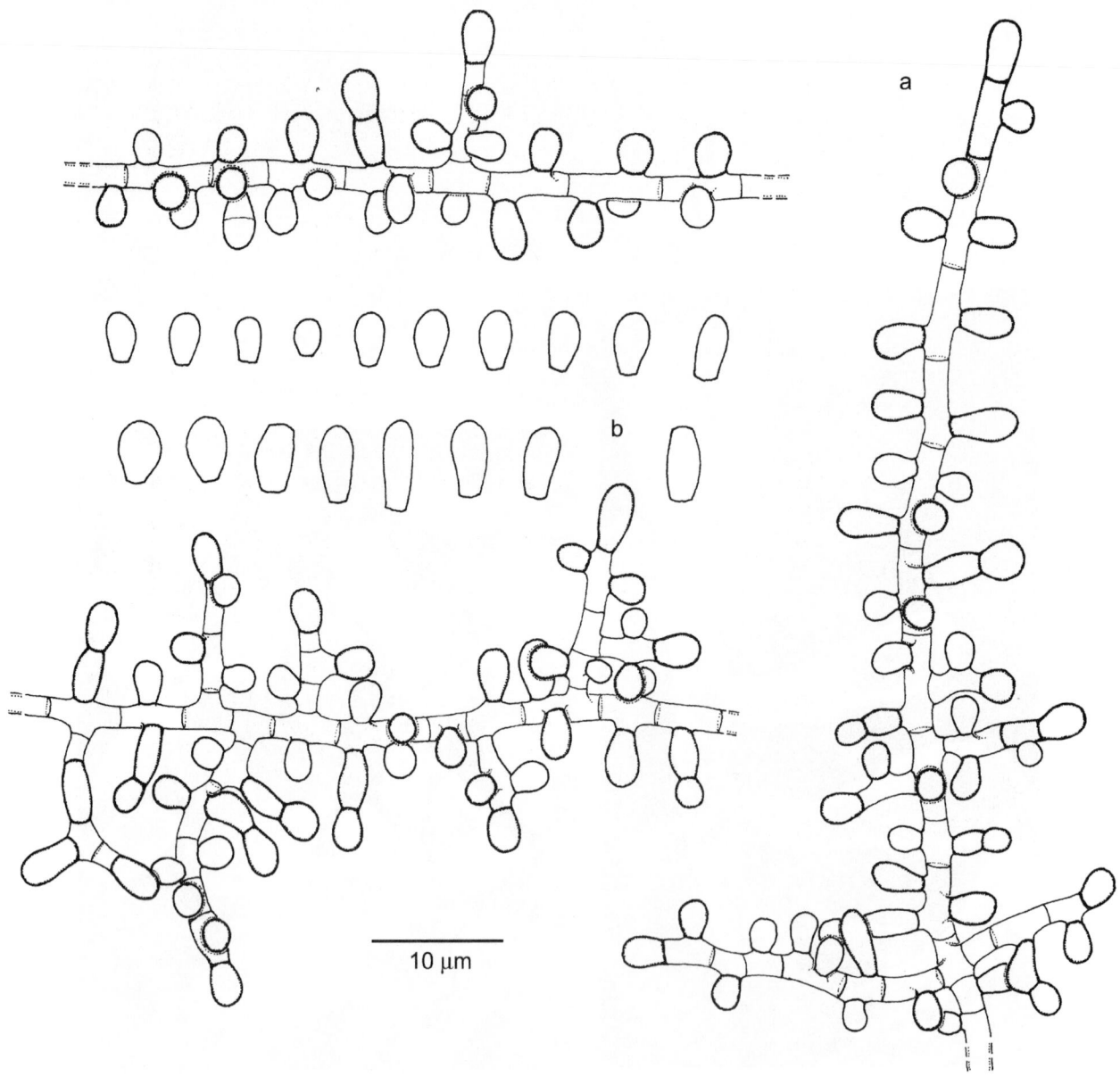

Nannizziopsis vriesii, CBS 407.71. a. Fertile hyphae bearing conidia; b. liberated conidia.

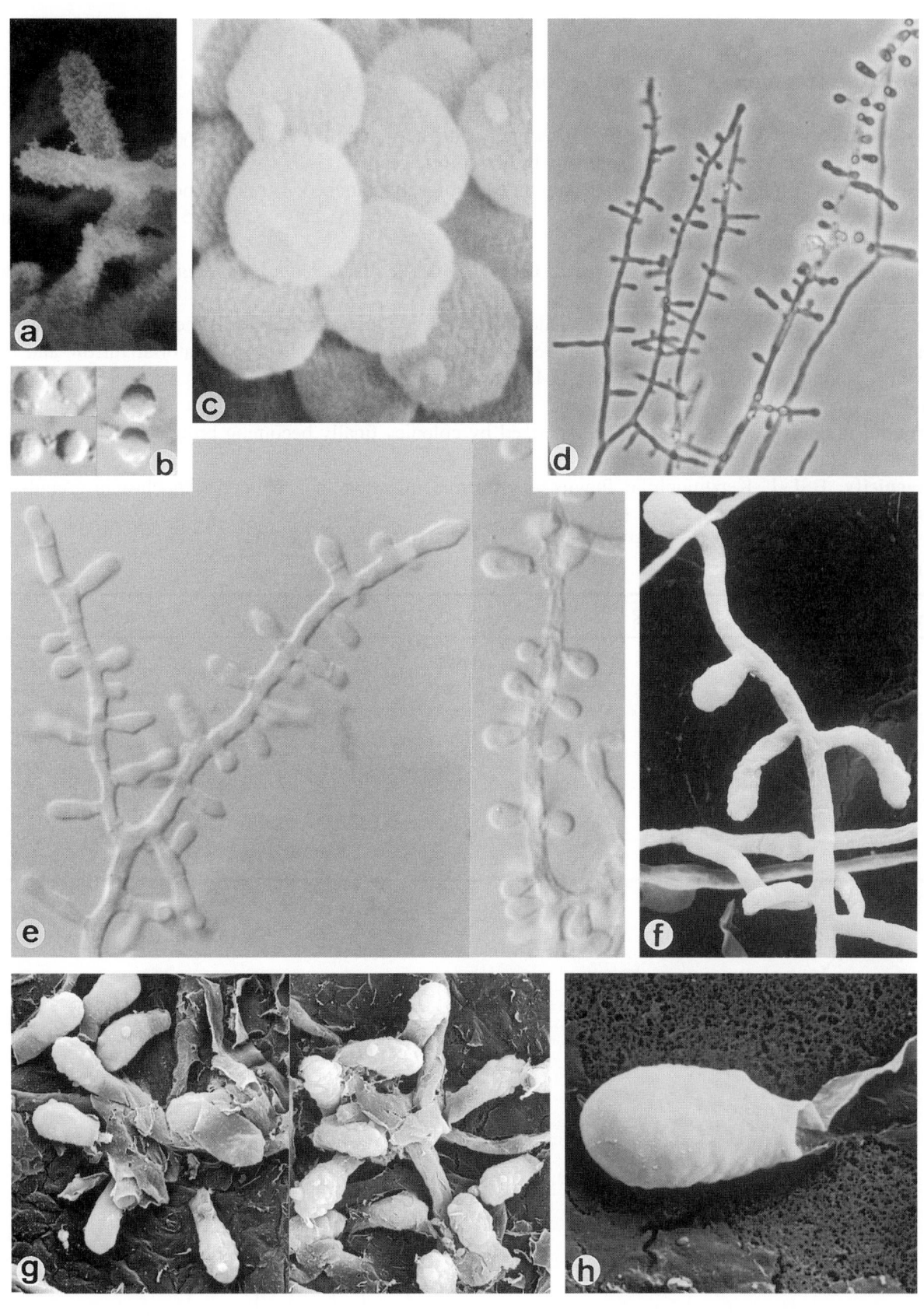

Nannizziopsis vriesii, RV 24806. a. Peridial hyphae; b, c. ascospores; d-h. fertile hyphae and conidia. a. ×1400; b. ×1800; c. ×13,200; d. ×640; c. ×1600; f, g. ×1900; h. ×5500.

Chrysosporium zonatum Al-Musallam & Tan

Colony characteristics. Colonies (PYE) growing moderately rapidly, dry, cottony, initially whitish, later becoming hazel to orange-brown; reverse finally brownish-orange.

Microscopy. Fertile hyphae hyaline, not differentiated, profusely branched. Terminal and lateral conidia on short, unswollen protrusions, subhyaline, thick-walled, verrucose at maturity, clavate, 5.0-7.5 × 3-5 μm, with truncate base. Intercalary conidia uncommon.

Teleomorph. *Uncinocarpus orisii* (Sur & G.R. Ghosh) Sigler & Flis (*Ascomycota, Euascomycetes, Onygenales: Onygenaceae*).

Ascomata spherical, 200-1100 μm in diam, reddish-brown; peridium composed of interwoven, undifferentiated hyphae. Asci subspherical, 8-14 × 7-10 μm. Ascospores oblate, with a shallow equatorial furrow and bipolar thickenings, smooth-walled but pitted under SEM, reddish-brown, 4.5-7.0 × 3.0-4.5 μm.

Physiology. Growth at 37°C better than at 25°C.

Differential diagnosis. The species is characterized by colonies finally becoming brownish, by good growth at 37°C and by conidia 3-5 μm in width.

Pathogenicity. BSL-1. Keratinophilic fungus. A systemic infection in a patient with chronic granulomatous disease was reported by Roilides *et al.* (1999).

References. Al-Musallam & Tan (1989), Sigler *et al.* (1998), Roilides *et al.* (1999).

Antifungal susceptibiliy.

Antifungal	MICs ranges	Mean MICs	Strains	Reference
AMB	<0.06-0.25	0.05	4	Roilides *et al.* (1999)
AMB	0.5	>128	3	Guarro *et al.* (2000d)
5FC	>128	>128	4	Roilides *et al.* (1999)
FLZ	32->128	106	4	Roilides *et al.* (1999)
FLZ	4-16	7.9	3	Guarro *et al.* (2000d)
ITZ	0.25-2	0.6	4	Roilides *et al.* (1999)
ITZ	0.5-1	0.63	3	Guarro *et al.* (2000d)
KTZ	0.25-1	0.5	3	Guarro *et al.* (2000d)
MCZ	1	1	3	Guarro *et al.* (2000d)

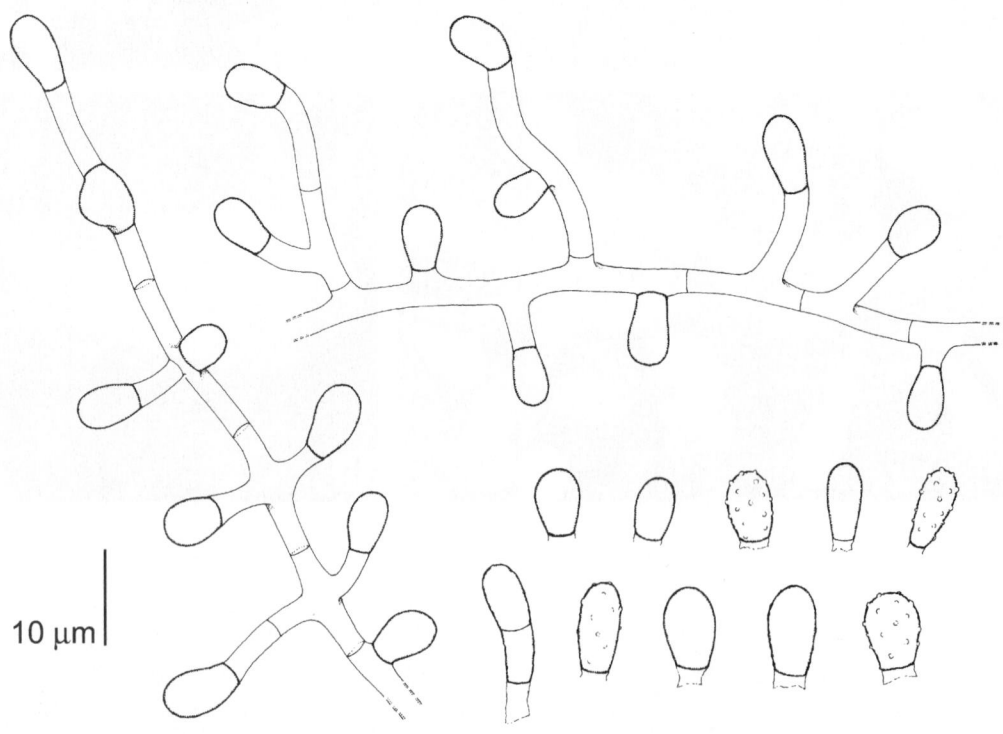

10 μm

Chrysosporium zonatum, CBS 340.89. Fertile hyphae and conidia.

Nomenclature. *Chrysosporium zonatum* Al-Musallam & Tan - Persoonia 14: 69, 1989.

Pseudoarachniotus orisii Sur & G.R. Ghosh, *in* Ghosh & Sur - Kavaka 12: 67, 1985 ≡ *Uncinocarpus orisii* (Sur & G.R. Ghosh) Sigler & Flis, *in* Sigler, Flis & Carmichael - Can. J. Bot. 76: 1627, 1998.

Gymnoascus arxii Cano & Guarro - Stud. Mycol. 31: 61, 1989.

Chrysosporium gourii P.C. Jain, Deshmukh & S.C. Agarwal - Mycoses 36: 77, 1993.

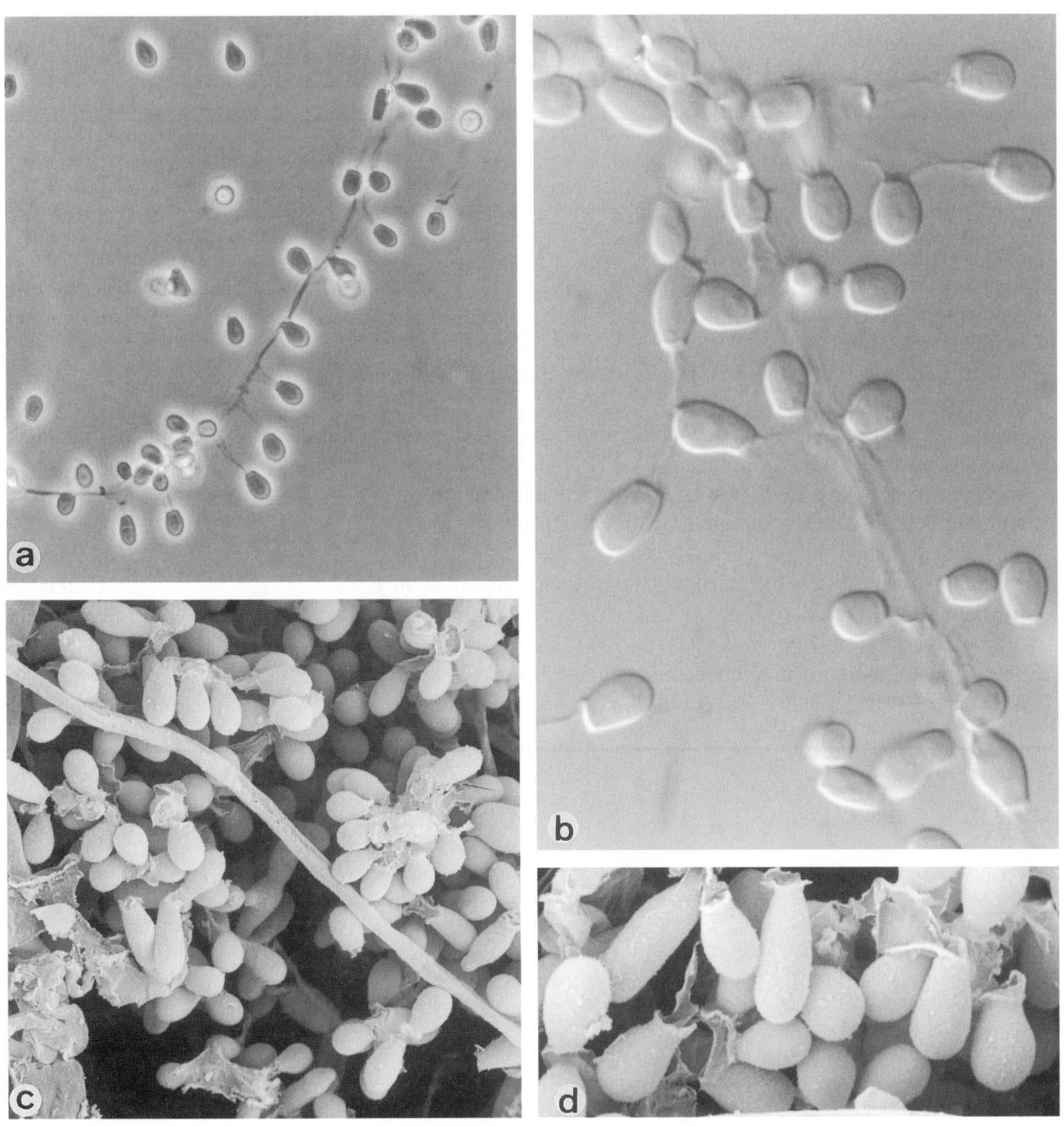

Chrysosporium zonatum, CBS 340.89. Fertile hyphae and conidia. a. ×640; b. ×1600; c. ×2700; d. ×5000.

Hyphomycetes. Genus: *CLADOPHIALOPHORA*

Generic description. Colonies restricted, powdery to woolly, greyish-green to olivaceous-green. Conidiophores absent or inconspicuous. Conidia one-celled, dry, arising in long, poorly branched, often coherent chains, with or without ramoconidia; conidial scars nearly unpigmented. Lactose and inositol mostly assimilated. Gelatin liquefaction absent. Tolerant to cycloheximide. Intolerant to benomyl.
Teleomorph. *Capronia* (*Ascomycota*, *Euascomycetes; Chaetothyriales: Herpotrichiellaceae*).
General remarks. The above description refers to the few human-pathogenic species previously classified in *Cladosporium* or *Xylohypha*. They differ from the saprobic *Cladosporium* species (p. 582) by having poorly differentiated conidiophores and nearly unpigmented scars. *Cladosporium* species have occasionally been found as opportunists in humans, having a wide clinical spectrum (de Hoog, 1983). In contrast, each *Cladophialophora* species provokes a mycosis which is characteristic for that species. *Cladophialophora* is related to the black yeasts; see Fig. 43 on p. 377. Phylogeny of the two groups based on 26S rRNA sequences was presented by Masclaux *et al.* (1995) and on 18S rDNA by Haase *et al.* (1999). An overview of physiological abilities was given by de Hoog *et al.* (1995), and ITS sequence data by Gerrits van den Ende & de Hoog (1999).
References. McGinnis & Borelli (1981), Kwon-Chung & de Vries (1983), Kwong-Chung *et al.* (1989), de Hoog *et al.* (1995), Masclaux *et al.* (1995), Gerrits van den Ende & de Hoog (1999).

Key to the treated species of Cladophialophora:

1a Growth on Mycosel agar (*Cladophialophora*) → **2**
1b. No growth on Mycosel agar . . . . ***Cladosporium elatum* (585)**
2a. Conidia nearly absent . . . . ***C. modesta* (578)**
2b. Conidia abundant → **3**
3a. Conidial chains sessile, without ramoconidia; conidia spherical; no assimilation of *meso*-erythritol . . . . ***C. boppii* (567)**
3b. Conidial chains ascending, with ramoconidia; conidia lemon-shaped, ellipsoidal or fusiform; assimilation of *meso*-erythritol → **4**
4a. Conidia formed on large denticles; no growth with ethanol → **5**
4b. Conidium-bearing denticles absent; growth with ethanol → **6**
5a. Conidial chains short; no growth at 40°C . . . . ***C. devriesii* (573)**
5b. Conidial system profusely branched with long chains; growth at 40°C . . . . ***C. arxii* (561)**
6a. Growth at 40°C . . . . ***C. bantiana* (564)**
6b. No growth at 40°C → **7**
7a. Conidia mostly shorter than 6 µm; conidial chains profusely branched . . . . ***C. carrionii* (570)**
7b. Conidia mostly longer than 6 µm; conidal chains poorly branched . . . . ***C. emmonsii* (576)**

Cladophialophora arxii Tintelnot

Colony characteristics. Colonies (PCA, 30°C) growing slowly, dry, felty, olivaceous grey; reverse olivaceous black.

Microscopy. Hyphae olivaceous brown, bearing lateral and terminal, profusely branched chains of acropetal conidia; lateral chains often inserted on outgrowths which are several μm in length. Conidia brown, smooth- and rather thick-walled, lemon-shaped to fusiform, 7-15 × 3-4 μm.

Physiology.

Growth:		Lactose	+	D-Galacturonate	+
D-Glucose	+	Raffinose	+	DL-Lactate	–
D-Galactose	+	Melezitose	+	Succinate	w
L-Sorbose	+	Inulin	–	Citrate	+
D-Glucosamine	–	Soluble starch	w	Methanol	–
D-Ribose	+	Glycerol	+	Ethanol	–
D-Xylose	+	*meso*-Erythritol	+	Nitrate	+
L-Arabinose	+	Ribitol	w	Nitrite	+
D-Arabinose	+	Xylitol	+	Ethylamine	+
L-Rhamnose	+	L-Arabinitol	w	L-Lysine	+
Sucrose	+	D-Glucitol	+	Cadaverine	+
Maltose	+	D-Mannitol	+	Creatine	+
α,α-Trehalose	+	Galactitol	+	Creatinine	+
methyl-α-D-glucoside	w	*myo*-Inositol	+	10% NaCl	–
Cellobiose	+	Glucono-δ-lactone	+	0.1% Cycloheximide	+
Salicin	+	5-Keto-D-Gluconate	+	Mycosel	+
Arbutin	+	D-Gluconate	+	Urease	+
Melibiose	+	D-Glucuronate	+	Growth at 40°C	+

Molecular diagnostics. SSU restriction map based on CBS 306.94 :

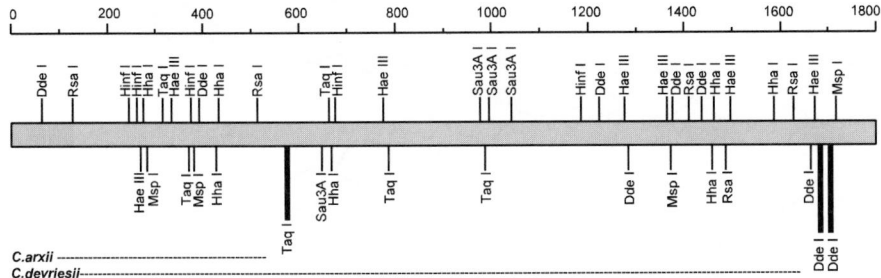

Pathogenicity. BSL-3. The species is known from a single human systemic infection in Germany (Tintelnot *et al.*, 1995).

Reference. Tintelnot *et al.* (1995).

Antifungal susceptibility.

Antifungal	Mean MICs	Strains	Reference
AMB	1.41	1	Tintelnot *et al.* (1999)
5FC	5.66	1	Tintelnot *et al.* (1999)
FLZ	17.66	1	Tintelnot *et al.* (1999)
ITZ	0.02	1	Tintelnot *et al.* (1999)
KTZ	0.18	1	Tintelnot *et al.* (1999)
MCZ	0.35	1	Tintelnot *et al.* (1999)

Nomenclature. *Cladophialophora arxii* Tintelnot, *in* Tintelnot, von Hunnius, de Hoog, Polak-Wyss, Guého & Masclaux - J. Med. Vet. Mycol. 33: 352, 1995.

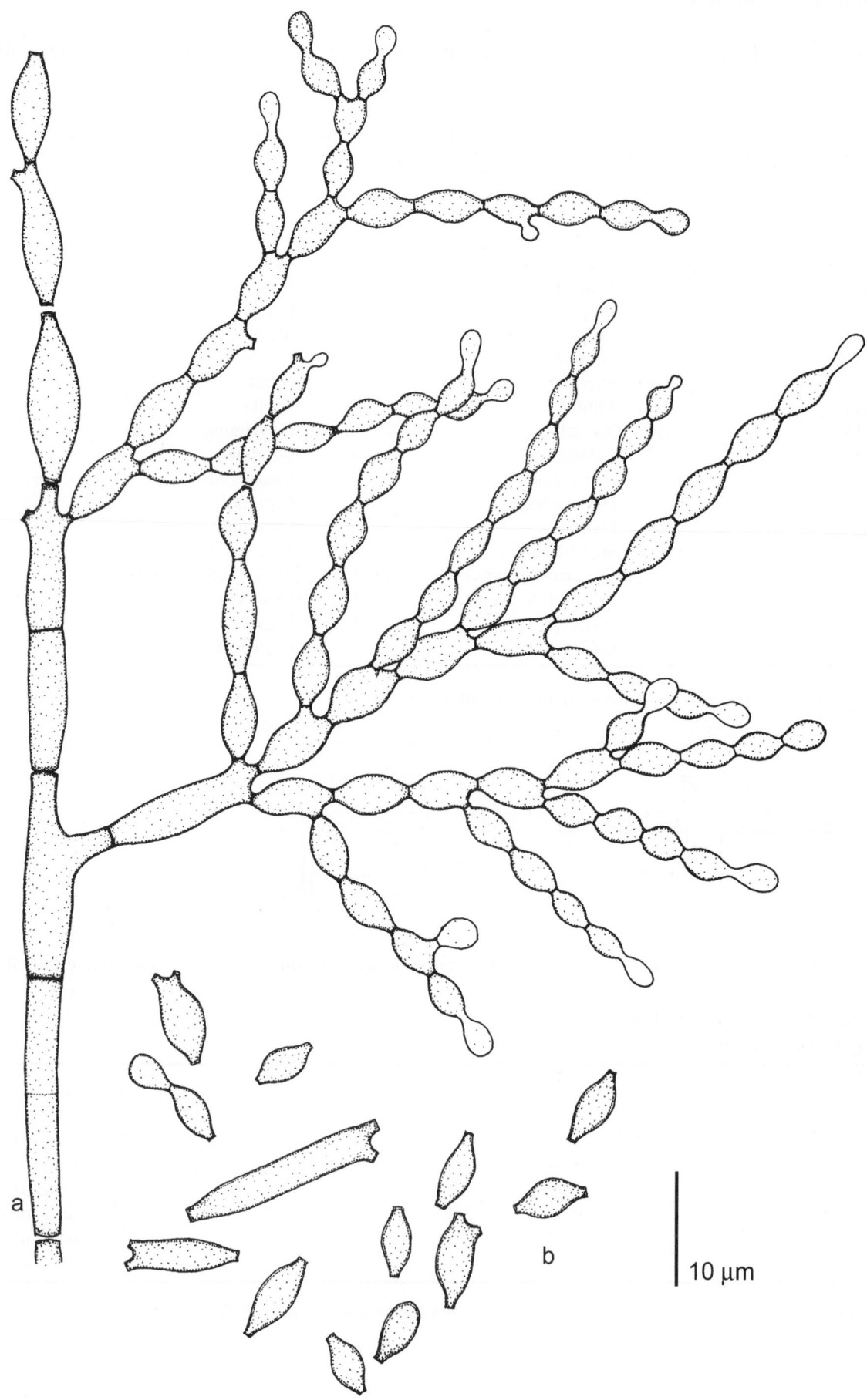

Cladophialophora arxii, CBS 306.94. a. Branched conidial system; b. liberated conidia.

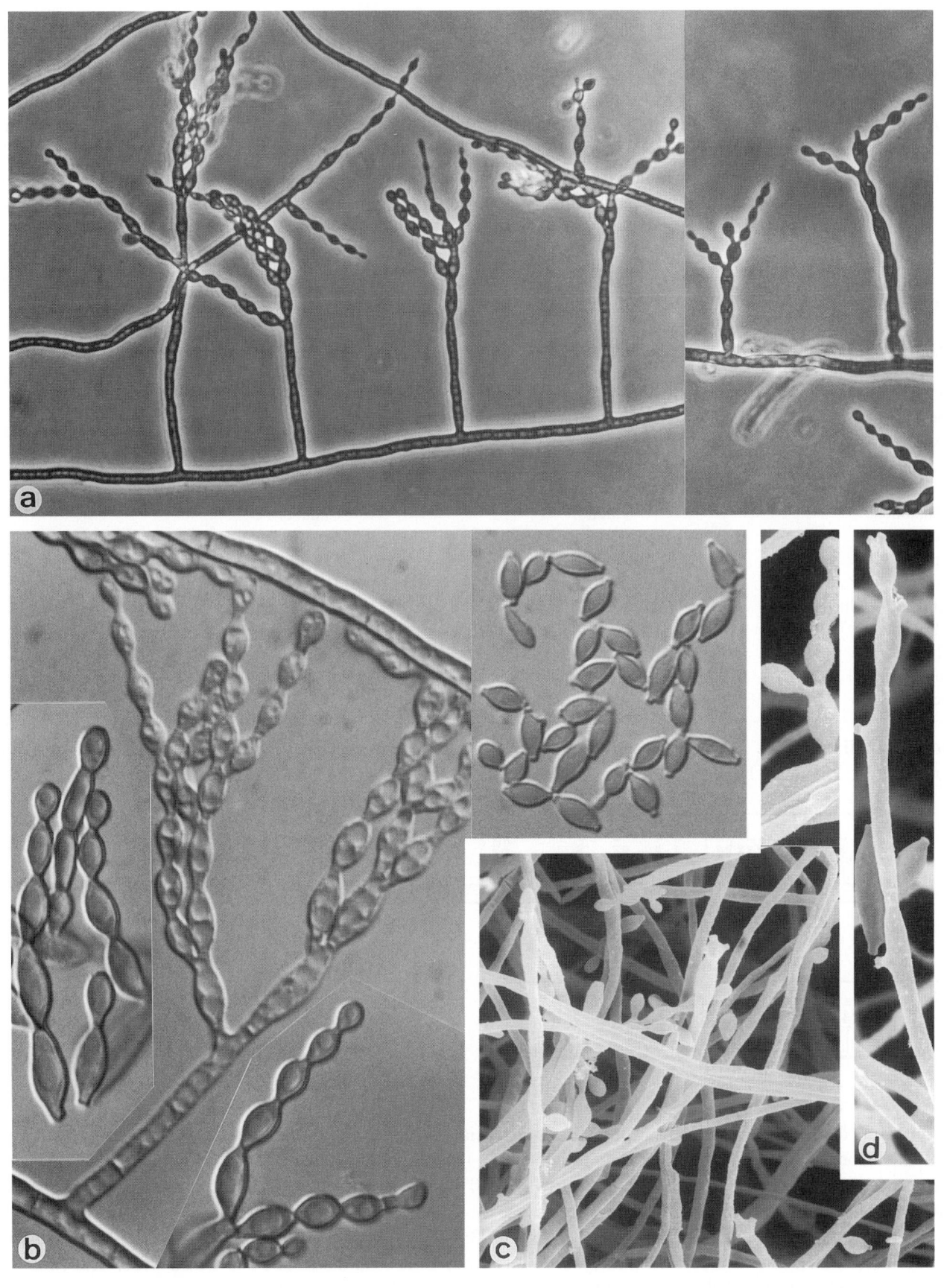

Cladophialophora arxii, CBS 306.94. Branched conidial system and liberated conidia. a. ×640; b. ×1600; c. ×1250; d. ×2000.

Cladophialophora bantiana (Sacc.) de Hoog *et al.*

Colony characteristics. Colonies (PCA, 30°C) moderately expanding, velvety to hairy, powdery when sporulating abundantly, olivaceous green; reverse olivaceous black.

Microscopy. Conidia formed in long, strongly coherent, poorly branched, sessile, lateral or terminal chains on undifferentiated hyphae, pale olivaceous, ellipsoidal to fusiform or nearly cylindrical, 6-11 × 2.5-5.0 µm. Large chlamydospores occasionally present.

Physiology.

Growth:					
D-Glucose	+	Raffinose	+	DL-Lactate	–,w
D-Galactose	+	Melezitose	+	Succinate	w
L-Sorbose	+	Inulin	–,w	Citrate	v
D-Glucosamine	–,w	Soluble starch	+	Methanol	–
D-Ribose	+,w	Glycerol	+	Ethanol	+
D-Xylose	+	*meso*-Erythritol	+	Nitrate	+
L-Arabinose	+	Ribitol	+	Nitrite	+
D-Arabinose	+,w	Xylitol	+	Ethylamine	+
L-Rhamnose	+	L-Arabinitol	–,w	L-Lysine	+
Sucrose	+	D-Glucitol	+	Cadaverine	+
Maltose	+	D-Mannitol	+	Creatine	+
α,α-Trehalose	+	Galactitol	+,w	Creatinine	+
methyl-α-D-Glucoside	+	*myo*-Inositol	+	10% NaCl	–
Cellobiose	+	Glucono-δ-lactone	+	0.1% Cycloheximide	+
Salicin	+	5-Keto-D-Gluconate	+,w	Mycosel	+
Melibiose	+	D-Gluconate	+,w	Urease	+
Lactose	+	D-Glucuronate	+	Growth at 40°C	+
		D-Galacturonate	+		

Molecular diagnostics. The species can be recognized by an intron-specific primer (Gerrits van den Ende & de Hoog, 1999). SSU and ITS restriction maps based on CBS 173.52 and CBS 364.80, resp.:

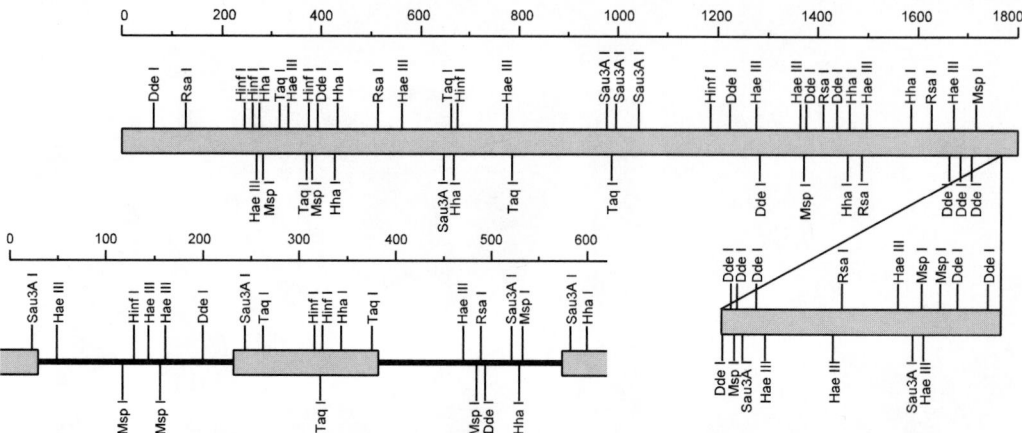

Differential diagnosis. With repeated transfer on artificial media, conidial chains tend to cohere more strongly, conidia becoming nearly cylindrical and rather pale. The very long, sparsely branched conidial chains are characteristic of the species, together with its growth at 40°C.

Pathogenicity. BSL-3. This neurotropic species causes cerebral phaeohyphomycosis in humans (Horré & de Hoog, 1999) and is usually fatal (Bennett *et al.,* 1973; Heney *et al.,* 1989; Sekhon *et al.,* 1992). Multiple, granulomatous abscesses are formed (Aravyski & Aronson, 1968). Its distribution is world-wide. The fungus is probably introduced via inhalation; a pulmonary case was described in an AIDS patient (Brenner *et al.,* 1996). Primarily young male patients are affected (Palaoglu *et al.,* 1993). The infection is often found in immunocompetent patients (Dixon *et al.,* 1989), but in several cases constitutional factors were reported such as solid organ transplant (Salama *et al.,* 1997; Silveira *et al.,* 2001) or drug abuse (Walz *et al.,* 1997). Occasionally the species occurs as an opportunist in immunocompromised patients (Emmens *et al.,* 1996). (Sub)cutaneous infections are rarely noted (Jacyk *et al.,* 1997). In skin tissue the morphology is irregular, with swollen, often yeast-like cells (Kwon-Chung *et al.,* 1989).

References. McGinnis & Borelli (1981), Kwon-Chung (1983), Kwon-Chung & de Vries (1983), McGinnis *et al.* (1986), Dixon *et al.* (1989), Kwon-Chung *et al.* (1989), de Hoog *et al.* (1995b), Gerrits van den Ende & de Hoog (1999).

Antifungal susceptibility.

Antifungal	Mean MICs	Strains	Reference
AMB	0.76	5	Tintelnot *et al.* (1999)
5FC	0.23	5	Tintelnot *et al.* (1999)
FLZ	40.5	5	Tintelnot *et al.* (1999)
ITZ	0.04	25	McGinnis & Pasarell (1998a)
KTZ	0.14	5	Tintelnot *et al.* (1999)
MCZ	1.41	5	Tintelnot *et al.* (1999)
TBF	0.08	25	McGinnis & Pasarell (1998a)

Nomenclature. *Torula bantiana* Saccardo - Annls Mycol. 10: 320, 1912 ≡ *Cladosporium bantianum* (Saccardo) Borelli - Riv. Anat. Patol. Oncol. 17: 620, 1960 ≡ *Xylohypha bantiana* (Saccardo) McGinnis, Padhye, Borelli & Ajello - J. Clin. Microbiol. 23: 1150, 1986 ≡ *Cladophialophora bantiana* (Saccardo) de Hoog, Kwon-Chung & McGinnis, *in* de Hoog, Guého, Masclaux, Gerrits van den Ende, Kwon-Chung & McGinnis - J. Med. Vet. Mycol. 33: 343, 1995.

   Cladosporium trichoides Emmons, *in* Binford, Thompson, Gorham & Emmons - Am. J. Clin. Path. 22: 535, 1952.

   Cladosporium trichoides Emmons var. *chlamydosporum* Kwon-Chung - Mycologia 75: 320, 1983.

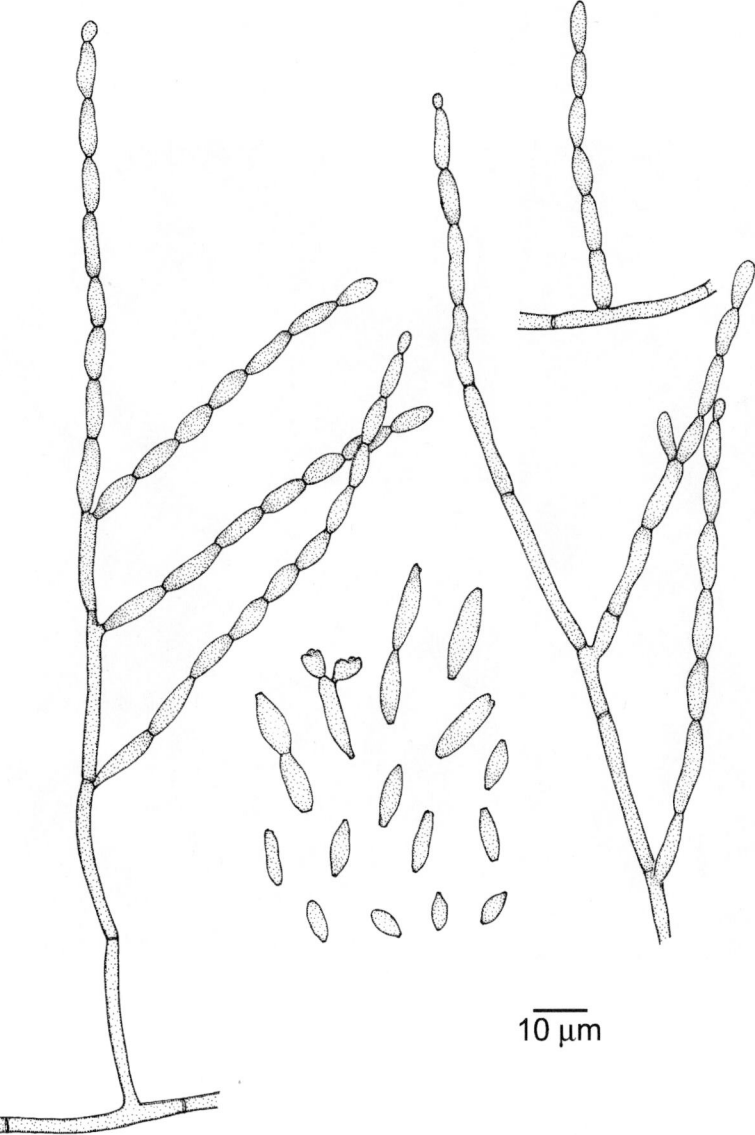

10 µm

Cladophialophora bantiana, CBS 328.65. Branching system with coherent chains of conidia.

565

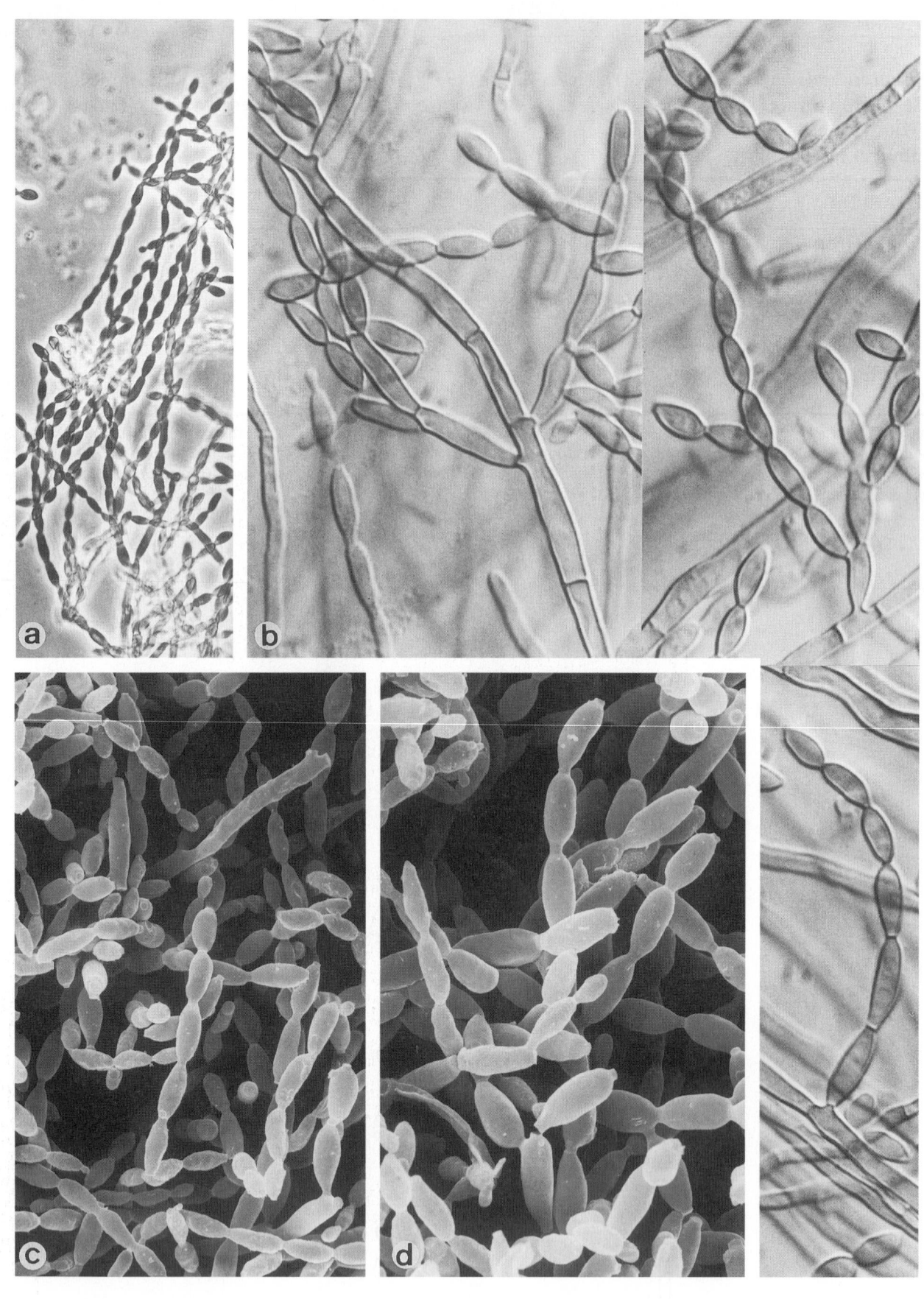

Cladophialophora bantiana, CBS 328.65. a-d. Branching system with coherent chains of conidia. a. ×640; b. ×1600; c. ×2050; d. ×2750.

Cladophialophora boppii (Borelli) de Hoog *et al.*

Colony characteristics. Colonies (PCA, 30°C) restricted, flat, with farinose centre and smooth margin, olivaceous grey to black.

Microscopy. Hyphae olivaceous, rather thick-walled, irregularly bent. Conidiogenous cells are undifferentiated, intercalary or terminal hyphal cells; conidia formed in long, mostly unbranched, acropetally elongating chains which are borne laterally on hyphae. Conidia spherical, 3-4 μm diam, smooth-walled, with moderately thick walls, olivaceous grey.

Physiology.

Growth:					
D-Glucose	+	Raffinose	+	DL-Lactate	–
D-Galactose	+	Melezitose	+	Succinate	–,w
L-Sorbose	+	Inulin	v	Citrate	–
D-Glucosamine	–,w	Soluble starch	w	Methanol	–
D-Ribose	w	Glycerol	+	Ethanol	–
D-Xylose	+	*meso*-Erythritol	–	Nitrate	+
L-Arabinose	+	Ribitol	+,w	Nitrite	v
D-Arabinose	+	Xylitol	+	Ethylamine	+
L-Rhamnose	+	L-Arabinitol	–,w	L-Lysine	+,w
Sucrose	+	D-Glucitol	+	Cadaverine	+
Maltose	+	D-Mannitol	+	Creatine	+
α,α-Trehalose	+	Galactitol	+,w	Creatinine	+
methyl-α-D-Glucoside	–	*myo*-Inositol	v	10% NaCl	–
Cellobiose	+	Glucono-δ-lactone	+	0.1% Cycloheximide	+
Salicin	+	5-Keto-D-Gluconate	+	Mycosel	+
Melibiose	+	D-Gluconate	+	Urease	+
Lactose	v	D-Glucuronate	+,w	Growth at 37°C	–
		D-Galacturonate	v		

Molecular diagnostics. SSU restriction map based on CBS 126.86:

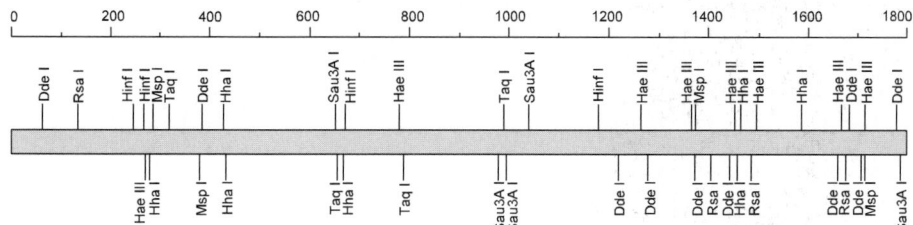

Pathogenicity. BSL-2. The species was described from skin lesions of the lower limbs of a South American female (Borelli, 1983). It was reported as causing chromoblastomycosis, but the characteristic muriform cells were not seen. It occurs occasionally on human skin, but may have been underdiagnosed.

Reference. De Hoog *et al.* (1995b).

Antifungal susceptibility.

Antifungal	Mean MICs	Strains	Reference
AMB	1	1	Tintelnot *et al.* (1999)
5FC	0.5	1	Tintelnot *et al.* (1999)
FLZ	128	1	Tintelnot *et al.* (1999)
ITZ	0.5	1	Tintelnot *et al.* (1999)
KTZ	0.5	1	Tintelnot *et al.* (1999)
MCZ	2	1	Tintelnot *et al.* (1999)

Nomenclature. *Taeniolella boppii* Borelli - Med. Cutan. Ibero Lat. Amer. Venez. 11: 232, 1983 ≡ *Cladophialophora boppii* (Borelli) de Hoog, Kwon-Chung & McGinnis, *in* de Hoog, Guého, Masclaux, Gerrits van den Ende, Kwon-Chung & McGinnis - J. Med. Vet. Mycol. 33: 345, 1995.

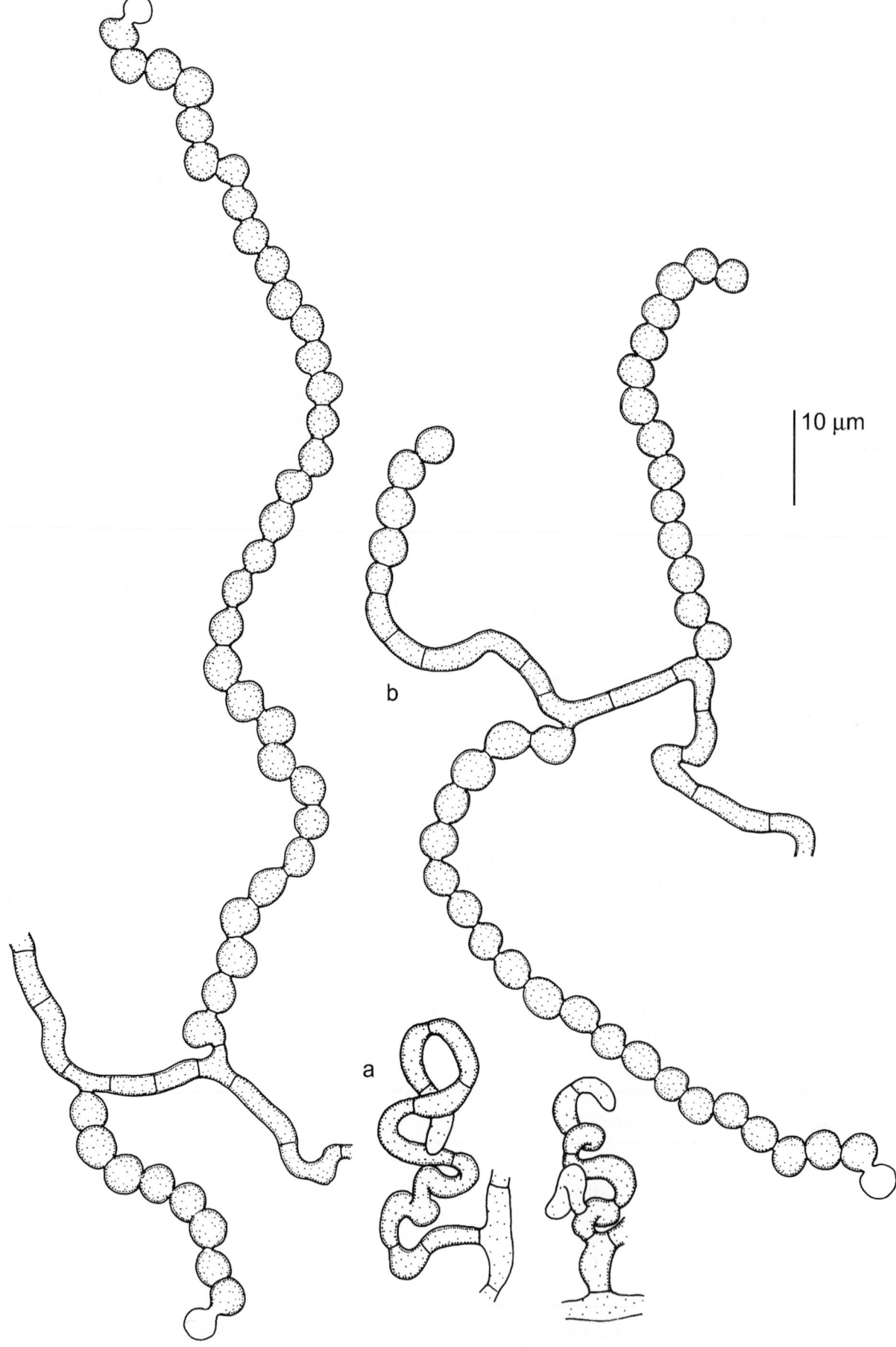

Cladophialophora boppii, CBS 126.86. a. Initial growth of conidial chains; b. full-grown conidial chains.

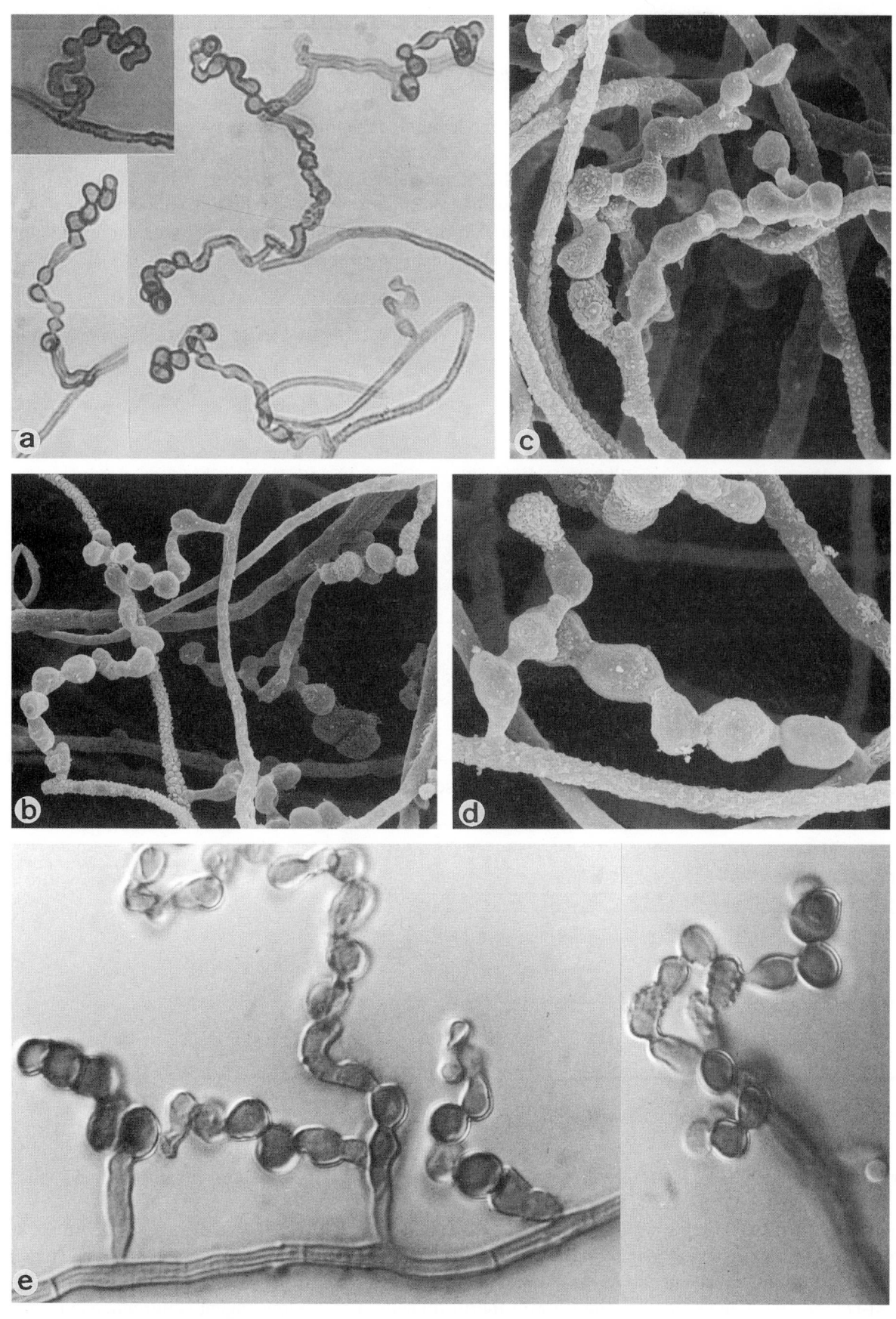

Cladophialophora boppii, FMR 3868. Conidial chains. a. ×640; b. ×1500; c. ×2700; d. ×3500; e. ×1600.

Cladophialophora carrionii (Trejos) de Hoog *et al.*

Colony characteristics. Colonies (PCA, 30°C) moderately expanding, powdery, olivaceous green; reverse olivaceous black.

Microscopy. Fertile hyphae ascending to erect, olivaceous-green, apically branched and bearing long, branched conidial chains. Conidia pale olivaceous, smooth-walled or slightly verrucose, limoniform to fusiform, 4.5-6.0 (- 8.5) × 2.2-2.6 µm; scars (2-4 per conidium) somewhat pigmented. Bulbous phialides with large collarettes and minute, hyaline conidia are occasionally formed on nutritionally poor media.

Physiology.

Growth:		Raffinose	w	DL-Lactate	–
D-Glucose	+	Melezitose	w	Succinate	w
D-Galactose	w	Inulin	w	Citrate	–
L-Sorbose	w	Soluble starch	+	Methanol	–
D-Glucosamine	–	Glycerol	+	Ethanol	+
D-Ribose	w	*meso*-Erythritol	+	Nitrate	+
D-Xylose	+	Ribitol	+	Nitrite	+
L-Arabinose	+	Xylitol	+	Ethylamine	w
D-Arabinose	+	L-Arabinitol	+	L-Lysine	+
L-Rhamnose	+	D-Glucitol	+	Cadaverine	+
Sucrose	+	D-Mannitol	w	Creatine	+
Maltose	+	Galactitol	–	Creatinine	+
α,α-Trehalose	+	*myo*-Inositol	w	10% NaCl	–
methyl-α-D-Glucoside	+	Glucono-δ-lactone	w	0.1% Cycloheximide	+
Cellobiose	+	5-Keto-D-Gluconate	+	Mycosel	+
Salicin	w	D-Gluconate	+	Urease	+
Melibiose	w	D-Glucuronate	+	Growth at 37°C	+
Lactose	w	D-Galacturonate	+	Growth at 40°C	–

Molecular diagnostics. SSU and ITS restriction maps based on CBS 260.83 and NCBI AF050262 resp.:

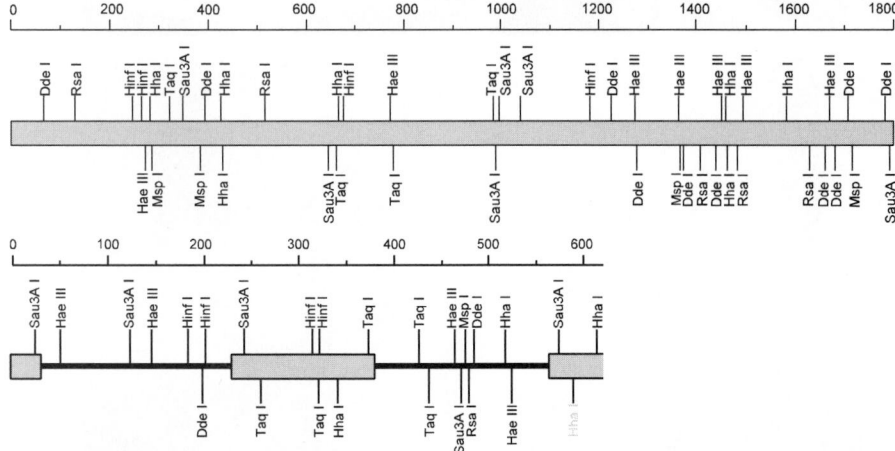

Differential diagnosis. Conidia are smaller and comprise heavily branched systems which fall apart much more easily than in the other *Cladophialophora* species.

Pathogenicity. BSL-2. Together with the two *Fonsecaea* species (p. 676) and *Phialophora verrucosa* (p. 864), *C. carrionii* is one of the relatively common causative agents of chromoblastomycosis. The disease is characterized by localized skin and subcutaneous lesions, leading to superficial, warty to cauliflower-like tumours (Simson, 1946; Al-Doory, 1972). In tissue, the fungus forms muriform cells. The etiologic agent is introduced via cutaneous injury. *Cladophialophora carrionii* is mainly reported from arid regions of tropical South America, from South Africa and from Australia (Lavelle, 1980).

References. Trejos (1954), de Hoog *et al.* (1995b).

Antifungal susceptibility.

Antifungal	Mean MICs	Strains	Reference
AMB	1.87	6	Tintelnot *et al.* (1999)
5FC	0.25	6	Tintelnot *et al.* (1999)
FLZ	26.7	6	Tintelnot *et al.* (1999)
ITZ	0.03	21	McGinnis & Pasarell (1998a)
KTZ	0.2	6	Tintelnot *et al.* (1999)
MCZ	0.71	6	Tintelnot *et al.* (1999)
TBF	0.03	21	McGinnis & Pasarell (1998a)

Nomenclature. *Cladophialophora ajelloi* Borelli - Pan Am. Health. Organ. Sci. Publ. 396: 335, 1980.

Cladosporium carrionii Trejos - Revta Biol. Trop. 2: 106, 1954 ≡ *Cladophialophora carrionii* (Trejos) de Hoog, Kwon-Chung & McGinnis, *in* de Hoog, Guého, Masclaux, Gerrits van den Ende, Kwon-Chung & McGinnis - J. Med. Vet. Mycol. 33: 345, 1995.

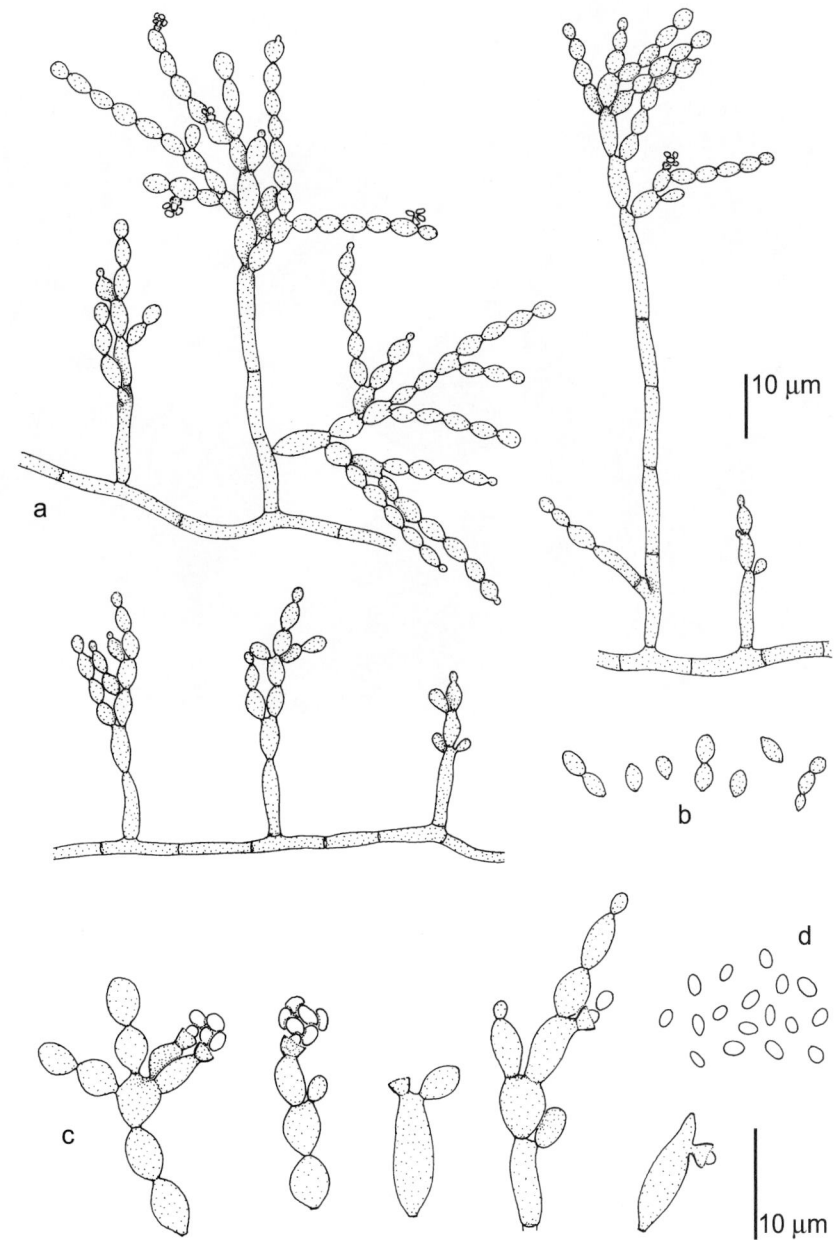

Cladophialophora carrionii, CBS 166.54. a. Conidial apparatus; b. conidia; c. phialides; d. phialoconidia.

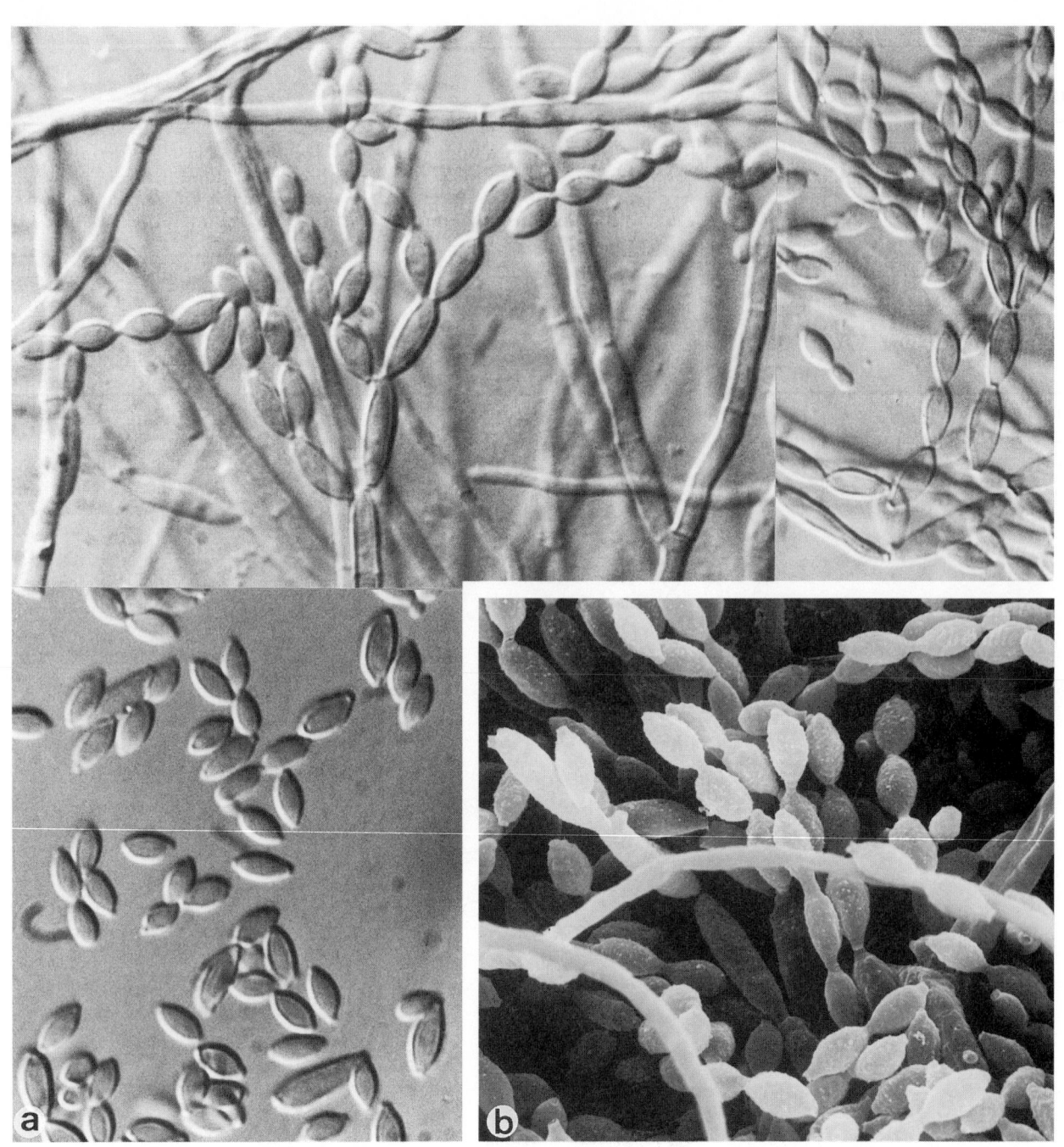

Cladophialophora carrionii, CBS 166.54. a, b. Conidial apparatus and conidia. a. ×1600; b. ×3450.

Cladophialophora devriesii (Padhye & Ajello) de Hoog *et al.*

Colony characteristics. Colonies (PCA, 30°C) growing slowly, dry, velvety, olivaceous black.

Microscopy. Conida formed in densely branched, acropetal chains which each contain up to 6 conidia; chains often located on distinct, unbranched or branched denticles. Conidia olivaceous brown, smooth-walled, lemon-shaped, narrowed towards both ends; scars pale pigmented.

Physiology.

Growth:					
D-Glucose	+	Raffinose	+	DL-Lactate	w
D-Galactose	+	Melezitose	+	Succinate	–
L-Sorbose	+	Inulin	–	Citrate	w
D-Glucosamine	–	Soluble starch	+	Methanol	–
D-Ribose	w	Glycerol	+	Ethanol	–
D-Xylose	+	*meso*-Erythritol	+	Nitrate	+
L-Arabinose	+	Ribitol	+	Nitrite	+
D-Arabinose	+	Xylitol	+	Ethylamine	+
L-Rhamnose	+	L-Arabinitol	+	L-Lysine	+
Sucrose	+	D-Glucitol	+	Cadaverine	+
Maltose	+	D-Mannitol	+	Creatine	+
α,α-Trehalose	+	Galactitol	+	Creatinine	+
methyl-α-D-Glucoside	+	*myo*-Inositol	+	10% NaCl	–
Cellobiose	+	Glucono-δ-lactone	+	0.1% Cycloheximide	+
Salicin	w	5-Keto-D-Gluconate	+	Mycosel	+
Melibiose	+	D-Gluconate	w	Urease	+
Lactose	+	D-Glucuronate	+	Growth at 37°C	+
		D-Galacturonate	+	Growth at 40°C	–

Molecular diagnostics. SSU restriction map based on CBS 147.84:

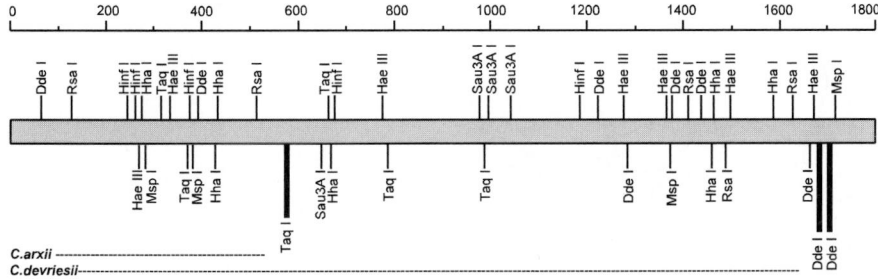

Pathogenicity. BSL-3. The species is known from a fatal, disseminated case in Puerto Rico (Gonzalez *et al.*, 1984; Mitchell *et al.* 1990). No neurotropism was reported.

References. Gonzalez *et al.* (1984), Mitchell *et al.* (1990), de Hoog *et al.* (1995b).

Antifungal susceptibility.

Antifungal	Mean MICs	Strains	Reference
AMB	1.41	1	Tintelnot *et al.* (1999)
5FC	5.66	1	Tintelnot *et al.* (1999)
FLZ	49.89	1	Tintelnot *et al.* (1999)
ITZ	0.12	1	Tintelnot *et al.* (1999)
KTZ	0.18	1	Tintelnot *et al.* (1999)
MCZ	0.25	1	Tintelnot *et al.* (1999)

Nomenclature. *Cladosporium devriesii* Padhye & Ajello, *in* Gonzalez, Alfonso, Seckinger, Padhye & Ajello - Sabouraudia 22: 430, 1984 ≡ *Cladophialophora devriesii* (Padhye & Ajello) de Hoog, Kwon-Chung & McGinnis, *in* de Hoog, Guého, Masclaux, Gerrits van den Ende, Kwon-Chung & McGinnis - J. Med. Vet. Mycol. 33: 344, 1995.

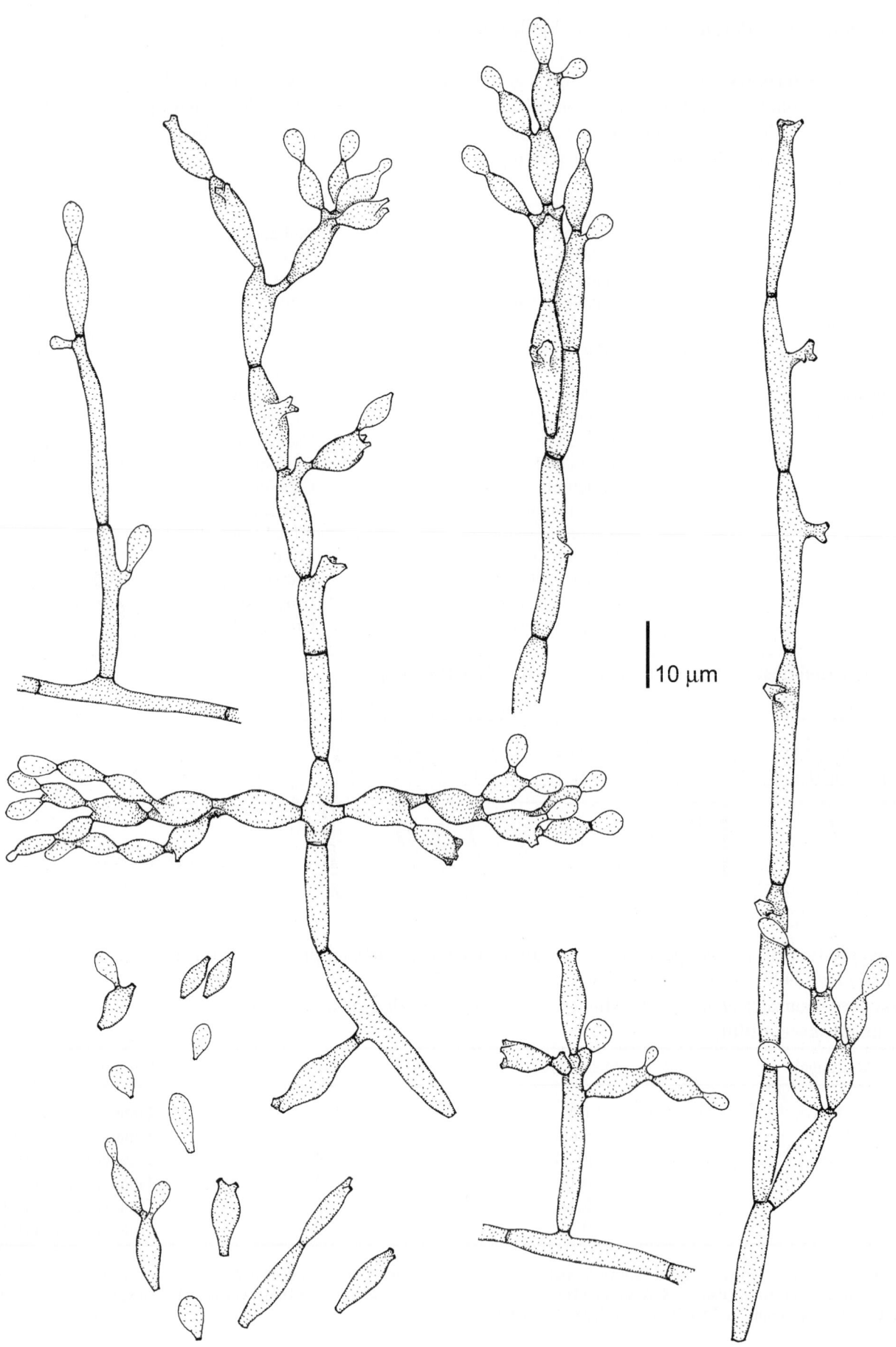

Cladophialophora devriesii, CBS 147.84. Conidial system and liberated conidia.

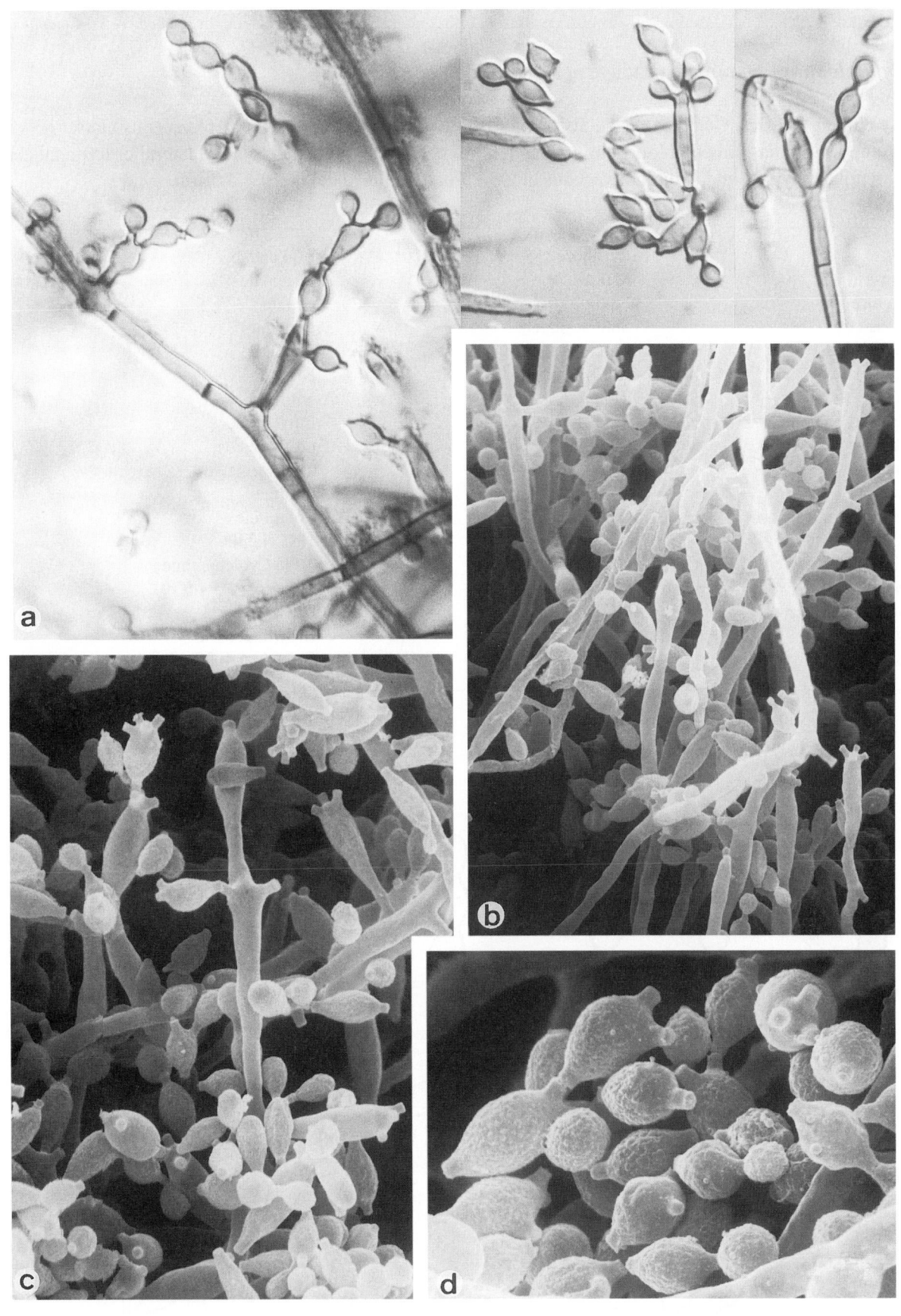

Cladophialophora devriesii, CBS 147.84. Conidial apparatus and conidia. a.×1600; b. ×2000; c. ×2850; d. ×5500.

Cladophialophora emmonsii (Padhye *et al.*) de Hoog & Padhye

Colony characteristics. Colonies (PCA, 30°C) moderately expanding, dry, velvety, olivaceous black.
Microscopy. Conidia formed in long, strongly coherent, moderately branched, sessile, lateral or terminal chains on undifferentiated hyphe, mid to dark olivaceous brown, broadly ellipsoidal, 3-10 × 3-5 μm.
Physiology.

Growth:					
D-Glucose	+	Raffinose	+	DL-Lactate	–
D-Galactose	+	Melezitose	+	Succinate	w
L-Sorbose	+	Inulin	–	Citrate	w
D-Glucosamine	–	Soluble starch	w	Methanol	–
D-Ribose	w	Glycerol	+	Ethanol	+
D-Xylose	+	*meso*-Erythritol	+	Nitrate	+
L-Arabinose	+	Ribitol	w	Nitrite	+
D-Arabinose	w	Xylitol	w	Ethylamine	+
L-Rhamnose	+	L-Arabinitol	–	L-Lysine	+
Sucrose	+	D-Glucitol	+	Cadaverine	+
Maltose	+	D-Mannitol	+	Creatine	+
α,α-Trehalose	+	Galactitol	w	Creatinine	+
methyl-α-D-Glucoside	+	*myo*-Inositol	+	10% NaC1	–
Cellobiose	+	Glucono-δ-lactone	+	0.1% Cycloheximide	+
Salicin	+	5-Keto-D-Gluconate	w	Mycosel	+
Melibiose	+	D-Gluconate	w	Urease	+
Lactose	w	D-Glucuronate	+	Growth at 37°C	+
		D-Galacturonate	+	Growth at 40°C	–

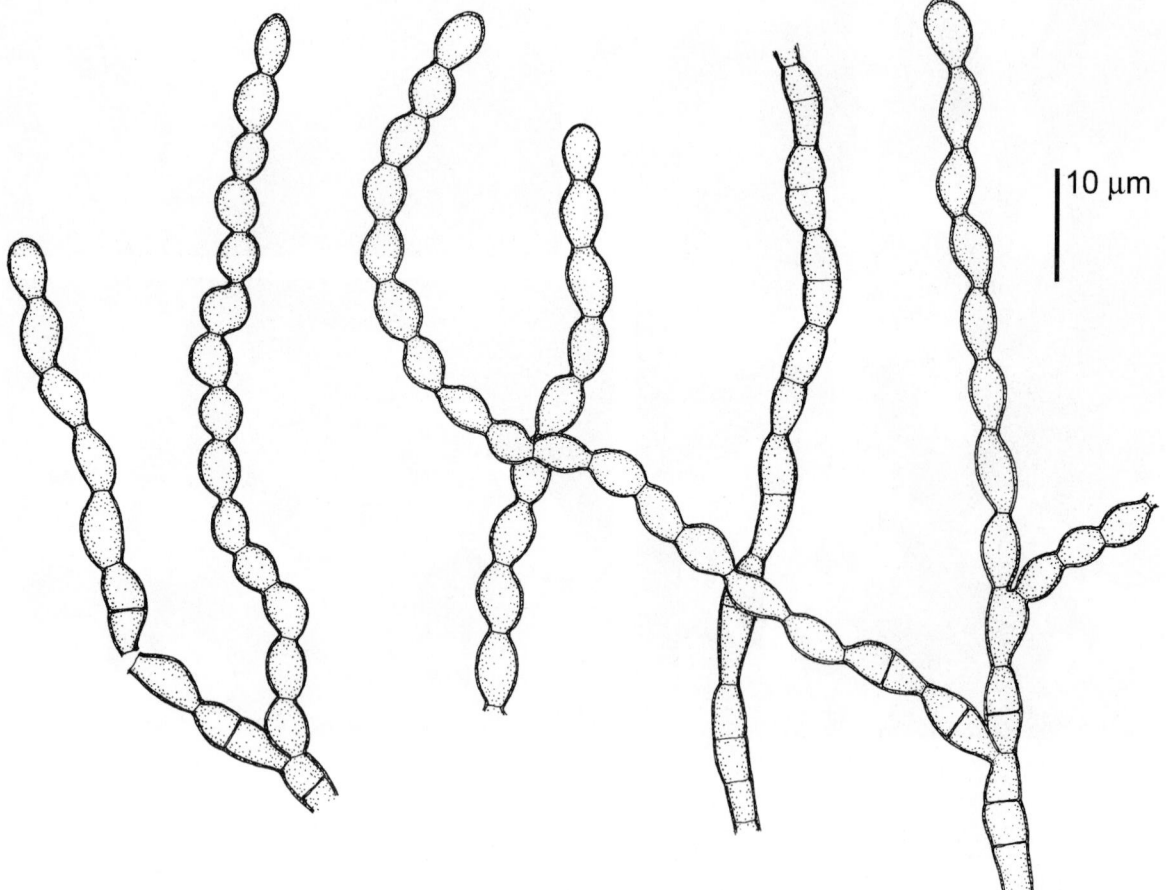

10 μm

Cladophialophora emmonsii, CBS 579.96. Conidial chains.

Differential diagnostics. The species has ellipsoidal conidia, as *C. bantiana*, but is unable to grow at 40°C.

Molecular diagnostics. ITS restriction map based on CBS 979.96:

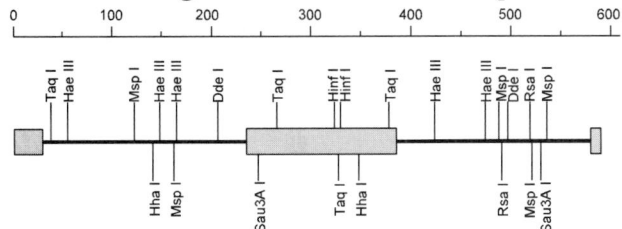

Pathogenicity. BSL-2. The species is particularly known from subcutaneous infections (Padhye *et al.*, 1988b).

References. Padhye *et al.* (1988a, b).

Nomenclature. *Xylohypha emmonsii* Padhye, McGinnis & Ajello, *in* Padhye, McGinnis, Ajello & Chandler - J. Clin. Microbiol. 26: 704, 1988 ≡ *Cladophialophora emmonsii* (Padhye, McGinnis & Ajello) de Hoog & Padhye, *in* Gerrits van den Ende & de Hoog - Stud. Mycol. 43: 160, 1999.

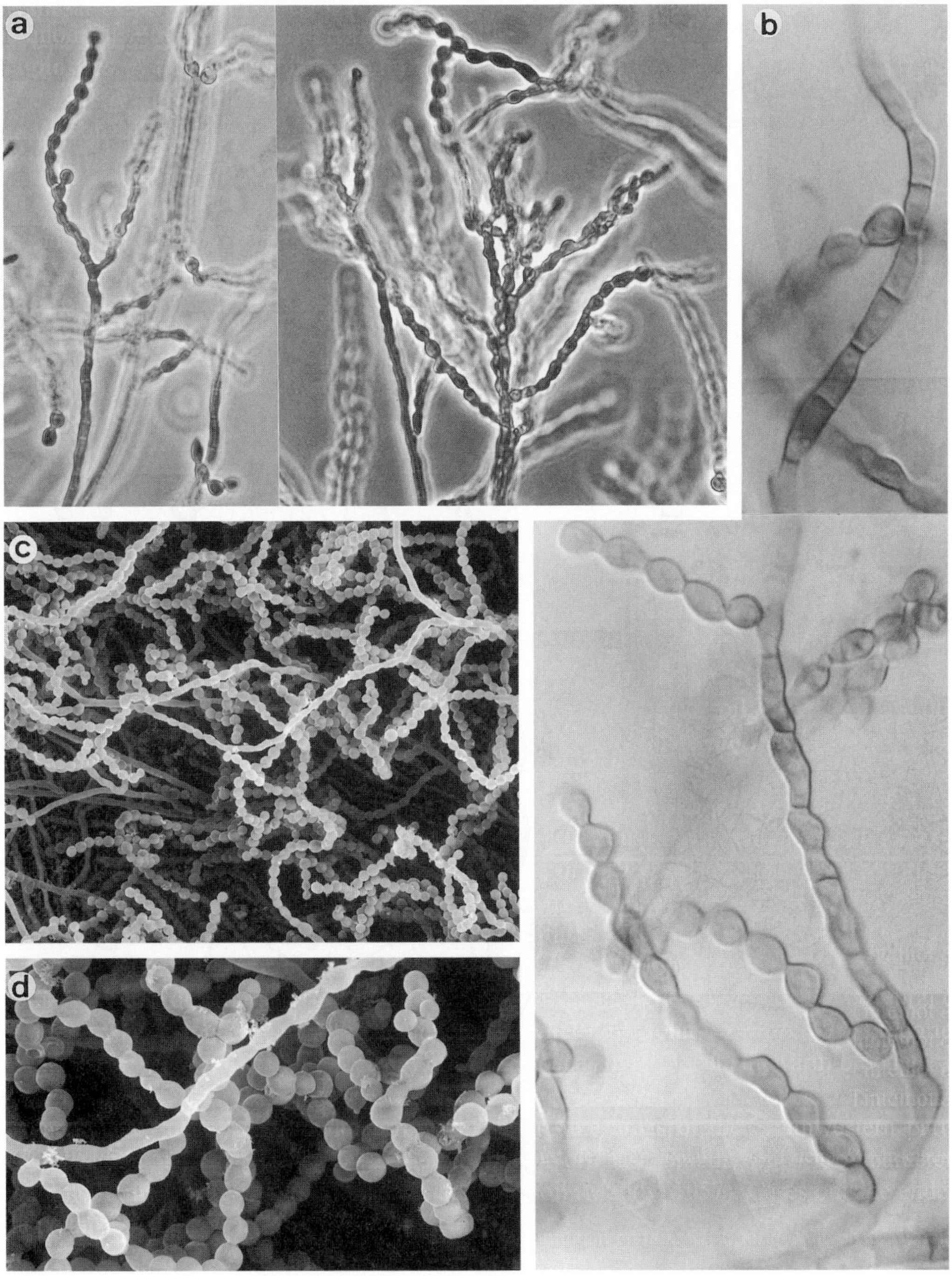

Cladophialophora emmonsii, CBS 979.96. Branched conidial system. a. ×640; b. ×1600; c. ×450; d. ×1150.

Cladophialophora modesta McGinnis *et al.*

Colony characteristics. Colonies (PCA, 30°C) growing moderately rapidly, dry, lanose, dark olivaceous grey.
Microscopy. Conidia formed in very low abundance in strongly coherent, mostly unbranched, lateral or terminal chains on undifferentiated hyphae, olivaceous brown, smooth- and rather firm-walled, broadly ellipsoidal to lemon-shaped, 6-10 × 4.5-7 μm, with wide conidial scars.
Molecular diagnostics. ITS restriction map based on CBS 985.96:

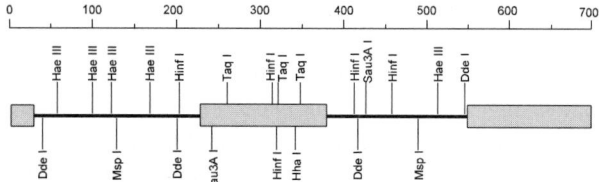

Differential diagnosis. The species grows at temperatures over 40°C. It shows very poor sporulation.
Pathogenicity. BSL-2. It is known from a single case of cerebral, possibly traumatically acquired mycosis (McGinnis *et al.*, 1999).
Reference. McGinnis *et al.* (1999).

Nomenclature. *Cladophialophora modesta* McGinnis, de Hoog & Haase, *in* McGinnis, Lemon, Walker, de Hoog & Haase - Stud. Mycol. 43: 170, 1999.

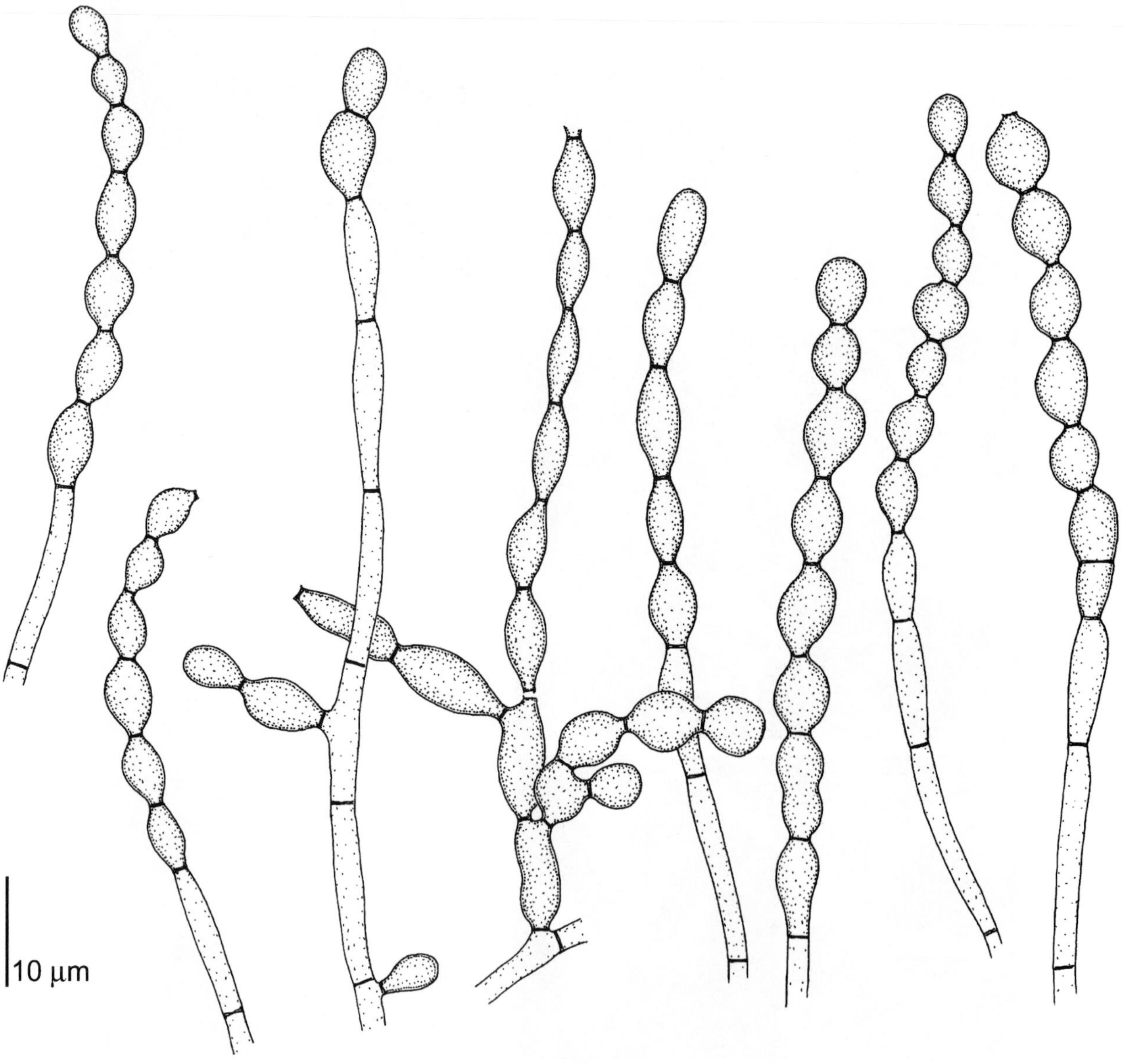

Cladophialophora modesta, CBS 985.96. Conidial chains.

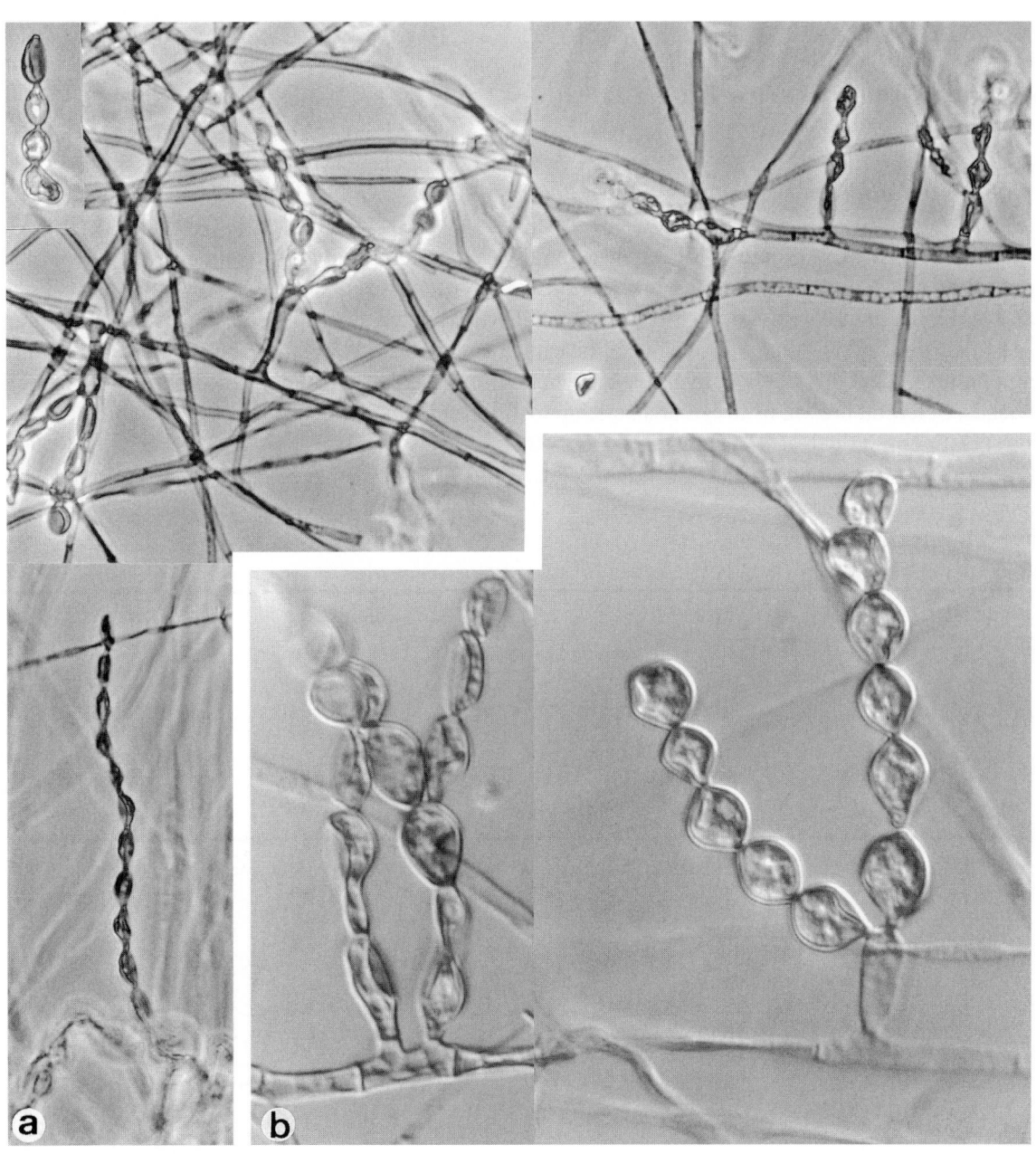

Cladophialophora modesta, CBS 985.96. Conidial apparatus. a. ×640; b. ×1600.

Hyphomycetes. Genus: *CLADORRHINUM*

Cladorrhinum bulbillosum W. Gams & Mouchacca

Cultural characteristics. Colonies (MEA 2%) rapidly expanding, initially hyaline, with local patches of sporulation, soon blackening from the margin due to formation of masses of sclerotia. Reverse greenish-black.

Microscopy. Hyphae wide, with narrow side branches. Fertile hyphae arising in clusters, repeatedly branched at right angles, (sub)hyaline, 2-4 µm wide, each cell bearing a single, distinct phialidic collarette near the middle on a 2-2.5 µm long neck. Conidia 1-celled, hyaline, subspherical to broadly ellipsoidal, 2.5-3.5 × 1.5-2.5 µm, adhering in slimy heads. Sclerotia brown, of irregular shape, up to 150 µm diam.

Physiology. Optimum growth at 30°C, good growth at 40°C.

Pathogenicity. BSL-1. Thermotolerant soil fungus. Two cases of keratitis have been reported (Zapater & Scattini, 1979; Chopin *et al.*, 1997).

Reference. Mouchacca & Gams (1993).

Antifungal susceptibility.

Antifungal	Mean MICs	Strains	Reference
AMB	32	1	Unpublished data
5FC	256	1	Unpublished data
FLZ	128	1	Unpublished data
ITZ	32	1	Unpublished data
KTZ	32	1	Unpublished data
MCZ	8	1	Unpublished data

Nomenclature. *Cladorrhinum bulbillosum* W. Gams & Mouchacca, *in* Mouchacca & Gams - Mycotaxon 48: 425, 1993.

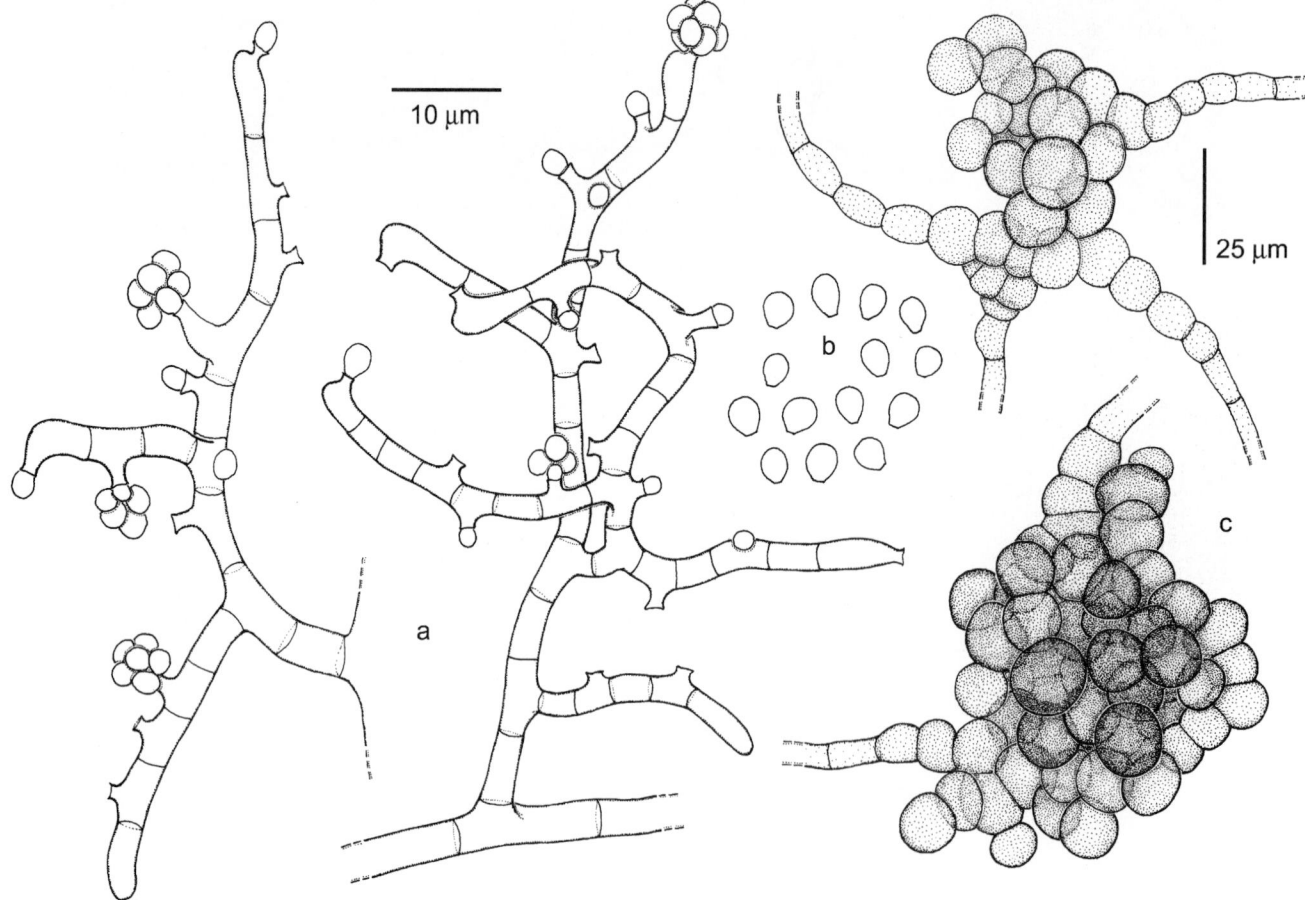

Cladorrhinum bulbillosum, CBS 604.75. a, b. Conidiophores and conidia; c. microsclerotia.

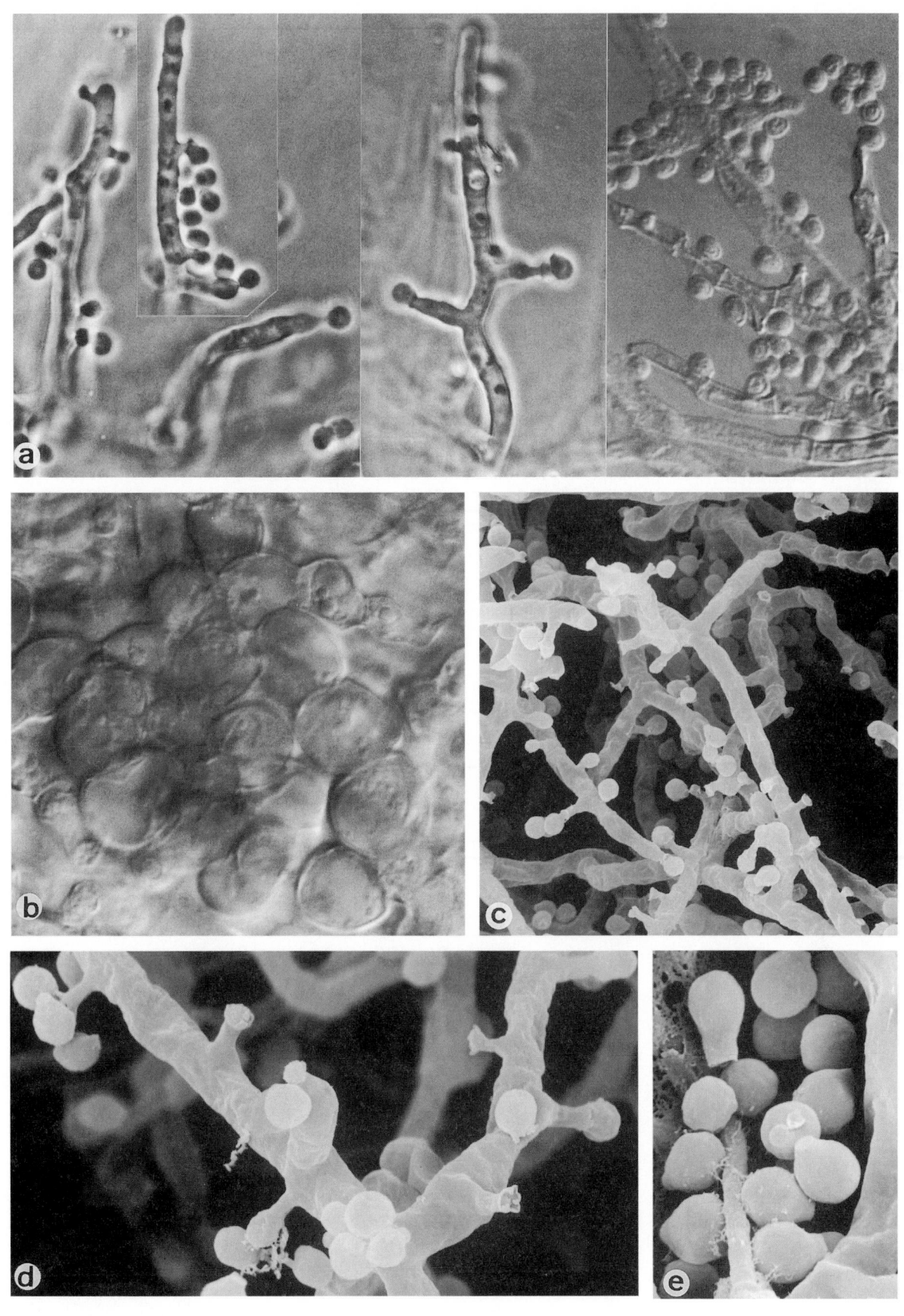

Cladorrhinum bulbillosum, CBS 604.75. a, c-e. Conidiophores and conidia; b. microsclerotium. a, b. ×1600; c. ×1900; d. ×3850; e. ×5500.

Hyphomycetes. Genus: *CLADOSPORIUM*

Generic description. Colonies spreading, powdery to woolly, greyish-green to olivaceous-green. Conidiophores usually present, erect, brown, sympodial, with blackish-brown conidial scars. Conidia dry, 1-(4)-celled, with blackish-brown scars at each end, pale to medium or dark brown, smooth-walled to verrucose, arising in branched chains which readily disarticulate; lower conidia are often septate ramoconidia. Gelatin liquified. Not thermo-tolerant. Intolerant to benomyl.

Teleomorph. *Mycosphaerella* (*Ascomycota, Euascomycetes, Dothideales: Mycosphaerellaceae*).

Differential diagnosis. *Cladophialophora* (p. 560) can be distinguished from *Cladosporium* by the absence of conidiophores, by unpigmented conidial scars and by being unable to liquify gelatin.

General remarks. Numerous saprophytic or plant-pathogenic *Cladosporium* species have been described. Some of the commonest species are often found as culture contaminants and very rarely as opportunists in humans (de Hoog, 1983). The truly human-pathogenic species, *C. bantianum*, *C. carrionii*, *C. devriesii* and *C. trichoides* have been reclassified in *Cladophialophora* (de Hoog *et al.*, 1995b; see p. 560).

References. Ellis (1971, 1976), de Hoog (1983), Ho *et al.* (1999).

Key to the treated species of Cladosporium:

1a. Conidiophores nodose   → **2**
1b. Conidiophores not nodose   → **3**
2a. Conidiophores longer than 500 µm; conidia smooth-walled .... ***C. oxysporum* (589)**
2b. Conidiophores shorter than 500 µm; conidia rough-walled .... ***C. herbarum* (587)**
3a. Conidia mostly spherical to subspherical .... ***C. sphaerospermum* (591)**
3b. Conidia ellipsoidal, limoniform or cylindrical → **4**
4a. Conidia with dark scars .... ***C. cladosporioides* (583)**
4b. Conidia with subhyaline scars .... ***C. elatum* (585)**

Cladosporium cladosporioides (Fres.) de Vries

Colony characteristics. Colonies (MEA 2%) expanding, velvety to powdery, olivaceous green to olivaceous brown; reverse olivaceous black.

Microscopy. Conidiophores of variable length, up to 350 μm long, 2-6 μm wide, without swellings, with terminal and lateral ramifications, bearing branched conidial chains, pale olivaceous brown. Conidia ellipsoidal to limoniform, smooth-walled or slightly verrucose, olivaceous brown, 1-celled, with dark scars, easily liberated.

Molecular diagnostics. ITS restriction map based on NCBI L25429:

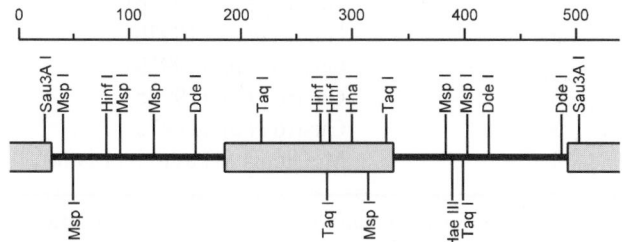

Pathogenicity. BSL-1. Very common air-borne saprobe, frequently encountered as a contaminant. The species was involved in a pulmonary infection (Kwon-Chung *et al.*, 1975) and cutaneous infections (Drabick *et al.*, 1990; Annessi *et al.*, 1992; López *et al.*, 1996). It was isolated from a case of keratitis (Polack *et al.*, 1976), from a dental

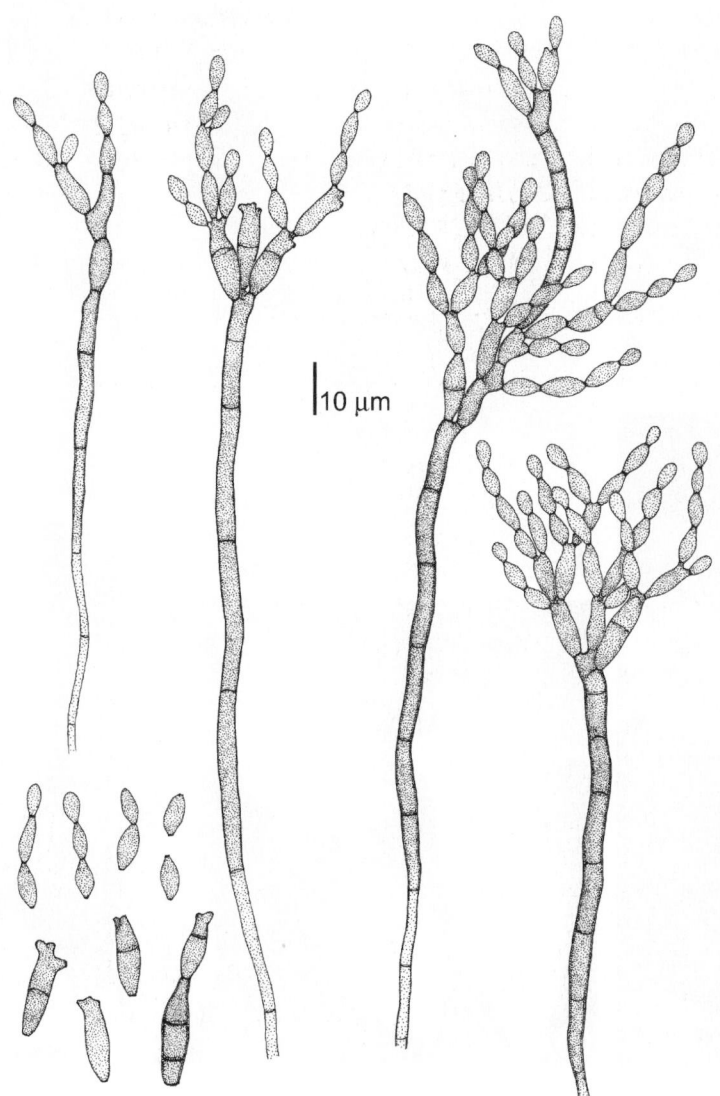

Cladosporium cladosporioides, CBS 171.54. Conidiophores and conidia.

granulome (Pepe & Bertolotto, 1991), from a phaeohyphomycosis (Gugnani *et al.*, 2000) and from a subcutaneous cyst (Pereiro *et al.*, 1998). It was the possible cause of a paranasal sinus fungus ball (Pritchard & Muir, 1987). A mixed disseminated infection was reported by Bentz & Sautter (1993).

References. De Vries (1967), Domsch *et al.* (1980), de Bièvre (1981, 1982).

Antifungal susceptibility.

Antifungal	Mean MICs	Strains	Reference
AMB	0.32	3	Guarro *et al.* (1997b)
5FC	0.5	3	Guarro *et al.* (1997b)
FLZ	25.39	3	Guarro *et al.* (1997b)
ITZ	0.25	3	Guarro *et al.* (1997b)
KTZ	0.25	3	Guarro *et al.* (1997b)
MCZ	0.63	3	Guarro *et al.* (1997b)
VCZ	0.08	3	McGinnis & Pasarell (1998b)

Nomenclature. *Penicillium cladosporioides* Fresenius - Beitr. Mykol. p. 22, 1863 ≡ *Cladosporium cladosporioides* (Fresenius) de Vries - Contrib. Knowl. Clad. p. 57, 1952.

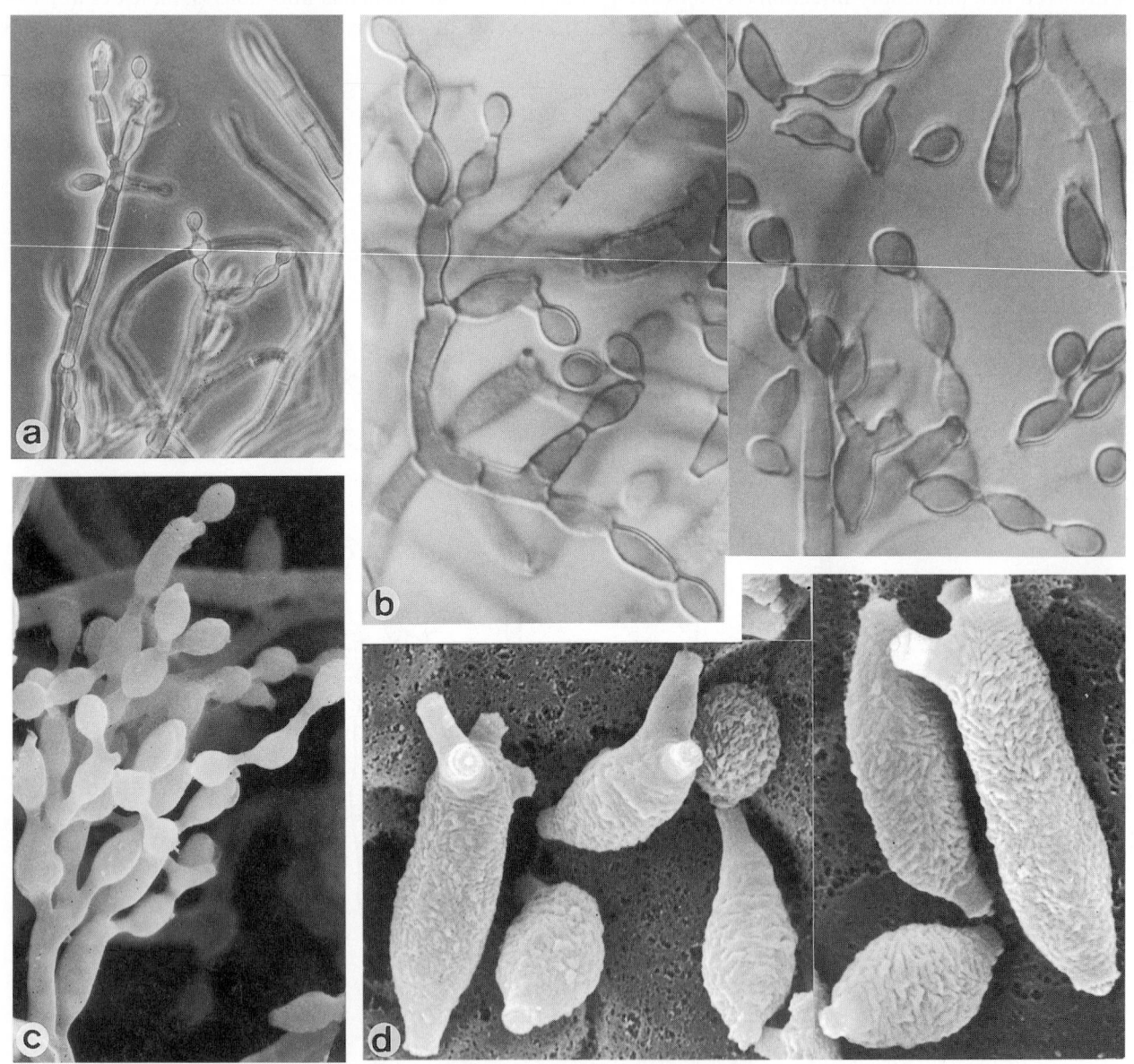

Cladosporium cladosporioides, CBS 171.54. Conidiophores and conidia. a. ×512; b. ×1600; c. ×1800; d. ×5900.

Cladosporium elatum (Harz) Nannf.

Colony characteristics. Colonies (MEA 2%) expanding, powdery, grey or olivaceous green; reverse olivaceous black.

Microscopy. Conidiophore stalks mostly absent. Conidia in very long chains, fusiform, lemon-shaped or subspherical, mostly 1-celled, tapered into a narrow cylindrical part at one or at both ends, smooth-walled, pale brown to olivaceous brown, 4-15 × 2-5 µm, with pale scars; branched, coherent chains of conidia forming wide, loose heads.

Physiology. Ethanol not assimilated; no growth at 37°C; cycloheximide intolerant.

Differential diagnosis. The species is morphologically indistinguishable from the human-pathogenic species of *Cladophialophora* (p. 560), but is different by being sensitive to cycloheximide.

Pathogenicity. BSL-1. The species was involved in a case of subcutaneous infection (Castro & Gomperta, 1984).

References. De Vries (1967), Ellis (1976), de Bièvre (1981, 1982).

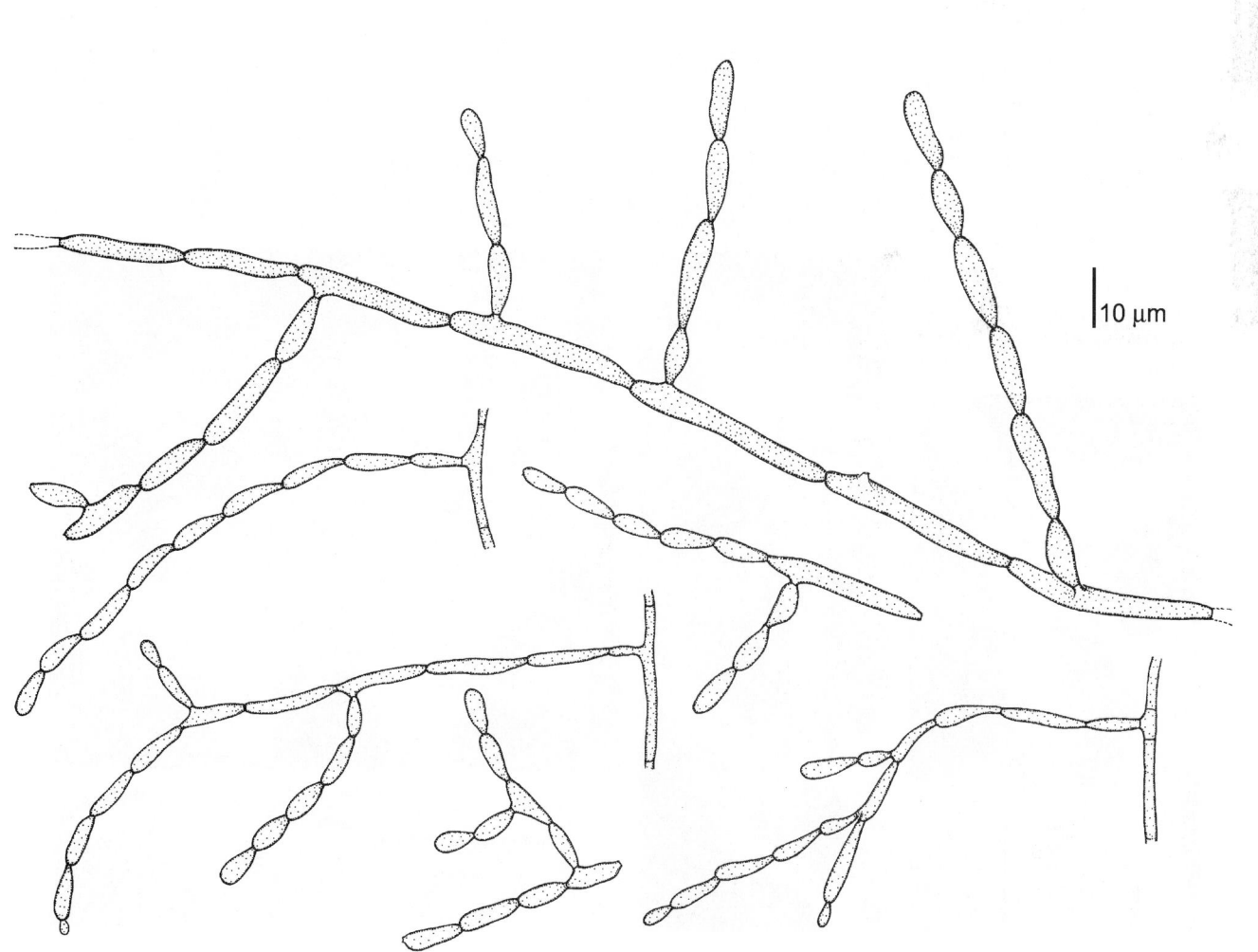

Cladosporium elatum, CBS 146.33. Branching system.

Antifungal susceptibility.

Antifungal	MICs	Strains	Reference
AMB	0.5	1	Guarro *et al.* (1997b)
5FC	64	1	Guarro *et al.* (1997b)
FLZ	32	1	Guarro *et al.* (1997b)
ITZ	0.5	1	Guarro *et al.* (1997b)
KTZ	1	1	Guarro *et al.* (1997b)
MCZ	2	1	Guarro *et al.* (1997b)

Nomenclature. *Hormodendrum elatum* Harz - Bull. Soc. Imp. Nat. Moscou 44: 140, 1871 ≡ *Cladosporium elatum* (Harz) Nannfeldt - Svenska Skogsvför. Tidskr. 32: 397, 1934.

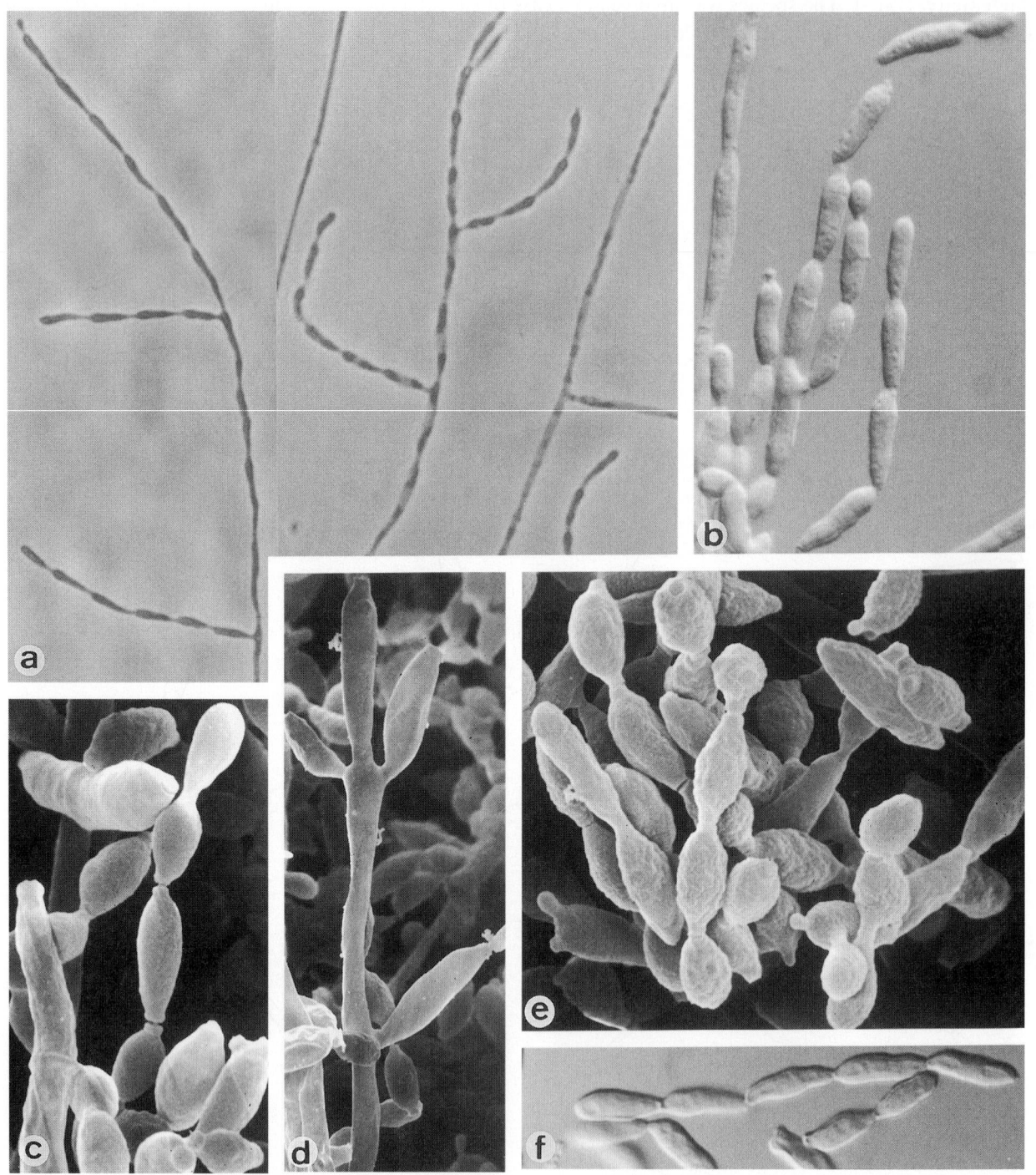

Cladosporium elatum, CBS 146.33. Branching system and conidia. a. ×512; b. ×1600; c. ×3600; d. ×2000; e. ×3600; f. ×1600.

Cladosporium herbarum (Pers.) Link: Fr.

Colony characteristics. Colonies (MEA 2%) moderately expanding, velvety, olivaceous green to olivaceous brown; reverse olivaceous black.

Microscopy. Conidiophores up to 250 µm long, 3-6 µm wide, with terminal and intercalary swellings and geniculate elongations, pale to mid olivaceous brown or brown, smooth-walled. Conidia ellipsoidal to cylindrical, with rounded ends, distinctly verrucose, with protuberant, brown scars, often 2- or more-celled, when 1-celled measuring 5.5-13.0 × 3.8-6.0 µm.

Teleomorph. *Mycosphaerella tassiana* (de Not.) Johanson (*Ascomycota, Euascomycetes, Dothideales: Mycosphaerellaceae*).

Molecular diagnostics. ITS sequences are identical to those of *Cyphellophora laciniata* (p. 621). ITS restriction map based on NCBI L25431:

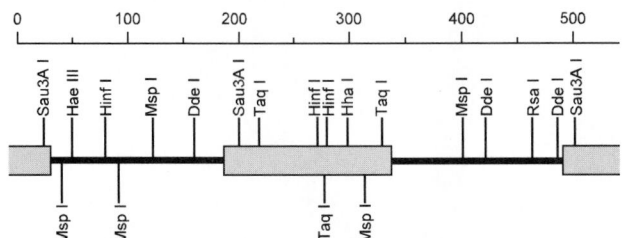

Pathogenicity. BSL-1. Very common saprobe on decaying plant material, frequently encountered in the clinical laboratory as a culture contaminant. This species was isolated from skin lesions (de Bièvre, 1982) and was reported as a possible agent of keratitis (Pritchard & Muir, 1987). Allergic pulmonary mycosis was reported by Muñoz-Ancillo *et al.* (1996).

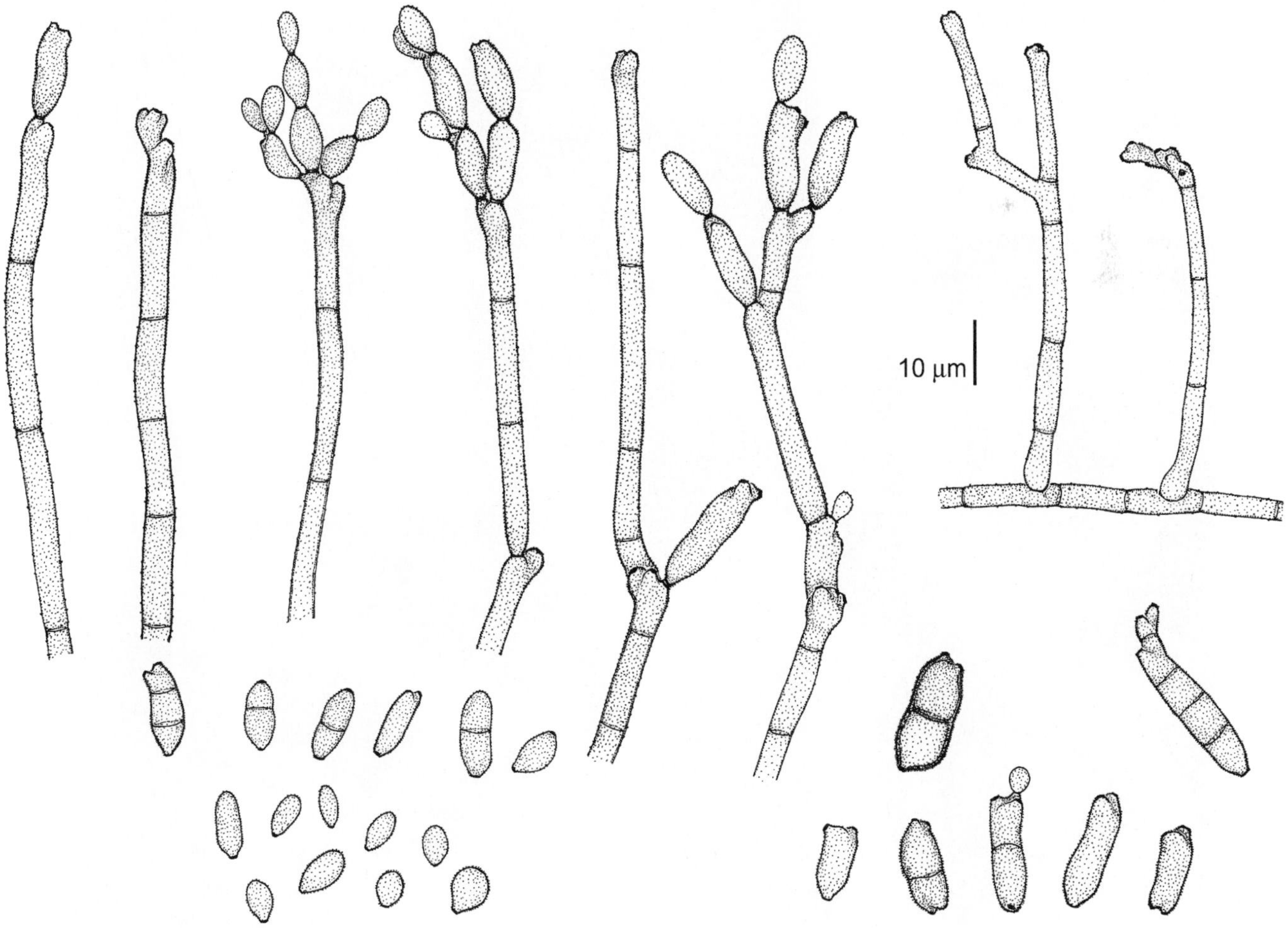

10 µm

Cladosporium herbarum, CBS 399.80. Conidiophores and conidia.

References. De Vries (1967), Ellis (1971), Domsch *et al.* (1980), Prašil & de Hoog (1988), David (1997).

Nomenclature. *Dematium herbarum* Persoon - Syn. Meth. Fung. p. 699, 1801 ≡ *Cladosporium herbarum* (Persoon) Link - Mag. Ges. Naturf. Freunde, Berlin 7: 37, 1815 ≡ *Cladosporium herbarum* (Persoon) Link: Fries - Syst. Mycol. 3: 370, 1829.

    Sphaerella tassiana de Notaris - Sferiac. Ital. 1: 87, 1863 ≡ *Mycosphaerella tassiana* (de Notaris) Johanson - Oefvers. Vet. Ak. Főrh. 41: 163, 1884.

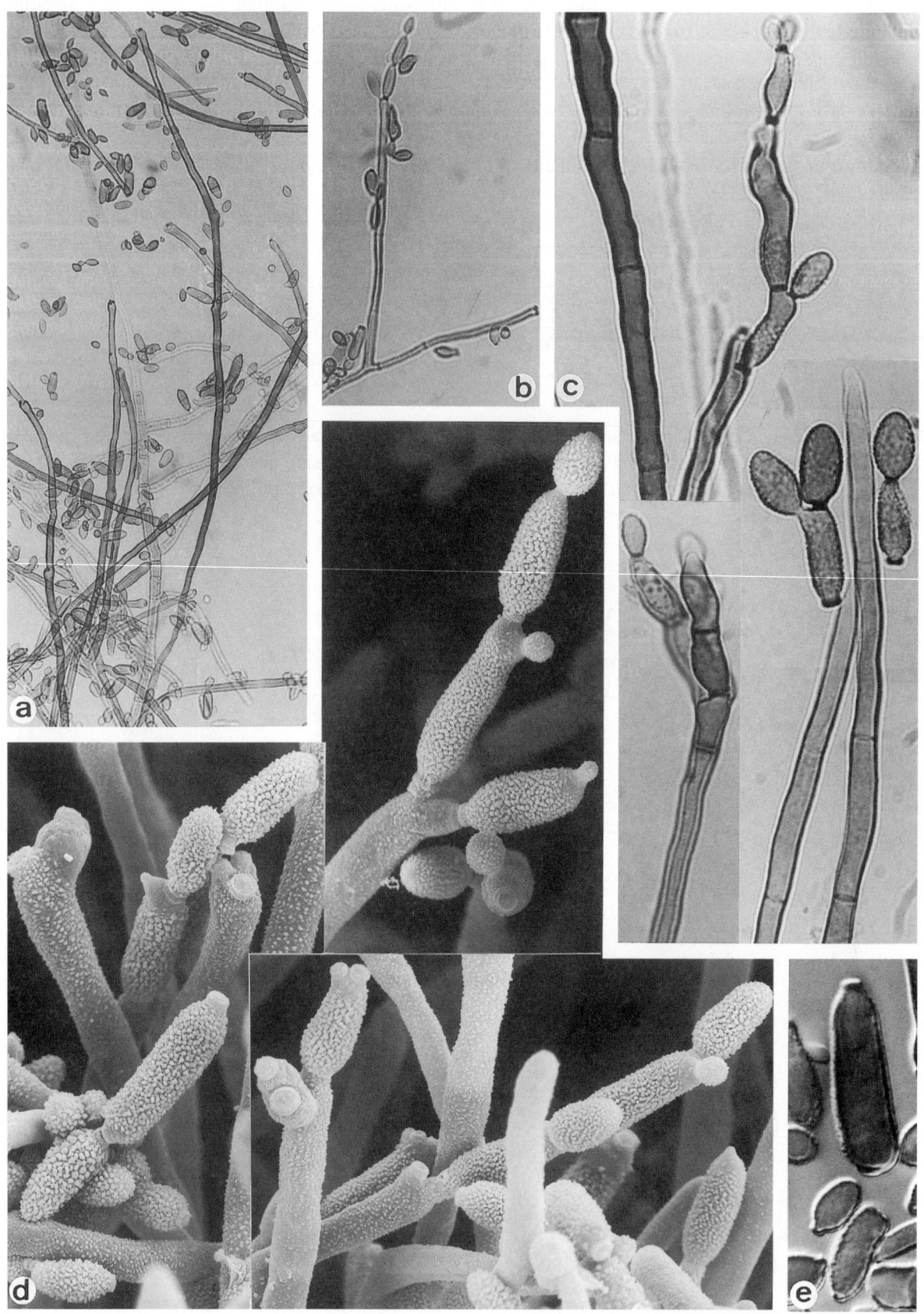

Cladosporium herbarum, CBS 399.80. Conidiophores and conidia. a. ×320; b. ×512; c. ×1280; d. ×2350; e. ×1600.

Cladosporium oxysporum Berk. & Curt.

Colony characteristics. Colonies (MEA 2%) moderately expanding, velvety and often floccose at the centre, olive to olive-green; reverse greenish-black.

Microscopy. Conidiophores straight or slightly flexuose, pale olive-green to light brown, up to 650 µm long, 4-5 µm wide, distinctly nodose with terminal and intercalary swellings. Ramoconidia cylindrical to clavate, 0-1-septate. Conidia spherical to subspherical, 3-5 µm diam, smooth-walled, pale olive-brown.

Molecular diagnostics. ITS restriction map based on NCBI L25432:

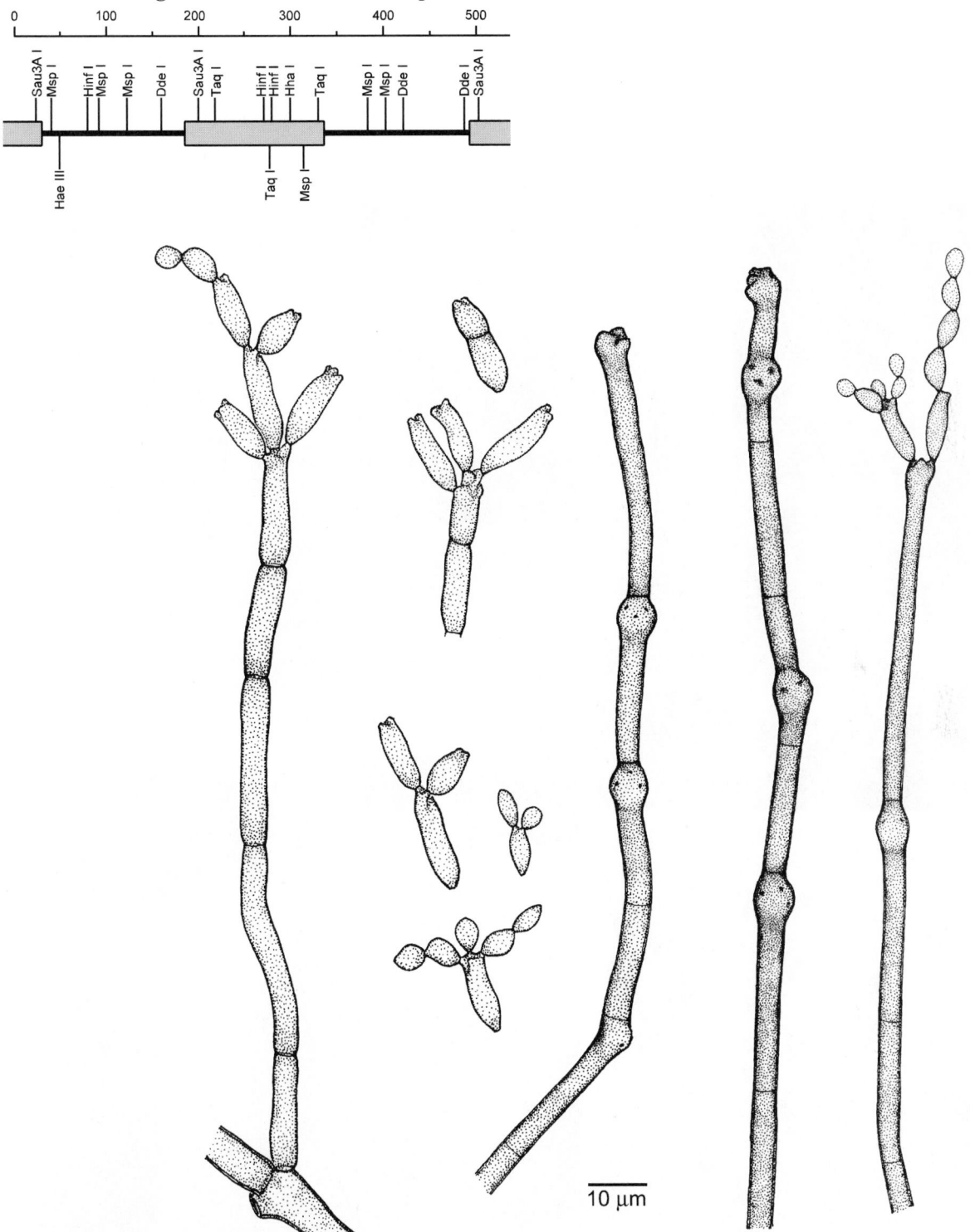

Cladosporium oxysporum, CBS 125.80. Conidiophores and conidia.

Pathogenicity. BSL-1. Saprobic species in warmer climates. Human cases: keratitis (Forster *et al.*, 1975b), cutaneous infection (Romano *et al.*, 1999).

References. De Vries (1967), Ellis (1971), de Bièvre (1981, 1982), McKemy & Morgan-Jones (1991).

Nomenclature. *Cladosporium oxysporum* Berkeley & Curtis - J. Linn. Soc. 10: 362, 1868.

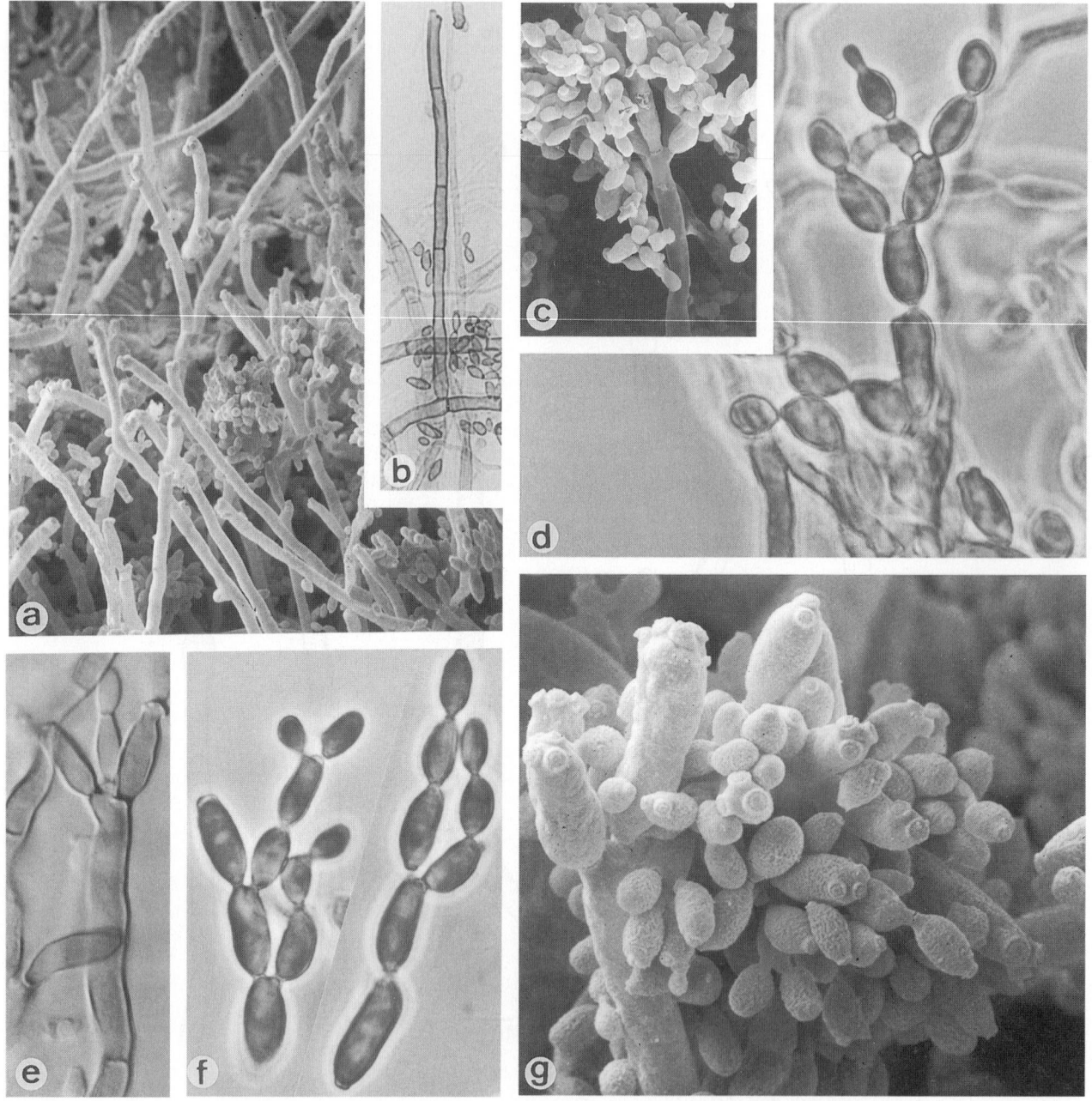

Cladosporium oxysporum, CBS 125.80. Conidiophores and conidia. a. ×620; b. ×512; c. ×1300; d. ×1280; e. ×1600; f. ×1280; g. ×3000.

Cladosporium sphaerospermum Penz.

Colony characteristics. Colonies (MEA 2%) moderately expanding, velvety to powdery, olive-green to olivaceous brown; reverse greenish-black.

Microscopy. Conidiophores of variable length, up to 300 µm long, 3-5 µm wide, pale to mid olivaceous brown, smooth-walled or verrucose, not nodose. Ramoconidia 0-3-septate, elongate, smooth-walled or verrucose. Conidia (sub)spherical, mid to dark olivaceous brown, verrucose, 3-5 µm diam.

Molecular diagnostics. ITS restriction map based on NCBI L25433:

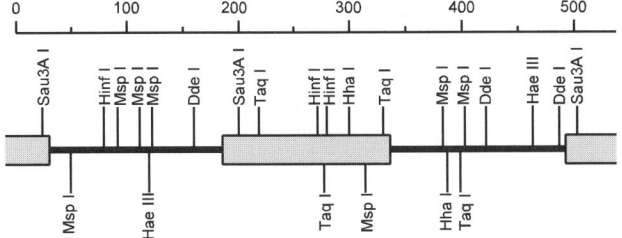

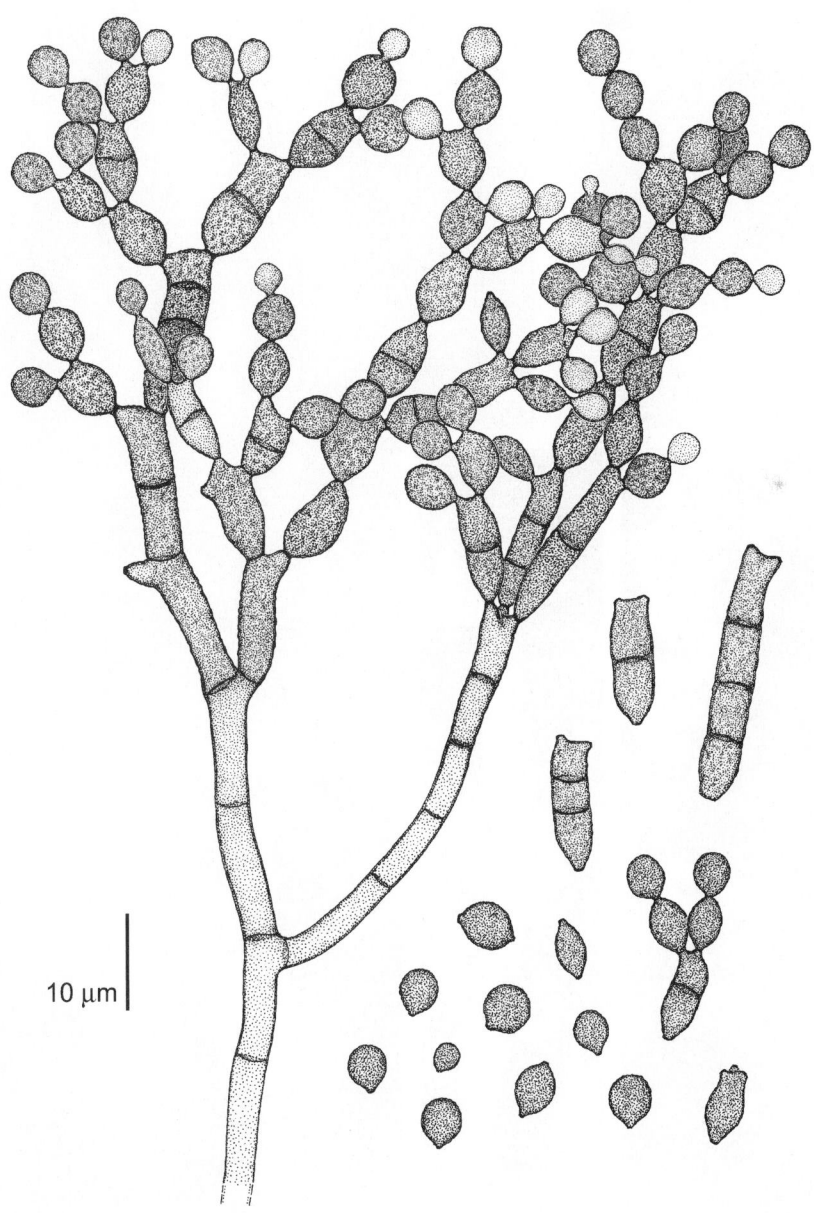

Cladosporium sphaerospermum, CBS 193.54. Conidiophores and conidia.

Pathogenicity. BSL-1. Common saprobe. This fungus was implicated in a case of human corneal ulcer (Shukla *et al.*, 1983) and has been isolated from skin lesions and onychomycoses (Badillet *et al.*, 1982).

References. De Vries (1967), Ellis (1971), Domsch *et al.* (1980), de Bièvre (1981, 1982).

Antifungal susceptibility.

Antifungal	Mean MICs	Strains	Reference
AMB	3.48	5	McGinnis & Pasarell (1998b)
5FC	5.66	2	Guarro *et al.* (1997b)
FLZ	128	2	Guarro *et al.* (1997b)
ITZ	0.76	5	McGinnis & Pasarell (1998b)
KTZ	2.83	2	Guarro *et al.* (1997b)
MCZ	2.83	2	Guarro *et al.* (1997b)
VCZ	0.87	5	McGinnis & Pasarell (1998b)

Nomenclature. *Cladosporium sphaerospermum* Penzig - Michelia 2: 473, 1882.

Hormodendron langeronii da Fonseca, de Area Leão & Peñido - Sciencia Med. 5: 563, 1927 ≡ *Cladosporium langeronii* (da Fonseca, de Area Leão & Peñedo) Vuillemin - Champ. Paras. Myc. Homme p. 78, 1931.

Cladosporium sphaerospermum, CBS 193.54. Conidiophores and conidia. a. ×512; b, c. ×1600; d. ×2000; e. ×3600; f. ×1800.

Hyphomycetes, dimorphic *Onygenales*. Genus: *COCCIDIOIDES*

Coccidioides immitis Rixford & Gilchrist

Colony characteristics. Colonies (SGA, 24°C) expanding, glabrous to felty, whitish to greyish, later becoming tan. Reverse cream-coloured, becoming brownish with age.

Microscopy. Fertile hyphae at 24°C usually arising at right angles, Arthroconidia hyaline, 1-celled, short cylindrical to barrel-shaped, smooth-walled, moderately thick-walled 3-8 × 3.5-4.5 µm, alternating with empty disjunctor cells, conidia after liberation with frill-like remains of the adjacent cells at both ends. At 37°C *in vitro* on BHI agar yeast-phase absent.

Physiology. Intolerant to benomyl.

Differential diagnosis. *In vivo:* after endogenous reactivation spherules are produced with endoconidia which are disseminated haematogenously. *In vitro:* enteroarthric conidia at 24°C and absence of budding at 37°C.

Serology. Immunodiffusion tests are commercially available.

Molecular diagnostics. A DNA probe for recognition of the species is commercially available (Beard *et al.* 1993; Stockman *et al.*, 1993; Padhye *et al.*, 1994d). A 26S rDNA-based specific probe was developed by Sandhu *et al.* (1995). A comparison of serological methods was presented by Martins *et al.* (1995). SSU and ITS restriction maps based on NCBI X58571 and NCBI U18360, resp.:

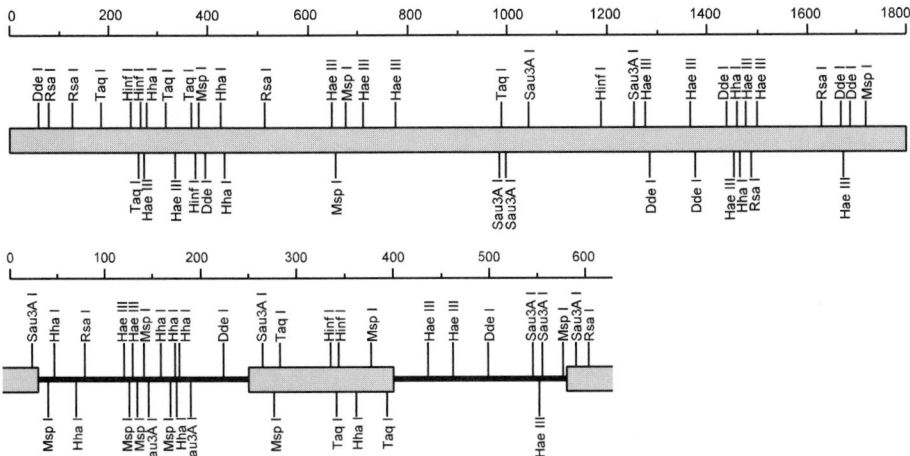

Pathogenicity. BSL-3. Agent of coccidioidomycosis (p. 25). The pathogenic form of growth consists of spherules, 30-60 µm diam, later liberating multinucleate endospores 2-5 µm diam, which arise through cleavage. There is a conspicuous association of infection and activities related to tillage of the soil, by, e.g., agricultural workers, telephone post diggers, archeologists or playing children. Dry arthroconidia are carried by dust storms and are highly infectious when inhaled. Initially, the infection is mild and is cured spontaneously in 95% of the immunocompetent hosts (Kirkland & Fierer, 1996). Infection rates differ with populations, East-Asians and Africans being more susceptible (Olivere *et al.*, 1999). Hematogenous dissemination to the skin, bones, joints, subcutaneous tissue and visceral organs may occur (Low *et al.*, 1996). The disseminated form may develop particularly during pregnancy (Powell *et al.*, 1983) and is frequently fatal. Polesky *et al.* (1999) described endotracheal lesions. Zepeda *et al.* (1998) reported a case with hyphae in the CSF. The species is emerging in patients with impaired acquired immunity (VandenBossche *et al.*, 1990; Jones *et al.*, 1995; Singh *et al.*, 1996), less frequently in other immune disorders such as bone marrow transplant (Riley *et al.*, 1993). Cerebral cases are rare (Mendel *et al.*, 1994; Mischel & Vinters, 1995; Banuelos *et al.*, 1996). Johnson *et al.* (1998) reported a case in a mandril and Adaska (1999) in a wild lion. The species is occasionally seen in warm-blooded water animals (Reidarson *et al.*, 1998).

Distribution. The species occurs in rodent burrows in desert-like areas of the Southwestern part of the USA (Ampel *et al.*, 1998; Standaert *et al.*, 1995), Argentina and Northeast Brazil, which are characterized by low rainfall, high summer temperatures, low altitude and alkaline soil (Maddy, 1957). Cases outside endemic areas are mostly imported (Jang *et al.*, 1999).

References. Emmons & Ashburn (1942), Ajello (1965), Sigler & Carmichael (1976), Cole & Sun (1985), Einstein & Catanzaro (1985), Sun *et al.* (1986), Pappagianis (1988), Chen *et al.* (1991), Batra (1992), Einstein & Johnson (1993), Raab *et al.* (1993), Pan *et al.* (1994), Wong *et al.* (1994), Lacaz *et al.* (1998).
Antifungal susceptibility.

Antifungal	MICs range	MIC90	Strains	Reference
AMB	0.5-2	1	100	Li *et al.* (2000)
ITZ	0.125-1	1	100	Li *et al.* (2000)
VCZ	≤0.03-0.5	0.25	100	Li *et al.* (2000)

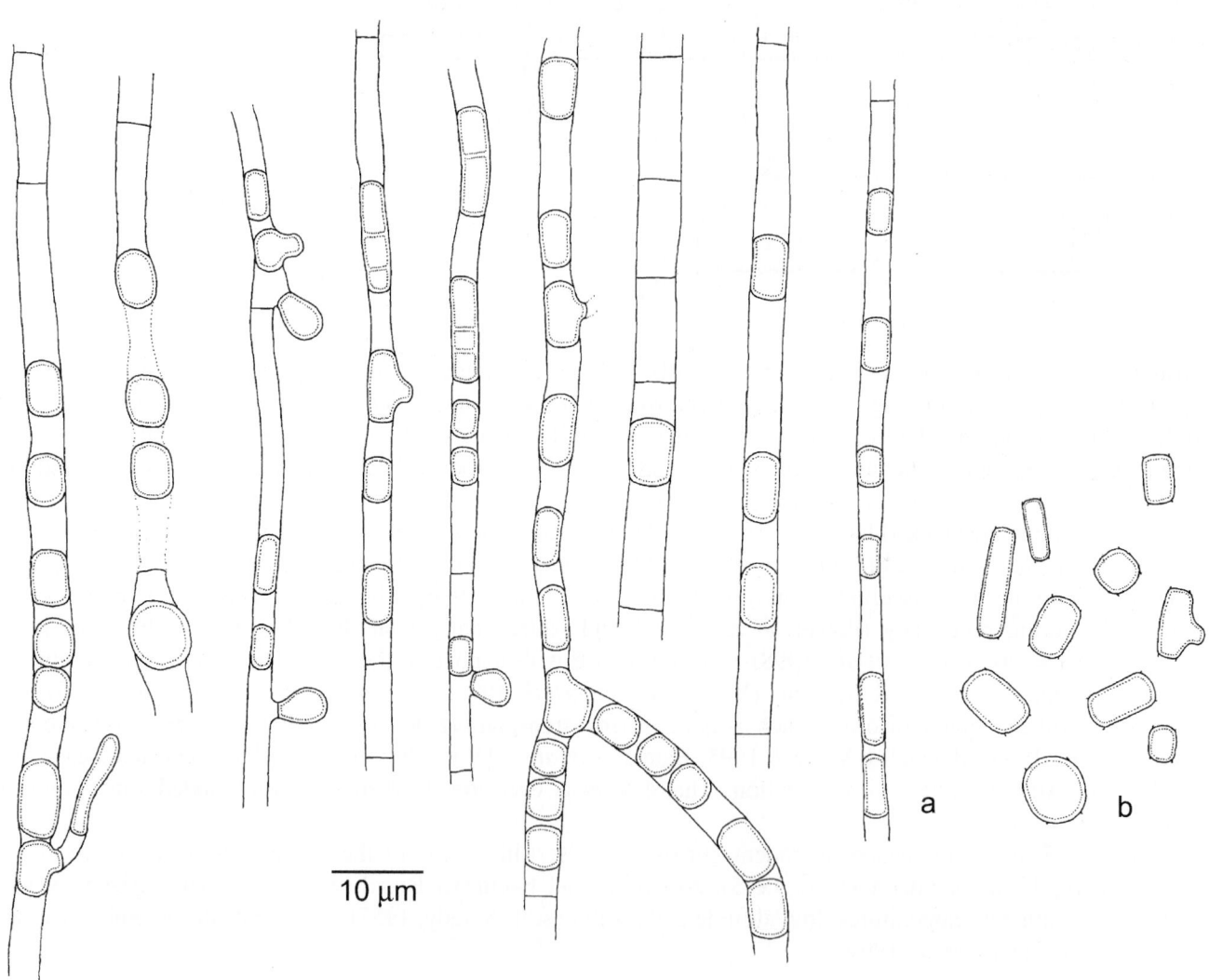

Coccidioides immitis, CBS 711.73. a. Hyphae with arthroconidium formation; b. liberated arthroconidia.

Nomenclature. *Coccidioides immitis* Rixford & Gilchrist - Johns Hopkins Hosp. Rep. 1: 209-268, 1896 = *Mycoderma immite* Rixford & Gilchrist) Verdun & Mandoul, *in* Verdun - Précis Parasitol., ed. 4, p. 769, 1924 = *Blastomycoides immitis* (Rixford & Gilchrist) Castellani - Am. J. Trop. Med. 8: 385, 1928 = *Zymonema immitis* (Rixford & Gilchrist) Froilano de Mello & Fernandes - Arq. Hig. Pat. Exot. 6, 1928 = *Geotrichum immitis* (Rixford & Gilchrist) Agostini - J. Trop. Med. Hyg. 35: 266, 1932 = *Aleurisma immite (*Rixford & Gilchrist) Bogliolo & Neves - Mycopath. Mycol. Appl. 6: 147, 1952.

Coccidioides pyogenes Rixford & Gilchrist - Johns Hopkins Hosp. Rep. 1: 209-268, 1896.

Coccidium neoplasicum Cantón - Trat. Zooparás. Cuerpo Um. p. 108-122, 1897.

Posadia esferiformis Cantón, *in* Posadas - Psorosperm. Infect. Gen., B. Aires, 1898 = *Coccidioides esferiformis* (Cantón) M. Moore - Annls Mo. Bot. Gdn 19: 421, 1932.

Pseudococcidioides mazzae da Fonseca, *in* Mazza & Parodi - Bol. Inst. Clín. Quirúr., B. Aires 4: 495, 1928.

Geotrichum louisianoideum Castellani - Med. Press. Circ. 136: 439, 1933; Castellani & Jacono - J. Trop. Med. Hyg. 36: 304, 1933.

Glenospora metaeuropea Castellani - J. Trop. Med. Hyg. 36: 309, 1933 = *Aleurisma metaeuropaea* (Castellani) C.W. Dodge - Med. Mycol. p. 790, 1935 = *Glenosporopsis metaeuropea* (Castellani) de Fonseca - Parasit. Méd. 1: 730, 1943.

Trichosporon proteolyticum Negroni & de Villafaña Lastra - Mycopath. Mycol. Appl. 2: 57, 1939 = *Geotrichum proteolyticum* (Negroni & de Villafaña Lastra) Coudert - Guide Prat. Mycol. Méd. p. 232, 1955.

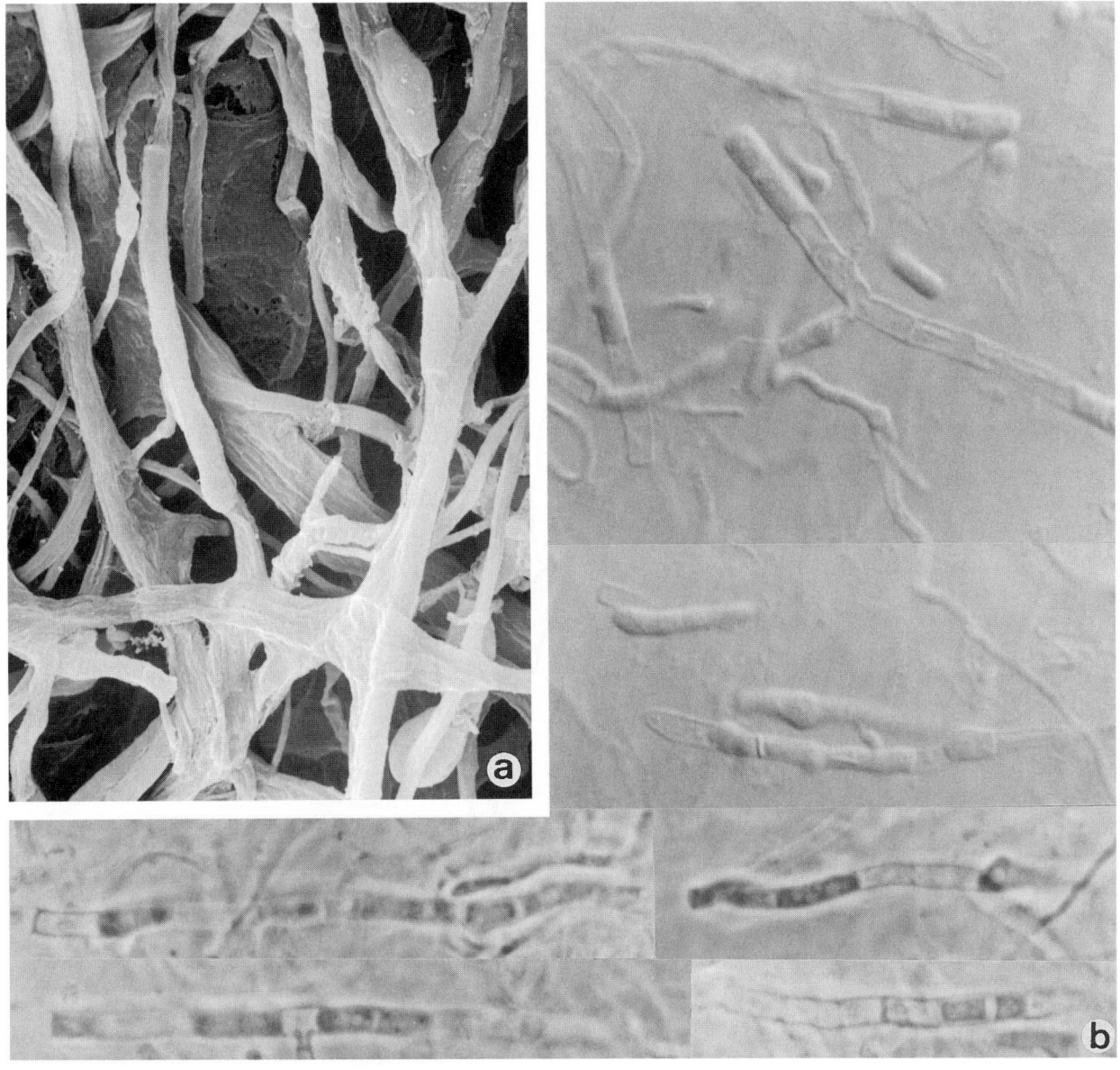

Coccidioides immitis, CBS 146.56. a. Spherules with endospores; b, c. hyphae with arthroconidium formation. a. ×500; b. ×3500; c. × 1600. a. Reprinted with permission from Lacaz *et al.* (1998).

Hyphomycetes, Dematiaceae. Genus: *CORYNESPORA*

Corynespora cassiicola (Berk. & Curt.) Wei

Colony characteristics. Colonies (PCA) with rapid growth, hairy or velvety, olive-green to greyish-black. **Microscopy.** Conidiophores erect, straight or flexuose, unbranched, septate, often nodose, pale brown, with several cylindrical proliferations, 100-800 × 4-11 µm. Conidiogenous cells integrated, elongating through the scars. Conidia solitary, occasionally catenate, obclavate or cylindrical, pale or brown when mature, with a thickened base, smooth-walled, with 4-20 distosepta, 40-500 × 9-22 µm.
Molecular diagnostics. ITS restriction map based on NCBI U95173:

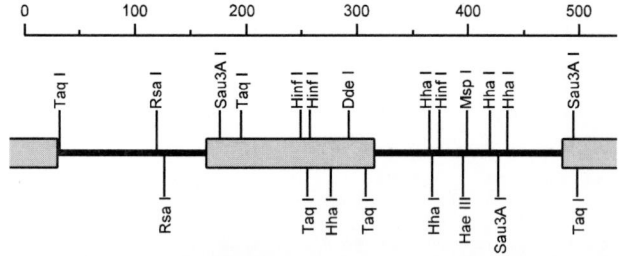

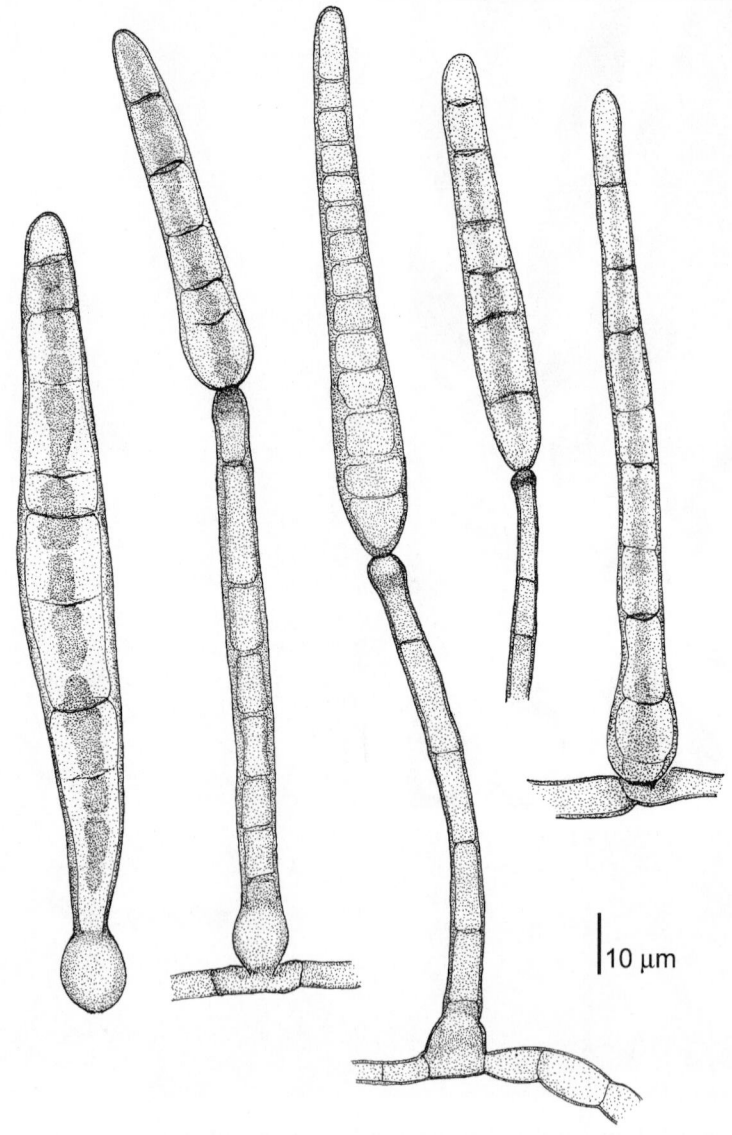

Corynespora cassiicola, UAMH 3480. Conidiophores and conidia.

Pathogenicity. BSL-1. Plant-pathogenic species. A human case of mycetoma was reported (Mahgoub, 1969).

Reference. Ellis (1971).

Nomenclature. *Helminthosporium cassiicola* Berkeley & Curtis, *in* Berkeley - J. Linn. Soc. Bot. 10: 361, 1869 ≡ *Corynespora cassiicola* (Berkeley & Curtis) Wei - Mycol. Pap 34: 5, 1950.

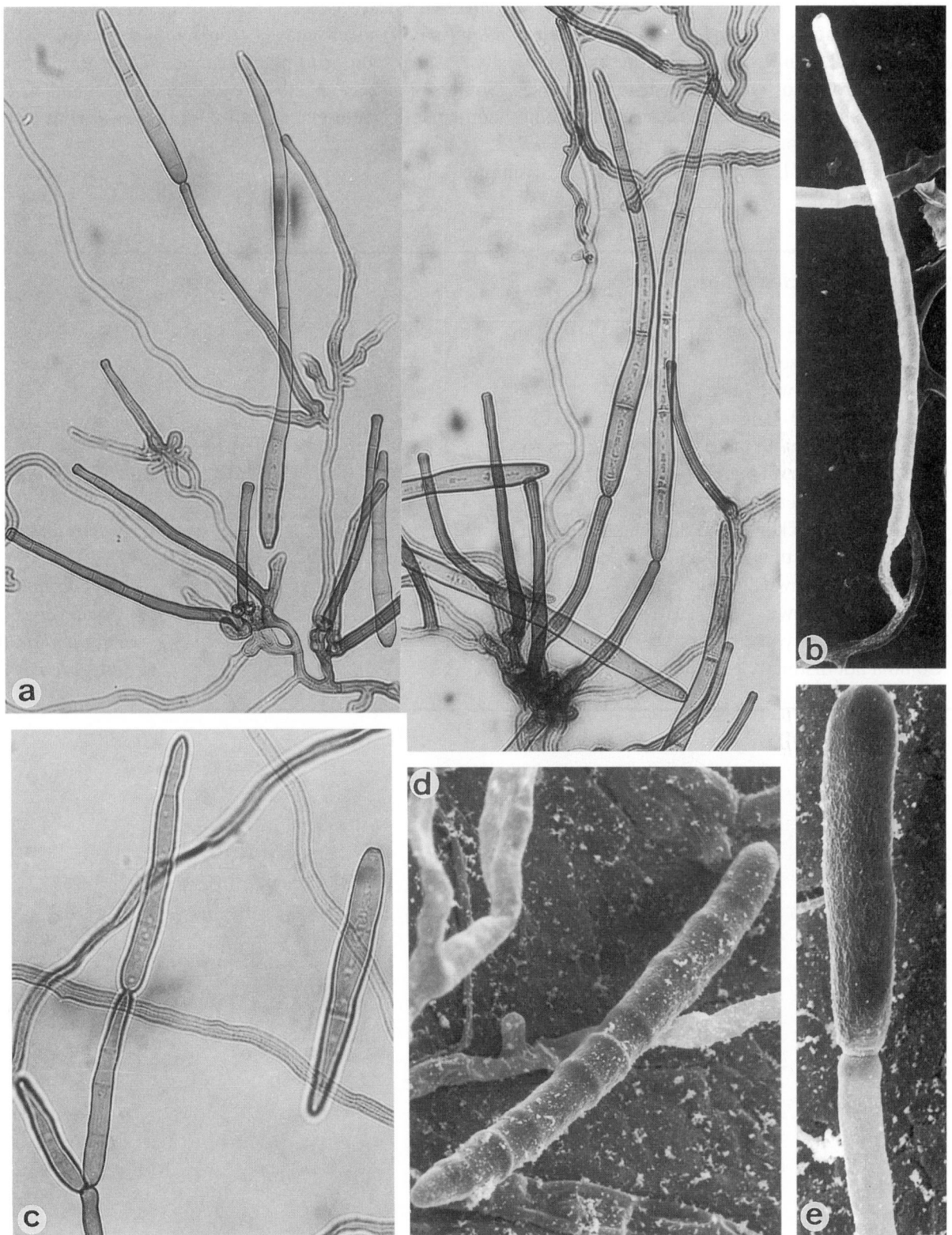

Corynespora cassiicola, UAMH 3480. Conidiophores and conidia. a. ×256; b. ×680; c. ×512; d. ×2800; e. ×3100.

Hyphomycetes, Dematiaceae. Genus: *CURVULARIA*

Generic description. Colonies black, hairy, expanding. Conidiophores erect, brown, multicellular, producing conidia in sympodial order; conidial scars dark, flat. Conidia ellipsoidal, often curved, with 3-4 true septa. Tolerant to benomyl.

Teleomorphs. *Cochliobolus, Pseudocochliobolus* (*Ascomycota, Euascomycetes, Pleosporales: Pleosporaceae*).

Differential diagnosis. *Bipolaris* (p. 526) and *Exserohilum* (p. 669) species have conidia with distosepta.

General remarks. Numerous species are known, mostly occurring on dead plant material. They are particularly common as saprobes or weak pathogens on grasses. Some ubiquitous species are occasionally found in cattle, rarely in humans, in both cases causing chronic, non-specific, allergic sinusitis, sometimes with cerebral involvement. In addition, traumatic infections are noted.

References. Ellis (1971, 1976), de Hoog (1983), Sivanesan (1987).

Key to the treated species of Curvularia:

1a.	Conidia verrucose, 20-40 × 12-17 µm	*C. verruculosa* (611)
1b.	Conidia smooth-walled → **2**	
2a.	Conidia mostly 3-septate → **3**	
2b.	Conidia usually with more than 3 septa → **6**	
3a.	Conidia mostly straight, clavate	*C. clavata* (601)
3b.	Conidia usually curved → **4**	
4a.	Conidia with thickened and dark median septum	*C. brachyspora* (599)
4b.	Conidia without thickened and dark median septum → **5**	
5a.	Conidia dark brown	*C. lunata* (605)
5b.	Conidia pale brown	*C. pallescens* (607)
6a.	Conidia distinctly geniculate, 18-37 × 8-14 µm	*C. geniculata* (603)
6b.	Conidia not distinctly geniculate, 19-30 × 10-14 µm	*C. senegalensis* (609)

Curvularia brachyspora Boedijn

Colony characteristics. Colonies (PCA) growing rapidly, velvety to floccose, greyish-brown to blackish-brown. Stromata black, cylindrical, in small groups.

Microscopy. Conidiophores arising singly or in groups, terminally and laterally on hyphae and on stromata, straight or flexuose, paler towards the apex, up to 700 μm long. Conidia usually slightly curved, ellipsoidal or broadly fusiform, 20-26 × 10-14 μm, asymmetrical, smooth-walled, dark brown, 3-septate, with the central septum truly meridional and darker than the other two.

Molecular diagnostics. ITS restriction map based on CBS 553.89:

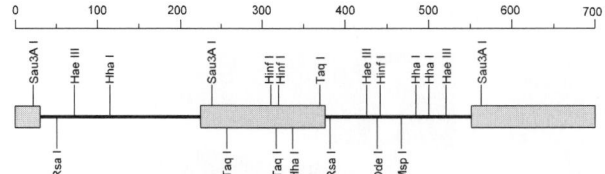

Pathogenicity. BSL-1. Saprobe on plant material. Human cases: keratitis (Marcus *et al.*, 1992) and cutaneous infection (Torda & Jones, 1997).

References. Ellis (1966), Sivanesan (1987).

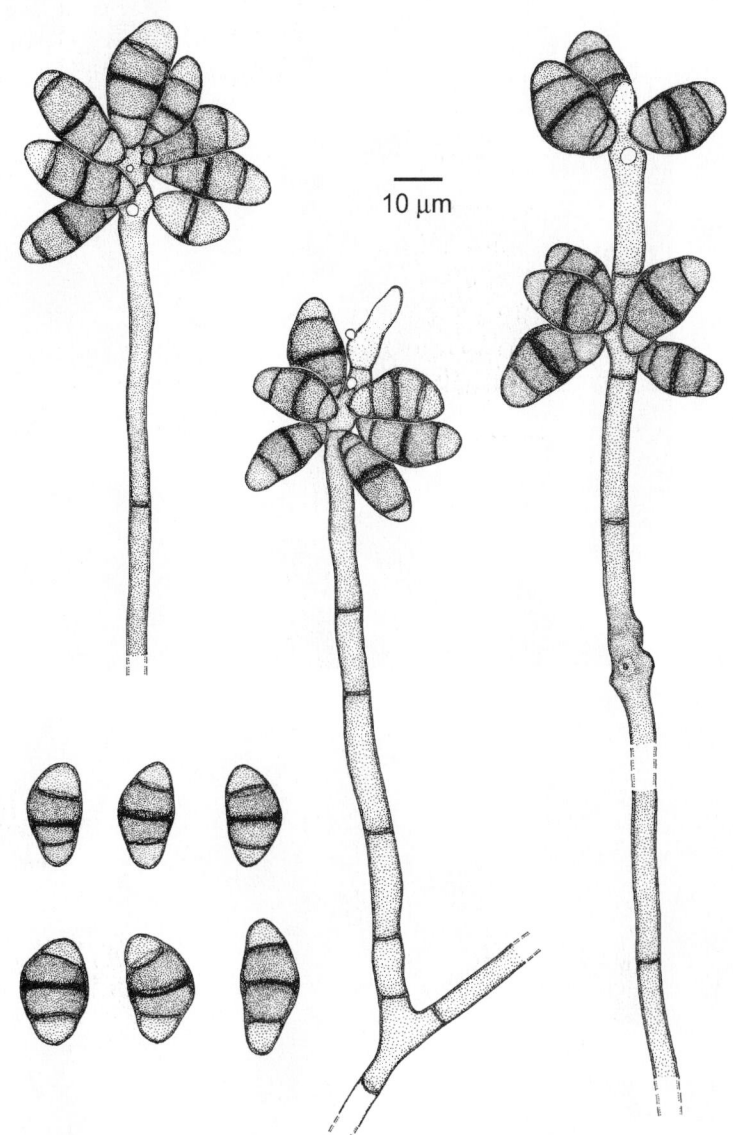

Curvularia brachyspora, CBS 553.89. Conidiophores and conidia.

Antifungal susceptibility.

Antifungal	Mean MICs	MICs range	Strains	Reference
AMB	0.29	0.06-2	4	Guarro *et al.* (1999b)
5FC	181	128-256	4	Guarro *et al.* (1999b)
FLZ	22.6	4-128	4	Guarro *et al.* (1999b)
ITZ	0.7	0.12-32	4	Guarro *et al.* (1999b)
KTZ	0.84	0.5-4	4	Guarro *et al.* (1999b)
MCZ	0.84	0.25-4	4	Guarro *et al.* (1999b)

Nomenclature. *Curvularia brachyspora* Boedijn - Bull. Jard. Bot. Buitenz., Ser. 3, 13: 126, 1933.

Curvularia brachyspora, CBS 553.89. Conidiophores and conidia. a. ×256; b. ×950; c. ×1200; d. ×2200; e. ×1280.

Curvularia clavata Jain

Colony characteristics. Colonies (PCA) spreading, greyish-brown or brown, cottony.

Microscopy. Conidiophores straight or flexuose, sometimes geniculate, brown, smooth-walled, up to 150 µm long 2.0-5.5 µm wide. Conidia brown, smooth-walled, straight or occasionally slightly curved, usually clavate, 3-septate, blackish-brown, basal cell paler, 17-29 × 7-13 µm.

Molecular diagnostics. ITS restriction map based on CBS 901.87:

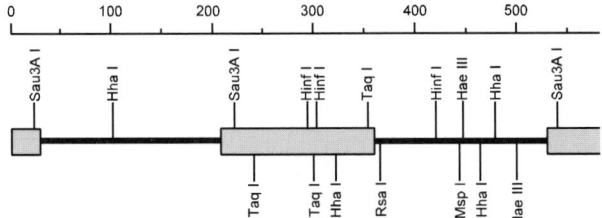

Pathogenicity. BSL-1. Skin infection (Gugnani *et al.*, 1990). An invasive sinusitis with cerebritis was described by Ebright *et al.* (1999).

References. Ellis (1966), Sivanesan (1987).

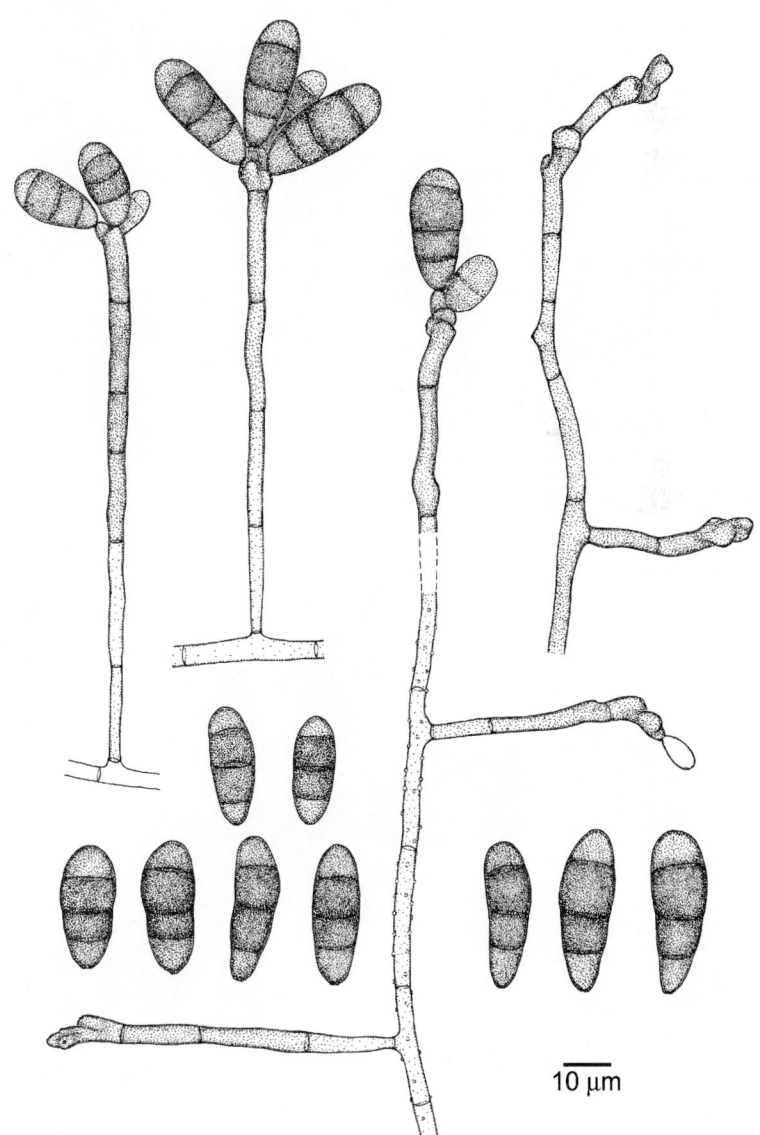

Curvularia clavata, CBS 539.70. Conidiophores and conidia.

Antifungal susceptibility.

Antifungal	Mean MICs	MICs range	Strains	Reference
AMB	0.31	0.25-0.5	3	Guarro *et al.* (1999b)
5FC	256	256	3	Guarro *et al.* (1999b)
FLZ	16	16	3	Guarro *et al.* (1999b)
ITZ	0.2	0.12-0.5	3	Guarro *et al.* (1999b)
KTZ	0.62	0.5-1	3	Guarro *et al.* (1999b)
MCZ	1	1-2	3	Guarro *et al.* (1999b)

Nomenclature. *Curvularia clavata* Jain - Trans. Br. Mycol. Soc. 45: 542, 1962.

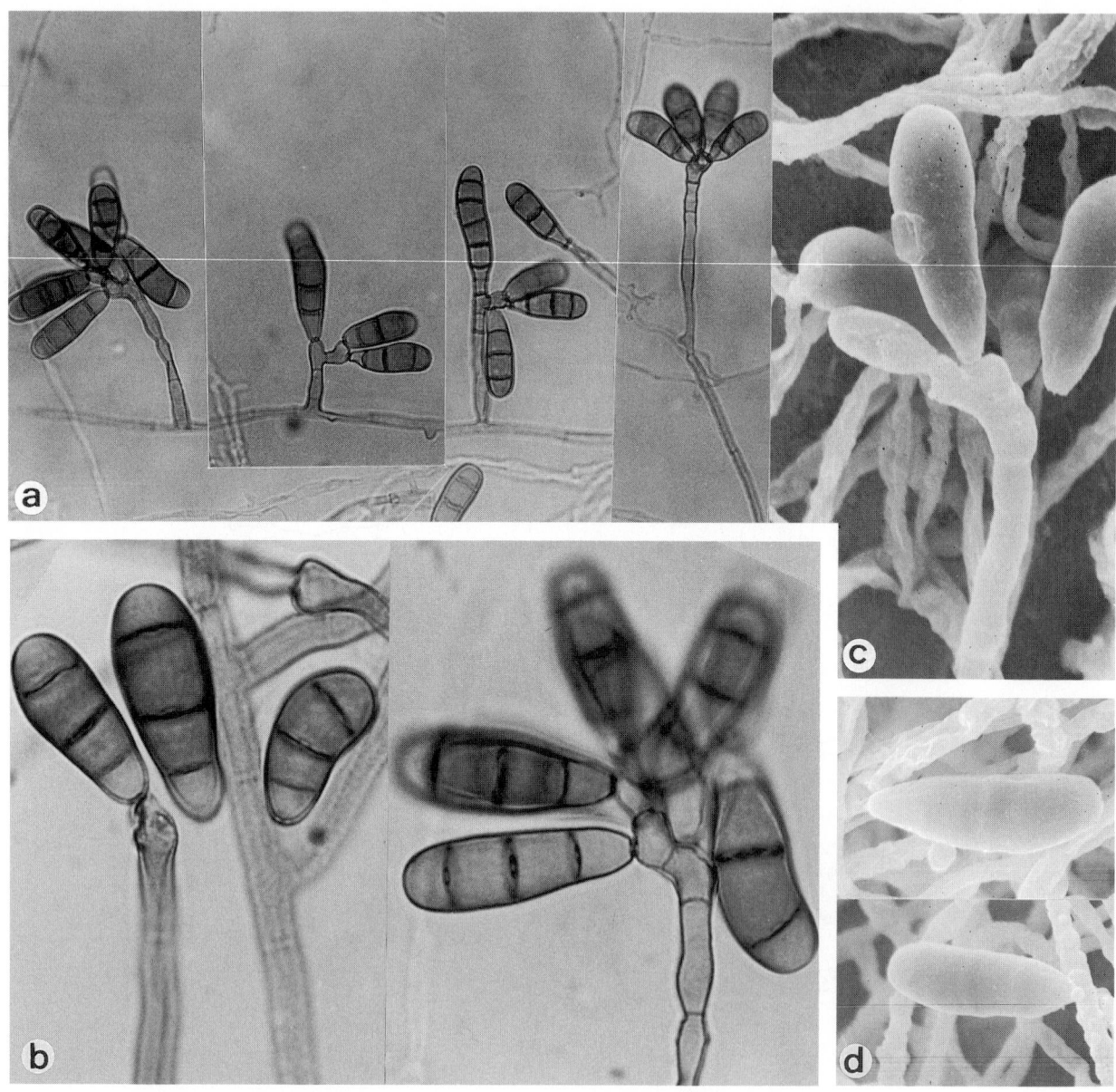

Curvularia clavata, CBS 539.70. Conidiophores and conidia. a. ×512; b. ×1280; c. ×1850; d. ×1900.

Curvularia geniculata (Tracy & Earle) Boedijn

Colony characteristics. Colonies (PCA) expanding, black, hairy.

Microscopy. Conidiophores erect, up to 600 µm long, unbranched, septate, flexuose in the apical part, with flat, dark brown scars. Conidia smooth-walled, dark brown, 4-septate, the central cell being the largest; conidia broadly ellipsoidal, unilaterally flattened to distinctly geniculate, 18-37 × 8-14 µm.

Teleomorph. *Cochliobolus geniculatus* Nelson (*Ascomycota, Euascomycetes, Pleosporales: Pleosporaceae*).

Molecular diagnostics. ITS restriction map based on CBS 534.70:

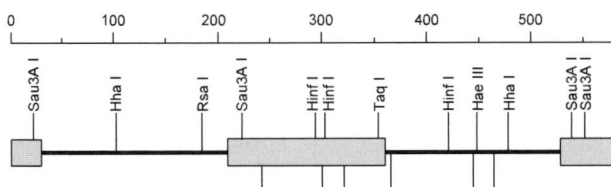

Differential diagnosis. The equally common *Curvularia lunata* (p. 605) has 3-septate conidia. Several reports in the literature ascribed to *C. geniculata* probably concerned *C. lunata* (de Hoog, 1983).

Pathogenicity. BSL-1. Occasionally found in humans, either after traumatic implantation in the eye (Nityananda *et al.*, 1964; Georg, 1964; Thomas *et al.*, 1988) or after colonizing sinus tracts, causing allergic sinusitis (Bartynski *et al.*, 1990). Invasive growth into the brain may occur (Killingsworth & Wetmore, 1990). Kaufman (1971) described a case of endocarditis after cardiac surgery. The fungus is relatively common in animals; the sinus

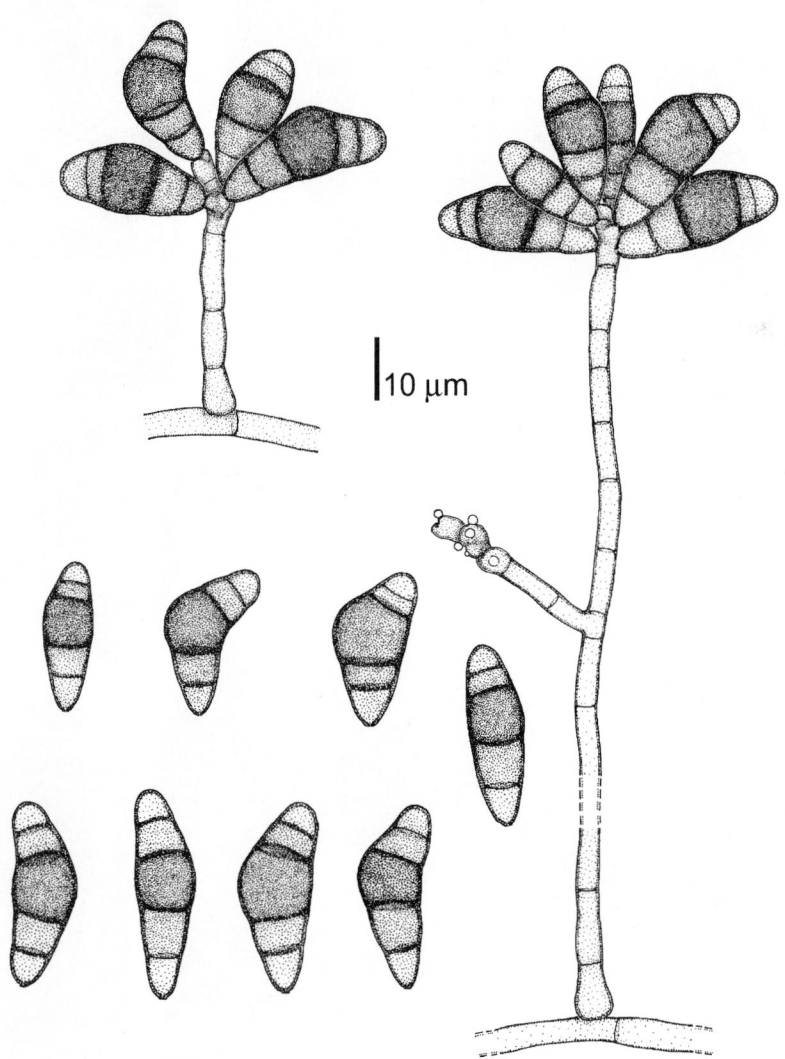

Curvularia geniculata, AAMRL 79.3. Conidiophores and conidia.

603

infection of cattle known in older literature as 'helminthosporiosis' may be caused by this and by related *Curvularia* and *Bipolaris* species. Further animal cases involve subcutaneous tumefactions in dogs (Bridges, 1957; Baylett *et al.*, 1959; Brodey *et al.*, 1967) and horses (Boomker *et al.*, 1977), osteomyelitis in a dog (Coyle *et al.*, 1984) and a central nervous infection in a parrot (Clark *et al.*, 1986).

References. Ellis (1966), de Hoog (1983), Sivanesan (1987).

Antifungal susceptibility.

Antifungal	Mean MICs	MICs range	Strains	Reference
AMB	0.12	0.06-0.25	4	Guarro *et al.* (1999b)
5FC	181	128-256	4	Guarro *et al.* (1999b)
FLZ	16	4-32	4	Guarro *et al.* (1999b)
ITZ	0.21	0.12-0.5	4	Guarro *et al.* (1999b)
KTZ	1	0.5-2	4	Guarro *et al.* (1999b)
MCZ	1.41	0.5-4	4	Guarro *et al.* (1999b)

Nomenclature. *Helminthosporium geniculatum* Tracy & Earle - Bull. Torrey Bot. Club 23: 207, 1896 ≡ *Curvularia geniculata* (Tracy & Earle) Boedijn - Bull. Jard. Bot. Buitenz., Ser. 3, 13: 129, 1933.

Cochliobolus geniculatus Nelson - Mycologia 56: 778, 1964.

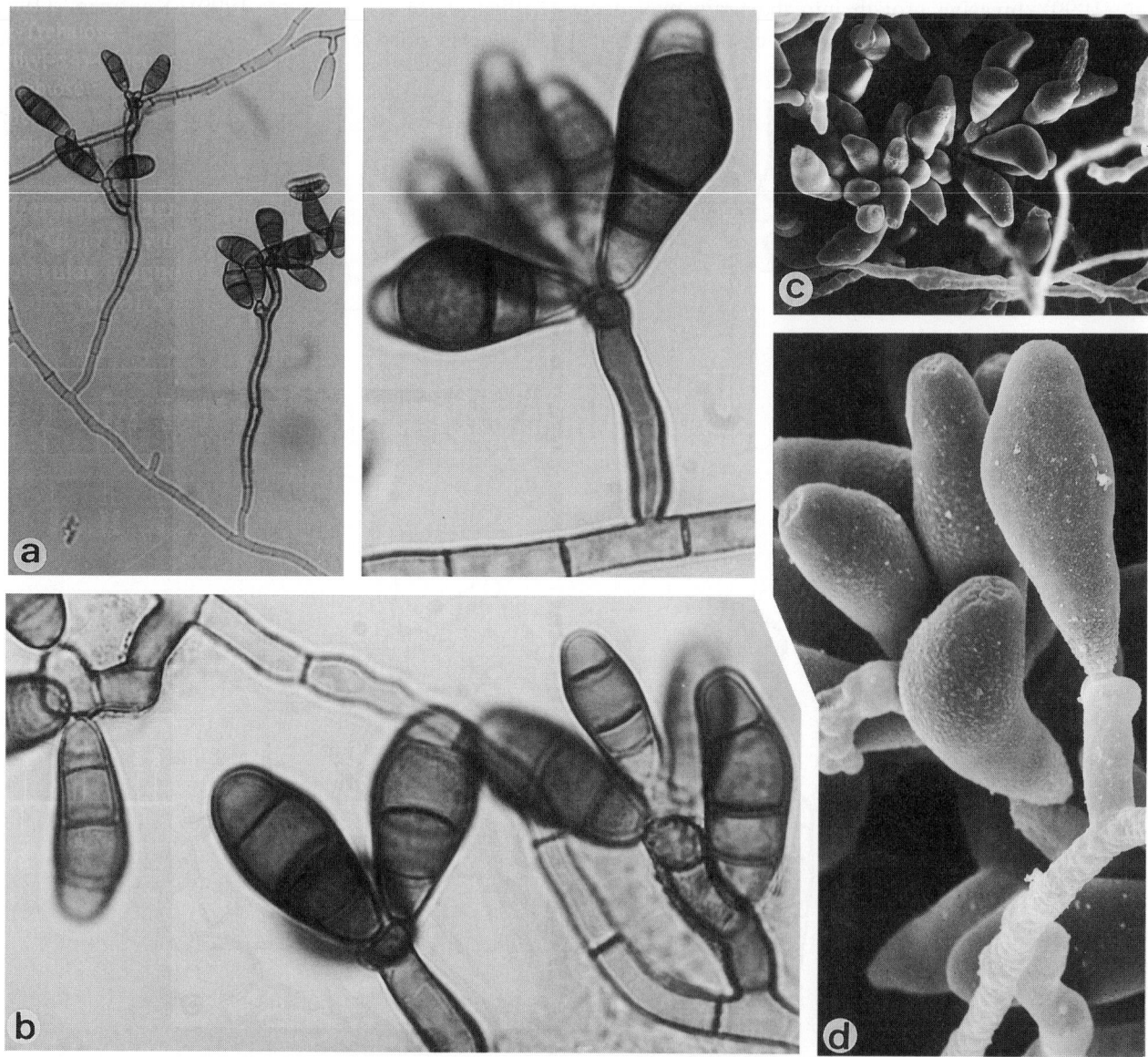

Curvularia geniculata, AAMRL 79.3. Conidiophores and conidia. a. ×320; b. ×1280; c. ×580; d. ×1600; e. ×2000.

Curvularia lunata (Wakker) Boedijn

Colony characteristics. Colonies (PCA) expanding, black, hairy.

Microscopy. Conidiophores erect, unbranched, septate, flexuose in the apical part, with flat, dark brown scars. Conidia smooth-walled, olivaceous brown, end cells somewhat paler; conidia obovoidal to broadly clavate, curved at the subterminal cell, 21-31 × 8.5-12.0 μm, 3-septate, the subterminal cell swollen and distinctly larger than the remaining cells.

Teleomorph. *Cochliobolus lunatus* Nelson & Haasis (*Ascomycota, Euascomycetes, Pleosporales: Pleosporaceae*).

Molecular diagnostics. ITS restriction map, based on UAMH 1349:

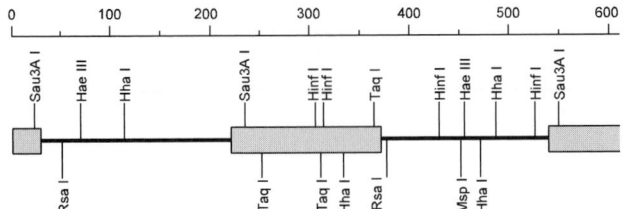

Pathogenicity. BSL-1. The species is an ubiquitous saprophyte on plant material. Human cases mainly concern sinusitis (Rinaldi *et al.*, 1987), occasionally with displacement of adjacent bone (Berry *et al.*, 1984) or intracranial extension (Ismael *et al.*, 1993). Pulmonary infiltrates are rarely reported (Halwig *et al.*, 1985). Monte & Hutchins (1985) described a disseminated case with cerebral involvement. Systemic infections other than in the brain are extremely rare (Bryan *et al.*, 1993). Cases of keratitis are relatively frequent (Nityananda *et al.*, 1962; Agrawal *et al.*, 1982; Luque *et al.*, 1985; Berger *et al.*, 1991). (Sub)cutaneous infections are noted, such as phaeohyphomycosis (Grieshop *et al.*, 1993), onychomycosis (Barde & Singh, 1983), cutaneous infections (Lopes & Jobim, 1998; Fernández *et al.*, 1999) or mycetoma (Mahgoub, 1973; Khan *et al.*, 1984; Janaki *et al.*, 1999). Robson & Craver (1994) reported a urinary tract infection. The species showed low virulence in a CAPD patient (DeVault *et al.*, 1985), though Lopes *et al.* (1994) and Guarner *et al.* (1989) reported peritonitis. Yau *et al.* (1994) reported infection after major surgery. A mycetoma in a dog was described by Elad *et al.* (1991).

Variety. *C. lunata* var. *aeria* was described for strains forming stromata in culture. It differs in *Hha*I and *Taq*I restriction sites in ITS rDNA (de Hoog *et al.*, 1997b).

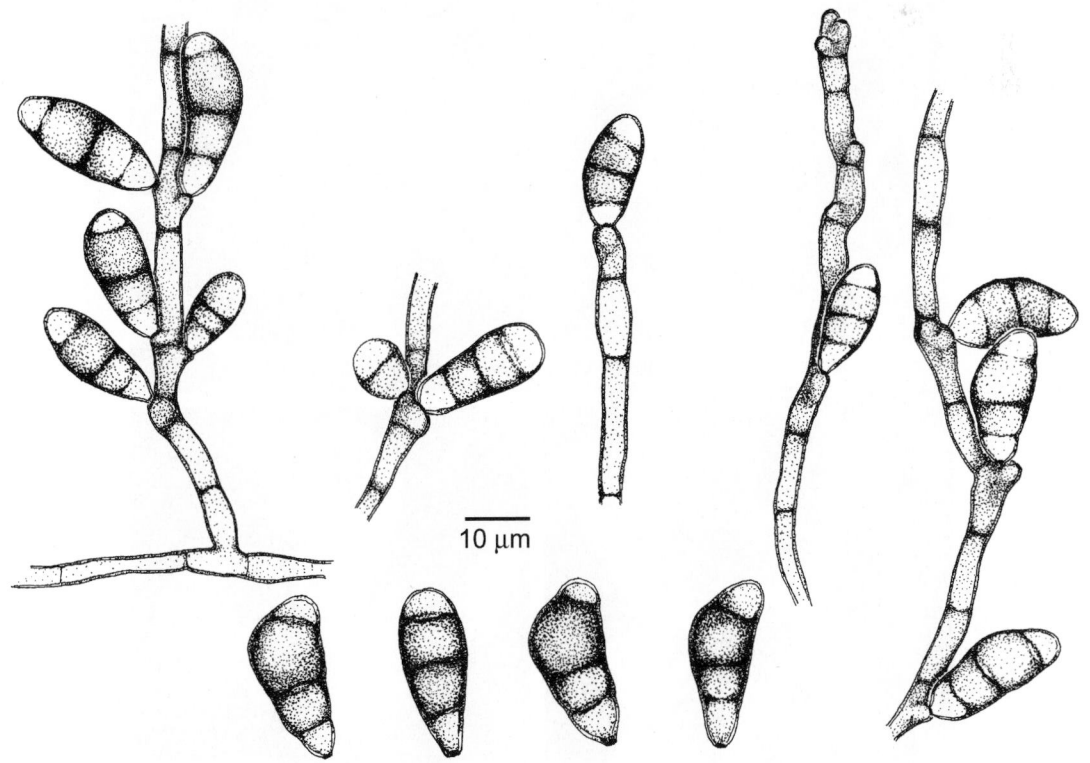

Curvularia lunata, UAMH 4388. Conidiophores and conidia.

References. Ellis (1966), de Hoog (1983), Rinaldi *et al.* (1987), Sivanesan (1987).
Antifungal susceptibility.

Antifungal	Mean MICs	MICs range	Strains	Reference
AMB	0.46	nd	17	McGinnis & Pasarell (1998b)
AMB	0.12	0.06-0.5	3	Guarro *et al.* (199911b)
5FC	181	128-256	4	Guarro *et al.* (1999b)
FLZ	26.9	16-128	4	Guarro *et al.* (1999b)
ITZ	0.19	nd	17	McGinnis & Pasarell (1998b)
ITZ	0.41	0.06-8	3	Guarro *et al.* (1999b)
KTZ	1	0.25-4	3	Guarro *et al.* (1999b)
MCZ	1	1-4	3	Guarro *et al.* (1999b)
VCZ	0.2	nd	17	McGinnis & Pasarell (1998b)

Nomenclature var. *lunata.* *Acrothecium lunatum* Wakker, *in* Wakker & Went - Ziekten Suikerr., Java p. 196, 1898 ≡ *Curvularia lunata* (Wakker) Boedijn - Bull. Jard. Bot. Buitenz., Ser. 3, 13: 127, 1933.
  Cochliobolus lunatus Nelson & Haasis - Mycologia 56: 316, 1964.
Nomenclature var. *aeria.* *Malustela aeria* Batista, Lima & Vasconcelos - Publçoes Inst. Micol., Recife 263: 7, 1960 ≡ *Curvularia lunata* (Wakker) Boedijn var. *aeria* (Batista, Lima & Vasconcelos) M.B. Ellis - Mycol. Pap. 106: 34, 1966.

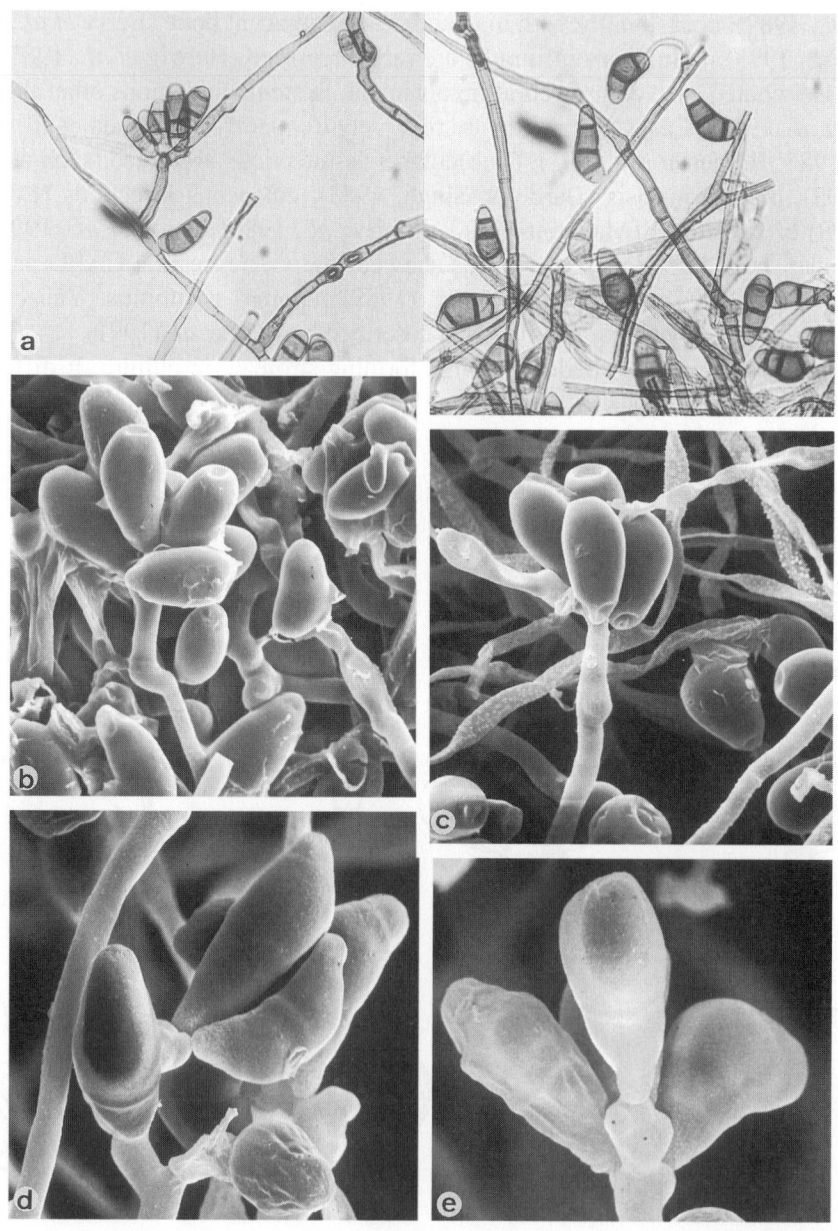

Curvularia lunata, UAMH 4388. Conidiophores and conidia. a. ×640; b. ×1050; c. ×1150; d. ×2000; e. ×2400.

Curvularia pallescens Boedijn

Colony characteristics. Colonies (PCA) spreading, woolly at the centre, often developing concentric zones.
Microscopy. Conidiophores simple, rarely branched, straight or sometimes geniculate near the apex, brown, variable in length, up to 5-6 µm wide. Conidia smooth-walled, pale brown, mostly 3-septate, ellipsoidal to fusiform, usually slightly curved, 17-32 × 7-12 µm.
Teleomorph. *Cochliobolus pallescens* (Tsuda & Ueyama) Sivanesan (*Ascomycota, Euascomycetes, Pleosporales: Pleosporaceae*).
Molecular diagnostics. ITS restriction map based on CBS 156.35:

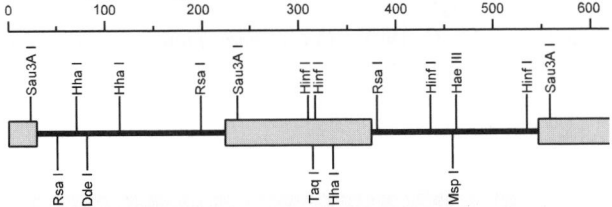

Pathogenicity. BSL-1. Disseminated infection, mainly affecting lungs and central nervous system (Lampert *et al.*, 1977; Friedman *et al.*, 1981); subcutaneous infection (Garcia *et al.*, 1992); cutaneous phaeohyphomycosis (Agrawal & Singh, 1995a).
References. Ellis (1966), Sivanesan (1987), Freire *et al.* (1998).

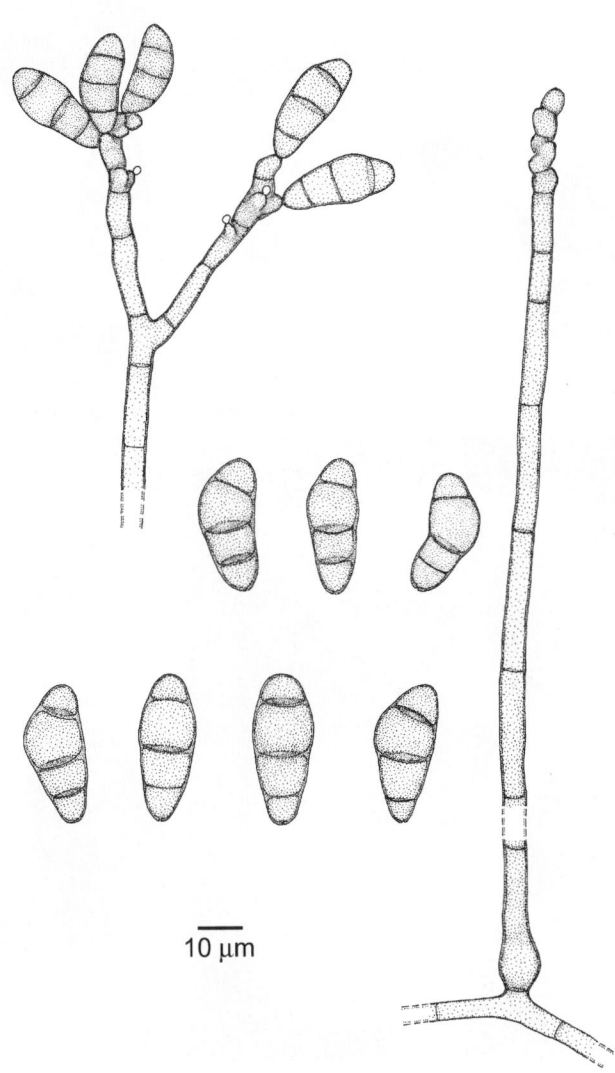

Curvularia pallescens, CBS 156.35. Conidiophores and conidia.

Antifungal susceptibility.

Antifungal	Mean MICs	MICs range	Strains	Reference
AMB	0.41	0.06-32	4	Guarro *et al.* (1999b)
5FC	215	128-256	4	Guarro *et al.* (1999b)
FLZ	76	16-128	4	Guarro *et al.* (1999b)
ITZ	2.82	0.5-16	4	Guarro *et al.* (1999b)
KTZ	4.7	0.66-32	4	Guarro *et al.* (1999b)
MCZ	1.99	0.55-4	4	Guarro *et al.* (1999b)

Nomenclature. *Curvularia pallescens* Boedijn - Bull. Jard. Bot. Buitenz., Ser. 3, 13: 127, 1933.

Pseudocochliobolus pallescens Tsuda & Ueyama - Mem. Coll. Agric., Kyoto Univ. 122: 86, 1983 ≡ *Cochliobolus pallescens* (Tsuda & Ueyama) Sivanesan - Mycol. Pap. 158: 188, 1987.

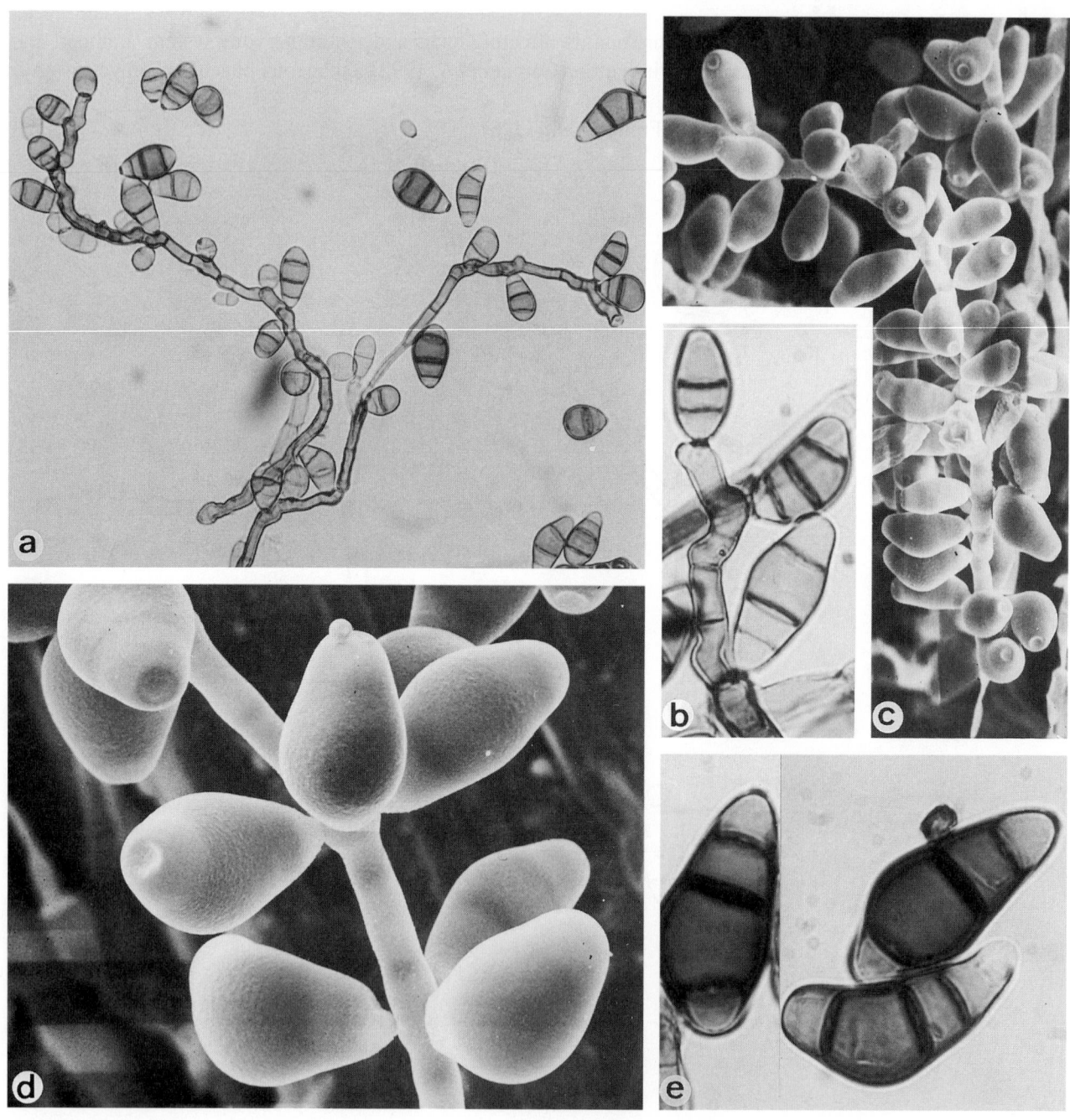

Curvularia pallescens, CBS 156.35. Conidiophores and conidia. a. ×1280; b. ×1400; c. ×2500.

Curvularia senegalensis (Speg.) Subram.

Colony characteristics. Colonies (PCA) spreading, velvety, dark blackish-brown to black.

Microscopy. Conidiophores simple or branched, straight or flexuose, brownish, smooth-walled, up to 150 µm long, 3-7 µm wide. Conidia smooth-walled, dark brown, terminal cells paler, usually curved, mostly 4-septate,19-30 × 10-14 µm.

Molecular diagnostics. ITS restriction map based on CBS 431.75:

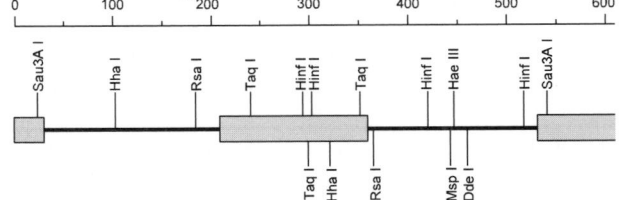

Pathogenicity. BSL-1. Keratitis (Forster *et al.*, 1975b; Guarro *et al.*, 1999b).

Reference. Sivanesan (1987).

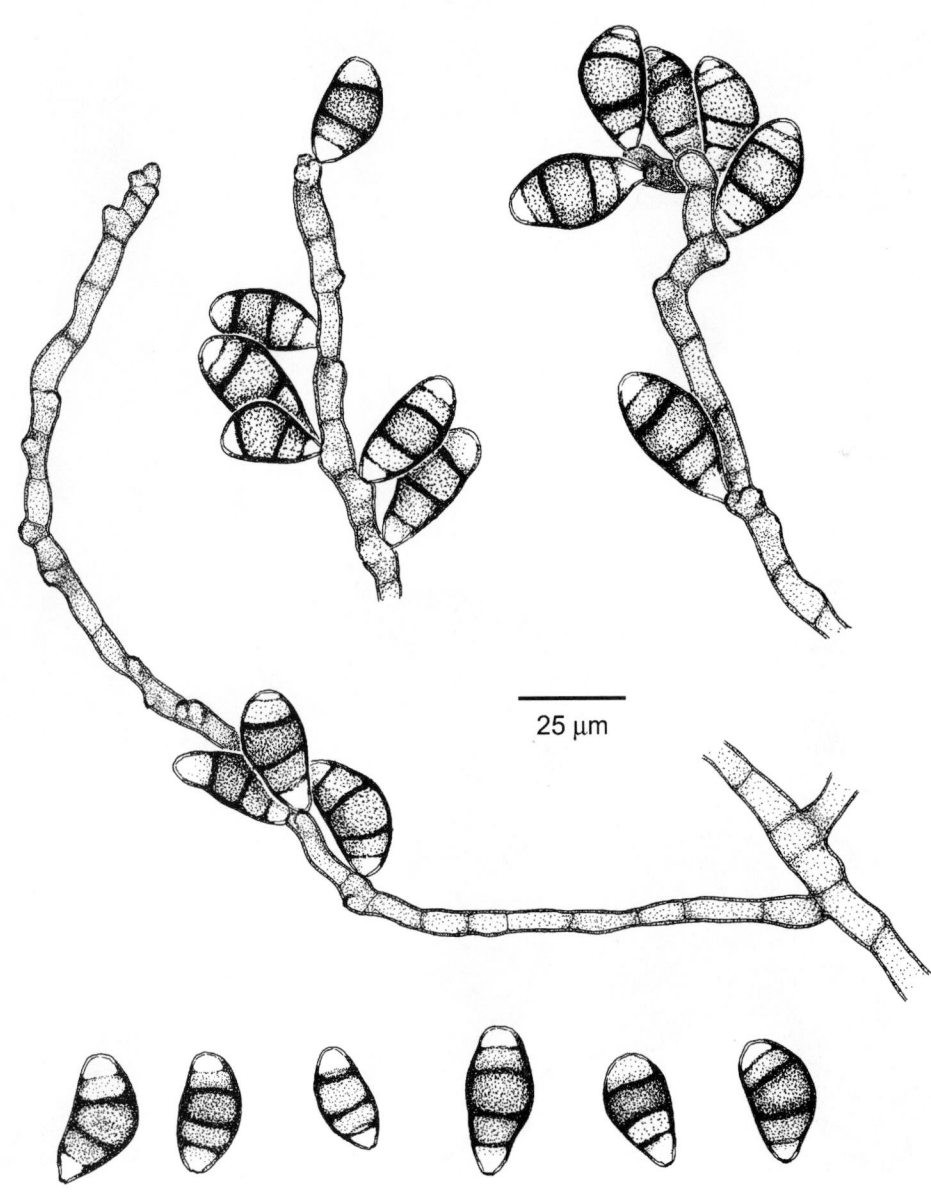

Curvularia senegalensis, UAMH 4373. Conidiophores and conidia.

Antifungal susceptibility.

Antifungal	Mean MICs	MICs range	Strains	Reference
AMB	0.5	0.25-2	3	Guarro *et al.* (1999b)
5FC	203	128-256	3	Guarro *et al.* (1999b)
FLZ	51	16-128	3	Guarro *et al.* (1999b)
ITZ	0.5	0.25-1	3	Guarro *et al.* (1999b)
KTZ	1.41	1-2	3	Guarro *et al.* (1999b)
MCZ	1	1-2	3	Guarro *et al.* (1999b)
VCZ	0.2	nd	3	McGinnis & Pasarell (1998b)

Nomenclature. *Brachysporium senegalense* Spegazzini - An. Mus. Nac. Hist. Nat., B. Aires 26: 133, 1914 ≡ *Curvularia senegalensis* (Spegazzini) Subramanian - J. Ind. Bot. Soc. 35: 467, 1956.

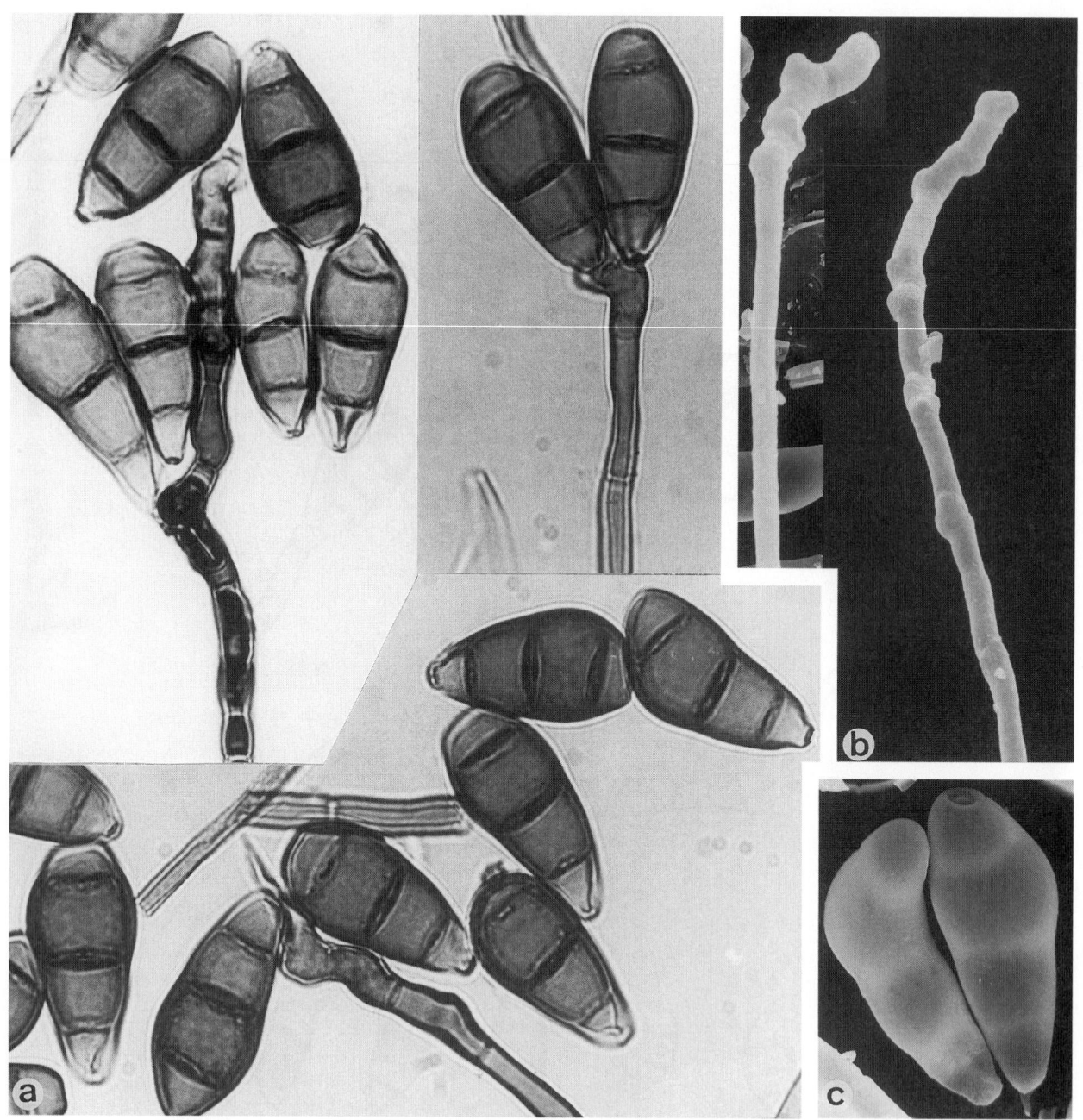

Curvularia senegalensis, UAMH 4373. Conidiophores and conidia. a. ×320; b. ×512; c. ×720; d. ×1900; e. ×1280.

Curvularia verruculosa Tandon & Bilgrami

Colony characteristics. Colonies (PCA) spreading, brown, lanose to felty; reverse dark brown to black.
Microscopy. Conidiophores simple, rarely branched, sometimes geniculate near the apex, brown to dark brown, variable in length, up to 5-6 μm wide. Conidia brown, terminal cells paler, verrucose, mostly 3-septate, ellipsoidal to fusiform, 20-40 × 12-17 μm.
Teleomorph. BSL-1. *Cochliobolus verruculosus* (Tsuda & Ueyama) Sivanesan (*Ascomycota, Euascomycetes, Pleosporales: Pleosporaceae*).
Molecular diagnostics. ITS restriction map based on CBS 148.63:

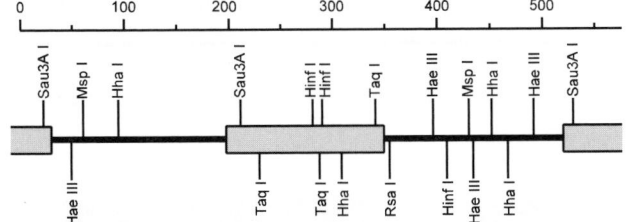

Pathogenicity. BSL-1. Keratitis (Forster *et al.*, 1975b).
Reference. Ellis (1966), Sivanesan (1987).

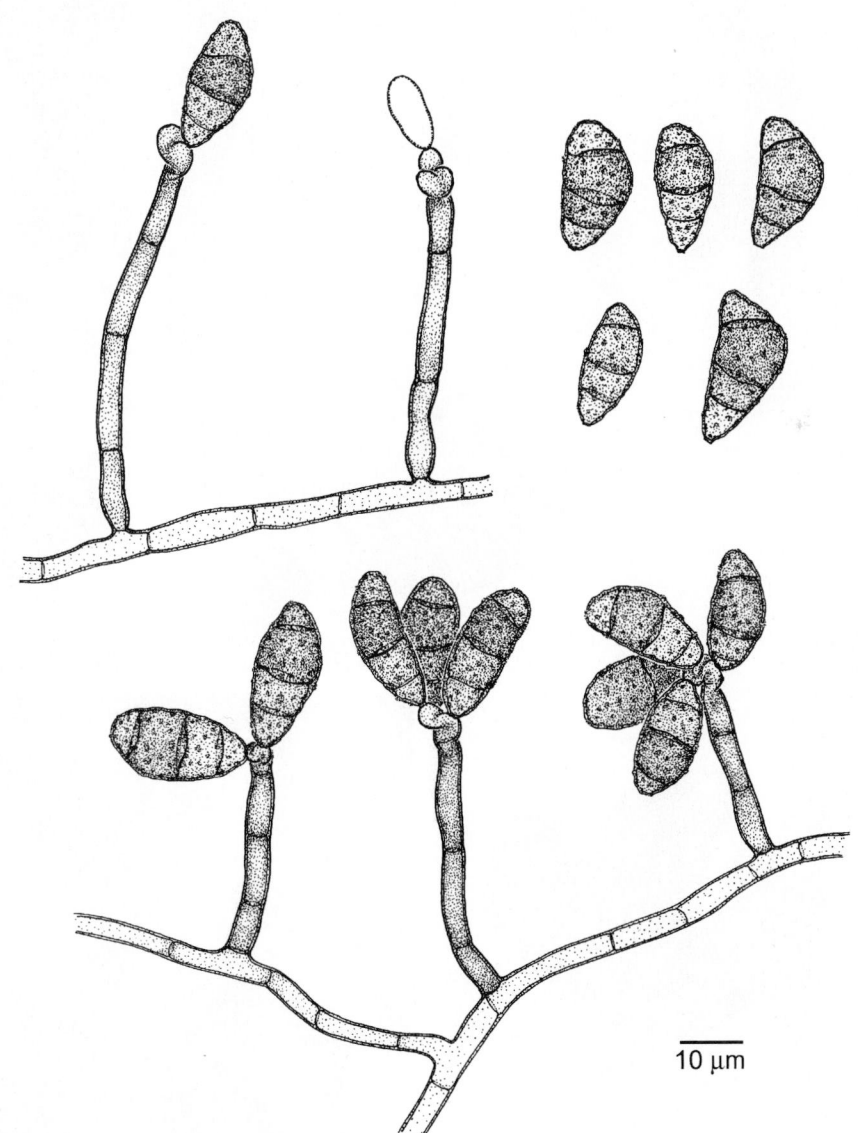

Curvularia verruculosa, CBS 148.63. Conidiophores and conidia.

Antifungal susceptibility.

Antifungal	Mean MICs	MICs range	Strains	Reference
AMB	0.25	0.125-0.5	4	Guarro *et al.* (1999b)
5FC	181	128-256	4	Guarro *et al.* (1999b)
FLZ	13.5	4-64	4	Guarro *et al.* (1999b)
ITZ	0.70	0.5-2	4	Guarro *et al.* (1999b)
KTZ	1.18	0.5-2	4	Guarro *et al.* (1999b)
MCZ	1.41	1-2	4	Guarro *et al.* (1999b)
VCZ	0.13	nd	3	McGinnis & Pasarell (1998b)

Nomenclature. *Curvularia verruculosa* Tandon & Bilgrami - Curr. Sci. 31: 254, 1962.

Pseudocochliobolus verruculosus Tsuda & Ueyama - Mycologia 74: 565, 1982 ≡ *Cochliobolus verruculosus* (Tsuda & Ueyama) Sivanesan - Bitun. Ascom. p. 366, 1984.

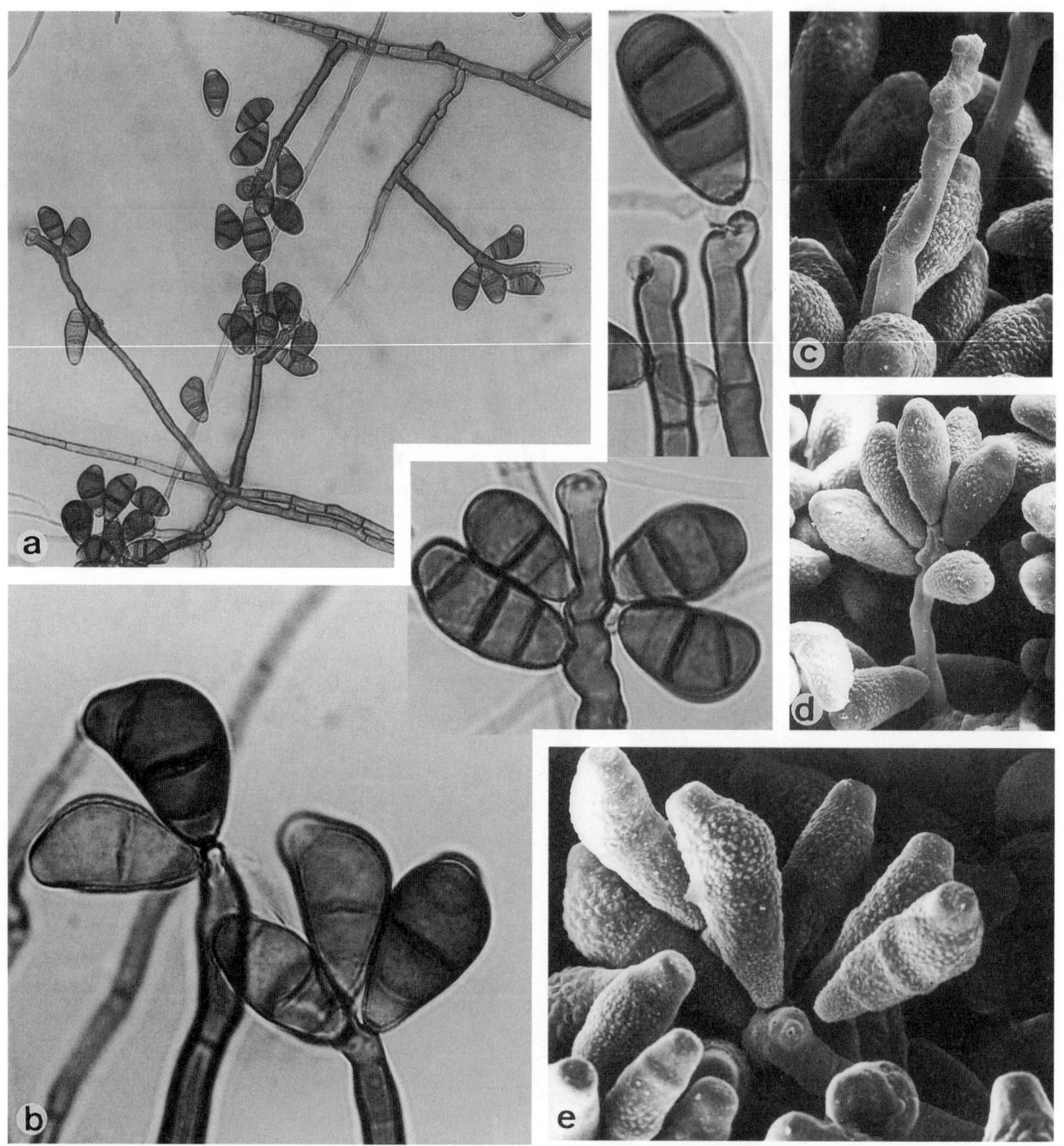

Curvularia verruculosa, CBS 148.63. Conidiophores and conidia. a. ×320; b. ×1280; c. ×1600; d. ×1000; e. ×1650.

Hyphomycetes. Genus: *CYLINDROCARPON*

Generic description. Colonies usually rather fast-growing, hyaline or bright-coloured, velvety, felty or woolly. Sporodochia occasionally present. Conidiophores consisting of simple or repeatedly verticillate phialides, arranged in brush-like structures. Phialides cylindrical to subulate, with a single apical opening with small collarette, producing hyaline, smooth-walled conidia which adhere in slimy masses; conidia often of two kinds: macroconidia hyaline, straight or curved, cylindrical to fusiform but with rounded apex and flat base, with one to several transverse septa; microconidia 1-celled, usually clearly distinct from the macroconidia. Chlamydospores present or absent, hyaline to brown, spherical, formed singly, in chains or in clumps, intercalary or terminal.

Teleomorph. *Nectria (Ascomycota, Euascomycetes, Hypocreales: Nectriaceae).*

Differential diagnosis. The genus differs from *Fusarium* by lacking an asymmetrical footcell on the macroconidia.

References. Booth (1966), Domsch *et al.* (1980), Samuels & Brayford (1990), Brayford & Samuels (1993).

Key to the treated species of Cylindrocarpon:

1a.	Conidia sparse, non-septate; clumps of chlamydospores prevalent	*C. cyanescens* (614)
1b.	Conidia abundant ➞ 2	
2a.	Non-septate conidia absent	*C. lichenicola* (618)
2b.	Non-septate conidia present in addition to septate conidia ➞ 3	
3a.	Microconidia small, mostly less than 10 µm long; macroconidia 1-3-septate	*C. destructans* (616)
3b.	Microconidia usually over 10 µm long; macroconidia 0-1(-3)-septate	see *Fusarium solani* (701)

Cylindrocarpon cyanescens (de Vries *et al.*) Sigler

Colony characteristics. Colonies (OA) restricted, compact, velutinous or somewhat cerebriform, cream-coloured, later becoming blackish. Fresh cultures exuding a blue pigment into the agar.

Microscopy. Thallus mainly consisting of clumps of thick-walled chlamydospores. Phialides sparse, cylindrical, tapering towards the apex, 0-3-septate, 10-40 µm long, with inconspicuous collarettes. Conidia hyaline, smooth-walled, broadly ellipsoidal, 3.5-6 × 2.5-4.0 µm.

Physiology. Maximum growth temperature 36°C. No assimilation of lactose and inositol. No gelatin liquefaction.

Differential diagnosis. The species is untypical for *Cylindrocarpon* because of its non-septate conidia. Phialides are rare, chlamydospores are prevalent.

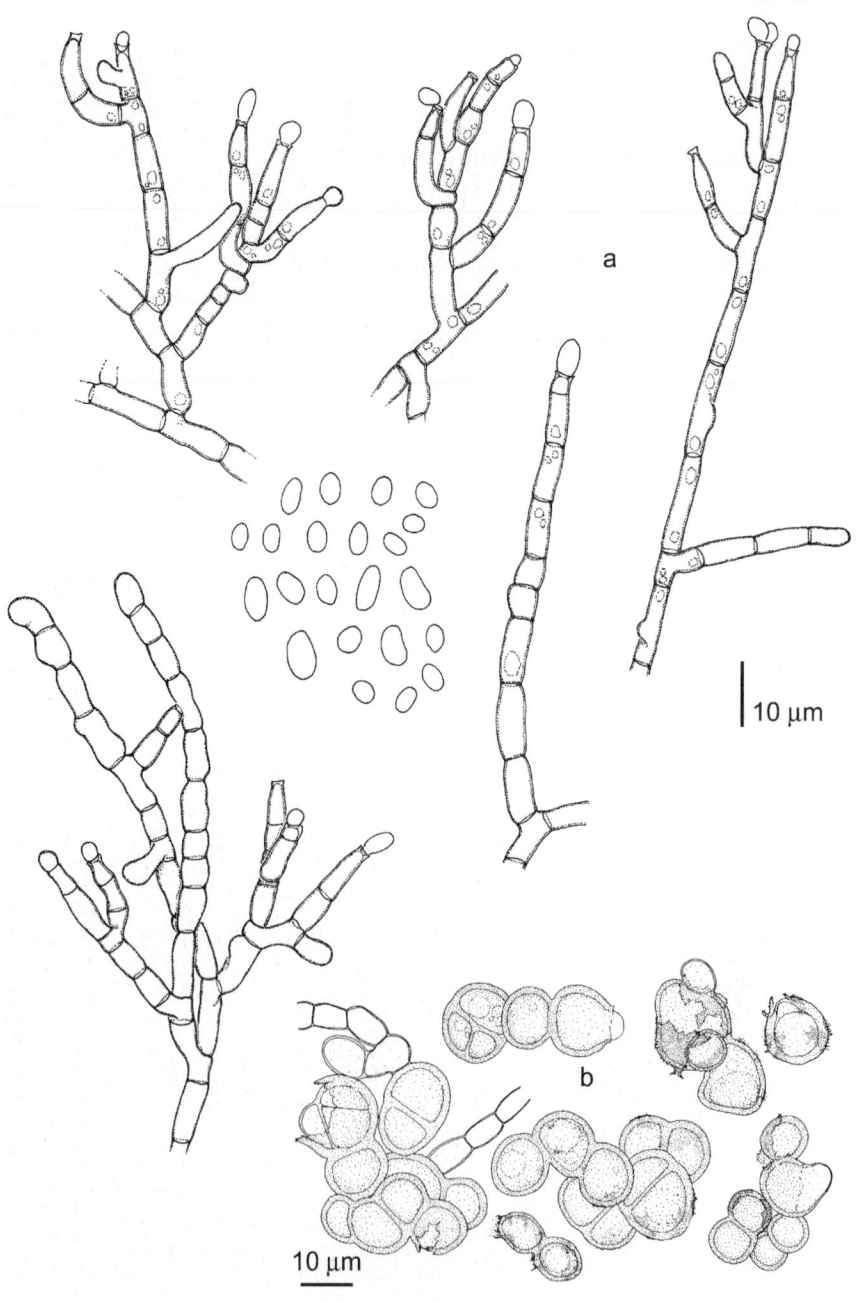

Cylindrocarpon cyanescens, CBS 518.82. a. Phialides and conidia; b. clumps of chlamydospores (b. reproduced with permission from *Antonie van Leeuwenhoek* 50: 152, 1984).

614

Pathogenicity. BSL-2. Described from a single strain from a human mycetome (de Bruyn *et al.,* 1985).

Reference. De Vries *et al.* (1984).

Nomenclature. *Phialophora cyanescens* de Vries, de Hoog & de Bruyn - Antonie van Leeuwenhoek 50: 150, 1984 ≡ *Cylindrocarpon cyanescens* (de Vries, de Hoog & de Bruyn) Sigler, *in* Zoutman & Sigler - Clin. J. Microbiol. 29: 1858, 1991.

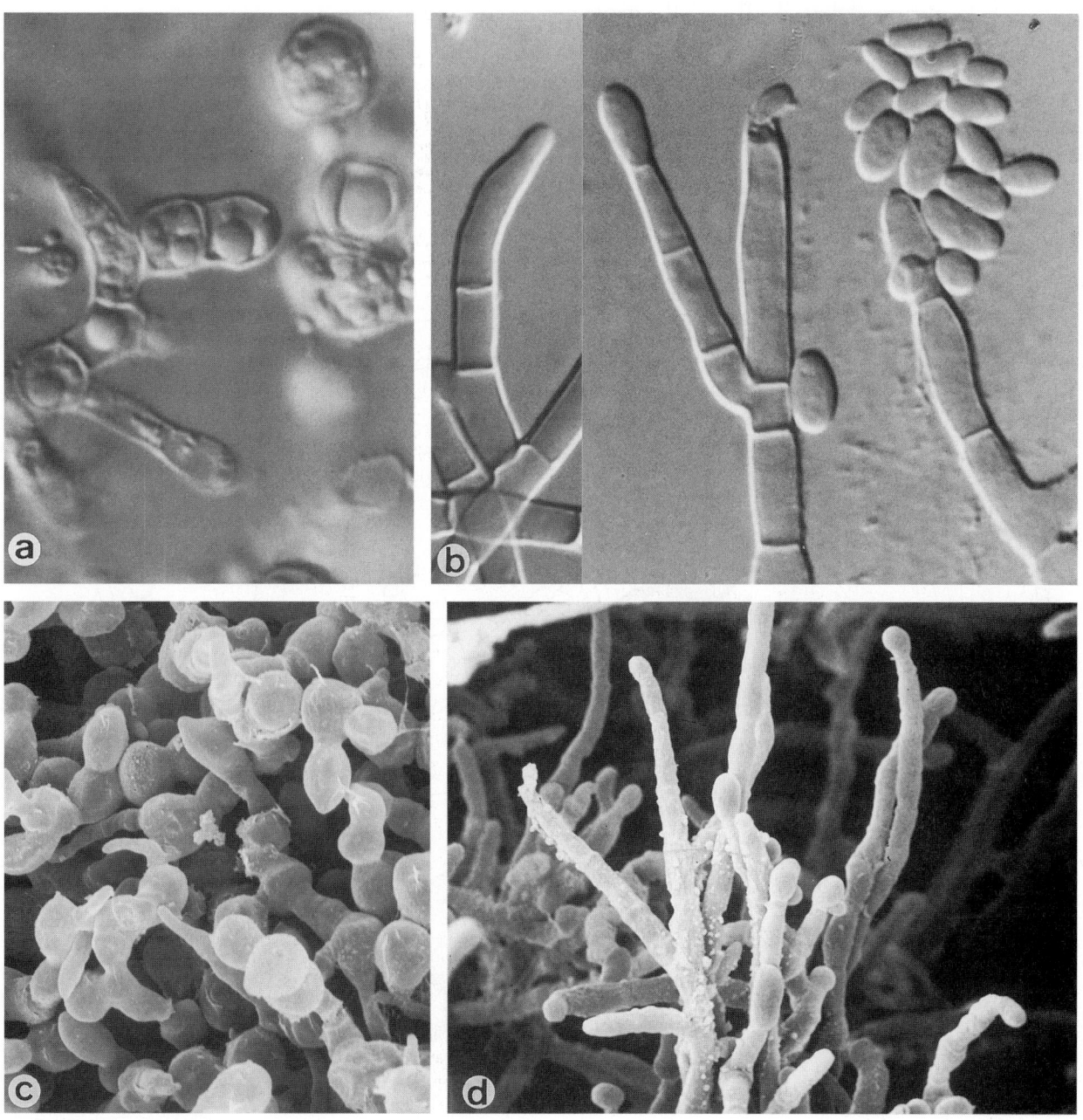

Cylindrocarpon cyanescens. a, c, d. CBS 518.82; b. CBS 637.82. a, c. Clumps of chlamydospores; b, d. young and mature phialides, and conidia. a, b. ×1600; c. ×2100; d. ×1500.

615

Cylindrocarpon destructans (Zins.) Scholten

Colony characteristics. Colonies (OA) moderately spreading, floccose, whitish, beige to pale brown or brown; reverse beige to deep brown.

Microscopy. Conidiophores repeatedly branched and bearing subulate phialides; sporodochia with cream-coloured to beige conidial slime commonly present. Macroconidia cylindrical with rounded ends, straight or curved and slightly truncate at the base, mostly 1-3-septate, 20-50 × 5.0-6.5 μm (though somewhat smaller in the clinical isolate of Zoutman & Sigler, 1991); microconidia not clearly differentiated from the macroconidia, 1-2-celled, 6-10 × 3.5-4.0 μm. Chlamydospores often abundant, intercalary or terminal, singly or in chains, spherical, 9-14 μm diam, brownish, smooth-walled or warted.

Teleomorph. *Nectria radicicola* Gerlach & Nilsson (*Ascomycota, Euascomycetes, Hypocreales: Nectriaceae*).

Molecular diagnostics. ITS restriction map based on NCBI AJ007352:

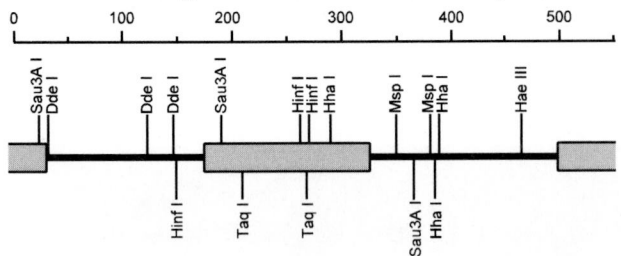

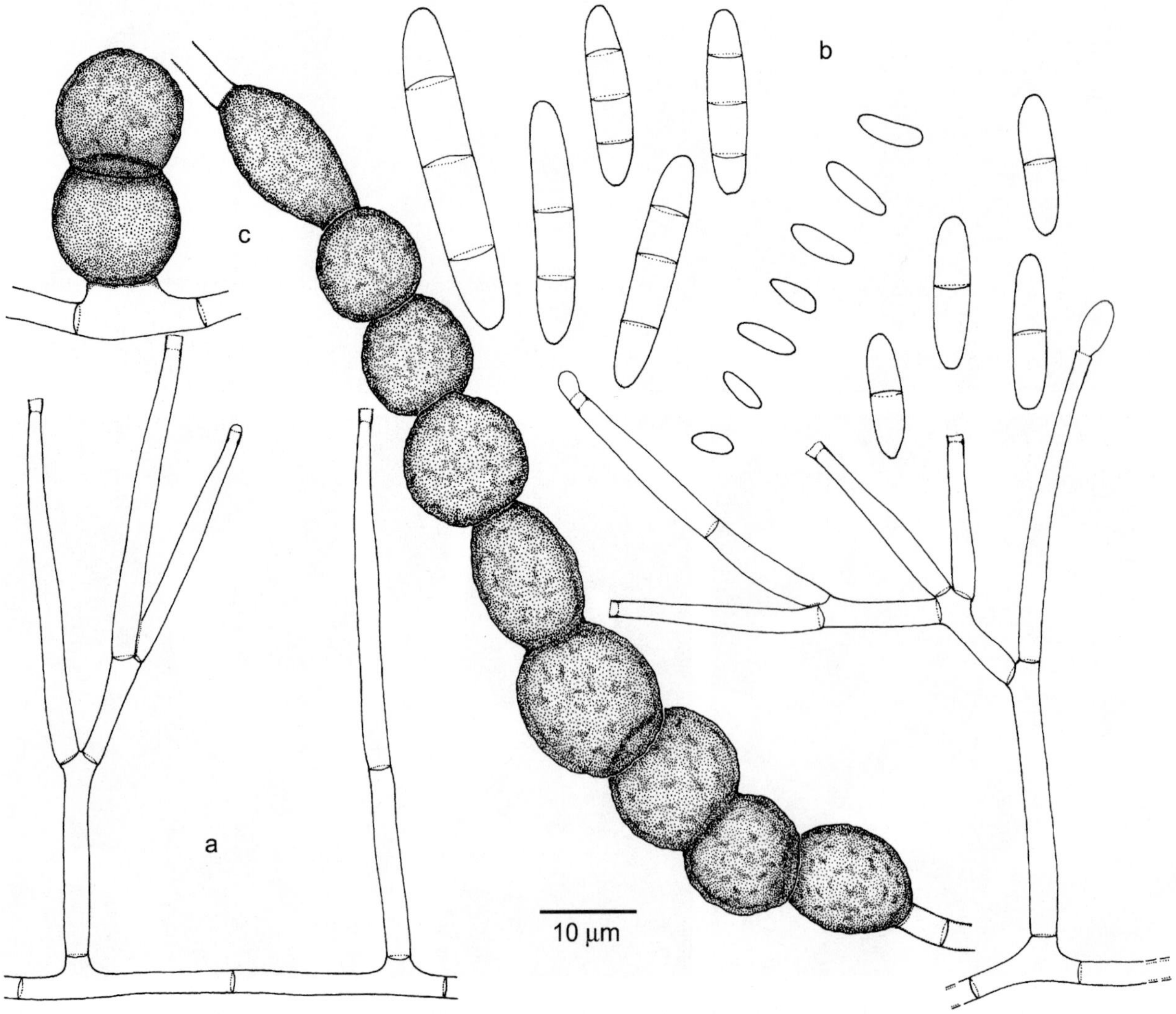

Cylindrocarpon destructans, CBS 264.65. a. Conidiophores; b. conidia; c. chlamydospores.

Pathogenicity. BSL-1. The species is commonly found in rhizosphere. A case of mycetoma of the foot has been reported (Zoutman & Sigler, 1991). Four cases of equine keratitis were described by Hendrix *et al.* (1996).

References. Booth (1966), Domsch *et al.* (1980), Samuels & Brayford (1990), Zoutman & Sigler (1991).

Nomenclature. *Ramularia destructans* Zinsmeister - Phytopathology 8: 570, 1918 ≡ *Cylindrocarpon destructans* (Zinsmeister) Scholten - Neth. J. Pl. Path. 70, Suppl. 2: 9, 1964.

Nectria radicicola Gerlach & Nilsson - Phytopath. Z. 48: 251, 1963.

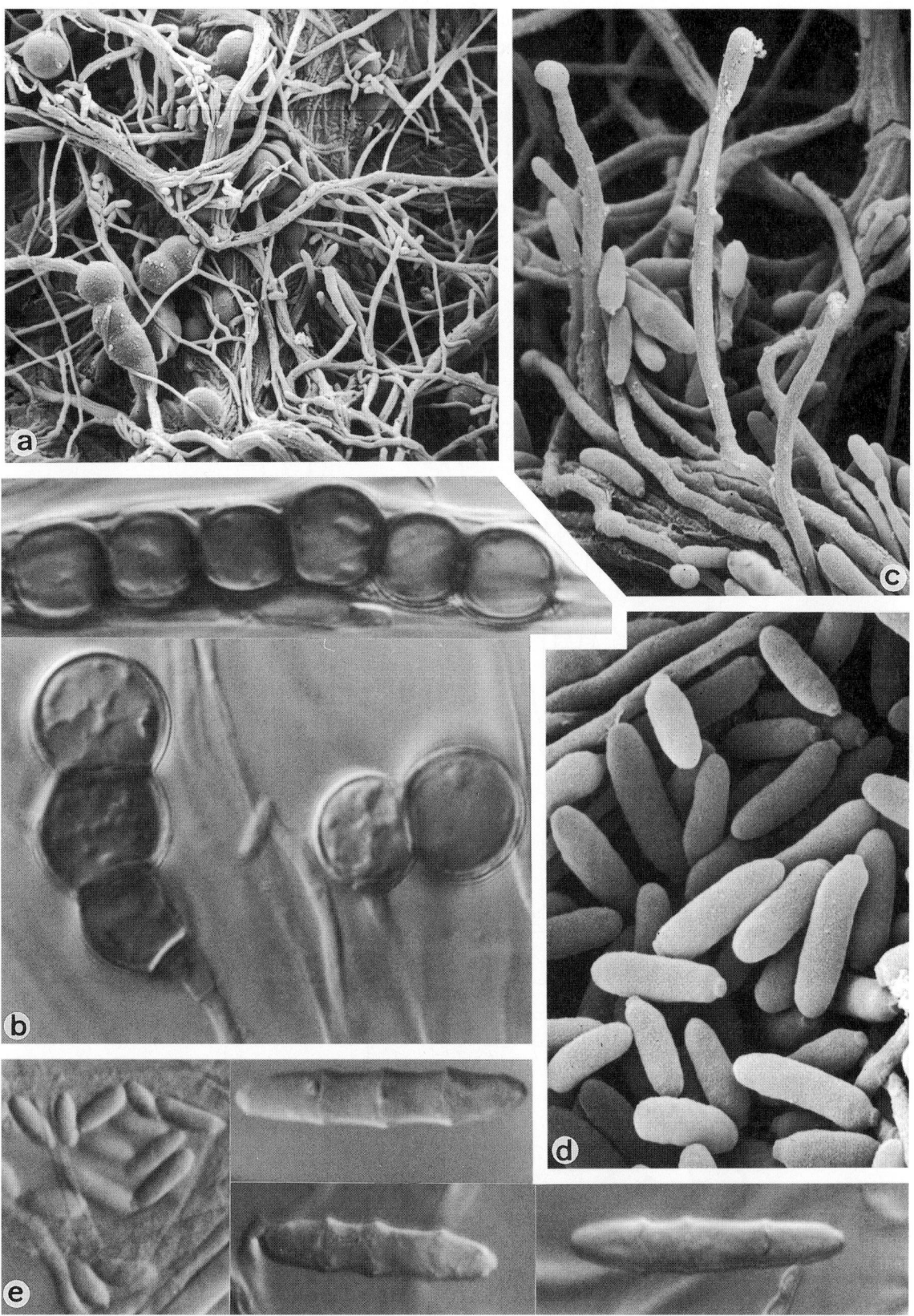

Cylindrocarpon destructans, CBS 264.65. a-c. Conidiophores, conidia and chlamydospores; d, e. macro- and microconidia. a. ×480; b. ×1600; c. ×2900; d. ×5200; e. ×1600.

Cylindrocarpon lichenicola (C. Massal.) D. Hawksw.

Colony characteristics. Colonies (OA) growing rapidly, purplish-red to pale brown, velvety to floccose; reverse dark brown.

Microscopy. Conidiophores rather long, simple or poorly branched. Phialides formed terminally on lateral branches which are forked once or twice, tapering towards the apex. Conidia straight, sometimes slightly curved, hyaline, smooth-walled, cylindrical to fusiform with rounded tip and distinctly truncate base, usually 3-septate but occasionally up to 5-septate, 18-40 × 5-7 μm; microconidia absent. Chlamydospores abundant, hyaline to brown, spherical, 6-9 μm diam, mostly terminally on short lateral branches but occasionally intercalary, single, in chains or in compact clusters.

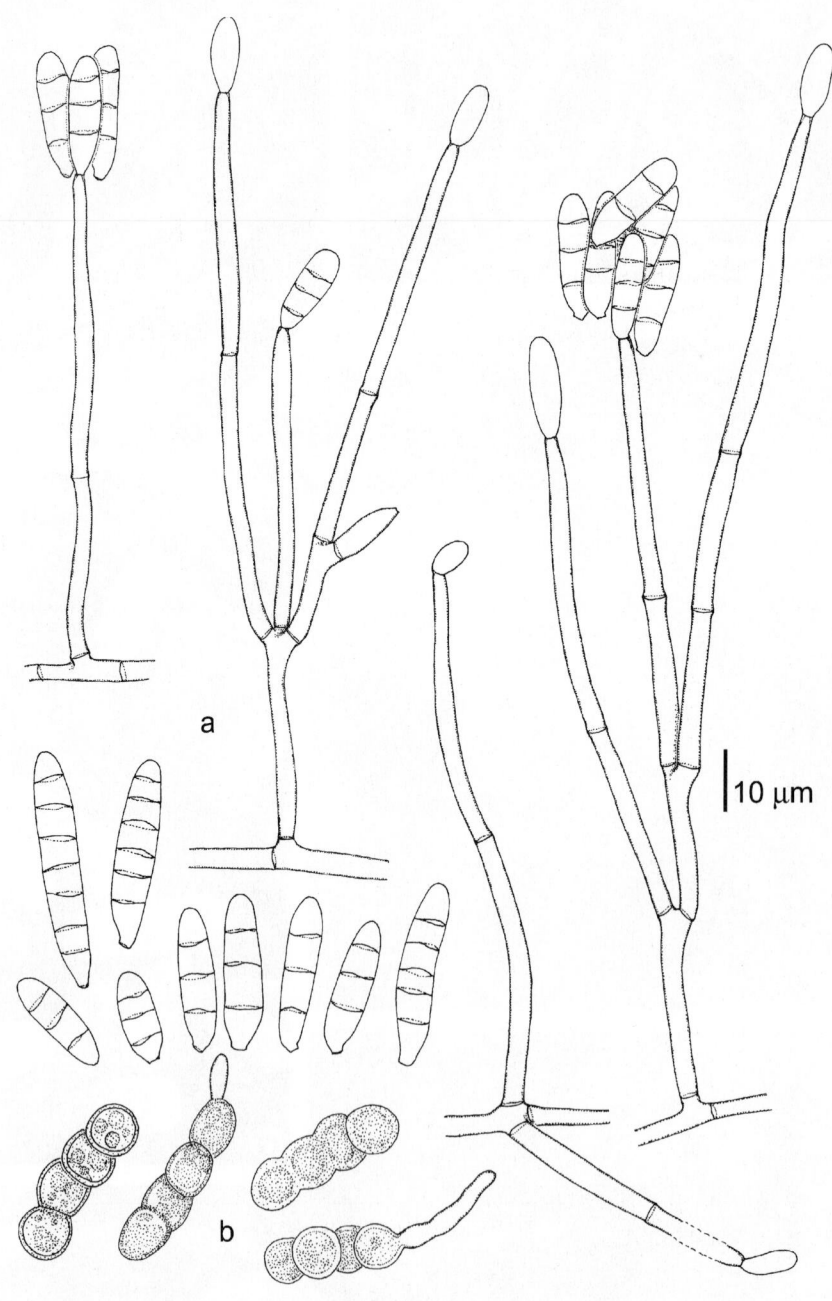

Cylindrocarpon lichenicola, IMI 55920. a. Conidiophores and conidia; b. chlamydospores.

Pathogenicity. BSL-1. Keratitis (Laverde *et al.*, 1973; Matsumoto *et al.*, 1979, Sharma *et al.*, 1998a, b: two cases). Iwen *et al.* (2000) reported a cutaneous case and Sharma *et al.* (1998) a CAPD-related infection. James *et al.* (1997) described a disseminated infection.

Reference. Booth (1966), Hawksworth (1979b).

Nomenclature. *Fusarium lichenicola* C. Massalongo, *in* Maire & Saccardo - Annls Mycol. 1: 223, 1903 ≡ *Cylindrocarpon lichenicola* (C. Massalongo) D. Hawksworth - Bull. Br. Mus., Bot. Ser. 6: 273, 1979.

  Cylindrocarpon tonkinense Bugnicourt - Encycl. Mycol. 11: 181, 1939.

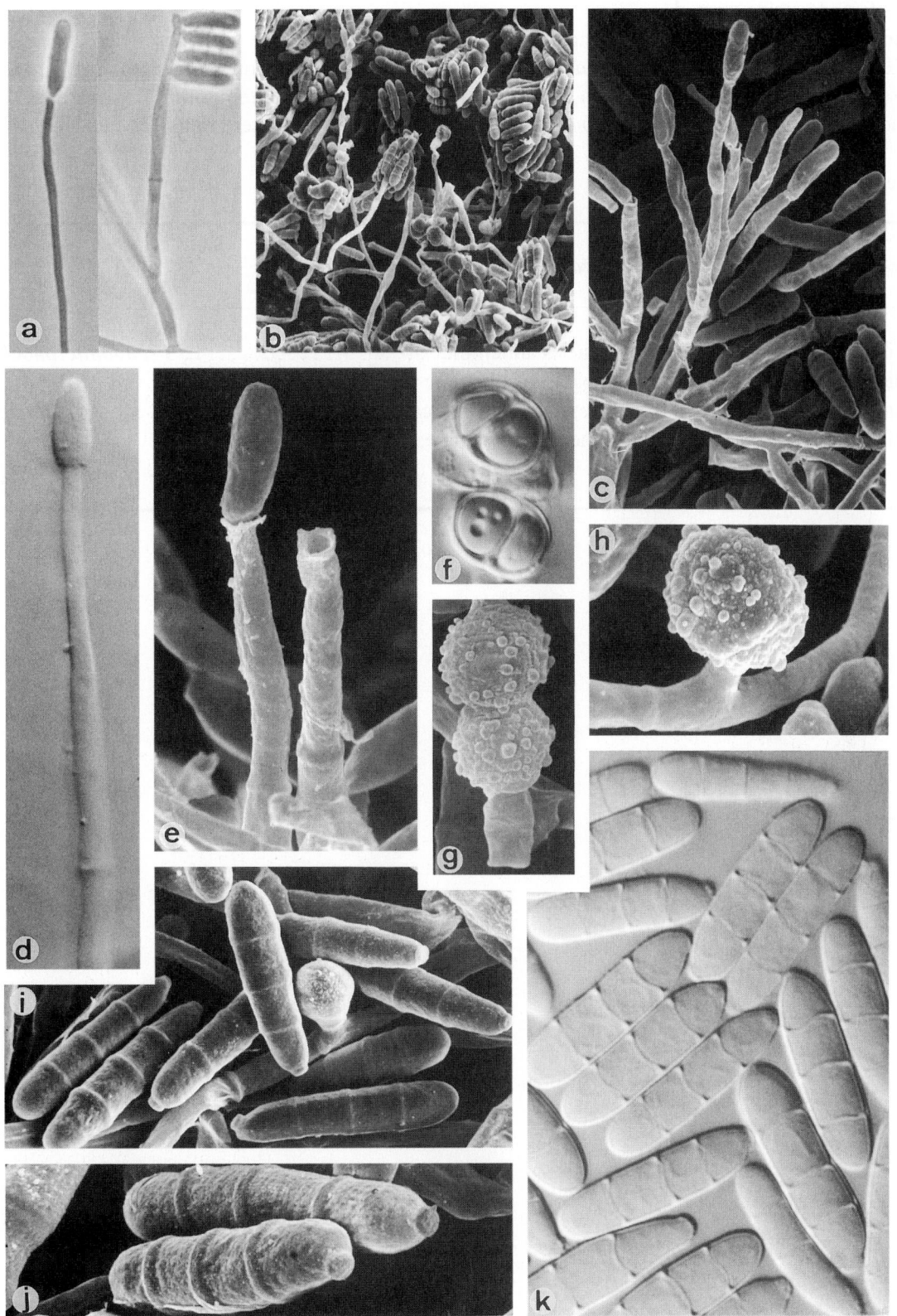

Cylindrocarpon lichenicola, IMI 55920. a-e. Conidiophores and conidia; f-h. chlamydospores; i-k. conidia. a. ×512; b. ×440; c. ×1350; d. ×1600; e. ×3800; f. ×1600; g. ×3700; h. ×4400; i. ×2650; j. ×3900; k. ×1600.

Hyphomycetes. Genus: *CYPHELLOPHORA*

Generic description. Colonies usually spreading, grey. Vegetative hyphae smooth-walled, olivaceous to fuscous. Conidiogenous cells are intercalary phialides, bearing thin lateral collarettes, occasionally laterally or terminally on hyphae with apical collarette. Conidia smooth-walled, at first hyaline, becoming brown, septate, straight or curved, fusiform to sickle-shaped.

Molecular diagnostics. Judging from unpublished ITS sequence data the genus is heterogeneous. The type species, *C. laciniata,* is very close to *Cladosporium herbarum* (p. 587; teleom.: *Mycosphaerella tassiana*, order *Dothideales*), but *C. pluriseptata* seems to have affinities to the order *Chaetothyriales* (compare p. 374). Part of the morphologically genus *Pseudomicrodochium* (p. 886) is related to the latter group (G.S. de Hoog, unpublished data).

Reference. De Vries (1962).

Key to the treated species of Cyphellophora:

1a. Conidia falcate, 3-6-septate; conidiogenesis annellidic .... *Pseudomicrodochium* (886)
1b. Conidia otherwise, 1-5-septate; conidiogenesis phialidic ➜ **2**
2a. Conidia sickle-shaped, mostly 1-septate; phialides intercalary .... *C. laciniata* (621)
2b. Conidia fusiform, sometimes curved, mostly 1-3-septate; phialides often lateral or
terminal .... *C. pluriseptata* (623)

Cyphellophora laciniata de Vries

Colony characteristics. Colonies (OA) growing moderately rapidly, velvety to lanose, in various shades of grey; reverse black.

Microscopy. Fertile hyphae pale brown, 1.5-3.0 μm wide, sometimes with constrictions at the septa. Phialides intercalary, sometimes on short side branches, each with a short, lateral or terminal collarette. Conidia sickle-shaped, brown, smooth-walled, 1-3-septate, 11-25 × 2-5 μm, adhering in small bundles.

Molecular diagnostics. Authentic CBS strains of this species proved to have very different ITS sequences (G.S. de Hoog, unpublished data). Part of the strains are similar to the environmental species *Cyphellophora vermispora* Walz & de Hoog (1987). ITS sequences of the type strain, however, are identical to those of *Cladosporium herbarum* (p. 587). Compare also *Pseudomicrodochium suttonii* (p. 886). ITS restriction map based on CBS 190.61:

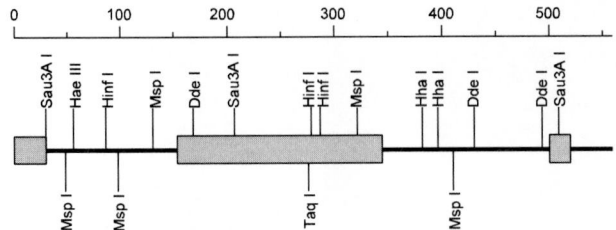

Pathogenicity. BSL-1. Isolated repeatedly from diseased human skin (de Vries *et al.*, 1986). The species has once been isolated from a bronchial lavage of a patient with a pulmonary infection (Sutton *et al.*, 1991).

Reference. De Vries (1962).

Nomenclature. *Cyphellophora laciniata* de Vries - Mycopath. Mycol. Appl. 16: 47, 1962.

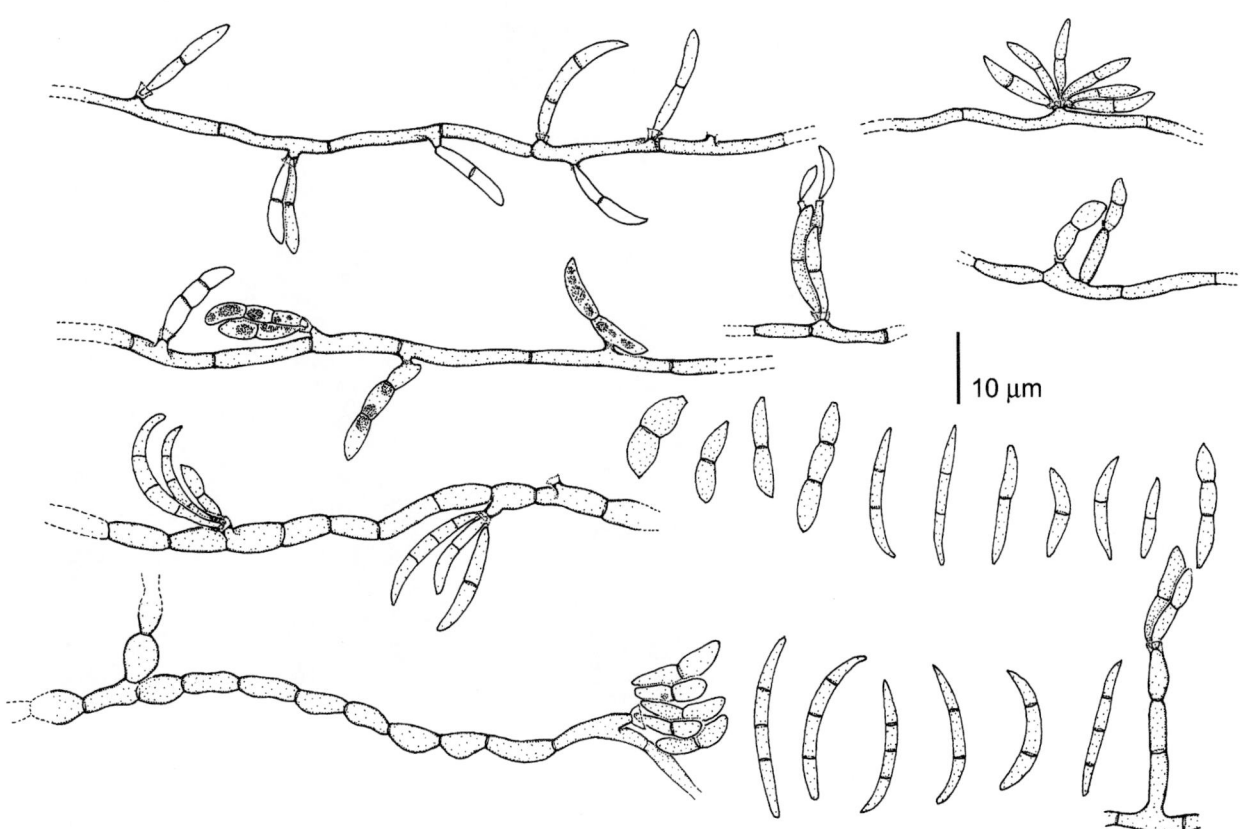

Cyphellophora laciniata, CBS 284.85. Hyphae with intercalary phialides and conidia.

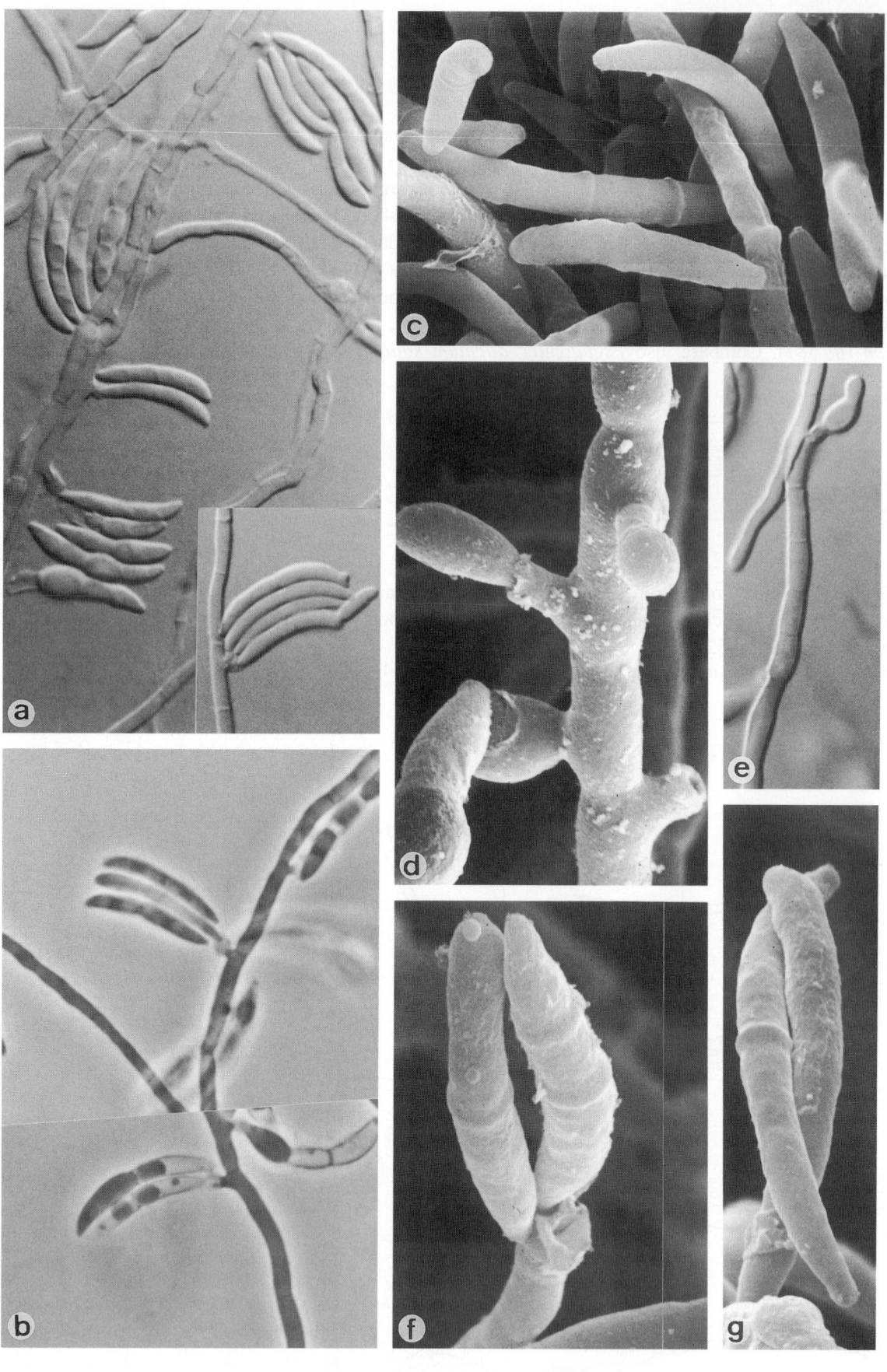

Cyphellophora laciniata, CBS 284.85. Phialides and conidia. a. ×1600; b. ×1280; c. ×6000; d. ×7100; e. ×1600; f. ×7000; g. ×10,000.

Cyphellophora pluriseptata de Vries *et al.*

Colony characteristics. Colonies (OA) growing rapidly, velvety to lanose, dark grey to brownish-grey.

Microscopy. Fertile hyphae pale brown, 1.5-5.0 μm wide. Phialides usually terminal or lateral, occasionally intercalary, pale brown, up to 24 μm long, 1-3 μm wide, basal part swollen, with a conspicuous collarette at the tip. Conidia cylindrical to fusiform, with rounded ends, straight or slightly curved, pale brown, (1-) 2-3 (-5)-septate, 7-18 × 1.7-2.3 μm.

Molecular diagnostics. ITS restriction map based on CBS 285.85:

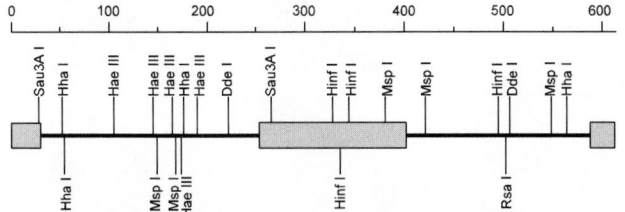

Pathogenicity. BSL-1. Isolated repeatedly from diseased human skin and nails (de Vries *et al.*, 1986).

Reference. De Vries *et al.* (1986).

Nomenclature. *Cyphellophora pluriseptata* de Vries, Elders & Luykx - Antonie van Leeuwenhoek 52: 141, 1986.

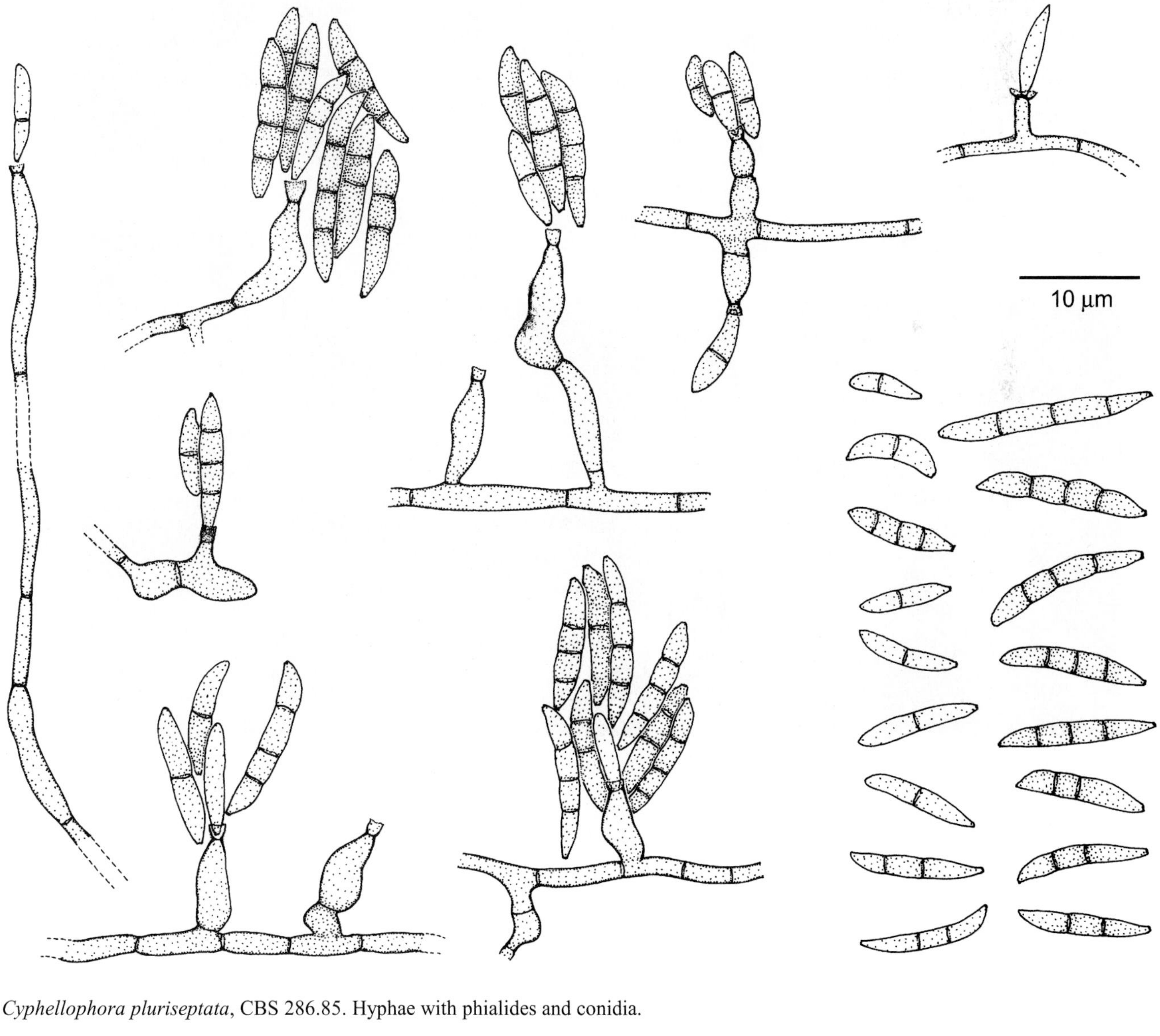

Cyphellophora pluriseptata, CBS 286.85. Hyphae with phialides and conidia.

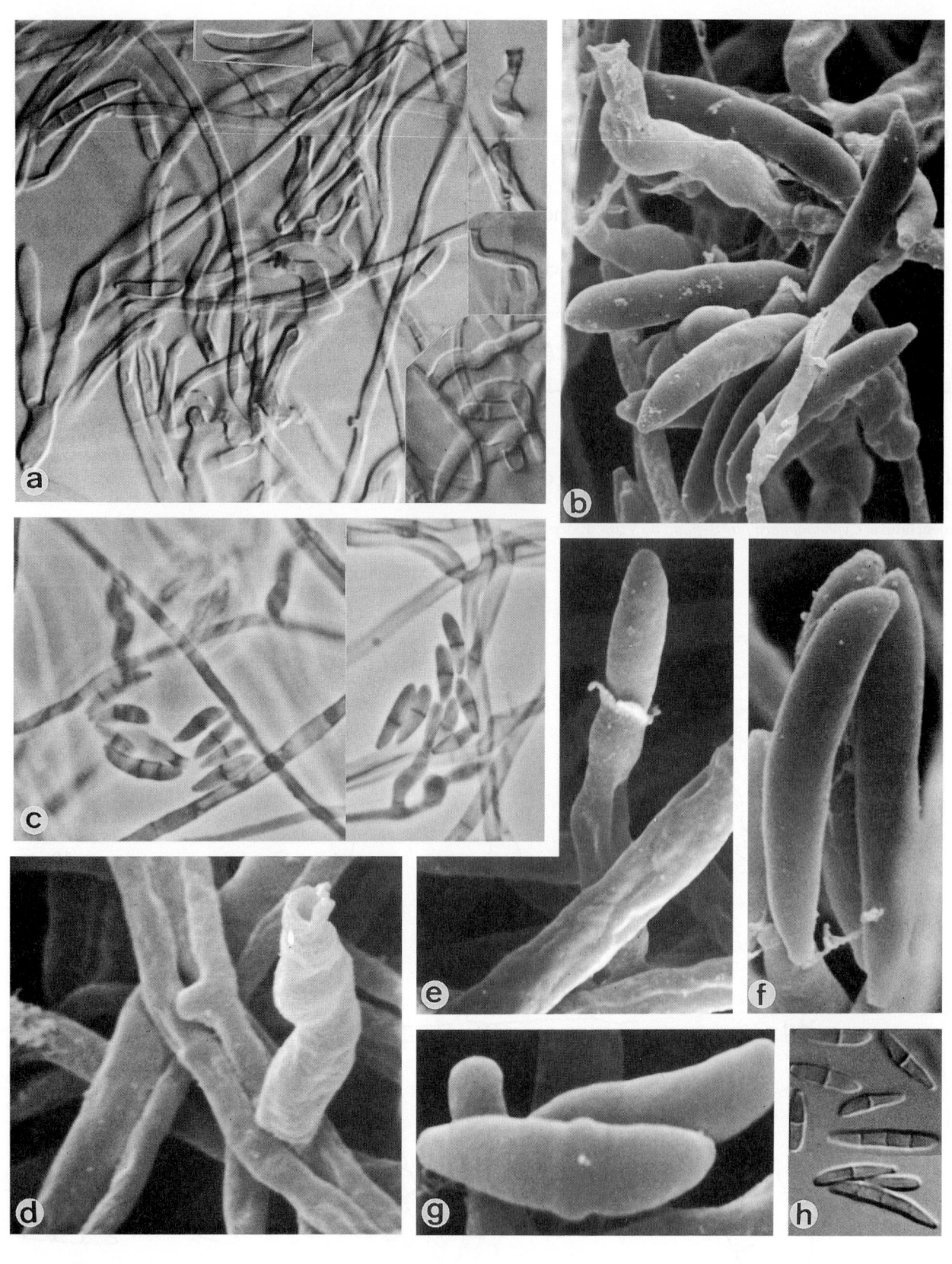

Cyphellophora pluriseptata, CBS 286.85. Phialides and conidia. a. ×1600; b. ×4800; c. ×1280; d. ×7600; e. ×8300; f. ×7400; g. ×11,000; h. ×1600.

Hyphomycetes. Genus: *DICHOTOMOPHTHORA*

Dichotomophthora portulacae Mehrlich & Fitzpatrick ex M.B. Ellis

Colony characteristics. Colonies (OA) growing rapidly, lanose to funiculose, olivaceous to dark brown.
Microscopy. Conidiophores ascending or erect, brown, septate, 100-280 × 4-5 µm, swollen and repeatedly dichotomously or trichotomously branched, lobed at the apex, branches usually short. Conidia solitary, terminally on the lobes, cylindrical or ellipsoidal, 1-7-celled, 15-90 × 6-15 µm, smooth-walled, brown. Sclerotial bodies often present.

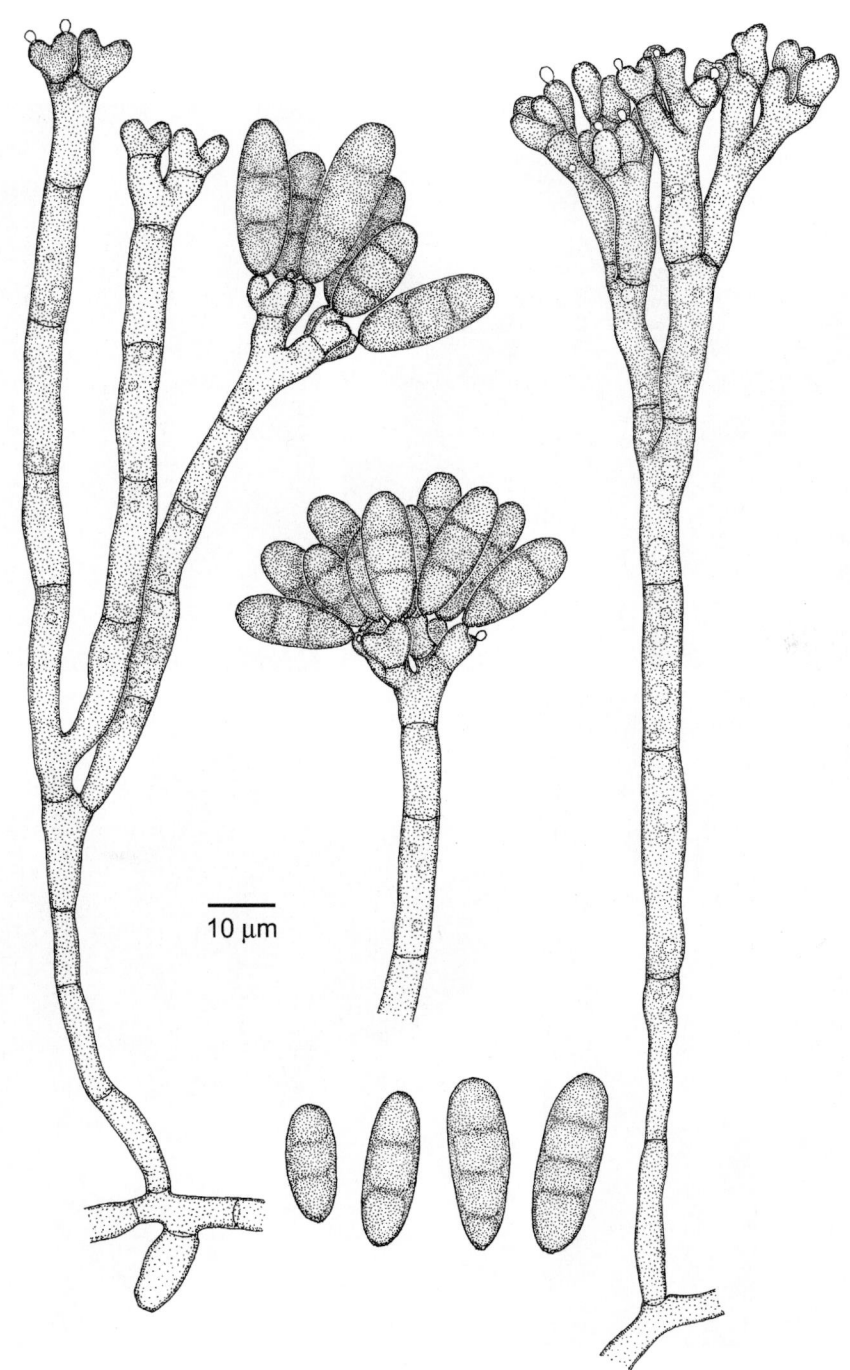

Dichotomophthora portulacae, IMI 260974. Conidiophores and conidia.

Pathogenicity. BSL-1. The species is a plant pathogen on purslane. A case of human keratitis was reported by Ormerod (1987).

References. Rao (1966), Ellis (1971), de Hoog & van Oorschot (1983).

Nomenclature. *Dichotomophthora portulacae* Mehrlich & Fitzpatrick - Mycologia 27: 550, 1935 ≡ *Dichotomophthora portulacae* Mehrlich & Fitzpatrick ex M.B. Ellis - Demat. Hyphom. p. 388, 1971.

Dichotomophthora portulacae, IMI 260974. Conidiophores and conidia. a. ×320; b. ×450; c. ×512; d. ×1000; e. ×1280.

Hyphomycetes. Genus: *DICHOTOMOPHTHOROPSIS*

Dichotomophthoropsis nymphaearum (Rand) M.B. Ellis

Colony characteristics. Colonies (OA) growing rapidly, velvety, olivaceous-brown.

Microscopy. Conidiophores erect, straight or flexuose, branched, often dichotomously or trichotomously forked, smooth-walled or verrucose, pale brown, up to 150 µm long, 3-7 µm wide. Conidiogenous cells terminal, integrated or discrete, lobed. Conidia coiled, borne singly at the tip of each lobe of the conidiogenous cell, multiseptate, often constricted at the septa, verruculose, straw-coloured or brown, 50-300 × 6-20 µm.

Pathogenicity. BSL-1. The species is a plant pathogen; a keratitis was reported by Wright *et al.* (1990).

Reference. Ellis (1976).

Nomenclature. *Helicosporium nymphaearum* Rand - J. Agric. Res. 8: 230, 1917 ≡ *Dichotomophthoropsis nymphaearum* (Rand) M.B. Ellis - Mycol. Pap. 125: 22, 1971.

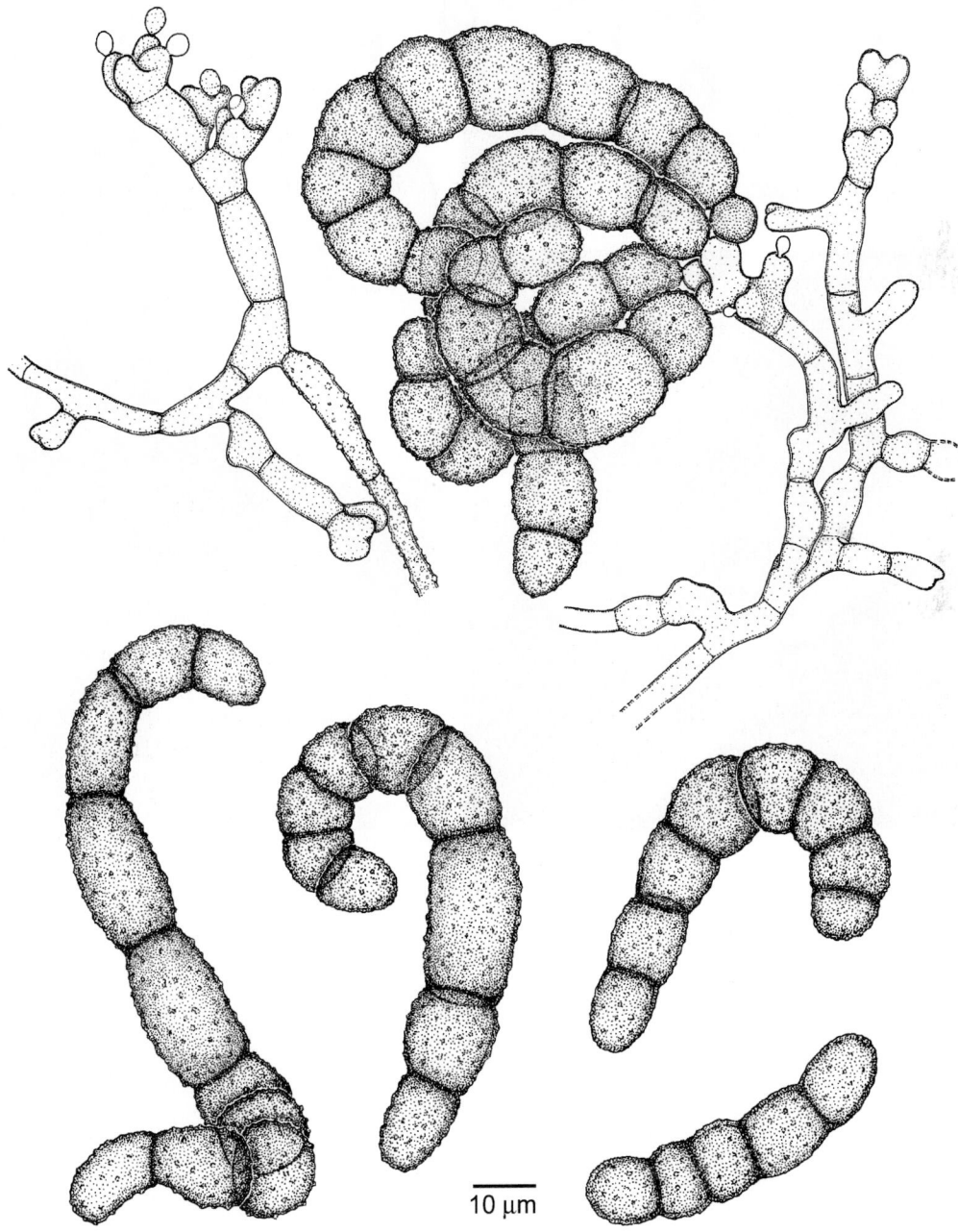

10 µm

Dichotomophthoropsis nymphaearum, IMI 146434. Conidiophores and conidia.

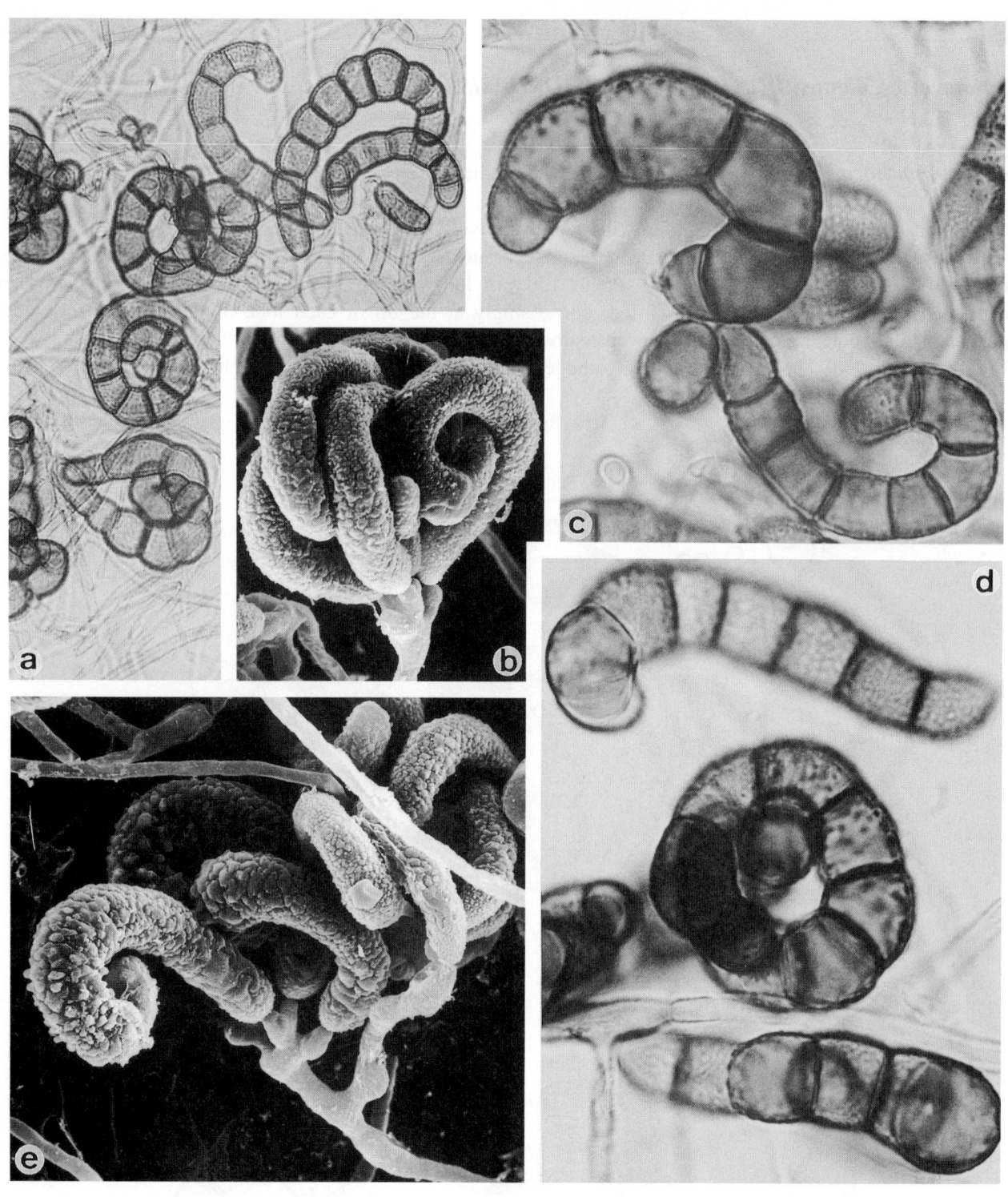

Dichotomophthoropsis nymphearum, IMI 146434. Conidiophores and conidia. a. ×512; b. ×1150; c. ×1280; d. ×1600; e. ×1250.

Hyphomycetes, Dematiaceae. Genus: *DISSITIMURUS*

Dissitimurus exedrus Simmons *et al.*

Colony characteristics. Colonies (PCA) growing slowly, aerial mycelium scarce, becoming velvety to cottony, blackish-brown.

Microscopy. Conidiophores erect, straight or flexuose, septate, brown, smooth-walled to verruculose, often geniculate after sympodial elongation, branched; each branch ending with a conidiogenous portion. Conidia with very variable morphology, ellipsoidal to ovoidal, sometimes narrow ovoidal with apical beak, pale brown, 1-5-celled, smooth-walled to verruculose, 7-33 × 5-10 µm, partly distoseptate, solitary or with secondary conidiophores.

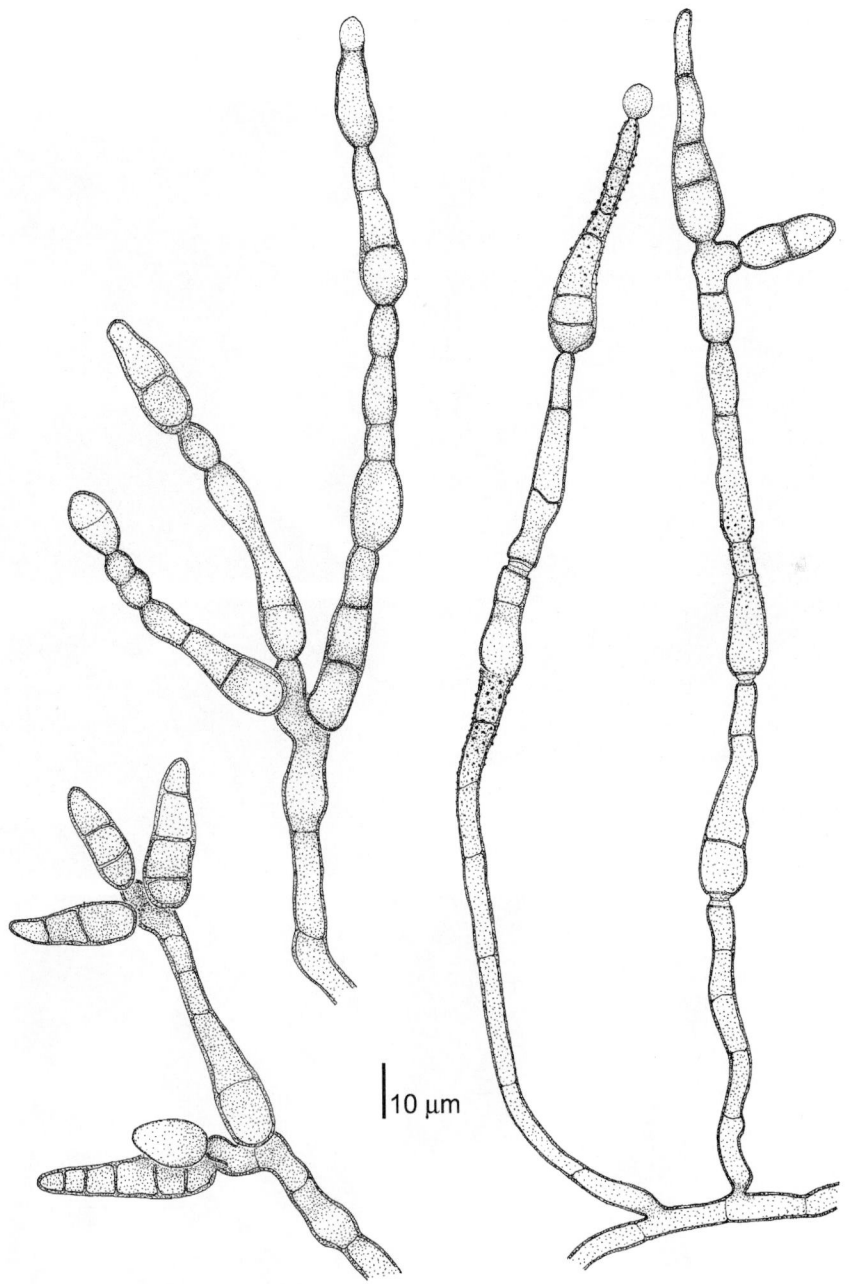

Dissitimurus exedrus, CBS 443.88. Conidiophores and conidia.

Differential diagnosis. The fungus resembles *Alternaria infectoria* (p. 430), but has diffusely branched conidiophores and pale brown, often distoseptate conidia. Probably a degenerate *Bipolaris*-like species is concerned; its ITS restriction profiles are close to those of *B. hawaiiensis* (p. 529).

Pathogenicity. BSL-2. The species is known from a single isolate from human nasopharynx and turbinate lesions (Simmons *et al.*, 1987).

Reference. Simmons *et al.* (1987).

Antifungal susceptibility.

Antifungal	Mean MICs	Strains	Reference
AMB	1	1	Unpublished data
5FC	4	1	Unpublished data
FLZ	128	1	Unpublished data
ITZ	0.5	1	Unpublished data
KTZ	2	1	Unpublished data
MCZ	1	1	Unpublished data

Nomenclature. *Dissitimurus exedrus* Simmons, McGinnis & Rinaldi - Mycotaxon 30: 248, 1987.

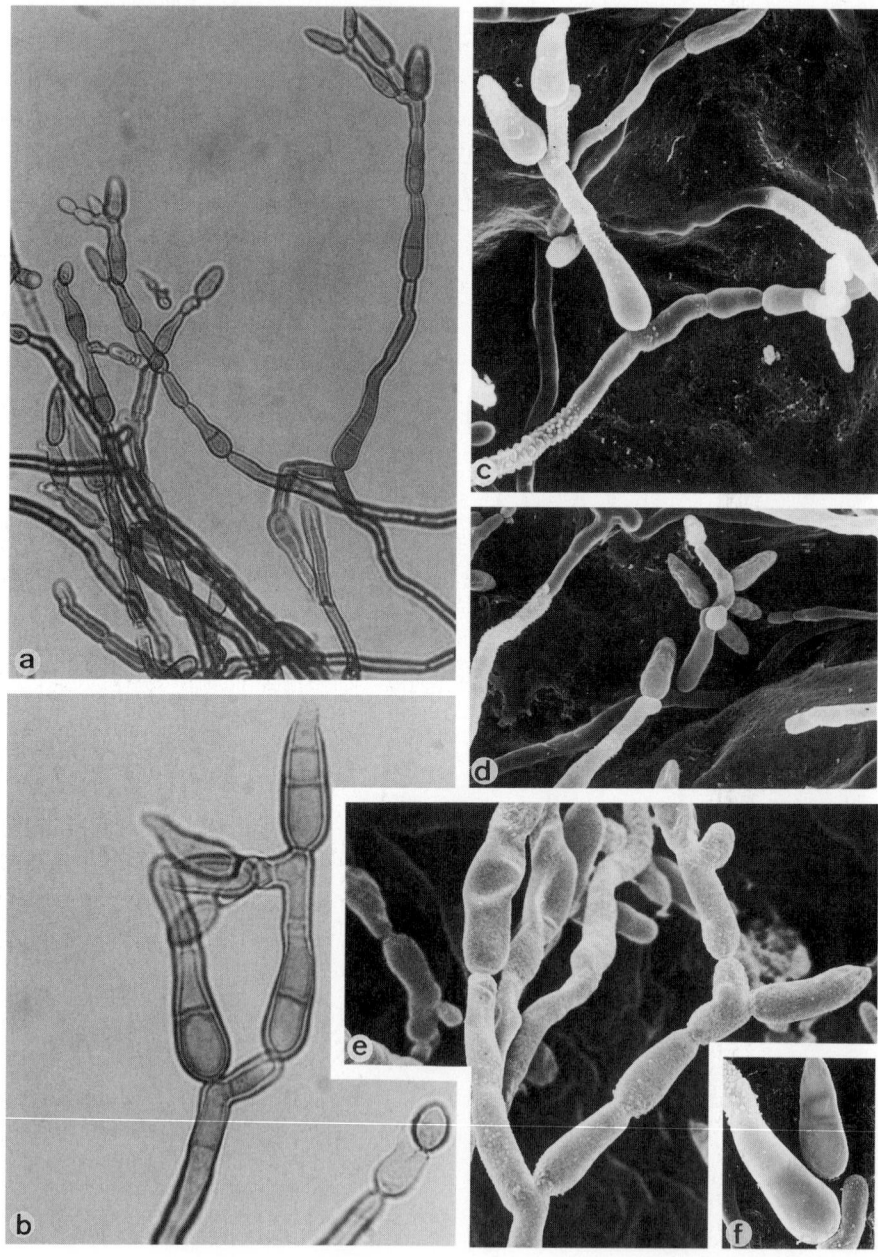

Dissitimurus exedrus, CBS 443.88. Conidiophores and conidia. a. ×350; b. ×870; c. ×925; d. ×640; e. ×1175; f. ×2310.

Hyphomycetes, Dematiaceae. Genus: *DRECHSLERA*

Drechslera biseptata (Sacc. & Roum.) Richardson & Fraser

Colony characteristics. Colonies (PCA) spreading, grey, velvety.
Microscopy. Conidiophores frequently arising from cushion-shaped, blackish-brown stromata, dark brown, erect, stiff, thick- and smooth-walled, 100-800 μm high, about 8-14 μm wide, somewhat narrower near the apex, intermingled with small, flexuose conidiophores. Conidia straight, broadly rounded at both ends, pale to mid brown, 2-3-distoseptate, broadly clavate, 23-33 × 14-17 μm, smooth-walled to finely roughened.
Differential diagnosis. *Drechslera*-like species with curved conidia are now classified in *Bipolaris* (p. 526). Conidia in *Bipolaris* germinate from polar cells, whereas in *Drechslera* germ tubes arise from all cells (p. 380).
Pathogenicity. BSL-1. Occurring on stored seeds. A case of sinusitis was reported by Washburn *et al.* (1988).
References. Ellis (1971), Mouchacca (1988), Sivanesan (1990).

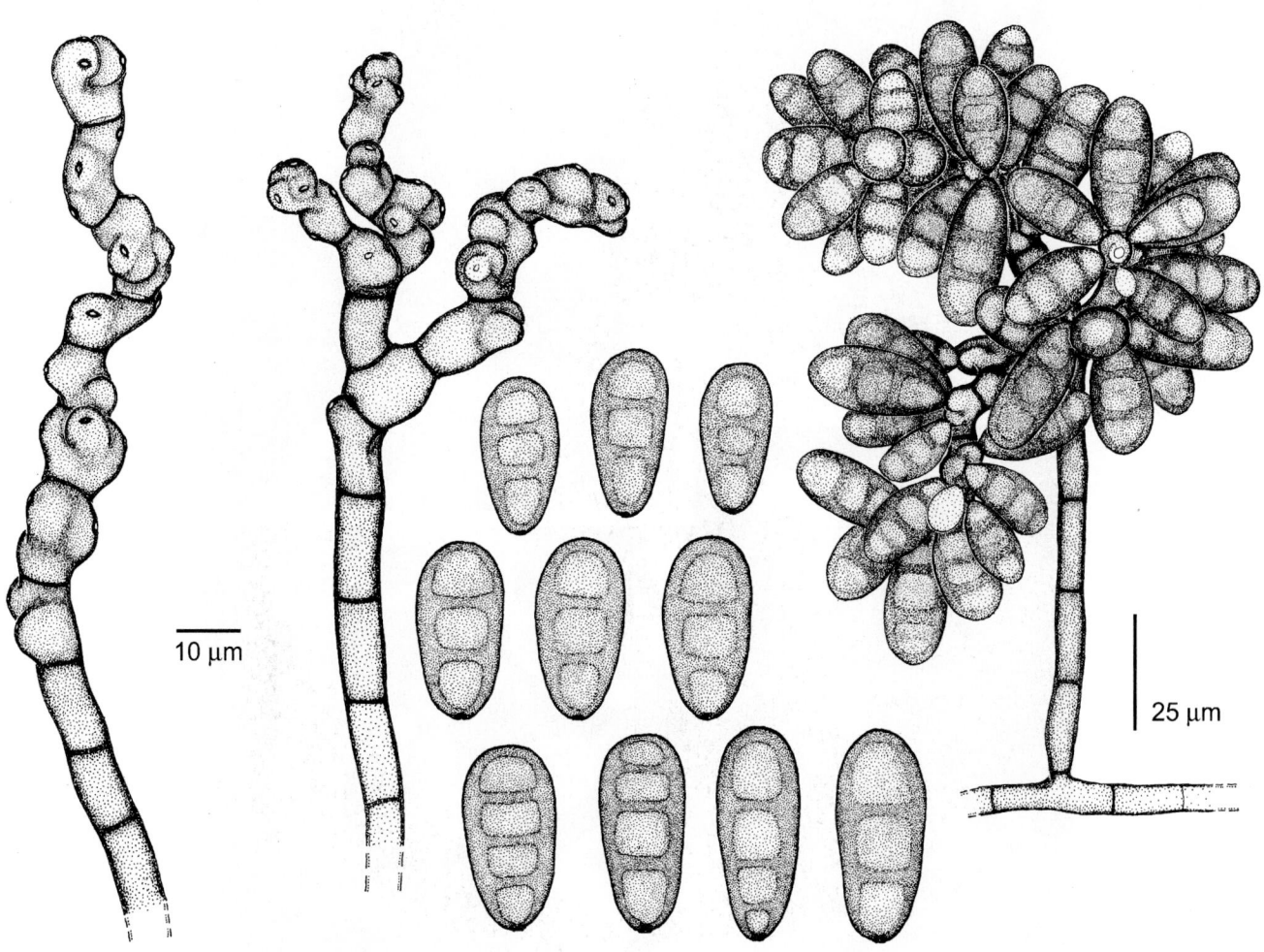

Drechslera biseptata, CBS 277.60. Conidiophores and conidia.

Antifungal susceptibility.

Antifungal	Mean MICs	Strains	Reference
AMB	0.25	2	Unpublished data
AMB	0.25	1	McGinnis & Pasarell (1998b)
5FC	256	2	Unpublished data
FLZ	64	2	Unpublished data
ITZ	2.8	2	Unpublished data
ITZ	0.25	1	McGinnis & Pasarell (1998b)
KTZ	2.8	2	Unpublished data
MCZ	4	2	Unpublished data
VCZ	0.06	1	McGinnis & Pasarell (1998b)

Nomenclature. *Helminthosporium biseptatum* Saccardo & Roumeguère - Revue Mycol. 3: 56, 1881 ≡ *Drechslera biseptata* (Saccardo & Roumeguère) Richardson & Fraser - Trans. Br. Mycol. Soc. 51: 148, 1968 ≡ *Marielliottia biseptata* (Saccardo & Roumeguère) Shoemaker - Can. J. Bot. 76: 1560, 1998.

Drechslera biseptata, CBS 277.60. a, b. Conidiophores bearing conidia; c. chlamydospores. a.× 640; b. ×1600; c. ×950; d. ×2300.

Hyphomycetes, dimorphic *Onygenales*. Genus: *EMMONSIA*

Generic description. Colonies white to tan, glabrous to cottony. Conidia one-celled, sessile or on stalks alongside hyphae or on ampulliform supportive cells. At 37°C either adiaspores or yeast cells are formed.
Teleomorph. *Ajellomyces* (*Ascomycota, Euascomycetes, Onygenales*: *Onygenaceae*).
Molecular diagnostics. Phylogeny based on ITS sequencing data was provided by Peterson & Sigler (1998). SSU rDNA restiction based on NCBI U29390, *E. parva:*

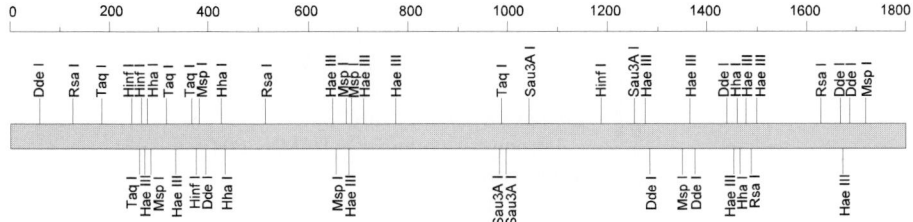

References. Sigler (1996), Peterson & Sigler (1998).

Key to the treated species of Emmonsia:

1a.   Growth at 40°C **→ 2**
1b.   Growth at 37°C but not at 40°C . . . . *E. crescens* **(635)**
2a.   Growth at 37°C and *in vivo* with yeast cells . . . . *E. pasteuriana* **(638)**
2b.   Growth at 37°C and in vivo with adiaspores . . . . *E. parva* **(637)**

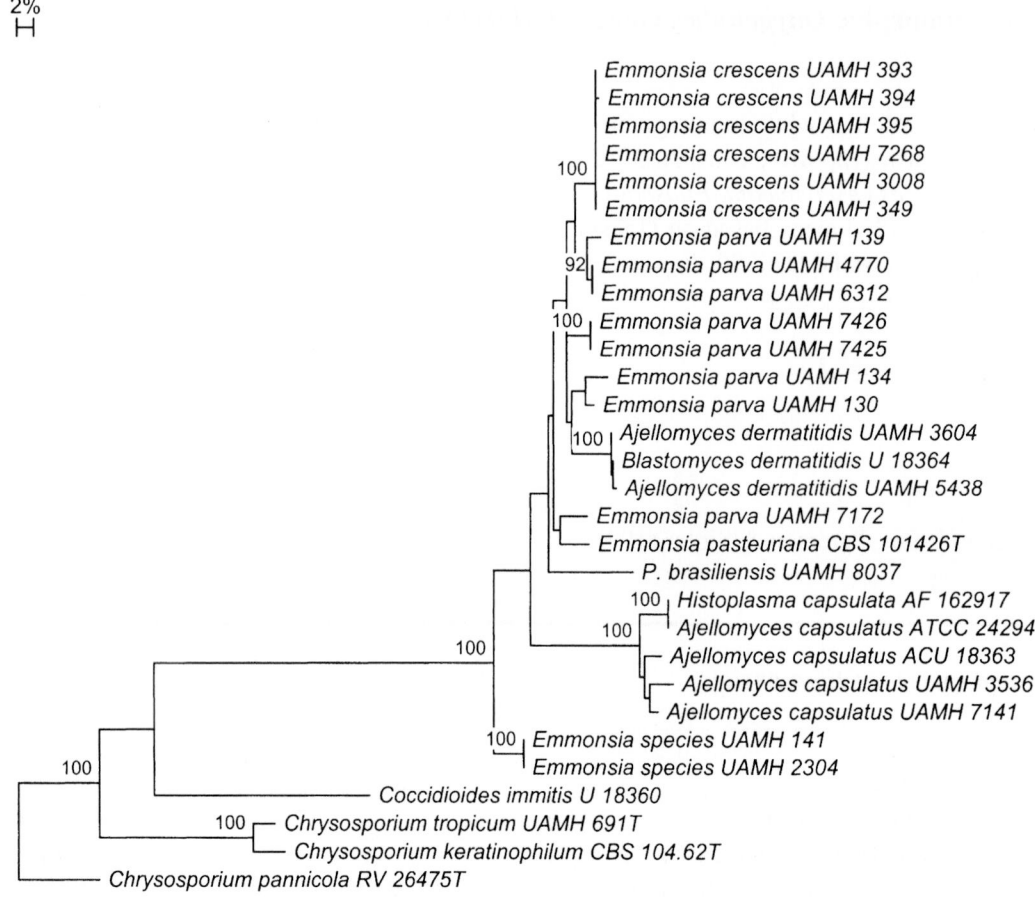

Fig. 56. Phylogenetic tree of *Emmonsia* and related systemic *Onygenales* based on the near-complete ITS rDNA domain using Neighbor joining algorithm with Kimura correction. Data are mainly those published by Peterson & Sigler (1998). Bootstrap values >90 from 100 resampled datasets are shown. *Chrysosporium pannicola* is chosen as outgroup. *Emmonsia parva* is more variable than *E. crescens*. *Coccidioides immitis* is found at a larger distance and shows more affinity to the geophilic *Chrysosporium* species.

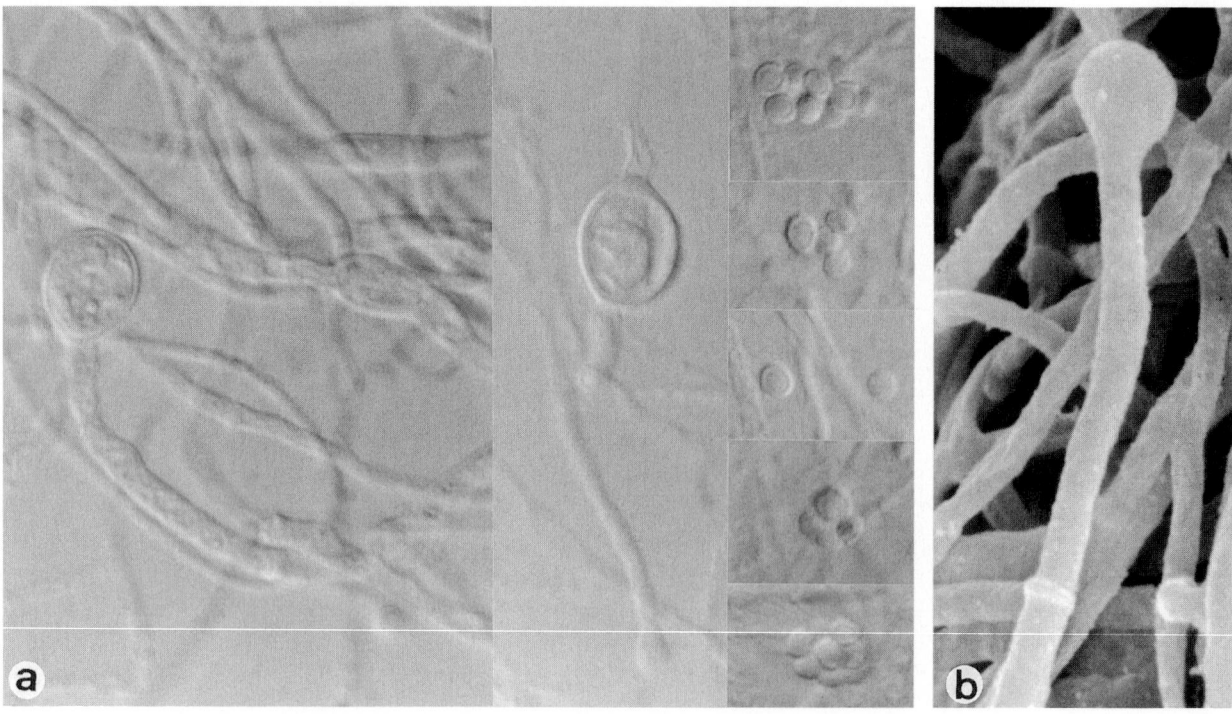

Emmonsia crescens, CBS 178.60. Adiaspores and clusters of conidia. a. ×1600; b. ×2500.

Emmonsia crescens Emmons & Jellison

Colony characteristics. Colonies (SGA, 24°C) with moderate growth, white to pale brown, often with yellowish or reddish tinge, glabrous, later cottony; reverse pale greyish-brown brown to reddish-brown. At 37°C on BHI colonies butryous, cerebriform.

Microscopy. Conidia at 24°C sessile or on slender stalks or terminal inflations, subhyaline, verrucose or smooth-walled, thin-walled, 1-celled, (sub)spherical, 2-5 × 2-4 μm, with narrow basal scars. At 37°C on BHI the blastic conidia swell and become spherical, thick-walled, 20-140 μm diam (adiaspores). Adiaspores produced *in vivo* multinucleate, 200-700 μm diam with walls up to 70 μm thick (Doby, 1986).

Teleomorph. *Ajellomyces crescens* Sigler (*Ascomycota, Euascomycetes, Onygenales: Onygenaceae*).

Physiology. Growth at 37°C but not at 40°C. Urase positive. Reduced growth on Mycosel agar.

Molecular diagnostics. ITS restriction map based on NCBI AF038335:

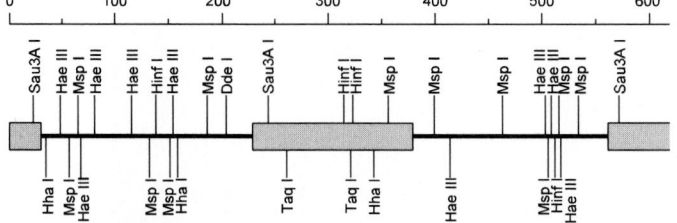

Pathogenicity. BSL-2. Agent of adiaspiromycosis (p. 25). Adiaspores are liberated conidia, which after inhalation enlarge in the alveoli of the host. The adiaspore remains localized and does not reproduce, causing no or only a mild disease (de Almeida Barbosa *et al.*, 1997). Each adiaspore is surrounded by a small granulome. In a later stage concentric layers of fibrous tissue are deposited; in these residual foci, fungal cells are often no longer visible. The fungus occurs in burrows of rodents (Emmons & Ashburn, 1942) and is inhaled by animals living in soil. In endemic areas, the percentage of infected animals is remarkably high (Hubálek *et al.*, 1993b, 1995).

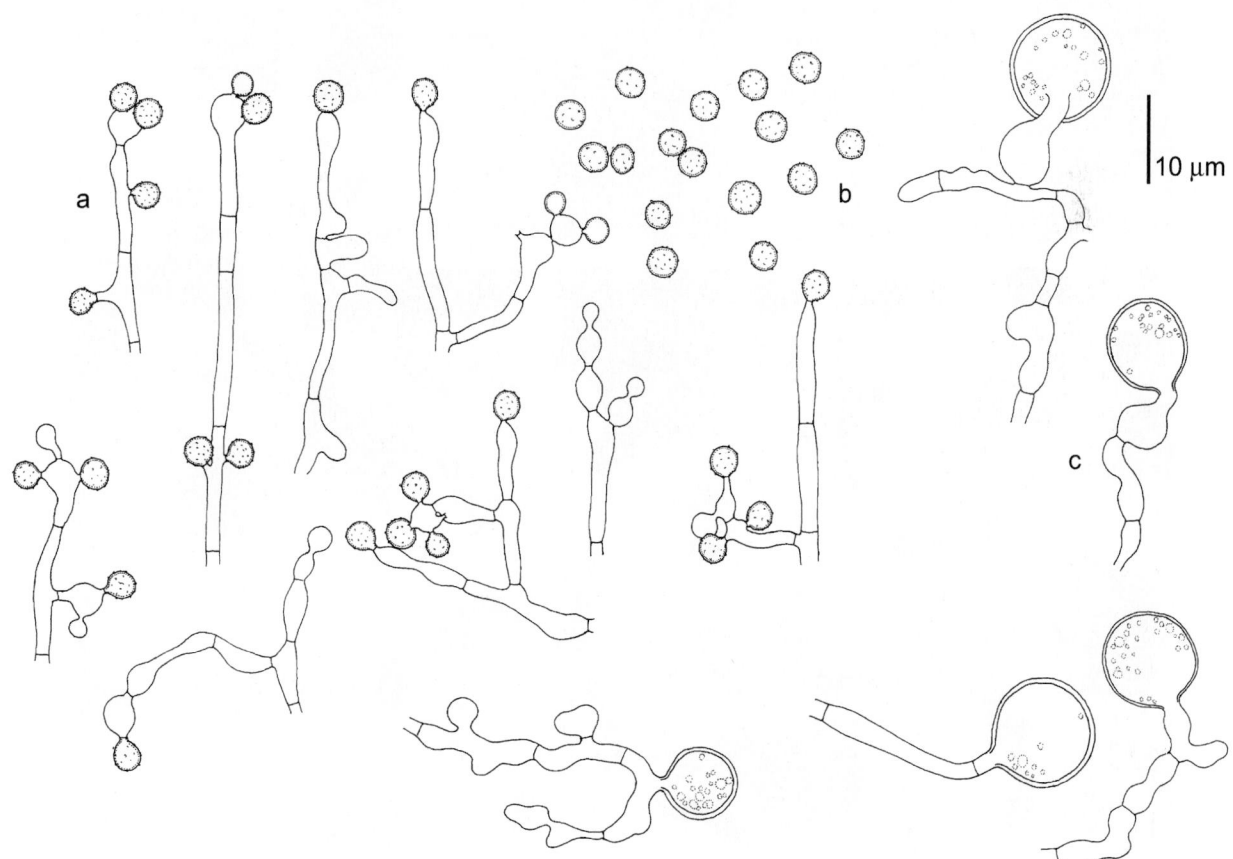

Emmonsia crescens, CBS 508.78. a. Conidiogenous cells; b. conidia; c. adiaspores in culture.

Odening *et al.* (1998) reported infections in zoo parrots. Mycoses in humans are rare (England & Hochholzer, 1993), then frequently in association with *Coccidioides immitis* (p. 593). Similar to the latter species, disseminated disease is particularly noted in patients with AIDS (Echavarria *et al.*, 1993).

Distribution. World-wide in temperate zones.

References. Emmons & Ashburn (1942), Emmons & Jellison (1960), Jellison (1969), van Oorschot (1980), Barbas-Filho *et al.* (1990), Peres *et al.* (1992), Sigler (1996).

Nomenclature. *Emmonsia crescens* Emmons & Jellison - Annls N.Y. Acad. Sci. 89: 96, 1960 ≡ *Chrysosporium parvum* (Emmons & Ashburn) Carmichael var. *crescens* (Emmons & Jellison) Carmichael - Can. J. Bot. 40: 1164, 1962 ≡ *Emmonsia parva* (Emmons & Ashburn) Ciferri & Montemartini var. *crescens* (Emmons & Jellison) van Oorschot - Stud. Mycol. 20: 59, 1980.

Ajellomyces crescens Sigler - J. Med. Vet. Mycol. 34: 305, 1996.

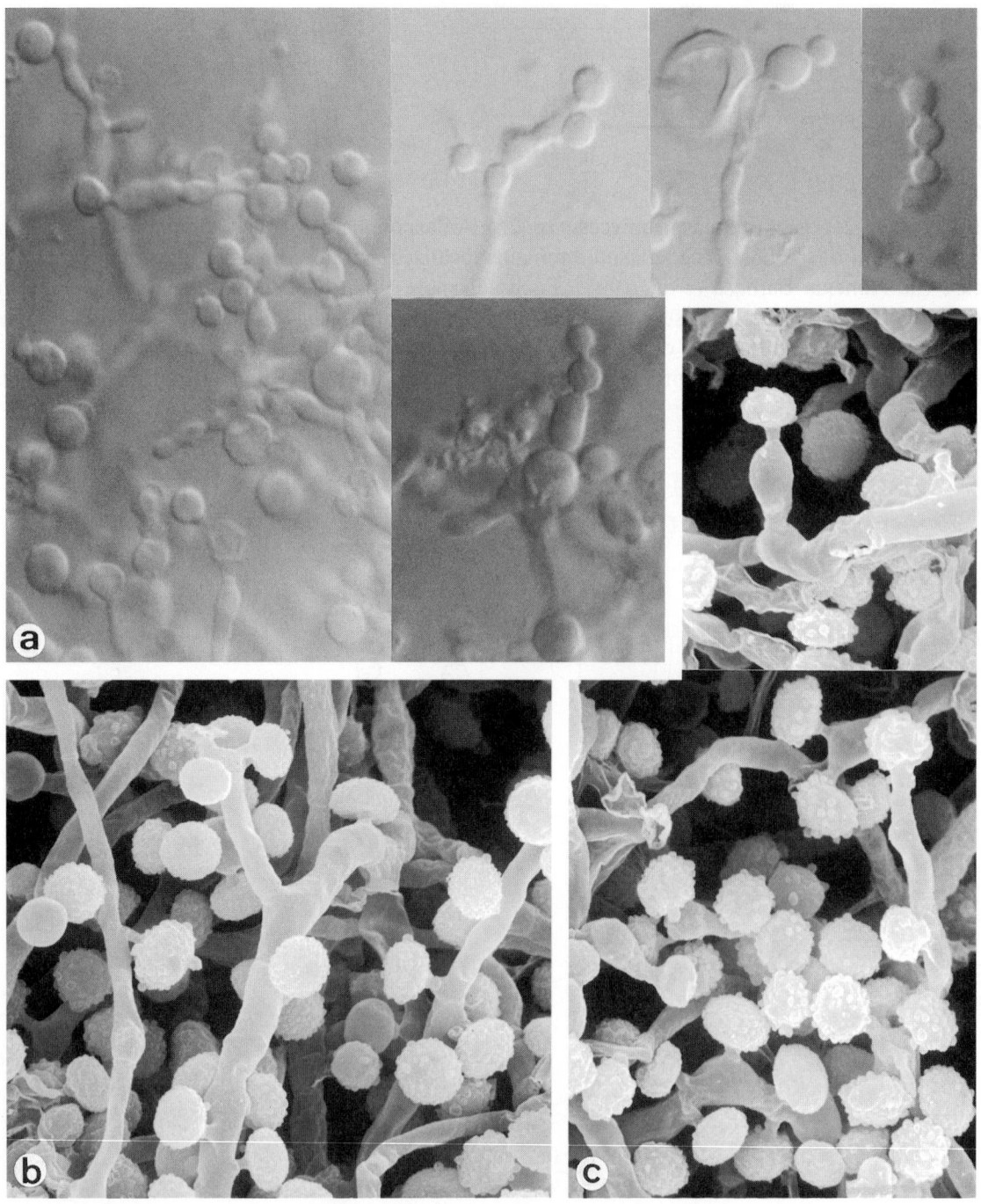

Emmonsia crescens, CBS 277.60. Conidiogenous cells and conidia. a. ×1600; b. ×3675; c. ×3875.

Emmonsia parva (Emmons & Ashburn) Ciferri & Montemartini

Cultural characteristics. Colonies (SGA, 24°C) with moderate growth, white to pale brown, glabrous, later cottony. At 37°C on BHI colonies butryous, cerebriform.

Microscopy. Conidia at 24°C sessile or on slender stalks or terminal inflations, subhyaline, verrucose or smooth-walled, thin-walled, 1-celled, (sub)spherical, 2-5 × 2-4 µm, with narrow basal scars. At 37°C on BHI the blastic conidia swell and become spherical, thick-walled, 8-20 µm diam (adiaspores). Adiaspores produced *in vivo,* oligonucleate, 10-40 µm diam.

Physiology. Growth at 40°C. Urease positive. Good growth on Mycosel agar.

Differential diagnosis. The filamentous form at 24°C is morphologically identical to that of *Emmonsia crescens* (p. 635). The species are distinguished by their adiaspores produced at 37°C. See also the comment on p. 634.

Molecular diagnostics. SSU and ITS restriction maps based on AF038324:

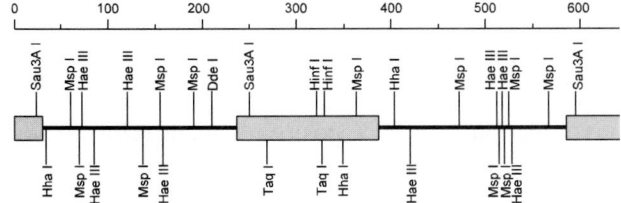

Pathogenicity. BSL-2. Agent of pulmonary adiaspiromycosis (p. 25) in rodents. A case in a beaver from Sweden was reported by Morner *et al.* (1999).

References. Sigler (1996), Peterson & Sigler (1998).

Distribution. Endemic zones in Southwestern U.S.A., Australia and Eastern Europe.

Nomenclature. *Haplosporangium parvum* Emmons & Ashburn - Publ. Health Rep., Washington 57, 1719, 1942 ≡ *Emmonsia parva* (Emmons & Ashburn) Ciferri & Montemartini - Mycopath. Mycol. Appl. 10: 314, 1959 ≡ *Chrysosporium parvum* (Emmonsia & Ashburn) Carmichael - Can. J. Bot. 40: 1164, 1962.

Emmonsia pasteuriana Drouhet *et al.*

Colony characteristics. Colonies (SGA, 24°C) with moderate growth, white, woolly to powdery with smooth sectors; reverse brownish. At 37°C on BHI butyrous.

Microscopy. Conidia at 24°C on slender stalks alongside hyphae or in groups on inflated cells, subhyaline, slightly verruculose, thin-walled, 1-celled, spherical, 2-3 × 3-4 µm. Additional sessile, broad-based, verrucose conidia 6-7 × 4.0-4.5 µm are present. At 37°C on BHI budding cells ellipsoidal, 2-4 µm in length.

Physiology. Good growth on Mycosel agar.

Molecular diagnostics. ITS restriction map based on CBS 101426.:

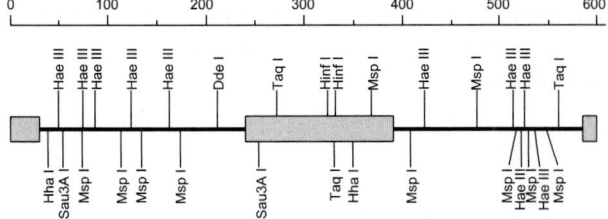

Pathogenicity. BSL-2. Described from an extensive cutaneous infection in a HIV positive patient (Gori *et al.*, 1998a). In tissue yeast cells are formed, intermingled with somewhat swollen cells.

References. Drouhet *et al.* (1998), Drouhet & Huerre (1999).

Nomenclature. *Emmonsia pasteuriana* Drouhet, Guého & Gori, *in* Drouhet, Guého, Gori, Huerre, Provost, Borgers & Dupont - J. Mycol. Méd. 8: 90, 1998.

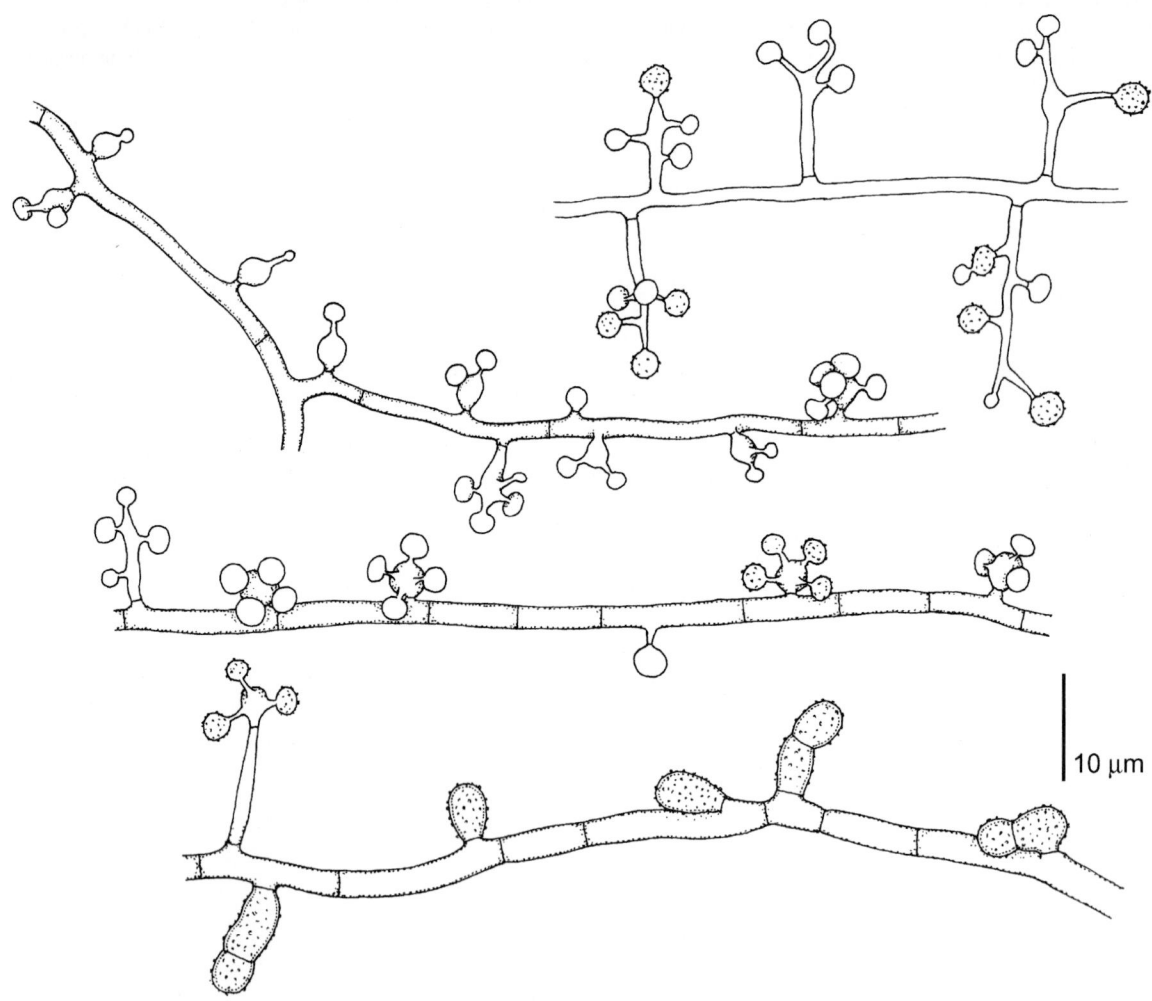

Emmonsia pasteuriana, CBS 101426. a. Hyphae with ampulliform cells and conidia; b. broad-based conidia.

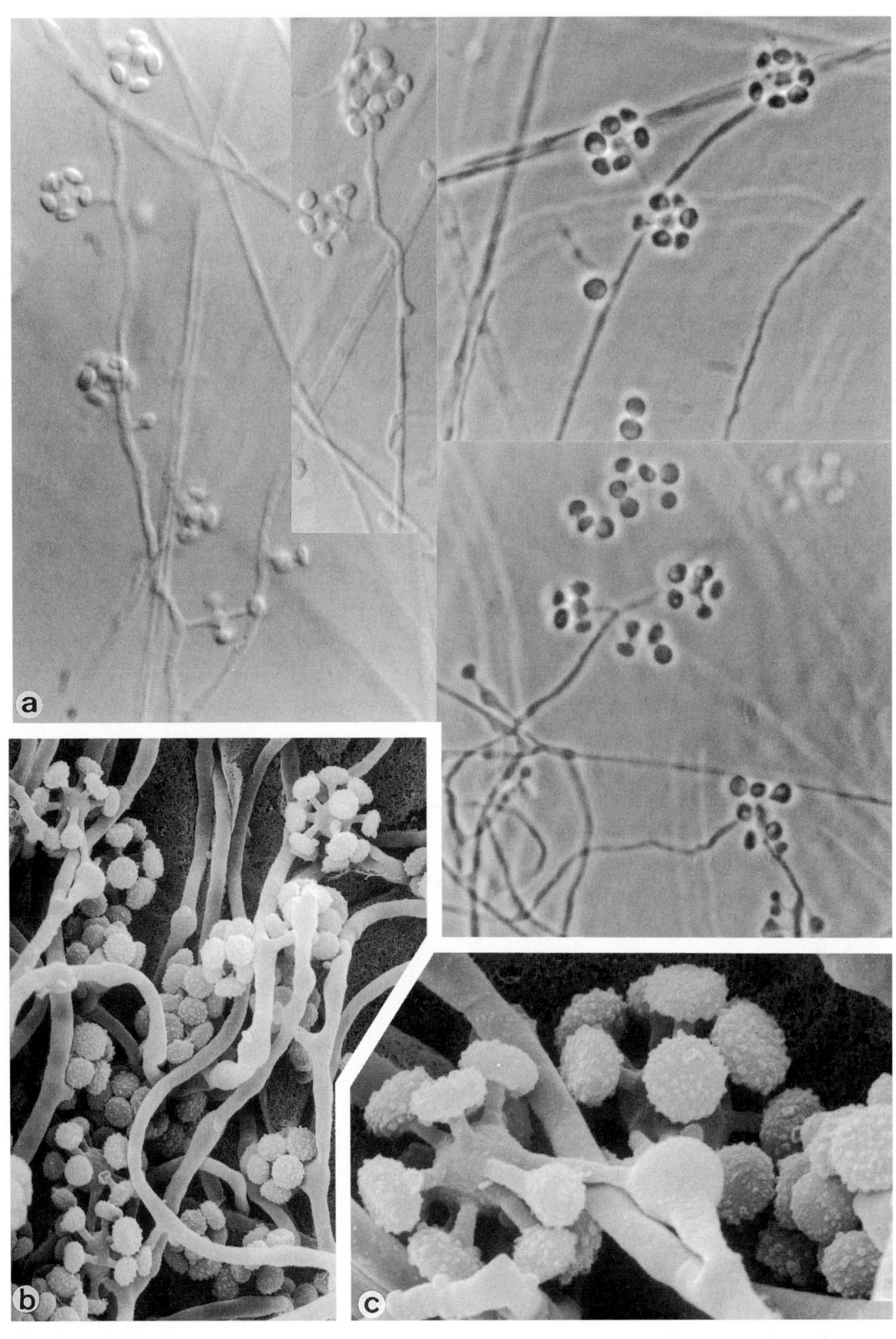

Emmonsia pasteuriana, CBS 101426. Conidiogenous cells and conidia. a. ×1600; b. ×3050; c. ×8000.

Hyphomycetes. Genus: *ENGYODONTIUM*

Engyodontium album (Limber) de Hoog

Colony characteristics. Colonies (MEA 2%) moderately expanding, white, appearing lanose to floccose, up to 2 mm high, sometimes zonate; reverse ochraceous-buff or uncoloured.

Microscopy. Conidiophores ascending, 2-4 µm wide, somewhat stiff, bearing conidiogenous cells in whorls of 1-3, at wide, often right angles. Conidiogenous cells consisting of an elongate to subcylindrical, tapering basal part, 10-24 (-30) × 1.5-2.5 µm, and a well developed rachis, up to 35 µm long and rather constantly 1 µm wide, geniculate, with up to 1 µm long denticles. Conidia hyaline, smooth-walled, (sub)spherical, 2-3 × 1.5-2.5 µm.

Molecular diagnostics. The species is related to the insect-inhabiting genus *Beauveria;* compare phylogenetic tree on p. 524. ITS restriction map based on NCBI Z54110:

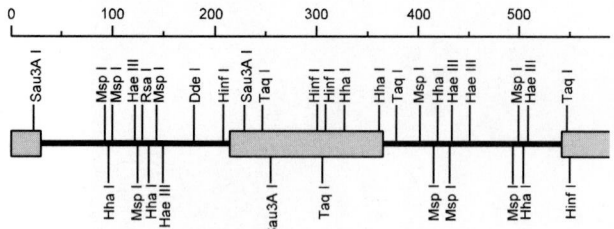

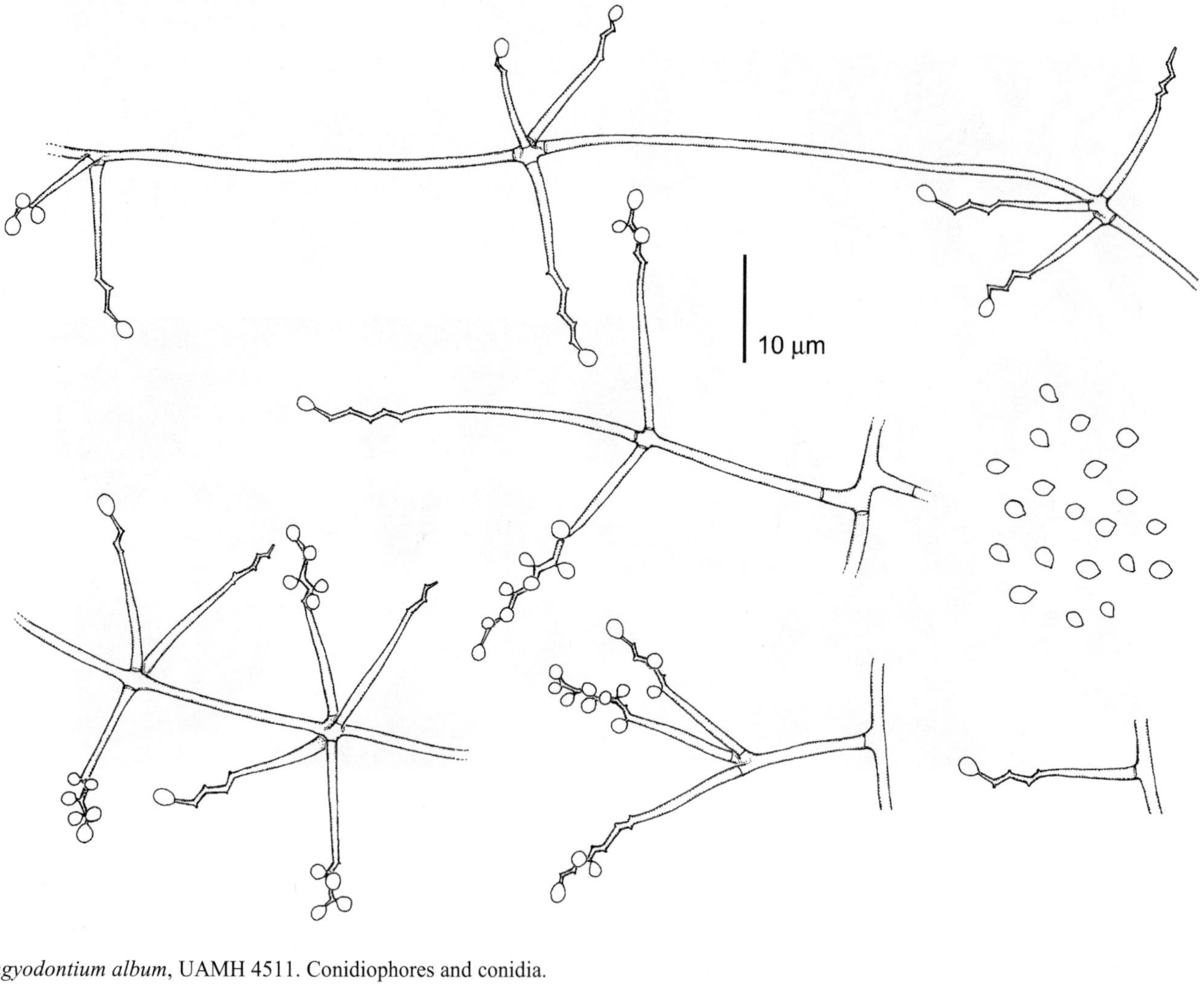

Engyodontium album, UAMH 4511. Conidiophores and conidia.

Pathogenicity. BSL-1. Air-borne saprobe. Cases include keratitis (McDonnell *et al.*, 1984), cerebritis (Seeliger, 1983) and endocarditis (Augustinsky *et al.*, 1990).

Reference. De Hoog (1972).

Nomenclature. *Tritirachium album* Limber - Mycologia 32: 27, 1940 ≡ *Beauveria alba* (Limber) Saccas - Revue Mycol. 13: 64, 1948 ≡ *Engyodontium album* (Limber) de Hoog - Persoonia 10: 53, 1978.

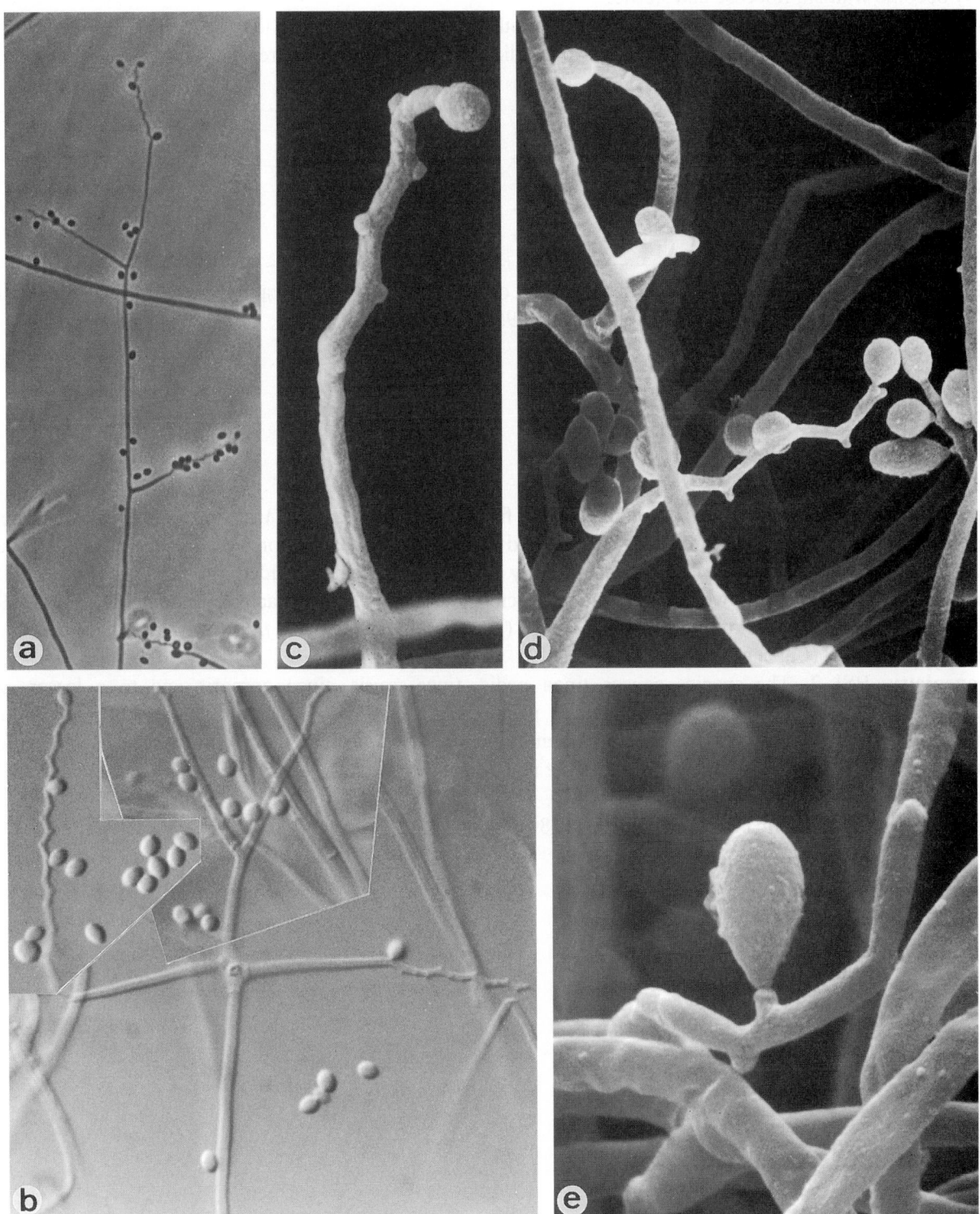

Engyodontium album, UAMH 4511. a, b. Conidiophores with conidiogenous cells; c-e. denticulate rachids with conidia. a. ×512; b. ×1600; c. ×7400; d. ×4200; e. ×12500.

Hyphomycetes, dermatophytes. Genus: *EPIDERMOPHYTON*

Epidermophyton floccosum (Harz) Langer. & Milochevitch

Colony characteristics. Colonies (SGA) velvety, sometimes powdery, felty, velvety, becoming woolly, gently folded, pale yellow to ochre or mustard-yellow; reverse yellowish-tan with yellow-brown centre. The colonies quickly become whitish, floccose and sterile.

Microscopy. Macroconidia arranged in clusters, smooth-walled, rather thin-walled, 2-5-celled, 10-40 × 6-12 µm, clavate with blunt tip. Chlamydospores and arthroconidia common in older cultures.

Physiology.

BCPCG	+	T-1	+	T-6	w	
Urease	w	T-2	+	T-7	w	
Hair perforation	−,+	T-3	+			
Rice (growth/sporulation)	+/+	T-4	+			
37°C growth	+	T-5	+			

Molecular diagnostics. The genus was compared with other dermatophyte genera by Kawasaki *et al.* (1996) using mtDNA RFLP. ITS restriction map based on CBS 970.95:

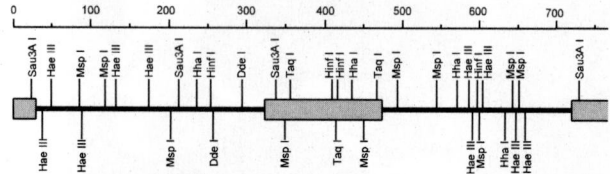

Differential diagnosis. The genus is distinguished from *Trichophyton* and *Microsporum* by the absence of microconidia. The macroconidia are rather thin-walled and develop as lateral or terminal outgrowths of mature hyphae, and initially lack a basal septum. Occasionally a reddish pigment is exuded into the agar (Frágner, 1968; Listemann, 1974). The dark brown, soil-inhabiting species *E. stockdaleae* Prochacki & Engelhardt-Zasada (1974), now known to be a synonym of *Trichophyton ajelloi* (p. 956), differs from *E. floccosum* by longer conidia and tolerance of 7% NaCl (Cabañes *et al.*, 1993).

Pathogenicity. BSL-2. The species causes tinea pedis, tinea cruris, tinea corporis and to a lesser extent onychomycosis. It may become epidemic among people using common shower or gym facilities. Seddon & Thomas (1997) described an invasive infection in an immunocompromised patient; Maruyama *et al.* (1999) found an association with genetic disorders. Cases are rare in animals (Stenwig & Taksdal, 1984; Terreni *et al.*, 1985).

Distribution. World-wide.

References. Rebell & Taplin (1970), Stockdale (1980).

Antifungal susceptibility.

Antifungal	Mean MICs	MIC 90	Strains	Reference
AMB	0.11	0.25	22	Fernández *et al.* (2000b)
CTZ	0.07	0.25	22	Fernández *et al.* (2000b)
FLZ	nd	4	3	Jessup *et al.* (2000)
FLZ	4.39	>64	13	Fernández *et al.* (2000b)
GRF	0.12	0.39	13	Wildfeuer *et al.* (1998)
GRF	nd	2	3	Jessup *et al.* (2000)
ITZ	<0.01	0.01	13	Wildfeuer *et al.* (1998)
ITZ	nd	<0.06	3	Jessup *et al.* (2000)
ITZ	0.03	0.125	22	Fernández *et al.* (2000b)
KTZ	0.01	0.03	13	Wildfeuer *et al.* (1998)
KTZ	0.08	0.125	22	Fernández *et al.* (2000b)
MCZ	0.06	1	22	Fernández *et al.* (2000b)
TBF	nd	0.015	3	Jessup *et al.* (2000)
TBF	0.02	0.06	22	Fernández *et al.* (2000b)
VCZ	0.03	0.1	13	Wildfeuer *et al.* (1998)
VCZ	0.03	0.125	22	Fernández *et al.* (2000b)

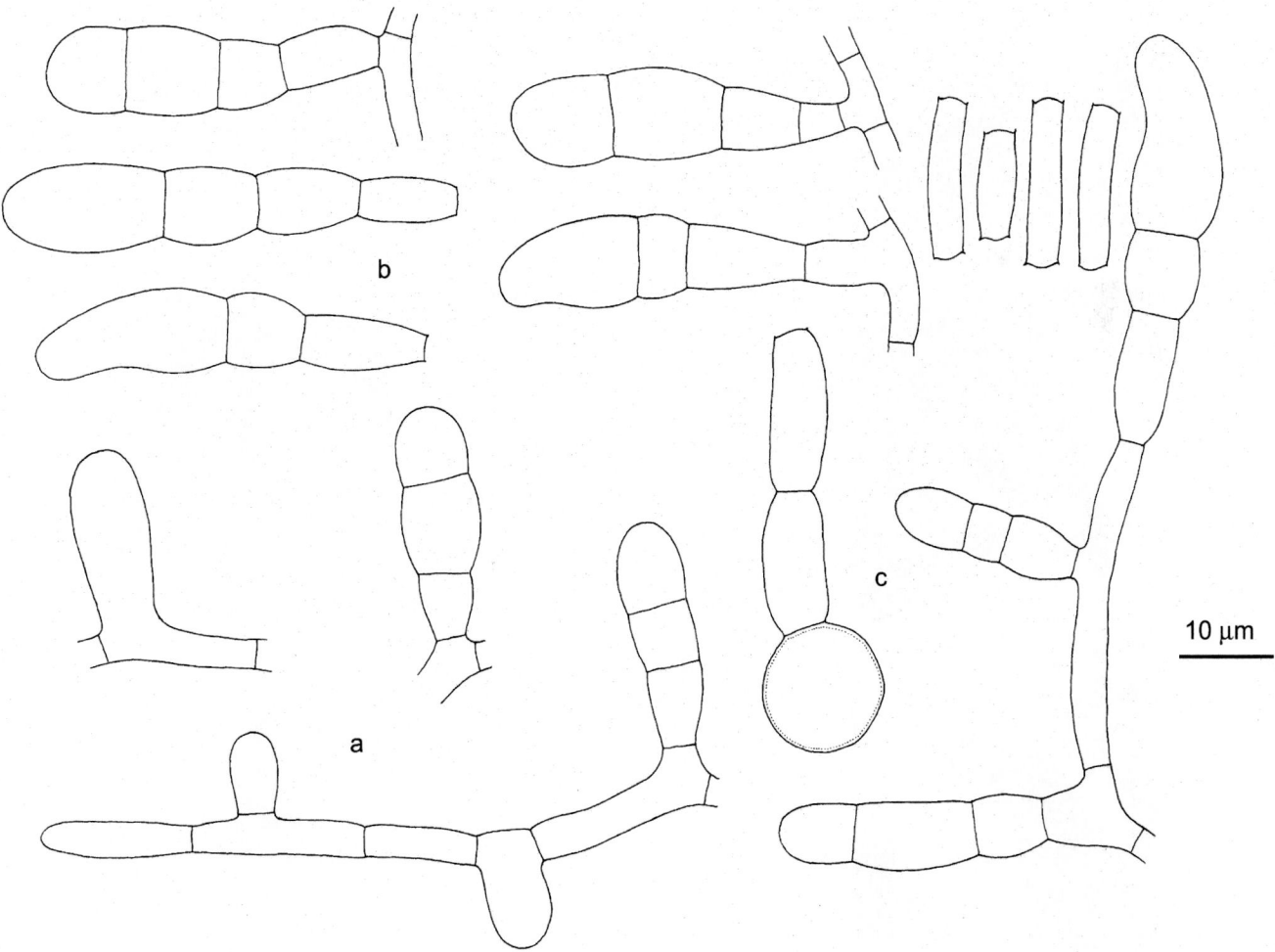

Epidermophyton floccosum, CBS 566.94. a. Conidial apparatus with developing macroconidia; b. macroconidia; c. chlamydospore-like cell; d. disarticulating hyphae.

Nomenclature. *Acrothecium floccosum* Harz - Bull. Soc. Imp. Nat. Moscou 44: 124, 1871 ≡ *Blastotrichum floccosum* (Harz) Belese & Voglino - Add. Syll. Nr. 3604, 1886 ≡ *Epidermophyton floccosum* (Harz) Langeron & Milochevitch - Annls Parasit. Hum. Comp. 8: 495, 1930 ≡ *Dactylium floccosum* (Harz) Sartory - Champ. Paras. Homme Anim. p. 871, 1923.

Trichophyton intertriginis Sabouraud - Dermatol. Topogr. 300: 51, 1905 ≡ *Trichophyton inguinale* Sabouraud - Arch. Méd. Expér. Anat. Pathol. 19: 565, 1907 (name change) ≡ *Epidermophyton inguinale* (Sabouraud) Sabouraud - Malad. Cuir Chev. 3: 420, 1910 ≡ *Closterosporia inguinalis* (Sabouraud) Grigorakis - Annls Sci. Nat., Bot., Sér. 10, 7: 411, 1925 ≡ *Microsporum inguinale* (Sabouraud) Guiart & Grigorakis - Lyon Méd. 141: 337, 1928.

Trichophyton cruris Castellani - J. Trop. Med. Hyg. 11: 262, 1908 ≡ *Epidermophyton cruris* (Castellani) Castellani & Chalmers - Man. Trop. Med. p. 609, 1910 ≡ *Fusoma cruris* (Castellani) Vuillemin - C.R. Hebd. Séanc. Acad. Sci., Paris 89: 450, 1929.

Epidermophyton plicarum Nicolau - Annls Derm. Syph., Sér. 5, 4: 65-87, 1913.

Epidermophyton clypeiforme MacCarthy - Annls Derm. Syph. 6: 30, 1925 ≡ *Epidermophyton floccosum* (Harz) Langeron & Milochevitch var. *clypeiforme* (MacCarthy) C.W. Dodge - Med. Mycol. p. 483, 1935.

Epidermophyton floccosum (Harz) Langeron & Milochevitch var. *nigricans* Frágner - Česká Mykol. 22: 203, 1968.

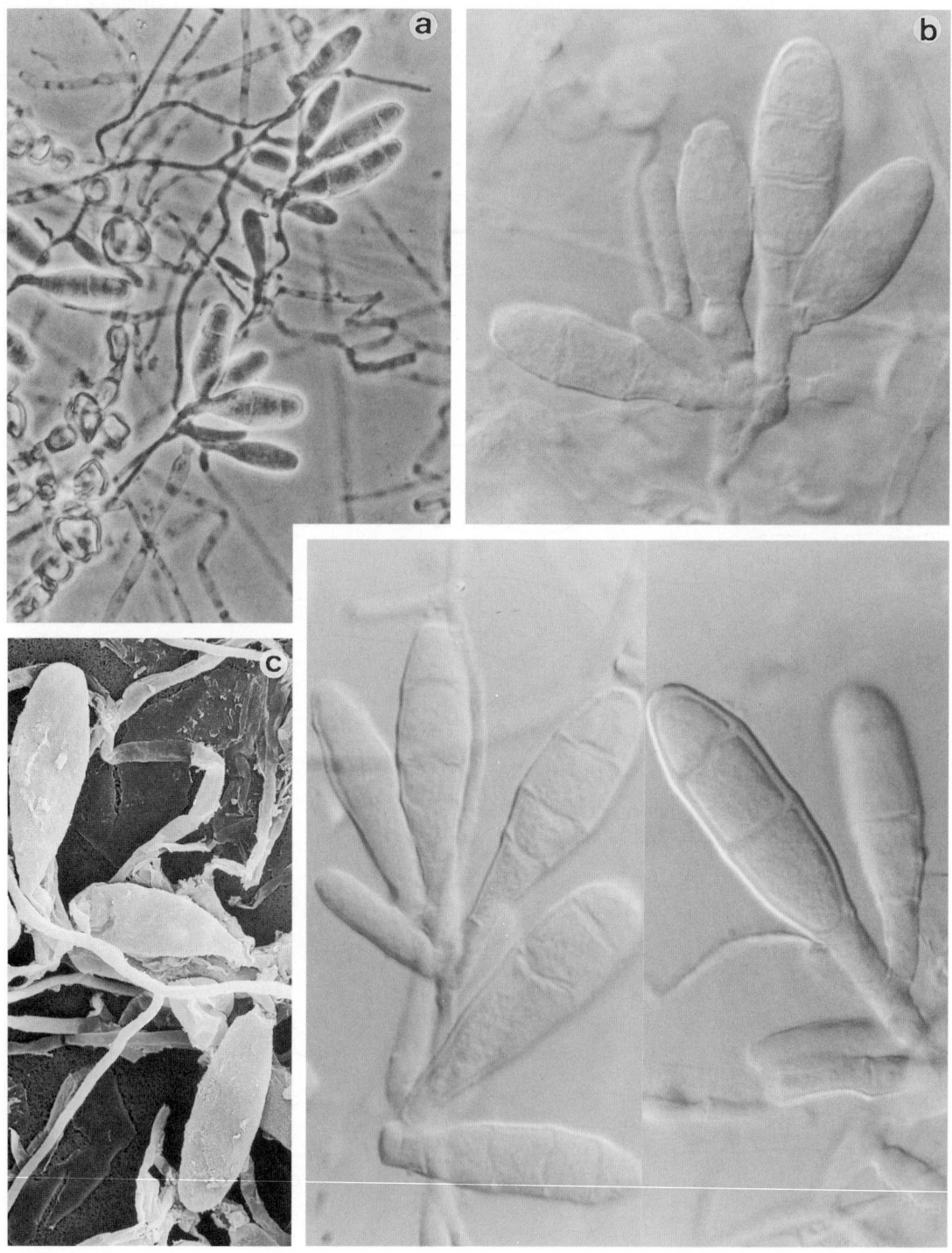

Epidermophyton floccosum, CBS 970.95. a. Fertile hyphae producing conidia and chlamydospore-like cells; b, c. conidial arrangement on fertile hyphae. a. ×640; b. ×1600; c. ×1400.

644

Hyphomycetes, black yeasts. Genus: *EXOPHIALA*

Generic description. Colonies mostly restricted, slimy at the centre due to yeast-like growth, smooth near the margin, later often becoming velvety or lanose, olivaceous-black. Conidiogenous cells intercalary, cylindrical, or free, flask-shaped or acicular, with relatively narrow, short or very short annellated zones. Conidia formed percurrently in slimy heads, (sub)hyaline, smooth-walled, 1-4-celled. Often chains of spherical cells are formed which bud profusely. Chains of barrel-shaped hyphal cells (torulose hyphae) may be present. Intolerant to 10% NaCl and benomyl. Lactate, inulin and citrate mostly not assimilated.

Teleomorph. *Capronia* (*Ascomycota, Euascomycetes, Chaetothyriales: Herpotrichiellaceae*).

Remarks. *Exophiala* is the main genus of black yeasts, characterised by annellidic conidiogenesis. Some cultures are entirely yeast-like (synanam. *Phaeococcomyces*), or form phialidic collarettes (synanam. *Phialophora*), sympodial conidiophores (synanam. *Rhinocladiella* without conidiophores, *Ramichloridium* with conidiophores) or dry conidial chains (synanam. *Cladophialophora*). Their taxonomic coherence has been proven by partial sequencing of the ribosomal gene (Haase *et al.*, 1995, 1999; Masclaux *et al.*, 1995; Spatafora *et al.*, 1995; see also tree on p. 377). Chlamydospores or sclerotial bodies may also be formed. The taxonomy of the genus is complicated; the number of taxa involved in vertebrate mycoses may actually be larger. Morphology is rather unreliable for species identification due to variable appearance.

References. De Hoog (1977, 1999), McGinnis (1978), de Hoog & McGinnis (1978), Matsumoto *et al.* (1987), Dixon & Polak-Wyss (1991), Uijthof & de Hoog (1995), de Hoog *et al.* (1995a), Uijthof (1996).

Key to treated species of Exophiala:

1a. Yeast cells absent or nearly absent → **2**
1b. Yeast cells present, at least at the colony centre → **3**
2a. Conidia (0-) 1 (-3)-septate; no assimilation of lactose, D-glucuronate and D-galacturonate **E. salmonis (664)**
2b. Conidia 0 (-1)-septate; assimilation of lactose, D-glucuronate and D-galacturonate . . . . . . **E. pisciphila (662)**
3a. Nitrite and creatine assimilated → **4**
3b. Nitrite and creatine not assimilated . . . . **E. dermatitidis (652)**
4a. Yeast cells with capsule visible in Indian Ink; erect, brown, multi-celled
    conidiophores may be present . . . . **E. spinifera (666)**
4b. Yeast cells without capsule; multi-celled conidiophores absent → **5**
5a. *meso*-Erythritol and ethanol not assimilated; conidiogenous cells with very short
    annellated zones, often in long chains . . . . **E. bergeri (648)**
5b. *meso*-Erythritol assimilated, ethanol assimilated or weakly assimilated; if the conidiogenous cells are
    in chains, they have longer annellations → **6**
6a. DL-Lactate assimilated; darkened, rocket-shaped conidiogenous cells mostly present . . . **E. jeanselmei (655)**
6b. DL-Lactate not assimilated; conidiogenous cells undifferentiated or ellipsoidal → **7**
7a. Salicin and xylitol not assimilated; mature conidiogenous cells have long annellated
    zones . . . . **E. moniliae (660)**
7b. Salicin and xylitol assimilated; conidiogenous cells always with very short annellated zones → **8**
8a. Lactose assimilated; L-arabinitol not assimilated . . . . **E. lecanii-corni (658)**
8b. Lactose not assimilated; L-arabinitol assimilated . . . . **E. castellanii (650)**

10%
H

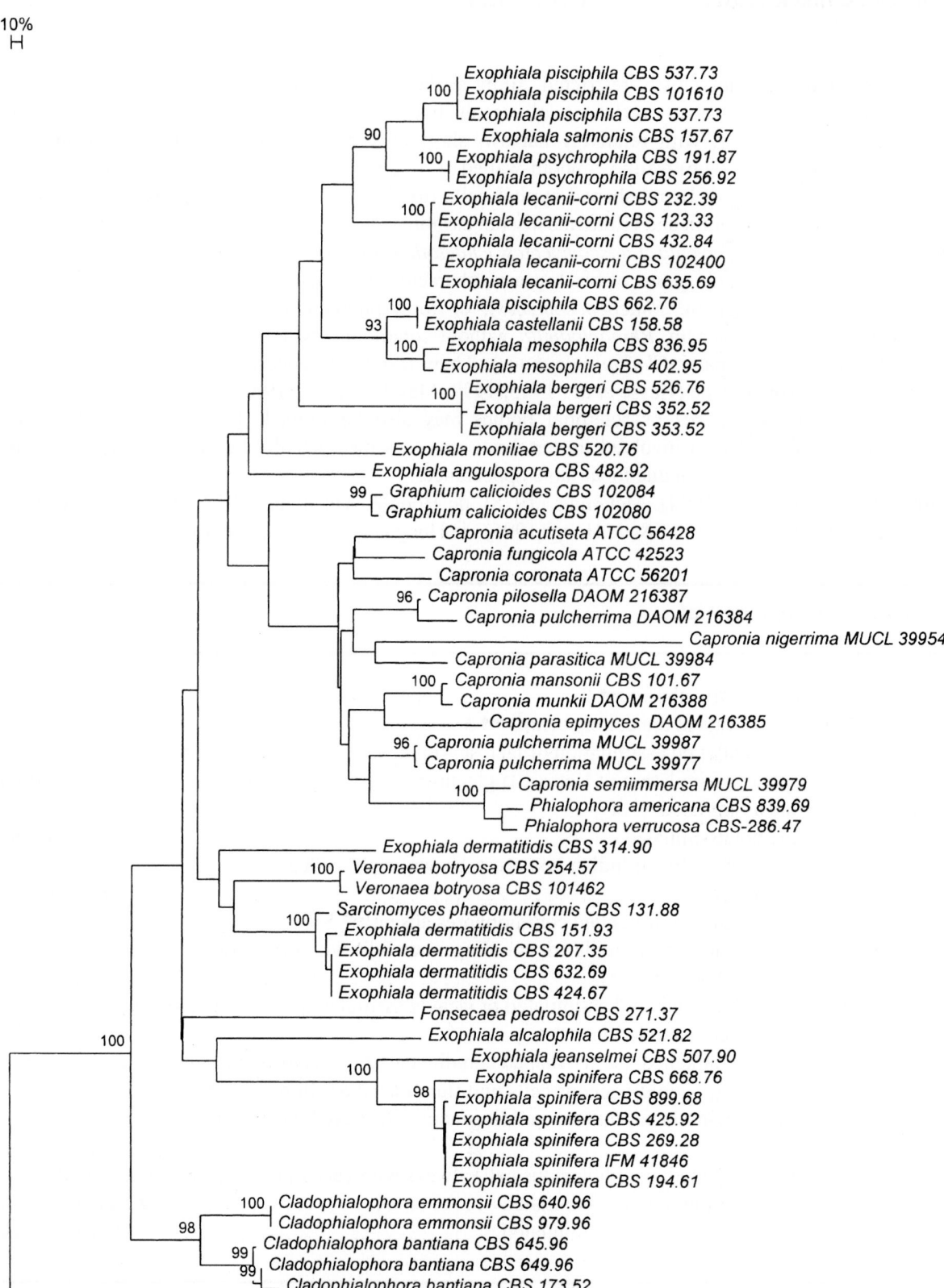

Fig. 57. Phylogenetic tree of *Exophiala* based on confidently aligned ITS rDNA sequences using Neighbor joining algorithm with Kimura correction. Bootstrap values >90 from 100 resampled datasets are shown. *Cladophialophora modesta* is chosen as outgroup. Within the genus, a number of clearly delimited species can be recognized, but overall the variability within the genus is large. In black yeasts and relatives, the molecular genera do not coincide with morphology-based taxonomy, as is also apparent from the SSU tree on p. 377.

Table 33. Salient physiological characteristics of *Exophiala* species.

	bergeri	castellanii	dermatitidis	jeanselmei	lecanii-corni	moniliae	pisciphila	salmonis	spinifera
D-Glucosamine	+	w	−	+	w	−	+	+	+
methyl-α-D-Glucoside	w,−	+	w	−	+	+	+	+	+
Salicin	w,−	+	+	w	+	−	+	+	+
Melibiose	+	+	+	−	+	+	+	+	+
Lactose	−	−	w	−	+	−	+	−	−
meso-Erythritol	−	+	+	+	+	+	+	+	+
Xylitol	+	+	+	+	+	−	+	+	+
L-Arabinitol	+,w	+	−	+	−	−	+	+	+
Galactitol	−	−	w	w	w	+	nd	w	v
myo-Inositol	+	−	+	w	−	−	w	+	+
D-Glucuronate	+	−	+	+	−	w	+	−	+
D-Galacturonate	+,w	−	w	+	−	−	+	−	+
DL-Lactate	v	−	−	+	−	−	−	−	+
Succinate	−	−	w	+	+	−	+	−	+
Ethanol	−	+	+	+	+	w	−	−	+
Nitrate	+	+	−	+	+	+	+	+	+
Nitrite	+	+	−	+	+	+	+	+	+
Creatine	+	+	−	+	+	+	+	+	+
Creatinine	+	+	−	+	+	+	+	+	+
0.1% Cycloheximide	−,w	+	+	−	+	−	+	+	+
Growth at 37°C	+	−	+	+	+	w,−	−	+	+
Growth at 40°C	−	−	+	−	w,−	−	−	−	−

Exophiala bergeri Haase & de Hoog

Colony characteristics. Colonies (SGA) restricted, black and slimy at the centre, with dry marginal zone. No brown diffusible pigment formed.

Microscopy. Yeast cells abudant. Hyphae dark olivaceous. Conidiogenous cells mostly arising as part of long, branched chains of subhyaline, ellipsoidal cells; annellated zones short, inconspicuous, only one on each conidiogenous cell, producing ellipsoidal conidia measuring about 3-4 × 2-3 μm.

Physiology.

Growth:					
D-Glucose	+	Lactose	–	D-Galacturonate	+,w
D-Galactose	+	Raffinose	+	DL-Lactate	v
L-Sorbose	+	Melezitose	+	Succinate	–
D-Glucosamine	+	Inulin	–	Citrate	–,w
D-Ribose	+,w	Soluble starch	–	Methanol	–
D-Xylose	+	Glycerol	+	Ethanol	–
L-Arabinose	+	*meso*-Erythritol	–	Nitrate	+
D-Arabinose	+,w	Ribitol	+	Nitrite	+
L-Rhamnose	+	Xylitol	+	Ethylamine	+
Sucrose	+	L-Arabinitol	+	L-Lysine	+
Maltose	+	D-Glucitol	+	Cadaverine	+
α,α-Trehalose	+	D-Mannitol	+	Creatine	+
methyl-α-D-Glucoside	w,–	Galactitol	–	Creatinine	+
Cellobiose	+	*myo*-Inositol	+	0.1% Cycloheximide	–,w
Salicin	w,–	Glucono-δ-lactone	+	Growth at 37°C	+
Melibiose	+	D-Gluconate	+,w	Growth at 40°C	–
		D-Glucuronate	+		

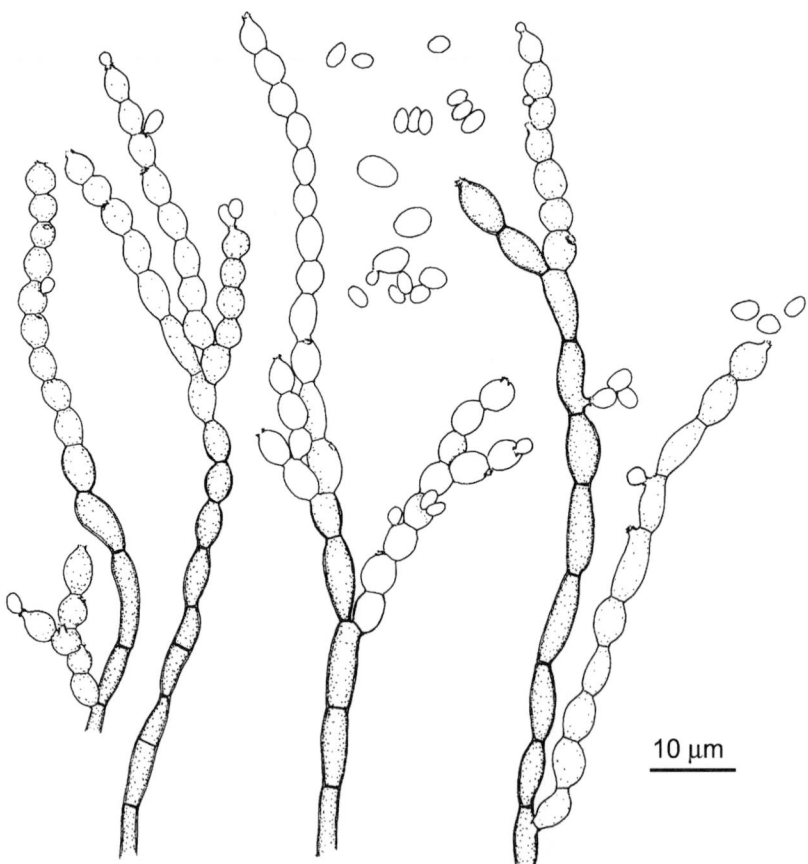

Exophiala bergeri, CBS 352.52. Conidial apparatus with conidia.

Differential diagnostics. *Exophiala castellanii* (p. 650) differs by absence of differentiated chains bearing conidiogenous cells, and by often having several annellated zones on a single cell. Numerous physiological criteria are available (Table 33).

Molecular diagnostics. ITS restriction map based on CBS 518.76:

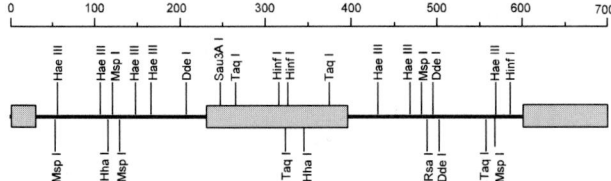

Pathogenicity. BSL-2. Rare species, known from (sub)cutaneous infections (Berger & Langeron, 1949). The species is identical to the neotype of *Sporotrichum gougerotii* (CBS 526.76) considered a *nomen confusum* (McGinnis & Ajello, 1982). *S. gougerotii* was originally described from a subcutaneous cyst (Matruchot, 1910).

Reference. Haase *et al.* (1999).

Nomenclature. *Candida nigra-canadensis* Berger, Beaudry & Gaumond, *in* Berger - Trans. R. Soc. Can. 38: 161, 1944 (invalid) ≡ *Torula bergeri* Langeron, *in* Berger & Langeron - Annls Parasit. Hum. Comp. 24: 597, 1949 ≡ *Pullularia bergeri* (Langeron) Seeliger, da Silva Lacaz & Ulson - Proc. Int. Congr. Trop. Med. Malar. 6: 641, 1959 ≡ *Exophiala bergeri* Haase & de Hoog, *in* Haase, Sonntag, Melzer-Krick & de Hoog - Stud. Mycol. 43: 91, 1999.

Exophiala castellanii Iwatsu *et al.*

Colony characteristics. Colonies (SGA) restricted, black and slimy at the centre, dry and flat, dark olivaceous towards the margin. No brown diffusible pigment formed.

Microscopy. Yeast cells abundant. Conidiogenous cells intercalary or free, young cells swollen, ellipsoidal, changing over into filaments; annellated zones inconspicuous, short, with 1-3 on a single conidiogenous cell, producing ellipsoidal to allantoid conidia of variable size.

Physiology.

Growth:					
D-Glucose	+	Lactose	–	D-Galacturonate	–
D-Galactose	+	Raffinose	+	DL-Lactate	–
L-Sorbose	+	Melezitose	+	Succinate	–
D-Glucosamine	w	Inulin	–	Citrate	–
D-Ribose	+	Soluble starch	w	Methanol	–
D-Xylose	+	Glycerol	+	Ethanol	+
L-Arabinose	+	*meso*-Erythritol	+	Nitrate	+
D-Arabinose	+	Ribitol	+	Nitrite	+
L-Rhamnose	+	Xylitol	+	Ethylamine	+
Sucrose	+	L-Arabinitol	+	L-Lysine	+
Maltose	+	D-Glucitol	+	Cadaverine	w
α,α-Trehalose	+	D-Mannitol	+	Creatine	+
methyl-α-D-Glucoside	+	Galactitol	–	Creatinine	+
Cellobiose	+	*myo*-Inositol	–	0.1% Cycloheximide	+
Salicin	+	Glucono-δ-lactone	w	Growth at 37°C	–
Melibiose	+	D-Gluconate	+		
		D-Glucuronate	–		

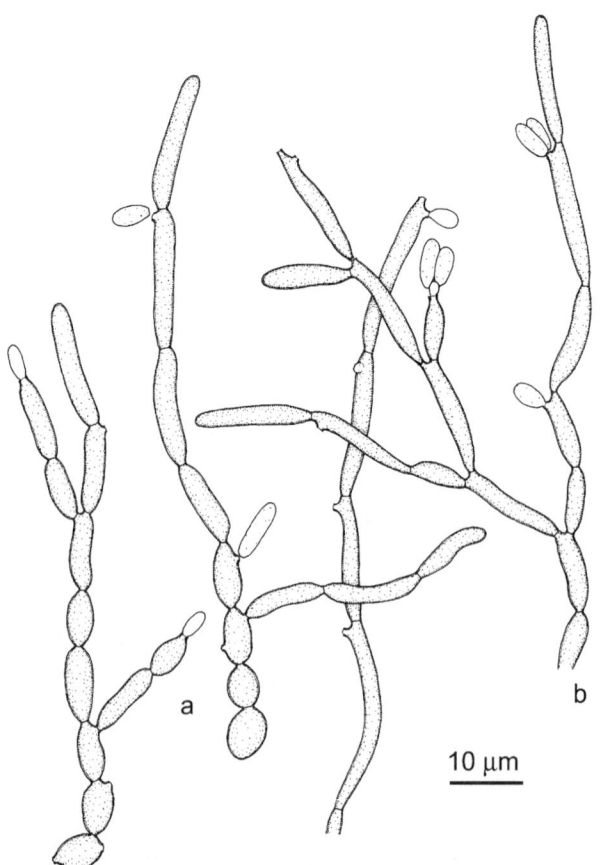

Exophiala castellanii, CBS 158.58. a. Young swollen cells; b. mature conidial apparatus.

Differential diagnosis. *Exophiala bergeri* is similar but produces conidiogenous cells in long chains; for physiological differences, see Table 33. *E. castellanii* differs from *E. dermatitidis* by assimilation of nitrate, nitrite, creatinine, absence of brown pigments and no growth at 40°C. *E. castellanii* differs from *E. jeanselmei* by having short annellated zones often with several on a cell, and by absence of growth with D-glucuronate, DL-lactate and succinate, and by tolerance of 0.1% cycloheximide.

Molecular diagnostics. SSU restriction map based on CBS 158.58:

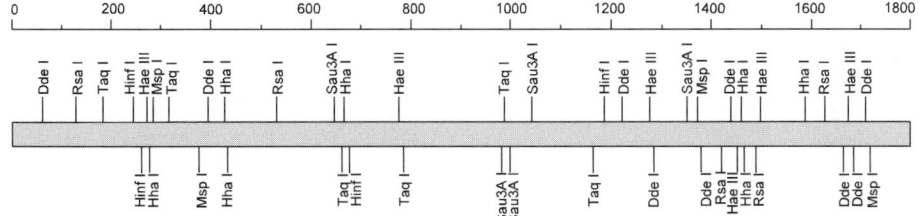

Pathogenicity. BSL-2. The species is occasionally isolated from clinical samples.

References. De Hoog (1977), Iwatsu *et al.* (1984).

Nomenclature. *Exophiala castellani* Iwatsu, Nishimura & Miyaji - Mycotaxon 20: 307, 1984 ≡ *Exophiala jeanselmei* (Langeron) McGinnis & Padhye var. *castellanii* (Iwatsu, Nishimura & Miyaji) Iwatsu & Udagawa - Mycotaxon 37: 292, 1990.

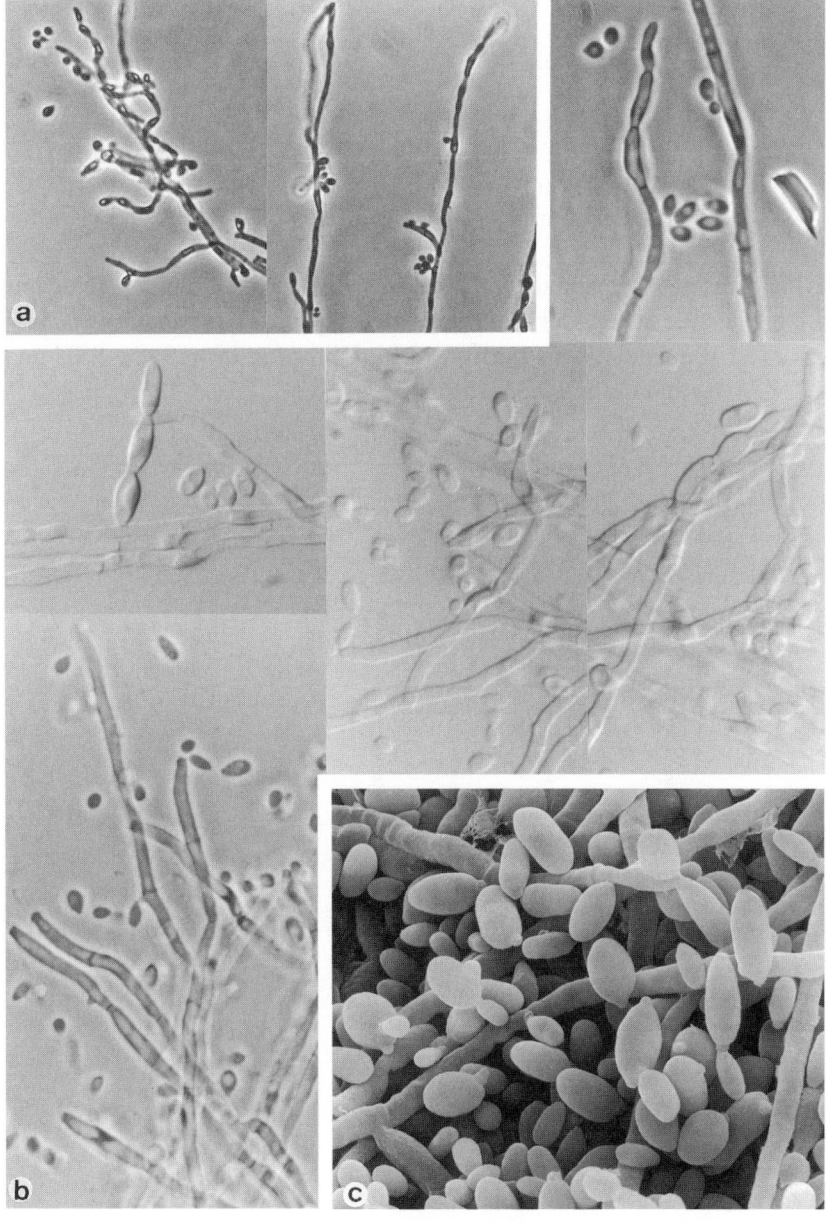

Exophiala castellanii, CBS 158.58. a, b. Conidial apparatus; c. yeast cells. a. ×640; b. ×1600; c. ×3350.

Exophiala dermatitidis (Kano) de Hoog

Colony characteristics. Colonies (SGA) restricted, smooth, waxy, often with brown pigment exuded into the agar.
Microscopy. Yeast cells abundant. Conidiogenous cells intercalary or free, then flask-shaped; annellated zones relatively wide, often in small groups, hardly elongating, producing broadly ellipsoidal conidia 2.5-4.0 × 2-3 μm. Spherical phialides with large, fragile collarettes may be present; phialoconidia small, ellipsoidal to subcylindrical. Sclerotial bodies sometimes formed.
Physiology.

Growth:		Lactose	w	D-Galacturonate	w
D-Glucose	+	Raffinose	+	DL-Lactate	–
D-Galactose	+	Melezitose	+	Succinate	w
L-Sorbose	+,w	Inulin	w	Citrate	–
D-Glucosamine	–	Soluble starch	w	Methanol	–
D-Ribose	w	Glycerol	+	Ethanol	+
D-Xylose	+	*meso*-Erythritol	+	Nitrate	–
L-Arabinose	+	Ribitol	+	Nitrite	–
D-Arabinose	+	Xylitol	+	Ethylamine	+
L-Rhamnose	+	L-Arabinitol	–	L-Lysine	+
Sucrose	+	D-Glucitol	+	Cadaverine	+
Maltose	+	D-Mannitol	w	Creatine	–
α,α-Trehalose	+	Galactitol	w	Creatinine	–
methyl-α-D-Glucoside	w	*myo*-Inositol	+	0.1% Cycloheximide	+
Cellobiose	+	Glucono-δ-lactone	+	Growth at 40°C	+
Salicin	+	D-Gluconate	+		
Melibiose	+	D-Glucuronate	+		

Differential diagnosis. The species is characterized by brown, flat, waxy colonies, wide annellated zones, growth at 40°C; no growth with nitrite. See also under the meristematic species *Sarcinomyces phaeomuriformis* (p. 897).
Molecular diagnostics. ITS1 sequencing was performed by Uijthof *et al.* (1998). SSU and ITS restriction maps based on NCBI X79312, Z75304 and CBS 207.35, resp.:

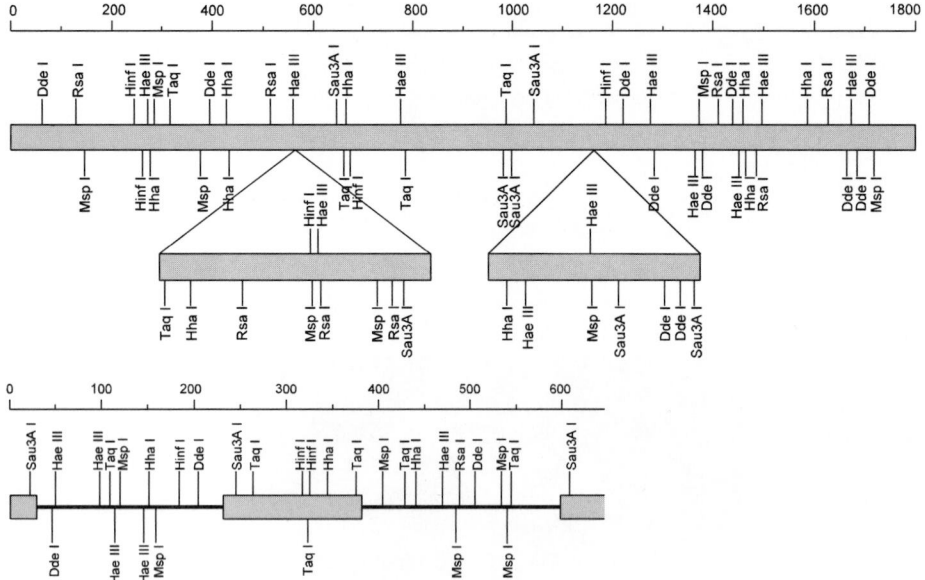

Pathogenicity. BSL-2. The species causes subcutaneous phaeohyphomycoses after traumatic implantation (Crosby *et al.*, 1989; Kim *et al.*, 1999). Keratitis was reported by Pospíšil *et al.* (1990) and Benoudia *et al.* (1999). It also is found subclinically in the lungs, occasionally causing pneumonia (Kusenbach *et al.*, 1990), and particularly occurs in patients with cystic fibrosis (Haase *et al.*, 1992). Otitis was reported by Kerkmann *et al.* (1999a). Infection through intravenous devices may occur (Kabel *et al.*, 1994; Nachman *et al.*, 1996). Disseminated, low virulence mycoses in leukemic patients were reported by Blaschke-Hellmessen *et al.* (1994) and Moissenet *et al.* (1995). Cerebral metastases may have a pulmonary portal of entry (Kenney *et al.*, 1992). In Asian patients chronic dissemination is often observed, the species showing a marked neurotropism (Shimazono *et al.*, 1963; Horré & de Hoog, 1999); its case-fatality rate is high (Matsumoto *et al.*, 1984b). The neurotropic clinical picture is particularly found in young, otherwise healthy persons (Hiruma *et al.*, 1993; Chang *et al.*, 2000).

References. Hohl *et al.* (1983), de Hoog & Haase (1993), Matsumoto *et al.* (1993), Uijthof *et al.* (1994, 1998), de Hoog *et al.* (1994c).

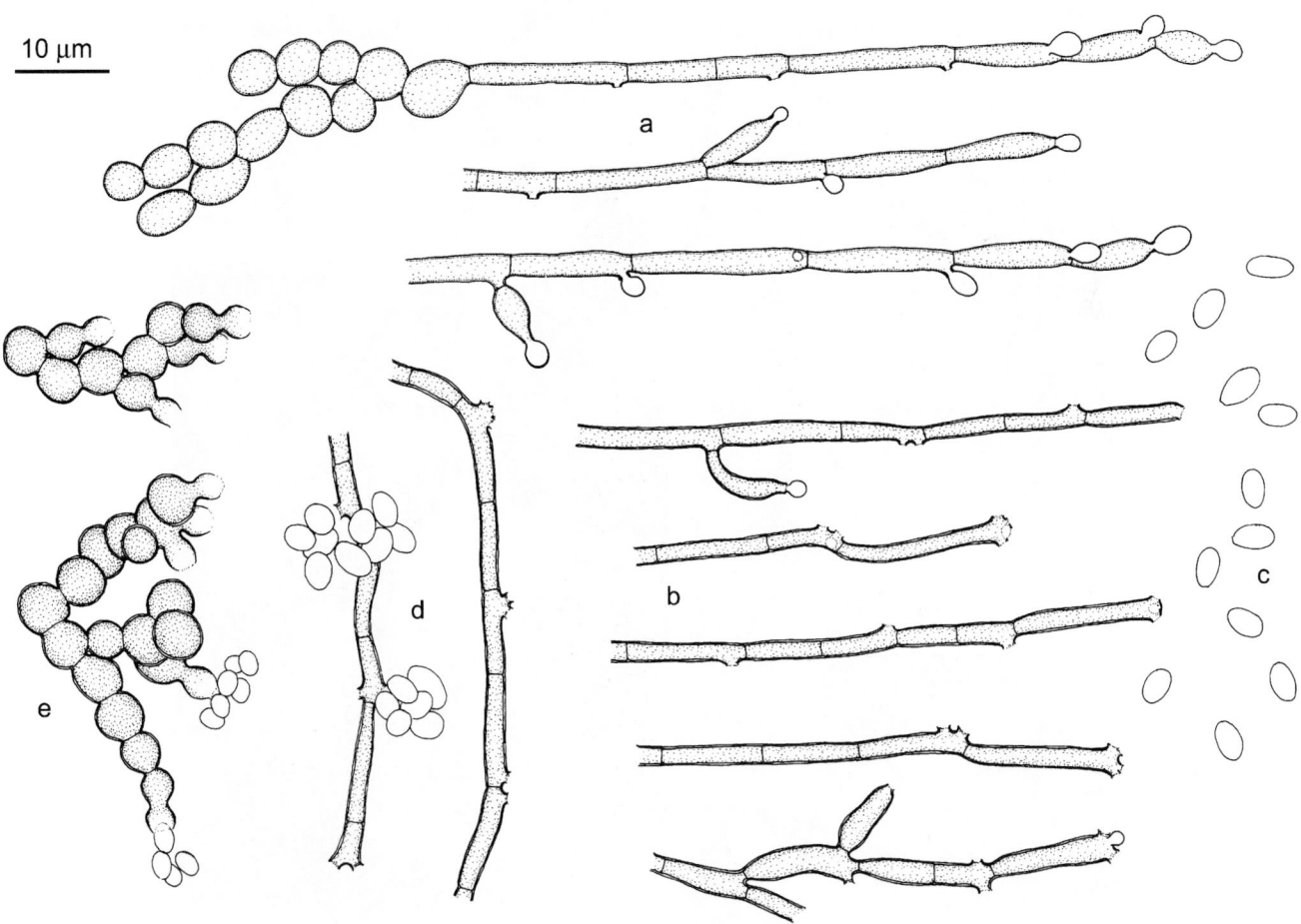

10 μm

Exophiala dermatitidis, CBS 748.88. a. Young conidial apparatus with conidia; b. mature conidiogenous cells; c. conidia; d. conidial heads; e. *Phialophora* synanamorph.

Antifungal susceptibility.

Antifungal	Mean MICs	Strains	Reference
AMB	0.25	3	Guarro *et al.* (1997)
5FC	4	3	Guarro *et al.* (1997)
FLZ	10.8	3	Guarro *et al.* (1997)
ITZ	0.31	8	McGinnis & Pasarell (1998a)
KTZ	0.25	3	Guarro *et al.* (1997)
MCZ	0.38	3	Guarro *et al.* (1997)
TBF	0.07	8	McGinnis & Pasarell (1998a)
VCZ	0.22	10	McGinnis & Pasarell (1998b)

Nomenclature. *Hormiscium dermatitidis* Kano - Aichi Igakukwai Zasshi 41: 11, 1934 ≡ *Fonsecaea dermatitidis* (Kano) Carrión - Arch. Derm. Syph. 61: 1008, 1950 ≡ *Hormodendrum dermatitidis* (Kano) Conant, *in* Conant, Smith, Baker, Callaway & Martin - Man. Clin. Mycol., ed. 2, p. 276, 1954 ≡ *Phialophora dermatitidis* (Kano) Emmons, Binford & Utz - Med. Mycol. p. 291, 1963 ≡ *Exophiala dermatitidis* (Kano) de Hoog - Stud. Mycol. 15: 118, 1977.

Mycotorula schawii Pereira-Filho - Revta Med. Rio Grande do Sul 5: 8, 1948.

Wangiella dermatitidis McGinnis - Mycotaxon 5: 355, 1977.

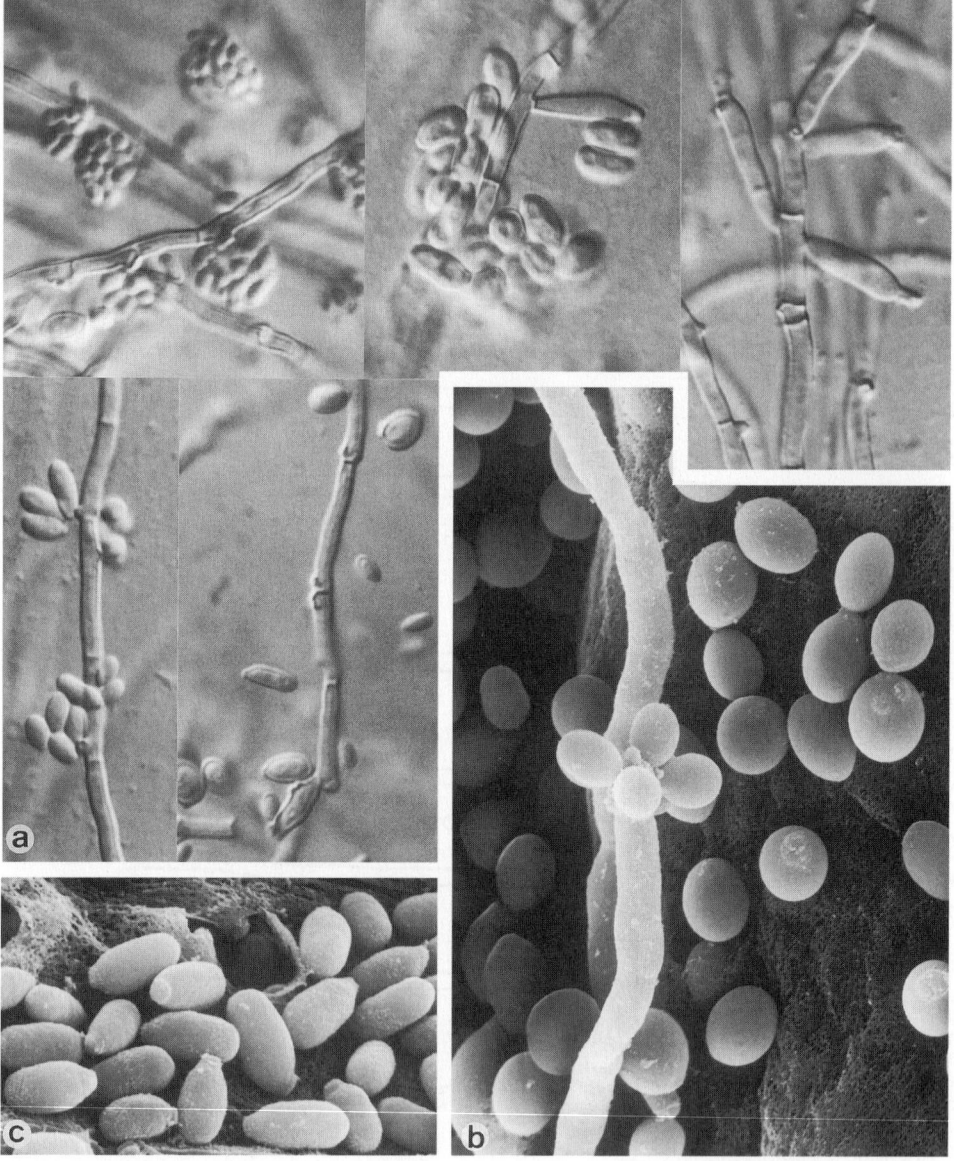

Exophiala dermatitidis, CBS 748.88. Conidial apparatus, conidia and yeast-like cells. a. ×1600; b. ×4000; c. ×6750.

Exophiala jeanselmei (Langer.) McGinnis & Padhye

Colony characteristics. Colonies (SGA) restricted, initially moist, soon forming velvety, olivaceous green aerial mycelium.

Microscopy. Yeast-like growth, when present, consisting of subspherical cells, each with an annellidic butt producing conidia, often cohering in long chains and changing over into hyphae. Conidiogenous cells on hyphae intercalary or rocket-shaped and brown, with inconspicuous annellated zones; the latter are slightly tapering, nearly smooth and produce narrow ellipsoidal conidia 2.6-5.9 × 1.2-2.5 μm.

Physiology.

Growth:					
D-Glucose	+	Lactose	–	D-Galacturonate	+
D-Galactose	+	Raffinose	–	DL-Lactate	+
L-Sorbose	+	Melezitose	+	Succinate	+
D-Glucosamine	+	Inulin	–	Citrate	–
D-Ribose	w	Soluble starch	–	Methanol	–
D-Xylose	+	Glycerol	+	Ethanol	+
L-Arabinose	+	*meso*-Erythritol	+	Nitrate	+
D-Arabinose	w	Ribitol	w	Nitrite	+
L-Rhamnose	+	Xylitol	w	Ethylamine	+
Sucrose	+	L-Arabinitol	+	L-Lysine	+
Maltose	+	D-Glucitol	+	Cadaverine	+
α,α-Trehalose	+	D-Mannitol	+	Creatine	+
methyl-α-D-Glucoside	–	Galactitol	w	Creatinine	+
Cellobiose	+	*myo*-Inositol	w	0.1% Cycloheximide	–
Salicin	w	Glucono-δ-lactone	w	Growth at 37°C	+
Melibiose	–	D-Gluconate	+	Growth at 40°C	–
		D-Glucuronate	+		

Differential diagnosis. *Exophiala lecanii-corni* (p. 658) does not assimilate L-arabinitol.

Molecular diagnostics. Heterogeneity of *E. jeanselmei* was found at the molecular level (Masuda *et al.*, 1989) and in the serology (Kaufman *et al.*, 1980). SSU and ITS restriction maps based on CBS 507.90:

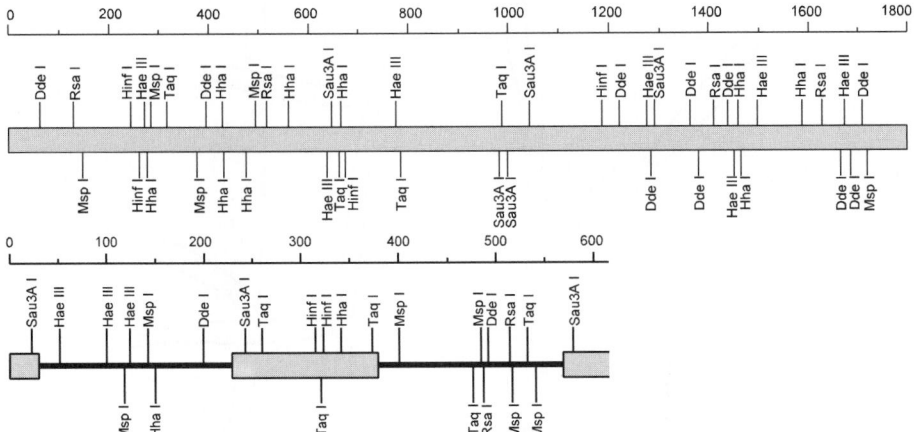

Pathogenicity. BSL-2. The clinical manifestations of the species are variable, but phaeohyphomycosis (Kim *et al.*, 1998), phaeohyphomycotic cysts (Hironaga *et al.*, 1982; Anzimlt, 1990; Schwinn *et al.*, 1993; Kawachi *et al.*, 1995) or black-grain mycetomes (Murray *et al.*, 1963; Nielsen *et al.*, 1968) have most frequently been reported. Most cases are associated with local or systemic immunosuppression (Chuan & Wu, 1995). The species may be associated with deficiencies of acquired immunity (Neumeister *et al.*, 1995). Chromoblastomycosis-like infections were reported by Naka *et al.* (1986) and Ishizawa & Kondo (1997).

Distribution. The species has world-wide occurrence, but infections are particularly noted in Asia (Hayashi *et al.*, 1994).

References. Wang (1966), de Hoog (1977), McGinnis & Padhye (1977).

Antifungal susceptibility.

Antifungal	Mean MICs	Strains	Reference
AMB	0.54	27	McGinnis & Pasarell (1998b)
5FC	2	2	Tintelnot *et al.* (1999)
FLZ	70.6	2	Tintelnot *et al.* (1999)
ITZ	0.09	27	McGinnis & Pasarell (1998b)
KTZ	0.7	2	Tintelnot *et al.* (1999)
MCZ	2	2	Tintelnot *et al.* (1999)
TBF	0.08	24	McGinnis & Pasarell (1998a)
VCZ	0.6	27	McGinnis & Pasarell (1998b)

Nomenclature. *Torula jeanselmei* Langeron - Annls Parasit. Hum. Comp. 6: 396, 1928 ≡ *Pullularia jeanselmei* (Langeron) C.W. Dodge - Med. Mycol. p. 675, 1935 ≡ *Phialophora jeanselmei* (Langeron) Emmons - Arch. Path. 39: 368, 1945 ≡ *Exophiala jeanselmei* (Langeron) McGinnis & Padhye - Mycotaxon 5: 345, 1977.

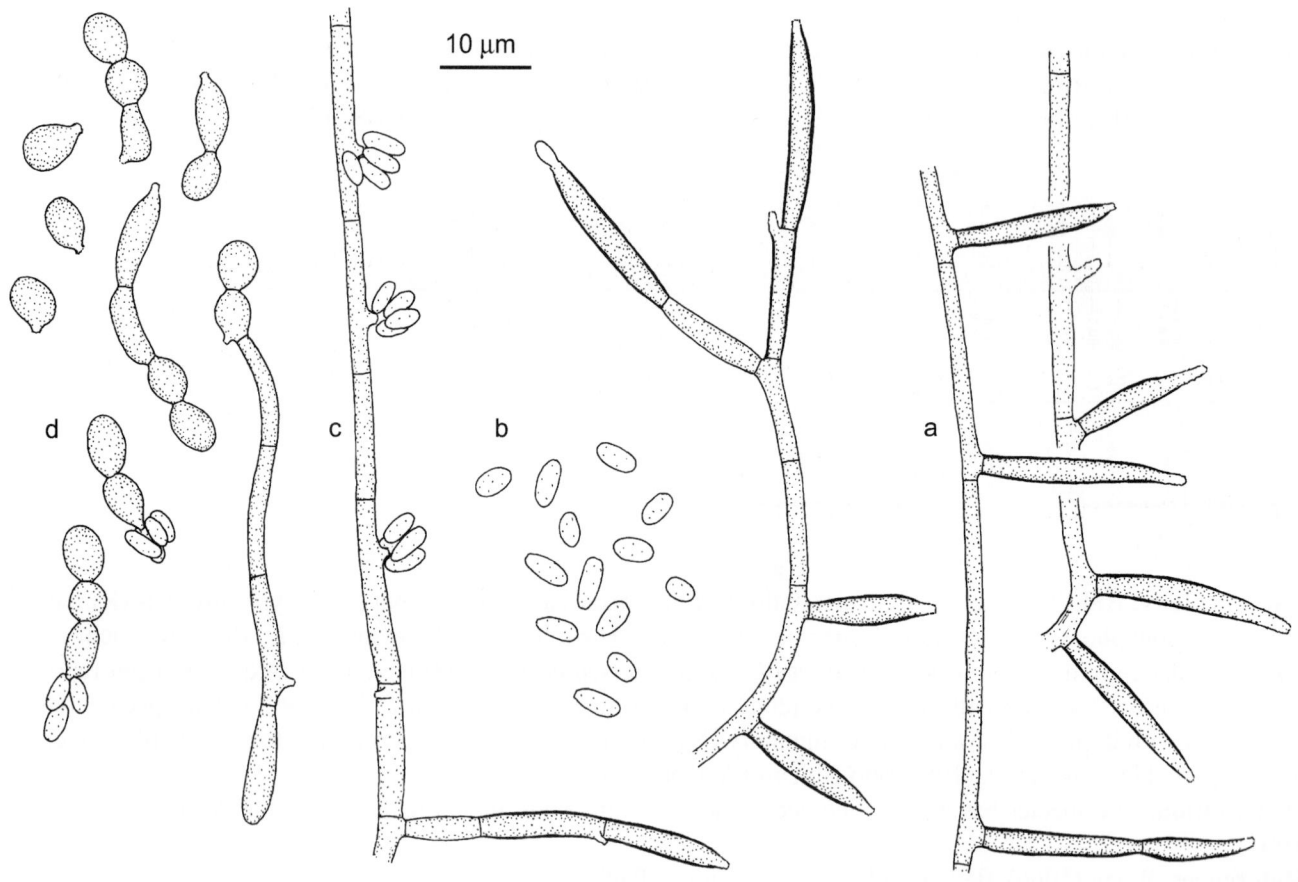

Exophiala jeanselmei, CBS 528.76. a. Conidiophores in aerial hyphae; b. conidia; c. submerged hyphae with conidial heads; d. yeast-like cells.

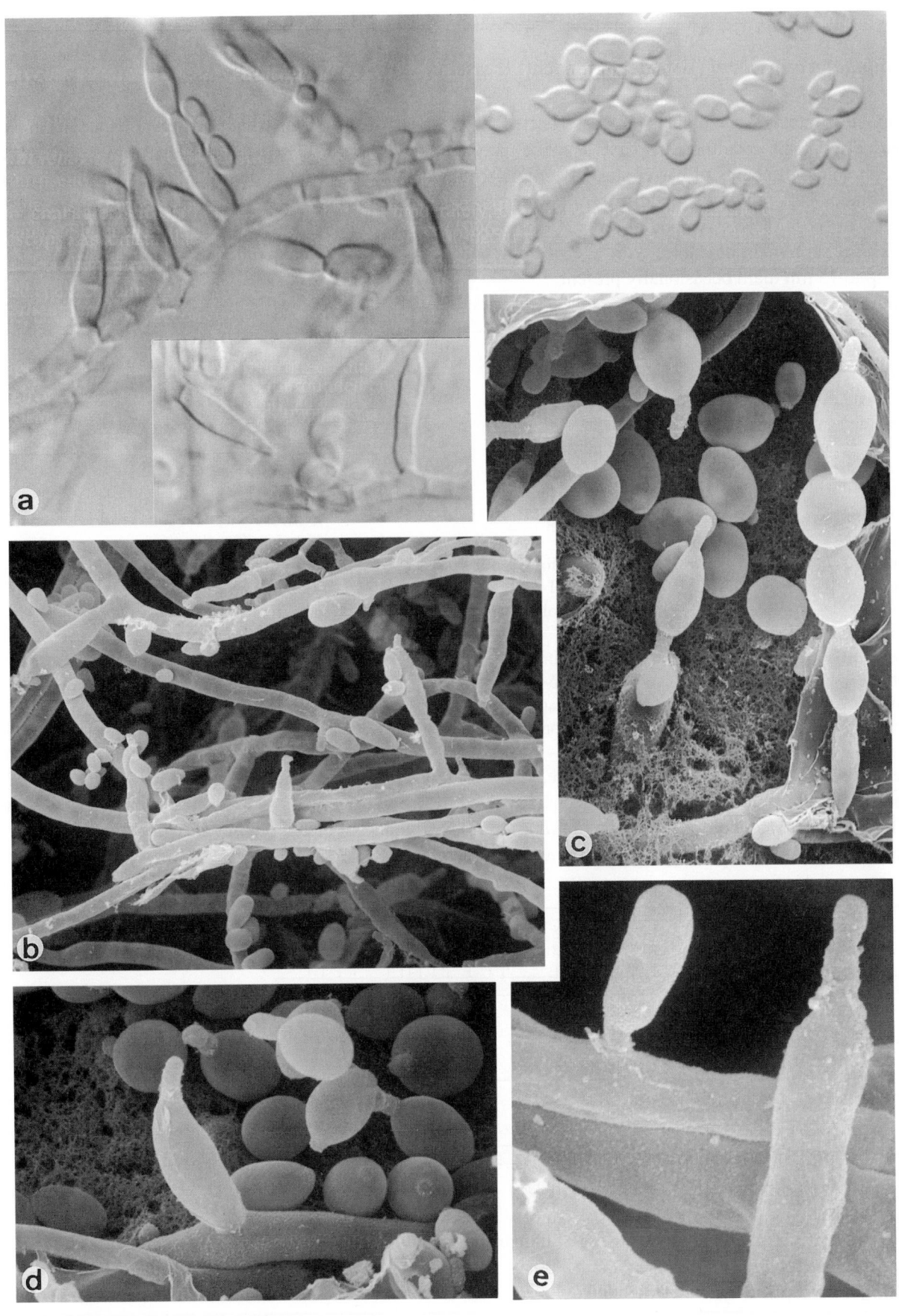

Exophiala jeanselmei, CBS 537.76. a, b, e. Conidiophores in aerial mycelium and conidia; c, d. torulose hyphae, conidia, and yeast-like cells. a. ×1600; b. ×2400; c. ×3650; d. ×4500; e. ×11,500.

Exophiala lecanii-corni (Benedek & Specht) Haase & de Hoog

Colony characteristics. Colonies (SGA) restricted, dry, velvety, locally somewhat powdery, olivaceous green.
Microscopy. Yeast cells mostly sparse. Hyphae pale olivaceous, profusely branched, in the apical part changing over into coherent chains of barrel-shaped conidiogenous cells up to 6 μm wide, interconnected with thin, broad septa. Chains elongating acropetally from a broad base. Conidial scars broad and flat, producing few conidia in basipetal or more or less sympodial order. Conidia broadly ellipsoidal, 5.5-9.0 × 3.0-4.5 μm wide, continuous or with a thin septum, subhyaline, with age becoming dark brown and thick-walled. Inflated chlamydospore-like cells, up to 15 μm diam occasionally present.
Physiology.

Growth:					
D-Glucose	+	Lactose	+	D-Galacturonate	–
D-Galactose	+	Raffinose	+	DL-Lactate	–
L-Sorbose	w	Melezitose	+	Succinate	+
D-Glucosamine	w	Inulin	–	Citrate	–
D-Ribose	+	Soluble starch	–	Methanol	–
D-Xylose	+	Glycerol	+	Ethanol	+
L-Arabinose	+	*meso*-Erythritol	+	Nitrate	+
D-Arabinose	w	Ribitol	+	Nitrite	+
L-Rhamnose	+	Xylitol	+	Ethylamine	+
Sucrose	+	L-Arabinitol	–	L-Lysine	+
Maltose	+	D-Glucitol	+	Cadaverine	+
α,α-Trehalose	+	D-Mannitol	+	Creatine	+
methyl-*α*-D-Glucoside	+	Galactitol	w	Creatinine	+
Cellobiose	+	*myo*-Inositol	–	0.1% Cycloheximide	+
Salicin	+	Glucono-δ-lactone	w	Growth at 37°C	w,–
Melibiose	+	D-Gluconate	+	Growth at 40°C	–
		D-Glucuronate	–		

Differential diagnosis. The relatively large, dark conidia and presence of barrel-shaped chains of cells are characteristic. Physiologically the taxon is characterized by growth with lactose and nitrate but absence of growth with L-arabinitol. Morphologically *E. pisciphila* (p. 662) is similar, but distinct by no growth with ethanol.
Molecular diagnostics. SSU and ITS restriction maps based on CBS 232.39:

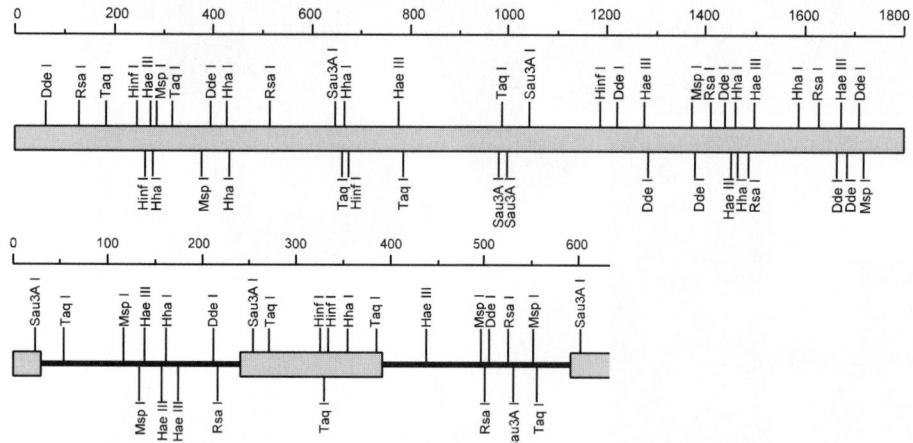

Pathogenicity. BSL-2. Some proven cases of human cutaneous skin-infections are known (de Hoog *et al.*, 1994b).
Reference. De Hoog *et al.* (1994b).
Antifungal susceptibility.

Antifungal	Mean MICs	Strains	Reference
AMB	1	2	Tintelnot *et al.* (1999)
5FC	0.18	2	Tintelnot *et al.* (1999)
KTZ	0.35	2	Tintelnot *et al.* (1999)

Nomenclature. *Torula lecanii-corni* Benedek & Specht - Zentbl. Bakt. Parasitkde, Abt. 1, 130: 86, 1933 ≡ *Pullularia fermentans* Wynne & Gott var. *benedekii* Wynne & Gott - J. Gen. Microbiol. 14: 518, 1956 (name change) ≡ *Exophiala jeanselmei* (Langeron) McGinnis & Padhye var. *lecanii-corni* (Benedek & Specht) de Hoog - Stud. Mycol. 15: 112, 1977 ≡ *Exophiala lecanii-corni* (Benedek & Specht) Haase & de Hoog, *in* Haase *et al.* - Stud. Mycol. 43: 93, 1999.

Hormodendrum negronii Pereira - Micol. Cromomyc., 1939 ≡ *Exophiala negronii* (Pereira) de Hoog - CBS List Cult., ed. 33, p. 124, 1994.

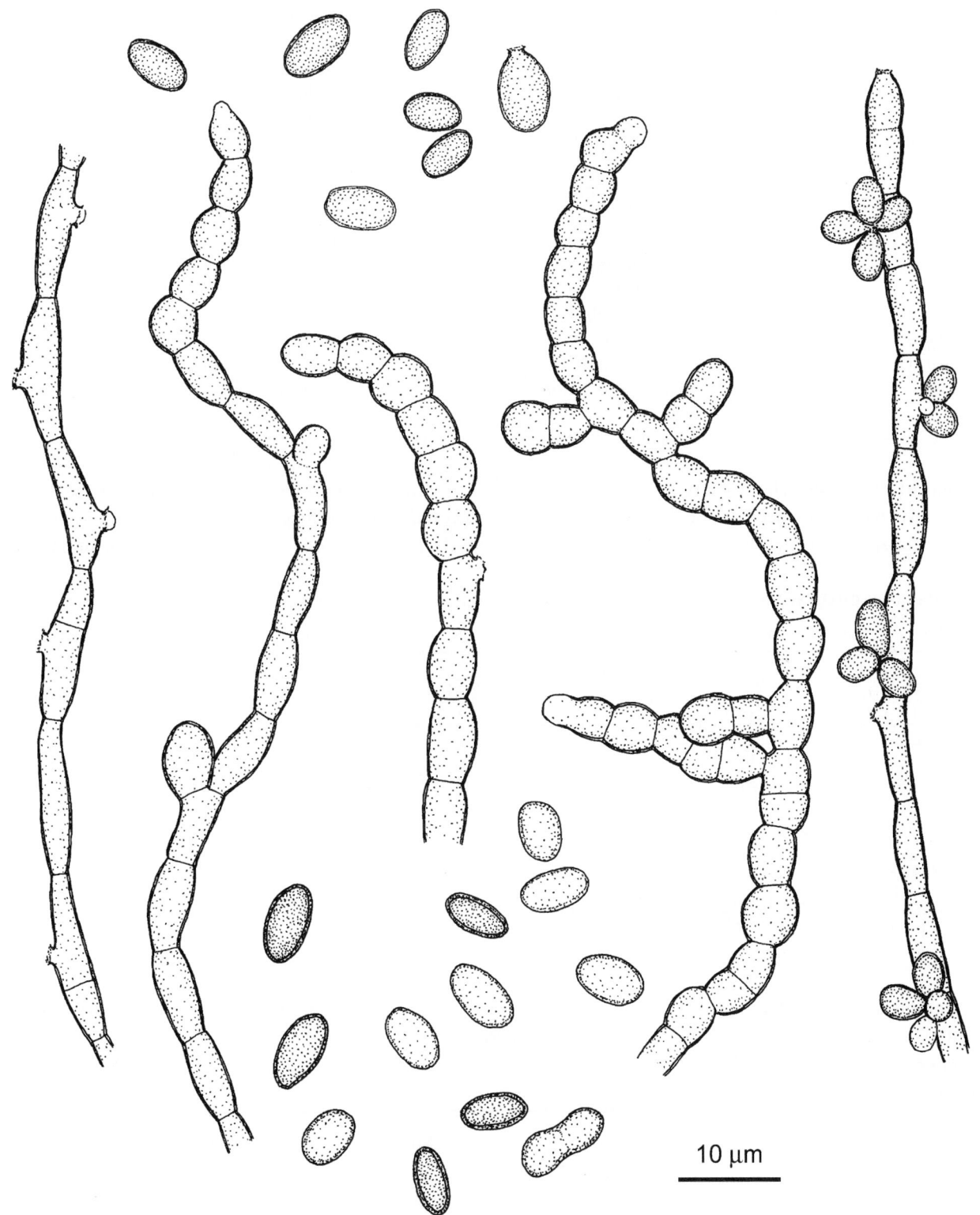

Exophiala lecanii-corni. a, c. CBS 232.39, more or less annellidic production of ellipsoidal conidia; b. CBS 464.81, mainly coherent chains of barrel-shaped cells.

Exophiala moniliae de Hoog

Colony characteristics. Colonies (SGA) restricted, black and moist, later becoming velvety.

Microscopy. Yeast cells mostly abundant. Hyphae consisting of inflated cells, producing short, inflated branchlets at the septa; conidiogenous cells ellipsoida, bearing a single annellated zone which may protrude and become rather long with age. Conidia broadly ellipsoidal, 2.5-4.0 × 1.5-2.5 µm. Additional reniform locally present.

Physiology.

Growth:					
D-Glucose	+	Lactose	–	D-Galacturonate	–
D-Galactose	+	Raffinose	+	DL-Lactate	–
L-Sorbose	+	Melezitose	+	Succinate	–
D-Glucosamine	–	Inulin	–	Citrate	–
D-Ribose	+	Soluble starch	–	Methanol	–
D-Xylose	+	Glycerol	+	Ethanol	w
L-Arabinose	+	*meso*-Erythritol	+	Nitrate	+
D-Arabinose	w	Ribitol	+	Nitrite	+
L-Rhamnose	+	Xylitol	–	Ethylamine	+
Sucrose	+	L-Arabinitol	–	L-Lysine	+
Maltose	+	D-Glucitol	+	Cadaverine	+
α,α-Trehalose	+	D-Mannitol	+	Creatine	+
methyl-α-D-Glucoside	+	Galactitol	+	Creatinine	+
Cellobiose	+	*myo*-Inositol	–	0.1% Cycloheximide	–
Salicin	–	Glucono-δ-lactone	+	Growth at 37°C	w,–
Melibiose	+	D-Gluconate	w		
		D-Glucuronate	w		

Differential diagnosis. The densely torulose, *Candida*-like branching system with small, nearly spherical conidia is characteristic for the species. Locally, additional reniform conidia are produced, which are characteristic for the species. The species is serologically different from the closely related species *E. jeanselmei* (Matsumoto *et al.*, 1984a).

Molecular diagnostics. SSU and ITS restriction maps based on CBS 520.76:

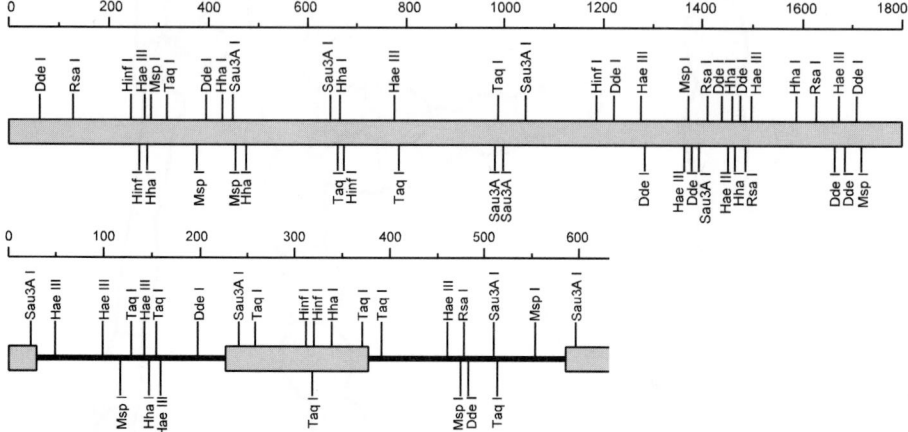

Pathogenicity. BSL-1. The species is found in the environment and is a rare agent of phaeohyphomycosis (McGinnis *et al.*, 1981). It may, however, have been underdiagnosed.

Reference. Matsumoto *et al.* (1984a).

Antifungal susceptibility.

Antifungal	Mean MICs	Strains	Reference
AMB	0.5	3	McGinnis & Pasarell (1998b)
ITZ	0.4	3	McGinnis & Pasarell (1998b)
VCZ	0.32	3	McGinnis & Pasarell (1998b)

Nomenclature. *Exophiala moniliae* de Hoog - Stud. Mycol. 15: 120, 1977.

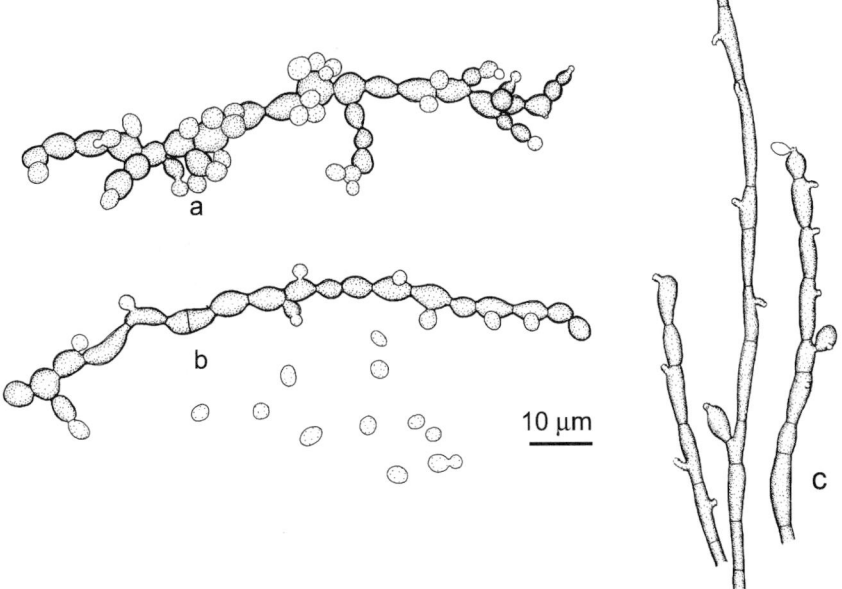

Exophiala moniliae, CBS 520.76. a. Young conidial apparatus with clustered conidiogenous cells; b. conidia; c. mature hyphal system with annellated zones.

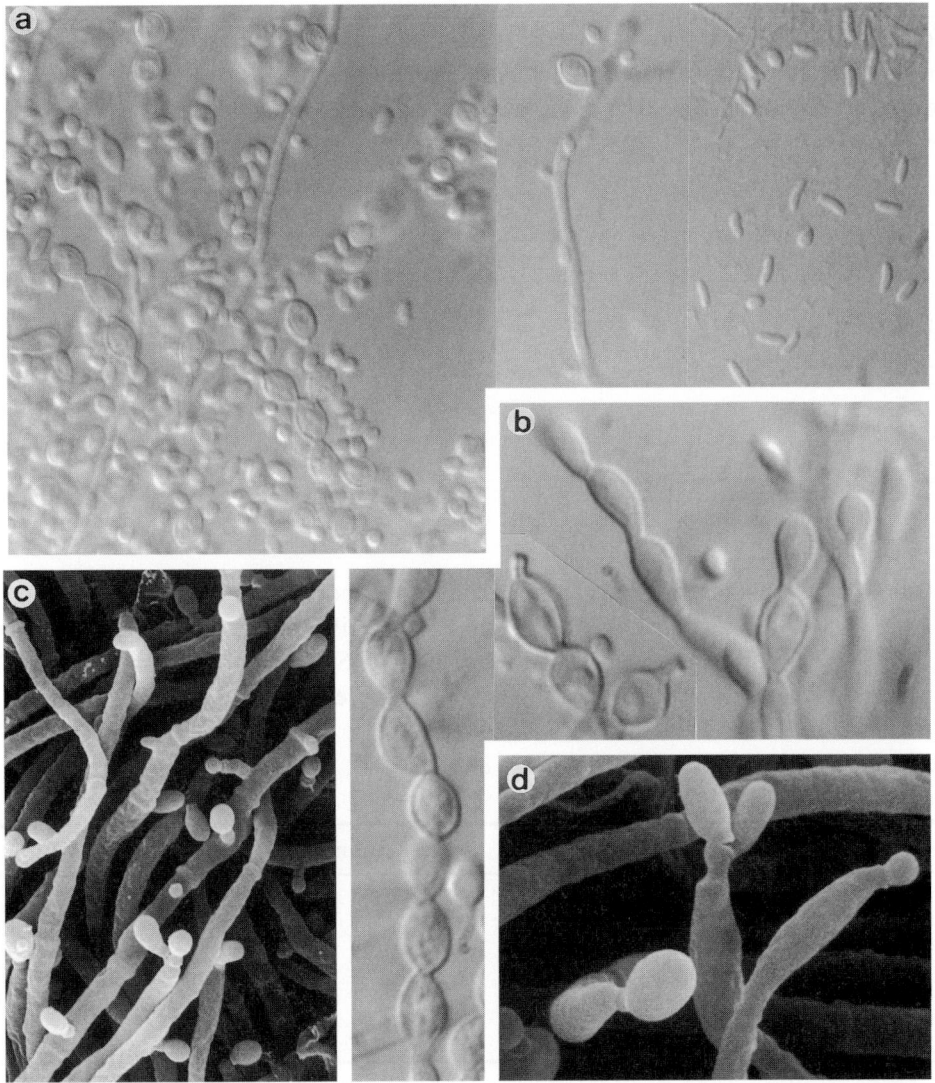

Exophiala moniliae, CBS 520.76. Conidial apparatus, torulose hyphae and conidia of two types. a, b. ×1600; c. ×4500; d. ×7500.

661

Exophiala pisciphila McGinnis & Ajello

Colony characteristics. Colonies (SGA) moderately expanding, dry, floccose, olivaceous black.
Microscopy. Yeast cells absent. Conidiogenous cells flask-shaped, mostly in loose clusters or branched systems, with inconspicuous annellated zones. Conidia 0 (-1)-septate, (sub)hyaline, ellipsoidal, 6-8 × 2.5-4.0 µm.
Physiology.

Growth:	+	Melibiose	+	D-Galacturonate	+
D-Glucose	+	Lactose	+	DL-Lactate	−
D-Galactose	+	Raffinose	+	Succinate	+
L-Sorbose	+	Melezitose	+	Citrate	−
D-Glucosamine	+	Inulin	+	Methanol	−
D-Ribose	+	Soluble starch	−	Ethanol	−
D-Xylose	+	Glycerol	+	Nitrate	+
L-Arabinose	+	*meso*-Erythritol	+	Nitrite	+
D-Arabinose	+	Ribitol	+	Ethylamine	+
L-Rhamnose	+	Xylitol	+	L-Lysine	+
Sucrose	+	L-Arabinitol	+	Cadaverine	w
Maltose	+	D-Mannitol	+	Creatine	+
α,α-Trehalose	+	*myo*-Inositol	w	Creatinine	+
methyl-α-D-Glucoside	+	Glucono-δ-lactone	+	0.1% Cycloheximide	+
Cellobiose	+	D-Gluconate	+	Growth at 37°C	−
Salicin		D-Glucuronate	+		

Differential diagnosis. Relatively expanding, woolly colonies without a yeast phase. Conidia broadly ellipsoidal, nearly all without septa. *E. salmonis* (p. 664) differs by no growth with lactose, D-gluconate and galacturonate.
Molecular diagnostics. SSU and ITS restriction maps based on CBS 537.73:

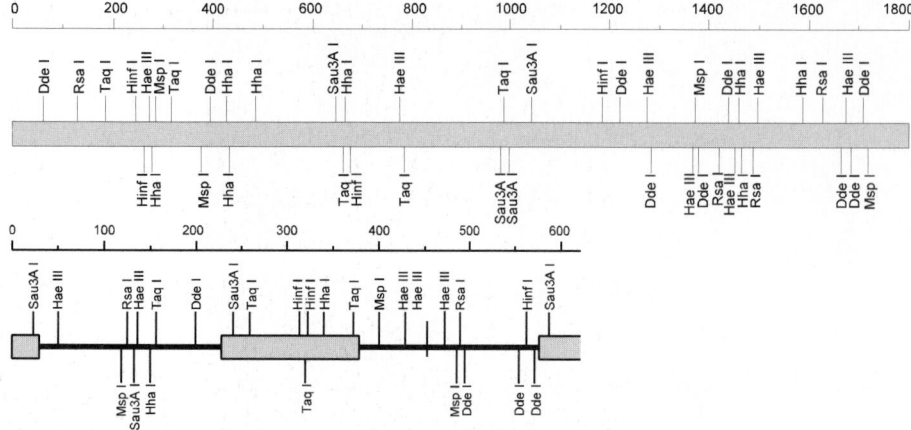

Pathogenicity. BSL-1. The species has been reported as an opportunistic invader in several species of fish (Gaskins & Cheung, 1986). Strains causing an epidemic in captive plaice, published as *Hormoconis resinae* (Strongman *et al.*, 1977) were also identified with this species (G.S. de Hoog, unpublished data). The species is also commonly found in soil. A skin infection in an immunosuppressed patient was reported by Sughayer *et al.* (1991).
References. McGinnis & Ajello (1974), Morgan-Jones *et al.* (1984), Iwatsu & Udagawa (1984).
Antifungal susceptibility.

Antifungal	Mean MICs	Strains	Reference
AMB	2.83	4	McGinnis & Pasarell (1998b)
5FC	0.7	1	Tintelnot *et al.* (1999)
ITZ	0.12	4	McGinnis & Pasarell (1998b)
KTZ	0.09	1	Tintelnot *et al.* (1999)
VCZ	0.59	4	McGinnis & Pasarell (1998b)

Nomenclature. *Exophiala pisciphila* McGinnis & Ajello - Mycologia 66: 518, 1974.

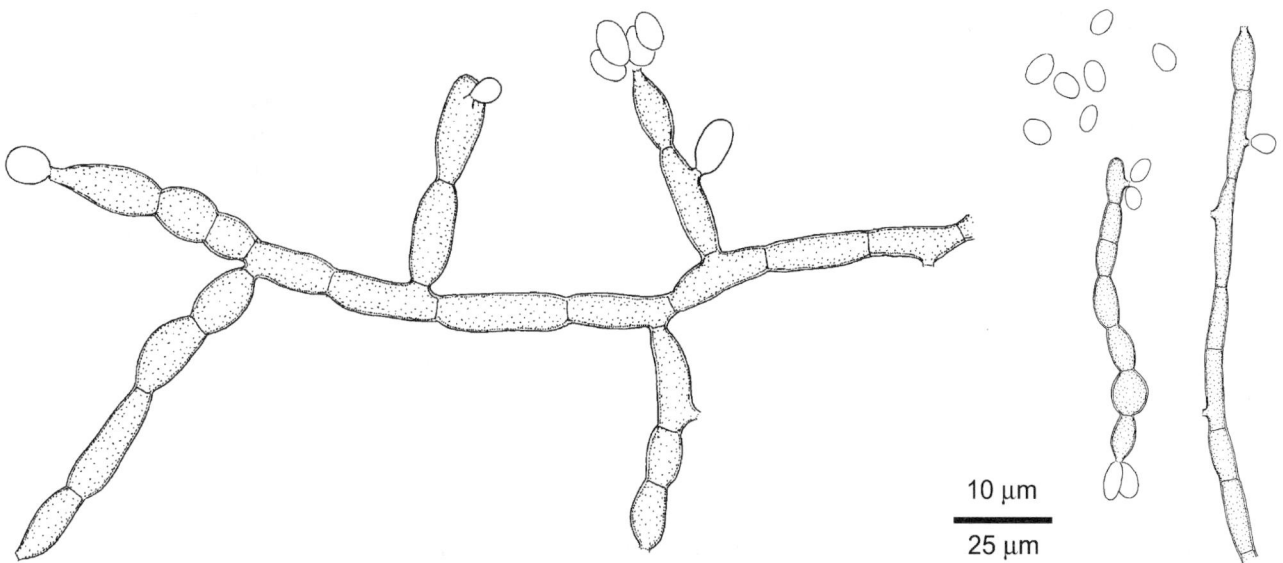

Exophiala pisciphila, CBS 537.73. Conidial apparatus with conidia.

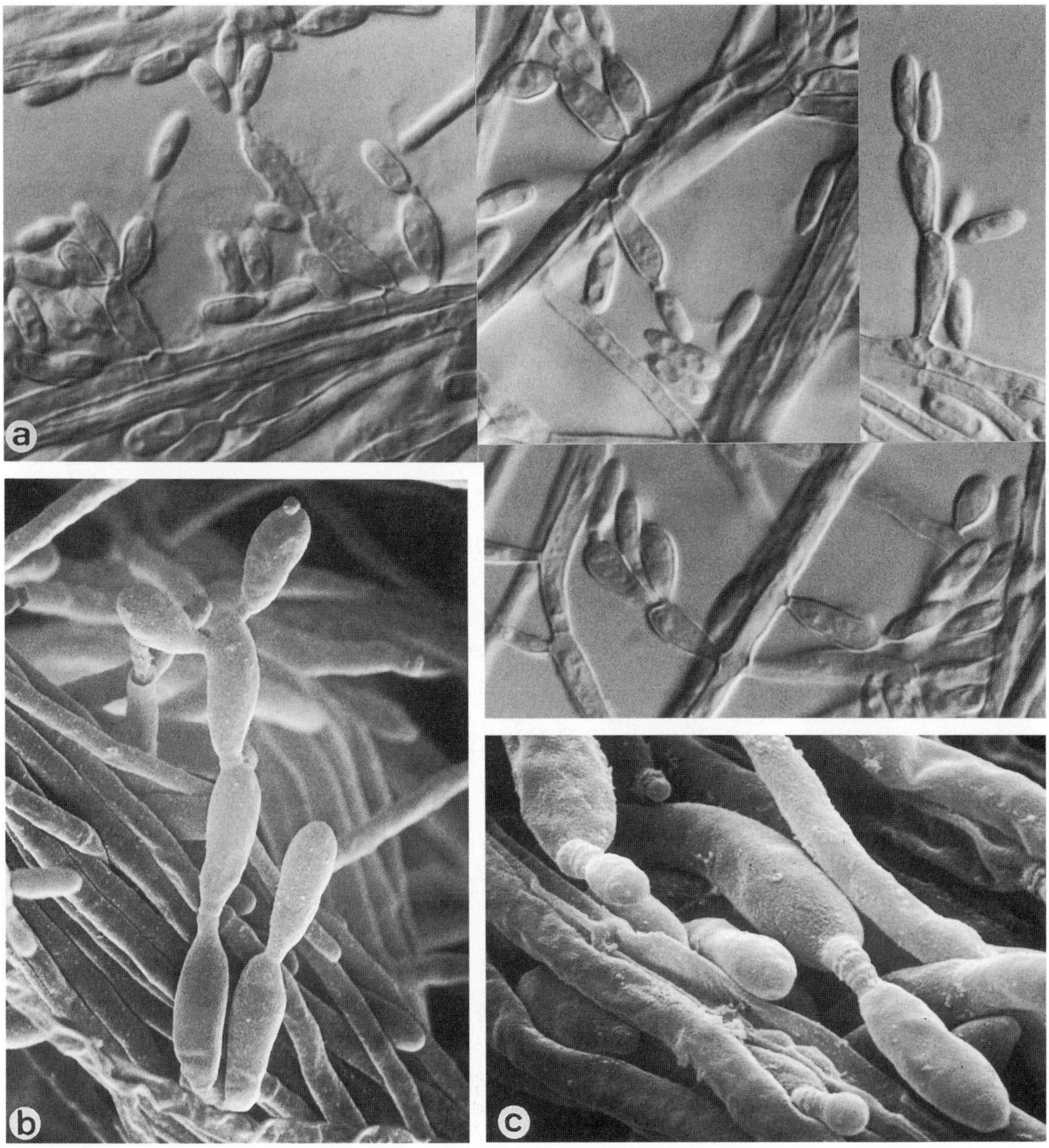

Exophiala pisciphila, CBS 160.89. Conidial apparatus with conidia. a. ×1600; b. ×3400; c. ×5800.

663

Exophiala salmonis Carmichael

Colony characteristics. Colonies (SGA) moderately expanding, dry, depressed, hairy, olivaceous-black.
Microscopy. Yeast cells nearly absent. Conidiogenous cells poorly differentiated, intercalary or flask-shaped; annellated zones short, inconspicuous. Conidia 0-3-septate, subhyaline to pale brown, ellipsoidal to short cylindrical, 5.5-8.5 × 2.0-3.5 μm.
Physiology.

Growth:					
D-Glucose	+	Melibiose	+	D-Galacturonate	−
D-Galactose	+	Lactose	−	DL-Lactate	−
L-Sorbose	+	Raffinose	+	Succinate	−
D-Glucosamine	+	Soluble starch	+	Citrate	−
D-Ribose	+	Glycerol	+	Methanol	−
D-Xylose	+	*meso*-Erythritol	+	Ethanol	−
L-Arabinose	+	Ribitol	+	Nitrate	+
D-Arabinose	+	Xylitol	+	Nitrite	+
L-Rhamnose	+	L-Arabinitol	+	Ethylamine	+
Sucrose	+	D-Glucitol	+	L-Lysine	+
Maltose	+	D-Mannitol	+	Cadaverine	+
α,α-Trehalose	+	Galactitol	w	Creatine	+
methyl-α-D-Glucoside	+	*myo*-Inositol	+	Creatinine	+
Cellobiose	+	Glucono-δ-lactone	w	0.1% Cycloheximide	+
Salicin	+	D-Gluconate	+	Growth at 37°C	+
		D-Glucuronate	−	Growth at 40°C	−

Molecular diagnostics. SSU and ITS restriction maps based on CBS 157.67:

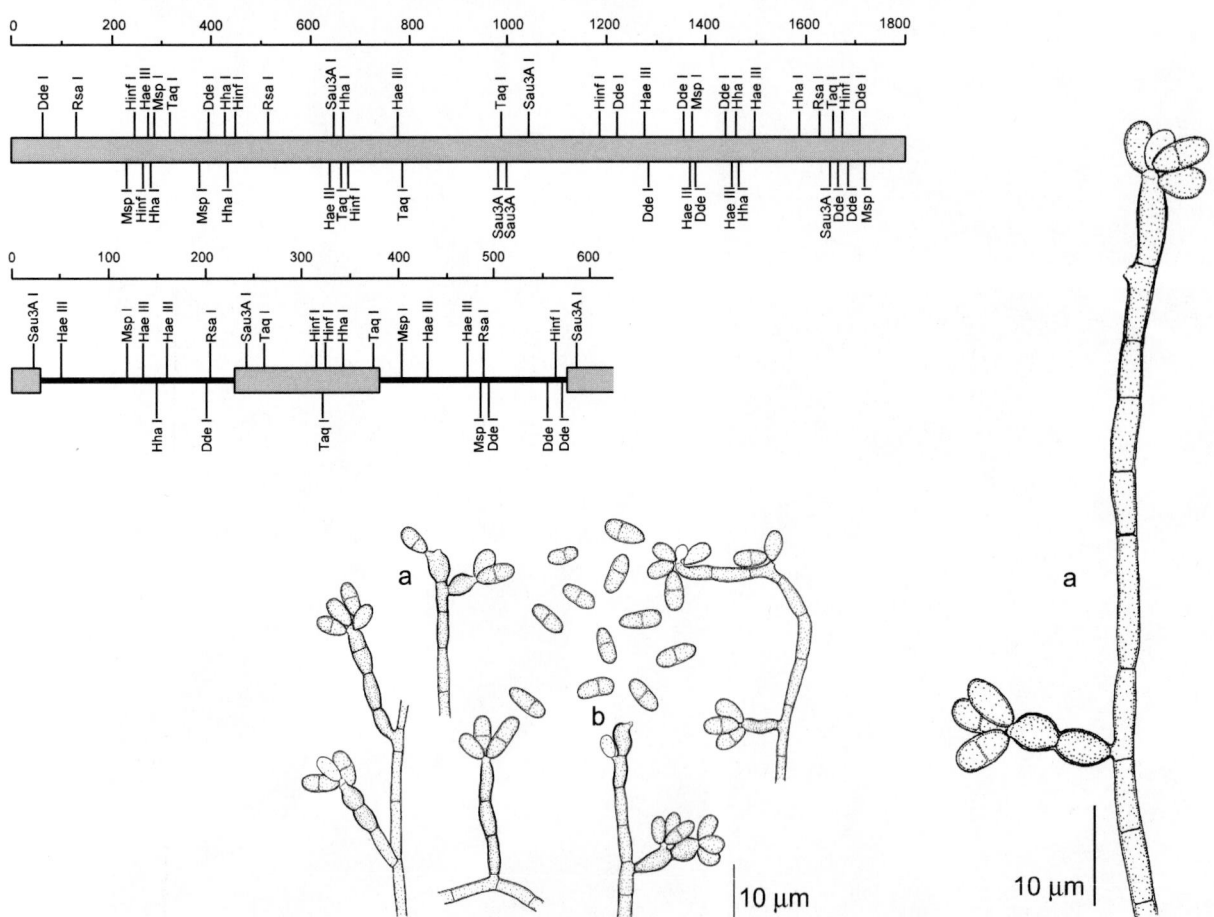

Exophiala salmonis, CBS 157.67. a. Conidial apparatus; b. conidia.

Differential diagnosis. Relatively expanding, woolly colonies without a yeast phase. Conidia frequently septate.

Pathogenicity. BSL-1. The species is a fish pathogen, causing cerebral phaeohyphomycoses in trout (Carmichael, 1966; Otis & Wolke, 1985).

References. Carmichael (1966), de Hoog (1977).

Nomenclature. *Exophiala salmonis* Carmichael - Sabouraudia 5: 122, 1966 ≡ *Aureobasidium salmonis* (Carmichael) Borelli - Med. Cafter. 3: 588, 1969.

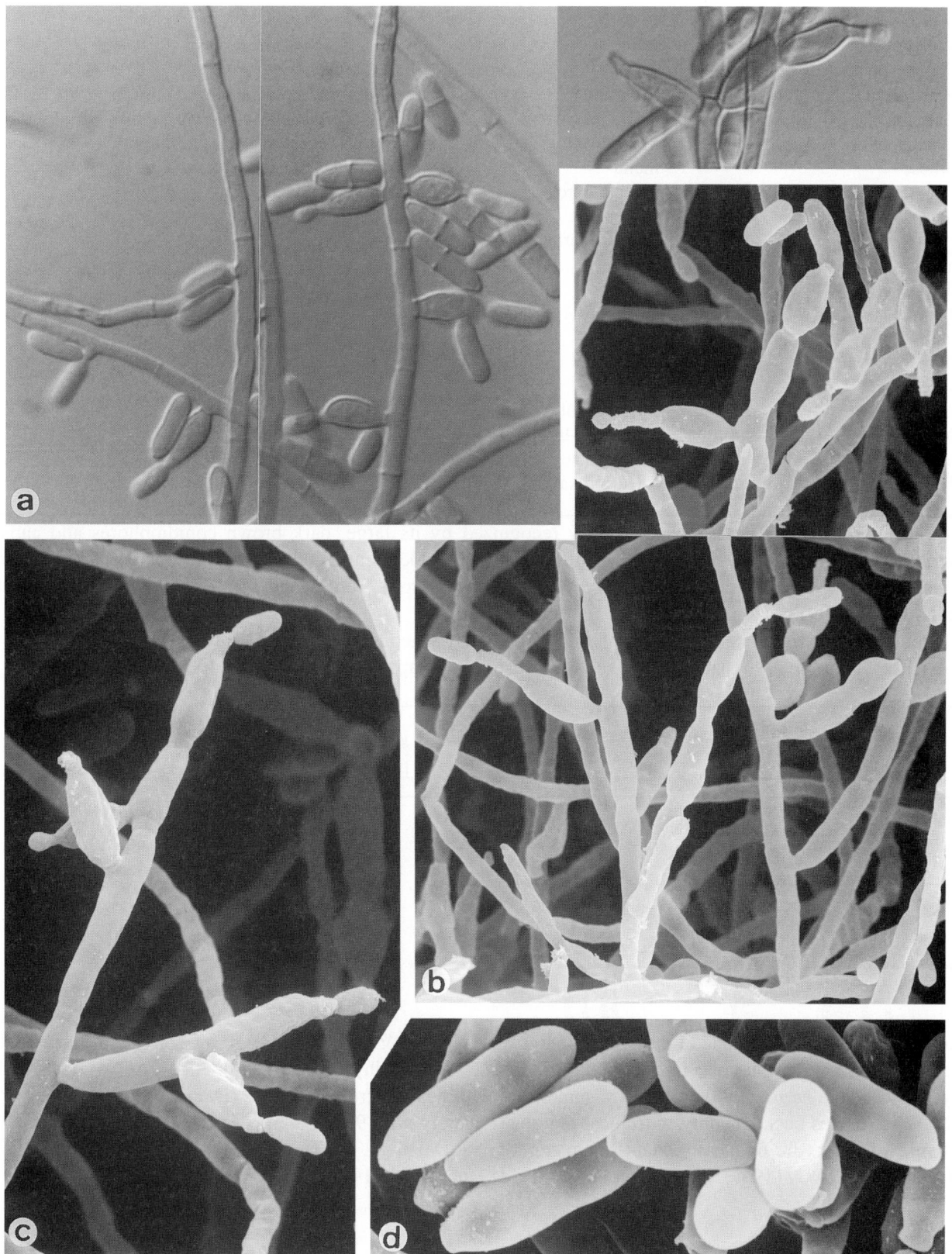

Exophiala salmonis, CBS 157.67. Conidial apparatus and conidia. a. ×1600; b. ×2000; c. ×2800; d. ×4700.

Exophiala spinifera (Nielsen & Conant) McGinnis

Colony characteristics. Colonies (SGA) restricted, olivaceous-black, with slimy centre due to yeast-like growth, and with floccose margin.

Microscopy. Yeast cells producing capsular material abundant. Conidiophores erect, brown, cylindrical, multi-celled, with apical and intercalary, narrow, cylindrical annellated zones up to several μm long. Conidia obovoidal, 1.8-2.8 × 2-4 μm.

Physiology.

Growth:					
D-Glucose	+	Lactose	–	D-Galacturonate	+
D-Galactose	+	Raffinose	+	DL-Lactate	+
L-Sorbose	+	Melezitose	+	Succinate	+
D-Glucosamine	+	Inulin	–	Citrate	v
D-Ribose	+	Soluble starch	–,w	Methanol	–
D-Xylose	+	Glycerol	+	Ethanol	+
L-Arabinose	+	*meso*-Erythritol	+	Nitrate	+
D-Arabinose	+	Ribitol	+	Nitrite	+
L-Rhamnose	+	Xylitol	+	Ethylamine	+
Sucrose	+	L-Arabinitol	+	L-Lysine	+
Maltose	+	D-Glucitol	+	Cadaverine	+
α,α-Trehalose	+	D-Mannitol	+	Creatine	+
methyl-α-D-Glucoside	+	Galactitol	v	Creatinine	+
Cellobiose	+	*myo*-Inositol	+	0.1% Cycloheximide	+
Salicin	+	Glucono-δ-lactone	+	Growth at 37°C	+
Melibiose	+	D-Gluconate	+	Growth at 40°C	–
		D-Glucuronate	+		

Differential diagnosis. The species is easily recognized by its large, dark brown conidiophores and capsular material around budding cells which are visible in Indian Ink. Occasionally the yeast phase is predominant. Some strains have mere budding cells (*Phaeococcomyces* synanamorph). Serological identification was introduced by Standard *et al.* (1991).

Molecular diagnostics. The species was found to be heterogeneous in mtDNA-RFLP (Ishizaki *et al.*, 1995) and DNA RFLP (Uijthof, 1996) patterns. SSU and ITS restriction maps based on CBS 899.68:

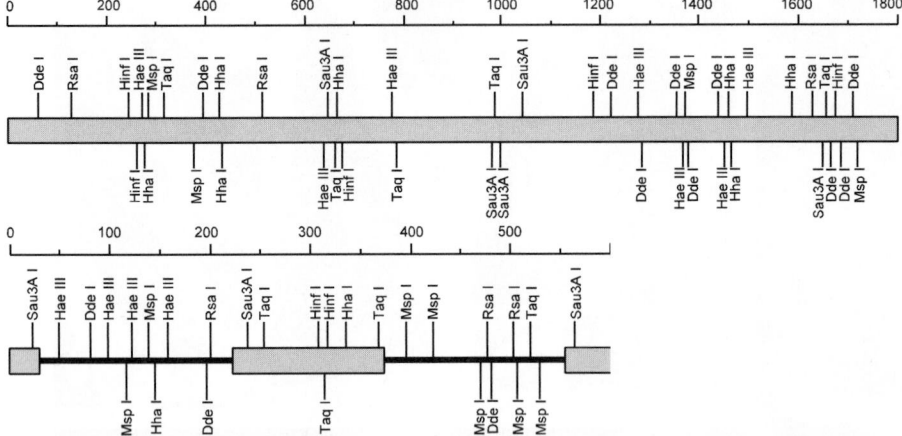

Pathogenicity. BSL-2. It is one of the agents of sinusitis (Nielsen & Conant, 1968) or phaeohyphomycotic cysts in (sub)cutaneous or intramuscular tissue (Kettlewell *et al.*, 1989; Padhye *et al.*, 1984). Also cases of chromoblast-omycosis (Padhye *et al.*, 1996) are known, sometimes with heavily crusted lesions (Dai *et al.*, 1987; Wang *et al.*, 1989). The species occurs mainly in tropical regions (Lacaz *et al.*, 1984; Campos-Takaki & Lobo Jardim, 1994). Systemic infections are frequently fatal (Rajam *et al.*, 1958; Dai *et al.*, 1987).

References. De Hoog (1977), Standard *et al.* (1991), de Hoog *et al.* (1999).

Antifungal susceptibility.

Antifungal	Mean MICs	Strains	Reference
AMB	0.7	9	McGinnis & Pasarell (1998b)
ITZ	0.11	9	McGinnis & Pasarell (1998b)
VCZ	0.21	9	McGinnis & Pasarell (1998b)

Nomenclature. *Phialophora spinifera* Nielsen & Conant - Sabouraudia 6: 228, 1968 ≡ *Rhinocladiella spinifera* (Nielsen & Conant) de Hoog - Stud. Mycol. 15: 93, 1977 ≡ *Exophiala spinifera* (Nielsen & Conant) McGinnis - Mycotaxon 5: 337, 1977.

Phaeococcus exophialae de Hoog - Stud. Mycol. 15: 127, 1977 ≡ *Phaeococcomyces exophialae* (de Hoog) de Hoog - Taxon 28: 348, 1979.

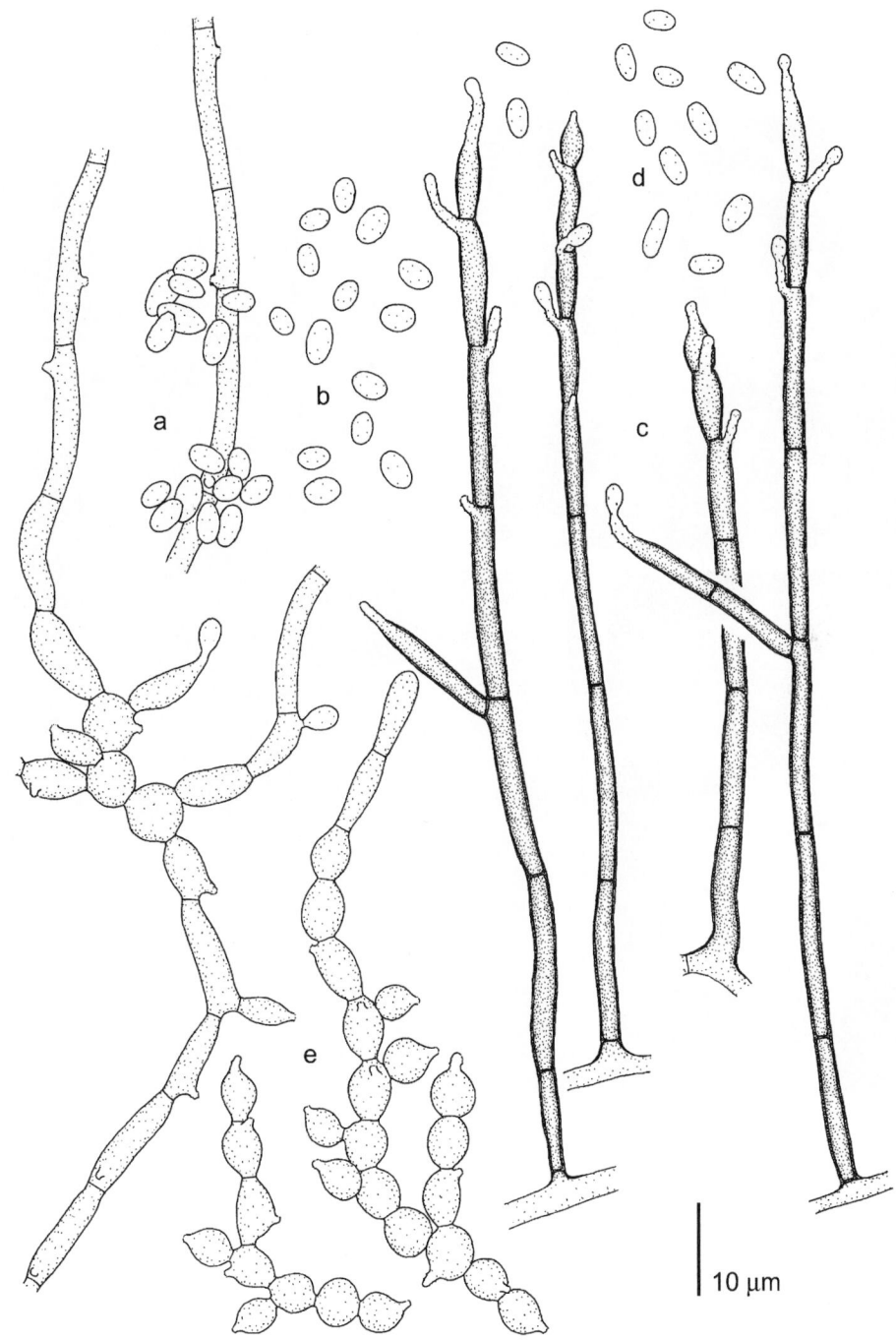

Exophiala spinifera, CBS 899.68. a, b. Submerged hyphae with conidia; c, d. hyphal conidiophores and conidia; e. torulose mycelium.

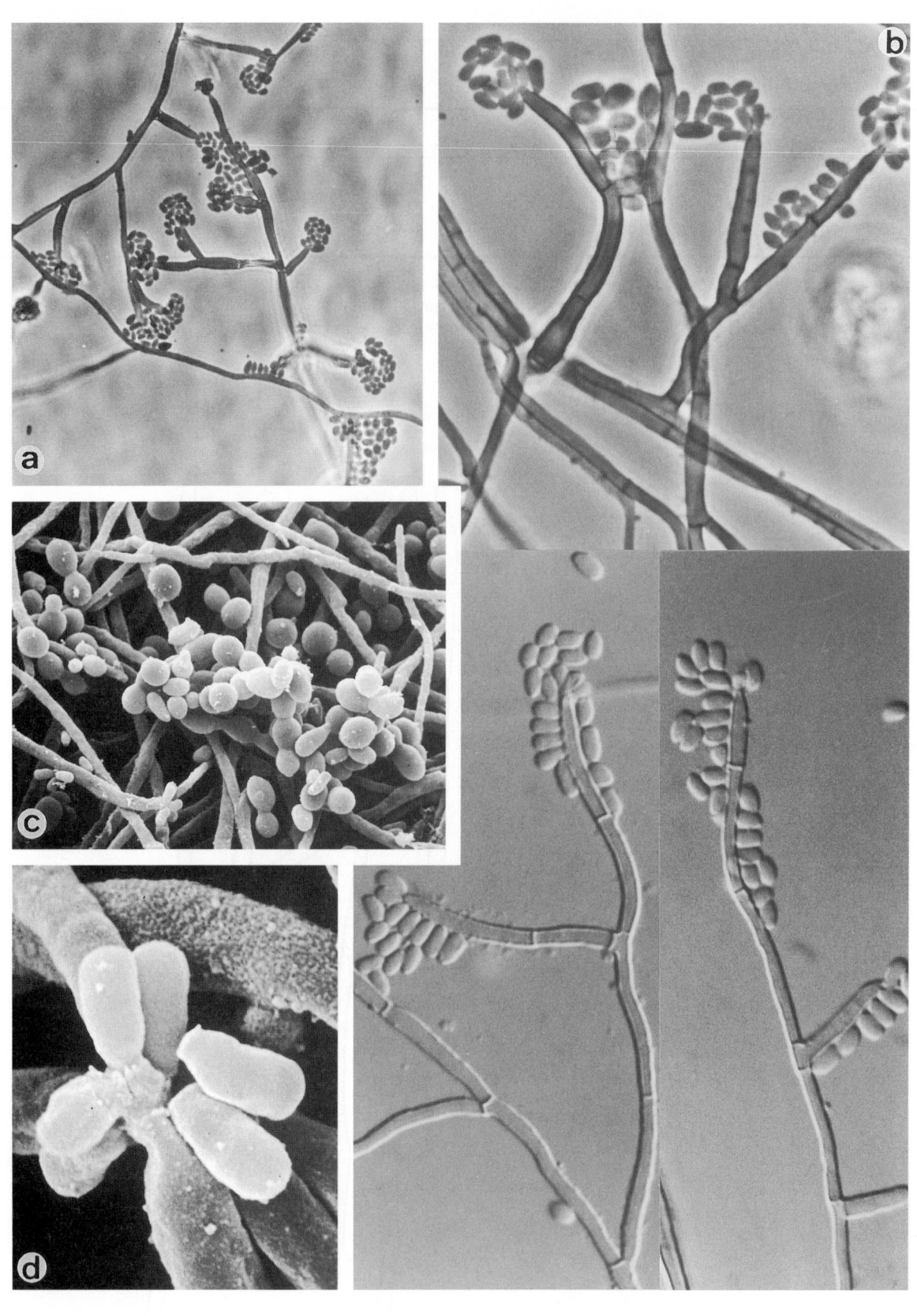

Exophiala spinifera, AMMRL 16.18. a, b, d. Conidial apparatus with conidia; c. torulose mycelium. a. ×640; b. ×1600; c. ×1500; d. ×10,970.

Hyphomycetes, Dematiaceae. Genus: *EXSEROHILUM*

Generic description. Colonies spreading, woolly or hairy, brown, grey or black. Conidiophores simple, thick- and smooth-walled, olivaceous brown to brown, with often geniculate terminal portion, sometimes nodose, producing conidia from conspicuous, darkened scars. Conidia fusiform, cylindrical or obclavate, straight to curved, compartmented with pseudosepta, brown, with a protuberant hilum at the base.

Teleomorph. *Setosphaeria* (*Ascomycota, Euascomycetes, Pleosporales: Pleosporaceae*).

Differential diagnosis. The genus is close to *Bipolaris* (p. 526), which differs by lacking protuberant conidial hila. For relationships with other genera, see the phylogenetic tree on p. 382.

General remarks. The genus comprises plant-pathogenic species, mainly occurring on grasses. Human mycoses mostly concern cases of sinusitis, eventually with cerebral involvement. The clinically relevant taxa are very close to each other; they all have identical ITS RFLP profiles.

Molecular diagnostics. SSU and ITS restriction maps based on NCBI U42487 and NCBI AF081453, resp.:

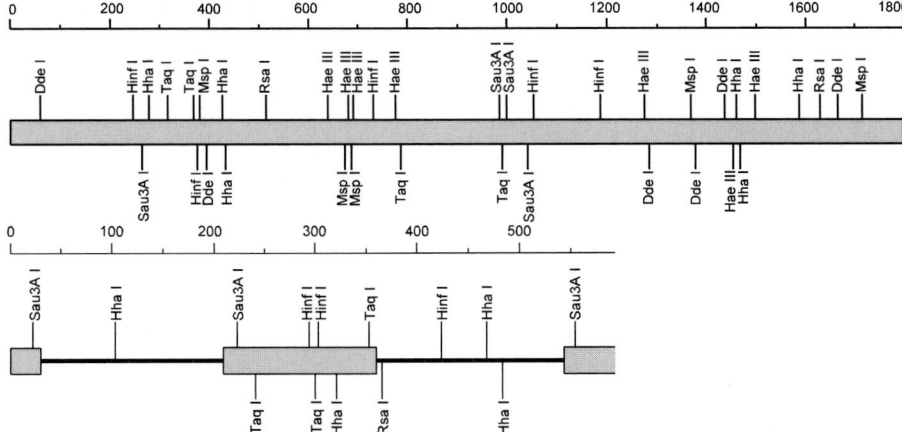

References. Ellis (1971, 1976), Alcorn (1983), Sivanesan (1987).

Key to the treated species of Exserohilum:

1a. Two types of conidia present: short and long, the latter measuring 196-260 × 13-16 μm, with 13-21 pseudosepta .... *E. longirostratum* **(670)**

1b. Only one type of conidia present ➞ **2**

2a. Conidia mostly with 9-11 pseudosepta without dark bands at the ends .... *E. mcginnisii* **(672)**

2b. Conidia mostly with 7-9 pseudosepta and darkly pigmented bands at the ends .... *E. rostratum* **(674)**

Exserohilum longirostratum (Subram.) Sivanesan

Colony characteristics. Colonies (PCA) spreading, hairy, brown.

Microscopy. Conidiophores up to 200 μm long, unbranched, flexuose, often geniculate. Conidia straight or slightly curved, widest near the base, gradually narrowing towards the apex into a long beak, with rounded ends, terminal cells often cut off by a dark, thick septum and with often paler end cells, smooth-walled to finely roughened, of two types: (1) long conidia 196-260 × 13-16 μm, 13-21-distoseptate; (2) shorter conidia 38-79 × 13-19 μm, 5-9-distoseptate.

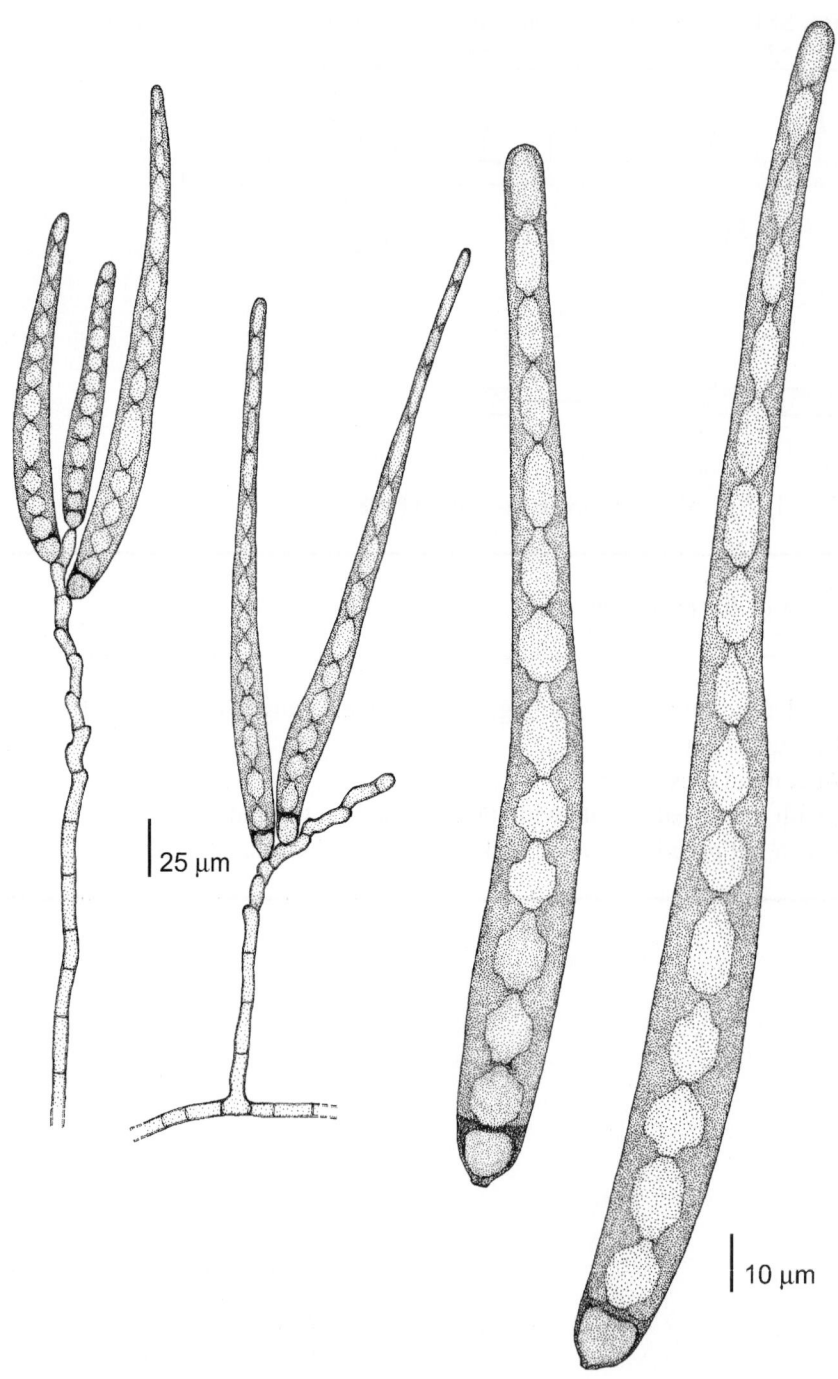

25 μm

10 μm

Exserohilum longirostratum, FMR 4283. Conidiophores and conidia.

670

Differential diagnosis. In culture the long conidia are often difficul to find; such cultures are closely similar to *E. rostratum*. Possibly both species are identical.

Pathogenicity. BSL-1. Several cases of disseminated infections have been reported (Drouhet *et al.*, 1985). Cutaneous cases were described by Lavoie *et al.* (1993) and Agrawal & Singh (1995b), and a corneal ulcer by Bouchon *et al.* (1994). Ziza *et al.* (1985) reported an arthritis. A clinical review was presented by Adam *et al.* (1986).

References. McGinnis *et al.* (1986b), Sivanesan (1987).

Nomenclature: *Helminthosporium longirostratum* Subramanian - J. Ind. Bot. Soc. 35: 463, 1956 ≡ *Drechslera longirostrata* (Subramanian) Subramanian - Proc. 1st. Int. Symp. Pl. Pathol. p. 195, 1970 ≡ *Exserohilum longirostratum* (Subramanian) Sivanesan - Trans. Br. Mycol. Soc. 83: 328, 1984.

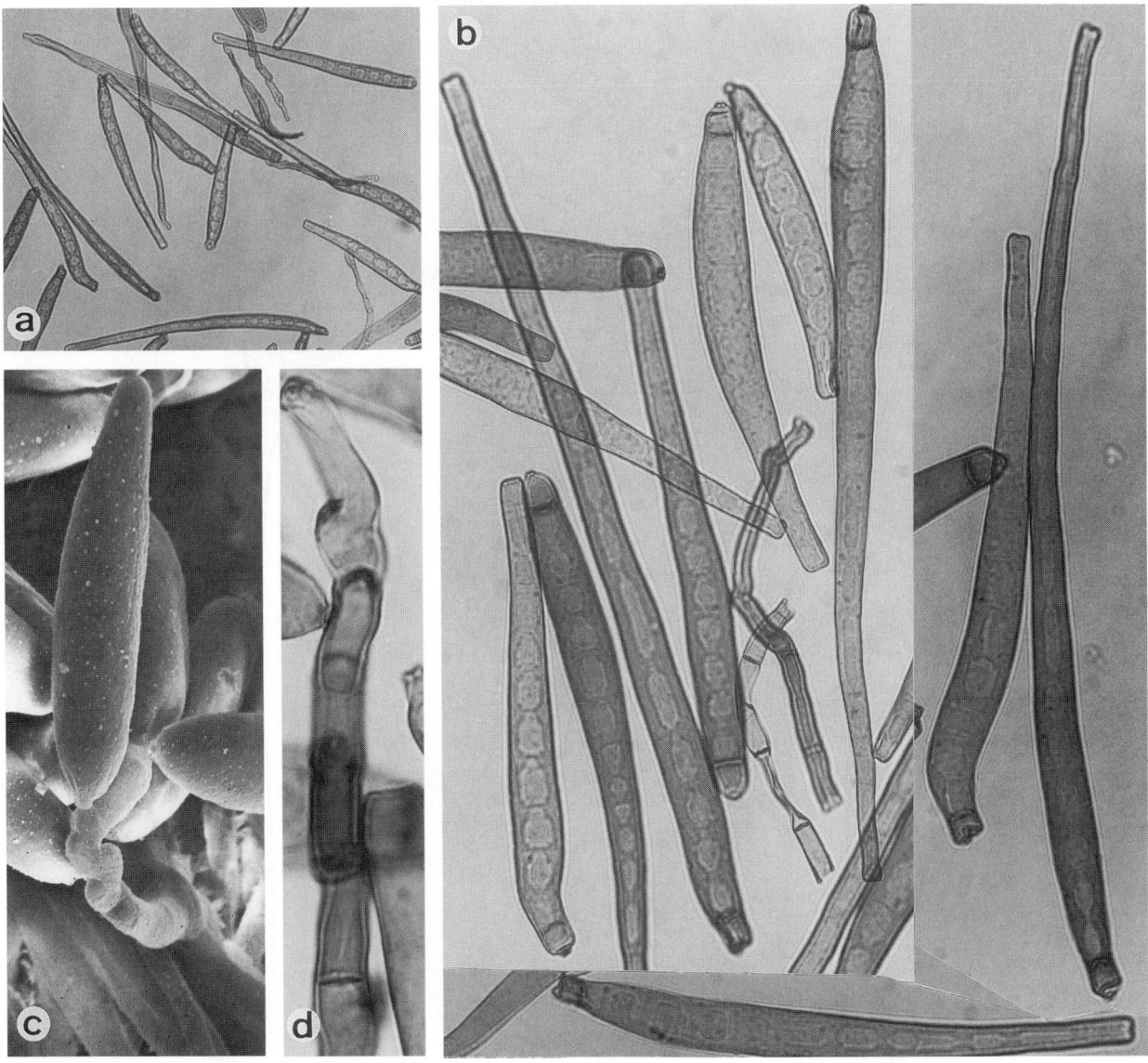

Exserohilum longirostratum, FMR 4283. Conidiophores and conidia. a. ×126; b. ×512; c. ×1150; d. ×1600.

Exserohilum mcginnisii Padhye & Ajello

Colony characteristics. Colonies (PCA) spreading, hairy or downy, olivaceous grey.
Microscopy. Conidiophores simple, up to 150 µm long, with flexuose apical part. Conidia smooth-walled, brown, straight, cylindrical to slightly clavate, 64-100 × 10-15 µm, with (4-) 9-11 (-13) pseudosepta and protuberant hilum.
Differential diagnosis. The species differs from *E. rostratum* (p. 674) and *E. longirostratum* (p. 670) by the absence of a true septum shutting off the end cells.
Pathogenicity. BSL-1. The species is known from a single strain from a case of human sinusitis with obstruction of the nasal cavity by polyps (Padhye *et al.*, 1986).
Reference. Padhye *et al.* (1986).
Nomenclature. *Exerohilum mcginnissi* Padhye & Ajello, *in* Padhye, Ajello, Wieden & Steinbronn - J. Clin. Microbiol. 24: 247, 1986.

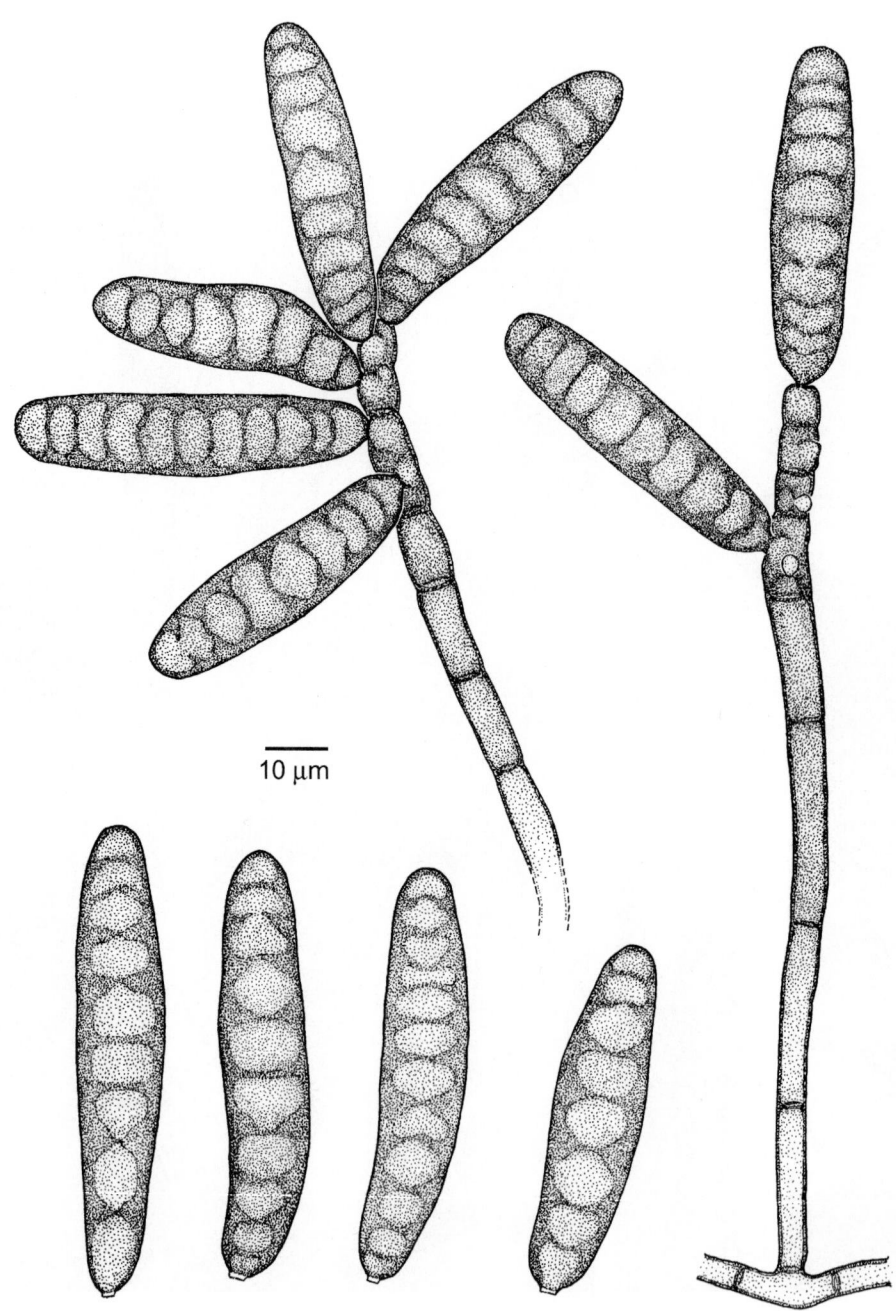

10 µm

Exserohilum mcginnisii, CBS 325.87. Conidiophores and conidia.

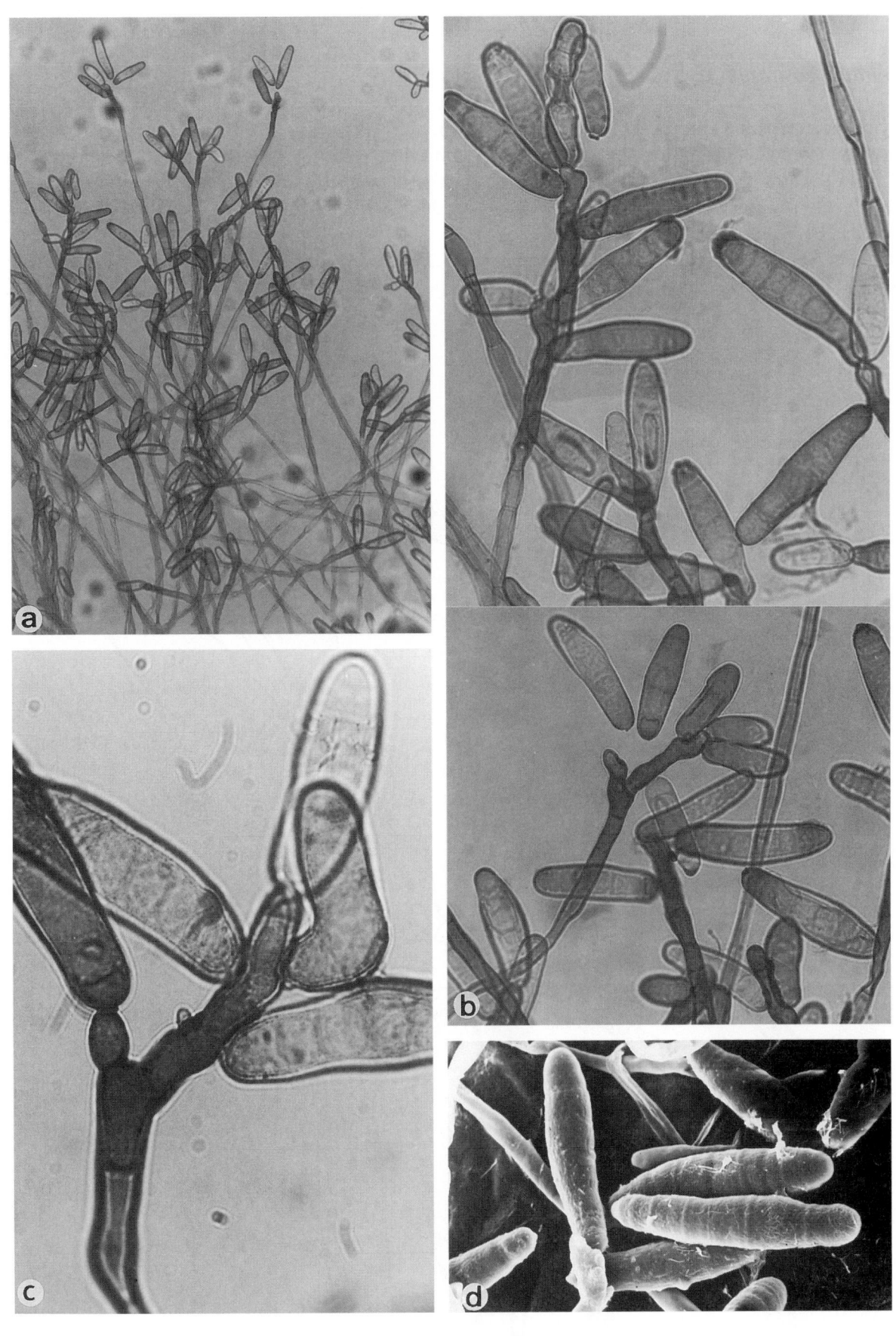

Exserohilum mcginnisii, CBS 325.87. Conidiophores and conidia. a. ×128; b. ×512; c. ×1280; d. ×880.

673

Exserohilum rostratum (Drechsler) Leonard & Suggs

Colony characteristics. Colonies (PCA) spreading, brown.

Microscopy. Conidiophores up to 230 μm long, 5-8 μm wide, erect to flexuose, often geniculate, olivaceous brown, pale towards the apex. Conidia straight or slightly curved, ellipsoidal to fusiform or rostrate, smooth-walled to finely roughened, brown to olivaceous brown, 30-128 × 9-23 μm, with (4-) 7-9 (-14) distosepta, with dark bands at both ends.

Teleomorph. *Setosphaeria rostrata* Leonard (*Ascomycota, Euascomycetes, Pleosporales: Pleosporaceae*).

Physiology. Tolerant to benomyl.

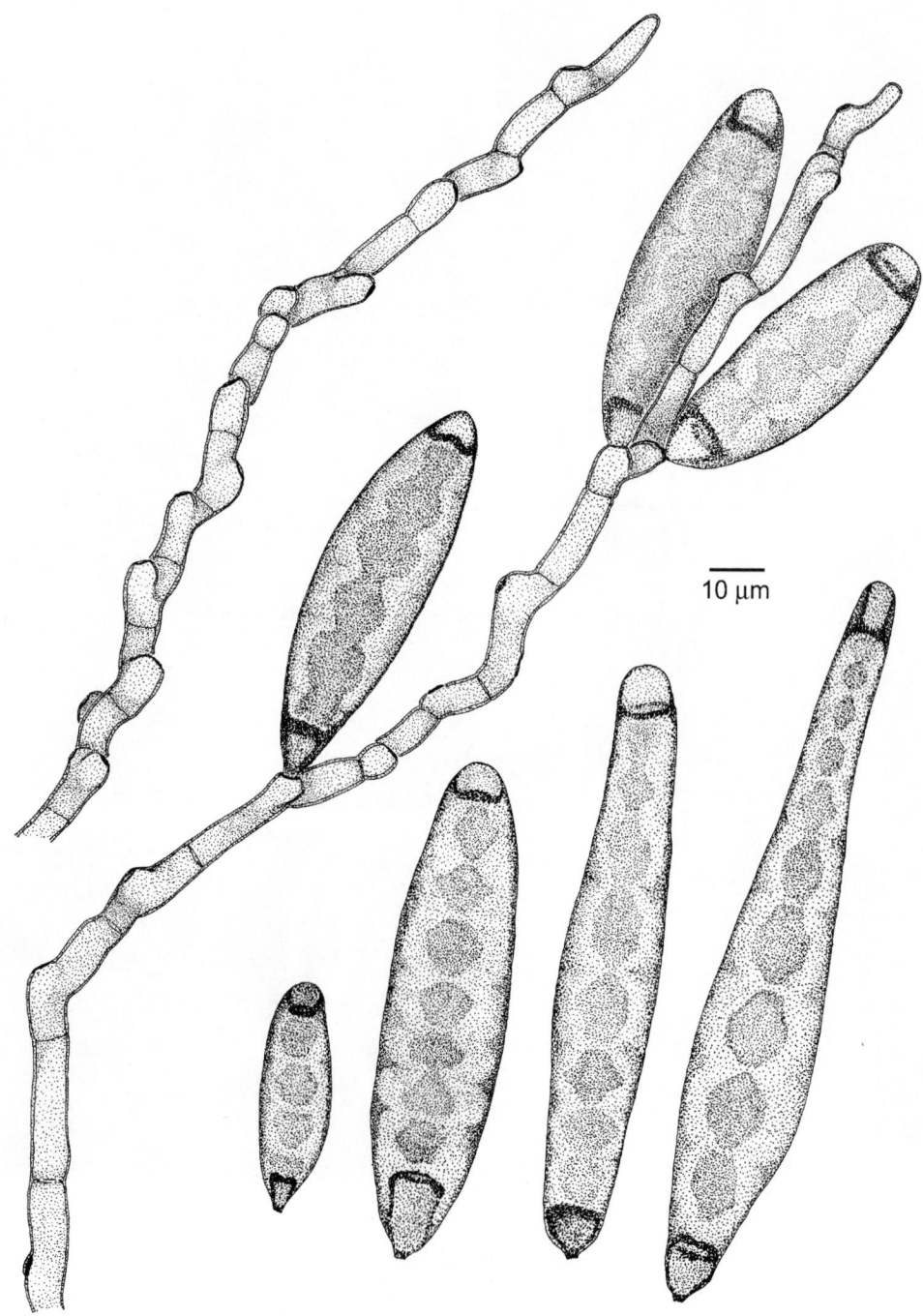

Exserohilum rostratum, CBS 732.96. Conidiophores and conidia.

Pathogenicity. BSL-1. Pathogen on grasses. Human cases: nasal phaeohyphomycosis (Douer *et al.*, 1987), eventually with bone erosion (Kinsilla *et al.*, 1996), keratitis (Jones, 1975; Forster *et al.*, 1975b; Kanungo & Srinivasan, 1996; Mathews & Maharajan, 1999), cutaneous (Agarwal & Singh, 1995) and subcutaneous (Moneymaker *et al.*, 1986; Burges *et al.*, 1987; Hsu & Lee, 1993) infections, invasive infections (Drouhet *et al.*, 1985; Adam *et al.*, 1986), fatal dissemination (Aquino *et al.*, 1995). Several animal infections have been reported (McKenzie & Connole 1977; Pritchard & Chick, 1977).

References. McGinnis *et al.* (1986), Sivanesan (1987).

Antifungal susceptibility.

Antifungal	Mean MICs	Strains	Reference
AMB	0.7	12	McGinnis & Pasarell (1998b)
ITZ	0.06	8	McGinnis & Pasarell (1998a)
TBF	0.1	8	McGinnis & Pasarell (1998a)
VCZ	0.17	12	McGinnis & Pasarell (1998b)

Nomenclature: *Helminthosporium rostratum* Drechsler - J. Agric. Res. 24: 724, 1923 ≡ *Drechslera rostrata* (Drechsler) Richardson & Fraser - Trans. Br. Mycol. Soc. 51: 148, 1968 ≡ *Exserohilum rostratum* (Drechsler) Leonard & Suggs - Mycologia 66: 290, 1974.

Helminthosporium halodes Drechsler - J. Agric. Res. 24: 709, 1923 ≡ *Drechslera halodes* (Drechsler) Subramanian & Jain - Curr. Sci. 35: 354, 1966.

Setosphaeria rostrata Leonard - Mycologia 68: 409, 1976.

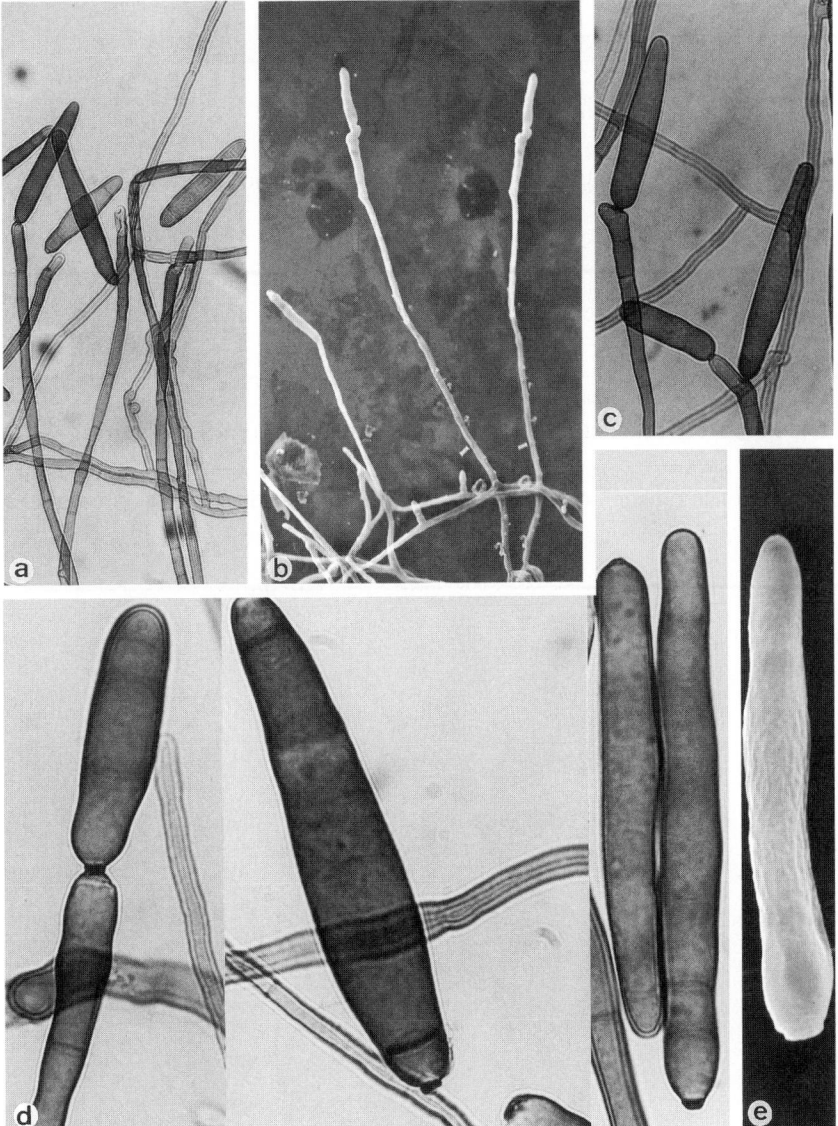

Exserohilum rostratum, AMMRL 106.9. Conidiophores and conidia. a. ×225; b. ×280; c. ×360; d. ×1120; e. ×1820.

Hyphomycetes. Genus: *FONSECAEA*

Generic description. Colonies restricted, velvety to cottony, dark olivaceous. Budding cells absent. Conidiophores poorly differentiated, pale to dark olivaceous, apically with a small cluster of cylindrical denticles or flat scars. Conidia 1-celled, arising singly or in short, branched chains. Dematiaceous, (sub)spherical phialides with collarettes may be present in low abundance on poor media.

Differential diagnosis. *Cladophialophora* species (p. 560) have longer conidial chains (>5 conidia/chain), conidia being fusiform to lemon-shaped. *Rhinocladiella* species (p. 893) have strictly non-catenate conidia and often have an additional *Exophiala* synanamorph. *Ramichloridium* (p. 888) has erect, brown conidiophores.

General remarks. Two species from humans are currently distinguished on the basis of morphology, *F. compacta* and *F. pedrosoi*, although their separation is not confirmed by molecular data (Attili *et al.*, 1998). Both species are agents of human chromoblastomycosis, and produce muriform cells in tissue. The phylogenetic affiliation of *Fonsecaea* to the ascomycete family *Herpotrichiellaceae* was presented by Masclaux *et al.* (1995) and Haase *et al.* (1999) on the basis of partial 26S rDNA and complete 18S rDNA sequencing, respectively. For relationships with the black yeasts, see p. 374.

Molecular diagnostics. SSU restriction map based on CBS 285.47:

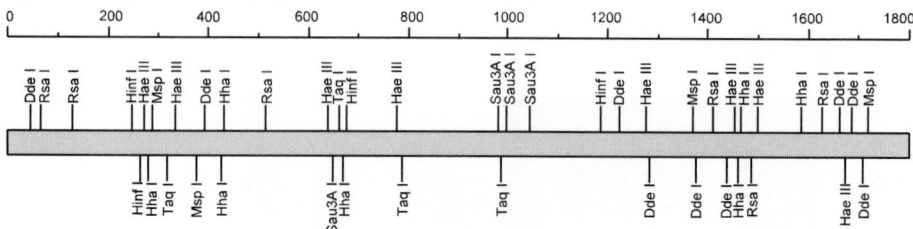

References. Carrión (1975), McGinnis & Schell (1980), Attili *et al.* (1998).

Key to the species of Fonsecaea:

1a. Conidiogenous cells in loosely branched systems, with prominent conidium-bearing denticles . . . . *F. pedrosoi* (679)

1b. Conidiogenous cells in dense clusters on dark conidiophores, with broad, flat conidial scars . . . . *F. compacta* (677)

Fonsecaea compacta Carrión

Colony characteristics. Colonies (OA) restricted, slightly heaped, powdery to velvety or hairy, olivaceous black; reverse olivaceous black.

Microscopy. Conidiophores suberect, olivaceous brown, apically densely branched. Conidiogenous cells in dense clusters, with broad, flat, pale pigmented scars. Conidia barrel-shaped, clustered, often remaining attached, smooth-walled, brown, or falling off as short branchlets, 2.5-4.5 × 2.0-3.5 μm. Phialides occasionally produced.

Molecular diagnostics. ITS restriction map based on CBS 102246:

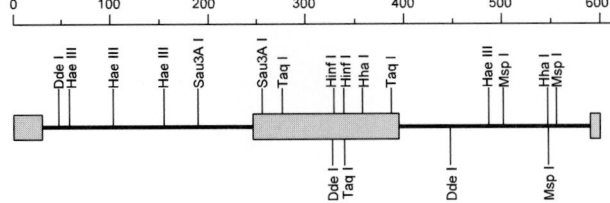

Pathogenicity. BSL-2. The species is a very rare agent of chromoblastomycosis in tropical Central and North America (Carrión, 1935, 1936; Emmons & Carrión, 1936).

References. Carrión (1935), Gezuele & Iraola (1970), Krempl-Lamprecht *et al.* (1987).

Antifungal susceptibility.

Antifungal	MICs	Strains	Reference
AMB	1.52	5	McGinnis & Pasarell (1998b)
5FC	0.25	1	Unpublished data
FLZ	64	1	Unpublished data
ITZ	0.08	5	McGinnis & Pasarell (1998b)
KTZ	0.5	1	Unpublished data
MCZ	0.5	1	Unpublished data
VCZ	0.014	5	McGinnis & Pasarell (1998b)

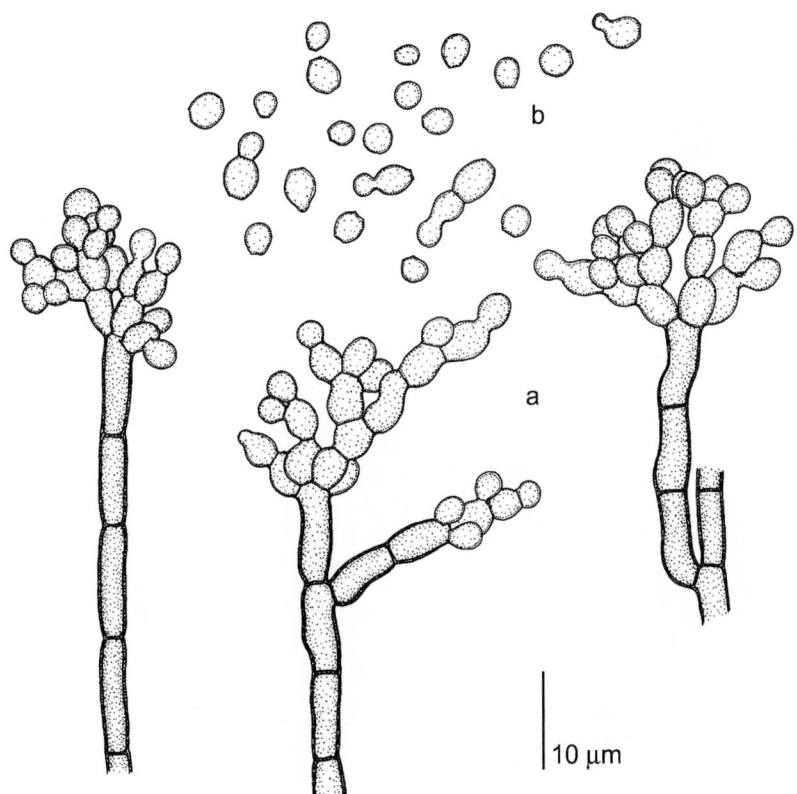

Fonsecaea compacta, CBS 285.47. a. Conidiophores; b. conidia.

Nomenclature. *Hormodendrum compactum* Carrión - Puerto Rico J. Publ. Health Trop. Med. 10: 543, 1935 (invalid) ≡ *Phialoconidiophora compacta* (Carrión) Moore & de Almeida - Annls Mo. Bot. Gdn 23: 549, 1936 ≡ *Fonsecaea compacta* Carrión ex Carrión - Puerto Rico J. Publ. Health Trop. Med. 15: 359, 1940 ≡ *Phialophora compacta* (Carrión) Redaelli & Ciferri - Granul. Fung. Uomo Reg. Trop. Subtrop. p. 595, 1942 ≡ *Rhinocladiella compacta* (Carrión) Schol-Schwarz - Antonie van Leeuwenhoek 34: 141, 1968.

Fonsecaea compacta, CBS 285.47. Conidiophores and conidia. a. ×640; b. ×1600; c. ×1650.

678

Fonsecaea pedrosoi (Brumpt) Negroni

Colony characteristics. Colonies (OA) spreading, lanose to velvety, olivaceous; reverse olivaceous black.

Microscopy. Conidiogenous cells pale olivaceous, in loosely branched systems, cylindrical, intercalary or terminal, with clusters of prominent denticles. Conidia pale olivaceous, in short chains, subhyaline, smooth- and thin-walled, clavate, 3.5-5.0 × 1.5-2.0 µm. Locally ampulliform, dark olivaceous brown phialides with deep, funnel-shaped collarettes may be present. Yeast cells are produced at low pH (Ibrahim-Granet *et al.*, 1985).

Physiology. Intolerant to benomyl.

Molecular diagnostics. Phylogeny based on chitin synthase genes was elaborated by Karuppayil *et al.* (1996) and mtDNA RFLP by Kawasaki *et al.* (1999), ITS restriction map based on CBS 271.37:

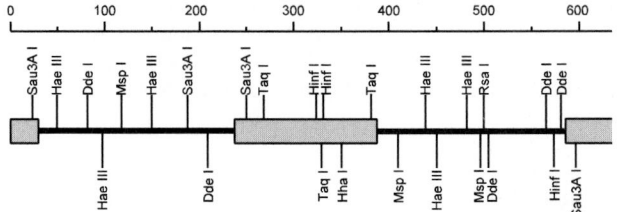

Pathogenicity. BSL-2. The species is one of the main agents of human chromoblastomycosis (Burks *et al.*, 1995) in humid tropical regions, particularly South America (Carrión & Silva-Hutner, 1971) and Japan (Fukushiro, 1983). Lesions are chronic (Usuki *et al.*, 1996). *Fonsecaea pedrosoi* has been reported from fatal brain abscesses after hematogenous dissemination (Fukushiro *et al.*, 1957; Fukushiro, 1983), but there may have been some misidentification. A case of nasal chromoblastomycosis was reported by Zaror *et al.* (1987), a paranasal sinusitis by Mehta *et al.* (1993) and keratitis by Fischman *et al.* (2000). It may infect preexisting lesions (Naka & Nishikawa, 1995). Its natural occurrence is in rotten wood and soil (Gezuele *et al.*, 1972). Cold-blooded animals living in swamps are frequently infected (Cicmanec, 1973; Beneke, 1978).

References. De Hoog (1977), Emmons *et al.* (1977), Attili *et al.* (1998).

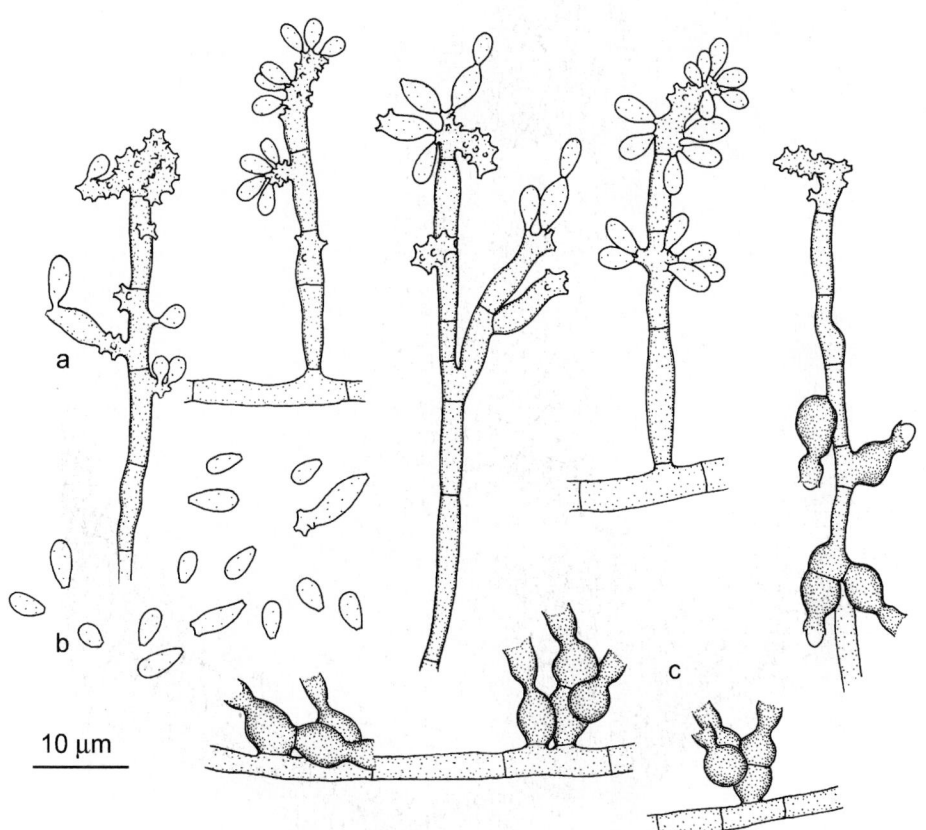

Fonsecaea pedrosoi, CBS 271.37. a. Sympodial conidiophores; b. conidia; c. *Phialophora* synanamorph.

Antifungal susceptibility.

Antifungal	Mean MICs	Strains	Reference
AMB	0.28	20	McGinnis & Pasarell (1998b)
5FC	2.83	2	Tintelnot *et al.* (1999)
ITZ	0.07	17	McGinnis & Pasarell (1998a)
KTZ	0.25	2	Tintelnot *et al.* (1999)
TBF	0.03	17	McGinnis & Pasarell (1998a)
VCZ	0.08	20	McGinnis & Pasarell (1998b)

Nomenclature. *Hormodendrum pedrosoi* Brumpt - Précis Parasitol., ed. 3, p. 1105, 1922 ≡ *Acrotheca pedrosoi* (Brumpt) da Fonseca & Aréa Leão - C.R. Soc. Biol., Paris 89: 762, 1923 ≡ *Carrionia pedrosoi* (Brumpt) Briceña-Iragorry - Revta Clín. Louis Razetti 1: 108, 1942 ≡ *Gomphinaria pedrosoi* (Brumpt) C.W. Dodge - Med. Mycol. p. 850, 1935 ≡ *Hormodendroides pedrosoi* (Brumpt) Moore & Almeida - Annls Mo. Bot. Gdn 23: 547, 1936 ≡ *Phialophora pedrosoi* (Brumpt) Redaelli & Ciferri - Tratt. Micopat. Um. 5: 592, 1942 ≡ *Trichosporium pedrosoi* (Brumpt) Brumpt - Précis Parasitol. 4: 1333, 1927 ≡ *Acrotheca pedrosiana* Ota - Jpn. J. Derm. Urol. 27: 916, 1927 (name change).

Hormodendrum rossicum Meriin - Arch. Derm. Syph. 162: 300-310, 1930.

Botrytoides monophora Moore & de Almeida - Annls Mo. Bot. Gdn 23: 546, 1936.

Phialoconidiophora guggenheimia Moore & de Almeida - Annls Mo. Bot. Gdn 23: 547, 1936 ≡ *Fonsecaea pedrosoi* (Brumpt) Negroni var. *cladosporioides* Carrión - Puerto Rico J. Publ. Health 15: 350, 1942 (name change) ≡ *Fonsecaea pedrosoi* (Brumpt) Negroni var. *cladosporium* O'Daly - Rev. Sanid., Caracas 8: 655, 1943 (name change) ≡ *Fonsecaea cladosporium* (O'Daly) Powell - Austr. J. Derm. 1: 218, 1952.

Fonsecaea pedrosoi (Brumpt) Negroni var. *communis* Carrión - Puerto Rico J. Publ. Health 15: 360, 1942.

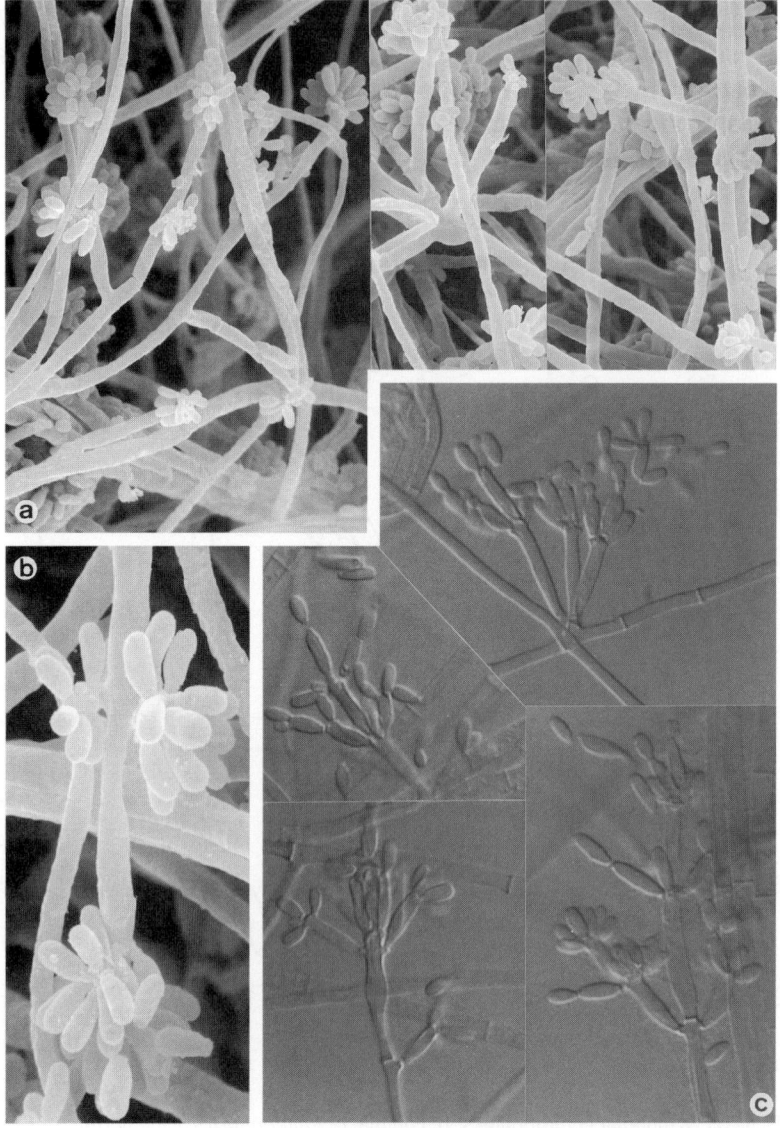

Fonsecaea pedrosoi, CBS 273.66. Sympodial conidiophores and conidia. a. ×1365; b. ×3100; c. ×990.

Hyphomycetes. Genus: *FUSARIUM*

Generic description. Colonies expanding, often coloured in pink, yellow, red or purple shades. Conidiogenous cells formed on aerial hyphae or on short, densely branched conidiophores, often clustered in sporodochia, ampulliform or cylindrical. Conidia often of three kinds: macroconidia, microconidia and blastoconidia. (1) Macroconidia falcate, with several transverse septa and an attenuated, often beaked apical cell and a pedicellate basal cell, produced in basipetal succession from monophialides or sporodochia and accumulating in slimy masses; (2) Microconidia ellipsoidal, ovoidal, subspherical, pyriform, clavate or allantoid, generally unicellular, with a rounded or truncate base, produced in basipetal series on mono- or polyphialides and accumulating in small slimy heads or in chains; (3). Blastoconidia produced singly on polyblastic cells, 0-3-septate. Chlamydospores often present, thick-walled, hyaline or pale, intercalary or terminal.

Teleomorphs. *Gibberella, Nectria (Ascomycota, Euascomycetes, Hypocreales: Hypocreaceae).*

General remarks. *Fusarium* species are very common plant pathogens or saprobes on plant debris and in soil. Some occur regularly on seeds, especially of cereals. They rather frequently occur as agents of various kinds of hyalohyphomycosis after traumatic inoculation (Anaissie *et al.*, 1988; Goldschmied-Reouven *et al.*, 1993; Vartivarian *et al.*, 1993; Rabodonirina *et al.*, 1994), particularly keratitis (Rosa *et al.*, 1994). They are emerging as opportunists in patients with impaired innate immunity (Helm *et al.*, 1990; Nucci *et al.*, 1992; Hennequin *et al.*, 1997). In leukemic patients sinusitis is often observed, eventually with complications (Lopes *et al.*, 1995). Clinical strains frequently have a reduced morphology and hence are difficult to identify with described plant-inhabiting taxa. Exoantigen tests for the recognition of clinical species were developed by Sekhon *et al.* (1995). *In vitro* susceptibility tests have been performed by Reuben *et al.* (1989), Sekhon *et al.* (1994), Pujol *et al.* (1997) and Arikan *et al.* (1999) and *in vivo* in mice by Guarro *et al.* (1999g). An experimental murine model of *Fusarium* infection was published by Legrand *et al.* (1991). Terminology used for *Fusarium* in particular is explained in the diagram below (Fig. 50). Species identification based on ITS RFLP was introduced by Edel *et al.* (1996). Molecular phylogeny based on mtDNA, ITS and β-tubulin sequences was introduced for some groups by O'Donnell *et al.* (1998, 2000) and taxonomic changes on the basis of these data were proposed by Nirenberg & O'Donnell (1998). Some 26S rDNA data were provided by Hennequin *et al.* (1999).

References. Booth (1971, 1977), Gerlach & Nirenberg (1982), Nelson *et al.* (1981, 1983), Joffe (1986), Nirenberg (1990). Several reviews of *Fusarium* infections in humans have been published (Anaissie *et al.*, 1992; Guarro & Gené, 1992, 1995; Nelson *et al.*, 1994; Boutati & Anaissie, 1997).

Key to the treated species of Fusarium:

1a. Colonies attaining less than 2 cm diam in 7-10 days; macroconidia, generally 1-2-septate ➡ **2**

1b. Colonies attaining over 2 cm, usually 4-8 cm diam, in 7-10 days ➡ **4**

2a. Macroconidia up to 55 μm long; chlamydospores absent .... *F. aquaeductuum* **(684)**

2b. Macroconidia shorter, usualy less than 25 μm long; chlamydospores absent or present ➡ **3**

3a. Macroconidia strongly curved and pointed at the apex; chlamydospores present .... *F. dimerum* **(688)**

3b. Macroconidia gently curved and with less pointed apex; chlamydospores absent ....
.... *Plectosporium tabacinum* **(880)**

4a. Microconidia rare or absent; colonies tan to brown; polyblastic conidiogenous cells abundant .... *F. incarnatum* **(690)**

4b. Microconidia numerous ➡ **5**

5a. Microconidia in chains ➡ **6**

5b. Microconidia not in chains ➡ **9**

6a. Microconidia exclusively on monophialides ➡ **7**

6b. Microconidia borne on polyphialides and monophialides ➡ **8**

7a. Napiform or lemon-shaped microconidia present .... *F. napiforme* **(692)**

7b. Napiform or lemon-shaped microconidia absent .... *F. verticillioides* **(704)**

8a. Chlamydospores absent .... *F. proliferatum* **(698)**

8b. Chlamydospores abundant .... *F. nygamai* **(694)**

9a. Polyblastic conidiogenous cells frequent ➡ **10**

9b.   Polyblastic conidiogenous cells absent ➞ **13**

10a.   Colonies red; chlamydospores abundant; blastoconidia predominant . . . . *F. chlamydosporum* **(686)**

10b.   Colonies pink or vinaceous to violet; chlamydospores absent; blastoconidia less abundant ➞ **11**

11a.   Microconidia produced from prostrate conidiophores on hyphae growing horizontally
above the agar surface, commonly aseptate, obovoidal to narrow ellipsoidal in shape . . *F. sacchari* **(703)**

11b.   Microconidia produced from erect conidiophores ➞ **12**

12a.   Pyriform microconidia present . . . . *F. anthophilum*[1]

12b.   Pyriform microconidia absent . . . . *F. subglutinans* **(702)**

13a.   Microconidia on short monophialides which are mostly lateral; macroconidia orange
in mass, without distinct foot cell; colonies white to purple, or tan to orange when
sporodochia are present . . . . *F. oxysporum* **(696)**

13b.   Microconidia on long monophialides; macroconidia with indistinct foot cell; colonies
white, cream coloured or blue to blue-green when sporodochia are present . . . . *F. solani* **(700)**

[1]This species was isolated together with *F. solani* from a disseminated infection in a leukemic patient (Okuda *et al.*, 1987).

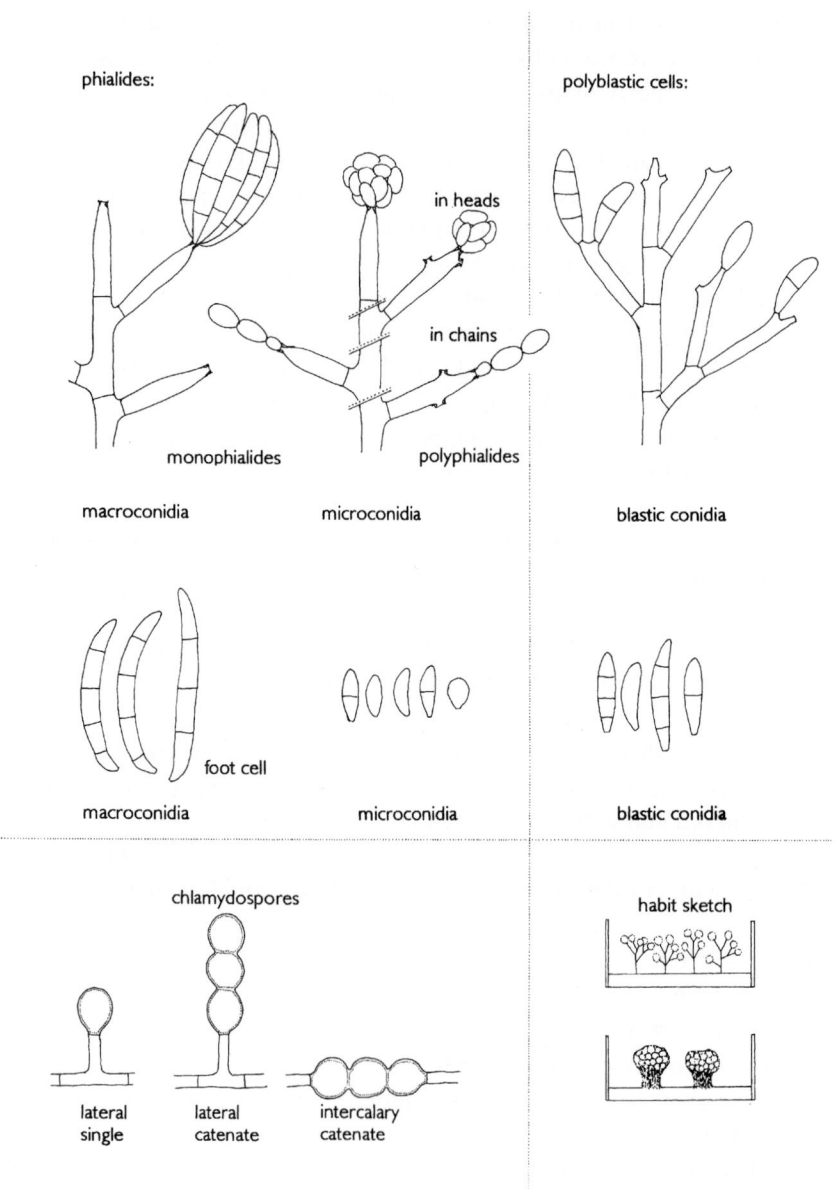

Fig. 50. Diagram of used terminology in *Fusarium*.

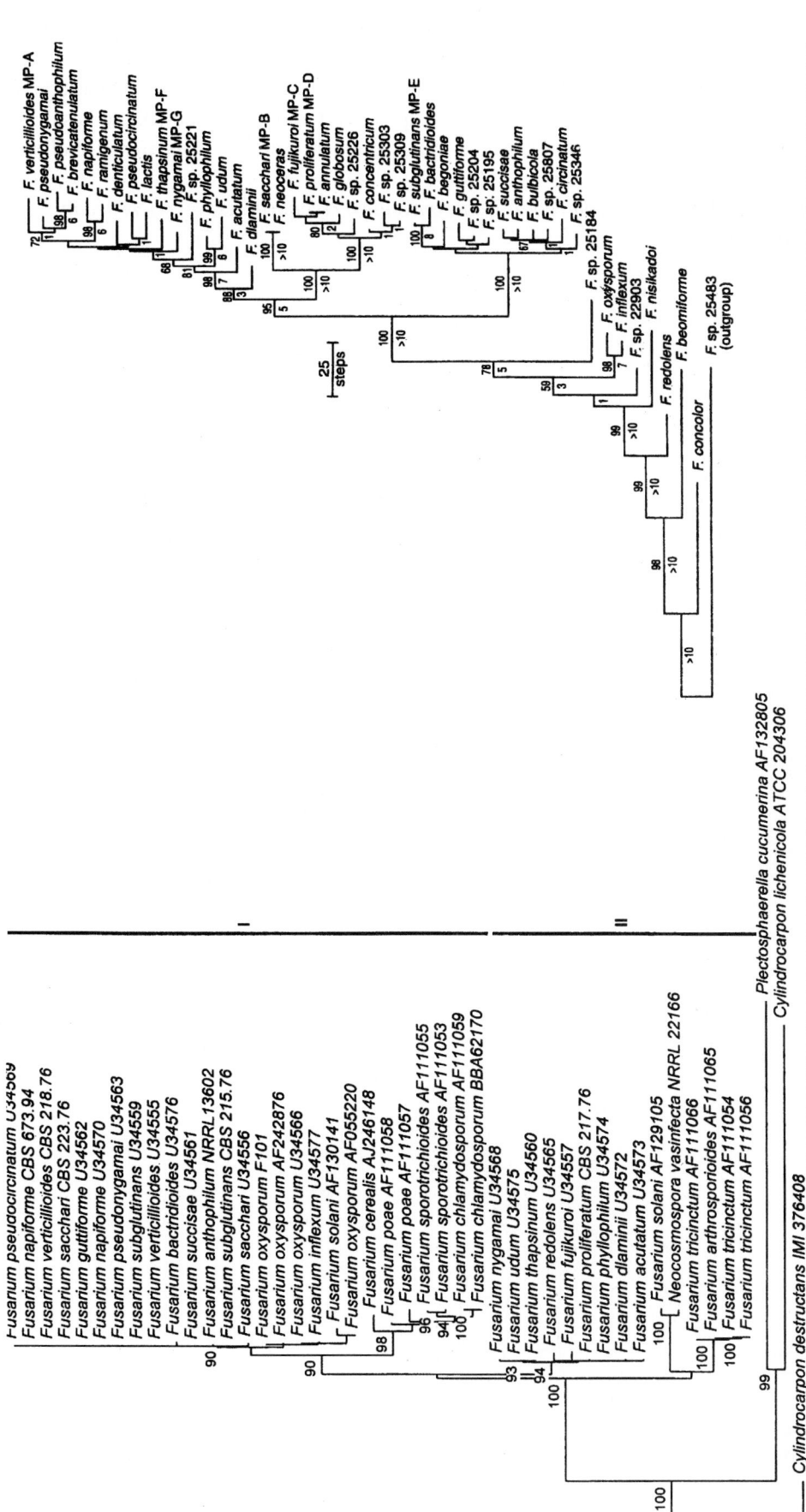

Fig. 51. (Left) Phylogenetic tree of *Fusarium* species based on the ITS rDNA domain using Neighbor joining alorithm with Kimura correction. Bootstrap values >90 from 100 resampled datasets are shown. Based on data taken from O'Donnell *et al.* (1998). Two types of ITS2 (I and II) may be present in a single strain, of which only the prevalent one is amplified in routine sequencing studies. This is shown e.g. by *Fusarium solani* appearing in both groups I and II. Within each of these groups, ITS sequences are very similar over numerous species. Hence, ITS is unsuitable for recognizing *Fusarium* species and therefore no restriction maps for this genus are included in the Atlas. O'Donnell *et al.* (1998) therefore published another, phylogenetically more informative tree based on a combination of 26S rDNA, mtSSU rDNA and β-tubulin sequence data; this is reproduced with permission on the right hand side.

Note that in the ITS tree the anamorph of *Neocosmospora* (p. 289), with an *Acremonium*-like anamorph, fits in *Fusarium*. In contrast, the anamorph of *Plectosphaerella* (p. 880), which was previously classified in *Fusarium*, is unrelated.

Fusarium aquaeductuum (Radlkofer & Rabenh.) Lagerh.

Colony characteristics. Colonies (OA) growing slowly, covered with orange slime from sporodochial conidia or partly covered by a white, floccose mycelium; reverse yellow-brown.

Microscopy. Conidiophores arising laterally from hyphae, later more or less loosely branched. Macroconidia abundant, strongly to moderately curved, falcate, with an often hooked apical cell and a non-pedicellate basal cell, (0-)1(-3)-septate, 14-55 × 2.0-3.5 µm. Microconidia absent, but abundant in var. *medium* Wollenw. Chlamydospores absent.

Teleomorph. *Nectria episphaeria* (Tode:Fr.) Fr. (*Ascomycota, Euascomycetes, Hypocreales: Hypocreaceae*).

Physiology. Weak tolerance to benomyl.

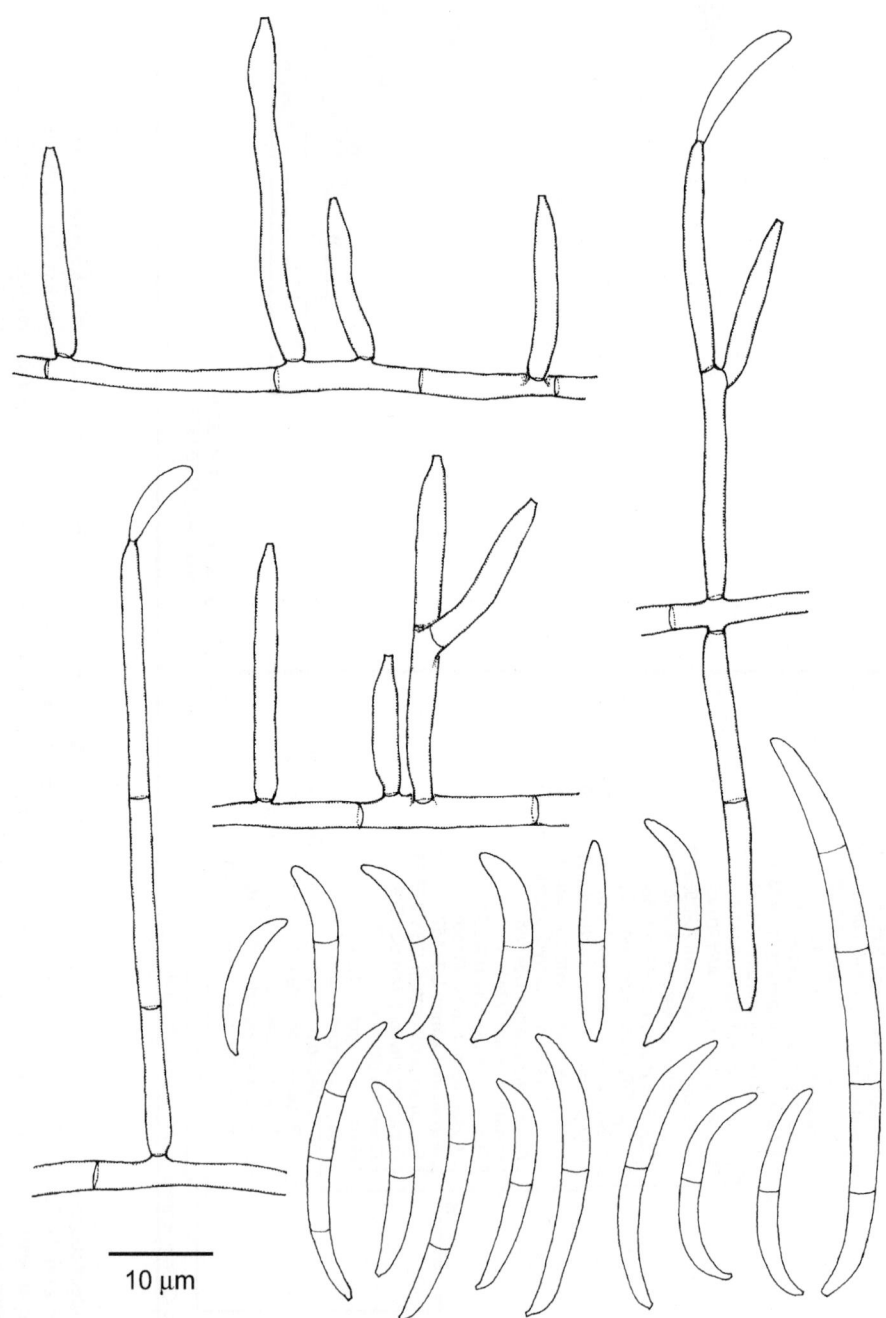

Fusarium aquaeductuum, CBS 837.85. Conidiophores, phialides and conidia.

Pathogenicity. BSL-1. Endophthalmitis (Pfulgfelder *et al.*, 1988).

References. Booth (1977), Domsch *et al.* (1980), Gerlach & Nirenberg (1982), Nelson *et al.* (1983).

Nomenclature. *Sphaeria episphaeria* Tode - Fung. Meckl. 2: 21, 1791 ≡ *Sphaeria episphaeria* Tode:Fries - Syst. Mycol. 2: 454, 1823 ≡ *Nectria episphaeria* (Tode:Fries) Fries - Summa Veg. Scand. 2: 388, 1849.

Selenosporium aquaeductuum Radlkofer & Rabenhorst - Hedwigia 2: 73, 1863 ≡ *Fusarium aquaeductuum* (Radlkofer & Rabenhorst) Lagerheim - Zentbl. Bakt. Parasitkde, Abt. 1, 9: 655, 1891.

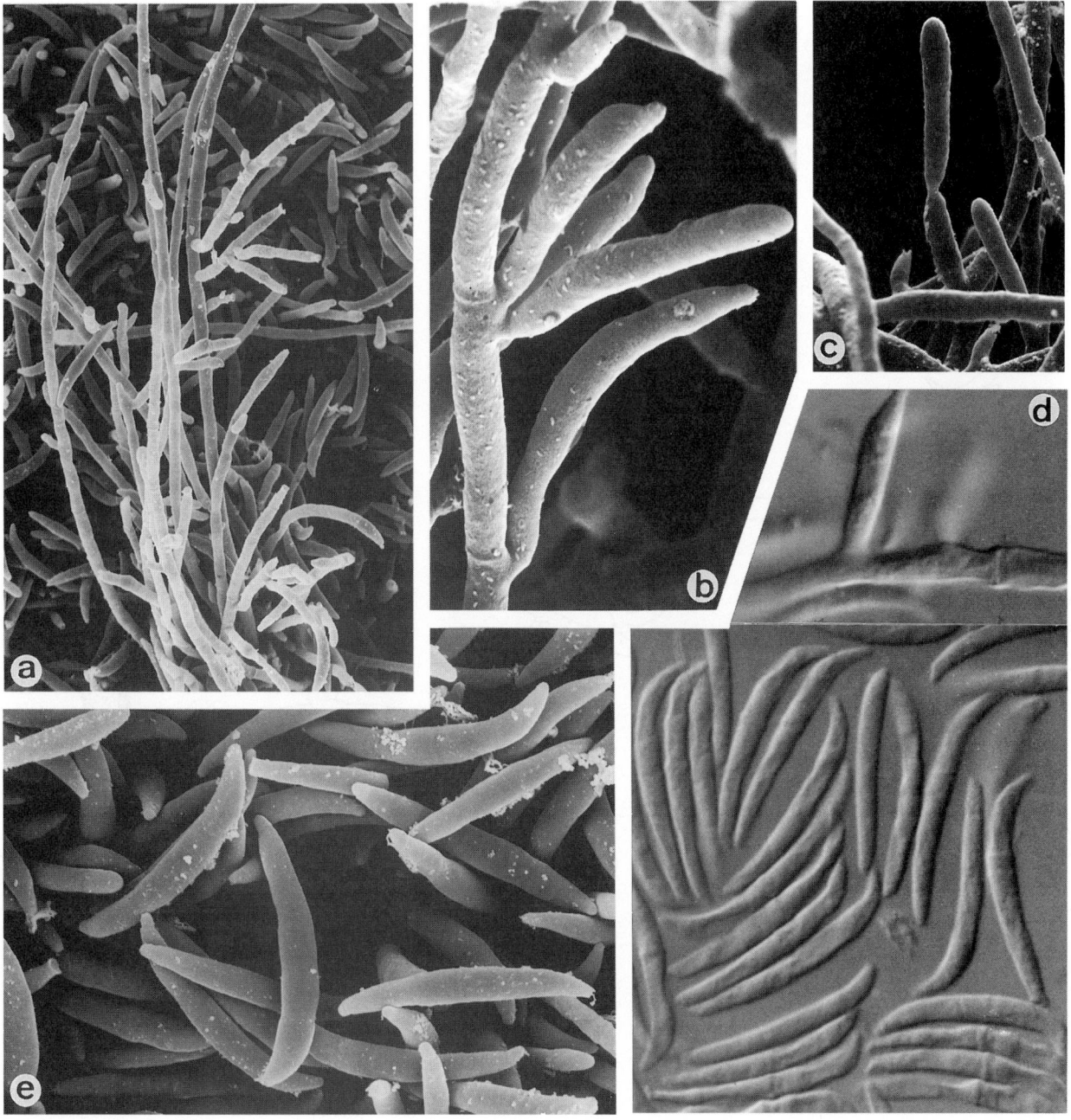

Fusarium aquaeductuum, CBS 837.85. Conidiophores and conidia. a. ×920; b. ×3900; c. ×2200; d. ×1600; e. ×2450.

Fusarium chlamydosporum Wollenweber & Reinking

Colony characteristics. Colonies (OA) growing rather rapidly, with abundant aerial mycelium, deep pink, red or ochraceous to brownish; reverse carmine red or tan to brown. Sporodochia orange, flesh-colour or ochraceous.
Microscopy. Conidiophores scattered over the aerial mycelium, branched; polyblastic conidiogenous cells numerous. Macroconidia rarely produced and appearing only on sporodochial phialides, 3-(5)-septate, slightly curved, 30-38 × 3.0-4.5 µm. Micro- and blastoconidia fusiform, rounded apically and tapered towards the base, 0-1 (-3)-septate, 6-26 × 2-4 µm. Chlamydospores abundant, intercalary, often roughened.

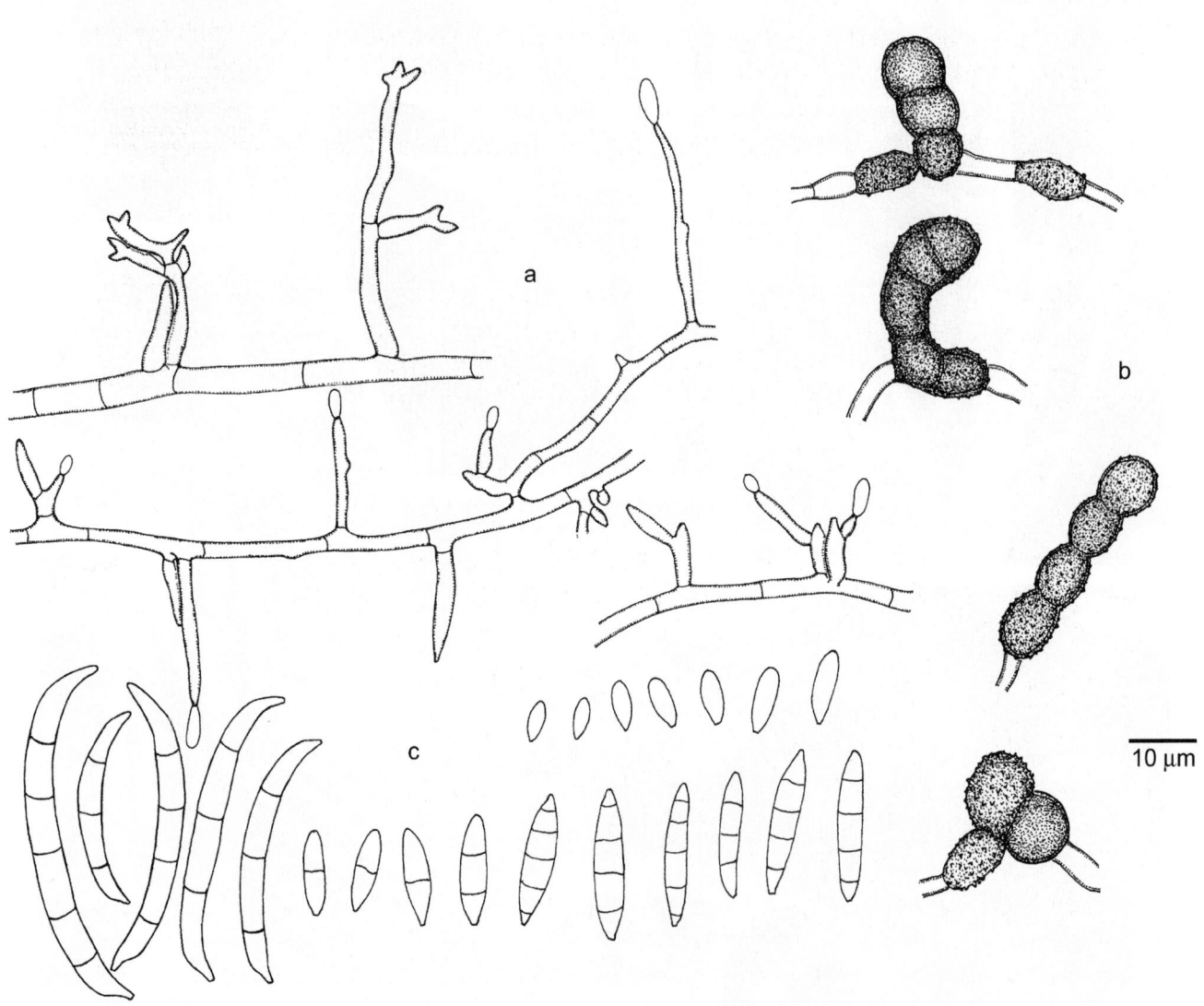

Fusarium chlamydosporum, CBS 615.87. a. Conidiophores; b. chlamydospores; c. macro-, micro- and blastoconidia.

Pathogenicity. BSL-1. Reported from a catheter-associated fungaemia in a patient with lymphocytic lymphoma (Kiehn *et al.*, 1985) and from an invasive infection of the turbinate in a neutropenic patient (Segal *et al.*, 1998).

References. Domsch *et al.* (1980), Gerlach & Nirenberg (1982), Nelson *et al.* (1983).

Nomenclature. *Fusarium chlamydosporum* Wollenweber & Reinking - Die Fusarien p. 47, 1935.

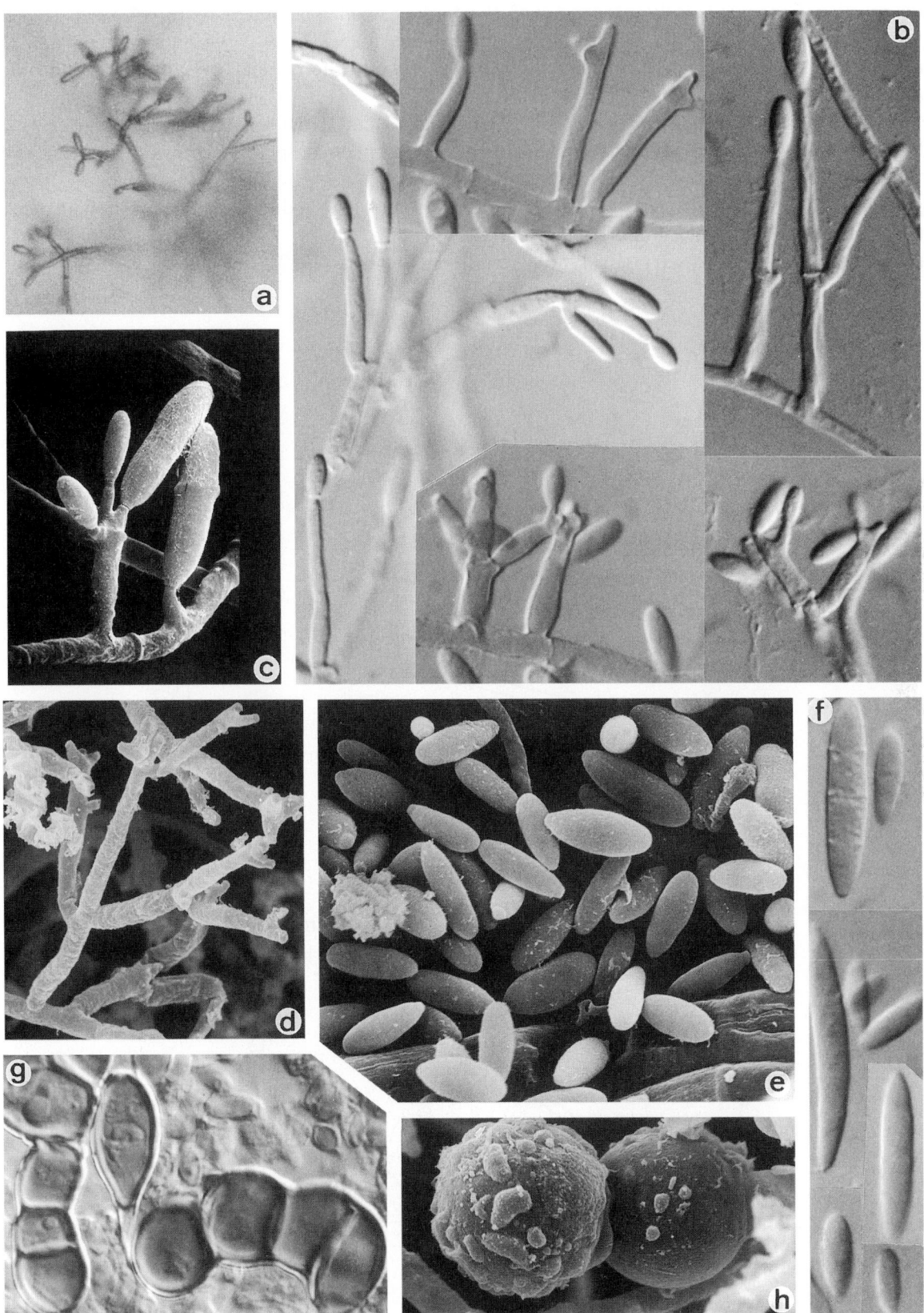

Fusarium chlamydosporum, CBS 615.87. a-d. Conidiophores with polyblastic conidiogenous cells and some monophialides; e, f. macro-, micro- and blastoconidia; g, h. chlamydospores. a. ×256; b. ×1600; c. ×2500; d. ×2300; e. ×3500; f,g. ×1600; h. ×3300.

Fusarium dimerum Penz.

Colony characteristics. Colonies (OA) growing slowly; surface usually orange to deep apricot due to confluent conidial slime; aerial mycelium sometimes floccose, whitish.

Microscopy. Conidiophores loosely branched, with short, often swollen phialides, 10-18 × 4-5 µm. Macroconidia strongly curved and pointed at the apex, mostly 1-(3)-septate, 5-25 (-32) × 1.5-4.2 µm. Microconidia absent. Chlamydospores mostly intercalary, exceptionally terminal, spherical to ovoidal, 6-12 µm diam, smooth-walled, single or in chains.

Differential diagnosis. *Microdochium nivale* (Fr.) Samuels & Hallett, previously known as *Fusarium nivale* (Fr.) Ces., differs by having conidia with rounded apex and by lacking chlamydospores. It grows and sporulates better at 18°C or below.

Pathogenicity. BSL-1. Keratitis (Zapater *et al.*, 1972; Zapater, 1986), disseminated infection in neutropenic patient (Poirot *et al.*, 1985), endocarditis (Camin *et al.*, 1999)

References. Booth (1977), Domsch *et al.* (1980), Gerlach & Nirenberg (1982), Nelson *et al.* (1983).

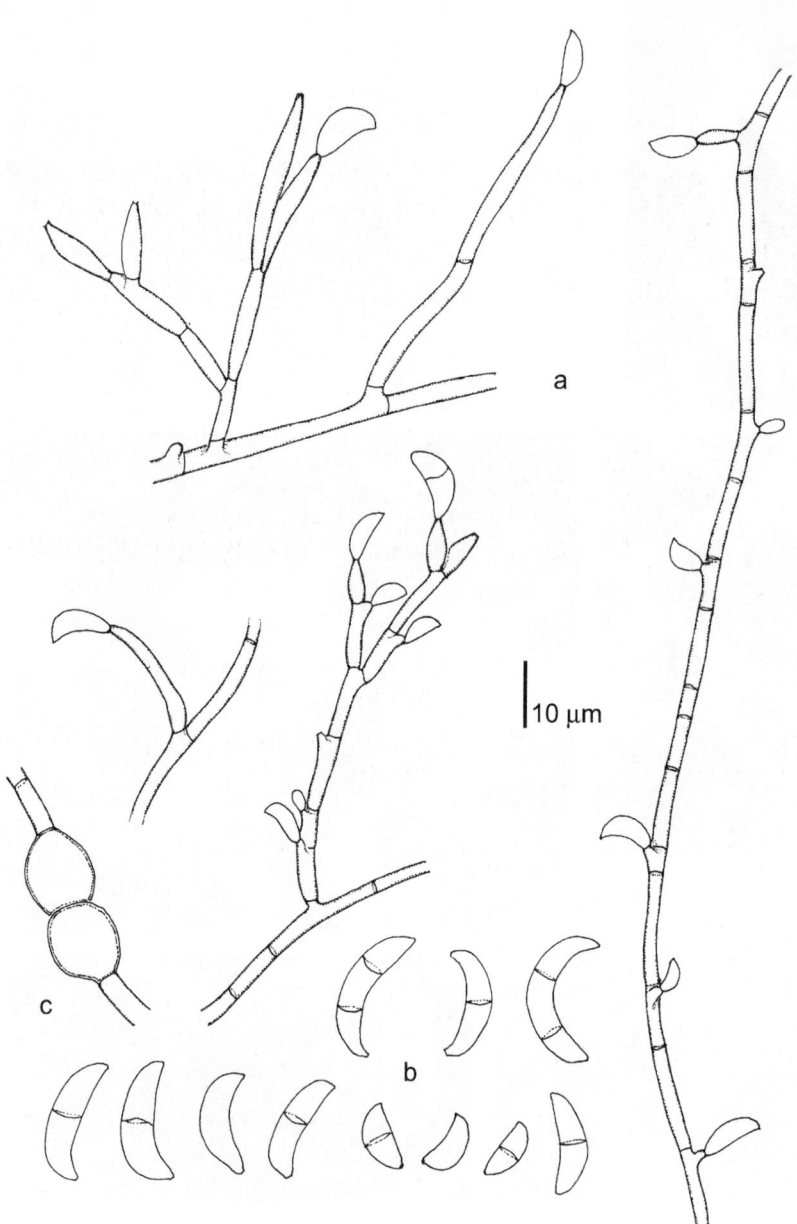

Fusarium dimerum, IMI 117087. a. Conidiophores; b. macroconidia; c. chlamydospores.

Antifungal susceptibility.

Antifungal	Mean MICs	Strains	Reference
AMB	0.48	4	Pujol *et al.* (1997)
5FC	>322.7	4	Pujol *et al.* (1997)
FLZ	>80	4	Pujol *et al.* (1997)
ITZ	>20	4	Pujol *et al.* (1997)
KTZ	30.5	4	Pujol *et al.* (1997)
MCZ	>40	4	Pujol *et al.* (1997)

Nomenclature. *Fusarium dimerum* Penzig, *in* Saccardo - Michelia 2: 484, 1882.
　Fusarium episphaeria (Tode) Snyder & Hansen - Am. J. Bot. 32: 662, 1945.

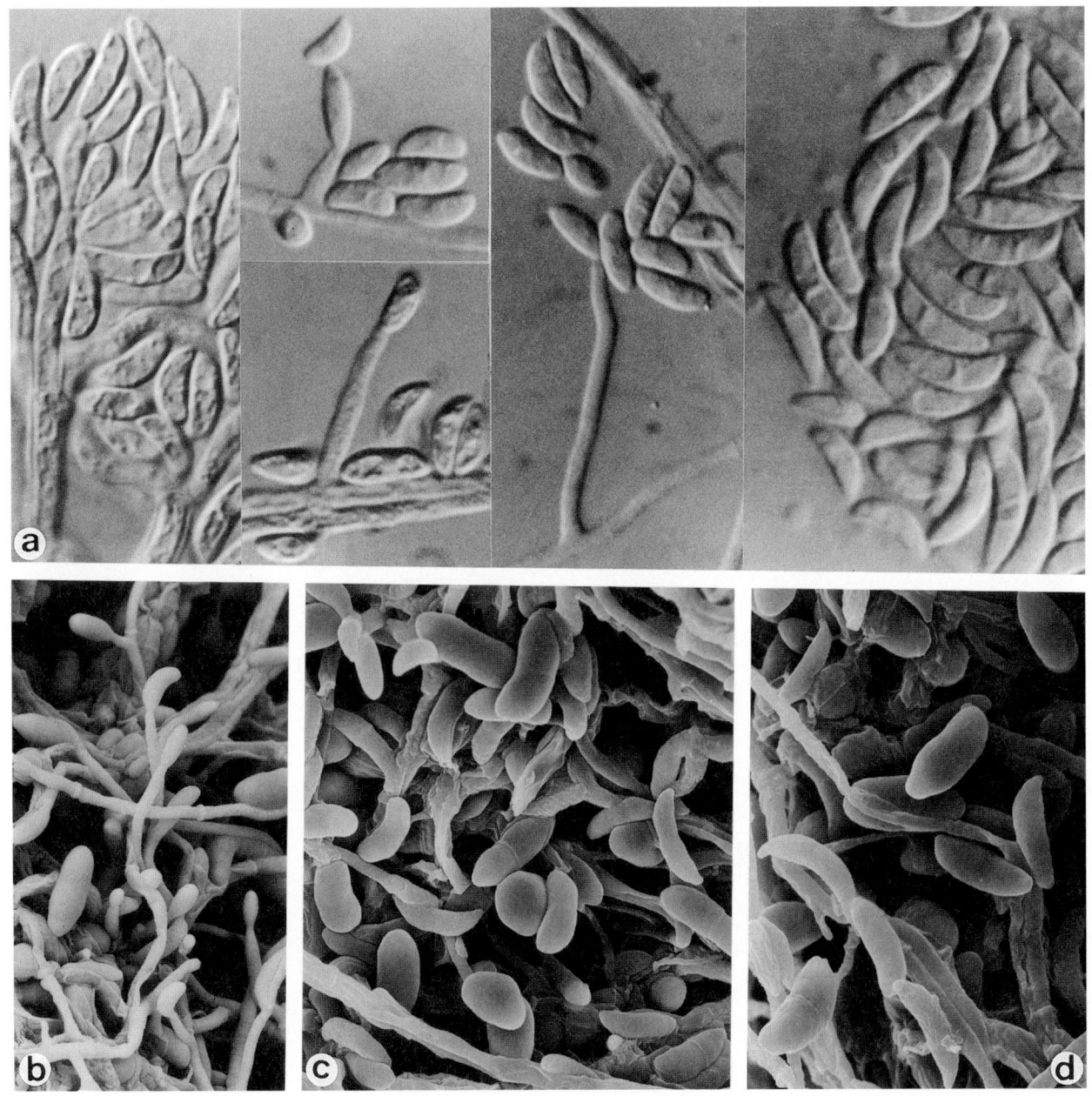

Fusarium dimerum, IMI 117087. Monophialides and conidia. a, b. ×1600; c. ×2000; d. ×2500.

Fusarium incarnatum (Rob.) Sacc.

Colony characteristics. Colonies (OA) growing rapidly; aerial mycelium floccose, at first whitish, later becoming avellaneous to buff-brown; reverse pale, becoming peach-coloured.

Microscopy. Conidiophores scattered in the aerial mycelium, loosely branched; polyblastic conidiogenous cells abundant. Sporodochial macroconidia slightly curved, with foot-cell, 3-7-septate, 20-46 × 3.0-5.5 µm. Conidia on aerial conidiophores (blastoconidia) usually borne singly on scattered denticles, fusiform to falcate, mostly 3-5-septate, 7.5-35 × 2.5-4.0 µm. Microconidia sparse or absent. Chlamydospores sparse, spherical, 10-12 µm diam, becoming brown, intercalary, single or in chains.

Pathogenicity. BSL-1. Isolated from skin burn of a patient who had suffered an electric shock (Imwidthaya *et al.*, 1984). A case of endocarditis after aortic valve replacement was reported (McGinnis *et al.*, 1994).

References. Booth (1977), Domsch *et al.* (1980), Gerlach & Nirenberg (1982), Nelson *et al.* (1983).

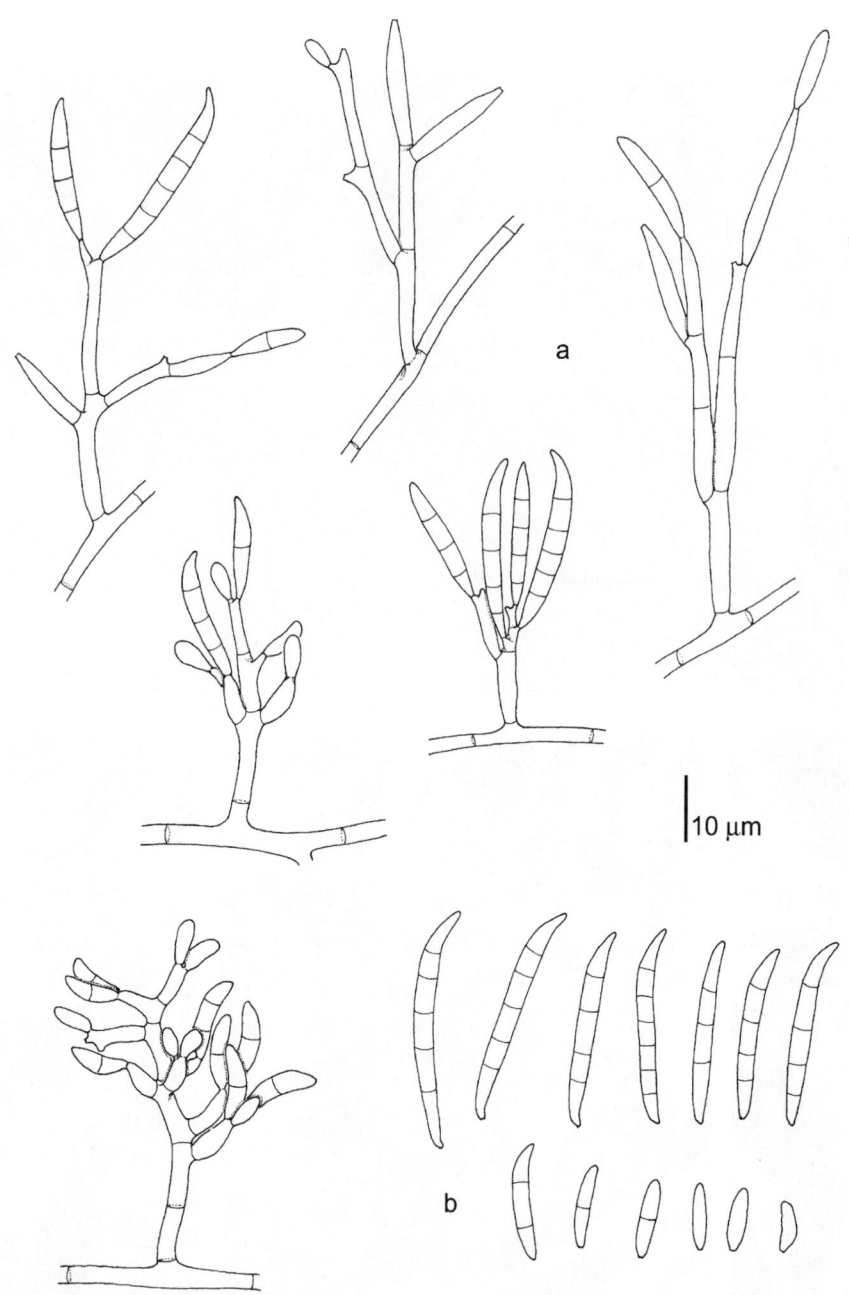

Fusarium incarnatum, CBS 131.73. a. Conidiophores; b. blasto- and microconidia.

Antifungal susceptibility.

Antifungal	Mean MICs	Strains	Reference
AMB	1.16	1	Pujol *et al.* (1997)
5FC	>322.7	1	Pujol *et al.* (1997)
FLZ	>80	1	Pujol *et al.* (1997)
ITZ	>10	1	Pujol *et al.* (1997)
KTZ	>51.2	1	Pujol *et al.* (1997)
MCZ	>40	1	Pujol *et al.* (1997)

Nomenclature. *Fusisporium incarnatum* Roberge - Annls Sci. Nat., Bot., Sér. 3, 11: 273-285, 1849 ≡ *Fusarium incarnatum* (Roberge) Saccardo - Syll. Fung. 4: 712, 1886.

Fusarium semitectum Berkeley & Ravenel, *in* Berkeley - Grevillea 3: 98, 1875.

Fusisporium pallidoroseum Cooke - Grevillea 6: 139, 1878 ≡ *Fusarium pallidoroseum* (Cooke) Saccardo - Syll. Fung. 4: 720, 1886.

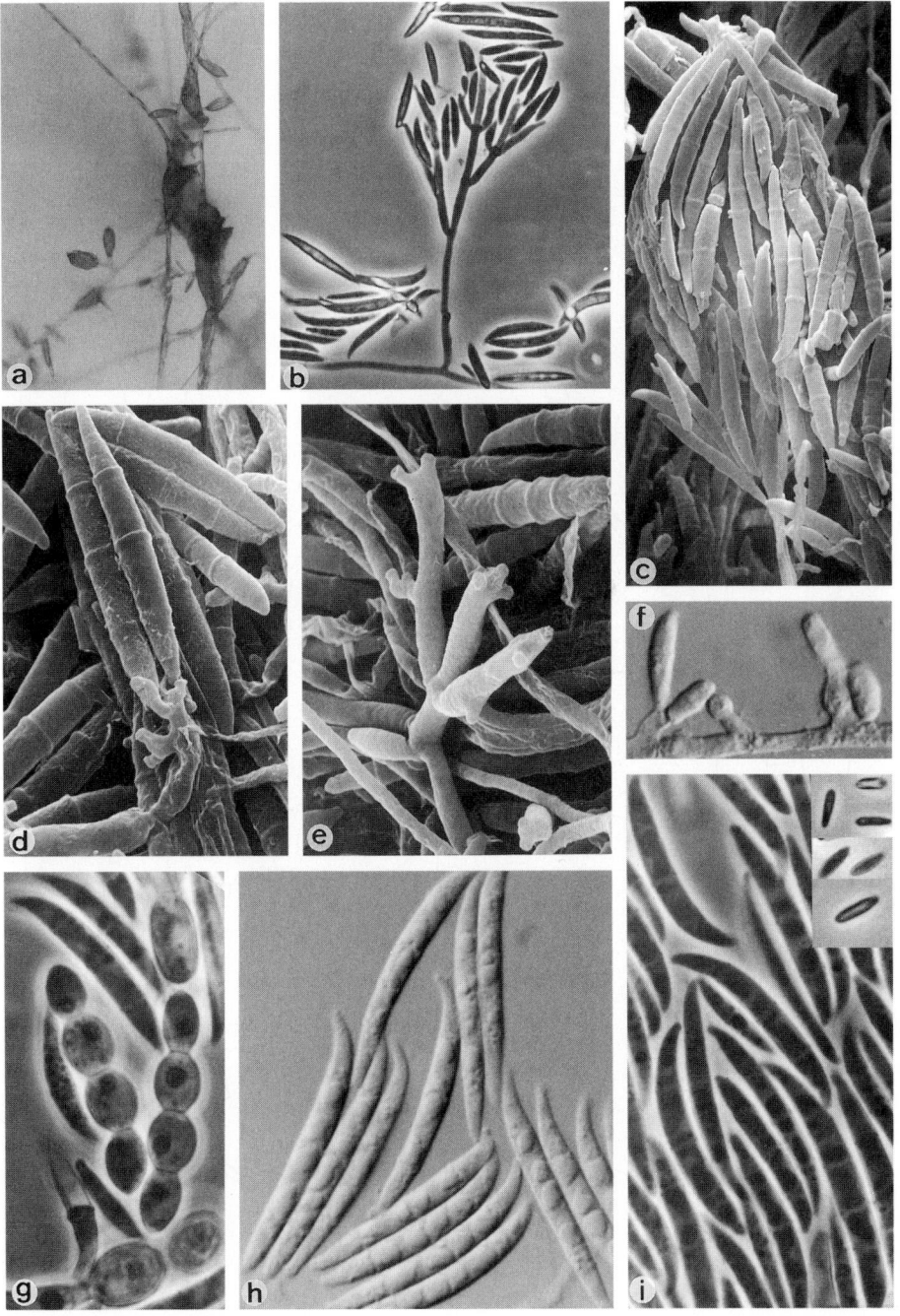

Fusarium incarnatum, CBS 131.73. a, b. Conidiophores; c-f, h, i. polyblastic conidiogenous cells, macro-, micro- and blastoconidia; g. chlamydospores. a. ×128; b. ×512; c. ×1450; d. ×2950; e. ×3050; f. ×1600; g. ×1280; h. ×1600; i. ×1280.

Fusarium napiforme Marasas *et al.*

Cultural characteristics. Colonies (OA) moderately expanding, floccose, white, with bright orange sporodochia; reverse purple.

Microscopy. Monophialides laterally on aerial hyphae or on branched conidiophores, subulate, up to 30 µm long. Macroconidia falcate, curved with inner wall nearly straight, mostly 5-septate, 29-90 × 3.5-5.0 µm. Microconidia in short chains, obovoidal with truncate base; some additional napiform or lemon-shaped conidia present. Chlamydospores clustered, hyaline to pale brown, 5-12 × 4-8 µm.

Differential diagnosis. Very similar to *F. verticillioides* (p. 704), differing by napiform microconidia and presence of chlamydospores in old cultures.

Pathogenicity. BSL-1. Disseminated infection in a leukemic patient (Melcher *et al.*, 1993).

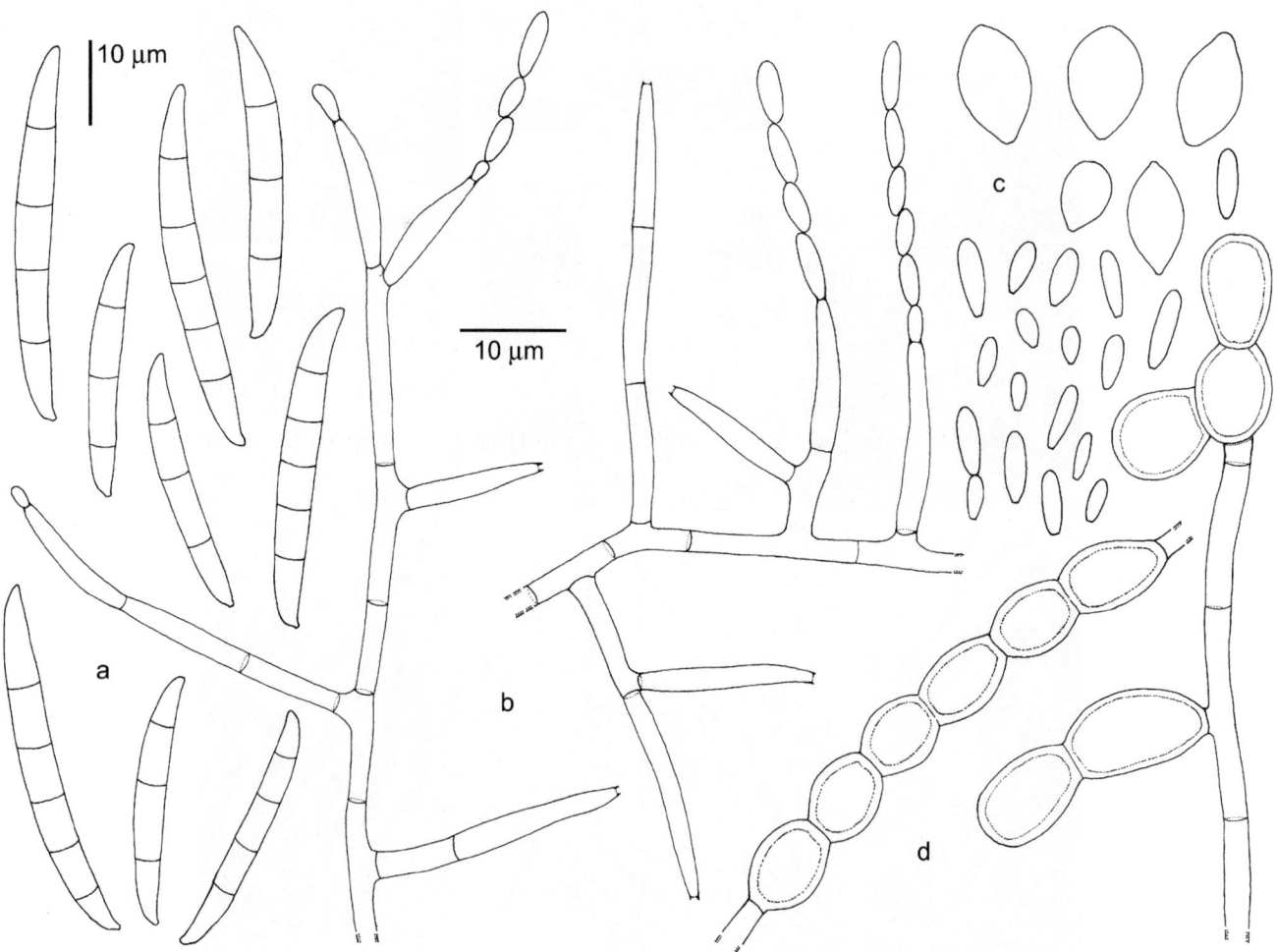

Fusarium napiforme, CBS 673.94. a, c. Liberated macro- and microconidia; b. monophialides producing false chains of microconidia; d. chlamydospores.

Antifungal susceptibility.

Antifungal	MICs	Strains	Reference
AMB	1	1	Unpublished data
5FC	256	1	Unpublished data
FLZ	128	1	Unpublished data
ITZ	32	1	Unpublished data
KTZ	32	1	Unpublished data
MCZ	32	1	Unpublished data

Nomenclature. *Fusarium napiforme* Marasas, Nelson & Rabie - Mycologia 79: 910, 1987.

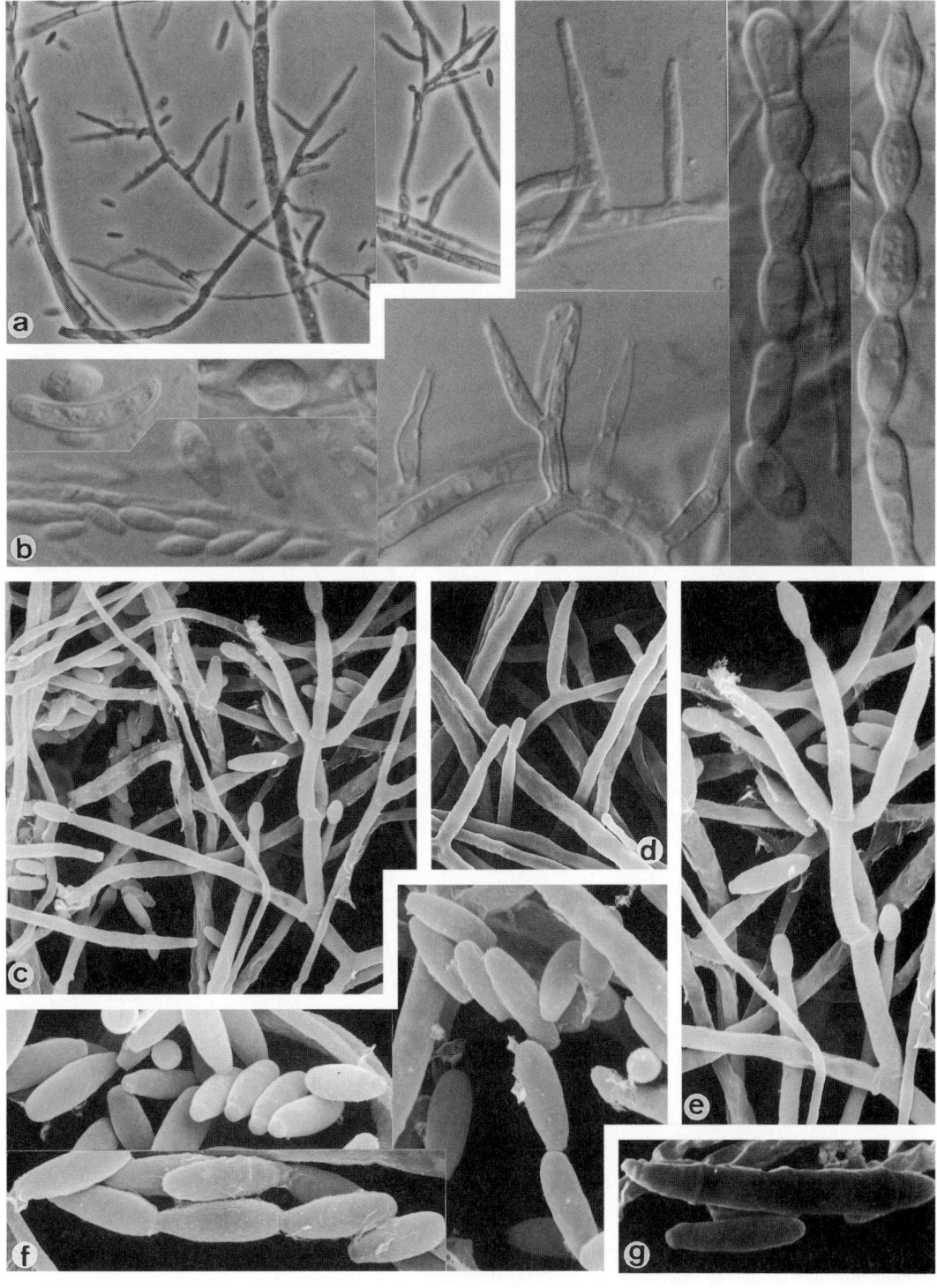

Fusarium napiforme, CBS 673.94. a-e. Monophialides with microconidia and chains of chlamydospores; f. microconidial chains; g. liberated macro- and microconidia. a. ×640; b. ×1600; c. ×1300; d. ×2000; e. 2400; f. ×2650; g. ×3875.

Fusarium nygamai Burgess & Trimboli

Colony characteristics. Colonies (OA) growing rapidly, floccose, white, becoming dark violet; reverse dark violet.

Microscopy. Macroconidial conidiophores arranged in dense sporodochia; phialides acicular. Macroconidia falcate, somewhat curved or nearly straight, with slightly curved pointed apex, 3-septate, basal cell pedicellate, 25-54 × 2-5 µm. Microconidial phialides single, lateral on hyphae, cylindrial. Microconidia abundant, in short chains or in heads, one-celled, fusiform, 5-15 × 2-3 µm. Chlamydospores mostly abundant, intercalary, single or in chains, hyaline, smooth- or rough-walled.

Teleomorph. *Gibberella nygamai* Klaasen & Nelson (*Ascomycota, Euascomycetes, Hypocreales: Hypocreaceae*).

Pathogenicity. BSL-1. In soil and on plant debris. A systemic infection in a leukemic patient was reported by Krulder *et al.* (1996).

References. Burgess & Trimboli (1986), Burgess *et al.* (1989), Klaasen & Nelson (1996).

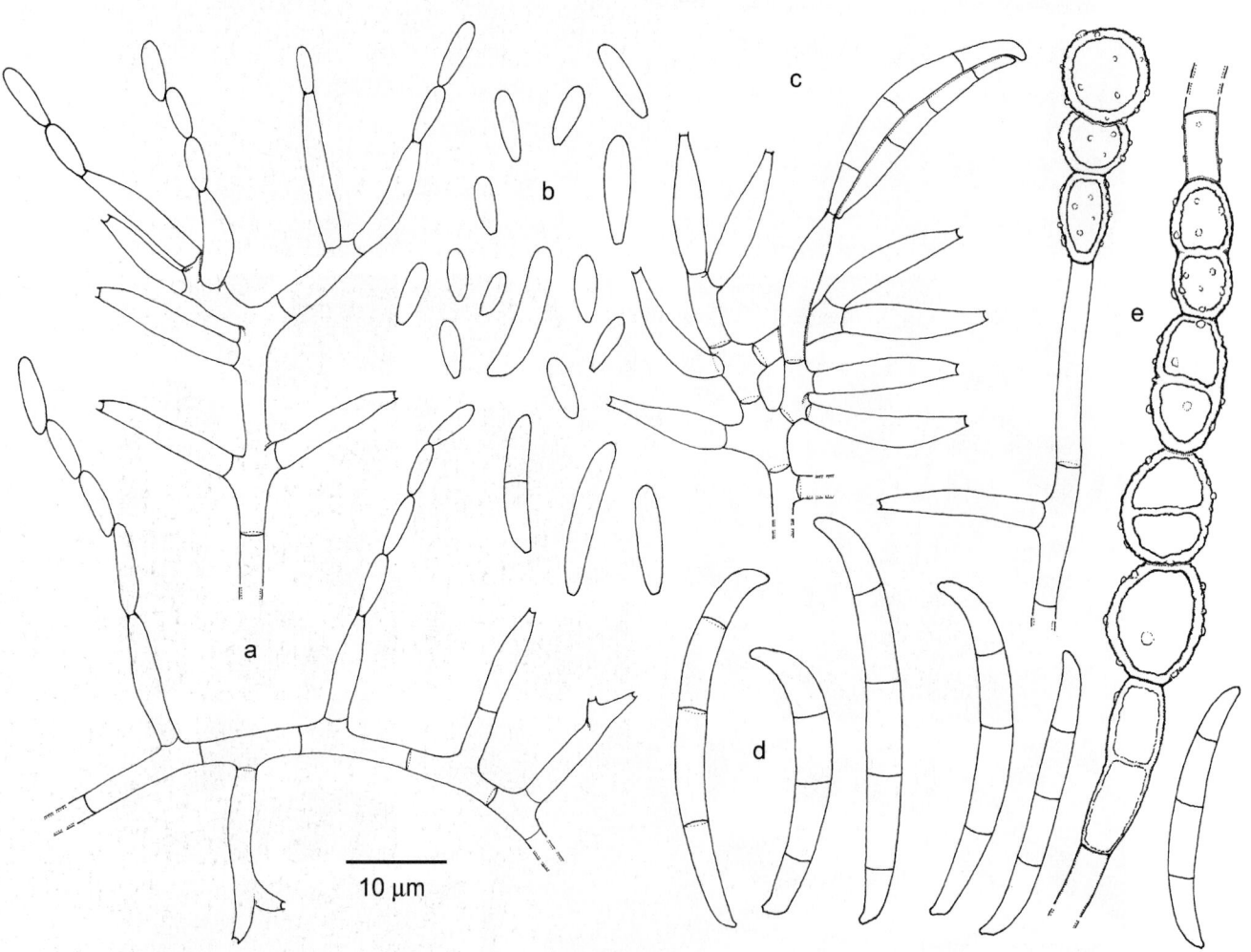

Fusarium nygamai, CBS 675.94. a. Mono- and polyphialides producing false chains of microconidia; b, d. liberated macro- and microconidia; c. monophialides producing macroconidia; e. chlamydospores.

Antifungal susceptibility.

Antifungal	MICs	Strains	Reference
AMB	1	1	Unpublished data
5FC	256	1	Unpublished data
FLZ	128	1	Unpublished data
ITZ	32	1	Unpublished data
KTZ	32	1	Unpublished data
MCZ	16	1	Unpublished data

Nomenclature. *Fusarium nygamai* Burgess & Trimboli - Mycologia 78: 223, 1986.
   Gibberella nygamai Klaasen & Nelson - Mycologia 88: 967, 1995.

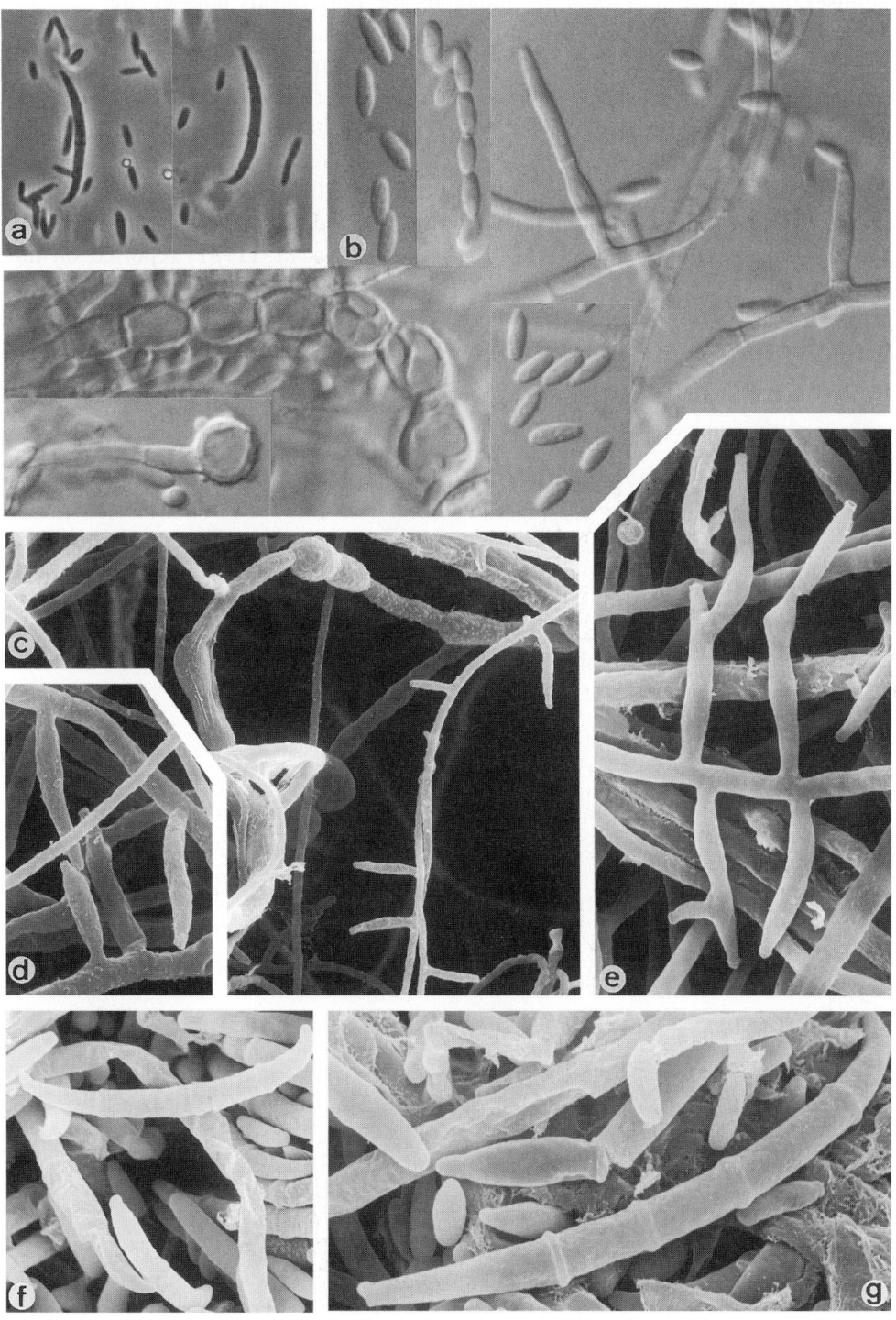

Fusarium nygamai, CBS 675.94. a, f, g. Liberated macro- and microconidia; b. microconidia, chlamydospores, and mono- and polyphialides. a. ×640; b. ×1600; c. ×1200; d. ×1900; e. ×2300; f. ×2500; g. ×3700.

Fusarium oxysporum Schlecht.:Fr.

Colony characteristics. Colonies (OA) growing rapidly; aerial mycelium white, usually becoming purple; discrete, erumpent, orange sporodochia are present in some strains; reverse hyaline to dark blue or dark purple. Some isolates have a characteristic odour suggesting lilac.

Microscopy. Conidiophores are short, single, lateral monophialides in the aerial mycelium, later arranged in densely branched clusters. Macroconidia fusiform, slightly curved, pointed at the tip, 3-(5)-septate, basal cells pedicellate, 23-54 × 3.0-4.5 µm. Microconidia abundant, never in chains, mostly non-septate, ellipsoidal to cylindrical, straight or often curved, 5-12 × 2.3-3.5 µm. Chlamydospores terminal or intercalary, hyaline, smooth- or rough-walled, 5-13 µm diam. Sclerotial pustules present in some isolates, pale to green or deep violet.

Physiology. Intolerant to benomyl.

Differential diagnoses. In contrast to *F. solani* the phialides are short, mostly non-septate.

Pathogenicity. BSL-2. The species is a common plant-pathogen; numerous host-specific variants have been described. Human cases involve: keratitis (Zapater, 1986), onychomycosis (Gianni *et al.,* 1997; Pereiro *et al.,* 1997; Romano *et al.,* 1998) disseminated and cutaneous infections (Steinberg *et al.,* 1983; Rippon *et al.,* 1988; Matsumoto & Matsuda, 1988; Sturm *et al.,* 1989; Agamolis *et al.,* 1991; Neumeister *et al.,* 1992). Peritoneal infection in a CAPD patient was reported by Farrell *et al.* (1994) and a central venous catheter infection by Raad & Hachem (1995). Disseminated infection in a HIV positive patient due to a port-a-cath was reported by Eljaschewitsch *et al.* (1996), in a transplant recipient by Mohammedi *et al.* (1995) and in an immunocompetent patient with respiratory distress by Sander *et al.* (1998). Wheeler *et al.* (1981) reported on infections of burn wounds.

References. Booth (1977), Domsch *et al.* (1980), Gerlach & Nirenberg (1982), Nelson *et al.* (1983), Gordon & Okamoto (1992), Brayford (1996).

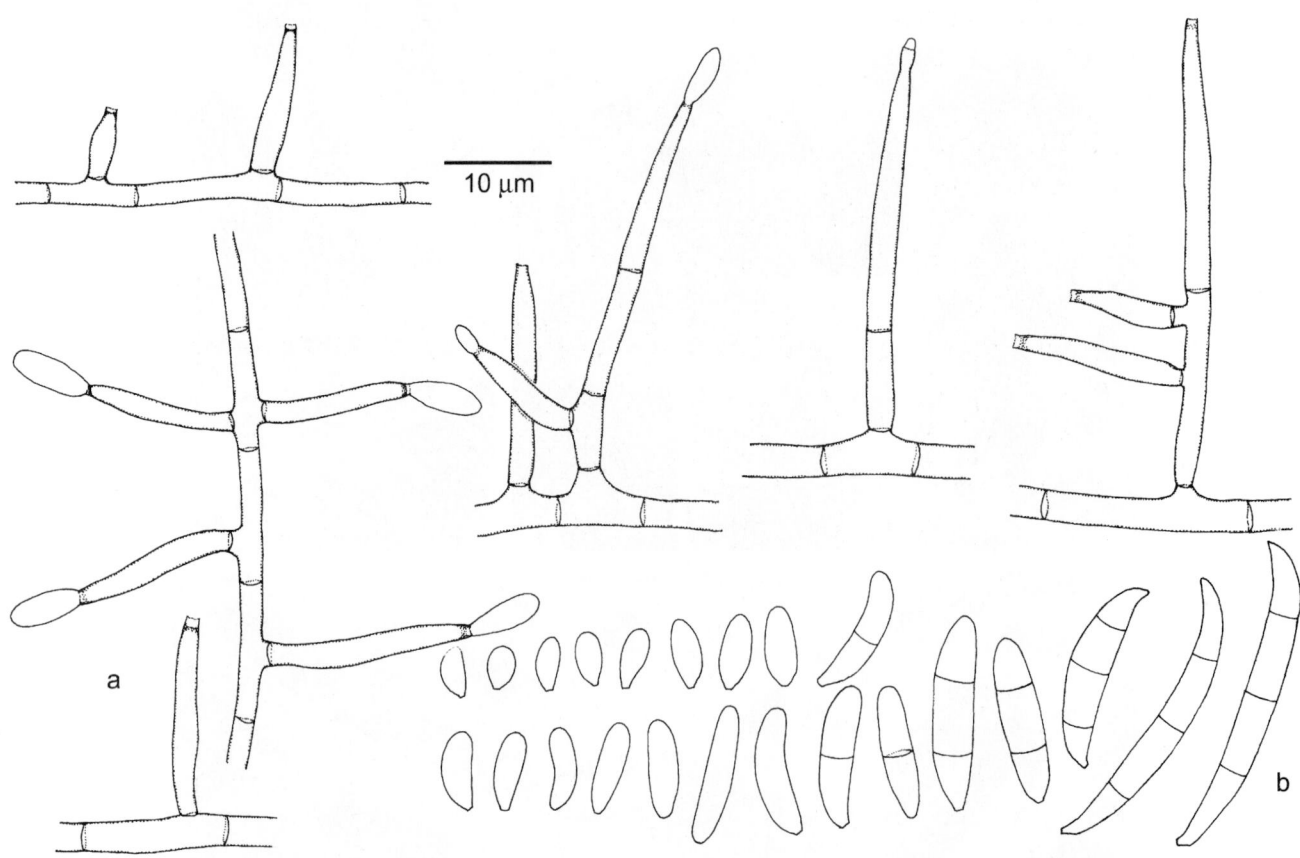

Fusarium oxysporum, CBS 620.87. a. Monophialides; b. macro- and microconidia.

Antifungal susceptibility.

Antifungal	Mean MICs	Strains	Reference
AMB	2.13	19	Pujol *et al.* (1997)
5FC	654	19	Pujol *et al.* (1997)
FLZ	160	19	Pujol *et al.* (1997)
ITZ	20	19	Pujol *et al.* (1997)
KTZ	33	19	Pujol *et al.* (1997)
MCZ	72	19	Pujol *et al.* (1997)

Nomenclature. *Fusarium oxysporum* Schlechtendahl - Fl. Berol. 2: 139, 1824 ≡ *Fusarium oxysporum* Schlechtendahl:Fries - Syst. Mycol. 3: 471, 1832.

Allantospora onychophila Vuillemin - Champ. Paras. Myc. Homme p. 63, 1931 ≡ *Hyalopus onychophilus* (Vuillemin) Ashieri - Atti Ist. Bot. Lab. Crittog. Univ. Paria, Ser. 4a, 3: 49, 1932.

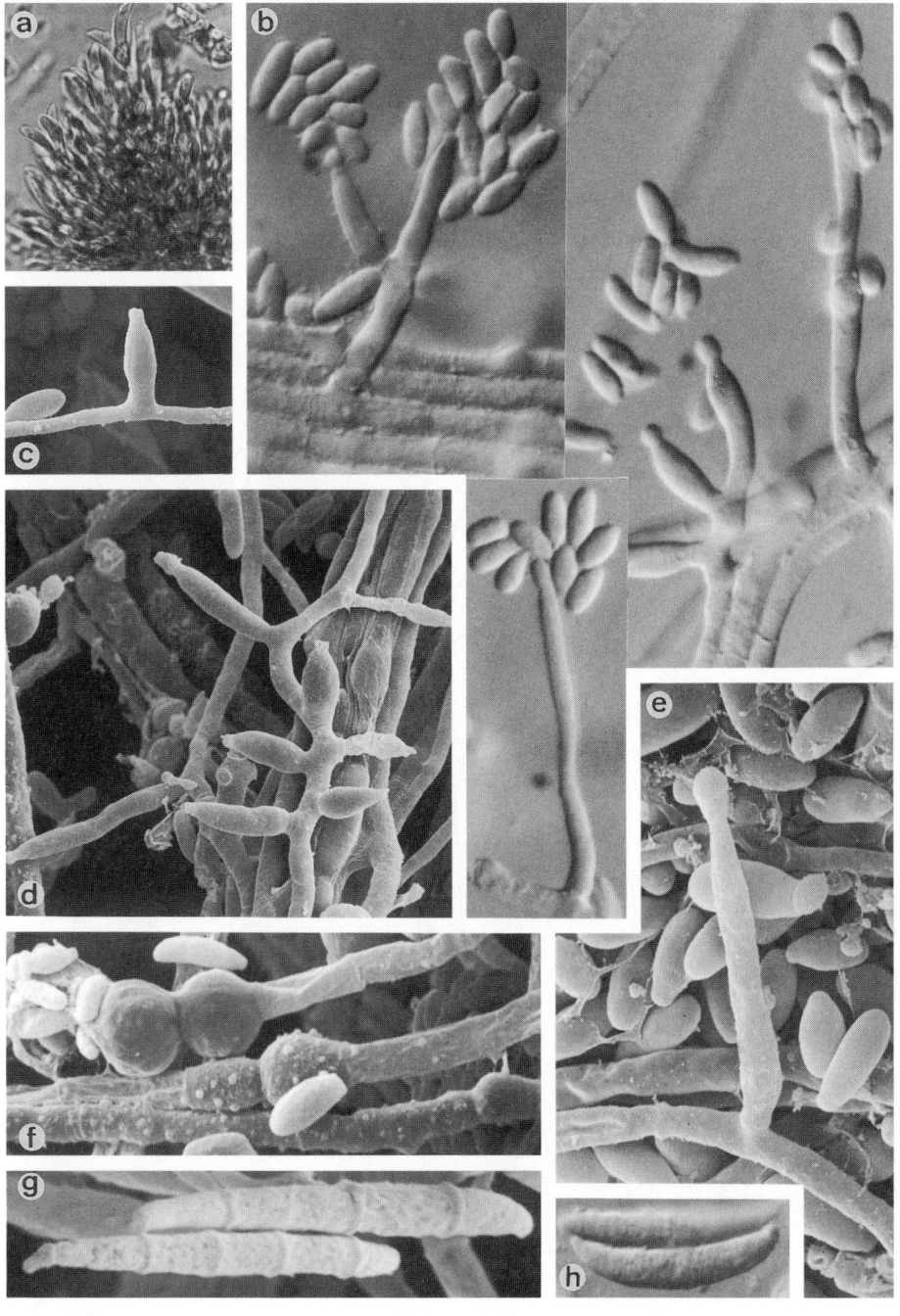

Fusarium oxysporum, CBS 620.87. a. Sporodochium; b-e. monophialides with microconidia; f. chlamydospores and microconidia; g, h. macroconidia. a. ×360; b. ×1120; c. ×1540; d. ×1295; e. ×1890; f. ×1540; g. ×1890; h. ×1120.

Fusarium proliferatum (Matsushima) Nirenberg

Colony characteristics. Colonies (OA) growing rapidly; aerial mycelium white to dark vinaceous or purple. Sporodochia present or absent; when present, they are tan to orange, discrete or confluent.

Microscopy. Conidiophores arising laterally from aerial hyphae, densely branched. Polyphialides abundant. Macroconidia abundant, falcate or nearly straight, with a distinct foot-cell, usually 3- or 5-septate, 27-58 × 2.6-5.0 µm. Microconidia clavate, with truncate base, 7-9 × 2.2-3.2 µm, in older cultures pyriform to ovoidal, 6.0-11.3 × 5-11 µm, in chains. Chlamydospores absent.

Differential diagnoses. Differs from *F. verticillioides* (p. 704) by microconidia being borne in chains on mono- and polyphialides. Leslie (1995) listed the species as one of the intermating populations of *Gibberella fujikuroi* (Sawada) Ito.

Pathogenicity. BSL-1. The species is a plant-pathogen. Strains from humans generally have a reduced morphology. The present description and drawing are of a clinical strain. Disseminated infections in immunocompromised patients (Summerbell *et al.*, 1988a; Helm *et al.*, 1990).

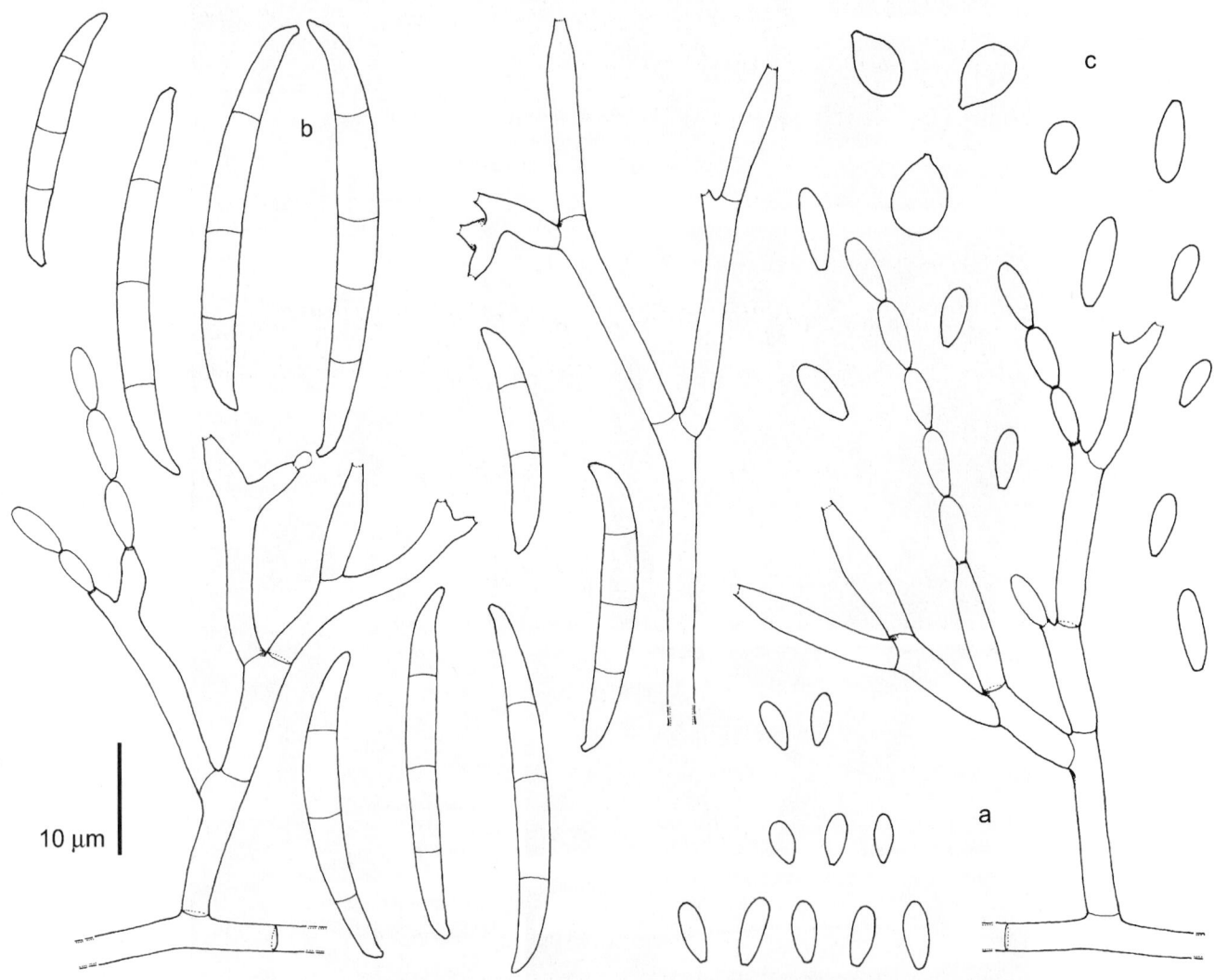

Fusarium proliferatum, CBS 579.78. a. Conidiophores with mono- and polyphialides; b. macroconidia; c. microconidia.

References. Gerlach & Nirenberg (1982), Nelson *et al.* (1983).

Nomenclature. *Cephalosporium proliferatum* Matsushima - Microf. Solomon Isl. Papua-New Guinea p. 11, 1971 ≡ *Fusarium proliferatum* (Matsushima) Nirenberg - Mitt. Biol. Bundesanst. Land- u. Forstw. 169: 38, 1976.

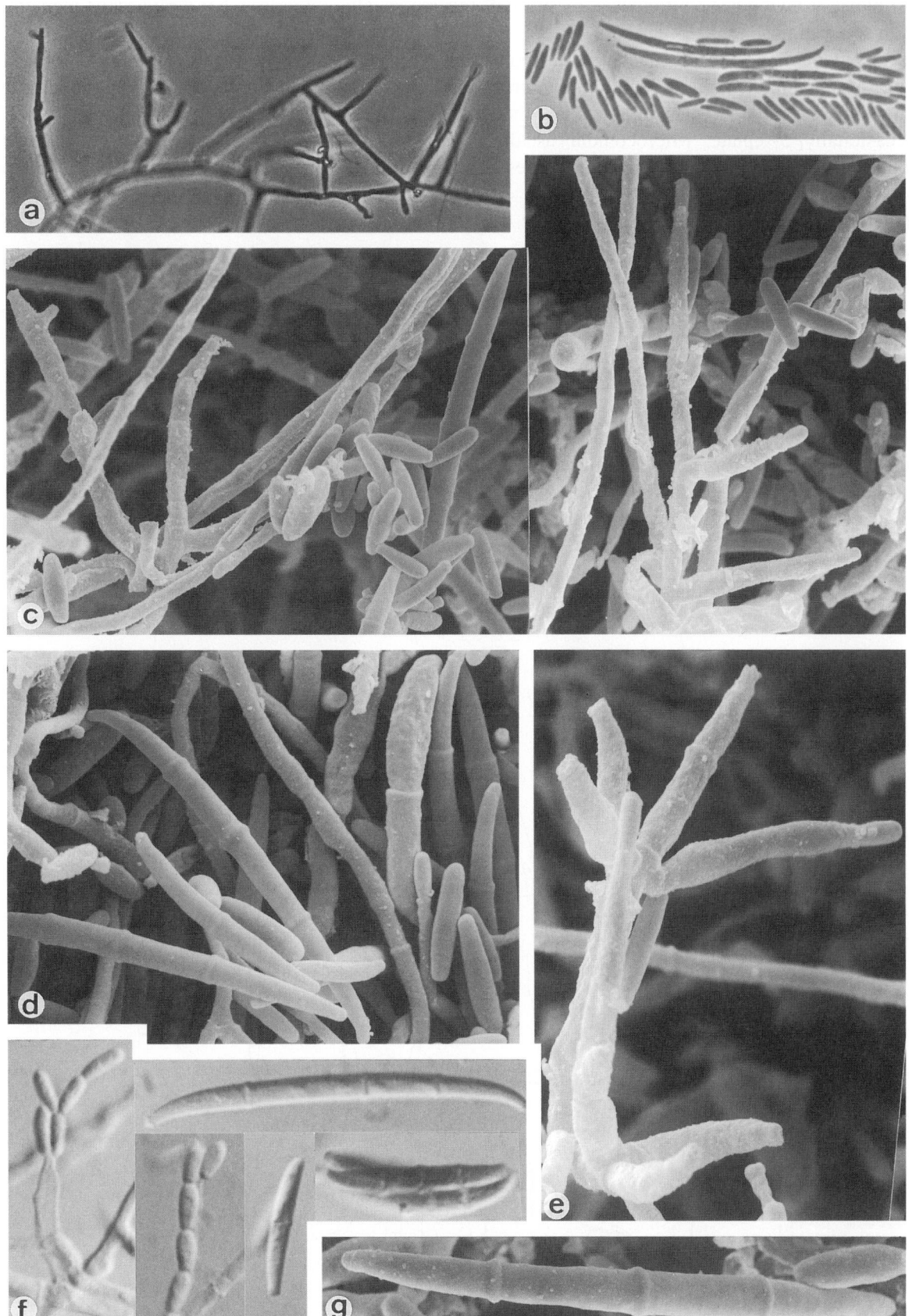

Fusarium proliferatum, CBS 597.78. a, c, e. Conidiophores; b, d, f, g. macro- and microconidia. a. ×512; b. ×640; c. ×2300; d. ×2700; e. ×3500; f. ×1600; g. ×3700.

Fusarium solani (Mart.) Sacc.

Colony characteristics. Colonies (OA) growing rapidly, with white to cream-coloured aerial mycelium, usually green to bluish-brown when sporodochia are present; reverse usually colourless, but a vinaceous pigmentation is present in some strains.

Microscopy. Conidiophores arising laterally from aerial hyphae. Monophialides mostly with a rather distinct collarette. Macroconidia produced on shorter, branched conidiophores which soon form sporodochia, usually moderately curved, with short, blunt apical and indistinctly pedicellate basal cells, mostly 3-septate, 28-42 × 4-6 µm, occasionally 5-septate. Microconidia usually abundant, produced on elongate, sometimes verticillate conidiophores, 8-16 × 2.0-4.5 µm. Chlamydospores frequent, singly or in pairs, terminal or intercalary, smooth- or rough-walled, 6-10 µm diam.

Teleomorph. *Nectria haematococca* Berk. & Br. var. *brevicona* (Wollenweber) Gerlach (*Ascomycota, Euascomycetes, Hypocreales: Hypocreaceae*).

Molecular diagnostics. PCR-diagnosis in keratitis was developed by Alexandrakis & Gloor (1994).

Pathogenicity. BSL-2. Most cases concern keratitis (Wahab *et al.*, 1979; Oji & Steele, 1982; Booth *et al.*, 1985; Zapater, 1986; Freidank, 1995; Hamdan *et al.*, 1995; Gari-Toussaint *et al.*, 1997). Ming & Yu (1966) claimed that a separate form of the s pecies would be concerned, but this is as yet unproven. Further clinical entities involve endophthalmitis (Louie *et al.*, 1994), subcutaneous (Hiemenz *et al.*, 1990; Leu *et al.*, 1995) and cutaneous infections (Datry *et al.*, 1988; Matsumoto & Matsuda, 1988; Rippon *et al.*, 1988; Caux *et al.*, 1993), septic arthritis (Jakle *et al.*, 1983) and mycetoma (Luque *et al.*, 1990). Kurien *et al.* (1992) reported two cases of sinusitis, and Ammani *et al.* (1993) a catheter-related infection. Disseminated cases were described in patients with leukemia (Matsuda & Matsumoto, 1986; Brint *et al.*, 1992; Patoux-Pibouin *et al.*, 1992), solid cancer (Bushelman *et al.*, 1995) and in transplant recipients (Arney *et al.*, 1997; Guinvarc'h *et al.*, 1998). Wolf & Blaschke-Hellmessen (1993), Cabañes *et al.* (1997) and Castellá *et al.* (1999) reported cases in turtles and Smith *et al.* (1989) fatal cases

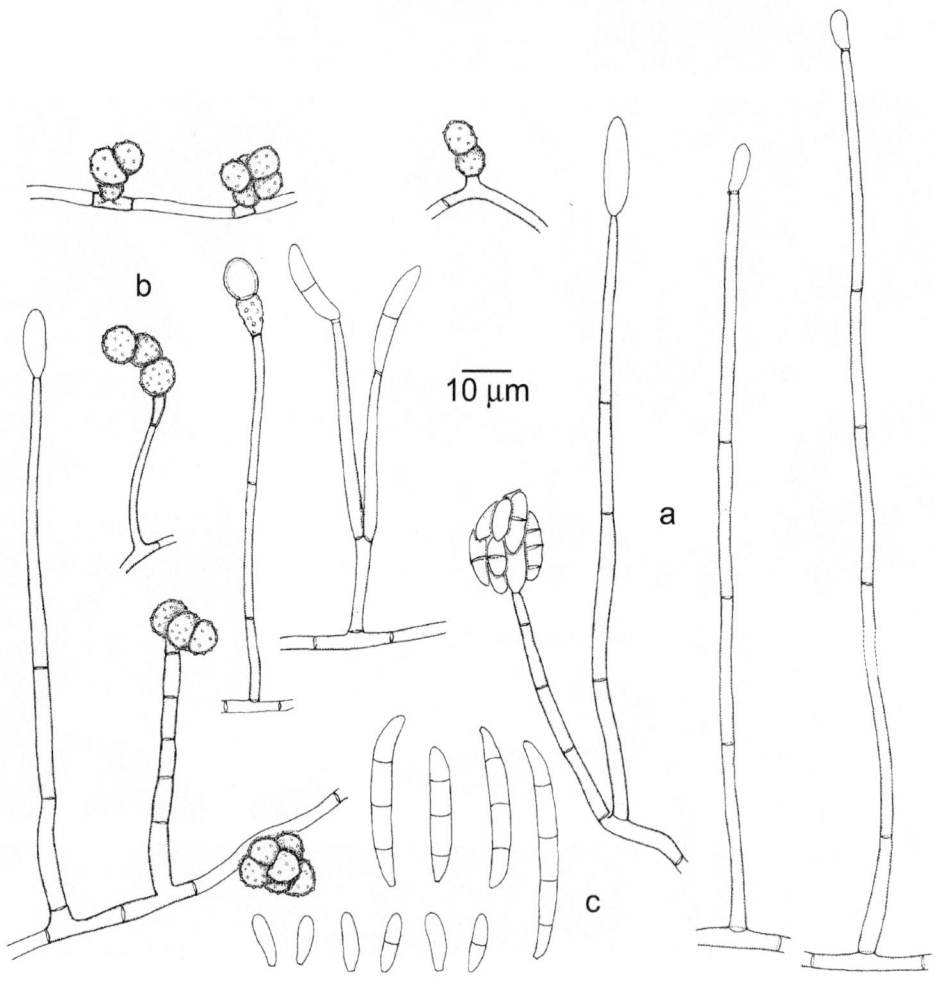

Fusarium solani, CBS 490.63. a. Conidiophores with monophialides; b. chlamydospores; c. macro- and microconidia.

700

in sharks. An AIDS-related sinusitis was reported by Bassi *et al.* (1995). The species was proven to be the most virulent opportunistic *Fusarium* species (Mayayo *et al.*, 1999).

References. Booth (1977), Domsch *et al.* (1980), Gerlach & Nirenberg (1982), Nelson *et al.* (1983), Guarro *et al.* (1999g).

Antifungal susceptibility.

Antifungal	Mean MIC	Strains	Reference
AMB	3.47	16	Pujol *et al.* (1997)
5FC	654	16	Pujol *et al.* (1997)
FLZ	160	16	Pujol *et al.* (1997)
ITZ	20	16	Pujol *et al.* (1997)
KTZ	88	16	Pujol *et al.* (1997)
MCZ	77	16	Pujol *et al.* (1997)

Nomenclature. *Fusisporium solani* Martius - Kartoff.-Epid. Letzt. Jahre p. 20, 1842 ≡ *Fusarium solani* (Martius) Saccardo - Michelia 2: 296, 1881.

Hypomyces haematococcus (Berkeley & Broome) Wollenweber var. *breviconus* Wollenweber - Fusar. Delin. No. 828, 1930 ≡ *Nectria haematococca* Berkeley & Broome var. *breviconia* (Wollenweber) Gerlach, *in* Nelson, Tousson & Cook - Fusar. Dis. Biol. Tax. p. 422, 1981.

Cephalosporium keratoplasticum Morikawa - Mycopathol. Mycol. Appl. 2: 66, 1939 ≡ *Hyalopus keratoplasticum* (Marikawa) Barbosa - Subs. Estud. Gen. Hyalopus p. 44, 1941.

Cylindrocarpon vaginae Booth, Clayton & Usherwood - Proc. Ind. Acad. Sci., Pl. Sci. 94: 436, 1985.

Fusarium solani (Martius) Saccardo f. *keratitis* Ming & Yu - Acta Microbiol. Sin. 12: 184, 1966.

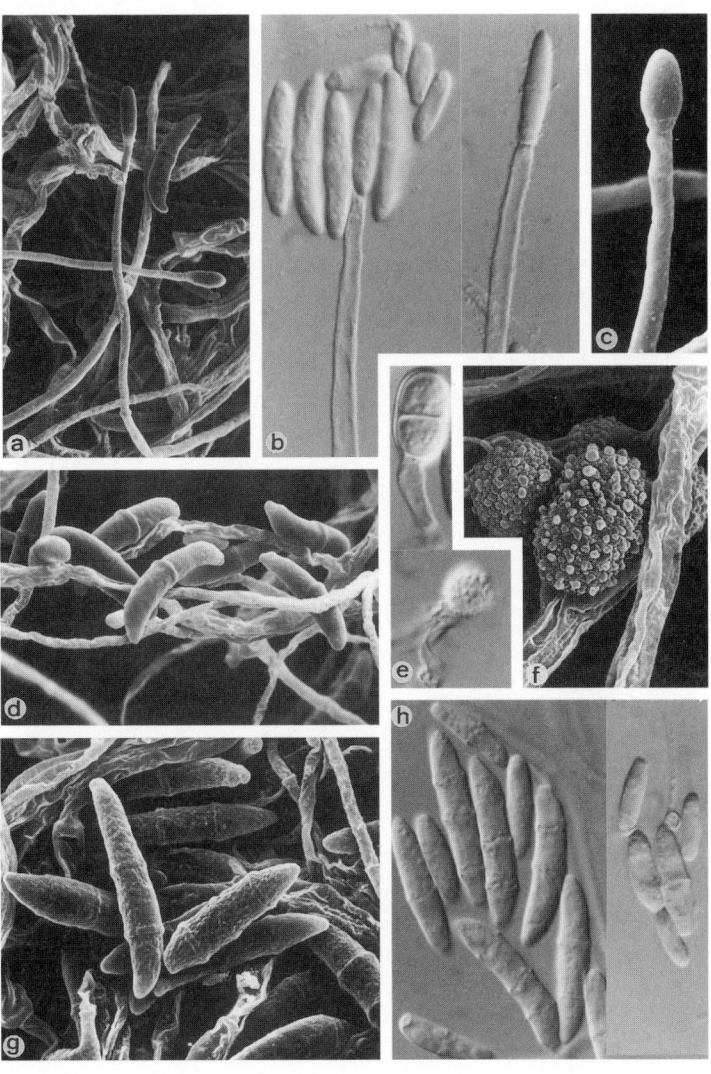

Fusarium solani, CBS 490.63. a-d, g, h. Conidiophores, macro- and microconidia; e, f. chlamydospores. a. ×1425; b. ×1600; c. ×4550; d. ×2175; e. ×1600; f. ×4300; g. ×2800; h. ×1600.

Fusarium subglutinans (Wollenw. & Reink) Nelson *et al.*

Colony characteristics. Colonies (OA) growing rapidly, pink or vinaceous to violet; aerial mycelium abundant. Sporodochia present or absent, when present they are tan to orange.

Microscopy. Conidiophores usually erect and branched. Macroconidia abundant, falcate to rather straight, 3-5-septate, with a distinct foot-cell, 27-73 × 3.4-5.2 μm. Blastoconidia straight or slightly curved, 2-3-septate, fusiform to lanceolate, with a somewhat pointed, often slightly asymmetrical apical cell and a truncate basal cell, 16-43 × 3.0-4.5 μm. Microconidia produced on polyphialides and aggregating in heads, usually unicellular, ovoidal, ellipsoidal or allantoid, 4-20 × 1.5-4.5 μm. Chlamydospores absent.

Teleomorph. *Gibberella fujikuroi* (Sawada) Wollenweber var. *subglutinans* Edwards (*Ascomycota, Euascomycetes, Hypocreales: Hypocreaceae*).

Differential diagnosis. Although molecular studies separate this species from *F. sacchari* (Butler) W. Gams, both are morphologically very close. The latter species differs by prostate conidiophores and a predominance of aseptate conidia. Both differ from *F. verticillioides* (p. 704) by microconidia borne in heads on microconidiophores. Leslie (1995) listed the species as intermating populations of *Gibberella fujikoroi*.

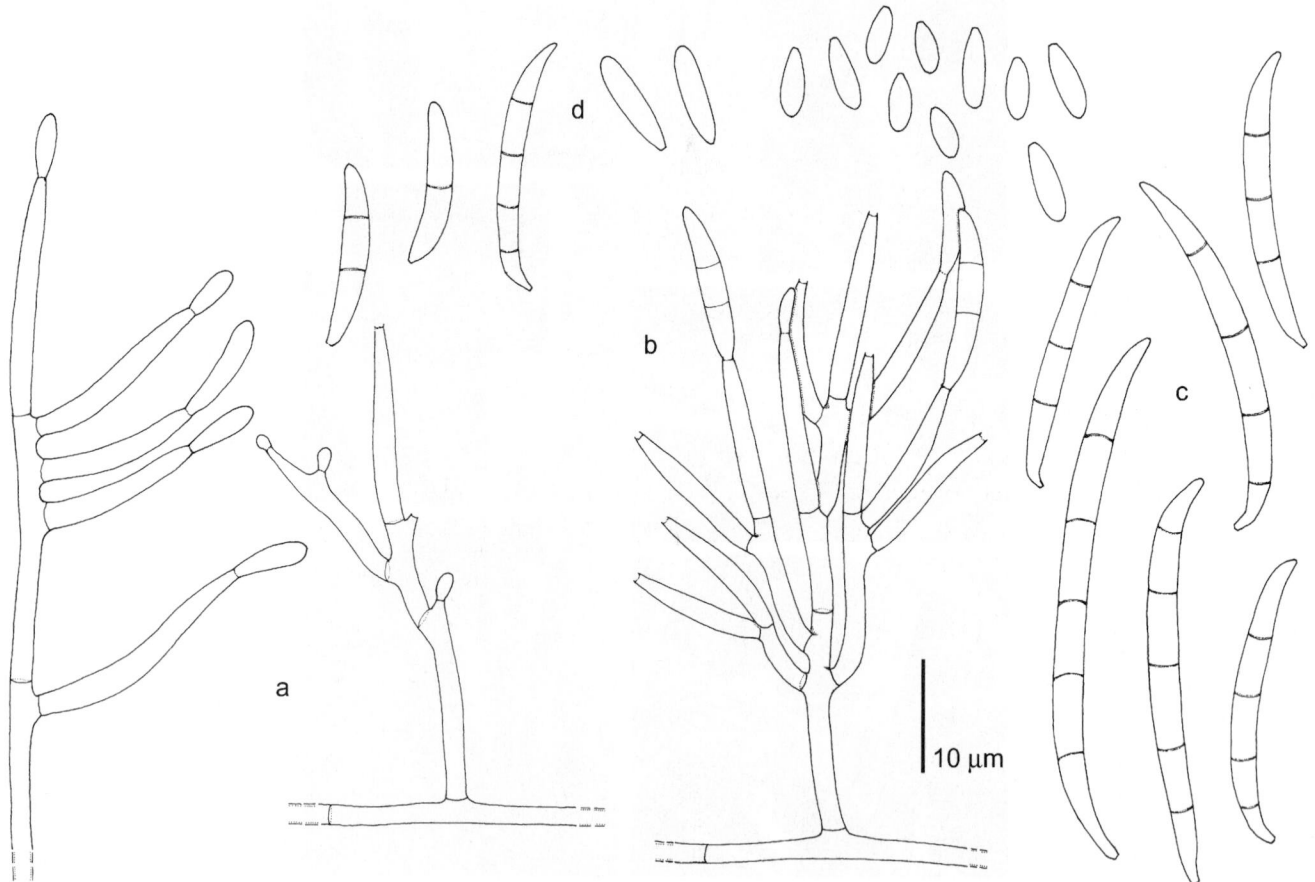

Fusarium subglutinans, CBS 215.76. a,b. Conidiophores; c. macroconidia; d. blasto- and microconidia.

Pathogenicity. BSL-1. *F. subglutinans* as well as *F. sacchari* were involved in cases of keratitis (Polenghi & Lasagni, 1976; Zapater, 1986; Guarro & Gené, 1995); a fungemia by *F. sacchari* was reported by Guarro *et al.* (2000c).

References. Gerlach & Nirenberg (1982), Nelson *et al.* (1983).

Nomenclature. *Fusarium moniliforme* Sheldon var. *subglutinans* Wollenweber & Reinking Phytopathology 15: 163, 1925 ≡ *Fusarium subglutinans* (Wollenweber & Reinking) Nelson, Toussoun & Marasas - Fus. Spec. Ill. Man. Ident. p. 135, 1983.

 Gibberella fujikuroi (Sawada) Wollenweber var. *subglutinans* Edwards - Agric. Gaz. N.S.W. 44: 896, 1933.

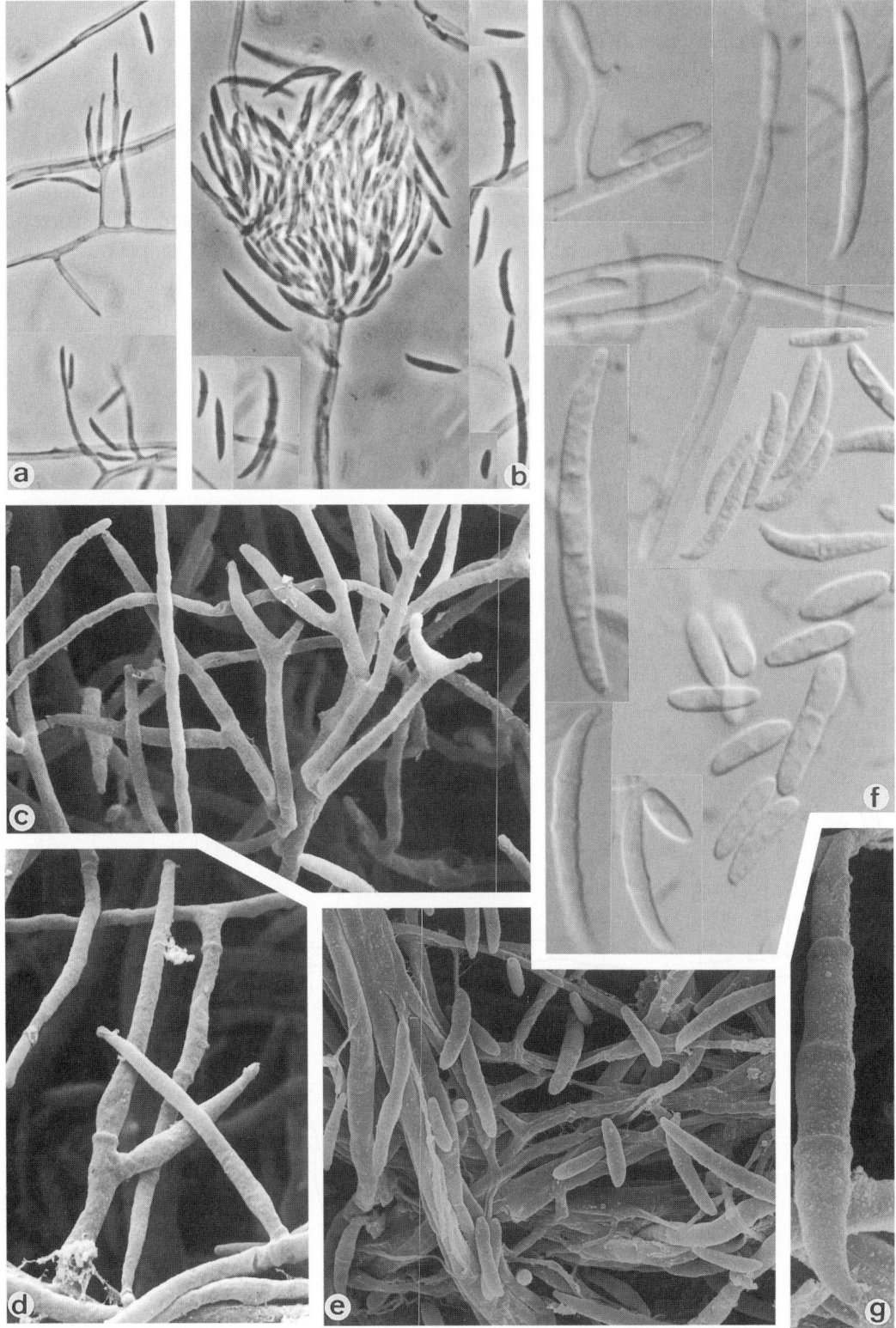

Fusarium subglutinans, CBS 215.76. a-d. Conidiophores bearing mon- and polyphialides; e-g. macro-, micro- and blastoconidia. a. ×512; b. ×640; c. ×2300; d. ×2800; e. ×1800; f. ×1600; g. ×4300.

Fusarium verticillioides (Sacc.) Nirenberg

Colony characteristics. Colonies (OA) very fast growing; aerial mycelium abundant, white and often becoming tinged with purple. Sporodochia may be present or absent; when present tan to orange, discrete or confluent.

Microscopy. Conidiophores arising laterally from hyphae in the aerial mycelium, sparsely branched. Conidiogenous cells strictly monophialidic. Macroconidia delicate, slender, falcate but rather straight, 3-5-septate, 31-58 × 2.7-3.6 µm, sometimes rare. Microconidia abundant, in chains, ovoidal to clavate, 7-10 × 2.5-3.2 µm. Sclerotia may be present, dark bluish, giving colonies a bluish appearance.

Teleomorph. *Gibberella moniliformis* Wineland (*Ascomycota, Euascomycetes, Hypocreales: Hypocreaceae*).

Physiology. Weakly tolerant to benomyl.

Pathogenicity. BSL-2. Keratitis (Zapater, 1986; Durán *et al.*, 1989; Pagliarusco *et al.*, 1995), endophthalmitis (Srdič *et al.*, 1993), disseminated infections (Young *et al.*, 1978; June *et al.*, 1986; Freidank, 1995), cutaneous infections (Veglia & Marks, 1987; Matsumoto & Matsuda, 1988; Rippon *et al.*, 1988; Gradon *et al.*, 1990; Castagnola *et al.*, 1993; Smith *et al.*, 1993; Shaoxi *et al.*, 1996; Pereiro *et al.*, 1999), subcutaneous nodules (Datry *et al.*, 1988), mycetoma (Ajello *et al.*, 1985), peritonitis (Piacentini *et al.*, 1984) and a fatal infection in a cold-blooded animal (Frelier *et al.*, 1985). A mixed infection by *F. verticillioides* and *F. solani* was reported by Guarro *et al.* (2000f).

References. Booth (1977), Domsch *et al.* (1980), Gerlach & Nirenberg (1982), Nelson *et al.* (1983), Nelson (1992).

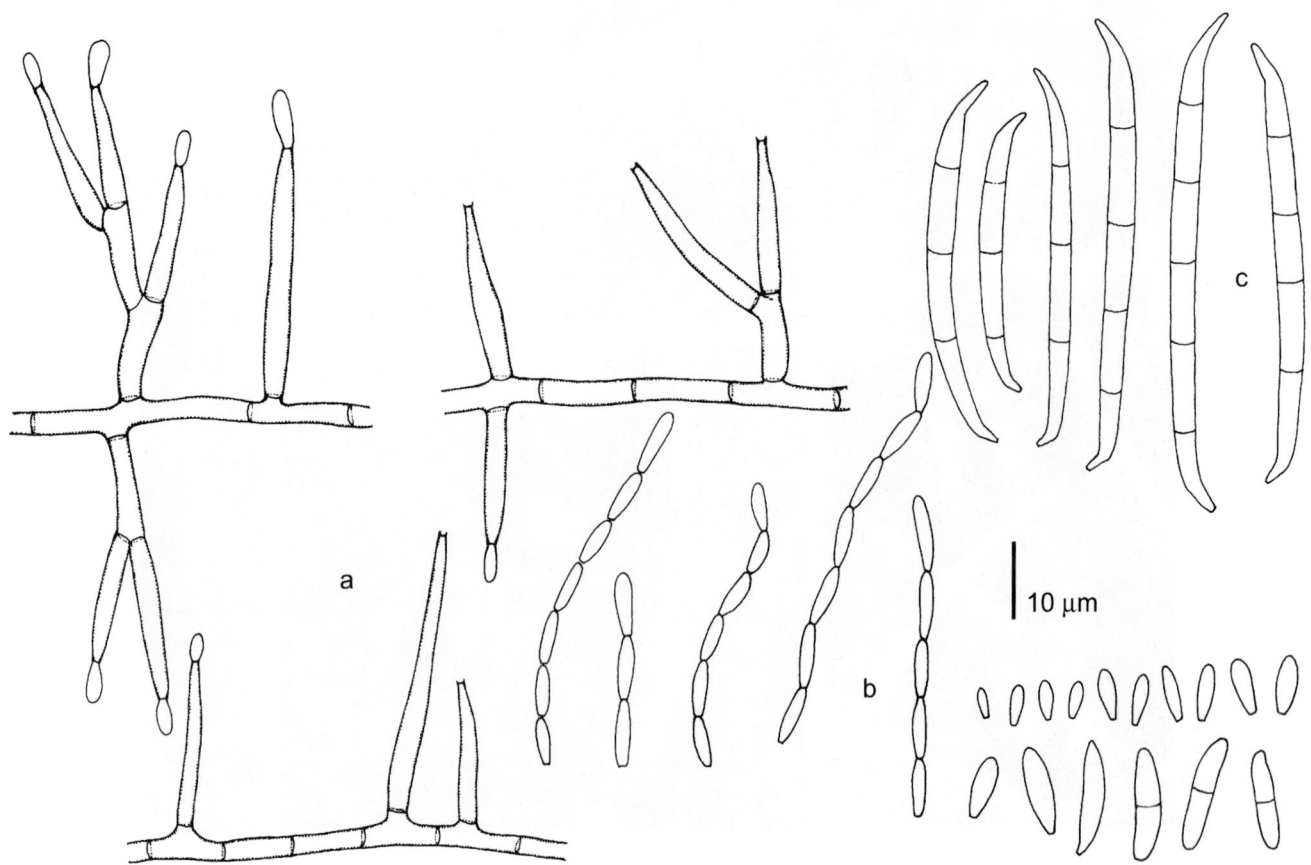

Fusarium verticillioides, CBS 624.87. a. Conidiophores with monophialides; b. microconidia; c. macroconidia.

Antifungal susceptibility.

Antifungal	Mean MIC	Strains	Reference
AMB	1.98	7	Pujol *et al.* (1997)
5FC	654.4	7	Pujol *et al.* (1997)
FLZ	160	7	Pujol *et al.* (1997)
ITZ	20	7	Pujol *et al.* (1997)
KTZ	38	7	Pujol *et al.* (1997)
MCZ	69	7	Pujol *et al.* (1997)

Nomenclature: *Oospora verticillioides* Saccardo - Fung. Ital. Fig. 789, 1881; Michelia 2: 546, 1882 ≡ *Fusarium verticillioides* (Saccardo) Nirenberg - Mitt. Biol. Bundesanst. Land- u. Forstw. 169: 26, 1976.
    Fusarium moniliforme Sheldon - Rep. Nebraska Agric. Exp. Stn. 17: 23, 1904.
    Gibberella moniliformis Wineland - J. Agric. Res. 28: 909, 1924.

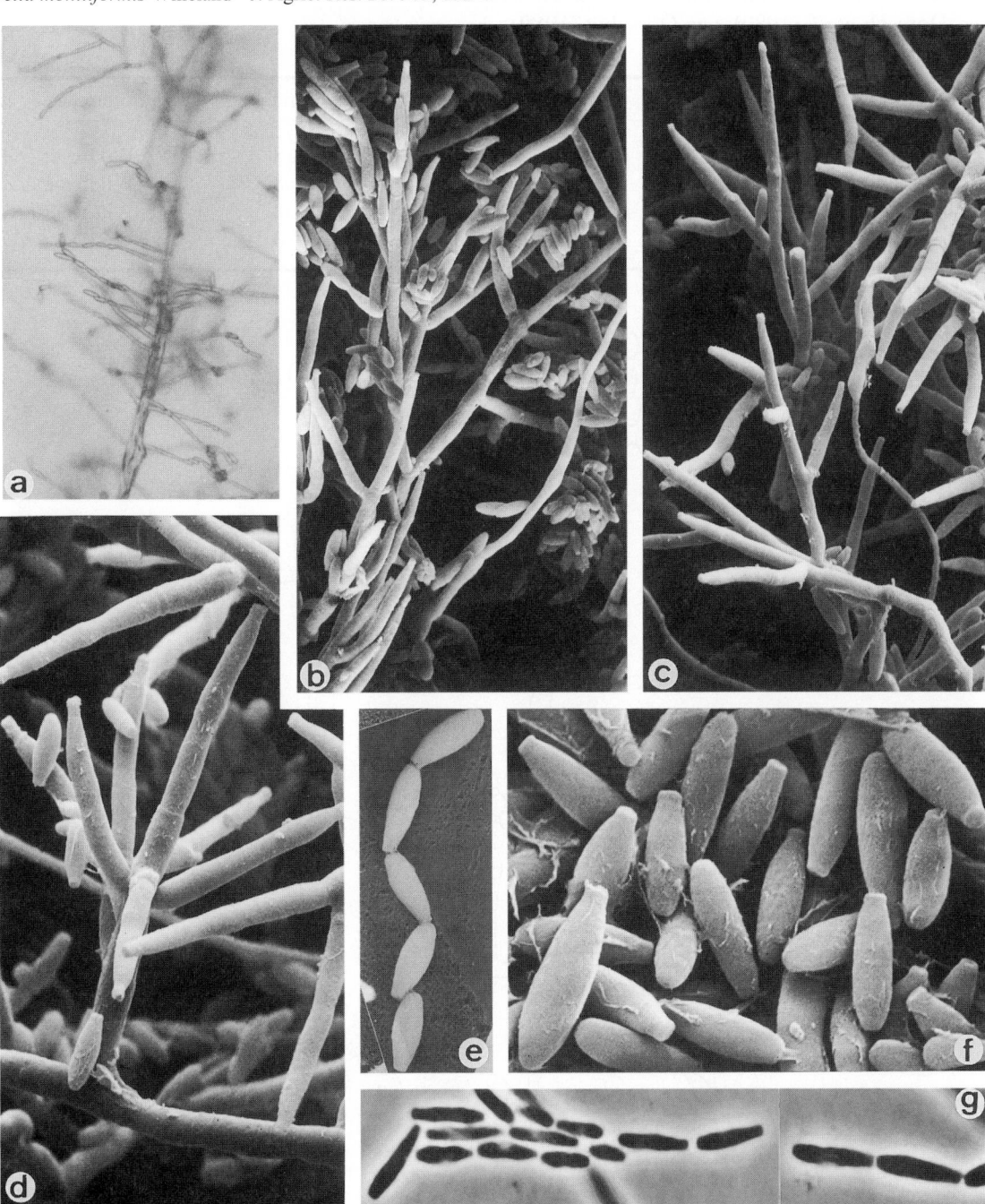

Fusarium verticillioides, CBS 624.87. a. Monophialides with conidial chains; b-d. conidiophores and microconidia; e-g. microconidia. a. ×256; b. ×920; c. ×1050; d. ×1800; e. ×2100; f. ×3900; g. ×1600.

Hyphomycetes. Genus: *GEOMYCES*

Geomyces pannorum (Link) Sigler & Carmichael

Colony characteristics. Colonies (PYE) growing slowly, felty to powdery or floccose, elevated at the centre, grey, buff or brown.

Microscopy. Conidiophores erect, arising from submerged or aerial hyphae, bearing short chains of 2-4 alternate arthroconidia. Conidia terminal or lateral, obovoidal, ellipsoidal, cuneiform or slightly clavate, pale greenish, echinulate or verrucose, thick-walled, 2-6 × 2-4 µm, with wide basal scars. Intercalary conidia barrel-shaped, echinulate or verruculose, 3-6 × 2.3-2.5 µm.

Pathogenicity. BSL-1. Reported from onychomycosis (Schönborn & Schmoranzer, 1970) and as the probable agent of infection in a dog (Gianni *et al.*, 2000; Zelenková, 2000)

References. Domsch *et al.* (1980), van Oorschot (1980).

Antifungal susceptibility.

Antifungal	Mean MICs	MIC 90	Strains	Reference
AMB	0.55	1.56	8	Wildfeuer *et al.* (1998)
ITZ	0.06	0.10	8	Wildfeuer *et al.* (1998)
KTZ	0.25	0.78	8	Wildfeuer *et al.* (1998)
VCZ	0.20	0.39	8	Wildfeuer *et al.* (1998)

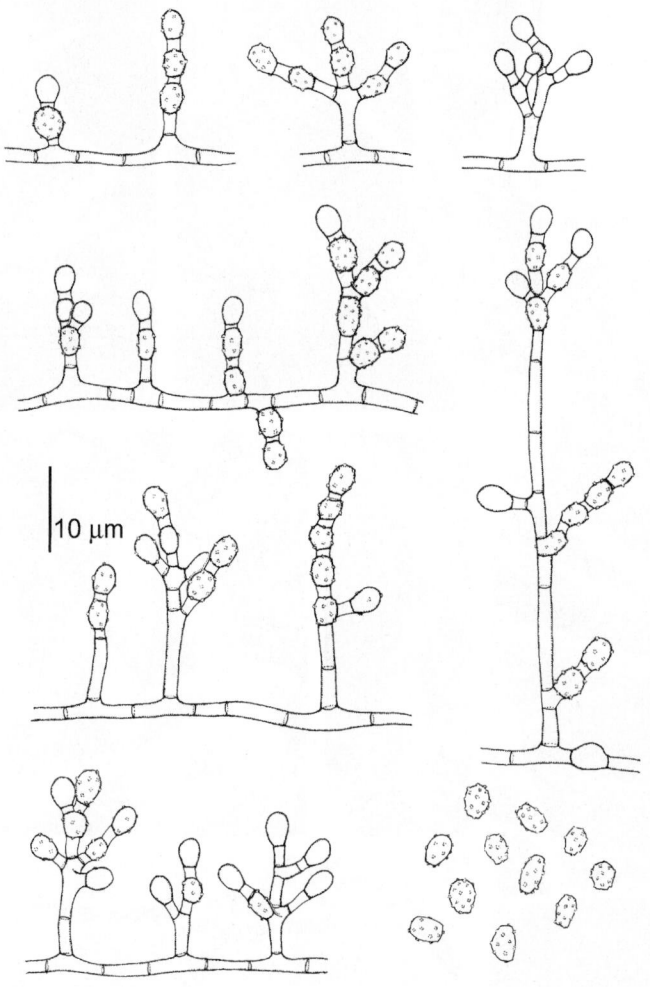

Geomyces pannorum, CBS 249.53. Conidiophores and conidia.

Nomenclature. *Sporotrichum pannorum* Link - Spec. Plant. 4-6: 13, 1824 ≡ *Geomyces pannorum* (Link) Sigler & Carmichael - Mycotaxon 4: 377, 1976.

Trichophyton albiscicans Nieuwenhuis - Tijdschr. Geneesk. Ned. Indië 48: 49, 1908 ≡ *Atrichophyton albiscicans* (Nieuwenhuis) Castellani & Chalmers - Man. Trop. Med., ed. 3, p. 1008, 1919 ≡ *Glenospora albiscicans* (Nieuwenhuis) Ota - Annls Parasit. Hum. Comp. 3: 84, 1925 ≡ *Glenosporella albiscicans* (Nieuwenhuis) Nannizzi, *in* Agostini - Atti Ist. Bot. Univ. Lab. Crittogam. Pavia 4: 98, 1931 ≡ *Aleurisma albiscicans* (Nieuwenhuis) C.W. Dodge - Med. Mycol. p. 788, 1935.

Corethropsis hominis Vuillemin, *in* Spillman & Jannin - Bull. Soc. Fr. Derm. Syph. 24: 227, 1913.

Sporotrichum carnis Brooks & Hansford - Trans. Br. Mycol. Soc. 8: 131, 1923 ≡ *Aleurisma carnis* (Brooks & Hansford) Bisby - Trans. Br. Mycol. Soc. 27: 111, 1944.

Aleurisma guilliermondii Grigorakis - C.R. Séanc. Soc. Biol. 91: 1381, 1924.

Aleurisma lugdunense Vuillemin, *in* Massia & Grigorakis - C.R. Séanc. Soc. Biol. 91: 1381, 1924.

Sporotrichum humanum Benedek - Derm. Wschr. 83: 1804, 1926.

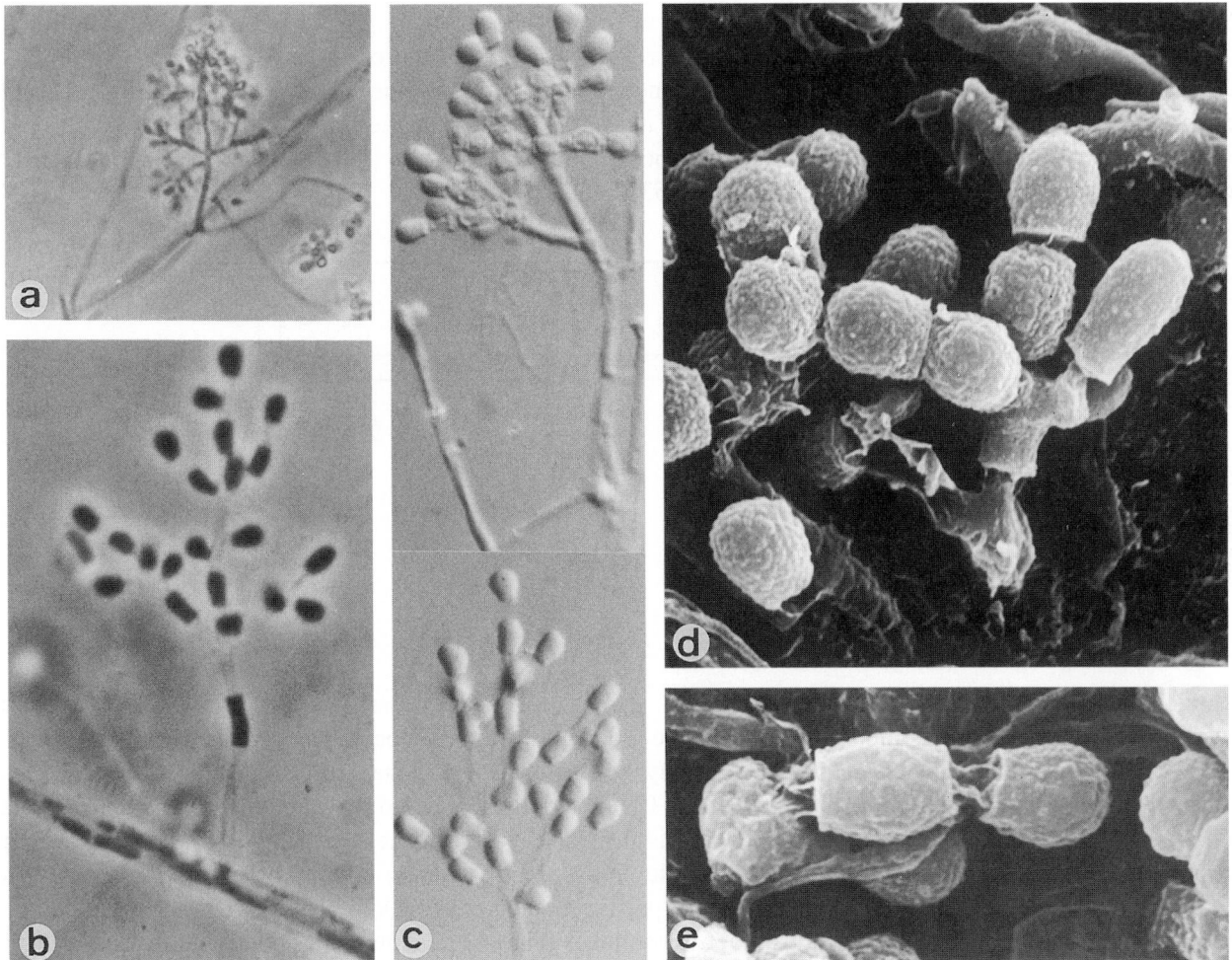

Geomyces pannorum, CBS 249.53. Conidiophores and conidia. a. ×512; b. ×1280; c. ×1600; d. ×6400; e. ×9600.

Hyphomycetes, dimorphic *Onygenales*. Genus: *HISTOPLASMA*

Histoplasma capsulatum Darling

Colony characteristics. Colonies (SGA, 24°C) expanding, granular to cottony, initially white, later becoming brownish. Reverse initially cream-coloured, becoming brownish with age.

Microscopy. Micro- as well as macro-conidia are produced. Microconidia (sub)hyaline, sessile or arising on short stalks from undifferentiated hyphae, hyaline, smooth- and thin-walled, 1-celled, pyriform to clavate, 1-4 × 2-6 µm. Chlamydospore-like macroconidia brown, arising from short conidiophores, thick-walled, tuberculate or with cylindrical projections, 1-celled, spherical, 8-14 µm diam. At 37°C on BHI, a yeast-phase with budding cells up to 7 µm is formed.

Teleomorph. *Ajellomyces capsulatus* (Kwon-Chung) McGinnis & Katz (*Ascomycota, Euascomycetes, Onygenales: Onygenaceae;* p. 251).

Physiology. Intolerant to benomyl.

Differential diagnosis. The species is recognized by its warted macroconidia. The warts sometimes arise in later stages of development. Similar structures are known in the *Chrysosporium* anamorph of *Arthroderma tuberculatum* Kuehn, but in that species the spines are sharp and small rather than blunt, and rarely also in *Duddingtonia* species (nematophagous fungi), but then chlamydospores are larger and nearly hyaline. *Histoplasma* is further characterized by the presence of a microconidial synanamorph.

Serology. Kits for the serological identification of *H. capsulatum* are commercially available. Extracts are assayed by immunodiffusion against specific H and M antibodies. Gómez *et al.* (1995) developed vaccination using recombinant heat shock proteins.

Molecular diagnostics. A probe for species recognition is commercially available (Hall *et al.*, 1992; Padhye *et al.*, 1992; Huffnagle & Gander, 1993; Stockman *et al.*, 1993b). Subspecific typing has been performed with mtDNA (see below) and RAPD (Woods *et al.*, 1993). A 26S rDNA-based specific probe was developed by Sandhu *et al.* (1995). A phylogeny based on highly conserved calmodulin genes was published by El-Rady & Shearer (1996). Okeke *et al.* (1998) sequenced the SSU intron of *H. capsulatum* and its varieties. Based on protein-coding gene phylogeny, Kasuga *et al.* (1999) suggested that the species should be split up into a number of separate taxa. SSU and ITS restriction maps based on NCBI X58572, Z75305 and AF038354:

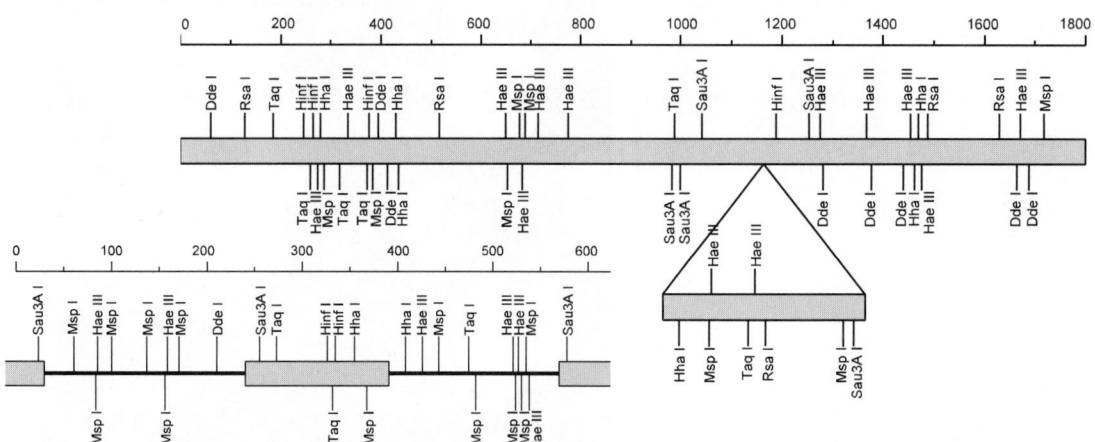

Pathogenicity. BSL-3. Agent of histoplasmosis (p. 25). The species causes an intracellular mycosis of the monocyte-macrophage system. Budding yeast cells are produced within phagocytosing cells and measure 3-4 × 2-3 µm. In stained tissue, the cells are recognizable by hyaline halos around a stained cytoplasm. Daughter cells are connected through narrow isthmi to the parent cell. Characteristically, histiocytes containing *Histoplasma* cells multiply in great numbers and replace original tissues. The infection is generally asymptomatic or mild, occasionally becoming systemic; eventually lymphatic tissues, lungs, kidney, visceral organs and nervous system can be involved. Infants in endemic areas are particularly at risk (Odio *et al.*, 1999), in addition to immunocompromised patients. Endogenous reactivation takes place in patients with underlying disease (Keath *et al.*, 1992), even 20 or 30 years after primary infection (Livas *et al.*, 1995). The fungus is particularly stimulated by defects in T-cell function, e.g., in AIDS (Taylor *et al.*, 1984; Wheat *et al.*, 1985, 1992; Machado *et al.*, 1991;

Conces *et al.*, 1992; Eidbo *et al.*, 1993). Also azathioprine therapy may be a risk factor (Poveda *et al.*, 1998). Disseminated cases may present with usual pictures, such as with myositis (Voloshin *et al.*, 1995). Oral pathology was reviewed by Ng & Siar (1996). In AIDS patients oral lesions are among the first signs of infection (Chinn *et al.*, 1995; Economopoulou *et al.*, 1998) and rather frequently lead to cerebral infections (Wheat, 1995). Origin of the disease is by inhalation of contaminated dust. The saprobic phase of *H. capsulatum* is particularly found in soil with high nitrogen content, i.e., enriched by chicken manure, or in bat or bird droppings under roosting sites (DiSalvo *et al.*, 1970). Local birds are generally not infected, but natural infections are common in canopy-dwelling mammals, particularly rodents, bats and opossums (Naiff *et al.*, 1996), occasionally other mammals (Bander *et al.*, 2000). A case in a pedigree dog was reported by Mackie *et al.* (1997) and in a dolphin by Jensen *et al.* (1998). Local outbreaks of the disease are often associated with the disturbance of roosting sites, e.g., in caves (Ashford *et al.*, 1999), so that humans are exposed to massive amounts of conidia (Suzaki *et al.*, 1995; Stobiersky *et al.*, 1996). Vincent *et al.* (1986) and Maresca & Kobayashi (1989) were able to differentiate several populations on the basis of mtDNA restriction patterns. One of these is a South American population, a second is the mainly North American population, of which soil and clinical isolates were proven to be identical (Spitzer *et al.*, 1989). A third population is a temperature-sensitive, AIDS-associated variant which has lower pathogenicity (Spitzer *et al.*, 1990).

Varieties. The African *Histoplasma capsulatum* var. *duboisii* (Vanbreuseghem) Ciferri (Carme *et al.*, 1993) differs by larger (7-15 µm) budding cells *in vivo*, which are thick-walled, uninucleate, and bud from a narrow base. Cultures grown at 24°C are identical to var. *capsulatum*, but var. *duboisii* differs by absence of urease activity. *In vivo* it resembles *Blastomyces dermatitidis*, but can be distinguished by the fact that the budding cells of this organism are multinucleate and have a broader base. It usually produces dissemination with secondary skin lesions and with involvement of cranial bones and thorax (Ndiaye *et al.*, 1998). An osteomyelitis of the radius was reported by Onwuasoigwe & Gugnani (1998). Despite these differences, Kwon-Chung (1973, 1975) demonstrated that the two varieties can be successfully intermated. Vincent *et al.* (1986) showed that var. *duboisii* had mtDNA restriction patterns identical to those of the majority of strains of var. *capsulatum*. The variety is the agent of human histoplasmosis in Central Africa (Gugnani & Muotoe-Okafor, 1997; Khalil *et al.*, 1998), and is rarely found in animals (Geffray *et al.*, 1994). Gugnani *et al.* (1994) demonstrated a natural association with bats, the fungus occurring in guano and showing a low infection rate in the animal. The variety is less AIDS-associated than the var. *capsulatum* (Eichmann & Schär, 1996).

Histoplasma capsulatum var. *farciminosum* (Rivolta) Weeks *et al.* is a closely related agent of lymphangitis in horses and mules (Weeks *et al.*, 1985; Soliman *et al.*, 1991). It may disseminate to several organs and causes characteristic ulcerative skin lesions (but compare Kasuga *et al.*, 1999).

Distribution. The var. *capsulatum* is common in Southeastern part of the United States and in Southeast Asia in warm regions with a high degree of humidity. In these areas 90% of the population may be asymptomatically infected, showing a positive skin reaction to histoplasmin, an antigen from the mycelial phase of the fungus. The Central and South American strains belong to another population as distinguished on the basis of mtDNA patterns (Vincent *et al.*, 1986). The taxon is infrequently isolated from other humid (sub)tropical areas world-wide. The different varieties are seen world-wide in AIDS patients as import mycoses (Manfredi *et al.*, 1994). The var. *duboisii* is endemic in tropical Africa. The var. *farciminosum* is particularly found in the Middle East (Al-Ani *et al.* (1998), but may have its evolutionary origin in South America.

References. Kwon-Chung (1973), Schwarz (1981), Kaji *et al.* (1987), Kritski *et al.* (1990), Salfelder (1990), Padhye *et al.* (1994c).

Antifungal susceptibility.

Antifungal	MICs range	MIC90	Strains	Reference
AMB	≤0.03-2	1	100	Li *et al.* (2000)
ITZ	≤0.03-0.5	0.06	100	Li *et al.* (2000)
VCZ	≤0.03-2	0.25	100	Li *et al.* (2000)

Nomenclature var. *capsulatum*. *Histoplasma capsulatum* Darling - J. Am. Med. Ass. 46: 1283, 1906 = *Cryptococcus capsulatus* (Darling) Castellani & Chalmers - Man. Trop. Med., ed. 3, p. 1076, 1919 = *Torulopsis capsulata* (Darling) de Almeida - Anais Fac. Med. S. Paulo 9: 76, 1933 = *Posadia capsulata* (Darling) M. Moore = Annls Mo. Bot. Gdn 21: 348, 1934.

Posadia pyriformis M. Moore - Annls Mo. Bot. Gdn 21: 347, 1934 = *Histoplasma pyriforme* (M. Moore) C.W. Dodge - Med. Mycol. p. 155, 1935.

Emmonsiella capsulata Kwon-Chung - Science 177: 368, 1972 = *Ajellomyces capsulatus* (Kwon-Chung) McGinnis & Katz - Mycotaxon 8: 158, 1979.

Nomenclature var. *duboisii*. *Histoplasma duboisii* Vanbreuseghem - Annls Soc. Belg. Méd. Trop. 32: 578, 1952 ≡ *Histoplasma capsulatum* Darling var. *duboisii* (Vanbreuseghem) Ciferri - Man. Micol. Med., ed. 2, p. 342, 1960.

Nomenclature var. *farciminosum*. *Cryptococcus farciminosus* Rivolta - Paras. Veg. p. 246, 1873 ≡ *Saccharomyces farciminosus* (Rivolta) Vuillemin - Revue Gén. Sci. Pures Appl. 12: 740, 1901 ≡ *Leishmania farciminosa* (Rivolta) Galli-Valerio - Centbl. Bakt. Parasitkde, Abt. 1, 44: 577-582, 1909 ≡ *Endomyces farciminosus* (Rivolta) Nègre & Bouquet - Bull. Soc. Path. Exot. 10: 274, 1917 ≡ *Parendomyces farciminosus* (Rivolta) Froilano de Mello & Fernandes - Arq. Hig. Pat. Exot. 6: 29, 1918 ≡ *Grubyella farciminosa* (Rivolta) Ota & Langeron - Annls Parasit. Hum. Comp. 3: 78, 1925 ≡ *Coccidioides farciminosus* (Rivolta) Vuillemin - Champ. Paras. Myc. Homme Anim. p. 140, 1931 ≡ *Torulopsis farciminosus* (Rivolta) de Almeida - Anais Fac. Med. S. Paulo 9: 76, 1933 ≡ *Histoplasma farciminosum* (Rivolta) Redaelli & Ciferri - Boll. Sez. Ital. Soc. Int. Microbiol. 6: 378, 1934 ≡ *Zymonema farciminosum* (Rivolta) C.W. Dodge - Med. Mycol. p. 169, 1935 ≡ *Histoplasma capsulatum* Darling var. *farciminosum* (Rivolta) Weeks, Padhye & Ajello - Mycologia 77: 969, 1985.

Saccharomyces equi Marcone - Atti R. Ist. Incoragg. Napoli, Ser. 5, 8-6: 1-19, 1895.

Cryptococcus rivoltae Fermi & Aruch - Centbl. Bakt. Parasitkde, Abt. 1, 17: 593-600, 1895.

Lymphosporidium equi Gasperini - Accad. Med. Fis. Fiorentina, Feb. 1905.

Leurocytozoon piroplasmoides Ducloux - C.R. Soc. Biol. 64: 593, 1908.

Monilia capsulata Lindner & Kreuth - Zeitschr. Infektkrankh. Haustiere 17: 299, 1916.

Parendomyces tokishigei Froilano de Mello & Fernandes - Arq. Hig. Pat. Exot. 6: 295, 1918.

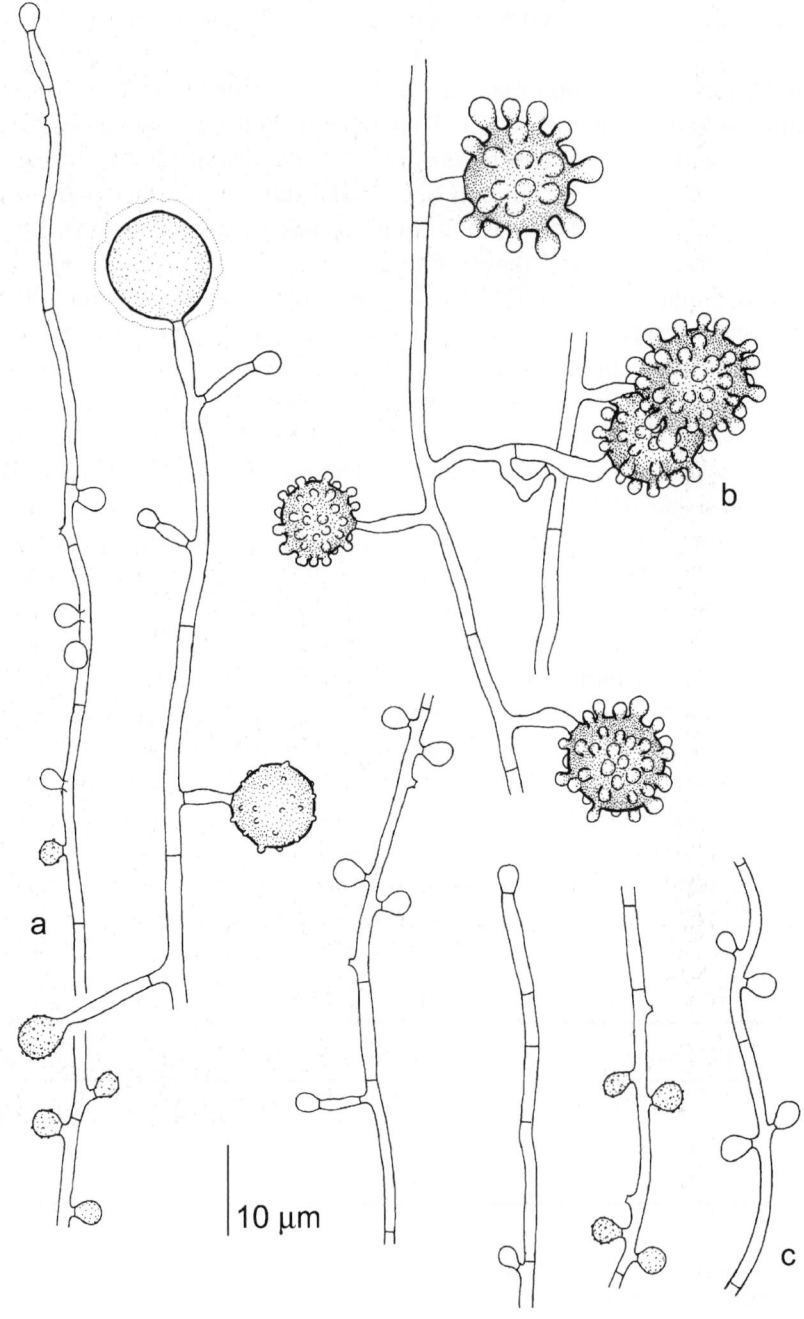

Histoplasma capsulatum, CBS 682.89. a. Microconidia and young, developing macroconidia; b. mature macroconidia with club-shaped ornamentation; c. smooth or finely warted microconidia.

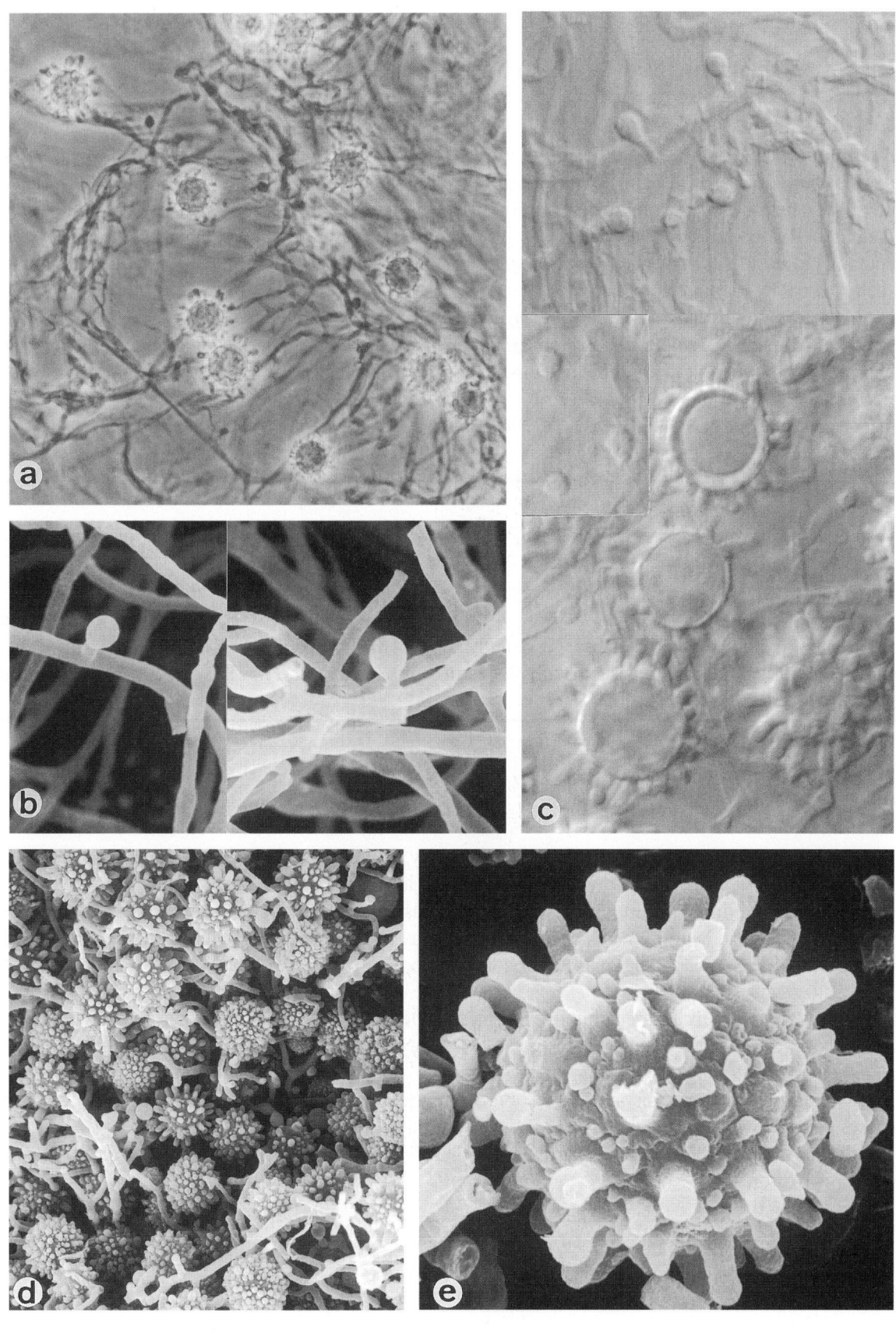

Histoplasma capsulatum, FMR 6518. Macro- and microconidia. a. ×640; b. ×2000; c. ×1600; d. ×800; e. ×4750.

Hyphomycetes, filamentous basidiomycetes. Genus: *HORMOGRAPHIELLA*

Generic description. Colonies growing rapidly, white or cream-coloured. Vegetative hyphae hyaline, septate, branched. Conidiophores erect, slightly or markedly differentiated, hyaline, with basal septum at a level often several µm higher than the supporting hypha; conidiophores in the apical part with repeated sympodial branching. Conidiogenous cells on conidiophores lateral or terminal, frequently branched or in whorls, falling apart into arthroconidia. Conidia usually catenate, hyaline, smooth- and thin-walled, cylindrical, with truncate ends but the terminal cell rounded at the tip.

Teleomorph. *Coprinus* (*Basidiomycota, Hymenomycetes, Agaricales: Coprinaceae*).

General remarks. Species of this genus, which are anamorphs of ink-caps (mushrooms), are frequent in compost and in sewage. Most species show optimal development at 37°C. Phylogeny of *Coprinus* was published by Park *et al.* (1999).

References. Guarro *et al.* (1992b), Gené *et al.* (1996), de Hoog & Gerrits van den Ende (1998).

Key to the treated species of Hormographiella:

1a. Conidiophores well-differentiated, with slimy conidial heads; conidia smooth-walled, 3.5-6.0 µm long
. . . . *H. aspergillata* (712)
1b. Conidiophores frequently undifferentiated, usually without distinct conidial heads;
conidia smooth-walled to verrucose, usually 4-15 µm long . . . . *H. verticillata* (715)

Hormographiella aspergillata Guarro *et al.*

Colony characteristics. Colonies (OA, 37°C) growing rapidly, white to cream-coloured, later becoming tan; aerial mycelium locally forming dense, white mycelial tufts; margin irregular.

Microscopy. Hyphae hyaline, 2-5 µm wide. Conidiophores usually differentiated, septate, simple or sympodially branched in the apical part, bearing a cluster of conidiogenous hyphae of variable length. Conidiogenous hyphae becoming septate and disarticulating into arthroconidia. Conidia smooth-walled, hyaline, cylindrical, (2.5-) 3.5-6.0 (-6.5) × (1.5-) 2.0-2.5 (-3.0) µm, aggregating in slimy heads. Occasionally sclerotial bodies are formed.

Teleomorph. *Coprinus cinereus* (Schaeff.:Fr.) S.F. Gray (*Basidiomycota, Hymenomycetes, Agaricales: Coprinaceae*).

Fruit bodies about 5 cm high, consisting of a slender stem and a greyish-brown cap, with purplish-black reverse. Basidiospores dark-brown, ellipsoidal, 9-12 × 6-7 µm. Cystidia present.

Molecular diagnostics. An RAPD comparison with related species was performed by Ito *et al.* (1998). SSU restriction map based on NCBI M92991:

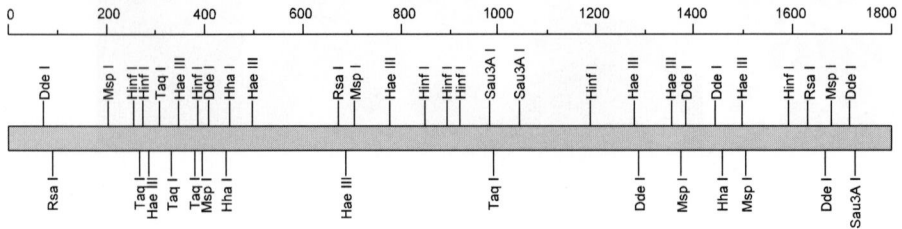

Pathogenicity. BSL-1. Isolated from human skin lesions on two occasions (Guarro *et al.*, 1992b) and from a fatal pulmonary infection in a leukemic patient (Verweij *et al.*, 1997).The species was isolated from sputum of a patient with chronic respiratory disease (Emmons, 1954). Otherwise it is found in compost and in sewage.

References. Guarro *et al.* (1992b), Gené *et al.* (1996a), Polak *et al.* (1997), de Hoog & Gerrits van den Ende (1998).

Antifungal susceptibility.

Antifungal	Mean MICs	Strains	Reference
AMB	1.05	7	Gené *et al.* (1996a)
5FC	53.8	7	Gené *et al.* (1996a)
FLZ	645	7	Gené *et al.* (1996a)
ITZ	0.49	7	Gené *et al.* (1996a)
KTZ	0.6	7	Gené *et al.* (1996a)
MCZ	1.8	7	Gené *et al.* (1996a)

Nomenclature. *Agaricus cinereus* Schaeffer - Fung. Bav. Palat. Ratisb. 1: 89, Tab. 100, 1821 ≡ *Agaricus cinereus* Schaeffer:Fries - Syst. Mycol. 1: 310, 1821 ≡ *Coprinus cinereus* (Schaeffer:Fries) S.F. Gray - Nat. Arr. Br. Plants 1: 634, 1821.

Hormographiella aspergillata Guarro, Gené & de Vroey, *in* Guarro, Gené, de Vroey & Guého - Mycotaxon 45: 182, 1992.

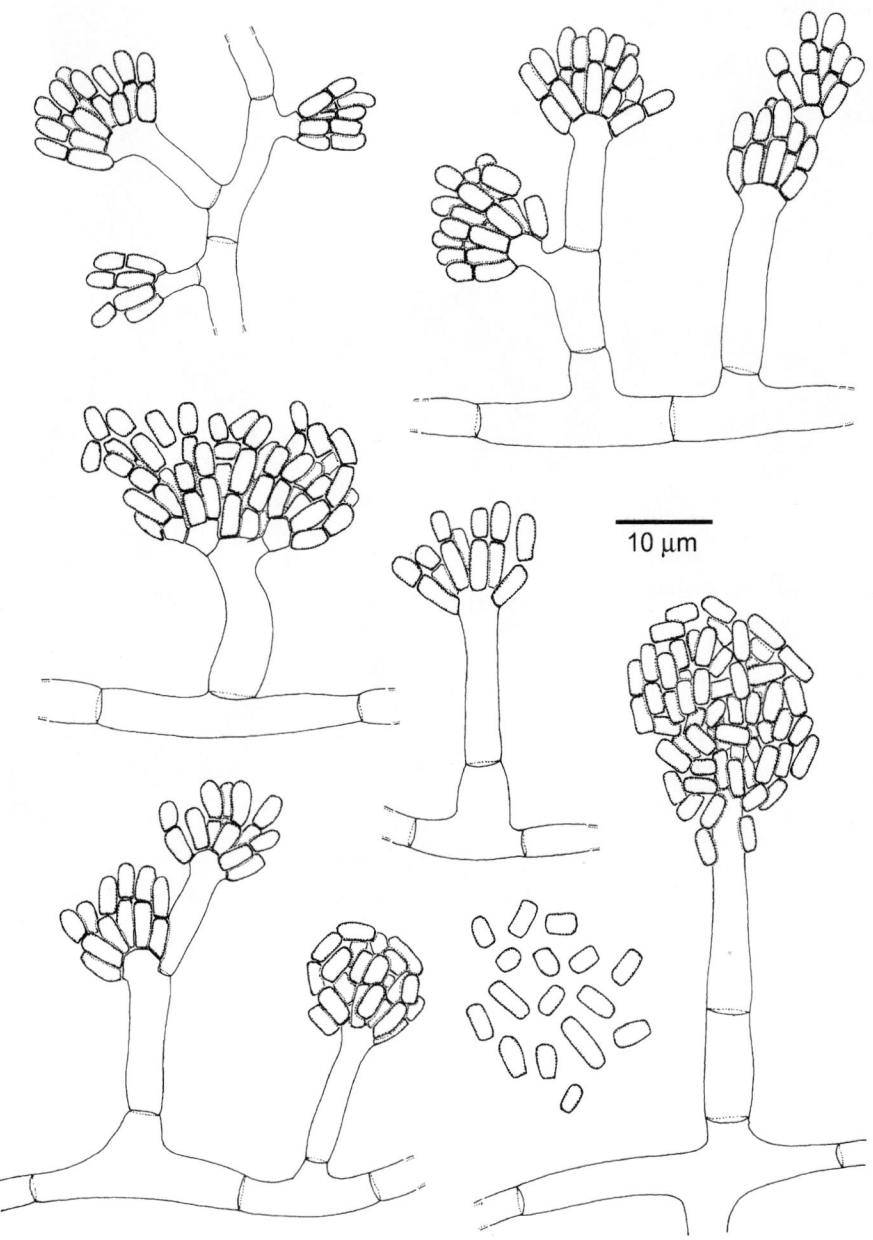

Hormographiella aspergillata, IMI 349604. Conidiophores and conidia.

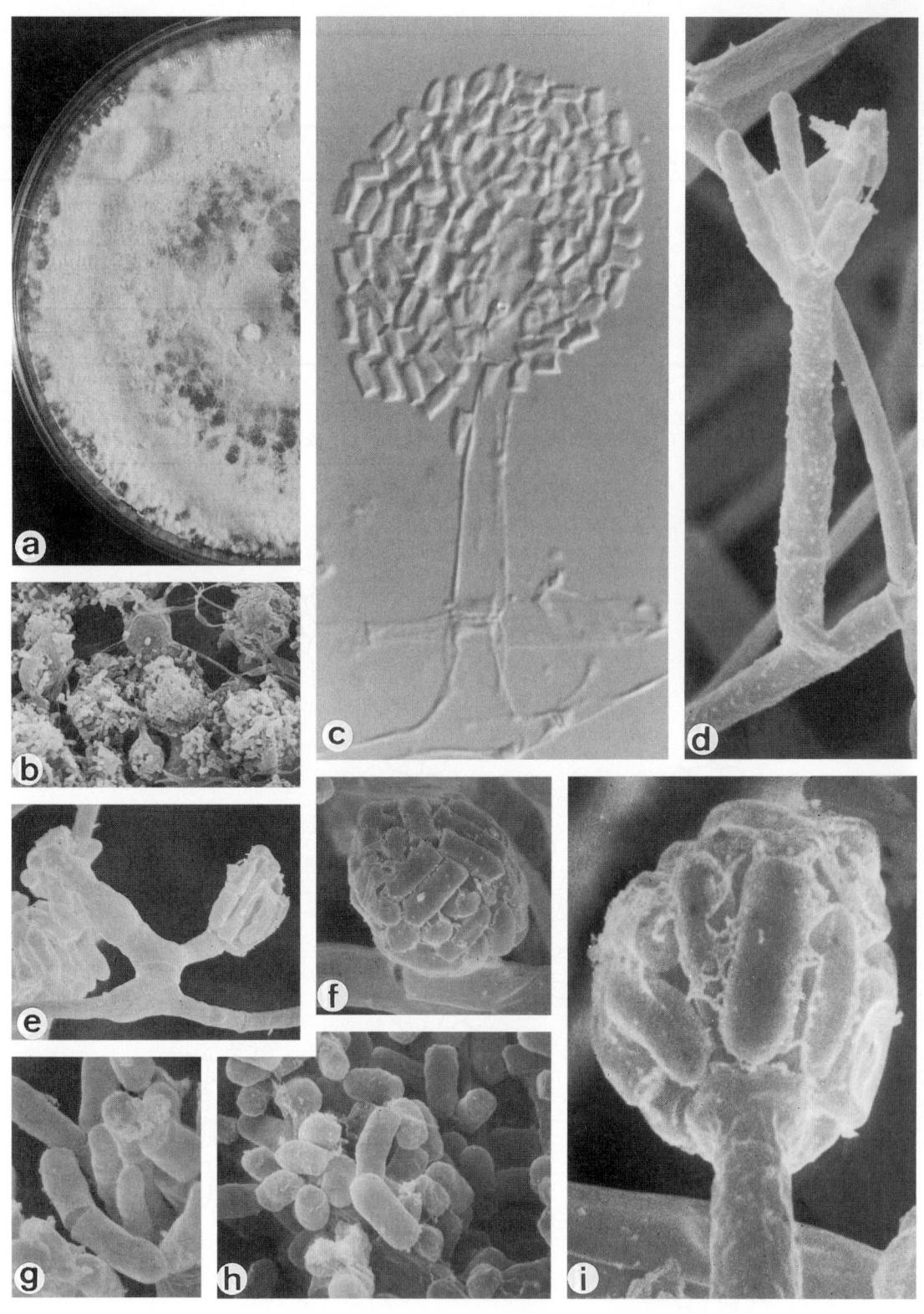

Hormographiella aspergillata. a-c. IMI 349604; d-i. IMI 349603. a, b. Colony morphology; c-e. conidiophores; f-i. conidial heads and conidia. b. ×520; c. ×1600; d. ×3800; e. ×2000; f. ×3600; g. ×4000; h. ×3600; i. ×5500. Reproduced with permission from Mycotaxon 45: 183, 1992.

Hormographiella verticillata Guarro *et al.*

Colony characteristics. Colonies (OA, 37°C) growing rapidly, white, flocccose to felty, with irregular patches, margin irregular.

Microscopy. Hyphae hyaline, 1.5-4.0 (-5.0) μm wide. Conidiophores usually undifferentiated, short, simple and aseptate, with a whorl of conidiogenous branches at the swollen apex, or septate and branched, then elongating repeatedly by sympodial proliferation, each branch producing a whorl of conidiogenous hyphae. Conidiogenous branches may also be borne directly on vegetative hyphae. Conidia hyaline, cylindrical (3-) 4-15 (-17) × (1.2-)1.5-2.0 (-3.0) μm, straight or slightly curved, smooth to slightly verrucose.

Pathogenicity. BSL-1. Isolated from a cerebro-spinal fluid catheter (Guarro *et al.*, 1992b).

References. Guarro *et al.* (1992b), Gené *et al.* (1996a), Küess (2000).

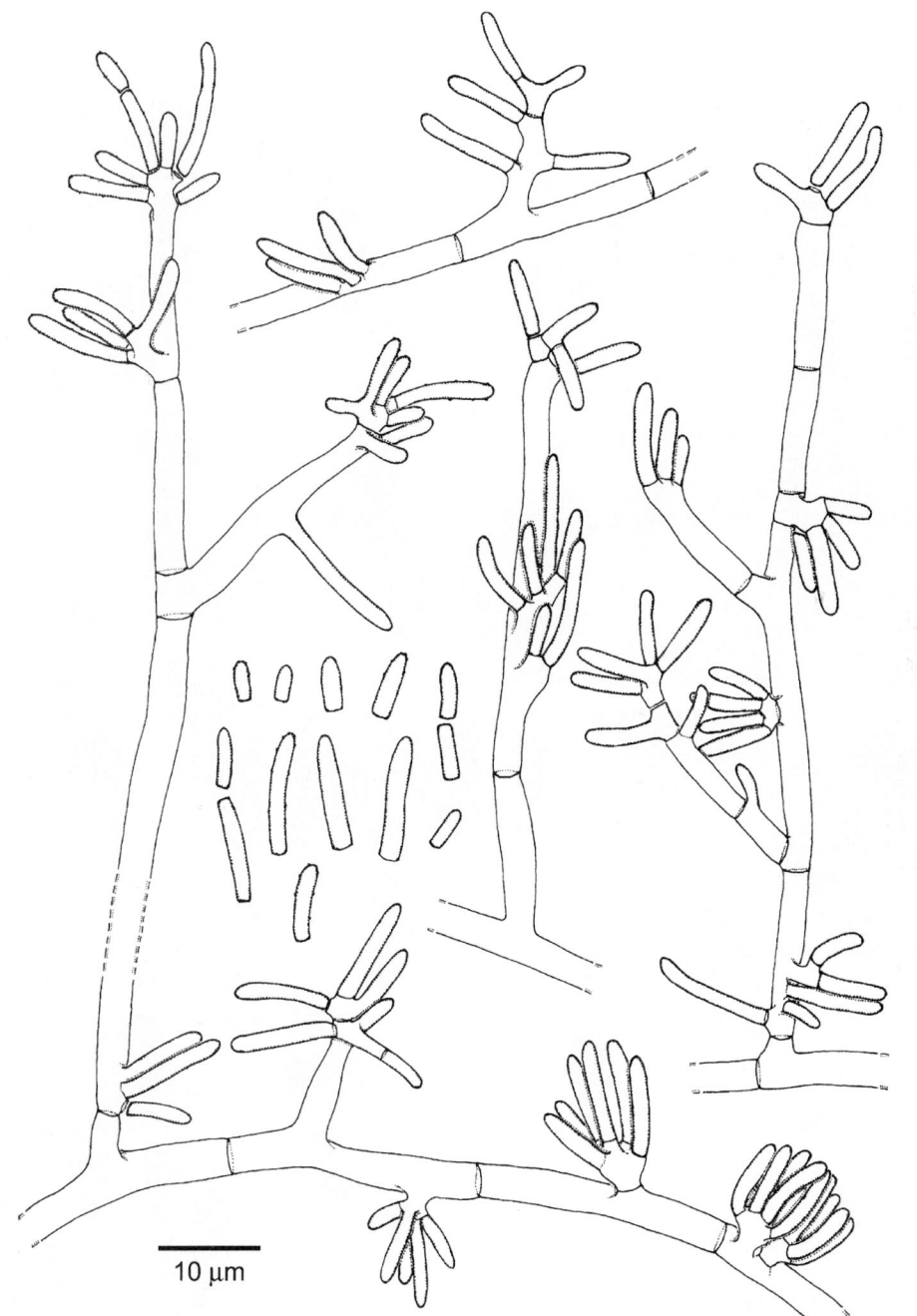

Hormographiella verticillata, FMR 3936. Conidiophores and conidia.

Antifungal susceptibility.

Antifungal	Mean MICs	Strains	Reference
AMB	0.1	5	Gené *et al.* (1996a)
5FC	488	5	Gené *et al.* (1996a)
FLZ	60.6	5	Gené *et al.* (1996a)
ITZ	0.25	5	Gené *et al.* (1996a)
KTZ	0.6	5	Gené *et al.* (1996a)
MCZ	1.06	5	Gené *et al.* (1996a)

Nomenclature. *Hormographiella verticillata* Guarro, Gené & Guého, *in* Guarro, Gené, de Vroey & Guého - Mycotaxon 45: 186, 1992.

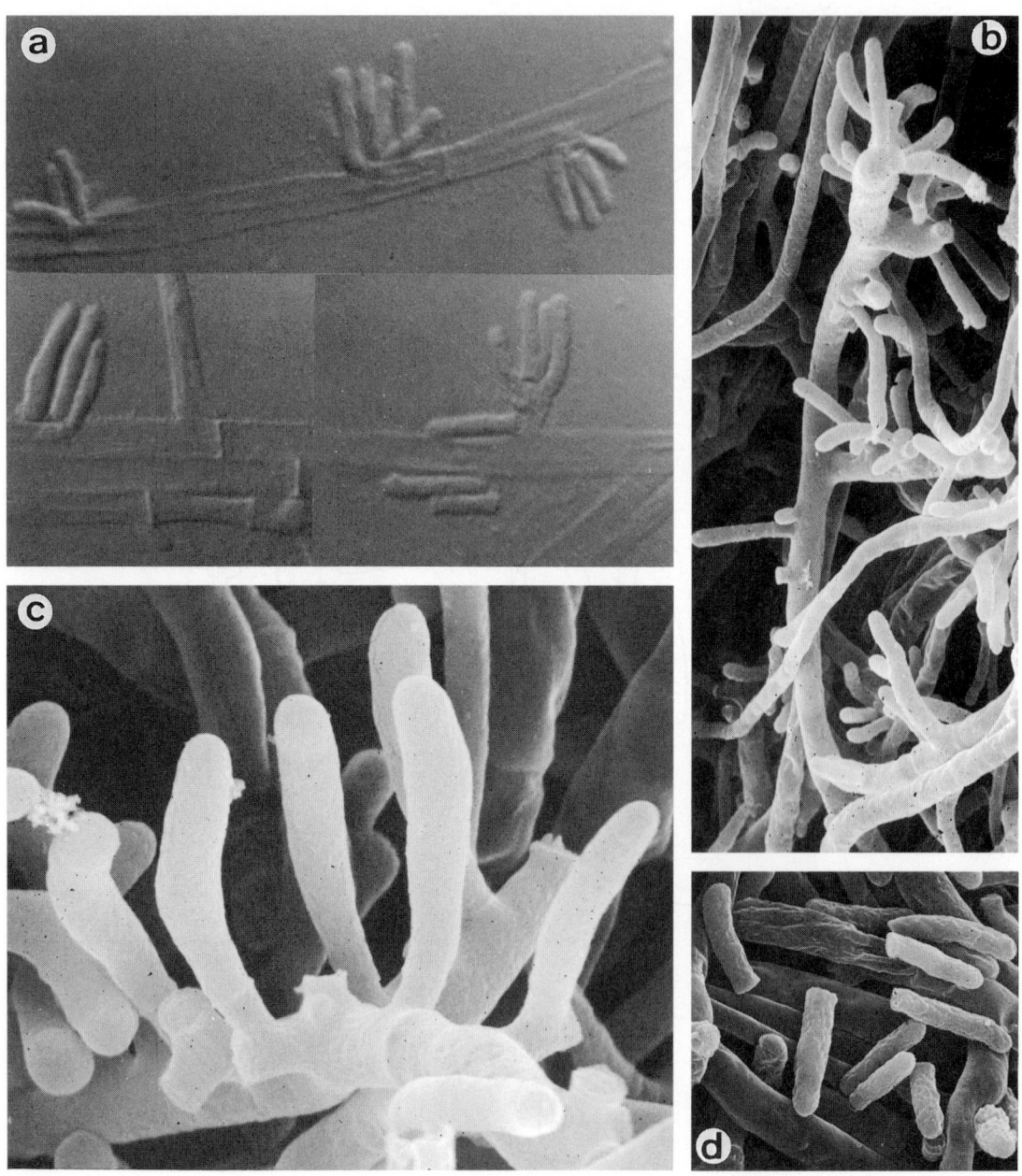

Hormographiella verticillata, FMR 3936. a-c. Conidiophores; d. conidia. a. ×1600; b. ×2200; c. ×6500; d. ×3200. Reproduced with permission from Mycotaxon 45: 188, 1992.

Hyphomycetes, black yeast. Genus: *HORMONEMA*

Hormonema dematioides Lagerberg & Melin

Colony characteristics. Colonies (MEA 2%) expanding, flat, smooth, moist, tough, initially with pinkish tinges, soon becoming olivaceous black.

Microscopy. Expanding hyphae non-septate, with irregularly dichotomous branching. Mature hyphae wide, densely septate, cells often becoming wider than long, locally converted into thick-walled chlamydospores. Conidia formed percurrently on undifferentiated hyphae from inconspicuous scars, hyaline, non-septate, smooth- and thin-walled, ellipsoidal, of variable size, often budding, finally becoming 1-septate and olivaceous brown.

Teleomorph. *Sydowia polyspora* (Brefeld & Tavel) E. Müller (*Ascomycota, Euascomycetes, Dothideales: Dothioraceae*).

Physiology.

Growth:		Lactose	v	Succinate	v
D-Glucose	+	Raffinose	+	Citrate	+,w
D-Galactose	w,–	Melezitose	+	Methanol	–
L-Sorbose	v	Inulin	–,w	Ethanol	+,w
D-Glucosamine	w,+	Soluble starch	+	Nitrate	+
D-Ribose	+	Glycerol	w,+	Nitrite	+
D-Xylose	+	*meso*-Erythritol	+,w	Ethylamine	+
L-Arabinose	+	Ribitol	+,w	L-Lysine	+
D-Arabinose	–,w	Xylitol	+	Cadaverine	+
L-Rhamnose	+	L-Arabinitol	+	Creatine	w,–
Sucrose	+	D-Glucitol	+	Creatinine	w,–
Maltose	+	D-Mannitol	+	10% NaCl	v
α,α-Trehalose	+	*myo*-Inositol	+	0.1% Cycloheximide	–
methyl-α-D-Glucoside	–,w	D-Gluconate	v	Mycosel	–
Cellobiose	+	D-Glucuronate	+,–	Urease	+
Salicin	+	D-Galacturonate	w,+	Growth at 37°C	–
Melibiose	+	DL-Lactate	w,+		

Molecular diagnostics. ITS restriction map based on AF013228:

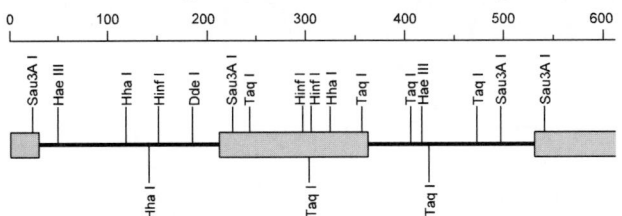

Differential diagnosis. The species differs from *Aureobasidium pullulans* (p. 520) by absence of synchronous conidiation and by absence of growth with methyl-α-D-glucoside and mostly also with D-gluconate.

Pathogenicity. BSL-1. A cutaneous phaeohyphomycosis was reported by Coldiron *et al.* (1990), and a fatal CAPD-associated peritonitis by Shin *et al.* (1998). Kent *et al.* (1998) reported fungemia after exposure to birds.

References. Hermanides-Nijhof (1977), de Hoog & Yurlova (1994).

Antifungal susceptibility.

Antifungal	Mean MICs	Strains	Reference
AMB	1.59	3	McGinnis & Pasarell (1998b)
ITZ	0.49	3	McGinnis & Pasarell (1998b)
VCZ	0.62	3	McGinnis & Pasarell (1998b)

Nomenclature. *Hormonema dematioides* Lagerberg & Melin, *in* Lagerberg, Lundberg & Melin - Svenska Skogsvfor. Tidskr. 2-4: 219, 1927.

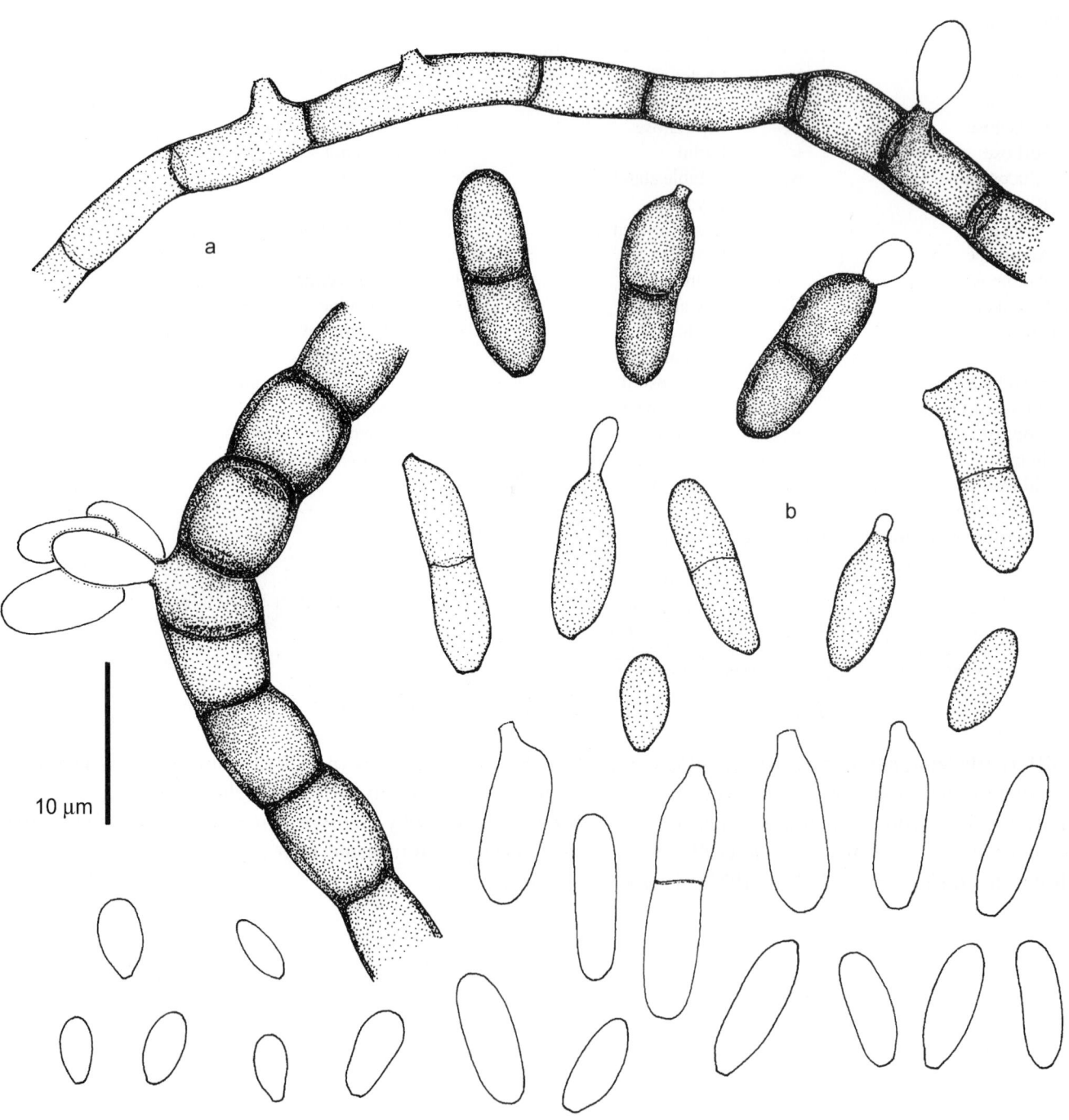

Hormonema dematioides, CBS 116.29. a. Hyphae with percurrent conidia, cells becoming chlamydospore-like; b. conidia with budding and changing over into thick-walled cells.

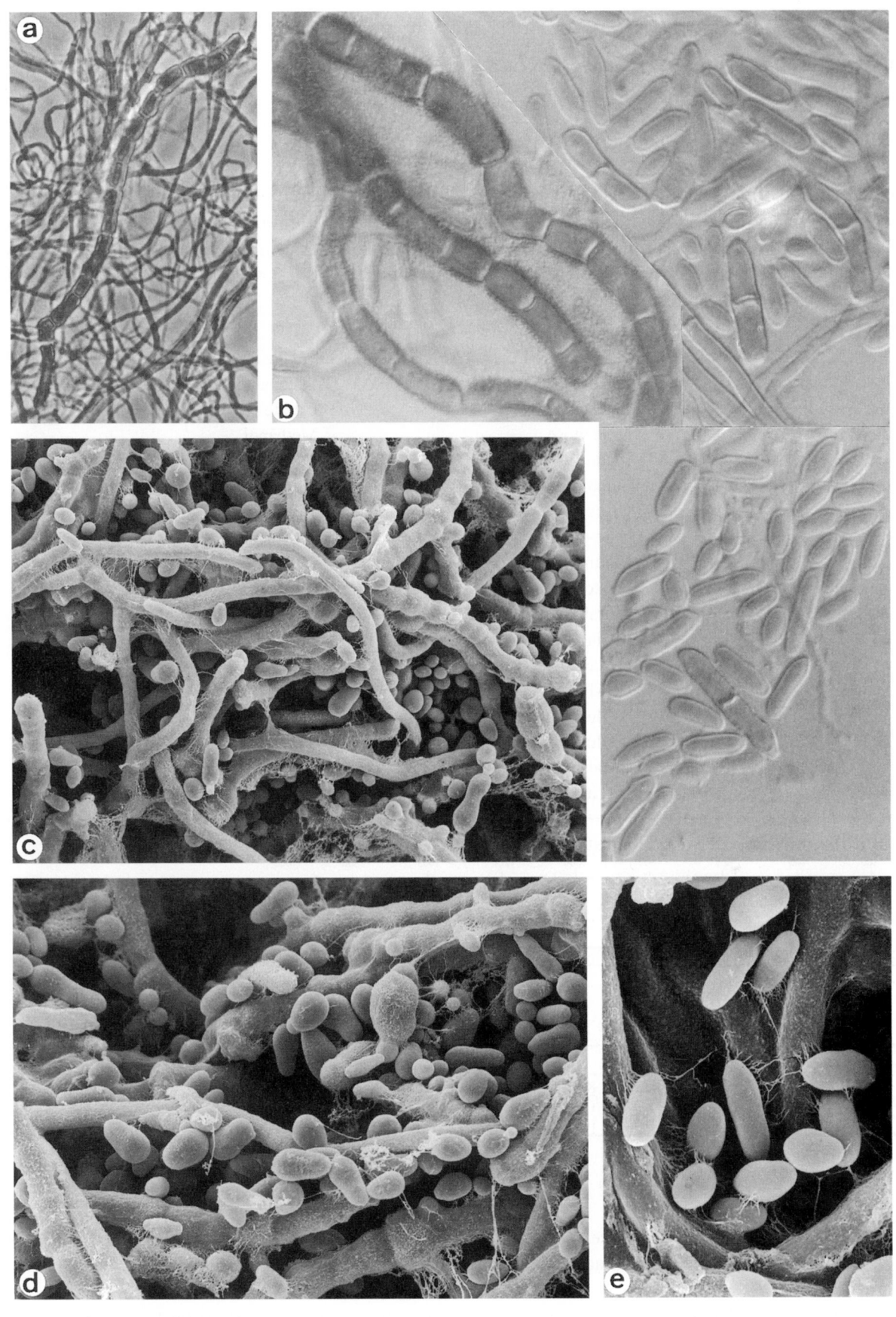

Hormonema dematioides, CBS 116.29. Hyphae with chlamydospore-like cells and conidia. a. ×640; b. ×1600; c. ×950; d. ×1500; e. ×3000.

Hyphomycetes, black yeasts. Genus: *HORTAEA*

Hortaea werneckii (Horta) Nishimura & Miyaji

Colony characteristics. Colonies (OA) restricted, smooth and slimy, olivaceous black.

Microscopy. Hyphae up to 6 µm wide, becoming densely septate in maturation, locally brown and thick-walled. Annellated zones conspicuous, 1-2 µm wide, formed on intercalary or lateral conidiogenous cells. Conidia hyaline, becoming pale olivaceous, broadly ellipsoidal, 1-, later often 2-celled, 7.0-9.5 × 3.5-4.5 µm, often budding, finally often converted into clumps of chlamydospores.

Physiology.

Growth:		Lactose	+	Succinate	+,w
D-Glucose	+	Raffinose	+	Citrate	v
D-Galactose	+	Melezitose	+	Methanol	–
L-Sorbose	+,w	Inulin	–,w	Ethanol	v
D-Glucosamine	–,w	Soluble starch	+,w	Nitrate	+
D-Ribose	+,w	Glycerol	+	Nitrite	+
D-Xylose	+	*meso*-Erythritol	+	Ethylamine	+,w
L-Arabinose	+	Ribitol	v	L-Lysine	–,w
D-Arabinose	+,w	Xylitol	v	Cadaverine	–,w
L-Rhamnose	+	L-Arabinitol	+,w	Creatine	–
Sucrose	+	D-Glucitol	v	Creatinine	–
Maltose	+	D-Mannitol	+	10% NaCl	+
α,α-Trehalose	+	*myo*-Inositol	+	0.1% Cycloheximide	–
methyl-α-D-Glucoside	–	D-Gluconate	v	Tyrosinase	+
Cellobiose	+	D-Glucuronate	v	Urease	+
Salicin	+,w	D-Galacturonate	–	Casein decomposition	–
Melibiose	+	DL-Lactate	–	Growth at 37°C	–

Molecular diagnostics. SSU and ITS restriction maps based on CBS 107.67:

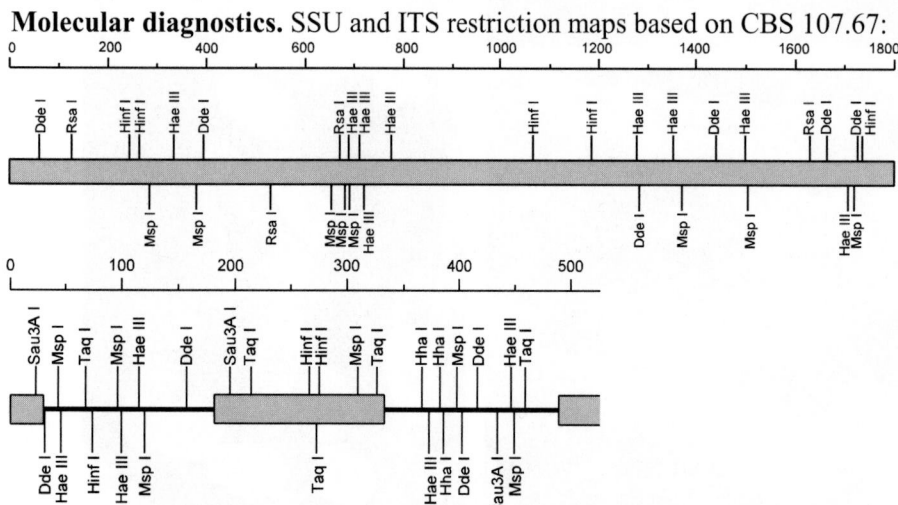

Differential diagnosis. Its broad, densely septate hyphae are reminiscent of *Aureobasidium* (p. 520), but the colonies are restricted and olivaceous black. The annellated zones differ from those of *Exophiala* (p. 645) by being much broader.

Pathogenicity. BSL-1. The species almost exclusively causes tinea nigra on one or on both hands, sometimes causing similar lesions on the sole. It is restricted to tropical or subtropical areas, where it occurs as a halophilic saprobe (Uijthof *et al.*, 1994) and in natural saltpans (Zalar *et al.*, 1999). Its presence on human hands concerns subclinical colonization (Göttlich *et al.*, 1995).

References. Miyaji & Nishimura (1985), McGinnis *et al.* (1985), de Hoog & Gerrits van den Ende (1992), Mittag (1993), Severo *et al.* (1994), de Cock (1994), Zalar *et al.* (1999).

Antifungal susceptibility.

Antifungal	Mean MICs	Strains	Reference
AMB	0.32	11	McGinnis & Pasarell (1998b)
5FC	32	1	Unpublished data
FLZ	2	1	Unpublished data
ITZ	0.07	11	McGinnis & Pasarell (1998a)
KTZ	0.03	1	Unpublished data
MCZ	2	1	Unpublished data
TBF	0.04	11	McGinnis & Pasarell (1998a)
VCZ	0.05	11	McGinnis & Pasarell (1998b)

Nomenclature. *Cladosporium werneckii* Horta - Revta Med.-Cirúrg. Braz. 22: 269, 1921 ≡ *Dematium werneckii* (Horta) C.W. Dodge - Med. Mycol. p. 676, 1935 / *Pullularia werneckii* (Horta) de Vries - Contrib. Knowl. Cladosp. p. 101, 1952 ≡ *Exophiala werneckii* (Horta) von Arx - Gen. Fungi. Sporul. Pure Cult. p. 180, 1970 ≡ *Hortaea werneckii* (Horta) Nishimura & Miyaji - Jpn. J. Med. Mycol. 26: 145, 1984 ≡ *Phaeoannellomyces werneckii* (Horta) McGinnis & Schell, *in* McGinnis, Schell & Carson - Sabouraudia 23: 182, 1985.

Cryptococcus metaniger Castellani - Arch. Derm. Syph. 16: 402, 1927 ≡ *Cladosporium metaniger* (Castellani) Ferrari - Atti Ist. Bot. Univ. Lab. Crittog. Pavia, Ser. 4, 3: 183, 1932 ≡ *Pullularia fermentans* Wynne & Gott. var. *castellanii* Wynne & Gott - J. Gen. Microbiol. 14: 518, 1956 (name change).

Cladosporium rietmanni A. Sartory - CBS List Cult. p. 16, 1931.

Pullularia fermentans Wynne & Gott var. *laeoi* Wynne & Gott - J. Gen. Microbiol. 14: 517, 1956.

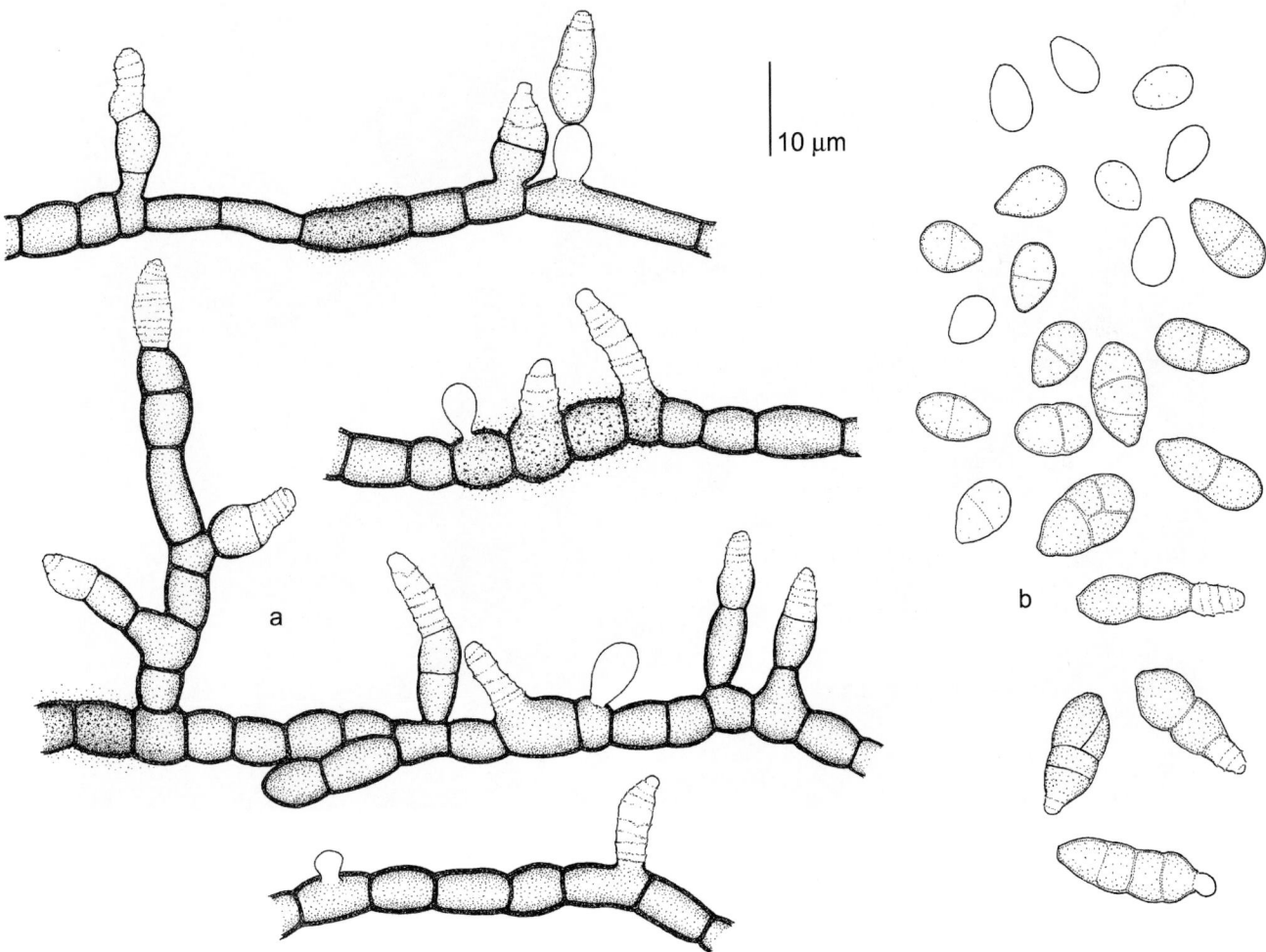

10 µm

Hortaea werneckii, CBS 410.51. a. Conidial apparatus; b. conidia, gradually developing into chlamydospore-like cells.

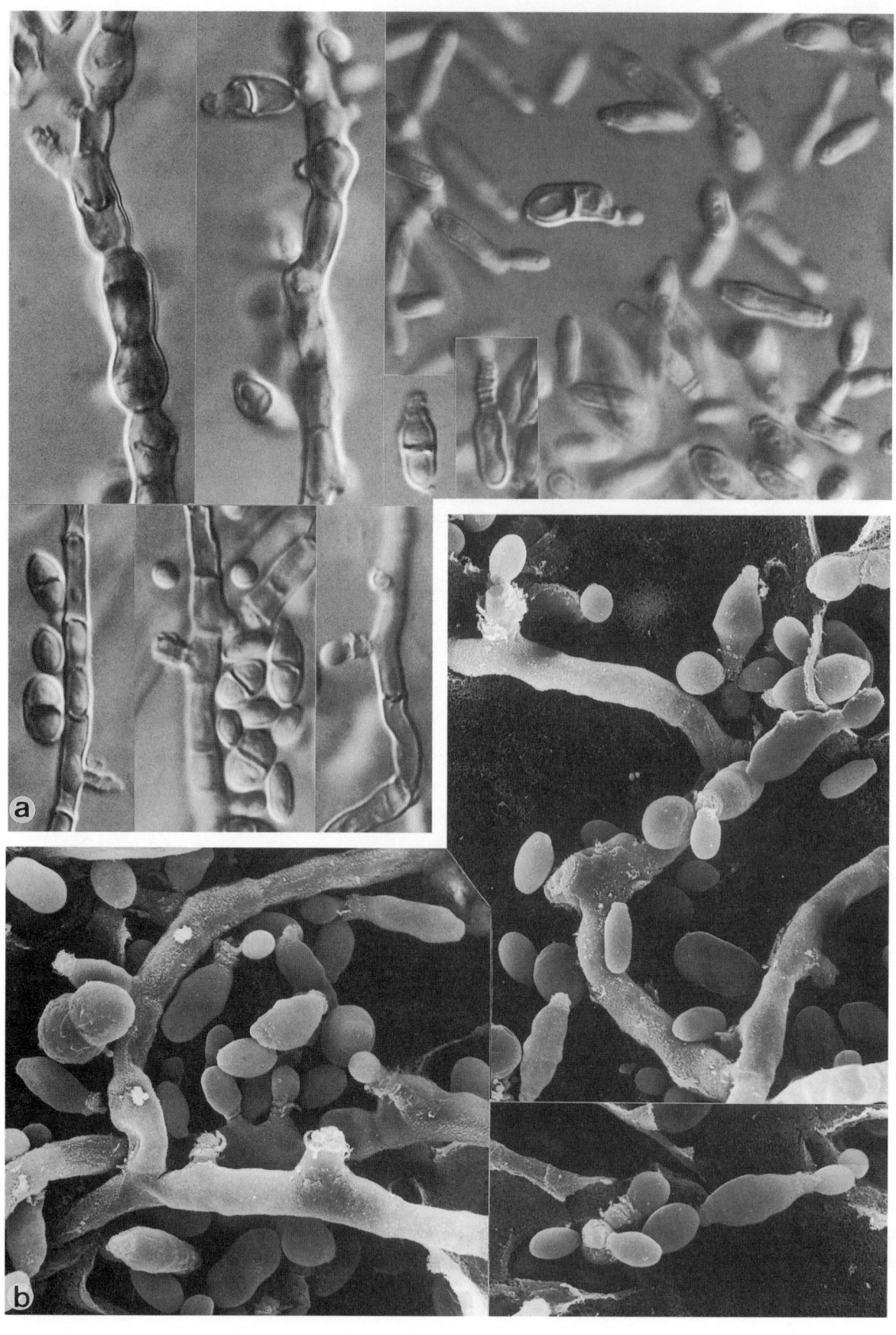

Hortaea werneckii, CBS 410.51. Conidial apparatus, conidia and chlamydospore-like cells producing conidia. a. ×1600; b. ×2650.

Hyphomycetes, dermatophytes. Genus: *KERATINOMYCES*

Keratinomyces ceretanicus Punsola & Guarro

Colony characteristics. Colonies (SGA, 15-17°C) restricted, flat, cottony, cream-coloured with elevated centre; reverse brown.

Microscopy. Macroconidia hyaline, smooth- and thick-walled, lanceolate to cylindrical (7-) 11-14 (-15)-celled, 50-78 × 4-5 µm, arranged in loose clusters which are borne laterally on creeping hyphae. Microconidia absent.

Physiology. The species is psychrophilic, with optimum growth below 20°C, and is sensitive to cycloheximide.

Pathogenicity. BSL-1. Geophilic dermatophyte.

Distribution. The species has been reported from Spain and Chile.

References. Punsola & Guarro (1984a), Cano & Sigler (1992).

Nomenclature. *Keratinomyces ceretanicus* Punsola & Guarro - Mycopathologia 85: 185, 1984.

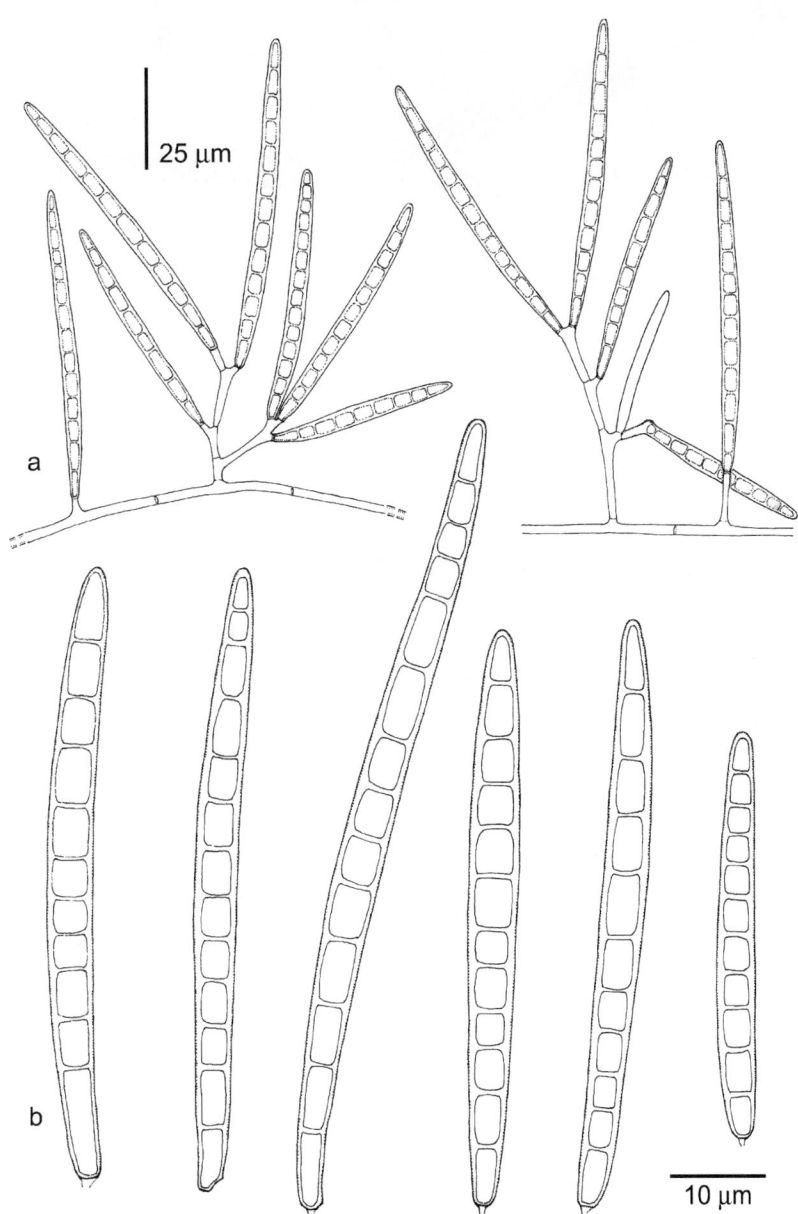

Keratinomyces ceretanicus, CBS 269.89. a. Arrangement of macroconidia; b. liberated macroconidia.

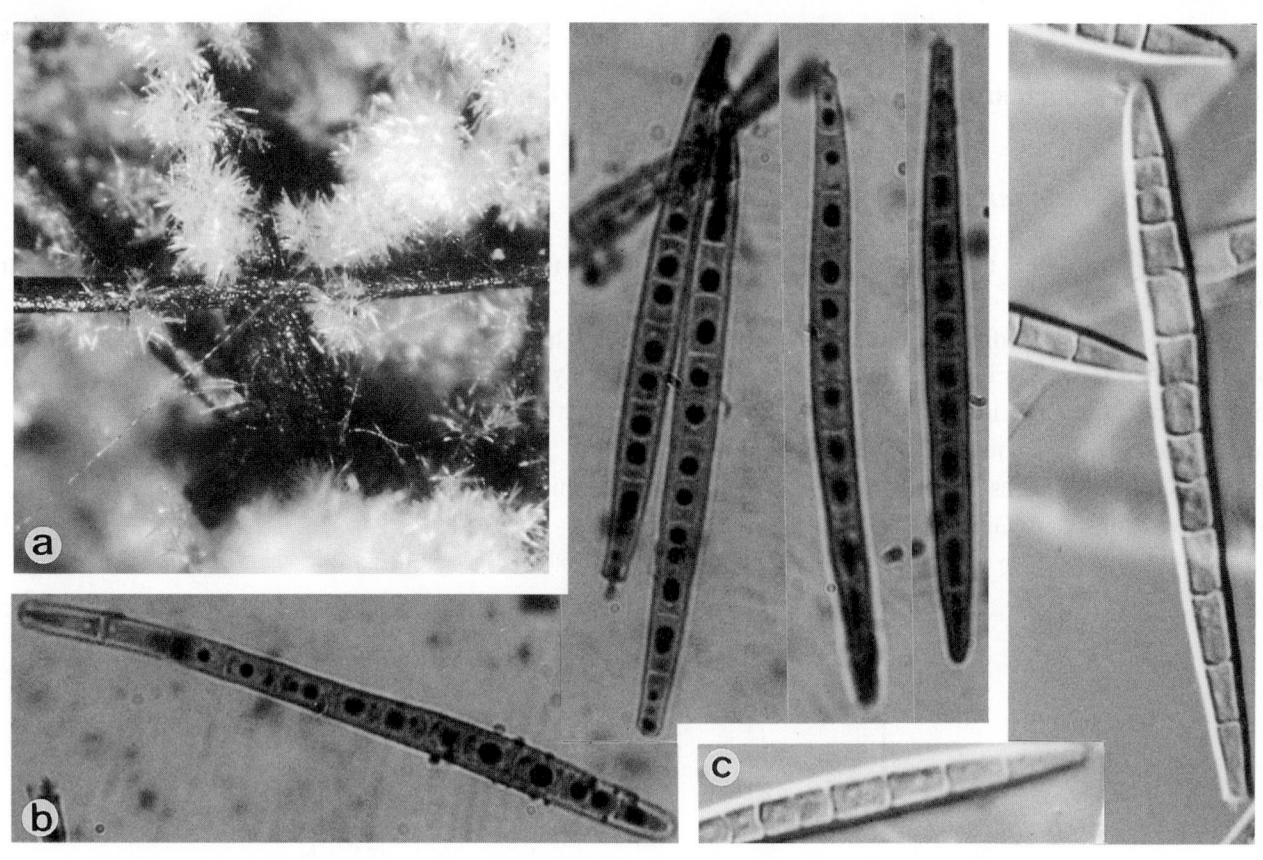

Keratinomyces ceretanicus, CBS 269.89. a. Growth on bait of horse hair; b, c. macroconidia. a. ×64; b. ×1400; c. ×1600.

Hyphomycetes. Genus: *LECYTHOPHORA*

Generic description. Colonies pink to salmon, later sometimes becoming blackish, smooth, often somewhat yeast-like. Hyphae hyaline to subhyaline, thin-walled. Conidiogenous cells poorly differentiated, usually lateral or intercalary, hyphae bearing one or several scattered, inconspicuous collarettes, either sessile or on small outgrowths. Conidia (sub)hyaline, smooth- and thin-walled, broadly ellipsoidal, cylindrical, reniform or allantoid. Chlamydospores may be present.

Teleomorph. *Coniochaeta* (*Ascomycota, Euascomycetes, Sordariales: Coniochaetaceae*).

References. De Hoog (1983), Gams & McGinnis (1983).

Key to the treated species of Lecythophora:

1a. Colonies remaining pink to orange; chlamydospores absent .... *L. hoffmannii* (726)
1b. Colonies darkening at the centre; chlamydospores present .... *L. mutabilis* (728)

Lecythophora hoffmannii (van Beyma) W. Gams & McGinnis

Colony characteristics. Colonies (MEA) with moderate growth, flat, smooth, moist, pink to orange, with regular and sharp margin; reverse pink.

Microscopy. Hyphae narrow, hyaline, producing conidia laterally from small collarettes directly on the hyphae, or from lateral cells which are sometimes arranged in dense groups; lateral cells flask-shaped or nearly cylindrical. Collarettes unpigmented, about 1.5 µm wide. Conidia hyaline, smooth- and thin-walled, broadly ellipsoidal to cylindrical or allantoid, 3.0-3.5 × 1.5-2.5 µm, produced in slimy heads.

Physiology. Intolerant to benomyl.

Teleomorph. *Coniochaeta ligniaria* (Grev.) Cooke (*Ascomycota, Euascomycetes, Sordariales: Coniochaetaceae*). Ascomata subspherical to pyriform, ostiolate, black, 200-350 µm in diam, setose. Asci cylindrical, 8-spored, 80-130 × 10-14 µm. Ascospores oblate, brownish-black, 11.5-17.0 × 10-12 × 7-8 µm, with a longitudinal germ slit.

Pathogenicity. BSL-1. Subcutaneous abscess (Rinaldi *et al.*, 1982), keratitis (McGinnis, 1978b). A sinusitis in an AIDS patient was described by Marriott *et al.* (1997).

Reference. De Hoog (1983)., Checa *et al.* (1988).

Antifungal susceptibility.

Antifungal	Mean MICs	Strains	Reference
AMB	0.25	5	McGinnis & Pasarell (1998b)
ITZ	0.38	5	McGinnis & Pasarell (1998b)
VCZ	0.29	5	McGinnis & Pasarell (1998b)

Nomenclature. *Sphaeria ligniaria* Greville - Scott. Crypt. Fl. 1: 82, 1828 ≡ *Coniochaeta ligniaria* (Greville) Cooke - Grevillea 16: 37, 1887.
 Margarinomyces hoffmannii van Beyma - Zentbl. Bakt. Parasitkde, Abt. 2, 99: 386, 1939 ≡ *Phialophora hoffmannii* (van Beyma) Schol-Schwarz - Persoonia 6: 79, 1970 ≡ *Lecythophora hoffmannii* (van Beyma) W. Gams & McGinnis - Mycologia 75: 985, 1983.

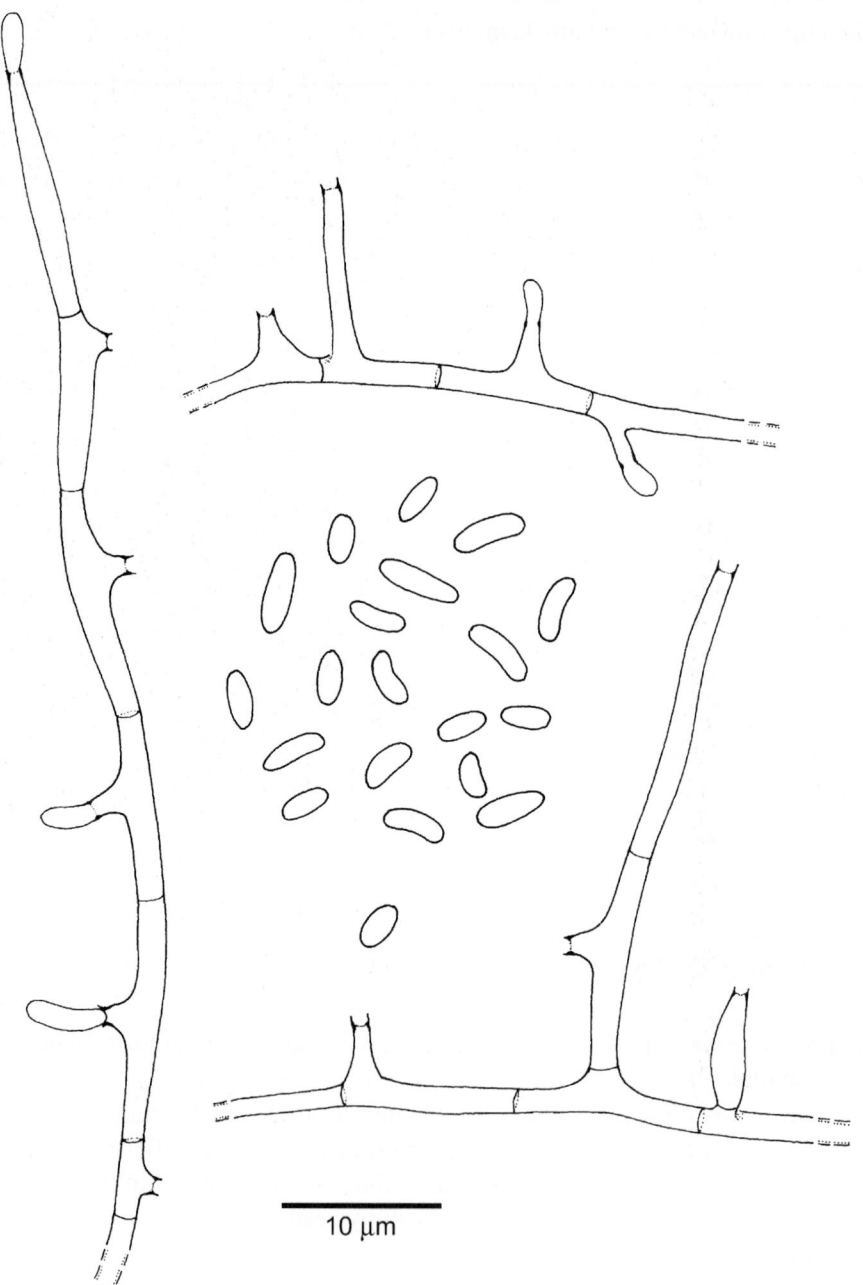

Lecythophora hoffmannii, CBS 140.41. Conidiogenous cells and conidia.

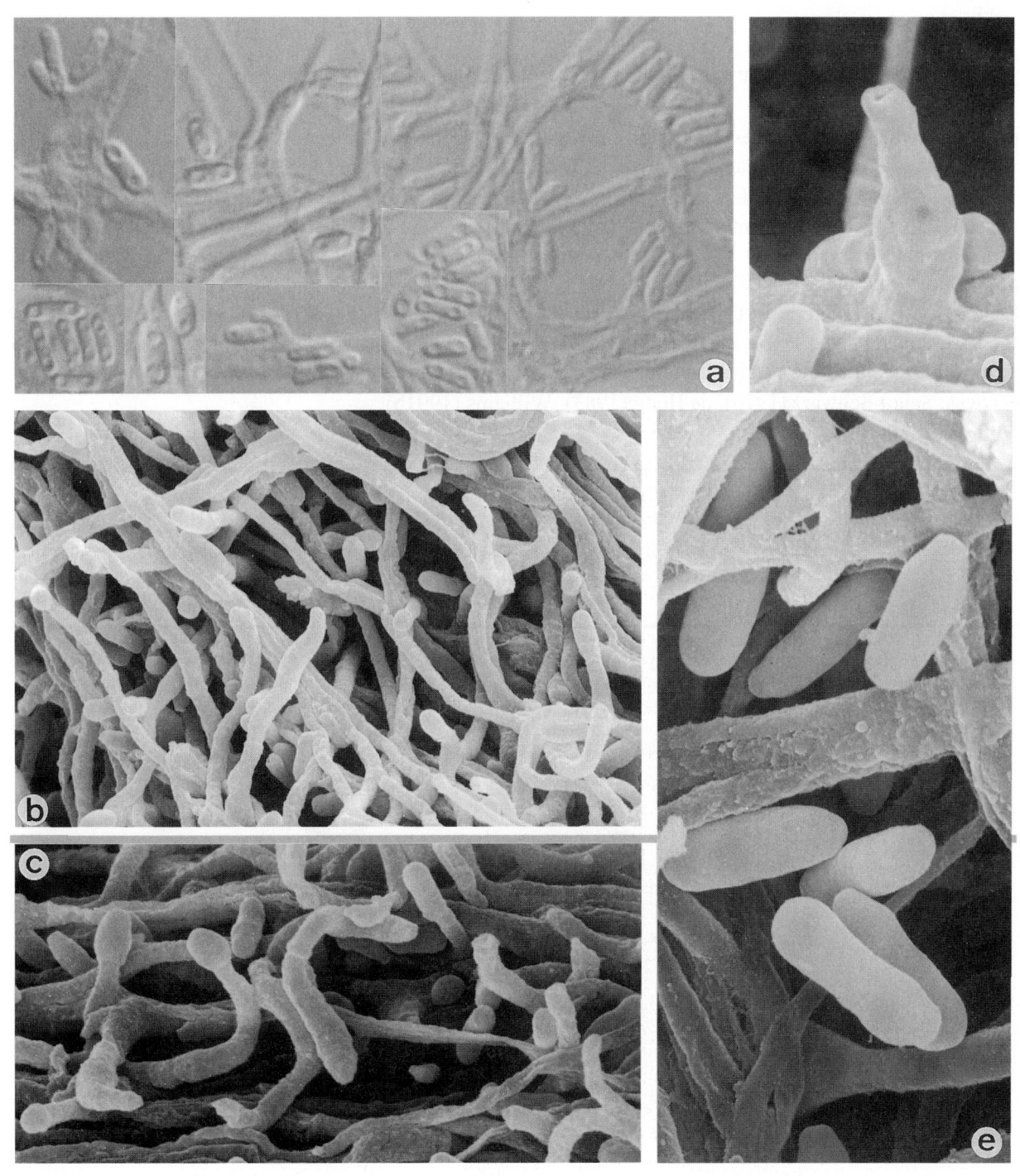

Lecythophora hoffmannii, CBS 140.41. Conidiogenous cells and conidia. a. ×1600; b. ×3500; c. ×4500; d. ×9500; e. ×8500.

Lecythophora mutabilis (v. Beyma) W. Gams & McGinnis

Colony characteristics. Colonies (OA) with moderate growth, flat, at the centre with ascending to erect, hyphal fascicles, pink, later becoming blackish-brown at the centre due to abundant formation of chlamydospores.

Microscopy. Conidiogenous cells emerging from hyphal strands, hyaline to subhyaline, flask-shaped to acicular, flexuose, 6-12 × 1.5-3.5 µm, or intercalary. Conidia hyaline, smooth- and thin-walled, subcylindrical to cylindrical, sometimes slightly curved, 4-6 × 1.8-2.5 µm. Chlamydospores broadly ellipsoidal to broadly clavate, about 7.0 × 4.5 µm, smooth- and thick-walled, brown.

Pathogenicity. BSL-1. Keratitis (Pritchard & Muir, 1987; Ho *et al.*, 1991), endophthalmitis (Marcus *et al.*, 1999), peritonitis (Ahmad *et al.*, 1985), endocarditis (Pierarch *et al.*, 1973; Slifkin & Bowers, 1975). Muotoe-Okafor & Gugnani (1993) isolated the species from internal organs of fruit-eating bats.

References. Schol-Schwarz (1970), de Hoog (1983), Williams (1991b).

Antifungal susceptibility.

Antifungal	Mean MICs	Strains	Reference
AMB	0.5	4	McGinnis & Pasarell (1998b)
ITZ	0.08	4	McGinnis & Pasarell (1998b)
VCZ	0.05	4	McGinnis & Pasarell (1998b)

Nomenclature. *Margarinomyces mutabilis* van Beyma - Antonie van Leeuwenhoek 10: 48, 1944 ≡ *Phialophora mutabilis* (van Beyma) Schol-Schwarz - Persoonia 6: 80, 1970 ≡ *Lecythophora mutabilis* (van Beyma) W. Gams & McGinnis - Mycologia 75: 985, 1983.

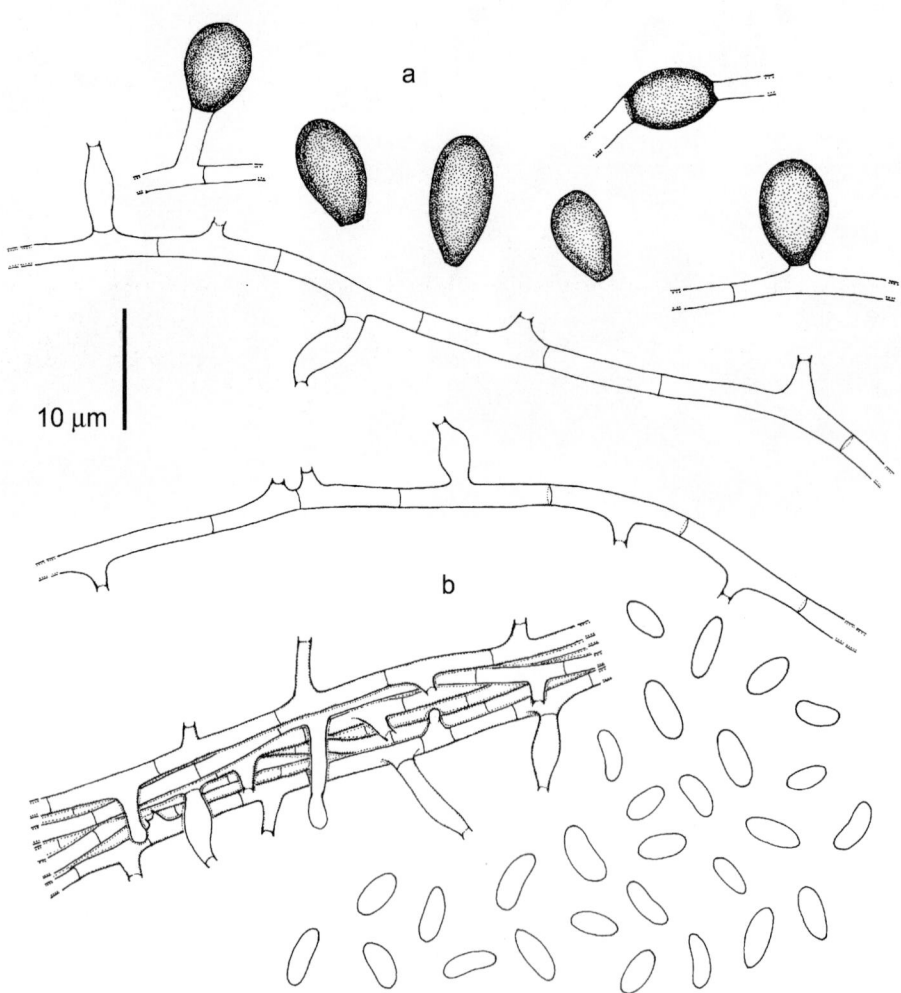

Lecythophora mutabilis, CBS 157.44. a. Chlamydospores; b. conidiogenous cells and conidia.

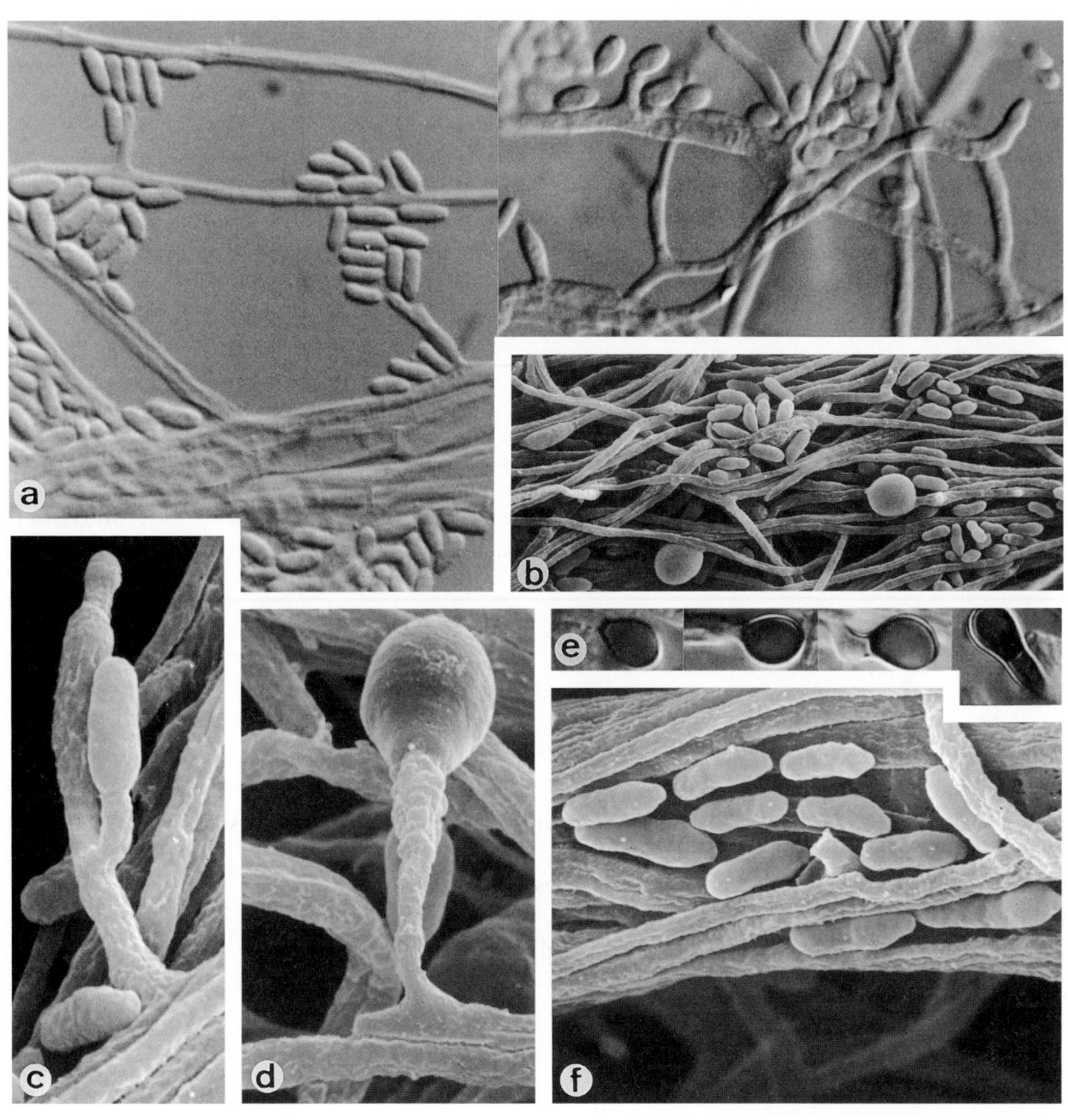

Lecythophora mutabilis, CBS 157.44. Conidiogenous cells, conidia and chlamydospores. a. ×1600; b. ×1675; c. ×7900; d. ×6600; e. ×1280; f. ×5400.

Hyphomycetes. Genus: *MADURELLA*

Generic description. Colonies growing slowly, dark, composed of a dense, melanized, mostly sterile mycelium.
General remarks. Isolates of *Madurella* are recovered from cases of black-grain mycetoma. In the laboratory they are dematiaceous and mostly sterile. *M. mycetomatis* has been reported to form occasional phialides with collarettes.
Reference. Kalamam *et al.* (1975).

Key to the treated species of Madurella:

1a.	No growth at 37°C; no brown pigment exuded into the medium; sucrose +, lactose –	*M. grisea* (730)
1b.	Growth at 37°C; brown pigment exuded into the medium; sucrose –, lactose + ...	*M. mycetomatis* (731)

Madurella grisea MacKinnon *et al.*

Colony characteristics. Colonies (OA) variable, growing slowly, raised to heaped, sometimes radially folded, felty, olivaceous to dark grey.
Microscopy. Colonies mostly sterile, composed of a dense, melanized mycelium.
Physiology. Sucrose assimilated but not lactose.
Molecular diagnostics. ITS rDNA restriction map based on CBS 331.50:

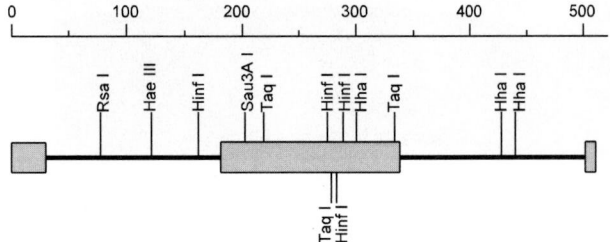

Pathogenicity. BSL-2. One of the main agents of human black-grain mycetoma (Southern, 1996; Severo *et al.*, 1999b).
Reference. Kwon-Chung & Bennett (1992).
Antifungal susceptibility.

Antifungal	MICs	Strains	Reference
AMB	0.25	1	McGinnis & Pasarell (1998b)
AMB	1	1	Unpublished data
ITZ	0.5	1	McGinnis & Pasarell (1998b)
ITZ	0.06	1	Unpublished data
KTZ	0.125	1	Unpublished data
MCZ	1	1	Unpublished data
VCZ	0.5	1	McGinnis & Pasarell (1998b)

Nomenclature. *Madurella grisea* MacKinnon, Ferrada & Montemayor - Mycopath. Mycol. Appl. 4: 389, 1949.

Madurella mycetomatis (Laveran) Brumpt

Colony characteristics. Colonies (OA) variable, with moderate growth, woolly, yellow or brown, generally producing a brownish diffusible pigment.

Microscopy. Colonies mostly sterile, composed of a dense, melanized mycelium; phialides with minute conidia in short chains occasionally present. Sclerotia frequently formed.

Physiology. Lactose assimilated but not sucrose. Intolerant to benomyl.

Molecular diagnostics. rDNA RFLP was used for species distinction by Al Ahmed *et al.* (1999) and Lopes *et al.* (2000). ITS rDNA restriction map based on CBS 247.48:

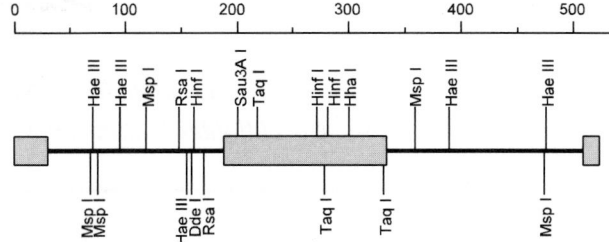

Pathogenicity. BSL-2. One of the main agents of human black-grain mycetoma (Chávez *et al.*, 1998). Host reactions were described by Fahal *et al.* (1995).

References. Mariat *et al.* (1977), Findlay *et al.* (1979), McGinnis (1996), Wethered *et al.* (1986).

Antifungal susceptibility.

Antifungal	Mean MICs	Strains	Reference
AMB	0.03	3	McGinnis & Pasarell (1998b)
ITZ	0.08	3	McGinnis & Pasarell (1998a)
TBF	0.1	3	McGinnis & Pasarell (1998a)
VCZ	0.05	3	McGinnis & Pasarell (1998b)

Nomenclature. *Streptothrix mycetomatis* Laveran - Bull. Acad. Méd. Paris, Sér. 3, 47: 773, 1902 ≡ *Madurella mycetomatis* (Laveran) Brumpt - C.R. Soc. Biol. 57: 997, 1905 (≡ *Madurella mycetomi Auctt.*).

 Madurella americana Gammel - Arch. Derm. Syph. 15: 263, 1927 ≡ *Acladium americanum* (Gammel) Ota - Jpn. J. Derm. Urol. 28, 1928.
 Madurella ikedae Gammel - Arch. Derm. Syph. 15: 281, 1927.

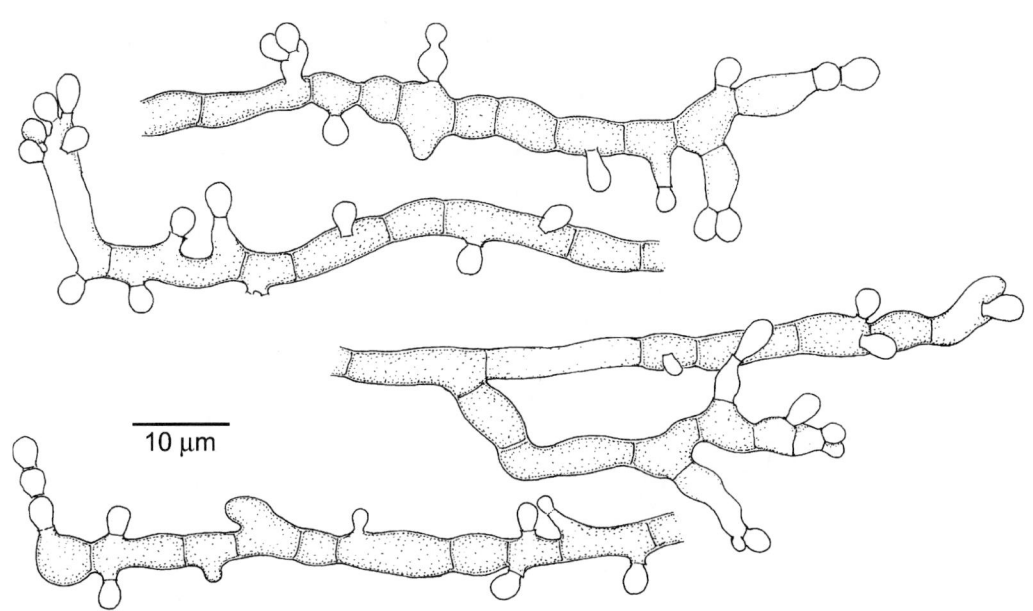

Madurella mycetomatis, dH 11952. Creeping hyphae with conidia in short chains.

Hyphomycetes. Genus: *MALBRANCHEA*

Generic description. Colonies growing moderately rapidly, white, yellow, tan orange or greenish. Hyphae hyaline, septate, branched; conidiophores absent. Conidia arising by disarticulation of fertile hyphae, which may be straight or spirally twisted. Conidia cylindrical, with truncate ends, separated from each other by sterile, empty cells.
Teleomorphs. *Auxarthron, Uncinocarpus* (*Ascomycota, Euascomycetes, Onygenaceae: Gymnoascaceae*).
Reference. Sigler & Carmichael (1976).

Key to the treated species of Malbranchea:

1a.	Fertile hyphae curved	***M. pulchella*** (732)
1b.	Fertile hyphae straight	see ***Coccidioides immitis*** (593)

Malbranchea pulchella Sacc. & Penz.

Colony characteristics. Colonies (MEA) restricted, cushion-shaped, felty, white, with tan to reddish-brown reverse.
Microscopy. Hyphae hyaline, 1-2 µm wide, bearing lateral, tightly coiled fertile branches which disarticulate into arthroconidia; conidia alternated with empty cells. Conidia cylindrical with truncate ends, pale yellowish, 2.5-5.5 × 1.5-2.5 µm.
Pathogenicity. BSL-1. A case of sinusitis was presented by Benda & Corey (1994).

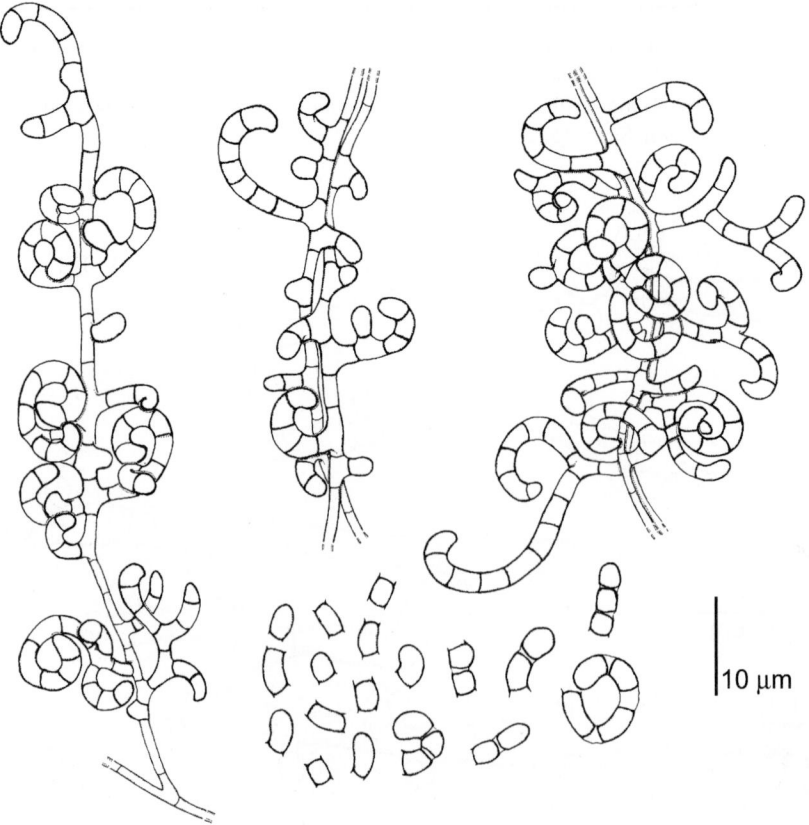

Malbranchea pulchella, FMR 4092. Fertile hyphae and arthroconidia.

Nomenclature. *Malbranchea pulchella* Saccardo & Penzig - Michelia 2: 638, 1882.

Malbranchea bolognesii-chiurcoi Vuillemin, Pollacci & Nannizzi, *in* Bolognesi & Chiurco - Archivi Biol. 1: 255, 1925 ≡ *Actinomyces bolognesii-chiurcoi* (Vuillemin, Pollacci & Nannizzi) C.W. Dodge - Med. Mycol. p. 766, 1935.

Malbranchia kambayashii Kambayashi - Arch. Derm. Syph. 170: 106, 1934.

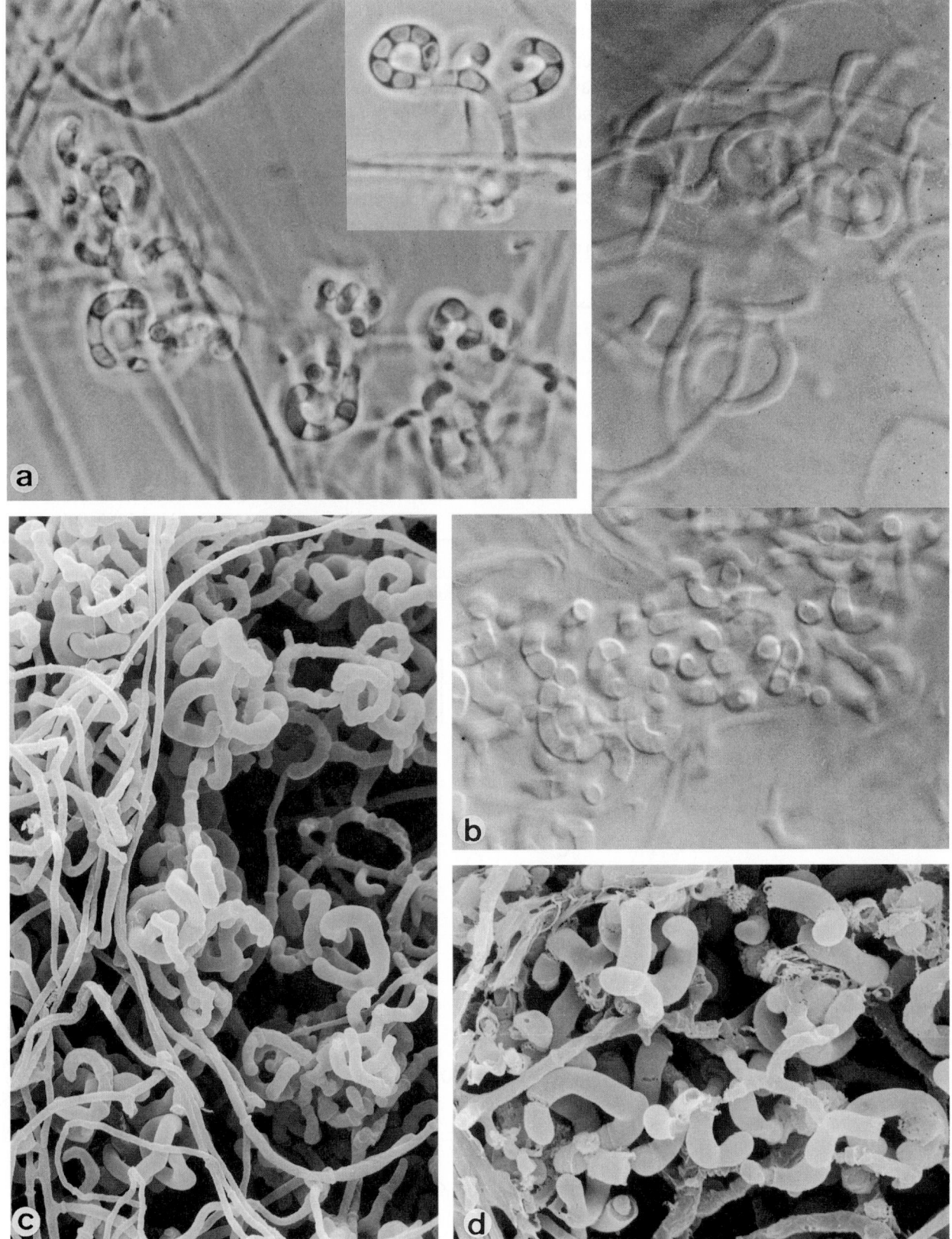

Malbranchea pulchella, FMR 4092. Fertile hyphae and arthroconidia. a. ×640; b. ×1600; c. ×2100; d. ×2800.

Hyphomycetes. Genus: *METARHIZIUM*

Metarhizium anisopliae (Metsch.) Sorok.

Colony characteristics. Colonies (OA) growing rather slowly, at first floccose, later becoming olivaceous green due to abundant conidiation; reverse yellowish to brownish.

Microscopy. Conidiophores aggregated in dense tufts, with repeated, more or less verticillate branching; phialides in dense, parallel arrangement. Phialides clavate, 9-14 µm long, with rounded apex. Conidia produced in long chains, cylindrical, 5-8 × 2.5-3.5 µm, thick-walled, yellowish-green in mass.

Molecular diagnostics. 26S rDNA sequence data were provided by Rakotonirainy *et al.* (1994).

Pathogenicity. BSL-1. This insect pathogen was reported from a human keratitis by de Cepero de García *et al.* (1997). Burgner *et al.* (1998) reported a disseminated infection in a leukemic child, and Revankar *et al.* (1999) two cases of sinusitis in immunocompetent hosts. An invasive rhinitis in a cat was described by Muir *et al.* (1998).

References. Tulloch (1976), Domsch *et al.* (1980), Bridge *et al.* (1993).

Antifungal susceptibility.

Antifungal	MICs	Strains	Reference
AMB	4	1	Unpublished data
5FC	256	1	Unpublished data
FLZ	128	1	Unpublished data
ITZ	4	1	Unpublished data
KTZ	4	1	Unpublished data
MCZ	8	1	Unpublished data

Nomenclature. *Entomophthora anisopliae* Metschnikov - Zap. Imp. Obschch. Khoz. Ross. p. 45, 1879 ≡ *Metarhizium anisopliae* (Metschnikov) Sorokin - Plant Paras. Man Anim. 2: 267, 1883 ≡ *Metarrhizium anisopliae Auctt.* (incorrect spelling).

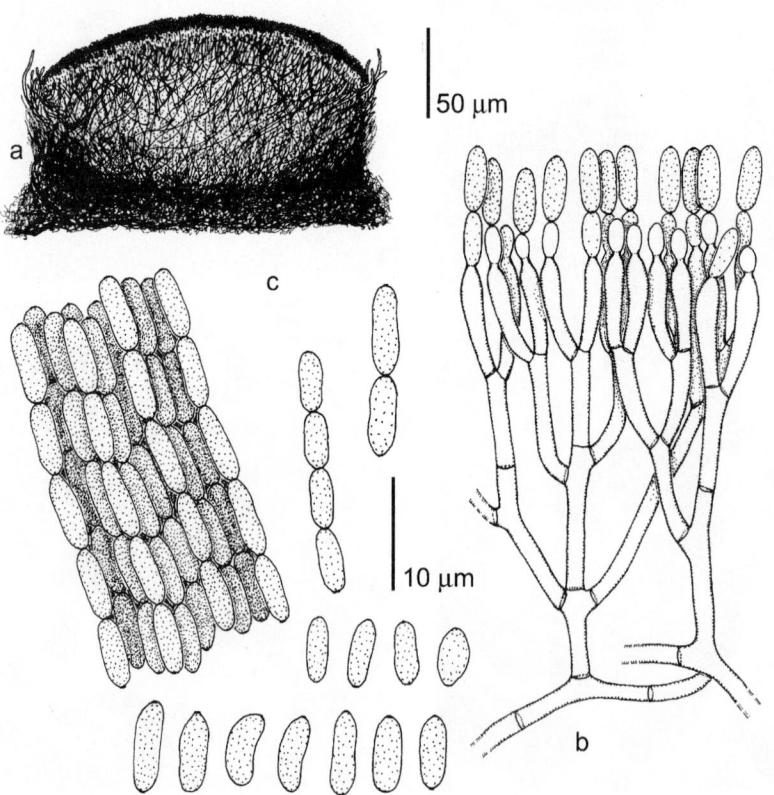

Metarhizium anisopliae, CBS 289.67. a. Sporodochium; b. conidiophores; c. conidia.

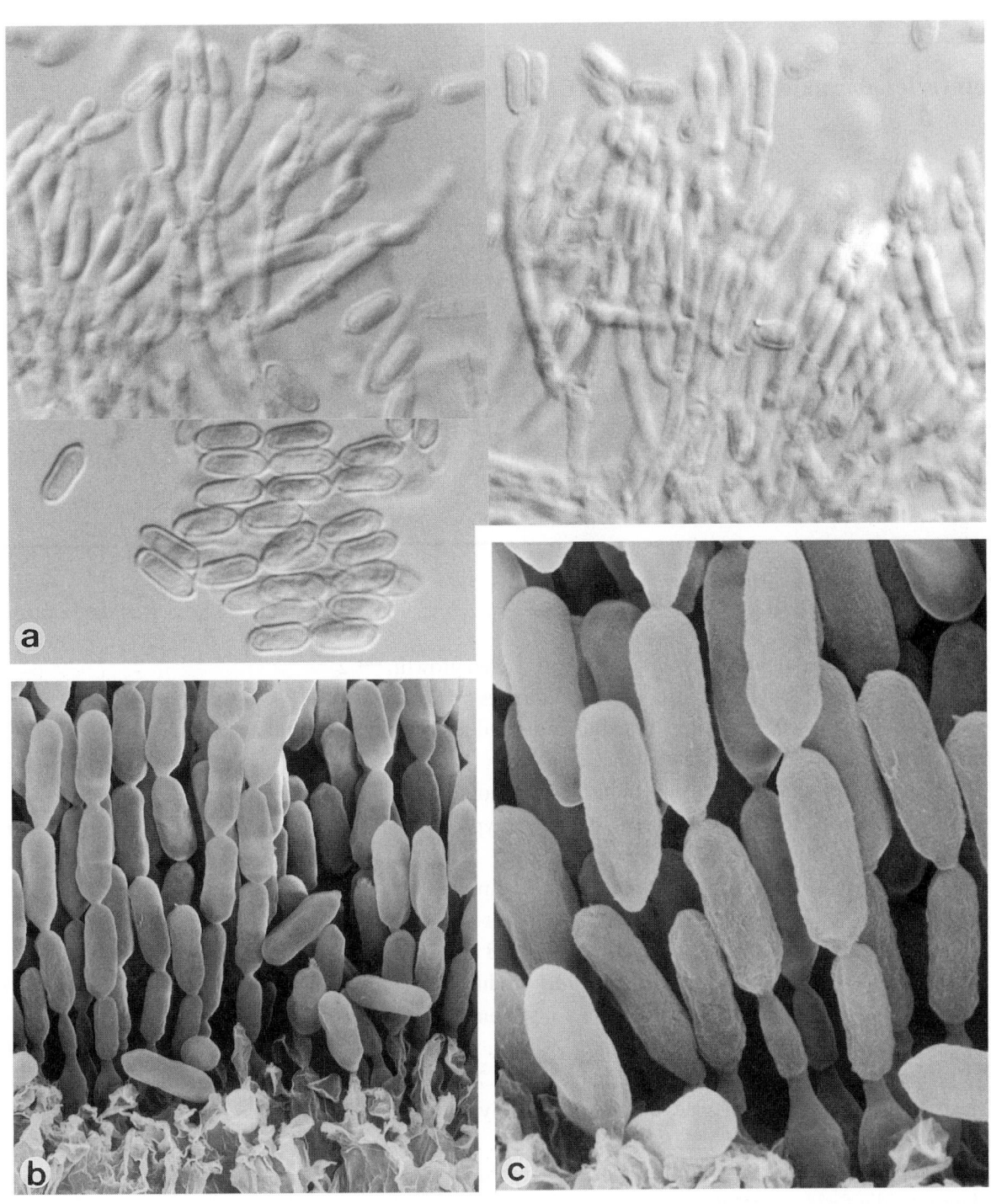

Metarhizium anisopliae, CBS 289.67. Conidiophores and chains of conidia. a. ×1600; b. ×2800; c. ×7000.

Hyphomycetes, dermatophytes. Genus: *MICROSPORUM*

Generic description. Colonies with slow or rapid growth, powdery, cottony to glabrous, white, buff to yellowish, sometimes with salmon tinges; reverse cream-coloured, reddish or yellowish. Macro- as well as microconidia present, solitary or in clusters at the end of or alongside undifferentiated hyphae. Macroconidia mostly arising in groups at acute angles, 2- to several-celled, thin- to thick-walled, echinulate to roughened, spindle- or cigar-shaped, hyaline, often with a frill at the base. Microconidia solitary, 1-celled, smooth- and thin-walled, hyaline, ovoidal to clavate, solitary.

Differential diagnosis. *Microsporum* differs from *Trichophyton* and *Epidermophyton* by having echinulate to roughened macroconidia, mostly with thick walls, while *Keratinomyces* has very thick, smooth walls.

Teleomorph. *Arthroderma (Ascomycota, Euascomycetes, Onygenales: Arthrodermataceae).*

References. Dvorák & Otcenásek (1969), Rebell & Taplin (1970), Badillet (1982), Kane *et al.* (1997), Gräser *et al.* (1999b, 2000a).

Key to the treated species of Microsporum:

1a. Sporulation reduced or absent; colonies with restricted growth → **2**
1b. Sporulation mostly abundant; faster growing colonies → **3**
2a. Colonies without aerial mycelium, often glabrous and leathery; macroconidia absent; bamboo hyphae present; pectinate hyphae absent .... *M. ferrugineum* (748)
2b. Colonies with aerial mycelium, not leathery; macroconidia, when present constricted near the middle; bamboo hyphae absent; pectinate hyphae present .... *M. audouinii* (740)
3a. Macroconidia mostly 2-celled, less than 18 µm long .... *M. nanum* (758)
3b. Macroconidia usually more than 2-celled, over 18 µm long → **4**
4a. Colony reverse reddish and/or exuding a red pigment into agar → **5**
4b. Colony reverse mostly yellowish, never reddish → **8**
5a. Macroconidia stalked, lanceolate; microconidia produced in clusters .... *M. racemosum* (765)
5b. Macroconidia not lanceolate; microconidia not produced in clusters → **6**
6a. Strawberry red pigment exuded into agar .... *M. gallinae* (752)
6b. No diffusion of strawberry red pigment into agar → **7**
7a. Macroconidia fusiform to clavate, thin- to moderately thick-walled, mostly less than 6-celled, 7-12 µm wide .... *M. fulvum* (749)
7b. Macroconidia fusiform, thick-walled, mostly more than 6-celled, 10-15 µm wide .... *M. cookei* (746)
8a. Macroconidia fusiform → **9**
8b. Macroconidia lanceolate → **11**
9a. Macroconidia with walls over 1.5 µm thick → **10**
9b. Macroconidia with walls less than 1.5 µm thick .... *M. gypseum* (755)
10a. Macroconidia less than 35 µm long, with a large oil droplet-like inclusion in each cell, with predominantly 3 thick septa and very small attachment scar .... *M. amazonicum* (738)
10b. Macroconidia over 35 µm long, without a large oil droplet-like inclusions, with more than 3 septa which have thinner walls .... *M. canis* (743)
11a. Colony with pinkish tinges; spiral hyphae mostly present; macroconidia with rounded apex .... *M. persicolor* (761)
11b. Colony without pinkish tinges; spiral hyphae absent; macroconidia with narrow apex ... *M. praecox* (763)

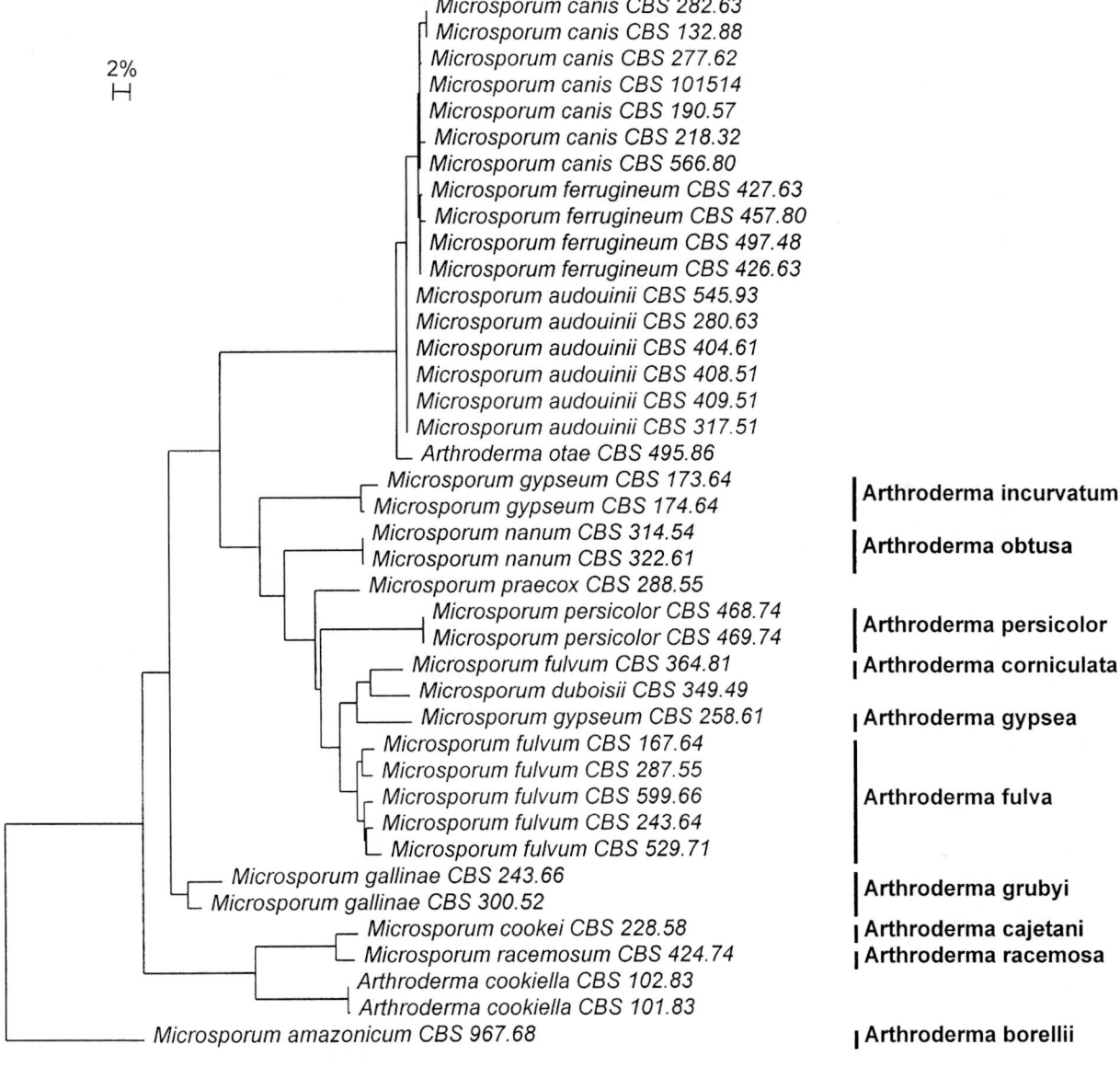

Fig. 58. Phylogenetic tree of *Microsporum.* and its teleomorphs based on confidently aligned complete ITS rDNA sequences using Neighbor joining algorithm with Kimura correction. Bootstrap values >90 from 100 resampled datasets are shown; data were provided by Y. Gräser. *Arthroderma borellii* is selected as outgroup.

Microsporum amazonicum Moraes *et al.*

Colony characteristics. Colonies (SGA) powdery to fluffy, grey olivaceous buff.

Microscopy. Macroconidia thin-walled, echinulate, spindle-shaped, 4-5 (-8)-celled, 13-35 × 3-10 μm, with 1.5 μm thick walls and with narrow scar. Microconidia sessile, clavate.

Teleomorph. *Arthroderma borellii* (Moraes *et al.*) Padhye *et al.* (*Ascomycota, Euascomycetes, Onygenales: Arthrodermataceae*).

Physiology.

Urease	+	T-1	+	T-4	+	T-7	+
Hair perforation	–	T-2	+	T-5	+		
		T-3	+	T-6	+		

Molecular diagnostics. ITS restriction map based on CBS 967.68:

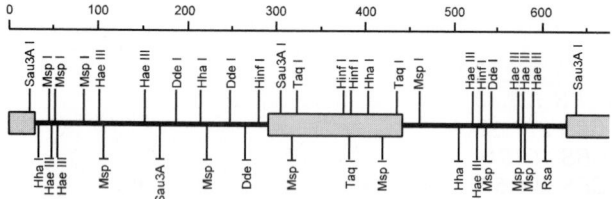

Pathogenicity. BSL-2. Zoophilic dermatophyte.

Distribution. Isolated from hairs of apparently normal rats in Brazil.

Reference. Moraes *et al.* (1967).

Nomenclature. *Microsporum amazonicum* Moraes, Borelli & Feo - Med. Cután. 11: 284, 1967.

Nannizzia borellii Moraes, Padhye & Ajello - Mycologia 67: 1112, 1975 ≡ *Arthroderma borellii* (Moraes, Padhye & Ajello) Padhye, Weitzman, McGinnis & Ajello, *in* Weitzman, McGinnis, Padhye & Ajello - Mycotaxon 25: 513, 1986.

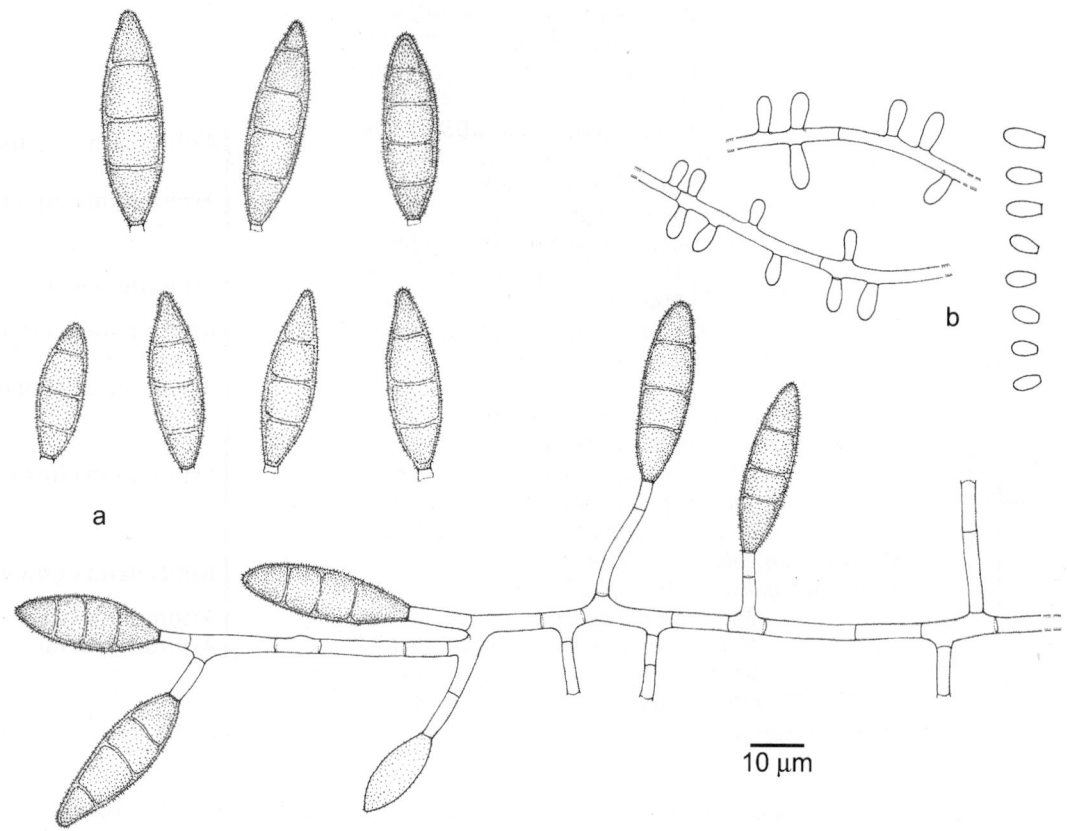

Microsporum amazonicum, CBS 967.68. a. Fertile hyphae and macroconidia; b. microconidia.

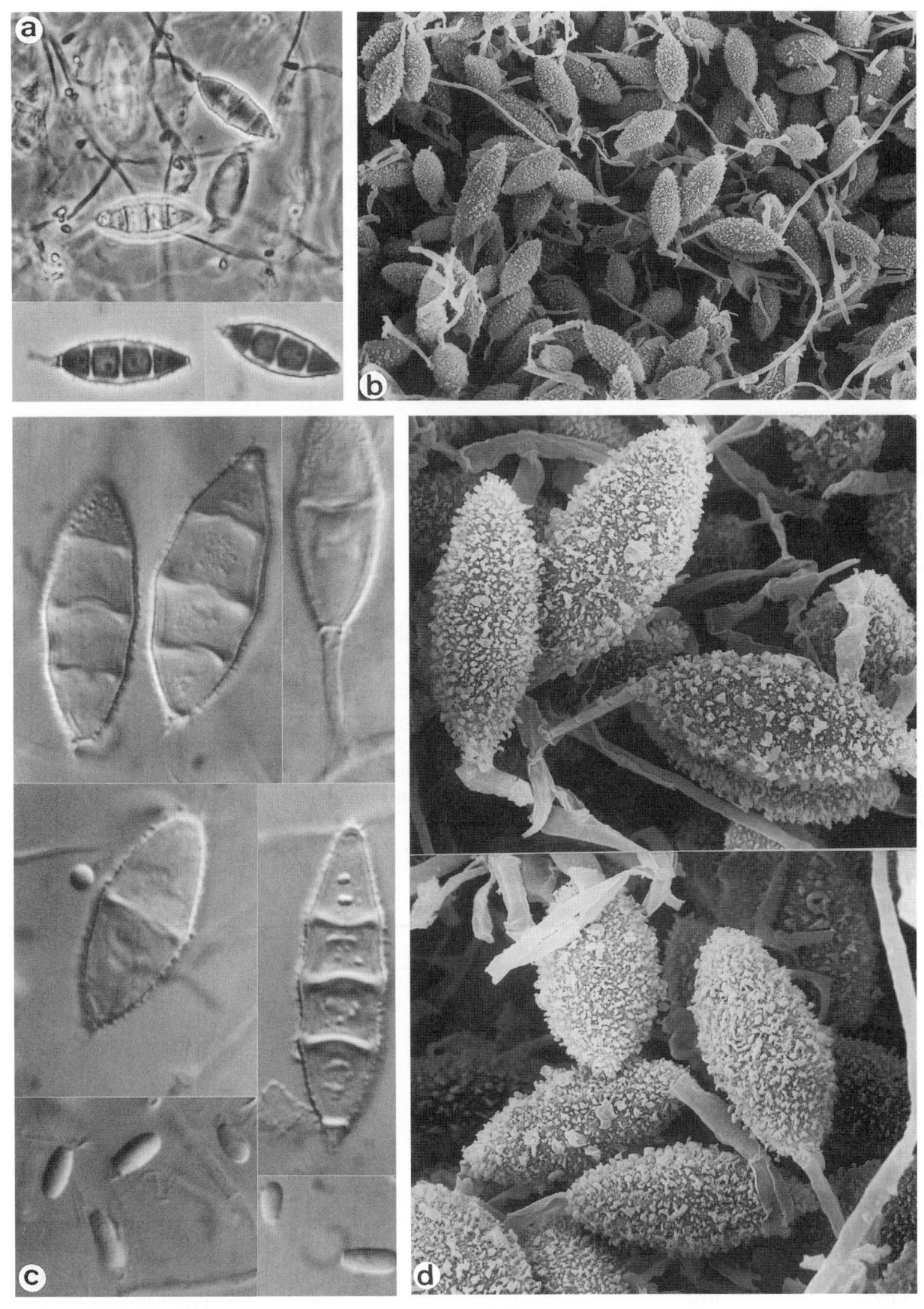

Microsporum amazonicum, CBS 967.68. Macro- and microconidia. a. ×640; b. ×600; c. ×1600; d. ×2200.

Microsporum audouinii Gruby

Colony characteristics. Colonies (SGA) woolly, flat, spreading, with radiating margin, greyish to tannish-white; reverse salmon-pink to rose-brown.

Microscopy. Pectinate hyphae and terminal chlamydospores may be present. Macroconidia rare, when present smooth-walled to sparsely echinulate, thick-walled, irregularly spindle-shaped, frequently somewhat isthmoid and rostrate, with constriction near the middle, of variable size and cell number, 30-82 × 8-14 μm, mostly with slightly bent, verrucose apex. Microconidia rare, ovoidal to clavate.

Physiology.

BCPCG	+	T-1	+	T-5	+
Urease	v	T-2	+	T-6	+
Hair perforation	–	T-3	+	T-7	+
Rice (growth/sporulation)	w/–	T-4	+		

Molecular diagnostics. ITS restriction map based on CBS 280.63:

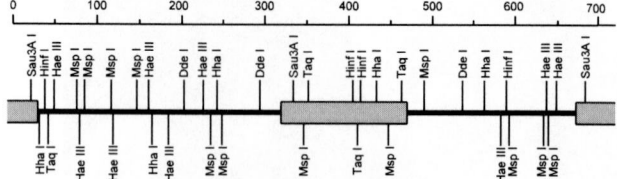

Differential diagnosis. *Microsporum audouinii* is closely related to *M. canis* (p. 743), differing physiologically by neither perforating hair *in vitro* nor sporulating on polished rice grains. The macroconidia resemble those of *M. canis* but are more irregularly shaped, mostly constricted near the middle.

Pathogenicity. BSL-2. Ectothrix with small cells, the hair usually brightly fluorescent when irradiated by UV light of Wood's lamp. The species causes human ringworm, primarily tinea capitis, more rarely tinea corporis. Occurs mainly in children and is less inflammatory and more chronic than *M. canis*. A recent case is that of Cabon *et al.* (1994) in a neonate originating from West Africa. Epidemics in France were reported by Viguie-Vallanet *et al.* (1997) and Weill *et al.* (1999).

Reference. Stockdale (1965), Whittle & Gresham (1970), Gräser *et al.* (1999b, 2000a).

Distribution. World-wide.

Antifungal susceptibility.

Antifungal	Mean MICs	MIC 90	Strains	Reference
AMB	0.13	0.125	8	Fernández *et al.* (2000b)
CTZ	0.04	0.125	8	Fernández *et al.* (2000b)
FLZ	8.65	16	8	Fernández *et al.* (2000b)
GRF	0.39	0.39	2	Wildfeuer *et al.* (1998)
ITZ	0.20	0.20	2	Wildfeuer *et al.* (1998)
ITZ	0.04	0.125	8	Fernández *et al.* (2000b)
KTZ	1.11	1.56	2	Wildfeuer *et al.* (1998)
KTZ	0.38	1	8	Fernández *et al.* (2000b)
MCZ	0.22	2	8	Fernández *et al.* (2000b)
STZ	0.41	2	8	Fernández *et al.* (2000b)
TBF	0.02	0.03	8	Fernández *et al.* (2000b)
VCZ	0.25	0.39	2	Wildfeuer *et al.* (1998)
VCZ	0.06	0.125	8	Fernández *et al.* (2000b)

Nomenclature. *Microsporum audouinii* Gruby - C.R. Hebd. Séanc. Acad. Sci., Paris 17: 301, 1843 ≡ *Sporotrichum audouinii* (Gruby) Saccardo - Syll. Fung. 4: 101, 1886 ≡ *Oidium microsporium* Kambayashi - Jpn. J. Derm. Urol. 21: 460, 1921 ≡ *Sabouraudites audouinii* (Gruby) Ota & Langeron - Annls Parasit. Hum. Comp. 1: 327, 1923 ≡ *Closteroaleurosporia audouinii* (Gruby) Grigorakis - Annls Sci. Nat., Bot., Sér. 10, 7: 412, 1925 ≡ *Veronaia audouinii* (Gruby) Benedek - Mycopath. Mycol. Appl. 14: 115, 1961.

Trichophyton decalvans Mamsten, *in* Creptin - Arch. Anat. Physiol. Wiss. Med. 1848: 19, 1848 ≡ *Trichomyces decalvans* (Malmsten) Malmsten - Arch. Anat. Physiol. Wiss. Med. 1848: 19, 1848.

Trichophyton microsporum Sabouraud - Annls Derm. Syph., Sér. 2, 3: 1061, 1892 ≡ *Martensella microspora* (Sabouraud) Vuillemin - Bull. Soc. Mycol. Fr. 11: 97, 1895.

Microsporum tomentosum Pelagatti, *in* Sabouraud - Malad. Cuir Chev. 3: 241, 1910.

Microsporum villosum Minne - Soc. Belge Derm. 7: 1906, 1907 ≡ *Sabouraudites villosus* (Minne) Lebasque - Champ. Teign. Cheval Bovidés p. 63, 1933.

Microsporum umbonatum Sabouraud - Annls Derm. Syph., Sér. 4, 8: 173, 1907 ≡ *Sabouraudites umbonatus* (Sabouraud) Ota & Langeron - Annls Parasit. Hum. Comp. 1: 328, 1923 ≡ *Closteroaleuriospora umbonata* (Sabouraud) Grigorakis - Annls Sci. Nat., Bot., Sér. 10, 7: 415, 1925.

Microsporum tardum Sabouraud - Malad. Cuir Chev. 3: 172, 1910 ≡ *Sabouraudia tardus* (Sabouraud) Ota & Langeron - Annls Parasit. Hum. Comp. 1: 328, 1923 ≡ *Closteroaleuriosporia tarda* (Sabouraud) Grigorakis - Annls Sci. Nat., Bot., Sér. 10, 7: 415, 1925 ≡ *Microsporum audouinii* Gruby var. *tardum* (Sabouraud) C.W. Dodge - Med. Mycol. p. 550, 1935.

Microsporum velveticum Sabouraud - Annls Derm. Syph., Sér. 4, 8: 178, 1907 ≡ *Aleurosporia velvetica* (Sabouraud) Grigorakis - Annls Sci. Nat., Bot., Sér. 10, 7: 413, 1925 ≡ *Microsporum audouinii* Gruby var. *velveticum* (Sabouraud) C.W. Dodge - Med. Mycol. p. 550, 1935.

Microsporum depauperatum Guéguen - Arch. Parasit. 14: 426 - 446, 1911 ≡ *Sabouraudites depauperatus* (Guéguen) Ota & Langeron - Annls Parasit. Hum. Comp. 1: 329, 1923 ≡ *Microsporum audouinii* Gruby var. *depauperatum* (Guéguen) C.W. Dodge - Med. Mycol. p. 551, 1935.

Sabouraudites langeronii Vanbreuseghem - Annls Parasit. Hum. Comp. 25: 516, 1950 ≡ *Microsporum langeronii* (Vanbreuseghem) Ciferri - Man. Micol. Med., ed. 2, 2: 416, 1960 ≡ *Microsporum audouinii* Gruby var. *langeronii* (Vanbreuseghem) Kane, Summerbell, Sigler, Krajden & Land - Lab. Handb. Dermatophytes p. 200, 1997.

Sabouraudites rivalieri Vanbreuseghem - Archs Belg. Derm. Syph. 7: 5, 1951 (invalid) ≡ *Microsporum rivalieri* Vanbreuseghem - Sabouraudia 2: 220, 1963 ≡ *Microsporum audouinii* Gruby var. *rivalieri* (Vanbreuseghem) Whittle & Gresham - Sabouraudia 8: 70, 1970.

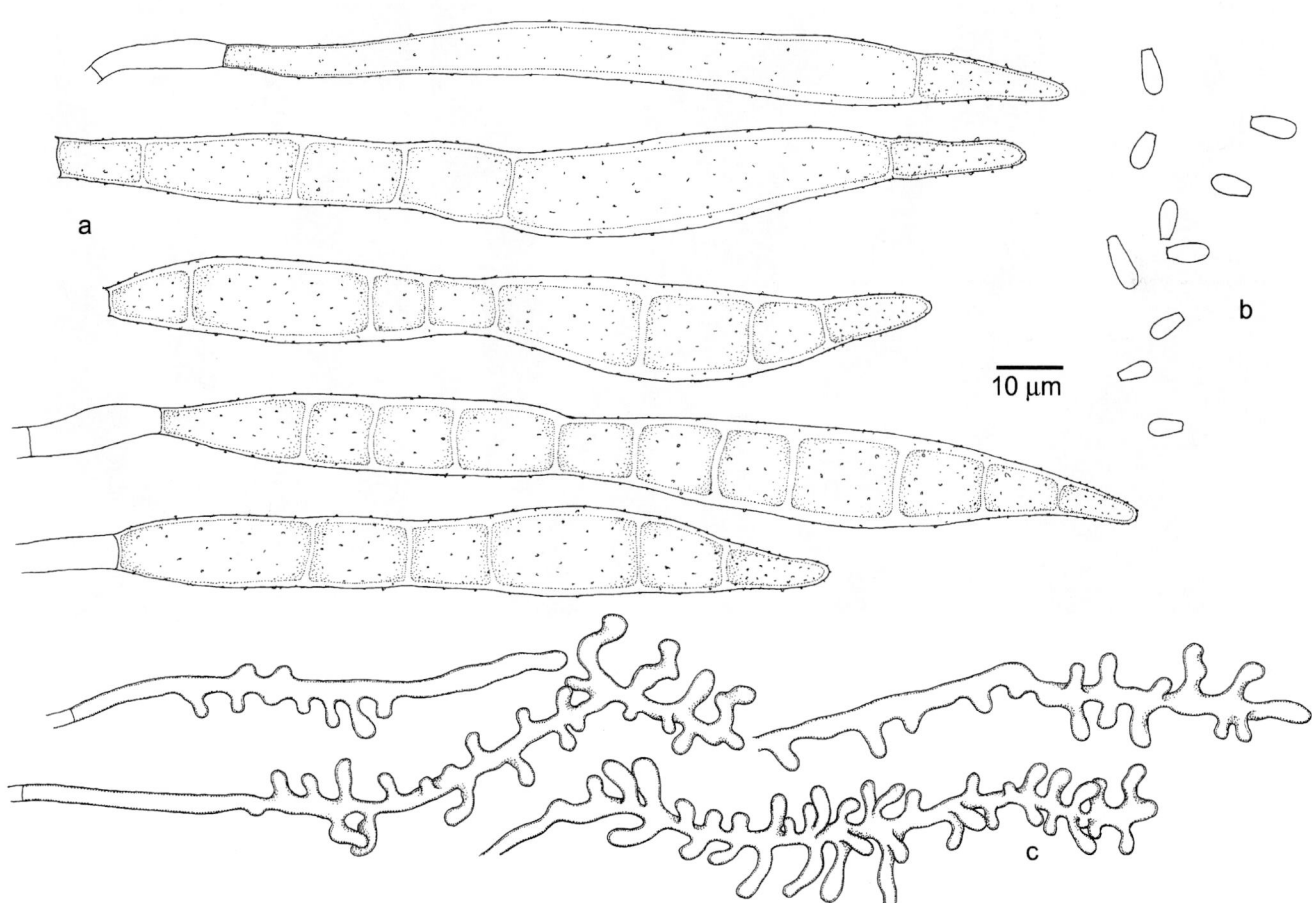

Microsporum audouinii, CBS 545.93. a. Macroconidia; b. microconidia; c. pectinate hyphae.

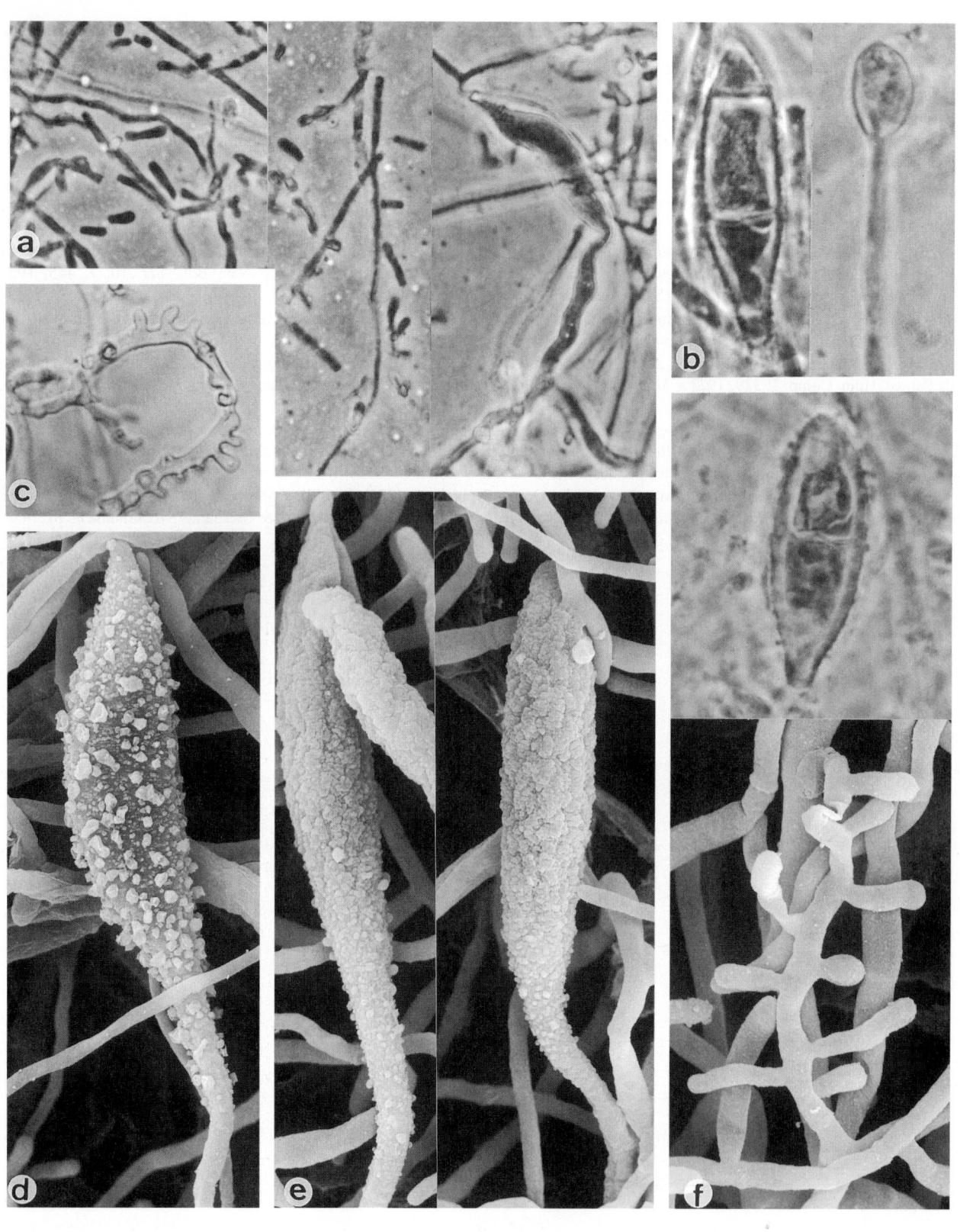

Microsporum audouinii, a, b, d-f. CBS 545.93; c. 99-053. a. Liberated microconidia and young macroconidium; b. liberated macroconidia and terminal chlamydospore; c. pectinate hypha; d-f. macro- and microconidia. a. ×640; b. ×1600; c. ×640; d. ×2050; e. ×2550; f. ×3050

Microsporum canis (Bodin) Bodin

Colony characteristics. Colonies (SGA) spreading, thin, woolly, strongly radiating, greyish- to tannish-white; reverse deep ochraceous-yellow.

Microscopy. Macroconidia 6-12-celled, rough-walled, with thick cell walls and thinner septa, 35-110 × 12-25 μm, spindle-shaped, with slightly bent, verrucose, rostrate apex. Microconidia clavate to pyriform, sessile alongside undifferentiated hyphae.

Teleomorph. *Arthroderma otae* (Hasegawa & Usui) McGinnis *et al.* (*Ascomycota, Euascomycetes, Onygenales: Arthrodermataceae*).

Physiology.

BCPCG	–	Rice (growth/sporulation)	+/+	T-2	++	T-5	++
Urease	+,w	Benomyl tolerance	–	T-3	++	T-6	++
Hair perforation	+	T-1		++	T-4	++	T-7	++

Molecular diagnostics. ITS restriction map based on CBS 496.86:

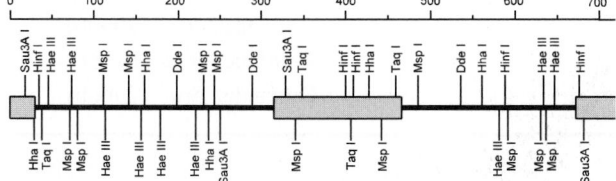

Differential diagnosis. Several *Microsporum* species, viz. *M. audouinii*, *M. canis* and *M. ferrugineum* are all closely related (Kawasaki *et al.*, 1995; Gräser *et al.*, 2000a), showing small-celled ectothrix, with bright fluorescence of affected hairs when irradiated by UV light of Wood's lamp. *M. distortum* is a dysgonic variant of *M. canis*.

Pathogenicity. BSL-2. Ectothrix with small cells. Fluorescence with Wood's lamp positive but has poor diagnostic value (Kefalidou *et al.*, 1997). The species causes ringworm in cats, dogs and monkeys, occasionally in other animals (Hörmansdorfer *et al.*, 1995). Quite frequently it resides in fur without causing symptoms (Sparkes *et al.*, 1994), but when transmitted to man, it leads to tinea capitis and tinea corporis, particularly children, sometimes reaching epizootic proportions (Brasch, 1989). Onychomycosis is rare (Piraccini *et al.*, 1996). Creach *et al.* (1995) presented a case involving the eyelashes and Margolis *et al.* (1998) of the vulva. In AIDS patients lesions may be extensive (Hevia *et al.*, 1991), whereas in patients undergoing immunosuppressive therapy invasion of the dermis is noted (King *et al.*, 1996; Virgili & Zampino, 1998; Voisard *et al.*, 1999).

Distribution. World-wide.

References. Klokke & de Vries (1963), Stockdale (1965), Matsumoto *et al.* (1983b), Butty *et al.* (1992), Vismer (1994), Gräser *et al.* (2000a).

Nomenclature. (?) *Microsporum audouinii* Gruby var. *equinum* Delacroix & Bodin, *in* Bodin - Teign. Tond. Cheval. p. 41, 1896 ≡ *Microsporum equinum* (Delacroix & Bodin) Guéguen - Champ. Paras. Homme Anim. p. 144, 1904 ≡ *Sabouraudites equinus* (Delacroix & Bodin) Ota & Langeron - Annls Parasit. Hum. Comp. 1: 329, 1923.

Microsporum audouinii Gruby var. *canis* Bodin, *in* Besnier, Brocq & Jacquet - Prat. Derm. p. 810, 1900 ≡ *Microsporum canis* (Bodin) Bodin - Champ. Paras. Homme p. 137, 1902 ≡ *Sabouraudites canis* (Bodin) Langeron - Précis Mycol. p. 534, 1945.

Microsporum lanosum Sabouraud - Annls Derm. Syph., Sér. 4, 8: 122, 1907 ≡ *Sabouraudites lanosus* (Sabouraud) Ota & Langeron - Annls Parasit. Hum. Comp. 1: 329, 1923 ≡ *Closterosporia lanosa* (Sabouraud) Grigorakis - Annls Sci. Nat., Bot., Sér. 10,7: 415, 1925.

Microsporum caninum Sabouraud - Annls Derm. Syph., Sér. 4, 9: 153, 1908.

Microsporum obesum Conant - Archs Derm. Syph. 36: 800, 1937.

Microsporum pseudolanosum Conant - Archs Derm. Syph. 36: 800, 1937.

Microsporum simiae Conant - Archs Derm. Syph. 36: 801, 1937.

Microsporum stilliansi Benedek - J. Trop. Med. Hyg. 41: 114, 1938.

Microsporum distortum di Menna & Marples - Trans. Br. Mycol. Soc. 37: 372, 1954 ≡ *Microsporum canis* Bodin var. *distortum* (di Menna & Marples) Matsumoto, Padhye & Ajello - Trans. Br. Mycol. Soc. 81: 649, 1983.

Nannizzia otae Hasegawa & Usui - Jpn. J. Med. Mycol. 16: 151, 1975 ≡ *Arthroderma otae* (Hasegawa & Usui) McGinnis, Weitzman, Padhye & Ajello, *in* Weitzman, McGinnis, Padhye & Ajello - Mycotaxon 25: 514, 1986.

Microsporum canis Bodin var. *album* van Cutsem, van Gerven, Geerts & Rochette - Mykosen 28: 401, 1985.

Antifungal susceptibility.

Antifungal	Mean MICs	MIC 90	Strains	Reference
AMB	0.29	1	34	Fernández *et al.* (2000b)
CLZ	0.09	0.125	34	Fernández *et al.* (2000b)
FLZ	nd	2	8	Jessup *et al.* (2000)
FLZ	4.58	≥16	34	Fernández *et al.* (2000b)
GRF	0.26	1.56	24	Wildfeuer *et al.* (1998)
GRF	nd	1	8	Jessup *et al.* (2000)
ITZ	0.05	0.20	24	Wildfeuer *et al.* (1998)
ITZ	nd	0.06	8	Jessup *et al.* (2000)
ITZ	0.05	0.125	34	Fernández *et al.* (2000b)
KTZ	0.22	0.78	24	Wildfeuer *et al.* (1998)
KTZ	0.18	0.5	34	Fernández *et al.* (2000b)
MCZ	0.08	0.25	34	Fernández *et al.* (2000b)
SCZ	0.22	0.5	34	Fernández *et al.* (2000b)
TBF	nd	0.03	8	Jessup *et al.* (2000)
TBF	0.02	0.06	34	Fernández *et al.* (2000b)
VCZ	0.12	0.39	24	Wildfeuer *et al.* (1998)
VCZ	0.04	0.125	34	Fernández *et al.* (2000b)

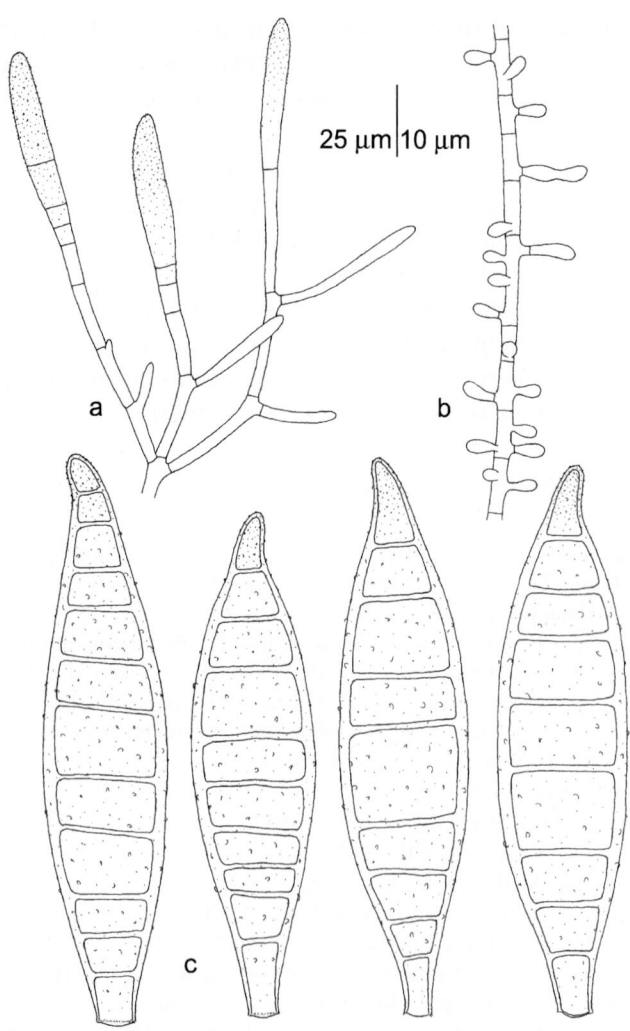

Microsporum canis, CBS 132.88. a. Conidial apparatus; b. microconidia; c. macroconidia.

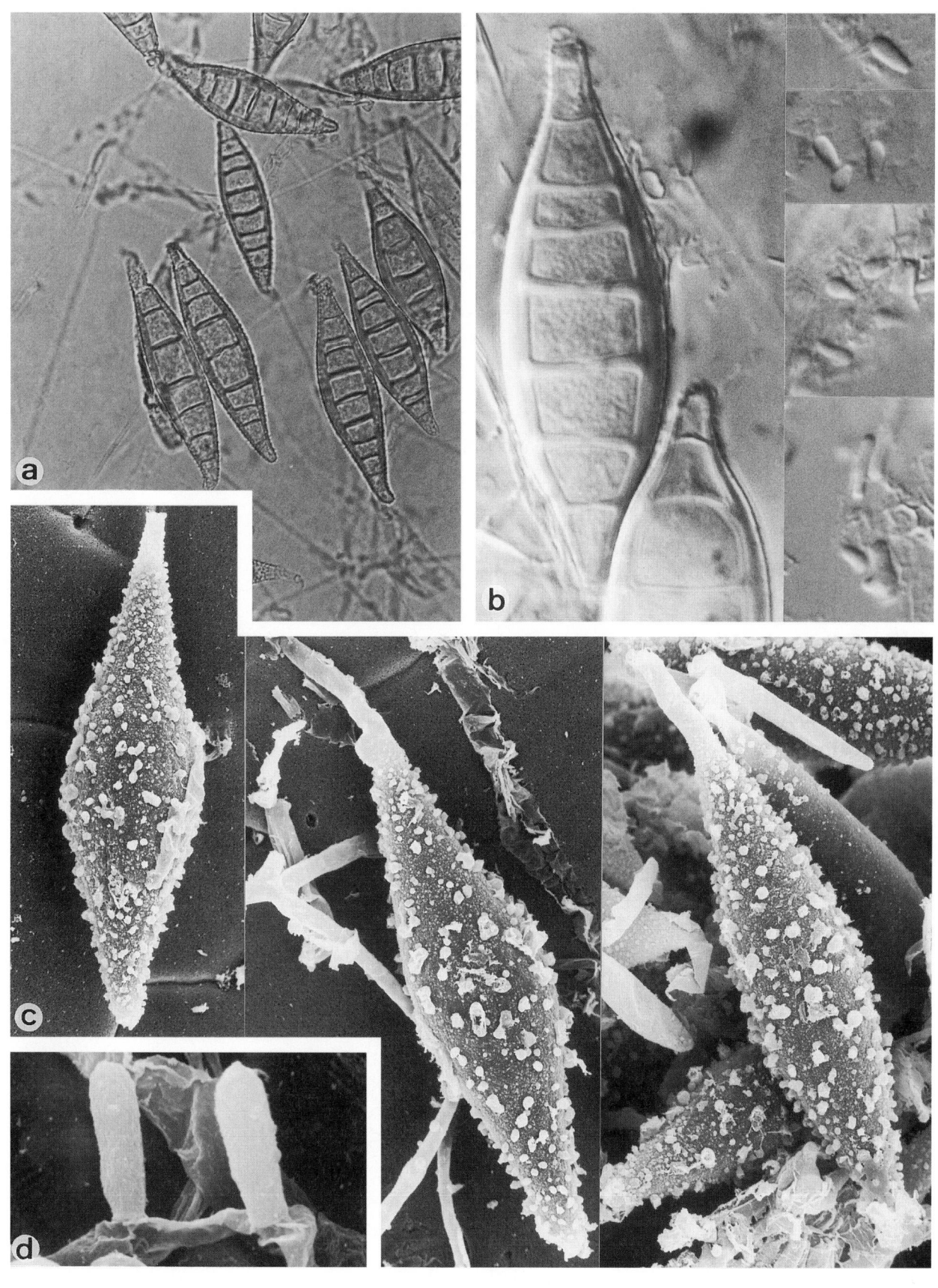

Microsporum canis, CBS132.88. Macro- and microconidia. a. ×512; b. ×1600; c. ×1350; d. ×8000.

Microsporum cookei Ajello

Colony characteristics. Colonies (SGA) spreading, powdery, becoming yellowish, greenish-buff, or dark brown. A deep grape-red pigment is exuded into the medium.

Microscopy. Macroconidia thick- and rough-walled, 6-7 (-10)-celled, broadly fusiform with rounded apex, 30-50 × 10-15 μm. Microconidia ovoidal to pyriform.

Teleomorph. *Arthroderma cajetani* (Ajello) Ajello *et al.* (*Ascomycota, Euascomycetes, Onygenales: Arthrodermataceae*).

Physiology.

BCPCG	–	T-1	+	T-4	+	T-7	+
Urease	+	T-2	+	T-5	+		
Hair perforation	+	T-3	+	T-6	+		

Molecular diagnostics. ITS restriction map based on CBS 228.58:

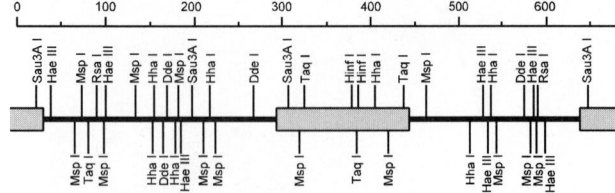

Differential diagnosis. *Microsporum racemosum* (p. 765) has longer, more slender macroconidia and the microconidia are produced in clusters.

Pathogenicity. BSL-1. The species is geophilic and has been reported from dogs, monkeys and squirrels (Caffara & Scagliarini, 1999). Superficial skin lesions in humans were reported by Schick (1966), Balogh *et al.* (1967), Lundell (1969) and Frey (1971).

Distribution. World-wide.

References. Ajello (1959, 1961).

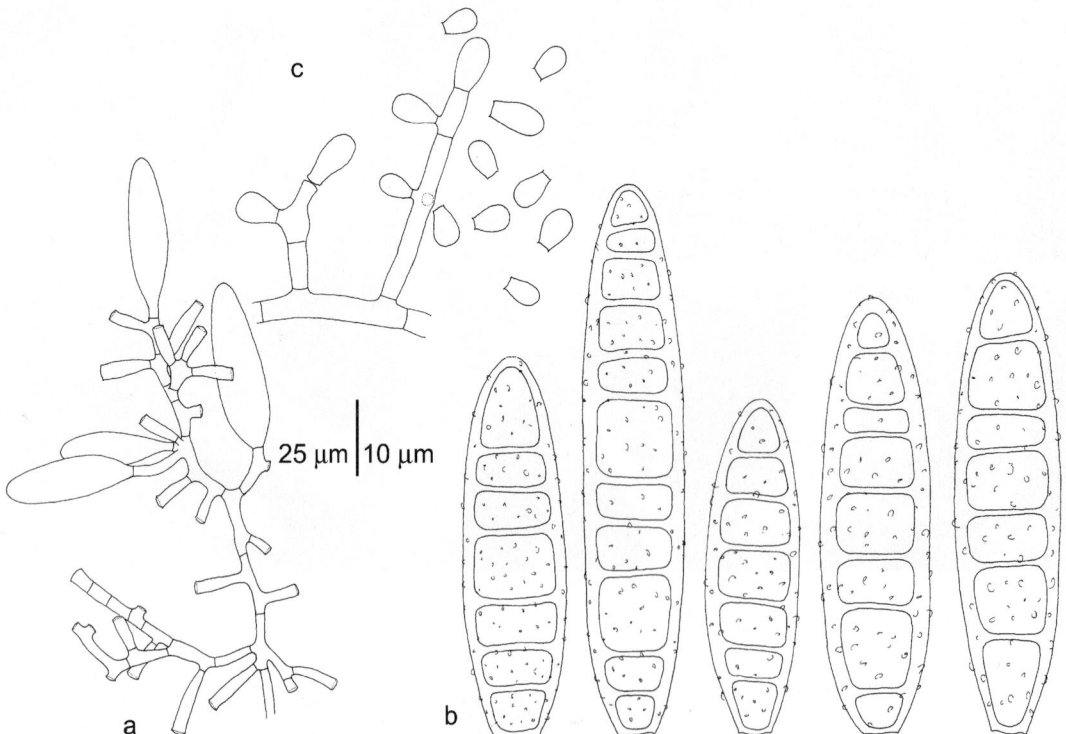

Microsporum cookei, CBS 495.70. a. Conidial apparatus; b. macroconidia; c. microconidia.

Antifungal susceptibility.

Antifungal	Mean MICs	Strains	Reference
AMB	>16	1	Fernández *et al.* (2000b)
CTZ	>16	1	Fernández *et al.* (2000b)
FLZ	4	1	Fernández *et al.* (2000b)
ITZ	0.03	1	Fernández *et al.* (2000b)
KTZ	8	1	Fernández *et al.* (2000b)
MCZ	>16	1	Fernández *et al.* (2000b)
TBF	>16	1	Fernández *et al.* (2000b)
VCZ	0.01	1	Fernández *et al.* (2000b)

Nomenclature. *Microsporum cookei* Ajello - Mycologia 51: 71, 1959.

Nannizzia cajetani Ajello - Sabouraudia 1: 175, 1961 ≡ *Arthroderma cajetani* (Ajello) Ajello, Weitzman, McGinnis & Padhye, *in* Weitzman, McGinnis, Padhye & Ajello - Mycotaxon 25: 513, 1986.

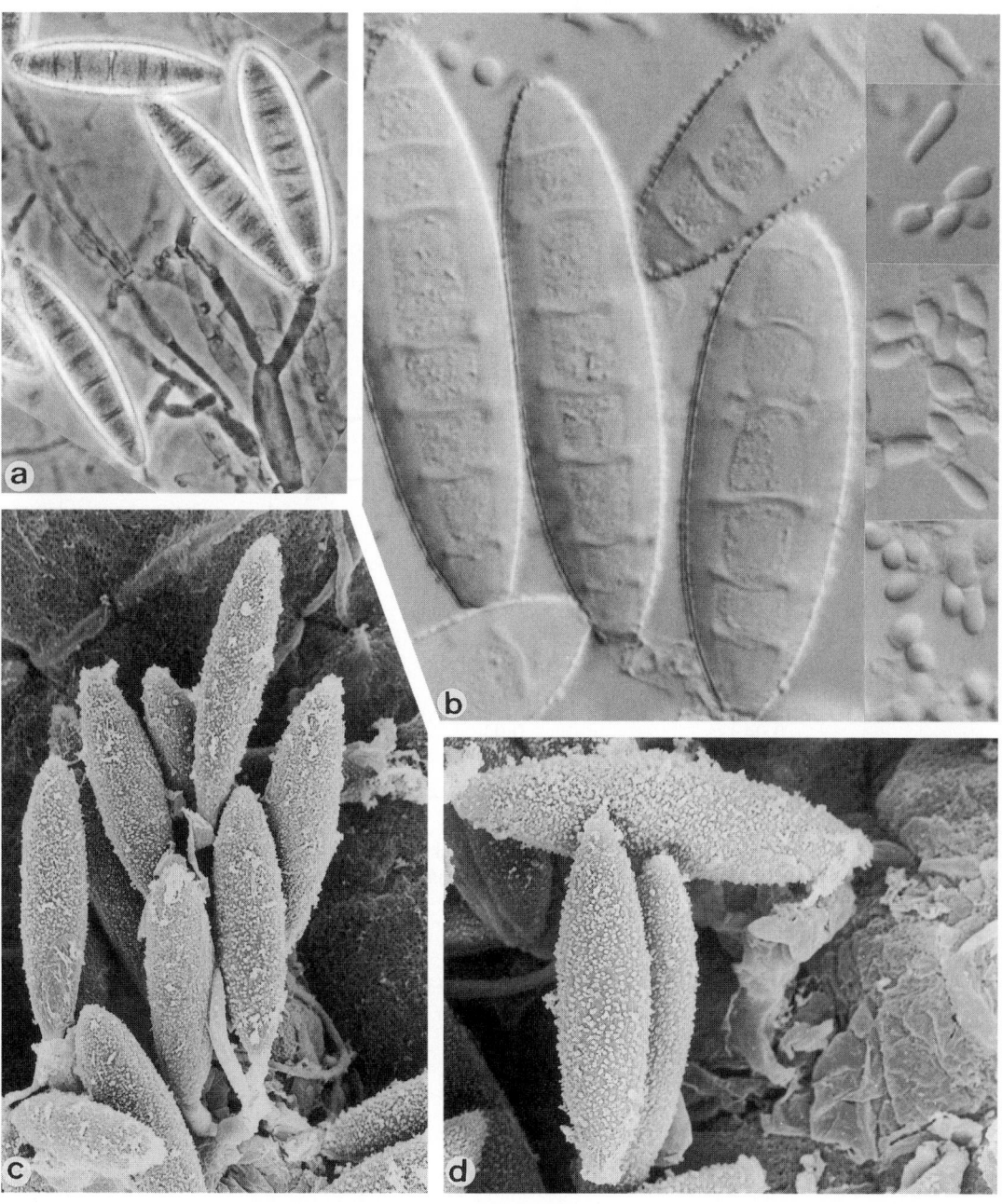

Microsporum cookei, CBS 495.70. Macro- and microconidia. a. ×640; b. ×1600; c. ×1000; d. ×1300.

Microsporum ferrugineum Ota

Colony characteristics. Colonies (SGA) glabrous, heaped, wrinkled, sometimes flat, yellow to cream-coloured; reverse cream-coloured to yellow.

Microscopy. Conidia usually absent. Long, straight, broad hyphae with prominent cross walls (bamboo hyphae) often present, showing a tendency to disarticulate. Spindle-shaped macroconidia similar to those of *M. canis* may be produced on dilute Sabouraud dextrose agar. The species is probably a sterile variant of *M. canis*.

Physiology.

BCPCG	–	Rice (growth/sporulation) w/–	T-3 +	T-6 +
Urease	–	T-1 +	T-4 ++	T-7 +
Hair perforation	–	T-2 +	T-5 +	

Molecular diagnostics. ITS restriction map based on AJ252335:

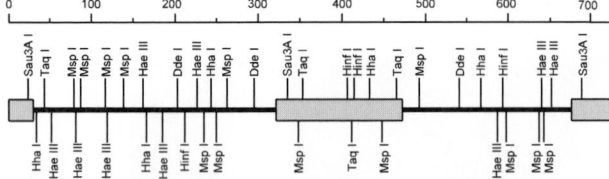

Pathogenicity. BSL-2. Ectothrix with small cells in chains; hairs fluorescent when irradiated by UV light of Wood's lamp. Causes human tinea capitis (Valencia Leon & Tio Polledo, 1989), primarily in children (Wisuthsarewong *et al.*, 1996). Chen *et al.* (1993) reported a subcutaneous infection.

Distribution. Asia, Eastern Europe and Africa. A slow growing type with heaped, yellow colonies is mostly isolated in Asia, while a faster growing type with flat, white colonies is prevalent in the Balkan (Kwon-Chung & Bennett, 1992).

Reference. Stockdale (1965).

Antifungal susceptibility.

Antifungal	Mean MICs	MIC90	Strains	Reference
AMB	0.14	0.125	4	Fernández *et al.* (2000b)
CTZ	0.05	0.03	4	Fernández *et al.* (2000b)
FLZ	4.73	4	4	Fernández *et al.* (2000b)
ITZ	0.06	0.125	4	Fernández *et al.* (2000b)
KTZ	0.04	0.03	4	Fernández *et al.* (2000b)
MCZ	0.06	nd	4	Fernández *et al.* (2000b)
STZ	0.12	0.125	4	Fernández *et al.* (2000b)
TBF	0.04	0.03	4	Fernández *et al.* (2000b)
VCZ	0.05	0.06	4	Fernández *et al.* (2000b)

Nomenclature. *Microsporum ferrugineum* Ota - Jpn. J. Derm. Urol. 21: 201, 1921 ≡ *Grubyella ferruginea* (Ota) Ota & Langeron - Annls Parasit. Hum. Comp. 1: 330, 1923 ≡ *Arthrosporia ferruginea* (Ota) Grigorakis - Annls Sci. Nat., Bot., Sér. 10, 7: 414, 1925 ≡ *Achorion ferrugineum* (Ota) Guiart & Grigorakis - Lyon Méd. 141: 377, 1928 ≡ *Trichophyton ferrugineum* (Ota) Talice - Annls Parasit. Hum. Comp. 9: 83, 1931.

Microporum aureum Takeya - Tohuko J. Exp. Med. 6: 93, 1925.

Oidium microsporium Kambayashi var. *japonicum* Kambayashi - Jpn. J. Derm. Urol. 21: 433, 1921 ≡ *Microsporum japonicum* (Kambayashi) Dohi & Kambayashi, *in* Kambayashi - Jpn. J. Derm. Urol. 21: 433, 1921.

Microsporum orientale Carol - Urol. Cutan. Rev. 32: 23, 1928.

Microsporum fulvum Uriburu

Colony characteristics. Colonies (SGA) spreading, flat, farinose to floccose, buff to pink buff; reverse red.

Microscopy. Macroconidia thin- or rather thick-walled, echinulate, broadly fusiform to clavate, (4-) 5-6 (-7)-celled, 25-60 × 7-12 μm. Microconidia sessile or short-stalked, clavate. Spiral hyphae often present. Rarely mutants occur with non-maturating conidia in strongly coherent, christmas tree-shaped clusters (*'T. longifusum'*).

Teleomorph. *Arthroderma fulvum* (Stockdale) Weitzman *et al.* (*Ascomycota, Euascomycetes, Onygenales: Arthrodermataceae*).

Physiology.

BCPCG	–	T-1	+	T-4	+	T-7	+	
Urease	+	T-2	+	T-5	+			
Hair perforation	+	T-3	+	T-6	+			

Molecular diagnostics. ITS restriction map based on CBS 287.55:

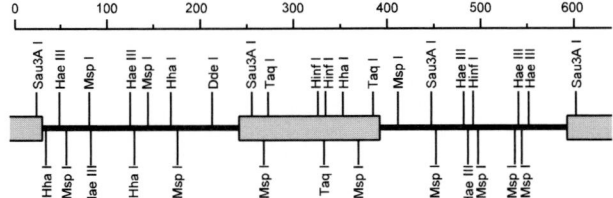

Differential diagnosis. Very similar to *M. gypseum* (Kawasaki *et al.*, 1990). *Microsporum boullardii* was proven to be a synonym (Gräser *et al.*, 1999b) but the purported teleomorph of that species, *Arthroderma corniculatum*, is a separate taxon.

Pathogenicity. BSL-1. Sparse ectothrix. Geophilic, rarely infecting man and animals.

Distribution. World-wide.

Reference. Stockdale (1963, 1965), Flórián & Galgóczy (1964), Dominik & Majchrowicz (1965).

Antifungal susceptibility.

Antifungal	Mean MICs	MIC 90	Strains	Reference
AMB	0.03	nd	1	Fernández *et al.* (2000b)
CTZ	0.06	nd	1	Fernández *et al.* (2000b)
FLZ	16	nd	1	Fernández *et al.* (2000b)
GRF	0.39	0.78	5	Wildfeuer *et al.* (1998)
ITZ	0.04	0.10	5	Wildfeuer *et al.* (1998)
ITZ	0.125	1	1	Fernández *et al.* (2000b)
KTZ	0.45	0.78	5	Wildfeuer *et al.* (1998)
KTZ	2	nd	1	Fernández *et al.* (2000b)
MCZ	0.5	nd	1	Fernández *et al.* (2000b)
TBF	0.03	nd	1	Fernández *et al.* (2000b)
VCZ	0.20	0.39	5	Wildfeuer *et al.* (1998)
VCZ	0.01	nd	1	Fernández *et al.* (2000b)

Nomenclature. *Microsporum fulvum* Uriburu - Argent. Med. 7, 1909 ≡ *Sabouraudites fulvus* (Uriburu) Ota & Langeron - Annls Parasit. Hum. Comp. 1: 329, 1923 ≡ *Closterosporia fulva* (Uriburu) Grigorakis - Annls Sci. Nat., Bot., Sér. 10, 7: 411, 1925.

Nannizzia fulva Stockdale - Sabouraudia 3: 120, 1963 ≡ *Nannizzia gypsea* (Nannizzi) Stockdale var. *fulva* (Stockdale) Apinis - Mycol. Pap. 96: 33, 1964 ≡ *Arthroderma fulvum* (Stockdale) Weitzman, McGinnis, Padhye & Ajello - Mycotaxon 25: 513, 1986.

Epidermophyton radiosulcatum Szathmáry var. *flavum* Szathmáry - Proc. Hung. Derm. Soc. 5: 85, 1941.

Keratinomyces longifusus Flórián & Galgóczy - Mycopath. Mycol. Appl. 24: 76, 1964 ≡ *Trichophyton longifusum* (Flórián & Galgóczy) Novák & Galgóczy - Acta Bot. Hung. 15: 130, 1969.

Microsporum boullardii Dominik & Majchrowicz - Ekol. Polska, Ser. A, 13: 426, 1965.

Microsporum ripariae Hubálek & Rush-Munro - Sabouraudia 11: 288, 1973.

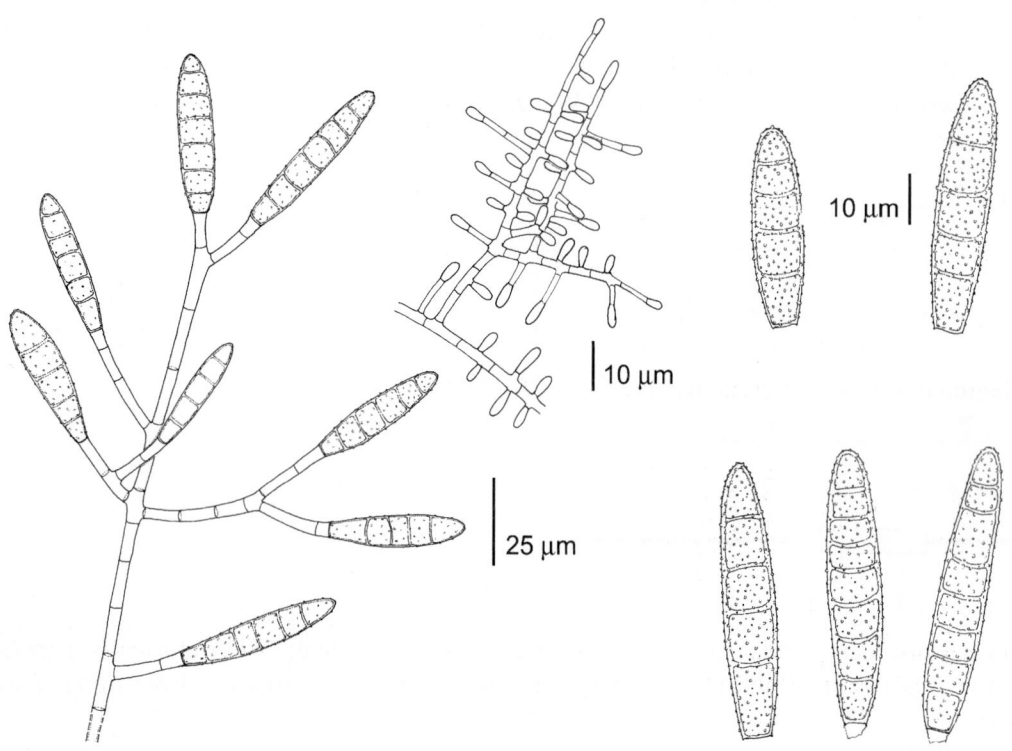

Microsporum fulvum, CBS 599.66. a. Fertile hyphae and macroconidia; b. fertile hyphae with microconidia; c. liberated conidia.

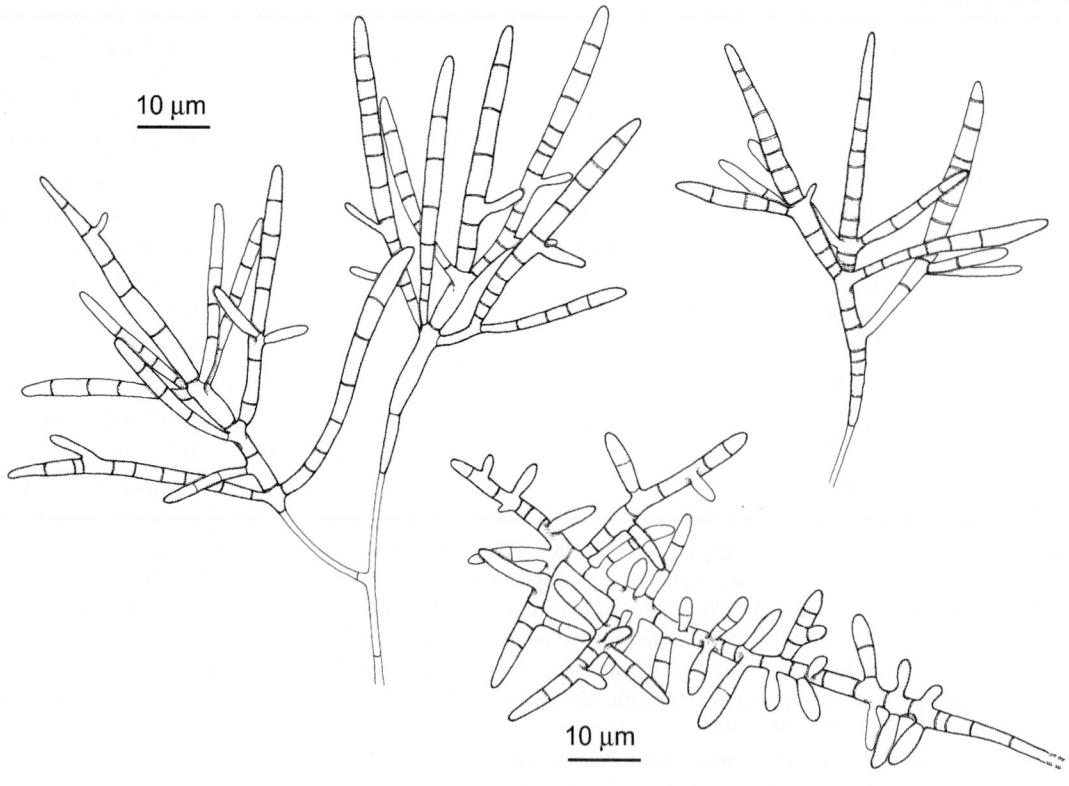

Microsporum fulvum, CBS 243.64. Mutant morphology showing christmas tree-like conidial clusters.

750

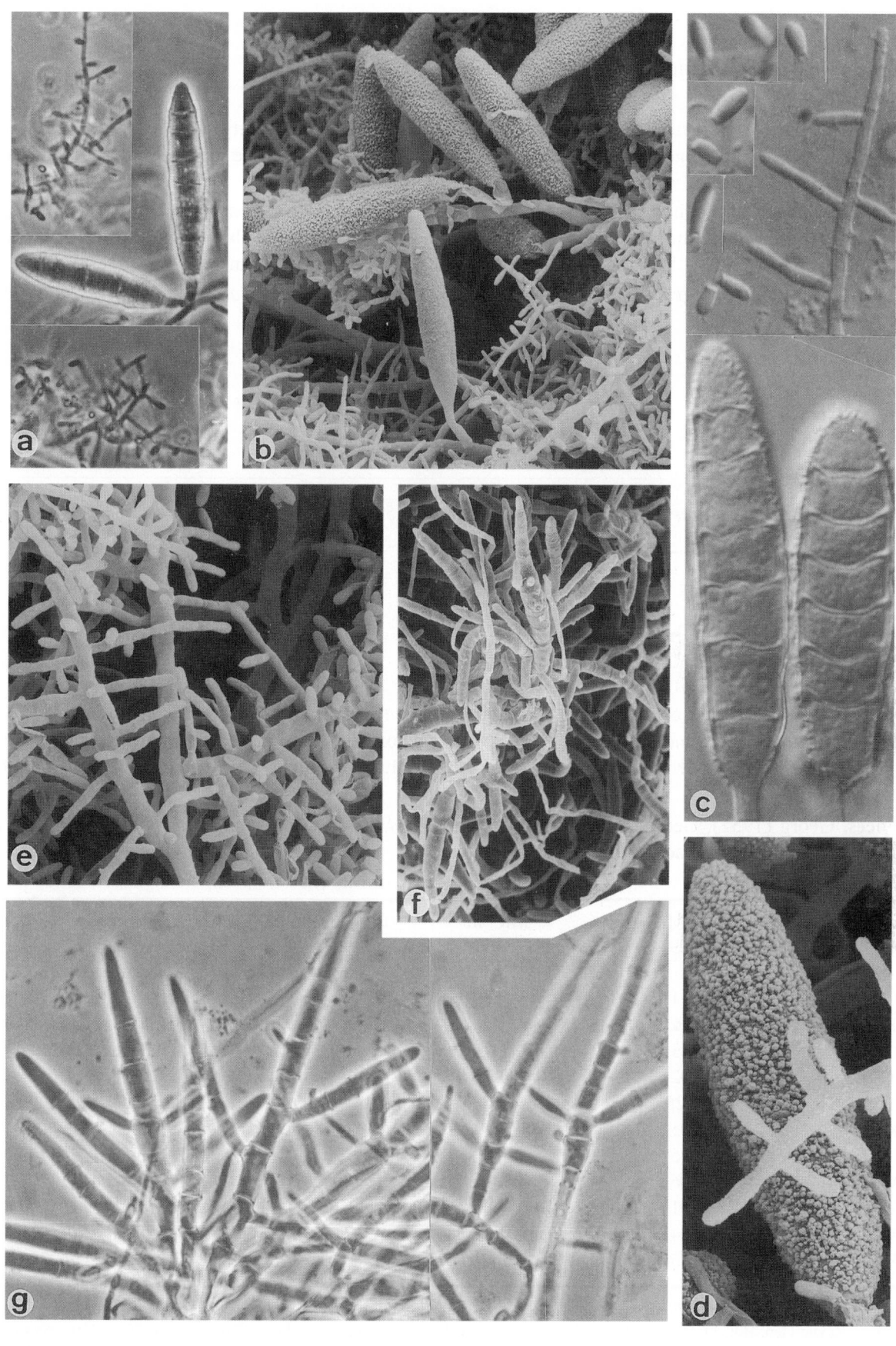

Microsporum fulvum, a-f. CBS 599.66; g. CBS 243.64. a-f. Fertile hyphae bearing conidia, and liberated macro- and microconidia; g. mutant showing coherent macroconidial system in christmastree-like arrangement. a. ×512; b. ×750; c. ×1600; d. ×2300; e. ×1550; f. ×550; g. × 640.

Microsporum gallinae (Mégnin) Grigorakis

Colony characteristics. Colonies (SGA) moderately fast growing, granular, velvety or satin, more or less wrinkled, white with pinkish or buff tinges. Reverse initially with a yellow, non-diffusible pigment, later a strawberry-red pigment diffuses into the agar.

Microscopy. Macroconidia, when present, arranged in unilateral clusters on pectinate hyphae, 2-12-celled (usually 5 to 6), thin- or thick-walled, smooth-walled to slightly echinulate, cylindrical to clavate with narrow base and blunt tip, sometimes slightly curved, 15-60 × 6-10 μm. Microconidia ovoidal to pyriform.

Teleomorph. *Arthroderma grubyi* (Georg *et al.*) Ajello *et al.* (*Ascomycota, Euascomycetes, Onygenales: Arthrodermataceae*).

Physiology.

Urease	w	T-1	+	T-4	++	T-7	++
Hair perforation	+	T-2	++	T-5	++		
		T-3	++	T-6	++		

Molecular diagnostics. ITS restriction map based on CBS 300.52:

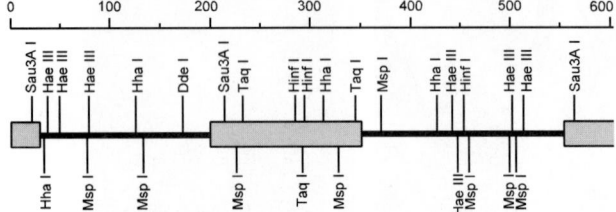

Pathogenicity. BSL-2. Sparse ectothrix with large cells. Causes ringworm in fowl, mainly on the comb and wattles, but infections in cats, dogs and squirrels have also been reported. Occasionally transmitted to man (Torres & Georg, 1956). A widespread dermatomycosis was reported in an AIDS patient (del Palacio-Hernanz *et al.*, 1992).

Distribution. World-wide.

References. Georg (1952), Georg *et al.* (1962), Londero *et al.* (1964), Morganti *et al.* (1975), Badillet (1982), Wawrkiewicz *et al.* (1987), Bradley *et al.* (1993).

Antifungal susceptibility.

Antifungal	Mean MICs	MIC 90	Strains	Reference
AMB	2	nd	1	Fernández *et al.* (2000b)
CTZ	0.01	nd	1	Fernández *et al.* (2000b)
FLZ	>16	nd	1	Fernández *et al.* (2000b)
GRF	7.43	12.5	4	Wildfeuer *et al.* (1998)
ITZ	0.02	0.03	4	Wildfeuer *et al.* (1998)
ITZ	0.125	nd	1	Fernández *et al.* (2000b)
KTZ	0.05	0.10	4	Wildfeuer *et al.* (1998)
KTZ	0.25	nd	1	Fernández *et al.* (2000b)
MCZ	0.06	nd	1	Fernández *et al.* (2000b)
TBF	0.01	nd	1	Fernández *et al.* (2000b)
VCZ	0.04	0.05	4	Wildfeuer *et al.* (1998)
VCZ	0.25	nd	1	Fernández *et al.* (2000b)

Nomenclature. *Epidermophyton gallinae* Mégnin - C.R. Soc. Biol. 33: 404, 1881 ≡ *Lophophyton gallinae* (Mégnin) Matruchot & Dassonville - Revue Gén. Bot. 11: 429, 1899 ≡ *Achorion gallinae* (Mégnin) Sabouraud - Malad. Cuir Chev. 3: 553, 1910 ≡ *Sabouraudites gallinae* (Mégnin) Ota & Langeron - Annls Parasit. Hum. Comp. 1: 327, 1923 ≡ *Closteroaleuriospora gallinae* (Mégnin) Grigorakis - Annls. Sci. Nat., Bot., Sér. 10, 7: 412, 1925 ≡ *Microsporum gallinae* (Mégnin) Grigorakis - Annls Derm. Syph., Sér. 6, 10: 42, 1929 ≡ *Trichophyton gallinae* (Mégnin) Georg - Mycologia 44: 486, 1952.

Microsporum vanbreuseghemii Georg, Ajello, Friedman & Brinkman - Sabouraudia 1: 191, 1962.

Nannizzia grubyi Georg, Ajello, Friedman & Brinkman - Sabouraudia 1: 194, 1962 ≡ *Arthroderma grubyi* (Georg, Ajello, Friedman & Brinkman) Ajello, Weitzman, McGinnis & Padhye, *in* Weitzman, McGinnis, Padhye & Ajello - Mycotaxon 25: 513, 1986.

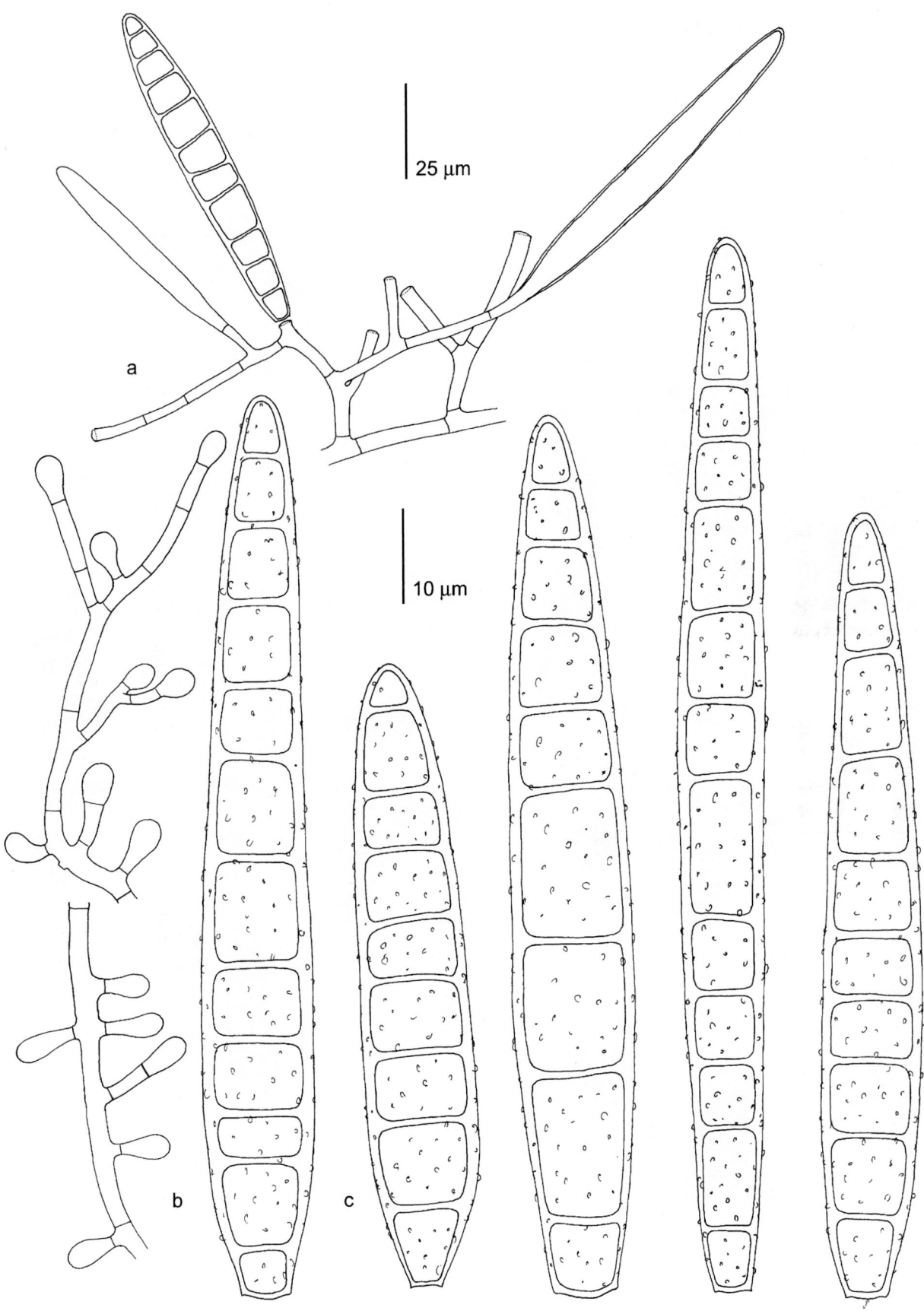

Microsporum gallinae, CBS 244.66. a. Conidial apparatus; b. microconidia; c. macroconidia.

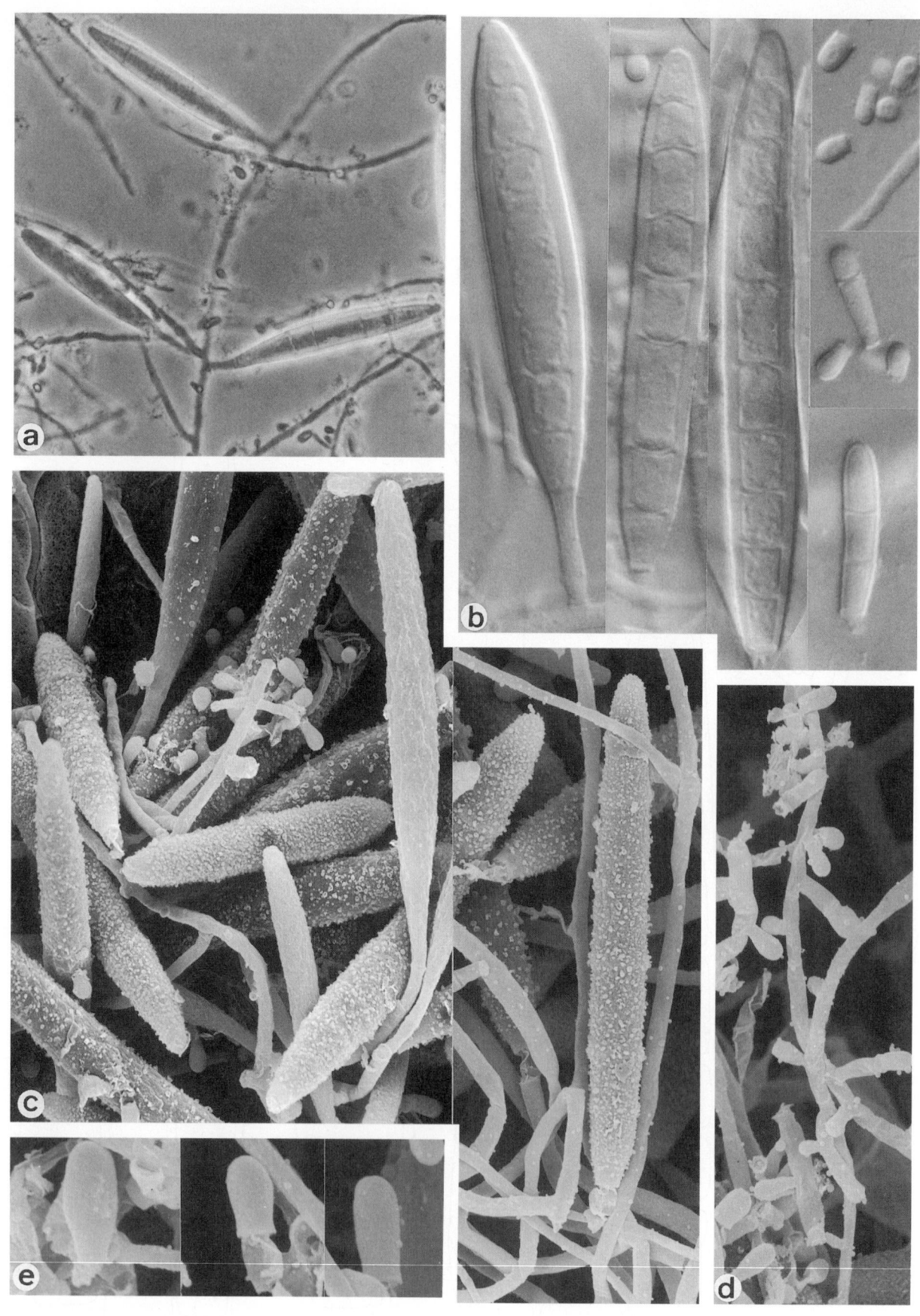

Microsporum gallinae, CBS 243.66. Macro- and microconidia. a.× 640; b. ×1600; c. ×1700; d. ×2050; e. ×3750.

Microsporum gypseum (Bodin) Guiart & Grigorakis

Colony characteristics. Colonies (SGA) growing rapidly, powdery, cinnamon-tan; reverse yellowish-buff, sometimes with pinkish tinges.

Microscopy. Macroconidia in large clusters, rather thin-walled, regularly verrucose, 3-6 (-8)-celled, fusiform, 25-60 × 8.5-15.0 µm. Microconidia sessile or stalked, smooth- and thin-walled, clavate, 3.5-8.0 × 2-3 µm.

Teleomorphs. *Arthroderma gypseum* (Nannizzi) Weitzman *et al.*, *A. incurvatum* (Stockdale) Weitzman *et al.* (*Ascomycota, Euascomycetes, Onygenales: Arthrodermataceae*); strains of the two species not isolated from soil are prevalent on animal and on human hosts, respectively (Demange *et al.*, 1992).

Physiology.

BCPCG	–	Rice (growth/sporulation)	+/+	T-2	+	T-5	+
Urease	+	Benomyl tolerance	–	T-3	+	T-6	+
Hair perforation	+	T-1		+	T-4	+	T-7	+

Molecular diagnostics. ITS restriction map based on CBS 161.69:

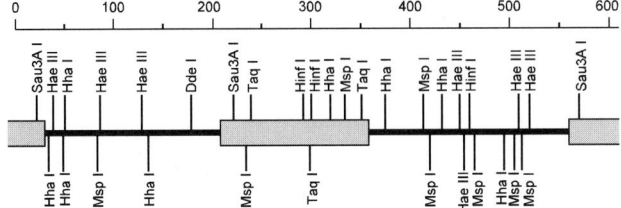

Differential diagnosis. The species differs from *Microsporum fulvum* (p. 749) by fusiform conidia.

Pathogenicity. BSL-1. Sparse ectothrix with large cells. Fluorescence, if any, dull. Geophilic species (Punsola & Guarro, 1984b), also reported from cats, dogs, rodents and horses (Pal *et al.*, 1994). In humans occasionally causing tinea corporis (Linhares *et al.*, 1998) or tinea capitis (Hayashi & Toshitani, 1983; Offidani *et al.*, 1998) were reported. Romano (1998) and Körte *et al.* (1998) described cases of onychomycosis. Infections in AIDS patients expand widely (Blanc *et al.*, 1994; Porro *et al.*, 1997; Giudice *et al.*, 1997).

Distribution. World-wide.

References. Stockdale (1963, 1965), Vismer *et al.* (1987), Ginter (1989), Singh *et al.* (1995), Mancianti & Papini (1997), Ranganathan & Balajee (2000).

Antifungal susceptibility.

Antifungal	Mean MICs	MIC 90	Strains	Reference
AMB	0.44	2	32	Fernández *et al.* (2000b)
FLZ	22.37	>16	32	Fernández *et al.* (2000b)
GRF	0.25	0.78	8	Wildfeuer *et al.* (1998)
ITZ	0.02	0.10	8	Wildfeuer *et al.* (1998)
ITZ	0.04	0.125	32	Fernández *et al.* (2000b)
KTZ	0.18	0.39	8	Wildfeuer *et al.* (1998)
KTZ	0.23	1	32	Fernández *et al.* (2000b)
MCZ	0.19	0.5	32	Fernández *et al.* (2000b)
TBF	0.04	0.06	32	Fernández *et al.* (2000b)
VCZ	0.11	0.39	8	Wildfeuer *et al.* (1998)
VCZ	0.06	0.5	32	Fernández *et al.* (2000b)

Nomenclature. *Trichophyton gypseum* Bodin - Champ. Paras. Homme p. 115, 1902 ≡ *Achorion gypseum* (Bodin) Bodin - Annls Derm. Syph. 4: 585, 1907 ≡ *Sabouraudites gypseus* (Bodin) Ota & Langeron - Annls Parasit. Hum. Comp. 1: 328, 1923 ≡ *Closterosporia gypsea* (Bodin) Grigorakis - Annls Sci. Nat., Bot., Sér. 10, 7: 411, 1925 ≡ *Microsporum gypseum* (Bodin) Guiart & Grigorakis - Lyon Méd. 141: 377, 1928 ≡ *Trichophyton mentagrophytes* (Robin) Blanchard var. *gypseum* (Bodin) Kamyszek - Med. Weteryn. 24: 146, 1945.

Microsporum flavescens Horta - Mem. Inst. Oswaldo Cruz 3: 301-308, 1912 ≡ *Sabouraudites flavescens* (Horta) Ota & Langeron - Annls Parasit. Hum. Comp. 1: 327, 1923.

Microsporum scorteum Priestley - Annls Trop. Med. Parasit. 8: 113, 1914 ≡ *Sabouraudites scorteus* (Priestley) Brumpt - Précis Parasitol., ed. 4, p. 1295, 1927 ≡ *Ecotrichophyton scorteum* (Priestley) C.W. Dodge - Med. Mycol. p. 503, 1935.

Microsporum xanthodes Fischer - Derm. Wschr. 66: 214-247, 1918 ≡ *Sabouraudites xanthodes* (Fischer) Ota & Langeron - Annls Parasit. Hum. Comp. 1: 328, 1923.

Gymnoascus gypseus Nannizzi - Atti Accad. Fisioscr. Siena Med.-Fis. 2: 93, 1927 ≡ *Nannizzia gypsea* (Nannizzi) Stockdale - Sabouraudia 3: 119, 1964 ≡ *Arthroderma gypseum* (Nannizzi) Weitzman, McGinnis, Padhye & Ajello - Mycotaxon 25: 514, 1986.

Nannizzia incurvata Stockdale - Sabouraudia 1: 46, 1961 ≡ *Nannizzia gypsea* (Nannizzi) Stockdale var. *incurvata* (Stockdale) Apinis Mycol. Pap. 96: 32, 1964 ≡ *Arthroderma incurvatum* (Stockdale) Weitzman, McGinnis, Padhye & Ajello - Mycotaxon 25: 514, 1986.

Favomicrosporon pinettii Benedek - Mycopath. Mycol. Appl. 31: 111, 1967.

Microsporum gypseum (Bodin) Guiart & Grigorakis var. *vinosum* Gordon & Lusick - Archs Derm. 91: 562, 1965.

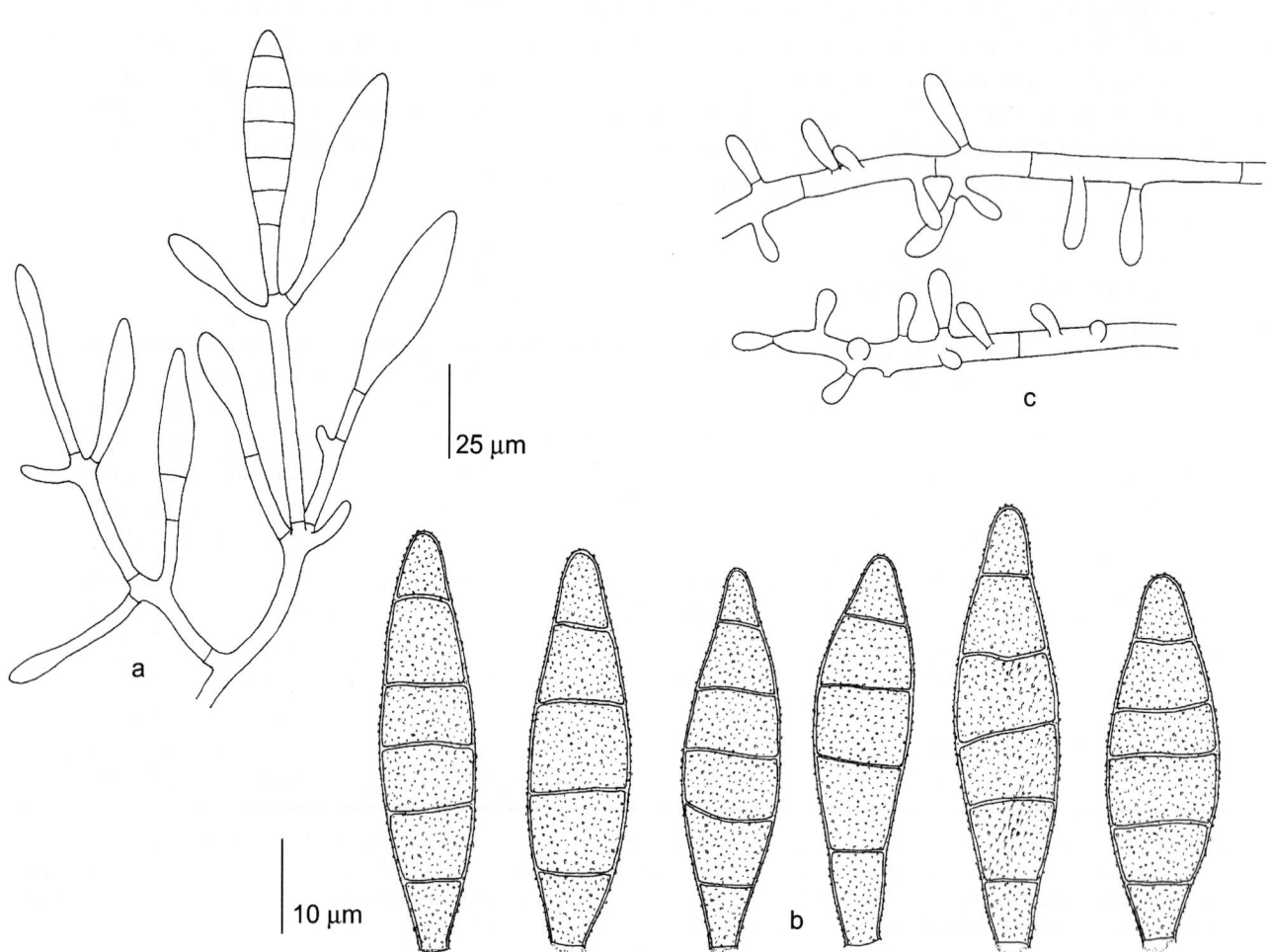

Microsporum gypseum, CBS 286.63. a. Conidial apparatus; b. macroconidia; c. microconidia.

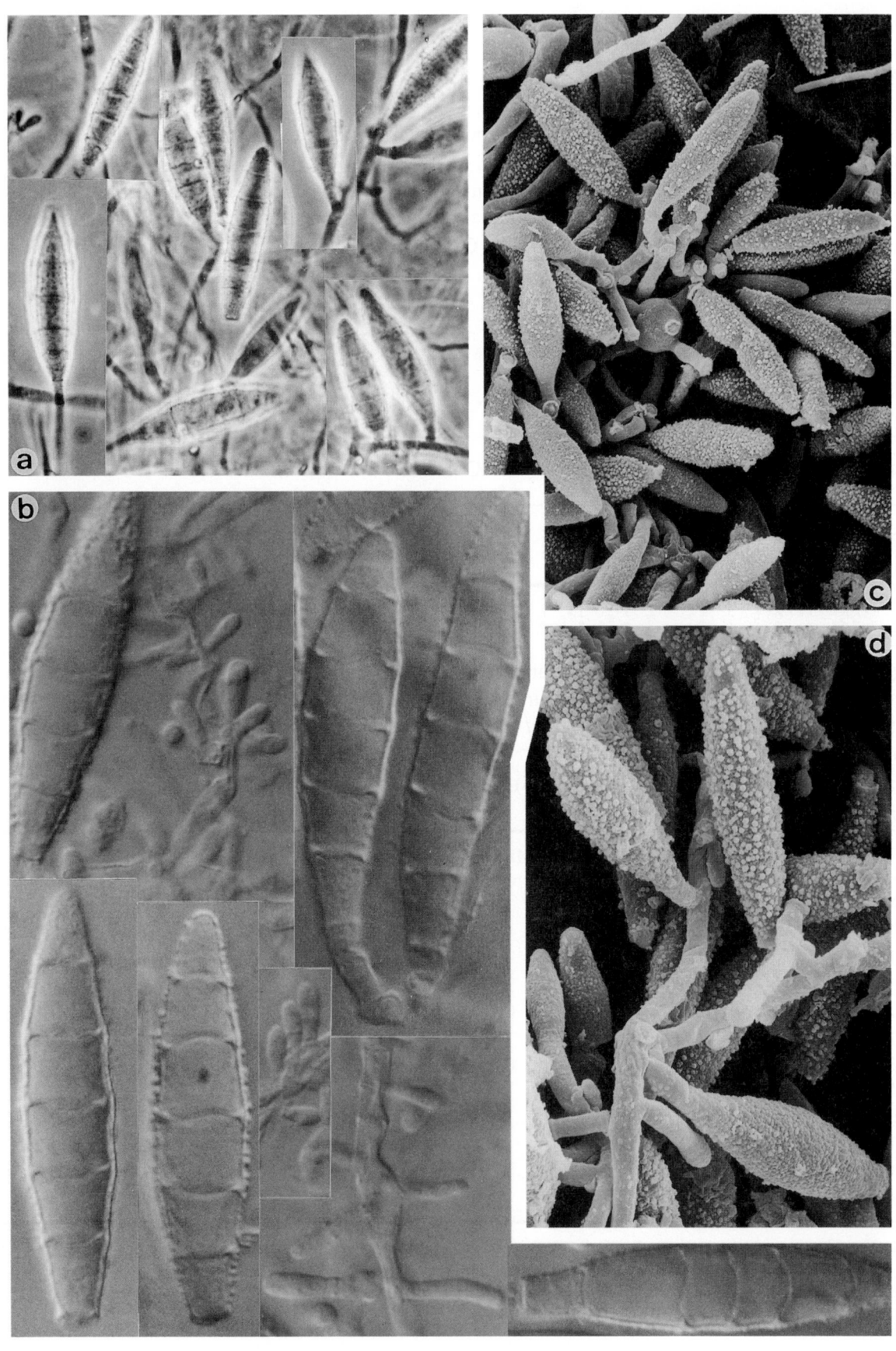

Microsporum gypseum, CBS 286.63. Fertile hyphae with macro- and microconidia. a. ×640; b. ×1600; c. ×1000; d. ×1700.

Microsporum nanum Fuentes

Colony characteristics. Colonies (SGA) spreading, powdery or cottony, often with some radial, shallow furrows, white, centrally buff; reverse reddish-brown.

Microscopy. Macroconidia mostly 2-celled, rather thin-walled, smooth-walled to verrucose, obovoidal to pyriform, 12-18 × 5.0-7.5 μm, with flat basal scars. Microconidia sessile alongside undifferentiated hyphae, clavate.

Teleomorph. *Arthroderma obtusum* (Dawson & Gentles) Weitzman *et al.* (*Ascomycota, Euascomycetes, Onygenales: Arthrodermataceae*).

Physiology.

Urease	+	T-1 +	T-4 +	T-7 +
Hair perforation	+	T-2 +	T-5 ++	
Rice (growth/sporulation)	+/+	T-3 +	T-6 +	

Molecular diagnostics. ITS restriction map based on CBS 322.61:

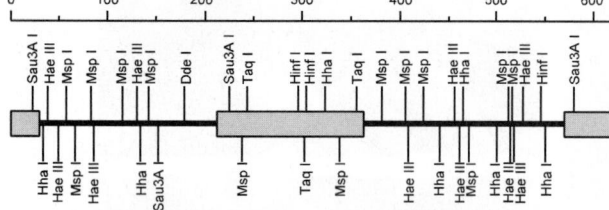

Pathogenicity. BSL-2. Sparse ectothrix with large cells. The species causes chronic, brownish, circular lesions on pig ears (Ginther & Bubash, 1966). Infections in man through direct contact may lead to tinea corporis and tinea capitis (Roller & Westblom, 1986; Pönnighaus *et al.*, 1995; Ranganathan *et al.*, 1997).

Distribution. World-wide.

References. Ajello *et al.* (1964), Stockdale (1965).

Antifungal susceptibility.

Antifungal	Mean MICs	MIC 90	Strains	Reference
AMB	0.125	nd	1	Fernández *et al.* (2000b)
CTZ	0.03	nd	1	Fernández *et al.* (2000b)
FLZ	1	nd	1	Fernández *et al.* (2000b)
GRF	0.98	3.13	3	Wildfeuer *et al.* (1998)
ITZ	0.10	0.78	3	Wildfeuer *et al.* (1998)
ITZ	0.03	nd	1	Fernández *et al.* (2000b)
KTZ	0.12	0.20	3	Wildfeuer *et al.* (1998)
KTZ	0.25	1	1	Fernández *et al.* (2000b)
MCZ	0.03	nd	1	Fernández *et al.* (2000b)
TBF	0.06	nd	1	Fernández *et al.* (2000b)
VCZ	0.20	0.78	3	Wildfeuer *et al.* (1998)
VCZ	0.01	nd	1	Fernández *et al.* (2000b)

Nomenclature. *Microsporum gypseum* (Bodin) Guiart & Grigorakis var. *nanum* Fuentes, Aboulafia & Vidal - J. Invest. Derm. 23: 56, 1954 (invalid) ≡ *Microsporum nanum* Fuentes - Mycologia 48: 614, 1956.

Nannizzia obtusa Dawson & Gentles - Sabouraudia 1: 56, 1961 ≡ *Arthroderma obtusum* (Dawson & Gentles) Weitzman, McGinnis, Padhye & Ajello - Mycotaxon 25: 514, 1986.

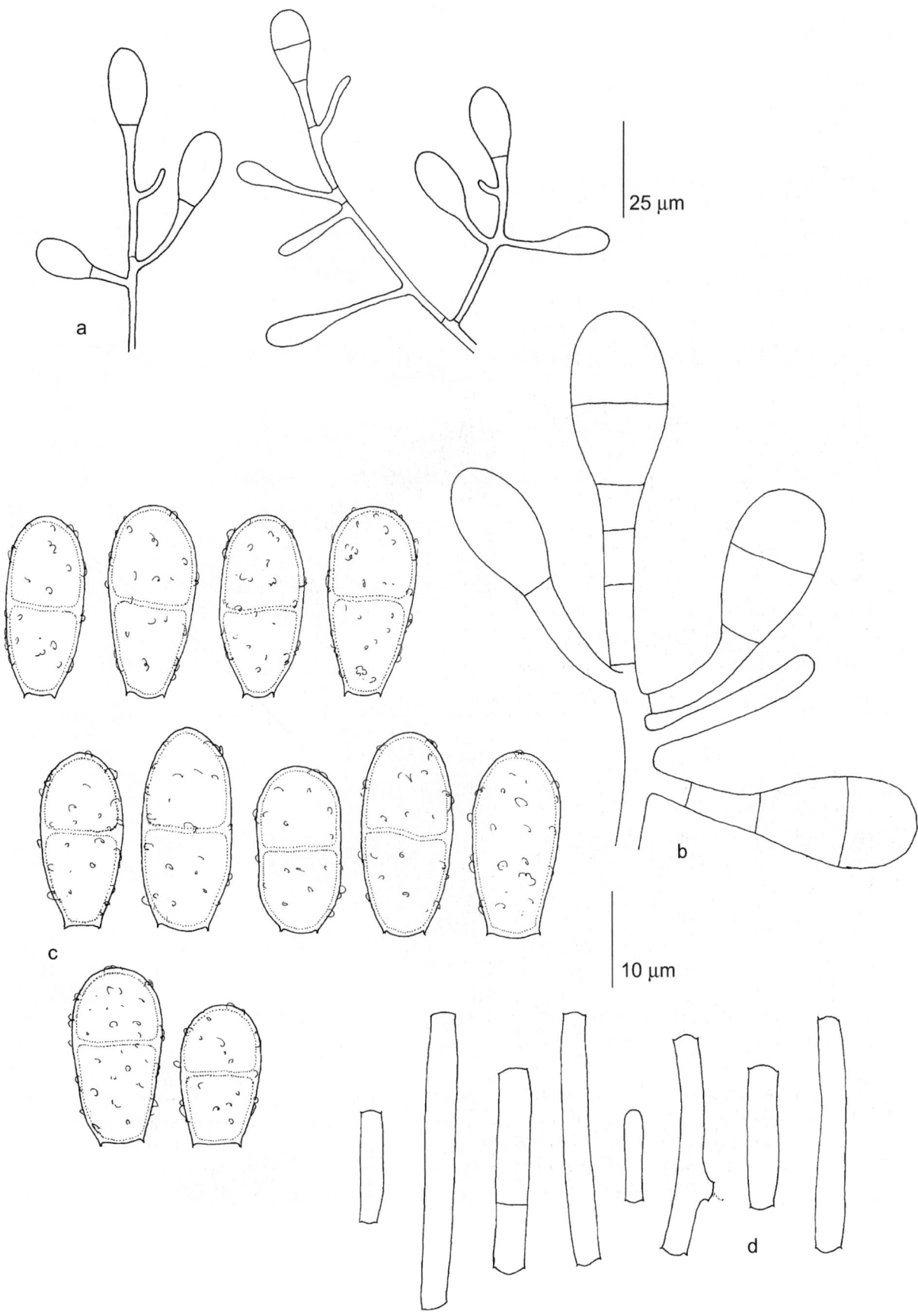

Microsporum nanum, CBS 727.88. a, b. Conidial apparatus; c. macroconidia; d. arthroconidia.

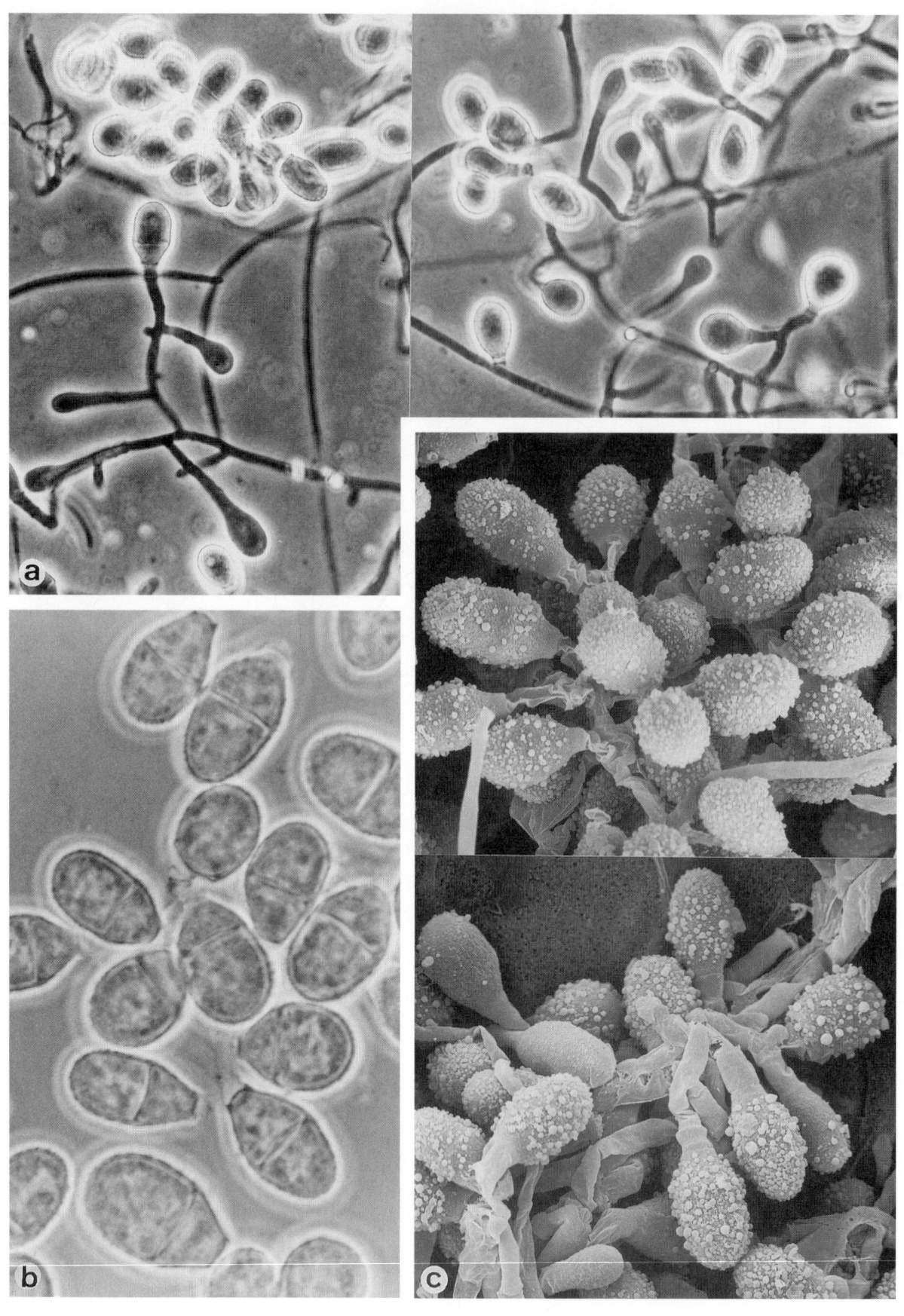

Microsporum nanum, CBS 727.88. a, c. Fertile hyphae producing macroconidia; b. liberated macroconidia. a. ×640; b. ×1600; c. ×1750.

Microsporum persicolor (Sabouraud) Guiart & Grigorakis

Colony characteristics. Colonies (SGA) expanding, powdery to fluffy, pale yellowish-buff to pinkish-buff; reverse ochraceous. On sugar-free media the colonies are rosaceous.

Microscopy. Macroconidia thin-walled, rough-walled at the tip, cigar-shaped, 4-7-celled, 40-60 × 4-8 μm. Microconidia in dense clusters, spherical. Spiral hyphae present.

Teleomorph. *Arthroderma persicolor* (Stockdale) Weitzman *et al.* (*Ascomycota, Euascomycetes, Onygenales: Arthrodermataceae*).

Physiology.

BCPCG	–	T-1	+	T-4	+	T-7 +
Urease	+	T-2	+	T-5	+	
Hair perforation	+	T-3	+	T-6	+	

Molecular diagnostics. ITS restriction map based on CBS 469.74:

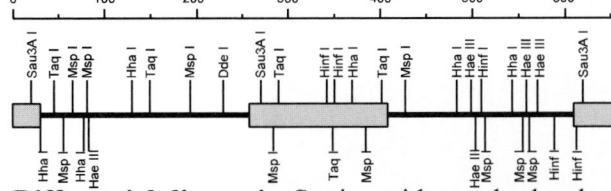

Differential diagnosis. Strains with poorly developed macroconidia may resemble *Trichophyton mentagrophytes* (p. 968), but young microconidia are pyriform and stalked.

Pathogenicity. BSL-2. Isolated from European bank and field voles (English, 1966). It is occasionally involved in cases of tinea corporis in humans (Onsberg, 1978; English *et al.*, 1978) and in dogs.

Distribution. Matches that of voles: Europe, Canada.

Reference. Stockdale (1967).

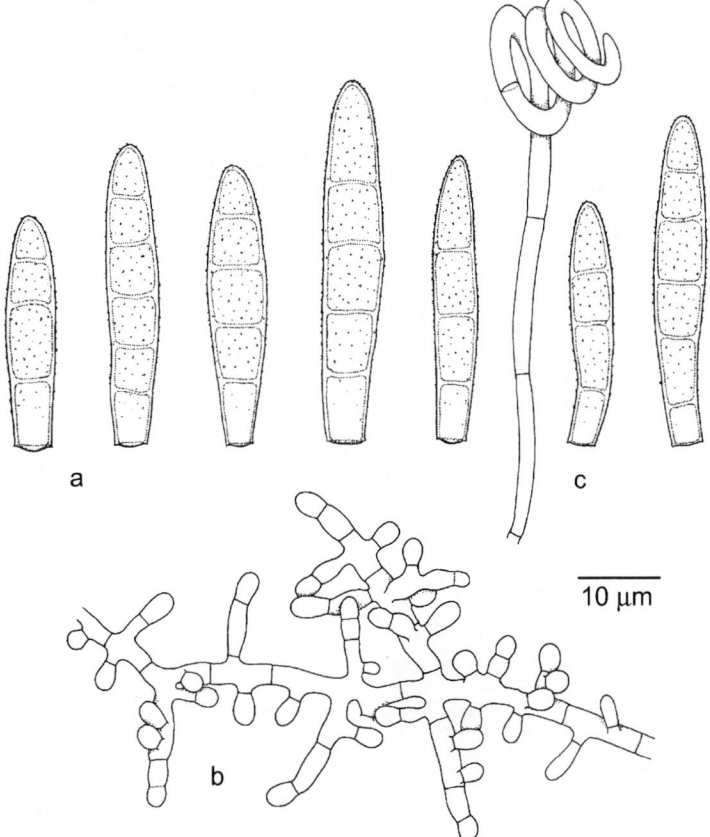

Microsporum persicolor, CBS 469.74. a. Macroconidia; b. microconidia; c. spiral hyphae.

Antifungal susceptibility.

Antifungal	Mean MICs	MIC 90	Strains	Reference
GRF	0.12	0.39	3	Wildfeuer *et al.* (1998)
ITZ	0.01	0.03	3	Wildfeuer *et al.* (1998)
KTZ	0.12	0.20	3	Wildfeuer *et al.* (1998)
VCZ	0.10	0.20	3	Wildfeuer *et al.* (1998)

Nomenclature. *Trichophyton persicolor* Sabouraud - Malad. Cuir Chev. 3: 632, 1910 ≡ *Ectotrichophyton persicolor* (Sabouraud) Castellani & Chalmers - Man. Trop. Med. p. 1005, 1918 ≡ *Sabouraudites persicolor* (Sabouraud) Ota & Langeron - Annls Parasit. Hum. Comp. 1: 329, 1923 ≡ *Closteroaleuriosporia persicolor* (Sabouraud) Grigorakis - Annls Sci. Nat., Bot., Sér. 10, 7: 412, 1925 ≡ *Microsporum persicolor* (Sabouraud) Guiart & Grigorakis - Lyon Méd. 141: 377, 1928 ≡ *Ctenomyces persicolor* (Sabouraud) Nannizzi - Tratt. Micopat. Um. 4: 154, 1934 ≡ *Epidermophyton persicolor* (Sabouraud) C.W. Dodge - Med. Mycol. p. 486, 1935 ≡ *Trichophyton mentagrophytes* (Robin) Blanchard var. *persicolor* (Sabouraud) Ueckert - Zentbl. Bakt. Parasitkde, Abt. 1, 176: 127, 1959.

Nannizzia persicolor Stockdale - Sabouraudia 5: 357, 1967 ≡ *Arthroderma persicolor* (Stockdale) Weitzman, McGinnis, Padhye & Ajello - Mycotaxon 25: 514, 1986.

Arthroderma benedekii Balogh, Liptovszky & Nagy-Peti - Mycopath. Mycol. Appl. 40: 73, 1970.

Nannizzia quinckeani Balabanov & Schick - Derm. Venerol. 9: 35, 1970.

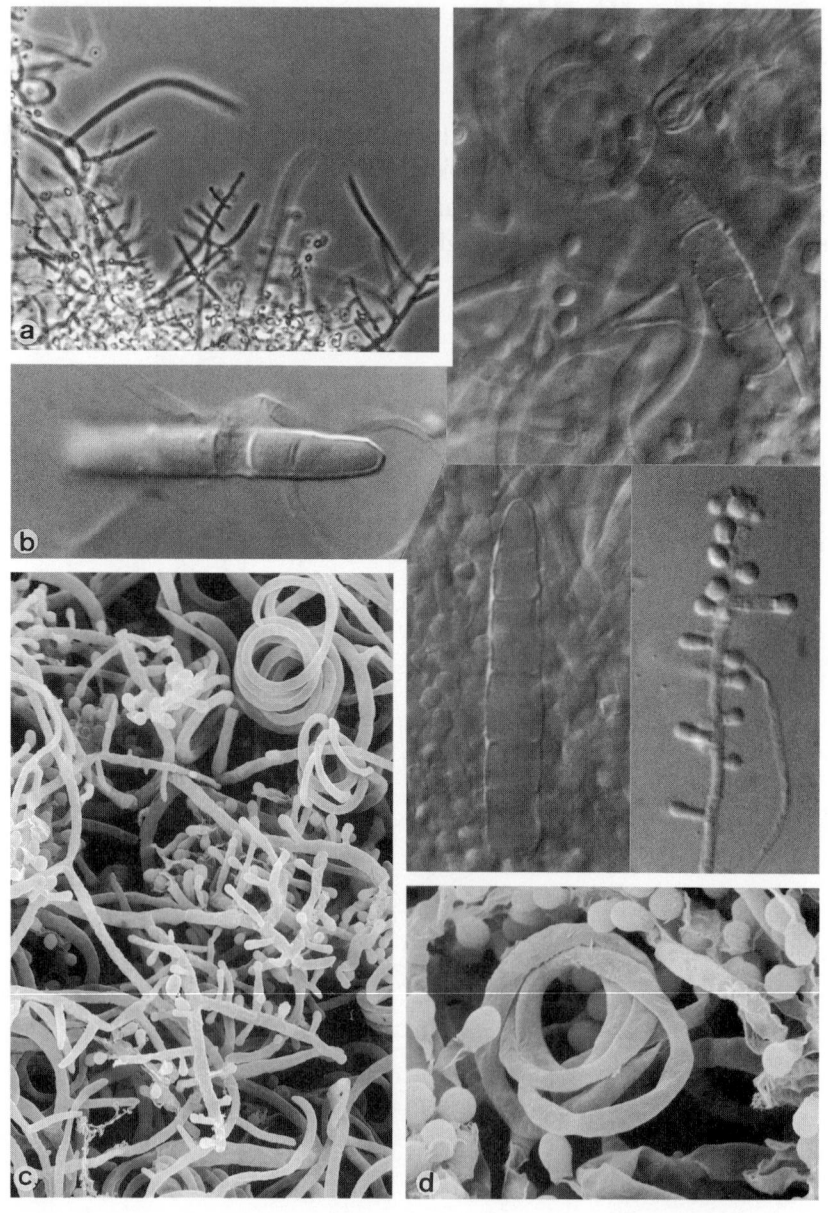

Microsporum persicolor, CBS 421.74. Microconidial branches, spiral hyphae, macro- and microconidia. a. ×490; b. ×1230; c. ×1040; d. ×2810.

Microsporum praecox Rivalier ex Padhye *et al.*

Colony characteristics. Colonies (SGA) moderately expanding, powdery, with concentric, cloudy growth waves, buff; reverse yellow-orange.

Microscopy. Macroconidia moderately thin-walled, echinulate, lanceolate with narrow apex, 6-9-celled, up to 65 × 9 μm. Microconidia, when present, in orthotropic arrangement, pyriform.

Physiology.

BCPCG	–	T-1	+	T-5	+
Urease	+	T-2	+	T-6	+
Hair perforation	–	T-3	+	T-7	+
		T-4	+		

Molecular diagnostics. ITS restriction map based on CBS 288.55:

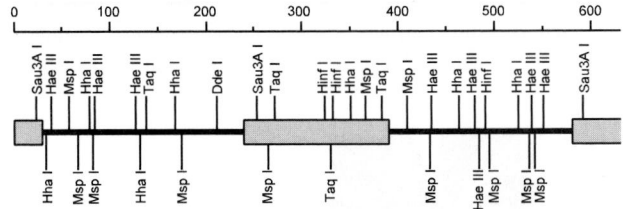

Pathogenicity. BSL-2. The species has been described as a rare cause of human skin lesions (de Vroey *et al.*, 1983a), causing tinea corporis and tinea capitis (Degeilh *et al.*, 1994). Infections are frequently acquired from the vicinity of horses (de Vroey *et al.*, 1983b).

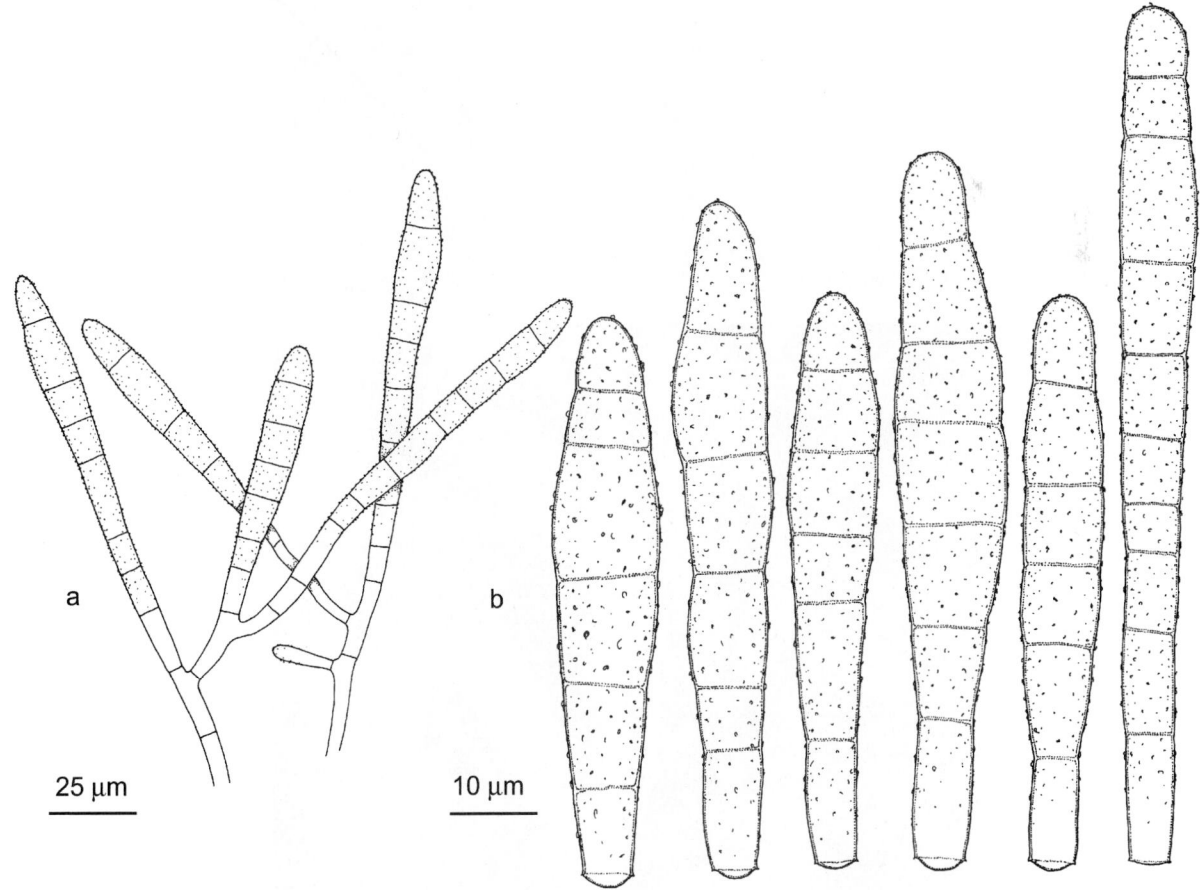

Microsporum praecox, CBS 671.89. a. Developing branching pattern; b. macroconidia.

763

Distribution. Rare species, hitherto isolated in France and the U.S.A.

References. Rivalier (1953), Avram *et al.* (1988), Phelippot *et al.* (1988), Padhye *et al.* (1989).

Antifungal susceptibility.

Antifungal	Mean MICs	Strains	Reference
AMB	1	1	Fernández *et al.* (2000b)
CTZ	0.01	1	Fernández *et al.* (2000b)
FLZ	>16	1	Fernández *et al.* (2000b)
ITZ	0.125	1	Fernández *et al.* (2000b)
KTZ	1	1	Fernández *et al.* (2000b)
MCZ	0.55	1	Fernández *et al.* (2000b)
TBF	0.01	1	Fernández *et al.* (2000b)
VCZ	0.03	1	Fernández *et al.* (2000b)

Nomenclature. *Sabouraudites praecox* Rivalier - Annls Inst. Pasteur 86: 276, 1954 (invalid) ≡ *Microsporum praecox* (Rivalier) Rivalier - Bull. Soc. Fr. Mycol. Méd. 7: 297, 1978 (invalid) ≡ *Microsporum praecox* Rivalier ex Padhye, Ajello & McGinnis, *in* Padhye, Detweiler, Frumkin, Bulmer, Ajello & McGinnis - J. Med. Vet. Mycol. 27: 316, 1989.

Microsporum praecox, CBS 671.89. Macroconidia. a. ×640; b. ×1600; c. ×470; d. ×1300; e. ×1650.

Microsporum racemosum Borelli

Colony characteristics. Colonies (MEA) spreading, powdery, often feathered, cream-coloured, later with reddish staining; reverse grape-red or brown.

Microscopy. Macroconidia stalked, cigar-shaped, thick- and rough-walled, 5-10-celled, 55-65 × 12-15 μm. Microconidia stalked, in clusters, clavate.

Teleomorph. *Arthroderma racemosum* (Rush-Munro *et al.*) Weitzman *et al.* (*Ascomycota, Euascomycetes, Onygenales: Arthrodermataceae*).

Physiology.

BCPCG	–	T-1	+	T-4	+	T-7	+
Urease	+	T-2	+	T-5	+		
Hair perforation	+	T-3	+	T-6	+		

Molecular diagnostics. ITS restriction map based on CBS 424.74:

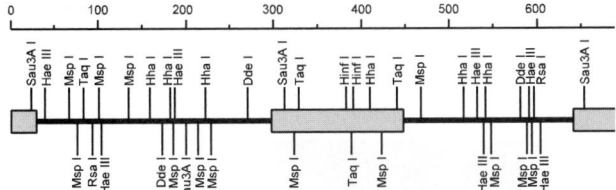

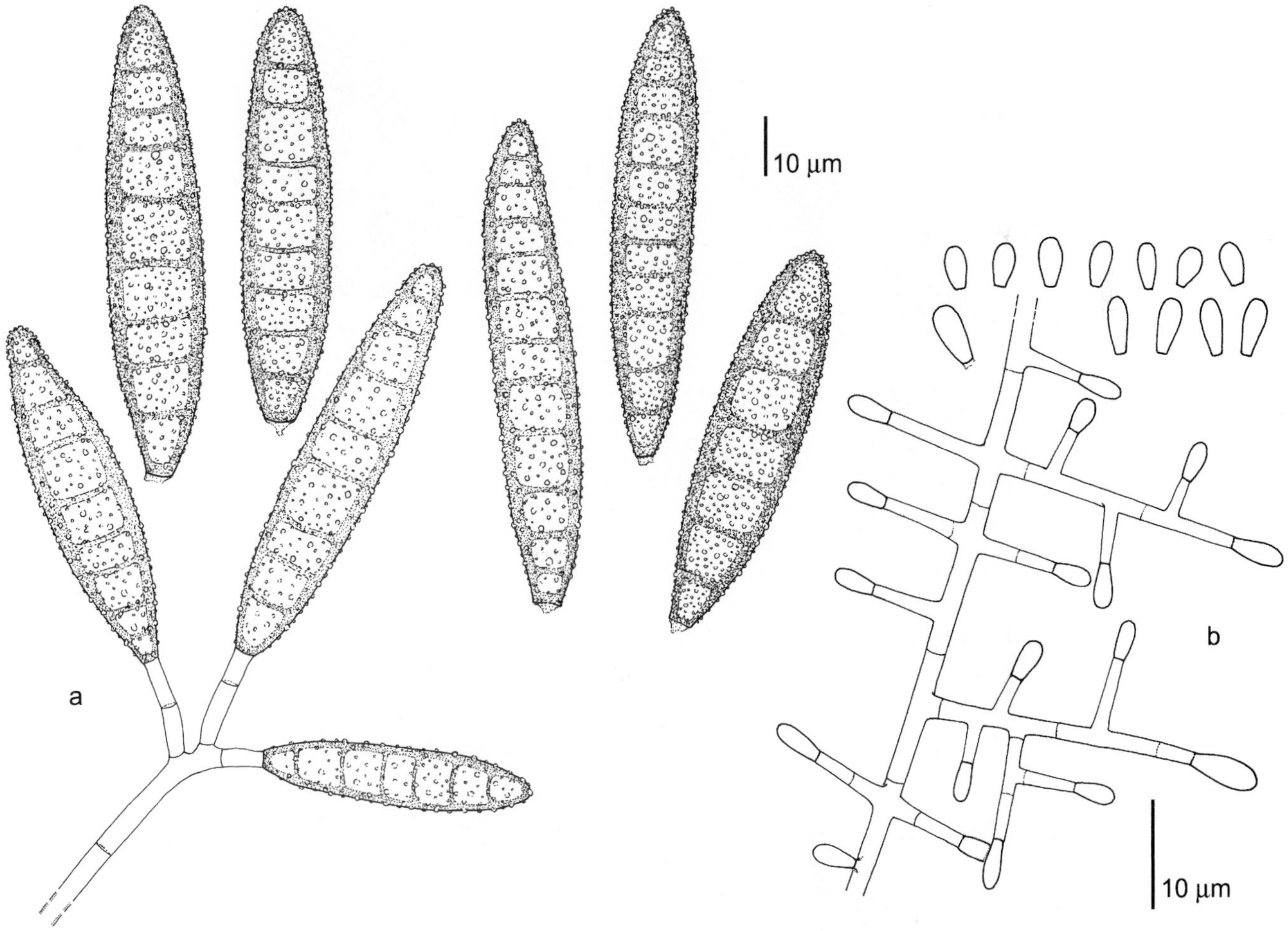

Microsporum racemosum, CBS 450.65. a. Macroconidia; b. microconidia.

Differential diagnosis. *Microsporum cookei* (p. 746) has shorter macroconidia with thicker walls.

Pathogenicity. BSL-1. The species is a geophilic dermatophyte. A human skin lesion was reported by Rippon & Andrews (1978) and onychomycosis by García-Martos *et al.* (1999).

References. Borelli (1965), Alteras & Evolceanu (1969a).

Antifungal susceptibility.

Antifungal	Mean MICs	Strains	Reference
AMB	8	1	Fernández *et al.* (2000b)
CLZ	>16	1	Fernández *et al.* (2000b)
FLZ	1	1	Fernández *et al.* (2000b)
ITZ	0.25	1	Fernández *et al.* (2000b)
KTZ	8	1	Fernández *et al.* (2000b)
MCZ	16	1	Fernández *et al.* (2000b)
TBF	>16	1	Fernández *et al.* (2000b)

Nomenclature. *Microsporum racemosum* Borelli - Acta. Méd. Venez. 12: 150, 1965.

Nannizzia racemosa Rush-Munro, J.M.B. Smith & Borelli - Mycologia 62: 658, 1970 ≡ *Arthroderma racemosa* (Rush-Munro, J.M.B. Smith & Borelli) Weitzman, McGinnis, Padhye & Ajello - Mycotaxon 25: 514, 1986.

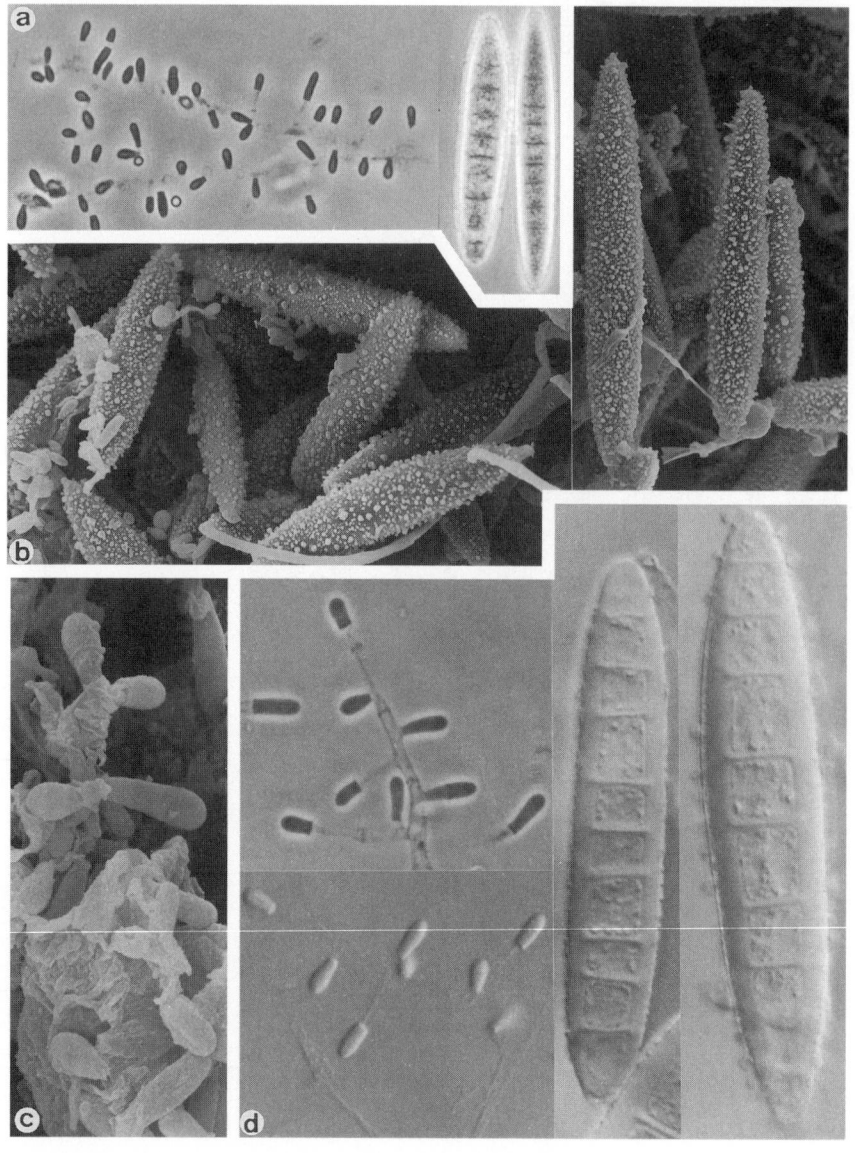

Microsporum racemosum, CBS 450.65. Macro- and microconidia. a. ×350; b. ×680; c. ×2515; d. ×1090.

Hyphomycetes, black yeasts. Genus: *MONILIELLA*

Moniliella suaveolens (Lindner ex Lindner) v. Arx

Colony characteristics. Colonies (PCA) restricted, at first cream-coloured, soon becoming olive to black, smooth, later cerebrifom; reverse pale yellow to dark olivaceous brown.

Microscopy. Hyphae hyaline, distally gradually changing over into conidial chains, which are often branched. Lower hyphal parts disarticulate into separate cells which often have protruding scars. Blastoconidia one-celled, arising in acropetal chains from undifferentiated hyphae, often also forming by disarticulation of supporting hyphae, smooth-walled, occasionally verrucose, ellipsoidal to broadly ellipsoidal; arthroconidia cylindrical, 12-27 (-40) × 4.0-5.8 µm, with truncate ends, subhyaline to brown. Blasto- and arthroconidia soon swell and become indistinguisable from each other.

Physiology.

Fermentation:		α,α-Trehalose	–	Ribitol	v
Glucose	+	Lactose	v	Galactitol	–
Galactose	v	Melibiose	–	D-Mannitol	+
Sucrose	+	Raffinose	v	D-Glucitol	v
Maltose	+	Melezitose	v	α-Methyl-D-glucoside	–
Lactose	v	Inulin	–	Salicin	v
Raffinose	–	Soluble starch	v	DL-Lactate	–
		D-Xylose	v	Succinate	+
Growth:		L-Arabinose	v	Citrate	v
Glucose	+	D-Arabinose	–	Inositol	–
Galactose	v	D-Ribose	+	Nitrate	+
L-Sorbose	–	L-Rhamnose	–	Vitamin-free	v
Sucrose	+	Glycerol	+	Urease	+
Maltose	+	*meso*-Erythritol	+	Growth at 37°C	–
Cellobiose	+				

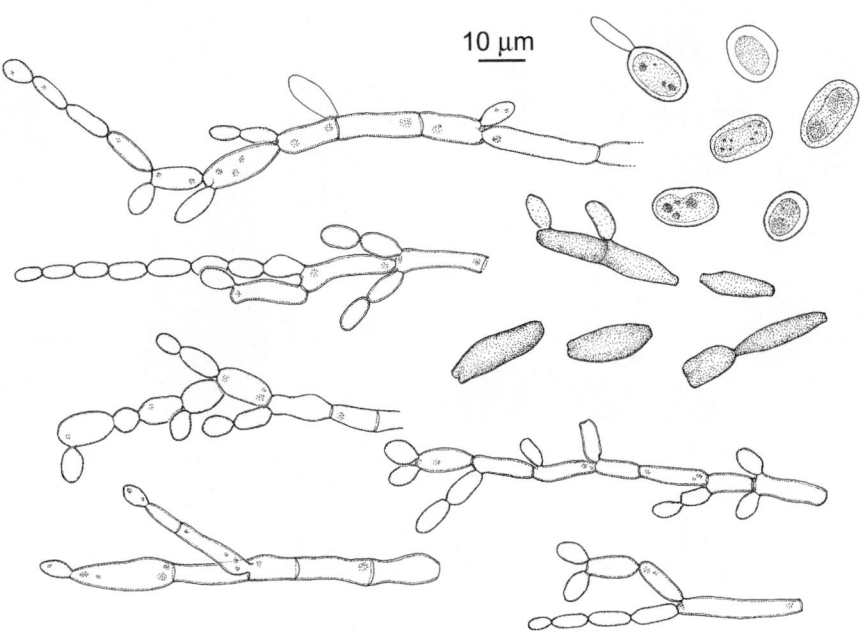

Moniliella suaveolens, CBS 120.63. Fertile hyphae and conidia.

767

Differential diagnosis. *Moniliella* is a basidiomycetous yeast-like genus phylogenetically close to *Trichosporon* (p. 164). Cultures are, however, melanized and hence not treated under the yeasts. In contrast to most other basidiomycetous yeasts, glucose is fermented.

Pathogenicity. BSL-1. Skin lesions on the palms (Kocková-Kratochvílová *et al.*, 1987). The species is an industrial, lipophilic organism.

Variety. *M. suaveolens* var. *nigra* (Burri & Staub) de Hoog is maintained for isolates that rapidly become melanized, but the two entities are genetically identical.

References. De Hoog (1979), de Hoog & Smith (1998b).

Nomenclature. *Sachsia suaveolens* Lindner - Mikrosk. Betriebskontr. Gärungsgew., ed. 5, p. 153, 1895 (nom. prov.) ≡ *Sachsia suaveolens* Lindner ex Lindner - Wschr. Brauerei 21: 1, 1906 ≡ *Candida suaveolens* (Lindner ex Lindner) Langeron & Guerra - Annls Parasit. Hum. Comp. 16: 505, 1938 ≡ *Moniliella suaveolens* (Lindner ex Lindner) von Arx - Antonie van Leeuwenhoek 38: 294, 1972.

Monilia nigra Burri & Staub - Landw. Jahrb. Schweiz 23: 512, 1909 ≡ *Moniliella suaveolens* (Lindner ex Lindner) von Arx var. *nigra* (Burri & Staub) de Hoog - Stud. Mycol. 19: 8, 1979.

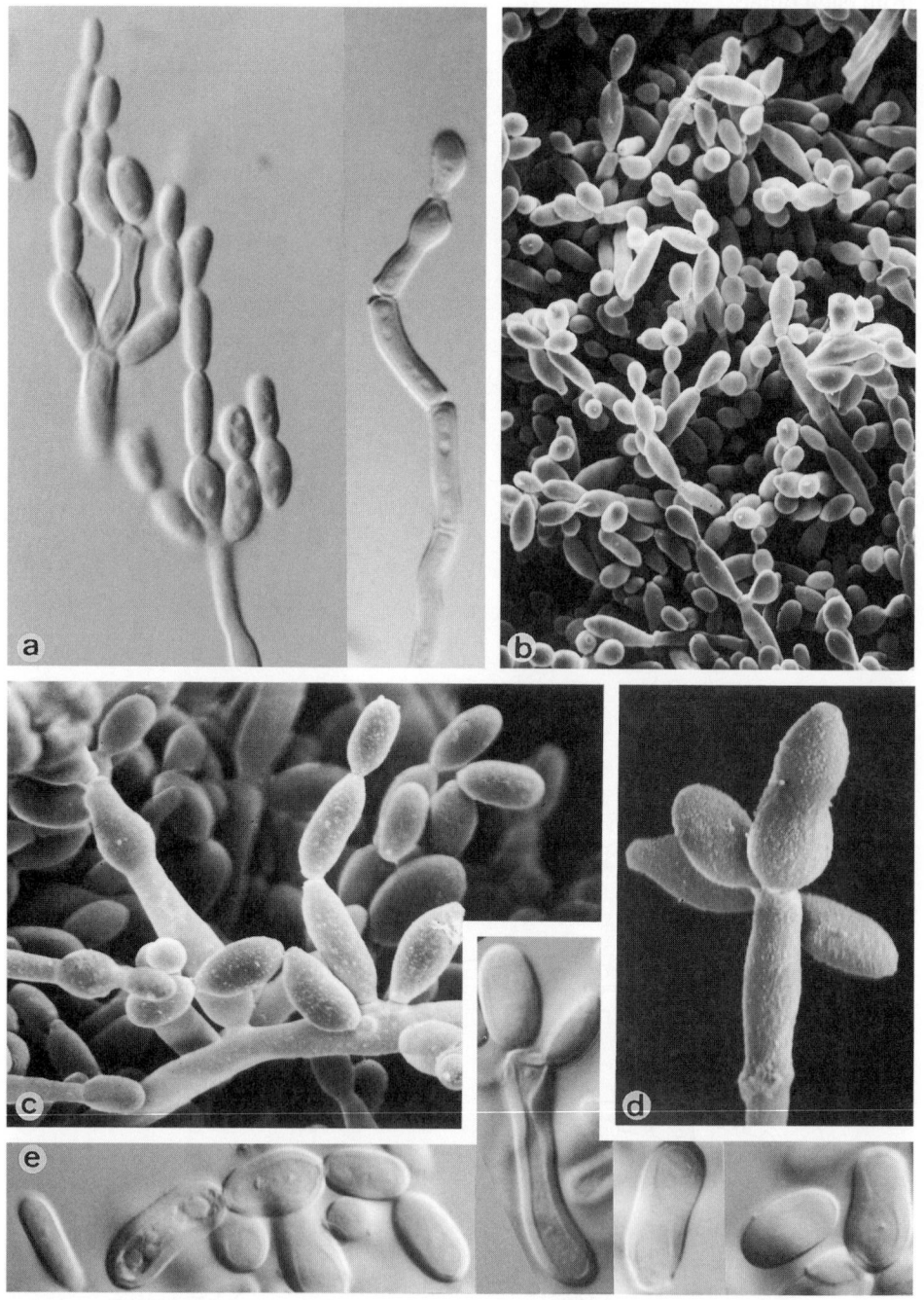

Moniliella suaveolens, CBS 120.63. a. Chains of blasto- and arthroconidia; b-d. fertile hyphae with blastoconidia; c. arthroconidia. a. ×1600; b. ×1300; c. ×2650; d. ×3700; e. ×1600.

Hyphomycetes. Genus: *MYCELIOPHTHORA*

Myceliophthora thermophila (Apinis) v. Oorschot

Colony characteristics. Colonies (MEA 2%, 30°C) growing rapidly, velvety to woolly, pale brown.

Microscopy. Conidiophores undifferentiated, hyaline, smooth-walled. Conidia borne on short protrusions, singly or in small groups on ampulliform swellings, obovoidal to pyriform, hyaline, 4.5-11.0 × 3.0-4.5 μm, initially rough-walled, becoming smooth-walled in old cultures, with thick walls.

Teleomorph. *Corynascus heterothallicus* (v. Klopotek) v. Arx (*Ascomycota, Euascomycetes, Sordariales: Chaetomiaceae*).

Ascomata spherical, non-ostiolate, glabrous. Peridium of *textura epidermoidea*. Asci 8-spored, spherical, 25-35 × 10-15 μm, with evanescent wall. Ascospores ellipsoidal to ovoidal, brown, 7.5-11.0 × 4.5-7.0 μm, with a distinct germ pore at both ends.

Physiology. The species is thermophilic.

Pathogenicity. BSL-1. Occurring in self-heated environments such as compost. Systemic infections in immunocompromised (McGough *et al.*, 1991; Bourbeau *et al.*, 1992) and major surgery patients (Farina *et al.*, 1998).

References. Van Oorschot (1977), Manoch *et al.* (1986).

Nomenclature. *Sporotrichum thermophilum* Apinis - Nova Hedwigia 5: 74, 1962 ≡ *Chrysosporium thermophilum* (Apinis) von Klopotek - Arch. Mikrobiol. 98: 366, 1974 / *Myceliophthora thermophila* (Apinis) van Oorschot - Persoonia 9: 403, 1977.

Thielavia heterothallica von Klopotek - Arch. Mikrobiol. 107: 223, 1976 ≡ *Corynascus heterothallicus* (von Klopotek) von Arx - Sydowia 34: 25, 1981.

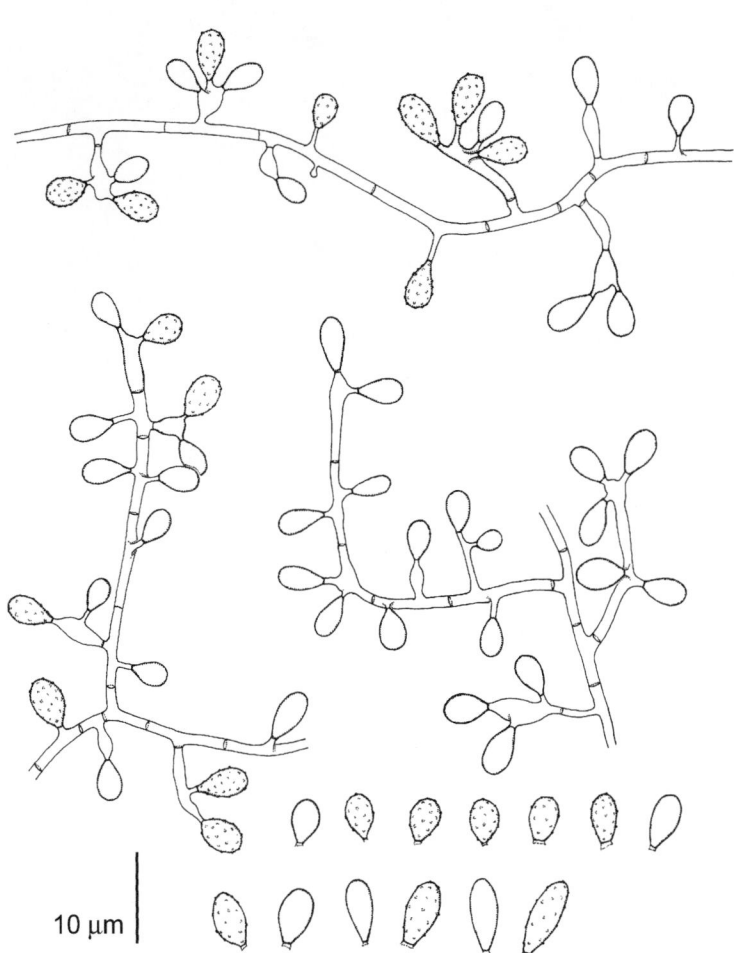

10 μm

Myceliophthora thermophila, CBS 117.65. Fertile hyphae and conidia.

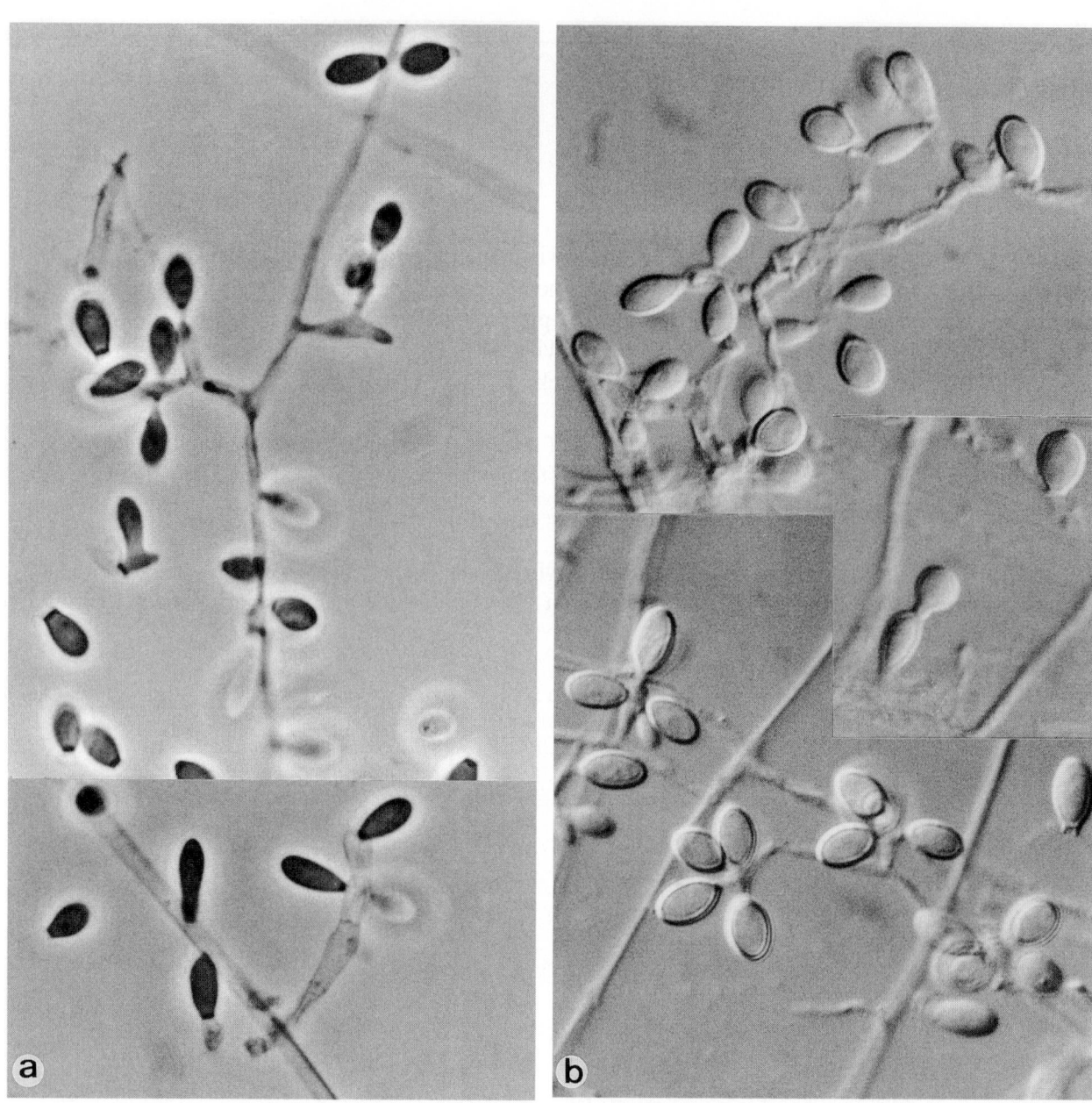

Myceliophthora thermophila, CBS 117.65. Fertile hyphae and conidia. a. ×1280; b. ×600.

Hyphomycetes. Genus: *MYCOCENTROSPORA*

Mycocentrospora acerina (Hartig) Deighton

Colony characteristics. Colonies (PCA) growing slowly, velvety, dark olivaceous to black; reverse olivaceous black.

Microscopy. Conidiogenous cells short, integrated or discrete, elongating sympodially, with unthickened conidial scars, pale olivaceous, smooth-walled. Conidia elongate to obclavate, apically rostrate, truncate at the base, smooth-walled, often with a lateral branch near the base, 4-25-septate, 50-300 × 5-7 µm, central cells slightly pigmented and with thickened walls, end cells pale or hyaline. Hyphal cells often locally become dark, swollen, thick-walled, chlamydospore-like.

Pathogenicity. BSL-1. Plant-pathogenic species. A case of chronic verrucose lesions on the body of an Indonesian patient (as *Cercospora apii*; Emmons *et al.*, 1957; Lie-Kian-Joe *et al.*, 1957; Deighton & Mulder, 1977).

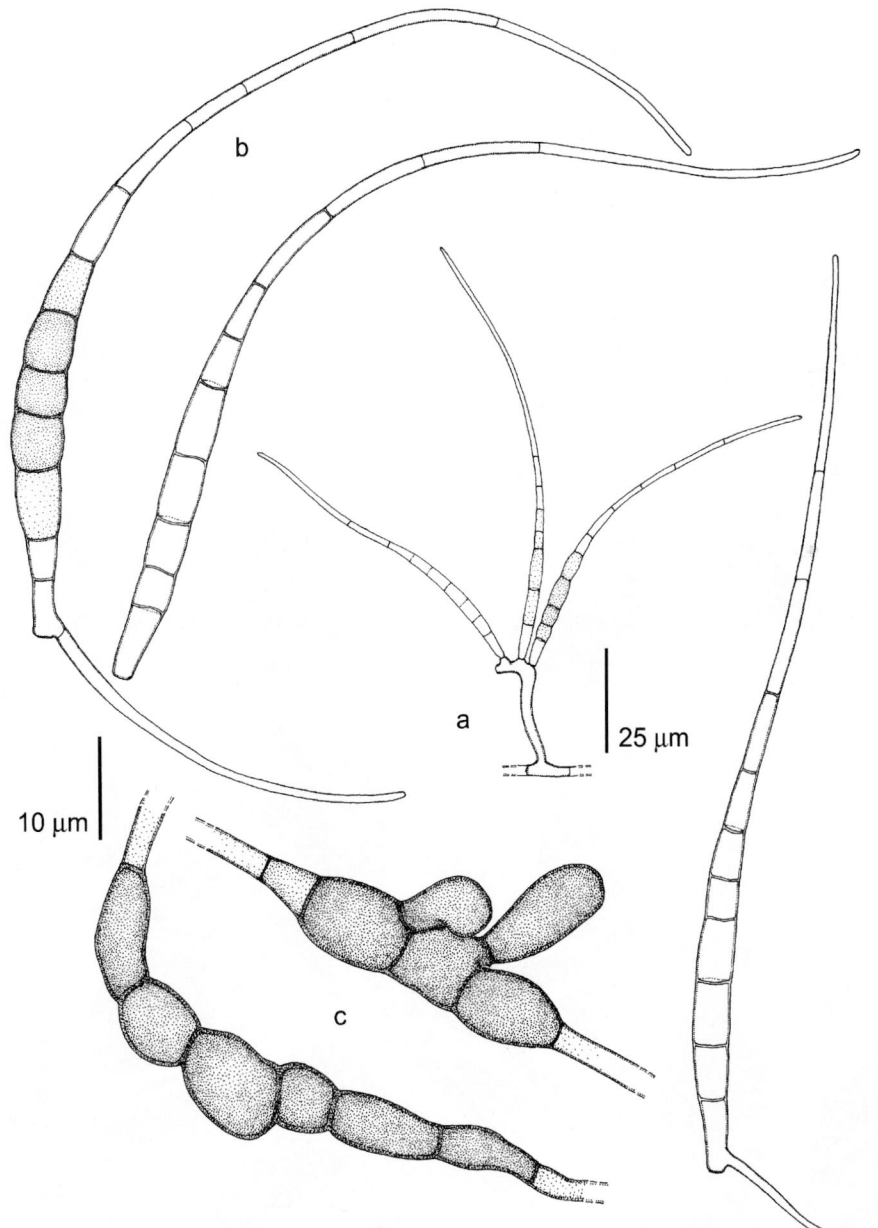

Mycocentrospora acerina, IMI 142050. a. Conidiophore and conidia; b. conidia; c. dark, swollen, chlamydospore-like cells.

References. Chupp (1957), Deighton (1971), Deighton & Mulder (1977).

Nomenclature. *Cercospora acerina* Hartig - Unters. Forstbot. Inst. Münch. 1: 58, 1880 ≡ *Mycocentrospora acerina* (Hartig) Deighton - Taxon 21: 716, 1972.

Mycocentrospora acerina, IMI 142050. Conidia. a. ×1280; b. ×750.

Hyphomycetes. Genus: *MYCOLEPTODISCUS*

Mycoleptodiscus indicus (Sahni) Sutton

Colony characteristics. Colonies (SGA) moderately expanding, flat, initially yellowish, finally greyish-black. Sporulation is obtained in slide cultures on PDA under near-UV light.

Microscopy. Hyphae pale brown, 3-7 µm wide. Sporodochia are flat, shield-like clusters of dark brown, thick-walled, compressed, ampulliform phialides which are 7-13 × 3.5-7.0 µm. Each phialide bears a single collarette up to 3 µm long. Conidia thin-walled, hyaline, curved, 10-16 × 4.5-7.0 µm, usually with thin, hyaline, hair-like appendage at each end.

Physiology. Growth at 37°C but not at 40°C; urease present; gelatin not liquified.

Pathogenicity. BSL-1. Plant-inhabiting species. It has been reported from a subcutaneous case in a gardener under prednisone treatment (Padhye *et al.*, 1995) and from an extended cutaneous lesion in a renal transplant recipient (B. Dupont & E. Guého, pers. comm.).

Reference. Sutton (1973).

Nomenclature. *Amerodiscosiella indica* Sahni - Mycopath. Mycol. Appl. 36: 277, 1968 ≡ *Mycoleptodiscus indicus* (Sahni) Sutton - Trans. Br. Mycol. Soc. 60: 528, 1973.

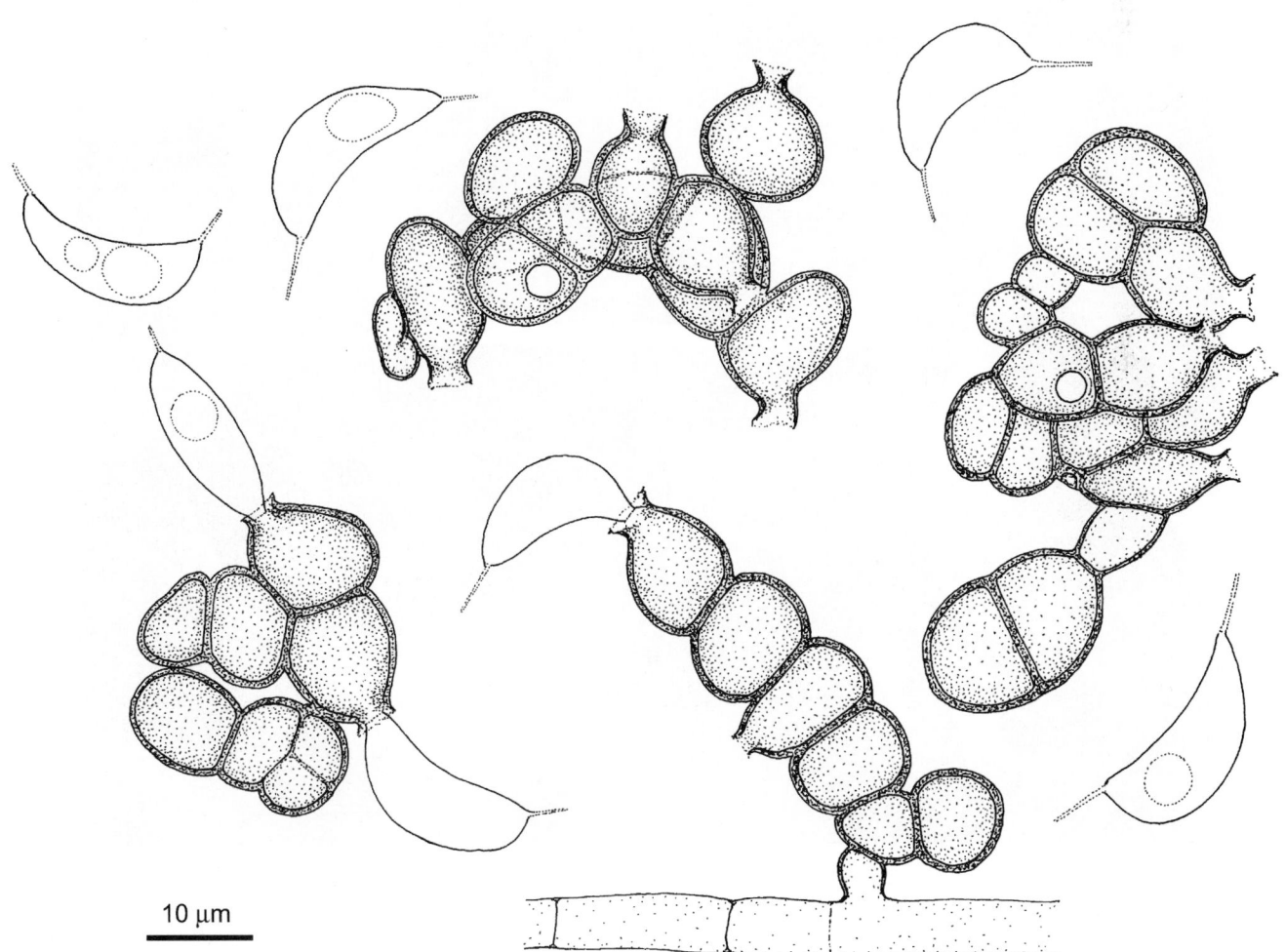

Mycoleptodiscus indicus, strain Padhye *et al.* (1995). Clusters of phialides with sessile collarettes, producing setose conidia.

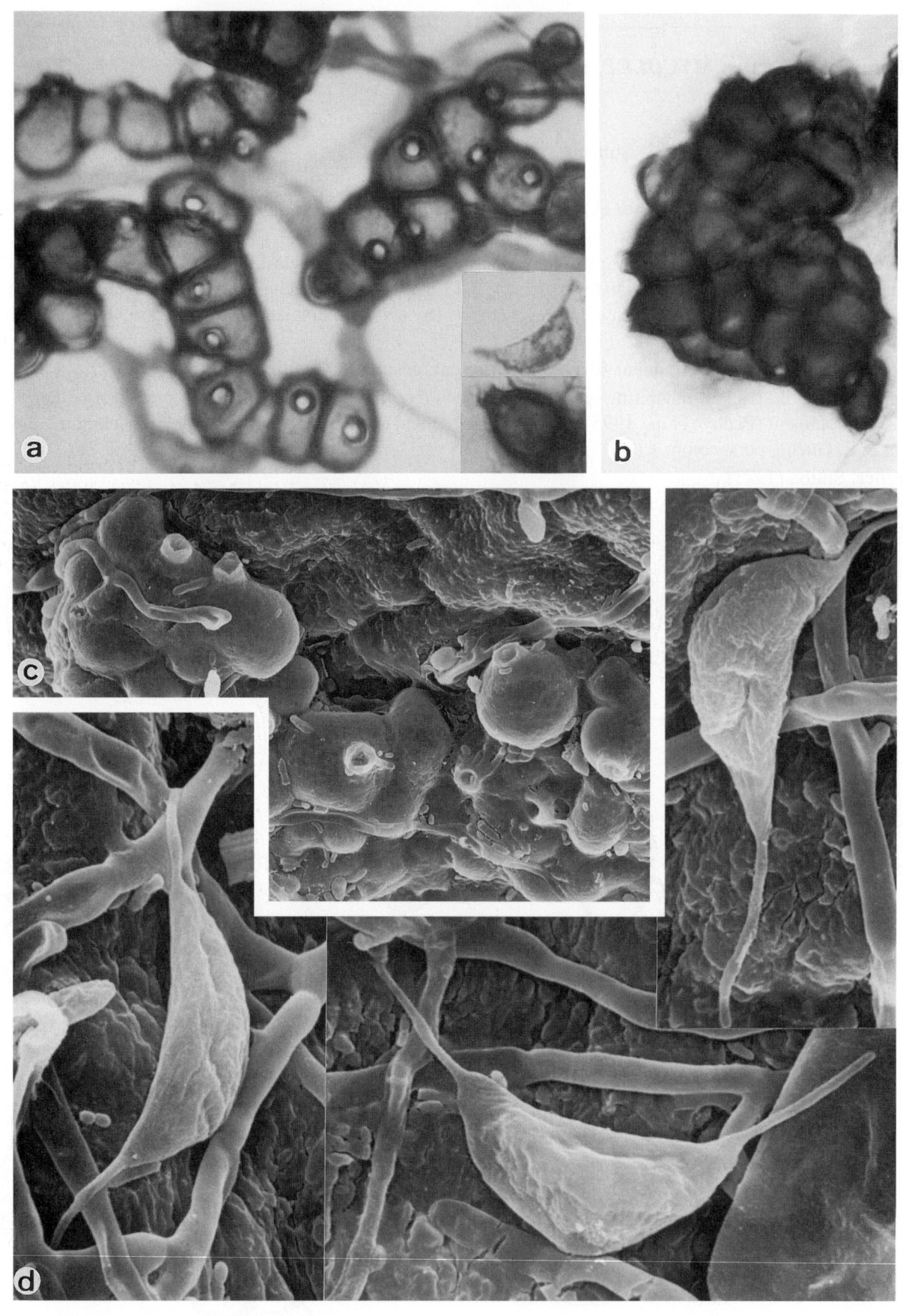

Mycoleptodiscus indicus, a. IP 2483.99. Courtesy E. Guého.; b-d. IMI 108220. a, c. Clusters of conidiogenous cells and conidium; b. sporodochium; d. conidia. a, b. ×1600; c. ×2050; d. ×4650.

Hyphomycetes. Genus: *MYRIODONTIUM*

Myriodontium keratinophilum Samson & Polonelli

Colony characteristics. Colonies (OA) growing moderately rapidly, white, floccose.

Microscopy. Fertile hyphae hyaline, smooth-walled, branched. Conidiogenous cells integrated, terminal or intercalary, polyblastic, denticulate. Conidia solitary, on long stalks, arising more or less synchronously, hyaline, subspherical to dacryoid, one-celled, 2-3 µm diam, smooth-walled.

Teleomorph. *Neoarachnotheca keratinophila* Ulfig *et al.* (*Ascomycota, Euascomycetes, Onygenales: Onygenaceae*).

Gymnothecia spherical, white to yellow, 150-700 µm diam. Peridium composed of a network of hyaline hyphae. Asci (sub)spherical, 8-spored, evanescent, 7-10 × 6-8 µm. Ascospores spherical, subhyaline, irregularly punctate-reticulate, with an irregular sheath, 3-4 × 2.7-3.8 µm.

Pathogenicity. BSL-1. The species has been isolated from soil and is probably keratinophilic (Chabasse, 1994). A case of frontal sinusitis in a Nigerian patient has been reported (Maran *et al.*, 1985).

References. Samson & Polonelli (1978), Cano *et al.* (1997).

Nomenclature. *Myriodontium keratinophilum* Samson & Polonelli - Persoonia 9: 505, 1978.

Neoarachnotheca keratinophila Ulfig, Cano & Guarro, *in* Cano, Ulfig, Guillamon, Vidal & Guarro - Antonie van Leeuwenhoek 72: 151, 1997.

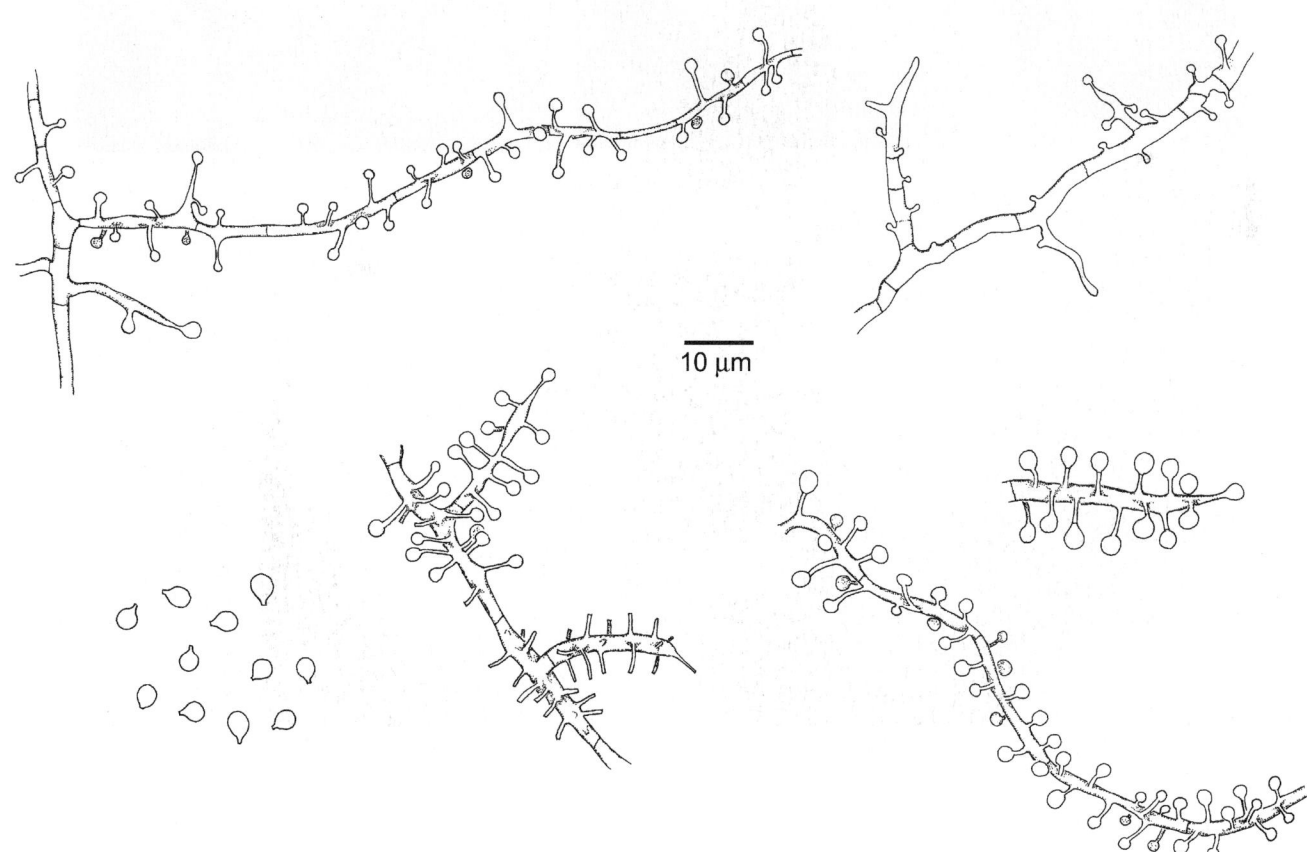

Myriodontium keratinophilum, FMR 3604. Fertile hyphae and conidia.

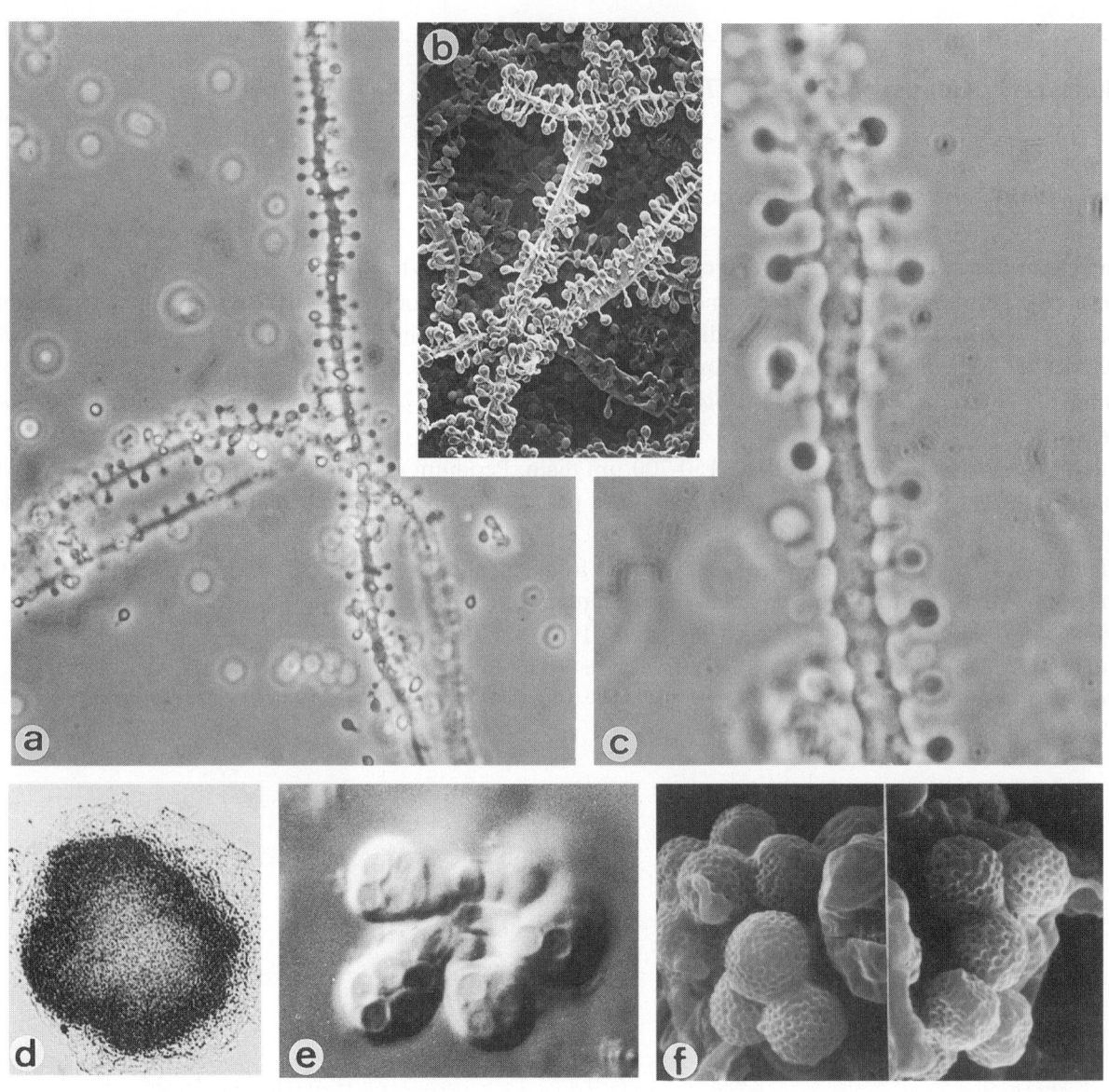

Myriodontium keratinophilum (anamorph), a-c. FMR 3604; d-f. *Neoarachnotheca keratinophila* (teleomorph), FMR 4017. a-c. Fertile hyphae and conidia; d. gymnothecium; e. asci; f. ascospores. a. ×640; b.×650; c, e. ×1600; d. ×60; f. ×4140.

Hyphomycetes. Genus: *NIGROSPORA*

Nigrospora sphaerica (Sacc.) Mason

Colony characteristics. Colonies (OA) expanding rapidly, lanose, at first white, later brown to black due to abundant sporulation.

Microscopy. Conidiogenous cells formed on superficial hyphae, lateral or terminal, swollen, ampulliform, 8-11 μm diam, hyaline, with a single conidium at the attenuated apex. Conidia spherical or oblate, solitary, black, opaque, smooth-walled, often with an equatorial furrow, 14-20 μm diam.

Pathogenicity. BSL-1. Plant-inhabiting species. Skin lesions in a leukemic patient were reported by Pritchard & Muir (1987). A case of keratitis probably caused by a *Nigrospora* sp. has been reported from India (Talwar & Sehgal, 1978).

References. Ellis (1971), Domsch *et al.* (1980).

Nomenclature. *Trichosporium sphaericum* Saccardo - Michelia 2: 579, 1882 ≡ *Nigrospora sphaerica* (Saccardo) Mason - Trans. Br. Mycol. Soc. 12: 158, 1927.

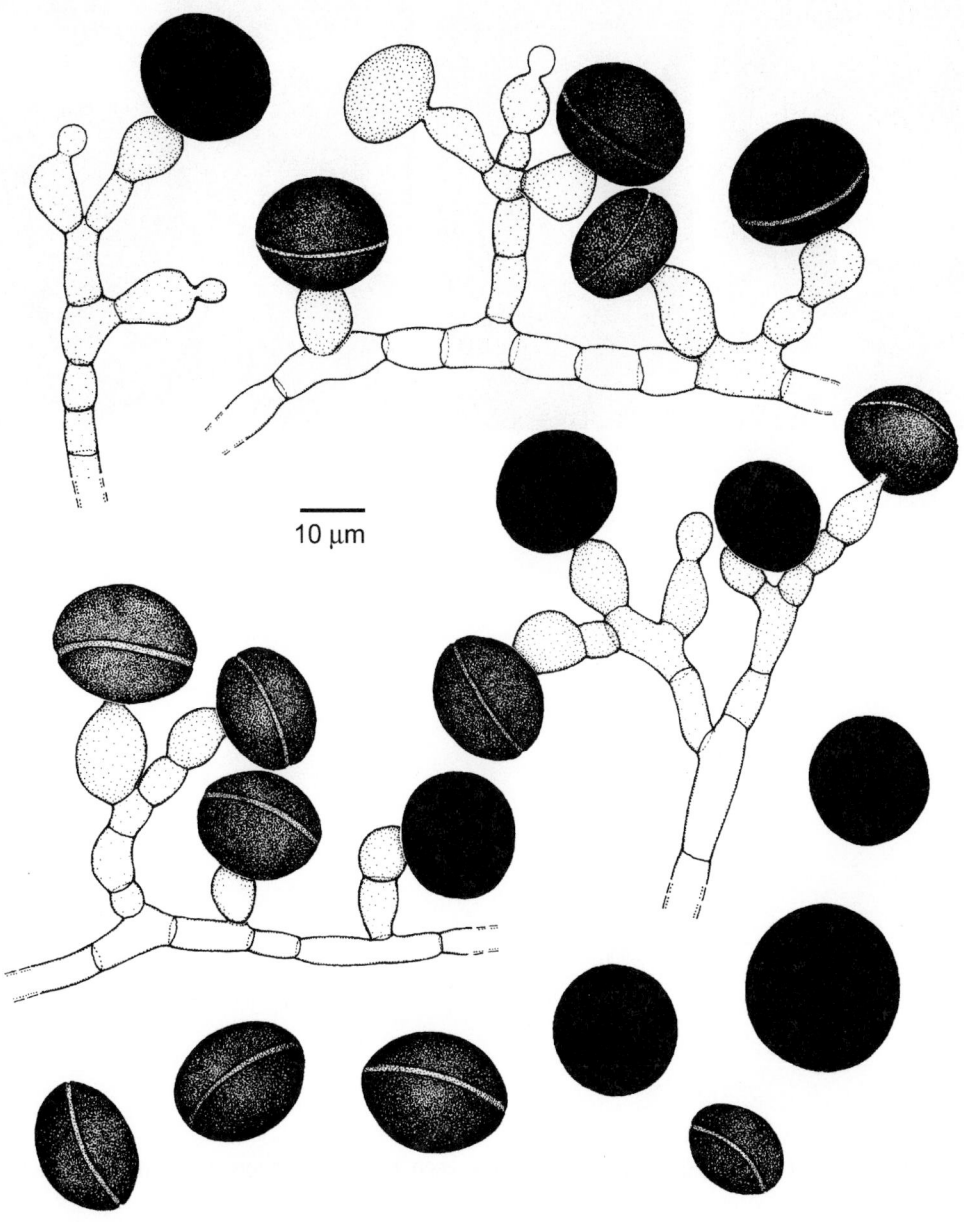

Nigrospora sphaerica, CBS 166.26. Conidiogenous cells and conidia.

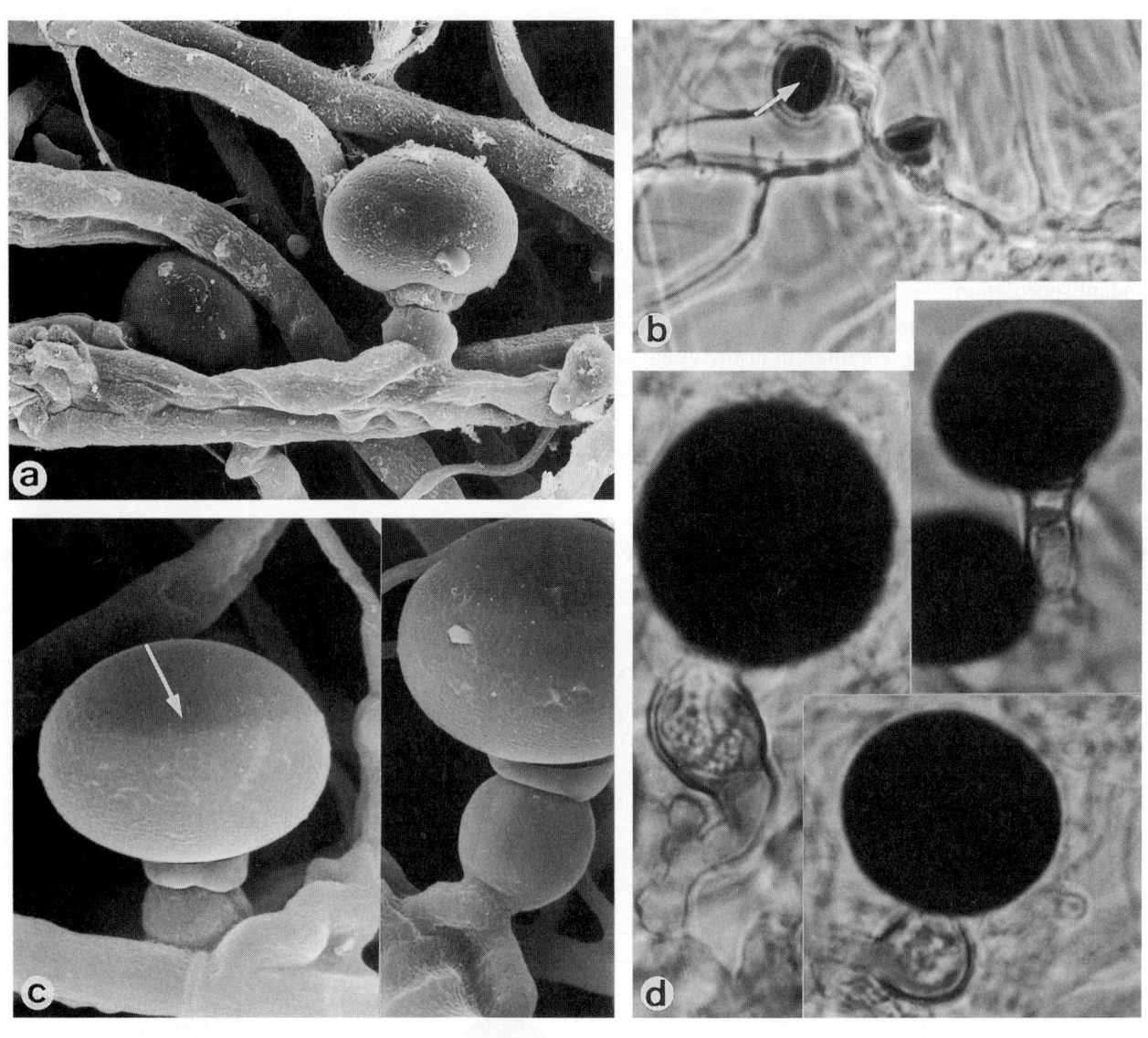

Nigrospora sphaerica, CBS 166.26. Conidiogenous cells and conidia. a. ×600; b. ×640; c. ×2700; d. ×1280. Arrows indicate germ slits.

Hyphomycetes. Genus: *OCHROCONIS*

Generic description. Colonies restricted, velvety to funiculose, brown to olivaceous, often with rust-brown reverse. Hyphae smooth- or somewhat rough-walled, pale olivaceous. Conidiophores slightly or conspicuously differentiated, cylindrical, often flexuose, producing conidia on scattered, cylindrical to conical denticles. After detachment an inconspicuous frill often remains both on the denticle and on the conidium base. Conidia 1- to 4-celled, pale olivaceous brown, smooth- or rough-walled, ellipsoidal, cylindrical, clavate or cuneiform.

Differential diagnosis. Colonies often exude rust-brown pigments into the agar. Conidia are liberated rhexolytically, i.e., the denticles are open, frills remaining on the conidial bases.

General remark. *Ochroconis gallopava* is a well-recognizable species, but the remaining taxa comprise a larger number of sibling species (Horré *et al.*, 1999).

References. Barron & Busch (1962), de Hoog (1983), Horré *et al.* (1999).

Key to the treated species of Ochroconis:

1a. Conidia usually 4-celled .... *O. tshawytschae* (**786**)
1b. Conidia usually 2-celled ➞ **2**
2a. Conidia distinctly clavate, upper cell being markedly wider than basal cell .... *O. gallopava* (**782**)
2b. Conidia broadly ellipsoidal or cylindrical, upper cell not markedly wider than basal cell ➞ **3**
3a. Conidia cylindrical, not constricted at the septum .... *O. humicola* (**784**)
3b. Conidia broadly ellipsoidal, constricted at the septum .... *O. constricta* (**779**)

Ochroconis constricta (Abbott) de Hoog & v. Arx

Colony characteristics. Colonies (OA) growing slowly, flat, velvety, brown, with rust-brown reverse.

Microscopy. Hyphae subhyaline to pale olivaceous, smooth- and rather firm-walled; aerial hyphae often strongly flexuose. Conidiophores arising laterally on hyphae, poorly differentiated, ovoidal to cylindrical, generally non-septate, bearing (1-) 3-6 conidia on long, open denticles. Conidia two-celled, verruculose, pale olivaceous brown, broadly ellipsoidal, constricted at the septum, with rounded ends, 5-12 × 2.5-4.0 μm.

Physiology. No growth on Mycosel agar; gelatin liquefaction positive.

Differential diagnosis. The species is close to *O. humicola*, but differs by shorter, non-septate conidiophores and shorter conidia which are constricted at the septum. *O. gallopava* has smooth-walled, nearly hyaline, clavate conidia.

Pathogenicity. BSL-1. The species was listed as a potentially neurotropic species by Dixon *et al.* (1986), but this report partly concerned *O. gallopava*, which had been synonymized with *O. constricta* by Dixon & Salkin (1986). *O. constricta* is regularly isolated from clinical specimens, but its pathogenic role has not yet been established.

Reference. Barron & Busch (1962).

Antifungal susceptibility.

Antifungal	Mean MICs	Strains	Reference
AMB	0.38	5	McGinnis & Pasarell (1998b)
ITZ	0.16	5	McGinnis & Pasarell (1998b)
VCZ	0.43	5	McGinnis & Pasarell (1998b)

Nomenclature. *Scolecobasidium constrictum* Abbott - Mycologia 19: 30, 1937 ≡ *Ochroconis constricta* (Abbott) de Hoog & von Arx - Kavaka 1: 57, 1973 ≡ *Dactylaria constricta* (Abbott) Dixon & Salkin - J. Clin. Microbiol. 24: 13, 1986.

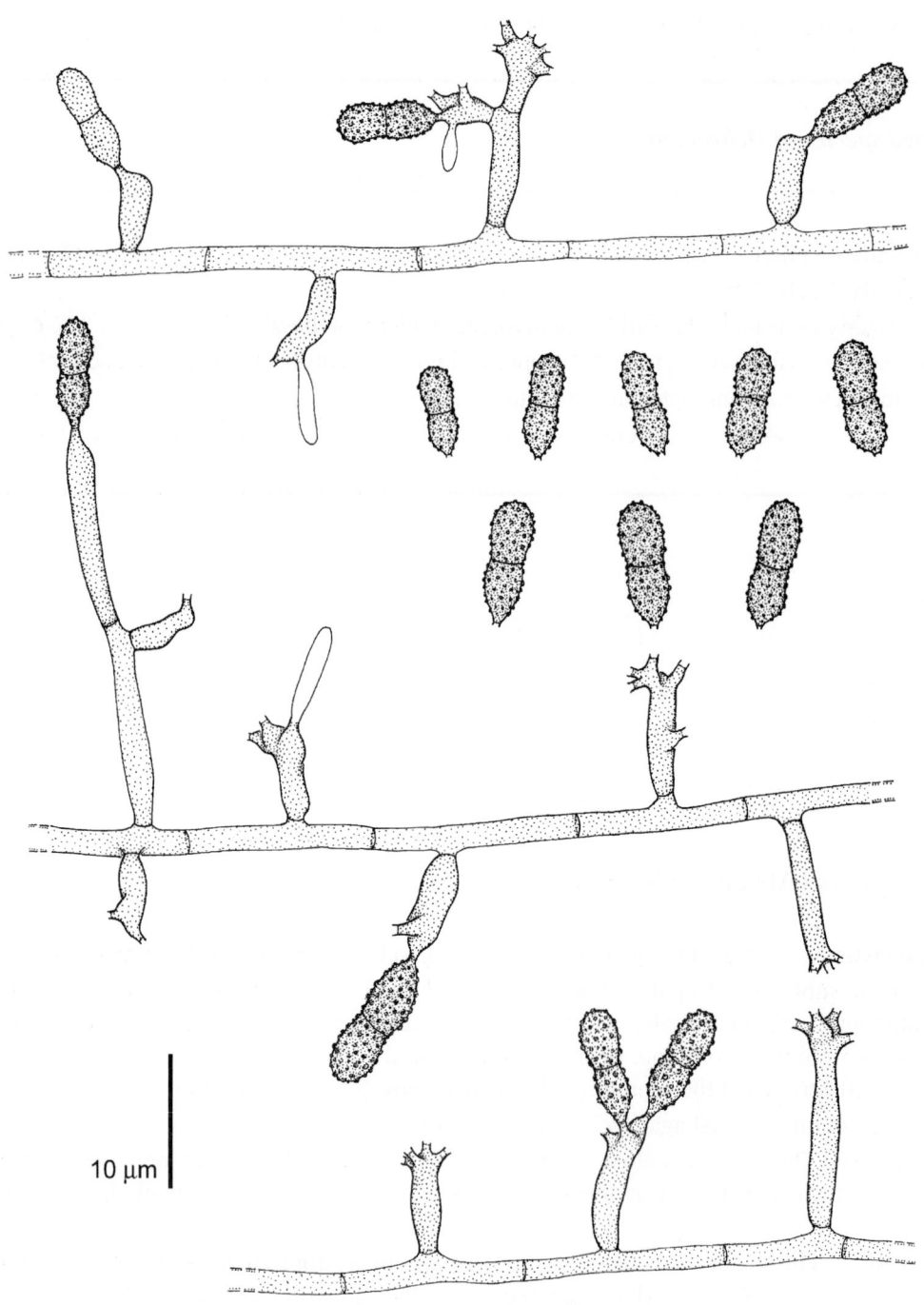

Ochroconis constricta, CBS 202.27. Conidiophores and conidia.

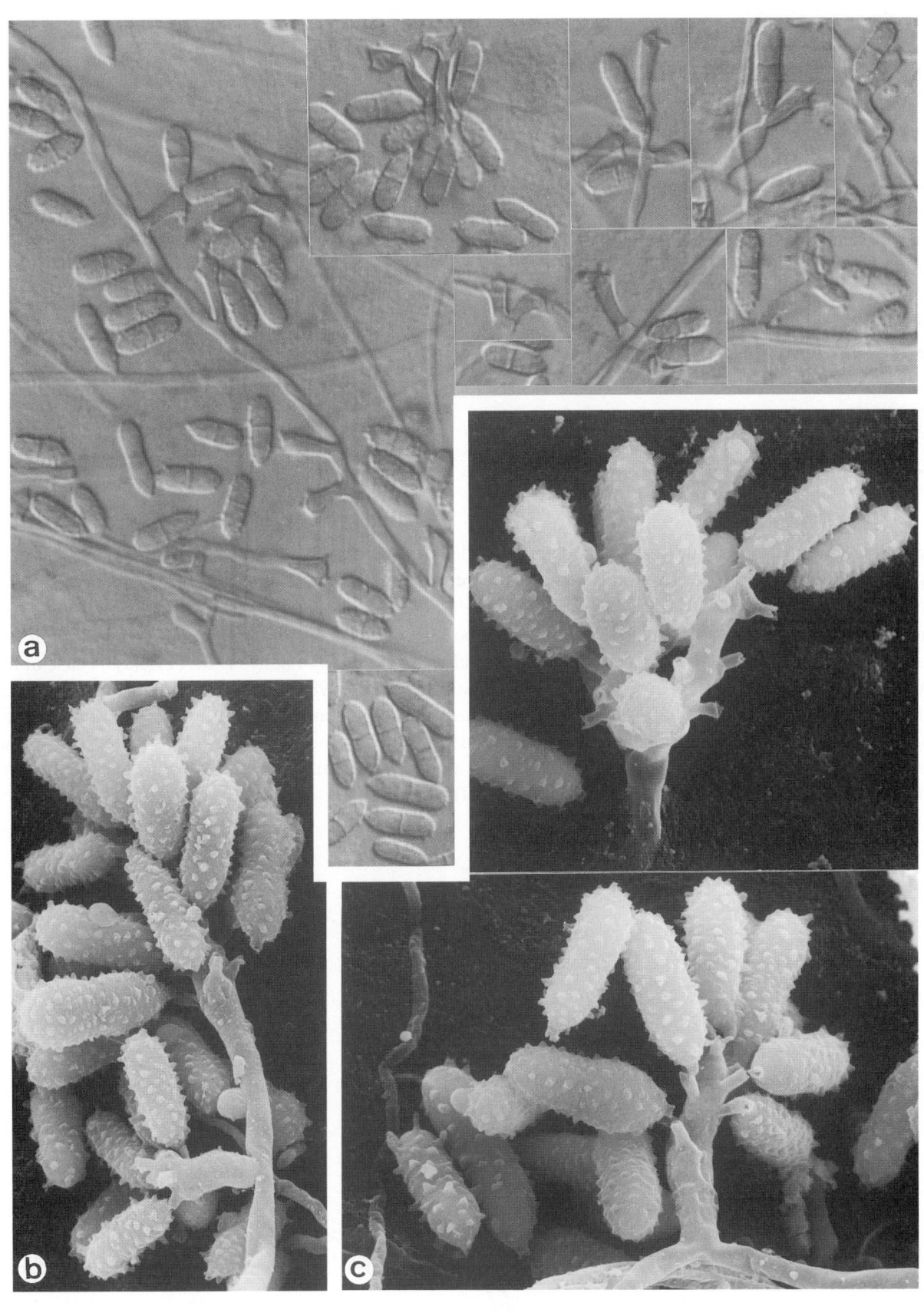

Ochroconis constricta, CBS 202.27. Conidiophores and conidia. a. ×1600; b. ×4300; c. ×500.

Ochroconis gallopava (W.B. Cooke) de Hoog

Colony characteristics. Colonies (OA) smooth to felty, dry, flat, tobacco-brown to brownish-black; a dark brown pigment is exuded into the agar.

Microscopy. Hyphae brown, with rather thick walls. Conidiophores flexible, mostly cylindrical to acicular, sometimes poorly differentiated, bearing few conidia at the tip. Denticles fragile, frills remaining on denticle and conidial base. Conidia two-celled, smooth-walled to verruculose, subhyaline to pale brown, clavate, constricted at septum, 11-18 × 2.5-4.5 μm, apical cell wider than the basal cell.

Physiology. No gelatin liquefaction. Optimum growth at 35°C, tolerant to 40°C. Intolerant to benomyl and cycloheximide.

Molecular diagnostics. ITS restriction map based on N759:

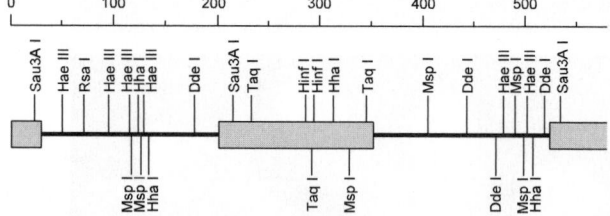

Differential diagnosis. The species is recognized by clavate, smooth- and thin-walled, nearly hyaline conidia. Antigenic relationships to other *Ochroconis* species were investigated by Sekhon *et al.* (1990), and rDNA RFLP diagnostics were developed by Horré *et al.* (1999).

Pathogenicity. BSL-2. The species is a relatively common agent of encephalitis in poultry (Blalock *et al.*, 1973) and other birds (Salkin *et al.*, 1990); a cerebral case in a cat was reported by Padhye *et al.* (1994a). A case of human subcutaneous phaeohyphomycosis was presented by Fukushiro *et al.* (1986), and a fatal endocarditis by Prevost-Smith *et al.* (1993). Sides *et al.* (1991) reported a case of phaeohyphomycotic brain abscess after treatment of malignant lymphoma with prednisone and Vukmir *et al.* (1994) and Kralovic & Rhodes (1995) after liver transplant; Jenney *et al.* (1998) reported a pulmonary infection in a heart transplant recipient. Systemic cases usually occur in immunocompromised patients (Horré & de Hoog, 1999).

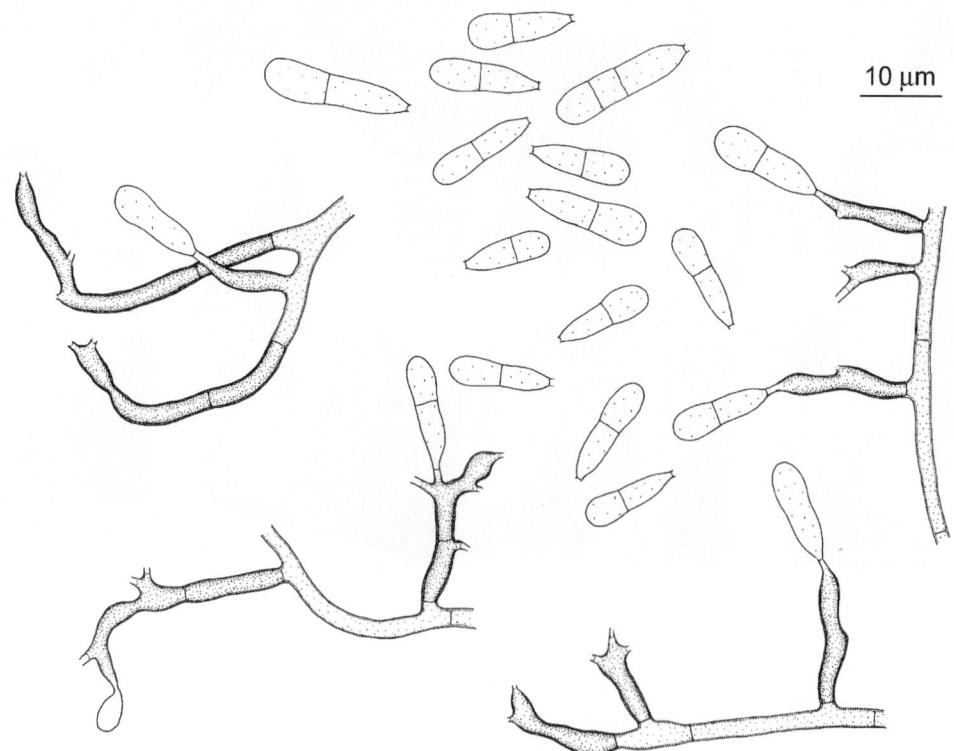

Ochroconis gallopava, CBS 437.64. a. Conidiogenous cells with frilled denticles; b. conidia.

References. Fukushiro *et al.* (1986), de Hoog (1983).

Antifungal susceptibility.

Antifungal	MICs range	MIC 90	Strains	Reference
AMB	0.06-0.5	0.5	13	Meletiadis *et al.* (1999)
AMB	0.25	nd	3	McGinnis & Pasarell (1998b)
5FC	4	nd	1	Unpublished data
FLZ	8	nd	1	Unpublished data
ITZ	0.01	0.01	3	McGinnis & Pasarell (1998b)
ITZ	0.03-0.5	0.25	13	Meletiadis *et al.* (1999)
KTZ	0.25	nd	1	Unpublished data
MCZ	0.5-2	2	13	Meletiadis *et al.* (1999)
TBF	0.03-0.06	0.06	13	Meletiadis *et al.* (1999)
VCZ	0.79	0.79	3	McGinnis & Pasarell (1998b)
VCZ	0.03-0.25	0.25	13	Meletiadis *et al.* (1999)

Nomenclature. *Diplorhinotrichum gallopavum* W.B. Cooke, *in* Georg, Bierer & W.B. Cooke - Sabouraudia 3: 241, 1964 ≡ *Dactylaria gallopava* (W.B. Cooke) Bhatt & Kendrick - Can. J. Bot. 46: 1257, 1968 ≡ *Ochroconis gallopava* (W.B. Cooke) de Hoog, *in* Howard - Fung. Path. Hum. Anim. B-II: 181, 1983 ≡ *Dactylaria constricta* (Abbott) Dixon & Salkin var. *gallopava* (W.B. Cooke) Salkin & Dixon - Mycotaxon 29: 379, 1987.

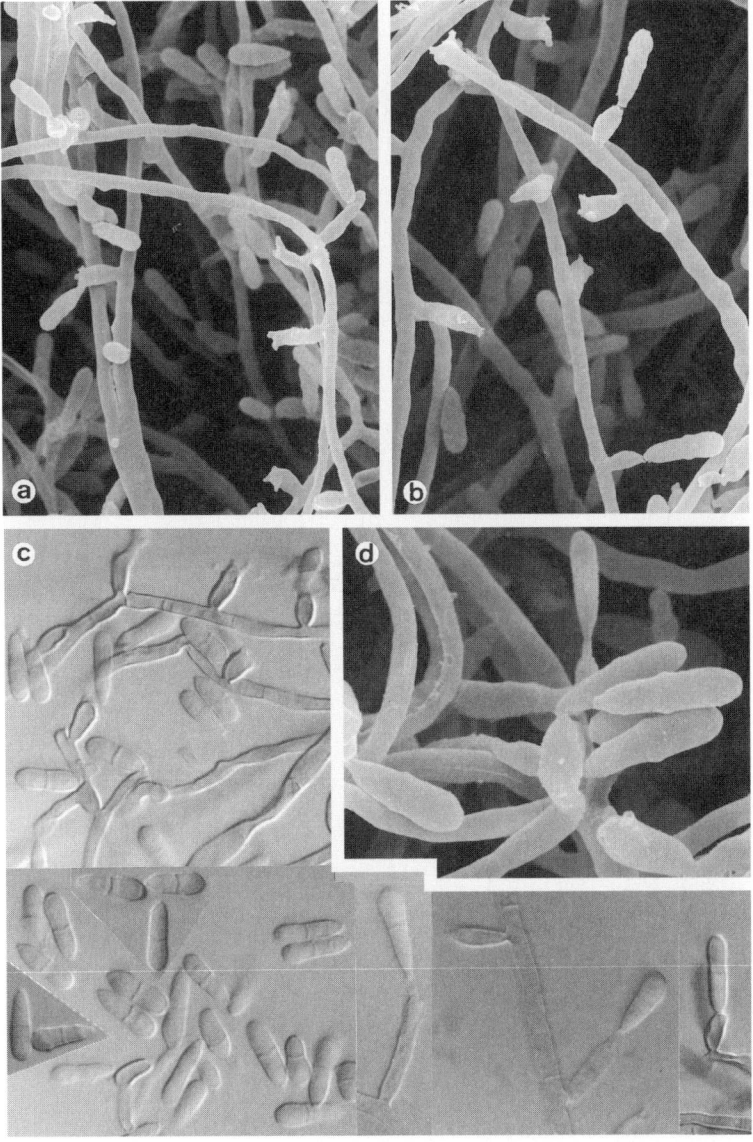

Ochroconis gallopava, CBS 265.97. Conidiogenous cells and conidia. a. ×2200; b. ×2700; c. ×1600; d. ×5000.

Ochroconis humicola (Barron & Busch) de Hoog & v. Arx

Colony characteristics. Colonies (OA) growing slowly, flat, velvety, brown, brownish-grey at the centre.

Microscopy. Hyphae subhyaline to pale olivaceous, smooth- and thick-walled, aerial hyphae often strongly flexouse; small sclerotial bodies may be present in the submerged mycelium. Conidiophores erect, flexuose, cylindrical, up to about 100 μm long, 2-3 μm wide; conidia produced on long denticles. Conidia mostly two-celled, smooth-walled or verruculose, pale olivaceous brown, cylindrical to slightly clavate, with rounded ends, 7-15 × 2.5-4.0 μm.

Physiology. No growth at 37°C; tolerant to cycloheximide; urease positive; tyrosine hydrolysis; no gelatin liquefaction.

Pathogenicity. BSL-2. The species is a rare pathogen of cold-blooded vertebrates, particularly fish, such as coho salmon (Ross & Yasutake, 1973), atlantic salmon (Schaumann & Priebe, 1994), rainbow trout (Ajello *et al.*, 1977), *Inimicus japonicus* (Wada *et al.*, 1995) and *Clarias batrachus* (Bhattacharya, 1988). It was also isolated from diseased frogs (Elkan & Philpot, 1973) and from a tortoise (Weitzman *et al.*, 1985). An ulcerative lesion in a human patient was reported by Goldschmied-Reouven *et al.* (1994) and in a cat by VanSteenhouse *et al.* (1988).

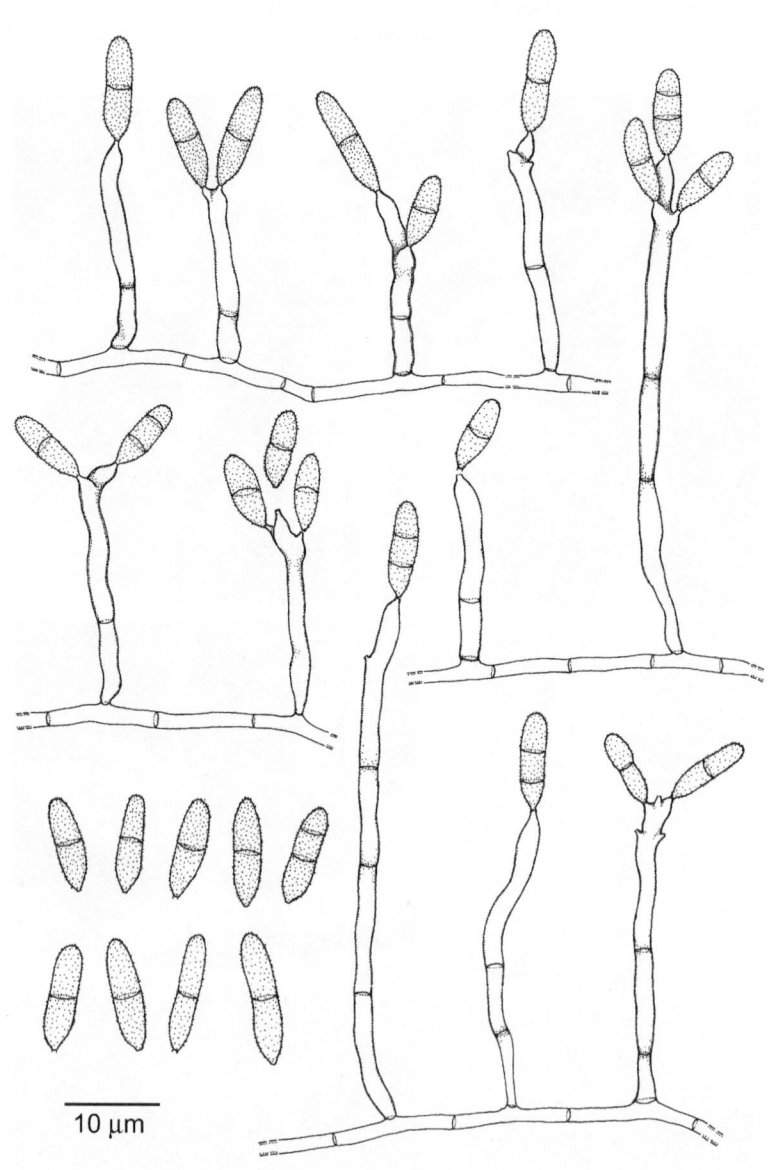

Ochroconis humicola, CBS 780.83. Conidiophores and conidia.

Antifungal susceptibility.

Antifungal	Mean MICs	Strains	Reference
AMB	0.7	4	McGinnis & Pasarell (1998b)
ITZ	0.07	4	McGinnis & Pasarell (1998b)
VCZ	0.21	4	McGinnis & Pasarell (1998b)

Reference. De Hoog (1983).

Nomenclature. *Scolecobasidium humicola* Barron & Busch - Can. J. Bot. 40: 83, 1962 ≡ *Ochroconis humicola* (Barron & Busch) de Hoog & von Arx - Kavaka 1: 57, 1973.

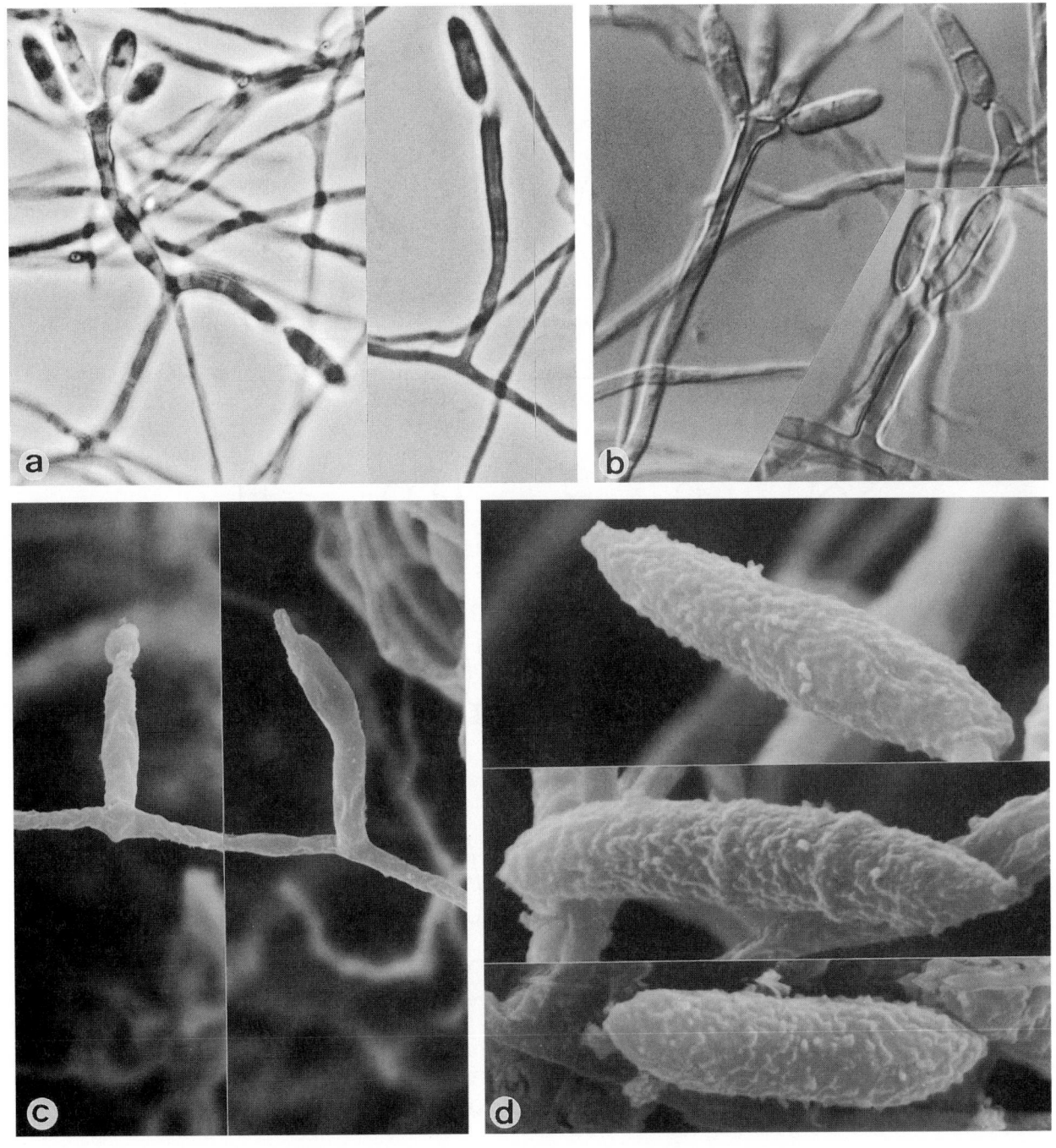

Ochroconis humicola, CBS 780.83. Conidiophores and conidia. a. ×1280; b. ×1600; c. ×2800; d. ×8500.

Ochroconis tshawytschae (Doty & Slater) Kirilenko & Al-Achmed

Colony characteristics. Colonies (OA) growing slowly, slightly domed, velvety, olivaceous brown to olive; reverse olivaceous black.

Microscopy. Hyphae subhyaline to pale olivaceous, smooth- and thick-walled, forming a compact mycelium. Conidiophores erect, straight, cylindrical, often somewhat inflated or flexuose, up to 30 μm long, 2.0-3.5 μm wide, conidia produced on denticles in the apical region. Conidia (2-)4-celled, verrucose, pale brown, cylindrical or slightly clavate, with rounded ends, 12-25 × 3.5-5.5 μm.

Pathogenicity. BSL-2. A case of infection in chinook salmon was reported (Doty & Slater, 1946).

Reference. De Hoog (1983).

Nomenclature. *Heterosporium tshawytschae* Doty & Slater - Am. Midl. Nat. 36: 663, 1946 ≡ *Scolecobasidium tshawytschae* (Doty & Slater) McGinnis & Ajello - Trans. Br. Mycol. Soc. 63: 202, 1974 ≡ *Ochroconis tshawytschae* (Doty & Slater) Kirilenko & Al-Achmed - Mykrobiol. Zh. 39: 305, 1977.

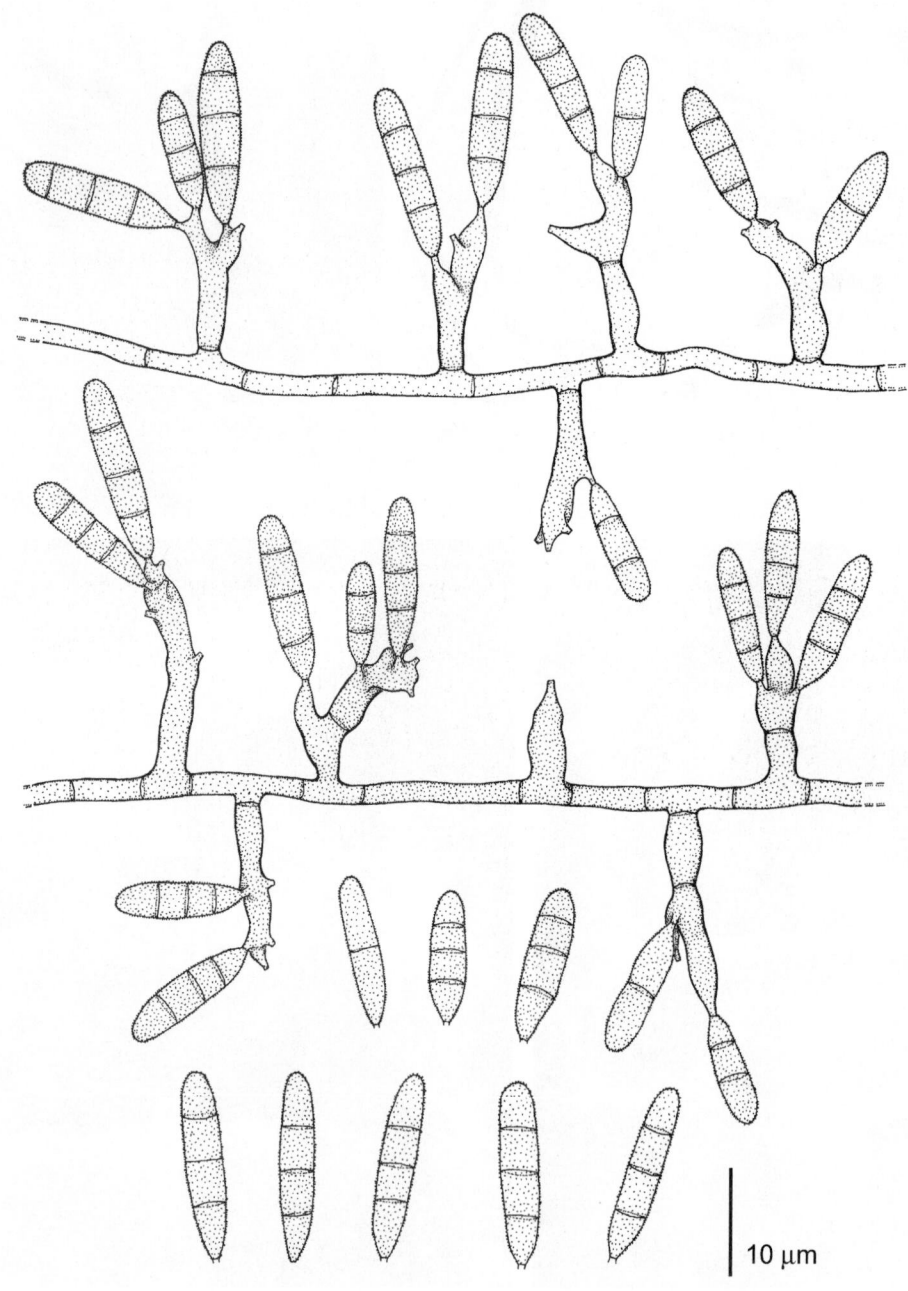

10 μm

Ochroconis tshawytschae, CBS 383.81. Conidiophores and conidia.

786

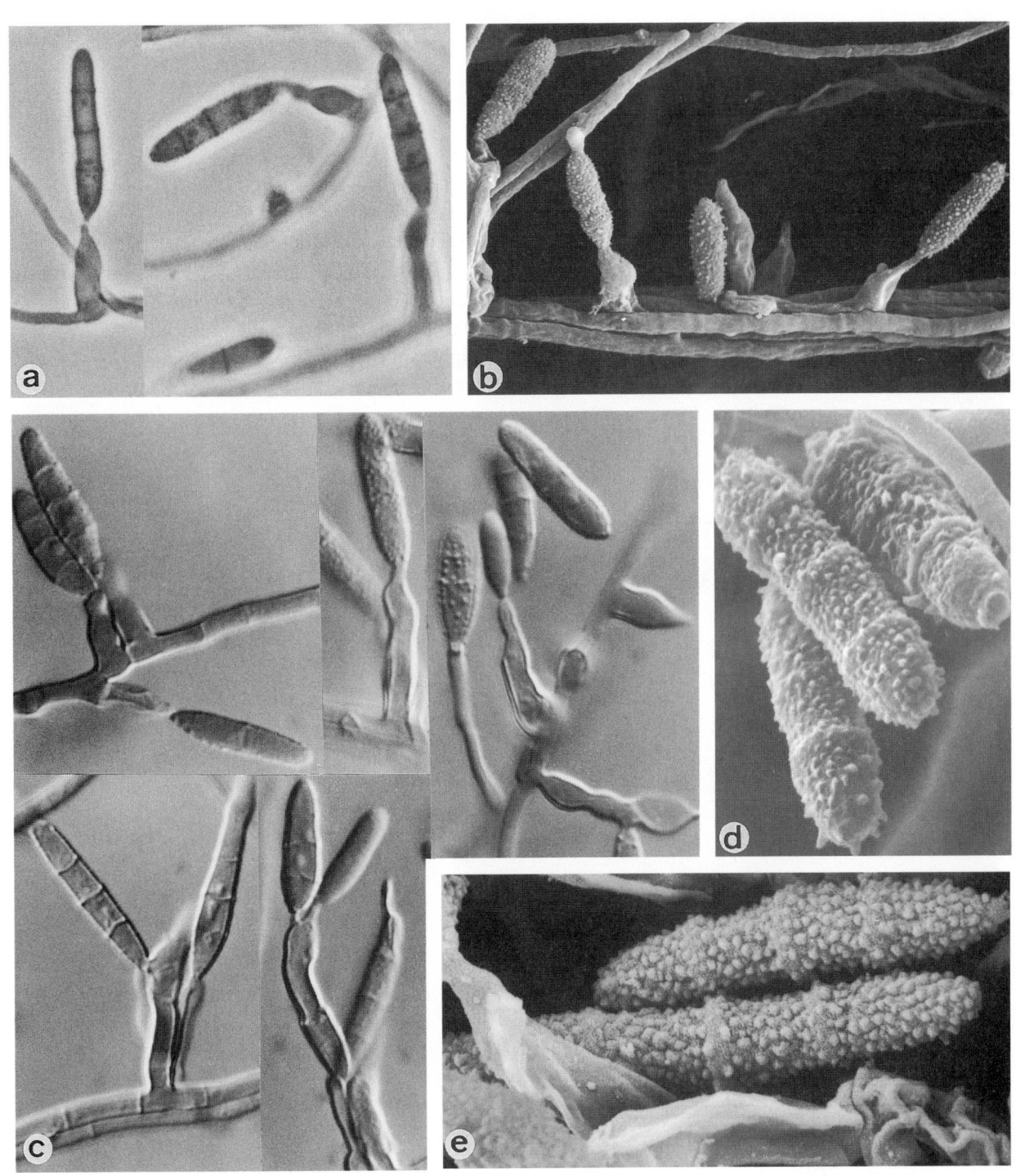

Ochroconis tshawytschae, CBS 383.81. Conidiophores and conidia. a. ×1280; b. ×2150; c. ×1600; d. ×6300; e. ×7000.

H yphomycetes. Genus: *OIDIODENDRON*

Oidiodendron cerealis (Thüm.) Barron

Colony characteristics. Colonies (OA) restricted, powdery, violaceous to purple-black; reverse purple-black.

Microscopy. Conidiophores erect, straight or flexuose, brown, up to 50 µm long, 1.5-2.5 µm wide, sometimes lacking, apically more or less verticillately repeatedly branched, the terminal branches forming chains of arthroconidia separated by disjunctors. Arthroconidia lenticular, brown, smooth-walled, unicellular, 3.5-5.5 × 2-3 µm, with a very distinctive equatorial band which protrudes slightly at each end.

Molecular diagnostics. ITS restriction map based on NCBI AF062788:

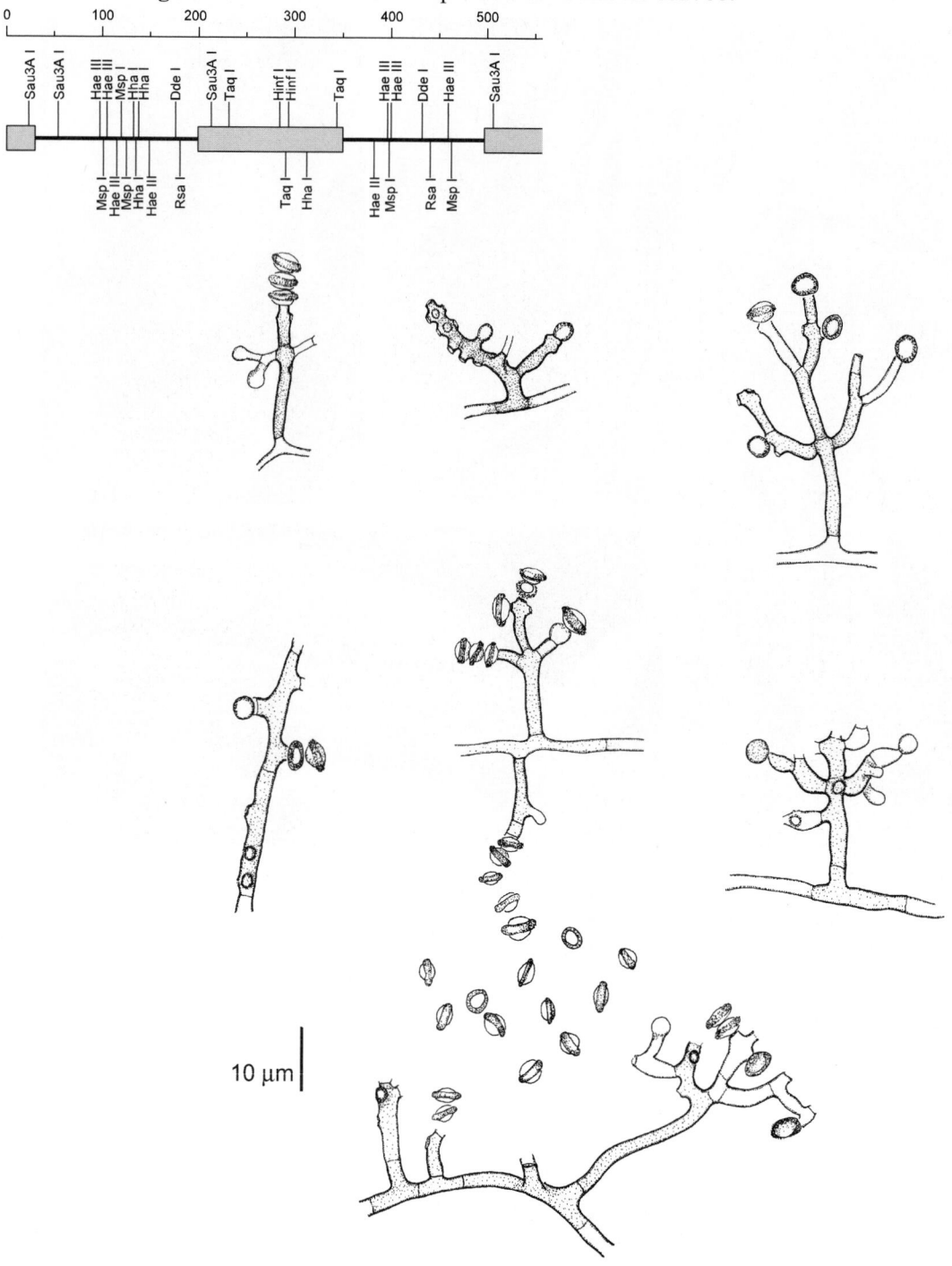

Oidiodendron cerealis, CBS 957.72. Conidiophores and conidia.

Pathogenicity. BSL-1. Isolated from a patient with neurodermitis nuchae (Blomquist & Salonen, 1969).

References. Ellis (1971), Domsch *et al.* (1980).

Nomenclature. *Sporotrichum cerealis* von Thümen - Hedwigia 19: 296, 1880 ≡ *Oidiodendron cerealis* (von Thümen) Barron - Can. J. Bot. 40: 594, 1962 ≡ *Stephanosporium cereale* (von Thümen) Swart - Trans. Br. Mycol. Soc. 48: 459, 1965.

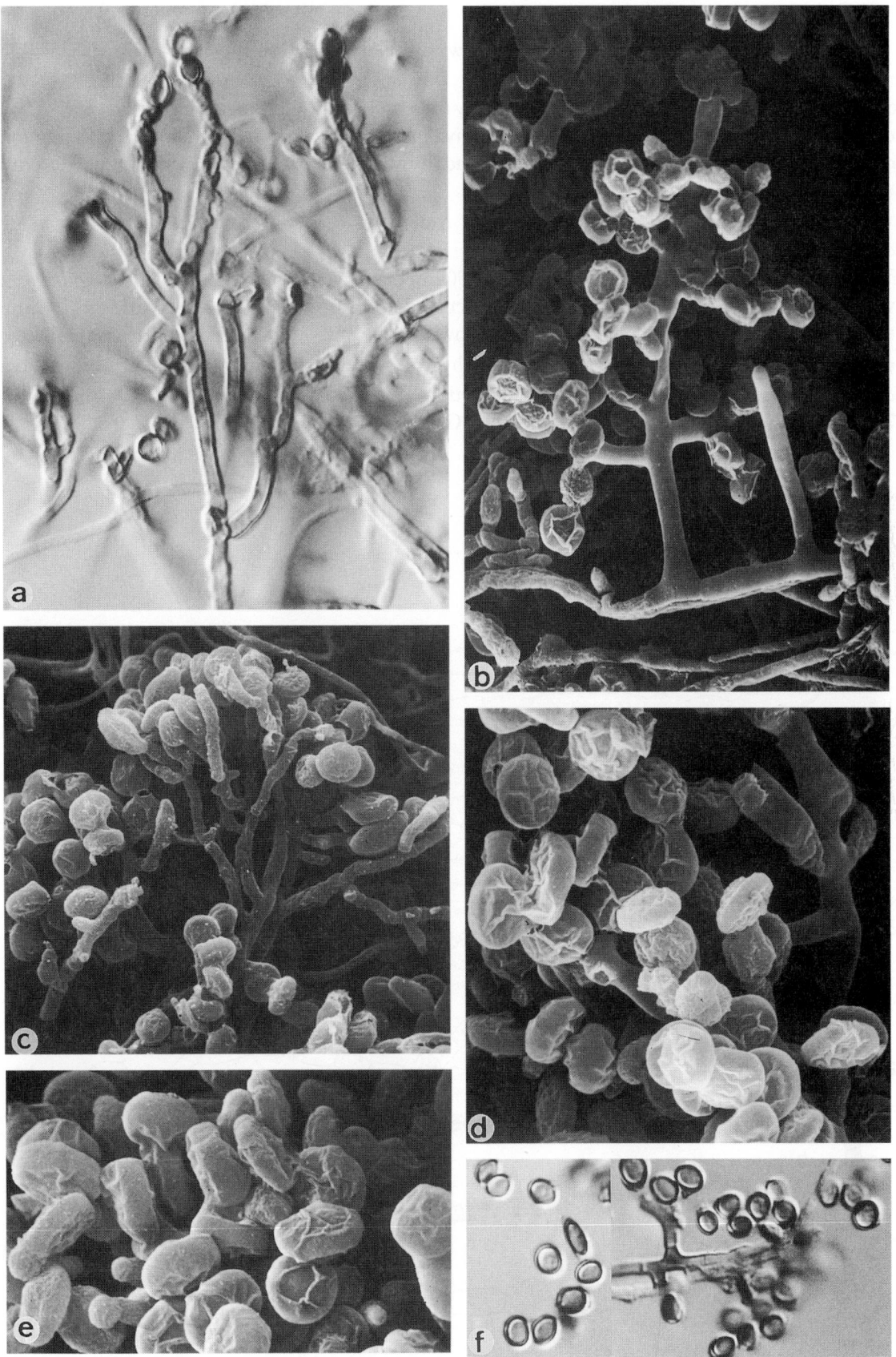

Oidiodendron cerealis, CBS 957.72. Conidiophores and arthroconidia. a. ×1600; b. ×3000; c. ×2550; d. ×4700; e. ×5400; f. ×1600.

Hyphomycetes. Genus: *ONYCHOCOLA*

Onychocola canadensis Sigler

Colony characteristics. Colonies (PDA) growing slowly, velvety to lanose, white to yellowish, with brownish reverse.

Microscopy. Fertile hyphae scarcely differentiated, hyaline, smooth-walled or slightly verrucose. Arthroconidia cylindrical to broadly ellipsoidal, one- or two-celled, hyaline to subhyaline, 4-16 × 2-5 µm, forming long chains which often secede with difficulty. In old cultures broad, brown, septate, sterile hyphae with darker brown knobs are present.

Teleomorph. *Arachnomyces nodososetosus* Sigler & Abbott (*Ascomycota, Euascomycetes, Onygenales: Onygenaceae*).

Ascomata spherical, non-ostiolate, dark brown, up to 450 µm diam, wall composed of *textura angularis*, bearing 3-8 thick-walled, about 1 µm long setae which are often curved at the tip. Asci (sub)spherical, 6.5-11.0 µm diam, 8-spored, evanescent. Ascospores smooth-walled, pale brown, oblate, 4.0-4.5 × 3.0-3.5 µm. Heterothallic.

Physiology. Intolerant to benomyl; casein and tyrosine hydrolysed; no assimilation of xanthine or hypoxanthine.

Pathogenicity. BSL-2. Known from nail and skin infections (Sigler & Congly, 1990; Sigler *et al.*, 1994; Koenig *et al.*, 1997; Chabasse *et al.*, 1997; Campbell *et al.*, 1997; Gupta *et al.*, 1998).

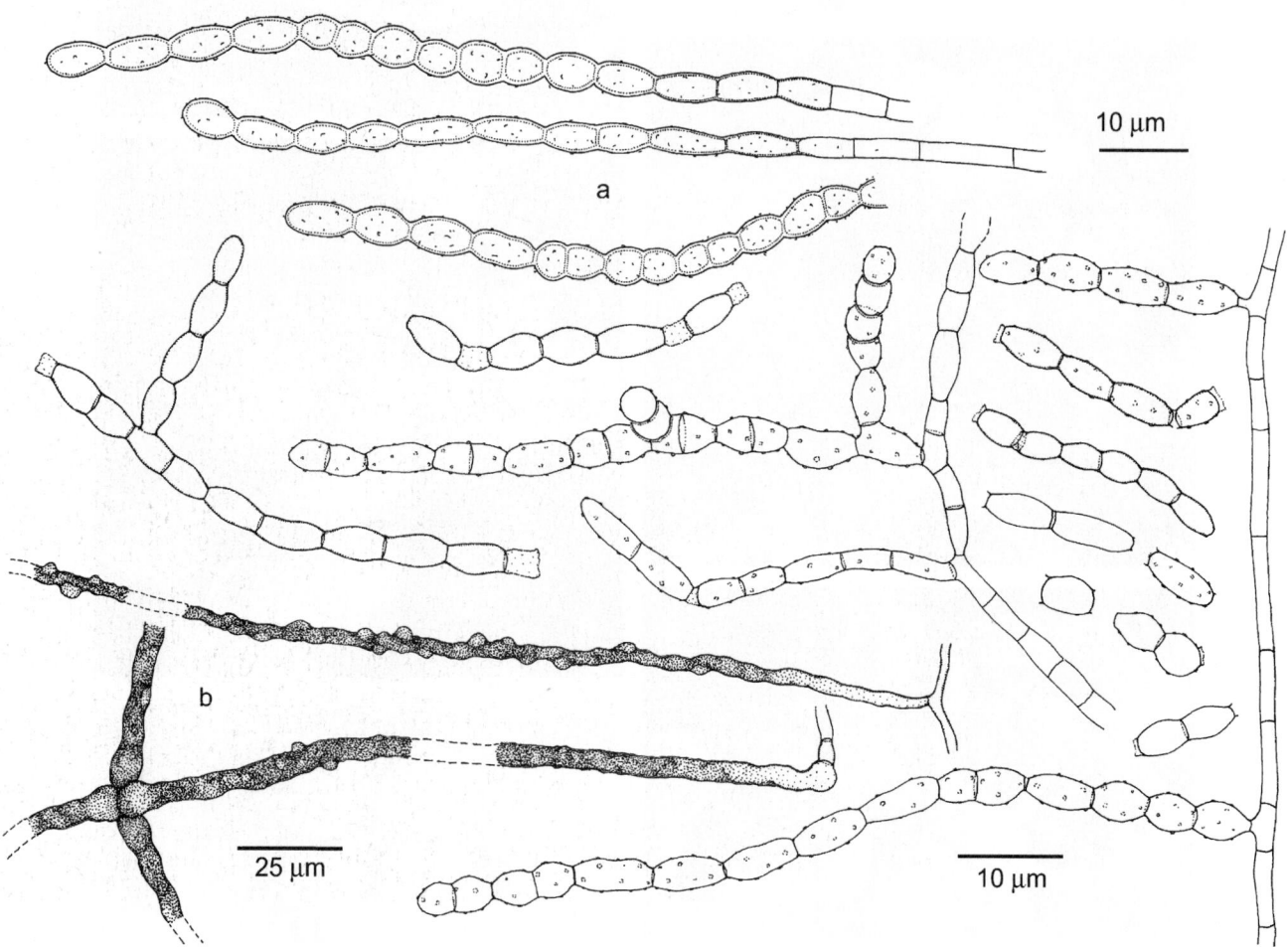

Onychocola canadensis, CBS 113.90. a. Fertile hyphae and arthroconidia; b. dark, sterile hyphae.

Reference. Sigler & Congly (1990), Sigler *et al.* (1994).

Nomenclature. *Onychocola canadensis* Sigler, *in* Sigler & Congly - J. Med. Vet. Mycol. 28: 409, 1990.

   Arachnomyces nodososetosus Sigler & Abbott, *in* Sigler, Abbott & Woodgyer - J. Med. Vet. Mycol. 32: 280, 1994.

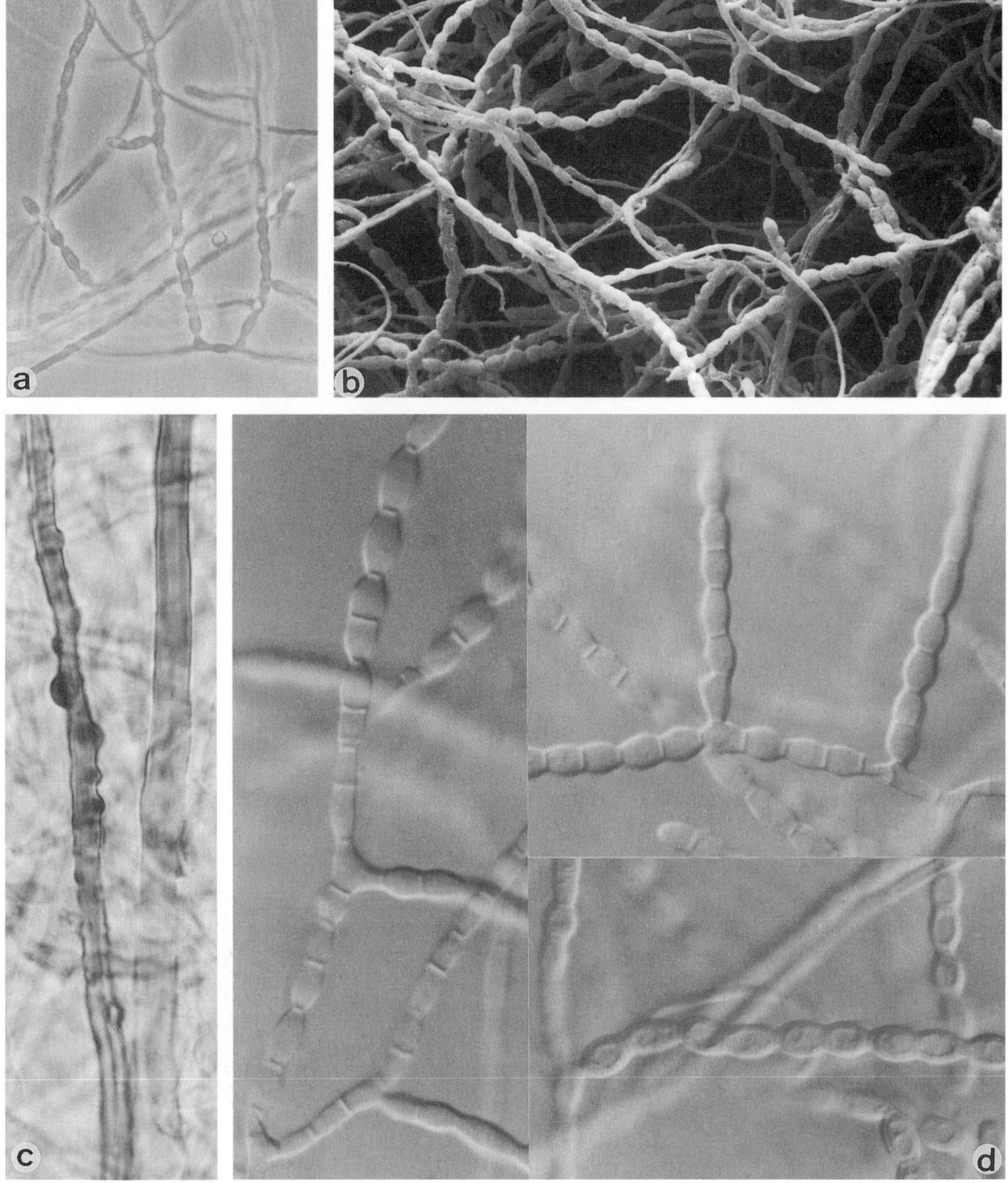

Onychocola canadensis, CBS 113.90. a, b, d. Fertile hyphae and arthroconidia; c. dark sterile hyphae. a. ×512; b. ×800; c. ×1280; d. × 1600.

Hyphomycetes. Genus: *OVADENDRON*

Ovadendron sulphureo-ochraceum (v. Beyma) Sigler & Carmichael

Colony characteristics. Colonies (MEA) rather restricted, white, floccose, with yellowish reverse.

Microscopy. Hyphae hyaline, septate, narrow. Fertile branches narrow, somewhat coiled, developing into a series of swollen arthroconidia and unswollen empty cells, maturating from the apex downwards. Conidia barrel-shaped, 2.5-4.0 × 2.0-2.5 µm.

Pathogenicity. BSL-1. The species was originally isolated from sputum of a patient with tuberculosis (van Beyma, 1933). A case of endophthalmitis after cataract extraction was reported by Lee *et al.* (1995). Further strains held in the CBS collection originate from sputum and from skin.

Reference. Sigler & Carmichael (1976).

Nomenclature. *Oospora sulphureo-ochracea* van Beyma - Zentbl. Bakt. Parasitkde, Abt. 2, 88: 134, 1933 ≡ *Ovadendron sulphureo-ochraceum* (van Beyma) Sigler & Carmichael - Mycotaxon 4: 392, 1976.

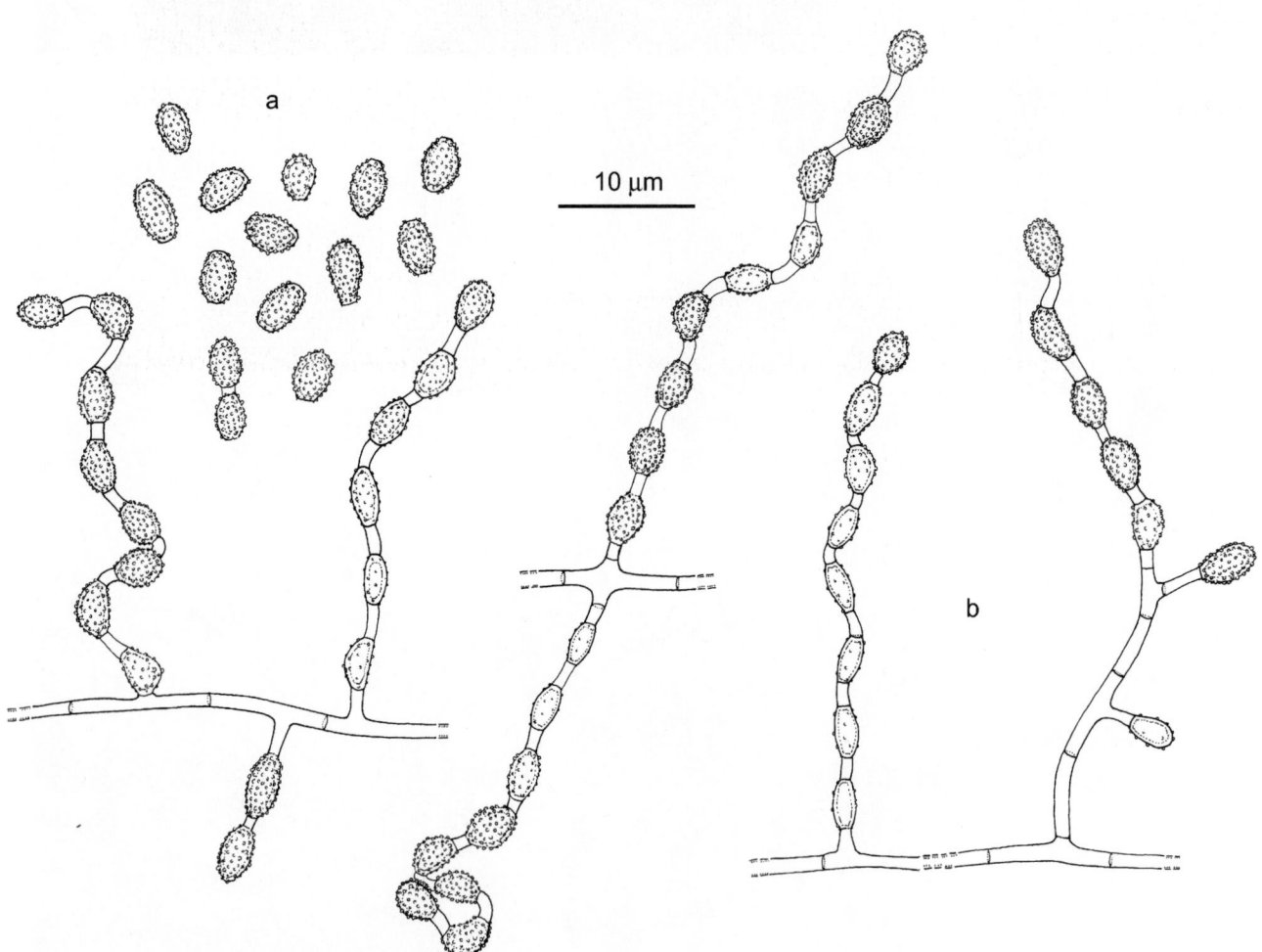

Ovadendron sulphureo-ochraceum, CBS 233.32. a. Liberated conidia; b. fertile hyphae bearing conidia.

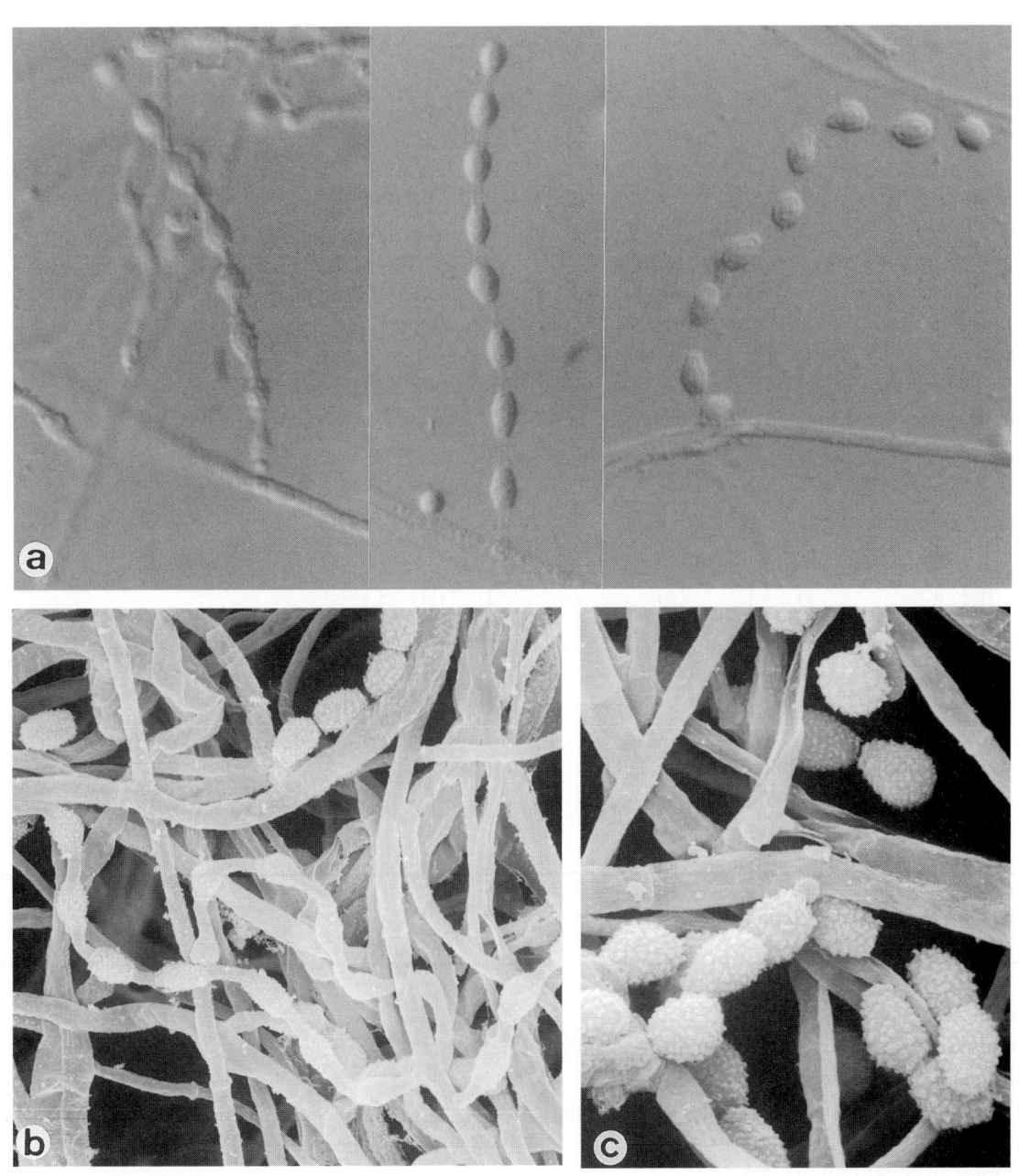

Ovadendron sulphureo-ochraceum, CBS 233.32. Fertile hyphae and liberated conidia. a. ×1600; b. ×3100; c. ×5500.

Hyphomycetes. Genus: *PAECILOMYCES*

Generic description. Colonies usually spreading broadly, white, brownish or in bright colours. Conidiophores simple, or irregularly or verticillately branched, bearing whorls of conidiogenous cells. Conidiogenous cells phialidic, swollen at the base and gradually narrowed into a long beak. Conidia formed in divergent chains, one-celled, hyaline or slightly pigmented, smooth-walled or echinulate, of various shapes. Chlamydospores, when present, usually thick-walled, smooth-walled or ornamented, borne singly or in short chains.

Teleomorphs. *Byssochlamys, Chromocleista, Talaromyces* (*Ascomycota, Euascomycetes, Eurotiales: Trichocomaceae*), *Thermoascus* (*Ascomycota, Euascomycetes, Eurotiales: Thermoascaceae*).

Molecular diagnostics. SSU restriction map based on NCBI M83263 and NCBI Y13996:

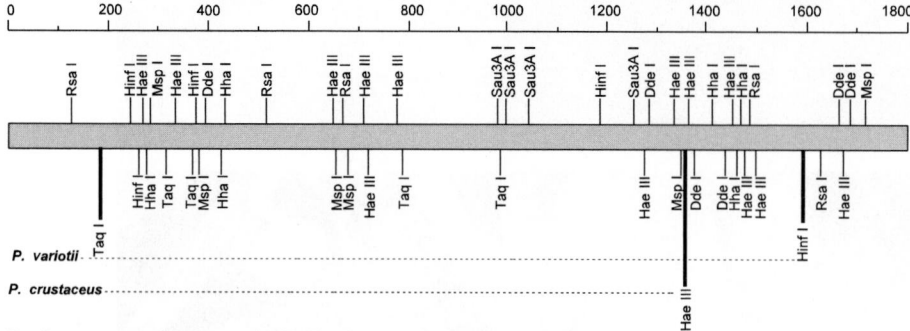

References. Samson (1974), Domsch *et al.* (1980).

Key to the treated species of Paecilomyces:

1a.	Ascomata present in culture; thermophilic	*P. crustaceus* (795)
1b.	Ascomata absent in culture; mesophilic or thermotolerant ➜ 2	
2a.	Colonies yellow-brown; with sweet odour	*P. variotii* (805)
2b.	Colonies white or in bright colours; non-aromatic ➜ 3	
3a.	Colonies vinaceous to violet ➜ 4	
3b.	Colonies in other colours ➜ 5	
4a.	Conidiophore stipes pigmented, rough-walled; chlamydospores absent	*P. lilacinus* (801)
4b.	Conidiophore stipes hyaline, smooth-walled; chlamydospores present	*P. marquandii* (803)
5a.	Mature colonies green or pink ➜ 6	
5b.	Mature colonies of other colour ➜ 7	
6a.	Colonies yellow-green; conidia spherical, 2.5-3.2 µm diam	*P. viridis* (808)
6b.	Colonies pink; conidia fusiform to cylindrical, 3-4 × 1-2 µm	*P. fumosoroseus* (797)
7a.	Chlamydospores absent; conidia fusiform, sometimes cylindrical, 4.0-7.4 × 1.2-1.7 µm	*P. javanicus* (799)
7b.	Chlamydospores present; conidia ellipsoidal to cylindrical, 3.5-4.0 × 1.7-2.0 µm	*P. puntonii* (1031)

Paecilomyces crustaceus Apinis & Chesters

Colony characteristics. Colonies (MEA 2%, 40°C) spreading broadly, white to pale orange.

Microscopy. Conidiophores emerging from the mycelial felt or trailing hyphae, up to 400 long μm, 3-10 μm wide, repeatedly branched. Phialides solitary or in whorls, cylindrical, tapering into a long tube, 17-24 × 5-6 μm. Conidia cylindrical to ellipsoidal, 7-8 × 5.0-6.5 μm, pale olivaceous brown, smooth-walled, in long, divergent chains.

Teleomorph. *Thermoascus crustaceus* (Apinis & Chesters) Stolk (*Ascomycota, Euascomycetes, Eurotiales: Thermoascaceae*).

Ascomata superficial, subspherical to hemispherical, non-ostiolate, yellowish-orange to reddish-orange, mostly 600-800 μm diam, surrounded by a loose cover of aerial hyphae. Peridium pseudoparenchymatous. Asci 8-spored, subspherical, irregularly arranged in the centrum. Ascospores pale yellow to pale brown, ovoidal to ellipsoidal, 7-8 × 4.5-6.0 μm, thick-walled, finely echinulate.

Physiology. The species is thermophilic.

Molecular diagnostics. SSU and ITS restriction maps based on NCBI M83263 and U18353, resp.:

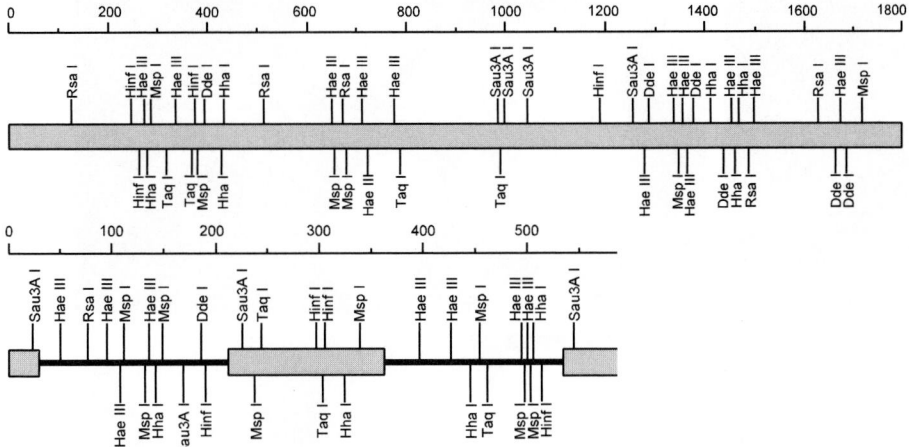

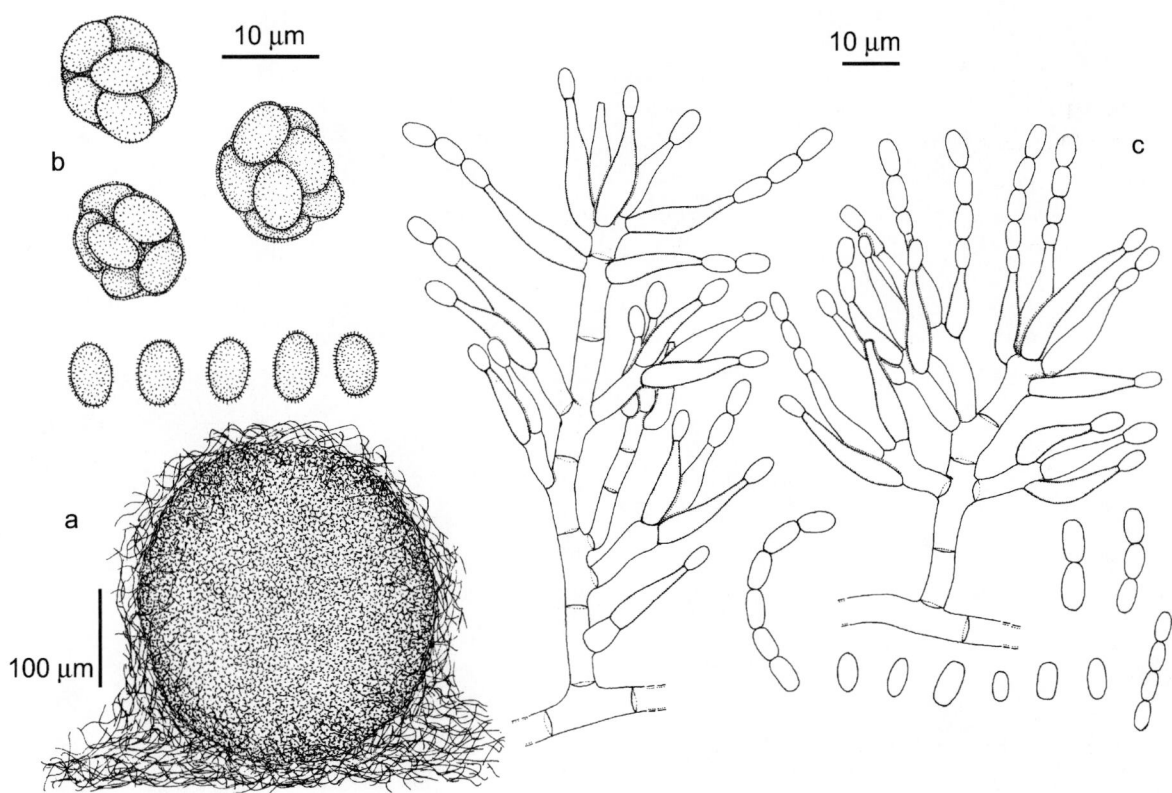

Paecilomyces crustaceus (*Thermoascus crustaceus*), IMI 158740. a. Ascomata; b. asci and ascospores; c. anamorph: conidiophores and conidia.

Pathogenicity. BSL-1. This fungus was isolated from the small intestine of a mouse and from a bronchial lavage of a patient with a pulmonary infarction (Pore & Larsh, 1967). It was grown repeatedly from monocyte cultures of patients suffering from AIDS (Kwon-Chung *et al.*, 1984). *Thermoascus taitungiacus* K.Y. Chen & Z.C. Chen, practically indistinguishable from *T. crustaceus* var. *verrucosus* Yaguchi *et al.*, has recently been reported as causing peritonitis (Korzets *et al.*, 2000).

References. Stolk (1965), Awao & Otsuka (1974).

Nomenclature. *Dactylomyces crustaceus* Apinis & Chesters - Trans. Br. Mycol. Soc. 49: 428, 1964 ≡ *Thermoascus crustaceus* (Apinis & Chesters) Stolk - Antonie van Leeuwenhoek 31: 272, 1965.

    Paecilomyces crustaceus Apinis & Chesters - Trans. Br. Mycol. Soc. 47: 428, 1964.

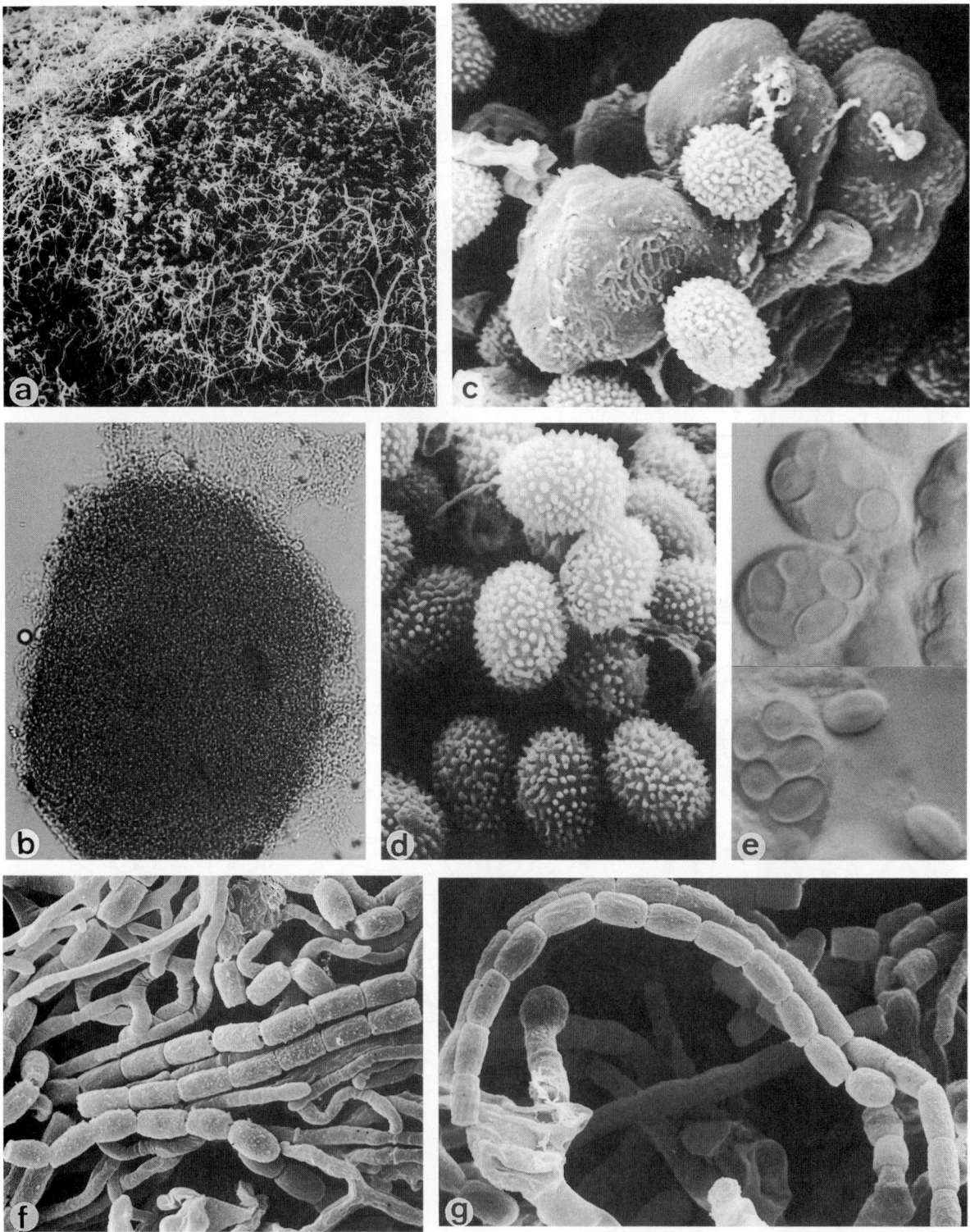

Paecilomyces crustaceus (*Thermoascus crustaceus*), IMI 158740. a, b. Ascomata; c-e. asci and ascospores; f, g. conidial chains. a. ×100; b. ×128; c. ×3400; d. ×4300; e. ×1600; f. ×2025; g. ×2150.

Paecilomyces fumosoroseus (Wize) A.H. Brown & G. Smith

Colony characteristics. Colonies (MEA 2%) rather slow growing, floccose, pink.
Microscopy. Conidiophores erect, sometimes forming synnemata, up to 100 μm in length, 1.5-2.0 μm wide, smooth-walled, hyaline, bearing several compact whorls of phialides which have a strongly inflated base tapering into a long and narrow neck. Conidia cylindrical to fusiform, smooth-walled, hyaline to pale pink, 3-4 × 1-2 μm.
Pathogenicity. BSL-1. Hyalohyphomycoses in a cat (Elliott *et al.*, 1984) and in a giant tortoise (Georg *et al.,* 1962b).
References. Samson (1974), Domsch *et al.* (1980).

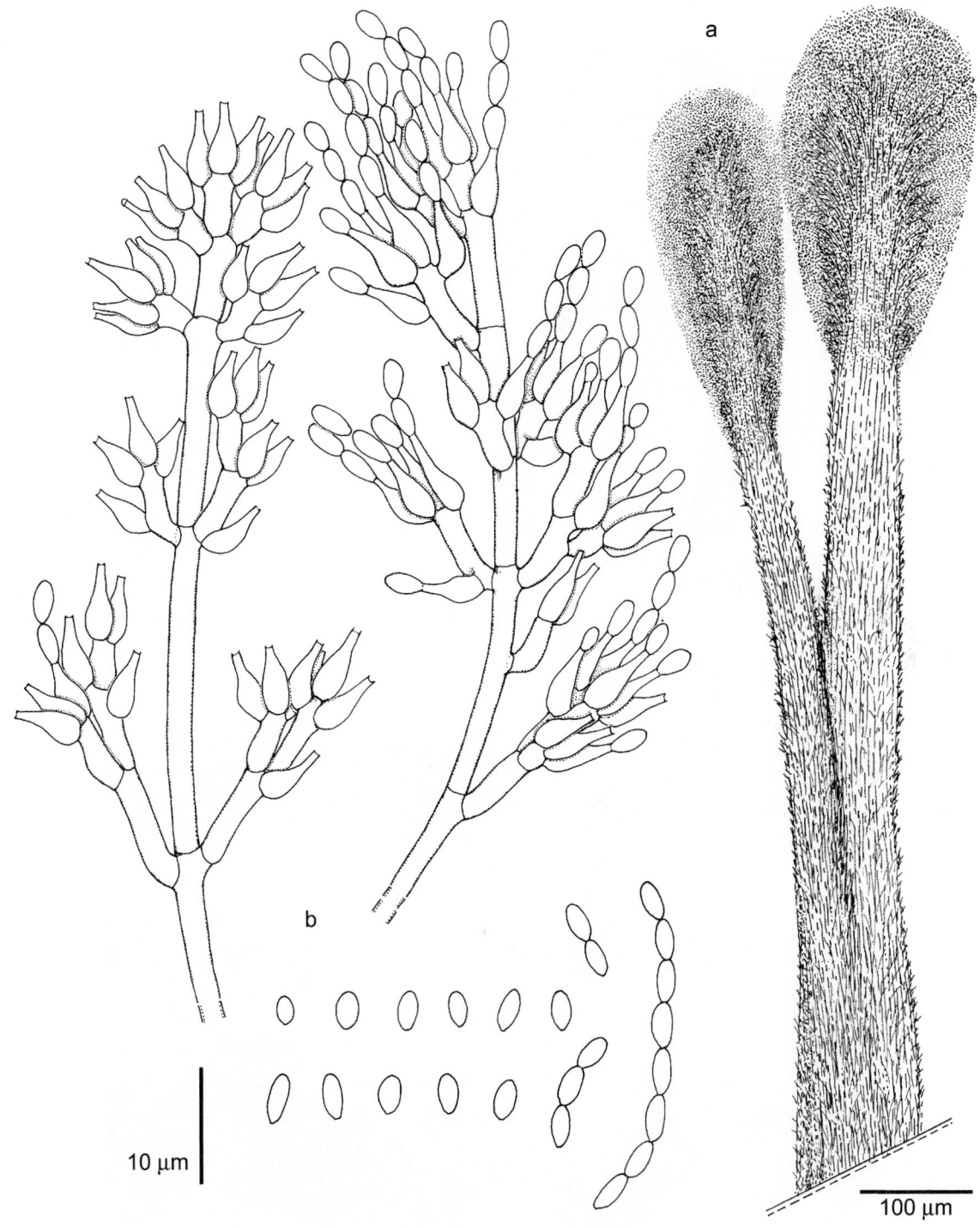

Paecilomyces fumosoroseus, CBS 101.73. a. Synnemata; b. conidiophores and conidia.

Antifungal susceptibility.

Antifungal	Mean MICs	Strains	Reference
AMB	0.63	6	Aguilar *et al.* (1998)
5FC	256	6	Aguilar *et al.* (1998)
FLZ	51	6	Aguilar *et al.* (1998)
ITZ	6.35	6	Aguilar *et al.* (1998)
KTZ	4	6	Aguilar *et al.* (1998)
MCZ	1.99	6	Aguilar *et al.* (1998)

Nomenclature. *Isaria fumosorosea* Wize - Bull. Int. Acad. Pol. Sci. Lett., Class Sci. Math. Nat. p. 72, 1904 ≡ *Paecilomyces fumosoroseus* (Wize) A.H. Brown & G. Smith - Trans. Br. Mycol. Soc. 40: 67, 1957.

Paecilomyces fumosoroseus, CBS 101.73. a, b. Synnemata; c-h. conidiophores and conidia. a. ×30; b. ×95; c. ×1480; d. ×1185; e. ×1035; f. ×2000; g. ×2440; h. ×2960.

Paecilomyces javanicus (Friederichs & Bally) A.H. Brown & G. Smith

Colony characteristics. Colonies (MEA 2%) growing slowly, powdery to floccose, at first white, becoming cream-coloured with age.

Microscopy. Conidiophores erect, arising from the aerial mycelium, up to 50 µm in length, 1.5-2.5 µm wide, bearing branches with phialides in whorls of 2 to 3. Phialides consisting of a cylindrical basal part, tapering into a thin neck. Conidia fusiform, sometimes cylindrical, hyaline, smooth-walled, 4.0-7.4 × 1.2-1.7 µm.

Pathogenicity. BSL-1. A case of endocarditis of native and prosthetic aortic valve (Allevato *et al.*, 1984).

Reference. Samson (1974).

Antifungal susceptibility.

Antifungal	Mean MICs	Strains	Reference
AMB	0.84	4	Aguilar *et al.* (1998)
5FC	256	4	Aguilar *et al.* (1998)
FLZ	45.3	4	Aguilar *et al.* (1998)
ITZ	1.66	4	Aguilar *et al.* (1998)
KTZ	2.37	4	Aguilar *et al.* (1998)
MCZ	2.37	4	Aguilar *et al.* (1998)

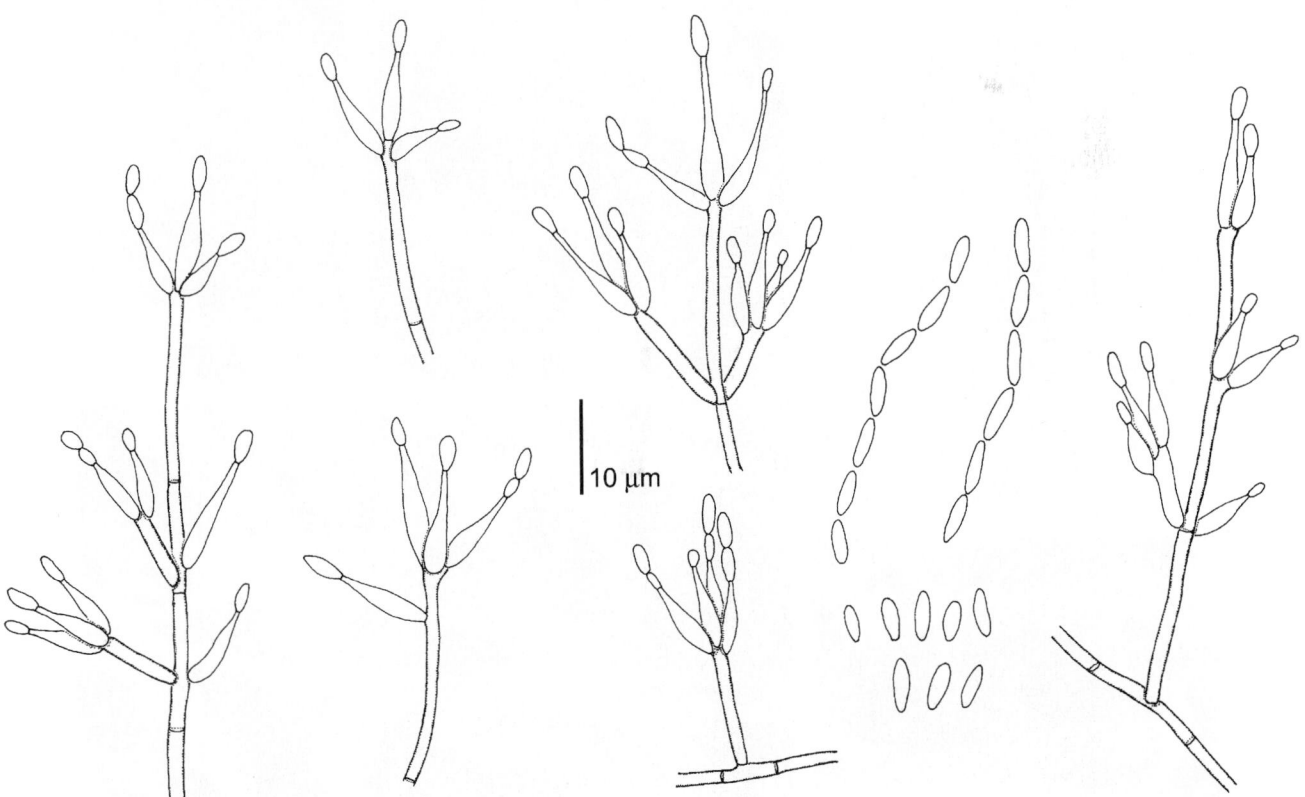

Paecilomyces javanicus, CBS 995.73. Conidiophores and conidia.

Nomenclature. *Spicaria javanica* Friederichs & Bally - Meded. Koffiebes. Fonds 6: 146, 1923 ≡ *Paecilomyces javanicus* (Friederichs & Bally) A.H. Brown & G. Smith - Trans. Br. Mycol. Soc. 40: 65, 1957.

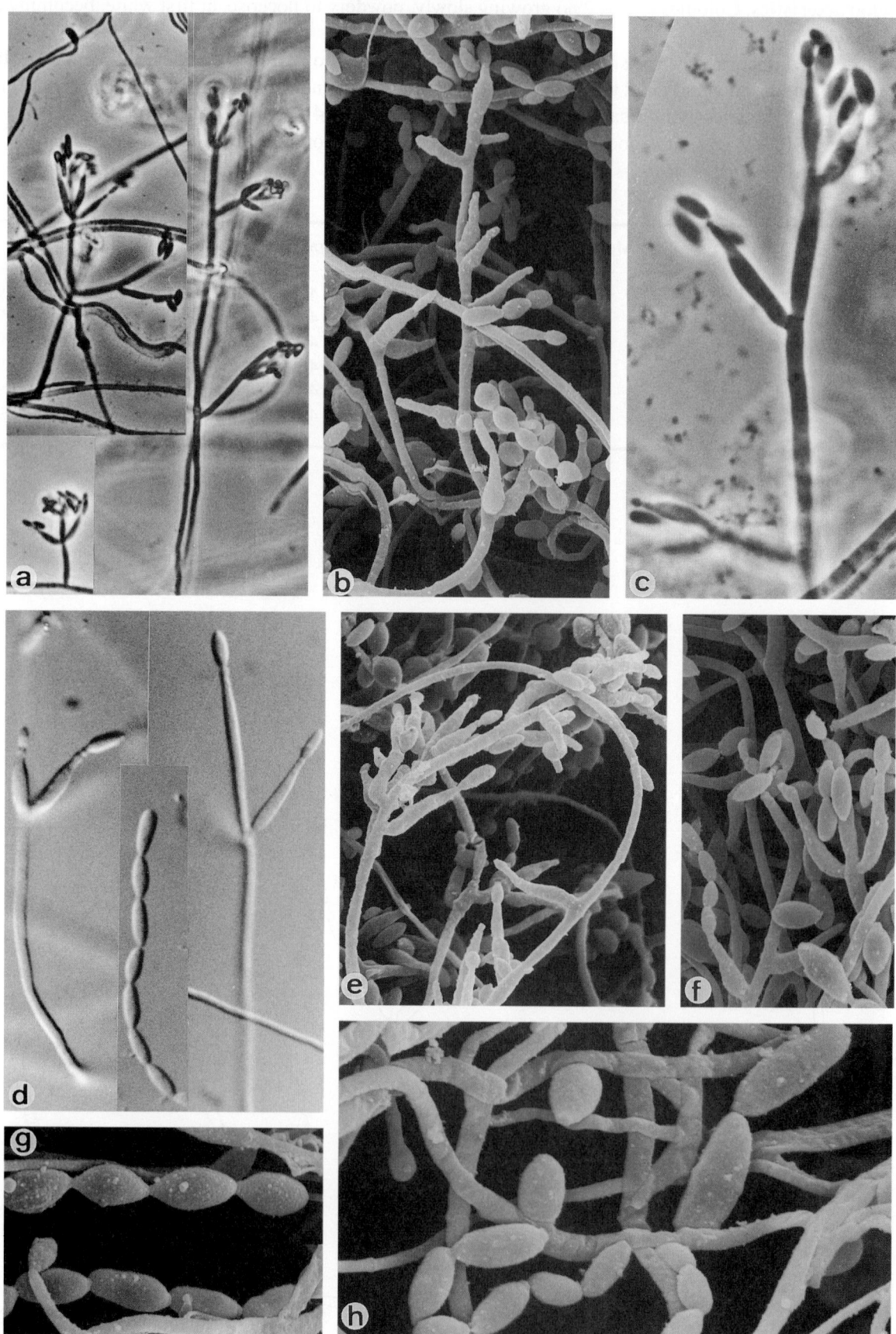

Paecilomyces javanicus, CBS 995.73. Conidiophores and conidia. a. ×640; b. ×1990; c, d. ×1600; e. ×1900; f. ×2300; g. ×2990; h. ×4000.

Paecilomyces lilacinus (Thom) Samson

Colony characteristics. Colonies (MEA 2%) growing rapidly, floccose, vinaceous to violet.

Microscopy. Conidiophores erect, 400-600 µm in length, mostly arising from submerged hyphae, occasionally forming tufts up to 2 mm high, bearing branches with densely clustered phialides; conidiophore stipes 3-4 µm wide, yellow to purple, rough-walled. Phialides consisting of a swollen basal part, tapering into a thin neck. Conidia ellipsoidal to fusiform, smooth-walled to slightly roughened, hyaline, purple in mass, 2.5-3.0 × 2.0-2.2 µm, in divergent chains.

Pathogenicity. BSL-1. Sinusitis is particularly common (Miller *et al.*, 1978; Theodore, 1978; Agrawal *et al.*, 1979; Rockhill & Klein, 1980; Rowley & Strom, 1982; Shukla *et al.*, 1984a; Gordon & Norton, 1985; Pflugfelder *et al.*, 1988; Gucalp *et al.*, 1996), eventually in diabetic patients (Saberhagens *et al.*, 1997). Other cases are ocular infections (O'Day, 1977; Watanabe *et al.*, 1985; Ohkubo *et al.*, 1994; Okhravi *et al.*, 1997), bursitis (Westenfeld *et al.*, 1996), cutaneous infections (Takayasu *et al.*, 1977; Diven *et al.*, 1996; Hecker *et al.*, 1997; Shin *et al.*, 1998) and onychomycoses (Fletcher *et al.*, 1998). Ono *et al.* (1999) described a pulmonary infection in an otherwise healthy patient and Clark (1999), but most deep infections are in immunocompromised or otherwise debilitated patients (Arai & Endo, 1977; Jade *et al.*, 1986; Castro *et al.*, 1990; Diven *et al.*, 1996; Marchese & Smoller, 1998). Several cases of catheter-related fungemia in immunocompromised patients have been reported (Bernacer *et al.*, 1992; Tan *et al.*, 1992;

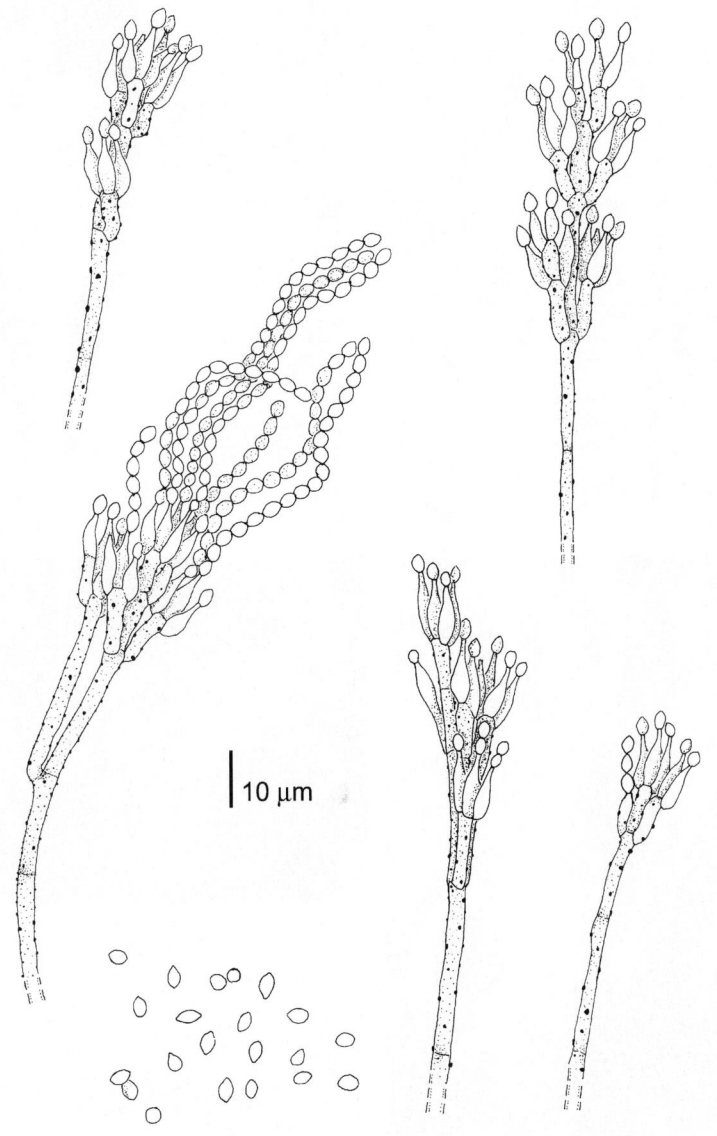

Paecilomyces lilacinus, FMR 2229. Conidiophores and conidia.

Chan-Tak *et al.*, 1999). The fungus is frequently found in high viscosity fluids; outbreaks of cutaneous mycosis in neutropenic patients due to contaminated skin lotion was described by Orth *et al.* (1996) and Itin *et al.* (1998). In animals: hyalohyphomycosis in turtles have been described (Heard *et al.*, 1986; Posthaus *et al.*, 1998).

References. Samson (1974), Domsch *et al.* (1980).

Antifungal susceptibility.

Antifungal	Mean MICs	Strains	Reference
AMB	10.3	11	Aguilar *et al.* (1998)
5FC	256	11	Aguilar *et al.* (1998)
FLZ	167	11	Aguilar *et al.* (1998)
ITZ	7.51	11	Aguilar *et al.* (1998)
KTZ	3.52	11	Aguilar *et al.* (1998)
MCZ	6.02	11	Aguilar *et al.* (1998)
VCZ	0.05	1	Wildfeuer *et al.* (1998)

Nomenclature. *Penicillium lilacinus* Thom - Bull. Bur. Anim. Ind. U.S. Dep. Agric. 118: 73, 1910 ≡ *Paecilomyces lilacinus* (Thom) Samson - Stud. Mycol. 6: 58, 1974.

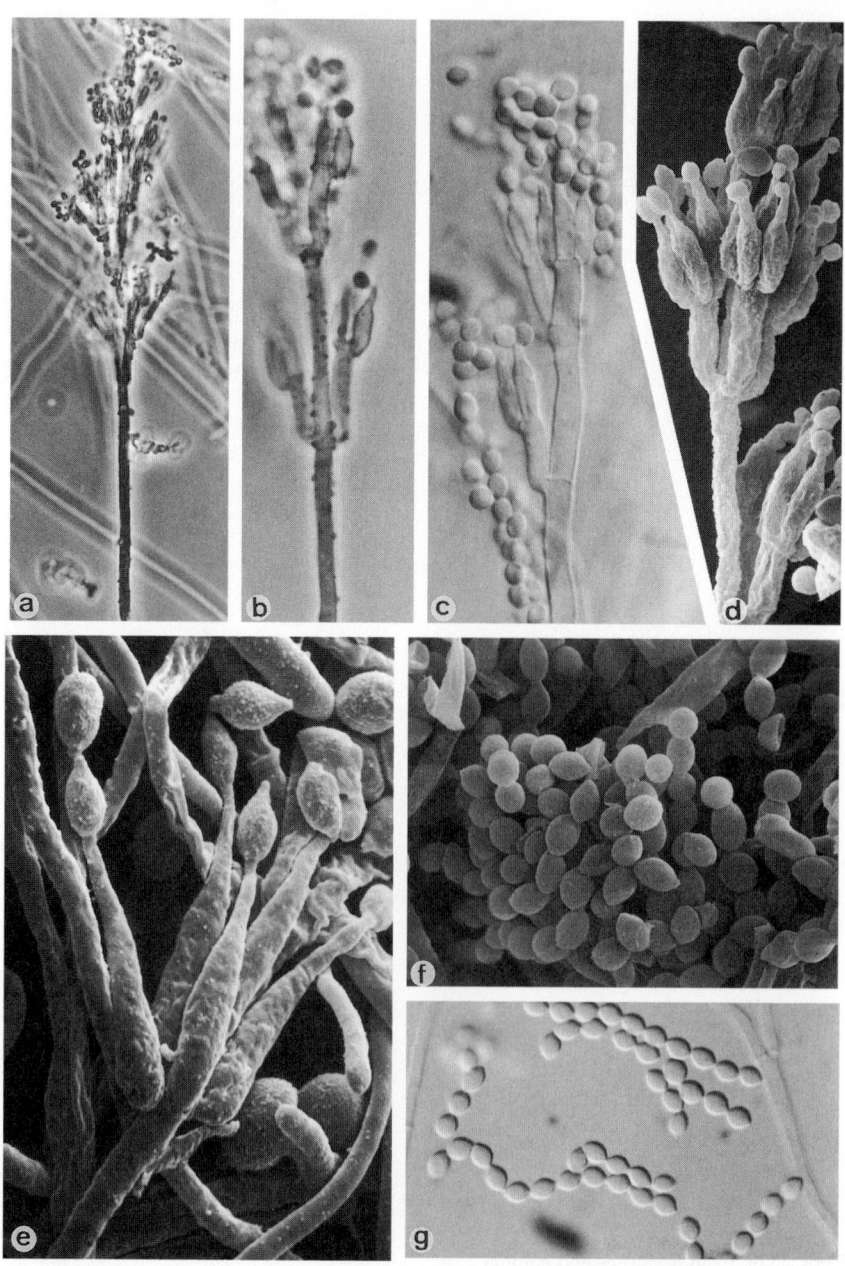

Paecilomyces lilacinus, FMR 2229. Conidiophores and conidia. a. ×360; b, c. ×1120; d. ×2170; e. ×4760; f. ×2730; g. ×1120.

Paecilomyces marquandii (Massee) Hughes

Colony characteristics. Colonies (MEA 2%) growing moderately rapidly, floccose, pale vinaceous to violet.
Microscopy. Conidiophores erect, arising from submerged hyphae, 50-300 μm in length, 2.5-3.0 μm wide, occasionally forming tufts up to 2 mm high, bearing loose whorls of branches and phialides, smooth-walled, hyaline. Phialides consisting of a swollen basal part, tapering into a thin, distinct neck. Conidia ellipsoidal to fusiform, smooth-walled to slightly roughened, hyaline, purple in mass, 2.5-3.0 × 2.0-2.2 μm. Spherical to ellipsoidal chlamydospores, 3-5 μm diam, present.
Pathogenicity. BSL-1. A case of cellulitis in an immunocompromised renal transplant patient (Harris *et al.*, 1979).
References. Samson (1974), Domsch *et al.* (1980).

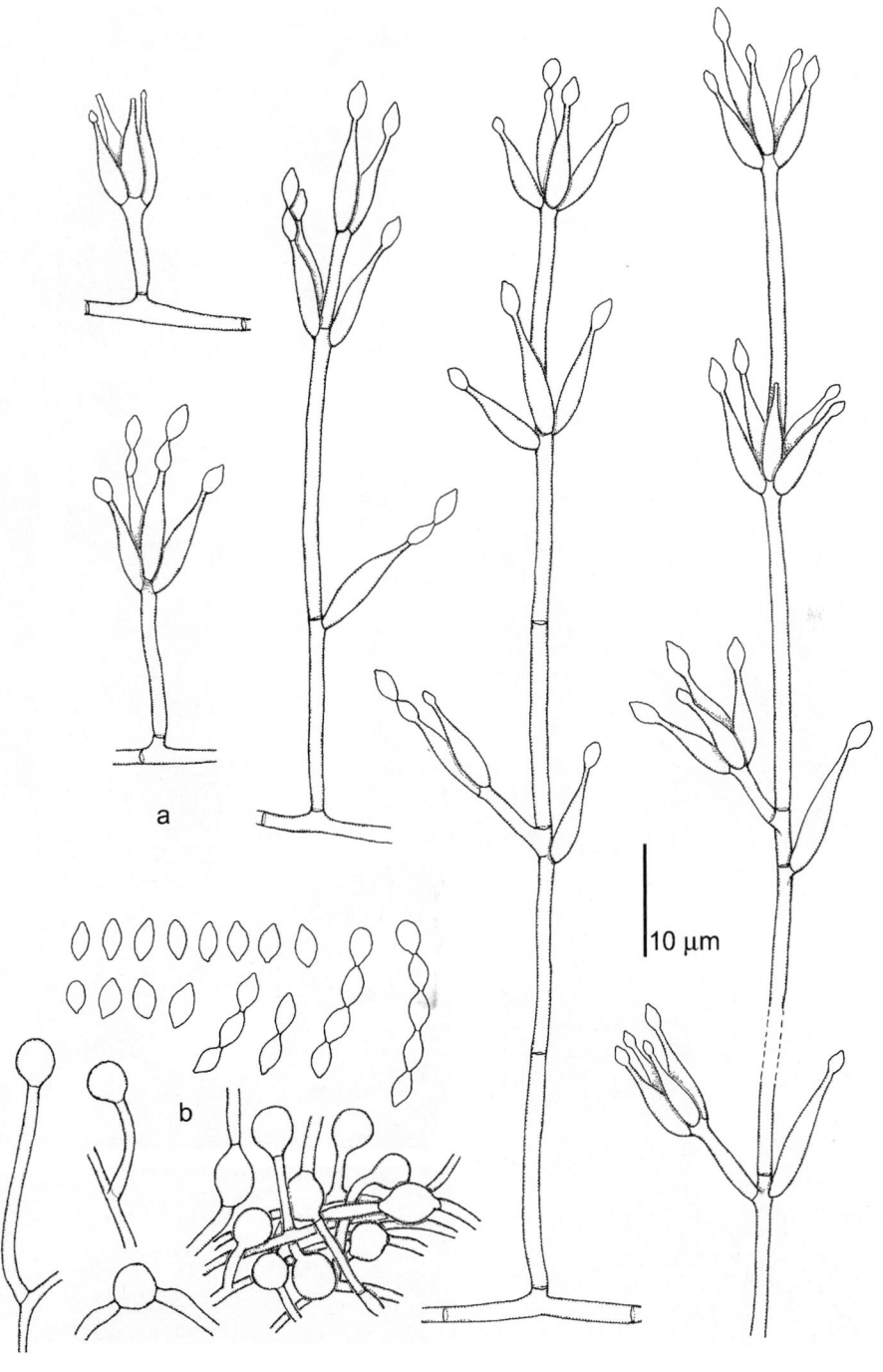

Paecilomyces marquandii, CBS 106.85. a. Conidiophores and conidia; b. chlamydospores.

Antifungal susceptibility.

Antifungal	Mean MICs	Strains	Reference
AMB	1.41	9	Aguilar *et al.* (1998)
5FC	256	9	Aguilar *et al.* (1998)
FLZ	123	9	Aguilar *et al.* (1998)
ITZ	5.88	9	Aguilar *et al.* (1998)
KTZ	3.17	9	Aguilar *et al.* (1998)
MCZ	5.03	9	Aguilar *et al.* (1998)

Nomenclature. *Verticillium marquandii* Massee - Br. Fung. Fl. 3: 24, 1898 ≡ *Paecilomyces marquandii* (Massee) Hughes - Mycol. Pap. 45: 30, 1951.

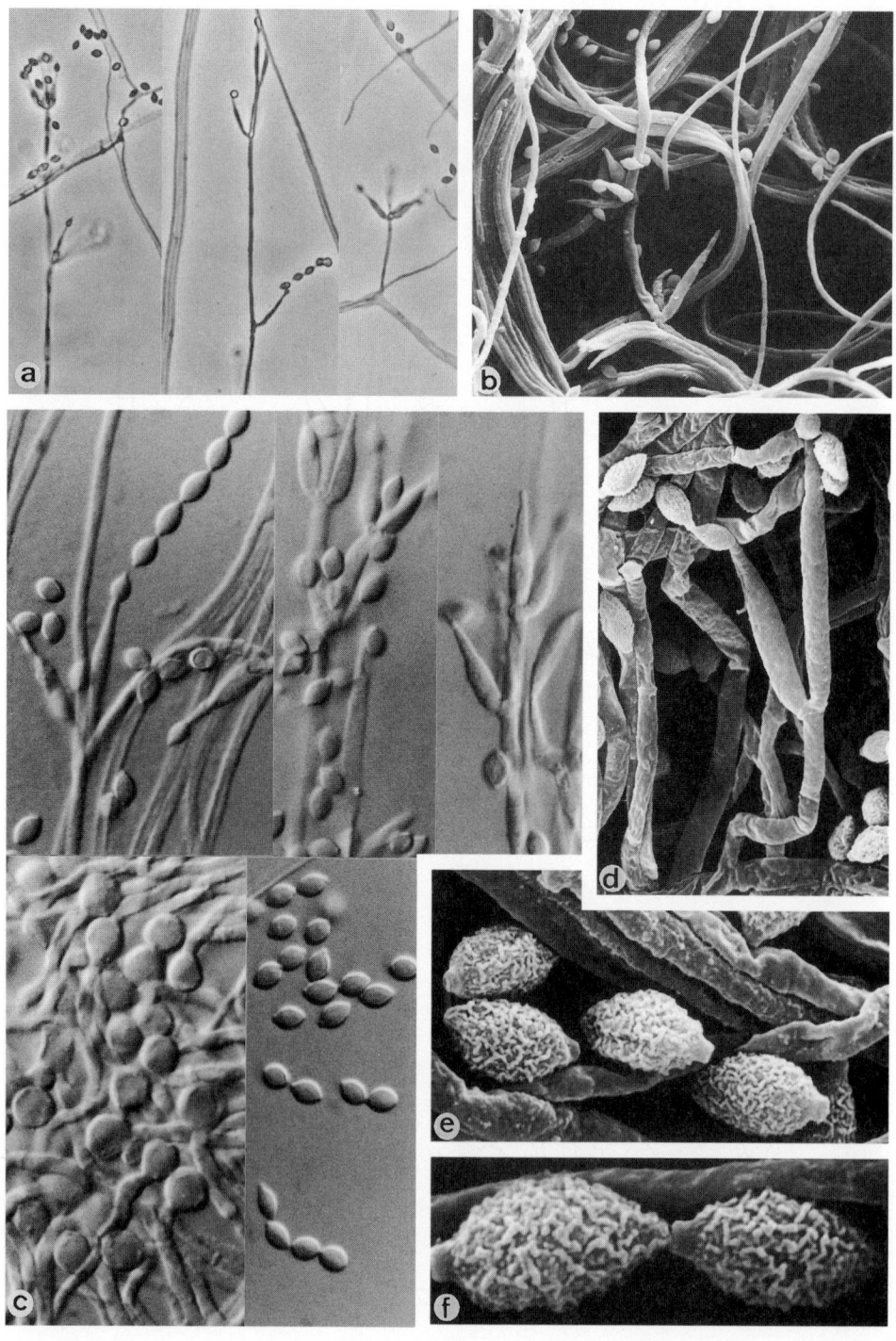

Paecilomyces marquandii, CBS 106.85. Conidiophores, conidia and chlamydospores. a. ×512; b. ×1300; c. ×1600; d. ×4000; e. ×9700; f. ×16,000.

Paecilomyces variotii Bain.

Colony characteristics. Colonies (MEA 2%) growing rapidly, powdery to floccose, funiculose or tufted, yellow-brown or sand colour. Odour sweet.

Microscopy. Conidiophores bearing dense, verticillately arranged branches, each bearing 2-7 phialides, up to 150 μm in length, 3.5-6.5 μm wide. Phialides cylindrical or ellipsoidal, tapering abruptly into a long, thin, cylindrical neck. Conidia subspherical, ellipsoidal to fusiform, hyaline to yellow, smooth-walled, 3-5 × 2-4 μm, arising in long, divergent chains. Chlamydospores usually present, singly or in short chains, brown, subspherical to pyriform, 4-8 μm diam, thick-walled, smooth-walled to slightly verrucose.

Physiology. The species is thermophilic. Intolerant to benomyl.

Molecular diagnostics. SSU and ITS restriction maps based on NCBI Y13996 and AF033395, resp.:

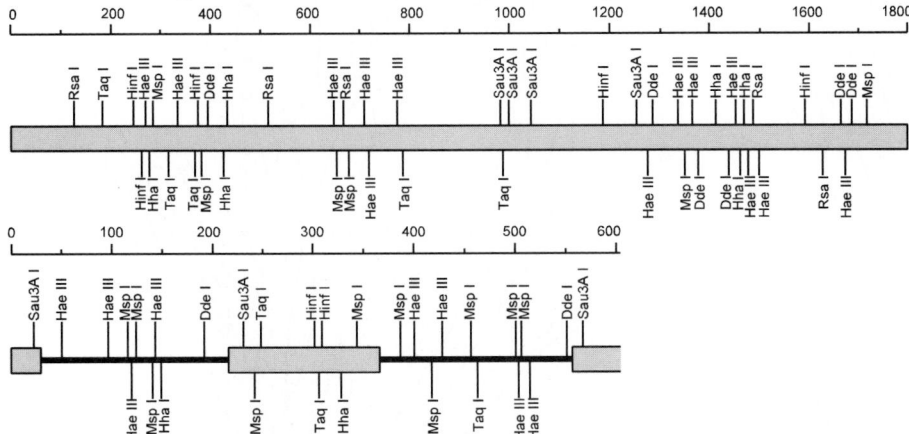

Pathogenicity. BSL-2. In humans: pneumonia (Byrd *et al.*, 1992), sinusitis (Otčenášek *et al.*, 1984; Thompson *et al.*, 1988; Eloy *et al.*, 1998), endophthalmitis (Lam *et al.*, 1999), otitis media (Dhindsa *et al.*, 1995), soft tissue infection of the heel (Williamson *et al.*, 1992), cutaneous hyalohyphomycoses (Jaishree & Singh, 1992; Athar *et al.*, 1996), onychomycosis (Arenas *et al.*, 1998) and osteomyelitis in a patient with granulomatous disorder (Cohen-Abbo & Edwards, 1995). Fungemia have been reported which were cleared after removal of catheters (Marzec *et al.*, 1993; Chan *et al.*, 1996; Kovac *et al.*, 1998), as well as a cerebrospinal fluid shunt obstruction (Fogerburg *et al.*, 1981). Also allergic alveolitis is known (Akhunova & Shustova, 1989). In animals: pulmonary and air sac infection in pigeons (Laclaire *et al.*, 1974), mastitis in a goat (Pepin & Pritchard, 1984), cutaneous outbreak in laboratory rats (Kunstýř *et al.*, 1997).

References. Samson (1974), Domsch *et al.* (1980), Naidu & Singh (1992), Samson *et al.* (1996).

Antifungal susceptibility.

Antifungal	Mean MICs	MIC90	Strains	Reference
AMB	0.08	nd	16	Aguilar *et al.* (1998)
AMB	0.35	0.78	7	Wildfeuer *et al.* (1998)
5FC	0.30	nd	16	Aguilar *et al.* (1998)
FLZ	24.75	nd	16	Aguilar *et al.* (1998)
ITZ	0.07	nd	16	Aguilar *et al.* (1998)
ITZ	0.20	0.39	7	Wildfeuer *et al.* (1998)
KTZ	0.47	nd	16	Aguilar *et al.* (1998)
KTZ	0.39	0.39	7	Wildfeuer *et al.* (1998)
MCZ	1.04	nd	16	Aguilar *et al.* (1998)
VCZ	0.58	1.56	7	Wildfeuer *et al.* (1998)

Nomenclature. *Paecilomyces variotii* Bainier - Bull. Trimest. Soc. Mycol. Fr. 23: 26, 1907.
Paecilomyces variotti Bainier var. *zaaminelli* Dehkan-Khodzhaeva, Shamsier & Shakirova - Pediatriya 9: 13, 1982.

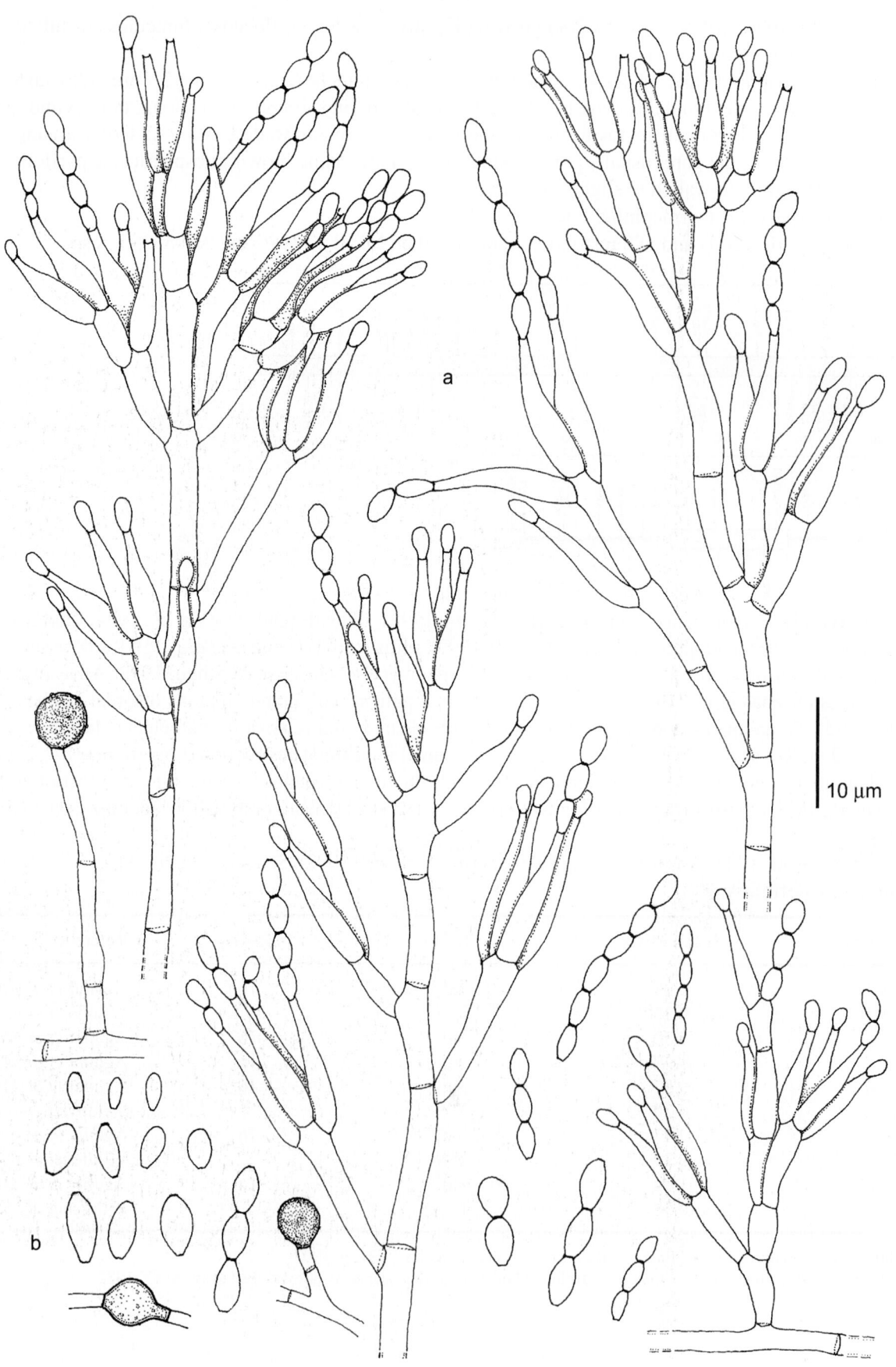

Paecilomyces variotii, FMR 3735. a. Conidiophores and conidia; b. chlamydospores and conidia.

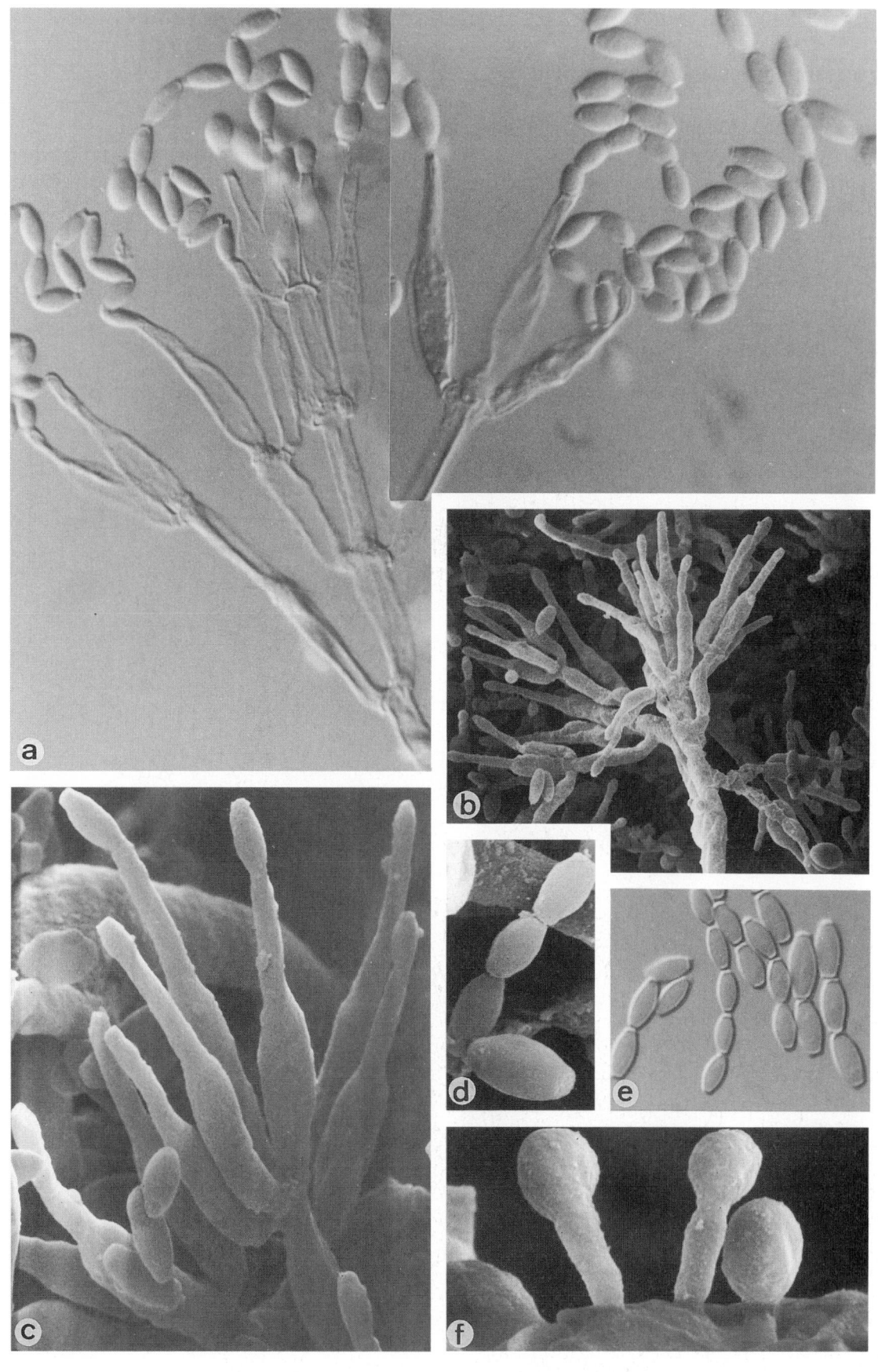

Paecilomyces variotii, FMR 3735. a-e. Conidiophores and conidia; f. chlamydospores. a. ×1600; b. ×1200; c. ×4100; d. ×5200; e. ×1600; f. ×4250.

Paecilomyces viridis Segretain *et al.* ex Samson

Colony characteristics. Colonies (MEA 2%) growing rapidly, velvety, yellow-green.

Microscopy. Conidiophores erect, up to 120 × 1.5-2.6 μm, consisting of verticillate branches bearing whorls of 2 to 4 phialides, or with verticillate phialides borne directly on the stipe. Phialides short, with a distinctly inflated base and a thin neck. Conidia spherical to subspherical, sometimes apiculate, hyaline, greenish in mass, 2.5-3.2 μm diam, smooth-walled.

Pathogenicity. BSL-1. The species was described from different organs of a camel (Segretain *et al.*, 1964). Akhunova (1991) reported an allergic bronchopulmonary infection in a human. A case of exogenous endophthalmitis published by Rodrigues & MacLeod (1975) probably concerned *P. variotii*.

Reference. Samson (1974).

Nomenclature. *Paecilomyces viridis* Segretain, Fromentin, Destombes, Brygoo & Bodin - C.R. Hebd. Séanc. Acad. Sci., Paris 259: 260, 1964 (invalid) ≡ *Paecilomyces viridis* Segretain, Fromentin, Destombes, Brygoo & Bodin ex Samson - Stud. Mycol. 6: 64, 1974.

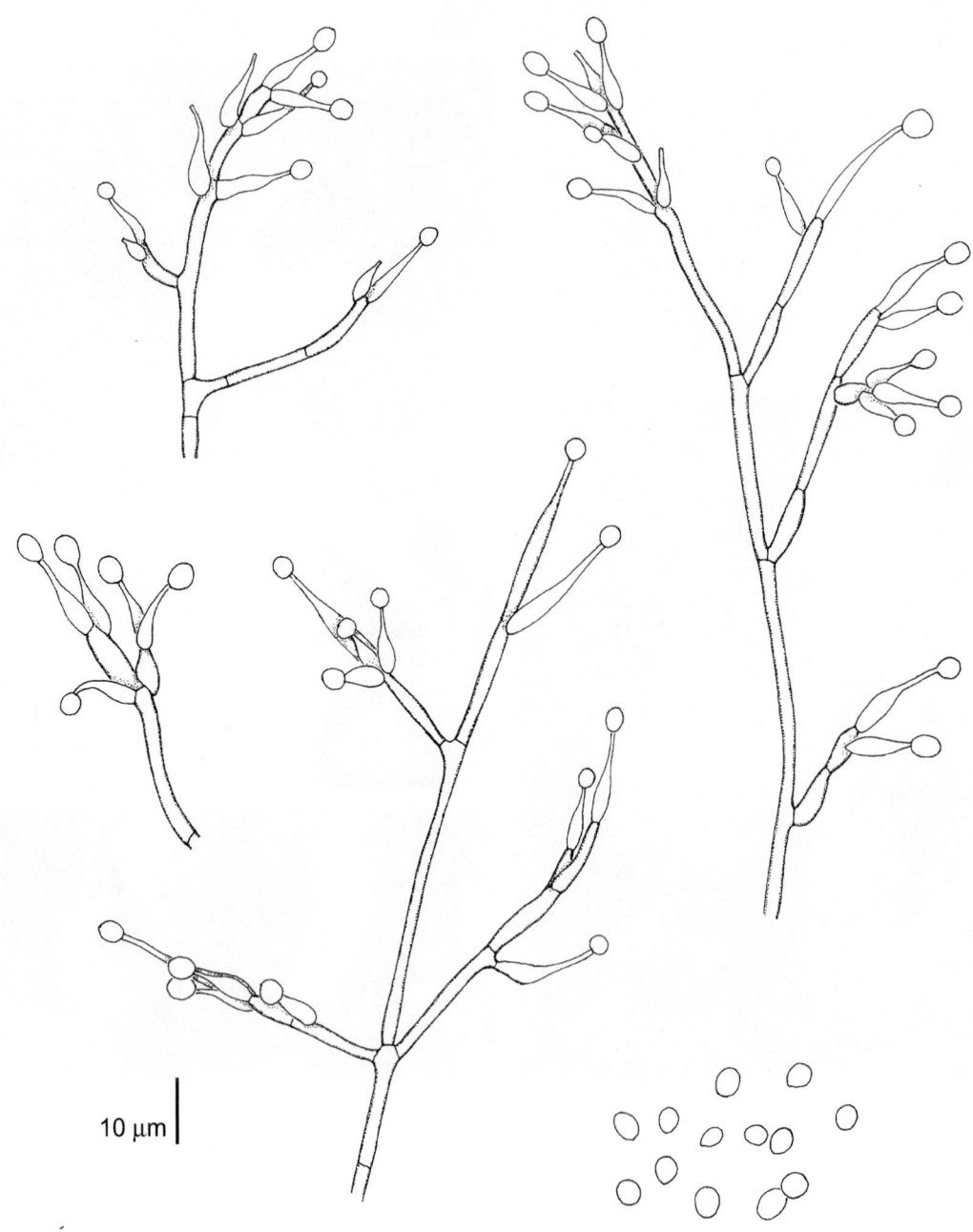

10 μm

Paecilomyces viridis, UAMH 2994. Conidiophores and conidia.

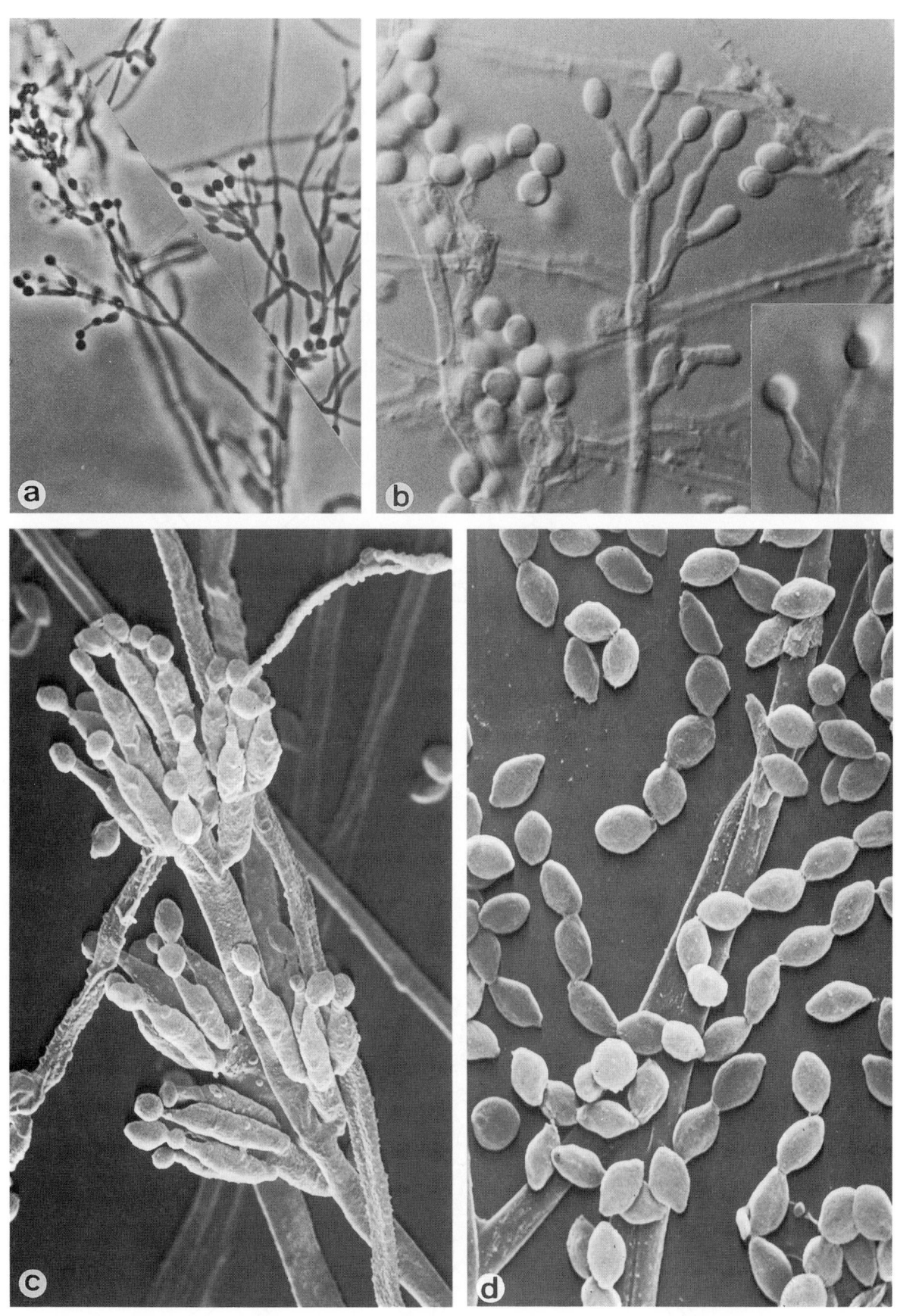

Paecilomyces viridis, UAMH 2994. Conidiophores and conidia. a. ×512; b. ×1600; c. ×3000; d. ×4500.

Hyphomycetes. Genus: *PAPULASPORA*

Papulaspora equi Shadomy & Dixon

Colony characteristics. Colonies (PDA) growing rapidly, white, green or greyish-brown.

Microscopy. Hyphae hyaline, of two sizes, narrow (1.0-2.5 µm diam) and broad (up to 8 µm diam). Bulbils at first hyaline, dark brown when mature, ovoidal to ellipsoidal, 170-300 × 80-120 µm, with an irregularly lobed surface; cells measuring 12-20 × 10-14 µm.

Pathogenicity. BSL-1. A case of eye infection in a horse has been reported (Shadomy & Dixon, 1989).

Reference. Shadomy & Dixon (1989).

Nomenclature. *Papulaspora equi* Shadomy & Dixon - Mycopathologia 106: 35, 1989.

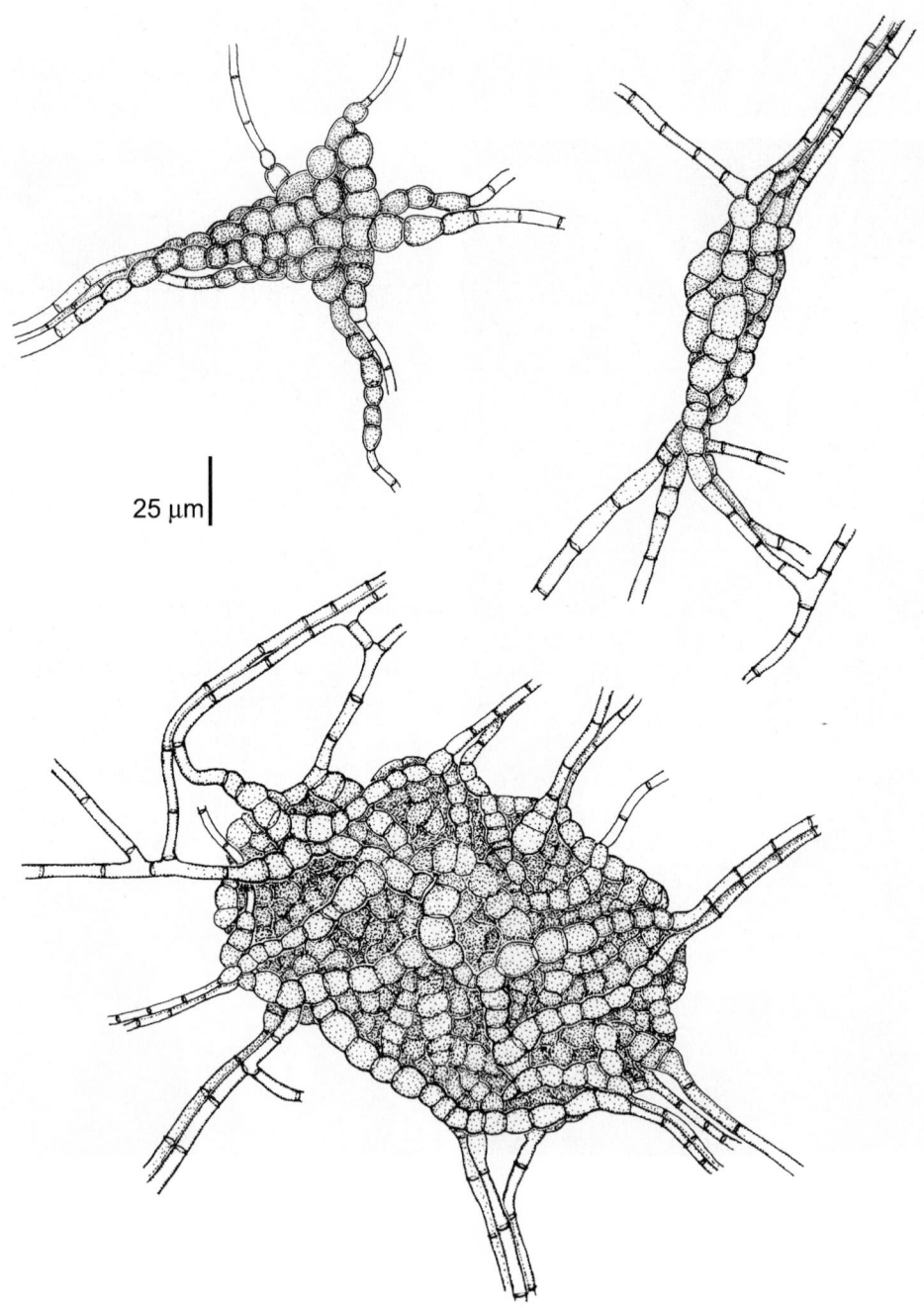

25 µm

Papulaspora equi, CBS 573.89. Young and mature bulbils.

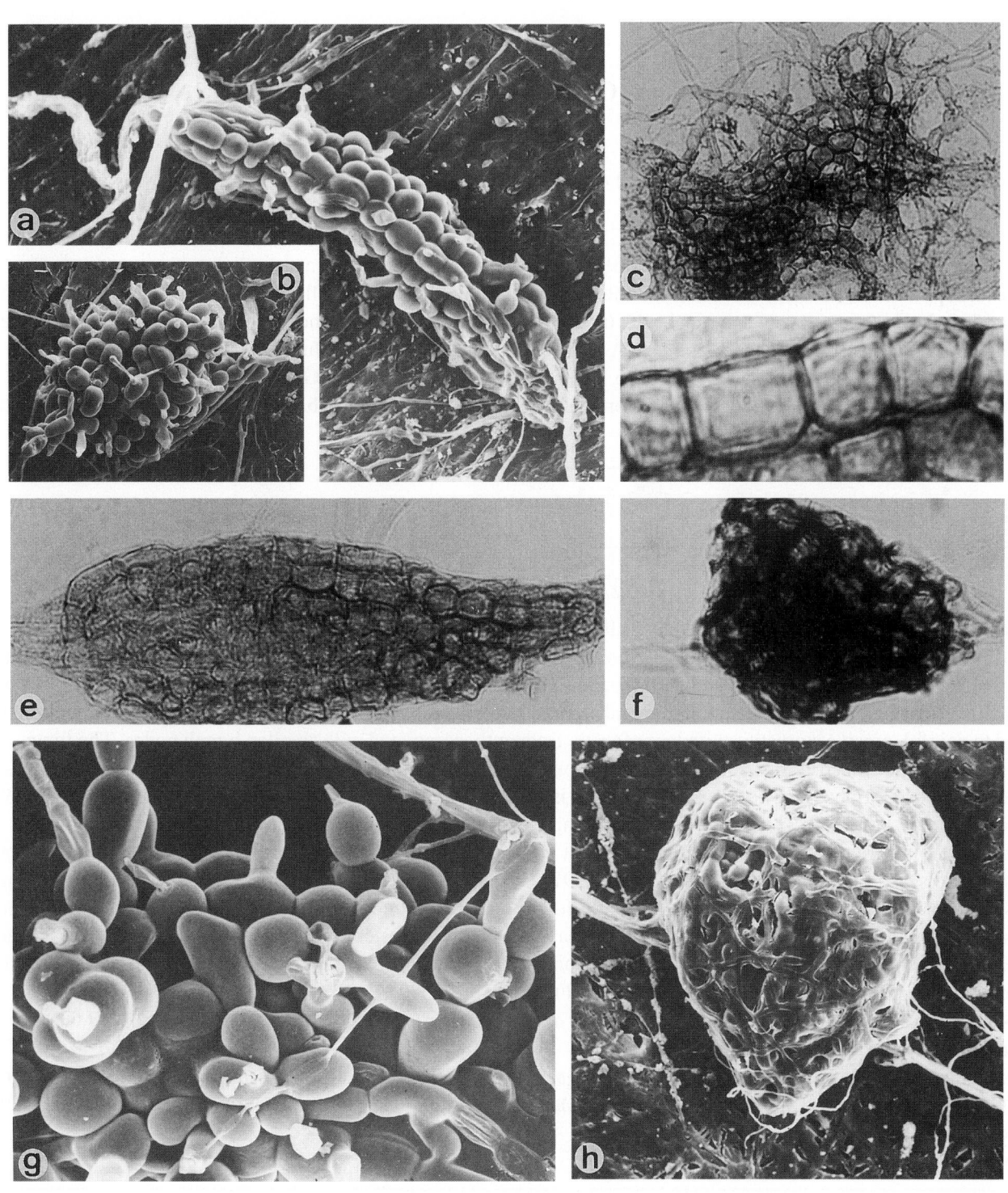

Papulaspora equi, CBS 573.89. a-c, e-h. Young and mature bulbils; d, g. detail of bulbil cells. a. ×495; b. ×378; c. ×512; d. ×1600; e. ×512; f. ×320; g. ×1250; h. ×360.

Hyphomycetes, dimorphic *Onygenales*. Genus: *PARACOCCIDIOIDES*

Paracoccidioides brasiliensis (Splendore) de Almeida

Colony characteristics. Colonies (SGA, 24°C) with slow growth, leathery, flat to wrinkled, becoming felty, whitish to brown with brownish reverse; at 37°C on BHI colonies butyrous, cerebriform.

Microscopy. Conidia at 24°C absent or are intercalary-like chlamydospore cells. At 37°C on BHI a yeast-like form is produced, which consists of spherical cells up to 30 μm diam, producing synchronous buds all over its surface.

Molecular diagnostics. A specific 26S rDNA primer was developed by Sandhu *et al.* (1997) and an ITS primer by Imai *et al.* (2000). A species-specific 110-bp DNA fragment was cloned by Goldani *et al.* (1995). De Brito *et al.* (1999) introduced *in situ* hybridization and Gomes *et al.* (2000) a specific PCR based on gp43 antigen. ITS restriction map based on NCBI AF038360:

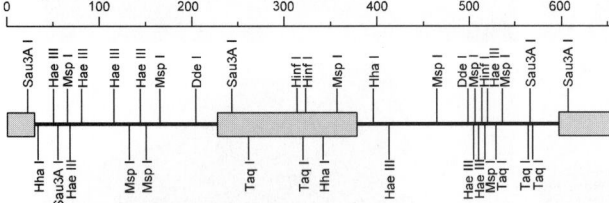

Serology. Serodiagnosis was described by Taborda & Camargo (1994).

Pathogenicity. BSL-3. Agent of paracoccidioidomycosis (p. 25). The species causes a systemic, chronic disease, occasionally occurring in otherwise healthy patients (Benard *et al.*, 1994). The portal of entry is not known with certainty. Possibly the fungus is acquired by inhalation, but it may also be introduced by local trauma. Infection may lead to painful, erosive stomatitis with loss of teeth, frequently associated with swollen lymph nodes. The fungus has a predilection for mucosa. It is abundantly present with yeast cells in pus and in tissue, allowing direct diagnosis by fine-needle aspiration (Drut, 1995). Bone invasion is often observed (Amstalden *et al.*, 1996; Fulciniti *et al.,* 1996). Eventually the lymphatic system, intestines, spleen, lungs and liver may become involved. Neurological cases are uncommon, but frequently have fatal outcome (Plá *et al.*, 1994). Infections are rarely superficial, with no signs of systemic disease (Tomimori-Yamashita *et al.*, 1997). Male patients are particularly affected. In contrast to most systemic *Onygenales*, it is relatively infrequently found in AIDS patients (Goldani & Sugar, 1995), although such cases do occur (Tobon *et al.*, 1998). Reactivation may take place long after the patient has left endemic areas (Manns *et al.*, 1996). Armadillos could be a natural reservoir of the fungus (Vidal *et al.*, 1995; Corredor *et al.*, 1999; Silva-Vergara *et al.*, 2000). The fungus is found in soil (Franco *et al.*, 2000).

Distribution. Central and South America (Blotta *et al.*, 1999), where it occurs in moist climates, mostly (sub)tropical rainforests (Restrepo, 1985).

References. San-Blas & San-Blas (1977), Köhler *et al.* (1988), Restrepo (1985), Brummer *et al.* (1993), Negroni (1993), San-Blas (1993), Sposto *et al.* (1993, 1994), Figueroa *et al.* (1994), Montenegro (1995), Sano *et al.* (1997), Lacaz *et al.* (1999).

Antifungal susceptibility.

Antifungal	MIC 90	Strains	Reference
AMB	0.16	11	Rodero *et al.* (1999)
FLZ	6.2	11	Rodero *et al.* (1999)
ITZ	0.31	11	Rodero *et al.* (1999)
KTZ	1.3	11	Rodero *et al.* (1999)

Nomenclature. *Zymonema brasiliense* Splendore - Bull. Soc. Path. Exot. 5: 317, 1912 ≡ *Mycoderma brasiliense* (Splendore) Brumpt - Précis Paras. Hum. Comp., ed. 5, p. 69, 1921 ≡ *Coccidioides brasiliensis* (Splendore) de Almeida - Anais Fac. Med., S. Paulo 4: 95, 1929 ≡ *Paracoccidioides brasiliensis* (Splendore) de Almeida - C.R. Soc. Biol. 105: 316, 1930 ≡ *Monilia brasiliensis* (Splendore) Vuillemin - Champ. Paras. Homme p. 86, 1931 ≡ *Blastomyces brasiliensis* (Splendore) Conant & Howell - J. Invest. Derm. 5: 357, 1942 ≡ *Aleurisma brasiliensis* (Splendore) Aroeia Neves & Bogliolo - Mycopath. Mycol. Appl. 5: 140, 1951.

Zymonema histosporocellulare Haberfeld - Gran. Gang. Maligno Blast., S. Paulo, p. 86, 1919 ≡ *Mycoderma histosporocellulare* (Haberfeld) Neveu-Lemaire - Précis Parasitol., ed. 5, p. 70, 1921 ≡ *Coccidioides histosporocellularis* (Haberfeld) Vuillemin - Champ. Paras. Myc. Homme p. 140, 1931 ≡ *Lutziomyces histosporocellularis* (Haberfeld) da Fonseca - An. Bras. Derm. Syph. 14: 108, 1939.

Paracoccidioides tenuis M. Moore, *in* de Almeida - Micol. Méd. p. 437, 1938.

Proteomyces faverae C.W. Dodge - Med. Mycol. p. 526, 1988.

Paracoccidioides antarcticus Gezuele - Encuent. Int. Paracocc. 4: B-2, 1989.

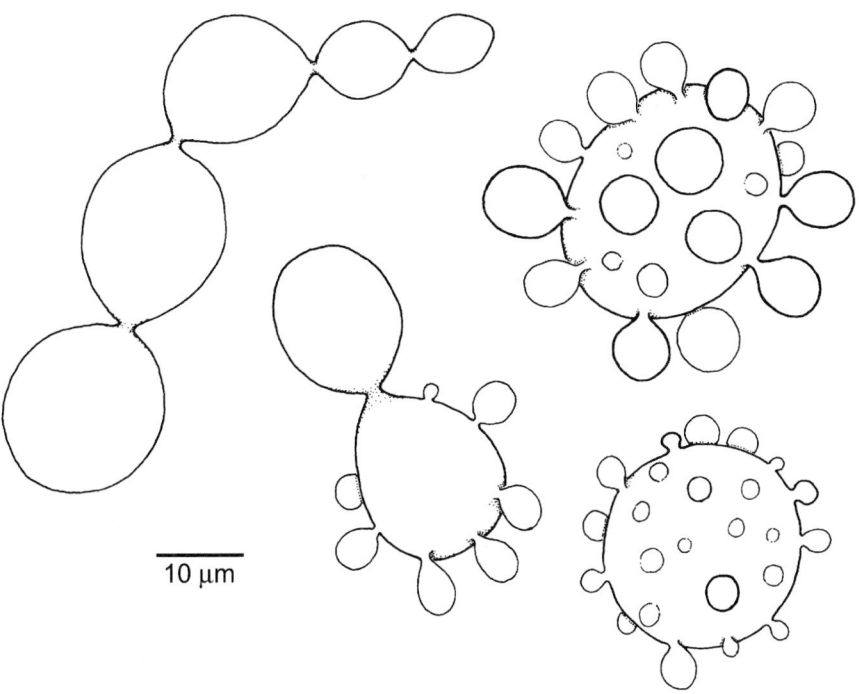

Paracoccidioides brasiliensis, FMR 5200. Yeast-like form *in vitro* at 37°C.

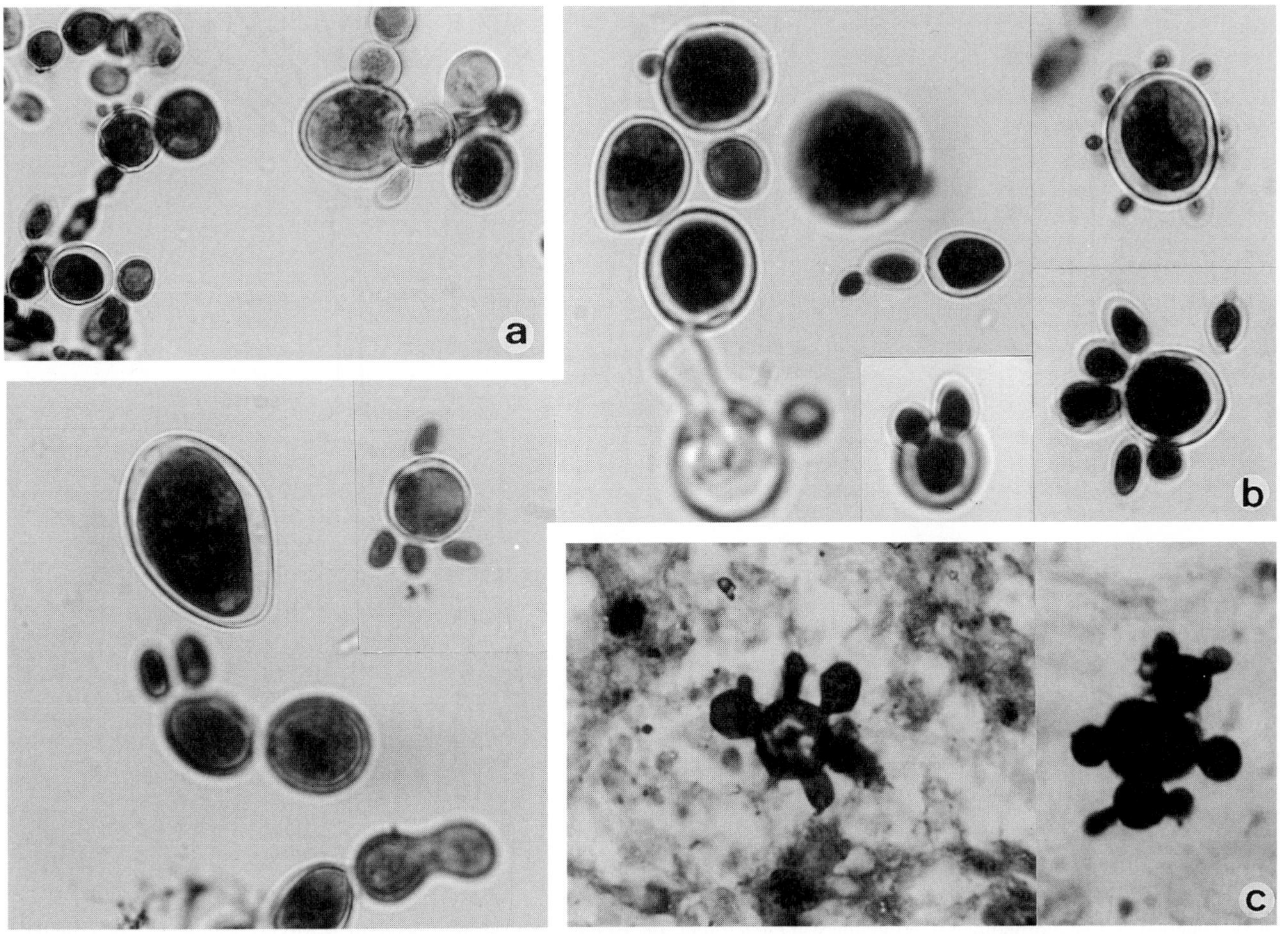

Paracoccidioides brasiliensis, a, b. FMR 5200. a, b. Yeast-like form *in vitro* at 37°C; c. yeast-like form *in vivo*. a. ×512; b. ×820; c. ×1200.

Hyphomycetes. Genus: *PENICILLIUM*

Generic description. Colonies usually growing rapidly, powdery, effuse, green, grey, yellow or white, rarely reddish. Conidiophores usually erect, simple or synnematous, hyaline, or pale pigmented, terminally bearing one or several whorls of upwardly directed, slender metulae which bear flask-shaped to acerose phialides. Conidia produced in dry basipetal chains, 1-celled, spherical, ellipsoidal, fusiform or short-cylindrical, often with truncate ends, hyaline or subhyaline, smooth-walled or ornamented. Sclerotia may be present.

Teleomorphs. *Eupenicillium, Hamigera, Talaromyces, Trichocoma* (*Ascomycota, Euascomycetes, Eurotiales: Trichocomaceae*).

General remarks. *Penicillium* is a very large and ubiquitous genus; it is the most abundant genus of mesophilic fungi in temperate soils. About 200 species are distinguishable. Nearly all are active producers of mycotoxins. Many species are airborne, and are therefore frequently encountered as laboratory contaminants. Most *Penicillium* species are hardly able to grow at 37°C, and are unlikely to be etiologic agents of systemic disease, with the complex around *P. marneffei* as the main exception. Review of *Penicillium* pathology were published by Mori *et al.* (1993) and Pitt (1994). Terminology used for *Penicillium* in particular is explained in the diagram below (Fig. 59). Phylogenetic sequencing data were presented by LoBuglio *et al.* (1993, 1994), LoBuglio & Taylor (1993) and Peterson (2000).

References. Raper & Thom (1949), Pitt (1979, 1988), Ramírez (1982), Samson & Pitt (1985, 1989), Samson *et al.* (1996), Samson & Pitt (2000).

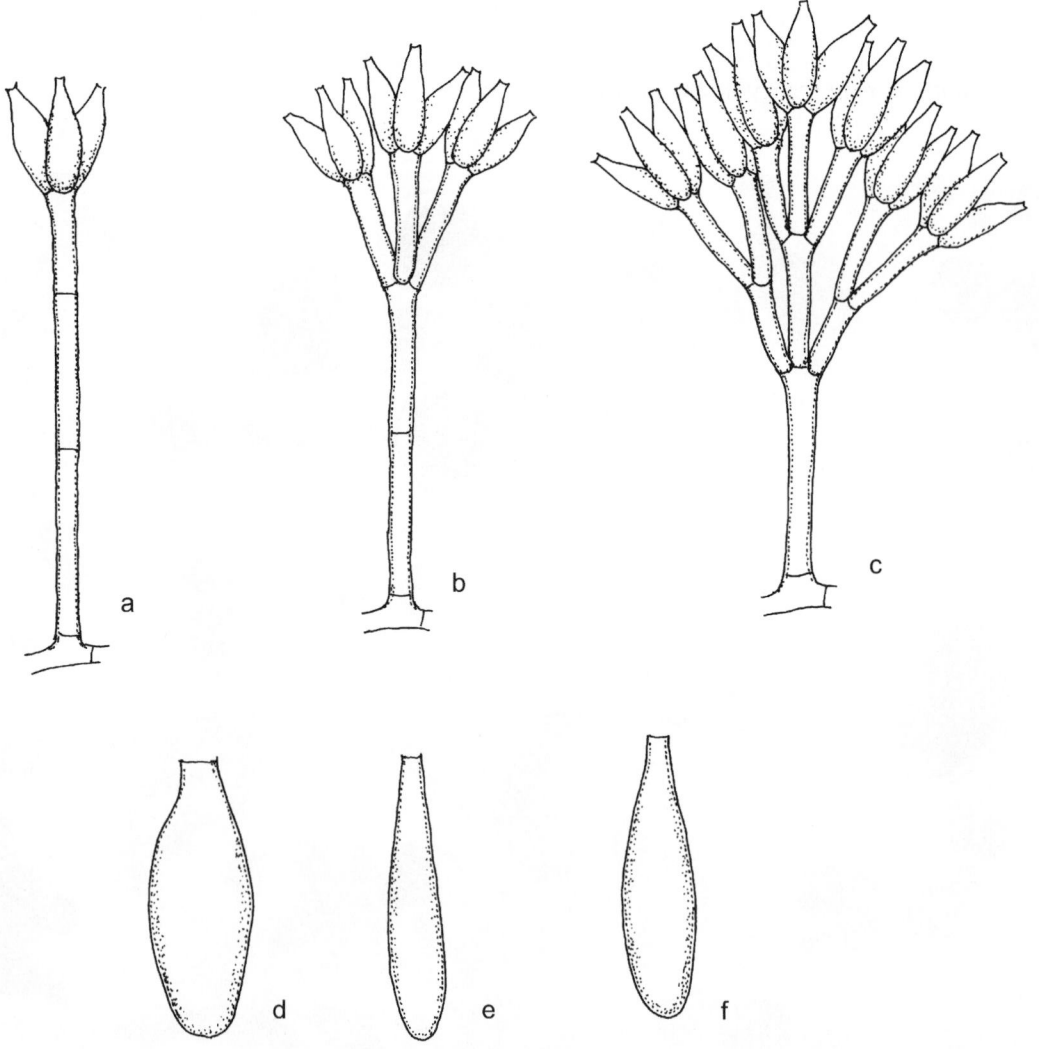

Fig 59. Diagram of used morphological terminology in *Penicillium*. a-c. Branching patterns of penicilli. a. Monoverticillate; b. biverticillate; c. terverticillate. d-f. Phialide shapes. d. Ampulliform; e. acerose; f. ampulliform-acerose, the intermediate type of *P. verruculosum*.

Key to the treated species of Penicillium:

1a. Conidiophores unbranched (monoverticillate) → **2**
1b. Conidiophores branched at least once (bi- or terverticillate) → **3**
2a. Colonies attaining over 30 mm diam in 12 days; conidia rough-walled . . . . *P. spinulosum* (**842**)
2b. Colonies attaining less than 25 mm diam in 12 days; conidia smooth-walled . . . . *P. decumbens* (**827**)
3a. Penicilli predominantly terverticillate → **4**
3b. Penicilli predominantly biverticillate → **9**
4a. Phialides less than 6 μm long, short-cylindrical, usually without a distinct neck; colonies
   greyish-green; conidiophores synnematous . . . . *P. griseofulvum* (**831**)
4b. Phialides longer than 6 μm, with distinct neck → **5**
5a. Colonies on MEA restricted (less than 20 mm diam in 7 days); penicilli broad and compact;
   conidia finely roughened . . . . *P. brevicompactum* (**819**)
5b. Colonies on MEA expanding (more than 20 mm diam in 7 days); penicilli otherwise;
   conidia smooth-walled → **6**
6a. Conidiophore stipes on MEA smooth-walled → **7**
6b. Conidiophore stipes on MEA rough-walled . . . . *P. aurantiogriseum* (**817**)
7a. Conidia blue to blue-green; exudate and soluble pigment bright yellow . . . . *P. chrysogenum* (**821**)
7b. Conidia green; exudate and soluble pigment pale to brown or absent → **8**
8a. Conidiophore stipes less than 200 μm long; exudate and pigment absent . . . . *P. commune* (**825**)
8b. Conidiophore stipes 200-500 μm long; exudate and pigment brown . . . . *P. expansum* (**829**)
9a. Penicilli biverticillate or more complex → **11**
9b. Penicilli biverticillate or irregularly monoverticillate → **10**
10a. Phialides acerose; conidia ellipsoidal, 3.5-5.0 × 2.5-4.0 μm, closely packed . . . . *P. oxalicum* (**1029**)
10b. Phialides ampulliform; conidia spherical to subspherical, 2-3 μm diam . . . . *P. citrinum* (**823**)
11a. Colonies attaining over 18 mm diam in 7 days → **12**
11b. Colonies attaining less than 12 mm diam in 7 days . . . . *P. rugulosum* (**840**)
12a. Conidia rough-walled; sporulation abundant → **13**
12b. Conidia mostly smooth-walled; sporulation scant → **14**
13a. Conidia ellipsoidal . . . . *P. purpurogenum* (**838**)
13b. Conidia spherical . . . . *P. verruculosum* (**844**)
14a. Colonies on MEA attaining 28-30 mm diam in 7 days, funiculose; red pigment
   exduded into the agar . . . . *P. marneffei* (**833**)
14b. Colonies on MEA attaining 17-25 mm diam in 7 days, flat to floccose; colony reverse
   yellow . . . . *P. piceum* (**836**)

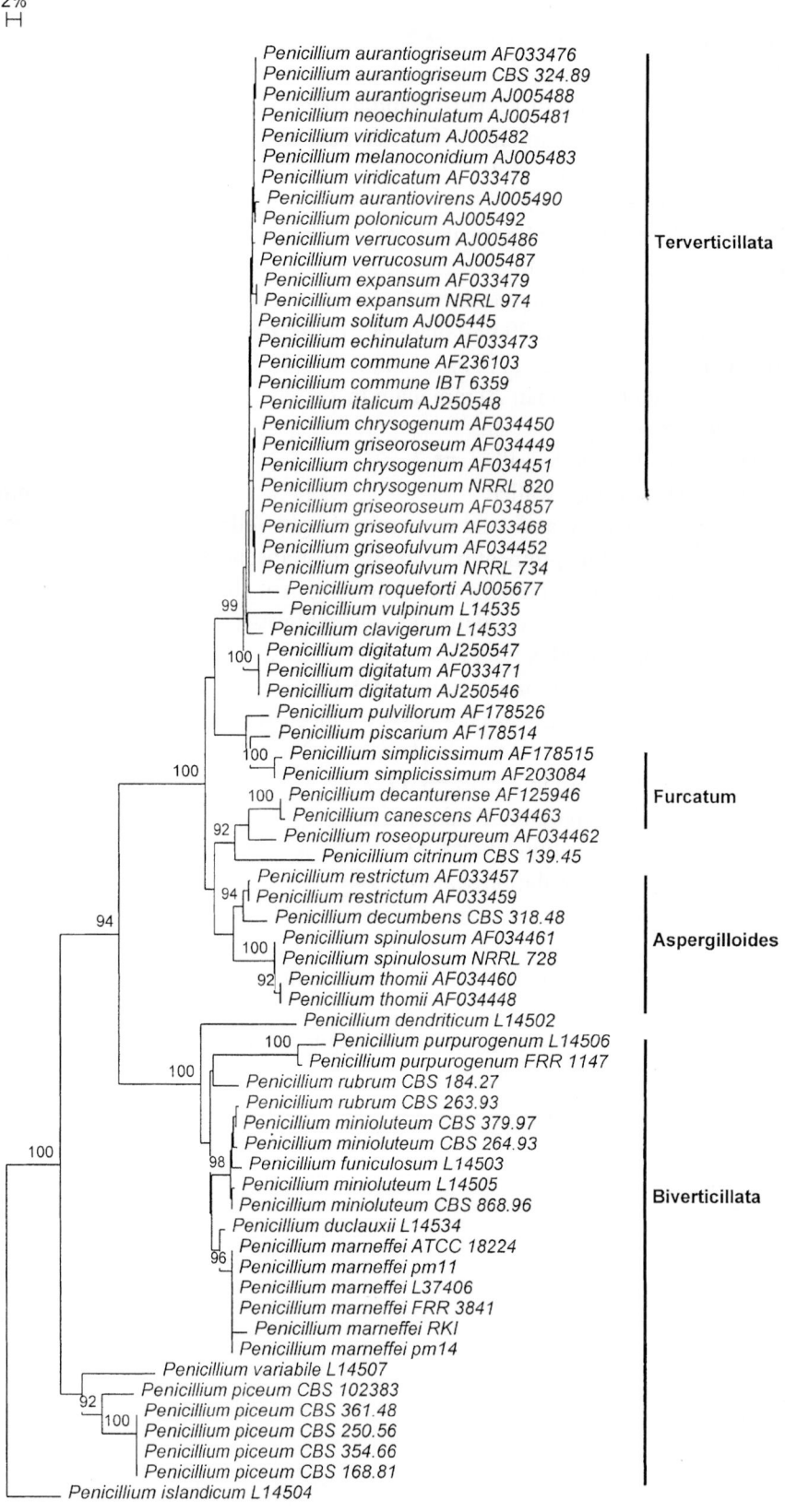

Fig. 60. Phylogenetic tree of *Penicillium*.species based on confidently aligned ITS rDNA sequences using Neighbor joining algorithm with Kimura correction. Bootstrap values >90 from 100 resampled datasets are shown. Part of the sequences were donated by N. Poonwan, who also made the alignment. Four subgenera of *Penicillium* are indicated. The subgenus *Biverticillium* contains several systemic, AIDS-associated species, among which is *P. marneffei*. The species in the subgenera *Terverticillata, Furcatum* and *Aspergilloides* are only exceptionally involved in mycoses, and these are mainly superficial or cutaneous. Hence they are likely to be saprobes.

Penicillium aurantiogriseum Dierckx

Colony characteristics. Colonies (CzA) growing moderately rapidly, granular, somewhat floccose, bright greyish-green, sometimes exuding a reddish-brown pigment into the medium; reverse orange to brown.

Microscopy. Conidiophore stipes mostly roughened, 200-400 μm long; penicilli usually terverticillate. Metulae 10-12 μm long. Phialides flask-shaped, 9-10 μm long. Conidia smooth-walled, subspherical to ellipsoidal, 3.5-4.0 μm long, bluish-grey.

Physiology. No or poor growth at 37°C.

Molecular diagnostics. ITS restriction map based on NCBI AF033476:

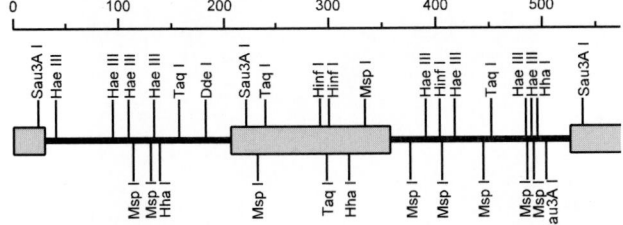

Pathogenicity. BSL-1. A ubiquitous air-borne species. An infection of the beak of a macaw was described by Bengoa *et al.* (1994).

References. Pitt (1979), Seifert & Louis-Seize (2000).

Nomenclature. *Penicillium aurantiogriseum* Dierckx - Annls Soc. Sci. Brux. 25: 88, 1901.
    Penicillium cyclopium Westling - Ark. Bot. 11: 90, 1911.

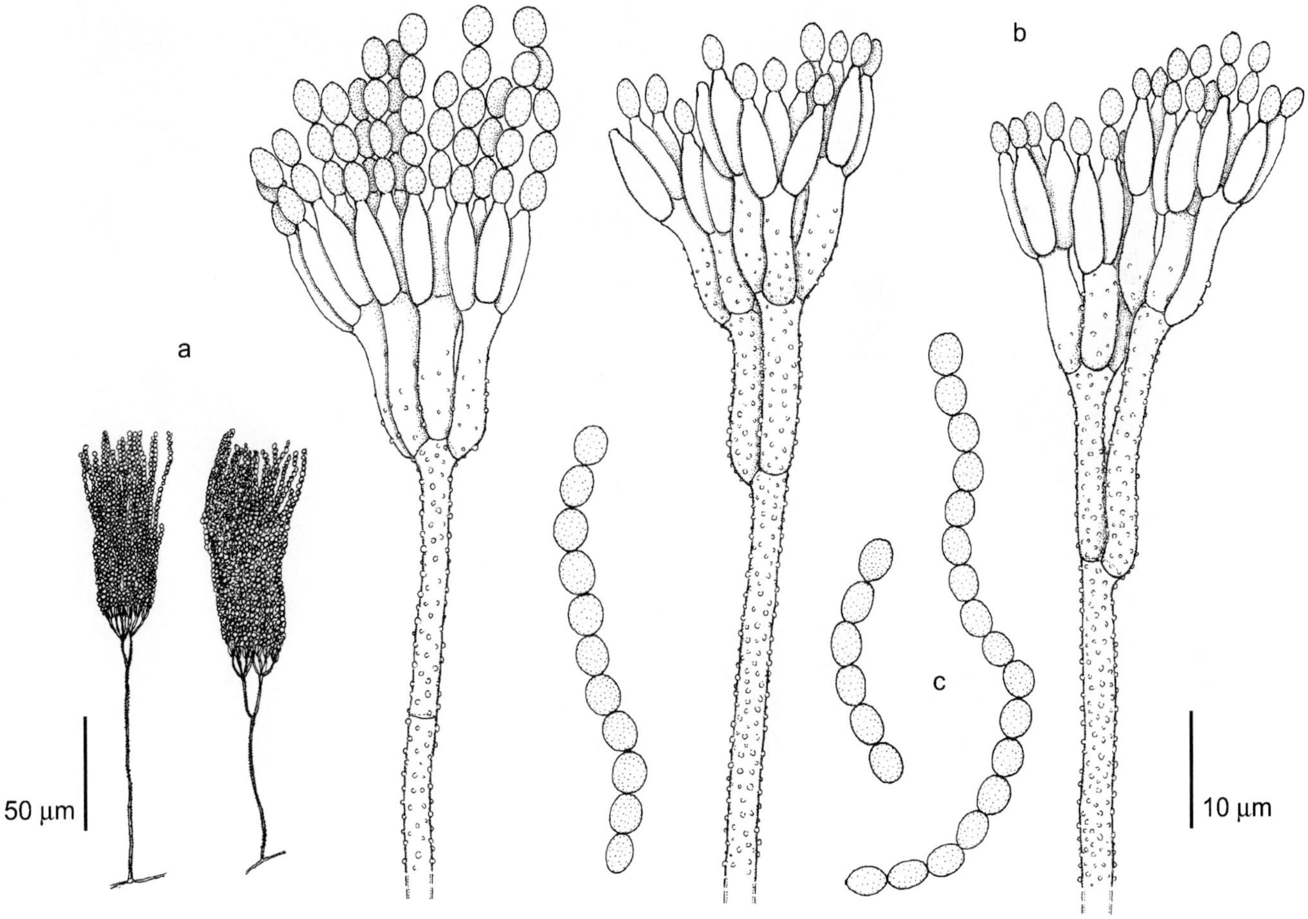

Penicillium aurantiogriseum, CBS 324.89. a. Habit sketch; b. conidiophores; c. conidia.

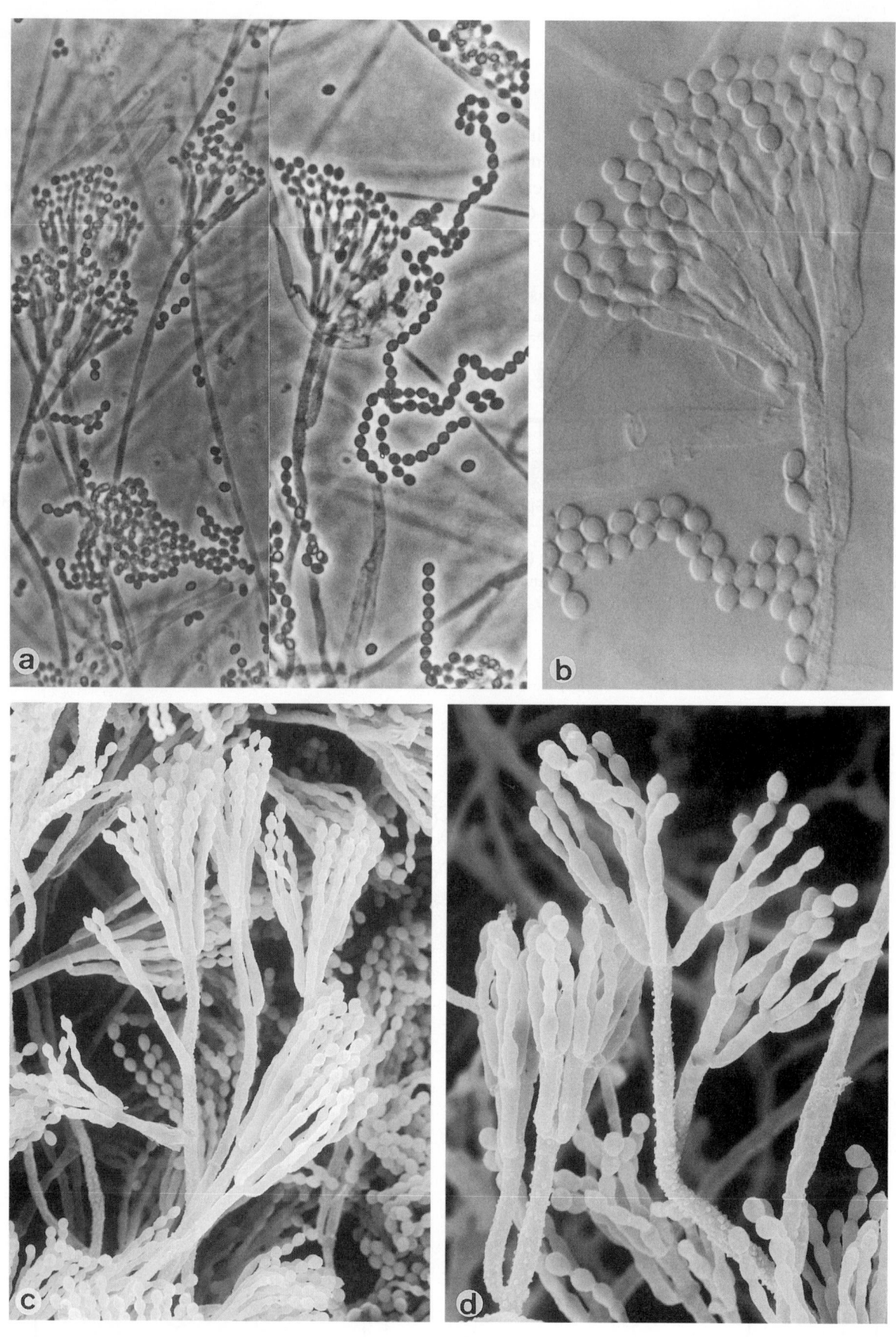

Penicillium aurantiogriseum, CBS 324.89. Conidiophores and conidia. a. ×640; b. ×1600; c. ×1150; d. ×2000.

Penicillium brevicompactum Dierckx

Colony characteristics. Colonies (CzA) growing moderately rapidly, dense cushion-shaped with radial fissures, granular, olivaceous-green, often with drops of exudate and mostly exuding a reddish-brown pigment into the medium; reverse yellowish to reddish-brown.

Microscopy. Conidiophore stipes smooth-walled, 500-800 μm long; penicillia usually terverticillate. Metulae compact, 9-12 μm long. Phialides flask-shaped, 6-9 μm long. Conidia smooth-walled or slightly verruculose, ellipsoidal, 2.5-3.5 μm long.

Physiology. No or poor growth at 37°C.

Pathogenicity. BSL-1. A pulmonary fungus ball in a transplant recipient was described by de la Camera *et al.* (1996); but note the species' inability to grow at 37°C.

Reference. Pitt (1979).

Nomenclature. *Penicillium brevicompactum* Dierckx - Annls Soc. Sci. Brux. 25: 88, 1901.

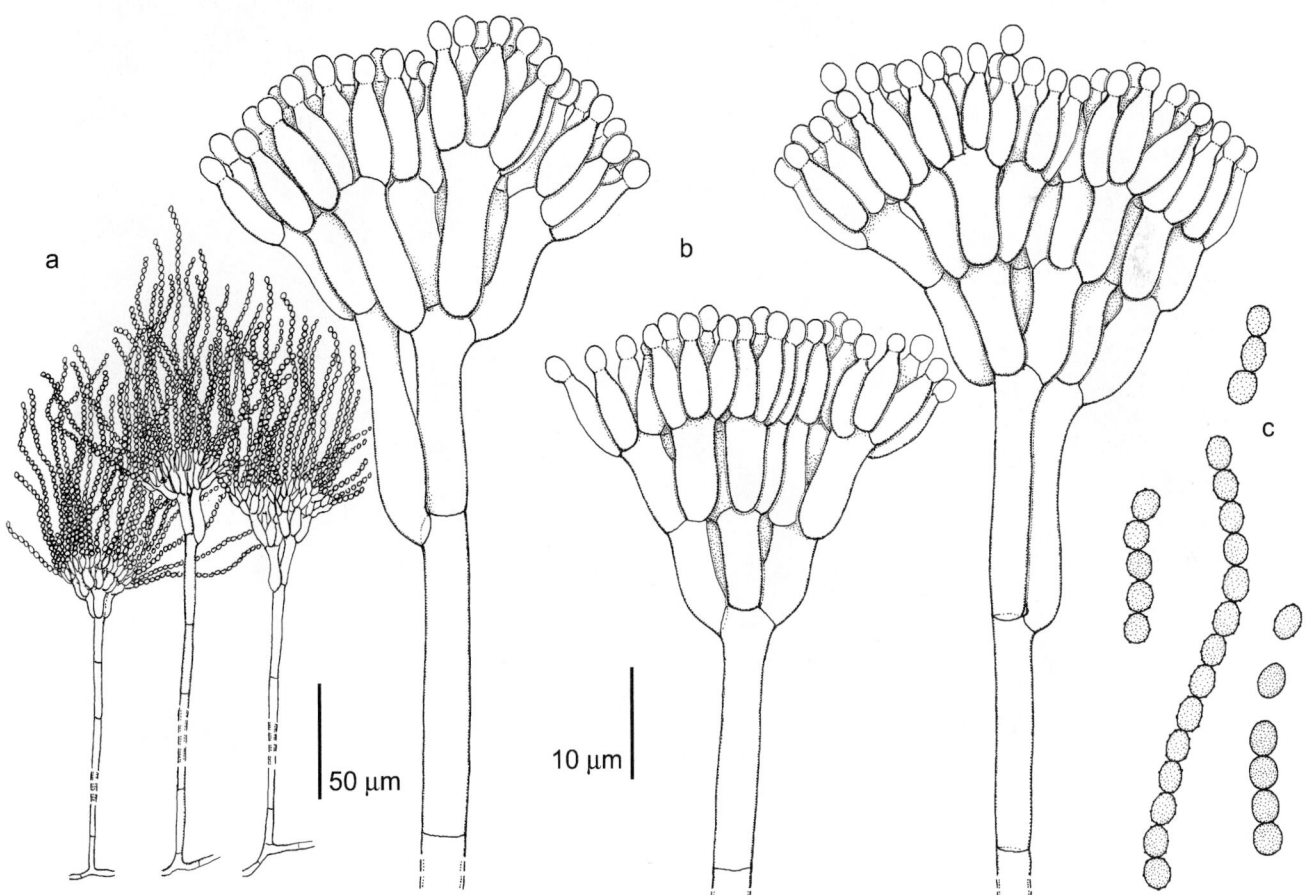

Penicillium brevicompactum, CBS 257.29. a. Habit sketch; b. conidiophores; c. conidia.

Penicillium brevicompactum, CBS 257.29. Conidiophores and conidia. a. ×1600; b. ×1450; c. ×5750.

Penicillium chrysogenum Thom

Colony characteristics. Colonies (CzA) growing rapidly, velutinous to floccose, exuding a bright yellow pigment into the medium; reverse yellow.

Microscopy. Conidiophore stipes smooth-walled, 200-300 µm long; penicilli usually terverticillate. Metulae 8-12 µm long. Phialides flask-shaped, 7-10 µm long. Conidia smooth-walled, ellipsoidal, 2.5-4.0 µm long, blue or bluish-green.

Physiology. No or limited growth at 37°C.

Molecular diagnostics. ITS restriction map based on NCBI AF034857:

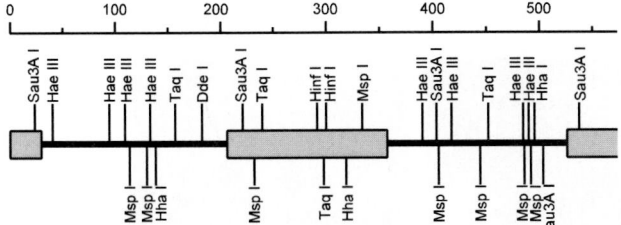

Pathogenicity. BSL-1. This species was implicated in cases of otomycoses (Yasin *et al.*, 1978), endophthalmitis (Eschete *et al.*, 1981), keratitis (Prasad & Nema, 1982), endocarditis (Upshaw, 1974) and a cutaneous infection (López-Martínez *et al.*, 1999). A fatal case of necrotizing esophagitis in an AIDS patient was reported by Hoffman *et al.* (1992) necrotizing pneumonia in a cancer patient by d'Antonio *et al.* (1997) and a systemic infection in an immunocompromised patient (Keung *et al.*, 1997).

References. Pitt (1979), Domsch *et al.* (1980).

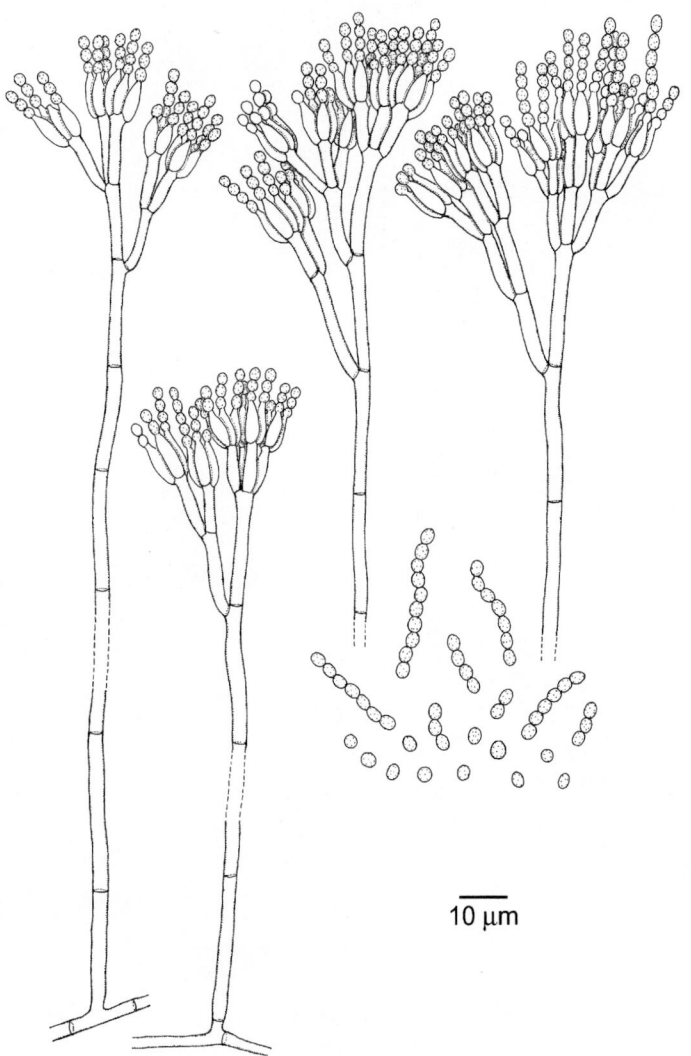

Penicillium chrysogenum, FMR 3088. Conidiophores and conidia.

Antifungal susceptibility.

Antifungal	Mean MICs	MIC90	Strains	Reference
AMB	0.28	0.39	2	Wildfeuer *et al.* (1998)
ITZ	<0.01	<0.01	2	Wildfeuer *et al.* (1998)
KTZ	0.05	0.05	2	Wildfeuer *et al.* (1998)
VCZ	0.10	0.20	2	Wildfeuer *et al.* (1998)

Nomenclature. *Penicillium chrysogenum* Thom - Bull. Bur. Anim. Ind. U.S. Dept Agric. 118: 58, 1910.

Penicillium chrysogenum, FMR 3088. Conidiophores and conidia. a. ×435; b. ×875; c. ×1090; d. ×1570; e. ×1090; f. ×2170; g. ×3015.

Penicillium citrinum Thom

Colony characteristics. Colonies (CzA) with slow to moderate growth, velutinous to floccose; mycelium white to greyish-orange. Conidial masses greyish-turquoise; frequently a pale yellow to reddish-brown soluble pigment is produced. Exudate on MEA greyish-blue.

Microscopy. Conidiophore stipes smooth-walled, 100-300 μm long; penicilli biverticillate. Metulae 12-15 μm long, divergent, in whorls of 3-5. Phialides flask-shaped, 7-12 μm long. Conidia spherical to subspherical, smooth-walled or finely roughened, 2.2-3.0 μm diam.

Physiology. Poor, whitish growth at 37°C.

Molecular diagnostics. ITS restriction map based on NCBI AF033422:

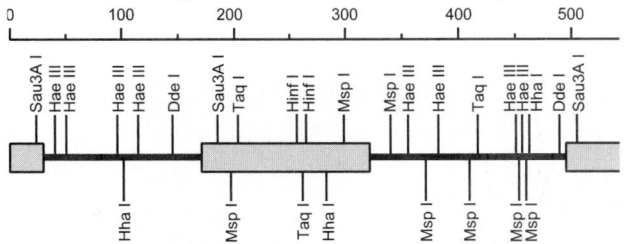

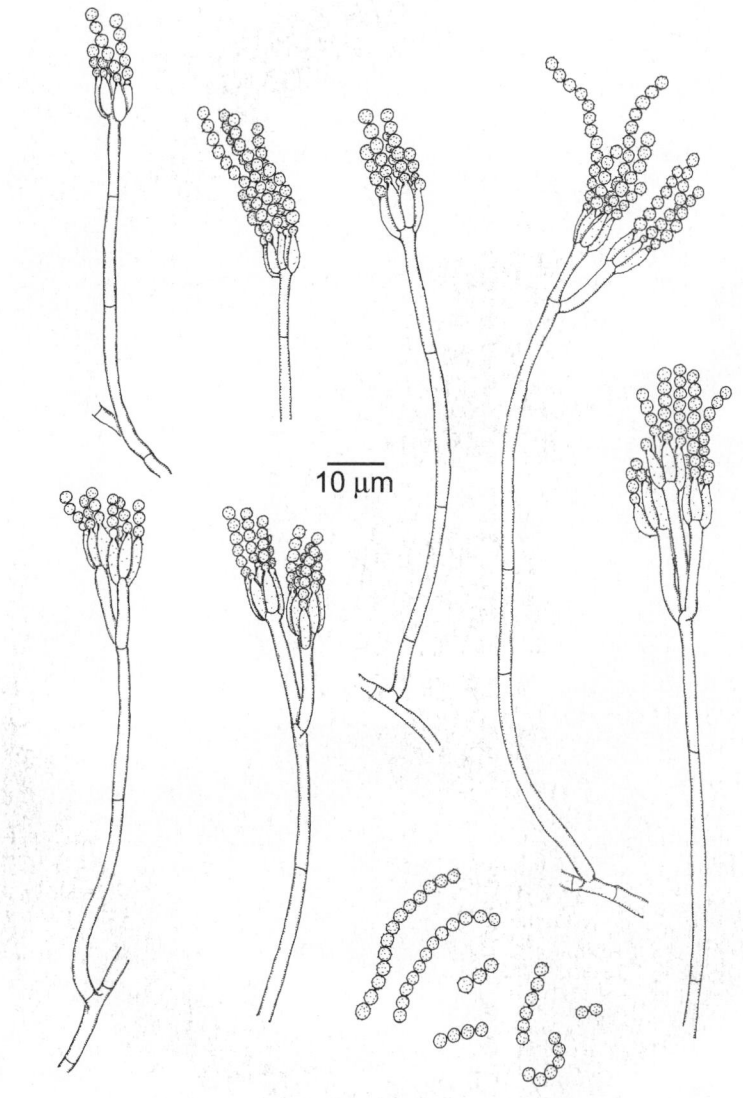

Penicillium citrinum, IFO 6352. Conidiophores and conidia.

Pathogenicity. BSL-1. The species was reported as being involved in infections of the urinary tract (Gilliam & Vest, 1951; Perry, 1964), lungs (Mori *et al.*, 1987) and in cases of keratitis (Jones *et al.*, 1970; Gugnani *et al.*, 1976). A fatal infection in a leukemic patient was reported by Mok *et al.* (1997).

Reference. Pitt (1979).

Nomenclature. *Penicillium citrinum* Thom - Bull. Bur. Anim. Ind. U.S. Dept Agric. 118: 61, 1910.

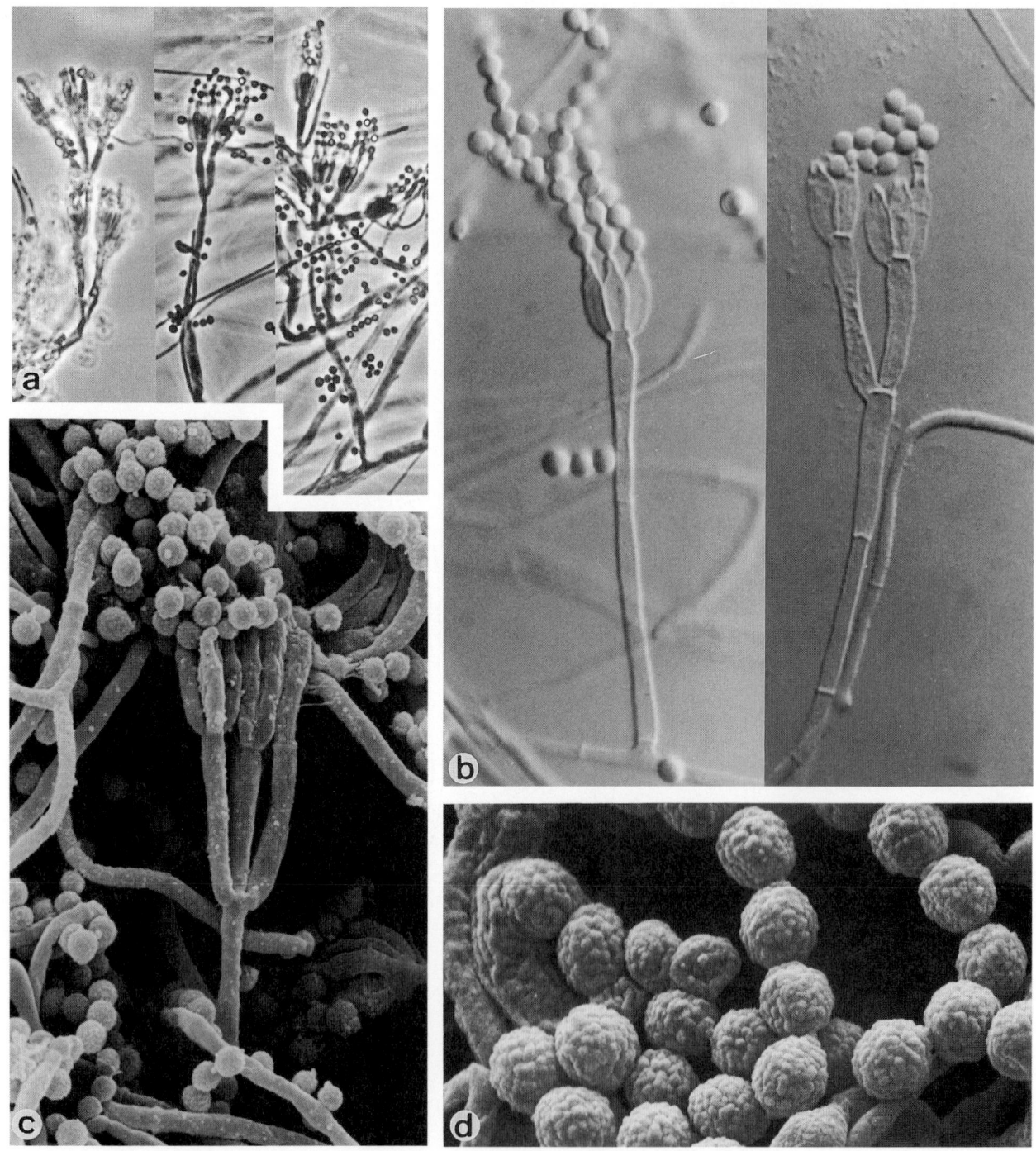

Penicillium citrinum, IFO 6352. Conidiophores and conidia. a. ×512; b. ×1600; c. ×2350; d. ×6900.

Penicillium commune Thom

Colony characteristics. Colonies (CzA) with moderate growth, velutinous to floccose; conidial mass dull green on MEA.

Microscopy. Conidiophore stipes rough-walled, 100-200 µm long; penicilli terverticillate. Metulae 8-15 µm long, in whorls of 2-5. Phialides flask-shaped, tapering into a narrow neck, 7-9 µm long. Conidia spherical to subspherical, smooth-walled, 3-4 µm diam.

Molecular diagnostics. ITS restriction map based on NCBI AJ004813:

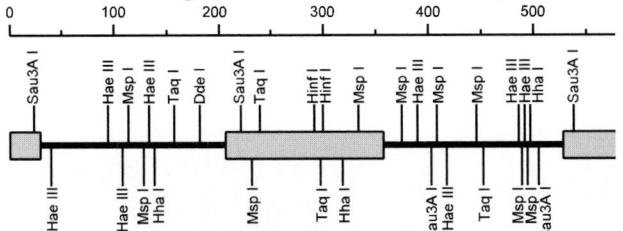

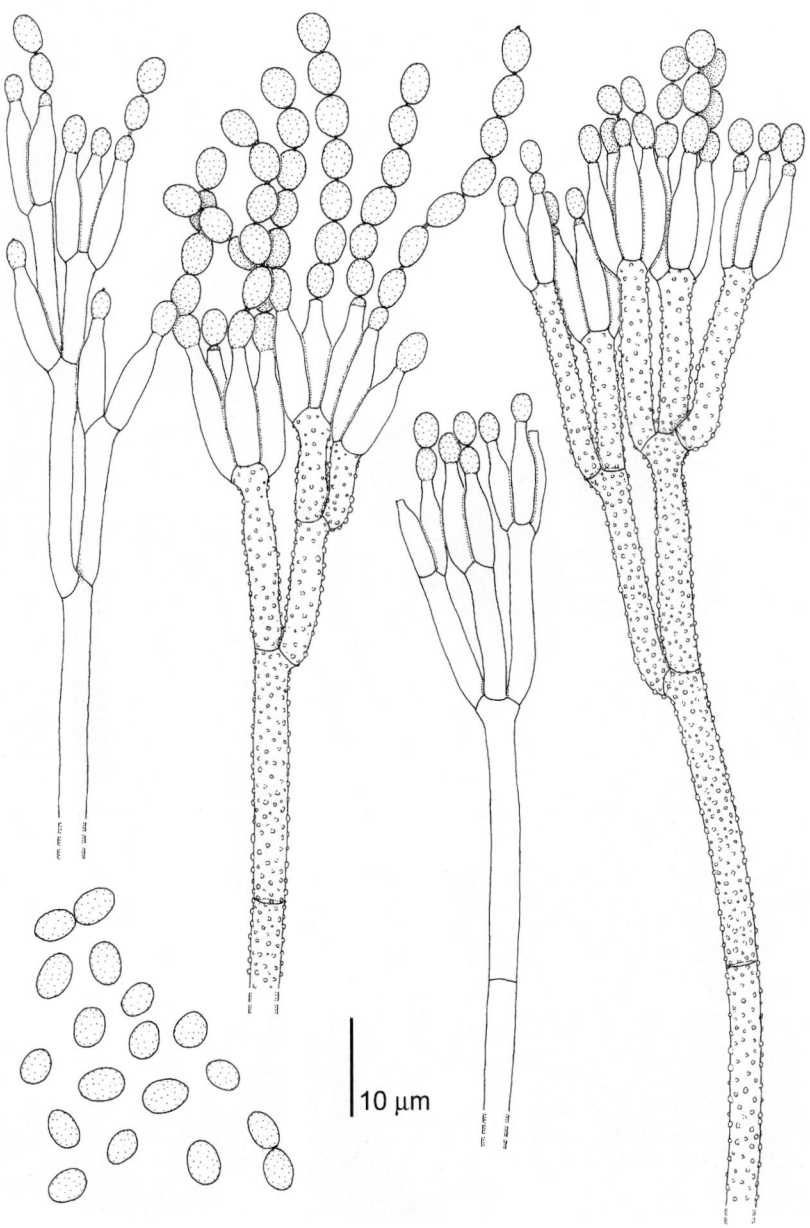

Penicillium commune, CBS 311.48. Conidiophores and conidia.

Pathogenicity. BSL-1. A case of acute, disseminated penicilliosis has been described (Huang & Harris, 1963).
Reference. Pitt (1979).
Nomenclature. *Penicillium commune* Thom - Bull. Bur. Anim. Ind. U.S. Dept Agric. 118: 56, 1910.

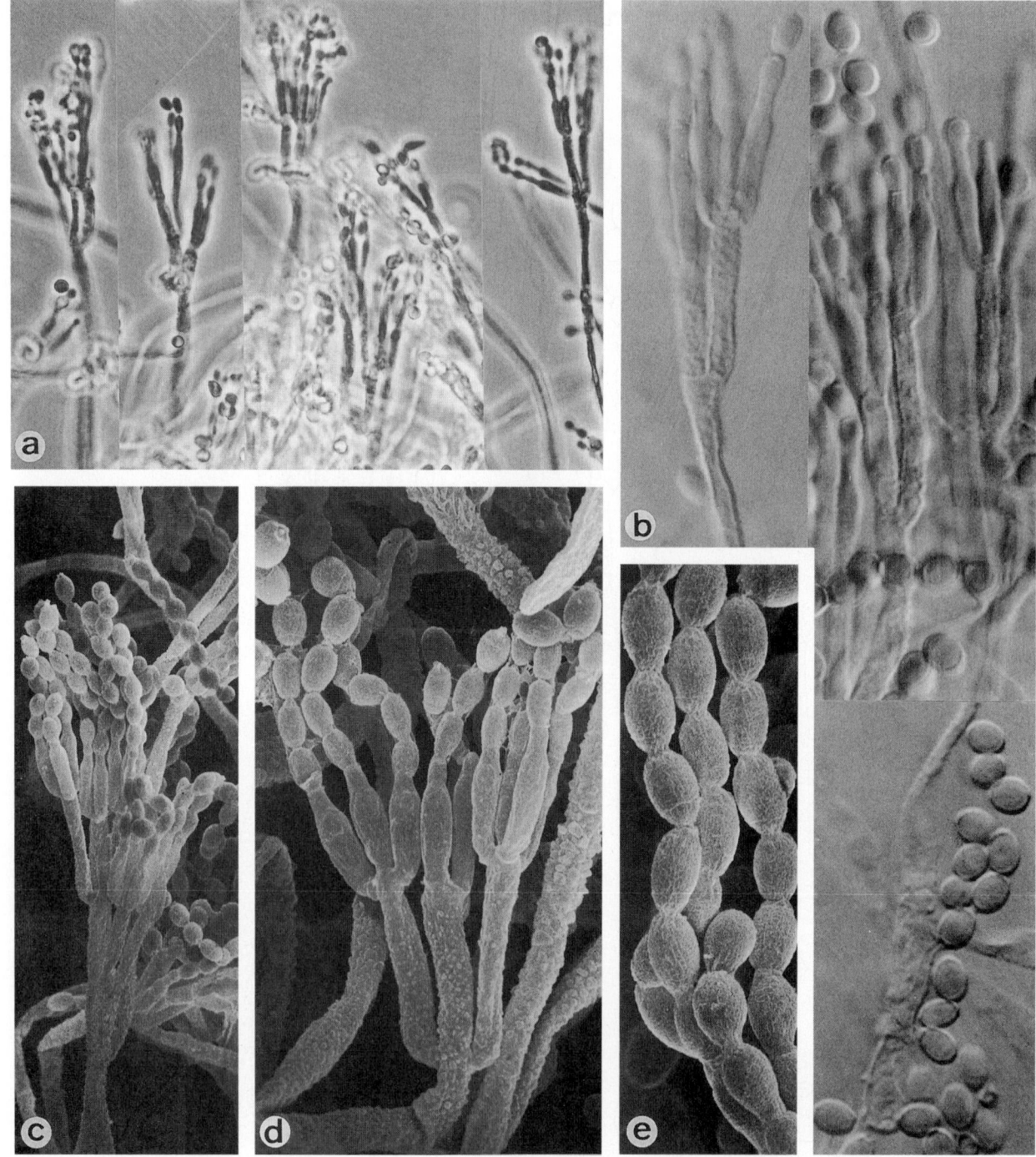

Penicillium commune, CBS 311.48. Conidiophores and conidia. a. ×512; b. ×1600; c. ×1100; d. ×2250; e. ×3450.

Penicillium decumbens Thom

Colony characteristics. Colonies (CzA) usually restricted, velutinous to somewhat floccose; mycelium white to cream-coloured; conidial mass dull blue-green.

Microscopy. Conidiophore stipes 50-100 µm long, smooth-walled; penicilli monoverticillate. Phialides flask-shaped, 8-14 µm long. Conidia ellipsoidal to subspherical, 2-3 µm long, smooth-walled.

Physiology. Restricted growth at 37°C.

Molecular diagnostics. ITS restriction map based on NCBI AF034458:

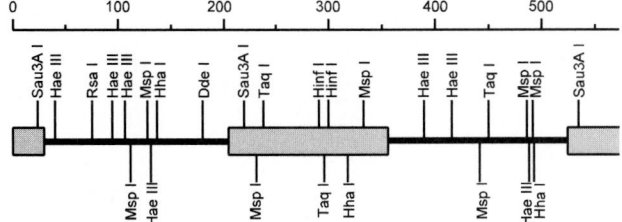

Pathogenicity. BSL-1. Cases known to date are a disseminated infection in a man with AIDS (Alvarez, 1990) and a case of fungus ball (Yoshida *et al.*, 1992).

References. Pitt (1979), Domsch *et al.* (1980).

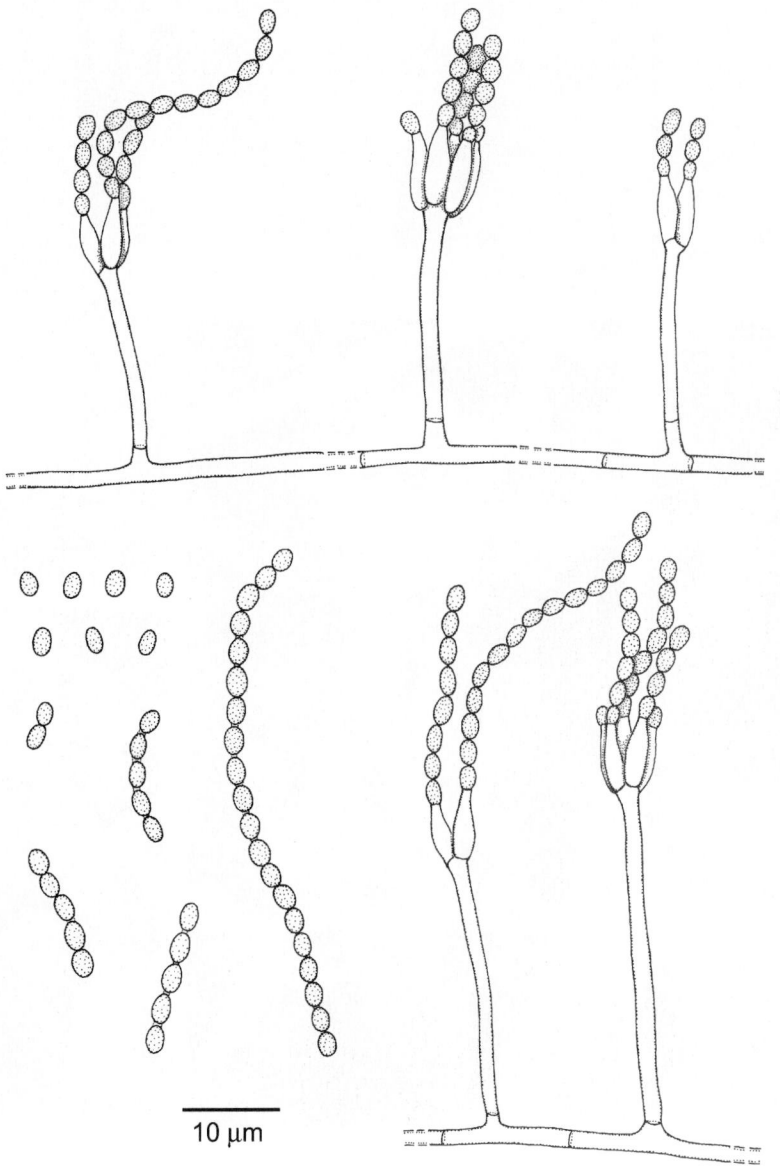

10 µm

Penicillium decumbens, CBS 230.81. Conidiophores and conidia.

Antifungal susceptibility.

Antifungal	Mean MICs	MIC90	Strains	Reference
AMB	0.55	1.56	2	Wildfeuer *et al.* (1998)
ITZ	1.11	1.56	2	Wildfeuer *et al.* (1998)
KTZ	0.28	0.39	2	Wildfeuer *et al.* (1998)
VCZ	0.39	0.39	2	Wildfeuer *et al.* (1998)

Nomenclature. *Penicillium decumbens* Thom - Bull. Bur. Anim. Ind. U.S. Dept Agric. 118: 71, 1910.

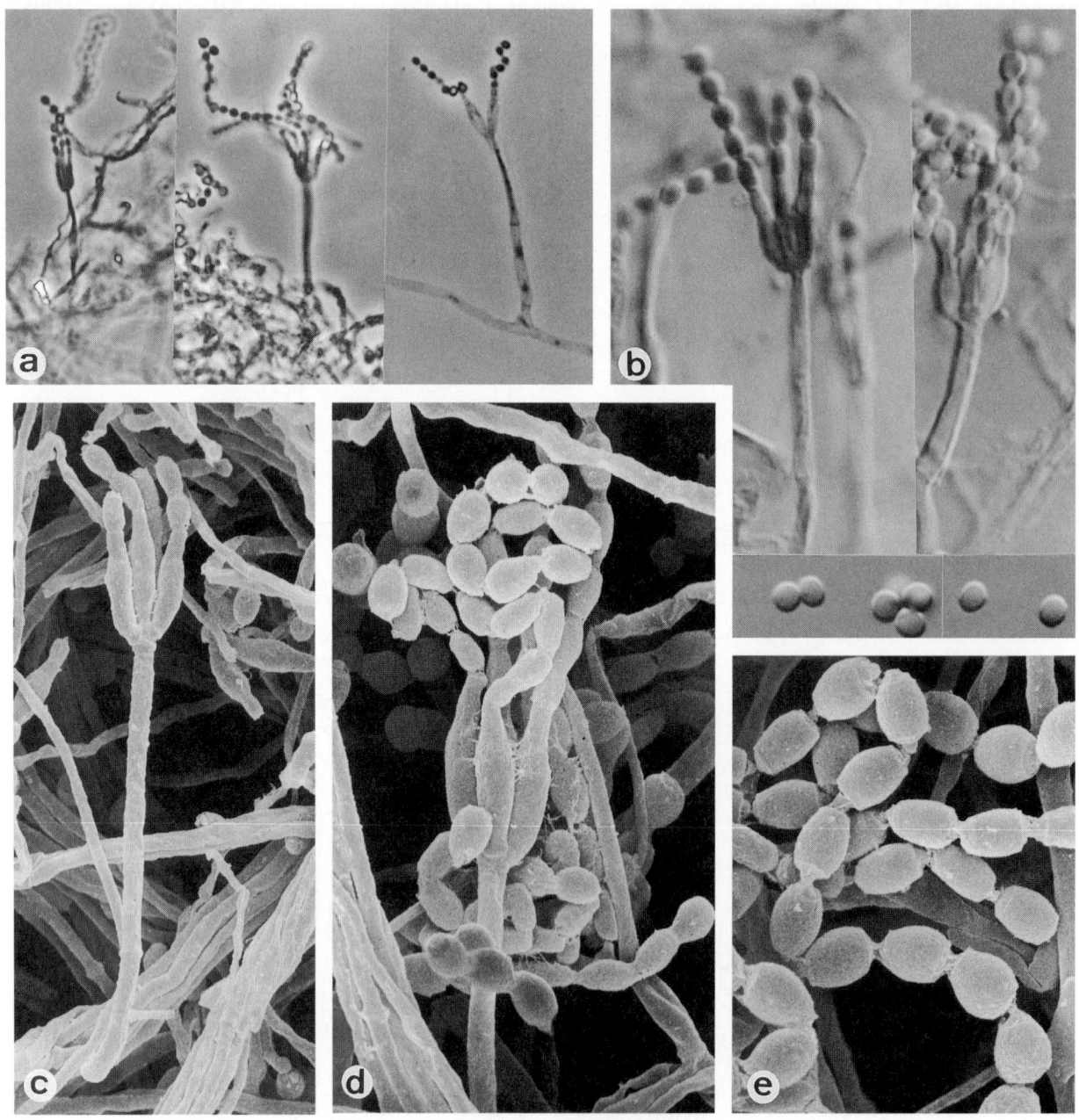

Penicillium decumbens, CBS 230.81. Conidiophores and conidia. a. ×512; b. ×1600; c. ×3300; d. ×3850; e. ×4800.

Penicillium expansum Link

Colony characteristics. Colonies (CzA) rapidly growing, fasciculate to synnematal; conidial mass dull green; exudate and soluble pigment brown.

Microscopy. Conidiophore stipes smooth-walled, 200-500 μm long; penicilli terverticillate. Metulae 12-18 μm long. Phialides closely packed, flask-shaped, tapering into a short, narrow neck, 8-11 μm long. Conidia ellipsoidal, smooth-walled, 3.0-3.5 μm long.

Physiology. No growth at 37°C.

Molecular diagnostics. ITS restriction map based on NCBI AF033479:

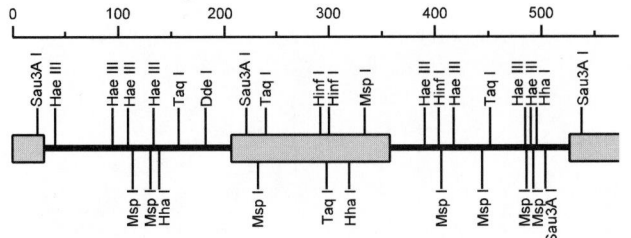

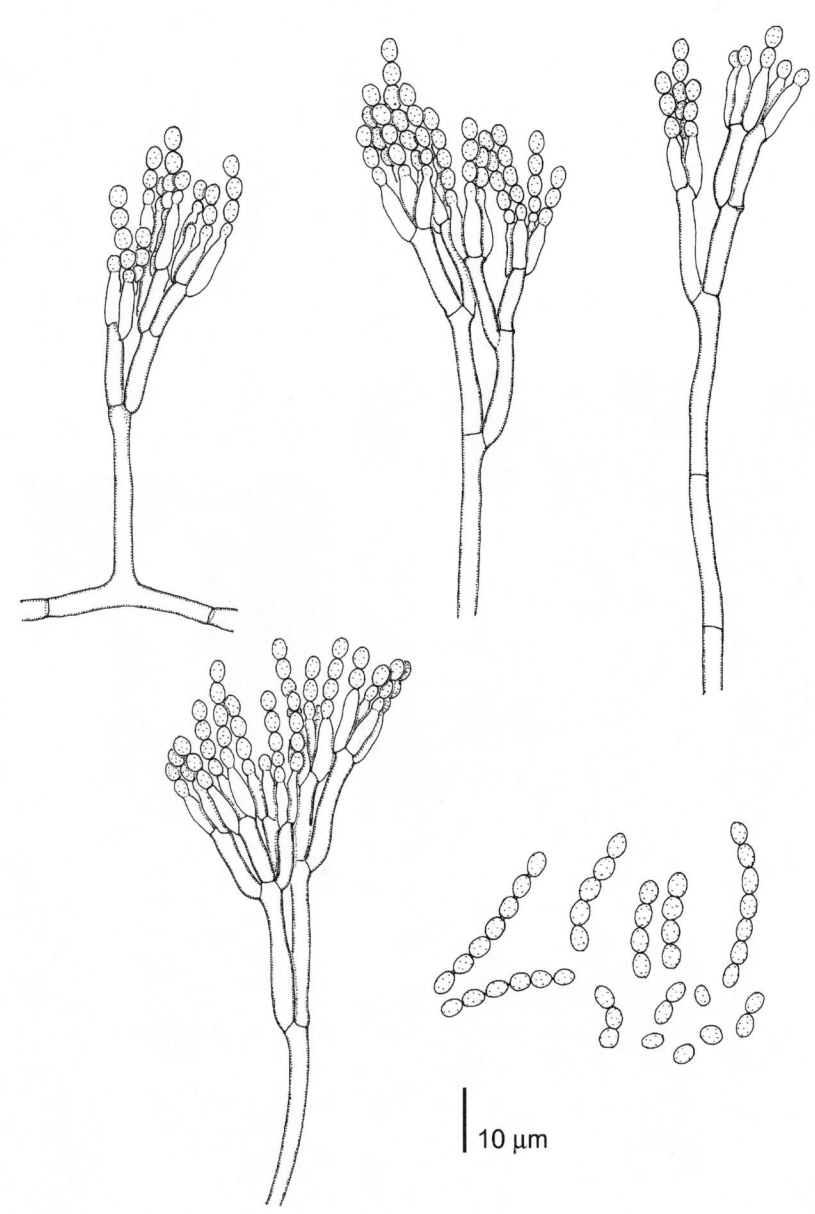

Penicillium expansum, IFO 6096. Conidiophores and conidia.

Pathogenicity. BSL-1. Keratitis (Gugnani *et al.*, 1976)
References. Pitt (1979), Kozakiewicz (1995).
Nomenclature. *Penicillium expansum* Link - Obs. Mycol. 1: 16, 1809.

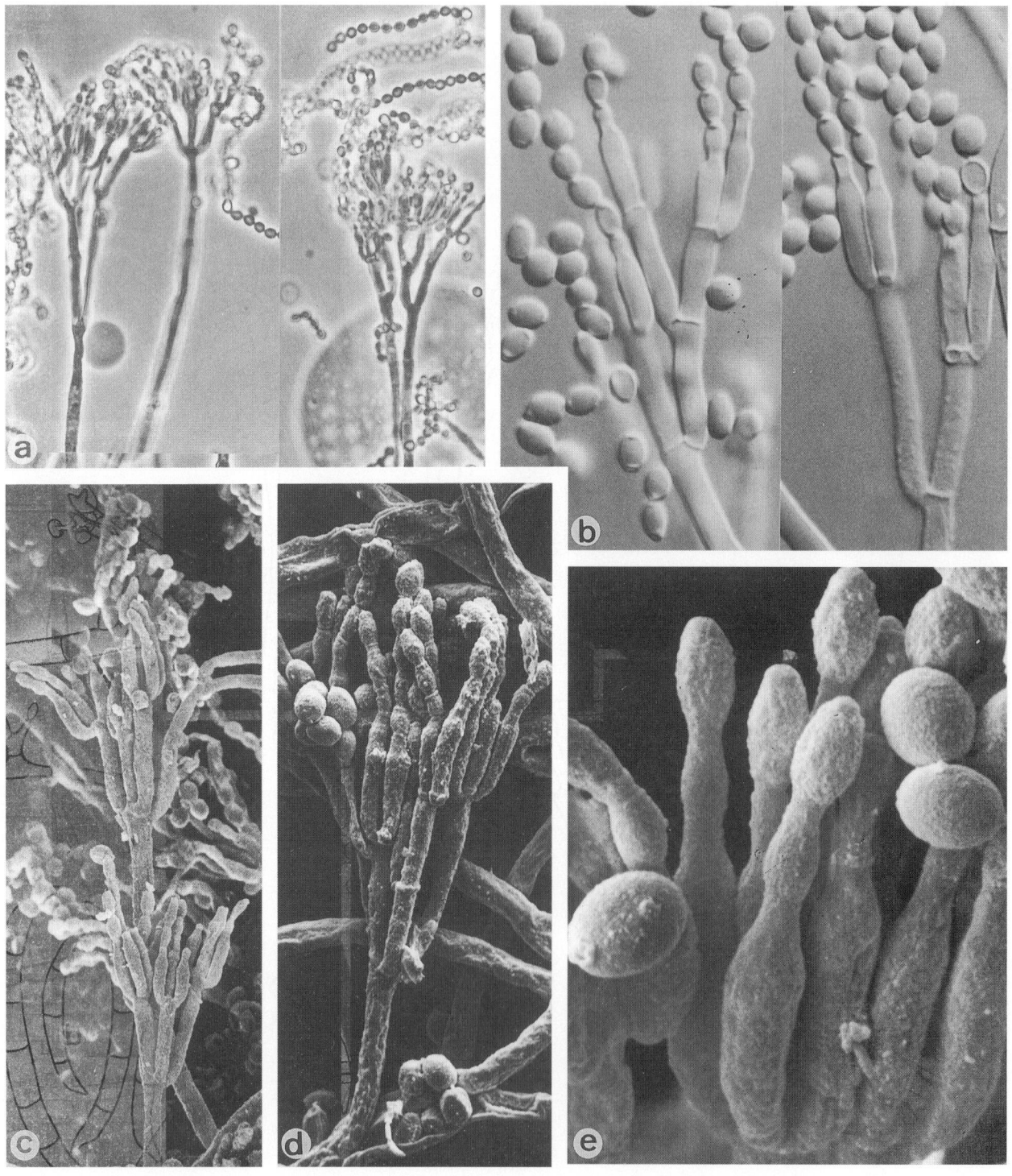

Penicillium expansum, IFO 6096. Conidiophores and conidia. a. ×512; b. ×1600; c. ×1200; d. ×1995; e. ×6600.

Penicillium griseofulvum Dierckx

Colony characteristics. Colonies (CzA) slowly growing, fasciculate to synnematal, greyish-green; soluble pigment reddish-brown.

Microscopy. Conidiophore stipes of very variable length, smooth-walled, brownish; penicilli terverticillate to quaterverticillate. Metulae 7-10 μm long, sometimes apically inflated. Phialides closely packed, very short, ampulliform, 4.5-6.0 μm. Conidia ellipsoidal, smooth-walled, 3.0-3.5 μm long.

Physiology. Intolerant to benomyl. No growth at 37°C.

Molecular diagnostics. ITS restriction map based on NCBI AF034452:

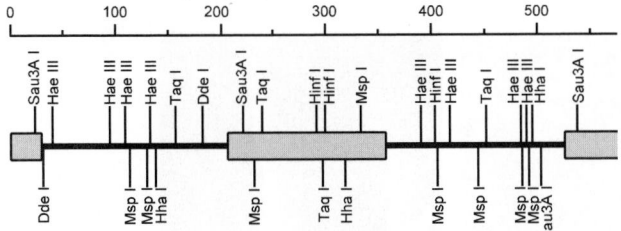

Pathogenicity. BSL-1. A case of systemic mycosis in a captive toucan (Aho *et al.*, 1990), though the species is unable to grow at 37°C (Pitt, 1979).

References. Pitt (1979), Domsch *et al.* (1980), Kozakiewicz (1995).

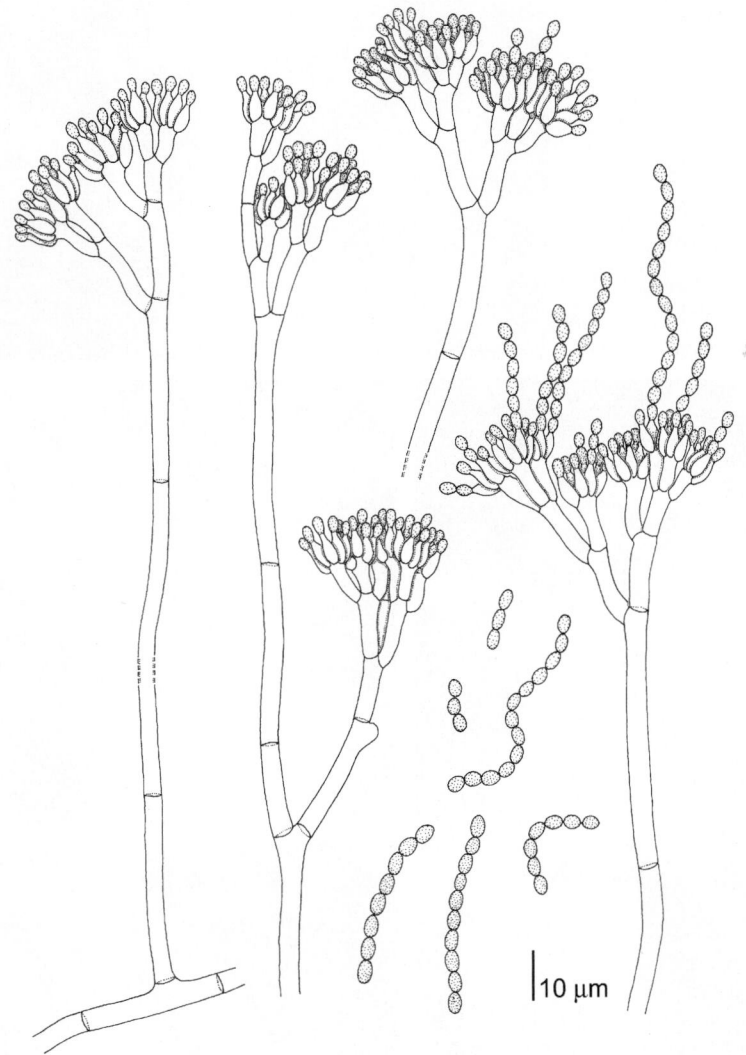

Penicillium griseofulvum, CBS 185.27. Conidiophores and conidia.

Antifungal susceptibility.

Antifungal	Mean MICs	Strains	Reference
AMB	3.13	1	Wildfeuer *et al.* (1998)
ITZ	1.56	1	Wildfeuer *et al.* (1998)
KTZ	0.20	1	Wildfeuer *et al.* (1998)
VCZ	0.78	1	Wildfeuer *et al.* (1998)

Nomenclature. *Penicillium griseofulvum* Dierckx - Annls Soc. Sci. Brux. 25: 88, 1901.

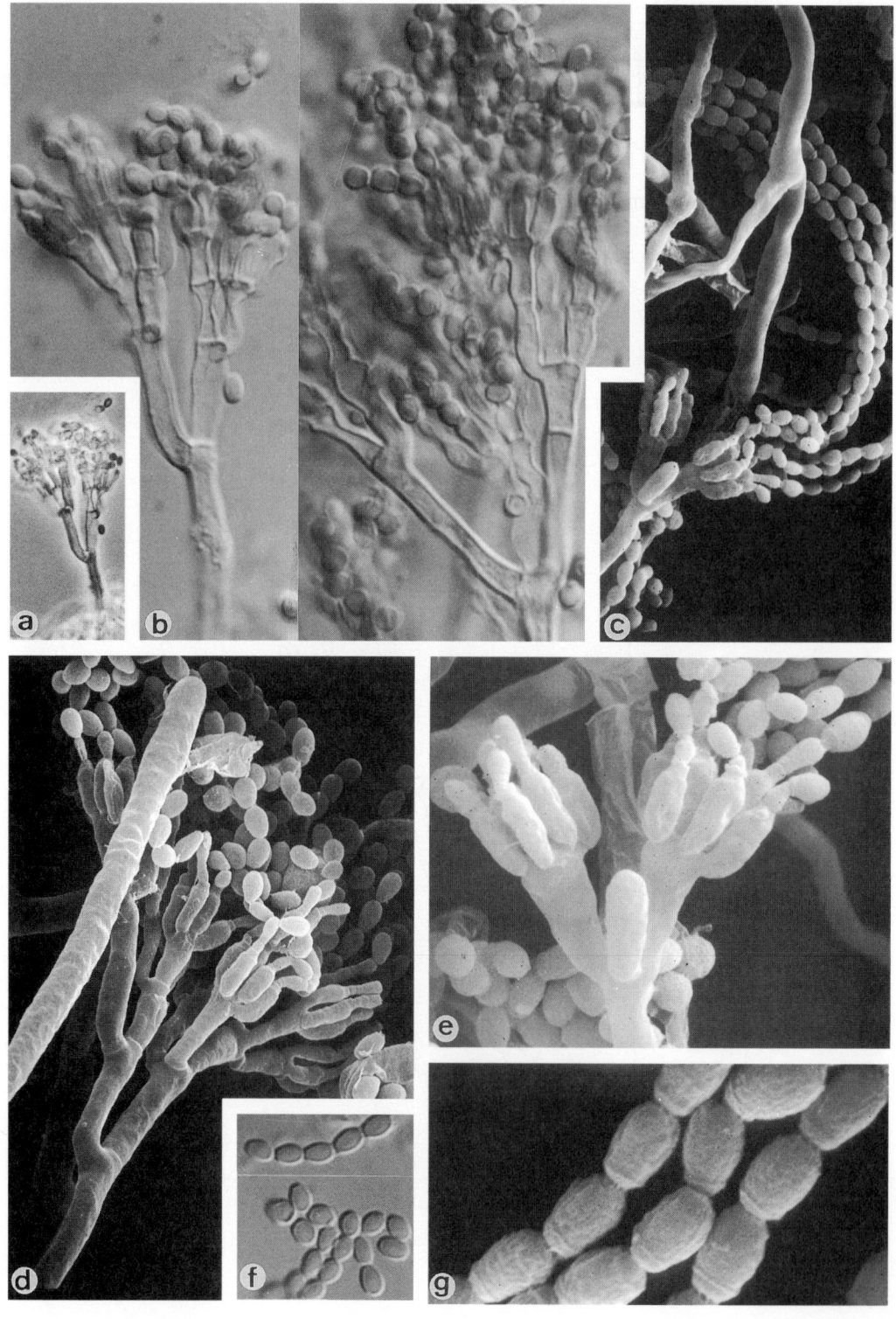

Penicillium griseofulvum, CBS 185.27. Conidiophores and conidia. a. ×512; b, c. ×1600; d. ×2450; e. ×3700; f. ×1600; g. ×7000.

Penicillium marneffei Segretain

Colony characteristics. Colonies (CzA, 30°C) flat, sparse, compact, greenish to purplish, on MEA exuding an orange or red pigment into the medium; primary cultures often canary yellow due to sterile aerial mycelium. At 37°C colonies restricted, whitish, yeast-like.

Microscopy. Hyphae in part spirally twisted. Conidiophores creeping or fasciculate, 70-150 × 2.5-3.0 µm; penicilli generally biverticillate but also irregularly monoverticillate or more complex. Metulae 7-11 µm long, in whorls of 3-5. Phialides in whorls of 4-7, ampulliform to acerose, 6-10 × 2.5-3.0 µm. Conidia smooth-walled, ellipsoidal, 2.5-4.0 × 2-3 µm, often with prominent scars, borne in short, disordered chains.

Physiology. An arthroconidial synanamorph ('yeast phase') is produced at 37°C.

Serology. Antigenic relationships were investigated by Sekhon *et al.* (1989). A specific fluorescent-antibody test was developed by Kaufman *et al.* (1995) and antigen detection by Kaufman *et al.* (1996) and Desakorn *et al.* (1998).

Molecular diagnostics. Species-specific ITS-based primers were developed by LoBuglio & Taylor (1995) and SSU probes by Vanittanakom *et al.* (1998). ITS restriction map based on NCBI L37406:

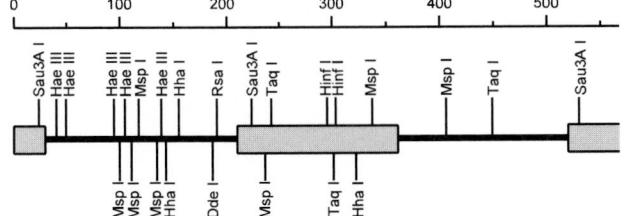

Differential diagnosis. *In vivo* the species differs from *Histoplasma* (p. 708) by the occurrence of some septate and slightly curved cells. Distinguished from *Pneumocystis* (p. 176) by positive Giemsa staining. *In vitro* the species shows poor growth and sporulation.

Pathogenicity. BSL-3. The species may disseminate in humans with AIDS which live in (Sirisanthana & Sirisanthana, 1995; Ukarapol *et al.*, 1998) or have visited (Kok *et al.*, 1994; Heath *et al.*, 1995; Depraetere *et al.*, 1998) endemic areas, rarely in healthy individuals (Louthrenoo *et al.*, 1994; Kwan *et al.*, 1997). The typical cutaneous lesions may be absent (Kantipong *et al.*, 1998). Dependence on acquired cellular immunity was proven in an animal experimental infection (Kudeken *et al.*, 1996). The mycosis is acquired by inhalation; first lesions are pulmonary (Breton *et al.*, 1998). It occurs as an intracellular mycosis, producing small, rounded arthroconidia (2-5 µm diam) inside macrophages. When untreated, the mycosis is mostly fatal. The species naturally occurs in bamboo rats and similar mammals in South East Asia (Chariyalertsak *et al.*, 1996); strains from different host animals show molecular differences (Vanittanakom *et al.*, 1996).

Distribution. Thailand and adjacent countries, Taiwan (Lin *et al.*, 1998), India (Singh *et al.*, 1999). Imported cases are common outside endemic areas. A possibly endemic case from Ghana (Africa) was reported by Lo *et al.* (2000)

References. Yayanetra *et al.* (1984), Deng *et al.* (1986), Mori *et al.* (1987), Qi *et al.* (1990), Chiewchanvit *et al.* (1991), Tsui *et al.* (1992), Drouhet (1993), Liu *et al.* (1994), Ajello *et al.* (1995), Wortman (1996), Vanittanakom & Sirisanthana (1997).

Antifungal susceptibility.

Antifungal	Mean MICs	Strains	Reference
AMB	32	1	Unpublished data
5FC	256	1	Unpublished data
FLZ	64	1	Unpublished data
ITZ	0.25	1	Unpublished data
ITZ	0.04	30	McGinnis *et al.* (2000)
KTZ	1	1	Unpublished data
MCZ	2	1	Unpublished data
TBF	0.09	30	McGinnis *et al.* (2000)

Nomenclature. *Penicillium marneffei* Segretain - Bull. Trim. Soc. Mycol. Fr. 75: 416, 1959.

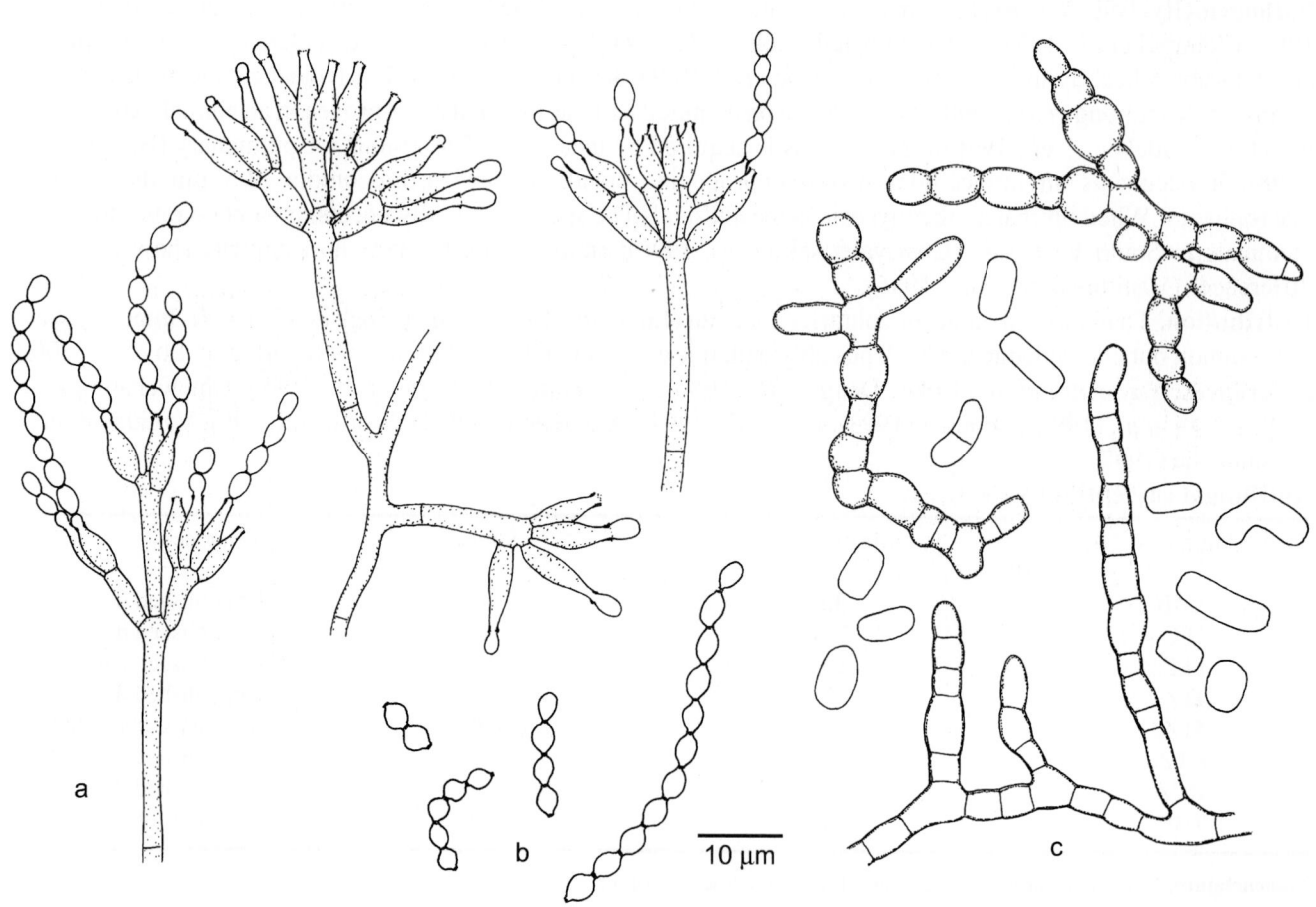

Penicillium marneffei, CBS 107.89. a. Mature conidiophores with conidial chains; b. liberated, coherent conidial chains; c. arthroconidial phase *in vitro* at 37°C.

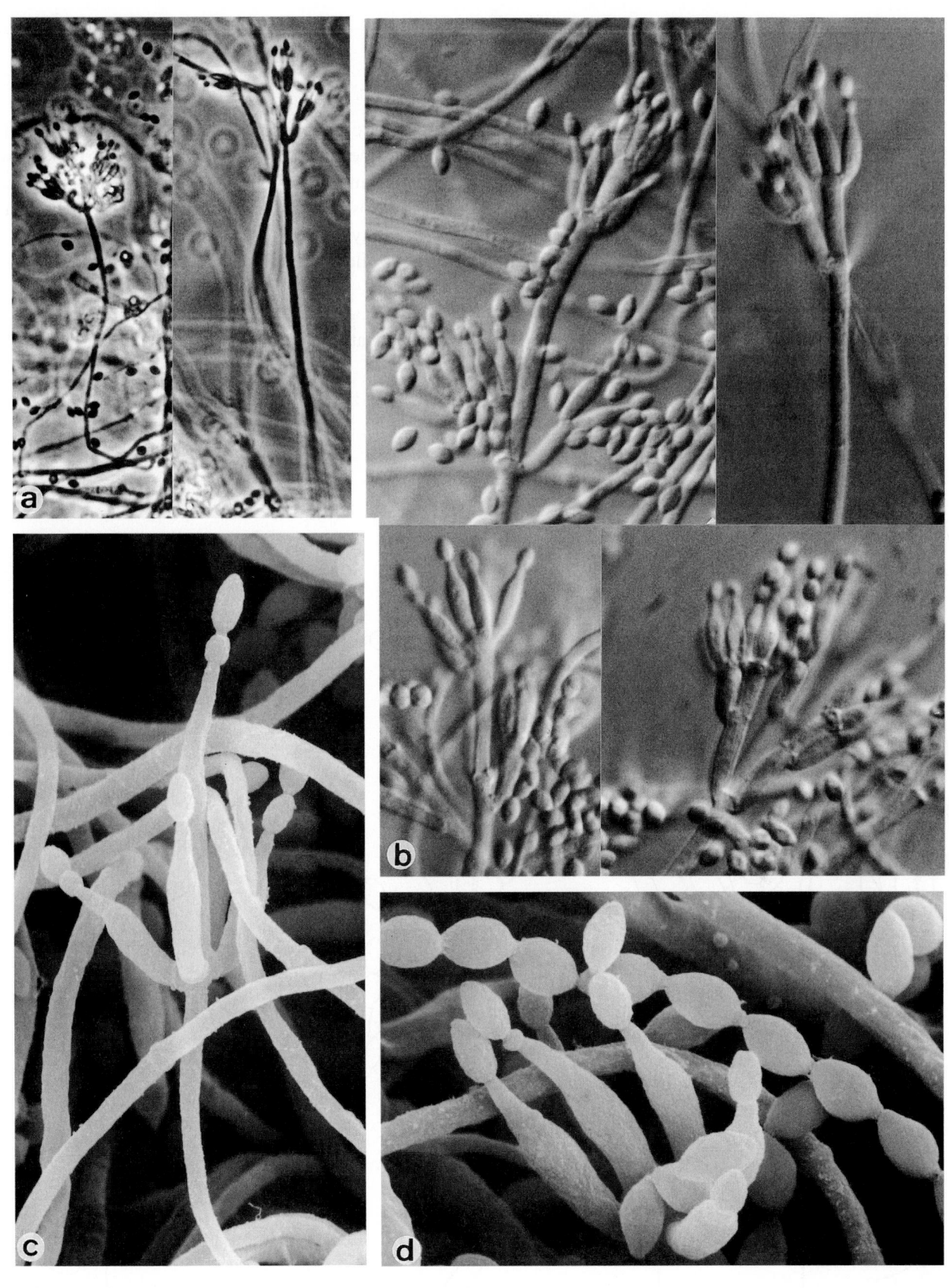

Penicillium marneffei, CBS 107.89. Conidiophores and conidia. a. ×640; b. ×1600; c. ×4750; d. ×6500.

Penicillium piceum Raper & Fennell

Colony characteristics. Colonies (CzA) growing moderatly rapidly, velutinous to floccose, straw yellow to bright yellow, with olivaceous green conidiation; no pigment exuded into the medium; reverse orange-brown to dark brown.
Microscopy. Conidiophore stipes thin- and smooth-walled, often somewhat flexuose, 15-22 μm long; penicilli biverticillate. Metulae in dense whorls, 7-12 × 3-4 μm. Phialides acerose, with of 3-8 on each metula, 7-9 × 2 μm. Conidia smooth-walled, broadly ellipsoidal, 3.0-3.5 × 2.2-2.5 μm, subhyaline.
Remark. Subcultures often lack sporulation, consisting of bright yellow, clumped aerial mycelium only. Older cultures and often degenerate, whitish, and nearly lacking conidia.
Physiology. Optimal growth at 37°C.
Pathogenicity. BSL-1. Horré *et al.* (2000) described a disseminated, fatal infection in a cancer patient. Pitt (1994) mentioned a strain from human nail, without a case report.
Reference. Pitt (1979).
Nomenclature. *Penicillium piceum* Raper & Fennell - Mycologia 40: 533, 1948.

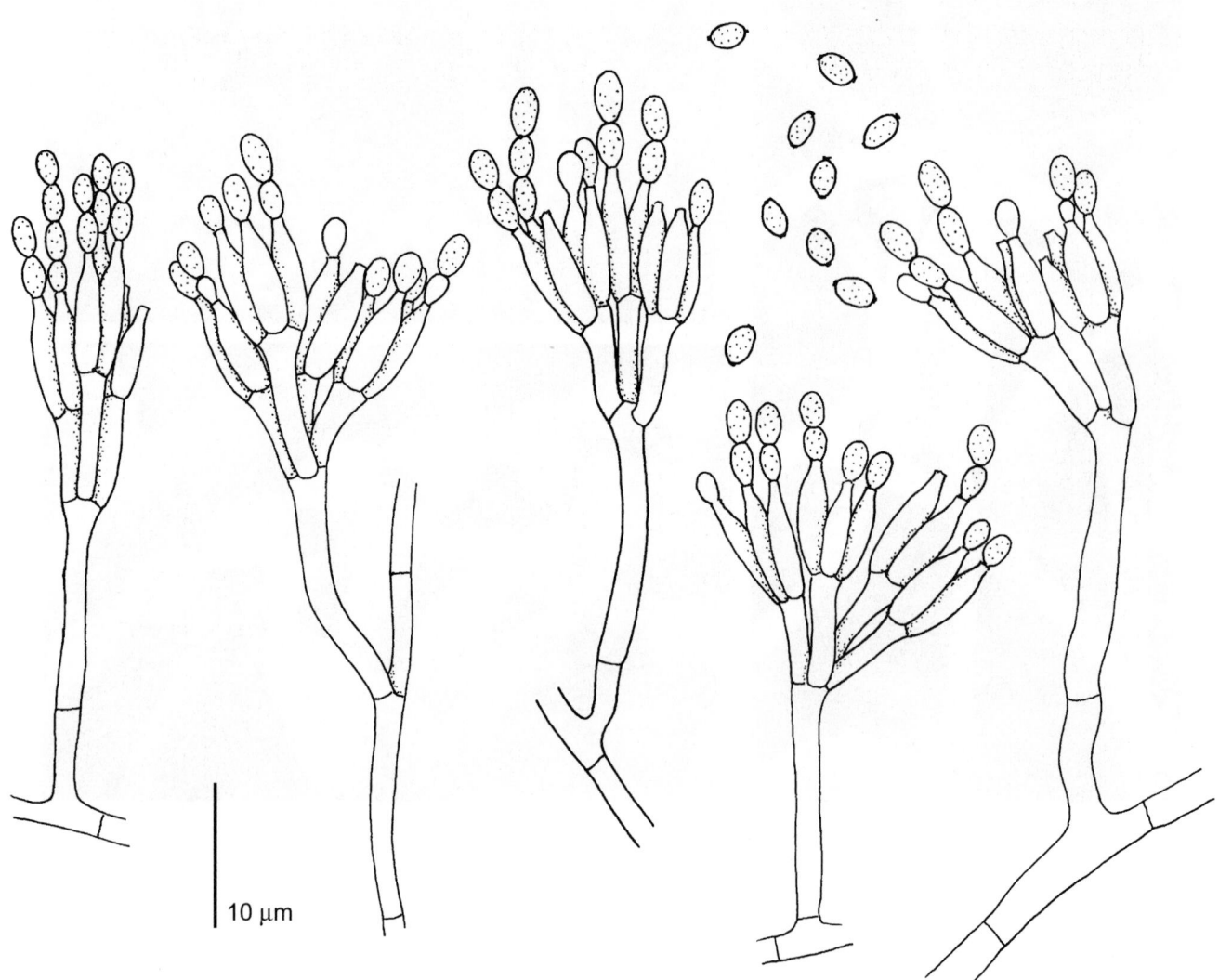

10 μm

Pennicillium piceum, CBS 361.48. Conidiophores and conidia.

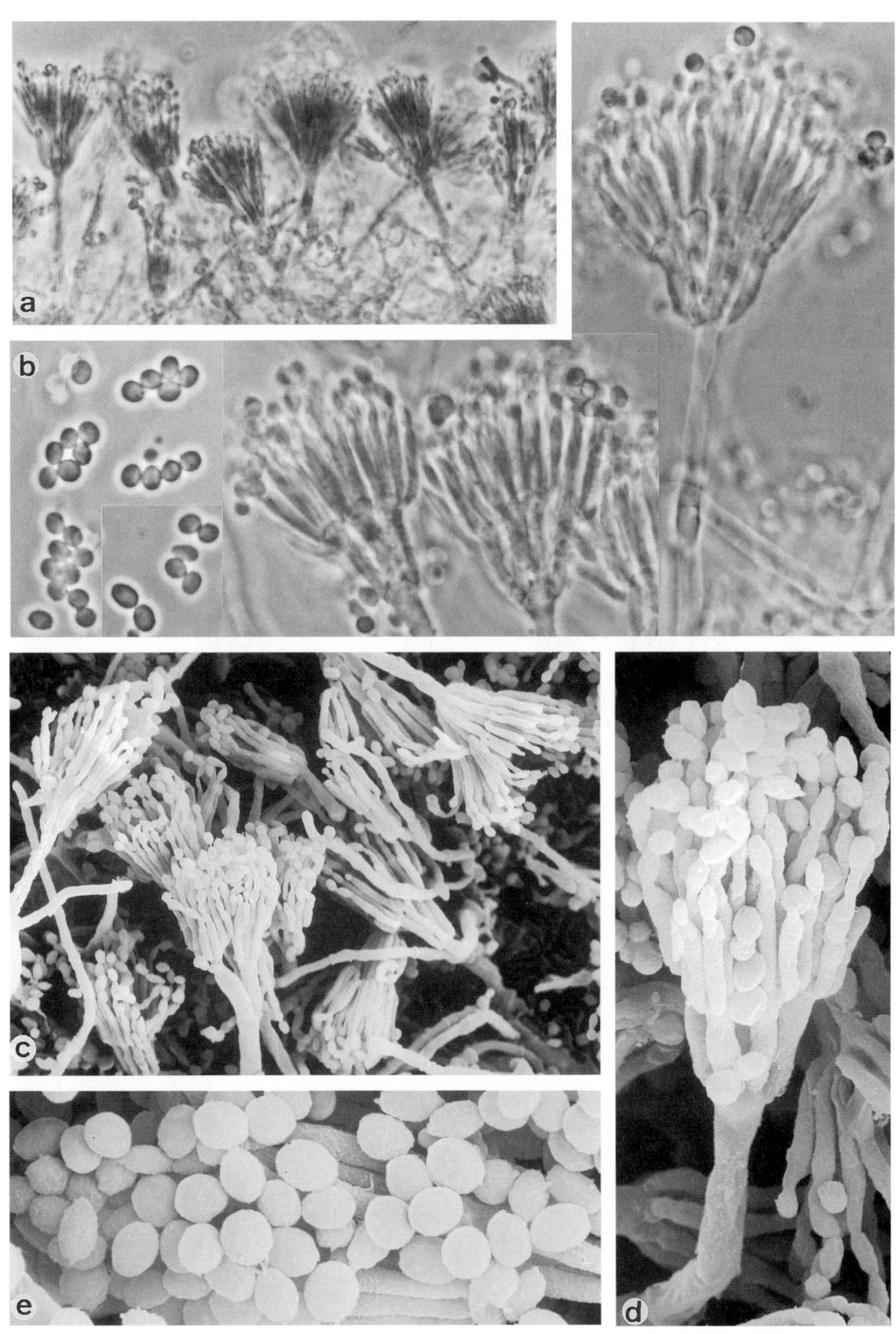

Penicillium piceum, CBS 102383. Conidiophores and conidia. a. ×640; b. ×1600; c. ×1250; d. ×3400; e. ×4750.

Penicillium purpurogenum Stoll

Colony characteristics. Colonies (CzA) restricted, velutinous; conidial masses deep yellow-green; soluble pigment and reverse deep red to dark reddish-purple.

Microscopy. Conidiophore stipes 70-300 μm long, smooth-walled, hyaline, conspicuously encrusted; penicilli biverticillate. Metulae and phialides 10-14 μm long. Phialides acerose. Conidia ellipsoidal, sometimes subspherical, apiculate, irregularly roughened, 3.0-3.5 × 2.5-3.0 μm.

Physiology. Growth at 37°C.

Molecular diagnostics. ITS restriction map based on NCBI L14506:

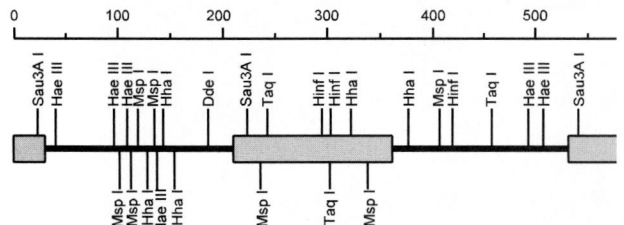

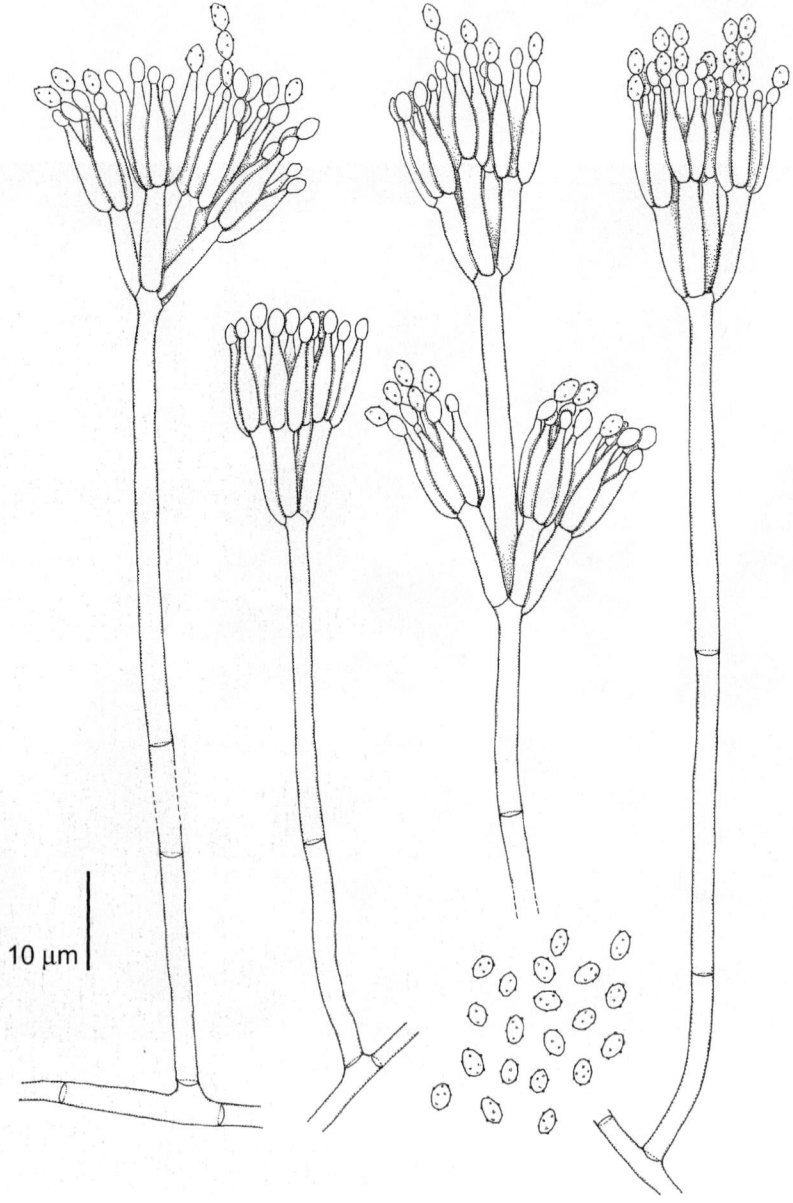

Penicillium purpurogenum, RV 38357. Conidiophores and conidia.

Pathogenicity. BSL-1. A case has been reported of a pulmonary infection in a patient with myeloblastic leukemia (Morin *et al.*, 1986). Pitt (1994) mentioned a strain from BAL.

References. Domsch *et al.* (1980), Pitt (1979).

Nomenclature. *Penicillium purpurogenum* Stoll - Beitr. Char. Penicill. p. 32, 1904.

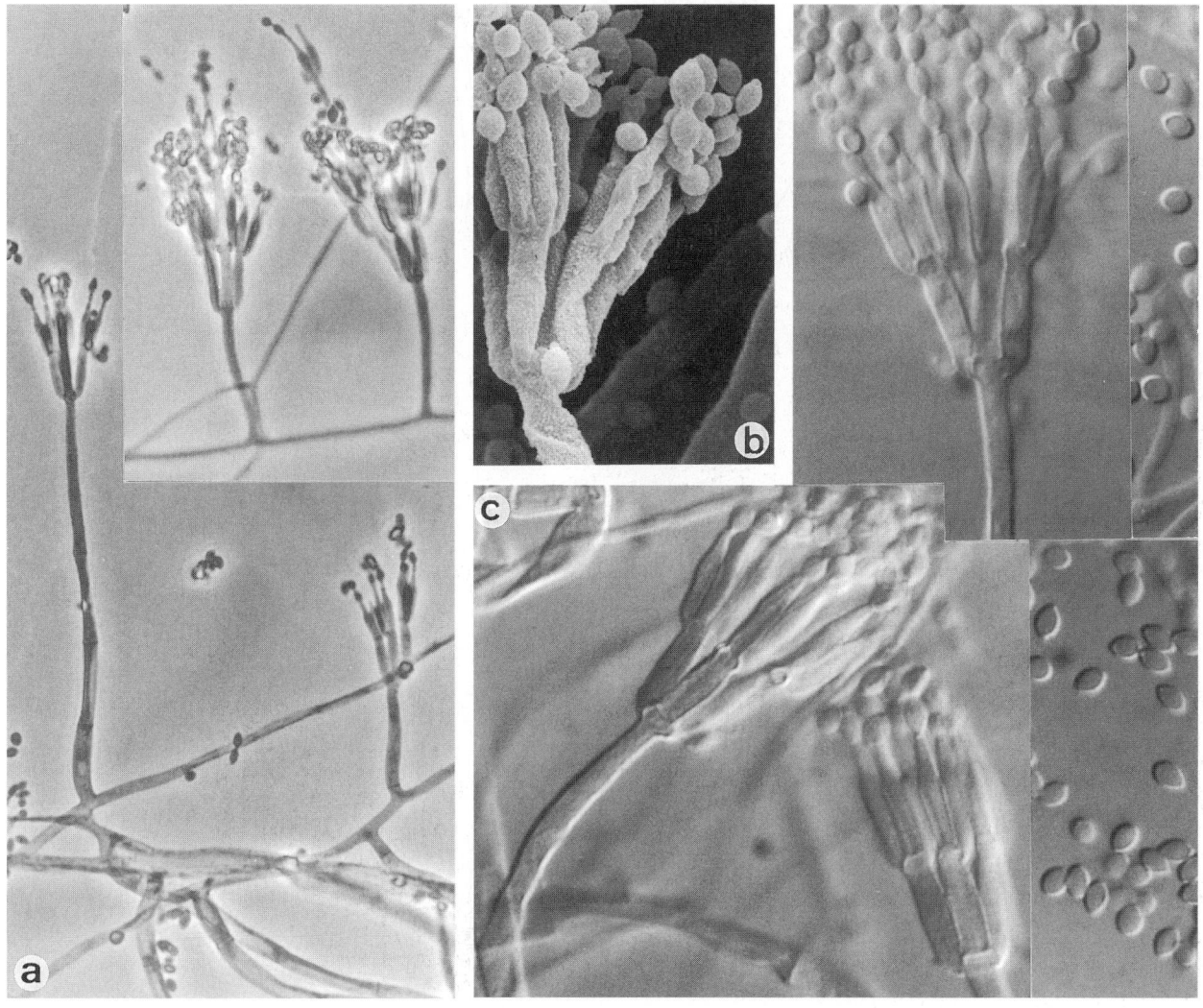

Penicillium purpurogenum, RV 38357. Conidiophores and conidia. a. ×512; b. ×2550; c. ×1600.

Penicillium rugulosum Thom

Colony characteristics. Colonies (CzA) restricted, velutinous, yellow-green to dark green.
Microscopy. Conidiophore stipes 70-100 µm long, with smooth walls; penicilli usually biverticillate. Metulae 10-15 µm long, appressed. Phialides acerose, 8-10 µm long. Conidia ellipsoidal, smooth- or rough-walled, 3.0-3.5 × 2.5-3.0 µm.
Pathogenicity. BSL-1. Corneal ulcers have been reported (Neuhann, 1976; Swietliczkowa *et al.*, 1984).
References. Pitt (1979), Domsch *et al.* (1980).
Nomenclature. *Penicillium rugulosum* Thom - Bull. Bur. Anim. Ind. U.S. Dept. Agr. 118: 60, 1910.

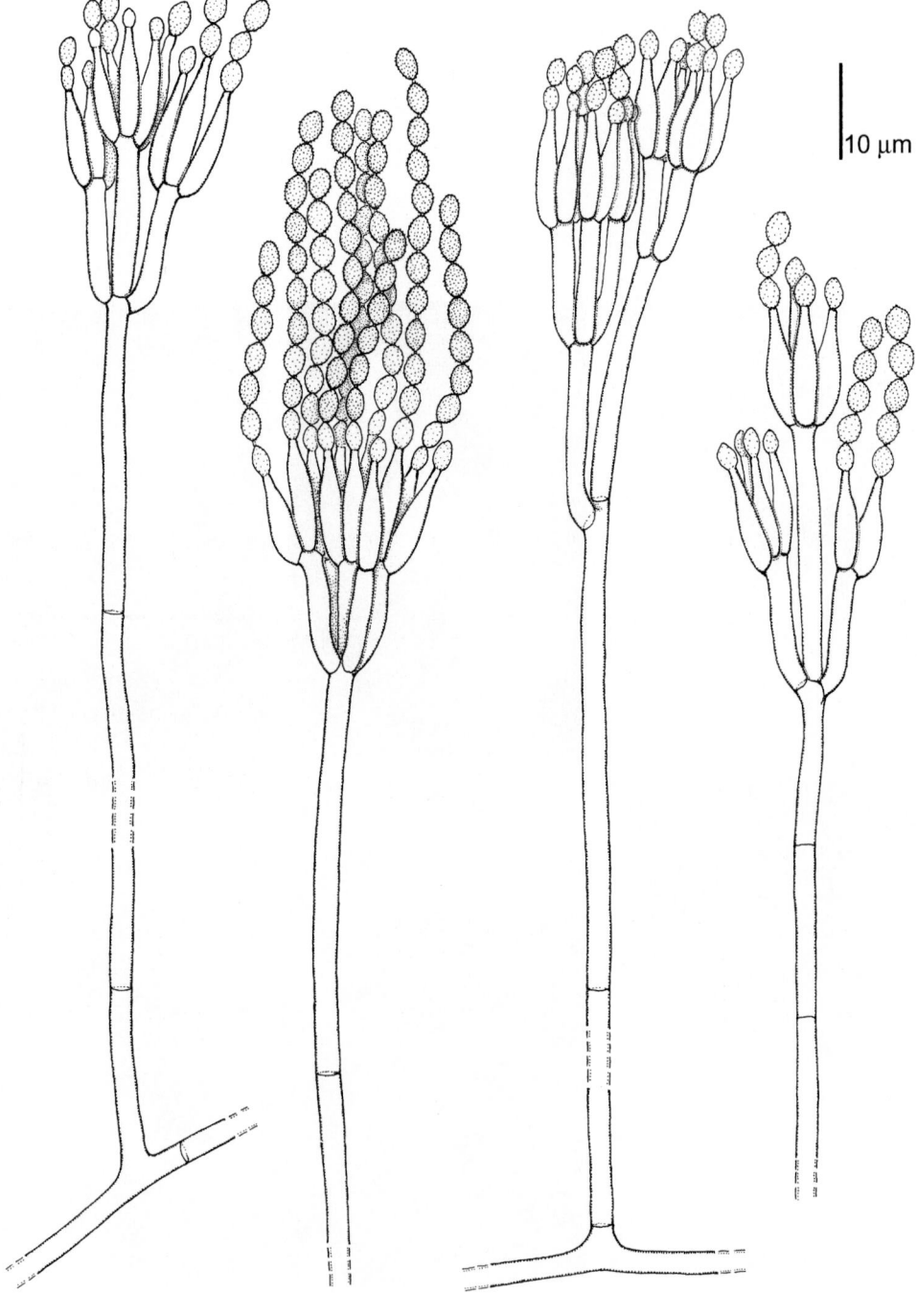

10 µm

Penicillium rugulosum, CBS 371.48. Conidiophores and conidia.

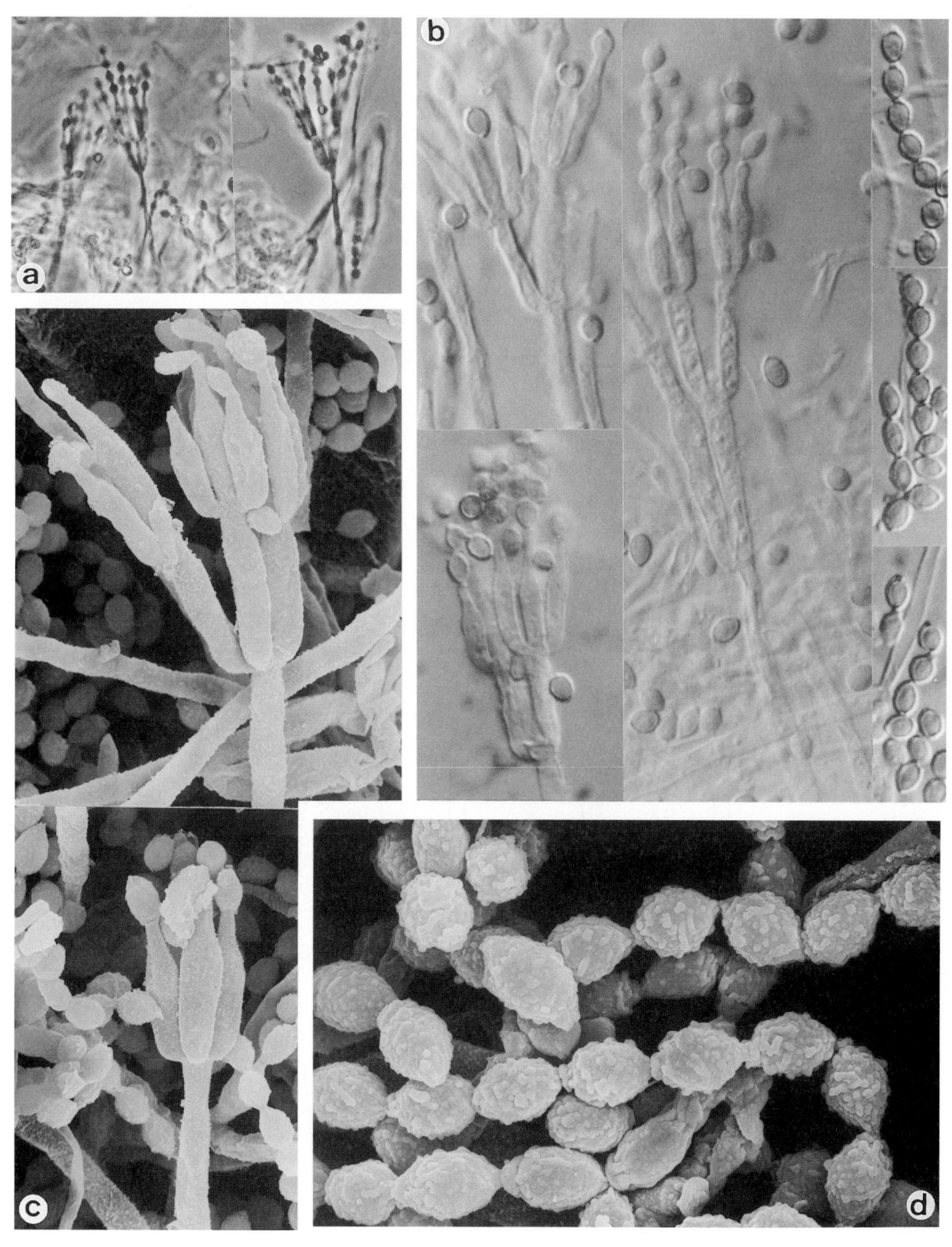

Penicillium rugulosum, CBS 371.48. Conidiophores and conidia. a. ×640; b. ×1600; c. ×2600; d. ×5500.

Penicillium spinulosum Thom

Colony characteristics. Colonies (CzA) spreading, velutinous to floccose; mycelium white; conidial mass dull green.
Microscopy. Conidiophore stipes 100-300 µm long, smooth- to rough-walled; penicilli monoverticillate. Phialides flask-shaped, 6-10 µm long. Conidia spherical, 3.0-3.5 µm diam, irregularly rough-walled to spinulose.
Physiology. No growth at 37°C.
Molecular diagnostics. ITS restriction map based on NCBI AF034461:

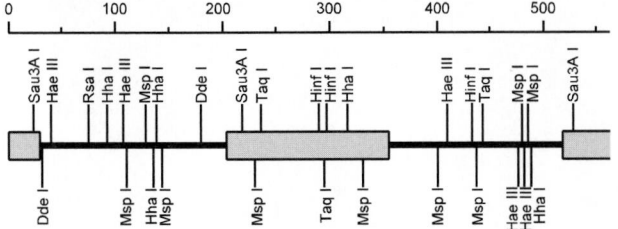

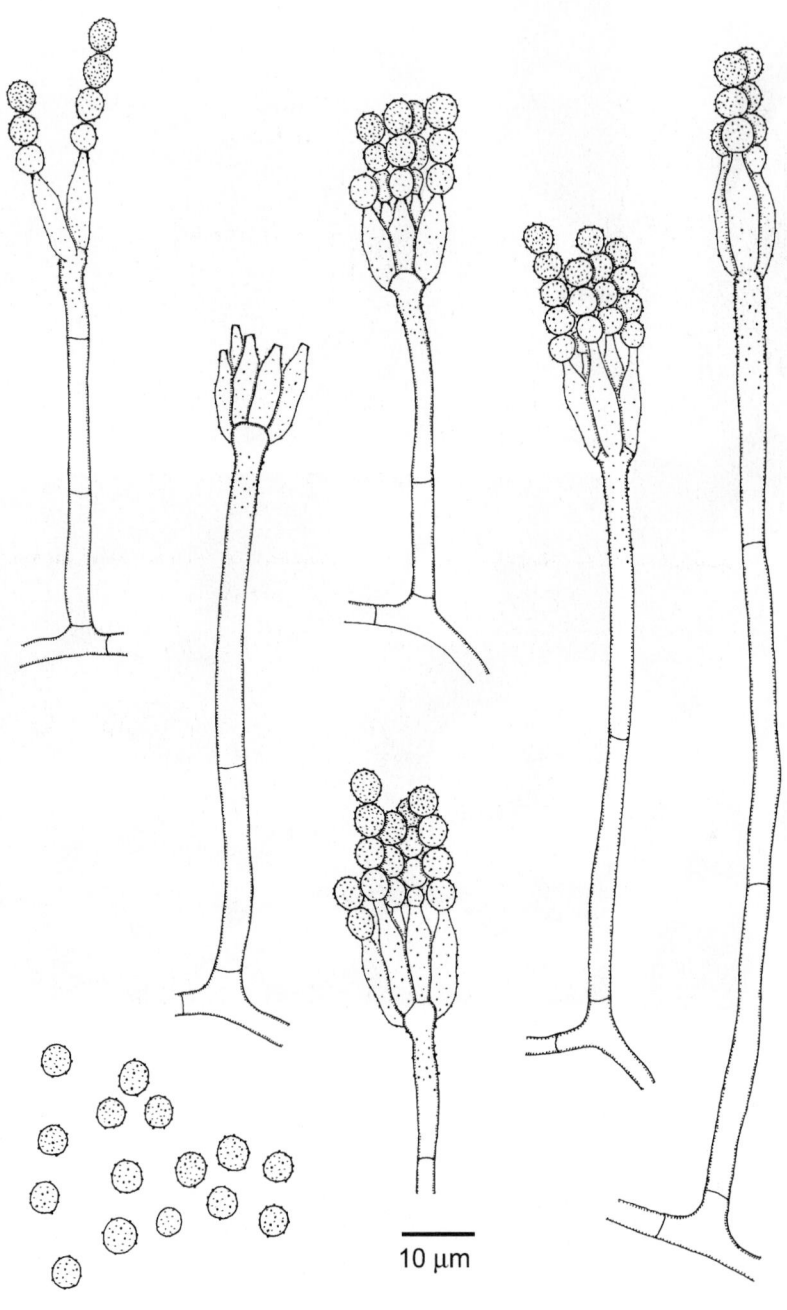

Penicillium spinulosum, IFO 6034. Conidiophores and conidia.

Pathogenicity. BSL-1. A case of broncho-pulmonary infection in a patient with chronic bronchitis (Delore *et al.*, 1955). The species was also involved in a case of keratitis (Anderson *et al.*, 1959) and has been isolated from cases of otomycosis (Senturia & Wolf, 1945).

References. Domsch *et al.* (1980), Pitt (1979).

Nomenclature. *Penicillium spinulosum* Thom - Bull. Bur. Anim. Ind. U.S. Dept Agric. 118: 76, 1910.

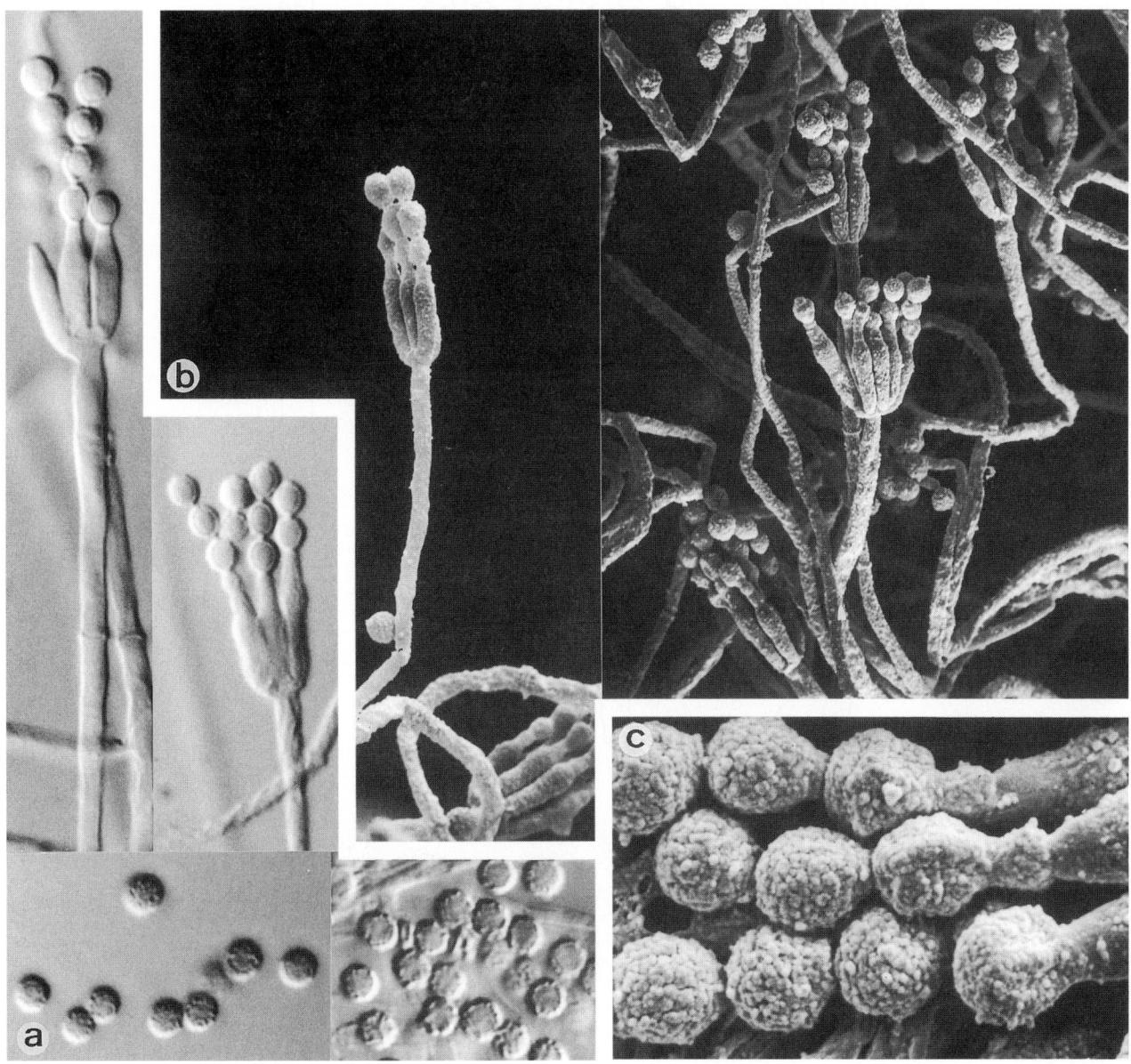

Penicillium spinulosum, IFO 6034. Conidiophores and conidia. a. ×1600; b. ×1800; c. ×7100.

Penicillium verruculosum Peyronel

Colony characteristics. Colonies (CzA) velutinous or somewhat floccose to funiculose; mycelium white to bright yellow; conidial mass green.

Microscopy. Conidiophore stipes 150-250 µm long, smooth-walled; penicilli usually biverticillate. Metulae in whorls of 7-10, 8-15 µm long. Phialides in whorls of 7-10, ampulliform to acerose, 7-10 µm long. Conidia spherical to subspherical, with roughened walls, 3.0-3.5 µm diam.

Physiology. Good growth at 37°C.

Pathogenicity. BSL-1. A case of osteomyelitis in a dog (Wigney *et al.*, 1990).

Reference. Pitt (1979).

Nomenclature. *Penicillium verruculosum* Peyronel - Germi Atmost. Fung. Micel. p. 22, 1913.

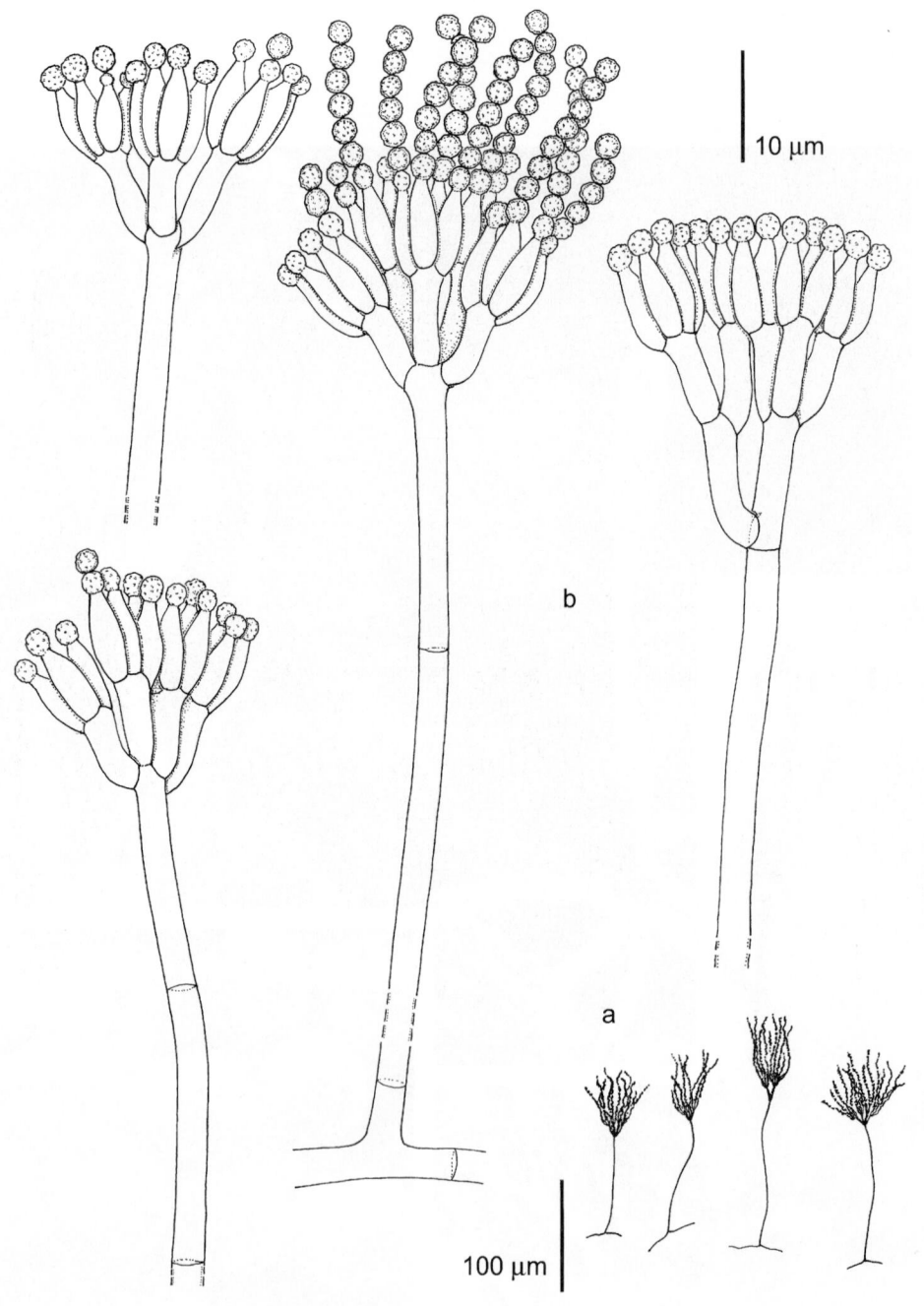

Penicillium verruculosum, CBS 603.74. a. Habit sketch; b. conidiophores and conidia.

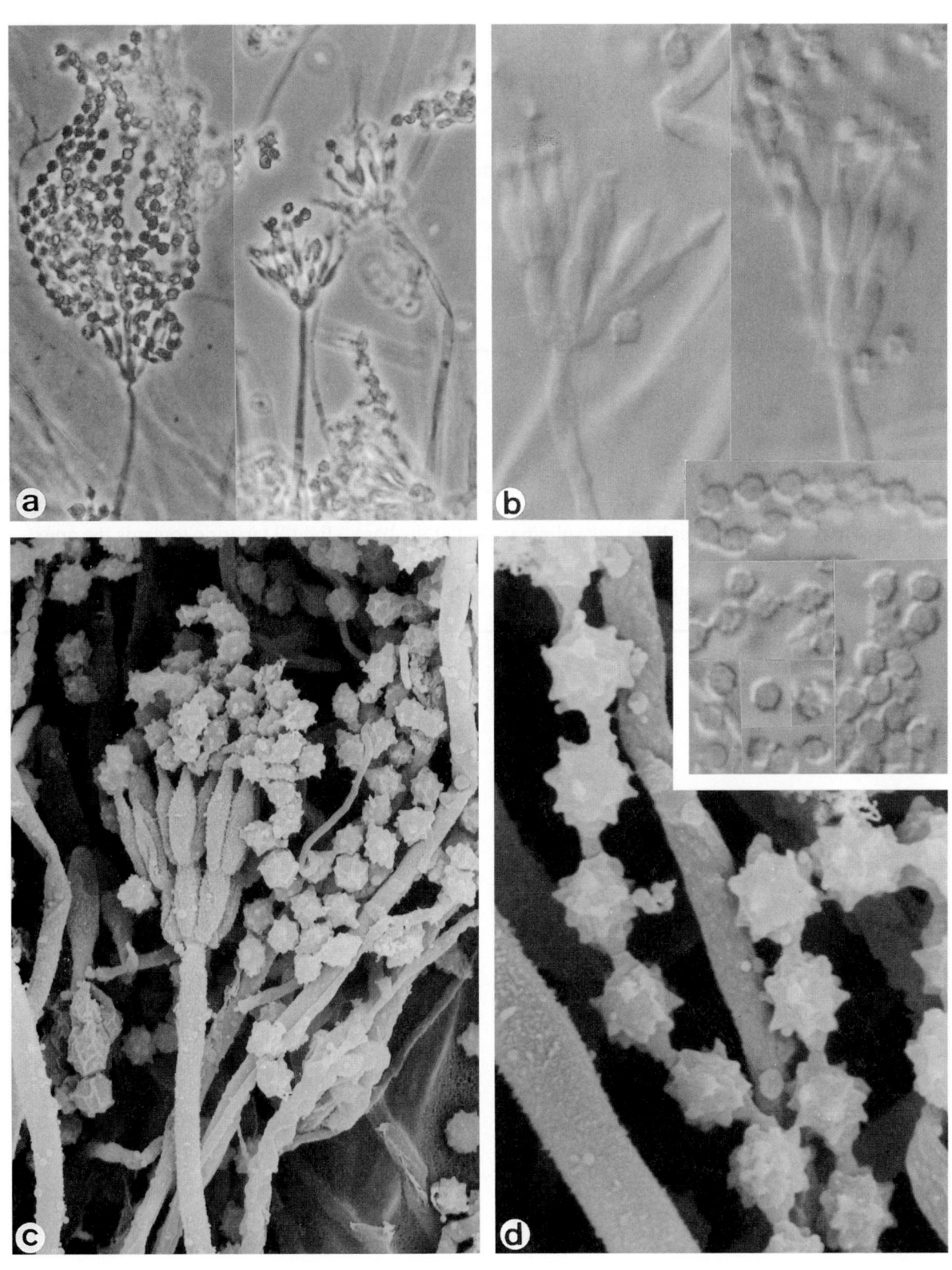

Penicillium verruculosum, CBS 312.59. Conidiophores and conidia. a. ×640; b. ×1600; c. ×2200; d. ×5500.

Hyphomycetes. Genus: *PHAEOACREMONIUM*

Generic description. Colonies expanding, woolly to cottony, greyish-olivaceous to greyish-brown. Conidiophores, when present, erect, stiff, cylindrical, irregularly branched. Phialides cylindrical, often slightly tapering towards the apex, with indistinct, more or less tubular collarettes. Conidia 1-celled, hyaline, smooth-walled, mostly allantoid.
Differential diagnosis. *Phialophora* differs by more or less funnel-shaped, splaying collarettes and mostly broadly ellipsoidal conidia.
References. Crous *et al.* (1996), Dupont *et al.* (1998).

Key to the treated species of Phaeoacremonium:

1a. Colonies with red reverse . . . . *P. rubrigenum* (851)
1b. Colonies with brownish reverse ➞ **2**
2a. Colonies olivaceous grey; phialides unswollen, widest at the base, slightly tapering towards the apex
. . . . *P. parasiticum* (849)
2b. Colonies grey to brown; phialides slightly swollen somewhat above the base . . . . *P. inflatipes* (847)

Phaeoacremonium inflatipes W. Gams *et al.*

Colony characteristics. Colonies (MEA) dull brownish-grey, locally with dark brown patches; reverse rust brown.
Microscopy. Hyphae becoming pale brown to brown, thick-walled, straight. Phialides subhyaline, slender, acicular, somewhat tapering towards the tip, often with one or several septa, 20-40 µm long, slightly swollen near the base, smooth- or slightly rough-walled, with narrowly funnel-shaped collarette which is fragile and may disappear in a later stage. Phialides may proliferate several times; they are often short, with no basal septum. Conidia hyaline, thin-walled, ellipsoidal or allantoid, rather variable in size, 3-5 × 1.5-2.0 µm.
Molecular diagnostics. ITS restriction map based on NCBI U31843 :

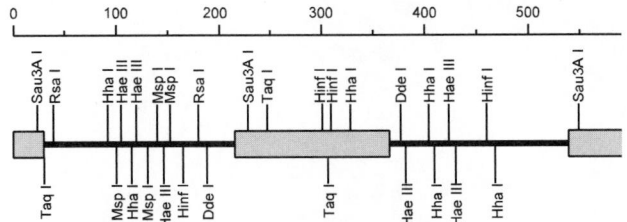

Pathogenicity. BSL-1. A mycetoma was described by de Albornoz (1974) under the name *Cephalosporium serrae*. In their list of material examined, Crous *et al.* (1996) included several strains from human subcutaneous infections. A phaeohyphomycosis was reported by Padhye *et al.* (1998).
Reference. Crous *et al.* (1996).
Antifungal susceptibility.

Antifungal	MICs	Strains	Reference
AMB	0.125	1	Unpublished data
5FC	0.25	1	Unpublished data
FLZ	128	1	Unpublished data
ITZ	0.06	1	Unpublished data
KTZ	0.25	1	Unpublished data
MCZ	2	1	Unpublished data

Nomenclature. *Phaeoacremonium inflatipes* W. Gams, Crous & Wingfield, *in* Crous, Gams, Wingfield & van Wijk - Mycologia 88: 786, 1996.

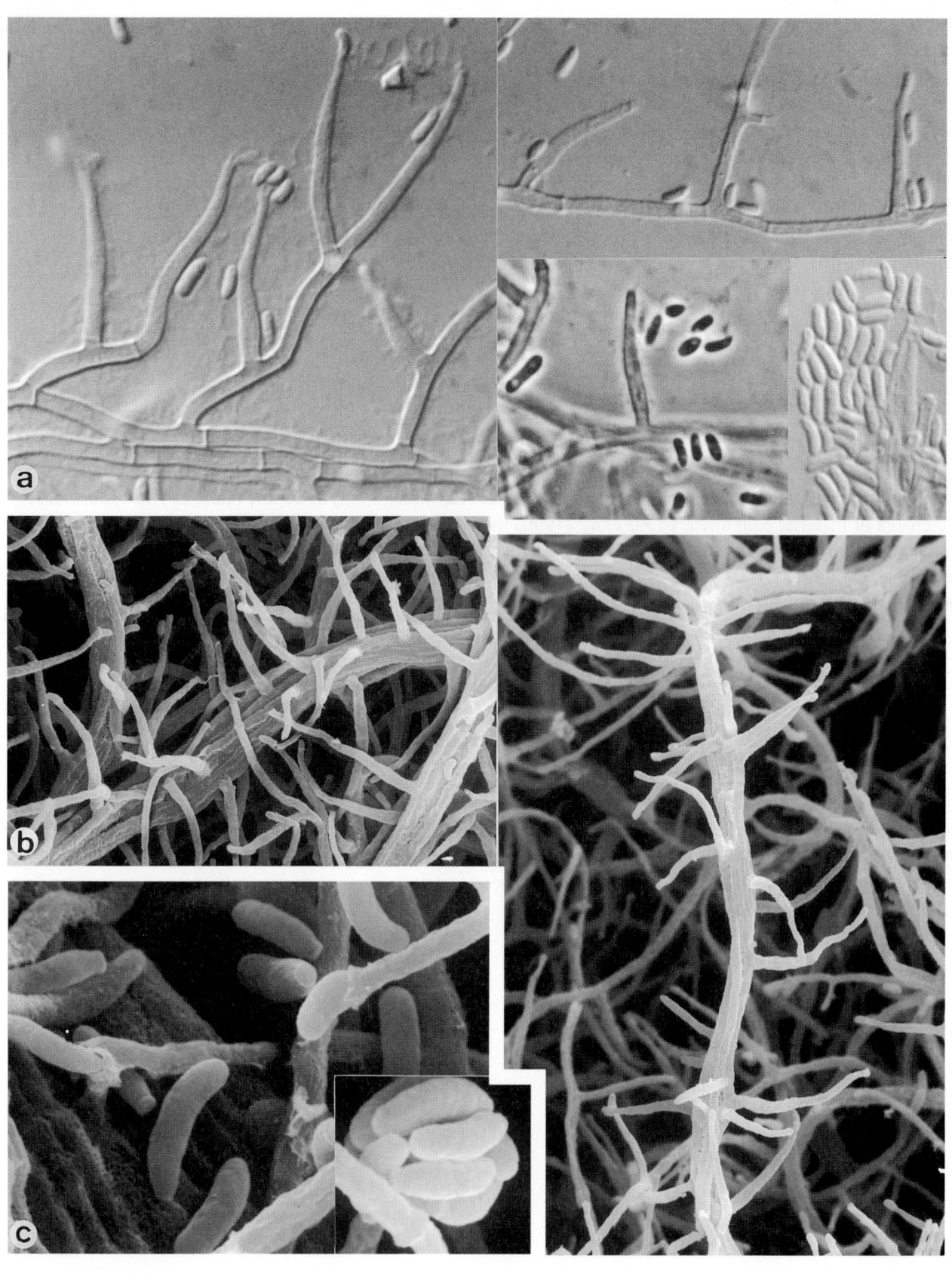

Phaeoacremonium inflatipes, CBS 729.97. Phialides and conidia. a. ×1600; b. ×1550; c. ×7750.

Phaeoacremonium parasiticum (Ajello *et al.*) W. Gams *et al.*

Colony characteristics. Colonies (MEA) expanding, with radiating furrows, initially velvety, later developing hyphal fascicles, olivaceous-grey with blackish reverse.

Microscopy. Hyphae hyaline, later brown. Phialides brown, thick-walled, slender, acicular to cylindrical, slightly tapering towards the tip, 15-50 µm long, often proliferating, with small, funnel-shaped collarettes. Conidia hyaline, thin-walled, cylindrical to sausage-shaped, 3-6 × 1-2 µm, later inflating.

Molecular diagnostics. ITS restriction map based on NCBI U31841:

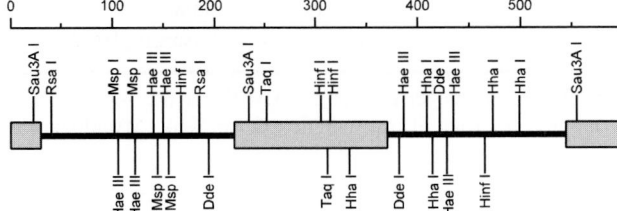

Pathogenicity. BSL-2. The species is a plant-pathogen (Hawksworth *et al.*, 1976) but is also known from cases of subcutaneous infection, such as phaeohyphomycoses (Fincher *et al.*, 1988), arthritis (Kaell & Weitzman, 1983; Ziza *et al.*, 1985) or mycetoma (Rowland & Farrar, 1987; Hood *et al.*, 1997). A case of fatal dissemination in an elderly patient was reported by Wong *et al.* (1989) and an endocarditis in a transplant recipient by Heath *et al.* (1997).

References. Ajello *et al.* (1974), Weitzman *et al.* (1984), Wang & Zabel (1990).

10 µm

Phaeoacremonium parasiticum, CBS 870.73. a. Branching system with phialides; b. conidia.

Antifungal susceptibility.

Antifungal	Mean MICs	Strains	Reference
AMB	3.08	8	McGinnis & Pasarell (1998b)
5FC	250	1	Unpublished data
FLU	128	1	Unpublished data
ITZ	6.17	8	McGinnis & Pasarell (1998b)
KTZ	32	1	Unpublished data
MCZ	32	1	Unpublished data
VCZ	0.55	8	McGinnis & Pasarell (1998b)

Nomenclature. *Phialophora parasitica* Ajello, Georg & Wang, *in* Ajello, Georg, Steigbigel & Wang - Mycologia 66: 493, 1974 ≡ *Phaeoacremonium parasiticum* (Ajello, Georg & Wang) W. Gams, Crous & Wingfield, *in* Crous, Gams, Wingfield & van Wijk - Mycologia 88: 794, 1996.

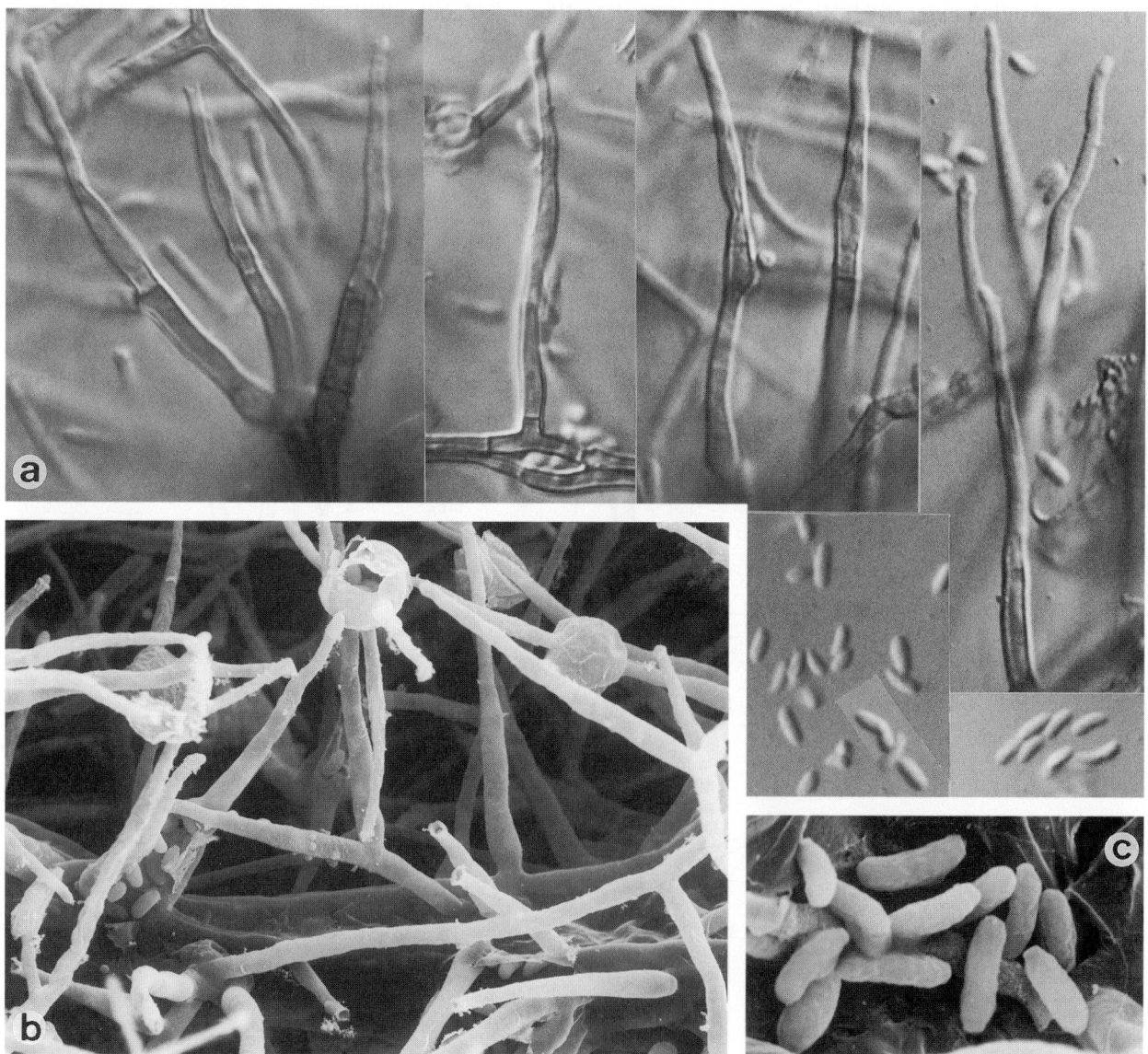

Phaeoacremonium parasiticum, CBS 483.94. Branching system with phialides and conidia. a. ×1600; b. ×1900; c. ×6000.

Phaeoacremonium rubrigenum W. Gams *et al.*

Colony characteristics. Colonies (MEA) restricted, flat, brownish-grey, with red reverse.
Microscopy. Hyphae pale brown, often rough-walled. Phialides subhyaline to pale brown, thick- and often rough-walled, acicular to cylindrical, slightly tapering towards the tip, 10-35 µm long, with narrow, funnel-shaped collarettes. Conidia hyaline, thin-walled, sausage-shaped, 3-4 × 1.5-2.0 µm, aggregating in slimy heads.
Physiology. Reduced growth at 37°C.
Pathogenicity. BSL-2. Matsui *et al.* (1999) reported a subcutaneous infection in an immunosuppressed patient.
Nomenclature. *Phaeoacremonim rubrigenum* W. Gams, Crous & Wingfield, *in* Crous, Gams, Wingfield & van Wijk, Mycologia 88: 795, 1996.

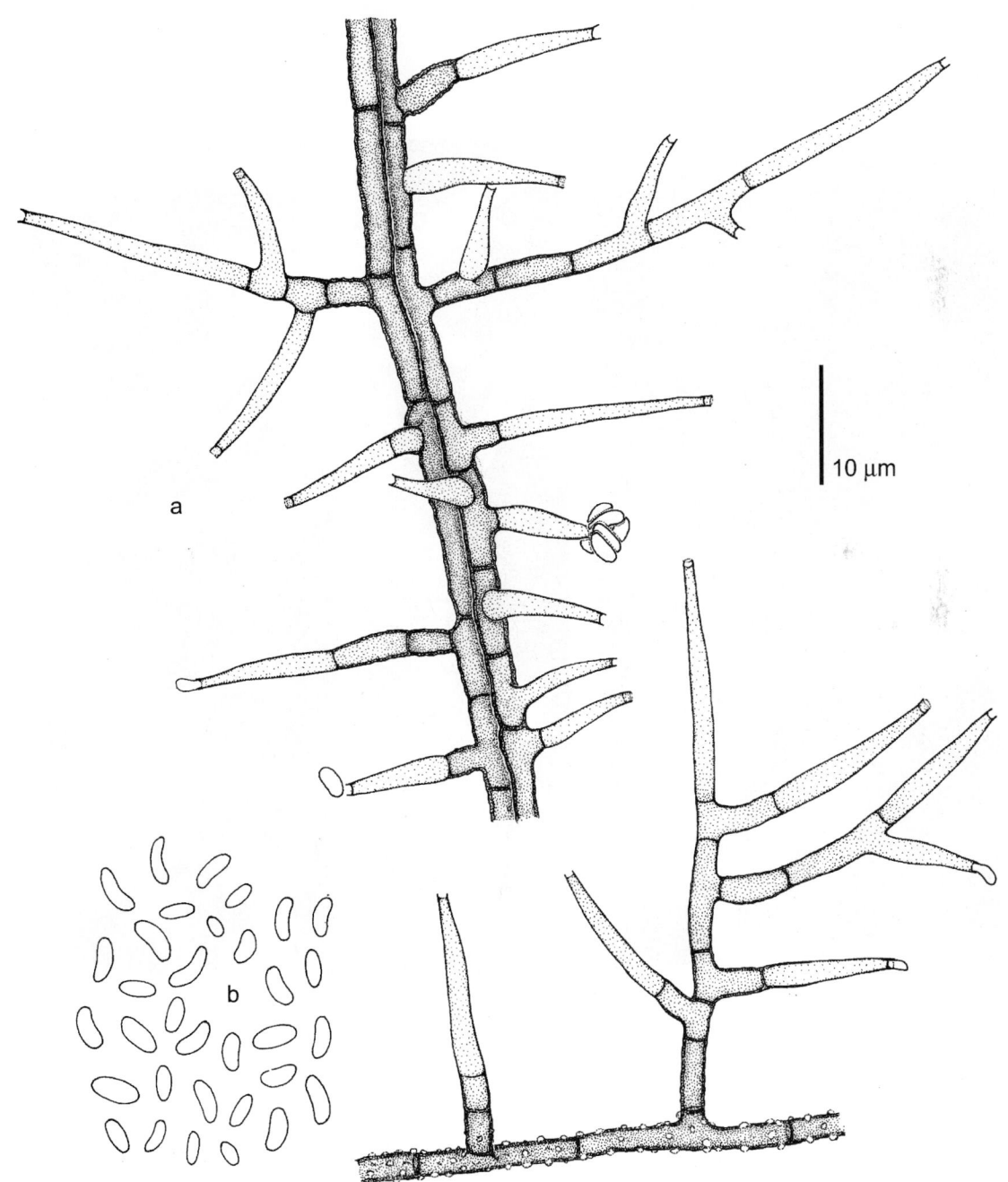

*Phaeoacremonium rubrigenum,*CBS 498.94. a. Hyphae with phialides; b. conidia.

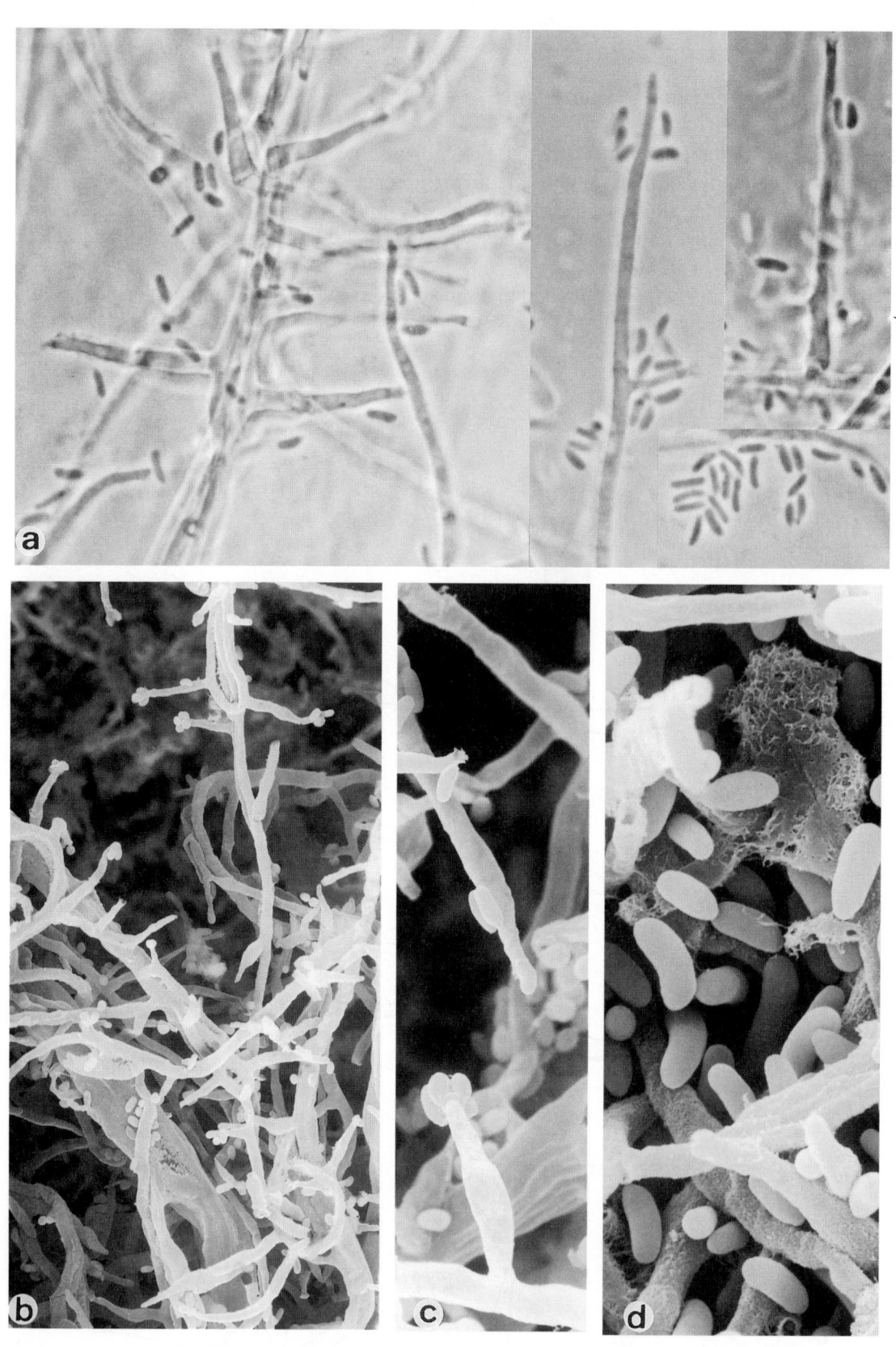

Phaeoacremonium rubrigenum, CBS 498.94. Phialides and conidia. a. ×1600; b. ×1250; c. ×3400; d. ×6000.

Hyphomycetes. Genus: *PHAEOISARIA*

Phaeoisaria clematitidis (Fuckel) S. Hughes

Colony characteristics. Colonies (OA) growing moderately rapidly, powdery, soon with hyphal fascicles which eventually develop into synnemata, olivaceous brown.

Microscopy. Hyphae pale olivaceous brown. Conidiogenous cells laterally on undifferentiated hyphae, often without basal septum; in a later stage slender synnemata are formed, which are composed of stiff, parallel hyphae with conidiogenous cells inserted laterally over the entire surface. Conidia borne on spine-like denticles of about 1 μm in length, inserted more or less randomly on the apical part of the conidiogenous cell. Conidia smooth- and thin-walled, subhyaline, fusiform, 6-8 × 1.5-4.0 μm. Large, spherical, dark brown chlamydospores occasionally present.

Pathogenicity. BSL-1. Common saprobe on rotten plant stems and wood. Guarro *et al.* (2000e) reported a keratitis after trauma.

Reference. De Hoog & Papendorf (1976).

Antifungal susceptibility.

Antifungal	Mean MICs	Strains	Reference
AMB	32	1	Guarro *et al.* (2000e)
5FC	256	1	Guarro *et al.* (2000e)
FLZ	128	1	Guarro *et al.* (2000e)
ITZ	32	1	Guarro *et al.* (2000e)
KTZ	32	1	Guarro *et al.* (2000e)
MCZ	32	1	Guarro *et al.* (2000e)

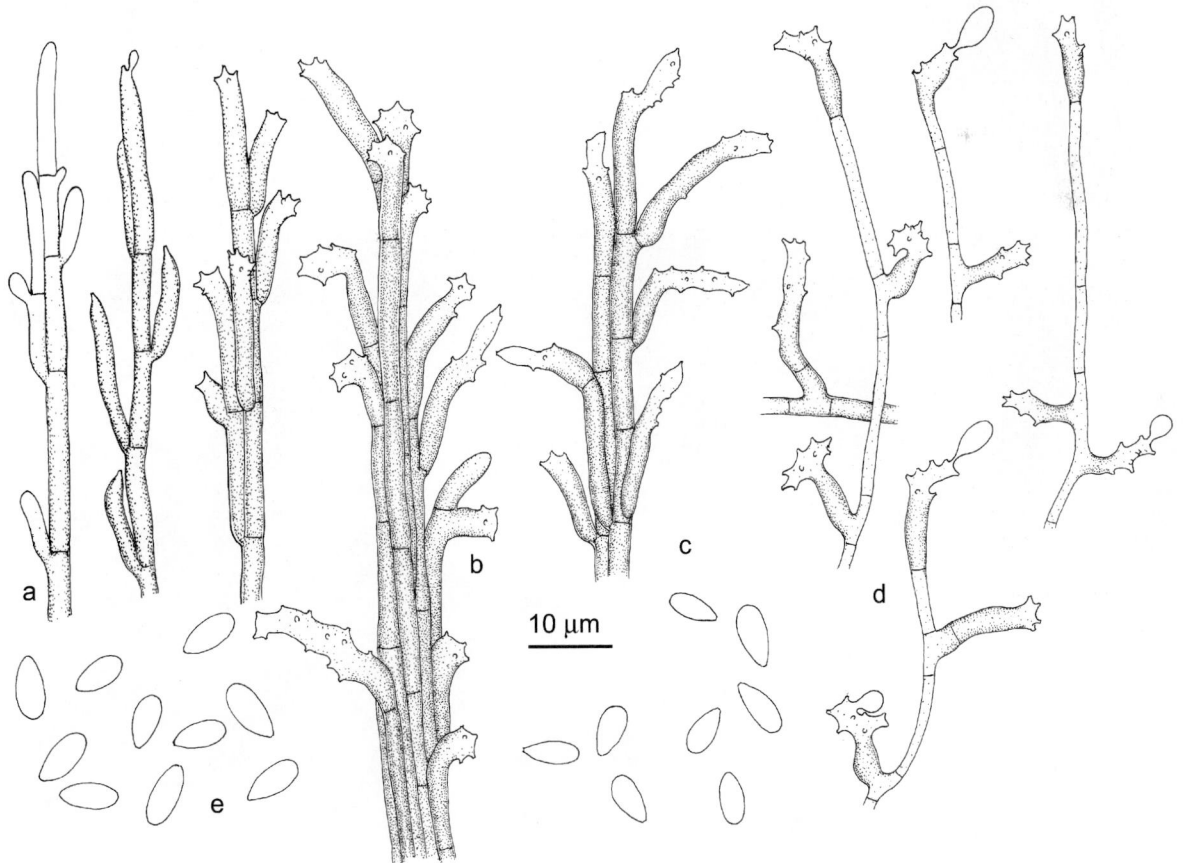

Phaeoisaria clematidis, CBS 429.73. a-c. Synnemata; b. conidiogenous cells borne directly on hyphae; e. conidia.

Nomenclature. *Stysanus clematitidis* Fuckel - Jb. Nassau. Ver. Naturk. 23-24: 365, 1870 ≡ *Phaeoisaria clematidis* (Fuckel) S. Hughes - Can. J. Bot. 36: 795, 1958

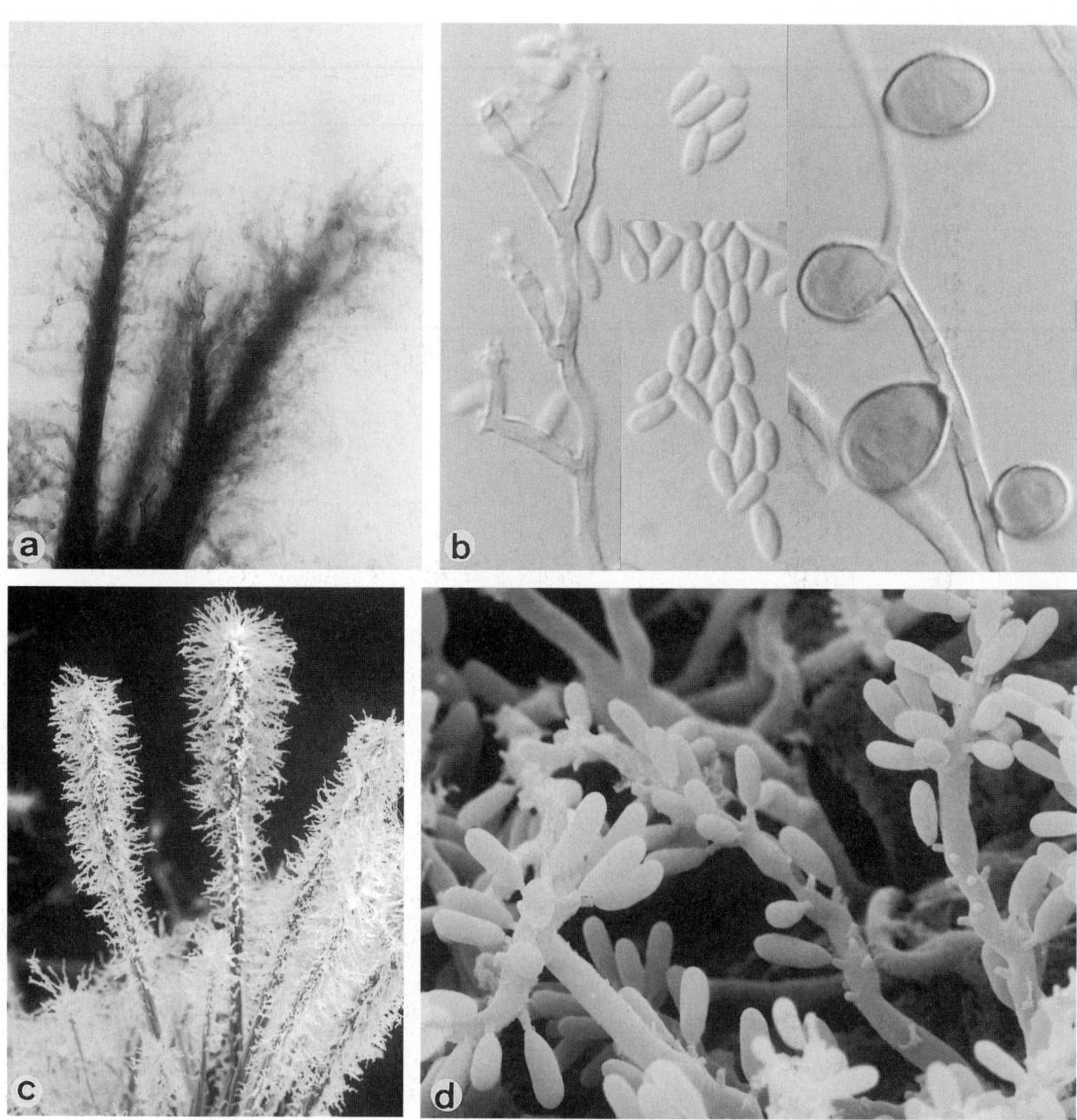

Phaeoisaria clematidis, CBS 102276. a, c. Synnemata; b, d. denticulate conidiogenous cells, chlamydospores and conidia. a. ×250; b.×1600; c. ×145; d. ×2600.

Hyphomycetes. Genus: *PHAEOSCLERA*

Phaeosclera dematioides Sigler *et al.*

Colony characteristics. Colonies (MEA 2%, 15°C for 1 wk, 20-22°C subsequently) restricted, dry, moriform to cerebriform, heaped, finally often cracking the agar medium, dull black.

Microscopy. Hyphae pale brown, initially in parallel arrangement, soon inflating and converting into clusters of thick-walled, blackish-brown cell clumps with thick longitudinal and transverse septation, strongly coherent, mostly smooth-walled, occasionally the outer wall layer is ruptured partially, releasing inner cells. Often local clumps of extracellular material present.

Molecular diagnostics. SSU and ITS restriction maps based on NCBI Y11716 and AJ244254:

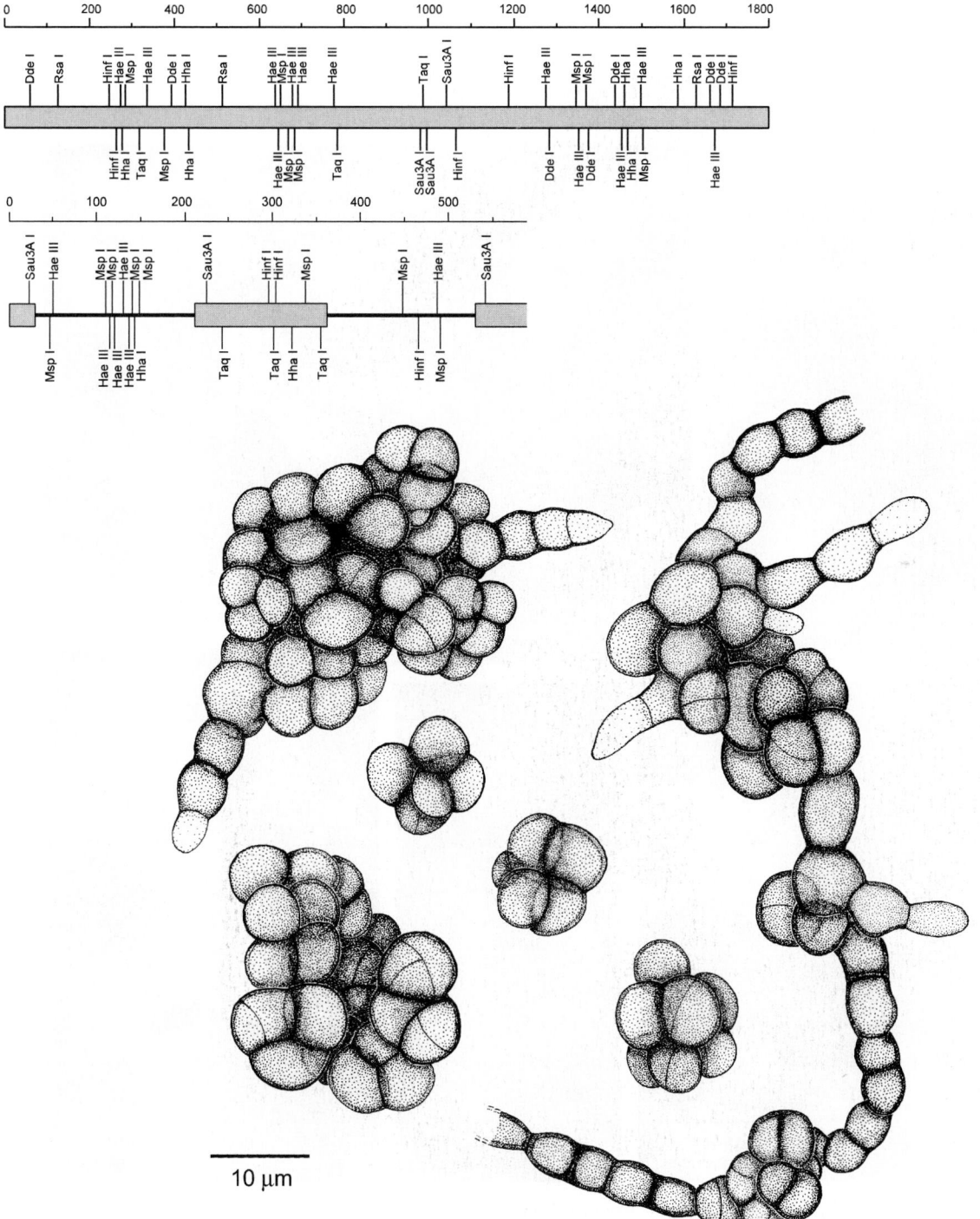

Phaeosclera dematioides, UAMH 4265. Clumps of meristematic cells on hyphae.

Pathogenicity. BSL-1. Reported from nasal granuloma in cattle (McGinnis *et al.*, 1985) and as a secondary invader in a case of human chromoblastomycosis (Krempl-Lamprecht *et al.*, 1987). A mixed skin infection with *Alternaria* has been reported (Palencarova *et al.*, 1995; Pec *et al.*, 1996).

Reference. Sigler *et al.* (1981).

Antifungal susceptibility.

Antifungal	Mean MICs	Strains	Reference
AMB	0.125	1	McGinnis & Pasarell (1998b)
ITZ	0.25	1	McGinnis & Pasarell (1998b)
VCZ	4	1	McGinnis & Pasarell (1998b)

Nomenclature. *Phaeosclera dematioides* Sigler, Tsuneda & Carmichael - Mycotaxon 12: 461, 1981.

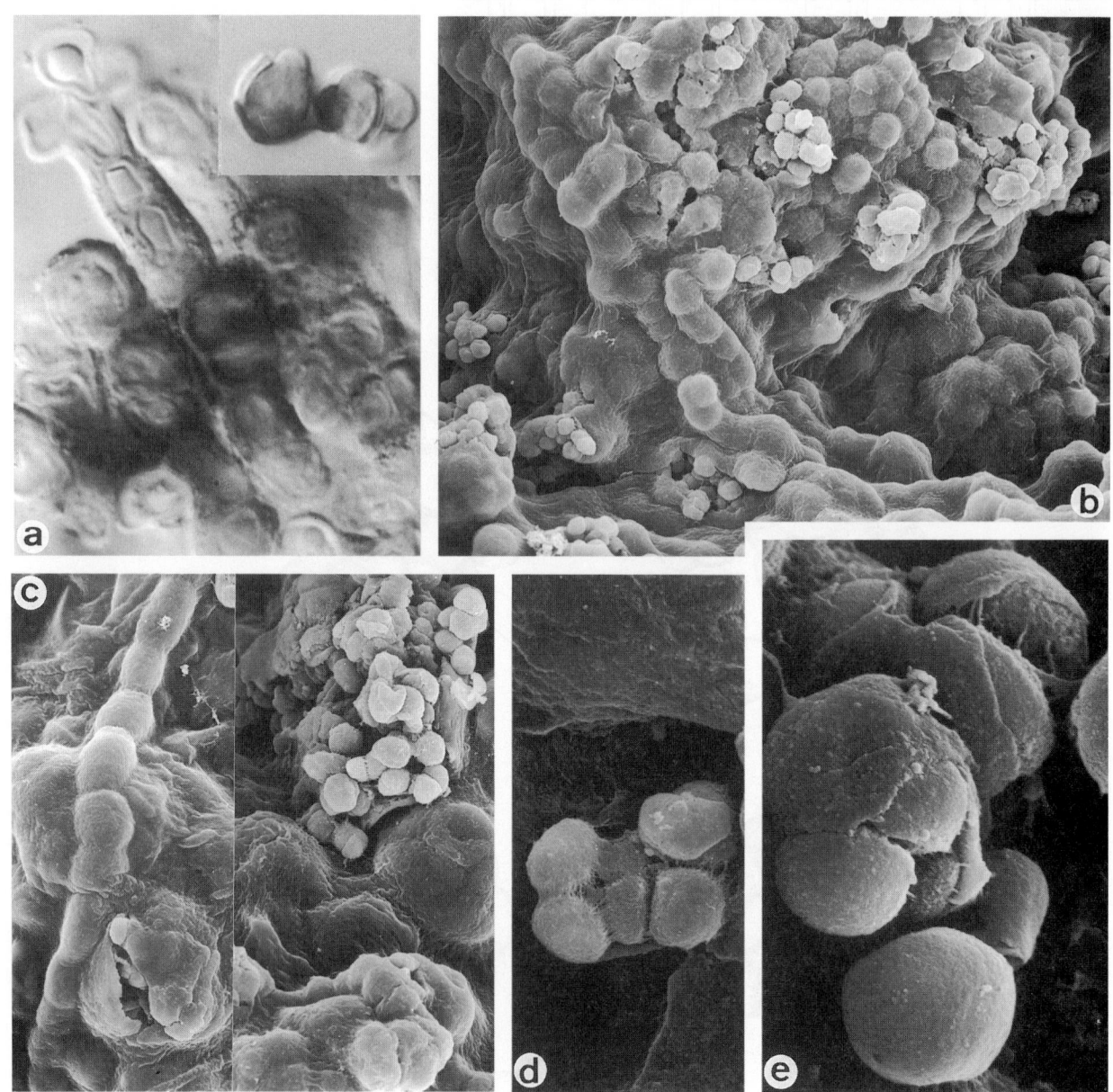

Phaeosclera dematioides, UAMH 4265. Meristematic cells. a. ×1640; b. ×700; c. ×1100; d. ×3500; e. ×6000.

Hyphomycetes. Genus: *PHAEOTRICHOCONIS*

Phaeotrichoconis crotalariae (Salam & Rao) Subram.

Colony characteristics. Colonies (PCA) growing rapidly, cottony, dark grey or greyish-brown. Sclerotia often present.

Microscopy. Conidiophores erect, straight or flexuose, often geniculate, brown, unbranched, up to 150 µm long, 4-8 µm wide. Conidiogenous cells integrated, sympodial, cylindrical, with flat scars. Conidia solitary, smooth- and thick-walled, obclavate, rostrate, 3-8-septate, brown, with a very long and narrow, hyaline beak; body 50-80 × 12-20 µm, beak 30-150 × 1.5-2.5 µm. Large, elongated, dark bron to black sclerotia often present.

Pathogenicity. BSL-1. Keratitis (Shukla *et al.*, 1984a, 1989).

Reference. Ellis (1971).

Nomenclature. *Trichoconis crotalariae* Salam & Rao - J. Ind. Bot. Soc. 33: 191, 1954 ≡ *Phaeotrichoconis crotalariae* (Salam & Rao) Subramanian - Proc. Ind. Acad. Sci., Sect. B, 44: 2, 1956.

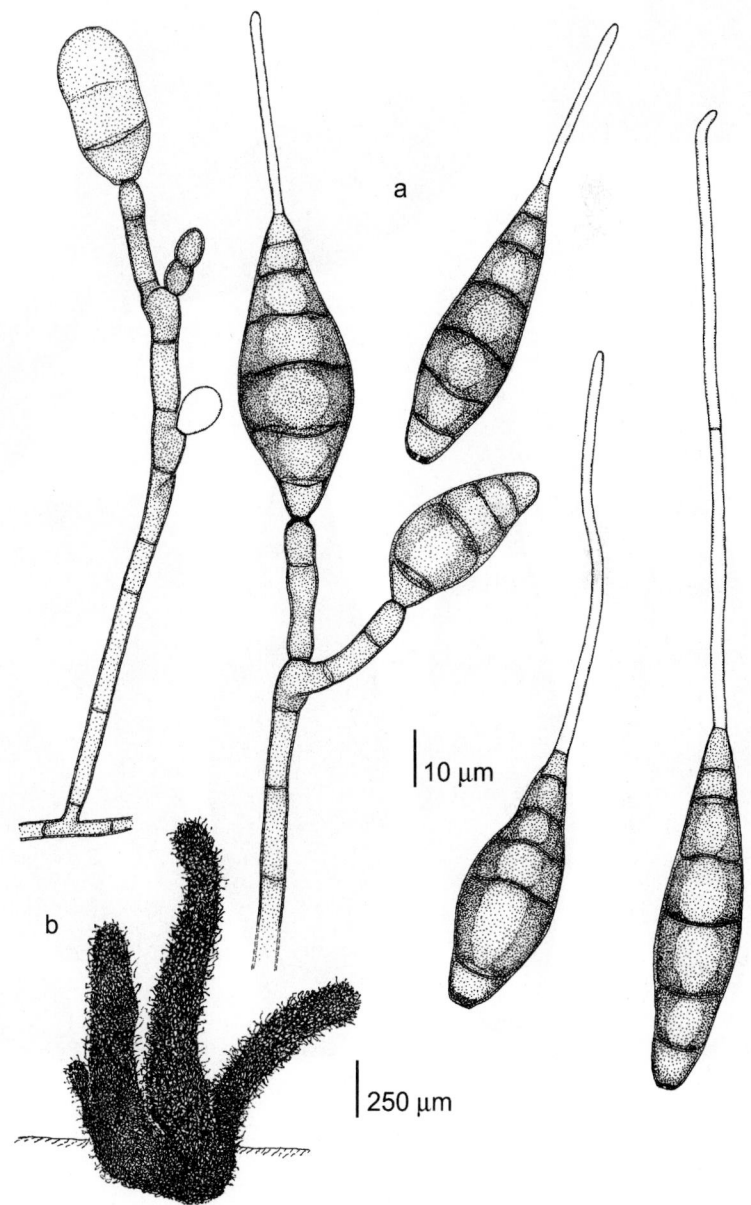

Phaeotrichoconis crotalariae, IMI 132040. a. Conidiophores and conidia; b. sclerotia.

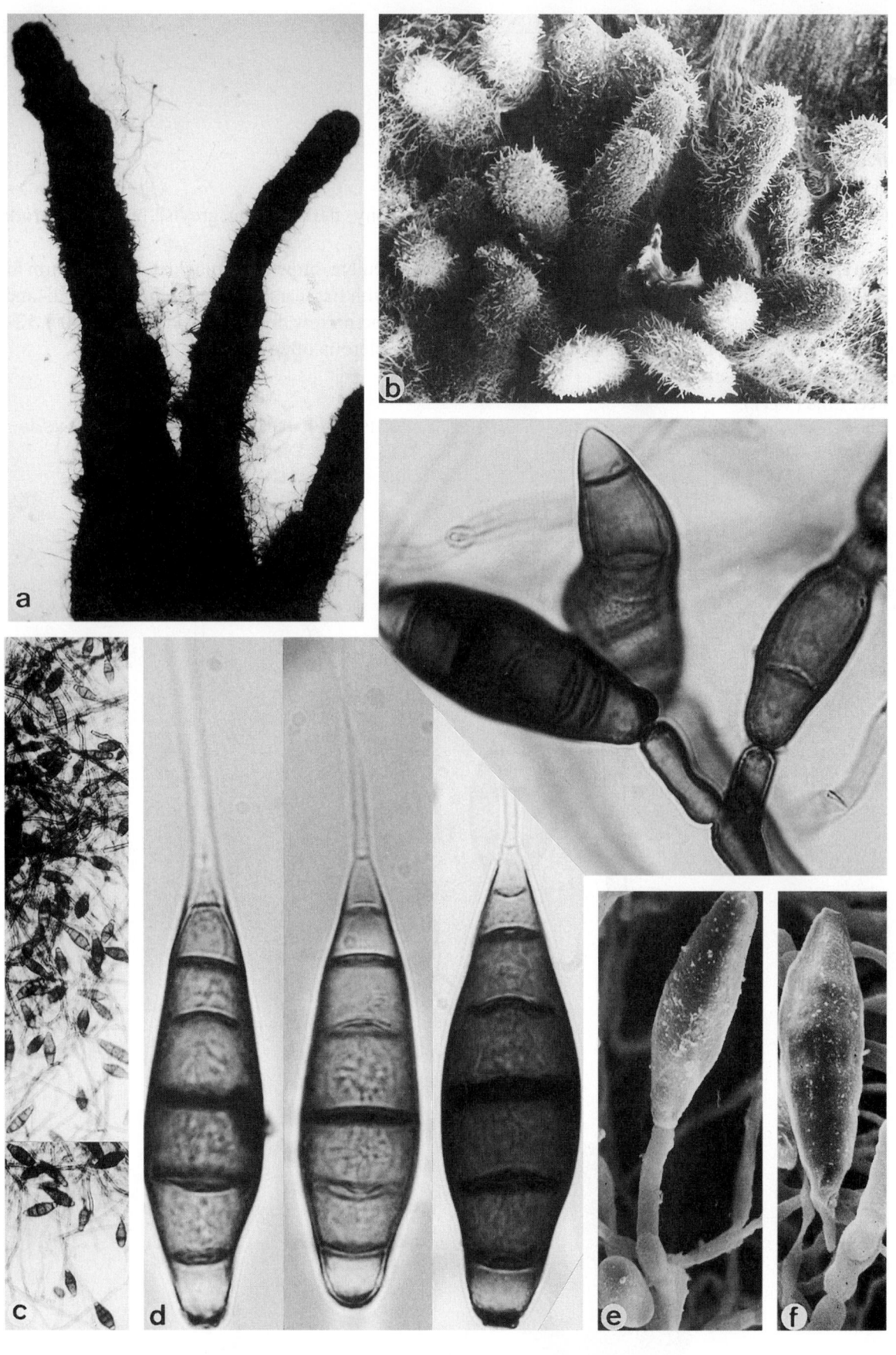

Phaeotrichoconis crotalariae, IMI 132040. a, b. Sclerotia; c-f. conidiogenous cells and conidia. a. ×52; b. ×112; c. ×256; d. ×1600; e. ×1120; f. ×1300.

Hyphomycetes Genus: *PHIALEMONIUM*

Generic description. Colonies spreading, moist, flat, without aerial mycelium, whitish. Phialides are generally intercalary cells with one, rarely several, large lateral protrusions. Discrete and lateral phialides may also be present. Collarettes inconspicuous. Conidia 1-celled, smooth-walled, hyaline, aggregating in slimy heads. Chlamydospores occasionally present in old cultures.

Differential diagnosis. Phialides of *Acremonium* species (p. 395) are predominantly lateral, separated from the supporting hyphae by a basal septum; colonies are usually pigmented more intensely. *Phialophora* species (p. 864) have more differentiated phialides or phialidic brushes. *Lecythophora* species (p. 725) have distinct collarettes. *Hormonema* (p. 717) has flat percurrent scars on undifferentiated hyphae.

Reference. Gams & McGinnis (1983).

Key to the treated species of Phialemonium:

Phialemonium curvatum W. Gams & W.B. Cooke

Colony characteristics. Colonies (PDA) expanding, white, becoming yellow or greyish, smooth or slightly floccose.
Microscopy. Conidia produced from inconspicuous, lateral collarettes or from discrete, lateral, tapering phialides, often without basal septa, up to 22 μm in length. Occasionally branched and clustered conidiophores are present. Conidia hyaline, smooth- and thin-walled, cylindrical to allantoid, 3.5-6.0 × 1.0-1.5 μm, aggregating in slimy heads.
Pathogenicity. BSL-2. The species is occasionally isolated from clinical specimens (Gams & McGinnis, 1983); catheter-related fungemia were described by Guarro *et al.* (1999f). A case of mycotic cyst produced by a fungus which resembles this species has been reported (King *et al.*, 1993). Endocarditis due to a mixed infection including a strain similar to *P. curvatum* was reported by Schønheyder *et al.* (1996).

References. Gams & McGinnis (1983), King *et al.* (1993).
Antifungal susceptibility.

Antifungal	MICs range	Mean MICs	Strains	Reference
AMB	nd	2.52	3	McGinnis & Pasarell (1998b)
AMB	0.125-0.5	0.3	5	Guarro *et al.* (1999f)
5FC	256	256	5	Guarro *et al.* (1999f)
FLZ	16	16	5	Guarro *et al.* (1999f)
ITZ	nd	0.4	3	McGinnis & Pasarell (1998b)
ITZ	0.06-0.25	0.08	5	Guarro *et al.* (1999f)
KTZ	0.25-0.5	0.29	5	Guarro *et al.* (1999f)
MCZ	0.5-2	0.66	5	Guarro *et al.* (1999f)
VCZ	nd	0.25	3	McGinnis & Pasarell (1998b)

Nomenclature. *Phialemonium curvatum* W. Gams & W.B. Cooke, *in* Gams & McGinnis - Mycologia 75: 980, 1983.
   Phialemonium dimorphosporum W. Gams & W.B. Cooke, *in* Gams & McGinnis - Mycologia 75: 981, 1983.

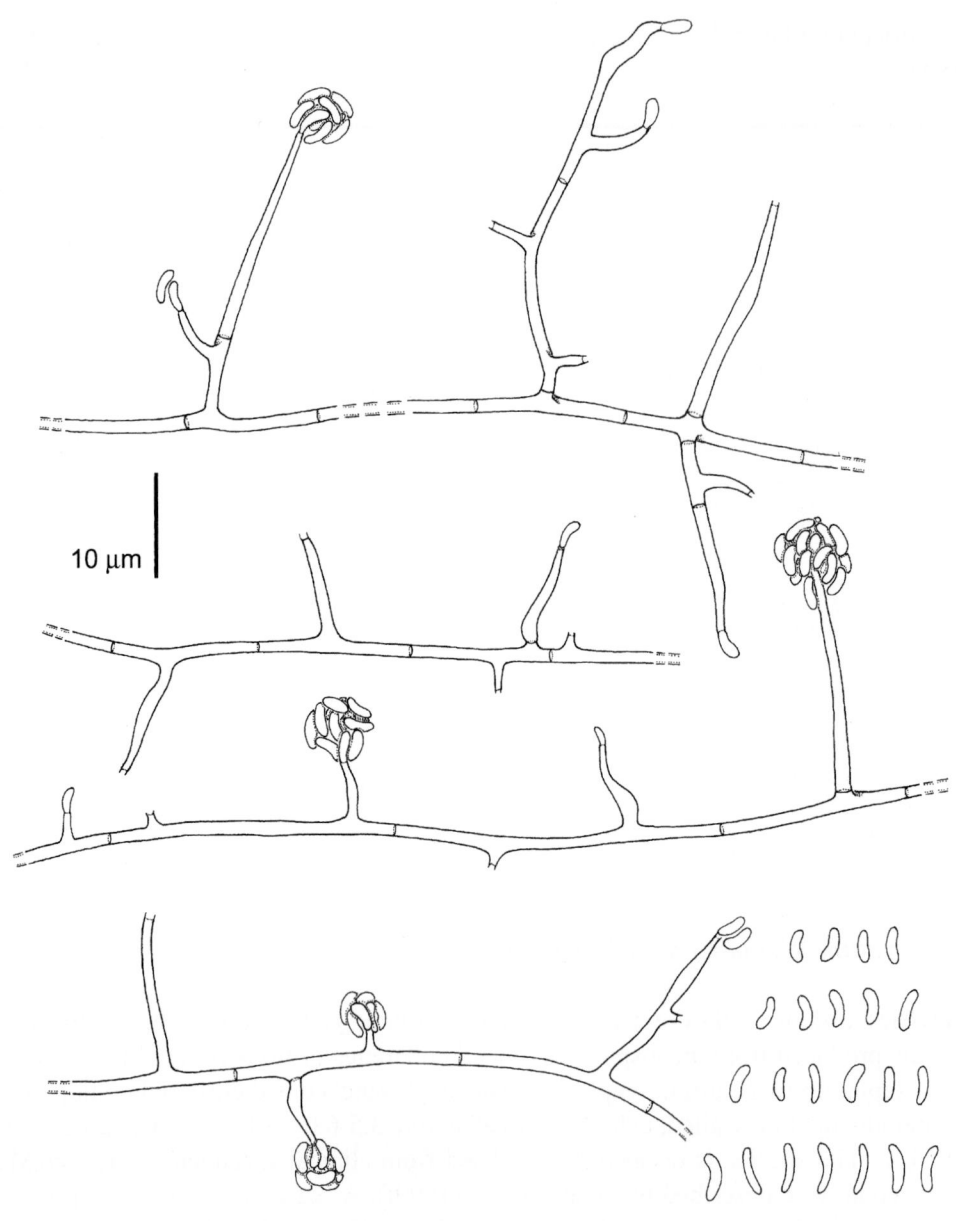

Phialemonium curvatum, CBS 490.82. Hyphae with lateral and intercalary phialides, and conidia.

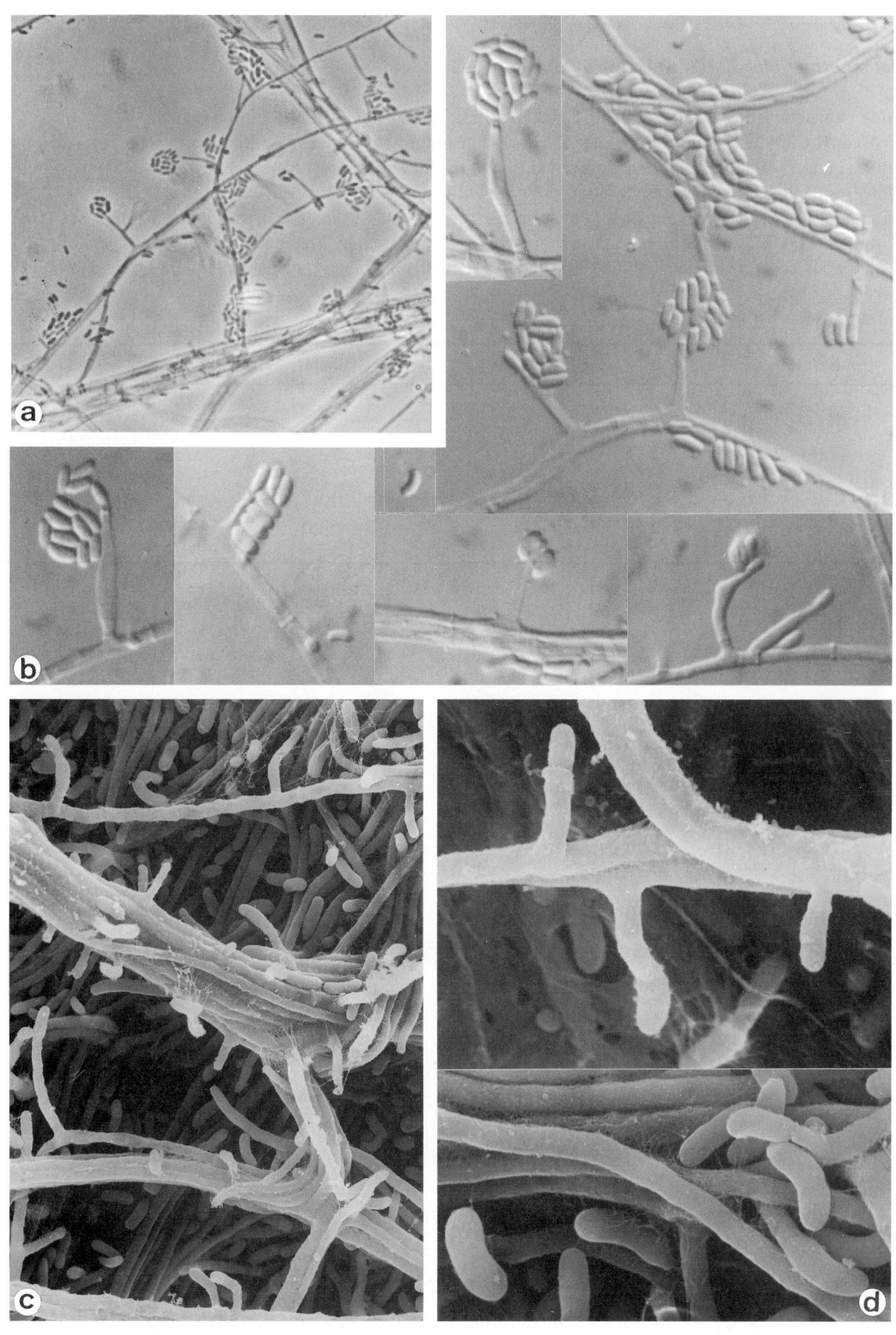

Phialemonium curvatum, CBS 490.82. Hyphae with lateral and intercalary phialides, conidia. a. ×640; b. ×1600; c. ×2500; d. ×7000.

Phialemonium obovatum W. Gams & McGinnis

Colony characteristics. Colonies (PDA) spreading, flat, smooth, pale ochraceous or greenish.

Microscopy. Conidia produced from lateral, non-septate outgrowths of creeping hyphae or from terminal, elongate phialides up to 15 μm in length. Conidia hyaline, smooth- and thin-walled, obovoidal, 3.5-6.0 × 1.5 μm, aggregating in slimy heads. Pale brown chlamydospores present in old cultures.

Pathogenicity. BSL-2. The species is known from strains isolated from several subcutaneous mycoses (Gams & McGinnis, 1983). A systemic infection in a burn patient was reported by McGinnis *et al.* (1986), a case of osteolytic phaeohyphomycosis by Magnon *et al.* (1993) and of peritonitis in a renal transplant recipient by King *et al.* (1993).

References. Gams & McGinnis (1983).

Antifungal susceptibility.

Antifungal	Mean MICs	Strains	Reference
AMB	6.96	3	McGinnis & Pasarell (1998b)
5FC	256	2	Guarro *et al.* (1999f)
FLU	22.6	2	Guarro *et al.* (1999f)
ITZ	0.67	3	McGinnis & Pasarell (1998b)
KTZ	1	2	Guarro *et al.* (1999f)
MCZ	0.71	2	Guarro *et al.* (1999f)
VCZ	0.5	3	McGinnis & Pasarell (1998b)

Nomenclature. *Phialemonium obovatum* W. Gams & McGinnis - Mycologia 75: 978, 1983.

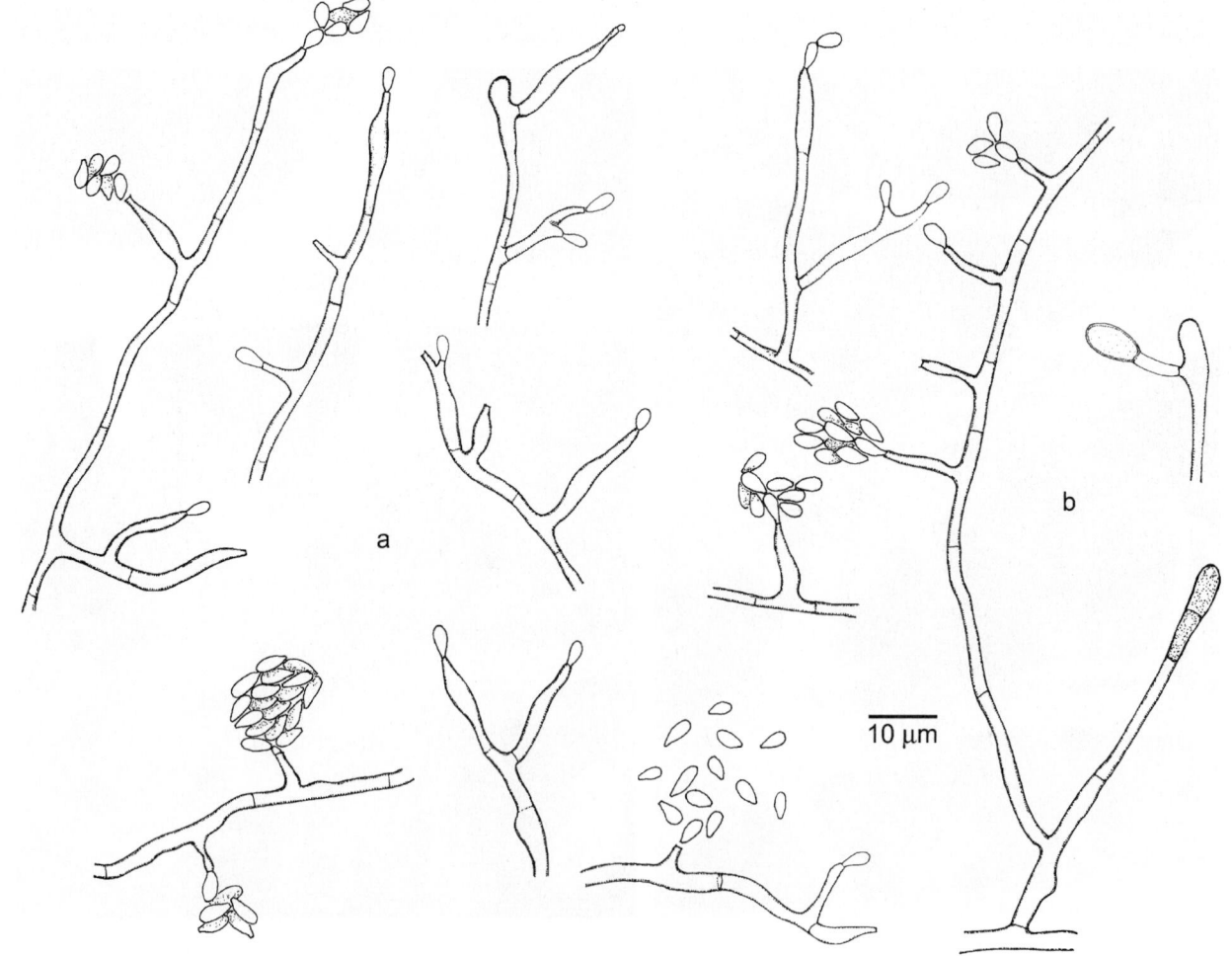

Phialemonium obovatum, CBS 279.76. a. Hyphae bearing phialides and conidia; b. chlamydospores.

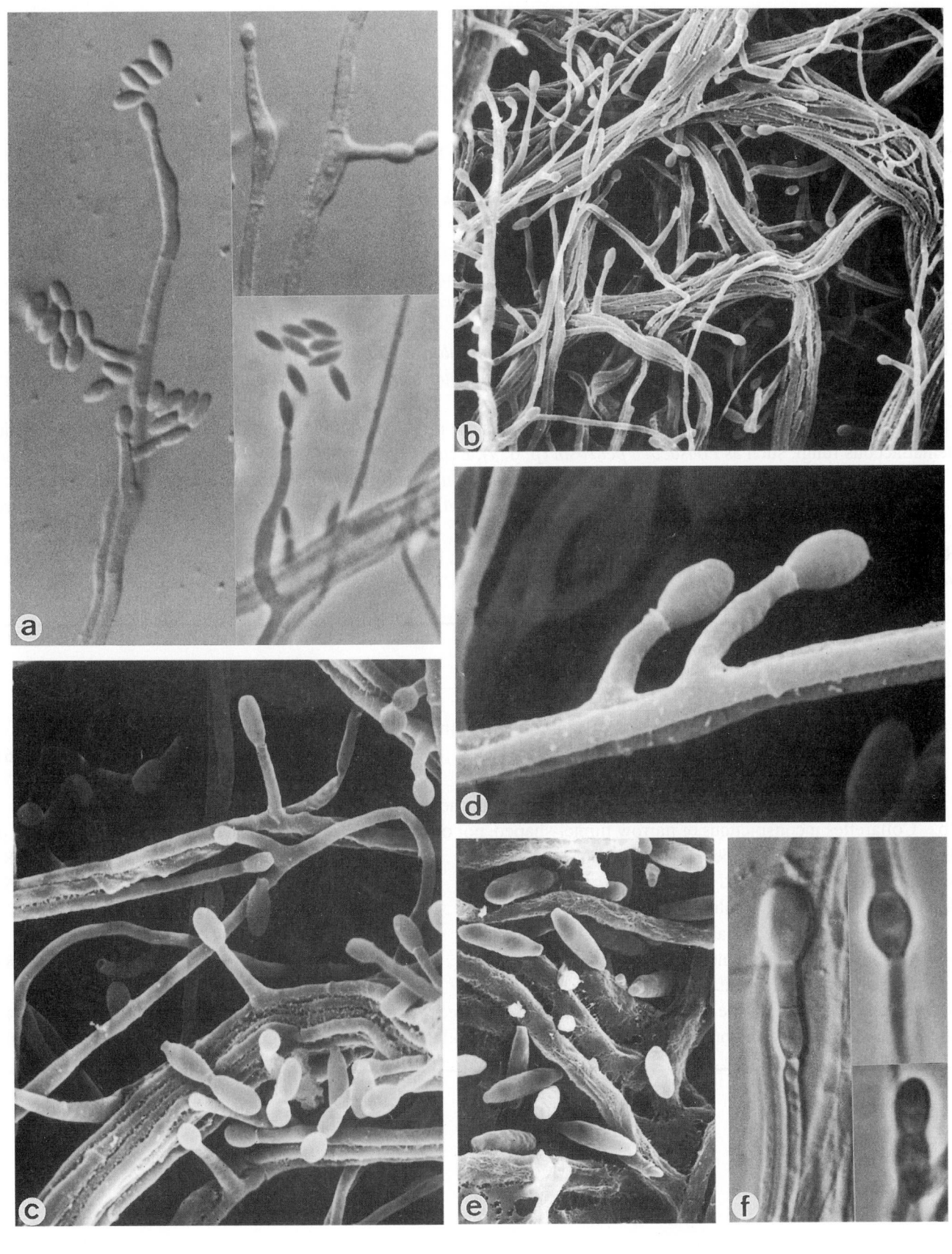

Phialemonium obovatum, CBS 279.76. a-e. Hyphae bearing phialides and conidia; f. chlamydospores. a. ×1600; b. ×1500; c. ×3600; d. ×8200; e. ×4400; f. ×1600.

Hyphomycetes. Genus: *PHIALOPHORA*

Generic description. Colonies expanding, woolly to cottony, grey, olivaceous to almost black; reverse grey to black. Conidiophores, when present, short, bearing phialides in brush-like arrangement. Phialides cylindrical to flask-shaped, with distinct collarettes. Conidia 1-celled, hyaline to brown, smooth-walled, ovoidal, cylindrical or allantoid, accumulating in slimy heads at the phialide tips. Intolerant to benomyl.

Teleomorphs. *Capronia* (*Ascomycota, Euascomycetes, Chaetothyriales: Herpotrichiellaceae*), *Gaeumannomyces* (*Ascomycota, Euascomycetes, Pleosporales: Magnaporthaceae*), *Mollisia* (*Ascomycota, Euascomycetes, Helotiales*: *Dermateaceae* and *Hyaloscyphaceae*), *Chaetosphaeria* (*Ascomycota, Euascomycetes, Sordariales: Chaetosphaeriaceae*).

Differential diagnosis. The genus is different from *Exophiala* (p. 645) by having distinct collarettes, usually more expanding, woolly growth and absence of yeast cells. *Phaeoacremonium* (p. 846) differs by narrow collarettes and usually allantoid conidia.

General remarks. *Phialophora*-like conidiogenous cells are produced in low abundance in a variety of pathogenic dematiaceous fungi, such as *Fonsecaea pedrosoi* (p. 679), *Exophiala dermatitidis* (p. 652), *E. spinifera* (p. 666) and *Cladophialophora carrionii* (p. 570). Otherwise, the genus *Phialophora* is heterogeneous and contains anamorphs of widely divergent, unrelated ascomycetes. Molecular studies of ribosomal genes of clinically significant species were published by Yan *et al.* (1995) and de Hoog *et al.* (1999b).

References. Schol-Schwarz (1970), Cole & Kendrick (1973), McGinnis (1978b), Wang & Zabel (1990), de Hoog *et al.* (1999), Gams (2000).

Key to the treated species of Phialophora:

1a.	Brown, subspherical conidia present in addition to hyaline, cylindrical or allantoid conidia . . . . *P. richardsiae* (876)
1b.	Conidia not obviously dimorphic → 2
2a.	Collarettes funnel- or vase-shaped clearly darker than the rest of the phialide → 3
2b.	Collarettes inconspicuous or narrow funnel-shaped, not or only slightly darker than the rest of the phialide→ 4
3a.	Collarettes funnel-shaped . . . . *P. verrucosa* (878)
3b.	Collarettes vase-shaped . . . . *P. americana* (865)
4a.	Conidia allantoid; phialides often penicillate . . . . *P. repens* (872)
4b.	Conidia otherwise; phialides scattered on hyphae → 5
5a.	Chlamydospores present . . . . *P. bubakii* (868)
5b.	Chlamydospores absent → 6
6a.	Collarettes often sessile on hyphae; melibiose assimilated . . . . *P. reptans* (874)
6b.	Collarettes terminally on phialides; melibiose not assimilated . . . . *P. europaea* (870)

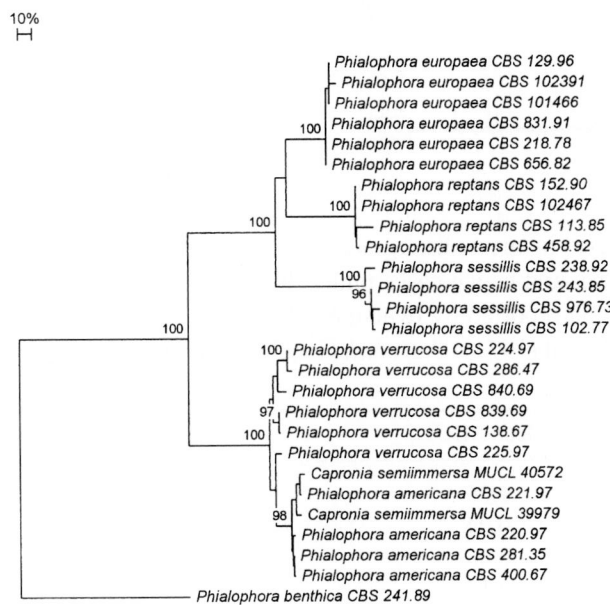

Fig. 60. Phylogenetic tree of *Phialophora* based on confidently aligned partial ITS rDNA sequences, using Neighor joining algorithm with Kimura correction. Bootstrap values >90 from 100 resampled datasets are shown. The species form well-delimited branches but are relatively far away from each other; the (groups of) species are only distantly related to each other despite similar morphology.

Phialophora americana (Nannf.) S.J. Hughes

Colony characteristics. Colonies (PDA) growing moderately rapidly, woolly, olivaceous grey; reverse olivaceous black.

Microscopy. Phialides scattered, flask-shaped to cylindrical, with darker, vase-shaped collarettes which may also be sessile, directly on hyphae. Conidia produced in slimy heads, subhyaline, broadly ellipsoidal, 2.5-4.0 × 1.5-3 µm.

Teleomorph. *Capronia semiimmersa* (Candousseau & Sulmont) Untereiner & Naveau (*Ascomycota, Euascomycetes, Chaetothyriales: Herpotrichiellaceae*).

Physiology.

Growth:					
D-Glucose	+	Lactose	–	DL-Lactate	+
D-Galactose	+	Raffinose	+	Succinate	+
L-Sorbose	+	Melezitose	+	Citrate	+,w
D-Glucosamine	+,w	Inulin	–,w	Methanol	–,w
D-Ribose	+,w	Soluble starch	+,w	Ethanol	+
D-Xylose	+	Glycerol	+	Nitrate	+
L-Arabinose	+	*meso*-Erythritol	+	Nitrite	+
D-Arabinose	+	Ribitol	+	Ethylamine	+
L-Rhamnose	+	Xylitol	+	L-Lysine	+
Sucrose	+	L-Arabinitol	+	Cadaverine	+
Maltose	+	D-Glucitol	+	Creatine	+
α,α-Trehalose	+	D-Mannitol	+	Creatinine	+
α-Methyl-D-glucoside	+	Galactitol	+	10% MgCl$_2$	+,w
Cellobiose	+	*myo*-Inositol	+	10% NaC1	–
Salicin	+	Glucono-δ-lactone	+	0.1% Cycloheximide	–
Arbutin	+	D-Gluconate	+,w	Urease	+
Melibiose	+	D-Glucuronate	+	Growth at 37°C	v
		D-Galacturonate	+	Growth at 40°C	–,w

Differential diagnosis. The species is distinct from *P. verrucosa* by vase-shaped rather than funnel-shaped collarettes, and by ITS × *Msp*I-profiles.

Molecular diagnostics. ITS restriction map based on NCBI U31838:

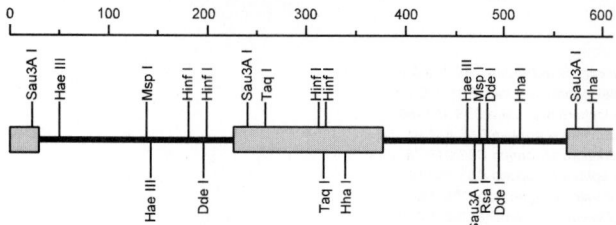

Pathogenicity. BSL-2. The species is one of the agents of chromoblastomycosis. In the medical literature it has not been separated from *P. verrucosa*, and hence cases under this name may actually be *P. americana* (de Hoog *et al.*, 1999b), such as a cystic granuloma reported by Iwatsu & Miyaji (1977).

References. Candousseau & Sulmont (1971), Yan *et al.* (1995), Untereiner & Naveau (1999), de Hoog *et al.* (1999b).

Antifungal susceptibility.

Antifungal	Mean MICs	Strains	Reference
AMB	1.26	3	McGinnis & Pasarell (1998b)
ITZ	0.04	3	McGinnis & Pasarell (1998a)
TBF	0.06	3	McGinnis & Pasarell (1998a)
VCZ	0.13	3	McGinnis & Pasarell (1998b)

Nomenclature. *Cadophora americana* Nannfeldt, *in* Melin & Nannfeldt - Svensk. Skogsvför. Tidskr. 32: 412, 1934 ≡ *Phialophora americana* (Nannfeldt) S.J. Hughes - Can. J. Bot. 36: 795, 1958.

Dictyotrichiella semiimmersa Candousseau & Sulmont - Revue Mycol. 36: 242, 1971 ≡ *Capronia semiimmersa* (Candousseau & Sulmont) Untereiner & Naveau - Mycologia 91: 73, 1999.

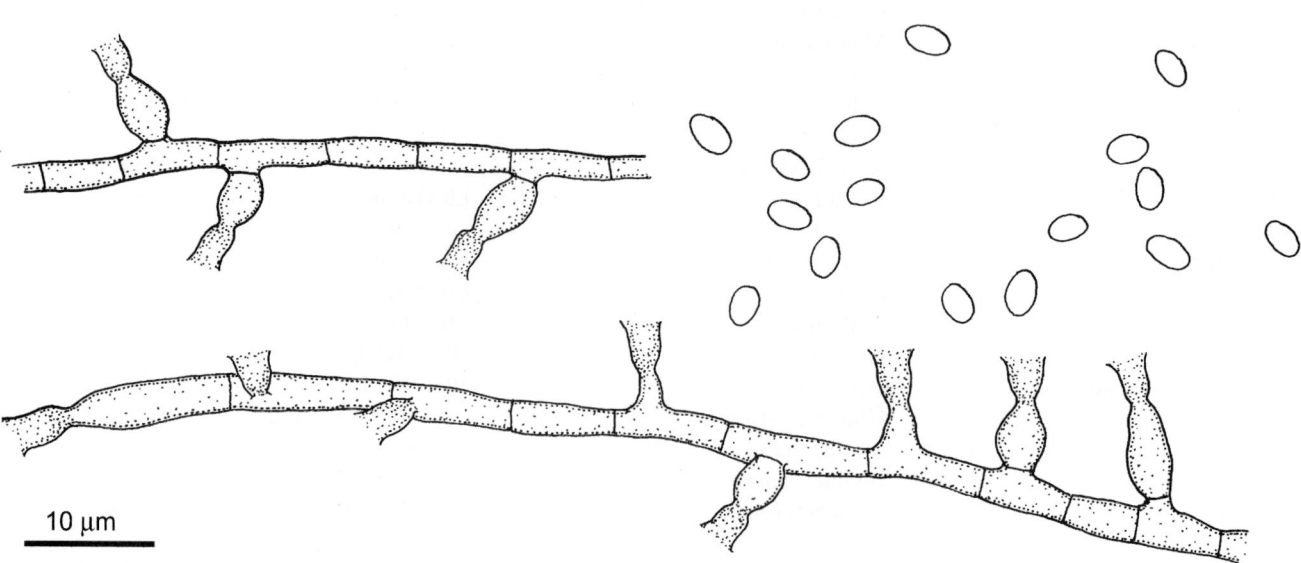

Phialophora americana, CBS 281.35. Hyphae with phialides and conidia.

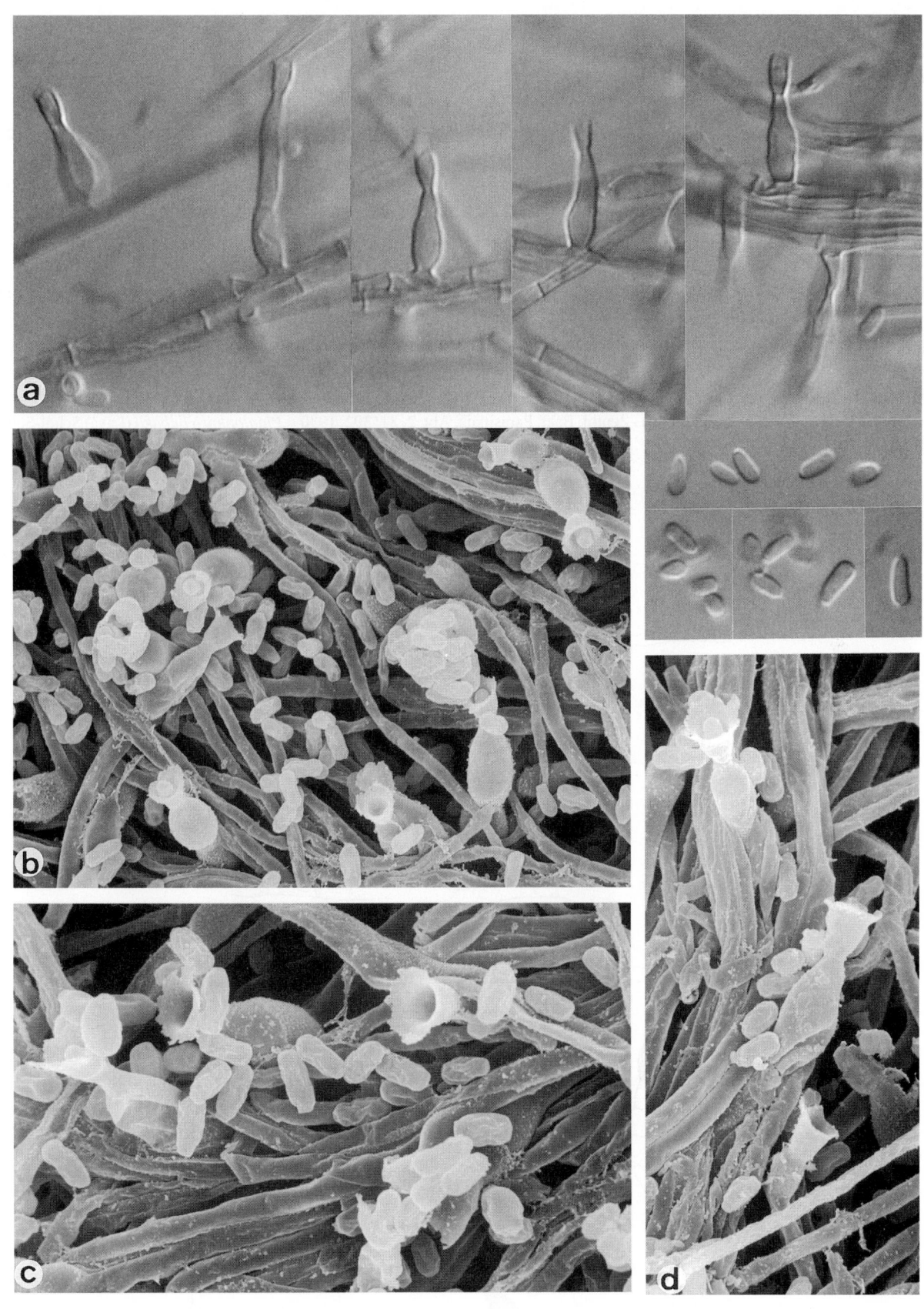

Phialophora americana, CBS 281.35. Hyphae with phialides and conidia. a.. ×1600; b. ×2800; c. ×4650; d. ×3800.

867

Phialophora bubakii (Laxa) Schol-Schwarz

Colony characteristics. Colonies (PDA) growing moderately rapidly, velvety, forming concentric rings, dark greyish-olive; reverse greenish-black.

Microscopy. Conidiophores undifferentiated, dark olivaceous, smooth-walled, occasionally verrucose. Phialides single or in groups, developing laterally on hyphae, flask-shaped, slender or swollen, with an indistinct collarette at the apex; occasionally phialides are intercalary cells. Conidia hyaline to subhyaline, in young cultures allantoid, 2.0-6.5 × 1-3 µm, in older cultures broadly ellipsoidal, 1.5-4.0 × 1-2 µm. Chlamydospores ovoidal, brown, 4.5-6.5 × 4-5 µm, sometimes with an irregularly thickened wall, developing laterally or terminally on the hyphae or intercalary. Torulose hyphae present.

Pathogenicity. BSL-1. Wood-inhabiting species. Subcutaneous infections (Porto, 1979), corneal ulcers (Eiferman *et al.*, 1983).

References. Schol-Schwarz (1970), Williams (1991a).

Nomenclature. *Margarinomyces bubakii* Laxa - Zentbl. Bakt. Parasitkde, Abt. 2, 81: 392, 1930 ≡ *Phialophora bubakii* (Laxa) Schol-Schwarz - Persoonia 6: 66, 1970.

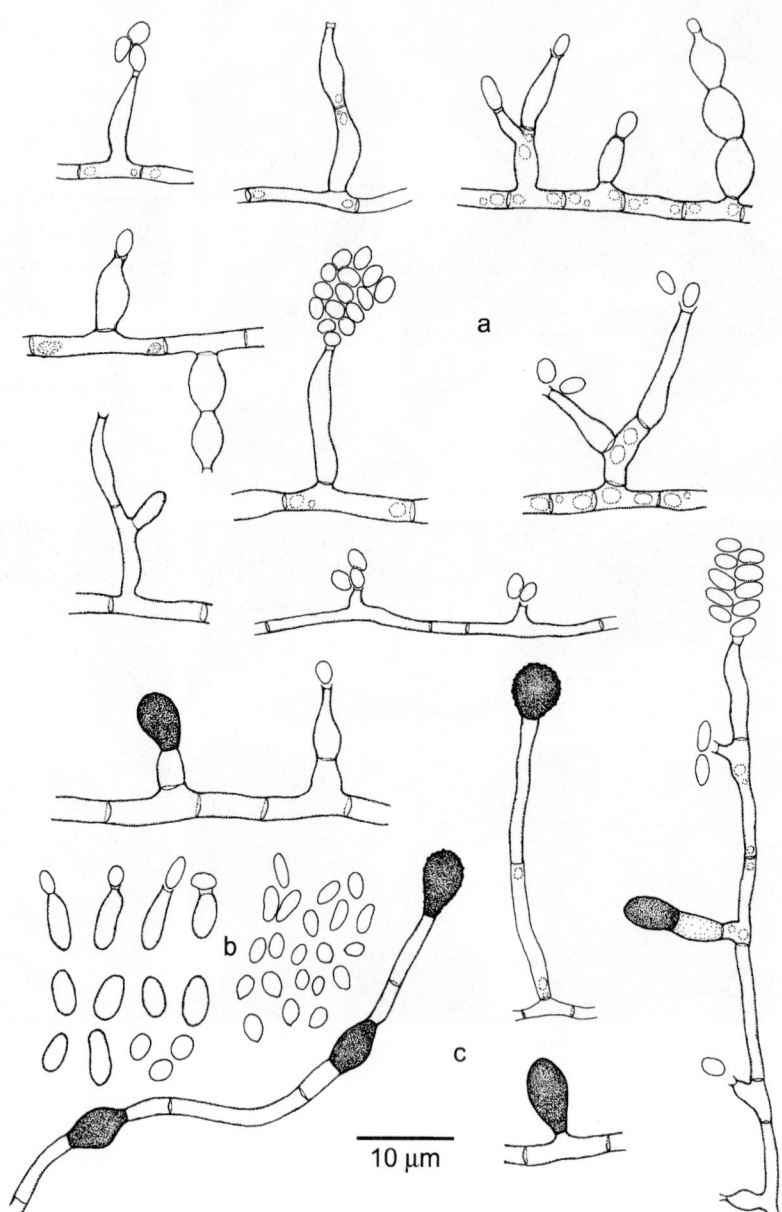

Phialophora bubakii, IMI 182533. a. Phialides producing conidia; b. budding cells and conidia; c. chlamydospores.

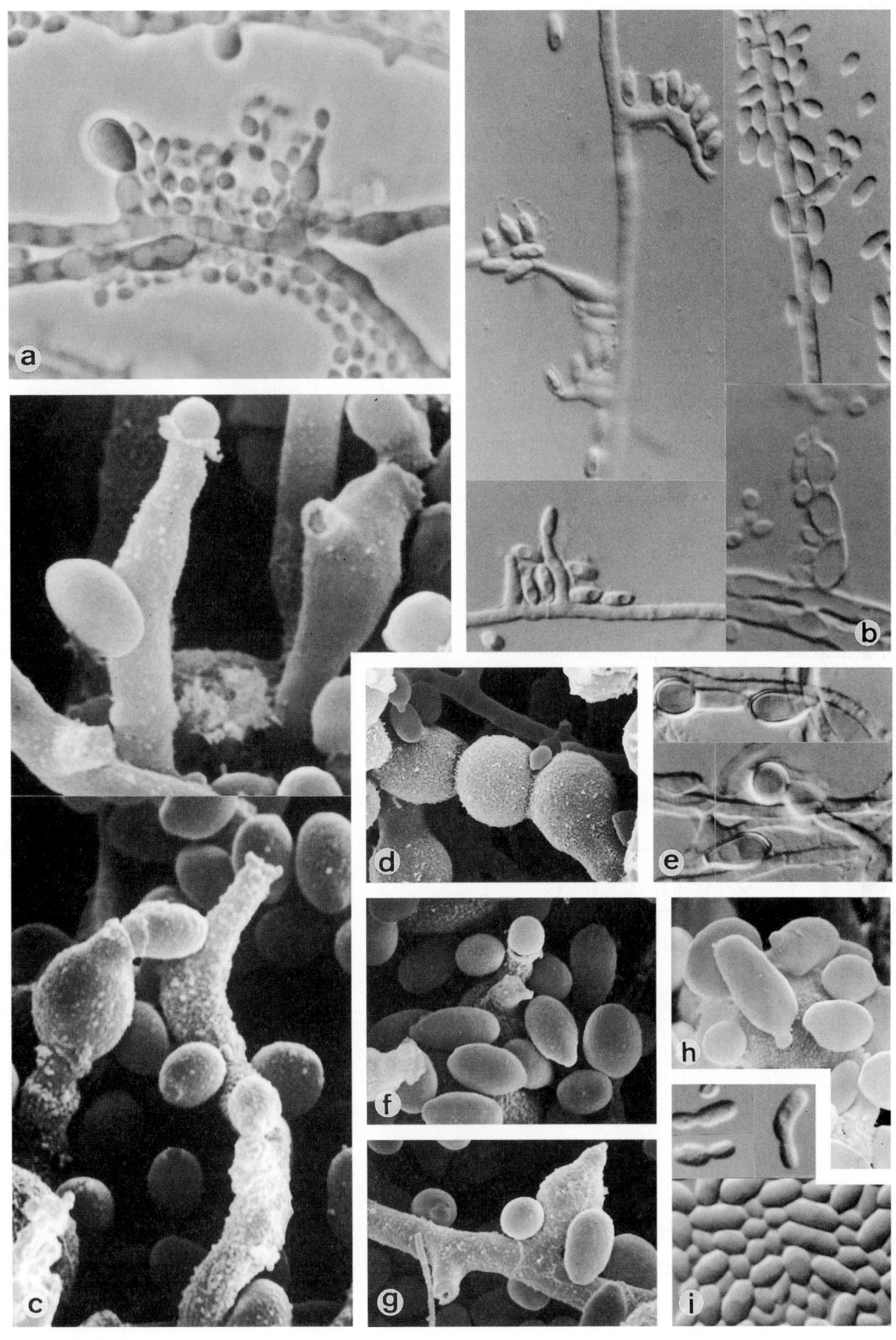

Phialophora bubakii, IMI 182533. a-c, f, g. Phialides and conidia; d. torulose hyphae; e. chlamydospores; h, i. budding cells and conidia. a. ×1280; b. ×1600; c. ×9600; d. ×2800; e. ×1600; f. ×5500; g. ×4700; h. ×5700; i. ×1600.

869

Phialophora europaea de Hoog *et al.*

Colony characteristics. Colonies (PDA) growing rather slowly, olivaceous black.
Microscopy. Hyphae pale olivaceous brown, 1.5-2.0 μm wide, with numerous anastomoses. Fertile hyphae bearing phialides either directly, or with 1-2 on slightly swollen subtending cells of about 2-3 μm wide; subtending cells occasionally arranged in chains containing 2-8 cells. Phialides flak-shaped to elongate, somewhat narrowed towards the tip, often with a nearly cylindrical apical portion, 6-9 μm long. Collarettes very short, flaring, producing conidia in fragile chains or in heads. Conidia subhyaline, (sub)spherical, 1.8-2.5 μm.
Physiology.

Growth:					
D-Glucose	+	Lactose	–,w	DL-Lactate	v
D-Galactose	+	Raffinose	+	Succinate	v
L-Sorbose	+,w	Melezitose	+	Citrate	–,w
D-Glucosamine	+,w	Inulin	–	Methanol	–
D-Ribose	+,w	Soluble starch	+,w	Ethanol	+,w
D-Xylose	+	Glycerol	+	Nitrate	+
L-Arabinose	+	*meso*-Erythritol	+	Nitrite	+
D-Arabinose	+,w	Ribitol	+	Ethylamine	+
L-Rhamnose	+	Xylitol	+	L-Lysine	+
Sucrose	+	L-Arabinitol	+	Cadaverine	+
Maltose	+	D-Glucitol	+	Creatine	+,w
α,α-Trehalose	+	D-Mannitol	+	Creatinine	–,w
α-Methyl-D-glucoside	+	Galactitol	v	10% MgCl$_2$	w
Cellobiose	+	*myo*-Inositol	v	10% NaCl	–
Salicin	+,w	Glucono-δ-lactone	+	0.1% Cycloheximide	+
Arbutin	+,w	D-Gluconate	+	Urease	+
Melibiose	–	D-Glucuronate	+,w	Growth at 37°C	+,w
		D-Galacturonate	v	Growth at 40°C	–

Molecular diagnostics. ITS restrictions map based on CBS 101466:

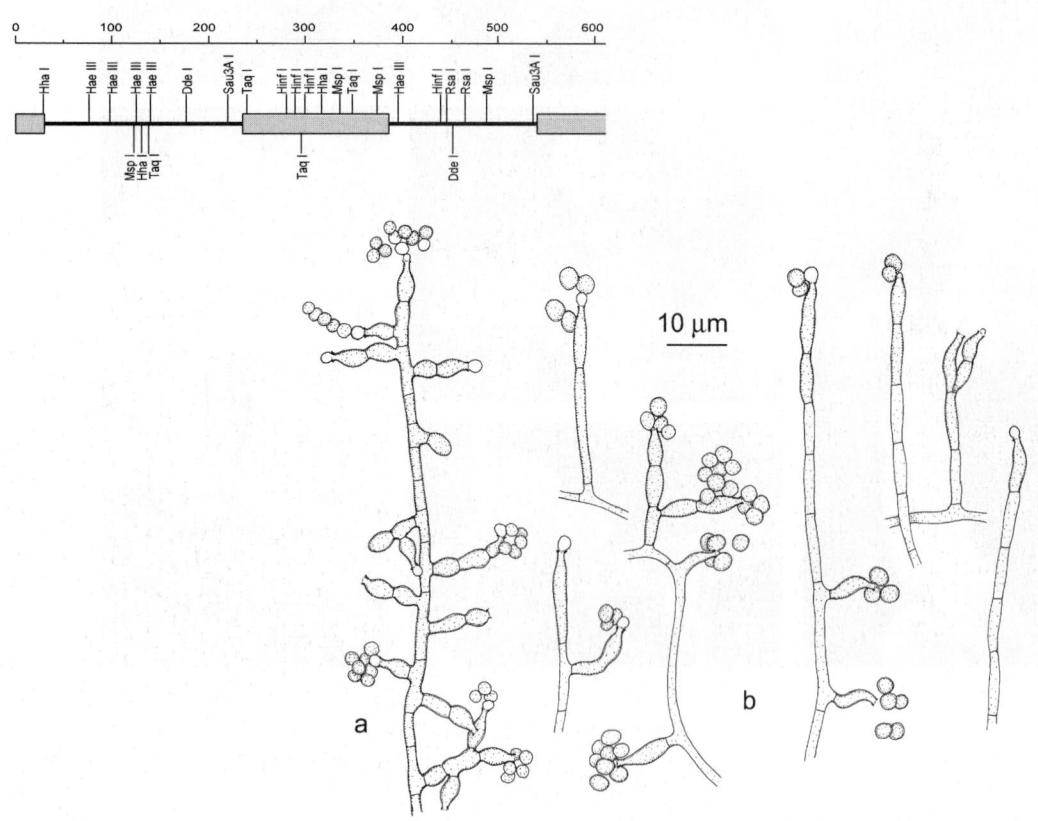

Phialophora europaea. a. CBS 101466, Conidial apparatus; b. CBS 129.96, poorly differentiated phialides.

Differential diagnosis. Phialides are usually borne on inflated cells but sometimes are poorly differentiated. The absence of melibiose assimilation is characteristic for the species.

Pathogenicity. BSL-2. Agent of cutaneous infections (de Hoog *et al.*, 2000).

Nomenclature. *Phialophora europaea* de Hoog, Mayser & Haase, *in* de Hoog, Mayser, Haase, Horré & Horrevorts - Mycoses (in press).

Phialophora europaea, CBS 101466. a-d. Conidial apparatus and conidia. a. ×1600; b. ×1750; c. ×2050; d. ×4750.

Phialophora repens (Davidson) Conant

Colony characteristics. Colonies (PDA) growing slowly, flat, moist and later more or less cottony, dark brown to olivaceous brown; reverse nearly black.

Microscopy. Conidiophores arising laterally on hyphae, bearing a dense, penicillately branched system of phialides. Phialides acicular to long-cylindrical, hyaline to pale brown, apically with a short, inconspicuous or slightly flared collarette. Conidia hyaline, thin- and smooth-walled, cylindrical to allantoid, 3-6 × 1.2-2.2 µm.

Differential diagnosis. *Acremonium atrogriseum* (p. 398) has phialides with slender necks and produces straight, ellipsoidal conidia. Gams (2000) recommends the use of the name *Cadophora repens* for *P. repens*.

Pathogenicity. BSL-1. Subcutaneous infections with granulomatous nodules (Meyers *et al.*, 1975; Hironaga *et al.*, 1989).

References. Schol-Schwarz (1970), de Hoog (1983).

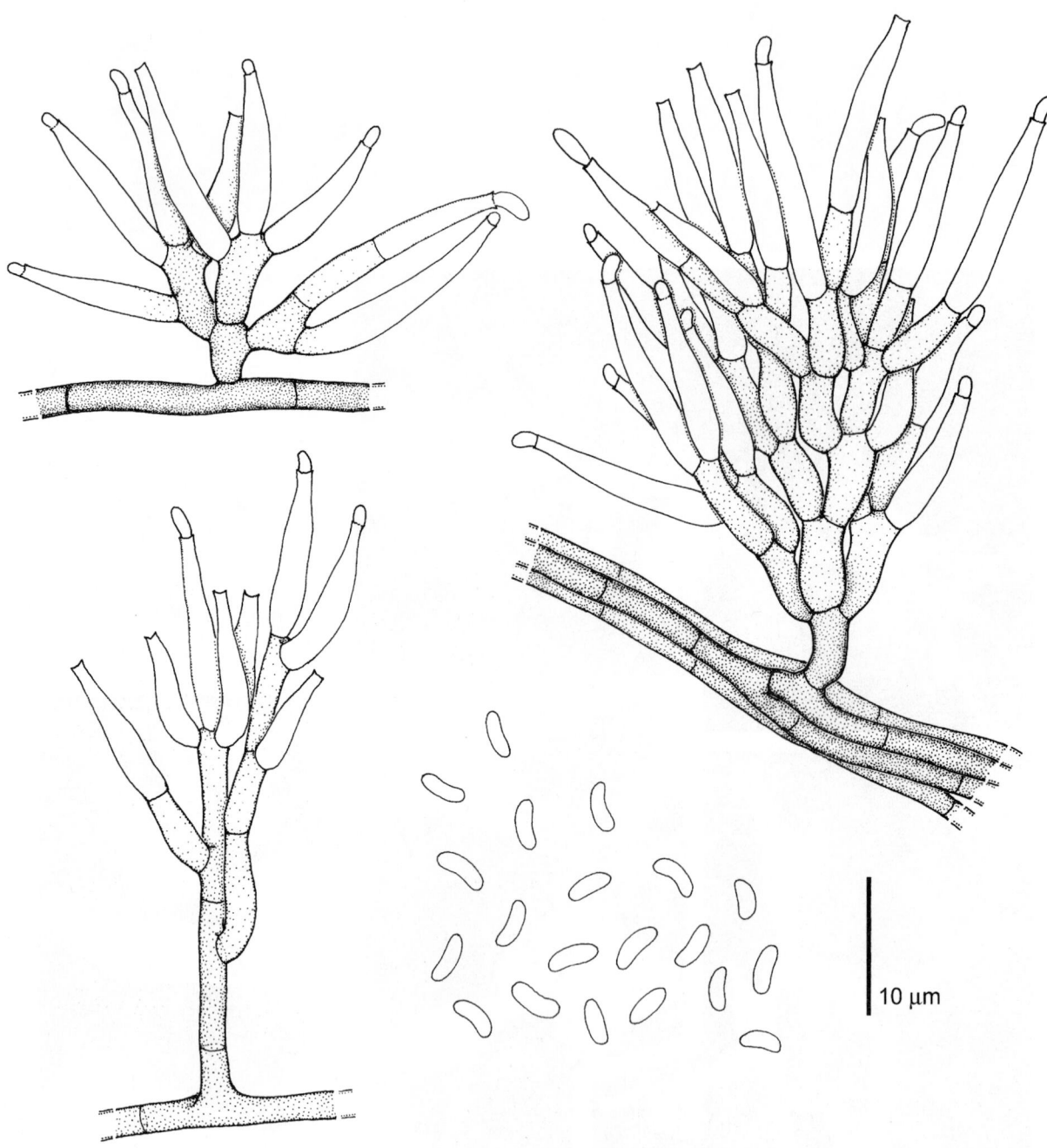

10 µm

Phialophora repens, CBS 294.39. Conidial apparatus and conidia.

Antifungal susceptibility.

Antifungal	Mean MICs	Strains	Reference
AMB	0.79	3	McGinnis & Pasarell (1998b)
ITZ	0.25	3	McGinnis & Pasarell (1998b)
VCZ	0.25	3	McGinnis & Pasarell (1998b)

Nomenclature. *Cadophora repens* Davidson - J. Agric. Res. 50: 803, 1935 ≡ *Phialophora repens* (Davidson) Conant - Mycologia 29: 598, 1937.

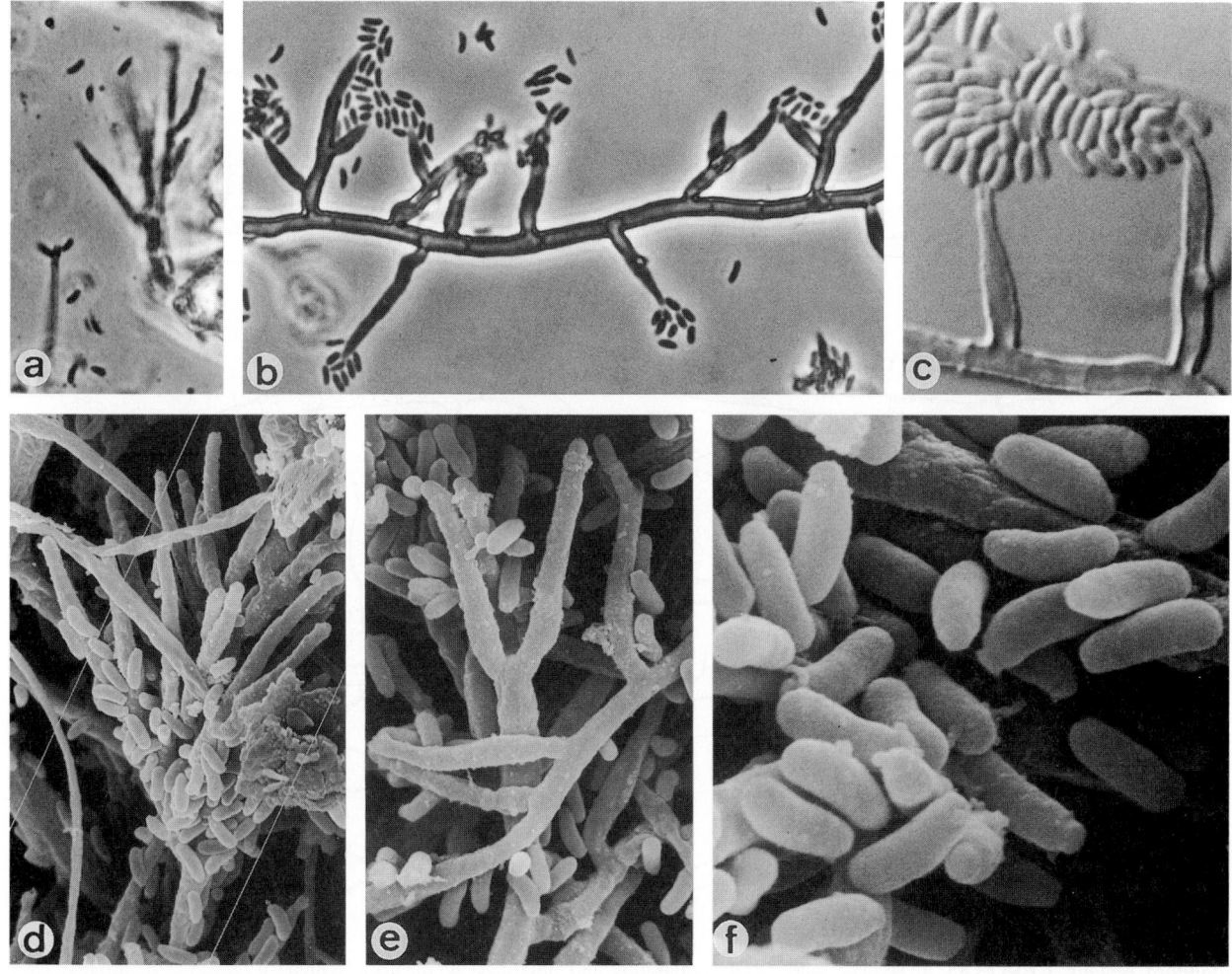

Phialophora repens, CBS 294.39. Conidial apparatus and conidia. a. ×512; b. ×640; c. ×1600; d. ×2200; e. ×3300; f. ×8500.

Phialophora reptans de Hoog

Colony characteristics. Colonies (PDA) restricted, olivaceous black, flat with slightly rough centre.
Microscopy. Torulose hyphae present at colony centre. Hyphae 1.5-2.2 μm wide, olivaceous. Phialidic collarettes mostly sessile on undifferentiated cells, sometimes on terminal cells; rarely lateral phialides present. Collarettes conspicuous, slightly darker than the rest of the phialide, narrow funnel-shaped to almost cylindrical, up to 2.5 μm long. Conidia subhyaline, obovoidal, about 3.0 × 1.8 μm, finally inflating and becoming melanized.
Physiology.

Growth:					
		Lactose	+	DL-Lactate	+
D-Glucose	+	Raffinose	+	Succinate	+,w
D-Galactose	+	Melezitose	+,w	Citrate	−,w
L-Sorbose	+	Inulin	−	Methanol	−
D-Glucosamine	v	Soluble starch	+,w	Ethanol	+,w
D-Ribose	+,w	Glycerol	+	Nitrate	+
D-Xylose	+	*meso*-Erythritol	+	Nitrite	+
L-Arabinose	+	Ribitol	+	Ethylamine	+
D-Arabinose	v	Xylitol	+	L-Lysine	+
L-Rhamnose	+	L-Arabinitol	+	Cadaverine	+,w
Sucrose	+	D-Glucitol	+	Creatine	+
Maltose	+,w	D-Mannitol	+	Creatinine	+
α,α-Trehalose	+	Galactitol	+,w	10% MgCl$_2$	+
α-Methyl-D-glucoside	+	*myo*-Inositol	+	10% NaCl	−,w
Cellobiose	+	Glucono-δ-lactone	+,w	0.1% Cycloheximide	+
Salicin	+,w	D-Gluconate	+	Urease	+
Arbutin	+	D-Glucuronate	+	Growth at 37°C	−
Melibiose	+	D-Galacturonate	+,w		

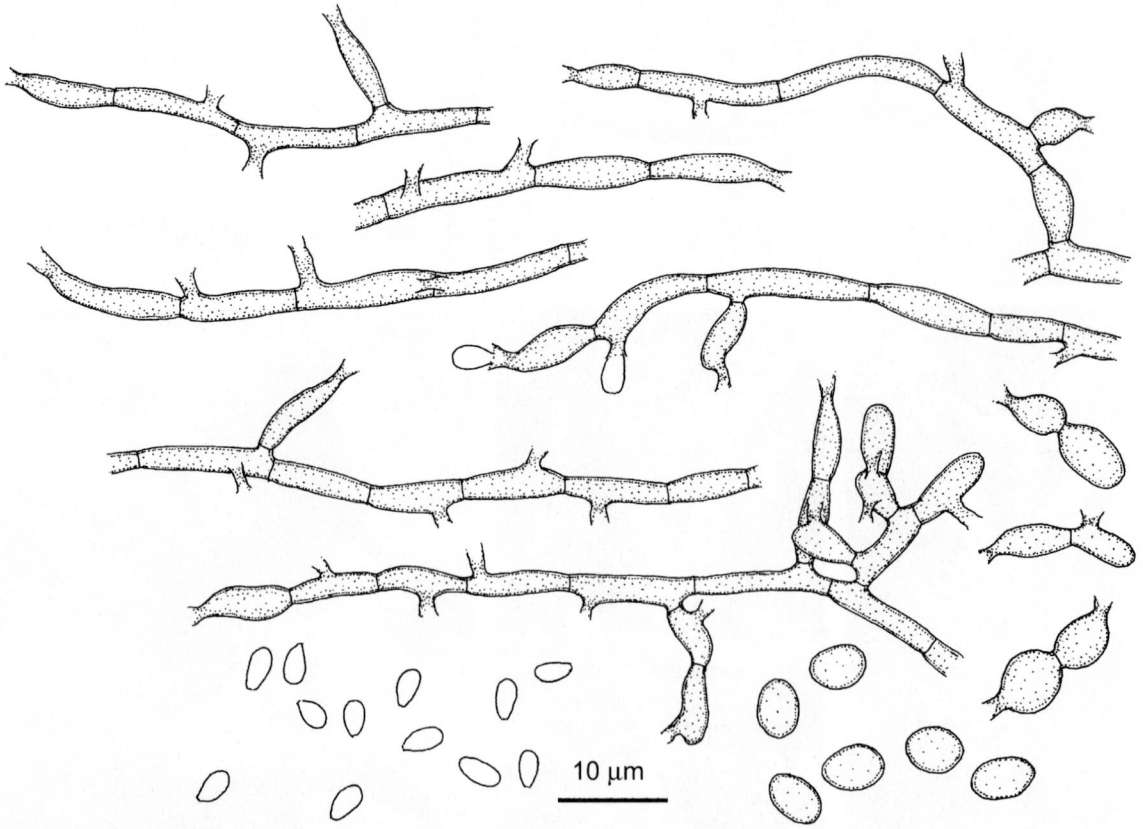

Phialophora reptans, CBS113.85. a. Hyphae with sessile collarettes and phialides; b. torulose hyphae; c. conidia; d. inflated conidia.. Reproduced with permission from Stud. Mycol. 43: 118, 1999.

Molecular diagnostics. ITS restriction map based on CBS 113.85:

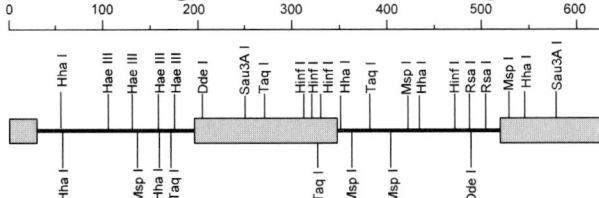

Pathogenicity. BSL-1. Reported from clinical samples of superficial infections (de Hoog *et al.*, 1999b).

Reference. De Hoog *et al.* (1999b).

Nomenclature. *Phialophora reptans* de Hoog, *in* de Hoog, Weenink & Gerrits van den Ende - Stud. Mycol. 43: 117, 1999.

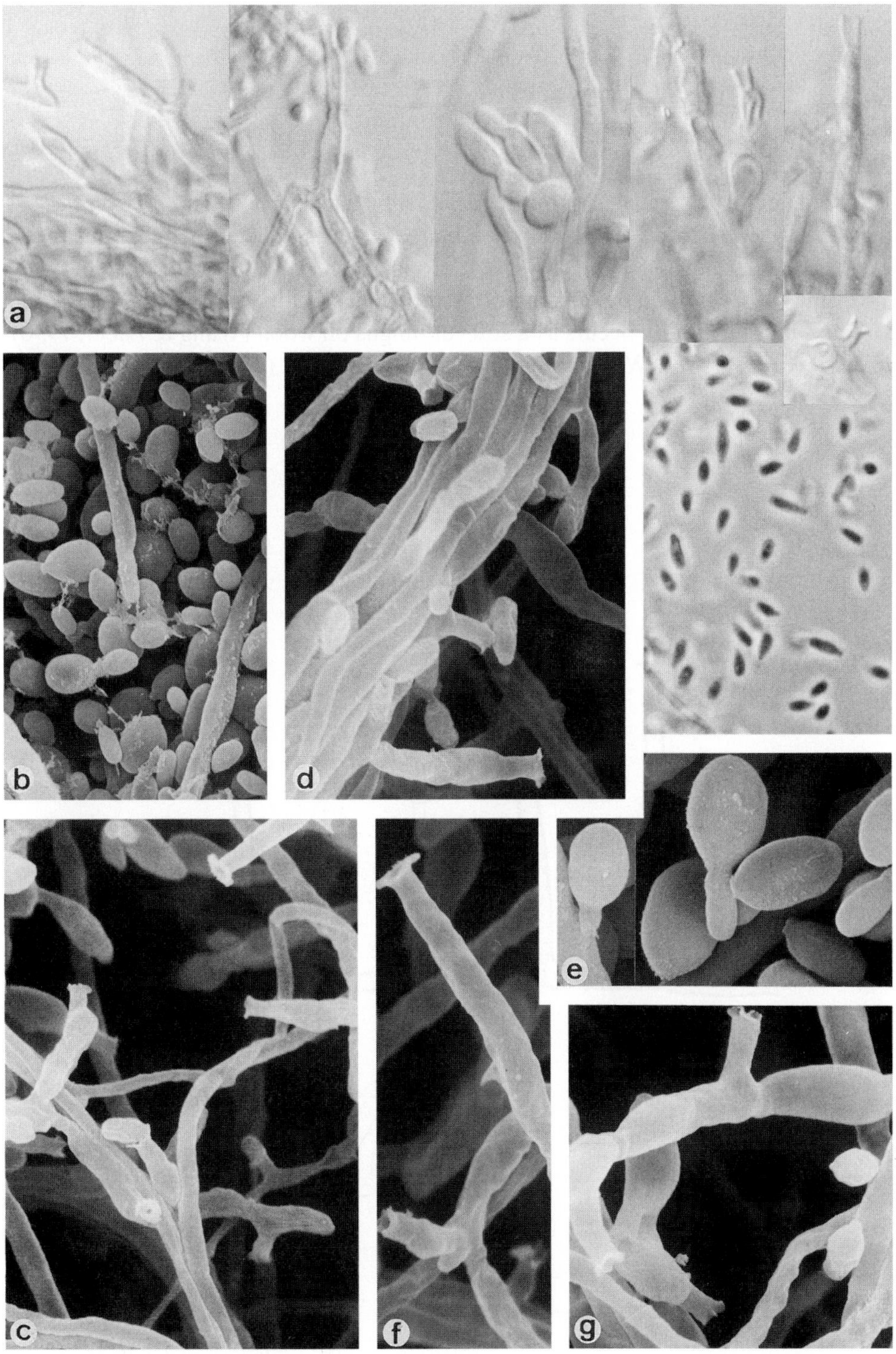

Phialophora reptans, CBS 113.85. Hyphae with sessile collarettes and phialides, and conidia. a. ×1600; b. ×2750; c. ×3800; d. ×4650; e. ×5000; f. ×5750; g. ×6500.

875

Phialophora richardsiae (Nannf.) Conant

Colony characteristics. Colonies (PDA) expanding, powdery, hairy in degenerate cultures, greyish-brown; reverse grey-brown to olivaceous black.

Microscopy. Two conidial types are produced: (1) hyaline conidia, which are allantoid or cylindrical, 3-6 × 1.5-2.5 μm are formed on inconspicuous, butt-shaped lateral outgrowths from thin-walled hyphae, having minute collarettes; (2) brown, thick-walled conidia which are (sub)spherical, 2.5-3.5 × 2-3 μm are formed on dark brown, slender, tapering phialides with flaring collarettes. Intermediate types are often present.

Molecular diagnostics. ITS restriction map based on NCBI U31844:

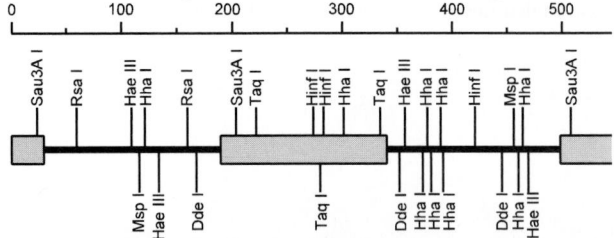

Pathogenicity. BSL-2. The species is an uncommon cause of subcutaneous phaeohyphomycotic cysts after traumatic implantation (Guého *et al.*, 1989), mostly in patients debilitated by, e.g., diabetes mellitus (Pitrak *et al.*, 1988). A case of osteomyelitis was reported in an AIDS patient by Uberti-Foppa *et al.* (1995). Localized cysts are mostly surrounded by a collagenous capsule. In severely immunocompromised patients recidives after surgery are frequently observed (Ikai *et al.*, 1988). The fungus naturally occurs as a soft-rot fungus on wood (Wang & Zabel, 1990).

References. Cole & Kendrick (1973), Wang & Zabel (1990), Singh *et al.* (1992).

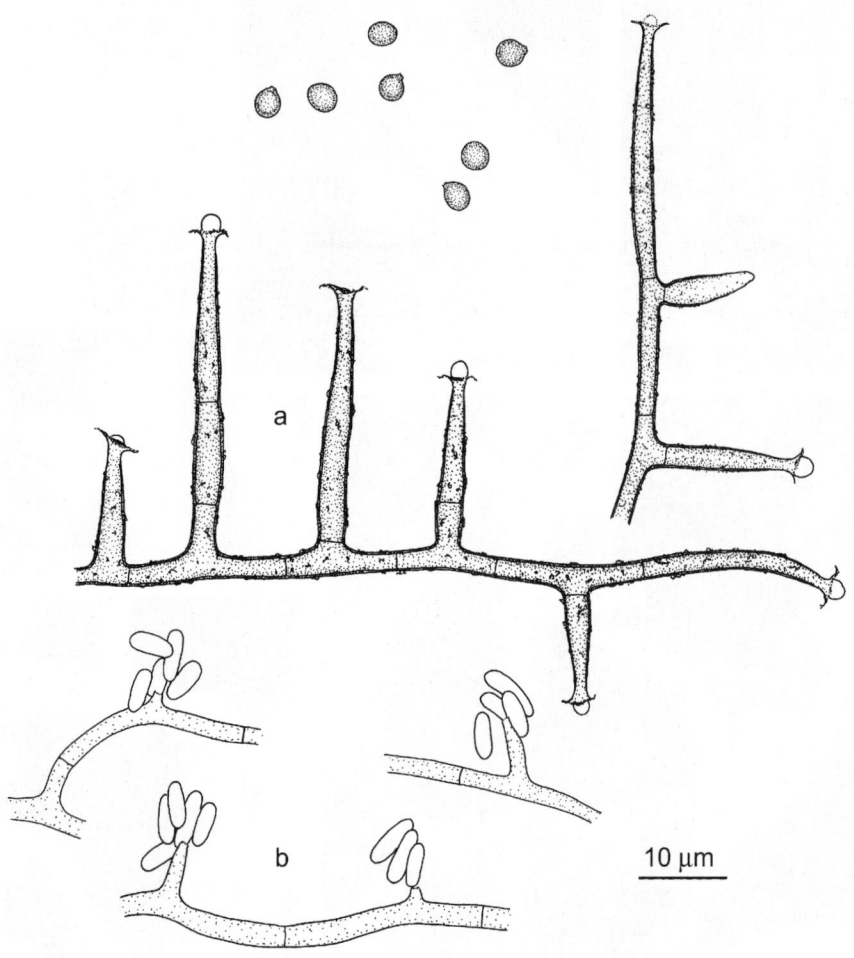

Phialophora richardsiae, CBS 506.90. a. Aerial phialides with brown, spherical conidia; b. submerged phialides with hyaline, cylindrical conidia.

Antifungal susceptibility.

Antifungal	MICs	Strains	Reference
AMB	0.73	11	McGinnis & Pasarell (1998b)
5FC	256	1	Unpublished data
FLZ	16	1	Unpublished data
ITZ	0.44	10	McGinnis & Pasarell. (1998a)
KTZ	0.25	1	Unpublished data
MCZ	2	1	Unpublished data
TBF	0.43	1	McGinnis & Pasarell (1998a)
VCZ	0.64	11	McGinnis & Pasarell (1998b)

Nomenclature. *Cadophora richardsiae* Nannfeldt, *in* Melin & Nannfeldt - Svenska Skogsvför. Tidskr. 32: 421, 1934 ≡ *Phialophora richardsiae* (Nannfeldt) Conant - Mycologia 29: 598, 1937.

Cadophora brunnescans Davidson - J. Agric. Res. 50: 803, 1955 ≡ *Phialophora brunnescens* (Davidson) Conant - Mycologia 29: 598, 1937.

Phialophora calyciformis G. Smith - Trans. Br. Mycol. Soc. 45: 391, 1962.

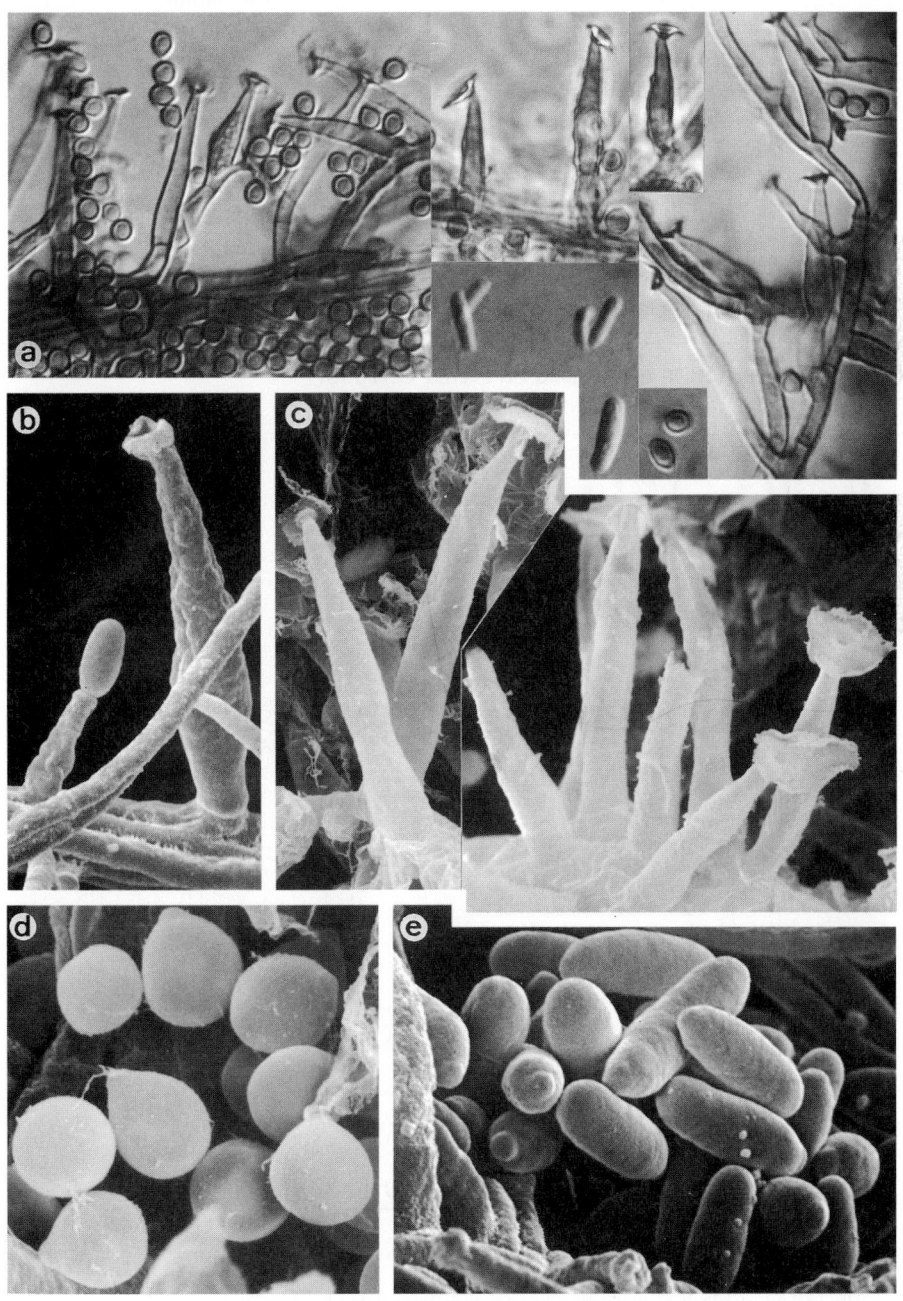

Phialophora richardsiae, CBS 506.90. Phialides, and hyaline and brown conidia. a. ×1150; b. ×6410; c. ×3240; d. ×6410; e. ×5040.

Phialophora verrucosa Medlar

Colony characteristics. Colonies (PDA) growing moderately rapidly, smooth, woolly or hairy, olivaceous black; reverse olivaceous black.

Microscopy. Phialides scattered, flask-shaped with darker, funnel-shaped collarettes. Conidia produced in slimy heads, subhyaline, smooth- and thin-walled, broadly ellipsoidal, 2.5-4.0 × 1.5-3.0 µm.

Physiology.

Growth:					
D-Glucose	+	Lactose	–	DL-Lactate	+
D-Galactose	+	Raffinose	+	Succinate	+
L-Sorbose	+	Melezitose	+	Citrate	v
D-Glucosamine	+,w	Inulin	–	Methanol	–,w
D-Ribose	v	Soluble starch	+,w	Ethanol	v
D-Xylose	+	Glycerol	+	Nitrate	+
L-Arabinose	+	*meso*-Erythritol	+	Nitrite	+
D-Arabinose	+,w	Ribitol	+	Ethylamine	+
L-Rhamnose	+	Xylitol	v	L-Lysine	+
Sucrose	+	L-Arabinitol	+,w	Cadaverine	+
Maltose	+	D-Glucitol	+	Creatine	+
α,α-Trehalose	+	D-Mannitol	+	Creatinine	+
α-Methyl-D-glucoside	+,w	Galactitol	v	10% MgCl$_2$	–,w
Cellobiose	+	*myo*-Inositol	+	10% NaCl	–
Salicin	+	Glucono-δ-lactone	+	0.1% Cycloheximide	+
Arbutin	+	D-Gluconate	+,w	Urease	v
Melibiose	+	D-Glucuronate	+	Growth at 37°C	v
		D-Galacturonate	+	Growth at 40°C	–,w

Molecular diagnostics. ITS restriction map based on NCBI U31848:

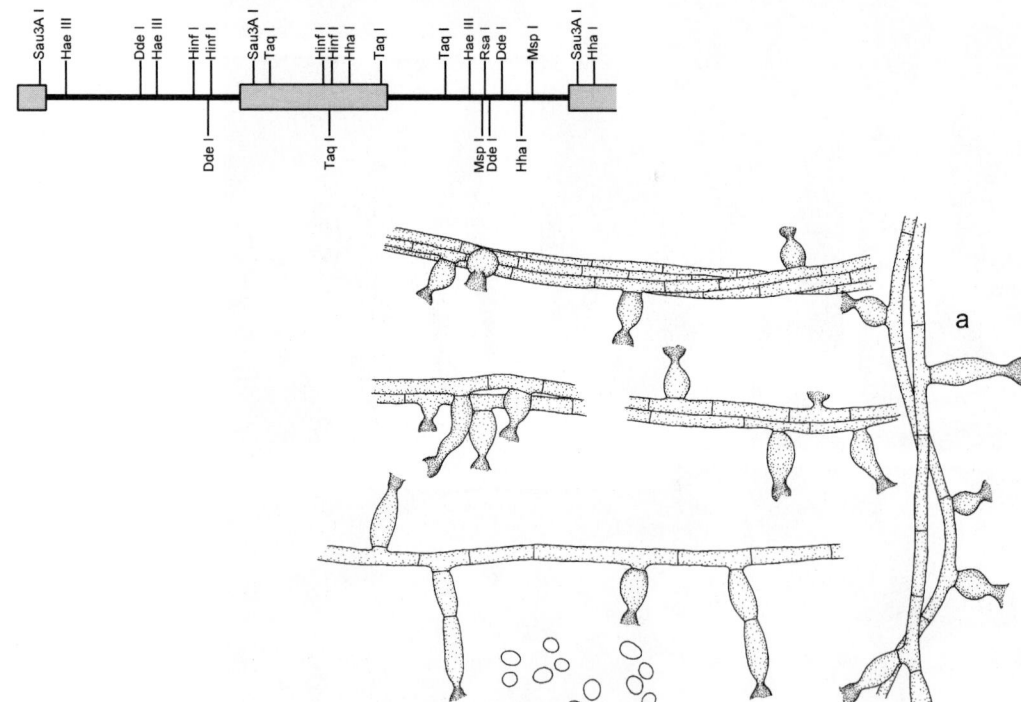

Phialophora verrucosa, CBS 138.67. a. Flask-shaped phialides on undifferentiated hyphae; b. conidia.

878

Pathogenicity. BSL-2. The species is one of the main agents of chromoblastomycosis in (sub)tropical regions, particularly in South-America (Velazquez *et al.*, 1976) and Japan (Fukushiro, 1983). Cases of mycetoma were reported by Pasarell *et al.* (1993) and Turiansky *et al.* (1995). Cases without muriform cells and hyperkeratosis were found in immunosuppressed or debilitated patients (Schnadig *et al.*, 1986). Phaeohyphomycoses were described by Duggan *et al.* (1995) and Tendolkar *et al.* (1998)

Reference. Cole & Kendrick (1973), Zweibel & Wang (1978), Yan *et al.* (1995), de Hoog *et al.* (1999b).

Antifungal susceptibility.

Antifungal	Mean MICs	Strains	Reference
AMB	0.36	25	McGinnis & Pasarell (1998b)
5FC	11.3	2	Tintelnot *et al.* (1999)
ITZ	0.07	17	McGinnis & Pasarell (1998a)
KTZ	1	2	Tintelnot *et al.* (1999)
TBF	0.03	17	McGinnis & Pasarell (1998a)
VCZ	0.12	25	McGinnis & Pasarell (1998b)

Nomenclature. *Phialophora verrucosa* Medlar - Mycologia 7: 203, 1915.

Phialophora macrospora Moore & Almeida - Ann. Mo. Bot. Gdn 23: 545, 1936 ≡ *Fonsecaea pedrosoi* (Brumpt) Negroni var. *phialophora* Carrión - Mycologia 34: 432, 1942 (name change).

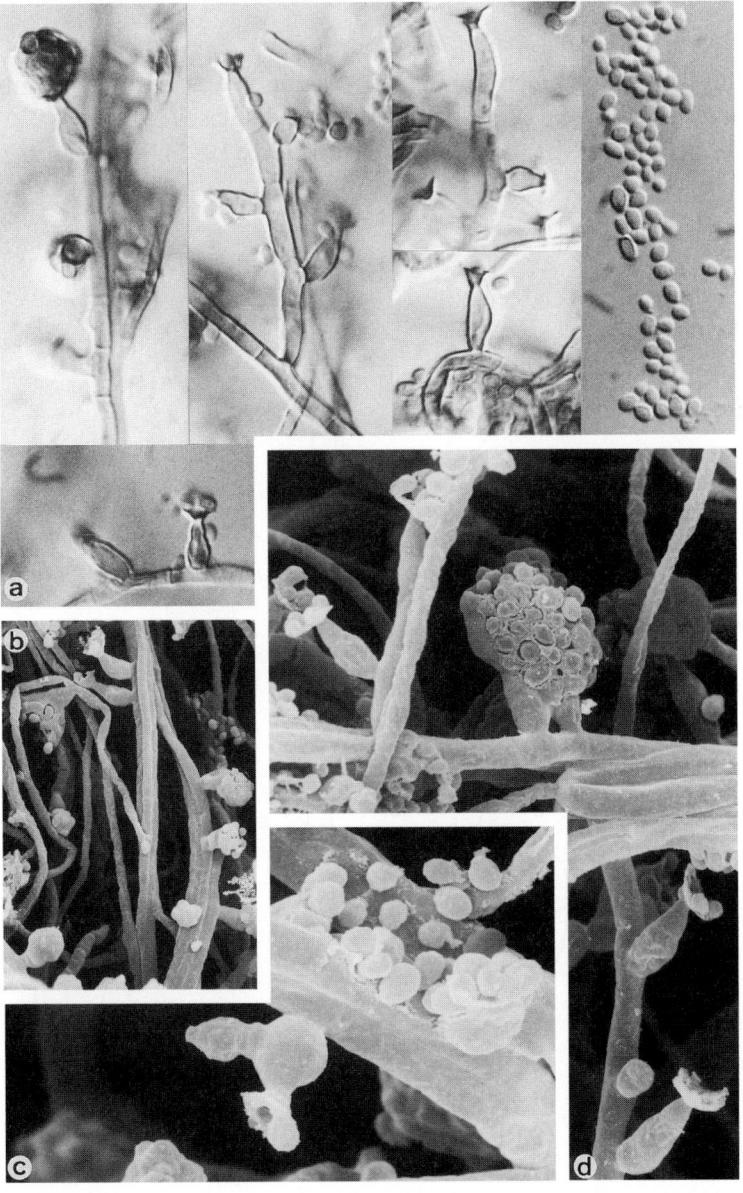

Phialophora verrucosa, CBS 138.67. Phialides and conidia. a. ×1055; b. ×955; c. ×3135; d. ×1780.

Hyphomycetes. Genus: *PLECTOSPORIUM*

Plectosporium tabacinum (v. Beyma) M.E. Palm *et al.*

Colony characteristics. Colonies (OA) growing rather slowly, whitish to beige, somewhat floccose; aerial mycelium generally sparse.

Microscopy. Conidiophores at first arising in the aerial mycelium as lateral phialides, later with sparse branching. Conidiogenous cells monophialidic. Macroconidia cylindrical, slightly curved with more or less pointed apex and wedge-shaped base, (0-) 1 (-3)-septate 12-16 × 3-4 μm. Microconidia absent.

Teleomorph. *Plectosphaerella cucumerina* (Lindfors) W. Gams (*Ascomycota, Euascomycetes, Polystigmatales: Phyllachoraceae*).

Molecular diagnostics. ITS restricton map based on NCBI L36640:

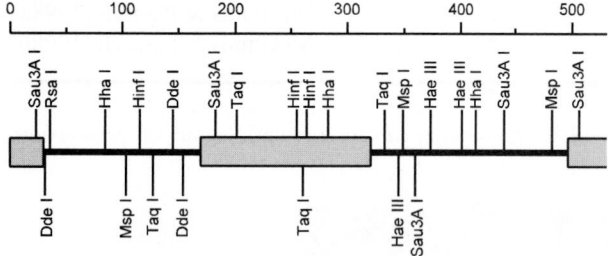

Pathogenicity. BSL-1. This is a very common soil fungus. A keratitis was reported by Simon *et al.* (2000) and black gill disease of crayfish from British fresh waters by Alderman & Polglase (1985).

References. Booth (1971), Gerlach & Nirenberg (1982), Palm *et al.* (1995), Rossman *et al.* (1999).

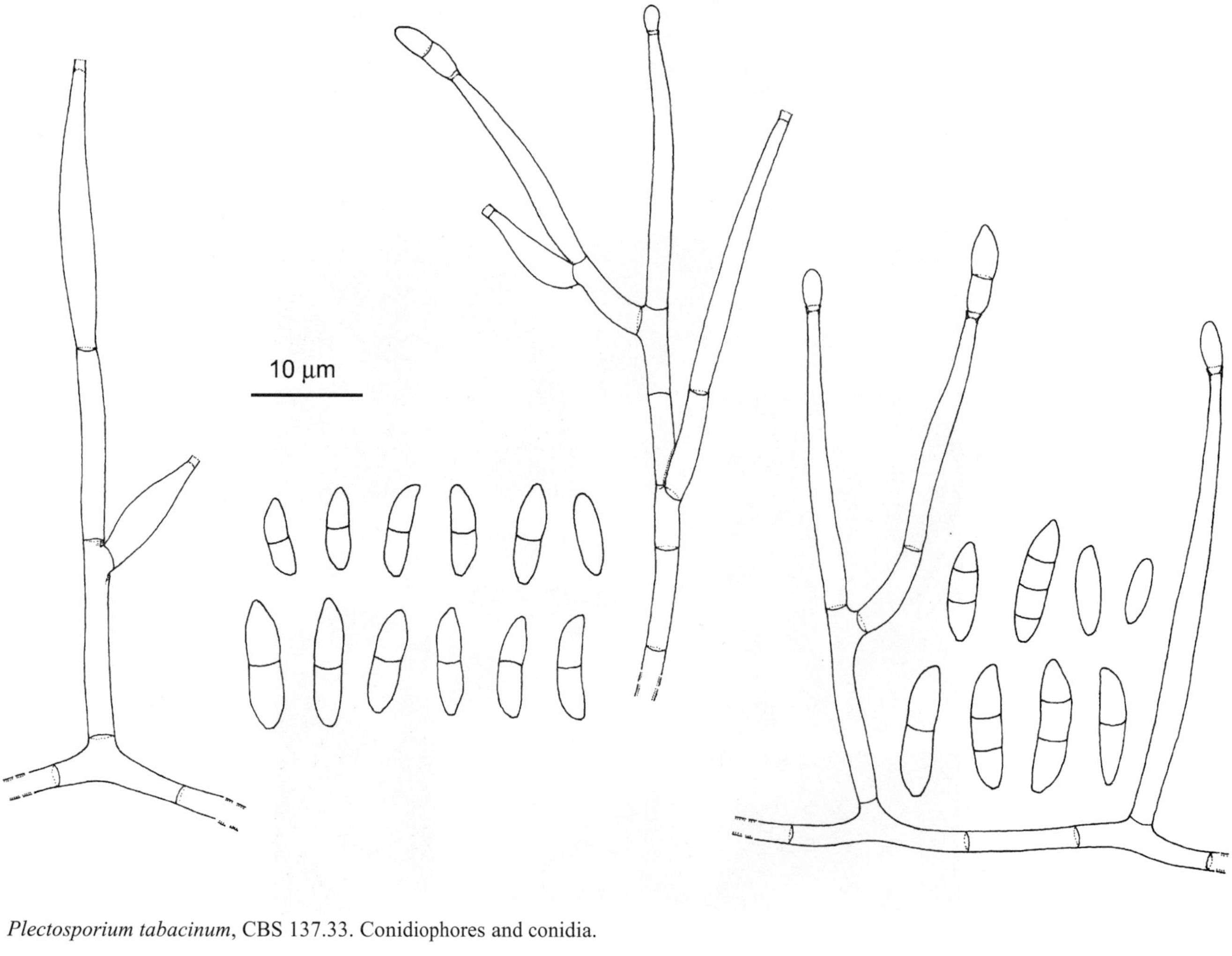

Plectosporium tabacinum, CBS 137.33. Conidiophores and conidia.

880

Nomenclature. *Cephalosporium tabacinum* van Beyma - Zentbl. Bakt., Abt. 2, 89: 240, 1933 =*Fusarium tabacinum* (van Beyma) W. Gams, *in* Gams & Gerlagh - Persoonia 5: 179, 1968 = *Plectosporium tabacinum* (van Beyma) M.E. Palm, W. Gams & Nirenberg - Mycologia 87: 399, 1995.

Venturia cucumerina Lindfors - Meddn. CentAnst. FörsVäs. JordbrOmråd, Stockholm 193-17: 1919 = *Plectosphaerella cucumerina* (Lindfors) W. Gams, *in* Domsch & Gams - Fungi Agric. Soils p. 160, 1972.

Plectosphaerella cucumeris Klebahn - Phytopath. Z. 1: 43, 1930.

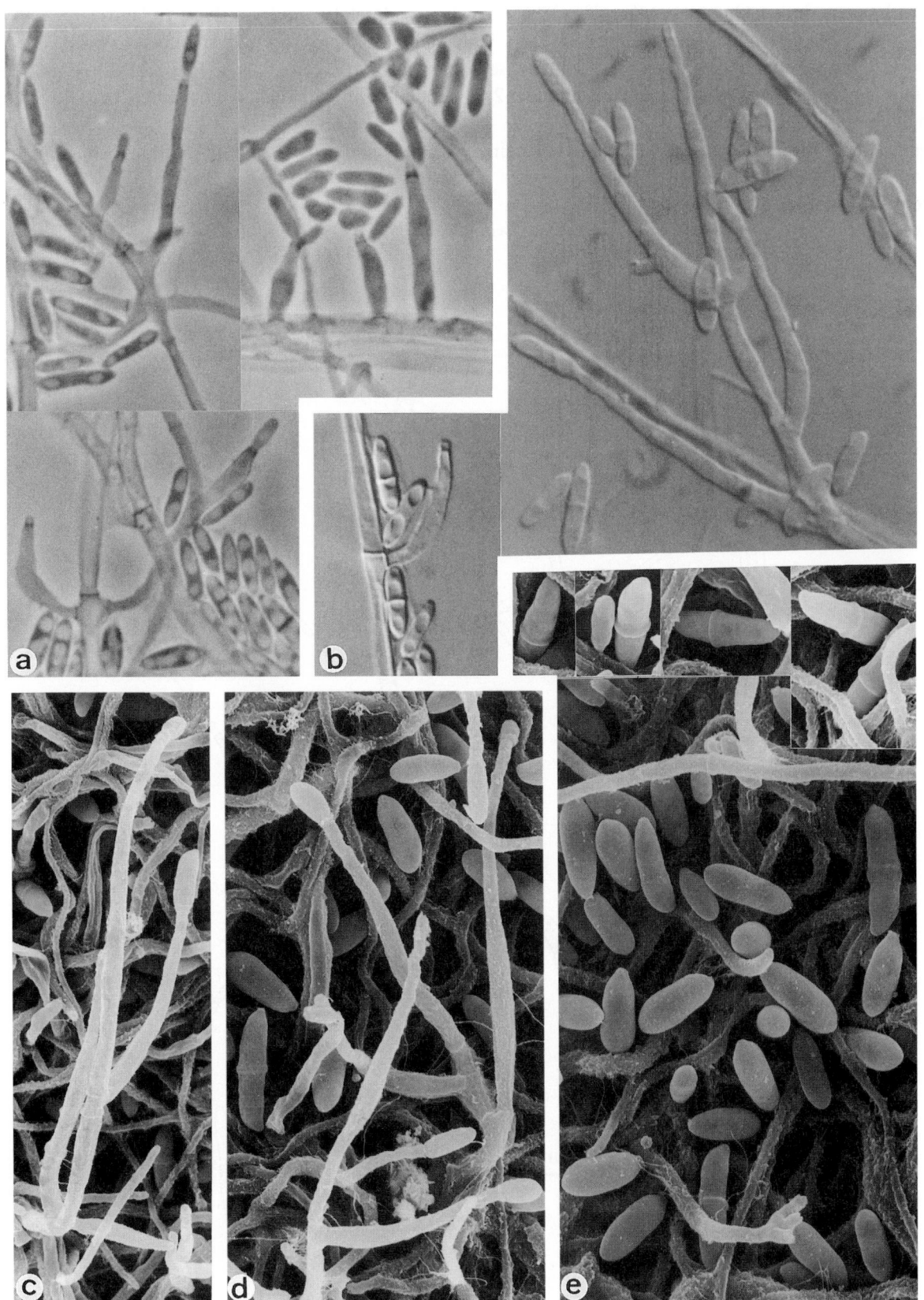

Plectosporium tabacinum, CBS 137.33. Conidiophores with phialides and conidia. a. ×1280; b. ×1600; c. ×2200; d. ×3000; e. ×3300.

Hyphomycetes. Genus: *POLYCYTELLA*

Polycytella hominis Campbell

Colony characteristics. Colonies (MEA 2%, 28°C) growing moderately rapidly, velvety to lanose.

Microscopy. Vegetative hyphae hyaline, 3-5 µm wide. Conidiophores undifferentiated. Conidiogenous cells integrated, terminal or intercalary. Conidia solitary, sessile or on short protrusions, cylindrical with a rounded apex and a funnel-shaped base, with a basal scar, hyaline, 2-13-septate, thin-walled except the apical cells which are pale brown and thick-walled, 22-92 × 3.2-5.5 µm.

Pathogenicity. BSL-2. The genus is known from a single strain which caused a white-grain mycetoma in an Indian patient (Campbell, 1987).

Reference. Campbell (1987).

Nomenclature. *Polycytella hominis* Campbell - J. Med. Vet. Mycol. 25: 302, 1987.

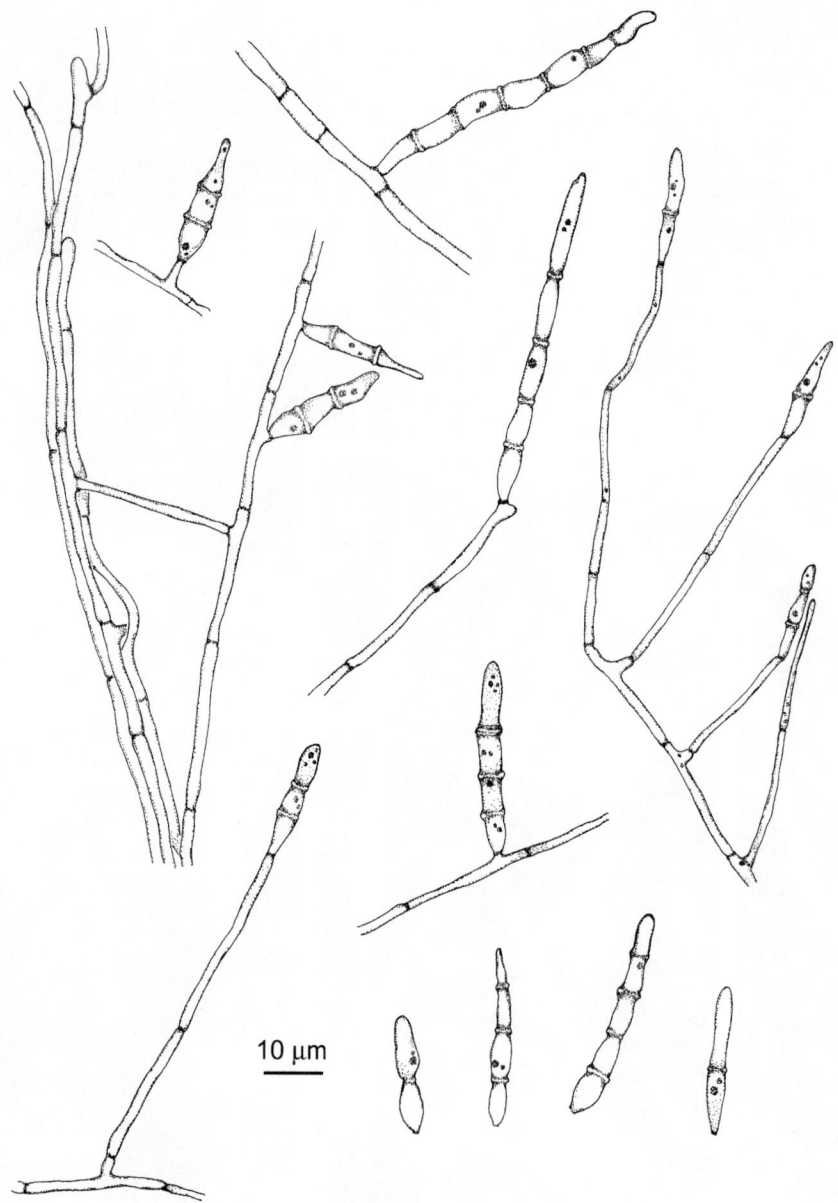

10 µm

Polycytella hominis, NCPF 2230. Conidiogenous cells and conidia.

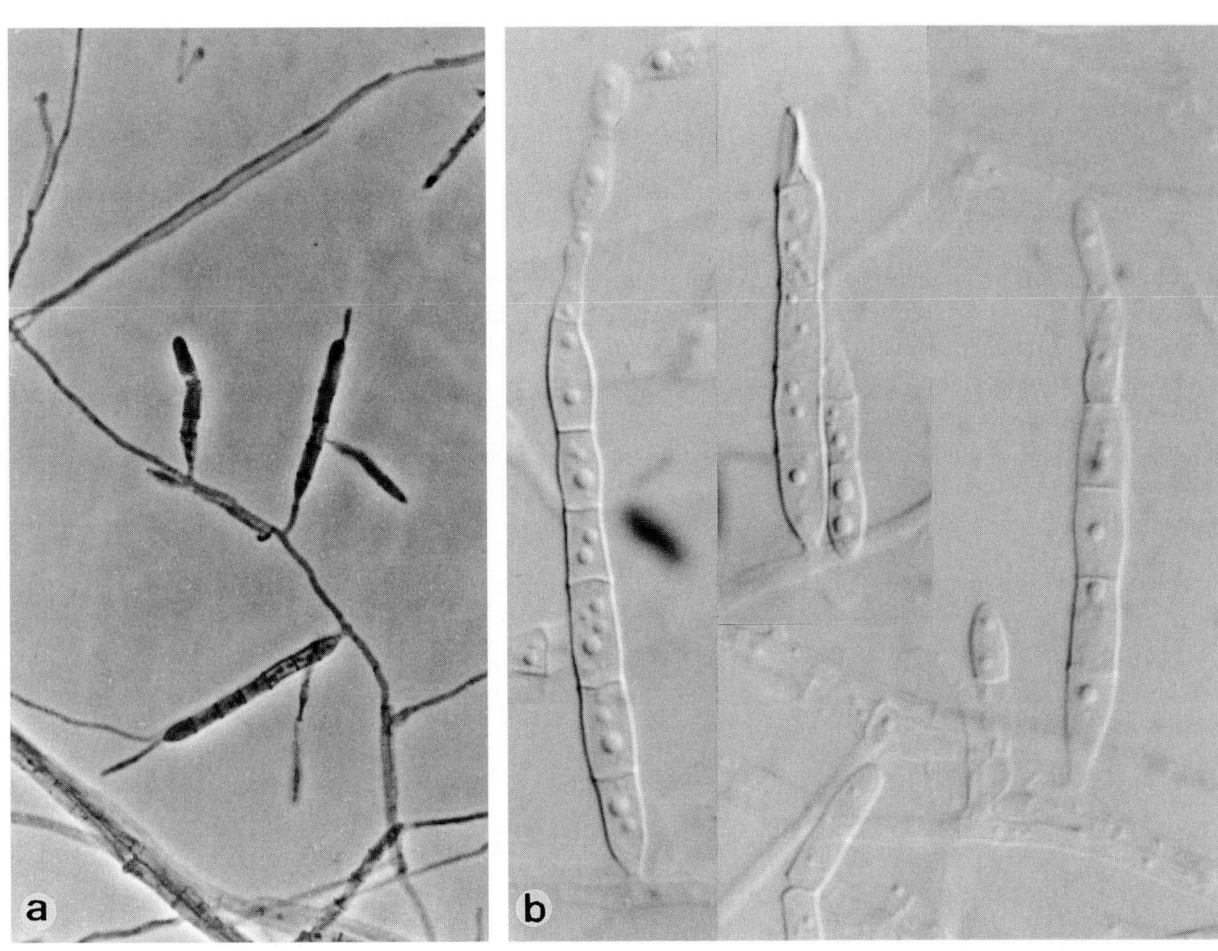

Polycytella hominis, NCPF 2230. Conidiogenous cells and conidia. a. ×640; b. ×1600.

Hyphomycetes. Genus: *POLYPAECILUM*

Polypaecilum insolitum G. Smith

Colony characteristics. Colonies (OA) with very slow growth, velvety, greyish-green; reverse blackish-olivaceous.
Microscopy. Conidiophores erect, straight or flexuous, pale brown, smooth-walled, up to 150 µm long, repeatedly di- or trichotomously branched; branches without septa, tube-like, swollen or tapered. Each terminal scar produces a series of conidia in basipetal order. Conidia spherical, subspherical or pyriform, with a truncate base, pale brown, unicellular, 4-7 µm diam, smooth-walled or verruculose, in chains. Terminal chlamydospores occasionally present.
Pathogenicity. BSL-1. Pulmonary infection (Coutelen *et al.*, 1955), otomycosis (Yamashita & Yamashita, 1972), nail infection (Piontelli & Toro, 1989).
Reference. Ellis (1971).

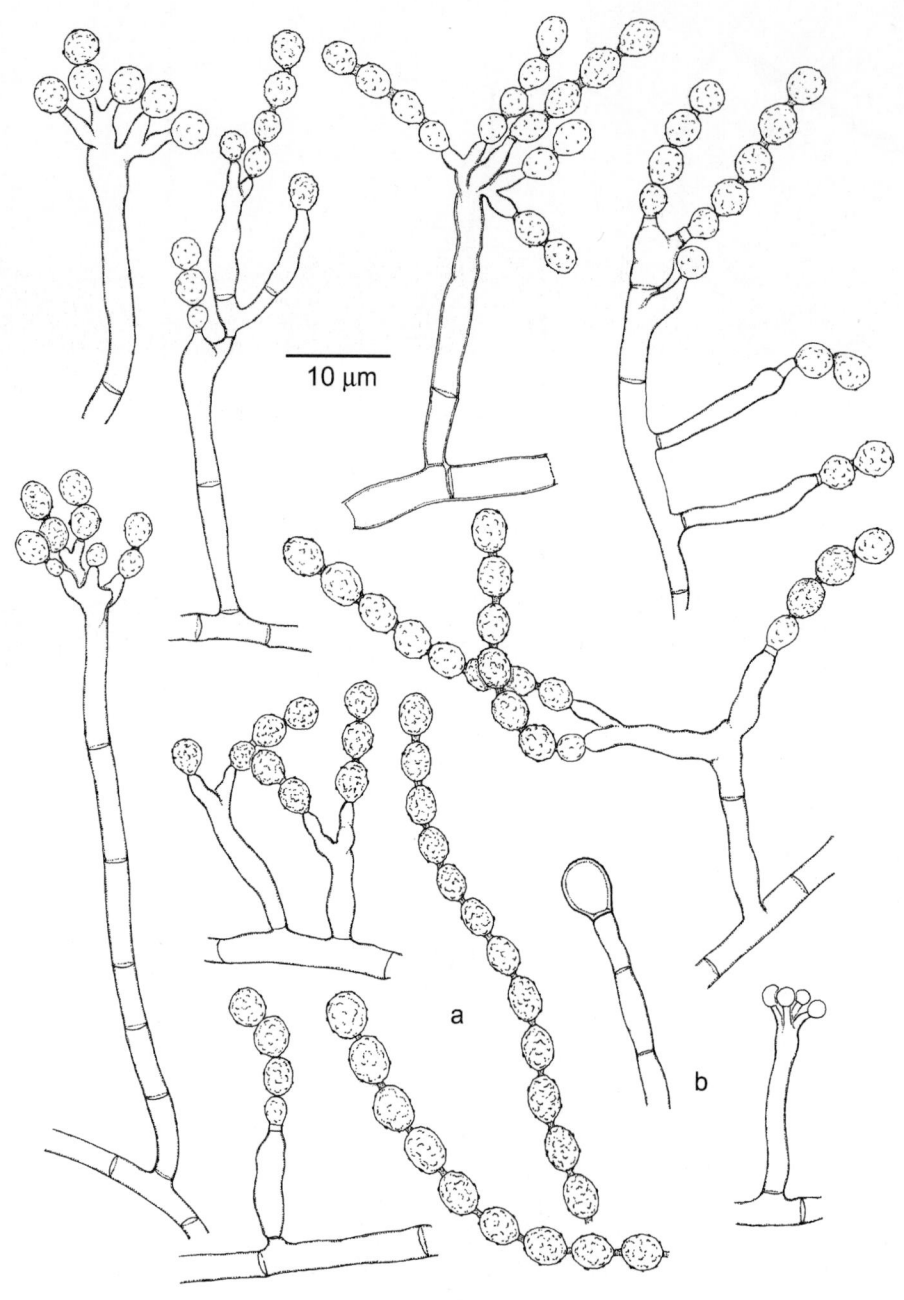

Polypaecilum insolitum, CBS 384.81. a. Conidiophores and conidia; b. chlamydospore.

Nomenclature. *Scopulariopsis insolita* Coutelen, Biguet, Cochet, Mullet & Doby-Dubois - Annls Parasit. Hum. Comp. 30: 410, 1955 (invalid).
  Scopulariopsis divaricata Yamashita - J. Otolaryngol. Jpn. 59: 129, 1956 (invalid).
  Scopulariopsis divaricata Yamashita var. *alba* Yamashita - J. Otolaryngol. Jpn. 59: 129, 1956 (invalid).
  Polypaecilum insolitum G. Smith - Trans. Br. Mycol. Soc. 44: 437, 1961.

Polypaecilum insolitum, CBS 384.81. Conidiophores and conidia. a. ×1600; b. ×2350; c. ×3100; d. ×5400.

Hyphomycetes. Genus: *PSEUDOMICRODOCHIUM*

Pseudomicrodochium suttonii Ajello *et al.*

Colony characteristics. Colonies (OA) restricted, grey to olivaceous black, somewhat moist.
Microscopy. Hyphae 1.5-2.0 µm wide, brownish. Conidiophores absent. Conidiogenous cells intercalary or lateral, then often without basal septum, cylindrical, with an indistinct collarette, producing subsequent conidia often in more or less sympodial order. Conidia straight to falcate, acicular, brown, smooth-walled, 18-30 × 1.0-1.2 µm, 3-8-septate.
Differential diagnosis. The genus *Pseudomicrodochium* is very close to the genus *Cyphellophora;* see p. 620.

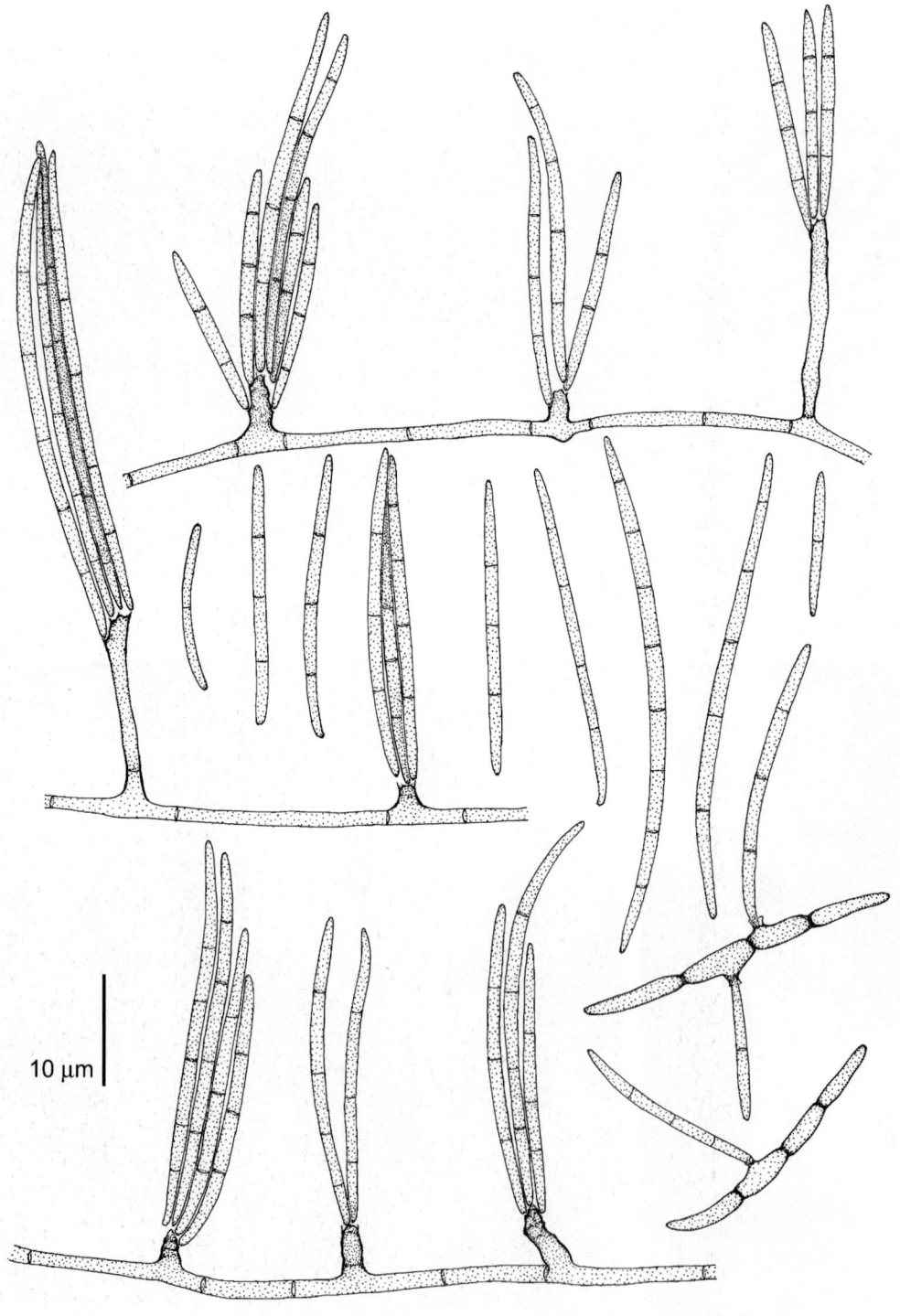

10 µm

Pseudomicrodochium suttonii, IMI 233463. Conidiogenous cells and conidia.

886

Pathogenicity. BSL-2. Subcutaneous lesion in a sarcoid patient (Schell & Perfect, 1995); ear lesion in a dog (Ajello *et al.*, 1980).

References. Ajello *et al.* (1980), Castañeda Ruíz *et al.* (1998), Torres Nunes *et al.* (1999).

Nomenclature. *Pseudomicrodochium suttonii* Ajello, Padhye & Payne - Mycotaxon 12: 133, 1980.

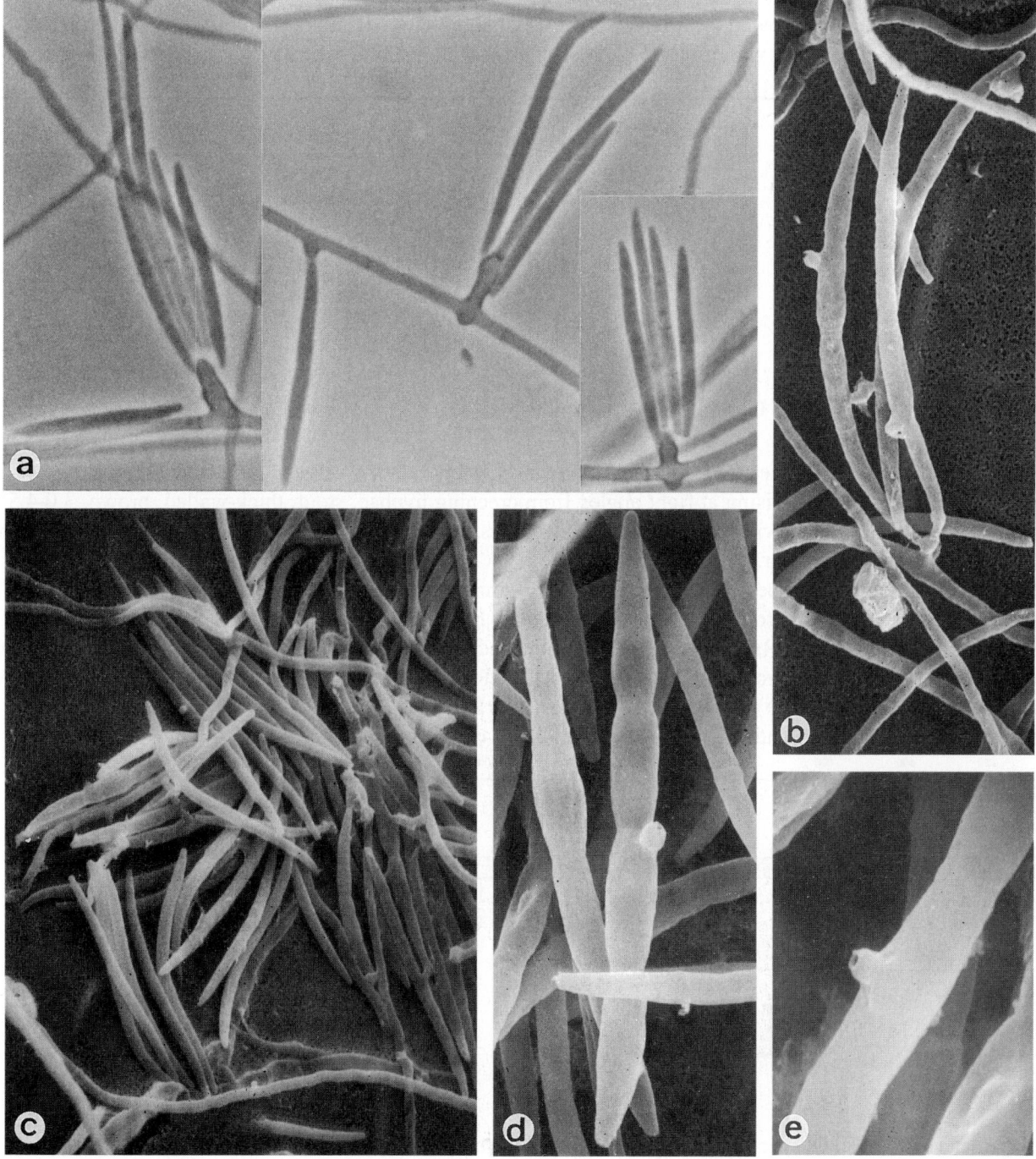

Pseudomicrodochium suttonii, IMI 233463. Conidiogenous cells and conidia. a. ×1600; b. ×3200; c. ×1935; d. ×4500; e. ×10000.

Hyphomycetes. Genus: *RAMICHLORIDIUM*

Generic description. Colonies growing rather rapidly, smooth, farinose or velvety, brown or olivaceous-green, often with orange or yellow soluble pigments. Conidiophores erect, brown, apically with small denticles on which conidia are produced in sympodial order. Conidia one-celled, hyaline to pale brown.
Teleomorph. *Capronia* (*Ascomycota, Euascomycetes, Chaetothyriales: Herpotrichiellaceae*).
Reference. De Hoog (1977).

Key to the treated species of Ramichloridium:

1a. Conidia (sub)hyaline, narrowed towards the base, on small, pointed denticles . . . . **R. schulzeri (891)**
1b. Conidia brown, ellipsoidal, with truncate basal scar 1 μm wide, on blunt denticles . . . . **R. mackenziei (888)**

Ramichloridium mackenziei Campbell & Al-Hedaithy

Colony characteristics. Colonies (OA, 30°C) growing moderately rapidly, velvety, olivaceous-brown.
Microscopy. Conidiophores arising at right angles from creeping hyphae, stout, thick-walled, brown, 3.0-4.5 μm wide, 10-25 μm long, apically with short-cylindrical denticles. Conidia brown, ellipsoidal, 8.5-12.0 × 4-5 μm, with prominent, 1 μm wide basal scar.
Differential diagnosis. The species is morphologically similar to *R. obovoideum* (Mats.) de Hoog, which has originally been described from forest humus and dead wood, but phylogenetically the two taxa are widely apart (E. Guého, pers. comm.). Naim-ur-Rahman *et al.* (1988) published a cerebral case ascribed to *R. obovoideum* which was identical to the cases reported from *R. mackenziei*. Its affiliation to the ascomycete family *Herpotrichiellaceae* was proven by Masclaux *et al.* (1995) on the basis of partial 26S rRNA sequencing.
Molecular diagnostics. SSU and ITS restriction maps based on CBS 650.93 and NCBI AF050288, resp.:

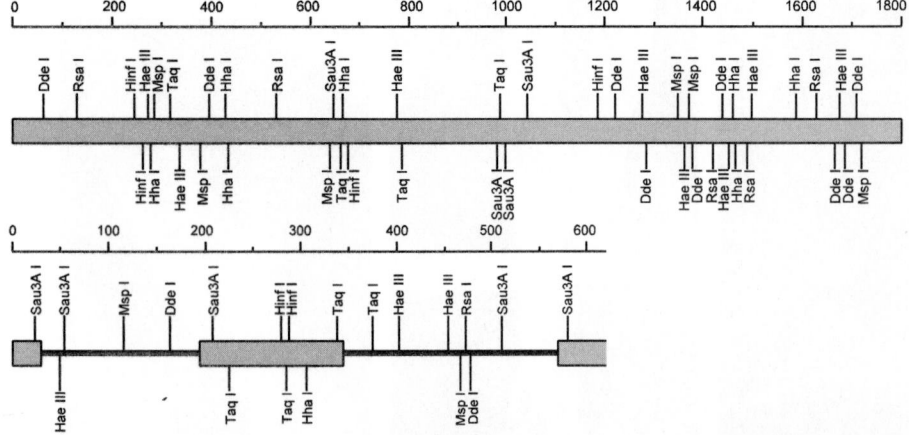

Pathogenicity. BSL-3. The species was reported to cause fatal brain lesions in the Middle-East (Campbell & Al-Hedaithy, 1993), where it is the most frequent cause of intracranial infections (Jamjoom *et al.*, 1995). Sutton *et al.* (1998) and Podnos *et al.* (1999) published American cases in residents from the Middle-East.

Antifungal susceptibility.

Antifungal	Mean MICs	Strains	Reference
AMB	2.8	2	Tintelnot *et al.* (1999)
AMB	0.5	1	Sutton *et al.* (2000)
AMB	3.16	2	Al-Abdely *et al.* (2000)
5FC	8	1	Sutton *et al.* (2000)
5FC	8	2	Tintelnot *et al.* (1999)
FLZ	22.3	2	Al-Abdely *et al.* (2000)
FLZ	16	1	Sutton *et al.* (2000)
ITZ	0.008	2	Al-Abdely *et al.* (2000)
ITZ	≤0.015	1	Sutton *et al.* (2000)
KTZ	0.086	2	Tintelnot *et al.* (1999)

Nomenclature. *Ramichlorium mackenziei* Campbell & Al-Hedaithy - J. Med. Vet. Mycol. 31: 330, 1993.

10 μm

Ramichloridium mackenziei, CBS 650.93. Conidial apparatus and conidia.

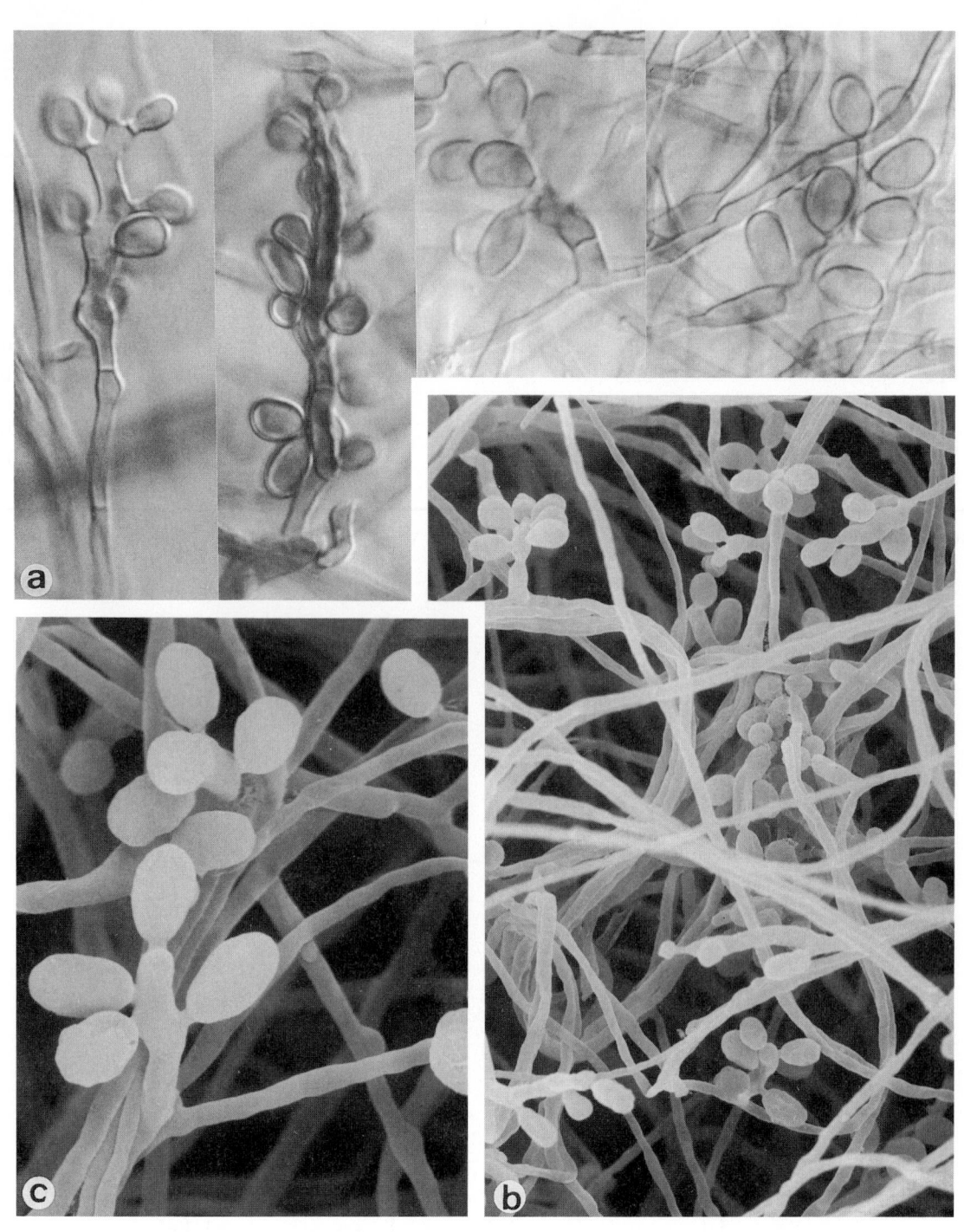

Ramichloridium mackenziei, CBS 650.93. Conidiophores and conidia. a. ×1600; b. ×1300; c. ×2600.

890

Ramichloridium schulzeri (Sacc.) de Hoog

Colony characteristics. Colonies (OA) growing moderately rapidly, consisting of a rather compact, flat, submerged mycelium, pale orange, locally with some powdery, brownish aerial mycelium; reverse pink to orange.

Microscopy. Conidiophores erect, straight, unbranched, thick-walled, reddish-brown, up to 250 μm high, gradually becoming paler towards the apex, of variable length, elongating sympodially during conidiogenesis, with scattered, pimple-shaped conidium-bearing denticles which have unpigmented scars. Conidia subhyaline, smooth-walled or slightly rough-walled, ellipsoidal, obovoidal or fusiform, 6.5-10.0 × 3-4 μm, usually with an acuminate base and unpigmented scars.

Pathogenicity. BSL-1. Saprobe. A case of 'golden tongue' syndrome in a leukemic patient (Rippon *et al.*, 1985).

Reference. De Hoog (1977).

Nomenclature. *Psilobotrys schulzeri* Saccardo, *in* Schulzer & Saccardo - Hedwigia 23: 126, 1884 ≡ *Ramichloridium schulzeri* (Saccardo) de Hoog - Stud. Mycol. 15: 64, 1977.

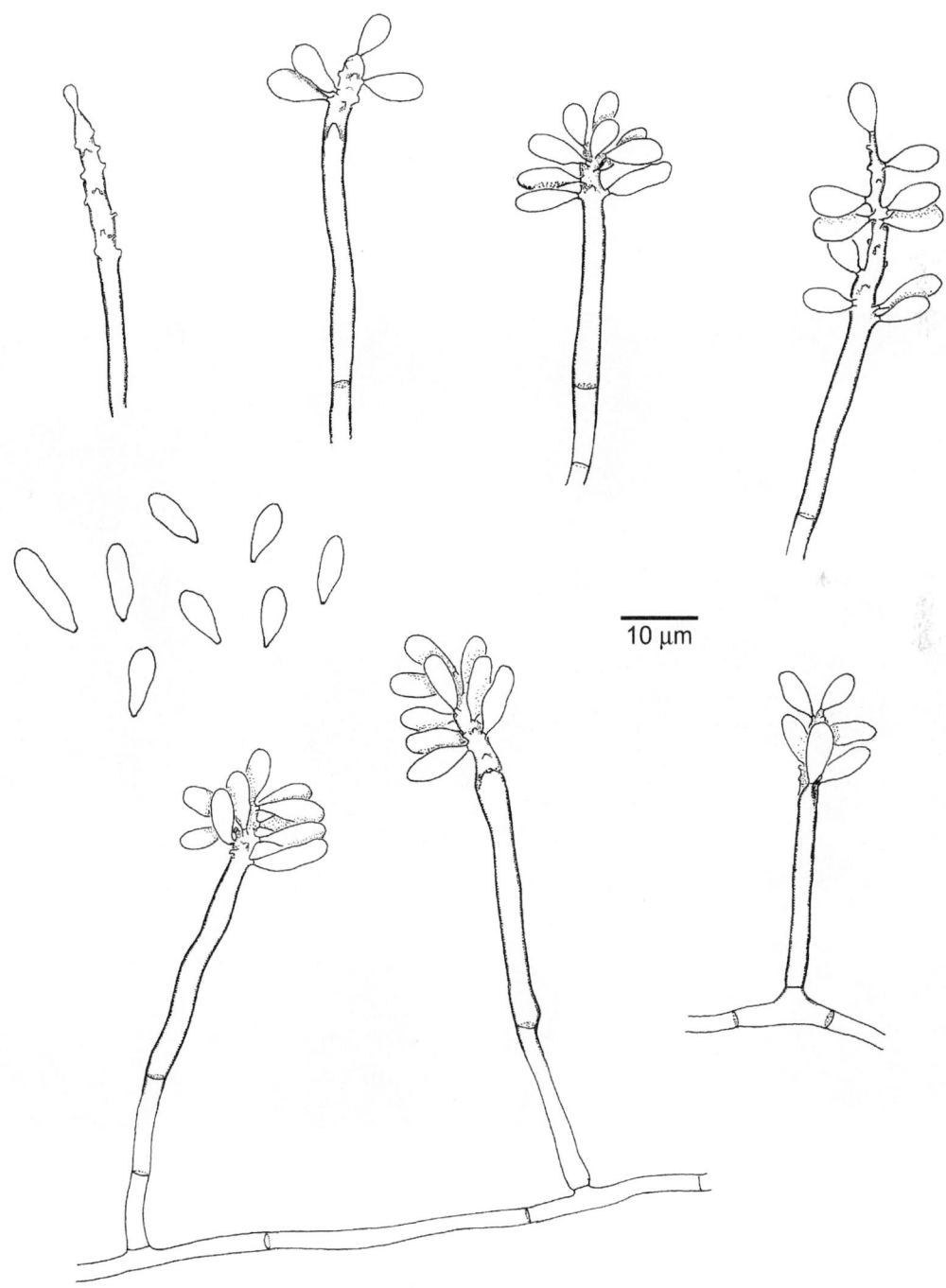

Ramichloridium schulzeri, ATCC 16310. Conidiophores and conidia.

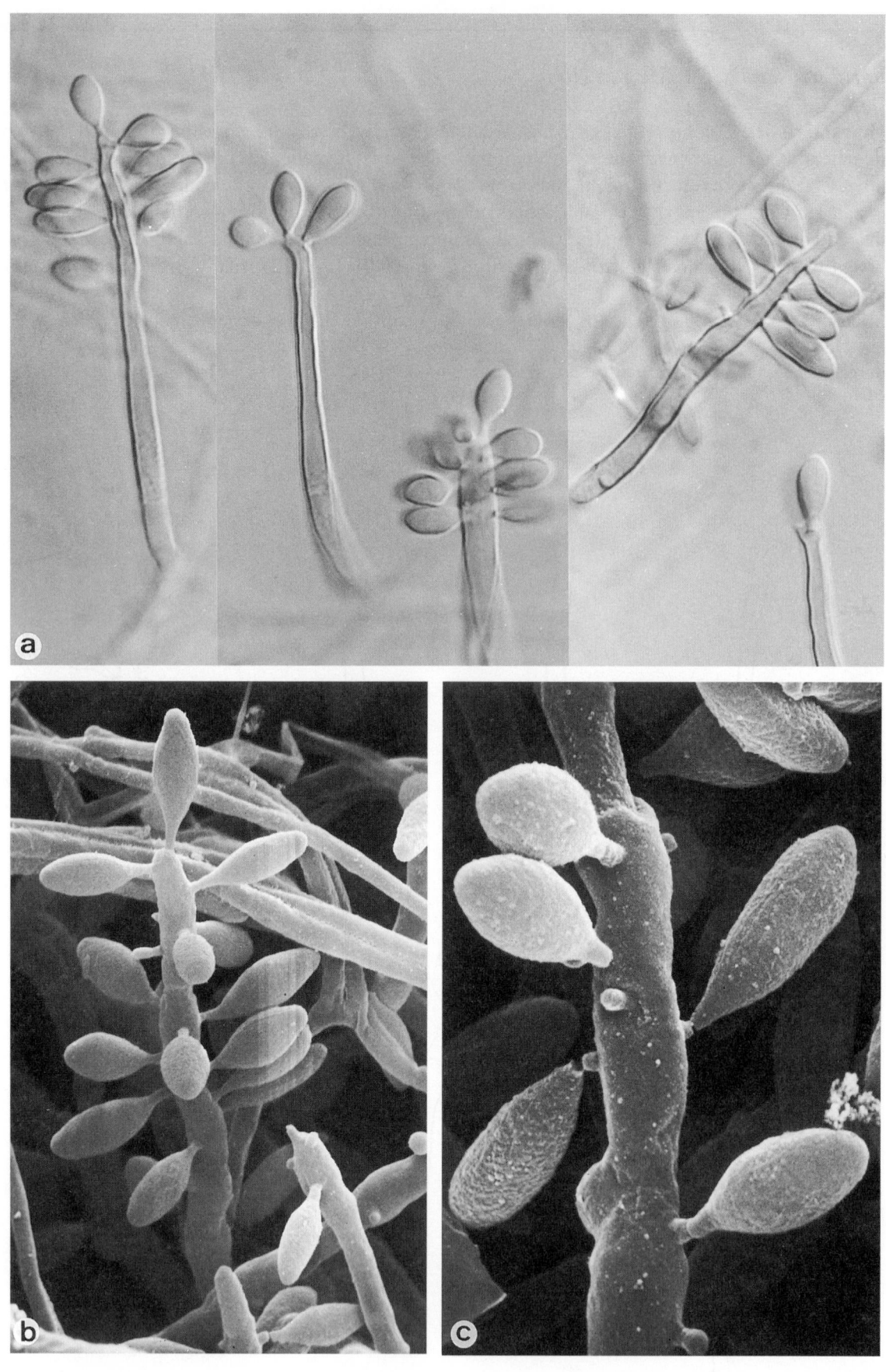

Ramichloridium schulzeri, ATCC 16310. Conidiophores and conidia. a. ×1600; b. ×3200; c. ×7000.

Hyphomycetes, black yeasts and relatives. Genus: *RHINOCLADIELLA*

Generic description. Colonies restricted, velvety, lanose or nearly smooth, grey, green or olivaceous brown. Hyphae pale olivaceous. Conidiophores slightly differentiated, suberect, usually branched, pale to dark brown. Conidiogenous cells intercalary or free, cylindrical, in the apical part with conidium-bearing denticles with unpigmented scars. Conidia hyaline to subhyaline, one-celled, smooth-walled, single or in short chains. Budding cells and an accompanying *Exophiala* state may be present.

References. De Hoog (1977, 1983), Schell *et al.* (1983).

Key to the treated species of Rhinocladiella:

1a. Conidia short-cylindrical, 1.2-1.8 µm wide .... ***R. atrovirens*** (895)
1b. Conidia ellipsoidal to clavate, 1.8-2.4 µm wide .... ***R. aquaspersa*** (893)

Rhinocladiella aquaspersa (Borelli) Schell *et al.*

Colony characteristics. Colonies (OA) restricted, velvety, elevated, olivaceous-black; reverse dark olivaceous.

Microscopy. Hyphae pale olivaceous, smooth- or slightly rough-walled. Conidiophores straight, unbranched, thick-walled, dark brown. Conidiogenous cells usually terminal, cylindrical, 9-23 µm long, with crowded, slightly prominent, unpigmented or faintly pigmented conidial scars. Conidia subhyaline, smooth- and thin-walled, one- or occasionally two-celled, ellipsoidal to clavate, 4.5-7.5 × 1.8-2.4 µm. Annellides occasionally present.

Pathogenicity. BSL-2. Rare cases of chromoblastomycosis (Borelli, 1972; Arango *et al.*, 1998; Pérez-Blanco *et al.* (1998).) have been published.

Reference. Schell *et al.* (1983).

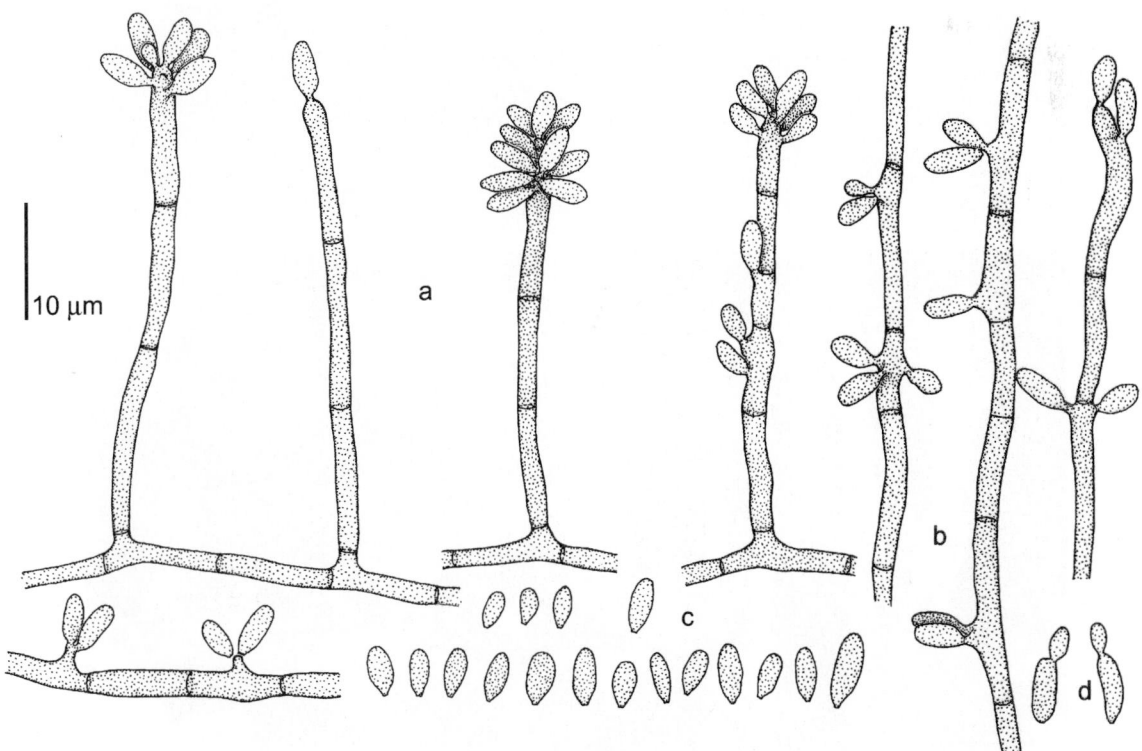

Rhinocladiella aquaspersa, FMC 295. a. Conidiophores; b. intercalary conidiogenous cells; c. conidia; d. budding cells.

Antifungal susceptibility.

Antifungal	Mean MICs	Strains	Reference
AMB	0.25	2	McGinnis & Pasarell (1998b)
ITZ	0.04	2	McGinnis & Pasarell (1998b)
VCZ	0.09	2	McGinnis & Pasarell (1998b)

Nomenclature. *Acrotheca aquaspersa* Borelli - Acta. Cient. Venez. 23: 195, 1972 ≡ *Rhinocladiella aquaspersa* (Borelli) Schell, McGinnis & Borelli - Mycotaxon 17: 343, 1983.

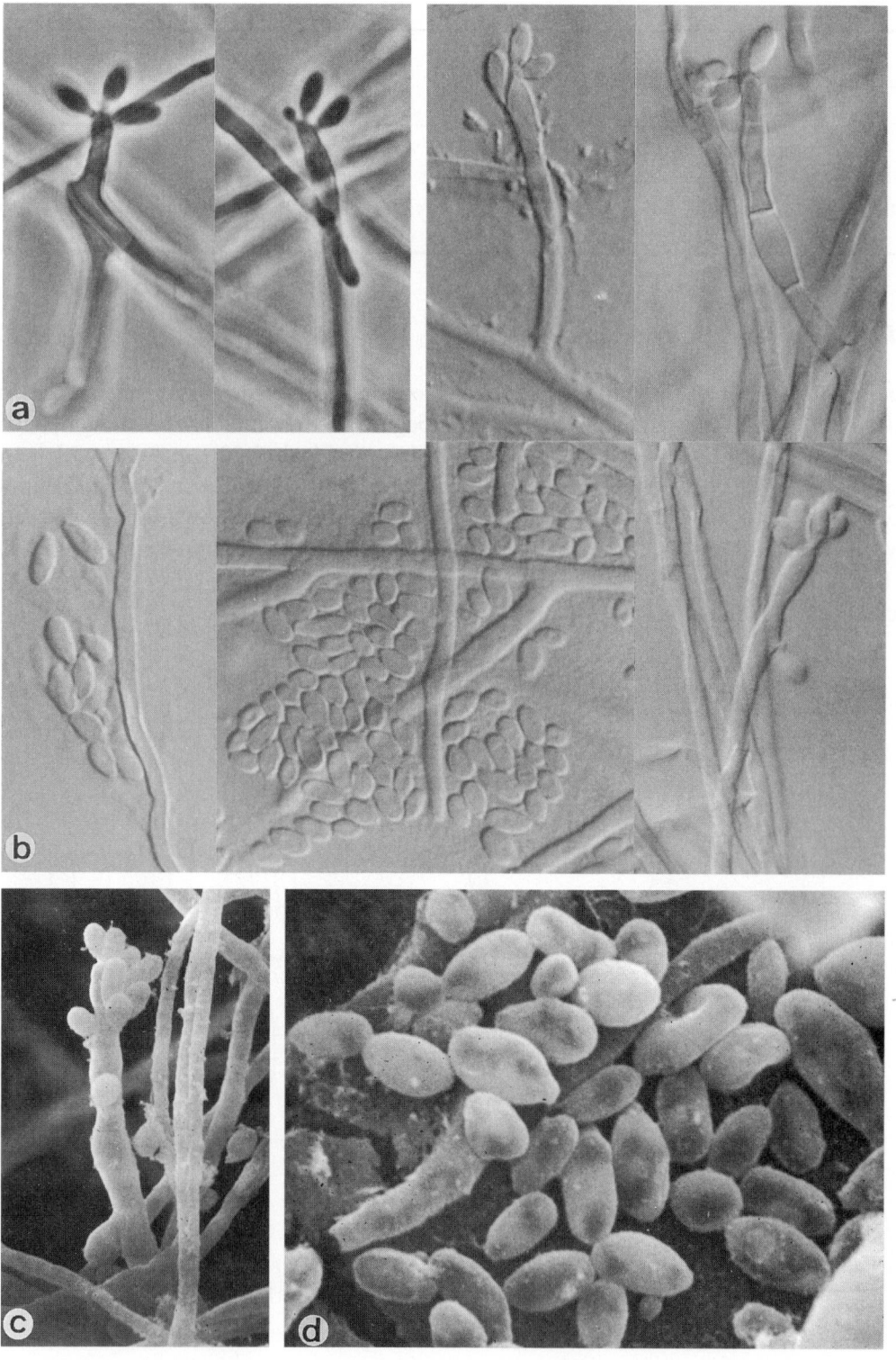

Rhinocladiella aquaspersa, FMC 295. Conidiophores and conidia. a. ×1280; b. ×1600; c. ×2800; d. ×5300.

Rhinocladiella atrovirens Nannf.

Colony characteristics. Colonies (OA) restricted, velvety or lanose, olivaceous, often slightly mucous at the centre; reverse dark olivaceous green to blackish.

Microscopy. Conidiophores short, brown, thick-walled. Conidiogenous cells cylindrical, intercalary or free, 9-19 × 1.6-2.2 µm; denticulate rachis up to 15 µm long, with crowded, flat or butt-shaped, unpigmented conidial denticles. Conidia hyaline, thin- and smooth-walled, short-cylindrical, with truncate basal scars, 3.7-5.5 × 1.2-1.8 µm. Budding cells, if present, hyaline, thin-walled, broadly ellipsoidal, 3.0-4.3 × 1.7-2.5 µm. Germinating cells inflated, spherical to subspherical, 4.5-6.0 (-7.0) µm. An annellidic *Exophiala* synanamorph may be present.

Physiology. Intolerant to benomyl.

Molecular diagnostics. SSU and ITS restriction maps based on CBS 688.76 and NCBI AF050289:

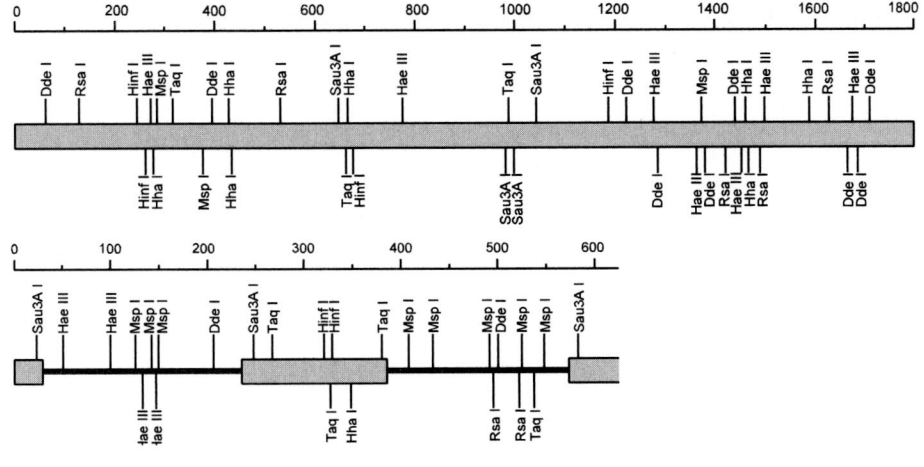

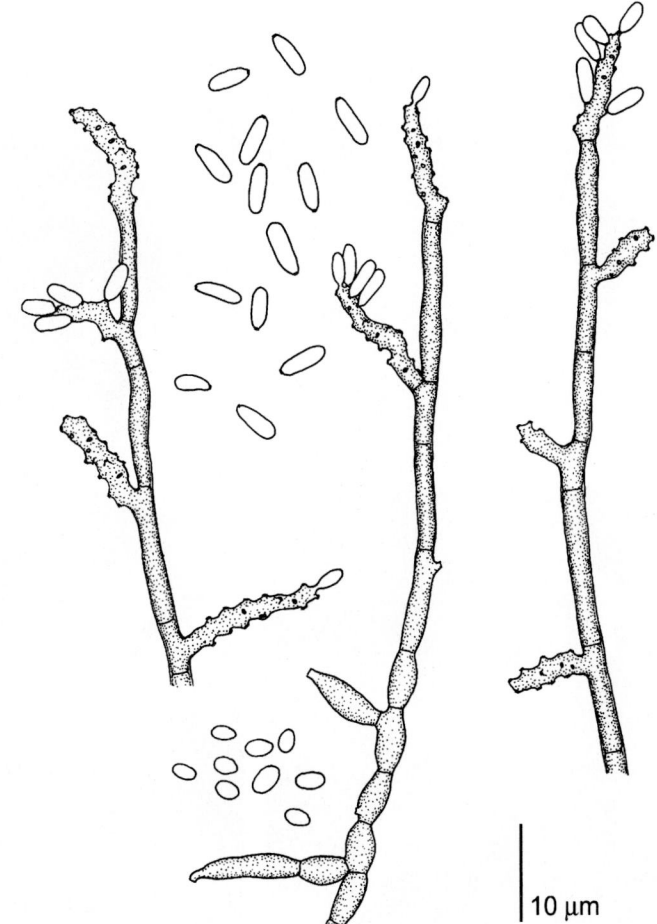

Rhinocladiella atrovirens, CBS 559.72. Conidiophores and conidia.

895

Pathogenicity. BSL-1. The species occurs on rotten wood. Documented clinical cases are an infection of the central nervous system (del Palacio-Hernanz *et al.*, 1989), mycetoma (Develoux *et al.*, 1994) and chromoblastomycosis (Resende *et al.*, 2000). The species is sometimes isolated from clinical specimens, such as cornea and sputum.
Reference. De Hoog (1977).
Antifungal susceptibility.

Antifungal	Mean MICs	Strains	Reference
AMB	0.12	3	McGinnis & Pasarell (1998b)
ITZ	0.04	3	McGinnis & Pasarell (1998b)
VCZ	0.01	3	McGinnis & Pasarell (1998b)

Nomenclature. *Rhinocladiella atrovirens* Nannfeldt, *in* Melin & Nannfeldt - Svenska Skogsvför. Tidskr. 32: 462, 1934.

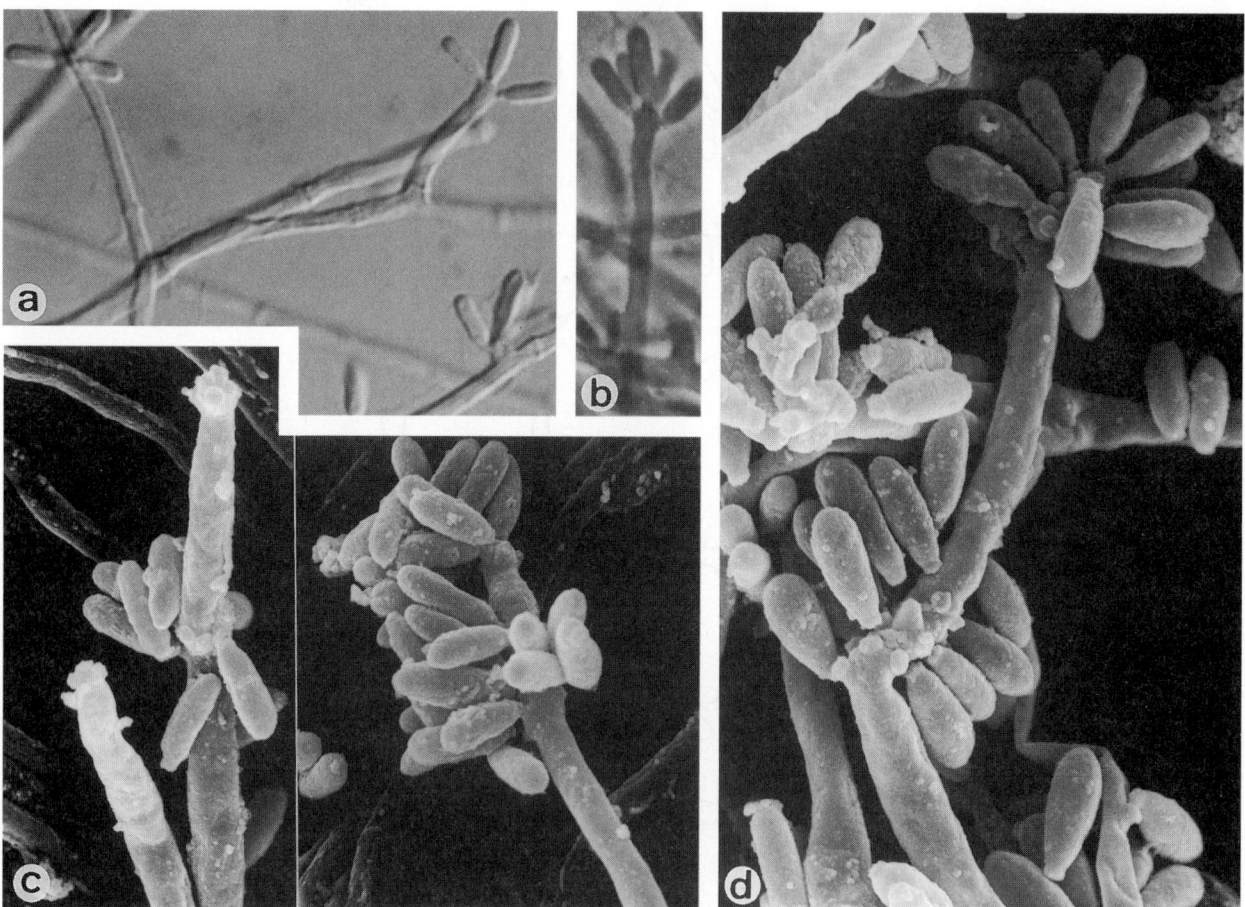

Rhinocladiella atrovirens, CBS 559.72. Conidiophores and conidia. a. ×1600; b. ×1280; c. ×4900; d. ×5200.

Hyphomycetes. Genus: *SARCINOMYCES*

Sarcinomyces phaeomuriformis Matsumoto *et al.*

Colony characteristics. Colonies (PDA 30°C) either with moderate, black, slimy growth, or growing very slowly, raised, forming an irregular, granular, folded, friable, mulberry-like mass.

Microscopy. Hyphae absent. Thallus initially yeast-like, either remaining yeast-like, or converting into aggregates of sclerotic cells, finally forming irregular bodies. Cells clumps thick-walled, muriform, spherical or irregular in shape, 12-20 μm diam. Chains of budding cells with a broad base are formed from unicellular elements. Budding multilateral.

Physiology.

Growth:						
D-Glucose	+	Melibiose	w	D-Gluconate	w
D-Galactose	+	Lactose	w	D-Glucuronate	w
L-Sorbose	+	Raffinose	w	D-Galacturonate	−
D-Glucosamine	−	Melezitose	+	DL-Lactate	−
D-Ribose	−	Inulin	−	Succinate	−
D-Xylose	+	Soluble starch	w	Citrate	−
L-Arabinose	+	Glycerol	+	Methanol	−
D-Arabinose	+	*meso*-Erythritol	+	Ethanol	w
L-Rhamnose	+	Ribitol	+	Nitrate	−
Sucrose	+	Xylitol	w	Nitrite	−
Maltose	+	L-Arabinitol	w	Ethylamine	+
α,α-Trehalose	+	D-Glucitol	w	L-Lysine	+
methyl-α-D-Glucoside	+	D-Mannitol	+	Cadaverine	+
Cellobiose	+	Galactitol	−	Creatine	−
Salicin	+	*myo*-Inositol	−	Creatinine	−
		Glucono-δ-lactone	−	0.1% Cycloheximide	+

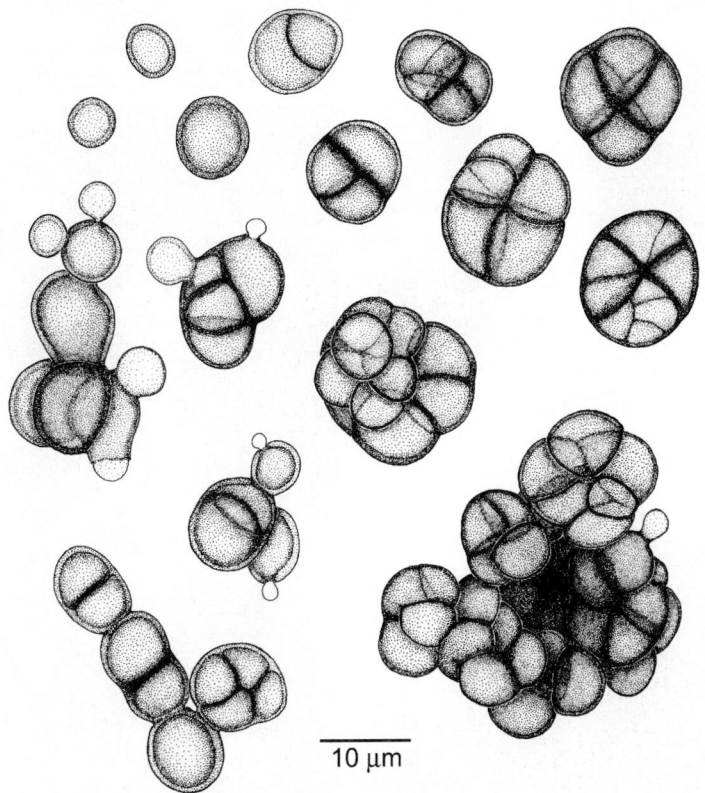

Sarcinomyces phaeomuriformis, CBS 131.88. Meristematic cells and budding cells.

897

Differential diagnosis. Yeast-like strains are culturally and morphologically indistinguishable from *Exophiala dermatitidis* (p. 652); also the physiological profiles of the two species are identical. Such cultures can be separated only by molecular methods (Uijthof *et al.*, 1998).

Molecular diagnostics. SSU and ITS restriction map based on NCBI X80710 and CBS 131.88:

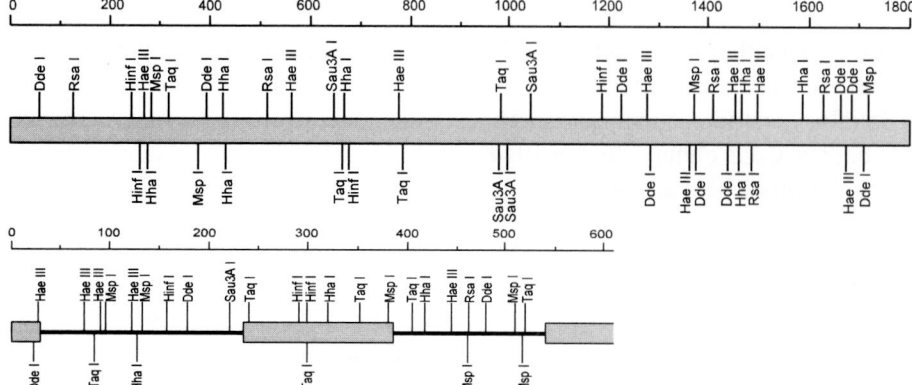

Pathogenicity. BSL-2. Skin infections (Mori *et al.*, 1961; Take, 1980; Matsumoto *et al.*, 1986).

References. Matsumoto *et al.* (1986), Uijthof *et al.* (1998).

Nomenclature. *Sarcinomyces phaeomuriformis* Matsumoto, Padhye, Ajello & McGinnis - J. Med. Vet. Mycol. 24: 396, 1986.

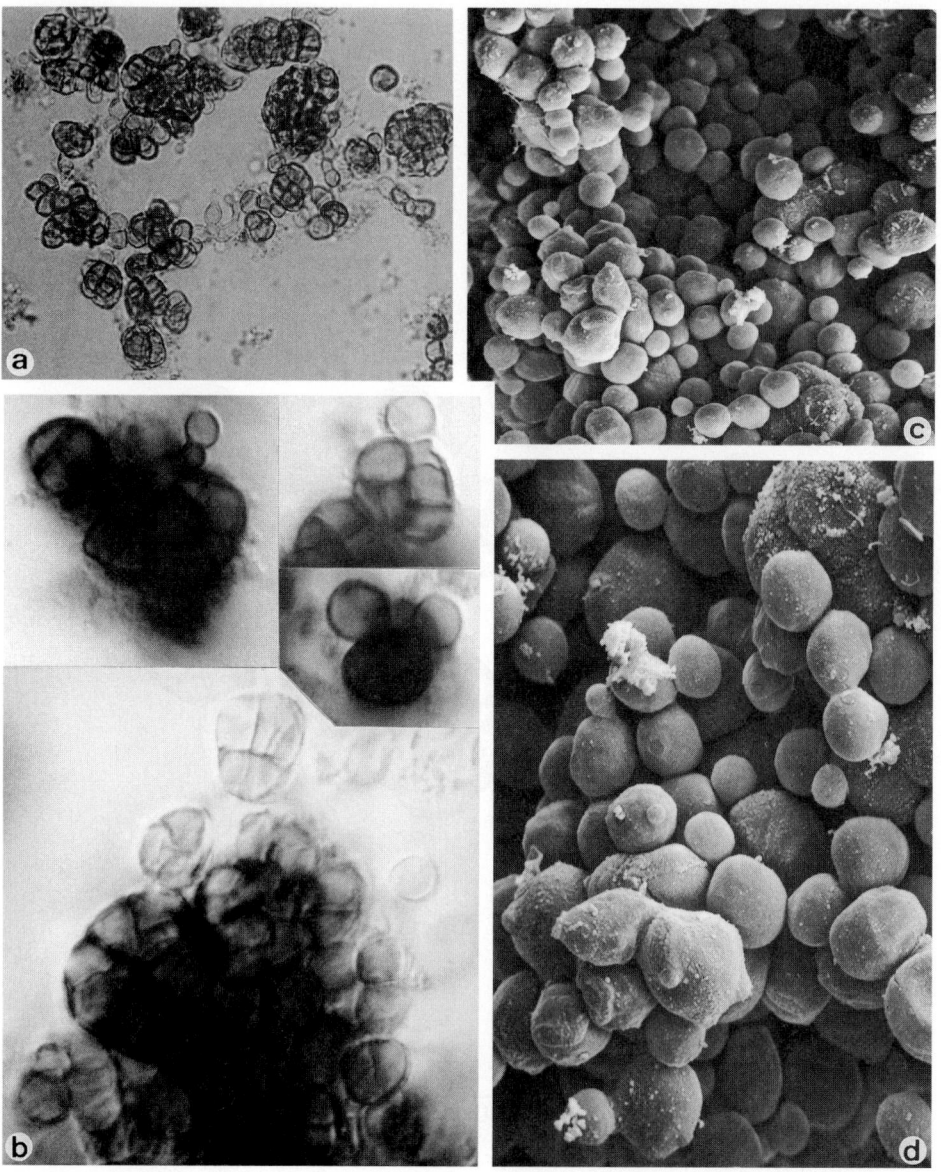

Sarcinomyces phaeomuriformis, CBS 131.88. Clusters of meristematic cells and budding cells. a. ×160; b. ×1600; c. ×1650; d. ×3400.

Hyphomycetes. Genus: *SCEDOSPORIUM*

Generic description. Colonies spreading, hairy, greyish-brown. Conidia sessile or on cylindrical or flask-shaped conidiogenous cells, either singly or formed percurrently. Annellated zones inconspicuous; conidial heads slimy. Conidia one-celled, subhyaline to brown. A *Graphium* synanamorph may be present.

Teleomorphs. *Pseudallescheria, Petriella* (*Ascomycota, Euascomycetes, Microascales: Microascaceae*).

Differential diagnosis. *Scopulariopsis* species (p. 902) are distinghuised by producing dry, hydrophobic conidia in chains rather than slimy conidia in heads.

References. Guého & de Hoog (1991), de Hoog *et al.* (1994a), Issakainen *et al.* (1999).

Key to the treated species of Scedosporium:

1a   Conidiogenous cells cylindrical, free or intercalary ➙ **2**
1b.   Conidiogenous cells with inflated base, then often in brush-like arrangement . . . . *S. prolificans* **(899)**
2a.   Conidia hyaline, thin-walled, often with papillate base . . . . *S. anam. of Petriella setifera* **(301)**
2b.   Conidia in part becoming pale brown, with truncate base . . . . *S. anam. of Pseudallescheria boydi* **(305)**

Scedosporium prolificans (Hennebert & Desai) Guého & de Hoog

Colony characteristics. Colonies (OA, 30°C) expanding, flat, moist with depressed, cobweb-like aerial mycelium, olivaceous grey to blackish.

Microscopy. Conidiogenous cells locally aggregated in small brushes, flask-shaped, often bearing a long, inconspicuously annellated zone. Darker and more inflated conidia may arise alongside hyphae. Conidia smooth-walled, aggregating in dense, slimy heads, soon becoming rather thick-walled and olivaceous brown, 3-7 × 2-5 µm.

Physiology.

Fermentation		Salicin	+,w	D-Galacturonate	–
D-Glucose	–	Melibiose	–,w	DL-Lactate	–
		Lactose	–,w	Succinate	–
Growth:		Raffinose	–	Citrate	–
D-Glucose	+	Melezitose	–	Methanol	–
D-Galactose	+	Inulin	–	Ethanol	+
L-Sorbose	–	Soluble starch	+	Nitrate	+
D-Glucosamine	w	Glycerol	+	Nitrite	+
D-Ribose	w	*meso*-Erythritol	–	Ethylamine	+
D-Xylose	+	Ribitol	–	L-Lysine	+
L-Arabinose	+	Xylitol	–	Cadaverine	v
D-Arabinose	–	L-Arabinitol	–	Creatine	–
L-Rhamnose	+	D-Glucitol	+	Creatinine	–
Sucrose	–	D-Mannitol	+	10% NaCl	–
Maltose	+	*myo*-Inositol	v	0.1% Cycloheximide	–
α,α-Trehalose	+	Glucono-δ-lactone	–	Mycosel	–
α-Methyl-D-glucoside	–	D-Gluconate	–	Urease	+
Cellobiose	+	D-Glucuronate	–	Growth at 40°C	+

Molecular diagnostics. SSU sequencing was performed by Issakainen *et al.* (1997, 1999). ITS-based diagnostic primers were developed by Wedde *et al.* (1998). SSU and ITS restriction maps based on NCBI U43910 and CBS 467.74, resp.:

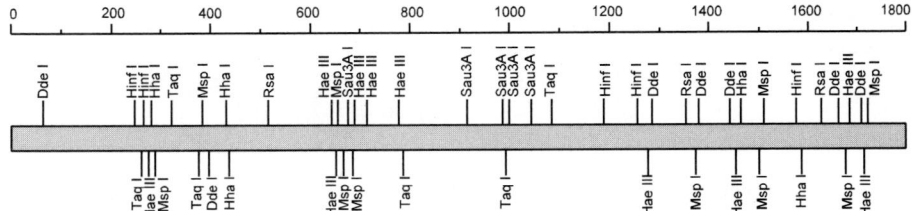

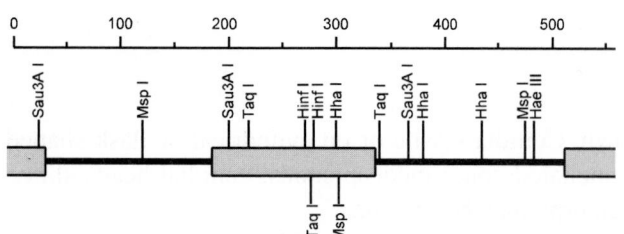

Antifungal susceptibility.

Antifungal	MICs range	MIC90	Strains	Reference
AMB	2-16	nd	5	Hennequin *et al.* (1997)
AMB	0.125->16	>8	33	Guarro *et al.* (2000)
AMB	1-16	nd	19	McGinnis & Pasarell (1998b)
AMB	8->16	>16	43	Cuenca-Estrella *et al.* (1999)
5FC	>16	>16	43	Cuenca-Estrella *et al.* (1999)
FLZ	32->64	>64	33	Guarro *et al.* (2000)
ITZ	2-16	nd	5	Hennequin *et al.* (1997)
ITZ	2-32	nd	19	McGinnis & Pasarell (1998b)
ITZ	8->16	>16	43	Cuenca-Estrella *et al.* (1999)
ITZ	8->16	>16	33	Guarro *et al.* (2000)
KTZ	>16	>16	43	Cuenca-Estrella *et al.* (1999)
KTZ	0.5-16	>16	33	Guarro *et al.* (2000)
MCZ	0.5->16	nd	5	Hennequin *et al.* (1997)
MCZ	0.25-16	>4	33	Guarro *et al.* (2000)
TBF	0.5-64	2	20	Meletiadis *et al.* (2000)
VCZ	1-32	>2	19	McGinnis & Pasarell (1998b)
VCZ	8-32	16	43	Cuenca-Estrella *et al.* (1999)
VCZ	0.06-4	nd	33	Guarro *et al.* (2000)

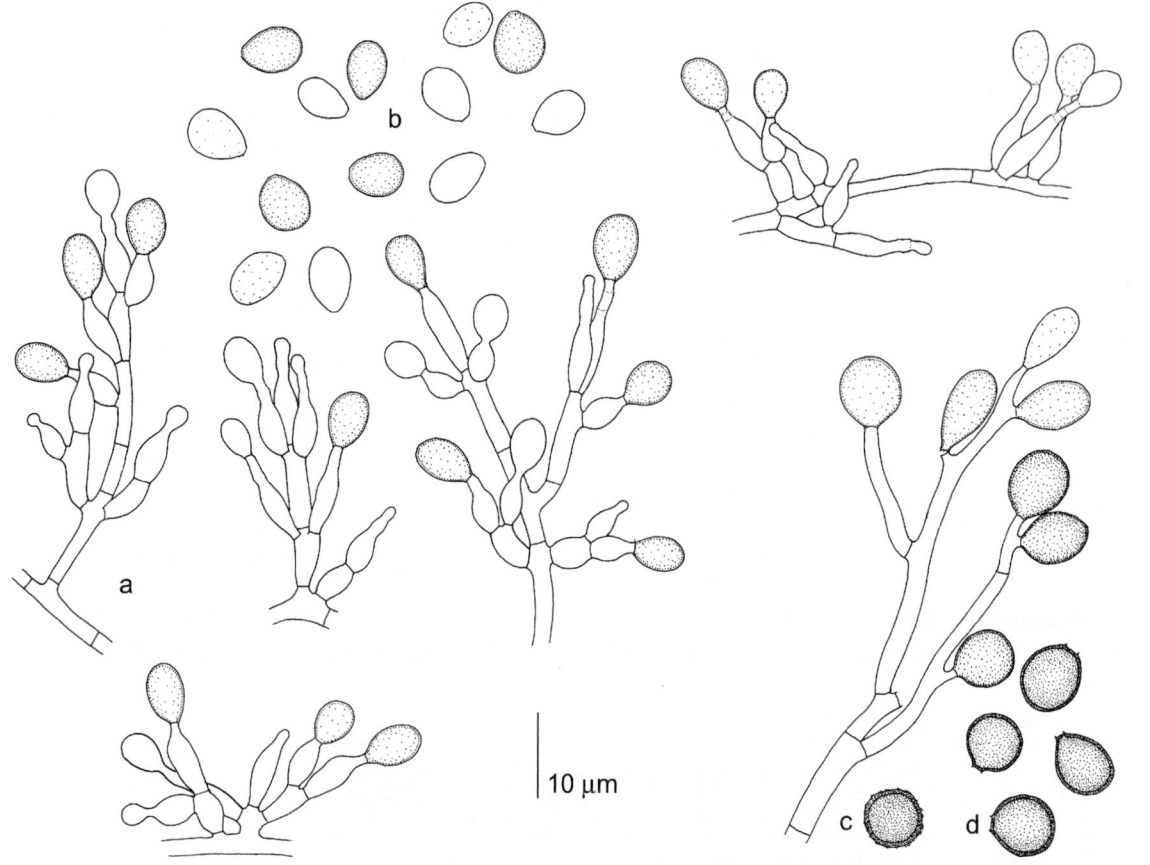

Scedosporium prolificans, CBS 114.90. a. Inflated conidiogenous cells; b. conidia; c, d. conidia sessile on hyphae.

900

Pathogenicity. BSL-2. The species occurs in relatively warm, indoor soil, such as in potted plants (Summerbell *et al.*, 1989b) or in glasshouses (Hennebert & Desai, 1974). It can be introduced traumatically into humans via thorns or splinters. It is frequently isolated from cutaneous (Gillum *et al.*, 1997) and subcutaneous lesions (Salkin *et al.*, 1988). In the body it has a predilection for cartilage and joint areas (Wilson *et al.*, 1990). Fatal dissemination may occur in immunocompromised patients (Guarro *et al.*, 1991b; Marin *et al.*, 1991; Perfect *et al.*, 1992; Wood *et al.*, 1992; Nielsen *et al.*, 1993; Wiswe *et al.*, 1993), leukemic patients (Feltkamp *et al.*, 1997; Gosbell *et al.*,1999) or transplant patients (Salesa *et al.*, 1993). On reversal of neutropenia such infections heal spontaneously (Bouza *et al.*, 1996). The species is one of the emerging opportunistic fungi (Berenguer *et al.*, 1997); Gossbell *et al.* (1999) reviewed recent cases. AIDS-related cases are rare. Nenoff *et al.* (1996) described fatal dissemination in an AIDS patient, but a Burkitt lymphoma was also present. A fatal case of endocarditis was reported by Prevost-Smith *et al.* (1993) and a meningoencephalitis by Madrigal *et al.* (1995). Hospital construction works are a major risk factor for these patients (Alvarez *et al.*, 1995). A canine osteomyelitis was reported by Salkin *et al.* (1992). The species' virulence was studied by Drouhet *et al.* (1991a) and Cano *et al.* (1992).

References. Malloch & Salkin (1984), Salkin *et al.* (1988), Sparrow *et al.* (1992), de Hoog *et al.* (1994a), Issakainen *et al.* (1999).

Nomenclature. *Lomentospora prolificans* Hennebert & Desai - Mycotaxon 1: 45, 1974 ≡ *Scedosporium prolificans* (Hennebert & Desai) Guého & de Hoog - J. Mycol. Méd. 1: 8, 1991.

Scedosporium inflatum Malloch & Salkin - Mycotaxon 21: 249, 1984.

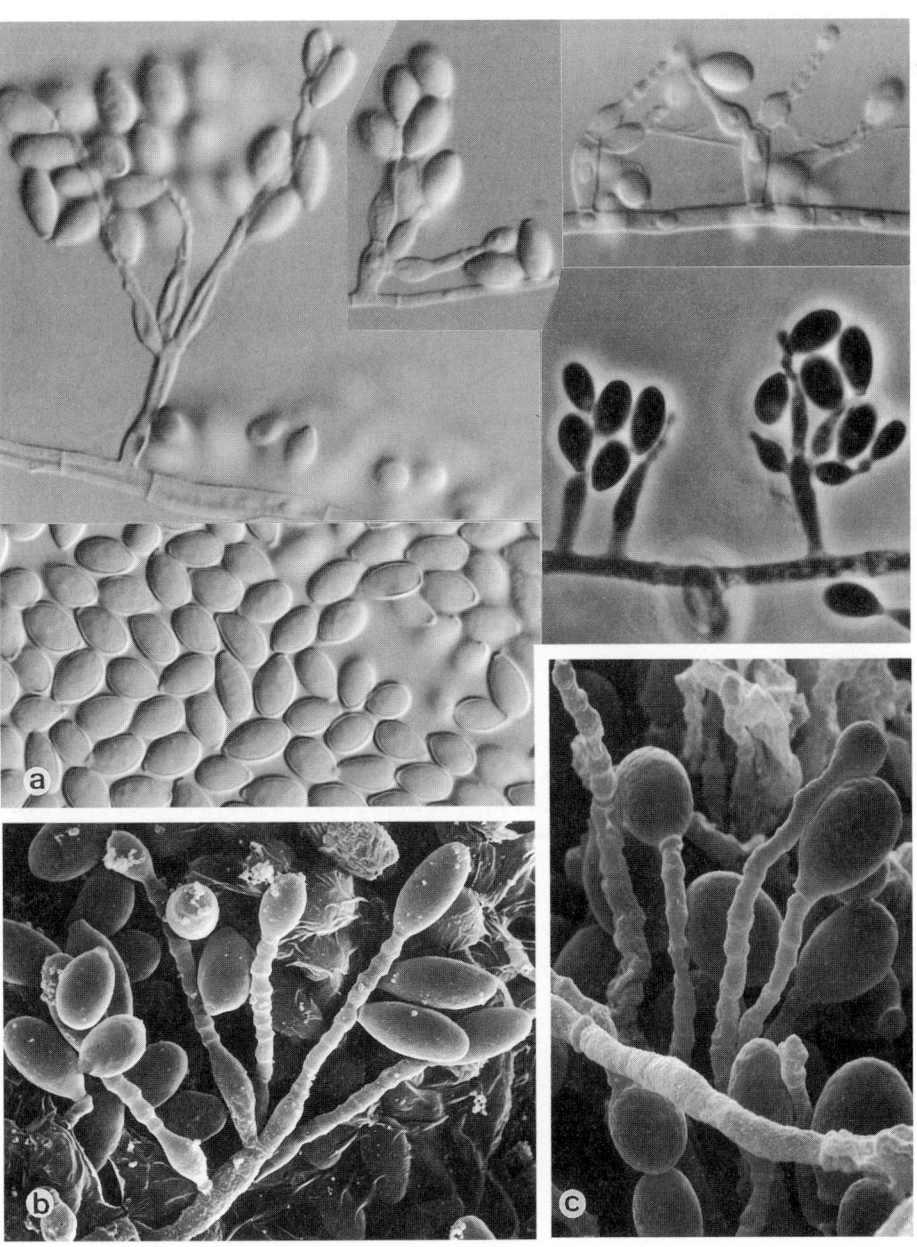

Scedosporium prolificans, CBS 114.90. Inflated conidiogenous cells and conidia. a. ×1185; b. ×3405; c. ×5625.

Hyphomycetes. Genus: *SCOPULARIOPSIS*

Generic description. Colonies expanding, white, buff, brown or black, powdery, velvety or somewhat funiculose; synnemata absent. Conidiophores erect, short, often branched, pale. Conidiogenous cells ampulliform to cylindrical, elongating with annellated zones on which conidia are produced. Conidia (sub)spherical, obovoidal, broadly clavate or bullet-shaped, truncate at the base, smooth-walled, punctate or verrucose, arranged in dry, basipetal chains, hyaline to pale brown.

Teleomorphs. *Microascus, Kernia* (*Ascomycota, Euascomycetes, Microascales: Microascaceae*).

Differential diagnosis. Species of *Scedosporium* (p. 899) have similar morphology but produce conidia in slimy heads rather than in dry chains.

General remarks. A large genus comprising mainly soil species which are frequently isolated from food, paper and other materials. They also occur as laboratory contaminants. Several species may be involved in human onychomycoses or pulmonary mycoses, and have recently been associated with invasive human infections (Wheat *et al.*, 1984; Neglia *et al.*, 1987; Anaissie *et al.*, 1989a, b).

References. Morton & Smith (1963), Domsch *et al.* (1980).

Key to the treated species of Scopulariopsis:

1a. Basal parts of conidiogenous cells less than 8 µm long; annellated zones less than 2 µm wide ➡ **2**

1b. Basal parts of conidiogenous cells usually longer than 8 µm; annellated zones 2.5-4.0 µm wide ➡ **4**

2a. Colonies pale buff . . . . **S. anam. of *Microascus cinereus* (277)**

2b. Colonies dark grey to black ➡ **3**

3a. Colonies (MEA) growing 10-16 mm in 7 days . . . . **S. brumptii (909)**

3b. Colonies (MEA) growing 4-6 mm in 7 days . . . . **S. chartarum (1018)**

4a. Conidia 8-14 µm long, ovoidal with pointed apex; colonies white to pale buff . . . . . **S. acremonium (903)**

4b. Not combining the above characteristics ➡ **5**

5a. Conidia smooth-walled ➡ **6**

5b. Conidia rough-walled ➡ **8**

6a. Colonies white . . . . **S. anam. of *Microascus manginii* (281)**

6b. Colonies avellaneous . . . . **S. koningii (915)**

6c. Colonies fuscous to nearly black ➡ **7**

7a. Colonies expanding . . . . **S. fusca (913)**

7b. Colonies restricted . . . . **S. anam. of *Microascus cirrosus* (279)**

8a. Colonies white . . . . **S. flava (911)**

8b. Colonies fuscous to nearly black . . . . **S. asperula (905)**

8c. Colonies avellaneous . . . . **S. brevicaulis (907)**

Scopulariopsis acremonium (Del.) Vuill.

Colony characteristics. Colonies (MEA 2%) growing slowly, white to pale buff, powdery to funiculose.
Microscopy. Conidiophores penicillate; conidiogenous cells cylindrical, 10-50 × 3.5-6.0 μm, basal part slightly swollen, and a long annellated zone of 2.5-4.0 μm wide. Conidia ovoidal, mostly with pointed apex and truncate base, 8-14 × 5-6 μm, hyaline to pale buff in mass.
Pathogencity. BSL-1. Onychomycoses (Krempl-Lamprecht, 1970). An invasive sinusitis in a leukemic patient was reported by Ellison *et al.* (1998).
References. Morton & Smith (1963), Domsch *et al.* (1980).
Antifungal susceptibility.

Antifungal	MICs	Strains	Reference
AMB	1	1	Aguilar *et al.* (1998)
5FC	256	1	Aguilar *et al.* (1998)
FLZ	128	1	Aguilar *et al.* (1998)
ITZ	32	1	Aguilar *et al.* (1998)
KTZ	1	1	Aguilar *et al.* (1998)
MCZ	2	1	Aguilar *et al.* (1998)

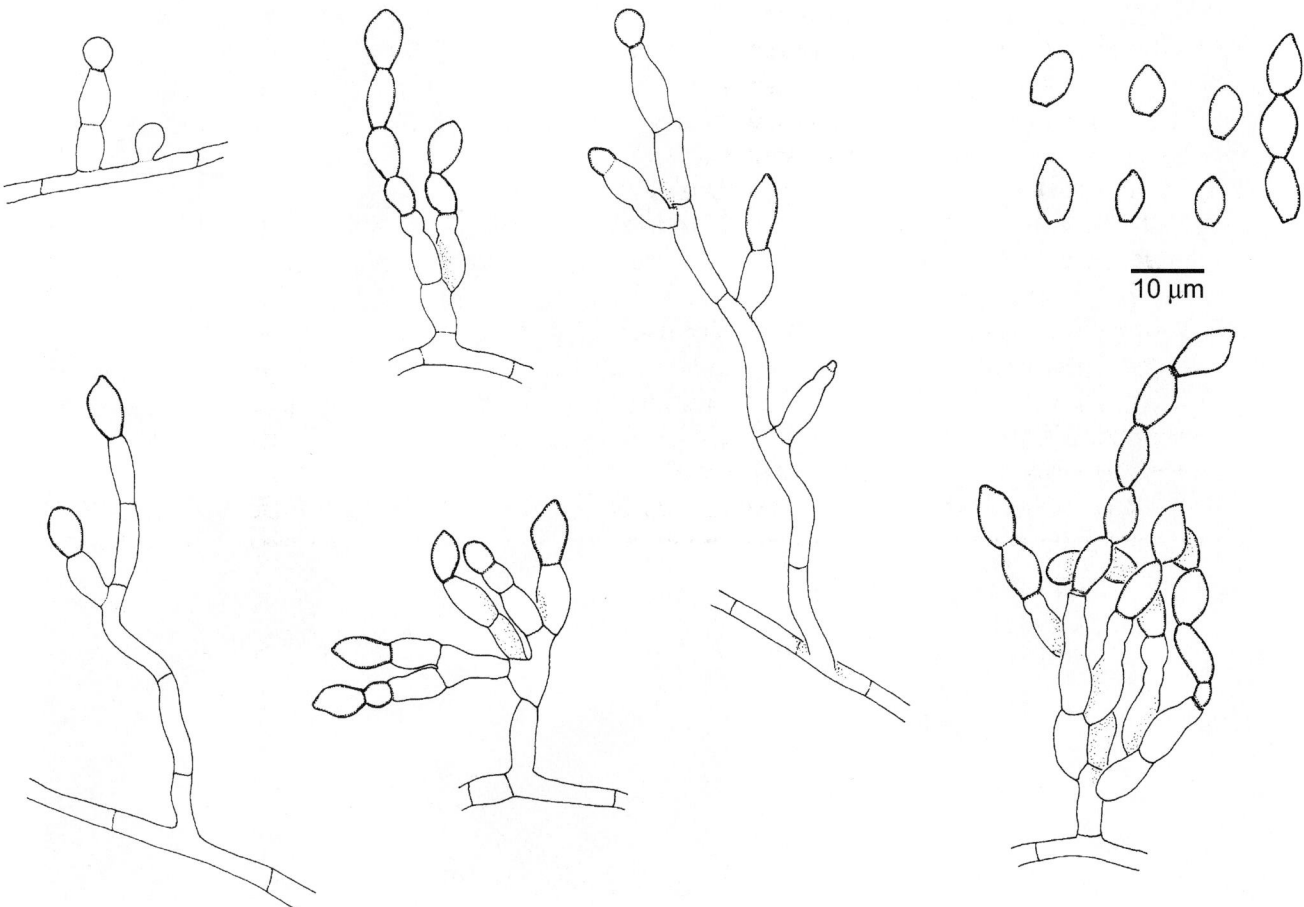

Scopulariopsis acremonium, IMI 250097. Conidiophores with annellated conidiogenous cells and conidia.

Nomenclature. *Monilia acremonium* Delacroix - Bull. Soc. Mycol. Fr. 13: 114, 1897 ≡ *Scopulariopsis acremonium* (Delacroix) Vuillemin - Bull. Soc. Mycol. Fr. 27: 148, 1911.

Scopulariopsis danica van Beyma - Zentbl. Bakt. Parasitkde, Abt. 2, 99: 390, 1938.

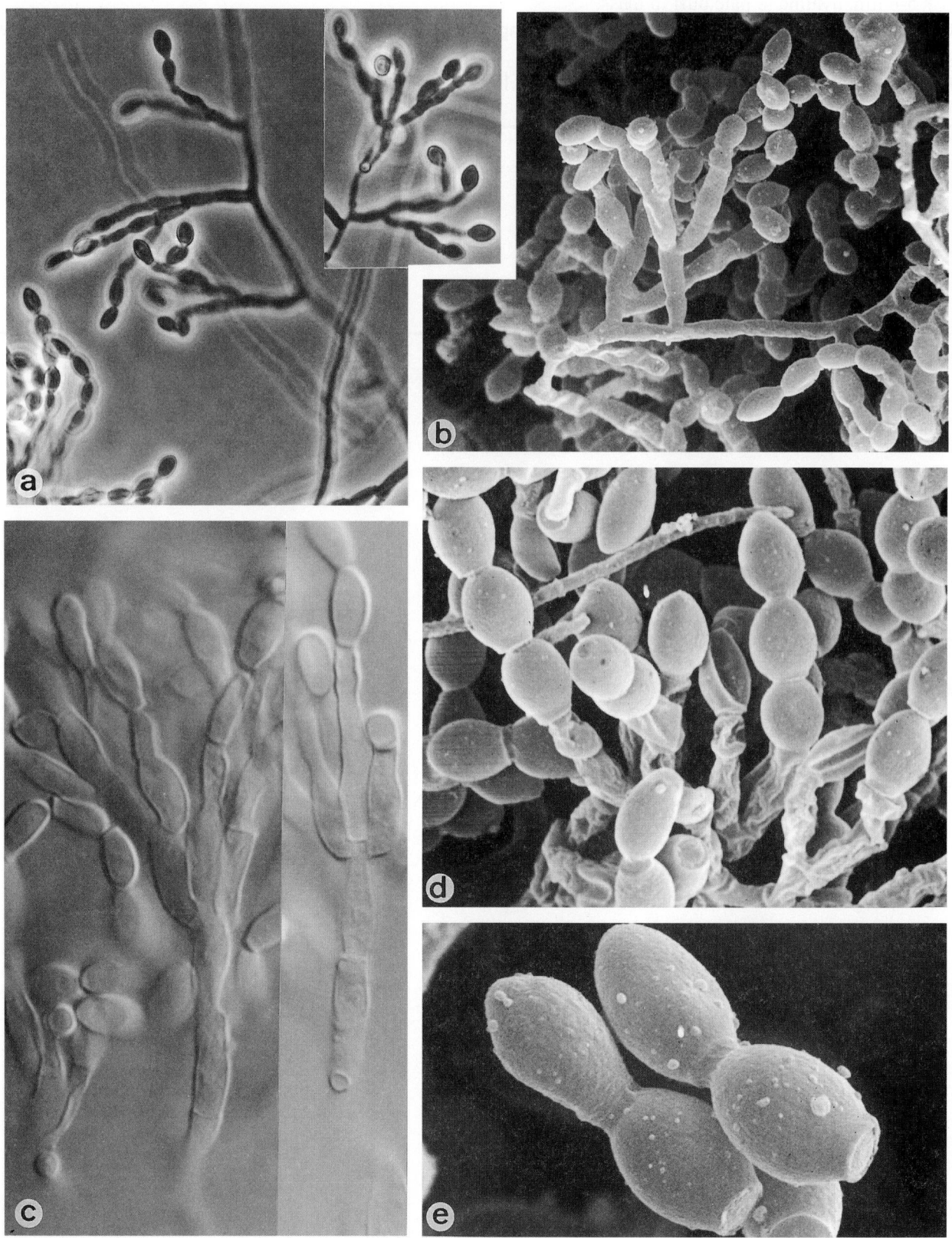

Scopulariopsis acremonium, IMI 250097. Conidiophores with annellated conidiogenous cells and conidia. a. ×512; b. ×1250; c. ×1600; d. ×2700; e. ×5300.

904

Scopulariopsis asperula (Sacc.) S.J. Hughes

Colony characteristics. Colonies (MEA 2%) growing rapidly, floccose to velvety or powdery, greyish-sepia or fuscous to nearly black.

Microscopy. Conidiogenous cells single, borne laterally on aerial hyphae, or in loose groups on short stalks, cylindrical, with slightly swollen basal part and a long annellated zone, 5-15 × 2.5-4.0 µm. Conidia rough-walled, spherical, occasionally ovoidal with rounded or pointed apex, 5-8 × 5-7 µm, sepia to fuscous in mass; base truncate with a distinct frill.

Pathogenicity. BSL-1. Onychomycoses (Krempl-Lamprecht, 1970).

Reference. Morton & Smith (1963).

Nomenclature. *Torula asperula* Saccardo - Michelia 2: 560, 1882 ≡ *Scopulariopsis asperula* (Saccardo) S.J. Hughes - Can. J. Bot. 36: 803, 1958.

  Scopulariopsis ivorensis Boucher - Bull. Soc. Pathol. Exot. 11: 313, 1918.

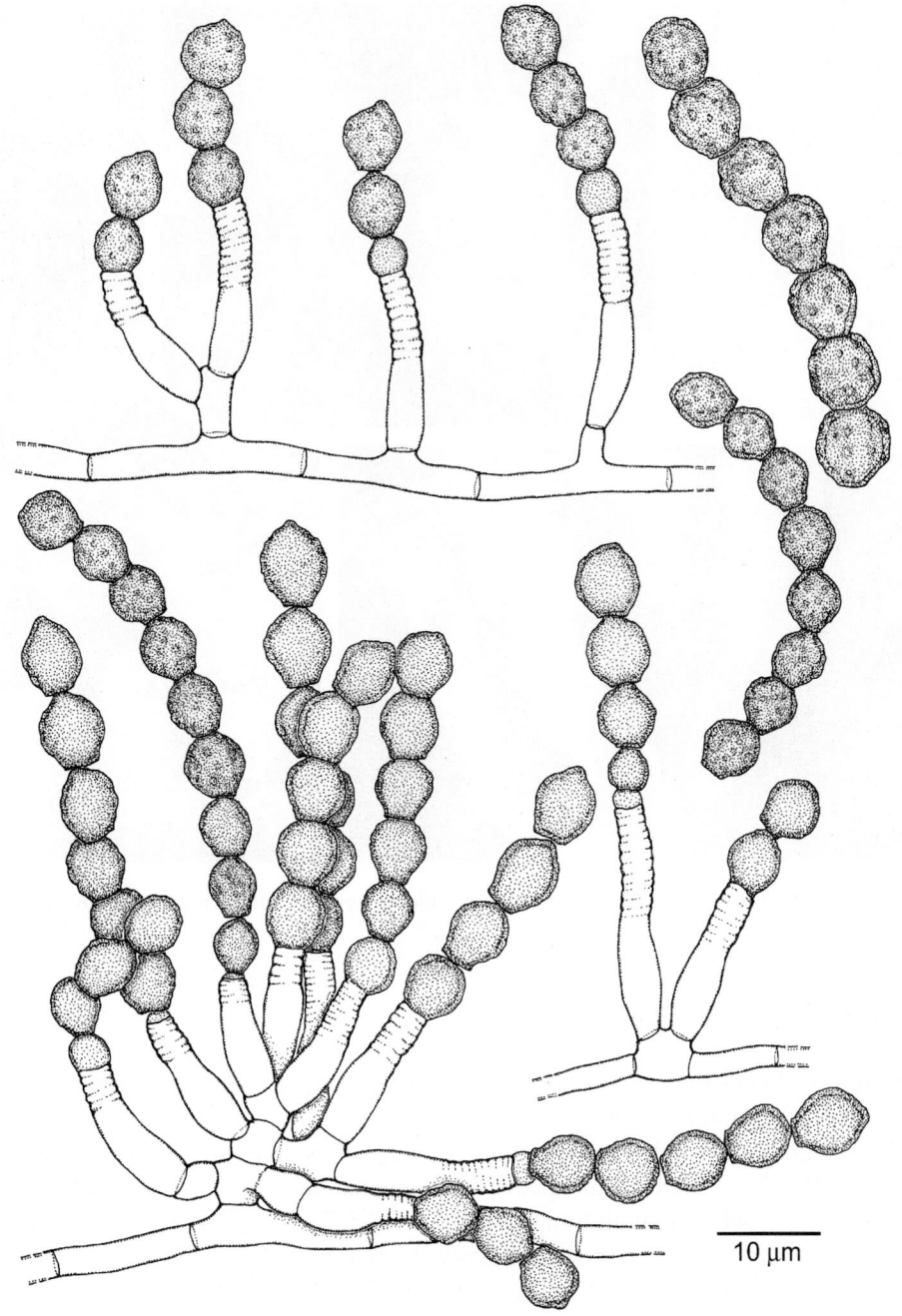

Scopulariopsis asperula, CBS 373.76. Conidiophores with annellated conidiogenous cells and conidia.

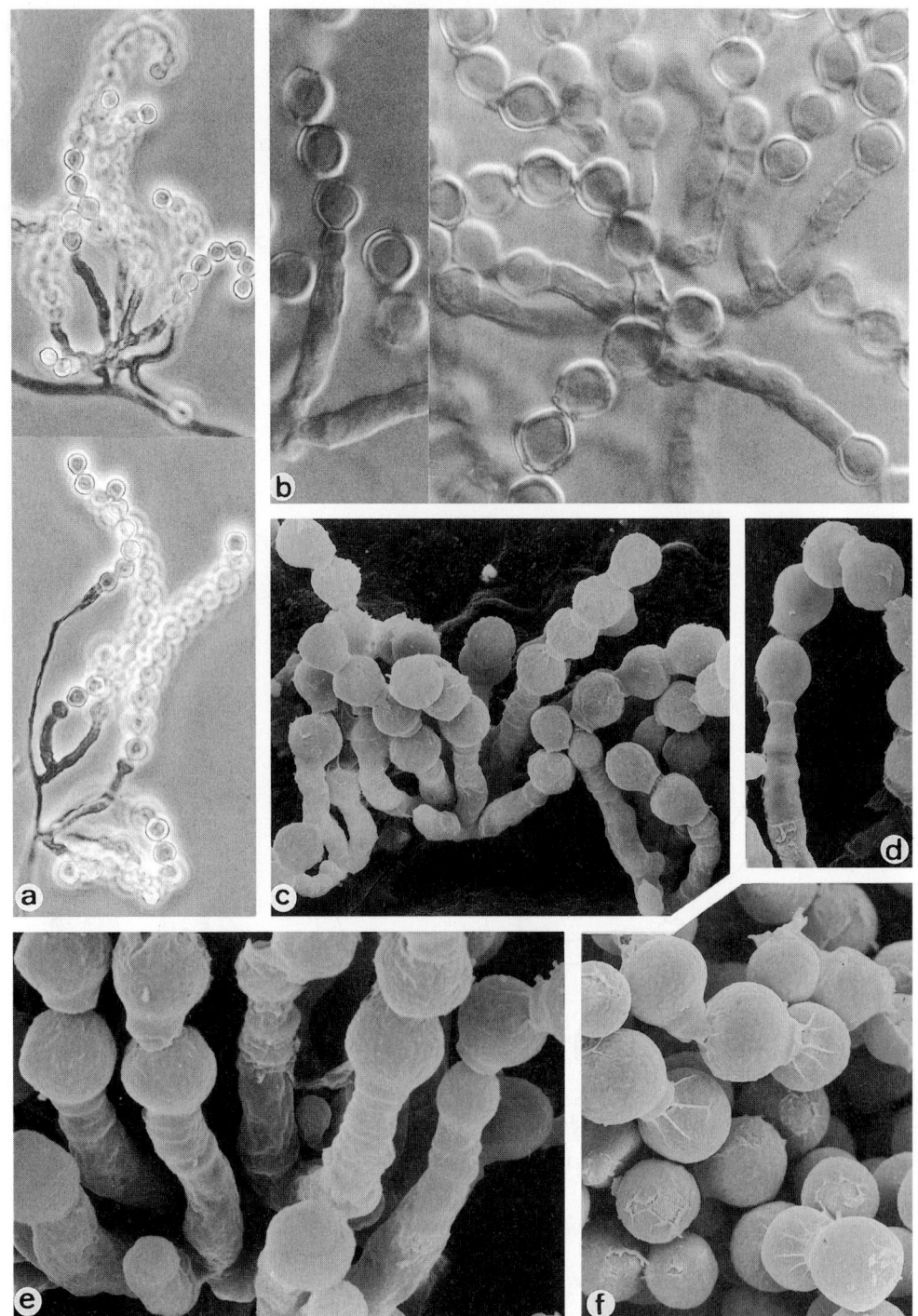

Scopulariopsis asperula, CBS 373.76. Conidiophores with annellated conidiogenous cells and conidia. a. ×385; b. ×1200; c. ×1500; d. ×1875; e. ×3000; f. ×2250.

Scopulariopsis brevicaulis (Sacc.) Bain.

Colony characteristics. Colonies (MEA 2%) expanding, whitish, powdery to felty, soon becoming avellaneous or pinkish-brown; reverse cream-coloured to brownish.

Microscopy. Conidiogenous cells single or in small brush-like groups on undifferentiated hyphae, cylindrical with slightly swollen base, 9-25 × 2.5-3.5 μm, with equally wide annellated zone of variable length. Conidia subhyaline, spherical to obovoidal or bullet-shaped, with truncate base, 5-8 × 5-7 μm, mostly rough-walled.

Teleomorph. *Microascus brevicaulis* Abbott (*Ascomycota, Euascomycetes, Microascales: Microascaceae*).

Ascomata spherical with a short neck, ostiolate, black, 80-150 × 70-130 μm. Peridium pseudoparenchymatous. Asci 8-spored, subspherical, 8-10 μm in diam, evanescent. Ascospores broadly reniform, 5-6 × 3.5-4.5 μm, smooth-walled, orange in mass.

Physiology. Tolerant to benomyl.

Pathogenicity. BSL-2. The species is frequently involved in onychomycoses (Tosti *et al.*, 1996); it is weakly keratinolytic (Filipello Marchisio, 2000). Rarely skin lesions are noted (Cox & Irving, 1993; Ginarte *et al.*, 1996; Arrese *et al.*, 1997), then mainly in immunocompromised patients (Dhar & Carey, 1993; Sellier *et al.*, 2000). A severe lesion in an otherwise healthy child was reported by Bruynzeel & Starink (1998). Invasive infections were described by Anaissie *et al.* (1989a, b) and Philips *et al.* (1989); endocarditis by Gentry *et al.* (1995) and Migrino *et al.* (1995); keratitis by Lotery *et al.* (1994) and del Prete *et al.* (1994) and endophthalmitis by Gariano & Kalina (1997). An animal case was reported by Purchio *et al.* (1980).

References. Morton & Smith (1963), Samson *et al.* (1996), Abbott *et al.* (1998).

Antifungal susceptibility.

Antifungal	Mean MICs	MIC 90	Strains	Reference
AMB	10.6	nd	5	Aguilar *et al.* (1998)
AMB	2.59	6.25	22	Wildfeuer *et al.* (1998)
5FC	256	nd	5	Aguilar *et al.* (1998)
FLZ	128	nd	5	Aguilar *et al.* (1998)
ITZ	32	nd	5	Aguilar *et al.* (1998)
ITZ	1.47	6.25	22	Wildfeuer *et al* (1998)
KTZ	1	nd	5	Aguilar *et al.* (1998b
KTZ	0.71	1.56	22	Wildfeuer *et al* (1998)
MCZ	12.12	nd	5	Aguilar *et al.* (1998)
VCZ	1.7	3.13	22	Wildfeuer *et al* (1998)

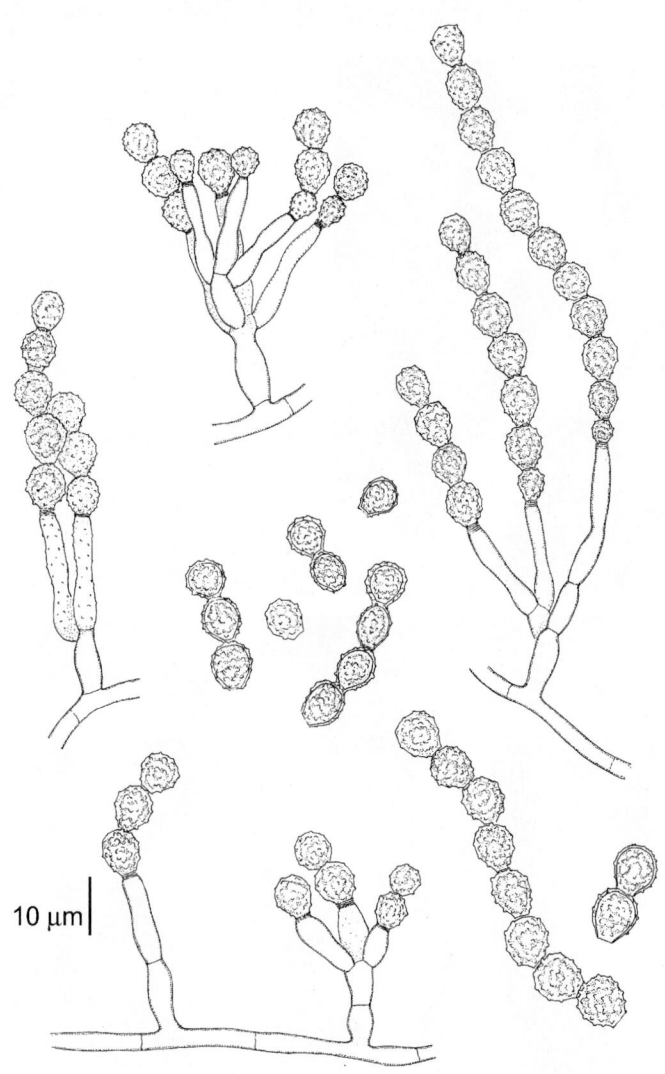

10 µm

Scopulariopsis brevicaulis, FMR 3503. Conidiogenous cells and catenate conidia.

Nomenclature. *Penicillium brevicaule* Saccardo - fung. Ital. No. 893, 1881 ≡ *Scopulariopsis brevicaulis* (Saccardo) Bainier - Bull. Soc. Mycol. Fr. 23: 99, 1907.

Penicillium brevicaule Saccardo var. *hominis* Brumpt & Langeron, *in* Brumpt - Précis Parasitol. p. 838, 1910 ≡ *Scopulariopsis brevicaulis* (Saccardo) Bainier var. *hominis* (Brumpt & Langeron) Brumpt & Langeron, *in* Brumpt - Précis Parasitol., ed. 2, p. 902, 1913 ≡ *Scopulariopsis hominis* (Brumpt & Langeron) Sartory - Champ. Paras. 8: 612, 1922.

Monilia penicillioides Delacroix - Bull. Soc. Mycol. Fr. 13: 114, 1897 ≡ *Penicillium penicillioides* (Delacroix) Vuillemin - Bull. Trim. Soc. Mycol. Fr. 27: 75, 1911.

Microascus brevicaulis Abbott, *in* Abbott, Sigler & Currah - Mycologia 90: 298, 1998.

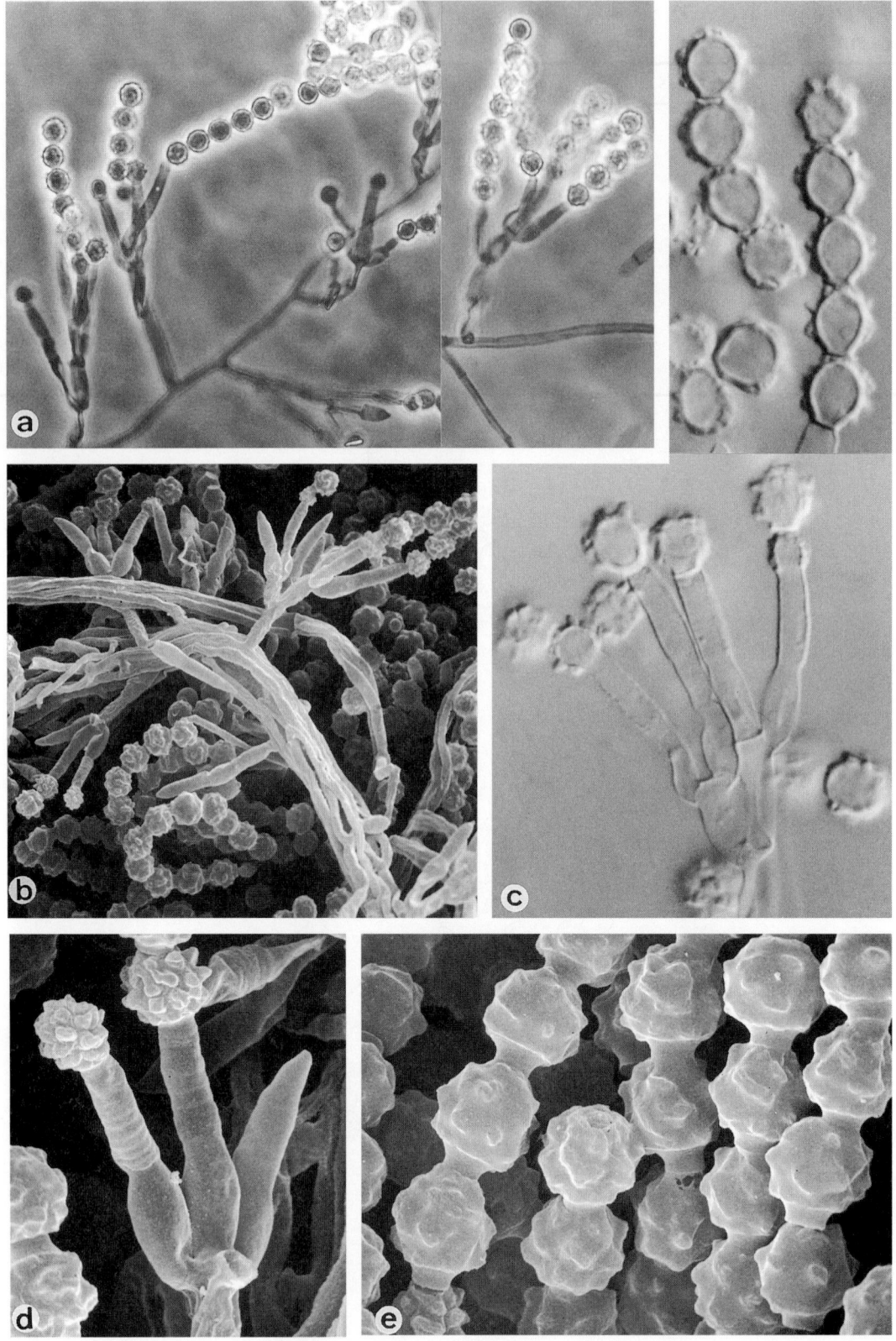

Scopulariopsis brevicaulis, FMR 3503. Annellated conidiogenous cells and catenate conidia. a.. ×530; b. ×645; c. ×1330; d. ×2240; e. ×2285.

Scopulariopsis brumptii Salvanet-Duval

Colony characteristics. Colonies (MEA 2%) moderately expanding, whitish, soon grey, finally sepia-grey to fuscous.
Microscopy. Conidiogenous cells single or in small brush-like groups on undifferentiated hyphae, flask-shaped, 5-10 × 2.5-3.5 µm, with cylindrical annellated zones of variable length, 0.8-1.5 µm wide; annellations inconspicuous. Conidia dark brown, obovoidal, with truncate base, 4.0-5.5 × 3.5-4.5 µm, mostly rough-walled.
Pathogenicity. BSL-2. The species is a common soil fungus. It is increasingly found as a pulmonary invader of patients with impaired cellular immunity (Grieble *et al.*, 1975; Salkin *et al.*, 1988).
References. Morton & Smith (1963), Salkin *et al.* (1988).

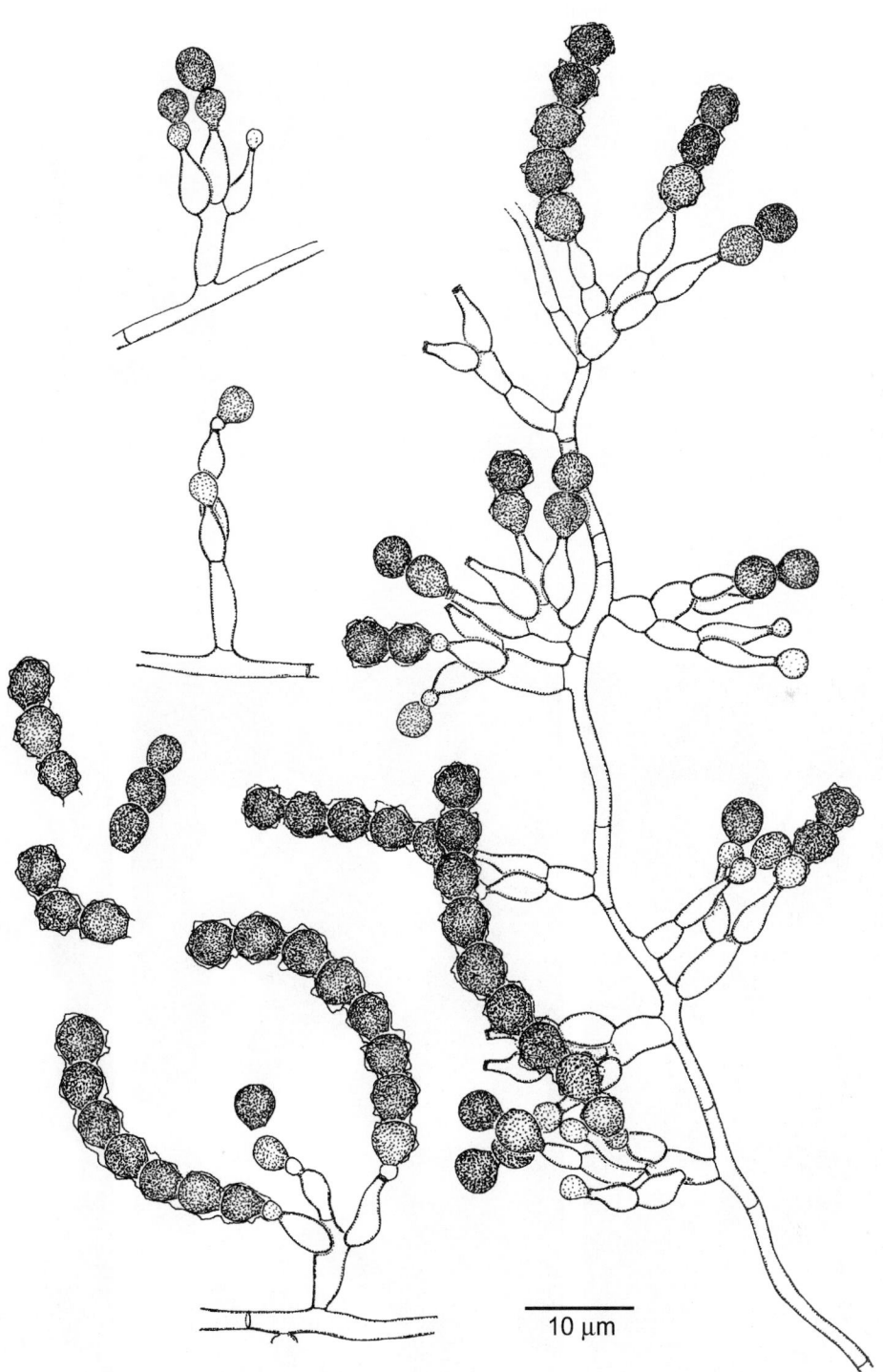

Scopulariopsis brumptii, FMR 3280. Conidial apparatus and chains of conidia.

Antifungal susceptibility.

Antifungal	Mean MICs	Strains	Reference
AMB	8	3	McGinnis & Pasarell (1998b)
5FC	256	1	Aguilar *et al.* (1998)
FLZ	128	1	Aguilar *et al.* (1998)
ITZ	32	3	McGinnis & Pasarell (1998b)
KTZ	8	1	Aguilar *et al.* (1998)
MCZ	4	1	Aguilar *et al.* (1998)
VCZ	4	3	McGinnis & Pasarell (1998b)

Nomenclature. *Scopulariopsis brumptii* Salvanet-Duval - Thèse Fac. Pharm. Paris 23: 58, 1935.

Scopulariopsis brumptii, FMR 3280. a-d. Conidiophores and conidia; e. detail of a conidial chain. a. ×512; b. ×1600; c. ×1900; d. ×4800; e. ×7200.

Scopulariopsis flava (Sopp) Morton & G. Smith

Colony characteristics. Colonies (MEA 2%) growing rapidly, floccose to fasciculate, white.
Microscopy. Conidiogenous cells single or in groups of 2-3, on short stalks or on larger, septate conidiophores, 5-25 × 2.5-3.5 μm, cylindrical, with a slightly swollen base. Conidia spherical to slightly obovoidal, with truncate base and a prominent frill, finely to coarsely roughened, 5-8 × 5-7 μm, hyaline, white in mass.
Pathogenicity. BSL-1. Onychomycosis (Piontelli & Toro, 1988).
Reference. Morton & Smith (1963).
Nomenclature. *Acaulium flavum* Sopp - VidenskSelsk. Christ. 11: 53, 1912 ≡ *Scopulariopsis flava* (Sopp) Morton & G. Smith - Mycol. Pap. 86: 43, 1963.

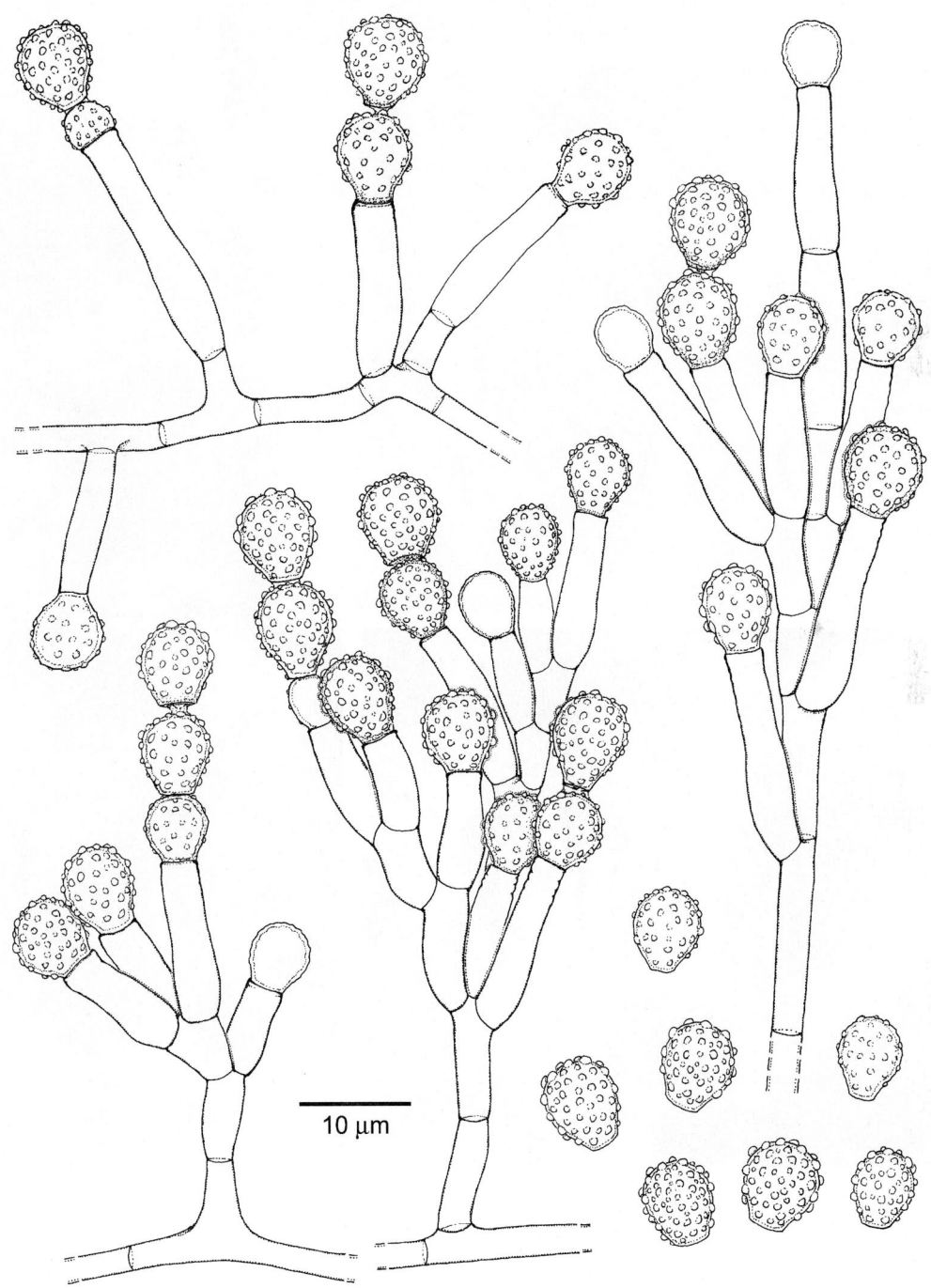

Scopulariopsis flava, CBS 207.61. Conidiophores with annellated conidiogenous cells and conidia.

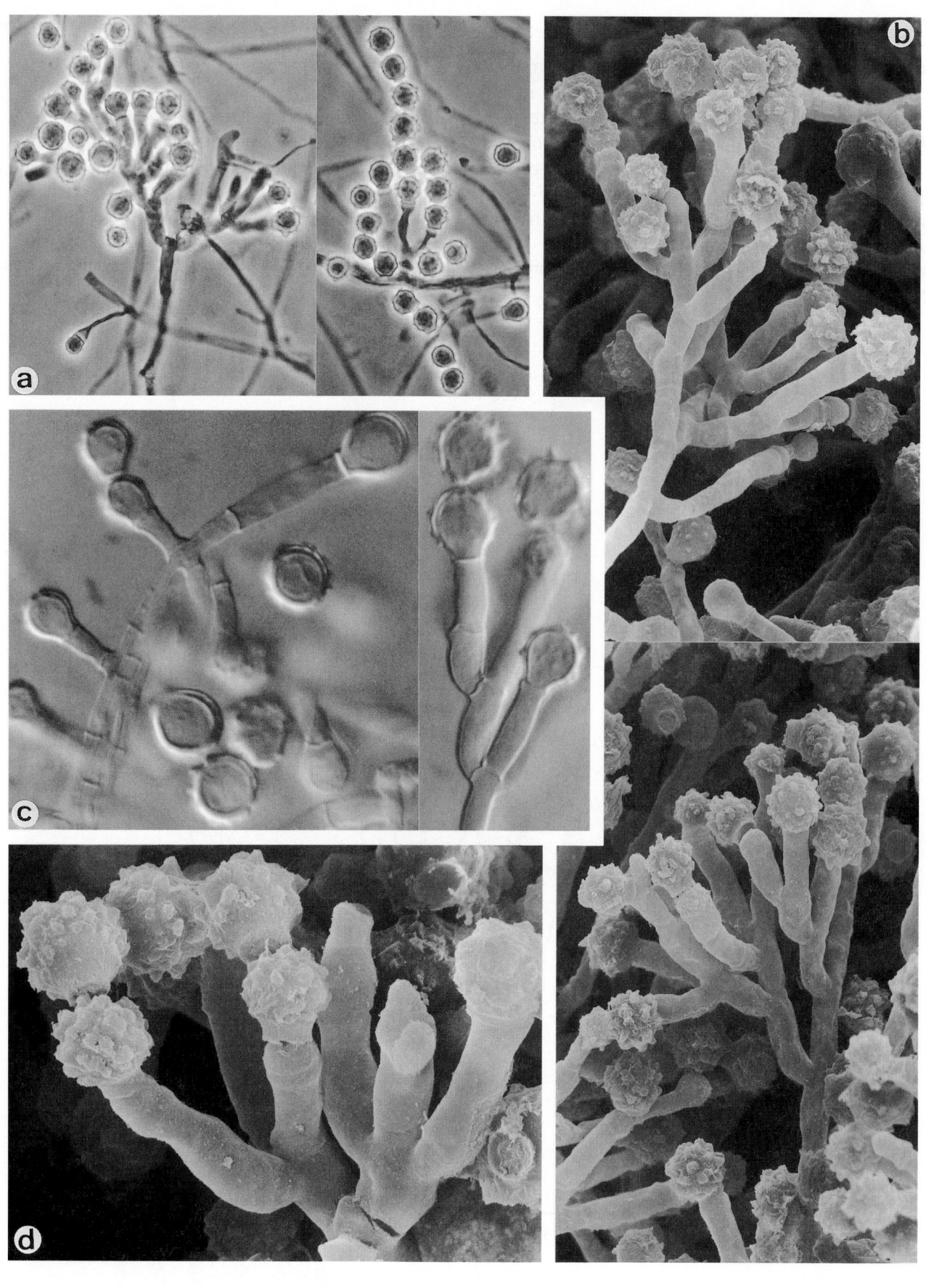

Scopulariopsis flava, CBS 207.61. Conidiophores with annellated conidiogenous cells and conidia. a. ×640; b. ×1750; c. ×1600; d. ×3300.

Scopulariopsis fusca Zach

Colony characteristics. Colonies (MEA 2%) growing rapidly, velvety to funiculose, avellaneous to vinaceous-brown or fuscous-black.

Microscopy. Conidiogenous cells arising from aerial hyphae, single or in groups on short stalks; basal parts 8-27 × 2.5-4.0 μm, cylindrical or somewhat swollen, annellated zones slightly narrower. Conidia spherical to broadly ovoidal, smooth-walled, 5-8 × 5-7 μm, olive to fuscous in mass, truncate, with prominent frill.

Pathogenicity. BSL-1. Onychomycoses (Krempl-Lamprecht, 1970; Schönborn & Schmoranzer, 1970).

References. Morton & Smith (1963), Domsch *et al.* (1980), Samson *et al.* (1996).

Antifungal susceptibility.

Antifungal	Mean MICs	Strains	Reference
AMB	32	1	Aguilar *et al.* (1998)
5FC	256	1	Aguilar *et al.* (1998)
FLZ	128	1	Aguilar *et al.* (1998)
ITZ	32	1	Aguilar *et al.* (1998)
KTZ	32	1	Aguilar *et al.* (1998)
MCZ	4	1	Aguilar *et al.* (1998)

Nomenclature. *Scopulariopsis fusca* Zach - Öst. Bot. Z. 83: 174, 1934.

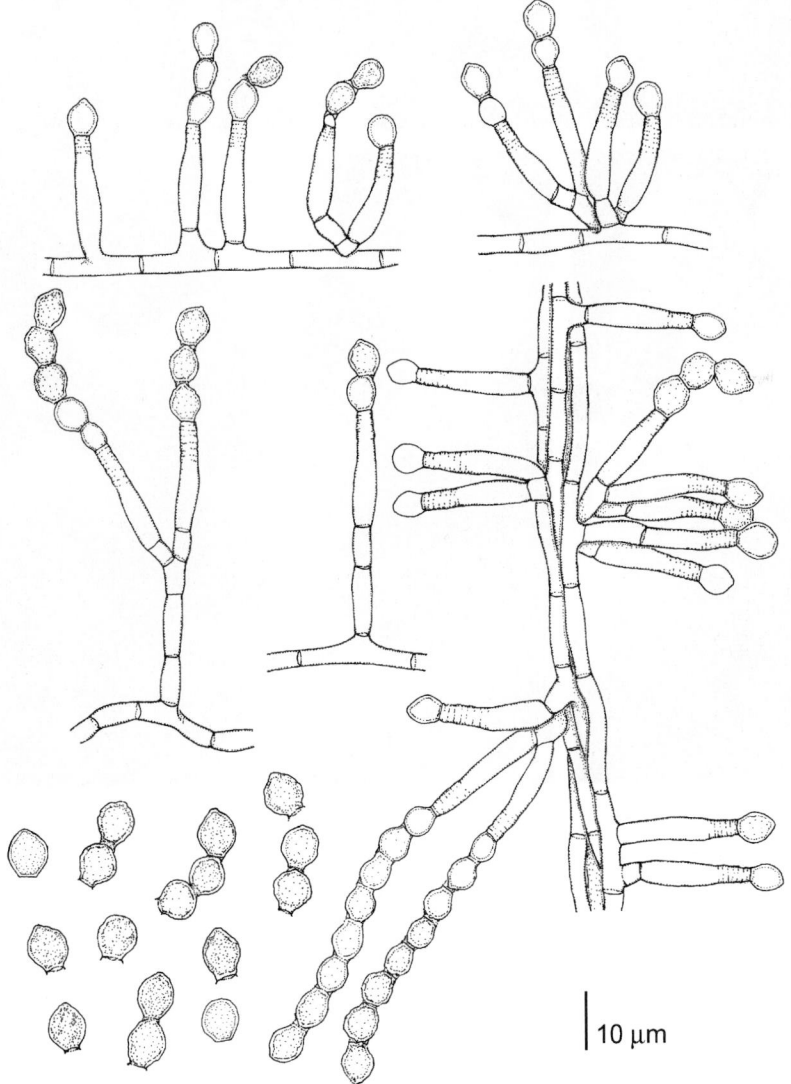

10 μm

Scopulariopsis fusca, CBS 334.53. Conidiophores with annellated conidiogenous cells and conidia.

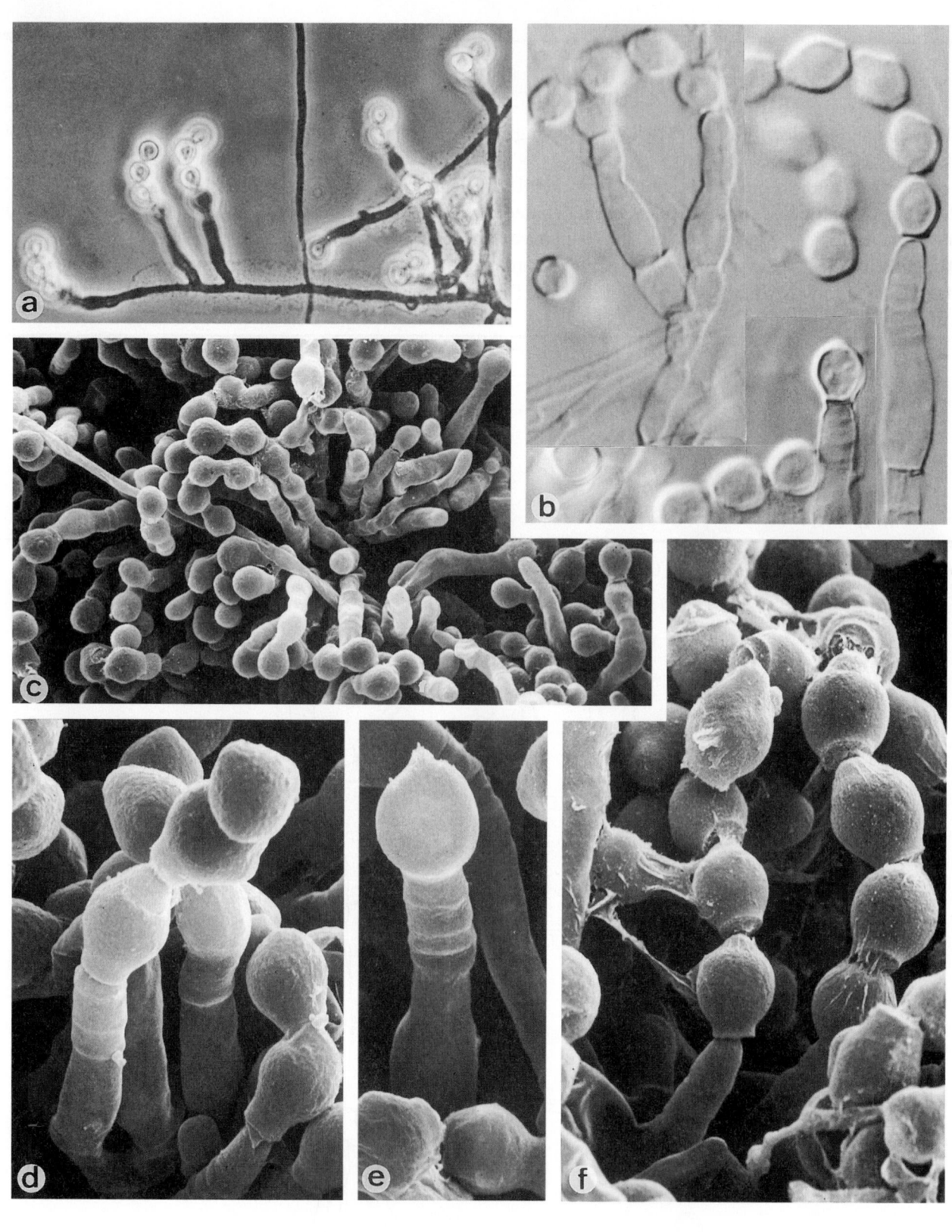

Scopulariopsis fusca, CBS 334.53. Conidiophores with annellated conidiogenous cells and conidia. a. ×512; b. ×1600; c. ×1350; d. ×3500; e. ×4250; f. ×3100.

Scopulariopsis koningii (Oudem.) Vuill.

Colony characteristics. Colonies (MEA 2%) growing rapidly, avellaneous, velvety to powdery.

Microscopy. Conidiogenous cells single or in groups, 10-28 × 2.5-3.5 µm, cylindrical or with slightly swollen basal parts; annellated zones 2.5-4.0 µm wide. Conidia spherical to ovoidal, 4.5-9.0 × 4-7 µm, smooth-walled, avellaneous in mass, truncate, usually with prominent frill. Chlamydospores occasionally present.

Pathogenicity. BSL-1. Onychomycoses (Schönborn & Schmoranzer, 1970; Krempl-Lamprecht, 1970).

Reference. Morton & Smith (1963).

Antifungal susceptibility.

Antifungal	Mean MICs	Strains	Reference
AMB	1.4	2	Aguilar *et al.* (1998)
5FC	256	2	Aguilar *et al.* (1998)
FLZ	128	2	Aguilar *et al.* (1998)
ITZ	32	2	Aguilar *et al.* (1998)
KTZ	2.8	2	Aguilar *et al.* (1998)
MCZ	2.8	2	Aguilar *et al.* (1998)

Nomenclature. *Monilia koningii* Oudemans, *in* Oudemans & Koning - Arch. Néerl. Sci., Ser. 2, 7: 287, 1902 ≡ *Scopulariopsis koningii* (Oudemans) Vuillemin - Bull. Soc. Mycol. Fr. 27: 143, 1911.

Torula bestae Pollacci - Riv. Biol. 4: 317, 1922 ≡ *Phaeoscopulariopsis bestae* (Pollacci) Ota - Jpn. J. Derm. Urol. 28: 405, 1928 ≡ *Scopulariopsis bestae* (Pollacci) Nannizzi - Tratt. Micopat. Um. 4: 254, 1934.

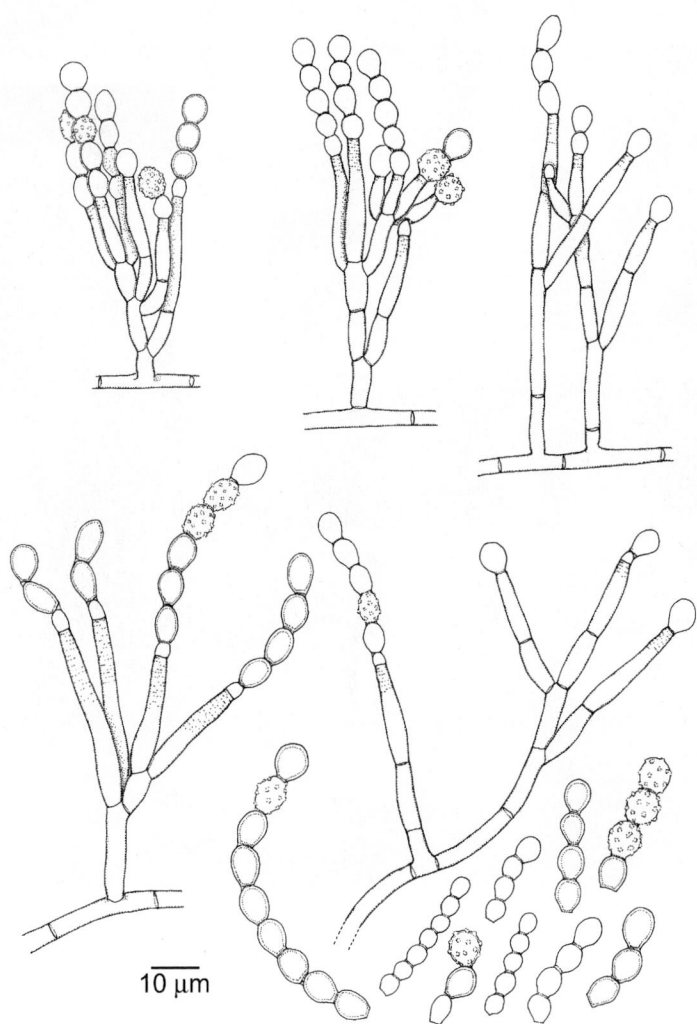

10 µm

Scopulariopsis koningii, CBS 208.61. Conidiophores with annellated conidiogenous cells and conidia.

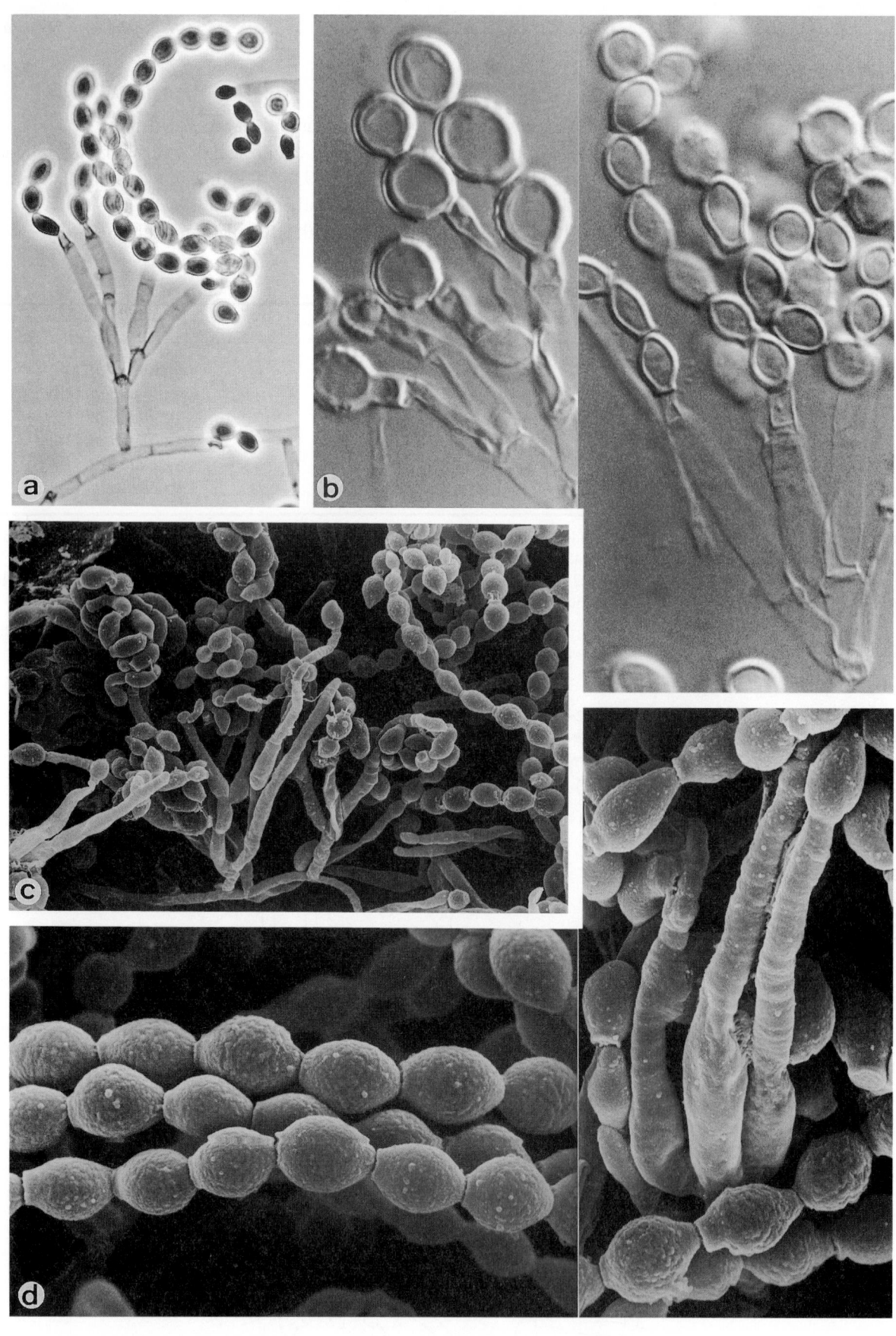

Scopulariopsis koningii, CBS 208.61. Conidiophores with annellated conidiogenous cells and conidia. a. ×512; b. ×1600; c. ×1100; d. ×3400.

Hyphomycetes. Genus: *SCYTALIDIUM*

Generic description. Colonies expanding, black or whitish. Hyphae melanized or hyaline, locally falling apart into arthroconidia. Intercalary chlamydospores present or absent.

Differential diagnosis. *Geotrichum* (p. 227) differs by purely white colonies with regular hyphae, entirely falling apart into rectangular arthroconidia. Many species of shelf fungi and mushrooms, such as *Bjerkandera* (p. 240) and *Coprinus* (described under the anamorph name *Hormographiella;* p. 712) in culture produce arthroconidia; such cultures are white, dry, fluffy. Basidiomycetes are urease positive.

General remarks. The coelomycete *Nattrassia mangiferae* (p. 329) has a *Scytalidium* synanamorph in culture. This species is black, but its melanin synthesis is easily blocked, leading to whitish colonies with an ochraceous-yellow colony reverse. This mutant is known in the medical literature as a separate *Scytalidium* species, *S. hyalinum,* but is treated here under *N. mangiferae.*

References. Campbell (1974), Campbell & Mulder (1977), Moore (1986, 1992), Sutton & Dyko (1989), Roeijmans *et al.* (1997).

Key to the treated species of Scytalidium:

1a. Conidia all hyaline; dark brown chlamydospores absent . . . .
. . . . ***S.* synanam. of *Nattrassia mangiferae* (hyaline mutant; 329)**

1b. Conida hyaline or pigmented; if hyaline, then additional, dark brown, chlamydospore-like conidia present ➙ **2**

2a. Conidia regularly rectangular, evenly brown pigmented . . . . ***S. infestans* (918)**

2b. Conidia of diverse shape and pigmentation ➙ **3**

3a. Hyaline and dark arthroconidia not clearly divided in two types . . . .
. . . . ***S.* synanam. of *Nattrassia mangiferae* (wild type; 329)**

3b. Hyaline, rectangular arthroconidia intermingled with swollen, dark brown chlamydospore-like conidia ➙ **4**

4a. Hyaline conidia measuring about 9-22 × 4.5-6.5 µm . . . . ***S. japonicum* (920)**

4b. Hyaline conidia measuring about 5-8 × 2 µm . . . . ***S. lignicola* (922)**

Scytalidium infestans Iwatsu *et al.*

Colony characteristics. Colonies (PDA) with restricted growth, umbonate, velvety, dark green.

Microscopy. Hyphae hyaline to pale brown, smooth-walled. Arthroconidia brown when mature, rather thick-walled, smooth-walled or verrucose, oblong, doliiform or broadly ellipsoidal, truncate at both ends, 0-1 (-3)-septate, 4-30 × 2.0-4.5 μm when non-septate or 7-23 × 4-5 μm when 1-septate, not easily detached.

Physiology. Ribose, melibiose, lactose, *myo*-inositol, creatine and creatinine assimilated; D-glucosamine not assimilated; tolerant to cycloheximide and 10% NaCl; urease present.

Molecular diagnostics. ITS restriction map based on CBS 161.91:

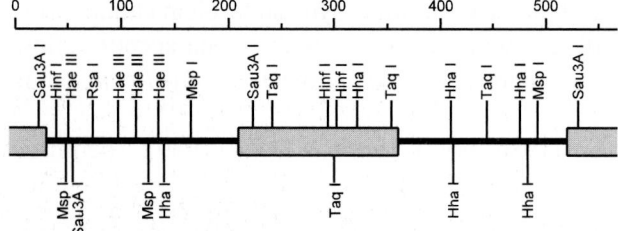

Pathogenicity. BSL-1. Systemic mycosis in fish (Iwatsu *et al.*, 1990).

Reference. Iwatsu *et al.* (1990).

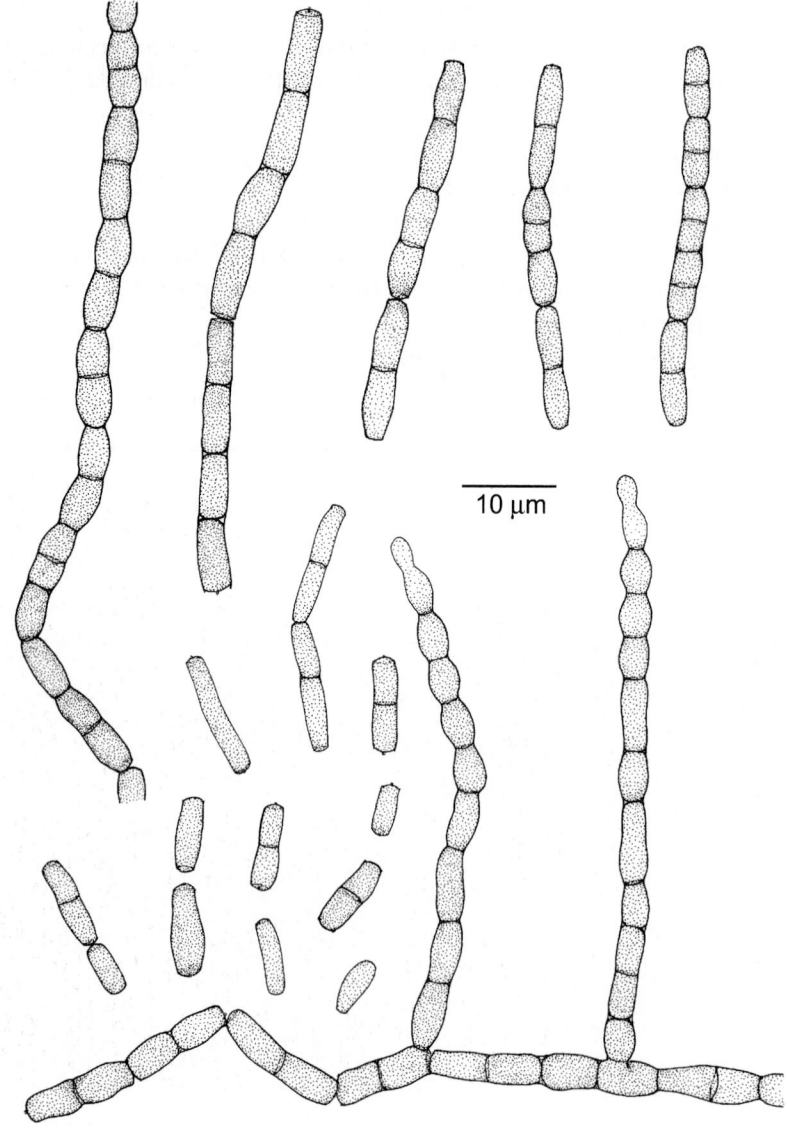

Scytalidium infestans, CBS 161.91. Fertile hyphae and arthroconidia.

Antifungal susceptibility.

Antifungal	Mean MICs	Strains	Reference
AMB	1	1	Guarro *et al.* (2000d)
5FC	2	1	Guarro *et al.* (2000d)
FLZ	8	1	Guarro *et al.* (2000d)
ITZ	0.12	1	Guarro *et al.* (2000d)
KTZ	0.25	1	Guarro *et al.* (2000d)
MCZ	1	1	Guarro *et al.* (2000d)

Nomenclature. *Scytalidium infestans* Iwatsu, Udagawa & Hatai - Trans. Mycol. Soc. Jpn. 31: 389, 1990.

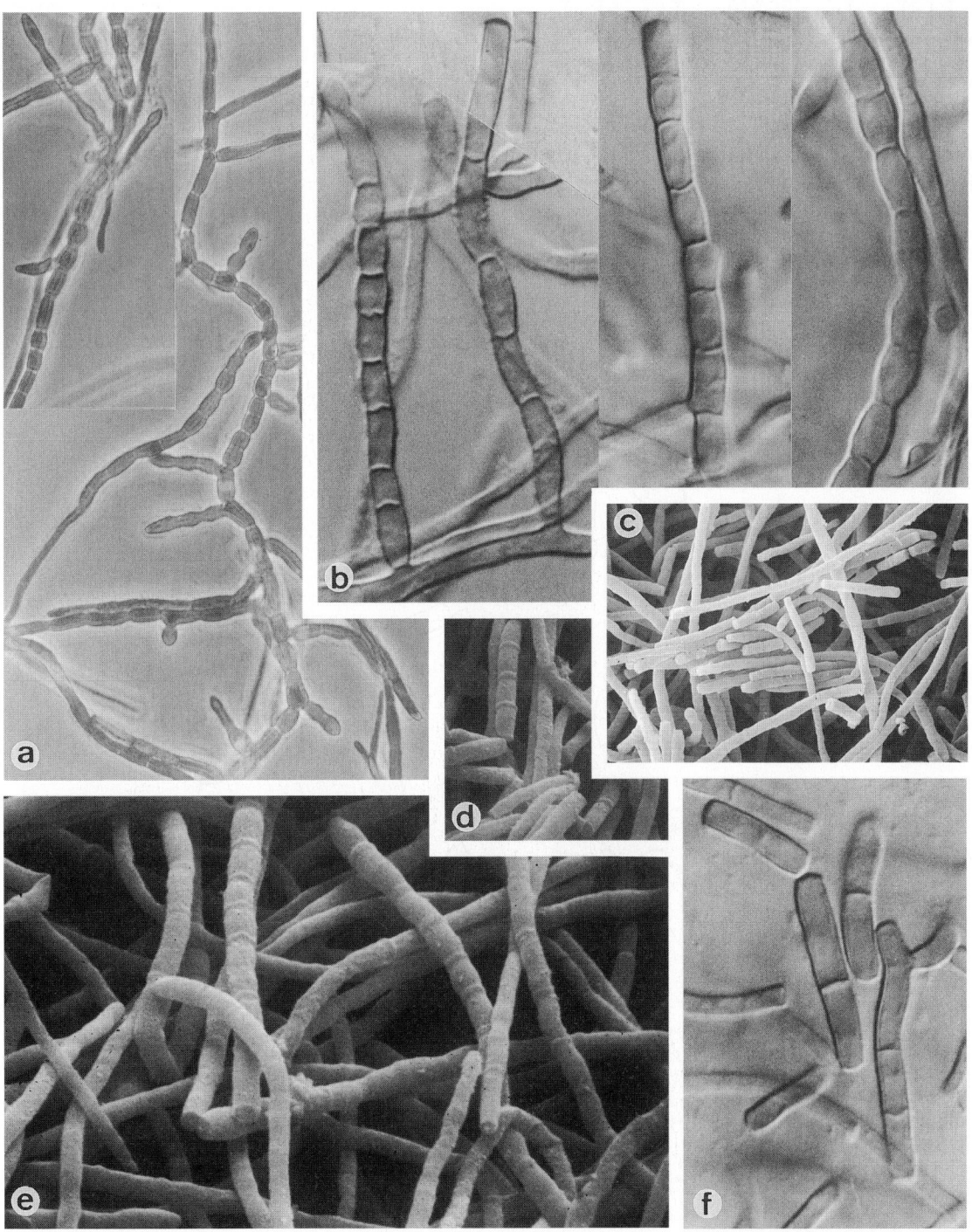

Scytalidium infestans, CBS 161.91. Fertile hyphae and arthroconidia. a. ×1280; b. ×1600; c. ×1015; d. ×1850; e. ×2600; f. ×1600.

Scytalidium japonicum Udagawa *et al.*

Colony characteristics. Colonies (PDA) growing rapidly, woolly, greenish-grey.

Microscopy. Hyphae hyaline to pale brown, smooth-walled to finely roughened, often in bundles or spirally twisted fascicles. Conidia of two kinds: (1) hyaline, 0(-1)-septate, thin-walled, cylindrical, truncate at both ends, 9-22 (-26) × 4.5-6.5 µm or 18-35 µm long when 1-septate, soon seceding, and (2) pale to dark brown, subspherical to ovoidal, 5.5-18.0 × 4-6 µm, thick- and smooth-walled, not easily seceding.

Molecular diagnostics. ITS restriction map based on CBS 494.88:

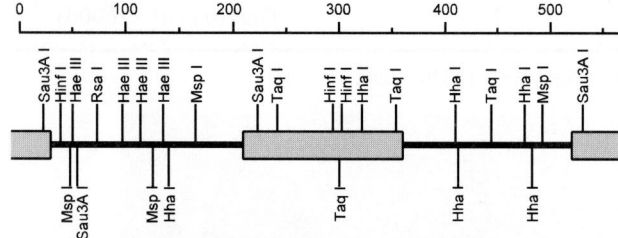

Pathogenicity. BSL-1. Bronchiolitis in cattle (Udagawa *et al.*, 1986).
Reference. Udagawa *et al.* (1986).

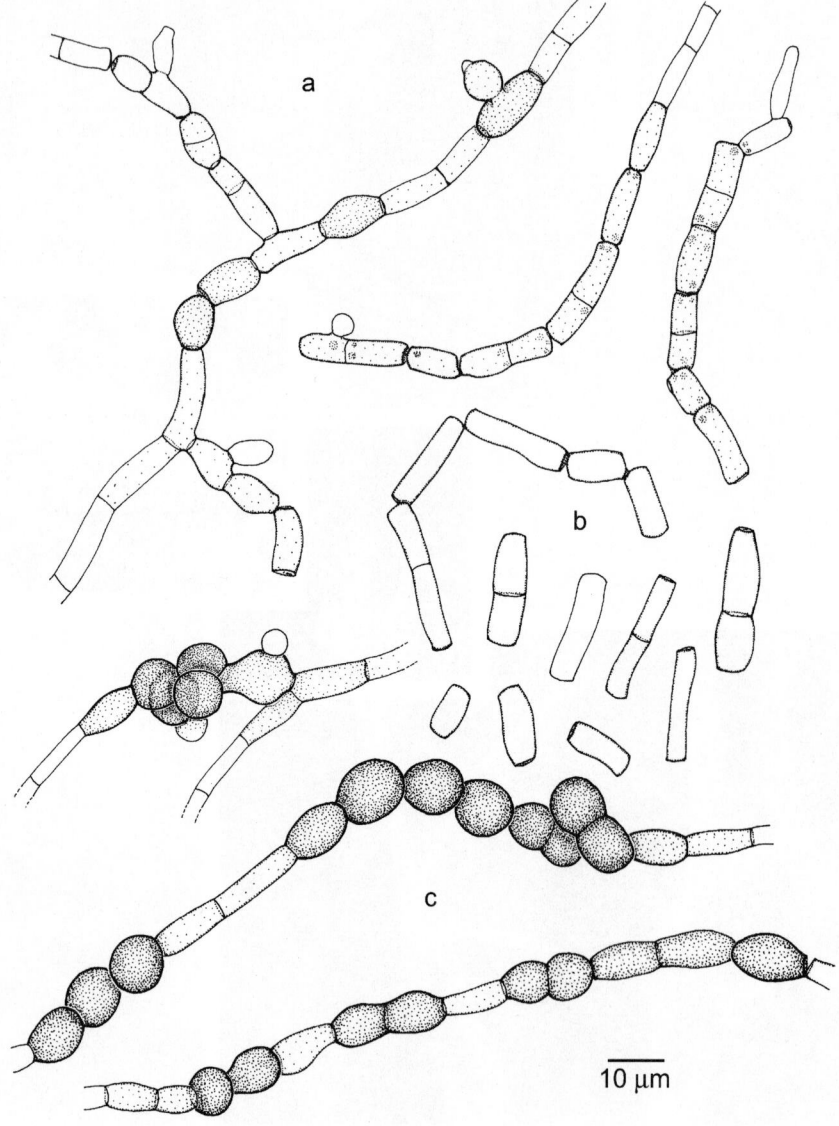

Scytalidium japonicum, CBS 494.88. a. Fertile hyphae; b. hyaline and cylindrical arthroconidia; c. pigmented and subspherical arthroconidia.

Antifungal susceptibility.

Antifungal	Mean MICs	Strains	Reference
AMB	0.06	1	Guarro *et al.* (2000d)
5FC	256	1	Guarro *et al.* (2000d)
FLZ	16	1	Guarro *et al.* (2000d)
ITZ	0.25	1	Guarro *et al.* (2000d)
KTZ	0.5	1	Guarro *et al.* (2000d)
MCZ	1	1	Guarro *et al.* (2000d)

Nomenclature. *Scytalidium japonicum* Udagawa, Tominga & Hamaoka - Mycotaxon 25: 281, 1986.

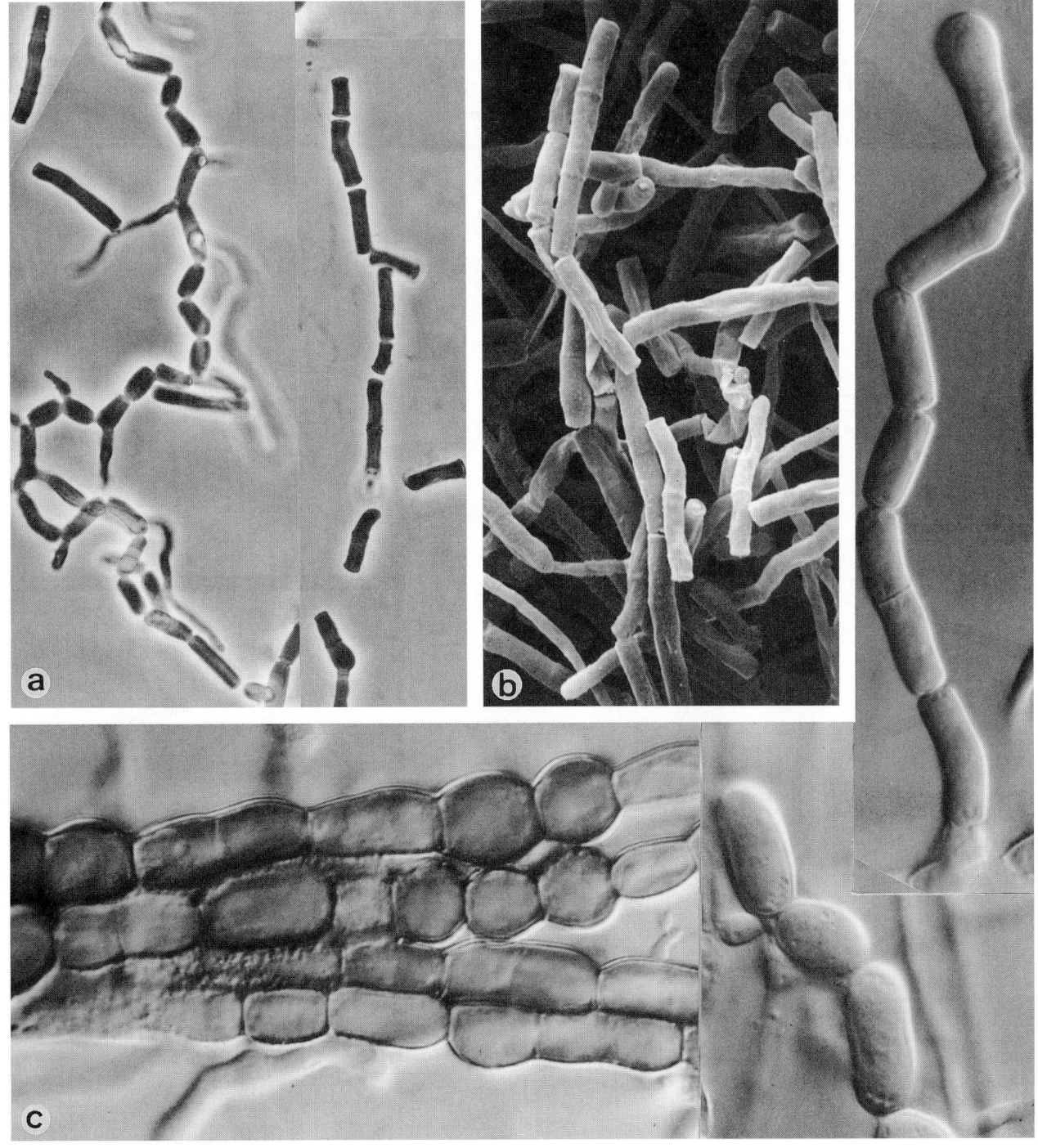

Scytalidium japonicum, CBS 494.88. Fertile hyphae with hyaline and dark arthroconidia. a. ×512; b. ×1050; c. ×1600.

Scytalidium lignicola Pesante

Colony characteristics. Colonies (PDA) effuse, flat with raised folds, cottony to woolly, initially whitish, finally becoming dark grey to black.

Microscopy. Hyphae hyaline at first, later becoming brown. Arthroconidia hyaline, thin-walled, rectangular, about 5-8 × 2 μm. Chlamydospore-like conidia single or in chains, dark brown, thick-walled, swollen up to 7 μm wide.

Molecular diagnostics. ITS restriction map based on CBS 387.59:

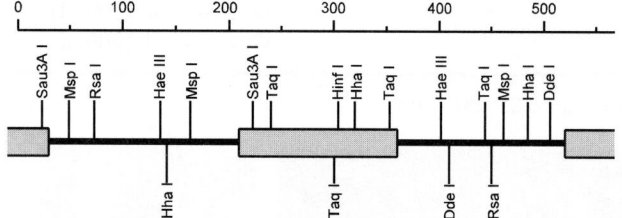

Pathogenicity. BSL-1. Strain IP 2311-95 caused a human a mycetoma (J. Waller, pers. comm). The interdigital mycosis reported by Costa *et al.* (1988a) concerned *Scytalidium dimidiatum* (p. 329); this might also hold true for a phaeohyphomycosis published by Dickinson *et al.* (1983).

Reference. Sigler & Carmichael (1976).

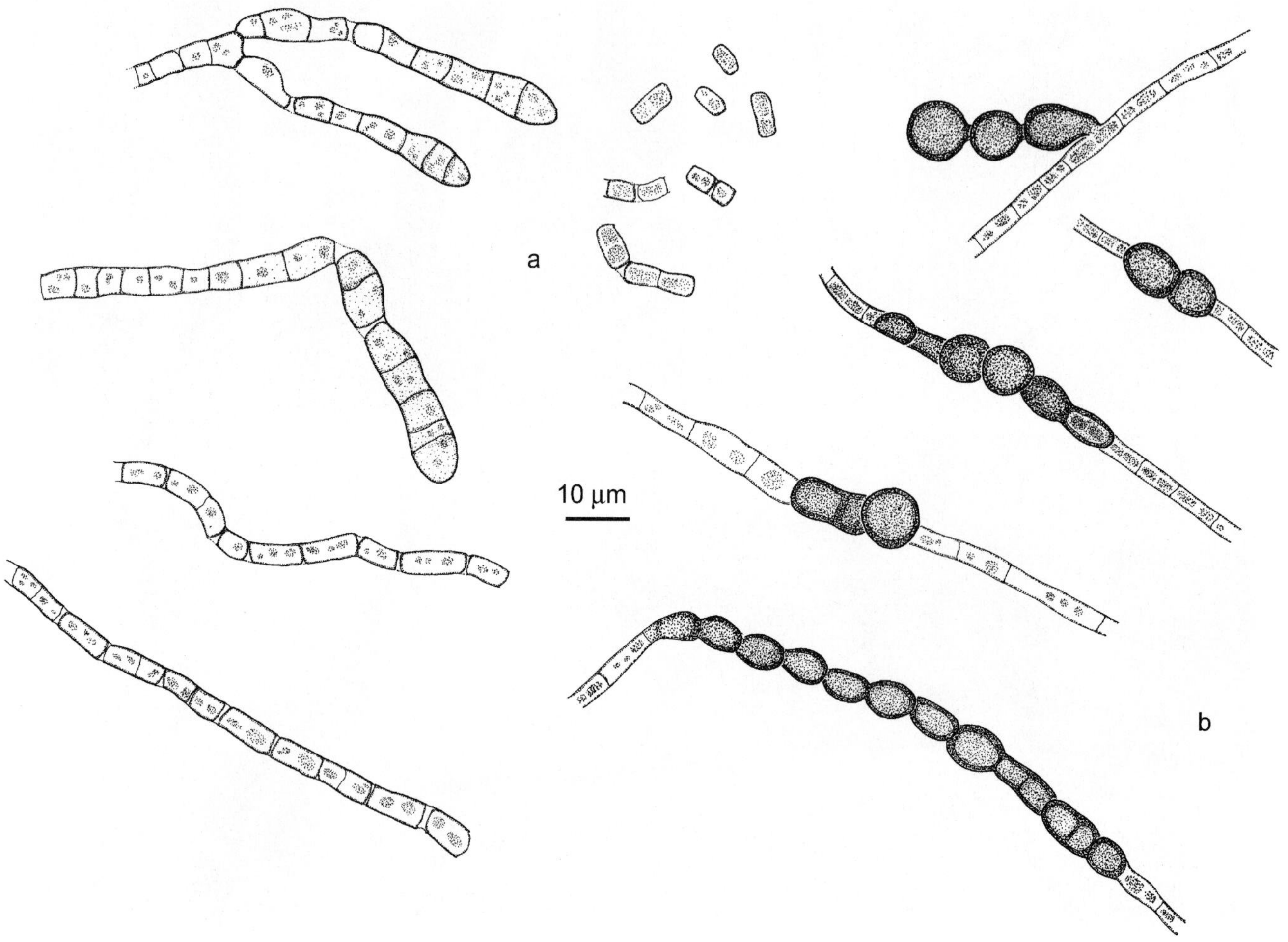

Scytalidium lignicola, CBS 233.57. a. Fertile hyphae and arthroconidia; b. hyphae with chlamydospores.

Antifungal susceptibility.

Antifungal	Mean MICs	Strains	Reference
AMB	0.5	2	Guarro *et al.* (2000d)
5FC	≤0.25	2	Guarro *et al.* (2000d)
FLZ	16	2	Guarro *et al.* (2000d)
ITZ	0.17	2	Guarro *et al.* (2000d)
KTZ	0.7	2	Guarro *et al.* (2000d)
MCZ	2	2	Guarro *et al.* (2000d)

Nomenclature. *Scytalidium lignicola* Pesante - Annali Spec. Agr., N. Ser., 11, Suppl., p. 261, 1957.

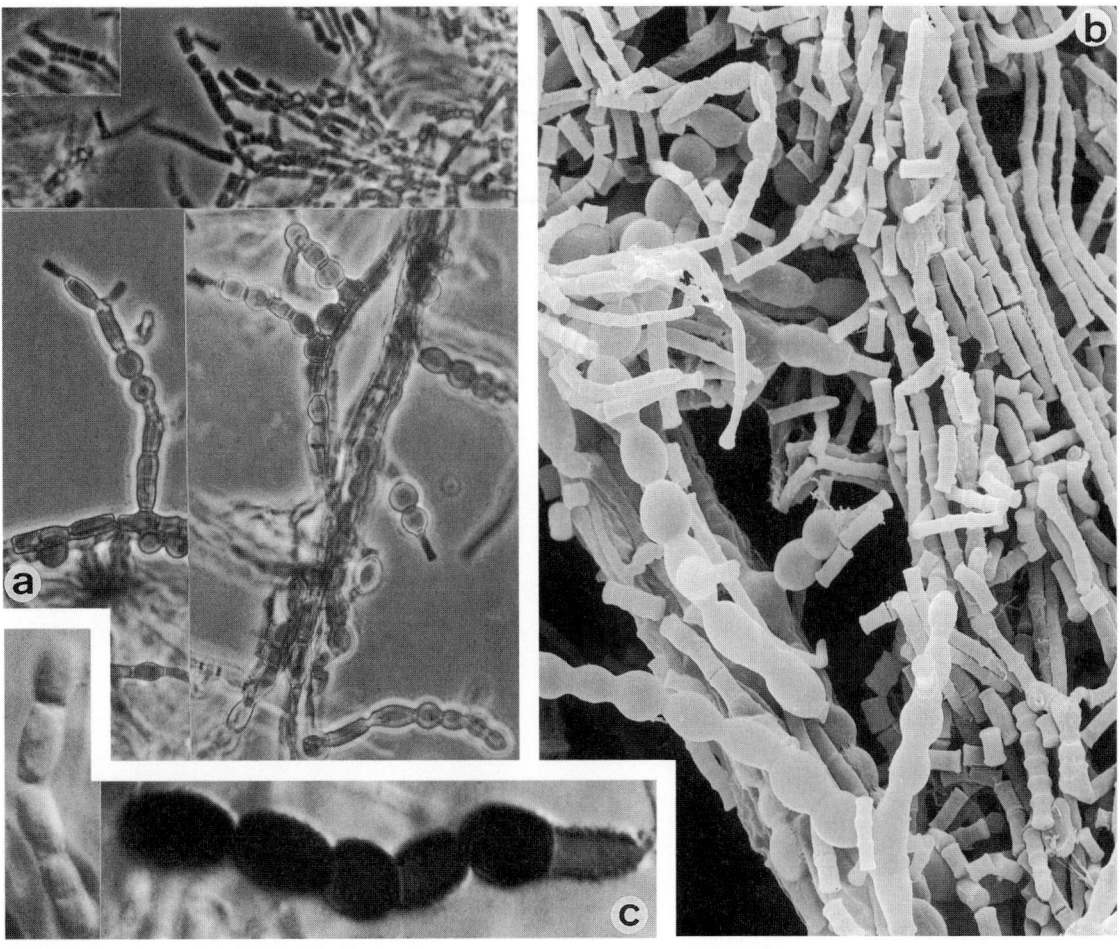

Scytalidium lignicola, CBS 233.57. Hyaline arthroconidia and intercalary dark brown chlamycospores. a. ×640; b. ×1400; c. ×1600.

Hyphomycetes. Genus: *SPOROTHRIX*

Generic description. Colonies rapidly growing, smooth and wrinkled or lanose and white. Conidiogenous cells arising from undifferentiated hyphae, forming conidia by sympodial growth on small, clustered denticles. Conidia 1-celled, tear-shaped to clavate, singly or in short chains. Often thin- or thick-walled, hyaline or brown conidia arise alongside the hyphae.

Teleomorph. *Ophiostoma* (*Ascomycota, Euascomycetes, Ophiostomatales: Ophiostomataceae*).

Molecular diagnostics. SSU restriction map based on NCBI M85053:

Differential diagnosis. The genus comprises ascomycetous hyphomycetes with 1-celled conidia on clustered denticles. A molecular comparison of *Sporothrix* species, using RFLP of mtDNA, was performed by Suzuki *et al.* (1988). Morphologically similar fungi are known in *Saccharomycetales* (*Candida*) and in *Exobasidiales* (*Cerinosterus*); for a comparison see under *S. schenckii* (p. 925).

References. De Hoog (1974, 1993), de Hoog *et al.* (1985), Wingfield *et al.* (1993).

Key to the treated species of Sporothrix:

1a. Colonies tough, often turning brown to black; sessile lateral conidia brown .... *S. schenckii* (925)
1b. Colonies floccose, remaining purely white; sessile lateral conidia
hyaline .... *S.* anam. of *Ophiostoma stenoceras* (299)

Sporothrix schenckii Hektoen & Perkins

Colony characteristics. Colonies (MEA) smooth and wrinkled, dirty whitish, tough; reverse grey to brownish-black.
Microscopy. Hyphae 1-2 μm wide. Conidiogenous cells arising from undifferentiated hyphae, forming conidia in groups on small, clustered denticles. Conidia 1-celled, tear-shaped to clavate, 2.5-5.5 × 1.5-2.5 μm. Often thin- or thick-walled, hyaline or brown conidia arise alongside the hyphae. A multilaterally budding yeast form may develop at 37°C.
Physiology.

Fermentation	–	Maltose	+	D-Mannitol	+
		α,α-Trehalose	+	Succinate	+
Growth:		methyl-α-D-Glucoside	+	Nitrate	+
D-Galactose	+	Cellobiose	+	Nitrite	+
D-Glucosamine	+	Salicin	+	Creatine	+
L-Sorbose	+	Melibiose	–	Creatinine	+
D-Ribose	+	Lactose	+	Cycloheximide	+
D-Xylose	+	Raffinose	v	Benomyl	–
L-Arabinose	+	Melezitose	+	Growth at 37°C	+
D-Arabinose	+	Soluble starch	+	Urease	–
L-Rhamnose	+	Glycerol	+		
Sucrose	+	*meso*-Erythritol	+		

Molecular diagnostics. Two 26S rDNA-based specific probes were developed by Sandhu *et al.* (1995). Infraspecific diversity based on mtDNA RFLP patterns was described by Suzuki *et al.* (1988) and Ishizaki *et al.* (1996).
Serology. An exoantigen test by immunodiffusion was described by Palmer *et al.* (1977).
Differential diagnosis. Conidium-bearing denticles unswollen, in contrast to those of the saprobe *Sporothrix inflata* de Hoog and the facultatively filamentous yeast *Candida ciferrii* (p. 192). The *Sporothrix* anamorph of *Ophiostoma stenoceras* (p. 299) has purely white, floccose, spreading colonies. The *Sporothrix*-like basidiomycete *Cerinosterus cyanescens* (p. 541) exudes a pH-dependent blue/purple pigment into the medium and has soft colonies.
Pathogenicity. BSL-2. The species is the agent of human sporotrichosis (p. 24), mostly developing characteristic lesions at regional lymph nodes (Restrepo *et al.*, 1986), occasionally in the face (Noguchi *et al.*, 1999). A localized cutaneous type of infection is also common (Itoh *et al.*, 1986). Occasionally arthritis (Purvis *et al.*, 1993), endophthalmitis (Cartwright *et al.*, 1993) or a combination of cutaneous and ocular mycosis (Vieira-Dias *et al.*, 1997) is noted. In tissue, mostly few tear-shaped yeast cells are seen, and sometimes characteristic asteroid bodies (Hiruma *et al.*, 1991) are present. Dissemination may occur in patients with impaired acquired immunity (Donabedian *et al.*, 1994; Morgan & Reves, 1996; Al-Tawfiq & Wools, 1998; Goldani *et al.*, 1999; Ware *et al.*, 1999) or after long-term use of corticosteroids (Severo *et al.*, 1999a). Exceptionally, the disorder has a fatal outcome in severely debilitated patients (Castrejón *et al.*, 1995). A pulmonary case in a HIV positive patient was presented by Gori *et al.* (1997) and a cutaneous infection in a patient with Cushing's disease by Kim *et al.* (1999). Werner & Werner (1994) surmised a significant role of cell-mediated immunity in disease outcome in cats. The fungus occurs in moist wood or plants, and develops in humans after puncture by a thorn or splinter, animal bites (Saravanakumar *et al.*, 1996) or occasionally without any apparent trauma (Cooper *et al.*, 1992). Epidemics can be traced back to common exposure to plant material, such as *Sphagnum* moss (Dixon *et al.*, 1991a) or hay (Dooley *et al.*, 1997). Transmission from animals is possible (Reed *et al.*, 1993; Bonifaz *et al.*, 1999) as the species frequently occurs in cats (Nakamura *et al.*, 1996), dogs (Marques *et al.*, 1993) and horses (Fishburn & Kelly, 1967; Greydanus-van der Putten *et al.*, 1994). Cases in armadillos were reported by Wenker *et al.* (1998).
Variety. The var. *luriei* (Ajello & Kaplan, 1969; Staib & Blisse, 1974) differs by the production of large, often septate budding cells. It does not assimilate creatine and creatinine. A fatal case by this variety was reported by Padhye *et al.* (1992a).
References. Summerbell *et al.* (1993), Padhye (1995), Vismer & Hull (1997).

Antifungal susceptibility.

Antifungal	MICs range	Mean MICs	Strains	Reference
AMB	0.5-16	2	5	Espinel-Ingroff *et al.* (1995)
FLZ	0.06-16	64	5	Espinel-Ingroff *et al.* (1995)
ITZ	0.06-16	5	5	Espinel-Ingroff *et al.* (1995)
KTZ	0.25-4	2	5	Espinel-Ingroff *et al.* (1995)

Nomenclature var. *schenckii*. - *Sporothrix schenckii* Hektoen & Perkins - J. Exp. Med. 5: 77, 1900 ≡ *Sporotrichum schenckii* (Hektoen & Perkins) de Beurmann & Gougerot - Archs Parasit. 15: 5, 1911 ≡ *Sporotrichum schenckii-beurmannii* Greco var. *schenckii* (Hektoen & Perkins) de Beurmann & Gougerot - Archs Parasit. 15: 38, 1911 ≡ *Rhinocladium schenckii* (Hektoen & Perkins) Verdun - Précis Parasitol., ed. 2, p. 677, 1913 ≡ *Rhinotrichum schenckii* (Hektoen & Perkins) Ota - Jpn. J. Derm. Urol. 27: 921, 1927 ≡ *Sporotrichum beurmannii* Matruchot & Ramond var. *schenckii* (Hektoen & Perkins) Redaelli & Ciferri - Tratt. Micopat. Um. 5: 452, 1942.

Sporotrichum beurmannii Matruchot & Ramond - C.R. Hebd. Séanc. Mém. Soc. Biol. 2: 380, 1905 ≡ *Trichosporium beurmanni* (Matruchot & Ramond) Lutz & Splendore - An. Ig. Sper. 17: 581, 1907 ≡ *Rhinocladium beurmannii* (Matruchot & Ramond) Vuillemin - Bull. Séanc. Soc. Sci. Nancy 11: 138, 1910 ≡ *Sporotrichum schenckii-beurmannii* Greco var. *beurmannii* (Matruchot & Ramond) de Beurmann & Gougerot - Archs Parasit. 15: 39, 1911≡ *Sporotrichopsis beurmannii* (Matruchot & Ramond) Guéguen, *in* de Beurmann & Gougerot - Archs Parasit. 15: 104, 1911 ≡ *Sporothrix beurmannii* (Matruchot & Ramond) Meyer & Aird - J. Infect. Dis. 16: 399, 1915 ≡ *Rhinotrichum beurmannii* (Matruchot & Ramond) Ota - Jpn. J. Derm. Urol. 28: 4, 1928 ≡ *Sporotrichum schenckii* (Hektoen & Perkins) de Beurmann & Gougerot var. *beurmannii* (Matruchot & Ramond) C.W. Dodge - Med. Mycol. p. 805, 1935.

Sporotrichum schenckii-beurmannii Greco - Argent. Med. 45: 699, 1907 ≡ *Sporothrix schenckii-beurmannii* (Greco) Meyer & Aird - J. Infect. Dis. 16: 407, 1915.

Sporotrichum indicum Castellani - J. Trop. Med. Hyg. 11: 261, 1908 ≡ *Sporothrix beurmannii* Matruchot & Ramond var. *indicum* (Castellani) de Beurmann & Gougerot - Revue Méd. Hyg. Trop. 7: 190, 1910 ≡ *Rhinocladium indicum* (Castellani) Verdun - Précis Parasit., ed. 2, p. 678, 1913 ≡ *Rhinotrichum indicum* (Castellani) Ota - Jpn. J. Derm. Urol. 27: 928, 1927.

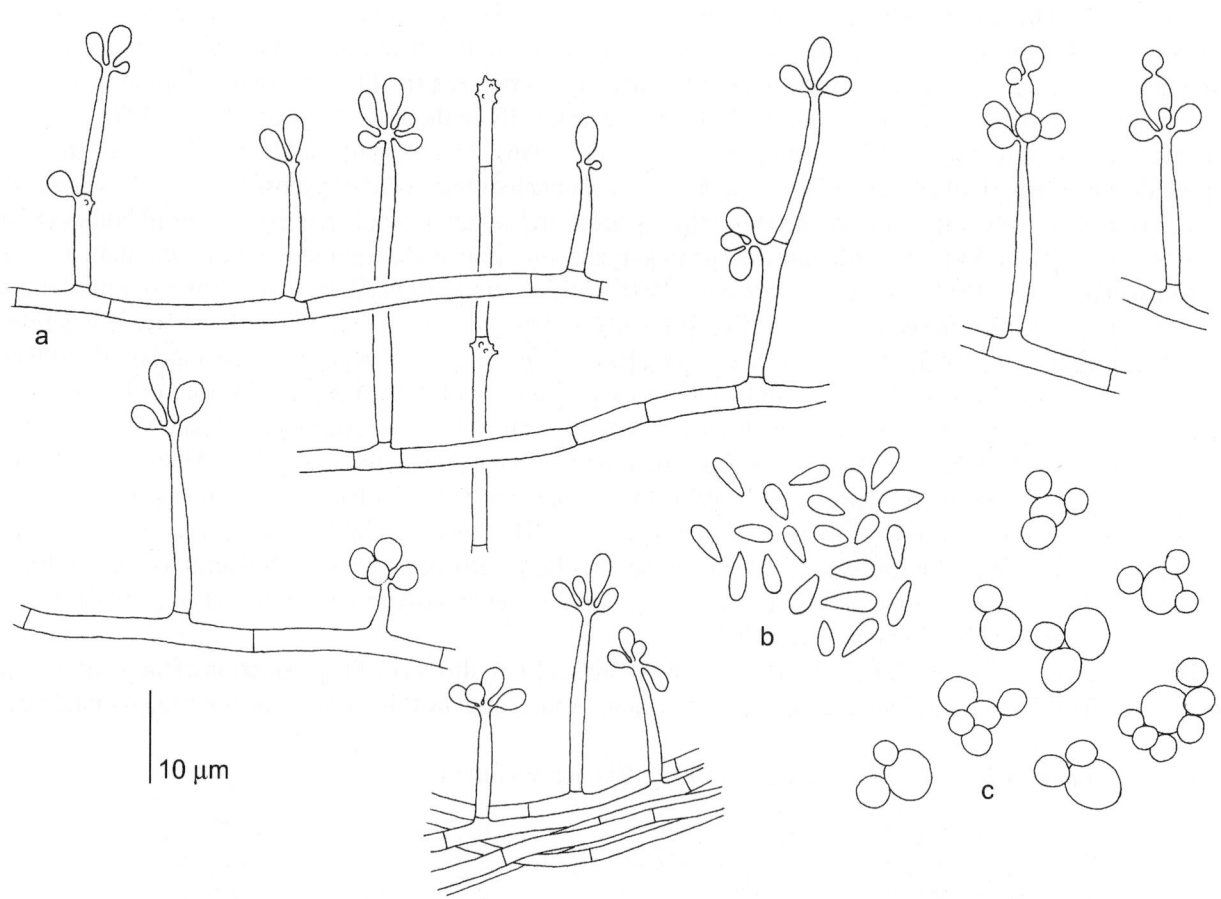

Sporothrix schenckii, CBS 359.36. a. Conidial apparatus; b. liberated conidia; c. yeast like cells produced *in vitro* at 37°C.

Sporotrichum asteroides Splendore - Revta Soc. Cient. S. Paulo 3: 62, 1908 ≡ *Sporotrichum beurmannii* Matruchot & Ramond var. *asteroides* (Splendore) de Beurmann & Gougerot - Les Sporotrichoses p. 179, 1912 ≡ *Rhinotrichum asteroides* (Splendore) Verdun - Précis Parasitol. 1912 ≡ *Rhinocladium asteroides* (Splendore) Verdun - Précis Parasit., ed. 2, p. 678, 1913 ≡ *Sporothrix asteroides* (Splendore) J. Davis - J. Infect. Dis. 12: 435, 1913 ≡ *Rhinocladium beurmannii* (Matruchot & Ramond) Vuillemin var. *asteroides* (Splendore) C.W. Dodge (as 'Vuillemin') - Med. Mycol. p. 802, 1935.

Sporotrichum equi Carougeau - J. Méd. Vét. Zootechn. 60: 80, 1909 ≡ *Rhinocladium equinum* Lurie - Mycologia 40: 107, 1948 (name change) ≡ *Rhinocladium equi* (Carougeau) Lurie - Mycologia 40: 112, 1948.

Sporotrichum jeanselmei Brumpt & Langeron, *in* Brumpt - Précis Parasit. p. 889, 1910 ≡ *Sporotrichum beurmannii* Matruchot & Ramond var. *jeanselmei* (Brumpt & Langeron) de Beurman & Gougerot - Archs Parasit. 15: 51, 1911 ≡ *Rhinocladium jeanselmei* (Brumpt & Langeron) Verdun - Précis Parasit., ed. 2, 1913 ≡ *Rhinotrichum jeanselmei* (Brumpt & Langeron) Ota - Jpn. J. Derm. Urol. 28: 5, 1928.

Sporotrichum fonsecae Pereira - Revta Med.-Cirurg. Brasil 37: 265, 1929 ≡ *Rhinocladium fonsecae* (Pereira) Pereira - Revta Med.-Cirurg. Brasil 38: 169, 1930.

Sporotrichum schenckii (Hektoen & Perkins) de Beurmann & Gougerot var. *greconis* (C.W. Dodge) - Med. Mycol. p. 808, 1935 ≡ *Sporotrichum greconis* (C.W. Dodge) Gougerot - Ann. N.Y. Acad. Sci. 50: 1348, 1950.

Sporotrichum grigsbyi C.W. Dodge - Med. Mycol. p. 801, 1935.

Sporotrichum verticilloides Sartory, R. Sartory & J. Meyer - C.R. Hebd. Séanc. Acad. Sci., Paris 201: 1502, 1935.

Sporotrichum tropicale Panja, Dey & Ghosh - Ind. Med. Gaz. 82: 202, 1947.

Nomenclature var. *luriei*. *Sporothrix schenckii* Hektoen & Perkins var. *luriei* Ajello & Kaplan - Mykosen 12: 642, 1969.

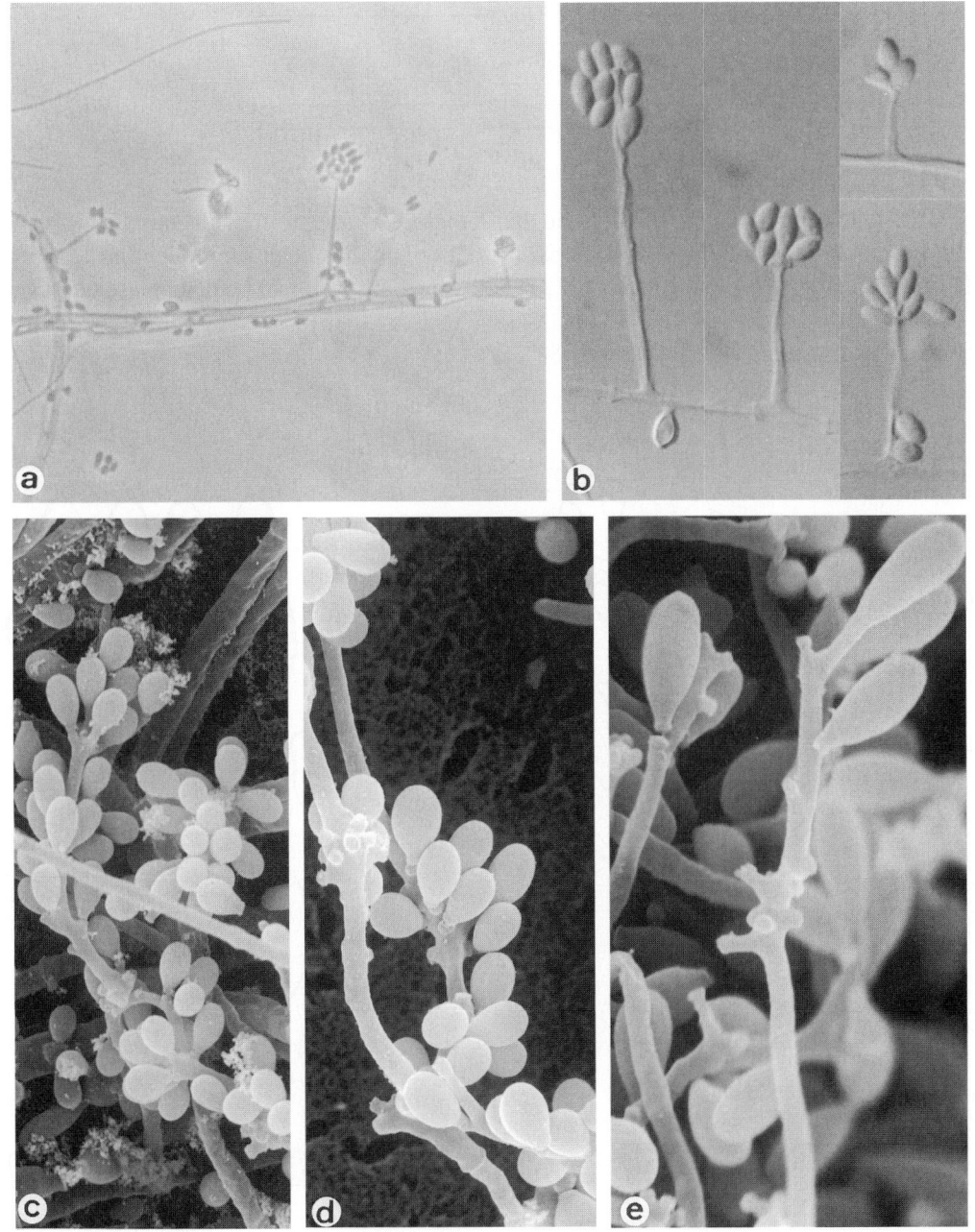

Sporothrix schenckii, a, b. IMI 77984, c-e. CBS 359.36. Denticulate conidiogenous cells and conidia. a. ×505; b. ×1265; c. ×3555; d. ×4345; e. ×5925.

Hyphomycetes, filamentous basidiomycetes. Genus: *SPOROTRICHUM*

Sporotrichum pruinosum Gilman & Abbott

Colony characteristics. Colonies (MEA) expanding, filling the Petri-dish within 14 days, flat, soft, farinose, dry, white to cream-coloured.

Microscopy. Marginal hyphae hyaline, 3-8 µm wide, with few septa, later developing more septation, soon with profuse branching at right angles, each branch terminating with a conidium. Conidia with broad base, ellipsoidal to nearly cylinderical, 5-10 × 3-5 µm, pale brown, with rather thick, smooth walls. Thick-walled, spherical chlamydospores 20-60 µm in diam, and hyaline arthroconidia frequently present.

Teleomorph. *Phanerochaete chrysosporium* Burdsall (*Basidiomycota, Hymenomycetes, Stereales: Corticiaceae*).

Physiology. Optimal growth at 36-40°C, maximum growth temperature 46-49°C.

Molecular diagnostics. SSU restriction map based on NCBI AF026593:

Pathogenicity. BSL-1. The fungus can be inhaled and then develops a tissue form with giant, thick-walled chlamydospores which are similar to those of *Emmonsia parva* (p. 636). This has been reproduced *in vivo* by experimental inoculation (Thirumalachar *et al.*, 1960; Singh *et al.*, 1992). The giant cells can be demonstrated in sputum (Khan *et al.*, 1988).

References. Burdsall & Eslyn (1974), Stalpers (1984), Wu (1998).

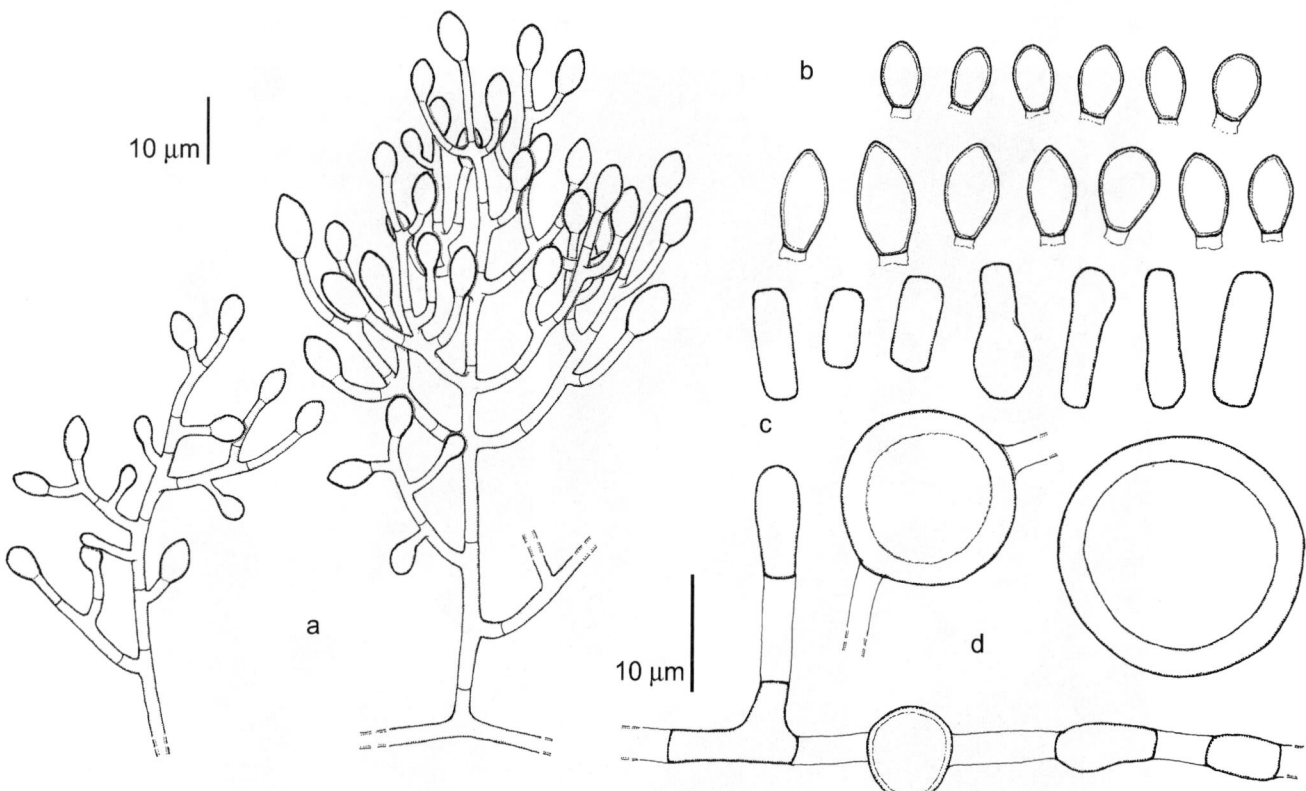

Sporotrichum pruinosum CBS 671.71. a. Conidiophores; b. conidia; c. arthroconidia; d. chlamydospores.

928

Antifungal susceptibility.

Antifungal	Mean MICs	Strains	Reference
AMB	0.03	1	Unpublished data
5FC	256	1	Unpublished data
FLZ	1	1	Unpublished data
ITZ	0.03	1	Unpublished data
KTZ	0.25	1	Unpublished data
MCZ	2	1	Unpublished data

Nomenclature. *Sporotrichum pruinosum* Gilman & Abbott - Iowa St. Coll. J. Sci. 1: 306, 1927 ≡ *Chrysosporium pruinosum* (Gilman & Abbott) Carmichael - Can. J. Bot. 40: 1166, 1962.

Emmonsia brasiliensis Batista, Lima, Pessoa & Shome - Revta Fac. Med. Univ. Ceará 3: 52, 1963.

Emmonsia ciferrina Thirumalachar, Padhye & Srinivasan - Mycopath. Mycol. Appl. 26: 330, 1960.

Phanerochaete chrysosporium Burdsall, *in* Burdsall & Eslyn - Mycotaxon 1: 124, 1974.

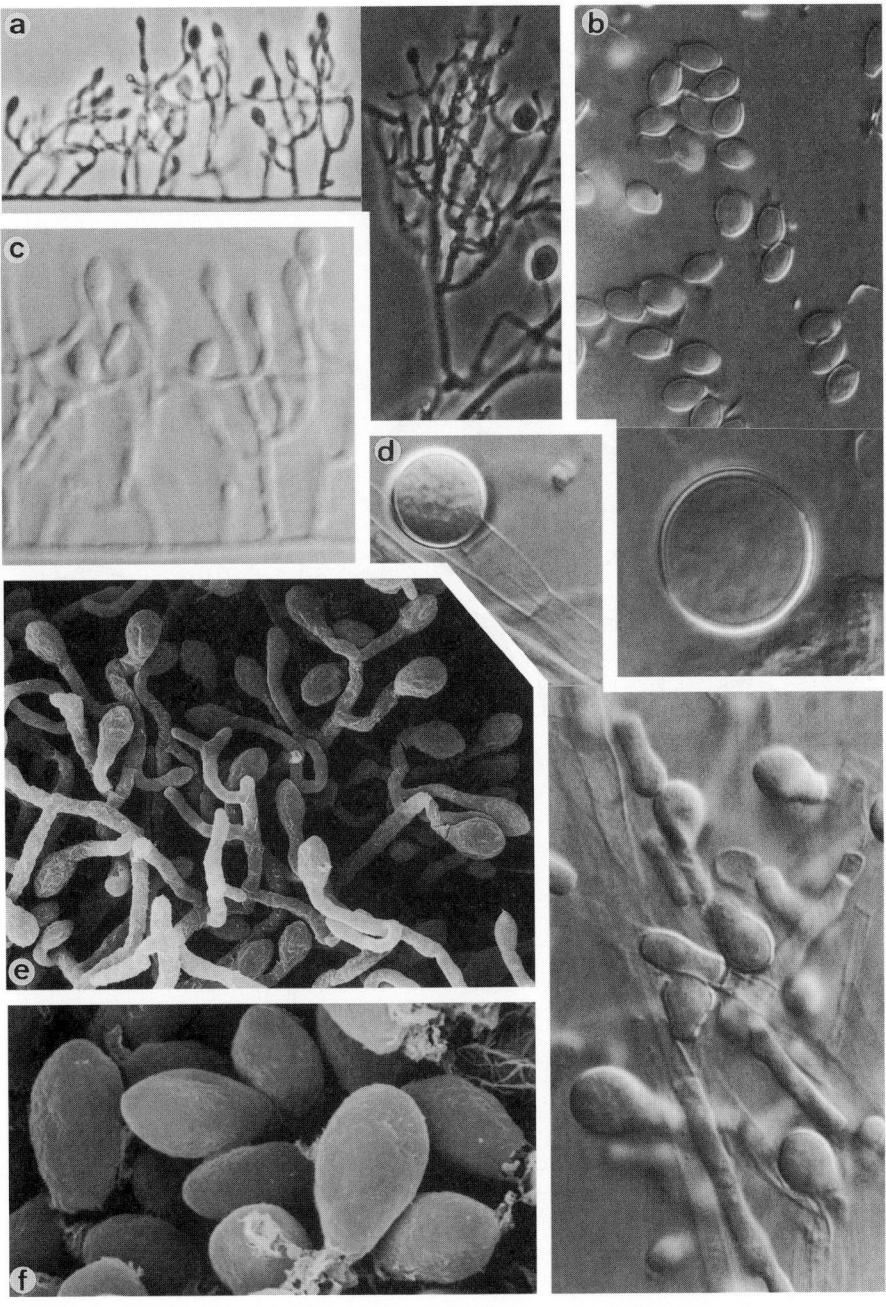

Sporotrichum pruinosum, a-c, e, f. CBS 671.71; d. CBS 129.27. Conidiophores, liberated conidia and chlamydospores. a. ×640; b-d. ×1600; e. ×2000; f. ×5750.

929

Hyphomycetes. Genus: *STAPHYLOTRICHUM*

Staphylotrichum coccosporum J. Meyer & Nicot

Colony characteristics. Colonies (PDA) expanding, cottony, yellowish-brown.
Microscopy. Hyphae hyaline. Conidiophores erect, up to 1200 μm long, thick-walled and dark brown in the lower part, thin-walled and hyaline near the apex, apically with irregular, hyaline side branches bearing sessile conidia. Conidia smooth- or nearly smooth-walled and thick-walled, pale to olivaceous brown, (sub)spherical, 10-12 μm diam.
Pathogenicity. BSL-1. Saprobe, found in soil. A subcutaneous infection mixed with *Microsporum canis* in a cat was reported by Fuchs *et al.* (1996).
References. Meyer & Nicot (1957), Maciejowska & Williams (1963).
Nomenclature. *Staphylotrichum coccosporum* J. Meyer & Nicot - Bull. Trim. Soc. Mycol. Fr. 72: 323, 1957.

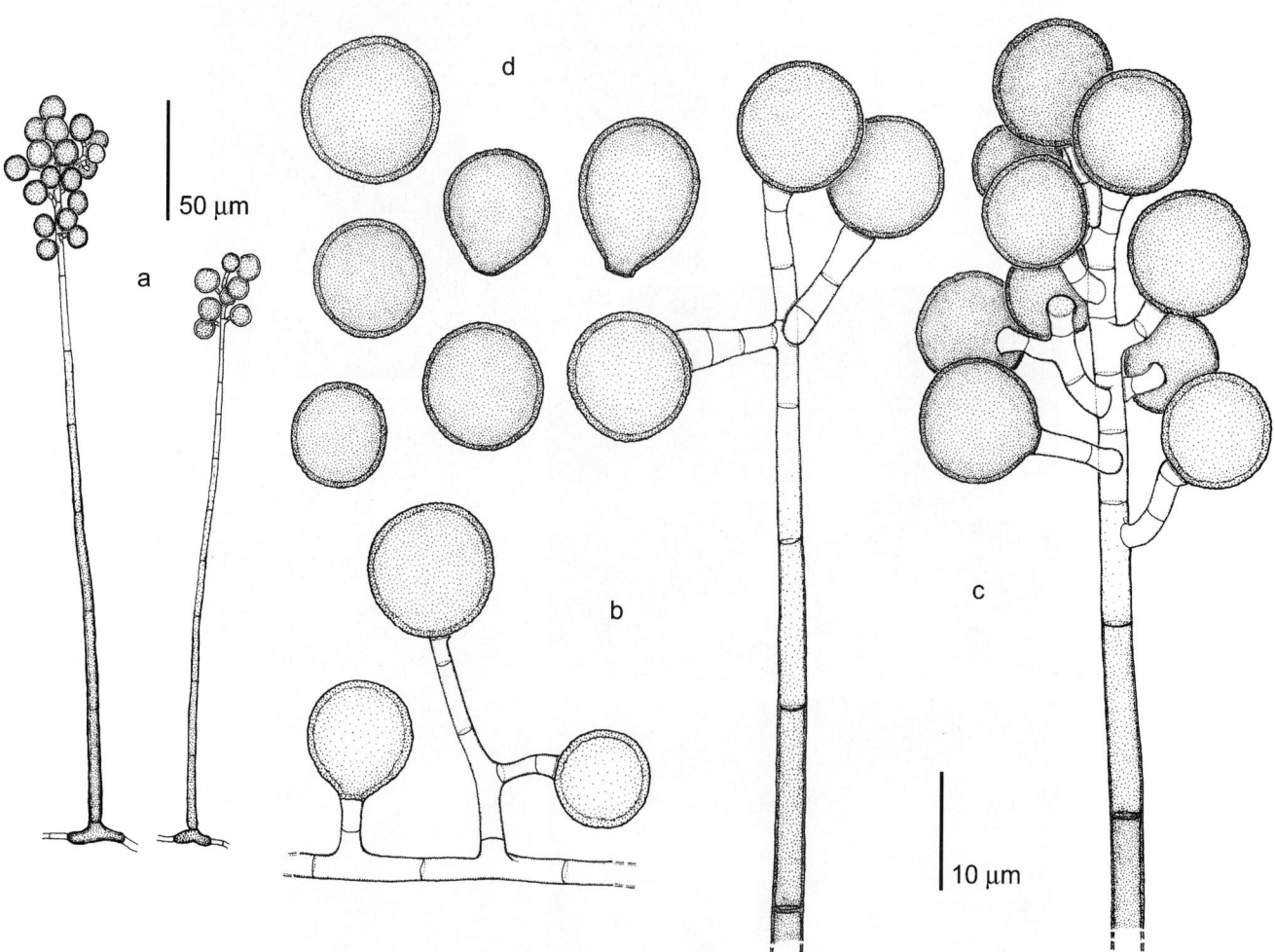

Staphylotrichum coccosporum, CBS294.55. a, c. Conidiophores; b. undifferentiated hypha producing conidia; d. liberated conidia.

930

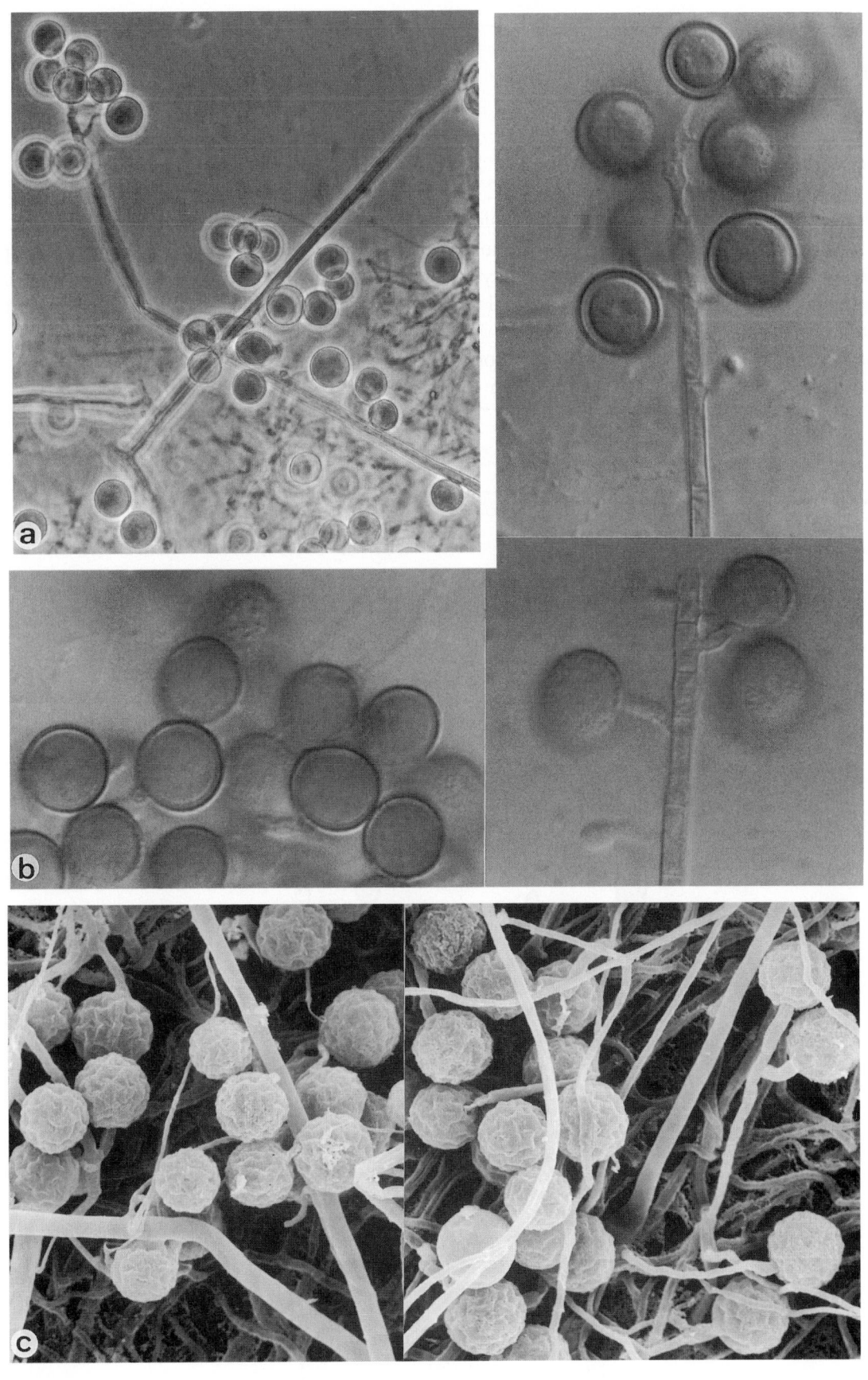

Staphylotrichum coccosporum, CBS 294.55. Conidiophores and conidia. a. ×640; b. ×1600; c. ×1800.

Hyphomycetes. Genus: *STENELLA*

Stenella araguata Sydow

Colony characteristics. Colonies (PDA) growing slowly, velvety, mouse grey to olive-green; reverse grey to black.
Microscopy. Conidiophores erect, straight to flexuose, simple or branched, pale brown to olivaceous, up to 65 μm long, 2-4 μm wide. Conidiogenous cells integrated, sympodial, often geniculate, with small, dark brown scars. Conidia solitary or in branched, acropetal chains, cylindrical to obclavate, verruculose, 0-4-septate, mostly 1-septate, pale to olivaceous-brown, 7-21 × 2.2-4.8 μm.
Molecular diagnostics. ITS restriction map based on CBS 486.80:

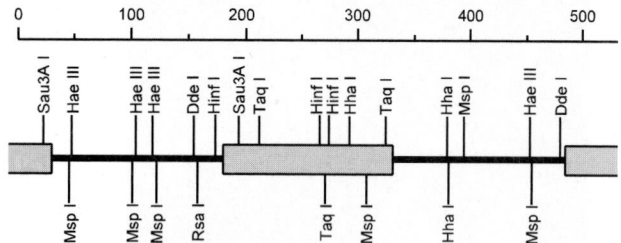

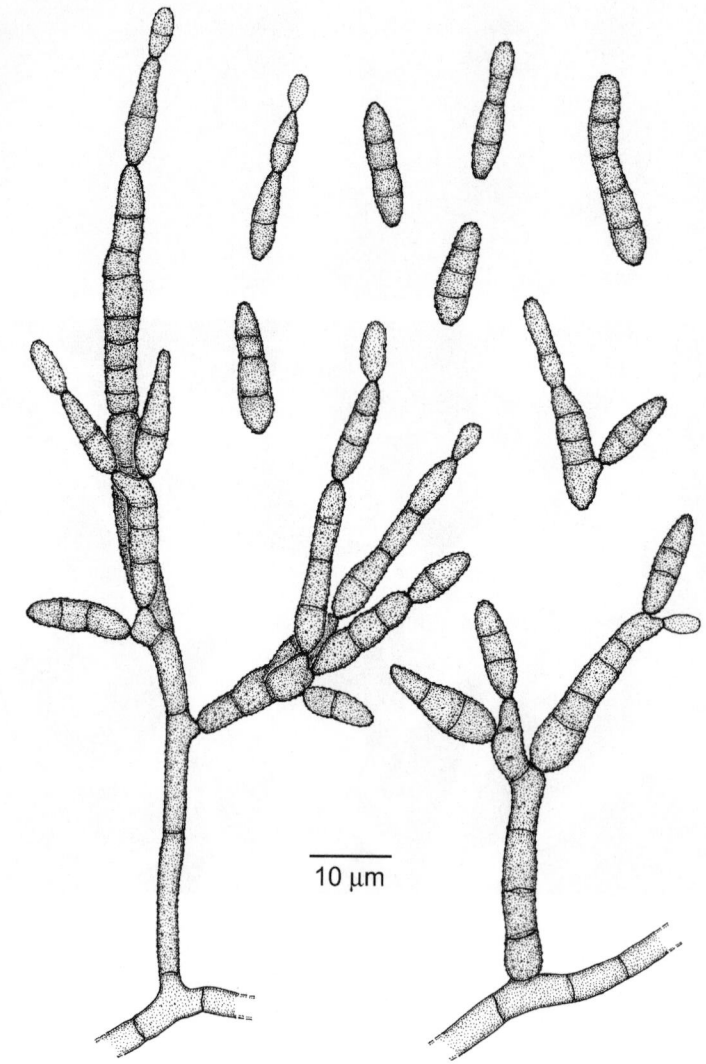

Stenella araguata, CBS 105.75. Conidiophores and conidia.

Pathogenicity. BSL-1. Several cases of a tinea nigra-like cutaneous infections were reported from Venezuela (di Prisco & Borelli, 1973; Marcano & Hutton, 1973; Borelli & Marcano, 1973; Reyes & Borelli, 1974).

References. Ellis (1971), McGinnis & Padhye (1978).

Nomenclature. *Stenella araguata* Sydow - Annls Mycol. 28: 205, 1930.
  Cladosporium castellanii Borelli & Marcano - Castellania 1: 151, 1973.

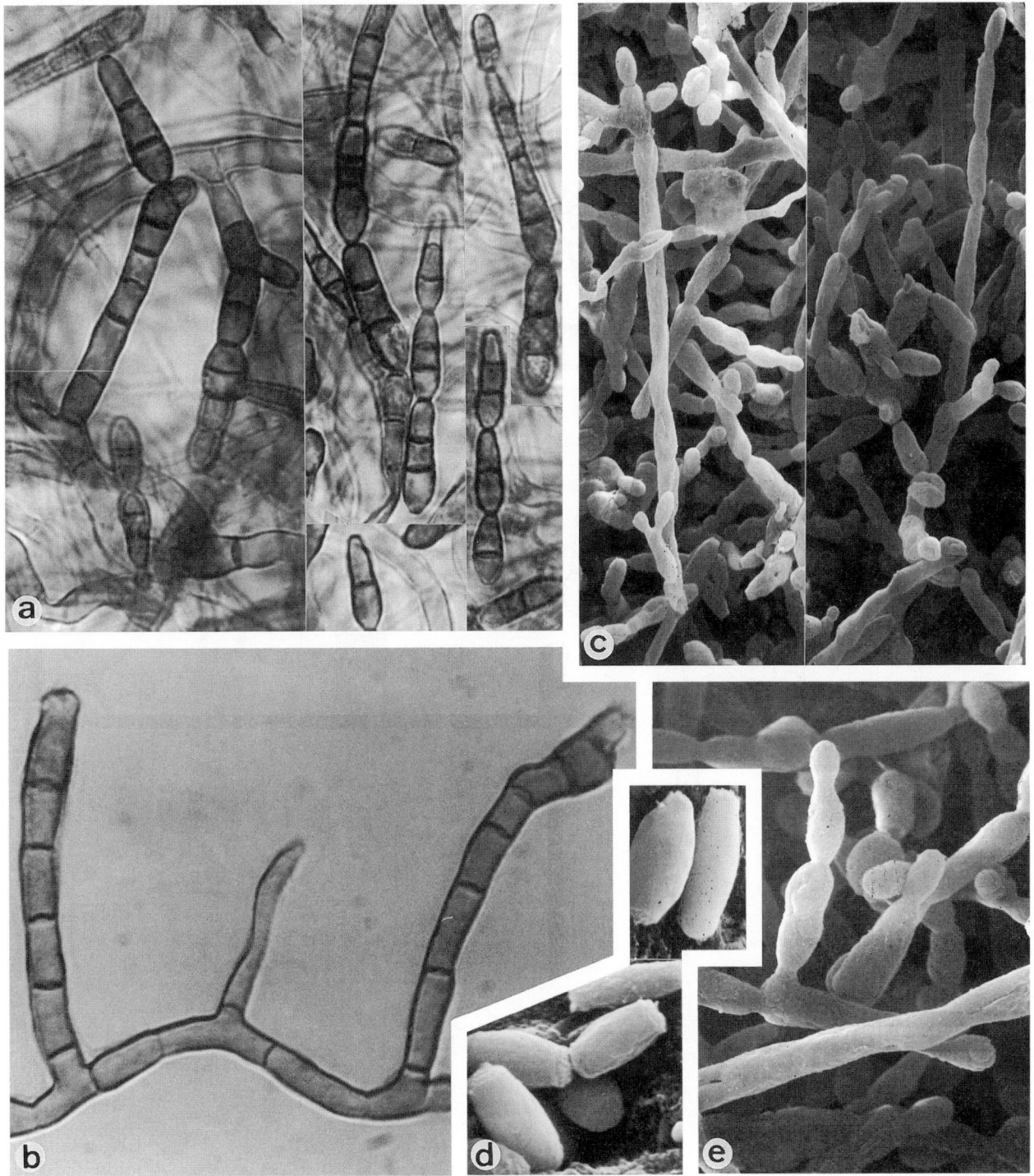

Stenella araguata, CBS 105.75. Conidiophores and conidia. a. ×1280; b. ×1600; c. ×1200; d. ×5200; e. ×2175.

Hyphomycetes. Genus: *TAENIOLELLA*

Taeniolella stilbospora (Corda) S.J. Hughes

Colony characteristics. Colonies (OA) restricted, dark olivaceous, velvety.

Microscopy. Conidiophores caespitose or scattered, 3-5 µm wide. Conidia straight or flexuose, cylindrical, rounded at the apex, often truncate at the base, brownish, smooth-walled, 3-24-septate, 25-140 × 7-11 µm.

Pathogenicity. BSL-1. Cutaneous lesions in the face of a human patient (Stewart *et al.,*1975; Pietrini & Stewart, 1977). A subcutaneous phaeohyphomycosis after thorn prick was caused by *T. exilis* (Alonso *et al.*, 1993). The latter species has conidia 12-15 µm in width.

Reference. Ellis (1971).

Nomenclature. *Torula stilbospora* Corda, *in* Sturm - Deutsch. Fl. 3, Pilze 2-8: 99, 1829 ≡ *Taeniolella stilbospora* (Corda) S.J. Hughes - Can. J. Bot. 36: 817, 1958.

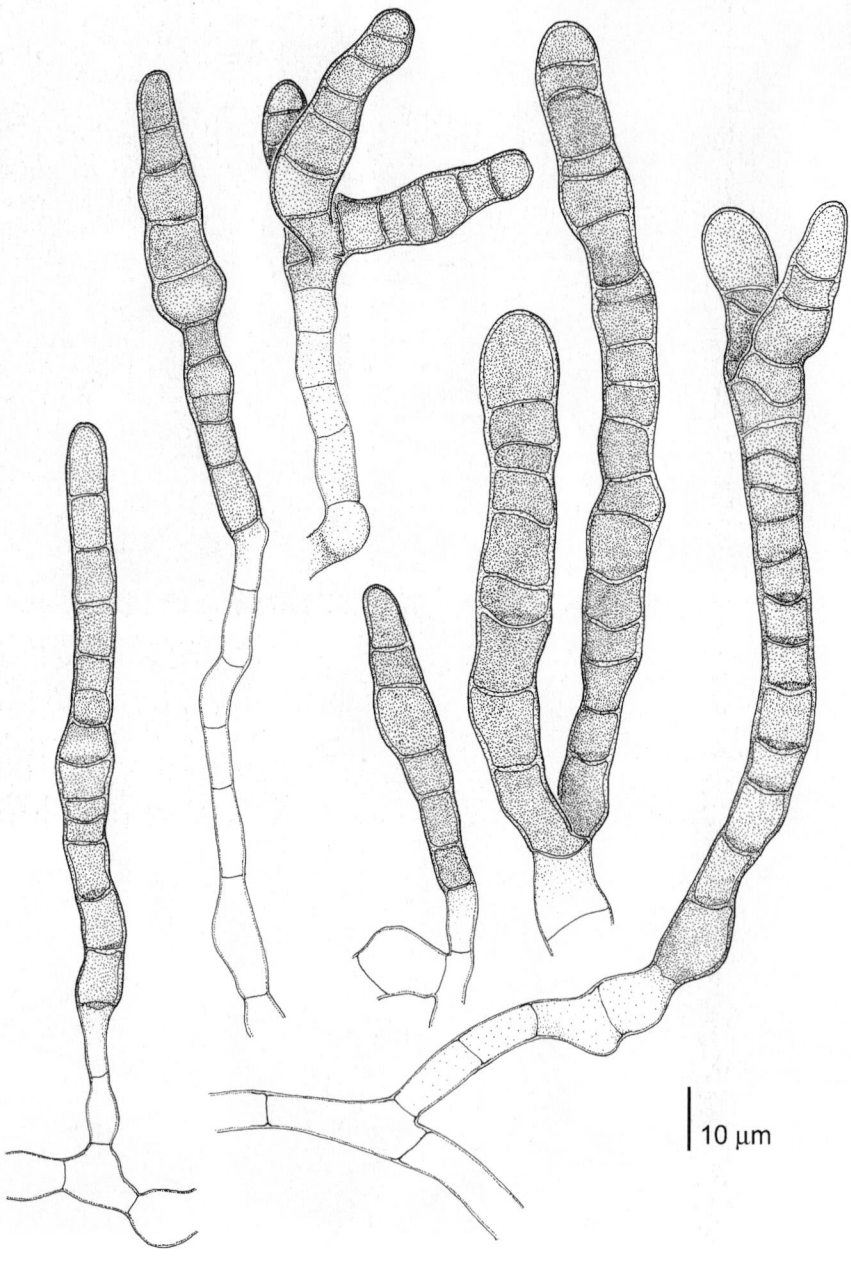

10 µm

Taeniolella stilbospora, CBS 307.75. Conidiophores and conidia.

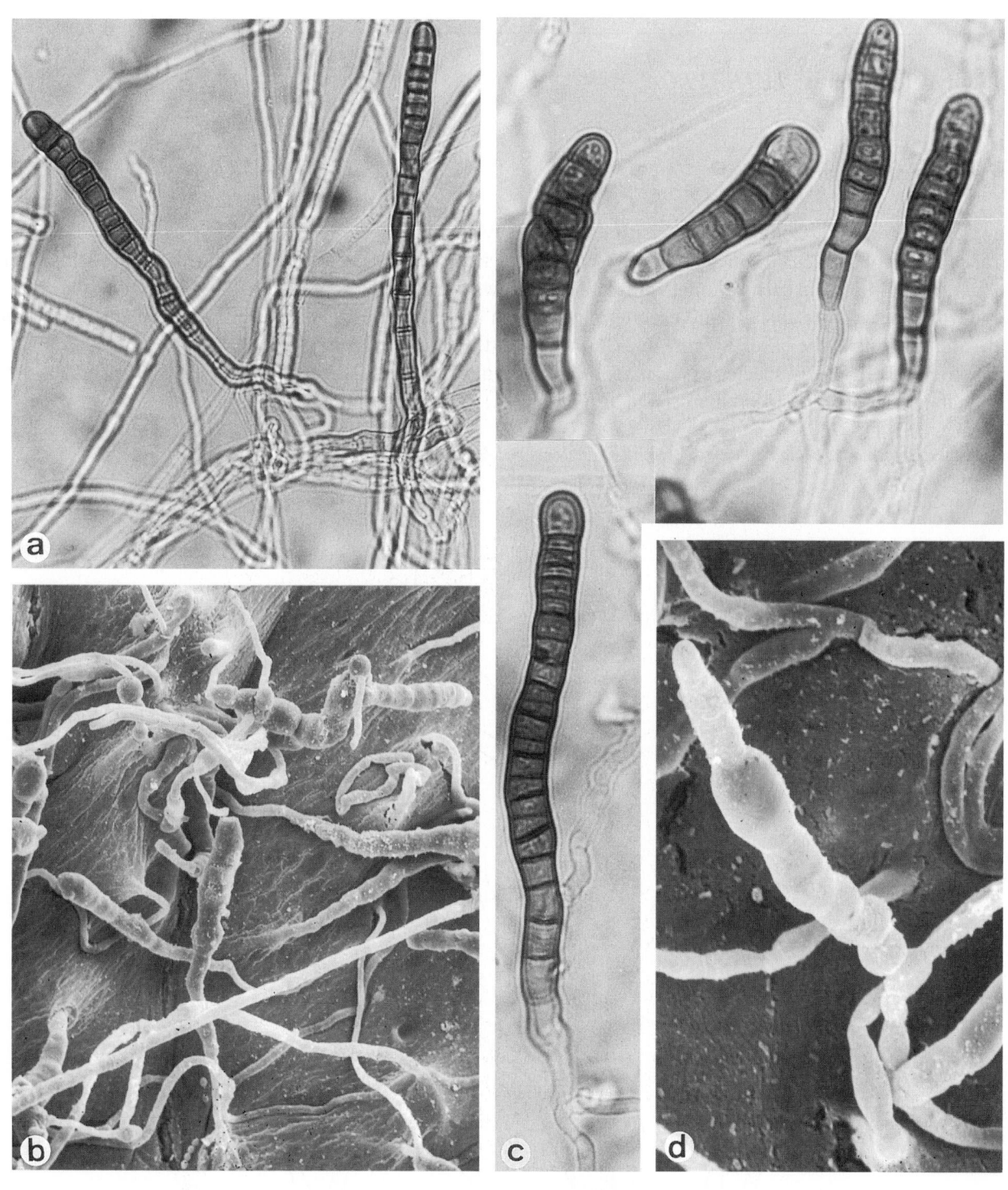

Taeniolella stilbospora, CBS 307.75. Conidiophores and conidia. a. ×400; b. ×620; c. ×640; d. ×1750.

Hyphomycetes. Genus: *TETRAPLOA*

Tetraploa aristata Berk. & Br.

Colony characteristics. Colonies (OA) restricted, dark greyish, velvety.

Microscopy. Hyphae 1.5-3.0 µm wide, pale to dark brown. Conidia sessile, solitary, brown, verruculose, tetra-columnar, with 3-4 cells in each column, with shallow furrows between the columns, 30-40 × 12-30 µm; the columns diverge from one another apically and each terminate in an appendage; appendages setiform, septate, 20-60 µm in length, 5-7 µm wide at the base, tapering to 2-3 µm wide at the tip.

Pathogenicity. BSL-1. The species has been reported as agent of keratitis (Newmark & Polack, 1970; Zapater, 1980) and subcutaneous infection (Markham *et al.*, 1990).

Reference. Ellis (1971).

Nomenclature. *Tetraploa aristata* Berkeley & Broome - Ann. Mag. Nat. Hist., Ser. 2, 5: 459, 1850.

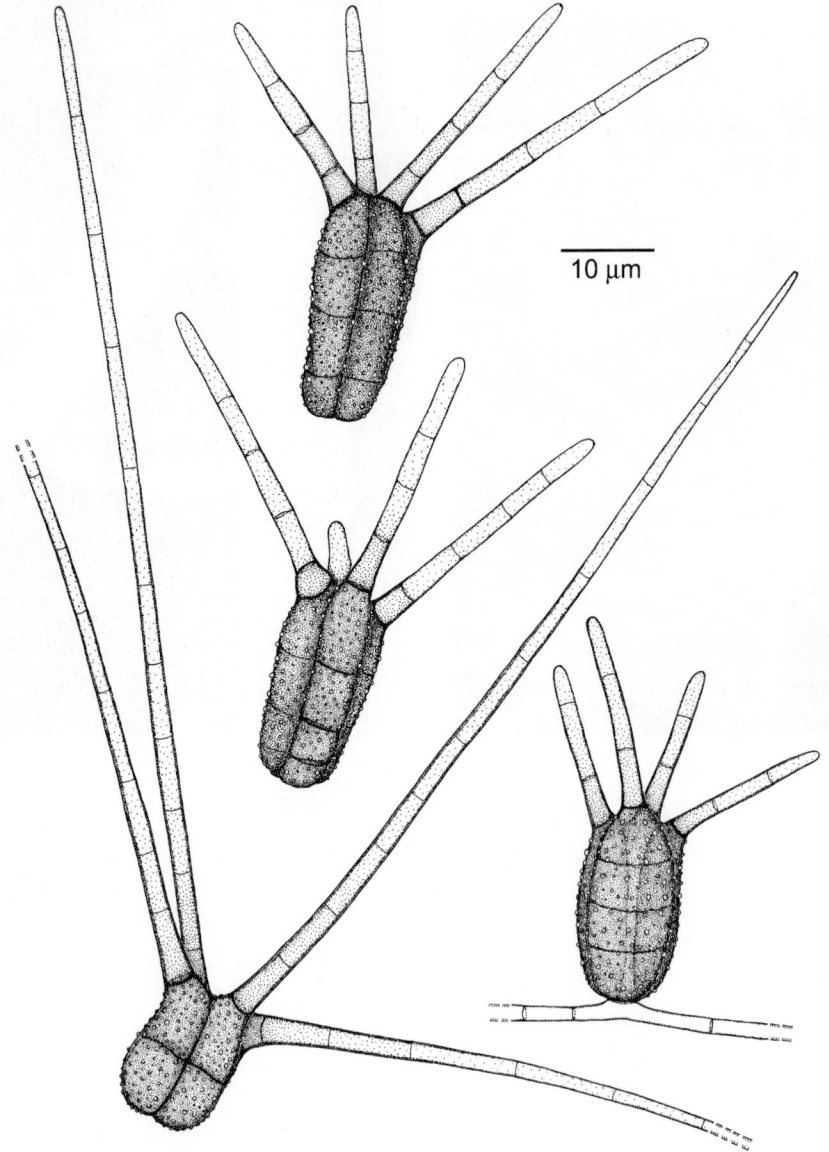

Tetraploa aristata, IMI 161260. Conidiogenous cells and conidia.

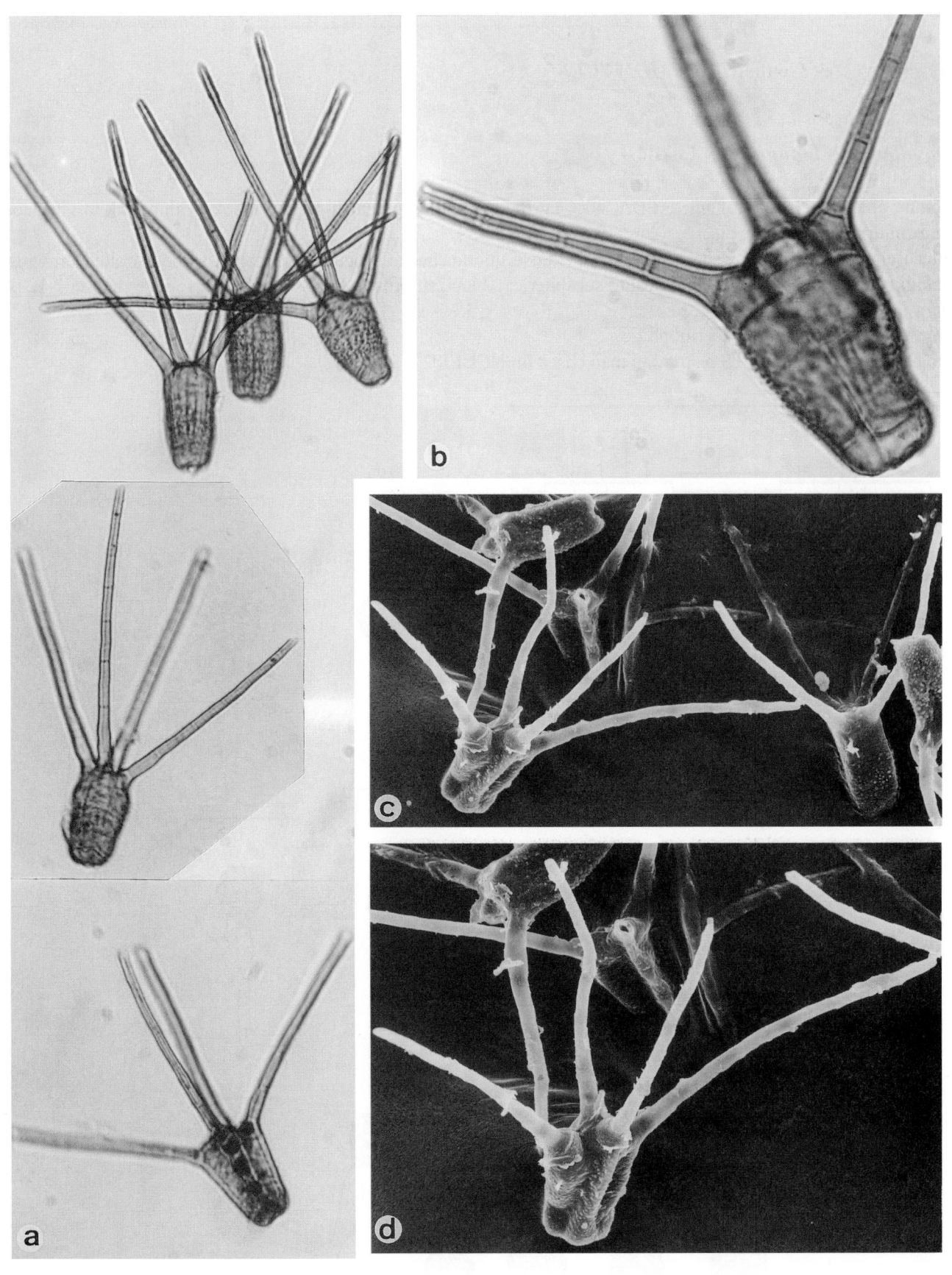

Tetraploa aristata, IMI 161260. Conidia. a. ×512; b. ×1280; c. ×690; d. ×890.

Hyphomycetes. Genus: *THERMOMYCES*

Thermomyces lanuginosus Tsiklinsky

Colony characteristics. Colonies (PDA, 40°C) growing rapidly, greenish-grey to black, with a pink to vinaceous pigment diffusing into the agar.

Microscopy. Conidiophores short, straight or flexuose, unbranched or irregularly branched, smooth-walled, brownish. Conidia solitary, dark brown, spherical to subspherical, blackish-brown, thick-walled, coarsely verrucose, 7-12 µm diam.

Physiology. The species is thermophilic.

Molecular diagnostics. ITS restriction map based on NCBI J02745:

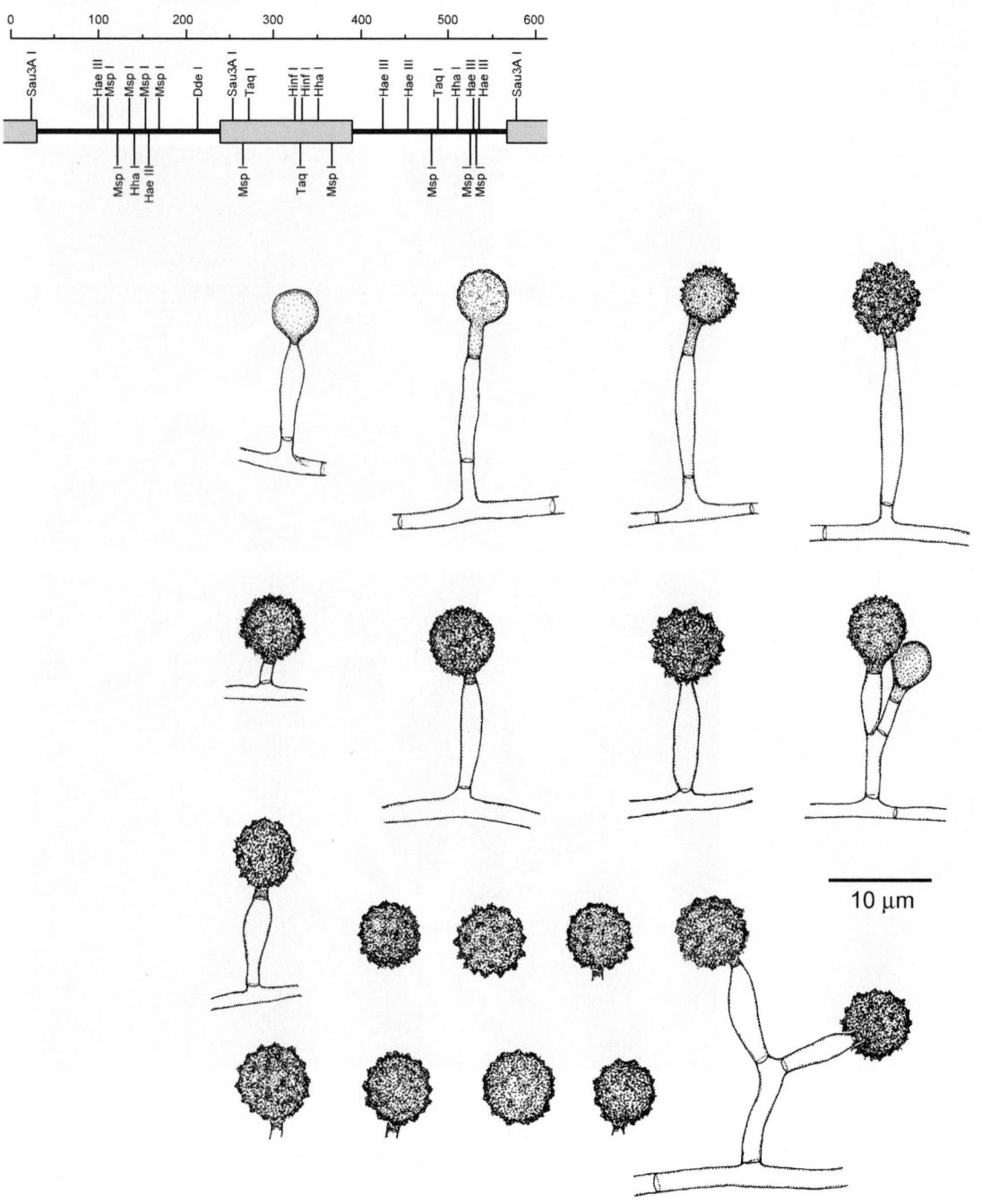

Thermomyces lanuginosus, CBS 224.63. Conidiophores and conidia.

Pathogenicity. BSL-1. Prosthetic valve endocarditis (Lesco-Bornet *et al.*, 1988).
References. Ellis (1971), Domsch *et al.* (1980).
Nomenclature. *Thermomyces lanuginosus* Tsiklinsky - Annls Inst. Pasteur 13: 500, 1899.

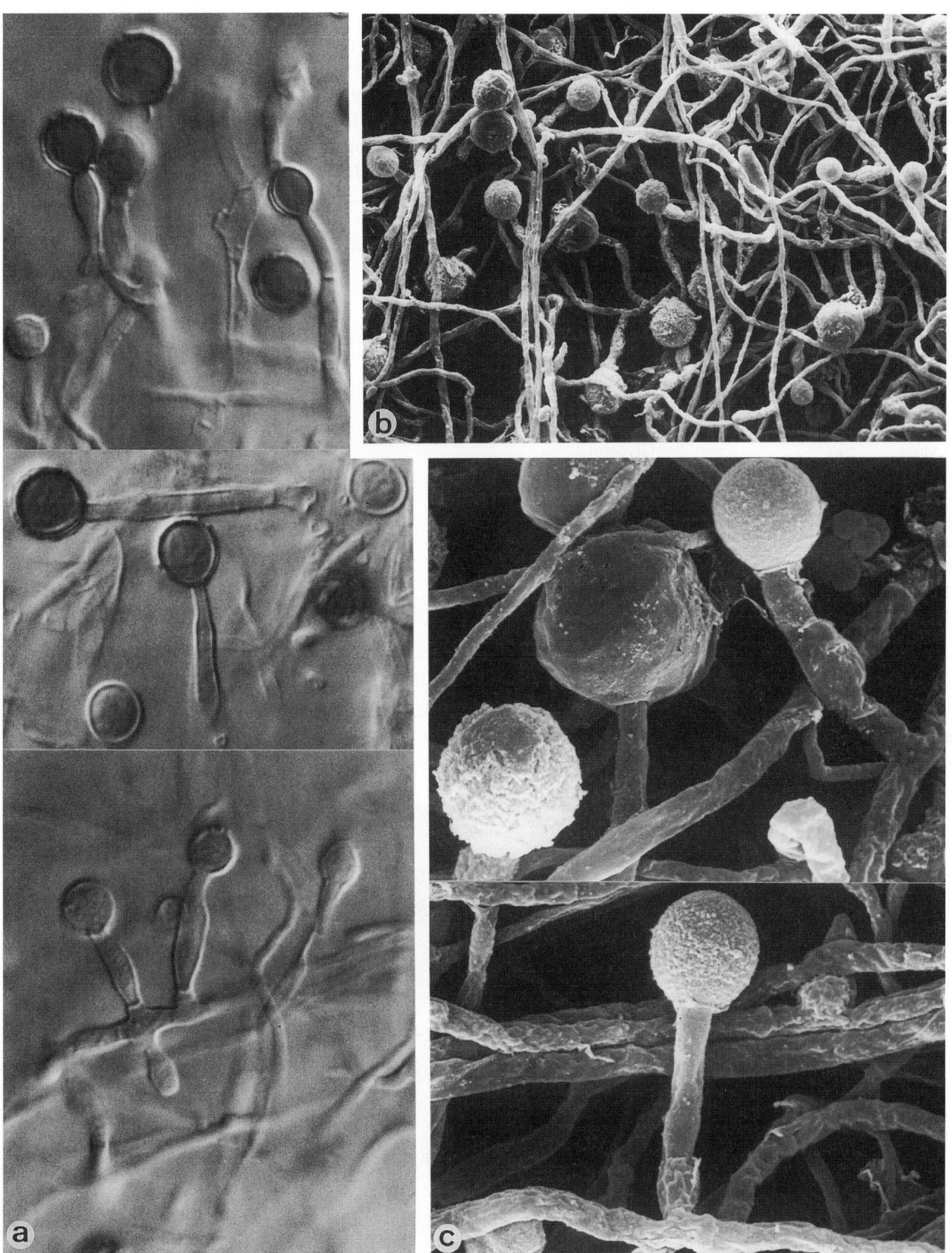

Thermomyces lanuginosus, CBS 224.63. Conidiophores and conidia. a. ×1600; b.×1350; c. ×5500.

Hyphomycetes, filamentous basidiomycetes. Genus: *TILLETIOPSIS*

Tilletiopsis minor Nyland

Colony characteristics. Colonies (MEYA) growing moderately rapidly, tough, smooth or farinose, later furrowed to cerebriform, dark reddish- to yellowish-brown; sometimes a brown pigment is exuded into the agar.

Microscopy. Budding cells usually absent. Hyphae hyaline, often partly lysed and locally producing densely spaced retraction septa. Conidia formed on unseptate lateral outgrowths of undifferentiated hyphae, fusiform to filiform, 6-25 × 1-2 µm. Additional ballistoconidia formed with one or more in sympodial order on sterigmata which are produced laterally on undifferentiated hyphae; ballistoconidia slender, somewhat curved, 6-16 × 1.0-2.5 µm. Intercalary chlamydospore-like cells present, pale brown.

Physiology.

Fermentation	–	Salicin	–	D-Gluconate	v
		Arbutin	+, w	D-Glucuronate	+
Growth:		Melibiose	+	DL-Lactate	+,w
D-Glucose	+	Lactose	+	Succinate	+
D-Galactose	+	Raffinose	+	Citrate	+,w
D-Glucosamine	–	Melezitose	+	Methanol	–
L-Sorbose	v	Inulin	+	Ethanol	–,w
D-Ribose	+	Soluble starch	+	Nitrate	+
D-Xylose	+	Glycerol	+	Nitrite	+
L-Arabinose	+	*meso*-Erythritol	+	Ethylamine	–,w
D-Arabinose	+	Ribitol	+	L-Lysine	v
L-Rhamnose	+	Xylitol	+	Cadaverine	–
Sucrose	+	L-Arabinitol	+	Creatine	–
Maltose	+	D-Glucitol	+	Creatinine	–
α,α-Trehalose	+	D-Mannitol	+	0.01% Cycloheximide	–
α-Methyl-D-glucoside	–	Galactitol	–	Growth at 37°C	–
Cellobiose	+	*myo*-Inositol	–	Urease	+

Pathogenicity. BSL-1. Phyllosphere fungus. Ramani *et al.* (1997) described a subcutaneous case. Also Boekhout (1991) listed some strains from human disorders. The species has a maximum growth temperature below 35°C.
Reference. Boekhout (1991).
Nomenclature. *Tilletiopsis minor* Nyland - Mycologia 42: 489, 1950.

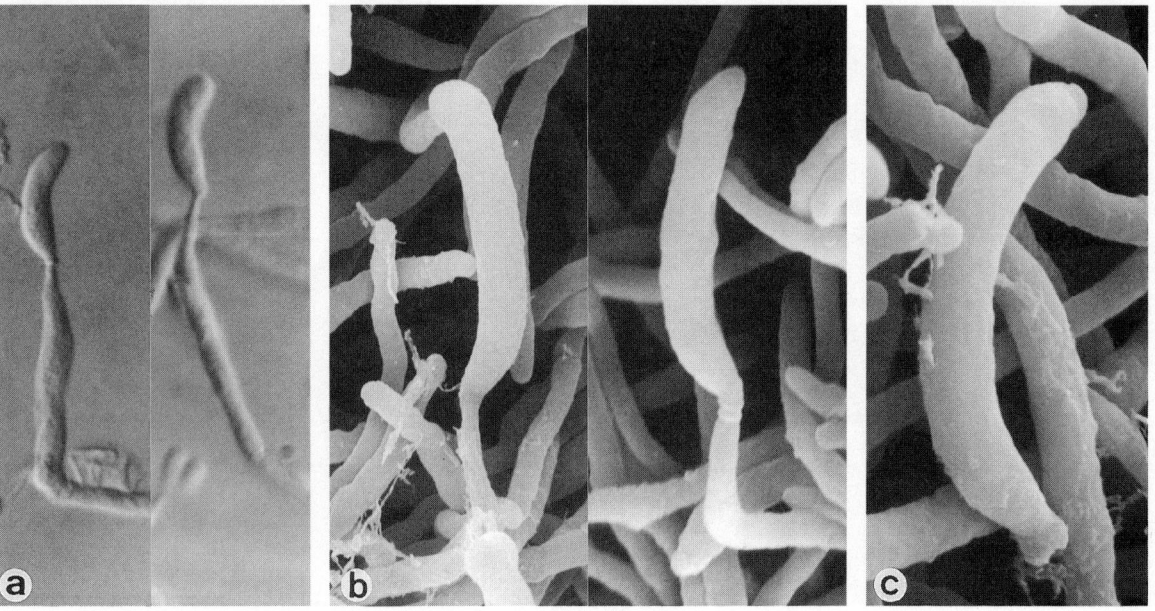

Tilletiopsis minor, CBS 335.32. a, b. Hyphae producing conidia, c. liberated conidium. a. ×1600; b. ×7500; c. ×10500.

940

Hyphomycetes. Genus: *TRICHOCLADIUM*

Trichocladium asperum Harz

Colony characteristics. Colonies (OA) spreading, cottony, at first white but later becoming grey; reverse olivaceous green.

Microscopy. Hyphae hyaline, smooth- and thin-walled, loosely branched. Conidiogenous cells undifferentiated, conidia frequently arising laterally from undifferentiated branches which are cylindrical, often slightly clavate, up to 20 μm long, non-septate or with some septa in the apical region. Conidia single, usually with one transverse septum, thick-walled, verrucose, dark brown, broadly cuneiform to ellipsoidal, with truncate base, constricted at the septum, 15-30 × 10-15 μm.

Pathogenicity. BSL-1. Saprobe on wood. Occasional cases of keratitis were reported (Bietii, 1922; Hoffmann, 1965).

References. Domsch *et al*. (1980), de Hoog (1983).

Nomenclature. *Trichocladium asperum* Harz - Bull. Soc. Impér. Moscou 44: 125, 1871.

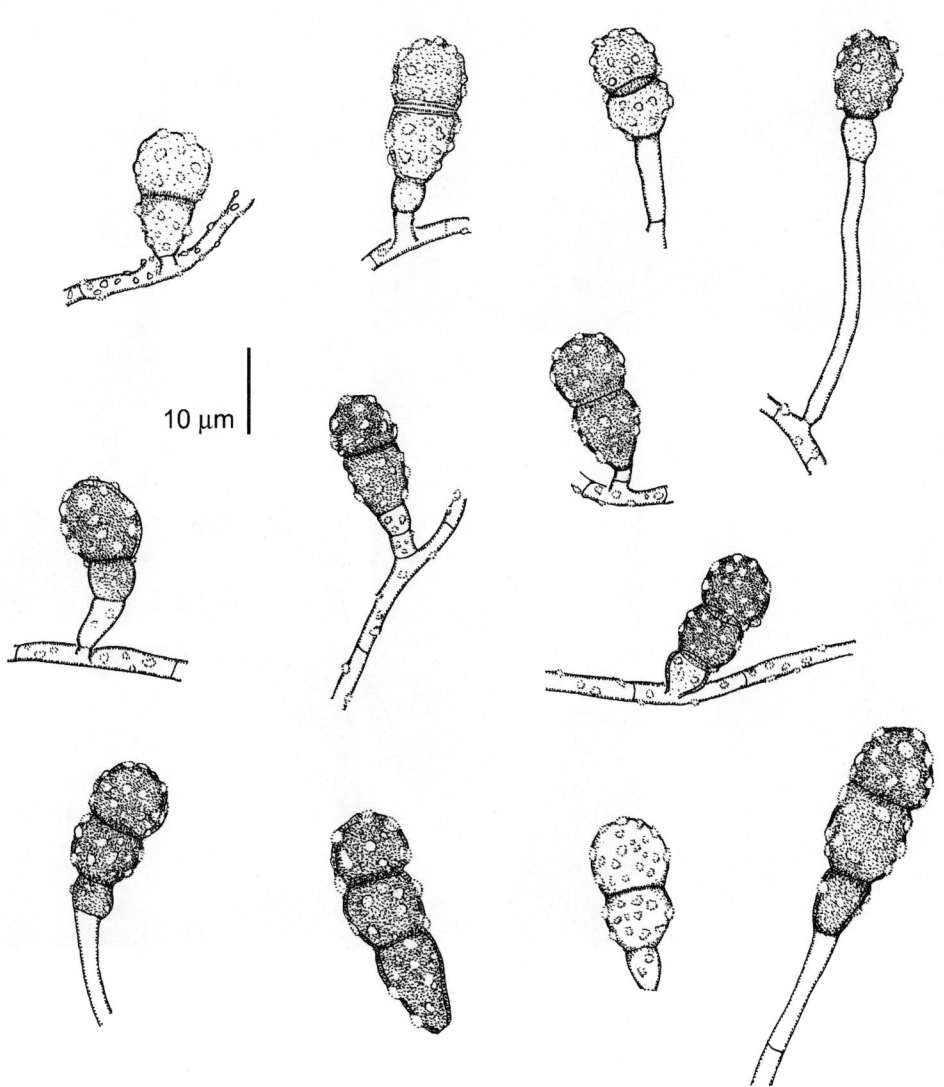

10 μm

Trichocladium asperum, CBS 903.85. Conidiogenous cells and conidia.

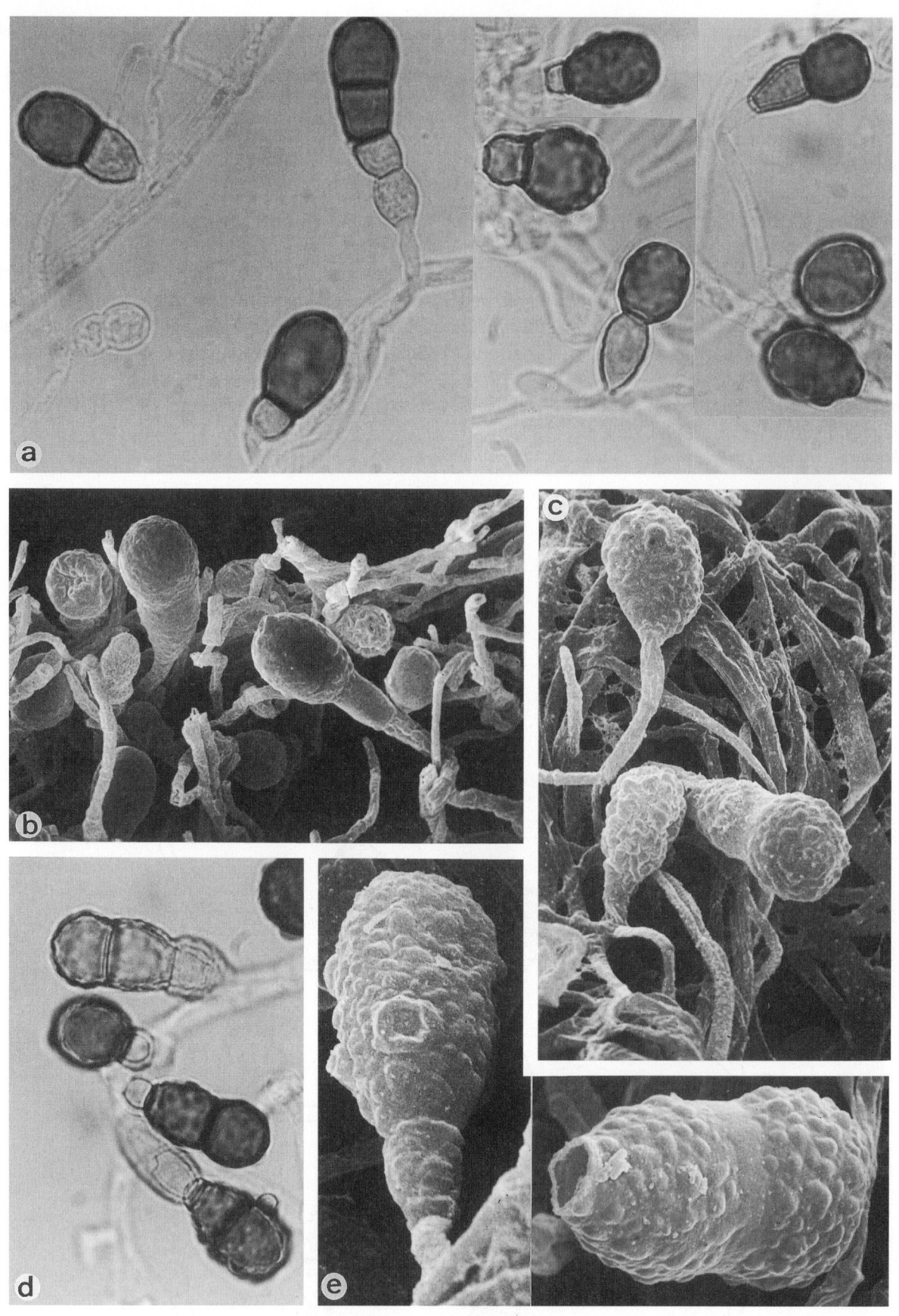

Trichocladium asperum, CBS 903.85. Conidiogenous cells and conidia. a. ×1280; b. ×1500; c. ×2058; d. ×1280; e. ×3900.

Hyphomycetes. Genus: *TRICHODERMA*

Generic description. Colonies with rapid growth, cream-coloured at first, soon becoming green due to abundant sporulation, starting at the edge of the colony. Conidiophores hyaline, loosely branched at right angles. Phialides flask-shaped, with very short, hardly visible collarettes. Conidia spherical to ellipsoidal, pale green, in slimy heads.

Teleomorph. *Hypocrea* (*Ascomycota, Euascomycetes, Hypocreales: Hypocreaceae*).

General remarks. Most *Trichoderma* species are soil-borne. Human cases are limited to severely weakened patients or emerge as complication of dialysis. The etiologic strains all belong to the section *Longibrachiatum; T. asperellum* (p. 1015) and *T. citrinoviride* (p. 1019) may be further clinically significant species.

References. Rifai (1969), Bissett (1984, 1991a, b), Kuhls *et al.* (1997), Gams & Bissett (1998), Gams & Meyer (1998), Kindermann *et al.* (1998), Kubicek & Harman (1998).

Key to the treated species of Trichoderma:

1a. Conidia rough-walled . . . . **T. viride (952)**
1b. Conidia smooth-walled ➝ **2**
2a. Phialides in regular whorls of 3-5 ➝ **3**
2b. Phialides more irregularly placed, often singly alongside conidiophore branches ➝ **4**
3a. Conidia (sub)spherical . . . . **T. harzianum (944)**
3b. Conidia ellipsoidal . . . . **T. koningii (946)**
4a. Phialides with more or less cylindrical base; conidia 3.5-7.0 × 2-3 µm . . . . **T. longibrachiatum (948)**
4b. Phialides swollen near the middle, narrower near the base;
conidia 3.5-4.5 × 2.0-2.5 µm . . . . **T. pseudokoningii (950)**

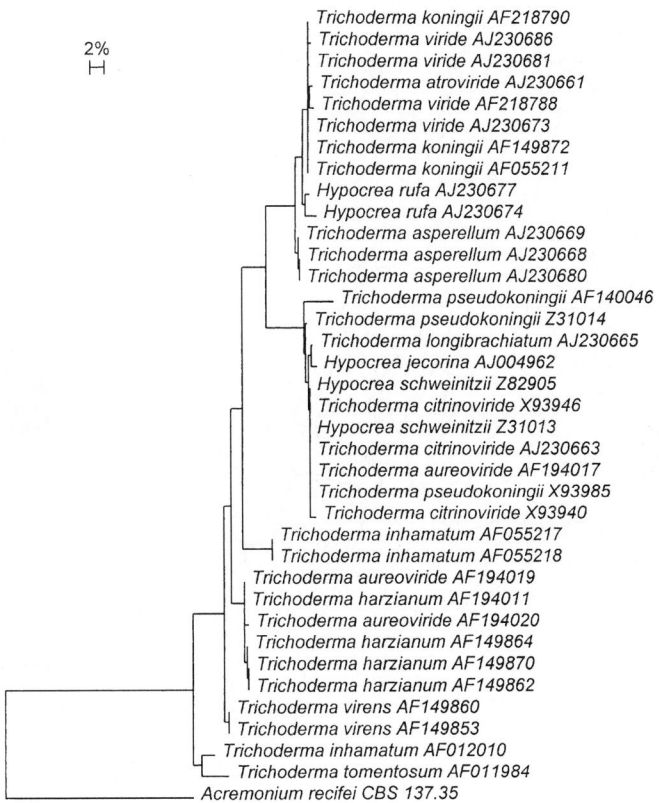

Fig. 62. Phylogenetic tree of *Trichoderma*.species based on confidently aligned ITS rDNA sequences using Neighbor joining algorithm with Kimura correction. Bootstrap values >90 from 100 resampled datasets are shown. *Acremonium recifei* was selected as outgroup. Some complexes of closely related taxa can be discerned. Within these complexes the individual species are insufficiently resolved. Hence, ITS is unsuitable for recognizing *Trichoderma* species .

Trichoderma harzianum Rifai

Colony characteristics. Colonies (OA) growing rapidly, initially glassy-white, soon with bright to dull green tufts of sporulation, first at the margin, later the entire colony.

Microscopy. Hyphae hyaline, 1.5-12.0 µm wide. Conidiophores branched in a pyramidal fashion, with short branches near the tip and longer branches in the lower part; branching at right angles. Phialides in whorls of 3-5, flask-shaped, abruptly attenuated near the ends, 4-7 × 2.5-3.5 µm. Conidia (sub)spherical, smooth-walled, subhyaline to pale green, 2.5-3.0 × 2.0-2.5 µm. Chlamydospores terminal or intercalary, smooth-walled, hyaline.

Differential diagnosis. The species differs from *T. koningii,* which also has a pyramidal branching system, by nearly spherical rather than elongate conidia.

Molecular diagnostics. Delimitation using UP-PCR was elaborated by Lübeck *et al.* (1999). ITS restriction map based on NCBI AF 194011:

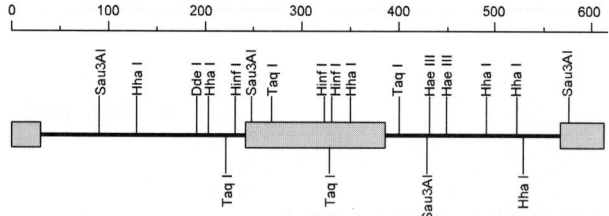

Pathogenicity. BLS-1. Common saprobe on wood. Guiserix *et al.* (1996) reported a fatal peritonitis in a patient undergoing peritoneal dialysis, and Guarro *et al.* (1999c) a fatal disseminated infection in a renal transplant recipient.

References. Rifai (1969), Gams & Meyer (1998).

Nomenclature. *Trichoderma harzianum* Rifai - Mycol. Pap. 116: 38, 1969.

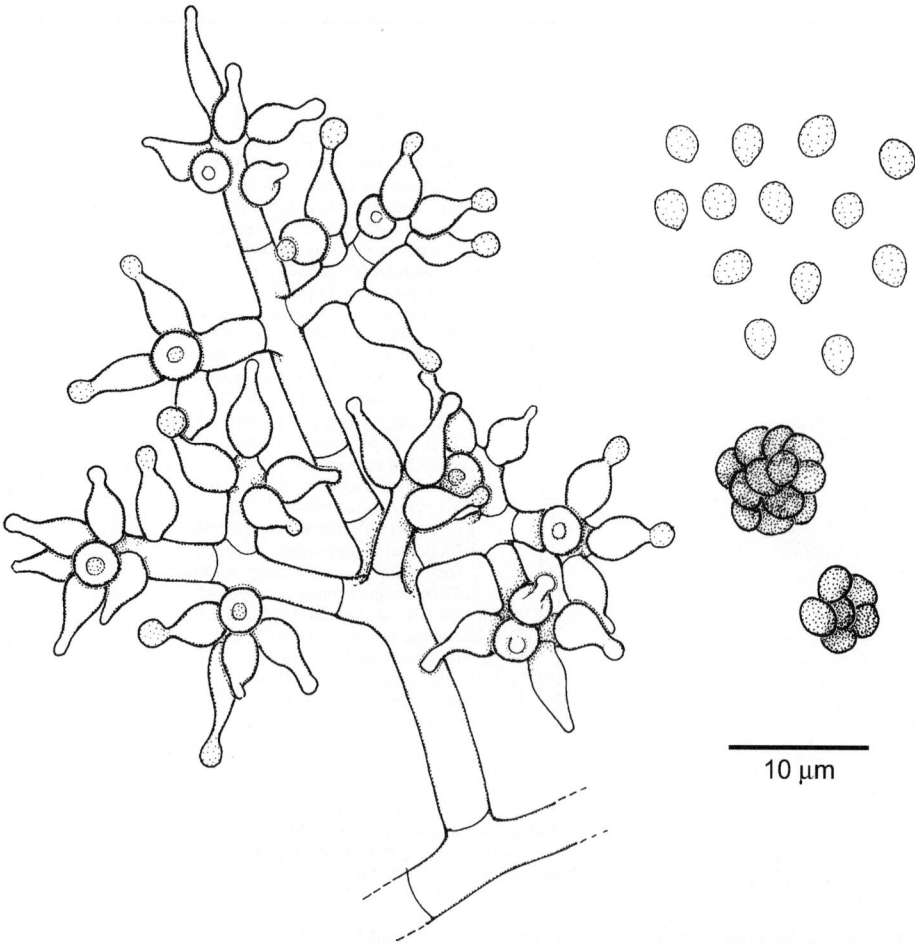

Trichoderma harzianum, CBS 102174. Conidiophore and conidia.

Antifungal susceptibility.

Antifungal	MICs	Strains	Reference
AMB	2	1	Guarro *et al.*, 1999b
5FC	256	1	Guarro *et al.*, 1999b
FLZ	128	1	Guarro *et al.*, 1999b
ITZ	32	1	Guarro *et al.*, 1999b
KTZ	8	1	Guarro *et al.*, 1999b
MCZ	8	1	Guarro *et al.*, 1999b

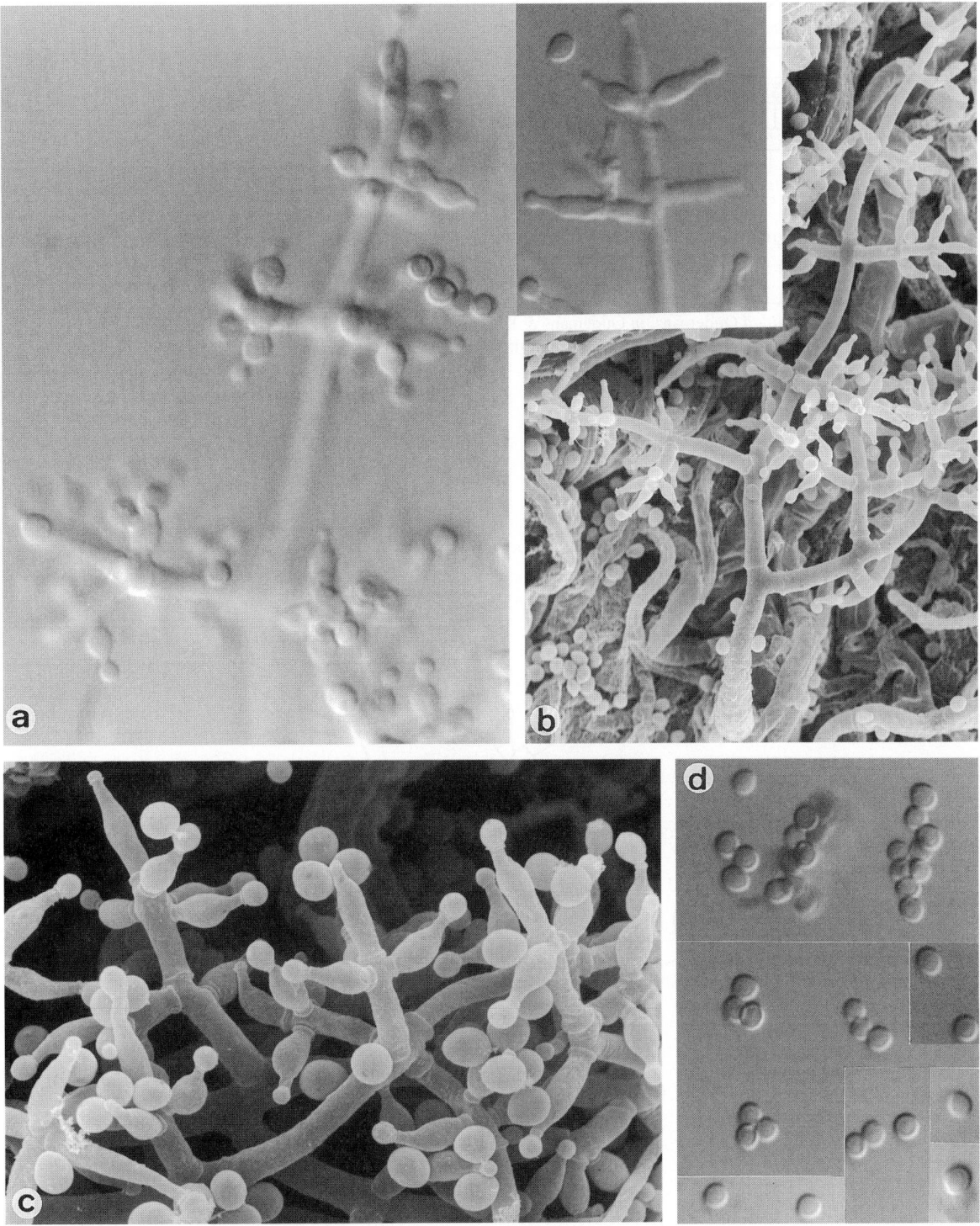

Trichoderma harzianum, CBS 102174. a, b. Pyramidal conidiophores; c, d. detail of phialides and conidia. a, d. ×1600; b. ×1300; c. ×3300.

Trichoderma koningii Oudem.

Colony characteristics. Colonies (OA) growing very rapidly, initially cream-coloured, gradually with greenish tufts of sporulation, first at the margin, later the entire colony.

Microscopy. Hyphae hyaline, up to 10 μm wide. Conidiophores branched in a pyramidal fashion, with short branches near the tip and longer branches in the lower part; branching at right angles. Phialides in whorls of 3-4, flask-shaped, abruptly attenuated near the ends, 7.5-8.2 × 2.5-3.0 μm, widely splaying out. Conidia broadly ellipsoidal, smooth-walled, green in mass, 3-5 × 2-3 μm. Chlamydospores terminal or intercalary, smooth-walled, pale brown, up to 12 μm wide.

Teleomorph. *Hypocrea koningii* Lieckfeldt *et al.* (*Ascomycota, Euascomycetes, Hypocreales: Hypocreaceae*).

Molecular diagnostics. ITS restriction map based on NCBI Z79628:

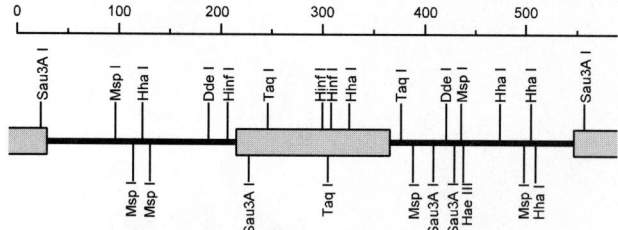

Pathogenicity. BSL-1. Ragnaud *et al.* (1984) and Campos-Herrero *et al.* (1996) reported infections in dialysis patients and McGough *et al.* (1994) in a transplant patient.

References. Rifai (1969), Lieckfeldt *et al.* (1998).

Nomenclature. *Trichoderma koningii* Oudemans, *in* Oudemans & Koning - Archs Néerl. Sci., Sér. 2, 7: 291, 1902.
   Hypocrea koningii Lieckfeldt, Samuels & W. Gams, *in* Lieckfeldt, Samuels, Börner & Gams - Can. J. Bot. 76: 1519, 1998.

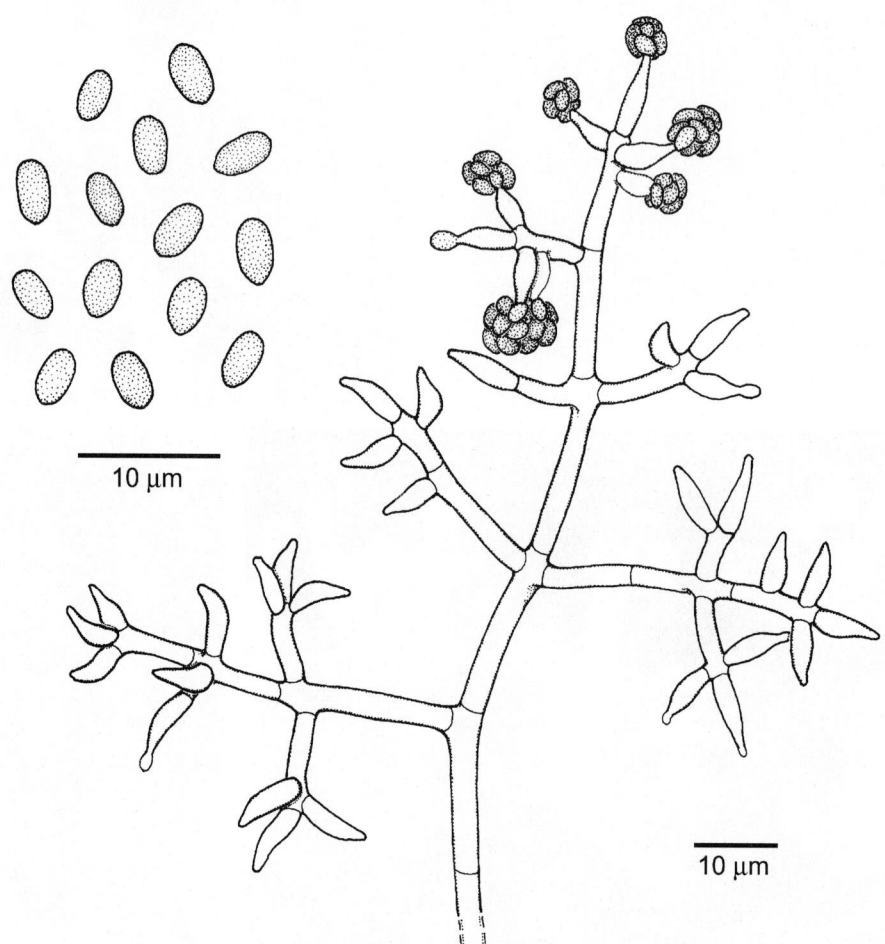

Trichoderma koningii, CBS 458.96. Conidiophore and conidia.

Antifungal susceptibility.

Antifungal	MICs	Strains	Reference
AMB	0.06	1	Guarro *et al.* (1999b)
5FC	256	1	Guarro *et al.* (1999b)
FLZ	64	1	Guarro *et al.* (1999b)
ITZ	0.5	1	Guarro *et al.* (1999b)
KTZ	2	1	Guarro *et al.* (1999b)
MCZ	2	1	Guarro *et al.* (1999b)

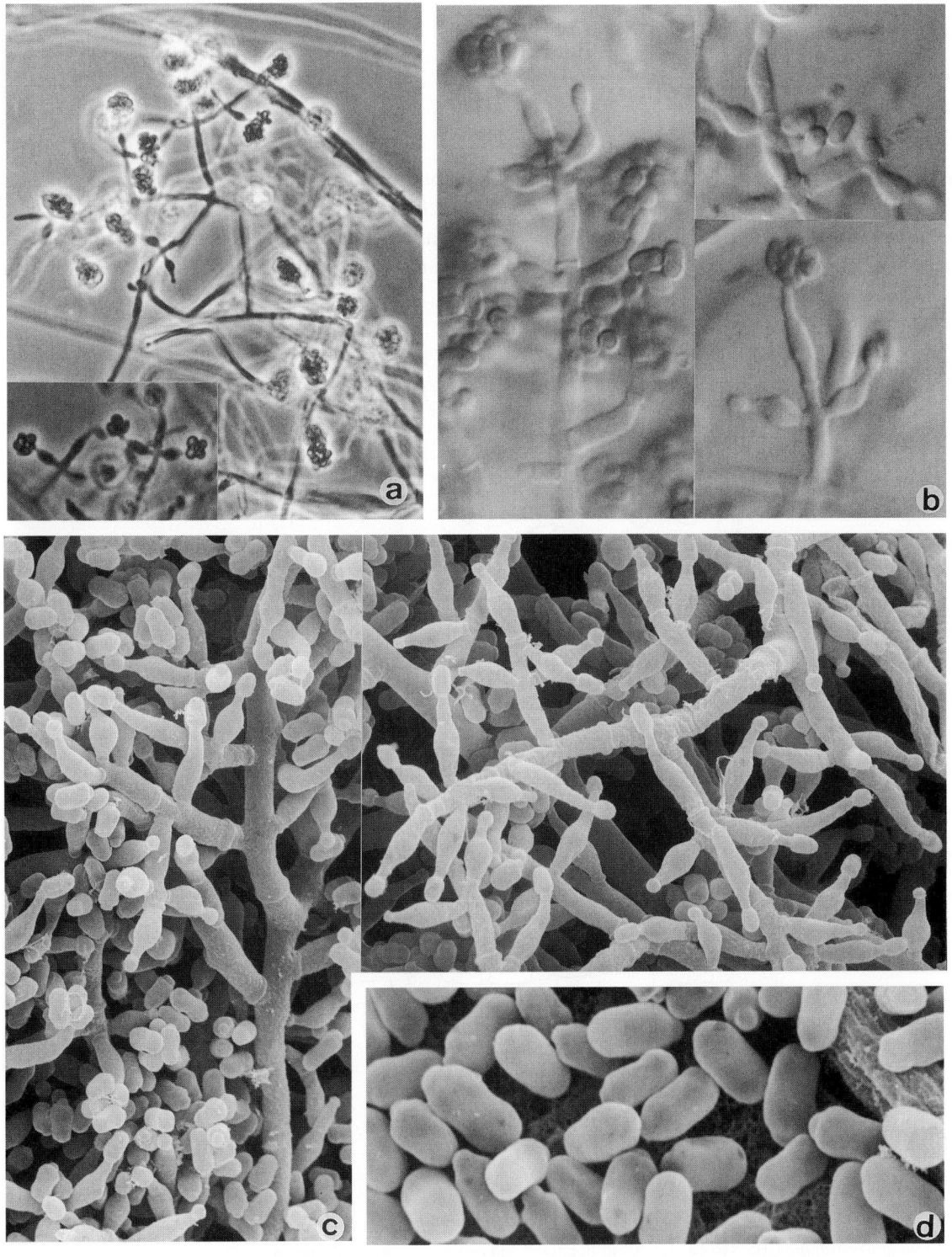

Trichoderma koningii, CBS 458.96. Conidiophores and conidia. a. ×640; b. ×1600; c. ×2750; d. ×5750.

Trichoderma longibrachiatum Rifai

Colony characteristics. Colonies (OA) growing very rapidly, initially off-white, soon with greyish-green tufts of sporulation, first at the margin, later the entire colony.

Microscopy. Hyphae hyaline, up to 10 μm wide. Conidiophores long, relatively poorly branched at right angles. Phialides mostly singly, flask-shaped with more or less cylindrical base and abruptly attenuated near the end, 6-14 × 2.5-3.0 μm, often curved, widely splaying out. Conidia broadly ellipsoidal, smooth-walled, green in mass, 3.5-7.0 × 2-3 μm. Chlamydospores terminal or intercalary, smooth-walled, hyaline, up to 10 μm wide.

Molecular diagnostics. ITS restriction map based on NCBI Z31019:

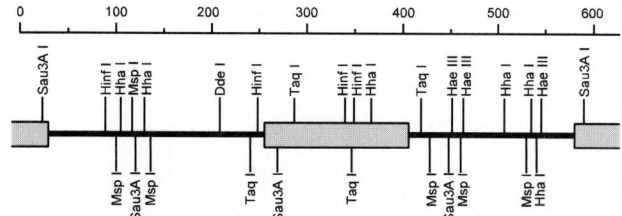

Pathogenicity. BSL-1. Tanis *et al.* (1995) described a fatal peritonitis in a patient on CAPD. Seguin *et al.* (1995) reported a cerebral and Muñoz *et al.* (1997) a cutaneous infection in a neutropenic patient. Furukawa *et al.* (1998) reported an invasive sinusitis and Richter *et al.* (1999) a disseminated infection in transplant recipients.

References. Rifai (1969), Bissett (1984, 1991b), Samuels *et al.* (1994, 1998), Turner *et al.* (1997), Kuhls *et al.* (1997, 1999).

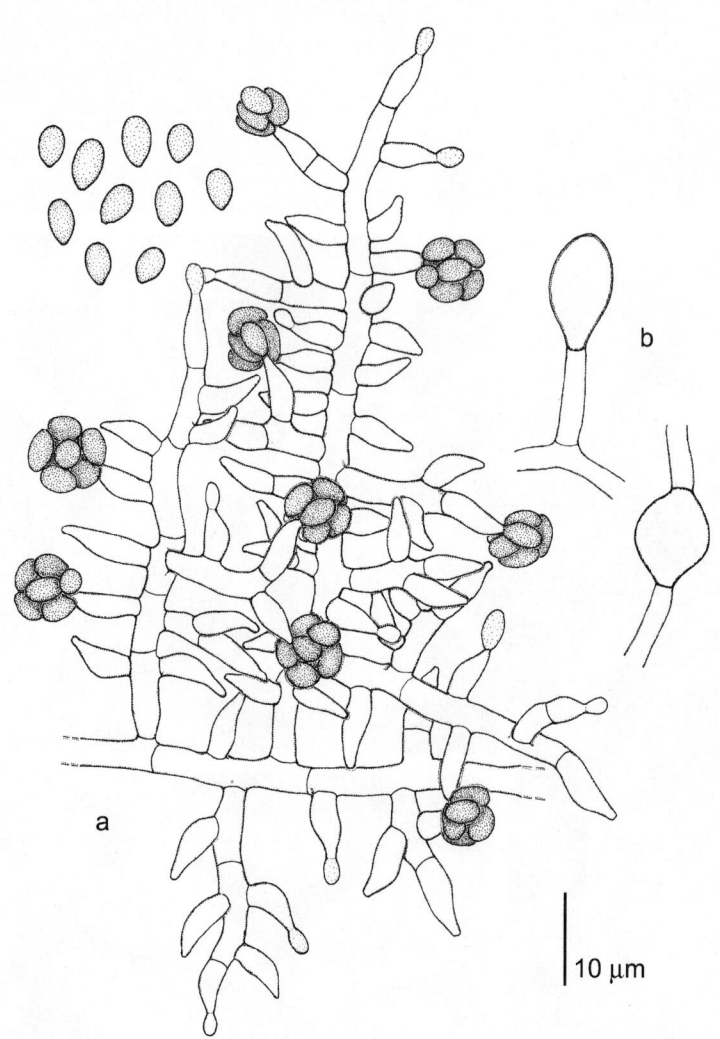

Trichoderma longibrachiatum, CBS 816.68. a, b. Conidiophores and conidia; c. chlamydospores.

Antifungal susceptibility.

Antifungal	MICs	Strains	Reference
AMB	0.5	1	Unpublished data
5FC	256	1	Unpublished data
FLZ	64	1	Unpublished data
ITZ	32	1	Unpublished data
KTZ	2	1	Unpublished data
MCZ	2	1	Unpublished data

Nomenclature. *Trichoderma longibrachiatum* Rifai - Mycol Pap. 116: 42, 1969.

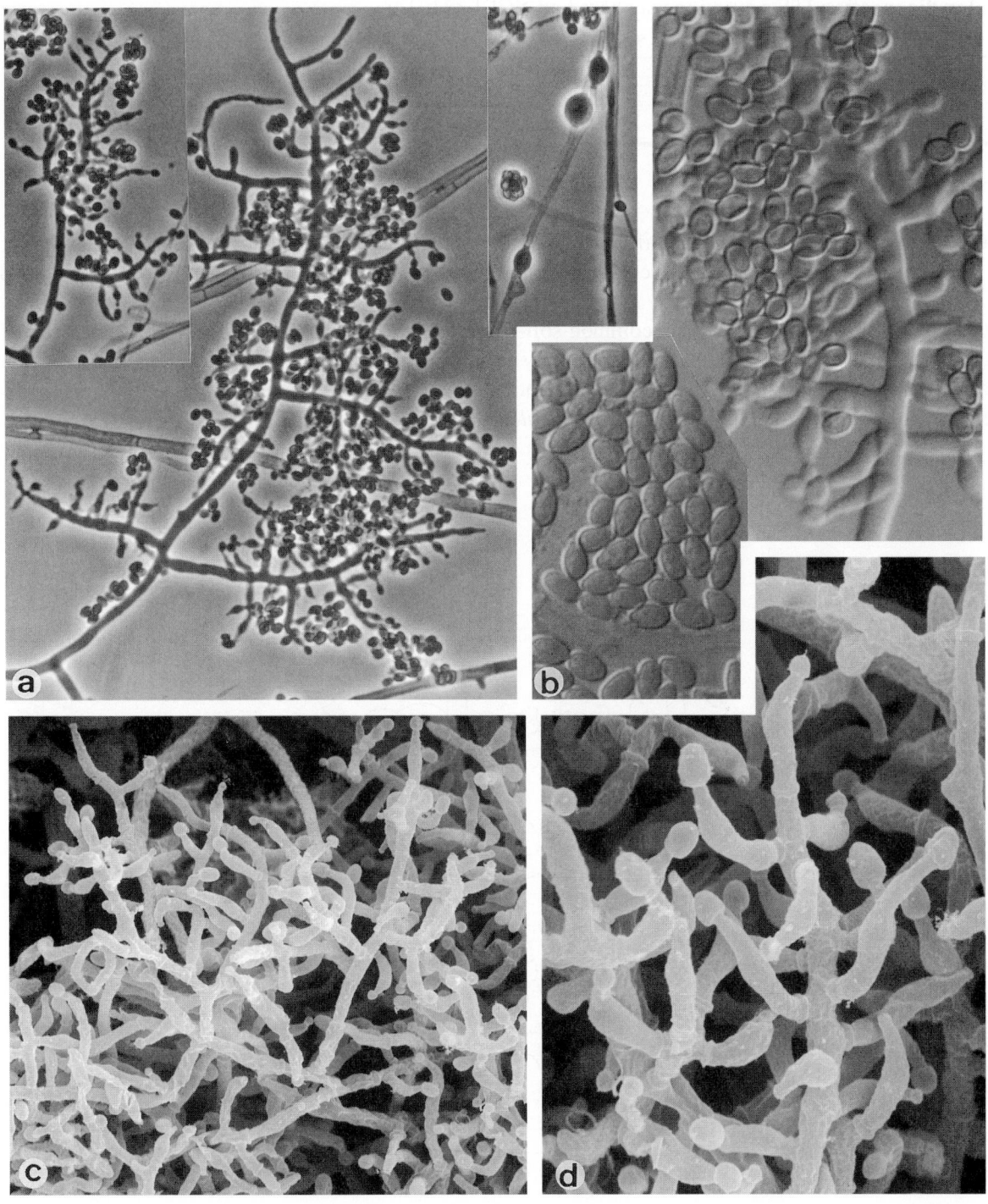

Trichoderma longibrachiatum, CBS 816.68. Conidiophores, conidia and chlamydospores. a. ×640; b. ×1600; c. ×1750; d. ×3800.

Trichoderma pseudokoningii Rifai

Colony characteristics. Colonies (OA) growing rapidly, with scant aerial mycelium, initially cream-coloured, gradually with greenish tufts of sporulation, first at the margin, later the entire colony; colony reverse often yellowish.
Microscopy. Hyphae hyaline, up to 10 μm wide. Conidiophores branched in a pyramidal fashion, with short branches near the tip and longer branches in the lower part; branching at right angles. Phialides in indistinct whorls of 1-4, slender, flask-shaped, swollen near the middle, with relatively long neck, 5.5-8.0 × 2.5-3.5 μm, widely splaying out. Conidia ellipsoidal to short cylindrical, smooth-walled, bluish-green in mass, 3.5-4.5 × 2.0-2.5 μm. Chlamydospores, when present, subspherical, smooth-walled, hyaline, up to 10 μm wide.
Teleomorph. *Hypocrea pseudokoningii* Samuels & Petrini (*Ascomycota, Euascomycetes, Hypocreales: Hypocreaceae*).
Molecular diagnostics. ITS restriction map based on NCBI Z31014:

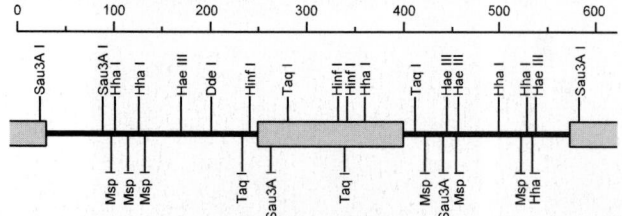

Differential diagnosis. Most strains identified in the literature may actually be *T. citrinoviride* Bissett (Turner *et al.*, 1997). That species differs by producing compact pustules and by conidia becoming dark green with age (Bissett, 1984).
Pathogenicity. BSL-1. Reported to be a common soil-borne species, but on the basis of RAPD data Turner *et al.* (1997) supposed that strains from outside Australia may have been misidentified. Systemic infections in patients with compromised innate immunity were reported by Gautheret *et al.* (1995) and Degeilh *et al.* (1993).
References. Rifai (1969), Bissett (1984), Samuels *et al.* (1994, 1998).

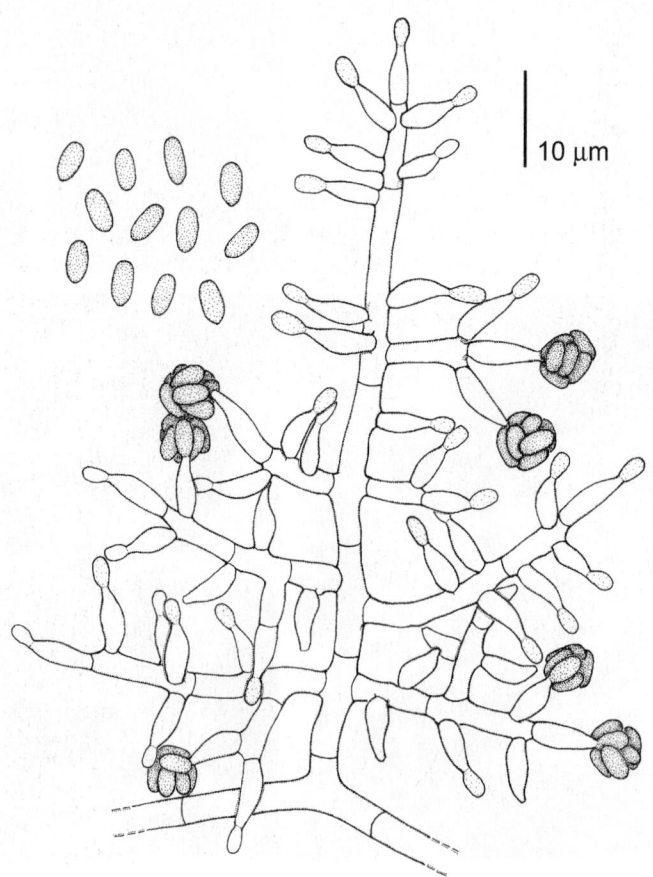

Trichoderma pseudokoningii, CBS 408.91. Conidiophore and conidia.

Antifungal susceptibility.

Antifungal	MICs	Strains	Reference
AMB	0.09	2	Guarro *et al.* (1999b)
5FC	256	2	Guarro *et al.* (1999b)
FLZ	90	2	Guarro *et al.* (1999b)
ITZ	4	2	Guarro *et al.* (1999b)
KTZ	2	2	Guarro *et al.* (1999b)
MCZ	2	2	Guarro *et al.* (1999b)

Nomenclature. *Trichoderma pseudokoningii* Rifai - Mycol. Pap. 116: 45, 1969.
   Hypocrea pseudokoningii Samuels & Petrini, *in* Samuels, Petrini, Kuhls, Lieckfeldt & Kubicek - Stud. Mycol. 41: 36, 1998.

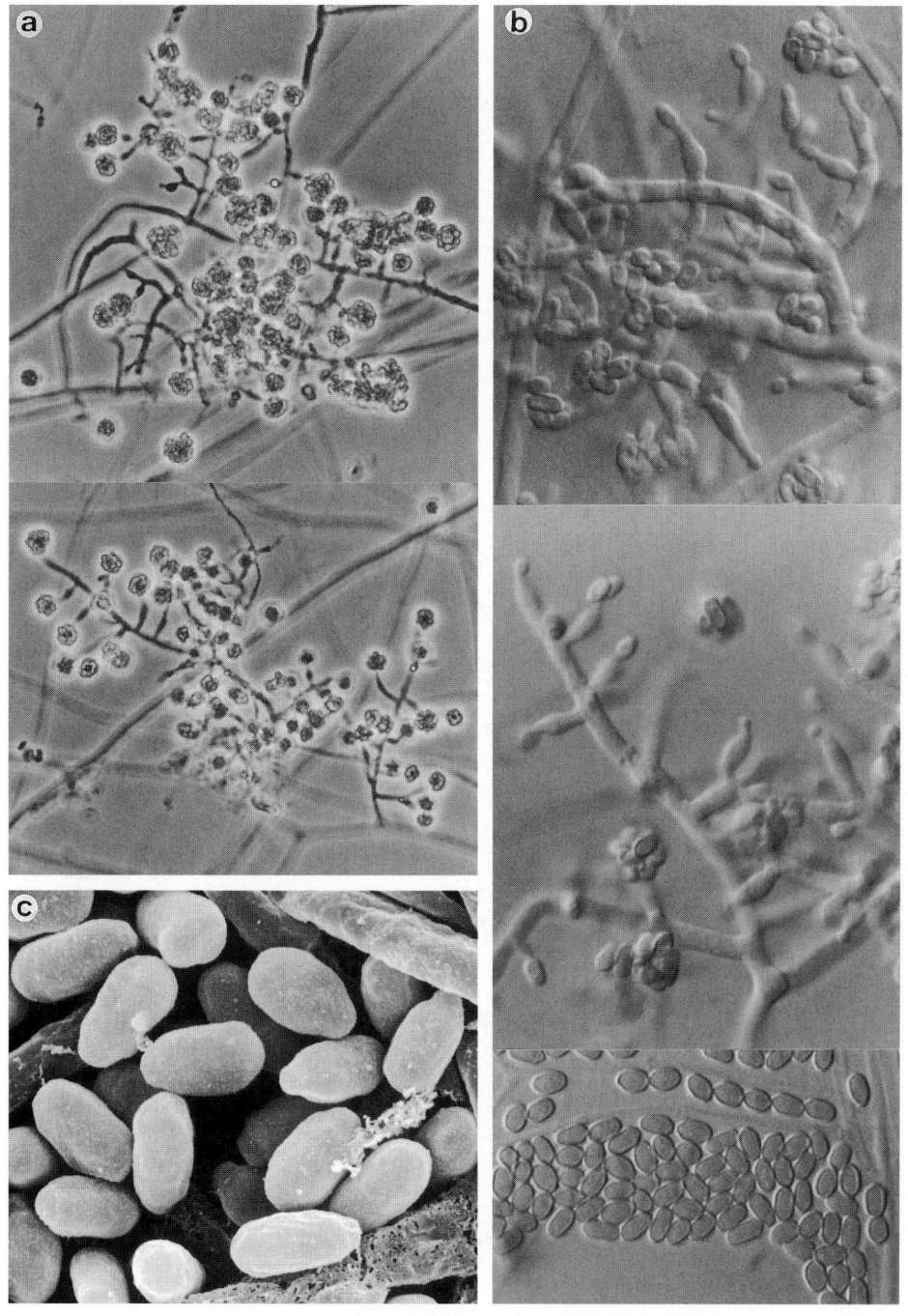

Trichoderma pseudokongii, CBS 408.81. Conidiophores and conidia. a.× 500; b. ×1250; c. ×5850.

Trichoderma viride Pers.: Fr.

Colony characteristics. Colonies (OA) growing very rapidly, initially more or less hyaline, soon becoming withish-green with tufted conidial areas in blue-green shades, first at the margin, later the entire colony.
Microscopy. Hyphae hyaline, up to 12 μm wide. Conidiophores branched in a pyramidal fashion, with short branches near the tip and longer branches in the lower part; branching at right angles. Phialides in whorls of 2-4, flask-shaped, abruptly attenuated near the ends, 8-14 × 2.5-3.0 μm, widely splaying out. Conidia (sub)spherical, hyaline, green in mass, 3.6-4.5 μm diam, roughened. Chlamydospores usually present in old cultures, intercalary, sometimes terminal, mostly spherical, hyaline, smooth-walled.
Physiology. Maximum growth temperature 30°C.
Teleomorph. *Hypocrea rufa* (Pers.:Fr.) Fr. (*Ascomycota, Euascomycetes, Hypocreales: Hypocreaceae*).

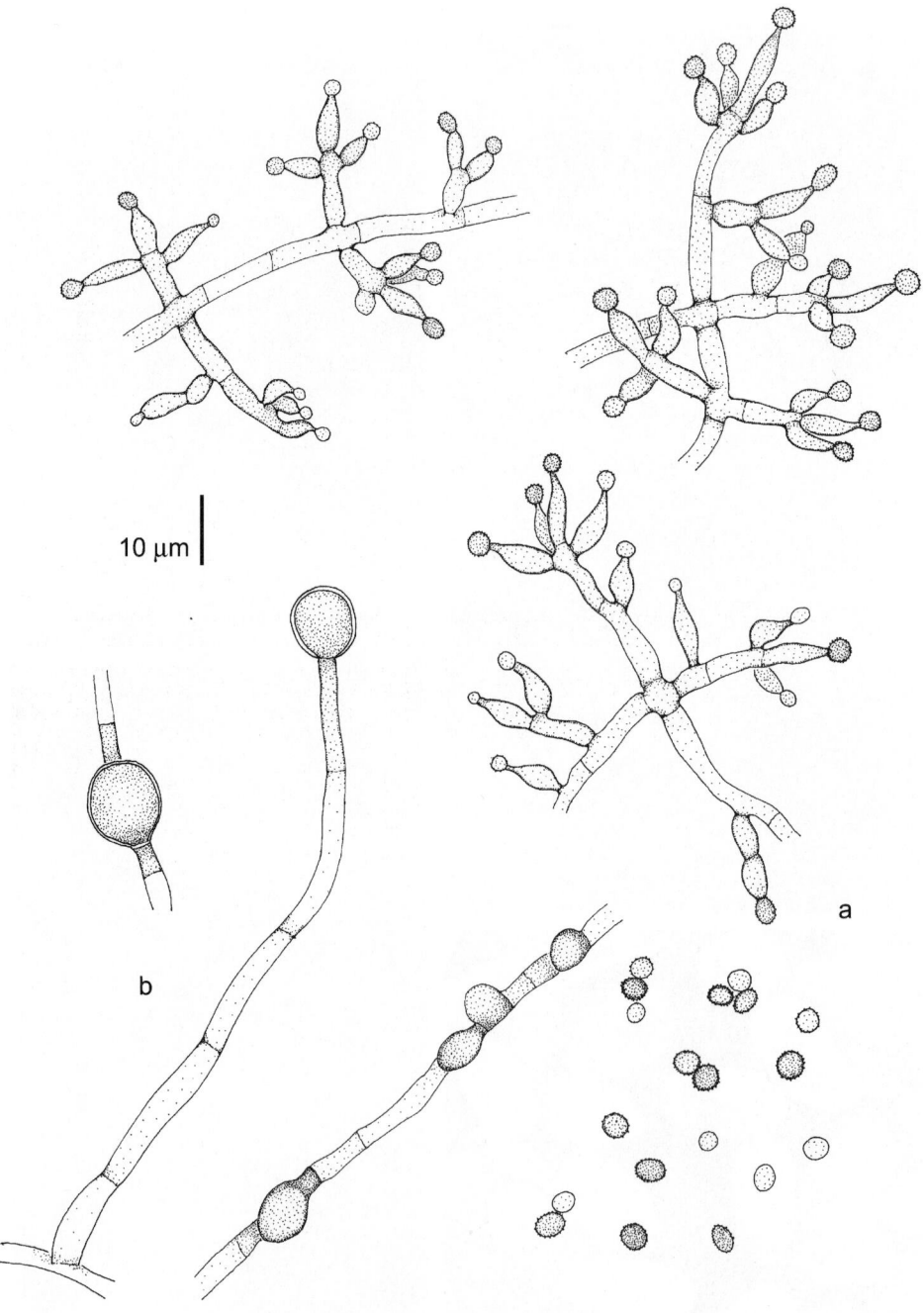

10 μm

b

a

Trichoderma viride, IFO 30498. a. Conidiophores and conidia; b. chlamydospores.

Molecular diagnostics. ITS restriction map based on NCBI X93979:

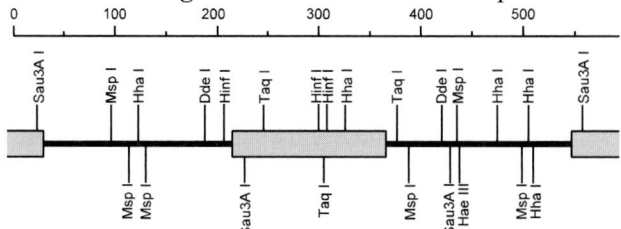

Pathogenicity. BSL-1. Pulmonary infection (Escudero *et al.*, 1976), peritonitis (Loeppky *et al.*, 1983) and a case of liver infection in an immunocompromised patient (Jacobs *et al.*, 1992). It should be noted, however, that the maximum growth temperature of typical *T. viride* is 30°C (Lieckfeldt *et al.*, 1999); clinical cases may have concerned *T. asperellum* Samuels *et al.* (1999) or a similar species.

References. Rifai (1969), Doi (1974), Domsch *et al.* (1980), Lieckfeldt *et al.* (1999).

Nomenclature. *Pyrenium lignorum* Tode var. *vulgarum* Tode - Fung. Mecklenb. 1: 33, 1790 ≡ *Trichoderma viride* Persoon - Römer's Neues Mag. Bot. 1: 92, 1794 (name change) ≡ *Trichoderma viride* Persoon:Fries - Syst. Mycol. 3: 215, 1829.

Sphaeria rufa Persoon - Syn. Meth. Fung. p. 13, 1801 ≡ *Sphaeria rufa* Persoon:Fries - Syst. Mycol. 2: 335, 1823 ≡ *Hypocrea rufa* (Persoon:Fries) Fries - Veg. Scand. p. 383, 1849.

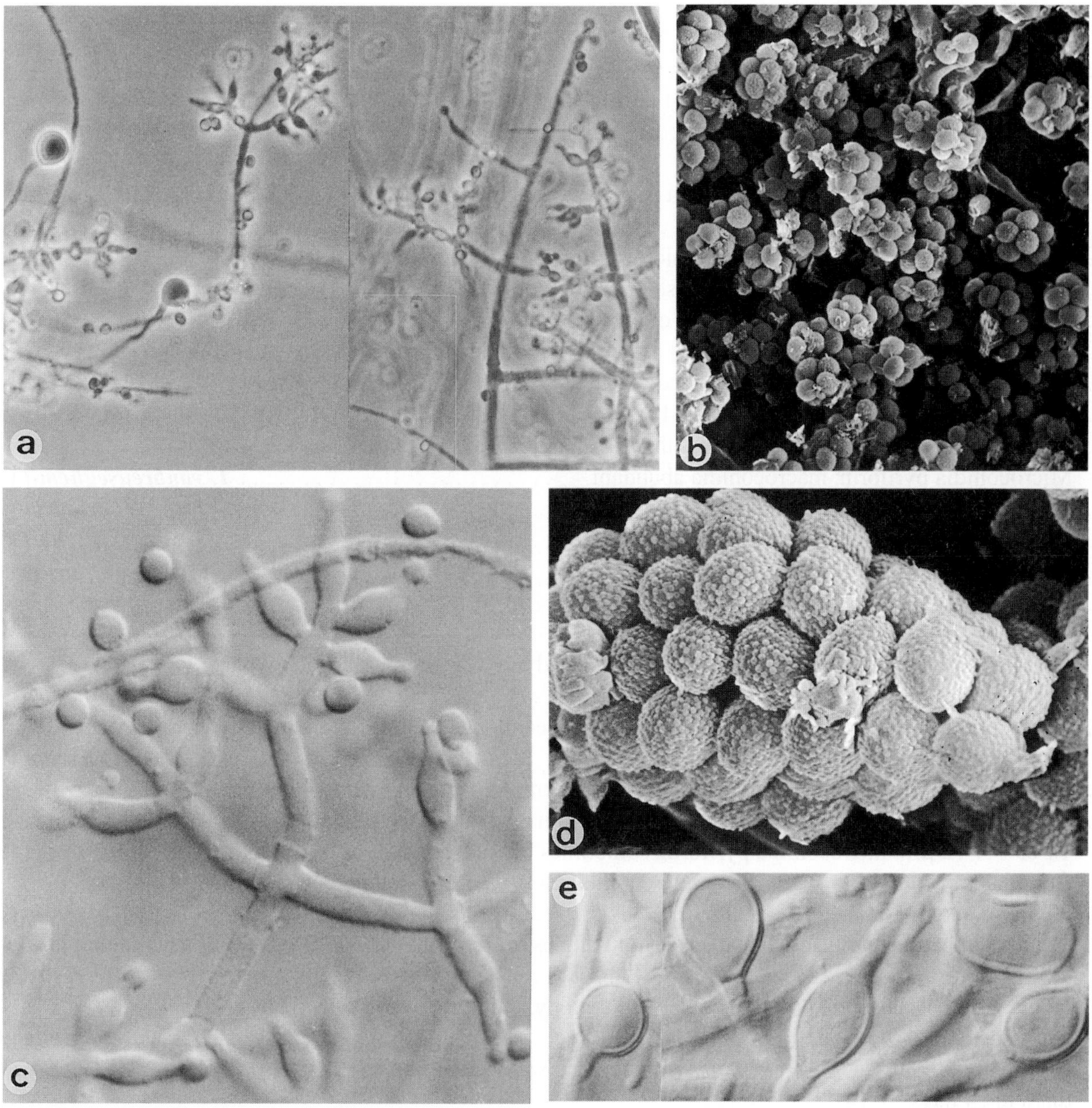

Trichoderma viride, IFO 30498. a-d. Conidiophores and conidia; e. chlamydospores. a. ×512; b. ×1000; c. ×1600; d. ×4400; e. ×1600.

953

Hyphomycetes, dermatophytes. Genus: *TRICHOPHYTON*

Generic description. Colonies waxy, glabrous or cottony, white, pinkish, yellowish or cream-coloured to brownish, with a cream-coloured, brown, red, violet or yellow colony reverse. Thallic macroconidia and microconidia, if present, terminally on or alongside undifferentiated hyphae. Macroconidia 2- or multi-celled, generally thin-walled, frequently absent, smooth-walled, hyaline, cylindrical, or clavate to cigar-shaped. Microconidia smooth- and thin-walled, hyaline, 1-celled, ovoidal, pyriform to clavate. The microconidia are often predominant. Intolerant to benomyl.

Teleomorph. *Arthroderma* (*Ascomycota, Euascomycetes, Onygenales: Arthrodermataceae*).

General remarks. *Trichophyton* differs from *Microsporum* by having smooth- and mostly thin-walled macroconidia and from *Epidermophyton* by the production of microconidia. *Keratinomyces* species have very thick cell walls.

References. Ajello (1968), Dvorák & Otcenásek (1969), Rebell & Taplin (1970), Badillet (1982), Kane *et al.* (1997), Gräser *et al.* (1999b).

Morphological and cultural key to the treated species of Trichophyton:

1a. Micro- and / or macro-conidia abundant → **2**
1b. Micro- and macro-conidia mostly absent or sparse → **13**
2a. Microconidia cashew nut-shaped . . . . *T. phaseoliforme* **(971)**
2b. Microconidia otherwise → **3**
3a. Colony in a later stage yellowish, with a dark purple pigment exuded into the agar; microconidia mostly absent . . . . *T. ajelloi* **(956)**
3b. Colony without dark purple pigment; microconidia mostly present → **4**
4a. Colony reverse of older culture yellowish or cream-coloured → **5**
4b. Colony reverse of older culture reddish or brown → **9**
5a. Macroconidia cylindrical, 2-7- celled; if absent, microconidia clavate → **6**
5b. Macroconidia of other shape, 5-11-celled → **8**
6a. Microconidia rare or absent, ovoidal, 5-16 μm in length . . . . *T. flavescens* **(961)**
6b. Microconidia present, not ovoidal, less than 7 μm in length → **7**
7a. Microconidia pyriform; macroconidia abundant . . . . *T. vanbreuseghemii* **(987)**
7b. Microconidia clavate; macroconidia rare or absent . . . . *T. erinacei* **(959)**
8a. Macroconidia fusiform, 30-85 × 6-11 μm . . . . *T. simii* **(979)**
8b. Macroconidia clavate, 9-60 × 3-7 μm . . . . *T. gloriae* **(963)**
9a. Microconidia spherical or pyriform → **10**
9b. Microconidia of variable size and shape → **11**
10a. Microconidia spherical, 2 μm diam, mostly produced in clusters on branched hyphae; spiral hyphae frequently present . . . . *T. mentagrophytes* **(968)**
10b. Microconidia pyriform, over 2 μm in length, produced alongside undifferentiated hyphae; spiral hyphae absent . . . . *T. rubrum* **(973)**
11a. Colony not furrowed; macroconidia not clearly differentiated from microconidia → **12**
11b. Colony radially or irregularly furrowed; macroconidia, when present, cigar-shaped, 2-6 celled, 10-65 × 4-12 μm; microconidia pyriform, sometimes balloon-shaped . . . . *T. tonsurans* **(984)**
12a. Conidial apparatus poorly branched; microconidia 3-5 × 2-3 μm; macroconidia up to 30 μm in length . . . . *T. thuringensis* **(983)**
12b. Conidial apparatus profusely branched at right angles; microconidia 4.0-6.5 × 1-5 μm; macroconidia 9-50 × 4-5 μm . . . . *T. terrestre* **(981)**
13a. Nodular bodies abundant; colony regular and cottony, attaining over 5 mm diam in 14 days . . . . *T. interdigitale* **(965)**
13b. Nodular bodies absent or rare; colony wrinkled and waxy, attaining less than 5 mm diam in 14 days → **14**
14a. Colony whitish to cream-coloured; reflexive hyphae absent → **15**

14b. Colony apricot-red, reddish to brown-purple; if yellowish with reflexive hyphae . . . . . . ***T. violaceum* (992)**
15a. Chlamydospores abundant, in chains; growth at 37°C better than at 24°C . . . . ***T. verrucosum* (989)**
15b. Chlamydospores rare or absent, not in chains; growth at 37°C absent or reduced ➜ **16**
16a. Favic chandeliers abundant . . . . ***T. schoenleinii* (977)**
16b. Favic chandeliers absent . . . . ***T. concentricum* (958)**

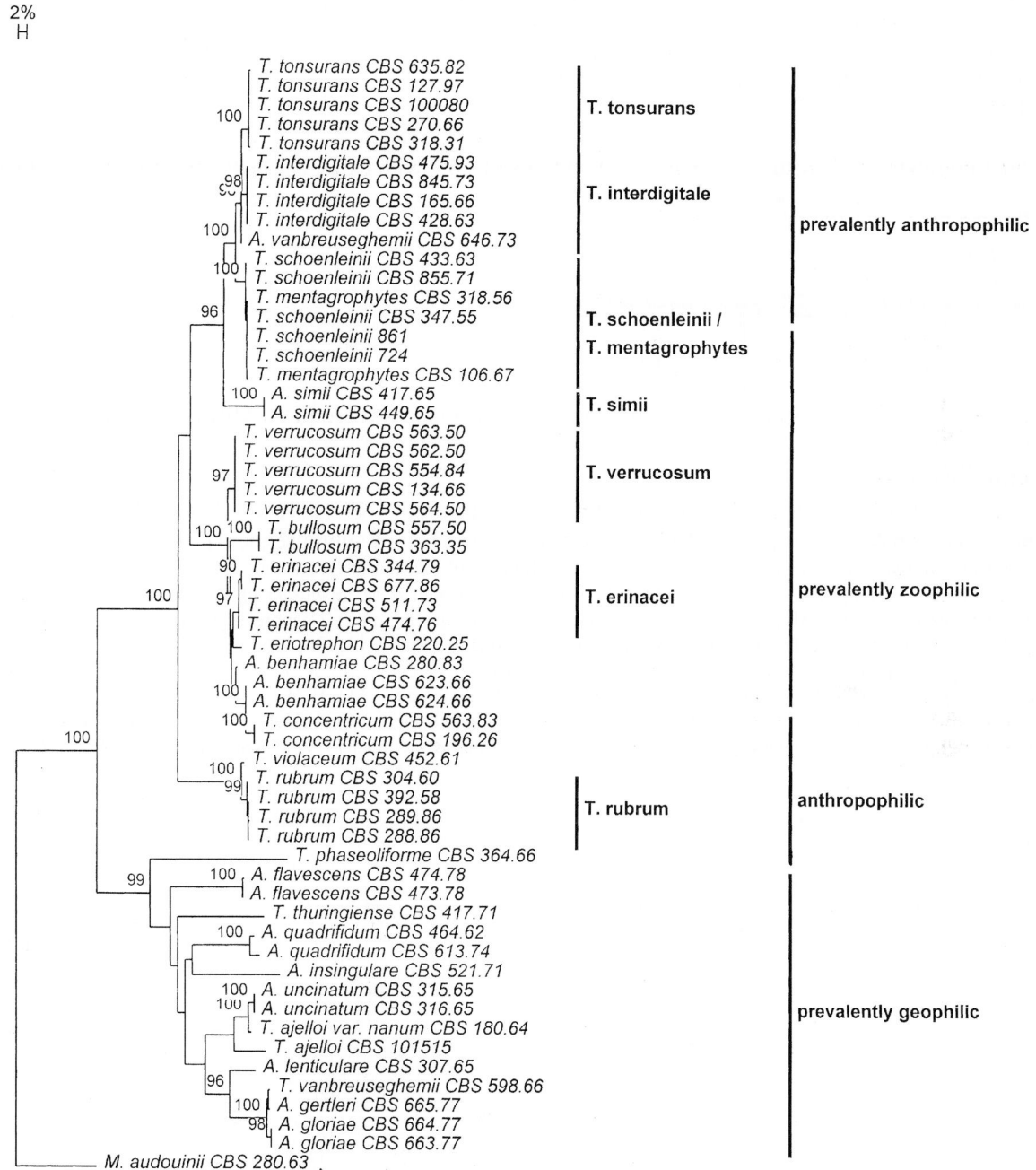

Fig. 63. Phylogenetic tree of *Trichophyton*.species based on confidently aligned complete ITS rDNA sequences using Neighbor joining algorithm with Kimura correction. Bootstrap values >90 from 100 resampled datasets are shown; *Microsporum audouinii* was selected as outgroup. Data were provided by Y. Gräser. In main traits an ecological subdivision is apparent. The geophilic species show a high degree of diversity and are remote from the core of *Trichophyton* (compare the tree of *Arthrodermataceae* on p. 390). The lowest degree of diversity is found among the prevalently anthropophilic species. *T. rubrum* and *T. violaceum* compose a second anthropophilic branch.
A = Arthroderma; M = Microsporum; T = Trichophyton.

Trichophyton ajelloi (Vanbreuseghem) Ajello

Colony characteristics. Colonies (SGA) expanding, flat, powdery to velvety, cream-coloured to ochraceous-buff; reverse yellowish, a dark purple pigment being exuded into the agar.
Microscopy. Macroconidia hyaline, smooth- and thick-walled, cigar-shaped, 8-12-celled, 40-70 × 9-12 µm. Microconidia sparse or absent, ovoidal to pyriform, 3-9 × 2-5 µm.
Teleomorph. *Arthroderma uncinatum* Dawson & Gentles (*Ascomycota, Euascomycetes, Onygenales: Arthrodermataceae*).
Physiology.

Urease	+	T1	+	T4	+	T7	+
Hair perforation	+	T2	+	T5	+		
Growth at 37°C	v	T3	+	T6	+		

Molecular diagnostics. ITS restriction map based on CBS 315.65, with extra *Msp*I site in var. *nanum*, CBS 180.64:

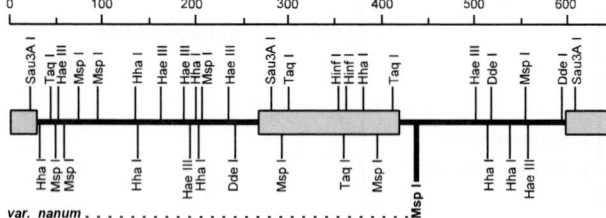

Pathogenicity. BSL-1. The species is geophilic; human cutaneous cases are rare (Alvarez & de Bracalenti, 1982; Diniz *et al.*, 1991).
Distribution. World-wide.
Variety. *T. ajelloi* var. *nanum* has 3-4-celled macroconidia measuring 14-40 × 8-18 µm (Kunert & Hejtmánek, 1964).
References. Dawson & Gentles (1961), Padhye & Carmichael (1971), Vanbreuseghem (1952, 1980).

Nomenclature var. *ajelloi*. *Keratinomyces ajelloi* Vanbreuseghem - Bull. Acad. R. Méd. Belg. 38: 1075, 1952 ≡ *Epidermophyton terrigenum* Evolceanu & Alteras - Mycopath. Mycol. Appl. 11: 202, 1959 (name change) ≡ *Microsporum ajelloi* (Vanbreuseghem) Arievitch & Stiepanishchewa - Proc. Int. Symp. Med. Mycol., Warsaw p. 43, 1965 ≡ *Trichophyton ajelloi* (Vanbreuseghem) Ajello - Sabouraudia 6: 148, 1966 ≡ *Epidermophyton ajelloi* (Vanbreuseghem) Novák & Galgóczy - Acta Bot. Hung. 15: 130, 1969.
 Arthroderma uncinatum Dawson & Gentles - Sabouraudia 1: 55, 1961.
 Epidermophyton stockdaleae Prochacki & Engelhardt-Zasada - Mycopathologia 54: 342, 1974.
Nomenclature var. *nanum*. *Keratinomyces ajelloi* Vanbreuseghem var. *nanum* Kunert & Hejtmánek - Česká Epid. Mikrobiol. Immunol. 13: 296, 1964 ≡ *Trichophyton ajelloi* (Vanbreuseghem) Ajello var. *nanum* (Kunert & Hejtmánek) Ajello - Sabouraudia 6: 148, 1966.

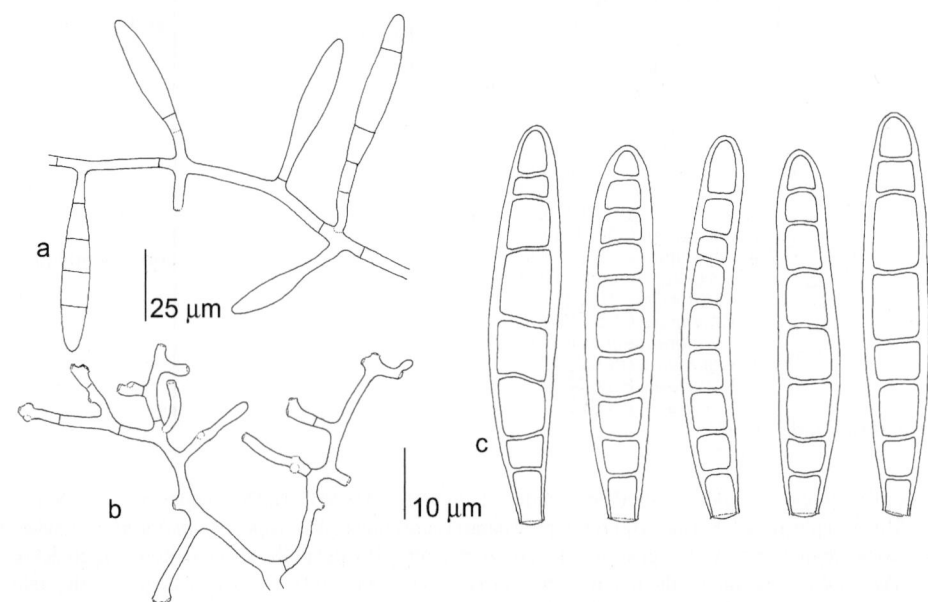

Trichophyton ajelloi, CBS 368.64. a. Young conidial apparatus; b. branches with detached conidia; c. mature conidia.

Antifungal susceptibility.

Antifungal	Mean MICs	MIC 90	Strains	Reference
AMB	0.35	nd	2	Fernández *et al.* (2000b)
CTZ	0.17	nd	2	Fernández *et al.* (2000b)
FLZ	2.81	nd	2	Fernández *et al.* (2000b)
GRF	4.42	25	6	Wildfeuer *et al.* (1998)
ITZ	0.04	nd	2	Fernández *et al.* (2000b)
ITZ	0.07	12.5	6	Wildfeuer *et al.* (1998)
KTZ	0.25	nd	2	Fernández *et al.* (2000b)
KTZ	0.07	6.25	6	Wildfeuer *et al.* (1998)
MCZ	0.12	nd	2	Fernández *et al.* (2000b)
TBF	0.01	nd	2	Fernández *et al.* (2000b)
VCZ	0.02	nd	2	Fernández *et al.* (2000b)
VCZ	0.05	0.78	6	Wildfeuer *et al.* (1998)

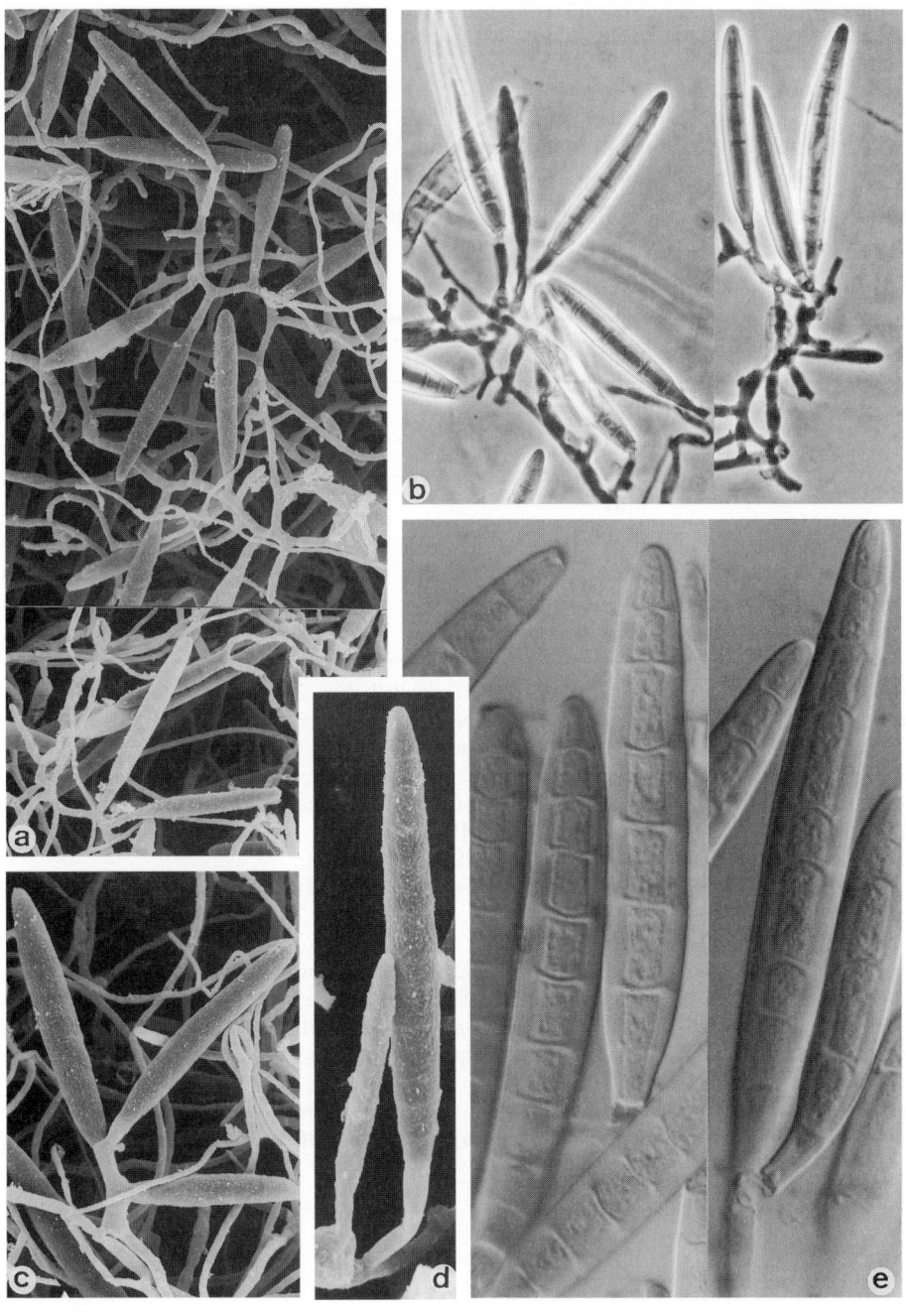

Trichophyton ajelloi, FMR 4564. Fertile hyphae and macroconidia. a. ×375; b. ×370; c. ×520; d. ×870; e. ×930.

Trichophyton concentricum Blanchard

Colony characteristics. Colonies (SGA) growing slowly, white, becoming cream-coloured, amber, honey-brown to orange, glabrous with a tendency to become slightly velvety, raised and irregularly, deeply folded, often cracking the agar; reverse cream-coloured to brown.

Microscopy. Tangled masses of branching hyphae present; conidia absent, or some tear-shaped microconidia formed.

Physiology. Growth at 37°C absent or weak.

Urease	–	T-1	–	T-5	+
Hair perforation	–	T-2	+	T-6	–
		T-3	++	T-7	w
		T-4	+		

Molecular diagnostics. ITS restriction map based on NCBI Z98012:

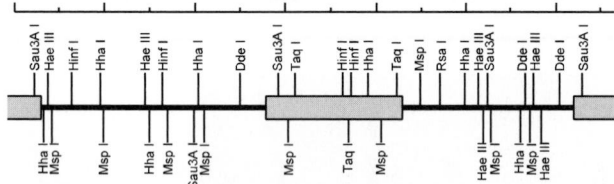

Pathogenicity. BSL-2. Causes human tinea imbricata ('tokelau'), a chronic skin disorder whereby fungal growth leads to concentric rings. Lesions may be located on all parts of the body, including the face.

Distribution. Major endemic areas are Pacific Islands, Southeast Asia and Central America.

Reference. Hay (1988).

Antifungal susceptibility.

Antifungal	Mean MICs	Strains	Reference
AMB	0.7	2	Fernández *et al.* (2000b)
CTZ	0.03	2	Fernández *et al.* (2000b)
FLZ	3.98	2	Fernández *et al.* (2000b)
ITZ	0.01	2	Fernández *et al.* (2000b)
KTZ	0.06	2	Fernández *et al.* (2000b)
MCZ	0.04	2	Fernández *et al.* (2000b)
TBF	0.007	2	Fernández *et al.* (2000b)
VCZ	0.04	2	Fernández *et al.* (2000b)

Nomenclature. *Trichophyton concentricum* Blanchard, *in* Bouchard - Traité Path. Gén. 2: 916, 1896 ≡ *Lepidophyton concentricum* (Blanchard) Gedoelst - Champ. Paras. Homme Anim. Domest. p. 147, 1902 ≡ *Aspergillus concentricum* (Blanchard) Castellani - Trans. Int. Derm. Congr. 6: 671, 1907 ≡ *Endodermophyton concentricum* (Blanchard) Castellani & Chalmers - Man. Trop. Med. p. 610, 1910 ≡ *Oospora concentrica* (Blanchard) Hanawa & Nagai - Jpn. J. Derm. Urol., Suppl., p. 47, 1917 ≡ *Achorion concentricum* (Blanchard) Guiart & Grigorakis - Lyon Méd. 141: 377, 1928 ≡ *Mycoderma concentricum* (Blanchard) Vuillemin - C.R. Hebd. Séanc. Acad. Sci., Paris 89: 405, 1929.

    Aspergillus lepidophyton Pinoy - Bull. Inst. Pasteur 1: 761-774, 1902.

    Aspergillus tokelau Wehmer - Centbl. Bakt. Parasitkde, Abt. 1, 35: 140, 1903.

    Trichophyton mansonii Castellani - Br. Med. J. 2: 1277, 1905 ≡ *Endodermophyton mansonii* (Castellani) Brumpt - Précis Parasitol., ed. 4, p. 1299, 1927.

    Endodermophyton castellanii Perry - Ceylon Med. Rep. 1907.

    Endodermophyton indicum Castellani - J. Trop. Med. Hyg. 14: 82, 1911 ≡ *Arthrosporia indica* (Castellani) Grigorakis - Annls Sci. Nat., Bot., Sér. 10, 7: 414, 1925 ≡ *Achorion indicum* (Castellani) Guiart & Grigorakis - Lyon Méd. 141: 377, 1928 ≡ *Mycoderma indica* (Castellani) Vuillemin - Champ. Paras. Myc. Homme p. 111, 1931 ≡ *Trichophyton indicum* (Castellani) Nannizzi - Tratt. Micopat. Um. 4: 186, 1934.

    Endodermophyton tropicale Castellani & Chalmers - Man. Trop. Med., ed. 3, p. 1017, 1919 ≡ *Arthrosporia tropicalis* (Castellani & Chalmers) Grigorakis - Annls Sci. Nat., Bot., Sér. 10, 7: 414, 1925.

    Endodermophyton roquettei da Fonseca - C.R. Soc. Biol. 92: 306, 1925 ≡ *Mycoderma roquettei* (da Fonseca) Vuillemin - C.R. Heb. Séanc. Acad. Sci., Paris 89: 405, 1929 ≡ *Trichophyton roquettei* (da Fonseca) Nannizzi - Tratt. Micopat. Um. 4: 196, 1934.

Trichophyton erinacei (J.M.B. Smith & Marbles) Quaife

Colony characteristics. Colonies (SGA) expanding, cottony or farinose, white; colony reverse becoming bright citron yellow.

Microscopy. Macroconidia, when present, cylindrical to clavate, variable in size, 2-6-celled. Microconidia abundant, slender, clavate, up to 6 μm, at right angles alongside hyphae, first widely interspaced, finally close together, liberated by deterioration of supporting hyphae. Arthroconidia common.

Differential diagnosis. The species differs from *T. mentagrophytes* (p. 968) by clavate, interspaced conidia and from *T. interdigitale* (p. 965) by abundant sporulation, and absence of nodular bodies and spiral hyphae.

Teleomorph. *Arthroderma benhamiae* Ajello & Cheng (*Ascomycota, Euascomycetes, Onygenales: Arthrodermataceae*).

Physiology.

Urease	+,w	T-1	++	T-4	++	T-6	++
Hair perforation	+,–	T-2	++	T-5	++	T-7	++
		T-3	++				

Molecular diagnostics. Species circumscription based on PCR-fingerprint, AFLP and ITS sequencing data was provided by Gräser *et al.* (1999b). ITS restriction map based on CBS 511.73:

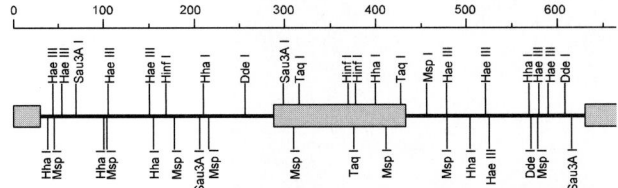

Pathogenicity. BSL-2. Zoophilic species on hedgehog (Contet-Andonneau *et al.*, 1991). Kano *et al.* (1998b) isolated the species from a rabbit. Human infections are rare (Pereiro *et al.*, 1989; Jury *et al.*, 1999).

Reference. Smith & Marples (1963).

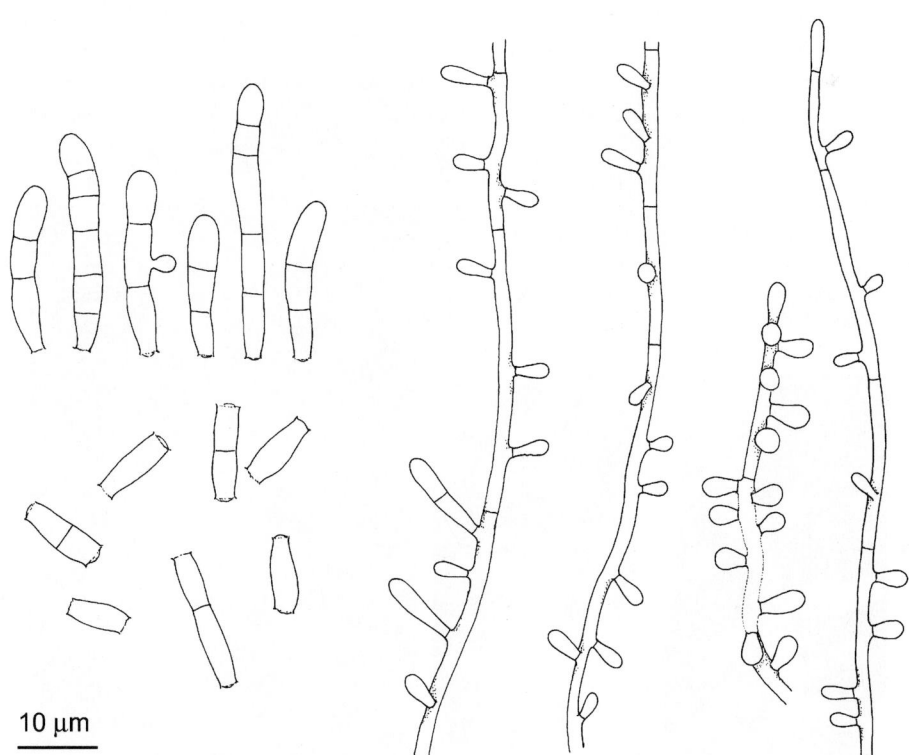

10 μm

Trichophyton erinacei, CBS 511.73. a, c. Young, slender micro- and macrononidia; b. mature conidia liberated by deterioration of supporting hypha; d. macroconidia; e. arthroconidia.

Antifungal susceptibility.

Antifungal	Mean MICs	MIC90	Strains	Reference
AMB	0.5	0.5	7	Fernández *et al.* (2000b)
FLZ	>16	>16	7	Fernández *et al.* (2000b)
ITZ	0.12	0.5	7	Fernández *et al.* (2000b)
KTZ	0.3	2	7	Fernández *et al.* (2000b)
MCZ	0.15	0.28	7	Fernández *et al.* (2000b)
TBF	0.02	0.06	7	Fernández *et al.* (2000b)
VCZ	0.09	0.125	7	Fernández *et al.* (2000b)

Nomenclature. *Trichophyton mentagrophytes* (Robin) Blanchard var. *erinacei* J.M.B. Smith & Marbles - Sabouraudia 3: 9, 1963 ≡ *Trichophyton erinacei* (J.M.B. Smith & Marbles) Quaife - J. Clin. Path. 19: 178, 1966 ≡ *Trichophyton benhamiae* Ajello & Cheng var. *erinacei* (J.M.B. Smith & Marbles) Takashio - Bull. Soc. Fr. Mycol. Méd. 4: 47, 1975.

    Arthroderma benhamiae Ajello & Cheng - Sabouraudia 5: 232, 1967.

    Trichophyton proliferans English & Stockdale - Sabouraudia 6: 267, 1968.

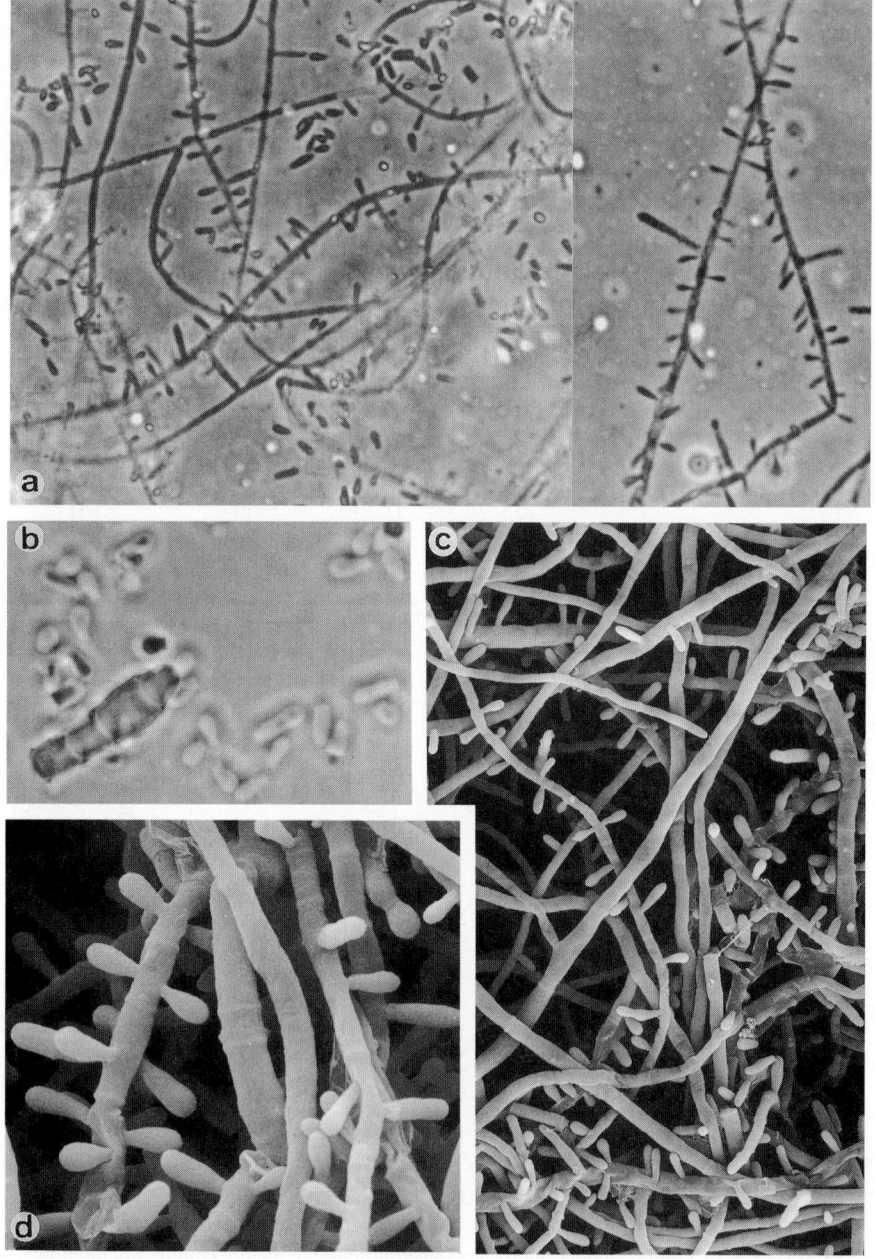

Trichophyton erinacei, CBS 677.86. Fertile hyphae producing microconidia, liberated microconidia and a macroconidium. a. ×640; b. ×1600; c. ×1250; d. ×3050.

Trichophyton flavescens Padhye & Carmichael

Colony characteristics. Colonies (SGA) spreading, pale yellow; reverse bright yellow to brown.
Microscopy. Macroconidia 2-6-septate, cylindrical, 25-80 × 8-14 μm, abruptly narrowed towards the base. Microconidia ovoidal, 1(-2)-celled, 5-16 × 4-8 μm, occasionally absent.
Teleomorph. *Arthroderma flavescens* Rees (*Ascomycota, Euascomycetes, Onygenales: Arthrodermataceae*).
Physiology.

Urease	w	T-1	++	T-5	++
Hair perforation	+	T-2	++	T-6	++
		T-3	++	T-7	++
		T-4	++		

Molecular diagnostics. ITS restriction map based on CBS 474.78:

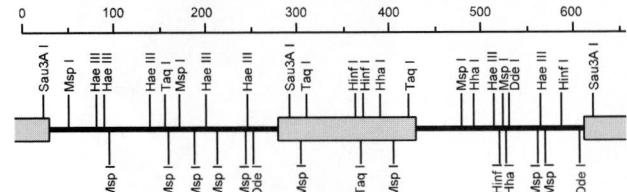

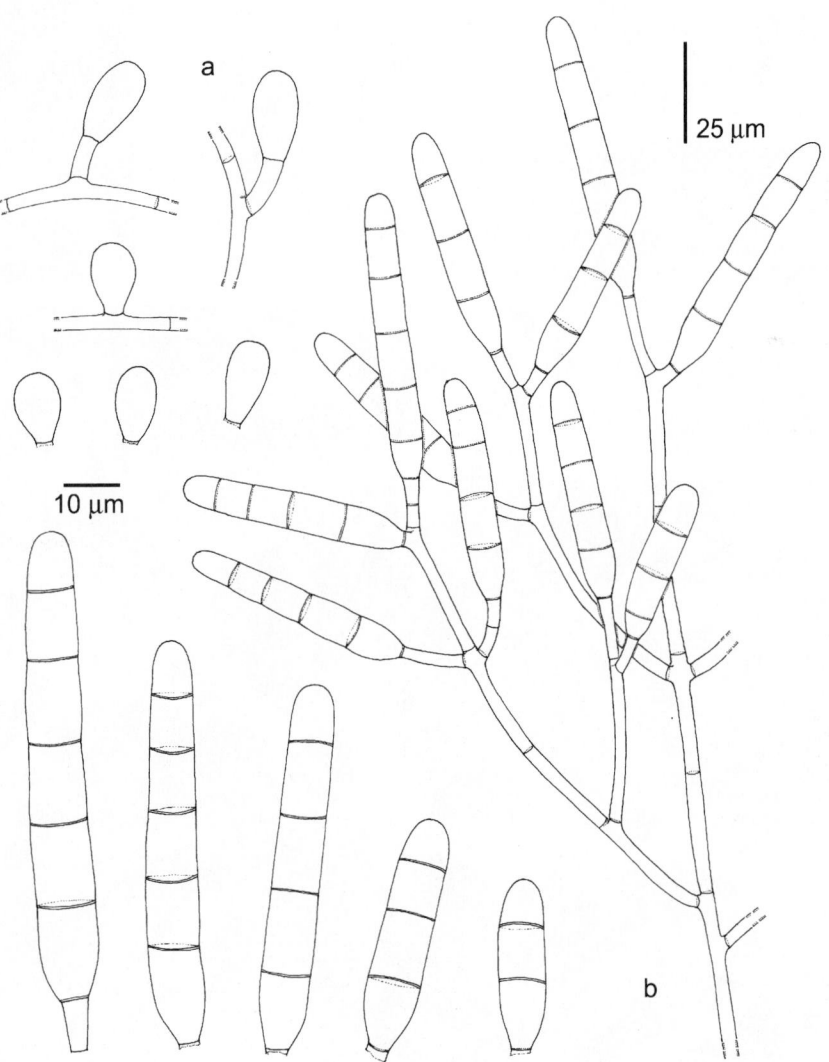

Trichophyton flavescens, CBS 473.78. a. Microconidia; b. fertile hyphae and macroconidia.

Pathogenicity. BSL-1. Geophilic dermatophyte, occurring on feathers.

Distribution. Australia.

Reference. Padhye & Carmichael (1971).

Nomenclature. *Trichophyton flavescens* Padhye & Carmichael - Can. J. Bot. 49: 1535, 1971.

    Arthroderma flavescens Rees - Sabouraudia 5: 206, 1967.

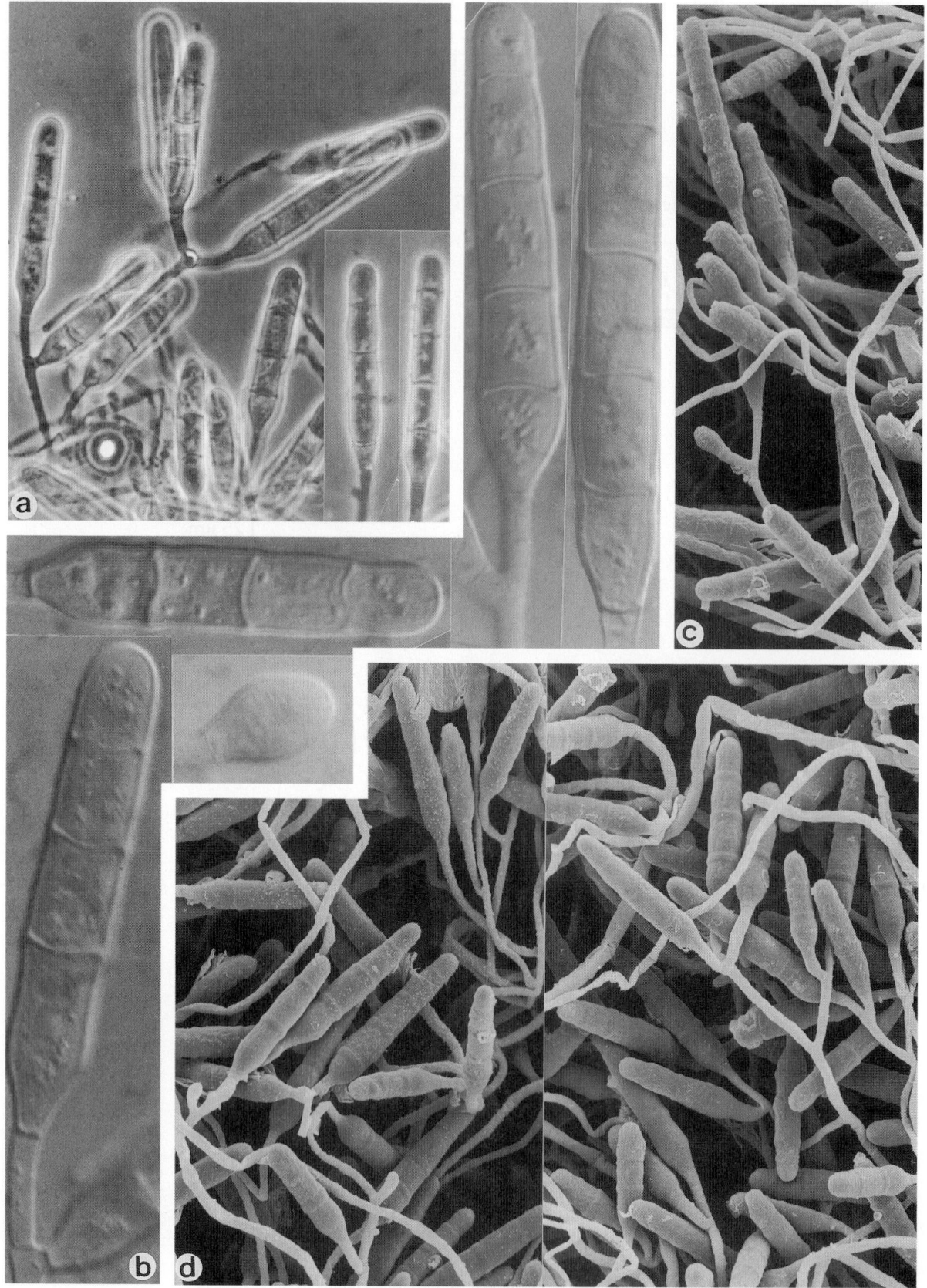

Trichophyton flavescens, CBS 473.78. a. Fertile hyphae with macroconidia; b-d. macro- and microconidia. a. ×512; b. ×1600; c. ×750; d. ×800.

Trichophyton gloriae Ajello

Colony characteristics. Colonies (SGA) flat, somewhat folded, powdery, whitish to cinnamon; reverse yellow.

Microscopy. Macroconidia in clusters, narrow clavate, mostly 5-11-celled, 9-60 × 3-7 µm, with walls up to 1 µm thick. Microconidia pyriform to clavate, 1.5-6.0 × 1.5-2.5 µm.

Teleomorph. *Arthroderma gloriae* Ajello (*Ascomycota, Euascomycetes, Onygenales: Arthrodermataceae*).

Physiology. *In vitro* hair perforation test positive.

Differential diagnosis. *Trichophyton vanbreuseghemii* (p. 987) is closely similar but has somewhat shorter macroconidia and a cream-coloured colony reverse.

Molecular diagnostics. ITS restriction map based on CBS 228.79:

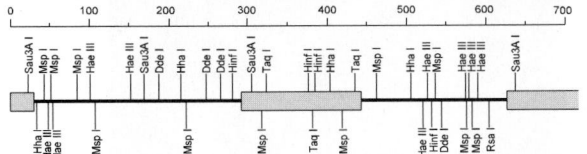

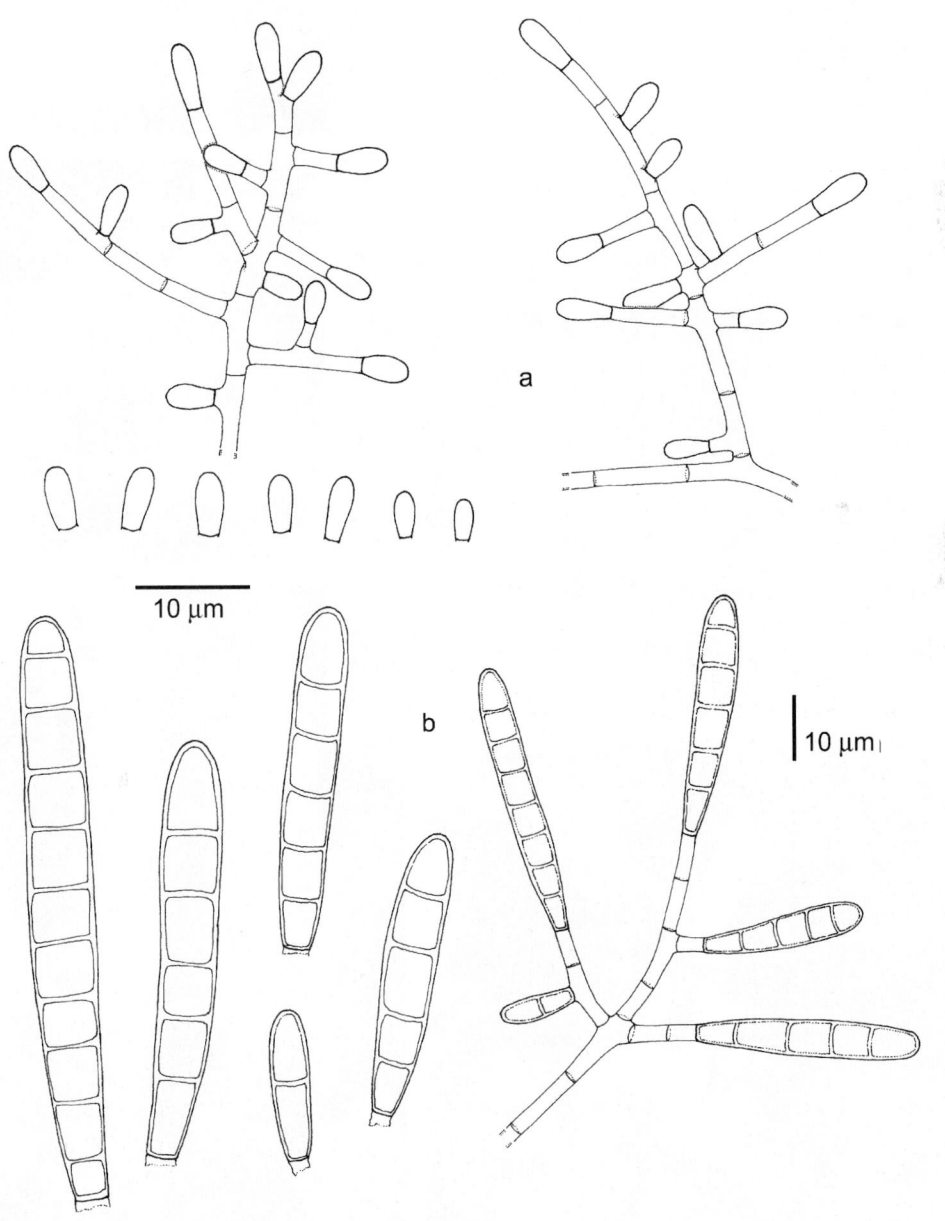

Trichophyton gloriae, CBS 228.79. a. Fertile hyphae and microconidia; b. fertile hyphae and macroconidia.

Pathogenicity. BSL-1. Geophilic dermatophyte.

Distribution. Arid regions in North America, occasionally in Italy (Caretta *et al.*, 1990).

Reference. Ajello & Cheng (1967).

Nomenclature. *Trichophyton gloriae* Ajello, *in* Ajello & Cheng - Mycologia 59: 257, 1967.
  Arthroderma gloriae Ajello & Cheng - Mycologia 59 : 257, 1967.

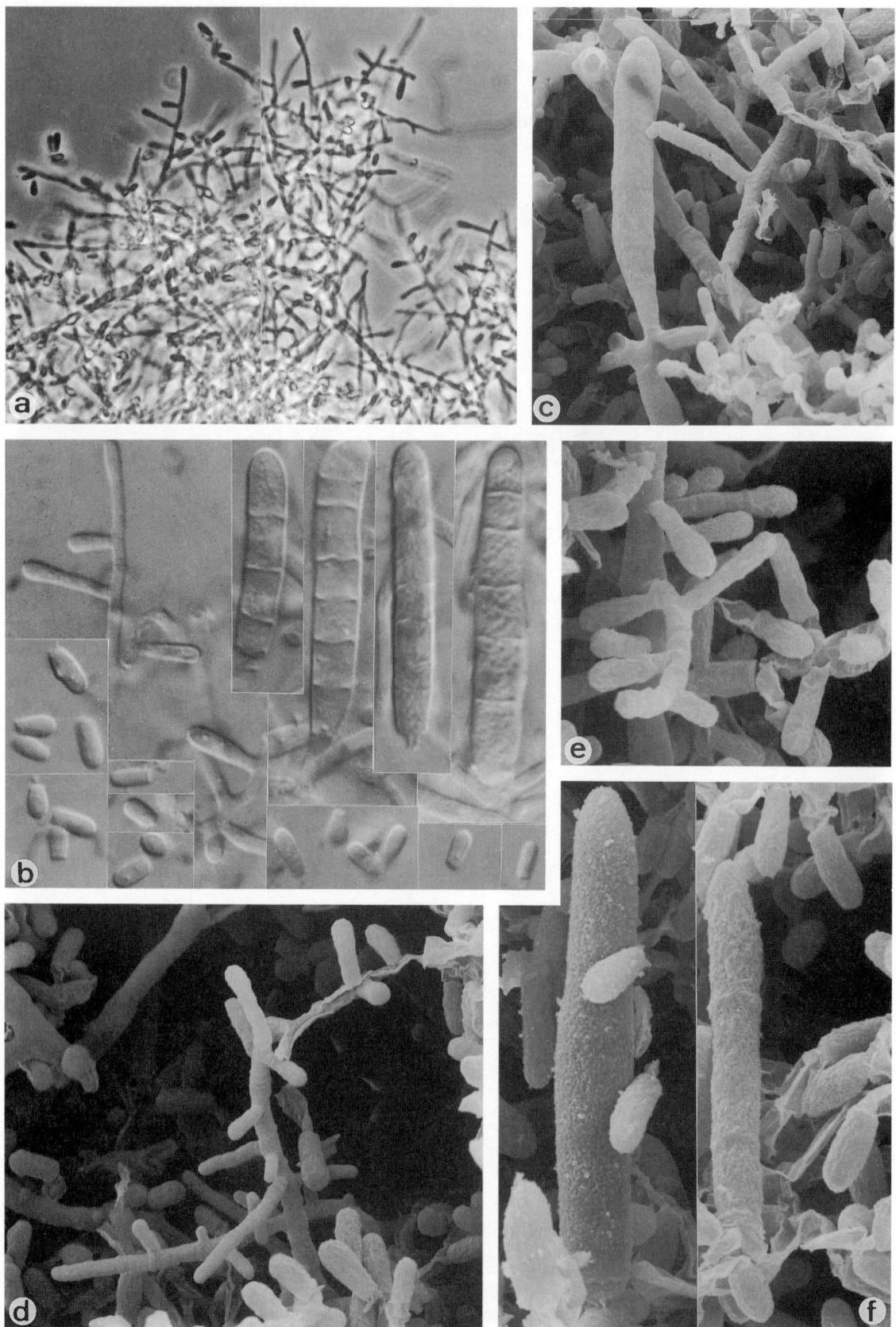

Trichophyton gloriae, CBS 228.79. Fertile hyphae with macro- and microconidia. a. ×640; b. ×1600; c. ×2300; d. ×3000; e. ×3700; f. ×3400.

Trichophyton interdigitale Priestley

Colony characteristics. Colonies (SGA) thin, floccose, white, or glabrous with deep yellow margin and orange reverse; the latter type of colonies do not produce any conidia but have yellowish hyphae with nodular bodies, and are known as var. *nodulare* (Kane *et al.*, 1992).

Microscopy. Conidia spherical to tear-shaped, frequently absent. Small nodular clumps of cells, surrounded by yellowish exudate, frequently present, particularly in non-sporulating strains.

Teleomorph. *Arthroderma vanbreuseghemii* Takashio (*Ascomycota, Euascomycetes, Onygenales: Arthrodermataceae*).

Physiology.

BCPCG	+	Sorbitol	–	T-4	++	
Urease	+	T-1	++	T-5	++	
Hair perforation	+	T-2	++	T-6	++	
Growth at 37°C	+	T-3	++	T-7	++	

Differential diagnosis. The species can be distinguished from *T. rubrum* by its positive urease test and positive *in vitro* hair perforation, absence of red colony reverse on cornmeal dextrose agar, inability to assimilate sorbitol and absence of green hue on Littman agar. On BCPCG a violet colour reaction is noted.

Molecular diagnostics. RAPD distinction from *T. rubrum* was described by Liu *et al.* (1996). Gräser *et al.* (1999) and Mochizuki *et al.* (1999) used fingerprinting and ITS sequencing. A specific primer based on chitin synthase 1 gene was introduced by Kano *et al.* (1999). ITS restriction map based on NCBI Z93001:

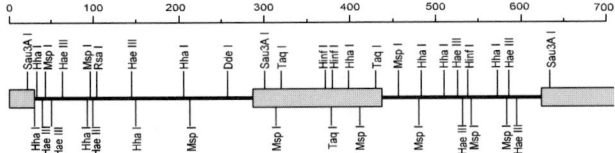

Pathogenicity. BSL-2. Hair affection undiagnostic, endo- as well as ectothrix being noted. Commonest European agent of dermatomycoses together with *T. rubrum*. Agent of tinea pedis, tinea corporis, tinea cruris, onychomycoses and infrequently tinea capitis (Urbanek *et al.*, 1995). The species was until recently known as *T. mentagrophytes* (p. 968), which is, however, a zoophilic species on rodents (Proulx & Onderka, 1997).

Distribution. World-wide.

References. Blank *et al.* (1969), Takashio (1975a), Pock-Steen (1968), Hantschke (1969), Gräser *et al.* (1998), Mochizuki *et al.* (1990, 1996a, b).

Antifungal susceptibility.

Antifungal	Mean MICs	MIC90	Strains	Reference
AMB	0.44	0.5	13	Fernández *et al.* (2000b)
CTZ	0.06	0.25	13	Fernández *et al.* (2000b)
FLZ	11.88	>64	13	Fernández *et al.* (2000b)
ITZ	0.06	0.25	13	Fernández *et al.* (2000b)
KTZ	0.38	1	13	Fernández *et al.* (2000b)
MCZ	0.23	1	13	Fernández *et al.* (2000b)
TBF	0.01	0.06	13	Fernández *et al.* (2000b)
VCZ	0.1	0.25	13	Fernández *et al.* (2000b)

Nomenclature. *Trichophyton interdigitale* Priestley - Med. J. Aust. 4: 475, 1917 ≡ *Sabouraudites interdigitalis* (Priestley) Ota & Langeron - Annls Parasit. Hum. Comp. 1: 328, 1923 ≡ *Epidermophyton interdigitale* (Priestley) MacCarthy - Archs Derm. Syph. 6: 24, 1925 ≡ *Trichophyton mentagrophytes* (Robin) Blanchard var. *interdigitale* (Priestley) Moraes - Anais Bras. Derm. Sif. 25: 230, 1950 ≡ *Kaufmannwolfia interdigitalis* (Priestley) Galgóczy & Novák, *in* Bakács - Az Orsz. Köz. Intéz. Mük. p. 224, 1962.

Trichophyton interdigitale Priestley var. *kaufmann-wolfii* Ota - Archs Derm. Syph. 5: 706, 1922 ≡ *Trichophyton gypseum* Bodin var. *kaufmann-wolfii* (Ota) Frágner - Česká Mykol. 10: 108, 1956.

Trichophyton rotundum MacCarthy - Annls Derm. Syph., Sér. 6, 6: 49, 1925.

Trichophyton batonrougei Castellani - J. Trop. Med. Hyg. 42: 373, 1939 ≡ *Trichophyton mentagrophytes* (Robin) Blanchard var. *batonrougei* (Castellani) de Vries & Cormane - Ned. Tijdschr. Geneeskd. 109: 1426, 1965.

Trichophyton mentagrophytes (Robin) Blanchard var. *nodulare* Georg & Meachling - J. Invest. Derm. 13: 349, 1949.

Trichophyton mentagrophytes (Robin) Blanchard var. *goetzii* Hantschke - Mykosen 12: 103, 1969.

Trichophyton candelabreum Listemann - Castellania 1: 53, 1973.

Arthroderma vanbreuseghemii Takashio - Annls Parasit. Hum. Comp. 48: 723, 1973.
Trichophyton krajdenii Kane, Scott, Summerbell & Diena - Mycotaxon 45: 309, 1992.

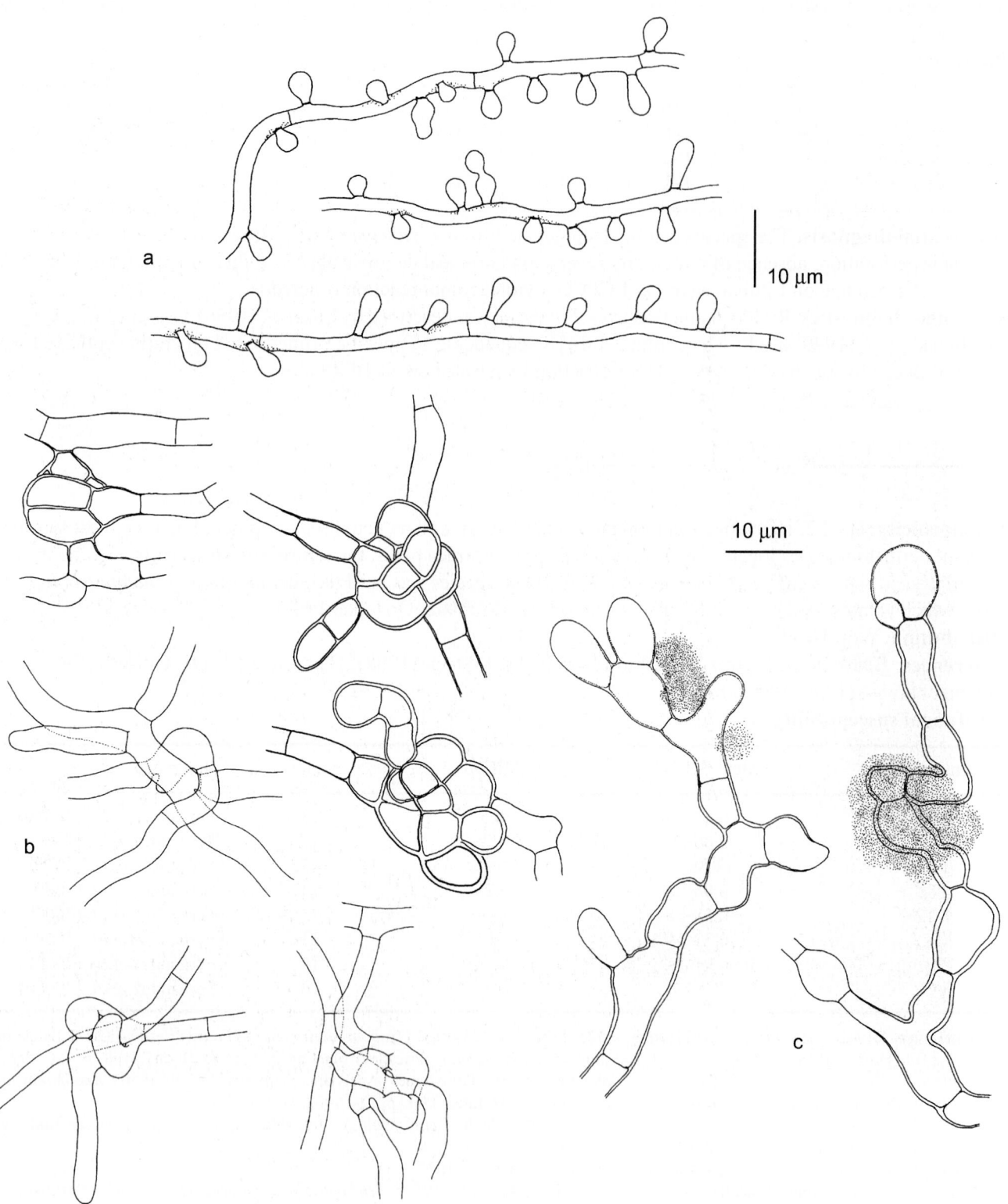

Trichophyton interdigitale, a. CBS 343.79; b, c. CBS 565.94. a. Microconidia; b. nodular bodies in aerial mycelium; c. pigment-exuding hyphae.

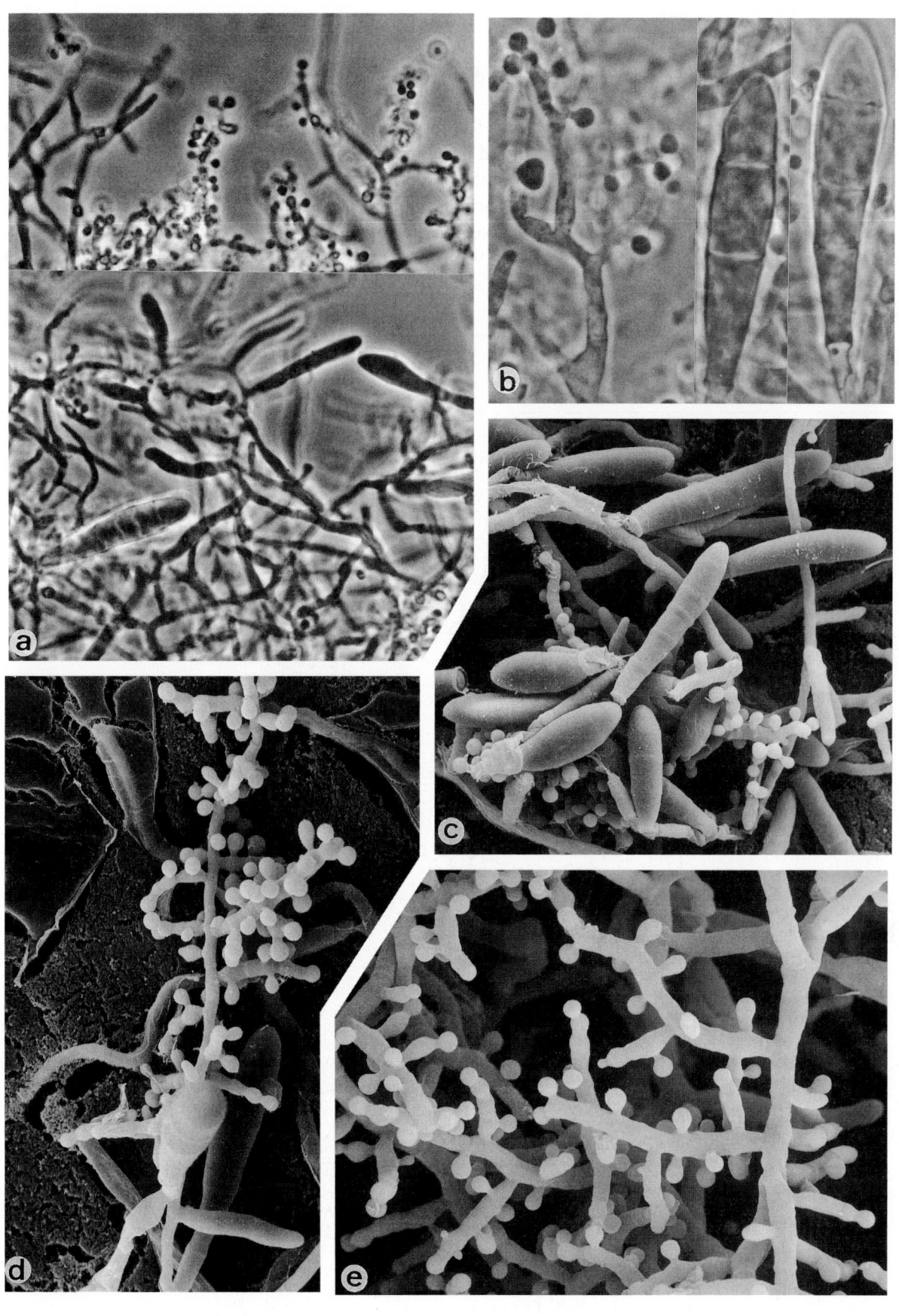

Trichophyton interdigitale, CBS 558.66. Fertile hyphae producing macro- and microconidia. a. ×640; b. ×1600; c. ×1250; d. ×1550; e. ×2050.

Trichophyton mentagrophytes (Robin) Blanchard

Colony characteristics. Colonies (SGA) powdery to floccose, cream-coloured to yellowish-buff; powdery colonies frequently somewhat star-shaped; reverse ochre to red-brown, occasionally yellow, or dark brown.

Microscopy. Macroconidia 3-8-celled, smooth- and thin-walled, clavate to cigar-shaped, 20-50 × 6-8 μm, usually sparse. Microconidia spherical, 2 μm diam, sessile, arranged in dense, grape-like clusters or alongside the hyphae. Spiral hyphae frequently present. Favic chandelier-like structures and chlamydospores occasionally present.

Differential diagnosis. *Trichophyton schoenleinii* (p. 977) is closely related but has exclusively favic chandeliers and chlamydospores, conidia being absent.

Physiology.

Urease	+	T-1	++	T-4	++	T-6	++
Hair perforation	v	T-2	++	T-5	++	T-7	++

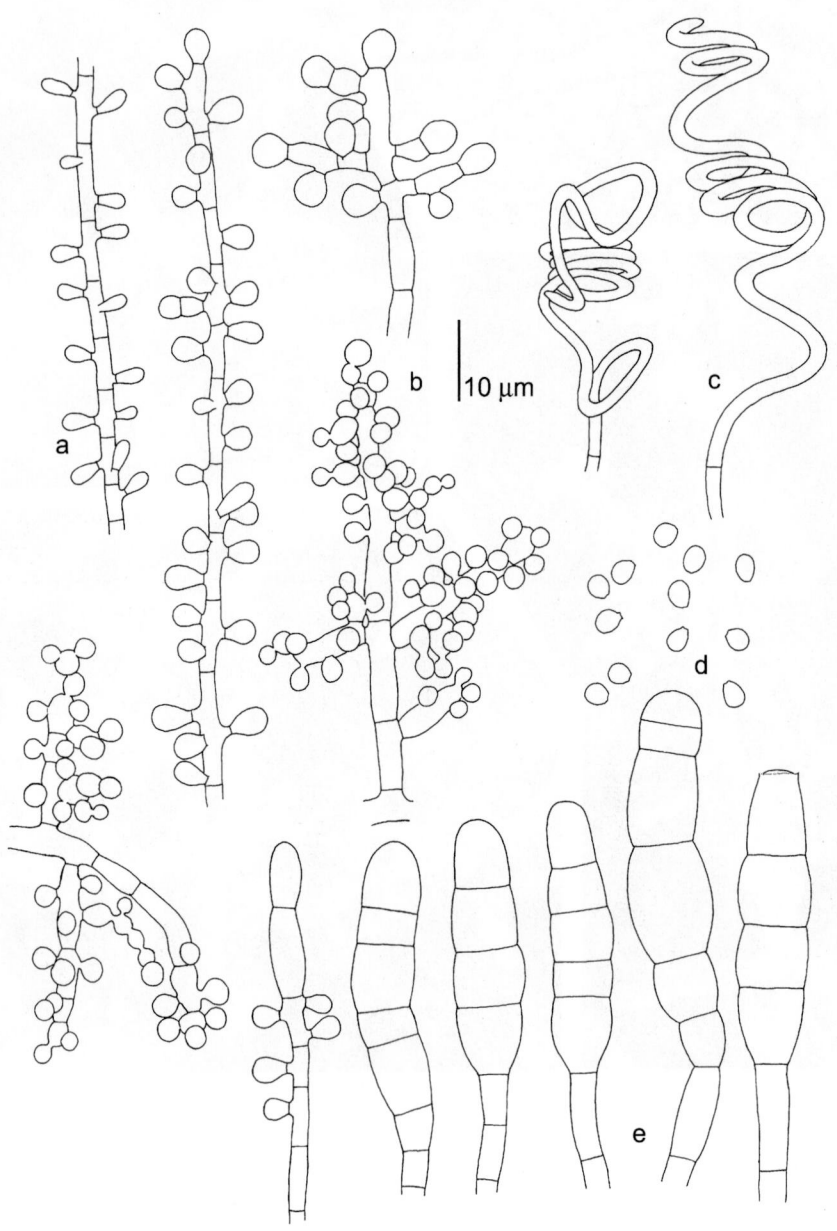

Trichophyton mentagrophytes, CBS 572.75. a. Young microconidia; b. mature microconidial branches; c. spiral hyphae; d. microconidia; e. macroconidia.

968

Molecular diagnostics. Species circumscription based on PCR-fingerprint, AFLP and ITS sequence data was provided by Gräser *et al.* (1999b). ITS restriction map based on NCBI Z97995:

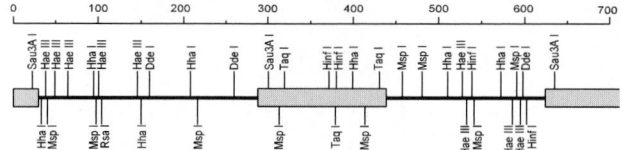

Pathogenicity. BSL-2. Produces favus on mice, rarely in humans (García-Sánchez *et al.*, 1997).
Reference. Ajello *et al.* (1968).
Antifungal susceptibility.

Antifungal	Mean MICs	MIC90	Strains	Reference
AMB	0.37	0.5	122	Fernández *et al.* (2000b)
CTZ	0.08	0.25	122	Fernández *et al.* (2000b)
FLZ	9.65	nd	6	Norris *et al.* (1999)
FLZ	nd	16	32	Jessup *et al.* (2000)
FLZ	15.08	nd	122	Fernández *et al.* (2000b)
GRF	nd	4	32	Jessup *et al.* (2000)
GRF	0.25	6.25	24	Wildfeuer *et al.* (1998)
GRF	nd	3	10	Niewerth *et al.* (1998)
GRF	0.15	nd	6	Norris *et al.* (1999)
ITZ	0.02	0.1	24	Wildfeuer *et al.* (1998)
ITZ	nd	1	10	Niewerth *et al.* (1998)
ITZ	0.1	nd	6	Norris *et al.* (1999)
ITZ	nd	0.5	32	Jessup *et al.* (2000)
ITZ	0.17	1	122	Fernández *et al.* (2000b)
KTZ	0.12	0.39	24	Wildfeuer *et al.* (1998)
KTZ	0.46	1	122	Fernández *et al.* (2000b)
MCZ	0.24	1	122	Fernández *et al.* (2000b)
TBF	<0.06	nd	6	Norris *et al.* (1999)
TBF	nd	0.001	32	Jessup *et al.* (2000)
TBF	0.04	0.06	122	Fernández *et al.* (2000b)
VCZ	0.1	0.2	24	Wildfeuer *et al.* (1998)
VCZ	0.09	1	122	Fernández *et al.* (2000b)

Nomenclature. *Microsporum mentagrophytes* Robin - Hist. Nat. Vég. Paras. Homme Anim. p. 430, 1853 ≡ *Sporotrichum mentagrophytes* (Robin) Saccardo - Syll. Fung. 4: 100, 1886 ≡ *Trichophyton mentagrophytes* (Robin) Blanchard - Traité Pathol. Gén. 2: 811, 1896 ≡ *Ectotrichophyton mentagrophytes* (Robin) Castellani & Chalmers - Man. Trop. Med., ed. 3. p. 1005, 1919 ≡ *Ctenomyces mentagrophytes* (Robin) Langeron & Milochevitch - Annls Parasit. Hum. Comp. 8: 484, 1930 ≡ *Spiralia mentagrophytes* (Robin) Grigorakis - C.R. Séanc. Soc. Biol. 109: 186, 1932.

Oidium quinckeanum Zopf - Die Pilze p. 481, 1890 ≡ *Achorion quinckeanum* (Zopf) Bodin - Archs Parasit. 5: 5-30, 1902 ≡ *Sabouraudites quinckeanus* (Zopf) Ota & Langeron - Annls Parasit. Hum. Comp. 1: 328, 1923 ≡ *Closteroaleuriospora quinckeana* (Zopf) Grigorakis - Annls Sci. Nat., Bot., Sér. 10, 12: 412, 1925 ≡ *Microsporum quinckeanum* (Zopf) Guiart & Grigorakis - Lyon Méd. 141: 377, 1928 ≡ *Trichophyton quinckeanum* (Zopf) MacLeod & Münde - Pract. Handb. Skin p. 361, 1940 ≡ *Trichophyton gypseum* Bodin var. *quinckeanum* (Zopf) Frágner - Česká Mykol. 10: 106, 1956 ≡ *Trichophyton mentagrophytes* (Robin) Blanchard var. *quinckeanum* (Zopf) J.M.B. Smith & Austwick, *in* Cotchin & Roe - Path. Lab. Rats Mice p. 684, 1967.

Trichophyton depressum MacCarthy - Annls Derm. Syph., Sér. 6, 6: 190, 1925.

Grubyella langeronii Baudet - Annls Parasit. Hum. Comp. 8: 417, 1930 ≡ *Trichophyton langeronii* (Baudet) Nannizzi - Tratt. Micopat. Um. 4: 210, 1934 ≡ *Favotrichophyton langeronii* (Baudet) C.W. Dodge - Med. Mycol. p. 517, 1935.

Trichophyton papillosum Lebasque - Champ. Teign. Cheval Bovidés p. 72, 1933; Annls Parasit. Hum. Comp. 12: 418, 1934.

Trichophyton sarkisovii Ivanova & Polyakov - Mikol. Fitopatol. 17: 364, 1983.

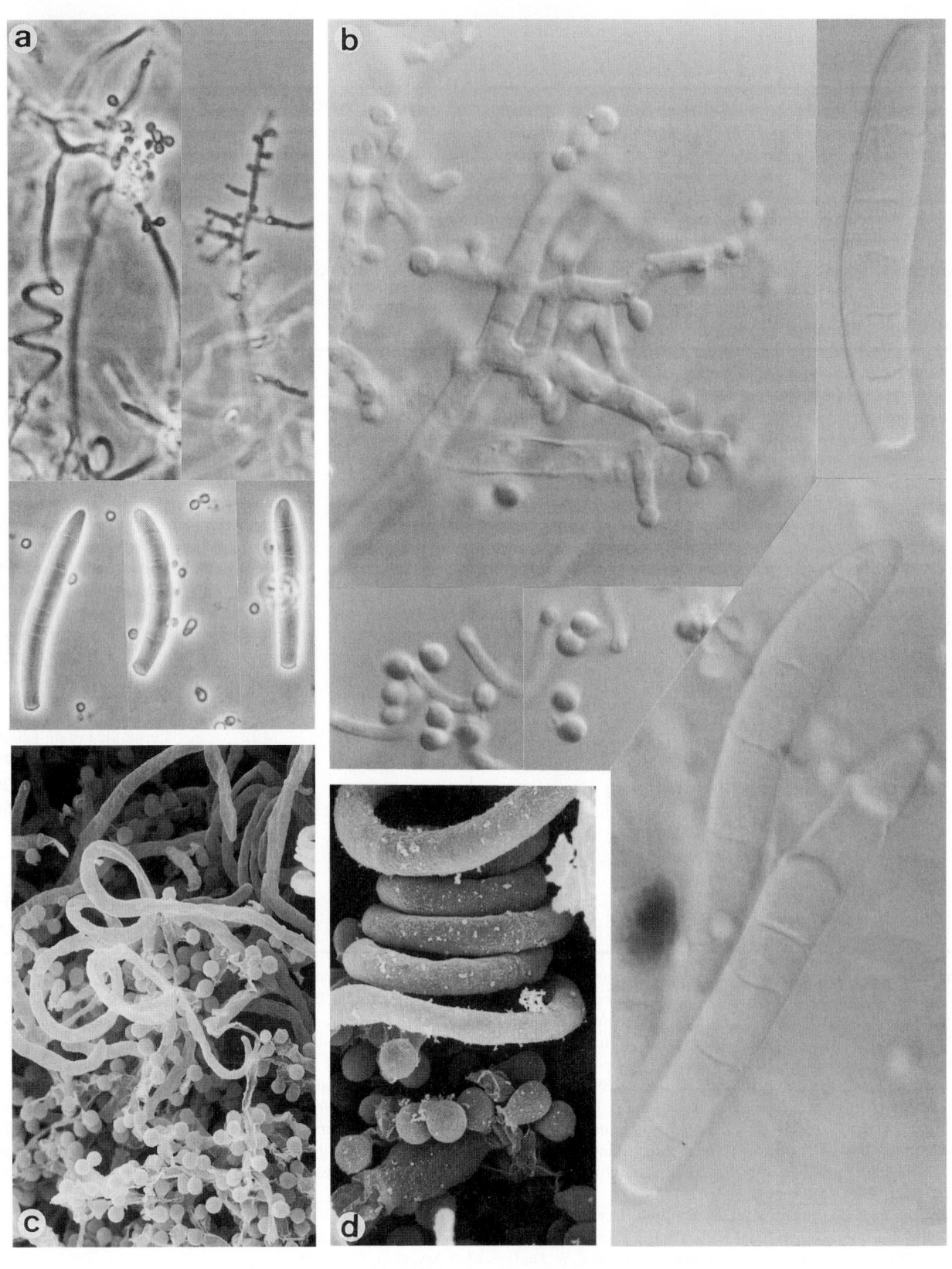

Trichophyton mentagrophytes, CBS 318.56. Fertile hyphae producing microconidia, liberated macro- and microconidia, and spiral hyphae. a. ×640; b. ×1600; c. ×1500; d. ×3200.

Trichophyton phaseoliforme Borelli & Feo

Colony characteristics. Colonies (SGA) spreading, powdery, white to bright cinnamon, with whitish pycnidium-like bodies.

Microscopy. Macroconidia usually absent; when present, in clusters, cigar-shaped, 2-5-celled. Microconidia curved, cashew nut-shaped, formed laterally on hyphae which locally are wider and densely aggregated in pycnidium-like bodies.

Physiology.

Urease	+	T-1	+	T-4	+	T-6		+
Hair perforation	+	T-2	+	T-5	+	T-7		+
		T-3	+					

Molecular diagnostics. ITS restriction map based on CBS 364.66:

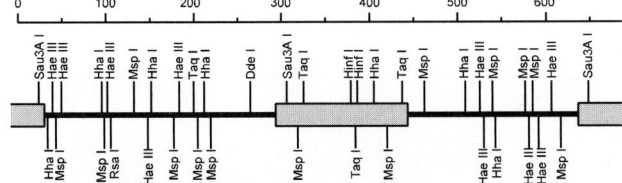

Pathogenicity. BSL-1. Geophilic dermatophyte.
Distribution. World-wide.
References. Borelli & Feo (1966), Alteras & Evolceanu (1969b).

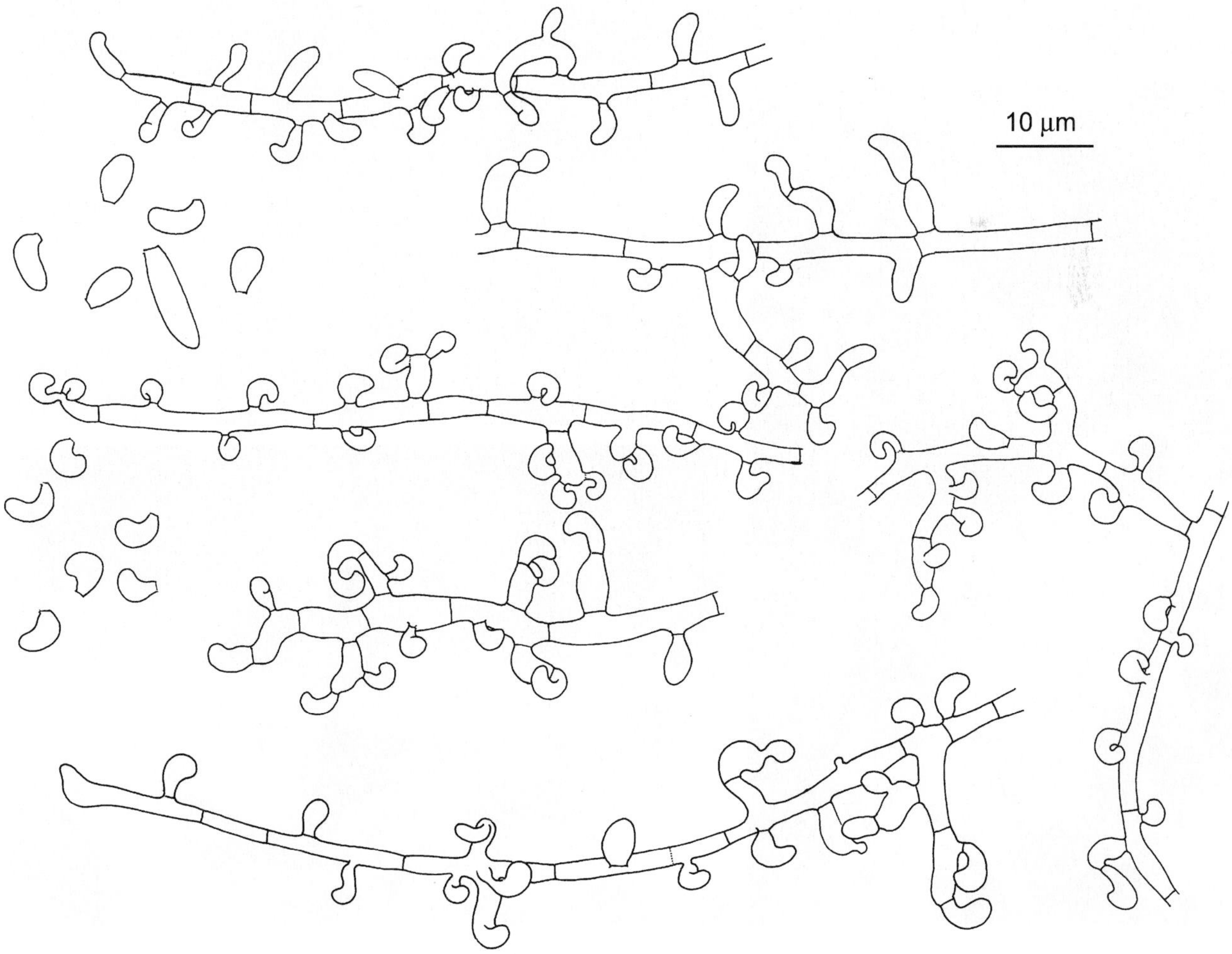

Trichophyton phaseoliforme, CBS 364.66. Branching system with microconidia.

971

Antifungal susceptibility.

Antifungal	Mean MICs	Strains	Reference
AMB	0.25	1	Fernández *et al.* (2000b)
CTZ	0.125	1	Fernández *et al.* (2000b)
FLZ	>16	1	Fernández *et al.* (2000b)
ITZ	0.5	1	Fernández *et al.* (2000b)
KTZ	2	1	Fernández *et al.* (2000b)
MCZ	0.06	1	Fernández *et al.* (2000b)
SCZ	8	1	Fernández *et al.* (2000b)
TBF	0.06	1	Fernández *et al.* (2000b)
VCZ	1	1	Fernández *et al.* (2000b)

Nomenclature. *Trichophyton phaseoliforme* Borelli & Feo - Acta Méd. Venez. 13: 176, 1966.

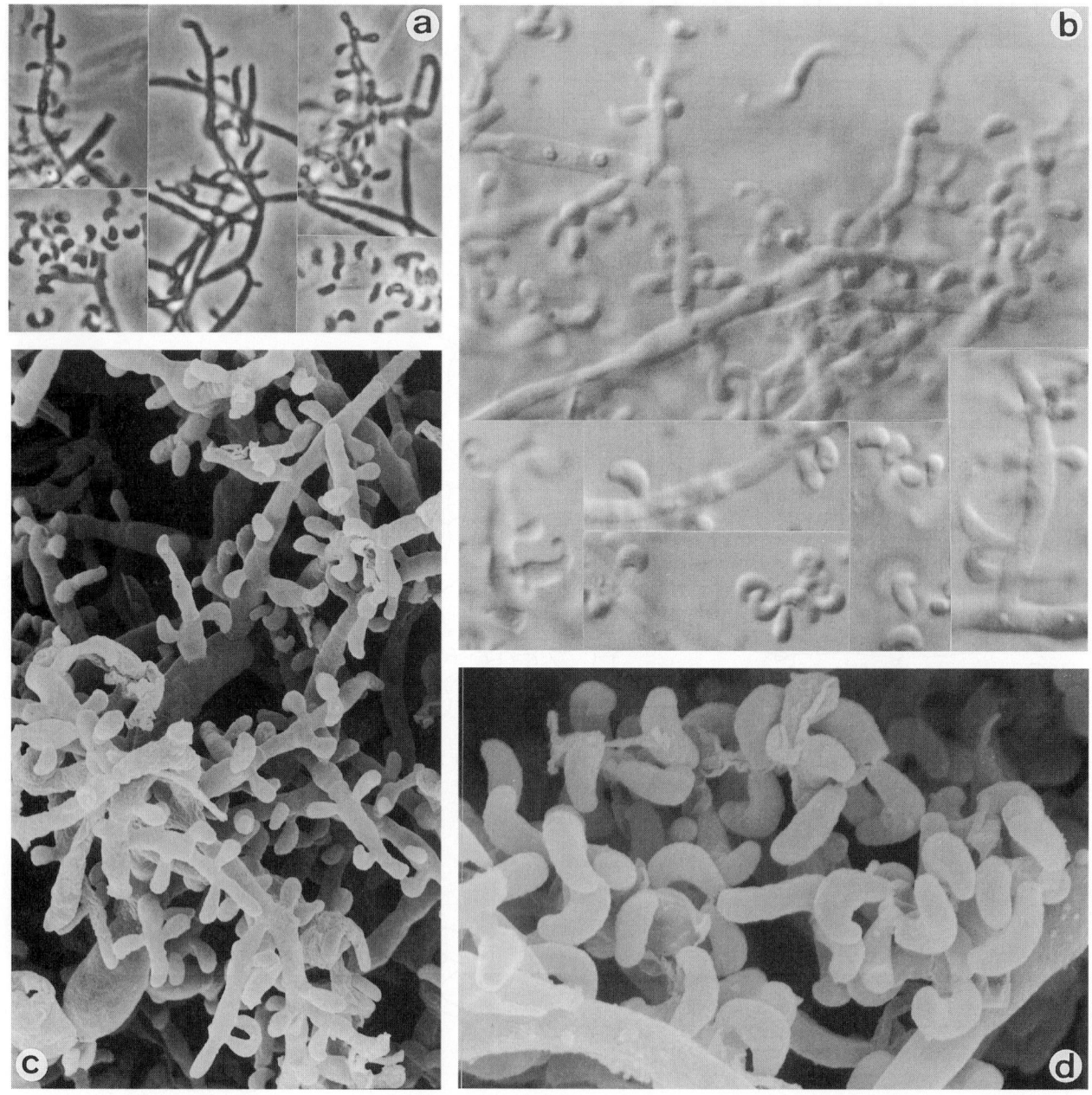

Trichophyton phaseoliforme, CBS 364.66. Fertile hyphae and microconidia. a. ×640; b. ×1600; c. ×2000; d. ×4500.

Trichophyton rubrum (Castell.) Semon

Colony characteristics. Colonies (SGA) fluffy to cottony, white, sometimes becoming rose when ageing; reverse wine-red to olive, sometimes yellow.

Microscopy. Macroconidia mostly absent, when produced thin-walled, poorly differentiated, of variable size, cylindrical to cigar-shaped, 40-55 × 6.0-7.5 μm, with a tendency to disarticulate. Microconidia peg-shaped to pyriform, 3.0-5.5 × 2.0-3.5 μm, sessile alongside undifferentiated hyphae. Occasionally only micro- or only macroconidia are present; cultures rarely sterile.

Physiology.

BCPCG	–(+)	T-1	++	T-5	++
Urease	v	T-2	++	T-6	+,–
Hair perforation	–	T-3	++	T-7	++
Growth at 37°C	+	T-4	++		
Sorbitol	+				

Differential diagnosis. The species can be distinguished from *T. interdigitale* (p. 965) by negative *in vitro* hair perforation test, ability to assimilate sorbitol (Kane & Fischer, 1971; Rezusta *et al.*, 1991) and by the production of a red pigment on cornmeal dextrose agar and a green hue on Littman agar. Colony occasionally yellow (Szilagyi & Reiss, 1968) or olivaceous brown (Frágner, 1966).

Molecular diagnostics. Mochizuki *et al.* (1996a) developed RFLP of total DNA to distinguish the species from *T. interdigitale*, whereas Liu *et al.* (1996) and Kac *et al.* (1999) used RAPD, Kano *et al.* (1999) chitin synthase genes, and de Bièvre & Rouffaud (1999), Summerbell *et al.* (1999) and Gräser *et al.* (2000X) ITS1 sequencing. Using fingerprinting and AFLP data, Gräser *et al.* (1999c) found that the species varies only little and is largely clonal. ITS restriction map based on NCBI Z97993:

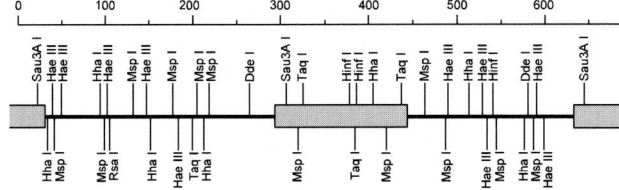

Pathogenicity. BSL-2. Affection of hair undiagnostic, endo- as well as ectothrix being noted. Most common agent of dermatomycoses. The species is anthropophilic, causing primarily tinea pedis, tinea corporis and onychomycosis (Ploysangam & Lucky, 1997); frequently isolated from tinea cruris and tinea manuum. Chronic infection is probably linked to genetic predisposition of the host (Zaias & Rebell, 1996), although these may be difficult to reveal (Pièrard-Franchimont *et al.*, 1996). Sentamilselvi *et al.* (1998) described a subcutaneous infection in an otherwise healthy patient; subcutaneous infections in non-AIDS patients are rare (Squeo *et al.*, 1998). Tinea capitis is uncommon (Schwinn *et al.*, 1995; Aoun *et al.*, 1998). Inflammatory cases are rare (Gupta *et al.*, 1998). Infections are more severe in HIV positive patients (Grossman *et al.*, 1995), occasionally with deep invasion (Tsang *et al.*, 1996). Occasionally found on cats and dogs.

Distribution. World-wide.

References. Young (1972), Blank & Mann (1975), Kane & Fischer (1975), Kaminsky (1985), Summerbell (1987), Pereiro *et al.* (1988).

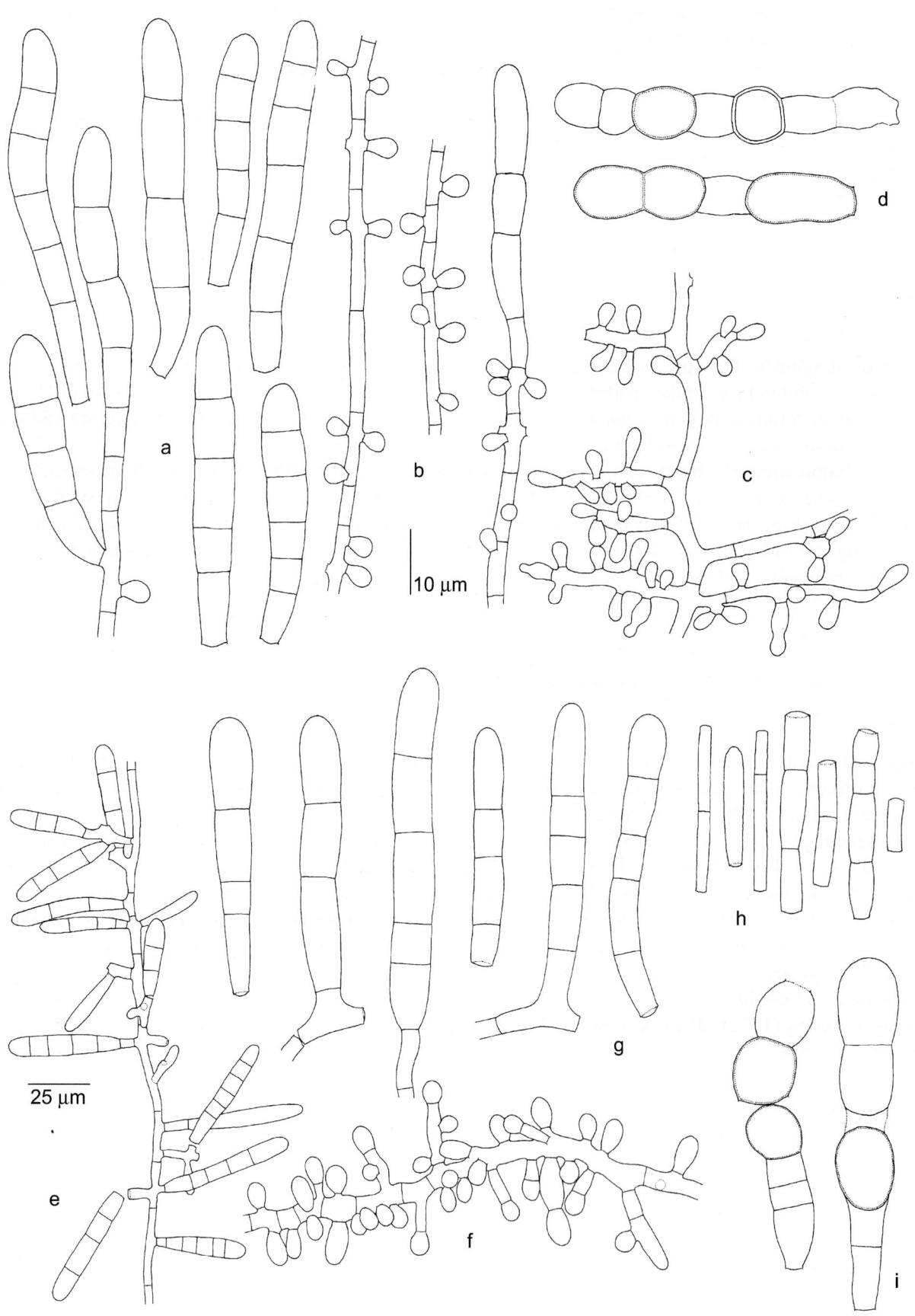

Trichophyton rubrum, a-e. CBS 304.60; f-i. CBS 289.86. a, e. Branching system with macroconidia; b, c, f. branching system with microconidia; g. macroconidia; d, i. macroconidia converted into strings of chlamydospores; h. arthric conidia.

Antifungal susceptibility.

Antifungal	Mean MICs	MIC 90	Strains	Reference
AMB	0.37	1	144	Fernández *et al.* (2000b)
CTZ	0.04	0.125	100	Fernández *et al.* (2000a)
FLZ	0.35	nd	6	Norris *et al.* (1999)
FLZ	nd	2	132	Jessup *et al.* (2000)
FLZ	0.25-2	nd	104	Bradley *et al.* (1999)
FLZ	2.80	16	144	Fernández *et al.* (2000b)
GRF	0.22	0.39	25	Wildfeuer *et al.* (1998)
GRF	nd	3	38	Niewerth *et al.* (1998)
GRF	0.35	nd	6	Norris *et al.* (1999)
GRF	nd	2	132	Jessup *et al.* (2000)
GRF	<0.125-2	nd	104	Bradley *et al.* (1999)
ITZ	0.03	0.1	25	Wildfeuer *et al.* (1998)
ITZ	nd	1	38	Niewerth *et al.* (1998)
ITZ	0.06	nd	6	Norris *et al.* (1999)
ITZ	nd	0.5	132	Jessup *et al.* (2000)
ITZ	<0.06-0.5	nd	104	Bradley *et al.* (1999)
ITZ	0.16	0.5	100	Fernández *et al.* (2000a)
KTZ	0.07	0.2	25	Wildfeuer *et al.* (1998)
KTZ	0.14	0.5	100	Fernández *et al.* (2000a)
MCZ	0.09	0.25	144	Fernández *et al.* (2000b)
SCZ	0.21	1	144	Fernández *et al.* (2000b)
TBF	0.01	0.01	100	Fernández *et al.* (2000a)
TBF	<0.06	nd	6	Norris *et al.* (1999)
TBF	nd	0.002	132	Jessup *et al.* (2000)
TBF	0.78	nd	104	Bradley *et al.* (1999)
VCZ	0.04	0.1	25	Wildfeuer *et al.* (1998)
VCZ	0.06	0.25	144	Fernández *et al.* (2000b)

Nomenclature. (?) *Trichophyton megninii* Blanchard, *in* Bouchard - Traité Pathol. Gén. 2: 915, 1896 ≡ *Ectotrichophyton megninii* (Blanchard) Castellani & Chalmers - Man. Trop. Med., ed. 3, p. 1007, 1919 ≡ *Megatrichophyton megninii* (Blanchard) Neveu-Lemaire - Précis Parasitol. Hum., ed. 5, p. 54, 1921.

(?) *Trichophyton roseum* Sabouraud, *in* Bodin - Champ. Paras. Homme p. 120, 1902 ≡ *Megatrichophyton roseum* (Bodin) C.W. Dodge - Med. Mycol. p. 510, 1935.

(?) *Trichophyton vinosum* Sabouraud - Malad. Cuir Chev. 3: 386, 1910 ≡ *Ectotrichophyton vinosum* (Sabouraud) Castellani & Chalmers - Man. Trop. Med., ed. 3, p. 1007, 1919 ≡ *Megatrichophyton vinosum* (Sabouraud) Neveu-Lemaire - Précis Parasitol. Hum., ed. 5, p. 55, 1921 ≡ *Megatrichophyton roseum* (Bodin) C.W. Dodge var. *vinosum* (Sabouraud) C.W. Dodge - Med. Mycol. p. 511, 1935.

Epidermophyton rubrum Castellani - Phil. J. Sci. 5: 203, 1910 ≡ *Trichophyton rubrum* (Castellani) Semon - Br. J. Derm. Syph. 34: 398, 1922 ≡ *Sabouraudites ruber* Ota & Langeron - Annls Parasit. Hum. Comp. 1: 328, 1923 ≡ *Sabouraudiella rubra* (Castellani) Boedijn - Mycopath. Mycol. Appl. 6: 125, 1951.

Trichophyton purpureum Bang - Annls Derm. Syph., Sér. 5, 1: 238, 1910 ≡ *Epidermophyton purpureum* (Bang) C.W. Dodge - Med. Mycol. p. 485, 1935 ≡ *Sabouraudiella purpurea* (Bang) Boedijn - Mycopath. Mycol. Appl. 6: 123, 1952.

Epidermophyton pernetii Castellani - Br. J. Derm. Syph. 22: 148, 1910 ≡ *Epidermophyton floccosum* (Harz) Langeron & Milochevitch var. *pernetii* (Castellani) Nannizzi - Tratt. Micopat. Um. 4: 173, 1934.

Trichophyton rubidum Priestley - Med. J. Aust. 4: 474, 1917 ≡ *Epidermophyton rubidum* (Priestley) C.W. Dodge - Med. Mycol. p. 486, 1935.

Trichophyton marginatum Muys - Ned. Tijdschr. Geneeskd. 65: 2205, 1921 = *Favotrichophyton violaceum* (Sabouraud) C.W. Dodge var. *marginatum* (Muys) C.W. Dodge - Med. Mycol. p. 524, 1935.

Trichophyton pedis Ota - Bull. Soc. Path. Exot. 15: 594, 1922 ≡ *Epidermophyton pedis* (Ota) C.W. Dodge - Med. Mycol. p. 480, 1935 ≡ *Kaufmanwolfia pedis* (Ota) Novák & Galgóczy - Acta Bot. Hung. 15: 132, 1969.

Epidermophyton plurizoniforme MacCarthy - Annls Derm. Syph., Sér. 6, 6: 37, 1925 ≡ *Sabouraudites plurizoniforme* (MacCarthy) Brumpt - Précis Parasitol., ed. 4, p. 1291, 1927 ≡ *Trichophyton plurizoniforme* (MacCarthy) Rippon - Med. Mycol. p. 256, 1988.

Epidermophyton lanoroseum MacCarthy - Annls Derm. Syph., Sér. 6, 6: 53, 1925 ≡ *Sabouraudites lanoroseus* (MacCarhy) Brumpt - Précis Parasitol., éd. 4, p. 1287, 1927.

Trichophyton coccineum Katoh - Trans. 6th Congr. Far East Assoc. Trop. Med., Tokyo p. 861, 1925 ≡ *Favotrichophyton coccineum* (Katoh) C.W. Dodge - Med. Mycol. p. 524, 1935.

Trichophyton multicolor de Magalhães & Neves - Mem. Inst. Oswaldo Cruz 20: 271-298, 1927 ≡ *Ectotrichophyton multicolor* (de Magalhães & Neves) C.W. Dodge - Med. Mycol. p. 500, 1935.

Trichophyton kagawaense Fujii - Jpn. J. Derm. Urol. 31: 305-357, 1931 ≡ *Ectotrichophyton kagawaense* (Fujii) C.W. Dodge - Med. Mycol. p. 506, 1935.

Trichophyton pervesii Catanei - Archs Inst. Pasteur Algér. 15: 267, 1937.

Trichophyton rodhainii Vanbreuseghem - Annls Parasit. Hum. Comp. 24: 244, 1949 ≡ *Trichophyton rubrum* (Castellani) Semon var. *rodhainii* (Vanbreuseghem) Armijo & Lachapelle - Annls Derm. Vénéréol. 108: 990, 1981.

Trichophyton kuryangei Vanbreuseghem & Rosenthal - Annls Parasit. Hum. Comp. 36: 802, 1961.

Trichophyton fluviomuniense Pereiro Miguens - Sabouraudia 6: 315, 1968.

Trichophyton rubrum (Castellani) Semon var. *nigricans* Frágner - Česká Mykol. 20: 27, 1966 ≡ *Trichophyton olexae* Watanabe - Jpn. J. Med. Mycol. 18: 77, 1977 (name change).

Trichophyton rubrum (Castellani) Semon var. *flavum* Szilagyi & Reiss - Mycopath. Mycol. Appl. 36: 193, 1968.

Trichophyton fischeri Kane - Sabouraudia 15: 239, 1977.

Trichophyton raubitschekii Kane, Salkin, Weitzman & Smitka - Mycotaxon 13: 260, 1981.

Trichophyton kanei Summerbell - Mycotaxon 28: 511, 1987.

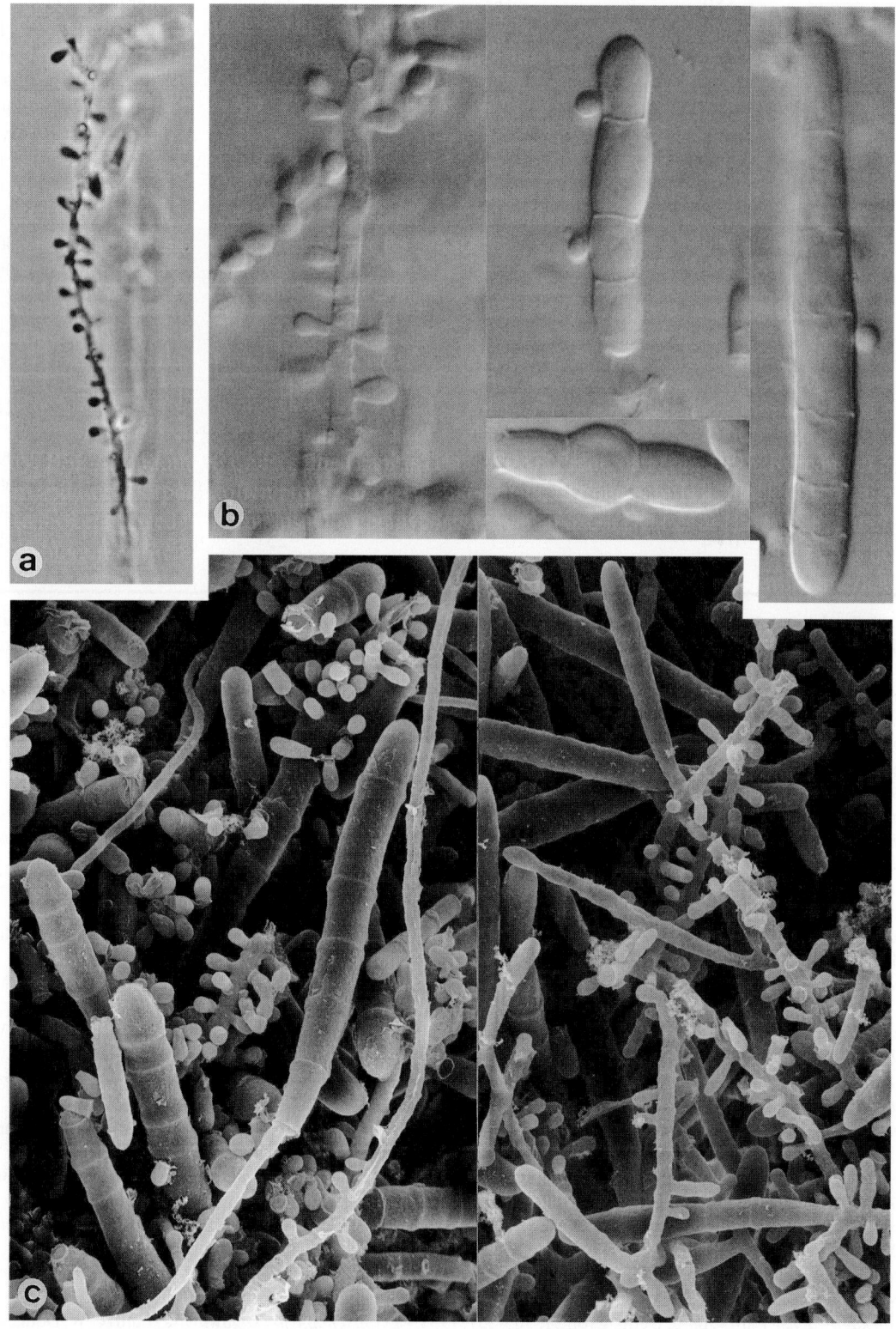

Trichophyton rubrum, CBS 304.60. Microconidial branches, macro- and microconidia. a. ×640; b. ×1600; c. ×1500.

976

Trichophyton schoenleinii (Lebert) Nannizzi

Colony characteristics. Colonies (SGA) growing rather slowly, waxy, later becoming velvety, folded, cerebriform and heaped with age, often cracking and splitting the agar, whitish to cream-coloured; margin sometimes feathered due to the presence of favic chandeliers; reverse unpigmented or pale yellow.

Microscopy. Macroconidia and microconidia usually absent. Antler-like hyphae, with dichotomously branched, swollen tips (favic chandeliers), present in submerged margin of fresh cultures. Chlamydospores abundant.

Physiology.

BCPCG	+	T-1	+	T-5	+
Urease	v	T-2	+	T-6	+
Hair perforation	–	T-3	+	T-7	+
Growth at 37°C	+	T-4	++		

Differential diagnosis. *Trichophyton mentagrophytes* (p. 968) may also produce favic chandelier-like structures and chlamydospores, but shows additional conidiation.

Molecular diagnostics. ITS restriction map based on NCBI Z98011:

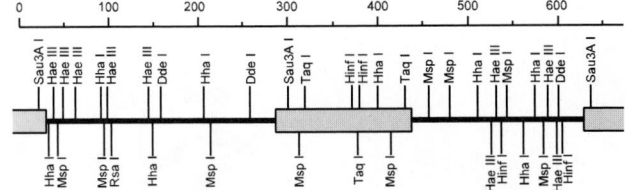

Pathogenicity. BSL-2. Producing characteristic air bubbles in the affected hair; hairs frequently with dull fluorescence when irradiated by a Wood's lamp. The species is anthropophilic, causing scarring or scaling forms of tinea capitis and tinea corporis (favus; Matte *et al.*, 1997); nails may also be affected. In typical favus, elevated, yellowish, cup-like crusts or mats of hyphae (scutella) occur on the scalp and on the body. The infection can cause permanent baldness. Because of its disfiguring effects it is one of the most severe kinds of dermatophyte infection. Contagious.

Distribution. Eurasia, North Africa.

Reference. Seeliger (1985).

Antifungal susceptibility.

Antifungal	Mean MICs	Strains	Reference
AMB	0.25	2	Fernández *et al.* (2000b)
CTZ	0.08	2	Fernández *et al.* (2000b)
FLZ	>16	2	Fernández *et al.* (2000b)
GRF	0.20	1	Wildfeuer *et al.* (1998)
ITZ	0.07	2	Fernández *et al.* (2000b)
ITZ	0.05	1	Wildfeuer *et al.* (1998)
KTZ	0.20	1	Wildfeuer *et al.* (1998)
KTZ	0.06	2	Fernández *et al.* (2000b)
MCZ	0.04	2	Fernández *et al.* (2000b)
TBF	0.007	2	Fernández *et al.* (2000b)
VCZ	0.02	2	Fernández *et al.* (2000b)
VCZ	0.14	1	Wildfeuer *et al.* (1998)

Nomenclature. *Oidium schoenleinii* Lebert - Physiol. Path. 2: 490, 1845 ≡ *Achorion schoenleinii* (Lebert) Remak - Diagn. Pathog. Unters. p. 13, 1845 ≡ *Schoenleinium achorion* Johan-Olsen - Zentbl. Bakt. Parasitkde, Abt. 2, 3: 276, 1897 (name change) ≡ *Grubyella schoenleinii* (Lebert) Ota & Langeron - Annls Parasit. Hum. Comp. 1: 320, 1923 ≡ *Arthrosporia schoenleinii* (Lebert) Grigorakis - Annls Sci. Nat., Bot., Sér. 7: 414, 1925 ≡ *Sporotrichum schoenleinii* (Lebert) Saccardo, *in* Vuillemin - Champ. Paras. Myc. Homme p. 69, 1931 ≡ *Trichophyton schoenleinii* (Lebert) Nannizzi - Tratt. Micopat. Um. 4: 198, 1934.

Oospora porriginis Saccardo - Syll. Fung. 4: 15, 1886.

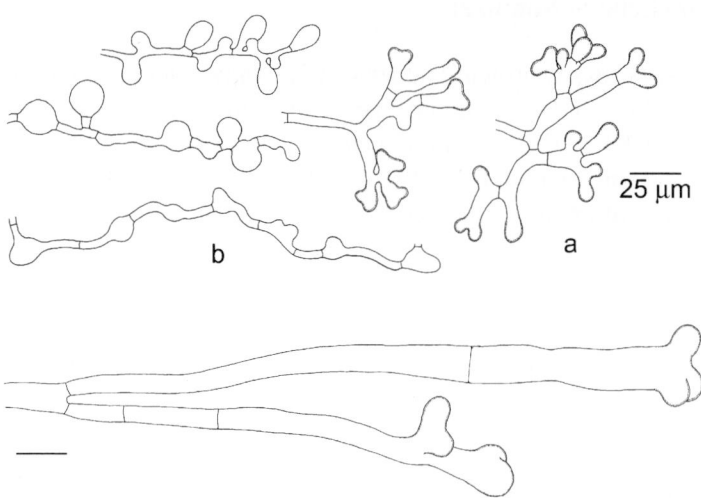

25 µm

Trichophyton schoenleinii, CBS 564.94. a. Antler-like hyphae; b. chlamydospore-like swellings.

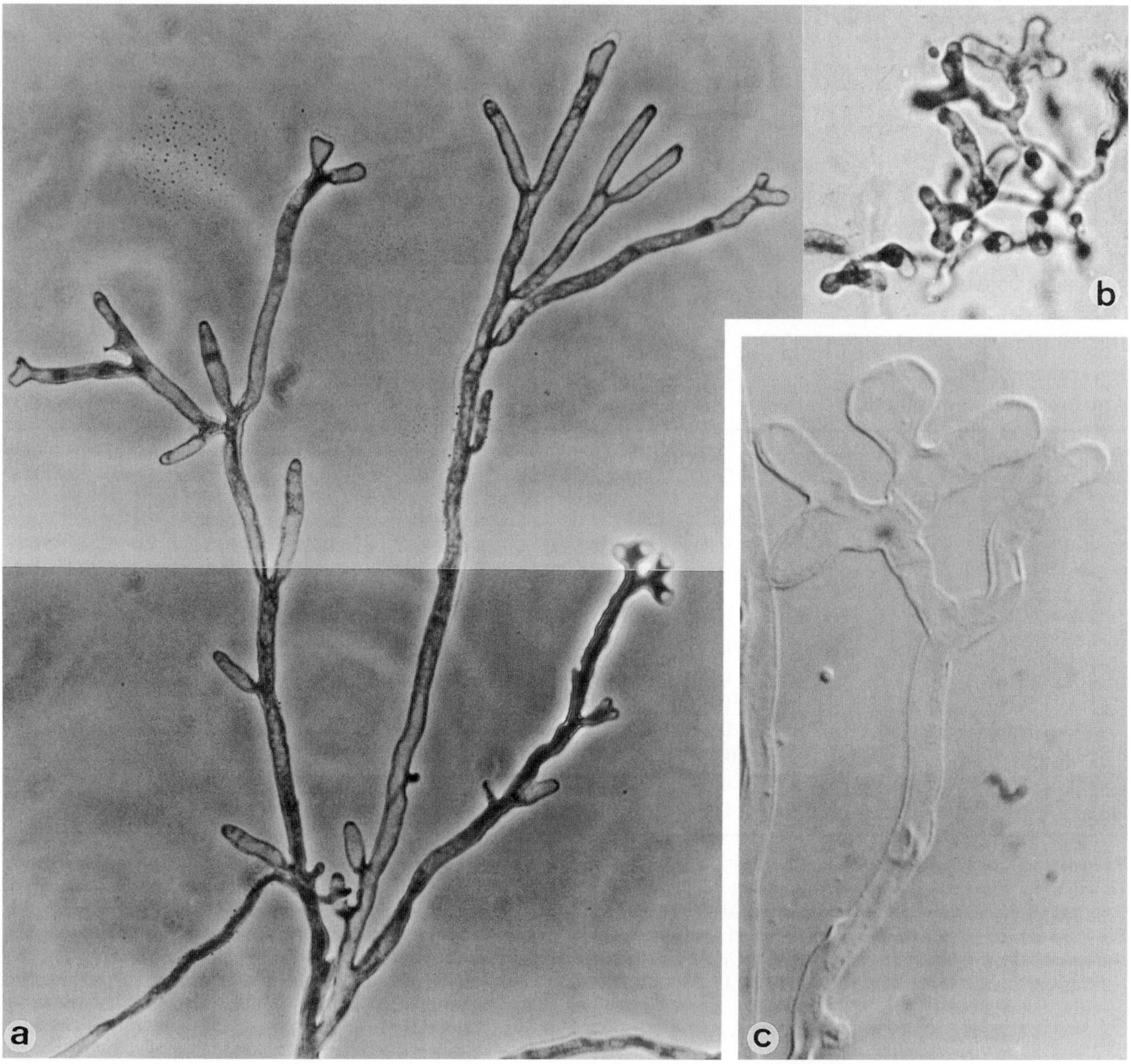

Trichophyton schoenleinii, CBS 335.32. Antler-like hyphae; c. chlamydospore-like swellings. a, b. ×640; c. ×1600.

Trichophyton simii (Pinoy) Stockdale *et al.*

Colony characteristics. Colonies (SGA) spreading, evenly granular with fluffy margin, whitish to pale buff; reverse yellowish to salmon, becoming vinaceous.

Microscopy. Macroconidia smooth-walled, fusiform, 30-85 × 6-11 µm, 5-10-celled; individual cells often swelling and becoming liberated as chlamydospores. Microconidia mostly sessile alongside undifferentiated hyphae, clavate to pyriform.

Teleomorph. *Arthroderma simii* Stockdale *et al.* (*Ascomycota, Euascomycetes, Onygenales: Arthrodermataceae;* p. 259).

Physiology.

BCPCG	+	T-1	++	T-5	++
Urease	+,s	T-2	++	T-6	++
Hair perforation	+	T-3	++	T-7	++
Growth at 37°C	+	T-4	++		

Molecular diagnostics. ITS restriction map based on NCBI Z98017:

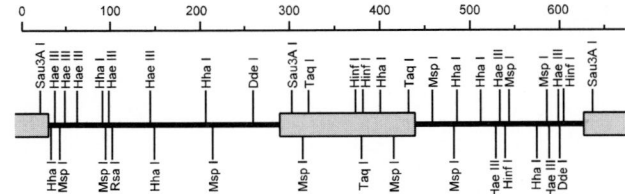

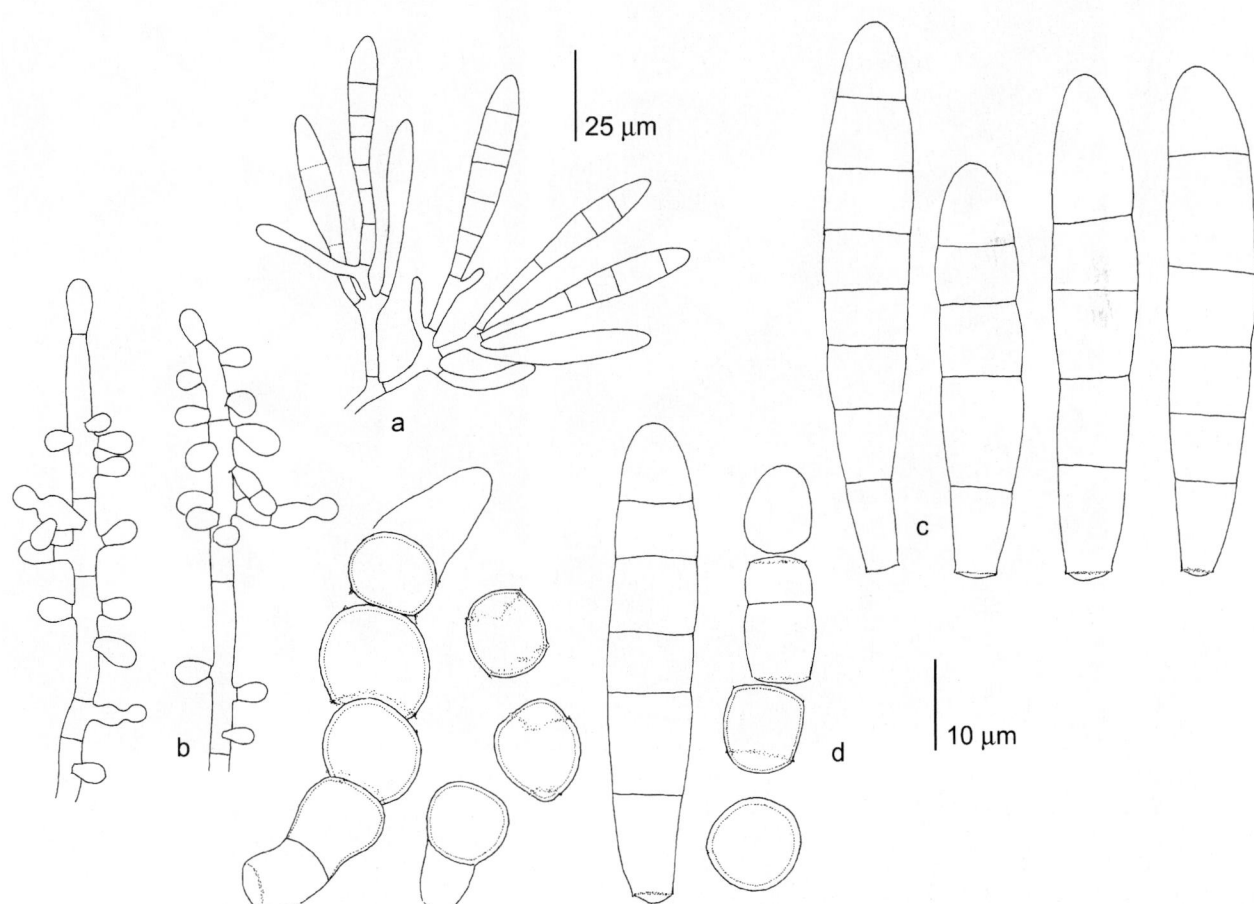

Trichophyton simii, CBS 449.65. a. Conidial apparatus; b. microconidial branches; c. macroconidia; d. macroconidia converted into disarticulating strings of chlamydospores.

Pathogenicity. BSL-2. Large ecto- as well as endothrix in chains, with bright green Fluorescense. Causes ringworm in monkeys (Boehringer *et al.*, 1999) and fowl, with occasional infections (tinea corporis) in humans.
Distribution. India.
References. Stockdale *et al.* (1965), Stockdale (1980).
Antifungal susceptibility.

Antifungal	Mean MICs	Strains	Reference
AMB	0.25	2	Fernández *et al.* (2000b)
CTZ	0.08	2	Fernández *et al.* (2000b)
FLZ	>16	2	Fernández *et al.* (2000b)
ITZ	0.12	2	Fernández *et al.* (2000b)
MCZ	1	2	Fernández *et al.* (2000b)
TBF	0.01	2	Fernández *et al.* (2000b)
VCZ	0.06	2	Fernández *et al.* (2000b)

Nomenclature. *Epidermophyton simii* Pinoy - C. R. Soc. Biol. 72: 59, 1912 ≡ *Pinoyella simii* (Pinoy) Castellani & Chalmers - Man. Trop. Med., ed. 3, p. 1023, 1919 ≡ *Trichophyton simii* (Pinoy) Stockdale, MacKenzie & Austwick - Sabouraudia 4: 114, 1965.
   Arthroderma simii Stockdale, MacKenzie & Austwick - Sabouraudia 4: 113, 1965.

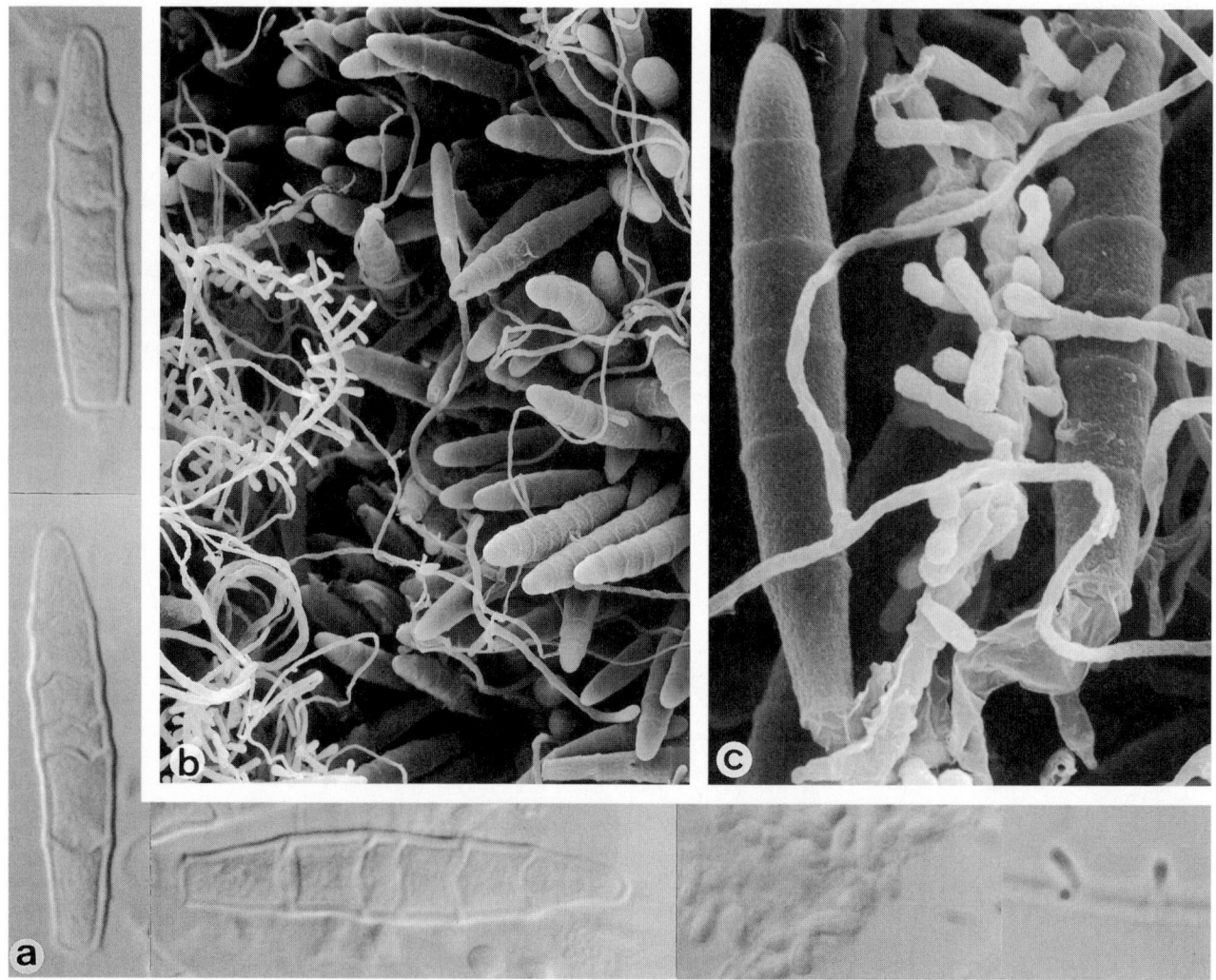

Trichophyton simii (Arthroderma simii), a. CBS 448.65; b, c. CBS 449.65. Macro and microconidia. a. ×1600; b. ×925; c. ×3000.

Trichophyton terrestre Durie & Frey

Colony characteristics. Colonies (SGA) expanding, felty, fluffy to powdery, whitish to pale cream-coloured or brownish; reverse greyish to yellowish or ochraceous.

Microscopy. Macroconidia not clearly differentiated from microconidia, 2-6-celled, smooth- and thin-walled, cylindrical or slightly clavate, 9-50 × 4-5 μm. Microconidia arising in densely arranged conidiophores with profuse orthotropic branching, short-cylindrical to short-clavate, 4.0-6.5 × 1.5 μm, truncate.

Teleomorphs. *Arthroderma insingulare* Padhye & Carmichael, *A. quadrifidum* Dawson & Gentles, *A. lenticulare* Pore *et al.* (*Ascomycota, Euascomycetes, Onygenales: Arthrodermataceae*).

Physiology.

BCPCG	+	T-1	+	T-5	+
Urease	+	T-2	+	T-6	+
Hair perforation	+	T-3	+	T-7	+
Growth at 37°C	–	T-4	+		

Differential diagnostics. The species is similar to *T. thuringiense* (p. 983), but *T. terrestre* differs by hyphae which are profusely branched at right angles.

Molecular diagnostics. ITS restriction map based on CBS 613.74:

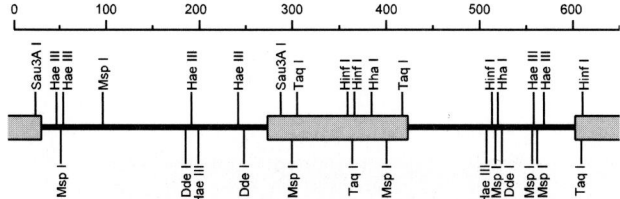

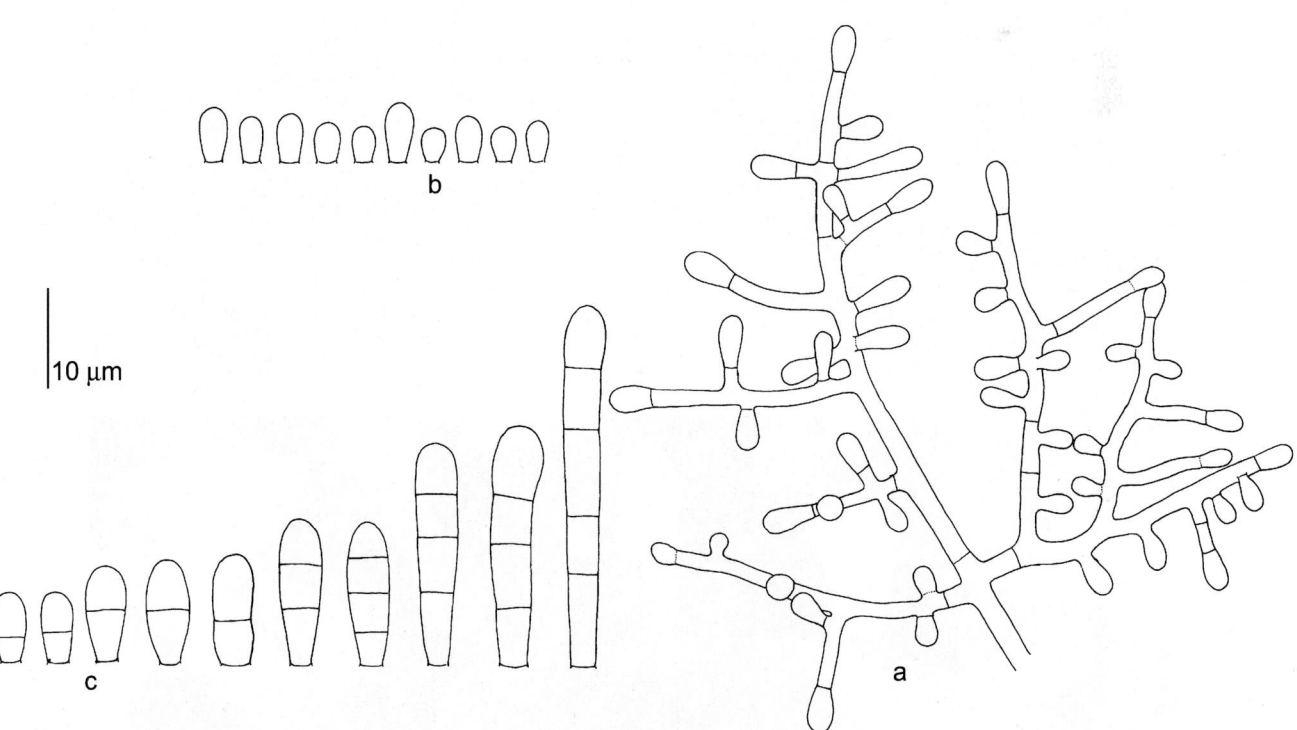

Trichophyton terrestre, CBS 567.94. a. Microconidial branches; b. microconidia; c. macroconidia.

Pathogenicity. BSL-1. The species is geophilic, mainly isolated from arid, alkaline soils.
Distribution. World-wide.
Reference. Durie & Frey (1957).
Nomenclature. *Trichophyton terrestre-primum* Szathmáry - Magya Orvosi Arch. 37-6: 1- 6, 1936 (invalid).
  Trichophyton terrestre Durie & Frey - Mycologia 49: 401, 1957.
  Arthroderma insingulare Padhye & Carmichael - Sabouraudia 10: 49, 1972.
  Arthroderma lenticulare Pore, Tsao & Plunkett - Mycologia 57: 970, 1965.
  Arthroderma quadrifidum Dawson & Gentles - Sabouraudia 1: 35, 1961.

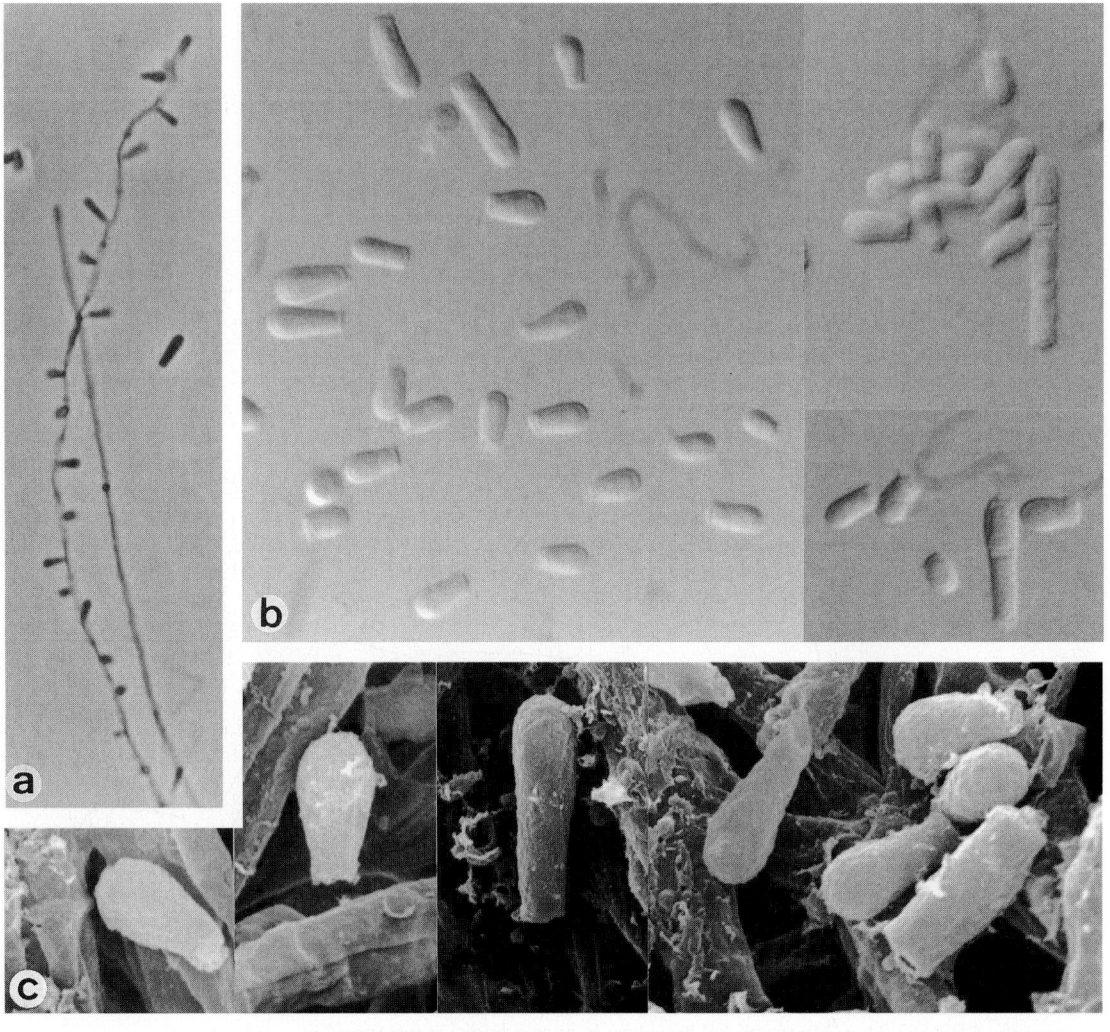

Trichophyton terrestre, CBS 464.62. a. Microconidial branche; b, c. macro- and microconidia. a. ×512; b. ×1600; c. ×6000.

Trichophyton thuringiense Koch

Colony characteristics. Colonies (SGA) expanding, cottony, white, becoming slightly brownish; reverse reddish-brown to buff.

Microscopy. Macroconidia hardly different from microconidia, cylindrical to clavate, 2-5-celled, up to 30 µm in length. Microconidia obovoidal to short-clavate with broad base, 3-5 ×2-3 µm, mostly sessile on thin, deteriorating hyphae. Arthroconidia present.

Physiology.

Urease	+	T-1	+	T-4	+	T-6	+
Hair perforation	–	T-2	+	T-5	+	T-7	+
		T-3	+				

Differential diagnosis. The species is morphologically close to *Trichopyton erinacei* (p. 959), but its young conidia are short-clavate with broad base. *T. terrestre* (p. 981) has a strongly, orthotropically branched conidial apparatus.

Molecular diagnostics. ITS restriction map based on CBS 417.71:

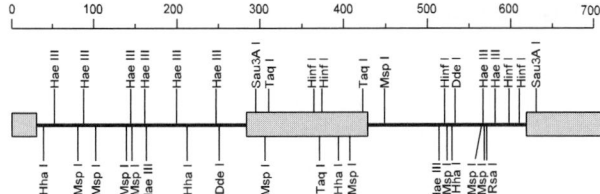

Pathogenicity. BSL-1. Known from a single strain from soil.

Reference. Koch (1969).

Nomenclature. *Trichophyton thuringiense* Koch - Mykosen 12: 288, 1969.

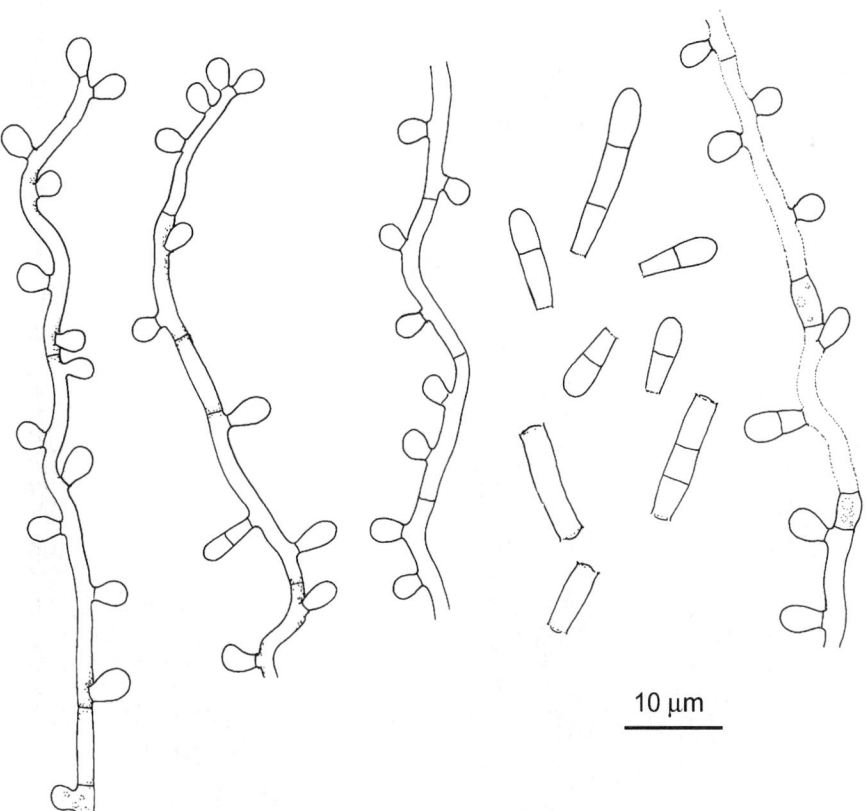

10 µm

Trichophyton thuringiense, CBS 417.71. a. Young micro- and macroconidia on hyphae; b. macro- and arthroconidia; c. mature conidia liberated by deterioration of supporting hypha.

Trichophyton tonsurans Malmsten

Colony characteristics. Colonies (SGA) rather variable; mostly suede-like, radially or irregularly furrowed, white to greyish, yellowish or brownish-buff, sometimes with pinkish or pale olivaceous centre; reverse mahogany-red, yellow or brown.

Microscopy. Microconidia of variable size, produced in abundance, formed on loosely clustered branches or thickened terminal hyphae, sessile, clavate to nearly cylindrical, sometimes inflating to balloon-shaped. Macroconidia, when present, variable, often somewhat thick-walled, 2-6-celled, cylindrical to cigar-shaped, 10-65 × 4-12 μm. Terminal and intercalary, swollen chlamydospores are formed in abundance.

Physiology.

BCPCG	v	Growth at 37°C	+	T-3	+(−)	T-6	v
Urease	v	T-1	+(−)	T-4	+(−)	T-7	w,−
Hair perforation	−	T-2	+(−)	T-5	+		

Molecular diagnostics. Gräser *et al.* (1999b) provided diagnostics based on ITS sequence data and PCR-fingerprinting, and Kim *et al.* (1999) RAPD. ITS restriction map based on NCBI Z98009:

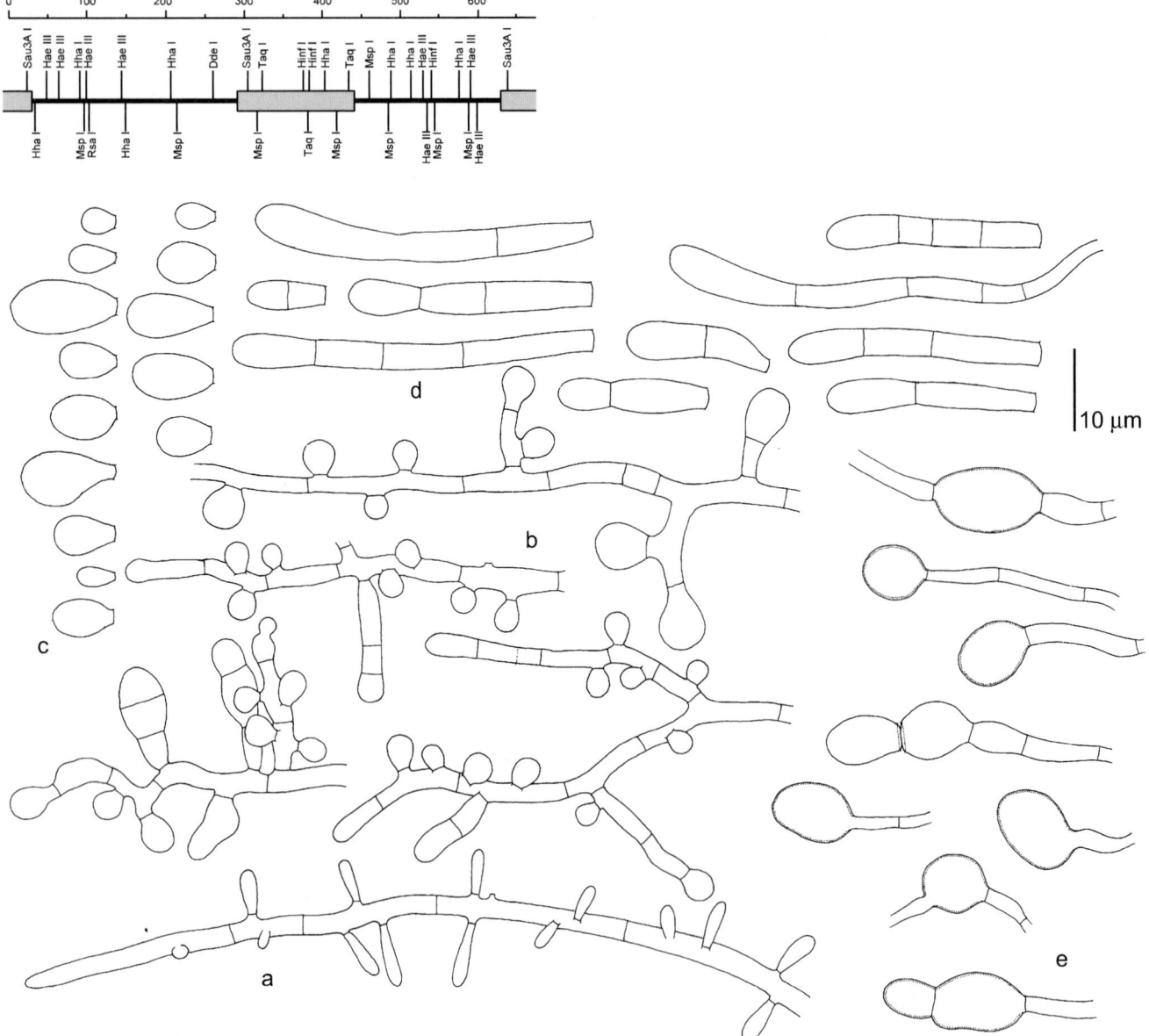

Trichophyton tonsurans, CBS 171.65. a. Young microconidia; b. mature microconidia with scattered macroconidia and inflated cells; c. microconidia; d. macroconidia; e. chlamydospore-like cell.

Differential diagnosis. The growth reactions on T-1 and T-4 are essential for recognition of the species. However, strains from horses are negative for most compounds.

Remark. Gräser *et al.* (1999b) noted that *T. balcaneum, T. radicosum* and *T. immergens*, listed as synonyms of *T. tonsurans* below, compose a separate subgroup.

Pathogenicity. BSL-2. Endothrix. Zoophilic, but frequently transmitted to man, then showing anthropophilic behaviour, primarily causing human scalp ringworm, frequently tinea corporis (Chao & Hsu, 1994). Increasingly found in tinea capitis in urban regions (Gupta & Summerbell, 1998). In tinea capitis a 'hen-skin' effect is noted due to broken, curled hair remains. Usually non-inflammatory, often occurring in elderly women (Derrick *et al.*, 1994) or in children (Leeming & Elliott, 1995). Contagious (Beller & Gessner, 1994; Hradil *et al.*, 1995), transmitted by sporting activities (El Fari *et al.*, 2000), also known from nosocomial epidemics (Lewis & Lewis, 1997; Wood & Rademaker, 1997). Clinical pictures may vary with the underlying disease (Maleszka & Ratka, 1998), eventually becoming invasive (Liao *et al.*, 1999). Strains on horse earlier were referred to as *T. equinum*, but this name was proven to be a synonym (Gräser *et al.*, 1999b). Transmission from horses to humans is known (Huovinen *et al.*, 1998).

Distribution. World-wide.

References. MacKenzie (1961), de Vries (1971), de Vries & Josephius Jitta (1973), Matsumoto *et al.* (1983a), Padhye *et al.* (1994e), Gräser *et al.* (1999b).

Antifungal susceptibility.

Antifungal	Mean MICs	MIC 90	Strains	Reference
AMB	0.16	0.5	18	Fernández *et al.* (2000b)
CTZ	0.05	0.125	18	Fernández *et al.* (2000b)
FLZ	7	nd	6	Norris *et al.* (1999)
FLZ	1.91	8	18	Fernández *et al.* (2000b)
GRF	1.13	nd	6	Norris *et al.* (1999)
GRF	nd	4	42	Jessup *et al.* (2000)
GRF	1.31	6.25	5	Wildfeuer *et al.* (1998)
ITZ	0.05	0.10	5	Wildfeuer *et al.* (1998)
ITZ	<0.06	nd	6	Norris *et al.* (1999)
ITZ	nd	0.06	42	Jessup *et al.* (2000)
ITZ	0.01	0.03	18	Fernández *et al.* (2000b)
KTZ	0.1	0.25	18	Fernández *et al.* (2000b)
KTZ	0.23	0.39	5	Wildfeuer *et al.* (1998)
MCZ	0.13	0.25	18	Fernández *et al.* (2000b)
TBF	<0.06	nd	6	Norris *et al.* (1999)
TBF	nd	0.008	42	Jessup *et al.* (2000)
TBF	0.09	0.01	18	Fernández *et al.* (2000b)
VCZ	0.04	0.125	18	Fernández *et al.* (2000b)
VCZ	0.10	0.20	5	Wildfeuer *et al.* (1998)

Nomenclature. *Trichophyton tonsurans* Malmsten - Arch. Anat. Physiol. Wiss. Med. 1848: 19, 1848 ≡ *Trichomyces tonsurans* (Malmsten) Malmsten - Arch. Anat. Physiol. Wiss. Med. 1848: 19, 1848 ≡ *Oidium tonsurans* (Malmsten) Zopf - Die Pilze p. 482, 1890.

Trichophyton depilans Mégnin - Arch. Gén. Méd. 141: 294-304, 1878 ≡ *Trichophyton epilans* Mégnin - Bull. Soc. Centr. Méd. Vét. 44: 183, 1890 (name change) ≡ *Favotrichophyton epilans* (Mégnin) C.W. Dodge - Med. Mycol. p. 526, 1935.

Trichophyton crateriforme Sabouraud, *in* Bodin - Champ. Paras. Homme p. 108, 1902 ≡ *Chlamydoaleuriospora crateriformis* (Sabouraud) Grigorakis - Annls Sci. Nat., Bot., Sér. 10, 7: 412, 1925.

Trichophyton equinum Gedoelst - Champ. Paras. Homme p. 88, 1902 ≡ *Ectotrichophyton equinum* (Gedoelst) Castellani & Chalmers - Man. Trop. Med., ed. 3, p. 1007, 1919 ≡ *Megatrichophyton equinum* (Gedoelst) Neveu-Lemaire - Précis Parasitol. Hum., ed. 5, p. 54, 1921 ≡ *Ctenomyces equinus* (Gedoelst) Nannizzi - Tratt. Micopat. Um. 4: 144, 1934.

Trichophyton acuminatum Bodin - Champ. Paras. Homme p. 110, 1902 = *Aleurosporia acuminata* (Bodin) Grigorakis - Annls Sci. Nat., Bot., Sér. 10, 7: 413, 1925 = *Trichophyton tonsurans* Malmsten var. *acuminatum* (Bodin) Drouhet & Robin - Bull. Soc. Fr. Derm. Syph. 76: 502, 1969.

Trichophyton flavum Bodin - Champ. Paras. Homme p. 119, 1902 = *Neotrichophyton flavum* (Bodin) Castellani & Chalmers - Man. Trop. Med., ed. 3, p. 1002, 1919.

Trichophyton regulare Sabouraud, *in* Dalla Favera - Annls Derm. Syph., Sér. 4, 10: 438, 1909.

Trichophyton fumatum Sabouraud, *in* Dalla Favera - Annls Derm. Syph., Sér. 4, 10: 442, 1909.

Trichophyton pilosum Sabouraud - Malad. Cuir Chev. 3: 314, 1910 = *Trichophyton sabouraudii* Blanchard var. *pilosum* (Sabouraud) C.W. Dodge - Med. Mycol. p. 535, 1935.

Trichophyton effractum Sabouraud - Malad. Cuir Chev. 3: 314, 1910 = *Aleurosporia effracta* (Sabouraud) Grigorakis - Ann. Sci. Nat., Bot., Sér. 10, 7: 413, 1925 = *Trichophyton tonsurans* Malmsten var. *effractum* (Sabouraud) C.W. Dodge - Med. Mycol. p. 532, 1935.

Trichophyton umbilicatum Sabouraud - Malad. Cuir Chev. 3: 315, 1910.

Trichophyton sulfureum Sabouraud - Malad. Cuir Chev. 3: 317, 1910 ≡ *Trichophyton tonsurans* Malmsten var. *sulfureum* (Sabouraud) MacKenzie - Sabouraudia 1: 58, 1961.

Trichophyton exsiccatum Uriburu, *in* Sabouraud - Malad. Cuir Chev. 3: 318, 1910 ≡ *Trichophyton tonsurans* Malmsten var. *exsiccatum* (Uriburu) C.W. Dodge - Med. Mycol. p. 532, 1935.

Trichophyton polygorum Uriburu, *in* Sabouraud - Malad. Cuir Chev. 3: 318, 1910.

Trichophyton cerebriforme Sabouraud - Malad. Cuir Chev. 3: 321, 1910.

Trichopyton plicatile Sabouraud - Malad. Cuir Chev. 3: 330, 1910 ≡ *Neotrichophyton plicatile* (Sabouraud) Castellani & Chalmers - Man. Trop. Med., ed. 3, p. 1002, 1919 ≡ *Aleurosporia plicatilis* (Sabouraud) Grigorakis - Annls Sci. Nat., Bot., Sér. 10, 7: 413, 1925.

Trichophyton balcaneum Castellani - J. Trop. Med. Hyg. 22: 174, 1919 ≡ *Bodinia balcanea* (Castellani) Ota & Langeron - Annls Parasit. Hum. Comp. 1: 329, 1923 = *Favotrichophyton balcanea* (Castellani) C.W. Dodge - Med. Mycol. p. 518, 1935.

Trichophyton ochropyrraceum Muijs, *in* Papegaaij - Path. Huidsch., Amsterdam p. 50, 1924 ≡ *Trichophyton crateriforme* Sabouraud var. *ochropyrraceum* (Muijs) Keller - Derm. Z. 49: 49, 1927.

Bodinia spadix Katoh - Trans. 6th Congr. Far East. Assoc. Trop. Med., Tokyo p. 865, 1925 = *Bodinia spadicea* Pollacci & Nannizzi - Miceti Pat. Uomo Anim. 10: 91, 1930 (name change) ≡ *Trichophyton spadiceum* (Pollacci & Nannizzi) Nannizzi - Tratt. Micopat. Um. 4: 201, 1934 ≡ *Favotrichophyton spadix* (Katoh) C.W. Dodge - Med. Mycol. p. 524, 1935 ≡ *Trichophyton spadix* (Katoh) Rippon - Med. Mycol., ed. 3, p. 256, 1988.

Trichophyton areolatum Negroni - Annls Parasit. Hum. Comp. 7: 438, 1929.

Trichophyton floriforme Beintema - Arch. Derm. 169: 575, 1934 = *Favotrichophyton floriforme* (Beintema) C.W. Dodge - Med. Mycol. p. 518, 1935.

Trichophyton immergens Milochevitch - C.R. Soc. Biol. 124: 469, 1937.

Trichophyton radicosum Catanei - Arch. Inst. Pasteur Algér. 15: 268, 1937.

Trichophyton equinum (Gedoelst) var. *autotrophicum* J.M.B. Smith, Jolly, Georg & Connole - Sabouraudia 6: 297, 1968.

Trichophyton tonsurans Malmsten var. *sulfureum* (Sabouraud) MacKenzie subvar. *perforans* Matsumoto, Padhye & Ajello - Mycotaxon 18: 240, 1983.

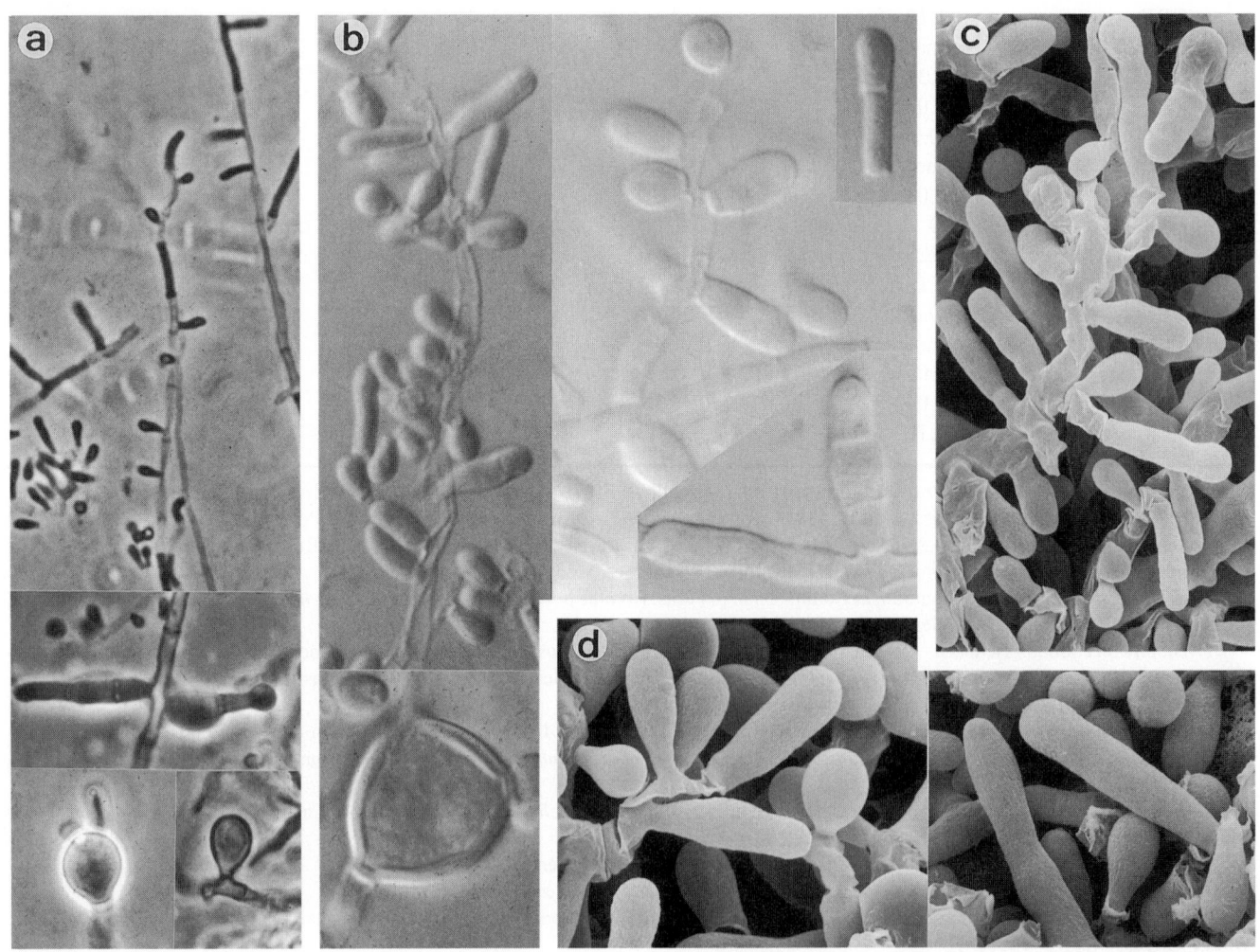

Trichophyton tonsurans, CBS 171.65. Fertile hyphae with macro- and microconidia and chlamydospore-like cells. a. ×640; b. ×1600; c. ×2400; d. ×3500.

Trichophyton vanbreuseghemii Rioux *et al.*

Colony characteristics. Colonies (SGA) velvety, folded, warm buff; reverse cream-coloured.
Microscopy. Macroconidia cylindrical, 4-7-celled, 30-55 × 6-8 µm. Microconidia pyriform, 2-7 × 1.5-2.5 µm.
Teleomorph. *Arthroderma gertleri* Böhme (*Ascomycota, Euascomycetes, Onygenales: Arthrodermataceae*).
Physiology.

Urease	+	T-1	++	T-4	++	T-6	++
Hair perforation	+	T-2	++	T-5	++	T-7	++
		T-3	++				

Differential diagnosis. *Trichophyton gloriae* (p. 963) is closely similar but has somewhat longer macroconidia and a yellow reverse.
Molecular diagnostics. ITS restriction map based on NCBI Z98013:

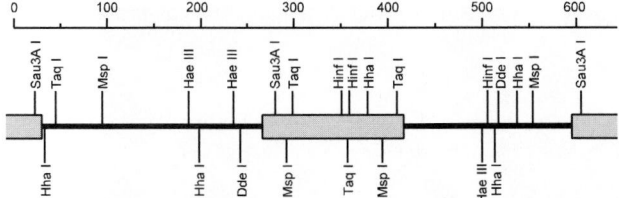

Pathogenicity. BSL-1. Ectothrix. The species rarely occurs on human skin (Lopes *et al.*, 1994c) and has also been reported from soil.
Distribution. Australia.
References. Rioux *et al.* (1964, 1966), Böhme (1967).
Nomenclature. *Trichophyton vanbreuseghemii* Rioux, Jarry & Juminez - Nat. Monspeliensia, Sér. Bot. 16: 158, 1964.
    Arthroderma gertleri Böhme - Mykosen 10: 251, 1967.

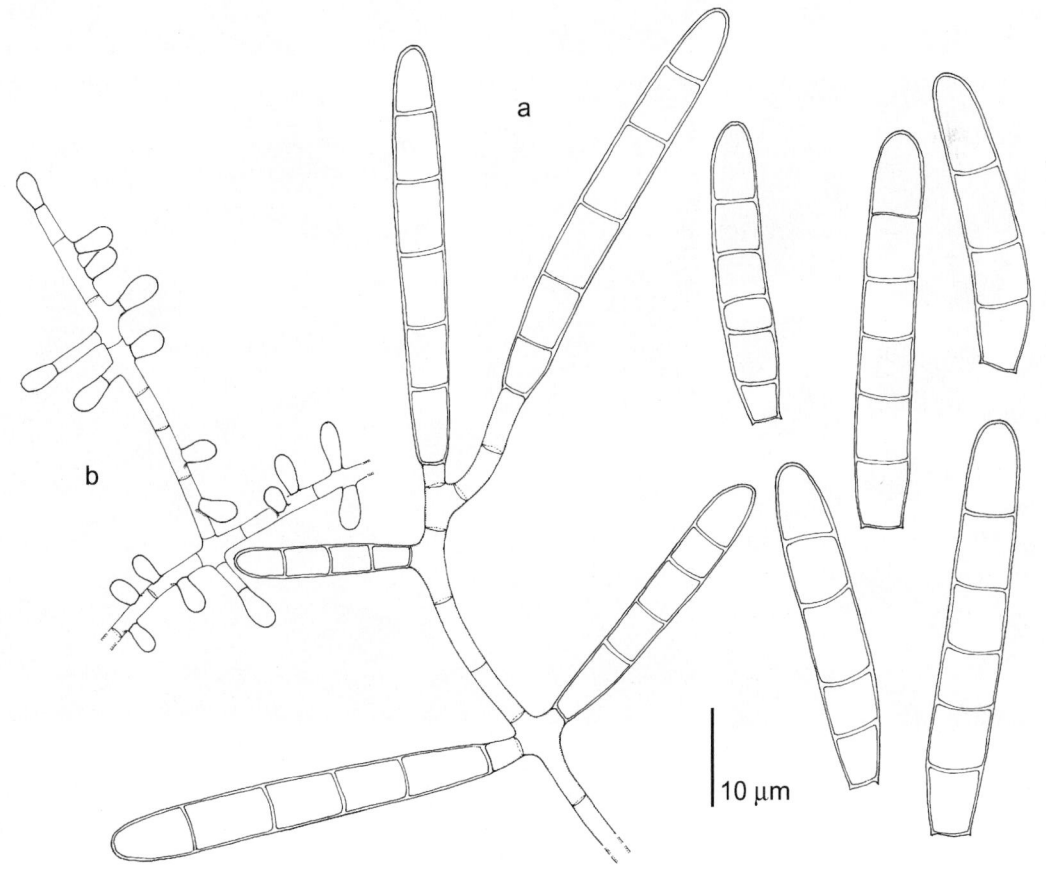

Trichophyton vanbreuseghemii, CBS 598.66. a. Conidial system and liberated macroconidia; b. microconidia.

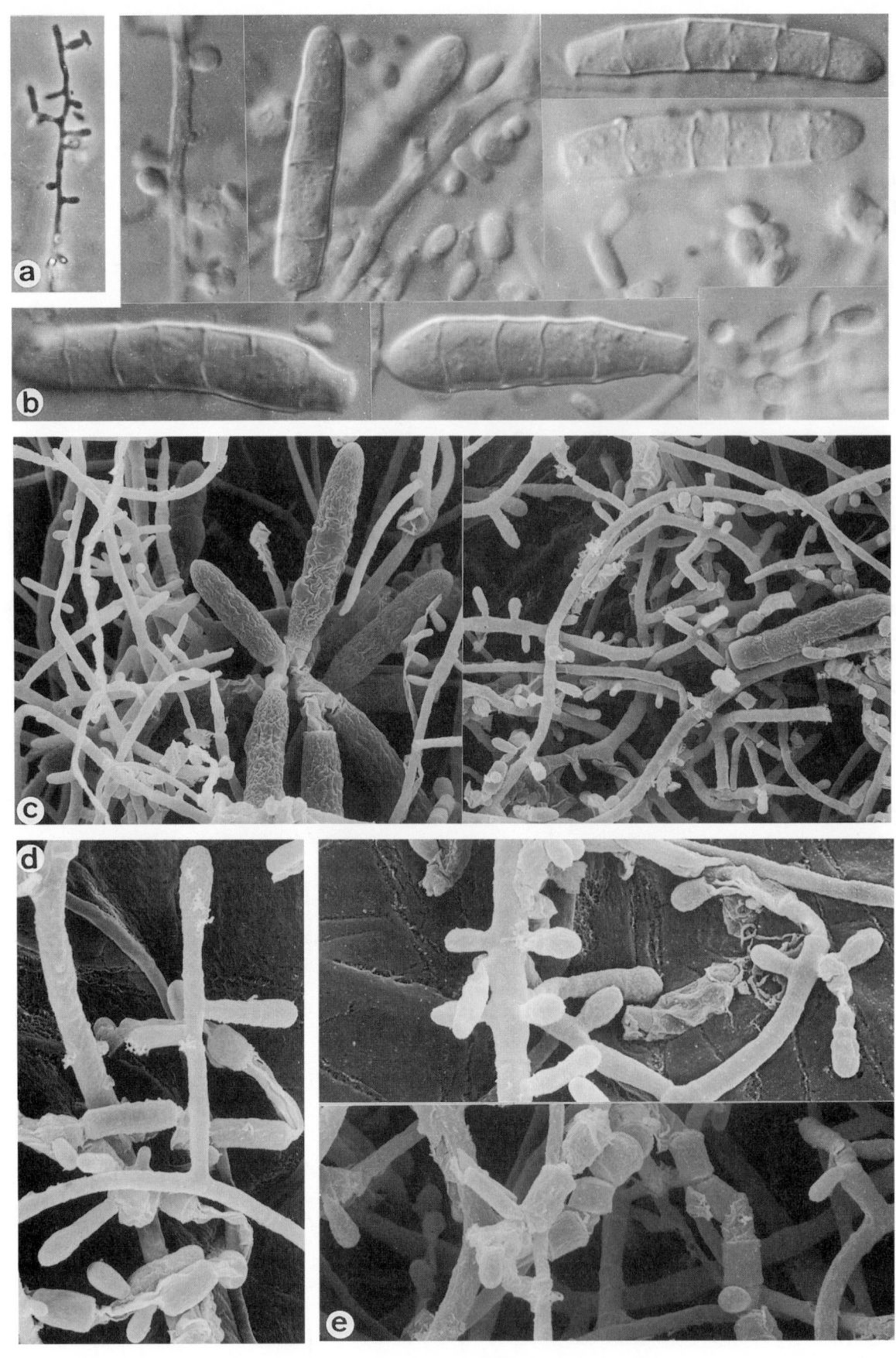

Trichophyton vanbreuseghemii, CBS 598.66. Macro- and microconidia. a. ×640; b. ×1600; c. ×1250; d. ×2500; e. ×2700.

Trichophyton verrucosum Bodin

Colony characteristics. Colonies (SGA) growing very slowly, heaped or button-like, feathered by a perimeter of submerged hyphae; colonies initially glabrous, later slightly velvety, cream-coloured or greyish-white, sometimes with salmon to yellow tinges; reverse pale cream- or salmon-coloured.

Microscopy. Sporulation absent or reduced. Macroconidia, when present (on T-3), 4-7-celled, smooth- and thin-walled, stringbean-shaped. Microconidia, when present, ovoidal to pyriform. Chlamydospores common in fresh isolates, often arranged in chains, sometimes ending with antler-like hyphal branching without swollen tips. Hyphal tips frequently swollen in fresh isolates.

Physiology.

BCPCG	+	T-1	–	T-5	–
Urease	–	T-2	w	T-6	–
Hair perforation	–	T-3	++	T-7	–
Growth at 37°C	++	T-4	+		

Differential diagnosis. It is the only dermatophyte that grows better at 37°C than at 24°C. Chlamydospores are typically produced at 37°C.

Molecular diagnostics. ITS restriction map based on NCBI Z98003:

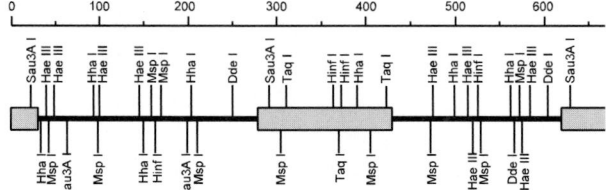

Pathogenicity. BSL-2. Large-celled ectothrix in chains. Fluorescence noted in cattle, when irradiated by a Wood's lamp (Yadav *et al.*, 1996). The species causes ringworm in cattle (Wabacha *et al.*, 1998) and, through direct transmission, in humans, then mostly in cattle breeders (Sabota *et al.*, 1996). Occasionally infections are from horses (Korman *et al.*, 1998). Human cases are highly inflammatory and particularly involve exposed parts of the body (tinea corporis, tinea capitis, tinea barbae; Janke & Newig, 1959). Non-occupational infections are rare (Bell *et al.*, 1996).

Distribution. World-wide.

References. Kane & Smitka (1978), Kielstein *et al.* (1998).

Antifungal susceptibility.

Antifungal	MICs	Strains	Reference
AMB	0.25	1	Fernández *et al.* (2000b)
CTZ	0.25	1	Fernández *et al.* (2000b)
FLZ	8	1	Fernández *et al.* (2000b)
ITZ	0.01	1	Fernández *et al.* (2000b)
KTZ	0.5	1	Fernández *et al.* (2000b)
MCZ	0.125	1	Fernández *et al.* (2000b)
TBF	0.007	1	Fernández *et al.* (2000b)
VCZ	0.06	1	Fernández *et al.* (2000b)

Nomenclature. *Trichophyton verrucosum* Bodin - Champ. Paras. Homme p. 121, 1902 ≡ *Favotrichophyton verrucosum* (Bodin) Neveu-Lemaire - Précis Parasitol. Hum., ed. 5, p. 55, 1921 ≡ *Ectotrichophyton verrucosum* (Bodin) Castellani & Chalmers - Man. Trop. Med., ed. 3, p. 1003, 1919.

Trichophyton mansonii Castellani - Br. Med. J. 2: 1277, 1905 ≡ *Endodermophyton mansonii* (Castellani) Brumpt - Précis Parasitol., ed. 4, p. 1299, 1927.

Trichophyton ochraceum Sabouraud - Annls Derm. Syph., Sér. 4, 9: 628-834, 1908 ≡ *Favotrichophyton ochraceum* (Sabouraud) Neveu-Lemaire - Précis Parasit. Hum., ed. 5, p. 55, 1921 ≡ *Grubyella ochracea* (Sabouraud) Ota & Langeron - Annls Parasit. Hum. Comp. 1: 330, 1923 ≡ *Arthrosporia ochracea* (Sabouraud) Grigorakis - Annls Sci. Nat., Bot., Sér. 10, 7: 414, 1925 ≡ *Achorion ochraceum* (Sabouraud) Guiart & Grigorakis - Lyon Méd. 141: 377, 1928 ≡ *Trichophyton faviforme* Bodin var. *ochraceum* (Sabouraud) Georg - Mycologia 42: 714, 1950 ≡ *Trichophyton verrucosum* Bodin var. *ochraceum* (Sabouraud) Georg, *in* Ainsworth & Georg - Mycologia 46: 10, 1954.

Trichophyton album Sabouraud - Annls Derm. Syph., Sér. 4, 9: 617, 1908 ≡ *Ectotrichophyton album* (Sabouraud) Castellani & Chalmers - Man. Trop. Med., ed. 3, p. 1004, 1919 ≡ *Favotrichophyton album* (Sabouraud) Neveu-Lemaire - Précis Parasitol. Hum., ed. 5, p. 55, 1921 ≡ *Grubyella alba* (Sabouraud) Ota & Langeron - Annls Parasit. Hum. Comp. 1: 330, 1923 ≡ *Arthrosporia alba* (Sabouraud) Grigorakis - Annls Sci. Nat., Bot., Sér. 10, 7: 414, 1925 ≡ *Achorion album* (Sabouraud) Guiart & Grigorakis - Lyon Méd. 141: 377, 1928 ≡ *Trichophyton verrucosum* Bodin var. *album* (Sabouraud) Georg, *in* Ainsworth & Georg - Mycologia 46: 10, 1954.

    Trichophyton discoides Sabouraud - Malad. Cuir Chev. 3: 408, 1910 ≡ *Favotrichophyton discoides* (Sabouraud) Neveu-Lemaire - Précis Parasit. Hum., ed. 5, p. 55, 1921 ≡ *Grubyella discoides* (Sabouraud) Ota & Langeron - Annls Parasit. Hum. Comp. 1: 330, 1923 ≡ *Trichophyton faviforme* Bodin var. *discoides* (Sabouraud) Georg - Mycologia 42: 714, 1950 ≡ *Trichophyton verrucosum* Bodin var. *discoides* (Sabouraud) Georg, *in* Ainsworth & Georg - Mycologia 46: 10, 1954.

    Trichophyton verrucosum Bodin var. *autotrophicum* D.B. Scott - Trans. Br. Mycol. Soc. 67: 343, 1976.

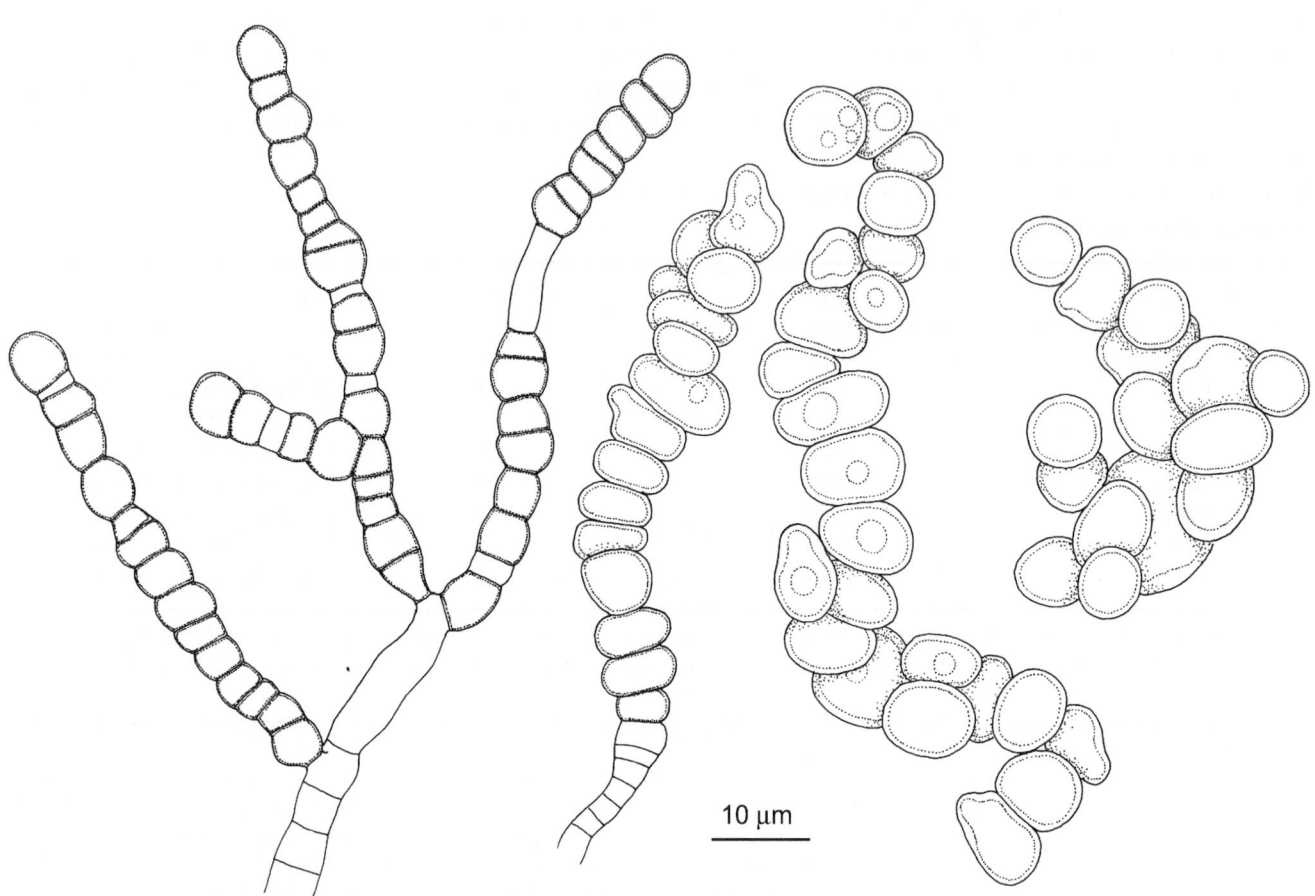

10 μm

Trichophyton verrucosum, M34-99. a. Developing hyphae, SGA; b, c. mature chlamydospors, blood agar; d. disarticulating chlamydospores, blood agar.

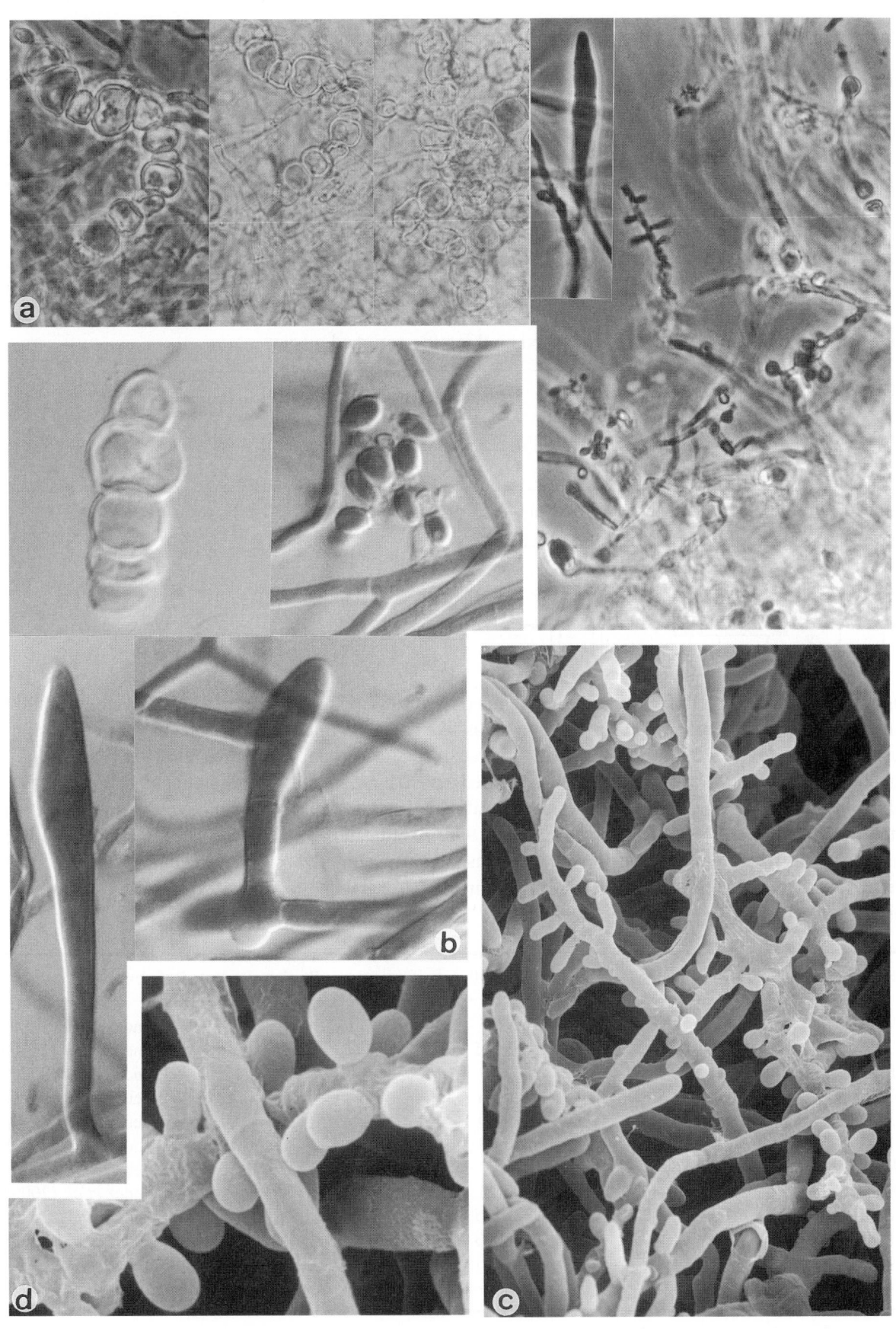

Trichophyton verrucosum, M34-99. Chlamydospores in chains, liberated microconidia, and fertile hyphae producing macro- and microconidia. a. ×640; b. ×1600; c. ×2250; d. ×6500.

Trichophyton violaceum Sab.

Colony characteristics. Colonies (SGA) growing slowly, glabrous, leathery, wrinkled, yellow, apricot-red or purple-red; reverse dark yellow (formerly *T. soudanense*), red-brown (formerly *T. gourvilii*), purple or violet, or becoming chocolate-brown with age and exuding a diffusable brown pigment into the agar (formerly *T. yaoundei*), or reddish with feathered margin (formerly *T. soudanense*). Strains easily loose their pigmentation by formation of white sectors. Colonies dark brown on Löwenstein's egg medium.

Microscopy. Hyphae highly distorted; they may have reflexive branching (formerly *T. soudanense*) and are often strongly septate, disarticulating into arthroconidia. Sporulation absent or reduced. Macroconidia very rare. Microconidia, when present, ovoidal, pyriform or clavate.

Physiology. Development of conidia is enhanced on thiamine-enriched media.

BCPCG	+	T-1	w	T-5	+	
Urease	–,+	T-2	+	T-6	+	
Hair perforation	–	T-3	+	T-7	+	
Growth at 37°C	+	T-4	++			

Molecular diagnostics. ITS restriction map based on CBS 374.92:

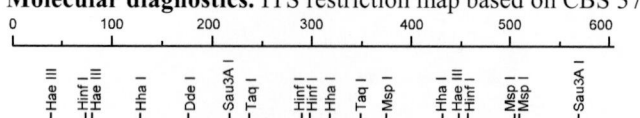
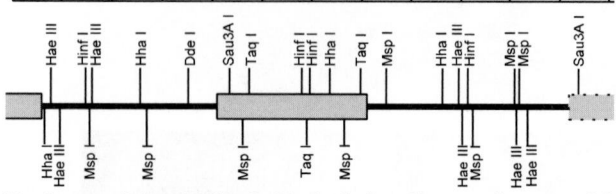

Pathogenicity. BSL-2. Endothrix. Causes tinea capitis (Haas, 1987; Figueroa *et al.*, 1997; Romano *et al.*, 2000), but also ringworm and tinea corporis, often producing favus-like crusts. In tinea capitis a 'black dot' or 'hen skin' effect is noted due to broken, curled hair remains. Onychomycosis may also occur (Kalter & Hay, 1988; Ricci *et al.*, 1998). Boudghène-Stambouli & Mérad-Boudia (1998) reported severe hyperkeratosis leading to cutaneous horns.

Distribution. Eastern Europe, Northern Africa, Central America. Cases in Europe and North America were mostly imported (Albanese *et al.*, 1995; Rübben & Krause, 1996; Armingaud *et al.*, 1998). The species is also known in Indian populations which originate from Africa (Hemashettar & Nadig, 1980; Hemashettar *et al.*, 1993).

References. Acton & McGuire (1929), Rebell & Taplin (1970), Ajello & Padhye (1987).

Antifungal susceptibility.

Antifungal	Mean MICs	MIC 90	Strains	Reference
AMB	0.18	0.25	7	Fernández *et al.* (2000b)
CTZ	0.05	0.125	7	Fernández *et al.* (2000b)
FLZ	2.43	>16	7	Fernández *et al.* (2000b)
GRF	1.11	1.56	2	Wildfeuer *et al.* (1998)
ITZ	0.07	0.10	2	Wildfeuer *et al.* (1998)
ITZ	0.05	0.03	7	Fernández *et al.* (2000b)
KTZ	0.78	0.78	2	Wildfeuer *et al.* (1998)
KTZ	0.125	0.25	7	Fernández *et al.* (2000b)
MCZ	0.08	0.125	7	Fernández *et al.* (2000b)
TBF	0.01	0.003	7	Fernández *et al.* (2000b)
VCZ	0.14	0.2	2	Wildfeuer *et al.* (1998)
VCZ	0.04	0.06	7	Fernández *et al.* (2000b)

Nomenclature. *Trichophyton violaceum* Sabouraud, *in* Bodin - Champ. Paras. Homme p. 113, 1902 ≡ *Achorion violaceum* (Sabouraud) Bloch - Derm. 18: 815, 1911 ≡ *Sabouraudites violaceum* (Sabouraud) Ota & Langeron - Annls Parasit. Hum. Comp. 1: 328, 1923 ≡ *Bodinia violacea* (Sabouraud) Ota & Langeron - Annls Parasit. Hum. Comp. 1: 329, 1923 ≡ *Arthrosporia violacea* (Sabouraud) Grigorakis - Annls Sci. Nat., Bot., Sér. 10, 7: 414, 1925 ≡ *Favotrichophyton violaceum* (Sabouraud) C.W. Dodge - Med. Mycol. p. 523, 1935.

   Trichophyton glabrum Sabouraud - Malad. Cuir Chev. 3: 312, 1910 ≡ *Bodinia glabra* (Sabouraud) Ota & Langeron - Annls Parasit. Hum. Comp. 1: 329, 1923 ≡ *Favotrichophyton glabrum* (Sabouraud) C.W. Dodge - Med. Mycol. p. 522, 1935.

   Trichophyton soudanense Joyeux - C.R. Soc. Biol. 72: 15, 1912 ≡ *Langeronia soudanensis* (Joyeux) Vanbreuseghem - Annls Soc. Belg. Méd. 30: 888, 1950 ≡ *Achorion soudanense* (Joyeux) Novák & Galgóczy - Acta Bot. Hung. 15: 131, 1969.

   Trichophyton violaceum Sabouraud var. *indicum* Acton & McGuire - Ind. Med. Gaz. 64: 245, 1929.

Trichophyton gourvilii Catanei - Bull. Soc. Path. Exot. 26: 377, 1933 ≡ *Favotrichophyton gourvilii* (Catanei) C.W. Dodge - Med. Mycol. p. 522, 1935.

    Trichophyton yaoundei Cochet & Doby-Dubois - Annls Parasit. Hum. Comp. 32: 585, 1957.

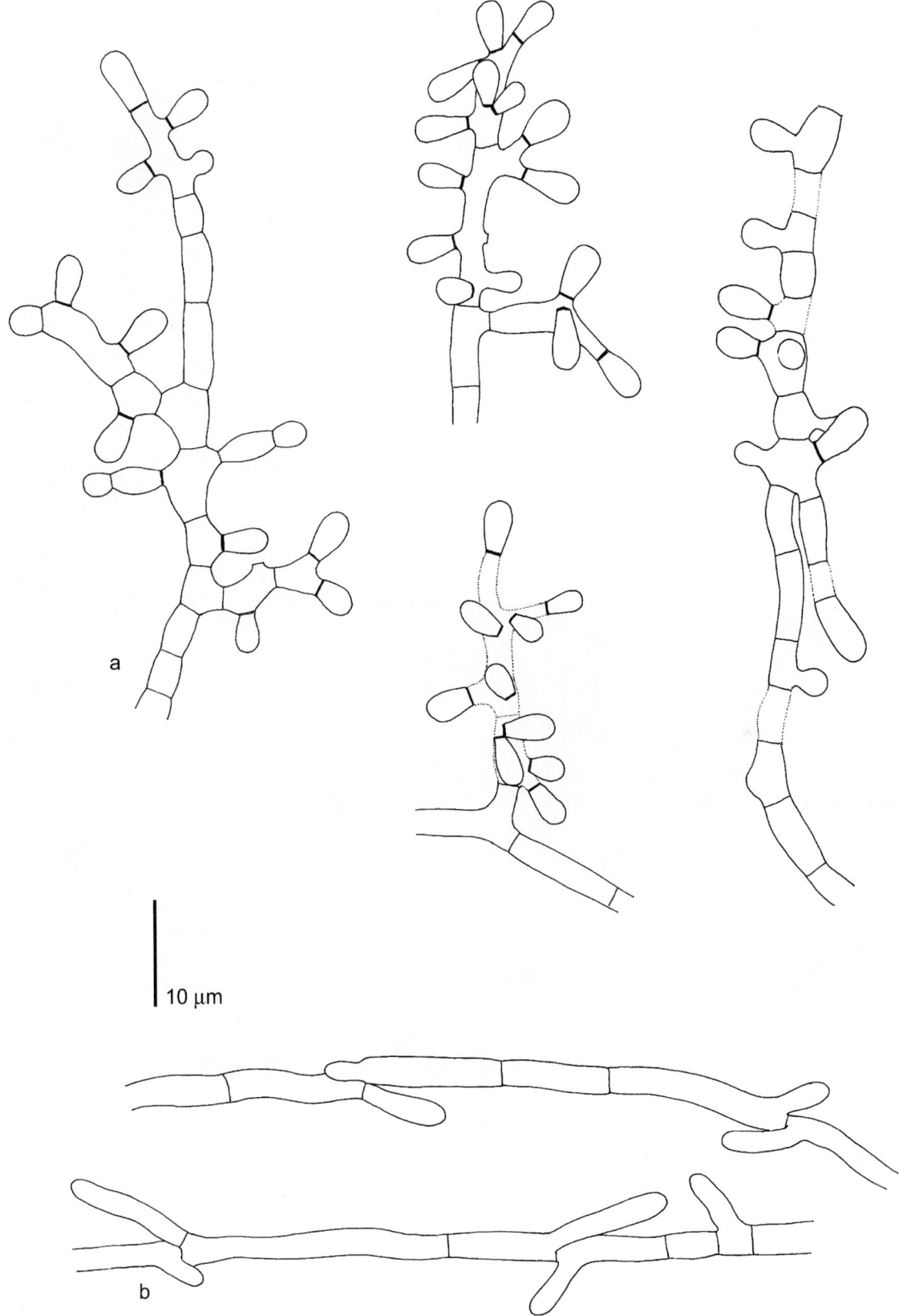

Trichophyton violaceum, CBS 411.88. a. Microconidial branches; b. reflexive branches.

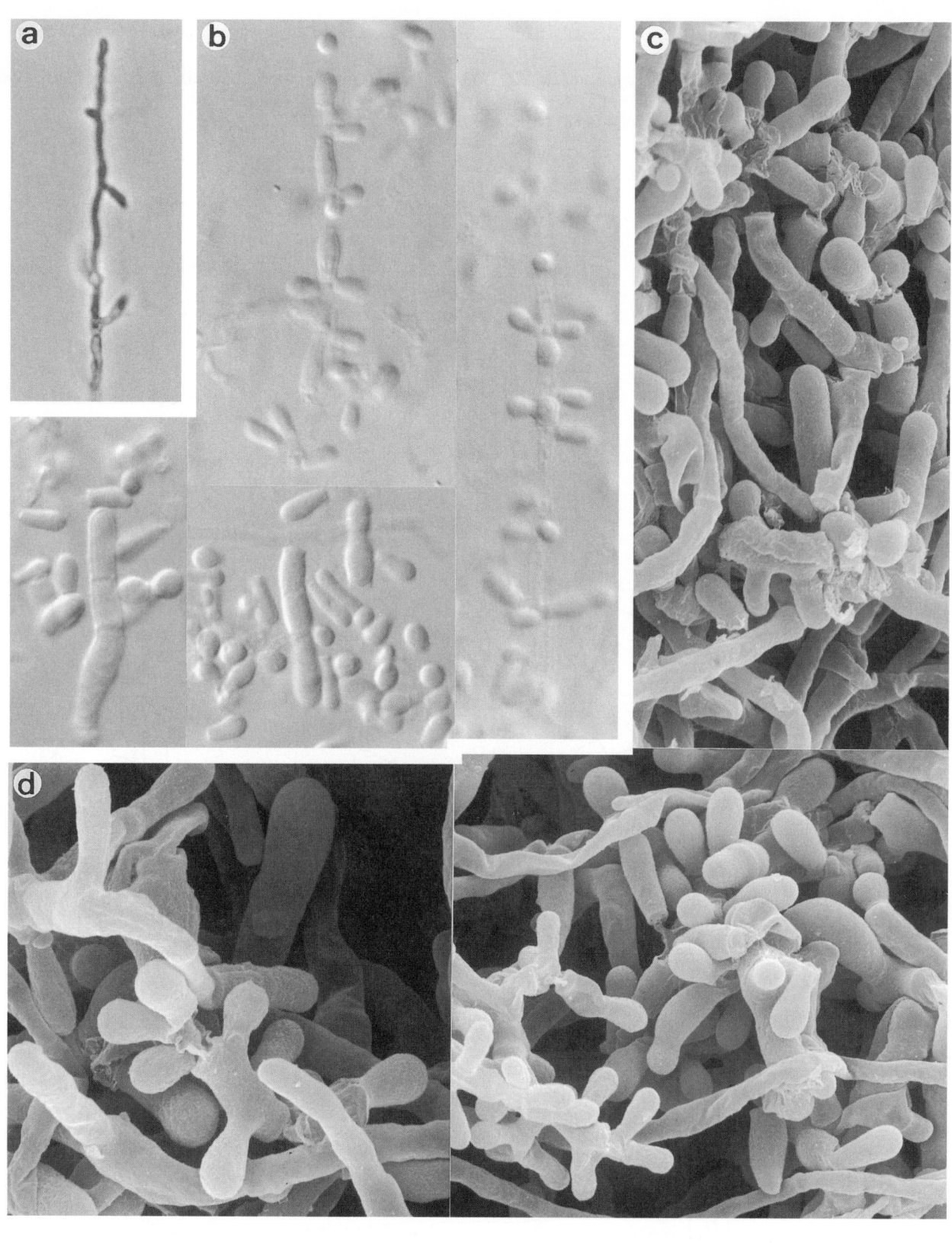

Trichophyton violaceum, CBS 411.88. a. Reflexive hypha; b-d. fertile hypha with microconidia, and liberated macro- and microconidia. a. ×640; b. ×1600; c. ×3100; d. ×4700.

Hyphomycetes. Genus: *TRITIRACHIUM*

Tritirachium oryzae (Vincens) de Hoog

Colony characteristics. Colonies (MEA 2%) growing slowly, velvety, rose vinaceous to lilac; reverse vinaceous-brown.

Microscopy. Hyphae hyaline to pale brown, smooth- and thin-walled, 1-2 µm wide, forming a compact submerged mycelium. Conidiophores suberect, reddish-brown, smooth- and thick-walled, 200-1500 µm long, in the upper part bearing several whorls of conidiogenous cells or verticillate side branches. Conidiogenous cells up to 200 µm, consisting of an elongate basal part, slightly swollen below the middle, and a narrow, regularly flexuose sympodial rachis, with flat scars. Conidia hyaline, smooth- and thin-walled, spherical to ellipsoidal, (1.5-) 2.0-3.0 (-3.5) × 1.5-3.0 µm.

Pathogenicity. BSL-1. Air-borne fungus. Agent of corneal ulcers (Baquis, 1905; Langeron, 1934, 1947; Srinivasa Rao Ramakrishnan, 1968; Kaplan, 1975; Rodrigues *et al.*, 1975). It has also been isolated from a case of otomycosis (Barale *et al.*, 1982).

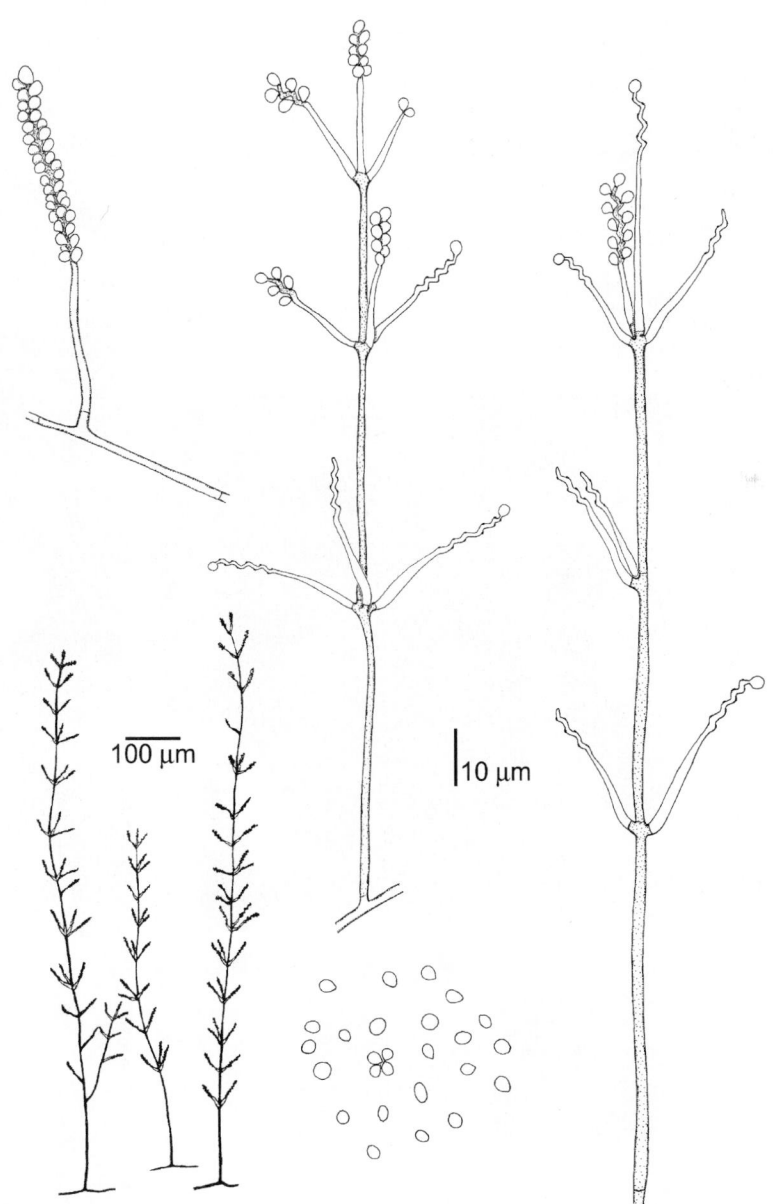

Tritirachium oryzae, UAMH 3334. Conidiophores and conidia.

Nomenclature. *Beauveria oryzae* Vincens - Revue Path. Vég. Ent. Agric. Fr. 10: 122, 1910 ≡ *Tritirachium oryzae* (Vincens) de Hoog - Stud. Mycol. 1: 22, 1972.

    Tritirachium roseum van Beyma - Antonie van Leeuwenhoek 8: 118, 1942.

    Beauveria brumptii Langeron & Lichaa - Bull. Acad. Méd., Paris 111: 133, 1934 ≡ *Tritirachium brumptii* - Annls Parasit. Hum. Comp. 22: 94, 1947.

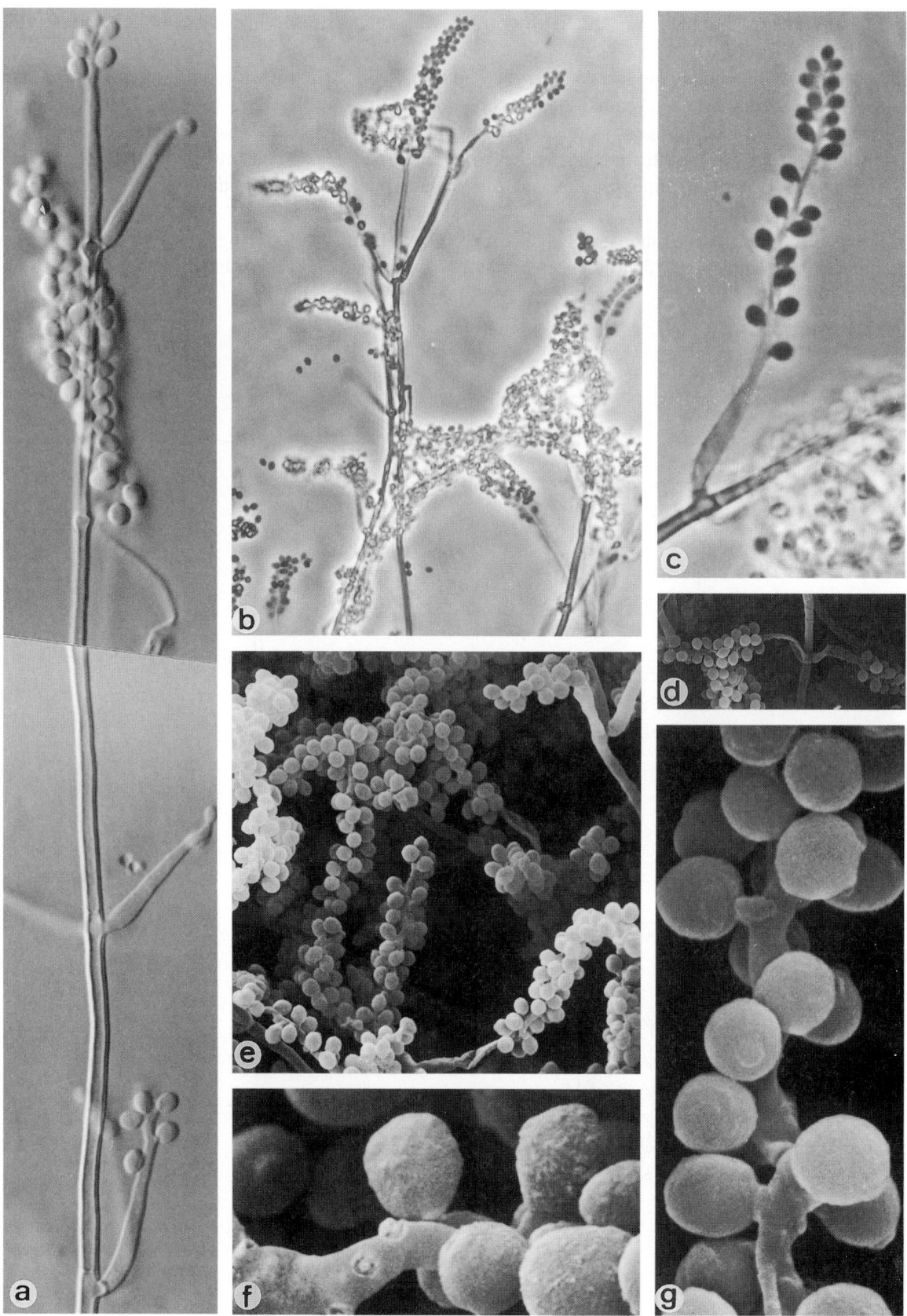

Tritirachium oryzae, UAMH 3334. Conidiophores and conidia. a. ×1600; b. ×512; c. ×1600; d. ×1150; e. ×1650; f. ×10500; g. ×9400.

Hyphomycetes, dematiaceae. Genus: *ULOCLADIUM*

Generic description. Colonies expanding, grey to olivaceous, powdery. Conidiophores erect, pale brown, multi-celled, producing conidia in sympodial order; conidial scar flat, pale brown or hyaline. Conidia black, rough-walled, with pointed base when young, with muriform septation, single or in very short chains.

Molecular diagnostics. SSU restriction map of *U. chartarum*, with *U. botrytis* extra restriction site indicated:

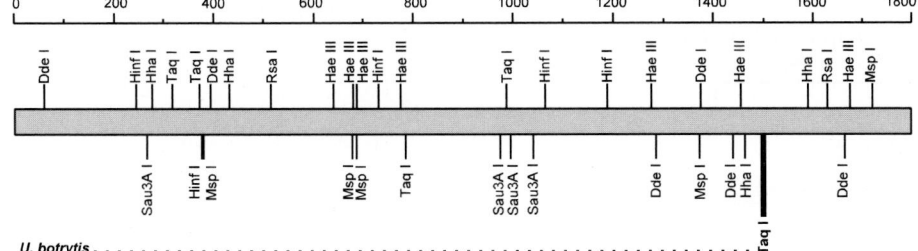

General remarks. *Ulocladium* species are common contaminants. They are closely similar to saprophytic *Alternaria* species (p. 422), but colonies sporulate abundantly and conidia arise in very short chains or are non-catenate, and become blackish-brown with coarse walls. Species closely resemble each other morphologically (compare Simmons, 1998).

Reference. Simmons (1967).

Key to the treated species of Ulocladium:

1a. Conidiophores often profusely branched with clavate ends; scars hyaline; conidia single . ***U. botrytis* (998)**
1b. Conidiophores flexuose, nearly unbranched; scars brownish; conidia commonly
    in chains . . . . ***U. chartarum* (1000)**

Ulocladium botrytis Preuss

Colony characteristics. Colonies (PCA) growing rapidly, powdery to lanose, olivaceous black.

Microscopy. Conidiophores erect, short, strongly branched, strongly geniculate with conidia on the nodes, up to 50 × 4.5 µm, somewhat swollen near the apices, subhyaline to pale brown; conidial scars hyaline. Conidia single, rarely in very short chains, obovoidal, without beaks, medium brown to olivaceous, finally black, verrucose to tuberculate, 19-25 × 7-12 µm, with (1-) 2-3 transverse and 0-2 oblique or longitudinal septa. Secondary conidiophores nearly absent.

Molecular diagnostics. ITS restriction map based on CBS 197.67:

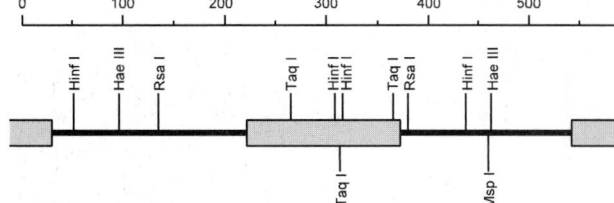

Pathogenicity. BSL-1. Common saprobe, rarely isolated from clinical specimens, with unproven pathogenicity.

Reference. Simmons (1967).

Antifungal susceptibility.

Antifungal	Mean MICs	Strains	Reference
AMB	2	3	Pujol *et al.* (2000)
5FC	256	3	Pujol *et al.* (2000)
FLZ	40.3	3	Pujol *et al.* (2000)
ITZ	0.99	3	Pujol *et al.* (2000)
KTZ	2	3	Pujol *et al.* (2000)
MCZ	4	3	Pujol *et al.* (2000)

Nomenclature. *Ulocladium botrytis* Preuss, *in* Sturm - Deutschl. Fl. 3, 30: 83, 1851.

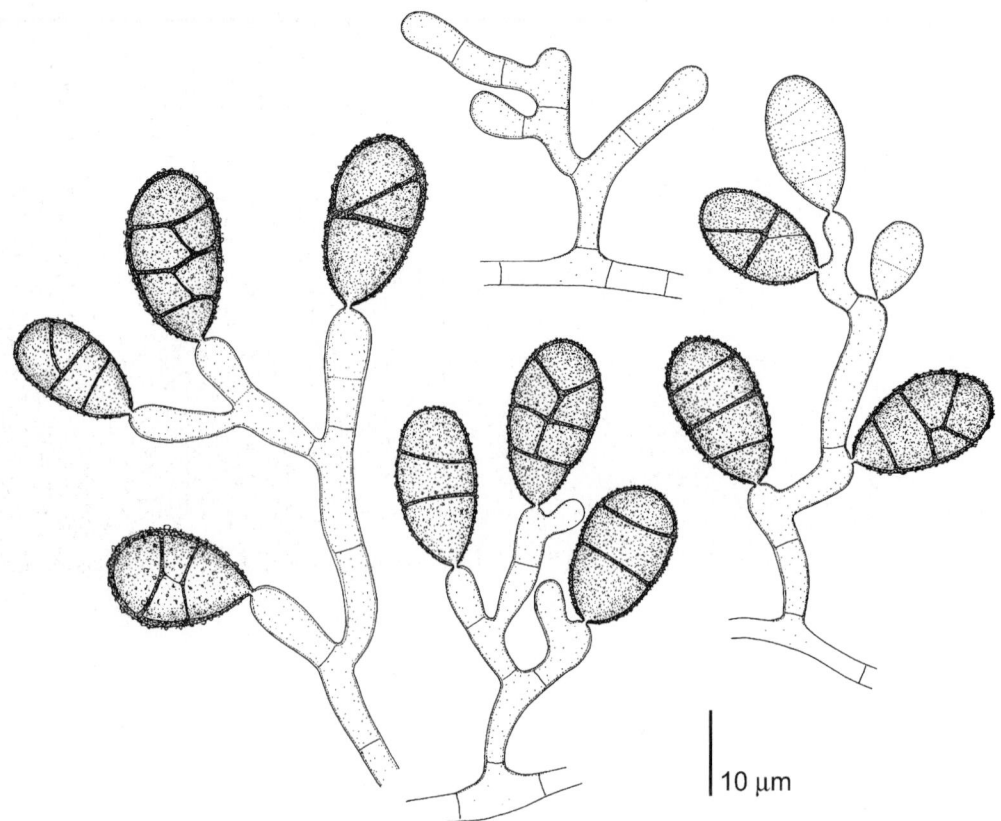

Ulocladium botrytis, CBS 197.67. Conidiophores and conidia.

Ulocladium botrytis, CBS 197.67. Conidiophores and conidia. a. ×640; b. ×1600; c. ×1300.

Ulocladium chartarum (Preuss) Simmons

Colony characteristics. Colonies (PCA) growing rapidly, powdery to lanose, black or olivaceous black.

Microscopy. Conidiophores erect, straight or flexuose, often somewhat geniculate, mostly unbranched, up to 50 × 4-5 µm, golden brown, smooth-walled; conidial scars brown. Conidia commonly in chains of 2-10, ellipsoidal or obovoidal, often with short beaks, medium brown to olivaceous, finally black, verrucose, 18-38 × 11-20 µm, with 1-5 (commonly 3) transverse and several oblique or longitudinal septa. Often secondary conidiophores present on conidia.

Molecular diagnostics. ITS restriction map based on CBS 105.32:

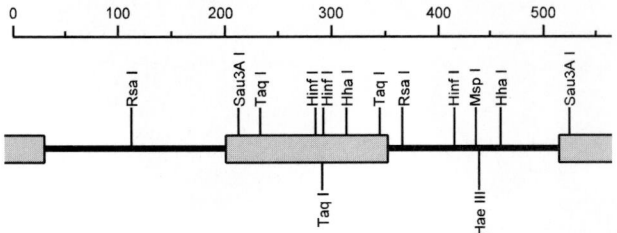

Differential diagnosis. Common contaminant. The species differs from *U. botrytis* (p. 998) by having conidia frequently in chains and dark conidiophore scars. It is distinct from *Alternaria alternata* (p. 423) by having conidia which are coarsely verrucose and have a short, spindle-shaped apex.

Pathogenicity. BSL-1. A case of extended infection of subcutaneous tissues was reported by Altmeyer & Schon (1981) and Seeliger (1983); further cutaneous infections by Blanc *et al.* (1984) in a transplant patient and by Verret *et al.* (1982) in a patient with Cushing's disease. Histologically, deeper infections may present with pale pigmented, ellipsoidal elements, as in *Alternaria* (p. 422).

References. Simmons (1967), Ellis (1976), David (1995).

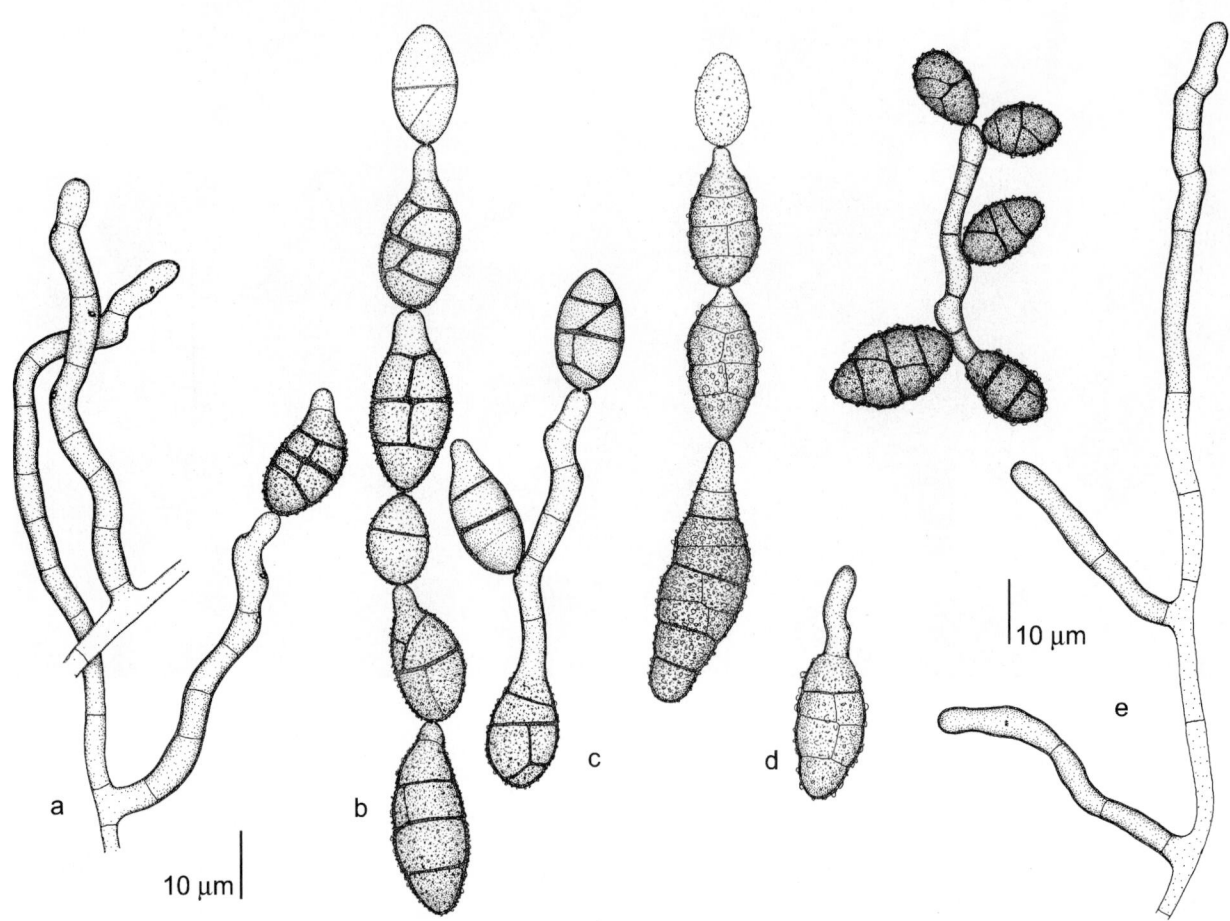

Ulocladium chartarum, a-c. CBS 108.41; d-f. CBS 200.67. a, f. Conidiophores; b, d. Conidial chains; c, f. conidia bearing short, secondary conidiophore.

Antifungal susceptibility.

Antifungal	Mean MICs	Strains	Reference
AMB	5.3	5	Pujol *et al.* (2000)
5FC	256	5	Pujol *et al.* (2000)
FLZ	111	5	Pujol *et al.* (2000)
ITZ	6.7	5	Pujol *et al.* (2000)
KTZ	6.1	5	Pujol *et al.* (2000)
MCZ	2.64	5	Pujol *et al.* (2000)

Nomenclature. *Alternaria chartarum* Preuss - Bot. Zeit. 6: 412, 1848 ≡ *Ulocladium chartarum* (Preuss) Simmons - Mycologia 59: 88, 1967. *Alternaria stemphylioides* Bliss - Mycologia 36: 538, 1944.

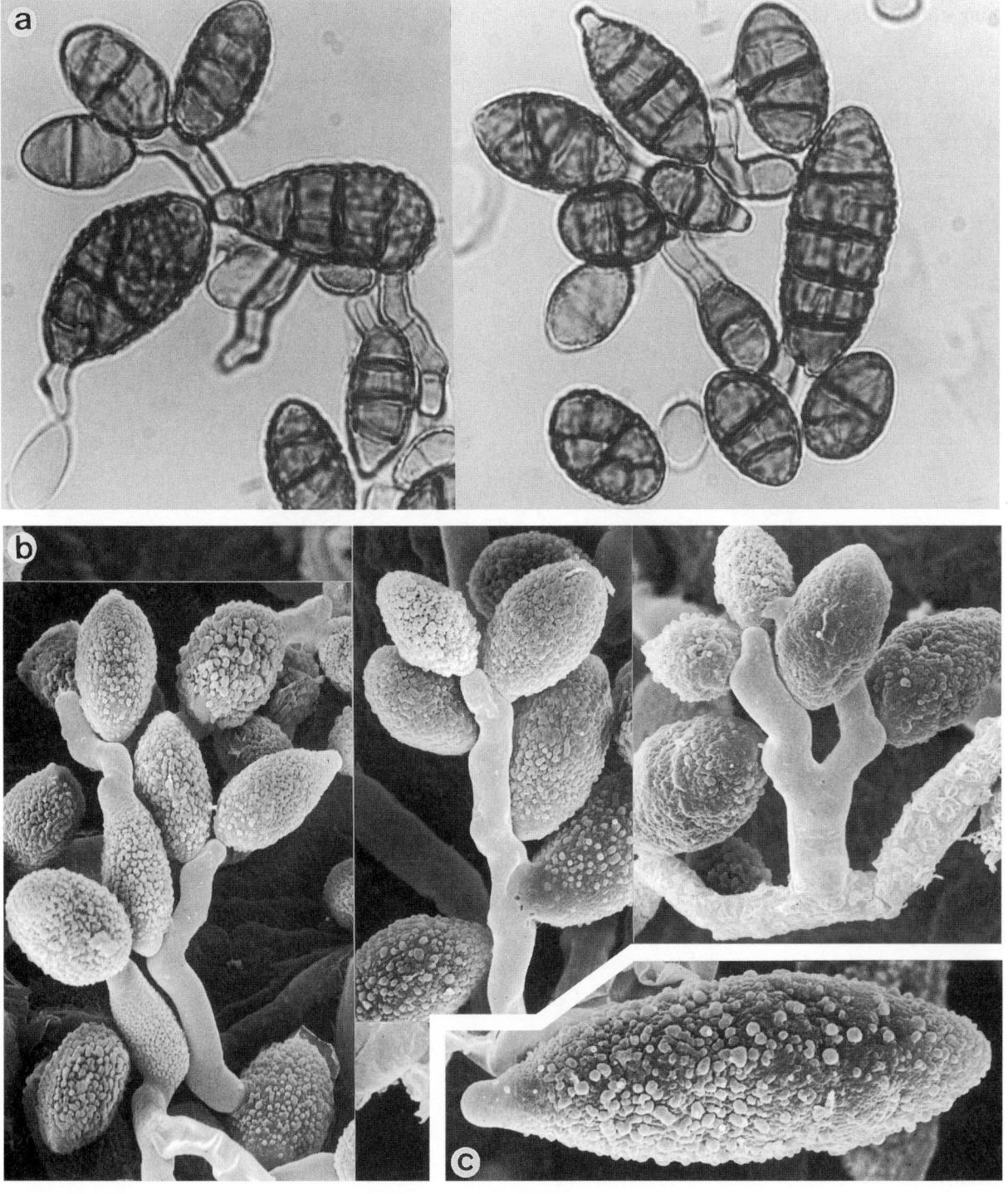

Ulocladium chartarum, CBS 108.41. Conidiophores and conidia. a. ×1600; b. ×2300; c. ×3200.

Hyphomycetes. Genus: *VERONAEA*

Veronaea botryosa Ciferri & Montemartini

Colony characteristics. Colonies (PDA) growing rapidly, velvety to lanose, greyish-brown or blackish-brown.

Microscopy. Conidiophores erect, straight or flexuose, unbranched or occasionally loosely branched, sometimes geniculate, smooth-walled, pale to medium or olivaceous brown, up to 250 μm long, 2-4 μm wide. Conidiogenous cells terminal or lateral, often becoming intercalary, cylindrical in the apical part with numerous flat scars. Conidia smooth-walled or slightly verrucose, sometimes cylindrical, rounded at the apex and truncate at the base, pale brown, usually 1-septate, 5-12 × 3-4 μm.

Physiology. Urease and nitrate positive, gelatin liquefied, starch hydrolyzed (Wang *et al.*, 1991).

Molecular diagnostics. ITS restriction map based on CBS 254.57:

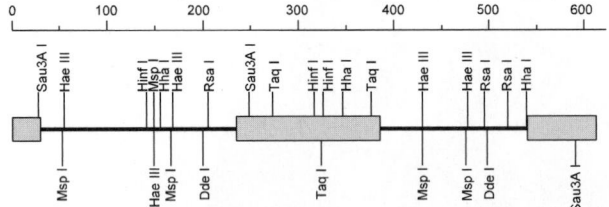

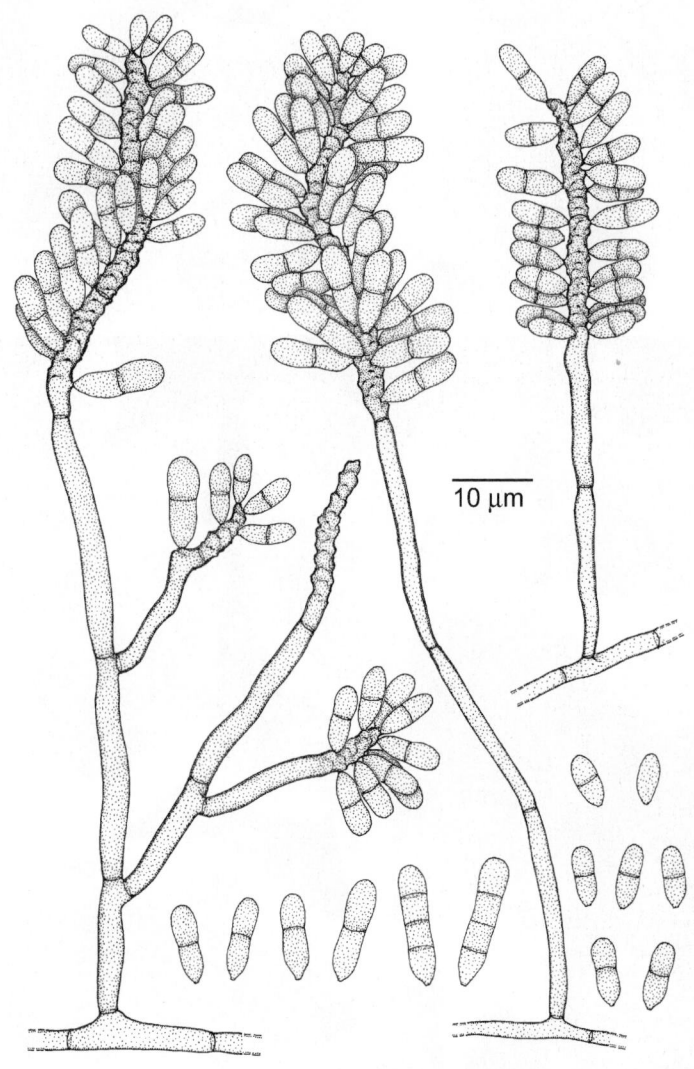

Veronaea botryosa, CBS 254.57. Conidiophores and conidia.

Pathogenicity. BSL-1. Cases of skin infection in China (Nishimura *et al.*, 1989), Libya (Ayadi *et al.*, 1995), France (Foulet *et al.*, 1999) and the Philippines (Medina *et al.*, 1998), mostly in immunocompromised patients.
Reference. Ellis (1971).
Nomenclature. *Veronaea botryosa* Ciferri & Montemartini - Atti Ist. Bot. Lab. Crittog. Univ. Pavia, Ser. 5, 15: 68, 1958.

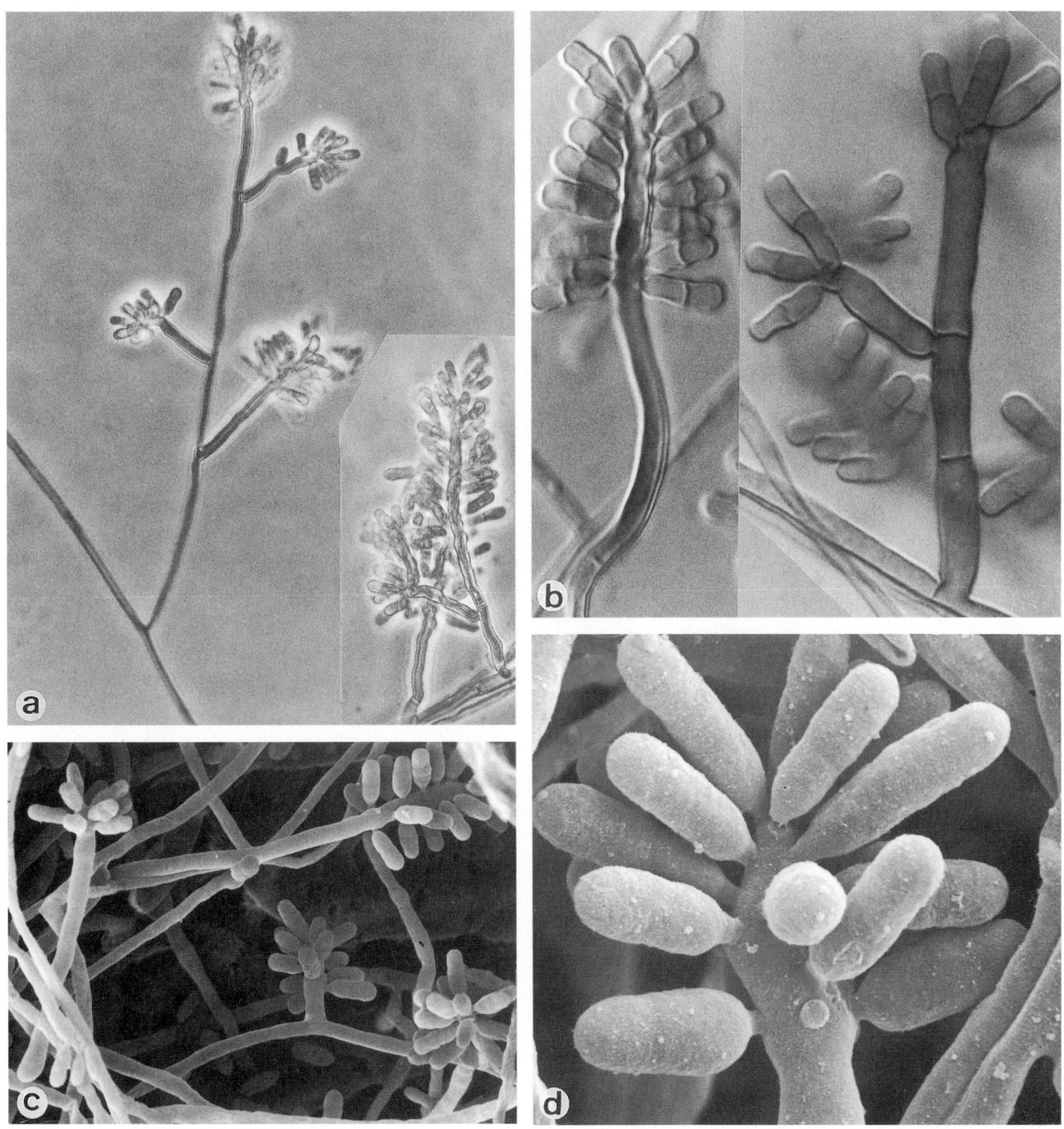

Veronaea botryosa, CBS 254.57. Conidiophores and conidia. a. ×512; b. ×1600; c. ×1350; d. ×5400.

1003

Hyphomycetes. Genus: *VOLUTELLA*

Volutella cinerescens (Ces.) Sacc.

Colony characteristics. Colonies (PDA) growing moderately rapidly, white to cream-coloured, pustulate due to formation of sporodochia.

Microscopy. Conidia produced mainly in sessile or somewhat stipitate sporodochia, which consist of a compact basal part and densely fasciculate, cylindrical phialides, 8.5-20.0 × 1-2 μm. Sterile setae present on the natural substrate, arising in variable numbers from a level somewhat below the phialides, hyaline, 5-10 μm wide, thick-walled, with pointed or blunt tips. Conidia ellipsoidal to cylindrical, hyaline, non-septate, smooth-walled, 2-4 × 0.8-1.2 μm.

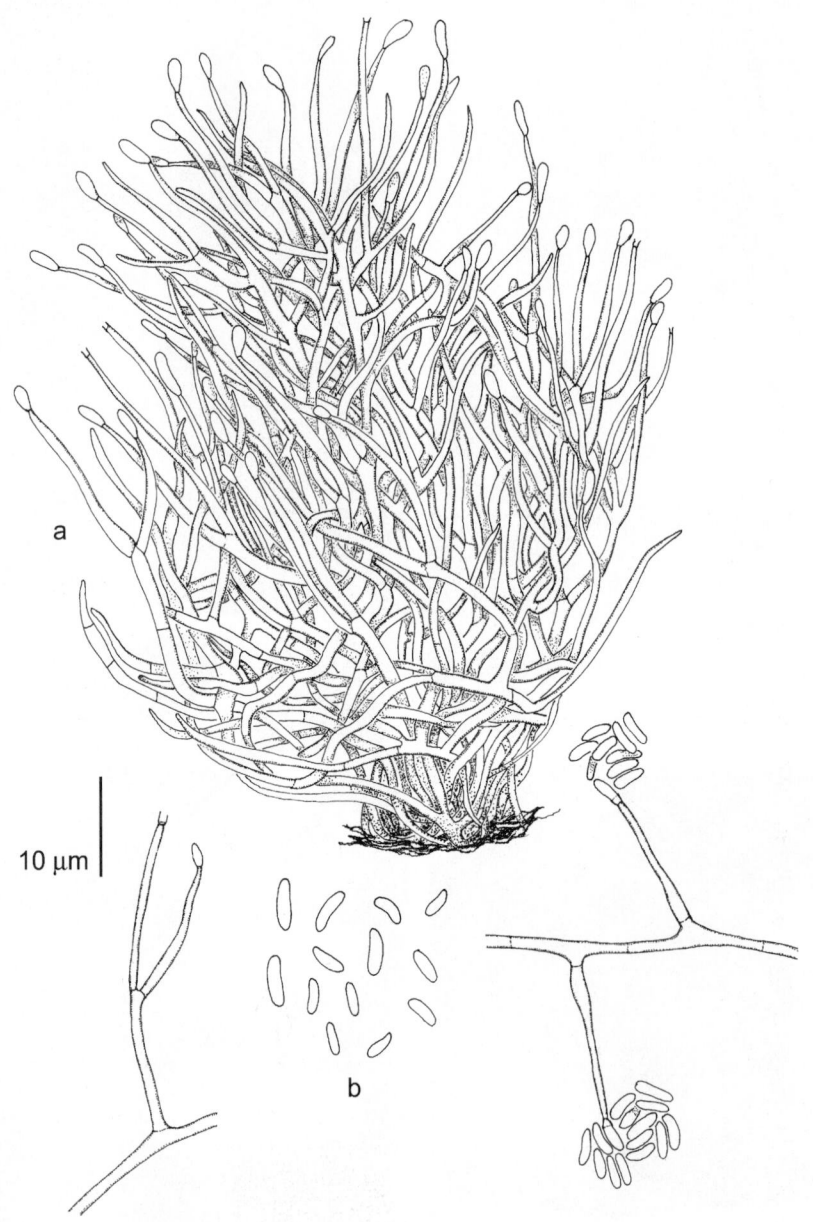

Volutella cinerescens, CBS 832.71. a. Sporodochium from culture, without setae; b. conidiogenous cells and conidia.

Pathogenicity. BSL-1. Occasionally causing endophthalmitis (Foster *et al.*, 1958; Theodore *et al.*, 1961; Rippon, 1988).

Reference. Theodore *et al.* (1961).

Nomenclature. *Psilonia cinerescens* Cesati - Flora p. 204, 1853 ≡ *Volutella cinerescens* (Cesati) Saccardo - Syll. Fung. 4: 688, 1886.

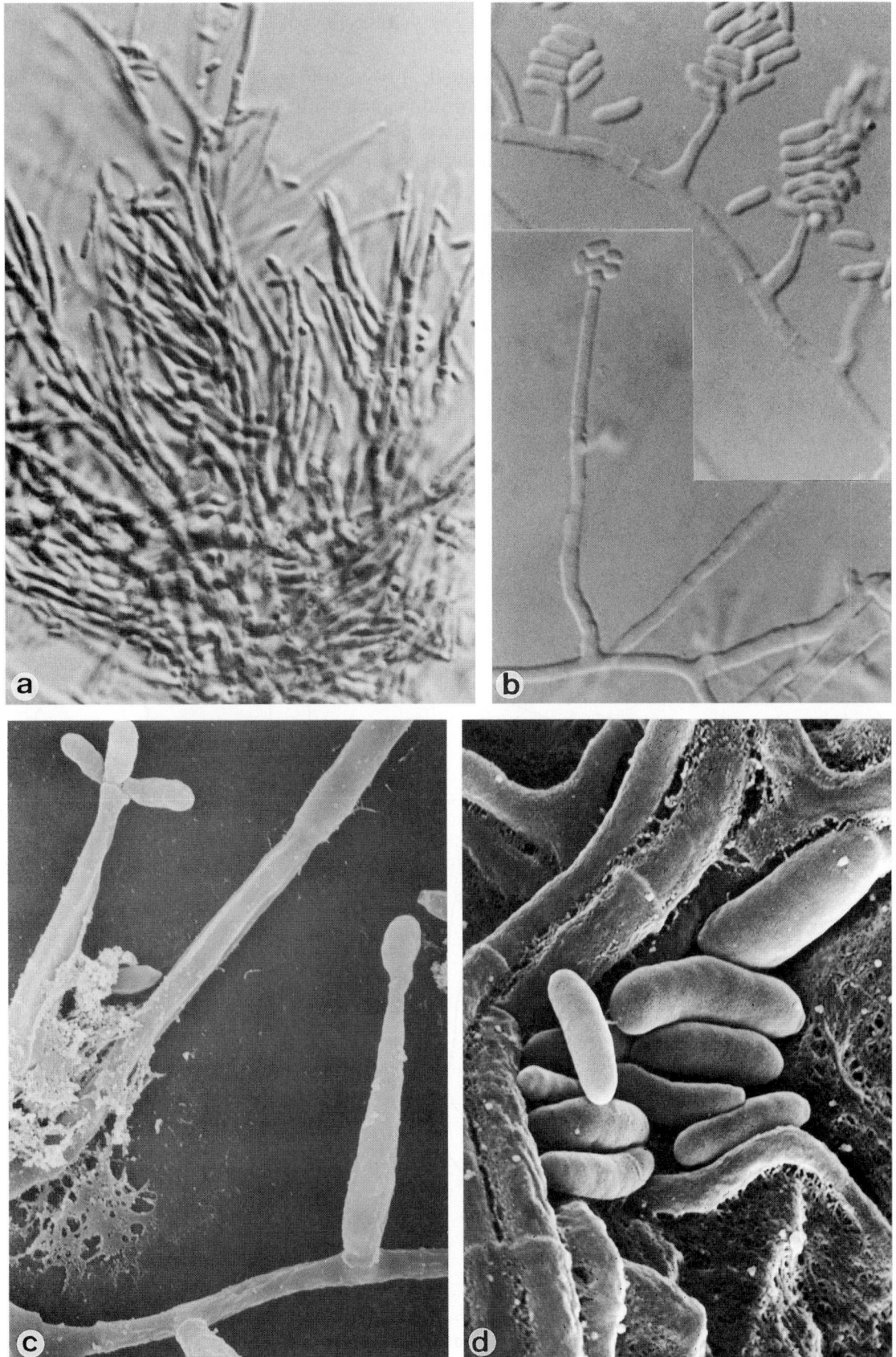

Volutella cinerescens, CBS 832.71. a. Sporodochium; b-d. conidiogenous cells and conidia. a, b. ×1600; c. ×2200; d. ×8000.

1005

Hyphomycetes. Genus: *WALLEMIA*

Wallemia sebi (Fr.) v. Arx

Colony characteristics. Colonies (MEA 4%) growing very slowly, punctiform, powdery in mass, rust brown to purplish-brown.

Microscopy. Hyphae hyaline, smooth- and thin-walled, 1.5-2.5 µm wide, forming a compact mycelium. Conidiophores erect, in dense, parallel arrangement, subhyaline, unbranched, smooth-walled, slightly constricted at the apex just below a brownish collarette which produces a cylindrical, verruculose meristematic cell disarticulating into arthric conidia. Conidia one-celled, pale brown, initially cubic, soon becoming spherical, 2.0-3.5 µm diam, finally with verrucose, rather thick walls.

Physiology. The species is xerophilic and should be grown on media with high sugar content.

Pathogenicity. BSL-1. Species occurring in dry food stuffs or in salty environments. Subcutaneous infections have been reported in old literature (Gougerot & Caraven, 1909; Auvrey, 1909; de Beurmann *et al.*, 1909; Janke 1950).

References. Domsch *et al.* (1980), de Hoog (1983).

Nomenclature. *Sporendonema sebi* Fries - Syst. Mycol. 3: 435, 1832 ≡ *Wallemia sebi* (Fries) von Arx - Gen. Fungi Sporul. Pure Cult. P. 166, 1970.

   Hemispora stellata Vuillemin - Bull. Soc. Mycol. Fr. 22: 125, 1906.

   Oospora d'agatae Saccardo, *in* d'Agata - Policlin. Sez. Chirurg. 25: 80, 1918 ≡ *Scopulariopsis d'agatae* (Saccardo) C.W. Dodge - Med. Mycol. p. 649, 1935.

   Sporendonema epizoum Ciferri & Redaelli - J. Trop. Med. Hyg. 37: 167, 1934.

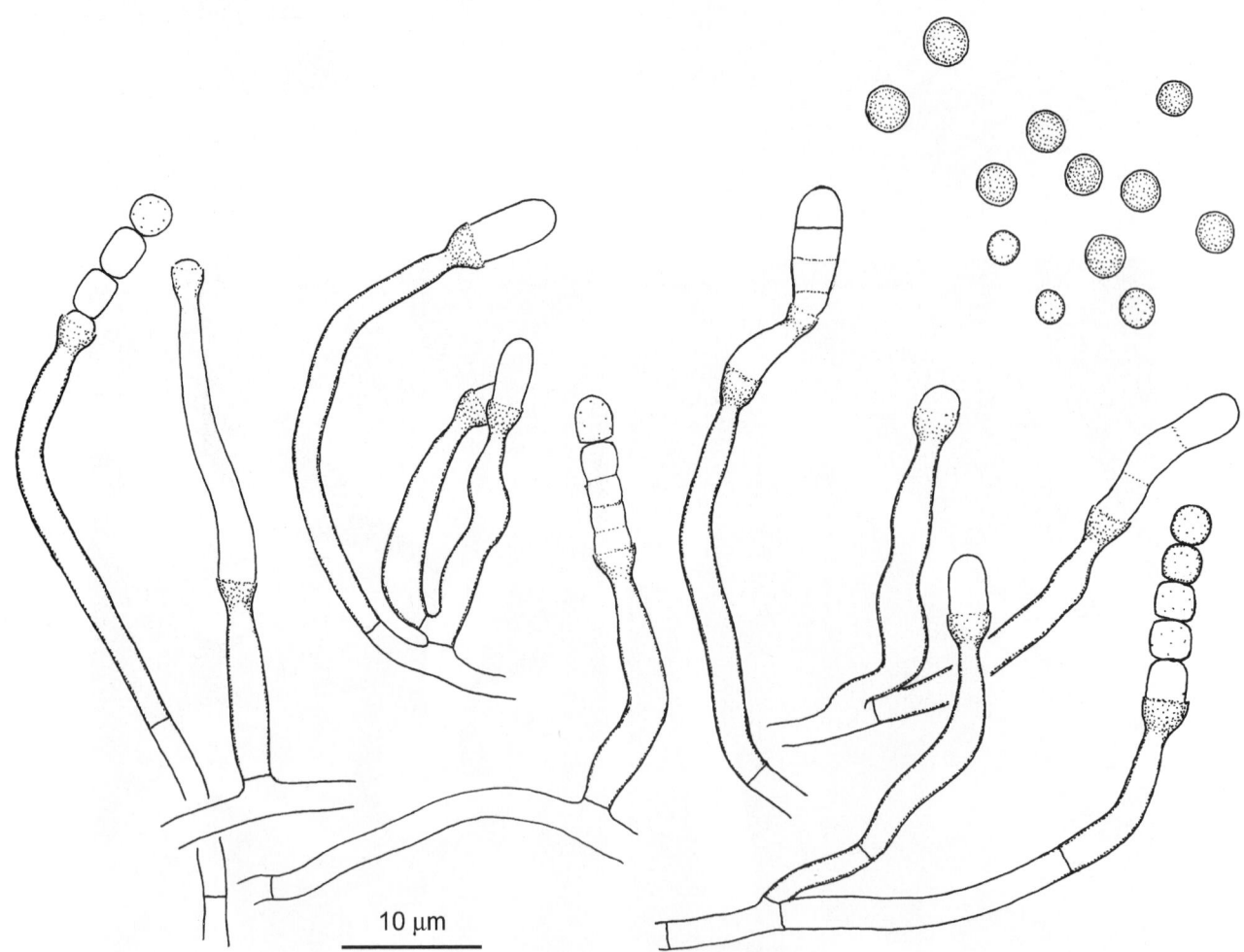

Wallemia sebi, CBS 196.56. Conidiogenous cells with collarettes, and conidia.

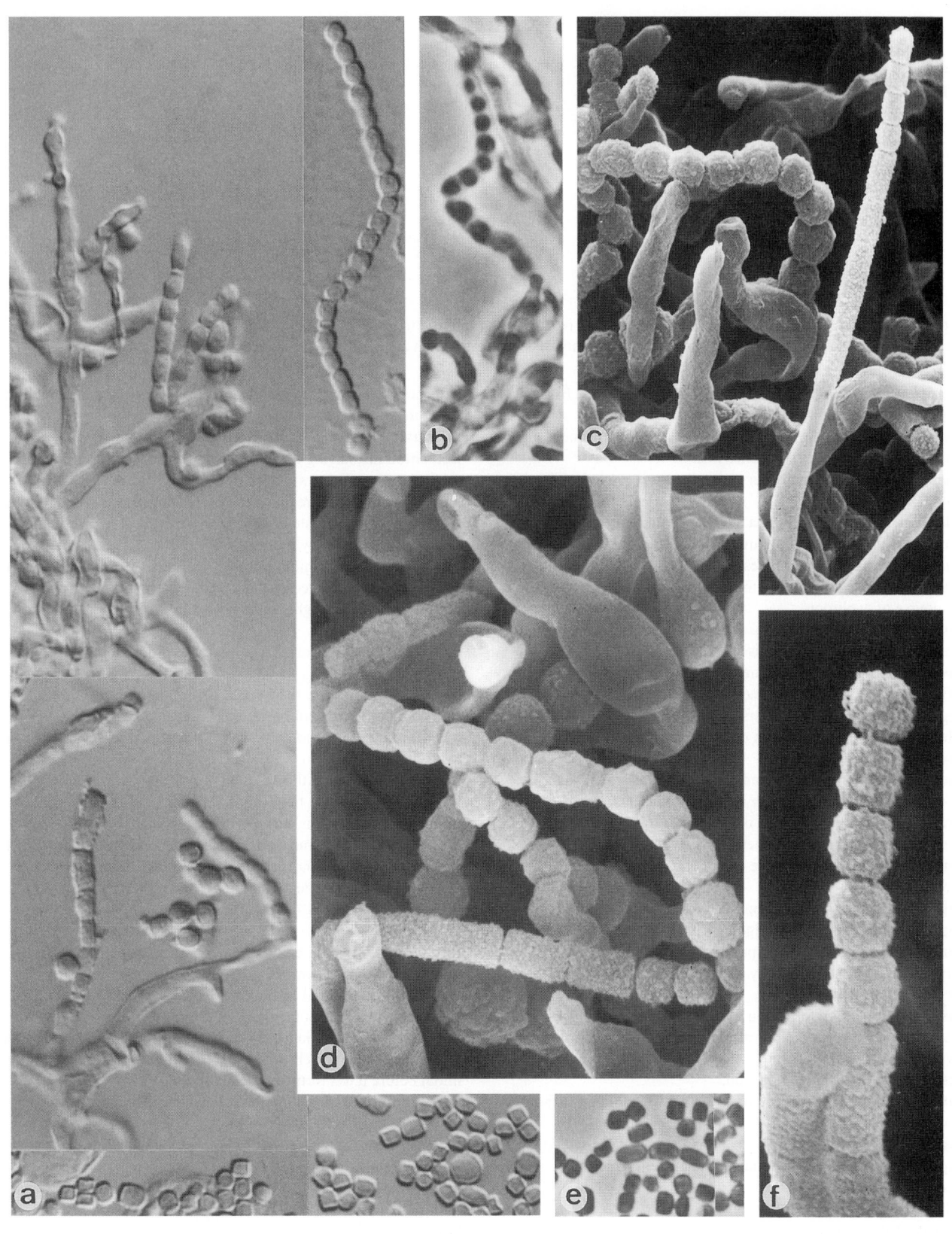

Wallemia sebi, CBS 453.80. Conidiogenous cells and conidia. a. ×1600; b. ×1280; c. ×3600; d. ×5400; e. ×1280; f. ×10,000.

GLOSSARY

Terminology used in this Atlas is explained below. The reader is also referred to the Dictionary of the Fungi (Hawksworth *et al.*, 1995). For descriptive terms in morphology, the Illustrated Dictionary of Mycology (Ulloa & Hanlin, 2000) is recommended.

acerose, rocket-shaped, slender, with somewhat swollen base and acute apex (Fig. 63-34); syn.: lanceolate.

acervulus, saucer-shaped conidial fruit body, conidiogenous cells being inserted on a pseudoparenchyma.

acicular, slender, gradually becoming narrower towards the apex (Fig. 63-36).

acladium, sessile microconidia alongside undifferentiated hyphae.

acropetal, chain of conidia growing at the tip.

aculeate, bearing narrow spines.

adaptation, acquiring factors to live in a particular environment in the course of evolution.

adiaspore, thick-walled fungal chlamydospore-like cell of *Emmonsia* species in human tissue.

adventitious sporulation, formation of morphologically differentiated reproductive structures in host tissue.

aggregated, in compact clusters.

allantoid, slightly curved with rounded ends; sausage-shaped (Fig. 63-44).

allochtonous, living on the mammal host because of availability of an essential factor, but actually having its niche elsewhere.

amplitude (ecological), range of conditions tolerated.

ampulliform, urn-shaped; spherical with short neck (Fig. 63-20).

amyloid, producing starch-like compounds, staining blue with Melzer's iodine.

anamorph, asexual form of sporulation.

angular, with several blunt edges.

annellated zone, pronounced part of the cell which comprises a series of scars having arisen by repetitive conidiogenesis through the previous scar.

annellidic, repetitive formation of conidia through a single scar, each subsequent scar being formed at a slightly or markedly higher level.

antheridium, male copulating cell.

antler-like hypha, (sub)erect hypha with swollen, blunt dichotomous lobes at the apex.

apophysis, swelling of sporangiophore just below sporangium.

appressorium, irregularly branched attachment structure.

arachnoid, cobweb-like.

arborescent, tree-shaped (bundles of hyphae).

arthric, thallic conidiogenesis by which a hyphal element is fragmented into conidia after transverse septation.

arthroconidium, conidium derived form a specialized hypha that has fallen apart into separate cells.

ascending, partly creeping with terminal part suberect.

ascogonium, female sex organ of *Ascomycota*.

ascoma, fruit body containing asci.

ascospore, sexually produced spore in ascus.

ascostroma, stroma in which cavities are formed containing asci.

ascus, cell in which ascospores are produced after karyogamy and meiosis.

assimilation, degradation of compounds for feeding.

asteroid body, fungal cell in human tissue on which antibodies have been precipitated, forming a star-shaped perimeter.

attenuated, narrowed towards the tip.

autochtonous, having its natural niche on the mammal body.

avellaneous, pale brown, hazelnut brown.

azygospore, zygospore on which one of the suspensors is lacking or shrivels in an early state of development.

bacilliform, narrow cylindrical (Fig. 63-27).

ballistoconidium, conidium formed on spine-like outgrowth of a cell, forcibly discharged by a droplet mechanism.

bamboo-hypha, straight hypha with darkened septa at regular intervals.

barrel-shaped, ellipsoidal with flattened poles, doliiform (Fig. 63-23).

basidiospore, forcibly discharged sexual spore in *Basidiomycota* formed on outgrowth of basidium.

basidium, cell in *Basidiomycota* in which karyogamy and meiosis takes place.

basipetal, growing at the base.

biflagellate, with two flagellae.

biguttulate, containing two droplets.

binomial, specific name, consisting of a first element indicating the generic relationship and a second element indicating the species.

bipolar, at the opposite extremities of a cell.

biseriate, phialides in *Aspergillus* being supported by a layer of metulae.

bitunicate, with double ascus wall, of which the innermost may bulge out at spore liberation.

biverticillate, conidial stipes in *Penicillium* which are once branched, each branch bearing whorl of phialides.

blastic, kind of conidiogenesis in which the conidium initial inflates from the supporting cell; it is already recognizable as a conidium initial before it has formed a basal septum.

blastoconidium, conidium formed by blastic conidiogenesis.

brick, brownish-red, colour of bricks.

budding cell, discrete cell reproducing by local inflation of the wall.

buff, yellowish-brown, pale leather coloured.

bulbil, multi-celled, clump-like propagule.

bullet-shaped, short, with pointed apex and flat base (Fig. 63-30).

butyrous, with buttery texture.

caespitose, arranged in groups together.

capilliconidium, slender propagule with adhesive knob on narrow erect hypha in *Entomophthorales*.

capillitium, mass of sterile, thread-like fibres forming the skeleton of the sporangium of *Mycetozoa*.

capsule, slimy excretion surrounding a cell.

cashewnut-shaped, strongly curved with rounded ends (Fig. 63-42).

catenate, in long chains > 3.

catenulate, in short chains of 2 or 3.

chlamydospore, swollen, thick-walled resting cell.

cerebriform, brain-like folded colony.

christmas tree-shaped, with short, right branches at the tip which become longer towards the base.

cigar-shaped, cylindrical but gradually becoming narrower towards both ends (Fig. 63-33).

cirrhus, curled slimy conidial mass forced out of pycnidium or perithecium.

clade, group with shared evolution and homologous characters.

clamp-connection, hyphal outgrowth in *Basidiomycota* which, at cell division, makes a connection between the two cells forming a by-pass around the septum to allow the migration of a nucleus.

clavate, club-shaped, slender becoming broader towards the rounded apex (Fig. 63-21).

cleistothecium, closed fruit body containing asci.

club-shaped, clavate, cylindrical and broadened towards the end.

coenocyte, multinuclear, non-septate mycelium.

collarette, cup- or funnel-shaped structure at the apex of a phialide, through which the conidia are produced.

colonization, growth on host without host response.

columella, sterile central axis of sporangium.

columnar, conidia in *Aspergillus* formed in dry chains which cohere in columns.

commensal, fungus living on products of the body without invading living tissue.

confluent, growing together, fusing at margins.

conical, triangular, becoming narrower towards the apex (Fig. 63-38).

conidiogenous cell, cell producing conidium.

conidioma, fruit body containing conidia.

conidiophore, specialized, differentiated hyphal structure bearing conidia.

conidium, asexual propagule in *Asco-* or *Basidiomycota* produced on or in hypha or specialized supporting structure.

coriaceous, with leathery texture.

cottony, loose, fragile aerial mycelium up to 5 mm high.

crest, longitudinal ridge.

cullulate, hat-shaped due to presence of slimy brim at the base (Fig. 63-14).

cuneiform, having the shape of an axe blade (Fig. 63-25).

cutaneous mycosis, fungal growth on exposed organs causing tissue damage and with host response.

cylindrical, elongate with parallel lateral walls (Fig. 63-26).

cyst, firm-walled resting cell derived from zoospore.

cystidium, large sterile cell in hymenium of *Basidiomycota*.

dematiaceous, pigmented in olivaceous or brown colours.

denticle, tooth-like conidium-bearing projection.

dermatomycosis, cutaneous infection of hair, skin or nails.

dermatophyte, fungus classified in *Epidermophyton*, *Microsporum* or *Trichophyton*.

dermatophytosis, cutaneous infection by a dermatophyte.

dextrioid, staining yellowish-brown with Melzer's iodine.

dichotomous, forked, with two symmetrical branches.

diffluent, breaking up in water.

dimorphic, displaying two morphological types, one environmental and one invasive.

diploid, containing a double set of chromosomes.

disarticulation, falling apart into separate cells.

discrete, separate; in morphology: single-celled; in taxonomy: clearly delimited group.

disease, damage to host tissue resulting from fungal growth.

disjunctor, structure between conidia to promote their disarticulation.

disseminated mycosis, fungal growth in various deep organ tissues located at some distance from each other.

distoseptate, conidia subdivided by inner wall layer only.

doliiform, barrel-shaped, ellipsoidal with flattened poles (Fig. 63-23).

dolipore, septal pore of *Basidiomycota* in which the walls of the pore canal are variously inflated.

dorsiventral, direction from upper side towards lower side.

downy, covered with short aerial hyphae.

echinate, covered with short spines.

echinulate, regularly rough-walled, covered with minute spines.

ecological strategy, life cycle optimised by vector/reservoir relations for survival of the species in a particular habitat.

ecology, science focussing on the natural behaviour of fungi.

effuse, thin and widely spreading.

ellipsoidal, three-dimensionally bilaterally symmetrical, l:b about 1:2 (Fig. 63-6); in broadly ellipsoidal l:b is about 1:1.5 (Fig. 63-5).

endoconidium, asexual propagule formed inside a cell.

endogenous, developmental process taking place inside a cell.

endoplasmic reticulum, cytoplasmic network of membranes bearing ribosomes.

endosaprobe, fungus feeding on the contents of the intestinal tract.

endosporulation, conidia formed inside a cell by wall formation around parts of the cytoplasm.

enteroarthric, disarticulation of hyphae into conidia by which the viable cells form an extra, internal wall layer and intermediate cells collaps after autolysis, promoting liberation of the conidia.

enteroblastic, inflation of conidium initial involving the inner wall layer through the outer wall layer of the mother cell.

epitheton, species-part of a fungus name.

euseptate, conidia subdivided by true septa involving all wall layers.

evanescent, rapidly dissolving.

exogenous, formed outside a cell, mostly on short branchlets.

expanding, with rapid growth rate, colony attaining at least 4 cm diam in 14 d.

extremophily, preferred growth under extreme environmental conditions.

exudate, droplets produced on the surface of a colony.

facultative pathogen, having a preference for invasion of the mammal body during a part of the life cycle, but also being able to reproduce in the environment, and thus being transmitted from host to environment and *vice versa*.

falcate, curved fusiform, narrow sickle-shaped (Fig. 63-40).

false chain, fragile conidial chain which grows at the base by repetitive formation of individual conidia through the same scar.

farinose, with minute granules like flour.

fascicle, small, irregular, compact bundle of hyphae.

favic chandelier, small aggregate of densely branched antler-like hyphae.

felty, with depressed, dense, irregular aerial mycelium.

filiform, thread-shaped (Fig. 63-29).

fission, a discrete cell is divided with one or more septa, each segment becoming liberated as a separate cell.

flagella, whip-like appendage of motile cell.

flaring, with irregular remains of cell walls.

flask-shaped, with obovoidal base and short neck (Fig. 63-19).

flexuose, with smooth bends.

floccose, with irregular tufts of aerial mycelium.

foot-cell, basal conidial cell in *Fusarium* macroconidium; hyphal element continuous with conidiophore stipe in *Aspergillus*.

frill, minute remains of outer cell wall after liberation of conidium.

fringed, with thin, often submerged, radial growth at the margin.

funiculose, aggregated into rope-like strands.

furrowed, with deep fissures.

fuscous, dark greyish-brown.

fusiform, spindle-shaped, swollen near the middle, strongly narrowed towards both ends (Fig. 63-32).

gametangium, cell in which gametes are formed.

gamete, cell liberated from gametangium to copulate with cell of opposite sex.

gemma, chlamydospore-like resting body in *Oomycota*.

geniculate, with repeated knee-like bents.

germ pore, circular thin-walled part of propagule through which germination takes place.

germ slit, elongate thin-walled part of a propagule through which germination takes place.

glabrous (of colony), without aerial mycelium; (of microscopic structure), smooth-walled.

granular (of colony), macroscopically coarsely powdery, (of cell), containing numerous oily droplets.

guttulate, containing a single droplet.

habitat, environment where the fungus resides.

hairy, with rather long, erect, sparse aerial mycelium.

haploid, containing a single set of chromosomes.

hat-shaped, hemispherical with a slimy brim at the base, cullulate (Fig. 63-14).

helical, spirally twisted.

heterokaryon, cell containing two nuclei, each derived from one of the parent strains.

heterothallic, sexual reproduction requiring the interaction of two different thalli.

hilum, slightly prominent basal scar.

holoarthric, falling apart of a hypha into separate cells, all cells remaining viable as a propagule.

holobasidium, basidium without septation.

holoblastic, inflation of conidium initial involving the entire wall of the mother cell.

holomorph, the total organism, comprising sexual and asexual life cycles.

holothallic, conversion of a pre-existing hyphal element into a single conidium.

homothallic, sexual reproduction requiring a single thallus.

hülle cell, thick-walled cell occurring in large numbers in association with the ascomata of *Aspergillus* species.

hyaline, colourless, transparent.

hymenium, layer of cells producing basidiospores.

hypha, septate, thread-like fungal element.

immersed, partly below the surface of the growth medium, upper part remaining exposed.

infection, multiplication of a fungus in viable tissue.

integrated, incorporated in the main axis of a conidiophore.

intercalary, incorporated in a mycelial filament, in between hyphal cells.

invasion, penetration into living host tissue.

isogamic, copulation of two morphologically indistinguishable cells.

isthmoid, with a constriction (Fig. 63-46).

karyogamy, fusion of two nuclei during a sexual process.

lanceolate, rocket-shaped, slender, with somewhat swollen base and acute apex (Fig. 63-34); syn.: acerose.

lanose, woolly, with abundant loose, regular aerial mycelium.

lenticular, lens-shaped, circular in face view and ellipsoidal in lateral view (Fig. 63-1).

limoniform, lemon-shaped, with small protrusions at the poles (Fig. 63-15).

linear, very narrow cylindrical (Fig. 63-28).

lobate, having small lobes or rounded projections.

longitudinal, of septation: parallel to length axis of hypha.

lunate, moon-shaped, curved with pointed ends (Fig. 63-39).

macroconidium, septate conidium, in fungi which potentially are able to produce additional 1-celled conidia.

mating, confrontation of two partners in a sexual process.

meiosis, division of a nucleus containing a double set of chromosomes, each daughter nucleus containing a single set.

meiospore, haploid spore formed after meiosis.

melanized, containing melanine, cells becoming olivaceous or brown.

membranaceous, skin-like fruit body texture, collapsing like a berry when a coverslip is applied.

meristematic, perpetual increase in biomass in all directions, and concordant septum formation.

merosporangium, cylindrical sporangium of *Mucorales* containing a chain of sporangiospores.

metula, cell or short branch on vesicle bearing phialides.

microconidium, one-celled conidium in fungi which potentially are able to produce additional septate conidia.

micropore, septal pore of about 10 nm width.

microspore, passively discharged secondary spore of *Entomophthorales.*

mitosis, nuclear division in which daughter nuclei contain the same number of chromosomes as the mother nucleus

moniliform, coherent chain of spherical cells.

monomorphic, having a single form of propagation.

monophialide, phialide with a single opening.

monopodial, with budding at the apical pole of the mother cell only.

monotypic, containing a single species.

monoverticillate, whorl of phialides directly on conidiophore stipe in *Penicillium.*

multilateral, budding from random positions on mother cell.

muriform, with septa in several directions, like a brick wall.

muriform cell, thick-walled cell which swells and becomes cruciately septate.

mycelium, fungal thallus composed of hyphal threads.

mycosis, symptomatic, clinically visible fungal growth on animal host.

myxamoeba, zoospore of *Mycetozoa* after becoming amoeba-like.

napiform, with the shape of a turnip.

navicular, elongate with pointed apex and flat base, matching the groundplan of a boat (Fig. 63-31).

niche, environment where fungus produces assimilative thallus and generates its progeny.

nodose, with swellings at irregular distances.

nodular body, knot of irregularly intertwined hyphae.

obclavate, cylindrical and widened towards the base (Fig. 63-22).

oblate, circular in face view and broadly ellipsoidal in lateral view (Fig. 63-2).

obligatory pathogen, having invasion of the mammal body as sole strategy, and thus being transmitted directly from one mammal host to the other.

oblique, septate at sharp angles to length axis of hypha.

obovoidal, inverted egg-shaped, with narrow end down (Fig. 63-8).

obpyriform, inverted pear-shaped, with narrow end down (Fig. 63-11).

obtuse, with rounded ends.

oidium, thin-walled, swollen resting cell in *Mucorales.*

oogonium, female sex organ of *Oomycota.*

opaque, vaguely transparent or non-transparent.

opportunist, fungus having its niche outside the mammal body, but, when inoculated coincidentally into an immunocompromised host, being able to cause systemic infection.

orthotropic, at right angles.

ossiform, bone-shaped, short cylindrical with swollen ends (Fig. 63-45).

ostiolum, pre-formed pore by which spores are freed from a fruit body.

ovoidal, egg-shaped with narrow end up (Fig. 63-7).

papillate, with small, rounded warts or denticles.

paraphysis, sterile upward growing hyphal element between asci in parallel arrangement inside a perithecium.

pathogen (in medicine), fungus causing disease; this concept is actual and refers to strains.

pathogen (in ecology), fungus adapted to take advantage of an animal host for its survival and evolution; this concept is potential and refers to species.

pathogenicity, ability of a fungus to cause disease.

pectinate hypha, hypha terminating with unilateral, comb-like outgrowths.

pimple-shaped, shaped like a small volcano.

peridial hypha, thick-walled hypha which composes a network around asci.

peridium, wall of fruit body.

periphysis, upward directed hypha around opening of pycnidium or perithecium.

periphysoid, hypha of tissue preceding asci and placed between developing asci.

perithecium, spherical fruit body, with apical opening, containing asci.

phialide, cell with terminal opening through which enteroblastic conidia are produced repetitively.

phialoconidium, propagule produced from a phialide.

phragmobasidium, basidium subdivided in individual cells by septation.

phylogeny, reconstruction of routes of evolution.

plagiotropic, at acute angles.

plasmodium, multinucleate, motile mass of protoplasm in *Mycetozoa*.

plasmogamy, fusion of cytoplasm during sexual process, the nuclei remaining separate.

pleomorphism, the occurrence of several asexual forms of reproduction, eventually in addition to the sexual one.

polyblastic, one cell producing several holoblastic conidia next to each other.

polyphialidic, one cell having several phialidic openings.

polyphyletic, having arisen in separate evolutionary lines.

progeny, offspring.

propagule, cell or cellular element serving dispersal.

pseudohypha, string of more or less coherent budding cells.

pseudomycelium, complex system composed of pseudohyphae.

pseudoparenchyma, fungal tissue composed of rounded cells.

pseudopodium, protoplasmic process which enables a myxomycete thallus to move.

pseudoseptate, cells compartimented but without septa.

pustulate, colony with local hyphal tufts.

pycnidium, spherical fruit body in which conidia are formed.

pycnostroma, stroma containing several conidial cavities.

pyriform, pear-shaped, with narrow end down (Fig. 63-11).

rachis, apical elongate part of a cell bearing conidia (plural: rachids).

radiate (*Aspergillus* morphology), diverging conidial chains.

radiate (taxonomic structure), containing a large number of closely related species.

ramoconidium, lower, branch-like conidium in a conidial chain of *Cladosporium*.

rectangular, broadly cylindrical, with parallel walls (Fig. 63-24).

recurved, curving backwards.

reflexive branching, backward branching at hyphal septa.

reniform, kidney- or bean-shaped (Fig. 63-43).

reservoir (in ecology), an environment where the fungus shelters in dormancy.

resistance factor, property of a fungus enhancing survival in host tissue.

restricted, with limited expansion growth, colony attaining less than 2 cm in 14 d.

reticulate, with net-like structure.

retraction septa, series of septa adjacent to lysing part of a hypha.

reverse, backside of colony in Petri-dish.

rhexolytic, conidial liberation in which an intermediate, supporting cell is sacrified.

rhizoid, root-like structure with which the organism grows into the substrate.

rhomboidal, kite-shaped with conically narrowed walls (Fig. 63-18).

rocket-shaped, lanceolate, slender, with somewhat swollen base and acute apex (Fig. 63-34).

rostrate, beaked, with rostrum (Fig. 63-47).

rostrum, beak-like elongation.

rugose, rough-walled with somewhat irregular ornamentation.

rugulose, finely rough-walled with somewhat irregular ornamentation.

saprobe, fungus feeding by dead organic matter.

sarcina, swollen, meristematic clump of cells.

sarcinic, thallic conidiogenesis by meristematic swelling of cells which become transversely and longitudinally septate and are liberated when a particular size is attained.

saturn-shaped, spherical with equatorial thickening (Fig. 63-16).

schizolytic, liberation of adjacent cells by fission of the septum.

sclerotial body, small sclerotium.

sclerotium, large, multi-cellular clump of cells which does not produce any spores or conidia.

secondary conidiophore, conidiophore produced as a longitudinal extension of a conidium.

sessile, sitting directly on supporting structure, without stalk or denticle.

seta, large, spine-like outgrowth.

single, not in chains or groups.

solitary, alone, only a single conidium being formed per conidiogenous cell.

spathulate, spoon-shaped, with inflated apical part (Fig. 63-13).

spherical, circular in outline (Fig. 63-3).

spherule, endoconidium-forming cell of *Coccidioides immites* formed in host.

spicula, curved outgrowth of hypha in filamentous basidiomycetes.

spindle-shaped, fusiform, swollen near the middle, strongly narrowed towards both ends (Fig. 63-32).

spinose, ornamented with rather large needles, like a hedgehog.

spinulose, ornamented with small needles.

sporangiole, sporangium of *Mucorales* containing one or very few spores only.

sporangiophore, stalk bearing sporangium.

sporangiospore, spore produced from sporangium.

sporangium, specialized organ in *Mucorales* producing asexual spores by cleavage.

sporophore, specialized stalk in *Entomophthorales* producing a single spore.

spore, general term for a reproductive propagule in fungi; in higher fungi it is meiotic (in contrast to conidium).

sporodochium, cushion-like, densely aggregated group of conidiophores.

sporophore, stalk bearing spore in *Zygomycota*.

stellate, star-shaped.

sterigma, slender, spine-like outgrowth of cell bearing a conidium.

sterigmatoconidium, conidium borne on spine-like outgrowth, not forcibly discharged.

stipe, conidiophore bearing vesicle in *Aspergillus*.

stipitate, elevated, with a small stalk.

stolon, creeping and locally rooting branch, bearing sporangiophores.

striate, ornamented with parallel stripes.

stringbean-shaped, linear and flexuose, sometimes with very narrow ends (Fig. 63-35).

stroma, large, irregular mass of vegetative hyphae.

stylospore, stalked chlamydospore of *Mucorales*.

subcutaneous mycosis, fungal growth in subcutaneous tissue.

submerged, below the surface of the growth medium.

subspherical, nearly spherical in shape (Fig. 63-4).

subulate, gradually becoming narrower towards the apex (Fig. 63-37).

superficial mycosis, growth of fungus on mammal host without breaking any line of host defence.

suspensor, hyphal element supporting a zygospore.

sympodial, continued, successive apical growth, each subsequent point of growth next to the previous one and terminating at a somewhat higher level.

synanamorph, one of the asexual forms of propagation of a fungus which forms several kinds of conidia.

synchronous, blastic conidium formation in which all conidium initials on a cell grow out at the same time.

synnema, bundle of conidiophores sporulating in the apical part.

systemic mycosis, localized fungal growth in deep organ tissue.

tadpole-shaped, subspherical with an extended basal part (Fig. 63-12).

tapering, gradually becoming narrower towards one of the ends.

taxonomic structure, kinds of species in a group, and their interrelationships.

tear-shaped, with rounded apex and pointed base (Fig. 63-9).

teleomorph, sexual form of sporulation.

teliospore, cell in smuts where karyogamy and meiosis takes place.

terverticillate, conidial stipes in *Penicillium* which are twice branched, each branch bearing whorl of phialides.

textura angularis, fruit body wall composed of angular cells.

textura epidermoidea, fruit body wall composed of jig-saw-shaped cells.

textura intricata, fruit body wall composed of interwoven hyphal elements.

thallic, kind of conidiogenesis in which a pre-existing hyphal element is converted into one or several conidia.

thallus, fungal organism as a more or less coherent structure, either one- or multi-celled.

tomentose, with a covering of soft hairs.

toruloid, with swollen cells like a chain of pearls.

transverse, of septation: at right angles to length axis of hypha.

trailing hyphae, hyphae resting on agar surface.

true chain, conidial chain produced by meristematic growth of the phialide tip, a hyphal element being extruded from the phialide

opening, later differentiating into conidia.

truncate, with flat base, the scar being nearly as wide as the conidium.

tuberculate, with blunt ornamentations.

umbonate, with a central swelling or thickening (Fig. 63-17).

uncinate, branching backward in a hooked fashion (Fig. 63-41).

undulate, of wavy appearance.

uniseriate, in *Aspergillus:* single layer of phialides on vesicle; of ascospores within an ascus: in a single row.

unitunicate, of asci: with a single wall.

vector (in ecology), an infected host transmitting the fungus to another suitable environment.

veined, of colony: with narrow and somewhat irregular bundles of hyphae.

velvety, with densely compacted, short, erect hyphae.

verrucose, distinctly and regularly rough-walled.

verruculose, finely rough-walled.

vesicle, swollen end of conidiophore of *Aspergillus* bearing conidiogenous cells, or similarly swollen end of sporangiophore of *Mucorales* bearing sporangioles.

villose, covered with slack hairs.

virulence, extent to which pathogenic ability is fulfilled.

virulence factor, specific property of a fungus enhancing multiplication in host tissue.

vitality factor, unspecific property of a non-pathogenic fungus enhancing its survival under hostile environmental conditions.

warted, with irregular clumps of extracellular material.

waxy, glabrous, somewhat elastic.

woolly, with abundant loose, regular aerial mycelium.

woronin body, electron-dense, circular organel at one or both sides of the septal pore of *Euascomycetes*, occluding the septum when the cell is damaged.

yeast, unicellular growth form, mostly belonging to the *Hemiascomycetes*, but also to the *Basidiomycota* and occasionally to the filamentous *Ascomycota* ('black yeasts').

zoodeme, mass of independently reproducing cells within a host tissue.

zoosporangium, sporangium producing zoospores.

zoospore, motile spore with flagella.

zygospore, resting spore resulting from the conjugation of hyphal tips in *Zygomycota*, in which karyogamy and meiosis takes place.

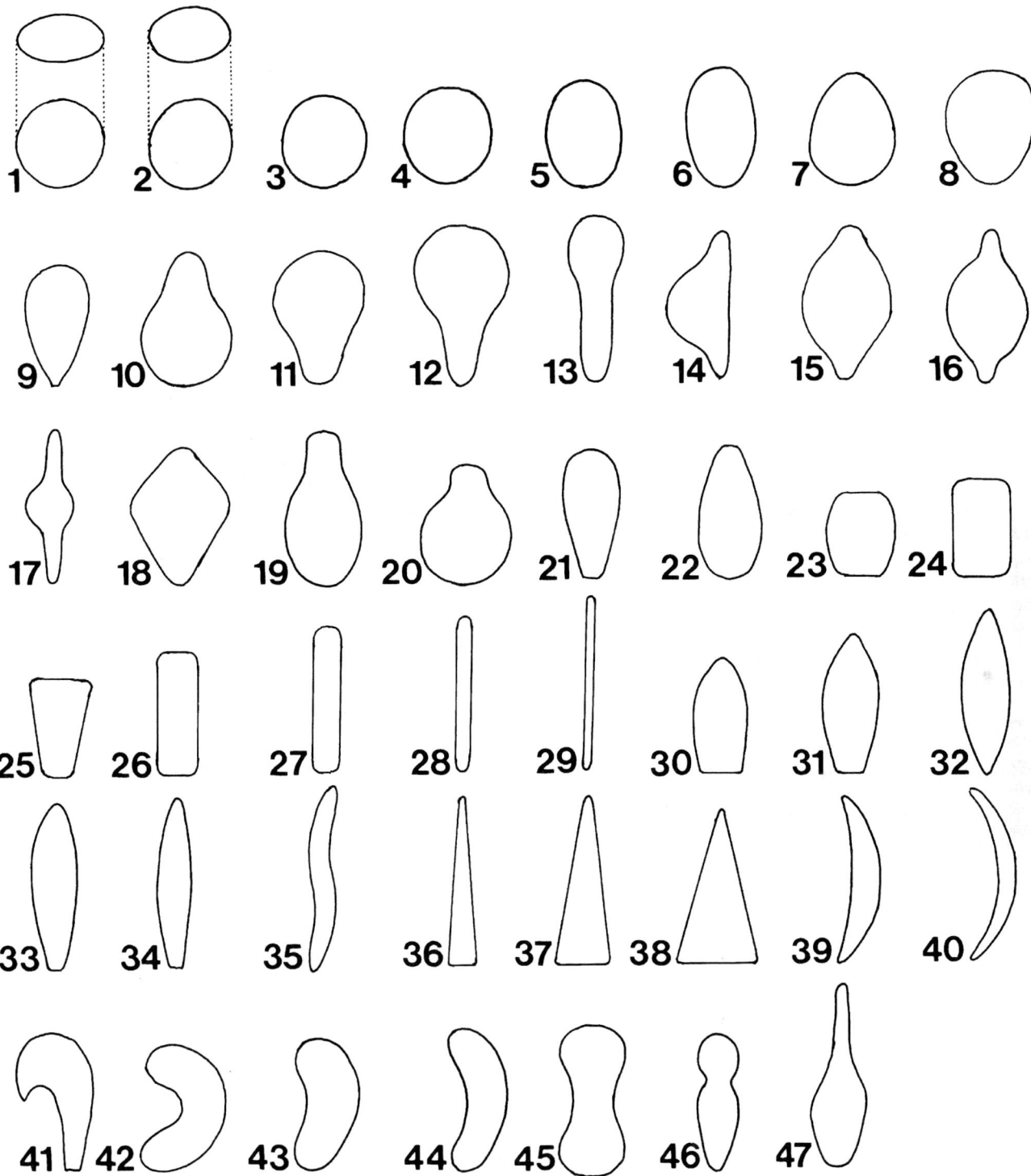

Fig. 63. Diagram of used shape terminology. 1. lenticular, lens-shaped; 2. oblate; 3. spherical; 4. subspherical; 5. broadly ellipsoidal; 6. ellipsoidal; 7. ovoidal, egg-shaped; 8. obovoidal; 9. tear-shaped; 10. obpyriform; 11. pyriform, pear-shaped; 12. tadpole-shaped; 13. spathulate, spoon-shaped; 14. cullulate, hat-shaped; 15. limoniform, lemon-shaped; 16. saturn-shaped; 17. umbonate; 18. rhomboid, kite-shaped; 19. flask-shaped; 20. ampulliform, urn-shaped; 21. clavate, club-shaped; 22. obclavate; 23. doliiform, barrel-shaped; 24. rectangular; 25. cuneiform, axeblade-shaped; 26. cylindrical; 27. bacilliform, narrow cylindrical; 28. linear, line-shaped; 29. thread-shaped; 30. bullet-shaped; 31. navicular, boat-shaped; 32. fusiform, spindle-shaped; 33. cigar-shaped; 34. lanceolate, rocket-shaped; 35. stringbean-shaped; 36. acicular; 37. subulate; 38. conical; 39. lunate, moon-shaped; 40. falcate, sickle-shaped; 41. Uncinate, hook-shaped; 42. cashewnut-shaped; 43. reniform, kidney-shaped; 44. allantoid, sausage-shaped; 45. ossiform, bone-shaped; 46. isthmoid, constricted; 47. rostrate, beaked.

INDEX OF DOUBTFUL NAMES AND UNCONFIRMED CLINICAL CASES

The list below contains names of fungi which have been reported as etiologic agents of disease, but whose identity is as yet uncertain. In addition, a number of names is included that are nomenclaturally correct but of which the case report was unconvincing. Further, some recently introduced taxa are listed which are nomenclaturally and clinically correct but were published to late for inclusion in the main part of the Atlas.

aaseri - *Candida aaseri* Dietrichson - Annls Parasit. Hum. Comp. 29: 476, 1954 ≡ *Candida aaseri* Dietrichson ex van Uden & Buckley, *in* Lodder - The Yeasts, ed. 2, p. 912, 1970.

The type strain of this species orginated from human sputum, but no etiology was proved.

abyssinica - *Bodinia abyssinica* Agostini - Atti Ist. Bot. Lab. Crittogam. Univ. Pavia, Ser. 4, 2: 123, 1931 ≡ *Trichophyton abyssinicum* (Agostini) Nannizzi - Tratt Micopat. Um 4: 174, 1934 ≡ *Favotrichophyton abyssinicum* (Agostini) C.W. Dodge - Med. Mycol. p. 517, 1935.

The type culture, CBS 126.34, intermediate between *T. interdigitale* and *T. tonsurans* (Gräser *et al.*, 2000b).

aceticum - *Oidium aceticum* Ricketts - J. Med. Res. 6: 516, 1901.

Doubtful species, authentic material lost. Possibly *Geotrichum candidum* was concerned (Carmichael, 1957).

acremonium - *Cephalosporium acremonium* Corda - Ic. Fung. 3: 11, 1839.

Doubtful *Acremonium* species. In older medical literature this name was frequently used for *Acremonium kiliense* (Gams, 1971).

actoni - *Achorion actoni* Dey & Marplestone - Ind. J. Med. Res. 23: 697, 1936.

Doubtful dermatophyte species.

acuminatum - *Sporotrichum acuminatum* Lurie - Mycologia 43: 120, 1951.

Possible synonym of *Sporothrix schenckii* (de Hoog, 1974); not intentionally introduced as a new species.

acutulum - *Trichophyton acutulum* da Veiga - Brasil. Med. 43: 837, 1929.

Doubtful dermatophyte species.

acutus - *Aspergillus acutus* Blaser - Sydowia 28: 33, 1975; teleom.: *Eurotium acutum* Blaser - Sydowia 28: 33, 1975.

The type is lost; Samson (1979) supposed *Eurotium herbariorum* was concerned.

africanum (Achorion) - *Achorion africanum* C.W. Dodge - Med. Mycol. p. 560, 1935.

Doubtful dermatophyte species.

africanum (Endodermophyton) - *Endodermophyton africanum* C.W. Dodge - Med. Mycol. p. 490, 1935.

Doubtful dermatophyte species.

alba (Monilia) - *Monilia alba* Castellani & Chalmers - Man. Trop. Med., ed. 3, p. 1089, 1919 ≡ *Castellania alba* (Castellani & Chalmers) C.W. Dodge - Med. Mycol. p. 262, 1935.

Doubtful yeast species.

alba (Scopulariopsis) - *Scopulariopsis alba* Szilvinyi - Zentbl. Bakt. Parasitkde, Abt. 2, 103: 172, 1941.

Doubtful *Scopulariopsis* species (Morton & Smith, 1963).

album (Epidermophyton) - *Epidermophyton album* Kesteren - Med. J. Austr. 1: 423, 1939.

Doubtful dermatophyte species.

album (Trichophyton) - *Trichophyton faviforme* Bodin var. *album* Gregorio - Actas Derm.-Sifil. 1934.

Doubtful dermatophyte species.

album (Trichophyton) - *Trichophyton ferrugineum* (Ota) Talice var. *album* Vanbreuseghem - Annls Parasit. Hum. Comp. 25: 489, 1950.

Dermatophyte species, probably synonymous with *Microsporum ferrugineum*.

album (Trichophyton) - *Trichophyton mentagrophytes* (Robin) Blanchard var. *album* Szathmáry - Mycopath. Mycol. Appl. 42: 53, 1970.

Doubtful dermatophyte species.

albus (Sabouraudites) - See this list under *blanche*.

alessandrina - *Monilia alessandrina* Panayotatou - J. Trop. Med. Hyg. 33: 17, 1930 ≡ *Pseudomonilia alessandrina* (Panayotatou) C.W. Dodge - Med. Mycol. p. 297, 1935.

Doubtful yeast species (Barnett *et al.*, 1990).

algeriense - *Hormodendrum algeriense* Montpellier & Catanei - Annls Derm. Syph., Sér. 6, 8: 634, 1927 ≡ *Cladosporium algeriense* (Montpellier & Catanei) Vuillemin - Champ. Paras. Myc. Homme p. 283, 1931.

Doubtful species close to or identical with *Fonsecaea pedrosoi* (de Hoog, 1977).

algiris - *Discomyces algiris* Blanchard & Nocard - Bull. Acad. Méd., Paris 60: 757, 1896 ≡ *Oospora algiris* (Blanchard & Nocard) Sartory - Champ. Paras. Homme Anim. p. 777, 1923 ≡ *Madurella algiris* (Blanchard & Nocard) C.W. Dodge - Med. Mycol. p. 682, 1935.

Agent of black-grain mycetoma; identity doubtful.

amazonica - *Glenosporopsis amazonica* da Fonseca - Parasitol. Med. 1: 711, 1943.

This species was reindentified by Raper & Fennell (1965) as *Aspergillus penicillioides* Speg.

angioformans - *Botryomyces angioformans* Greco - Orig. Tum. Obs. Myc. Argent. p. 493, 1916.

Doubtful species; no material has been preserved.

anglicum - *Sporotrichum anglicum* Castellani - J. Trop. Med. Hyg. 40: 313, 1937 ≡ *Sporotrichum beurmannii* Matruchot & Ramond var. *anglicum* (Castellani) Redaelli & Ciferri - Tratt. Micopat. Um. 5: 459, 1942.

The species was reidentified as *Hyphopichia burtonii* (Boidin *et al*.) v. Arx & v.d. Walt (Barnett *et al.*, 1990).

angolense - *Favotrichophyton angolense* Froilano de Mello & Paes - Congr. Méd. Trop. Afr. Occ., 1923; Bull. Inst. Pasteur 22: 410, 1924 ≡ *Grubyella angolensis* (Froilano de Mello & Paes) Brumpt - Précis Parasitol., ed. 4, p. 1304, 1927 ≡ *Trichophyton angolensis* (Froilano de Mello & Paes) Nannizzi - Tratt. Micopat. Um. 4: 175, 1934.

Doubtful; possibly *Trichophyton concentricum* or a similar species was concerned.

angustisporum - *Epidermophyton angustisporum* Boedijn - Mycopath. Mycol. Appl. 6: 129, 1951.

Doubtful dermatophyte species.

annamensis - *Rhizopus equinus* Costantin & Lucet var. *annamensis* Bernard - Bull. Trimest. Soc. Mycol. Fr. 30: 230, 1914.

Doubtful *Rhizopus* species, possibly belonging to the *R. microsporus* complex.

annulosum - *Achorion annulosum* Cazalbou - Rev. Pathol. Comp. 14: 329, 1923 ≡ *Bodinia annulosum* (Cazalbou) Ota & Langeron - Annls Parasit. Hum. Comp. 1: 329, 1923 ≡ *Trichophyton annulosum* (Cazalbou) Nannizzi - Tratt. Micopat. Um. 4: 210, 1934.

Doubtful dermatophyte species.

anoma - See this list under *anomus*.

anomaeon - *Microsporum anomaeon* Vidal - C.R. Soc. Biol. 31: 3, 1879.

Possibly bacterial organism was concerned (Vuillemin, 1895).

anomus - *Hyalopus anomus* Boucher - Bull. Soc. Path. Exot. 11: 334, 1918 (as '*anoma*').

Material lost; possibly a *Fusarium* species was concerned (Gams, 1971).

antacustica - *Sterigmatocystis antacustica* Cramer - Vjschr. Naturf. Ges. Zürich 4: 325, 1859.

The original strain from human ear probably concerned *Aspergillus niger* (Raper & Fennell, 1965).

anthophilum - *Fusisporium anthophilum* A. Braun, *in* Rabenhorst - Fung. Europ. No. 1964, 1875 ≡ *Fusarium anthophilum* (A. Braun) Wollenweber - Fus. Autogr. Del. No. 176, 1916.

Recognized *Fusarium* species. Okuda *et al.* (1987) reported an extended cutaneous infection in a leukemic patient.

apiculata - See this list under *uvarum*.

apii - *Cercospora apii* Fresenius - Beitr. Mykol. 3: 91, 1863.

The strain from a clinical case published under this name was reidentified as *Mycocentrospora acerina* (p. 771; Chupp, 1957; Deighton & Mulder, 1971).

arloini - *Achorion arloini* Busquet - Ann. Microgr. 3: 9, 1891 ≡ *Aleurima arloini* (Busquet) Vuillemin - Champ. Paras. Myc. Homme p. 114, 1931.

Doubtful dermatophyte species.

arrhizus - *Rhizopus arrhizus* Fischer, *in* Rabenhorst - Krypt. Fl. 1: 233, 1892 ≡ *Mucor arrhizus* (Fischer) Hagem - Unters. Norw. Mucorin. p. 37, 1907/08.

Doubtful, possibly identical to *Rhizopus oryzae* (Schipper, 1984).

artagaveytiae - *Debaryomyces artagaveytiae* Batista, J. Silveira & Coêlho - Mycopath. Mycol. Appl. 14: 22, 1961.

Doubtful yeast species (Barnett *et al.*, 1990).

articulata - *Sporormia articulata* Viégas - Bragantia 3: 155, 1943.

The species was described from a human skin lesion (Viégas, 1943). A *Sporormiella* is concerned, but the epithet is not listed in the extensive reviews of *Sporormia* and *Sporormiella* by Ahmed & Cain (1972).

arxii - *Calyptrozyma arxii* Boekhout & Spaay, *in* Boekhout, Roeijmans & Spaay - Mycol. Res.. 99: 1244, 1995.

Discomycete-like fungus described from the human lower oesophagus but without a convincing case report.

asperellum - *Trichoderma asperellum* Samuels, Lieckfeldt & Nirenberg - Sydowia 41: 81, 1999.

It has been suggested to be the main clinical *Trichoderma* species, but as yet no case report has been published.

asteroides - *Trichophyton asteroides* Sabouraud - Malad. Cuir Chev. 3: 347, 1910 ≡ *Trichophyton mentagrophytes* (Robin) Blanchard var. *asteroides* (Sabouraud) Neveu-Lemaire - Précis Parasitol. Anim. Domest. p. 69, 1912 ≡ *Sabouraudites asteroides* (Sabouraud) Ota & Langeron - Annls Parasit. Hum. Comp. 1: 327, 1923 ≡ *Spiralia asteroides* (Sabouraud) Grigorakis - Annls Sci. Nat., Bot., Sér. 10, 7: 409, 1925 ≡ *Microsporum asteroides* (Sabouraud) Guiart & Grigorakis - Lyon Méd. 141: 377, 1928 ≡ *Trichophyton gypseum* Bodin var. *asteroides* (Sabouraud) Frágner - Česká Mykol. 10: 106, 1956.

Doubtful dermatophyte species.

atra - *Stachybotrys atra* Corda - Ic. Fung. 1: 21, 1837.

A toxigenic fungus. It was isolated from the lungs of a child with pulmonary hemisiderosis (Elidemir *et al.*, 1999).

atramentum - Misspelling for *atramentarium*; see under *Colletotrichum coccodes* (p. 315).

atro-olivaceum - *Spondylocladium atro-olivaceum* Neves - Mem. Inst. Oswaldo Cruz 25: 323, 1931.

Doubtful dematiaceous fungus; possibly a *Curvularia* species.

aurantiacum - *Microsporum aurantiacum* Conant - Archs Derm. Syph. 36: 792, 1937.

Doubtful dermatophyte species.

aurea (Sterigmatocystis) - *Sterigmatocystis aurea* Greco - Orig. Tum. Obs. Myc. Argent. p. 685, 1916 ≡ *Aspergillus greconis* C.W. Dodge - Med. Mycol. p. 624, 1935 (name change).

Doubtful *Aspergillus* species.

aureus (Scopulariopsis) - *Scopulariopsis aureus* Sartory - Champ. Paras. Homme Anim. 9: 650, 1922.

Doubtful *Scopulariopsis* species (Morton & Smith, 1963).

avaricum - *Trichophyton avaricum* Szathmáry - Mykosen 13: 551, 1970 ≡ *Arthroderma avaricum* Szathmáry - Česká Mykol. 31: 98, 1977.

Doubtful geophilic dermatophyte.

avium - *Favotrichophyton avium* Neveu-Lemaire - Précis Parasitol. Hum., ed. 5: 55, 1921 ≡ *Grubyella avium* (Neveu-Lemaire) Brumpt - Précis Parasitol., ed. 4, p. 1304, 1927 ≡ *Trichophyton avium* (Neveu-Lemaire) Nannizzi - Tratt. Micopat. Um. 4: 176, 1934.

Doubtful dermatophyte species from bird.

avramii - *Plenedomus avramii* Borelli - Acta Symp. Intern. Micetomas, Venezuela 1978.

Doubtful *Phoma*-like coelomycete, reported from black-grain mycetoma (Borelli, 1978).

azymatica - *Mycotorula azymatica* Serrano - Z. Parasitkde 12: 18, 1940.

Doubtful yeast species.

balcaneum - *Trichophyton balcaneum* Castellani - J. Trop. Med. Hyg. 22: 174, 1919 ≡ *Bodinia balcanea* (Castellani) Ota & Langeron - Annls Parasit. Hum. Comp. 1: 329, 1923 ≡ *Favotrichophyton balcaneum* (Castellani) C.W. Dodge - Med. Mycol. p. 518, 1935.

The type strain, CBS 359.62, is a dermatophyte which could as yet not be identified with certainly.

balzeri - *Parendomyces balzeri* Gougerot & Burnier, *in* Balzer, Gougerot & Burnier - Annls Derm. Syph., Sér. 5, 3: 286, 1912 ≡ *Saccharomyces balzeri* (Gougerot & Burnier) Castellani & Chalmers - Man. Trop. Med., ed. 3, p. 982, 1919 ≡ *Monilia balzeri* (Gougerot & Burnier) Brumpt - Précis Parasitol., ed. 34, 1922 ≡ *Trichosporon balzeri* (Gougerot & Burnier) Ota - Annls Parasit. Hum. Comp. 4: 12, 1926 ≡ *Trichosporium balzeri* (Gougerot & Burnier) Bolognesi & Chiurco - Micosi Chirurg. p. 597, 1927 ≡ *Geotrichoides balzeri* (Gougerot & Burnier) Langeron & Talice - Annls Parasit. Hum. Comp. 10: 68, 1932 ≡ *Candida balzeri* (Gougerot & Burnier) de Almeida - Anais Fac. Med. Univ. S. Paulo 9: 77, 1933 ≡ *Proteomyces balzeri* (Gougerot & Burnier) C.W. Dodge - Med. Mycol. p. 212, 1935.

This species was listed by Barnett *et al.* (1990) as a synonym of *Trichosporon cutaneum*, but as a doubtful species by Guého *et al.* (in Kutzman & Fell, 1998).

batrachea - *Monilia batrachea* H.H. Scott - Proc. Zool. Soc., Lond. 45: 692, 1926.

Doubtful yeast species.

beauveriei - *Saccharomyces beauveriei* Froilano de Mello, *in* Froilano de Mello & Fernandes - Arq. Hig. Pat. Exot. 6: 246, 1918.

Doubtful yeast species (Barnett *et al.*, 1990).

beemeri - *Trichosporon beemeri* Kuttin & J. Müller - Antonie van Leeuwenhoek 47: 257, 1981.

See this list under *jirovecii*; synonym of that species (Guého *et al.*, 1992).

behrendii - *Trichosporon behrendii* Lodder & Kreger-van Rij - The Yeasts p. 673, 1952.

The type strain, CBS 2552, has been reidentified as *Hyphopichia burtonii* (Barnett *et al.*, 1990). The type strain was described from human sputum, without a proven case.

beigeliana - See this list under *beigelii*.

beigelii - *Pleurococcus beigelii* Küchenmeister & Rabenhorst, *in* Rabenhorst - Hedwigia 6: 49, 1867 ≡ *Sclerotium beigelianum* Hallier - Parasitol. Unters. p. 75, 1868 (name change) ≡ *Zoogloea beigelina* (Hallier) Ebert - Zentbl. Med. Wiss. 11: 307, 1873 ≡ *Chlamydatomus beigelii* (Küchenmeister & Rabenhorst) Trevisan - Rc. Ist. Lomb. Sci. Lett., Ser. 2, 12: 146, 1879 ≡ *Hyalococcus beigelii* (Küchenmeister & Rabenhorst) Schroeter - Kryptfl. Schlesien 3: 152, 1886 ≡ *Micrococcus beigelii* (Küchenmeister & Rabenhorst) Migula - Syst. Bakt. 1: 193, 1900 ≡ *Trichosporon beigelii* (Küchenmeister & Rabenhorst) Vuillemin - Archs Parasit. 5: 59, 1902 ≡ *Sporotrichum beigelii* (Küchenmeister & Rabenhorst) Saccardo & Traverso - Syll. Fung. 20: 871, 1911 ≡ *Geotrichum beigelii* (Küchenmeister & Rabenhorst) Coudert - Guide Prat. Mycol. Méd. p. 233, 1955.

Guého *et al.* (1992a) listed it as a doubtful *Trichosporon* species.

beijingensis - *Aspergillus beijingensis* D.-M. Li, Horie, Y.-X. Wang & R.-Y. Li - Mycoscience 39: 299, 1998.

Described from human maxillary sinusitis but without case report.

bernasconii - *Cryptococcus bernasconii* Fontoynont & Boucher - Annls Derm. Syph., Sér. 6, 4: 318, 1923 ≡ *Torulopsis bernasconii* (Fontoynont & Boucher) de Almeida - Anais Fac. Med. Univ. S. Paulo 9: 10, 1933 ≡ *Eutorula bernasconii* (Fontoynont & Boucher) C.W. Dodge - Med. Mycol. p. 355, 1935.

Doubtful yeast species (Barnett *et al.*, 1990).

bertaccinii - *Scopulariopsis bertaccinii* Redaelli, *in* Bertaccini - Giorn. Ital. Derm. Syph. 75: 783-828, 1934.

Doubtful species (Morton & Smith, 1963).

bertai - *Penicillium bertai* Talice & McKinnon - Annls Parasit. Hum. Comp. 7: 97, 1929.

Pitt (1979) supposed the species to be *Penicillium citreonigrum* Dierckx, but no authentic material is known to be preserved. *P. bertai* was described from a pulmonary mycosis (Talice & McKinnon, 1929), but *P. citreonigrum* is not known from humans or animals.

bettencourti - *Octomyces bettencourti* Froilano de Mello & Fernandes - Arq. Hig. Pat. Exot. 6: 237, 1918.

Doubtful yeast species (Barnett *et al.*, 1990).

bicolor (Penicillium) - *Penicillium bicolor* Fries - Syst. Mycol. 3: 408, 1829.

Doubtful *Penicillium* species (Pitt, 1979).

bicolor (Trichophyton) - *Trichophyton bicolor* da Veiga - Brasil. Med. 43: 830, 1929.

Doubtful dermatophyte species.

biseptatum - *Helminthosporium biseptatum* Saccardo & Roumeguère - Revue Mycol. 3: 56, 1881 ≡ *Drechslera biseptata* (Saccardo & Roumeguère) Richardson & Fraser - Trans. Br. Mycol. Soc. 51: 148, 1968.

This species, now known as *D. biseptata*, was reported from a mixed rhinocerebral infection (Washburn *et al.*, 1988) with *Curvularia lunata*; the etiologic role of *D. biseptata* was insufficiently proved.

bisporus - *Saccharomyces bisporus* Mattlet - Annls Soc. Belge Méd. Trop. 6: 32, 1926 ≡ *Hansenula bispora* (Mattlet) Nannizzi - Tratt. Micopat. Um. 4: 134, 1934.

Barnett *et al.* (1990) listed *H. bispora* as a synonym of *Pichia anomala* but *S. bispora* as a doubtful species. No type material is known to be preserved.

blanchardii - See this list under *sabouraudii*.

blanche - *Sabouraudites ruber* Ota & Langeron var. *blanche* Hashimoto, Irizawa & Ota - Jpn. J. Derm. Urol. 30: 1, 1930 ≡ *Sabouraudites ruber* (Castellani) Ota & Langeron var. *albus* Nannizzi - Tratt. Micopat. Um. 4: 168, 1934 (name change) ≡ *Ectotrichophyton otae* C.W. Dodge - Med. Mycol. p. 500, 1935 (name change).

Doubtful dermatophyte species.

blanci - *Candida blanci* Blanc & Catanei - Archs Inst. Pasteur Algér. 22: 161, 1944.

Doubtful yeast species.

blankii - *Candida blankii* Buckley & van Uden - Mycopath. Mycol. Appl. 36: 259, 1968.

The species was described from a systemic infection in a mink, but no case report was provided.

bodinii - *Montoyella bodinii* Castellani - J. Trop. Med. Hyg. 10: 65, 1907. Doubtful species.

bogolepoffii - *Tilachlidium bogolepoffii* Vuillemin - Bull. Trimest. Soc. Mycol. Fr. 28: 119, 1912 - *Mycoderma bogolepoffii* (Vuillemin) Jannin - Les Mycod., Nancy p. 102, 1913 ≡ *Geotrichum bogolepoffii* (Vuillemin) Basgal - Contrib. Estudo Blast. Pulmon. p. 48, 1931 ≡ *Cephalosporium bogolepoffii* (Vuillemin) C.W. Dodge - Med. Mycol. p. 830, 1935 ≡ *Hyalopus bogolepoffii* (Vuillemin) Barbosa - Subs. Estud. Hyalopus, Recife, 1941.

The authentic description refers to *Acremonium* aff. *kiliense* (Gams, 1971), but other authors had arthroconidial specimens (de Hoog *et al.*, 1986).

bonaërensis - *Endomyces bonaërensis* Greco - Tum. Obs. Myc. Argent. p. 123, 1916 ≡ *Zymonema bonaërense* (Greco) C.W. Dodge - Med. Mycol. p. 175, 1935.

Material lost; doubtful yeast species.

bonordenii - See this list under *candida*.

bossae - *Ctenomyces bossae* Milochevitch - Annls Parasit. Hum. Comp. 13: 559, 1935.

Doubtful dermatophyte species.

bostonense - *Geotrichum bostonense* C.W. Dodge - Med. Mycol. p. 223, 1935.

Doubtful species, possibly *Aureobasidium* or *Hormonema* (de Hoog *et al.*, 1986).

botrytis - *Monilia botrytis* Vuillemin - Champ. Paras. Myc. Homme p. 87, 1931.

Doubtful, yeast-like species.

bouffardii - *Aspergillus bouffardii* Brumpt - Archs Parasit. 10: 525, 1906 ≡ *Madurella bouffardii* (Brumpt) C.W. Dodge - Med. Mycol. p. 685, 1935.

Doubtful species; Vuillemin (1931) supposed that a mixed infection was concerned.

bovoi - *Madurella bovoi* Brumpt - Précis Parasitol., ed. 2, p. 944, 1913.

Agent of black-grain mycetoma which has not been cultured; identity doubtful.

brachytomum - *Microsporum brachytomum* Oho - Kyoto Med. J. 16: 20, 1919.

Neither material nor literature was available for study.

brasiliense (Oidium) - *Oidium brasiliense* de Magalhães - Mems Inst. Oswaldo Cruz 10: 20, 1914 ≡ *Mycoderma brasiliense* (de Magalhães) Neveu-Lemaire - Précis Parasitol. Hum. p. 69, 1921 ≡ *Monilia brasiliensis* (de Magalhães) Brumpt - Précis Parasitol., ed. 3, p. 1100, 1922 ≡ *Myceloblastanon brasiliense* (de Magalhães) Ota - Jap. J. Derm. Urol. 28: 171, 1928 ≡ *Candida brasiliensis* (de Magalhães) Basgal - Contrib. Estudo Blast. Pulmon. p. 50, 1931 ≡ *Trichosporon brasiliense* (de Magalhães) Puntoni - Man. Microbiol., ed. 2, 1935 ≡ *Geotrichum brasiliense* (de Magalhães) Brumpt - Précis Parasitol., ed. 6, p. 1781, 1949.

Doubtful species, possibly *Trichosporon* (de Hoog *et al.*, 1986).

brasiliense (Trichophyton) - See this list under *griseum (Trichophyton)*.

brassicicola - *Helminthosporium brassicicola* Schweinitz - Trans. Am. Phil. Soc., N. Ser., 4: 279, 1832 ≡ *Alternaria brassicicola* (Schweinitz) Wiltshire - Mycol. Pap. 20: 8, 1947.

Alternaria brassicicola, mentioned by de Bièvre (1990) from unconfirmed cases of human alternariosis, is an obligate plant pathogen on cabbage.

brocianum - *Mycoderma brocianum* Pinoy, *in* Anderson, Colas-Belcour & Brocq - Archs Inst. Pasteur Afr. N. 19: 317, 1930 ≡ *Proteomyces brocianus* (Pinoy) C.W. Dodge - Med. Mycol. p. 209, 1935 ≡ *Geotrichum brocianum* (Pinoy) Brumpt - Précis Parasitol., ed. 6, p. 1742, 1949.

Doubtful, possibly a *Lecythophora* species (de Hoog *et al.*, 1986).

brocquii - *Parendomyces brocquii* Beintema - Pseudom. Brocq, Pautrier et Fernet p. 171, 1930; Annls Derm. Syph., Sér. 7, 4: 421, 1933 ≡ *Geotrichum brocquii* (Beintema) Castellani & Jacono - J. Trop. Med. Hyg. 36: 317, 1933 ≡ *Trichosporon brocquii* (Beintema) Nannizzi -

Tratt. Micopat. Um. 4: 288, 1934 ≡ *Proteomyces brocquii* (Beintema) C.W. Dodge - Med. Mycol. p. 210, 1935.

Doubtful, possibly a *Lecythophora* species (de Hoog *et al.*, 1986).

brumptii (Achorion) - *Achorion brumptii* Langeron & Baeza - Annls Parasit. Hum. Comp. 14: 399, 1936 ≡ *Trichophyton brumptii* (Langeron & Baeza) Asani & Faghih - Annls Parasit. Hum. Comp. 26: 250, 1951.

Doubtful dermatophyte species.

brumptii (Indiella) - *Indiella brumptii* Silva Piraja - Sobre Madur. Graõs Branc. Prod. Ind., Thesis, Bahia, 1922 ≡ *Madurella brumptii* (Silva Piraja) Ciferri & Redaelli - Mycopathologia 3: 197, 1941.

Doubtful species, from mycetoma; it was not cultured.

brunaudii - *Sporotrichum brunaudii* Nannizzi - Tratt. Micopat. Um. 4: 436, 1934.

Doubtful species (Stalpers, 1984).

brunneum - *Sporotrichum brunneum* Schenk - Verh. Phys.-Med. Ges. Würzb. 1: 73, 1850 ≡ *Trichosporum brunneum* (Schenk) Saccardo - Syll. Fung. 4: 294, 1886.

Doubtful species (Stalpers, 1984).

bullosum - *Trichophyton bullosum* Lebasque - Champ. Teign. Cheval Bovidés p. 53, 1933.

Doubtful dermatophyte species.

burnieri - *Debaryomyces burnieri* Ota - Annls Parasit. Hum. Comp. 2: 35, 1924 ≡ *Cryptococcus burnieri* (Ota) Nannizzi - Tratt. Micopat. Um. 4: 305, 1934.

Doubtful yeast species (Barnett *et al.*, 1990).

caballinum - *Favotrichophyton caballinum* Neveu-Lemaire - Précis Parasitol. Hum., ed. 5, p. 55, 1921 ≡ *Grubyella caballina* (Neveu-Lemaire) Brumpt - Précis Parasitol., ed. 4, p. 1304, 1927 ≡ *Trichophyton caballinum* (Neveu-Lemaire) Nannizzi - Tratt. Micopat. Um. 4: 177, 1934.

Doubtful dermatophyte species, probably *Trichophyton verrucosum*.

cactivorum - *Helminthosporium cactivorum* Petrak - Gartenbau-Wiss. 5: 226, 1931 ≡ *Drechslera cactivora* (Petrak) M.B. Ellis - Demat. Hyphom. p. 432, 1970.

Strains from onychomycoses published by Negroni (1985) were reidentified by McGinnis *et al.* (1986) as *Exserohilum rostratum*.

calistegicum - *Trichophyton mentagrophytes* (Robin) Blanchard var. *calistegicum* Szathmáry - Mycopath. Mycol. Appl. 42: 53, 1970.

Doubtful dermatophyte species.

camerounensis - *Grubyella camerounensis* Ota & Gaillard - Annls Parasit. Hum. Comp. 4: 14, 1926 ≡ *Favotrichophyton camerounense* (Ota & Gaillard) C.W. Dodge - Med. Mycol. p. 515, 1935 ≡ *Trichophyton camerounense* (Ota & Gaillard) Nannizzi - Tratt. Micopat. Um. 4: 210, 1934.

Doubtful dermatophyte species.

candida (Monilia) - *Monilia candida* Bonorden - Handb. Allgem. Mykol. p. 76, 1851 ≡ *Monilia bonordenii* Vuillemin - Bull. Trimest. Soc. Mycol. Fr. 27: 140, 1911 (name change) ≡ *Endomyces candidus* (Bonorden) Castellani - Lancet 1912: 13, 1912 ≡ *Myceloblastanon candidus* (Bonorden) Ota - Jpn. J. Derm. Urol. 27: 169, 1927 ≡ *Candida bonordenii* (Vuillemin) Basgal - Contrib. Estud. Blastom. Pulmon. p. 49, 1931.

Barnett *et al.* (1990) listed this species as a synonym of *Candida tropicalis*, but since Bonorden's material has not been preserved it is regarded here as of doubtful identity.

candida (Thailandia) - *Thailandia candida* Vardhanabuthi - Sydowia 13: 102, 1959.

Onygenalean ascomycete; material was not available.

candida - *Torula candida* Saito - Jpn. J. Bot. 1: 1-54, 1922 ≡ *Torulopsis candida* (Saito) Lodder - Anaskosp. Hefen 2: 163, 1934 ≡ *Candida saitoana* Nakase & Suzuki - J. Gen. Appl. Microbiol. 31: 85, 1985 (name change)

The species is presently known as *Candida saitoana*, a soil-inhabiting saprophyte. The agent of a cathether-related fungemia (St.-Germain & Laverdière, 1986) has been reidentified as *C. palmioleophila;* see under that species name.

candidus (Endomyces) - *Endomyces candidus* Castellani - Lancet 1912:

15, 1912.

Doubtful species (de Hoog *et al.*, 1986).

canina - *Oospora canina* Costantin & Sabrazès - C.R. Séanc. Soc. Biol., Paris 45: 511, 1893 ≡ *Achorion caninum* (Costantin & Sabrazès) Brumpt - Précis Parasitol., ed. 2, p. 875, 1913 ≡ *Bodinia canina* (Costantin & Sabrazès) Brumpt - Précis Parasitol., ed. 4, p. 1297, 1927 ≡ *Trichophyton caninum* (Costantin & Sabrazès) Nannizzi - Tratt. Micopat. Um. 4: 210, 1934.

Doubtful dermatophyte species.

caninum - *Trichophyton caninum* Matruchot & Dassonville - Bull. Soc. Méd. Vét. Deps Cent., N. Sér., 20: 54, 1902 ≡ *Sabouraudites caninus* (Matruchot & Dassonville) Ota & Langeron - Annls Parasit. Hum. Comp. 1: 327, 1923 ≡ *Megatrichophyton caninum* (Matruchot & Dassonville) C.W. Dodge - Med. Mycol. p. 509, 1935.

Doubtful dermatophyte species.

caoi - *Monilia caoi* Verdun - Précis Parasitol. Hum., ed. 2, 1913 ≡ *Mycoderma caoi* (Verdun) Basgal - Contrib. Estud. Blastom. Pulmon. p. 49, 1931.

Doubtful species, possibly *Geotrichum capitatum* (de Hoog *et al.*, 1986).

carateum - *Trichophyton carateum* Brumpt - Précis Parasitol., ed. 2, p. 847, 1913 ≡ *Mycoderma carateum* (Brumpt) Vuillemin - Champ. Parasit. Myc. Homme p. 112, 1931.

Doubtful dermatophyte species.

carlsbergensis - *Saccharomyces carlsbergensis* Hansen - C.R. Trav. Lab. Carlsberg 7: 184, 1908.

This is a synonym of the non-pathogenic species *S. pastorianus*; see under that species.

carougeaui - *Sporotrichum carougeaui* Langeron in Brumpt - Précis Parasitol., ed. 2, p. 921, 1913 ≡ *Rhinocladium carougeaui* (Langeron) Neveu-Lemaire - Précis Parasitol. Hum. p. 85, 1921 ≡ *Sporotrichum beurmannii* Matruchot & Ramond var. *carougeaui* (Langeron) Redaelli & Ciferri - Tratt. Micopat. Um. 5: 458, 1942.

This is supposed to be a synonym of *Hyphopichia burtonii* (Stalpers, 1984).

castanea - *Grubyella castanea* Taniguchi - Jpn. J. Derm. Urol. 25: 38, 1925 ≡ *Trichophyton schoenleinii* (Lebert) Nannizzi var. *castaneum* (Taniguchi) Nannizzi - Tratt. Micopat. Um. 4: 177, 1934.

Doubtful dermatophyte species.

castellanii (Acladium) - *Acladium castellanii* Pinoy, *in* Castellani - Br. Med. J. 2910: 486, 1916 - *Pseudomicrosporon castellanii* (Pinoy) Craik - J. Trop. Med. Hyg. 26: 186, 1916 ≡ *Aleurisma castellanii* (Pinoy) Vuillemin - Champ. Paras. Myc. Homme p. 115, 1931 ≡ *Sporotrichum schenckii* (Hektoen & Perkins) de Beurmann & Gougerot var. *castellanii* (Pinoy) E.J. Butler - Parasitology 29: 259, 1937 ≡ *Sporotrichum beurmannii* (Pinoy) Redaelli & Ciferri - Tratt. Micopat. Umana 5: 481, 1942 ≡ *Raffaelea castellanii* (Pinoy) de Hoog - Stud. Mycol. 7: 44, 1974.

The identity of this species (type strain: CBS 100.26) has not yet been established (Rainer *et al.*, 2000).

castellanii (Scopulariopsis) - *Scopulariopsis castellanii* Ota & Komaya - Derm. Wschr. 78: 163, 1924.

Doubtful species (Morton & Smith, 1963).

castellanii (Trichophyton) - *Trichophyton castellanii* Brooke - Aids Trop. Med. p. 163, 1908.

Doubtful dermatophyte species (Dodge, 1935).

catanei - *Atelosaccharomyces catanei* C.W. Dodge - Med. Mycol. p. 343, 1935.

Doubtful yeast species (Barnett *et al.*, 1990).

catarrhalis - *Oospora catarrhalis* Sartory & Bailly - Contrib. Myc. Pulmon. 4: 57, 1921 ≡ *Actinomyces catarrhalis* (Sartory & Bailly) Brumpt - Précis Parasitol., ed. 4, p. 1195, 1927.

Doubtful Actinomyces species.

catenata - *Oospora catenata* Schaede - Derm. Wschr. 98: 523, 1934.

No material was preserved; it may have been *Geotrichum candidum* (de Hoog *et al.*, 1986).

cavarae - *Cryptococcus cavarae* Pollacci & Turconi, *in* Bendici & Federici - Atti R. Accad. Fisiocrit. Siena, Ser. 10, 3: 746-766, 1928 ≡ *Syringospora cavarae* (Pollacci & Turconi) C.W. Dodge - Med. Mycol.

p. 280, 1935.

Doubtful yeast species (Barnett *et al.*, 1990).

caviae - *Trichophyton mentagrophytes* (Robin) Blanchard var. *caviae* Takashio - Bull. Soc. Fr. Mycol. Méd. 4: 47, 1975.

Invalid name introduced by Takashio (1975) for the americo-european race of *T. erinacei*.

cavicola - *Cryptococcus cavicola* Artault - Archs Parasit. 1: 259-265, 1898 ≡ *Torula cavicola* (Artault) de Almeida - Anais Fac. Med. Univ. S. Paulo 9: 76, 1933.

Doubtful red yeast species (Barnett *et al.*, 1990).

ceratophagus - *Achorion ceratophagus* Ercolani - Mem. R. Accad. Sci. Ist. Bologna 3: 363-381, 1875 ≡ *Oospora porriginis* (Montagne) Saccardo var. *ceratophagus* (Ercolani) Saccardo - Syll. Fung. 4: 15, 1886.

Doubtful dermatophyte species.

cerebriforme (Epidermophyton) - *Epidermophyton cerebriforme* C.W. Dodge - Med. Mycol. p. 479, 1935.

Doubtful dermatophyte species.

cerebriforme (Sporotrichum) - *Sporotrichum cerebriforme* de Vries & Kleine-Natrop - Mycopath. Mycol. Appl. 8: 159, 1957 ≡ *Streptomyces cerebriformis* (de Vries & Kleine-Natrop) G. Müller - Wiss. Z. Humboldt-Univ., Berlin 13: 637, 1964 ≡ *Trichosporiella cerebriformis* (de Vries & Kleine-Natrop) W. Gams, *in* von Arx - Persoonia 6: 184, 1971.

The identity of the type strain, CBS 344.53, is uncertain.

cerebriformis (Oospora) - *Oospora cerebriformis* Kambayashi - Jpn. J. Derm. Urol. 23: 483-511, 1922 ≡ *Trichosporon cerebriforme* (Kambayashi) Ota - Jpn. J. Derm. Urol. 28: 202, 1928.

Doubtful *Trichosporon* species (Guého *et al.*, 1992a).

cerebriformis (Paraccocccidioides) - *Paracoccidioides cerebriformis* M. Moore - Arch. Urug. Med. Cir. Espec. 8: 224, 1936.

Rippon (1988) listed the species as a synonym of *P. brasiliensis*, but Lacaz *et al.* (1997) claimed that a different fungus was concerned.

ceroton - *Trichophyton ceroton* Cazalbou - Rev. Gén. Méd. Vét. 24: 2, 1914 ≡ *Favotrichophyton ceroton* (Cazalbou) C.W. Dodge - Med. Mycol. p. 519, 1935.

Doubtful dermatophyte species.

ceylonense - *Trichophyton ceylonense* Castellani - J. Trop. Med. Hyg. 11: 267, 1908.

Doubtful dermatophyte species; material probably lost.

chalmersii - *Endomyces chalmersii* Castellani - Archs Parasit. 16: 187, 1913 ≡ *Monilia chalmersii* (Castellani) Castellani & Chalmers - Man. Trop. Med., ed. 2, p. 831, 1913 ≡ *Myceloblastanon chalmersii* (Castellani) Ota - Jpn. J. Derm. Urol. 27: 176, 1927 ≡ *Candida chalmersii* (Castellani) Basgal - Contrib. Estud. Blastom. Pulmon. p. 49, 1931 ≡ *Castellania chalmersii* (Castellani) C.W. Dodge - Med. Mycol. p. 257, 1935 ≡ *Candida lodderi* Langeron & Guerra - Annls Parasit. Hum. Comp. 16: 472, 1938 (name change) ≡ *Mycotorula chalmersii* (Castellani) Redaelli & Ciferri - Atti. Ist. Bot. Univ. Lab. Crittogam. Pavia 5: 47, 1943.

De Hoog *et al.* (1986) listed the species as a possible synonym of *Candida parapsilosis*, depending on the authenticity of the strain CBS 2193. Barnett *et al.* (1990) regarded this as doubtful.

chartarum - *Masonia chartarum* G. Smith - Trans. Br. Mycol. Soc. 35: 150, 1952 ≡ *Scopulariopsis chartarum* (G. Smith) Morton & G. Smith - Mycol. Pap. 63: 64, 1963.

This fungus, with the currently accepted name *Scopulariopsis chartarum*, was reported by Wesh & Ely (1999) from a disseminated infection in a dog.

chibaensis - *Sabouraudites asteroides* (Sabouraud) Ota & Langeron var. *chibaensis* Ogata - Jpn. J. Derm. Urol. 29: 1200, 1929 ≡ *Ectotrichophyton mentagrophytes* (Robin) Castellani & Chalmers var. *chibaense* - (Ogata) C.W. Dodge - Med. Mycol. p. 506, 1935.

Doubtful dermatophyte species.

chlamydeum - *Cladosporium chlamydeum* Ciferri & Redaelli - Mycopathologia 8: 184, 1957.

Material probably lost; judging from the description this was *Moniliella suaveolens*.

chosenica - *Grubyella chosenica* Takahashi - Jpn. J. Derm. Urol. 25: 284, 1925 ≡ *Trichophyton chosenica* (Takahashi) Takahashi - Jpn. J. Derm. Urol. 25: 285, 1925.

Doubtful dermatophyte species; no material has been preserved.

ciferrii - *Saccharomyces ciferrii* Froilano de Mello & de Faria - Mycopathologia 1: 207, 1938.

Doubtful yeast species (Barnett *et al.*, 1990).

cineraceum - *Trichophyton cineraceum* da Veiga - Brasil Med. 43: 837, 1929.

Doubtful dermatophyte species.

cinereum - *Trichophyton cinereum* Cazalbou - Rev. Gén. Méd. Vét. 24: 5, 1914 ≡ *Favotrichophyton cinereum* (Cazalbou) C.W. Dodge - Med. Mycol. p. 519, 1935.

Doubtful dermatophyte species close to *Trichophyton verrucosum*.

cinnabarina - *Sphaeria cinnabarina* Tode - Fungi Mecklenb. 2: 9, 1791 ≡ *Sphaeria cinnabarina* Tode:Fries - Syst. Mycol. 2: 412, 1823 ≡ *Nectria cinnabarina* (Tode:Fries) Fries - Summa Veg. Scand. 2: 388, 1849.

Nectria cinnabarina is a very common fungus on bark. An ophthalmic case was published by Affeldt *et al.* (1987) under its anamorph name, *Tubercularia vulgaris*, but the identity of this strain is unconfirmed.

cinnabarinum - *Acremonium cinnabarinum* A. & R. Sartory - C.R. Hebd. Séanc. Acad. Sci., Paris 216: 391, 1943.

Acremonium species, probably *A. kiliense* (Gams, 1971).

cinnamomina - *Sterigmatocystis cinnamomina* Weiss - Annls Parasit. Hum. Comp. 8: 183, 1930 ≡ *Aspergillus cinnamominus* (Weiss) C.W. Dodge - Med. Mycol. p. 627, 1935.

This was an *Aspergillus* species close or identical to *A. terreus* (Raper & Fennell, 1965).

circonvolutum - *Trichophyton circonvolutum* Sabouraud - Malad. Cuir Chev. 3: 320, 1010.

Doubtful dermatophyte species (Gräser *et al.*, 2000b).

circuluscentricus - See this list under *circuluscentum*.

circuluscentum - *Microsporum circuluscentum* de Magalhães - Mem. Inst. Oswaldo Cruz 21: 178, 1928 ≡ *Sabouraudites circuluscentricus* Nannizzi - Tratt. Micopat. Um. 4: 159, 1934 (name change) ≡ *Ectotrichophyton circuluscentricum* (Nannizzi) C.W. Dodge - Med. Mycol. p. 503, 1935.

Doubtful dermatophyte species.

citricarpa - *Phoma citricarpa* McAlpine - Dis. Citrus Austral. p. 21, 1899 ≡ *Phyllosticta citricarpa* (McAlpine) van der Aa - Stud. Mycol. 5: 40, 1973; teleom.: *Guignardia citricarpa* Kiely - Proc. Linn. Soc. N.S.W. 73: 259, 1948.

This pathogen on *Citrus* was once mentioned from a human traumatic infection, but the etiology was not ascertained (Punithalingam, 1979).

citrinoviride - *Trichoderma citrinociride* Bissett - Can. J. Bot. 62: 926, 1984.

Kuhls *et al.* (1999) identified a strain from a human source, but no convincing case report has as yet been published.

clamitans - *Endomycopsis clamitans* Batista & Coelho - Publções Inst. Micol. Recife 242: 3, 1960.

Doubtful yeast species.

clericus - *Atelosaccharomyces clericus* C.W. Dodge - Med. Mycol. p. 345, 1935.

Doubtful yeast species (Barnett *et al.*, 1990).

closterosporiger - See this list under *closterosporigerum*.

closterosporigerum - *Epidermophyton niveum* MacCarthy var. *closterosporigerum* Kesteren - Med. J. Austr. 1: 422, 1939 (as '*closterosporiger*').

Doubtful dermatophyte species.

closterosporum - *Trichophyton tonsurans* Malmsten var. *closterosporum* Maeke - An. Inst. Biol. Univ. Mex. 12: 547, 1941.

Doubtful dermatophyte species.

coccinatum - *Favotrichophyton violaceum* (Sabouraud) C.W. Dodge var. *coccinatum* C.W. Dodge - Med. Mycol. p. 524, 1935.

Doubtful dermatophyte, probably identical to *Trichophyton violaceum*.

coeruleum - *Fusarium coeruleum* Saccardo - Syll. Fung. 4: 705, 1886 ≡

Fusarium solani (Martius) Saccardo var. *coeruleum* (Saccardo) C. Booth - Gen. Fus. p. 51, 1971.

The mycetoma published by Thianprasit & Sivayathorn (1984) showed a fungus with microconidia, whereas *Fusarium coeruleum* typically has macroconidia. Possibly *F. oxysporum* or *F. solani* was concerned.

coeruleum-cuticulare - *Oidium coeruleum-cuticulare* Greco - Orig. Tum. Obs. Myc. Argent. p. 54, 1916.

Neither literature nor material was available for study.

colombiana - See this list under *giganteum*.

colorans - *Oospora colorans* van Beyma - Antonie van Leeuwenhoek 6: 285, 1939.

The type culture, CBS 140.40, was an *Acremonium* species; it is now sterile (W. Gams, pers. comm.).

communis (Endomyces) - *Endomyces communis* Castellani - Arch. Parasit. 16: 187, 1913.

Doubtful species (de Hoog *et al.*, 1986).

communis (Monilia) - *Monilia communis* Castellani - Annls Inst. Pasteur 30: 152, 1916 ≡ *Parendomyces communis* (Castellani) C.W. Dodge - Med. Mycol. p. 245, 1935.

Doubtful yeast species (Barnett *et al.*, 1990).

conglobata - *Torulopsis conglobata* Redaelli - Micet. Ass. Microb. Tuberc. Pulmon. Cavit. p. 58, 1925 ≡ *Cryptococcus conglobatus* (Redaelli) Pollacci & Nannizzi - Micet. Pat. Uomo Anim., Fasc. 7, no. 63, 1928 ≡ *Candida albicans* (Robin) Berkhout var. *conglobata* (Redaelli) Ciferri - Man. Micol. Med., ed. 2, 2: 245, 1960 ≡ *Candida conglobata* (Redaelli) van Uden & Buckley, *in* Meyer & Ahearn - Mycotaxon 17: 297, 1983.

This rare species was originally reported from human lungs, but no modern case report has been published.

congolensis - *Discomyces congolensis* Baerts - Bull. Med. Katanga 2: 67, 1925 ≡ *Actinomyces congolensis* (Baerts) Brumpt - Précis Parasitol., ed. 4, p. 1206, 1927 ≡ *Sporotrichum congolense* (Baerts) C.W. Dodge - Med. Mycol. p. 809, 1935.

Doubtful species (Stalpers, 1984); probably an actinomycete was concerned.

conicum - *Trichophyton conicum* Cazalbou - Rev. Gén. Méd. Vét. 24: 4, 1914 ≡ *Favotrichophyton conicum* (Cazalbou) C.W. Dodge - Med. Mycol. p. 521, 1935.

Doubtful dermatophyte species.

conori - *Atelosaccharomyces conori* C.W. Dodge - Med. Mycol. p. 344, 1935.

Doubtful yeast species (Barnett *et al.*, 1990).

convoluta (Fonsecaea) - *Fonsecaea pedrosoi* (Brumpt) Negroni var. *convoluta* Borelli - PAHO Sci. Publ. 396: 334, 1980.

Proposed for dysplastic form of *F. pedrosoi*, but without proper description.

convolutum (Trichophyton) - *Trichophyton circonvolutum* Sabou-raud - Malad. Cuir Chevelu 3: 320, 1910.

Doubtful dermatophyte species; material lost.

cooperi - *Cryptococcus cooperi* C.W. Dodge - Med. Mycol. p. 338, 1935.

Doubtful yeast species (Barnett *et al.*, 1990).

coralloides - *Rhodotorula minuta* (Saito) Harrison var. *coralloides* Ruiz - An. Inst. Biol. Univ. Mex. 14: 121, 1943.

Doubtful yeast species.

cordoniforme - *Cephalosporium cordoniforme* Barbosa -Mycopathologia 3: 93, 1941 ≡ *Hyalopus cordoniformis* (Barbosa) Barbosa - Subs. Estud. Hyalopus, Recife p. 46, 1941.

Doubtful species, possibly a *Fusarium* (Gams, 1971).

coremigerum - *Epidermophyton niveum* MacCarty var. *coremigerum* Kesteren - Med. J. Austr. 1: 426, 1939 (as '*coremiger*').

Doubtful dermatophyte species.

coremiiforme - *Hemispora coremiiformis* M. Moore - Ann. Mo. Bot. Gdn 22: 328, 1935 ≡ *Geotrichum coremiiforme* (M. Moore) Brumpt - Précis Parasitol., ed. 6, p. 1783, 1949 ≡ *Trichosporon coremiiforme* (M. Moore) Guého & M. Th. Smith, *in* Guého, Smith, de Hoog, Billon-Grand, Christen & Batenburg-van der Vegte - Antonie van Leeuwenhoek 61:

309, 1992 ≡ *Trichosporon asahii* Akagi var. *coremiiformis* (M. Moore) Shinoda & Sugita, *in* Sugita, Nishikawa & Shinoda - J. Gen. Appl. Microbiol., Tokyo 40: 405, 1994.

This *Trichosporon* species was described from a single strain, CBS 2482, originating from a human head lesion (Moore, 1935).

cornua-bovis - *Microsporum cornua-bovis* Galli & Conforti - Annali Fac. Med. Vet. Torino 4: 177, 1954.

Doubtful dermatophyte species.

corniculatum - *Nannizzia corniculata* Takashio & de Vroey - Mycotaxon 14: 384, 1982 ≡ *Arthroderma corniculatum* (Takashio & de Vroey) Weitzman, McGinnis, Padhye & Ajello - Mycotaxon 25: 513, 1986.

This is a geophilic dermatophyte. It has been reported to be the teleomorph of *Microsporum boulardii*, but this is incorrect.

coronatum - *Trichophyton coronatum* Cazalbou - Revue Gén. Méd. Vét. 24: 3, 1914 ≡ *Favotrichophyton coronatum* (Cazalbou) C.W. Dodge - Med. Mycol. p. 520, 1935.

Doubtful dermatophyte species.

corsellii - *Cryptococcus corsellii* Neveu-Lemaire - Précis Parasitol. Anim. Domest. p. 60, 1912 ≡ *Torula corsellii* (Neveu-Lemaire) Vuillemin - Champ. Paras. Myc. Homme p. 113, 1931.

Doubtful yeast species (Barnett *et al.*, 1990).

corticale - *Coniosporium corticale* Ellis & Everhart in Ellis, Everhart & Benjamin - J. Mycol. 5: 69, 1889 ≡ *Cryptostroma corticale* (Ellis & Everhart) Gregory & Waller - Trans. Br. Mycol. Soc. 34: 594, 1951.

Cryptostroma corticale is a plant pathogen, agent of maple-bark disease (Gregory & Waller, 1951). The specimen from human lung described by Emanuel *et al.* (1962, 1966) produced mycetoma-like lesion in experimental inoculation (Bulman & Stretton, 1974). The strain was not available for study.

cracoviense - *Sporotrichum cracoviense* Lipinski - Medycyna Dosw. Spol. 2: 153, 1924 ≡ *Monilia cracoviensis* (Lipinski) Vuillemin - Champ. Paras. Myc. Homme p. 83, 1931 ≡ *Rhinocladium cracoviense* (Lipinski) Coudert - Guide Prat. Mycol. Méd. p. 296, 1955.

Doubtful species; has been supposed to be a *Trichosporon* (Stalpers, 1984) or a *Sporothrix* species (Rippon, 1988).

craikii - See this list under *depauperatum*.

crateriformis - *Endomyces crateriformis* Hudelo, Sartory & Montlaur - C.R. Hebd. Séanc. Acad. Sci., Paris 170: 1086, 1920 ≡ *Sporotrichum crateriforme* (Hudelo, Sartory & Montlaur) Vuillemin - Champ. Paras. Myc. Homme p. 70, 1931 ≡ *Zymonema crateriforme* (Hudelo, Sartory & Montlaur) C.W. Dodge - Med. Mycol. p. 174, 1935.

The authentic material is lost. Possibly *Saccharomycopsis capsularis* Schiönning was concerned (Stalpers, 1984).

crustaceum - *Penicillium crustaceum* Fries - Syst. Mycol. 3: 407, 1930.

Doubtful *Penicillium* species (Thom, 1930).

cucurbeticum - *Trichophyton mentagrophytes* (Robin) Blanchard var. *cucurbeticum* Szathmáry - Mycopath. Mycol. Appl. 42: 53, 1970.

Doubtful dermatophyte species.

currii - *Trichophyton currii* Chalmers & Marshall - J. Trop. Med. Hyg. 17: 262, 1914 ≡ *Ateleothylax currii* (Chalmers & Marshall) Ota & Langeron - Annls Parasit. Hum. Comp. 1: 333, 1923 ≡ *Matruchotiella currii* (Chalmers & Marshall) Grigorakis - C.R. Hebd. Séanc. Acad. Sci., Paris 179: 1424, 1924.

Doubtful dermatophyte species; material probably lost.

debellae-marengoi - *Haplographium debellae-marengoi* Pollacci - Atti Ist. Bot. Univ. Lab. Crittogam. Pavia 18: 125, 1921.

Doubtful; possibly the anamorph of *Pseudallescheria boydii* (p. 305) was concerned.

debuenii - *Achorion debuenii* Langeron & Baeza - Annls Parasit. Hum. Comp. 14: 399, 1936.

Doubtful dermatophyte species.

decipiens - *Favotrichophyton decipiens* Boedijn & Verbunt - Mycopathologia 1: 194, 1938 ≡ *Achorion decipiens* (Boedijn & Verbunt) Boedijn - Mycopath. Mycol. Appl. 6: 133, 1951.

Doubtful dermatophyte species; material probably not preserved.

degenerans - *Cryptococcus degenerans* Vuillemin - Rev. Gén. Sci. Pures Appl. 12: 740, 1901 ≡ *Zymonema degenerans* (Vuillemin) Froilano de

Mello, Paes & Sousa - Arq. Hig. Patol. Exot. 6: 33, 1918 ≡ *Torulopsis degenerans* (Vuillemin) de Almeida - Anais Fac. Med. Univ. S. Paulo 9: 76, 1933.

Doubtful yeast species (Barnett *et al.*, 1990).

delbrueckii - *Saccharomyces delbrueckii* Lindner - Mikrosk. Betriebskontr. Gärungsgew., 1895. ≡ *Torulaspora delbrueckii* (Lindner) Lindner - Jahrb. Versuchs- u. Lehranst. Berlin 7: 441-464, 1904.

Torulaspora delbrueckii is a yeast inhabiting lipid-rich food stuffs (Barnett *et al.*, 1990).

dendrobatidis - *Batrachochytrium dendrobatidis* Longcore, Pessier & Nichols - Mycologia 91: 220, 1999.

This is a member of *Chytridiomycota* known as a pathogen on *Amphibia.*.

denticulatum - *Trichophyton denticulatum* Sabouraud - Malad. Cuir Chev. 3: 374, 1910 ≡ *Trichophyton niveum* Sabouraud var. *denticulatum* (Sabouraud) Sabouraud - Malad. Cuir Chev. 3: 374, 1910 ≡ *Ectotrichophyton denticulatum* (Sabouraud) Castellani & Chalmers - Man. Trop. Med., ed. 3, p. 1006, 1919 ≡ *Microtrichophyton denticulatum* (Sabouraud) Neveu-Lemaire - Précis Parasit. Hum., ed. 5, p. 54, 1921 ≡ *Aleurosporia denticulata* (Sabouraud) Grigorakis - Annls Sci. Nat., Bot., Sér. 10, 7: 413, 1925 ≡ *Ctenomyces denticulatus* (Sabouraud) Nannizzi - Tratt. Micopat. Um. 4: 144, 1934 ≡ *Ectotrichophyton felineum* (Blanchard) Castellani & Chalmers var. *denticulatum* (Sabouraud) C.W. Dodge - Med. Mycol. p. 498, 1935.

Doubtful dermatophyte species.

demonbreunii - *Gymnoascus demonbreunii* Ajello & Cheng - Mycologia 59: 692, 1967 ≡ *Neogymnomyces demonbreunii* (Ajello & Cheng) Orr - Can. J. Bot. 48: 1061, 1970.

The species was originally described as the teleomorph of *Histoplasma capsulatum*, but it has been proven to be unrelated to this taxon (Currah, 1985).

depauperatum - *Verticillium depauperatum* Bruhns & Alexander, *in* Jadassohn - Handb. Haut- u. Geschlechtskrankh. 11: 129, 1928 ≡ *Microsporum iris* Pasini var. *craikii* C.W. Dodge - Med. Mycol. p. 540, 1935 (name change).

Doubtful dermatophyte species.

dermatensis - *Endomycopsis dermatensis* Batista, Campos & Coelho - Publções Inst. Micol. Recife 178: 3, 1960.

The authentic strain, CBS 5154, has been reidentified as *Candida diddensiae* (Fell *et al.*) Fell & S.A. Meyer. The etiology of the cutaneous lesions mentioned by Batista *et al.* (1960) was insufficiently proven.

dermatodes - *Sporotrichum dermatodes* Kane, *in* Castellani & Chalmers - Man. Trop. Med., ed. 3, p. 1032, 1919.

Mentioned as a synonym of *Ustilago hypodytes* Schlecht.; no original diagnosis is known to exist.

dermatosum - *Sporotrichum dermatosum* Schwartz, Tulipan & Birmingham - Occupat. Dis. Skin, ed. 3, 1957.

Doubtful species (Stalpers, 1984).

dermo-unguis - *Chaetophoma dermo-unguis* S.M. Singh & Barde - Mykosen 29: 276, 1986.

The dried type specimen, IMI 254387, was too poor to establish its identity. Possibly *Phoma sorghina* (p. 347) was concerned (H.A. van der Aa, pers. comm.).

dessyi - *Aspergillus dessyi* Spegazzini - Revta Soc. Argent. Cienc. Nat. 8: 115, 1925 ≡ *Thomiella dessyi* (Spegazzini) C.W. Dodge - Med. Mycol. p. 834, 1935.

Doubtful *Aspergillus* species (Thom & Church, 1928).

diplocyste - *Eurotium diplocyste* Sartory, R. Sartory, Hufschmitt & J. Meyer - C.R. Soc. Biol. 104: 883, 1930 ≡ *Aspergillus diplocystis* (Sartory, R. Sartory, Hufschmitt & J. Meyer) C.W. Dodge - Med. Mycol. p. 625.

Doubtful *Aspergillus* species (Raper & Fennell, 1965).

dombrayi - *Mycoderma dombrayi* Vuillemin - C.R. Hebd. Acad. Sci., Paris 189: 406, 1929 ≡ *Geotrichum dombrayi* (Vuillemin) Nannizzi - Tratt. Micopat. Um. 4: 393, 1934.

Doubtful species (de Hoog *et al.*, 1986).

dori - *Sporotrichum dori* de Beurmann & Gougerot - Annls Derm. Syph.

7: 996, 1906 ≡ *Discomyces dori* (de Beurmann & Gougerot) de Beurmann & Gougerot - Nouv. Myc. p. 59, 1909 ≡ *Rhinocladium dori* (de Beurmann & Gougerot) Neveu-Lemaire - Précis Parasitol. Hum., ed. 5, p. 84, 1921 ≡ *Oospora dori* (de Beurmann & Gougerot) Sartory - Champ. Parasitol. Homme Anim. p. 770, 1923 ≡ *Actinomyces dori* (de Beurmann & Gougerot) Brumpt ≡ Précis Parasitol., ed. 4, p. 1206, 1927 ≡ *Nocardia dori* (de Beurmann & Gougerot) Vuillemin - Encyl. Mycol. 2: 123, 1931.

The identity of this species is doubtful (Stalpers, 1984).

doukourei - *Cephalosporium doukourei* Boucher - Bull. Soc. Path. Exot. 11: 330, 1918 ≡ *Hyalopus doukourei* (Boucher) Ota - Jpn. J. Derm. Urol. 27: 938, 1927.

Doubtful *Acremonium* species close to *A. kiliense* (Gams, 1971).

duboisii - *Sabouraudites duboisii* Vanbreuseghem - Annls Parasit. Hum. Comp. 24: 254, 1949 ≡ *Microsporum duboisii* (Vanbreuseghem) Ciferri - Man. Micol., ed. 2, p. 414, 1960.

The type strain, CBS 349.49, represents a hitherto unidentified *Microsporum* species.

dupontii - *Penicillium dupontii* Griffon & Maublanc - Bull. Trimest. Soc. Mycol. Fr. 27: 73, 1911; teleom.: *Talaromyces thermophilus* Stolk - Antonie van Leeuwenhoek 31: 268, 1965.

Species from compost once isolated from an endocarditis (Pitt, 1994), but without a case report.

eguttulatus - *Cryptococcus eguttulatus* Pollacci & Nannizzi in Nannizzi - Tratt. Micopat. Um. 4: 309, 1934.

Doubtful yeast species (Barnett *et al.*, 1990).

ehimeënse - *Trichophyton ehimeënse* Fujii - Jpn. J. Derm. Urol. 31: 305-357, 1931 ≡ *Ectotrichophyton ehimeënse* (Fujii) C.W. Dodge - Med. Mycol. p. 506, 1935.

Doubtful dermatophyte species.

elegans - *Cunninghamella elegans* Lendner - Bull. Herb. Boissier 7: 250, 1907.

In older literature mentioned as a pathogen, but now it is assumed that these cases were all due to *C. bertholletiae* (p. 74).

ellisii - *Chalara ellisii* Nag Raj & Kendrick - Monogr. Chalara Allied Gen. p. 113, 1975.

A mixed infection in the femur of a human patient was reported by Cimerman *et al.* (1999).

elongatum - *Blastodendrion elongatum* Mattlet - Annls Soc. Belge Med. Trop. 6: 16, 1926 ≡ *Monilia elongata* (Mattlet) Brumpt - Précis Parasitol., ed. 4, p. 1241, 1927 ≡ *Cryptococcus elongatus* (Mattlet) Nannizzi - Tratt. Micopat. Um. 4: 324, 1924.

Doubtful yeast species.

emersonii - *Penicillium emersonii* Stolk - Antonie van Leeuwenhoek 31: 262, 1965; teleom: *Talaromyces emersonii* Stolk - Antonie van Leeuwenhoek 31 : 262, 1965.

For extended desription, see Stolk & Samson (1972). The species has been reported from a chronically colonized respiratory tract of a CF patient (Cimon *et al.* , 1999).

epidermidis - *Cryptococcus epidermidis* Castellani & Chalmers - Man. Trop. Med., ed. 3, p. 1075, 1919 ≡ *Torulopsis epidermidis* (Castellani & Chalmers) de Almeida - Anais Fac. Med. Univ. S. Paulo 9: 76, 1933.

Doubtful yeast species (Barnett *et al.*, 1990).

epigaeum (Penicillium) - *Penicillium epigaeum* Motta - Atti Clin. Oto-Rino-Laringol. 25: 305, 1927 ≡ *Scopulariopsis mottai* Vuillemin - Champ. Paras. Myc. Homme p. 62, 1931 (name change) ≡ *Scopulariopsis epigaeus* (Motta) Coudert - Guide Prat. Mycol. Méd. p. 217, 1955.

Doubtful species (Morton & Smith, 1963).

epigaeum (Sporotrichum) - *Sporotrichum epigaeum* Brunaud - Annls Soc. Sci. Nat. Rochelle 24: 81, 1888 ≡ *Sporotrichum schenckii* (Hektoen & Perkins) de Beurmann & Gougerot var. *fioccoi* C.W. Dodge - Med. Mycol. p. 808, 1935 (name change) ≡ *Beauveria epigaea* (Brunaud) Langeron, *in* Brumpt - Précis Parasitol., ed. 5, p. 1839, 1936 ≡ *Sporotrichum beurmannii* Matruchot & Ramond var. *fioccoi* (C.W. Dodge) Redaelli & Ciferri - Tratt. Micopat. Um. 5: 481, 1942 ≡ *Tritirachium epigaeum* (Brunaud) Langeron - Annls Parasit. Hum. Comp. 22: 98, 1947.

Doubtful species, possibly *Beauveria brougniartii* (Saccardo) Petrak (de Hoog, 1972).

equinum (Fusarium) - *Fusarium equinum* Növegaard - Sci., N.S. 14: 899, 1901.

Doubtful *Fusarium* species (Wollenweber & Reinking, 1935).

equinum (Trichosporon) - *Trichosporon equinum* Fambach - Z. Infektkrankh. Haustiere 29: 124, 1926.

Doubtful *Trichosporon* species, agent of animal white piedra (Guého et al., 1992a).

erioton - *Trichophyton erioton* Cazalbou - Bull. Soc. Centr. Méd. Vét. 91: 118, 1914 = *Ecotrichophyton erioton* (Cazalbou) C.W. Dodge - Med. Mycol. p. 500, 1935.

Doubtful dermatophyte species.

eriotrephon - *Trichophyton eriotrephon* Papegaaij - Ned. Tijdschr. Geneeskd. 69: 885, 1925 = *Ctenomyces eriotrephon* (Papegaaij) Nannizzi - Tratt. Micopat. Um. 4: 146, 1934 = *Ectotrichophyton eriotrephon* (Papegaaij) C.W. Dodge - Med. Mycol. p. 497, 1935.

The type strain, CBS 220.25, shows reduced sporulation and can as yet not be identified.

ernobii - *Torulopsis ernobii* Lodder & Kreger-van Rij - The Yeasts p. 671, 1952 = *Candida ernobii* (Lodder & Kreger-van Rij) S.A. Meyer & Yarrow, *in* Yarrow & Meyer - Int. J. Syst. Bact. 28: 612, 1978.

The species was mentioned as causing sepsis (Majoret, 1986) and vaginitis (Tóth et al., 1973). However, since it is a bark beetle associated fungus (Barnett et al., 1990) misidentifications may have been concerned.

estorilensis - *Torulopsis estorilensis* Castellani - Mykosen 12: 508, 1969.

Doubtful yeast species..

etiennei - *Saccharomyces etiennei* Proton - Revue Méd. Est 45: 814- 826, 1913 = *Octomyces etiennei* (Proton) C.W. Dodge - Med. Mycol. p. 180, 1935.

Doubtful yeast species (Barnett et al., 1990).

exiguus - *Saccharomyces exiguus* Reess - Bot. Unters. Alkoholgär. 1-87: 22, 1870.

Teleomorph of *Candida holmii*; for discussion of pathogenicity, see this list under *holmii*.

exilis - *Septonema exile* P. Karsten - Meddel. Soc. Flora Fauna Fennica 14: 98, 1887 = *Taeniolella exilis* (P. Karsten) S.J. Hughes - Can. J. Bot. 36: 817, 1958.

Taeniolella exilis normally occurs on wood (Ellis, 1971). A traumatically implanted mycosis was reported by Alonso et al. (1993).

expansum - *Trichophyton expansum* Cazalbou - Revue Gén. Méd. Vét. 24: 7, 1914 = *Favotrichophyton expansum* (Cazalbou) C.W. Dodge - Med. Mycol. p. 520, 1935.

Doubtful dermatophyte species.

farinulentum - *Trichophyton farinulentum* Sabouraud - Malad. Cuir Chev. 3: 368, 1910 = *Ectotrichophyton farinulentus* (Sabouraud) Castellani & Chalmers - Man. Trop. Med., ed. 3, p. 1005, 1919 = *Sabouraudites farinulentus* (Sabouraud) Ota & Langeron - Annls Parasit. Hum. Comp. 1: 327, 1923 = *Closteroaleuriosporia farinulenta* (Sabouraud) Grigorakis - Ann. Sci. Nat., Bot., Sér. 10, 7: 415, 1925 = *Microsporum farinulentum* (Sabouraud) Guiart & Grigorakis - Lyon Méd. 141: 377, 1928 = *Ctenomyces farinulentus* (Sabouraud) Nannizzi - Tratt. Micopat. Um. 4: 147, 1934.

Doubtful dmatophyte species.

faviforme-aviare - See this list under *avium*.

felineum (Microsporum) - *Microsporum felineum* Fox & Blaxall - Br. J. Derm. Syph. 10: 354, 1896 = *Sabouraudites felineus* (Fox & Blaxall) Ota & Langeron - Annls Parasit. Hum. Comp. 1: 327, 1923 = *Closterosporia felinea* (Fox & Blaxall) Grigorakis - Ann. Sci. Nat., Bot., Sér. 10, 7: 411, 1925.

No material was preserved; doubtful dermatophyte species. Rippon (1988) listed it as a synonym of *Microsporum canis*.

felineum - *Trichophyton felineum* Blanchardm, *in* Bouchard - Traité Pathol. Gén 2: 914, 1895 = *Ctenomyces felineus* (Blanchard) Nannizzi - Tratt. Micopat. Um. 4: 148, 1934.

Doubtful dermatophyte species.

fennelliae - *Neosartorya fennelliae* Kwon-Chung & Kim - Mycologia 66: 629, 1974; anam.: *Aspergillus fennelliae* Kwon-Chung & Kim - Mycologia 66: 632, 1974.

This species was isolated from homogenized eye balls of rabbits, but proved non-pathogenic in experimental inoculation (Kwon-Chung & Kim, 1974).

fermentans - *Pichia fermentans* Lodder - Zentbl. Bakt. Parasitkde, Abt. 2, 86: 242, 1932.

This is the teleomorph of *Candida lambica*. It was reported as the cause of arthritis in an alcoholic (Trowbridge et al., 1999).

filiformis - *Torulopsis inconspicua* Lodder & Kreger-van Rij var. *filiformis* Dietrichson - Annls Parasit. Hum. Comp. 29: 474, 1954.

Now known as *Candida inconspicua*. Pathogenicity is doubtful; see this list under *inconspicua*.

finckii - *Scopulariopsis finckii* Vuillemin - Champ. Paras. Myc. Homme p. 65, 1931.

Doubtful hyphomycete species (Morton & Smith, 1963).

fioccoi - See this list under *epigaeum (Sporotrichum)*.

flavovirens - *Trichophyton flavovirens* da Veiga - Brasil Med. 43: 836, 1929 = *Favotrichophyton flavovirens* (da Veiga) C.W. Dodge - Med. Mycol. p. 518, 1935.

Doubtful dermatophyte species.

flavum - *Epidermophyton flavum* Kesteren - Med. J. Austr. 1: 424, 1939.

Doubtful dermatophyte species.

floreale - *Trichophyton floreale* Cazalbou - Revue Gén. Méd. Vét. 24: 6, 1914 = *Favotrichophyton floreale* (Cazalbou) C.W. Dodge - Med. Mycol. p. 517, 1935.

Doubtful dermatophyte species.

fontoynontii (Aspergillus) - *Aspergillus f ontoynontii* Guéguen - C.R. Soc. Biol. 46: 1052, 1909.

Aspergillus species close to *A. penicillioides* Speg. (Raper & Fennell, 1965)

fontoynontii (Cryptococcus) - *Cryptococcus fontoynontii* Vuillemin - Champ. Paras. Myc. Homme p. 97, 1931.

Doubtful yeast species (Barnett et al., 1990).

fontoynontii (Hormodendrum) - *Hormodendrum fontoynontii* Langeron - Archs Parasit. 17, 1913.

Doubtful; possibly a *Cladosporium* species (de Hoog, 1977).

formosense - *Grubyella formosensis* Hasegawa - Jpn. J. Derm. Urol. 27: 140, 1927 = *Achorion formoseum* Hasegawa - Jpn. J. Derm. Urol. 27: 8, 1927 (name change) = *Trichophyton formosense* (Hasegawa) Nannizzi - Tratt. Micopat. Um. 4: 184, 1934.

Doubtful dermatophyte species.

formoseum - See this list under *formosense*.

foxii - *Trichosporum foxii* Castellani - J. Trop. Med. Hyg. 11: 261, 1908 = *Trichosporon foxii* (Castellani) Nannizzi - Tratt. Micopat. Um. 4: 295, 1934.

On the basis of the very scant original description its identity cannot be established.

fragrans - *Cephaloascus fragrans* Hanawa - Jpn. J. Derm. Urol. 20: 14, 1920.

This species was originally isolated from a human ear, but its prevalent ecological niche is on wood in association with bark insects.

fresenii - *Aspergillus fresenii* Subramanian - Hyphomycetes p. 552, 1971.

Synonym of *A. sulphureus*; see this list under *sulphureus*.

fructicola - *Graphium fructicola* Él. & Ém. Marchal - Bull. Soc. R. Bot. Belge, N. Sér., 4: 26, 1921.

Clinical cases under this name (e.g. Käufer & Weber, 1976) mostly concern the synnematous anamorph of *Pseudallescheria boydii* (p. 305). Whether or not the two taxa are identical remains to be established.

fujii - *Blastodendrion cutaneum* (Ota) C.W. Dodge var. *fujii* C.W. Dodge - Med. Mycol. p. 287, 1935.

Doubtful yeast species.

fuliginea - *Bodinia fuliginea* Ogata - Jpn. J. Derm. Urol. 29: 1192, 1929 = *Trichophyton fuligineum* (Ogata) Ogata - Jpn. J. Derm. Urol. 29: 1192, 1929 = *Favotrichophyton fuligineum* (Ogata) C.W. Dodge - Med. Mycol. p. 526, 1935.

Doubtful dermatophyte species.

funiculosum - *Penicillium funiculosum* Thom - Bull. Bur. Anim. Ind. US dept. Agric. 118: 69, 1910.

Pitt (1994) mentioned a strain from a systemic infection but did not provide a case report.

fusarioides - *Pseudomicrodochium fusarioides* B.C. Sutton & C.K. Campbell, *in* Sutton, Campbell & Goldschmied-Reouven - Mycopathologia 114: 160, 1991.

This species, isolated from human brochial lavage (Sutton *et al.*, 1991), is very similar to *Cyphellophora laciniata* (p. 621); the type strain was currently not available for study.

fuscinum - *Trichophyton glabrum* Sabourand var. *fuscinum* Catanei - Archs Inst. Pasteur Algér. 15: 270, 1937.

Doubtful dermatophyte species.

fuscosulcatum - *Trichophyton fuscosulcatum* Neuber - Derm. Wschr. 80: 861-872, 1925 ≡ *Trichophyton fuscum* (Neuber) C.W. Dodge - Med. Mycol. p. 531, 1935 (name change).

Doubtful dermatophyte species.

fuscum - See this list under *fuscosulcatum*.

fuscus - *Cryptococcus fuscus* Fontoynont & Boucher - Annls Derm. Syph., Sér. 6, 4: 328, 1923 ≡ *Torulopsis fusca* (Fontoynont & Boucher) de Almeida - Anais Fac. Med. S. Univ. Paulo 9: 76, 1933 ≡ *Eutorula fusca* (Fontoynont & Boucher) C.W. Dodge - Med. Mycol. p. 355, 1935.

Doubtful yeast species (Barnett *et al.*, 1990).

gallicum - *Geotrichum rotundatum* (Castellani) de Almeida var. *gallicum* Castellani - J. Trop. Med. Hyg. 43: 81, 1940.

Doubtful species (de Hoog *et al.*, 1986).

gallopavum - *Trichophyton gallopavum* Metianu, Lucas & Drouhet - Mycopath. Mycol. Appl. 30: 29, 1966.

Doubtful dermatophyte species.

gamelleirae - *Trichophyton gamelleirae* Magalhães - Mem. Inst. Oswaldo Cruz 30: 36, 1935.

Doubtful dermatophyte species.

gandavensis - *Glenospora gandavensis* Vuillemin - C.R. Hebd. Séanc. Acad. Sci., Paris 173: 378, 1921 ≡ *Glenosporopsis gandavensis* (Vuillemin) da Fonseca - Parasit. Méd. 1: 727, 1943.

Doubtful, no material was preserved. Perhaps *Pseudallescheria boydii* (p. 305).

gemmulolaniferum - *Trichophyton gypseum* Bodin var. *gemmulolaniferum* von Berde - Arch. Derm. Syph. 176: 4, 1937.

Doubtful dermatophyte species.

georgiae - *Trichophyton georgiae* Varsavsky & Ajello - Riv. Patol. Veg., Padova 4: 357, 1964 ≡ *Chrysosporium georgiae* (Varsavsky & Ajello) van Oorschot - Riv. Patol. Veg., Padova 4: 357, 1964.

This is a geophilic *Chrysosporium* species, the anamorph of *Arthroderma ciferii* (van Oorschot, 1980).

giganteum (Trichosporon) - *Trichosporon giganteum* Unna - Dt. Med. Ztg 10: 255, 1895 ≡ *Geotrichum giganteum* (Unna) Coudert - Guide Prat. Mycol. Méd. p. 234, 1955.

Doubtful *Trichosporon* species (Guého *et al.*, 1992a).

giganteum (Trichosporon) - *Trichosporon giganteum* Vuillemin - Revue Gén. Sci. 12: 732-751, 1901 ≡ *Piedraia colombiana* C.W. Dodge - Med. Mycol. p. 133, 1935.

Probably a *Trichosporon* species was concerned. Dodge (1935) explicitly stated that the meant species was different from *T. giganteum* Unna and cited some papers describing ascospores. The whitish colonies indicate that the attribution to the genus *Piedraia* is probably incorrect.

giordanoi - *Penicillium giordanoi* Vuillemin - Champ. Paras. Myc. Homme p. 62, 1931.

Doubtful *Penicillium* species from a systemic mycosis; material not preserved.

glaucum - *Penicillium glaucum* Link - Mag. Ges. Naturf. Freunde, Berlin 3: 17, 1809.

Doubtful *Penicillium* species.

glycophile - *Trichosporon glycophile* DuBois - Annls Derm. Syph., Sér. 5, 1: 456, 1910.

Doubtful *Trichosporon* species (de Hoog *et al.*, 1997).

goënsis - *Proteomyces goënsis* C.W. Dodge - Med. Mycol. p. 215, 1935.

Fungus of doubtful identity (Barnett *et al.*, 1990).

gotoi - *Cryptococcus gotoi* C.W. Dodge - Med. Mycol. p. 333, 1935.

Doubtful yeast species (Barnett *et al.*, 1990).

gougerotii (Arthrosporia) - *Arthrosporia gougerotii* Grigoraki - C.R. Soc. Biol. 114: 258, 1933.

Dermatophyte species, possibly *Trichophyton schoenleinii*.

gougerotii (Sporotrichum) - *Sporotrichum gougerotii* Matruchot - C.R. Hebd. Séanc. Acad. Sci., Paris 150: 545, 1910 ≡ *Rhinocladium gougerotii* (Matruchot) Verdun - Précis Parasitol. Hum., ed. 2, p. 677, 1913 ≡ *Dematium gougerotii* (Matruchot) Grigoraki - Bull. Trim. Soc. Mycol. Fr. 40: 274, 1924 ≡ *Torula gougerotii* (Matruchot) Brumpt - Précis Parasitol., ed. 5, 1791, 1936 ≡ *Oospora gougerotii* (Matruchot) Janke - Arch. Derm. Syphil., Berlin 187: 693, 1949 ≡ *Phialophora gougerotii* (Matruchot) Borelli - Acta Cient. Venez. 6: 81, 1955 ≡ *Cladosporium gougerotii* (Matruchot) Carrión & da Silva Lacaz - Arch. Derm. 72: 532, 1955.

The name has been applied to different fungi and has therefore been abandoned (McGinnis & Ajello, 1982).

gougerotti-ganceae - *Cryptococcus gougerotii-ganceae* Nannizzi - Tratt. Micopat. Um. 4: 309, 1934.

Doubtful yeast species (Barnett *et al.*, 1990).

gracile - *Pythium gracile* Schenk - Verh. Phys.-Med. Ges. Würzb. 9: 13, 1859.

Doubtful *Pythium* species (van der Plaats-Niterink, 1981). The two cases from horse published under this name (Hutchins & Johnston, 1972; Ichitani & Amemiya, 1980) probably concerned *P. insidiosum* (p. 55).

gracilioides - *Cryptococcus gracilioides* Castellani - J. Trop. Med. Hyg. 28: 222, 1925.

Doubtful yeast species (Barnett *et al.*, 1990).

granulatus - *Saccharomyces granulatus* Vuillemin & Legrain - Arch. Parasit. 3: 237-268, 1900.

Doubtful yeast species.

granulomatis - *Cephalosporium granulomatis* Weidman & Kligman - J. Bact. 1: 491, 1945.

Doubtful *Acremonium* species, possibly *A. kiliense* (Gams, 1971).

granulomatogenes - *Saccharomyces granulomatogenes* Sanfelice - Z. Hyg. 29: 499, 1898 (as '*granulomatosus*' in heading) ≡ *Cryptococcus granulomatogenes* (Sanfelice) Vuillemin - Revue Gén. Sci. Pures Appl. 12: 741, 1901 ≡ *Torulopsis granulomatogenes* (Sanfelice) de Almeida - Anais Fac. Med. S. Univ. Paulo 9: 76, 1933.

Doubtful yeast species (Barnett *et al.*, 1990).

granulomatosus - See this list under *granulomatogenes*.

granulosa (Oospora) - *Oospora granulosa* Kambayashi - Jpn. J. Derm. Urol. 23: 509, 1922.

Doubtful *Trichosporon* species.

granulosum (Trichophyton) - *Trichophyton granulosum* Sabouraud, *in* Pécus - Rev. Gén. Méd. Vét. 15: 561, 1909 ≡ *Trichophyton mentagrophytes* (Robin) Blanchard var. *granulosum* (Sabouraud) Neveu-Lemaire - Précis Parasitol. Anim. Domest. p. 71, 1912 ≡ *Ectotrichophyton granulosum* (Sabouraud) Castellani & Chalmers - Man. Trop. Med., ed. 3, p. 1006, 1919 ≡ *Sabouraudites granulosus* (Sabouraud) Ota & Langeron - Annls Parasit. Hum. Comp. 1: 328, 1923 ≡ *Chlamydoaleuriospora granulosa* (Sabouraud) Grigorakis - Annls Sci. Nat., Bot., Sér. 10, 7: 412, 1925 ≡ *Trichophyton gypseum* Bodin var. *granulosum* (Sabouraud) Frágner - Česká Mykol. 10: 108, 1956.

Doubtful dermatophyte species.

greconis - See this list under *aurea (Sterigmatomyces)*.

griewankii - *Mycoderma griewankii* Neveu-Lemaire - Précis Parasitol. Hum., ed. 5, p. 70, 1921 ≡ *Proteomyces griewankii* (Neveu-Lemaire) C.W. Dodge - Med. Mycol. p. 214, 1935.

Doubtful yeast species (Barnett *et al.*, 1990).

griseum (Cephalosporium) - *Cephalosporium griseum* Gougerot, Burnier & Duché - Bull. Soc. Fr. Derm. Syph. 40: 417, 1933 ≡ *Hyalopus griseus* (Gougerot, Burnier & Duché) Barbosa - Subs. Estud. Hyalopus, Recife, p. 46, 1941.

Doubtful, possibly a *Fusarium* species (Gams, 1971).

griseum (Epidermophyton) - *Epidermophyton griseum* Kesteren - Med. J. Austr. 1: 426, 1939.

Doubtful dermatophyte species.

griseum (Trichophyton) - *Trichophyton griseum* F. Fischer - Verh. Berl. Derm. Ges. 1909/10: 112, 1910 = *Ectotrichophyton griseum* (F. Fischer) C.W. Dodge - Med. Mycol. p. 499, 1935.

Doubtful dermatophyte close to *Trichophyton mentagrophytes*.

griseum (Trichophyton) - *Trichophyton griseum* Vasconcellos - Mem. Inst. Oswaldo Cruz 6: 11, 1914 = *Sabouraudites griseus* (Vasconcellos) Ota & Langeron - Annls Parasit. Hum. Comp. 1: 28, 1923 = *Ectotrichophyton griseum* (F. Fischer) C.W. Dodge var. *brasiliense* C.W. Dodge - Med. Mycol. p. 499, 1935 (name change).

Doubtful dermatophyte close to *Trichophyton mentagrophytes*.

guayaquilense - *Rhinocladium guayaquilense* Valenzuela, *in* Brumpt - Précis Parasitol., ed. 4, p. 1329, 1927.

Fungus of doubtful identity; material not preserved.

guggenheimii - *Parasaccharomyces oosporoides* (Zach) C.W. Dodge var. *guggenheimii* C.W. Dodge - Med. Mycol. p. 267, 1935.

Doubtful yeast species (Barnett *et al.*, 1990).

guilliermondii - *Hanseniaspora guilliermondii* Pijper - Proc. K. Ned. Akad. Wet. 31: 989, 1928 = *Willia guilliermondii* (Pijper) Vuillemin - Champ. Paras. Myc. Homme p. 274, 1931.

Reported by Vuillemin (1931) from a human nail without proven clinical significance. The current name of the organism is *Hanseniaspora guilliermondii*.

guzzonii - *Trichophyton guzzonii* Castellani - J. Trop. Med. Hyg. 42: 374, 1939.

Doubtful dermatophyte species.

gypseum - *Epidermophyton gypseum* MacCarthy - Annls Derm. Syph., Sér. 6, 6: 41, 1925 = *Ctenomyces gypseus* (MacCarthy) Nannizzii - Tratt. Micopat. Um. 4: 149, 1934.

Doubtful dermatophyte species.

hagiwarae - *Saccharomyces hagiwarae* C.W. Dodge - Med. Mycol. p. 319, 1935.

Doubtful yeast species (Barnett *et al.*, 1990).

harteri - *Cryptococcus harteri* Verdun - Précis Parasitol., 1912 = *Atelosaccharomyces harteri* (Verdun) de Beurmann & Gougerot in Guilliermond - Les Levures p. 491, 1912 = *Monilia harteri* (Verdun) Castellani & Chalmers - Man. Trop. Med. p. 831, 1913 = *Parasaccharomyces harteri* (Verdun) Froilano de Mello, Paes & Sousa - Arq. Hig. Patol. Exot. 6: 33, 1918 = *Myceloblastanon harteri* (Verdun) Ota - Jpn. J. Derm. Urol. 27: 173, 1927 = *Torulopsis harteri* (Verdun) de Almeida - Anais Fac. Med. Univ. S. Paulo 9: 76, 1933 = *Zymonema harteri* (Verdun) C.W. Dodge - Med. Mycol. p. 178, 1935.

Doubtful yeast species (Barnett *et al.*, 1990).

hashimotoi - *Castellania hashimotoi* C.W. Dodge - Med. Mycol. p. 260, 1935.

Doubtful yeast species (Barnett *et al.*, 1990).

hessleris - *Blastomyces hessleris* Rettger - Centbl. Bakt. Parasitkde, Abt. 1, 36: 527, 1904 = *Cryptococcus hessleris* (Rettger) Castellani & Chalmers - Man. Trop. Med., ed. 2, p. 771, 1913 = *Torulopsis hessleris* (Rettger) de Almeida - Anais Fac. Med. Univ. S. Paulo 9: 76, 1933 = *Parendomyces hessleris* (Rettger) C.W. Dodge - Med. Mycol. p. 242, 1935.

Doubtful yeast species (Barnett *et al.*, 1990).

heteromorpha - *Trichosporium heteromorphum* Nannfeldt, *in* Melin & Nannfeldt - Svenska Skogsvför. Tidskr. 22: 467, 1934 = *Margarinomyces heteromorpha* (Nannfeldt) Mangenot - Revue Gén. Bot. 59: 391, 1952 = *Phialophora heteromorpha* (Nannfeldt) Wang - Can. J. Bot. 42: 1015, 1964 = *Exophiala jeanselmei* (Langeron) McGinnis & Padhye var. *heteromorpha* (Nannfeldt) de Hoog - Stud. Mycol. 15: 110, 1977 = *Exophiala heteromorpha* (Nannfeldt) Rinaldi - Infect. Dis. Clin. N. Amer. 3: 67, 1989.

The species was listed as an emerging opportunist by Rinaldi (1989), but no confirmed case has been published yet (Uijthof, 1996).

hiulcum - *Trichophyton gypseum* Bodin var. *hiulcum* von Berde - Arch. Derm. Syph. 176: 4, 1937.

Doubtful dermatophyte species.

holmii - *Torula holmii* Jörgensen - Mikroorg. Gärungsind., ed. 5, p. 395,

1909 = *Cryptococcus holmii* (Jörgensen) Skinner - Amer. Midl. Nat. 43: 248, 1950 = *Candida holmii* (Jörgensen) S.A. Meyer & Yarrow, *in* Yarrow & Meyer - Int. J. Syst. Bact. 28: 612, 1978.

The correct name is *Candida holmii*; it is the anamorph of *Saccharomyces exiguus*. This fungus occurs in nutrient-rich, fermenting substrates. A human osteomyelitis was published by Murdock *et al.* (1983).

hominis (Acrothecium) - *Acrothecium hominis* Oláh - Atti Ist. Bot. Univ. Lab. Crittogam. Pavia, Ser. 4a, 9: 176. 1937.

Authentic material is lost; probably *Curvularia lunata* was concerned.

hominis (Aspergillus) - *Aspergillus awamori* Nakazawa var. *hominis* Batista & Maia - Anais Soc. Biol. Pernamb. 15: 186, 1957.

The type strain, CBS 107.55 from human skin lesion, is *Aspergillus phoenicis* (Corda) Thom. The variety was published without convincing case report.

hominis (Basidiobolus) - *Basidiobolus hominis* Casagrandi - Riv. Biol. 13: 8, 1931.

Doubtful zygomycete species (Dodge, 1935).

hominis (Saccharomyces) - *Saccharomyces hominis* Klein & Gordon - Centbl. Bakt. Parasitkde, Abt. 1, 35: 138, 1904 = *Monilia hominis* (Klein & Gordon) Vuillemin - Champ. Paras. Myc. Homme p. 85, 1931 = *Torulopsis hominis* (Klein & Gordon) de Almeida - Anais Fac. Med. Univ. S. Paulo 9: 76, 1933.

Doubtful yeast species (Barnett *et al.*, 1990).

hominis (Saccharomyces) - *Saccharomyces hominis* Castellani & Chalmers - Man. Trop. Med., ed. 2, p. 769, 1913 = *Castellania hominis* (Castellani & Chalmers) C.W. Dodge - Med. Mycol. p. 250, 1935.

The name refers to the case described by Klein & Gordon (1904). Its identity is doubtful (Barnett *et al.*, 1990).

hominis (Schizosaccharomyces) - *Schizosaccharomyces hominis* Benedek - Zentbl. Bakt. Hyg. Parasitkde, Abt. 1, 104: 291, 1927 = *Mycoderma hominis* (Benedek) Vuillemin - C.R. Hebd. Séanc. Acad. Sci., Paris 89: 408, 1929.

According to Dodge (1935) a *Bacillus* species was concerned.

hominis (Schwanniomyces) - *Schwanniomyces hominis* Bastista, Vieira & Coelho - Publções Inst. Micol. Recife 175: 3, 1960.

The type strain, CBS 4865, has been reidentified as *Torulaspora delbrueckii* (Lindner) Lindner (Barnett *et al.*, 1990). This species normally occurs in food stuffs. The case report, from human skin lesion (Batista *et al.*, 1960) is unconfirmed.

homopilatum - *Chaetomium homopilatum* Omvik - Mycologia 47: 749, 1953.

A mixed infection with *Aspergillus fumigatus* in a leukemic patient was reported by Schulze *et al.* (1997).

humahuaquense - *Trichosporon humahuaquense* Mazza & Niño - Reun. Soc. Argent. Patol. Reg. Norte Santiago 1: 239-317, 1933.

Doubtful *Trichosporon* species, agent of human white piedra (Guého *et al.*, 1992a).

inconspicua - *Torulopsis inconspicua* Lodder & Kreger-van Rij - The Yeasts p. 436, 1952 = *Candida inconspicua* (Lodder & Kreger-van Rij) S.A. Meyer & Yarrow, *in* Yarrow & Meyer - Int. J. Syst. Bact. 28: 612, 1978.

This species, very closely related to *C. krusei* (p. 206), is repeatedly isolated from sputum and blood (Bailey *et al.,* 1997). Tchang & Gilardi (1973) reported an osteomyelitis, and d'Antionio *et al.* (1998) a series of nosocomial infections.

indica - *Monodictys indica* S.M. Singh & Barde - Curr. Sci. 54: 1001, 1985.

This species was reported from a mixed superficial infection primarily due to *Nattrassia mangiferae* (Singh & Barde, 1985).

indicae-seudatiae - *Thermomucor indicae-seudatiae* Subrahmanyam, Mehrotra & Thirumalachar - Geor. J. Sci. 35: 1, 1977.

The species was reported from human skin, but without true pathology (Subrahmanyam & Lakshmi, 1993).

indicum (Epidermophyton) - *Epidermophyton indicum* Castellani & Chalmers - Man. Trop. Med., ed. 3, p. 2327, 1919.

The name was listed in the index only, referring to page 2027 where a causative agent of tinea intersecta was described but no fungal name was mentioned.

indicum (Trichophyton) - *Trichophyton indicum* Randhawa & R.S. Sandhu - Mycopath. Mycol. Appl. 20: 227, 1963 ≡ *Chrysosporium indicum* (Randhawa & R.S. Sandhu) Garg - Sabouraudia 4: 262, 1966.

The correct name is *Chrysosporium indicum;* it is the anamorph of *Aphanoascus terreus.* The species is geophilic.

infiltrans - *Saccharomyces infiltrans* Casagrandi & Santori - Annali Ig. Sper., N. Ser., 9: 141, 1899 ≡ *Cryptococcus infiltrans* (Casagrandi & Santori) Nannizzi - Tratt. Micopat. Um. 4: 345, 1934.

Doubtful yeast species (Barnett *et al.*, 1990).

interdigitalis - *Endomycopsis interdigitalis* Batista & Coelho - Publções Inst. Micol. Recife 241: 3, 1960.

Doubtful yeast species.

interposita - *Monilia interposita* Sutejew, Utenkow & Zeitlin - Fortschr. Geb. Röntgenst. 40: 475, 1929.

Doubtful yeast species (Barnett *et al.*, 1990).

iris - *Microsporum iris* Pasini - G. Ital. Mal. Vener. Pelle 52: 614, 1911 ≡ *Sabouraudites iris* (Pasini) Ota & Langeron - Annls Parasit. Hum. Comp. 1: 329, 1923.

Doubtful dermatophyte species.

irritans - *Cryptococcus irritans* Mattlet - Annls Soc. Belge Méd. Trop. 6: 15, 1926 ≡ *Eurotorula irritans* (Matlett) C.W. Dodge - Med. Mycol. p. 356, 1935.

Doubtful yeast species (Barnett *et al.*, 1990).

irritum - *Blastodendrion irritum* Mattlet - Annls Soc. Belge Méd. Trop. 6: 16, 1926.

Doubtful yeast species.

issavi - *Mycoderma issavi* Mattlet - Annls Soc. Belge Méd. Trop. 6: 26, 1926 ≡ *Geotrichum issavi* (Mattlet) Nannizzi - Tratt. Micopat. Um. 4: 394, 1934.

Possibly *Geotrichum candidum* was concerned (Carmichael, 1957); authentic material is lost.

janthinellum - *Penicillium janthinellum* Biourge - Cellule 33: 258, 1923.

A mixed infection of burnt skin was reported by Sandner & Schönborn (1973). Pitt (1994) mentioned a strain from sputum.

japonicum - *Hormodendrum japonicum* Takahashi - Jpn. J. Derm. Urol. 41: 53, 1937.

Rippon (1988) listed this species as a synonym of *Fonsecaea pedrosoi*, but authentic material for confirmation is lost (de Hoog, 1977).

jeanselmei - *Cryptococcus jeanselmei* Burnier & Langeron - Congr. Derm. Syph. Langue Franc. 1: 40, 1922 ≡ *Torulopsis jeanselmei* (Burnier & Langeron) de Almeida - Anais Fac. Med. Univ. S. Paulo 9: 76, 1933.

Doubtful, no material was preserved.

jirovecii - *Trichosporon jirovecii* Frágner - Česká Mykol. 23: 160, 1969 ≡ *Trichosporon cutaneum* (de Beurmann, Gougerot & Vaucher) Ota var. *jirovecii* (Frágner) Frágner - Česká Mykol. 24: 160, 1970.

The species was described from diseased human nails, but without a convincing case report. A second strain, described as *T. beemeri* Kuttin & J. Müller (see this list under *beemeri*), originated from a crocodile.

karafutoënse - *Microsporum karafutoënse* Takatsuki - Jpn. J. Derm. Urol. 40: 150, 1936.

Doubtful dermatophyte species.

keratitidis - *Periconia keratitidis* Oláh - Atti Ist. Bot. Univ. Lab. Crittogam. Pavia, Ser. 4a, 9: 158, 1937.

Original material is lost; the description does not allow the establishment of its identity.

khartoumense (Trichophyton) - *Trichophyton violaceum* Sabouraud var. *khartoumense* Chalmers & MacDonald - J. Trop. Med. Hyg. 18: 145, 1915 ≡ *Favotrichophyton violaceum* (Sabouraud) C.W. Dodge var. *khartoumense* (Chalmers & MacDonald) C.W. Dodge - Med. Mycol. p. 523, 1935.

Doubtful dermatophyte, probably identical to *Trichophyton violaceum.*

khartoumensis (Glenospora) - *Glenospora khartoumensis* Chalmers & Archibald - Ann. Trop. Med. Parasit. 10: 169, 1916 ≡ *Trichosporium khartoumense* (Chalmers & Archibald) C.W. Dodge - Med. Mycol. p. 793, 1935 ≡ *Madurella khartoumensis* (Chalmers & Archibald) Ciferri &

Redaelli - Mycopath. Mycol. Appl. 3: 196, 1941 ≡ *Glenosporopsis khartoumensis* (Chalmers & Archibald) da Fonseca - Parasitol. Méd. 1: 728, 1943.

Doubtful species; no material was preserved.

kieta - *Mycoderma kieta* Mattlet - Annls Soc. Belge Méd. Trop. 6: 29, 1926 ≡ *Geotrichum kieta* (Mattlet) Nannizzi - Tratt. Micopat. Um. 4: 394, 1934.

Doubtful, possibly *Geotrichum candidum* (Carmichael, 1957).

kluyveri - *Saccharomyces kluyveri* Phaff, M.W. Miller & Shifrine - Antonie van Leeuwenhoek 22: 151, 1956.

The species is known from plant exudates. (Vaughan-Martini & Martini, 1998). A disseminated case in an AIDS patient reported by Pynka *et al.* (1998) may be misidentification.

kochii - *Rhodomyces kochii* von Wettstein - Sitzber. K. Akad. Wiss. Wien, Abt. 1, 9: 51, 1885 ≡ *Candida kochii* (von Wettstein) Basgal - Contrib. Estud. Blastom. Pulmon. p. 49, 1931.

Doubtful red yeast, possibly *Sporobolomyces salmonicolor* (Kreger-van Rij, 1984).

kuehnii - *Thallomicrosporon kuehnii* Benedek - Mycopath. Mycol. Appl. 23: 96, 1964.

The type strain CBS 119.64 is a hitherto unidentified dermatophyte species.

lacrymeatus - *Cryptococcus lacrymeatus* Jeaume - Blastom. Voies Lacrym. Ane Maroc., Thèse, Avignon, 1926.

Doubtful yeast species (Barnett *et al.*, 1990).

lacteus - *Endomyces lacteus* Castellani - Lancet 1912: 15, 1912.

Doubtful species (de Hoog *et al.*, 1986).

lacticolor (Endomyces) - *Endomyces lacticolor* Castellani - Lancet 1912: 15, 1912.

Doubtful species (de Hoog *et al.*, 1986).

lacticolor (Trichophyton) - *Trichophyton lacticolor* Sabouraud - Malad. Cuir Chev. 3: 362, 1910 ≡ *Ectotrichophyton lacticolor* (Sabouraud) Castellani & Chalmers - Man. Trop. Med., ed. 3, p. 1006, 1919 ≡ *Sabouraudites lacticolor* (Sabouraud) Ota & Langeron - Annls Parasit. Hum. Comp. 1: 329, 1923 ≡ *Chlamydoaleuriosporia lacticolor* (Sabouraud) Grigorakis - Annls Sci. Nat., Bot., Sér. 10, 7: 412, 1925 ≡ *Trichophyton gypseum* Bodin var. *lacticolor* (Sabouraud) Ballagi - Derm. Wschr. 83: 1155, 1926 ≡ *Ctenomyces lacticolor* (Sabouraud) Langeron & Milochevitch - Annls Parasit. Hum. Comp. 8: 489, 1930 ≡ *Trichophyton mentagrophytes* (Robin) Blanchard var. *lacticolor* (Sabouraud) López-Martinez, Mier & Quirarte - Mycopathologia 88: 112, 1984.

Doubtful dermatophyte species.

lactis - *Mycoderma lactis* Dombrowski - Centbl. Bakt. Parasitkde, Abt. 2, 28: 377, 1928 ≡ *Kluyveromyces lactis* (Dombrowski) van der Walt - Bothalia 10: 417, 1971.

Reported from the buccal cavity of HIV-positive patients, but without clinical significance (Millon *et al.*, 1997).

lackanawa - *Madurella lackanawa* Hanan & Zurett - Arch. Derm. Syph. 37: 947-966, 1938.

Possibly a synonym of *Madurella mycetomatis* (p. 731; MacKinnon, 1954).

lambaranensis - *Geotrichoides lambaranensis* A. Sartory, R. Sartory, Rietmann & J. Meyer - Annls Inst. Pasteur 57: 544, 1936.

Doubtful, possibly *Arthrographis kalrae* (p. 440; de Hoog *et al.*, 1986).

lambica - *Mycoderma lambica* Lindner & Genoud - Wschr. Brauerei 30: 366, 1913 ≡ *Candida lambica* (Lindner & Genoud) S.A. Meyer & Ahearn - Mycotaxon 17: 297, 1983.

This is the anamorph of *Pichia fermentans.* A case of arthritis in an alcoholic was reported by Trowbridge *et al.* (1999).

lanatus - *Sabouraudites lanatus* Lebasque - Champ. Teign. Cheval Bovid. p. 42, 1933.

Doubtful zoophilic dermatophyte species.

langeronii (Rubromadurella) - *Rubromadurella langeronii* Talice - Annls Parasit. Hum. Comp. 13: 584-590, 1935 ≡ *Madurella langeronii* (Talice) Ciferri & Redaelli - Mycopathologia 3: 197, 1941.

Doubtful, probably an actinomycete.

langeronii (Trichophyton) - *Trichophyton langeronii* Milochevitch - Annls Parasit. Hum. Comp. 9: 460, 1931 ≡ *Favotrichophyton avellaneum* C.W. Dodge - Med. Mycol. p. 525, 1935 (name change).

Doubtful dermatophyte species.

lanuginosus - *Blastomycoides lanuginosus* Castellani - Br. J. Derm. 42: 372, 1930 ≡ *Glenospora lanuginosa* (Castellani) Agostini - Atti Ist. Bot. Univ. Lab. Crittogam. Pavia 3: 67, 1931 ≡ *Trichosporium gammelii* (Pollacci & Nannizzi) C.W. Dodge var. *lanuginosum* (Castellani) C.W. Dodge - Med. Mycol. p. 792, 1935 ≡ *Glenosporopsis lanuginosa* (Castellani) da Fonseca - Parasit. Méd. 1: 729, 1943.

Doubtful, possibly *Blastomyces dermatitidis.*

laryngitidis - *Saccharomyces laryngitidis* Sartory & Poitier - C.R. Hebd. Séanc. Acad. Sci., Paris 228: 1970, 1949.

Doubtful yeast species.

lecante - *Sporotrichum lecante* de Beurmann & Gougerot, *in* Sartory - Champ. Paras. Homme Anim. 1922.

Possibly an *Exophiala* species, as it was unintentionally introduced as a synonym of *Sporotrichum gougerotii* (Müller, 1965).

leproides - *Scopulariopsis leproides* Leger & Nogue - Bull. Soc. Pathol. Exot. 15: 656, 1922 ≡ *Cladosporium leproides* (Leger & Nogue) Nannizzi - Tratt. Micopat. Um. 4: 409, 1954 ≡ *Hormodendrum leproides* (Leger & Nogue) C.W. Dodge - Med. Mycol. p. 848, 1935.

Doubtful *Cladosporium* species (de Hoog, 1977).

leproticus - *Zygosaccharomyces leproticus* Froilano de Mello & de Faria - Mycopathologia 1: 203, 1938.

Doubtful yeast species (Barnett *et al.,* 1990).

lesnei - *Rhinocladium lesnei* Vuillemin - Bull. Séanc. Soc. Sci. Nancy 11: 143, 1910 ≡ *Sporotrichum lesnei* (Vuillemin) Castellani & Chalmers - Man. Trop. Med., ed. 3, p. 1121, 1919 ≡ *Graphium lesnei* (Vuillemin) Mason - Mycol. Pap. 4: 94, 1937.

Doubtful; Stalpers (1984) supposed it may have been *Sporothrix schenckii,* but the clinical picture (mycetoma) and the presence of a *Graphium* anamorph make *Pseudallescheria boydii* (p. 305) much more likely. Judging from unpublished ITS sequence data the type strain, CBS 108.10, may be a degenerate culture of *Graphium putredinis.*

lignicola - *Hyalodendron lignicola* Diddens - Zentbl. Bakt. Parasitkde, Abt. 2, 90: 317, 1934.

This is a basidiomycetous, *Trichosporon*-like species mainly isolated from woodpulp (de Hoog, 1979). A strain described by Batista *et al.* (1959) from vaginitis, CBS 315.79, is similar to *Tilletiopsis.*

lignieresii - *Aspergillus lignieresii* Costantin & Lucet - Annls Sci. Nat., Sér. Bot., 9: 137, 1905.

Aspergillus species close or identical to *A. fumigatus* (Raper & Fennell, 1965).

lilacunis - *Paecilomyces lilacunis,* misspelling (Marchese & Smoller, 1998) for *P. lilacinus* (p. 801).

lipoferum - *Trichophyton lipoferum* Kominami - Jpn. J. Med. Mycol. 2: 40, 1961.

Doubtful dermatophyte species

lipsiense - *Sporotrichum lipsiense* Benedek - Derm. Wschr. 83: 1695, 1926 ≡ *Rhinotrichum lipsiense* (Benedek) Ota - Jpn. J. Derm. Urol. 27: 929, 1927 ≡ *Rhinocladium lipsiense* (Benedek) Brumpt - Précis Parasitol., ed. 6, p. 1836, 1949.

The authentic material is lost; a secondary specimen was found to be *Geomyces pannorum* (Stalpers, 1984).

loboi (Glenosporella) - *Glenosporella loboi* O.O.R. da Fonseca & de Area Leão - Revta Méd. Cirurg. Brasil 48: 154, 1940 ≡ *Paracoccidioides loboi* (O.O.R. da Fonseca & de Area Leão) de Almeida & da Silva Lacaz - Anais Fac. Med. Univ. S. Paulo 24: 16, 1949 ≡ *Blastomyces loboi* (O.O.R. da Fonseca & de Area Leão) Langeron & Vanbreuseghem - Précis Mycol. p. 491, 1952 ≡ *Loboa loboi* (O.O.R. da Fonseca & de Area Leão) Ciferri, de Azevedo, Campos & Carneiro - Pubļcões Inst. Micol. Univ. Recife 53: 12, 1956 ≡ *Lobomyces loboi* (O.O.R. da Fonseca & de Area Leão) Borelli - Derm. Venez. 1: 286, 1958.

The type specimen probably concerned *Paracoccidioides brasiliensis* (Artagaveytia-Allende & Montmayor, 1949). The real agent of Lobo's disease has been re-introduced as *Lacazia loboi.*

loboi (Lacazia) - *Paracoccidioides loboi* O.M. Fonseca & da Silva Lacaz - Revta Inst. Med. Trop. Sao Paulo 13: 247, 1971 (invalid) ≡ *Paracoccidioides loboi* O.M. Fonseca & da Silva Lacaz ex da Silva Lacaz - Revta Inst. Med. Trop. S. Paulo 38: 230, 1996 ≡ *Lacazia loboi* (O.M. Fonseca & da Silva Lacaz ex da Silva Lacaz) P.R. Taborda, V.A. Taborda & McGinnis - J. Clin. Microbiol. 37: 2031, 1999.

The species, presently known as *Lacazia,* cannot be cultured on routine artificial media. For clinical description, see Kwon-Chung & Bennett (1992) and Haubold *et al.* (2000), and for molecular data, see Haubold *et al.* (1998). The latter case concerned a cutaneous infection in a dolphin. Note that a similar dolphin case was attributed by de Vries & Laarman (1974) to *Petriella setifera* (p. 301).

lodderae - See this list under *chalmersii.*

louisianicum - *Trichophyton louisianicum* Castellani - New Orleans Med. Surg. J. 79: 897, 1927.

Doubtful dermatophyte species; material lost. Dodge (1935) listed it as a possible synonym of *T. soudanense.*

luteolus - *Torula luteola* Saito - Jpn. J. Bot. 1: 44, 1922 ≡ *Cryptococcus luteolus* (Saito) Skinner - Am. Midl. Nat. 43: 249, 1950.

This *Cryptococcus* species (Barnett *et al.,* 1990) was reported from a pulmonary case by Binder *et al.* (1956), but the identity of the agent was questioned by Krajden *et al.* (1991). The species was found to occur on human skin by Malkina & Agarunova (1973).

luxurians - *Trichophyton luxurians* Brault & Vignier - C.R. Soc. Biol. 77: 243, 1914 ≡ *Ectotrichophyton luxurians* (Brault & Vignier) Castellani & Chalmers - Man. Trop. Med., ed. 3, p. 1004, 1919 ≡ *Favotrichophyton luxurians* (Brault & Vignier) Neveu-Lemaire - Précis Parasitol. Hum., ed. 5, p. 55, 1921 ≡ *Grubyella luxurians* (Brault & Vignier) Brumpt - Précis Parasitol., ed. 4, p. 1305, 1927.

Doubtful dermatophyte species.

macfiei - *Aspergillus macfiei* C.W. Dodge - Med. Mycol. p. 629, 1935.

Doubtful *Aspergillus* species.

mackinnonii - *Pyrenochaeta mackinnonii* Borelli - Castellania 4: 230, 1976.

Pyrenochaeta species known as rare agent of mycetoma; see p. 360..

macrosporium (Epidermophyton) - *Epidermophyton macrosporium* Kesteren - Med. J. Austr. 1: 425, 1939.

Doubtful dermatophyte species.

macrosporium (Microsporum) - *Microsporum audouinii* Gruby var. *macrosporium* Craik - Br. Med. J. 1: 672, 1921 ≡ *Sabouraudites audouinii* (Gruby) Ota & Langeron var. *macrosporium* (Craik) Nannizzi - Tratt. Micopat. Um. 4: 158, 1934.

Doubtful dermatophyte species.

macrosporoideum - *Epochnium macrosporoideum* Berkeley & Broome - Ann. Mag. Nat. Hist., Ser. 1, 13: 346, 1844 ≡ *Stemphylium macrosporoideum* (Berkeley & Broome) Saccardo - Syll. Fung. 4: 518, 1886.

Neergaard (1945) treated this species as of doubtful identity.

macrosporus - *Talaromyces flavus* (Klöcker) Stolk & Samson var *macrosporus* Stolk & Samson - Stud. Mycol. 2: 15, 1972 ≡ *Talaromyces macrosporus* (Stolk & Samson) Frisvad, Samson & Stolk, *in* Frisvad, Filtenborg, Samson & Stolk - Antonie van Leeuwenhoek 57: 186, 1990.; anam.: *Penicillium macrosporum* Frisvad, Filtenborg, Samson & Stolk - Antonie van Leeuwenhoek 57: 186, 1990.

Preferred name is *Talaromyces macrosporus.* Mentioned by Pitt (1994) from a vulval ulcer but without a case report.

madagascariense - *Cladosporium madagascariense* Verdun - Précis Parasitol. Hum. 1912 ≡ *Hormodendron madagascariense* (Verdun) C.W. Dodge - Med. Mycol. p. 845, 1935.

No material preserved; doubtful species.

madurae - *Sporobolomyces roseus* Kluyver & van Niel var. *madurae* Janke - Zentbl. Bakt. Parasitkde, Abt. 1, 161: 1954.

The type strain, CBS 2646, reidentified as *Sporobolomyces roseus,* was isolated from a case of human mycetoma. Its etiology is unconfirmed.

maduromycosis - *Scopulariopsis maduromycosis* Chen, *in* Wang, Xu & Wang - Chin. Med. J. 99: 378, 1986.

The species was invalidly described; probably *Scopulariopsis brevicaulis* was concerned.

magalhaesii - *Scedosporium magalhaesii* Fróes - Mycet. Pedis Brazil p. 49, 1930 ≡ *Monosporium magalhaesii* (Fróes) C.W. Dodge - Med. Mycol. p. 841, 1935.

Probable synonym of *Scedosporium apiospermum,* the anamorph of *Pseudallescheria boydii* (p. 305)

maltosa - *Candida maltosa* Komagata, Nakase & Katsuya - J. Gen. Appl. Microbiol., Tokyo 10: 327, 1964.

The species was described by Kitamura *et al.* (1991) from a chronic, progressive infection in a cow.

mandschuricus - *Aspergillus mandschuricus* Hanawa - Jpn. J. Derm. Urol. 20: 13, 1920.

Doubtful *Aspergillus* species.

mansonii (Indiella) - *Indiella mansonii* Brumpt - Les Mycétomes p. 63, 1906 ≡ *Madurella mansonii* (Brumpt) Vuillemin - Champ. Paras. Myc. Homme p. 226, 1931.

Sterile, hyaline fungus; cannot be identified.

mansonii (Microsporum) - *Microsporum mansonii* Castellani - Br. Med. J. 2: 1271, 1905 ≡ *Foxia mansonii* (Castellani) Castellani - J. Trop. Med. Hyg. 11: 261, 1908 ≡ *Malassezia mansonii* (Castellani) Verdun - Précis Paras. Hum., ed. 2, p. 698, 1912 ≡ *Cladosporium mansonii* (Castellani) Castellani & Chalmers - Man. Trop. Med. p. 1100, 1919 ≡ *Torula mansonii* (Castellani) Vuillemin - C.R. Hebd. Séanc. Acad. Sci., Paris 89: 406, 1919 ≡ *Sporotrichum mansonii* (Castellani) Toro - Sci. Survey Porto Rico Virgin Isl., N.Y. Acad. Sci. 8: 222, 1932 ≡ *Dematium mansonii* (Castellani) C.W. Dodge - Med. Mycol. p. 678, 1935 ≡ *Aureobasidium mansonii* (Castellani) W.B. Cooke - Mycopath. Mycol. Appl. 17: 34, 1962 ≡ *Rhinocladiella mansonii* (Castellani) Schol-Schwarz - Antonie van Leeuwenhoek 34: 122, 1968 ≡ *Wangiella mansonii* (Castellani) de Bièvre & Mariat - Bull. Soc. Fr. Mycol. Méd. 8: 127, 1979.

This name has been abandoned because of confusion (McGinnis, 1978). The neotype strain, CBS 158.58, is also the type strain of *Exophiala castellanii* (p. 649).

mantegazzae - *Trichosporium mantegazzae* Pollacci - Riv. Biol. 4: 327, 1922 ≡ *Glenospora mantegazzae* (Pollacci) Vuillemin - Champ. Paras. Myc. Homme p. 117, 1931.

The type culture, CBS 364.35 is sterile. Possibly *Pseudallescheria boydii* (p. 305) is concerned.

margaritiferum - *Oosporidium margaritiferum* Stautz - Phytopath. Z. 3: 196, 1931 ≡ *Trichosporon margaritiferum* (Stautz) Buchwald - Fung. Imp. 1939.

The maximum growth temperature of *Oosporidium margaritiferum* lies around 30°C (Barnett *et al.,* 1990). Consequently the case reported by Serpa (1958) probably concerned a misidentification.

marginatum - *Microsporum marginatum* Cazalbou - C.R. Congr. Int. Pathol. Comp. 2: 307, 1914.

Doubtful dermatophyte species.

mariatii - *Trichophyton mariatii* Tapia de Fossart, Mizrachi, Padhye & Ajello - PAHO Sci. Publ. 396: 154, 1980.

Doubtful dermatophyte species.

mautneri - See this list under *pulmonalis* (*Parendomyces*).

maydis - *Uredo maydis* De Candolle - Fl. Franc., ed. 3, 6: 77, 1815 ≡ *Ustilago maydis* (De Candolle) Corda - Ic. Fung. 5: 3,1842.

Ustilago maydis is a smut fungus of corn. Reports of this species in older literature probably were misidentifications.

mekundu - *Cryptococcus mekundu* Mattlet - Annls Soc. Belge Méd. Trop. 6: 14, 1926 ≡ *Torulopsis mena* (Fontoynont & Boucher) C.W. Dodge var. *mekundu* (Mattlet) C.W. Dodge - Med. Mycol. p. 350, 1935.

Doubtful yeast species (Barnett *et al.,* 1990).

melanographioides - *Ascosubramania melanographioides* Rajendran - J. Med. Vet. Mycol. 35: 336, 1997.

Described as the teleomorph of *Fonsecaea pedrosoi* (p. 679), but the connection between the two taxa has as yet not been confirmed. Guarro (1998) supposed that a *Microascus* species might be concerned.

melibiosica - *Candida melibiosica* Buckley & van Uden - Mycopath. Mycol. Appl. 36: 264, 1968

The species was reported from human sputum but without a convincing case report.

melinii (Candida) - *Candida melinii* Diddens & Lodder - Anaskosp. Hefen 2: 367, 1942.

This is the anamorph of *Pichia canadensis* (Wickerham) Kurtzman. A case of orchitis of the testis (Kastl & Horáček, 1980) is unconfirmed.

melinii (Penicillium) - *Penicillium melinii* Thom - The Penicillia p. 273, 1930.

Reported from a non-invasive antromycosis (Bassiouny *et al.,* 1982).

melis - *Arthroderma melis* Křivanec, Janečková & Otčenášek - Česká Mykol. 31: 92, 1977.

Geophilic dermatophyte teleomorph; type strain: CBS 669.80.

membranifaciens - *Candida guilliermondii* (Castellani) Langeron & Guerra var. *membranifaciens* Lodder & Kreger-van Rij - The Yeasts p. 518, 1952.

This is the anamorph of *Pichia ohmeri*. An endocarditis secondary to a bacterial infection was reported by Lavarde *et al.* (1985).

membranogenes (Geotrichum) - See this list under *pulmonalis-membranogenes.*

membranogenes (Saccharomyces) - *Saccharomyces membranogenes* Steinhaus - Centbl. Bakt. Parasitkde, Abt. 1, 43: 64, 1907 ≡ *Cryptococcus membranogenes* (Steinhaus) Vuillemin - Champ. Paras. Myc. Homme p. 94, 1931 ≡ *Atelosaccharomyces membranogenes* (Steinhaus) C.W. Dodge - Med. Mycol. p. 344, 1935.

Doubtful yeast species (Barnett *et al.,* 1990).

mencieri - *Aspergillus mencieri* Sartory & Flamant, *in* Sartory - Champ. Paras. Homme Anim. p. 578, 1922.

Doubtful *Aspergillus* species.

messanensis - *Mycotorula messanensis* Ciferri & Redaelli, *in* Redaelli & Ciferri - Atti Ist. Bot. Univ. Lab. Crittogam. Pavia, Ser. 5, 3: 46, 1943.

Doubtful yeast species (Barnett *et al.,* 1990).

metamericana - *Glenospora metamericana* Castellani - J. Trop. Med. Hyg. 36: 310, 1933 ≡ *Aleurisma metamericanum* (Castellani) C.W. Dodge - Med. Mycol. p. 790, 1935 ≡ *Glenosporopsis metamericana* (Castellani) da Fonseca - Parasitol. Méd. 1: 730, 1943.

Doubtful species; no material has been preserved.

micetogenum - See this list under *mycetogenum.*

microthecius - *Aspergillus microthecius* Samson & W. Gams, *in* Samson & Pitt - Adv. Pen. Asp. Syst. p. 48, 1985.

This is the anamorph of *Emericella parvathecia* (Raper & Fennell) Malloch & Cain. It was described from human skin (Raper & Fennell, 1965), but without a convincing case report.

milochevitchii - *Achorion milochevitchii* Langeron & Baeza - Annls Parasit. Hum. Comp. 14: 399, 1936 ≡ *Trichophyton milochevitchii* (Langeron & Baeza) Ansari & Faghih - Annls Parasit. Hum. Comp. 36: 218, 1951.

Doubtful dermatophyte species.

minima - *Scopulariopsis minima* Sartory, Hufschmitt & J. Meyer - Bull. Acad. Méd. 103: 606, 1930.

Doubtful hyphomycete (Morton & Smith, 1963).

minioluteum - *Penicillium minioluteum* Dierckx - Annls Soc. Sci. Brux. 25: 87, 1901.

Pitt (1994) mentioned strains from clinical sources, but did not provide a case report.

minutissimum - *Microsporum minutissimum* Burchardt - Ueber Chloasma Vorkomm. Pilzf. 1859 ≡ *Sporotrichum minutissimum* (Burchardt) Saccardo - Syll. Fung. 4: 100, 1886 ≡ *Microsporoides minutissimus* (Burchardt) Neveu-Lemaire - Précis Parasitol. Hum. p. 607, 1907 ≡ *Actinomyces minutissimus* (Burchardt) Brumpt - Précis Parasitol., ed. 4, p. 1199, 1927 ≡ *Oospora minutissima* (Burchardt) Ridet - Les Oosp. Oosporoses p. 68, 1911 ≡ *Nocardia minutissima* (Burchardt) Castellani & Chalmers - Man. Trop. Med., ed. 2, p. 819, 1913 ≡ *Proactinomyces minutissimus* (Burchardt) Krassilnikov - Opred. Actinom. 1941 ≡ *Corynebacterium minutissimum* (Burchardt) Sarkany, Taplin & Blanc - Archs Derm. 85: 578, 1962.

The species is the bacterial agent of human erythrasma; the correct name is *Corynebacterium minutissimum* (Sneath *et al.,* 1986).

mirandei - *Cryptococcus mirandei* Velu - Bull. Soc. Pathol. Exot. 17: 545, 1924 ≡ *Torulopsis mirandei* (Velu) de Almeida - Anais Fac. Med. Univ. S. Paulo 9: 76, 1933 ≡ *Castellania mirandei* (Velu) C.W. Dodge - Med. Mycol. p. 251, 1935.

Doubtful yeast species (Barnett *et al.*, 1990).

mitis - *Cryptococcus mitis* Mazza, de Nucci & Feijóo - Bol. Inst. Clin. Quirúrg. Univ. B. Aires 5: 293, 1930 ≡ *Torulopsis mitis* (Mazza, de Nucci & Feijóo) C.W. Dodge - Med. Mycol. p. 350, 1935.

Doubtful yeast species (Barnett *et al.*, 1990).

mongolica - *Grubyella schoenleinii* (Lebert) Ota & Langeron var. *mongolica* Hashimoto & Ota - Jpn. J. Derm. Urol. 27: 386, 1927 ≡ *Trichophyton schoenleinii* (Lebert) Nannizzi var. *mongolicum* (Hashimoto) Nannizzi - Tratt. Micopat. Um. 4: 200, 1934 ≡ *Achorion schoenleinii* (Lebert) Remak var. *mongolicum* (Hashimoto & Ota) C.W. Dodge - Med. Mycol. p. 559, 1935.

Doubtful dermatophyte species.

montevideënse - *Aspergillus montevideënsis* Talice & Mackinnon - C.R. Séanc. Soc. Biol. Argent. 108: 1007, 1931 ≡ *Eurotium montevideënse* (Talice & Mackinnon) Malloch & Cain - Can. J. Bot. 50: 64, 1965.

The authentic strain, CBS 491.65 was isolated from a case of human otomycosis, but no convincing case report was published.

montoyae - *Monilia montoyae* Castellani & Chalmers - Man. Trop. Med. p. 622, 1910 ≡ *Myceloblastanon montoyae* (Castellani & Chalmers) Ota - Jpn. J. Derm. Urol. 27: 178, 1927.

Doubtful yeast species (Barnett *et al.*, 1990).

mottae - See this list under *epigaeum* (*Penicillium*).

mugera - *Cryptococcus mugera* Mattlet - Annls Soc. Belge Méd. Trop. 6: 12, 1926 ≡ *Asporomyces mugera* (Mattlet) C.W. Dodge - Med. Mycol. p. 356, 1935.

Doubtful yeast species (Barnett *et al.*, 1990).

muisa - *Torula muisa* Mattlet - Annls Soc. Belge Méd. Trop. 6: 25, 1926 ≡ *Monilia muisa* (Mattlet) Nannizzi - Tratt. Micopat. Um. 4: 369, 1934 ≡ *Geotrichum muisa* (Mattlet) C.W. Dodge - Med. Mycol. p. 220, 1935.

Doubtful species (de Hoog *et al.*, 1986).

multifermentans - *Geotrichum multifermentans* Castellani, *in* Castellani & Jacono - J. Trop. Med. Hyg. 36: 303, 1936 ≡ *Mycoderma multifermentans* (Castellani) C.W. Dodge - Med. Mycol. p. 228, 1935.

Doubtful species (de Hoog *et al.*, 1986).

muris - *Microsporum muris* Gluge & d'Udekem - Bull. Acad. R. Belg., Cl. Sci. 3: 338, 1857 ≡ *Sporotrichum muris* (Gluge & d'Udekem) Saccardo - Syll. Fung. 10: 533, 1893 ≡ *Malassezia muris* (Gluge & d'Udekem) Escomel - Bull. Soc. Path. Exot. 12: 350, 1924 ≡ *Achorion muris* (Gluge & d'Udekem) C.W. Dodge - Med. Mycol. p. 555, 1935.

Doubtful dermatophyte species (Stalpers 1984).

murorum - *Chaetomium murorum* Corda - Ic. Fung. 1: 24, 1837.

Cellulolytic fungus. The case described by Liu *et al.* (1995) is unconfirmed.

muthuonii - *Acremonium muthuonii* Fontoynont & Boucher - Annls Derm. Syph., Sér. 6, 4: 343, 1923.

This is an *Acremonium* species close to *A. kiliense* (Gams, 1971); authentic material is lost.

muyaga - *Mycoderma muyaga* Mattlet - Annls Soc. Belge Méd. Trop. 6: 28, 192 ≡ *Geotrichum muyaga* (Mattlet) Nannizzi - Tratt. Micopat. Um. 4: 395, 1934.

Type material is lost; the species is regarded as a synonym of *Geotrichum candidum* (Carmichael, 1957).

mycetogenum - *Penicillium mycetogenum* Mantelli & Negri - Giorn. Accad. Med. Torino, Ser. 4, 21: 165, 1915 ≡ *Penicillium mycetomagenum* Negri - Atti R. Accad. Sci. Torino 56: 33, 1921 (name change).

Doubtful *Penicillium* species, possibly *P. aurantiogriseum* (Pitt, 1979).

mycetomagenum - See this list under *mycetogenum*.

mycetomi (*Aspergillus*) - *Aspergillus mycetomi* Villabruzzi & Gelonesi - Ann. Med. Nav. Colon., Roma 33: 283, 1927.

Doubtful, possibly *Madurella*-like (Raper & Fennell, 1965).

mycetomi (*Mucor*) - *Mucor mycetomi* Gelonesi - Ann. Med. Nav. Colon., Roma 33: 303, 1927.

Doubtful zygomycete species.

mycetomi (*Penicillium*) - *Penicillium mycetomi* Neveu-Lemaire - Précis Parasitol. Hum. p. 123, 1908.

Doubtful *Penicilllium* species (Pitt, 1979).

myrmeciae - *Cryptococcus myrmeciae* Chalmers & Christopherson - J. Trop. Med. Hyg. 17: 129, 1914 ≡ *Torulopsis myrmeciae* (Chalmers & Christopherson) de Almeida - Anais Fac. Med. Univ. S. Paulo 9: 76, 1922.

Doubtful yeast species (Barnett *et al.*, 1990).

nakamurae - *Ectotrichophyton nakamurae* C.W. Dodge - Med. Mycol. p. 502, 1935.

Doubtful dermatophyte species.

natricis - *Cephalosporium acremonium* Corda var. *natricis* Frágner - Paras. Pilze Mensch p. 199, 1958.

Invalidly published *Acremonium* species (Gams, 1971).

neveuxii - *Cryptococcus neveuxii* Vuillemin & Lasseur, *in* Lasseur & Servet - Revue Méd. Est. 50: 357-370, 1922 ≡ *Torulopsis neveuxii* (Vuillemin & Lasseur) de Almeida - Anais Fac. Med. Univ. S. Paulo 9: 76, 1933.

Doubtful yeast species (Barnett *et al.*, 1990).

niger (*Hyalopus*) - See this list under *nigrum* (*Cephalosporium*).

niger (*Saccharomyces*) - *Saccharomyces niger* Maffuci & Sirleo - Policlinico, Sez. Chirurg. 2: 143, 1895 ≡ *Cryptococcus niger* (Maffuci & Sirleo) Gedoelst - Champ. Paras. Homme Anim. Dom. p. 59, 1902 ≡ *Torulopsis nigra* (Maffuci & Sirleo) de Almeida - Anais Fac. Med. Univ. S. Paulo 9: 76, 1933.

Doubtful species (Barnett *et al.*, 1990).

nigra (*Periconia*) - *Periconia tenuissima* Peck var. *nigra* Redaelli - Micet. Assoc. Microb. Tuberc. Polmon. Cavit. p. 39, 1925.

Doubtful hyphomycete; original material not known to be preserved

nigra (*Torulopsis*) - See this list under niger (*Saccharomyces*).

nigrescens - *Aspergillus nigrescens* Robin - Hist. Nat. Veg. Paras. p. 518, 1853.

Aspergillus species close to, or identical with, *A. fumigatus* (Raper & Fennell, 1965).

nigricans - *Penicillium nigricans* Bainier, *in* Thom - The Penicillia p. 351, 1930.

This species (type strain: CBS 354.48) is a synonym of the soil fungus *P. janczewskii* Zaleski. It was mentioned as an agent of pulmonary aspergilloma by Mori *et al.* (1993), but without a case report.

nigrum (*Acrothecium*) - *Acrothecium nigrum* Ciferri - Annls Parasit. Hum. Comp. 7: 532, 1929.

The type strain, CBS 105.28 is now sterile. Judging from the original description, *Curvularia lunata* or a similar species may have been concerned.

nigrum (*Cephalosporium*) - *Cephalosporium nigrum* Kambayashi - Bot. Mag., Tokyo 51: 443, 1937 ≡ *Hyalopus niger* (Kambayashi) Barbosa - Thesis, 1941.

Doubtful; material lost. Gams (1971) supposed that a *Phialophora* species may have been concerned.

nivalis - *Lanosa nivalis* Fries - Syst. Orb. Veg. p. 317, 1825 ≡ *Fusarium nivale* (Fries) Sorauer - Z. Pflkrankh. 11: 220, 1901 ≡ *Microdochium nivale* (Fries) Samuels & Hallett - Trans. Br. Mycol. Soc. 81: 479, 1983.

The species was mentioned in several lists of pathogenic taxa (Otčenášek & Dvořák, 1985; Thomas, 1994), but no convincing case has been published. This somewhat psychrophilic taxon is now known under the name *Microdochium nivale*, anamorph of *Monographella nivalis* (Schaffnit) E. Müller.

niveo-velutina - *Mortierella niveo-velutina* Ciferri & Ashford - Puerto Rico J. Publ. Health Trop. Med. 5: 142, 1929.

Doubtful species, probably not a *Mortierella* (W. Gams, pers. comm.).

niveum (*Epidermophyton*) - *Epidermophyton niveum* MacCarthy - Annls Derm. Syph., Sér. 6, 6: 44, 1925.

Doubtful dermatophyte species, close to *Trichophyton mentagrophytes*.

niveum (*Microsporum*) - *Microsporum tomentosum* Pelagatti var. *niveum* Truffi - Arch. Ital. Derm. Sif. Venereol. 1: 210, 1925 ≡ *Microsporum niveum* (Truffi) Caruso - G. Ital. Derm. Sif. 67: 280, 1926 ≡ *Microsporum felineum* Fox & Blaxall var. *niveum* (Truffi) C.W. Dodge - Med. Mycol. p. 542, 1935.

Doubtful dermatophyte species.

nobile - *Mycoderma nobile* Sartory, R. Sartory, J. Meyer & Charles - Annls Mycol. 29: 325, 1931 ≡ *Geotrichum nobile* (Sartory, R. Sartory, J. Meyer & Charles) Nannizzi - Tratt. Micopat. Um. 4: 395, 1934.

No material is known to be preserved. It may have been a *Lecythophora* species (de Hoog *et al.*, 1986).

nodiformans - *Trichophyton nodiformans* Castellani - Proc. R. Soc. Med., Derm. Sect., 6: 35, 1912 ≡ *Ectotrichophyton nodiformans* (Castellani) Castellani & Chalmers - Man. Trop. Med., ed. 3, p. 1007, 1919 ≡ *Megatrichophyton nodiformans* (Castellani) Neveu-Lemaire - Précis Parasitol. Hum., éd. 5, p. 55, 1921.

Doubtful dermatophyte species.

noelting - *Aspergillus noelting* Hallier - Z. Parasitkde 2: 259-284, 1870.

Doubtful *Aspergillus* species, possibly *A. flavus* (Raper & Fennell, 1965).

nogamii - *Castellania nogamii* C.W. Dodge - Med. Mycol. p. 248, 1935.

Doubtful yeast species (Barnett *et al.*, 1990).

nyabisii - *Mycoderma nyabisii* Mattlet - Annls Soc. Belge Méd. Trop. 6: 30, 1926 ≡ *Geotrichum nyabisii* (Mattlet) Basgal - Contrib. Estudo Blastom. Pulmon. p. 49, 1931.

No material is preserved; the name is doubtful (Carmichael, 1957).

obovoideum - *Rhinocladiella obovoidea* Matsushima - Ic. Microf. Matsushima Lect. p. 123, 1975 ≡ *Ramichloridium obovoideum* (Matsushima) de Hoog - Stud. Mycol. 15: 73, 1977.

This is a litter-inhabiting fungus, morphologically resembling the neurotropic *R. mackenziei* (p. 888) but in its 26S rDNA sequences clearly distinct from that taxon (E. Guého, pers. comm.).

oceanicum - *Trichophyton oceanicum* Dompmartin, Drouhet & F. Moreau - Bull. Soc. Fr. Derm. Syph. 82: 426, 1976.

Invalidly described for slow-growing strains similar to *Trichophyton rubrum*.

ochoterenai - *Grubyella ochoterenai* Ciferri - Annls Parasit. Hum. Comp. 7: 511, 1929 ≡ *Trichophyton ochoterenai* (Ciferri) Nannizzi - Tratt. Micopat. Um. 4: 210, 1934.

Doubtful dermatophyte species.

ohmeri - *Endomycopsis ohmeri* Etchells & Bell - Food Technol. 4: 82, 1950 ≡ *Pichia ohmeri* (Etchells & Bell) Kreger-van Rij - Tax. Stud. Yeast Gen. Endom. Pichia Debar. p. 98, 1964.

Pichia ohmeri is the teleomorph of *Candida guilliermondii* var. *membranaefaciens*. An endocarditis secondary to a bacterial infection was reported by Lavarde *et al.* (1985) and a disseminated infection by Bergman *et al.* (1998).

oláhii - *Arthoderma oláhii* Balogh, Liptovsky & Nagy-Peti - Mycopath. Mycol. Appl. 40: 76, 1970.

Invalidly described dermatophyte teleomorph; no material is known to be preserved.

olivispora - *Acremoniella olivispora* Ciferri & Ashford - Mycologia 22: 67, 1930.

Humicola-like species; the authentic material is probably lost.

onychophila (Allantospora) - *Allantospora onychophila* Vuillemin - Champ. Paras. Myc. Homme p. 63, 1931 ≡ *Hyalopus onychophilus* (Vuillemin) Aschieri - Atti Ist. Bot. Univ. Lab. Crittogam. Pavia., Ser. 4, 3: 49, 1932.

The type strain, CBS 224.34, is a *Fusarium* species close to *F. oxysporum* (Gams, 1971).

onychophilum (Coniosporium) - *Coniosporium onychophilum* Agostini - Boll. Sez. Ital. Soc. Internaz. Microbiol. 3: 214, 1931.

Neither literature nor material was available for study.

onychis - *Pichia onychis* Yarrow - Antonie van Leeuwenhoek 31: 465, 1965.

This yeast species was described from human nails, but without a convincing case report (Yarrow, 1965).

orbiculare - *Pityrosporum orbiculare* Gordon - Mycologia 43: 532, 1951.

Malassezia species; in absence of authentic material it cannot be

assigned to any of the recently distinguished species (Guého *et al.*, 1996).

orbiculata - *Torulopsis orbiculata* Ruiz - An. Inst. Biol. Mex. 14: 369, 1943.

Doubtful yeast species.

osloënsis - *Saccharomyces veronae* Lodder & Kreger-van Rij var. *osloënsis* Dietrichson - Annls Parasit. Hum. Comp. 29: 466, 1954.

The type strain, CBS 2919 isolated from human sputum, has been reidentified as *Saccharomyces cerevisiae*.

osloensis - *Torulopsis osloensis* Dietrichson - Annls Parasit. Hum. Comp. 29: 475, 1954.

Type strain CBS 2752 has been reidentified as the non-pathogenic species *Zygosaccharomyces rouxii* (Boutroux) Yarrow (Barnett *et al.*, 1990). Originally described from sputum, without case report.

oswaldoi - *Madurella oswaldoi* Horta - Patol. Geral 4: 1, 1919.

Authentic material lost; doubtful fungus.

otae (Cryptococcus) - *Cryptococcus otae* Nannizzi - Tratt. Micopat. Um. 4: 329, 1934 ≡ *Syringospora otae* (Nannizzi) C.W. Dodge - Med. Mycol. p. 281, 1935.

Doubtful yeast species (Barnett *et al.*, 1990).

otae (Ectotrichophyton) - See this list under *blanche*.

otomycosis - *Fusarium otomycosis* Ming & Yu - Acta Microbiol. Sin. 12: 178, 1966.

Doubtful *Fusarium* species.

ovale (Trichosporon) - *Trichosporon ovale* Unna, *in* Vuillemin - Archs Parasit. 5: 38-66, 1902.

Doubtful *Trichosporon* species (Guého *et al.*, 1992a).

ovale (Trichosporon) - *Trichosporon ovale* Paoli - Giorn. Ital. Malad. Ven. Pelle 54: 566, 1913.

Doubtful *Trichosporon* species.

oxalicum - *Penicillium oxalicum* Currie & Thom - J. Biol. Chem. 22: 289, 1915.

An ocular infection by this species was reported by Rodriguez de Kopp & Vidal (1998).

palatii - *Blastodendrion palatii* Fröhlich, Zach & Piringer - Arch. Derm. Syph. 179: 521, 1939.

Doubtful yeast species (Barnett *et al.*, 1990).

palmioleophila - *Candida palmioleophila* Nakase & Itho, *in* Nakase, Itoh, Suzuki, Komagata & Kodama - J. Gen. Appl. Microbiol., Tokyo 34: 496, 1988.

The agent of a catheter-associated fungemia, reported as *Torulopsis candida* (St.Germain & Laverdière, 1986) was reidentified by Sugita *et al.* (1999) as *Candida palmioleophila*, a palm-oil degrading species (Nakase *et al.*, 1988).

parachalmersii - *Monilia parachalmersii* Castellani & Chalmers - Man. Trop. Med., ed. 3, p. 1087, 1919 ≡ *Myceloblastanon parachalmersii* (Castellani & Chalmers) Ota - Jpn. J. Derm. Urol. 27: 178, 1927 ≡ *Parasaccharomyces parachalmersii* (Castellani & Chalmers) C.W. Dodge - Med. Mycol. p. 267, 1935.

Doubtful yeast species (Barnett *et al.*, 1990).

paraferrugineum - *Microsporum paraferrugineum* de Magalhães - Mem. Inst. Oswaldo Cruz 30: 46, 1935.

Doubtful dermatophyte species.

pararugosa - *Hemispora pararugosa* Castellani & Douglas - J. Trop. Med. Hyg. 24: 129, 1921 ≡ *Trichosporon pararugosum* (Castellani & Douglas) Nannizzi - Tratt. Micopat. Um. 4: 292, 1934 ≡ *Mycoderma pararugosum* (Castellani & Douglas) C.W. Dodge - Med. Mycol. p. 225, 1935 ≡ *Geotrichum pararugosum* (Castellani & Douglas) de Almeida & Lacaz - Folia Clin. Biol., S. Paulo 12: 46, 1940.

An authentic strain, CBS 2478, is a *Trichosporon* species close to *T. coremiiforme* (M. Moore) Guého & M.Th. Smith.

parasiticus - *Hyalobyssus moniliformis* Zukal var. *parasiticus* Maffei - Atti Ist. Bot. Lab. Crittog. Univ. Pavia, Ser. 4a, 3: 27, 1933.

No material is known to be preserved. Chains of somewhat rough-walled chlamydospores were illustrated; the identity of this fungus is doubtful.

parvathecius - *Aspergillus parvathecius* Raper & Fennell - Gen. Asp. p. 509, 1965 ≡ *Emericella parvathecia* (Raper & Fennell) Malloch & Cain -

Can. J. Bot. 50: 62, 1972.

This species was described from human skin (Raper & Fennell, 1965) but without a convincing case report. The correct name is *Emericella parvathecia;* the anamorph is *Aspergillus microthecius.*

pastorianus - *Saccharomyces pastorianus* Hansen - Centbl. Bakt. Parasitkde, Abt. 2, 12: 350, 1904.

The species is doubtfully distinct from *S. cerevisiae* (Barnett *et al.*, 1990). It was implicated in a gastritis by Ahlund *et al.* (1967).

pedis - *Trichophyton pedis* Ota - Bull. Soc. Path. Exot. 15: 594, 1922 ≡ *Epidermophyton pedis* (Ota) C.W. Dodge - Med. Mycol. p. 480, 1935 ≡ *Kaufmanwolfia pedis* (Ota) Novák & Galgóczy - Acta Bot. Hung. 15: 132, 1969.

Doubtful dermatophyte species, probably *Trichophyton rubrum.*

pedrosoi - *Microascus pedrosoi* Fuentes & Wolf - Mycologia 48: 63, 1956.

Described as the teleomorph of *Fonsecaea pedrosoi*, but it was later found to be a mixed culture. The ascomycete was reidentified as *Microascus cinereus* (Fuentes & Wolf, 1956).

peltata - *Torulopsis peltata* Yarrow - Antonie van Leeuwenhoek 35: 418, 1969 ≡ *Selenozyma peltata* (Yarrow) Yarrow, *in* von Arx, Rodrigues de Miranda, Smith & Yarrow - Stud. Mycol. 14: 29, 1977 ≡ *Candida peltata* (Yarrow) S.A. Meyer & Ahearn - Mycotaxon 17: 298, 1983.

Candida peltata is known from a single strain, CBS 5576, from milk of a mastitic cow. It was described without a convincing case report (Yarrow, 1968).

penicillioides - *Aspergillus penicillioides* Spegazzini - Revta Agr. Vet. La Plata 1896: 246, 1896.

This is an *Aspergillus* species (Thom & Raper, 1945) close to *A. restrictus* (p. 500). A case was reported by Marsálek *et al.* (1960). The CBS collection holds a strain from human skin.

peralbida - *Glenosporella peralbida* Castellani - Mycopathologia 23: 166, 1964.

Original material is probably lost; from the description it cannot be identified.

pereirae - *Sporotrichum pereirae* Miranda - Um Novo Espor. 1936 (invalid) ≡ *Sporotrichum pereirae* Miranda ex Redaelli & Ciferri - Tratt. Micopat. Um. 5: 460, 1942.

Doubtful species; possibly *Sporothrix schenckii* was concerned (Stalpers, 1984).

perieri - *Oospora perieri* Matruchot & Antoine - Annls Inst. Pasteur 32: 202, 1918 ≡ *Monilia perieri* (Matruchot & Antoine) Castellani & Chalmers - Man. Trop. Med., ed. 3, p. 1092, 1919 ≡ *Trichosporon perieri* (Matruchot & Antoine) Ota - Jpn. J. Derm. Urol. 28: 16, 1928 ≡ *Candida perieri* (Matruchot & Antoine) de Almeida - Anais Fac. Med. Univ. S. Paulo 9: 11, 1933 ≡ *Proteomyces perieri* (Matruchot & Antoine) C.W. Dodge - Med. Mycol. p. 213, 1935.

Doubtful yeast species (Barnett *et al.*, 1990).

perpulchrum - *Chaetomium perpulchrum* L. Ames - Monogr. Chaetom. p. 33, 1963.

The species was listed as doubtful by von Arx *et al.* (1986). A strain from onychomycosis published by Costa *et al.* (1988b) was reidentified by Abbott *et al.* (1995) and Guarro *et al.* (1995) as *C. globosum.*

pertenue - *Microsporum pertenue* Plaut, *in* Klehmet - Dt. Med. Wschr. 45: 1188, 1919 ≡ *Sabouraudites pertenuis* (Plaut) Nannizzi - Tratt. Micopat. Um. 4: 165, 1934 ≡ *Microsporum audouinii* Gruby var. *depauperatum* (Guéguen) C.W. Dodge f. *pertenue* (Plaut) C.W. Dodge - Med. Mycol. p. 551, 1935.

Dermatophyte species probably identical to *Microsporum audouinii.*

pettiti - *Torula pettiti* Hranova - (publication not located) ≡ *Cryptococcus pettiti* (Hranova) Nannizzi - Tratt. Micopat. Um. 4: 318, 1934.

Doubtful yeast species (Barnett *et al.*, 1990).

phoenicis - *Ustilago phoenicis* Corda - Ic. Fung. 4: 9, 1840 ≡ *Aspergillus phoenicis* (Corda) Thom - The Aspergilli p. 175, 1926.

A doubtful case of cutaneous mycosis by this *Aspergillus* species is listed under *hominis.*

pictor (Monilia) - *Monilia pictor* Neveu-Lemaire - Précis Parasitol. Hum., ed. 4, 1908.

Doubtful yeast species (Barnett *et al.*, 1990).

pictor (Trichophyton) - *Trichophyton pictor* Blanchard, *in* Bouchard - Traité Pathol. Gén. 2: 916, 1895.

Doubtful dermatophyte species.

pittalugae - *Achorion pittalugae* Langeron & Baeza - Annls Parasit. Hum. Comp. 14: 399, 1936.

Doubtful dermatophyte species.

planum - *Epidermophyton planum* Kesteren - Med. J. Austr. 1: 426, 1939.

Doubtful dermatophyte species.

platensis - *Monilia platensis* Peruchena - Semana Méd. 36: 527, 1929 ≡ *Castellania platensis* (Peruchena) C.W. Dodge - Med. Mycol. p. 262, 1935.

Doubtful yeast species (Barnett *et al.*, 1990).

pleurospora - *Phoma pleurospora* Saccardo - Michelia 2: 97, 1880 ≡ *Dendrophoma pleurospora* (Saccardo) Saccardo - Syll. Fung. 3: 178, 1884 ≡ *Pleurophoma pleurospora* (Saccardo) von Höhnel - Sber. Akad. Wiss. Wien 123: 117, 1914.

We were unable to obtain sporulation in cultures of this *Pleurophoma* species; for a description, see Sutton (1980). Cutaneous lesions in a diabetic patient were described by Dooley *et al.* (1980).

pluriseptatum - *Sporidesmium mucosum* Saccardo var. *pluriseptatum* Karsten & Hariot - J. Bot., Paris 4: 363, 1890 ≡ *Alternaria pluriseptata* (Karsten & Hariot) Jørstad - Meld. Stat. Plantepat. Inst. 1: 95, 1945.

The species is occasionally mentioned (e.g., Wadhwani & Srivastava, 1985), but all medical stains we have seen were reidentified with other *Alternaria* species.

poikilospora - *Torula poikilospora* Takahashi - Jpn. J. Derm. Urol. 41: 31, 1937.

Borelli (1980) supposed this was a dysplastic variant of *Fonsecaea pedrosoi;* no authentic material has been preserved.

pollaccii - *Sporobolomyces pollaccii* Verona & Ciferri - Atti Ist. Bot. Univ. Lab. Crittogam. Pavia 4: 248, 1939.

The type strain, CBS 485, from a skin lesion, was reidentified as *Sporobolomyces roseus* (Barnett *et al.*, 1990), which is otherwise known as a saprobe.

polymorpha - *Hansenula polymorpha* de Morais & Maria - Anais Esc. Sup. Quím. Univ. Recife 1: 15, 1959.

This is a synonym of *Pichia angusta* (Theunisson *et al.*) Kurtzman, wich is a food-borne fungus. The fungus was isolated from lymph nodes of human patient (McGinnis *et al.*, 1980).

porriginis - *Oidium porriginis* Montagne, *in* Berkeley & Broome - Annls Mag. Nat. Hist., Ser. 2, 7: 178, 1851 ≡ *Oospora porriginis* (Montagne) Saccardo - Syll. Fung. 4: 15, 1886.

Doubtful dermatophyte species, possibly *Trichophyton schoen-leinii* (Rippon, 1986).

pratense - *Trichophyton pratense* Szathmáry - Z. Haut- u. Geschlkrankh. 67: 675, 1941.

Dermatophyte close to *Trichophyton mentagrophytes.*

pruinosum - *Trichophyton pruinosum* Catanei - Bull. Soc. Path. Exot. 24: 296, 1931 ≡ *Favotrichophyton pruinosum* (Catanei) C.W. Dodge - Med. Mycol. p. 520, 1935.

Doubtful dermatophyte species.

pseudo-umbonatum - *Microsporum pseudo-umbonatum* Sartory, R. Sartory, J. Meyer & Malgras - C.R. Hebd. Séanc. Acad. Sci., Paris 229: 138, 1949.

Doubtful dermatophyte species.

pseudotumoralis - *Candida pseudotumoralis* Morquer, Puget & Bazex - Revue Mycol. 19: 81, 1954.

Doubtful yeast species (Barnett *et al.*, 1990).

pubescens - *Microsporum pubescens* Sabouraud - Malad. Cuir Chevelu 3: 243, 1910 ≡ *Sabouraudites pubescens* (Sabouraud) Ota & Langeron - Annls Parasit. Hum. Comp. 1: 328, 1923 ≡ *Closterosporia pubescens* (Sabouraud) Grigorakis - Ann. Sci. Nat., Bot., Sér. 10, 7: 411, 1925.

Doubtful dermatophyte species.

pullulans - *Oidium pullulans* Lindner - Mikrosk. Betriebskontr. Gärungsgew., ed. 2, p. 286, 1901 ≡ *Trichosporon pullulans* (Lindner)

Diddens & Lodder - Anaskosp. Hefen p. 413, 1942.

Guého *et al.* (1992b) in their revision of *Trichosporon* listed some particularly psychrophilic strains under *T. pullulans*. If thus seems likely that case reports under this name (e.g., Catellani, 1960; Hughes *et al.*, 1988; Kunovó *et al.*, 1995) are misidentifications.

pulmonale (Haplographium) - *Haplographium debellaemarengoi* Pollacci var. *pulmonale* Redaelli - Micet. Assoc. Microb. Tuberc. Pulmon. Cavit. Pavia p. 65, 1925.

Hyphomycete strongly reminiscent of *Cladosporium sphaerospermum* (p. 591).

pulmonale (Rhinosporidium) - *Rhinosporidium pulmonale* Kirschenblatt - C.R. Hebd. Séanc. Acad. Sci., Paris 33: 406, 1939.

No material is preserved; this species probably was *Emmonsia crescens* (Sigler, 1996).

pulmonalis (Antennaria) - *Antennaria pulmonalis* Redaelli - Micet. Assoc. Microb. Tuberc. Pulmon. Cavit. Pavia p. 29, 1925 ≡ *Antennularia pulmonalis* (Redaelli) Nannizzi - Tratt. Micopat. Um. 4: 281, 1934.

Doubtful coelomycete species.

pulmonalis (Parendomyces) - *Parendomyces pulmonalis* Plaut, *in* Mautner - Centbl. Bakt. Parasitkde, Abt. 1, 74: 207, 1914 ≡ *Monilia pulmonale* (Plaut) Sartory - Champ. Parasit. Homme Anim. p. 709, 1922 ≡ *Myceloblastanon pulmonale* (Plaut) Ota - Jpn. J. Derm. Urol. 27: 178, 1927 ≡ *Candida pulmonalis* (Plaut) de Almeida - Anais Fac. Med. Univ. S. Paulo 9: 77, 1933 ≡ *Monilia mautneri* Nannizzi - Tratt. Micopat. Um. 4: 366, 1934 (name change).

Doubtful yeast species.

pulmonalis-membranogenes - *Zymonema pulmonalis- membranogenes* Martins - C.R. Soc. Biol. 98: 1162, 1928 ≡ *Geotrichum membranogenes* C.W. Dodge - Med. Mycol. p. 222, 1935 (name change).

Doubtful *Geotrichum* species, possibly *G. clavatum* (de Hoog *et al.*, 1986).

pulmoneum (Oidium) - *Oidium pulmoneum* Bennett - Trans. R. Soc. Edinburgh 15: 277, 1842 ≡ *Oospora pulmoneum* (Bennett) Saccardo - Syll. Fung. 4: 16, 1886 ≡ *Mycoderma pulmoneum* (Bennett) Vuillemin - Bull. Mens. Séanc. Soc. Sci. Nancy 3: 19, 1891 ≡ *Monilia pulmonea* (Bennett) Castellani & Chalmers - Man. Trop. Med., ed. 2, p. 829, 1913 ≡ *Geotrichum pulmoneum* (Bennett) Basgal - Contrib. Est. Blastom. Pulmon. p. 49, 1931.

The original material is lost; it may have been a *Geotrichum* species (de Hoog *et al.*, 1986).

pulmoneum (Oidium) - *Oidium pulmoneum* Magalhães - Brasil. Med. 28: 313, 1914 ≡ *Neogeotrichum pulmoneum* (Magalhães) Magalhães - Mem. Inst. Oswaldo Cruz 26: 151-167, 1932.

Doubtful, possibly a *Trichosporon*-like species.

pulmonum-hominis - *Apergillus pulmonum-hominis* Welcker, *in* Küchenmeister - Lebend. Mensch. Vork. Paras., Z. Pfl. Paras. p. 144, 1855.

Aspergillus species close or identical to *A. fumigatus* (Raper & Fennell, 1965).

pulverulentum - *Microsporum canis* Bodin var. *pulverulentum* Rivalier & Badillet - Annls Parasit. Hum. Comp. 44: 265-272, 1969.

Doubtful dermatophyte species.

puntonii - *Corethropsis puntonii* Vuillemin - C.R. Hebd. Séanc. Acad. Sci., Paris 190: 1334, 1930 ≡ *Paecilomyces puntonii* (Vuillemin) Nannizzi - Tratt. Micopat. Um. 4: 245, 1934.

The strain is only known from a single isolate from human skin, without a convincing case report (Vuillemin, 1930). A taxonomic description was given by Samson (1974) under the name *Paecilomyces puntonii*.

pyogenes - *Cryptococcus pyogenes* Mattlet - Annls Soc. Belge Méd. Trop. 6: 13, 1926 ≡ *Atelosaccharomyces pyogenes* (Mattlet) C.W. Dodge - Med. Mycol. p. 345, 1935.

Doubtful yeast species (Barnett *et al.*, 1990).

qizutongii - *Aspergillus qizutongii* D.-M. Li, Horie, Y.-X. Wang & R.Y. Li - Mycoscience 39: 301, 1998.

Described from a human maxillary sinusitis, but without case report.

quasilinguae-pilosae - *Cryptococcus quasilingae-pilosae* Cooper - Am.

J. Trop. Med. 12: 97, 1932.

Doubtful yeast species (Barnett *et al.*, 1990).

quintanilhae - *Piedraia quintanilhae* van Uden & de Barros-Machado - Revta Bras. Portug. Biol. Ger. 3: 271, 1963.

This species differs from *P. hortae* (p. 303) by absence of ascospore appendages. No material was available for study.

rabesalama - *Mycoderma rabesalama* Fontoynont & Boucher - Annls Derm. Syph., Sér. 4, 6: 330, 1923 ≡ *Geotrichum rabesalama* (Fontoynont & Boucher) Basgal - Contrib. Estudo Blastom. Pulmon. p. 49, 1931 ≡ *Pseudomycoderma rabesalama* (Fontoynont & Boucher) C.W. Dodge - Med. Mycol. p. 236, 1935.

Doubtful arthroconidial yeast. The description is strongly reminiscent of *Trichosporon inkin*, but the clinical picture does not correspond with that species.

racovitzae - *Myxotrichum racovitzae* Lagarde - Arch. Zool. Exp. Gén. 53: 280, 1913 ≡ *Gymnoascus racovitzae* (Lagarde) Lagarde - Arch. Zool. Exp. Gén. 53: 281, 1913 ≡ *Kuehniella racovitzae* (Lagarde) Orr - Mycotaxon 4: 171, 1976.

The current name of this species is *Kuehniella racovitzae*. It was described from a dog lesion but without convincing case report (Orr, 1976).

radians - *Trichophyton radians* Sabouraud - Malad. Cuir Chev. 3: 374, 1910 ≡ *Trichophyton niveum* Sabouraud var. *radians* (Sabouraud) Sabouraud - Malad. Cuir Chev. 3: 338, 1910 ≡ *Aleurosporia radians* (Sabouraud) Grigorakis - Annls Sci. Nat., Bot., Sér. 10, 7: 413, 1925 ≡ *Trichophyton gypseum* Bodin var. *radians* (Sabouraud) Frágner - Česká Mykol. 10: 106, 1956 ≡ *Trichophyton mentagrophytes* (Robin) Blanchard ssp. *radians* (Sabouraud) Badillet - Bull. Soc. Fr. Mycol. Méd., N. Sér., 6: 6, 1977.

Doubtful dermatophyte species.

radicicola - *Fusarium javanicum* Koorders var. *radicicola* Wollenweber - Z. Parasitkde 3: 286, 1931 ≡ *Fusarium solani* (Martius) Saccardo f. sp. *radicicola* (Wollenweber) Snyder & Hansen - Am. J. Bot. 28: 740, 1942.

The forma specialis of *F. solani* pathogenic to potato was reported as an agent of keratitis by Matsumoto & Soejima (1976).

radicosum - *Trichophyton radicosum* Catanei - Archs Inst. Pasteur Algér. 15: 268, 1937.

The type strain, CBS 304.38, could as yet not be identified.

radiolatum - *Trichophyton radiolatum* Sabouraud - Malad. Cuir Chev. 4: 355, 1910 ≡ *Ectotrichophyton radiolatum* (Sabouraud) Castellani & Chalmers - Man. Trop. Med., ed. 3, p. 1006, 1919 ≡ *Sabouraudites radiolatus* (Sabouraud) Ota & Langeron - Annls Parasit. Hum. Comp. 1: 328, 1923 ≡ *Spiralia radiolata* (Sabouraud) Grigorakis - Annls Sci. Nat., Bot., Sér. 10, 7: 409, 1925 ≡ *Microsporum radiolatum* (Sabouraud) Guiart & Grigorakis - Lyon Méd. 141: 377, 1928 ≡ *Trichophyton gypseum* (Bodin) Guiart & Grigorakis var. *radiolatum* (Sabouraud) Bruhns & Alexander, *in* Jadassohn - Handb. Haut- u. Geschlechtskrankh. 11: 174, 1928 ≡ *Ectotrichophyton mentagrophytes* (Robin) Castellani & Chalmers var. *radiolatum* (Sabouraud) C.W. Dodge - Med. Mycol. p. 505, 1935.

Doubtful dermatophyte species.

radioplicatum - *Trichophyton gypseum* Bodin var. *radioplicatum* W. Fischer - Derm. Wschr. 57: 1402, 1913 ≡ *Trichophyton radioplicatum* (W. Fischer) Bruhns & Alexander, *in* Jadassohn - Handb. Haut- u. Geschlkrankh. 11: 173, 1928 ≡ *Ctenomyces radioplicatus* (W. Fischer) Nannizzi - Tratt. Micopat. Um. 4: 154, 1934 ≡ *Ectotrichophyton radioplicatum* (W. Fischer) C.W. Dodge - Med. Mycol. p. 498, 1935.

Doubtful dermatophyte species.

ramiroi - *Madurella ramiroi* Pirajá da Silva - Mem. Inst. Butantan 1: 187, 1918; Brasil Med. 38: 81, 1919.

Doubtful agent of human mycetoma.

ramos - *Microsporon ramos* Horta - Brasil Med. 38: 59, 1924; Bull. Inst. Pasteur 22: 410, 1924.

Doubtful dermatophyte species.

repens - *Lepocolla repens* Eklund - Annls Derm. Syph., Sér. 1, 2: 197, 1883 ≡ *Epidermophyton repens* (Eklund) Sartory - Champ. Paras. Homme Anim. p. 516, 1922.

Doubtful species; possibly an actinomycete was concerned (Dodge, 1935).

resinae - _Hormodendrum resinae_ Lindau, _in_ Rabenhorst - Kryptfl. 8: 699, 1907 ≡ _Cladosporium resinae_ (Lindau) de Vries - Antonie van Leeuwenhoek 21: 167, 1955 ≡ _Hormoconis resinae_ (Lindau) von Arx & de Vries, _in_ von Arx - Verh. K. Ned. Akad. Wet., Ser. 2, 61: 62, 1973; teleom.: _Amorphotheca resinae_ Perbery - Aust. J. Bot. 17: 340, 1969.

This fungus typically occurs in kerosene, creosote and resins. The agent reported in an epidemic in captive plaices (Strongman _et al._, 1997) was reidentified (unpublished data) as _Exophiala pisciphila_ (p. 662).

reukaufii - _Anthomyces reukaufii_ Grüss - Ber. Dt. Bot. Ges. 35: 746-761, 1918 ≡ _Candida reukaufii_ (Grüss) Diddens & Lodder - Anaskosp. Hefen 2: 289, 1942; teleom.: _Metschnikowia reukaufii_ Pitt & M.W. Miller - Mycologia 60: 671, 1968.

This species is associated with honey and bees. It was implicated in a pulmoneous disorder of a patient with tuberculosis by Raab (1964).

reynieri - _Indiella reynieri_ Brumpt - Les Mycétomes p. 70, 1906 ≡ _Madurella reynieri_ (Brumpt) Vuillemin - Champ. Paras. Myc. Homme p. 326, 1931.

Hyaline sterile fungus; cannot be identified.

rhoi - _Endomyces rhoi_ Castellani - Man. Trop. Med. p. 601, 1910 ≡ _Monilia rhoi_ (Castellani) Castellani & Chalmers - Man. Trop. Med., ed. 2, p. 828, 1913 ≡ _Myceloblastanon rhoi_ (Castellani) Ota - Jpn. J. Derm. Urol. 27: 178, 1927 ≡ _Candida rhoi_ (Castellani) de Almeida - Anais Fac. Med. Univ. S. Paulo 9: 11, 1933.

Doubtful yeast species (Barnett _et al._, 1990).

ribeiroi - _Trichosporon ribeiroi_ Moraes - Cultura Méd., Rio de J. 3: 50, 1941.

Doubtful _Trichosporon_ species (de Hoog _et al._, 1997).

rifanum - _Madurella rifanum_ Gastaminza - Med. Paises Calidos 2: 445-449, 1929.

Neither material nor literature were available for study.

ripariae - _Microsporum ripariae_ Hubálek & Rush-Munro - Sabouraudia 11: 288, 1973.

Zoophilic dermatophyte (type strain: CBS 529.71), isolated from bird nests (Hubálek & Rush-Munro, 1973).

rogeri - _Cryptococcus rogeri_ Sartory & Demanche - Bull. Soc. Mycol. Fr. 23: 179, 1907 ≡ _Torulopsis rogeri_ (Sartory & Demanche) de Almeida - Anais Fac. Med. Univ. S. Paulo 9: 76, 1933 ≡ _Castellania rogeri_ (Sartory & Demanche) C.W. Dodge - Med. Mycol. p. 254, 1935.

Doubtful yeast species (Barnett _et al._, 1990).

rohii - See this list under _rhoi_.

rosaceum - _Trichophyton rosaceum_ Sabouraud - Trichoph. Hum. F. 192, 1894 ≡ _Aleurosporia rosacea_ (Sabouraud) Grigorakis - Annls Sci. Nat., Bot., Sér. 10, 7: 413, 1925.

No type material has been preserved. Rippon (1988) supposed that the species was identical to _Trichophyton megninii_ which is now regarded as a synonym of _T. rubrum_ (p. 973).

roseum (Epidermophyton) - _Epidermophyton interdigitale_ (Priestly) MacCarthy var. _roseum_ Kesteren - Med. J. Austr. 1: 425, 1939.

Doubtful dermatophyte species.

roseum (Fusisporium) - _Fusisporium roseum_ Link - Mag. Ges. Naturf. Freunde, Berlin 3: 10, 1809 ≡ _Fusarium roseum_ Link:Fries - Syst. Mycol. 3: 471, 1832.

This name has been used in different ways and is a source of error; it should be abandoned (Gerlach & Nirenberg, 1982; Gams _et al._, 1997).

roseum (Oidium) - _Oidium roseum_ Zenoni - Sperimentale 66: 33-66, 1912 ≡ _Monilia rosea_ (Zenoni) Castellani & Chalmers - Man. Trop. Med. p. 829, 1913 ≡ _Myceloblastanon roseum_ (Zenoni) Ota - Jpn. J. Derm. Urol. 27: 176, 1927 ≡ _Candida rosea_ (Zenoni) de Almeida - Anais Fac. Med. Univ. S. Paulo 9: 77, 1923 ≡ _Mycocandida rosea_ (Zenoni) C.W. Dodge - Med. Mycol. p. 293, 1935.

Doubtful red yeast species (Barnett _et al._, 1990).

roseum (Trichoderma) - _Trichoderma roseum_ Persoon - Römer's Neues Mag. Bot. 1: 92, 1794 ≡ _Trichoderma roseum_ Persoon:Fries - Syst. Mycol. 3: 427, 1832 ≡ _Trichothecium roseum_ (Persoon:Fries) Link - Mag. Ges. Naturf. Freunde 3: 18, 1809.

Trichothecium roseum is a saprobe. The case presented by Földvári & Polgar (1951) probably concerned a dermatophyte.

rotundatus - _Endomyces rotundatus_ Castellani - Br. Med. J. 2: 1209, 1912 ≡ _Monilia rotundata_ (Castellani) Castellani & Chalmers - Man. Trop. Med., ed. 2, p. 828, 1913 ≡ _Oidium rotundatum_ (Castellani) Castellani & Chalmers - Man. Trop. Med., ed. 3, p. 1095, 1919 ≡ _Mycoderma rotundatum_ (Castellani) Brumpt - Précis Parasitol., ed. 3, p. 1084, 1922 ≡ _Torulopsis rotundata_ (Castellani) Redaelli - Micet. Ass. Microb. Tuberc. Pulmon. Cavit. p. 47, 1925 ≡ _Myceloblastanon rotundatum_ (Castellani) Ota - Jpn. J. Derm. Urol. 27: 178, 1927 ≡ _Candida rotundata_ (Castellani) Basgal - Contrib. Estud. Blastom. Pulmon. p. 40, 1931 ≡ _Geotrichum rotundatum_ (Castellani) de Almeida - Anais Fac. Med. Univ. S. Paulo 9: 12, 1934 ≡ _Cryptococcus rotundatus_ (Castellani) Nannizzi - Tratt. Micopat. Um. 4: 320, 1934.

An authentic strain, CBS 2468, was listed in the CBS List of Cultures (ed. 34, 1996) as a hitherto unidentified _Trichosporon_ species.

Doubtful dermatophyte species; authentic material lost.

rouxii - _Amylomyces rouxii_ Calmette - Annls Inst. Pasteur 6: 604, 1892 ≡ _Mucor rouxii_ (Calmette) Wehmer - Centbl. Bakt. Parasitkde, Abt. 2, 6: 364, 1900.

The original material is lost; the species is of doubtful identity. A strain described by Wehmer (1900) was reidentified by Schipper (1978a) as _Mucor indicus_ Lendner.

rubrobrunneum - See this list under _rubrobrunneum-cerebriforme_.

rubrobrunneum-cerebriforme - _Cephalosporium rubrobrunneum-cerebriforme_ Benedek - Arch. Derm. Syph. 154: 166, 1928 ≡ _Cephalosporium rubrobrunneum_ Nannizzi - Tratt. Micopat. Um. 4: 455, 1934 (name change).

Acremonium-like species close to _A. kiliense_; type strain: CBS 124.29.

rubrum (Microsporum) - _Microsporum rubrum_ Cazalbou - Bull. Soc. Cent. Méd. Vét. 67: 77, 1913 ≡ _Sabouraudites rubrum_ (Cazalbou) Nannizzi - Tratt. Micopat. Um. 4: 171, 1934.

Doubtful zoophilic dermatophyte.

rubrum (Verticillium) - _Verticillium rubrum_ Baquis - Annali Ottal. Clin. Ocul. 34: 947, 1905 ≡ _Beauveria rubra_ (Baquis) Langeron - Bull. Acad. Méd., Paris 111: 133, 1934 ≡ _Spicaria rubra_ (Baquis) C.W. Dodge - Med. Mycol. p. 843, 1935 ≡ _Tritirachium rubrum_ (Baquis) Langeron - Annls Parasit. Hum. Comp. 22: 98, 1947.

Doubtful species, possibly _Tritirachium oryzae_ (de Hoog, 1972).

rugosus - _Endomyces rugosus_ Castellani - Br. Med. J. 2: 1209, 1912 ≡ _Monilia rugosa_ (Castellani) Castellani & Chalmers - Man. Trop. Med., ed. 2, p. 828, 1913 ≡ _Hemispora rugosa_ (Castellani) Castellani & Chalmers - Man. Trop. Med., ed. 3, p. 1108, 1919 ≡ _Parendomyces rugosus_ (Castellani) Ota - Derm. Wschr. 78: 236, 1924 ≡ _Trichosporium rugosum_ (Castellani) Ota - Annls Parasit. Hum. Comp. 4: 12, 1926 ≡ _Geotrichum rugosum_ (Castellani) C.W. Dodge - Med. Mycol. p. 219, 1935.

An authentic strain, CBS 2469, was listed in the CBS List of Cultures (ed. 34, 1996) as a hitherto unidentified _Trichosporon_ species.

rugulosa - _Acremoniella rugulosa_ de Area Leão - C.R. Soc. Biol. 116: 1160, 1934.

The type strain, CBS 110.35, is now sterile; possibly it is _Pseudallescheria boydii_ (p. 305). It was described from a traumatic subcutaneous lesion with fistels (de Area Leão, 1934).

sabouraudii - _Trichophyton sabouraudii_ Blanchard, _in_ Bouchard - Traité Pathol. Gén. 2: 907, 1896 ≡ _Trichophyton blanchardii_ Castellani - J. Trop. Med. Hyg. 11: 264, 1908 (name change) ≡ _Atrichophyton blanchardii_ (Castellani) Castellani & Chalmers - Man. Trop. Med., ed. 3, p. 1008, 1919.

Doubtful species; material lost. Dodge (1935) supposed that _Malassezia_ was concerned, but Rippon (1988) listed it under _Trichophyton tonsurans_.

sakuranei - _Cryptococcus sakuranei_ Gedoelst - Synops. Parasitol. 1911 ≡ _Mycoderma sakuranei_ (Gedoelst) Brumpt - Précis Parasit., ed. 2, p. 933, 1913 ≡ _Monilia sakuranei_ (Gedoelst) Vuillemin - Champ. Parasit. Myc. Homme p. 86, 1931 ≡ _Geotrichum sakuranei_ (Gedoelst) de Almeida - Anais Fac. Med. Univ. S. Paulo 9: 78, 1933 ≡ _Dematium sakuranei_

(Gedoelst) C.W. Dodge - Med. Mycol. p. 679, 1935.

No material has been preserved; possibly *Exophiala dermatitidis* or a similar black yeast was concerned.

salmoneus - *Cryptococcus salmoneus* Sartory - C.R. Soc. Biol. 61: 650, 1906 ≡ *Torulopsis salmoneus* (Sartory) de Almeida - Anais Fac. Med. Univ. S. Paulo 9: 76, 1933.

Doubtful yeast species (Barnett *et al.*, 1990).

saltense - *Trichophyton saltense* Mazza, Niño & Cornejo - Reun. Soc. Argent. Pat. Reg. 8: 322, 1934.

Doubtful dermatophyte species; neither literature nor material were available fur study.

sangiorgii - *Torulopsis sangiorgii* C.W. Dodge - Med. Mycol. p. 350, 1935.

Doubtful yeast species (Barnett *et al.*, 1990).

sapporense - *Microsporum sapporense* Komuro, Aoki & Odashima - Sapporo Med. 3: 1-13, 1952.

Doubtful dermatophyte species.

sartoryi - *Lichtheimia sartoryi* Sartory & Bailly, *in* Sartory - Champ. Parasit. Homme Anim. 2: 5, 1927.

Doubtful mucoralean species.

scandinavicus - *Saccharomyces scandinavicus* Dietrichson - Annls Parasit. Hum. Comp. 29: 461, 1954.

The type strain, CBS 4611 isolated from human sputum, has been re-identified as the non-pathogenic yeast *Pichia fermentans* Lodder.

sarlatinosus - *Mucor scarlatinosus* Hallier - Z. Parasitkde 1: 117, 1869.

Doubtful mucoralean species (Dodge, 1935).

schweitzeri - *Blastodendrion schweitzeri* Sartory, R. Sartory, Meyer & Weiss - Bull. Acad. Méd. 113: 486, 1935.

Doubtful yeast species (Barnett *et al.*, 1990).

sehnsuchta - *Scopulariopsis sehnsuchta* Froilano de Mello - Bull. Soc. Pathol. Exot. 25: 296, 1932.

Doubtful hyphomycete (Morton & Smith, 1963).

semonii - *Glenospora semonii* Chalmers & Archibald - New Orl. Med. Surg. J. 70: 455, 1917 ≡ *Glenosporopsis semonii* (Chalmers & Archibald) da Fonseca - Parasitol. Méd. 1: 730, 1943.

Described from red-grain mycetoma; an Actinomycete species must have been concerned.

septicoides - *Clitopilus septicoides* Kummer - Führer Pilzk. 1871.

A case described by Batista *et al.* (1955) was reidentified by Rihs *et al.* (1996) as *Schizophyllum commune*.

septicus - *Saccharomyces septicus* Gaetano - Rif. Med. 13: 590, 1897 ≡ *Cryptococcus septicus* (Gaetano) C.W. Dodge - Med. Mycol. p. 340, 1935.

Doubtful yeast species (Barnett *et al.*, 1990).

serisei - *Achorion serisei* Cazalbou - Bull. Soc. Path. Exot. 6: 302, 1913 ≡ *Trichophyton serisei* (Cazalbou) Nannizzi - Tratt. Micopat. Um. 4: 210, 1934.

Dermatophyte species close to *Microsporum gypseum*; material lost.

serrae - *Cephalosporium serrae* Maffei - Att. Ist. Bot. Univ. Pavia, Ser. 4, 1: 196, 1930 ≡ *Verticillium serrae* (Maffei) van Beyma - Antonie van Leeuwenhoek 6: 40, 1939 ≡ *Hyalopus nigrescens* (Maffei) Barbosa - Subs. Estud. Hyalopus, Recife, 1941.

The type strain, CBS 290.30 is close to *Verticillium nigrescens* Pethybridge. No convincing case report has been published on this species. The strain described by de Albornoz (1974) under this name, CBS 651.85, has been reidentified as *Phaeoacremonium inflatipes*; see under that species (p. 847).

shiotae - *Isaria shiotae* Kuru - Jpn. J. Med. Sci. Trans. Abstr., Ser. 9, 2: 351, 1931 ≡ *Beauveria shiotae* (Kuru) Langeron, *in* Brumpt - Précis Parasitol., ed. 5, p. 1839, 1936 ≡ *Tritirachium shiotae* (Kuru) Langeron - Annls Parasit. Hum. Comp. 22: 98, 1947.

This species was listed by de Hoog (1972) as a synonym of *Beauveria bassiana*, but the authentic culture, CBS 337.32 is a *Tolypocladium*-like species.

silvae - *Candida silvae* Vidal-Leiria & van Uden - Antonie van Leeuwenhoek 29: 261, 1963.

This species was described from the intestinal tract of man and

animals and from human sputum (Vidal-Leiria & van Uden, 1963), but as yet no proven clinical case has been published.

singulare - *Trichophyton singulare* Cazalbou - Revue Gén. Méd. Vét. 24: 9, 1914 ≡ *Favotrichophyton singulare* (Cazalbou) C.W. Dodge - Med. Mycol. p. 515, 1935 — *Favotrichophyton album* (Sabouraud) Neveu-Lemaire var. *singulare* (Cazalbou) Gammel & Work - Arch. Derm. Syph. 38: 756, 1938.

Doubtful dermatophyte species.

skutetzkyi - *Myceloblastanon skutetzkyi* Ota - Jpn. J. Derm. Urol. 27: 177, 1927 ≡ *Cryptococcus skutetzkyi* (Ota) Nannizzi - Tratt. Micopat. Um. 4: 331, 1934 ≡ *Mycocandida skutetzkyi* (Ota) C.W. Dodge - Med. Mycol. p. 293, 1935.

Barnett *et al.* (1990) listed the *Cryptococcus* as doubtful and the remaining names as synonyms of *Candida albicans*. No material is preserved to verify which option is the right one.

slovacum - *Aureobasidium slovacum* Svobodová, Chmel & Bojanovská - Proc. 1st Congr. Soc. Int. Derm. Trop. p. 116, 1966 ≡ *Chmelia slovaka* (Svobodová, Chmel & Bojanovská) Svobodová - Biológia, Bratislava 21: 83, 1966.

The type strain, CBS 567.66, is sterile; it could as yet not be identified.

somaliensis - *Indiella somaliensis* Brumpt - Arch. Parasit. 10: 560, 1906 (These p. 71) ≡ *Discomyces somaliensis* (Brumpt) Brumpt - Précis Parasit. p. 967, 1913 ≡ *Nocardia somaliensis* (Brumpt) Chalmers & Christopherson - Annls Trop. Med. Parasit. 10: 223-282, 1916 ≡ *Streptothrix somaliensis* (Brumpt) Mischer (cited by Ciferri & Redaelli, 1942) ≡ *Actinomyces somaliensis* (Brumpt) Brumpt - Précis Parasit., ed. 5, 1927 ≡ *Indiellopsis somaliensis* (Brumpt) Ciferri & Redaelli (as 'Brumpt') - Mycopathologia 3: 194, 1942.

Doubtful actinomycete species causing mycetoma.

sphaerica - *Torula sphaerica* Hammer & Cordes - Res. Bull. Agric. Exp. Stn Iowa State Coll. Agric. Mech. Arts 61: 14, 1920 ≡ *Candida sphaerica* (Hammer & Cordes) S.A. Meyer & Yarrow, *in* Yarrow & Meyer - Int. J. Syst. Bact. 28: 614, 1971.

Reported from the buccal cavity of HIV-positive patients, but without clinical significance (Millon *et al.*, 1997).

sphaericus - *Saccharomyces sphaericus* Bizzozero - Virchows Arch. Path. Anat. 98: 441, 1884.

Probably *Malassezia furfur* was concerned (Ingham & Cunningham, 1993).

sphaeroides - *Geotrichum sphaeroides* Gauvreau & Grandbois - Laval Méd. 19: 66, 1954 (as '*spheroides*').

Doubtful species. De Hoog *et al.* (1986) supposed that a sterile hyphomycete was concerned.

spinulosum - *Chrysosporium spinulosum* Negroni - Boln Acad. Nac. Med., B. Aires 41: 534, 1963.

Doubtful *Chrysosporium* species (van Oorschot, 1980).

sputicola - *Monilia sputicola* Galippe - J. Anat. 21: 538, 1885 ≡ *Scopulariopsis sputicola* (Galippe) C.W. Dodge - Med. Mycol. p. 648, 1935.

Doubtful hyphomycete (Morton & Smith, 1963).

steatolytica - *Candida steatolytica* Yarrow - Antonie van Leeuwenhoek 35: 24, 1969.

This taxon, now known as *Candida hellenica* (Verona & Picci) King & Jong, the anamorph of *Zygoascus hellenicus* M.Th. Smith, was originally isolated from mastitic cow udder, but its prevalent ecological niche is grape must (Smith, 1986).

stromatoides - *Aspergillus stromatoides* Raper & Fennell - Gen. Aspergillus p. 421, 1965; teleom.: *Chaetosartorya stromatoides* Wiley & Simmons - Mycologia 65: 935, 1973.

A human orbit infection under this name was published by Sachs *et al.* (1987), but the identity of this strain was doubted by Kwong-Chung (1987). Another sclerotial strain of the orbit, CBS 616.94 has been identified as *A. flavus*.

subcapitulatum - *Acrothecium obovatum* Cooke & Ellis var. *subcapitulatum* Ciferri & Ashford - Mycologia 22: 184, 1930.

The authentic specimen from human skin is probably lost. Judging from the description a *Diplococcium*-like fungus was concerned.

subcutaneus- tumefaciens - *Saccharomyces subcutaneus-tumefaciens* Curtis - Annls Inst. Pasteur 10: 449, 1896.

Doubtful yeast species.

subfuscum - *Trichophyton gypseum* Bodin var. *subfuscum* von Berde - Arch. Derm. Syph. 176: 1, 1937 ≡ *Trichophyton subfuscum* (von Berde) Szathmáry - Zentbl. Haut- u. Geschlkrankh. 67: 675, 1941.

Doubtful dermatophyte species.

subtile (Oidium) - *Oidium subtile* Kotliar - Vratche No. 12: 2055, 1892.

Doubtful; Rippon (1988) listed the name as a synonym of *Malassezia furfur*.

subtile (Oidium) - See this list under *subtile-cutis*.

subtile-cutis - *Oidium subtile-cutis* Babès - Biol. Centbl. 18: 569, 1882 ≡ *Oidium subtile* Blanchard - Malad. Parasit. 1895 (name change) ≡ *Endomyces subtilis* (Blanchard) Castellani & Chalmers - Man. Trop. Med. p. 601, 1910 ≡ *Mycoderma subtile* (Blanchard) Brumpt - Précis Parasitol., ed. 2, p. 932, 1913 ≡ *Monilia subtilis* (Blanchard) Castellani & Chalmers - Man. Trop. Med., ed. 2, p. 829, 1913 ≡ *Oospora subtilis* (Blanchard) Saccardo - Syll. Fung. 22: 1243, 1913 ≡ *Parasaccharomyces subtilis* (Blanchard) Froilano de Mello & Fernandes - Arq. Hig. Pat. Exot. 6: 273, 1918 ≡ *Geotrichum subtile* (Blanchard) de Almeida - Anais Fac. Med. S. Univ. Paulo 9: 78, 1933.

Doubtful species, possibly *Geotrichum candidum* (de Hoog *et al.*, 1986).

sulphureus - *Sterigmatocystis sulphurea* Fresenius - Beitr. Mykol. 3: 83, 1863 ≡ *Aspergillus sulphureus* (Fresenius) Wehmer - Pilzgatt. Asp. p. 113, 1901.

Mentioned by Durie & Brown (1953) from pulmonary empyema, but without proven case report.

suomiense - *Zygozyma suomiense* M.Th. Smith, van der Walt & Yamada, *in* Smith, van der Walt, Yamada & Batenburg-van der Vegte - Antonie van Leeuwenhoek 56: 284, 1989.

This yeast species is known from a single strain from a bovine skin lesion, but a case report has not been published.

symbioticum - *Cephalosporium symbioticum* Pinkerton - Ann. Mo. Bot. Gdn 23: 42, 1936.

Authentic material lost. Gams (1971) supposed that a species close to *Lecythophora hoffmannii* may have been concerned.

syncephalis - *Aspergillus syncephalis* Guéguen - Champ. Paras. Homme Anim. p. 163, 1904.

Aspergillus species close or identical to *A. fumigatus* (Raper & Fennell, 1965).

talicei - *Achorion talicei* Langeron & Baeza - Annls Parasit. Hum. Comp. 14: 399, 1936.

Doubtful dermatophyte species.

tanakae - *Pseudomycoderma tanakae* C.W. Dodge - Med. Mycol. p. 237, 1935.

Doubtful yeast species (Barnett *et al.*, 1990).

teheraniense - *Trichophyton teheraniense* Ożegović - Med. Arch. 24: 37-44, 1970.

Doubtful dermatophyte species.

tenuishyphum - *Trichophyton tenuishyphum* Castellani - J. Trop. Med. Hyg. 42: 376, 1939.

Doubtful dermatophyte species.

terrestre - *Trichosporon terrestre* van der Walt & E. Johannsen - Antonie van Leeuwenhoek 41: 361, 1975 ≡ *Arxula terrestris* (van der Walt & E. Johannsen) van der Walt, M.Th. Smith & Yamada - Antonie van Leeuwenhoek 57: 60, 1990.

An acute pneumonia caused by this species, now known as *Arxula terrestris*, was reported by Miyazaki *et al.* (1995). However, the identity of the human strain is unconfirmed.

terreum - *Keratinophyton terreum* Randhawa & R.S. Sandhu - Sabouraudia 3: 253, 1964 ≡ *Aphanoascus terreus* (Randhawa & R.S. Sandhu) Apinis - Mycopath. Mycol. Appl. 35: 99, 1968 ≡ *Anixiopsis terrea* (Randhawa & R.S. Sandhu) de Vroey & Recacochea - Bull. Soc. Fr. Mycol. Méd. 6: 209, 1977.

The current name is *Aphanoascus terreus*; it is the teleomorph of

Chrysosporium indicum. The species is geophilic.

terricola - *Fusidium terricola* J.H. Miller, Giddens & Foster - Mycologia 49: 796, 1957.

The type strain, CBS 243.59, is *Acremonium implicatum* (Gilman & Abbot) W. Gams. Mentioned by Thomas (1994) as agent of keratitis, but without case report.

thermophilus - *Talaromyces thermophilus* Stolk - Antonie van Leeuwenhoek 31: 268, 1965; anam.: *Penicillium dupontii* Griffon & Maublanc - Bull. Trimest. Soc. Mycol. Fr. 27: 73, 1911.

Species from compost once isolated from an endocarditis (Pitt, 1994), but without a case report.

tonkinense - *Cylindrocarpon tonkinense* Bugnicourt - Encycl. Mycol. 11: 181, 1939.

The type strain, CBS 115.40, is identical to *C. lichenicola* (C.B. Massalongo) D. Hawksworth.

tonkinii - *Blastomyces tonkinii* Legendre - Bull. Soc. Path. Exot. 4: 24, 1911 ≡ *Cryptococcus tonkinii* (Legendre) Castellani & Chalmers - Man. Trop. Med., ed. 2, p. 771, 1913 ≡ *Torulopsis tonkinii* (Legendre) de Almeida - Anais Fac. Med. Univ. S. Paulo 9: 76, 1933.

Fungus of doubtful identity; no material was preserved.

tonsillae (Torulopsis) - *Torulopsis tonsillae* Carnevale-Ricci - Ferment. Pseudoferment. Paras. Tonsill., Pavia p. 46, 1926 ≡ *Cryptococcus tonsillae* (Carnevale-Ricci) Nannizzi - Tratt. Micopat. Um. 4: 340, 1934.

Doubtful yeast species (Barnett *et al.*, 1990).

tonsillae (Mycotorula) - *Mycotorula tonsillae* Carnevale-Ricci - Ferment. Pseudoferment. Paras. Tonsill., Pavia p. 22, 1926 ≡ *Cryptococcus tonsillarum* Nannizzi - Tratt. Micopat. Um. 4: 341, 1934 (name change).

Doubtful yeast species (Barnett *et al.*, 1990).

tonsillarum - See list under *tonsillae (Mycotorula)*.

tozeuri - *Oospora tozeuri* Nicolle & Pinoy - Bull. Soc. Path. Exot. 1: 99, 1908 ≡ *Madurella tozeuri* (Nicolle & Pinoy) Pinoy - Annls Derm. Syph., Sér. 4, 3: 341, 1912.

Doubtful filamentous fungus.

trachomatosum - *Microsporum trachomatosum* Noizewski - Gaz. Lekarska 25: 998, 1890.

Doubtful dermatophyte species; literature was not available.

tremoniensis - *Debaryomyces tremoniensis* Ota - Derm. Wschr. 78: 289, 1924 ≡ *Debaryomyces fabrii* Ota var. *tremoniensis* (Ota) C.W. Dodge - Med. Mycol. p. 313, 1935.

Doubtful yeast species (Barnett *et al.*, 1990).

trichophyticus - *Ctenomyces trichophyticus* Szathmáry - Mykosen 3: 77, 1960.

Hejtmánek (1963) regards the species as identical to one of the teleomorphs of *Trichophyton terrestre* (p. 981). Szathmáry (1960), however, treats *T. gypseum* var. *asteroides* as the anamorph of *C. trichophyticus*, and that name was listed by Rippon (1988) as a synonym of *T. mentagrophytes* (p. 968). No material was available to confirm either supposition.

tropicale - *Cladosporium tropicale* Sartory, R. Sartory, J. Meyer & Weiss - Bull. Acad. Méd. 113: 890, 1935.

Doubtful; probably an *Exophiala* species was concerned.

tuberculatum - *Arthroderma tuberculatum* Kuehn - Mycopath. Mycol. Appl. 13: 190, 1960; anamorph: *Chrysosporium tuberculatum* Dominik - Zesz. Nauk. Wyzsz. Szk. Roln. Szczec. 24: 59, 1968.

Geophilic dermatophyte..

tumefaciens - *Monilia tumefaciens* Pollacci & Nannizzi - Miceti Patog. No. 55, 1927 ≡ *Geotrichoides tumefaciens* (Pollacci & Nannizzi) Langeron & Talice - Annls Parasit. Hum. Comp. 10: 68, 1932 ≡ *Candida tumefaciens* (Pollacci & Nannizzi) C.W. Dodge - Med. Mycol. p. 232, 1935.

Doubtful yeast species (Barnett *et al.*, 1990).

tunetana - *Sterigmatocystis tunetana* Langeron - Bull. Soc. Path. Exot. 17: 345, 1924 ≡ *Aspergillus tunetanus* (Langeron) C.W. Dodge - Med. Mycol. p. 635, 1935.

Doubtful *Aspergillus* species.

uncinulosa - *Endospora uncinulosa* Szathmáry - Mykosen 17: 38, 1974.

Doubtful onygenalean fungus with peridial hyphae resembling those of _Uncinocarpus._

unguis (Acremoniella) - _Acremoniella unguis_ Sartory & R. Sartory - C.R. Hebd. Séanc. Acad. Sci., Paris 218: 808, 1944.

Doubtful filamentous fungus, perhaps a _Scopulariopsis_ species.

unguis (Spicaria) - _Spicaria unguis_ Weill & Gaudin - Arch. Méd. Exp. Anat. Path. 28: 452, 1918 = _Monilia unguis_ (Weill & Gaudin) Vuillemin - Champ. Parasit. Myc. Homme p. 85, 1931 = _Mycotoruloides unguis_ (Weill & Gaudin) Langeron & Talice - Annls Parasit. Hum. Comp. 10: 50, 1932.

Type material lost; doubtful species (Samson, 1974).

unguium - _Saccharomyces unguium_ Bourgeois - Derm. Z. 22: 411, 1915 = _Castellania unguium_ (Bourgeois) C.W. Dodge - Med. Mycol. p. 259, 1935.

Doubtful yeast species (Barnett _et al._, 1990).

urenae - _Sabouraudites urenae_ Ochoterena - Revta Mex. Biol. 4: 94, 1924 = _Favotrichophyton urenae_ (Ochoterana) C.W. Dodge - Med. Mycol. p. 517, 1935.

Doubtful dermatophyte species.

urinae - _Candida urinae_ C.W. Dodge - Med. Mycol. p. 230, 1935.

Doubtful yeast, either _Candida albicans_ or _C. krusei_ (Barnett _et al._, 1990).

uruguayense - _Trichophyton ferrugineum_ (Ota) Langeron & Milochevitch var. _uruguayense_ Talice - Annls Parasit. Hum. Comp. 9: 85, 1931.

Doubtful dermatophyte species; type material probably lost.

uvarum - _Kloeckeraspora uvarum_ Niehaus - Zentbl. Bakt. ParasitKde, Abt. 2, 87: 146, 1932 = _Hanseniaspora uvarum_ (Niehaus) Shehata, Mrak & Phaff - Mycologia 42: 807, 1955.

This yeast, known as _Hanseniaspora uvarum,_ teleomorph of _Kloeckera apiculata,_ was reported from clinical specimens by Garcia-Martos _et al._ (1999), but without a proper case report.

vaginae - _Diplosporium vaginae_ Nannizzi - Atti R. Accad. Fisiocrit., Siena, Ser. 9, 17: 491, 1926.

No material has been preserved. Booth _et al._ (1985) supposed it might be _Cylindrocarpon vaginae,_ which is now known as a synonym of _Fusarium solani_ (p. 703).

vaginalis - _Monilia vaginalis_ Escomel - Bull. Soc. Pathol. Exot. 17: 922, 1924.

Doubtful yeast species.

valbyensis - _Hanseniaspora valbyensis_ Klöcker - Centbl. Bakt. Parasitkde, Abt. 2, 35: 385, 1912.

This is a food-inhabiting yeast. The identity of the strain causing skin lesions (Batista _et al._, 1960) is unconfirmed.

valeriana - _Castellania valeriana_ C.W. Dodge - Med. Mycol. p. 259, 1935.

Doubtful yeast species (Barnett _et al._, 1990).

vanderburghii - _Oidium vanderburghii_ Kohlbrugge - Arch. Schiffs-Tropenhyg. 5: 394, 1931 = _Parendomyces vanderburghii_ (Kohlbrugge) C.W. Dodge - Med. Mycol. p. 244, 1935.

Doubtful yeast species (Barnett _et al._, 1990).

variabile (Epidermophyton) - _Epidermophyton variabile_ Karrenberg - Arch. Derm. Syph. 17: 523, 1928.

Doubtful dermatophyte close to _Trichophyton mentagrophytes._

variabilis (Proteomyces) - _Proteomyces variabilis_ Boedijn & Verbunt - Mycopath. Mycol. Appl. 1: 189, 1938 = _Geotrichum variabile_ (Boedijn & Verbunt) Brumpt - Précis Parasitol., ed. 6, p. 1744, 1949.

This may have been a _Trichosporon_ species (de Hoog _et al._, 1986). The authentic strain is lost.

varians - _Torulopsis varians_ Carnevale-Ricci - Ferment. Pseudoferment. Paras. Tonsill., Pavia p. 47, 1926 = _Cryptococcus varians_ (Carnevale-Ricci) Nannizzi - Tratt. Micopat. Um. 4: 331, 1934.

Doubtful yeast species.

velutinum - _Penicillium velutinum_ van Beyma - Zentbl. Bakt. Parasitkde, Abt. 2, 91: 352, 1935.

This species (teleom.: _Eupenicillium stolkiae_ D.B. Scott) was described by van Beyma (1935) from human sputum, but without a proven case. Other strains of the species are primarily soil-borne.

venerei - _Scopulariopsis venerei_ Greco - Orig. Tum. Obs. Myc. Argent. p. 716, 1916.

No material preserved; doubtful species (Morton & Smith, 1963).

venezuelense - _Monosporium venezuelense_ Castellani, _in_ Castellani & Jacono - J. Trop. Med. Hyg. 36: 307, 1933.

Doubtful hyphomycete, possibly identical to _Scedosporium prolificans_ (p. 899).

vexans - _Cercosporella vexans_ Massee, _in_ Russ - Lancet 204: 77, 1923.

Doubtful filamentous fungus.

viannae - _Trichophyton viannae_ de Mello - Ind. J. Med. Res. 5: 228, 1917 (as '_vianai_') = _Sabouraudites viannae_ (de Mello) Ota & Langeron - Annls Parasit. Hum. Comp. 1: 329, 1923 = _Beauveria viannae_ (de Mello) Langeron - Bull. Acad. Méd. 111: 133, 1934 = _Ctenomyces viannae_ (de Mello) Nannizzi - Tratt. Micopat. Um. 4: 155, 1934 = _Ateleothylax viannae_ (de Mello) C.W. Dodge - Med. Mycol. p. 431, 1935 = _Tritirachium viannae_ (de Mello) Langeron - Annls Parasit. Hum. Comp. 22: 98, 1947.

Doubtful species; de Hoog (1972) supposed that _Tritirachium oryzae_ may have been concerned.

vignolo-lutatii - _Acaulium vignolo-lutatii_ Matruchot, _in_ Vignolo-Lutati - Arch. Derm. Syph. 118: 681-698, 1913 = _Scopulariopsis vignolo-lutatii_ (Matruchot) C.W. Dodge - Med. Mycol. p. 650, 1935.

Doubtful _Scopulariopsis_ species (Morton & Smith, 1963).

vinosum - _Fusarium vinosum_ Greco - Orig. Tum. Obs. Myc. Argent. p. 670, 1916.

Doubtful, no material has been preserved.

violacea - _Allantospora violacea_ Ambrosini - Riv. Paras. 2: 151-158, 1938.

Doubtful _Cylindrocarpon_ species.

viridobrunnea - _Glenospora viridobrunnea_ Redaelli & Ciferri - Atti Ist. Bot. Univ. Lab. Crittogam. Pavia, Ser. 5, 3: 25, 1943.

Doubtful; possibly the _Scedosporium_ anamorph of _Pseudallescheria boydii_ (p. 305) was concerned.

virulens (Geotrichum) - _Geotrichum virulens_ de Almeida & Lacaz - Folia Clin. Biol., S. Paulo 12: 46, 1940.

Doubtful species.

virulens (Mycoderma) - _Mycoderma virulens_ C.W. Dodge - Med. Mycol. p. 224, 1935.

Introduced as a provisional name; yeast of doubtful identity.

vuilleminii - _Endomyces vuilleminii_ Froilano de Mello & Fernandes - Arq. Hig. Patol. Exot. 6: 230, 1918 = _Parendomyces vuilleminii_ (Froilano de Mello & Fernandes) C.W. Dodge - Med. Mycol. p. 245, 1935.

Doubtful yeast species (Barnett _et al._, 1990).

vulgaris - _Tubercularia vulgaris_ Tode - Fungi Mecklenb. 1: 18, 1790 = _Tubercularia vulgaris_ Tode:Fries - Syst. Mycol. 3: 464, 1832.

This is the anamorph of the common twig-inhabiting ascomycete _Nectria cinnabarina_ (Tode:Fr.) Fr. An ophthalmic mycosis was listed by Affeldt _et al._ (1987), but without a convincing case report.

wangduanlii - _Aspergillus wangduanlii_ D.-M. Li, Horie, Y.-X. Wang & R.-Y. Li - Mycoscience 39: 302, 1998.

Described from a human maxillary sinusitis, but without case report.

wentii - _Aspergillus wentii_ Wehmer - Centbl. Bakt. Parasitkde, Abt. 2, 2: 150, 1896.

The species was listed as an agent of keratitis by Thomas (1994) but without a proper case report.

zeae - _Ustilago zeae_ Schweinitz - Syn. Fung. Carol. Sup. p. 71, 1822.

This is _Ustilago maydis,_ a smut fungus of corn. Reports under the name _Ustilago_ in older literature probably were misidentifications; see this list under _maydis._

zeylanicus - _Endomyces zeylanicus_ Castellani - Arch. Parasit. 16: 187, 1913 = _Monilia zeylanica_ (Castellani) Castellani & Chalmers - Man. Trop. Med., ed. 2, p. 826, 1913 = _Parasaccharomyces zeylanicus_ (Castellani) Froilano de Mello, _in_ Froilano de Mello & Fernandes - Arq. Hig. Patol. Exot. 6: 273, 1918 = _Myceloblastanon zeylanicum_ (Castellani) Ota - Jpn. J. Derm. Urol. 27: 178, 1927 = _Candida zeylanica_ (Castellani) Basgal - Contrib. Estud. Blastom. Pulmon. p. 50, 1931 = _Parendomyces zeylanicus_ (Castellani) C.W. Dodge - Med. Mycol. p. 241, 1935.

Doubtful yeast species (Barnett *et al.*, 1990).

zignoëllae - *Acrotheca zignoëllae* Höhnel - Sitzber. K. Akad. Wiss. Wien, Abt. 1, 118: 332, 1909 ≡ *Cylindrotrichum zignoëllae* (Höhnel) W. Gams & Holubová-Jechová - Stud. Mycol. 13: 53, 1996.

Cylindrotrichum zignoëllae is a wood-inhabiting species (Gams & Holubová-Jechová, 1976). A strain from a human superficial infection, CBS 103.44 (Ciferri *et al.*, 1943) was *Pseudallescheria boydii* (p. 305).

zimmeri - *Saccharomyces zimmeri* C.W. Dodge - Med. Mycol. p. 321, 1935.

Doubtful yeast species (Barnett *et al.*, 1990).

REFERENCES

Anon., 1994. The *Pneumocystis* Workshop. Revised Nomenclature for *Pneumocystis carinii*. J Eukaryot Microbiol 41, Suppl.. 1: 121-122.

Aa, H.A. van der, 1973. Studies in *Phyllosticta* I. Stud. Mycol. 5: 1-110.

Abbott, S.P., Sigler, L., McAleer, R., McGough, D.A., Rinaldi, M.G. & Mizell, G., 1995. Fatal cerebral mycoses caused by the ascomycete *Chaetomium strumarium*. J. Clin. Microbiol. 33: 2692-2698.

Abbott, S.P., Sigler, L. & Currah, R.S., 1998. *Microascus brevicaulis* sp. nov., the teleomorph of *Scopulariopsis brevicaulis* supports placement of *Scopulariopsis* with the *Microascaceae*. Mycologia 90: 297-302.

Achten, G., Want-Rouard, J., Wiame, L. & Hoof, F. van, 1979. Les onychomycoses à moisissures. Dermatologica 159: 128-140.

Acland, K.M., Hay, R.J. & Groves, R., 1998. Cutaneous infection with *Alternaria alternata* complicating immunosuppression: successful treatment with itraconazole. Br. J. Derm. 138: 354-356

Acton, H.W. & McGuire, C., 1929. 'Cooly itch'. A purulent folliculitis due to *Trichophyton violaceum* variety *indicum*. Ind. Med. Gaz. 64: 241-246.

Adam, B.A., Soo-Hoo, T.S. & Chong, K.C., 1977. Black piedra in West Malaysia. Aust. J. Derm. 18: 45-47.

Adam, R.D., Paquin, M.L., Persen, E.A., Saubolle, M.A., Rinaldi, M.G., Corcoran, J.C., Galgiani, J.N. & Sobonya, R.E., 1986. Phaeohyphomycosis caused by the fungal genera *Bipolaris* and *Exserohilum*. A report of 9 cases and review of the literature. Medicine, Baltimore 65: 203-217.

Adaska, J.M., 1999. Peritoneal coccidioidomycosis in a Mountain Lion in California. J. Wildlife Dis. 35: 75-77.

Addrizzo-Harris, D.J., Harkin, T.J., McGuinness, G., Naidich, D.P. & Rom, W.N., 1997. Pulmonary aspergilloma and AIDS. A comparison of HIV-infected and HIV-negative individuals. Chest 111: 612-618.

Ader, P. & Dodd, J., 1979. Mucormycosis and entomophthoromycosis. A bibliography. Mycopathologia 68: 67-99.

Adler, D.E., Milhorat, T.H. & Miller, J.I., 1998. Treatment of rhinocerebral mucormycosis with intravenous, interstitial, and cerebrospinal fluid administration of amphotericin B: case report. Neurosurgery 42: 644-649.

Affeldt, J.C., Flynn, H.W., Forster, R.K., Mandelbaum, S., Clarkson, J.G. & Jarns, G.D., 1987. Microbial endophthalmitis resulting from ocular trauma. Ophthalmology 94: 407-413.

Agamamolis, D.P., Kalwinsky, D.K., Krill, C.E., Dasu, S., Halasa, B. & Galloway, P.G., 1991. *Fusarium* meningoencephalitis in a child with acute leukemia. Neuropediatrics 22: 110-112.

Agarwal, A. & Singh, S.M., 1995. A case of cutaneous phaeohyphomycosis caused by *Exserohilum rostratum*, its *in vitro* susceptibility and review of the literature. Mycopathologia 131: 9-12.

Agostini, A., 1930. *Glenosporella dermatitidis* n. sp. causa di dermatomicosi umana. Atti Ist. Bot. Univ. Lab. Crittog. Pavia 4: 93-101.

Agostini, A., 1932. *Coniosporium onychophilum* n.sp. causa di onicomicosi. Atti Ist. Bot. Crittog. Lab. Univ. Pavia, Ser. 4a, 3: 29-36.

Agrawal, A. & Singh, S.M., 1995a. Two cases of cutaneous phaeohyphomycosis caused by *Curvularia pallescens*. Mycoses 38: 301-303.

Agrawal, A. & Singh, S.M., 1995b. A case of cutaneous phaeohyphomycosis caused by *Exserohilum rostratum*, its *in vitro* sensitivity and review of literature. Mycopathologia 131: 9-12.

Agrawal, P.K., Lal, B., Shukla, P.K., Khan, Z.A. & Srivastava, O.P., 1982. Clinical and experimental keratitis due to *Curvularia lunata* (Wakker) Boedijn var. *aeria* (Batista, Lima and Vasconcelos) Ellis. Sabouraudia 20: 225-232.

Agrawal, P.K., Lal, B., Wahab, S., Srivastava, O.P. & Misra, S.C., 1979. Orbital paecilomycosis due to *Paecilomyces lilacinus* (Thom) Samson. Sabouraudia 17: 363-369.

Aguilar, C., Pujol, I., Sala, J. & Guarro, J., 1998. Antifungal susceptibilities of *Paecilomyces* species. Antimicrob. Agents Chemother. 42: 1601-1604.

Aguilar, C., Pujol, I. & Guarro, J., 1999. *In vitro* antifungal susceptibilities of *Scopulariopsis* isolates. Antimicrob. Agents Chemother. 43: 1520-1522.

Ahlund, H.O., Pallin, B., Peterhoff, R. & Schönebeck, J., 1967. Mycosis of the stomach. Acta Chir. Scand. 133: 555-562.

Ahmad, S., Johnson, R.J., Hillier, S., Shelton, W.R. & Rinaldi, M.G., 1985. Fungal peritonitis caused by *Lecythophora mutabilis*. J. Clin. Microbiol. 22: 182-186.

Ahmed, S.I. & Cain, R.F., 1972. Revision of the genera *Sporormia* and *Sporormiella*. Can. J. Bot. 50: 419-477.

Ahluwalia, K.B., 1999. Culture of the organism that causes rhinosporidiosis. J. Laryngol. Otol. 113: 523-528.

Ahmed, A.O.A., Mukhtar, M.M., Kools-Sijmons, M., Fahal, A.H., Hoog, G.S. de, Gerrits van den Ende, A.H.G., Zijlstra, E., Verbrugh, H., Abugroun, A.M., Elhassan, A.M. & Belkum, A. van, 1999. Development of a species-specific PCR-RFLP procedure for the identification of *Madurella mycetomatis*. J. Clin. Microbiol. 37: 3175-3178

Aho, R., 1987. Mycological studies on *Microsporum equinum* isolated in Finland, Sweden and Norway. J. Med. Vet. Mycol. 25: 255-260.

Aho, R., Westerling, R., Ajello, L., Padhye, A.A. & Samson, R.A., 1990. Avian penicilliosis caused by *Penicillium griseofulvum* in a captive toucanet. J. Med. Vet. Mycol. 28: 349-354.

Ajello, L., 1952. The isolation of *Allescheria boydii* Shear, an etiologic agent of mycetomas, from soil. Am. J. Trop. Med. Hyg. 1: 227-238.

Ajello, L., 1959. A new *Microsporum* and its occurrence on man and animals. Mycologia 51: 69-76.

Ajello, L., 1961. The ascigerous state of *Microsporum cookei*. Sabouraudia 1: 173-177.

Ajello, L. (ed.), 1965. Coccidioidomycosis. Proc. 2nd Cocc. Symp., Phoenix, 434 pp.

Ajello, L., 1968. A taxonomic review of the dermatophytes and related species. Sabouraudia 6: 147-159.

Ajello, L., 1980. Natural habitats of the fungi that cause pulmonary mycoses. In H.-J. Preusser (ed.): Medical Mycology, pp. 31-42. Gustav Fischer, Stuttgart.

Ajello, L. & Cheng, S.Y., 1967. A new geophilic *Trichophyton*. Mycologia 59: 255-263.

Ajello, L., Bostick, L. & Cheng, S.-L., 1968. The relationship of *Trichophyton quinckeanum* to *Trichophyton mentagrophytes*. Mycologia 60: 1185-1189.

Ajello, L., Georg, L.K., Steigbigel, R.T. & Wang, C.J.K., 1974. A case of phaeohyphomycosis caused by a new species of *Phialophora*. Mycologia 66: 490-498.

Ajello, L. & Kaplan, W., 1969. A new variant of *Sporothrix schenckii*. Mykosen 12: 633-644.

Ajello, L., McGinnis, M.R. & Camper, J., 1977. An outbreak of phaeohyphomycosis in rainbow trout caused by *Scolecobasidium humicola*. Mycopathologia 62: 15-22.

Ajello, L. & Padhye, A.A., 1987. Macroconidial formation in *Trichophyton soudanense* Joyeux 1912. Mykosen 30: 258-262.

Ajello, L., Padhye, A.A., Chandler, F.W., McGinnis, M.R., Morganti,

L. & Alberici, F., 1985. *Fusarium moniliforme*, a new mycetoma agent. Restudy of a European case. Eur. J. Epidemiol. 1: 5-10.

Ajello, L., Padhye, A.A. & Payne, M., 1980. Phaeohyphomycosis in a dog caused by *Pseudomicrodochium suttonii* sp. nov. Mycotaxon 12: 131-136.

Ajello, L., Padhye, A.A., Sukroongreung, S., Nilakul, C.H. & Tantimavanic, S., 1995. Occurrence of *Penicillium marneffei* infections among wild bamboo rats in Thailand. Mycopathologia 131: 1-8.

Ajello, L., Varsovosky, E., Ginther, O.J. & Bubash, G., 1964. The natural history of *Microsporum nanum*. Mycologia 56: 873-884.

Akagi, M., Fujino, T., Hayashi, S., Yamamoto, T. & Akabe, T., 1958. A case of pulmonary infection caused by black yeast-like fungus. Med. J. Osaka Univ. 8: 585-599.

Akhunova, A.M., 1991. Infectious allergic bronchopulmonary paecilomycosis. Terap. Ceskij Arch., Moscow 63: 19-24.

Akhunova, A.M. & Shustova, V.I., 1989. Paecilomycosis. Probl. Tuberk. 1989: 38-42.

Akova, M., Akalin, H.E., Uzun, Ö. & Gür, D., 1991. Emergence of *Candida krusei* infections after therapy of oropharyngeal candidiasis with fluconazole. Eur. J. Clin. Microbiol. Infect. Dis. 10: 598-599.

Al-Abdely, H.M., Najvar, L., Bocanegra, R., Fothergill, A., Loebenberg, D., Rinaldi, M.G. & Graybill, J.R., 2000. SCH 56592, amphotericin B, or itraconazole therapy of experimental murine cerebral phaeohyphomycosis due to *Ramichloridium obovoideum* ('*Ramichloridium mackenziei*'). Antimicrob. Agents Chemother. 44: 1159-1162.

Al-Ani, F.K., Ali, A.H. & Banna, H.B., 1998. *Histoplasma farciminosum* infection in horse in Iraq. Vet. Arhiv 68: 101-107.

Albanese, G., Cintio, R. di, Crippa, D. & Galbiati, G., 1995. *Trichophyton soudanense* in Italy. Mycoses 38: 229-230.

Albornoz, M.B. de, 1974. *Cephalosporium serrae*, agente etiologico de micetomas. Mycopath. Mycol. Appl. 54: 485-498.

Alcorn, J.L., 1983. Generic concepts in *Drechslera*, *Bipolaris* and *Exserohilum*. Mycotaxon 17: 1-86.

Alderman, D.J. & Polglase, J.L., 1985. *Fusarium tabacinum* (Beyma) Gams as a gill parasite in the crayfish *Austropotamobius pallipes* Lereboullet. J. Fish Dis. 8: 249-252.

Al-Doory, Y., 1972. Chromomycosis. Mountain Press, Missoula, 203 pp.

Al-Doory, Y. & DiSalvo, A.F. (eds), 1992. Blastomycosis. Plenum, New York, 268 pp.

Alexandrakis, G. & Gloor, P., 1994. Diagnosis of *Fusarium* keratitis in an animal model using the polymerase chain reaction. Invest. Ophthalmol. Vis. Sci. 35: 1676.

Al-Hajjar, S., Perfect, J., Hashem, F., Tufenkeji, H. & Kayes, S., 1996. Orbitofacial conidiobolomycosis in a child. Pediatr. Infect. Dis. J. 15: 1130-1132.

Al-Hedaikthy, M., 1998. Cutaneous zygomycosis due to *Saksenaea vasiformis:* case report and literature review. Annls Saudi Med. 18: 428-431.

Allevato, P.A., Ohorodnik, J.M., Mezger, E. & Eisses, J.F., 1984. *Paecilomyces javanicus* endocarditis of native and prosthetic aortic valve. Am. J. Clin. Pathol. 82: 247-252.

Almeida Barbosa, A. de, Moreira Lemos, A.C. & Severo, L.C., 1997. Acute pulmonary adiaspiromycosis. Report of three cases and review of 16 other cases collected from the literature. Revta Iberoam. Micol. 14: 177-180.

Almeida, H.L., Salebian, A. & Rivitti E.A., 1991. Ultrastructure of black piedra. Mycoses. 34: 447-451.

Al-Musallam, A., 1980. Revision of the black *Aspergillus* species. Thesis, Univ. Utrecht, 91 pp.

Al-Musallam, A. & Tan, C.S., 1989. *Chrysosporium zonatum*, a new keratinophilic fungus. Persoonia 14: 69-71.

Alomar, Y., Linas, M.D., Lloveras, J.J., Becasse, F., Durand, D. & Seguela, J.P., 1989. Mycose systémique à *Candida lusitaniae* chez un malade nephrectomisé: problèmes thérapeutiques. Bull. Soc. Fr. Mycol. Méd. 18: 113-116.

Alonso i Tarrés, C., Heures, J.K., Guého, E. & Cuatrecases, M., 1993. Subcutaneous phaeohyphomycosis caused by *Taeniolella exilis*. Proc. CEMM, Paris, p. 124.

Al-Rajhi, A.A., Awad, A.H., Al-Hedaithy, S.S.A., Forster, R.K. &

Caldwell, K.C., 1993. *Scytalidium dimidiatum* fungal endophthalmitis. Br. J. Ophthalmol. 77: 388-390.

Alsina, A., Mason, M., Uphoff, R.A., Riggsby, W.S., Becker, J.M. & Murphy, D., 1988. Catheter-associated *Candida utilis* fungemia in a patient with acquired immunodeficiency syndrome: species verification with a molecular probe. J. Clin. Microbiol. 26: 621-624.

Al-Tawfiq, J.A. & Wools, K.K., 1998. Disseminated sporotrichosis and *Sporothrix schenckii* fungemia as the initial presentation of human immunodeficiency virus infection. Clin. Infect. Dis. 26: 1403-1406.

Alter, S.J. & Farley, J., 1994. Development of *Hansenula anomala* infection in a child receiving fluconazole therapy. Pediatr. Infect. Dis. J. 13: 158-159.

Alteras, I. & Evolceanu, R., 1969a. First isolation of *Microsporum racemosum* Dante Borelli 1965 from Romanian soil. Mykosen 12: 223-230.

Alteras, I. & Evolceanu, R., 1969b. *Trichophyton phaseoliforme* (Dante Borelli & Feo - 1966) in Romanian soil. Mykosen 12: 421-426.

Altmeyer, P. & Schon, K., 1981. Kutane Schimmelpilzgranulome durch *Ulocladium chartarum*. Hautarzt 32: 36-38.

Alvarez, D.P. & Bracalenti, B.T. de, 1982. Dermatofitosis causada por *Trichophyton ajelloi*. Mycopathologia 77: 27-29.

Alvarez, M., Lopez Ponga, B., Rayon, C., Garcia Gala, J., Consolacion, M.R.P., Gonzalez, M., Martínez-Suarez, J. & Rodríguez-Tudela, J.L., 1995. Nosocomial outbreak caused by *Scedosporium prolificans (inflatum)*: four fatal cases in leukemic patients. J. Clin. Microbiol. 33: 3290-3295.

Alvarez, O.A., Marples, J.A., Tio, F.O. & Lee, M., 1995. Severe diarrhea due to *Cokeromyces recurvatus* in a bone marrow transplant recipient. Am. J. Gastroenterol. 90: 1350-1351.

Alvarez, S., 1990. Systemic infections caused by *Penicillium decumbens* in a patient with acquired immunodeficiency syndrome. J. Infect. Dis. 162: 283.

Amann, R.I., Ludwig, W. & Schleifer K.-H., 1995. Phylogenetic identification and *in situ* detection of individual microbial cells without cultivation. Microb. Rev. 59: 143-169.

Ames, L.M., 1963. A monograph of the *Chaetomiaceae*. US Army Res. Div. Ser. 2: 1-125.

Amft, N., Miadonna, A., Viviani, M.A. & Tedeschi, A., 1996. Disseminated *Geotrichum capitatum* infection with predominant liver involvement in a patient with non-Hodgkin's lymphoma. Haematologica 81: 352-355.

Amitani, R., Nishimura, K., Niimi, A., Kobayashi, H., Nawada, R., Murayama, T., Taguchi, H. & Kuze, F., 1996. Bronchial mucoid impaction due to the monokaryotic mycelium of *Schizophyllum commune*. Clin. Infect. Dis. 22: 146-148.

Ammari, L.K., Puck, J.M. & McGowan, K.L., 1993. Catheter-related *Fusarium solani* fungemia and pulmonary infection in a patient with leukemia in remission. Clin. Infect. Dis. 16: 148-150.

Ampel, N.M., 1999. Delayed-type hypersensitivity, *in vitro* T-cell responsiveness and risk of active coccidioidomycosis among HIV-infected patients living in the coccidioidal area. Med. Mycol. 37: 245-250.

Ampel, N.M., Mosley, D.G., England, B., Vertz, P.D., Komatsu, K. & Hajjeh, R.A., 1998. Coccidioidomycosis in Arizona: increase in incidence from 1990 to 1995. Clin. Infect. Dis. 27: 1528-1530.

Amstalden, E.M.I., Xavier, R., Kattapuram, S.V., Bertolo, M.B., Swartz, M.N. & Rosenberg, A.E., 1996. Paracoccidioidomycosis of bones and joints: a clinical, radiologic, and pathologic study of 9 cases. Medicine, Baltimore 75: 212-225.

Amstead, G.M., Sutton, D.A., Thompson, E.H., Weitzman, I., Otto, R.A. & Ahuja, S.K., 1999. Disseminated zygomycosis due to *Rhizopus schipperae* after heatstroke. J. Clin. Microbiol. 37: 2656-2667.

Amstrong, D., 1989. Problems in management of opportunistic fungal diseases. Rev. Infect. Dis. 11: 1591-1599.

Anaissie, E.J., Bodey, G.P., Kantarjian, H., Ro, J., Vartivarian, S.E., Hopfer, R., Hoy, J. & Rolston, K., 1989a. New spectrum of fungal infections in patients with cancer. Rev. Infect. Dis. 11: 369-378.

Anaissie, E., Bodey, G.P. & Rinaldi, M.G., 1989b. Emerging fungal pathogens. Eur. J. Clin. Microbiol. Infect. Dis. 8: 323-330.

Anaissie, E., Kantarjian, H., Ro, J., Hopfer, R., Rolston, V., Fainstein, V. & Bodey, G., 1988. The emerging role of *Fusarium* infections in patients with cancer. Medicine 67: 77-83.

Anaissie, E., Nelson, P., Beremand, M., Komtoyiannis, M. & Rinaldi, M., 1992. *Fusarium*-caused hyalohyphomycosis: an overview. Curr. Topics Med. Mycol. 4: 231-249.

Anandi, V., Jhon, T.J., Walter, A., Shastry, J.M., Lalitha, M.K., Padhye, A.A., Ajello, L. & Chandler, F.W., 1989. Cerebral phaeohyphomycosis caused by *Chaetomium globosum* in a renal transplant recipient. J. Clin. Microbiol. 27: 2226-2229.

Anandi, V., Suryawanshi, N.B., Koshi, G., Padhye, A.A. & Ajello, L., 1988. Corneal ulcer caused by *Bipolaris hawaiiensis*. J. Med. Vet. Mycol. 26: 301-306.

Andersen, B. & Thrane, U., 1996. Differentiation of *Alternaria infectoria* and *Alternaria alternata* based on morphology, metabolite profiles, and cultural characteristics. Can. J. Microbiol. 42: 685-689.

Anderson, B., Roberts, S.S., Gonzalez, C. & Chick, E.W., 1959. Mycotic ulcerative keratitis. Arch. Ophthalmol. 62: 169-179.

Anderson, K., Morris, G., Kennedy, H., Croall, J., Michie, J., Richardson, M.D. & Gibson, B., 1996. Aspergillosis in immuno-compromised pediatric patients: assocations with building hygiene, design, and outdoor air. Thorax 51: 256-261.

Annessi, G., Cimitan, A., Zambruno, G. & Disilverio, A., 1992. Cutaneous phaeohyphomycosis due to *Cladosporium cladosporioides*. Mycoses 35: 243-246.

Antonio, D. d', Fioritoni, G., Iacone, A., Betti, S., Fazii, P., Isola, M. dell', Gianfilippo, R. di, Silverio, A. di, Ubezio, S., Zeccara, C., Mosca, M. & Torlontano, G., 1992. Hepatosplenic infection caused by *Candida parapsilosis* in patients with acute leukemia. Mycoses 35: 311-313.

Antonio, D. d', Mazzoni, A., Iacone, A., Violante, B., Capuani, M.A., Schioppa, F. & Romano, F., 1996. Emergence of fluconazole-resistant strains of *Blastoschizomyces capitatus* causing nosocomial infections in cancer patients. J. Clin. Microbiol. 34: 753-755.

Antonio, D. d', Piccolomini, R., Fioritoni, G., Iacone, A., Betti, S., Fazii, P. & Mazzoni, A., 1994. Osteomyelitis and intervertebral discitis caused by *Blastoschizomyces capitatus* in a patient with acute leukemia. J. Clin. Microbiol. 32: 224-227.

Antonio, D. d', Romano, F., Iacone, A., Violante, B., Fazii, P., Pontieri, E., Staniscia, T., Caracciolo, C., Bianchini, S., Sferra, R., Vetuschi, A., Gaudio, E. & Carruba, G., 1999. Onychomycosis caused by *Blastoschizomyces capitatus*. J. Clin. Microbiol. 37: 2927-2930.

Antonio, D. d', Violante, B., Farina, C., Sacco, R., Angelucci, D., Masciulli, M., Iacone, A. & Romano, F., 1997. Necrotizing preumonia caused by *Penicillium chrysogenum*. J. Clin. Microbiol. 35: 3335-3337.

Antonio, D. d', Violante, B., Mazzoni, A., Bonfini, T., Assunta Capuani, M., Aloia, F. d', Iacone, A., Schioppa, F. & Romano, F., 1998. A nosocomial cluster of *Candida inconspicua* infections in patients with hematological malignancies. J. Clin. Microbiol. 36: 792-795.

Anzimlt, B.J.A., 1990. Subcutaneous phaeohyphomycosis due to *Exophiala jeanselmei* in an immunosuppressed patient: case report. N.Z. Med. J. 103: 321-322.

Aoun, K., Bouratbine, A., Mokni, M., Chatti, S., Ben Ismail, R. & Ben Osman, A., 1998. Teignes du cuir chevelu causées par *Trichophyton rubrum* chez deux enfants atteints de dermatophytie extensive. J. Mycol. Méd. 8: 200-202.

Apinis, A.E. & Rees, R.G., 1976. An undescribed keratinophilic fungus from southern Queensland. Trans. Br. Mycol. Soc. 67: 522-524.

Aquino, V.M., Norvell, J.M., Krisher, K. & Mustafa, M.M., 1995. Fatal disseminated infection due to *Exserohilum rostratum* in a patient with a plastic anemia: case report and review. Clin. Infect. Dis. 20: 176-178.

Arab, H.C., Yilmaz, H., Ucar, I., Yildirim, E. & Ozkul, M., 1995. A chronic cavity pulmonary histoplasmosis case from Turkey. J. Trop. Med. Hyg. 98: 190-191.

Arai, H. & Endo, T., 1977. Deep mycosis *(Fonsecaea pedrosoi* and *Paecilomyces lilacinus)* after renal transplantation. Rinsho Hifuka 31: 481.

Arango, M., Jaramillo, C., Cortés, A. & Restrepo, A., 1998. Auricular chromoblastomycosis caused by *Rhinocladiella aquaspersa*. J. Med. Vet. Mycol. 36: 43-45.

Aravysky, R.A. & Aronson, V.B., 1968. Comparative histopathology of chromomycosis and cladosporiosis in the experiment. Mycopath. Mycol.

Appl. 36: 322-340.

Arceneaux, K.A., Tabaoda, J. & Hosgood, G., 1998. Blastomycosis in dogs: 115 cases (1980-1995). J. Am. Vet. Med. Ass. 213: 658-664.

Arêa Leão, A.-E. de, 1934. Sur une mycose osseuse par *Acremoniella*. Nouvelle espèce de champignon trouvée chez l'homme: *Acremoniella rugulosa* n.sp. C.R. Soc. Biol, Paris. 116: 1158-1160.

Arêa Leão, A.-E. & Lobo, J., 1934. Mycétome du pied à *Cephalosporium recifei* var. sp. Mycétome à grains blancs. C.R. Soc. Biol., Paris 107: 303-305.

Arenas, R., Arce, M., Muñoz, H. & Ruiz-Esmenjaud, J., 1998. Onychomycosis due to *Paecilomyces variotii*. Case report and review. J. Mycol. Méd. 8: 32-33.

Arievich, A.M., Stepanisheva, Z.G., Tjufilina, O.V., Kukoleva, Z.I., Malkina, A.J. & Teplitz, V.V., 1966. Gummato-ulcerative cephalosporiosis of the skin. Mycopathol. Mycol. Appl. 28: 113-121.

Arikan, S., Lazono-Chiu, M., Paetznick, V., Nangia, S. & Rex, J.H., 1999. Microdilution susceptibility testing of amphotericin B, itraconazole, and voriconazole against clinical isolates of *Aspergillus* and *Fusarium* species. J. Clin. Microbiol. 37: 3946-3951.

Arker, S. van den, 1952. Een schimmelinfectie *(Cephalosporium potronii)* in de mondholte van een kat. Tijdschr. Diergeneeskd. 77: 515-516.

Armingand, P., Toledano, C., Barthez, J.P., Aubry, A.M. & Esteve, E., 1998. Papules purpuriques péri-orbitaires révélant une tinea faciei à *Trichophyton soudanense*. J. Mycol. Méd. 8: 203-204.

Arney, K.L., Tiernan, R. & Judson, M.A., 1997. Primary pulmonary involvement of *Fusarium solani* in a lung transplant recipient. Chest 112: 1128-1130.

Arrese, J.E., Pierard-Franchimont, C. & Pierard, G.E., 1997. Unusual mould infection in the human stratum corneum. J. Med. Vet. Mycol. 35: 225-227.

Artagaveytia-Allende, R.C. & Montemayor, L., 1949. Estudio comparativo de varias cepas de *Paracoccidioides brasiliensis* y especies afines. Mycopath. Mycol. Appl. 4: 356-366.

Artis, W.M., Fountains, J.A., Deldher, H.K. & Jones, H.E., 1982. A mechanism of susceptibility to mucormycosis in diabetic ketoacidosis: transferrin and iron availability. Diabetes 31: 1109-1114.

Aru, A., Munk-Nielsen, L. & Federspiegel, B.M., 1997. The soil fungus *Chaetomium* in the human paranasal sinuses. Eur. Arch. Oto-Rhino-Laryngol. 254: 350-352.

Arx, J.A. von, 1957. Die Arten der Gattung *Colletotrichium* Cda. Phytopath. Z. 29: 413-468.

Arx, J.A. von, 1970. A revision of the fungi classified as *Gloeosporium*. Bibl. Mycol. 24: 1-203.

Arx, J.A. von, 1975. On *Thielavia* and some similar genera of Ascomycetes. Stud. Mycol. 8: 1-29.

Arx, J.A. von, 1981a. The genera of fungi sporulating in pure culture. J. Cramer, Vaduz, 424 pp.

Arx, J.A. von, 1981b. On *Monilia sitophila* and some families of Ascomycetes. Sydowia 34: 13-29.

Arx, J.A. von, Guarro, J. & Figueras, M.J., 1986. The ascomycete genus *Chaetomium*. Beih. Nova Hedwigia 84, 162 pp.

Arx, J.A. von, Figueras, M.J. & Guarro, J., 1988. Sordariaceous Ascomycetes without ascospore ejaculation. Beih. Nova Hedwigia 94, 104 pp.

Arx, J.A. von & Müller, E., 1954. Die Gattungen der amerosporen Pyrenomyceten. Beitr. Kryptfl. Schweiz 11-1: 7-434.

Ashford, D.A., Hajjeh, R.A., Kelley, M.F., Kaufman, L., Hutwagner, L. & McNeil, M.M., 1999. Outbreak of histoplasmosis among cavers attending the National Speleological Society Annual Convention, Texas, 1994. Am. J. Trop. Med. Hyg. 60: 899-903.

Ashikaga, R., 1920. A new kind of keratomycosis. Jpn. J. Ophthalmol. 24: 554-561 (In Japanese).

Athar, M.A., Sekhon, A.S., Megrath, J.V. & Malone, R.M., 1996. Hyalohyphomycosis caused by *Paecilomyces variotii* in an obstetrical patient. Eur. J. Epidemiol. 12: 33-35.

Atlas, R.M. & Parks, L.S., 1993. Handbook of Microbiological Media. CRC, Boca Raton, 1079 pp.

Attapattu, M.C., 1995. Acute rhinocerebral mucormycosis caused by *Rhizopus arrhizus* from Sri Lanka. J. Trop. Med. Hyg. 98: 355-358.

Attili, D.S., Hoog, G.S. de & Pizzirani-Kleiner, A.A., 1998. rDNA-RFLP and ITS1 sequencing of species of the genus *Fonsecaea*, agents of chromoblastomycosis. Med. Mycol. 36: 219-225.

Aue, R., Müller, E. & Stoll, C., 1969. *Pseudophaeotrichum sudanense* nov. gen. et nov. spec. Nova Hedwigia 17: 83-91.

Augustinsky, J., Kammeyer, P., Husain, A., Hoog, G.S. de & Libertin, C.R., 1990. *Engyodontium album* endocarditis. J. Clin. Microbiol. 28: 1479-1481.

Aulakh, H.S., Straus, S.E. & Kwon-Chung, K.J., 1981. Genetic relatedness of *Filobasidiella neoformans (Cryptococcus neoformans)* and *Filobasidiella bacillispora (Cryptococcus bacillisporus)* as determined by deoxyribonucleic acid base composition and sequence homology studies. Int. J. Syst. Bact. 31: 97-103.

Austwick, P.K.C., 1983. Some mycoses of reptiles. Proc. 8th Congr. ISHAM. Palmerton, pp. 383-384.

Austwick, P.K.C. & Venn, J.A.J., 1962. Mycotic abortion in England and Wales 1954-1960. Proc. 4th Int. Congr. Anim. Repr., The Hague.

Auvrey, M., 1909. A propos d'une nouvelle mycose observée chez l'homme. Suppuration cervicale due à l'*Hemispora stellata*. Bull. Mem. Soc. Chir. Paris 20: 686.

Avram, A., 1964. Étude clinique et mycologique concernant le premier cas Européen de mycétome déterminé par *Cephalosporium* sp. Mycopathol. Mycol. Appl. 24: 177-194.

Avram, A., 1967. Grains expérimentaux maduromycosiques et actinomycosiques à *Cephalosporium falciforme, Monosporium apiospermum, Madurella mycetomi* et *Nocardia asteroides*. Mycopathol. Mycol. Appl. 32: 319-336.

Avram, A., Gauchy, O., Blanchet, P. & Buot, G., 1988. Trois cas de Dermatophytie à *Microsporum praecox*. Bull. Soc. Fr. Mycol. Méd. 17: 329-334.

Awao, T. & Otsuka, S.-i., 1974. Notes on thermophilic fungi in Japan (3). Trans. Mycol. Soc. Japan 15: 7-22.

Ayadi, A., Huerre, M.-R. & Bièvre, C. de, 1995. Phaeohypho-mycosis caused by *Veronaea botryosa*. Lancet 346: 1703-1704.

Azadeh, B., Baghoumian, N. & El-Bakri, O.T., 1994. Rhinosporidiosis: immunohistochemical and electron microscopic studies. J. Laryngol. Otol. 108: 1048-1054.

Aznar, C., Bièvre, C. de & Guiguen, C., 1989. Maxillary sinusitis from *Microascus cinereus* and *Aspergillus repens*. Mycopathologia 105: 93-97.

Baddley, J.M.L., Moser, S.A., Sutton, D.A. & Pappas, G., 2000. *Microascus cinereus* (anamorph *Scopulariopsis*) brain abscess in a bone marrow transplant recipient. J. Clin. Microbiol. 38: 395-397.

Badillet, G., 1982. Les Dermatophytes. Atlas Clinique et Biologique. Varia, Paris, 219 pp.

Badillet, G., 1991. Les alternarioses cutanées. Revue de la littérature. J. Mycol. Méd. 2: 59-71.

Badillet, G., Bièvre, C. de & Guého, E., 1987. Champignons contaminants des cultures, champignons opportunistes. Atlas clinique et biologique. Varia, Paris, 2 vols.

Badillet, G., Bièvre, C. de, Kafando, C., Sene, S., Blanchet-Bardon, C. & Puissant, A., 1986. Un cas de mycose à *Anthopsis deltoidea*. Annls Derm. Vénéréol. 113: 916-918.

Badillet, G., Bièvre, C. de & Spizajzen, S., 1982. Isolement de dématiées à partir d'ongles et de squames. Bull. Soc. Fr. Mycol. Méd. 11: 69-72.

Baily, J.A. & Jeger, M.J., 1992. *Colletotrichum*: biology, pathogenicity, and control. CAB International, Wallingford, 388 pp.

Baily, G.G., Moore, C.B., Essayag, S.M., De Wit, S., Burnie, J.P. & Denning, D.W., 1997. *Candida inconspicua*, a fluconazole-resistant pathogen in patients infected with human immunodeficiency virus. Clin. Infect. Dis. 25: 161-163.

Baird, D.R., Harris, M., Menon, R. & Stoddart, R.W., 1985. Systemic infection with *Trichosporon capitatum* in two patients with acute leukemia. Eur. J. Clin. Microbiol. 4: 62-64.

Baker, J.G., Nadler, H.L., Forgacs, P. & Kurtz, S.R., 1984. *Candida lusitaniae*: a new opportunistic pathogen of the urinary tract. Diagn. Microbiol. Infect. Dis. 2: 145-149.

Baker, J.G., Salkin, I.F., Forgacs, P., Haines, J.H. & Kemma, M.E., 1987. First report of subcutaneous phaeohyphomycosis of the foot caused by *Phoma minutella*. J. Clin. Microbiol. 25: 2395-2397.

Bakerspigel, A., 1970. The isolation of *Phoma hibernica* from a lesions on a leg. Sabouraudia 7: 261-264.

Bakerspigel, A., Lowe, D. & Rostas, A., 1981. The isolation of *Phoma eupyrena* from a human lesion. Arch. Derm. 117: 362-363.

Balogh, M., Liptovsky, J. & Nagy-Peti, Z., 1967. *Microsporum cookei* als Krankheitserreger der Mikrosporia superficialis. Derm. Wschr. 153: 1426-1428.

Bambule, G., Savary, M., Grigoriu, D. & Delacretaz, J., 1982. Les otomycoses. Annls Oto-Laryngol. 99: 537-540.

Banerji, S., Lugli, E.B., Miller, R.F. & Wakefield, A.E., 1995. Analysis of genetic diversity at the arom locus in isolates of *Pneumocystis carinii*. J. Eukaryot. Microbiol. 42: 675-679.

Banno, I., 1967. Studies on the sexuality of *Rhodotorula*. J. Gen. Appl. Microbiol. 13: 167-196.

Banuelos, A.F., Williams, P.L., Johnson, R.H., Bibi, S., Fredricks, D.N., Gilroy, S.A., Bhatti, S.U., Aguet, J. & Stevens, D.A., 1996. Central nervous system abscess due to *Coccidioides* species. Clin. Infect. Dis. 22: 240-250.

Baquis, E., 1905. Una nuova forma di cheratomicosi. Annls Ottalmol. Clin. Ocul. 34: 945-947.

Barale, T., Bièvre, C. de, Chobaut, C. & Cantenot, A., 1982. A propos de la première souche de *Tritirachium oryzae* isolée au C.H.U. de Besançon. Bull. Soc. Fr. Mycol. Méd. 11: 243-245.

Barbas-Filho, J.V., Amato, M.B., Deheinzelin, D., Saldiva, P.H. & Carvalho, C.R. de, 1990. Respiratory failure caused by adiaspiromycosis. Chest 97: 1171-1175.

Barber, G.R., Brown, A.E., Kiehn, T.E., Edwards, F.F. & Amstrong, D., 1993. Catheter-related *Malassezia furfur* fungemia in immunocompromised patients. Am. J. Med. 95: 365-370

Barbor, P.R.H., Rotimi, V.O. & Fatani, H., 1995. Paravertebral abscess caused by *Trichosporon capitatum* in a child with acute lymphoblastic leukemia. J. Infect. 31: 251-252.

Barde, A.K. & Singh, S.M., 1983. A case of onychomycosis caused by *Curvularia lunata* (Wakker) Boedijn. Mykosen 26: 311-316.

Barnett, J.A., Payne, R.W. & Yarrow, D., 1990. Yeasts: Characteristics and Identification, 2nd ed. Cambridge Univ. Press, Cambridge, 1002 pp.

Barnett, J.A., Payne, R.W. & Yarrow, D., 2000. Yeasts: Characteristics and Identification, 3rd ed. Cambridge Univ. Press, Cambridge, 1152 pp.

Barr, D.J.S., Warwick, S.I. & Désaulniers, N.L., 1997. Isoenzyme variation in heterothallic species and related asexual isolates of *Pythium*. Can. J. Bot. 75: 1927-1935.

Barrios, N.J., Kirkpatrick, D.V., Murciano, A., Stine, K., Dyke, R.B. van & Humbert, J.R., 1990. Successful treatment of disseminated *Fusarium* infection in an immunocompromised child. Am. J. Hematol. Oncol. 12: 319-324.

Barron, G.L. & Busch, L.V., 1962. Studies on the soil hyphomycete *Scolecobasidium*. Can. J. Bot. 40: 77-84.

Barron, G.L., Cain, R.F. & Gilman, J.C., 1961a. The genus *Microascus*. Can. J. Bot. 39: 1609-1631.

Barron, G.L., Cain, R.F. & Gilman, J.C., 1961b. A revision of the genus *Petriella*. Can. J. Bot. 39: 837-845.

Barson, W.J. & Ruymann, F.B., 1986. Palmar aspergillosis in immunocompromised children. Pediatr. Infect. Dis. 5: 264-268.

Barth, G. & Gaillardin, C., 1997. Physiology and genetics of the dimorphic fungus *Yarrowia lipolytica*. FEMS Microbiol. Rev. 19: 219-237.

Barthez, J.P., Pierre, D., Bièvre, C. de & Arbeille, M., 1984. Peritonite à *Chaetomium globosum* chez un insuficient rénal traité par D.P.C.A. Bull. Soc. Fr. Mycol. Méd. 13: 205-208.

Bartolome, B., Valks, R., Fraga, J., Buendia, V., Fernandez-Herrera, J. & Garcia-Diez, A., 1999. Cutaneous alternariosis due to *Alternaria chlamydospora* after bone marrow transplantation. Acta Derm. Vener. 79: 244.

Barton, J.T., Daft, B.M., Read, D.H., Kinde, H. & Bickford, A.A., 1992. Tracheal aspergillosis in 61 2-week-old chickens caused by *Aspergillus flavus*. Avian Dis. 36: 1081-1085.

Bartynski, J.M., McCaffrey, T.V. & Frigas, E., 1990. Allergic fungal sinusitis secondary to dematiaceous fungi - *Curvularia lunata* and *Alternaria*. Otolaryngology 103: 32-39.

Baselski, V.S., Robison, M.K., Pifer, L.W. & Woods, D.R., 1990. Rapid detection of *Pneumocystis carinii* in bronchoalveolar lavage samples by using cellufluor staining. J. Clin. Microbiol. 28: 393-394.

Bassetti, S., Frei, R. & Zimmerli, W., 1998. Fungemia with *Saccharomyces cerevisiae* after treatment with *Saccharomyces boulardii*. Am. J. Med. 105: 71-72.

Bassiouny, A., Maher, A., Bucci, T.J., Moawad, M.K. & Hendawy, D.S., 1982. Non invasive antromycosis (diagnosis and treatment). J. Laringol. Otol. 96: 215-228.

Batista, A.C., Campos, S.T.C. & Coelho, R.P., 1960a. *Endomycopsis dermatensis* n.sp., isolado de lesões epidermicas do homem. Publçoes Inst. Micol. Recife 178: 3-13.

Batista, A.C., Guedes, M.D.C. & Silveira, J.S., 1959. Vaginite humana produzida por *Hyalodendron lignicola* Diddens. Publçoes Inst. Micol. Recife 205: 1-18.

Batista, A.C., Maia, H. da S., Alecrim, I. da C., 1955. Onicomicose produzida por '*Aspergillus clavato-nanica*', n. sp. An. Fac. Méd. Recife 15: 197-203.

Batista, A.C., Maia, H. da S. & Cavalcanti, W., 1960b. Otomicose produzida por *Waldemaria pernambucensis* n. gen. n. sp. Atas Inst. Micol. Recife 1960: 5-12.

Batista, A.D., Maia, J.A. & Singer, R., 1955. Basidioneuromycosis on man. An. Soc. Biol. Pernambuco 13: 52-60.

Batista, A.C., Vieira, J.R. & Coêlho, R.P., 1960c. *Schwanniomyces hominis* n.sp. como agente de lesão epidermica em recem-nascido. Publçoes Inst. Micol. Recife 175: 3-14.

Batra, P., 1992. Pulmonary coccidioidomycosis. J. Thorac. Imag. 7: 29-38.

Bauder, B., Kübber-Heiss, A., Steineck, T., Kuttin, E.S. & Kaufman, L., 2000. Granulomatous skin lesions due to histoplasmosis in a badger (*Meles meles*) in Austria. Med. Mycol. 38: 249-253.

Baumgardner, D.J. & Paretsky, D.P., 1999. The *in vitro* isolation of *Blastomyces dermatitidis* from a woodpile in north central Wisconsin, USA. Med. Mycol. 37: 163-168.

Baumgardner, D.J., Paretsky, D.P. & Yopp, A.C., 1995. The epidemiology of blastomycosis in dogs: north central Wisconsin, USA. J. Med. Vet. Mycol. 33: 171-176.

Baylet, J., Camain, R. & Segretain, G., 1959. Identification des agents des maduromycoses du Sénégal et de la Mauritanie. Description d'une espèce nouvelle. Bull. Soc. Pathol. Exot. 52: 448-477.

Bayó, M., Agut, M. & Calvo, M.A., 1994. Otitis externas infecciosas: etiologia en el área de Terrassa, métodos de cultivo y consideraciones sobre la otomicosis. Microbiol. SEM 10: 279-284.

Beaman, L., Pappagianis, D. & Benjamini, E., 1977. Significance of T cells in resistance to experimental murine coccidioidomycosis. Infect. Immun. 17: 580-585.

Beard, J.S., Benson, P.M. & Skillman, L., 1993. Rapid diagnosis of coccidioidomycosis with a DNA probe to ribosomal RNA. Arch. Derm. 129: 1589-1593.

Bearer, E.A., Nelson, P.R., Chowers, M.Y. & Davis, C.E., 1994. Cutaneous zygomycosis caused by *Saksenaea vasiformis* in a diabetic patient. J. Clin. Microbiol. 32: 1823-1824.

Beer, P.L.R & Taine, J., 1990. Otomycoses dans le territoire fédéral Amazonien (Venezuela). Bull. Soc. Fr. Mycol. Méd. 19: 257-264.

Begerow, D., Bauer, R. & Boekhout, T., 2000. Phylogenetic placements of ustilaginomycetous anamorphs as deduced from nuclear rDNA sequences. Mycol. Res. 104: 53-60.

Behar, S.M. & Chertow, G.M., 1998. Olecranon bursitis caused by infection with *Candida lusitaniae*. J. Rheumatol. 25: 598-600.

Belay, T., Cherniak, R., O'Neill, E.B. & Kozel, T.R., 1996. Serotyping of *Cryptococcus neoformans* by dot enzyme assay. J. Clin. Microbiol. 34: 466-470.

Belhadj, S.E., Daoud, A., Gastli, M., Idir, L., Cherif, R., Kallel, K., Zribi, A. & Chaker, E., 1997. Mucormycose rhino-orbitaire et diabète. À propos de 6 cas Tunisiens. J. Mycol. Méd. 7: 159-161.

Belkum, A. van, Quint, W.G.V., Pauw, B.E. de, Melchers, W.J.G. & Meis, J.F., 1993. Typing of *Aspergillus* species and *Aspergillus fumigatus* isolates by interrepeat polymerase chain reaction. J. Clin. Microbiol. 31: 2502-2505.

Bell, R.F., 1976. The development in beef cattle manure of *Petriellidium boydii* (Shear) Malloch, a potential pathogen for man and cattle. Can. J.

Microbiol. 22: 552-556.

Bell, R.G., 1978. Comparative virulence and immunodiffusion analysis of *Petriellidium boydii* (Shear) Malloch strains isolated from feedlot manure and a human mycetoma. Can. J. Microbiol. 24: 856-863.

Bell, S.A., Röcken, M. & Korting, H.C., 1996. Tinea axillaris, a variant of intertriginous tinea, due to non-occupational infection with *Trichophyton verrusosum*. Mycoses 39: 471-474.

Beller, M. & Gessner, B.D., 1994. An outbreak of tinea corporis gladiatorum on a high school wrestling team. J. Am. Acad. Derm. 31: 197-201.

Benard, G., Orii, N.M., Marques, H.H.S., Mendoça, M., Aquino, M.Z., Campeas, A.E., Negro, G.B. del, Durandy, A. & Duarte, A.J.S., 1994. Severe acute paracoccidioidomycosis in children. Pediatr. Infect. Dis. J. 13: 510-515.

Benda, T. & Corey, J.P., 1994. *Malbranchea pulchella* fungal sinusitis. Otolaryngol. Head Neck Surg. 110: 501-504.

Beneke, E.S., 1978. Dematiaceous fungi in laboratory-housed frogs. PAHO Sci. Publ. 356: 101-108.

Benevent, D., Peyronnet, P., Lagarde, C. & Leroux-Robert, C., 1985. Fungal peritonitis in patients on continuous ambulatory peritoneal dialysis. Three recoveries in 5 cases without catheter removal. Nephron 41: 203-206.

Bengoa, A., Briones, V., López, M.B. & Payá, M.J., 1994. Beak infection by *Penicillium cyclopium* in a macaw (*Ara aranaura*). Avian Dis. 38: 922-927.

Ben Hamidi, F., Achard, J.M., Westeel, P.F., Chandenier, J., Bouzernidj, M., Petit, J., Carme, B. & Fournier, A., 1993. Leg granuloma due to *Neocosmospora vasinfecta* in a renal graft recipient. Transpl. Proc. 25: 2292.

Benjamin, R.K., 1962. A new *Basidiobolus* that forms microspores. Aliso 5: 223-233.

Benne, C.A., Neeleman, C., Bruin, M., Hoog, G.S. de & Fleer, A., 1993. Disseminating infection with *Scytalidium dimidiatum* in a granulocytopenic child. Eur. J. Clin. Microbiol. Infect. Dis. 12: 118-121.

Bennett, J.E., Bonner, H., Jennings, A.E. & Lopez, R.I., 1973. Chronic meningitis caused by *Cladosporium trichoides*. Am. J. Clin. Pathol. 59: 398-407.

Benoldi, D., Alinovi, A., Polonelli, L., Conti, S., Gerloni, M., Ajello, L., Padhye, A.A. & Hoog, G.S. de, 1991. *Botryomyces caespitosus* as an agent of cutaneous phaeohyphomycosis. J. Med. Vet. Mycol. 29: 9-13.

Benoudia, F., Assouline, M., Pouliquen, Y., Bouvet, A. & Guého, E., 1999. *Exophiala (Wangiella) dermatitidis* keratitis after keratoplasty. Med. Mycol. 37: 53-56.

Bentz, M.S. & Sautter, R.L., 1993. Disseminated infection with *Aspergillus fumigatus* and *Cladosporium cladosporioides* in an immunocompromised host. Abstr. Gen. Meet. ASM 93: 533.

Ben-Ze'ev, J. & Kenneth, R.G., 1982. Features - criteria of taxonomic value in the Entomophthorales: I. A revision of the Batkoan classification. Mycotaxon 14: 393-455.

Berardinelli, S. & Opheim, D.J., 1985. New germ tube induction medium for the identification of *Candida albicans*. J. Clin. Microbiol. 22: 861-862.

Berbee, M.L., Pirseyedi, M. & Hubbard, S., 1999. *Cochliobolus* phylogenetics and the origin of known, highly virulent pathogens, inferred from ITS and glyceraldehyde-3-phosphate dehydrogenase gene sequences. Mycologia 91: 964-977.

Berenguer, J., Daz-Mediavilla, J., Urra, D. & Muñoz, P., 1989. Central nervous system infection caused by *Pseudallescheria boydii*: case report and review. Rev. Infect. Dis. 11: 890-896.

Berenguer, J., Moreno, S., Muñoz, P., Parras, F., Barros, C. & Bouza, E., 1988. Mucormycosis, clinical manifestations, diagnosis, treatment, and outcome of 12 cases. Revta Iber. Micol. 5, Suppl. 1: 100.

Berenguer, J., Rodríguez-Tudela, J.L., Richard, C., Alvarez, M., Sanz, M.A., Gaztelurrutia, L., Ayats, J. & Martínez-Suárez, J.V., 1997. Deep infections caused by *Scedosporium prolificans*: a report of 16 cases in Spain and a review of the literature. Medicine, Baltimore 76: 256-265.

Bereston, E.S. & Waring, W.S., 1946. *Aspergillus* infection of the nails. Arch. Derm. Syph., Chicago 54: 552-557.

Berger, L. & Langeron, M., 1949. Sur un type nouveau de

chromomycose observé au Canada (*Torula bergeri* n. sp.). Annls Parasit. Hum. Comp. 24: 574-599.

Berger, L., Speare, R., Daszak, P., Green, D.E., Cunningham, A.A., Goggin, C.L., Slocombe, R., Ragan, M.A., Hyatt, A.D., McDonald, K.R., Hines, H.B., Lips, K.R., Marantelli, G. & Parkes, H., 1998. Chytridiomycosis causes amphibian mortality associated with population declines in the rain forests of Australia and Central America. Proc. Nat. Am. Acad. Sci. U.S.A. 95: 9031-9036.

Berger, L., Speare, R. & Humphrey, J., 1997. Mucormycosis in a free-ranging tree frog from Australia. J. Wildlife Dis. 33: 903-907.

Berger, S.T., Katser, D.A., Mondino, B.J. & Pettit, T.H., 1991. Macroscopic pigmentation in a dematiaceous fungal keratitis. Cornea 10: 272-276.

Bergman, A.G. & Kauffman, C.A., 1984. Dermatitis due to *Sporobolomyces* infection. Arch. Derm. 120: 1059-1060.

Bergman, M.M., Gagnon, D. & Doem, G.V., 1998. *Pichia ohmeri* fungemia. Diagn. Microbiol. Infect. Dis. 30: 229-231.

Berkowitz, I.D., Robboy, S.J., Karchmer, A.W. & Kunz, L.J., 1979. *Torulopsis glabrata* fungemia: a clinical pathological study. Medicine 58: 430-440.

Bernacer, M., Gadea, I., Esteban, J., Gegundez, M.I., Kamal, K. & Soriano, F., 1992. Catheter-related fungemia due to *Paecilomyces lilacinus* in a leukemic child. Med. Microbiol. Lett. 1: 207-212.

Berner, R., Sauter, S., Michalski, Y. & Niemeyer, C.M., 1996. Central venous catheter infection by *Aspergillus fumigatus* in a patient with B-type non-Hodgkin lymphoma. Med. Pediatr. Oncol. 27: 202-204.

Bernhardt, H., Wellmer, A., Zimmermann, K. & Knoke, M., 1995. Growth of *Candida albicans* in normal and altered faecal flora in the model of continuous flow culture. Mycoses 38: 265-270.

Berry, A.J., Kerkering, T.M., Giordano, A.M. & Chiancone, J., 1984. Phaeohyphomycotic sinusitis. Pediatr. Infect. Dis. J. 3: 150-152.

Berry, W.L. & Leisewitz, A.L., 1996. Multifocal *Aspergillus terreus* discospondylitis in two German shepherd dogs. J. S. Afr. Vet. Ass. 67: 222-228.

Bertoli, G., Rivasi, F. & Fabio, U., 1992. *Rhodotorula glutinis* keratitis. Int. Ophthalmol. 16: 187-190.

Beurmann, L. de, Clair, M. & Gougerot, H., 1909. Une nouvelle mycose, l'hémisporose. Un cas d'hémisporose de la verge. Bull. Mém. Soc. Méd. Hôp., Paris 3: 917-911.

Beyma, F.H. van, 1931. Über einen neuen *Rhizopus, Rhizopus bovinus* nov. spec. Verh. K. Ned. Akad. Wet., Ser. C, 29: 38-40.

Beyma, F.H. van, 1933. Beschreibung einiger neuen Pilzarten aus dem Centraalbureau voor Schimmelcultures - Baarn (Holland). Zentbl. Bakt. Parasitkde, Abt. 2, 88: 132-141.

Beyma, F.H. van, 1935. Beschreibung einiger neuen Pilzarten aus dem Centraalbureau voor Schimmelcultures, Baarn (Holland). III. Mitteiling. Zentbl. Bakt. Parasitkde, Abt. 2, 91: 345-355.

Bhattacharya, U., 1988. *Scolecobasidium humicola*, a new fungal fish infection from India. Environ. Ecol. 6: 532-533.

Biasoli, M.S. & Bracalenti, B.J.C. de, 1986. Endoftalmitis por *Aspergillus nidulans*. Revta Argent. Micol. 9: 35-38.

Bibashi, E., Papagianni, A., Kelesidis, A., Antoniadou, R. & Papadimitriou, M., 1993. Peritonitis due to *Aspergillus niger* in a patient on continuous ambulatory peritoneal dialysis shortly after kidney graft rejection. Nephrol. Dial. Transpl. 8: 185-187.

Bietti, A., 1922. Tre casi di forma benigna di cheratomicosi con ifomiceti non ancora rinvenuti nella cornea. Annls Ottalmol. Clin. Ocul. 50: 301.

Bièvre, C. de, 1981. Détermination des espèces de *Cladosporium* Link ex Fr. pathogènes pour l'homme et de quelques espèces saprophytes. Bull. Soc. Fr. Mycol. Méd. 10: 277-283.

Bièvre, C. de, 1982. Étude comparative des *Cladosporium* isolés de diverses lésions humaines. Bull. Soc. Path. Exot. 75: 390-399.

Bièvre, C. de, 1991. Les *Alternaria* pathogènes pour l'homme: mycologie épidémiologique. J. Mycol. Méd. 1: 50-58.

Bièvre, C. de & Rouffaud, M.A., 1999. Analyse des séquences de l'ADN codant pour les ITS1 (séquences internes transerites-1) de *Trichophyton rubrum* Sabouraud (Castellani). Comparaison avec principales espèces de *Trichophyton*. J. Mycol. Méd. 9: 1-8.

Biggs, P.J., Allen, R.L., Powers, J.M. & Holley, H.P., 1986. Phaeohyphomycosis complicating compound skull fracture. Surg. Neurol.

25: 393-396.

Bikandi, J., San Millán, R., Moragues, M.D., Cebas, G., Clarke, M., Coleman, D.C., Sullivan, D.J., Quindós, G. & Pontón, J., 1998. Rapid identification of *Candida dubliniensis* by indirect immunofluorescence based on differential localization of antigens on *C. dubliniensis* blastospores and *Candida albicans* germ. tubes. J. Clin. Microbiol. 36: 2428-2433.

Binder, L., Csillag, A. & Toth, G., 1956. Diffuse infiltration of lungs due to *Cryptococcus luteolus*. Lancet 270: 1043-1045.

Bissett, J., 1984. A revision of the genus *Trichoderma*. I. Section *Longibrachiatum* sect. nov. Can. J. Bot. 62: 924-931.

Bisset, J., 1991a. A revision of the genus *Trichoderma*. II. Infrageneric classification. Can. J. Bot. 69: 2357-2372.

Bisset, J., 1991b. A revision of the genus *Trichoderma*. III. Additional notes on section *Longibrachiatum*. Can. J. Bot. 69: 2418-2420.

Bissonnette, K.W., Sharp, N.J.H., Dyksta, M.H., Robertson, I.R., Davis, B., Padhye, A.A. & Kaufman, L., 1991. Nasal and retrobulbar mass in a cat caused by *Pythium insidiosum*. J. Med. Vet. Mycol. 29: 39-44.

Bittencourt, A.L., Londero, A.T., Aranjo, M.G.S., Mendoça, N. & Bastos, J.L.A., 1979. Occurrence of subcutaneous zygomycosis caused by *Basidiobolus haptosporus* in Brazil. Mycopathologia 68: 101-104.

Blake, J.S., Dahl, M.V., Herron, M.J. & Nelson, R.D., 1991. An immunoinhibitory cell wall glycoprotein (mannan) from *Trichophyton rubrum*. J. Invest. Derm. 96: 657-661.

Blalock, H.G., Georg, L.K. & Durieux, W.T., 1973. Encephalitis in turkey poults due to *Dactylaria (Diplorhinotrichum) gallopava*, a case report and its experimental reproduction. Avian Dis. 17: 179-204.

Blanc, V., Cremer, G., Benkhraba, F., Ochonisky, S., Houin, R. & Revuz, J., 1994. Dermatophytie à *Microsporum gypseum* chez un patient atteint du syndrome d'immunodéficience acquisé (SIDA). J. Mycol. Méd. 4: 172-174.

Blanc, C., Lamey, B. & Lapalu, J., 1984. Alternariose cutanée chez un transplante rénal. Bull. Soc. Fr. Mycol. Méd. 13: 213-216.

Blank, H. & Mann, S.J., 1975. *Trichophyton rubrum* infections according to age, anatomical distribution and sex. Br. J. Derm. 92: 171-174.

Blank, H., Taplin, D. & Zaias, N., 1969. Cutaneous *Trichophyton mentagrophytes* infections in Vietnam. Arch. Derm. 99: 135-144.

Blaschke-Hellmessen, R., Lauterbach, I., Paul, K.-D., Tintelnot, K. & Weißbach, G., 1994. Nachweis von *Exophiala dermatitidis* (Kano) de Hoog 1977 bei Septikämie eines Kindes mit akuter lymphatischer Leukämie und bei Patienten mit Mukoviszidose. Mycoses 37, Suppl. 1: 89-96.

Blaser, P., 1975. Taxonomische und physiologische Untersuchungen über die Gattung *Eurotium* Link ex Fr. Sydowia 28: 1-49.

Blechschmidt, J. & Meinhof, W., 1989. *Candida*-Mykosen in der Praxis. Diagnostik und Therapie. Diesbach, Berlin, 69 pp.

Blinkhorn, R.J., Adelstein, D. & Spagnuolo, P.J., 1989. Emergence of a new opportunistic pathogen, *Candida lusitaniae*. J. Clin. Microbiol. 27: 236-240.

Block, R.L. & McCormick, S.A., 1994. Granulomatous dacryoadenitis secondary to *Rhodotorula*: a case mimicking sarcoidosis. Invest. Ophthalmol. Vis. Sci. 35: 1447.

Blomquist, K. & Salonen, A., 1969. *Oidiodendron cerealis* isolated from neurodermitis nuchae. Dermatologica 139: 158-160.

Bloom, J.D., Hamor, R.E. & Gerding, P.A., 1996. Ocular blastomycosis in dogs: 73 cases, 108 eyes (1985-1993). J. Am. Vet. Med. Ass. 209: 1271-1274.

Blotta, M.H.S.L., Mamoni, R.L., Oliveira, S.J., Nouer, S.A., Papaiordanou, P.M.O., Goveia, A. & Pires de Camargo, Z., 1999. Endemic regions of paracoccidioidomycosis in Brazil: a clinical and epidemiologic study of 584 cases in the southeast region. Am. J. Trop. Med. Hyg. 61: 390-394.

Blumenfeld, W., McCook, O., Holodniy, M. & Katzenstein, D.A., 1992. Correlation of morphologic diagnosis of *Pneumocystis carinii* with the presence of *Pneumocystis* DNA amplified by polymerase chain reaction. Mod. Pathol. 5: 103-106.

Bock, M., Maiwald, M., Kappe, R., Nickel, P. & Näher, H., 1994. Polymerase chain reaction-based detection of dermatophyte DNA with a fungus-specific primer system. Mycoses 37: 79-84.

Bodey, G.P. (ed.), 1993. Candidiasis. Raven Press, New York.

Bodey, G.P. & Fainstein, V., 1985. Candidiasis. Raven, New York, 281 pp.

Bodey, G.P. & Vartivarian, S., 1989. Aspergillosis. Eur. J. Clin. Microbiol. Infect. Dis. 8: 413-437.

Boehringer, S.I., Cicuta, M.E., Santa Cruz, A., Gómez, L., Patiño, E.M. & Bordia, J.T., 1999. *Trichophyton simii* en una colonia de monos 'Caí' (*Cebus apella*), en la provincia de Corrientes, Argentina. Revta Iberoam. Micol. 15: 300-301.

Boekhout, T., 1991. A revision of ballistoconidia-forming yeasts and fungi. Stud. Mycol. 33: 1-193.

Boekhout, T., Belkum, A. van, Leenders, A.C.A.P., Verbrugh, H.A., Mukamurangwa, P., Swinne, D. & Scheffers, W.A., 1997. Molecular typing of *Cryptococcus neoformans*: taxonomic and epidemiological aspects. Int. J. Syst. Bact. 47: 432-442.

Boekhout, T., Kamp, M. & Guého, E., 1998. Molecular typing of *Malassezia* species with PFGE and RAPD. Med. Mycol. 36: 365-372.

Boekhout, T., Roeijmans, H.J. & Spaay, F., 1995. A new pleomorphic ascomycete, *Calyptrozyma arxii* gen. et. sp. nov., isolated from the human lower oesophagus. Mycol. Res. 99: 1239-1246.

Boekhout, T. & Scorzetti, G., 1997. Differential killing toxin sensitivity patterns of varieties of *Cryptococcus neoformans*. J. Med. Vet. Mycol. 35: 147-149.

Boelaert, J.R., Locht, M. de, Cutsem, J. van, Kerrels, V., Cantinieaux, B., Verdonck, A., Landuyt, H.W. van & Schneider, Y.-J., 1993. Mucormycosis during deferoxamine therapy is a siderophore-mediated infection: *in vitro* and *in vivo* animal studies. J. Clin. Invest. 91: 1979-1986.

Boerema, G.H., 1993. Contributions towards a monograph of *Phoma* (Coelomycetes). II. Section *Peyronellaea*. Persoonia 15: 197-221.

Boerema, G.H. & Dorenbosch, M.M.J., 1973. The *Phoma* and *Ascochyta* species described by Wollenweber and Hochapfel in their study on fruit-rotting. Stud. Mycol. 3: 1-50.

Boerema, G.H., Dorenbosch, M.M.J. & Kesteren, H.A. van, 1965. Remarks on the species of *Phoma* referred to *Peyronellaea*. Persoonia 4: 47-68.

Boerema, G.H., Gruyter, J. de & Noordeloos, M.E., 1997. Contributions towards a monograph of *Phoma* (Coelomycetes) - IV. Section *Heterospora*: taxa with large sized conidial dimorphs, *in vivo* sometimes as *Stagonosporopsis* synanamorphs. Persoonia 16: 335-371.

Boerema, G.H., Loerakker, W.M. & Hamers, M.E.C., 1996. Contributions towards a monograph of *Phoma* (Coelomycetes) - III. 2. Misapplications of the type species name and the generic synonyms of section *Plenodomus* (excluded species). Persoonia 16: 141-190.

Boerlin, P., Boerlin-Petzold, F., Durussel, C., Addo, M., Pagani, J.-L., Chave, J.-P. & Bille, J., 1995. Cluster of atypical *Candida* isolates in a group of human immunodeficiency virus-positive drug users. J. Clin. Microbiol. 33: 1129-1135.

Böhme, H., 1967. *Arthroderma gertleri* sp. nov., die perfekte Form von *Trichophyton vanbreuseghemii* Rioux, Jarry et Juminer. Mykosen 10: 247-252.

Boltansky, H., Kwon-Chung, K.J., Macher, A.M. & Gallin, J.I., 1984. *Acremonium strictum*-related pulmonary infection in a patient with granulomatous disease. J. Infect. Dis. 149: 653.

Bond, R., Anthony, R.M., Dodd. M. & Llloyd, D.H., 1996. Isolation of *Malassezia sympodialis* from feline skin. J. Med. Vet. Mycol. 34: 145-147.

Bonifaz, A., Saul, A., Montes de Oca, G. & Mercadillo, P., 1999. Superficial cutaneous sporotrichosis in specific anergic patient. Int. J. Derm. 38: 700-703.

Bonomo, R.A., Strauss, M., Blinkhorn, R. & Salata, R.A., 1996. *Torulopsis (Candida) glabrata*: a new pathogen found in spinal epidural abscess. Clin. Infect. Dis. 22: 588-589.

Boomker, J., Coetzer, J.A.W. & Scott, D.B., 1977. Black grain mycetoma (maduromycosis) in horses. Onderstepoort J. Vet. Res. 44: 249-252.

Booth, C., 1966. The genus *Cylindrocarpon*. Mycol. Pap. 104: 1-56.

Booth, C., 1971. The genus *Fusarium*. Commonwealth Mycological Institute, Kew, 257 pp.

Booth, C., 1977. *Fusarium* Laboratory Guide to the Identification of the Major Species. Commonweath Mycological Institute, Kew, 58 pp.

Booth, C., Clayton, M. & Usherwood, M., 1985. *Cylindrocarpon* species associated with mycotic keratitis. Proc. Indian Acad. Sci. (Plant Sci.) 94: 433-436.

Borderie, V.M., Bourcier, T.M., Poirot, J.-L.P., Baudrimont, M., Prudhomme de Saint-Maur, P. & Laroche, L., 1997. Endophthalmitis after *Lasiodiplodia theobromae* corneal abscess. Graefe's Arch. Clin. Exp. Ophthalmol. 235: 259-261.

Borelli, D., 1965. *Microsporum racemosum* nova species. Acta Méd. Venez. 12: 148-151.

Borelli, D., 1972. *Acrotheca aquaspersa* nova species agente de cromomicosis. Acta Cient. Venez. 23: 193-196.

Borelli, D., 1976. *Pyrenochaeta mackinnonii* nova species agente de micetoma. Castellania 4: 227-234.

Borelli, D., 1978. *Plenodomus avramii*, nova species, agente de micetoma. In: Acta I Simp. Intern. Micetomas. Barquísimeto, Venezuela.

Borelli, D., 1980. Causal agents of chromoblastomycosis (chromomycetes). PAHO Sci. Publ. 396: 334-335.

Borelli, D., 1983. *Taeniolella boppii*, nova species, agente de cromomicosis. Med. Cut. I.L.A. 11: 227-232.

Borelli, D., 1995. *Botryodiplodia theobromae* agente de onicomicosis podal. Revta Iberoam. Micol. 12: 2-5.

Borelli, D. & Feo, M., 1966. *Trichophyton phaseoliforme*, nova species. Med. Cutánea 2: 165-172.

Borelli, D. & Marcano, C., 1973. *Cladosporium castellanii* nova species agente de tinea nigra. Castellania 1: 151-154.

Borelli, D. & Zamora, R., 1973. *Chaetosphaeronema larense* nova species agente de micetoma - presentacion del tipo. Bol. Mens. Soc. Venez. Derm. 6: 17-18.

Borelli, D., Zamora, R. & Senabre, G., 1976. *Chaetosphaeronema larense* nova species agente de micetoma. Gaceta Med. Caracas 84: 307-318.

Borne, M.J., Elliott, J.H. & O'Day, D.M., 1993. Ocular fluconazole treatment of *Candida parapsilosis* endophthalmitis after failed intravitreal amphotericin B. Arch. Ophthalmol. 111: 1326-1327.

Bossi, P., Mortier, E., Michon, C., Gaudin, H., Simonpoli, A.M., Pouchot, J. & Vinceneux, P., 1995. Sinusite à *Fusarium solani* chez un patient atteint de SIDA. J. Mycol. Méd. 5: 56-57.

Bottone, E.J., Weitzman, I. & Hanna, B.A., 1979. *Rhizopus rhizopodiformis:* emerging etiological agent of mucormycosis. J. Clin. Microbiol. 9: 530-537.

Bouchon, C.L., Greer, D.L. & Genre, C.F., 1994. Corneal ulcer due to *Exserohilum longirostratum*. Am. J. Clin. Pathol. 101: 452-455.

Boudghène-Stambouli, O. & Mérad-Boudia, A., 1998. Maladie dermatophytique: hyperkératose exubérante avec des cornes cutanées. Annls. Derm. Vénéréol. 125: 705-707.

Bougnoux, M.-E., Guého, E. & Potocka, A.-C., 1993. Resolutive *Candida utilis* fungaemia in a non-neutropenic patient. J. Clin. Microbiol. 31: 1644-1645.

Bouhry, H., Kissi, B., Audry, L., Smreczak, M., Sadkowska-Todys, M., Kulonen, K., Tordo, N., Zmudzinski, J.F. & Holmes, E.C., 1999. Ecology and evolution of rabies virus in Europe. J. Gen. Virol. 80: 2545-2557.

Bourbeau, P., McGough, D.A., Fraser, H., Shah, N. & Rinaldi, M.G., 1992. Fatal disseminated infection caused by *Myceliophthora thermophila*, a new agent of mycosis: case history and laboratory characteristics. J. Clin. Microbiol. 30: 3019-3023.

Bouza, E., Muñoz, P., Vega, L., Rodríguez-Creixems, M., Berenguer, J. & Escudero, A., 1996. Clinical resolution of *Scedosporium prolificans* fungemia associated with reversal of neutropenia following administration of granulocyte colony-stimulating factor. Clin. Infect. Dis. 23: 192-193.

Boutati, E.I. & Anaissie, E.J., 1997. *Fusarium*, a significant emerging pathogen in patients with hematologic malignancy: ten years' experience at a cancer center and implications for management. Blood 90: 999-1008.

Bowman, B.H. & Taylor, J.W., 1993. Molecular phylogeny of pathogenic and non-pathogenic *Onygenales*. In D.R. Reynolds & J.W. Taylor (eds): The Fungal Holomorph, pp. 169-178.

Bowman, B.H., Taylor, J.W., Brownlee, A.G., Lee, J., Lu, S.D. & White, T.J., 1992. Molecular evolution of the fungi: relationship of the

Basidiomycetes, Ascomycetes, and *Chytridiomycetes*. Molec. Biol. Evol. 9: 285-296.

Bowman, B.H., White, T.J. & Taylor, J.W., 1995. Human pathogenic fungi and their close nonpathogenic relatives. Molec. Phyl. Evol. 6: 89-96.

Boyce, J.M., Lawson, L.A., Lockwood, W.R. & Hughes, J.L., 1981. *Cunninghamella bertholletiae* wound infection of probable nosocomial origin. South. Med. J. 74: 1132-1135.

Bradley, F.A., Bickford, A.A. & Walker, R.L., 1993. Diagnosis of favus (avian dermatophytosis) in oriental breed chickens. Avian Dis. 37: 1147-1150.

Bradley, M.C., Leidich, S., Isham, N., Elewski, B.E. & Ghannoum, M.A., 1999. Antifungal suscepbitibilies and genetic relatedness of serial *Trichophyton rubrum* isolates from patients with onychomycosis of the toenail. Mycoses 42: 105-110.

Brakhage, A.A., Jahn, B. & Schmidt, A., 1999. *Aspergillus fumigatus*: biology, clinical aspects and molecular approaches to pathogenicity. Karger, New York, 221 pp.

Branchini, M.L., Pfaller, M.A., Rhine-Chalberg, J., Frempong, T. & Isenberg, H.D., 1994. Genotypic variation and slime production among blood and catheter isolates of *Candida parapsilosis*. J. Clin. Microbiol. 32: 453-456.

Brandt, M.E., Harrison, L.H., Pass, M., Sofair, A.N., Huie, S., Li, R.-K., Morrison, C.J., Warnock, D.W. & Hajjeh, R.A., 2000. *Candida dubliniensis* fungemia: the first four cases in North America. Emerg. Infect. Dis. 6: 46-49.

Brasch, J., 1989. *Microsporum canis* with polymorphous macroconidia. Mycoses 32: 33-38.

Braun, D.K. & Kauffmann, C.A., 1992. *Rhodotorula* fungaemia: a life-threatening complication of indwelling central venous catheters. Mycoses 35: 305-308.

Brawner, D.L., Anderson, G.L. & Yuen, K.Y., 1992. Serotype prevalence of *Candida albicans* from blood culture isolates. J. Clin. Microbiol. 30: 149-153.

Brayford, D., 1996. IMI descriptions of fungi and bacteria Nos. 1261-1270. Mycopathologia 133: 35-63.

Brayford, D. & Samuels, G.J., 1993. Some didymosporous species of *Nectria* with non-microconidial *Cylindrocarpon* anamorphs. Mycologia 85: 612-637.

Breitenbach, J. & Kränzlin, F., 1986. Pilze der Schweiz, Band 2: Nichtblätterpilze. Mykologia, Luzern.

Brenier-Pinchart, M.P., Pelloux, H., Lebeau, B., Pinel, C., Ambroise-Thomas, P. & Grillot, R., 1999. Towards a molecular diagnosis of invasive aspergillosis? A review of the literature. J. Mycol. Méd. 9: 16-23.

Brennan, R.O., Crain, B.J., Proctor, A.M., Durack, D.T. & Phil, D., 1983. *Cunninghamella*: a newly recognized cause of rhinocerebral mucormycosis. Am. J. Clin. Pathol. 80: 98-102.

Brenner, S.A., Morgan, J., Rickert, P.D. & Rimland., D., 1996.*Cladophialophora bantiana* isolated from an AIDS patient with pulmonary infiltrates. J. Med. Vet. Mycol. 34: 427-429.

Bretagne, S., Bart-Delabesse, E., Kuentz, M., Dhenin, N. & Cordonnier, C., 1997. Fatal primary cutaneous aspergillosis in a bone marrow transplant recipient: nosocomial acquisition in a laminar-air flow room. J. Hosp. Infect. 36: 235-239.

Bretagne, S., Costa, J.-M., Marmorat-Khuong, A., Poron, F., Cordonnier, C., Vidaud, M. & Fleury-Feith, J., 1995. Detection of *Aspergillus* species DNA in bronchoalveolar lavage samples by competitive PCR. J. Clin. Microbiol. 33: 1164-1168.

Breton, P., Bani Sadr, F., Germaud, P., Leautez, S., Morin, O. & Raffi, F., 1998. *Penicillium marneffei* infection: a rare cause of pulmonary mycosis. Revue Pneumol. Clin. 54: 85-87.

Bridge, P.D., Williams, M.A.J., Prior, C. & Paterson, R.R.M., 1993. Morphological, biochemical and molecular characteristics of *Metarhizium anisopliae* and *M. flavoviride*. J. Gen. Microbiol. 139: 1163-1169.

Bridges, C.H., 1957. Maduromycotic mycetomas in animals. *Curvularia geniculata* as an etiologic agent. Am. J. Pathol. 33: 411-427.

Brint, J.M., Flynn, P.M., Pearson, T.A. & Pui, C.-H., 1992. Disseminated fusariosis involving bone in an adolescent with leukemia. Pediatr. Infect. Dis. J. 11: 965-968.

Brito, T. de, Sandhu, G.S., Kline, B.C., Aleff, R.A., Sandoval, M.P., Santos, R.T., Brandão, A.A.H. & Lacaz, C.S., 1999. *In situ* hybridization in paracoccidioidomycosis. Med. Mycol. 37: 207-211.

Brodey, R.S., Schrijver, H.F., Deubler, M.J., Kaplan, W. & Ajello, L., 1967. Mycetoma in a dog. J. Am. Vet. Med. Ass. 151: 442-451.

Brummer, E., Castañeda, E. & Restrepo, A., 1993. Paracoccidioidomycosis: an update. Clin. Microbiol. Rev. 6: 89-117.

Bruyn, H.P. de, Broekman, J.M., Vries, G.A. de, Klokke, A.H. & Greep, J.M., 1985. Een patiënt met eumycetoma in Nederland. Ned. Tijdschr. Geneeskd. 129: 1099-1101.

Bruynzeel, I. & Starink, T.M., 1998. Granulomatous skin infection caused by *Scopulariopsis brevicaulis*. J. Am. Acad. Derm. 39: 365-367.

Bryan, C.S., Smith, C.W., Berg, D.E. & Karp, R.B., 1993. *Curvularia lunata* endocarditis treated with terbinafine: case report. Clin. Infect. Dis. 16: 30-32.

Bryce, E.A., Walker, M., Scharf, S., Lim, A.T., Walsh, A., Sharp, N. & Smith, J.A., 1996. An outbreak of cutaneous aspergillosis in a tertiary-care hospital. Infect. Contr. Hosp. Epid. 17: 170-172.

Buchanan, K.L.- & Murphy, J.W., 1998. What makes *Cryptococcus neoformans* a pathogen ? Emerg. Infect. Dis. 4: 71-83.

Buchman, T.G., Rossier, M., Mertz, W.G. & Charache, P., 1990. Detection of surgical pathogens by *in vitro* DNA amplification. Part I. Rapid identification of *Candida albicans* by *in vitro* amplification of a fungus-specific gene. Surgery 108: 338-347.

Buddie, A.G., Martínez-Culebras, P., Bridge, P.D., García, M.D., Querol, A., Cannon, P.F. & Monte, E., 1999. Molecular characterization of *Colletotrichum* strains derived from strawberry. Mycol. Res. 103: 385-394.

Bullock, J.D., Jampol, L.M. & Fezza, A.J., 1974. Two cases of orbital phycomycosis with recovery. Am. J. Ophthalmol. 78: 811-815.

Bulman, R.A. & Stretton, R.J., 1974. Lesions experimentally produced by fungi implicated in extrinsic allergic alveolitis. J. Hyg., Camb. 73: 369-374.

Buracco, P. & Gallo, M.G., 1990. Bone lesions caused by *Cryptococcus laurentii* associated with subclinical visceral leishmaniasis in a dog. Canine Pract. 15: 5-11.

Burdsall, H.H. & Eslyn, W.E., 1974. A new *Phanerochaete* with a *Chrysosporium* imperfect state. Mycotaxon 1: 123-133.

Burgener-Kairuz, P., Zuber, J.-P., Jaunin, P., Buchman, T.G., Bille, J. & Rossier, M., 1994. Rapid detection and identification of *Candida albicans* and *Torulopsis (Candida) glabrata* in clinical specimens by species-specific nested PCR amplification of a cytochrome P-450 lanosterol-α-demethylase (L1A1) gene fragment. J. Clin. Microbiol. 32: 1902-1907.

Burges, G.E., Walls, C.T. & Maize, J.C., 1987. Subcutaneous phaeohyphomycosis caused by *Exserohilum rostratum* in an immunocompetent host. Arch. Derm. 123: 1346-1350.

Burgess, L.W. & Trimboli, D., 1986. Characterization and distribution of *Fusarium nygamai*, sp. nov. Mycologia 78: 223-229.

Burgess, L.W., Nelson, P.E. & Toussoun, T.A., 1989. Stability of morphological characters of *Fusarium nygamai*. Mycologia 81: 480-482.

Burgner, D., Eagles, G., Burgess, M., Procopis, P., Rogers, M., Muir, D., Pritchard, R., Hocking, A. & Priest, M., 1998. Disseminated invasive infection due to *Metarrhizium anisopliae* in an immunocompromised child. J. Clin. Microbiol. 36: 1146-1150.

Burkitt, D.P., Wilson, A.M.M. & Jelliffe, D.B., 1964. Subcutaneous phycomycosis: review of 31 cases seen in Uganda. Br. Med. J. 1964: 1669-1672.

Burks, J.B., Wakabongo, M. & McGinnis, M.R., 1995. Chromoblastomycosis. A fungal infection primarily observed in the lower extremity. J. Am. Podiatr. Med. Ass. 85: 260-264.

Burrell, S.R., Ostlie, D.J., Saubolle, M., Dimler, M. & Barbour, S.D., 1998. *Apophysomyces elegans* infection associated with cactus spine injury in an immunocompetent pediatric patient. Pediatr. Infect. Dis. J. 17: 663-664.

Busapakum, R., Youngchaiyud, U., Sriumpal, S., Segretain, G. & Fromentin, H., 1983. Disseminated infection with *Conidiobolus incongruus*. Sabouraudia 21: 323-330.

Bushelman, S.J., Callen, J.P., Roth, D.N. & Cohen, L.M., 1995. Disseminated *Fusarium solani* infection. J. Am. Acad. Derm. 32: 346-351.

Butty, P., Gorenflot, A., Mallié, M. & Bastide, J.-M., 1992. Low voltage scanning electron microscopy study of naftifine activity on *Microsporum canis*. Mycoses 35: 335-342.

Byard, R.N., Bonin, R.A. & Haq, A.U., 1986. Invasion of paranasal sinuses by *Aspergillus oryzae*. Mycopathologia 96: 41-43.

Byrd, R.P., Roy, T.M., Fields, C.L. & Lynch, J.A., 1992. *Paecilomyces varioti* pneumonia in a patient with diabetes mellitus. J. Diabetes Complic. 6: 150-153.

Cabañes, F.J., Abarca, M.L., Bragulat, M.R. & Castella, G., 1993. Sodium chloride tolerance in strains of *Epidermophyton floccosum* and *Epidermophyton stockdaleae*. Mycopathologia 124: 153-156.

Cabañes, F.J., Alonso, J.M., Castella, G., Alegre, F., Domingo, M. & Pont, S., 1997. Cutaneous hyalohyphomycosis caused by *Fusarium solani* in a loggerhead sea turtle (*Caretta caretta* L.). J. Clin. Microbiol. 35: 3343-3345.

Cabon, N., Moulinier, C., Taieb, A. & Maleville, J., 1994. Dermatophytie à *Microsporum langeronii* chez un nouveau-né contaminé en France. Acta Derm. Vénéréol. 121: 247-248.

Cáceres, A.M., Sardiñas, C., Marcano, C., Guevara, R., Barros, J., Bianchi, G., Rosario, V., Balza, R., Silva, M., Redondo, M.C. & Nuñez, M., 1997. *Apophysomyces elegans* limb infection with a favorable outcome: case report and review. Clin. Infect. Dis. 25: 331-332.

Caffara, M. & Scagliarini, A., 1999. Study of diseases of the grey squirrel (*Sciurus carolinensis*) in Italy. First isolation of the dermatophyte *Microsporum cookei*. Med. Mycol. 37: 75-77.

Cailliez, J.C., Séguy, N., Denis, C.M., Aliouat, E.M., Mazars, E., Polonelli, L., Camus, D. & Dei-Cas, E., 1996. *Pneumocystis carinii*: an atypical micro-organism. J. Med. Vet. Mycol. 34: 227-239.

Cairoli, R., Marenco, P., Perego, R. & Cataldo, F. de, 1995. *Saccharomyces cerevisiae* fungemia with granulomas in the bone marrow in a patient undergoing BMT. Bone Marrow Transpl. 15: 785-786.

Caiuby, M.J., Monteiro, P.C.F. & Nishikawa, M.M., 1996. Isolation of *Trichophyton raubitschekii* in Rio de Janeiro (Brazil). J. Med. Vet. Mycol. 34: 361-363.

Callejas, A., Ordoñez, N., Rodriguez, M.C. & Castañeda, E., 1998. First isolation of *Cryptococcus neoformans* var. *gattii*, serotype C, from the environment in Colombia. Med. Mycol. 36: 341-344.

Cambon, M., Ourgaud, D., Rigal, D., Nizzoli, J.M., Kemeny, J.L., Bièvre, C. de & Coulet, M., 1992. Keratomycose à *Chlamydoabsidia padenii* Hesseltine et Ellis. J. Mycol. Méd. 2: 103-105.

Camenen, I., Closets, F. de, Vaillant, L., Muret, A. de, Pilette, M, Fouquet, B. & Lorette, G., 1988. Alternariose cutanée à *Alternaria tenuissima*. Annls Derm. Vénéréol. 115: 839-842.

Camera, R. de la, Pinilla, I., Muñoz, E., Buendia, B., Steegmann, J.L. & Fernandez-Ranada, J.M., 1996. *Penicillium brevicompactum* as the cause of a necrotic lung ball in an allogeneic bone marrow transplant recipient. Bone Marrow Transpl. 18: 1189-1193.

Cameron, J.A., Badawi, E.M., Hoffman, P.A. & Tabbara, K.F., 1996. Chronic endophthalmitis caused by *Acremonium falciforme*. Can. J. Ophthalmol. 31: 367-368.

Camin, A.-M., Michelet, C., Langanay, T., Place, C. de, Chevrier, S., Guého, E. & Guiguen, C., 1999. Endocarditis due to *Fusarium dimerum* four years after coronary artery bypass grafting. Clin. Infect. Dis. 28: 150.

Campbell, C.K., 1974. Studies on *Hendersonula toruloidea* isolated from human skin and nail. Sabouraudia 12: 150-156.

Campbell, C.K., 1987. *Polycytella hominis* gen. et sp. nov., a cause of human pale grain mycetoma. J. Med. Vet. Mycol. 25: 301-305.

Campbell, C.K. & Al-Hedaithy, S.S., 1993. Phaeohyphomycosis of the brain caused by *Ramichloridium mackenziei* sp. nov. in Middle Eastern countries. J. Med. Vet. Mycol. 31: 325-332.

Campbell, C.K., Johnson, E.M. & Warnock, D.W., 1997. Nail infection caused by *Onychocola canadensis*: report of the first four British cases. J. Med. Vet. Mycol. 35: 423-425.

Campbell, C.K. & Mulder, J.L., 1977. Skin and nail infection by *Scytalidium hyalinum* sp. nov. Sabouraudia 15: 161-166.

Campbell, C.K. & Smith, M.D., 1982. Conidiogenesis in *Petriellidium boydii (Pseudallescheria boydii)*. Mycopathologia 78: 145-150.

Campos-Herrero, M.I., Bordes, A., Perera, A., Ruiz, M.C. & Fernandez, A., 1996. *Trichoderma koningii* peritonitis in a patient undergoing peritoneal dialysis. Clin. Microbiol. Newsl. 18: 150-152.

Campos-Takaki, G.M. & Lobo Jardim, M., 1994. Report of chronic

subcutaneous abscess caused by *Exophiala spinifera*. Mycopathologia 127: 73-76.

Camprin, A.C. & Matthews, R.C., 1993. Application of the polymerase chain reaction to the diagnosis of candidosis by amplification of an HSP 90 gene fragment. J. Med. Microbiol. 39: 233-238.

Candousseau, F. & Sulmont, P., 1971. *Dictyotrichiella semiimmersa* nov. spec. Revue Mycol. 36: 238-242.

Cannon, P.F. & Hawksworth, D.L., 1984. A revision of the genus *Neocosmospora (Hypocreales)*. Trans. Br. Mycol. Soc. 82: 637-688.

Cano, J. & Guarro, J., 1990. The genus *Aphanoascus*. Mycol. Res. 94: 355-377.

Cano, J., Guarro, J., Mayayo, E. & Fernández-Ballart, J., 1992. Experimental infection with *Scedosporium inflatum*. J. Med. Vet. Mycol. 30: 413-420.

Cano, J. & Sigler, L., 1992. Re-evaluation of the synonymy between *Keratinomyces ceretanicus* and *Trichophyton ajelloi*. J. Med. Vet. Mycol. 30: 327-331.

Cano, J., Ulfig, K., Guillamon, J.M., Vidal, P. & Guarro, J., 1997. Studies on keratinophilic fungi. IX: *Neoarachnotheca* gen. nov. and new species of *Nannizziopsis*. Antonie van Leeuwenhoek 72: 149-158.

Caporale, N.E., Calegari, L., Perez, D. & Gezuele, E., 1996. Peritoneal catheter colonization and peritonitis with *Aureobasidium pullulans*. Periton. Dial. Int. 16: 97-98.

Caretta, G., Ajello, L. & Padhye, A.A., 1990. Occurrence of *Arthroderma gloriae*, a geophilic keratinophilic ascomycete in Italy. J. Med. Vet. Mycol. 28: 99-102.

Carlotti, A., Couble, A., Domingo, J., Miroy, K. & Villard, J., 1996. Species-specific identification of *Candida krusei* by hybridization with the CkF1,2 DNA probe. J. Clin. Microbiol. 34: 1726-1731.

Carman, W.F., Frean, J.A., Crewe-Brown, H.H., Culligan, G.A. & Young, C.N., 1989. Blastomycosis in Africa. A review of known cases diagnosed between 1951 and 1987. Mycopathologia 107: 25-32.

Carme, B., Hayette, M.P., Ngaporo, A.I., Ngoler, A., Darozzin, F., Moyikona, A., Ndeli, D., Lehenaff, Y.M., Obengui, M. & Mbalawa, C.G., 1993. Histoplasmose Africaine à *Histoplasma duboisii (Histoplasma capsulatum* var. *duboisii):* Quatorze cas Congolais observés en 10 ans (1981-1990). J. Mycol. Méd. 3: 67-73.

Carme, B., Itoua Ngaporo, A., Ngolet, A., Ibara, J.R. & Ebikili, B., 1992. Disseminated African histoplasmosis in a Congolese patient with AIDS. J. Med. Vet. Mycol. 30: 245-248.

Carmichael, J.W., 1957. *Geotrichum candidum*. Mycologia 49: 820-830.

Carmichael, J.W., 1962. *Chrysosporium* and some other aleuriosporic hyphomycetes. Can. J. Bot. 40: 1137-1173.

Carmichael, J.W., 1966. Cerebral mycetoma of trout due to a *Phialophora*-like fungus. Sabouraudia 5: 120-123.

Carmichael, J.W., Kendrick, W.B., Conners, I.L. & Sigler, L., 1980. Genera of Hyphomycetes. Univ. Alberta Press, Edmonton, 386 pp.

Carrega, G., Riccio, G., Santoriello, L., Pasqualini, M. & Pellicci, R., 1997. *Candida famata* fungemia in a surgical patient successfully treated with fluconazole. Eur. J. Clin. Microbiol. Infect. Dis. 16: 698-699.

Carrière, J., Bagnis, C., Guého, E., Bitker, M.O., Danis, M. & Datry, A., 1997. Subcutaneous phaeohyphomycosis caused by *Phoma cruris-hominis* in renal transplant patient. Abstr. 13th Congr. ISHAM p. 90.

Carrión, A.L., 1935. Chromoblastomycosis: preliminary report on a new clinical type of the disease caused by *Hormodendrum compactum*, nov. sp. Puerto Rico J. Publ. Health Trop. Med. 10: 543-545.

Carrión, A.L., 1936. Chromoblastomycosis: a new clinical type caused by *Hormodendrum compactum*. Puerto Rico J. Publ. Health Trop. Med. 11: 663-682.

Carrión, A.L., 1939. Estudio micologico de un caso de micetoma por *Cephalosporium* en Puerto Rico. Mycopath. Mycol. Appl. 2: 165-170.

Carrión, A.L., 1951. *Cephalosporium falciforme* sp. nov., a new etiologic agent of maduromycosis. Mycologia 43: 522-523.

Carrión, A.L., 1975. Chromoblastomycosis and related infections: new concepts, differential diagnosis and nomenclatural implications. Int. J. Derm. 14: 27-32.

Carrión, A.L. & Silva-Hutner, M., 1971. Taxonomic criteria for the fungi of chromoblastomycosis with reference to *Fonsecaea pedrosoi*. Int. J. Derm. 10: 35-43.

Cartwright, M.J., Promersberger, M. & Stevens, G.A., 1993.

Sporothrix schenckii endophthalmitis presenting as granulomatous uveitis. Br. J. Ophthalmol. 77: 61-62.

Casadevall, A. & Perfect, J.R., 1998. *Cryptococcus neoformans.* ASM Press, Washington, 541 pp.

Casolari, C., Nanetti, A., Cavallini, C.M., Rivasi, F., Fabio, U. & Mazoni, A., 1992. Keratomycosis with an unusual etiology *(Rhodotorula glutinis):* a case report. Microbiologica 15: 83-88.

Castagnola, E., Garaventa, A., Conte, M., Baretta, A., Faggi, E. & Viscoli, C., 1993. Survival after fungemia due to *Fusarium moniliforme* in a child with neuroblastoma. Eur. J. Clin. Microbiol. Infect. Dis. 12: 308-309.

Castañeda Ruíz, R.F., Kendrick, B. & Guarro, J., 1998. Notes on conidial fungi. XVIII. New species of *Pseudomicrodochium* and *Refractohilum* from rainforest litter. Mycotaxon 68: 23-32.

Castanet, J., Lacour, L.P., Toussaint-Gary, M., Perrin, C., Rodot, S. & Ortonne, J.P., 1995. Infection cutanée pluri-focale à *Alternaria tenuissima.* Annls Derm. Véneréol. 122: 115-118.

Castellá, G., Cano, J., Guarro, J. & Cabañes, F.J., 1999. DNA fingerprinting of *Fusarium solani* isolates related to a cutaneous infection in a sea turtle. Med. Mycol. 37: 223-226.

Castellani, A., 1963. Balanoprosthitis chronica ulcerativa- cryptococcia. Derm. Trop. 2: 137-141.

Castillo, A. del, Lucas, A.N.S. & Santandreu, R., 1993. A method of taxonomic determination of *Candida albicans* with DNA probes. Curr. Microbiol. 26: 57-60.

Castrejón, O.V., Robles, M. & Zubieta Arroyo, O.E., 1995. Fatal fungaemia due to *Sporothrix schenckii.* Mycoses 38: 373-376.

Castro, L.G.M., Ito, E., Nunes, R.S., Heins-Vaccari, E.M., Lacaz, C. da Silva & Guarro, J., 2000. Subcutaneous hyalohyphomycosis by *Colletotrichum gloeosporioides* in a renal transplant patient. Proc. 14th ISHAM Congr., B. Aires p. 264.

Castro, L.G.M., Salebian, A. & Sotto, M.N., 1990. Hyalohyphomycosis by *Paecilomyces lilacinus* in a renal transplant patient and a review of human *Paecilomyces* species infections. J. Med. Vet. Mycol. 28: 15-26.

Castro, R.M. & Gomperta, O.F., 1984. Feohifomicose subcutânea por *Cladosporium elatum.* Relato de un caso. An. Bras. Derm. 59: 235-237.

Catalano, P., Lawson, W., Bottone, E. & Lebenger, J., 1990. Basidiomycetous (mushroom) infection of the maxillary sinus. Otolarhyngol. Head Neck Surg. 102: 183-185.

Catanei, A., 1933. Description de *Trichophyton gourvili* n. sp., agent d'une teigne de l'homme. Bull. Soc. Path. Exot. 26: 377-381.

Catellani, G., 1960. Un caso di micosi in un bufalo. Isolamento di un micete della specie *Trichosporon pullulans.* Acta Med. Vet. 6: 359-370.

Caux, F., Aractingi, S., Baurmann, H., Reygagne, P., Dombret, H., Romand, S. & Dubertret, L., 1993. *Fusarium solani* cutaneous infection in a neutropenic patient. Dermatology 186: 232-235.

Celard, M., Dannaoui, E., Piens, M.A., Guého, E., Kirkorian, G., Greenland, T., Vandenesch, F. & Picot, S., 1999. Early *Microascus cinereus* endocarditis of a prosthetic valve implanted after *Staphylococcus aureus* endocarditis of the native valve. Clin. Infect. Dis. 29: 691-692.

Cepero de García, M.C., Arboleda, M.L., Barraquer, F. & Grose, E., 1997. Fungal keratitis caused by *Metarhizium anisopliae* var. *anisopliae.* J. Med. Vet. Mycol. 35: 361-363.

Chabasse, D., 1994. Les nouveaux champignons opportunistes apparus en médicine. Revue générale. J. Mycol. Méd. 4: 9-28.

Chabasse, D., Avenel-Audran, M., Bouchara, J.P., Cimon, B., Verret, J.L., Pichard, E. & Bièvre, C. de, 1997. Deux nouveaux cas Français d'onyxis à *Onychocola canadensis.* J. Mycol. Méd. 7: 43-46.

Chabasse, D., Bièvre, C. de, Legrand, E., Saint-André, J.P., Gentile, L. de, Cimon, B. & Bouchara, J.P., 1995a. Subcutaneous abscess caused by *Pleurophomopsis lignicola* Petr.: first case. J. Med. Vet. Mycol. 33: 415-417.

Chabasse, D. & Contet-Audonneau, N., 1994. Du saprophytisme au parasitisme épidémiologique des champignons kératinophiles isolés en France. J. Mycol. Méd. 4: 80-89.

Chabasse, D., Loison, J., Guého, E., Gentile, L. de, Cimon, B., Chennebault, J.M. & Bouchara, J.P., 1995b. Septicemia caused by *Cryptococcus albidus* in an HIV patient. Abstr. 2nd ECMM, Brussels.

Chabasse, D., Verret, J.L., Smulevici, A., Robert, R., Hocquet, P., 1982. Pathologie à *Alternaria.* A propos d'un cas d'alternariose cutanée.

Bull. Soc. Fr. Mycol. Méd. 11: 247-250.

Chaffin, M.K., Schumacher, J. & McMullan, W.C., 1995. Cutaneous pythiosis in the horse. Vet. Clinics North Am., Eq. Pract. 11: 91-103.

Chai, F.C., Auret, K.A., Christiansen, K., Yuen, P.W. & Gardam, D., 2000. Malignant otitis externa caused by *Malassezia sympodialis.* Head Neck 22: 87-89.

Chaiprasert, A., Samerpitak, K., Wanachiwanawin, W. & Thasnakorn, P., 1990. Induction of zoospore formation in Thai isolates of *Pythium insidiosum.* Mycoses 33: 317-323.

Chakrabarti, A., Kumar, P., Padhye, A.A., Chatha, L., Singh, S.K., Das, A., Wig, J.D. & Kataria, R.N., 1997a. Primary cutaneous zygomycosis due to *Saksenaea vasiformis* and *Apophysomyces elegans.* Clin. Infect. Dis. 24: 580-583.

Chakrabarti, A., Panda, N., Varma, S.C., Singh, K., Das, A., Sharma, S.C. & Padhye, A.A., 1997b. Craniofacial zygomycosis caused by *Apophysomyces elegans.* Mycoses 40: 419-421.

Chalet, M., Howard, D.H., McGinnis, M.R. & Zapatero, I., 1986. Isolation of *Bipolaris australiensis* from a lesion of viral vesicular dermatitis on the scalp. J. Med. Vet. Mycol. 24: 461-465.

Chan, T.H., Koehler, A. & Li, P.K., 1996. *Paecilomyces variotii* peritonitis in patients on continuous ambulatory peritoneal dialysis. Am. J. Kidney Dis. 27: 138-142.

Chandenier, J., Hayette, M.P., Bièvre, C. de, Westeal, P.F., Petit, J., Achard, J.M., Bove, N. & Carme, B., 1993. Tuméfaction de la jambe à *Neocosmospora vasinfecta* chez un tranplanté rénal. J. Mycol. Méd. 3: 165-168.

Chandler, F.W., Kaplan, W. & Ajello, L., 1980. A colour atlas and textbook of histopathology of mycotic diseases. Wolfe, London, 333 pp.

Chang, C.L., Kim, D.-S., Park, D.J., Kim, H.J., Lee, C.H. & Shin, J.H., 2000. Acute cerebral phaeohyphomycosis due to *Wangiella dermatitidis* accompanied by cerebrospinal fluid eosinophilia. J. Clin. Microbiol. 38: 1965-1966.

Chang, W.J., Shields, C.L., Shields, J.A., Depotter, P.V., Schiffman, R., Eagle, R.C. & Nelson, L.B., 1996. Bilateral orbital involvement with massive allergic fungal sinusitis. Arch. Ophthalmol. 114: 767-768.

Chan-Tak, K.M., Thio, C.L., Miller, N.S., Karp, C.L., Ho, C. & Merz, W.G., 1999. *Paecilomyces lilacinus* fungemia in an adult bone marrow transplant recipient. Med. Mycol. 37: 57-60.

Chao, S.C. & Hsu, M.M.L., 1994. *Trichophyton tonsurans* infection in Tainan area. J. Formosan Med. Ass. 93: 697-201.

Chapman, S.W., Rogers, P.D., Rinaldi, M.G. & Sullivan, D.C., 1998. Susceptibilities of clinical and laboratory isolates of *Blastomyces dermatitidis* to ketoconazole, intraconazole, and fluconazole. Antimicrob. Agents Chemother. 42: 978-980.

Chariyalertsak, S., Vanittanakom, P., Nelson, K.E., Sirisanthana, V. & Vanittanakom, N., 1996. *Rhizomys sumatrensis* and *Cannomys badius,* new natural animal hosts of *Penicillium marneffei.* J. Med. Vet. Mycol. 34: 105-110.

Charles, C., 1989. Disseminated aspergillosis in a German Shepherd. Vet. Pathol. Rep. Aust. Soc. Vet. Pathol. 23: 25-26.

Chartois-Léauté, A.G., Wolfrom, E., Bièvre, C. de, Geniaux, M. & Couprie, B., 1995. Alternariose cutanée à *Alternaria chlamydospora.* J. Mycol. Méd. 5: 182-183.

Chatterjee, A., Mitra, K. & Saha, G.R., 1987. Isolation of *Acremonium blochii* (Matr.) W. Gams from cutaneous lesions in Asian buffaloes. Indian J. Anim. Health 26: 171-172.

Chaumentin, G., Boibieux, A., Piens, M.A., Douchet, C., Buttard, P., Bertrand, J.L. & Peyramond, D., 1996. *Trichosporon inkin* endocarditis: short-term evolution and clinical report. Clin. Infect. Dis. 23: 396-397.

Chavanet, P., Lefranc, T., Bonnin, A., Waldner, A. & Portier, H., 1990. Unusual cause of pharyngeal ulcerations in AIDS. Lancet 8711: 383-384.

Chávez, G., Arenas, R., Pérez-Polito, A., Torres, B. & Estrada, R., 1998. Micetomas eumicéticos por *Madurella mycetomatis.* Informe de seis casos. Revta Iberoam. Micol. 15: 90-93.

Checa, J., Barrasa, J.M., Moreno, G., Fort, F. & Guarro, J., 1988. The genus *Coniochaeta* (Sacc.) Cooke *(Coniochaetaceae, Ascomycotina)* in Spain. Crypt. Mycol. 9: 1-34.

Chechani, V. & Kamholz, S.L., 1990. Pulmonary manifestations of disseminated cryptococcosis in patients with AIDS. Chest 98: 1060-1066.

Chen, A.W., Kuo, J.W., Chen, J.S., Sun, C.C. & Huang, S.F., 1993. Dermatophyte pseudomycetoma: a case report. Br. J. Derm. 129: 729-732.

Chen, C.H., Shih, J.F., Hsu, Y.T. & Perng, R.P., 1991. Disseminated coccidioidomycosis with lung, skin and lymph node involvement: report of a case. J. Formos. Med. Ass. 90: 788-792.

Chen, Y.C., Eisner, J.D., Kattar, M.M., Rassoulian-Barrett, S.L., Lafe, K., Yarfitz, S.L., Limaye, A.P. & Cookson, B.T., 2000. Identification of medically important yeasts using PCR-based detection of DNA sequence polymorphisms in the internal transcribed spacer 2 region of the rRNA genes. J. Clin. Microbiol. 38: 2302-2310.

Chermette, R., Ferreiro, L., Bièvre, C. de, Camadro, J.P., Mialot, M. & Vauzelle, P., 1997. *Exophiala spinifera* nasal infection in a cat and literature review of feline phaeohyphomycoses. J. Mycol. Méd. 7: 149-158.

Chien, C.-Y., Bhat, D.J. & Kendrick, W.B., 1992. Mycological observations on *Saksenaea vasiformis* (*Saksenaceae, Mucorales*). Trans. Mycol. Soc. Jpn. 33: 443-448.

Chiewchanvit, S., Mahanupab, P., Hirunsri, P. & Vanittanakom, N., 1991. Cutaneous manifestations of disseminated *Penicillium marneffei* mycosis in five HIV-infected patients. Mycoses 34: 245-249.

Chim, C.S., Ho, P.L. & Yuen, K.Y., 1998. Simultaneous *Aspergillus fischeri* and herpes simplex pneumonia in a patient with multiple myeloma. Scand. J. Infect. Dis. 30: 190-191.

Chinn, H., Chernoff, D.N., Migliorati, C.A., Silverman, S. & Green, T.L., 1995. Oral histoplasmosis in HIV-infected patients. Oral Surg. Oral Med. Oral Path. 79: 710-714.

Chodorowska, G. & Lecewicz-Torún, B., 1996. Blastomycosis opis przypadku własnego. Mikol. Lek. 3: 35-40.

Chopin, J.B., Sigler, L., Connole, M.D., O'Boyle, D.A., Mackay, B. & Goldstein, L., 1997. Keratomycosis in a Percheron cross horse caused by *Cladorrhinum bulbillosum*. J. Med. Vet. Mycol. 35: 53-55.

Chouaid, C., Roux, P., Lavard, I., Poirot, J.L. & Housset, B., 1995. Use of the polymerase chain reaction technique on induced-sputum samples for the diagnosis of *Pneumocystis carinii* pneumonia in HIV-infected patients. A clinical and cost analysis study. Am. J. Clin. Path. 104: 72-75.

Christenson, J.C., Guruswamy, A., Mukwaya, G. & Rettig, P.J., 1987. *Candida lusitaniae:* an emerging human pathogen. Pediatr. Infect. Dis. J. 6: 755-757.

Chuan, M.-T. & Wu, M.-C., 1995. Subcutaneous phaeohypho-mycosis caused by *Exophiala jeanselmei:* successful treatment with itraconazole. Int. J. Derm. 34: 563-566.

Chugh, K.S., Padhye, A.A., Chakrabarti, A., Sakhuja, V., Gupta, K.L. & Kathuria, P., 1996. Renal zygomycosis in otherwise healthy hosts. J. Mycol. Méd. 6: 22-25.

Chung, Y.L., Chang, S.-N., Hann, S.-K., Cho, H.-J., Suh, S.-B. & Lee, K.H., 1999. Spontaneously healed primary cutaneous alternariosis: reports of 2 cases. Kor. J. Med. Mycol. 4: 137-142.

Chupp, C., 1957. The possible infection of the human body with *Cercospora apii*. Mycologia 49: 773-774.

Cicmanec, J.L., 1973. Spontaneous occurrence and experimental transmission of the fungus *Phialophora pedrosoi* in the marine toad, *Bufo marinus*. Lab. Anim. Sci. 23: 43-47.

Ciferri, R., Batista, A.C. & Campos, S., 1956. Isolation of *Schizophyllum commune* from a sputum. Atti Ist. Bot. Lab. Crittogam. Univ. Pavia 14: 3-5.

Ciferri, R. & Redaelli, P., 1925. Monografia delle Torulopsidacee a pigmento rosso. Atti Ist. Bot. R. Univ. Pavia, Ser. 3, 2: 147-303.

Ciferri, R. & Redaelli, P., 1942. Sulle affinitá e sulla posizione sistematica dei generi *Madurella* e *Indiella*. Mycopathologia 3: 182-202.

Cimerman, M., Gunde-Cimerman, N., Zalar, P. & Perkovic, T., 1999. Femur osteomyelitis due to a mixed fungal infection in a previously healthy man. J. Clin. Microbiol. 37: 1532-1535.

Cimon, B., Carrere, J., Chazalette, J.P., Vinatier, J.F., Chabasse, D. & Bouchara, J.P., 1999. Chronic airway colonization by *Penicillium emersonii* in a patient with cystic fibrosis. Med. Mycol. 37: 291-293.

Clancy, C.J. & Nguyen, M.H., 1998. *In vitro* efficacy and fungicidal activity of voriconazole against *Aspergillus* and *Fusarium* species. Eur. J. Clin. Microbiol. Infect. Dis. 17: 573-575.

Clancy, C.J., Wingard, J.R. & Nguyen, M.H., 2000. Subcutaneous phaeohyphomycosis in transplant recipients: review of the literature and demonstration of *in vitro* synergy between antifungal agents. Med. Mycol. 38: 169-175.

Clark, E.C., Silver, S.M., Hollick, G.E. & Rinaldi, M.G., 1995. Continuous ambulatory peritoneal dialysis complicated by *Aureobasidium pullulans* peritonitis. Am. J. Nephrol. 15: 353-355.

Clark, F.D., Jones, L.P. & Panigrahy, B., 1986. Mycetoma in a grand eclectus *(Eclectus rotatus rotatus)* parrot. Avian Dis. 30: 441-443.

Clark, N.M., 1999. *Paecilomyces lilacinus* infection in a heart transplant recipient and succesful treatment with terbinafine. Clin. Infect. Dis. 28: 1169-1170.

Cochet, G., 1939. Sur un nouveau champignon arthrosporé *(Arthrographis langeroni* n. g., n. sp.) agent pathogène d'une onychomycose humaine. Annls Parasit. Hum. Comp. 17: 97-102.

Cock, A.W.A.M. de, 1994. Population biology of *Hortaea werneckii* based on restriction patterns of mitochondrial DNA. Antonie van Leeuwenhoek 65: 21-28.

Cock, A.W.A.M. de, Mendoza, L., Padhye, A.A., Ajello, L. & Kaufman, L., 1987. *Pythium insidiosum* sp. nov., the etiologic agent of pythiosis. J. Clin. Microbiol. 25: 344-349.

Cofrancesco, E., Viviani, M.A., Boschetti, C., Tortorano, A.M., Balzani, A. & Castagnone, D., 1995. Treatment of chronic disseminated *Geotrichum capitatum* infection with high cumulative dose of colloidal amphotericin B and itraconazole in a leukaemic patient. Mycoses 38: 377-384.

Cohen, M. & Montgomerie, J.Z., 1993. Hematogenous endophthalmitis due to *Candida tropicalis:* report of two cases and review. Clin. Infect. Dis. 17: 270-272.

Cohen-Abbo, A., Bozeman, P.M. & Patrick, C.P., 1993. *Cunninghamella* infections: review and report of two cases of *Cunninghamella* pneumonia in immunocompromised children. Clin. Infect. Dis. 17: 173-177.

Cohen-Abbo, A. & Edwards, K.M., 1995. Multifocal osteomyelitis caused by *Paecilomyces variotii* in a patient with chronic granulomatous disease. Infection 23: 55-57.

Coimbra, C.E.A. & Santos, R.V., 1984. Black piedra among Zoro Indians from Amazonia (Brasil). Mycopathologia 107: 57-60.

Coldiron, B.M., Wiley, E.L. & Rinaldi, M.G., 1990. Cutaneous phaeohyphomycosis caused by a rare fungal pathogen, *Hormonema dematioides:* successful treatment with ketoconazole. J. Am. Acad. Derm. 23: 363-367.

Cole, G.T. & Kendrick, W.B., 1968. Conidium ontogeny in hyphomycetes. The imperfect state of *Monascus ruber* and its meristem arthrospores. Can. J. Bot. 46: 987-992.

Cole, G.T. & Kendrick, B., 1973. Taxonomic studies of *Phialophora*. Mycologia 65: 661-688.

Cole, G.T. & Samson, R.A., 1979. Patterns of development in conidial fungi. Pittman, London, 190 pp.

Cole, G.T., Seshan, K.R., Lynn, K.T. & Franco, M., 1993. Gastrointestinal candidiasis: histopathology of *Candida*-host interactions in a murine model. Mycol. Res. 97: 385-408.

Cole, G.T. & Sun, S.H., 1985. Arthroconidium-spherule-endospore transformation in *Coccidioides immitis*. In P.J. Szaniszlo (ed.): Fungal Dimorphism, pp. 281-333. Plenum, New York.

Coleman, D.C., Bennett, D.E., Sullivan, D.J., Gallagher, P.J., Henman, M.C., Shanlay, D.B. & Russell, R.J., 1993. Oral *Candida* in HIV infection and AIDS: new perspectives, new approaches. Crit. Rev. Microbiol. 19: 61-82.

Coleman, D.C., Sullivan, D.J., Bennett, D.E., Moran, G.P., Barry, H.J. & Shanley, D.B., 1997. Candidiasis: the emergence of a novel species. AIDS, London 11: 557-567.

Comiter, C.V., Mcdonald, M., Minton, J. & Yalla, S.V., 1996. Fungal bezoar and bladder rupture secondary to *Candida tropicalis*. Urology 47: 439-441.

Compton, J., 1991. Nucleic acid sequence-based amplification. Nature 350: 91-92.

Conces, D.J., Stockberger, S.M., Tarver, R.D. & Wheat, L.J., 1992. Disseminated histoplasmosis in AIDS: findings on chest radiographs. Am. J. Roentgen. 160: 15-19.

Confalonieri, M., Nanetti, A., Gandola, L., Colavecchio, A., Aiolfi, S., Cannatelli, G., Parigi, P., Scartabellati, A., Della Porte, R. & Mazzoni, A., 1995. Histoplasmosis capsulati in Italy: autotochthonous or imported? Eur. J. Epidemiol. 10: 435-439.

Connolly, J.H., Krockenberger, M.B., Malik, R., Canfield, P.J., Wigney, D.I. & Muir, D.B., 1999. Asymptomatic carriage of *Cryptococcus neoformans* in the nasal cavity of the koala (*Phascolarctos cinereus*). Med. Mycol. 37: 331-338.

Conte, P. le, Blanloeil, Y., Germaud, P., Morin, O. & Moreau, P., 1995. Aspergillose invasive en réanimation. Annls Fr. Anest. Réanim. 14: 198-208.

Contet-Audonneau, N., Barbaud, A., Guérin, V., Basile, A.M., Guiguen, C., Rassemusse, C. & Percebois, G., 1991a. Alternariose cutanée et syndrome de Cushing: nouvelle observation. J. Mycol. Méd. 1: 82-83.

Contet-Audonneau, N., Saboureau, M. & Percebois, G., 1991b. *Trichophyton erinacei* chez le trévission. J. Mycol. Méd. 1: 29-32.

Cooke, W.B. & Kahler, P.G., 1955. Isolation of potentially pathogenic fungi from polluted water and sewage. Publ. Health Rep. 170: 689-694.

Cooper, C.R., Dixon, D.M. & Salkin, I.F., 1992. Laboratory-acquired sporotrichosis. J. Med. Vet. Mycol. 30: 169-171.

Cooter, R.D., Lim, I.S., Ellis, D.H. & Leitch, I.O.W., 1990. Burn wound zygomycosis caused by *Apophysomyces elegans*. J. Clin. Microbiol. 28: 2151-2153.

Coriglione, G., Stella, G., Gafa, L., Spata, G., Oliveri, S., Padhye, A.A. & Ajello, L., 1990. *Neosartorya fischeri* var. *fischeri* (Wehmer) Malloch and Cain 1972 (Anamorph: *Aspergillus fischerianus* Samson and Gams 1985) as a cause of mycotic keratitis. Eur. J. Epidemiol. 6: 382-385.

Corredor,G.G. Castaño, J.H., Peralta, L.A., Díez, S., Arango, M., McEwen, J. & Restrepo, A., 1999. Isolation of *Paracoccidioides brasiliensis* from the nine-banded armadillo *Dasypus novemcinctus*, in an endemic area for paracoccidioidomycosis in Colombia. Revta Iberoam. Micol. 16: 216-220.

Costa, A.R., Pires, M.C., Porto, E., Silva Lacaz, C. da, HeinsVaccari, E.-M. & Maranhâo, W.M., 1988a. Interdigital, cutaneous phaeohyphomycosis due to *Scytalidium lignicola* Pesante 1957. A case report. Mycoses 31: 604-612.

Costa, A.R., Porto, E., Lacaz, C.S. da Silva, Melo, N.T. de, Calux, M.J.F. de & Valente, N.Y.S., 1988b. Cutaneous and ungual phaeohyphomycosis caused by species of *Chaetomium* Kunze (1817) ex Fresenius, 1829. J. Med. Vet. Mycol. 26: 261-268.

Costa, A.R., Porto, E., Tayah, M., Valente, N.Y.S., Lacaz, C. da Silva, Maranhão, W.M. & Rodrigues, M.C., 1990. Subcutaneous mucormycosis caused by *Mucor hiemalis* Wehmer f. *luteus* (Linnemann) Schipper 1973. Mycoses 33: 241-246.

Costa, A.R., Porto, J.R., Pegas, J.R.P., Reis, V.M.S. dos, Pires, M.C., Lacaz, C. da S., Rodrigues, M.C., Müller, H. & Cucé, L.C., 1991. Rhinofacial zygomycosis caused by *Conidiobolus coronatus*. A case report. Mycopathologia 115: 1-8.

Coutelen, F., Biguet, J., Cochet, G., Muller, S. & Doby-Dubois, M., 1955. Étude d'un champignon nouveau isolé d'un tumeur mycotique pulmonaire. Annls Parasit. Hum. Comp. 30: 395-419.

Cox, G.H., Schell, W.A., Scherr, R.L. & Perfect, J.R., 1994. First report of involvement of *Nodulisporium* species in human disease. J. Clin. Microbiol. 32: 2301-2304.

Cox, G.H., Rude, T.H., Dijkstra, C.C. & Perfect, J.R., 1995. The actin gene from *Cryptococcus neoformans*: structure and phylogenetic analysis. J. Med. Vet. Mycol. 33: 261-266.

Cox, N.H. & Irving, B., 1993. Cutaneous 'ringworm' lesions of *Scopulariopsis brevicaulis*. Br. J. Derm. 129: 726-728.

Coyle, V., Isaacs, J.P. & O'Boyle, D.A., 1984. Canine mycetoma: a case report and review of the literature. J. Small Anim. Pract. 25: 261-268.

Creach, P., Auffret, N., Buot, G. & Binet, O., 1995. Tinea ciliaris and blepharitis by *Microsporum canis*. Annls Derm. Véneréol. 122: 773-774.

Crespo Erchiga, V., Ojeda Martos, A., Vera Canaño, A., Crespo Erchiga, A., Sanchez Fajardo, F. & Guého, E., 1999. Mycology of pityriasis versicolor. J. Mycol. Méd. 9: 143-148.

Crosby, J.H., O'Quinn, M.H., Steele, J.C.H. & Rao, R.N., 1989. Fine-needle aspiration of subcutaneous phaeohyphomycosis caused by *Wangiella dermatitidis*. Diagn. Cytopathol. 5: 293-297.

Crous, P.W., Gams, W., Wingfield, M.J. & Wijk, P.S. van, 1996. *Phaeoacremonium* gen. nov. associated with wilt and decline diseases of woody hosts and human infections. Mycologia 88: 786-796.

Crozier, W.J. & Coats, H., 1977. A case of onychomycosis due to *Candida ravautii*. Aust. J. Derm. 18: 139-140.

Cuenca-Estrella, M., Ruiz-Diez, B., Martinez-Suarez, J.V., Monzon, A. & Rodriguez-Tudela, J.L., 1999. Comparative *in-vitro* activity of voriconazole (UK-109,496) and six other antifungal agents against clinical isolates of *Scedosporium prolificans* and *Scedosporium apiospermum*. J. Antimicrob. Chemother. 43: 149-151.

Cunha, T.D. & Lusins, J., 1973. *Cryptococcus albidus* meningitis. South. Med. J. 66: 1230-1243.

Cunningham, A.C., Leeming, J.P, Ingham, E. & Gowland, G., 1990. Differentiation of three serovars of *Malassezia furfur*. J. Appl. Bact. 68: 439-446.

Cuozzo, D.W., Aaronson, B., Benson, P.M. & Sau, P., 1995. *Candida krusei* abdominal wall abscess presenting as ecchymosis. Diagnosis with ultrasound. Arch. Derm. 131: 275-277.

Currah, R.S., 1985. Taxonomy of the *Onygenales: Arthrodermataceae, Gymnoascaceae, Myxotrichaceae* and *Onygenaceae*. Mycotaxon 24: 1-216.

Curran, M.P. & Lemke, L.G., 1996. *Torulopsis glabrata* spinal osteomyelitis involving two contiguous vertebrae. A case report. Spine 21: 866-870.

Cushion, M.T. & Ebbets, D., 1990. Growth and metabolism of *Pneumocystis carinii* in axenic culture. J. Clin. Microbiol. 28: 1385-1394.

Cutsem, J. van, Franssen, J. & Janssen, P.A.J., 1988. Experimental zygomycosis due to *Rhizopus* spp. infection by various routes in guinea-pigs, rats and mice. Mycoses 31: 563-578.

Czwerwiec, F.S., Bilsker, M.S., Kamerman, M.L. & Bisno, A.L., 1993. Long-term survival after fluconazole therapy of candidal prosthetic valve endocarditis. Am. J. Med. 94: 545-546.

Dai, W.L., Wang, D.L., Ren, Z.F., Li, R.Y. & Wang, J.Z., 1987. First case of systemic phaeohyphomycosis caused by *Exophiala spinifera* in China. Chin. J. Derm. 20: 13-15.

Dannemiller, S., Watson, J.R. & Rozmiarek, H., 1995. Fluconazole therapy in a rhesus monkey *(Macaca mulatta)* with epidural *Trichosporon beigelii* in a cephalic recording cylinder. Lab. Anim. Sci. 45: 31-35.

Dannaoui, E., Persat, F., Monier, M.-F., Borel, E., Piens, M.-A. & Picot, S., 1999. *In-vitro* susceptibility of *Aspergillus* spp. Isoaltes to amphotericin B and itraconazole. J. Antimicrob. Chemother. 44: 553-555.

Das, T., Vyas, P. & Sharma, S., 1993. *Aspergillus terreus* postoperative endophthalmitis. Br. J. Ophthalmol. 77: 386-387.

Dasgupta, L.R., Agarwal, S.C., Varma, R.A., Bedi, B.M.S. & Chatterjee, H., 1976. Subcutaneous phycomycosis from Pondicherry, South India. Sabouraudia 14: 123-127.

Date, A., Ramakrishna, B., Lee, V.N. & Sundararaj, G.D., 1995. Tumoral rhinosporidiosis. Histopathology 27: 288-290.

Datry, A., Leblond, V., Feger, C., Gabarre, J., Guého, E., Lecso, G., Danis, M., Binet, J.L. & Gentilini, M., 1988. A propos de 3 cas de mycose à *Fusarium* sp. Bull. Soc. Fr. Mycol. Méd. 17: 137-142.

Davey, K.G., Holmes, A.D., Johnson, E.M., Szekely, A. & Warnock, D.W., 1998. Comparative evaluation of FUNGITEST and broth microdilution methods for antifungal drug susceptibility testing of *Candida* species and *Cryptococcus neoformans*. J. Clin. Microbiol. 36: 926-930.

David, J.C., 1991. IMI Descriptions of Fungi and Bacteria 1073. *Alternaria dianthicola*. Mycopathologia 116: 49-50.

David, J.C., 1995. IMI Descriptions of Fungi and Bacteria 1224. *Ulocladium atrum*. Mycopathologia 129: 47-48.

David, J.C., 1997. A contribution to the systematics of *Cladosporium*. Revision of the fungi previously referred to *Heterosporium*. Mycol. Pap. 172: 1-157.

David, M., Charlin, M. & Naudascher, 1951. Infiltration mycosique à *Aspergillus amstelodami* du lobe temporal simulant un abcès encapsulé. Ablation en masse. Guérison operatoire. Rev. Neurol. 85: 121-124.

Davis, S.R., Ellis, D.H., Goldwater, P., Dimitriou, S. & Byard, R., 1994. First human culture-proven Australian case of ento-mophthoromycosis caused by *Basidiobolus ranarum*. J. Med. Vet. Mycol. 32: 225-230.

Davison, F.D. & MacKenzie, D.W.R., 1984. DNA homology studies in the taxonomy of dermatophytes. Sabouraudia 22: 117-123.

Davison, F.D., MacKenzie, D.W.R. & Owen, R.J., 1980. Deoxyribonucleic acid base composition of dermatophytes. J. Gen. Microbiol. 118: 465-470.

Dawkins, R., 1990. Parasites, desiderata lists and the paradox of the organism. Parasitology 100: S63-S73.

Dawson, C.O. & Gentles, J.C., 1961. The perfect states of *Keratinomyces ajelloi* Vanbreuseghem, *Trichophyton terrestre* Durie & Frey and *Microsporum nanum* Fuentes. Sabouraudia 1: 49-57.

Dawson, C.O. & Lepper, A.W.D., 1970. *Peyronella glomerata* infection of the ear pinna in goats. Sabouraudia 8: 145-148.

Degeilh, B., Contet-Audonneau, N., Chevrier, S. & Guiguen, S., 1994. À propos de trois noveaux cas de dermatophytie à *Microsporum praecox*. J. Mycol. Méd. 4: 175-178.

Degeilh, B., Seguin, P., Brasy, P., Maugendre, S. & Guignen, C., 1993. Abces cérébral à *Trichoderma pseudokoningii* chez une jeune leucemique. Abstr. 1st Congr. ECMM, Paris p. 121.

Degos, R., Segretain, G., Badillet, G. & Maisler, A., 1970. Aspergillose de la paupière. Bull. Soc. Fr. Derm. Syph. 77: 732-734.

Degrave, B., Joujoux, J.M., Danduraud, M. & Guillot, B., 1997. First report of mycetoma caused by *Arthrographis kalrae*: successful treatment with itraconazole. J. Am. Acad. Derm. 37: 318-320.

Dei-Cas, E., Cailliez, J.C., Palluault, F., Aliouat, E.M., Mazars, E., Soulez, B., Suppin, J. & Camus, D., 1992. Is *Pneumocystis carinii* a deep mycosis-like agent? Eur. J. Epidemiol. 8: 460-470.

Dei-Cas, E., Mazars, E., Aliouat, E.M., Nevez, G., Cailliez, J.C. & Camus, D., 1998. The host specificity of *Pneumocystis carinii*. J. Mycol. Méd. 8: 1-6.

Deicke, P. von & Gemeinhardt, H., 1980. Embolisch-metastatische Pilz-Enzephalitis durch *Trichosporon capitatum* nach Infusionstherapie. Dtsch. GesundhWes. 35: 673-677.

Deighton, F.C., 1971. Studies on *Cercospora* and allied genera. III. *Centrospora*. Mycol. Pap. 124: 1-13.

Deighton, F.C. & Mulder, J.L., 1977. *Mycocentrospora acerina* as a human pathogen. Trans. Br. Mycol. Soc. 69: 326-327.

Deka, P.N. & Rao, A.T., 1988. Aspergillosis caused by *Aspergillus nidulans*. Cheiron 17: 232-234.

Delage, A., Lauraire, M.C., Eglin, G., Degrave, B., Dandurand, M. & Guillot, B., 1998. Deux observations de mycoses dues à *Arthrographis langeronii* chez des malades immunocompromis du Sud de la France. J. Mycol. Méd. 8: 205-208.

Delore, P., Coudert, L.R. & Fayolle, J., 1955. Un cas de mycose bronchique avec localisations musculaires septicémiques. Presse Méd. 63: 1580-1582.

Demange, C., Contet-Audonneau, N., Kombila, M., Miegeville, M., Berthonneau, M., Vroey. C. de & Percebois, G., 1992. *Microsporum gypseum* complex in man and animals. J. Med. Vet. Mycol. 30: 301-308.

Deng, Z., Yun, M. & Ajello, L., 1986. Human penicilliosis marneffei and its relation to the bamboo rat. J. Med. Vet. Mycol. 24: 383-389.

Denis, C.M., Mazars, E., Guyot, K., Ödberg-Ferragut, C., Viscogliosi, E., Dei-Cas, E. & Wakefield, A.E., 2000. Genetic divergence at the *SODA* locus of six different *formae speciales* of *Pneumocystis carinii*. Med. Mycol. 38: 289-300.

Depraetere, K., Colebunders, R., Ieven, M., De Droogh, E., Pelgrom, Y., Hauben, E., Van Marck, E. & Vroey, C. de, 1998. Two imported cases of *Penicillium marneffei* infection in Belgium. Acta Clin. Belg. 53: 255-258.

Dermoumi, H., 1993. A rare zygomycosis due to *Cunninghamella bertholletiae*. Mycoses 36: 293-294.

Derrick, E.K., Voyce, M.E. & Proce, M.L., 1994. *Trichophyton tonsurans* kerion in an elderly woman. Br. J. Derm. 130: 683.

Desakorn, V., Smith, M.D., Walsh, A.L., Simpson, A.J.H., Sahassananda, D., Rajanuwong, A., Wuthiekanun, V., Howe, P., Angus, B.J., Suntharasamai, P. & White, N.J., 1998. Diagnosis of *Penicillium marneffei* infection by quantitation of urinary antigen using an enzyme immunoassay. J. Clin. Microbiol. 37: 117-121.

Destombes, P., Mariat, F., Rosati, L., Segretain, G. 1977. Les mycetomes en Somalie - conclusions d'une enquete menée de 1959 à 1964. Acta Trop. 34: 355-373.

Develoux, M., Ndiaye, B., Kane, A., Ndir, O., Huerre, M. & Bièvre, C. de, 1994. Madura foot caused by *Rhinocladiella atrovirens*. Abstr. 12th ISHAM Congr., Adelaide, p. D79.

Deventer, A.J.M. van, Goessens, W.H.F., Belkum, A. van, Vliet, H.J.A. van, Etten, E.W.M. van & Verbrugh, H.A., 1995. Improved detection of *Candida albicans* by PCR in blood of neutropenic mice with systemic candidiasis. J. Clin. Microbiol. 33: 625-628.

Deventer, A.J.M. van, Vliet, H.J.A. van, Hop, W.C.J. & Goessens, W.H.F., 1994. Diagnostic value of anti-*Candida* enolase antibodies. J. Clin. Microbiol. 32: 17-23.

DeVault, G.A., Brown, S.T., King, J.W., Fowler, M. & Oberle, A., 1985. Tenckhoff catheter obstruction resulting from invasion by *Curvularia lunata* in the absence of peritonitis. Am. J. Kidney Dis. 6: 124-127.

Dewhurst, A.G., Cooper, M.J., Khan, S.M., Pallett, A.P. & Dathan, J.R.E., 1990. Invasive aspergillosis in immunosuppressed patients: potential hazard of building work. Br. Med. J. 301: 802-804.

Dhar, J. & Carey, P.B., 1993. *Scopulariopsis brevicaulis* skin lesions in an AIDS patient. AIDS, Philad. 7: 1283-1284.

Dhindsa, M.K., Naidu, J., Singh, S.M. & Jain, S.K., 1995. Chronic suppurative otitis media caused by *Paecilomyces variotii*. J. Med. Vet. Mycol. 33: 59-61.

Diaz, M.R. & Fell, J.W., 2000. Molecular analyses of the IGS & ITS regions of rDNA of the psychrophilic yeasts in the genus *Mrakia*. Antonie van Leeuwenhoek 77: 7-12.

Diaz Guerra, T.M., Mellado, E., Cuenca Estrella, M., Laguna, F. & Rodriguez Tudela, J.L., 1999. Molecular characterization by PCR-fingerprinting of *Candida dubliniensis* strains isolated from two HIV-positive patients in Spain. Diagn. Microbiol. Infect. Dis. 35: 113-119.

Dick, J.D., Rosengard, B.R., Merz, W.G., Stuart, R.K., Hutchins, G.M. & Saral, R., 1985. Fatal disseminated candidiasis due to amphotericin-B-resistant *Candida guilliermondii*. Annls Int. Med. 102: 67-68.

Dickensheets, D.L., 1989. *Hansenula anomala* infection. Rev. Infect. Dis. 11: 507-508.

Dickinson, G.M., Cleary, T.J., Sanderson, T. & McGinnis, M.R., 1983. First case of subcutaneous phaeohyphomycosis caused by *Scytalidium lignicola* in a human. J. Clin. Microbiol. 17: 155-158.

Diekema, D.J., Messer, S.A., Hollis, R.J., Wenzel, R.P. & Pfaller, M.A., 1997. An outbreak of *Candida parapsilosis* prosthetic valve endocarditis. Diagn. Microbiol. Infect. Dis. 29: 147-153.

Dietrichson, E., 1954. Étude d'une collection norvégienne de levures (2e partie). Annls Parasit. Hum. Comp. 29: 460-498.

Dilorenzo, D.J., Wong, G. & Ludmir, J., 1997. *Candida lusitaniae* chorioamnionitis in a bone marrow transplant patient. Obstet. Gynec. 90: 702-703.

Diniz, L., Souza Filho, J.B. de & Vitória, L.C. da, 1991. Dermatofitose por *Trichophyton ajelloi* em ser humano. Anais Bras. Derm. 66: 275-276.

Dion, W.M. & Dukes, T.W., 1985. *Candida rugosa*: experimental infection in a dairy cow. Sabouraudia 20: 95-100.

Dion, W.M., Pukay, B.P. & Bundaza, A., 1982. Feline cutaneous phaeohyphomycosis caused by *Phialophora verrucosa*. Can. Vet. J. 23: 48-49.

Disalvo, A.F., Bigler, W.J., Ajello, L., Johnson, J.E. & Palmer, J., 1970. Bat and soil studies for sources of histoplasmosis in Florida. Publ. Health Rep. USDA 85: 1063-1069.

Diven, D.G., Newton, R.C., Sang, J.L., Beightler, E.L., McGinnis, M.R. & Macdonald-Davidson, E., 1996. Cutaneous hyalohyphomycosis caused by *Paecilomyces lilacinus* in a patient with lymphoma. J. Am. Acad. Derm. 35: 779-781.

Dix, N. & Webster, J., 1995. Fungal ecology. Chapman & Hall, London, 549 pp.

Dixon, D.M. & Polak-Wyss, A., 1991. The medically important dematiaceous fungi and their identification. Mycoses 34: 1-18.

Dixon, D.M. & Salkin, I.F., 1986. Morphologic and physiologic studies of three dematiaceous pathogens. J. Clin. Microbiol. 24: 12-15.

Dixon, D.M., Salkin, I.F., Duncan, R.A., Hurd, N.J., Haines, J.H., Kemna, M.E. & Coles, F.B., 1991a. Isolation and characterization of *Sporothrix schenckii* from clinical and environmental sources associated with the largest U.S. epidemic of sporotrichosis. J. Clin. Microbiol. 29: 1106-1113.

Dixon, D.M., Szaniszlo, P.J. & Polak, A., 1991b. Dihydroxy-naphthalene (DHN) melanin and its relationship with virulence in the early stages of phaeohyphomycosis. In G.T. Cole & H.C. Hoch (eds): The Fungal Spore and Disease Initiation in Plants and Animals, pp. 297-318. Plenum, New York.

Dixon, D.M., Walsh, T.J., Merz, W.G. & McGinnis, M.R., 1989. Infections due to *Xylohypha bantiana (Cladosporium trichoides)*. Rev. Infect. Dis. 11: 515-525.

Dixon, D.M., Walsh, T.J., Salkin, I.F. & Polak, A., 1986. *Dactylaria constricta:* another dematiaceous fungus with neurotropic potential in animals. J. Med. Vet. Mycol. 25: 55-58.

Doby, J.M., 1986. l'Adiaspiromycose humaine par *Emmonsia crescens* dans le monde. Bilan 20 ans après la découverte du premier cas à Rennes en 1963. Bull. Soc. Fr. Mycol. Méd. 15: 221-226.

Doby, J.M. & Kombila-Favry, M., 1978. Présence de formes sexuées (cleistothèces et Hülle cells) dans un cas humain d'aspergillose du sinus maxillaire chez *Aspergillus nidulans* associé à *Aspergillus fumigatus*. Mycopathologia 64: 157-163.

Dodge, C.W., 1935. Medical Mycology. Mosby, St. Louis, 900 pp.

Doi, Y., 1974. *Hypocrea rufa* (Pers. ex Fr.) Fr., *Hypomyces aurantius* (Pers. per S.F. Gray) Tul., and their allies in Japan. Jpn. J. Bot. 20: 403-412.

Dominik, T. & Majchrowicz, I., 1965. Second contribution to the knowledge of keratinolytic and keratinophilic soil fungi in the region of Szczecin. Ecol. Polska, Ser. A, 13: 415-447.

Domsch, K.H., Gams, W. & Anderson, T.-H., 1980. Compendium of soil fungi. Academic Press, London, 2 vols.

Donabedian, H., O'Donnell, E., Olszewski, C., MacArthur, R.D. & Budd, N., 1994. Disseminated cutaneous and meningeal sporotrichosis in an AIDS patient. Diagn. Microbiol. Infect. Dis. 18: 111-115.

Donnelly, S.M., Sullivan, D.J., Shanley, D.B. & Coleman, D.C., 1999. Phylogenetic analysis and rapid identification of *Candida dubliniensis* based on analysis of ACT1 intron and exon sequences. Microbiology, Reading 145: 1871-1882.

Dooley, D.P., Beckius, M.L., Jeffery, B.S., McAllister, C.K., Radentz, W.H., Feldman, A.R., Rinaldi, M.G., Bailey, S.R. & Keeling, J.H., 1989. Phaeohyphomycotic cutaneous disease caused by *Pleurophoma* in a cardiac transplant patient. J. Infect. Dis. 159: 503-507.

Dooley, D.P., Bostic, P.S. & Beckius, M.L., 1997. Spook house sporotrichosis: a point-source outbreak of sporotrichosis associated with hay bale props in a halloween haunted house. Arch. Intern. Med. 157: 1885-1887.

Dorenbosch, M.J., 1970. Key to nine ubiquitous *Phoma*-like fungi. Persoonia 6: 1-14.

Doty, M.S. & Slater, D.W., 1946. A new species of *Heterosporium* pathogenic on young chinook salmon. Am. Midl. Nat. 36: 663-665.

Douchet, C., Thérizol-Ferly, M., Kombila, M., Duong, T.H., Gomez de Diaz, M., Barrabes, A. & Richard-Lenoble, D., 1994. White piedra and *Trichosporon* species in equatorial Africa. III. Identification of *Trichosporon* species by slide agglutination test. Mycoses 37: 261-264.

Douer, D., Goldschmied-Reouven, A., Segev, S. & Ben-Bassat, I., 1987. Human *Exserohilum* and *Bipolaris* infections: report of *Exserohilum* nasal infection in a neutropenic patient with acute leukemia and review of the literature. J. Med. Vet. Mycol. 25: 235-241.

Douvin, D., Lefichoux, Y. & Huguet, C., 1975. Gastric phycomycosis. Early anatomo-pathological and mycological diagnosis. Favourable course under medical then surgical treatment. Arch. Anat. Path., Paris 23: 133-138.

Drabick, J.J., Gomatos, P.J. & Solis, J.B., 1990. Cutaneous cladosporiosis as a complication of skin testing in a man positive for human immunodeficiency virus. J. Am. Acad. Derm. 22: 135-136.

Drakos, P.E., Nagler, A., Or, R., Naparstek, E., Kapelushnik, J., Engelhard, D., Rahav, G., Ne'emean, D. & Slavin, S., 1993. Invasive fungal sinusitis in patients undergoing bone marrow transplantation. Bone Marrow Transpl. 12: 203-208.

Drechsler, C., 1953. Three new species of *Conidiobolus* isolated from leaf mold. J. Wash. Acad. Sci. 43: 29-43.

Dreizen, S., Bodey, G.P., McCredie, K.B. & Keating, M.J., 1985. Orofacial aspergillosis in acute leukemia. Oral Surg. 59: 499-504.

Dromer, F., Guého, E., Improvisi, L., Provost, F. & Dupont, B., 1993a. *Cryptococcus curvatus*: a new pathogen during HIV infection? Abstr. 1st Congr. ECMM, Paris p. 79.

Dromer, F., Guého, E., Ronin, O. & Dupont, B., 1993b. Serotyping *Cryptococcus neoformans* by using a monoclonal antibody specific for capsular polysaccharide. J. Clin. Microbiol. 31: 359-363.

Dromer, F., Mathoulin, S., Dupont, B., Letenneur, L. & Ronin, O., 1996. Individual and environmental factors associated with infection due to *Cryptococcus neoformans* serotype D. Clin. Infect. Dis. 23: 91-96.

Drouhet, E., 1993. Penicilliosis due to *Penicillium marneffei:* a new emerging systemic mycosis in AIDS patients travelling or living in Southeast Asia. J. Mycol. Méd. 4: 195-224.

Drouhet, E., Bièvre, C. de & Wahalo, S., 1985. *Drechslera longirostrata* (Subramanian) and other *Drechslera* species pathogenic to humans and animals. Proc. Indian Acad. Sci. (Plant Sci.) 94: 453-463.

Drouhet, E., Dupont, B. & Ravisse, P., 1991a. Étude expérimentale d'une souche hautement virulente de *Scedosporium inflatum* isolée d'une arthrite du genou. J. Mycol. Méd. 1: 16-20.

Drouhet, E., Guého, E., Gori, S., Huerre, M., Provost, F., Borgers, M. & Dupont, B., 1998. Mycological, ultrastructural and experimental aspects of a new dimorphic fungus *Emmonsia pasteuriana* sp. nov. isolated from a cutaneous disseminated mycosis in AIDS. J. Mycol. Méd. 8: 64-77.

Drouhet, E. & Huerre, M., 1999. Yeast tissue phase of *Emmonsia pasteuriana* inoculated in golden hamster by intratesticular way. Mycoses 42, Suppl. 2: 11-18.

Drouhet, E., Luciani, J., Frantz, P., Chomette, G., Ravisse, P. & Dupont, B., 1991b. Alternariose cutanée et sarcome de Kaposi chez un greffe rénal. J. Mycol. Méd. 1: 84-87.

Drut, R., 1995. Paracoccidioidomycosis: diagnosis by fine-needle aspiration cytology. Diagn. Cytopathol. 13: 52-53.

Dubé, M.P., Heseltine, P.N.R., Rinaldi, M.G., Evans, S. & Zawacki, B., 1994. Fungemia and colonization with nystatin-resistant *Candida rugosa* in a burn unit. Clin. Infect. Dis. 18: 77-82.

Dufait, R., Velho, R. & Vroey, C. de, 1987. Rapid identification of the two varieties of *Cryptococcus neoformans* by D-proline assimilation. Mykosen 30: 483.

Duffner, F., Brandner, S., Opitz, H., Klier, R. & Grote, E.H., 1997. Primary *Candida albicans* empyema associated with epidural hematomas in craniocervical junction. Clin. Neuropathol. 16: 143-146.

Duggan, J.M., Wolf, M.D. & Kauffman, C.A., 1995. *Phialophora verrucosa* infection in an AIDS patient. Mycoses 38: 215-218.

Dupont, J., Laloui, W. & Roquebert, M.F., 1998. Partial ribosomal DNA sequences show an important divergence between *Phaeoacremonium* species isolated from *Vitis vinifera*. Mycol. Res. 102: 631-637.

Durán, J.A., Malvar, A., Pereiro, M. & Pereiro, M., 1989. *Fusarium moniliforme* keratitis. Acta Ophthalmol. 67: 710-713.

Durand-Joly, I., Wakefield, A.E., Palmer, R.J., Denis, C.M., Creusy, C., Fleurisse, L., Ricard, I., Gut, J.P. & Dei-Cas, E., 2000. Ultrastructural and molecular characterization of *Pneumocystis carinii* isolated from a rhesus monkey (*Macaca mulatta*). Med. Mycol. 38: 61-72.

Durie, E.B. & Brown, S., 1953. Fungous infections in hospital practice. Med. J. Aust. 40: 813-814.

Durie, E.B. & Frey, D., 1957. A new species of *Trichophyton* from New South Wales. Mycologia 49: 401-411.

Dvorák, J. & Otcenásek, M., 1969. Mycological Diagnosis of Animal Dermatophytoses. Academia, Praha, 213 pp.

Dworzack, D.L., Clark, R.B., Borkowski, W.J., Smith, D.L., Dykstra, M., Pugsley, M.P., Horowitz, E.A., Connolly, T.L., McKinney, D.L., Hostetler, M.K., Fitzgibbons, J.F. & Galant, M., 1989. *Pseudallescheria boydii* brain abscess: association with near-drowning and efficacy of high-dose, prolonged miconazole therapy in patients with multiple abscesses. Medicine 68: 218-224.

Dykstra, M.J., Salkin, I.F. & McGinnis, M.R., 1989. An ultrastructural comparison of conidiogenesis in *Scedosporium apiospermum, Scedosporium inflatum* and *Scopulariopsis brumptii*. Mycologia 81: 896-904.

Dykstra, M.J., Sharp, N.J.H., Olivry, T., Hillier, A., Murphy, K.M., Kaufman, L., Kunkle, G.A. & Pucheu-Haston, C., 1999. A description of cutaneous-subcutaneous pythiosis in fifteen dogs. Med. Mycol. 37: 427-433.

Eaton, M.E., Padhye, A.A., Schwartz, D.A. & Steinberg, J.P., 1994. Osteomyelitis of the sternum caused by *Apophysomyces elegans*. J. Clin. Microbiol. 32: 2827-2828.

Ebright, J.R., Chandrasekar, P.H., Marks, S., Fairfax, M., Aneziokoro, A. & McGinnis, M.R., 1999. Invasive sinusitis and cerebritis due to *Curvularia clavata* in an immunocompetent adult. Clin. Infect. Dis. 28: 687-689.

Echavarria, E., Cano, E.L. & Restrepo, A., 1993. Disseminated adiaspiromycosis in a patient with AIDS. J. Med. Vet. Mycol. 31: 91-97.

Economopoulou, P., Laskaris, G. & Kittas, K., 1998. Oral histoplasmosis as an indicator of HIV infection. Oral Med. Oral Path. Oral Radiol. Endodonth. 86: 203-206.

Edel, V., Steinberg, C., Gautheron, N. & Alabouvette, C., 1997. Evaluation of restriction analysis of polymerase chain reaction (PCR)-amplified ribosomal DNA for the identification of *Fusarium* species. Mycol. Res. 101: 179-187.

Eeftinck Schattenkerk, J.K.M., 1997. Pneumocystosis and microsporidiosis in HIV-infection. Clinical, diagnostic and therapeutic aspects. Thesis, Amsterdam, 183 pp.

Egan, J.J., Yonan, N., Carroll, K.B., Deiraniya, A.K., Webb, A.K. & Woodcock, A.A., 1996. Allergic bronchopulmonary aspergillosis in lung allograft recipients. Eur. Respir. J. 9: 169-171.

Eichmann, A. & Schär, G., 1996. Afrikanische Histoplasmose bei Patient mit HIV-2-Infektion. Schweiz. Med. Wschr. 126: 765-769.

Eidbo, J., Sanchez, R.L., Tschen, J.A. & Ellner, K.M., 1993. Cutaneous manifestations of histoplasmosis in the acquired immune deficiency syndrome. Am. J. Surg. Path. 17: 110-116.

Eiferman, R.A., Snyder, J.W. & Barbee, J.W., 1983. Corneal chromomycosis. Am. J. Ophthalmol. 95: 255-256.

Eiff, M. von, Fahrenkamp, A., Roos, N., Fegeler, W. & Loo, J. van de, 1990. Hepatosplenic candidosis, a late manifestation of *Candida* septicemia. Mycoses 33: 283-290.

Einstein, H.E. & Catanzaro, A. (eds), 1985. Coccidioidomycosis. Proc. 4th Int. Conf., Nat. Found. Infect. Dis., Washington.

Einstein, H.E. & Johnson, R.H., 1993. Coccidioidomycosis: new aspects of epidemiology and therapy. Clin. Infect. Dis. 16: 349-354.

Elad, D., Orgad, U., Yakobson, B., Perl, S., Golomb, P., Trainin, R., Tsur, I., Shenkler, S. & Bor, A., 1991. Eumycetoma caused by *Curvularia lunata* in a dog. Mycopathologia 116: 113-118.

El-Ani, A.S., 1966. A new species of *Leptosphaeria*, an etiologic agent of mycetoma. Mycologia 58: 406-411.

El-Ani, A.S. & Gordon, M.A., 1965. The ascospore sheath and taxonomy of *Leptosphaeria senegalensis*. Mycologia 57: 275-278.

El Fari, M., Gräser, Y., Presber, W. & Tietz, H.-J., 2000. An epidemic of tinea corporis caused by *Trichophyton tonsurans* among children (wrestlers) in Germany. Mycoses 43: 191-196.

Elidemir, O., Colasurdo, G.N., Rossmann, S.N. & Fan, L.L., 1999. Isolation of *Stachybotrys* from the lung of a child with pulmonary hemisiderosis. Pediatrics 104: 964-966.

Elewski, B.E., 1996. Onychomycosis caused by *Scytalidium dimidiatum*. J. Am. Acad. Derm. 35: 336-338.

Elewski, B.E. & Greer, D.L., 1991. *Hendersonula toruloidea* and *Scytalidium hyalinum*. Review and update. Arch. Derm. 127: 1041-1044.

Eljaschewitsch, J., Sandfort, J., Tintelnot, K., Horbach, I. & Ruf, B., 1996. Port-a-cath-related *Fusarium oxysporum* infection in an HIV-infected patient: treatment with liposomal amphotericin B. Mycoses 39: 115-119.

Elkan, E. & Philpot, C.M., 1973. Mycotic infections in frogs due to a *Phialophora*-like fungus. Sabouraudia 11: 99-105.

Elliot, G.S., Whitney, M.S., Reed, W. & Tuite, S.F., 1984. Antemorte diagnosis of paecilomycosis in a cat. J. Am. Vet. Med. Ass. 1: 93-94.

Ellis, D.H., 1994. Clinical Mycology. The Human Opportunistic Mycoses. Pfizer, New York, 166 pp.

Ellis, D.H. & Pfeiffer, T., 1990. Natural habitat of *Cryptococcus neoformans* var. *gattii*. J. Clin. Microbiol. 28: 1642-1644.

Ellis, D.H. & Pfeiffer, T., 1992. The ecology of *Cryptococcus neoformans*. Eur. J. Epidemiol. 8: 321-325.

Ellis, J.J., 1985. Species and varieties in the *Rhizopus arrhizus-Rhizopus oryzae* group as indicated by their DNA complementarity. Mycologia 77: 243-247.

Ellis, J.J., 1986. Species and varieties in the *Rhizopus microsporus* group as indicated by their DNA complementarity. Mycologia 78: 508-510.

Ellis, J.J. & Hesseltine, C.W., 1965. The genus *Absidia:* globose-spored species. Mycologia 57: 222-235.

Ellis, M.B., 1965. Dematiaceous hyphomycetes. VI. Mycol. Pap. 103: 1-46.

Ellis, M.B., 1966. Dematiaceous hyphomycetes. VII. *Curvularia, Brachysporium*, etc. Mycol. Pap. 106: 1-57.

Ellis, M.B., 1971. Dematiaceous hyphomycetes. Commonwealth Mycological Institute, Kew, 608 pp.

Ellis, M.B., 1976. More dematiaceous hyphomycetes. Commonwealth Mycological Institute, Kew, 507 pp.

Ellison, M.D., Hung, R.T., Harris, K. & Campbell, B.H., 1998. Report of the first case of invasion fungal sinusitis caused by *Scopulariopsis acremonium*: review of *Scopulariopsis* infections Arch. Otolar. Head Neck Surg. 124: 1014-1016.

Eloy, P., Collet, S., Bertrand, B. & Delos, M., 1996. Allergic *Aspergillus* sinusitis presenting as a mucocele. Case report. Mikol. Lek. 3: 103-111.

Eloy, O., Favre, A., Decousser, J.W., Legendre, C., Lesca, C., Badot, F., Sentilhes, C. & Ghnassia, J.C., 1998. Sinusite à *Paecilomyces variotii*. À propos d'un cas. J. Mycol. Méd. 8: 30-31.

El-Rady, J. & Shearer, G., 1996. Isolation and characterization of a calmodulin-encoding cDNA from the pathogenic fungus *Histoplasma capsulatum*. J. Med. Vet. Mycol. 34: 163-169.

El-Shoura, S., 1993. Ultrastructural interaction between multinucleate giant cells and the fungus in aspergillomas of human paranasal sinuses. Virchovs Arch. B-64: 395-400.

Emanuel, D.A., Lawton, B.R. & Wenzel, F.J., 1962. Maple-bark disease. Pneumonitis due to *Coniosporium corticale*. New Engl. J. Med. 266: 333-337.

Emanuel, D.A., Wenzel, F.J. & Lawton, B.R., 1966. Pneumonitis due to *Cryptostroma corticale* (maple-bark disease). New Engl. J. Med. 274: 1413-1418.

Emmens, R.K., Richardson, D., Thomas, W., Hunter, S., Hennigar, R.A., Wingard, J.R. & Nolte, F.S., 1996. Necrotizing cerebritis in an allogeneic bone marrow transplant recipient due to *Cladophialophora bantiana*. J. Clin. Microbiol. 34: 1330-1332.

Emmons, C.W., 1942. Biology of *Coccidioides*. In W.J. Nickerson (ed.): Biology of Pathogenic Fungi, pp. 71-82. Chron. Bot., Waltham.

Emmons, C.W., 1954. Isolation of *Myxotrichum* and *Gymnoascus* from lungs of animals. Mycologia 46: 334-338.

Emmons, C.W. & Ashburn, L.L., 1942. The isolation of *Haplosporangium parvum* n.sp. and *Coccidioides immitis* from wild rodents. Their relationship to coccidioidomycosis. Publ. Health Rep. 57: 1715-1727.

Emmons, C.W., Binford, C.H., Utz, J.P. & Kwon-Chung, K.J., 1977. Medical Mycology, 3rd ed. Lea & Febiger, Philadelphia, 592 pp.

Emmons, C.W. & Bridges, C.H., 1961. *Entomophthora coronata*, the etiologic agent of a phycomycosis of horses. Mycologia 53: 307-312.

Emmons, C.W. & Carrión, A.L., 1936. The *Phialophora* type of sporulation in *Hormodendrum pedrosoi* and *Hormodendrum compactum*. Puerto Rico J. Publ. Health Trop. Med. 11: 703-710.

Emmons, C.W. & Jellison, W.L., 1960. *Emmonsia crescens* sp. n. and adiaspiromycosis (haplomycosis) in mammals. Annls N.Y. Acad. Sci. 89: 91-101.

Emmons, C.W., Lie-Kian-Joe, Njo-Injo Tjoei Eng, Pohan, A., Kertopati, S. & Meulen, A. van der, 1957. *Basidiobolus* and *Cercospora* from human infections. Mycologia 49: 1-10.

England, D.M. & Hochholzer, L., 1993. Adiaspiromycosis: an unusual fungal infection of the lung. Report of 11 cases. Am. J. Surg. Pathol. 17: 876-886.

English, M.P., 1966. *Trichophyton persicolor* infection in the field vole and pipistrelle bat. Sabouraudia 4: 219-222.

English, M.P., Harman, R.R.M. & Turvey, J.W.J., 1967. *Pseudeurotium ovalis* in toenails. Some problems of mycological diagnosis of nail infections. Br. J. Derm. 79: 553-556.

English, M.P., Kapica, L. & Maciejewska, J., 1978. On the occurrence of *Microsporum persicolor* in Montreal, Canana. Mycopathologia 64: 35-37.

Escalante-Glorsky, S., Youssef, A.I. & Chen, Y.K., 1995. *Torulopsis glabrata* infected pseudocystis. Diagnosis and treatment. J. Clin. Gastroent. 21: 230-232.

Eschete, M.L., King, J.M., Nest, B.C. & Oberle, A., 1981. *Penicillium chrysogenum* endophthalmitis. First reported case. Mycopathologia 74: 125-127.

Escudero, M.R., Pino, E. & Muñoz, R., 1976. Micoma pulmonar causado por *Trichoderma viride*. Actas Dermo-Sifil. 9-10: 673-680.

Espinel-Ingroff, A., 1996. Antifungal susceptibility testing. Clin. Microbiol. News. 18: 161-167.

Espinell-Ingroff, A., 1998. *In vitro* activity of the new triazole voriconazole (UK-109,496) against opportunistic filamentous and dimorphic fungi and common and emerging yeast pathogens. J. Clin. Microbiol. 36: 198-202.

Espinel-Ingroff, A., Barchiesi, F., Hazen, K.C., Martínez-Suárez, J.V. & Scalise, G., 1998. Standardization of antifungal susceptibility testing and clinial relevance. Med. Mycol. 36, Suppl. 1: 68-78.

Espinel-Ingroff, A., Dawson, K., Pfaller, M., Anaissie, E., Brisllus, B., Dixon, D., Fothergill, A., Paetznick, V., Peter, J., Rinaldi, M. & Walsh, T., 1995. Comparative and collaborative evaluation of standardization of antifungal susceptibility testing for filamentous fungi. Antimicrob. Agents Chemother. 39: 314-319.

Espinel-Ingroff, A., Oakley, L.A. & Kerkering, T.M., 1987. Opportunistic zygomycotic infections. A literature review. Mycopathologia 97: 33-41.

Estes, S.A., Merz, W.G. & Maxwell, L.G., 1977. Primary cutaneous phaeohyphomycosis caused by *Drechslera spicifera*. Arch. Derm. 113: 813-815.

Estrader, F., Longefait, H., Lalaveé, G., Coury, C. & Constans, P., 1972. *Aspergillus restrictus* forme rare d'aspergillome intra-cavitaire associée à une tuberculose active. J. Fr. Med. Chir. Thor. 26: 241-249.

Etta, L.L. van, Peterson, L.R. & Gerding, D.N., 1983. *Acremonium falciforme* (*Cephalosporium falciforme*) mycetoma in a renal transplant patient. Arch. Derm. 119: 707-708.

Ewald, P.W., 1996. Guarding against the most dangerous emerging pathogens: insights from evolutionary biology. Emerg. Infect. Dis. 2: 245-257.

Fadl, F.A., Gugnani, H.C. & Perera, D.J.B., 1995. Rhinosporidiosis in Saudi Arabia: report of four cases. Mycoses 38: 219-221.

Fahal, A.H., El Toum, E.A., El Hassan, A.M., Mahgoub, E.S. & Gumaa, S.A., 1995. The host tissue reaction to *Madurella mycetomatis*: new classification. J. Med. Vet. Mycol. 33: 15-17.

Failla, P.J., Cerise, F.P., Karam, G.H. & Summer, W.R., 1995. Blastomycosis: pulmonary and pleural manifestations. South. Med. J. 88: 405-410.

Fakih, M.G., Barden, G.E., Oakes, C.A. & Berenson, C.S., 1995. First reported case of *Aspergillus granulosus* infection in a cardiac transplant patient. J. Clin. Microbiol. 33: 471-473.

Falser, N., 1983. Pilzbefall des Ohres. Harmloser Saprophyt oder pathognomonische Risikofaktor? Laryngol. Rhinol. Otol. 62: 140-146.

Fan, M., Currie, B.P., Gutell, R.R., Ragan, M.A. & Casadevall, A., 1994. The 16S-like, 5.8S and 23S-like rRNAs of the two varieties of *Cryptococcus neoformans:* sequence, secondary structure, phylogenetic analysis and restriction length polymorphisms. J. Med. Vet. Mycol. 32: 163-180.

Fanci, R., Pecile, P., Martinez, R.L., Fabbri, A. & Nicoletti, P., 1997. Amphotericin B treatment of fungemia due to unusual pathogens in neutropenic patients: report of two cases. J. Chemotherap. 9: 427-430.

Farina, C., Gamba, A., Tambini, R., Beguin, H. & Trouillet, J.L., 1998. Fatal aortic *Myceliophthora thermophila* infection in a patient affected by cystic medial necrosis. Med. Mycol. 36: 113-118.

Farina, C., Punithalingam, E., Ruggenenti, P. & Goglio, A., 1997. Phaeohyphomycotic soft tissue disease caused by *Pleurophomopsis lignicola* in a kidney transplant patient. J. Med. Microbiol. 46: 699-703.

Farrell, D., Abbey, L. & Payne, C., 1994. *Fusarium oxysporum* peritonitis as a complication of continuous peritoneal dialysis (CAPD): a case report and review. Abstr. 12th ISHAM Congr., Adelaide p. D38.

Faure-Fontenla, M.A., Bracho-Blanchet, E., Yañez-Molina, C. & Barragan-Tame, L., 1997. Gastric perforation with *Candida tropicalis* invasion in a previously healthy girl. Mycoses 40: 175-177.

Favel, A., Michel-Nguyen, A., Chastin, C., Trousson, F., Penaud, A. & Regli, P., 1997. *In-vitro* susceptibility pattern of *Candida lusitaniae* and evaluation of the Etest mehod. J. Antimicrob. Chemother. 39: 591-596.

Febré, N., Silva, V., Medeiros, E.A.S., Godoy, P., Reyes, E., Halker, E. & Fischman, O., 1999. Contamination of peritoneal dialysis fluid by filamentous fungi. Revta Iberoam. Micol. 16: 238-239.

Feio, C.L., Bauwens, L., Swinne, D. & Meurichy, W. de, 1999. Isolation of *Basidiobolus ranarum* from ectotherms in Antwerp Zoo with special reference to characterization of the isolated strains. Mycoses 42: 291-296.

Fell, J.W. & Statzell-Tallman, A., 1998. *Cryptococcus* Vuillemin. In C.P. Kurtzman & J.W. Fell (eds.): The Yeasts, a Taxonomic Study, 4th ed., pp. 742-767. Elsevier, Amsterdam.

Fell, J.W. & Statzell-Tallman, A., 1998. *Rhodotorula* F.C. Harrison. In Kurtzman, C.P. & Fell, J.W. (eds): The Yeasts, a taxonomic study, 4th ed., pp. 800-827. Elsevier, Amsterdam.

Fell, J.W., Statzell-Tallman, A., Hetz, M.S. & Kurtzman, C.P., 1992. Partial rRNA sequences in marine yeasts: a model for identification of marine eukaryotes. Mol. Mar. Biol. Biotech. 1: 175-186.

Feltkamp, M.C.W., Kersten, M.J., Lelie, J. van der, Burggraaf, J.D., Hoog, G.S. de & Kuijper, E.J., 1997. Fatal *Scedosporium prolificans* infection in a leukemic patient. Eur. J. Clin. Microbiol. Infect. Dis. 16: 460-464.

Fenn, J.P., Billetdeaux, E., Segal, H., Skodack-Jones, L., Padilla, P.E., Bale, M. & Carroll, K., 1999. Comparison of four methodologies for rapid and cost-effective identification of *Candida glabrata*. J. Clin. Microbiol. 37: 3387-3389.

Fernández, B., Vázquez-Veiga, H., Llovo, X., Pereiro, M. & Guarro, J., 2000a. *In vitro* susceptibility to itraconazole, clotrimazole, ketoconazole and terbinafine of 100 isolates of *Trichophyton rubrum*. Chemotherapy (in press).

Fernández, B., Pereiro, M. & Guarro, J., 2000b. Unpublished data.

Fernandez, M., Noyola, D.E., Rossmann, S.N. & Edwards, M.S., 1999. Cutaneous phaeohyphomycosis caused by *Curvularia lunata* and a review of *Curvularia* infections in pediatrics. Pediatr. Infect. Dis. J. 18: 727-731.

Ferra, C., Doebbeling, B.N., Hollis, R.J., Pfaller, M.A., Lee, C.K. & Gingrich, R.D., 1994. *Candida tropicalis* vertebral osteomyelitis: a late sequela of fungemia. Clin. Infect. Dis. 19: 697-703.

Ferreiro, J.A., Carlson, B.A. & Cody, D.T., 1997. Paranasal sinus fungus balls. Head Neck 19: 481-486.

Ferry, A.P. & Abedi, S., 1983. Diagnosis and management of rhino-orbitocerebral mucormycosis (phycomycosis). Ophthalmology 90: 1096-1104.

Fetchick, R.J., Rinaldi, M.G. & Sun, S.H., 1986. Zygomycosis due to *Mucor circinelloides*, a rare agent of human fungal disease: clinical and mycological aspects. Bact. Proc. 1986, F-42.

Feuilhade de Chauvin, M. & Bièvre, C. de, 1985. Onyxis et perionyxis à *Aspergillus sclerotiorum*. Bull. Soc. Fr. Mycol. Méd. 14: 77-79.

Fidel, P.L., Vazquez, J.A. & Sobel, J.D., 1999. *Candida glabrata:* review of epidemiology, pathogenesis, and clinical disease with comparison to *C. albicans*. Clin. Microbiol. Rev. 12: 80-96.

Figueras, M.J. & Guarro, J., 1997. X-ray microanalysis of black piedra. Antonie van Leeuwenhoek 72: 275-281.

Figueras, M.J. & Guarro, J., 2000. Ultrastructural aspects of keratinolytic activity of piedra. In R.K.S. Kushwaha & J Guarro (eds): Biology of dermatophytes and other keratinophilic fungi, pp. 136-141.

Figueras, M.J., Guarro, J. & Zaror, L., 1996. New findings in black piedra infection. Br. J. Derm. 135: 157-158.

Figueras, M.J., Guarro, J. & Zaror, L., 1997. Ultrastructural aspects of hair digestion in black piedra infection. J. Med. Vet. Mycol. 35: 1-6.

Figueroa, J.I., Hamilton, A., Allen, M. & Hay, R., 1994. Immunohistochemical detection of a novel 22- to 25-kilodalton glycoprotein of *Paracoccidioides brasiliensis* in biopsy material and partial characterization by using species-specific monoclonal antibodies. J. Clin. Microbiol. 32: 1566-1574.

Figueroa, J.I., Hawranek, T., Abraha, A. & Hay, R.J., 1997. Tinea capitis in south-western Ethiopia: a study of risk factors for infection and carriage. Int. J. Derm. 36: 661-666.

Filipello Marschisio, V., Fontana, A. & Luppi Mosca, A.M., 1976. *Anthopsis deltoidea*, a new genus and species of Dematiaceae from soil. Can. J. Bot. 55: 115-112.

Filipello Marchisio, V., Fuscuni, A. & Queirio, F.L., 2000. *Scopulariopsis brevicaulis:* a keratinophilic or a keratinolytic fungus? Mycoses 43: 281-292.

Fincher, R.M., Fisher, J.F., Lovell, R.D., Newman, C.L., Espinel-Ingroff, A. & Shadomy, H.J., 1991. Infection due to the fungus *Acremonium (Cephalophorium)*. Medicine, Baltimore 70: 398-409.

Fincher, R.M.E., Fisher, J.F., Padhye, A.A., Ajello, L. & Steele, J.C.H., 1988. Subcutaneous phaeohyphomycotic abscess caused by *Phialophora parasitica* in a renal allograft recipient. J. Med. Vet. Mycol. 26: 311-314.

Findlay, G.H., Vismer, H.F. & Liebenberg, N. van den, 1979. Black grain mycetoma: the ultrastructure of *Madurella mycetomi* Mycopathologia 67: 51-54.

Finkelstein, R., Reinhertz, G., Hashman, N. & Merbach, D., 1993. Outbreak of *Candida tropicalis* fungemia in a neonatal intensive care unit. Infect. Control Hosp. Epidemiol. 14: 587-590.

Fishburn, F. & Kelly, D., 1967. Sporotrichosis in a horse. J. Am. Vet. Med. Ass. 151: 45-46.

Fisher, J.F., Shadomy, S., Teabeaut, J.R., Woodward, J., Michaels, G.E., Newman, M.A., White, E., Cook, P., Seagraves, A., Yaghmai, F. & Rissing, J.P., 1982. Near-drowning complicated by brain abscess due to *Petriellidium boydii*. Arch. Neurol. 39: 511-513.

Fischman, O., Godoy, P., Guarro, J., Zorat, M.C. & Zaror, L., 2000. *Fonsecaea pedrosoi* as an agent of corneal infection. Proc. 14th ISHAM Congr, B. Aires, p. 276.

Flahaut, M., Sanglard, M., Monod, M., Bille, J. & Rossier, M., 1997. Rapid detection of *Candida albicans* in clinical samples by DNA amplification of common regions from *C. albicans*-secreted aspartic proteinase genes. J. Clin. Microbiol. 36: 395-401.

Flanagan, K.L. & Bryceson, A.D.M., 1997. Disseminated infection due to *Bipolaris australiensis* in an young immunocompetent man: case report and review. Clin. Infect. Dis. 25: 311-313.

Fletcher, C.L., Hay, R.J., Midgley, G. & Moore, M., 1998. Onychomycosis caused by infection with *Paecilomyces lilacinus*. Br. J. Derm. 139: 1133-1135.

Flórián, E & Galgoczy, J., 1964. *Keratinomyces longifusus* sp. nov. from Hungary. Mycopathol. Mycol. Appl. 24: 73-80.

Flynn, P.M., Williams, B.G., Hethrington, S.V., Williams, B.F., Giannini, M.A. & Pearson, T.A., 1993. *Aspergillus terreus* during hospital renovation. Infect. Control Hosp. Epidemiol. 14: 363-365.

Fogerburg, R., Suh, B., Buckley, R., Lorber, B. & Karian, J., 1981. Cerebrospinal fluid shunt colonization and obstruction by *Paecilomyces variotii*. J. Neurosurg. 54: 257-260.

Földvári, F. & Polgar, F., 1951. À propos de divers symptômes dûs au *Trichothecium roseum*. Dermatologica 102: 135-139.

Fonseca, A., Scorzetti, G. & Fell, J.W., 2000. Diversity in the yeast *Cryptococcus albidus* and related species as revealed by ribosomal DNA sequence analysis. Can. J. Microbiol. 46: 7-27.

Forster, R.K. & Rebell, G., 1975. Therapeutic surgery in failure of medical treatment for fungal keratitis. Br. J. Ophthalmol. 59: 336-371.

Forster, R.K., Rebell, G. & Stiles, W., 1975a. Recurrent keratitis due to *Acremonium potronii*. Am. J. Ophthalmol. 79: 121-125.

Forster, R.K., Rebell, G. & Wilson, L.A., 1975b. Dematiaceous fungal keratitis. Clinical isolates and management. Br. J. Ophthalmol. 59: 372-376.

Fortun, J., Lopez-San Roman, A., Velasco, J.J., Sanchez-Sousa, A., Vicente, E. de, Nuno, J., Quereda, C., Barcena, R., Monge, G., Candela, A., Honrubia, A. & Guerrero, A., 1997. Selection of *Candida glabrata* strains with reduced susceptibility to azoles in four liver transplant patients with invasive candidiasis. Eur. J. Clin. Microbiol. Infect. Dis. 16: 314-318.

Foster, J.B.T., Almeda, E. & Littman, M.L., 1958. Intraocular and conjunctival effects of amphotericin B in man and in the rabbit. Arch. Ophthalmol. 60: 555-564.

Foulet, F., Duvoux, C., Bièvre, C. de, Hézode, C. & Bretagne, S., 1999. Cutaneous phaeohyphomycosis caused by *Veronaea bothryosa* in a liver transplant recipient successfully treated with itraconazole. Clin. Infect. Dis. 29: 689-690.

Fournier, S., Dupont, B., Begue, P., Improvisie, L., Saint Martin, L. de & Huerre, M., 1995. Infection rhino-faciale à *Conidiobolus coronatus* avec lyse osseuse et adénomégalie. Difficultés thérapeutiques. J. Mycol. Méd. 5, Suppl. 1: 35-39.

Frágner, P., 1966. *Trichophyton rubrum* (Cast.) Sabouraud var. *nigricans* var. nova. Česká Mykol. 20: 27-28.

Frágner, P., 1968. *Epidermophyton floccosum* (Harz) Langeron et Milochevitch variatio *nigricans*, var. nov. Česká Mykol. 22: 202-205.

Frank, U.K., Nishimura, S.L., Li, N.C., Sugal, K., Yajko, D.M., Hadley, W.K. & Ng, V.L., 1993. Evaluation of an enzyme immunoassay for detection of cryptococcal capsular polysaccharide antigen in serum and cerebrospinal fluid. J. Clin. Microbiol. 31: 97-101.

Frank, U.K., Nishimura, S.L., Li, N.C., Sugai, K., Yajko, Frágner, P., 1966. Mykoflora der Onychomykosen. Mykosen 9: 29-34.

Franco, M., Bagagli, E., Scapolio, S. & Lacaz, C. da Silva, 2000. A critical analysis of isolation of *Paracoccidioides brasiliensis* from soil. Med. Mycol. 38: 185-191.

Frank, W., Roester, U. & Scholer, H.J., 1974. Sphaerulen-Bilddung bei einer *Mucor*-Spezies in inneren Organen von Amphibien. Zentralbl. Bakt. Parasitenkd., Abt. 1, 226: 405-417.97-101.

Frankel, D.H. & Rippon, J.W., 1989. *Hendersonula toruloidea* infection in man. Index cases in the non-endemic North American host, and a review of the literature. Mycopathologia 105: 175-186.

Franzot, S.P., Mukherjee, J., Cherniak, R., Chen, L.C., Hamdan, J.S. & Casadevall, A., 1998. Micro-evolution of a standard strain of *Cryptococcus neoformans* resulting in differences in virulence and other phenotypes. Infect. Immun. 66: 89-97.

Franzot, S.P., Salkin, I.F. & Casadevall, A.F., 1999. *Cryptococcus neoformans* var. *grubii:* separate varietal status for *Cryptococcus neoformans* serotype A isolates. J. Clin. Microbiol. 37: 838-840.

Frappaz, D., Aubert, G., Bouteille, M., Dorche, G. & Freycon, F., 1990. Fungemie à *Candida rugosa* compliquant une allogreffe de moëlle osseuse. Méd. Malad. Infect. 20: 42-44.

Fraser, V.J., Keath, E.J. & Powderly, W.G., 1991. Two cases of blastomycosis from a common source: use of DNA restriction analysis to identify strains. J. Infect. Dis. 163: 1378-1381.

Frean, J., Blumberg, L. & Woolf, M., 1993. Disseminated blastomycosis masquerading as tuberculosis. J. Infect. 26: 203-206.

Fredenucci, I., Chomarat, M., Boucaud, C. & Flandrois, J.P., 1998. *Saccharomyces boulardii* fungemia in a patient receiving ultra-levure therapy. Clin. Infect. Dis. 27: 222-223.

Freidank, H., 1995. Hyalohyphomycoses due to *Fusarium* spp. Two case reports and review of the literature. Mycoses 38: 69-74.

Freire, S.V., Paiva, L.M., Luna-Alves, E.A. de & Costa Maia, L., 1998. Morphological. cytological, and cultural aspects of *Curvularia pallescens*. Revta Microbiol. 29: 197-201.

Frelier, P.F., Sigler, S. & Nelson, P.E., 1985. Mycotic pneumonia caused by *Fusarium moniliforme* in an alligator. Sabouraudia 23: 399-402.

Freour, P., Lahourcade, M. & Chomy, P., 1966. Les champignons 'Beauveria' en pathologie humaine. A propos d'un cas à localisation pulmonaire. Presse Méd. 74: 3317-2320.

Frey, D., 1971. Isolation of *Microsporum cookei* from a human case. Sabouraudia 9: 146-148.

Fridkin, S.K., Kremer, F.B., Bland, L.A., Padhye, A., Mcneil, M.M. & Jarvis, W.R., 1996. *Acremonium kiliense* endophthalmitis that occurred after cataract extraction in an ambulatory surgical center and was traced to an environmental reservoir. Clin. Infect. Dis. 22: 222-227.

Friedman, A.D., Campos, J.M., Rorke, L.B., Bruce, D.A. & Arbeter, A.M., 1981. Fatal recurrent *Curvularia* brain abscess. J. Pediatr. 99: 413-415.

Fromentin, H., 1982. Enzymatic characterization with the API ZYM system of *Entomophthorales* potentially pathogenic to man. Curr. Microbiol. 7: 315-318.

Fromentin, H. & Ravisse, P., 1977. Les entomophthoromycoses tropicales. Acta Trop. 34: 375-394.

Fromentin, H., Segretain, G.L. & Segretain, G.M., 1981. Identification d'un *Conidiobolus incongruus* agent d'une mycose profonde en Thailande. Bull Soc. Fr. Mycol. Méd. 10: 77-80.

Fromtling, R.A., Kosanke, S.D., Jensen, J.M. & Bulner, G.S., 1979. Fatal *Beauveria bassiana* infection in a captive American alligator. J. Am. Vet. Med. Ass. 175: 934-936.

Frye, C.B. & Reinhardt, J., 1993. Characterization of groups of the zygomycete genus *Rhizopus*. Mycopathologia 124: 139-147.

Fryen, A., Mayser, P., Glanz, H., Füssle, R., Breithaupt, H. & Hoog, G.S. de, 1999. Allergic fungal sinusitis caused by *Bipolaris (Drechslera) hawaiiensis*. Eur. Arch. Otorhinolaryngol. 256: 330-334.

Fuchs, A., Axmann, H., Zuckermann, A. & Kuttin, E.S., 1996. Subcutaneous mycosis in a cat due to *Staphylotrichum coccosporum*. Mycoses 39: 381-385.

Fuentes, C.A. & Wolf, F.A., 1956. *Microascus pedrosoi* is *M. cinereus* - a correction. Mycologia 48: 446-448.

Fujita, S.-i., Lasker, B.A., Lott, T.J., Reiss, E. & Morrison, C.J., 1995. Microtitration plate enzyme immunoassay to detect PCR amplified DNA from *Candida* species in blood. J. Clin. Microbiol. 33: 962-967.

Fukushiro, R., 1983. Chromomycosis in Japan. Int. J. Derm. 22: 221-229.

Fukushiro, R., Kagawa, S., Nishiyama, S. & Takahashi, H., 1957. Un cas de chromoblastomycose cutanée avec métastase cérébrale mortelle. Presse Méd. 65: 2142.

Fukushiro, R., Udagawa, S., Kawashima, Y. & Kawamura, Y., 1986. Subcutaneous abscesses caused by *Ochroconis gallopavum*. J. Med. Vet. Mycol. 24: 175-182.

Fulciniti, F., Troncone, G., Fazioli, F., Vetrani, A., Zeppa, P., Manco, A. & Palombini, L., 1996. Osteomyelitis by *Paracoccidioides brasiliensis* (South American blastomycosis): cytologic diagnosis on fine-needle aspiration biopsy smears. A case report. Diagn. Cytopath. 15: 442-446.

Furman, R.M. & Ahearn, D.-G., 1983. *Candida ciferrii* and *Candida chiropterorum* isolated from clinical specimens. J. Clin. Microbiol. 18: 1252-1255.

Furukawa, H., Kusne, S., Sutton, D.A., Manez, R., Carrau, R., Nicols, L., Abu-Elmagd, K., Skedros, D., Todo, S. & Rinaldi, M.G., 1998. Acute invasive sinusitis due to *Trichoderma longibrachiatum* in a liver and small bowel transplant recipient. Clin. Infect. Dis. 26: 487-489.

Fuste, F.J., Ajello, L., Threlkeld, R. & Henry, E.R., 1973. *Drechslera hawaiiensis*: causative agent of fatal fungal meningoencephalitis. Sabouraudia 11: 59-63.

Gadallah, M.F., White, R., El-Shahawy, M.A., Abreo, F., Oberlec, A. & Work, J., 1995. Peritoneal dialysis complicated by *Bipolaris hawaiiensis*: successful therapy with catheter removal and itraconazole without the use of amphotericin-B. Am. J. Nephrol. 15: 348-352.

Gales, A.C., Pfaller, M.A., Houston, A.K., Joly, S., Sullivan, D.J., Coleman, D.C. & Soll, D.R., 1999. Identification of *Candida dubliniensis* based on temperature and utilization of xylose and α-methyl-D-glucoside as determined with the API 20C AUX and Vitek YBC systems. J. Clin. Microbiol. 37: 3804-3808.

Galetta, S.L., Wulc, A.E., Goldberg, H.I., Nichols, C.W. & Glaser J.S., 1990. Rhinocerebral mucormycosis: management and survival after carotid occlusion. Annls Neurol. 28: 103-107.

Gams, W., 1971. *Cephalosporium*-artige Schimmelpilze (Hyphomycetes). Gustav Fischer, Stuttgart, 262 pp.

Gams, W., 2000. *Phialophora* and some similar morphologically little-differentiated anamorphs of divergent ascomycetes. Stud. Mycol. 45: 187-199.

Gams, W. & Bissett, J., 1998. Morphology and identification of *Trichoderma*. In C.P. Kubicek & G.E. Harman (eds): *Trichoderma* and *Gliocladium* 1: 3-34. Taylor & Francis, London.

Gams, W. & Holubová-Jechová, V., 1976. *Chloridium* and some other dematiaceous hyphomycetes growing on decaying wood. Stud. Mycol. 13: 1-99.

Gams, W. & McGinnis, M.R., 1983. *Phialemonium*, a new anamorphic genus intermediate between *Phialophora* and *Acremonium*. Mycologia 75: 977-987.

Gams, W. & Meyer, W., 1998. What exactly is *Trichoderma harzianum*? Mycologia 90: 904-915.

Gams, W., Nirenberg, H.I., Seifert, K.A., Brayford, D. & Thrane, U., 1997. Proposal to conserve the name *Fusarium sambucinum* (Hyphomycetes). Taxon 46: 111-113.

Garcia, J., Berg, D., Murray, J., Schell, W. & Perfect, J., 1992. Chronic leg ulcer due to *Curvularia pallescens*. J. Cutan. Pathol. 19: 523.

Garcia-Arata, M.I., Otero, M.J., Zomeno, M., Figuera, M.A. de la, Cuevas, M.C. de las & Lopez-Brea, M., 1996. *Scedosporium apiospermum* pneumonia after autologous bone marrow transplatation. Eur. J. Clin. Microbiol. Infect. Dis. 15: 600-603.

García-Martos, P., Gené, J., Solé, M., Mira, J., Ruíz-Henestrosa, R. & Guarro, J., 1999a. Case of onychomycosis caused by *Microsporum racemosum*. J. Clin. Microbiol. 37: 258-260.

García-Martos, P., Hernández-Molina, J.M., Carillo-Muñoz, A.J., Arroyo, F., Marquez, J. & Romero, C., 1994. Fungal sepsis caused by *Pichia anomala*. Abstr.12th ISHAM Congr., Adelaide p. D72.

García-Martos, P., Hernández-Molina, J.M., Galan, F., Ruiz-Henestrosa, J.R., Garcia-Agudo, R., Palomo, M.J. & Mira, J., 1999b. Isolation of *Hanseniaspora uvarum (Kloeckera apiculata)* in humans. Mycopathologia 144: 73-75.

García-Martos, P., Rubia, F. de la, Paloma, M.J., Mar Alvarez, M. del, Marín, P. & Mira, J., 1993. *Candida lipolytica*, un nuevo patógeno oportunista. Enferm. Infecc. Microbiol. 11: 163.

Garcia-Sanchez, M.S., Pereiro, M., Pereiro, M.M. & Toribio, J., 1997. Favus due to *Trichophyton mentagrophytes* var. *quinckeanum*. Dermatology, Basel 194: 177-179.

Gargas, A. & DePriest, P.T., 1996. A nomenclature for fungal PCR primers with examples from intron-containing SSU rDNA. Mycologia 88: 745-748.

Gargas, A. & Taylor, J.W., 1992. Polymerase chain reaction (PCR) primers for amplifying and sequencing nuclear 18S rDNA from lichenized fungi. Mycologia 84: 589-592.

Gargeya, I.B., Pruitt, W.R., Meyer, S.A. & Ahearn, D.G., 1991. *Candida haemulonii* from clinical specimens in the USA. J. Med. Vet. Mycol. 29: 335-338.

Gargeya, I.B., Pruitt, W.R., Simmons, R.B., Meyer, S.A. & Ahearn, D.G., 1990. Occurrence of *Clavispora lusitaniae*, the teleomorph of *Candida lusitaniae*, among clinical isolates. J. Clin. Microbiol. 28: 2224-2227.

Gariano, R.F. & Kalina, R.E., 1997. Posttraumatic fungal endophthalmitis resulting from *Scopulariopsis brevicaulis*. Retina 17: 256-258.

Gari-Toussaint, M., Leguay, J.M., Zur, C., Michiels, J.F., Ferrara, L., Nègre, F. & Fichoux, Y. le, 1997. Kératite à *Fusarium solani* chez une patiente diabétique. J. Mycol. Méd. 7: 227-231.

Garma-Avina, A., 1995. Cytologic findings in 43 cases of blastomycosis diagnosed *ante-mortem* in naturally-infected dogs. Mycopathologia 131: 87-91.

Gaskins, J.E. & Cheung, P.J., 1986. *Exophiala pisciphila*. A study of its development. Mycopathol. Mycol. Appl. 93: 173-184.

Gautheret, A., Dromer, F., Bourhis, J.-H. & Andremont, A., 1995. *Trichoderma pseudokoningii* as a cause of fatal infection in a bone marrow transplant recipient. Clin. Infect. Dis. 20: 1063-1064.

Gautret, P., Kauffmann-Lacroix, C., Rodier, M.H., Kull, E., Charron, M., Silvain, C. & Jacquemin, J.L., 1998. Molecular typing of *Candida parapsilosis* isolated from a patient undergoing a fungal pancreatitis. J. Mycol. Méd. 8: 188-191.

Gebhard, F., Chastagner, P., Maillot, D., Kures, L., Georges, J.L., Schmitt, C., Bordigoni, P. & Sommelet, D., 1995. Évolution favorable d'une mucormycose orbitonasosinusienne compliquant le traitement d'induction d'une leucémie aiguë lymphoblastique. Arch. Pédiatr. 2: 47-51.

Geffray, L., Veyssier, P., Cevallos, R., Baud, B., Mayolle, J., Nogier, C., Ray, E. & Thouvenot, D., 1994. Histoplasmose africaine: aspects cliniques et thérapeutiques - relations avec le SIDA. Annls Méd. Interne 145: 424-428.

Geis, P.A. & Szaniszlo, P.J., 1984. Carotenoid pigments of the dematiaceous fungus *Wangiella dermatitidis*. Mycologia 76: 268-273.

Geiser, D.M., Frisvad, J.C. & Taylor, J.W., 1998. Evolutionary relationships in *Aspergillus* section *Fumigati* inferred from partial ß-tubulin and hydrophobin sequences. Mycologia 90: 831-845.

Gemeinhardt, H., 1965. Lungenpathogenität von *Trichosporon capitatum* beim Menschen. Zentbl. Bakteriol. Parasitenkd. Infektkrankh., Abt. 1, 196: 121-133.

Gené, J., Azon-Masoliver, A., Guarro, J., Ballester, F., Pujol, I., Llovera, M. & Ferrer, C., 2000. Cutaneous phaeohyphomycosis caused by *Alternaria longipes* in an immunosuppressed patient. J. Clin. Microbiol. 33: 2774-2776.

Gené, J., Guillamón, J.M., Guarro, J., Pujol, I. & Ulfig, K., 1996a. Molecular characterization, relatedness and antifungal susceptibility of the basidiomycetous *Hormographiella* species and *Coprinus cinereus* from clinical and environmental sources. Antonie van Leeuwenhoek 70: 49-57.

Gené, J., Guillamón, J.M., Ulfig, K. & Guarro, J., 1996b. Studies on keratinophilic fungi. X. *Arthrographis alba* sp. nov. Can. J. Microbiol. 42: 1185-1189.

Genet, P., Pulik, M., Lionnet, F., Pettididier, C., Touahri, T. & Jonghe, B. de, 1995. Severe *Candida krusei* infection in an AIDS patient receiving long-term treatment with fluconazole. AIDS, Philad. 9: 661.

Gentile, L. de, Bouchara, J.P., Cimon, B. & Chabasse, D., 1991. *Candida ciferrii*: clinical and microbiological features of an emerging pathogen. Mycoses 34: 125-128.

Gentile, L. de, Bouchara, J.-P., Cler'h, C. le, Cimon, B., Symoens, F. & Chabasse, D., 1995. Prevalence of *Candida ciferrii* in elderly patients with trophic disorders of the legs. Mycopathologia 131: 99-102.

Gentry, L.O., Nasser, M.M. & Kielhofner, M., 1995. *Scopulariopsis* endocarditis with Duran ring valvuloplasty. Texas Heart Inst. J. 22: 81-85.

Georg, L.K., 1952. Cultural and nutritional studies of *Trichophyton gallinae* and *Trichophyton megninii*. Mycologia 44: 470-492.

Georg, L.K., 1964. *Curvularia geniculata*, a cause of mycotic keratitis. J. Med. Ass. State Ala. 31: 234-236.

Georg, L.K., Ajello, L., Friedman, L. & Brinkman, S.A., 1962a. A new species of *Microsporum* pathogenic to man and animals. Sabouraudia 1: 189-196.

Georg, L.K., Kaplan, W. & Camp, L.B., 1957. *Trichophyton equinum*. A re-evaluation of its taxonomic status. J. Invest. Derm. 29: 27-37.

Georg, L.K., Williamson, W.M., Tilden, E.B. & Getty, R.E., 1962b. Mycotic pulmonary disease of captive giant tortoises due to *Beauveria bassiana* and *Paecilomyces fumosoroseus*. Sabouraudia 2: 80-86.

Gerber, J., Chomicki, J., Brandsberg, J.W., Jones, R. & Hammerman, K.J., 1973. Pulmonary aspergillosis caused by *Aspergillus fischeri* var. *spinosus*. Am. J. Clin. Pathol. 60: 861-866.

Gerlach, W. & Nirenberg, H., 1982. The genus *Fusarium*, a pictorial atlas. Mitt. Biol. Bundesanst. Land-Forstw. Berlin-Dahlem 209: 1-406.

Gerrits van den Ende, A.H.G. & Hoog, G.S. de, 1999. Variability and molecular diagnostics of the neurotropic species *Clado-phialophora bantiana*. Stud. Mycol. 43: 151-162.

Gerritsen, J., Dissel, J.T. van & Verwey, H.F., 1998. *Candida tropicalis* endocarditis. Circulation 98: 90-91.

Gershan, W.M., Rusakow, L.S., Henrickson, K.J. & Splaingard, M.L., 1994. Brain abscess caused by *Blastomyces dermatitidis* in a child with cystic fibrosis. Chest 106: 601-603.

Gezuele, E. & Iraola, M.L., 1970. Cromomicosis humana experimental por *Phialophora compacta* tratada con anfotericina B. Revta Urug. Patol. Clin. 8: 134-140.

Gezuele, E., Mackinnon, J.E. & Conti-Diaz, I.A., 1972. The frequent isolation of *Phialophora verrucosa* and *Phialophora pedrosoi* from natural sources. Sabouraudia 10: 266-273.

Ghosh, G.R., Sur, B. & Roy, K., 1976. Dermatomycoses in Cuttack, Orissa and report on pseudocleistothecia formation by strains of *Trichophyton rubrum* Castellani. Kavaka 4: 25-30.

Gianni, C., Cerri, A. & Crosti, C., 1997. Unusual clinical features of finger nail infection by *Fusarium oxysporum*. Mycoses 40: 455-459.

Gigase, P. & Kastelyn, P., 1993. Further African cases of rhinosporidiosis. Annls Soc. Belge Méd. Trop. 73: 149-152.

Gianni, C., Caretta, G., Romano, C., Braidotti, P. & Crosti, C., 2000. *Geomyces pannorum* var. *pannorum* as a rare agent of tinea corporis. Abstr. 14th ISHAM Congr., B. Aires, p. 276.

Gilbert, H.M., Peters, E.D., Lang, S.J. & Hartman, B.J., 1996. Successful treatment of fungal prosthetic valve endocarditis: case report and review. Clin. Infect. Dis. 22: 348-354.

Gilfillan, G.D., Sullivan, J.D., Haynes, K., Parkinson, T., Coleman, D. & Gow, N.A.R., 1998. *Candida dubliniensis*: phylogeny and putative virulence factors. Microbiology, Reading 144: 829-838.

Gilliam, J.S. & Vest, S.A., 1951. *Penicillium* infection of the urinary tract. J. Urol. 65: 484-489.

Gillum, P.S., Gurswami, A. & Taira, J.W., 1997. Localized cutaneous infection by *Scedosporium prolificans (inflatum)*. Int. J. Derm. 36: 297-299.

Ginarte, M., Pereiro, M., Fernandez-Redondo, V. & Toribio, J., 1996. Plantar infection by *Scopulariopsis brevicaulis*. Dermatology, Basel 193: 149-151.

Ginter, G., 1989. Oekologie, Epidemiologie und klinische Symptomatik von *Microsporum gypseum* Infektionen. Mykosen 32: 531-535.

Ginter, G., Hoog, G.S. de, Pschaid, A., Fellinger, M., Bogiatzis, A., Berghold, C., Reich, E.M. & Odds, F.C., 1995. Arthritis without grains caused by *Pseudallescheria boydii*. Mycoses 38: 369-371.

Ginter, G., Petutschnig, B., Pierer, G., Soyer, H.P., Reischle, S., Kern, T. & Hoog, G.S. de, 1999. Atypical cutaneous pseudallescheriosis refractory to antifungal agents. Mycoses 42: 507-511.

Ginther, O.J. & Bubash, G.R., 1966. Experimental *Microsporum nanum* infection in swine. J. Am. Vet. Med. Ass. 148: 1034-1037.

Giorgi, W., Genovez, M.E., Porto, E. & Heins, E.M., 1986. Metrite purulenta em égua puro-sangue inglés por *Candida rugosa*. Revta Microbiol., S. Paulo 17: 225-227.

Gip, L. & Paldrok, H., 1967. Onychomycosis caused by *Phyllostictina* Sydow. Acta Derm.-Venerol. 47: 186-189.

Girardi, L., Malowitz, R., Tortora, G.I. & Spitzer, E.D., 1993. *Aureobasidium pullulans* septicaemia. Clin. Infect. Dis. 16: 338-339.

Girardin, H., Monod, M. & Latgé, J.-P., 1995. Molecular characterization of the food-borne fungus *Neosartorya fischeri* (Malloch and Cain). Appl. Environm. Microbiol. 61: 1378-1383.

Girmenia, C., Castaldi, R. & Martino, P., 1995. Catheter-related cutaneous aspergillosis complicated by fungemia and fatal pulmonary infection in a HIV-positive patient with acute lymphocytic leukemia. Clin. Microbiol. Infect. Dis. 14: 524-526.

Girmenia, C., Martino, P., Bernardis, F. de, Gentile, G., Boccanera, M., Monaco, M., Antonucci, G. & Cassone, A., 1996. Rising incidence of *Candida parapsilosis* fungemia in patients with hematologic malignancies: clinical aspects, predisposing factors, and differential pathogenicity of the causative strains. Clin. Infect. Dis. 23: 506-514.

Giudice, M.C., Szeszs, M.W., Sarpini, R.L., Ninomyia, A., Oliveira Trifilio, M. de, Pereira Pinto, W. & Sousa Carvalho Melhem, M. de, 1997. Clinical and epidemiological study in an AIDS patient with *Microsporum gypseum* infection. Revta Iberoam. Micol. 14: 184-187.

Glenn, A.E., Bacon, C.W., Price, R. & Hanlin, R.T., 1996. Molecular phylogeny of *Acremonium* and its taxonomic implications. Mycologia 88: 369-383.

Glick, C., Graves, G.R. & Feldman, S., 1993. *Torulopsis glabrata* in the neonate: an emerging fungal pathogen. South. Med. J. 86: 969-970.

Glick, A.D. & Kwon-Chung, K.J., 1973. Ultrastructural comparison of coils and ascospores of *Emmonsiella capsulata* and *Ajellomyces dermatitidis*. Mycologia 65: 216-220.

Glowacka, A., Jeske, J., Lupa, S. & Ochęcka-Szymańska, A., 2000. *Candida lipolytica* infection - a clinical study. Mycoses 43: 226.

Göttlich, E., 1996. Untersuchungen zur Pilzbelastung der Luft an Arbeitsplätzen in Betrieben zur Abfallbehandlung. Thesis, Erich Schmidt Verlag, 244 pp.

Göttlich, E., Hoog, G.S. de, Yoshida, S., Takeo, K., Nishimura, K. & Miyaji, M., 1995. Cell-surface hydrophobicity and lipolysis as essential factors in human tinea nigra. Mycoses 38: 489-494.

Goldani, L.Z., Aquino, V.R. & Dargel, A.A., 1999. Disseminated cutaneous sporotrichosis in an AIDS patient receiving maintenance therapy with fluconazole for previous cryptococcal meningitis. Clin. Infect. Dis. 28: 1337-1338.

Goldani, L.Z., Craven, D.E. & Sugar, A.M., 1995a. Central venous catheter infection with *Rhodotorula minuta* in a patient with AIDS taking suppressive doses of fluconazole. J. Med. Vet. Mycol. 33: 267-270.

Goldani, L.Z., Maia, A.L. & Sugar, A.M., 1995b. Cloning and nucleotide sequence of a specific DNA fragment from *Paracoccidioides brasiliensis.* J. Clin. Microbiol. 33: 1652-1654.

Goldani, L.Z. & Sugar, A.M., 1995. Paracoccidioidomycosis and AIDS. Clin. Infect. Dis. 21: 1275-1281.

Goldman, M., Pottage, J.C. & Weaver, D.C., 1993. *Candida krusei* fungemia. Report of 4 cases and review of the literature. Medicine, Baltimore 72: 143-150.

Goldschmied-Reouven, A., Friedman, J. & Block, C.S., 1993. *Fusarium* species isolated from non-ocular sites: a 10 year experience at an Israeli general hospital. J. Mycol. Méd. 3: 99-102.

Goldschmied-Reouven, A., Hassin, D., Schneiderman, J., Block, C., Padhye, A.A. & Keller, N., 1994. *Ochroconis humicola* isolated from ischemic ulcer in a diabetic patient. Abstr. 12th ISHAM Congr., Adelaide, p. D76.

Gomes, G.M., Cisalpino, P.S., Taborda, C.P. & Camargo, Z.P. de, 2000. PCR for diagnosis of paracoccidioidomycosis. J. Clin. Microbiol. 38: 3478-3480.

Gomez, F.J., Allendoerfer, R. & Deepe, G.S., 1995. Vaccination with recombinant heat shock protein 60 from *Histoplasma capsulatum* protects mice against pulmonary histoplasmosis. Infect. Immun. 63: 2587-2595.

Gonis, G. & Starr, M., 1997. Fatal rhinoorbital mucormycosis caused by *Saksenaea vasiformis* in an immunocompromised child. Pediatr. Infect. Dis. J. 16: 714-716.

Gonzalez, M.S., Alfonso, B., Seckinger, D., Padhye, A.A. & Ajello, L., 1984. Subcutaneous phaeohyphomycosis caused by *Cladosporium devriesii,* sp. nov. Sabouraudia 22: 427-432.

González-Cabo, J.F., Espejo Serrano, J. & Bárcena Asensio, M.C., 1995. Mycotic pulmonary disease by *Beauveria bassiana* in a captive tortoise. Mycoses 38: 167-169.

González-Escalada, Palgcio, A., Blacio, A. Del, Calvo, M.T., Gené, J. & Guarro, J., 2000. A propósito de dos casos de colonización por hongos filamentosos en secreciones respiratorias y en herida traumática de cuero cabelludo. Revta Iberoam. Micol. 17: 140-151.

Gordon, M.A., Salkin, I.F. & Stone, W.B., 1975. *Phoma (Peyronellaea)* as a zoopathogen. Sabouraudia 13: 329-333.

Gordon, M.A., Holzman, R.S., Senter, H., Lapa, E.W. & Kupersmith, M.J., 1976. *Aspergillus oryzae* meningitis. J. Am. Med. Ass. 235: 2122-2123.

Gordon, M.A. & Norton, S.W., 1985. Corneal transplant infection by *Paecilomyces lilacinus.* Sabouraudia 23: 295-301.

Gordon, M.A., Simmons, B.P., Appelbaum, P.C. & Aber, R.C., 1980. Intra-abdominal abscess and fungemia caused by *Candida krusei.* Arch. Intern. Med. 140: 1239-1240.

Gordon, M.A. & Weitzman, I., 1972. Pulmonary cryptococcosis. A case due to *Cryptococcus albidus.* Am. Rev. Respir. Dis. 106: 786-787.

Gordon, T.R. & Okamoto, D., 1992. Variation within and between populations of *Fusarium oxysporum* based on vegetative compatibility and mitochondrial DNA. Can. J. Bot. 70: 1211-1217.

Gori, S., Drouhet, E., Guého, E., Huerre, M., Lofaro, A., Parenti, M. & Dupont, B., 1998a. Cutaneous disseminated mycosis in a patient with AIDS due to a new dimorphic fungus. J. Mycol. Méd. 8: 57-63.

Gori, S., Lupetti, A., Moscato, G., Parenti, M. & Lofaro, A., 1997. Pulmonary sporotrichosis with hyphae in a human immunodeficiency virus-infected patient: a case report. Acta Cytol. 41: 519-521.

Gori, S., Pellegrini, G., Filipponi, F., Capanna, S. delle, Biancofiore, G., Mosca, F. & Lofaro, A., 1998b. Pulmonary aspergillosis caused by *Neosartorya fischeri (Aspergillus fischerianus)* in a liver transplant recipient. J. Mycol. Méd. 8: 105-107.

Gori, S. & Scasso, A., 1994. Cytology and differential diagnosis of rhinosporidiosis. Acta Cytol. 38: 167-169.

Gortel, K., McKierman, B.C., Johnson, J.K. & Campbell, K.L., 1999. Calcinosis associated with systemic blastomycosis in three dogs. J. Anim. Hosp. Ass. 35: 368-374.

Gosbell, I.B., Morris, M.L., Gallo, J.H., Weeks, K.A., Neville, S.A., Rogers, A.H., Andrews, R.H. & Ellis, D.H., 1999. Clinical, pathologic and epidemiologic features of infection with *Scedosporium prolificans:* four cases and review. Clin. Microbiol. Infect. 5: 672-686.

Goss, G., Grigg, A., Rathbone, P. & Slavin, M., 1994. *Hansenula anomala* infection after bone marrow transplantation. Bone Marrow Transpl. 14: 995-997.

Gottlieb, J.L., McAllister, I.L., Guttman, F.A. & Vine, A.K., 1995. Choroidal blastomycosis: a report of two cases. Retina 15: 248-252.

Gougerot, H. & Caraven, M., 1909. Mycose nouvelle: l'hémisporose, ostette humaine primitive du tibia due à l'*Hemispora stellata* (note préliminaire). C.R. Soc. Biol., Paris 11: 474.

Gradon, J.D., Lerman, A. & Lutwick, L.I., 1990. Septic arthritis due to *Fusarium moniliforme.* Rev. Infect. Dis. 12: 716-717.

Graham, D.R. & Frost, H.M., 1973. *Candida guilliermondii* infection of the knee complicating rheumatoid arthritis: a case report. Arthr. Rheum. 16: 272-277.

Gräser, Y., El Fari, M., Presber, W., Sterry, W. & Tietz, H.-J., 1998. Identification of common dermatophytes *(Trichophyton, Microsporum, Epidermophyton)* using polymerase chain reactions. Br. J. Derm. 138: 576-582.

Gräser, Y., El Fari, M., Vilgalys, R., Kuijpers, A.F.A., Hoog, G.S. de, Presber, W. & Tietz, H.-J., 1999a. Phylogeny and taxonomy of the family *Arthrodermataceae* (dermatophytes) using sequence analysis of the ribosomal ITS region. Med. Mycol. 37: 105-114.

Gräser, Y., Kühnisch, J. & Presber, W., 1999b. Molecular markers reveal exclusively clonal reproduction in *Trichophyton rubrum.* J. Clin. Microbiol. 37: 3713-3717.

Gräser, Y., Kuijpers, A.F.A., El Fari, M., Presber, W. & Hoog, G.S. de, 2000a. Molecular and conventional taxonomy of the *Microsporum canis* complex. Med. Mycol. 38: 143-153.

Gräser, Y., Kuijpers, A.F.A., Presber, W. & Hoog, G.S. de, 1999c. Molecular taxonomy of *Trichophyton mentagrophytes* and *T. tonsurans.* Med. Mycol. 37: 315-330.

Gräser, Y., Kuijpers, A.F.A., Presber, W. & Hoog, G.S. de, 2000b. Molecular taxonomy of the *Trichophyton rubrum* complex. J. Clin. Microbiol. 38: 3329-3336.

Grauer, M.E., Bokemeyer, C., Bautsch, W., Freund, M. & Link, H., 1994. Successful treatment of a *Trichosporon beigelii* septicemia in a granulocytic patient with amphotericin B and granulocyte colony-stimulating factor. Infection 22: 283-286.

Grau Salvat, C., Pont Sanjuan, V., Sanchez-Carazo, L., Vilata Corell, J.J. & Boniche, A.A., 1998. Tiña inflamatoria diseminada: presentación inusual. Revta Iberoam. Micol. 15: 100-102.

Green, D., Still, J.M. & Law, E.J., 1994. *Candida parapsilosis* sepsis in patients with burns: report of six cases. J. Burn Care Rehab. 15: 240-243.

Green, W.R., Font, R.I. & Zimmerman, L.E., 1969. Aspergillosis of the orbit. Report of ten cases and review of the literature. Arch. Ophthalmol. 82: 302-313.

Greer, D.L. & Friedman, L., 1966. Studies on the genus *Basidiobolus* with reclassification of the species pathogenic for man. Sabouraudia 4: 231-241.

Gregory, J.K. & Haller, J.A., 1992. Chronic postoperative *Rhodotorula* endophthalmitis. Arch. Ophthalmol. 110: 1686-1687.

Gregory, P.H. & Waller, S., 1951. *Cryptostroma corticale* and sooty bark disease of Sycamore *(Acer pseudoplatanns).* Trans. Br. Mycol. Soc. 34: 579-596.

Gregory, R.K., Powles, R.L., Treleaven, J.G., Smith, M.L., Mortimer, P.S., Watherspoon, A. & Riley, U., 1999. Systemic candidiasis with *Candida* vasculitis due to *Candida kruzei* in a patient with acute myeloid leukaemia. Bone Marrow Transpl. 23: 103-104.

Greydanus-van der Putten, S.W.M., Klein, W.R., Blankenstein, B., Hoog, G.S. de & Koeman, J.P., 1994. Sporotrichosis bij een paard. Tijdschr. Diergeneeskd. 119: 500-502.

Grieble, H.G., Rippon, J.W., Maliwan, N. & Daun, V., 1975. *Scopulariopsis* and hypersensitivity pneumonitis in an addict. Annls Intern. Med. 83: 326-329.

Grieshop, T.J., Yarbrough, D. & Farrar, W.E., 1993. Case report: phaeohyphomycosis due to *Curvularia lunata* involving skin and subcutaneous tissue after an explosion at a chemical plant. Am. J. Med. Sci. 305: 387-389.

Grigg, A. & Clouston, D., 1995. Disseminated fungal infection and early onset of microangiopathy after allogenic bone marrow transplantation. Bone Marrow Transpl. 15: 795-797.

Grigoriu, D. & Grigoriu, A., 1975. Les onychomycoses. Rev. Méd. Suisse Romande 95: 839-849.

Groff, J.M., Mughannam, A., McDowell, T.S., Wong, A., Dykstra, M.J., Frye, F.L. & Hedrick, R.P., 1991. An epizootic of cutaneous zygomycosis in cultured dwarf African clawed frogs *(Hymenochirus curtipes)* due to *Basidiobolus ranarum.* J. Med. Vet. Mycol. 29: 215-223.

Grose, E.S. & Marinkelle, C.J., 1968. A new species of *Candida* from Colombian bats. Mycopath. Mycol Appl. 36: 225-227.

Grossman, M.E., Pappert, A.S., Garzon, M.C. & Silvers, D.N., 1995. Invasive *Trichophyton rubrum* infection in the immmuno-compromised host: report of three cases. J. Am. Acad. Derm. 33: 315-318.

Gruyter, J. de & Noordeloos, M.E., 1992. Contributions towards a monograph of *Phoma* (Coelomycetes). 1. Section *Phoma:* taxa with very small conidia in vitro. Persoonia 15: 71-92.

Gruyter, J. de, Noordeloos, M.E. & Boerema, G.H., 1993. Contributions towards a monograph of *Phoma* (Coelomycetes). I, 2. Section *Phoma:* additional taxa with very small conidia and taxa with conidia up to 7 μm long. Persoonia 15: 369-400.

Guadet, J., Julien, J., Lafay, J.F. & Brygoo, Y., 1995. Phylogeny of some *Fusarium* species, as determined by large-subunit rRNA sequence comparison. Mol. Biol. Evol. 6: 227-242.

Guarner, J., Rio, C. del, Williams, P. & McGowan, J.E., 1989. Fungal peritonitis caused by *Curvularia lunata* in a patient undergoing peritoneal dialysis. Am. J. Med. Sci. 298: 320-323.

Guarro, J., 1998. Comments on recent human infections caused by ascomycetes. Med. Mycol. 36: 349.

Guarro, J., Aguilar, C. & Pujol, I., 1999a. *In vitro* antifungal susceptibilities of *Basidiobolus* and *Conidiobolus* strains. J. Antimicrob. Chemother. 44: 557-560.

Guarro, J., Akiti, T., Almada-Horta, R., Leite-Filho, L.A.M., Gené, J., Ferreira-Gomes, S., Aguilar, C. & Ortoneda, M., 1999b. Mycotic keratitis due to *Curvularia senegalensis* and *in vitro* antifungal susceptibilities of *Curvularia* spp. J. Clin. Microbiol. 37: 4170-4173.

Guarro, J., Antolín-Ayala, M.I., Gené, J., Gutiérrez-Calzada, J., Nieves-Díez, C. & Ortoneda, M., 1999c. Fatal case of *Trichoderma harzianum* infection in a renal transplant recipient. J. Clin. Microbiol. 37: 3751-3755.

Guarro, J., Cano, J., Gené, J., Solé, M. & Carrillo-Muñoz, A.J., 2000a. *Scedosporium prolificans* as an emerging opportunistic human pathogen. Proc. 14th ISHAM Congr., B. Aires, p. 84.

Guarro, J., Cano, J. & Vroey, C. de, 1991a. *Nannizziopsis (Ascomycotina)* and related genera. Mycotaxon 42: 193-200.

Guarro, J., Gams, W., Pujol, I. & Gené, J. 1997a. *Acremonium* species: new emerging fungal opportunists. *In vitro* antifungal susceptibilities and review. Clin. Infect. Dis. 25: 1222-1229.

Guarro, J., Gaztelurrutia, L., Marín, J. & Bàrcena, J., 1991b. *Scedosporium inflatum,* un nuevo hongo patògeno. A propósito de dos casos con desenlace fatal. Enf. Infec. Microbiol. Clin. 9: 557-560.

Guarro, J. & Gené, J., 1992. *Fusarium* infections. Criteria for the identification of the responsible species. Mycoses 35: 109-114.

Guarro, J. & Gené, J., 1995. Opportunistic fusarial infections in humans. Eur. J. Clin. Microbiol. Infect. Dis.14: 741-754.

Guarro, J., Gené, J. & Stchigel, A.M., 1999d. Developments in fungal taxonomy. Clin. Microbiol. Rev. 12: 454-500.

Guarro, J., Gené, J., Vroey, C. de & Guého, E., 1992. *Hormographiella,* a new genus of hyphomycetes from clinical sources. Mycotaxon 45: 179-190.

Guarro, J., Llop, C., Aguilar, C. & Pujol, I., 1997b. Comparison of *in vitro* antifungal susceptibilities of conidia and hyphae of filamentous fungi. Antimicrob. Agents. Chemother. 41: 2760-2762.

Guarro, J., Mayayo, E., Tapiol, J., Aguilar, C. & Cano, J., 1999e. *Microsphaeropsis olivacea* as an etiological agent of human skin infection. Med. Mycol. 37: 133-137.

Guarro, J., Nucci, M., Akiti, T., Gené, J., Cano, J., Gloria, M. da & Aguilar, C., 2000b. *Phialemonium* fungemia: two documented nosocomial cases. J. Clin. Microbiol. 37: 2493-2497.

Guarro, J., Nucci, M., Akiti, T., Gené, J., Cano, J., Barreiro, M.D.C. Gonçalves, C.T., 2000c. Fungemia due to *Fusarium sacchari* in an immunosuppressed patient. J. Clin. Microbiol. 38: 419-421.

Guarro, J., Nucci, M., Akiti, T., Gené, J., Cano, J., Barreiro, M.G.C. & Aguilar, C., 1999f. *Phialemonium* fungemia: two documented nosocomial cases. J. Clin. Microbiol. 37: 2483-2407.

Guarro, J., Nucci, M., Akiti, T. & Gené, J., 2000d. Mixed infection caused by two species of *Fusarium* in a human immunodeficiency virus-positive patient. J. Clin. Microbiol. 38: 3460-3462.

Guarro, J., Pujol, I., Aguilar, C. & Ostaneda, M., 2000e. Antifungal *in vitro* susceptibility of non dermatophyte keratinophilic fungi. Revta Iberoam. Mycol. 17: 144-149.

Guarro, J., Pujol, I. & Mayayo, 1999g. *In vitro* and *in vivo* experimental activities of antifungal agents against *Fusarium solani.* Antimicrob. Agent Chemother. 43: 1256-1257.

Guarro, J., Soler, L. & Rinaldi, M.G., 1995. Pathogenicity and antifungal susceptibility of *Chaetomium* species. Eur. J. Clin. Microbiol. Infect. Dis. 14: 613-618.

Guarro, J. & Stchigel, A.M., 1999. Sobre la implicación de *Trichophyton simii* en un brote de micosis superficiales en simios. Revta Iberoam. Micol. 16: 118.

Guarro, J., Svidzinski, T.E., Zaror, L., Forjaz, M.H., Gené, F. & Fischman, O., 1998. Subcutaneous hyalohyphomycosis caused by *Colletotrichum gloeosporioides.* J. Clin. Microbiol 36: 3060-3065.

Guarro, J., Vieira, L.A., Freitas, D. de, Gené, J., Zaror, L., Hofling-Lima, A.L., Fischman, O., Zorat-Yu, C. & Figueras, M.J., 2000f. *Phaeoisaria clematidis* as a cause of keratomycosis. J. Clin. Microbiol. 38: 2434-2437.

Guarro, J., Vroey, C. de & Gené, J., 1991c. Concerning the implication of *Arthrobotrys oligospora* in a case of keratitis. J. Med. Vet. Mycol. 29: 349-352.

Gucalp, R., Carlisle, P., Gialanella, P., Mitsudo, S., Mckitrick, J. & Dutcher, J., 1996. *Paecilomyces* sinusitis in an immunocompromised adult patient: case report and review. Clin. Infect. Dis. 23: 391-393.

Guccion, J.G., Rohatgi, P.K., Saini, N.B., French, A., Tavaloki, S. & Barr, S., 1996. Disseminated blastomycosis and acquired immunodeficiency syndrome: a case report and ultrastructural study. Ultrastruct. Pathol. 20: 429-435.

Guého, E., Bonnefoy, A., Luboinski, J., Petit, J.-C. & Hoog, G.S. de, 1989. Subcutaneous granuloma caused by *Phialophora richardsiae:* case report and review of the literature. Mycoses 32: 219-223.

Guého, E., Faegermann, J., Lyman, C. & Anaissie, E.J., 1995. *Malassezia* and *Trichosporon:* two emerging pathogenic basidiomycetous yeast-like fungi. J. Med. Vet. Mycol. 32, Suppl. 1: 367-378.

Guého, E. & Guillot, J., 1999. Comments on *Malassezia* species from dogs and cats. Mycoses 42: 673-674.

Guého, E. & Hoog, G.S. de, 1991. Taxonomy of the medical species of *Pseudallescheria* and *Scedosporium.* J. Mycol. Méd. 1: 3-9.

Guého E., Hoog, G.S. de & Smith, M.Th., 1992a. Neotypification of the genus *Trichosporon.* Antonie van Leeuwenhoek 61: 285-288.

Guého, E., Hoog, G.S. de, Smith, M.Th. & Meyer, S.A., 1987a. DNA relatedness, taxonomy, and medical significance of *Geotrichum capitatum.* J. Clin. Microbiol. 25: 1191-1194.

Guého, E., Improvisi, L., Christen, R. & Hoog, G.S. de, 1993. Phylogenetic relationships of *Cryptococcus neoformans* and some related basidiomycetous yeasts determined from partial large subunit rRNA sequences. Antonie van Leeuwenhoek 63: 175-189.

Guého, E., Improvisi, L., Hoog, G.S. de & Dupont, B., 1994. *Trichosporon* on humans: a practical account. Mycoses 37: 3-10.

Guého, E., Leclerc, M.C., Hoog, G.S. de & Dupont, B., 1997. Molecular taxonomy and epidemiology of *Blastomyces* and *Histoplasma* species. Mycoses 40: 69-81.

Guého, E. & Meyer, S.A., 1989. A reevaluation of the genus *Malassezia* by means of genome comparison. Antonie van Leeuwenhoek 55: 245-251.

Guého, E., Midgley, G. & Guillot, J., 1996. The genus *Malassezia* with description of four new species. Antonie van Leeuwenhoek 69: 337-355.

Guého, E., Simmons, R.B. & Ahearn, D.G., 1987b. *Malassezia* spp. et infections systémiques. Bull. Soc. Fr. Mycol. Méd. 16: 329-332.

Guého, E., Smith, M.Th. & Hoog, G.S. de, 1998. *Trichosporon* Behrend. In C.P. Kurtzman & J.W. Fell (eds): The Yeasts, a Taxonomic Study, 4th ed., pp. 854-872. Elsevier, Amsterdam.

Guého, E., Smith, M.Th., Hoog, G.S. de, Billon-Grand, G., Christen, R. & Batenburg-van der Vegte, W.H., 1992b. Contributions to a revision of the genus *Trichosporon*. Antonie van Leeuwenhoek 61: 289-316.

Guého, E., Villard, J. & Guinet, R., 1985. A new human case of *Anixiopsis stercoraria* mycosis: discussion of its taxonomy and pathogenicity. Mykosen 28: 430-436.

Guerra, R., Cavallini, G.M., Longanesi, L., Casolari, C., Bertoli, G., Rivasi, F. & Fabio, U., 1992. *Rhodotorula glutinis* keratitis. Int. J. Ophthalmol. 16: 187-190.

Gugnani, H.C., 1992. Entomophthoromycosis due to *Conidiobolus*. Eur. J. Epidemiol. 8: 391-396.

Gugnani, H.C. & Muotoe-Okafor, F.A., 1997. African histo- plasmosis: a review. Revta Iberoam. Micol. 14: 155-159.

Gugnani, H.C., Muotoe-Okafor, F.A., Kaufman, L. & Dupont, B., 1994. A natural focus of *Histoplasma capsulatum* var. *duboisii* is a bat cave. Mycopathologia 127: 151-157.

Gugnani, H.C. & Okafor, J.I., 1980. Mycotic flora of the intestine and other internal organs of certain reptiles an amphibians with special reference to characterization of *Basidiobolus* isolates. Mykosen 23: 260-268.

Gugnani, H.C., Okeke, C.N. & Sivanesan, A., 1990. *Curvularia clavata* as an aetiological agent of human skin infection. Lett. Appl. Microbiol. 10: 47-49.

Gugnani, H.C. & Oyeka, C.A., 1989. Foot infections due to *Hendersonula toruloidea* and *Scytalidium hyalinum* in coal miners. J. Med. Vet. Mycol. 27: 169-179.

Gugnani, H.C., Sood, N., Singh, B. & Makkar, R., 2000. Subcutaneous phaeohyphomycosis due to *Cladosporium cladosporioides*. Mycoses 43: 85-87.

Gugnani, H.C., Telwar, R.J., Njoku-obi, A.N.U. & Kodilinye, H., 1976. Mycotic keratitis in Nigeria. A study of 21 cases. Br. J. Ophthalmol. 60: 607-613.

Guillamón, J.M., Cano, J., Ramón, D. & Guarro, J., 1996. Molecular differentiation of *Keratinomyces (Trichophyton)* species. Antonie van Leeuwenhoek 69: 223-227.

Guillot, J. & Bond, R., 1999. *Malassezia pachydermatis:* a review. Med. Mycol. 37: 295-306.

Guillot, J., Chermette, R. & Guého, E., 1994. Prévalence du genre *Malassezia* chez les mammifères. J. Mycol. Méd. 4: 72-79.

Guillot, J., Collobert, C., Guého, E., Mialot, M. & Lagarde, E., 1997. *Emericella nidulans* as an agent of guttural pouch mycosis in a horse. J. Med. Vet. Mycol. 35: 433-435.

Guillot, J. & Guého, E., 1995. The diversity of *Malassezia* yeasts confirmed by rRNA sequence and nuclear DNA comparisons. Antonie van Leeuwenhoek 67: 297-314.

Guinet, R., Chanas, J., Gouillier, A., Bonnefoy, G. & Ambroise-Thomas, P., 1983. Fatal septicemia due to amphotericin B-resistant *Candida lusitaniae*. J. Clin. Microbiol. 18: 443-444.

Guinvarc'h, A., Guilbert, L., Marmorat-Khuong, A., Lavarde, V., Chevalier, P., Amrein, C., Guillemain, R. & Berrebi, A., 1998. Disseminated *Fusarium solani*-infection with endocarditis in a lung transplant recipient. Mycoses 41: 59-61.

Guiserix, J., Ramdane, R., Finielz, P., Michault, A. & Rajaonarivelo, P., 1996. *Trichoderma harzianum* peritonitis in peritoneal dialysis. Nephron 74: 473-474.

Guppy, K.H., Thomas, C., Thomas, K. & Anderson, D., 1998. Cerebral fungal infections in the immunocompromised host: a literature review and a new pathogen - *Chaetomium atrobrunneum*: case report. Neurosurgery, Baltimore 43: 1463-1469.

Gupta, A.K., Horgan-Bell, C.B. & Summerbell, R.C., 1998. Onychomycosis associated with *Onychocola canadensis*: ten case reports and a review of the literature. J. Am. Acad. Derm. 39: 410-417.

Gupta, A.K., Kohli, Y. & Summerbell, R.C., 2000. Molecular differentiation of seven *Malassezia* species. J. Clin. Microbiol. 38: 1869-1875.

Gupta, A.K. & Summerbell, R.C., 1998. Increased incidence of *Trichophyton tonsurans* tinea capitis in Ontario, Canada between 1985 and 1996. Med. Mycol. 36: 55-60.

Gupta, G., Burden, A.D. & Roberts, D.T., 1999. Acute suppurative ringworm (kerion) caused by *Trichophton rubrum*. Br. J. Derm. 140: 369-370.

Gyaurgieva, O.H., Bogomolova, T.S. & Gorshkova, G.I., 1996. Meningitis caused by *Rhodotorula rubra* in an HIV-infected patient. J. Med. Vet. Mycol. 34: 357-359.

Haas, N., 1987. Tinea capitis caused by *Trichophyton soudanense* in black schoolchildren. Mykosen 30: 226-228.

Haase, G., Skopnik, H., Groten, T., Kusenbach, G. & Posselt, H.-G., 1991. Long-term fungal cultures from patients with cystic fibrosis. Mycoses 34: 373-376.

Haase, G., Sonntag, L., Melzer-Krick, B. & Hoog, G.S. de, 1999. Phylogenetic interference by SSU-gene analysis of members of the *Herpotrichiellaceae* with special reference to human pathogenic species. Stud. Mycol. 43: 80-97.

Haase, G., Sonntag, L., Peer, Y. van de, Uijthof, J.M.J., Podbielski, A. & Melzer-Krick, B., 1995. Phylogenetic analysis of ten black yeast species using unclear small subunit rRNA gene sequences. Antonie van Leeuwenhoek 68: 19-33.

Hadfield, T.L., Smith, M.B., Winn, R.E., Rinaldi, M.G. & Guerra, C., 1987. Mycoses caused by *Candida lusitaniae*. Rev. Infect. Dis. 9: 1006-1012.

Hagan, M.E., Potter, L., Klotz, S., Bartholomew, W.R. & Nelson, M., 1994. *Rhodotorula rubra* isolated from bronchoscopy specimens: a pseudoepidemic. Abstr. Gen. Meet. ASM 94: 602.

Hajjeh, R.A. & Blumberg, H.M., 1995. Bloodstream infection due to *Trichosporon beigelii* in a burn patient: case report and review of therapy. Clin. Infect. Dis. 20: 913-916.

Hajsig, M., Vries, G.A. de, Sertić, V. & Naglić, T., 1974. *Chrysosporium evolceanui* from pathologically changed dogs skin. Veterinarski Arhiv 44: 209-211.

Halde, C., Padhye, A.A., Haley, L.D., Rinaldi, M.G., Kay, D. & Leeper, R., 1976. *Acremonium falciforme* as a cause of mycetoma in California. Sabouraudia 14: 319-326.

Hall, G.S., Pratt-Rippin, K. & Washington, J.A., 1992. Evaluation of a chemiluminescent probe assay for identification of *Histoplasma capsulatum* isolates. J. Clin. Microbiol. 30: 3003-3004.

Hall, J.E., 1965. Multiple maduromycotic mycetoma in a dog caused by *Helminthosporium*. Southwest Vet. 18: 233-234.

Halwig, J.M., Brueske, D.A., Greenberger, P.A., Dreisin, R.B. & Sommers, H.M., 1985. Allergic bronchopulmonary curvulariosis. Am. Rev. Respir. Dis. 132: 186-188.

Hambleton, S., Egger, K.N. & Currah, R.S., 1998. The genus *Oidiodendron*: species delimitation and phylogenetic relationships based on nuclear ribosomal DNA analysis. Mycologia 90: 854-869.

Hamdan, J.S., Resende, M.A. de, Piancastelli, S., Vieira Dias, D., Viana, E.M. & Kiesling Casali, A., 1995. A case of mycotic keratitis caused by *Fusarium solani*. Revta Inst. Med. Trop. S. Paulo 37: 181-183.

Hamilton, A.J. & Goodley, J., 1993. Purification of the 115-kilodalton exoantigen of *Cryptococcus neoformans* and its recognition by immune sera. J. Clin. Microbiol. 31: 335-339.

Hanlin, R.T., 1973. Keys to the families, genera and species of the *Mucorales*. J. Cramer, Vaduz, 49 pp.

Hantschke, D., 1969. Morphologie und Biologie des *Trichophyton mentagrophytes* (Robin) Blanchard var. *goetzii* var. nova. Mykosen 12: 97-104.

Hantschke, D., 1989. Die Bedeutung zentraler Venenkatheter bei der Entstehung von *Candida*-Endomykosen. Mycoses 32: 235-238.

Harley, W.B., Dummer, J.S., Anderson, T.L. & Goodman, S., 1995. Malignant external otitis due to *Aspergillus flavus* with fulminant dissemination to the lungs. Clin. Infect. Dis. 20: 1052-1054.

Harmon, C.B., Daniel, W.P., & Peters, M.S., 1993. Cutaneous aspergillosis complicating pyoderma gangrenosum. J. Am. Acad. Derm. 29: 656-658.

Harmsen, D., Schwinn, A., Weig, M., Bröcker, E.-B. & Heesemann, J., 1995. Phylogeny and dating of some pathogenic keratinophilic fungi using small subunit ribosomal RNA. J. Med. Vet. Mycol. 33: 299-303.

Haron, E., Anaissie, E., Dumphy, F., McCredie, K. & Fainstein, V., 1988. *Hansenula anomala* fungaemia. Rev. Infect. Dis. 10: 1182-1186.

Haron, E., Vartivarian, S., Anaissie, E., Dekmezian, R. & Bodey, G.P., 1993. Primary *Candida* pneumonia. Experience at a large cancer center and review of the literature. Medicine 72: 137-142.

Harpster, W.H., González, C. & Opal, S.M., 1985. Pansinusitis caused by the fungus *Drechslera*. Otolaryngol.-Head Neck Surg. 93: 683-685.

Harris, L.F., Dan, B.M., Lefkowitz, L.B. & Alford, R.H., 1979. *Paecilomyces* cellulitis in a renal transplant patient: successful treatment with intravenous miconazole. South. Med. J. 72: 897-898.

Hattori, N., Adachi, M., Kaneko, T., Shimozuma, M., Ichinohe, M. & Iozumi, K., 2000. Onychomycosis due to *Chaetomium globosum* successfully treated with itraconazole. Mycoses 43: 89-92.

Haubold, E.M., Aronson, J.F., Cowan, D.F., McGinnis, M.R. & Cooper, C.R., 1998. Isolation of fungal rDNA from bottlenose dolphin skin infected with *Loboa loboi*. Med. Mycol. 36: 263-267.

Haubold, E.M., Cooper, J.R., Wen, J.W., McGinnis, M.R. & Cowan, D.F., 2000. Comparative morphology of *Lacazia loboi* (syn. *Loboa loboi*) in dolphins and humans. Med. Mycol. 38: 9-14.

Hawksworth, D.L., 1971. A revision of the genus *Ascotricha* Berk. Mycol. Pap. 126: 1-28.

Hawksworth, D.L., 1979a. Ascospore sculpturing and generic concepts in the *Testudinaceae* (syn. *Zopfiaceae*). Can. J. Bot. 57: 91-99.

Hawksworth, D.L., 1979b. The lichenicolous hyphomycetes. Bull. Br. Mus., Bot. Ser. 6: 183-300.

Hawksworth, D.L., 1991. The fungal dimension of biodiversity: magnitude, significance, and conservation. Mycol. Res. 95: 641-655.

Hawksworth, D.L. & Booth, C., 1974. A revision of the genus *Zopfia* Rabenh. Mycol. Pap. 135: 1-38.

Hawksworth, D.L., Gibson, I.A.S. & Gams, W., 1976. *Phialophora parasitica* associated with disease conditions in various trees. Trans. Br. Mycol. Soc. 66: 427-431.

Hawksworth, D.L., Kirk, P.M., Sutton, B.C. & Pegler, D.N., 1995. Ainsworth & Bisby's Dictionary of the Fungi, 8th ed. CAB International, Wallingford, 616 pp.

Hawksworth, D.L. & Pitt, J.I., 1983. A new taxonomy for *Monascus* species based on cultural and microscopical characters. Aust. J. Bot. 31: 51-61.

Hay, C.E.M., Loveday, R.K., Spencer, B.M.T. & Scott, B., 1978. Bilateral mycotic myositis, osteomyelitis and nephritis in a dog caused by a *Cephalosporium*-like hyphomycete. J. S. Afr. Vet. Ass. 49: 359-361.

Hay, R.J., 1988. Tinea imbricata. Curr. Topics Med. Mycol. 2: 55-72.

Hay, R.J. & Moore, M.K., 1984. Clinical features of superficial fungal infections caused by *Hendersonula toruloidea* and *Scytalidium hyalinum*. Br. J. Derm. 110: 677-683.

Hayashi, M., Kiryu, H., Suenaga, Y. & Asahi, M., 1994. A case of cutaneous infection by *Exophiala jeanselmei*. J. Derm. 21: 971-973.

Hayashi, N. & Toshitani, S., 1983. Human infections with *Microsporum gypseum* in Japan. Mykosen 26: 527-530.

Haynes, K.A. & Westerneng, T.J., 1996. Rapid identification of *Candida albicans*, *C. glabrata*, *C. parapsilosis* and *C. krusei* by species-specific PCR of large subunit ribosomal DNA. J. Med. Microbiol. 44: 390-395.

Haynes, K.A., Westerneng, T.J., Fell, J.W. & Moens, W., 1995. Rapid detection and identification of pathogenic fungi by polymerase chain reaction amplication of large subunit ribosomal DNA. J. Med. Vet. Mycol. 33: 319-325.

Hazen, K.C., Theisz, G.W. & Howell, S.A., 1999. Chronic urinary tract infection due to *Candida utilis*. J. Clin. Microbiol. 37: 824-827.

Head, C.B. & Ratnam, S., 1988. Comparison of API ZYM system with API AN-indent, API 20A, Minitek Anaerobe II, and RapID-ANA systems for identification of *Clostridium difficile* J. Clin. Microbiol. 26: 144-146.

Heard, D.J., Cantor, G.H., Jacobson, E.R., Purich, B., Ajello, L. & Padhye, A.A., 1986. Hyalohyphomycosis caused by *Paecilomyces lilacinus* in an Aldabra tortoise. J. Am. Vet. Med. Ass. 189: 1143-1145.

Heath, C.H., Lendrum, J.L., Wetherall, B.L., Wesselingh, S.L. & Gordon, D.L., 1997. *Phaeoacremonium parasiticum* endocarditis following liver transplantation. Clin. Infect. Dis. 25: 1251-1252.

Heath, T.C.B., Patel, A., Fisher, D., Bowden, F.J. & Currie, B., 1995. Disseminated *Penicillium marneffei* presenting illness of advanced HIV infection: a clinicopathological review, illustrated by a case report. Pathology 27: 101-105.

Hecker, M.S., Weinberg, J.M., Bagheri, B., Tangoren, I.A., Rudikoff, D., Bottone, E., Bilodeau-McCarthy, E., Rudolph, R.I. & Phelps, R.G., 1997. Cutaneous *Paecilomyces lilacinus* infection: report of two novel cases. J. Am. Acad. Derm. 37: 270-271.

Hegedus, D.D. & Khachatourians, G.G., 1996. Identification and differentiation of the entomopathogenic fungus *Beauveria bassiana* using polymerase chain reaction and single-strand conformation polymorphism analysis. J. Invert. Pathol. 67: 289-299.

Hejtmánek, M., 1963. *Trichophyton terrestre* Durie et Frey - izolace konidiového a perfektního stádia. Česká Mykol. 17: 195-199.

Hek, L.G. van 't, Verweij, P.E., Weemaes, C.M., Dalen, R. van Yntema, J.-B. & Meis, J.F.G.M., 1998. Successful treatment with voriconazole of invasive aspergillosis in chronic granulomatous disease. Am. J. Resp. Crit. Care Med. 157: 1694-1696.

Hellman, E. & Raethel, S., 1964. *Trichosporon capitatum* als Ursache eines Abortes beim Rind. Berl. Münch. Tierärztl. WochenSchr. 77: 380-381.

Helm, K.F. & Lookingbill, D.P., 1993. *Pityrosporum* folliculitis and severe pruritus in two patients with Hodgkin's disease. Arch. Derm. 129: 380-381.

Helm, T.N., Longworth, D.L., Hall, G.S., Bolwell, B.J., Fernandez, B. & Tomecki, K.J., 1990. Case report and review of resolved fusariosis. J. Am. Acad. Derm. 23: 393-398.

Hemashettar, B,.M. & Nadig, V.S., 1980. First isolation of *Trichophyton soudanense* in India. Ind. J. Path. Microbiol. 23: 53-54.

Hemashettar, B.M., Patil, C.S., Yenni, V.V., Malur, P.R. & Campbell, C.K., 1993. Isolation of *Trichophyton yaoundei* in India. J. Med. Vet. Mycol. 31: 333-336.

Hendriks, L., Goris, A., Neefs, J.-M., Peer, Y. van de, Hennebert, G. & Wachter, R. de, 1989. The nucleotide sequence of the small ribosomal subunit RNA of the yeast *Candida albicans* and the evolutionary position of the fungi amongst the Eukaryotes. Syst. Appl. Microbiol. 12: 223-229.

Hendrix, D.V.H., Chmielewski, N.T., Smith, P.J., Brooks, D.E., Gelatt, K.N. & Whittaker, C., 1996. Keratomycosis in four horses caused by *Cylindrocarpon destructans*. Vet. Compar. Ophthalmol. 6: 252-256.

Heney, C., Song, E., Kellen, A., Raal, F., Miller, S.D. & Davis, V., 1989. Cerebral phaeohyphomycosis caused by *Xylohypha bantiana*. Eur. J. Clin. Microbiol. Infect. Dis. 8: 984-988.

Hennebert, G.L. & Desai, B.G., 1974. *Lomentospora prolificans*, a new hyphomycete from greenhouse soil. Mycostaxon 1: 45-50.

Hennequin, C., Abachin, E., Symoens, F., Lavarde, V., Reboux, G., Nolard, N. & Berche, P., 1999. Identification of *Fusarium* species involved in human infection by 28S rRNA gene sequencing. J. Clin. Microbiol. 37: 3586-3589.

Hennequin, C., Benailly, N., Silly, C., Sorin, M., Scheinmann, P., Lenoir, G., Gaillard, J.L. & Berche, P., 1997a. *In vitro* susceptibilities to amphotericin B, itraconazole, and miconazole of filamentous fungi isolated from patients with cystic fibrosis. Antimicrob. Agents Chemother. 41: 2064-2066.

Hennequin, C., Lavarde, V., Poirot, J.L., Rabodonirina, M., Datry, A., Aractingi, S., Dupouy-Camet, J., Caillot, D., Grange, F., Kures, L., Morin, O., Lebeau, B., Bretagne, S., Guigen, C., Basset, D. & Grillot, R., 1997b. Invasive *Fusarium* infections: a retrospective survey of 31 cases. J. Med. Vet. Mycol. 35: 107-114.

Herbrecht, R., Koenig, H., Waller, J., Liu, K.L. & Guého, E., 1993. *Trichosporon* infections: clinical manifestations and treatment. J. Mycol. Méd. 3: 129-136.

Herceg, M., Maržan, B., Hajsig, M., Naglić, T. & Huber, I., 1977. Pathomorphological observations of spontaneous candidal encephalitis in a monkey. Vet. Arhiv 47: 183-187.

Hermanides-Nijhof, E.J., 1977. *Aureobasidium* and allied genera. Stud. Mycol. 14: 144-176.

Hernandez-Molina, J.M., 1993. Bibliographic review on *Malassezia (Pityrosporum)*: taxonomy and its role in systemic infections. Revta Iberoam. Micol. 10: 24-28.

Hernandez-Molina, J.M., Garcia-Martos, P., Mira, J., Carillo-Muñoz, A.J., Bueno, M.J. & Alonso, J., 1994. Vaginitis due to *Kluyveromyces marxianus*. Abstr. 12th ISHAM Congr., Adelaide p. D72.

Herr, R.A., Ajello, L., Taylor, J.W., Arseculeratne, S.N. & Mendoza, L., 1999. Phylogenetic analysis of *Rhinosporidium seeberi*'s 18S small-subunit ribosomal DNA groups of this pathogen among members of the protoctistan *Mycetozoa* clade. J. Clin. Microbiol. 37: 2750-2754.

Hesseltine, C.W. & Ellis, J.J., 1966. Species of *Absidia* with ovoid sporangiospores I. Mycologia 58: 761-785.

Hesseltine, C.W. & Ellis, J.J., 1973. *Mucorales.* In G.G Ainsworth *et al.* (eds): The Fungi IVb: 187-217. Academic, New York.

Hevia, O., Kligman, D., & Penneys, N.S., 1991. Nonscalp hair infection caused by *Microsporum canis* in a patient with acquired immunodeficiency syndrome. Am. Acad. Derm. 24: 789-790.

Hickley, W.F., Sommerville, L.H. & Schoen, F.J., 1983. Disseminated *Candida glabrata:* report of a unique severe infection and a literature review. Am. J. Clin. Pathol. 80: 724-727.

Hiemenz, J.W., Kennedy, B. & Kwon-Chung, K.J., 1990. Invasive fusariosis associated with an injury by a stingray barb. J. Med. Vet. Mycol. 28: 209-213.

Hironaga, M., Mochizuki, T. & Watanabe, S., 1982. Cutaneous phaeohyphomycosis of the sole caused by *Exophiala jeanselmei* and its susceptibility to amphothericin B, 5-FC and ketoconazole. Mycopathologia 79: 101-104.

Hironaga, M., Nakano, K., Yokoyama, I. & Kitajima, J., 1989. *Phialophora repens*, an emerging agent of subcutaneous phaeohyphomycoses in humans. J. Clin. Microbiol. 27: 394-399.

Hironaga, M. & Watanabe, S., 1980. Annellated conidiogenous cells in *Petriellidium boydii (Scedosporium apiospermum).* Sabouraudia 18: 261-268.

Hirsch, B.E., Farber, B.F., Shapiro, J.F. & Kemelly, S., 1996. Successful treatment of *Aureobasidium pullulans* prosthetic hip infection. Infect. Dis. Clin. Pract. 5: 205-207.

Hiruma, M., Kawada, A. & Ishibashi, A., 1991. Ultrastructure of asteroid bodies in sporotrichosis. Mycoses 34: 103-107.

Hiruma, M., Kawada, A., Ohata, H., Ohnishi, Y., Takahashi, H., Yamazaki, M., Ishibashi, A., Hatsuse, K., Kakihara, M. & Yoshida, M., 1993. Systemic phaeohyphomycosis caused by *Exophiala dermatitidis.* Mycoses. 36: 1-7.

Ho, M.H.-M., Castañeda, R.F., Dugan, F.M. & Jong, S.C., 1999. *Cladosporium* and *Cladophialophora* in culture: description and expanded key. Mycotaxon 72: 115-158.

Ho, R.H.T., Bernard, P.J. & McClellan, K.A., 1991. *Phialophora mutabilis* keratomycosis. Am. J. Ophthalmol. 112: 728-729.

Hoffmann, D.H., 1965. Pilzinfektionen des Auges. Systematik, Klinik, Erkennung und Behandlung. Fortschr. Augenheilkd. 16: 63-217.

Hoffman, M., Bash, E., Berger, S.A., Burke, M. & Yust, I., 1992. Fatal necrotizing esophagitis due to *Penicillium chrysogenum* in a patient with acquired immunodeficiency syndrome. Eur. J. Clin. Microbiol. Infect. Dis. 11: 1158-1160.

Hofling Lima, A.L., Freitas, D., Fischman, O., Yu, M.C.Z., Roizenblatt, R. & Belfort, R., 1999. *Exophiala jeanselmei* causing late endophtalmitis after cataract surgery. Am. J. Ophthalmol. 128: 512-514.

Hohl, P.E., Holley, H.P., Prevost, E., Ajello, L. & Padhye, A.A., 1983. Infections due to *Wangiella dermatitidis* in humans: report of the first documented case from the United States and a review of the literature. Rev. Infect. Dis. 5: 854-864.

Holland, J., 1997. Emerging zygomycoses of humans: *Saksenaea vasiformis* and *Apophysomyces elegans.* Curr. Topics Med. Mycol. 8: 27-34.

Holmes, A.R., Cannon, R.D., Shepherd, M.G. & Jenkinson, H.F., 1994. Detection of *Candida albicans* and other yeasts in blood by PCR. J. Clin. Microbiol. 32: 228-231.

Holmes, A.R., Lee, Y.C., Cannon, R.D., Jenkinson, H.F. & Shepherd, M.G., 1992. Yeast-specific DNA probes and their application for the detection of *Candida albicans.* J. Med. Microbiol. 37: 346-351.

Holschu, D.L., Presley, H.L., Miranda, M. & Phaff, H.J., 1979. Identification of *Candida lusitaniae* as an opportunistic yeast in humans. J. Clin. Microbiol. 10: 202-205.

Hood, S.V., Moore, C.B. & Denning, D.W., 1996. Isolation of *Candida norvegensis* from clinical specimens: four case reports. Clin. Infect. Dis. 23: 1185-1187.

Hood, S.V., Moore, C.B., Cheesbrough, J.S., Mene, A. & Denning, D.W., 1997. Atypical eumycetoma casued by *Phialophora parasitica* successfully treated with itraconazole and flucytosine. Br. J. Derm. 136: 953-956.

Hoog, G.S. de, 1972. The genera *Beauveria, Isaria, Tritirachium* and *Acrodontium* gen. nov. Stud. Mycol. 1: 1-41.

Hoog, G.S. de, 1974. The genera *Blastobotrys, Sporothrix, Calcarisporium* and *Calcarisporiella* gen. nov. Stud. Mycol. 7: 1-84.

Hoog, G.S. de, 1977. *Rhinocladiella* and allied genera. Stud. Mycol. 15: 1-140.

Hoog, G.S. de, 1979. The black yeasts, II: *Moniliella* and allied genera. Stud. Mycol. 19: 1-90.

Hoog, G.S. de, 1983. On the potentially pathogenic dematiaceous Hyphomycetes. In: D.H. Howard (ed.): The Fungi Pathogenic to Humans and Animals, A: 149-216. Marcel Dekker, New York.

Hoog, G.S. de, 1993. *Sporothrix*-like anamorphs of *Ophiostoma* species and other fungi. In M.J. Wingfield *et al.* (eds): *Ceratocystis* and *Ophiostoma*, pp. 53-60. APS, St. Paul.

Hoog, G.S. de, 1996. Risk assessment of fungi reported from humans and animals. Mycoses 39: 407-417.

Hoog, G.S., de, 1997. Significance of fungal evolution for the understanding of their pathogenicity, illustrated with agents of phaeohyphomycosis. Mycoses 40, Suppl. 2: 5-8.

Hoog, G.S. de (ed.), 1999. Ecology and evolution of black yeasts and their relatives. Stud. Mycol. 43: 1-208.

Hoog, G.S. de, Buiting, A., Tan, C.S., Stroebel, A.B., Ketterings, C., Boer, E.J. de, Naafs, B., Brimicombe, R., Nohlmans-Paulssen, M.K.E., Fabius, G.T.J., Klokke, A.H. & Visser, L.G., 1993. Diagnostic problems with imported cases of mycetoma in The Netherlands. Mycoses 36: 81-87.

Hoog, G.S. de & Gerrits van den Ende, A.H.G., 1992. Nutritional pattern and eco-physiology of *Hortaea werneckii*, agent of human tinea nigra. Antonie van Leeuwenhoek 62: 321-329.

Hoog, G.S. de & Gerrits van den Ende, A.H.G., 1998a. Molecular diagnostics of clinical strains of filamentous Basidiomycetes. Mycoses 41: 183-189.

Hoog, G.S. de, Gerrits van den Ende, A.H.G., Uijthof, J.M.J. & Untereiner, W.A., 1995a. Nutritional physiology of type isolates of currently accepted species of *Exophiala* and *Phaeococcomyces.* Antonie van Leeuwenhoek 68: 43-49.

Hoog, G.S. de, Guého, E., Masclaux, F., Gerrits van den Ende, A.H.G., Kwon-Chung, K.J. & McGinnis, M.R., 1995b. Nutritional physiology and taxonomy of human-pathogenic *Cladosporium (Xylohypha)* species. J. Med. Vet. Mycol. 33: 339-347.

Hoog, G.S. de, Guého, E. & Smith, M.Th., 1997a. Nomenclatural notes on some arthroconidial yeasts. Mycotaxon 63: 345-347.

Hoog, G.S. de & Haase, G., 1993. Nutritional physiology and selective isolation of *Exophiala dermatitidis.* Antonie van Leeuwenhoek 64: 17-26.

Hoog, G.S. de, Marvin-Sikkema, F.D., Lahpor, G.A., Gottschall, J.C., Prins, R.A. & Guého, E., 1994a. Ecology and physiology of the emerging opportunistic fungi *Pseudallescheria boydii* and *Scedosporium prolificans.* Mycoses 37: 71-78.

Hoog, G.S. de, Matsumoto, T., Matsuda, T. & Uijthof, J.M.J., 1994b. *Exophiala jeanselmei* var. *lecanii-corni*, an etiologic agent of human phaeohyphomycosis, with report of a case. J. Med. Vet. Mycol. 32: 373-380.

Hoog, G.S. de, Mayser, P., Haase, G., Horré, R. & Horrevorts, A.M., 2000. A new species of *Phialophora* from human skin. Mycoses (in press).

Hoog, G.S. de & McGinnis, M.R., 1987. Ascomycetous black yeasts. In G.S. de Hoog *et al.* (eds): The expanding realm of yeast-like fungi, pp. 187-199. Elsevier Sci. Publ., Amsterdam, 510 pp.

Hoog, G.S. de & Oorschot, C.A.N. van, 1983. Taxonomy of the *Dactylaria* complex. I. Notes on the genus *Dichotomophthora.* Proc. K. Ned. Akad. Wet., Ser. C, 86: 55-61.

Hoog, G.S. de, Poonwan, N. & Gerrits van den Ende, A.H.G., 1999a. Taxonomy of *Exophiala spinifera* and its relationship to *E. jeanselmei.* Stud. Mycol. 43: 133-142.

Hoog, G.S. de, Rantio-Lehtimäki & Smith, M.Th., 1985. *Blastobotrys, Sporothrix* and *Trichosporiella*: generic delimitation, new species, and a *Stephanoascus* teleomorph. Antonie van Leeuwenhoek 51: 79-109.

Hoog, G.S. de & Rubio, C., 1982. A new dematiaceous fungus from human skin. Sabouraudia 20: 15-20.

Hoog, G.S. de & Smith, M.Th., 1998b. *Moniliella* Stolk & Dakin. In C.P. Kurtzman & J.W. Fell (eds): The Yeasts, a Taxonomic Study, ed. 4, pp. 785-788. Elsevier Sci. Publ., Amsterdam.

Hoog, G.S. de, Smith, M.Th. & Guého, E., 1986. A revision of the genus *Geotrichum* and its teleomorphs. Stud. Mycol. 29: 1-131.

Hoog, G.S. de, Smith, M.Th. & Weijman, A.C.M. (eds), 1987. The Expanding Realm of Yeast-like Fungi. Elsevier Sci. Publ., Amsterdam, 510 pp.

Hoog, G.S. de, Takeo, K., Yoshida, S., Göttlich, E., Nishimura, K. & Miyaji, M., 1994c. Pleoanamorphic life cycle of *Exophiala (Wangiella) dermatitidis*. Antonie van Leeuwenhoek 65: 143-153.

Hoog, G.S. de, Uijthof, J.M.J., Gerrits van den Ende, A.H.G., Figge, M.J. & Weenink, X.O., 1997b. Comparative rDNA diversity in medically significant fungi. Microbiol. Cult. Coll. 13: 39-48.

Hoog, G.S. de & Vries, G.A. de, 1973. Two new species of *Sporothrix* and their relation to *Blastobotrys nivea*. Antonie van Leeuwenhoek 39: 515-520.

Hoog, G.S. de, Weenink, X.O. & Gerrits van den Ende, A.H.G., 1999b. Taxonomy of the *Phialophora verrucosa* complex with the description of four new species. Stud. Mycol. 43: 107-142.

Hoog, G.S. de & Yurlova, N.A., 1994. Conidiogenesis, nutritional physiology and taxonomy of *Aureobasidium* and *Hormonema*. Antonie van Leeuwenhoek 65: 41-54.

Hoog, G.S. de, Zalar, P., Urzì, C., De Leo, F., Yurlova, N.A. & Sterflinger, K., 1999c. Relationships of dothideaceous black yeasts and meristematic fungi based on 5.8S and ITS2 rDNA sequences Stud. Mycol. 43: 31-37.

Hopfer, R.L., Walden, P., Setterquist, S. & Highsmith, W.E., 1993. Detection and differentiation of fungi in clinical specimens using polymerase chain reaction (PCR) amplification and restriction enzyme analysis. J. Med. Vet. Mycol. 31: 65-75.

Hoppe, J.E., Klingebiel, T. & Niethammer, D., 1994. Selection of *Candida glabrata* in pediatric bone marrow transplant recipients receiving fluconazole. Pediatr. Hematol. Oncol. 11: 207-210.

Hoppin, E.C., McCoy, E.L. & Rinaldi, M.G., 1983. Opportunistic mycotic infection caused by *Chaetomium* in a patient with acute leukemia. Cancer 52: 555-556.

Hopwood, V., Hicks, D.A., Thomas, S. & Evans, E.G.V., 1992. Primary cutaneous zygomycosis due to *Absidia corymbifera* in a patient with AIDS. J. Med. Vet. Mycol. 30: 399-402.

Hörmansdorfer, S., Heinritzi, K. & Bauer, J., 1995. *Microsporum canis* als Ursache einer Bestandsenzootie beim Schwein. Ein Fallbericht. Tierärztl. Praxis 23: 465-468.

Horowitz, I.D. & Blumberg, E.A., 1993. *Cryptococcus albidus* and mucormycosis empyema in a patient receiving hemodialysis. South Med. J. 86: 1070-1072.

Horré, R., Crecelius, A., Yassin, A.F., Marklein, G., Stratmann, H., Wadelmann, E., Gilges, S., Tintelnot, K., Hoog, G.S. de & Schaal, K.P., 2000a. Mycetoma due to *Pseudallescheria boydii* and *Nocardia abscessus* after a road accident. J. Clin. Microbiol. (in press).

Horré, R. & Hoog, G.S. de, 1999. Primary cerebral infections by melanized fungi: a review. Stud. Mycol. 43: 176-193.

Horré, R., Hoog, G.S. de, Kluczny, C., Marklein, G. & Schaal, K.P., 1999. rDNA diversity and physiology of *Ochroconis* and *Scolebasidium* species reported from humans and other vertebrates. Stud. Mycol. 43: 194-205.

Horré, R., Kupfer, K., Marklein, G., Evert, M., Breig, P., Hoog, G.S. de, Hoekstra, E. & Schaal, K.P., 2000b. Human fungemia due to *Penicillium piceum*. Biospectrum, Suppl. 2000: 126.

Hradil, E., Hersle, K., Nordin, P. & Faegermann, J., 1995. An epidemic of tinea corporis caused by *Trichophyton tonsurans* among wrestlers in Sweden. Acta Derm.-Venereol. 75: 305-306.

Hrdy, D.B., Nassar, N.N. & Rinaldi, M.G., 1995. Traumatic joint infection due to *Geotrichum candidum*. Clin. Infect. Dis. 20: 468-469.

Hsu, C.-F., Wang, C.-C., Hung, C.-S., Cheng, S.-N., Chen, Y.-H. & Chu, M.-L., 1998. *Trichosporon beigelii* causing oral mucositis and fungemia: report of one case. Acta Pediatr. Sin. 39: 191-194.

Hsu, M.M.L. & Lee, J.Y.Y., 1993. Cutaneous and subcutaneous phaeohyphomycosis caused by *Exserohilum rostratum*. J. Am. Acad. Derm. 28: 340-344.

Huang, S. & Harris, L.S., 1963. Acute disseminated penicilliosis. Am. J. Clin. Path. 39: 167-174.

Huang, H.-P., Little, C.J.L. & Fixter, L.M., 1993. Effects of fatty acids on the growth and composition of *Malassezia pachydermatis* and their relevance to canine otitis externa. Res. Vet. Sci. 55: 119-123.

Huang, J.-L., Yang, C.-P. & Hung, I.-J., 1993. *Candida tropicalis* fungemia in children with leukemia and lymphoma. Acta Paed. Sin. 34: 257-263.

Huang, Y.-C., Lin, T.-Y., Leu, H.-S., Peng, H.-L., Wu, J.-H. & Chang, H.-Y., 1999. Outbreak of *Candida parapsilosis* fungemia in neonatal intensive care units: clinical implications and genotyping analysis. Infection 27: 97-102.

Huang, Y.-C., Lin, T.-Y., Peng, H.-L., Wu, J.-H., Chang, H.-Y. & Leu, H.-S., 1998. Outbreak of *Candida albicans* fungemia in a neonatal intensive care unit. Scand. J. Infect. Dis. 30: 137-142.

Hubálek, Z., Nesvadbová, J. & Rychnovsky, B., 1995. A heterogeneous distribution of *Emmonsia parva* var. *crescens* in an agro-ecosystem. J. Med. Vet. Mycol. 33: 197-200.

Hubálek, Z. & Rush-Munro, F.M., 1993a. A dermatophyte from birds: *Microsporum ripariae* sp. nov. Sabouraudia 11: 287-292.

Hubálek, Z., Zejda, J., Svobodová, S. & Kucera, J., 1993b. Seasonality of rodent adiasporomycosis in a lowland forest. J. Med. Vet. Mycol. 31: 359-366.

Huffnagle, K.E. & Gander, R.M., 1993. Evaluation of Gen-Probe's *Histoplasma capsulatum* and *Cryptococcus neoformans* AccuProbes. J. Clin. Microbiol. 31: 419-421.

Huffnagle, K.E., Southern, P.M., Byrd, L.T. & Gander, R.M., 1992. *Apophysomyces elegans* as an agent of zygomycosis in a patient following trauma. J. Med. Vet. Mycol. 30: 83-86.

Hughes, C.E., Serstock, D., Wilson, B.D. & Payne, W., 1988. Infection with *Trichosporon pullulans*. Annls Intern. Med. 108: 772-773.

Humber, R.A., 1989. Synopsis of a revised classification for the *Entomophthorales (Zygomycotina)*. Mycotaxon 34: 441-460.

Humber, R.A., Brown, C.C. & Kornegay, R.W., 1989. Equine zygomycosis caused by *Conidiobolus lamprauges*. J. Clin. Microbiol. 27: 573-576.

Hung, C.-C., Chang, S.-C., Chen, Y.-C., Tien, H.-F & Hsieh, W.-C., 1995. *Trichosporon beigelii* fungemia in patients with acute leukemia: report of three cases. J. Formos. Med. Ass. 94: 127-131.

Huovinen, S., Tunnela, E., Huovinen, P., Kuijpers, A.F.A. & Suhonen, R., 1998. Human onychomycosis caused by *Trichophyton equinum* transmitted from a racehorse. Br. J. Derm. 138: 1082-1084.

Hurle, A., Campos-Herrero, M.I., Rodriguez, H., Elcuaz, R., Arroyo, J., Floriano, P. & Abad, C., 1996. Cutaneous mucormycosis of the thoracic wall. Clin. Infect. Dis. 22: 373-374.

Huss, V.A. & Sogin, M.L. 1990. Phylogenetic position of some *Chlorella* species within the *Chlorococcales* based upon complete small-subunit ribosomal RNA sequences J. Molec. Evol. 31: 432-442.

Hutchins, D.R. & Johnston, K.G., 1972. Phycomycosis in the horse. Aust. Vet. J. 48: 269-278.

Ibanez, R. & Serrano Heranz, R., 1999. Pancreatic infection with *Candida parapsilosis*. Scand. J. Infect. Dis. 31: 415-416.

Ibrahim-Granet, O., Guého, E. & Bièvre, C. de, 1985. Induction of yeast-like cells in a strain of *Fonsecaea pedrosoi*, cultured under very acidic conditions. Mycopathologia 90: 35-39.

Ichitani, T. & Amemiya, J., 1980. *Pythium gracile* isolated from the foci of granular dermatitis in the horse (*Equus caballus*). Trans. Mycol. Soc. Jpn. 21: 263-265.

Iemmolo, R.M., Rossanese, A., Rotilio, A., Mattisi, G., Gerunda, G.E., Merenda, R., Neri, D., Crepaldi, G. & Strazzabosco, M., 1998. Cerebral aspergillosis in a liver transplant recipient: a case report of long-term survival after combined treatment with liposomal amphotericin B and surgery. J. Hepat. 28: 518-522.

Ikai, K., Tomono, H. & Watanabe, S., 1988. Phaeohyphomycosis caused by *Phialophora richardsiae*. J. Am. Acad. Derm. 19: 478-481.

Imai, T., Sano, A., Mikami, Y., Watanabe, K., Aoki, F.H., Branchini, M.L.M., Negroni, R., Nishimura, K. & Miyaji, M., 2000. A new PCR primer for the identification of *Paracoccidioides brasiliensis* on rRNA sequences coding the internal transcribed spacers (ITS) and 5.8S regions. Med. Mycol. 38: 323-326.

Imwidthaya, P., 1994. Human pythiosis in Thailand. Postgrad. Med. J. 70: 558-560.

Imwidthaya, S., Chantrasakul, C. & Chantarakul, N., 1984. Opportunistic fungal infection of the burn wound. J. Med. Ass. Thailand 67: 242-248.

Ingham, E. & Cunningham, A.C., 1993. *Malassezia furfur*. J. Med. Vet. Mycol. 31: 265-288.

Irokanulo, E.O.A., Makinde, A.A., Akuesgi, C.O. & Ekwonu, M., 1997. *Cryptococcus neoformens* var. *neoformans* isolated from droppings of captive birds in Nigeria. J. Wildlife Dis. 33: 343-345.

Isenberg, H.D. (ed.), 1992. Clinical Microbiology Procedures Handbook. ASM, Washington, 2 vols.

Isenberg, H.D., Tucci, V., Cintron, F., Singer, C., Weinstein, G.S. & Tyras, D.H., 1989. Single-source outbreak of *Candida tropicalis* complicating coronary bypass surgery. J. Clin. Microbiol. 27: 2426-2428.

Ishibashi, Y., Kaufman, H.E., Ichinoe, M. & Kagawa, S., 1987. The pathogenicity of *Beauveria bassiana* in the rabbit cornea. Mykosen 30: 115-126.

Ishizaki, H., Kawasaki, M., Aoki, M., Miyaji, M., Nishimura, K. & Garcia Fernandez, J.A., 1996. Mitochondrial DNA analysis of *Sporothrix schenckii* in Costa Rica. J. Med. Vet. Mycol. 34: 71-74.

Ishizaki, H., Kawasaki, M., Nishimura, K. & Miyaji, M., 1995. Mitochondrial DNA analysis of *Exophiala spinifera*. Mycopathologia 131: 67-70.

Ishizawa, T. & Kondo, S., 1997. A case of chromomycosis caused by *Exophiala jeanselmei*. Acta Derm., Kyota 92: 337-344.

Ismail, Y., Johnson, R.H., Wells, M.V., Pusavat, J., Douglas, K. & Arsura, E.L., 1993. Invasive sinusitis with intracranial extension caused by *Curvularia lunata*. Arch. Intern. Med. 153: 1604-1606.

Issakainen, J., Jalava, J., Eerola, E. & Campbell, C.K., 1997. Relatedness of *Pseudallescheria, Scedosporium* and *Graphium pro parte* based on SSU rDNA sequences. J. Med. Vet. Mycol. 35: 389-398.

Issakainen, J., Jalava, J., Saari, J. & Campbell, C.K., 1999. Relationship of *Scedosporium prolificans* with *Petriella* confirmed by partial LSU rDNA sequences. Mycol. Res. 103: 1179-1184.

Itin, P.H., Frei, R., Lautenschlager, S., Buechner, S.A., Surber, C., Gratwohl, A. & Widmer, A.F., 1998. Cutaneous manifestations of *Paecilomyces lilacinus* infection induced by a contaminated skin lotion in patients who are severely immunosuppressed. J. Am. Acad. Derm. 39: 401-409.

Ito, Y., Fushimi, T. & Yanagi, S.O., 1998. Discrimination of species and strains of basidiomycete genus *Coprinus* by random amplified polymorphic DNA (RAPD) analysis. Mycoscience 39: 361-365.

Itoh, T., Hosokawa, H., Kohdera, U., Toyazaki, N. & Asada, Y., 1996. Disseminated infection with *Trichosporon asahii*. Mycoses 39: 195-199.

Itoh, M., Okamoto, S. & Kariya, H., 1986. Survey of 200 cases of Sporotrichosis. Dermatologica 172: 209-213.

Iwatsu, T. & Miyaji, M., 1977. Subcutaneous cystic granuloma caused by *Phialophora verrucosa*. Mycopathologia 64: 165-168.

Iwatsu, T., Nishimura, K. & Miyaji, M., 1984. *Exophiala castellanii* sp. nov. Mycotaxon 20: 307-314.

Iwatsu, T. & Udagawa, S.-I., 1984. Materials for the fungus flora of Japan (36). Trans. Mycol. Soc. Jpn. 25: 389-394.

Iwatsu, T., Udagawa, S. & Hatai, K., 1990. *Scytalidium infestans* sp. nov. isolated from striped jack *(Pseudocaranx dentex)* as a causal agent of systemic mycosis. Trans. Mycol. Soc. Jpn. 31: 389-397.

Iwatsu, T., Udagawa, S., Norizuki, K., Chiba, N. & Miki, R., 1990. *Cunninghamella bertholletiae* recovered from human disseminated zygomycosis in Japan. Trans. Mycol. Soc. Jpn. 31: 259-270.

Iwen, P.C., Kelly, D.M., Reed, E.C. & Hinrichs, S.H., 1993. *Candida krusei* causing invasive candidiasis in compromised patients not treated with fluconazole. Abstr. Gen. Meet. ASM 93: 536.

Iwen, P.C., Kelly, D.M., Reed, E.C. & Hinrichs, S.H., 1995. Invasive infection due to *Candida krusei* in immunocompromised patients not treated with fluconazole. Clin. Infect. Dis. 20: 342-347.

Iwen, P.C., Rupp, M.E., Bishop, M.R., Rinaldi, M.G., Sutton, D.A., Tarantolo, S. & Hinrichs, S.H., 1998a. Disseminated aspergillosis caused by *Aspergillus ustus* in a patient following allogeneic peripheral stem cell transplantation. J. Clin. Microbiol. 361: 3713-3717.

Iwen, P.C., Rupp, M.E., Langnas, A.N., Reed, E.C. & Hinrichs, S.H., 1998b. Invasive pulmonary aspergillosis due to *Aspergillus terreus:* 12-year experience and review of the literature. Clin. Infect. Dis. 26: 1092-1097.

Iwen, P.C., Tarantolo, S.R., Sigler, L., Sutton, D.A., Rinaldi, M.G., Lackner, R.P., McCarthy, D.I. & Hinrichs, S.H., 1999. Pulmonary infection caused by *Gymnascella hyalinospora* in a patient with acute myelogenous leukemia. Abstr. Gen Meet. ASM p. 26F-6.

Iwen, P.C., Tarantolo, S.R., Sutton, D.A., Rinaldi, M.G. & Hinrichs, S.H., 2000. Cutaneous infection caused by *Cylindrocarpon lichenicola* in a patient with acute myelogenous leukemia. J. Clin. Microbiol. 38: 3375-3378.

Jabado, N., Casanova, J.-L., Haddad, E., Dulieu, F., Fournet, J.-C., Dupont, B., Fischer, A., Hennequin, C. & Blanche, S., 1998. Invasive pulmonary infection due to *Scedosporium apiospermum* in two children with chronic granulomatous disease. Clin. Infect. Dis. 27: 1437-1441.

Jabra-Rizk, M.A., Baqui, A.A.M.A., Kelley, J.I., Falkler, W.A., Merz, W.G. & Meiller, T.F., 1999. Identification of *Candida dubliniensis* in a prospective study of patients in the United States. J. Clin. Microbiol. 37: 321-326.

Jabra-Rizk, M.A., Falkler, W.A., Merz, W.G., Baqui, A.A.M.A., Kelley, J.I. & Meiller, T.F., 2000. Retrospective identification and characterization of *Candida dubliniensis* isolates among *Candida albicans* clinical laboratory isolates from Human Immunodeficiency Virus (HIV)-infected and non-HIV-infected individuals. J. Clin. Microbiol. 38: 2423-2426.

Jackson, L., Klotz, S.A. & Normand, R.E., 1990. A pseudoepidemic of *Sporothrix cyanescens* pneumonia occurring during renovation of a bronchoscopy suite. J. Med. Vet. Mycol. 28: 455-459.

Jacob, M. & Bhat, D.J., 2000. Two new endophytic conidial fungi from India. Crypt. Mycol. 21: 81-88.

Jacobs, F., Byl, B., Bourgeois, N., Coremans-Pelseneer, J., Florquin, S., Depre, G., Vandestadt, J., Adler, M., Gelim, M & Thys, J.P., 1992. *Trichoderma viride* infection in a liver transplant recipient. Case report. Mycoses 35: 301-303.

Jacyk, W.K., Bruyn, J.H. du, Holm, N., Gryffenberg, H. & Karusseit, V.O., 1997a. Cutaneous infection due to *Cladophialo-phora bantiana* in a patient receiving immunosuppressive therapy. Br. J. Derm. 136: 428-430.

Jacyk, W.K., Shah, A.H. & Pillay, M.K., 1997b. Rhinosporidiosis. Report of two South African cases. Mikol. Lek. 4: 115-117.

Jade, K.B., Lyons, M.F. & Gnann, J.W., 1986. *Paecilomyces lilacinus* cellulitis in an immunocompromised patient. Arch. Derm. 122: 1169-1170.

Jaffay, P.B., Haque, A.K., El-Zaatari, M., Pasarell, L. & McGinnis, M.R., 1990. Disseminated *Conidiobolus* infection with endocarditis in a cocaine abuser. Arch. Pathol. Lab. Med. 114: 1276-1278.

Jaishree, N. & Singh, S.M., 1992. Hyalohyphomycosis caused by *Paecilomyces variotii*: a case report, animal pathogenicity and *'in vitro'* sensitivity. Antonie van Leeuwenhoek 62: 225-230.

Jakle, C., Leek, J.C., Olson, D.A. & Robbins, D.L., 1983. Septic arthritis due to *Fusarium solani*. J. Rheumatol. 10: 151-153.

James, E.A., Orchard, K., McWhinney, P.H.W., Warnock, D.W., Johnson, E.M., Mehta, A.B. & Kibbler, C.C., 1997. Disseminated infection due to *Cylindrocarpon lichenicola* in a patient with acute myeloid leukaemia. J. Infect. 34: 65-67.

Jamjoom, A.B., Al-Hedaithy, S.A.S., Jamjoom, Z.A.B. & Al-Hedaithy, S.F., 1995. Intracranial mycotic infections in neurosurgical practice. Acta Neurochir. 137: 78-84.

Janaki, C., Sentamilselvi, G., Janaki, V.R., Devesh, S. & Ajithados, K., 1999. Eumycetoma due to *Curvularia lunata*. Mycoses 42: 345-346.

Jandourek, A., Brown, P. & Vazquez, J.A., 1999. Community-acquired fungemia due to multiple-azole-resistant strain of *Candida tropicalis*. Clin. Infect. Dis. 29: 1583-1584.

Jang, J., Lee, H.J., Lee, I., Cho., Y.K., Kim, H.J. & Sohn, K.-H., 1999. The first imported case of pulmonary coccidioidomycosis in Korea. J. Korean Med. Sci. 14: 206-209.

Jang, S.S., Dorr, T.E., Biberstein, E.I. & Wong, A., 1986. *Aspergillus deflectus* infections in four dogs. J. Med. Vet. Mycol. 24: 95-104.

Janke, D., 1949. Zur Klinik und Mykologie der Cephalosporiose. Ein Beitrag zur Kenntnis seltener Mykosen. Arch. Derm. Syph. 188: 357-373.

Janke, D., 1950. Zur Kenntniss der Hemisporose. Arch. Derm. Syph. 190: 95-113.

Janke, D., 1954. Kasuistik seltener Mykosen. Hautarzt 4: 387-390.

Janke, D. & Newig, H., 1959. *Trichophyton verrucosum* als Erreger von Trichophytien bei Mensch und Tier in Oberhessen. Mykosen 2: 75-89.

Jantunen, E., Kolho, E., Ruutu, P., Koukila-Kahkola, P., Virolainen, M., Juvonen, E. & Volin, L., 1996. Invasive cutaneous mucormycosis caused by *Absidia corymbifera* after allogeneic bone marrow transplantation. Bone Marrow Transpl. 18: 229-230.

Jellison, W.L., 1969. Adiaspiromycosis (= haplomycosis). Mountain Press Publ., Missoula, 99 pp.

Jenney, A., Maslen, M., Bergin, P., Tang, S.-K., Esmore, D. & Fuller, A., 1998. Pulmonary infection due to *Ochroconis gallopavum* treated successfully after orthopedic heart transplantation. Clin. Infect. Dis. 26: 236-237.

Jensen, E.D., Lipscomb, T., Bonn, B. van, Miller, G., Fradkin, J.M. & Ridgway, S.H., 1998. Disseminated histoplasmosis in an Atlantic bottlenose dolphin (*Tursiops truncatus*). J. Zoo Wildlife Med. 29: 456-460.

Jensen, H.E., 1992. Murine subcutaneous granulomatous zygomycosis induced by *Absidia corymbifera*. Mycoses 35: 261-268.

Jensen, H.E., Aalbaek, B. & Han, J., 1995. Induction of systemic zygomycosis in pregnant mice by *Absidia corymbifera*. Lab. Anim. Sci. 45: 254-257.

Jesenska, Z., Durkovsky, J., Rosinski, I., Polak, M., Zamboova, E. & Baca, B., 1992. Filamentous micromycetes in otitis. Česk Epidemiol. Mikrobiol. Imunol. 41: 337-341.

Jessup, C.J., Ryder, N.S., Ghannoum, M.A., 2000. An evaluation of the *in vitro* activity of terbinafine. Med. Mycol. 38: 155-159.

Jessup, C.J., Warner, J., Isham, N., Hasan, I. & Ghannoum, M.A., 2000. Antifungal susceptibility testing of dermatophytes: establishing a medium for inducing conidial growth and evaluation of susceptibility of clinical isolates. J. Clin. Microbiol. 38: 341-344.

Joffe, A.Z., 1986. *Fusarium* Species: their Biology and Toxicology. Wiley, New York, 588 pp.

Johnson, A.S., Ranson, M., Scarffe, J.H., Morgenstern, G.R., Shaw, A.J. & Oppenheim, B.A., 1993. Cutaneous infection with *Rhizopus oryzae* and *Aspergillus niger* following bone marrow transplantation. J. Hosp. Infect. 25: 293-296.

Johnson, E.M., Szekely, A. & Warnock, D.W., 1998. *In-vitro* activity of voriconazole, itraconazole and amphotericin B against filamentous fungi. J. Antimicr. Chemother. 42: 741-745.

Johnson, E.M., Szekely, A. & Warnock, D.W., 1999. *In vitro* activity of Syn-2869, a novel triazole agent, against emerging and less common mold pathogens. Antimicrob. Agents Chemother. 43: 1260-1263.

Johnson, J.H., Wolf, A.M., Edwards, J.F., Walker, M.A., Homco, L., Jensen, J.M., Simpson, B.R. & Taliaferro, L., 1998. Disseminated coccidioidomycosis in a mandrill baboon (*Mandrillus sphinx*): a case report. J. Zoo Wildl. Med.29: 208-213.

Johnson, J.S., 1987. Pulmonary aspergillosis. Semin. Respir. Med. 9: 187-199.

Johnson, L.B., Bradley, S.F. & Kauffman, C.A., 1998. Fungemia due to *Cryptococcus laurentii* and a review of non-*neoformans* cryptococcaemia. Mycoses 41: 277-280.

Johnston, B.L., Schlech, W.F. & Marrie, T.J., 1994. An outbreak of *Candida parapsilosis* prosthetic valve endocarditis following cardiac surgery. J. Hosp. Infect. 28: 103-112.

Johnston, P.R. & Jones, D., 1997. Relationship among *Colletotrichum* isolates from fruit-rots assessed using rDNA sequences. Mycologia. 89: 420-430.

Joly, P., 1964. Le genre *Alternaria*. Encycl. Mycol. 33: 1-250.

Joly, S., Pujol, C., Rysz, M., Vargas, K. & Soll, D.R., 1999. Development and characterization of complex DNA fingerprinting probes for the infectious yeast *Candida dubliniensis*. J. Clin. Microbiol. 37: 1035-1044.

Jones, B.R., 1975. Principles in the management of oculomycosis. Trans. Am. Acad. Ophthalmol. Otolaryngol. 79: 15-53.

Jones, D.B., 1977. Therapy of postsurgical fungal endophthalmitis. Ophthalmology 85: 357-373.

Jones, D.B., Sexton, R. & Rebell, G., 1970a. Mycotic keratitis in South Florida: a review of thirty-nine cases. Trans. Ophthalmol. Soc. U.K. 89: 781-797.

Jones, D.B., Wilson, L., Sexton, R. & Rebell, G., 1970b. Early diagnosis of mycotic keratitis. Trans. Ophthalmol. Soc. U.K. 89: 805-813.

Jones, F.R. & Christensen, G.R., 1974. *Pullularia* corneal ulcer. Arch. Ophthalmol. 92: 529-530.

Jones, J.L., Fleming, P.L., Ciesielski, C.A., Hu, D.J., Kaplan, J.E. & Ward, J.W., 1995. Coccidioidomycosis among persons with AIDS in the United States. J. Infect. Dis. 171: 961-966.

Jonnalagadda, S., Veerabagu, M.P., Rakela, J., Kusne, S., Randhawa, P. & Rabinovitz, M., 1996. *Candida albicans* osteomyelitis in a liver transplant recipient: a case report and review of the literature. Transplantation 62: 1182-1184.

Jordan, J.A., 1994. PCR Identification of four medically important *Candida* species by using a single primer pair. J. Clin. Microbiol. 32: 2962-2967.

June, C., Beatty, P.G., Shulman, H.M. & Rinaldi, M.G., 1986. Disseminated *Fusarium moniliforme* infection after allogenic marrow transplantation. South. Med. J. 79: 513-515.

Jury, C.S., Lucke, T.W. & Bilsland, D., 1999. *Trichophyton erinacei*: an unusual case of kerion. Br. J. Derm. 141: 606-607.

Kabel, P.J., Illy, K.E., Holl, R.A., Buiting, A.G.M. & Wintermans, R.G.F., 1994. Nosocomial intravascular infection with *Exophiala dermatitidis*. Lancet 344: 1167-1168.

Kac, G., Bougnoux, M.E., Feuilhade, M., Sené, S. & Derouin, F., 1999a. Différentiation d'isolats de *Trichophyton rubrum* et de *Trichophyton interdigitale* par la technique de Random Amplification of Polymorphic DNA (RAPD). J. Mycol. Méd. 9: 39-41.

Kac, G., Piriou, P., Guého, E., Roux, P., Trémoulet, J., Denis, M. & Judet, T., 1999b. Osteoarthritis caused by *Neocosmospora vasinfecta*. Med. Mycol. 37: 213-217.

Kaczmarski, E.B., Liu Yin, J.A., Tooth, J.A., Love, E.M. & Delamore, I.W., 1986. Systemic infection with *Aureobasidium pullulans* in a leukaemic patient. J. Infect. 13: 289-291.

Kaell, A.T. & Weitzman, I., 1983. Acute monoarticular arthritis due to *Phialophora parasitica*. Am. J. Med. 74: 519-522.

Kahler, J.S., Leach, M.W., Jang, S. & Wong, A., 1990. Disseminated aspergillosis attributable to *Aspergillus deflectus* in a Springer spaniel. J. Am. Vet. Med. Ass. 197: 871-874.

Kaji, T., Udagawa, S., Inaoki, M., Miwa, A., Miyaji, M. & Fukushiro, R., 1987. Cutaneous histoplasmosis - report of a case, with special reference to the mycology of the isolated *Histoplasma capsulatum*. Jpn. J. Med. Mycol. 28: 366-372.

Kalter, D.C. & Hay, R.J., 1988. Onychomycosis due to *Trichophyton soudanense*. Clin. Exp. Derm. 13: 221-227.

Kamal, M.M., Luley, A.S., Mundhada, S.G. & Bobhate, S.K., 1995. Rhinosporidiosis. Diagnosis by scrape cytology. Acta Cytol. 39: 931-935.

Kamalam, A. & Thambiah, A.S., 1980. Cutaneous infection by *Syncephalastrum*. Sabouraudia 18: 19-20.

Kamalam, A., Yesudia, P. & Thambiah, A.S., 1975. Black grain mycetoma - 2 case reports. Mycopathologia 57: 27-29.

Kamei, K., McCullough, M.J. & Stevens, D.A., 2000. Initial case of *Candida dubliniensis* infection from Asia: non-mucosal infection. Med. Mycol 38: 81-83.

Kamei, K., Unno, H., Nagao, K., Kuriyama, T., Nishimura, K. & Miyaji, M., 1994. Allergic bronchopulmonary mycosis caused by the basidiomycetous fungus *Schizophyllum commune*. Clin. Infect. Dis. 18: 305-309.

Kaminski, G.W., 1985. The routine use of modified Borelli's Lactitmel agar (MBLA). Mycopathologia 91: 57-59.

Kan, V.L., 1993. Polymerase chain reaction for the diagnosis of candidemia. J. Infect. Dis. 168: 779-783.

Kanaizuka, I., Sugita, Y., Takahashi, Y. & Nakajima, H., 1992. Polymerase chain reaction for the diagnosis of candidiasis. Detection of *Candida* specific DNA from twenty-five-year-old paraffin-embedded tissue. Jpn. J. Derm. 102: 1243-1247.

Kane, J., 1977. *Trichophyton fischeri* sp. nov.: a saprophyte resembling *Trichophyton rubrum*. Sabouraudia 15: 231-241.

Kane, J. & Fischer, J.B., 1971. The differentiation of *Trichophyton rubrum* and *T. mentagrophytes* by use of Christensen's urea broth. Can. J. Microbiol. 17: 911-913.

Kane, J. & Fischer, J.B., 1975. Occurrence of *Trichophyton megninii* in Ontario: identification with a simple procedure. J. Clin. Microbiol. 12: 111-114.

Kane, J., Krajden, S., Summerbell, R.C. & Sibbald, R.G., 1990. Infections caused by *Trichophyton raubitschekii*: clinical and epidemiological features. Mycoses 33: 499-506.

Kane, J., Padhye, A.A. & Ajello, L., 1982. *Microsporum equinum* in North America. J. Clin. Microbiol. 16: 943-947.

Kane, J., Salkin, I.F., Weitzman, I. & Smitka, C., 1981. *Tricho-phyton raubitschekii*, sp. nov. Mycotaxon 13: 259-266.

Kane, J., Scott, J.A., Summerbell, R.C. & Diena, B., 1992. *Trichophyton krajdenii*, new species: an anthropophilic dermatophyte. Mycotaxon 45: 307-316.

Kane, J., Sigler, L. & Summerbell, R.C., 1987. Improved procedures for differentiating *Microsporum persicolor* from *Trichophyton mentagrophytes*. J. Clin. Microbiol. 25: 2449-2452.

Kane, J. & Smitka, C., 1978. Early detection and identification of *Trichophyton verrucosum*. J. Clin. Microbiol. 8: 740-747.

Kane, J., Summerbell, R., Sigler, L., Krajden, S. & Land, G., 1997. Laboratory Handbook of Dermatophytes. Star Publ., Belmont, 344 pp.

Kano, R., Nakamura, Y., Watari, T., Watanabe, S., Takahashi, H., Tsujimoto, H. & Hasegawa, A., 1998a. Molecular analysis of chitin synthase (CHS1) gene sequences of *Trichophyton mentagrophytes* and *T. rubrum*. Curr. Microbiol. 37: 236-239.

Kano, R., Nakamura, Y., Watari, T., Watanabe, S., Takahashi, H., Tsujimoto, H. & Hasegawa, A., 1999. Species-specific primers of chitin synthase 1 gene for the differentiation of the *Trichophyton mentagrophytes* complex. Mycoses 42: 71-74.

Kano, R., Nakamura, Y., Yasuda, K., Watari, T., Watanabe, S., Takahashi, H., Tsujimoto, H. & Hasegawa, A., 1998b. The first isolation of *Arthroderma benhamiae* in Japan. Microbiol. Immunol. 42: 575-578.

Kantipong, P., Panich, V., Pongsurachet, V. & Watt, G., 1998. Hepatic penicilliosis in patients without skin lesions. Clin. Infect. Dis. 26: 1215-1217.

Kanungo, R. & Srinivasan, R., 1996. Corneal ulcer phaeohyphomycosis due to *Exserohilum rostratum*: a case report and brief review. Acta Ophthalmol. Scand. 74: 197-199.

Kaplan, W., 1959. The occurrence of black piedra in primate pelts. Trop. Geogr. Med. 11: 115-126.

Kaplan, W., 1975. Exogenous corneal ulcer caused by *Tritirachium roseum*. Am. J. Ophthalmol. 80: 804-806.

Kaplan, W., Chandler, F.W., Ajello, L., Gauthier, R., Higgins, R. & Cayouette, P., 1975. Equine phaeohyphomycosis caused by *Drechslera spicifera*. Can. Vet. J. 16: 205-208.

Karim, M., Alam, M., Shah, A.A., Ahmed, R. & Sheikh, H., 1997. Chronic invasive aspergillosis in apparently immunocompetent hosts. Clin. Infect. Dis. 24: 723-733.

Karim, M., Hizbullah, S., Mehboob, A. & Yunus, S., 1993. Disseminated *Bipolaris* infection in an asthmatic patient: case report. Clin. Infect. Dis. 17: 248-253.

Karlowsky, J.A., Zhanel, G.G., Balko, T.V., Zelenitsky, S.A., Kabani, A.M. & Hoban, D.J., 1997. *In vitro* antifungal activity of BMS-181184 against systemic isolates of *Candida, Cryptococcus,* and *Blastomyces* species. Diagn. Microbiol. Infect. Dis. 28: 179-182.

Karuppayil, S.M., Peng, M., Mendoza, L., Levins, T.A. & Szaniszlo, P.J., 1996. Identification of the conserved coding sequences of three chitin synthase genes in *Fonsecaea pedrosoi*. J. Med. Vet. Mycol. 34: 117-125.

Kassamili, H., Anaissie, E., Ro, J., Rolston, K., Kantarjian, H., Fainstein, V. & Bodey, G., 1987. Disseminated *Geotrichum candidum* infection. J. Clin. Microbiol. 25: 1782-1783.

Kastl, J. & Horáček, J., 1980. Kvasinková orchitida imitující tumorózní proces. Česká Derm. 55: 111-114.

Kasuga, T., Taylor, J.W. & White, T.J., 1999. Phylogenetic relationships of varieties and geographical groups of the human pathogenic fungus *Histoplasma capsulatum* Darling. J. Clin. Microbiol. 37: 653-663.

Kataoka-Nishimura, S., Akiyama, H., Saku, K., Kashiwa, M.M., Mori, S., Tanikawa, S., Sakamaki, H. & Onozawa, Y., 1998. Invasive infection due to *Trichosporon cutaneum* in patients with hematologic malignancies. Cancer 82: 484-487.

Katz, G., Winchester, K. & Lam, S., 1993. Ocular aspergillosis isolated in the anterior chamber. Ophthalmology 100: 1815-1818.

Käufer, I. & Weber, A., 1977. *Graphium fructicola* als Ursache einer Systemmykose beim Hund. Mykosen 20: 39-46.

Kauffman, C.A. & Tan, J.S., 1974. *Torulopsis glabrata* renal infection. Am. J. Med. 57: 217-224.

Kauffman, C.A. & Zarins, L.T., 1998. *In vitro* activity of voriconazole against *Candida* species. Diagn. Microbiol. Infect. Dis. 31: 297-300.

Kaufman, L., 1992. Immunohistologic diagnosis of systemic mycoses: an update. Eur. J. Epidemiol. 8: 377-382.

Kaufman, L. & Lopez, R.B., 1980. Immunodiffusion Studies of Morphologically Similar Dermatophyte Species. Publ. Health, Washington, 15 pp.

Kaufman, L., Mendoza, M. & Standard, P.G., 1990. Immunodiffusion test for serodiagnosing subcutaneous zygomycosis. J. Clin. Microbiol. 28: 1887-1890.

Kaufman, L., Padhye, A.A. & Parker, S., 1988. Rhinocerebral zygomycosis of *Saksenaea vasiformis*. J. Med. Vet. Mycol. 26: 237-241.

Kaufman, L., Standard, P.G., Anderson, S.A., Jalbert, M. & Swisher, B.L., 1995. Development of a specific fluorescent-antibody test for tissue form of *Penicillium marneffei*. J. Clin. Microbiol. 33: 2136-2138.

Kaufman, L., Standard, P.G., Jalbert, M., Kantipong, P., Limpakarnjanarat, K. & Mastro, T.D., 1996. Diagnostic antigenemia tests for penicilliosis marneffei. J. Clin. Microbiol. 34: 2503-2505.

Kaufman, L., Standard, P.G. & Padhye, A., 1980. Serologic relationships among isolates of *Exophiala jeanselmei (Phialophora jeanselmei, P. gougerotii)* and *Wangiella dermatitidis*. PAHO Sci. Publ. 396: 252-258.

Kaufman, S.M., 1971. *Curvularia* endocarditis following cardiac surgery. Am. J. Clin. Pathol. 56: 466-470.

Kawachi, Y., Tateishi, T., Shojima, K., Iwata, M. & Otsuka, F., 1995. Subcutaneous phaeomycotic cyst of the finger caused by *Exophiala jeanselmei*: association with a wooden splinter. Cutis 56: 41-43.

Kawasaki, H., Yoshimura, K., Kohdera, U., Horio, T., Toyasaki, N. & Kobayashi, Y., 1996. Primary cutaneous zygomycosis due to *Absidia corymbifera* in a child with acute leukemia. Int. J. Clin. Oncol. 1: 118-120.

Kawasaki, M., Aoki, M. & Ishizaki, H., 1995. Phylogenetic relationships of some *Microsporum* and *Arthroderma* species inferred from mitochondrial DNA analysis. Mycopathologia 130: 11-21.

Kawasaki, M., Aoki, M., Ishizaki, H., Nishimura, K. & Miyaji, M., 1996. Phylogeny of *Epidermophyton floccosum* and other dermatophytes. Mycopathologia 134: 121-128.

Kawasaki, M., Aoki, M., Ishizaki, H., Miyaji, M., Nishimura, K., Nishimoto, K., Matsumoto, T., Vroey, C. de, Negroni, R., Mendoça, M., Andriantsimahavandy, A. & Esterre, P., 1999. Molecular epidemiology of *Fonsecaea pedrosoi* using mitochondrial DNA analysis. Med. Mycol. 37: 435-440.

Kawasaki, M., Ishizaki, H., Aoki, M. & Watanabe, S., 1990. Phylogeny of *Nannizzia incurvata, N. gypsea, N. fulva* and *N. otae* by restriction enzyme analysis of mitochondrial DNA. Mycopathologia 112: 173-177.

Keath, E.J., Kobayashi, G.S. & Medoff, G., 1992. Typing of *Histoplasma capsulatum* by restriction fragment length polymorphisms in a nuclear gene. J. Clin. Microbiol. 30: 2104-2107.

Keay, S., Denning, D.W. & Stevens, D.A., 1991. Endocarditis due to *Trichosporon beigelii*: *in vitro* susceptibility of isolates and review. Rev. Infect. Dis. 13: 383-386.

Keeling, P.J., Luker, M.A. & Palmer, J.D., 2000. Evidence from β-tubulin phylogeny that microsporidia evolved from within the fungi. Molec. Biol. Evol. 17: 23-31.

Kefalidou, S., Odia, S., Gruseck, E., Schmidt, T., Ring, J. & Abeck, D., 1997. Wood's light in *Microsporum canis* positive patients. Mycoses 40: 461-463.

Keller, S., 1987. Arthropod-pathogenic *Entomophthorales* of Switzerland. I. *Conidiobolus, Entomophaga* and *Entomophthora.* Sydowia 34: 122-167.

Kelly, S.E., Shaw, S.E. & Clark, W.T., 1995. Long-term survival of four dogs with disseminated *Aspergillus terreus* infection treated with itraconazole. Austr. Vet. J. 72: 311-313.

Kemna, M.E., Cooper, C.E., Neri, R.C. & Salkin, I.F., 1993. *Cokeromyces recurvatus* isolated from an endo-cervical specimen. Abstr. Gen. Meet. ASM 93: 533.

Kemna, M.E., Neri, R.C., Ali, R. & Salkin, I.F., 1994. *Cokeromyces recurvatus*, a mucoraceous Zygomycete rarely isolated in clinical laboratories. J. Clin. Microbiol. 32: 843-845.

Kennedy, F.A., Buggage, R.R. & Ajello, L., 1995. Rhinosporidiosis: a description of an unprecedented outbreak in captive swans (*Cygnus* sp.) and a proposal for revision of the ontogenic nomenclature of *Rhinosporidium seeberi.* J. Med. Vet. Mycol. 33: 157-165.

Kennedy, M.J. & Volz, P.A., 1985. Ecology of *Candida albicans* gut colonization: inhibition of *Candida* adhesion, colonization, and dissemination from the gastrointestinal tract by bacterial antagonism. Infect. Immun. 49: 654-663.

Kenney, R.T., Kwon-Chung, K.J., Waytes, A.T., Melnick, D.A., Pass, H.I., Merino, M.J. & Gallin, J.I., 1992. Successful treatment of systemic *Exophiala dermatitidis* infection in a patient with chronic granulomatous disease. Clin. Infect. Dis. 14: 235-242.

Kent, D., Wong, T., Osgood, R., Kosinski, K., Coste, G. & Bor, D., 1998. Fungemia due to *Hormonema dematioides* following intense avian exposure. Clin. Infect. Dis. 26: 759-760.

Kerkmann, M.-L., Blaschke-Hellmessen, R. & Mikulin, H.-D., 1994. Successful treatment of cerebral aspergillosis by stereotactic operation and antifungal therapy. Mycoses 37: 123-126.

Kerkmann, M.-L., Piontek, K., Mitze, H. & Haase, G., 1999a. Isolation of *Exophiala (Wangiella) dermatitidis* in a case of otitis externa. Clin. Infect. Dis. 29: 929-930.

Kerkmann, M.-L., Schuppler, M., Paul, K.-D., Schoenian, G. & Smith, M. Th., 1999b. Red-pigmented *Candida albicans* in patients with cystic fibrosis. J. Clin. Microbiol. 37: 278.

Kern, M.E. & Uecker, F.A., 1986. Maxillary sinus infection caused by the homobasidiomycetous fungus *Schizophyllum commune.* J. Clin. Microbiol. 23: 1001-1005.

Kerr, P.G., Turner, H., Davidson, A., Bennett, C. & Maslen, M., 1988. Zygomycosis requiring amputation of the hand: an isolated case in a patient receiving haemodialysis. Med. J. Aust. 148: 258-259.

Kerwin, S.C., McCarthy, R.J., VanSteenhouse, J.L., Partington, B.P. & Taboada, J., 1998. Cervical spinal cord compression caused by cryptococcosis in a dog: successful treatment with surgery and fluconazole. J. Am. Anim. Hosp. Ass. 34: 523-526.

Kettlewell, P., McGinnis, M.R. & Wilkinson, G.T., 1989. Phaeohyphomycosis caused by *Exophiala spinifera* in two cats. J. Med. Vet. Mycol. 27: 257-264.

Keung, Y.-K., Kimbrough, R., Yuan, K.-Y., Wong, W.-C. & Lobos, E., 1997. *Penicillium chrysogenum* infection in a cotton farmer with acute myeloid leukemia. Infect. Dis. Clin. Pract. 6: 482-483.

Khalil, M.A., Hassan, A.W. & Gugnani, H.C., 1998. African histoplasmosis: report of four cases from northeastern Nigeria. Mycoses 41: 293-295.

Khan, K.A., Khan, A.F., Masih, M., Farooqi, A.H. & Ansari, A.M., 1984. Clinical and pathological findings of mycetoma with special reference to the etiology. Asian Med. J. 27: 250-257.

Khan, Z.U., Gopalakrishanan, G., Al-Awadi, K., Gupta, R.K., Moussa, S.A., Chugh, T.D. & Krajci, D., 1995. Renal aspergilloma due to *Aspergillus flavus.* Clin. Infect. Dis. 21: 210-212.

Khan, Z.U., Prakash, B., Kapoor, M.M., Madda, J.P. & Chandy, R., 1998. Basidiobolomycosis of the rectum masquering as Crohn's disease: case report and review. Clin. Infect. Dis. 26: 521-523.

Khan, Z.U., Randhawa, H.S., Towshik, T., Gaur, S.N. & Vries, G.A. de, 1988. The pathogenic potential of *Sporotrichum pruinosum* isolated from the human respiratory tract. J. Med. Vet. Mycol. 26: 145-151.

Khan, Z.U., Sanyal, S.C., Mokaddas, E., Vislocky, I., Anim, J.T., Salama, A.L. & Shuhaiber, H., 1997. Endocarditis due to *Aspergillus flavus.* Mycoses 40: 213-217.

Khardori, N., 1989. Host-parasite interaction in fungal infection. Eur. J. Clin. Microbiol. Infect. Dis. 8: 331-351.

Khashnobish, A. & Shearer, C.A., 1996. Phylogenetic relationships in some *Leptosphaeria* and *Phaeosphaeria* species. Mycol. Res. 100: 1355-1363.

Kiehn, T.E., Edwards, F., Amstrong, D., Rosen, P.P. & Weitzman, I., 1979. Pneumonia caused by *Cunninghamella bertholletiae* complicating chronic lymphatic leukemia. J. Clin. Microbiol. 10: 374-379.

Kiehn, T.E., Gorey, E., Brown, A.E., Edwards, F.F. & Amstrong, D., 1992. Sepsis due to *Rhodotorula* related to use of indwelling central venous catheters. Clin. Infect. Dis. 14: 841-846.

Kiehn, T.E., Nelson, P.E., Bernard, E.M., Edwards, F.F., Koziner, B. & Armstrong, G.D., 1985. Catheter-associated fungemia caused by *Fusarium chlamydosporum* in a patient with lymphocytic lymphoma. J. Clin. Microbiol. 21: 501-504.

Kiehn, T.E., Polsky, B., Punithalingam, E., Edwards, F.F., Brown, A.E. & Armstrong, D., 1987. Liver infections caused by *Coniothyrium fuckelii* in a patient with acute myelogenous leukemia. J. Clin. Microbiol. 25: 2410-2412.

Kielstein, P., Wolf, H., Gräser, Y., Buzina, W. & Blanz, P., 1998. Zur Variabilität von *Trichophyton verrucosum.* Isolaten aus Impfbeständen mit Rindertrichophytie. Mycoses 41, Suppl. 2: 58-64.

Killingsworth, S.M. & Wetmore, S.J., 1990. *Curvularia / Drechslera* sinusitis. Laryngoscope 100: 932-937.

Kilpatrick, W.R., Revankar, S.G., McAtee, R.K., Lopez-Ribot, J.I., Fothergill, A.W., McCarthy, D.I., Sanche, S.E., Cantu, R.A., Rinaldi, M.G. & Patterson, T.F., 1998. Detection of *Candida dubliniensis* in oropharyngeal samples from human immunodeficiency virus-infected patients in North America by primary CHROMagar Candida screening and susceptibility testing of isolates. J. Clin. Microbiol. 36: 3007-3012.

Kim, D.G., Hong, S.C., Kim, H.J., Chi, J.G., Han, M.H., Choi, K.S. & Han, D.H., 1993. Cerebral aspergillosis in immunologically competent patients. Surg. Neurol. 40: 326-331.

Kim, D.S., Yoon, Y.M. & Kim, S.W., 1999. Phaeohyphomycosis due to *Exophiala dermatitidis* successfully treated with itraconazole. Korean J. Med. Mycol. 4: 79-83.

Kim, H.U. & Kang, S.H. & Matsumoto, T., 1998. Subcutaneous phaeohyphomycosis caused by *Exophiala jeanselmei* in a patient with advanced tuberculosis. Br. J. Derm. 138: 351-353.

Kim, J.A., Buarque de Gusmao, N., Okada, K., Campos Takaki, G.M. de, Fukushima, K., Nishimura, K. & Miyaji, M., 1999. Identification of *Trichophyton tonsurans* by random polymorphic DNA. Annls Derm. 11: 135-141.

Kim, S., Rusk, M.H. & James, W.D., 1999. Erysipeloid sporotrichosis in a woman with Cushing's disease. J. Am. Acad. Derm. 40: 272-274.

Kimura, M., Smith, M.B. & McGinnis, M.R., 1999. Zygomycosis due to *Apophysomyces elegans:* report of 2 cases and review of the literature. Archs Path. Lab. Med. 123: 386-390.

Kimura, M., Udagawa, S., Toyazaki, N., Jimori, M. & Hashimoto, S., 1995. Isolation of *Rhizopus microsporus* var. *rhizopodiformis* in the ulcer of human gastric carcinoma. J. Med. Vet. Mycol. 33: 137-139.

Kinderlerer, J.L., 1995. Czapek casein 50% glucose (CZC50G): a new medium for the identification of foodborne *Chrysosporium* spp. Lett. Appl. Microbiol. 21: 131-136.

Kindermann, J., El-Ayouti, Y., Samuels, G.J. & Kubicek, C.P., 1998. Phylogeny of the genus *Trichoderma* based on sequence analysis of the internal transcribed spacer region 1 of the rDNA cluster. Fung. Gen. Biol. 24: 298-309.

King, D., Cheever, L.W., Hood, A., Horn, T., Rinaldi, M.G. & Merz, W.G., 1996. Primary invasive cutaneous *Microsporum canis* infectious in immunocompromised patients. J. Clin. Microbiol. 34: 460-462.

King, D., Pasarell, L., Dixon, D.M., McGinnis, M.R. & Merz, W.G., 1993. A phaeohyphomycotic cyst and peritonitis caused by *Phialemonium* species and a reevaluation of its taxonomy. J. Clin. Microbiol. 31: 1804-1810.

King, D.S., 1977. Systematics of *Conidiobolus (Entomophthorales)* using numerical taxonomy. III. Descriptions of recognized species. Can. J. Bot. 55: 718-729.

King, D.S. & Jong, S.C., 1975. *Sarcinosporon*: a new genus to accommodate *Trichosporon inkin* and *Prototheca filamenta*. Mycotaxon 3: 89-94.

King, D.S & Jong, S.C., 1976. Identity of etiological agent of the first deep entomophthoraceous infection of man in the United States. Mycologia 68: 181-183.

Kinsella, J.B., Rassekh, C.H., Bradfield, J.L., Chaljub, G., Mcnees, S.W., Gourley, W.K. & Calhoun, K.H., 1996. Allergic fungal sinusitis with cranial bone erosion. Head Neck 18: 211-217.

Kiraz, N., Gülbas, Z. & Akgün, Y., 2000. *Rhodotorula rubra* fungaemia due to use of indwelling catheters. Mycoses 43: 209-210.

Kirk, P.M., 1991a. IMI Descriptions of Fungi and Bacteria 1051. *Acrophialophora fusispora*. Mycopathologia 115: 131-132.

Kirk, P.M., 1991b. IMI Descriptions of Fungi and Bacteria 1053. *Arthrinium phaeospermum*. Mycopathologia 115: 135-136.

Kirkland, T.N. & Fierer, J., 1996. Coccidioidomycosis: a reemer-ging infectious disease. Emerg. Infect. Dis. 2: 192-199.

Kirkness, C.M., Seal, D.V. & Clayton, Y.M., 1991. *Sphaeropsis subglobosa* keratomycosis. First reported case. Cornea 10: 85-89.

Kitada, K., Oka, S., Kimura, S., Shimada, K., Serikawa, T., Yamada, J., Tsunoo, H., Egawa, K. & Nakamura, Y., 1991. Detection of *Pneumocystis carinii* sequences by polymerase chain reaction: animal models and clinical application to non invasive specimens. J. Clin. Microbiol. 29: 1985-1990.

Kitamura, H., Anri, A., Fuse, K., Seo, M. & Itakura, C., 1990. Chronic mastitis caused by *Candida maltosa* in a cow. Vet. Pathol. 27: 465-466.

Klaasen, J.A. & Nelson, P.E., 1996. Identification of a mating population, *Gibberella nygamai* sp. nov., within the *Fusarium nygamai* anamorph. Mycologia 88: 965-969.

Klein, A.S., Tortora, G.T., Malowitz, R. & Greene, W.H., 1988. *Hansenula anomala*: a new fungal pathogen. Arch. Intern. Med. 148: 1210-1213.

Klein, E. & Gordon, M., 1904. Über die Herkunft einer Rosahefe. Centbl. Bakt. Parasitkde, Abt. 1, 35: 138-139.

Klich, M.A. & Pitt, J.I., 1988. A Laboratory Guide to the Common *Aspergillus* Species and their Teleomorphs. CSIRO, North Ryde, Australia, 116 pp.

Kligman, A., 1950. A basidiomycete probably causing onychomycosis. J. Invest. Derm. 14: 67-70.

Klokke, A.H. & Vries, G.A. de, 1963. Tinea capitis in chimpanzees caused by *Microsporum canis* 1902 resembling *M. obesum* Conant 1937. Sabouraudia 2: 268-270.

Klossek, J.-M., Peloquin, L., Fourcroy, P.-J., Ferrie, J.-C. & Fontanel, J.-P., 1996. Aspergillomas of the sphenoid sinus: a series of 10 cases treated by endoscopic sinus surgery. Rhinology, Utrecht 34: 179-183.

Koc, A.N., Utas, C., Oymak, O. & Sehman, E., 1998. Peritonitis due to *Acremonium strictum* in a patient on continuous ambulatory peritoneal dyalisis. Nephron 79: 357-358.

Koch, H.A., 1969. *Trichophyton thuringiense* spec. nov. Mykosen 12: 287-290.

Koch, H.A. & Haneke, H., 1965. *Chaetomium funicolum* Cooke als möglicher Erreger einer tiefen Mykose. Mykosen 9: 23-28.

Kocková-Kratochvílová, A., Simordová, M. & Sternbersky, S., 1987. *Moniliella suaveolens* var. *nigra*. Mykosen 30: 544-547.

Koenig, H., Ball, C. & Bièvre, C. de, 1997. First European cases of onychomycosis caused by *Onychocola canadensis*. J. Med. Vet. Mycol. 35: 71-72.

Koenig, H., Bièvre, C. de, Waller, J. & Conraux, C., 1985. *Aspergillus alliaceus*, agent d'otorrhée chronique. Bull. Soc. Fr. Mycol. Méd. 14: 85-87.

Koenig, H., Warter, A., Bièvre, C. de, Waller, J., Weitzeblum, E. & Morand, G., 1984. Mycose pulmonaire à *Drechslera hawaiiensis*. Bull. Soc. Fr. Mycol. Méd. 13: 373-376.

Köhler, C., Klotz, M., Daus, H., Schwarze, G. & Dette, S., 1988. Viszerale Paracoccidioidomykose bei einem Goldgräber aus Brasilien. Mycoses 31: 395-403.

Kok, I., Veenstra, J., Rietra, P.J.G.M., Dirks-Go, S., Blaauwgeers, J.L.G. & Weigel, H.M., 1994. Disseminated *Penicillium marneffei* infections as an imported disease in HIV-1 infected patients. Neth. J. Med. 44: 18-22.

Kolbeck, P.C., Mahkoul, R.G., Bollinger, R.R. & San Filippo, F., 1985. Widely disseminated *Cunninghamella* mucormycosis in an adult renal transplant recipient. Case report and review of literature. Am. J. Clin. Pathol. 83: 747-753.

Kombila, M., Martz, M., Gomez de Diaz, M., Bièvre, C. de & Richard-Lenoble, D., 1990. *Hendersonula toruloidea* as an agent of mycotic foot infection in Gabon. J. Med. Vet. Mycol. 28: 215-223.

Komoda, M., Itoi, Y., Ozai, Y., Kimura, Y., Koizumi, S. & Takatori, K., 1988. An infection of cow with *Mortierella wolfii*. Mycopathologia 101: 89-93.

Komshian, S.V., Uwaydah, A.K., Sobel, J.D. & Crane, L.R., 1989. Fungemia caused by *Candida* species and *Torulopsis glabrata* in the hospitalized patient: frequency, characteristics, and evaluation of factors influencing outcome. Rev. Infect. Dis. 11: 379-390.

Kontou-Kastellanou, C., Leonardopoulos, J., Botriasi, L., Kastellanos, S. & Toutouzas, P., 1990. A case of *Candida parapsilosis* endocarditis. Mycoses 33: 427-429.

Kontoyianis, D.P., Vartivarian, S., Anaissie, E.J., Samonis, G., Bodey, G.P. & Rinaldi, M., 1994. Infections due to *Cunninghamella bertholletiae* in patients with cancer: report of three cases and review. Clin. Infect. Dis. 18: 925-928.

Koppang, H.S., Olsen, I., Stuge, V. & Sandven, P., 1991. *Aureobasidium* infection of the jaw. J. Oral Path. Med. 20: 191-195.

Kordossis, T., Avlami, A., Velegraki, A., Stefanou, I., Georgakopoulos, G., Papalambrou, C. & Legakis, N.J., 1998. First report of *Cryptococcus laurentii* meningitis and a fatal case of *Cryptococcus albidus* cryptococcaema in AIDS patients. Med. Mycol. 36: 335-339.

Korfel, A., Menssen, H.D., Schwartz, S. & Thiel, E., 1998. Cryptococcosis in Hodgkin's disease: description of two cases and review of the literature. Annls Hematol. 76: 283-286.

Korman, T.M., Fuller, A., Dowling, J.P., 1998. Inflammatory tinea corporis due to *Trichophyton verrucosum*. Clin. Infect. Dis. 26: 220-221.

Körte, C., Edo, D., Canteros, C., Braga, M.E., Leitner, R.M. & Rusiñol, J., 1998. Onicomicosis por *Microsporum gypseum*. Observación de un caso y actualización. Prensa Med. Arg. 85: 352-356.

Korzenioswka-Kosela, M., Halweg, H., Bestry, I., Podsiadlo, B. & Krakowka, P., 1990. Pulmonary aspergilloma caused by *Aspergillus niger*. Pneumon. Pol. 58: 328-333.

Korzets, A., Weinberger, M., Chagnak, A., Goldschmied-Reouven, A. & Sutton, D.A., 2000. Fungal peritonitis due to *Thermoascus crustaceus* (anamorph *Paecilomyces crustaceus*) in a CAPD patient. Abstr. 14th ISHAM Congr. p. 267.

Koshi, G., Anandi, V., Kurien, M., Kirubakaran, M.G., Padhye, A.A. & Ajello, L., 1987. Nasal phaeohyphomycosis caused by *Bipolaris hawaiiensis*. J. Med. Vet. Mycol. 25: 397-402.

Koshi, G., Kurien, T., Sudarsanam, D., Selvapandian, A.J. & Mammen, K.E., 1972. Subcutaneous phycomycosis caused by *Basidiobolus*. A report of three cases. Sabouraudia 10: 237-243.

Koshi, G., Padhye, A.A., Ajello, L. & Chandler, F.W., 1979. *Acremonium recifei* as an agent of mycetoma in India. Am. J. Trop. Med. Hyg. 28: 692-696.

Kovac, D., Lindic, J., Lejko-Zupanc, T., Bren, A.F., Knap, B., Lesnik, M., Gucek, A. & Ferluga, D., 1998. Treatment of severe *Paecilomyces varioti* peritonitis in a patient on continuous peritoneal dialysis. Nephrol. Dial. Transpl. 13: 2943-2946.

Kozakiewicz, Z., 1989. *Aspergillus* species on stored products. Mycol. Pap. 161: 1-188.

Kozakiewicz, Z., 1995. IMI Descriptions of Fungi and Bacteria, Set 126. Mycopathologia 132: 41-62.

Krachmer, J.H., Anderson, R.L., Binder, P.S., Waring, G.O., Rowsey, J.J. & Meck, S., 1978. *Helminthosporium* corneal ulcers. Am. J. Ophthalmol. 85: 666-670.

Krajden, S., Summerbell, R.C., Kane, J., Salkin, I.F., Kemna, M.E., Rinaldi, M.G., Fuksa, M., Spratt, E., Rodrigues, C. & Choe, J., 1991.

Normally saprobic cryptococci isolated from *Cryptococcus neoformans* infections. J. Clin. Microbiol. 29: 1883-1887.

Kralovic, S.M. & Rhodes, J.C., 1995. Phaeohyphomycosis caused by *Dactylaria* (human dactylariosis): report of a case with review of the literature. J. Infect. 31: 107-113.

Kramer, B.S., Hernandez, A.D., Reddick, R.L. & Levine, A.S., 1977. Cutaneous infarction. Manifestation of disseminated mucormycosis. Arch. Derm. 113: 1075-1076.

Krasinski, K., Holzman, R.S., Hanna, B., Greco, M.A., Graff, M. & Bhogal, M., 1985. Nosocomial fungal infection during hospital renovation. Infect. Control 6: 278-282.

Krcmery, V., Spanik, S., Grausova, S., Trupl, J., Krupova, I., Roidova, A., Salek, T., Sulfiarsky, J. & Mardiak, J., 1998. *Candida parapsilosis* fungemia in cancer patients - incidence, risk factors and outcome. Neoplasma, Bratislava 45: 336-342.

Kreger-van Rij, N.J.W. (ed.), 1984. The Yeasts, a Taxonomic Study, 3rd ed. Elsevier Sci. Publ., Amsterdam, 1082 pp.

Kreger-van Rij, N.J.W. & Veenhuis, M., 1970. An electron microscope study of the yeast *Pityrosporum ovale*. Arch. Mikrobiol. 71: 123-131.

Kreger-van Rij, N.J.W. & Veenhuis, M., 1971. A comparative study of the cell wall structure of basidiomycetous and related yeasts. J. Gen. Microbiol. 68: 87-95.

Krempl-Lamprecht, L., 1970. *Scopulariopsis*-Arten bei Onychomykosen. Proc. 2th Int. Symp. Med. Mycol. Poznan, 1967, pp. 45-86.

Krempl-Lamprecht, L., Luderschmidt, C. & Wehrmann, W., 1987. Chromomykose durch *Fonsecaea compacta* (Carrion 1940) mit phaeohyphomykotischem Sekundärbefall durch *Phaeoscler dema-tioides* (Sigler 1981). Mykosen 30: 454-467.

Krisher, K.K., Holdridge, N.B., Mustafa, M.M., Rinaldi, M.G. & McGough, D.A., 1995. Disseminated *Microascus cirrosus* infection in pediatric bone marrow transplant recipient. J. Clin. Microbiol. 33: 735-737.

Krishnan, S.G.S., Sentamilselvi, G., Kamalam, A., Das, K.A. & Janaki, C., 1998. Entomophthoromycosis in India - a 4-year study. Mycoses 41: 55-58.

Kritsky, A.L., Lemle, A., Souza, G.R. de, Sousa, R.V. de, Nogueira, S.A., Pereira, N.G. & Bethlem, N.M., 1990. Pulmonary function changes in the acute stage of histoplasmosis,with follow-up. Chest 97: 1244-1245.

Krueger, W., Sobottka, I., Stockschlader, M., Mross, K., Hoffknecht, M., Ruessmann, B., Horstmann, M., Betker, R. & Zander, A., 1996. Fatal outcome of disseminated candidosis after allogeneic bone marrow transplantation under treatment with liposomal and conventional amphotericin-B. A report of 4 cases with determination of the MIC values. Scand. J. Infect. Dis. 28: 313-316.

Krulder, J.W.M., Brimicombe, R.W., Wijermans, P.W. & Gams, W., 1996. Systemic *Fusarium nygamai* infection in a patient with lymphoblastic non-Hodgkin's lymphoma. Mycoses 39: 121-123.

Krumholz, R.A., 1972. Pulmonary cryptococcosis. A case due to *Cryptococcus albidus*. Am. Rev. Respir. Dis. 105: 421-424.

Krzystolik, M.G., Ciulla, T.A., Topping, T.M. & Baker, A.S., 1997. Exogenous *Aspergillus niger* endophthalmitis in a patient with a filtering bleb. Retina 17: 461-462.

Kubicek, C.P. & Harman, G.E. (eds), 1998. *Trichoderma* and *Gliocladium*. 2 vols. Taylor & Francis, London.

Kudeken, N., Kawakami, K., Kusano, N. & Saito, A., 1996. Cell-mediated immunity in host resistance against infection caused by *Penicillium marneffei*. J. Med. Vet. Mycol. 34: 371-378.

Kuehn, H.H., Orr, G.F. & Ghosh, G.R., 1961. A new and widely distributed species of *Pseudoarachniotus*. Mycopath. Mycol. Appl. 14: 215-229.

Kuehnert, M.J., Clark, E., Lockhart, S.R., Soll, D.R., Chia, J. & Jarvis, W.R., 1998. *Candida albicans* endocarditis associated with a contaminated aortic valve allograft: implications for regulation of allograft processing. Clin. Infect. Dis. 27: 688-691.

Kuemmerle, S. & Wedler, H., 1998. Renal abscess in an AIDS patient caused by *Aspergillus fumigatus*. Urol. Int. 61: 52-54.

Kües, U., 2000. Life history and developmental process in the basidiomycete *Coprinus cinereus*. Microbiol. Mol. Biol. Rev. 64: 316-353.

Kuhls, K., Lieckfeldt, E., Börner, T. & Guého, E., 1999. Molecular reidentification of human pathogenic *Trichoderma* isolates as *Trichoderma longibrachiatum* and *T. citrinoviride*. Med. Mycol. 37: 25-33.

Kuhls, K., Lieckfeldt, E., Samuels, G.J., Meyer, W., Kubicek, C.P. & Börner, T., 1997. Revision of *Trichoderma* sect. *Longibrachiatum* including related teleomorphs based on analysis of ribosomal DNA internal transcribed spaces sequences. Mycologia 89: 442-460.

Kujath, P., Lerch, K. & Dämmrich, J., 1990. Fluconazole monitoring in *Candida* peritonitis based on histological control. Mycoses 33: 441-448.

Kunert, J. & Hejtmánek, M., 1964. Izolace nového dermatofyta rodu *Keratinomyces* Vanbreuseghem 1952. Česka Epid. Mikrobiol. Immunol. 13: 293-297.

Kunová, A., Sorkovská, D., Sufliarsky, J., Helpienska, L. & Krcmery, V., 1995. Report of catheter-associated *Trichosporon pullulans* breakthrough fungemia in a cancer patient. Eur. J. Clin. Microbiol. Infect. Dis. 14: 729-730.

Kunová, A., Spanik, S., Kollar, T. & Krcemery, V., 1996. Breakthrough fungemia due to *Hansenula anomala* in a leukemia patient successfully treated with amphotericin B. J. Chemother. 8: 85-86.

Kunstÿř, I., Jelinek, F., Bitzenhofer, U. & Pittermann, W., 1997. Fungus *Paecilomyces:* a new agent in laboratory animals. Lab. Anim. 31: 45-51.

Kunstÿř, I., Niculescu, E., Naumann, S. & Lippert, E., 1980. *Torulopsis pintolopesii* - an opportunistic pathogen in guineapigs? Lab. Anim. 14: 43-45.

Kurien, M., Anandi, V., Raman, R. & Brahmadathan, K.N., 1992. Maxillary sinus fusariosis in immunocompetent hosts. J. Laryngol. Otol. 106: 733-736.

Kurtzman, C.P., 1973. Formation of hyphae and chlamydospores by *Cryptococcus laurentii*. Mycologia 65: 388-395.

Kurtzman C.P., 1984. Synonymy of the yeast genera *Hansenula* and *Pichia* demonstrated through comparisons of deoxyribonucleic acid relatedness. Antonie van Leeuwenhoek 50: 209-217.

Kurtzman, C.P., 1998. *Pichia* E.C. Hansen *emend*. Kurtzman. In C.P. Kurtzman & J.W. Fell (eds.): The Yeasts, a Taxonomic Study. 4th ed., pp. 273-352. Elsevier, Amsterdam.

Kurtzman, C.P. & Fell. J.W. (eds), 1998. The Yeasts, a Taxonomic Study. 4th ed. Elsevier, Amsterdam, 1055 pp.

Kurtzman, C.P. & Robnett, C.J., 1997. Identification of clinically important ascomycetous yeasts based on nucleotide divergence in the 5' end of the large-subunit (26S) ribosomal DNA gene. J. Clin. Microbiol. 35: 1216-1223.

Kurtzman, C.P. & Robnett, C.J., 1998. Identification and phylogeny of ascomycetous yeasts from analysis of nuclear large subunit (26S) ribosomal DNA partial sequences. Antonie van Leeuwenhoek 73: 331-371.

Kurtzman, C.P., Smiley, M.J., Robnett, C.J. & Wicklow, D.T., 1986. DNA relatedness among wild and domestic species in the *Aspergillus flavus* group. Mycologia 78: 955-959.

Kurzai, O., Heinz, W.J., Sullivan, D.J., Coleman, D.C., Frosch, M. & Mühlschlegel, F.A., 1999. Rapid PCR test for discriminating between *Candida albicans* and *Candida dubliniensis* isolates using primers derived from the pH-regulated PHR1 and PHR2 genes of *C. albicans*. J. Clin. Microbiol. 37: 1587-1590.

Kusenbach, G., Skopnik, J., Haase, G., Friedrichs, F. & Doehmen, H., 1992. *Exophiala dermatitidis* pneumonia in cystic fibrosis. Eur. J. Pediatr. 151: 344-346.

Kushawa, R.K.S. & Guarro, J. (eds), 2000. Biology of dermatophytes and other keratinophilic fungi. Revta Iberoam. Micol. 17: 1-176.

Kuzo, R.S. & Goodman, L.R., 1996. Blastomycosis. Semin. Roentgenol. 31: 45-51.

Kwan, E.Y.W., Lan, Y.L., Yuen, K.Y., Jones, B.M. & Low, L.C.K., 1997. *Penicillium marneffei* infection in a non-HIV infected child. J. Pediatr. Child Health 33: 267-271.

Kwochka, K.W., Mays, M.B.C., Ajello, L. & Padhye, A.A., 1984. Canine phaeohyphomycosis caused by *Drechslera spicifera:* a case report and literature review. J. Am. Anim. Hosp. Ass. 20: 625-633.

Kwon-Chung, K.J., 1972. Genetic study on the incompatibility system in *Arthroderma simii*. Sabouraudia 10: 74-78.

Kwon-Chung, K.J., 1973. Studies on *Emmonsiella capsulata*. I. Heterothallism and development of the ascocarp. Mycologia 65: 109-121.

Kwon-Chung, K.J., 1975. Perfect state *(Emmonsiella capsulata)* of the fungus causing large-form African histoplasmosis. Mycologia 67: 980-990.

Kwon-Chung, K.J., 1977. Perfect state of *Cryptococcus uni-guttulatus*. Int. J. Syst. Bacteriol. 27: 293-299.

Kwon-Chung, K.J., 1983. A new variety of *Cladosporium trichoides*. Mycologia 75: 320-323.

Kwon-Chung, K.J., 1987a. *Filobasidiaceae* - a taxonomic survey. In G.S. de Hoog *et al.* (eds): The Expanding Realm of Yeast-like Fungi, pp. 75-85.

Kwon-Chung, K.J., 1987b. Infection of the human orbit by *Aspergillus stromatoides*. Mycopathologia 100: 113.

Kwon-Chung, K.J., 1994. Phylogenetic spectrum of fungi that are pathogenic to humans. Clin. Infect. Dis. 19, Suppl. 1: 1-7.

Kwon-Chung, K.J. & Bennett, J.W., 1992. Medical Mycology. Lea & Febiger, Philadelphia, 861 pp.

Kwon-Chung, K.J. & Chang, Y.C., 1994. Gene arrangement and sequence of the 5S rRNA in *Filobasiella neoformans (Cryptococcus neoformans)* as a phylogenetic indicator. Int. Syst. Bact. 44: 209-213.

Kwon-Chung, K.J. & Droller, D.D., 1984. Infection of the olecranon bursa by *Anthopsis deltoidea*. J. Clin. Microbiol. 20: 221-223.

Kwon-Chung, K.J., Folks, T. & Sell, K.W., 1984. Unusual isolates of *Thermoascus crustaceus* from three monocyte cultures of AIDS patients. Mycologia 76: 375-379.

Kwon-Chung, K.J. & Kim, S.J., 1974. A second heterothallic *Aspergillus*. Mycologia 66: 628-638.

Kwon-Chung, K.J., Polacheck, I. & Bennett, J., 1982. Improved diagnostic medium for separation of *Cryptococcus neoformans* var. *neoformans* (serotypes A and D) and *Cryptococcus neoformans* var. *gattii* (serotypes B and C). J. Clin. Microbiol. 15: 535-537.

Kwon-Chung, K.J., Schwartz, I.S. & Ryback, B.J., 1975. A pulmonary fungus ball produced by *Cladosporium cladosporioides*. Am. J. Clin. Path. 64: 564-568.

Kwon-Chung, K.J., Varma, A. & Howard, D.H., 1990. Ecology of *Cryptococcus neoformans* and prevalence of its two varieties in AIDS and non-AIDS associated cryptococcosis. In H. Vanden Bossche *et al.* (eds.): Mycoses in AIDS Patients, pp. 103-113. Plenum, New York.

Kwon-Chung, K.J. & Vries, G.A. de, 1983. Comparative study of an isolate resembling Banti's fungus with *Cladosporium trichoides*. Sabouraudia 21: 59-72.

Kwon-Chung, K.J., Wickes, B.L. & Plaskowitz, J., 1989. Taxonomic clarification of *Cladosporium trichoides* Emmons and its subsequent synonyms. J. Med. Vet. Mycol. 27: 413-426.

Laakkonen, J. & Sukura, A., 1997. *Pneumocystis carinii* of the common shrew, *Sorex araneus,* shows a discrete phenotype. J. Eukaryot. Microbiol. 44: 117-121.

Lacaz, C. da Silva, Heins-Vaccari, E.M., Takahashi de Melo, N. & Hernandez-Arriagada, G.L., 1996. Basidiomycosis: a review of the literature. Revta Inst. Med. Trop. S. Paulo 38: 379-390.

Lacaz, C. da Silva & Netto, C.F., 1954. Contribuçao para o estudo dos agentes etiológicos da maduromicose. Folia Clin. Biol., S. Paulo 21: 331-352.

Lacaz, C. da Silva, Porto, E., Andrade, J.G. & Telles Filho, F.Q., 1984. Feohifomicose disseminada por *Exophiala spinifera*. An. Bras. Derm. 59: 238-243.

Lacaz, C. da Silva, Porto, E., Carneiro, J.J., Paziami, I.O. & Pimenta, W.P., 1981. Endocardite em prótese de dura-mater provocada pelo *Acremonium kiliense*. Revta Inst. Med. Trop. S. Paulo 23: 274-279.

Lacaz, C. da Silva, Porto, E., Cucé, L.C. & Salebian, A., 1979. Maduromicose por *Cephalosporium acremonium*. Registro de um caso. Revta Inst. Med. Trop. S. Paulo 21: 56-61.

Lacaz, C. da Silva, Porto, E., Heins-Vaccari, E.H. & Melo, N.T. de, 1998. Guia para Identificação Fongos Actinomicolos de Interesse Médico. Sarvier, S. Paulo, 445 pp.

Lacaz, C. da Silva, Vidal, M.S.M., Heins-Vaccari, E.M., Melo, N.T. de, Negro, G.M.B. del, Arriagada, G.L.H. & Freitas, R. dos Santos, 1999. *Paracoccidioides brasiliensis*. A mycologic and immunochemical study of two strains. Revta Inst. Med. Trop. S. Paulo 41: 79-86.

Lacaz, C. da Silva, Vidal, M.S.M., Pereira, C.N., Heins-Vaccari, E.M., Melo, N.T. de, Sakai-Valente, N. & Arriagada, G.L.H., 1997. *Paracoccidioides cerebriformis* Moore, 1935. Mycologic and immunochemical study. Revta Inst. Med. Trop. S. Paulo 39: 141-144.

Lacey, J., 1986. *Microascus cinereus* (Émile-Weil & Gaudin) Curzi, a human pathogen? Mycopathologia 96: 137-142.

Laclaire, M.C., Morand, M. & Euzeby, J., 1974. Investigation mycologique. Note II. Recherche sur quelques syndromes pseudo-aspergillaires; paecilomycose des sacs aériens des poumons. Bull. Soc. Sci. Vét. Méd. Comp. Lyon 76: 317-319.

Lacroix, C., Jacquemin, J.L., Guilhot, F., Rabot, M.H., Burucoa, C. & Bièvre, C. de, 1988. Septicémie à *Acremonium kiliense* avec dissémination sécondaire chez une patiente atteinte d'un myelome à forte masse tumorale. Bull. Soc. Fr. Mycol. Méd. 17: 93-98.

Lakshmi, V., Rani, T.S., Savitri, S., Mohan, V.S., Sundaram, C., Rao, R.R. & Satyanarayana, G., 1993. Zygomycotic necrotizing fasciitis caused by *Apophysomyces elegans*. J. Clin. Microbiol. 31: 1368-1369.

Lalla, F. de, Vaglia, A., Franzetti, M., Manfrin, V., Pellizzer, G.P. & Fabris, P., 1993. Cryptococcal pleural effusion as first indicator of AIDS: a case report. Infection 21: 192.

Lam, D.S.C., Koehler, A.P., Fan, D.S.P., Cheuk, W., Leung, A.S. & Ng, J.S.K., 1999. Endogenous fungal endophthalmitis caused by *Paecilomyces variotii*. Eye, London 13: 113-116.

Lampert, R.P., Hutto, J.H., Donnelly, W.H. & Shulman, S.T., 1977. Pulmonary and cerebral mycetoma caused by *Curvularia pallescens*. J. Pediatr. 91: 603-605.

Landry, M. & Parkins, C.W., 1993. Calcium oxalate crystal deposition in necrotizing otomycosis caused by *Aspergillus niger*. Mod. Pathol. 6: 493-496.

Langeron, M., 1934. Mycose oculaire primitive due à '*Beauveria brumpti*'. Bull. Acad. Méd., Paris 111: 133-137.

Langeron, M., 1947. *Tritirachium brumpti* (Langeron et Lichaa 1934) Langeron 1947 et le genre *Tritirachium* Limber 1940. Annls Parasit. Hum. Comp. 22: 94-99.

Larsson, L., Pehrson, C., Wiebe, T. & Christensson, B., 1994. Gas chromatographic determination of D-arabinitol / L-arabinitol ratios in urine: a potential method for diagnosis of disseminated candidiasis. J. Clin. Microbiol. 32: 1855-1859.

Lascaux, A.S., Bouscarat, F., Descamps, V., Casalino, E., Picard-Dahan, C., Crickx, B. & Belaich, S., 1998. Manifestations cutanées au cours d'une trichosporonose disséminée chez un malade sidéen. Derm. Vénéréol. 125: 111-113.

Latgé, J.-P., 1999. *Aspergillus fumigatus* and aspergillosis. Clin. Microbiol. Rev. 12: 310-350.

Latgé, J.P., Verweij, P.E. & Bretagne, S., 2000. Molecular typing of *Aspergillus fumigatus*. In R.A. Samson & J.I. Pitt (eds): Integration of Modern Taxonomic Methods for *Penicillium* and *Aspergillus* Classification, pp. 471-482.

Lautenschlager, I., Lyytikäinen, O., Jokipii, L., Maiche, A., Ruutu, T., Tukiainen, P. & Ruutu, R., 1996. Immunodetection of *Pneumocystis carinii* in bronchoalveolar lavage specimens compared to methenamine silver stain. J. Clin. Microbiol. 34: 728-730.

Lavarde, V., Daniel, F., Saëz, H., Arnold, M. & Faguer, B., 1984. Peritonite mycosique à *Torulopsis haemulonii*. Bull. Soc. Fr. Mycol. Méd. 13: 173-176.

Lavarde, V., Saez, H. & Jouffre, F., 1985. Endocardite fongique à *Pichia ohmeri*. Bull. Soc. Fr. Mycol. Méd. 14: 265-268.

Lavelle, P., 1980. Chromoblastomycosis in Mexico. PAHO Sci. Publ. 396: 235-247.

Laverde, S., Moncada, L.H., Restrepo, A. & Vera, C.L., 1973. Mycotic keratitis, 5 cases caused by unusual fungi. Sabouraudia 11: 119-123.

Lavoie, S.R., Espinel-Ingroff, A. & Kerkering, T., 1993. Mixed cutaneous phaeohyphomycosis in a cocaine user. Clin. Infect. Dis. 17: 114-116.

Lawrence, R.M., Snodgrass, W.T., Reichel, G.W., Padhye, A.A., Ajello, L. & Chandler, F.W., 1986. Systemic zygomycosis caused by *Apophysomyces elegans*. J. Med. Vet. Mycol. 24: 57-65.

Lazéra, M.S., Cavalcanti, M.A.S., Trilles, L., Nishikawa, M.M. & Wanke, B., 1998. *Cryptococcus neoformans* var. *gattii* - evidence for a natural habitat related to decaying wood in a pottery tree hollow. Med. Mycol. 36: 119-122.

Lazéra, M.S., Pires, F.D.A., Camillo-Coura, L., Nishikawa, M.M., Bezerra, C.C.F., Trilles, L. & Wanke, B., 1996. Natural habitat of *Cryptococcus neoformans* var. *neoformans* in decaying wood forming hollows in living trees. J. Med. Vet. Mycol. 34: 127-131.

Lazéra, M.S., Wanke, B. & Nishikawa, M.M., 1993. Isolation of both varieties of *Cryptococcus neoformans* from saprophytic sources in the city of Rio de Janeiro. J. Med. Vet. Mycol. 31: 449-454.

Leask, B.G.S. & Yarrow, D., 1976. *Pichia norvegensis* sp. nov. Sabouraudia 14: 61-63.

Leclerc, M.C., Philippe, H. & Guého, E., 1994. Phylogeny of dermatophytes and dimorphic fungi based on large subunit ribosomal RNA sequence comparisons. J. Med. Vet. Mycol. 32: 331-341.

Lecso-Bornet, M., Guého, E., Barbier-Boehm, G., Berthelot, G., Gaildrat, M., Taravella, D. & Bergogne-Berezin, E., 1991. Prostatic valve endocarditis due to *Thermomyces lanuginosus* Tsiklinsky. First case report. J. Med. Vet. Mycol. 29: 205-209.

Lederberg, J., 1999. Paradoxes of the host-parasite relationship. ASM News 65: 811-816.

Lee, B.L., Grossniklaus, H.E., Capone, A., Padhye, A.A. & Sekhon, A.S., 1995. *Ovadendron sulphureo-ochraceum* endophthalmitis after cataract surgery. Am. J. Ophthalmol. 119: 307-312.

Lee, E.-J., Lee, M.-Y., Huang, Y.-C. & Wang, L.-C., 1998. Orbital rhinocerebral mucormycosis associated with diabetic ketoacidosis: report of survival of a 10-year-old boy. J. Formosan Med. Ass. 97: 720-723.

Lee, M.W., Kim, J.C., Choi, J.S., Kim, K.H. & Greer, D.L., 1995. Mycetoma caused by *Acremonium falciforme*. Successful treatment with itraconazole. J. Am. Acad. Derm. 32: 897-900.

Lee, S. & Hanlin, R.T., 1999. Phylogenetic relationships of *Chaetomium* and similar genera based on ribosomal DNA sequences. Mycologia 91: 434-442.

Lee, S.C., Dickson, D.W. & Casadevall, A., 1996. Pathology of cryptococcal meningoencephalitis: analysis of 27 patients with pathogenetic implications. Hum. Pathol. 27: 839-847.

Leeming, J.G. & Elliott, T.S.J., 1995. The emergence of *Trichophyton tonsurans* tinea capitis in Birmingham, U.K. Br. J. Derm. 133: 929-931.

Leeming, J.P., Sutton, T.M. & Fleming, P.J., 1995. Neonatal skin as a reservoir of *Malassezia* species. Pediatr. Infect. Dis. J. 14: 719-721.

Legrand, C., Anaissie, E., Hashem, R., Nelson, P., Bodey, G.J. & Ro, J., 1991. Experimental fusarial hyalohyphomycosis in a murine model. J. Infect. Dis. 164: 944-948.

Lellouche, B., Serrano, E., Linas, M.D., Massip, P., Nevez, G., Yardéni, E. & Pessey, J.J., 1993. À propos d'un cas de mucormycose rhino-cérébrale. J. Mycol. Méd. 3: 114-117.

Lemairé, S.L., Bauer, R.W., Foil, C.S. & Roy, A.F., 1994. Conidiobolomycosis in a dog. Vet. Derm. 5: 144.

Lemerle, E., Bastien, M., Demolliens-Dreux, G., Forest, J.-L., Boyer, E., Chabasse, D. & Celerier, P., 1998. Scédosporiose cutanée révélée par un purpura bullo-nécrotique. Annls Derm. Véneréol. 125: 711-714.

Lennon, P.A., Cooper, C.R., Salkin, I.F. & Lee, S.B., 1994. Ribosomal DNA internal transcribed spacer analysis supports synonymy of *Scedosporium inflatum* and *Lomentospora prolificans*. J. Clin. Microbiol. 32: 2413-2416.

Leong, K.W., Crowley, B., White, B., Crotty, G.M., O'Brien, D.S., Keane, C. & Mccann, S.R., 1997. Cutaneous mucormycosis due to *Absidia corymbifera* occurring after bone marrow transplantation. Bone Marrow Transpl. 19: 513-515.

Lesire, V., Hazouard, E., Dequin, P.F., Delain, M., Thérizol-Ferly, M. & Legras, A., 1999. Possible role of *Chaetomium globosum* in infection after autologous bone marrow transplantation. Intens. Care Med. 25: 124-125.

Leslie, J.F., 1995. *Gibberella fujikuroi:* available populations and variable traits. Can. J. Bot. 73: S282-S291.

Leu, H.S., Lee, A.Y.S. & Kuo, T.T., 1995. Recurrence of *Fusarium solani* abscess formation in an otherwise healthy patient. Infection 23: 303-305.

Levin, A.S., Costa, S.F., Mussi, N.S., Basso, M., Sinto, S.I., Machado, C., Geiger, C., Villares, M.C.B., Schreiber, A.Z., Barone, A.A. & Branchini, M.L.M., 1998. *Candida parapsilosis* fungemia associated with implantable and semi-implantable central venous catheters and the hands of healthcare workers. Diagn. Microbiol. Infect. Dis. 30: 243-249.

Levy, I., Rubin, L.G., Vasishtha, S., Tucci, V. & Sood, S.K., 1998. Emergence of *Candida parapsilosis* as the predominant species causing candidemia in children. Clin. J. Infect. Dis. 26: 1086-1088.

Levy-Klotz, B., Badillet, G., Cavelier-Balloy, B., Chemaly, P., Leverger, G. & Civatte, J., 1985. AIDS associated cutaneous alternariosis. Am. Derm. Véneréol., Paris 112: 739-740.

Lewis, S.J. & Freedman, A.R., 1998. The use of biotherapeutic agents in the prevention and treatment of gastrointestinal disease. Alim. Pharmacol. Therap. 12: 807-822.

Lewis, S.M. & Lewis, B.G., 1997. Nosocomial transmission of *Trichophyton tonsurans* tinea corporis in a rehabilitation hospital. Infect. Contr. Hosp. Epid. 18: 322-325.

Li, R.K., Ciblak. M.A., Nordoff, N., Pasarell, L., Warnock, D.W. & McGinnis, M.R., 2000. *In vitro* activities of voriconazole, itraconazole and amphotericin B against *Blastomyces dermatitidis, Coccidioides immitis,* and *Histoplasma capsulatum.* Antimicrob. Agents Chemother. 44: 1734-1736.

Li, S., Cullen, D., Hjort, M., Spear, R. & Andrews, J.H., 1996. Development of an oligonucleotide probe for *Aureobasidium pullulans* based on the small-subunit rRNA gene. Appl. Environm. Microbiol. 62: 1514-1518.

Li, S., Perlman, J.I., Edward, D.P. & Weiss, R., 1998. Unilateral *Blastomyces dermatitidis* endophthalmitis and orbital cellulitis. A case report and literature review. Ophthalmology 105: 1466-1470.

Li, S., Spear, R.N. & Andrews, J.H., 1997. Quantitative fluorescense *in situ* hybridization of *Aureobasidium pullulans* on microscope slides and leaf surfaces. Appl. Environm. Microbiol. 63: 3261-3267.

Li, R.K., Ciblak, M.A., Nordoff, N., Pasarell, L., Warnock, D.W. & McGinnis, M.R. 2000. *In vitro* activities of voriconazole, itraconazole and amphotericin B against *Blastomyces dermatitidis, Coccidioides immitis,* and *Histoplasma capsulatum.* Antimicrob. Agents Chemother. 44: 1734-1736.

Liao, W.-Q., Li, Z.-G., Guo, M. & Zhang, J.-Z., 1993. *Candida zeylanoides* causing candidiasis as tinea cruris. Chin. Med. J. 106: 542-545.

Liao, W.-Q., Shao, J.Z., Li, S.Q., Li, T.Z., Wu, S.X., Zhang, U.Z. & Chen, Q.T., 1983. *Colletotrichum dematium* caused keratitis. Chin. Med. J. 96: 391-394.

Liao, W.-Q., Yao, Z.R., Li, Z.Q., Xu, H. & Zhao, J., 1995. Pyoderma gangraenosum caused by *Rhizopus arrhizus.* Mycoses 38: 75-77.

Liao, Y.-H., Chu, S.-H., Hsiao, G.-H., Chou, N.-K. Wang, S.-S. & Chiu, H.-C., 1999. Majocchi's granuloma caused by *Trichophyton tonsurans* in a cardiac transplant recipient. Br. J. Derm. 140: 1194-1196.

Libertin, C.R., Wilson, W.R. & Roberts, G.D., 1985. *Candida lusitaniae* - an opportunistic pathogen. Diagn. Microbiol. Infect. Dis. 3: 69-71.

Lie-Kian-Joe, Njo-Injo Tjoei Eng & Sartono Kertopati, 1957. A new verrucous mycosis caused by *Cercospora apii.* Arch. Derm. 75: 864-870.

Lieckfeldt, E., Samuels, G.J., Börner, T. & Gams, W., 1998. *Trichoderma koningii:* neotypification and *Hypocrea* teleomorph. Can. J. Bot. 76: 1507-1522.

Lieckfeldt, E., Samuels, G.J., Nirenberg, H.I. & Petrini, O., 1999. A morphological and molecular perspective of *Trichoderma viride:* is it one or two species? Appl. Environm. Microbiol. 65: 2418-2428.

Lieckfeldt, E., & Seifert, K.A., 2000. An evaluation of the use of ITS sequences in the taxonomy of the *Hypocreales.* Stud. Mycol. 45: 35-44.

Liesegang, T.J. & Forster, R.K., 1980. Spectrum of microbial keratitis in South Africa. Am. J. Ophthalmol. 90: 38-47.

Lin, W.-C., Dai, Y.-S., Tsai, M.-J., Huang, L.-M. & Chiang, B.-L., 1998. Systemic *Penicillium marneffei* infection in a child with common variable immunodeficiency. J. Formosan Med. Ass. 97: 780-783.

Lin, Y. & Li, X., 1995. First case of phaeohyphomyccosis caused by *Chaetomium murorum* in China. Chin. J. Derm. 28: 367-369.

Lin, Y., Xiangyin, L. & Al, E., 1995. First case of phaeo-hyphomycosis caused by *Chaetomium murorum* in China. Zhonghua Pifuke Zazhi 28: 367-369.

Lindahl, K.J. & Limbird, T.J., 1987. *Torulopsis glabrata* vertebral osteomyelitis: case report and review of the literature. Spine 12: 593-595.

Linder, N., Keller, N., Huri, C., Kuint, J., Goldschmied-Reouven, A. & Barzilai, A., 1998. Primary cutaneous mucormycosis in a premature infant: case report and review of the literature. Am. J. Perinat. 15: 35-38.

Lindsberg, P.J., Pieninkeroinen, I. & Valtonen, M., 1997. Meningoencephalitis caused by *Cryptococcus macerans*. Scand. J. Infect. Dis. 29: 430-433.

Linhares, A.C., Miranda, M.F.R., Brito, A.C. de, Zaitz, C., Carvalho, T. de & Carneiro, F.R.O., 1998. *Microsporum gypseum* infection showing a white-paint-dot appearance. Int. J. Derm. 37: 956-957.

Lipperheide, V., Andraka, L., Pontón, J. & Quindós, G., 1993. Evaluation of the Albicans IC^R-systems plate method for the rapid identification of *Candida albicans*. Mycoses 36: 417-420.

Lischewski, A., Amann, R.I., Harmsen, D., Merkert, H., Hacker, J. & Morschhäuser, J., 1996. Specific detection of *Candida albicans* and *Candida tropicalis* by fluorescent *in situ* hybridization with an 18S rRNA-targeted oligonucleotide probe. Microbiology 142: 2731-2740.

Lischewski, A., Kretschmar, M., Hof, H., Amann, R., Hacker, J. & Morschhäuser, J., 1997. Detection and identification of *Candida* species in experimentally infected tissue and human blood by rRNA-specific fluorescent *in situ* hybridization. J. Clin. Microbiol. 35: 2943-2948.

Listemann, H., 1974. Erstmaliges Auftreten von *Epidermophyton floccosum* (Harz 1870) Langeron et Milochevitch 1930 var. *nigricans* Frágner 1968 in Deutschland. Castellania 2: 23.

Listemann, H., Schönrock-Nabulsi, P., Kuse, R. & Meigel, W., 1996. Geotrichosis of oral mucosa. Mycoses 39: 289-291.

Listemann, H., Schulz, K.D., Wasmuth, R., Begemann, F. & Meigel, W., 1998. Oesophagitis caused by *Candida kefyr*. Mycoses 41: 343-344.

Liu, D., Coloe, S., Baird, R. & Pedersen, J., 1997. PCR identification of *Trichophyton mentagrophytes* var. *interdigitale* and *T. mentagrophytes* var. *mentagrophytes* dermatophytes with a random primer. J. Med. Microbiol. 46: 1043-1046.

Liu, D., Coloe, S., Pedersen, J. & Baird, R., 1996. Use of arbitrarily primed polymerase chain reaction to differentiate *Trichophyton* dermatophytes. FEMS Microbiol Lett. 136: 147-150.

Liu, M.-T., Wong, C.-K. & Fung, C.-P., 1994. Disseminated *Penicillium marneffei* infection with cutaneous lesions in a HIV-positive patient. Br. J. Derm. 131: 280-283.

Liu, Z., Hou, T., Shen, Q., Liao, W. & Xu, H., 1995. Osteomyelitis of sacral spine caused by *Aspergillus versicolor* with neurologic deficits. Clin. Med. J. 108: 472-475.

Liv, K., Howell, D.N., Perfect, J.R. & Schell, W.A., 1998. Morphologic criteria for the preliminary identification of *Fusarium, Paecilomyces*, and *Acremonium* species by histopathology. Microbiol. Infect. Dis. 109: 45-54.

Livas, I.C., Nechay, P.S. & Nauseef, W.M., 1995. Clinical evidence of spinal and cerebral histoplasmosis twenty years after renal transplantation. Clin. Infect. Dis. 20: 692-695.

Lo, Y., Tintelnot, K., Lippert, U. & Hoppe, T., 2000. Disseminated *Penicillium marneffei* infection in an African AIDS patient. Trans. R. Soc. Trop. Med. Hyg. 94: 187.

LoBuglio, K.F., Pitt, J.I. & Taylor, J.W., 1993. Phylogenetic analysis of two ribosomal DNA regions indicate multiple losses of a sexual *Talaromyces* state among asexual *Penicillium* species in subgenus *Biverticillium*. Mycologia 85: 592-604.

LoBuglio, K.F., Pitt, J.I. & Taylor, J.W., 1994. Independent origins of the synnematous *Penicillium* species, *P. duclauxii, P. clavigerum*, and *P. vulpinum,* as assessed by two ribosomal DNA regions. Mycol. Res. 98: 250-256.

LoBuglio, K.F. & Taylor, J.W., 1995. Phylogeny and PCR identification of the human pathogenic fungus *Penicillium marneffei*. J. Clin. Microbiol. 33: 85-89.

Loeppky, C.B., Sprouse, R.F., Carlson, J.V. & Everett, E.D., 1983. *Trichoderma viride* peritonitis. South. Med. J. 76: 789-799.

Lombardi, G., Padhye, A.A. & Ajello, L., 1988. *In vitro* conversion of African isolates of *Blastomyces dermatitidis* to their yeast form. Mycoses 31: 447-450.

Lombardi, G., Padhye, A.A., Standard, P.G., Kaufman, L. & Ajello, L., 1989. Exoantigen tests for the rapid and specific identification of *Apophysomyces elegans* and *Saksenaea vasiformis*. J. Med. Vet. Mycol. 27: 113-120.

Londero, A.T., Fischman, O. & Ramos, C.D., 1964. *Trichophyton gallinae* in Brazil. Sabouraudia 3: 233-234.

Longcore, J.E., Pessier, A.P. & Nichols, D.K., 1999. *Batrachochytrium dendrobatidis* gen. et sp. nov., a chytrid pathogenic to amphibians. Mycologia 91: 219-227.

Longley, R.E. & Cozad, G.C., 1979. Thymosin restoration of cellular immunity of *Blastomyces dermatitidis* in T cell depleted mice. Infect. Immun. 26: 187-192.

Lonial, S., Williams, L., Carum, G., Ostrowski, M. & McCarthy, P., 1997. *Neosartorya fischeri*: an invasive fungal pathogen in an allogeneic bone narrow transplant patient. Bone Marrow Transpl. 19: 753-755.

Lopes, J.O., Alves, S.H., Benevenga, J.P., Brauner, F.B., Castro, M.S. & Melchiors, E., 1994b. *Curvularia lunata* peritonitis complicating peritoneal dialysis. Mycopathologia 127: 65-67.

Lopes, J.O., Alves, S.H., Benevenga, J.P. & Encarnação, C.S., 1994a. Nodular infection of the hair caused by *Malassezia furfur*. Mycopathologia 125: 149-152.

Lopes, J.O., Alves, S.H., Rosa, A.C., Silva, C.B., Sarturi, J.C. & Souza, C.A.R., 1995a. *Acremonium kiliense* peritonitis complicating continuous ambulatory peritoneal dialysis: report of two cases. Mycopathologia 131: 83-85.

Lopes, J.O., Alves, S.H., Klock, C., Oliveira, T.L.O. & Forno, N.R.F. dal, 1997. *Trichosporon inkin* peritonitis during continuous ambulatory peritoneal dialysis with bibliography review. Mycopathologia 139: 15-18.

Lopes, J.O. & Jobim, N.M., 1998. Dermatomycosis of the toe web caused by *Curvularia lunata*. Revta Inst. Med. Trop. S. Paulo 40: 327-328.

Lopes, J.O., Kolling, L.C. & Neumaier, W., 1995b. Kerionlike lesion of the scalp due to *Acremonium kiliense* in a noncompromised boy. Revta Inst. Med. Trop. S. Paulo 37: 365-368.

Lopes, J.O., Mello, E.S. de & Klock, C., 1995a. Mixed intranasal infection caused by *Fusarium solani* and a zygomycete in a leukaemic patient. Mycoses 38: 281-284.

Lopes, J.O., Mello, E.S. de, Villalba, B., Alves, S.H. & Klock, C., 1996a. Mixed invasive *Candida kefyr* and *Aspergillus flavus* pulmonary infection in a leukemic patient. Revta Iberoam. Micol. 13: 3-5.

Lopes, J.O., Pereira, D.V., Alves, S.H., Castro, M.S. & Benevenga, J.P., 1996. Pulmonary zygomycosis due to *Cunninghamella bertholletiae*: report of the first case in Brazil and review. Revta Iberoam. Micol. 13: 29-30.

Lopes, J.O., Pereira, D.V., Streher, L.A., Fenalte, A.A., Alves, S.H. & Benevanga, J.P., 1995b. Cutaneous zygomycosis caused by *Absidia corymbifera* in a leukemic patient. Mycopathologia 130: 89-92.

Lopes, J.O., Salebian, A., Alves, S.H. & Benevanga, J.P., 1994c. First case of human infection by *Trichophyton vanbreuseghemii* in Brazil. Revta Inst. Med. Trop. Sao Paulo 36: 379-380.

Lopes, M.M., Freitas, G. & Boiron, P., 2000. Potential utility of random polymorphic DNA (RAPD) and restriction endonuclease assay (REA) as typing systems for *Madurella mycetomatis*. Curr. Microbiol. 40: 1-5.

López, C., Ramos, L., Weisburd, G., Margasin, S. & Ramirez, R., 1996. Feohifomicosis causada por *Cladosporium cladosporioides*. Revta Iberoam. Micol. 13: 87-89.

Lopez, F.A., Crowley, R.S., Wastila, L., Valentine, H.A. & Remington, J.S., 1998. *Scedosporium apiospermum (Pseudallescheria boydii)* infection in a heart transplant recipient: a case of mistaken identity. J. Heart Lung Transpl. 17: 321-324.

Lopez-Jiménez, J., Cabezudo, E., Sousa, A., García-Larana, J., Velasco, J., Villalón, L., Perez-Oteyza, J., Odriozola, J., Cancelas, J. & Navarro, J.L., 1994. *Candida parapsilosis:* a yeast whose frequencey is increasing in bone marrow transplant recipients. Bone Marrow Transpl. Abstr. 184.

López-Martínez, R., Neumann, L. & González-Mendoza, A., 1999. Cutaneous penicilliosis due to *Penicillium chrysogenum*. Mycoses 42: 347-349.

López-Martínez, R., Soto-Hernández, J.L., Ostrosky-Zeichner, L., Castañon-Olivares, L.R., Angeles-Morales, V. & Sotelo, J., 1996. *Cryptococcus neoformans* var. *gattii* among patients with cryptococcal meningitis in Mexico. First observations. Mycopathologia 134: 61-64.

Lotery, A.J., Kerr, J.R. & Page, B.A., 1994. Fungal keratitis caused by *Scopulariopsis brevicaulis*: successful treatment with topical amphotericin

B and chloramphenicol without the need of surgical debridement. Br. J. Ophthalmol. 78: 730.

Lott, T.J., Kuykendall, R.J. & Reiss, E., 1993. Nucleotide sequence analysis of the 5.8S rDNA and adjacent ITS2 region of *Candida albicans* and related species. Yeast 9: 1199-1206.

Loudon, K.W., Coke, A.P., Burnie, J.P., Oppenheim, B.A. & Morris, C.Q., 1996. Kitchens as a source of *Aspergillus niger* infection. J. Hosp. Infect. 32: 191-198.

Louie, T., Baba, F.El, Shulman, M. & Jiminez-Lucho, V., 1994. Endogenous endophthalmitis due to *Fusarium:* case report and review. Clin. Infect. Dis. 18: 585-588.

Louthrenoo, W., Thampraser, K. & Sirisanthana, T., 1994. Osteoarticular penicilliosis marneffei. A report of eight cases and review of the literature. Br. J. Rheumatol. 33: 1145-1150.

Low, W.S., Seid, A.B., Pransky, S.M. & Kearns, D.B., 1996. *Coccidioides immitis* subperiosteal abscess of the temporal bone in a child. Arch. Otolaryngol. Head Neck Surg. 122: 189-192.

Lowinger-Seoane, M., Torres-Rodriguez, J.M., Madrenys-Brunet, N., Aregallfuste, S. & Saballs, P., 1992. Extensive dermatophytoses caused by *Trichophyton mentagrophytes* and *Microsporum canis* in a patient with AIDS. Mycopathologia 120: 143-146.

Lu, J.-J., Chen, C.-H., Bartlett, M.S., Smith, J.W. & Lee, C.-H., 1995. Comparison of six different PCR methods for detection of *Pneumocystis carinii.* J. Clin. Microbiol. 33: 2785-2788.

Lucas, G.M., Tucker, P. & Merz, W.G., 1999. Primary cutaneous *Aspergillus nidulans* infection associated with a Hickman catheter in a patient with neutropenia. Clin. Infect. Dis. 29: 1594-1596.

Lübeck, M., Alekhina, I.A., Lübeck, P.S., Jensen, D.F. & Bulat, S.A., 1999. Delimitation of *Trichoderma harzianum* into two different genetic groups by a higly robust fingerprint method, UP-PCR, and UP-PCR product cross-hybridization. Mycol. Res. 103: 289-298.

Lui, A.Y., Turett, G.S., Karter, D.L., Bellman, P.C. & Kislak, J.W., 1998. Amphotericin B lipid complex therapy in an AIDS patient with *Rhodotorula rubra* fungemia. Clin. Infect. Dis. 27: 892-893.

Lund, O.E., Miño de Kaspar, H. & Klauss, V., 1993. Strategie der Untersuchung und Therapie bei mykotischer Keratitis. Klin. MonatsBl. Augenheilk. 202: 188-195.

Lundell, E., 1969. *Microsporum cookei* Ajello in an eczematous skin lesion. Mykosen 12: 123-126.

Lundqvist, N., 1972. Nordic *Sordariaceae s. lat.* Symb. Bot. Upsal. 20: 1-374.

Luque, A.G., Mujica, M.T., Anna, M.L. d' & Alvarez, D.P., 1991. Micetoma podal por *Fusarium solani* (Mart.) Appel & Wollenweber. Bol. Micol. Valparaiso 6: 55-57.

Luque, A.G., Nanni, R. & Bracalenti, B.J.C. de, 1985. Mycotic keratitis caused by *Curvularia lunata* var. *aeria.* Mycopathologia 93: 9-12.

Lutwick, L.I., Phaff, H.J. & Stevens, D.A., 1980. *Kluyveromyces fragilis* as an opportunistic fungal pathogen in man. Sabouraudia 18: 69-73.

Lye, G.R., Wood, G. & Nimmo, G., 1996. Subcutaneous zygomycosis due to *Saksenaea vasiformis:* rapid isolate identification using a modified sporulation technique. Pathology 28: 364-365.

Lynch, J.P., Schaberg, D.R., Kissner, D.G. & Kauffman, C.A., 1981. *Cryptococcus laurentii* lung abscess. Am. Rev. Respir. Dis. 123: 135-138.

Maaten, J.C. ter, Golding, R.P., Schijndel, R.J.M.S. van & Thijs, L.G., 1995. Disseminated aspergillosis after near-drowning. Neth. J. Med. 47: 21-24.

MacGregor, R.R., Schimmer, B.M. & Steinberg, M.E., 1979. Results of combined amphotericin B 5-fluorocytosine therapy for prosthetic knee joint infected with *Candida parapsilosis.* J. Rheumatol. 6: 451-455.

Machado, A.A., Coelho, I.C.B., Roselino, A.M.F., Trad, E.S., Figueire-do, J.F. de C., Martinez, R. & Costa, J.C. de, 1991. Histoplasmosis in individuals with acquired immunodefiency syndrome (AIDS): report of six cases with cutaneous mucosal involvement. Mycopathologia 115: 13-18.

Maciejowska, Z. & Williams, E.B., 1963. Studies on morphological forms of *Staphylotrichum coccosporum.* Mycologia 55: 221-225.

Mackenzie, D.W.R., 1961. The extra-human occurrence of *Trichophyton tonsurans* in a residential school. Sabouraudia 1: 58-64.

Mackie, J.T., Kaufman, L. & Ellis, D., 1997. Confirmed histoplasmosis in an australian dog. Aust. Vet. J. 75: 362-363.

MacKinnon, J.E., 1951. Los agentes de maduromicosis de los géneros *Monosporium, Allescheria, Cephalosporium,* y otros de dudosa identidad. An. Fac. Med. Montevideo, Sec. Cient., 36: 153-180.

MacKinnon, J.E., 1954. A contribution to the study of the causal organisms of maduromycosis. Trans. R. Soc. Trop. Med. Hyg. 48: 470-480.

MacKinnon, J.E., 1970. On the importance of South American blastomycosis. Mycopathol. Mycol. Appl. 41: 187-183.

Maddy, K., 1957. Ecological factors possibly relating to the geographic distribution of *Coccidioides immitis.* Public Health Serv. Publ. 575: 144-157.

Madrigal, V., Alonso, J., Bureo, E., Figols, F.J. & Salesa, R., 1995. Fatal meningoencephalitis caused by *Scedosporium inflatum (Scedosporium prolificans)* in a child with lymphoblastic leukemia. Eur. J. Clin. Microbiol. Infect. Dis. 14: 601-603.

Magalhaes, O.M.C., Queiroz, L.A. de, Fernando, M.J., Souza, C.M. de & Rorres, L., 1996. *Aspergillus sydowii* isolated from two bronchial lavage samples. Bol. Micol. 11: 95-97.

Magnon, K.C., Jalbert, M. & Padhye, A.A., 1993. Osteolytic phaeohyphomycosis caused by *Phialemonium obovatum.* Arch. Pathol. Lab. Med. 117: 841-843.

Mahgoub, E.S., 1969. *Corynespora cassiicola*, a new agent of maduromycetoma. J. Trop. Med. Hyg. 7: 217-221.

Mahgoub, E.S., 1971. Mycological and serological studies on *Aspergillus flavus* isolated from paranasal aspergilloma in Sudan. J. Trop. Med. Hyg. 74: 162-165.

Mahgoub, E.S., 1973. Mycetomas caused by *Curvularia lunata, Madurella grisea, Aspergillus nidulans* and *Nocardia brasiliensis* in Sudan. Sabouraudia 11: 179-182.

Mahrous, M., Sawant, A.D., Pruitt, W.R., Lott, T., Meyer, S.A. & Ahearn, D.G., 1992. DNA relatedness, karyotyping and gene probing of *Candida tropicalis, Candida albicans* and its synonyms *Candida stellatoidea* and *Candida claussenii.* Eur. J. Epidemiol. 8: 444-451.

Mahul, P., Piens, M.-A., Guyotat, D., Godard, J., Archimbaud, E., Bui-Xuan, B. & Motin, J., 1989. Disseminated *Geotrichum capitatum* infection in a patient with acute myeloid leukemia. Mycoses 32: 573-577.

Maiwald, M., Kappe, R. & Sonntag, H.G., 1994. Rapid presumptive identification of medically relevant yeasts to the species level by polymerase chain reaction and restriction enzyme analysis. J. Med. Vet. Mycol. 32: 115-122.

Majoret, M., 1986. *Torulopsis ernobii, Torulopsis haemulonii:* levures opportunistes chez l'immunodéprimé? Bull. Soc. Fr. Mycol. Méd. 15: 143-146.

Makimura, K., Murayama, S.Y., Gotoh, M., Oguchi, A., Kamakura, M., Kinoshita, T., Yamanaka, M., Mitsuya, M., Wada, K., Suzuki, M., Uchida, K. & Yamaguchi, H., 1993. *Hansenula anomala* fungemia in a patient with acute myelocytic leukemia. Jpn. J. Med. Mycol. 34: 451-457.

Makimura, K., Murayama, S.Y. & Yamaguchi, H., 1994. Detection of a wide range of medically important fungi by the polymerase chain reaction. J. Med. Microbiol. 40: 358-364.

Makimura, K., Tamura, Y., Kudo, M., Uchida, K., Saito, H. & Yamaguchi, H., 2000. Species identification and strain typing of *Malassezia* species stock strains and clinical isolates based on the DNA sequences of nuclear ribosomal internal transcribed spacer 1 regions. J. Med. Microbiol. 49: 29-35.

Male, O. & Pehamberger, H., 1985. Die kutane Alternariose. Fallberichte und Literaturübersicht. Mykosen 28: 278-305.

Maleszka, R. & Ratka, P., 1999. Clinical and epidemiological aspects of various forms of fungal infections caused by *Trichophyton tonsurans.* Revta Iberoam. Micol. 15: 286-289.

Malkina, A.Y. & Agarunova, Y.S., 1973. Comparative studies of fungi belonging to *Cryptococcus* Vuill. genus isolated from human organisms and environment. Mikol. Fitopatol. 7: 194-197.

Malloch, D. & Salkin, I.F., 1984. A new species of *Scedosporium* associated with osteomyelitis in humans. Mycotaxon 21: 247-255.

Malloch, D. & Sigler, L., 1988. The *Eremomycetaceae (Ascomycotina).* Can. J. Bot. 66: 1929-1932.

Maloisel, F., Dufour, P., Waller, J., Herbrecht, R., Marcellin, L., Koenig, H., Liu, K.L., Weber, J.-C., Bergerat, J.-P. & Oberling, F., 1991. *Cunninghamella bertholletiae:* an uncommon agent of opportunistic fungal infection. Case report and review. Nouv. Rev. Fr. Hematol. 33: 311-315.

Manavathu, E.K., Vakulenko, S.B., Obedeanu, N. & Lerner, S.A., 1996. Isolation and characterization of a species-specific DNA probe for the detection of *Candida krusei*. Curr. Microbiol. 33: 147-151.

Mancianti, F. & Papini, R., 1997. Mating types of *Microsporum gypseum* complex isolated from animals in Italy. J. Mycol. Méd. 7: 87-89.

Manfredi, R., Mazzoni, A., Nanetti, A. & Chiodo, F., 1994. Histoplasmosis capsulati and duboisii in Europe: the impact of the HIV pandemic, travel and immigration. Eur. J. Epidemiol. 10: 675-681.

Manfredi, R., Salfi, N., Alampi, G., Mazzoni, A., Nanetti, A., Cillia, C. de & Chiodo, F., 1998. AIDS-related visceral aspergillosis: an underdiagnosed disease during life? Mycoses 41: 453-460.

Manns, B.J., Baylis, B.W., Urbanski, S.J., Gibb, A.P. & Rabin, H.R., 1996. Paracoccidioidomycosis: case report and review. Clin. Infect. Dis. 23: 1026-1032.

Manoch, L., Payapanon, A., Tubaki, K. & Sato, S., 1986. Taxonomic study of thermophilic fungi in Thailand. Trans. Mycol. Soc. Jpn. 27: 257-269.

Manso, E., Montillo, M., Frongia, G., Centurioni, R. & Murer, B., 1994. Rhinocerebral mucormycosis caused by *Absidia corymbifera*: an unusual localization in a neutropenic patient. J. Mycol. Méd. 4: 104-107.

Mansur, A.J., Safi, J., Ricardo, M., Markus, P., Aiello, V.D., Grinberg, M. & Pomerantzeff, P.M.A., 1996. Late failure of surgical treatment for bioprosthetic valve endocarditis due to *Candida tropicalis*. Clin. Infect. Dis. 22: 380-381.

Maran, A.G.D., Kwong, K., Milne, L.J.R. & Lamb, D., 1985. Frontal sinusitis caused by *Myriodontium keratinophilum*. Br. Med. J. 20: 207.

Marcano, C. & Hutton, 1973. Tinea nigra plantaris por *Clado-sporium* sp., segundo caso. Castellania 1: 129-131.

Marcelin-Little, D.J., Sellon, R.K., Kyles, A.E., Lemons, C.L. & Kaufman, L., 1996. Chronic localized osteomyelitis caused by an atyprical infection with *Blastomyces dermatitidis* in a dog. J. Am. Vet. Med. Ass. 209: 1877-1879.

Marchese, S.M. & Smoller, B.R., 1998. Cutaneous *Paecilomyces lilacunis* infection in a hospitalized patient taking corticosteroids. Int. J. Derm. 37: 433-453.

Marco, F., Pfaller, M.A., Messer, S. & Jones, R.N., 1998. *In vitro* activities of voriconazole (UK-109,496) and four other antifungal agents against 394 clinical isolates of *Candida* spp. Antimicrob. Agents Chemother. 42: 161-163.

Marcus, D.M., Hull, D.S., Rubin, R.M. & Newman, C.L., 1999. *Lecythophora mutabilis* endophthalmitis after long-term corneal cyanoacrylate. Retina 19: 351-353.

Marcus, L., Visner, H.F., Hoven, H.J. van der, Gove, E. & Meewes, P., 1992. Mycotic keratitis caused by *Curvularia brachyspora*. Mycopathologia 119: 29-33.

Maresca, B. & Kobayashi, G., 1989. Dimorphism in *Histoplasma capsulatum:* a model for the study of cell differentiation in pathogenic fungi. Microbiol. Rev. 53: 186-209.

Maresca, B. & Kobayashi, G.S. (eds), 1994. Molecular Biology of Pathogenic Fungi. A Laboratory Manual. Telos Press, New York, 577 pp.

Margolis, D.J., Weinberg, J.M., Tangoren, I.A., Cheney, R.T. & Johnson, B.L., 1998. Trichophytic granuloma of the vulva. Dermatology, Basel 197: 69-70.

Mariat, F., 1971. Adaption de *Ceratocystis stenoceras* (Robak) Moreau à la vie parasitaire chez l'animal. Étude de la souche sauvage et des mutants pathogènes. Comparaison avec *Sporothrix schenckii* Hektoen et Perkins. Revue Mycol. 36: 3-24.

Mariat, F., Destombes, P. & Segretain, G., 1977. The mycetomas: clinical features, pathology, etiology and epidemiology. Contrib. Microbiol. Immunol. 4: 1-39.

Marín, G. & Campos, R., 1984. Dermatofitosis por *Aphanoascus fulvescens*. Sabouraudia 22: 311-314.

Marin, J., Sanz, M.A., Sanz, G.F., Guarro, J., Martínez, M.L., Prieto, M., Guého, E. & Menezo, J.L., 1991. Disseminated *Scedosporium inflatum* in a patient with acute myeloblastic leukemia. Eur. J. Clin. Microbiol. Infect. Dis. 10:15-19.

Marjanková, K., Křivanek, K. & Zajiček, J., 1978. Mass occurrence of necrotic inflammation of the penis in ganders caused by phycomycetes. Mycopathologia 66: 21-26.

Markham, W.D., Key, R.D., Padhye, A.A. & Ajello, L., 1990. Phaeohyphomycotic cyst caused by *Tetraploa aristata*. J. Med. Vet. Mycol. 28: 147-150.

Marot-Leblond, A., Grimaud, L., Nail, S., Bouterige, S., Apaire-Marchais, V., Sullivan, D.J. & Robert, R., 2000. New monoclonal antibody specific for *Candida albicans* germ tube. J. Clin. Microbiol. 38: 61-67.

Marques, A.R., Kwon-Chung, K.J., Holland, S.M., Turner, M.L. & Gallin, J.I., 1995. Suppurative cutaneous granulomata caused by *Microascus cinereus* in a patient with chronic granulomatous disease. Clin. Infect. Dis. 20: 110-114.

Marques, S.A., Franco, S.R., Camargo, R.M. de, Dias, L.D., Haddad, V. & Fabris, V.E., 1993. Sporotrichosis of the domestic cat (*Felis catus*): human transmission. Rev. Inst. Med. Trop. Sao Paulo 35: 327-330.

Marriot, D.J.E., Wong, K.H., Aznar, E., Harkness, J.L., Cooper, D.A. & Muir, D., 1997. *Scytalidium dimidiatum* and *Lecythophora hoffmannii*: unusual causes of fungal infections in a patient with AIDS. J. Clin. Microbiol. 35: 2949-2952.

Maršálek, E., Zika, Z., Riha, V., Dušek, J. & Dvorácek, C., 1960. Plicni aspergilóza s generalizací vyvolaná druhem *Aspergillus restrictus*. Lék. Ces. 99: 1285-1292.

Marshall, D.H., Brownstein, S., Jackson, W.B., Mintsioulis, G., Gilberg, S.M. & Al-Zeerah, B.F., 1997. Post-traumatic corneal mucormycosis caused by *Absidia corymbifera*. Ophthalmology 104: 1107-1111.

Martino, P., Girmenia, C., Micozzi, A., Raccah, R., Gentile, G., Venditti, M. & Mandelli, F., 1993. Fungemia in patients with leukemia. Am. J. Med. Sci. 306: 225-232.

Martino, P., Venditti, M., Micozzi, A., Morace, G., Polonelli, L., Mantovani, M.P., Petti, M.C., Burgio, V.L., Santini, C. & Serra, P., 1990. *Blastoschizomyces capitatus:* an emerging cause of invasive fungal disease in leukemic patients. Rev. Infect. Dis. 12: 570-582.

Martins, T.B., Jaskowski, T.D., Mouritsen, C.L. & Mill, H.R., 1995. Comparison of commercially available enzyme immunoassay with traditional serological tests for detection of antibodies to *Coccidioides immitis*. J. Clin. Microbiol. 33: 940-943.

Maruyama, R., Katoh, T. & Nishioka, K., 1999. A case of Unna-Thost disease accompanied by *Epidermophyton floccosum* infection. J. Derm., Tokyo 26: 63-66.

Marzec, A., Heron, L.G., Pritchard, R.C., Butcher, R.H., Powell, H.R., Disney, A.P.S. & Tosolini, F.A., 1993. *Paecilomyces variotii* in peritoneal dialysate. J. Clin. Microbiol. 31: 2392-2395.

Masclaux, F., Guého, E., Hoog, G.S. de & Christen, R., 1995. Phylogenetic relationships of human-pathogenic *Cladosporium (Xylohypha)* species inferred from partial LS rRNA sequences. J. Med. Vet. Mycol. 33: 327-338.

Maslen, M.M., Collis, T. & Stuart, R., 1996. *Lasiodiplodia theobromae* isolated from a subcutaneous abscess in a Cambodian immigrant to Australia. J. Med. Vet. Mycol. 34: 279-283.

Masuda, M., Naka, W., Tajima, S., Harada, T., Nishikawa, T., Kaufman, L. & Standard, P., 1989. Deoxyribonucleic acid hybridization studies of *Exophiala dermatitidis* and *Exophiala jeanselmei*. Microbiol. Immunol. 33: 631-639.

Mateev, G., Kantjardjiev, T., Vassileva, S. & Tsankov, N., 1993. Chronic mucocutaneous candidosis with osteomyelitis of the frontal bone. Int. J. Derm. 32: 888-889.

Mathews, M.S. & Kuriakose, T., 1995. Keratitis due to *Cephaliophora irregularis* Thaxter. J. Med. Vet. Mycol. 33: 359-360.

Mathews, M.S. & Maharajan, S.V., 1999. *Exserohilum rostratum* causing keratitis in India. Med. Mycol. 37: 131-132.

Mathews, M.S., Mukundan, U., Lalitha, M.K., Aggarwal, S., Chandy, S.M., Padhye, A.A. & Ewing, E.P., 1993. Subcutaneous zygomycosis caused by *Saksenaea vasiformis* in India. A case report and review of the literature. J. Mycol. Méd. 3: 95-98.

Mathews, M.S. & Prabhakar, S., 1995. Chronic meningitis caused by *Trichosporon beigelii* in India. Mycoses 38: 125-126.

Mathews, M.S., Raman, A. & Nair, A., 1997. Nosocomial zygomycotic post-surgical necrotizing fasciitis in a healthy adult caused by *Apophysomyces elegans* in South India. J. Med. Vet. Mycol. 35: 61-63.

Mathur, P.N., 1967. Studies on members of *Sphaeropsidales* among Indian soil fungi and morphological and cultural studies on some members of *Plectascales* from soils in India. Thesis, University of Agra, Agra.

Matruchot, L., 1910. Sur un nouveau champignon pathogène pour l'homme. C.R. Hebd. Séanc. Acad. Sci., Paris 150: 543-545.

Matsuda, T. & Matsumoto, T., 1986. Disseminated hyalo-hyphomycosis in a leukemia patient. Arch. Derm. 122: 1171-1175.

Matsui, T., Nishimoto, K., Udagawa, S.-I., Ishihara, H. & Ono, T., 1999. Subcutaneous phaeohyphomycosis caused by *Phaeoacremo-nium rubrigenum* in an immunosuppressed patient. Jpn. J. Med. Mycol. 40: 99-102.

Matsumoto, T., Masaki, J. & Okabe, T., 1979. *Cylindrocarpon tonkinense* as a cause of keratomycosis. Trans. Br. Mycol. Soc. 72: 503-504.

Matsumoto, T. & Matsuda, T., 1988. Critical review of hyalo-hyphomycosis caused by *Fusarium* species. Proc. 10th Congr. ISHAM, Barcelona, pp. 292-296.

Matsumoto, T., Matsuda, T., McGinnis, M.R. & Ajello, L., 1993. Clinical and mycological spectra of *Wangiella dermatitidis* infections. Mycoses 36: 145-155.

Matsumoto, T., Nishimoto, K., Kimura, K., Padhye, A.A., Ajello, L., & McGinnis, M.R., 1984a. Phaeohyphomycosis caused by *Exophiala moniliae*. Sabouraudia 22: 17-26.

Matsumoto, T., Padhye, A.A. & Ajello, L., 1983a. *In vitro* hair perforation by a new subvariety of *Trichophyton tonsurans* var. *sulfureum*. Mycotaxon 18: 235-242.

Matsumoto, T., Padhye, A.A. & Ajello, L., 1983b. Successful mating of *Microsporum distortum* and *Nannizzia otae*. Trans. Br. Mycol. Soc. 81: 645-650.

Matsumoto, T., Padhye, A.A. & Ajello, L., 1987. Medical significance of the so-called black yeasts. Eur. J. Epidemiol. 3: 87-95.

Matsumoto, T., Padhye, A.A., Ajello, L. & McGinnis, M.R., 1986. *Sarcinomyces phaeomuriformis:* a new dematiaceous hyphomycete. J. Med. Vet. Mycol. 24: 395-400.

Matsumoto, T., Padhye, A.A., Ajello, L. & Standard, P.G., 1984b. Critical review of human isolates of *Wangiella dermatitidis*. Mycologia 76: 232-249.

Matsumoto, T. & Soejima, N., 1976. Keratomycosis. Mykosen 19: 217-222.

Matsuo, T., Nakagawa, H. & Matsuo, N., 1995. Endogenous *Aspergillus* endophthalmitis associated with peridontitis. Ophthalmologica 209: 109-111.

Matsuzaki, O., Yasuda, M. & Ichinohe, M., 1988. Keratomycosis due to *Glomerella cingulata*. Revta Ibér. Micol. 5, Suppl. 1: 30.

Matte, S.M.W., Lopes, J.O., Melo, I.S. & Costa Beber, A.A., 1997. A focus of favus due to *Trichophyton schoenleinii* in Rio Grande do Sul, Brasil. Revta Inst. Med. Trop. S. Paulo 39: 1-3.

Matthews, R.C., 1996. Comparative assessment of the detection of candidal antigens as a diagnostic tool. J. Med. Vet. Mycol. 34: 1-10.

Mayayo, E., Pujol, I. & Guarro, J., 1999. Experimental pathogenicity of four opportunist *Fusarium* species in a murine model. J. Med. Microbiol. 48: 363-366.

Mayser, P., Haze, P., Papavassilis, C., Pickel, M., Gruender, K. & Guého, E., 1997. Differentiation of *Malassezia* species: selectivity of cremophor EL, castor oil and ricinoleic acid for *M. furfur*. Br. J. Derm. 137: 208-213.

Mayser, P., & Pape, B., 1998. Decreased susceptibility of *Malassezia furfur* to UV light by synthesis of tryptophane derivatives. Antonie van Leeuwenhoek 73: 315-319.

Mayser, P., Wille, P., Imkampe, A., Arnold, N. & Monsees, T., 1998. Synthesis of fluorochromes and pigments in *Malassezia furfur* by use of tryptophan as the single nitrogen source. Mycoses 41: 265-271.

Mazade, M.A., Margolin, M.F., Rossman, S.N. & Edwards, M.S., 1998. Survival from pulmonary infection with *Cunninghamella bertholletiae:* case report and review of the literature. Pediatr. Infect. Dis. J. 17: 835-839.

Mazars, E. & Dei-Cas, E., 1998. Epidemiological and taxonomic impact of *Pneumocystis* biodiversity. FEMS Immunol. Med. Microbiol. 22: 75-80.

McAleer, R., 1988. Four opportunistic fungal brain lesions. Revta Ibér. Micol. 5, Suppl. 1: 36.

McCormack, J.G., McIntyre, P.B., Tilse, M.H. & Ellis, D.H., 1987. Mycetoma associated with *Acremonium falciforme* infection. Med. J. Aust. 147: 187-188.

McCray, E., Rampell, N., Solomon, S.L., Bond, W.W., Martone, W.J. & O'Day, D., 1986. Outbreak of *Candida parapsilosis* endophthalmitis after cataract extraction and intraocular lens implantation. J. Clin. Microbiol. 24: 625-628.

McCullough, M.J., Clemons, K.V., McCusker, J.H. & Stevens, D.A., 1998a. Intergenic transcribed spacer PCR ribotyping for differentiation of *Saccharomyces* species and interspecific hybrids. J. Clin. Microbiol. 36: 1035-1038.

McCullough, M.J., Clemons, K.V., McCusker, J.H. & Stevens, D.A., 1998b. Species identification and virulence attributes of *Saccharomyces boulardii* (nom. inval.). J. Clin. Microbiol. 36: 2613-2617.

McDonald, J.A. & Saulsbury, F.T., 1997. Chronic *Candida albicans* otitis media in children with immunodeficiency. Pediatr. Infect. Dis. J. 16: 529-531.

McDonnell, M. & Isaacs, D., 1995. Neonatal systemic candidiasis. J. Pediatr. Child Health 31: 490-492.

McDonnell, P.J., Werblin, T.P., Sigler, L. & Green, W.R., 1984. Mycotic keratitis due to *Beauveria alba*. Cornea 3: 213-216.

McDonough, E.S. & Lewis, A.L., 1968. The ascigerous state of *Blastomyces dermatitidis*. Mycologia 60: 76-83.

McGinnis, M.R., 1978a. Taxonomy of *Exophiala jeanselmei* (Langeron) McGinnis & Padhye. Mycopath. Mycol. Appl. 65: 79-87.

McGinnis, M.R., 1978b. Human pathogenic species of *Exophiala, Phialophora*, and *Wangiella*. PAHO Sci. Publ. 356: 37-59.

McGinnis, M.R., 1996. Mycetoma. Dermatol. Clinics 14: 97-104.

McGinnis, M.R. & Ajello, L., 1974. A new species of *Exophiala* isolated from channel catfish. Mycologia 66: 518-520.

McGinnis, M.R. & Ajello, L., 1982. A note on *Sporotrichum gougerotii* Matruchot 1910. Mycotaxon 16: 232-238.

McGinnis, M.R. & Borelli, D., 1981. *Cladosporium bantianum* and its synonym *Cladosporium trichoides*. Mycotaxon 13: 127-136.

McGinnis, M.R., Borelli, D., Padhye, A.A., & Ajello, L., 1986a. Reclassification of *Cladosporium bantianum* in the genus *Xylohypha*. J. Clin. Microbiol. 23: 1148-1151.

McGinnis, M.R., Campbell, G., Gourley, W.K. & Lucia, H.L., 1992. Phaeohyphomycosis caused by *Bipolaris spicifera:* an informative case. Eur. J. Epidemiol. 8: 383-386.

McGinnis, M.R., Gams, W. & Goodwin, M.N., 1986b. *Phialemonium obovatum* infection in a burned child. J. Med. Vet. Mycol. 24: 51-55.

McGinnis, M.R., Lemon, S.M., Walker, D.H., Hoog, G.S. de & Haase, G., 1999. Fatal cerebritis caused by a new species of *Cladophialophora*. Stud. Mycol. 43: 166-171.

McGinnis, M.R., McKenzie, R.A. & Connole, M.D., 1985. *Phaeosclera dematioides*, a new etiologic agent of phaeohyphomycosis in cattle. Sabouraudia 23: 133-135.

McGinnis, M.R., Midez, J., Pasarell, L. & Haque, A., 1993. Necrotizing fasciitis caused by *Apophysomyces elegans*. J. Mycol. Méd. 3: 175-179.

McGinnis, M.R., Nordoff, N.G., Ryder, N.S. & Nunn, G.B., 2000. *In vitro* comparison of terbinafine and itraconazole against *Penicillium marneffei*. Antimicrob. Agents Chemother. 44: 1407-1408.

McGinnis, M.R. & Padhye, A.A., 1977. *Exophiala jeanselmei*, a new combination for *Phialophora jeanselmei*. Mycotaxon 5: 341-352.

McGinnis, M.R. & Padhye, A.A., 1978. *Cladosporium castellanii* is a synonym of *Stenella araguata*. Mycotaxon 7: 415-418.

McGinnis, M.R., Padhye, A.A. & Ajello, L., 1982a. *Pseudallescheria* Negroni et Fischer, 1943 and its later synonym *Petriellidium* Malloch, 1970. Mycotaxon 14: 94-102.

McGinnis, M.R. & Pasarell, L., 1998a. *In vitro* evaluation of terbinafine and itraconazole against dematiaceous fungi. Med. Mycol. 36: 243-246.

McGinnis, M.R. & Pasarell, L., 1998b. *In vitro* testing of susceptibilities of filamentous ascomycetes to voriconazole, itra-conazole, and amphotericin B, with consideration of phylogenetic implication. Antimicrob. Agents Chemother. 36: 2353-2355.

McGinnis, M.R., Rinaldi, M.G. & Winn, R.E., 1986c. Emerging agents of phaeohyphomycosis: pathogenic species of *Bipolaris* and *Exserohilum*. J. Clin. Microbiol. 24: 250-259.

McGinnis, M.R. & Schell, W.A., 1980. The genus *Fonsecaea* and its relationship to the genera *Cladosporium, Phialophora, Rami-chloridium,* and *Rhinocladiella.* PAHO Sci. Publ. 396: 215-234.

McGinnis, M.R., Schell, W.A. & Carson, J., 1985. *Phaeoannellomyces* and the *Phaeococcomycetaceae,* new dematiaceous blastomycete taxa. Sabouraudia 23: 179-188.

McGinnis, M.R., Severo, L.C., Kalil, R. & Falleiro, P.T., 1994. Endocarditis caused by *Fusarium pallidoroseum.* J. Mycol. Méd. 4: 45-47.

McGinnis, M.R., Sorell, D.F., Miller, R.L. & Kaminski, G.W., 1981. Subcutaneous phaeohyphomycosis caused by *Exophiala moniliae.* Mycopathol. Mycol. Appl. 73: 69-72.

McGinnis, M.R., Walker, D.H., Dominy, I.E. & Kaplan, W., 1982b. Zygomycosis caused by *Cunninghamella bertholletiae.* Arch. Path. Lab. Med. 106: 282-286.

McGinnis, M.R., Walker, D.H. & Foids, J.D., 1980. *Hansenula polymorpha* infection in a child with chronic granulomatous disease. Arch. Pathol. Lab. Med. 104: 290-292.

McGough, D.A., 1993. Clinical and laboratory aspects of the 'black yeasts'. Clin. Microbiol. Newsl. 15: 145-152.

McGough, D.A., Fothergill, A.W., Kusne, S., Furukukawa, H. & Rinaldi, M.G., 1994. *Trichoderma koningii:* yet another new agent of contemporary mycoses. Abstr. Gen. Meet. ASM 94: 602.

McGough, D.A., Fothergill, A.W. & Rinaldi, M.G., 1990. *Cokeromyces recurvatus* Poitras, a distinctive zygomycete and potential pathogen: criteria for identification. Clin. Microbiol. News 12: 113-117.

McGough, D.A., Praser, H., Bourbeau, P., Naumovita, D., Shah, N., Fothergill, A.W. & Rinaldi, M.G., 1991. Fatal disseminated infection in an acute myeloblastic leukemic caused by *Myceliophthora thermophila* (Apinis) van Oorschot, a rare agent of mycosis. Proc. 10th Congr. ISHA-M, Montreal, p. 115.

McKemy, J.M. & Morgan-Jones, G., 1991. Studies in the genus *Cladosporium sensu lato.* IV. Concerning *Cladosporium oxysporum,* a plurivorous, predominantly saprophytic species in warm climates. Mycotaxon 51: 397-405.

McKenzie, D.W.R., 1988. Host resistance and predisposing factors in human mycoses. Quaderni Coop. Sanit. 8: 25-34.

McKenzie, D.W.R., 1990. *Pneumocystis carinii:* a nomadic taxon. In H. VandenBossche *et al.* (eds): Mycoses in AIDS Patients, pp. 55-63. Plenum, New York.

McKenzie, R.A. & Connole, M.D., 1977. Mycotic nasal granuloma in cattle. Am. Vet. J. 53: 260-270.

McNamee, C.J., Wang, S. & Modry, D., 1998. Purulent pericarditis secondary to *Candida parapsilosis* and *Peptostreptococccus* species. Can. J. Cardiol. 14: 85-86.

Medina, A.L., Redondo, J.A.D. & Nebrida, L.M., 1998. Two unusual cases of mycoses in the Philippines. Proc. 4th China Japan Int. Congr. Mycol. p. 88.

Mehrotra, B.S. & Baijal, U., 1963. Species of *Mortierella* from India - III. Mycopath. Mycol. Appl. 20: 50-54.

Mehta, S.A., Kaul, S., Mehta, M.S. Kelkar, R.S. & Mehta, A.R., 1993. Phaeohyphomycosis of the paranasal sinuses masquerading as a neoplasm: a case report. Head Neck Surg. 15: 59-61.

Meiller, T.F., Jabra-Rizk, M.A., Baqui, A.A.M.A., Kelley, J.I., Meeks, V.I., Merz, W.G. & Falkler, W.A., 1999. Oral *Candida dubliniensis* as a clinically important species in HIV-seropositive patients in the United States. Oral Surg. Oral Path. Oral Rad. Endodont. 88: 573-580.

Meireles, M.C.A., Riet-Correa, F., Fischman, O., Zambrano, A.F.H., Zambrano, M.S. & Ribeiro, G.A., 1993. Cutaneous pythiosis in horses from Brazil. Mycoses 36: 139-142.

Meis, J.F.G.M., Ruhnke, M., Pauw, B.E. de, Odds, F.C., Siegert, W. & Verweij, P.E., 1999. *Candida dubliniensis* candidemia in patients with chemotherapy-induced neutropenia and bone marrow transplantation. Emerg. Infect. Dis. 5: 150-153.

Meis, J.F.G.M., Kullberg, B.-J., Pruszczynski, M. & Veth, R.P.H., 1994. Severe osteomyelitis due to the zygomycete *Apophysomyces elegans.* J. Clin. Microbiol. 32: 3078-3081.

Melcher, G.P., McGough, D.A., Fothergill, A.W., Norris, C. & Rinaldi, M.G., 1993. Disseminated hyalohyphomycosis caused by a novel human pathogen, *Fusarium napiforme.* J. Clin. Microbiol. 31: 1461-1467.

Melchers, W.J.G., Verweij, P.E., Hurk, P. van den, Belkum, A. van, Pauw, B.E. de, Hoogkamp-Korstanje, J.A.A. & Meis, J.F.G.M., 1994. General primer-mediated PCR for detection of *Aspergillus* species. J. Clin. Microbiol. 32: 1710-1717.

Meletiadis, J., Meis, J.F.G.M., Horré, R. & Verweij, P.E., 1999. *In vitro* antifungal activity of six drugs against 13 clinical isolates of *Ochroconis gallopava.* Stud. Mycol. 43: 206-208.

Meletiadis, J., Mouton, J.W., Rodriguez-Tudela, J.L., Meis, J.F.G.M. & Verweij, P.E., 2000. *In vitro* interaction of terbinafine with itraconazole against clinical isolates of *Scedosporium prolificans.* J. Clin. Microbiol. 44: 470-472.

Melo, J.C., Srinivasan, S., Scott, M.L. & Raff, M., 1980. *Cryptococcus albidus* meningitis. J. Infect. 2: 79-82.

Mencl, K., Otcenásek, M., Spacek, J. & Rehulová, E., 1985. *Aspergillus restrictus* and *Candida parapsilosis.* Erreger von Endokarditiden nach Herzklappentransplantationen. Mykosen 28: 127-134.

Mendel, E., Milefchik, E.N., Ahmadi, J. & Gruen, P., 1994. Coccidioidomycotic brain abscess. J. Neurosurg. 80: 140-142.

Mendoza, L., Ajello, L. & McGinnis, M.R., 1996. Infections caused by the oomycetous pathogen *Pythium insidiosum.* J. Mycol. Méd. 6: 151-164.

Mendoza, L., Alfaro, A.A. & Villalobos, J., 1988. Bone lesions caused by *Pythium insidiosum* in a horse. J. Med. Vet. Mycol. 26: 5-12.

Mendoza, L., Hernandez, F. & Ajello, L., 1993. Life cycle of the human and animal Oomycete pathogen *Pythium insidiosum.* J. Clin. Microbiol. 31: 2967-2973.

Mendoza, L. & Marin, G., 1989. Antigenic relationship between *Pythium insidiosum* de Cock et al. 1987 and its synonym *Pythium destruens* Shipton 1987. Mycoses 32: 73-77.

Mercano, C. & Hutton, B., 1973. Tinea nigra plantaris por *Cladosporium* sp., segundo caso. Castellania 1: 129-131.

Mercantini, R., Marsella, R., Moretto, D., Mercantini, P., Balus, L., Mastroianni, A. & Ferraro, C., 1995. Macroscopic and microscopic characteristics of an African *Blastomyces dermatitidis* strain. Mycoses 38: 477-480.

Merz, W.G., 1984. *Candida lusitaniae:* frequency of recovery, colonization, infection and amphotericin B resistance. J. Clin. Microbiol. 20: 1194-1195.

Merz, W.G., Karp, J.E., Schron, D. & Saral, R., 1986. Increased incidence of fungemia caused by *Candida krusei.* J. Clin. Microbiol. 24: 581-584.

Meyer, K.C., McManus, E.J. & Maki, D.G., 1993. Overwhelming pulmonary blastomycosis associated with the adult respiratory distress syndrome. N. Engl. J. Med. 329: 1231-1236.

Meyer, S.A., Payne, R.W. & Yarrow, D., 1998. Candida Berkhout. In C.P. Kurtzman & J.W. Fell (eds): The Yeasts, a Taxonomic Study, 4th ed., pp. 454-573.

Meyers, W.M., Dooley, J.R. & Kwon-Chung, K.J., 1975. Mycotic granuloma caused by *Phialophora repens.* Am. J. Clin. Path. 64: 549-555.

Meyohas, M.-C., Roux, P., Bollens, D., Chouaid, C., Rozenbaum, W., Meynard, J.-L., Poirot, J.-L., Frottier, J. & Mayaud, C., 1995. Pulmonary cryptococcosis: localized and disseminated infections in 27 patients with AIDS. Clin. Infect. Dis. 21: 628-633.

Michel-Nguyen, A., Favel, A., Penaud, A., Trousson, F. & Carlotti, A., 1996. Unusual case of *Candida lusitaniae* infection in a severely immunocompromised patient: multiple localizations of a single infective strain documented by restriction endonuclease analysis. J. Mycol. Méd. 6: 172-177.

Mickelsen, P.A., Viano-Paulson, M.C., Stevens, D.A. & Diaz, P., 1988. Clinical and microbiological features of infection with *Malassezia pachydermatis* in high-risk infants. J. Infect. Dis. 157: 1163-1168.

Middelhoven, W.J., Guého, E. & Hoog, G.S. de, 2000. Phylogenetic position and physiology of *Cerinosterus cyanescens.* Antonie van Leeuwenhoek 77: 313-320.

Midgley, G., 1993. Morphological variation in *Malassezia* and its significance in pityriasis versicolor. In H. VandenBossche *et al.* (eds): Dimorphic Fungi in Biology and Medicine, pp. 267-277. Plenum, New York.

Midha, N.K., Mirzanejad, Y. & Soni, M., 1996. *Colletotrichum* sp.: plant or human pathogen? Antimicr. Inf. Dis. Newsl. 15: 26-27.

Migrino, R.Q., Hall, G.S. & Longworth, D.L., 1995. Deep tissue infections caused by *Scopulariopsis brevicaulis*: report of a case of prosthetic valve endocarditis and review. Clin. Infect. Dis. 21: 672-674.

Miller, G.R., Rebell, G., Magoon, R.C., Kulvin, S.M. & Forster, R.K., 1978. Intravitreal antimycotic therapy and the cure of mycotic endophthalmitis caused by a *Paecilomyces lilacinus* contaminated pseudophakos. Ophthalmic Surg. 9: 54-63.

Miller, W.T., Sais, G.J., Gefter, W.B., Aronchick, J.M. & Miller, W.T., 1994. Pulmonary aspergillosis in patients with AIDS: clinical and radiographic correlations. Chest 105: 37-44.

Millon, L., Reboux, G. & Barale, T., 1994. Emergence de *Candida glabrata* et *Candida krusei* chez des patients séropositifs pour le VIH atteints de candidose oropharyngée, traités de façon prolongée par le fluconazole. J. Mycol. Méd. 4: 90-92.

Millon, L., Reboux, G., Drobacheff, C., Claude, B., Comparot, S., Monod, M. & Barale, T., 1997. Isolements répétés de *Kluyveromyces lactis* de la cavité buccale de patients VIH positives. J. Mycol. Méd. 7: 28-32.

Min, Y.-G., Kim, S.H., Lee, K.-S., Kang, M.-K. & Han, M.H., 1996. *Aspergillus* sinusitis: clinical aspects and treatment outcomes. Otolaryngol. Head Neck Surg. 115: 49-52.

Ming, Y.-n. & Yu, T.-f., 1966. Identification of a *Fusarium* species, isolated from corneal ulcer. Acta Microbiol. Sin. 12: 180-186.

Miro, O., Sacanella, E., Nadal, P., Lluch, M.M., Nicolas, J.M., Milla, J. & Urbano-Marquez, A., 1994. *Trichosporon beigelii:* fungemia and metastatic pneumonia in a trauma patient. Eur. J. Clin. Microbiol. Infect. Dis. 13: 604-606.

Mirza, S.H., 1996. Cryptococcal meningitis in non-immunocompromised patients: report of two cases. Saudi Med. J. 16: 569-571.

Mischel, P.S. & Vinters, H.V., 1995. Coccidioidomycosis of the central nervous system: neuropathological and vasculopathic manifestations and clinical correlates. Clin. Infect. Dis. 20: 400-405.

Misra, P.C., Srivastava, K.J. & Latas, K., 1979. *Apophysomyces*, a new genus of the *Mucorales*. Mycotaxon 8: 377-382.

Misra, V.C. & Randhawa, H.S., 1976. *Sporobolomyces salmonicolor* var. *fischerii*, a new yeast. Arch. Microbiol. 108: 141-143.

Mitchell, A.J., Salomon, A.R., Beneke, E.S. & Anderson, T.F., 1983. Subcutaneous alternariosis. J. Am. Acad. Derm. 8: 673-676.

Mitchell, D.H., Sorell, T.C., Allworth, A.M., Heath, C.H., McGregor, A.R., Papansoum, K., Richards, M.J. & Gottlieb, T., 1995. Cryptococcal disease of the CNS in immunocompetent hosts: influence of cryptococcal variety on clinical manifestations and outcome. Clin. Infect. Dis. 20: 611-616.

Mitchell, D.M., Fitz-Henley, M. & Horner-Bryce, J., 1990. A case of disseminated phaeohyphomycosis caused by *Cladosporium devriesii*. West Ind. Med. J. 39: 118-123.

Mitchell, R.G., Chaplin, A.J. & Mackenzie, D.W.R., 1987. *Emericella nidulans* in a maxillary sinus fungal mass. J. Med. Vet. Mycol. 25: 339-341.

Mitchell, T.G., Freedman, E.Z., White, T.J. & Taylor, J.W., 1994. Unique oligonucleotide primers in PCR for identification of *Cryptococcus neoformans*. J. Clin. Microbiol. 32: 253-255.

Mitchell, T.G. & Perfect, J.R., 1995. Cryptococcosis in the era of AIDS: 100 years after the discovery of *Cryptococcus neoformans*. Clin. Microbiol. Rev. 8: 515-548.

Mitchell, T.G., White, T.J. & Taylor, J.W., 1992. Comparison of 5.8S ribosomal DNA sequences among the basidiomycetous yeast genera *Cystofilobasidium, Filobasidium* and *Filobasidiella*. J. Med. Vet. Mycol. 30: 207-218.

Mittag, H., 1993. The fine structure of *Hortaea werneckii*. Mycoses 36: 343-350.

Miyaji, M. & Nishimura, K., 1985. Conidial ontogenesis of pathogenic black yeasts and their pathogenicity for mice. Proc. Ind. Soc. Acad. Sci., Plant Sci. 94: 437-451.

Miyakawa, Y. & Mabuchi, T., 1994. Characterization of a species-specific DNA fragment originating from the *Candida albicans* mitochondrial genome. J. Med. Vet. Mycol. 32: 71-75.

Miyakawa, Y., Mabuchi, T. & Fukazawa, Y., 1993. New method for detection of *Candida albicans* in human blood by polymerase chain reaction. J. Clin. Microbiol. 31: 3344-3347.

Miyamoto, T., Sasaoka, R., Kawaguchi, M., Ishioka, S., Inoue, T., Yamada, N. & Mihare, M., 1998. *Scedosporium apiospermum* skin infection: a case report and review of the literature. J. Am. Acad. Derm. 39: 498-500.

Miyazaki, E., Sugisaki, K., Shigenaga, T., Matsumoto, T., Kita, S., Inobe, Y. & Tsuda, T., 1995. A case of acute eosinophilic pneumonia caused by inhalation of *Trichosporon terrestre*. Am. J. Respir. Crit. Care Med. 151: 541-543.

Mizuki, M., Chikuba, K. & Tanaka, K., 1994. A case of chronic necrotizing pulmonary aspergillosis due to *Aspergillus nidulans*. Mycopathologia 128: 75-79.

Mochizuki, T., Kawasaki, M., Ishizaki, H. & Makimura, K., 1999. Identification of several clinical isolates of dermatophytes based on the nucleotide sequence of internal transcribed spacer 1 (ITS1) in nuclear ribosomal DNA. J. Derm., Tokyo 26: 276-281.

Mochizuki, T., Takada, K., Watanabe, S., Kawasaki, M. & Ishizaki, H., 1990. Taxonomy of *Trichophyton interdigitale* (*Trichophyton mentagrophytes* var. *interdigitale*) by restriction enzyme analysis of mitochondrial DNA. J. Med. Vet. Mycol. 28: 191-196.

Mochizuki, T., Uehara, M., Menon, T. & Ranganathan, S., 1996a. Minipreparation of total cellular DNA is useful as an alternative molecular marker of mitochondrial DNA for the identification of *Trichophyton mentagrophytes* and *T. rubrum*. Mycoses 39: 31-35.

Mochizuki, T., Watanabe, S. & Uehara, M., 1996b. Genetic homogeneity of *Trichophyton mentagrophytes* var. *interdigitale* isolated from geographically distant regions. J. Med. Vet. Mycol. 34: 139-143.

Mohammedi, I., Gachot, B., Grossin, M., Marchie, C., Wolff, M. & Vachon, F., 1995. Overwhelming myocarditis due to *Fusarium oxysporum* following bone marrow transplantation. Scand. J. Infect. Dis. 27: 643-644.

Mohl, W., Lerch, M.M., Klotz, M., Freidank, H. & Zeitz, M., 1998. Infection of an intravenous port system with *Metschnikowia pulcherrima* Pitt *et* Miller. Mycoses 41: 425-426.

Moissenet, D., Marsol, P. & Thieu, H.V., 1995. Isolement répété d'*Exophiala dermatitidis* dans des hémocultures au cathéter. J. Mycol. Méd. 5: 179-181.

Mok, T., Koehler, A.P., Yu, M.Y., Ellis, D.H., Johnson, P.J. & Wickham, N.W.R., 1997. Fatal *Penicillium citrinum* pneumonia with pericarditis in a patient with acute leukemia. J. Clin. Microbiol. 35: 2654-2656.

Moll, H.D., Schumacher, J. & Hoover, T.R., 1992. Entomophthoromycosis conidiobolae in a llama. J. Am. Vet. Med. Ass. 200: 969-970.

Moneymaker, C.S., Shenep, J.L. & Pearson, T.A., 1986. Primary cutaneous phaeohyphomycosis due to *Exserohilum rostratum (Drechslera rostrata)* in a child with leukemia. Pediatr. Infect. Dis. J. 5: 380-382.

Monod, M., Togni, G., Hube, B. & Sanglard, D., 1994. Multiplicity of genes encoding secreted aspartic proteinases in *Candida* species. Molec. Microbiol. 13: 357-368.

Montagna, M.T., Tortorano, A.M., Fiore, L., Ingletti, A.M. & Barbuti, S., 1997a. *Cryptococcus neoformans* var. *gattii* en Italie. Note I. Premier cas autochtone de méningite à sérotype B chez un sujet VIH positif. J. Mycol. Méd. 7: 90-92.

Montagna, M.T., Viviani, M.A., Pulito, A., Aralla, C., Tortorano, A.M., Fiore, L. & Barbati, S., 1997b. *Cryptococcus neoformans* var. *gattii* in Italy. Note II. Environmental investigation related to an autochtonous clinical case in Apulia. J. Mycol. Méd. 7: 93-96.

Monte, S.M. de la & Hutchins, G.M., 1985. Disseminated *Curvularia* infection. Arch. Path. Lab. Med. 109: 872-874.

Montenegro, J., Aguirre, R., González, O., Martinez, I. & Saracho, R., 1995. Fluconazole treatment of *Candida* peritonitis with delayed removal of the peritoneal dialysis catheter. Clin. Nephrol. 44: 60-63.

Montenegro, M.R., 1995. Host parasite relationship in paracoccidioidomycosis. Jpn. J. Med. Mycol. 36: 209-213.

Montero, A., Cohen, J.E., Fernandez, M.A., Mazzolini, G., Gomez, R. & Perugini, J., 1998. Cerebral pseudallescheriasis due to *Pseudallescheria boydii* as the first manifestation of AIDS. Clin. Infect. Dis. 26: 1476-1477.

Montes, M.A., DiNisco, S., Dry, S. & Galvanek, E., 1998. Fine needle aspiration cytology of primary isolated splenic *Blastomyces dermatitidis:* a case report. Acta Cytol. 42: 396-398.

Mooney, J.E. & Wanger, A., 1993. Mucormycosis of the gastrointestinal tract in children: report of a case and review of the literature. Pediatr. Infect. Dis. J. 12: 872-876.

Moore, B.R., Reed, S.M., Kowalski, J.J. & Bertone, J.J., 1993. Aspergillosis granuloma in the mediastrinum of a non-immunocompromised horse. Cornell Vet. 83: 97-104.

Moore, C.B., Walls, C.M., Denning, D.W., 2000. *In vitro* activity of the new triazole BMS-207147 against *Aspergillus* species in comparison with itraconazole and amphotericin B. Antimicrob. Agents Chemother. 44: 441-443.

Moore, C.K., Hellreich, M.A., Coblentz, C.L. & Roggli, V.L., 1988. *Aspergillus terreus* as a cause of invasive pulmonary aspergillosis. Chest 94: 889-891.

Moore, M., 1935. Head infection caused by a new *Hemispora: H. coremiformis.* Annls Bot. Gdn 22: 317-334.

Moore, M.K., 1986. *Hendersonula toruloidea* and *Scytalidium hyalinum* infections in London, England. J. Med. Vet. Mycol. 24: 219-230.

Moore, M.K., 1988. Morphological and physiological studies of isolates of *Hendersonula toruloidea* Nattrass cultured from human skin and nail samples. J. Med. Vet. Mycol. 26: 25-39.

Moore, M.K., 1992. The infection of human skin and nail by *Scytalidium* species. Curr. Topics Med. Mycol. 4: 1-42.

Moore, R.T., 1987. Micromorphology of yeasts and yeast-like fungi and its taxonomic implications. Stud. Mycol. 30: 203-226.

Morace, G., Sanguinetti, M., Posterado, B., Cascio, G. & Fadda, G., 1997. Identification of various medically important *Candida* species in clinical specimens by PCR-restriction enzyme analysis. J. Clin. Microbiol. 35: 667-672.

Moraes, M., Borelli, D. & Feo, M., 1967. *Microsporum amazonicum* nova species. Med. Cutánea 11: 281-286.

Mordue, J.E.M., 1967. CMI Descriptions of Pathogenic Fungi and Bacteria 131, 132, 133.

Mordue, J.E.M., 1971. CMI Descriptions of Pathogenic Fungi and Bacteria 315, 316, 317.

Moreau, P., Zahar, J.-R., Milpied, N., Baron, O., Mahe, B., Wu, D., Germaud, P., Despins, P., Delajarte, A.-Y. & Harousseau, J.-L., 1993. Localized invasive pulmonary aspergillosis in patients with neutropenia: effectiveness of surgical resection. Cancer, Philad. 72: 3223-3226.

Moreno-Ancillo, A., Diaz-Pena, J.-M., Ferrer, A., Martin-Munoz, F., Martin-Barroso, J.-A., Martin-Esteban, M. & Ojeda, J.-A., 1996. Allergic bronchopulmonary cladosporiosis in a child. J. All. Clin. Immunol. 97: 714-715.

Morenz, J., 1963. *Geotrichum candidum* Link. Taxonomie, Diagnose und medizinische Bedeutung. Mykol. SchrReihe 1: 1-79.

Morenz, J., 1970. Geotrichosis. Handb. Spez. Path. Anat. Hist. 3: 919-952.

Morgan, M. & Reves, R., 1996. Invasive sinusitis due to *Sporothrix schenckii* in a patient with AIDS. Clin. Infect. Dis. 23: 1319-1320.

Morgan-Jones, G., 1974. Notes on Hyphomycetes, V. A new thermophilic species of *Acremonium.* Can. J. Bot. 52: 429-431.

Morgan-Jones, G., Culbreath, A.K. & Rodriguez-Kabana, R., 1984. Notes on Hyphomycetes. XLIX. *Xenokylindria obovata*, a new species isolated from diseased eggs of the nematode *Meloidogyne arenaria*, and *X. prolifera.* Mycotaxon 20: 599-606.

Morganti, L., Padhye, A.A. & Ajello, L., 1975. Recovery of *Nannizzia grubyia* from a stray Italian cat *(Felis catus).* Mycologia 67: 434-436.

Morhart, M., Rennie, R., Ziola, B., Bow, E. & Louie, T.J., 1994. Evaluation of enzyme immunoassay for *Candida* cytoplasmic antigens in neutropenic cancer patients. J. Clin. Microbiol. 32: 766-776.

Mori, A., Moriwaki, H. & Kagi, M., 1961. A case of chromoblastomycosis due to *Hormiscium dermatitidis.* Skin Res., Osaka 3: 158.

Mori, T., Ebe, T., Takahashi, M., Kohara, T., Isomuma, H. & Matsumura, M., 1993. Clinical aspects of penicilliosis, a rare infection. Jpn. J. Med. Mycol. 34: 145-153.

Mori, T., Matsumura, M., Kohara, T., Wantabe, Y., Ishiyama, T., Wakabayashi, Y., Ikemoto, H., Watanabe, A., Tanno, M., Shirai, T. & Ichinoe, M., 1987. A fatal case of pulmonary penicilliosis. Jpn. J. Med. Mycol. 28: 341-348.

Mori, T., Matsumura, M., Yamada, K., Irie, S., Oshimi, K., Suda, K., Oguri, T. & Ichinoe, M., 1998. Systemic aspergillosis caused by an aflatoxin-producing strain of *Aspergillus flavus.* Med. Mycol. 36: 107-112.

Morin, O., Germaud, P., Miegeville, M. & Milpied, N., 1986. Mycose pulmonaire à *Penicillium purpurogenum*, à propos d'une observation chez un malade immunodéprimé. Bull. Soc. Fr. Mycol. Méd. 15: 441-448.

Morin, O., Milpield, N., Audovin, A.F. & Maillot, M., 1988. Mycose opportuniste invasive à *Acremonium strictum* chez un malade atteint de myelome. Bull. Soc. Fr. Mycol. Méd. 17: 357-362.

Morner, T., Avenas, A. & Mattson, R., 1999. Adiaspiromycosis in a European beaver from Sweden. J. Wildlife Dis. 35: 367-370.

Morquer, R. & Enjalbert, L., 1957. Étude morphologique et physiologique d'un *Aspergillus* nouvellement isolé au cours d'une affection pulmonaire de l'homme. C.R. Acad. Sci., Paris 244: 1405-1408.

Morquer, R., Lombard, C., Berthelon, M. & Lacoste, L., 1965. Pouvoir pathogène des *Mucorales* dans le règne animal. Une nouvelle mycose chez les bovidés et les porcins. C.R. Acad. Sci., Paris 260: 6173-6176.

Morris, A., Schell, W.A., McDonagh, D., Chaffee, S. & Perfect, J.R., 1995. Pneumonia due to *Fonsecaea pedrosoi* and cerebral abscesses due to *Emericella nidulans* in a bone marrow transplant recipient. Clin. Infect. Dis. 21: 1346-1348.

Morris, J.T., Beckius, M.L. & McAllister, C.K., 1991. *Sporobolomyces* infection in an AIDS patient. J. Infect. Dis. 164: 623-624.

Morris, J.T., Beckius, M.L., Jeffery, B.S., Longfeld, R.N., Heaven, R.F. & Baker, W.J., 1995. Lung mass caused by *Phoma* species. Infect. Dis. Clin. Pract. 4: 58-59.

Morris, J.T. & McAllister, K.C., 1993. Fungemia due to *Torulopsis glabrata.* South. Med. J. 86: 356-357.

Morrison, V.A. & Weisdorf, D.J., 1993. *Alternaria:* a rhinonasal pathogen of immunocompromised hosts. Clin. Infect. Dis. 16: 265-270.

Morschhäuser, J., Köhler, G. & Hacker, J., 1996. Gibt es Pathogenitätsfaktoren bei Pilzen? Mycoses 39, Suppl. 1: 51-54.

Morschhäuser, J., Ruhnke, M., Michel, S. & Hacker, J., 1999. Identification of CARE-2-negative *Candida albicans* isolates as *Candida dubliniensis.* Mycoses 42: 29-32.

Morton, F.J. & Smith, G., 1963. The genera *Scopulariopsis* Bainier, *Microascus* Zukal, and *Doratomyces* Corda. Mycol. Pap. 86: 1-96.

Morton, S.J., Midthun, K. & Merz, W.G., 1986. Granulomatous encephalitis caused by *Bipolaris hawaiiensis.* Arch. Pathol. Lab. Med. 110: 1183-1185.

Mós, E., Birgel, E.H., Araújo, W. & Mendes, M.J., 1978. Mamite bovina devida a levedura do gênero *Candida.* Revta Fac. Med. Vet. Zootec. Univ. S. Paulo 15: 161-164.

Moses, J.S., Balachandron, C., Sandhanam, S., Ratnasamy, N., Thanappan, S., Rajaswar, J. & Moses, D., 1990. Ocular rhinosporidiosis in Tamil Nadu, India. Mycopathologia 111: 5-8.

Mostaza, J.M., Barrado, F.J., Fernández-Martin, J., Peña-Yañez, J. & Vazquez-Rodriguez, J.J., 1989. Cutaneoarticular mucor-mycosis due to *Cunninghamella bertholletiae* in a patient with AIDS. Rev. Infect. Dis. 11: 316-318.

Mouchacca, J., 1973a. Deux *Alternaria* des sols arides d'Égypte: *A. chamydosporum* sp. nov. et *A. phragmospora* van Emden. Mycopath. Mycol. Appl. 50: 217-225.

Mouchacca, J., 1973b. Espècies nouvelles et communes de *Drechslera*, isolées de sols de régions arides. Revue. Mycol. 38: 102-108.

Mouchacca, J., 1987. Quelques micromycètes intéressants observés sur des feuilles vivantes ou mortes de *Carpinus betulus* L. Crypt. Mycol. 8: 141-158.

Mouchacca, J. & Gams, W., 1993. The hyphomycete genus *Cladorrhinum* and its teleomorph connections. Mycotaxon 48: 415-440.

Moulias, S, Hazouard, E., Delain, M., Barrabes, A., Thérizol-Ferly, M. & Legras, A., 1998. Fongémie prolongée a _Acremonium recifei_ après autogreffe de moelle. J. Mycol. Méd. 8: 26-29.

Moulsdale, M.T., Harper, J.M. & Thatcher, G.N., 1981. Fungal peritonitis: complication of continuous ambulatory peritoneal dialysis. Med. J. Aust. 1: 88.

Moussongo, J. & Miegeville, M., 1998. Teignes à _Trichophyton soudanense_. Enquête familiale à partir de plusieurs cas isolés au Centre Hospitalier Universitaire de Nantes. Enquête scolaire dans le district de Nantes. J. Mycol. Méd. 8: 18-20.

Mouy, R., Ropert, J.C., Donadieu, J., Hubert, P., Blic, J. de, Revillon, Y., Brunelle, F., Schollet-Martin, S., Descamps, B., Debré, M., Griscelli, C. & Fischer, A., 1995. Granulomatose septique chronique révélée par une aspergillose néonatale. Arch. Pediatr. 2: 861-864.

Mouzin, E.K. & Beilke, M.A., 1996. Female genital blastomycosis: case report and review. Clin. Infect. Dis. 22: 718-719.

Müller, E., 1950. Die schweizerischen Arten der Gattung _Leptosphaeria_ und ihrer Verwandten. Sydowia 4: 185-319.

Müller, G., 1964. Die Gattung _Sporotrichum_ Link. I. Wiss. Z. Humboldt-Univ. Berlin 13: 843-860.

Müller, G., 1965. Die Gattung _Sporotrichum_ Link. II. Wiss. Z. Humboldt-Univ. Berlin 14: 753-798.

Müller, G.H., Kaplan, W., Ajello, L. & Padhye, A.A., 1975. Phaeohyphomycosis caused by _Drechslera spicifera_ in a cat. J. Am. Vet. Med. Ass. 166: 150-154.

Müller, J., 1994. Pathogenese, Immunbiologie und Epidemiologie der Cryptococcose. Mycoses 37, Suppl. 1: 34-42.

Muhm, M., Zuckermann, A., Prokesch, R., Pammer, J., Hiesmayr, M. & Haider, W., 1996. Early onset of pulmonary mucormycosis with pulmonary vein thrombosis in a heart transplant recipient. Transplantation 62: 1185-1187.

Muir, D., Martin, P., Kendall, K. & Malik, R., 1998. Invasive hyphomycotic rhinitis in a cat due to _Metarhizium anisopliae_. J. Med. Vet. Mycol. 36: 51-54.

Mukhopadhyay, D., Ghosh, L.M., Thammayya, A. & Sanyal, M., 1995. Entomophthoromycosis caused by _Conidiobolus coronatus_: clinicomycological study of a case. Auris Nasus Larynx 22: 139-142.

Mukhtar, A.U., 1999. Rhinosporidiosis of the conjunctiva. Centr. Afr. J. Med. 45: 20-21.

Munipalli, B., Rinaldi, M.G. & Greenberg, S.B., 1996. _Cokeromyces recurvatus_ isolated from pleural and peritoneal fluid: case report. J. Clin. Microbiol. 34: 2601-2603.

Muñoz, F., Demmler, G.J., Travis, W.R., Ogden, A.K., Rossmann, S.N. & Rinaldi, M.G., 1997. _Trichoderma longibrachiatum_ infection in a pediatric patient with aplastic anemia. J. Clin. Microbiol. 35: 499-503.

Muñoz, M.C., González, M. & Alvárez, E., 1989. Placentitis micótica y aborto por _Aspergillus ochraceus_ en una vaca. Rev. Salud Animal 11: 190-195.

Muñoz, P., Garcia-Leoni, M.E., Berenguer, J., Bernaldo de Queiros, J.C.L. & Bouza, E., 1989. Catheter-related fungemia by _Hansenula anomala_. Arch. Intern. Med. 143: 709-713.

Muotoe-Okafor, F.A. & Gugnani, H.C., 1993. Isolation of _Lecythophora mutabilis_ and _Wangiella dermatitidis_ from the fruit eating bat, _Eidolon helvum_. Mycopathologia 122: 95-100.

Murakawa, G.J., Kerschmann, R. & Berger, T., 1996. Cutaneous _Cryptococcus_ infection and AIDS. Arch. Derm. 132: 545-548.

Muralidhar, S. & Sulthana, C.M., 1995. _Rhodotorula_ causing chronic dacryocystitis: a case report. Ind. J. Ophthalmol. 43: 196-198.

Murdock, C.B., Fisher, J.F., Loebl, D. & Chew, W.H., 1983. Osteomyelitis of the hand due to _Torulopsis holmii_. South. Med. J. 76: 1460-1461.

Murphy, A. & Kavanagh, K., 1999. Emergence of _Saccharomyces cerevisiae_ as a human pathogen: implications for biotechnology. Microb. Techn. 25: 551-557.

Murphy, N., Damjanovic, V., Hart, C.A., Buchanan, C.R., Whitaker, R. & Cooke, R.W.I., 1986. Infection and colonization of neonates by _Hansenula anomala_. Lancet 8476: 291-293.

Murphy, J.W., 1999. Immunological down-regulation of host defenses in fungal infections. Mycoses 42, Suppl. 2: 37-43.

Murray, I.G., Dunkerley, G.E. & Hughes, K.E.A., 1963. A case of Madura foot caused by _Phialophora jeanselmei_. Sabouraudia 3: 175-177.

Mussa, A.Y., Singh, V.K., Randhawa, H.S. & Khan, Z.U., 1998. Disseminated fatal trichosporonosis: first case due to _Trichosporon inkin_. J. Mycol. Méd. 8: 196-199.

Mylonakis, E., Barlam, T.F., Flanigan, T. & Rich, J.D., 1998. Pulmonary aspergillosis and invasive disease in AIDS: review of 342 cases. Chest 114: 251-262.

Mylonakis, E., Rich, J.D., Flanigan, T., Kwakwa, H., Orchis, D.F. de, Boyce, J. & Mileno, M.D., 1996. Muscle abscess due to _Aspergillus fumigatus_ in a patient with AIDS. Clin. Infect. Dis. 23: 1323-1324.

Nachman, S., Alpan, O., Malowitz, R. & Spitzer, E.D., 1996. Catheter-associated fungemia due to _Wangiella (Exophiala) dermatitidis_. J. Clin. Microbiol. 34: 1011-1013.

Nagahama, T., Sato, H., Shimazu, M. & Sugiyama, J., 1995. Phylogenetic divergence of the entomophthoralean fungi: evidence from nuclear 18S ribosomal RNA gene sequences. Mycologia 87: 203-209.

Nagai, H., Yamakami, Y., Hashimoto, A., Tokimatsu, I. & Nasu, M., 1999. PCR detection of DNA specific for _Trichosporon_ species in serum of patients with disseminated trichosporonosis. J. Clin. Microbiol. 37: 694-699.

Naglić, T., Hajsig, D., Herceg, M., Hajsig, M. & Huber, I., 1986. Systemic mycoses in domestic and wild ruminants. III. Gastric phycomycosis in domestic and zebu cattle, red deer and reindeer. Vet. Arhiv. 56: 255-260.

Naguib, M.T., Huycke, M.M., Pederson, J.A., Pennington, L.R., Burton, M.E. & Greenfield, R.A., 1995. _Apophysomyces elegans_ infection in a renal transplant recipient. Am. J. Kidney Dis. 26: 381-384.

Naidu, J. & Singh, S.M., 1992. Hyalohyphomycosis caused by _Paecilomyces variotii:_ a case report, animal pathogenicity and 'in vitro' sensitivity. Antonie van Leeuwenhoek 62: 225-230.

Naidu, J. & Singh, S.M., 1994. _Aspergillus chevalieri_ (Mangin) Thom and Church: a new opportunistic pathogen of human cutaneous aspergillosis. Mycoses 37: 271-274.

Naidu, J., Singh, S.M. & Pouranik, M., 1991. Onychomycosis caused by _Chaetomium globosum_ Kunze. Mycopathologia 113: 31-34.

Naiff, R.D., Barrett, T.V., Farias Naiff, M. de, Lima Ferreira, L.C. de & Arias, J.R., 1996. New records of _Histoplasma capsulatum_ from wild animals in the Brazilian Amazon. Revta Inst. Med. Trop. S. Paulo 38: 273-277.

Naim-ur-Rahman, Jamjoom, A., Al-Hedaithy, S.S.A., Jamjoom, Z.A.B., Al-Sohaibani, O. & Aziz, S.A., 1996. Cranial and intracranial aspergillosis of sinonasal origin. Acta Neurochir. 138: 944-950.

Naim-ur-Rahman, Mahgoub, E.S. & Chagla, A.H., 1988. Fatal brain abscesses caused by _Ramichloridium obovoideum:_ report of three cases. Acta Neurochirurg. 93: 92-95.

Naka, W., Harada, T., Nishikawa, T. & Fukushiro, R., 1986. A case of chromoblastomycosis with special reference to the mycology of the isolated _Exophiala jeanselmei_. Mykosen 29: 445-452.

Naka, W., Masuda, M., Konahana, A., Shinoda, T. & Nishikawa, T., 1995. Primary cutaneous cryptococcosis and _Cryptococcus neoformans_ serotype D. Clin. Exper. Derm. 20: 221-225.

Naka, W. & Nishikawa, T., 1995. _Fonsecaea pedrosoi_ isolated from skin crusts of Bowen's disease. Mycoses 38: 127-129.

Nakamura, B., Weil, W.B. & Kaufman, D.B., 1989. Fatal fungal peritonitis in an adolescent on continuous ambulatory peritoneal dialysis: association with deferoxamine. Pediatr. Nephrol. 3: 80-82.

Nakamura, Y., Sato, H., Watanabe, S., Takahashi, H., Koide, K. & Hasegawa, A., 1996. _Sporothrix schenckii_ isolated from a cat in Japan. Mycoses 39: 125-128.

Nakase, T., Itoh, M., Suzuki, M., Komagata, K. & Kodama, T., 1998. _Candida palmioleophila_ sp. nov., a yeast capable of assimilating crude palm oil, formerly identified as _Torulopsis candida_. J. Gen. Appl. Microbiol., Tokyo 34: 493-498.

Nakase, T. & Suzuki, M., 1985. Taxonomic studies on _Debaryomyces hansenii_ (Zopf) Lodder _et_ Kreger-van Rij and related species. II. Practical discrimination and nomenclature. J. Gen. Appl. Microbiol. 31: 71-86.

National Committee of Clinical and Laboratory Standards, 1997. Reference method for broth dilution antifungal susceptibility of yeasts. Approved standard M27-A. Wayne.

National Committee of Clinical and Laboratory Standards, 1998. Reference method for broth dilution antifungal susceptibility of conidium-forming filamentous fungi; proposed standard. NCCLS document M38-P. Wayne.

Navarro, D., Monzonis, E., Lopez-Ribot, J.L., Sepulveda, P., Casanova, M., Nogueira, J.M. & Martinez, J.P., 1993. Diagnosis of systemic candidiasis by enzyme immunoassay detection of specific antibodies to mycelial phase cell wall and cytoplasmic candidal antigens. J. Clin. Microbiol. Infect. Dis. 12: 839-846.

Nayak, B.C., Rao, A.G., Ray, S.K. & Chanda, S.K., 1975. Mycotic ruminitis in a calf. Indian Vet. J. 52: 56-57.

Nayeri, F., Cameron, R., Chryssanthou, E., Johansson, L. & Soderstrom, C., 1997. *Candida glabrata* prosthesis infection following pyelonephritis and septicaemia. Scand. J. Infect. Dis. 29: 635-638.

Nazir, Z., Hasan, R., Pervaiz, S., Alam, M. & Moazam, F., 1997. Invasive retroperitoneal infection due to *Basidiobolus ranarum* with response to potassium iodide - case report and review of the literature. Annls Trop. Paediatr. 17: 161-164.

Ndiaye, B., Develoux, M., Dieng, M.T., Ndir, O., Huerre, M. & Raphenon, G., 1998. l'Histoplasmose à *Histoplasma capsulatum* var. *duboisii* au Sénégal. À propos de deux cas dont un associé à l'infection à HLTV1. J. Mycol. Méd. 8: 163-166.

Neame, P. & Rayner, D., 1960. Mucormycosis. A report of twenty-two cases. Arch. Pathol., Chicago 70: 261-268.

Neergaard, P., 1945. Danish Species of *Alternaria* and *Stemphylium*. Munksgaard, Copenhagen, 560 pp.

Neglia, J.P., Hurd, D.D., Ferrieri, P. & Snover, D.C., 1987. Invasive *Scopulariopsis* in the immunocompromised host. Am. J. Med. 83: 1163-1166.

Negroni, P., 1933. Onychomycose par *Cephalosporium spinosus* n. sp. Negroni 1933. C.R. Séanc. Soc. Biol., B. Aires 113: 478-480.

Negroni, P., 1943. El problema de las onixis micóticas no específicas. Revta Argent. Dermatosif. 24: 194-199.

Negroni, P., 1985. *Drechslera cactivora*, nuevo agente de micosis oportunista de las uñas. Revta Argent. Micol. 8: 17-20.

Negroni, P., 1993. Paracoccidioidomycosis (South American blastomycosis, Lutz's mycosis). Int. J. Derm. 32: 847-859.

Neijens, H.J., Frenkel, J., Muinck Keizer-Schrama, S.M.P.F. de, Dzoljic-Danilovic, G., Meradji, M. & Dongen, J.J.M. van, 1990. Invasive *Aspergillus* infection in chronic granulomatous disease: treatment with itraconazole. J. Pediatr. 115: 1016-1019.

Nelson, P.E., 1992. Taxonomy and biology of *Fusarium moniliforme*. Mycopathologia 117: 29-36.

Nelson, P.E., Dignani, M.C. & Anaissie, E.J., 1994. Taxonomy, biology, and clinical aspects of *Fusarium* species. Clin. Microbiol. Rev. 7: 479-504.

Nelson, P.E., Tousson, T.A. & Cook, R.J., 1981. *Fusarium*: diseases, biology, and taxonomy. Pennsylvania State Univ. Press, London, 457 pp.

Nelson, P.E., Tousson, T.A. & Marasas, W.F.O., 1983. *Fusarium* species. An illustrated manual for identification. Pennsylvania State Univ. Press, London, 192 pp.

Nenoff, P., Gütz, U., Tintelnot, K., Bosse-Henck, A., Mierzwa, M., Hofmann, J., Horn, L.-C. & Haustein, U.-F., 1996. Disseminated mycosis due to *Scedosporium prolificans* in an AIDS patient with Burkitt lymphoma. Mycoses 39: 461-465.

Nenoff, P., Kellermann, S., Schober, R., Nenning, H., Kubel, M., Winkler, J. & Haustein, U.-F., 1998. Rhinocerebral zygomycosis following bone marrow transplantation in chronic myelogenous leukaemia. Report of a case and review of the literature. Mycoses 41: 365-372.

Neuhann, T., 1976. Clotrimazol in der Behandlung von Keratomykosen. Klin. Monatsbl. Augenheilkd. 169: 459-462.

Neumeister, B., Bartmann, P., Gaedicke, G. & Marre, R., 1992. A fatal infection due to *Fusarium oxysporum* in a child with Wilms' tumour: case report and review of the literature. Mycoses 35: 115-119.

Neumeister, B., Rockemann, M. & Marre, R., 1992. Fungaemia due to *Candida pelliculosa* in a case of acute pancreatitis. Mycoses 35: 309-310.

Neumeister, B., Zollner, T.M., Krieger, D., Sterry, W. & Marre, R., 1995. Mycetoma due to *Exophiala jeanselmei* and *Mycobacterium chelonae* in a 73-year-old man with idiopathic CD4[+] T lymphocytopenia. Mycoses 38: 271-276.

Newmark, E. & Polack, F.M., 1970. *Tetraploa* keratomycosis. Am. J. Ophthalmol. 70: 1013-1015.

Ng, K.H., Chin, C.S., Jalleh, R.D., Siar, C.H., Ngui, C.H. & Singaram, S.P., 1991. Nasofacial zygomycosis. Oral Surg. Oral Med. Oral Pathol. 72: 685-688.

Ng, K.H. & Siar, C.H., 1996. Review of oral histoplosmosis in Malaysians. Oral Surg. Oral Med. Oral Pathol. Oral Rad. Endodont. 81: 303-307.

Ng, V.L., Yajko, D.M., McPhaul, L.W., Gartner, I., Byford, B., Goodman, C.D., Nassos, P.S., Sanders, C.A., Howes, E.L., Leoung, G., Hopewell, P.C. & Hadley, W.K., 1990. Evaluation of an indirect fluorescent-antibody stain for detection of *Pneumocystis carinii* in respiratory samples. J. Clin. Microbiol. 28: 975-979.

Nguyen, M.H., Peacock, J.E., Morris, A.J., Tanner, D.C., Nguyen, M.L., Snijdman, D.R., Wagener, M.M., Rinaldi, M.G. & Yu, V.L., 1996. The changing face of candidemia: emergence of non-*Candida albicans* species and antifungal resistance. Am. J. Med. 100: 617-623.

Nho, S., Anderson, M.J., Moore, C.B. & Denning, D.W., 1997. Species differentiation by internally trancribed spacer PCR and HhaI digestion of fluconazole-resistant *Candida krusei, Candida inconspicua*, and *Candida norvegensis* strains. J. Clin. Microbiol. 35: 1036-1039.

Niamba, P., Weill, F.X., Sarlangue, J., Labreze, C., Couprie, B. & Taieb, A., 1998. Is common neonatal pustulosis (neonatal acne) triggered by *Malassezia sympodialis*? Arch. Derm. 134: 995-998.

Nicand, E., Buisson, Y., Auzanneau, G., Improvisi, L. & Dupont, B., 1993. Septicémie à *Debaryomyces hansenii (Candida famata)*, levure pathogène opportuniste. J. Mycol. Méd. 3: 242-244.

Nicot, J. & Meyer, J., 1956. Un hyphomycète nouveau des sols tropicaux: *Staphylotrichum coccosporum* nov. gen., nov. sp. Bull. Soc. Mycol. Fr. 72: 318-323.

Nielsen, H., Bentsen, K.D., Hojtved, L., Willemoes, E.H., Scheutz, F., Schiodt, M., Stoltze, K. & Pindborg, J.J., 1994. Oral candidiasis and immune status of HIV-infected patients. J. Oral Path. Med. 23: 140-143.

Nielsen, H., Stenderup, J., Bruun, B. & Ladefoged, J., 1990. *Candida norvegensis* peritonitis and invasive disease in a patient on continuous ambulatory peritoneal dialysis. J. Clin. Microbiol. 28: 1664-1665.

Nielsen, H. & Stenderup, J., 1996. Invasive *Candida norvegensis* infection in immunocompromised patients. Scand. J. Infect. Dis. 28: 311-312.

Nielsen, H.S. & Conant, N.F., 1968. A new human pathogenic *Phialophora*. Sabouraudia 6: 228-231.

Nielsen, H.S., Conant, N.F., Weinberg, T. & Reback, J.F., 1968. Report of a mycetoma due to *Phialophora jeanselmei*. Sabouraudia 6: 330-333.

Nielsen, K., Lang, H., Chum, A.C., Woodruff, K. & Cherry, J.D., 1993. Disseminated *Scedosporium prolificans* infection in an immunocompromised adolescent. Pediatr. Infect. Dis. J. 12: 882-884.

Niesters, H.G.M., Goessens, W.H.F., Meis, J.F.M.G & Quint, W.G.V., 1993. Rapid, polymerase chain reaction-based identification assays for *Candida* species. J. Clin. Microbiol. 31: 904-910.

Niewerth, M., Splanemann, V., Korting, H.C., Ring, J. & Abeck, D., 1998. Antimicrobial susceptibility testing of dermatophytes. Comparison of the agar macrodilution and broth microdilution test. Chemotherapy 44: 31-35.

Nimmo, G.R., Whiting, R.F. & Strong, R.W., 1988. Disseminated mucormycosis due to *Cunninghamella bertholletiae* in a liver transplant recipient. Postgrad. Med. J. 64: 82-84.

Ninin, E., Morin, O., Tortorec, S. le, Milpied, N., Moreau, P. & Harousseau, J.L., 1997. Infection invasive à *Candida lipolytica* après allogreffe de moelle osseuse. J. Mycol. Méd. 7: 212-214.

Niño, F.L. & Freire, R.S., 1964. Existencia de un foco endemico de rhinosporidioisis en la provincia del Chaco. V. Estudio de nuevas observaciones y consideraciones finales. Mycopath. Mycol. Appl. 24: 92-102.

Nirenberg, H.I., 1990. Recent advances in the taxonomy of *Fusarium*. Stud. Mycol. 32: 91-101.

Nirenberg, H.I. & O'Donnell, K., 1998. New *Fusarium* species and combinations within the *Gibberella fujikuroi* species complex. Mycologia 90: 434-458.

Nishida, H., Ando, K., Ando, Y., Hirata, A. & Sugiyama, J., 1995. *Mixia osmundae:* transfer from the **Ascomycota** to the *Basidiomycota* based on evidence from molecules and morphology. Can. J. Bot. 73: S660-S666.

Nishida, T., Mayumi, H., Kawachi, Y., Tokunaga, S., Murayama, A., Yasui, H. & Tokunaga, K., 1994. The efficacy of fluconazole in treating prosthetic valve endocarditis caused by *Candida glabrata*: report of a case. Surg. Today, Tokyo 24: 651-654.

Nishikawa, A., Tomomatsu, H., Sugita, T., Ikeda, R. & Shinoda, T., 1996. Taxonomic position of clinical isolates of *Candida famata*. J. Med. Vet. Mycol. 34: 411-419.

Nishimura, K., Miyaji, M., Taguchi, H., Wang, D.L., Li, R.Y. & Meng, Z.H., 1989. An ecological study on pathogenic dematiaceous fungi in China. Proc. 4th Int. Symp. Res. Center of Path. Fungi Microbiol. Toxic, Tokyo, pp. 17-20.

Nityananda, K., Sivasubramaniam, P. & Ajello, L., 1962. Mycotic keratitis caused by *Curvularia lunata.* Case report. Sabouraudia 2: 35-39.

Nityananda, K., Sivasubramaniam, P. & Ajello, L., 1964. A case of mycotic keratitis caused by *Curvularia geniculata*. Arch. Ophthalmol. 71: 456-458.

Nitzulescu, V. & Niculescu, M., 1975. Ophtalmopathie déterminée par *Candida humicola*. Arch. Roum. Pathol. Exp. Microbiol. 34: 357-361.

Noble, R.C., Salgado, J., Newell, S.W. & Goodman, N.L., 1997. Endophthalmitis and lumbar diskitis due to *Acremonium falciforme* in a splenectomized patient. Clin. Infect. Dis. 24: 277-278.

Noguchi, H., Hiruma, M. & Kawada, A., 1999. Sporotrichosis succesfully treated with itraconazole in Japan. Mycoses 42: 571-576.

Nohinek, B., Zee-Cheng, C.S., Barnes, W., Dall, L. & Gibbs, H.R., 1987. Infective endocarditis of a bicuspid aortic valve caused by *Hansenula anomala*. Am. J. Med. 82: 165-168.

Norris, H.A., Elewski, B.E. & Ghannoum, M.A., 1999. Optimal growth conditions for the determination of the antifungal susceptibility of three species of dermatophytes with the use of a microdilution method. J. Am. Acad. Derm. 40: 9-13.

Nucci, M., Pulcheri, W., Spector, N., Bueno, A.P., Bacha, P.C., Caiuby, M.J., Derossi, A., Costa, R., Morais, J.C. & Oliviera, H.P. de, 1995. Fungal infections in neutropenic patients: an 8-year prospective study. Revta Inst. Med. Trop. S. Paulo 37: 397-406.

Nucci, M., Spector, N., Lucena, S., Bacha, P.C., Pulcheri, W., Lamosa, A., Derossi, A., Caiuby, M.J., Macieira, J. & Oliveira, H.P., 1992. Three cases of infection with *Fusarium* species in neutropenic patients. Eur. J. Clin. Microbiol. Infect. Dis. 11: 1160-1162.

Nussbaum, E.S. & Hall, W.A., 1994. Rhinocerebral mucormycosis: changing patterns of disease. Surg. Neurol. 41: 152-156.

Nyirjesy, P., Vazquez, J.A., Ufberg, D.D., Sobel, J.D., Boikov, D.A. & Buckley, H.R., 1995. *Saccharomyces cerevisiae* vaginitis: transmission from yeast used in baking. Obstetrics Gynecol. 86: 326-329.

Obayashi, T., Yoshida, M., Mori, T., Goto, H., Yasuoka, A., Iwasaki, H., Teshima, H., Kohno, S., Horiuchi, A., Ito, A., Yamaguchi, H., Shimada, K. & Kawai, T., 1995. Plasma (1→3)-ß-D-glucan measurement in diagnosis of invasive deep mycosis and fungal febrile episodes. Lancet 345: 17-20.

Obendorf, D.L., Peel, B.F. & Munday, B.L., 1993. *Mucor amphibiorum* infection in platypus *(Ornithorhynchus anatinus)* from Tasmania. J. Wildlife Dis. 29: 485-487.

O'Day, D.M., 1977. Fungal endophthalmitis caused by *Paecilomyces lilacinus* after intraocular lens implantation. Am. J. Opthalmol. 83: 130-131.

Odds, F.C., 1979. *Candida* and Candidosis. Leicester Univ. Press, Leicester, 382 pp.

Odds, F.C., 1987. *Candida* infections: an overview. CRC Crit. Rev. Microbiol. 15: 1-115.

Odds, F.C., 1996. Epidemiological shifts in opportunistic and nosocomial *Candida* infections: mycological aspects. Int. J. Antimicrob. Agents 6: 141-144.

Odds, F.C., Arai, T., Disalvo, A.F., Evans, E.G.V., Hay, R.J., Randhawa, H.S., Rinaldi, M.G. & Walsh, T.J., 1992. Nomenclature of fungal diseases: a report and recommendations from a sub-committee of the International Society for Human and Animal Mycology (ISHAM). J. Med. Vet. Mycol. 30: 1-10.

Odds, F.C. & Bernaerts, R., 1994. CHROMagar *Candida*, a new differential isolation medium for presumptive identification of clinically important *Candida* species. J. Clin. Microbiol. 32: 1923-1929.

Odds, F.C., Gerven, F., van, Espinel-Ingroff, A., Bartlett, M.S., Ghannoum, M.A., Lancaster, M.V., Pfaller, M.A., Rex, J.H., Rinaldi, M.G. & Walsh, T.J., 1998a. Evaluation of possible correlations between antifungal susceptibilities of filamentous fungi *in vitro* and antifungal treatment outcomes in animal infection models. Antimicrob. Agents Chemother. 42: 282-288.

Odds, F.C., Nuffel, L. van & Dams, G., 1998b. Prevalence of *Candida dubliniensis* isolates in a yeast stock collection. J. Clin. Microbiol. 36: 2869-2873.

Odening, K., Aue, A., Ochs, A. & Stolte, M., 1998. *Emmonsia crescens (Ascomycotina)* and *Sarcocystis ochotonae* n. sp. *(Sporozoa)* in picas *(Ochotona)* from China in the Berlin Zoological Garden. Zool. Gart. 68: 80-94.

Odio, C.M., Navarrete, M., Carrillo, J.M., Mora, J. & Carranza, A., 1999. Disseminated histoplasmosis in infants. Pediatr. Infect. Dis. J. 18: 1065-1068.

O'Donnell, K.L., 1979. Zygomycetes in Culture. Dept. Botany, Univ. Georgia, Athens, 257 pp.

O'Donnell, K.L., Cigelnik, E. & Nirenberg, H.I., 1998. Molecular systematics and phylogeography of the *Gibberella fujikuroi* species complex. Mycologia 90: 465-493.

O'Donnell, K.L., Nirenberg, H.I., Aoki, T. & Cigelnik, E., 2000. A multigene phylogeny of the *Gibberella fujikuroi* species complex: detection of additional phylogenetically distinct species. Mycoscience 41: 61-78.

Oelz, O., Schaffner, A., Frick, P. & Schärr, G., 1983. *Trichosporon capitatum:* thrush-like oral infection, local invasion, fungaemia and metastatic abscess formation in leukaemic patient. J. Infect. 6: 183-185.

Offidani, A., Simoncini, C., Arzeni, D., Cellini, A., Amerio, P. & Scalise, G., 1998. Tinea capitis due to *Microsporum gypseum* in an adult. Mycoses 41: 239-241.

Ohashi, Y., 1960. On a rare disease due to *Alternaria tenuis* (alternariasis). Tohoku J. Exp. Med. 72: 78-82.

Ohkubo, S., Torisaki, M., Higashide, T., Mochizuki, K. & Ishibashi, Y., 1994. Endophthalmitis caused by *Paecilomyces lilacinus* after cataract surgery: a case report. Nippon Ganka Gakkai Zasshi 98: 103-110.

Oji, E.O. & Steele, D.M., 1982. *Fusarium solani* keratitis. East Afr. Med. J. 59: 632-638.

Okafor, J.I., Testrake, D., Mushinsky, H.R. & Yangco, B.G., 1984. *Basidiobolus* sp. and its association with reptiles and amphibians in Southern Florida. Sabouraudia 22: 47-51.

Okeke, C.N., Kappe, R., Zakikhani, S., Nolte, O. & Sonntag, H.-G., 1998. Ribosomal genes of *Histoplasma capsulatum* var. *duboisii* and var. *farciminosum*. Mycoses 41: 355-362.

Okhravi, N., Dart, J.K., Towler, H.M. & Lightman, S., 1997. *Paecilomyces lilacinus* endophthalmitis with secondary keratitis: a case report and literature review. Arch. Ophthalm. 115: 1320-1324.

Okuda, C., Ito, M., Sato, M., Oka, K. & Hotchi, M., 1987. Disseminated cutaneous *Fusarium* infection with vascular invasion in a leukemic patient. J. Med. Vet. Mycol. 15: 177-186.

Olive, L.S., 1968. An unusual new Heterobasidiomycete with *Tilletia*-like basidia. J. Elisha Mitchell Sci. Soc. 84: 261-266.

Olivere, J.W., Meier, P.A., Fraser, S.L., Morrison, W.B., Parsons, T.W. & Drehner, D.M., 1999. Coccidioidomycosis - the airborne assault continues: an unusual presentation with a review of the history, epidemiology, and military relevance. Aviation Space Environm. Med. 70: 790-796.

Oliveri, S., Cammarata, E., Augello, G., Mancuso, P., Tropea, R., Ajello, L. & Padhye, A.A., 1988. *Rhizopus arrhizus* in Italy as the causative agent of primary cerebral zygomycosis in a drug addict. Eur. J. Epidemiol. 4: 284-288.

Ollert, M.W., Wende, C., Görlich, M., McMullan-Vogel, C.G., Borg-von Zepelin, M., Vogel, C.-W. & Korting, H.C., 1995. Increased expression of *Candida albicans* secretory proteinase, a putative virulence factor, in isolates from human immunodeficiency virus-positive patients. J. Clin. Microbiol. 33: 2543-2549.

Olsson, M., Elvin, K., Löfdahl, S. & Linder, E., 1993. Detection of *Pneumocystis carinii* DNA in sputum and bronchoalveolar lavage samples by polymerase chain reaction. J. Clin. Microbiol. 31: 221-226.

Ono, N., Sato, K., Yokomise, H. & Tamura, K., 1999. Lung abscess caused by *Paecilomyces lilacinus* Respiration 66: 85-87.

Onsberg, P., 1978. Human infections with *Microsporum persicolor* in Denmark. Br. J. Derm. 99: 531-536.

Onwuasoigwe, O. & Gugnani, H.C., 1998. African histoplasmosis: osteomyelitis of the radius. Mycoses 41: 105-107.

Oo, M.M., Kutteh, L.A., Koc, O.N., Strauss, M. & Lazarus, H.M., 1998. Mucormycosis of petrous bone in an allogeneic stem cell recipient. Clin. Infect. Dis. 27: 1546-1547.

Oorschot, C.A.N. van, 1977. The genus *Myceliophthora*. Persoonia 9: 401-408.

Oorschot, C.A.N. van, 1980. A revision of *Chrysosporium* and allied genera. Stud. Mycol. 20: 1-89.

Opal, S.M., Asp, A.A., Cannady, P.B., Morse, P.L., Burton, L.J. & Hammer, P.G., 1986. Efficacy of infection control measures during a nosocomial outbreak of disseminated aspergillosis associated with hospital construction. J. Infect. Dis. 153: 634-637.

Orem, J., Mpanga, L., Habyara, E., Nambuya, A., Aisu, T., Wamukota, W. & Otim, M.A., 1998. Disseminated *Aspergillus fumigatus* infection: case report. East Afr. Med. J. 75: 436-438.

Oriol, A., Ribera, J.-M., Arnal, J., Milla, F., Battle, M. & Feliu, E., 1993. *Saccharomyces cerevisiae* septicemia in a patient with myelodysplastic syndrome. Am. J. Hematol. 43: 325-326.

Ormerod, L.D., 1987. Causation and management of microbial keratitis in subtropical Africa. Ophthalmology 94: 1662-1668.

Orr, G.F., 1976. *Kuehniella*, a new genus of the *Gymnoascaceae*. Mycotaxon 4: 171-178.

Orr, G.F., Kuehn, H.H. & Plunkett, O.A., 1963. The genus *Myxotrichum* Kunze. Can J. Bot. 41: 1457-1480.

Orr, P.H., Safneck, J.R. & Napier, L.B., 1993. *Monosporium apiospermum* endophthalmitis in a patient without risk factors for infection. Can. J. Ophthalmol. 28: 187-190.

Orth, B., Frei, R., Itin, P.H., Rinaldi, M.G., Speck, B., Gratwohl, A. & Widmer, A.F., 1996. Outbreak of invasive mycoses caused by *Paecilomyces lilacinus* from a contaminated skin lotion. Annls Int. Med. 125: 799-806.

Ortiz, A.M., Sanz-Rodriguez, C., Culebras, J., Buendia, B., Gonzalez-Alvaro, I., Ocon, E. & Camara, R. de la, 1998. Multiple spondylodiscitis caused by *Blastoschizomyces capitatus* in an allogeneic bone marrow transplantation recipient. J. Rhematol. 25: 2276-2278.

Ota, M., 1923. Sur une nouvelle espèce d'*Aspergillus* pathogène: *Aspergillus jeanselmei* n. sp. Annls Parasit. Hum. Comp. 1: 137-146.

Otčenášek, M. & Dvořák, J., 1985. Houby infikující člověka. Taxonomie původů humánních mykóz v abecedním přehledu. Česká Mykol. 39: 155-164.

Otčenášek, M., Janeckova, V., Kaupa, R., Medek, B. & Nevludová, D., 1976. Ketiologii plicnich aspergilomú. Cesk. Epidemiol. 25: 263-268.

Otčenášek, M., Jirousek, Z., Nožička, Z. & Mencl, K., 1984. Paecilomycosis of the maxillary sinus. Mykosen 27: 242-251.

Otis, E.J. & Wolke, R.E., 1985. Infection of *Exophiala salmonis* in Atlantic salmon (*Salmo salar* L.). J. Wildlife Dis. 21: 61-64.

Owen, P.G., Willis, B.K. & Benzel, E.C., 1992. *Torulopsis glabrata* vertebral osteomyelitis. J. Spinal Disord. 5: 370-373.

Owens, W.R., Miller, R.I., Haynes, P.F. & Snider, T.G., 1984. *Basidiobolus haptosporus* in two horses. J. Am. Vet. Med. Ass. 186: 703-705.

Oyeka, C.A. & Gugnani, H.C., 1991. Physiological characteristics of clinical isolates of *Hendersonula toruloidea* and *Scytalidium* species. Mycoses 34: 369-371.

Padhye, A.A., 1995. Sporotrichosis - an occupational mycosis. In Landis, T.D. & Cregg, B.: USDA Techn. Rep. PNW-GTR-365.

Padhye, A.A., Ajello, L., Wieden, M.A. & Steinbronn, K.K., 1986. Phaeohyphomycosis of the nasal sinuses caused by a new species of *Exserohilum*. J. Clin. Microbiol. 24: 245-249.

Padhye, A.A., Amster, R.L., Browning, M. & Ewing, E.P., 1994a. Fatal encephalitis caused by *Ochroconis gallopavum* in a domestic cat *(Felis domesticus)*. J. Med. Vet. Mycol. 32: 141-145.

Padhye, A.A. & Carmichael, J.W., 1971. The genus *Arthroderma* Berkeley. Can. J. Bot. 49: 1525-1540.

Padhye, A.A., Davis, M.S., Baer, D., Reddick, A., Sinha, K.K. & Ott, J., 1998. Phaeohyphomycosis caused by *Phaeoacremonium inflatipes*. J. Clin. Microbiol. 36: 2763-2765.

Padhye, A.A., Davis, M.S., Reddick, A., Bell, M.F., Gearhart, E.D. & Moll, L. von, 1995. *Mycoleptodiscus indicus:* a new etiologic agent of phaeohyphomycosis. J. Clin. Microbiol. 33: 2796-2797.

Padhye, A.A., Detweiler, J.G., Frumkin, A., Bulmer, G.S., Ajello, L. & McGinnis, M.R., 1989. Tinea capitis caused by *Microsporum praecox* in a patient with sickle cell anaemia. J. Med. Vet. Mycol. 27: 313-317.

Padhye, A.A., Godfrey, J.H., Chandler, F.W. & Peterson, S.W., 1994b. Osteomyelitis caused by *Neosartorya pseudofischeri*. J. Clin. Microbiol. 32: 2832-2836.

Padhye, A.A., Gutekunst, R.W., Smith, D.J. & Punithalingam, E., 1997. Maxillary sinusitis caused by *Pleurophomopsis lignicola*. J. Clin. Microbiol. 35: 2136-2141.

Padhye, A.A., Hampton, A.A., Hampton, M.T., Hutton, N.W., Prevost-Smith, E. & Davis, M.S., 1996. Chromoblastomycosis caused by *Exophiala spinifera*. Clin. Infect. Dis. 22: 331-335.

Padhye, A.A., Helwig, W.B., Warren, N.G., Ajello, L., Chandler, F.W. & McGinnis, M.R., 1988a. Subcutaneous phaeohyphomycosis caused by *Xylohypha emmonsii*. J. Clin. Microbiol. 26: 709-712.

Padhye, A.A., Kaplan, W., Neuman, M.A., Case, P. & Radcliffe, G.N., 1984. Subcutaneous phaeohyphomycosis caused by *Exophiala spinifera*. Sabouraudia 22: 493-500.

Padhye, A.A., Kaufman, L., Durry, E., Banerjee, C.K., Jindal, S.K., Talwar, P. & Chakrabarti, A., 1992a. Fatal pulmonary sporotrichosis caused by *Sporothrix schenckii* var. *luriei* in India. J. Clin. Microbiol. 30: 2492-2494.

Padhye, A.A., Koshi, G., Anandi, V., Ponniah, J., Sitaram, V., Jacob, M., Mathai, R., Ajello, L. & Chandler, F.W., 1988b. First case of subcutaneous zygomycosis caused by *Saksenaea vasiformis* in India. Diagn. Microbiol. Infect. Dis. 9: 69-77.

Padhye, A.A., McGinnis, M.R., Ajello, L. & Chandler, F.W., 1988c. *Xylohypha emmonsii*, sp. nov., a new agent of phaeohyphomycosis. J. Clin. Microbiol. 26: 702-708.

Padhye, A.A., Pathak, A.A., Katkar, V.J., Hazare, V.K. & Kaufman, L., 1994c. Oral histoplasmosis in India: a case report and an overview of cases reported during 1968-92. J. Med. Vet. Mycol. 32: 93-103.

Padhye, A.A., Smith, G., McLaughlin, D., Standard, P.G. & Kaufman, L., 1992b. Comparative evaluation of a chemiluminescent DNA probe and exoantigen test for rapid identification of *Histoplasma capsulatum*. J. Clin. Microbiol. 30: 3108-3111.

Padhye, A.A., Smith, G., Standard, P.G., McLaughlin, D. & Kaufman, L., 1994d. Comparative evaluation of chemiluminescent DNA probe assays and exoantigen tests for rapid identification of *Blastomyces dermatitidis* and *Coccidioides immitis*. J. Clin. Microbiol. 32: 867-870.

Padhye, A.A., Weitzman, I. & Domenech, E., 1994e. An unusual variant of *Trichophyton tonsurans* var. *sulphureum*. J. Med. Vet. Mycol. 32: 147-150.

Padhye, A.A., Young, C.N. & Ajello, L., 1980. Hair perforation as a diagnostic criterion in the identification of *Epidermophyton, Microsporum* and *Trichophyton* species. PAHO Sci. Publ. 396: 115-120.

Pagano, L., Morace, G., Ortu-la Barbera, E., Sanguinetti, M. & Leone, G., 1996a. Adjuvant therapy with rhGM-CSF for the treatment of *Blastoschizomyces capitatus* systemic infection in a patient with acute myeloid leukemia. Annls Hematol. 73: 33-34.

Pagano, L., Ricci, P., Montillo, M., Cenacchi, A., Nosari, A., Tonso, A., Cudillo, L., Chierichini, A., Savignano, C., Buelli, M., Melillo, L., Barbera, E.O. la, Sica, S., Hohaus, S., Bobini, A., Bucaneve, G. & Favero, A. del, 1996b. Localization of aspergillosis to the central nervous system among patients with acute leukemia: report of 14 cases. Clin. Infect. Dis. 23: 628-630.

Pagliarusco, A., Tomazzoli, L., Amalfitano, G., Polonelli, L. & Bonomi, L. 1995. Mycotic keratitis by *Fusarium moniliforme*. Acta Ophtholmol. Scand. 73: 560-562.

Pal, M., 1992. Disseminated *Aspergillus terreus* infection in a caged pigeon. Mycopathologia 119: 137-139.

Pal, M., 1995a. Nasal rhinosporidiosis in a bullock in Gujarat (India). Revta Iberoam. Micol. 12: 61-62.

Pal, M., 1995b. *Aphanoascus fulvescens*: first report of its isolation from canine dermatitis in India. Revta IberoAm. Micol. 12: 1.

Pal, M., 1997. First report of isolation of *Cryptococcus neoformans* var. *neoformans* from avian excreta in Kathmandu, Nepal. Revta Iberoam. Micol. 14: 181-183.

Pal, M., Matsusaka, N. & Lee, C.W., 1994. Clinical and mycological observations on equine ringworm due to *Microsporum gypseum*. Korean J. Vet. Clin. Med. 11: 5-8.

Palacio, A. del, Garau, M., Tena, D., Sainz, J., Arribi, A. & Carillo, A., 1999a. Otitis externa por *Scedosporium apiospermum*. Revta Iberoam. Micol. 16: 161-163.

Palacio, A. del, Gomez-Hernando, C., Revenga, F., Carabias, E., Gonzalez, A., Cuetara, M.S. & Johnson, E.M., 1996. Cutaneous *Alternaria alternata* successfully treated with itraconazole. Clin. Exp. Derm. 21: 241-243.

Palacio, A. del, Ramos, M.J., Pérez, A., Arribi, A., Amondarain, I., Alonso, S. & Cruz Ortíz, M., 1999b. Zigomicosis. A propósito de cinco casos. Revta Iberoam. Micol. 16: 50-56.

Palacio-Hernanz, A. del, Moore, M.K., Campbell, C.K., Palacio-Medel, A. del, & Castillo, R. del, 1989. Infection of the central nervous system by *Rhinocladiella atrovirens* in a patient with acquired immunodeficiency syndrome. J. Med. Vet. Mycol. 27: 127-130.

Palacio-Hernanz, A. del, Moore, M.K., Campbell, C.K., Palacio, A. del, Pereiro-Miguens, M., Gimeno, C., Cuetara, M.S., Rubio, R., Costa, R. & Romero, G., 1992. Widespread dermatophytosis due to *Microsporum gallinae* in a patient with AIDS. Clin. Exp. Derm. 17: 449-453.

Palaoglu, S., Sav, A., Basak, T., Yalcinlar, Y. & Scheithauer, B.W., 1993. Cerebral phaeohyphomycosis. Neurosurgery 33: 894-897.

Paldrok, H., 1965. Report on a case of subcutaneous dissemination of *Aspergillus niger*, type *awamori*. Acta Derm.-Venereol. 45: 275-282.

Palencarova, E., Jesenska, Z., Plank, L., Straka, S., Baska, T., Hajtman, A. & Pec, J., 1995. Phaeohyphomycosis caused by *Alternaria* species and *Phaeosclera dematioides* Sigler, Tsuneda and Carmichael. Clin. Exper. Derm. 20: 419-422.

Palm, M.E., Gams, W. & Nirenberg, H., 1995. *Plectosporium*, a new genus for *Fusarium tabacinum*, the anamorph of *Plectosphaerella cucumerina*. Mycologia 87: 397-406.

Palmer, D.F., Kaufman, L., Kaplan, W. & Cavallaro, J.J., 1977. Preparation of *Sporothrix schenckii* culture filtrate antigen. In A. Balows (ed.): Serodiagnosis of Mycotic Diseases, pp. 138-139.

Pan, S., Sigler, L. & Cole, G.T., 1994. Evidence for a phylogenetic connection between *Coccidioides immitis* and *Uncinocarpus reesii* (Onygenaceae). Microbiology 140: 1481-1494.

Panagiotidou, D., Kapetis, E., Chryssomalis, F., Karakatsanis, G. & Badillet, G., 1991. Deux cas d'alternariose cutanée en Grèce. J. Mycol. Méd. 1: 88-89.

Papendorf, M.C. & Hoog, G.S. de, 1976. The genus *Phaeoisaria*. Persoonia 8: 407-414.

Pappagianis, D., 1988. Epidemiology of coccidioidomycosis. Curr. Topics Med. Mycol. 2: 199-238.

Pappas, P.G., Pottage, J.C., Powderly, W.G., Fraser, V.J., Stratton, C.W., McKenzie, S., Trapper, M.L., Chmel, H., Bonebrake, F.C., Blum, R., Shafer, R.W., King, C. & Dismukes, W.E., 1992. Blastomycosis in patients with the acquired immonodeficiency syndrome. Annls Intern. Med. 116: 847-853.

Pařenicová, L., Skouboe, P., Frisvad, J., Samson, R.A., Rossen, L., Hoor-Suykerbuyk, M. ten & Visser, J., 2000. Combined molecular and biochemical approach identifies *Aspergillus japonicus* and *Aspergillus aculeatus* as two species. Appl. Envir. Microbiol. (in press).

Paré, J.A., Sigler, L., Hunter, D.B., Summerbell, R.C., Smith, D.A. & Machin, L., 1997. Cutaneous mycoses in chameleons caused by the *Chrysosporium* anamorph of *Nannizziopsis vriesii* (Apinis) Currah. J. Zoo Wild Med. 28: 443-453.

Park, D.S., Go, S.J., Kim, Y.S., Seok, S.J., Ryu, J.C. & Sung, J.M., 1999. Phylogenetic relationships of the genera *Coprinus* and *Psathyrella* on the basis of ITS region sequences. Korean J. Mycol. 27: 274-279.

Pasarell, L., Garcia, M.C.C. de, Baraquer, F. & McGinnis, M.R., 1994. Keratitis caused by *Lasiodiplodia theobromae*: case report and review of the human pathogenic *Sphaeropsidales*. Abstr. Gen. Meet. ASM 94: 601.

Pasarell, L., Kemna, M.E. & McGinnis, M.R., 1993. Mycetoma caused by *Phialophora verrucosa*. Abstr. Gen. Meet. ASM 93: 532.

Pasarell, L., McGinnis, M.R. & Standard, P.G., 1990. Differentiation of medically important isolates of *Bipolaris* and *Exserohilum* with exoantigens. J. Clin. Microbiol. 28: 1655-1657.

Pasha, T.M., Leighton, J.A., Smilack, J.D., Heppell, J., Colby, T.V. & Kaufman, L., 1997. Basidiobolomycosis: an unusual fungal infection mimicking bowell disease. Gastroenterology 112: 250-254.

Pastor, J., Pumarola, M., Cuenca, R. & Lavin, S., 1993. Systemic aspergillosis in a dog. Vet. Rec. 132: 412-413.

Paterson, D.L., Robson, J.M.B., Ridley, M.F., Sullivan, J.J. & Nicollaides, N.J., 1997. Five cases of infection with the dematiaceous fungus, *Coniothyrium fuckelii*. Clin. Infect. Dis. 25: 383.

Patoux-Pibouin, M., Couatarmanach, A., Gall, F. le, Bergeron, C., Bièvre, C. de, Guiguen, C. & Chevrant-Breton, J., 1992. Fusariose à *Fusarium solani* chez un adolescent leucémique. J. Ann. Derm. Vénérol. 119: 377-380.

Patton, C.S., 1977. *Helminthosporium spiciferum* as the cause of dermal and nasal maduromycosis in a cow. Cornell. Vet. 67: 236-244.

Pauzner, R., Goldschmied-Reouven, A., Hay, I., Vered, Z., Ziskind, Z., Hassin, N. & Farfel, Z., 1997. Phaeohyphomycosis following cardiac surgery: case report and review of serious infection due to *Bipolaris* and *Exserohilum* species. Clin. Infect. Dis. 25: 921-923.

Pavan, P.R. & Margo, C.E., 1993. Endogenous endophthalmitis caused by *Bipolaris hawaiiensis* in a patient with acquired immunodeficiency syndrome. Am. J. Ophthalmol. 116: 644-645.

Pec, J., Palencarova, E., Plank, L., Straka, S., Pec, M., Jesenska, Z. & Filo, V., 1996. Phaeohyphomycosis due to *Alternaria* spp. and *Phaeosclera dematioides*: a histopathological study. Mycoses 39: 217-221.

Pecarrere, J.L., Huerre, M., Lafond, P., Esterre, P., Raharisolo, C. & Rotalier, P. de, 1994. Les entomophthoromycoses à Madagascar. Arch. Inst. Pasteur Madagascar 61: 99-102.

Pe'er, J., Gnessin, H., Levinger, S., Averbukh, E., Levy, Y. & Polachek, I., 1996. Conjunctival oculosporidosis in East Africa caused by *Rhinosporidium seeberi*. Arch. Path. Lab. Med. 120: 854-858.

Peltroche-Llacsahuanga, H., Schnitzler, N., Lütticken, R. & Haase, G., 1999. Rapid identification of *Candida glabrata* by using a dipstick to detect trehalase-generated glucose. J. Clin. Microbiol. 37: 202-205.

Penn, P., Degasne, I., Asfar, P., Chennebault, J.M., Foussard, C., Legrand, E., Gentile, L. de & Chabasse, D., 1998. Spondylodiscites aspergillaires. À propos de deux observations. J. Mycol. Méd. 8: 167-171.

Pepe, R.R. & Bertolotto, C., 1991. Primo isolamento di *Cladosporium cladosporioides* (Fres.) de Vries da gramulomi dentali. Minerva Stomatol. 40: 781-785.

Pepin, G.A. & Pritchard, G.C., 1984. Fungal mastitis in a goat due to infection with *Paecilomyces variotii*. Vet. Med. J. 5: 12.

Pereiro, M., Jo-Chu, J. & Toribio, J., 1998. Phaeohyphomycotic cyst due to *Cladosporium cladosporioides*. Dermatology, Basel 197: 90-92.

Pereiro, M., Labandeira, J. & Toribio, J., 1999. Plantar hyperkeratosis due to *Fusarium verticillioides* in a patient with malignancy. Clin. Exp. Derm. 24: 175-178.

Pereiro, M., Pereiro, M., Pereiro-Miguens, M. & Toribio, J., 1988. Las micosis por *Trichophyton megninii* en Galicia (con revisión de la taxonomia de este dermatofito). J. Med. Vet. Mycol. 26: 93-100.

Pereiro, M., Pereiro Ferreirós, M.M., Pereiro-Miguens, M. & Toribio, J., 1989. Some remarks concerning *Trichophyton proliferans*. Mycoses 32: 87-92.

Pereiro, M., Pereiro, E., Toribo, J. & Pereiro-Miguens, M., 1997. Superficial white toenail onychomycosis due to *Fusarium oxysporum*. A case reported and review of the literature. J. Mycol. Méd. 7: 219-222.

Perelman, B., & Kuttin, E., 1992. Zygomycosis in ostriches. Avian Pathol. 21: 675-680.

Peres, L.C., Figueiredo, F., Peinada, M. & Soares, F.A., 1992. Fulminant disseminated pulmonary adiaspiromycosis in humans. Am. J. Trop. Med. Hyg. 46: 146-150.

Pérez-Blanco, M., Fernández-Zeppenfeldt, G., Hernández, R., Yegres, F. & Borelli, D., 1998. Cromomycosis por *Rhinocladiella aquaspersa:* descripción del primer caso en Venezuela. Revta Iberoam. Micol. 15: 51-54.

Perfect, J.R., 1996. Fungal virulence genes as targets for antifungal chemotherapy. Antimicrob. Agents Chemother. 40: 1577-1583.

Perfect, J.R., Schell, W.A. & Rinaldi, M.G., 1992. Uncommon invasive fungal pathogens in the acquired immunodeficiency syndrome. J. Med. Vet. Mycol. 31: 175-179.

Perlman, E.M. & Binns, L., 1997. Intense photophobia caused by *Arthrographis kalrae* in a contact lens-wearing patient. Am. J. Ophthalmol. 123: 547-549.

Perry, J.E., 1964. Opportunistic fungal infections of the urinary tract. Texan J. Med. 60: 146-148.

Perttala, Y., Peltokallio, P., Leiviskä, T. & Sipponen, J., 1975. Yeast bezoar formation following gastric surgery. Am. J. Roentgenol. Radium Therap. Nuclear Med. 125: 365-373.

Perzigian, R.W. & Faiz, R.G., 1993. Primary cutaneous aspergillosis in a preterm infant. Am. J. Perinatol. 10: 269-271.

Peterson, S.W., 1992. *Neosartorya pseudofischeri* sp. nov. and its relationship to other species in *Aspergillus* section *fumigati*. Mycol. Res. 96: 547-554.

Peterson, S.W., 2000. Phylogenetic analysis of *Penicillium* species based on ITS and LSU-rDNA nucleatide sequences. In R.A. Samson & J.I. Pitt (eds): Integration of Modern Taxonomic Methods for *Penicillium* and *Aspergillus* Classification, pp. 163-178.

Peterson, S.W., 2000. Phylogenetic relationships in *Aspergillus* based on rDNA sequence analysis. In R.A. Samson & J.I. Pitt (eds): Integration of Modern Taxonomic Methods for *Penicillium* and *Aspergillus* Classification, pp. 323-355.

Peterson, S.W. & Sigler, L., 1998. Molecular genetic variation in *Emmonsia crescens* and *Emmonsia parva*, etiologic agents of adiaspiromycosis, and their phylogenetic relationship to *Blastomyces dermatitidis (Ajellomyces dermatitidis)* and other systemic fungal pathogens. J. Clin. Microbiol. 36: 2918-2925.

Petrak, F., 1924. Mykologische Notizen. VII. Annls Mycol. 22: 1-182.

Pfaller, M.A., Jones, R.N., Messer, S.A., Edmond, M.B., Wenzel, R.P. & SCOPE Participant Group, 1998a. National surveillance of nosocomial blood stream infection due to species of *Candida* other than *Candida albicans:* frequency of occurrence and antifungal susceptibility in the SCOPE Program. Diagn. Microbiol. Infect. Dis. 30: 121-129.

Pfaller, M.A., Jones, R.N., Messer, S.A., Edmond, M.B. & Wenzel, R.P., 1998b. National surveillance of nosocomial blood stream infection due to *Candida albicans:* frequency of occurrence and antifungal susceptibility in the SCOPE Program. Diagn. Microbiol. Infect. Dis. 31: 327-332.

Pfaller, M.A., Messer, S.A. & Coffman, S., 1997. *In vitro* susceptibilities of clinical yeast isolates to a new echinocandin derivative, LY303366, and other antifungal agents. Antimicrob. Agents Chemother. 41: 763-766.

Pfaller, M.A., Messer, S.A., Gee, S., Joly, S., Pujol, C., Sullivan, D.J., Coleman, D.C. & Soll, D.R., 1999. *In vitro* susceptibilities of *Candida dubliniensis* isolates tested against the new triazole and echinocandin antifungal agents. J. Clin. Microbiol. 37: 870-872.

Pfeiffer, T.J. & Ellis, D.H., 1993. Serotypes of Australian environmental and clinical isolates of *Cryptococcus neoformans*. J. Med. Vet. Mycol. 31: 401-404.

Phelippot, R., Feuilhade de Chauvin, M., Michel, Y., Pietrini, P., Teillac, D., Boniatsi, L. & Badillet, G., 1988. *Microsporum praecox:* à propos de 4 cas. Annls Dermatol. Véneréol. 115: 1154-1156.

Pflugfelder, S.C., Flynn, H.W., Zwickey, T.A., Forster, R.K., Tsiligianni, A., Culbertson, W.W. & Mandelbaum, S., 1988. Exogenous fungal endophthalmitis. Ophthalmology 95: 19-30.

Philips, P., Wood, W.S., Phillips, G. & Rinaldi, M.G., 1989. Invasive hyalohyphomycosis caused by *Scopulariopsis brevicaulis* in a patient undergoing allogeneic bone marrow transplant. Diagn. Microbiol. Infect. Dis. 12: 429-432.

Phillips, R., 1981. Mushrooms and other fungi of Great Britain and Europe. Pan Books, London, 288 pp.

Piacentini, I., Biasioli, S., Chiaramonte, S., Fabris, A., Feriani, M., Pisani, E., Ronco, C. & Greca, G. la, 1984. *Fusarium verticillioides* nuovo patogeno opportunistica. Giorn. Malat. Infet. Paras. 36: 64-67.

Piens, M.A., Dannaoui, E., Sauzet, C., Chouvet, B., Monier, M.F. & Picot, S., 1999. Primary mucormycosis due to *Absidia corymbifera* in the absence of risk factors. J. Mycol. Méd. 9: 230-232.

Pier, A.C., Hodges, A.B., Lauze, J.M. & Raisbeck, M., 1995. Experimental immunity to *Microsporum canis* and cross reactions with other dermatophytes of veterinary importance. J. Med. Vet. Mycol. 33: 93-97.

Pieracci, F. & Colosi, A., 1957. Micosi polmonaire in equino: studio eziologico ed istopatologiso. Atti Soc. Ital. Sci. Vet. 11: 759-762.

Pierarch, C.A., Gulman, C., Dhar, G.J. & Kiser, J.C., 1973. *Phialophora mutabilis* endocarditis. Annls Intern. Med. 79: 900-901.

Piérard-Franchimont, C., Deleixhe, F. & Piérard, G.E., 1996. Dermatophytose cutanée chronique, onychomycose occulte et hamartome épidermolytique linéaire. Une impasse thérapeutique. J. Mycol. Méd. 6: 190-194.

Pietrini, P. & Stewart, W.M., 1977. Granulome péri-narinaire dû à *Taeniolella stilbospora* (Corda) Hughes. Bull. Soc. Fr. Mycol. Méd., N. Sér., 6: 97-99.

Pincus, D.H., Coleman, D.C., Pruitt, W.R., Padhye, A.A., Salkin, I.F., Geimer, M., Bassel, A., Sullivan, D.J., Clarke, M. & Hearn, V., 1999. Rapid identification of *Candida dubliniensis* with commercial yeast identification systems. J. Clin. Microbiol. 37: 3533-3539.

Pinjon, E., Sullivan, D., Salkin, I., Shanley, D. & Coleman, D., 1998. Simple, inexpensive, reliable method for differentiation of *Candida dubliniensis* from *Candida albicans*. J. Clin. Microbiol. 36: 2093-2095.

Piontelli, E. & Toro, M.A., 1988. Comentarios biomorfológicos y clínicos sobre el género *Scopulariopsis* Bainier. Hialohifomicosis en uñas y piel. II. Bol. Micol. 3: 259-273.

Piontelli, E. & Toro, M.A., 1989. Un raro caso hialohifomicosis en uñas por *Polypaecilum insolitum* G. Smith. Bol. Micol. 4: 155-159.

Piraccini, B.M., Morelli, R., Stinchi, C. & Tosti, A., 1996. Proximal, subungual onychomycosis due to *Microsporum canis*. Br. J. Derm. 134: 175-177.

Pitrak, D.L., Koneman, E.W., Estupinan, R.C. & Jackson, J., 1988. *Phialophora richardsiae* infection in humans. Rev. Infect. Dis. 10: 1195-1203.

Pitt, J.I., 1979. The genus *Penicillium* and its teleomorphic states *Eupenicillium* and *Talaromyces*. Academic Press, London, 634 pp.

Pitt, J.I., 1985. A Laboratory Guide to Common *Penicillium* Species. North Ryde, Australia, 184 pp.

Pitt, J.I., 1994. The current role of *Aspergillus* and *Penicillium* in human and animal health. J. Med. Vet. Mycol. 32, Suppl. 1: 17-32.

Pitt, J.I. & Hocking, A.D., 1997. Fungi and Food Spoilage. Blackie Acad., London.

Plá, M.-P., Hartung, C., Mendoza, P., Stukanoff, A. & Moreno, M.-J., 1994. Neuroparacoccidioidomycosis: case reports and review. Mycopathologia 127: 139-144.

Plaats-Niterink, A.J. van der, 1981. Monograph of the genus *Pythium*. Stud. Mycol. 21: 1-242.

Pletincx, M., Legein, J. & Vandenplas, Y., 1995. Fungemia with *Saccharomyces boulardii* in a 1-year-old girl with protacted diarrhea. J. Pediatr. Gastroenterol. Nutr. 21: 113-115.

Plouffe, J.F., Brown, D.G., Silva, J., Eck, T., Stricof, R.L. & Fekety, F.R., 1977. Nosocomial outbreak of *Candida parapsilosis* fungemia related to intravenous infusions. Arch. Intern. Med. 137: 1686-1689.

Ploysangam, T. & Lucky, A.W., 1997. Childhood white superficial onychomycosis caused by *Trichophyton rubrum:* report of seven cases and review of the literature. J. Am. Acad. Derm. 36: 29-32.

Plum, G., Scheid, C., Franzen, C., Schuett-Gerowitt, H., Seifert, H. & Wickramanayake, P.D., 1996. Empirical liposomal amphotericin-B therapy in a neutropenic patient: breakthrough of disseminated *Blastoschizomyces capitatus* infection. Zentbl. Bakt. 284: 361-366.

Plzas, J., Portilla, J., Boix, V. & Perez-Mateo, M., 1994. *Sporobolomyces salmonicolor* lymphadenitis in an AIDS patient. Pathogen or passenger? AIDS, Philad. 8: 387-388.

Pock-Steen, B., 1968. Isolation of a strain of *Trichophyton mentagrophytes* producing a brown diffusable pigment. Acta Derm. Venereol. 48: 313-315.

Podnos, Y.D., Anastasio, P., Maza, L. de la & Kim, R.B., 1999. Cerebral phaeohyphomycosis caused by *Ramichloridium obovoideum (Ramichloridium mackenziei):* case report. Neurosurgery 45: 372-375.

Podsiadlo, B. & Halweg, H., 1988. *Absidia coerulea* jako przyczyna grzybicy pluc. Acta Mycol. 24: 71-76.

Poelma, F.G., Vries, G.A. de, Blythe-Russell, E.A. & Luykx, M.H.F., 1974. Lobomycosis in an Atlantic bottle-nosed dolphin in the Dolphinarium Harderwijk. Aquatic Mammals 13: 11-15.

Poirot, J.L., Laporte, J.P., Guého, E., Verny, A., Gorin, N.C., Najman, A., Marteau, M. & Roux, P., 1985. Mycose profonde à *Fusarium*. Presse Méd. 14: 45.

Polacheck, I., Nagler, A., Okon, E., Drakos, P., Plaskowitz, J. & Kwon-Chung, K.J., 1992. *Aspergillus quadrilineatus*, a new causative agent of fungal sinusitis. J. Clin. Microbiol. 30: 3290-3293.

Polacheck, I., Salkin, I.F., Kitzes-Cohen, R. & Raz, R., 1992. Endocarditis caused by *Blastoschizomyces capitatus* and taxonomic review of the genus. J. Clin. Microbiol. 30: 2318-2322.

Polachek, I., Strahilevitz, J., Sullivan, D., Donnelly, S., Salkin, I.F. & Coleman, D.C., 2000. Recovery of *Candida dubliniensis* from non-human immunodeficiency virus-infected patients in Israel. J. Clin. Microbiol. 38: 170-174.

Polack, F.M., Silverio, C. & Bresky, R.H., 1976. Corneal chromoblastomycosis double infection to *Phialophora verrucosa* (Medlar) and *Cladosporium cladosporioides* Fresenius. Annls Ophthalmol. 8: 139-144.

Polak, E., Hermann, R., Kües, U. & Aebi, M., 1997. Asexual sporulation in *Coprinus cinereus:* structure and development of oidiophores and oidia in an *Amut Bmut* homokaryon. Fung. Genet. Biol. 22: 112-126.

Polenghi, F. & Lasagni, A., 1976. Observation on a case of mycokeratitis and its treatment with BAY b 5097 (Canesten). Mykosen 19: 223-226.

Polesky, A., Kirsch, C.M., Snyder, L.S., LoBue, P., Kagawa, F.T., Dykstra, B.J., Wehner, J.H., Catanzaro, A., Ampel, N.M. & Stevens, D.A., 1999. Airway coccidioidomycosis - report of cases and review. Clin. Inefct. Dis. 28: 1273-1280.

Polonelli, L., Dettori, G., Morace, G., Rosa, R., Castagnola, M. & Schipper, M.A.A., 1988. Antigenic studies on *Rhizopus microsporus, Rh. rhizopodiformis*, progeny and intermediates *(Rh. chinensis)*. Antonie van Leeuwenhoek 54: 5-18.

Pönninghaus, J.M., Warndorff, D. & Port, G., 1995. *Microsporum nanum* - a report from Malawi (Africa). Mycoses 38: 149-150.

Pore, R.S., d'Amato, R.F. & Ajello, L., 1977. *Fissuricella* gen. nov.: a new taxon for *Prototheca filamenta*. Sabouraudia 15: 69-78.

Pore, R.S. & Larsh, H.W., 1967. First occurrence of *Thermoascus aurantiacus* from animal and human sources. Mycologia 59: 927-928.

Porges, N., Muller, J.F. & Lockwood, L.B., 1935. A *Mucor* found in fowl. Mycologia 27: 330-331.

Pore, R.S. & Larsh, H.W., 1968. Experimental pathology of *Aspergillus terreus-flavipes* group species. Sabouraudia 6: 89-93.

Porro, A.M., Yoshioka, M.C.N., Kaminski, S.K., Carmo, M. do, Fischman, O. & Alchorne, M.M.A., 1997. Disseminated dermatophytosis caused by *Microsporum gypseum* in two patients with the acquired immunodeficiency syndrome. Mycopathologia 137: 13-16.

Porto, E.C., 1979. *Phialophora bubakii*, isolaments de abscessu subcutaneo, em transplantado renal. Revta Inst. Med. Trop., S. Paulo 21: 106-109.

Pospíšil, L., 1989. The significance of *Candida pulcherrima* findings in human clinical specimens. Mycoses 32: 581-583.

Pospíšil, L., Skorkovská, Š. & Moster, M., 1990. Corneal phaeohyphomycosis caused by *Wangiella dermatitidis*. Ophthalmologica 201: 128-132.

Posthaus, H., Krampe, M., Pagan, O., Guého, E., Suter, C. & Bacciarini, L., 1997. Systemic paecilomycosis in a hawksbill turtle *(Eretmochelys imbricata)*. J. Mycol. Méd. 7: 223-226.

Potron, M. & Noisette, G., 1911. Un cas de mycose. Revue Méd. Est 43: 132-149.

Poveda, F., Garcia-Alegria, J., Nieves, M.A. de las, Villar, E., Montiel, N. & Arco, A. del, 1998. Disseminated histoplasmosis successfully treated with liposomal amphotericin B following azathioprine therapy in a patient from a nonendemic area. Eur. J. Clin. Microbiol. Infect. Dis. 17: 357-359.

Powderly, W.G., 1993. Cryptococcal meningitis and AIDS. Clin. Infect. Dis. 17: 837-842.

Powell, B.L., Drutz, D.J., Huppert, M. & Sun, S.H., 1983. Relationship of progesterone- and estradiol-binding proteins in *Coccidioides immitis* to coccidioidal dissemination in pregnancy. Infect. Immun. 40: 478-485.

Powers, C.N., 1998. Diagnosis of infectious diseases: a cytopathologist's perspective. Clin. Microbiol. Rev. 11: 341-365.

Pracharktam, R., Changtrakool, P., Sathapatayavongs, B., Jayanetra, P. & Ajello, L., 1991. Immunodiffusion test for diagnosis and monitoring of human Pythiosis insidiosi. J. Clin. Microbiol. 29: 2661-2662.

Pracharktam, R., Siriporn, S. & Jayanetra, P., 1992. Morphological variation in pathogenic strains of *Penicillium marneffei*. J. Med. Ass. Thail. 75: 172-179.

Prariyachatigul, C., Chaiprasert, A., Meevotisom, V. & Pattanakitsakul, S., 1996. Assessment of a PCR technique for the detection and identification of *Cryptococcus neoformans*. J. Med. Vet. Mycol. 34: 251-258.

Prasad, S. & Nema, H.V., 1982. Mycotic infection of cornea (drug sensitivity studies). Indian J. Ophthalmol. 30: 81-85.

Prasil, K. & Hoog, G.S. de, 1988. Variability in *Cladosporium herbarum*. Trans. Br. Mycol. Soc. 90: 49-54.

Prats, E., Sans, J., Valldeperas, J., Ferrer, J.E. & Manresa, F., 1995. Pulmonary mycetoma-like lesion caused by *Candida tropicalis*. Respir. Med. 89: 303-304.

Prete, A. del, Sepe, G., Ferrante, M., Loffredo, C., Masciello, M. & Sebastiani, A., 1994. Fungal keratitis due to *Scopulariopsis brevicaulis* in an eye previously suffering from herpetic keratitis. Ophthalmologica 208: 333-335.

Prevoo, R.L.M.A., Starink, T.M. & Haan, P. de, 1991. Primary cutaneous mucormycosis in a heathly young girl. Report of a case caused by *Mucor hiemalis* Wehmer. J. Am. Acad. Derm. 24: 882-885.

Prevost-Smith, E., Hutton, N., Padhye, A.A., Upshur, J.E. & Bakel, A.B. van, 1993. Fatal phaeohyphomycotic infection due to *Dactylaria gallopava* and *Scedosporium prolificans* in a cardiac transplant patient. Abstr. Gen. Meet. ASM 93: 533.

Prisco, J. di & Borelli, D., 1973. Tinea nigra por *Cladosporium* species. Castellania 1: 97-100.

Pritchard, D. & Chick, B.F., 1977. Mycotic mycetoma due to *Drechslera rostrata* infection in a cow. Aust. Vet. J. 53: 241-244.

Pritchard, R.C. & Muir, D.B., 1987. Black fungi: a survey of dematiaceous hyphomycetes from clinical specimens identified over a five year period in a reference laboratory. Pathology 19: 281-284.

Prochacki, H. & Engelhardt-Zasada, C., 1974. *Epidermophyton stockdaleae* sp. nov. Mycopathologia 54: 341-345.

Proulx, G. & Onderka, D.D., 1997. *Trichophyton mentagrophytes* ringworm infection in a Northern Pocket Gopher, *Thomomys talpoides*. Can. Field-Nat. 111: 633-634.

Pujol, I., Aguilar, C., Gené, J. & Guarro, J., 2000. *In vitro* antifungal susceptibility of *Alternaria* spp. and *Ulocladium* spp. J. Antimicrob. Chemother. 46 (in press).

Pujol, F., Angirekula, M., Weiner, M., Jindrak, K. & Pachter, B.R., 1995. Parotitis due to *Torulopsis glabrata*. Clin. Infect. Dis. 21: 1342-1343.

Pujol, I., Guarro, J., Gené, J. & Sala, J., 1997. *In-vitro* antifungal susceptibility of clinical and environmental *Fusarium* spp. strains. J. Antimicrob. Chemother. 39: 163-167.

Punithalingam, E., 1969. Studies on *Sphaeropsidales* in culture. Mycol. Pap. 119: 1-24.

Punithalingam, E., 1979. Sphaeropsidales in culture from humans. Nova Hedwigia 31: 119-158.

Punsola, L. & Guarro, J., 1984a. *Keratinomyces ceretanicus* sp. nov., a psychrophilic dermatophyte from soil. Mycopathologia 85: 185-190.

Punsola, L. & Guarro, J., 1984b. Distribution of mating types of the *Microsporum gypseum* complex in Spanish soils. Mykosen 27: 191-193.

Purchio, A., Macado, A., Gambale, W., Paula, C.R. & Mariano, M., 1980. *Scopulariopsis brevicaulis:* a possible etiologic agent of pityriasis rosea in piglets. PAHO Sci. Publ. 396: 104-111.

Purcell, K.L., Johnson, P.J., Kreeger, J.M. & Wilson, D.A., 1994. Jejunal obstruction caused by a *Pythium insidiosum* granuloma in a mare. J. Am. Vet. Med. Ass. 205: 337-339.

Purvis, R.S., Diven, D.G., Drechsel, R.D., Calhoun, J.H. & Tyring, S.K., 1993. Sporotrichosis presenting as arthritis and subcutaneous nodules. J. Am. Acad. Derm. 28: 879-884.

Pynka, M., Wnuk, A., Bander, D., Syczewska, M., Boroń, A., Prost, B. & Wrzecion, S., 1998. Disseminated infection with *Saccharomyces kluyveri* in a patient with AIDS. Infection 26: 184-186.

Qi, Z.T., Sun, Z.-M., Wang, D.-L., Li, J.-S. & Wei, X.-G., 1990. Taxonomic notes on *Penicillium marneffei*. Mycosystema 3: 9-18.

Quindos, G., Cabrera, F., Arilla, M.D., Burgos, A., Ortiz-Vigon, R., Canon, J.L. & Ponton, J., 1994. Fatal *Candida famata* peritonitis in a patient undergoing continuous ambulatory peritoneal dialysis who was treated with fluconazole. Clin. Infect. Dis. 18: 658-660.

Quindos, G., San Millan, R., Robert, R., Bernard, C. & Ponton, J., 1997. Evaluation of Bichro-latex Albicans, a new method for rapid identification of *Candida albicans*. J. Clin. Microbiol. 35: 1263-1265.

Raab, S.S., Silverman, J.F. & Zimmerman, K.G., 1993. Fine-needle aspiration biopsy of pulmonary coccidioidomycosis. Spectrum of cytologic findings in 73 patients. Am. J. Clin. Pathol. 99: 582-587.

Raab, W., 1964. Die gegenseitige Beeinflüssung von Tuberkulose und systematisierter Moniliasis. Derm. Wschr. 149: 401-410.

Raabe, P., Mayser, P. & Weiß, R., 1998. Demonstration of *Malassezia furfur* and *M. sympodialis* together with *M. pachydermatis* in veterinary specimens. Mycoses 41: 493-500.

Raad, I. & Hachem, R., 1995. Treatment of central venous catheter-related fungemia due to *Fusarium oxysporum*. Clin. Infect. Dis. 20: 709-711.

Rabodonirina, M., Piens, M.A., Monier, M.F., Guého, E., Fière, D. & Mojon, M., 1994. *Fusarium* infections in immunocompromised patients: case reports and literature review. Eur. J. Clin. Microbiol. Infect. Dis. 13: 152-161.

Radix, A.E., Bieluch, V.M. & Graeber, C.W., 1996. Peritonitis caused by *Monilia sitophila* in a patient undergoing peritoneal dialysis. Int. J. Artif. Organs 19: 218-220.

Radner, A.B., Witt, M.D. & Edwards, J.E., 1995. Acute invasive rhinocerebral zygomycosis in an otherwise healthy patient: case report and review. Clin. Infect. Dis. 20: 163-166.

Radosavljevic, M., Koenig, H., Letschner-Bru, V., Waller, J., Maloisel, F., Lioure, B. & Herbrecht, R., 1999. *Candida catenulata* fungemia in a cancer patient. J. Clin. Microbiol. 37: 475-477.

Ragnaud, J.M., Marceau, C., Roche-Bezian, M.C. & Wone, C., 1984. Infection péritonéale à *Trichoderma koningii* sur dialyse péritonéale continué ambulatoire. Méd. Malad. Infect. 7: 402-405.

Rai, M.K., 1989a. *Phoma sorghina* infection in human being. Mycopathologia 105: 167-170.

Rai, M.K., 1989b. Mycosis in man due to *Arthrinium phaeospermum* var. *indicum*. First case report. Mycoses 32: 472-475.

Rainer, J., Hoog, G.S. de, Wedde, M., Gräser, Y. & Gilges, S., 2000. Molecular variability of *Pseudallescheria boydii*, a neurotropic opportunist. J. Clin. Microbiol. 38: 3267-3273.

Rajam, R.V., Kandhari, K.C. & Thirumalachar, M.J., 1958. Chromoblastomycosis caused by a rare yeast-like dematiaceous fungus. Mycopath. Mycol. Appl. 9: 5-19.

Rakotonirainy, M.S., Cariou, M.L., Brygoo, Y. & Riba, G., 1994. Phylogenetic relationships with the genus *Metarhizium* based on 28S rRNA sequences and isozyme comparison. Mycol. Res. 98: 225-230.

Ramanan, C. & Ghorpade, A., 1996. Giant cutaneous rhinosporidiosis. Int. J. Dermatol 35: 441-442.

Ramani, R., Kahn, B.T. & Chaturvedi, V., 1997. *Tilletiopsis minor*: a new etiologic agent of human subcutaneous mycosis in an immunocompromised host. J. Clin. Microbiol. 35: 2992-2995.

Ramírez, C., 1982. Manual and Atlas of the Penicillia. Elsevier, Amsterdam, 847 pp.

Randhawa, H.S., Charturvedi, V.P., Kini, S. & Khan, Z.U., 1985. *Blastomyces dermatitidis* in bats, first report of its isolation from the live of *Rhinopoma hardwickei hardwickei* Gray. Sabouraudia 23: 69-76.

Ranganathan, S. & Balajee, S.A.M., 2000. *Microsporum gypseum* complex in Madras, India. Mycoses 43: 177-180.

Ranganathan, S., Menon, T. & Balajee, S.A.M., 1996. Isolation of *Microsporum nanum* from a patient with tinea corporis in Madras, India. Mycoses 40: 229-230.

Rao, P.N., 1966. A new species of *Dichotomophthora* on *Portulaca oleracea* from Hyderabad (India). Mycopathologia 28: 137-140.

Rao, N.A., Nerenberg, A.V. & Forster, D.J., 1991. *Torulopsis candida (Candida famata)* endophthalmitis simulating *Propionibacterium acnes* syndrome. Arch. Ophthalmol. 109: 1718-1721.

Rapelanoro, R., Mortureux, P., Couprie, B., Maleville, J. & Taieb, A., 1996. Neonatal *Malassezia furfur* pustulosis. Arch. Derm. 132: 190-193.

Raper, J.R. & Krongelb, G.S., 1958. Genetic and environmental aspects of fruiting in *Schizophyllum commune* Fr. Mycologia 50: 707-740.

Raper, K.B. & Fennell, D.I., 1965. The Genus *Aspergillus*. Williams & Wilkins, Baltimore, 686 pp.

Raper, K.B. & Thom, C., 1949. A Manual of the Penicillia. Williams & Wilkins, Baltimore, 875 pp.

Ravisse, P., 1987. Les entomophthoromycoses. Bull. Soc. Fr. Mycol. Méd. 16: 51-60.

Rebell, G. & Foster, R.K., 1976. *Lasiodiplodia theobromae* as a cause of keratomycoses. Sabouraudia 14: 155-170.

Rebell, G. & Taplin, D., 1970. Dermatophytes: their recognition and identification. 2nd. ed., Univ. Miami Press, Miami, 124 pp.

Reboux, G., Comparot, S., Kirchgesner, V. & Barale, T., 1995. À propos de 19 souches de *Chrysosporium* isolées au Centre Hospitalier Universitaire de Besançon. Bilan de 10 années: 1984-1994. J. Mycol. Méd. 5: 105-110.

Redkar, R.J., Dubé, M.P. McCleskey, F.K., Rinaldi, M.G. & DelVecchio, V.G., 1996. DNA fingerprinting of *Candida rugosa* via repetitive sequence-based PCR. J. Clin. Microbiol. 34: 1677-1681.

Redmond, A., Carré, I.J., Biggart, J.D. & Mackenzie, D.W.R., 1965. Aspergillosis *(Aspergillus nidulans)* involving bone. J. Path. Bact. 89: 391-395.

Redondo-Lopez, V., Lynch, M., Schmitt, C., Cook, R. & Sobel, J.D., 1990. *Torulopsis glabrata* vaginitis: clinical aspects and susceptibility to antifungal agents. Obstet. Gynecol. 76: 651-654.

Reed, K.D., Moore, F.M., Geiger, G.E. & Stemper, M.E., 1993. Zoonotic transmission of sporotrichosis: case report and review. Clin. Infect. Dis. 16: 384-387.

Reed, P.A., Girgis, R.M., Ashok, P.S., Ravi, J., Agrawal, R.M. & Brodmerkel, G.J., 1993. A pseudocyst infected by *Torulopsis glabrata*: a unique problem. J. Gastroenterol. 88: 1962-1963.

Rehbinder, C. & Mattson, R., 1994. Mycotic skin lesions in an adult reindeer (*Rangifer tarandus tarandus* L.): a case report. Rangifer 14: 131-132.

Reich, J.D., Huddleston, K., Jorgensen, D. & Berkowitz, F.E., 1997. Neonatal *Torulopsis glabrata* fungemia. South. Med. J. 90: 246-248.

Reichard, U., Margraf, S. & Rüchel, R., 1993. Anwendung der Polymerase-Kettenreaktion in der Diagnostik tiefer Mykosen im Tiermodell. Abstr. MYK, Greifswald, p. 22.

Reidarson, T.H., Griner, L.A., Pappagiansis, D. & McBain, J., 1998. Coccidioidomycosis in a bottlenose dolphin. J. Wildlife Dis. 34: 629-631.

Reimund, E. & Ramos, A., 1994. Disseminated neonatal gastrointestinal mucormycosis: a case report and review of the literature. Pediatr. Pathol. 14: 385-389.

Reinhardt, J.F., Ruane, P.J., Walker, L.J. & George, W.L., 1985. Intravenous catheter-associated fungemia due to *Candida rugosa*. J. Clin. Microbiol. 22: 1056-1057.

Reiss, E. & Morrison, C.J., 1993. Nonculture methods for diagnosis of disseminated candidiasis. Clin. Microbiol. Rev. 6: 311-323.

Resende, M.A., Caligiorne, R.B., Aguilar, C.R. & Gontijo, M.M., 2000. Case of chromoblastomycosis due to *Rhinocladiella atrovirens*. Abstr. 14th ISHAM Congr., B. Aires, p. 274.

Restrepo, A., 1985. The ecology of paracoccidioidomycosis brasi-liensis: a puzzle still unsolved. Sabouraudia 23: 323-334.

Restrepo, A., Arango, M., Velez, H. & Uribe, L., 1976. The isolation of *Botryodiplodia theobromae* from a nail lesion. Sabouraudia 14: 1-4.

Restrepo, A.D., Greer, D.L., Robledo, M., Osorio, O. & Mondragon, H., 1973. Ulceration of the palate caused by a basidiomycete *Schizophyllum commune*. Sabouraudia 11: 201-204.

Restrepo, A., Greer, D.L. & Vasconcellos, M., 1973. Paracoccidioidomycosis: a review. Rev. Med. Vet. Mycol. 8: 97-123.

Restrepo, A., McGinnis, M.R., Malloch, D., Porras, A., Giraldo, N., Villegas, A. & Herrera, J., 1984. Fungal endocarditis caused by *Arnium leporinum* following cardiac surgery. J. Med. Vet. Mycol. 22: 225-234.

Restrepo, A., Robledo, J., Gomez, I., Tabares, A.M. & Gutierrez, R., 1968. Itraconazole therapy in lymphangitic and cutaneous sporotrichosis. Arch. Derm. 122: 413-417.

Retamal, C., Díaz, C., Salamanca, L., Ferrada, L. & Alvarez de Oro, R., 1984. Aspergilosis pulmonar en Chile. Enfoque immunológico. Bol. Micol. 2: 11-16.

Reuben, A., Anaissie, E., Nelson, P.E., Hashem, R., Legrand, C., Ho, D.H. & Bodey, G.P., 1989. Antifungal susceptibility of 44 clinical isolates of *Fusarium* species determined by using a broth microdilution method. Antimicrob. Agents Chemother. 33: 1647-1649.

Revankar, S.G., Sutton, D.A., Sanche, S.E., Rao, J., Zervos, M., Dashti, F. & Rinaldi, M.G., 1999. *Metarhizium anisopliae* as a cause of sinusitis in immunocompetent hosts. J. Clin. Microbiol. 37: 195-198.

Reyes, O. & Borelli, D., 1974. Caso de tiña negra por una cepa peculiar de *Cladosporium castellanii*. Revta Derm. Venez. 13: 20-28.

Rex, J.H., Ginsberg, A.M., Fries, L.F., Pass, H.I. & Kwon-Chung, K.J., 1988. *Cunninghamella bertholletiae* infection associated with deferoxamine therapy. Rev. Infect. Dis. 10: 1187-1194.

Rezusta, A., Rubio, M.C. & Alejandre, M.C., 1991. Differentiation between *Trichophyton mentagrophytes* and *T. rubrum* by sorbitol assimilation. J. Clin. Microbiol. 29: 219-220.

Rhyan, J.C., Stackhouse, L.L. & Davis, E.G., 1990. Disseminated geotrichosis in two dogs. J. Am. Vet. Med. Ass. 197: 358-360.

Ribes, J.A., Limper, A.H., Espy, M.J. & Smith, T.F., 1997. PCR detection of *Pneumocystis carinii* in bronchoalveolar lavage specimens: analysis of sensitivity and specificity. J. Clin. Microbiol. 35: 830-835.

Ribes, J.A., Vanover-Sams, C.L. & Baker, D.J., 2000. Zygomycetes in human disease. Clin. Microbiol. Rev. 13: 236-301.

Ricci, C., Monod, M. & Baudraz-Rosselet, F., 1998. Onychomycosis due to *Trichophyton soudanense* in Switzerland. Dermatology, Basel 197: 297-298.

Ricci, R.M., Evans, J.S., Meffert, J.J., Kaufman, L. & Sadkowski, L.C., 1998. Primary cutaneous *Aspergillus ustus* infection: second reported case. J. Am. Acad. Derm. 38: 797-798.

Richardson, M.D. & Warnock, D.W., 1997. Fungal infection. Diagnosis and management. 2nd ed. Blackwell, Oxford, 272 pp.

Richardson, M.D. & Johnson, E.M., 2000. The Pocket Guide to Fungal Infection. Blackwell, Oxford, 114 pp.

Richter, S., Cormican, M.G., Pfaller, M.A., Lee, C.K., Gingrich, R., Rinaldi, M.G. & Sutton, D.A., 1999. Fatal disseminated *Trichoderma longibrachiatum* infection in an adult bone marrow transplant patient: species identification and review of the literature. J. Clin. Microbiol. 37: 1154-1160.

Rieske, K., Handrick, W., Müller, H. & Sterker, I., 1998. Therapy of sinuorbital aspergillosis with amphotericin B colloidal dispersion. Mycoses 41: 287-292.

Rifai, M.A., 1969. A revision of the genus *Trichoderma*. Mycol. Pap. 116: 1-56.

Rihs, J.D., Padhye, A.A. & Good, C.B., 1996. Brain abscess caused by *Schizophyllum commune*: an emerging basidiomycete pathogen. J. Clin. Microbiol. 34: 1628-1632.

Riley, D.K., Galgiani, J.N., O'Donnell, M.R., Ito, J.I., Beatty, P.G. & Evans, T.G., 1993. Coccidioidomycosis in bone marrow transplant recipients. Transplantation 56: 1531-1533.

Rimek, D., Zimmermann, T., Hartmann, M., Prariyachatigul, C. & Kappe, R., 1999. Disseminated *Penicillium marneffei* infection in an HIV-positive female from Thailand in Germany. Mycoses 42, Suppl. 2: 25-28.

Rinaldi, M.G., 1989. Emerging opportunists. Infect. Dis. Clin. N. Am. 3: 65-76.

Rinaldi, M.G., Inderlied, C.B., Mahnovski, V., Monforte, H., Lam, G.L., Fothergill, A.W. & McGough, D.A., 1991. Fatal *Chaetomium atrobrunneum* Ames, 1949, systemic mycosis in a patient with acute lymphoblastic leukemia. Proc. 11th Congr. ISHAM, Montreal, p. 107.

Rinaldi, M.G., McCoy, E.L. & Winn, D.F., 1982. Gluteal abscess caused by *Phialophora hoffmannii* and review of the role of this organism in human mycoses. J. Clin. Microbiol. 16: 181-185.

Rinaldi, M.G., Phillips, P., Schwartz, J.G., Winn, R.E., Holt, G.R., Shagets, F.W., Elrod, J., Nishiok, G. & Audfemorte, T.B., 1987. Human *Curvularia* infections. Report of five cases and review of the literature. Diagn. Microbiol. Infect. Dis. 6: 27-39.

Rioux, J.A., Jarry, D.T., Jarry, D.M. & Juminer, B., 1966. *Trichophyton vanbreuseghemii* Rioux, Jarry *et* Juminer, 1964. Observations complémentaires. Annls Parasit. 41: 195-201.

Rioux, J.A., Jarry, D.T. & Juminer, B., 1964. Un nouveau Dermatophyte isolé du sol. Natur. Montpell., Sér. Bot. 16: 153-162.

Rippon, J.W., 1985. The changing epidemiology and emerging patterns of dermatophyte species. Curr. Topics Med. Mycol. 1: 208-234.

Rippon, J.W., 1988. Medical Mycology. The Pathogenic Fungi and the Pathogenic Actinomycetes, 3rd ed. Saunders, Philadelphia, 797 pp.

Rippon, J.W. & Andrews, T.W., 1978. *Microsporum racemosum*. Second clinical isolation from the United States and the Chicago area. Mycopathologia 64: 187-190.

Rippon, J.W., Arnow, P.M., Larson, R.A. & Zang, K.L., 1985. 'Golden tongue' syndrome caused by *Ramichloridium schulzeri*. Arch. Derm. 121: 892-894.

Rippon, J.W. & Carmichael, J.W., 1976. Petriellidiosis (allescheriosis): four unusual cases and review of the literatur. Mycopathologia 58: 117-124.

Rippon, J.W., Larson, R.A., Rosenthal, D.M. & Clayman, J., 1988. Disseminated cutaneous and peritoneal hyalohyphomycosis caused by *Fusarium* species: three cases and review of the literature. Mycopathologia 101: 105-111.

Rischin, M., 1921. Ueber einen Fall von bisher noch nicht beschriebener *Parendomyces*. Erkrankung, die unter dem Bilde der tiefen (Sycosis parasitaria) und oberflächlichen Trichophytie verlief. Arch. Derm. Syph. 134: 232-242.

Ritterband, D.C., Shah, M. & Seedor, J.A., 1997. *Colletotrichum graminicola*: a new corneal pathogen. Cornea 16: 362-364.

Rivalier, E., 1953. Description de *Sabouraudites praecox* nova species suivie de remarques sur le genre *Sabouraudites*. Annls Inst. Pasteur 86: 276-284.

Robinson, L.A., Reed, E.C., Galbraith, T.A., Alonso, A., Moulton, A.L. & Fleming, W.H., 1995. Pulmonary resection for invasive *Aspergillus* infections in immunocompromised patients. J. Thorac. Cardiovasc. Surg. 109: 1182-1197.

Robson, A.M. & Craver, R.D., 1994. *Curvularia* urinary tract infection: a case report. Pediatr. Nephrol. 8: 83-84.

Rockhill, R.C. & Klein, M.D., 1980. *Paecilomyces lilacinus* as the cause of chronic maxillary sinusitis. J. Clin. Microbiol. 11: 737-739.

Roder, B.L., Sonnenschein, C. & Hartzen, S.H., 1991. Failure of fluconazole therapy in *Candida krusei* fungemia. Eur. J. Clin. Microbiol. Infect. Dis. 10: 173.

Rodero, L., Canteros, C.E., Rivas, C., Lee, W. & Davel, G., 1999. Sensibilidad *in vitro* de *Paracoccidioides brasiliensis* frente a los antifúngicos de un sistémico. Revta Argent. Microbiol. 31: 78-81.

Rodrigues, M.M., Laibson, P. & Kaplan, W., 1975. Exogenous corneal ulcer caused by *Tritirachium roseum*. Am. J. Ophthalmol. 80: 804-806.

Rodrigues, M.M. & Macleod, D., 1975. Exogenous fungal endophthalmitis caused by *Paecilomyces*. Am. J. Ophthalmol. 79: 687-690.

Rodrigues de Kopp, N. & Vidal, G., 1998. Micosis ocular postraumática por *Penicillium oxalicum*. Revta Iberoam. Micol. 15: 103-106.

Roeijmans, H.J., Hoog, G.S. de, Tan, C.S. & Figge, M.J., 1997. Molecular taxonomy and GC/MS of metabolites of *Scytalidium hyalinum* and *Nattrassia mangiferae* (*Hendersonula toruloidea*). J. Med. Vet. Mycol. 35: 181-188.

Roilides, E., Sigler, L., Bibashi, E., Katsifa, H., Flaris, N. & Panteliadis, C., 1999. Disseminated infection due to *Chrysosporium zonatum* in a patient with chronic granulomatous disease and review of non-*Aspergillus* fungal infections with this disease. J. Clin. Microbiol. 37: 18-25.

Roller, J.A. & Westblom, T.U., 1986. *Microsporum nanum* infection in hay farmers. J. Am. Acad. Derm. 15: 935-939.

Rolston, K.V.I., Hopfer, R.L. & Larson, D.L., 1985. Infections caused by *Drechslera* species: case report and review of the literature. Rev. Infect. Dis. 7: 525-529.

Romano, C., 1998. Onychomycosis due to *Microsporum gypseum* Mycoses 41: 349-351.

Romano, C., Bilenchi, R., Alessandrini, C. & Miracco, C., 1999. Cutaneous phaeohyphomycosis caused by *Cladosporium oxysporum*. Mycoses 42: 111-115.

Romano, C., Fimiani, M., Pellegrino, M., Valenti, L., Casini, L., Miracco, C. & Faggi, E., 1996. Cutaneous phaeohyphomycosis due to *Alternaria tenuissima*. Mycoses 39: 211-215.

Romano, C., Massai, L. & Difonzo, E.M., 2000. Dermatophytosis due to *Trichophyton violaceum* in Tuscany from 1985 to 1997. Mycoses 43: 169-172.

Romano, C., Miracco, C. & Difonzo, E.M., 1998. Skin and nail infections due to *Fusarium oxysporum* in Tuscany, Italy. Mycoses 41: 433-437.

Roosje, P.J., Hoog, G.S. de, Koeman, J.P. & Willemse, T., 1993. Phaeohyphomycosis in a cat caused by *Alternaria infectoria* Simmons. Mycoses 36: 451-454.

Rosa, R.H., Miller, D. & Alfonso, E.C., 1994. The changing spectrum of fungal keratitis in South Florida. Ophthalmology 101: 1005-1013.

Rosenberger, R.S., West, B.C. & King, J.W., 1983. Survival from sino-orbital mucormycosis due to *Rhizopus rhizopodiformis*. Am. J. Med. Sci. 286: 25-30.

Rosenthal, J., Katz, R., Dubois, D.B., Morrissey, A. & Machicao, A., 1992. Chronic maxillary sinusitis associated with the mushroom *Schizophyllum commune* in a patient with AIDS. Clin. Infect. Dis. 14: 46-48.

Ross, A.J. & Yasutake, W.T., 1973. *Scolecobasidium humicola*, a fungal pathogen of fish. J. Fish. Res. Board Can. 30: 994-995.

Rossman, A.Y., Samuels, G.J., Rogerson, C.T. & Lowen, R., 1999. Genera of *Bionectriaceae, Hypocreaceae* and *Nectriaceae (Hypocreales, Ascomycetes)*. Stud. Mycol. 42: 1-248.

Rowe-Jones, J., 1993. Paranasal aspergillosis - a spectrum of disease. J. Laryngol. Otol. 107: 773-774.

Rowland, M.D. & Farrar, W.E., 1987. Case report: thorn-induced *Phialophora parasitica* arthritis treated successfully with synovectomy and ketaconazole. Am. J. Med. Sci. 30: 393-395.

Rowley, S.D. & Strom, C.G., 1982. *Paecilomyces* fungus infection of the maxillary sinus. Laryngoscope 92: 332-334.

Rübben, A. & Krause, H., 1996. Tinea superficialis capitis due to *Trichophyton soudanense* in African immigrants. Mycoses 39: 397-398.

Ruben, S.J., Scott, T.E. & Seltzer, H.M., 1987. Intracranial and paranasal sinus infection due to *Drechslera*. South. Med. J. 80: 1057-1058.

Rubin, M.M. & Sanfilippo, R.J., 1990. Osteomyelitis of the hyoid caused by *Torulopsis glabrata* in a patient with acquired immunodeficiency syndrome. J. Oral Max. Surg. 48: 1217-1219.

Romand, S., Bourée, P. & Taburet, A.M., 1995. Inoculation accedentelle de *Aureobasidium pullulans:* à propos d'un cas. J. Mycol. Méd. 5: 259-260.

Rummelt, V., Ruprecht, K.W., Boltze, H.J. & Naumann, G.O.H., 1991. Chronic *Alternaria alternata* endophthalmitis following intraocular lens implantation. Arch. Ophthalmol. 109: 178.

Rusthoven, J.J., Feld, R. & Tuffnell, P.G., 1984. Systemic infection by *Rhodotorula* spp. in the immunocompromised host. J. Infect. 8: 241-246.

Ryu, S.L. Murooka, Y. & Kaneko, Y., 1998. Reciprocal translocation at duplicated RPL2 loci might cause speciation of *Saccharomyces bayanus* and *Saccharomyces cerevisiae*. Curr. Genet. 33: 345-351.

Ryder, N.S., 1999. Activity of terbinafine against serious fungal pathogens. Mycoses 42, Suppl. 2: 115-119.

Ryvarden, L. & Gilbertson, R.L., 1993. European Polypores, part 1: *Abortiporus - Lindtneria*. Fungiflora, Oslo.

Sa'adah, M.A., Araj, G.F., Diab, S.M. & Nazzal, M., 1995. *Cryptococcal* meningitis and confusional psychosis: a case report and literature review. Trop. Georgr. Med. 47: 224-226.

Saberhagens, C., Klotz, S.A., Bartholomew, W., Drews, D. & Dixon, A., 1997. Infection due to *Paecilomyces lilacinus:* a challenging clinical identification. Clin. Infect. Dis. 25: 1411-1413.

Sabota, J., Brodell, R., Rutecki, G.W. & Hoppes, W.L., 1996. Severe tinea barbae due to *Trichophyton verrucosum* infection in dairy farmers. Clin. Infect. Dis. 23: 1308-1310.

Saccente, M., Abernathy, R.S., Pappas, P.G., Shah, H.R. & Bradsher, R.W., 1998. Vertebral blastomycosis with paravertebral abscess: report of eight cases and review of the literature. Clin. Infect. Dis. 26: 413-418.

Sacho, H., Stead, K.J., Klugman, K.P. & Lawrence, Z., 1987. Infection of the human orbit by *Aspergillus stromatoides*. Case report. Mycopathologia 97: 97-100.

Sachs, S.W., Baum, J. & Mies, J., 1985. *Beauveria bassiana* keratitis. Br. J. Ophthalmol. 69: 548-550.

Saëz, H., 1970. Champignons isolés du poumon et du tube digestif de quelques Psittacidés. Anim. Compagn. 15: 27-41.

Saffele, J.K. van den & Boelaert, J.R., 1996. Zygomycosis in HIV-positive patients: a review of the literature. Mycoses 39: 77-84.

Saidi, S.A., Bhatt, S., Richard, J.L., Sikdar, A. & Ghosh, G.R., 1994. *Chrysosporium tropicum* as a probable cause of mycosis of poultry in India. Mycopathologia 125: 143-147.

Sakata, T., Fukuda, M., Hirashima, K. & Inoue, Y., 1995. A case of nodular subcutaneous abscesses due to *Candida tropicalis* complicated by acute myeloid leukemia. Jpn. J. Med. Mycol. 36: 251-257.

Salama, A.D., Rogers, T., Lord, G.M., Lechler, R.I. & Mason, P.D., 1997. Multiple *Cladosporium* brain abscesses in a renal transplantation recipient. Transplantation 63: 160-162.

Salesa, R., Burgos, A., Fernandez-Mazarrasa, C., Quindos, G. & Ponton, J., 1991. Transient fungaemia due to *Candida pelliculosa* in a patient with AIDS. Mycoses 34: 327-329.

Salesa, R., Burgos, A., Ondiviela, R., Richard, C., Quindos, G. & Ponton, J., 1993. Fatal disseminated infection by *Scedosporium inflatum* after bone marrow transplantation. Scand. J. Infect. Dis. 25: 389-393.

Salfelder, K., 1990. Atlas of Fungal Pathology. Kluwer, Dordrecht, 200 pp.

Salkin, I.F., Cooper, C.R., Bartges, J.W., Kemna, M.E. & Rinaldi, M.G., 1992. *Scedosporium inflatum* osteomyelitis in a dog. J. Clin. Microbiol. 30: 2797-2800.

Salkin, I.F., Dixon, D.M., Kemna, M.E., Danneman, P.J. & Griffith, J.W., 1990. Fatal encephalitis caused by *Dactylaria constricta* var. *gallopava* in a snowy owl chick *(Nyctea scandiaca)*. J. Clin. Microbiol. 28: 2845-2847.

Salkin, I.F., Gordon, M.A. & Stone, W.B., 1976. Cutaneous infection of a porcupine *(Erethizon dorsatum)* by *Aureobasidium pullulans*. Sabouraudia 14: 47-49.

Salkin, I.F., Martinez, J.A. & Kemna, M.E., 1986. Opportunistic infection of the spleen caused by *Aureobasidium pullulans*. J. Clin. Microbiol. 23: 828-831.

Salkin, I.F., McGinnis, M.R., Dykstra, M.J. & Rinaldi, M.G., 1988. *Scedosporium inflatum*, an emerging pathogen. J. Clin. Microbiol. 26: 498-503.

Saltarelli, C.G., 1989. *Candida albicans*, the pathogenic fungus. Hemisphere Publ. Co., New York, 290 pp.

Samaranayake, L.P. & MacFarlane, T.W., 1990. Oral Candidosis. Butterworth, London, 265 pp.

Samaranayake, Y.H. & Samaranayake, L.P., 1994. *Candida krusei:* biology, epidemiology, pathogenicity and clinical manifestations of an emerging pathogen. J. Med. Microbiol. 41: 295-310.

Samson, R.A., 1969. Revision of the genus *Cunninghamella* (Fungi, *Mucorales*). Proc. K. Ned. Akad. Wet., Ser. C, 72: 322-335.

Samson, R.A., 1974. *Paecilomyces* and some allied hyphomycetes. Stud. Mycol. 6: 1-119.

Samson, R.A., 1979. A compilation of the Aspergilli described since 1965. Stud. Mycol. 18: 1-40.

Samson, R.A., Hoekstra, E. van, Frisvad, J.C. & Filtenborg, O., 1996. Introduction to food-borne fungi, 5th ed. Centraalbureau voor Schimmelcultures, Baarn, 322 pp.

Samson, R.A. & Mahmood, T., 1970. The genus *Acrophialophora* (Fungi, *Moniliales*). Acta Bot. Neerl. 19: 804-808.

Samson, R.A., Nielsen, P.V. & Frisvad, J.C., 1990. The genus *Neosartorya*: differentiation by scanning electron microscopy and mycotoxin profiles. In R.A. Samson & J.I. Pitt (eds): Modern Concepts in *Penicillium* and *Aspergillus* Classification, pp. 455-467. Plenum, New York.

Samson, R.A. & Pitt, J.I. (eds), 1985. Advances in *Penicillium* and *Aspergillus* Systematics. NATO ASI Ser. 102. Plenum, New York, 483 pp.

Samson, R.A. & Pitt, J.I. (eds), 1990. Modern concepts in *Penicillium* and *Aspergillus* classification. NATO ASI Ser. 185. Plenum, New York, 478 pp.

Samson, R.A. & Pitt, J.I. (eds), 2000. Integration of Modern Taxonomic Methods for *Penicillium* and *Aspergillus* Classification. Harwood, Amsterdam, 510 pp.

Samson, R.A. & Polonelli, L., 1978. *Myriodontium keratinophilum*, gen. et sp. nov. Persoonia 9: 505-509.

Samuels, G.J. & Brayford, D., 1990. Variation in *Nectria radicicola* and its anamorph, *Cylindrocarpon destructans*. Mycol. Res. 94: 433-442.

Samuels, G.J., Lieckfeldt, E. & Nirenberg, H.I., 1999. *Trichoderma asperellum*, a new species with warted conidia, and redescription of *T. viride*. Sydowia 51: 71-88.

Samuels, G.J., Petrini, O., Kuhls, K., Lieckfeldt, E. & Kubicek, C.P., 1998. The *Hypocrea schweinitzii* complex and *Trichoderma* sect. *Longibrachiatum*. Stud. Mycol. 41: 1-54.

Samuels, G.J., Petrini, O. & Manguin, S., 1994. Morphological and macromolecular characterization of *Hypocrea schweinitzii* and its *Trichoderma* anamorph. Mycologia 86: 421-435.

San-Blas, G., 1993. Paracoccidioidomycosis and its etiologic agent *Paracoccidioides brasiliensis*. J. Med. Vet. Mycol. 31: 99-113.

San-Blas, G. & San-Blas, F., 1977. *Paracoccidioides brasiliensis*: cell wall structure and virulence. A review. Mycopathologia 62: 77-86.

Sanchez, P.J. & Cooper, B.H., 1987. *Candida lusitaniae*: sepsis and meningitis in a neonate. Pediatr. Infect. Dis. J. 6: 758-759.

Sanchez, M.R., Ponge-Wilson, I., Moy, J.A. & Rosenthal, S., 1994. Zygomycosis in HIV infection. J. Am. Acad. Derm. 30: 904-908.

Sanchez, V., Vazquez, J.A., Barth-Jones, D., Dembry, L., Sobel, J.D. & Zervos, M.J., 1992. Epidemiology of nosocomial acquisition of *Candida lusitaniae*. J. Clin. Microbiol. 30: 3005-3008.

Sanchez-Portocarrero, J., Martín-Rabadán, P., Saldaña, C.J. & Pérez-Cecilia, E., 1994. *Candida* cerebrospinal fluid shunt infection. Diagn. Microbiol. Infect. Dis. 20: 33-40.

Sander, A., Beyer, U. & Amberg, R., 1998. Systemic *Fusarium oxysporum* infection in an immunocompetent patient with an adult respiratory distress syndrome (ARDS) and extracorporal membrane oxygenation (ECMO). Mycoses 41: 109-111.

Sandhu, D.K. & Sandhu, R.S., 1972. A new variety of *Aspergillus nidulans*. Mycologia 55: 297-299.

Sandhu, D.K., Sandhu, R.S. & Misra, V.C., 1976. Isolation of *Candida viswanathii* from cerebral spinal fluid. Sabouraudia 14: 251-254.

Sandhu, G.S., Aleff, R.A., Kline, B.C. & Lacaz, C. da Silva, 1997. Molecular detection and identification of *Paracoccidioides brasiliensis*. J. Clin. Microbiol. 35: 1894-1896.

Sandhu, G.S., Kline, B.C., Stockman, L. & Roberts, G.D., 1995. Molecular probes for the diagnosis of fungal infections. J. Clin. Microbiol. 33: 2913-2919.

Sandhu, R.S. & Randhawa, H.S., 1962. On the reisolation and taxonomic study of *Candida viswanathii* Viswanathan and Randhawa. Mycopath. Mycol. Appl. 18: 179-183.

Sandin, R.L., Fang, T.-T., Hiemenz, J.W., Greene, J.N., Kalik, A. & Szakacs, J.E., 1993a. *Malassezia furfur* folliculitis in cancer patients: the need for interaction of microbiologist, surgical pathologist, and clinician in facilating identification by the clinical microbiology laboratory. Annls Clin. Lab. Sci. 23: 377-384.

Sandin, R.L., Meier, C.S., Crowder, M.L. & Greene, J.N., 1993b. Concurrent isolation of *Candida krusei* and *Candida tropicalis* from multiple blood cultures in a patient with acute leukemia. Arch. Pathol. Lab. Med. 117: 521-523.

Sandler, B., Potter, T.S. & Hashimoto, K., 1996. Cutaneous *Pneumocystis carinii* and *Cryptococcus neoformans* in AIDS. Br. J. Derm. 134: 159-163.

Sandler, R., Tallman, C.B., Keamy, D.G. & Irving, W.R., 1971. Successfully treated rhinocerebral phycomycosis in well controlled diabetes. New Engl. J. Med. 18: 1180-1182.

Sandner, V.K. & Schönborn, C., 1973. Schimmelpilzinfektion der Haut bei ausgedehnter Verbrennung. Deutsch. GesWesen 28: 125-128.

Sands, J.M., Macher, A.M., Ley, T.J. & Nienhuis, A.W., 1985. Disseminated infection caused by *Cunninghamella bertholletiae* in a patient with beta-thalassemia. Case report and review of the literature. Annls Intern. Med. 102: 59-63.

Sandven, P., Nilsen, K., Digranes, A., Tjade, T. & Lasseu, J., 1997. *Candida norvegensis*: a fluconazole-resistant species. Antimicrob. Agents Chemother. 41: 1375-1376.

Sano, A., Tanaka, R., Nishimura, K., Okurokawa, C.S., Iabuki, K., Coelho, R., Franco, M., Montenegro, M.R. & Miyaji, M., 1997. Characteristics of 17 *Paracoccidioides brasiliensis* strains. Mycoscience 38: 117-122.

Santurio, J.M., Bardemaker, A. & Trindade, A., 1998. Cutaneous pythiosis insidiosi in calves from the Pantanal region of Brazil. Mycopathologia 141: 123-125.

Sanyal, M., Thammaya, A. & Basu, N., 1971. Studies on *Gymnoascaceae*. I. *Arachniotus flavoluteus* in cases of dermatophytoses. Bull. Calcutta School Trop. Med. 19: 86-87.

Saravanakumar, P.S., Eslami, P. & Zar, F.A., 1996. Lymphocutaneous sporotrichosis associated with a squirrel bite: case report and review. Clin. Infect. Dis. 23: 647-648.

Sarma, P.S.A., Durairaj, P. & Padhye, A.A., 1993. *Candida lusitaniae* causing fatal meningitis. Postgrad. Med. J. 69: 878-880.

Sarosi, G.A., Eckman, M.R., Davies, S.F. & Laskey, W.K., 1979. Canine blastomycosis as a harbinger of human disease. Annls Intern. Med. 91: 733-735.

Sathapatayavongs, B.P., Leelachaikul, P., Prachaktum, R., Atichartakarn, V., Sriphojanart, S., Trairatvorakul, P., Jirasiritham, S., Montasut, S., Euvilaichit, C. & Flagel, T., 1989. Human pythiosis associated with thalassemia hemoglobinopathy. J. Infect. Dis. 159: 274-280.

Sato, T., Koseki, S., Takahashi, S. & Maie, O., 1990. Localised cutaneous cryptococcosis successfully treated with itraconazole. Review of medication in 18 cases reported in Japan. Mycoses 33: 455-463.

Savelkoul, P.H.M., Aarts, H.J.M., Haas, J. de, Dijkshoorn, L., Duim, B., Otsen, M., Rademaker, J.L.W., Schouls, L. & Lenstra, J.A., 1999. Amplified-fragment length polymorphism analysis: the state of an art. J. Clin. Microbiol. 37: 3083-3091.

Saxen, H., Virtanen, M., Carlson, P., Hoppu, K., Pohjavuori, M., Vaara, M., Vuopio-Varkila, J. & Peltola, H., 1995. Neonatal *Candida parapsilosis* outbreak with a high case fatality rate. Pediatr. Infect. Dis. J. 14: 776-781.

Scalarone, G.M., Legendre, A.M., Clark, K.A. & Pusater, K., 1992. Evaluation of a commercial DNA probe assay for the identification of clinical isolates of *Blastomyces dermatitidis* from dogs. J. Med. Vet. Mycol. 30: 43-49.

Schaffner, A., 1989. Experimental basis for the clinical epidemiology of fungal infections. A review. Mycoses 32: 499-515.

Schauffer, A.F., 1972. Maduromycotic mycetoma in an aged mare. J. Am. Vet. Med. Ass. 160: 998-1000.

Schaumann, K. & Priebe, K., 1994. *Ochroconis humicola* causing muscular black spot disease of Atlantic salmon *(Salmo salar)*. Can. J. Bot. 72: 1629-1634.

Schechtman, R.C., Midgley, G. & Hay, R.J., 1995. HIV disease and *Malassezia* yeasts: a quantitative study of patients presenting with seborrhoeic dermatitis. Br. J. Derm. 133: 694-698.

Schell, W.A., 1995. New aspects of emerging fungal pathogens: a multifaceted challenge. Clin. Lab. Med. 15: 365-387.

Schell, W.A., McGinnis, M.R. & Borelli, D., 1983. *Rhinocladiella aquaspersa* a new combination for *Acrotheca aquaspersa*. Mycotaxon 17: 341-348.

Schell, W.A. & Perfect, J.R., 1993. *Coniothyrium fuckelii*, a species in need of a genus. Second case of human infection. Abstr. Gen. Meet. ASM 93: 533.

Schell, W.A. & Perfect, J.R., 1996. Fatal, disseminated *Acremonium strictum* infection in a neutropenic host. J. Clin. Microbiol. 34: 1333-1336.

Schell, W.A. & Perfect, J.R., 1995. *Pseudomicrodochium suttonii* isolated from a subcutaneous lesion in a sarcoid patient. Abstr. Gen. Meet. ASM 95: 108.

Schepelmann, K., Müller, F. & Dichgans, J., 1993. Cryptococcal meningitis with severe visual and hearing loss and radiculopathy in a patient without immunodeficiency. Mycoses 36: 429-432.

Schiemann, R., Glasmacher, A., Bailly, E., Horré, R., Molitor, E., Leutner, C., Smith, M.T., Kleinschmidt, R., Marklein, G. & Sauerbruch, T., 1998. *Geotrichum capitatum* septicaemia in neutropenic patients: case report and review of the literature. Mycoses 41: 113-116.

Schick, G.W., 1966. Drei positive Befunde von *Microsporum cookei* Ajello 1959 in Hautläsionen beim Menschen. Derm. Wschr. 152: 177-183.

Schipper, M.A.A., 1973. A study on variability in *Mucor hiemalis* and related species. Stud. Mycol. 4: 1-40.

Schipper, M.A.A., 1976. On *Mucor circinelloides, Mucor racemosus* and related species. Stud. Mycol. 12: 1-40.

Schipper, M.A.A., 1978a. On certain species of *Mucor* with a key to all accepted species. Stud. Mycol. 17: 1-52.

Schipper, M.A.A., 1978b. On the genera *Rhizomucor* and *Parasitella*. Stud. Mycol. 17: 53-71.

Schipper, M.A.A., 1984. A revision of the genus *Rhizopus*. I. The *Rhizopus stolonifer*-group and *Rhizopus oryzae*. Stud. Mycol. 25: 1-19.

Schipper, M.A.A., 1990. Notes on *Mucorales* - I. Observations on *Absidia*. Persoonia 14: 133-148.

Schipper, M.A.A., Maslen, M.M., Hogg, G.G., Chow, C.W. & Samson, R.A., 1996. Human infection by *Rhizopus azygosporus* and the occurrence of azygospores in the Zygomycetes. J. Med. Vet. Mycol. 34: 199-203.

Schipper, M.A.A. & Stalpers, J.A., 1984. A revision of the genus *Rhizopus* II. The *Rhizopus microsporus* group. Stud. Mycol. 25: 30-34.

Schlitzer, R.L. & Ahearn, D.G., 1982. Characterization of atypical *Candida tropicalis* and other uncommon clinical yeast isolates. J. Clin. Microbiol. 15: 511-516.

Schmidt, G., Calanni, L., Iacono, M. & Negroni, R., 2000. *Cerinosterus cyanescens* fungemia: report of a case. Proc. 14th ISHAM Congr., B. Aires, p. 272.

Schnadig, V.J., Long, E.G., Washington, J.M., McNeely, M.C. & Troum, B.A., 1986. *Phialophora verrucosa*-induced subcutaneous phaeohyphomycosis. Acta Cytol. 30: 425-429.

Schneider, R., 1979. Die Gattung *Pyrenochaeta* de Notaris. Mitt. Biol. Bundesanst. Land- u. Forstw. 189: 1-73.

Schönborn, C. & Schmoranzer, H., 1970. Untersuchungen über Schimmelpilzinfektionen der Zehennägel. Mykosen 13: 253-272.

Schoepfer, C., Carla, H., Bezou, M.J., Cambon, M., Girault, D., Deméocq, F. & Malpuech, G., 1995. Septicémie à *Malassezia furfur* au décours d'une greffe de moelle. Arch. Pédiatr. 2: 245-248.

Scholer, H.J., Müller, E. & Schipper, M.A.A., 1983. *Mucorales.* In D.H. Howard (ed.): Fungi Pathogenic for Humans and Animals, A: 9-59. Marcel Dekker, New York.

Schol-Schwarz, M.B., 1970. Revision of the genus *Phialophora* (*Moniliales*). Persoonia 6: 59-94.

Scholtens, R.E.M. & Harrison, S.M., 1994. Subcutaneous phycomycosis. Trop. Geogr. Med. 46: 371-373.

Scholz, H.D. & Meyer, L., 1965. *Mortierella polycephala* as a cause of pulmonary mycosis in cattle. Berl. Münch. Tieräztl. Wochenschr. 78: 27-30.

Schønheyder, H.C., Jensen, H.E., Gams, W., Nyvad, O., Nga, P. van, Aalbaek, B. & Stenderup, J., 1996. Late bioprosthetic valve endocarditis caused by *Phialemonium aff. curvatum* and *Streptococcus sanguis*: a case report. J. Med. Vet. Mycol. 34: 209-214.

Schoofs, A., Odds, F.C., Colebunders, R., Ieven, M. & Goossens, H., 1997. Use of specialized isolation media for recognition and identification of *Candida dubliniensis* isolates from HIV-infected patients. J. Clin. Microbiol. Infect. Dis. 16: 296-300.

Schrank, J.H. & Dooley, D.-P., 1995. Purulent pericarditis caused by *Candida* species: case report and review. Clin. Infect. Dis. 21: 182-187.

Schroeder, H., Jardine, J.E. & Davis, V., 1994. Systemic phaeohyphomycosis caused by *Xylohypha bantiana* in a dog. J.S. Afr. Vet. Ass. 65: 175-178.

Schubert, M.S. & Goetz, D.W., 1998. Evaluation and treatment of allergic fungal sinusitis. J. Allerg. Clin. Immunol. 102: 387-402.

Schulze, H., Aptroot, A., Grote-Metke, A. & Balleisen, L., 1997. *Aspergillus fumigatus* und *Chaetomium homopilatum* bei einem Leukämiepatienten. Pathogene Bedeutung von *Chaetomium*-arten. Mycoses 40, Suppl. 1: 104-109.

Schwartz, J., 1981. Histoplasmosis. Praeger, New York, 472 pp.

Schwinn, A. Ebert, J. & Bröcker, E.-B., 1995. Frequency of *Trichophyton rubrum* in tinea capitis. Mycoses 38: 1-7.

Schwinn, A., Strohm, S., Helgenberger, M., Rank, C. & Bröcker, E.-B., 1993. Phaeohyphomycosis caused by *Exophiala jeanselmei* treated with itraconazole. Mycoses 36: 445-448.

Scully, C., El-Kabir, M. & Samaranayake, L.P., 1994. *Candida* and oral candidosis: a review. Crit. Rev. Oral Biol. Med. 5: 125-157.

Seddon, M.E. & Thomas, M.G., 1997. Invasive disease due to *Epidermophyton floccosum* in an immunocompromised patient with Behcet's syndrome. Clin. Infect. Dis. 25: 153-154.

Seebacher, C. & Blaschke-Hellmessen, R., 1990. Mykosen. Epidemiologie - Diagnostik - Therapie. Gustav Fischer, Jena.

Seeliger, H.P.R., 1983. Infections of man by opportunistic molds. Their identification and nomenclature of their diseases. Mykosen 26: 587-598.

Seeliger, H.P.R., 1985. The discovery of *Achorion schoenleinii*: facts and 'stories'. Mykosen 28: 161-182.

Seeliger, H.P.R. & Heymer, T., 1981. Diagnostik Pathogener Pilze des Menschen und seiner Umwelt. Thierme, Stuttgart.

Seeverens, H.J.J., Tijhuis, G.J., Ruijs, G.J.H.M., Kazzaz, B.A. & Kauffmann, R.H., 1992. Dialysis associated mucormycosis and desferrioxamine treatment: a case report with review of the role of oxygen radicals. Neth. J. Med. 41: 275-279.

Segal, B.H., Walsh, T.J., Liu, J.M., Wilson, J.D. & Kwon-Chung, K.J., 1998. Invasive infection with *Fusarium chlamydosporum* in a patient with aplastic anemia. J. Clin. Microbiol. 36: 1772-1776.

Segretain, G., Baylet, J., Darasse, H. & Camain, R., 1959. *Leptosphaeria senegalensis* n. sp., agent de mycétome à grains noirs. C.R. Acad. Sci., Paris 248: 3730-3732.

Segretain, G. & Destombes, P., 1961. Description d'un nouvel agent de maduromycose, *Neostudina rosatii*, n. gen., n. sp., isolé en Afrique. C.R. Séanc. Acad. Sci., Paris 253: 2577-2579.

Segretain, G., Fromentin, H., Destombes, P., Brygoo, E.-R. & Dodin, E., 1964. *Paecilomyces viridis* n.sp., champignon dimorphique, agent d'une mycose généralisée de *Chameleo lateralis* Gray. C.R. Hebd. Séanc. Acad. Sci., Paris 259: 258-261.

Segretain, G. & Vieu, M., 1947. Formes parasitaires des *Aspergillus* dans l'aspergillome bronchique: diagnostic biologique des aspergilloses broncho-pulmonaires. Arch. Biol. Méd. 33: 1281-1289.

Seguin, P., Degeilh, B., Grulois, I., Gacouin, A., Maugendre, S., Dufour, T., Dupont, B. & Camus, C., 1995. Succesful treatment of a brain abscess due to *Trichoderma longibrachiatum* after surgical resection. Eur. J. Clin. Microbiol. Inf. Dis. 14: 445-448.

Seifert, K.A., 2000. ASP45, a synoptic key to common species of *Aspergillus*. In R.A. Samson & J.I. Pitt (eds): Integration of Modern Taxonomic Methods for *Penicillium* and *Aspergillus* Classification, pp. 139-146.

Seifert, K., Kendrick, B. & Murase, G., 1983. A key to the Hyphomycetes on dung. Univ. Waterloo Biol. Ser. 27: 1-62.

Seifert, K.A. & Louis-Seize, G., 2000. Phylogeny and species concepts in the *Penicillium aurantiogriseum* complex as inferred from partial β-tubulin gene DNA sequences. In R.A. Samson & J.I. Pitt (eds): Integration of Modern Taxonomic Methods for *Penicillium* and *Aspergillus* Classification, pp. 189-198.

Sekhon, A.S., Galbraith, J., Mielke, B.W., Garg, A.K. & Sheehan, G., 1992. Cerebral phaeohyphomycosis caused by *Xylohypha bantiana*, with a review of the literature. Eur. J. Epidemiol. 8: 387-390.

Sekhon, A.S., Garg, A.K., Padhye, A.A., Standard, P.G., Kaufman, L. & Ajello, L., 1989. Antigenic relationship of *Penicillium marneffei* to *P. primulinum*. J. Med. Vet. Mycol. 27: 105-112.

Sekhon, A.S., Kaufman, L., Moledina, N., Summerbell, R.C., Padhye, A.A., Ambrosie, E.A. & Panter, T., 1995. An exoantigen test for the rapid identification of medically significant *Fusarium* species. J. Med. Vet. Mycol. 33: 287-289.

Sekhon, A.S., Padhye, A.A., Garg, A.K., Ahmad, H. & Moledina, N., 1994. *In vitro* sensitivity of medically significant *Fusarium* species to various antimycotics. Chemotherapy 40: 239-244.

Sekhon, A.S., Padhye, A.A., Kaufman, L., Garg, A.K., Ajello, L., Ambrosie, E. & Panter, T., 1997. Antigenic relationships among pathogenic *Beauveria bassiana* with *Engyodontium album (B. alba)* and non-pathogenic species of the genus *Beauveria*. Mycopathologia 138: 1-4.

Sekhon, A.S., Padhye, A.A., Standard, P.G., Kaufman, L., Ajello, L. & Garg, A.K., 1990. Antigenic relationship of *Dactylaria gallopava* to *Scolecobasidium constrictum*. J. Med. Vet. Mycol. 28: 59-66.

Sellier, P., Monsvery, J.J., Lacroix, C., Feray, C., Evans, J., Minozzi, C., Vayre, F., Giudice, P. Del, Feuilhade, M., Pinel, C., Vittecoq, D. & Passeron, J., 2000. Recurrent subcutaneous infection due to *Scopulariopsis brevicaulis* in a liver transplant recipient. Clin. Infect. Dis. 38: 820-823.

Senneville, E., Ajana, F., Gerard, Y., Bourez, J.-M., Alfandari, S., Chidiac, C. & Mouton, Y., 1996. Bilateral ureteral obstruction due to *Saccharomyces cerevisiae* fungus balls. Clin. Infect. Dis. 23: 636-637.

Sentamilselvi, G., Janaki, C., Kamalam, A. & Thambiah, A.S., 1998. Deep dermatophytosis caused by *Trichophyton rubrum* - a case report. Mycopathologia 142: 9-11.

Senturia, B.H. & Wolf, F.T., 1945. Treatment of external otitis II. Action of sulfonamide compounds in fungi isolated from cases of otomycoses. Arch. Otolaryngol. 41: 56-63.

Serody, J.S., Mill, M.R., Detterbeck, F.C., Harris, D.T. & Cohen, M.S., 1993. Blastomycosis in transplant recipients: report of a case and review. Clin. Infect. Dis. 16: 54-58.

Serpa, J., 1958. Tricosporonose cutânea curada com biflorina. Anais Fac. Med. Univ. Recife 18: 33-37.

Serrano, J.A., Pisano, I.D. & Lopez, F.A., 1998. Black grain minimycetoma caused by *Pyrenochaeta mackinnonii*. The first clinical case of eumycetoma reported in Barinas State, Venezuela. J. Mycol. Méd. 8: 34-39.

Sethi, N. & Mandell, W., 1988. *Saccharomyces* fungemia in a patient with AIDS. N.Y. State J. Med. 88: 278-279.

Severo, L.C., Bassanesi, M.C. & Londero, A.T., 1994. Tinea nigra: report of four cases observed in Rio Grande do Sul (Brazil) and a review of Brazilian literature. Mycopathologia 126: 157-162.

Severo, L.C., Festugato, M., Bernardi, C. & Londero, A.T., 1999a. Widespread cutaneous lesions due to *Sporothrix schenckii* in a patient under long term steroids therapy. Revta Inst. Med. Trop. S. Paulo 41: 59-62.

Severo, L.C., Vetoratto, G., Mattos Oliveira, F. de & Thomaz Londero, A., 1999b. Eumycetoma by *Madurella grisea*. Report of the first case observed in the Southern Brazilian region. Revta Inst. Med. Trop. S. Paulo 41: 139-142.

Severo, L.C., Resin Geyer, G., Silva Porto, N. da, Wagner, M.B. & Londero, A.T., 1997. Pulmonary *Aspergillus niger* intracavitary colonization. Report of 23 cases and a review of the literature. Revta Iberoam. Micol. 14: 104-110.

Shadomy, H.J. & Dixon, D.M., 1989. A new *Papulaspora* species from the infected eye of a horse: *Papulaspora equi* sp. nov. Mycopathologia 106: 35-39.

Shadomy, H.J. & Philpot, C.M., 1980. Utilization of standard laboratory methods in the laboratory diagnosis of problem dermatophytes. Am. J. Clin. Pathol. 74: 197-201.

Shao, J.Z., Liao, W.Q., Li, S.O., Wu, S.X., Zhang, J.Z. & Huang, J.J., 1983. Mycologic identification of *Emericella nidulans* and *Aspergillus flavus* caused pulmonary infection. Chin. Med. J. 96: 306-308.

Shaoxi, W., Ningro, G. & Guixia, L., 1996. A rare case of *Fusarium verticillioides* facial granuloma successfully treated by itraconazole. J. Mycol. Méd. 6: 88-90.

Sharma, R., Farmer, C.K.T., Grandsen, W.R & Ogg, C.S., 1998a. Peritonitis in continuous ambulatory peritoneal dialysis due to *Cylindrocarpon lichenicola* infection. Nephr. Dial. Transpl. 13: 2662-2664.

Sharma, R., Prem, R.R., Padhye, A.A. & Carzoli, R.P., 1994. Disseminated septic arthritis due to *Mucor ramosissimus* in a premature infant: a rare fungal infection and its successful management. Pediatr. Res. 35: 303A.

Sharma, R., Premachandra, B.R. & Carzoli, R.P., 1995. Disseminated septic arthritis due to *Mucor ramosissimus* in a premature infant: a rare fungal infection. J. Mycol. Méd. 5: 167-174.

Sharma, S., Singh, S.M. & Chatterjee, P.K., 1998b. Clinical, experimental and therapeutic aspects of mycotic corneal ulcer caused by *Cylindrocarpon lichenicola*. In S.M. Singh & J. Naidu (eds): Current Advances in Medical Mycology pp 157-166.

Sharma, V.D., Sethi, M.S. & Negi, S.K., 1971. Fungal flora of the respiratory tract of fowls. Poult. Sci. 50: 1041-1044.

Shek, Y.H., Tucker, M.C., Viciana, A.L., Manz, H.J. & Connor D.H., 1989. *Malassezia furfur*, disseminated infection in premature infants. Am. J. Clin. Pathol. 92: 595-603.

Shenep, J.L., English, B.K., Kaufman, L., Pearson, T.A., Thompson, J.W., Kaufman, R.A., Frisch, G. & Rinaldi, M.G., 1998. Successful medical therapy for deeply invasive facial infection due to *Pythium insidiosum* in a child. Clin. Infect. Dis. 27: 1388-1393.

Shenoy, S., Samuga, M., Anuradha, K.M., Kurian, M.M., Augustine, A., Anand, A.R. & Prasad, A., 1996. Intravenous catheter-related *Candida rugosa* fungaemia. Trop. Doctor 26: 31.

Shih, M.-H., Sheu, M.-M., Chen, H.-Y, & Lin, S.-R., 1999. Fungal keratitis caused by *Candida utilis* Kaohsiung J. Med. Sci. 15: 171-174.

Shimazono, Y., Isaki, K., Torii, H. & Otsuka, R., 1963. Brain abscess due to *Hormodendrum dermatitidis* (Kano) Conant, 1953. Report of a case and review of the literature. Folia Psych. Neurol. Jpn. 17: 80-96.

Shimizu, K., Tanaka, C., Peng, Y.-L. & Tsuda, M., 1998. Phylogeny of *Bipolaris* inferred from nucleotide sequences of *Brn1*, a reductase gene involved in melanin biosynthesis. J. Gen. Appl. Microbiol. 44: 251-258.

Shin, J.H., Lee, S.K., Suh, S.P., Ryang, D.W., Kim, N.H., Rinaldi, M.G. & Sutton, D.A., 1998. Fatal *Hormonema dematioides* peritonitis in a patient on continuous ambulatory peritoneal dialysis: criteria for organism identification and review of other known fungal agents. J. Clin. Microbiol. 36: 2157-2163.

Shin, S.B., Lee, H.N., Kim, S.W., Park, G.S., Cho, B.K. & Kim, H.J., 1998. Cutaneous abscess caused by *Paecilomyces lilacinus* in a renal transplant patient. Korean J. Med. Mycol. 3: 185-189.

Shing, M.M.K., Ip, M., Li, C.K., Chik, K.W. & Yuen, P.M.P., 1996. *Paecilomyces variotii* fungemia in a bone marrow transplant patient. Bone Marrow Transpl. 17: 281-283.

Shitara, T., Yugami, S.-I., Sotomatu, M., Oshima, Y., Ijima, H., Kuroume, T. & Matsumoto, T., 1993. Invasive aspergillosis in leukemic children. Pediatr. Hematol. Oncol. 10: 169-174.

Shlosberg, A., Zadikov, I., Perl, S., Yakobson, B., Varod, Y., Elad, D., Rapoport, E. & Handji, V., 1991. *Aspergillus clavatus* as the probable cause of a lethal mass neurotoxicosis in sheep. Mycopathologia 114: 35-39.

Shmuely, H., Kremer, I., Sagie, A. & Pitlik, S., 1997. *Candida tropicalis* multifocal endophthalmitis as the only initial manifestation of pacemaker endocarditis. Am. J. Ophthalmol. 123: 559-560.

Shoemaker, R.A., 1984. Canadian and some extralimital *Leptosphaeria* species. Can. J. Bot. 62: 2688-2729.

Shparago, N.I., Bruno, P.P. & Bennett, J., 1995. Systemic *Malassezia furfur* infection in an adult receiving total parenteral nutrition. J. Am. Osteopat. Ass. 95: 375-377.

Shrestha, S., Hennig, A. & Parija, S.C., 1998. Prevalence of rhinosporidiosis of the eye and its adnexa in Nepal. Am. J. Trop. Med. Hyg. 59: 231-234.

Shukla, P.K., Jain, M., Lal, B., Agrawal, P.K. & Srivastava, O.P., 1985. A study on keratomycoses caused by the species of *Aspergillus*. Biol. Mem. 11: 161-167.

Shukla, P.K., Jain, M., Lal, B., Agrawal, P.K. & Srivastava, O.P., 1989. Mycotic keratitis caused by *Phaeotrichoconis crotalariae*. New report. Mycoses 32: 230-232.

Shukla, P.K., Khan, Z.A., Lal, B., Agrawal, P.K. & Srivastava, O.P., 1983. Clinical and experimental keratitis caused by *Colletotrichum* state of *Glomerella cingulata* and *Acrophialophora fusispora*. Sabouraudia 21: 137-147.

Shukla, P.K., Khan, Z.A., Lal, B., Agrawal, P.K. & Srivastava, O.P., 1984a. A study on the association of fungi in human corneal ulcers and their therapy. Mykosen 27: 385-390.

Shukla, N.P., Rajak, R.K., Agrawal, G.P. & Gupta, D.K., 1984b. *Phoma minutispora* as a human pathogen. Mykosen 27: 260-263.

Shwayder, T., Andreae, M. & Babel, D., 1994. *Trichophyton equinum* from riding bareback: first reported U.S. case. J. Am. Acad. Derm. 30: 785-787.

Siddiqi, S.U. & Freedman, J.D., 1994. Isolated central nervous system mucormycosis. South. Med. J. 87: 997-1000.

Sides, E.H., Benson, J.D. & Padhye, A.A., 1991. Phaeohyphomycotic brain abscess due to *Ochroconis gallopavum* in a patient with malignant lymphoma of a large cell type. J. Med. Vet. Mycol. 29: 317-322.

Sigler, L., 1996. *Ajellomyces crescens* sp. nov., taxonomy of *Emmonsia* spp., and relatedness with *Blastomyces dermatitidis* (teleomorph *Ajellomyces dermatitidis*). J. Med. Vet. Mycol. 34: 303-314.

Sigler, L. & Abbott, S.P., 1997. Characterizing and conserving diversity of filamentous basidiomycetes from human sources. Microbiol. Cult. Coll. 13: 21-27.

Sigler, L., Abbott, S.P. & Woodgyer, A.J., 1994. New records of nail and skin infection due to *Onychocola canadensis* and description of its teleomorph *Arachnomyces nodosetosus* sp. nov. J. Med. Vet. Mycol. 32: 275-285.

Sigler, L., Bartley, J.R., Parr, D.H. & Morris, A.J., 1999a. Maxillary sinusitis caused by a medusoid form of *Schizophyllum commune*. J. Clin. Microbiol. 37: 3395-3398.

Sigler, L. & Carmichael, J.W., 1976. Taxonomy of *Malbranchea* and some other hyphomycetes with arthroconidia. Mycotaxon 4: 349-488.

Sigler, L. & Carmichael, J.W., 1983. Redisposition of some fungi referred to *Oidium microspermum* and a review of *Arthrographis*. Mycotaxon 18: 495-507.

Sigler, L. & Congly, H., 1990. Toenail infection caused by *Onychocola canadensis* gen. et. sp. nov. J. Med. Vet. Mycol. 28: 405-417.

Sigler, L., Estrada, S., Montealegre, N.A., Jaramillo, E., Arango, M., Bedout, C. de & Restrepo, A., 1997a. Maxillary sinusitis caused by *Schizophyllum commune* and experience with treatment. J. Med. Vet. Mycol. 35: 365-370.

Sigler, L., Flis, A.L. & Carmichael, J.W., 1998. The genus *Uncinocarpus* (Onygenaceae) and its synonym *Brunneospora*: new concepts, combinations and connections to anamorphs in *Chrysosporium*, and further evidence of relationship with *Coccidioides immitis*. Can. J. Bot. 76: 1624-1636

Sigler, L., Harris, J.L., Dixon, D.M., Flis, A.L., Salkin, I.F., Kemna, M. & Duncan, R.A., 1990. Microbiology and potential virulence of *Sporothrix cyanescens*, a fungus rarely isolated from blood and skin. J. Clin. Microbiol. 28: 1009-1015.

Sigler, L., Kibsey, P.C., Sutton, D.A., Abbott, S.P., Zilkie, E., McCarthy, D.I. & Fothergill, A., 1999b. *Monascus ruber*, causing renal infection. Abstr. ASM 26 F-8.

Sigler, L., Maza, L. de la, Tan, G., Egger, K.N. & Sherburne, R.K., 1995. Diagnostic difficulties caused by a nonclamped *Schizophyllum commune* isolate in a case of fungus ball of the lung. J. Clin. Microbiol. 33: 1979-1983.

Sigler, L., Summerbell, R.C., Poole, L., Wieden, M., Sutton, D.A., Rinaldi, M.G., Aguirre, M., Estes, G.W. & Galgiani, J.N., 1997b. Invasive *Nattrassia mangiferae* infections: case report, literature review, and therapeutic and taxonomic appraisal. J. Clin. Microbiol. 35: 433-440.

Sigler, L., Tsuneda, A. & Carmichael, J.W., 1981. *Phaeotheca* and *Phaeosclera*, two new genera of dematiaceous hyphomycetes and a redescription of *Sarcinomyces* Linder. Mycotaxon 12: 449-467.

Silva-Vergara, M.L., Martinez, R., Camargo, Z.P., Malta, M.H.B., Maffei, C.M.L. & Chadu, J.B., 2000. Isolation of *Paracoccidioides brasiliensis* from armadillos (*Dasypus novemcinctus*) in an area where the fungus was recently isolated from soil. Med. Mycol. 38: 193-199.

Silverira, E.R., Caligiorne, R.B., Mariano, V.S., Coura, W.A., Alkmim, L.D.A., Starling, C.E., Cruz, G.G., Benício, L.H.A., Paula, A.M., Gomes, J.A., Santos, G.D., Macedo, M.A.M., Salum, K.E., Gontijo, M.M. & Resende, M.A., 2001. Brain abscess in a renal transplant patient due to *Cladophialophora bantiana*. Transplantation (in press).

Simmons, E.G., 1967. Typification of *Alternaria, Stemphylium* and *Ulocladium*. Mycologia 59: 67-92.

Simmons, E.G., 1981. *Alternaria* themes and variations. Mycotaxon 13: 16-34.

Simmons, E.G., 1990. *Alternaria* themes and variations (27-53). Mycotaxon 37: 79-119.

Simmons, E.G., 1995a. *Alternaria* themes and variations (106-111). Mycotaxon 50: 409-427.

Simmons, E.G., 1995b. *Alternaria* themes and variations (112-144). Mycotaxon 55: 55-163.

Simmons, E.G., 1998. Multiplex conidium morphology in species of the *Ulocladium atrum* group. Can. J. Bot. 76: 1533-1539.

Simmons, E.G., McGinnis, M.R. & Rinaldi, M.G., 1987. *Dissitimurus*, a new dematiaceous genus of Hyphomycetes. Mycotaxon 30: 247-252.

Simmons, R.B. & Guého, E., 1990. A new species of *Malassezia*. Mycol. Res. 94: 1146-1149.

Simon, G., Rákóczy, G., Galgóczy, J., Verebély, T. & Bókay, J., 1991. *Acremonium kiliense* in oesophagus stenosis. Mycoses 34: 257-260.

Simon, G., Tóth, J. & Török, J., 2000. Fungal keratitis caused by *Plectosporium tabacinum*: first reported case. Abstr. 14th ISHAM Congr. B. Aires p. 266.

Simson, F.W., 1946. Chromoblastomycosis. Some observations on the types of the disease in South Africa. Mycologia 38: 432-449.

Singh, B.S., Singh, D.V., Singh, B.S. & Sinha, R.J., 1995. A variety of unusual strains of *Microsporum gypseum* in agra soils (India). Mycotaxon 54: 465-478.

Singh, M.P. & Singh, C.M., 1970. Fungi associated with superficial mycoses of cattle and sheep in India. Indian J. Anim. Health. 9:75-77.

Singh, M.S., Singh, M. & Mukherjee, S., 1992. Pathogenicity of *Sporotrichum pruinosum* and *Cladosporium oxysporum*, isolated from the bronchial secretions of a patient, for laboratory mice. Mycopathologia 117: 145-152.

Singh, N., Gayowski, T., Singh, J. & Yu, V.L., 1995. Invasive gastrointestinal zygomycosis in a liver transplant recipient: case report and review of zygomycosis in solid-organ transplant recipients. Clin. Infect. Dis. 20: 617-620.

Singh, P.N., Ranjana, K., Singh, Y.I., Singh, K.P., Sharma, S.S., Kulachandra, M., Nabakumar, Y., Chakrabarti, A., Padhye, A.A., Kaufman, L. & Ajello, L., 1999. Indigenous disseminated *Penicillium marneffei* infection in the state of Manipur, India: report of four autochtonous cases. J. Clin. Microbiol. 37: 2699-2702.

Singh, S.M., Agrawal, A., Naidu, J., Hoog, G.S. de & Figueras, M.J., 1992. Cutaneous phaeohyphomycosis caused by *Phialophora richardsiae* and the effect of topical clotrimazole in its treatment. Antonie van Leeuwenhoek 61: 51-55.

Singh, S.M. & Barde, A.K., 1985. *Monodictys indica* sp. nov. as a saprophyte but transitory fungus on human skin. Curr. Sci. 54: 1001-1003.

Singh, S.M. & Barde, A.K., 1986. Opportunistic infections of skin and nails by non-dermatophytic fungi. Mykosen 29: 272-277.

Singh, S.M., Naidu, J., Jain, S., Nawange, S.R. & Dhindsa, M.K., 1996. Maxillary sinusitis caused by *Ascotricha chartarum* Berk. (anamorph *Dicyma ampullifera* Boul.): a new phaeoid opportunistic human pathogen. J. Med. Vet. Mycol. 34: 215-218.

Singh, S.M., Naidu, J. & Pouranik, M., 1990a. Ungual and cutaneous phaeohyphomycosis caused by *Alternaria alternata* and *Alternaria chlamydospora*. J. Med. Vet. Mycol. 28: 275-278.

Singh, S.M., Sharma, S. & Chatterjee, P.K., 1990b. Clinical and experimental mycotic keratitis caused by *Aspergillus terreus* and the effect of subconjunctival oxiconazole treatment in the animal model. Mycopathologia 112: 127-137.

Singh, V.R., Smith, D.K., Lawrence, J., Kelly, P.C., Thomas, A.R., Spitz, B. & Sarosi, G.A., 1996. Coccidioidomycosis in patients infected with human immunodeficiency virus: review of 91 cases at a single institution. Clin. Infect. Dis. 23: 563-568.

Sinnott, J.T., Rodnite, J., Emmanuel, P.J. & Campos, A., 1989. *Cryptococcus laurentii* infection complicating peritoneal dialysis. Pediatr. Infect. Dis. J. 8: 803-805.

Sirisanthana, V. & Sirisanthana, T., 1995. Disseminated *Penicillium marneffei* infection in human immunodeficiency virus-infected children. Pediatr. Infect. Dis. J. 14: 935-940.

Sivanesan, A., 1987. Graminicolous species of *Bipolaris, Curvularia, Drechslera, Exserohilum* and their teleomorphs. Mycol. Pap. 158: 1-261.

Sivanesan, A., 1990. CMI Descriptions of fungi and bacteria 1007. Mycopathologia 111: 123-124.

Sivanesan, A., 1991. IMI Descriptions of Fungi an Bacteria 1038. Mycopathologia 114: 59-60.

Slifkin, M. & Bowers, H.M., 1975. *Phialophora mutabilis* endocarditis. Am. J. Clin. Pathol. 63: 120-130.

Slomovic, A.R., Forster, R.K. & Gelender, H., 1985. *Lasiodiplodia theobromae* panophthalmitis. Can. J. Ophthalmol. 20: 225-228.

Slunt, J.B., Taketomi, E.A., Woodfolk, J.A., Hayden, M.L. & Platt-Mills, T.A.E., 1996. The immune response to *Trichophyton tonsurans*. J. Immunol. 157: 5192-5197.

Smith, A.G., Bustamante, C.I. & Gilmor, G.D., 1989. Zygomycosis (absidiomycosis) in an AIDS patient. Absidiomycosis in AIDS. Mycopathologia 105: 7-10.

Smith, A.G., Bustamante, C.I. & Ood, C.W., 1993. Disseminated cutaneous and vascular invasion by *Fusarium moniliforme* in a fatal case of acute lymphocytic leukemia. Mycopathologia 122: 15-20.

Smith, A.G., Muhvich, A.G., Muhvich, K.H. & Wood, C., 1989. Fatal *Fusarium solani* infections in baby sharks. J. Med. Vet. Mycol. 27: 83-91.

Smith, G., 1943. Two new species of *Aspergillus*. Trans. Br. Mycol. Soc. 26: 25-27.

Smith, J.M.B. & Griffin, J.F.T., 1995. Strategies for the development of a vaccine against ringworm. J. Med. Vet. Mycol. 33: 87-91.

Smith, J.M.B., Jolly, R.D., Georg, L.K. & Connole, M.D., 1968. *Trichophyton equinum* var. *autotrophicum*. Its characteristics and geographical distribution. Sabouraudia 6: 296-304.

Smith, J.M.B. & Marples, M.J., 1963. *Trichophyton mentagrophy-tes* var. *erinacei*. Sabouraudia 3: 1-10.

Smith, M.Th., 1986. *Zygoascus hellenicus* gen. nov., sp. nov., the teleomorph of *Candida hellenica* (= *C. inositophila* = *C. steatolytica*). Antonie van Leeuwenhoek 52: 25-37.

Smith, M.Th. & Batenburg-van der Vegte, W.H., 1985. Ultrastructure of septa in *Blastobotrys* and *Sporothrix*. Antonie van Leeuwenhoek 51: 121-128.

Smith, M.Th., Cock, A.W.A.M. de, Poot, G. & Steensma, H.Y., 1995. Genome comparisons in the yeast-like fungal genus *Galactomyces* Redhead et Malloch. Int. J. Syst. Bact. 45: 826-831.

Smith, M.Th. & Poot, G.A., 1998. *Dipodascus capitatus, Dipodascus spicifer and Geotrichum clavatum:* genomic characterization. Antonie van Leeuwenhoek 74: 229-235.

Smith, M.Th., Poot, G.A. & Cock, A.W.A.M. de, 2000. Re-examination of some species of the genus *Geotrichum* Link: Fr. Antonie van Leeuwenhoek 77: 71-81.

Smith, M.Th., Walt, J.P. van der & Johannsen, E., 1976. The genus *Stephanoascus* gen. nov. (*Ascoideaceae*). Antonie van Leeuwenhoek 42: 119-127.

Smolyanskaya, A.Z., Dronova, O.M., Dmitrieva, N.V., Sokolova, V.I., Zhabina, M.I. & Sokolova, E.N., 1996. Infections of fungal etiology in cancer patients. Klinich. Lab. Diagn. 1: 30-32.

Sneath, P.H.A., Mair, N.S., Sharpe, M.E. & Holt, J.G. (eds), 1986. Bergey's Manual of Systematic Bacteriology, vol. 2: 965-1599. Williams & Wilkins, Baltimore.

Snidvongs, M.L.K., Supanakorn, S. & Supiyaphun, P., 1998. Severe epitaxis from rhinosporidiosis: a case report. J. Med. Ass. Thail. 81: 555-558.

Sobol, S.M., Love, R.G., Stutman, H.R. & Pysher, T.J., 1984. Phaeohyphomycosis of the maxilloethmoid sinus caused by *Drechslera spicifera:* a new fungal pathogen. Laryngoscope 94: 620-627.

Sobottka, I., Daneke, J., Pothman, W., Heinemann, A. & Mack, D., 1999. Fatal native valve endocarditis due to *Scedosporium apiospermum (Pseudallescheria boydii)* following trauma. Eur. J. Clin. Microbiol. Infect. Dis. 18: 387-389.

Soliman, R., Ebeid, M., Essa, M., Abd El-Hamid, M.A., Khamis, Y. & Said, A.H., 1991. Ocular histoplasmosis due to *Histoplasma farciminosum* in Egyptian donkeys. Mycoses 34: 261-266.

Solomon, S.L., Khabbaz, R.F., Parker, R.H., Anderson, R.I., Geraghty, M.A., Furman, R.M. & Martone, W.J., 1984. An outbreak of *Candida parapsilosis* bloodstream infections in patients receiving parenteral nutrition. J. Infect. Dis. 149: 98-102.

Southern, P.M., 1996. Mycetoma due to *Madurella grisea* acquired in Mexico. Trop. Doctor 26: 187-188.

Spapen, H., Fiasse, M., Diltoer, M., Deuvaert, F.E. & Huyghens, L., 1995. Catheter-related intracardiac thrombosis: a rare complication of *Candida glabrata* sepsis. Acta Clin. Belg. 50: 314-317.

Sparkes, A.H., Werrett, G., Stokes, C.R. & Gruffydd-Jones, T.J., 1994. Inapparent carriage by cats and the viability of arthrospores. J. Small Anim. Pract. 35: 397-401.

Sparrow, S.A., Hallam, L.A., Wild, B.E. & Baker, D.L., 1992. *Scedosporium inflatum:* first case report of disseminated infection and review of the literature. Pediatr. Haematol. Oncol. 9: 293-295.

Spatafora, J.W., Mitchell, T.G. & Vilgalys, R., 1995. Analysis of genes coding for small-subunit rRNA sequences in studying phylogenetics of dematiaceous fungal pathogens. J. Clin. Microbiol. 33: 1322-1326.

Speed, B. & Dunt, D., 1995. Clinical and host differences between infections with the two varieties of *Cryptococcus neoformans*. Clin. Infect. Dis. 21: 28-34.

Speers, D.-J., Cole, C.H. & Wild, B.E., 1995. *Candida tropicalis* infection in childhood leukaemia. Aust. N. Zeal. J. Med. 25: 545.

Speller, D.C.E. & MacIver, A.C., 1971. Endocarditis caused by a *Coprinus* species: a fungus of toadstool group. J. Med. Microbiol. 4: 370-374.

Spinillo, A., Capuzzo, E., Egbe, T.O., Baltaro, F., Nicola, S. & Piazzi, G., 1995. *Torulopsis glabrata* vaginitis. Obstet. Gynecol. 85: 993-998.

Spitzer, E.D., Keath, E.J., Travis, S.J., Painter, A.A., Kobayashi, G.S. & Medoff, G., 1990. Temperature-sensitive variants of *Histoplasma capsulatum* isolated from patients with acquired immunodeficiency syndrome. J. Infect. Dis. 162: 258-261.

Spitzer, E.D., Lasker, B.A., Travis, S.J., Kobayashi, G.S. & Medoff, G., 1989. Use of mitochondrial and ribosomal DNA polymorphisms to classify clinical and soil isolates of *Histoplasma capsulatum*. Infect. Immun. 57: 1409-1412.

Spitzer, E.D. & Spitzer, S.G., 1992. Use of a dispersed repetitive DNA element to distinguish clinical isolates of *Cryptococcus neoformans*. J. Clin. Microbiol. 30: 1094-1097.

Sposto, M.R., Mendes-Giannini, M.J., Moraes, R.A., Branco, F.C. & Scully, C., 1994. Paracoccidioidomycosis manifesting as oral lesions: clinical, cytological and serological investigation. J. Oral. Pathol. Med. 23: 85-87.

Sposto, M.R., Scully, C., Almeida, O.P. de, Jorge, J., Graner, E. & Bozzo, L., 1993. Oral paracoccidioidomycosis: a study of 36 South American patients. Oral Surg. Oral Med. Oral Pathol. 75: 461-465.

Spreadbury, C., Holden, D., Aufauvre-Brown, A., Bainbridge, B. & Cohen, J., 1993. Detection of *Aspergillus fumigatus* by polymerase chain reaction. J. Clin. Microbiol. 31: 615-621.

Squeo, R.F., Beer, R., Silvers, D., Weitzman, I. & Grossman, M., 1998. Invasive *Trichophyton rubrum* resembling blastomycosis infection in the immunocompromised host. J. Am. Acad. Derm. 39: 379-380.

Srdić, N., Radulović, Š., Nonković, Z., Velimirović, S., Cvetković, L. & Vico, I., 1993. Two cases of exogenous endophthalmitis due to *Fusarium moniliforme* and *Pseudomonas* species as associated aetiological agents. Mycoses 36: 441-444.

Srinivasa Rao, P.N. & Ramakrishnan, T.S., 1968. Studies on the fungus diseases affecting man in and around Manipal. III. Fungi and diseases of the eye. Indian J. Pathol. Bacteriol. 11: 53-60.

Srivastava, O.P., Lal, B., Agrawal, P.K., Agarwal, S.C., Chandra, B. & Mathur, I.S., 1977. Mycotic keratitis due to *Rhizoctonia* sp. Sabouraudia 15: 125-131.

Srivastava, S., Kleinman, G. & Manthous, C.A., 1996. *Torulopsis* pneumonia: a case report and review of the literature. Chest 110: 858-861.

Staib, F., 1987. Kryptokokkose bei AIDS aus mykologisch-diagnostischer und -epidemiologischer Sicht. AIDS-Forsch. 2: 363-382.

Staib, F. & Blisse, A., 1974. Stellungnahme zur *Sporothrix schenckii* var. *luriei*. Ein Beitrag zum diagnostischen Wert der Assimilation von Kreatinin, Kreatin und Guanidinessigsäure durch *Sporothrix schenckii*. Ztbl. Bakt. Parasitkde A229: 261-263.

Staib, P. & Morschhäuser, J., 1999. Chlamydospore formation on Staib agar as a species-specific characteristic of *Candida dubliniensis*. Mycoses 42: 521-524.

Staley, J.T., Palmer, F. & Adams, J.B., 1982. Microcolonial fungi: common inhabitants of desert rocks? Science 215: 1093-1095.

Stalpers, J.A., 1978. Identification of wood-inhabiting *Aphyllo-phorales* in pure culture. Stud. Mycol. 16: 1-248.

Stalpers, J.A., 1984. A revision of the genus *Sporotrichum*. Stud. Mycol. 24: 1-105.

Standaert, S.M., Schaffner, W., Galgiani, J.N., Pinner, R.W., Kaufman, L., Durry, E. & Hutcheson, R.H., 1995. Coccidioido-mycosis among visitors to a *Coccidioides immitis* endemic area: an outbreak in a military reserve unit. J. Infect. Dis. 171: 1672-1675.

Standard, P.G. & Kaufman, L., 1985. Exoantigen test. Rapid identification of pathogenic mould isolates by immunodiffusion. CDC, Atlanta, 61 pp.

Standard, P.G., Padhye, A.A. & Kaufman, L., 1991. Exoantigen test for the rapid identification of *Exophiala spinifera*. J. Med. Vet. Mycol. 29: 273-277.

Stchigel, A.M., Cano, J. & Guarro, J., 1999. A new species of *Emericella* and a rare morphological variant of *E. quadrilineata*. Mycol. Res. 103: 1057-1064.

Steele, P.E., 1992. Current concepts of virulence. Adv. Path. Lab. Med. 4: 107-119.

Steinberg, G.K., Britt, R.H., Enzmann, D.R., Finlay, J.L. & Arvim, A.M., 1983. *Fusarium* brain abscess. J. Neurosurg. 56: 598-601.

Steinfeld, P., Durez, P., Hauzeur, J.-P., Motte, S. & Appelboom, T., 1997. Articular aspergillosis: two case reports and review of the literature. Br. J. Rheumatol. 36: 1331-1334.

Stenson, J., Brookner, A. & Rosenthal, S., 1982. Bilateral endogenous necrotizing scleritis due to *Aspergillus oryzae*. Annls Ophthalmol. 14: 67-72.

Stenwig, H. & Taksdal, T., 1984. Isolation of *Epidermophyton floccosum* from a dog in Norway. Sabouraudia 22: 171-172.

Stephens, C.P. & Gibson, J.A., 1997. Disseminated zygomycosis caused by *Conidiobolus incongruus* in a deer. Austr. Vet. J. 75: 358-359.

Sterflinger, K., 1998. Temperature and NaCl-tolerance of rock-inhabiting meristematic fungi. Antonie van Leeuwenhoek 74: 271-281.

Sterflinger, K., De Baere, R., Hoog, G.S. de, De Wachter, R., Krumbein, W. & Haase, G., 1997. *Coniosporium perforans* and *C. apollinis*, two new rock-inhabiting fungi isolated from marble in the Sanctuary of Delos (Cyclades, Greece). Antonie van Leeuwenhoek 72: 349-363.

Sterflinger, K. & Hain, M., 1999. *In situ* hybridization with rRNA targeted probes as a new tool for the detection of black yeasts and meristematic fungi. Stud. Mycol. 43: 23-30.

Sterflinger, K., Hoog, G.S. de & Haase, G., 1999. Phylogeny and ecology of meristematic ascomycetes. Stud. Mycol. 43: 5-22.

Stevens, D.A., 1989. The interface of mycology and endocrinology. J. Med. Vet. Mycol. 27: 133-140.

Stewart, N.J., Munday, B.L. & Hawkesford, T., 1999. Isolation of *Mucor circinelloides* from a case of ulcerative mycosis of platypus (*Ornithorhynchus anatinus*), and a comparison of *Mucor circinelloides* and *Mucor amphibiorum* to different culture temperatures. Med. Mycol. 37: 201-206.

Stewart, W.M., Pietrini, P., Lauret, P., Thomine, E. & Leroy, D., 1975. Granulome mycosique péri-narinaire dû à *Taeniolella stilbospora* (Corda) Hughes. Bull. Soc. Fr. Derm. Syph. 82: 396-397, 1975.

St.-Germain, G. & Laverdière, M., 1986. *Torulopsis candida*, a new opportunistic pathogen. J. Clin. Microbiol. 24: 884-885.

St.-Germain, G., Robert, A., Ishak, M., Tremblay, C. & Claveau, S., 1993. Infection due to *Rhizomucor pusillus*: report of four cases in patients with leukemia and review. Clin. Infect. Dis. 16: 640-645.

Still, J.M., Belcher, K. & Law, E.J., 1995. Management of *Candida* septicaemia in a regional burn unit. Burns 21: 594-596.

Still, J.M., Orlet, K. & Law, E.J., 1994. *Trichosporon beigelii* septicaemia in a burn patient. Burns 20: 467-468.

Stiller, M., Gordon, M., Rosenthal, S., Potter, J., Tepperman, L. & Schupack, J., 1994. Primary cutaneous infection by *Aspergillus ustus* in a liver-transplant recipient. Abstr. Gen. Meet. ASM 94: 603.

Stiller, M.J., Rosenthal, S., Summerbell, R.C., Pollack, M.S. & Chan, A., 1992. Onychomycosis of the toenails caused by *Chaetomium globosum*. J. Am. Acad. Derm. 26: 775-776.

Stobiersky, M.G., Hospedales, C.J., Hall, W.N., Robinson-Dunn, B., Hoch, D. & Sheill, D.A., 1996. Outbreak of histoplasmosis among employees in a paper factory - Michigan 1993. J. Clin. Microbiol. 34: 1220-1223.

Stockdale, P.M., 1963. The *Microsporum gypseum* complex *(Nannizzia incurvata* Stockd., *N. gypsea* (Nann.) comb. nov., *N. fulva* sp. nov.). Sabouraudia 3: 114-126.

Stockdale, P.M., 1965. CMI Descriptions of Pathogenic Fungi and Bacteria 61-70.

Stockdale, P.M., 1967. *Nannizzia persicolor* sp. nov., the perfect state of *Trichophyton persicolor*. Sabouraudia 5: 355-359.

Stockdale, P.M., 1980. CMI Descriptions of Pathogenic Fungi and Bacteria 641-650.

Stockdale, P.M., MacKenzie, D.W.R. & Austwick, P.K.C., 1965. *Arthroderma simii* sp. nov., the perfect state of *Trichophyton simii* (Pinoy) comb. nov. Sabouraudia 4: 112-123.

Stockman, L., Clark, K.A., Hunt, J.M. & Roberts, G.D., 1993. Evaluation of commercially available acridinium ester-labeled chemiluminescent DNA probes for culture identification of *Blastomyces dermatitidis, Coccidioides immitis, Cryptococcus neoformans*, and *Histoplasma capsulatum*. J. Clin. Microbiol. 31: 845-850.

Stolk, A.C., 1955. The genera *Anixiopsis* Hansen and *Pseudeurotium* van Beyma. Antonie van Leeuwenhoek 21: 65-79.

Stolk, A.C., 1965. Thermophilic species of *Talaromyces* Benjamin and *Thermoascus* Miehe. Antonie van Leeuwenhoek 31: 262-276.

Stolk, A.C. & Samson, R.A., 1992. The genus *Talaromyces*. Studies on *Talaromyces* and related genera II. Stud. Mycol. 2: 1-65.

Stringer, J.R., 1996. *Pneumocystis carinii*: what is it, exactly? Clin. Microbiol. Rev. 9: 489-498.

Sturm, A.W., Grave, W. & Kwee, W.S., 1989. Disseminated *Fusarium oxysporum* infection in a patient with heatstroke. Lancet 29: 968.

Su, Y.-C., Huang, H., Liu, X.-Y. & Zheng, R.-Y., 1999. Systematic relationship of several controversial *Cunninghamella* taxa inferred from sequence comparisons of ITS2 of rDNA. Mycol. Res. 103: 805-810.

Subrahmanyam, A. & Lakshmi, B.V., 1993. *Thermomucor indicae-seudaticae* on human skin. Mycoses 36: 201-202.

Sugar, A.M., 1991. Cryptococcosis in the patient with AIDS. Mycopathologia 114: 153-157.

Sugar, A.M. & Stevens, D.A., 1985. *Candida rugosa* in immunocompromised infection. Case reports, drug susceptibility, and review of the literature. Cancer 56: 318-320.

Sughayer, M., DeGirolami, P.C., Khettry, U., Korzeniowski, D., Grumney, A., Pasarell, L. & McGinnis, M.R., 1991. Human infection caused by *Exophiala piscyphila*: case report and review. Rev. Infect. Dis. 13: 379-382.

Sugita, T., Kagaya, K., Takashima, M., Suzuki, M., Fukazawa, Y. & Nakase, T., 1999. A clinical isolate of *Candida palmioleophila* formerly identified as *Torulopsis candida*. Jpn. J. Med. Mycol. 40: 21-25.

Sugita, T. & Nakase, T., 1998. Molecular phylogenetic study of the basidiomycetous anamorphic yeast genus *Trichosporon* and related taxa based on small subunit ribosomal DNA sequences. Mycoscience 39: 7-13.

Sugita, T., Nishikawa, A., Ikeda, R., Shinoda, T., Sakashita, H., Sakai, Y. & Yoshizawa, Y., 1998a. First report of *Trichosporon ovoides* isolated from the home of a summer-type hypersensitivity pneumonitis patient. Microbiol. Immunol. 42: 475-478.

Sugita, T., Nishikawa, A. & Shinoda, T., 1994. Reclassification of *Trichosporon cutaneum* by DNA relatedness by the spectrophotometric method and the chemiluminometric method. J. Gen. Appl. Microbiol. 40: 397-408.

Sugita, T., Nishikawa, A. & Shinoda, T., 1998b. Rapid detection of species of the opportunistic yeast *Trichosporon* by PCR. J. Clin. Microbiol. 36: 1458-1460.

Sugita, T., Nishikawa, A., Shinoda, T. & Kume, H., 1995. Taxonomic position of deep-seated, mucosa-associated, and superficial isolates of *Trichosporon cutaneum* from trichosporonosis patients. J. Clin. Microbiol. 33: 1368-1370.

Sugita, Y., Kanaizuka, I., Nakajima, H., Ibe, M., Yokota, S. & Matsuyama, S., 1993. Detection of *Candida albicans* DNA in cerebrospinal fluid. J. Med. Vet. Mycol. 31: 353-358

Sugiyama, M., Ohara, A. & Mikawa, T., 1999. Molecular phylogeny of onygenalean fungi based on small subunit ribosomal DNA (SSU rDNA) sequences. Mycoscience 40: 251-258.

Sukroongreung, S., Kitiniyom, K., Nilakul, C. & Tantimavanich, S., 1998. Pathogenicity of basidiospores of *Filobasidiella neoformans* var. *neoformans*. Med. Mycol. 36: 419-424.

Sullivan, D.J., Bennett, D., Henman, M., Harwood, P., Flint, S., Mulcahy, F., Shanley, D. & Coleman, D., 1993. Oligonucleotide fingerprinting of isolates of *Candida* species other than *C. albicans* and of a typical *Candida* species from human immunodeficiency virus-positive and AIDS patients. J. Clin. Microbiol. 31: 2124-2133.

Sullivan, D. & Coleman, D., 1997. *Candida dubliniensis*: an emerging opportunistic pathogen. Curr. Topics Med. Mycol. 8: 15-25.

Sullivan, D.J., Westerneng, T.J., Haynes, K.A., Bennett, D.E. & Coleman, D.C., 1995. *Candida dublinensis* sp. nov.: phenotypic and molecular characterization of a novel species associated with oral candidosis in HIV-infected individuals. Microbiology, Reading 141: 1507-1521.

Summerbell, R.C., 1987. *Trichophyton kanei*, sp. nov., a new anthropophilic dermatophyte. Mycotaxon 38: 509-523.

Summerbell, R.C., 2000. Form and function in the evolution of dermatophytes. Revta Iberoam. Micol. 17: 32-45.

Summerbell, R.C., Haugland, R.A., Li, A. & Gupta, A.K., 1999. rRNA gene internal transcribed spacer 1 and 2 sequences of a sexual, anthropophilic dermatophytes related to *Trichophyton rubrum*. J. Clin. Microbiol. 37: 4005-4011.

Summerbell, R.C., Kane, J. & Krajden, S., 1989a. Onychomycosis, tinea pedis and tinea manuum caused by non-dermatophytic filamentous fungi. Mycoses 32: 609-619.

Summerbell, R.C., Kane, J., Krajden, S. & Duke, E.E., 1993a. Medically important *Sporothrix* species and related ophiostomatoid fungi. In M.J. Wingfield *et al.* (eds): *Ceratocystis* and *Ophiostoma*, pp. 185-192. APS, St. Paul.

Summerbell, R.C., Krajden, S. & Kane, J., 1989b. Potted plants in hospitals as reservoirs of pathogenic fungi. Mycopathologia 106: 13-22.

Summerbell, R.C., Krajden, S., Kane, J., Levine, R. & Fuksa, M., 1993b. Subcutaneous phaeohyphomycosis caused by *Lasiodiplodia theobromae*. Abstr. Gen. Meet. ASM 93: 534.

Summerbell, R.C., Repentigny, L. de, Chartrand, C. & St. Germain, G., 1992. Graft-related endocarditis caused by *Neosartorya fischeri* var. *spinosa*. J. Clin. Microbiol. 30: 1580-1582.

Summerbell, R.C., Richardson, S.E. & Kane, J., 1988a. *Fusarium proliferatum* as an agent of disseminated infection in an immunosuppressed patient. J. Clin. Microbiol. 7: 589-594.

Summerbell, R.C., Rosenthal, S.A. & Kane, J., 1988b. Rapid method for differentiation of *Trichophyton rubrum*, *Trichophyton mentagrophytes*, and related dermatophyte species. J. Clin. Microbiol. 26: 2279-2282.

Sun, S.H., Cole, G.T., Drutz, D.J. & Harrison, J.L., 1986. Electron-microscopic observations of the *Coccidioides immitis* parasitic cycle *in vivo*. J. Med. Vet. Mycol. 24: 183-192.

Sutton, B.C., 1973. *Pucciniopsis, Mycoleptodiscus* and *Amerodiscosiella*. Trans. Br. Mycol. Soc. 60: 525-536.

Sutton, B.C., 1980. The Coelomycetes. Fungi with Pycnidia, Acervuli and Stromata. CAB, Kew, 696 pp.

Sutton, B.C., Campbell, C.K. & Goldschmied-Rouven, A., 1991. *Pseudomicrodochium fusarioides* sp. nov., isolated from human bronchial fluid. Mycopathologia 114: 159-161.

Sutton, B.C. & Dyko, B.J., 1989. Revision of *Hendersonula*. Mycol. Res. 93: 466-488.

Sutton, D.A., 1999. Coelomycetous fungi in human disease. A review: clinical entities, pathogenesis, identification and therapy, Revta Iberoam. Micol. 16: 171-179.

Sutton, D.A., Fothergill, A.W. & Rinaldi, M.G., 1998. Guide to Clinically Significant Fungi. Williams & Wilkins, Baltimore.

Sutton, D.A., Sanche, S.E., Revankar, S.G., Fothergill, A.W. & Rinaldi, M.G., 1999a. *In vitro* amphotericin B resistance in clinical isolates of *Aspergillus terreus*, with a head-to-head comparison to voriconazole. J. Clin. Microbiol. 37: 2343-2345.

Sutton, D.A., Sigler, L., Kalassian, K.G., Fothergill, A.W. & Rinaldi, M.G., 1997. Pulmonary *Acrophialophora fusispora*: case history, literature review and mycology. Abstr. 13th ISHAM Congr., Parma, p. 160.

Sutton, D.A., Slifkin, M., Yakulis, R. & Rinaldi, M.G., 1998. U.S. Case report of cerebral phaeohyphomycosis caused by *Ramichloridium obovoideum (R. mackenziei)*: criteria for identification, therapy, and review of other known dematiaceous neurotropic taxa. J. Clin. Microbiol. 36: 708-715.

Sutton, D.A., Timm, W.D., Morgan-Jones, G. & Rinaldi, M.G., 1999b. Human phaeohyphomycotic osteomyelitis caused by the coelomycete *Phomopsis* Saccardo 1905: criteria for identification, case history, and theraphy. J. Clin. Microbiol. 37: 807-811.

Suzaki, A., Kimura, M., Kimura, S., Shimada, K., Miyaji, M. & Kaufman, L., 1995. An outbreak of histoplasmosis contacted by Japanese visitors to a bat-inhabited cave near Manaus, Brazil. J. Mycol. Méd. 5: 40-43.

Suzuki, K., Kawazaki, M. & Ishizaki, H., 1988. Analysis of restriction profiles of mitochondrial DNA from *Sporothrix schenckii* and related fungi. Mycopathologia 103: 147-151.

Suzuki, Y., Udagawa, S., Wakita, H., Yamada, N., Ichikawa, H., Furukawa, F. & Takigawa, M., 1998. Subcutaneous phaeohyphomycosis caused by *Geniculosporium* species; a new fungal pathogen. Br. J. Derm. 138: 346-350.

Swerdloff, J.N., Filler, S.G. & Edwards, J.E., 1993. Severe candidal infections in neutropenic patients. Clin. Infect. Dis., Suppl. 17: S457-S467.

Swietliczkowa, I., Szusterowska-Martinowa, E. & Braciak, W., 1984. Ocena kliniczna 1% ma´sci Clotrimazol w leczeniu grzzybic rogówki. Klin. Oczna 86: 221-223.

Swinne, D. & Kayembe, K., 1987. Cryptococcosis. Health Coop. Pap. 8: 175-180.

Switchenko, A.C., Miyada, C.G., Goodman, T.C., Walsh, T.J., Wong, B., Becker, M.J. & Ullman, E.F., 1994. An automated enzymatic method for measurement of D-arabinitol, a metabolite of pathogenic *Candida* species. J. Clin. Microbiol. 32: 92-97.

Szathmáry, S., 1960. Die Cleistothecium-Bildung von *Ctenomyces trichophyticus - Trichophton gypseum* var. *asteroides* - auf sterili-siertem Heu. Mykosen 3: 77-83.

Szilagyi, G. & Reiss, F., 1968. *Trichophyton rubrum* (Castellani) var. *flava,* var. nova. A yellow pigment forming *Trichophyton rubrum*. Mycopath. Mycol. Appl. 36: 193-198.

Taborda, C.P. & Camargo, Z.P., 1994. Diagnosis of paracoccidioidomycosis by dot immunobinding assay for antibody detection using the purified and specific antigen gp43. J. Clin. Microbiol. 32: 554-556.

Tack, K.J., Rhame, F.S., Brown, B. & Thompson, R.C., 1982. *Aspergillus* osteomyelitis: report of four cases and review of the literature. Am. J. Med. 73: 295-300.

Tadros, T.S., Workowski, K., Siegel, R.J., Hunter, S. & Schwartz, D.A., 1998. Pathology of hyalohyphomycosis caused by *Scedosporium apiospermum (Pseudallesheria boydii)*: an emerging mycosis. Human Path. 29: 1266-1272.

Taillandier, J., Alemanni, M., Cerrina, J., Roy Ladurie, F. le & Dartevelle, P., 1997. *Aspergillus* osteomyelitis after heart-lung transplantation. J. Heart Lung Transpl. 16: 436-438.

Takashio, M., 1973. Étude des phenomènes de réproduction liés au viellissement et au rajeumissement des cultures de champignons. Annls Soc. Belge. Méd. Trop. 53: 480-492.

Takashio, M., 1975a. The *Trichophyton mentagrophytes* complex. In K. Iwata (ed.): Recent Advances in Medical and Veterinary Mycology, pp. 271-276. Univ. Tokyo, Tokyo.

Takashio, M., 1975b. Contribution à l'étude d'*Arthroderma benhamiae* var. *erinacei* et de souches du hérisson Africain *Erinaceus albiventris* Wagner. Bull. Soc. Fr. Mycol. Méd. 4: 47-50.

Takashio, M. & Vanbreuseghem, R., 1971. Production of ascospores by *Piedraia hortai in vitro*. Mycologia 63: 612-618.

Takashio, M. & Vroey, C. de, 1975. Piedra noire chez des chimpanzes du Zaire. Sabouraudia 13: 58-62.

Takayasu, S., Akagi, M. & Shimizu, Y., 1977. Cutaneous mycosis caused by *Paecilomyces lilacinus*. Arch. Derm. 113: 1687-1690.

Take, M., 1980. A case of chromomycosis. Jpn. J. Derm. 90: 1039.

Takeo, K. & Hoog, G.S. de, 1991. Karyology and hyphal characters as taxonomic criteria in ascomycetous black yeasts and related fungi. Antonie van Leeuwenhoek 60: 35-42.

Takeo, K. & Nakai, K., 1986. Mode of cell growth of *Malassezia (Pityrosporum)* as revealed by using plasma membrane configurations as natural markers. Can. J. Microbiol. 32: 389-394.

Talice, R.-V. & Mackinnon, J.-E., 1929. *Penicillium bertai* n.sp. agent d'une mycose broncho-pulmonaire de l'homme. Annls Parasit. Hum. Comp. 7: 97-106.

Talwar, P. & Sehgal, S.C., 1978. Mycotic infections of the eye in Chandigarh and neighbouring areas. Ind. J. Med. Res. 67: 929-933.

Tambini, R., Farina, C., Fiocchi, R., Dupont, B., Guého, E., Delvecchio, G., Mamprin, F. & Gavazzeni, G., 1996. Possible pathogenic role for *Sporothrix cyanescens* isolated from a lung lesion in a heart transplant patient. J. Med. Vet. Mycol. 34: 195-198.

Tamura, M., Watanabe, K., Imai, T., Mikami, Y. & Nishimura, K., 2000. New PCR primer pairs specific for *Candida dubliniensis* and detection of the fungi from the *Candida albicans* clinical isolates in Japan. Clin. Lab. 46: 33-40.

Tan, G., Kaufman, L., Peterson, E.M. & Mada, L.M. de la, 1993. Disseminated atypical blastomycosis in two patients with AIDS. Clin. Infect. Dis. 16: 107-111.

Tan, H.P., Wahlstrom, H.E., Zamora, J.U. & Hassanein, T., 1997. *Aureobasidium* pneumonia in a post liver transplant recipient: a case report. Hepat. Gastr. 44: 1215-1218.

Tan, T.Q., Ogden, A.K., Tillman, J., Demmler, G.J., Rinaldi, M.G., 1992. *Paecilomyces lilacinus* catheter-related fungemia in an immunocompromised pediatric patient. J. Clin. Microbiol. 30: 2479-2483.

Tanabe, Y., Nagahama, T., Saikawa, M. & Sugiyama, J., 1999. Phylogenetic relationship of *Cephaliophora* to nematophagous hyphomycetes including taxonomic and nomenclatural emendations of the genus *Lecophagus*. Mycologia 91: 830-835.

Tanaka, K.-i., Miyazaki, T., Maesaki, S., Mitsutake, K., Kakeya, H., Yamamoto, Y., Yanagihara, K., Hossain, M.A., Tashiro, T. & Kohno, S., 1996. Detection of *Cryptococcus neoformans* gene in patients with pulmonary cryptococcosis. J. Clin. Microbiol. 34: 2826-2828.

Tang, C.M., Holden, D.W., Aufauvre-Brawn, A. & Cohen, J., 1993. The detection of *Aspergillus* spp. by the polymerase chain reaction and its evaluation in bronchoalveolar lavage fluid. Am. Rev. Respir. Dis. 148: 1313-1317.

Tanis, B.C., Pijl, H. van der, Ogtrop, M.L. van, Kibbelaar, R.E. & Chang, P.C., 1995. Fatal fungal peritonitis by *Trichoderma longibrachiatum* complicating peritoneal dialysis. Nephrol. Dial. Transpl. 10: 114-116.

Tashiro, T., Nagai, H., Goto, Y., Kamberi, P. & Nasu, M., 1995. *Trichosporon beigelii* pneumonia in patients with hematologic malignancies. Chest 108: 190-195.

Tauphaichitr, V.S., Chaiprasert, A., Suvatte, V. & Thasnakorn, P., 1990. Subcutaneous mucormycosis caused by *Saksenaea vasiformis* in a thallassaemic child: first case report in Thailand. Mycoses 33: 303-309.

Tawara, S., Ikeda, F., Maki, K., Morishita, Y., Otomo, K., Teratani, N., Goto, T., Tomishima, M., Ohki, H., Yamada, A., Kawabata, K., Takasugy, H., Sakane, K., Tanaka, H., Matsumoto, F., Kuwahara, S., 2000. *In vitro* activities of a new lipopeptide antifungal agent, FK 463 against a variety of clinically important fungi. Antimicrob. Agents Chemother. 44: 57-62.

Tawfik, O.W., Papasian, C.J., Dixon, A.Y. & Potter, L.M., 1989. *Saccharomyces cerevisiae* pneumonia in a patient with acquired immune deficiency syndrome. J. Clin. Microbiol. 27: 1689-1691.

Taylor, J.W. & Bowman, B.H., 1993. *Pneumocystis carinii* and the ustomycetous red yeast fungi. Molec. Microbiol. 8: 425-426.

Taylor, M.N., Baddour, L.M. & Alexander, J.R., 1984. Disseminated histoplasmosis associated with the acquired immune deficiency syndrome. Am. J. Med. 77: 579-580.

Tchang, F.K.M. & Gilardi, G.L., 1973. Osteomyelitis due to *Torulopsis inconspicua*. J. Bone Joint Surg. 55: 1739-1743.

Tedder, M., Spratt, J.A., Anstadt, M.P., Hedge, S.S., Tedder, S.D. & Lowe, J.E., 1994. Pulmonary mucormycosis: results of medical and surgical therapy. Annls Thorac. Surg. 5: 1044-1050.

Tell, L.A., Nichols, D.K., Fleming, W.P. & Bush, M., 1997. Cryptococcosis in tree shrews (*Tupaia tana* and *Tupaia minor*) and elephant shrews (*Macroscelides proboscides*). J. Zoo Wildlife Med. 28: 175-181.

Tendolkar, V.M., Kerkar, P., Jerajani, U., Gogate, A. & Padhye, A.A., 1998. Phaeohyphomycotic ulcer caused by *Phialophora verrucosa*: successful treatment with itraconazole. J. Infect. 36: 122-125.

Terreni, A.A., Gregg, W.B., Morris, P.R. & DiSalvo, A.F., 1985. *Epidermophyton floccosum* infection in a dog from the United States. Sabouraudia 23: 141-142.

Testa, J., Vuillecard, E., Ravisse, P., Dupont, B., Gonzalez, J.P. & Georges, A.J., 1987. À propos de deux nouveaux cas de rhinoentomophthoromycose diagnostiques en R.C.A. (revue de la literature). Bull. Soc. Path. Exot. Fil. 80: 787-791.

Tewari, R.P. & Macpherson, C.R., 1968. Pathogenicity and neurologic effects of *Oidiodendron kalrai* for mice. J. Bact. 95: 1130-1139.

Tewari, R.P & Macpherson, C.R., 1971. A new dimorphic fungus, *Oidiodendron kalrai:* morphological and biochemical characteristics. Mycologia 63: 602-611.

Thanos, M., Schönian, G., Meyer, W., Schweynoch, C., Gräser, Y., Mitchell, T.G., Presber, W. & Tietz, H.-J., 1996. Rapid identification of *Candida* species by DNA fingerprinting with PCR. J. Clin. Microbiol. 34: 615-621.

Thappa, D.M., Venkatesan, S., Sirka, C.S., Jaisankar, T.J. & Ratnakar, C., 1998. Disseminated cutaneous rhinosporidiosis. J. Derm., Tokyo 25: 527-532.

Theodore, F.H., 1978. Etiology and diagnosis of fungal postoperative endophthalmitis. Ophthalmology 85: 327-340.

Theodore, F.H., Littman, M.L. & Almeda, E., 1961. The diagnosis and management of fungus endophthalmitis following cataract extraction. Arch. Ophthalmol. 66: 39-51.

Thérizol-Ferly, M., Kombila, M., Gomez de Diaz, M., Douchet, C., Salaum, Y., Barrabes, A., Duong, T.H. & Richard-Lenoble, D., 1994. White piedra and *Trichosporon* species in equatorial Africa II. Clinical and mycological associations: an analysis of 449 superficial inguinal specimens. Mycoses 37: 255-260.

Thianprasit, M. & Sivayathorn, A., 1984. Black dot mycetoma. Mykosen 27: 219-226.

Thianprasit, M. & Thagerngpol, K., 1989. Rhinosporidiosis. Curr. Topics Med. Mycol. 3: 64-85.

Thirumalachar, M.J., Padhye, A.A. & Srinivasan, M.C., 1960. *Emmonsia ciferrina*, a new species from India. Mycopath. Mycol. Appl. 26: 323-332.

Thom, C., 1930. The Penicillia. Williams & Wilkins, Baltimore, 644 pp.

Thom, C. & Church, M.B., 1926. The Aspergilli. Williams & Wilkins, Baltimore, 272 pp.

Thom, C. & Raper, K.B., 1945. A Manual of the Aspergilli. Williams & Wilkins, Baltimore, 373 pp.

Thomas, C., Mileusnic, D., Carey, R.B., Kampert, M. & Anderson, D., 1999. Fatal *Chaetomium* cerebritis in a bone marrow transplant patient. Hum. Path. 30: 874-879.

Thomas, I., 1993. Superficial and deep candidosis. Int. J. Derm. 32: 778-783.

Thomas, P.A., 1994. Mycotic keratitis - an underestimated mycosis. J. Med. Vet. Mycol. 32: 235-256.

Thomas, P.A., Abraham, D.J. & Kalawathy, C.M., 1988. Oral itraconazol therapy for mycotic keratitis. Mykosen 31: 271-279.

Thomas, P.A., Garrison, R.G. & Jansen, T., 1991. Intrahyphal hyphae in corneal tissue from a case of keratitis due to *Lasiodiplodia theobromae*. J. Med. Vet. Mycol. 29: 263-267.

Thomas, P.A. & Kuriakose, T., 1990. Keratitis due to *Arthrobotrys oligospora* Fres. 1850. J. Med. Vet. Mycol. 28: 47-50.

Thompson, R.F., Bode, R.B., Rhodes, J.C. & Gluckman, J.L., 1988. *Paecilomyces variotii:* an unusual cause of isolated sphenoid sinusitis. Arch. Otolaryngol. Head Neck Surg. 114: 567-569.

Thuler, L.C.S., Faivichenco, S., Velasco, E., Martins, C.A., Nascimento, C.R.G. & Castilho, I.A.M.A., 1997. Fungemia caused by

Hansenula anomala - an outbreak in a cancer hospital. Mycoses 40: 193-196.

Tietz, H.-J., Czaika, V. & Sterry, W., 1999. Osteomyelitis caused by high resistant *Candida guilliermondii.* Mycoses 42: 577-580.

Tintelnot, K., Hunnius, P. von, Hoog, G.S. de, Polak-Wyss, A., Guého, E. & Masclaux, F., 1995. Systemic mycosis caused by a new *Cladophialophora* species. J. Med. Vet. Mycol. 33: 349-354.

Tintelnot, K., Llop, C., Hoog, G.S. de, Raddatz, B. & Guarro, J., 1999. Unpublished data.

Tintelnot, K. & Nitsche, B., 1989. *Rhizopus oligosporus* as a cause of mucormycosis in man. Mycoses 32: 115-118.

Tiwari, S., Singh, S.M. & Jain, S., 1995. Chronic bilateral suppurative otitis media caused by *Aspergillus terreus.* Mycoses 38: 297-300.

Tobon, A.M., Orozco, B., Estrada, S., Jaramillo, E., Bedout, C. de, Arango, M. & Restrepo, A., 1998. Paracoccidioidomycosis and AIDS: report of the first two Colombian cases. Revta Inst. Med. Trop. S. Paulo 40: 377-381.

Todorova, S.I., Côté, J.-C. & Coderre, D., 1998. Distinction between *Beauveria* and *Tolypocladium* by carbohydrate utilization. Mycol. Res. 102: 81-87.

Tomimori-Yamashita, J., Tagliolatto, S., Porro, A.M., Ogawa, M.M., Michalany, N.S. & Camargo, Z.P., 1997. Paracoc-cidioidomycosis: an uncommon localisation in the scrotum. Mycosis 40: 415-418.

Torda, A.J. & Jones, P.D., 1992. Necrotizing cutaneous infection caused by *Curvularia brachyspora* in an immunocompetent host. Austr. J. Derm. 38: 85-87.

Toro, C., Palacio, A. del, Alvarez, C., Rodriguez-Peralto, J.L., Carabias, E., Cuetara, M.S., Carpintero, Y. & Gomez, C., 1998. Cutaneous zygomycosis caused by *Rhizopus arrhizus* in a surgical wound. Revta Iberoam. Micol. 15: 94-96.

Torres, G. & Georg, L.K., 1956. A human case of *Trichophyton gallinae* infection. Disease contracted from chickens. Arch. Derm. 74: 191-197.

Torres Nunes, A., Cavalcanti, M.A. de & Queiroz, L.A. de, 1999. Occurrence of *Pseudomicrodochium suttonii* in Brazil. Revta Microbiol. 30: 52-53.

Torres-Rodriguez, J.M., Madrenys-Brunet, N., Siddat, M., López-Jodra, J. & Jiménez, T., 1998. *Aspergillus versicolor* as cause of onychomycosis: report of 12 cases and susceptibility testing to antifungal drugs. J. Eur. Acad. Derm. Venereol. 11: 25-31.

Torssander, J., Chryssanthou, E. & Petrini, B., 1996. Increased prevalence of oral *Candida albicans* serotype B in homosexual men: a comparative and longitudinal study in HIV-infected and HIV-negative patients. Mycoses 39: 353-356.

Tosti, A. & Piraccini, B.M., 1998. Proximal subungual onycho-mycosis due to *Aspergillus niger:* report of two cases. Br. J. Derm. 139: 156-157.

Tosti, A., Piraccini, B.M., Stinchi, C. & Lorenzi, S., 1996. Onychomycosis due to *Scopulariopsis brevicaulis:* clinical features and response to systemic antifungals. Br. J. Derm. 135: 799-802.

Tóth, B., Horvath, A. & Palos, H., 1973. Auswertung der mit Amphotericin-B vaginal Tabletten erzielten Heilungsergebnisse bei vaginaler Sprosspilzinfektionen. Z. GesKrankh. 48: 47-58.

Toth, J., Bausz, M. & Imre, L., 1996. Unilateral *Malassezia furfur* blepharitis after perforating keratoplasty. Br. J. Ophthalmol. 80: 488.

Toubas, D., Vilgre, J.P., Himberlin, C., Delepine, G., Aubert, D., Pignon, B. & Pinon, J.M., 2000. Pneumonia in an immunocompromised patient caused by *Microascus cirrosus.* Proc. 14th ISHAM Congr., B. Aires, p. 272.

Towler, H.M.A., Ligtman, S. & Matheson, M., 1995. *Candida* endophthalmitis. Br. J. Ophthalmol. 79: 1141-1142.

Toy, E.C., Scerpella, E.C. & Riggs, J.W., 1995. Tuboovarian abscess associated with *Candida glabrata* in a woman with an intrauterine device: a case report. J. Reprod. Med. 40: 223-225.

Travis, L.B., Roberts, G.D. & Wilson, W.R., 1985. Clinical significance of *Pseudallescheria boydii:* a review of 10 years' experience. Mayo Clin. Proc. 60: 531-537.

Trejos, A., 1954. *Cladosporium carrionii* n. sp. and the problem of Cladosporia from chromoblastomycosis. Revta Biol. Trop. 2: 75-112.

Tritz, D.M. & Woods, G.L., 1993. Fatal disseminated infection with *Aspergillus terreus* in immunocompromised hosts. Clin. Infect. Dis. 16: 118-122.

Trowbridge, J., Ludmer, L.M., Riddle, V.D., Levy, C.S. & Barth, W.F., 1999. *Candida lambica* polyarthritis in a patient with chronic alcoholism. J. Rheumatol. 26: 1846-1848.

Tsai, T.W., Hammond, L.A., Rinaldi, M., Martin, K., Tio, F., Maples, C.O., Freytes, C.O. & Roodman, G.D., 1997. *Cokeromyces recurvatus* infection in a bone marrow transplant recipient. Bone Marrow Transpl. 19: 301-302.

Tsang, P., Hopkins, T. & Jimenez-Lucho, V., 1996. Deep dermatophytosis caused by *Trichophyton rubrum* in a patient with AIDS. J. Am. Acad. Derm. 34: 1090-1091.

Tsui, W.M., Ma, K.F. & Tsang, D.N., 1992. Disseminated *Penicillium marneffei* infection in HIV-infected subject. Histopathology 20: 287-293.

Tulloch, M., 1976. The genus *Metarhizium.* Trans. Br. Mycol. Soc. 66: 407-411.

Tumbarello, M., Bevilacque, N., Federico, G., Morace, G., Cauda, R. & Tacconelli, E., 1996. Fluconazole-resistant *Candida parapsilosis* fungemia in a patient with AIDS. Clin. Infect. Dis. 22: 179-180.

Tumbarello, M., Ventura, G., Caldarola, G., Morace, G., Cauda, R. & Ortana, L., 1994. An emerging opportunistic infection in HIV patients: a retrospective analysis of 11 cases of pulmonary aspergillosis. Eur. J. Epidemiol. 9: 638-644.

Tümbay, E., Seeliger, H.P.R. & Anğ, Ö. (eds), 1991. *Candida* and Candidamycosis. Plenum, New York, 301 pp.

Tumietto, F., Raimondi, C., Frasca, G.M., Martello, M., Bari, M.A. di, Costigliola, P., Bonomini, V. & Chiodo, F., 1995. Disseminated cryptococcosis in a lupus nephritis patient under long-term immunosuppression. Nephrol. Dial. Transpl. 10: 896-899.

Turiansky, G.W., Benson, P.M., Sperling, L.C., Sau, P., Salkin, I.F., McGinnis, M.R. & James, W.D., 1995. *Phialophora verrucosa*: a new cause of mycetoma. J. Am. Acad. Derm. 32: 311-315.

Turner, D., Kovacs, W., Kuhls, K., Lieckfeldt, E., Peter, B., Arisan-Atac, I., Strauss, J., Samuels, G.J., Börner, T. & Kubicek, C.P., 1997. Biogeography and phenotypic variation in *Trichoderma* sect. *Longibrachiatum* and associated *Hypocrea* species. Mycol. Res. 101: 449-459.

Turner, P.D., 1964. *Syncephalastrum* associated with bovine mycotic abortion. Nature 204: 309.

Uberti-Foppa, C., Fumagalli, L., Gianotti, N., Viviani, A.M., Vaiani, R. & Guého, E., 1995. First case of osteomyelitis due to *Phialophora richardsiae* in a patient with HIV infection. AIDS, Philad. 9: 975-976.

Udagawa, S.-i., Horie, Y. & Cannon, P.F., 1989. Two new species of *Neocosmospora* from Japan, with a key to the currently accepted species. Sydowia 41: 349-359.

Udagawa, S.-i, Tominaga, K. & Hamaoka, T., 1986. *Scytalidium japonicum,* a new species, the causal agent of cattle bronchiolitis. Mycotaxon 25: 279-286.

Uden, N. van, 1952. Zur Kenntniss von *Torulopsis pintolopesii* sp. nov. Mit Beobachtungen über die parasitäre Phase von *Acladium castellanii* Pinoy. Arch. Mikrobiol. 17: 199-208.

Uden, N. van & Kolipinski, M.C., 1962. *Torulopsis haemulonii* nov. spec., a yeast from the Atlantic Ocean. Antonie van Leeuwenhoek 28: 78-80.

Uijthof, J.M.J., 1996. Taxonomy and phylogeny of the human pathogenic black yeast genus *Exophiala* Carmichael. Thesis, Utrecht, 120 pp.

Uijthof, J.M.J., Belkum, A. van, Hoog, G.S. de & Haase, G., 1998. *Exophiala dermatitidis* and the related taxon *Sarcinomyces phaeomuriformis:* physiology, ITS1-sequencing and the development of a molecular probe. J. Med. Vet. Mycol. 36: 143-151.

Uijthof, J.M.J., Cock, A.W.A.M. de, Hoog, G.S. de, Quint, W.G.V. & Belkum, A. van, 1994. Polymerase chain reaction mediated genotyping of *Hortaea werneckii,* causative agent of tinea nigra. Mycoses 37: 307-312.

Uijthof, J.M.J., Figge, M.J. & Hoog, G.S. de, 1997. Molecular and physiological investigations of *Exophiala* species described from fish. System. Appl. Microbiol. 20: 585-594.

Uijthof, J.M.J. & Hoog, G.S. de, 1995. PCR-ribotyping of type isolates of currently accepted *Exophiala* and *Phaeococcomyces* strains. Antonie van Leeuwenhoek 68: 35-42.

Uijthof, J.M.J., Hoog, G.S. de, Cock, A.W.A.M. de, Takeo, K. & Nishimura, K., 1994. PCR-based evaluation of pathology of strains of the black yeast *Exophiala (Wangiella) dermatitidis.* Mycoses 37: 235-242.

Ukarapol, N., Sirisanthana, V. & Wongsawasdi, L., 1998. *Penicillium marneffei* mesenteric lymphadenitis in human immunodeficiency virus-infected children. J. Med. Ass. Thail. 81: 637-640.

Ulloa, M. & Hanlin, R.T., 2000. Illustrated Dictionary of Mycology. APS, St. Paul, 448 pp.

Untereiner, W.A. & Naveau, F., 1999. Molecular systematics of the *Herpotrichiellaceae* with an assessment of the phylogenetic positions of *Exophiala dermatitidis* and *Phialophora americana*. Mycologia 91: 67-83.

Upadhyay, H.P., 1981. A Monograph of *Ceratocystis* and *Ceratocystiopsis*. Univ. Georgia Press, Athens, 176 pp.

Upshaw, C.B., 1974. *Penicillium* endocarditis of aortic valve prosthesis. J. Thorac. Cardiovasc. Surg. 68: 428-431.

Urbanek, M., Neill, S.M. & Miller, J.A., 1995. Kerion: a case report. Clin. Experim. Derm. 20: 413-414.

Usuki, K., Yotsumoto, S.-i., Hamada, H., Shimada, T., Fukumitsu, K. & Kanzaki, T., 1996. A case of chromomycosis with tumor-like growth. J. Derm. Tokyo 23: 643-647.

Valencia Leon, G. & Tio Polledo, L., 1989. Aislamiento de *Microsporum ferrugineum* en Cuba. Presentación de un caso. Revta Cub. Med. Trop. 41: 290-298.

Valenton, M.J., Rinaldi, M.G. & Butler, E.E., 1975. A corneal abscess due to the fungus *Botryodiplodia theobromae*. Can. J. Ophthalmol. 10: 416-418.

Vanbreuseghem, R., 1952. Intérêt théorique et pratique d'un nouveau dermatophyte isolé du sol: *Keratinomyces ajelloi* gen. nov., sp. nov. Bull. Acad. Roy. Méd. Belg. 38: 1068-1077.

Vanbreuseghem, R. (ed.), 1972. Les Mycoses Profondes des Régions Tropicales. 2me Coll. Int. Mycol. Méd., Bruxelles; Annls Soc. Belge Méd. Trop. 52: 243-495.

Vanbreuseghem, R., 1980. *Keratinomyces ajelloi* ou *Trichophyton ajelloi?* Bull. Soc. Fr. Mycol. Méd. 10: 257-260.

Vanbreuseghem, R. & Vroey, C. de, 1966. Gymnoascacées isolées de lesions cutanées en République de Somalie. Annls Soc. Belge Méd. Trop. 46: 451-456.

Vanbreuseghem, R. & Vroey, C. de, 1979. Dermatophyte infection by *Anixiopsis stercoraria* in a wild boar *(Sus scrofa)*. Mykosen 23: 183-187.

Vanbreuseghem, R., Vroey, C. de & Takashio, M., 1978. Practical Guide to Medical and Veterinary Mycology. 2nd ed. Masson, Paris.

VandenBossche, H., Mackenzie, D.W.R. & Cauwenbergh, G. (eds.), 1988. *Aspergillus* and Aspergillosis. Plenum, New York, 322 pp.

VandenBossche, H., Mackenzie, D.W.R., Cauwenberg, G., Cutsem, J. van, Drouhet, E. & Dupont, B. (eds), 1990. Mycoses in AIDS Patients. Plenum, New York, 337 pp.

Vanittanakom, N., Cooper, C.R., Chariyalertsak, S., Youngchim, S., Nelson, K.E. & Sirisanthana, T., 1996. Restriction endonuclease analysis of *Penicillium marneffei*. J. Clin. Microbiol. 34: 1834-1836.

Vanittanakom, N., Merz, W.G., Sittisombut, N., Khamwam, C., Nelson, K.E. & Sirisanthana, T., 1998. Specific identification of *Penicillium marneffei* by a polymerase chain reaction/hybridization technique. Med. Mycol. 36: 169-175.

Vanittanakom, N. & Sirisanthana, T., 1997. *Penicillium marneffei* infection in patients infected with human immunodeficiency virus. Curr. Topics Med. Mycol. 8: 35-42.

VanSteenhouse, J.L., Padhye, A.A. & Ajello, L., 1988. Subcutaneous phaeohyphomycosis caused by *Scolecobasidium humicola* in a cat. Mycopathologia 102: 123-127.

Varga, J., Vida, Z., Tóth, B., Debets, F. & Horie, Y., 2000. Phylogenetic analysis of newly described *Neosartorya* species. Antonie van Leeuwenhoek 77: 235-239.

Vartivarian, S.E., Anaissie, E.J. & Bodey, G.P., 1993. Emerging fungal pathogens in immunocompromised patients: classification, diagnosis and management. Clin. Infect. Dis., Suppl. 17: S487-S491.

Vasei, M. & Imanieh, M.H., 1999. Duodenal colonization by *Geotrichum candidum* in a child with transient low serum levels of IgA and IgM: a case report. APMIS 107: 681-684.

Vastag, M., Papp, T., Kaska, Z. & Vágvögyi, C., 1998. Differentiation of *Rhizomucor* species by carbon source utilization and isoenzyme analysis. J. Clin. Microbiol. 36: 2153-2156.

Vaughan, L.M., 1993. Allergic bronchopulmonary aspergillosis. Clin. Pharm. 12: 24-33.

Vaughan-Martini, A. & Martini, A., 1998. *Saccharomyces* Meyer ex Reess. In C.P. Kurtzman, & J.W. Fell (eds): The Yeasts, a Taxonomic Study, 4th ed. pp. 358-371. Elsevier, Amsterdam.

Vazquez, J.A., Lundstrom, T., Dembry, L., Chandrasekar, P., Boikov, D., Perri, M.B. & Zervos, M.J., 1995. Invasive *Candida guilliermondii* infection: *in vitro* susceptibility studies and molecular analysis. Bone Marrow Transpl. 16: 849-853.

Vazquez, J., Lundstrom, T., Dembry, L., Perry, M.B. & Zervos, M., 1993. Disseminated *Torulopsis candida (Candida famata)* infection: an unusual human pathogen. Abstr. Gen. Meet. ASM 93: 534.

Vecchiarelli, A., Retini, C., Monari, C., Tascini, C., Bistoni, F. & Kozel, T.R., 1996. Purified capsular polysaccharide of *Cryptococcus neoformans* induces interleukin-10 secretion by human monocytes. Infect. Immun. 64: 2846-2849.

Veglia, K.S. & Marks, V.J., 1987. *Fusarium* as a pathogen. A case report of *Fusarium* species and review of the literature. J. Am. Acad. Derm. 16: 260-263.

Velagraki, A. & Logotheti, M., 1998. Presumptive identification of an emerging yeast pathogen: *Candida dubliniensis* (sp. nov.) reduces 2,3,5-triphenyltetrazolium chloride. Immunol. Med. Microbiol. 20: 239-241.

Velazquez, L.F., Restrepo, A. & Calle, G., 1976. Cromomicosis: experiencia de doce años. Acta Med. Colomb. 1: 165-171.

Velez, A., Fernandez-Roldan, J.-C., Linares, M. & Casal, M., 1996. Melanonychia due to *Candida humicola*. Br. J. Derm. 134: 375-376.

Vennewald, I., Henker, M., Klemm, E. & Seebacher, C., 1999. Fungal colonization of the paranasal sinuses. Mycoses 42, Suppl. 2: 33-36.

Vennewald, I., Seebacher, C. & Roitzsch, E., 1998. *Post-mortem* findings in patients with repeatedly mycological demonstration of *Candida glabrata*. Mycoses 41: 125-132.

Verduyn Lunel, F.M., Voss, A., Neeleman, C. & Meis, J.F.G.M., 1996. Candidemie met *Candida krusei* bij een niet-neutropene intensive care-patiënt. Ned. Tijdschr. Med. Microbiol. 4: 28-29.

Verma, S. & Graham, E.M., 1995. *Cryptococcus* presenting as cloudy choroiditis in an AIDS patient. Br. J. Ophthalmol. 79: 618-619.

Vermeil, C., Cordeff, A., Leroux, M.J., Morin, O. & Bouc, M., 1971. Blastomycose chéloidienne à *Aureobasidium pullulans* (de Bary) Arnaud en Bretagne. Mycopathol. Mycol. Appl. 43: 35-39.

Verret, J.-L., Gaborieau, F., Chabasse, D., Rohmer, V., Avenel, M. & Smûlevici, A., 1982. Alternariose cutanée révélatrice d'une maladie de Cushing, un cas avec étude ultrastructurale. Annls Derm. Vénéréol. 109: 841-846.

Verweij, P.E., Bergh, M.F.Q. van den, Rath, P.M., Pauw, B.E. de, Voss, A. & Meis, J.F.G.M., 1999. Invasive aspergillosis caused by *Aspergillus ustus:* case report and review. J. Clin. Microbiol. 37: 1606-1609.

Verweij, P.E., Kasteren, M. van, Nes, J. van de, Hoog, G.S. de, Pauw, B.E. de & Meis, J.F.G.M., 1997. Fatal pulmonary infection caused by the basidiomycete *Hormographiella aspergillata*. J. Clin. Microbiol. 35: 2675-2678.

Verweij, P.E., Meis, J.F.G.M., Hurk, P. van den, Zoll, J., Samson, R.A. & Melchers, W.J.G., 1995. Phylogenetic relationships of five species of *Aspergillus* and related taxa as deduced by comparison of sequences of small subunit ribosomal RNA. J. Med. Vet. Mycol. 33: 185-190.

Verweij, P.E., Mensink, M., Rijs, A.J.M.M., Donnelly, J.P., Meis, J.F.G.M. & Denning, D.W., 1998. *In-vitro* activities of amphotericin B, itraconazole and voriconazole against 150 clinical and environmental *Aspergillus fumigatus* isolates. J. Antimicrob. Chemother. 42: 389-392.

Vicek, T.J., Oliver, J.L. & Reese, K.W., 1995. Systemic trichosporonosis caused by *Trichosporon beigelii* in a white-handed gibbon *(Hylobates lar)*. J. Zoo Wildlife Med. 26: 115-118.

Vidal, M.S.M., Melo, N.T. de, Garcia, N.M., Negro, G.M.B. del, Assis, C.M. de, Heins-Vaccari, E.M., Naiff, R.D., Mendes, R.P. & Lacaz, C. da Silva, 1995. *Paracoccidioides brasiliensis*. A mycologic and immunochemical study of a sample isolated from an armadillo *(Dasipus novencinctus)*. Revta Inst. Med. Trop. S. Paulo 37: 43-49.

Vidal, P., Vinuesa, M. de los, Sánchez-Puelles, J.M. & Guarro, J., 2000. Phylogeny of the anamorphic genus *Chrysosporium* and related taxa based on rDNA internal transcribed spacer sequences. Revta Iberoam. Micol. 17: 24-31.

Vidal-Leiria, M. & Uden, N. van, 1963. *Candida silvae* sp.n., a yeast from humans and horses. Antonie van Leeuwenhoek 29: 261-264.

Viégas, A.P., 1943. Notas sôbre uma nova espécie de *Sporormia*. Bragantia 3: 155-164.

Vieira, R., Martins, L., Alfonso, A., Rego, F. & Cardoso, J., 1998. Cutaneous alternariosis. Revta Iberoam. Micol. 15: 97-99.

Vieira-Dias, D., Sena, C.M., Oréfice, F., Tanure, M.A.G. & Hamdan, J.S., 1997. Ocular and concomittant cutaneous sporotrichosis. Mycoses 40: 197-201.

Vignale, R., Mackinnon, J.E., Casella de Vilaboa, E. & Burgoa, F., 1964. Chronic, destructive, mucocutaneous phycomycosis in man. Sabouraudia 3: 143-147.

Vigui-Vallanet, C., Savaglio, N., Piat, C. & Tourte-Schaefer, C. 1997. Epidémiologie des teignes à *Microsporum langeronii* en région Parisienne. Resultats de deux enquetes scolaires et familiales. Annls Derm. Véneréol. 124: 696-699.

Vilgalys, R. & Hester, M., 1990. Rapid genetic identification and mapping of enzymatically amplified ribosomal DNA from several *Cryptococcus* species. J. Bact. 172: 4238-4246.

Vincent, R.D., Goewert, R., Goldman, W.E., Kobayashi, G.S., Lambowitz, A.M. & Medoff, G., 1986. Classification of *Histoplasma capsulatum* isolates by restriction fragment polymorphisms. J. Bact. 165: 813-818.

Virgile, R., Perry, H.D., Pardanani, B., Szabo, K., Rahm, E.K., Stone, J., Salkin, I.F. & Dixon, D.M., 1993. Human infectious corneal ulcer caused by *Pythium insidiosum*. Cornea 12: 81-83.

Virgili, A. & Zampino, M.R., 1998. Relapsing tinea capitis by *Microsporum canis* in an adult female renal transplant recipient. Nephron 80: 61-62.

Vismer, H.F., 1994. *Microsporum canis* scalp ringworm: its primary or secondary ectothrix character. Scan. Microb. 7: 671-676.

Vismer, H.F., Beer, H.A. de & Dreyer, L., 1980. Subcutaneous phycomycosis caused by *Basidiobolus haptosporus* (Drechsler 1947). S. Afr. Med. J. 58: 644-647.

Vismer, H.F., Findley, G.H. & Eicker, A., 1987. The septal ontogeny, germination and electron microscopy of *Microsporum gypseum* macroaleurioconidia. Mycopathologia 98: 149-164.

Vismer, H.F. & Hull, P.R., 1997. Prevalence, epidemiology and geographical distribution of *Sporothrix schenckii* infections in Gauteng, South Africa. Mycopathologia 137: 137-143.

Vissiennon, T., Schüppel, K.-F., Ullrich, E. & Kuijpers, A.F.A., 1999. A disseminated infection due to *Chrysosporium queenslandi-cum* in a garter snake (*Thamnophis*). Mycoses 42: 107-110.

Viswanathan, R. & Randhawa, H.S., 1959. *Candida viswanathii* sp. nov., isolated from a case of meningitis. Sci. Cult. 25: 86-87.

Viviani, M.A., Tortorano, A.M., Laria, G., Giannetti, A. & Bignotti, G., 1986. Two new cases of cutaneous alternariosis with a review of the literature. Mycopathologia 96: 3-12.

Vivas, C., 1998. Endocarditis caused by *Aspergillus niger:* case report. Clin. Infect. Dis. 27: 1322-1323.

Vogelaers, D., Petrovic, M., Deroo, M., Verplancke, R., Claessens, Y., Naeyaert, J.M. & Afschrift, M., 1997. A case of primary cutaneous cryptococcosis. Eur. J. Clin. Microbiol. Infect. Dis. 16: 150-152.

Vogeser, M., Haas, A., Aust, D. & Ruckdeschel, G., 1997. *Postmortem* analysis of invasive aspergillosis in a tertiary care hospital. Eur. J. Clin. Microbiol. Infect. Dis. 16: 1-6.

Voigt, K., Cigelnik, E. & O'Donnell, K., 1999. Phylogeny and PCR identification of clinically important Zygomycetes based on nuclear ribosomal. DNA sequence data. J. Clin. Microbiol. 37: 3957-3964.

Voigt, K., Schleier, S. & Brückner, B., 1995. Genetic variability in *Gibberella fujikuroi* and some related species of the genus *Fusarium* based on random amplification of polymorphic DNA (RAPD). Curr. Genet. 27: 528-535.

Voisard, J.J., Weil, F.X., Beylot-Barry, M., Vergier, B., Dromer, C. & Beylot, C., 1999. Dermatophytic granuloma caused by *Microsporum canis* in a heart-lung recipient. Dermatology, Basel 198: 317-319.

Voloshin, D.K., Lacomis, D. & Mcmahon, D., 1995. Disseminated histoplasmosis presenting as myositis and fasciitis in a patient with dermatomyositis. Muscle Nerve 18: 531-535.

Vries, G.A. de, 1962. *Cyphellophora laciniata* nov. gen., nov. sp. and *Dactylium fusarioides* Fragoso et Ciferri. Mycopathol. Mycol. Appl. 16: 47-54.

Vries, G.A. de, 1967. Contribution to the Knowledge of the Genus *Cladosporium* Link ex Fr. Bibl. Mycol. 3: 1-121.

Vries, G.A. de, 1971. Observations on *Trichophyton tonsurans*. Sabouraudia 9: 1-5.

Vries, G.A. de, Elders, M.C.C. & Luykx, M.H.F., 1986. Description of *Cyphellophora pluriseptata* sp. nov. Antonie van Leeuwenhoek 52: 141-143.

Vries, G.A. de, Hoog, G.S. de & Bruyn, H.P. de, 1984. *Phialophora cyanescens* sp. nov. with *Phaeosclera*-like synanamorph, causing white-grain mycetoma in man. Antonie van Leeuwenhoek 50: 149-153.

Vries, G.A. de & Josephius Jitta, C.R., 1973. An epizootic in horses in The Netherlands caused by *Trichophyton equinum* var. *equinum*. Sabouraudia 11: 137-139.

Vries, G.A. de, Kemp, R.F.O. & Speller, D.C.E., 1971. Endocarditis caused by *Coprinus delicatulus*. Proc. 5th Congr. ISHAM, Paris, pp. 185-186.

Vries, G.A. de & Kleine-Natrop, H.E., 1957. *Sporotrichum cerebriforme* nov. spec. Mycopath. Mycol. Appl. 8: 154-160.

Vries, G.A. de & Laarman, J.J., 1973. A case of Lobo's disease in the dolphin *Sotalia guianensis*. Aquatic Mammals 13: 1-8.

Vroey, C. de, 1976. Sur quelques Ascomycètes isolés de lésions cutanées chez l'homme. Bull. Soc. Fr. Mycol. Méd. 5: 161-163.

Vroey, C. de, Lasagni, A., Tossi, E., Schroeder, F. & Song, M., 1992. Onychomycoses due to *Microascus cirrosus* (syn. *M. desmosporus)*. Case report. Mycoses 35: 7-8.

Vroey, C. de, Song, M., Wiame, L. & Achten, G., 1983a. Infections cutanées par *Microsporum praecox*. Bull. Soc. Fr. Mycol. Méd. 12: 71-74.

Vroey, C. de, Wuytack-Raes, C. & Fossoul, F., 1983b. Isolation of saprophytic *Microsporum praecox* Rivalier from sites associated with horses. Sabouraudia 21: 255-257.

Vuillemin, P., 1895. Structure et affinités de *Microsporum*. Bull. Trim. Soc. Mycol. Fr. 11: 94-103.

Vuillemin, P., 1931. Les champignons parasites et les mycoses de l'homme. Encycl. Mycol. 2: 1-290.

Vukmir, R.B., Kusne, S., Linden, P., Pasculle, W., Fothergill, A.W., Schaeffer, J., Nieto, J., Segal, R., Merhav, H., Martinez, A.J. & Rinaldi, M.G., 1994. Successful therapy for cerebral phaeohyphomycosis due to *Dactylaria gallopava* in a liver transplant recipient. Clin. Infect. Dis. 19: 714-719.

Vukovic, Z., Bobic-Radovanovic, A., Latkovic, Z. & Radovanovic, Z., 1995. An epidemiological investigation of the first outbreak of rhinosporidiosis in Europe. J. Trop. Med. Hyg. 98: 333-337.

Wabacha, J.K., Gitau, G.K., Bebora, L.C., Bwanga, C.O., Wamuri, Z.M. & Mbithi, P.M.F., 1998. Occurrence of dermatomycosis (ringworm) due to *Trichophyton verrucosum* in dairy calves and its spread to animal attendants. J.S. Afr. Vet. Ass. 69: 172-173.

Wada, S., Nakamura, K. & Hatai, K., 1995. First case of *Ochroconis humicola* infection in marine cultured fish in Japan. Fish Path. 30: 125-126.

Wada, Y., Nakaoka, Y., Matsui, T. & Ikeda, T., 1994. Candidiasis caused by *Candida glabrata* in the forestomachs of a calf. J. Compar. Path. 111: 315-319.

Wadhwani, K. & Srivastava, A.K., 1984. Fungi from otitis media of agricultural field workers. Mycopathologia 88: 155-159.

Wadhwani, K. & Srivastava, A.K., 1985. Some cases of onycho-mycosis from North India in different working environments. Mycopathologia 92: 149-155.

Wahab, S., Lal, B., Jacob, Z., Pandey, V.C. & Srivastava, O.P., 1979. Studies on a strain of *Fusarium solani* (Mart.) Sacc. isolated from a case of mycotic keratitis. Mycopatholgia 68: 31-38.

Wakefield, A.E., Peters, S.E., Banerji, S., Bridge, P.D., Hall, G.S., Hawksworth, D.L., Guiver, L.A., Allen, A.G. & Hopkin, J.M., 1992. *Pneumocystis carinii* shows DNA homology with the ustomycetous red yeast fungi. Mol. Microbiol. 6: 1903-19011.

Walker, S.D., Clark, R.V., King, C.T., Humphries, K.E., Lytle, L.S. & Butkus D.E., 1992. Fatal disseminated *Conidiobolus coronatus* infection in a renal transplant patient. J. Clin. Path. 98: 559-564.

Waller, J., Lutun, P., Jaegle, M.-L., Marcellin, L., Koenig, H., Altieri, M., Heldt, N. & Robles, G., 1991a. Aspergillose invasive à *Aspergillus fumigatus* chez une transplantée hépatique. J. Mycol. Méd. 1: 225-230.

Waller, J., Lutun, P., Koenig, H., Marcellin, L., Boudjama, K., Robles, G. & Altieri, M., 1991b. Aspergillose disséminée à *Aspergillus terreus* chez une transplantée hépatique. J. Mycol. Méd. 1: 168-171.

Waller, J., Woehl-Jaegle, M.-L., Guého, E., Koenig, H., Marcellin, L. & Ellero, B., 1993. Mucormycose abdominale nosocomiale à *Rhizopus rhizopodiformis* chez un transplanté hépatique. Revue de la Littérature. J. Mycol. Méd. 3: 180-186.

Walsh, T.J., Lee, J.W., Sien, T., Schaufele, R., Bacher, J., Switchenko, A.C., Goodman, T.C. & Pizzo, P.A., 1994a. Serum D-arabinitol measured by automated quantitative enzymatic assay for detection and therapeutic monitoring of experimental disseminated candidiasis: correlation with tissue concentrations of *Candida albicans*. J. Med. Vet. Mycol. 32: 205-215.

Walsh, T.J., Peter, J., McGough, D., Fothergill, A.W., Rinaldi, M.G. & Pizzo, P. A., 1995. Activities of amphotericin B and antifungal azoles alone and in combination against *Pseudallescheria boydii*. Antimicrob. Agents Chemother. 39: 1361-1364.

Walsh, T.J., Renshaw, G., Andrews, J., Kwon-Chung, K.J., Cunnion, R.C., Pass, H.J., Taubenberger, J., Wilson, W. & Pizzo, P.A., 1994b. Invasive zygomycosis due to *Conidiobolus incrongruus*. Clin. Infect. Dis. 19: 423-430.

Walsh, T.J., Salkin, I.F., Dixon, D.M. & Hurd, N.J., 1989. Clinical, microbiological and experimental animal studies of *Candida lipolytica*. J. Clin. Microbiol. 27: 927-931.

Walshe, M.M. & English, M.P., 1966. Fungi in nails. Br. J. Derm. 78: 198-207.

Walz, A. & Hoog, G.S. de, 1987. A new species of *Cyphellophora*. Antonie van Leeuwenhoek 53: 143-146.

Walz, R., Bianchin, M., Chaves, M.L., Cerski, M.R., Severo, L.C. & Londero, A.T., 1997. Cerebral phaeohyphomycosis caused by *Cladophialophora bantiana* in a Brazilian drug abuser. J. Med. Vet. Mycol. 35: 427-431.

Wanachiwanawim, W., Thiamaprasit, M., Fucharoem, S., Chaiprasert, A., Ayvdhya, N. Sirithamaratkul, S.N. & Piankijagum, A., 1993. Fatal arteritis due to *Pythium insidiosum* infection in patients with Thalassaemia. Trans. R. Soc. Trop. Med. Hyg. 87: 296-298.

Wang, H.Y. & Lin, J.L., 1999. *Trichosporon beigelii* fungemia in a patient with haemodialysis. Nephrol. Dial. Transpl. 14: 2017-2018.

Wang, C.J.K., 1966. Annellophores in *Torula jeanselmei*. Mycologia 58: 614-621.

Wang, C.J.K. & Zabel, R.A., 1990. Identification Manual for Fungi from Utility Poles in the Eastern United States. ATCC, Rockville, 356 pp.

Wang, D.L., Li, R.Y. & Wang, Z.X., 1989. One case of phaeo-hyphomycosis caused by *Exophiala spinifera*. Chin. J. Derm. 22: 262-263.

Wang, D.L., Li, R.Y., Wang, X.H. & Zhang, H.E., 1991. Studies on *Veronaea botryosa*, agent of the first human case. Acta Mycol. Sin. 10: 159-165.

Wang, J., Qiangqiang, Z. & Li, L., 1998. Phaeohyphomycosis caused by *Chaetomium globosum:* first case report in China. Zhonghua Pifuke Zazhi 31: 273-275.

Wang, J.-J., Satoh, H., Takahashi, H. & Hasegawa, A., 1990. A case of cutaneous mucormycosis in Shanghai, China. Mycoses 33: 311-315.

Wang, L., Yokoyama, K., Miyaji, M. & Nishimura, K., 1998. The identification and phylogenetic relationship of pathogenic species of *Aspergillus* based on the mitochondrial cytochrome b gene. Med. Mycol. 36: 153-164.

Wan-Qing, L., 1988. Mycological identification of pulmonary aspergil-loma caused by *Aspergillus oryzae* with proliferating heads. Revta Ibér. Micol. 5, Suppl. 1: 51.

Ward, H.P., Martin, W.J., Ivins, J.C. & Weed, L.A., 1961. *Cephalosporium* arthritis. Proc. Mayo Clin. 36: 337-343.

Ware, A.J., Cockerell, C.J., Skiest, D.J. & Kussman, H.M., 1999. Disseminated sporotrichosis with extensive cutaneous involvement in a patient with AIDS. J. Am. Acad. Derm. 40: 350-355.

Warnock, D.W. & Johnson, E.M., 1991. Clinical manifestations and management of hyalohyphomycosis, phaeohyphomycosis and other uncommon forms of fungal infection in the compromised patient. In D.W. Warnock & M.D. Richardson (eds): Fungal Infection in the Compromised Patient, 2nd. ed., pp. 247-310. Wiley, Chichester.

Washburn, R.G., Kennedy, D.W., Begley, M.G., Henderson, D.K. & Bennett, J.E., 1988. Chronic fungal sinusitis in apparently normal hosts. Medicine 67: 231-247.

Watanabe, J.-i., Hori, H., Tanabe, K. & Nakamura, Y., 1989. Phylogenetic association of *Pneumocystis carinii* with the *Rhizopoda / Myxomycota / Zygomycota* group indicated by comparison of 5S ribosomal RNA sequences. Molec. Biochem. Parasit. 32: 163-168.

Watanabe, K., Yamana, T. & Inomata, H., 1985. Postoperative fungal endophthalmitis caused by *Paecilomyces lilacinus*. Jpn. J. Clin. Ophthalmol. 39: 1141-1144.

Watling, R. & Sweeney, J., 1971. Observations on *Schizophyllum commune* Fries. Sabouraudia 12: 214-226.

Wawrzkiewicz, K., Lobarzewski, J. & Wolski, T., 1987. Intra-cellular keratinase of *Trichophyton gallinae*. J. Med. Vet. Mycol. 25: 261-268.

Weber, A. & Kolb, S., 1986. Über die mehrmalige Isolierung von *Candida pulcherrima* (Lindner) Windisch aus Blutkulturen eines parenteral ernährten Patienten. Mykosen 29: 127-131.

Webster, J., 1970. Coprophilous fungi. Trans. Br. Mycol. Soc. 54: 161-180.

Wedde, M., Müller, D., Tintelnot, K., Hoog, G.S. de & Stahl, U., 1998. PCR-based identification of clinically relevant *Pseudallescheria / Scedosporium* strains. Med. Mycol. 36: 61-67.

Weeks, R.J., Padhye, A.A., & Ajello, L., 1985. *Histoplasma capsulatum* variety *farciminosum:* a new combination for *Histoplasma farciminosum*. Mycologia 77: 964-970.

Weers-Pothof, G., Havermans, J.F., Kamphuis, J., Sinnige, H.A.M. & Meis, J.F.G.M., 1997. *Candida tropicalis* arthritis in a patient with acute myeloid leukemia successfully treated with fluconazole: case report and review of the literature. Infection 25: 109-111.

Wehmer, C., 1900. Studien über technische Pilze. VII. Die 'Chinesische Hefe' und der sogenannte Amylomyces (= *Mucor rouxii*). Centbl. Bakt. Parasitkde, Abt. 2, 6: 353-365.

Wehrspann, P. & Füllbrandt, U., 1985. *Yarrowia lipolytica* van der Walt und von Arx isolated from blood culture. Mykosen 28: 217-222.

Weijman, A.C.M. & Golubev, W.I., 1987. Carbohydrate patterns and taxonomy of yeasts and yeast-like fungi. Stud. Mycol. 30: 361-371.

Weiler, H., Staib, F., Keller, H. & Stäcker, W., 1991. Luftsackmykose beim Pferd. Ein Beitrag zur Pathologie und Aetiologie. Pferdeheilkunde 7: 179-187.

Weising, K., Nybom, H., Wolff, K. & Meyer, W., 1995. DNA Fingerprinting in Plants and Fungi. CRC Press, Boca Raton, 323 pp.

Weiss, L.M. & Thiemke, W.A., 1983. Disseminated *Aspergillus ustus* infection following cardiac surgery. Am. J. Clin. Pathol. 80: 408-411.

Weissgold, D.J., Maguire, A.M. & Brucker, A.J., 1996. Management of postoperative *Acremonium* endophthalmitis. Ophthalmology 103: 745-756.

Weissman, Z., Berdicevsky, I. & Cavari, B., 1995. Molecular identification of *Candida albicans*. J. Med. Vet. Mycol. 33: 205-207.

Weitzman, I. & Crist, M.Y., 1979. Studies with clinical isolates of *Cunninghamella*. I. Mating behavior. Mycologia 71: 1024-1033.

Weitzman, I. & Crist, M.Y., 1980. Studies with clinical isolates of *Cunninghamella*. II. Physiological and morphological studies. Mycologia 72: 661-669.

Weitzman, I., Della-Latta, P., Housey, G. & Rebatta, G., 1993. *Mucor ramosissimus* Samutsevitsch isolated from a thigh lesion. J. Clin. Microbiol. 31: 2523-2525.

Weitzman, I., Gordon, M.A., Henderson, R.W. & Lapa, E.W., 1984. *Phialophora parasitica*, an emerging pathogen. J. Med. Vet. Mycol. 22: 331-339.

Weitzman, I., McGough, D.A., Rinaldi, M.G. & Della-Latta, P., 1996. *Rhizopus schipperae*, sp. nov., a new agent of zygomycosis. Mycotaxon 59: 217-225.

Weitzman, I., McGinnis, M.R., Padhye, A.A. & Ajello, L., 1986. The genus *Arthroderma* and its synonym *Nannizzia*. Mycotaxon 25: 505-518.

Weitzman, I., Rosenthal, S.A. & Shupack, J.L., 1985. A comparison between *Dactylaria gallopava* and *Scolecobasidium humicola:* first report of an infection in a tortoise caused by *S. humicola.* Sabouraudia 23: 287-293.

Weitzman, I. & Padhye, A.A., 1996. Dermatophytes, gross and microscopic. Derm. Clinics 14: 9-22.

Weitzman, I. & Summerbell, R.C., 1995. The dermatophytes. Clin. Microbiol. Rev. 8: 240-259.

Welbel, S.F., McNeil, M.M., Pramanik, A., Silberman, R., Oberle, A.D., Midgley, G., Crow, S. & Jarvis, W.R., 1994. Nosocomial *Malassezia pachydermatis* bloodstream infections in a neonatal intensive care unit. Pediatr. Infect. Dis. J. 13: 104-108.

Welbel, S.F., McNeil, M.M., Kuykendall, R.J., Lott, T.J., Pramanik, A., Silberman, R., Oberle, A.D., Bland, L.A., Aguero, S., Arduino, M., Crow, S. & Jarvis, W.R., 1996. *Candida parapsilosis* bloodstream infections in neonatal intensive care unit patients: epidemiologic and laboratory confirmation of a common source outbreak. Pediatr. Infect. Dis. J. 15: 998-1002.

Wells, G.M., Gajjar, A., Pearson, T.A., Hale, K.L. & Shenep, J.L., 1998. Pulmonary cryptosporidiosis and *Cryptococcus albidus* fungemia in a child with acute lymphatic leukemia. Med. Pediatr. Oncol. 31: 544-546.

Welsh, R.A. & Buchness, J.M., 1955. *Aspergillus* endocarditis, myocarditis and lung abscesses. Report of a case. Am. J. Clin. Pathol. 25: 782-786.

Welsh, R.D. & Ely, R.W., 1999. *Scopulariopsis chartarum* systemic mycosis in a dog. J. Clin. Microbiol. 37: 2102-2103.

Welti, C.V., Weiss, S.D., Cleary, T.J. & Gyori, F., 1984. Fungal cerebritis from intravenous drug abuser. J. Forensic Sci. 29: 260-268.

Wendisch, J., Blaschke-Hellmessen, R., Kaulen, F., Schwarze, R. & Kabus, M., 1996. Tödliche meningoenzephalitis durch *Cryp-tococcus neoformans* var. *neoformans* bei einem Mädchen ohne schwerwiegende Immundefekte. Mycoses 39, Suppl. 1: 97-101.

Wendt, B., Haglund, L., Razavi, A. & Rath, R., 1998. *Candida lusitaniae:* an uncommon cause of prosthetic valve endocarditis. Clin. Infect. Dis. 26: 769-770.

Wenker, C.J., Kaufman, L., Bacciarini, L.N. & Robert, N., 1998. Sporotrichosis in a nine-banded armadillo (*Dasypus novemcinctus*). J. Zoo Wildlife Med. 29: 474-478.

Werff, P.J. van der, 1951. Longaandoeningen veroorzaakt door schimmels. Ned. Tijdschr. Geneesk. 95: 1682-1690.

Werner, A.H. & Werner, B.E., 1994. Sporotrichosis in man and animal. Int. J. Derm. 33: 692-700.

West, B.C., Kwon-Chung, K.J., King, J.W., Grafton, W.D. & Rohr, M.S., 1983. Inguinal abscess caused by *Rhizopus rhizopodiformis:* successful treatment with surgery and amphotericin B. J. Clin. Microbiol. 18: 1384-1387.

West, B.C., Oberle, A.D. & Kwon-Chung, K.J., 1995. Mucor-mycosis caused by *Rhizopus microsporus* var. *microsporus:* cellulitis in the leg of a diabetic patient cured by amputation. J. Clin. Microbiol. 33: 3341-3344.

Westenfeld, F., Alston, W.K. & Winn, W.C., 1996. Complicated soft tissue infection with prepatellar bursitis caused by *Paecilomyces lilacinus* in an immunocompetent host: case report and review. J. Clin. Microbiol. 34: 1559-1562.

Wethered, D.-B., Markey, M.A., Hay, R.J., Mahgoub, E.S. & Gumaa, S.A., 1986. Ultrastructural and immunogenic changes in the formation of mycetoma grains. J. Med. Vet. Mycol. 25: 39-46.

Wheat, J., 1995. Endemic mycoses in AIDS: a clinical review. Clin. Microbiol. Rev. 8: 146-159.

Wheat, L.J., Bartlett, M., Cicarelli, M. & Smith, J.W., 1984. Opportunistic *Scopulariopsis* pneumonia in an immunocompromised host. South. Med. J. 77: 1608-1609.

Wheat, L.J., Connolly-Stringfield, P., Williams, B., Connolly, K., Blair, R., Bartlett, M. & Durkin, M., 1992. Diagnosis of histoplasmosis in patients with the acquired immunodeficiency syndrome by detection of *Histoplasma capsulatum* polysaccharide antigen in bronchoalveolar lavage fluid. Am. Rev. Respir. Dis. 145: 1421-1424.

Wheat, L.J., Slama, T.G. & Zeckel, M.L., 1985. Histoplasmosis in the acquired immune deficiency syndrome. Am. J. Med. 78: 203-209.

Wheeler, M.S., McGinnis, M.R., Schell, W.A., 1981. *Fusarium* infection in burned patients. Am. J. Clin. Pathol. 74: 304-311.

Whitby, S., Madu, E.C. & Bronze, M.S., 1996. Case report: *Candida zeylanoides* infective endocarditis complicating infection with the human immunodeficiency virus. Am. J. Med. Sci. 312: 138-139.

White, T.J., Bruns, T., Lee, S. & Taylor, J., 1990. Amplification and direct sequencing of fungal ribosomal RNA genes for phylogenetics. In M.A Innis *et al.* (eds): PCR Protocols, pp. 315-322. Academic, San Diego.

White, A. & Goetz, M.B., 1995. *Candida parapsilosis* prosthetic joint infection unresponsive to treatment with fluconazole. Clin. Infect. Dis. 20: 1068-1069.

Whittle, C.H. & Gresham, G.A., 1970. *Microsporum rivalieri* isolated from tinea capitis in East Anglia (England). Sabouraudia 8: 65-71.

Wickes, B.L., Hicks, J.B., Merz, W.G. & Kwon-Chung, K.J., 1992. The molecular analysis of synonymy among medically important yeasts within the genus *Candida.* J. Gen. Microbiol. 138: 901-907.

Wickline, C.L., Cornitius, T.C. & Butler, T., 1989. Cellulitis caused by *Rhizomucor pusillus* in a diabetic patient receiving continuous insulin infusion pump therapy. South. Med. J. 82: 1432-1434.

Widjojoatmodjo, M.N., Borst, A., Schukkink, A.F., Box, A.T.A., Tacken, N.M.M., Gemen, B. van, Verhoef, J., Top, B. & Fluit, A.C., 1999. Nucleic acid sequence-based amplification (NASBA) detection of medically important *Candida* species. J. Microbiol. Meth. 38: 81-90.

Wieden, M.A., Steinbronn, K.K., Padhye, A.A., Ajello, L. & Chandler, F.W., 1985. Zygomycosis caused by *Apophysomyces elegans.* J. Clin. Microbiol. 22: 522-526.

Wieser, H.G., 1974. Pathogenicity of *Cryptococcus albidus.* Schweiz. Med. WSchr. 103: 475-481.

Wiest, P.M., Wiese, K., Jacobs, M.R., Morrissey, A.B., Abelsom, T.I., Witt, W. & Lederman, M.M., 1987. *Alternaria* infection in a patient with acquired immunodeficiency syndrome: case report and review of invasive *Alternaria* infections. Rev. Infect. Dis. 9: 799-803.

Wigney, D.I., Allan, G.S., Hay, L.E. & Hocking, A.D., 1990. Osteomyelitis associated with *Penicillium verruculosum* in a German shepherd dog. J. Small. Anim. Pract. 31: 449-452.

Wildfeuer, A., Schlenk, R. & Friedrich, W., 1996. Detection of *Candida albicans* DNA with a yeast-specific primer system by polymerase chain reaction. Mycoses 39: 341-346.

Wildfeuer, A., Seidl, H.P., Paule, I. & Habereiter, A., 1998. *In vitro* evaluation of voriconazole against clinical isolates of yeasts, moulds and dermatophytes in comparison with itraconazole, ketoconazole, amphotericin B and griseofulvin. Mycoses 41: 309-319.

Wiley, B.J. & Simmons, E.G., 1973. New species and a new genus of Plectomycetes with *Aspergillus* states. Mycologia 65: 934-938.

Willemsen, R., Schots, R., Shahabpour, M. & Pierard, D., 1997. *Pseudallescheria boydii* tenosynovitis. J. Mycol. Méd. 7: 100-105.

Williams, B., Popoola, B. & Ogundana, K., 1984. A possible new pathogenic *Aspergillus*, isolation and general mycological properties of the fungus. Afr. J. Med. Sci. 13: 111-115.

Williams, D.W., Wilson, M.J., Lewis, M.A.O. & Potts, A.J.C., 1995. Identification of *Candida* species by PCR and restriction fragment length polymorphism analysis of intergenic spacer regions of ribosomal DNA. J. Clin. Microbiol. 33: 2476-2479.

Williams, J.C., Schned, A.R., Richardson, J.R., Heaney, J.A., Curtis, M.R., Rupp, I.P. & Fordham von Reyn, C., 1995. Fatal genitourinary mucormycosis in a patient with undiagnosed diabetes. Clin. Infect. Dis. 21: 682-684.

Williams, M.A.J., 1987. CMI Descriptions of Pathogenic Fungi and Bacteria 934. Mycopathologia 100: 175-176.

Williams, M.A.J., 1991a. IMI Descriptions of Fungi and Bacteria 1085. Mycopathologia 116: 137-138.

Williams, M.A.J., 1991b. IMI Descriptions of Fungi and Bacteria 1088. Mycopathologia 116: 143-144.

Williamson, P.R., Kwon-Chung, K.J. & Gallin, J.I., 1992. Successful treatment of *Paecilomyces variotii* infection in a patient with chronic granulomatous disease and a review of *Paecilomyces* species infections. Clin. Infect. Dis. 14: 1023-1026.

Wilson, C.M., O'Rourke, E.J., McGinnis, M.R. & Salkin, I.F., 1990. *Scedosporium inflatum:* clinical spectrum of a newly recognized pathogen. J. Infect. Dis. 161: 102-107.

Wilson, M., Robson, J., Pyke, C.M. & McCormack, J.G., 1998. *Saksenaea vasiformis* breast abscess related to gardening injury. New.. Zeal. J. Med. 28: 845-846.

Winer-Muram, H.T. & Rubin, S.A., 1992. Pulmonary blastomycosis. J. Thorac. Imag. 7: 23-28.

Wingard, J.R., Merz, W.G., Rinaldi, M.G., Johnson, T.R., Karp, J.E. & Saral, R., 1991. Increase in *Candida krusei* infection among patients with bone marrow transplantation and neutropenia treated prophylactically with fluconazole. N. Engl. J. Med. 325: 1274-1277.

Wingard, J.R., Merz, W.G. & Saral, R., 1979. *Candida tropicalis:* a major pathogen in immunocompromised patients. Annls Intern. Med. 91: 539-543.

Wingfield, M.J., Seifert, K.A. & Webber, J.F. (eds), 1993. *Ceratocystis* and *Ophiostoma*. Taxonomy, Ecology and Pathogenicity. APS, St. Paul, 293 pp.

Winkler, S., Stanek, G., Hubsch, P., Willinger, B., Susani, S., Rosenkranz, A.R. & Pohanka, E., 1996. Pneumonia due to *Blastomyces dermatitidis* in a European renal transplant recipient. Nephrol. Dial. Transpl. 11: 1376-1379.

Winkler, S., Susani, S., Willinger, B., Apsner, R., Rosenkranz, A.R., Pötzi, R., Berlakovich, G.A. & Pohanka, E., 1996. Gastric mucormycosis due to *Rhizopus oryzae* in a renal transplant recipient. J. Clin. Microbiol. 34: 2585-2587.

Winquist, E.W., Walmsley, S.L. & Berinstein, N.L., 1993. Reactivation and dissemination of blastomycosis complicating Hodgkin's disease: a case report and review of the literature. Am. J. Hematol. 43: 129-132.

Winston, D.J., Balsley, G.E., Rhodes, J. & Linné, S.R., 1977. Disseminated *Trichosporon capitatum* infection in an immunosuppressed host. Arch. Intern. Med. 137: 1192-1195.

Wisuthsarewong, W., Chaiprasert, A. & Viravan, S., 1996. Outbreak of tinea capitis by *Microsporum ferrugineum* in Thailand. Mycopathologia 135: 157-161.

Wiswe, K.A., Speed, B.R., Ellis, D.H. & Andrew, J.H., 1993. Two fatal infections in immunocompromised patients caused by *Scedosporium inflatum*. Pathology 25: 187-189.

Witzig, R.S., Greer, D.L. & Hyslop, N.E., 1996. *Aspergillus flavus* mycetoma and epidural abscess successfully treated with itraconazole. J. Med. Vet. Mycol. 34: 133-137.

Wolf, H. & Blaschke-Hellmessen, R., 1993. Animale Fusariosen. *Fusarium solani*-Infektion bei zwei Schildkröten. Abstr. MYK, Greifswald, P-15.

Wolfram, S. & Zach, F., 1934. Über durch niedere Pilze verursachte Nagelerkrankungen beim Menschen. Arch. Derm. Syph. 170: 681-694.

Wollenweber, H.W. & Hochapfel, H., 1937. Beiträge zur Kenntniss parasitärer und saprophytischer Pilze. 4. *Coniothyrium* und seine Beziehung zur Fruchtfäule. Z. Parasitkde 9: 600-637.

Wong, B., Kiehn, T.E., Edwards, F., Bernard, E.M., Marcove, R.C., Harven, E. de, & Amstrong, D., 1982. Bone infection caused by *Debaryomyces hansenii* in a normal host: a case report. J. Clin. Microbiol. 16: 545-548.

Wong, J.S., Herman, S.J., Hoyos, A. de & Weisbrod, G.L., 1994. Pulmonary manifestations of coccidioidomycosis. Can. Ass. Radiol. J. 45: 87-92.

Wong, P.K., Ching, W.T.W., Kwon-Chung, K.J. & Meyer, R.D., 1989. Disseminated *Phialophora parasitica* infection in humans: case report and review. Rev. Infect. Dis. 11: 770-775.

Wong, V.K.W., Tasman, W., Eagle, R.C. & Rodriguez, A., 1997. Bilateral *Candida parapsilosis* endophthalmitis. Arch. Ophthalmol. 115: 670-672.

Wood, B. & Rademaker, M., 1997. Nosocomial *Trichophyton tonsurans* in a long stay ward. New Zeal. Med. J. 110: 277-278.

Wood, G.M., McCormack, J.G., Muir, D.B., Ellis, D.H., Ridley, M.F., Pritchard, R. & Harrison, M., 1992. Clinical features of human infection with *Scedosporium inflatum*. Clin. Infect. Dis. 14: 1027-1033.

Woods, J.P., Kersulyte, D., Goldman, W.E. & Berg, D.E., 1993. Fast DNA isolation from *Histoplasma capsulatum:* methodology for arbitrary primer polymerase chain reaction-based epidemiological and clinical studies. J. Clin. Microbiol. 31: 463-464.

Wortman, P.D., 1996. Infection with *Penicillium marneffei*. Int. J. Derm. 35: 393-401.

Wright, E.D., Clayton, Y.M., Howlader, A., Nazrul, I. & Husain, R., 1990. Keratomycosis caused by *Dichotomophthoropsis nymphea-rum*. Mycoses 33: 477-481.

Wu, S.-H., 1998. Nine new species of *Phanerochaete* from Taiwan. Mycol. Res. 102: 1126-1132.

Wylen, E.L. & Nanda, A., 1999. *Blastomyces dermatitidis* occurring as an isolated cerebellar mass. Neurosurg. Rev. 22: 152-154.

Yadav, J.S., Singh, N. & Singh, A.P., 1996. Haematological studies on bovine dermatomycosis. Ind. Vet. J. 73: 616-619.

Yaguchi, T., Someya, A. & Udagawa, S., 1994. *Fennellia flavipes* and *Neosartorya stramenia*, two new records from Japan. Mycoscience 35: 175-178.

Yamada, S., Maruoka, T., Nagai, K., Tsumura, N., Yamada, T., Sakata, Y., Tominaga, K., Motohiro, T., Kato, H., Makimura, K. & Yamaguchi, H., 1995. Catheter-related infections by *Hansenula anomala* in children. Scand. J. Infect. Dis. 27: 85-87.

Yamada, Y., Banno, I., Arx, J.A. von & Walt, J.P. van der, 1987. Taxonomic significance of the coenzyme Q system in yeasts and yeast-like fungi. Stud. Mycol. 30: 299-308.

Yamada, Y., Nagahama, T., Kawasaki, H. & Banno, I., 1990. Significance of the co-enzyme Q-system in the classification of yeasts and yeast-like organisms. 36. The phylogenetic relationship of the genera *Phaffia* Miller, Yoneyama *et* Soneda and *Cryptococcus* Kützing *emend*. Phaff et Spencer (*Cryptococcaceae*) based on partial sequences of 18S and 26S ribosomal ribonucleic acids. J. Gen. Appl. Microbiol. 36: 403-414.

Yamakami, Y., Hashimoto, A., Tokimatsu, I. & Nasu, M., 1996. PCR detection of DNA specific for *Aspergillus* species in serum of patients with invasive aspergillosis. J. Clin. Microbiol. 34: 2464-2468.

Yamashita, K. & Yamashita, T., 1972. *Polypaecilum insolitum* (= *Scopulariopsis divaricata*) isolated from cases of otomycosis. Sabouraudia 10: 128-131.

Yan, Z.H., Rogers, S.O. & Wang, C.J.K., 1995. Assessment of *Phialophora* species based on ribosomal DNA internal transcribed spacers and morphology. Mycologia 87: 72-83.

Yangco, G.G., Nettlow, A., Okafor, J.I., Park, J. & Strake, D. te, 1986. Comparative antigenic studies of species of *Basidiobolus* and other medically important fungi. J. Clin. Microbiol. 23: 679-682.

Yarrish, R., Sepulveda, J., Torres, R. & Britton, D., 1991. Invasive *Aureobasidium* infection in a patient with acquired immunodeficiency syndrome. Abstr. MB 2214, 7th Int. Conf. AIDS, Florence.

Yarrow, D., 1965. *Pichia onychis* sp. n. Antonie van Leeuwenhoek 31: 465-467.

Yarrow, D., 1968. *Torulopsis peltata* sp. n. Antonie van Leeuwen-hoek 34: 81-84.

Yasin, A., Maher, A. & Moawad, M.H., 1978. Otomycosis: a survey in the eastern province of Saudi Arabia. J. Laringol. Otol. 92: 869-876.

Yasin, M.S., Abdul-Samad, S., Misngi, S. & Arumugham, G., 1988. Laboratory isolation of *Cryptococcus albidus* from two cases of meningitis. Trop. Biomed. 5: 145-148.

Yau, Y.C.W., Nanassy, J. de, Summerbell, R.C., Matlow, A.G. & Richardson, S.E., 1994. Fungal sternal wound infection due to *Curvularia lunata* in a neonate with congenital heart disease: case report and review. Clin. Infect. Dis. 19: 735-740.

Yayanetra, P., Nitiyanant, P., Ajello, L., Padhye, A.A., Lolekha, S., Atichartakarn, V., Vathesatogit, P., Sathaphatayavongs, B. & Prajaktam, R., 1984. Penicilliosis marneffei in Thailand: report of five human cases. Am. J. Trop. Med. Hyg. 33: 637-644.

Yeghen, T., Fenelon, L., Campbell, C.K., Warnock, D.W., Hoffbrand, A.V., Prentice, H.G. & Kibbler, C.C., 1996. *Chaetomium* pneumonia in patient with acute myeloid leukaemia. J. Clin. Path., London 49: 184-186.

Yeldandi, V., Laghi, F., McCabe, M.A., Larson, R., O'Keefe, P., Husain, A., Montoya, A. & Garrity, E.R., 1995. *Aspergillus* and lung transplantation. J. Heart Lung Transpl. 14: 883-890.

Yen, A., Knipe, R.C. & Tyring, S.K., 1994. Primary cutaneous blastomycosis: report of a case acquired by direct inoculation of a bullous pemphigoid lesion. J. Am. Acad. Derm. 31: 277-278.

Yinnon, A.M., Woodin, K.A. & Powell, K.R., 1992. *Candida lusitaniae* infection in the newborn: case report and review of the literature. Pediatr. Infect. Dis. J. 11: 878-880.

Yohai, R.A., Bullock, J.D., Aziz, A.A. & Markert, R.J., 1994. Survival factors in rhino-orbital-cerebral mucormycosis. Survey Ophthalmol. 39: 3-22.

Yokoi, S., Iizasa, T., Yoshida, S., Kamei, K., Hiroshima, K., Ohwada, O. & Fujisawa, T., 1999. Localized pulmonary zygomycosis without preexisting immunocompromised status. Mycoses 42: 675-677.

Yoshida, K., Hiraoka, T., Ando, M., Uchida, K. & Mohsenin, V., 1992. *Penicillium decumbens*. A new case of fungus ball. Chest 102: 1152-1153.

Yoss, B.S., Sautter, R.L. & Brenker, H.J., 1997. *Trichosporon beigelii*. A new neonatal pathogen. Am. J. Perinatol. 14: 113-117.

Young, C.N., 1972. Range of variation among isolates of *Trichophyton rubrum*. Sabouraudia 10: 164-170.

Young, C.N., Swart, J.G., Ackerman, D. & Davidge-Pitts, K., 1978. Nasal obstruction and bone erosion caused by *Drechslera hawaiiensis*. J. Laryngol. Otol. 92: 137-143.

Young, L.S., 1984. *Pneumocystis carinii* pneumonia. Pathogenesis, diagnosis, treatment. Marcel Dekker, New York, 244 pp.

Young, N.A., Kwon-Chung, K.J., Freeman, J., 1973. Subcutaneous abscess caused by *Phoma* sp. resembling *Pyrenochaeta romeroi*. Unique fungal infection occurring in immunosuppressed recipient of renal allograft. Am. J. Clin. Path. 59: 810-816.

Young, N.A., Kwon-Chung, K.J., Kubota, T.T., Jennings, A.E. & Fisher, R.I., 1978. Disseminated infection by *Fusarium moniliforme* during treatment of malignant lymphoma. J. Clin. Microbiol. 7: 589-594.

Young, R.C., Jennings, A. & Bennett, J.E., 1972. Species identification of invasive aspergillosis in man. Am. J. Clin. Pathol. 58: 554-557.

Yu, R.-Y., 1996. Cutaneous cryptococcosis. Mycoses 39: 207-210.

Yurlova, N.A., Uijthof, J.M.J. & Hoog, G.S. de, 1996. Distinction of species in *Aureobasidium* and related genera by PCR-ribotyping. Antonie van Leeuwenhoek 69: 323-329.

Yurlova, N.A., Hoog, G.S. de & Gerrits van den Ende, A.H.G., 1999. Taxonomy of *Aureobasidium* and allied genera. Stud. Mycol. 43: 63-69.

Zahid, M.A., Klotz, S.A. & Hinthorn, D.R., 1994. Medical treatment of recurrent candidemia in a patient with probable *Candida parapsilosis* prosthetic valve endocarditis. Chest 105: 1597-1598.

Zaias, N. & Rebell, G., 1996. Chronic dermatophytosis caused by *Trichophyton rubrum*. J. Am. Acad. Derm. 35: S17-S20.

Záitz, C., Heins-Vaccari, E.M., Santos de Freitas, R., Hernandez Arriagada, G.L., Ruiz, L., Totoli, S.A.S., Maques, A.C., Rezze, G.G., Mueller, H., Valente, N.S. & Silva Lacaz, C. da, 1997. Subcutaneous phaeohyphomycosis caused by *Phoma cava*. Report of a case and review of the literature. Revta Inst. Med. Trop. S. Paulo 39: 43-48.

Záitz, C., Lacaz, C. da Silva, Salebian, A., Ruiz, L.R., HeinsVaccari, E.M. & Melo, N.T. de, 1988. Eumicetoma podal por *Acremonium falciforme*. Registro de um caso. Anais Bras. Derm. 63: 413-418.

Záitz, C., Porto, E., Heins-Vaccari, E.M., Sadahrio, A., Rangel, Ruiz, B., Muller, H. & Lacaz, C. da Silva, 1995. Subcutaneous hyalohyphomycosis caused by *Acremonium recifei*. Revta Inst. Med. Trop. S. Paulo 37: 267-270.

Zalar, P., Hoog, G.S. de & Gunde-Cimerman, N., 1999. Ecology of halotolerant dothideaceous black yeasts. Stud. Mycol. 43: 38-48.

Zamos, D.T., Schumacher, J. & Loy, J.K., 1996. Nasopharyngeal conidiobolomycosis in a horse. J. Am. Vet. Med. Ass. 208: 100-101.

Zapater, R.C., 1980. Keratomycosis caused by dematiaceous fungi. PAHO Sci. Publ. 396: 82-87.

Zapater, R.C., 1986. Opportunistic fungus infection - *Fusarium* infections - Keratomycosis by *Fusarium*. Jpn. J. Med. Mycol. 27: 68-69.

Zapater, R.C., Albesi, E.J. & Garcia, G.H., 1975. Mycotic keratitis by *Drechslera spicifera*. Sabouraudia 13: 295-298.

Zapater, R.C., Arrechea, A. de & Guevara, V.H., 1972. Queratomicosis por *Fusarium dimerum*. Sabouraudia 10: 274-275.

Zapater, R.C. & Scattini, F., 1979. Mycotic keratitis by *Cladorrhinum*. Sabouraudia 17: 65-69.

Zaror, L., Fischman, O., Pereira, C.A., Felipe, R.G., Gregório, L.C. & Castelo, A., 1987. A case of primary nasal chromoblastomycosis. Mykosen 30: 468-471.

Zaror, L. & Moreno, M.I., 1980. Onicomicosis por *Aspergillus candidus* Link. Revta Argent. Micol. 3: 13-15.

Zelenková, H., 2000. *Chrysosporium pannorum* as a probable mycotic infection in a chow-chow dog. Abstr. 14th ISHAM, B. Aires, p. 186.

Zepeda, M., Kobayashi, G.K., Appleman, M.D. & Navarro, A., 1998. *Coccidioides immitis* presenting as a hyphal form in cerebrospinal fluid. J. Nat. Med. Ass. 90: 435-436.

Zeppenfeldt, G., Richard-Yegres, N., Yegres, F. & Hernández, R., 1994. *Cladosporium carrionii*: hongo dimorfo en cactáceas de la zona endémica para la cromomicosis en Venezuela. Revta Iberoam. Micol. 11: 61-63.

Zerva, L., Hollis, R.J. & Pfaller, M.A., 1998. *In vitro* susceptibility testing and DNA typing of *Saccharomyces cerevisiae* clinical isolates. J. Clin. Microbiol. 34: 3031-3034.

Zhanei, G.G., Karlowsky, J.A., Zelenitsky, S.A., Turik, M.A. & Hoban, D.J., 1998. Susceptibilities of *Candida* species isolated from the lower gastrointestinal tracts of high-risk patients to the new semisynthetic echinocandin LY303366 and other antifungal agents. Antimicrob. Agents Chemother. 42: 2446-2448.

Zhao, Y.M., Deng, C.R. & Chen, X., 1990. *Arthrinium phaeospermum* causing dermatomycosis, a new record of China. Acta Mycol. Sin. 9: 232-235.

Zheng, R.-y. & Chen, C.-q., 1991. A non-thermophilic *Rhizomucor* causing human primary cutaneous mucormycosis. Mycosystema 4: 45-57.

Zheng, R.-y. & Chen, C.-q., 1993. Another non-thermophilic *Rhizomucor* causing human primary cutaneous mucormycosis. Mycosystema 6: 1-12.

Ziskind, J., Pizzolato, P. & Buff, E., 1958. Aspergillosis of the brain: report of a case. Am. J. Clin. Path. 29: 554-559.

Ziza, J.M., Dupont, B., Boissonnas, A., Meyniard, O., Bedrossian, J., Drouhet, E. & Cremer, G.A., 1985. Ostéoarthrites à champignons noirs (dématiés). À propos de 3 observations. Annls Med. Interne 136: 393-397.

Zoutman, D.E. & Sigler, L., 1991. Mycetoma of the foot caused by *Cylindrocarpon destructans*. J. Clin. Microbiol. 29: 1855-1859.

Zweibel, S.M. & Wang, C.J.K., 1978. Reexamination of *Phialophora verrucosa*. PAHO Sci. Publ. 356: 91-100.

Zycha, H., Siepmann, R. & Linnemann, G., 1969. *Mucorales*, eine Beschreibung aller Gattungen und Arten dieser Pilzgruppe. Cramer, Lehre, 355 pp.

INDEX